WATER ENCYCLOPEDIA

OCEANOGRAPHY; METEOROLOGY; PHYSICS AND CHEMISTRY; WATER LAW; AND WATER HISTORY, ART, AND CULTURE

WATER ENCYCLOPEDIA

WATER ENCYCLOPEDIA

OCEANOGRAPHY; METEOROLOGY; PHYSICS AND CHEMISTRY; WATER LAW; AND WATER HISTORY, ART, AND CULTURE

Jay Lehr, Ph.D.
Editor-in-Chief

Jack Keeley
Senior Editor

Janet Lehr
Associate Editor

Thomas B. Kingery III
Information Technology Director

The *Water Encyclopedia* is available online at
http://www.mrw.interscience.wiley.com/eow/

WILEY-INTERSCIENCE

A John Wiley & Sons, Inc., Publication

Published by John Wiley & Sons, Inc., Hoboken, New Jersey.
Published simultaneously in Canada.

Library of Congress Cataloging-in-Publication Data is available.

Lehr, Jay
Water Encyclopedia: Oceanography; Meteorology; Physics and Chemistry; Water Law; and Water History, Art, and Culture

ISBN 0-471-73684-8
ISBN 0-471-44164-3 (Set)

Printed in the United States of America

10 9 8 7 6 5 4 3 2 1

CONTENTS

Physics and Chemistry of Water

Water Law and Economics

Water History, Art, and Culture

PREFACE

No natural molecule on the planet is more fascinating than water. It has unique properties ranging from the unusual angles formed between its two hydrogen ions and its single oxygen molecule, to the fact that unlike most substances, it expands when it freezes rather than shrinks and reaches a maximum density as a liquid 4°F above its freezing point. These and many other aspects of the special physics and chemistry are described in this volume, including the impact of a wide variety of chemicals occurring in water, osmosis, diffusion, hydration, isotope exchange, along with the fun physics of the mariotte bottle.

Equally fascinating are the many unusual physical and chemical encounters in both the ocean and the atmosphere. Although oceanography and meteorology are frequently considered separate sciences from hydrology, their limited inclusion in the *Water Encyclopedia* was deemed necessary to tell the complete story. Tidal changes, benthic nutrients, the sea floor, el nino, sea level, and ocean/climate relationships make up but a few of the oceans fascinating stories, whereas water spouts, hurricanes, monsoons, droughts, sublimation, and barometric efficiency just touch the tip of what this volume has in store in the area of meteorology.

But water has a wonderful human face as well. We have reached around the world to describe the history of water and its role in the development of civilizations and the many beliefs held about it. As society developed, the distribution of water needed supervision, which lead to a wide variety of water laws we have attempted to categorize and describe in an interesting and meaningful way.

We are equally proud of our open-minded effort to describe the role that water has played in art and culture. We have attempted not be judgmental, with stories of water forms and water intelligence along with some medical theories and, of course, the wonderful descriptions of early water clocks. This volume is a true intellectual cornucopia of water in the life of humankind on a personal level.

We are confident that in the coming years and editions of the *Water Encyclopedia*, this volume will expand with more participation from individuals working in unusual fields relating to water.

Jay Lehr
Jack Keeley

CONTRIBUTORS

M. Abad, *Universidad de Huelva. Avda. de las Fuerzas Armadas, Huelva,* A Statistical Approach to Critical Storm Period Analysis

Edinara Adelaide Boss, *Artur-Nogueira-SP, Brazil,* Sublimation

Joseph H. Aldstadt, III, *Genetic Technologies, Inc. Testing Institute, Waukesha, Wisconsin, In Situ* Chemical Monitoring

Alaska Biological Science Center—U.S. Geological Survey, Oceanographic Environment of Glacier Bay

Sergio Alonso, *University of the Balearic, Palma de Mallorca, Spain,* Relative Humidity

Marie de Angelis, *Humboldt State University, Arcata, California,* Major Ions in Seawater

Carolyn Ann Koh, *Colorado School of Mines, Golden, Colorado,* Clathrate Hydrates, Hydration

Yoseph Negusse Araya, *Open University, Milton Keynes, United Kingdom,* Hydrosphere

Ann Azadpour-Keeley, *National Risk Management Research Laboratory, ORD, U.S.EPA, Ada, Oklahoma,* Water and the History of Man

W.D. Bach, Jr., *(from The Handbook of Weather, Climate, and Water: Dynamics, Climate, Physical Meteorology, Weather Systems, and Measurements, Wiley 2003),* Basic Research for Military Applications

Patrick L. Barry, *NASA,* Water on the Space Station

F. Batmanghelidj, *Global Health Solutions, Falls Church, Virginia,* The Myth of Bad Cholesterol: Why Water Is A Better Cholesterol-Lowering Medication, Water: The Key to Natural Health And Healing, The Mistake of Waiting to Get Thirsty

M. Eric Benbow, *Michigan State University, East Lansing, Michigan,* A History of Hawaiian Freshwater Resources

Mike Bettwy, *Goddard Space Flight Center, NASA,* African Monsoons

David Biggs, *University of Washington, Seattle, Washington,* Canals in the Mekong Delta: A Historical Overview from 200 C.E. to the Present

Jessica Black, *NOAA/CREST CCNY, New York, New York,* Remote Sensing of Applications in Hydrology

Peter E. Black, *SUNY ESF, Syracuse, New York,* Water Resource Organizations

Julián Blasco, *Institute of Marine Sciences of Analucia, Cadiz, Spain,* Marine and Estuarine Microalgal Sediment Toxicity Tests

M. Boström, *Linköping University, Linköping, Sweden,* Hofmeister Effects, Dissolved Gases

Kristofor R. Brye, *University of Arkansas, Fayetteville, Arkansas,* Nitrogen

Bureau of Labor Statistics, U.S. Department of Labor, Water Transportation Occupations

Bureau of Reclamation—U.S. Department of the Interior, Hoover Dam History

Robert M. Burgess, *U.S. Environmental Protection Agency, Narragansett, Rhode Island,* Ammonia

Albert J. Burky, *University of Dayton, Dayton, Ohio,* A History of Hawaiian Freshwater Resources

James Butler, *Climate Monitoring and Diagnostics Laboratory,* NOAA and University Scientists Study Methyl Bromide Cycling in the North Pacific

Bradford Butman, *U.S. Geological Survey,* Mapping the Sea Floor of the Historic Area Remediation Site (HARS) Offshore of New York City

Pietro Chiavaccini, *Universitá di Pisa, Pisa, Italy,* Shallow Water Waves, Breakwaters

G.G. Clarke, *University of Hawaii, Hilo, Hawaii,* Climate and Water Balance on the Island of Hawaii

Jay Clausen, *AMEC Earth and Environmental, Inc., Westford, Massachusetts,* Technetium in Water, Beryllium in Water

Aldo Conti, *Frascati (RM), Italy,* Floating Ice, Condensation, Desertification, Dew, Distilled Water, Isotopes, Rain Forests, Dew Point

Giuseppe Cortese, *Alfred Wegener Institute for Polar and Marine Research (AWI), Bremerhaven, Germany,* Seawater Temperature Estimates in Paleoceanography

Sara Cotton, *University of Miami,* NOAA and University Scientists Study Methyl Bromide Cycling in the North Pacific

L.M. Cáceres, *Universidad de Huelva. Avda. de las Fuerzas Armadas, Huelva,* A Statistical Approach to Critical Storm Period Analysis

Eli Dahi, *Environmental Development Corporation, Søborg, Denmark,* Defluroidation

Carl W. David, *University of Connecticut, Storrs, Connecticut,* The Hydronium Ion

Jana Davis, *Smithsonian Environmental Research Center, Edgewater, Maryland,* Marine Stock Enhancement Techniques

Christine Dickenson, *Florida Institute of Technology, Melbourne, Florida,* An Analysis of the International Maritime Organization–London Convention Annual Ocean Dumping Reports

Robert E. Dickinson, *(from The Handbook of Weather, Climate, and Water: Dynamics, Climate, Physical Meteorology, Weather Systems, and Measurements, Wiley 2003),* Overview: The Climate System

Priyanka K. Dissanayake, *Moratuwa, Sri Lanka,* Estuarian Waters

Iver W. Duedall, *Florida Institute of Technology, Melbourne, Florida,* An Analysis of the International Maritime Organization–London Convention Annual Ocean Dumping Reports

Sandra Dunbar, *Napier University, Edinburgh, Scotland, United Kingdom,* Partitioning and Bioavailability

Antonín Dvořák, *University of Economics, Prague, Czech Republic,* Negotiating between Authority and Polluters: An Approach to Managing Water Quality

David R. Easterling, *(from Handbook of Weather, Climate, and Water: Dynamics, Climate, Physical Meteorology, Weather Systems, and Measurements, Wiley 2003),* Observations of Climate and Global Change from Real-Time Measurements

Theodore A. Endreny, *SUNY-ESF, Syracuse, New York,* Remote Sensing of Applications in Hydrology, Evaporation, Radar Use in Rainfall Measurements, Rainfall and Runoff

Environment Canada, Water—Nature's Magician

Timothy Erickson, *National Weather Service New Orleans/Baton Rouge Forecast Office, Slidell, Louisiana,* Coastal Fog Along the Northern Gulf of Mexico

Kirsten Exall, *National Water Research Institute, Burlington, Ontario, Canada,* Coagulation and Flocculation

David L. Feldman, *University of Tennessee, Knoxville, Tennessee,* Water Supply Planning—Federal, Water Transfers

Montserrat Filella, *University of Geneva, Geneva, Switzerland,* Antimony in Aquatic Systems

Sylwester Furmaniak, *N. Copernicus University, Toruń, Poland,* Mechanisms of Water Adsorption on Carbons

Piotr A. Gauden, *N. Copernicus University, Toruń, Poland,* The Effect of Carbon Surface Chemical Composition on the Mechanism of Phenol Adsorption from Aqueous Solutions, Mechanisms of Water Adsorption on Carbons

Geophysical Fluid Dynamics Laboratory—NOAA, El Niño: The Interannual Prediction Problem, Hurricanes: Modeling Nature's Fury, Global Climate Change

Andrea K. Gerlak, *Columbia University, New York, New York,* Wetlands Policy in the United States: From Drainage to Restoration

Michael H. Glantz, *(from The Handbook of Weather, Climate, and Water: Atmospheric Chemistry, Hydrology, and Societal Impacts, Wiley 2003),* Climate and Society

Global Change Research Program—U.S. Geological Survey, Flattop Mountain Snotel Snowpack: Water Year 2004

Jyotsna Goel, *Centre for Fire, Explosives, and Environmental Safety, Timarpur, India,* Ion Exchange and Inorganic Adsorption

Kelly Goodwin, *University of Miami,* NOAA and University Scientists Study Methyl Bromide Cycling in the North Pacific

David M. Gray, *Mettler-Toledo Thornton Inc., Bedford, Massachusetts,* Conductivity-Electric

Great Lakes Environmental Research Laboratory (NOAA), Technology Development: Hardware Development—Marine Instrumentation Laboratory (MIL)

Seth I. Gutman, *NOAA Forecast Systems Laboratory,* Ground-Based GPS Meteorology at FSL

Janusz Guziur, *University of Warmia and Mazury in Olsztyn, Olsztyn, Poland,* History of Pond Fisheries in Poland

David Haley, *Picacadilly, Ulverston, Cumbria, United Kingdom,* Water and Well-Being, Evolution

Jon Hare, *NOAA Center for Coastal Fisheries and Habitat Research,* NOS/NMFS Cooperative Research on Coastal Fisheries and Habitats at the Beaufort Laboratory

Michael J. Hayes, *Climate Impacts Specialist—National Drought Mitigation Center,* Drought Indices

William Henderson, *University of Waikato, Hamilton, New Zealand,* Analysis of Aqueous Solutions Using Electrospray Ionization Mass Spectrometry (esi ms)

Warwick Hillier, *The Australian National University, Canberra, Australia,* Isotope Exchange in Gas-Water Reactions

Joseph Holden, *University of Leeds, Leeds, United Kingdom,* Mariotte Bottle—Use in Hydrology, Rain Simulators

S. Holgate, *Proudman Oceanographic Laboratory, Birkenhead, United Kingdom,* The Permanent Service for Mean Sea Level

Arthur M. Holst, *Philadelphia Water Department, Philadelphia, Pennsylvania,* Frost Damage, Droughts, Water Cycle, Monsoon, Permanent Frost, United States Weather Bureau, Waterspout, Cloud Seeding, Cyclones, Fog, Frost, Chinook, Flood Control Act of 1944, Jacob's Well, Gordon and Franklin Rivers and the Tasmanian Wilderness World Heritage Area

Kirk L. Holub, *NOAA Forecast Systems Laboratory,* Ground-Based GPS Meteorology at FSL

Paul R. Houser, *(from The Handbook of Weather, Climate, and Water: Atmospheric Chemistry, Hydrology, and Societal Impacts, Wiley 2003),* Infiltration and Soil Moisture Processes

S.A. Hsu, *Louisiana State University, Baton Rouge, Louisiana,* Air–Sea Interaction

Jason A. Hubbart, *University of Idaho, Moscow, Idaho,* Hydrologic Cycle, Water Resources, and Society, Hydrologic History, Problems, and Perspectives

Basia Irland, *University of New Mexico, Albuquerque, New Mexico,* A Concise Glimpse of Water in the History of Photography

Mohammed Riajul Islam, *University of Idaho, Moscow, Idaho,* Rain and Rocks: The Recipe for River Water Chemistry

James A. Jacobs, *Environmental Bio-Systems, Inc., Mill Valley, California,* Fenton's Reaction and Groundwater Remediation, Regulatory Issues and Remediation: Risk, Costs, and Benefits, Water, Bacteria, Life on Mars, and Microbial Diversity

Sharad K. Jain, *National Institute of Hydrology, Roorkee, India,* Isohyetal Method, Hydrologic Cycle

Purnima Jalihal, *National Institute of Ocean Technology, Chennai, India,* Renewable Energies from the Ocean

S. Jevrejeva, *Proudman Oceanographic Laboratory, Birkenhead, United Kingdom,* The Permanent Service for Mean Sea Level

Alicia Jimenez, *Michigan State University, East Lansing, Michigan,* Transboundary Waters in Latin America: Conflicts and Collaboration

Anne Jones-Lee, *G. Fred Lee & Associates, El Macero, California,* Clean Water Act, Water Quality Criteria/Standards, TMDLs, and Weight-of-Evidence Approach for Regulating Water Quality

Andrew Juhl, *Lamont–Doherty Earth Observatory of Columbia University, Palisades, New York,* Food Chain/Foodweb/Food Cycle

James O. Juvik, *University of Hawaii, Hilo, Hawaii,* Climate and Water Balance on the Island of Hawaii

K. Kadirvelu, *Centre for Fire, Explosives, and Environmental Safety, Timarpur, India,* Ion Exchange and Inorganic Adsorption

Th.D. Karapantsios, *Aristotle University, Thessaloniki, Greece,* Sorption Kinetics

Thomas R. Karl, *(from Handbook of Weather, Climate, and Water: Dynamics, Climate, Physical Meteorology, Weather Systems, and Measurements, Wiley 2003),* Observations of Climate and Global Change from Real-Time Measurements

Melinda R. Kassen, *Trout Unlimited, Colorado Water Project, Boulder, Colorado,* Legal Protection for In-stream Flow, Interface between Federal Water Quality Regulation and State Allocation of Water Quantity

Gholam A. Kazemi, *Shahrood University of Technology, Shahrood, Iran,* Age Dating Old Groundwater, Chlorine-36 and Very Old Groundwaters, Deuterium, Freshwater, Hard Water, Heavy Water, Hydrogen Ion, Isotope Fractionation, Soft Water, Strontium Isotopes in Water and Rock, Chlorofluorocarbons (CFCs)

Jack W. Keeley, *Environmental Engineer, Ada, Oklahoma,* Water and the History of Man

Giora J. Kidron, *The Hebrew University of Jerusalem, Jerusalem, Israel,* Dew Deserts

Daniel King, *University of Colorado,* NOAA and University Scientists Study Methyl Bromide Cycling in the North Pacific

Piotr Kowalczyk, *Faculty of Science, Chiba University, Chiba, Japan,* The Effect of Carbon Surface Chemical Composition on the Mechanism of Phenol Adsorption from Aqueous Solutions, Mechanisms of Water Adsorption on Carbons

Upadhyayula V. K. Kumar, *Choa Chu kang Ave-4, Singapore,* Coastal Water Pollutants

W. Kunz, *University of Regensburg, Regensburg, Germany,* Hofmeister Effects

Frederic Lasserre, *Laval University, Quebec City, Quebec, Canada,* Great Lakes Governors' Agreement

Marshall Lawson, *Land Conservation Legal Services, LLC, Columbia, South Carolina,* Quantitative GroundWater Law

Jamie R. Lead, *University of Birmingham, Birmingham, United Kingdom,* Dissolved Organic Carbon

G. Fred Lee, *G. Fred Lee & Associates, El Macero, California,* Clean Water Act, Water Quality Criteria/Standards, TMDLs, and Weight-of-Evidence Approach for Regulating Water Quality

Maggie Lee, *Santa Fe, New Mexico,* Free Flowing Water: A Source of Wisdom

Nai Kuang Liang, *National Taiwan University, Taipei, Taiwan,* Water Waves

Sharon L. Lien, *The Groundwater Foundation, Lincoln, Nebraska,* Effective Water Education Strategies in a Nontraditional Setting

Clive D. Lipchin, *Ann Arbor, Michigan,* Water Between Arabs and Israelis: Researching Twice-Promised Resources, Conflict and Water Use in the Middle East

Eileen Loiseau, *Bigelow Laboratory for Ocean Sciences,* NOAA and University Scientists Study Methyl Bromide Cycling in the North Pacific

M.X. Loukidou, *Aristotle University, Thessaloniki, Greece,* Sorption Kinetics

Helen H. Lou, *Lamar University, Beaumont, Texas,* Solubility of Hydrocarbons and Sulfur Compounds in Water, Solubility of Chemicals in Water

Nancy A. Lowery, *San Diego, California,* United States-Mexico Border Waters: Conventions, Treaties, and Institutions

Luís M. Lubián, *Institute of Marine Sciences of Analucia, Cadiz, Spain,* Marine and Estuarine Microalgal Sediment Toxicity Tests

Paolo Magni, *IMC—International Marine Centre, Torregrande-Oristano, Italy,* Laboratory Experiments On Bivalve Excretion Rates of Nutrients, Temporal Scaling of Benthic Nutrient Regeneration in Bivalve-Dominated Tidal Flat, Physical and Chemical Variability of Tidal Streams, Tidally Mediated Changes in Nutrient Concentrations

Babs A. Makinde-Odusola, *Riverside, California,* Radon in Water

William E. Marks, *Water Consciousness, Inc., Martha's Vineyard, Massachusetts,* Benjamin Franklin: From Kite to Lightning Rod, Benjamin Franklin's Armonica: A Water Music Instrument, Our Evolving Water Consciousness, Water Clocks, Ben Franklin's Gulf Stream Weather and Swim Fins

D.L. Marrin, *Hanalei, Hawaii,* Cosmic Water, Molecular Network Dynamics, Sound in Water, Water Symbolism

K.A. Matis, *Aristotle University, Thessaloniki, Greece,* Sorption Kinetics

Patricia Matrai, *Bigelow Laboratory for Ocean Sciences,* NOAA and University Scientists Study Methyl Bromide Cycling in the North Pacific

Mollie D. McIntosh, *Michigan State University, East Lansing, Michigan,* A History of Hawaiian Freshwater Resources

Kevin S. McLeary, Physical Properties

Lucius O. Mendis, *Colombo, Sri Lanka,* Ancient Water and Soil Conservation Ecosystems of Sri Lanka

Carlos D. Messina, *University of Florida, Gainesville, Florida,* Degree Day Method

Paulette Middleton, *(from The Handbook of Weather, Climate, and Water: Atmospheric Chemistry, Hydrology, and Societal Impacts, Wiley 2003),* Acid Rain and Society

Elsie F. Millano, *ERM, Inc., Vernon Hills, Illinois,* Great Lakes Water Quality Initiative

A. Trent Millet, *Newark, Vermont,* The Medicinal Properties of the Waters of Saratoga Springs

Koichi Miyashita, *Okayama University, Kurashiki, Japan,* Transpiration

Eric P. Mollard, *Institut de Recherche pour le Développement, France,* Curious Uses of Agricultural Water in the World, Jaubert de Passa: The First World History of Irrigation in 1846

Shigeru Montani, *Hokkaido University, Hakodate, Japan,* Physical and Chemical Variability of Tidal Streams, Laboratory Experiments On Bivalve Excretion Rates of Nutrients, Tidally Mediated Changes in Nutrient Concentrations, Temporal Scaling of Benthic Nutrient Regeneration in Bivalve-Dominated Tidal Flat, Seasonal Coupling Between Intertidal Macrofauna and Sediment Column Porewater Nutrient Concentrations

Robert M. Moore, *Dalhousie University, Halifax, Nova Scotia, Canada,* Marine Sources of Halocarbons

Ignatio Moreno-Garrido, *Institute of Marine Sciences of Analucia, Cadiz, Spain,* Marine and Estuarine Microalgal Sediment Toxicity Tests

Suparna Mukherji, *IIT Bombay, Powai, Mumbai, India,* Adsorption of Organic Compounds

J.M. Muñoz, *Universidad de Sevilla. Avda. Reina Mercedes, Sevilla,* A Statistical Approach to Critical Storm Period Analysis

Prasad K. Narasimhan, *Lamar University, Beaumont, Texas,* Adsorption Capacity of Activated Carbon for Water Purification, Solubility of Hydrocarbons and Sulfur Compounds in Water, Solubility of Chemicals in Water, Solubility of Hydrocarbons in Salt Water

NASA—Goddard Space Flight Center, Black Water Turns the Tide on Florida Coral, The Water Cycle, Weather Forecasting Through The Ages, The Earth Observing System: Aqua

NASA—Langley Research Center, CERES: Understanding The Earth's Clouds and Climate

NASA Marshall Space Flight Center, Where Water Floats

National Drought Mitigation Center, What is Climatology?

National Oceanographic and Atmospheric Administration (NOAA), NOAA's Atlantic Oceanographic and Meteorological Laboratory

Natural Resources Conservation Service, Conservation and the Water Cycle

John W. Nielsen-Gammon, *(from Handbook of Weather, Climate, and Water: Dynamics, Climate, Physical Meteorology, Weather Systems, and Measurements, Wiley 2003),* Overview of Weather Systems

Robert Y. Ning, *King Lee Technologies, San Diego, California,* Silica in Natural Waters, Carbonate in Natural Waters

B.W. Ninham, *University of Regensburg, Regensburg, Germany and Australian National University, Canberra, Australia,* Hofmeister Effects, Dissolved Gases

NOAA Coral Reef Information System (CORIS), Deep Water Corals

NOAA National Ocean Service, NOS Sanctuaries Protect Nation's Maritime History

NOAA—Pacific Marine Environmental Laboratory, Pacific Marine Environmental Laboratory—30 Years of Observing The Ocean

Northeast Fisheries Science Center—NOAA, Woods Hole: The Early Years

Gertrud K. Nürnberg, *Freshwater Research, Baysville, Ontario, Canada,* Quantification of Anoxia and Hypoxia in Water Bodies

Pacific Fisheries Environmental Laboratory, NOAA, Physical Oceanography

Pacific Northwest National Laboratory—Shrub-Steppe Ecology Series, What About Meteorology?

Stefano Pagliara, *Universitá di Pisa, Pisa, Italy,* Shallow Water Waves, Breakwaters

Magni Paolo, *IMC-International Marine Centre, Torregrande-Oristano, Italy,* Seasonal Coupling Between Intertidal Macrofauna and Sediment Column Porewater Nutrient Concentrations

Jose O. Payero, *University of Nebraska-Lincoln, North Platte, Nebraska,* Evapotranspiration

Mauricio Peredo, *NASA Goddard Space Flight Center,* Electricity as a Fluid

Naraine Persaud, *Virginia Polytechnic Institute and State University, Blacksburg, Virginia,* Adiabatic Cooling, Humidity—Absolute, Heat of Vaporization, Vapor Pressure

Geoff Petts, *University of Birmingham, Birmingham, United Kingdom,* Water in History

Tony Phillips, *NASA,* Water on the Space Station

Jim Philp, *Napier University, Edinburgh, Scotland, United Kingdom,* Partitioning and Bioavailability

Laurel Phoenix, *Green Bay, Wisconsin,* Water as a Human Right

Physics Laboratory, National Institute of Standards and Technology, Early Clocks

Ralph W. Pike, *Louisiana State University, Baton Rouge, Louisiana,* Solubility of Hydrocarbons and Sulfur Compounds in Water, Solubility of Chemicals in Water

R. Pino, *Universidad de Sevilla. Avda. Reina Mercedes, Sevilla,* A Statistical Approach to Critical Storm Period Analysis

Richard Z. Poore, *U.S. Geological Survey, Reston, Virginia,* Sea Level and Climate

Bobby J. Presley, *Texas A&M University, College Station, Texas,* Trace Element Pollution

Nitish Priyadarshi, *Ranchi University, Ranchi, Jharkhand, India,* Sodium in Natural Waters

Jian-Wen Qiu, *Hong Kong Baptist University, Kowloon Tong, Kowloon, Hong Kong,* Larvae and Small Species of Polychaetes in Marine Toxicological Testing

Climent Ramis, *University of the Balearic, Palma de Mallorca, Spain,* Relative Humidity

Jorge Ramírez-Vallejo, *Harvard University, Cambridge, Massachusetts,* Economic Value of Water: Estimation

Todd Rasmussen, *The University of Georgia, Athens, Georgia,* Barometric Efficiency

Howard Reminick, *Ohno Institute on Water and Health, Willoughby, Ohio,* An Analysis of The Impact of Water on Health and Aging: Is All Water The Same?

Martin Reuss, Ph.D, *Office of History Headquarters, U.S. Army Corps of Engineers,* The Constitution and Early Attempts at Rational Water Planning

J. Rodríguez Vidal, *Universidad de Huelva. Avda. de las Fuerzas Armadas, Huelva,* A Statistical Approach to Critical Storm Period Analysis

A. Rodríguez-Ramírez, *Universidad de Huelva. Avda. de las Fuerzas Armadas, Huelva,* A Statistical Approach to Critical Storm Period Analysis

Romualdo Romero, *University of the Balearic, Palma de Mallorca, Spain,* Relative Humidity

William R. Roy, *Illinois State Geological Survey, Champaign, Illinois,* Iron

F. Ruiz, *Universidad de Huelva. Avda. de las Fuerzas Armadas, Huelva,* A Statistical Approach to Critical Storm Period Analysis

David L. Russell, *Global Environmental Operations, Inc., Lilburn, Georgia,* A Brief History of the Water Pollution Control Act in the U.S.

Timothy J. Ryan, *Ohio University, Athens, Ohio,* The Clean Water Act

Gerhard Rychlicki, *N. Copernicus University, Toruń, Poland,* The Effect of Carbon Surface Chemical Composition on the Mechanism of Phenol Adsorption from Aqueous Solutions, Mechanisms of Water Adsorption on Carbons

S. Sabri, *University of Malaya, Petaling Jaya Selangor, Malaysia,* Islamic Water Law

Basu Saha, *Loughborough University, Loughborough, United Kingdom,* Removal of Organic Micropollutants and Metal Ions from Aqueous Solutions by Activated Carbons

Eric Saltzman, *University of Miami,* NOAA and University Scientists Study Methyl Bromide Cycling in the North Pacific

Peter H. Santschi, *Texas A&M University, Galveston, Texas,* Marine Colloids

Edward S. Sarachik, *(from The Handbook of Weather, Climate, and Water: Dynamics, Climate, Physical Meteorology, Weather Systems, and Measurements, Wiley 2003),* The Ocean in Climate

Eva Saroch, *Panjab University, Chandigarh, India,* Representing Geopolitics of (Hydro) Borders in South Asia

Petr Šauer, *University of Economics, Prague, Czech Republic,* Negotiating between Authority and Polluters: An Approach to Managing Water Quality

Donald Savage, *NASA Headquarters, Washington,* Mars Exploration Rover Mission

Reuel Shinnar, *The City College of the CUNY, New York,* The Mirage of The H_2 Economy

Melissa A. Singer Pressman, *Genetic Technologies, Inc. Testing Institute, Waukesha, Wisconsin, In Situ* Chemical Monitoring

Pratap Singh, *National Institute of Hydrology, Roorkee, India,* Snow Density, Snow and Snowmelt, Snow Surveys

Vijay P. Singh, *Louisiana State University, Baton Rouge, Louisiana,* Entropy Theory For Hydrologic Modeling, Unit Hydrograph Theory, Hydrologic Cycle, Isohyetal Method

D.C. Singleton, *University of Hawaii, Hilo, Hawaii,* Climate and Water Balance on the Island of Hawaii

Bellie Sivakumar, *University of California, Davis, California,* Hydropsychology

E. Dendy Sloan, *Colorado School of Mines, Golden, Colorado,* Clathrate Hydrates

James A. Smith, *(from The Handbook of Weather, Climate, and Water: Atmospheric Chemistry, Hydrology, and Societal Impacts, Wiley 2003),* Rainfall

Fiona Stainsby, *Napier University, Edinburgh, Scotland, United Kingdom,* Partitioning and Bioavailability

Susan-Marie Stedman, *Silver Spring, Maryland,* Coastal Waters

Kenneth F. Steele, *University of Arkansas, Fayetteville, Arkansas,* Nitrogen, Carbonate Geochemistry

David P. Stern, *Goddard Space Flight Center, NASA,* Electricity as a Fluid, Weather and the Atmosphere

Bradley A. Striebig, *Gonzaga University, Spokane, Washington,* Chemical Precipitation

Georgina Sturrock, *Commonwealth Scientific and Industrial Research Organization—Australia,* NOAA and University Scientists Study Methyl Bromide Cycling in the North Pacific

Artur P. Terzyk, *N. Copernicus University, Toruń, Poland,* The Effect of Carbon Surface Chemical Composition on the Mechanism of Phenol Adsorption from Aqueous Solutions, Mechanisms of Water Adsorption on Carbons

Ryszard Tokarczyk, *University of Miami,* NOAA and University Scientists Study Methyl Bromide Cycling in the North Pacific

Marta E. Torres, *Oregon State University, Corvallis, Oregon,* Distribution and Dynamics of Gas Hydrates in the Marine Environment

Christopher Tracey, *U.S. Geological Survey, Reston, Virginia,* Sea Level and Climate

Anne M. Trehu, *Oregon State University, Corvallis, Oregon,* Distribution and Dynamics of Gas Hydrates in the Marine Environment

Robert C. Upstill-Goddard, *University of Newcastle upon Tyne, Newcastle upon Tyne, United Kingdom,* The Role of Oceans in the Global Cycles of Climatically-Active Trace-Gases

David B. Vance, *ARCADIS, Midland, Texas,* Regulatory Issues and Remediation: Risk, Costs, and Benefits, Water, Bacteria, Life on Mars, and Microbial Diversity

Javier Velez-Arocho, *U.S. Environmental Protection Agency, Washington, District of Columbia,* Marine Debris Abatement

Roger C. Viadero Jr., *West Virginia University, Morgantown, West Virginia,* Henry's Law

Rolf R. von Oppenfeld, *The TESTLaw Practice Group, Phoenix, Arizona,* The National Pollution Discharge Elimination System, The Safe Drinking Water Act

Linda Voss, *U.S. Centennial of Flight Commission,* Ballooning and Meteorology in the Twentieth Century

U.S. Department of Labor, Bureau of Labor Statistics, Atmospheric Scientists

U.S. Environmental Protection Agency—Oceans and Coastal Protection Division, Coral Reefs and Your Coastal Watershed

U.S. Geological Survey, Water Quality

U.S. Global Change Research Program, The Global Water Cycle

Guy Webster, *Jet Propulsion Laboratory, Pasadena, California,* Mars Exploration Rover Mission

Radosław P. Wesołowski, *N. Copernicus University, Toruń, Poland,* Mechanisms of Water Adsorption on Carbons

Scott Whiteford, *Michigan State University, East Lansing, Michigan,* Transboundary Waters in Latin America: Conflicts and Collaboration

Patrick Willems, *Hydraulics Laboratory, Leuven, Belgium,* Uncertainties in Rainfall–Runoff Modeling

Richard S. Williams, Jr., *U.S. Geological Survey, Woods Hole, Massachusetts,* Sea Level and Climate

Marek Wiśniewski, *N. Copernicus University, Toruń, Poland,* Mechanisms of Water Adsorption on Carbons

Ming Hung Wong, *Hong Kong Baptist University, Kowloon Tong, Kowloon, Hong Kong,* Larvae and Small Species of Polychaetes in Marine Toxicological Testing

Eve Woods, *Denver, Colorado,* Reserved Water Rights for Indian and Federal Lands

P.L. Woodworth, *Proudman Oceanographic Laboratory, Birkenhead, United Kingdom,* The Permanent Service for Mean Sea Level

Tom Wydrzynski, *The Australian National University, Canberra, Australia,* Isotope Exchange in Gas-Water Reactions

Carl L. Yaws, *Lamar University, Beaumont, Texas,* Adsorption Capacity of Activated Carbon for Water Purification, Solubility of Hydrocarbons and Sulfur Compounds in Water, Solubility of Chemicals in Water, Solubility of Hydrocarbons in Salt Water

Brian Yocis, *Bigelow Laboratory for Ocean Sciences,* NOAA and University Scientists Study Methyl Bromide Cycling in the North Pacific

David W. Yoskowitz, *Texas A&M University—Corpus Christi, Laredo, Texas,* U.S./Canadian Boundary Waters Treaty and the Great Lakes Water Quality Agreement

Mark A. Young, *University of Iowa, Iowa City, Iowa,* Environmental Photochemistry in Surface Waters

Shari Yvon-Lewis, *Atlantic Oceanographic and Meteorological Laboratory,* NOAA and University Scientists Study Methyl Bromide Cycling in the North Pacific

Karl Erik Zachariassen, *Norwegian University of Science and Technology, Trondheim, Norway,* Freezing and Supercooling of Water, Osmosis-Diffusion of Solvent or Caused by Diffusion of Solutes?

A. Zaharuddin, *University of Malaya, Petaling Jaya Selangor, Malaysia,* Islamic Water Law

OCEANOGRAPHY

AIR–SEA INTERACTION

S.A. Hsu
Louisiana State University
Baton Rouge, Louisiana

INTRODUCTION

Air–sea interaction is, according to Geer (1), the interchange of energy (e.g., heat and kinetic energy) and mass (e.g., moisture and particles) that takes place across the active surface interface between the top layer of the ocean and the layer of air in contact with it and vice versa. The fluxes of momentum, heat, moisture, gas, and particulate matter at the air–water interface play important roles, for example, in environmental hydraulics and water–environment–health interactions, during low wind speeds before the onset of wave breaking, the exchange of air bubbles is limited. If this situation persists for a long time, algal blooms may develop, ultimately affecting water quality. On the other hand, during typhoon/hurricane conditions, the storm surge may affect the sewerage outflow at a greater depth than normal because of shoaling. In the area on the right-hand side of the storm track (in the Northern Hemisphere), runoff may also be blocked due to the surge, thus increasing the flood potential and saltwater intrusion.

Air–sea interaction encompasses vast scales in both spatial and temporal viewpoints, so only a few basic and applied topics are summarized here, including the parameterization of stability length, determination of friction velocity, wind–wave interaction, and the estimation of shoaling depth during storms. For more detailed laws and mechanisms in air–sea interaction, see Donelan (2) and recently Csanady (3); for air–sea exchange of gases and particles, see Liss (4) and most recently Donelan et al. (5); for the role of air–sea exchange in geochemical cycling, see Buat-Menard (6) and recently Liss (7); for larger scale air–sea interaction by La Niña and its impacts, see Glantz (8); for more physics, chemistry, and dynamics related to air–sea exchange, see Geernaert (9); and for wind–wave interaction, see Janssen (10).

PARAMETERIZATION OF THE STABILITY LENGTH

In the atmospheric boundary layer, the buoyancy length scale, L, also known as the Obukhov (or Monin–Obukhov) length, is a fundamental parameter that characterizes the "stability" of the surface layer (11). L describes the relative importance between the buoyancy effect (or thermal turbulence) and wind shear (or mechanical turbulence). According to Hsu and Blanchard (12), L can be parameterized as follows.

For unstable conditions (i.e., when $T_{sea} > T_{air}$),

$$\frac{z}{L} = -\frac{1000(T_{sea} - T_{air})(1 + \frac{0.07}{B})}{(T_{air} + 273.2)U_z^2} \tag{1}$$

For stable conditions ($T_{air} > T_{sea}$),

$$\frac{z}{L} = \frac{620(T_{air} - T_{sea})}{(T_{air} + 273.2)U_z^2} \tag{2}$$

According to Hsu (13,14),

$$B = 0.146(T_{sea} - T_{air})^{0.49} \tag{3}$$

where z is the height normally set to 10 m; T_{air} and T_{sea} stand for the air and sea temperatures, respectively; U_z is the wind speed at height z; and B is the Bowen ratio.

For operational and engineering applications, $z/L \leq -0.4$ is unstable, $|z/L| < 0.4$ is neutral, and $z/L \geq 0.4$ is stable.

PARAMETERIZATION OF THE ROUGHNESS LENGTH

The roughness parameter Z_0 can be computed based on the formula provided in Taylor and Yelland (15) that

$$\frac{Z_0}{H_s} = 1200\left(\frac{H_s}{L_p}\right)^{4.5} \tag{4}$$

and, for deep water waves,

$$L_p = \frac{gT_p^2}{2\pi} \tag{5}$$

where g is gravitational acceleration, H_s and L_p are the significant wave height and peak wavelength for the combined sea and swell spectrum, and T_p is its corresponding wave period. Note that H_s is defined as the average of the highest one-third of all wave heights during the 20-minute sampling period.

ESTIMATION OF THE FRICTION VELOCITY

The friction velocity (u_*) can be obtained as follows:

$$u_* = U_{10}C_d^{1/2} \tag{6}$$

where U_{10} is the wind speed at 10 m and C_d is the drag coefficient.

According to Amorocho and DeVries (16), the WAMDI Group (17), and Hsu (18), one may classify the air–sea interaction into three broad categories based on wave breaking conditions:

In light winds, $U_{10} < 7.5$ m/s, prior to the onset of wave breakers,

$$C_d = \left(\frac{u_*}{U_{10}}\right)^2 = 1.2875 * 10^{-3} \tag{7}$$

Both thermal and mechanical turbulence are important.

In moderate winds, $7.5 \leq U_{10} \leq 20$ m/s, the range after the onset but before the saturation of wave breakers,

$$C_d = (0.8 + 0.065U_{10}) * 10^{-3} \tag{8}$$

Mechanical turbulence is more important than thermal effects.

In strong winds, $U_{10} > 20$ m/s, after the saturation of wave breakers,

$$C_d = 2.5 * 10^{-3} \tag{9}$$

Mechanical turbulence dominates.

ESTIMATING LATENT HEAT FLUX (OR EVAPORATION)

Using the parameter of the Bowen ratio supplied by Hsu (13,14), as shown in Eq. 3, the latent heat flux (H_{latent}) can be estimated as

$$H_{\text{latent}}(\text{W m}^{-2}) = \frac{1}{B}H_{\text{sensible}} = \frac{1}{B}\rho_a C_p C_T (T_{\text{sea}} - T_{\text{air}})U_{10} \tag{10}$$

where H_{sensible} is the sensible heat flux, ρ_a (=1.2 kg m^{-3}) is the air density, C_p (=1004 J kg^{-1} K^{-1}) is the specific heat at constant pressure for dry air, C_T ($=1.1 * 10^{-3}$) is the transfer coefficient for heat, $(T_{\text{sea}} - T_{\text{air}})$ is in K, and U_{10} in m s^{-1}.

A latent heat flux of 1 W m^{-2} is equivalent to an evaporation rate of $3.56 * 10^{-3}$ cm day^{-1}, so Eq. 10 can be used to estimate the evaporation rate.

ESTIMATING MAXIMUM SUSTAINED WIND SPEED DURING A HURRICANE

Under hurricane/typhoon conditions, intense air–sea interaction occurs. Beach erosion, engineering structures, storm surge, and sewerage outflow can all be affected, so this topic should deserve more attention than the deepwater environment. The very first subject related to a tropical cyclone is to estimate its maximum sustained wind speed at the standard height of 10 m (i.e., U_{10}). This is accomplished as follows.

From the cyclostrophic equation (i.e., centrifugal force = pressure gradient) (11),

$$\frac{U_a^2}{\gamma} = \frac{1}{\rho_a}\frac{\partial P}{\partial \gamma} = \frac{1}{\rho_a}\frac{\Delta P}{\Delta \gamma} = \frac{1}{\rho_a}\frac{P_n - P_0}{\gamma - 0} \tag{11}$$

where U_a is the maximum sustained wind speed above the surface boundary layer, γ is the radius of the hurricane, $\partial P/\partial \gamma$ is the radial pressure gradient P_n is the pressure outside the hurricane effect (1013 mb, the mean sea level pressure), and P_0 is the hurricane's minimum central pressure. Because $\rho_a = 1.2$ kg m^{-3}, $\Delta P = (1013 - P_0)$ mb, and 1 mb = 100 N m^{-2} = 100 kg m^{-1} s^{-2}, Eq. 11 becomes

$$U_a = \left[\frac{100\ \text{kg m}^{-1}\,\text{s}^{-2}}{1.2\ \text{kg m}^{-3}}\right]^{1/2} \sqrt{\Delta P} = 9\sqrt{\Delta P} \tag{12}$$

According to Powell (19), $U_{10} = 0.7U_a$; therefore

$$U_{10} = 6.3\sqrt{\Delta P} = 6.3(1013 - P_0)^{1/2} \tag{13}$$

where U_{10} is in m s^{-1} and ΔP in mb.

Equation 13 has been verified by Hsu (18). In 1985, during Hurricane Kate over the Gulf of Mexico, the U.S. National Data Buoy Center (NDBC) buoy #42003, located on the right-hand-side of the storm track near the radius of maximum wind, recorded a minimum sea-level pressure (P_0) of 957.1 mb. Therefore, $\Delta P = (1013 - 957.1) = 55.9$ mb. Substituting this value in Eq. 13, $U_{10} = 47.1$ m s^{-1} which is in excellent agreement with the measured $U_{10} = 47.3$ m s^{-1}.

Another verification is provided in Fig. 1. According to Anthes (20, p. 22 and Fig. 2.8),

$$U_{10\gamma} = U_{10\,\text{max}}\left(\frac{R}{\gamma}\right)^{0.5} \tag{14}$$

where $U_{10\gamma}$ is the sustained wind speed at a distance 10 m away from the storm center and $U_{10\,\text{max}}$ is the maximum sustained wind at 10 m at the radius of maximum wind, R.

According to Hsu et al. (21), for operational applications,

$$\frac{R}{\gamma} = \ln\left(\frac{1013 - P_0}{P_\gamma - P_0}\right) \tag{15}$$

where P_γ is the pressure at a point located at a distance from the storm center and ln is the natural logarithm.

Substituting Eq. 15 in Eq. 14,

$$U_{10\gamma} = U_{10\,\text{max}}\left[\ln\left(\frac{1013 - P_0}{P_\gamma - P_0}\right)\right]^{0.5} \tag{16}$$

During Hurricane Lili in 2002, the NDBC had two buoys, #42001 located near R, and #42003 located due east along 26 °N, approximately 280 km from 42001. The wind speed measurement at both buoys was 10 m. From the NDBC website (www.ndbc.noaa.gov/), at 20Z 2 October 2002 at #42001, $P_0 = 956.1$ mb. Substituting this P_0 in Equation 13, $U_{10\,\text{max}} = 47.5$ m s^{-1}, in excellent agreement with the measured value of 47.2 m s^{-1} (=106 mph). Therefore, Eq. 13 is further verified. At the same time, $P_\gamma = 1011.1$ mb was measured at #42003. Substituting this P_γ in Eq. 16, we obtain $U_{10\gamma} = 8.8$ m s^{-1}. The measured $U_{10\gamma}$ at 42003 was 9.2 m s^{-1}. The difference is only about 4%, so we conclude that Eqs. 13 and 16 can be used for nowcasting using the pressure measurements at P_γ and P_0 which are normally available via the official "Advisory" during a hurricane.

ESTIMATING MAXIMUM SIGNIFICANT WAVE HEIGHT DURING A HURRICANE

According to the USACE (22, p. 3–85, Eq. 3–64),

$$T_p = 12.1\sqrt{\frac{H_s}{g}} \tag{17}$$

$$\therefore \frac{H_s}{gT_p^2} = 0.0068 \tag{18}$$

More verification of Eq. 18 is provided in Hsu (23).

According to Hsu et al. (21), based on the evaluation of nine fetch-limited wind–wave interaction formulas, that provided by Donelan et al. (24) ranked best as follows:

$$\frac{gH_s}{U_{10}^2} = 0.00958\left(\frac{gT_p}{U_{10}}\right)^{1.65} \tag{19}$$

Figure 1. Satellite data (visible channel from NOAA-16) received and processed at the Earth Scan Lab, Louisiana State University, during Hurricane Lili (2002) in the Gulf of Mexico. The solid line represents the storm track. Data from NDBC buoys 42001 and 42003 are employed in this study. Note that the anemometers for both bouys were located at the standard 10 m height.

From Eqs. 18 and 19,

$$H_s = 0.00492\ U_{10}^2 \tag{20}$$

Substituting Eq. 13 in Eq. 20,

$$H_{s\ max} = 0.20\ \Delta P \tag{21}$$

where $H_{s\ max}$ is in meters and ΔP is in mb. Equation 21 is verified in Fig. 1. Buoy 42001, located near the radius of maximum wind, measured $P_0 = 956.1$ mb at 20Z 2 October 2002, so that $\Delta P = (1013 - 956.1) = 56.9$ mb. Substituting this value in Eq. 21, $H_{s\ max} = 11.38$ m, which is in excellent agreement with that of 11.22 m measured at 21Z 2 October 2002 (within 1 hour after the measured minimal P_0).

ESTIMATING STORM SURGE AND SHOALING DEPTH

To estimate a typhoon/hurricane—generated storm surge (ΔS), and shoaling depth (D_{shoaling}), the following formulas are useful operationally, provided that the storm's minimum (or central) pressure near the sea surface (P_0) is known.

According to Hsu (23), for the storm surge in deep water before shoaling (i.e., when the waves feel the sea floor),

$$S_I = 0.070(1010 - P_0) \tag{22}$$

where S_I is the initial peak storm surge before shoaling. For the peak surge at the coast,

$$S_P = 0.070\,(1010 - P_0) * F_S * F_M \tag{23}$$

where F_S is a shoaling factor dependent on shelf topography and width and F_M is a correction factor for storm motion. Both F_S and F_M for certain areas are included in Hsu (23). A verification of Eq. 23 during Hurricane Georges in 1998 is also available in Hsu (23).

The shoaling depth is computed as follows:

From Taylor and Yelland (15), $D_{shoaling} = 0.2\ L_p$, and from Eq. 5,

$$D_{shoaling} = 0.2\,\frac{gT_p^2}{2\pi} = \frac{0.2\,H_s}{2\pi\left(\dfrac{H_s}{gT_p^2}\right)}$$

where $H_s/gT_p{}^2$ is called wave steepness, a useful parameter in coastal engineering.

From Hsu et al. (21) and under hurricane conditions from Eq. 18, $H_s/gT_p{}^2 = 0.0068$. Thus, from Eq. 21,

$$D_{shoaling} = 4.7\,H_s = 4.7 * 0.2(1013 - P_0)$$

$$\therefore \text{shoaling depth (meters)} \approx (1013 - P_0) \qquad (24)$$

CONCLUDING REMARKS

Although all formulas presented in this article are based on the open literature, they may need some verification before being applied to site-specific conditions. For example, Eq. 22 for the storm surge is for an open coast before shoaling. It needs to be adjusted for flooding at the coast due to different storm speeds and local bathymetry, as needed in Eq. 23.

BIBLIOGRAPHY

1. Geer, I.W. (Ed.). (1996). *Glossary of Weather and Climate With Related Oceanic and Hydrologic Terms*. American Meteorological Society, Boston, MA.
2. Donelan, M.A. (1990). Air-sea interaction. In: *The Sea*. Wiley, Vol. 9. Hoboken, NJ.
3. Csanady, G.T. (2001). *Air-Sea Interaction: Laws and Mechanisms*. Cambridge University Press, Cambridge, UK.
4. Liss, P.S. (1983). *Air-Sea Exchange of Gases and Particles*. Kluwer Academic, Springer, New York.
5. Donelan, M.A., Drennan, W.M., Saltman, E.S., and Wanninkhof, R. (Eds.). (2002). *Gas Transfer at Water Surfaces. Geophysical Monograph* #127, American Geophysical Society, Washington, DC.
6. Buat-Menard, P. (1986). *Role of Air-Sea Exchange in Geochemical Cycling*. Kluwer Academic, Springer, New York.
7. Liss, P.S. (1997). *Sea Surface and Global Change*. Cambridge University Press, Cambridge, UK.
8. Glantz, M.H. (2002). *La Niña and Its Impacts: Facts and Speculation*. Brookings Inst. Press, Washington, DC.
9. Geernaert, G.L. (1999). *Air-Sea Exchange: Physics, Chemistry & Dynamics*. Kluwer Academic, Springer, New York.
10. Janssen, P. (2004). *The Interaction of Ocean Waves and Wind*. Cambridge University Press, Cambridge, UK.
11. Hsu, S.A. (1988). *Coastal Meteorology*. Academic Press, San Diego, CA.
12. Hsu, S.A. and Blanchard, B.W. (2004). On the estimation of overwater buoyancy length from routine measurements. *Environ. Fluid Mech.*, in press.
13. Hsu, S.A. (1998). A relationship between the Bowen ratio and sea-air temperature difference under unstable conditions at sea. *J. Phys. Oceanogr.* **28**: 2222–2226.
14. Hsu, S.A. (1999). On the estimation of overwater Bowen ratio from sea-air temperature difference. *J. Phys. Oceanogr.* **29**: 1372–1373.
15. Taylor, P.K. and Yelland, M.J. (2001). The dependence of sea surface roughness on the height and steepness of the waves. *J. Phys. Oceanogr.* **31**: 572–590.
16. Amorocho, J. and DeVries, J.J. (1980). A new evaluation of the wind stress coefficient over water surfaces. *J. Geophys. Res.* **85**(C1): 433–442.
17. The WAMDI Group. (1988). The WAM model—a third generation ocean wave prediction model. *J. Phys. Oceanogr.* **18**: 1775–1810.
18. Hsu, S.A. (2003). Estimating overwater friction velocity and exponent of the power-law wind profile from gust factor during storms. *J. Waterways Port Coastal Ocean Eng.* **129**(4): 174–177.
19. Powell, M.D. (1982). The transition of the Hurricane Frederic boundary-layer wind field from the open Gulf of Mexico to landfall. *Mon. Weather Rev.* **110**: 1912–1932.
20. Anthes, R.A. (1982). *Tropical Cyclones, Their Evolution, Structure, and Effects. Meteorological Monographs Number* 41, American Meteorological Society, Boston, MA.
21. Hsu, S.A., Martin, M.F., Jr., and Blanchard, B.W. (2000). An evaluation of the USACE's deepwater wave prediction techniques under hurricane conditions during Georges in 1998. *J. Coastal Res.* **16**(3): 823–829.
22. U.S. Army Corps of Engineers. (1984). *Shore Protection Manual*. Vicksburg, MS.
23. Hsu, S.A. (2004). A wind-wave interaction explanation for Jelesnianski's open-ocean storm surge estimation using Hurricane Georges (1998) measurements. *Natl. Weather Dig.*, in press.
24. Donelan, M.A., Hamilton, J., and Hui, W.H. (1985). Directional spectra of wind-generated waves. *Philoso. Transa. Roy. Soc. London, Ser. A* **315**: 509–562.

NOAA'S ATLANTIC OCEANOGRAPHIC AND METEOROLOGICAL LABORATORY

National Oceanographic and
Atmospheric Administration
(NOAA)

June 18, 1999—The Atlantic Oceanographic and Meteorological Laboratory (AOML) in Miami, Florida, is one of 12 environmental research laboratories that work on environmental issues for NOAA's Office of Oceanic and Atmospheric Research (OAR). OAR research advances NOAA's ability to predict weather, helps monitor and provides understanding of climate and global change, as well as improve coastal ocean health.

AOML's mission is to conduct a basic and applied research program in oceanography, tropical meteorology, atmospheric and oceanic chemistry, and acoustics. The programs seek to understand the physical characteristics and processes of the ocean and the atmosphere, both individually and as a coupled system.

The principal focus of these investigations is to provide knowledge that may ultimately lead to improved prediction and forecasting of severe storms, better use and management of marine resources, better understanding of the factors affecting both climate and environmental quality, and improved ocean and weather services for the nation.

Schematic Hurricane Structure

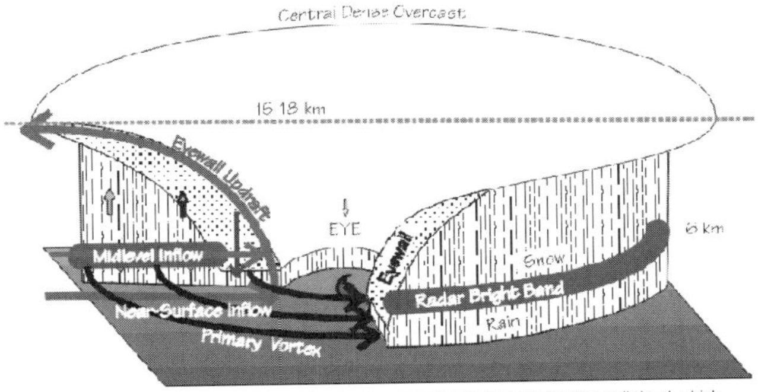

The inner core of a mature hurricane, showing the clear eye surrounded by the wall cloud, which is in turn surrounded by the overhanging Central Dense Overcast (CDO). The primary circulation is the air orbiting horizontally around an axis of rotation inside the eye. Superimposed upon the primary circulation is a secondary, in-up-out, circulation. The updrafts slope outward so that rain and snow or graupel fal from them.

Originally under the jurisdiction of the Environmental Science Services Administration (ESSA), the forerunner of NOAA, AOML was founded in Miami, Florida, in 1967. Several months after NOAA was established in 1970, groundbreaking began on a new 12-acre federally funded research facility on Virginia Key. AOML dedicated its new location on Feb. 9, 1973. It celebrated its 25th anniversary in 1998.

AOML has four main research divisions: Hurricane Research, Ocean Acoustics, Ocean Chemistry, and Physical Oceanography.

To learn more about AOML visit: http://www.aoml.noaa.gov/

HURRICANE RESEARCH DIVISION

The Hurricane Research Division (HRD) is NOAA's primary component for research on hurricanes. Its highest priority is improving the understanding and prediction of hurricane motion and intensity change. A key aspect of this work is the annual hurricane field program, supported by the NOAA Aircraft Operation's Center research/reconnaissance aircraft. Research teams analyze data from field programs, develop numerical hurricane models, conduct theoretical studies of hurricanes, prepare storm surge atlases, and study the tropical climate.

HRD works with the National Hurricane Center/Tropical Prediction Center in all phases of its research, the National Meteorological Center and the Geophysical Fluid Dynamics Laboratory—another of OAR's research labs—in research related to numerical modeling of hurricanes, and the National Severe Storms Laboratory—yet another OAR lab—in the study of landfalling hurricanes,

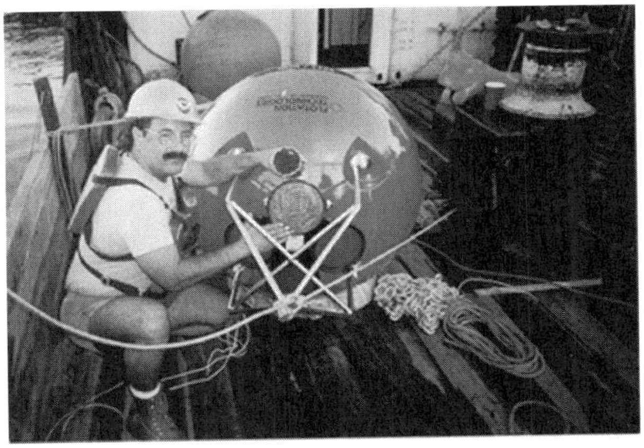

An AOML/OAD Oceanographer prepares a 150 kHZ upward looking Acoustic Doppler Current Profiler (ADCP) for deployment 3 miles east of the Port of Miami. The 4 red transducers are covered with a mixture of cayenne pepper and silicone grease to prevent biofouling. The ADCP is used to obtain water column current profiles and provides such information as current speed and direction, information used in this project to determine optimal conditions for disposal of dredge materials.

This image depicts the manual deployment of a global surface Lagrangian drifter (buoy). These devices measure sea surface temperature, surface currents, bathymetric pressure and wind. There are currently 756 of these instruments transmitting data back to AOML/PhOD from the oceanic regions all over the globe. The data re used for a variety of projects and programs.

as well as other NOAA groups, federal agencies, and universities in a variety of basic and applied research.

OCEAN ACOUSTICS DIVISION

The Ocean Acoustics Division (OAD) gathers, analyzes and reports coastal ocean data on human-related discharges and their potential environmental impacts. Additionally, OAD has an ongoing research program on the use of acoustics to measure coastal and deep ocean rainfall, an important element in calculating the global energy balance for climate monitoring and prediction. The Division works in cooperation with other federal, state, and local authorities to maximize research knowledge for use in economically and environmentally important projects in the coastal ocean.

AOML/OCD and RSMAS scientists install multi-sensor optical/acoustic moorings off the Florida Keys in support of an ERL funded Pilot Study investigating the biology, physics and chemistry of Gulf Stream topographic and frontal eddies. The moorings provide time series data within an Ocean Surface Current Radar (OSCR) field.

OCEAN CHEMISTRY DIVISION

With a diverse scientific staff of marine, atmospheric, and geological chemists, as well as chemical, biological, and geological oceanographers, the Ocean Chemistry Division (OCD) is able to use multidisciplinary approaches to solve scientific research questions. The Division's work includes projects that are important in assessing the current and future effects of human activities on our coastal to deep ocean and atmospheric environments.

PHYSICAL OCEANOGRAPHY DIVISION

The Physical Oceanography Division (PhOD) provides and interprets oceanographic data and conducts research relevant to decadal climate change and coastal ecosystems. This research includes the dynamics of the ocean, its interaction with the atmosphere, and its role in climate and climate change. Data is collected from scientific expeditions, both in the deep ocean and in coastal regions. Satellite data is processed and incorporated into the analyses. PhOD manages the Global Ocean Observing (GOOS) Center, which manages the global collection, processing, and distribution of drifting buoy data and the information collected from ocean temperature profilers. This information is crucial to understanding and predicting shifts in weather patterns and the relationship of the ocean and the atmosphere as a coupled system.

LABORATORY EXPERIMENTS ON BIVALVE EXCRETION RATES OF NUTRIENTS

Paolo Magni
IMC—International Marine Centre
Torregrande-Oristano, Italy

Shigeru Montani
Hokkaido University
Hakodate, Japan

BACKGROUND

Benthic nutrient regeneration may be referred to as a new availability to the water column of significant amounts of nitrogen, phosphorus, and other nutrients, as a consequence of the metabolism of organic matter by the benthos (1). The processes of benthic nutrient regeneration in coastal marine systems are strongly influenced by the presence of abundant macrofauna (2–5). Correct evaluation of the biogenic flux of nutrients due to the excretory activity of infaunal species is therefore an important background of information to investigate the cycling of biophilic elements (nitrogen, phosphorus, and silicon).

In field studies, major drawbacks include the difficulty to discern between nutrient upward flux due to animal excretion and a number of local effects, such as microbial mineralization (4,6–8) and uptake (9–14), animal bioturbation and irrigation currents (15–19), tidal currents and wind-generated waves (20–22). Laboratory experiments on the animal excretion rates of nutrients under more controlled conditions represent a useful tool for quantifying the actual biogenic contribution by macrofauna to the total upward flux of nutrients from sediments. Nevertheless, these experiments have often restricted their investigations to ammonium (23–28) or, in a few cases, to ammonium and phosphate (29).

CASE STUDY

This study was conducted within a multidisciplinary project on the cycling of nutrients and organic matter in a tidal estuary in the Seto Inland Sea (30–37). Laboratory experiments were carried out on the excretion rates of ammonium, phosphate, and silicate by different size classes of the bivalves, *Ruditapes philippinarum* and *Musculista senhousia*. These species were selected as they were dominant on a sandflat of the estuary under investigation. An extrapolation of these results to a field community is presented in Temporal Scaling of Benthic Nutrient Regeneration in Bivalve-Dominated Tidal Flat. Both studies will be the basis of a third companion paper on the relationship between the temporal scaling of bivalve nutrient excretion and the seasonal change of nutrient concentrations in the porewater (Seasonal Coupling Between Intertidal Macrofauna and Sediment Column Porewater Nutrient Concentrations).

In these experiments, 2.5 L aquaria with and without (control) animals were employed and run on two different occasions for 24 hours. Each experiment consisted of a 10 h day (light) and a 10 h night (dark) treatment in which the changes in nutrient concentrations were measured every 2 h. Between the two (light and dark) treatments, a low tide lasting 2 h (like that approximately on the flat where animals were collected) was created, during which the experimental animals were not removed from sediments to keep the experiment continuous. The experimental setup and procedure are detailed in our associated paper where the bivalve excretion rates of ammonium and phosphate have been presented and discussed (35). We will extend this study to silicate, a nutrient species whose regeneration through animal excretory activity has been less investigated, either *in situ* (38,39) or in the laboratory (5,40). Table 1 includes some characteristics of the experimental animals, as well as the field-relevant (33) amount of algal food (*Thalassiosira* sp.) offered in four spikes during each experiment (35). In all treatments, there was a marked increase in all three nutrient concentrations, in the control (no animals), the increase was much more limited (i.e., silicate) or not observed (i.e., ammonium) (Fig. 1).

Based on the differences in nutrient concentrations between treatments and controls, relevant linear regression lines of five to six measurements were used to calculate the nutrient excretion rates for each size class of *R. philippinarum* and *M. senhousia* (Table 2). This approach may be a more reliable way to quantify the daily bivalve excretion, whereas previous similar experiments have been based on shorter incubations and/or the sole difference between initial and final values of nutrient concentrations (4,7,22,28,29). The data sets were subjected to ANOVA in a two-factor randomized complete block design, using the day/night variable as factor A, the time (hour) variable as factor B, and the size classes of each bivalve species as replicates (35). As found for ammonium in *R. philippinarum*, but not in *M. senhousia*, silicate excretion was significantly higher (57%, $p < 0.001$, $n = 36$) during the day than during the night, suggesting a possible effect of light on the excretory activity of this bivalve species. A comparison of nutrient excretion rates ($\mu mol\ g^{-1}\ DW\ h^{-1}$) of bivalve species obtained through *in situ* or laboratory experiments is given in Table 3.

According to the excretion rates of silicate found in our laboratory experiments, this study points to the importance of the excretory activity of these bivalve species to the biogenic regeneration of silicate. This aspect

Table 1. Animals Employed in the Laboratory Experiments and Experimental Conditions. Ind.: Number of Individuals[a,b]

Species		Size, mm	Ind., n	TOT, mg	DW, mg	Temp, °C	Chl*a*, (μg L⁻1)	Expt, date
				Ruditapes philippinarum				
Size class I	AVG	9.4	12	197	9.9	19.6	26.3	May 1996
	SD	±1.4	–	±79	±4.0	±1.5	±8.7	–
Size class II	AVG	15.5	15	830	37.0	19.6	24.6	May 1996
	SD	±1.0	–	±174	±9.2	±1.5	±8.8	–
Size class III	AVG	18.9	9	1520	63.6	21.6	38.9	Sep 1996
	SD	±0.8	–	±149	±10.4	±0.3	±12.8	–
				Musculista senhousia				
Size class I	AVG	16.7	14	431	27.6	19.6	24.5	May 1996
	SD	±1.3	–	±117	±7.7	±1.5	±9.1	–
Size class II	AVG	23.5	8	1264	52.4	21.6	47.4	Sep 1996
	SD	±1.7	–	±172	±8.0	±0.3	±22.8	–

[a] Reproduced from Reference 35.

[b] TOT: mean (live) weight for each size class of the experimental animals; DW: mean dry soft-body weight for each size class of the experimental animals; Temp: experimental temperature; Chl *a* (Chlorophyll *a*) is the mean (AVG) ± standard deviation (SD) of four spikes of cultures of *Thalassiosira* sp. (Chl $a = 0.01 \times$ *Thalassiosira*sp. cell + 3.6, $r^2 = 0.908$; $p < 0.001$, $n = 40$) for each day/night treatment

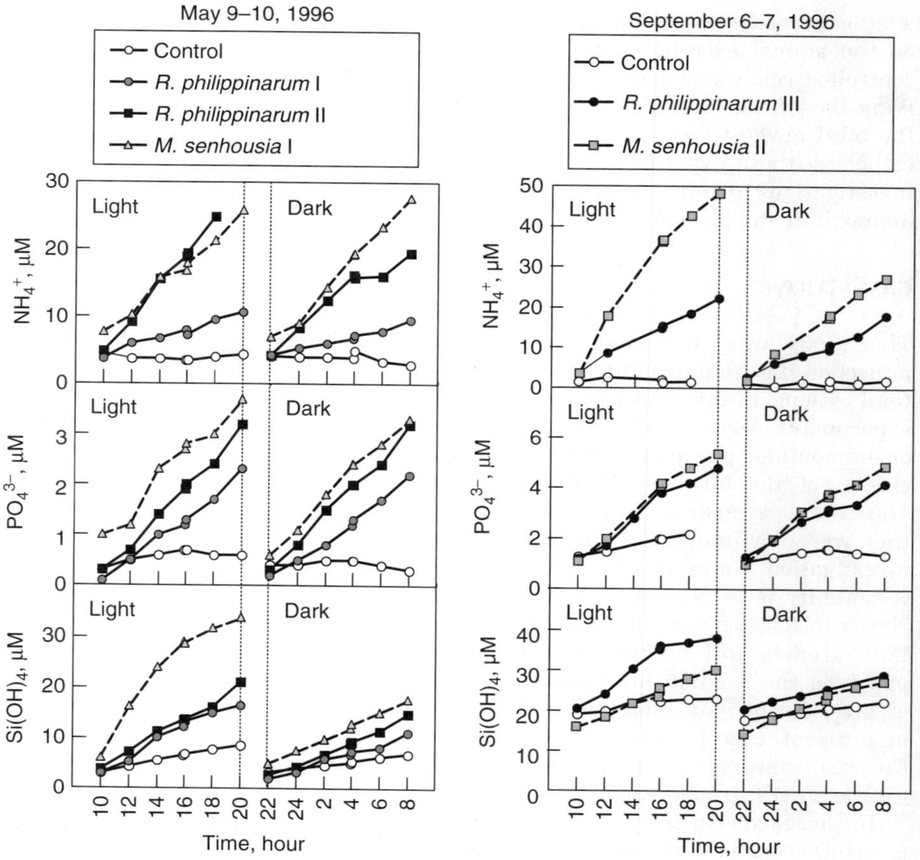

Figure 1. Laboratory experiments: changes of nutrient concentrations [ammonium, NH_4^+–N); phosphate, PO_4^{3-}–P; and silicate $Si(OH)_4$–Si] during the day (light) and night (dark) treatments. Vertical dashed lines: left line (time 20:00) indicates the end of the day (light) treatment; right line (time 22:00) indicates the start of the night (dark) treatment; between lines: low tide between treatments (based on Ref. 35).

Table 2. Nutrient Excretion Rate for Each Size class of *Ruditapes philippinarum* and *Musculista senhousia* During day/night Treatments (Experimental Temperature as in Table 1)[a]

	Nutrient Excretion Rate, μmol g^{-1}DW h^{-1}					
	NH_4^+		PO_4^{3-}		$Si(OH)_4$	
Species	Light	Dark	Light	Dark	Light	Dark
Ruditapes philippinarum						
Size class I	10.6	7.9	3.4	3.9	15.8	10.9
Size class II	9.6	5.8	1.0	1.1	4.1	3.3
Size class III	5.0	3.8	0.9	0.7	4.0	1.1
Light/dark mean	8.4[b]	5.8	1.8	1.9	8.0[b]	5.1
Total mean	7.1		1.9		6.6	
Musculista senhousia						
Size class I	9.3	11.4	1.2	1.5	14.5	4.8
Size class II	16.9	9.7	1.6	1.3	4.2	5.5
Light/dark mean	13.1	10.6	1.4	1.4	9.4	5.2
Total mean	11.8		1.4		7.3	

[a]Based on Reference 35.
[b]Mean day (light) excretion significantly higher (ANOVA $p < 0.001$) than night (dark) excretion (based on Reference 35).

has been more controversial than the contribution of bivalves to the regeneration of ammonium and phosphate. The extent of silicate excretion also varies considerably, depending on the bivalve species investigated and the environmental/experimental conditions employed.

In *in situ* experiments, Prins & Small (4) found no significant excretion of silicate by *Mytilus edulis* beds on an intertidal zone of the Westerschelde (The Netherlands). The occurrence of silicate fluxes was attributed to the possible increased rate of dissolution of silicate at higher temperature. Asmus et al. (38) in the eastern Wadden Sea (Germany) and Dame et al. (39) in the Easterschelde and the western Wadden Sea (The Netherlands) found high fluxes of silicate from mussel beds of *M. edulis*. Although, in both studies, the actual excretion rate of silicate by *M. edulis* was not estimated per biomass unit (e.g., μmol g^{-1} DW h^{-1}), Dame et al. (39) suggested that silicate release from mussel beds results from phytoplankton cells breaking down as they are metabolized by the mussels. Dame et al. (39) argued that the longer turnover time for silicate, compared to phosphate and ammonium, implies a lesser role for the mussel beds in recycling this nutrient species in the two estuaries under investigation.

In contrast, Asmus et al. (38) found rapid silicate release in the Sylt-flume study and suggested that the mussels are an accelerator in recycling biogenic silica. Similarly, a study on the nutrient excretion of *R. philippinarum* in core incubation experiments found that silicate regeneration was, on average, 9.2 times faster in the site farmed with clams (5). In mesocosm experiments using large tanks, Doering et al. (40) also found that the level of flux was elevated in the presence of another clam, *Mercenaria mercenaria*, by 86% and 57% for silicate and ammonium, respectively. Our results indicate

Table 3. Comparison of Nutrient Excretion Rates (μmol g^{-1} DW $hour^{-1}$) for Different Species of Mussels (m), Clams (c) and Oysters (o)[a]

Species and Study Area	Method[b]	NH_4^+	PO_4^{3-}	$Si(OH)_4$	T, °C	Reference
Mytilus edulis (m)						
Narragansett Bay, USA	Lab	3.1	nd[c]	nd[c]	15	41
Linher River, U.K.	Lab	4.9–34.6	nd	nd	11–21	23
Sound, DK	*In situ*	0.14–3.1	0.10–0.53	nd	0.7–18	42
Western Scheldt, NL	*In situ*	1.1	nd	nd	12	43
Musculista senhousia (m)						
Seto Inland Sea, JPN	Lab	9.3–16.9	1.2–1.6	–	18–22	35
Seto Inland Sea, JPN	Lab	9.3–16.9	1.2–1.6	4.2–14.5	18–22	This study
Modiolus demissus (m)						
Narragansett Bay, USA	Lab	3.58 ± 1.73	nd	nd	21	41
Great Sippewissett, USA	Lab	2.5	nd	nd	annual	2
Donax serra (m)						
Maitland River, S. Africa	*In situ*	0.35–8.1	nd	nd		44
Sundays River, S. Africa	Lab/*In situ*	2.2	nd	nd	15–25	26
Donax sordidus (m)						
Sundays River, S. Africa	Lab/*In situ*	2.9	nd	nd	15–25	26
Aspatharia wahlbergi (m)						
Lake Kariba, Zimbabwe	Lab	6.1	0.48	nd	25.2	29
Corbicula africana (c)						
Lake Kariba, Zimbabwe	Lab	12.9	nd	nd	25.2	29
Corbicula japonica (c)						
Lake Shinji, JPN	Lab	14.3	nd	nd	27	28
Mercenaria mercenaria (c)						
Delaware Bay, USA	Lab	0.9–1.5	nd	nd	20	45
Macoma balthica (c)						
Wadden Sea, DK	Lab	0.1[d]	nd	nd	13–15	16
Ruditapes philippinarum (c)						
Virgin Islands, USA	Lab	1.9–4.9	nd	nd	20.1	24
Moss Landing, USA	Lab	1–2.3	nd	nd	27.1	46
Moss Landing, USA	Lab	0.6–0.9	nd	nd	12, 15, 18	47
Marennes-Oléron, F	Lab	0.5–13	nd	nd	5–25	25
Hatchery, Ireland	Lab	0.16–1	nd	nd	18.8	27
Seto Inland Sea, JPN	Lab	3.8–10.6	0.7–3.9	—	18–22	35
Seto Inland Sea, JPN	Lab	3.8–10.6	0.7–3.9	1.1–15.8	18–22	This study
Crassostrea virginica (o)						
Delaware Bay, USA	Lab	0.5–0.9	nd	nd	20.1	24
Crassostrea gigas (o)						
North Brittany, F	*In situ*	0.28–6.6	nd	nd	27.1	48

[a]Based on Reference 35.
[b]Lab: laboratory experiments.
[c]nd: not determined.
[d]Excretion rate calculated as a wet soft-body weight.

well-balanced stoichiometric ratios among the nutrient species excreted by the bivalves (35), which can also be related to the use of the diatom *Thalassiosira* sp. as a food (35). This was aimed to approximate the actual field situation on the tidal flat, where abundant microalgal biomass, including resuspended benthic diatoms, is available to filter-feeders such as in *R. philippinarum* and *M. senhousia* (31,33). The high excretion rates of silicate of these two dominant bivalve species found in our laboratory experiments, together with those of ammonium and phosphate, suggest a major contribution of bivalve nutrient excretion to the upward flux of nutrients from sediments in our relevant study area. This will be the subject of the subsequent paper where we will apply these rates to the actual bivalve standing stock found in the field (Temporal Scaling of Benthic Nutrient Regeneration in Bivalve-Dominated Tidal Flat).

BIBLIOGRAPHY

1. Nixon, S.W., Oviatt, C.A., and Hale, S.S. (1976). *Nitrogen Regeneration and the Metabolism of Coastal Marine Bottom Communities. The Role of Terrestrial and Aquatic Organisms in Decomposition Process*. J. Anderson and A. MacFayden (Eds.). Blackwell, Oxford, pp. 269–283.
2. Jordan, T.E. and Valiela, I. (1982). A nitrogen budget of the ribbed mussel, *Geukensia demissa* and its significance in nitrogen flow in a New England salt marsh. *Limnol. Oceanogr.* **27**: 75–90.
3. Murphy, R.C., and Kremer, J.N. (1985). Bivalve contribution to benthic metabolism in a California lagoon. *Estuaries* **8**: 330–341.
4. Prins, T.C. and Smaal, A.C. (1994). The role of the blue mussel *Mytilus edulis* in the cycling of nutrients in Oosterschelde estuary (the Netherlands). *Hydrobiol.* **282/283**: 413–429.

5. Bartoli, M. et al. (2001). Impact of *Tapes philippinarum* on nutrient dynamics and benthic respiration in the Sacca di Goro. *Hydrobiol.* **455**: 203–212.
6. Falcao, M. and Vale, C. (1990). Study of the Rio Formosa ecosystem: benthic nutrient remineralization and tidal variability of nutrients in the water. *Hydrobiol.* **207**: 137–146.
7. Gardner, W.S., Briones, E.E., Kaegi, E.C., and Rowe, G.T. (1993). Ammonium excretion by benthic invertebrates and sediment–water nitrogen flux in the Gulf of Mexico near the Mississipi river outflow. *Estuaries* **16**: 799–808.
8. Smaal, A.C. and Zurburg, W. (1997). The uptake and release of suspended and dissolved material by oysters and mussels in Marennes-Oléron Bay. *Aquat. Living Resour.* **10**: 23–30.
9. Rizzo, W.M. (1990). Nutrient exchanges between the water column and a subtidal benthic microalgal community. *Estuaries* **13**: 219–226.
10. Sundbäck, K., Enoksson, V., Granéli, W., and Pettersson, K. (1991). Influence of sublittoral microphytobenthos on the oxygen and nutrient flux between sediment and water: a laboratory continuous-flow study. *Mar. Ecol. Prog. Ser.* **74**: 263–279.
11. van Duyl, F.C., van Raaphorst, W., and Kop, A.J. (1993). Benthic bacterial production and nutrient sediment-water exchange in sandy North Sea sediments. *Mar. Ecol. Prog. Ser.* **100**: 85–95.
12. Risgaard-Petersen, N., Rysgaard, S., Nielsen, L.P., and Revsbech, N.P. (1994). Diurnal variation of denitrification and nitrification in sediments colonized by benthic microphytes. *Limnol. Oceanogr.* **39**: 573–579.
13. Feuillet-Girard, M., Gouleau, D., Blanchard, G., and Joassard, L. (1997). Nutrient fluxes on an intertidal mudflat in Marennes-Oléron Bay, and influence of the emersion period. *Aquat. Living Resour.* **10**: 49–58.
14. Trimmer, M., Nedwell, D.B., Sivyer, D.B., and Malcolm, S.J. (1998). Nitrogen fluxes through the lower estuary of the river Great Ouse, England: the role of the bottom sediments. *Mar. Ecol. Prog. Ser.* **163**: 109–124.
15. Callender, E. and Hammond, D.E. (1982). Nutrient exchange across the sediment–water interface in the Potomac river estuary. *Estuar. Coast. Shelf Sci.* **15**: 395–413.
16. Henriksen, K., Rasmussen, M.B., and Jensen, A. (1983). Effect of bioturbation on microbial nitrogen transformations in the sediment and fluxes of ammonium and nitrate to the overlaying water. *Environ. Biogeochem.* **35**: 193–205.
17. Yamada, H. and Kayama, M. (1987). Liberation of nitrogenous compounds from bottom sediments and effect of bioturbation by small bivalve, *Theora lata* (Hinds). *Estuar. Coast. Shelf Sci.* **24**: 539–555.
18. Aller, R.C. (1988). Benthic fauna and biogeochemical processes: the role of borrow structures. In: *Nitrogen Cycling in Coastal Marine Environments*. T.H. Blackburn and J. Sørensen (Eds.). Wiley, pp. 301–338.
19. Luther, I.II. et al. (1998). Simultaneous measurements of O_2, Mn, Fe, I^-, and S(-II) in marine pore waters with a solid-state voltammetric microelectrode. *Limnol. Oceanogr.* **43**: 325–333.
20. Nakamura, Y. (1994). Effect of flow velocity on phosphate release from sediment. *Water Sci. Technol.* **30**: 263–272.
21. Miller-Way, T. and Twilley, T.T. (1996). Theory and operation of continuous flow systems for the study of benthic–pelagic coupling. *Mar. Ecol. Prog. Ser.* **140**: 257–269.
22. Asmus, R.M. et al. (1998). The role of water movement and spatial scaling for measurements of dissolved inorganic nitrogen fluxes in intertidal sediments. *Estuar. Coast. Shelf Sci.* **46**: 221–232.
23. Bayne, B.L. and Scullard, C. (1977). Rates of nitrogen excretion by species of *Mytilus* (Bivalvia: Mollusca). *J. Mar. Biol. Ass. U.K.* **57**: 355–369.
24. Langton, R.W., Haines, K.C., and Lyon, R.E. (1977). Ammonia-nitrogen production by the bivalve mollusc *Tapes japonica* and its recovery by the red seaweed *Hypnea musciformis* in a tropical mariculture system. *Helgol. Wiss Meeresunters* **30**: 217–229.
25. Goulletquer, P. et al. (1989). Ecophysiologie et bilan énergétique de la palourde japonaise d' élevage *Ruditapes philippinarum*. *J. Exp. Mar. Biol. Ecol.* **132**: 85–108.
26. Cockcroft, A.C. (1990). Nitrogen excretion by the surf zone bivalves *Donax serra* and *D. sordidus*. *Mar. Ecol. Prog. Ser.* **60**: 57–65.
27. Xie, Q. and Burnell, G.M. (1995). The effect of activity on the physiological rates of two clam species, *Tapes philippinarum* (Adams & Reeve) and *Tapes decussatus* (Linnaeus). *Biol. Environ. Proc. Royal Irish Acad.* **95B**: 217–223.
28. Nakamura, M., Yamamuro, M., Ishikawa, M., and Nishimura, H. (1988). Role of the bivalve *Corbicula japonica* in the nitrogen cycle in a mesohaline lagoon. *Mar. Biol.* **99**: 369–374.
29. Kiibus, M., and Kautsky, N. (1996). Respiration, nutrient excretion and filtratin rate of tropical freshwater mussels and their contribution to production and energy flow in Lake Kariba, Zimbabwe. *Hydrobiol.* **331**: 25–32.
30. Magni, P. (2000). Distribution of organic matter in a tidal estuary of the Seto Inland Sea, Japan, and its relationship with the macrozoobenthic communities. In: *Ad hoc Benthic Indicator Group—Results of Initial Planning Meeting*. IOC Technical Series No. 57, UNESCO, pp. 20–25.
31. Magni, P. and Montani, S. (1997). Development of benthic microalgal assemblages on an intertidal flat in the Seto Inland Sea, Japan: effects of environmental variability. *La mer* **35**: 137–148.
32. Magni, P. and Montani, S. (1998). Responses of intertidal and subtidal communities of the macrobenthos to organic load and oxygen depletion in the Seto Inland, Japan. *J. Rech. Océanogr.* **23**: 47–56.
33. Magni, P. and Montani, S. (2000). Physical and chemical variability in the lower intertidal zone of an estuary in the Seto Inland Sea, Japan: seasonal patterns of dissolved and particulate compounds. *Hydrobiol.* **432**: 9–23.
34. Magni, P., Abe, N., and Montani, S. (2000). Quantification of microphytobenthos biomass in intertidal sediments: layer-dependent variation of chlorophyll *a* content determined by spectrophotometric and HPLC methods. *La mer* **38**: 57–63.
35. Magni, P., Montani, S., Takada, C., and Tsutsumi, H. (2000). Temporal scaling and relevance of bivalve nutrient excretion on a tidal flat of the Seto Inland Sea, Japan. *Mar. Ecol. Progr. Ser.* **198**: 139–155.
36. Magni, P., Montani, S., and Tada, K. (2002). Semidiurnal dynamics of salinity, nutrients and suspended particulate matter in an estuary in the Seto Inland Sea, Japan, during a spring tide cycle. *J. Oceanogr.* **58**: 389–402.
37. Montani, S. et al. (1998). The effect of a tidal cycle on the dynamics of nutrients in a tidal estuary in the Seto Inland Sea, Japan. *J. Oceanogr.* **54**: 65–76.
38. Asmus, H., Asmus, R.M., and Reise, K. (1990). Exchange processes in an intertidal mussel bed: A Sylt-flume study in the Wadden Sea. *Berichte Biolog. Anstalt Helgol.* **6**: 78.
39. Dame, R.F. et al. (1991). The influence of mussel beds on nutrients in the western Wadden Sea and Eastern Scheldt estuaries. *Estuaries* **14**: 130–138.

40. Doering, P.H., Kelly, J.R., Oviatt, C.A., and Sowers, T. (1987). Effect of the hard clam *Mercenaria mercenaria* on benthic fluxes of inorganic nutrients and gases. *Mar. Biol.* **94**: 377–383.
41. Nixon, S.W., Oviatt, C.A., Garber, J., and Lee, V. (1976). Diel metabolism and nutrient dynamics in a salt marsh embayment. *Ecology* **57**: 740–750.
42. Schlüter, L. and Josefsen, S.B. (1994). Annual variation in condition, respiration and remineralisation of *Mytilus edulis* L. in the Sound, Denmark. *Helgol. Meeres.* **48**: 419–430.
43. Smaal, A.C., Vonck, A.P.M.A., and Bakker, M. (1997). Seasonal variation in physiological energetics of *Mytilus edulis* and *Cerastoderma edule* of different size classes. *J. Mar. Biol. Ass. U.K.* **77**: 817–838.
44. Prosh, R.M. and McLachlan, A. (1984). The regeneration of surf-zone nutrients by the sand mussel, *Donax Serra* Röding. *J. Exp. Mar. Biol. Ecol.* **80**: 221–233.
45. Snra, R.F. and Baggaley, A. (1976). Rate of excretion of ammonia by the hard clam *Mercenaria mercenaria* and the American oyster *Crassostrea virginica*. *Mar. Biol.* **36**: 251–258.
46. Mann, R. and Glomb, S.J. (1978). The effect of temperature on growth and ammonia excretion of the Manila clam *Tapes japonica*. *Estuar. Coast. Shelf Sci.* **6**: 335–339.
47. Mann, R. (1979). The effect of temperature on growth, physiology, and gametogenesis in the Manila clam *Tapes philippinarum* (Adams & Reeve, 1850). *J. Exp. Mar. Biol. Ecol.* **38**: 121–133.
48. Boucher, G. and Boucher-Rodoni, R. (1988). In situ measurements of respiratory metabolism and nitrogen fluxes at the interface of oyster beds. *Mar. Ecol. Prog. Ser.* **44**: 229–238.

TEMPORAL SCALING OF BENTHIC NUTRIENT REGENERATION IN BIVALVE-DOMINATED TIDAL FLAT

Paolo Magni
IMC—International Marine Centre
Torregrande-Oristano, Italy

Shigeru Montani
Hokkaido University
Hakodate, Japan

Beside light and temperature (1–5), nutrients such as ammonium ($NH_4{}^+$-N), phosphate ($PO_4{}^{3-}$-P), and silicate [$Si(OH)_4$-Si] are a key factor in controlling the growth, abundance, and structure of primary producers in the ocean (6,7). Hence, it is important to investigate the availability, sources, and distribution of these nutrient species, as well as their spatial and temporal scaling. Biological processes strongly influence nutrient regeneration in different marine systems. In the open ocean, an important portion of reduced N-forms (e.g., $NH_4{}^+$-N), for instance, is made available *in situ* from waste products of plankton metabolism (8,9) and supports the so-called "regenerated" primary production (10). In coastal marine ecosystems, benthic nutrient regeneration is a major driving force in cycling biophilic elements (e.g., N, P, and Si) (11–13) and abundant macrofauna, for example, dense assemblages of bivalves, it has been shown, play a major role in these processes (14–16).

BACKGROUND

The contribution of benthic macrofauna to the total upward flux of nutrients has been investigated (mostly for ammonium and, to a lesser extent, for phosphate) in many coastal and estuarine areas using several approaches. They include laboratory and mesocosm experiments (17–19). *In situ* benthic chambers and sediment core incubations (20–22), and open flow/tunnel systems (16,23–26). Measurements of macrofauna-influenced nutrient flux, however, often have temporal limitations as they are based on one or relatively few sampling occasions, and thus seasonal patterns are in most cases not known. In this article, we evaluate the magnitude and temporal scaling of biogenic flux of nutrients from intertidal sediments densely populated by bivalves, based on extrapolating nutrient excretion rates of dominant bivalves to a field community. In particular, we show that the seasonal pattern of nutrient fluxes can be strongly influenced by the animal standing stock and its temporal distribution. This is beside the effect and importance of variation in excretion rate due to animal physiological factors, such as seasonal cycles of gametogenesis, storage and use of body reserves, and water temperature (27,28). The nutrient species considered in this study include ammonium, phosphate, and silicate, for which we quantified in associated laboratory experiments Laboratory Experiments On Bivalve Excretion Rates of Nutrients the excretion rates of different size classes of two dominant bivalve species. The relevance of macrofaunal excretion in regenerating the inorganic forms of three major bioelements such as N, P, and Si is discussed.

MACROFAUNAL COMMUNITIES

We present here the macrofaunal composition and distribution at an individual station (Stn B5) of a transect line selected in a sandflat of the Seto Inland Sea of Japan (29). At this station, the porewater nutrient concentrations (ammonium, phosphate, and silicate) in the uppermost 10 cm of sediments were also investigated in parallel from January 1995 to April 1996. They will be the subject of a subsequent associated paper focusing on the relationship between the seasonal pattern of bivalve nutrient excretion, described here, and the seasonal variation of porewater nutrient concentrations Seasonal Coupling Between Intertidal Macrofauna and Sediment Column Porewater Nutrient Concentrations. The total density and biomass of macrofauna varied from 7,400 (July 1995) to 22,050 ind.m^{-2} (October 1995), and from 70.9 (July 1995) to 244 g DW m^{-2} (September 7, 1995), respectively (Fig. 1). The bivalves *Ruditapes philippinarum* and *Musculista senhousia* and the polychaetes *Ceratonereis erithraeensis* and *Cirriformia tentaculata* were dominant; they accounted for 60.5% and 94.7% of the total density and biomass, respectively. Remarkably, *R. philippinarum* and *M. senhousia* alone accounted for up to 83.3 ± 6.7% of the total biomass when this exceeded

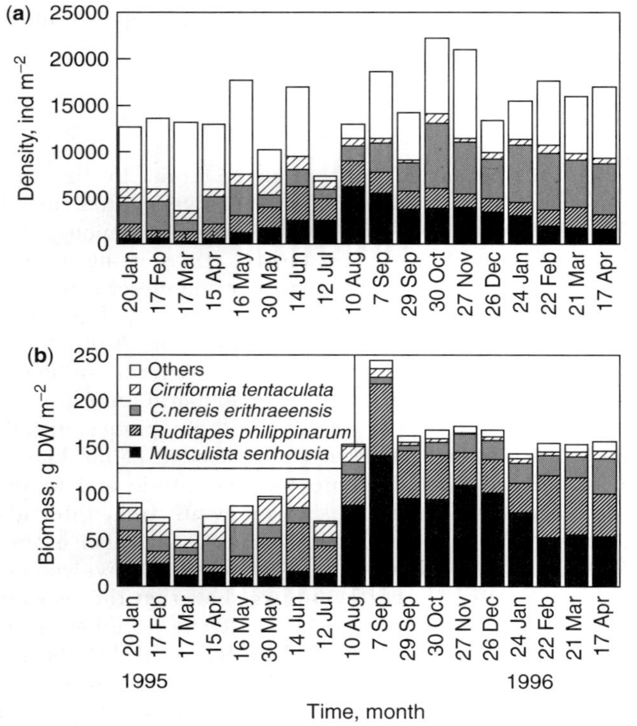

Figure 1. Seasonal variation of density (**a**) and biomass (**b**) of dominant macrozoobenthic species at Stn B5 (29). Note that in May 1995 and September 1995, sampling was carried out fortnightly.

120 g DW m^{-2}, from August 1995 till the end of the investigations. The high values of total macrofaunal and bivalve biomass may be regarded as a typical feature of many estuarine and intertidal areas, which are amongst the most productive systems in the ocean (30). In addition, biomass was markedly lower during the first half of the year (January 1995 to July 1995) than between late summer and winter; yet values progressively increased from early spring (March 1995) to early summer (June 1995). These marked temporal changes of macrofaunal communities reflect the high variability of these ecosystems. To estimate the magnitude and temporal scaling of biogenic nutrient excretion, we used an indirect approach. The mean excretion rates of ammonium, phosphate, and silicate for the two dominant bivalves *R. philippinarum* and *M. senhousia*, which were obtained in laboratory experiments Laboratory Experiments On Bivalve Excretion Rates of Nutrients, were applied to the relevant monthly biomass values found in the field. The ammonium and phosphate excretion rates of each bivalve species and their scaling to a field community have been extensively reported in Magni et al. (29). In this article, we applied these excretion rates to the bivalve biomass found at Stn B5 and extended this scaling to silicate, whose size-class dependent excretion rates are presented in Laboratory Experiments On Bivalve Excretion Rates of Nutrients. For silicate, we adopted the same temperature-dependent excretion rate factors as those used for ammonium and phosphate (Table 1).

BIVALVE NUTRIENT EXCRETION

The highest excretion rates of nutrients were estimated in September 7, 1995, up to a total of 50.2, 7.5, and 34.1 mmol m^{-2} d^{-1} for ammonium, phosphate, and silicate, respectively (Fig. 2). This corresponded to the period of highest biomass of both *R. philippinarum* and *M. senhousia*, which also accounted for the highest bivalve percentage (91.8%) of the total macrofaunal biomass. The lowest excretion rates occurred in April 1995 for *R. philippinarum* (lowest biomass on the same occasion), in March 1995 for *M. senhousia* (lowest biomass on May 16, 1995), and in February 1995 as the sum of the two bivalve species excretion rates. These latter rates were 4.1, 0.64, and 2.9 mmol m^{-2} d^{-1} for ammonium, phosphate, and silicate, respectively. The upward flux rates of nutrients obtained through this extrapolation of laboratory experiments on bivalve nutrient excretion to a field community are comparable to the highest biogenic releases reported for dense assemblages of bivalves such as oyster reefs (34) and mussel beds (16,35). This study also points to the importance of bivalve excretion to the biogenic regeneration of silicate, as previously suggested by field measurements indicating evidence of increased levels of silicate flux in the

Table 1. Adopted Temperature-Dependent Excretion Rates of Ammonium (NH_4^+-N), Phosphate (PO_4^{3-}-P), and Silicate [$Si(OH)_4$-Si] for *Ruditapes philippinarum* and *Musculista senhousia*

Period		Temperature, °C			Excretion Rate, μmol g^{-1}DW h^{-1}							
Month		Station			*R. philippinarum*				*M. senhousia*			
		B5-B1	H1	Y3	f^a	NH_4^+	PO_4^{3-}	$Si(OH)_4$	f^a	NH_4^+	PO_4^{3-}	$Si(OH)_4$
Dec, Jan,	AVG	4.7	5.9	10.4	0.6	4.3	1.1	4.0	0.5	5.9	0.7	3.7
Feb	SD	±1.8	±1.8	±2.1	–	–	–	–	–	–	–	–
Nov, Mar	AVG	11.7	11.9	15.2	0.9	6.4	1.7	5.9	0.7	8.3	1.0	5.1
–	SD	±0.9	±0.2	±3.0	–	–	–	–	–	–	–	–
Apr, May,	AVG	19.6	19.9	21.7	1	7.1	1.9	6.6	1	11.8	1.4	7.3
Jun, Oct	SD	±2.8	±4.0	±3.6	–	–	–	–	–	–	–	–
Jul, Aug,	AVG	27.7	28.0	26.9	0.9	6.4	1.7	5.9	1.2	14.2	1.7	8.8
Sep	SD	±2.3	±3.1	±1.1	–	–	–	–	–	–	–	–

[a] A factor (f) of 1 is used for the mean of the excretion rates obtained in laboratory experiments (see Laboratory Experiments On Bivalve Excretion Rates of Nutrients and based on Ref. 29).

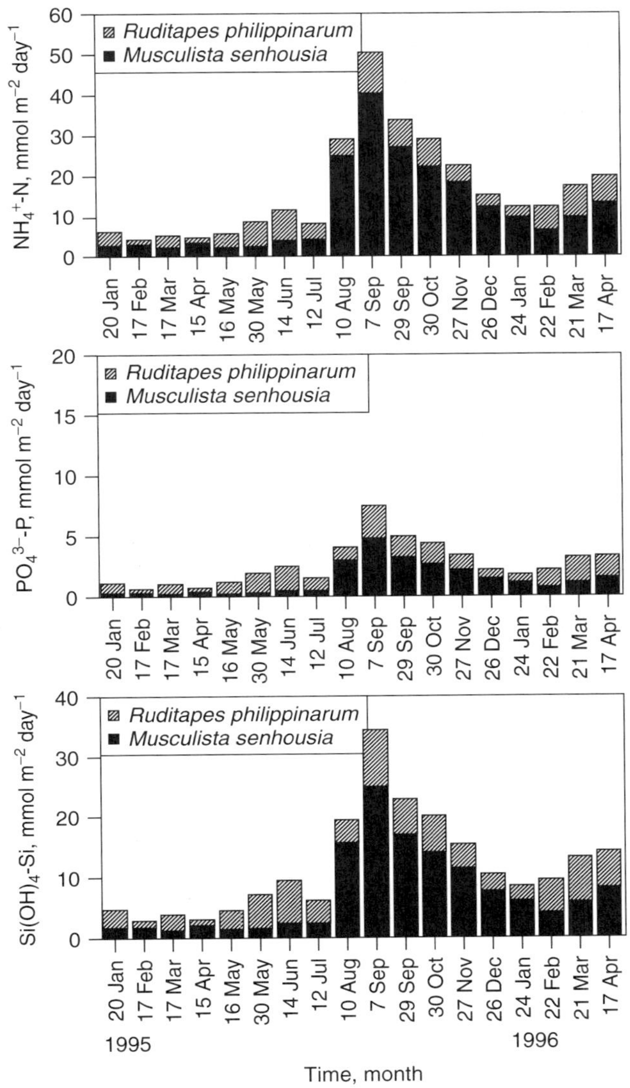

Figure 2. Magnitude and temporal scaling of ammonium (NH_4^+-N), phosphate (PO_4^{3-}-P), and silicate [$Si(OH)_4$-Si] excretion by *Ruditapes philippinarum* and *Musculista senhousia* in a field community.

presence of bivalves (17,24,36). The temporal scaling of bivalve nutrient excretion showed a marked seasonal pattern of large variations of nutrient flux, up to ca. 10-fold (*R. philppinarum*) and 20-fold (*M. senhousia*) between March–April 1995 and September 7, 1995, and a progressive decrease from late summer through winter. This approach may involve some limitations, such as the effect of differences between the bivalve performance in controlled laboratory experiments and that in the field and a relative approximation in adopting different excretion rates at temperatures other than those actually employed in the laboratory experiments (29). However, it indicates the strong influence of animal distribution on the magnitude and temporal scaling of biogenic nutrient regeneration due to bivalve excretion.

The great potential of this biogenic source of nutrients in cycling biophilic elements can also be highlighted by a comparison with the extent of benthic nutrient regeneration through diffusive flux. In particular, nutrient flux measured from nutrient concentrations in the porewater in adjacent intertidal and coastal areas was more than one order of magnitude lower; it varied from 0.2 to 1.5 mmol NH_4^+-N m^{-2} d^{-1} and from 0.01 to 0.05 mmol PO_4^{3-}-P m^{-2} d^{-1}. It can be inferred that a marked increase in biogenic nutrient regeneration is importantly controlled by the animal biomass increase (36) and has a major impact, acting as a positive feedback, on primary producers (41). These results indicate that abundant macrofauna and its excretory products play a primary role in benthic nutrient regeneration, are well balanced in their stoichiometric ratios, and thus act as a major factor to support primary production within the intertidal zone.

BIBLIOGRAPHY

1. Admiraal, W. and Peletier, H. (1980). Influence of seasonal variations of temperature and light on the growth rate of culture and natural population of intertidal diatoms. *Mar. Ecol. Progr. Ser.* **2**: 35–43.
2. Harrison, W.G. and Platt, T. (1980). Variations in assimilation number of coastal marine phytoplankton: effects of environmental co-variates. *J. Plank. Res.* **2**: 249–260.
3. Verhagen, J.H.G. and Nienhuis, P.H. (1983). A simulation model of production, seasonal changes in biomass and distribution of eelgrass (*Zoostera marina*) in Lake Grevelingen. *Mar. Ecol. Progr. Ser.* **10**: 187–195.
4. Giesen, W.B.J.T., van Katwijk, M.M., and den Hartog, C. (1990). Eelgrass condition and turbidity in the Dutch Wadden Sea. *Aquat. Bot.* **37**: 71–85.
5. Agustí, S. et al. (1994). Light harvesting among photosynthetic organisms. *Funct. Ecol.* **8**: 273–279.
6. Justic, D., Rabalais, N.N., and Turner, T. (1995). Stoichiometric nutrient balance and origin of coastal eutrophication. *Mar. Poll. Bull.* **30**: 41–46.
7. Valiela, I. et al. (1997). Macroalgal blooms in shallow estuaries: controls and ecophysiological and ecosystem consequences. *Limnol. Oceanogr.* **5**: 1105–1118.
8. Park, Y.C. and Carpenter, E.J. (1987). Ammonium regeneration and biomass of macrozooplankton and Ctenophores in Great South Bay, New York. *Estuaries* **4**: 316–320.
9. Wen, Y.H. and Peters, R.H. (1994). Empirical models of phosphorus and nitrogen excretion rates by zooplankton. *Limnol. Oceanogr.* **39**: 1669–1679.
10. Harrison, W.G. et al. (1992). Nitrogen dynamics at the VERTEX time-series site. *Deep-Sea Res.* **39**: 1535–1552.
11. Rowe, G.T., Clifford, C.H., Smith, K.L., and Hamilton, P.L. (1975). Benthic nutrient regeneration and its coupling to primary productivity in coastal waters. *Nature* **225**: 215–217.
12. Nixon, S.W. (1981). Remineralization and nutrient cycling in coastal marine ecosystems. In: *Nutrient Enrichment in Estuaries*. B. Nelson and L.E. Cronbin (Eds.). Humana Press, Clifton, NJ, pp. 111–138.
13. Klump, J.V. and Martens, C.S. (1983). Benthic nitrogen regeneration. In: *Nitrogen in the Marine Environment*. E.J. Carpenter and D.G. Capone (Eds.). Academic Press, New York, pp. 411–457.
14. Jordan, T.E. and Valiela, I. (1982). A nitrogen budget of the ribbed mussel, *Geukensia demissa*, and its significance in

nitrogen flow in a New England salt marsh. *Limnol. Oceanogr.* **27**: 75–90.
15. Murphy, R.C. and Kremer, J.N. (1985). Bivalve contribution to benthic metabolism in a California lagoon. *Estuaries* **8**: 330–341.
16. Prins, T.C. and Smaal, A.C. (1994). The role of the blue mussel *Mytilus edulis* in the cycling of nutrients in Oosterschelde estuary (the Netherlands). *Hydrobiol.* **282/283**: 413–429.
17. Doering, P.H., Kelly, J.R., Oviatt, C.A., and Sowers, T. (1987). Effect of the hard clam *Mercenaria mercenaria* on benthic fluxes of inorganic nutrients and gases. *Mar. Biol.* **94**: 377–383.
18. Nakamura, M., Yamamuro, M., Ishikawa, M., and Nishimura, H. (1988). Role of the bivalve *Corbicula japonica* in the nitrogen cycle in a mesohaline lagoon. *Mar. Biol.* **99**: 369–374.
19. Kiibus, M. and Kautsky, N. (1996). Respiration, nutrient excretion and filtration rate of tropical freshwater mussels and their contribution to production and energy flow in Lake Kariba, Zimbabwe. *Hydrobiol.* **331**: 25–32.
20. Yamada, H. and Kayama, M. (1987). Liberation of nitrogenous compounds from bottom sediments and effect of bioturbation by small bivalve, *Theora lata* (Hinds). *Estuar. Coast. Shelf Sci.* **24**: 539–555.
21. Gómez-Parra, A. and Forja, J.M. (1993). Benthic fluxes in Cadiz Bay (SW Spain). *Hydrobiol.* **252**: 23–34.
22. Yamamuro, M. and Koike, I. (1993). Nitrogen metabolism of the filter-feeding bivalve *Corbicula japonica* and its significance in primary production of a brackish lake in Japan. *Limnol. Oceanogr.* **35**: 997–1007.
23. Dame, R.F., Zingmark, R.G., and Haskin, E. (1984). Oyster reefs as processors of estuarine materials. *J. Exp. Mar. Biol. Ecol.* **83**: 239–247.
24. Asmus, H., Asmus, R.M., and Reise, K. (1990). Exchange processes in an intertidal mussel bed: a Sylt-flume study in the Wadden Sea. *Berichte Biolog. Anstalt Helgol.* **6**: 78.
25. Prins, T.C. and Smaal, A.C. (1990). Benthic-pelagic coupling: the release of inorganic nutrients by an intertidal bed of *Mytilus edulis*. In: *Trophic relationships in the marine environment. Proc. 24th Europ. Mar. Biol. Symp.*, Aberdeen Univ. Press, pp. 89–103.
26. Asmus, R.M., Asmus, H., Wille, A., Zubillaga, G.F., and Reise, K. (1994). Complementary oxygen and nutrient fluxes in seagrass beds and mussels banks? In: *Changes in Fluxes in Estuaries: Implications from Science to Management.* K.R. Dyer and R.J. Orth (Eds.). Olsen & Olsen, Fredensborg, pp. 227–238.
27. Bayne, B.L. and Scullard, C. (1977). Rates of nitrogen excretion by species of *Mytilus* (Bivalvia: Mollusca). *J. Mar. Biol. Ass. U.K.* **57**: 355–369.
28. Schlüter, L. and Josefsen, S.B. (1994). Annual variation in condition, respiration and remineralisation of *Mytilus edulis* L. in the Sound, Denmark. *Helgol. Meeres.* **48**: 419–430.
29. Magni, P., Montani, S., Takada, C., and Tsutsumi, H. (2000). Temporal scaling and relevance of bivalve nutrient excretion on a tidal flat of the Seto Inland Sea, Japan. *Mar. Ecol. Progr. Ser.* **198**: 139–155.
30. Heip, C.H.R. et al. (1995). Production and consumption of biological particles in temperate tidal estuaries. *Oceanogr. Mar. Biol. Ann. Rev.* **33**: 1–149.
31. Magni, P. and Montani, S. (1997). Development of benthic microalgal assemblages on a tidal flat in the Seto Inland Sea, Japan: effects of environmental variability. *La mer* **35**: 109–120.
32. Magni, P. and Montani, S. (1998). Responses of intertidal and subtidal communities of the macrobenthos to organic load and oxygen depletion in the Seto Inland Sea, Japan. *J. Rech. Océanogr.* **23**: 47–56.
33. Magni, P. and Montani, S. (2000). Physical and chemical variability in the lower intertidal zone of an estuary in the Seto Inland Sea, Japan: seasonal patterns of dissolved and particulate compounds. *Hydrobiol.* **432**: 9–23.
34. Dame, R.F., Wolaver, T.G., and Libes, S.M. (1985). The summer uptake and release of nitrogen by an intertidal oyster reef. *Neth. J. Sea Res.* **19**: 265–268.
35. Dame, R.F. et al. (1991). The influence of mussel beds on nutrients in the western Wadden Sea and Eastern Scheldt estuaries. *Estuaries* **14**: 130–138.
36. Bartoli, M. et al. (2001). Impact of *Tapes philippinarum* on nutrient dynamics and benthic respiration in the Sacca di Goro. *Hydrobiol.* **455**: 203–212.
37. Kuwae, T., Hosokawa, Y., and Eguchi, N. (1998). Dissolved inorganic nitrogen cycling in Banzu intertidal sand-flat, Japan. *Mangroves Salt Marshes* **2**: 167–175.
38. Matsukawa, Y., Sato, Y., and Sasaki, K. (1987). Benthic flux of nutrient salts on an intertidal flat. *Nippon Suisan Gakkaishi* **53**: 985–989.
39. Takayanagi, K. and Yamada, H. (1999). Effects of benthic flux on short term variations of nutrients in Aburatsubo Bay. *J. Oceanogr.* **55**: 463–469.
40. Yamamoto, T., Matsuda, O., Hashimoto, T., Imose, H., and Kitamura, T. (1998). Estimation of benthic fluxes of dissolved inorganic nitrogen and phosphorus from sediments of the Seto Inland Sea. *Umi to Kenkyu (Oceanogr. Soc. Japan)* **7**: 151–158 (in Japanese).
41. Peterson, B.J. and Heck, K.L. Jr. (2001). Positive interaction between suspension-feeding bivalves and seagrass—a facultative mutualism. *Mar. Ecol. Progr. Ser.* **213**: 143–155.

BREAKWATERS

Stefano Pagliara
Pietro Chiavaccini
Università di Pisa
Pisa, Italy

Breakwaters are coastal structures used to protect harbor and shore areas by dissipating and reflecting wave energy. They are built to

— reduce wave disturbance in coastal and harbor areas and preserve related activities;
— protect ships and boats from wave forces;
— when located near shore, in the same direction as the coastline, they can stabilize the coastline, modifying cross-shore and long-shore sediment transport.

The choice of the type of structure depends on the availability of materials used, the characteristics of the incident wave, the bottom morphology, the geotechnical parameters of the soil, and the necessity of obtaining a flexible or rigid structure. Breakwaters can be classified as rubble-mound structures, vertical breakwaters, and floating breakwaters.

RUBBLE-MOUND STRUCTURES

The typical cross section of a rubble-mound breakwater is sketched in Fig. 1. It consists of different layers of stones. The center core is made up of quarry run. The external layer (armor) consists of large armor units, that can be either rock or specially designed concrete units (cubes, tetrapods, dolos). The breakwater crest is generally 1–2 m over the still water level (SWL). The crest width should be large enough to allow transport and installation of material during construction and when a repair is made (1).

Because of the relative dimensions of the units of the armor and those of the core, in some cases, it is necessary to build the breakwater as a filter of three or four layers (underlayers), so that the finer material of the core cannot be removed by the waves through the voids of the armor layer. To prevent removal of finer material, the filter must satisfy the following relations:

$$D_{15}(upper\ layer) < 4 \div 5D_{85}(lower\ layer) \quad (1)$$

$$D_{15}(upper\ layer) < 20 \div 25D_{15}(lower\ layer) \quad (2)$$

where D_{15} = nominal size that is exceeded by the 85% of the sample
D_{85} = nominal size that is exceeded by the 15% of the sample

A toe filter is necessary if the breakwater is built on erodible material. The toe filter prevents breaking waves from removing material from the base of the structure. If the breakwater is located in shallow water, the filter toe is exposed to extreme wave action. To avoid, or just limit, wave overtopping, it is possible to use a concrete structure (crown wall) located over the crest (Fig. 1).

When the breakwater is small and not high, it is possible to avoid using a center core. This kind of structure, of single sized stones, is called a "reef breakwater," and it is normally used for small submerged breakwaters (2).

Stability

The rubble-mound breakwater causes the dissipation of wave energy by generating eddies due to the breaking. The voids and the roughness of the structural material, as well as the permeability of the structure, are very important in the dissipation process. The wave energy entering through the structure creates shear stresses that can move the masses causing loss of stability.

Armor Stability. The stability formulas are based on experiments carried out on hydraulic models. One of the most used stability formulas is Hudson's (3) determined for a nonovertopping structure:

$$\frac{H}{\Delta D_{n50}^2} = (K \cot \alpha)^{1/3} \quad (3)$$

where H = characteristic height of the wave (H_s to $H_{1/10}$);
D_{n50} = equivalent cube length of median rock;
α = slope angle;
$\Delta = (\rho_s/\rho_w - 1)$ where ρ_s and ρ_w are rock density and water density, respectively;
K = stability coefficient (Tables 1 and 2).

The damage D represents the measure of the modification of the structure's profile under wave action. The damage can be defined by counting the number of rocks moved or by measuring the variation of the armor layer

Table 1. Values of the Stability Coefficient K for $H = H_s$[a]

		Damage D 0–5%	
Stone Shape	Placement	Breaking Waves	Nonbreaking Waves
Smooth, rounded	Random	2.1	2.4
Rough, angular	Random	3.5	4.0
Rough, angular	Special	4.8	5.5

[a]Slope $1.5 \leq \cot\alpha \leq 3.0$.

Table 2. Values of the Stability Coefficient K for $H = H_{1/10}$

		Damage D 0–5%			
		Breaking Waves		Non Breaking Waves	
Stone shape	Placement	Trunk	Head	Trunk	Head
Smooth, rounded	Random	1.2	1.1	2.4	1.9
Rough, angular	Random	2.0	1.3^a–1.9^b	4.0	2.3^a–3.2^b
Rough, angular	Special	5.8	5.3	7.0	6.4
Tetrapods	Random	7.0	3.5^a–5^b	8.0	4.0^a–6.0^b

[a]$\cot\alpha = 3$.
[b]$\cot\alpha = 1.5$.

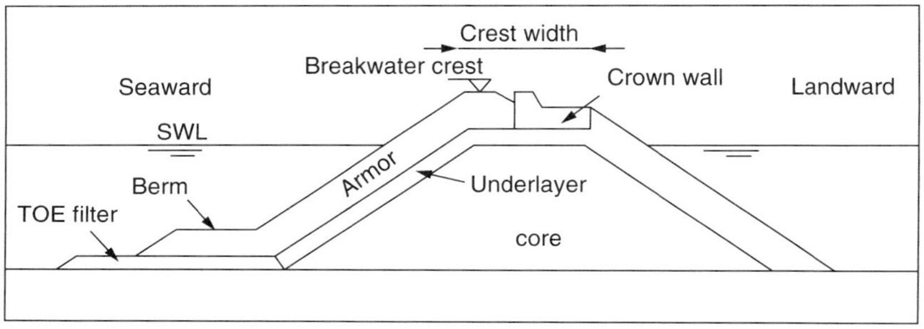

Figure 1. Typical section of a rubble-mound breakwater.

area (eroded area A of the damaged section). For this second case, Broderick (4) introduced a parameter (relative eroded area) defined as

$$S = \frac{A}{D_{n50}^2} \tag{4}$$

where D_{n50} is the nominal diameter, corresponding to 50% of the weight of the sample. The damage can be considered the number of masses of dimension equal to D_{n50} eroded in a strip of section of the same length. Zero damage means that there is nominally no removal of the armor units from the breakwater face.

The K value of Hudson's formula is different for the trunk and the head of the structure. The stones will be less stable on the head than on the trunk. In this case, K must be decreased by about 20%.

Van der Meer (5) derives expressions that include some additional parameters for an incident wave:

$$\frac{H_s}{\Delta D_{n50}} = 6.2 S^{0.2} P^{0.18} N_z^{-0.1} \xi_m^{-0.5}$$
$$\text{plunging waves } \xi_m < \xi_{cm} \tag{5}$$

$$\frac{H_s}{\Delta D_{n50}} = 1.0 S^{0.2} P^{-0.13} N_z^{-0.1} (\cot\alpha)^{0.5} \xi_m^P$$
$$\text{surging waves } \xi_m > \xi_{cm} \tag{6}$$

$$\xi_m = s_m^{-0.5} \tan\alpha$$

$$\xi_{mc} = [6.2 P^{0.31} (\tan\alpha)^{0.5}]^{1/(P+0.5)}$$

where S = relative eroded area (normally equal to 2);
P = notional permeability;
N_z = number of waves;
s_m = wave steepness $s_m = H_s/L_{\text{om}}$;
L_{om} = deepwater wavelength corresponding to the mean period.

For a homogeneous structure (no core, no filter, and stones of the same size), $P = 0.6$; a rock armor layer with a permeable core gives $P = 0.5$; an armor layer with filter on a permeable core gives $P = 0.4$. For a breakwater with an impermeable core, $P = 0.1$.

For overtopped and low crested structures, Van der Meer suggests multiplying D_{n50} by a reduction factor f_i defined by

$$f_i = \left(1.25 - 4.8 \frac{R_c}{H_s} \sqrt{\frac{s_{\text{op}}}{2\pi}}\right) \tag{7}$$

R_c is the freeboard, and $s_{\text{op}} = H_s/L_{\text{op}}$ (L_{op} is the deep water wavelength referred to the peak period). Expression (7) can be used in the range $0 < \left(\frac{R_c}{H_s}\sqrt{\frac{s_{\text{op}}}{2\pi}}\right) < 0.052$.

For submerged breakwaters, the following expression can be used (6):

$$\frac{h_c}{h} = (2.1 + 0.1S) \exp(0.14 N_s^*) \tag{8}$$

where h = water depth;
h_c' = height of the structure from the base;
S = relative eroded area;
$N_s^* = \frac{H_s}{\Delta D_{n50}} s_p^{-1/3}$ = spectral stability number.

Run Up and Overtopping

Run up is a phenomenon in which the incident crest wave runs up along a sloping structure to a level higher then the original wave crest. Together with overtopping, it plays, a very important role in the design of a rubble-mound structure because it depends on the characteristics of the structure (slope roughness, berm length, permeability).

Run up is expressed by $R_{u,x\%}$ that represents the level reached by the wave exceeded in x% of the cases by the incident wave. The run up level is referred to the SWL. For rubble-mound structures, Van der Meer's (6) formula is used:

$$\frac{R_{u,x\%}}{H_s} = a\xi_m, \qquad \xi_m < 1.5 \tag{9}$$

$$\frac{R_{u,x\%}}{H_s} = b\xi_m^c, \qquad \xi_m > 1.5 \tag{10}$$

These formulas are valid for breakwaters that have an impermeable or almost impermeable core ($P < 0.1$). If the breakwaters have a permeable core ($0.1 < P < 0.4$) Equations 9 and 10 become

$$\frac{R_{u,x\%}}{H_s} = d \tag{11}$$

It is usual in breakwater design to consider that $x\% = 2$; this means that $R_{u,2\%}$ is the run up exceeded by 2% of the waves. In this case, the values of the parameters of Equations (9,10), and 11 are $a = 0.96$, $b = 1.17$, $c = 0.46$, and $d = 1.97$.

ξ_m is the breaker parameter for deep water, corresponding to the mean period ($\xi_m = s_m^{-0.5} \tan\alpha$ where the symbols are explained in Equations 5 and 6).

In a low crest elevation, overtopping is allowed. Overtopping is the quantity of water passing over the crest of a structure per unit time, and it has the same dimensions of a discharge $Q(m^3/s)$, often expressed for unit length $q[m^3/(sm)]$.

A knowledge of overtopping is important in defining the necessary protection of the splash area and in assessing the risk to people or installations behind the breakwater. The amount of overtopping varies considerably from wave to wave; the overtopping discharge changes in time and space, and the greatest quantity is due to a small number of the incident waves. Wave overtopping for an impermeable rock armored slope structure with a crown wall can be expressed by the equation of Bradbury and Allsop (7), using the parameters of Aminti and Franco (8):

$$\frac{q}{gH_s T_{\text{om}}} = a \left[\left(\frac{R_c}{H_s}\right)^2 \sqrt{\frac{S_{\text{om}}}{2\pi}} \right]^{-b} \tag{12}$$

where S_{om} = deepwater wave steepness, based on mean period;
H_s = significant wave height;
R_c = crest freeboard relative to SWL in m;
T_{om} = deepwater wave mean period;
a, b = parameters as specified in Table 3.

G is the width (seaward) of the armor crest till the crown wall, and α is the slope of the armor layer.

Table 3. Coefficients for Equation 15 from Experimental Results

Armor Units	Cotα	G/H_s	a	b
Rock	2	1.10	17×10^{-8}	2.41
	2	1.85	19×10^{-8}	2.30
	2	2.60	2.3×10^{-8}	2.68
	1.33	1.10	5.0×10^{-8}	3.10
	1.33	1.85	6.8×10^{-8}	2.65
	1.33	2.60	3.1×10^{-8}	2.69
Tetrapods	2	1.10	1.9×10^{-8}	3.08
	2	1.65	1.3×10^{-8}	3.80
	2	2.6	1.1×10^{-8}	2.86
	1.33	1.10	5.6×10^{-8}	2.81
	1.33	1.85	1.7×10^{-8}	3.02
	1.33	2.60	0.92×10^{-8}	2.98

For rock armored permeable slopes, that have a theoretical permeability $P = 0.4$ and a berm in front of a crown wall, we can use the Pedersen and Burcharth formula (9):

$$\frac{qT_{\text{om}}}{L_{\text{om}}^2} = 3.2 \times 10^{-5} \left(\frac{H_s}{R_c}\right)^3 \frac{H_s^2}{A_c B \cot \alpha} \qquad (13)$$

where A_c = level of the berm from SWL;
B = the width of the berm;
R_c = the level of the crest of the crown wall from SWL;
α = slope of armor.

Wave Reflection

Each coastal structure causes a wave reflection. Reflection plays a very important role because of the interaction between reflected and incident waves that can create a very confused sea, increasing the wave steepness. It is a problem especially at the entrance of an harbor because the steepness makes ship and boat maneuver very difficult. Besides, strong reflection increases the erosive force in front of the structure.

Rubble-mound breakwaters, which are permeable, rough, and sloping structures and structure of limited crest level, absorb a significant portion of the wave energy. For these structures, the reflection coefficient is small. For nonovertopped structures, that have a theoretical permeability P, we can use the following equation (10):

$$K_r = 0.071 P^{-0.082} (\cot \alpha)^{-0.62} S_{\text{op}}^{-0.46} \qquad (14)$$

where K_r is the ratio of the reflected wave height and the incident wave height.

Wave Transmission

When energy passes over and through a breakwater, there is a wave transmission. The wave action in the landward side of the structure is smaller than in the seaward side. A wave is transmitted when a considerable amount of water overtops the structure and when the breakwater is very permeable and the wave period is relatively long. We define the coefficient of transmission as the ratio of transmitted to incident characteristic wave height or the ratio of the square of transmitted mean energy to incident mean wave energy:

$$K_t = \frac{H_{\text{st}}}{H_s} = \left(\frac{E_{\text{st}}}{E_s}\right)^{0.5} \qquad (15)$$

For rock armored, low crested, submerged and reef breakwaters, we can use the Van der Meer and d'Angremond formula (11):

$$K_t = \left(0.031 \frac{H_s}{D_{n50}} - 0.24\right) \frac{R_c}{D_{n50}} + b \qquad (16)$$

where $b = -5.42 s_{\text{op}} + 0.0323 \frac{H_s}{D_{n50}} - 0.0017 \left(\frac{B}{D_{n50}}\right)^{1.84} + 0.51$ for a conventional structure,
$b = -2.6 s_{\text{op}} - 0.05 \frac{H_s}{D_{n50}} + 0.85$ for a reef type structure,
H_s = significant wave height;
s_{om} = deepwater wave steepness, based on peak period;
R_c = crest freeboard relative to SWL, negative for submerged breakwaters;
B = width of crest;
D_{n50} = median of nominal diameter of rock for design conditions.

For conventional structures, K_t has a maximum of 0.75 and a minimum of 0.075, and for reef type structures, K_t varies between 0.15 and 0.6.

This formula can be used in the following range:

$$1 < \frac{H_s}{D_{n50}} < 6, \quad 0.01 < s_{\text{op}} < 0.05, \quad -2 < \frac{R_c}{D_{n50}} < 6$$

VERTICAL BREAKWATERS

Vertical breakwaters are rigid and not as flexible as rock armored ones. It is necessary to know very well the wave climate (direction, intensity, frequency, and duration of waves) to obtain a good design. A typical cross section of a vertical breakwater is sketched in Fig. 2a.

Forces on Vertical Structures

Nonbreaking Wave. A nonbreaking wave is perfectly reflected (coefficient of reflection = 1.0) and can be considered static because the period is longer than the natural period of oscillation of the structure. This stationary and static wave is called "clapotis." Assuming perfect reflection, the seaward wave has a height double the incident wave.

If the vertical breakwater is overtopped, a reduction of wave pressure occurs because it is truncated at the crest level. According to Sainflou (12), the pressures are (Fig. 2a)

$$p_1 = p_2 + \rho_w g h_s \left(\frac{H + \delta_0}{h_s + H + \delta_0}\right) \qquad (17)$$

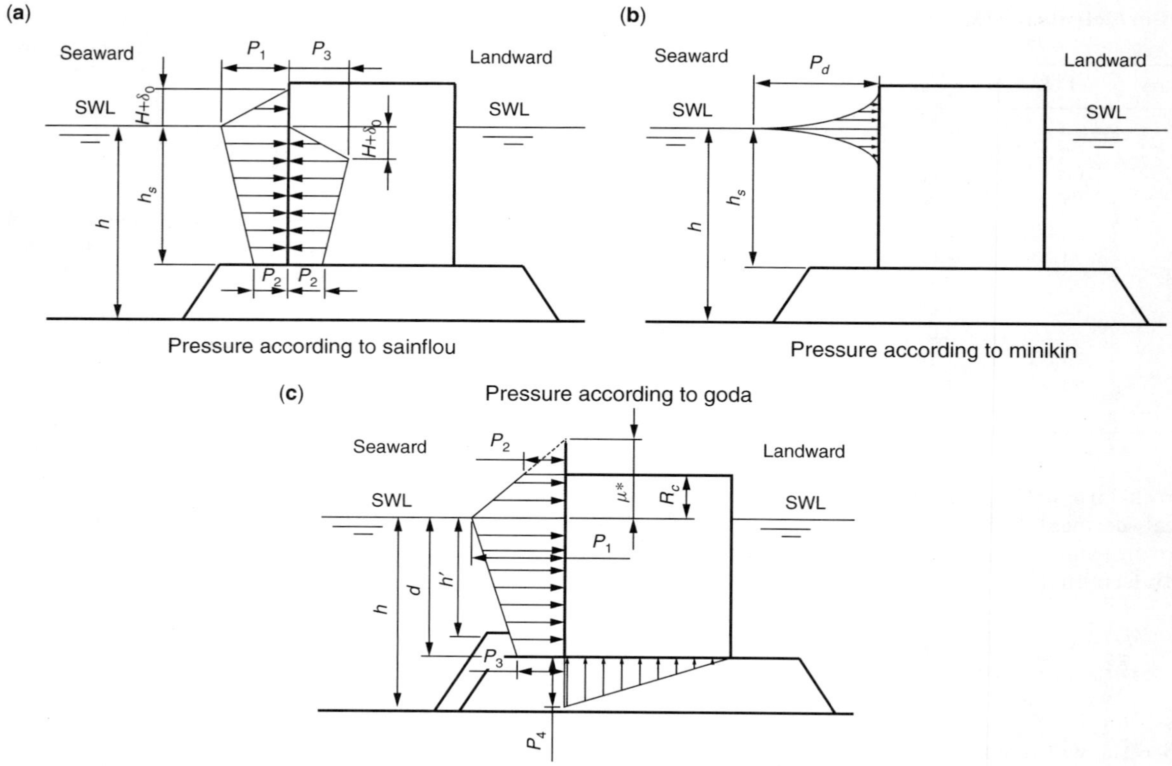

Figure 2. Pressure on vertical breakwaters.

$$p_2 = \frac{\rho_w g H}{\cosh(2\pi h_s/L_0)} \tag{18}$$

$$p_3 = \rho_w g(H - \delta_0) \tag{19}$$

$$\delta_0 = \frac{\pi H^2}{L_0}\coth(2\pi h_s/L_0) \tag{20}$$

where H = wave height;
p_1 = pressure at the still water level (SWL), corresponding to wave crest;
p_2 = pressure at the base of a vertical wall;
p_3 = pressure at the still water level, corresponding to a wave trough;
δ_0 = vertical shift in the wave crest and trough at the wall;
ρ_w = water density;
h_s = water depth at the foot of the structure;
L_0 = wavelength in deep water.

Breaking Waves. The worst condition for a vertical breakwater occurs when the water depth in front of the structure causes the wave to break. In this case, the dynamic component of the pressure must be considered, especially if the breaking is sudden, as happens for plunging waves. The forces generated by breaking are very high and a very short duration. According to Minikin (13), the dynamic pressure (Fig. 2b) is

$$p_d = 100\frac{h_t}{h_s L_0}(h_s + h_t)\rho_w g H \tag{21}$$

where H = wave height;
p_d = dynamic pressure at the still water level, corresponding to the wave crest;
ρ_w = water density;
h_s = water depth at the foot of the structure;
h_t = height of the caisson between the SWL and the berm;
L_0 = wavelength in deep water.

Another expression that can be used to determine the total pressure for breaking waves is (14)

$$p_1 = 0.5(1 + \cos\beta)(\alpha_1 + \alpha_2\cos^2\beta)\gamma H \tag{22}$$

$$p_2 = p_1\frac{\eta^* - h_c}{\eta^*} \tag{23}$$

$$p_3 = \alpha_3 p_1 \tag{24}$$

$$p_4 = 0.5(1 + \cos\beta)\alpha_1\alpha_3\gamma H \tag{25}$$

$$\eta^* = 0.75(1 + \cos\beta)H \tag{26}$$

$$\alpha_1 = 0.6 + 0.5\left(\frac{2kh}{\sinh 2kh}\right)^2 \tag{27}$$

$$\alpha_2 = \min\left\{\frac{h_b - d}{3h_b}\left(\frac{H}{d}\right)^2, \frac{2d}{H}\right\} \tag{28}$$

$$\alpha_3 = 1 - \frac{h}{h}\left(1 - \frac{1}{\cosh kh}\right) \tag{29}$$

$$H = \min(1.8H_s, H_f) \tag{30}$$

$$H_f = 0.18\frac{gT^2}{2\pi}\left(1 - \exp\left\{\frac{3\pi^2 h_b}{gT^2}[1 + 15(\tan\vartheta)]^{4/3}\right\}\right) \quad (31)$$

where h_b = water depth at a distance of $5H_s$ from the vertical breakwater (seaward);
β = angle of incidence of the wave;
ϑ = bottom angle within the horizontal.

The other symbols are explained in Fig. 2c.

Ice Force. Ice exerts a force on both the landward and seaward sides. Ice force acts at the water level, and it can be as high as the crushing strength of the ice (1.5 Mpa) multiplied by the ice thickness. Even if ice is present in both sides of the structure, the worst case (ice acting only on one side) represents the design condition. Ice and waves do not occur simultaneously.

Wave Transmission

Wave transmission for a vertical breakwater is mainly the result of overtopping. Goda (15) has investigated the wave transmission of a vertical wall breakwater placed over a rock berm. He related $K_t = H_t/H_i$ to the relative freeboard $R_c/H(-1 < R_c/H < 1)$ and found this equation:

$$K_t = 0.2\left(\frac{h_v}{h_s}\right)^2 - 0.4\left(\frac{h_v}{h_s}\right) + 0.58 - 0.32\left(\frac{R_c}{H}\right) \quad (32)$$

where h_s = water depth at the foot of the structure;
h_v = height of the wall from the berm;
R_c = freeboard crest;
H = design wave height.

Wave Overtopping

The overtopping of a vertical breakwater can be determined by using Franco's formula (16):

$$\frac{q}{\sqrt{gH_s^3}} = a\exp\left(-b\frac{Rc}{\gamma_g\gamma_{b\sigma}H_s}\right) \quad (33)$$

where q =unit discharge (m^3/s per m);
a, b =coefficient depending on structure shape and on the water behavior at the seaward face; for a vertical plain wall, $a = 0.082$, $b = 3$;
$\gamma_{b\sigma}$ =depends on the direction of wave attack. For long crested waves,

$\gamma_{b\sigma} = \cos(\beta)$ for $0 \le \beta \le 37°$,
$\gamma_{b\sigma} = 0.79$ for $37° \le \beta$.

For short crested waves,

$\gamma_{b\sigma} = 0.83$ for $0 \le \beta \le 20°$
$\gamma_{b\sigma} = 0.83\cos(20 - \beta)$ for $20° \le \beta$.

γ_g =depends on the type of the wall: for a plain wall $\gamma_g = 1$; for a plain wall with round hole perforations, $\gamma_g = 0.792$; for a plain wall with rectangular hole perforations, $\gamma_g = 0.722$; for a plain wall with rectangular hole perforations and open deck, $\gamma_g = 0.577$.

Concrete Caissons

The main component of a vertical breakwater is a cellular concrete caisson, a hollow box that is floated and *in situ*is filled with granular material to be sunk and placed on a berm prepared with sand or rock to form a stable structure (Fig. 3b).

Sometimes, the berm is a partial rubble-mound structure forming a composite breakwater. In some locations, the seaward face of the caisson is protected by sloping armor to avoid the breaking of incident waves in the front of the structure. The types of failure of a vertical caisson are shown in Fig. 3c. Normally, the loss of stability occurs from

— sliding of the caisson over the base;
— overturning of the caisson around its landward corner;
— loss of geotechnical stability of the base.

The forces acting are total wave pressure seaward (different for breaking or nonbreaking waves), hydrostatic force landward due to the still water depth, the buoyancy and the uplift force under the caisson, and the weight of the caisson. In some locations, ice force must be considered.

Sliding is verified if the total normal force multiplied by the coefficient of friction between the caisson and the berm is greater then 1.3 times the horizontal sliding force. Overturning is verified if the total stabilizing moment is greater then 1.5 times the overturning total moment. The point of overturning is the landward corner of the caisson. The last verification is to see if the resistance of the base (its critical value) is greater than the pressure due to the resultant of the acting forces. Stress is transmitted from the bottom of the caisson to the soil at an angle of 45°; this is ensured if the berm is large enough with respect to the caisson.

Sheet Piling Breakwater

Sheet piling structures are a particular type of vertical breakwater that consist of a series of piles engaged to form a continuous structure. The material most used is steel, but it possible to use reinforced concrete and timber. The choice depends on economic considerations and the availability of material. A vertical breakwater can be made of a single line of sheet piling, a double wall of sheet piles, or a cellular unit that has its the cavity filled with sand or rock. The head of a sheet piling is generally of rock, concrete slab, or paving. If steel piling is used, protection against corrosion should be provided.

The forces acting on sheet piling are the same as on caissons. Generally the worst conditions happen when the trough of the waves passes. In this case, the saturated backfill soil could cause the wall to fail seaward. The penetration of the sheet pile in to the soil has to develop the necessary resistance to cantilever action. The double wall ensures more resistance; the walls are as far apart as the depth of the water, and can be connected by cross-walls to form boxes. These boxes can be filled with sand or rocks to increase stability. A cellular sheet piling is necessary in deep water to stiffen the structure. Timber sheet pilings

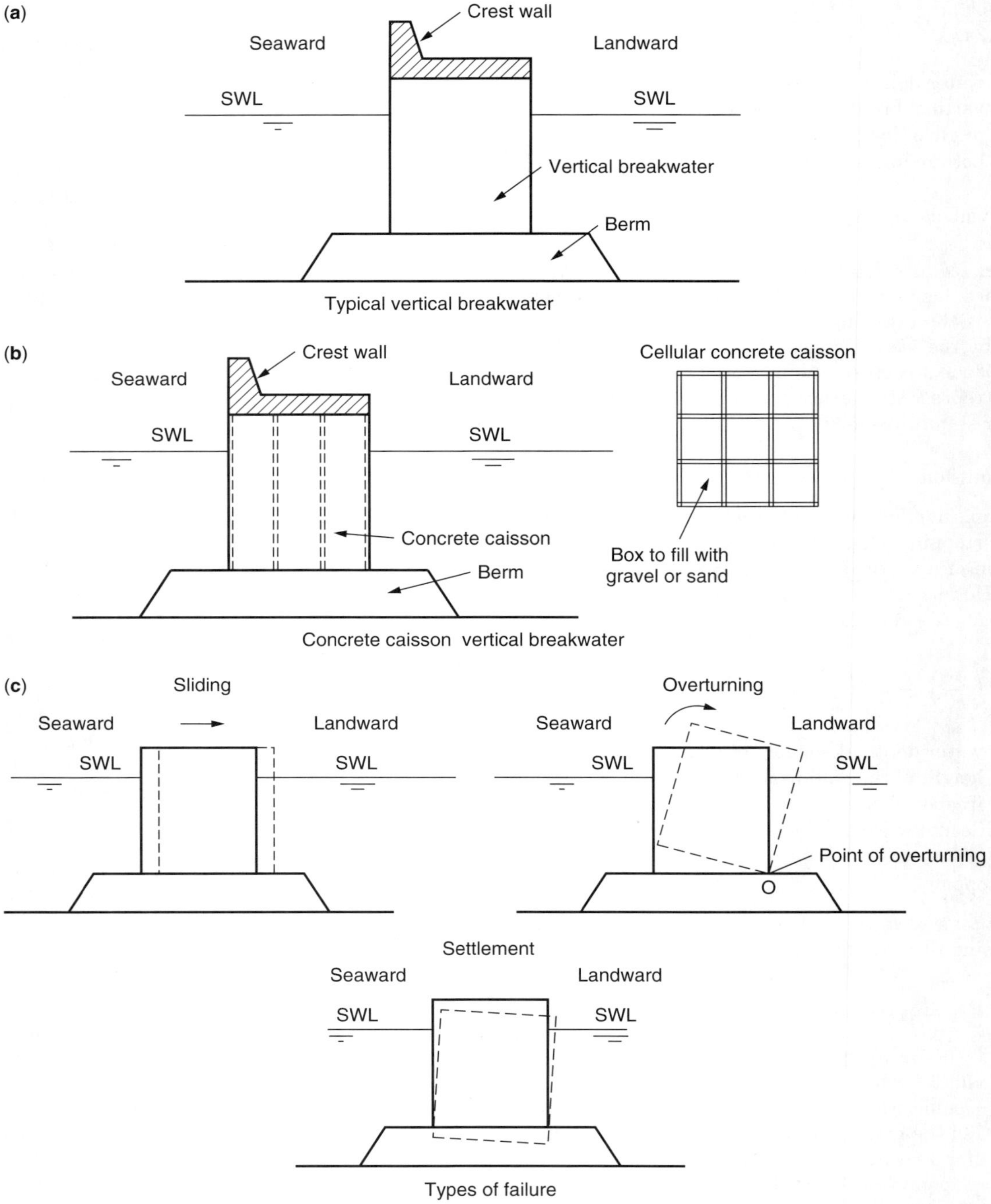

Figure 3. Vertical breakwater sections and types of failure.

are used in shallow water locations where wave conditions are moderate.

FLOATING BREAKWATERS

Floating breakwaters consist of composite structures that float on water (17). They can be used where waves are moderate and have a very short period. These structures are low in cost, insensitive to water depth and are easily portable, even if they provide less protection than fixed breakwaters. Generally, they consist of a box, pontoons, tethered floats, and floating docks affixed to piles. The floating breakwater reduces wave energy by reflection, dissipation interference, and conversion of the energy into a mono-oscillatory motion. The dimensions of floating breakwaters (the penetration depth, in particular) depend on the amount of energy that is necessary to dissipate and the mooring force on the anchor of the structure. The more the structure extends to the bottom the greater the reflection and the dissipation, but the mooring force also increases.

BIBLIOGRAPHY

1. CERC U.S. Corps of Eng. (1984). *Shore Protection Manual.* Vicksburg, MS.
2. U.S. Corps of Eng. (1992). EM110-2-2904 *Coastal Groins and Nearshore Breakwaters.* Washington, DC.
3. Hudson, R.Y. (1959). Laboratory investigation of rubble mound breakwaters. *J. of the Waterways and Harbors Division* **85**: 93–121.
4. Broderick, L.L. (1983). Riprap stability: A progress report. *Proceedings of costal Structures ASCE* **83**: 320–330.
5. Van der Meer, J.W. (1987). *Stability of Breakwaters Armor Layer, Co. Eng.* **11**.
6. Van der Meer, J.W. (1993). *Conceptual design of rubble mound breakwaters*, Rep. 483. Delft University, the Netherlands.
7. Bradbury, A.P. and Allsop, N.W. (1988). Hydraulic effects of breakwater crown ways. In: *Proceedings of the Breakwater '88 Conference, Institution of Civil Eng.* Thomas Telford Publishing, London, UK, pp. 385–396.
8. Aminti, P. and Franco, L. (1988). Wave overtopping on rubble mound breakwaters. *Proceedings of the 21st International Costal Engineering Conference ASCE* **1**: 770–781.
9. Pedersen, J. and Burchart, H.F. (1992). wave forces on crown walls. In: *Proceedings of the 23rd International Coastal Eng. Conference ASCE.* **2**: 1489–1502.
10. Postma, G.M. (1989). *Wave Reflection from Rock Scopes Under Random Wave Attack.* M.S Thesis Delft University of Technology, the Netherlands.
11. D'Angremond, K. and Van Roode, F.C. (2001). *Breakwaters and Closure Dams.* Delft University Press, the Netherlands, pp. 139–149.
12. Sainflou, M. (1928). *Treatise on Vertial Breakwaters.* Annal Des Ponts Et Chaussee, Paris, France.
13. Minikin, R.R. (1950). *Winds, Waves and Maritime Structures.* Griffin, London, pp. 38–39.
14. Goda, Y. (2000). *Random Seas and Design of Maritime Structure.* World Scientific, Singapore, pp. 134–139.
15. Goda, Y. (1969). *Reanalysis of Laboratory Data on Wave Transmission Over Breakwaters.* Report of the port and harbor research institute Vol. 8, no.3.
16. Franco, C. and Franco, L. (1999). Overtopping formulas for caisson breakwaters with nonbreaking 3D waves. *J. of Waterway, Port, Coastal and Ocean Eng.* **125**: 98–108.
17. U.S. Corps of Eng. (1996). EM110-2-2904 *Design of Breakwater and Jetties.* Washington, DC.

THE OCEAN IN CLIMATE

EDWARD S. SARACHIK

(from *The Handbook of Weather, Climate, and Water: Dynamics, Climate, Physical Meteorology, Weather Systems, and Measurements*, Wiley 2003)

INTRODUCTION

Earth's present climate is intrinsically affected by the ocean—the climate without the ocean would be different in many essential ways: Without the evaporation of water from the sea surface, the hydrological cycle would be different; without ocean heat transport and uptake, the temperature distribution of the globe would be different; and without the biota in the ocean, the total amount of carbon in the atmosphere would be many time its current value. Yet, while we may appreciate the role of the ocean in climate, the difficulty and expense of making measurements below the ocean's surface has rendered the vast volume of the ocean a sort of *mare incognita*. Why is the ocean so important in Earth's climate, and which of its properties are of special significance for climate? How have we learned about the ocean and its role in climate, and what more do we need to know?

PROPERTIES OF THE OCEAN AND PROCESSES IN THE OCEAN

The ocean covers 70% of Earth's surface to an average depth of about 4 km. The mass of the ocean is 30 times and its heat capacity 120 times that of the atmosphere, and the ocean contains 80 times the carbon dioxide stored in the atmosphere.

The density of the ocean is controlled both by its temperature and by its salt content. Ocean density increases with salinity and decreases with temperature. Unlike fresh water, which has a maximum density at 4 °C (so that colder water and ice float on 4 °C water and the temperature at the bottom of a nonfrozen lake is 4 °C), water saltier than 26 parts per thousand of water is continuously denser as the temperature is lowered, and the temperature at the bottom of the ocean is closer to 1 °C.

Since heat always diffuses from warm to cool temperatures, why does not the temperature of the deep ocean eventually adjust and become the same as its surface temperature? Cold water constantly sinks at high latitudes (both in the Northern and Southern Hemispheres) and fills the deeper parts of the oceans with cold water so that the water at depth is always cold, even when the surface temperature is very warm. This circulation is called the thermohaline circulation.

About 7% of the ocean surface is covered by sea ice. Growth of sea ice radically changes the nature of the ocean surface: Sea ice reflects solar radiation, thereby preventing it from being absorbed by the surface and blocks the transfer of heat and moisture from the surface of the ocean to the atmosphere.

The average salinity in the global oceans is 34.7 parts per thousand salt to water by weight. As the total amount of salt in the ocean is constant, changes in salinity only occur because of additions and subtractions of fresh water. Salinity decreases as rain falls on the ocean or river water enters the ocean, and it increases as water evaporates from the surface of the ocean. As sea ice grows, it rejects salt into the ocean thereby increasing its salinity. Similarly, as ice melts, it dilutes the surrounding ocean and lowers its salinity. A specific parcel of water can either increase or decrease its salinity by mixing with parcels with different salinities.

HOW THE OCEAN INTERACTS WITH THE ATMOSPHERE TO AFFECT THE CLIMATE

The ocean interacts with the atmosphere at (or very near) the sea surface where the two media meet. Visible light can penetrate into the ocean several tens of meters, but heat, moisture, and momentum, carbon dioxide, and other gases exchange directly at the surface. Sea ice forms at the surface and helps to determine the local exchanges. The basic problem of the ocean in climate is to explain these interchanges and to determine those characteristics of the ocean that affect these exchanges.

The ocean may be considered to interact with the atmosphere in two distinct ways: passively and actively. It interacts passively when the ocean affects the atmosphere but does not change the essential manner in which the atmosphere is operating. An example of a passive interaction is the oceanic response to adding CO_2 to the atmosphere where the ocean simply delays the greenhouse warming of the atmosphere as heat and CO_2 enters the ocean.

Active interaction with the atmosphere produces results that would not otherwise be there—an example is where the warming of the atmosphere reduces the thermohaline circulation and produces a climate reaction that could not have been obtained without the essential interaction of the ocean. In particular, since the northern branch, say, of the thermohaline circulation brings cold water from high latitudes toward the equator, and since the water must be replaced by warm water that is cooled as it moves northward, the net effect of the thermohaline circulation is to transport heat northward and thereby warm the higher latitudes. As the atmosphere warms, the water becomes less dense both by the effect of temperature and by increased rainfall, a necessary concomitant of global warming. Since the atmosphere sees a reduced north–south temperature gradient at the sea surface, it reacts fundamentally differently than if the thermohaline circulation were at full strength.

Our present models of greenhouse warming have the ocean acting in both the active and passive modes—active when warming leads to a slowed thermohaline circulation and passive when heat and CO_2 simply enter the ocean surface and is therefore lost to the atmosphere. Another example of active interaction is El Niño, a phenomenon that would not exist were it not for the active interaction of the atmosphere and the ocean (see the chapter by Trenberth). The ocean also has been inferred (by examining the composition of ancient ice stored in the Greenland and Antarctic ice sheets) to have, and probably actively take part in causing, climatic variability on time scales anywhere from decades to a few thousand years, a type of variability not seen on Earth since it emerged from the last glacial maximum some 18,000 years ago.

MEASURING THE OCEAN

The ocean is remarkably poorly measured. While the global atmosphere is constantly probed and analyzed for the purposes of weather prediction, until recently no such imperative existed for the ocean. Our ability to measure the ocean is severely limited basically by the inability of radiation to penetrate very far into the ocean—this requires direct *in situ* measurements of the interior of the ocean. The world's oceanographic research fleet is small and incapable of monitoring the breadth and depth of the world's ocean, although valuable research measurements are constantly being taken at selected places. As a result of ocean observations, we know the basic pathways of water in the ocean, we have a good idea of the transports by the ocean, we have some idea of the basic mechanisms for much of the major ocean currents, and we have a good idea of how the exchanges at the surface are accomplished. Yet we cannot measure vertical velocity (it is far too small), and so we remain completely ignorant of the processes by which the interior of the ocean affects the surface. Similarly, we are ignorant of the basic processes of mixing and friction in the ocean, both basic to being able to model the ocean for climate purposes.

Major oceanographic programs have been conducted in the last decade (the World Ocean Circulation Experiment, WOCE) and (the Tropical Ocean–Global Atmosphere, TOGA), and while they have taught us much about the ocean circulation and El Niño, respectively, the basic lesson is that, unless we can make continuous long-term measurements beneath the surface of the ocean, it will forever remain unknown territory.

MODELING THE OCEAN

Because the ocean is poorly measured, and because much of what we need to know about the past and predict about the future cannot be directly known, only inferred, models have played a particularly important role in the development of oceanography and, in particular, the role of the ocean in climate.

The basic tool of climate studies is the coupled model, where the various components of the climate system—the atmosphere, ocean, cryosphere, biosphere, and chemosphere—are simultaneously and consistently coupled. The ocean component of such a coupled model exchanges its heat, fresh water, and momentum with the atmosphere at the sea surface. The test of the successful coupling of the atmosphere and ocean is the correct simulation of the time-varying sea surface temperature and surface winds, both of which are relatively easy to measure: Directly by ship or mooring, remotely by satellite, or by a combination of the two.

The development of off-line ocean-only models requires the heat, momentum, and freshwater forcing from the atmosphere to be known. Since precipitation and evaporation, in particular, are so poorly measured over the ocean, it is a continual struggle to know whether errors in the ocean model are due to errors in the model itself or errors in the forcing of the ocean by the atmosphere.

Ocean models themselves are relatively simple in concept: The known equations of water and salt are discretized and time stepped into the future. The discretization process requires a trade-off between fine resolution for accuracy and the need to simulate over long periods of time, which, because of limited computer resources, requires coarser resolution. While the equation of state of seawater relating density to salt, temperature, and pressure cannot be written down simply, it has, over the course of time, become known to high accuracy.

What makes ocean modeling difficult is the specification of those mixing processes that unavoidably cannot be resolved by whatever resolution is chosen. We are beginning to understand that enhanced small-scale mixing occurs near bottom topography and near boundaries: Purposeful release experiments, where a dye is released and then followed in time to see how the dye cloud evolves, has revealed this to us. Larger scale mixing, where parcels are interchanged because of the large-scale circulation (but still unresolved by the ocean models) itself is more problematic, but recent advances in parameterizing these unresolved mixing effects have shown promise.

THE FUTURE OF THE OCEAN IN CLIMATE

It is clear that the ocean is a crucial component of the climate system. Since so much of what is not known about the past and future of the climate system depends on active interactions with the ocean, it is clear that we have to learn more about its essential processes. How to go about learning about the ocean is the difficult question.

Direct measurements are very expensive, and satellites, while giving a global look at the entire ocean, see only its surface. Designs are currently underway for a Global Ocean Observing System (GOOS), but the cost of implementing such a system in toto is prohibitive, even if shared among the wealthier countries of the world.

It is likely that a combination of studies, perhaps conducted for entirely different purposes, will advance the field most rapidly. In particular, the advent of the El Niño–Southern Oscillation (ENSO) prediction, which requires subsurface ocean data as initial conditions, has made almost permanent the Tropical Atmosphere-Ocean (TAO) array in the tropical Pacific, giving an unprecedented and continuous view of a significant part of the tropical ocean. We may extend the reasoning to say that, where predictability is indicated and shows societal or economic value, the measurement systems to produce the initial data will almost certainly be implemented. The promise of predicting climate from seasons to a few years will expand the ocean-observing system considerably. Additional expansions will come from resource monitoring, pollution monitoring, and various types of monitoring for national security purposes. While monitoring for security has traditionally meant the data is classified, once taken, data can eventually reach the research arena—the vast amount of Soviet and U.S. data that was declassified after the end of the cold war has shown this.

Observations can be also combined with models to give "value-added" observations. Data at individual points in the ocean exist without reference to neighboring points unless models are used to dynamically interpolate the data to neighboring points using the equation of motion of a fluid. This so-called four-dimensional data assimilation is in the process of development and shows promise as a powerful way of optimally using the ocean data that can be taken.

Models can also be compared with other models. While this might seem sterile, fine-resolution models can be used to develop parameterizations of large-scale mixing for use in coarse-resolution ocean models that can be run the long times needed to participate in coupled model simulations of climate. Advances in computer power will ultimately allow successive refinements in resolution so that finer scale resolution models can be run directly.

We close by reemphasizing the crucial role that the ocean plays in climate and climate variability and the necessity to know more about the ocean for all aspects of the climate problem.

COASTAL WATERS

Susan-Marie Stedman
Silver Spring, Maryland

Coastal waters are usually defined as those waters in a zone whose landward boundary is the limit of tidal influence and whose seaward boundary is the edge of the continental shelf (or a depth of 200 m if there is no continental shelf). Salinity, temperature, turbidity, and other characteristics of coastal waters vary considerably in this transitional and very dynamic zone.

The effects of waves, wind, and other processes create two distinct layers in coastal waters: an upper surface layer known as the "mixed layer," which varies in thickness from tens to hundreds of meters thick, and a lower "deep layer." The mixed layer is warmer and less saline than the deep layer. The deep layer generally contains more nutrients due to the decomposition of organic matter that has fallen to the ocean bottom. The transition between the two layers, called the pycnocline, is usually very abrupt, occurring over a space of a meter or less (1).

The majority of coastal waters occur in three types of environments: the relatively low energy environments of estuaries and bays, the high-energy environments of waters along rocky shores and beaches, and the deeper waters of the continental shelf. All coastal waters are subject to the effects of tides, waves, and currents, to varying degrees. Coastal waters are also often affected by activities in coastal terrestrial areas, especially in parts of the world where urban and suburban development in coastal areas is intense.

PHYSICAL PROCESSES IN COASTAL WATERS

Waves

Waves in coastal waters are produced by the wind and modified by their interaction with the coastline itself. There are two general types of waves: chop, which are the short steep waves seen on windy days; and swell, which are the long gentle waves that are present even on calm days. Chop is produced by local wind conditions, whereas swell is produced by storms very distant from the coastline. Waves are often described by their period (the time between two successive crests passing the same point), length (the distance from one crest to another), height (the distance from the trough to the crest of the wave), and their speed, which is determined using the equation

$$C(\text{speed}) = \Lambda(\text{length})/T(\text{period})$$

As waves move from the deep ocean to shallow coastal waters their speed is affected by the depth of the water as defined by the equation $c = \sqrt{gH}$, where g is the acceleration due to gravity and H is the water depth. As waves approach the shore, the speed of the wave crests becomes greater than the speed of the wave troughs, creating the characteristic "breakers" that crash onto the shore.

If waves approach a shoreline at an angle they will be refracted: that is, they will bend so that they become more parallel to the shore. Wave refraction affects where wave energy is concentrated; for example, a protrusion of the shoreline will cause waves to bend toward it, concentrating wave energy in that area. Areas of wave convergence are usually subject to intense erosion, whereas areas of wave divergence tend to be depositional; thus the overall effect of waves is to straighten the coastline.

Tides

Tides are created by complex interactions between astronomical forces and the earth's geomorphology. One can begin to understand tides by imagining that the earth is covered with a layer of water with uniform depth. Gravitational and centrifugal forces associated with the moon and, to a lesser extent, the sun produce a bulge in the water at points both closest to the moon and on the opposite side of the earth. As the earth completes one rotation, a hypothetical point on the earth's surface will pass through a bulge twice, generating two high tides (and two low tides), which is known as a semidiurnal tide. Because the moon is rotating around the earth, each high tide theoretically occurs a bit later (approximately 30 min) than 12 h after the previous one. The sun's effect on tides produces the difference in tidal range seen during the lunar cycle. When the earth, moon, and sun are aligned, the tidal bulge is largest and the coast sees the highest high tides and the lowest low tides (both are known as "spring" tides). When the sun and moon are aligned at right angles with respect to each other and the earth, the tidal bulge is decreased, resulting in reduced tides (know as "neap" tides).

The actual dynamics of tides is much more complex than this simple theory alone can explain. Tides are affected not only by astronomical factors but also by the sizes, shapes, and depths of water bodies, by winds, by the earth's rotation, and by changes in barometric pressure. A wide range of tidal cycles and tidal ranges occurs around the world. Although most parts of the world have semidiurnal tides, some have only one high tide and one low tide per day (diurnal tides), and some experience a phenomenon known as "double low water," where two extreme low tides occur between high tides occurring 24 h apart (2). Tidal ranges vary not only with the lunar cycle but geographically, from a few centimeters in the Gulf of Mexico to the famous 12-m tides in the Bay of Fundy. However, at any given location along the coast, the tidal cycle is fairly constant, allowing prediction of tidal periodicity and height based on data from past years. As physical oceanographers become more adept at modeling tidal dynamics, predictions also can be made for areas without such records.

Currents

Both waves and tides, as well as the wind, generate currents in the ocean that circulate water between numerous and sometimes disparate oceanic and coastal environments. Where these currents are close to shore, they strongly influence not only the coastal waters but the shape of the coastline itself. Of most importance, however, is the way currents transport nutrients, sediments, and marine life along the coasts. Currents generally fall into three categories: tidal currents, drift currents, and a phenomenon known as upwelling.

Tidal Currents. Among the strongest forces affecting coastal waters are tides. Tidal currents are strongest where a constriction (such as a narrow tidal inlet) or other topographic feature (a valley in the continental shelf) concentrates the tidal flow. Tidal current velocities range between zero at slack tide (exact high or low tide) to midtide maximums that are generally in the range of 1–5 km/h. Under the Golden Gate Bridge in San Francisco, tidal currents sometimes reach 9 km/h, and at Seymour Narrows, British Columbia, they can reach 24 km/h (3). Tidal currents have long been considered a source of renewable energy and are being used to generate power in Canada, Russia, and France. These projects involve damming a portion of an estuary and using the force of the tide leaving the estuary to turn turbines. More recently, proposals have been made to place undersea "windmills" in areas of strong tidal currents to generate power without incurring the environmental cost of damming an estuary.

Drift Currents. Drift currents are caused by winds moving the surface layer of water. The strength of a drift current depends on wind speed, duration (how long the wind has been blowing), and fetch (how wide the body of water is). The velocity of drift currents rarely exceeds 1–2% of wind speed (1). Drift currents tend to move in the direction the wind is blowing but are also affected by the rotation of the earth, which creates a phenomenon known as the Coriolis effect. The Coriolis effect is best known for describing the direction in which a whirlpool or hurricane will rotate (clockwise in the northern hemisphere, counterclockwise in the southern hemisphere), but it also affects currents the same way. Thus, drift currents in the northern hemisphere will tend to curve to the right (and in the southern hemisphere to the left), giving rise to what is known as "Ekman drift." Finally, when a drift current encounters a coastline, it will tend to parallel the coastline in what is known as "littoral drift."

Upwelling. Upwelling refers to the vertical movement of water that occurs when the surface layer is pushed offshore

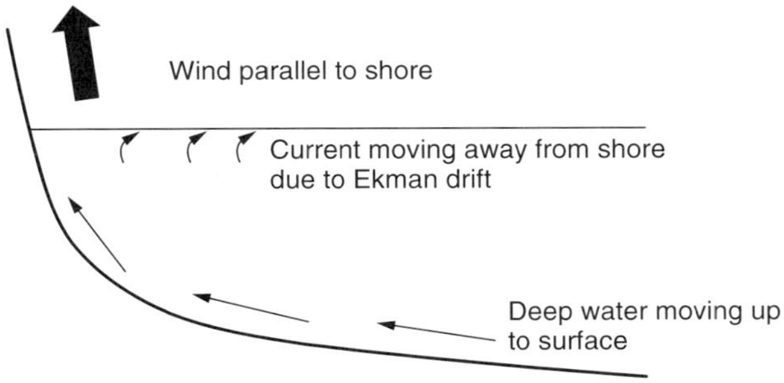

Figure 1. Upwelling along a coast.

by wind and Ekman drift, causing deeper waters to rise up and take its place (Fig. 1). While upwelling can be localized and transitory, prevailing winds create major upwelling areas along the coasts of Oregon in the United States, Peru in South America, northwest Africa, southwest Africa, and the western Indian Ocean. Where winds are steady, the upwelling is likewise fairly constant, for example, the Peruvian upwelling system; but where winds fluctuate (northwest Africa) or are only seasonal (the monsoon winds of the eastern Indian Ocean), upwelling is also periodic or seasonal. Upwelling introduces deep waters rich in nutrients into the coastal ecosystems, creating areas of high productivity and high concentration of marine life (4).

CHEMICAL CHARACTERISTICS OF COASTAL WATERS

Although fresh water is an important component of estuaries, ocean water is the primary component of most coastal waters. The primary dissolved constituents of ocean water (in greatest to lowest concentration) are calcium, magnesium, sodium, potassium, bicarbonate, sulfide, chloride, and bromide. These molecules combine in various ways to create salts, of which sodium chloride (table salt) is the most common. Salt content, or salinity, is measured in parts per thousand (ppt). The average salinity of ocean water is 35 ppt (5), but in coastal waters salinity can be reduced by the addition of fresh water from rivers, overland runoff, and/or groundwater discharge.

Ocean water also contains nutrients in the form of nitrogen and phosphate salts. The concentration of these nutrients is very low in the surface layer of the open ocean, only 0.001–0.08 mg/L N and 0.002–0.003 mg/L P, but can be 10–100 times higher in the deep layer. Oxygen, a very important constituent for biological activity, is dissolved in ocean water at concentrations from about 8 cm^3/L (for cold waters near the poles) to 4.5 cm^3/L (for warm waters at the equator). Other dissolved gases in ocean water include nitrogen, carbon dioxide, and hydrogen sulfide. The pH of ocean water varies between 7.9 and 8.3 (1).

With 44% of the world's population living within 150 km of the coast (6), human activities have a significant influence on the chemistry of coastal waters. Garbage, sewage, effluent from manufacturing plants, runoff from industrial and agricultural areas, as well as many other sources contribute nutrients, sediment, metals, pesticides, hydrocarbons, and other pollutants to coastal waters. Clumps of floating garbage, dead fish, and oiled beaches are some of the most visible effects of humans on coastal waters, but more recent concern has centered on "dead zones," areas where excess nutrients have led to overproduction of plant material, which in turn leads to oxygen depletion in a process known as "eutrophication" (7). Nutrient enrichment of coastal waters is a global problem that continues to grow as more countries increase their use of fossil fuels and inorganic fertilizers, the two principal sources of anthropogenic phosphorus and nitrogen (8).

TYPES OF COASTAL WATERS

Estuaries and Bays

An estuary is "a semi-enclosed coastal body of water which has a free connection with the open sea and within which sea water is measurably diluted with fresh water derived from land drainage" (9). Although the "land drainage" is not specified in Pritchard's classic definition, it is generally assumed that the freshwater source is a river, as is mentioned in many other definitions of an estuary (10). The tidal portion of streams that drain into estuaries is also considered to be part of the estuary. Bays are also semienclosed bodies of water, normally open to the ocean, but they differ from estuaries in not being associated with a river. Estuaries are much more common along the world's coasts than bays, although the term "bay" is often used in the name of an estuary, for example, Chesapeake Bay. Both estuaries and bays provide protected aquatic environments for marine life, but estuaries are more diverse and often more productive due to the range of habitats created by the salinity gradient and input of nutrients from streams. Dyer (11) described four physiographic types of estuaries: coastal plain estuaries (such as the drowned river valleys along the southeast coast of the United States), fjords (such as the submerged glacial valleys of Canada and New Zealand), lagoons (estuaries created by spits or bars), and tectonic estuaries (such as San Francisco Bay).

Rivers bring fresh water into an estuary, and tides bring saline water into the estuary. Where mixing is minimal, a layered water body is created with the less dense fresh water on top and the more dense saline water on the bottom (Fig. 2).

Turbulence in the water causes mixing of the two layers, which increases the thickness of the fresher upper layer as it moves from the landward part of the estuary toward

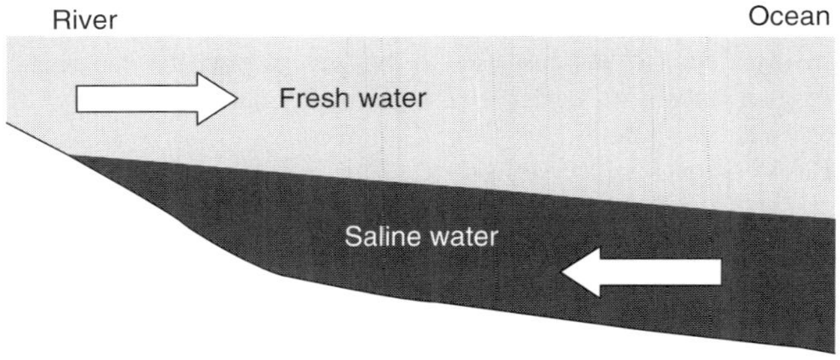

Figure 2. Layers of fresh water and saline water in a stratified estuary.

the ocean (4). The degree of mixing between the layers depends on the amount of freshwater input, the strength of tidal flow, and the effects of wind. Estuaries are often described as being one of three types: salt-wedge, partially mixed, or vertically homogeneous. When an estuary has a large freshwater input relative to tidal flow, the saline water tends to remain at the bottom with minimal mixing and the tip of this "salt wedge" changes location with the seasonal variation of freshwater flow. In partially mixed estuaries, the effects of tides and/or wind create a transition zone between the saline and fresh water layers. The mixing of the layers results in resuspension of sediments and nutrients brought into the estuary by the river and tide. In a vertically homogeneous estuary, there are no longer two water layers, but rather a single mixed body of water with a salinity gradient from low at the river mouth to high at the estuary's outlet to the ocean. In fjords, a sill at the mouth of the estuary traps saline water behind it, and mixing of fresh and saline water only occurs at the surface in water lying above the sill. The deeper waters behind the sill are replaced only when storm events disrupt this balance.

River discharge and tidal exchange both bring sediment and nutrients into an estuary. Estuaries tend to act as sediment traps due to two changes that occur when river discharge meets seawater. First, when clay particles in fresh water encounter saline water, the particles tend to flocculate (creating what is called a "turbidity maximum") and ultimately sink to the bottom of the estuary. Second, as rivers discharge fresh water at the surface of the estuary and tides bring saline water into the bottom of the estuary, current velocities decrease, causing sand particles to settle out. Complex hydrodynamics determine how much sediment stays in the estuary and how much is transported through the estuary to the ocean. Ultimately, sediments in estuaries undergo many cycles of transport by tides, currents, and waves before they find a permanent resting place (12).

Nutrients can be brought into estuaries in particular form, as dissolved constituents, or even via the atmosphere. Phosphorus, one important nutrient, has natural sources in soils and rocks. Weathering creates small particles with phosphorus bound to them that are carried into estuaries. Nitrogen, another important nutrient, is also found naturally in soils, but it is much more soluble than phosphorus and so tends to enter an estuary predominantly in dissolved form. Natural sources of nitrogen and phosphorus are generally small when compared to human-related sources such as agriculture (inorganic fertilizers), wastewater, and fossil fuel. Atmospheric deposition of nitrogen derived from fossil fuel combustion or animal waste is thought to be an important source of excess nutrients to estuaries in the United States (8).

Rocky Shores and Beaches

Rocky shores and beaches are high energy environments. Waves and currents keep the water almost constantly in motion and well aerated. The shoreface is a transition zone where waves become influenced by the bottom and change shape, increase in height, and realign to become more parallel to the shoreline. In the surf zone, currents are driven by the interaction of waves with the shoreface and with other waves. In some places the water draining off the shoreface creates very strong seaward-moving currents known as "rip currents." Offshore from the surf zone, currents are dominated by wind and tide energy but still tend to parallel the coastline.

Plants and animals must have special adaptations to thrive in these environments. Kelp and seaweed have strong attachment systems to keep them in place, and the dominant invertebrates have protective shells. Furthermore, the vigorous movement of water along rocky shores and beaches is essential to the survival of these organisms because they feed by filtering the water or snatching debris as it floats by. Sandy areas are the most dynamic because not only is the water constantly moving, but the substrate itself tends to move as well. Although many beaches seem devoid of plant or animal life, that appearance occurs only because most beach inhabitants are buried in the sand.

Shelf Waters

Seaward of estuaries and the coastline are the waters overlying the continental shelf. Tides continue to influence these waters, creating turbulence that results in a mixed water layer that can extend for tens of kilometers outward from the coast (2). Beyond this mixed layer, coastal waters become stratified into the surface and deep layers described earlier. Areas of high productivity tend to occur at the transition between mixed and stratified waters.

Coastal shelf waters are an important component of ocean circulation. Prevailing winds and the earth's rotation create a system of major currents that circle the ocean basins and in some cases move water from one basin to another. The strongest currents occur on the western sides of ocean basins, where they carry water toward the poles, and weaker currents occur on the eastern sides of ocean basins, where they carry water toward the equator. These major currents, which transport volumes ranging from 10 to 100 m^3/s (2), serve many important biological functions related to migration, dispersal of juvenile organisms, and transportation of nutrients. One of the best-known oceanic currents is the Gulf Stream, which occurs off the east coast of North America. Coastal waters also contain coastal boundary currents that trend parallel to the coastline, their overall direction determined by the shape of the coastline and the way winds tend to reorient perpendicular to the coast as they approach the coastline. For example, landward of the Gulf Stream in the western north Atlantic Ocean, a coastal boundary current flows south—opposite to the direction of Gulf Stream flow. Coastal boundary currents transport lesser volumes of water than oceanic currents—about 10 m^3/s on the average (2).

IMPORTANCE OF COASTAL WATERS

Coastal waters are extremely important for fisheries. With the exception of a few deep-water migratory species such as tuna, most other commercial and recreational marine fish (and shellfish) feed, reproduce, and are harvested in coastal waters. In the United States, coastal waters support a commercial fishing industry that in 2002 landed 4.3 million metric tons of fish and shellfish valued at $3.1 billion. The commercial fishing industry as a whole contributed $28.4 billion to the U.S. gross national product (GNP) in 2002 (13). Recreational fishing also produces billions of dollars of economic benefits per year (14). Coastal waters are also a very important habitat for marine mammals such as seals, manatees, and sea lions, some of which are listed as threatened or endangered under the U.S. Endangered Species Act.

Coastal waters are also very important to interstate and international commerce. Transport by water is generally the most economical and efficient means to move goods. U.S. ports and waterways transport more than 2 billion tons of domestic and foreign goods each year. The marine transportation industry contributes $742 billion to the U.S. GNP and provides approximately 13 million jobs (15).

Given the importance of coastal waters to fisheries, marine transportation, and numerous other uses that are significant from either an economic or ecological perspective, conflicts over coastal waters are common. Laws and regulations governing environmental protection and management of natural resources in the United States and other countries often require a cost–benefit analysis of proposed activities in coastal waters, or a determination that a proposal is in the best interests of the public. However, uncertainties about quantifying ecological values and predicting the response of coastal systems to change make it difficult to conduct these kinds of evaluations. The future health of coastal waters depends on a better understanding of their processes and their response to human activities.

BIBLIOGRAPHY

1. Barnabé, G. and Barnabé-Quet, R. (2000). *Ecology and Management of Coastal Waters: The Aquatic Environment*. Springer-Verlag, New York.
2. Beer, T. (1997). *Environmental Oceanography*. 2nd Edn. CRC Press, Boca Raton, FL.
3. National Imagery and Mapping Agency (NIMA). (1995). Tides and tidal currents. In: *The American Practical Navigator, Publ. No. 9*. NIMA, Bethesda, MD, pp. 143–164.
4. Mann, K.H. (2000). *Ecology of Coastal Waters with Implications for Management*. 2nd Edn. Blackwell Science, Malden, MA.
5. Dittmar, W. (1884). Report on the researches into the composition of ocean water collected by *H.M.S. Challenger* during the years 1873–1876. In: *Report on the Scientific results of the Voyage of the H.M.S. Challenger: Physics and Chemistry*. J. Murray (Ed.). Vol. 1. HMSO, London.
6. Cohen, J.E. et al. (1998). Estimates of coastal populations. *Science* **278**(5341): 1211–1212.
7. Nixon, S.W. (1995). Coastal marine eutrophication: a definition, social causes, and future concerns. *Ophelia* **41**: 199–219.
8. National Research Council (NRC). (2000). *Clean Coastal Waters: Understanding and Reducing the Effects of Nutrient Pollution*. National Academy Press, Washington, DC.
9. Pritchard, D.W. (1967). What is an estuary: physical viewpoint. In: *Estuaries*. G.H. Lauf (Ed.). *AAAS* Publ. No. 83, Washington, DC, pp. 3–5.
10. Perillo, G.M.E. (1995). Geomorphology and sedimentology of estuaries: definitions and geomorphologic classifications of estuaries. *Dev. Sedimentol.* **53**.
11. Dyer, K. (1973). *Estuaries: A Physical Introduction*. Wiley-Interscience, Hoboken, NJ.
12. Nichols, M.M. and Biggs, R.B. (1985). Estuaries. In: *Coastal Sedimentary Environments*. 2nd Edn. R.A. Davis (Ed.). Springer Verlag, New York, pp. 77–186.
13. U.S. Department of Commerce (DOC). (2003). *Fisheries of the United States, 2002*. DOC, Silver Spring, MD.
14. American Sportfishing Association (ASA). (2002). *Sportfishing in America: Values of Our Traditional Pastime*. Alexandria, VA.
15. U.S. Department of Transportation (DOT). (1999). *An Assessment of the U.S. Marine Transportation System—A Report to Congress*. DOT, Washington, DC.

MARINE COLLOIDS

Peter H. Santschi
Texas A&M University
Galveston, Texas

Aquatic colloids, as nanoparticles and macromolecules in the 1 nm to 1 μm size range (1), differ fundamentally from solutions. In a solution, all the relevant interactions are comparable to the thermal energy, kT. Thus, even if water 'prefers' water, the tendency toward disorder—entropy—is dominant. In contrast, the colloidal state is unstable because of large interfacial energies and because particle–particle interactions are stronger than kT. In aquatic systems, physical techniques such as diffusion, permeation, light scattering, and viscosity can be used to characterize colloidal systems, chemical techniques can help to unravel their compositional complexities, and biological techniques can give insights into degradability, bioavailability, and stability. New insights into colloids in marine systems require application of state-of-the-art techniques and also paradigm shifts.

Most of the organic matter in the ocean is in the dissolved state, and 20–30% of the marine DOC is colloidal and subject to rapid turnover (2). Marine colloidal matter (COM) contains biologically resistant heteropolycondensations, such as humic-type molecules (geopolymers) that act as soaps, and products of carbon fixation and consumption, such as exopolymeric substances (EPS) that self-assemble (3,4) and thus act as glues (5,6, and references therein). Both geopolymers and biopolymers also contain strong metal-binding ligand groups, rendering them as metal-sequestering agents. Therefore, colloids in the ocean can act as vectors of metal removal and of solubilization, depending on the biomolecule, as well as modifiers of metal bioavailability. In addition to questions of metal bioavailability in the delicately balanced marine ecosystem, EPS in the marine colloidal pool are also important in their response to global change, as they provide some of the sinking material, that is, one of the conveyor belts of global carbon cycling. Last but not least, man-made nanoparticles, such as buckeyballs, carbon fibers, and nanotubes, which are potentially toxic to man, will make their way into the environment, and ultimately into the ocean, where they have unknown behavior and fate.

Colloids in seawater can be small and globular or long and fibrillar, as shown in Figure 1. Aquagenic fibrillar gel-forming microbial exudates are classified as "rigid", other colloids, such as terrestrially derived (pedogenic) nanometer-sized humic acids, are more globular and are called "flexible" polymers (8–11). Both types of colloids are stabilized by metal ions such as Ca^{2+}, which holds individual humic acid molecules together as

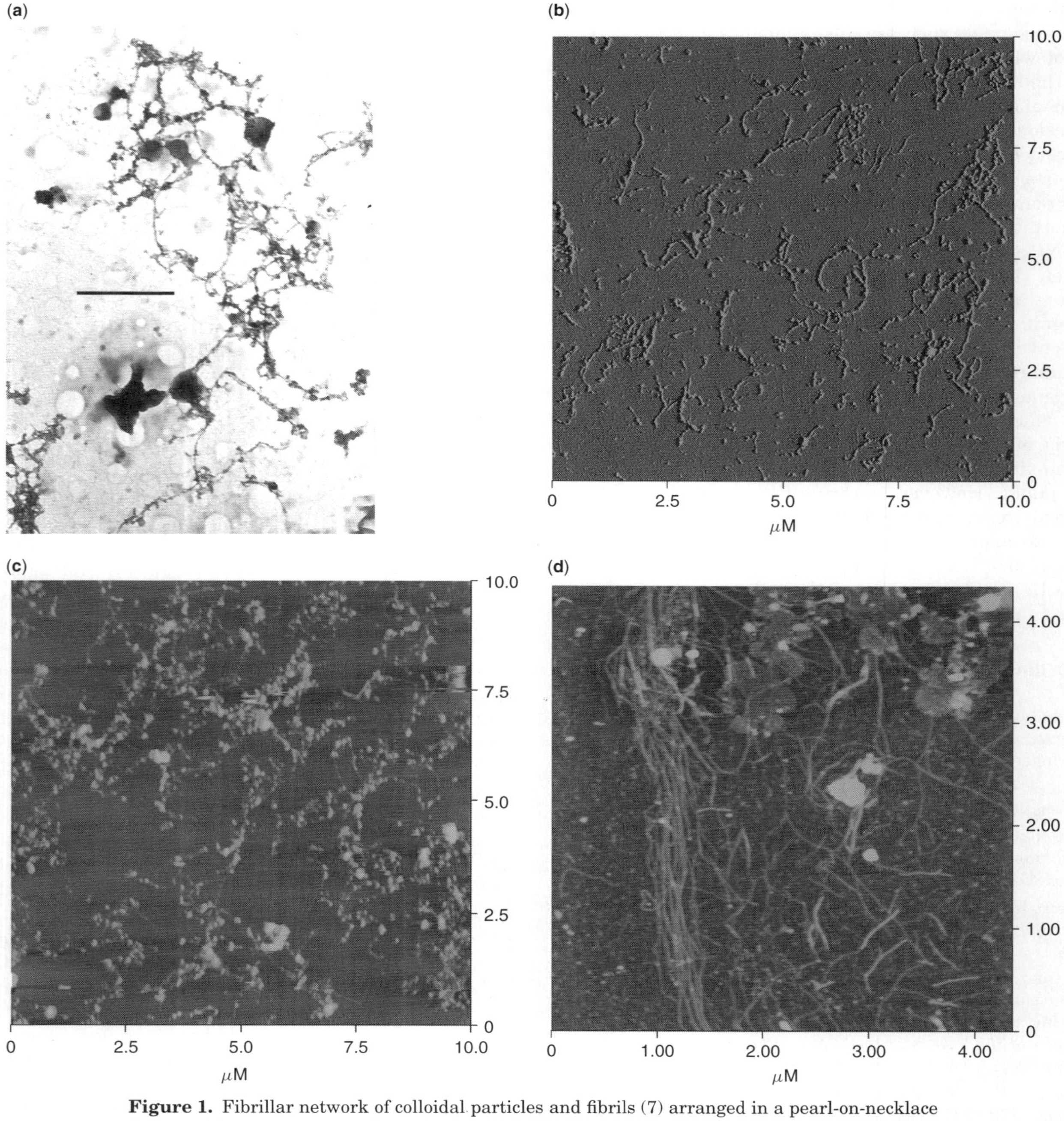

Figure 1. Fibrillar network of colloidal particles and fibrils (7) arranged in a pearl-on-necklace fashion. (**a**). Transmission electron microscopy (TEM) of stained fibrillar network, embedded in a hydrophilic resin, which is part of colloidal organic matter, COM (ultrafilter size range of 2–400 nm), sampled from Middle Atlantic Bight surface waters (courtesy of K. Wilkinson); (**b**). Atomic force microscopy (AFM) of sample from Middle Atlantic Bight surface water COM; Fibrils were also identified by AFM, albeit at lower abundance, in nepheloid layers at 2500 m but not in midwaters; (**c**). AFM of COM from Galveston Bay waters, TX. (**d**). AFM image of COM from surface waters of the Gulf of Mexico, which shows the potential for gel formation of fibrillar macromolecules at the edge of the mica support plate, where the gel-like aggregates likely formed as a consequence of drying.

aggregates (12) and gives rigidity to alpha-helical regions within exopolymeric fibrils (13). These exopolymeric fibrils are up to several 1000 nm in length, but only 1–3 nm wide, and can exist abundantly attached to cell surfaces or as free colloids (7,14–16). The degree of rigidity of fibrillar biopolymers can be studied using atomic force microscopy (AFM) or SANS (small angle neutron scattering). These fibrils can also be characterized by persistence length (defines a kind of stiff-chain-segment length for a linear polymer chain), end-to-end distance, and contour length (11). These measures allow predicting their physicochemical behavior in solution.

Marine colloids are commonly sampled for characterization using cross-flow ultrafiltration (17,18, and references therein) or field-flow fractionation techniques (19). Recent applications of ultrafiltration methods to the sampling of aquatic colloids have advanced our understanding of the abundance, distribution, physical conformation, chemical composition, and environmental behavior of colloids in aquatic systems, including freshwater, seawater, and groundwater. Applications of ultrafiltration to marine systems date back to the 1970s and 1980s, when mostly stirred cell ultrafiltration units with membrane disks were used. Applications of large volume cross-flow ultrafiltration systems to sampling marine colloids started in the early 1990s, and they are now commonly used in marine science (20,21).

In estuarine waters, many metals such as Cu, Ag, Fe, Hg, and Zn, have been found largely in colloidal form (5,18,22,23, and references therein). In marine environments, studies indicated that high molecular weight (HMW) COM has relatively high biological and chemical reactivity and short residence times in the water column (2,24). Therefore, colloids are also important intermediaries in organic carbon cycling in aquatic environments. Recent advances in sampling techniques now enable the isolation of large quantities (e.g., 100s of mg) of environmental colloids from large volumes of natural waters for elemental, isotopic, and molecular characterization (2,7,20,25–27), although there is still a debate about optimal sampling procedures (21,28, and references therein).

Due to their amphiphilic (or amphipathic) nature, aquatic colloids are not only charged, thus providing potentially strong chelating binding sites for trace metals and radionuclides, but also contain hydrophobic moieties for binding more hydrophobic organic trace compounds (5,29,30, and references therein). There are many potential metal binding sites in aquatic biomacromolecules, but most of them are occupied by major seawater cations, and much fewer with the much rarer trace metal ions. Binding sites include carboxyl, hydroxyl, phosphate, sulfate, amino, and sulfhydryl groups; A (or hard) metals prefer O over N and S binding sites, and B (or soft) metals prefer S (or N) over O binding sites (31). One would also expect to find that metal enrichments follow the Irving–Williams series, which has been documented for marine colloids by Guo et al. (32), and for bacterial EPS by Rudd et al. (33). Many of these sites are not exceptionally strong when acting alone, but the particular steric environment of the macromolecule can provide strongly chelating binding sites for trace metals when acting in concert (34,35). In addition, in the "egg-box" chelating environment (13,36), where Ca^{2+} stabilizes the alpha helix of acid polysaccharides (APS) such as alginic acid, Ca^{2+} can be replaced by other cations with similar radii.

EPS are usually rich in APS, and because much more is known about biopolymers such as APS that are studied in food science or biomedical science, where they are used as emulsifiers, stabilizers, coagulants (or anticoagulants) through their amphiphilic properties (37), the concepts and approaches in these scientific disciplines are very useful here. APS include alginates, carrageenans, xanthans, and hyaluronic acids, which contain building blocks that are also found in EPS and transparent exopolymeric particles (TEP) (14,38,39). One may therefore apply knowledge from these scientific disciplines to the biomolecular complexants in the marine colloidal pool, which are not as well studied. The physicochemical properties of EPS molecules are important because they can provide "hideouts" for micronutrients such as trace metals. EPS molecules contain APS and also proteins or lipids, which makes these molecules amphiphilic (1). The amphipilicity is caused by the generally hydrophilic properties of their acid polysaccharidic regions and the more hydrophobic properties of their protein or lipid-rich regions (37). Many of these EPS biopolymers also form gels (3,6) and thus, have properties of hydrocolloids.

Colloidal ligands that have surface-active properties and are found in the filter-passing fraction can be the cause of particle concentration effects on particle–water partition coefficients (K_d, the concentration ratio in particles and in solution) and kinetic constants (k_i) of trace metals or radionuclides, which have been documented in the literature for many years. Colloidal ligands can often be removed by coagulation (40,41) and flocculation (5). Such particle-concentration effects disappear, however, when one corrects for the presence of colloidal ligands (29). Values of the partition coefficients of the A-metal ^{234}Th(IV) to colloids (K_c) are generally similar to K_d values for particles (42), but this is not necessarily the case for B-metals such as Cu (43,44) where generally K_c is larger than K_d.

Aggregation rates of colloidally complexed metals show two distinctive reaction rates; a larger rate constant is consistent with Brownian "colloidal pumping" that occurs at a uniform rate and is faster at higher particle concentrations and a smaller rate constant that is different for different metals (22,45).

^{234}Th, which is generated *in situ* from the radioactive decay of ^{238}U and has a half-life of 24 days, can be used to derive residence times of Th(IV)-binding colloids. Average colloidal residence times (with respect to coagulation) range from fractions of days in estuaries to a few weeks in the surface ocean (2,20). Similar results are obtained in controlled laboratory coagulation experiments using radioactive metals bound to estuarine COM (23,45). These laboratory results also confirmed particle concentration effects on removal rate constants, previously demonstrated for Th(IV) (34,40,41).

Residence times that reflect colloidal organic carbon ages are longer, as much of the carbon is recycled. For example, Santschi et al. (2,7) and Guo et al. (42,46) obtained mostly "modern" (i.e., younger than 1950) apparent radiocarbon ages for high molecular weight colloids $\geq$10 kDa, but mostly older ages, for example, 400 to 4500 years, for low molecular weight COM $\geq$1 kDa in surface waters containing mostly biopolymers (2). In contrast, if the COM $\geq$1 kDa is enriched to 100% polysaccharides through repeated alcohol precipitation, modern radiocarbon ages are obtained for this material (7). In contrast to colloids from the euphotic zone, colloids from benthic nepheloid layers of estuaries and continental shelf or slope areas of the Gulf of Mexico and the Middle Atlantic Bight, which contain mostly geopolymers generated by hydrodynamic and physicochemical fractionation processes near the sediment–water interface during sediment resuspension (46,47), this age distribution is reversed. For example, apparent radiocarbon ages are up to 10 ky for $\geq$10 kDa colloids, and younger for $\geq$1 kDa colloids, indicating fundamental differences between biopolymers and geopolymers.

It thus appears that whenever colloids are generated by microorganisms in surface and bottom waters, their importance in trace metal binding and removal might be disproportionately higher than would be expected from the observed colloidal fractions of trace metals.

TEP-like biopolymers control the scavenging of metals and radionuclides and the coagulation/flocculation of particles in marine systems (5, and references therein) and also the early development of biofilms (15), which is largely due to their surface-active nature. These biopolymers can also initiate or modify precipitation of MnO_2 and FeOOH (48,49), SiO_2 (50), and $CaCO_3$ (51). Moreover, APS rich polymers in the extracellular milieu form flocs (52,53) and also bind extracellular enzymes in their active forms. This enzymatic activity allows trace metal binding (24,34) and modifies the solubility of associated molecules (14).

What do we know about the physicochemical parameters of colloidal organic matter in the ocean, for example, proton reactive sites, their pK_a values, electrophoretic mobilities, and relative hydrophobicities? Santschi et al. (2) reported a total concentration of 1.4 meq/g of proton reactive sites in marine colloids from the Gulf of Mexico, which is intermediate between that of humic and fulvic acids (10–16 meq/g) and polysaccharides (0.3–0.8 meq/g) (1). These proton-reactive sites are often assumed to have pK_a values of 2, 4, 6, 8, and 10 (54). Actual values of $pK_{a1} \leq 3$ have been determined for aquatic colloids (7,34,54–56), which could suggest sulfate or polyphosphate as possible ligands. Values for pK_{a2} are more variable and increase from about 4 to 9 across the salinity gradient in the Firth of Clyde estuary (55,56), suggesting that the nature of colloids changes dramatically across an estuarine salinity gradient. Interestingly, log K values, as well as fractions of metal-specific ligands (C_L) in the colloidal fraction (colloidal C_L/total C_L) of selected metals in coastal waters were significantly related to the pK_{a2} of the colloids; Cu showed a positive relationship for both log K and % colloidal C_L with pK_{a2}, whereas for Pb, a negative relationship was found (55). This is possibly due to the different affinities of these two metals to the two predominant forms of freshwater colloids (19), that is, the smaller organic carbon rich (enriched in Cu) and larger Fe-rich (enriched in Pb) colloids, both of which are also found in coastal waters (23).

In agreement with the pKa values, measured electrophoretic mobilities of aquatic colloids are negative. Electrophoretic mobilities, u_E, determined by dynamic light scattering in a ZetaSizer instrument at pH 8, were measured at about -2.5×10^{-8} m^2 V^{-1} s^{-1}, ranging from about (-2 to -3.4) $\times 10^{-8}$ m^2 V^{-1} s^{-1} (55–57). Their values were weakly related to the colloidal fraction of metal-specific ligands (colloidal C_L/total C_L); Cu and Pb again related in opposite ways.

In accordance with the relative high concentration of moderately strong acidic sites, for example, pK_{a1} values of ~3 (55,56) or below for marine colloids (9), isoelectric points of about 2 to 4 for marine colloids have been determined by Quigley et al. (34) and Alvarado-Quiroz (35), suggesting that the surface charge is dominated by relatively strong acidic groups. Conversely, positively charged ligand groups such as amino groups are considerably rarer.

Marine colloids are negatively charged, suggesting hydrocolloid-like behavior, but that does not preclude coexistence with more hydrophobic sites, as has been said before. EPS and APS have been described as amphiphilic (1,37,58). Furthermore, it is likely that more hydrophobic ligands would be more particle-reactive and thus, have a shorter residence time in water. However, this has not yet been demonstrated in the field. Laboratory studies of bacterial attachment, however, show a clear relationship between attachment probability and relative hydrophobicity of bacterial surfaces (58,59), which are composed of EPS.

There are many things we do not know yet about marine colloids. For example, more than 50% of the organic matter in this pool remains uncharacterized at the molecular level (26,27,60), our knowledge about their tendency toward gel formation is still in its infancy (6), and we still do not have a quantitative assessment of the relative role of biodegradation (61) versus self-assembly (3) or aggregation contributing to the observed size distribution of marine colloids and particles (2,6,24). Even the molecular weight distribution of DOC is not as certain as one would conclude from the application of just one single method. For example, results from the application of state-of-the-art electrospray mass spectrometry techniques (62,63) indicate average molecular weights of DOC, normalized to charge, of less than 1 kDa, possibly due to fragmentation of carboxylic acid-rich compounds during electrospray ionization. On the other hand, much higher average molecular weights of 10^5 to 10^6 Da can be obtained from images of freshwater (9) or marine colloids (7) when using transmission electron microscopy (TEM) of stained specimens consisting of abundant fibrillar macromolecules 100s to 1000s of nm long, when embedded in hydrophilic resins, or AFM techniques of the same specimens mounted by adsorption or centrifugation onto mica surfaces that are flat on the atomic scale (Fig. 1). Thus, scientists interested in the properties of marine colloidal matter still have a long way to go until the role of this abundant material in seawater and self-assembles on the nanoscale (3) become better understood.

Acknowledgments

This review was supported, in parts, also by grants to the author from the U.S. National Science Foundation (NIRT-0210865 and OCE-0351559) and the Texas Institute of Oceanography.

BIBLIOGRAPHY

1. Buffle, J. (1990). *Complexation Reactions in Aquatic Systems: An Analytical Approach*. Ellis Horwood, New York.
2. Santschi, P.H. et al. (1995). Isotopic evidence for contemporary origin of high molecular weight organic matter in oceanic environments. *Geochim. Cosmochim. Acta* **59**: 625–631.
3. Chin, W.C., Orellana, M.V., and Verdugo, P. (1998). Spontaneous assembly of marine dissolved organic matter into polymer gels. *Nature* **391**: 568–572.
4. Kerner, M. et al. (2003). Self-organization of dissolved organic matter to micelle-like microparticles in river water. *Nature* **422**: 150–154.
5. Santschi, P.H., Burd, A.B., Gaillard, J-F., and Lazarides, A.A. (2004). Transport of materials and chemicals by nano-scale colloids and micro to macro-scale flocs in marine, freshwater and engineered systems. In: *Flocculation in Natural and Engineered Environmental Systems*. I.G. Droppo, G.G. Leppard, S.N. Liss, and T.G. Milligan (Eds.). CRC Press, London, UK in press.
6. Verdugo, P. et al. (2004). The oceanic gel phase: A bridge in the DOM-POM continuum. *Mar. Chem.*, in press.
7. Santschi, P.H. et al. (1998). Fibrillar polysaccharides in marine macromolecular organic matter as imaged by atomic force microscopy and transmission electron microscopy. *Limnol. Oceanogr.* **43**: 896–908.
8. Buffle et al. (1998).
9. Wilkinson, K.J., Balnois, E., Leppard, G.G., and Buffle, J. (1999). Characteristic features of the major components of freshwater colloidal organic matter revealed by transmission electron and atomic force microscopy. Colloids *Surf. A: Physicochemical Eng. Aspects* **155**: 287–310.
10. Balnois, E., Wilkinson, K.J., Lead, J.R., and Buffle, J. (1999). Atomic force microscopy of humic substances: Effects of pH and ionic strength. *Environ. Sci. Technol.* **33**: 3911–3917.
11. Balnois, E. et al. (2000). Conformations of succinoglycan as observed by atomic force microscopy. *Macromolecules* **33**: 7440–7447.
12. Simpson et al. (2000).
13. Moe, S., Draget, K., SkjÅk-Bræk, G., and Smidrød, O. (1995). Alginates. In: *Food Polysaccharides and their Application*. A. Stephen (Ed.). Marcel Dekker, New York, pp. 245–286.
14. Leppard, G. (1995). The characterization of algal and microbial mucilages and their aggregates in aquatic ecosystems. *Sci. Total Environ.* **165**: 103–131.
15. Leppard, G. (1997). Colloidal organic fibrils of acid polysaccharides in surface water: Electron-optical characteristics, activities and chemical estimates of abundance. *Colloids Surf. A: Physicochemical Eng. Aspects* **120**: 1–15.
16. Leppard, G.G. et al. (1997). A classification scheme for marine organic colloids in the Adriatic Sea: Colloid speciation by transmission electron microscopy. *Can. J. Fish. Aquat. Sci.* **54**: 2334–2349.
17. Guo, L., Wen, L., Tang, D., and Santschi, P.H. (2000). Re-examination of cross flow ultrafiltration for sampling aquatic colloids: evidence from molecular probes. *Mar. Chem.*
18. Guo et al. (2004).
19. Lyven, B. et al. (2003). Competition between iron- and carbon-based colloidal carriers for trace metals in a freshwater assessed using flow field- flow fractionation coupled to ICPMS, *Geochim. Cosmochim. Acta* **67**(20): 3791–3802.
20. Guo, L. and Santschi, P.H. (1997). Composition and cycling of colloids in marine environments. *Rev. Geophys.* **35**: 17–40.
21. Wells, M.L. (2002). Marine colloids and trace metals. In: *Biogeochemistry of Marine Dissolved Organic Matter*. D.A. Hansell and C.A. Carlson (Eds.). Academic Press, New York, pp. 367–404.
22. Wen, L-S., Santschi, P.H., Gill, G., and Paternostro, C. (1997). Colloidal and particulate silver in river and estuarine waters of Texas. *Environ. Sci. Technol.* **31**: 723–731.
23. Wen, L.S., Santschi, P.H., Paternostro, C., and Gill, G. (1999). Estuarine trace metal distributions in Galveston Bay I: Importance of colloidal forms in the speciation of the dissolved phase. *Mar. Chem.*, **63** (3–4): 185–212.
24. Santschi, P.H. et al. (2003). Control of acid polysaccharide production, and ^{234}Th and POC export fluxes by marine organisms. *Geophysical Res. Lett.* **30**(2), Art. No. 1044, dio 10.1029/2002GL016046.
25. Guo, L., Tanaka, N., Schell, D.M., and Santschi, P.H. (2003). Nitrogen and carbon isotopic composition of high-molecular-weight dissolved organic matter in marine environments. *Mar. Ecol. Prog. Ser.* **252**: 51–60.
26. Benner, R. (2002). Chemical composition and reactivity. In: *Biogeochemistry of Marine Dissolved Organic Matter*. D.A. Hansell and C.A. Carlson (Eds.). Academic Press, New York, pp. 59–90.
27. Aluwihare, L., Repeta, D., and Chen, R. (2002). Chemical composition and cycling of dissolved organic matter in the Mid-Atlantic Bight. *Deep Sea Res.* **II**(49): 4421–4437.
28. Guo, L., Santschi, P.H., and Warnken, K.W. (2000). Trace metal composition of colloidal organic matter in marine environments. *Mar. Chem.* **70**(4): 257–275.
29. Santschi, P.H., Lenhart, J.J., and Honeyman, B.D. (1997). Heterogeneous process affecting trace contaminant distribution in estuaries: the role of natural organic matter. *Mar. Chem.* **58**: 99–125.
30. Santschi, P.H., Guo, L., and Bianchi, T.S. (1999). Dissolved organic matter in estuaries of the Gulf of Mexico. In: *Biogeochemistry of Gulf of Mexico Estuaries*. T.S. Bianchi (Ed.). John Wiley & Sons, New York, pp. 269–299.
31. Stumm, W. and Morgan, J. (1996). *Aquatic Chemistry: Chemical Equilibria and Rates in Natural Waters*. 3rd Edn. John Wiley & Sons, New York.
32. Guo et al. (2002).
33. Rudd, T., Sterritt, R., and Lester, J. (1984). Formation and conditional stability constants of complexes formed between heavy metals and bacterial extracellular polymers. *Water Res.* **18**: 379–384.
34. Quigley, M.S. et al. (2002). Importance of polysaccharides for ^{234}Th complexation to marine organic matter. *Limnol. Oceanogr.* **47**: 367–377.
35. Alvarado-Quiroz, N. (2004). Ph.D. Dissertation, Dept. of Oceanography, Texas A&M University, College Station, TX.
36. Kohn, R. (1972). Ion binding on polyuronides—alginate and pectin. *Pure Appl. Chem.* **42**: 371–397.
37. Dickinson, E. (2003). Hydrocolloids at interfaces and the influence on the properties of dispersed systems. *Food Hydrocolloids* **17**: 25–39.
38. Decho, A.W. (1990). Microbial exopolymer secretions in the ocean environments their role(s) in food webs and marine processes. *Oceanogr. Mar. Biol. Annu. Rev.* **28**: 73–153.

39. Passow, U. (2002). Transparent exopolymer particles (TEP) in aquatic environments. *Prog. Oceanogr.* **55**: 287–333.
40. Honeyman, B.D. and Santschi, P.H. (1989). Brownian-pumping model for oceanic trace metal scavenging: Evidence from Th isotopes. *J. Mar. Res.* **47**: 951–992.
41. Honeyman, B. and Santschi, P. (1991). Coupling of trace metal adsorption and particle aggregation: Kinetic and equilibrium studies using ^{59}Fe-labeled hematite. *Environ. Sci. Technol.* **25**: 1739–1747.
42. Guo, L., Santschi, P.H., and Baskaran, M. (1997). Interaction of thorium isotopes with colloidal organic matter in oceanic environments. *Colloids Surf. A* **120**: 255–272.
43. Muller, F. (1996). Interactions of copper, lead and cadmium with the dissolved, colloidal and particulate components of estuarine and coastal waters. *Mar. Chem.* **52**: 245–268.
44. Wells et al. (1998).
45. Stordal, M.C., Santschi, P.H., and Gill, G.A. (1996). Colloidal pumping: Evidence for the coagulation process using natural colloids tagged with ^{203}Hg. *Environ. Sci. Technol.* **30**: 3335.
46. Guo, L. et al. (1996). Cycling of high molecular weight dissolved organic matter in the Middle Atlantic Bight as revealed by carbon isotopic (^{13}C and ^{14}C) signatures. *Limnol. Oceanogr.* **41**: 1242–1252.
47. Guo and Santschi. (2000).
48. Cowen, J. and Bruland, K. (1985). Metal deposits associated with bacteria—implications for Fe and Mn marine biogeochemistry. *Deep Sea Res.* **32**: 253–272.
49. Taillefert et al. (2000).
50. Kinrade, S., Gillson, A., and Knight, C. (2002). Silicon-29 NMR evidence of a transient hexavalent silicon complex in the diatom *Navicula pelliculosa*. *J. Chem. Soc., Dalton Trans.* **3**: 307–309.
51. Leveille, R., Fyfe, W., and Longstaffe, F. (2000). Geomicrobiology of carbonate-silicate microbialites from Hawaiian basaltic sea caves. *Chem. Geol.* **169**: 339–355.
52. Alldredge, A.L., Passow, U., and Logan, B.E. (1993). The abundance and significance of a class of large, transparent organic particles in the ocean. *Deep-Sea Res.* **I**(40): 1131–1140.
53. Mopper et al. (1995).
54. Murphy, R.J., Lenhart, J.J., and Honeyman, B.D. (1999). The sorption of thorium (IV) and uranium (VI) to hematite in the presence of natural organic matter. *Colloids Surf. A: Physicochemical Eng. Aspects* **157**: 47–62.
55. Muller, F.L.L. (1998). Colloid/solution partitioning of metal-selective organic ligands, and its relevance to Cu, Pb and Cd cycling in the Firth of Clyde. *Estu. Coast. Shelfish Sci.* **46**: 419–437.
56. Muller, F. (1999). Evaluation of the effects of natural dissolved and colloidal organic ligands on the electrochemical lability of Cu, Pb, and Cd in the Arran Deep, Scotland. *Mar. Chem.* **67**: 43–60.
57. Lead, J., Hamilton-Taylor, J., Davison, W., and Harper, M. (1999). Trace metal sorption by natural particles and coarse colloids. *Geochim. Cosmochim. Acta* **63**: 1661–1670.
58. Stenstörm, T.A. (1989). Bacterial hydrophobicity, an overall parameter for the measurement of adhesion potential to soil particles. *Appl. Environ. Microbiol.* **55**: 142.
59. Ahimou, F. et al. (2001). Influence of electrical properties on the evaluation of the surface hydrophobicity of Bacillus subtilis. *J. Microbiol. Methods* **45**: 119–126.
60. Repeta et al. (2002).
61. Amon, R.M.W. and Benner, R. (1994). Rapid cycling of high molecular weight dissolved organic matter in the ocean. *Nature* **369**: 549–552.
62. Leenheer, J.A. et al. (2001). Molecular resolution and fragmentation of fulvic acid by electrospray ionization/multistage tandem mass spectrometry. *Anal. Chem.* **73**: 1461–1471.
63. Kujawinski, B. et al. (2002). The application of electrospray ionization mass spectrometry (ESI MS) to the structural characterization of natural organic matter. *Org. Geochem.* **33**: 171–180.

READING LIST

Gustafsson, O. and Gschwend, P. (1997). Aquatic colloids: concepts, definitions, and current challenges. *Limnol. Oceanogr.* **42**: 519–528.

DEEP WATER CORALS

NOAA Coral Reef Information System (CORIS)

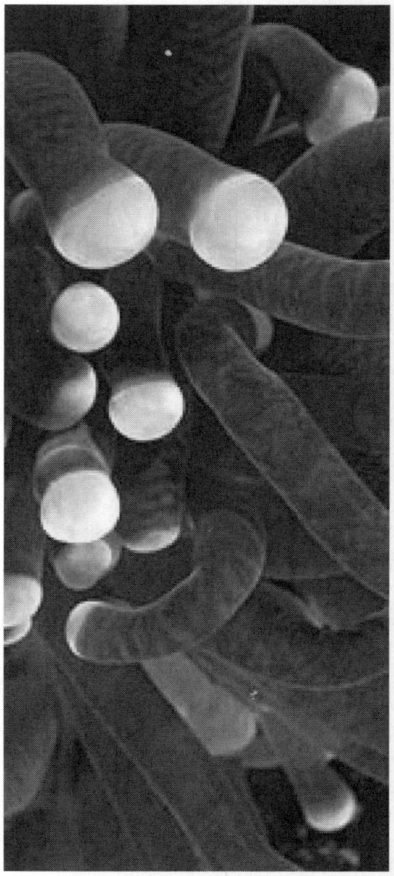

Many people are familiar with the coral reefs that thrive in shallow, well lighted, clear tropical waters where

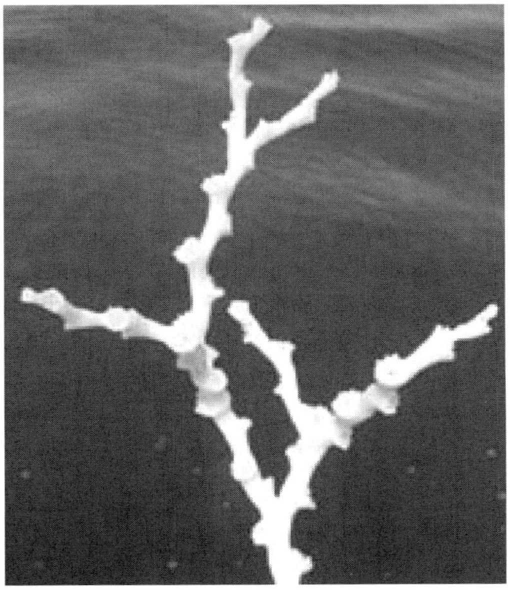

This healthy branch of *Lophelia* coral was sampled from deep ocean reefs off the coast of South Carolina. Unlike tropical species of coral, *Lophelia* possesses no symbiotic algae (zooxanthellae).

myriad colorful hard and soft corals provide habitat "infrastructure" for numerous invertebrates and fishes. The corals provide protection and cover, sources of nutrition, and sites for reproduction. Corals, however, also grow in the deep, cold sea. Although the existence of some of these deep-sea coral thickets has been known for several centuries, initially from pieces of broken corals brought up with fishing gear, scientists know little about their distribution, biology, behavior, and function as essential habitats for fishes and invertebrates.

Some deep-water corals (also called cold-water corals) do not form reefs exactly like those in tropical waters. Often, they form colonial aggregations called patches, mounds, banks, bioherms, massifs, thickets or groves. These aggregations are often still referred to as "reefs." Deep-water corals also provide crucial habitat and reproductive grounds for commercially important fisheries including sea bass, snapper, porgy, rock shrimp and calico shrimp, thus drawing the commercial fishing industry to these fragile areas.

Human activities constitute the most serious threat to these fragile corals. Destructive bottom fishing, as well as oil and gas exploration and exploitation have the potential to destroy large areas of coral habitat in a relatively short time. These activities create coral rubble, which is not a suitable habitat for fishes and invertebrates. In recent years, scientists have begun to study deep-water corals more closely, and some countries with deep-water corals in their territorial waters have begun to implement fishing restrictions in sensitive coral areas.

WHAT ARE DEEP-WATER CORALS?

Far fewer species of corals exist in cold, deep water than in the tropics. Three main groups of corals make up deep-water coral communities: hard (stony) corals of the Order Scleractinia, which form hard, ahermatypic reefs; black and horny corals of the Order Antipatharia; and soft corals of the order Alcyonacea, which includes the gorgonians (sea fans) (1). Deep-water corals are similar in some ways to the more familiar corals of shallow, tropical seas. Like their tropical equivalents, the hard corals develop sizeable reef structures that host rich and varied invertebrate and fish fauna. However, unlike their tropical cousins, which are typically found in waters above 70 m depth and at temperatures between 23° and 29 °C, deep-water corals live at depths just beneath the surface to the abyss (2000 m), where water temperatures may be as cold as 4 °C and utter darkness prevails.

At these depths, corals lack zooxanthellae. These symbiotic algae provide food for many shallow-water corals through photosynthesis. They also assist in the formation of the calcareous skeleton, and give most tropical corals their coloration. By contrast, the polyps of deep-water corals appear to be suspension feeders. They capture and consume organic detritus and plankton that are transported by strong, deep-sea currents. These corals are commonly found along bathymetric highs such as seamounts, ridges, pinnacles and mounds (2).

Deep-water corals range in size from small solitary colonies to huge, massive reef structures, which appear as oases of teeming life surrounded by more barren bathymetry. The gorgonians (sea fans) also range from small individuals to those with tree-like dimensions. The gorgonian, *Paragorgia arborea,* may grow in excess of three meters in length (3). Growth rates of branching deep-water coral species, such as *Lophelia* and *Oculina,* range

This thicket of *Paragorgia* corals was viewed by the deep-sea submersible *Alvin* at 1,043 m depth. *Photo*: Barbara Hecker.

from ~1.0–2.5 cm/yr, whereas branching shallow-water corals, such as *Acropora,* may exceed 10–20 cm/yr. Using coral age-dating methods, scientists have estimated that some living deep-water corals date back at least 10,000 years (4). However, little is known of their basic biology, including how they feed or their methods and timing of reproduction.

GENERAL DISTRIBUTION OF COLD-WATER CORALS

Deep-water corals are found globally, from coastal Antarctica to the Arctic Circle. In northern Atlantic waters, the principal coral species that contribute to reef formation are *Lophelia pertusa, Oculina varicosa, Madrepora oculata, Desmophyllum cristagalli, Enallopsammia rostrata, Solenosmilia variabilis*, and *Goniocorella dumosa*. Four of those genera (*Lophelia, Desmophyllum, Solenosmilia*, and *Goniocorella*) constitute the majority of known deep-water coral banks at depths of 400 to 700 m (5).

Madrepora oculata occurs as deep as 2,020 m and is one of a dozen species that occur globally and in all oceans, including the Subantarctic (6). Colonies of *Enallopsammia* contribute to the framework of deep-water coral banks found at depths of 600 to 800 m in the Straits of Florida (5).

TWO IMPORTANT DEEP-WATER CORALS

Two of the more significant deep-sea coral species are *Lophelia pertusa* and *Oculina varicosa*. These species form extensive deep-water communities that attract commercially important species of fishes, making them susceptible to destructive bottom trawling practices (7). Increased sedimentation places additional stress on corals.

Oil and gas exploration structures and activities, particularly in the North Sea and adjacent areas, also damage *Lophelia* communities. Subsequent oil and gas production activities may also introduce noxious substances into these areas.

LOPHELIA PERTUSA DISTRIBUTION

Lophelia pertusa is the most common aggregate-forming deep-water coral. Typically, it is found at depths between 200 and 1,000 m in the northeast Atlantic, the Mediterranean Sea, along the mid-Atlantic Ridge, the West African and Brazilian coasts, and along the eastern shores of North America (e.g., Nova Scotia, Blake Plateau, Florida Straits, Gulf of Mexico) as well as in parts of the Indian and Pacific Oceans. Like tropical coral reefs, *Lophelia* communities support diverse marine life, such as sponges, polychaete worms, mollusks, crustaceans, brittle stars, starfish, sea urchins, bryozoans, sea spiders, fishes, and many other vertebrate and invertebrate species.

Lophelia has been found most frequently on the northern European continental shelves between 200 and 1000 m, where temperatures range from 4° to 12 °C, but it has also been found at depths greater than 2,000 m. Once a colonial patch is established, it can spread over a broad area by growing on dead and broken pieces of coral (coral rubble). *Lophelia* has a linear extension of the polyps of about 10 mm per year. The reef structure has been estimated to grow about 1 mm per year (8). Scientists have also found *Lophelia* colonies on oil installations in the North Sea (9).

Lophelia pertusa can occur in a variety of structures and forms. DNA-based sequencing tests conducted at the University of Southampton Oceanography Centre, UK, have indicated that different morphological varieties of *Lophelia* all belong to the same species (10).

NOTED LOPHELIA AREAS

The world's largest known deep-water *Lophelia* coral complex is the Røst Reef. It lies in depths between 300 and 400 m west of Røst Island in the Lofoten archipelago, Norway. Discovered during a routine survey in May 2002, the reef is still largely intact. It covers an area approximately 40 km long by 3 km wide.

Relatively close by is the Sula Reef, located on the Sula Ridge, west of Trondheim on the mid-Norwegian Shelf,

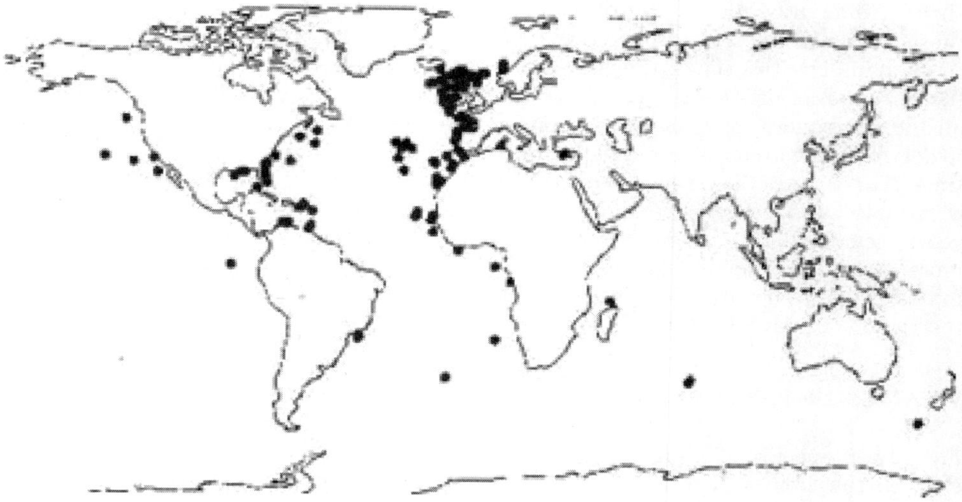

Global distribution of *Lophelia pertusa*. Image: Southampton Oceanography Centre, UK.

Reefs of *Lophelia pertusa* have been recorded on raised offshore seabeds in the north Atlantic off Britain and west Ireland. Image: Dr. Brian Bett, Southampton Oceanography Centre, UK.

at 200 to 300 m depth. It is estimated to be 13 km long, 700 m wide, and up to 35 m high (11), an area one-tenth the size of the 100 km^2 Røst Reef.

Discovered and mapped in 2002, Norway's 1,000-year-old and 2-km-long Tisler Reef lies in the Skagerrak—the submarine border between Norway and Sweden at a depth of 74 to 155 m. The Tisler Reef contains the world's only known yellow *Lophelia pertusa* corals. At present, Norway is the only European nation to enact laws protecting its *Lophelia* reefs from trawling, pollution, and oil and gas exploration. However, the European Commission has introduced an interim ban on trawling in the Darwin Mounds area, west of Scotland, in August 2003. A permanent ban on trawling is expected to follow.

Elsewhere in the northeastern Atlantic, *Lophelia* is found around the Faroe Islands, an island group between the Norwegian Sea and the Northeast Atlantic Ocean. At depths from 200 to 500 m, *Lophelia* is chiefly on the Rockall Bank and on the shelf break north and west of Scotland (12).

One of the most researched deep-water coral areas in the United Kingdom are the Darwin Mounds. The mounds were discovered in 1998 by the Atlantic Frontier Environmental Network (AFEN) while conducting large-scale regional surveys of the sea floor north of Scotland. They discovered two areas of hundreds of sand and cold-water coral mounds at depths of about 1,000 m, in the northeast corner of the Rockall Trough, approximately 185 km northwest of the northwest tip of Scotland. Named after the research vessel *Charles Darwin,* the Darwin Mounds have been extensively mapped using low-frequency side-scan sonar. They cover an area of approximately 100 km^2 and consist of two main fields—the Darwin Mounds East, with about 75 mounds, and the Darwin Mounds West, with about 150 mounds. Other mounds are scattered in adjacent areas. Each mound is about 100 m in diameter and 5 m high.

The tops of the mounds are covered with *Lophelia* corals and coral rubble, which attract other marine life. Side-scan sonar images show that the mounds appear to be sand volcanoes, each with a unique feature—a "tail." The tails are up to several hundred m long, all oriented downstream (13). The tails and the mounds are uniquely characterized by large congregations of deep-sea organisms called xenophyophores (*Syringammina fragilissima*), which are giant unicellular organisms that can grow up to 25 cm in diameter (14). Scientists are uncertain why these interesting organisms congregate in these areas. In addition, the *Lophelia* of the Darwin Mounds are growing on sand rather than hard substrate, an unusual condition unique to this area. Usually, coral larvae almost always settle and grow on hard substrates, such as dead coral skeletons or rock.

Lophelia corals exist in Irish waters as well (15). The Porcupine Seabight, the southern end of the Rockall Bank, and the shelf to the northwest of Donegal all exhibit large, mound-like *Lophelia* structures. One of them, the Theresa Mound, is particularly noted for its *Lophelia pertusa* and *Madrepora oculata* colonies. *Lophelia* reefs are also found along the U.S. East Coast at depths of 500 to 850 m along the base of the Florida-Hatteras slope. South of Cape Lookout, NC, rising from the flat sea bed of the Blake Plateau, is a band of ridges capped with thickets of *Lophelia*. These are the northernmost East Coast *Lophelia pertusa* growths. The coral mounds and ridges here rise as much as 150 m from the plateau plain. These *Lophelia* communities lie in unprotected areas of potential oil and gas exploration and cable-laying operations, rendering them vulnerable to future threats (16).

Finally, *Lophelia* is known to exist around the Bay of Biscay, the Canary Islands, Portugal, Madeira, the Azores, and the western basin of the Mediterranean Sea (13).

OCULINA VARICOSA DISTRIBUTION

Oculina varicosa is a branching ivory coral that forms giant but slow-growing, bushy thickets on pinnacles up to 30 m in height. The Oculina Banks, so named because they consist mostly of *Oculina varicosa*, exist in 50 to 100 m of water along the continental shelf edge about 26 to 50 km off of Florida's central east coast.

Discovered in 1975 by scientists from the Harbor Branch Oceanographic Institution conducting surveys of the continental shelf, *Oculina* thickets grow on a series of pinnacles and ridges extending from Fort Pierce to Daytona, Florida (7,17–19).

Like the *Lophelia* thickets, the Oculina Banks host a wide array of macroinvertebrates and fishes. They also are significant spawning grounds for commercially important species of food fishes including gag, scamp, red grouper, speckled hind, black sea bass, red porgy, rock shrimp, and the calico scallop (20).

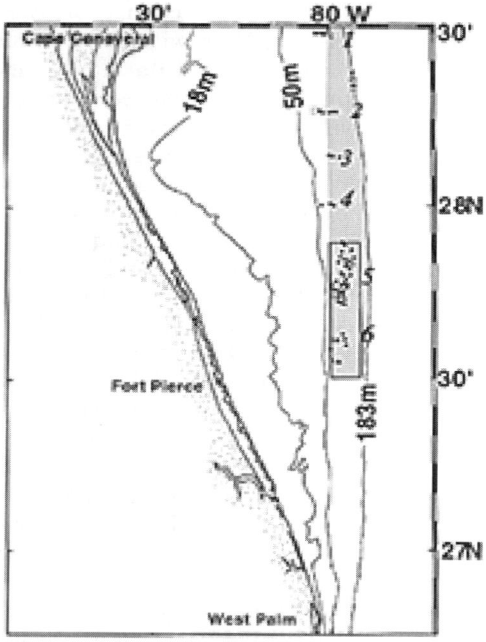

Chart of the Deep-water Oculina Coral Banks Marine Protected Area (MPA). The shaded area is the entire Oculina Habitat Area of Particular Concern (HAPC); the Experimental Oculina Research Reserve (EORR) is the smaller inset box. Recent dive sites from 2001 to 2003 include: 1) Cape Canaveral, 2) Cocoa Beach, 3) Eau Gallie, 4) Malabar, 5) Sebastian, and 6) Chapmans Lumps/Jeff's Reef. (Courtesy of: Dr. John K. Reed, Harbor Branch Oceanographic Institution).

THREATS TO LOPHELIA AND OCULINA CORALS

Both *Lophelia* and *Oculina* corals face uncertain futures. Until recently, *Lophelia* habitats remained undisturbed by human activity. However, as traditional fish stocks are depleted from northern European waters, bottom trawling has moved into deeper waters, where the gear has affected the coral beds. The towed nets break up the reef structure, kill the coral polyps and expose the reef to sediment by altering the hydrodynamic and sedimentary processes around the reef (21).

The fishes and invertebrates that depend on the coral structure lose their habitat and move out of the area. Damage may range from a decrease in the size of the coral habitat with a corresponding decrease in the abundance and biodiversity of the associated invertebrate and fish species, to the complete destruction of the coral habitat. The trawls also may resuspend sediments that, in turn, may smother corals growing downstream of the current. In addition, oil and gas exploration and extraction operations have begun to move into these deep-water areas, further threatening the resident corals.

Scientists estimate that within the Norwegian Exclusive Economic Zone, 30 to 50 percent of the total coral area of the Norwegian shelf has been damaged or destroyed by trawling (8). Scientists from the International Council for the Exploration of the Sea, the main provider of scientific advice to the European Commission on fisheries and environmental issues in the northeast Atlantic, have recommended that to protect the remaining deep-water coral groves, all of Europe's deep corals must be accurately mapped and then closed to fishing trawlers (22).

In 1999, the first complete mapping of the Sula Reef was carried out by the Norwegian Hydrographic Society, which used the latest available multibeam echosounder equipment to record both depth and backscatter data. The mapping was the product of joint cooperation between the Geological Survey of Norway and the Institute of Marine Research. That same year, the Norwegian Ministry of Fisheries issued regulations for the protection of the *Lophelia* reefs. An area of 1,000 km^2 at Sula, including the large reef, is now closed to bottom trawling. In 2000, an additional area, about 600 km^2 was closed. The Røst Reef, an area of about 300 km^2, was closed to bottom trawling in 2002.

Florida's Oculina Banks, once teeming with commercially important fish, now appear to be severely depleted of fish stocks (23). Much of the *Oculina* coral has been reduced to rubble, probably the result of a combination of destructive bottom trawling and natural causes like bioerosion and episodic die-offs. Regardless of the cause, the Oculina Banks now support a drastically reduced fishery because most of the significant spawning grounds have been destroyed (7,19).

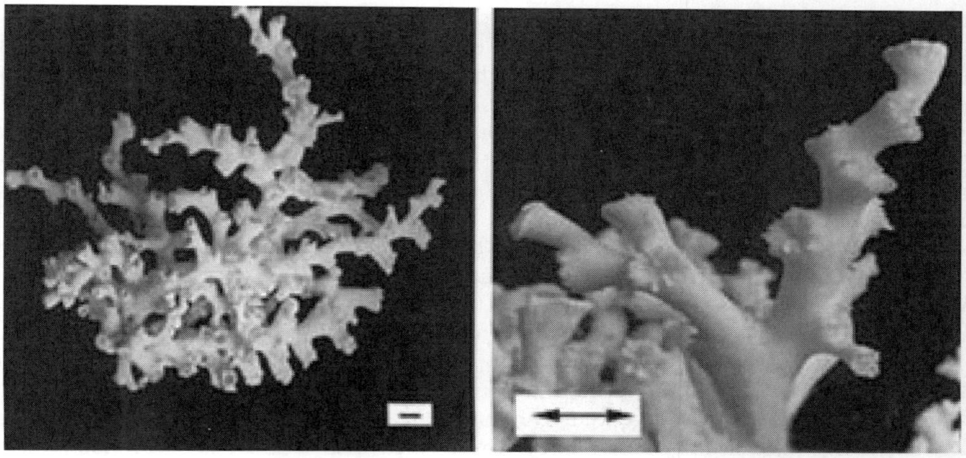

Two images of the deep sea coral *Lophelia pertusa*. The image on the left shows the complex growth structure of a small colony. On the right is a closeup of individual polyps. *Photo*: Ref. 7.

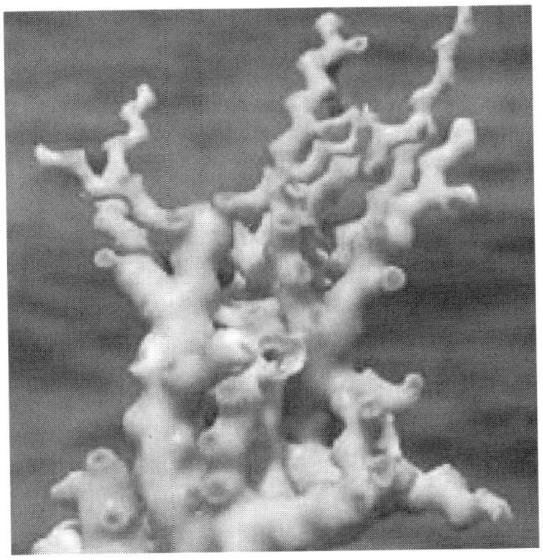

This beautiful species of *Lophelia* coral was collected off the coast of South Carolina.

Efforts to protect the Oculina Banks began in 1980, when scientists from Harbor Branch Oceanographic Institution initiated a call to implement protective measures for the area. Since then, progressively stricter legal protections have been implemented to facilitate a recovery. In 1984, a 315 km^2 portion of deep-water Oculina reef system was designated as a Habitat Area of Particular Concern by the South Atlantic Fishery Management Council (7,19), a designation that categorized it as an area of special ecological significance worthy of stricter regulatory and enforcement activity.

In 1994, after the area showed no significant recovery, part of the Oculina Banks was completely closed to bottom fishing in an area called the Experimental *Oculina* Research Reserve. In 1996, the South Atlantic Fisheries Management Council implemented additional protections in the reserve by prohibiting fishing vessels from dropping anchors, grapples, or attached chains there. In 1998, the council also designated the reserve as an Essential Fish Habitat. In 2000, the deep-water Oculina Marine Protected Area was extended to 1029 km^2 (7,19,24).

The Oculina Banks remain a hot spot for research and efforts to rehabilitate the coral (23). Scientists recently deployed concrete "reef balls" in the area in an attempt to attract fish and provide substrate for coral attachment, settlement and growth. They are cautiously optimistic about their initial restorative efforts in the reserve (23).

BIBLIOGRAPHY

1. Williams, G.C. 2001. Octocoral Research Center Web Site. Available on-line at http://www.calacademy.org/research/izg/orc_home.html
2. Southampton Oceanographic Centre. Coral ecosystems in deep water. Available on-line at http://www.soc.soton.ac.uk/GDD/DEEPSEAS/coralecosystems.html
3. Watling, L. 2001. Deep sea coral. Available on the NOAA Ocean Explorer Web site at http://oceanexplorer.noaa.gov/explorations/deepeast01/background/corals/corals.html.
4. Mayer, T. 2001. 2000 Years Under the Sea. Available on-line at http://exn.ca/Stories/2001/10/16/52.asp
5. Cairns, S. and G. Stanley. 1982. *Ahermatypic coral banks: Living and fossil counterparts*. Proceedings of the Fourth International Coral Reef Symposium, Manila (1981), 1: 611–618.
6. Cairns, S.D. 1982. *Antarctic and Subantarctic Scleractinia. Antarctic Research Series* 34: 74 pp.
7. Reed, J.K. 2002. *Comparison of deep-water coral reefs and lithoherms off southeastern U.S.A.* Hydrobiologia 471: 43–55.
8. Fossa, J.H., P.B. Mortensen, and D.M. Furevic. 2002. The deep-water coral *Lophelia pertusa* in Norwegian waters: distribution and fishery impacts. Hydrobiologia 417: 1–12. Available on-line at: http://www.imt.no/dokumenter/fossa.pdf
9. Bell, N. and J. Smith. 1999. Coral growing on North Sea oil rigs. *Nature* **402**: 601.
10. Rogers, A., M. Le Goff, and B. Stockley. Ecology of deep-water stoney corals (the Darwin Mounds). Available on-line at http://www.soton.ac.uk/~merg/corals.htm
11. Bellona Foundation. 2001. Coral reefs in Norwegian Waters. Available on-line at http://www.bellona.no/en/b3/biodiversity/coral_reefs/12784.html.
12. Tyler-Walters, H. 2003. *Lophelia* reefs. Marine Life Information Network: Biology and Sensitivity Key Information Sub-programme. Plymouth: Marine Biological Association of the United Kingdom. Available on-line at http://www.marlin.ac.uk/Bio_pages/Bio_BasicInfo_COR.Lop.htm

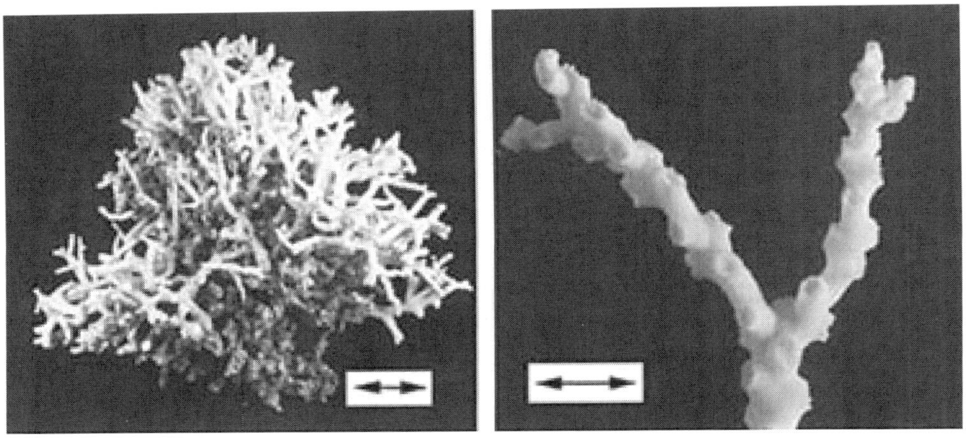

Two images of the deep-sea copral *Oculina varicosa*. The image on the left shows the complex growth structure of a small colony. On the right is a closeup of an individual branch. *Photo*: Ref. 7.

13. International Council for Exploration of the Sea (ICES). 2001a. Marine World, article on deep-sea coral. Available on-line at http://www.ices.dk/marineworld/deepseacoral.asp
14. International Council for Exploration of the Sea (ICES). 2002. *Advisory Committee on Ecosystems*. Report of the study group on mapping the occurrence of cold-water corals: 1–17.
15. Rogers, A.D. 1999. The biology of *Lophelia pertusa* (Linnaeus 1758) and other deep-water reef-forming corals and impacts from human activities. *International Review of Hydrobiology* **84**: 315–406.
16. Sulak, K. and S. Ross. 2001. A profile of the *Lophelia* reefs. Available on the NOAA Ocean Explorer Web site at http://oceanexplorer.noaa.gov/explorations/islands01/background/islands/sup10_lophelia.html.
17. Avent, R.M., M.E. King, and R.M. Gore. 1977. Topographic and faunal studies of shelf-edge prominences off the central eastern Florida coast. *Int. Revue ges.Hydrobiol.* **62**: 185–208.
18. Reed, J.K. 1981. *In Situ growth rates of the scleractinian coral Oculina varicosa occurring with zooxanthellae on 6-m reefs and without on 80-m banks*. pp. 201–206. in: W.J. Richards (ed.), Proceedings of Marine Recreational Fisheries Symposium.
19. Reed, J.K. 2002. Deep-water Oculina reefs of Florida: biology, impacts and management. *Hydrobiologia* **471**: 43–55.
20. Koenig, C., C.F. Coleman, and S. Brooke. Coral restoration in the Experimental *Oculina* Research Reserve of the South Atlantic. Available on-line at http://www.bio.fsu.edu/ifre/ifre_research_oculina.html
21. Hall-Spencer, J. et al. 2002. *Trawling damage to Northeast Atlantic ancient coral reefs*. Proc. R. Soc. London B.
22. International Council for Exploration of the Sea (ICES). 2001b. Close Europe's cold-water coral reefs to fishing. Available on-line at http://www.ices.dk/aboutus/pressrelease/coral.asp
23. Marine Protected Areas of the United States (MPA). 2002. Experimental *Oculina* Research Reserve. Available on-line at http://mpa.gov/mpadescriptive/cs_obrr.html
24. National Oceanic and Atmospheric Administration. 2000. *Final Rule, Amendment 4 to the Fishery Management Plan for Coral, Coral Reefs, and Live/Hard Bottom Habitats of the South Atlantic Regions (Coral FMP)*. Federal Register, Vol. 65, No. 115, June 14, 2000.

READING LIST

Atlantic Coral Ecosystem Study (ACES), 2000. Consortium of European scientists completed an environmental assessment of the status of Europe's deep-water corals, available on-line at: http://www.cool-corals.de

Hoskin, C.M., J.K. Reed, and D.H. Mook. 1987. *Sediments from a living shelf-edge reef and adjacent area off central eastern Florida*. Pp. 42–77, In: F, JMR, Maurrasse (ed.), Symposium on south Florida geology. Miami Geological Society. Memoirs 3.

International Council for Exploration of the Sea (ICES). 2001c: *Extract of the report of the Advisory Committee on Ecosystems 2002 on identification of areas where cold-water corals may be affected by fishing*.

National Oceanic and Atmospheric Administration. 1982. *Fishery Management Plan for Coral and Coral Reefs of the Gulf of Mexico and South Atlantic*. Gulf of Mexico and South Atlantic Fishery Management Councils.

Reed, J.K. and C.M. Hoskin. 1984. Studies of geological and biological processes at the shelf-edge with use of submersibles. In: *Undersea research and technology—Scientific application and future needs—abstracts, Symposia Series for Undersea Research*, Vol. 2, NOAA.

Reed, J.K., R.H. Gore, L.E. Scotto and K.A. Wilson. 1982. Community composition, structure, aerial and trophic relationships of decapods associated with shallow and deep-water *Oculina varicosa* coral reefs. *Bull. Mar. Sci.* **32**: 761–786.

MARINE DEBRIS ABATEMENT

JAVIER VELEZ-AROCHO
U.S. Environmental Protection Agency
Washington, District of Columbia

TRASH IN OUR OCEANS—YOU CAN BE PART OF THE SOLUTION

Marine debris, often called litter, has become a problem along shorelines, coastal waters, estuaries, and oceans throughout the world. It is defined as any man-made, solid material that enters our waterways directly (e.g., by dumping) or indirectly (e.g., washed out to sea via rivers, streams, storm drains, etc.). Objects ranging from detergent bottles, hazardous medical wastes, and discarded fishing line all qualify as marine debris. In addition to being unsightly, it poses a serious threat to everything with which it comes into contact. Marine debris can be life-threatening to marine organisms and humans and can wreak havoc on coastal communities and the fishing industry.

Sources of Marine Debris

There are two different sources from which debris pollutes our oceans. The first is from the land and includes users of

the beach, storm water-runoff, landfills, solid waste, rivers, and streams, floating structures, ill maintained garbage bins and dumps and litterbugs. Marine debris also comes from combined sewer overflows, and storm drains. Typical debris from these sources includes medical waste, street litter and sewage. Land-based sources cause 80% of the marine debris found on our beaches and waters.

The second source of marine debris is from ocean sources, and this type of debris includes galley waste and other trash from ships, recreational boaters and fishermen and offshore oil and gas exploration and production facilities.

Adding to this problem is the population influx along our nation's shores. More people means more paved area and wastes generated in coastal areas. These factors; combined with the growing demand for manufactured and packaged goods, have led to an increase in non-biodegradable solid wastes in our waterways.

The Beaches Environmental Assessment and Coastal Health Act (BEACH) of 2000

The BEACH Act was enacted on October 10, 2000, and it is designed to reduce the risk of disease to users of the Nation's coastal recreation waters. The act authorizes the EPA to award program development and implementation grants to eligible states, territories, tribes, and local governments to support microbiological testing and monitoring of coastal recreational waters, including the Great Lakes, that are adjacent to beaches or similar points of access used by the public. BEACH Act grants provide support for developing and implementing programs to notify the public of the potential for exposure to disease-causing microorganisms in coastal recreation waters. The act also authorizes EPA to provide technical assistance to States and local governments for the assessment and monitoring of floatable materials. In partially fulfilling that obligation, EPA has compiled the most current information to date on assessing and monitoring floatable materials in the document Assessing and Monitoring Floatable Debris.

The International Coastal Cleanup

The Ocean Conservancy, formerly known as the Center for Marine Conservation, established and maintains the annual International Coastal Cleanup (ICC) with support from EPA and other stakeholders. The first cleanup was in 1986 in Texas, and the campaign currently involves all of the states and territories of the United States and more than 100 countries around the world. The ICC is the largest volunteer environmental data-gathering effort and associated cleanup of coastal and underwater areas in the world. It takes place every year on the third Saturday in September. In 2001, over 140,000 people across the U.S. participated in the ICCC. They removed about 3.6 million pounds of debris from more than 7,700 miles of coasts, shorelines, and underwater sites.

National Marine Debris Monitoring Program

EPA along with other federal agencies helped to design the National Marine Debris Monitoring Program (NMDMP), and EPA is supporting The Ocean Conservancy's implementation of the study. NMDMP is designed to gather scientifically valid marine debris data following a rigorous statistical protocol. The NMDMP is designed to identify trends in the amounts of marine debris affecting the U.S. coastline and to determine the main sources of the debris. This scientific study is conducted every 28 days by teams of volunteers at randomly selected study sites along the U.S. coastline. The NMDMP requires, at a maximum, that 180 monitoring sites located along the coast of contiguous U.S. States and Alaska, Hawaii, Puerto Rico, and the U.S. Virgin Islands be fully operational. The program began in 1996 with the establishment of 40 monitoring sites ranging from the Texas/Mexico border to Port Everglades, Florida and included Puerto Rico and the U.S. Virgin Islands. To date 163 study sites have been designated and 128 sites are collecting data. The program will run for a 5-year period once all of the study sites have been established. For more information please visit: http://www.oceanconservancy.org/dynamic/learn/issues/debris/debris.htm.

Other Floatable Debris-Related Legislation

In response to a growing concern over floatable debris, governments have taken actions nationally, as well as internationally, to reduce discharges at their source. For example, intentional at-sea dumping of garbage generated on land became subject to international control in 1972 through the Convention on the Prevention of Marine Pollution by Dumping of Wastes and Other Matter,

commonly called the London Convention. Similarly, at-sea disposal of garbage generated during the routine operation of ships (e.g., garbage not deliberately carried to sea for the purpose of disposal) was addressed through a 1978 Protocol to the 1973 International Convention for the Prevention of Pollution by Ships (commonly called the MARPOL Convention).

In the Shore Protection Act of 1989 provides for controls on operations relating to the vessel transport of certain solid wastes (municipal or commercial waste) so that those wastes are not deposited in coastal waters. This act has provisions that reduce wastes being deposited in coastal waters during the transfer of waste from the waste source to the transport vessel to the waste receiving station.

As amended by the Water Quality Act of 1987, the Clean Water Act requires EPA to establish regulations that treat storm water and combined sewer overflows (CSOs) as point source discharges that must be regulated under the National Pollutant Discharge Elimination System (NPDES). The permits will prohibit non-storm water discharges into storm sewers while leading to improved control techniques and best management practices.

Debris Hazardous for Marine Animals

Each year millions of seabirds, sea turtles, fish, and marine mammals become entangled in marine debris or ingest plastics which they have mistaken for food. As many as 30,000 northern fur seals per year get caught in abandoned fishing nets and either drown or suffocate. Whales mistake plastic bags for squid, and birds may mistake plastic pellets for fish eggs. At other times, animals accidentally eat the plastic while feeding on natural food. According to recent estimates, 267 marine species have been reported entangled in or having ingested marine debris (Marine Mammal Commission, 1998). The plastic constricts the animals' movements, or kills the marine animals through starvation, exhaustion, or infection from deep wounds caused by tightening material. The animals may starve to death, because the plastic clogs their intestines preventing them from obtaining vital nutrients. Toxic substances present in plastics can cause death or reproductive failure in the fish, shellfish, and wildlife that use the habitat.

Polluted Waters Make Trouble for Humans, Too

Humans can also be directly affected by marine debris. Swimmers and divers can become entangled in abandoned netting and fishing lines like marine organisms. Beach users can be injured by stepping on broken glass, cans, needles or other litter. Appearance of debris, such as plastic, can also result in economic consequences. Floating debris, either as a floating slick or as dispersed items, is visually unappealing and can result in lost tourism revenues. New Jersey now spends $1,500,000 annually to clean up its beaches, and $40,000 to remove debris from the New York/New Jersey Harbor (YOTO Discussion Papers, 1998).

Fishing Industries and Communities Hit Hard by Marine Debris

Marine debris also acts as a navigational hazard to fishing and recreational boats by entangling propellers and clogging cooling water intake valves. Repairing boats damaged by marine debris are both time consuming and expensive. Fixing a small dent in a large, slow-moving vessel can take up to 2 days, costing the shipping company $30,000–40,000 per day in lost carrying fees, as well as up to $100,000 for the repair itself (Peter Wallace, personal contact). According to Japanese estimates, the Japanese fishing industry spent $4.1 billion (U.S.) on boat repairs in 1992 (YOTO Discussion Papers, 1998). Lost lobster traps cost New England fishing communities $250 million in

1978. These traps continue to catch lobsters and other marine organisms that are never harvested and sold; the communities' economies are therefore adversely affected.

Beach Raking

Mechanical beach raking, which is accomplished with a tractor and is used to remove debris from the shoreline, can help to remove floatable material from beaches and marine shorelines. However, it can also be harmful to aquatic vegetation, nesting birds, sea turtles, and other types of aquatic life. A study in Maine compared a raked beach and an adjacent natural beach to determine the effects of beach raking on vegetation. Beach raking not only prevents the natural re-vegetation process, but it reduces the integrity of the sand root mat just below the surface that is important in slowing beach erosion. Other problems include disturbance of vegetation if raking is conducted too, close to a dune. By removing seaweed, beach erosion can also be caused. Sand compaction is reduced when seaweed is removed, resulting in suspension of the sand in the water during high tides and contributing to loss of sand and erosion of the beach. Beach cleaning machines are harmful to nest birds and can destroy potential nesting sites, crush plover nests and chicks, and remove the plovers' natural wrack-line feeding habitat. To reduce the effects on nesting birds, beach raking should not be done during the nesting season.

Unseen Consequences

Once debris reaches coastal and ocean bottom, especially in areas with little current, it may continue to cause environmental problems. When plastic film and other debris settle on the bottom, it can suffocate immobile plants and animals, producing areas essentially devoid of life. In areas with some currents, such as coral reefs, debris can wrap around living coral, smothering the animals and breaking up their coraline structures.

Trash Indicator of Bigger Problems

The typical floatable debris from Combined Sewer Overflows includes street litter, sewage (e.g., condoms, tampons, applicators), and medical items (e.g., syringes), resin pellets, and other material that might have washed into the storm drains or from land runoff. These materials or objects can make it unsafe to walk on the beaches, and pathogens or algae's blooms can make it unsafe to swim. Pollutants, such as toxic substances, can make it unsafe to eat the fish caught from the waters. Swimming in or ingesting waters which are contaminated with pathogens can result in human health problems such as, sore throat, gastroenteritis, meningitis or even encephalitis. Pathogens can also contaminate shellfish beds.

What Needs to Be Done

Reducing marine debris means reducing the amount of waste generated on land and disposing of it properly. Recycling can significantly reduce the amounts of litter reaching marine and coastal waters. Volunteer coastal cleanups and public education efforts can also help reduce the amount of debris reaching our waterways.

What You Can Do

- Dispose of trash properly. This will help reduce the amount of trash that is washed into our waterways from storm drains.
- Reduce, reuse and recycle. Visit the EPA's solid waste web site for more information.
- Cut the rings of plastic six-pack holders. This lowers the risk of entanglement to marine animals if the holders do make it out to sea.
- Participate in local beach, river or stream clean ups.
- Practice good housekeeping.
- Look for alternative materials or avoid excessive packaging when deciding on purchases.
- Educate others about marine debris.
- Get involved in your local area.

What the EPA and Other Partners Are Doing

- Providing resources to educate the public on the impact of marine debris and guidance for how to eliminate it.
- Providing resources and technical support for the development of studies to address the impact of marine debris.
- Providing guidance for the implementation of regulations addressing marine debris.
- Organizing and sponsoring cleanup efforts, like the U.S. and International Coastal Cleanups.

Whether directly or indirectly, many of our land-based activities ultimately impact and adversely affect the oceans. At the heart of the problem lie the great quantities of trash generated in our daily lives. Reusing and recycling efforts have helped, but a great deal more can be done. You can be part of the solution.

LARVAE AND SMALL SPECIES OF POLYCHAETES IN MARINE TOXICOLOGICAL TESTING

Jian-Wen Qiu
Ming Hung Wong
Hong Kong Baptist University
Kowloon Tong, Kowloon, Hong Kong

The ultimate goal of ecotoxicological testing is to provide data to predict potential impacts of pollutants on the ecosystem. To fulfill this goal, the tests should meet several criteria: the species should be ecologically relevant and sensitive to pollutants; the species should be readily available through field collection or laboratory culture; and the tests should yield reasonably rapid results at low cost. Although rarely consumed as seafood, polychaetes are a diverse group in marine ecosystems, typically consisting 35% to 75% of the total number of species in soft bottom coastal areas (1). They serve as prey for demersal fish and are important in the transfer of organic material and nutrients between the sediment and the overlying water. The use of polychaetes as bioassay organisms began in the 1960s, stimulated by the awareness that many pollutants were threatening the ecosystems and the surge in studies of polychaete life history. Reish and Gerlinger (2) is the most updated comprehensive review on the use of polychaetes in bioassays. Among the 8000 described species worldwide, only 49 polychaete species have been used in bioassays, of which all inhabit shallow waters. Most tests have been conducted with small species (*Neanthes arenaceodentata, Capitella capitata, Dinophilus gyrociliatus, Ophryotrocha labronica*, and *O. diadema*), which can be obtained from laboratory culture. Large polychaetes, i.e., *Nereis virens* and *Neanthes diversicolor*, have also been used. These species are usually obtained from field populations and have been used mostly in pollutant accumulation studies.

Polychaetes have been employed in acute, chronic, and life-cycle tests to determine the toxicity of metals, contaminated sediment, sediment pore water, municipal effluents, petroleum hydrocarbons, pesticides, and other synthetic organic chemicals. Although several species, such as *Capitella capitata, Dinophilus gyrociliatus, Ophryotrocha labronica*, and *O. diadema*, have been used in all test types, many other species have been used only in acute or chronic tests. The acute tests can be conducted with larvae, juveniles, and adults. During the test, the worms are usually not fed. When juveniles and adults are used, the acute test is usually run for 96 h. The test can be static, static renewal, or flow-through. In larval tests, however, the tests are usually static, with shorter exposure periods (24 or 48 h). Survivorship is usually the endpoint of acute tests. Chronic tests are usually conducted using juveniles or adults. The test protocol is the same as in acute test except the exposure duration is 10 to 28 days, during which the polychaetes are fed. The bioassay usually adopts a static renewal design. In addition to survival, sublethal effects such as cessation of feeding, defecation, or reduced growth are the endpoints. The life-cycle tests have been employed infrequently in the past. The test protocol is similar to that of the chronic test. In addition to survival and growth effects, reproductive parameters, such as eggs or juveniles produced, are the endpoints. Only small species have been used in life-cycle tests because their life cycle can be completed in less than one month. In recent years, many published toxicity studies have used small polychaete species in chronic or life-cycle tests. Examples are Bridges and Farrar (3), Anderson et al. (4), Linke-Gamenick et al. (5), McPherson et al. (6), and Nipper and Carr (7). There has also been new development in the use of polychaete larvae (8–10).

Below are several general conclusions regarding the use of polychaetes in bioassays. These conclusions are supported by examples.

1. *Polychaete Larvae are Very Sensitive to Pollutants.* In *Capitella capitata,* bifurcated abnormal larvae occurred at 10 ppb copper (11). In *Arenicola cristata*, abnormal larvae occurred at 0.75 ppb TBTO (12). In *Galeolaria caespitosa*, the 48-h EC50 for larval development ranged from 20 to 28 ppb copper (9).
2. *Polychaete Larvae are More Sensitive than the Adults to Pollutants.* In *Capitella capitata,* the 96-h LC50 for trochophores and adults was 0.22 and 7.5 ppm cadmium, respectively (13). In *Hydroides elegans*, the EC50 for trochophores was 29 ppb copper, whereas mortality of the adults occurred only when copper concentrations reached 500 ppb (10).
3. *Test Sensitivity Varies with Polychaete Species and Toxicant.* Reish and LeMay (14) compared the toxicity of seven metals with five polychaete species. They found that, for the same metal, different polychaete species gave very different results. For cadmium, the 96-h LC50 ranged from 2.6 ppm in *Pectinaria californiensis* to 14.1 ppm in *Neanthes arenaceodentata*. For the same species, sensitivity varies with toxicant. *Ophryotrocha labronica* was the most sensitive to copper, but it was the second most tolerant to chromium, and the most tolerant to mercury.
4. *The Chronic Test may not be More Sensitive than the Acute Test.* Reish and Gerlinger (2) noted that very limited data on the comparison of acute and chronic tests existed. After exposure of *Neanthes arenaceodentata* to 5.1 ppm cadmium for 20 days, no mortality existed, but the growth rate was significantly lower than that of the control. In another study with the same polychaete species, however, the 28-d LC50 values for mercury and copper were not lower than the 4-day values (15).
5. *Life-Cycle Tests may not be more Sensitive than Acute Tests.* At 100 ppb pentachlorophenol, *Dinophilus gyrociliatus* adults survived well, but the total number of eggs plus juveniles was significantly reduced (16). Similarly, at 50 ppb mercury, *Ophryotrocha diadema* survived well, but the number of offspring was reduced; in contrast, the 96-h LC50 for mercury was very close to the concentration that resulted in declined reproductive output (17).

BIBLIOGRAPHY

1. Qian, P.Y., Qiu, J.W., Kennish, R., and Reid, C.A. (2003). Recolonization of benthic infauna subsequent to capping of contaminated dredged material in East Sha Chau, Hong Kong. *Estuarine Coastal Shelf Sci.* **56**: 819–831.
2. Reish, D.J. and Gerlinger, T.V. (1997). A review on the toxicological studies with polychaetous annelids. *Bull Mar Sci.* **60**: 584–607.
3. Bridges, T.S. and Farrar, J.D. (1997). The influence of worm age, duration of exposure and endpoint selection on bioassay sensitivity for *Neanthes arenaceodentata* (Annelida: Polychaeta). *Environ Toxicol Chem.* **16**: 1650–1658.
4. Anderson, B.S. et al. (1998). Comparison of marine sediment toxicity test protocols for the amphipod *Rhepoxynius abronius* and the polychaete worm *Nereis (Neanthes) arenaceodentata*. *Environ Toxicol Chem.* **17**: 859–866.
5. Linke-Gamenick, I., Forbes, V.E., and Sibly, R.M. (1999). Density-dependent effects of a toxicant on life-history traits and population dynamics of a capitellid polychaete. *Mar Ecol Prog Ser.* **184**: 139–148.
6. McPherson, C.A. et al. (2002). Toxicity of 1,4-dichlorobenzene in sediments to juvenile polychaete worms. *Mar Pollut Bull.* **44**: 1405–1414.
7. Nipper, M. and Carr, R.S. (2003). Recent advances in the use of meifaunal polychaetes for ecotoxicological assessments. *Hydrologia.* **496**: 347–353.
8. Palau-Casellas, A. and Hutchinson, T.H. (1998). Acute toxicity of chlorinated organic chemicals to the embryos and larvae of the marine worm *Platynereis dumerilii* (Polychaeta: Nereidae). *Environ Toxicol Water Qual.* **13**: 149–155.
9. Ross, K.E. and Bidwell, J.R. (2001). A 48-h larval development toxicity test using the marine polychaete *Galeolaria caespitosa* Lamarck (Fam. Serpulidae). *Arch Environ Contam Toxicol.* **40**: 489–496.
10. Xie, Z.C., Wong, N.C., Qian, P.Y., and Qiu, J.W. (2004). Responses of polychaete *Hydroides elegans* life stages to copper stress. *Mar Ecol Prog Ser.* (in press).
11. Reish, D.J., Plitz, F.M., and Martin, J.M. (1974). Induction of abnormal polychaete worms. *Mar Poll Bull.* **51**: 125–126.
12. Walsh, G.E., Louie, M.K., McLaughlin, L.L., and Lores, E.M. (1986). Lugworm *Arenicola cristata* larvae in toxicity tests: survival and development when exposed to organotins. *Environ Toxicol Chem.* **5**: 749–754.
13. Reish, D.J., Martin, J.M., Plitz, F.M., and Word, J.Q. (1976). The effect of heavy metals on laboratory populations of two polychaetes with comparisons to the water quality conditions and standards in Southern California marine waters. *Water Res.* **10**: 299–302.
14. Reish, D.J. and LeMay, J.A. (1991). Toxicity and bioconcentration of metals and organic compounds by polychaetes. *Ophelia Suppl.* **5**: 653–660.
15. Reish, D.J. (1980). *The Effect of Different Pollutants on Ecologically Important Polychaete Worms*. U.S. Environmental Protection Agency. Environmental Research Laboratory, Narragansett, RI, EPA 600/3-80-053, p. 138.
16. Carr, R.S., Curran, M.D., and Mazurkiewicz, M. (1986). Evaluation of the archiannelid *Dinophilus gyrociliatus* for use in short-term life-cycle tests. *Environ Toxicol Chem.* **5**: 703–712.
17. Reish, D.J. and Carr, R.S. (1978). The effect of heavy metals on survival, reproduction, development and life cycles of two species of polychaetous annelids. *Mar Pollut Bull.* **9**: 24–27.

EL NIÑO: THE INTERANNUAL PREDICTION PROBLEM

Geophysical Fluid Dynamics
Laboratory—NOAA

Once every few years, the normally cool waters of the eastern tropical Pacific become unusually warm in a phenomenon known as El Niño. During El Niño years, disruptive weather patterns often occur over wide regions of the globe, including North America. For example, flooding rains can strike California, producing coastal erosion, mudslides, and crop damage, while droughts may occur in Australia and other regions. Along the South American coast, the local fishing industry is severely disrupted by the unusually warm ocean waters.

GFDL and Princeton University maintain active collaborative efforts aimed at understanding and predicting El Niño. George Philander of Princeton University is a pioneer in the field of El Niño simulation and theory. His work with GFDL's Ron Pacanowski established the viability of simulating El Niño using coupled ocean-atmosphere models and set the stage for attempts to use such models for El Niño prediction. At GFDL, Kikuro Miyakoda and Tony Rosati have led efforts to build a system to predict both El Niño and its impact on large-scale weather

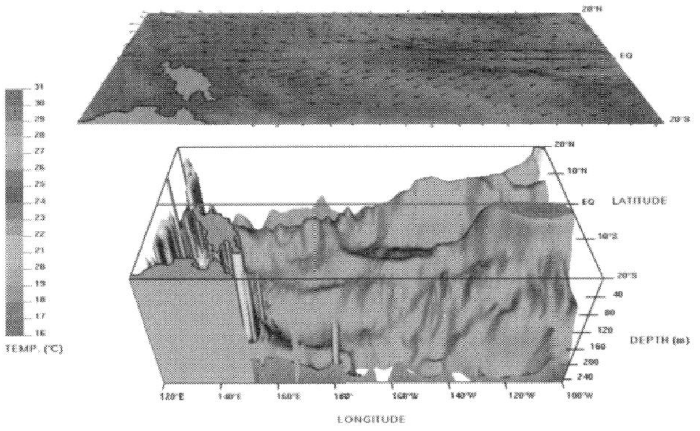

Modeled ocean temperature and surface current distribution over the tropical Pacific Ocean region obtained from a GFDL coupled ocean-atmosphere model used to predict El Niño/Southern Oscillation (ENSO). Lower (blue surface): three-dimensional depiction of the surface on which the ocean temperature is 20 °C. Undulations of this surface can be used to monitor ocean heat content changes associated with ENSO. During the cold phase of ENSO, this surface is deep in the warmer western tropical Pacific but rises toward the sea surface in the cooler eastern equatorial Pacific, as in the example shown. Upper: corresponding distribution of sea surface temperature and surface ocean currents. Gray regions depict land areas of New Guinea and northern Australia.

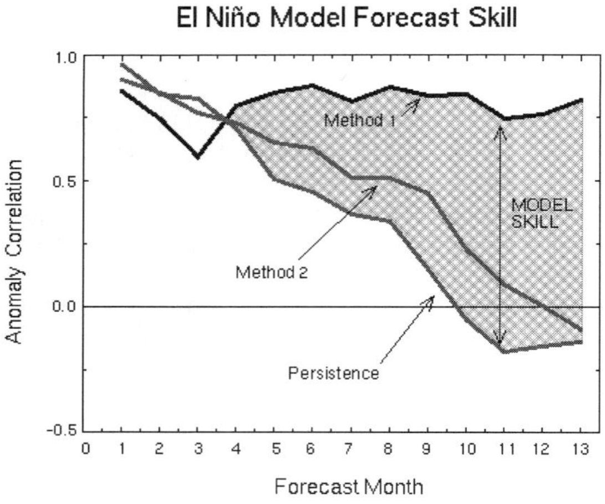

Model skill in the prediction of eastern tropical Pacific sea surface temperature anomalies for different forecast lead times. Higher values indicate greater skill. Based on an ensemble of 14 retrospective forecasts for 1982–1988 using a GFDL coupled ocean-atmosphere model. Forecast Method 1 uses both surface and sub-surface ocean data for initialization, whereas Forecast Method 2 uses only surface data. The results show the crucial importance of subsurface data for the success of the forecasts with this prediction system. [*Source*: Anthony Rosati, et al., Monthly Weather Review, submitted.]

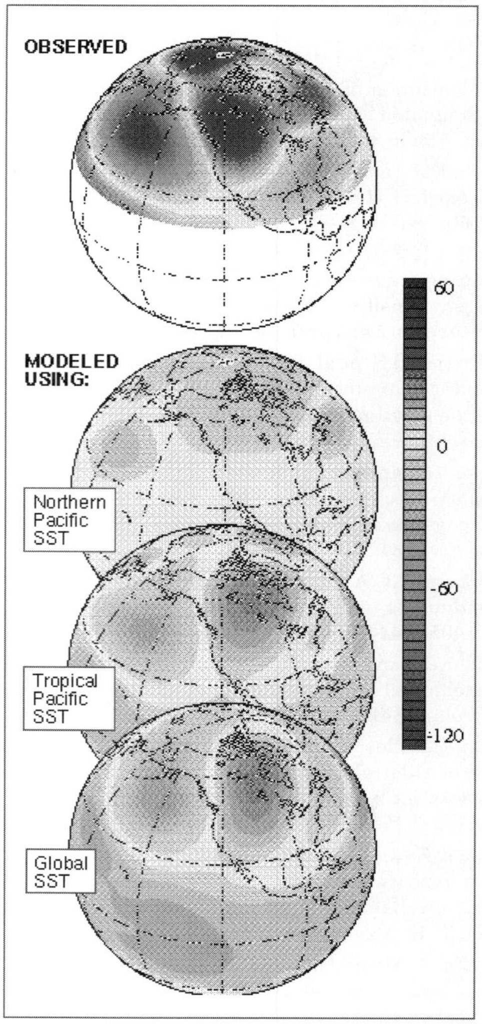

Observed and modeled atmospheric circulation anomalies over the Pacific Ocean and North America during six El Niño winters. The model simulations illustrate how long-range (1 year) predictions of tropical Pacific sea surface temperature may ultimately lead to improved long-range predictions of seasonal weather anomalies, even in regions remote from the tropical Pacific such as the continental United States. Shown are composite 500 mb geopotential height anomalies in meters. The three bottom diagrams are from atmospheric model experiments in which observed sea surface temperatures have been specified for different domains. [*Source*: Ngar-Cheung Lau and Mary Jo Nath, Journal of Climate, August 1994.]

patterns. Miyakoda has long been a leader in efforts to extend the limits of useful weather forecasts. Many of his research group's earlier modeling innovations had been incorporated into operational weather forecasting models at the National Meteorological Center, leading to significant improvements in the 3–5 day forecasts now available to the public.

Using coupled ocean-atmosphere models developed at GFDL and other research centers, scientists now have begun to make physically based predictions of El Niño conditions, with lead times of a year or more. These models have the potential to predict El Niño's effects on weather patterns over North America and other regions far removed from the tropical Pacific. Farmers, energy planners, and water resource managers are examples of those who would benefit from improved extended-range weather and short-term climate predictions.

GFDL'S MODULAR OCEAN MODEL

Detailed predictions of El Niño are made possible by the "Modular Ocean Model," a product of extensive research at GFDL. Through the efforts and foresight of Mike Cox, Keith Dixon, Ron Pacanowski, and Tony Rosati at GFDL, this ocean model is now a shared resource, used by hundreds of researchers worldwide. The model has become the mainstay of the oceanographic modeling community, not just in new operational El Niño predictions at the National Centers for Environmental Prediction, but for global climate change research and other applications as well.

RENEWABLE ENERGIES FROM THE OCEAN

Purnima Jalihal
National Institute of Ocean Technology
Chennai, India

INTRODUCTION

Today renewable energies are the focus of attention because of the realization that fossil fuels are on the verge of extinction. Renewables like solar and wind energy have been studied thoroughly, and a lot of standard energy conversion devices have also been designed, tested, and proven. The ocean has a large potential in terms of the energy availability. However, harnessing the ocean energy forms is still in the infancy stage.

The oceans have several untapped features like varying temperature, waves, and tides. The mechanical motions and the thermal variation caused by solar radiation cause the oceans to have entrapped in them tremendous amounts of energy, which can be tapped. The energy tapped from these phenomena occurring in the oceans is renewable and therefore needs exploration. Although several attempts have been made to harness these energies and convert them into electricity, none of the methods are yet commercially viable. However, further studies in these areas and performance evaluation of pilot plants can lead to improvements in the economic and technical prospects of such plants. The three energy sources in the oceans are waves, tides, and thermal variation along the depth. This chapter gives a brief description of each type of energy and gives the Indian case studies for wave energy and ocean thermal energy conversion.

WAVE ENERGY

Waves result when wind blows over the ocean surface and energy gets transferred from the wind to the sea. The amount of energy transferred depends on the wind speed, the distance along which it affects the water, and the amount of time during which the wind blows. When a wave breaks at seashore, one can see the tremendous amount of energy released.

Wave energy is an irregular and oscillating low-frequency energy source that can be converted into usable electric energy. It varies from location to location and from season to season.

Harnessing wave energy involves complicated devices in which the mechanical motion of the pressure waves has to be converted to electrical energy. These devices can be generally categorized as follows:

1. Floating or pitching devices
2. Oscillating water columns (OWC)
3. Surge or focusing devices

The first type consists of mechanical linkages between floating or fixed objects. The oscillating water column type of devices transfer the energy from the waves to pneumatic power by alternate compression and decompression of air in a duct or chamber. This pneumatic power is converted to mechanical power and thereby to electric power through a generator. The surge devices channel large waves into a small area, which forces the height of waves to increase, and when this water passes through hydroelectric turbines on its way back to sea level, electricity is generated.

Wave Energy Plants Around the World

In Japan, research on wave energy began during the late 1970s. Several tests were conducted on OWCs, and the biggest focus has been on what is known as the "Mighty Whale" developed by Japan Marine Science and Technology (JAMSTEC). This OWC is probably the world's largest. However, Japan has been active in the area of the floating OWC and a concept called the Backward Bent Ducted Buoy has been developed. This is a floating conventional OWC but with the opening away from the wave direction. Norway has had research being carried out in wave energy for the past 25 years mainly by the Norwegean University of Science and Technology. They have been working on the Tapered Channel and OWC schemes. Portugal also is an active player in wave energy research. The Archimedes wave swing was invented here, and the concept uses the periodic wave pressure to cause the upper part of a buoy to move while the lower part stays in position. The United Kingdom has conducted research on the largest number of varied type of wave energy devices. The Shoreline OWC on Islay, the LIMPET OWC with Wells turbine, and the large OSPREY OWC are all the U.K.'s contributions to wave energy research. Some large prototypes like OSPREY failed before installation was completed. In Sweden, work has been carried out on a floating buoy with a submerged tube underneath. The tube is open to the sea at both ends, and there are some components for the power take off mechanism in the buoy. However, until now, no single plant has been completely commercialized and made repeatable. One must note that the Indian wave energy plant has been functional for the past 10 years, but it generates very low average powers. Details of this plant follow in a later section. Before we move on to the case study, a few details on the nature of waves will be discussed briefly.

Wave Theory

The average power P (W/m) in a regular sine wave per meter wave front of waves with height H and period T can be expressed as

$$P = \rho g^2 H^2 T / 8\pi$$

where ρ is the density of water and g is the acceleration caused by gravity. However, in practice, waves are far from ideal. In nature, waves are irregular and can be described by statistical models. If the wave conditions are measured, over 20 minutes, for example, the mean wave height H_m and the significant wave height H_s can be calculated. The significant wave height is defined as the average of the highest 33% of the waves. Under such circumstances, the wave power can be stated as:

$$P = 0.55 H^2{}_s T_z \text{ kW/m length of wave crest}$$

where T_z is the zero crossing period (1). It is very clear from the above equations that the energy in waves is dependent directly on the wave height and period, which makes it a variable and intermittent source of energy.

CASE STUDY

The Indian Wave Energy program

Indian wave energy research began in 1983 under the sponsorship of the Government's Department of

Figure 1. Wave energy plant.

Ocean Development. After about 6 years of research in the laboratory, the wave energy group and the Ocean Engineering Center, Indian Institute of Technology, Madras, coordinated a project, which involved the fabrication, and construction of the OWC-based wave energy system, which is shown in Fig. 1.

The structure consists of a heavy concrete caisson. The site selected was 45 m in front of the Vizhinjam Fisheries Harbor breakwater, off the Trivandrum Coast at the southern tip of the Indian Peninsula. The water depth at this location is 10 m, typical for fisheries breakwaters, and the average annual wave power potential is of the order of 15 kW/m. The already existing breakwater gives the approach needed for this structure. Figure 2 shows the sectional view of the caisson.

Stages of Research

Several stages of research were carried out on the power module by IIT, Chennai, and subsequently by the National Institute of Ocean Technology, India.

Stage I (1991–1994). The Constant chord Wells turbine coupled to a 150-kW slip ring-type induction generator was used. The major defect found in the system was the stalling behavior of the turbine at higher flow rate. Hence, efforts were made to carry out further studies on turbine and to reduce the no-load losses in the system. The studies resulted in the development of the stage II turbine.

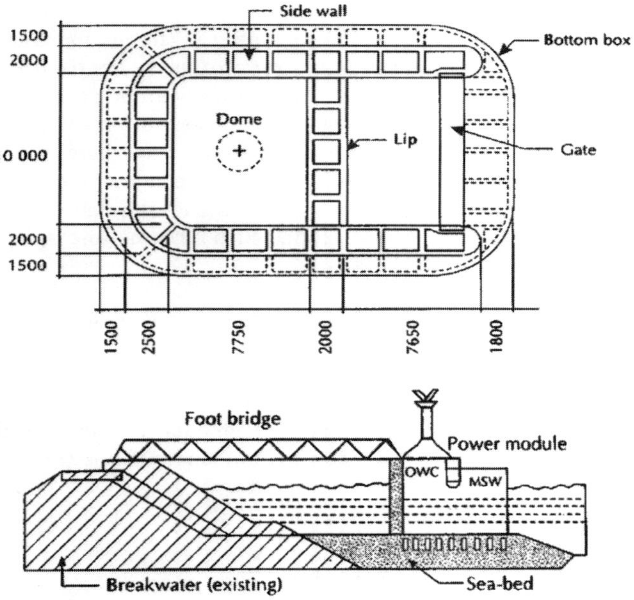

Figure 2. Cross-sectional view and elevation of wave energy device.

Stage II (1994–1997). A varying chord Wells turbine that has characteristics of self-pitch control, which can reduce the stalling effect, was installed in 1995. The data studies made on the system revealed that the efficiency of the turbine has been improved because of the reduced stalling and the selection of proper generator rating.

Stage III (1997–2000). An Impulse turbine with self-governing guide vanes coupled to a 55-kW generator was installed, and the studies revealed that the system was operable for all periods of the year. The problem with the impulse turbine is that the flipping guide vanes deteriorate the efficiency with time because of the corrosion of bearings in the marine environment, and a change in the geometry of vanes can result because of many cycles of operation. Hence, it was decided to change to fixed guide vanes.

Stage IV (2000–Present). A special variable-speed permanent magnet brushless alternator with constant power output is now used along with the impulse turbine with fixed guide vanes. The system can deliver power to charge a battery bank of 300 Ah capacity between 400 rpm and 1000 rpm. The objective of using a battery bank is to deliver a constant power requirement of 7 hp to a desalination system already installed at the site. These studies helped to understand the wave-to-wire conversion process and the pumping of power to the grid. The impulse turbine proved to give the highest efficiency, and this finding has led other wave energy researchers around the world to also consider using impulse turbine in place of wells turbine for wave energy conversion. It was decided to use the average power being generated by the plant to run a reverse osmosis-based desalination plant.

Wave-Powered Desalination System

The wave-powered desalination system consists of the impulse turbine, a special variable-speed alternator, and a battery backup. As the waves are random in nature, the alternator is a device that gives a constant DC voltage with varying input. When the wave heights are low, the turbine speed is low and may not be in the range within which the alternator generates voltage. In this case, the battery discharges and powers up the load. When more power is generated than is required by the load, the batteries are charged. This system was put in place after rigorous computer simulations and laboratory testing. The system is now operational and is now supplying water to the local village community. The system is self-sustaining in that energy from the sea generates freshwater out of seawater.

TIDAL ENERGY

Tides are caused by the gravitational attraction of the moon and the sun acting on the rotation of Earth. The relative motions of these bodies cause the surface of the sea to be raised and lowered periodically, according to several interacting cycles.

The tide moves a huge amount of water twice each day, and if we could harness it, large amounts of energy could be generated. However, converting it into useful electrical power is not easy. Tidal power works in a manner similar to a hydroelectric plant except that the dam is much larger. A huge dam or barrage is built across a river or a bay with a large difference between its low and high tides. When the tide goes in and out, the water flows through tunnels in the dam. The ebb and flow of the tides can turn a turbine. A major drawback of tidal power stations is that they can only generate when the tide is flowing in or out, i.e., only for 10 hours each day. However, some advantages are that there are no wastes like greenhouse gases, and production of electricity is reliable because tides are totally predictable.

Tidal energy potential has been investigated by several countries, notably, France, where a 240-MW demonstration plant was built on the Rance estuary during the 1960s and has now completed 30 years of successful operation. The Russians have built a small 400-kW device near Murmansk, which was later followed by a 17.4-MW experimental device, built by the Canadians at Annapolis on a small inlet off the Bay of Fundy. Small plants have also been installed in China. None of these countries have developed these plants any further. In the United Kingdom, a series of industrial consortia in collaboration with the Government have investigated the prospects for tidal energy on the Severn, Mersey and several smaller estuaries (2).

Every type of renewable energy project has its drawbacks, and tidal power plants have their problems. Obviously these stations would not be effective as the sole supplier of required energy (3). Also, some environmental issues develop with tidal energy. One issue is that the flooding of surrounding land can bring about much destruction. It could mean the relocation of settlement and agriculture plots. It can also change the composition of the habitat and threaten local inhabitants. Vegetation and marine life may not to adapt to the new water levels and may cease to grow or live where they previously existed. The main advantage of tidal energy is that it is nonpolluting. Hence, when designing tidal energy plants, these factors can be built into the design to make it economical.

India is also trying to venture into tidal energy in the Gulf of Khambhat region of Gujarat on the western side of India. Once the Gulf is closed, water levels within the reservoir can be controlled while the tidal fluctuation outside the reservoir continues, and hence, they can be harnessed for the generation of tidal energy.

OCEAN THERMAL ENERGY CONVERSION

The method of harnessing the energy by using the temperature difference between the surface of the ocean and at depths of around 1000 m is called ocean thermal energy conversion (OTEC) and is again a nonpolluting renewable energy source. It is capital intensive, but the unit cost comes down drastically as the rating becomes higher.

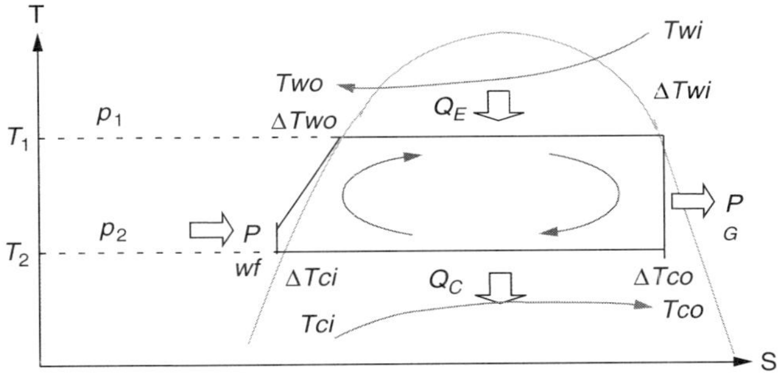

Figure 3. T-S Diagram for the closed Rankine cycle.

OTEC—Principle and Systems

OTEC uses the temperature-difference existing between warm surface seawater of around 27–29 °C and the cold deep-sea water of around 5–7 °C, which is available at a depth of 800 to 1000 m (4). Based on the power cycle, the OTEC system can be classified in to two open cycle and closed cycle systems. In the open cycle system, the warm surface water is flash-evaporated in a chamber maintained under high vacuum and the generated vapor drives a low-pressure turbine connected with the generator. The exhaust steam is condensed with cold seawater taken from the ocean depth. For an open cycle system, a condenser can be a means for freshwater production. The closed cycle system working on Rankine cycle uses a low boiling point liquid like R 134a or ammonia as the working fluid. The T-S diagram for the closed Rankine cycle is shown in Fig. 3.

The fluid is evaporated in a heat exchanger with the warm seawater from the ocean surface. After the vapor is expanded and drives the turbine, it is condensed by cold seawater from the depth of the ocean. This condensate is pumped back to the evaporator and recycled with the working fluid pump. A simplified flow diagram of the OTEC power cycle along with the Rankine cycle is shown below in Fig. 4.

Design Requirements of OTEC Plants

Choice of Working Fluid. The selection of equipment, sizing of piping, working conditions, and so on depend on the working fluid, which should have low-temperature boiling and condensation, high heat transfer coefficient, be environment friendly, safe, nontoxic, and so on. Some working fluids are ammonia, fluro carbon, tetra fluro ethane (HFL-1349), and propane. If safety is the most important issue, HFL-134a is the choice; however, ammonia is less expensive, easily available, and has superior thermodynamic properties. To handle ammonia, certain safety guidelines need to be followed carefully and strictly.

Choice of Material for Heat Exchangers. For an OTEC plant, the heat exchanger material should have corrosion resistance to seawater and compatibility with working

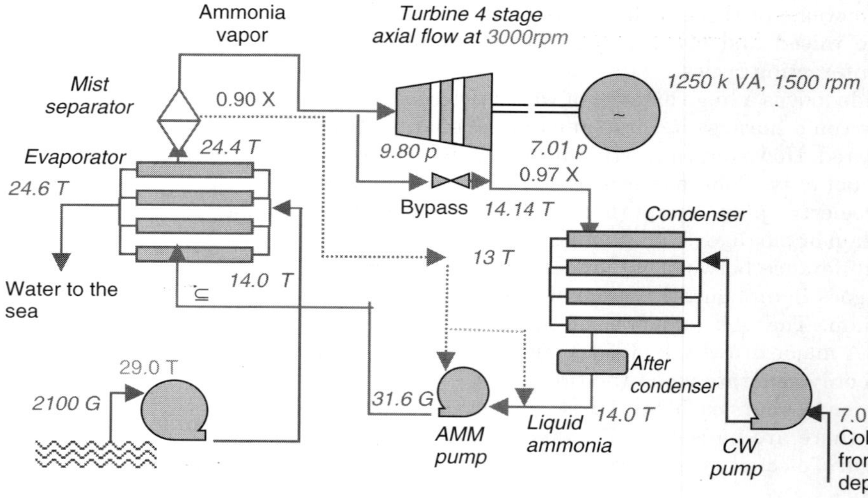

Figure 4. Energy balance diagram for 1-MW OTEC plant.

fluid. Materials that can be considered are aluminum, nickel, stainless steel, and titanium. Combinations like titanium with ammonia and aluminum with propane are workable.

Equipment Selection

Heat Exchangers. Heat exchangers are critical components, and the shell-tube, plate-fin, plate type of exchangers need to be compared. Considering volume required, fouling rates, heat transfer area, pressure drop, maximum working pressure, and other criteria, plate heat exchangers may have an edge over other types.

Turbine. The turbine generator system is another critical component of an OTEC plant, and the net power production depends on the turbine efficiency. Axial flow turbines give better adiabatic efficiency compared with radial flow turbines.

Sea Water Pumps. Seawater pumps for the warm and cold seawater consume about 30–40% of the gross power produced by the plant. The choice of material and characteristics to suit large discharge and low head should be considered for the selection.

Cold Water Pipe Material. The cold water pipe is usually around 1 km long and is subjected to varying hydrostatic pressures, temperatures, and forces caused by current drag and waves. In addition to the conventional steel pipes, other materials that can be considered are concrete, high-density polyethylene (HDPE), glass reinforced plastics (GRP), and so on. HDPE has some additional advantages in that it is buoyant and can float in water, which helps during towing and installation.

Other Criteria to be Considered

Station Keeping Systems. For floating OTEC plants, either the platform should be self-propelled or moored offshore. Design of the mooring depends on various environmental factors and the life expectancy of the plant.

Biofouling Control. In warm seawater, any solid surface begins to acquire a film in the form of a slime layer in a few hours. This process is called fouling, and it affects the heat exchanger performance when it grows within. It also increases drag on the pipes and vessels if not cleared for extended periods of time.

Some means of regular cleaning and addition of antifoulants like chlorine are measures to be taken to reduce the effects.

Power Transmission. For offshore plants, the power generated needs to be transmitted to shore and this requires underwater cables, which should have minimum transmission losses, resistance to water pressure, and such other characteristics. These requirements and more need to be catered to for designing an OTEC plant.

The most complex part of setting up a floating OTEC plant offshore is the deployment of the long cold water pipe and its connection to the barge. A HDPE pipe is buoyant, and it floats easily when towed on the surface. However, the lowering and the upending of the pipe requires several craft and tugs with cranes and other handling equipment. The mooring system should be designed well for operating as well as for storm conditions.

Case Study: The Indian 1-MW OTEC Plant

The National Institute of Ocean Technology is currently in the process of installing a floating barge-mounted OTEC plant off the Southern coast of India of gross rating 1 MW. A 1-km-long pipe made of high-density polyethylene will draw the cold water. The heat exchangers are of plate type and made of titanium. The turbine is a 1-MW ammonia turbine designed and fabricated for the first time. The working fluid is ammonia, and the Rankine cycle is being used. The power plant is housed on a non-self-propelled barge, which needs to be moored in 1000-m water depth. The configuration of the pipe/mooring system is shown in Fig. 5.

The motions of the barge under the environmental forces caused by wind, wave, and current affect the pipe and mooring behavior. Several analyses have been carried

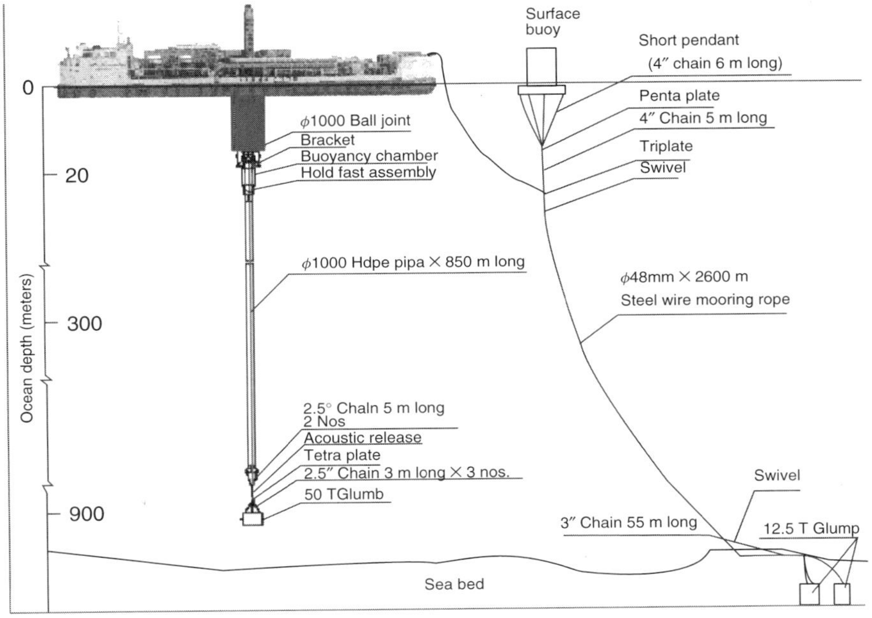

Figure 5. Pipe mooring configuration for 1-MW OTEC plant.

out to study the behavior of the system and arrive at the final design. The barge was specially designed for the OTEC plant, and some special features have been introduced in it, such as:

- A large 14-m-high retractable sump supporting the cold water pumps to meet the pump net positive suction head (NPSH) requirements.
- A quick-release mechanism to disconnect the mooring from the barge.
- A quick-release mechanism to disconnect the cold water pipe.

When commissioned, this will be the first megawatt-rating floating OTEC plant in the world.

CONCLUSIONS

Harnessing ocean energies in the various forms described in this article is challenging and needs several engineering and technological innovations. Most methods are still under study. However, with a focus toward overcoming the engineering complexities, a day will dawn when these renewable sources of energy will be harnessed and will replace the fossil fuels. The oceans are vast storehouses of energy, and man's endeavors and ingenuity will capture these ocean energy forms in such a way as to be environment friendly and convert energy-starved areas in the world to areas self-sufficient in energy.

Acknowledgment
The author expresses her gratitude to Ms. Sharmila N and Ms. Sri Lakshmi I of the National Institute of Ocean Technology, for helping with information gathering and word processing of this work.

BIBLIOGRAPHY

1. Shaw, R. (1982). *Wave Energy a Design Challenge*. Ellis Horwood Publishers, New York.
2. Available: http://europa.eu.int/comm/energy_transport/atlas/htmlu/tidalintro.html.
3. Available: http://ririessortment.com/tidal energy-rdrx.htm.
4. Avery, W.H. (1994). *Renewable Energy from the Ocean*. Oxford University Press, Oxford, UK.

ESTUARIAN WATERS

Priyanka K. Dissanayake
Moratuwa, Sri Lanka

IMPORTANCE OF ESTUARIES

Estuaries are in many ways the most complicated and variable of aquatic ecosystems. An estuary is a place where salt water from an ocean mixes with freshwater from the land and creates a unique and special environment in which marine species live, feed, and reproduce. Estuaries and intertidal regions are naturally exposed to stressful conditions.

An estuary and its immediate catchments form a complex system of ecological, physical, chemical, and social processes that interact in a highly involved and, at times, unexpected fashion. Estuaries are the end point for the detrimental effects of many upstream catchment activities. Materials carried from the land by rivers vary in quality and quantity, sometimes with strong seasonal patterns of high biological significance. Tidal oscillations cause vigorous reversals of flow. Inherent hydrographic patterns can lead to accumulation of materials and to upstream transport from the point of addition. The estuaries as recipients of wastes from rivers entering them and cities and industries along their shores are obviously more immediately susceptible to pollution damage than any other part of the marine system. Increased loads of nutrients and sediments from activities in the surrounding catchments are the two key threats to marine and estuarine water quality.

Estuaries are often used as disposal sites for unwanted biological or industrial products, usually referred to as waste. If such waste enters the estuarine ecosystem and is potentially harmful to life or has potentially detrimental effects on its health it is usually called a pollutant.

The usual way of controlling negative impacts of pollutant dispersal in estuaries is by monitoring its concentration in the water and making sure that the concentration does not exceed a certain level above which the pollutant is considered harmful. This problem consists of two parts, the *near field* and the *far field*. The near field problem considers technical details of pollutant outfall design and ways to inject the pollutant into the estuary from a point source so that its concentration is reduced as rapidly as possible. This is an engineering task, and therefore, the near field problem is not considered in oceanography. The far field problem studies the distribution of the pollutant through the entire estuary, starting from the disposal site not as a point source but as an extended source of uniform concentration across the estuary depth and width. The far field problem thus assumes that the engineering design of the pollutant outfall achieves reasonably uniform distribution over a sizeable section of the estuary and uses this situation as the starting point for the analysis.

PHYSICOCHEMICAL PROCESSES IN ESTUARIES

If the concentration of a pollutant where it is introduced into the estuary is known, its distribution through the estuary can be predicted from theoretical considerations. To understand how this is done, it is useful to begin with a look at the distribution of salt and freshwater in the estuary. Let us consider salt as a "pollutant" and follow its path through the estuary. Its source is at the estuary mouth, where its concentration is given by the oceanic salinity S_0. The salt enters the estuary with the net water movement of the lower layer. As it moves up the estuary, its concentration decreases through turbulent diffusion; in other words, the salt "pollutant" is diluted with freshwater that comes down from the river. We can normalize the salt "pollutant" distribution along the estuary by making it independent of its original "concentration" S_0 and define the salt–water fraction s as $s = S/S_0$, where S is the local depth averaged salinity at any arbitrary location in the estuary. This normalized salt concentration or saltwater fraction decreases from 1 at the mouth to 0 at the inner end of the estuary.

Similarly, we can consider the freshwater introduced by the river as a "pollutant" and follow its path through the estuary. Its source is at the inner end of the estuary, where its normalized concentration or freshwater fraction f is equal to 1. The freshwater then moves down through the estuary with the net water movement in the upper layer and reaches a concentration of zero at the estuary mouth. The estuary contains only saltwater and freshwater, so the saltwater fraction and the freshwater fraction always add up to 1 locally, which means that $f + s = 1$.

Not all pollutants are introduced into the estuary through the river or from the ocean. There are three classes of marine pollutants. Conservative pollutants are substances that are inert in the marine environment; their concentrations change only as a result of turbulent diffusion. (Salt and freshwater can be considered such substances.) Nonconservative pollutants undergo natural decay; their concentration depends on turbulent diffusion and also on the time elapsed since their introduction into the environment. The concentration of coupled nonconservative pollutants depends on turbulent diffusion and natural decay, both of which act to decrease their concentration, and on the availability of other substances in the environment, which allow the concentration to increase over time.

Conservative Pollutants

A conservative pollutant released at a point between the inner end and the mouth of the estuary spreads by turbulent diffusion in both directions, downstream and upstream. Its diffusive behavior is no different from the diffusive behavior of salt or freshwater. In the upstream direction, it diffuses in the same way as the salt water diffuses upstream; in the downstream direction, it follows the freshwater diffusion. If its concentration at the outlet is c_{out}, its concentration C in the estuary upstream from the release point is thus proportional to the salt fraction; downstream from the release point, it is proportional to the freshwater fraction:

$$\text{Upstream:} \quad C = c_{out}(S/S_{out})$$

$$\text{Downstream:} \quad C = c_{out}(f/f_{out})$$

Here, S_{out} is the vertically averaged salinity at the outlet location, and f_{out} is the freshwater fraction at the outlet. This shows that it is possible to predict the distribution of a pollutant in the entire estuary if the salinity distribution is known. The equation assumes steady-state conditions, continuous release of the pollutant at the constant concentration c_{out}.

Nonconservative Pollutants

The concentration of nonconservative pollutants decreases even in the absence of diffusion, through either biochemical or geochemical reaction. An example is the concentration of coliform bacteria released from a sewage outlet. The relationship between the concentration C of such a pollutant, the salinity S, and the fresh water fraction f is not as straightforward as for conservative pollutants, but concentration levels are always lower than those derived from the conservative case.

The concentration can be estimated by subdividing the estuary into compartments. If the compartments are chosen so that the ratio r of the freshwater volume V_f in the compartment and the volume of freshwater R introduced by the river into the estuary in one tidal cycle is constant ($r = V_f/R$ = constant), it can be shown that the concentration of a nonconservative pollutant can be approximated by an equation for the concentration in each compartment:

$$\text{Upstream:}\quad C_p = C_{\text{out}}\frac{S}{S_{\text{out}}}\left[\frac{r}{1-(1-r)e^{-kT}}\right]^{n+1-p}$$

$$\text{Downstream:}\quad C_p = C_{\text{out}}\frac{f}{f_{\text{out}}}\left[\frac{r}{1-(1-r)e^{-kT}}\right]^{p+1-n}$$

In this equation, compartments are numbered from the inner end of the estuary, n is the compartment containing the outfall, and p is the compartment where the concentration is evaluated. T is the tidal period, and k is the decay constant for the pollutant. The larger k, the faster the decrease of the concentration over time. $k = 0$ represents the case where there is no independent decrease without diffusion, or the case of a conservative pollutant. Figure 1 compares this behavior with that of a conservative pollutant. Note that for $k = 0$, the equation for the concentration reduces to

$$\text{Upstream:}\quad C = c_{\text{uot}}(S/S_{\text{out}})$$

$$\text{Downstream:}\quad C = c_{\text{out}}(f/f_{\text{out}})$$

which is the same equation that was derived for the conservative pollutant.

Coupled Nonconservative Pollutants

Biochemical or geochemical processes reduce the concentration of a nonconservative pollutant, and they often also give rise to an increase in the concentration of another substance, which in turn may produce a third substance as a result of its own decay, and so on. Whether this chain of events constitutes pollution or not depends on the degree of harmfulness of each substance. Ammonium nitrogen, for example, is used as a fertilizer on land. Under natural conditions, it converts to nitrate (Fig. 2), which is also a nutrient.

The introduction of nutrients into the marine environment does not automatically constitute pollution. But when nitrate and other nutrients are again mineralized in seawater, this process requires oxygen, which is not in unlimited supply in the ocean. Too much nutrient can lead to such a reduction of oxygen levels that the lack of oxygen can become a threat to marine life.

The indicators of nutrients and sediments are influenced by human activities from both diffuse inputs (urbanization and agriculture) and point discharges to the environment. The recommended indicators need to be biologically meaningful, yet readily collected and interpreted in relation to ecological conditions.

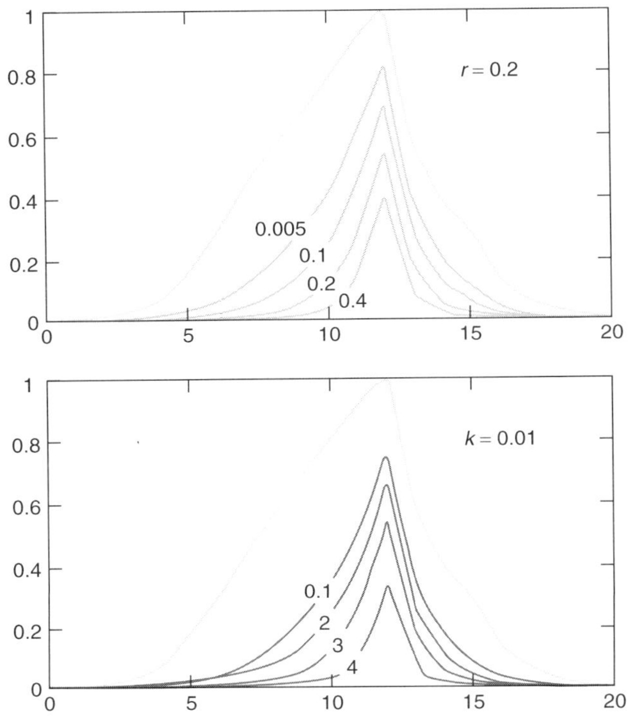

Figure 1. Behavior of conservative and nonconservative pollutants.

ROLE OF NUTRIENTS IN ESTUARIES

Nutrient Status

Six indicators of nutrients are recommended for assessing the status of marine and estuarine waters: dissolved inorganic nitrogen (DIN), total nitrogen (TN), dissolved inorganic phosphorous (DIP), total phosphorous (TP), chlorophyll a (Chl a), and dissolved oxygen (DO). Individual indicators have their own limitations, but when all six are used together, they provide a strong indication of nutrient status.

1. *DIN.* DIN is the combined concentration of nitrate, nitrite, and ammonia. DIN is the bioavailable fraction of the nitrogen in the water column.
2. *TN.* DIN may rapidly disappear as algae assimilate it, and so TN provides a better indication of the nitrogen cycling through the algal community and how much is dissolved in the water. TN is the sum of the concentrations of bioavailable (dissolved) and particulate forms of inorganic and organic nitrogen.

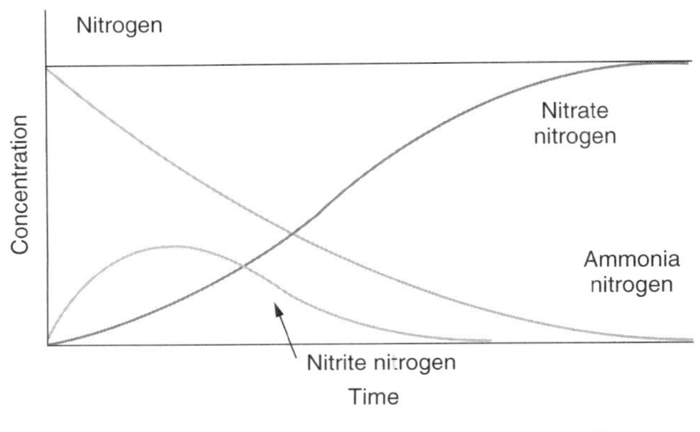

Figure 2. Behavior of a coupled nonconservative pollutant.

3. *DIP.* DIP is typically measured as the concentration of reactive orthophosphates, which is the form of phosphorous most readily bioavailable to aquatic plants.
4. *TP.* Like DIN, algae may rapidly assimilate DIP and so TP provides a better indication of the phosphorous cycling through the algal community and how much is dissolved in the water. TP is the sum of the concentrations of dissolved and particulate forms of inorganic and organic phosphorous.
5. *Chl a.* Chl a is a crude measure of phytoplankton biomass. Apart from light, phytoplankton are primarily limited by nutrient availability; Chl a is often used as an integrated, surrogate measure of nutrient status.
6. *DO.* DO, measured preferably as percent saturation (%), is the net balance of the amount of oxygen produced by photosynthesis, relative to that consumed by respiration and other forms of biological and chemical oxygen demand. Low DO (hypoxia) can arise following blooms of phytoplankton stimulated by nutrient enrichment. As the large mass of plant tissue decays, DO is lowered in the water column, and this can result in fish kills and nutrient inputs from the sediments.

Nutrients That Are Limiting Factors for Estuarine Phytoplankton

A limiting nutrient is a nutrient or trace element that is essential for phytoplankton to grow but is available in quantities smaller than required by the plants or algae to increase in abundance. Therefore, if more of a limiting nutrient is added to an aquatic ecosystem, larger algal populations develop until nutrient limitation or another environmental factor (e.g., light or water temperature) curtails production, although at a higher threshold than previously.

It is often said that nitrogen is the limiting nutrient in marine and coastal waters; however, the general assumption is often incorrect (1). Phosphorous, silica, and iron can also limit production in marine and coastal waters, and different trophic groups within the same ecosystem can be limited by different elements and nutrients (1).

Redfield Ratios. The recommended ratios of nutrients in estuaries are given by Redfield ratios, which are C:N:P = 106:16:1 for algae and Si:N:P = 20:106:16:1 for diatoms (1). Recently, Harris has suggested that a C:N:P ratio of 166:20:1 is probably more representative of the average phytoplankton composition (1). However, a review of the literature by Hecky and Kilham showed that the optimum N:P ratios for 14 freshwater and marine phytoplankton, for which data were available, ranged from 7–87 (1).

If the concentration of Si or N relative to P falls below these ratios (N:P<16:1, Si:P<20:1, these may be limiting nutrients (carbon is never limiting because it is available as a gas (CO_2); carbon in organic matter comes from dissolved CO_2 in water. The nitrogen levels control the rate of primary production. If the system is provided with high levels of nitrogen, algal blooms occur. Systems may be phosphorous limited, however, or become so when nitrogen concentrations are high and N:P>20:1(1). In such cases, excess phosphorous triggers eutrophic conditions. For optimum alagal growth, the N:P ratio should be in the range of 16–20:1(1). Increased reliance on the N:P ratio as a management tool is widely accepted by marine scientists and engineers to control nuisance algal blooms.

Another limitation on using N:P ratios is the considerable uncertainty over what measure of nitrogen and phosphorous should be used in calculating the N:P ratio. The ratio of TN:TP generally overestimates the algal available nitrogen to algal available phosphorous ratio aN:aP because organic phosphorous is more readily available than organic nitrogen and DIN:DRP underestimates aN:aP because DRP includes some organic phosphorous compounds in addition to orthophosphates (1). Therefore, the ratio of DIN:DIP gives a better estimate of the N:P ratio (1).

The recommended levels of total phosphorous in estuaries and coastal ecosystems to avoid algal blooms is 0.01–0.10 mg/*l* and 0.1–1 mg/l nitrogen (a 10:1 ratio of N:P) (1). It is recommended that total phosphorous levels in waterbodies should range within –0.005–0.15 mg/L to prevent eutrophication in estuarine waters; guidelines for protecting aquatic ecosystems recommend that phosphate values should not exceed 0.015 mg/L in estuaries (1).

ANZECC guidelines recommend the following concentration ranges to protect aquatic ecosystems: total N range from 0.01–0.10 mg/L for estuaries and 0.01–0.06 mg/L for coastal waters. The nitrate value should not exceed 0.1 mg/L for estuaries (1).

Regeneration of Nutrients (cycling)

Nitrogen Cycle in An Estuary (Fig. 3)

Phosphorous Cycle in An Estuary

$$\text{Organic P} \rightarrow \text{Inorganic P} \rightarrow \text{Organic P}$$

BUDGETARY MODELS FOR ESTUARIES

In science, models are tools that help us conceptualize, integrate, and generalize knowledge. "Budget models" are simple mass balance calculations of specific variables (such as water, salt, nitrogen, phosphorous, etc.) within defined geographic areas and defined periods.

The fluxes of materials to and from a system may be budgeted by many different procedures, but there are inherent similarities among these procedures. A budget describes the rate of material delivery to the system ('inputs"), the rate of material removal from the system ("outputs"), and the rate of change of material mass within the system ("storage"). Some materials may undergo internal transformations. Such changes are sometimes referred to as "internal sources and sinks." Figure 4 depicts the budget for any material.

Without further interpretation, budgets do not provide information on the processes, which account for the summed sources minus sinks (2). Some inferences about the actual processes can be drawn by comparisons among

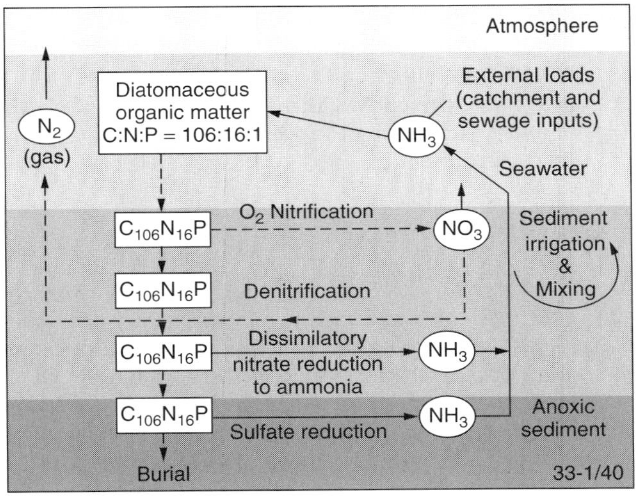

Figure 3. Nitrogen cycle in an estuary.

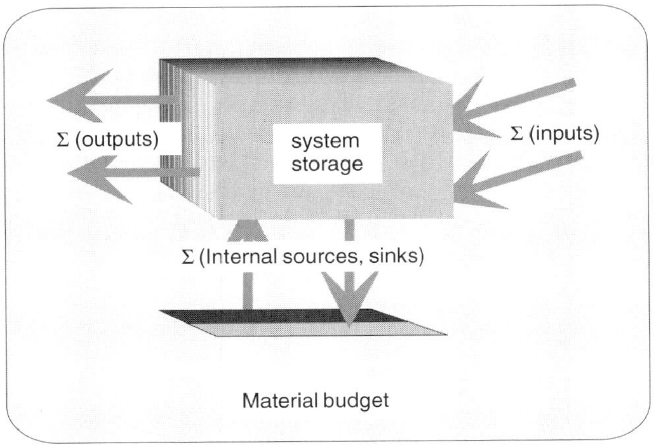

Figure 4. Budget for any material.

budgets for different materials. This approach is referred to as "stoichiometric analysis" of the fluxes.

It is also useful to describe such a budget in terms of this simple equation:

$$\frac{dM}{dt} = \sum \text{inputs} - \sum \text{outputs} + \sum (\text{sources} - \text{sin ks})$$

where dM/dt represents the change in mass of any particular material in the system with respect to time (2).

In broad summary, water and salt budgets are used to estimate water exchange in coastal systems. Nutrient budgets (as a minimum, dissolved inorganic phosphorus and dissolved inorganic nitrogen) are also developed, and departure of the nutrient budgets from conservative behavior is a measure of net system biogeochemical fluxes.

Stoichiometric Analysis of Fluxes

In chemistry, "stoichiometry" is the study of the combination of elements in chemical reactions. Therefore, we use the term "stoichiometric analysis" to describe this approach to budgetary analysis. It should be kept in mind that the discussion to follow is based on the ΔY's for C, N, and P, as derived from nutrient budgets. The following is generally computed using the ΔY's.

- Net ecosystem metabolism, that is, the difference between primary production and respiration
- Net nitrogen fixation minus denitrification
- Net sulfate reduction
- Net $CaCO_3$ precipitation
- Net CO_2 gas flux across the air–sea interface.

NONCONSERVATIVE DISSOLVED INORGANIC P AND N FLUXES IN NEGOMBO ESTUARY, SRI LANKA, USING THE STANDARD LAND–OCEAN INTERACTION IN THE COASTAL ZONE (LOICZ) METHODOLOGY (3)

The Negombo estuary (7°4′–7° 12′ N: 79°47′–79°51′ E) is situated on the west coast of Sri Lanka. The estuary receives water from the Attanagalu Oya (ja-Ela and Dandugam Oya) drainage basin and is a dominant morphological feature of the watershed. The brackish water area is 3239 ha and is considered the estuarine part of the contiguous wetland system of the Muthurajawela Marsh Negombo Estuary. The main freshwater source, Attanagalu Oya, empties at Ja-Ela and Dandugam Oya at the southern tip of the estuary. In addition, the Hamilton canal is the connecting watercourse of the Kelani Estuary and the Negombo Estuary running parallel to the west coast from the north to the south along the Muthurajawela Marsh.

LOICZ, a one-box, one-layer model was used for this study, and the data available are applied on a seasonal basis (dry and wet). The data used are mentioned below and are the averages for dry and wet seasons:

1. river discharge
2. precipitation
3. evaporation
4. average salinity outside the system, within the system, and in the river discharge
5. dissolved inorganic phosphate (DIP) outside the system, within the system, and in the river discharge
6. dissolved inorganic nitrogen (DIN) outside the system, within the system, and in the river discharge

The nonconservative DIN and DIP fluxes are given in (Table 1).

$$[p - r] = -\Delta \text{DIP} \times \left(\frac{C}{P}\right)_{\text{part}}$$

$$[\text{nfix} - \text{denit}] = \Delta N_{\text{obs}} - \Delta N_{\text{exp}}$$

$$(\text{nfix} - \text{denit}) = \Delta N - \Delta P \times (N:P)_{\text{part}}$$

The system is an internal source for DIP and an internal sink for DIN in both the wet and dry seasons. In the dry season, the N:P ratio is less than the Redfield ratio of 16:1; therefore, the estuary is N limiting. In the wet season, the N:P ratio is greater than the Redfield ratio

Table 1. Nonconservative Dissolved Inorganic P and N Fluxes in Negombo Estuary (Sri Lanka) Considering a one–box, one Layer System[a]

Process	Dry Season Rate, ($mmolm^{-2}d^{-1}$)	Wet season Rate, ($mmolm^{-2}d^{-1}$)
Area	32.39 km^2	32.39 km^2
DDIP	+0.052	+0.008
DDIN	−0.104	−0.837
DIN:DIP	13	18
Stoichiometric Analysis		
(nfix-denit)phytoplankton[b]	−0.936	−0.965
(*p-r*) phytoplankton[c]	−5.512	−0.848
(nfix-denit)sea grass[b]	−1.664	−1.077
(*p-r*)sea grass[c]	−28.6	−4.4
nfix-denit)mangrove[b]	−0.676	−0.925
(*p-r*)mangrove[c]	−67.6	−10.4

[a]Reference (3).
[b]$(N:P)_{part}$ assumed to be 16 (plankton), 11(mangrove), and 30 (sea grass).
[c]$(C:P)_{part}$ assumed to be 106(plankton), 1300 (mangrove), and 550 (sea grass).

of 16:1 but less than 20:1; therefore, the estuary has optimum nutrient concentrations, which support optimum phytoplankton growth.

The DDIP>0 in both seasons. This system with DDIP>0 is producing DIC via net respiration ($p - r$) and is heterotrophic. The negative signs of ($p - r$) calculated for phytoplankton, sea grass, and mangrove support this assumption if phytoplankton, sea grass, and mangrove dominate the input of organic matter from outside the system respectively. The (nfix−denit) is negative for phytoplankton as well as for sea grass and mangroves, indicating that denitrification is the main process taking place. The denitrificate rates are in the range of −0.676 to −1.664 for the dry season and −0.925 to −1.077 in the wet season.

Nitrogen is typically the nutrient that controls primary production in marine coastal ecosystems and, as such, is commonly implicated in the eutrophication of coastal waters. Denitrification is probably the most important nitrogen cycling pathway because it is one of the few natural processes that can counteract eutrophication. Up to 60 to 80% of the external nitrogen load delivered to coastal ecosystems may be lost to coupled sediment nitrification–denitrification. Despite its importance, sediment denitrification has rarely been measured in Sri Lankan coastal ecosystems.

Worldwide denitrification studies in tropical and subtropical estuaries where N_2 fluxes have been directly measured are also rare; most work has been carried out in temperate West European and North American systems. There are distinct biogeochemical differences between tropical/subtropical and temperate estuaries. Differences in tropical/subtropical estuaries most likely to influence denitrification rates include

generally low water column nitrogen concentrations,
different temperature and light regimes,
differences in benthic infauna,
shallower, so benthic processes that interact with denitrification such as benthic productivity are more important, and
episodically driven resulting in differences in the quality, timing, and delivery rates of carbon inputs to the sediments.

WATER QUALITY GOALS

In developing an effective strategy for mitigating the effects of nutrient overenrichment, one must understand the physical and ecological relationships that determine the extent and causes of nutrient overenrichment, along with societal objectives and behavioral responses. Societal objectives determine goals that a management strategy strives to achieve and is the bench mark against which it is evaluated.

The appropriate set of policies for any given estuary depends on the nutrient sources for that estuary. For example, if the main nutrient source is agriculture, a set of policies designed to promote the adoption of best management practices is required. These can be implemented at the local, regional, or national level. Alternatively, if atmospheric deposition is the main source of nutrients, policies that reduce atmospheric emissions of nitrogen are needed. The source of atmospheric nitrogen is often outside the local jurisdiction governing the estuary, so policies to combat this nutrient source must be implemented at the regional or national level.

Establishing Criteria and Standards

A water quality criterion is that concentration, quality, or intensive measure that, if achieved or maintained, allows or makes possible a specific water use. Water quality criteria are often the starting point in deriving standards, but criteria do not have a direct regulatory impact because they relate to the effect of pollution rather than its causes. A water quality standard is the translation of a water quality criterion into a legally enforceable ambient concentration, mass discharge, or effluent limitation expressed as a definite rule, measure, or limit for a particular water quality parameter. A standard may or may not be based on a criterion.

Traditionally, water quality standards have been absolute numbers as a concentration or discharge of a toxic substance that may not be exceeded or an oxygen concentration that must be maintained. However, water quality in a given waterbody can fluctuate as a result of random factors, such as weather and uncertainties in hydrologic processes. Eutrophication impacts include increased primary productivity; increased phytoplankton biomass; reduction in water clarity; increased incidents of low oxygen; changes in trophic structure, trophic interactions, and trophodynamics of phytoplankton, zooplankton, and benthic communities; damage to coral reefs; fish kills; reduced fisheries production; and decreased biotic diversity.

In general, the public must accept the need for improved receiving water quality and prevention measures. If the public does not perceive a problem, it is unlikely that elected officials will pursue this issue or that agency staff will have the resources or authority to implement solutions.

When setting up water quality standards, many other factors must also be considered (4)

- Are damages subject to threshold effects or are they continuous?
- During which season are impacts the greatest?
- Are sudden discharges controlled?
- What are the flushing and mixing conditions in the estuary?
- What timescales are involved?

BIBLIOGRAPHY

1. ANZECC/ARMCANZ. (October 2000). *Australian and New Zealand Guidelines for Fresh and Marine Water Quality*. Available: http://www.ea.gov.au/water/quality/nwqms/#quality.
2. Gorden, D.C., Jr., Bondrean, P.R., Mann, K.H., Ong, J.E., Silnert, W.L., Smith, S.V., Wattayakern, G., Wulff, F., and Yangi, T. (1996). *LOICZ Biogeochemical Modelling Guidelines*. LOICZ/R & S/95-5, vit p. 96 LOICZ, Texel, the Netherlands.
3. Dissanayake, U.A.P.K. (2003). *Strategies for the Improved Management of the Coastal Zone of Sri Lanka*. M.Sc. Thesis Environmental Management, University of Moratuwa, Sri Lanka.
4. Committee on the Causes and Management of Eutrophication. (2000). Ocean Studies Board, Water Science and Technology Board, National Research Council. *Clean Coastal Waters: Understanding and Reducing the Effects of Nutrient Pollution*. National Academy Press, Washington, DC.

NOS/NMFS COOPERATIVE RESEARCH ON COASTAL FISHERIES AND HABITATS AT THE BEAUFORT LABORATORY

Jon Hare
NOAA Center for Coastal Fisheries and Habitat Research

EARLY LIFE HISTORY OF FISHES

Laboratory researchers have studied fish early life history since the early 1900's. Work is conducted on egg, larval and juvenile stages of species ranging from anadromous shad to highly migratory tuna and billfishes. Topics of study include taxonomy, distribution, growth, predation, physical transport, behavior and stock identification.

Research on early life history stages at the Laboratory is supported by a fish rearing facility. A variety of species have been reared ranging from summer flounder, to black sea bass, to spotted sea trout. Fish from this facility are used in ecological, physiological, and behavioral studies.

FISHERIES OCEANOGRAPHY

Laboratory staff have been active in fisheries oceanography since the late-1970's. The goal of this research is to better understand the linkages between biological and physical oceanographic processes and fish population distribution, abundance and dynamics.

Recently, the Laboratory has been involved in the South Atlantic Bight Recruitment Experiment (SABRE), the purpose of which is to understand the effect of environmental variation on young menhaden survival, and ultimately, recruitment to the fishery. Employees have conducted egg and larval investigations on the continental shelf, have studied ingressing larvae at inlets, and have examined juveniles in estuarine nursery habitats. Coupling field observations, behavioral models and physical models has provided insights into the role that physical transport plays in controlling larval supply to inlets.

Current research in fisheries oceanography ranges from the effect of Langmuir circulation on plankton distributions, to the exchange of larval fishes across the Cape Hatteras zoogeographic boundary, to the role of the Charleston Bump as essential spawning and nursery habitat. The scales studied range from meters to hundreds of kilometers and from hours to decades, thereby allowing Laboratory researchers to link processes-oriented work with studies examining annual and decadal variation in fish populations.

HARMFUL ALGAL BLOOMS

Laboratory employees have been studying the "red tide" dinoflagellate, *Gymnodinium breve*, since a bloom in coastal North Carolina waters in 1987 caused both fish kills and shellfish closures. The source of *G. breve* cells is the shelf off the west coast of Florida, where blooms have occurred in 21 of the last 22 years. Local eddy circulation in the northeastern Gulf of Mexico and in the vicinity of the Dry Tortugas affects the retention and coastal distribution of blooms on the western Florida shelf as well as influences the entrainment of cells into the Loop Current-Florida Current-Gulf Stream systems. All blooms off the Florida Keys and in the South Atlantic Bight can be traced to the inoculation of shelf water with Gulf Stream water, as well as the proper physical conditions that support blooms. Monitoring and research of *G. breve* is continuing, as is work with other species that potentially cause harmful algal blooms.

The Laboratory's remote sensing capabilities contribute to the study of harmful algal blooms. The Laboratory serves as the Southeast Node of NOAA's CoastWatch program and researchers at the Laboratory are involved in the development of regional ocean color algorithms for the SeaWIFS satellite.

FLORIDA BAY-EVERGLADES NATIONAL PARK

Laboratory scientists initiated fisheries and habitat research in Florida Bay and adjacent waters in 1983. Past research has examined a variety of topics including larval fish distributions, ecology of juvenile spotted seatrout and gray snapper, evaluation of habitat use by juvenile fishes and invertebrates, nutrient limitation of seagrass growth and responses of both plant and animal communities to the die-off of seagrasses first identified in 1987.

Building on this past work, investigators at the Laboratory are conducting laboratory and field research as part of the Florida Bay Study. The overall goal of this work is to evaluate the relation between environmental and habitat change and the recruitment, growth and survivorship for fishery organisms in Florida Bay. Current research includes: examination of the influence of environment and habitat on larval and juvenile fish growth and survival; development of an individual-based energetic model for larval spotted seatrout; evaluation of the effect of habitat on the distribution of fishes and invertebrates and determination of the origin of elevated mercury levels in upper trophic levels.

SEAGRASS RESTORATION

Research on various aspects of seagrass ecology at the Laboratory has been continuous for 20 years. Areas of investigation have included development and dissemination of planting techniques monitoring protocols, and success criteria, as well as studies to determine the light requirements of seagrasses, functional equivalency of restored beds as compared to natural, undisturbed systems, and landscape scale studies regarding the temporal dynamics of seagrass bed pattern and distribution. Emphasis has been placed on research information transfer to managers, active participation in research projects, and litigation. The research approach has been to sustain a broad-based program covering a variety of ecological processes which allows the scientists to quickly adapt and respond to changing management concerns and issues.

MANAGEMENT AND RESEARCH USES OF GIS

Laboratory employees use Geographic Information Systems for various research and management activities. Menhaden researchers are examining the spatial and environmental aspects of menhaden catch and marine mammal researchers are studying the spatial interaction between fisheries and marine mammals.

As a more in depth example, researchers at the Laboratory are using GIS to map and to understand the distribution of aquatic seagrasses throughout the southeastern United States. Aquatic seagrass beds are important habitat to many fish and shellfish species. Linkage with the NOAA Coastal Assessment Framework associates the aquatic beds with coastal drainage basins. Within North Carolina, the vast majority of seagrass beds are inside the barrier islands of the Pamlico-Albemarle Estuarine Complex. Distribution of seagrass beds is related to annual extremes of salinity but other, more-local factors are also involved. The restricted abundance of seagrass beds along mainland shores reflects salinity stress, water quality, and water use problems from coastal development. These large-scale studies of seagrass are being linked with the smaller scale seagrass dynamics studies conducted by Laboratory employees.

MENHADEN RESEARCH

Scientists at the Beaufort Laboratory have been studying Atlantic menhaden since the 1950's and Gulf menhaden since the 1960's. In 1996, catch of both species totaled 1.8 million pounds and was valued at $94.2 million. Research has included studies of life history, movements, juvenile abundance, and patterns in the fishery. Current efforts include examination of the environmental factors that contribute to recruitment variability and continuation of the monitoring of fishery patterns.

Beginning in the late 1970's, menhaden boasts were asked to complete logbooks of daily purse-seine activities. For each purse-seine set, time, location, catch, and weather conditions are enumerated. Approximately 6,800 forms describing 23,000 purse seine sets are collected annually. Laboratory personnel also collect data on landings, fishing effort, and size and age of catch. These data are used for stock assessments, are helpful to industry personnel and are used by researchers to address various scientific questions. For example, these data are being used to examine the effects of hypoxia on Gulf menhaden distribution. Preliminary analyses indicate that catches during the summer of 1995 were low in hypoxic areas near shore. These data on adult fish distribution allow examination of the effects of hypoxia on higher tropic levels.

REEF FISH RESEARCH

Laboratory employees have studied reef fish off the southeastern United States since the late 1960's and have addressed both applied fishery issues and basic reef fish ecology. The Laboratory's Southeast Region Headboat Survey collects fisheries and biological data to support management activities. Approximately 165 vessels are monitored throughout the southeast, and in 1996, biological samples were collected from 37, 435 fishes of 124 species. These data are also used to examine patterns in the fishery and to study the structure and distribution of reef fish communities.

A variety of studies of reef fish ecology have been conducted by Laboratory employees using SCUBA. A temporal comparison off North Carolina suggests an increase in reef fish diversity and abundance possibly owing to climate change. The impact of removing piscivorous predators from reef areas is also being studied. Employees are continuing to examine changes in fish assemblages on managed and unmanaged reefs at Key Largo and participate in annual dives to assess reef fish populations at the Grays Reef National Marine Sanctuary off Sapelo Island, Georgia. Future research includes working with fisheries and GIS data to map essential reef fish habitat, and studying recruitment of larval fishes to reef habitats.

STOCK ASSESSMENT

Stock assessments are conducted by Laboratory employees for a variety of species in support of fisheries management. Stock assessments involve estimating population size-at-age, data which are collected by laboratory employees. From estimates of population size at age, age-specific fishing mortality for different levels of natural mortality can be calculated. Information on reproduction, such as age-specific maturity and fecundity, are used to calculate spawning potential ratios and to provide an indication of the level of fishing pressure experienced by the stock. The results of these assessments are provided to regional fishery management councils and commissions and include recommendations on how to either build or sustain healthy fish stocks along the Atlantic and Gulf coasts.

Laboratory employees conduct annual stock assessments for Atlantic menhaden and five year assessments for Gulf menhaden. Stock assessments are also completed for various reef fishes; recent assessments include red porgy, black sea bass, red snapper, vermilion snapper and scamp. Staff also conduct or provide advice on stock assessments for other fishery species in the southeast including blue crab, red drum, weakfish and wreckfish.

LINKING STOCK ASSESSMENTS WITH OCEANOGRAPHIC AND ECOLOGICAL PROCESSES

The ability to conduct stock assessment work and process-oriented work has allowed Laboratory researchers to further understand the processes that influence recruitment and stock abundance. For example, Gulf menhaden recruitment has been examined relative to Mississippi River discharge, a relation suggested by work done at the Laboratory during the GOMEX project. As another example, Atlantic menhaden recruitment has been related to the ingress of menhaden larvae into Beaufort Inlet. This time series of wintertime larval ingress at Beaufort Inlet has been used by participants in the SABRE program and data collection will continue.

MARINE MAMMALS

The Laboratory's research on marine mammals focuses on their interactions with fisheries, as well as life history and stock identification. Bottlenose dolphins are the primary species studied. Activities include recovery of stranded animals, live captures for tagging and data collection, photo-identification, and analysis of life-history samples. Results contribute to basic biological knowledge and to effective management and conservation.

As an example of some of the research conducted, dolphin 717 was freeze-branded during a live capture in 1995, and has since been regularly observed in the Beaufort area. Laboratory scientists know that she was a 3.5 year old female at the time of capture. Monitoring continues to determine when she will have her first calf as part of life history and recovery studies.

SEA TURTLES

Laboratory scientists continue an active program of sea turtle research. Surveys conducted since 1988 have underscored the importance of North Carolina's inshore waters, particularly the Pamlico-Albemarle Estuarine Complex, to juvenile loggerhead (*Caretta caretta*), green (*Chelonia mydas*), and Kemp's ridley (*Lepidochelys kempii*) sea turtles. Sea turtles are present in the Complex from April-December and during their emigration in early winter, are vulnerable to capture in pound nets set behind barrier islands. North Carolina pound nets are a passive gear that allow turtles to feed and to surface, to breathe.

Parts of the Pamlico-Albemarle Estuarine Complex were established as a pilot index area for sea turtles in 1995, using catch rates in pound nets as an index of abundance. The feasibility and methodology was established and the fishery was sampled again during fall 1996 and 1997. The goal of this project is to establish an index-abundance-area to monitor, assess, and predict the status of and impacts to sea turtles and their ecosystems. The cooperation of pound net fishermen in the area provides data not only on abundance, but also on movement and demography of these populations. In addition, blood samples are taken for analyses of health status, sex, and genetics.

PHRAGMITES RESEARCH

Researchers at the Laboratory are studying the use of *Phragmites* marsh by fish and shellfish species. Throughout the eastern United States many *Spartina alterniflora* salt marsh systems are being altered through *Phragmites australis* invasion. While information is lacking on the use of *P. australis* marsh by fish and shellfish species, increase in spatial distribution of *Phragmites* has concerned resource managers, who view such a shift as a degradation in habitat quality. Research conducted by Laboratory staff shows no significant differences in fish abundance or biomass between tidal *P. australis* and *S. alterniflora* marsh. Further, species diversity was higher in *P. australis* marsh. Stable isotope data from marsh fauna indicate that *P. australis* is an integral part of the estuarine food web.

MARINE CHEMISTRY

The Laboratory has a long history and continues research in marine chemistry. Although not highlighted here, the work bridges the gap between chemistry and habitat and fisheries issues. Studies include metals in the environment and their transfer through the food chain, the relation between iron and primary production, and otolith microchemistry as an indicator of past environmental exposure.

DISTRIBUTION AND DYNAMICS OF GAS HYDRATES IN THE MARINE ENVIRONMENT

Anne M. Trehu
Marta E. Torres
Oregon State University
Corvallis, Oregon

Gas hydrate is an ice-like substance that forms at low temperature and high pressure when adequate amounts

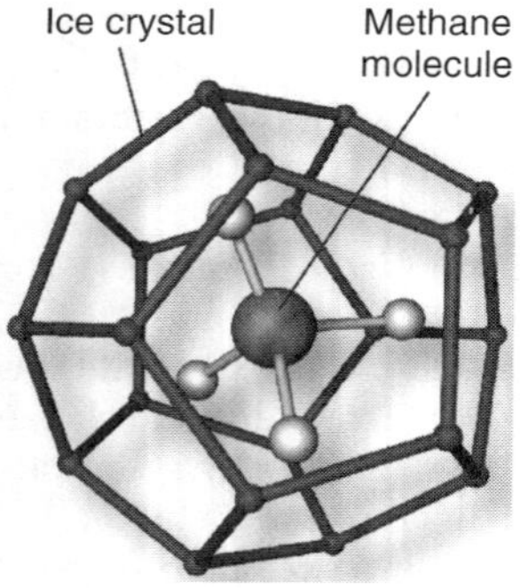

Figure 1. Schematic diagram of a gas hydrate molecule.

of water and low molecular weight gases, such as carbon dioxide, methane, and other hydrocarbons, are present (Fig. 1). In nature, gas hydrates containing primarily methane are widespread in the Arctic and in the shallow sediments on continental margins. Under certain geologic conditions, enough methane hydrate may be present for these deposits to represent a fossil fuel resource for the future. Concentrated deposits of gas hydrate near the seafloor form the substratum for unique assemblages of microbes that are only now being discovered; these microbes constitute the base of a foodweb for clams and other organisms, similar to those found at hydrothermal vent sites. Destabilization of gas hydrate in marine sediment may occur when artificial structures that heat up the surrounding sediments are installed; this may, in turn, result in catastrophic structural collapse. Because methane is a powerful greenhouse gas, sudden destabilization of gas hydrate on a large scale can have an impact on global climate. Indeed, past episodes of climate change linked to gas hydrate destabilization have been documented in the geologic record. For all of these reasons, researchers in many countries are trying to better understand the distribution and dynamics of gas hydrates in the marine environment.

NECESSARY CONDITIONS FOR FORMATION OF GAS HYDRATE IN NATURE

The temperature and pressure conditions required for gas hydrate to be chemically stable are found at the seafloor everywhere the water depth is greater than 300–500 m. (The minimum depth for gas hydrate stability decreases as the water temperature decreases.) As temperature increases beneath the seafloor at a rate that is controlled by the regional geothermal gradient, the effect of increasing temperature will eventually overtake the effect of increasing pressure, and gas hydrate will no longer be stable (Fig. 2). This balance between increasing pressure and temperature results in a gas hydrate stability zone beneath the seafloor that ranges in thickness from a few centimeters to several hundred meters. Based on these criteria, gas hydrate has the potential of forming almost everywhere beneath the continental slope and ocean basins.

Another critical condition that must be met for gas hydrate to form is that the concentration of gas present

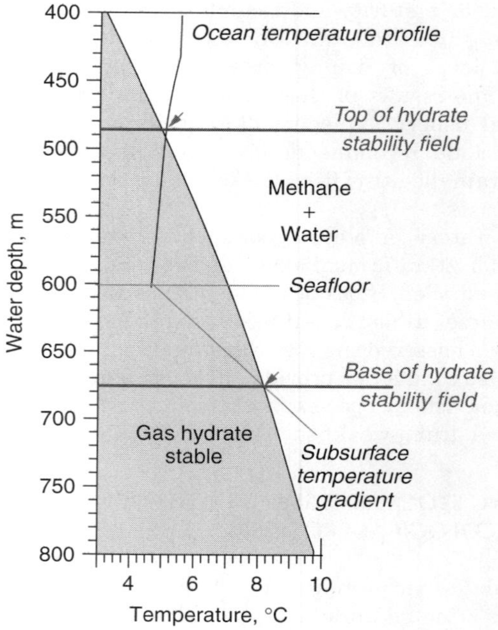

Figure 2. The temperature and pressure conditions under which gas hydrate is stable are constrained thermodynamically as shown by the stability boundary (shown here for pure methane hydrate in seawater). Hydrate is stable at low temperature and high pressure (which increases as water depth increases). This is shown by the light gray field. In this example, which depicts a site located at 600 m depth, the water temperature profile and the subseafloor temperature gradient are typical for the continental margin offshore Oregon. The top of the gas hydrate stability zone here is located within the water column at 485 m. Gas hydrates are stable in the sediments from the seafloor to a depth of 665 m, where the subsurface temperature becomes too high for methane hydrate to be stable.

must exceed that required to stabilize the gas hydrate structure. Not enough methane is present in the sediments throughout most of the ocean basins. Gas hydrate is restricted to regions where organic material, which decomposes to form methane, is unusually abundant or where fluid flow transports gas into the gas hydrate stability zone from greater depth. These conditions are met on many continental margins (Fig. 3).

Other factors affecting gas hydrate stability are the salinity of the pore waters and the composition of the hydrate-forming gases that are present. Moreover, temperature and pressure conditions can change locally as a result of focused fluid flow, tectonic activity, or the introduction of artificial structures. It can change globally as a result of climatic perturbations. The distribution and abundance of gas hydrate in the marine environment are therefore constantly changing. New discoveries are being made every year, increasing the number of verified gas hydrate sites.

ORIGIN OF THE GAS IN GAS HYDRATE

The gas in gas hydrate has two possible origins. Methane can be generated in the upper few hundred meters by microbes (biogenic gas), or it can migrate into the

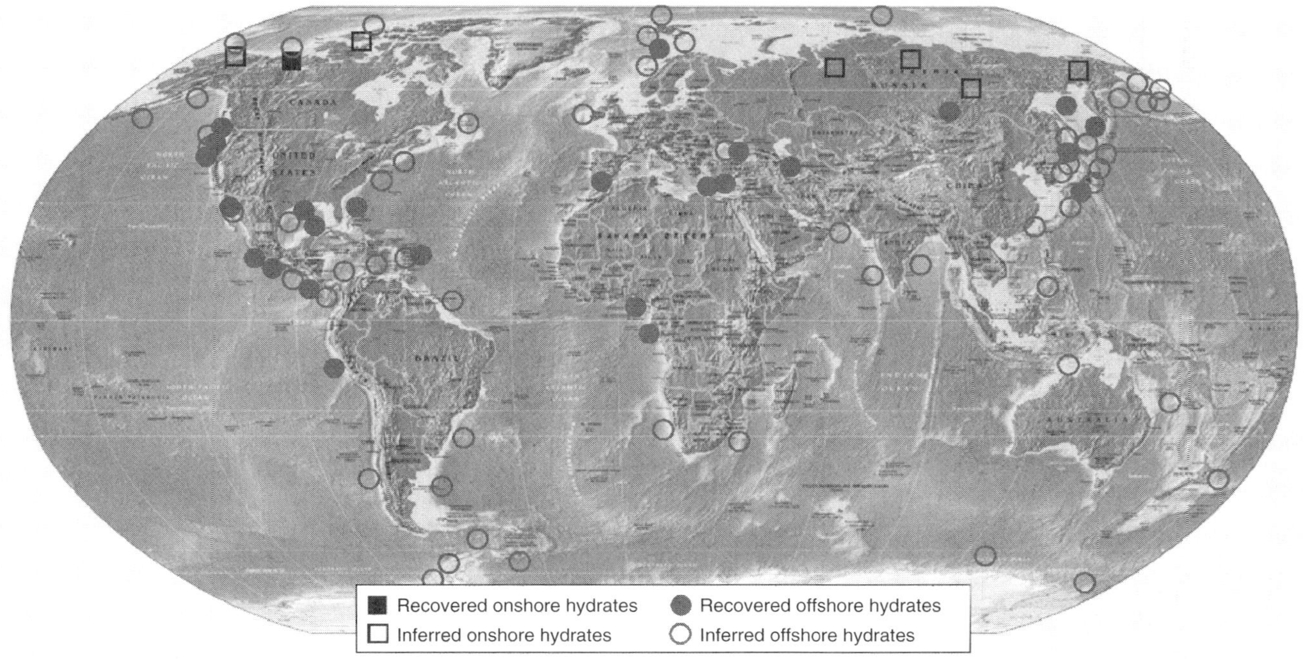

Figure 3. Global map showing locations where gas hydrate has been identified either through sample recovery or remote sensing. Figure courtesy of Alexei Milkov (BP America, Exploration and Production Technology Corp).

gas hydrate stability zone at greater depth. If the gas comes from subseafloor depths greater than about 2 km, it probably contains higher order hydrocarbon gases such as ethane and propane and has a distinctive carbon isotopic signature (thermogenic gas). Both sources of gas contribute to gas hydrate formation in marine sediments. Thermogenic gas is important in focused gas hydrate deposits, whereas biogenic gas dominates in the gas hydrates that are broadly distributed throughout large regions.

REMOTE SENSING FOR SUBMARINE GAS HYDRATES

Because methane hydrate is not stable at atmospheric pressure unless the temperature is below $-60\,^{\circ}$C, gas hydrate samples are rare. Most of the gas hydrate in a sediment sample is probably lost because the gas hydrate decomposes as the sample is recovered. Only large chunks of gas hydrate survive the trip from the seafloor to the earth's surface. Even these large chunks decompose rapidly after recovery, and the methane that is released by this decomposition can be burned, leading to the dramatic phenomenon of "burning ice" (Fig. 4). Special tools are required to sample and preserve natural gas hydrates for detailed studies.

Several geochemical and geophysical techniques are available for indirectly identifying the presence of gas hydrate *in situ*. One of the most widespread techniques is the use of seismic reflection data to detect the subseafloor boundary between sediments containing gas hydrate and sediments that contain free gas (Fig. 5). These data are analogous to an echo sounding of the earth beneath the seafloor. A sharp contrast in the elastic properties of sediments across this boundary results in a distinctive seismic reflection known as the bottom-simulating reflection (BSR). This reflection can be mapped over large areas of the ocean. Similarly, gas hydrates near the seafloor appear as strong reflections at the seafloor, although many other geologic features also produce seafloor "bright spots." Geophysical surveys are therefore a primary tool for determining whether gas hydrate is present in a "frontier" area. In many places, the presence of gas hydrate has been inferred from observations of a BSR in data that were originally

Figure 4. Burning the methane that is released as a gas hydrate sample decomposes. Hands belong to Robert Collier (College of Oceanic and Atmospheric Science, Oregon State University). Picture taken by one of the authors (M. Torres).

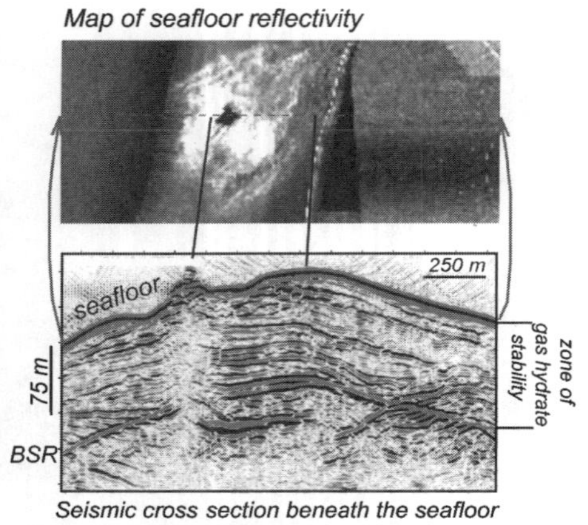

Figure 5. Data showing how the seafloor and subseafloor reflect high frequency sound waves. Such remote sensing strategies are used to survey the seafloor for the presence of gas hydrate. Bright (white) spots in the upper image indicate the presence at the seafloor of gas hydrate and associated carbonate. The presence of the reflection labeled BSR in the subsurface (lower image) indicates that gas hydrate is widespread deep in the sediments, even though massive gas hydrate at the seafloor is restricted to the summit of this structure. The seismic data can also be used to image the subsurface plumbing system that led to development of the hydrate deposit. Seafloor reflectivity data were acquired in 1999 by Chris Goldfinger (College of Oceanic and Atmospheric Science, Oregon State University). Seismic data were acquired by one of the authors (A. Trehu) in collaboration with Nathan Bangs (Institute for Geophysics, University of Texas).

acquired to address other problems in marine geology. These reflective patterns, however, are of limited use for determining how much gas hydrate is present and how it is distributed within the sediments. Researchers are developing new geophysical techniques to better quantify the distribution of gas hydrate from seismic and other remote-sensing techniques.

OBSERVATIONS OF GAS HYDRATE MOUNDS ON THE SEAFLOOR

Mounds of gas hydrates have been directly observed at the seafloor during submersible dives and by using deep-towed cameras and remotely operated vehicles (ROVs). Although many of these gas hydrate occurrences were originally identified on seafloor imaging data, some have been found by serendipity. For example, massive gas hydrate mounds offshore Vancouver Island, Canada, were first discovered when fishermen dragging for ground fish accidentally dredged some up in their nets (Fig. 6). Because gas hydrate is buoyant, their net rose rapidly to the sea surface when the gas hydrate broke free from the seafloor. Someone on board recognized that the bubbling, decomposing "ice" in the net was gas hydrate. Fortunately, no one was smoking!

Although these seafloor gas hydrate deposits may constitute only a small percentage of the total amount of gas hydrate present in marine sediments, they represent the most accessible and best studied deposits and are usually accompanied by complex and as yet poorly understood faunal assemblages. In all cases that have been studied in detail, these deposits contain a mixture of biogenic and thermogenic gas and result from geological structures that focus gas to the seafloor from a deep-seated source. They represent gas hydrate 'sweet spots" that punctuate widespread diffuse deposits inferred from seismic data.

Vigorous expulsion of gas bubbles into the ocean is often observed above seafloor hydrate mounds, even when the mound lies within the thermodynamic gas hydrate stability field. The gas discharge, which is often episodic, provides evidence for the dynamic processes that lead to the formation of massive hydrate near the seafloor. It is likely that these deposits are ephemeral and that they are subject to destruction when fluid flow patterns change or when the seafloor is perturbed. It is possible that

Figure 6. A thinly sedimented mound of gas hydrate exposed on the seafloor. White spots around the hydrate mound are clams, which are often associated with seafloor methane seeps. Photo was taken in August 2002 approximately 50 km offshore Vancouver Island during a survey by ROPOS (Remotely Operated Platform for Ocean Science), a Canadian research vehicle. Courtesy of Ross Chapman (University of Victoria).

large blocks of gas hydrate can spontaneously break away and rise to the sea surface, transporting methane from the seafloor to the atmosphere. Long-term observations of these deposits are needed to evaluate whether they represent a useable resource and whether they are an important factor in transferring methane from the ocean to the atmosphere.

SAMPLING GAS HYDRATES THROUGH DRILLING

More detailed information on the fine-scale distribution of gas hydrate in marine sediments can be obtained by drilling into the subsurface. Logging techniques permit measurement of geophysical and geochemical parameters throughout the entire borehole; coring allows for recovery of subsurface samples. Logging and coring to study gas hydrates is very expensive and has only been undertaken in a handful of places worldwide (Fig. 3). The Ocean Drilling Program (ODP), an international endeavor that enables the research community to sample up to a few kilometers beneath the Earth's surface to address problems of environmental change, earth history, and mantle dynamics, has dedicated several expeditions to study gas hydrates offshore Chile (Leg 141), the eastern United States (Leg 164), Japan (Leg 196), and the Pacific Northwest of the United States (Legs 146 and 204). The Japanese government has recently supported several expeditions to drill gas hydrates in the Nankai Trough offshore Japan. New programs are being planned to sample submarine gas hydrates in the Gulf of Mexico and elsewhere.

These drilling projects provide critical information needed to "ground truth" estimates of gas hydrate abundance and distribution inferred from remote sensing techniques. One conclusion from recent experiments is that the total amount of gas hydrate present in the globe, while still huge, may be less than previously thought. More importantly, these projects have revealed the "patchiness" of gas hydrate distribution beneath the seafloor and provide information on factors, such as sediment grain size, that may be controlling the heterogeneity in their distribution. They also provide unique information of the chemistry of the gases that form the gas hydrate and on the origin of these gases. This information is essential for understanding how gas hydrate deposits develop.

QUANTIFYING THE AMOUNT OF GAS HYDRATE IN BOREHOLES

To estimate the amount of gas hydrate *in situ*, several different types of measurements must be used because different techniques have different sensitivity and spatial resolution. The only way to directly measure the amount of gas in a core is to maintain the *in situ* pressure as it is recovered. This is technically difficult and can only be done for a limited number of samples in each drill hole. Another common technique is to measure the salinity or chloride concentration in the pore waters. Because salts are excluded when gas hydrate forms, the water that is released when it decomposes is fresh. Estimating the amount of hydrate present in the sediments using this technique requires knowledge of the salinity of the *in situ*pore water. This technique is limited by the need to extract pore water from the core, which is time consuming and disrupts the sample for other studies. Yet another technique is to measure the temperature of cores using a scanning infrared camera. Because gas hydrate decomposition is endothermic (takes up heat), locations in the core where gas hydrate is decomposing are up to several degrees colder than the rest of the core. This technique can efficiently and noninvasively sample the entire core but must be calibrated through comparison with other methods. Because core recovery is often incomplete, geophysical logging methods (which measure electrical resistivity, elastic wave velocity, density, and other physical parameters) are needed to sample the entire borehole. Models for interpreting these geophysical parameters obtained through logging must also be calibrated.

CONCLUSION

Gas hydrate is widespread in shallow sediments on the margins of the continents and represents an important, but often overlooked, component of the global carbon cycle. The distribution of gas hydrate is very heterogeneous. While gas hydrate probably represents only ~1–2% of the total volume of continental margin sediments that lie within the gas hydrate stability zone at the present time, concentrated deposits containing 10–30% gas hydrate have been found. These generally result from geologic conditions that lead to focusing of gas from a large volume in the subsurface into the gas hydrate stability zone.

The amount and distribution of gas hydrate in the seafloor, however, are probably not static. Instead, they change with time in response to changes in temperature, pressure, and fluid flow, which are, in turn, driven by tectonics (e.g., earthquakes and underwater mountain building) and global climate change. Rapid release of methane from gas hydrate in response to changes in ocean temperature and depth may affect global climate.

OCEANOGRAPHIC ENVIRONMENT OF GLACIER BAY

Alaska Biological Science
Center—U.S. Geological Survey

Oceanography describes one of the most fundamental physical aspects of a marine ecosystem. Glacier Bay exhibits a highly complex oceanographic regime within a small area. An understanding of many of the resource and research issues in Glacier Bay will not be possible without an understanding of the underlying oceanographic processes causing the large spatial and annual variation found within the Bay.

The Bay is a recently (300 years ago) deglaciated fjord located within Glacier Bay National Park in Southeast Alaska. Glacier Bay is a fjord estuarine system that has multiple sills. These sills are often associated with contractions and are backed by very deep basins with tidewater glaciers and many streams. Glacier Bay experiences a large amount of runoff, high sedimentation, and large tidal variations. Melting occurs year-round, which is thought to fuel the estuarine circulation even through the winter. This runoff, and the presence of the tidewater glaciers makes the bay extremely cold. There are many small- and large-scale mixing and upwelling zones at sills, glacial faces, and streams. The complex topography and strong currents lead to highly variable salinity, temperature, sediment, productivity, light penetration, and current patterns within a small area. This complexity defies simple characterization or modeling based on other areas in Southeast Alaska. While several oceanographic studies have been conducted in Glacier Bay, these studies are contradictory and were of short duration and limited coverage, missing much of the spatial, seasonal and annual variation. In addition, some assumptions based on past studies have been contradicted by recent results. The constantly changing nature of the Bay may contribute to contradictions among past studies and between recent and historical results.

The primary data used in this study (see the Fjord Oceanographic Monitoring Handbook) were oceanographic surveys consisting of samples of 24 central-channel stations from the mouth of the bay to the heads of both the East and West Arm taken between 1992 and 2000. A oceanographic instrument (CTD) capable of recording depth, temperature, salinity, light penetration, amount of sediment, and amount of phytoplankton was used to obtain measurements at one-meter intervals throughout the water column to a depth of 300 m at each station. Surveys were conducted up to seven times a year in such a manner as to encompass the primary annual variation in oceanographic patterns.

Results from the current work indicate several shifts in the dominant paradigm of oceanographic understanding for this area. Primarily, deep-water renewal, and with it increased nutrient availability, is not limited to the winter months but can and probably does occur regularly in the spring/summer/fall at least during spring-tide series. We believe that Glacier Bay is not a traditional silled fjord estuary nor a plain fjord estuary like most estuaries in Southeast Alaska, but a combination of a stratified deep basin estuary and a tidally mixed estuary. This results in a complete blockage of estuarine circulation at the mouth of the bay; instead, mixing occurs by turbulent diffusion rather than buoyancy-driven entrainment as seen in a traditional estuary. In addition, where this turbulent water meets the deep stratified basin, a tidally-mixed front is created. This front is ideal for encouraging high primary productivity by phytoplankton. Mixing phenomena have now been observed to be much more extensive in Glacier Bay than previously thought. This, as with the increased deep-water renewal, almost certainly results in a more nutrient enhanced system. The Bay is now known to exhibit phenomenally high primary productivity, which is sustained throughout the summer in sharp contrast to other interior waters of Southeast Alaska and adjoining shelf and oceanic areas. The limits on this productivity appear to be caused not by grazing by zooplankton or nutrient limitation, but by light penetration. Light penetration in Glacier Bay is significantly reduced by both sedimentation phenomena related to the young nature of the terrestrial environment and to the high levels of the primary productivity itself. In addition to these major changes in understanding, we have identified several new phenomena that will need further study to understand their nature and significance.

As a result of the oceanographic monitoring program and the synthesis effort funded by the NPS there is now an extensive body of oceanographic and weather data integrated into the Glacier Bay information management system and available on CD-ROM. The Oceanographic Analyst Extension, a Geographic Information System (GIS) tool, has been created to allow viewing, analysis, and manipulation of these complex data in 3 and 4 dimensions.

NOS SANCTUARIES PROTECT NATION'S MARITIME HISTORY

NOAA National Ocean Service

Marine archaeology allows us to unravel the mysteries and riddles of shipwrecks and clarify our history. Each fragile wreck is a time capsule of our past. The remains of sunken vessels and their contents provide a distinctive window into the history of the ships and the lives of its passengers and crew. Using state-of-the-art technology, such as sonar and remotely operated vehicles, marine archaeologists are now able to conduct advanced explorations and create accurate depictions of wreck sites.

The National Ocean Service's National Marine Sanctuary Program plays a pivotal role in the exploration and preservation of the nation's maritime history. By protecting oceanic treasures within a sanctuary, marine archaeologists are able to uncover and preserve invaluable information and artifacts. Shipwrecks discovered in the sanctuaries, such as the *USS Monitor, Portland*, and those in Thunder Bay, are essential to understanding the importance of maritime history.

USS MONITOR

Undiscovered for more than 100 years, the *USS Monitor* was found in 1973 off the coast of North Carolina by scientists from Duke University using sidescan sonar technology. In 1862, this famous Civil War ship's battle with the Confederate ironclad *Virginia* revolutionized the

The crew on the deck of the *USS Monitor*. Members of the crew were lost when the *Monitor* sank in 1862. Human remains recently were found when the newly recovered turret was excavated last summer. (Monitor Collection, NOAA)

war at sea, and marked the end of an era of wooden-hulled sailing warships. The *Monitor* was the first of a class of low-freeboard, turreted war ships developed during the Civil War. Its revolving turret revolutionized naval warfare, and other features of the *Monitor* foreshadowed the future of naval technology.

The *Monitor* is more than an era frozen in time. Its dedication in 1975 as the first national marine sanctuary opened the door to advances in maritime archaeology and historic preservation. Since its discovery, numerous research expeditions have been conducted to study and recover the wreck. One of the most recent was in the *summer of 2002*. With the help of a claw-like structure known as the spider, the turret of the *Monitor* was recovered. Human remains also were found during the excavation of the turret in 2002. More than 100 artifacts also have been retrieved from the wreck, including the ship's anchor, condiment bottles, dinnerware, and a portion of leather bookbinding. These artifacts are now part of The Monitor Collection, which is housed in The Mariners' Museum Research Library. More information about the *USS Monitor* is available *from the Monitor National Marine Sanctuary*.

PORTLAND AT STELLWAGEN BANK

The coastal steam ship *Portland*, which rests on the sea floor of the *Stellwagen Bank National Marine Sanctuary*, sank Nov. 27, 1898, during the infamous "Portland Gale of 1898." All 192 passengers and crew were lost. With the help of state-of-the-art technology, the sanctuary obtained three new side-scan sonar images of the wreck. The images clearly show the side-by-side smoke stacks and the diamond-shaped metal walking beam that provided power to the side paddle wheels. The *Portland* has yet to be fully explored, but another mission is scheduled for September 2003. During this missions, scientists will

The *Monitor's* turret broke the surface during the summer 2002 expedition—the first time it had done so in 140 years

use state-of-the-art technology including side-scan sonar and remote operating vehicles for exploration. View the *side-scan sonar images of the Portland* that were collected last year.

Stellwagen Bank, first discovered by Captain Henry Stellwagen in 1854, was especially important to navigators when they sailed at night or in poor weather. Mariners could gauge the distance to the sea floor and, when they saw the bank, would know that they were entering the Massachusetts Bay. The wealth of sea life present in the area also allowed for centuries of prosperous fishing fleets and other economic gains.

Stellwagen Bank was designated on Nov. 4, 1992, as the nation's 11th national marine sanctuary. Stretching 842 square miles, more than 1 million people each year visit the sanctuary. Many of them are intent on experiencing an encounter with the well-known Humpback whale.

A sonar image of the *Portland*, taken in 2002

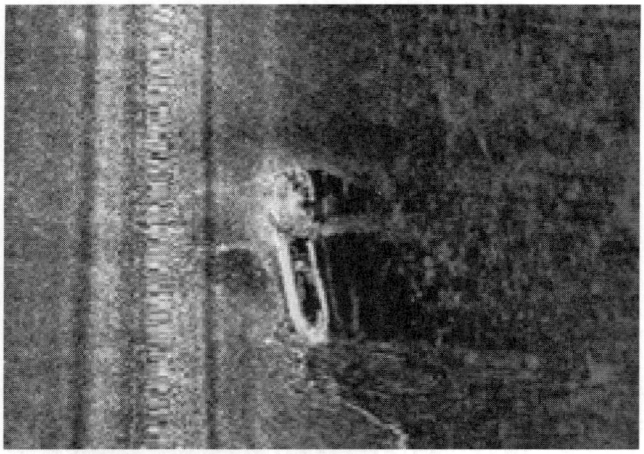

On June 30, 1906, the *New Orleans* was carrying coal when she collided with the *William R. Linn*. Al of the crew boarded the *Linn* safely before the pilothouse of he *New Orleans* blew off and the vessel sank. This sonar image of the wreck was taken in summer 2001

THUNDER BAY

The *Thunder Bay National Marine Sanctuary and Underwater Preserve*, dedicated on Oct. 7, 2000, is the first freshwater and Great Lakes sanctuary. More than 100 19th and 20th century shipwrecks rest in Thunder Bay and the surrounding waters of Lake Huron. For this reason, it has earned the nickname "Shipwreck Alley." Located off the coast of Alpena, Mich., Thunder Bay's collection of shipwrecks represents the diversity of vessels that navigated the Great Lakes, from wooden schooners to steel-hulled steamers.

These preserved vessels still have stories to tell of Great Lakes maritime history and commerce. Some of the wrecks have been broken apart and washed away, and others have recently been found and are waiting to be explored. Many wrecks are yet to be discovered but likely will be found as explorations continue. For more information, read more about *surveying Thunder Bay's Shipwreck Alley* in June 2001.

QUANTIFICATION OF ANOXIA AND HYPOXIA IN WATER BODIES

Gertrud K. Nürnberg
Freshwater Research
Baysville, Ontario, Canada

Hypoxic conditions (dissolved oxygen, DO, concentration below saturation) are widely spread in freshwater and saline environments. Especially in recent years, anthropogenic impacts have led to severe increases in estuarine and coastal anoxia [e.g., Gulf of Mexico (1) and European coast-lines (2)]. Diaz (3) describes 44 marine areas of moderate to severe hypoxia worldwide. Hypoxia and anoxia (lack of oxygen or zero concentration of DO) may even be more established in the freshwater environment, where it can occur naturally in lakes (4) or not (5) and where recent increases in organic and nutrient loading and changes in water flow have increased oxygen depletion in lakes (6), reservoirs (7), and large rivers (8,9).

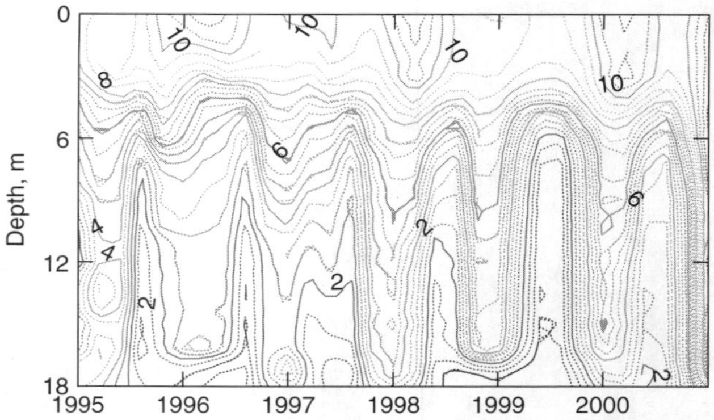

Figure 1. Dissolved oxygen isopleths for small urban Lake Wilcox, Ontario. The beginning of each year is indicated.

OCCURRENCE OF DO DEPLETION

An example isopleth (Fig. 1) for several years of DO data for an urban lake demonstrates DO variability with respect to space and time. The lake experiences severe anoxia in the hypolimnion because the bottom water is stagnant during the summer and fall. Entrainment of hypolimnetic anoxic water may lead to hypoxic conditions in the epilimnion in late summer. Such variation can be quantified by computing anoxic and hypoxic factors.

DEFINITION

The *anoxic* and *hypoxic factor* [AF (4), HF (10)] are calculating methods that quantify the extent and duration of anoxia and hypoxia based on oxygen profiles and morphometric (hypsographic) information. They are based on a series of measured oxygen profiles and morphometric data and can be computed for any water body (lake, reservoir, river, marine area). To render this index comparable across waters like other areal measures, for example, areal nutrient loads and fish yields, AF is corrected for (lake) surface area by simple division. Expressed this way, AF or HF is a ratio that represents the number of days in a year or season that a sediment area equal to the (lake) surface area is anoxic or hypoxic. Hence, its units are d/yr and d/season, summer or winter.

The exact steps in computing the specific factors are described in Fig. 2. The difference between the HF and the AF lies in the different threshold DO concentration used. The AF is based on the oxycline depth or its approximation; the HF is based on a certain threshold that indicates the hypoxic conditions of choice for that particular project, for example, 6.5 mg/L DO. To consider the mixed layer anoxia or hypoxia separately, epilimnetic factors (AF_{epi}, HF_{epi}) can be computed, as defined in Fig. 2.

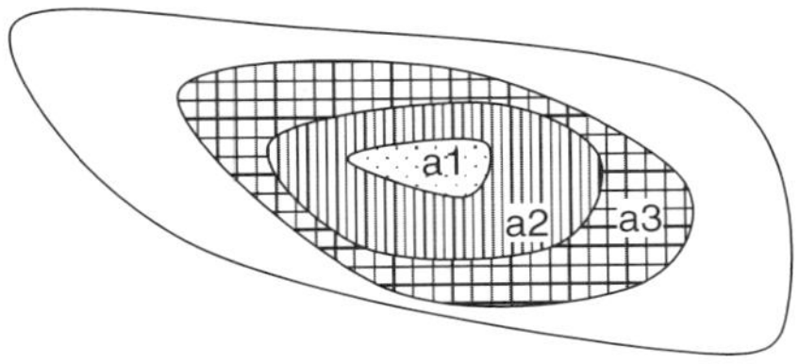

Figure 2. Computation and definition of anoxic and hypoxic factors. The sketch indicates different lake areas that become anoxic sequentially with time, where a_1 becomes anoxic first, then a_2, etc.

Determination of Anoxic or Hypoxic Factor

1. For the anoxic factor, the oxycline is determined at 1 or 2 mg/L dissolved oxygen (DO) concentration from (DO) profiles. The choice of the threshold values depends on the method of DO measurement. For example, 2 mg/L measured by a DO probe may actually coincide with anoxic conditions at the sediment surfaces located at that depth.
   For the hypoxic factor, the depth of the DO concentration threshold in question, for example, 6.5 mg/L, is determined.
2. The periods (total number n), for which the DO levels according to the chosen threshold are at approximately the same depth, are established.
3. The duration of each period of constant DO levels (t_i, in days) is multiplied by the corresponding area (a_i) and divided by the total surface area (A_{oi}) for the period (i). Especially in reservoirs where volumes and therefore areas change, is it important to use the surface area specific for the period. In water bodies without large volume changes, the average surface area A_o can be used instead.
4. These n terms, numbers of periods at different oxyclines, are then added up over the season or year, according to

$$\frac{t_1 \cdot a_1}{A_{o1}} + \frac{t_2 \cdot a_2}{A_{o2}} + \frac{t_3 \cdot a_3}{A_{o3}} + \cdots + \frac{t_n \cdot a_n}{A_{on}} = \sum_{i=1}^{n} \frac{t_i \cdot a_i}{A_{oi}}$$

Therefore AF and HF are defined as follows, where a lower DO threshold is used for AF (1 or 2 mg/L) than for HF (e.g., 6.5 mg/L):

$$\text{AF or HF} = \sum_{i=1}^{n} \frac{t_i \cdot a_i}{A_{oi}}$$

End of Summer Stratification. In stratified lakes, much of the oxygen depletion occurs in the fall when DO profiles may not always be available. The duration of the last period of anoxia or hypoxia can be estimated from the average summer hypolimnetic temperature according to a model that predicts the fall turnover date (or end of summer stratification) from the following equation, based on 92 worldwide lakes, $R^2 = 0.47$, $p < .0001$ (11):

$$\text{Log(fall turnover date)} = 2.62 - 0.116 \log(T) + 0.042 \log(z) - 0.002\ \text{latitude}$$

where the fall turnover date is the Julian day of the year; T is the average July, August temperature at circa 1 m above the bottom at the deepest location of the lake (C°); and z is mean depth, lake volume/lake surface area (m).

Seasonal Factors. The AF and HF can be determined for the stratification periods separately as winter (e.g., AF_{win} in d/winter) or summer factor (e.g., AF_{sum} in d/summer). When there is no winter anoxia, AF_{win} is zero, and AF_{sum} is equal to AF in d/yr.

Epilimnetic Factors. In stratified water bodies, it can be interesting to separate hypolimnetic from epilimnetic hypoxia, so that guidelines can be established specifically for the mixed zone. Therefore, epilimnetic factors (AF_{epi}, HF_{epi}) can be defined. These factors are determined by subtracting, for each period i, terms corresponding to anoxia or hypoxia in the seasonal stratified area below the thermocline ($a_{\text{thermo } i}$) from those for the whole water column before summation, according to

$$AF_{epi} \text{ or } HF_{epi} = \sum_{i=1}^{n} \left(\frac{t_i \cdot a_i}{A_{o\ i}} - \frac{t_i \cdot a_{\text{thermo } i}}{A_{o\ i}} \right)$$

Application

Internal Load Calculation. The anoxic factor was originally developed to determine the internal phosphorus load in lakes as the product of anoxic areal release rates (experimentally determined or predicted from sediment phosphorus concentration) and AF according to Eq. 1 (12).

$$\text{Internal load} = \text{Release rate} \times AF_{sum} \tag{1}$$

where (summer) internal load is in mg TP per lake surface area (m^2) per summer, and release rate represents a summer average in mg TP released per anoxic sediment area per day of the anoxic summer period. If there is phosphorus release under ice in the winter, a release rate representative for the winter and AF_{win} are substituted in Eq. 1 to estimate the winter internal load.

Lake to Lake Variation. Anoxia can vary extensively between lakes. Even in geographically near areas like the Muskoka—Haliburton region on the Precambrian shield in southern Central Ontario, the AF ranged from 0 to 70 d/yr in relatively pristine lakes that were often small, but deep, whereas at least 50% had an AF > 10 d/yr. Lakes in more fertile regions, such as in the St. Lawrence–Great Lakes region, had consistently high AFs between 40 and 70 d/yr. Such differences in lakes and their AF indicate the need for different management strategies (5).

Quantification of anoxia by the AF has been proven useful in explaining lake to lake variation in anoxia. In particular, dependency of anoxia on nutrient concentration and morphometry was found and multiple regressions of the AF on average lake TP and TN concentrations and a morphometric ratio were highly significant in North American lakes (4) (Table 1, Eqs. A to C). These relationships can be used to model (or predict) the AF

Table 1. Relationships and Models Involving the Anoxic Factor (AF) that Demonstrate Between Lake Variation of Anoxia in North American Stratified Lakes[a]

Equation	Predicted	Regression Equation[b]	N	R^2	Source
A	AF_{sum}	$-35.4\ (5.1) + 44.2\ (4.3) \log(TP_{annual}) + 0.950\ (0.187)\ z/A^{0.5}$	73	0.65	(4)
B	AF_{sum}	$-36.2\ (5.2) + 50.1\ (4.4) \log(TP_{summer}) + 0.762\ (0.196)\ z/A^{0.5}$	70	0.67	(13)
C	AF_{sum}	$-173\ (25) + 72.5\ (9.7) \log(TN_{summer}) + 0.789\ (0.219)\ z/A_o^{0.5}$	61	0.54	unpubl. data
D	AF_{sum}	$-39.9\ (9.7) + 27.0\ (4.0) \log(\text{TP-Load})$	17	0.76	(4)
E	Fish species number	$0.97\ (2.42) - 1.53\ (0.49) \log(AF_{sum}+1) + 5.38\ (1.02) \log(A_o)$	52	0.51	(14)
F	Fish species number	$4.92\ (1.38) - 6.12\ (1.44) \log(AF_{win}+1)^b + 0.56\ (0.11)\ z$	32	0.71	(14)
G	Winterkill AF_{win}	$-1 + 10^{(0.091z+0.804)}$	–	–	Based on Eq. F

[a] AF_{sum}, summer AF (d/summer); AF_{win}, winter AF (d/winter); Winterkill-AF_{win}, value of AF_{win} above which a fish winterkill is likely for a given z; z, mean depth (m); A_o, lake surface area (km^2); TP_{annual}, annual water-column average TP (μg/L); TP_{summer}, epilimnetic summer TP (μg/L); fish species number, also called fish species richness is based on Central Ontario lakes.

[b] All regressions are highly significant at $p < .001$, except as noted by[b], $p < .01$. Standard errors of the regression coefficients are given in parenthesis.

in lakes where DO data are not available or difficult to obtain, as in polymictic lakes. The AF describes the sediment surface that is overlain by anoxic water in stratified lakes, so it can be hypothesized that its predicted value resembles anoxic sediment surfaces in polymictic lakes. Such a modeled variable is useful in determining internal load according to Eq. 1 for shallow lakes, where release rates may be available but not the extent of anoxic sediment areas.

Trophic State Classification. The concept of the anoxic factor was used to establish limits for trophic state classification with respect to anoxia (13). When classified with respect to the trophic state, below 20 d/yr, the AF indicates oligotrophic conditions, 20 to 40 d/yr are usually found in mesotrophic lakes, 40–60 d/yr represents eutrophic conditions, and above 60 d/yr is typical of hypereutrophic conditions (13). A more exact measure of the trophic state would include lake morphometry, measured as $z/A^{0.5}$, because the deeper the lake compared to its area (the larger $z/A^{0.5}$), the larger the AF. Based on the relationship of the AF with the summer average TP and the morphometric ratio of Eq. B in Table 1, a chart was drawn to assist in evaluating a specific lake's trophic state (Fig. 3).

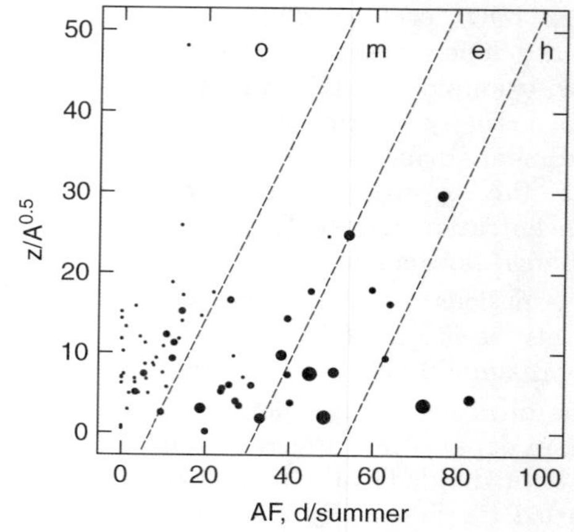

Figure 3. Chart to determine the lake trophic state from the anoxic factor (AF) and the morphometric variable $z/A_o^{0.5}$. Observed values from 70 North American lakes are presented. The symbol size indicates the four trophic states, oligotrophy, mesotrophy, eutrophy, and hypereutrophy (from small to large), and lines indicate the trophic state boundaries based on TP_{summer} in Equation B, Table 1).

Fish species richness correlated with the AF in Central Ontario lakes after variability due to lake area was taken into account, and winter kill could be predicted by using a seasonal AF that quantifies anoxia under ice, and mean depth (14) (Table 1). In particular, cold water species including *Salmonidae*, *Coregonidae*, and *Gadidae* were sensitive to summer and winter anoxia, and they occurred only when AF_{sum} was below 32 d/summer or AF_{win} was below 4.4 d/winter (Fig. 4).

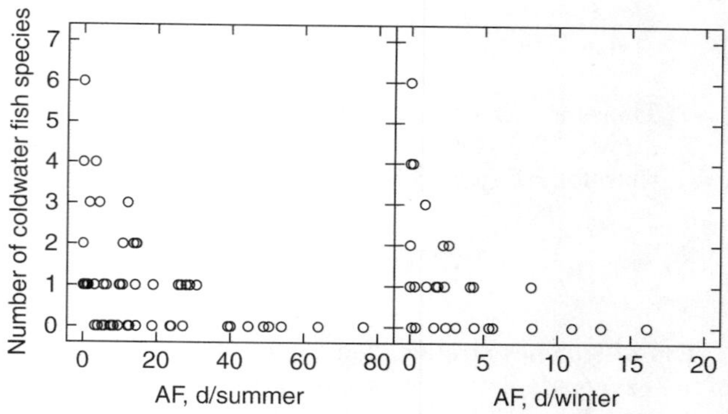

Figure 4. Number of South Central Ontario coldwater fish species in relation to AF_{sum} (left panel) and AF_{win} (right panel).

Spatial Within-Lake Variability. In large and morphometrically distinct water bodies, the factors can be estimated separately for different sections. For example, the AF was quite different between the two basins in small (77 ha) Cinder Lake, Ontario, at 12 and 40 d/summer (4). Similarly pronounced were differences for the 100-km long, narrow (maximum 1 km wide) reservoir, Brownlee Reservoir, on the Snake River in Idaho and Oregon (15). The total reservoir had an AF of 68 d/yr on average between 1962 and 2000, ranging from 53 to 90 d/yr. But the AF of the 48-km long, deep lacustrine section was higher, 88 d/yr on average, ranging from 72 to 114 d/yr, because this section is stratified all summer long. The AF for the 52-km long, shallow riverine section was only 10 d/yr in 1999 and 14 d/yr in 2000 because it is often mixed and reaerated from atmospheric oxygen. Epilimnetic factors were smaller than factors for the whole water column, as expected. Severe anoxia in this reservoir is apparent because the long-term average lacustrine AF_{epi} was 10 d/yr, indicating that overall, an area in the lacustrine epilimnion equal to the lacustrine surface area is overlain by water below 2 mg/L DO for 10 days per summer. The average HF_{epi} of 58 d/yr indicates that the Oregon state criterion of 6.5 mg/L DO for the water column was exceeded in the lacustrine epilimnion for an equivalent of circa 2 months, on average.

Annual Variability. Year-to-year variability can be quite large and ranged from 72 to 114 d/yr AF and 102 to 215 d/yr HF in lacustrine Brownlee Reservoir between 1962 and 2000, from 45 to 68 d/yr AF in a small urban lake, Lake Wilcox, between 1987 and 2000, and from 0 to 10, 9 to 30, and 10 to 50 d/yr AF in three small glacial lakes on the Precambrian shield [Nürnberg, unpublished data and (4,5,10)].

Although the average AF can be predicted from TP and morphometry according to the equations in Table 1, TP concentration cannot always explain the variability in the AF between years, as the variation between annual average TP is often small. Instead, physical characteristics, especially hydrology in flow-managed reservoirs have been found to control the residual variance of DO depletion between years.

For example, several annual and seasonal hydrologic variables were significantly ($p < .01$) correlated with AF, HF, AF_{epi}, and HF_{epi} in the Snake River Brownlee Reservoir. In particular, spring (April–May), summer inflow (July–Sept) (Fig. 5), and late fall (Oct–Dec) flushing rate or its inverse, the residence time, were strongly correlated with the factors, so that they were smallest during periods of high inflow or flushing (10). Such a good inverse relationship of the fall outflow with anoxia and hypoxia can be expected because severe DO depletion happens mostly in the summer and fall; high flow rates out of the dam at that time would destabilize stratification in the lacustrine section, leading to decreased hypoxia. On the other hand, a strong inverse correlation for spring inflow with hypoxia can be explained by a delay in the onset of stratification.

Climatic Effects. The factors quantify annual trends, so climatic effects on DO depletion can be investigated. For example, the Pacific Northwest Index (PNI) was significantly correlated with HF and HF_{epi} [Fig. 6 (10)]. The PNI is a normalized index based on three terrestrial climate variables for the western North American continent: the air temperature at San Juan Islands, the total precipitation at Cedar Lake in the Cascade Mountains, and the snowpack depth on Mount Rainier (16).

Establishment of Criteria and Guidelines. The factors are particularly useful in setting DO criteria and guidelines because the quantification of hypoxia is flexibly based on individual threshold values and can be adapted to various DO levels. Typically, a threshold DO concentration is set by governmental agencies under which a water body is considered "impaired" and a plan to attain the guidelines, including the process of determining a TMDL (total maximum total load) (17), is initiated. These thresholds or acceptable levels vary for different water bodies, agencies, and intended usage but can all be accommodated by various definitions of the hypoxic factor.

Additional Applications. Furthermore, AF helped to compare methods to measure denitrification rates in lakes and determine the most appropriate one (18) and reveal the dependency of hypolimnetic anoxia in tropical African reservoirs on stage height and volume (19). The anoxic

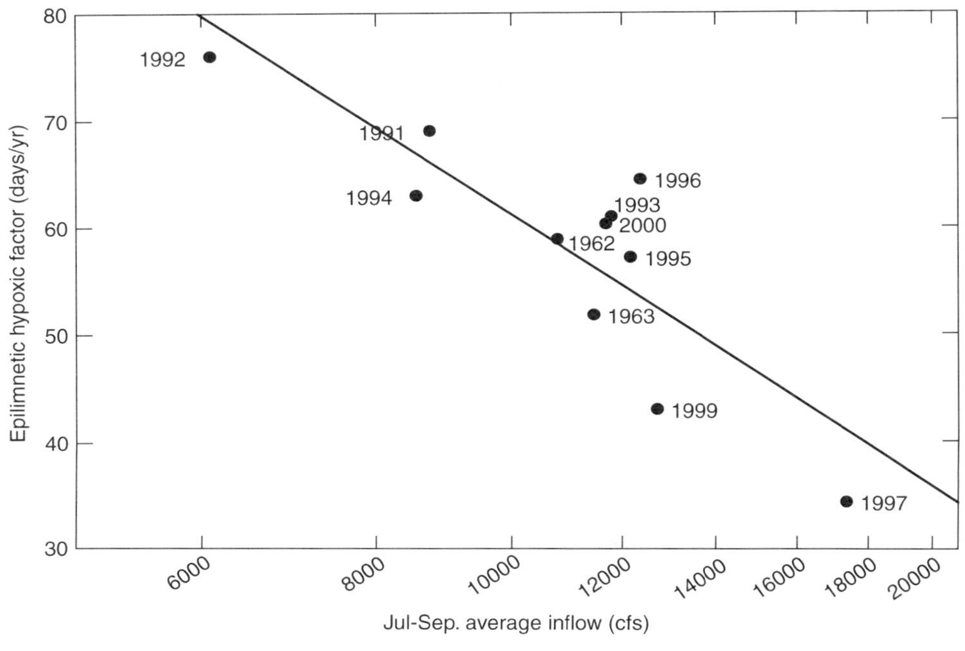

Figure 5. Lacustrine HF_{epi} of the Snake River Brownlee Reservoir, versus average summer inflow (July–Sept, computed from daily average flow values in cubic feet per second) between 1962 and 2000. The regression line is shown for $n = 11$, $p < .001$, $R^2 = 0.72$.

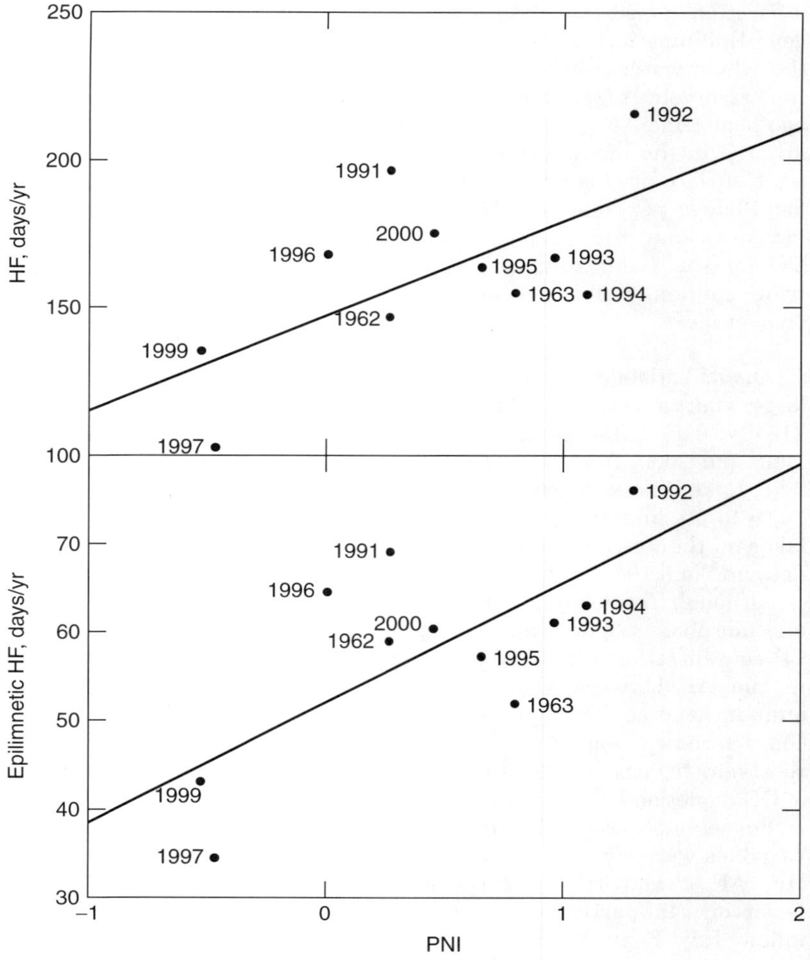

Figure 6. Lacustrine HF (top) and HF_{epi} (bottom) of Brownlee Reservoir versus the Pacific Northwest Climate Index (PNI). Regression lines are shown for $n = 11$, HF: $p < .05$, $R^2 = 0.41$; HF_{epi}: $p < .02$, $R^2 = 0.49$.

factor and its relationship with TP and morphometry were combined with information on fossil midge (*Chironomidae*) assemblages in paleolimnological studies to establish training sets and hindcast hypolimnetic anoxia for the past (20–22).

CONCLUSIONS

In general it can be concluded that the quantification of anoxia and hypoxia leads to increased knowledge of factors that control oxygen depletion in any freshwater and should be useful in estuarine and marine systems as well.

BIBLIOGRAPHY

1. Rabalais, N.N., Turner, R.E., and Wiseman, W.J. (2001). Hypoxia in the Gulf of Mexico. *J. Environ. Qual.* **30**: 320–329.
2. Johannessen, T. and Dahl, E. (1996). Declines in oxygen concentrations along the Norwegian Skagerrak Coast, 1927–1993: A signal of ecosystem changes due to eutrophication? *Limnol. Oceanogr.* **41**: 766–778.
3. Diaz, R.J. (2001). Overview of hypoxia around the world. *J. Environ. Qual.* **30**: 275–281.
4. Nürnberg, G.K. (1995). Quantifying anoxia in lakes. *Limnol. Oceanogr.* **40**: 1100–1111.
5. Nürnberg, G.K. (1997). Coping with water quality problems due to hypolimnetic anoxia in Central Ontario Lakes. *Water Quality Research Journal of Canada* **32**: 391–405.
6. Schertzer, W.M. and Sawchuk, A.M. (1990). Thermal structure of the Lower Great Lakes in a warm year: implications for the occurrence of hypolimnion anoxia. *Trans. Am. Fish. Soc.* **119**: 195–209.
7. Gordon, J.A. (1993). Dissolved oxygen in streams and reservoirs. *Water Environ. Res.* **65**: 571–573.
8. Quinn, J.M. and McFarlane, P.N. (1989). Epilithon and dissolved oxygen depletion in the Manawuta River, New Zealand: Simple models and management implications. *Water Res.* **23**: 825–832.
9. Chambers, P.A., Brown, S., Culp, H.M., and Lowell, R.B. (2000). Dissolved oxygen decline in ice-covered rivers of northern Alberta and its effects on aquatic biota. *J. Aquat. Ecosys. Stress Recov.* **8**: 27–38.
10. Nürnberg, G.K. (2002). Quantification of oxygen depletion in lakes and reservoirs with the hypoxic factor. *Lake and Reservoir Manage* **18**: 000–000.
11. Nürnberg, G.K. (1988). A simple model for predicting the date of fall turnover in thermally stratified lakes. *Limnol. Oceanogr.* **33**: 1190–1195.
12. Nürnberg, G.K. (1987). A comparison of internal phosphorus loads in lakes with anoxic hypolimnia: laboratory incubations

versus hypolimnetic phosphorus accumulation. *Limnol. Oceanogr.* **32**: 1160–1164.
13. Nürnberg, G.K. (1996). Trophic state of clear and colored, soft- and hardwater lakes with special consideration of nutrients, anoxia, phytoplankton and fish. *Lake and Reservoir Management* **12**: 432–447.
14. Nürnberg, G.K. (1995). Anoxic factor, a quantitative measure of anoxia and fish species richness in Central Ontario lakes. *Trans. Am. Fish. Soc.* **124**: 677–686.
15. Freshwater Research, Brown and Caldwell. (2001). *Assessment of Brownlee Reservoir Water Quality.* 2001, Technical Report for the City of Boise, ID.
16. Petersen, J.H. and Kitchell, J.F. (2001). Climate regimes and water temperature changes in the Columbia River: bioenergetic implications for predators of juvenile salmon. *Can. J. Fish. Aquat. Sci.* **58**: 1831–1841.
17. EPA. (1999). *Protocol for Developing Nutrient TMDLs.* U.S. EPA, Washington, DC.
18. Molot, L.A. and Dillon, P.J. (1993). Nitrogen mass balances and denitrification rates in Central Ontario lakes. *Biogeochem.* **20**: 195–212.
19. Townsend, S.A. (1999). The seasonal pattern of dissolved oxygen, and hypolimnetic deoxygenation, in two tropical Australian reservoirs. *Lake & Reservoirs: Research and Management* **4**: 41–53.
20. Quinlain, R., Smol, J.P., and Hall, R.I. (1998). Quantitative inferences of past hypolimnetic anoxia in south-central Ontario lakes using fossil midges (Diptera: Chironomidae). *Can. J. Fish. Aquat. Sci.* **55**: 587–596.
21. Little, J.L. and Smol, J.P. (2001). A chironomid-based model for inferring late-summer hypolimnetic oxygen in southeastern Ontario lakes. *J. Paleolimnol.* **26**: 259–270.
22. Clerk, S., Hall, R., Quinlain, R., and Smol, J.P. (2000). Quantitative inferences of past hypolimnetic anoxia and nutrient levels from a Canadian Precambrian Shield lake. *J. Paleolimnol.* **23**: 319–336.

FLOATING ICE

Aldo Conti
Frascati (RM)
Frascati, Italy

Floating ice, or sea ice, is very important because it regulates the surface exchange of heat, moisture, and salt between the atmosphere and the oceans. It is a high latitude phenomenon and can be found only in the Arctic ocean and in the oceans around Antarctica. Although the local amount of sea ice changes with the season, at any given time sea ice covers, globally, an area larger than the North American continent.

In the Arctic ocean, sea ice covers an average 14 to 16 million square kilometers in late winter and 7 to 9 million square kilometers at the end of summer. Here, the ice cover circulation is characterized by two large-scale motions. A strong transpolar drift current transports sea ice from the Bering Strait over the North Pole down to the east coast of Greenland. Then, much of the Arctic winter is characterized by a persistent high-pressure center over the western Arctic region, which stimulates a clockwise rotation of surface winds and waters, which accumulate a large amount of drifting sea ice that floats around the Beaufort Gyre.

In Antarctica, sea ice covers from 17 to 20 million square kilometers in late winter, but then the seasonal decrease is much larger, with only about 3 to 4 million square kilometers left at the end of summer. Here, sea ice is continually stirred by winds and ocean currents. Ice moves at an average speed of 10 to 20 kilometers per day, and it is divergent northward with divergences of up to 10% per day.

Floating ice can be divided into two categories, namely ice shelf and proper sea ice, which have a completely different origin. Ice shelves are floating masses of ice that originate from the land and are normally attached to the glacier that produced them, which is, in fact, terrestrial ice that has recently seeped onto the surface of the ocean. Icebergs are fragments of these ice shelves. The largest is the Ross Ice Shelf, in Antarctica, that covers an area of about 540,000 square kilometers, about the size of Texas. It contains a third of all the floating ice in the Antarctic oceans and is 90 meters thick at its hinge point and 200 meters at the front.

Sea ice is formed from ocean water that freezes. As the oceans are salty, freezing occurs at $-1.8\,^{\circ}$C. Sea ice itself is largely fresh, because the ocean salt, by a process called brine rejection, is expelled from the forming and consolidating ice. The resulting highly saline (and hence dense) water has an important influence on the ocean, overturning circulation. Fresh sea ice is formed by the cooling of the ocean as heat is lost into the atmosphere. The uppermost layer of the ocean is supercooled to slightly below the freezing point, at which time tiny ice platelets, known as frazil ice, form. Water movements push the frazil crystals against each other, causing the fragile arms of the crystals to break. Soon, a mixture of damaged crystals and crystal remnants occurs that results in an oily-looking ice soup on the surface of water, called grease ice. The name comes from its dark, oily appearance. In time, grease ice become frozen into a thin, transparent layer of ice. Still dark, like the grease ice, the first transparent stage is called dark nilas. It then grows thicker, thick enough to reflect light, and goes from dark to light. This solid white layer is referred to as light nilas. Water starts freezing at the bottom of the ice, increasing its thickness. When nilas becomes 30 cm thick, it is called first-year ice. In a single winter, first-year ice can grow up to 2.5 meters in thickness.

Sea ice influences the climate system in a number of ways, by altering the heat and chemical exchanges between the atmosphere and oceans. It has a high albedo; that is, sea ice is a strong reflector of solar radiation, and therefore energy. The growth and decay of sea ice are positive feedback processes: freezing and melting processes are self-promoting. When ice melts, for example, more solar energy is absorbed into the ocean surface waters, thereby promoting further melting. Alternately, more energy is reflected as ice growth occurs, restricting the solar radiation that warms the surface of the ocean. Less sea ice during one season may allow surface waters to store enough energy to inhibit sea ice formation in the following season. Sea ice also acts as an insulator, inhibiting exchanges of mass, heat, and momentum across the air-sea interface. The perpetual drift of sea ice extends these influences to regional and global scales. It influences the chemistry and circulation of the world's oceans by redistributing salt and freshwater and stimulating dense water formation, which facilitates overturn of the global ocean.

The seasonal sea ice cycle affects both human activities and biological habitats. For example, shipping vessels have to work quickly during periods of low ice concentration. Sea ice is the natural habitat of many mammals, like polar bears and seals, that hunt, feed, and breed on the ice. Scientists are worried that global warming might reduce ice formation, affecting the very limited food chain these animals depend on.

TECHNOLOGY DEVELOPMENT: HARDWARE DEVELOPMENT—MARINE INSTRUMENTATION LABORATORY (MIL)

Great Lakes Environmental
Research Laboratory (NOAA)

MIL's goal is to provide high quality instrumentation packages for oceanographic, biological, chemical, and water resources data collection. The highly skilled engineers and technicians of the Marine Instrumentation Laboratory (MIL) design electronics, develop schematics, manufacture circuit boards and develop computer software to support major systems design. MIL supports and maintains existing GLERL instrumentation, provides the hardware and software necessary for data acquisition systems, and provide various facility and ancillary support services.

OCEANOGRAPHY

The purpose of this equipment is the measurement of water currents, temperature, water quality parameters, wave conditions and the sampling of water and sediments for subsequent analysis.

Focus on Current Measurements

MIL maintains a strong commitment to maintaining the current state of the art in this area. We currently maintain an inventory consisting of 8 RD ADCP's and several Sontek point current meters. We are currently also investigating the 3DACM current meter produced by Falmouth Scientific. MIL has also developed in house two series of satellite reporting Lagrangian drifters using Loran and GPS position sensing systems. Additionally MIL maintains a large active inventory of Vector Averaging Current Meters (VACM), Marsh-McBirney EM current meters and Neal Brown smart acoustic current meters (SACM). These aging systems are in the sunset phase of their utilization.

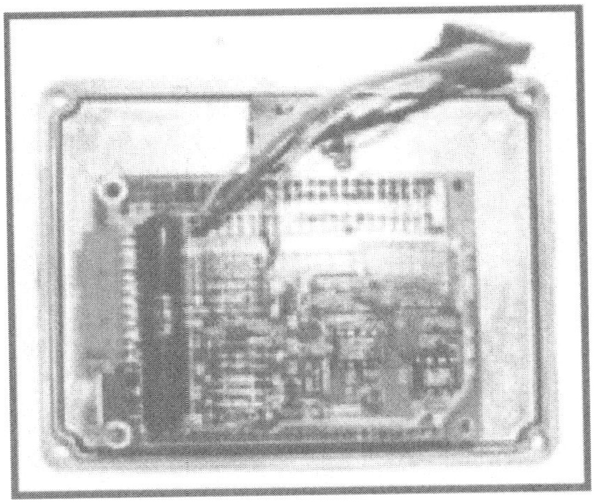

Autonomous CTD Profiler

The ACP was programmed to mechanically drive a CTD package up and down a cable through 40 meters of the water column at the rate of once per hour, taking measurements at sub-meter intervals. The ACP uses an electric motor to drive the CTD package up and down a wire rope cable. Rubber end stops attached to the cable determine the limits of travel and can be adjusted to meet the specific requirements of the experiment. The operation of the ACP is controlled by an embedded microprocessor and data is stored onto a compact flash disk. The flexibility of the programmable data logger allows specifying the interval of the cycles as well as the sample rate. The unit can sample and record data from the sensors in intervals as small as 10 cm.

Microcontrollers and Dataloggers

Recent advances in electronics have produced very small and economical computer driven controllers. These powerful tools have made it possible for instrument developers to design powerful measurement platforms with a minimum investment of time and capital. The controller recently developed for the MIL programmable sequential sediment sampler is shown.

Lagrangian Drifters

An important area of work for MIL has been the design and development of drifting current measurement systems. An example of an MIL designed drifter which used the GPS system to report locations to the ARGOS series satellites is shown. This unit is controlled with a Tattletale 2B datalogger. Currently we have 11 of these units in inventory. These units were produced significantly in advance of the availability of similar commercial units.

Sequential Sediment Samplers

MIL has designed in house a programmable 24 bottle suspended sediment sampler. This instrument is currently in it's third version. Early examples utilized an entirely in house designed controller. The current version uses a Tattletale TFX-11 datalogger. Currently there 22 units in active use with 14 committed to the EEGLE project in

FY99. Several of these instruments have been distributed both nationally and internationally.

Plankton Survey System (PSS)

The Plankton Survey System (PSS) was assembled by GLERL's Marine Instrumentation Lab (MIL) in 1997 to measure distribution and abundance of zooplankton, tiny waterborne crustaceans that serve as a food base for various species of larval fish. It measures turbidity, chlorophyll a, photosynthetically active radiation (PAR), conductivity, temperature, and zooplankton distribution. The PSS and its onboard instruments has been successfully used on an ongoing basis to survey physical, chemical and biological conditions before, during and after Lake Michigan sediment resuspension events as part of the 5-year Episodic Events Great Lakes Experiment (EEGLE) program.

METEOROLOGY

WebCams and Met Stations

GLERL's Marine Instrumentation Laboratory has deployed and is maintaining a real-time network of shore-based meteorological instrument packages in Wisconsin, Michigan, Illinois, and Indiana. Stations in Illinois and Michigan also include web cameras. All meteorological observation stations measure and record wind speed, wind gust, wind direction, and air temperature. In addition, instruments in Chicago measure water levels, and instruments at Muskegon measure dew point, relative humidity, atmospheric pressure, and light level. This data as well as camera images are updated at 5-30 minute intervals and displayed on GLERL's website.

Web Cams:	Met Stations:
Muskegon, MI Alpena, MI Chicago, IL	Alpena, MI Muskegon, MI Saugatuck, MI Michigan City, IN Chicago, IL Kenosha, WI Milwaukee, WI

Lake Michigan Wireless Environmental Observatory

MIL is developing the first environmental observatory using wireless internet technology on the Great Lakes at GLERL's Lake Michigan Field Station in Muskegon, MI. When fully developed, the observatory will provide chemical, physical and biological data to support long-term research on the Great Lakes. These data will be available to the scientific and educational community via

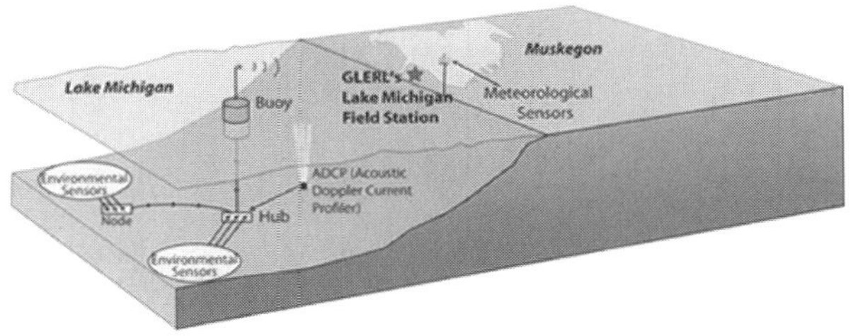

the internet through real-time observations. The goal of this project is to put in place the infrastructure needed to simplify sensor deployment and data acquisition to allow information access by scientific researchers, educators and the public. This is an important contribution to GLERL's leadership in supporting and promoting observation system development among Great Lakes universities and non-governmental organizations.

The environmental observatory consists of an offshore buoy connected to a hub that receives data from various environmental sensors such as an acoustic doppler current profiler. The data are then sent through a wireless link to an onshore receiver connected to the internet.

SEASONAL COUPLING BETWEEN INTERTIDAL MACROFAUNA AND SEDIMENT COLUMN POREWATER NUTRIENT CONCENTRATIONS

Paolo Magni
IMC-International Marine Centre, Torregrande-Oristano, Italy

Shigeru Montani
Hokkaido University
Hakodate, Japan

BACKGROUND

It is well established that dense bivalve assemblages play a major role in the processes of benthic nutrient regeneration in coastal marine ecosystems (1–7). In recent years, there is increasing evidence of the close relationship between nutrient concentrations in sediments and macrofauna-influenced upward flux of nutrients that enhances benthic primary productivity. This coupling has been shown, to varying degree, for a variety of benthic communities dominated by both suspension and deposit feeders. The former, including semi-infaunal mussel species, such as *Musculista senhousia* and *Modiolus americanus* (8,9), filter particulate matter from the water column; the latter, such as holothurians (10), consume high amounts of sediment bacteria, microalgae, and dead organic matter (11). Due to the complexity and multiplicity of mechanisms that govern nutrient cycling in coastal ecosystems, these studies highlight the need for integrated investigations on the variability of porewater nutrient concentrations and the composition, biology, and behavior of *in situ* macrofaunal species (12). There is still a noticeable lack of knowledge about the relationship between macrofauna and the temporal distribution and spatial variability of nutrients in sediments. Some works have conducted a seasonal study of porewater chemistry focusing on nitrogen organic compounds, such as dissolved free amino acids (13), and ammonium profiles and production (13–15). These studies have indicated that temperature has a pronounced effect on the porewater ammonium production rate and seasonal variation. In addition to the effect of environmental parameters that exhibit seasonal patterns and mineralization processes in sediments by bacteria, macrofauna also contribute to the total benthic metabolism by feeding, assimilation, and respiration. Recently, we have shown that the excretory activity of macrofauna strongly influences the magnitude and seasonal variability of the biogenic upward flux of nutrients (16; Temporal Scaling of Benthic Nutrient Regeneration in Bivalve-Dominated Tidal Flat). However, field studies on the relationship between porewater nutrient concentrations and macrofaunal communities come often from isolated surveys (17) or from transplantation/manipulation experiments (8,9). In particular, evidence of a coupling between the seasonal variability of nutrient concentrations in porewater and macrofauna-influenced upward flux of nutrients is still lacking.

SEASONALITY OF POREWATER NUTRIENT CONCENTRATIONS

We present here relevant results obtained from a multidisciplinary study carried out on a sandflat in the Seto Inland Sea (18–20). Extended information on the environmental characteristics and variability of this flat are reported in our associated papers (21–23). Emerged sediment temperature ranges widely through the year, from 2.8 °C (December) to 30.0 °C (July) in 1995 (Fig. 1).

At an individual sampling station, we investigated the macrofaunal communities (Temporal Scaling of Benthic Nutrient Regeneration in Bivalve-Dominated Tidal Flat) in parallel with the distribution of ammonium, phosphate, and silicate in porewater, monthly from January 1995 to April 1996. All these nutrient species showed marked spatial and temporal patterns and had a 10 to >30-fold variability between layers and months (Fig. 2). Ammonium varied from 75.1 (0–0.5 cm, January 1995) to

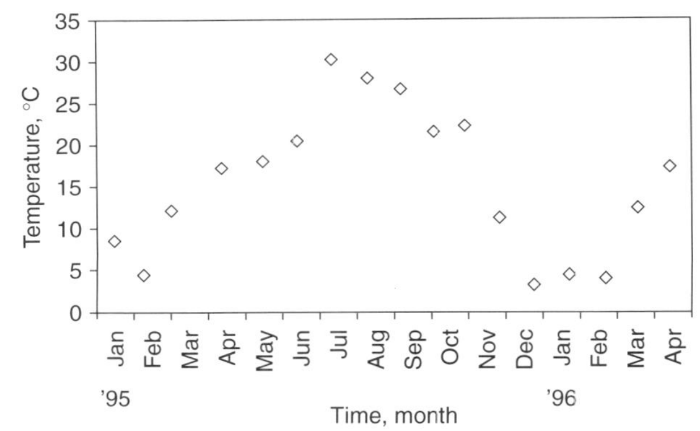

Figure 1. Seasonal variation of sediment temperature.

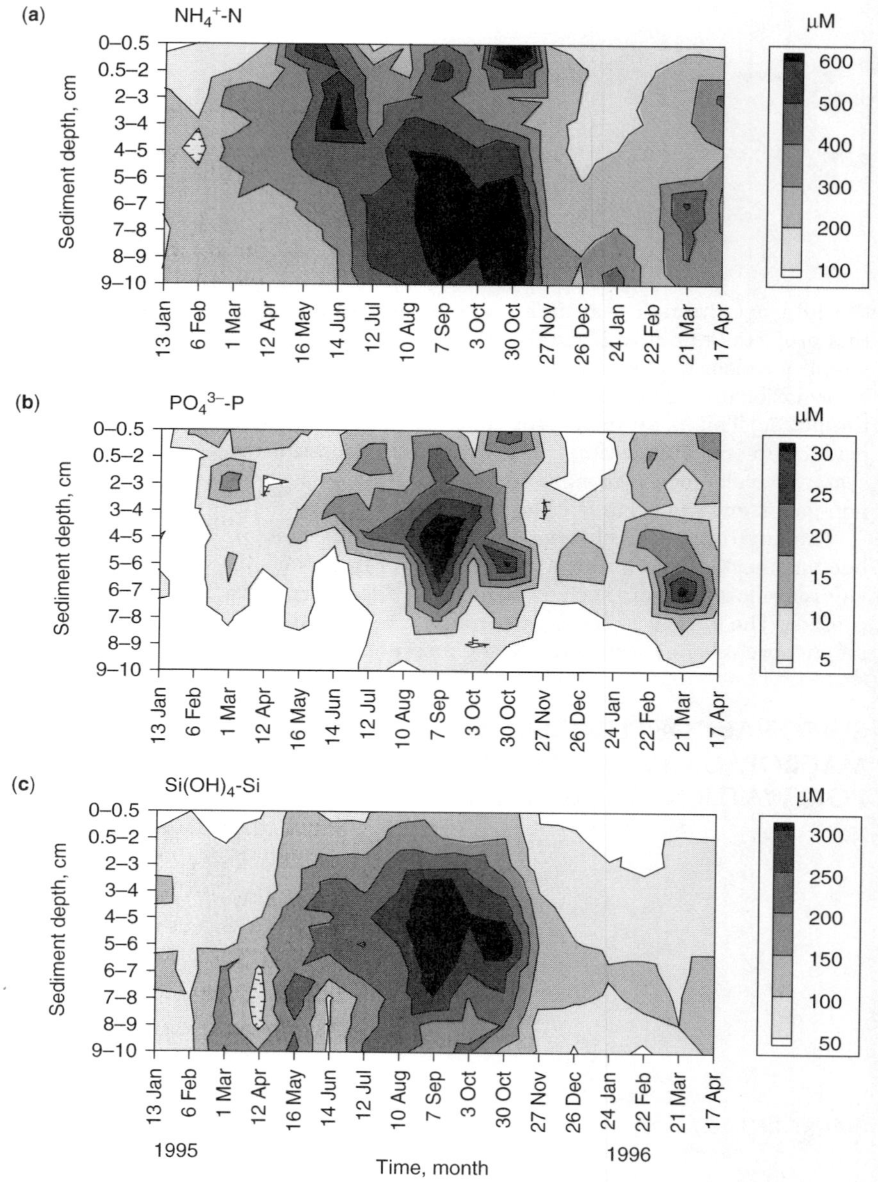

Figure 2. Spatial and seasonal variation of porewater ammonium (NH_4^+-N) (**a**), phosphate (PO_4^{3-}-P) (**b**), and silicate [$Si(OH)_4$-Si] (**c**) concentrations.

908 µM (0–0.5 cm, 30 October 1995), phosphate from 0.9 (0–0.5 cm, January 1995) to 36.9 µM (4–5 cm, September 1995), and silicate from 17.1 (0–0.5 cm, February 1996) to 379 µM (5–6 cm, 30 September 1995).

The spatial and temporal distribution of ammonium, phosphate, and silicate concentrations were consistent with each other. They were lowest in winter, progressively increased through spring and summer in the uppermost layers, and were highest between September and October 1995; a major increase occurred in intermediate layers (i.e., between 3 and 8 cm). Subsequently, minor but noticeable peaks of ammonium and phosphate concentrations were also found in March 1996, up to 518 and 32.7 µM at 6–7 cm, respectively.

For each sampling occasion and nutrient species, we summed the concentrations measured in each layer of the sediment column to be representative of an all-layer monthly pool expressed on a square meter basis. This was obtained by calculating the sediment porosity of each layer from the water content (weight loss on drying at 105°C for 20 h), assuming the bulk density of sediment particles as 2.5 g cm^{-3}. Each volume of porewater was subsequently multiplied by the relevant nutrient concentrations, which were finally expressed as areal depth-integrated values (mmol m^{-2}). Ammonium, phosphate, and silicate concentrations showed a strong correlation with each other (Fig. 3). These results suggest that similar and/or coincident processes may govern the spatial and seasonal variability of major inorganic forms of N, P, and Si in sediment porewater.

We were thus interested in assessing the existence of common environmental factors (i.e., temperature) and/or biological (i.e., macrofauna-influenced) processes that influence the variability of ammonium, phosphate, and silicate in porewater. Ammonium and silicate were highly correlated with temperature (Fig. 4). This could

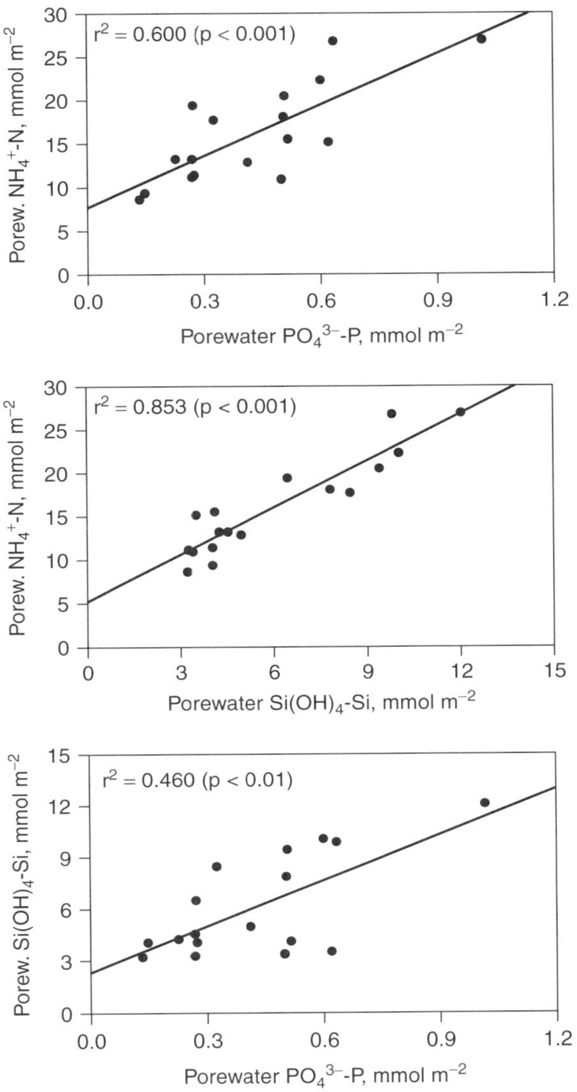

Figure 3. Relationship between porewater ammonium (NH_4^+-N), phosphate (PO_4^{3-}-P), and silicate [$Si(OH)_4$-Si] concentrations in the uppermost 10 cm of sediments.

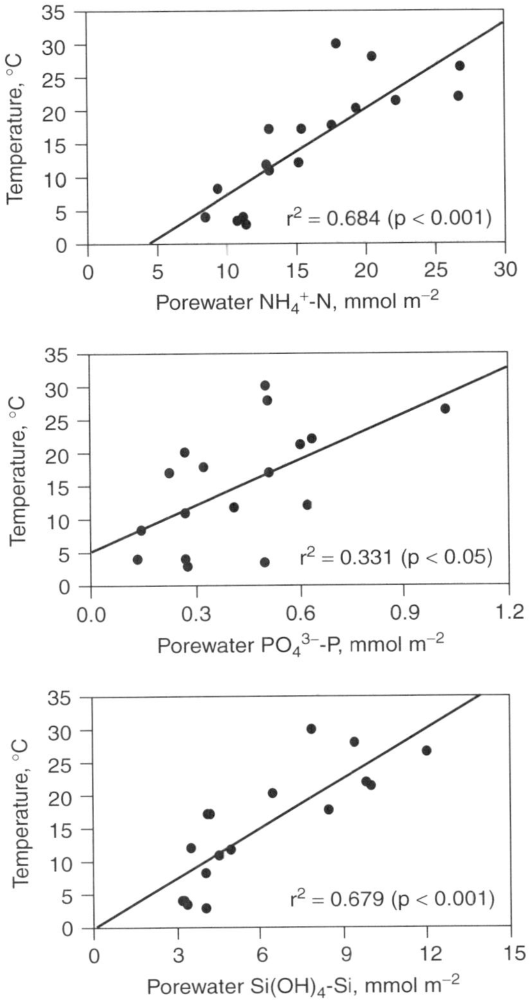

Figure 4. Relationship between porewater ammonium (NH_4^+-N), phosphate (PO_4^{3-}-P), and silicate [$Si(OH)_4$-Si] concentrations in the uppermost 10 cm sediments and temperature.

be consistent with previous studies that have focused on the distribution of ammonium and found that it was strongly dependent on seasonal variations in temperature (13,24,25), whereas little combined information is available on the distribution of ammonium, phosphate, and silicate in the sediment column and relevant influencing factors. In our study, for instance, the correlation between temperature and phosphate was significant, yet rather weak. We then tested the hypothesis that the seasonal variation of all three nutrient species could be related to the activity of *in situ* benthic macrofauna.

COUPLING WITH THE EXCRETORY ACTIVITY OF DOMINANT BIVALVES

We based our considerations on previous physiological measurements (Laboratory Experiments On Bivalve Excretion Rates of Nutrients) and scaling up (Temporal Scaling of Benthic Nutrient Regeneration in Bivalve-Dominated Tidal Flat) of nutrient excretion rates by two bivalve species dominant on this flat, *Ruditapes philippinarum* and *Musculista senhousia*. The plots of bivalve excretion rates of ammonium, phosphate, and silicate versus their relevant pool in the porewater showed a highly significant positive correlation in all cases (Fig. 5). These results suggest the importance of the physiological activity of the benthos on the seasonal variability of porewater chemistry. We conclude that the seasonal patterns of nutrient concentrations in the porewater are strongly coupled with the extent of biogenic regeneration of nutrients due to bivalve excretory activity. This study thus provides evidence of the influence of biological processes on the seasonal patterns of porewater nutrient distribution, suggesting a major role of macrofauna not only at the sediment–water interface, but also in the year-round processes that occur within sediments.

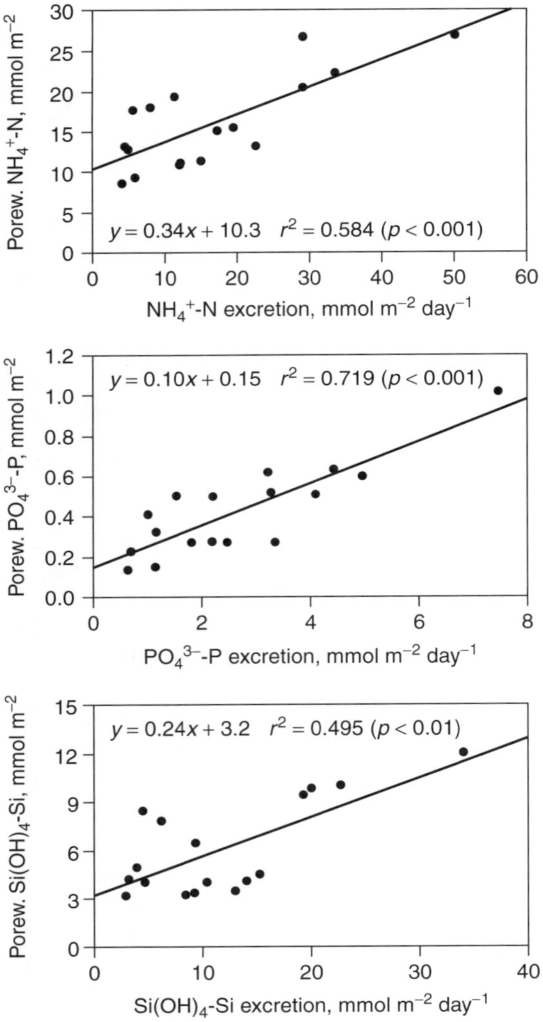

Figure 5. Relationship between porewater ammonium (NH_4^+-N), phosphate (PO_4^{3-}-P), and silicate [$Si(OH)_4$-Si] concentrations in the uppermost 10 cm of sediments and bivalve-influenced upward flux of those nutrients.

BIBLIOGRAPHY

1. Jordan, T.E. and Valiela, I. (1982). A nitrogen budget of the ribbed mussel, *Geukensia demissa*, and its significance in nitrogen flow in a New England salt marsh. *Limnol. Oceanogr.* **27**: 75–90.
2. Dame, R.F., Zingmark, R.G., and Haskin, E. (1984). Oyster reefs as processors of estuarine materials. *J. Exp. Mar. Biol. Ecol.* **83**: 239–247.
3. Murphy, R.C. and Kremer, J.N. (1985). Bivalve contribution to benthic metabolism in a California lagoon. *Estuaries* **8**: 330–341.
4. Boucher, G. and Boucher-Rodoni, R. (1988). In situ measurements of respiratory metabolism and nitrogen fluxes at the interface of oyster beds. *Mar. Ecol. Prog. Ser.* **44**: 229–238.
5. Nakamura, M., Yamamuro, M., Ishikawa, M., and Nishimura, H. (1988). Role of the bivalve *Corbicula japonica* in the nitrogen cycle in a mesohaline lagoon. *Mar. Biol.* **99**: 369–374.
6. Prins, T.C. and Smaal, A.C. (1990). Benthic-pelagic coupling: the release of inorganic nutrients by an intertidal bed of *Mytilus edulis*. In: *Trophic Relationships in the Marine Environment*. Proc. 24th Europ. Mar. Biol. Symp. Aberdeen, Univ. Press, pp. 89–103.
7. Yamamuro, M. and Koike, I. (1993). Nitrogen metabolism of the filter-feeding bivalve *Corbicula japonica* and its significance in primary production of a brackish lake in Japan. *Limnol. Oceanogr.* **35**: 997–1007.
8. Reusch, T.B.H. and Williams, S.L. (1998). Variable responses of native eelgrass *Zostera marina* to a non-indigenous bivalve *Musculista senhousia*. *Oecologia* **113**: 428–441.
9. Peterson, B.J. and Heck, K.L., Jr. (2001). Positive interaction between suspension-feeding bivalves and seagrass—a facultative mutualism. *Mar. Ecol. Progr. Ser.* **213**: 143–155.
10. Uthike, S. (2001). Interaction between sediment-feeders and microalgae on coral reefs: grazing losses versus production enhancement. *Mar. Ecol. Progr. Ser.* **210**: 125–138.
11. Uthike, S. (1999). Sediment bioturbation and impact of feeding activity of *Holothuria (Halodeima) atra* and *Stichopus chloronotus*, two sediment feeding holothurians, at Lizard Island, Great Barrier Reef. *Bull. Mar. Sci.* **64**: 129–141.
12. Christensen, B., Vedel, A., and Kristensen, E. (2000). Carbon and nitrogen fluxes in sediments inhabited by suspension-feeding (*Nereis diversicolor*) and non-suspension-feeding (*N. virens*) polychaetes. *Mar. Ecol. Prog. Ser.* **192**: 203–217.
13. Landén, A. and Hall, P.O.J. (1998). Seasonal variation of dissolved and adsorbed amino acids and ammonium in a near-shore marine sediment. *Mar. Ecol. Prog. Ser.* **170**: 67–84.
14. Blackburn, T.H. (1980). Seasonal variation in the rate of organic-N mineralization in oxic marine sediments. In: *Biogechimie de la matière organique à l'interface eau-sediment marine*. Édition du CNRS, Paris, pp. 173–183.
15. Laima, M.J.C. (1992). Extraction and seasonal variation of NH_4^+ pools in different types of coastal marine sediments. *Mar. Ecol. Prog. Ser.* **82**: 75–84.
16. Magni, P., Montani, S., Takada, C., and Tsutsumi, H. (2000). Temporal scaling and relevance of bivalve nutrient excretion on a tidal flat of the Seto Inland Sea, Japan. *Mar. Ecol. Progr. Ser.* **198**: 139–155.
17. Lomstein, B.A., Blackburn, T.H., and Henriksen, K. (1989). Aspects of nitrogen and carbon cycling in the northern Bering Shelf sediment: I. The significance of urea turnover in the mineralization of ammonium ion. *Mar. Ecol. Prog. Ser.* **57**: 237–248.
18. Magni, P. (1998). A multidisciplinary study on the dynamics of biophilic elements (C, N, P, Si) in a tidal estuary of the Seto Inland Sea, Japan: physico-chemical variability and macrozoobenthic communities. Ph.D. Thesis, The United Graduate School of Ehime University, Japan, p. 258.
19. Montani, S. et al. (1998). The effect of a tidal cycle on the dynamics of nutrients in a tidal estuary in the Seto Inland Sea, Japan. *J. Oceanogr.* **54**: 65–76.
20. Magni, P., Montani, S., and Tada, K. (2002). Semidiurnal dynamics of salinity, nutrients and suspended particulate matter in an estuary in the Seto Inland Sea, Japan, during a spring tide cycle. *J. Oceanogr.* **58**: 389–402.
21. Magni, P. and Montani, S. (1997). Development of benthic microalgal assemblages on an intertidal flat in the Seto Inland Sea, Japan: effects of environmental variability. *La mer.* **35**: 137–148.
22. Magni, P. and Montani, S. (1998). Responses of intertidal and subtidal communities of the macrobenthos to organic

load and oxygen depletion in the Seto Inland Sea, Japan. *J. Rech. Océanogr.* **23**: 47–56.

23. Magni, P. and Montani, S. (2000). Physical and chemical variability in the lower intertidal zone of an estuary in the Seto Inland Sea, Japan: seasonal patterns of dissolved and particulate compounds. *Hydrobiol.* **432**: 9–23.
24. Blackburn, T.H. and Henriksen, K. (1983). Nitrogen cycling in different types of sediments from Danish waters. *Limnol. Oceanogr.* **28**: 477–493.
25. Klump, J.V. and Martens, C.S. (1989). Seasonality of nutrient regeneration in an organic-rich coastal sediment: kinetic modeling of changing pore-water nutrient and sulfate distribution. *Limnol. Oceanogr.* **34**: 559–577.

MAPPING THE SEA FLOOR OF THE HISTORIC AREA REMEDIATION SITE (HARS) OFFSHORE OF NEW YORK CITY

Bradford Butman
U.S. Geological Survey

Repeated surveys using a multibeam mapping system document changes in the topography and distribution of sediments on the sea floor caused by placement of dredged material, remedial capping, and natural processes.

INTRODUCTION

The area offshore of New York City has been used for the disposal of dredged material for over a century. The area has also been used for the disposal of other materials such as acid waste, industrial waste, municipal sewage sludge, cellar dirt, and wood. Between 1976 and 1995, the New York Bight Dredged Material Disposal Site, also known as the Mud Dump Site (MDS), received on average about 6 million cubic yards of dredged material annually. In September 1997 the MDS was closed as a disposal site, and it and the surrounding area were designated as the Historic Area Remediation Site (HARS) (Figs. 1 and 2). The sea floor of the HARS, approximately 9 square nautical miles in area, currently is being remediated by placing a minimum 1-m-thick cap of clean dredged material on top of the surficial sediments that are contaminated from previous disposal of dredged and other materials. The U.S. Geological Survey (USGS) is working cooperatively with the U.S. Army Corps of Engineers (USACE) to map the sea floor geology of the HARS and changes in the characteristics of the surficial sediments over time.

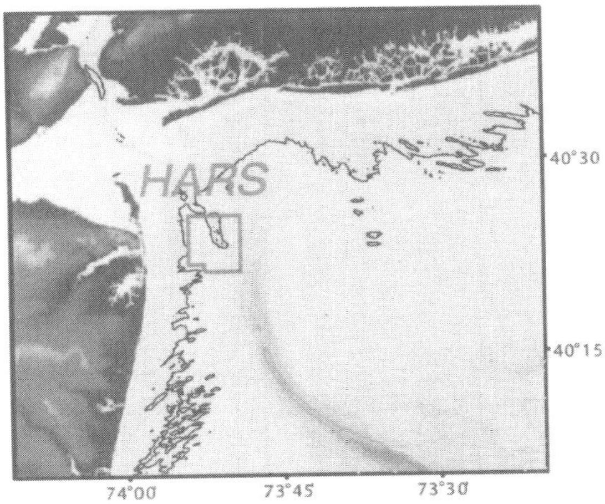

Figure 1. Map showing the area offshore of New York and New Jersey that has been used for the disposal of dredged materials and other wastes since the late 1800's. The Historic Area Remediation Site (HARS) is outlined in red.

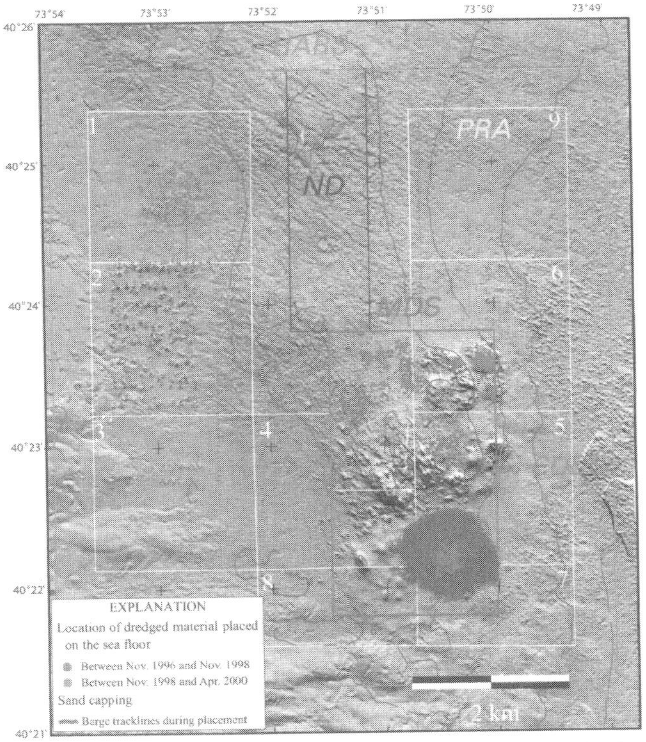

Figure 2. Shaded relief image of the Historic Area Remediation Site (HARS) in April 2000 showing the Primary Remediation Area (PRA, divided into nine cells), the no discharge zone (ND), the former Mud Dump Site (MDS), and the discontinued Cellar Dirt Disposal Site (CDDS). Companion images are shown in Figures 4 and 5. See text for a description of this image and major features. Bathymetric contour interval is 5 m (red lines).

HIGH-RESOLUTION SURVEYS OF THE SEA FLOOR OF THE HARS

Surveys of the HARS were conducted in November 1996 (prior to the closing of the Mud Dump Site), November 1998 (during early remediation of the HARS), and April 2000 (during continued remediation of the HARS) using a Simrad EM1000 multibeam mapping system (Fig. 3). Survey lines were run approximately 100 m apart to provide full coverage of the sea floor. The EM1000 measured the depth of water (to an accuracy of

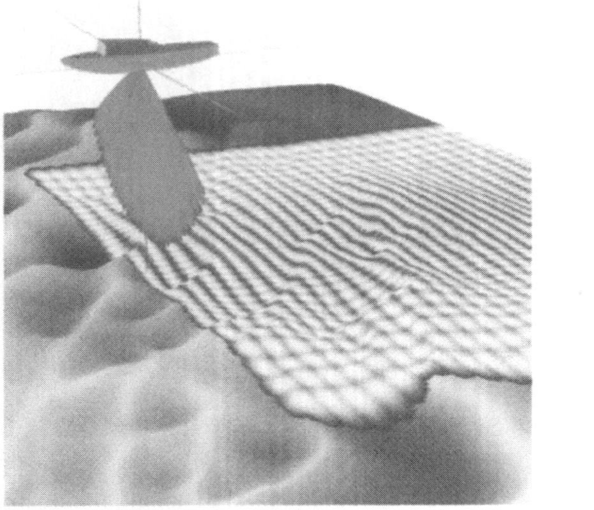

Figure 3. High-resolution multibeam mapping systems use sound from an array of transducers to measure water depth and sediment characteristics of the sea floor. The horizontal resolution of the maps is a few meters, providing an image of the sea floor topography and sediment properties somewhat comparable to an aerial photograph.

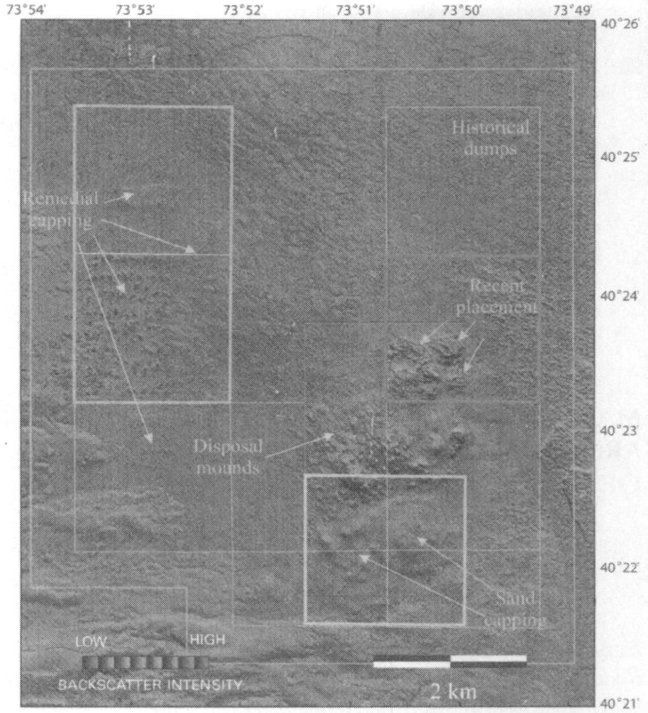

Figure 4. Pseudo-colored backscatter intensity and shaded relief map of the entire HARS in April 2000. The faint north-trending stripes run parallel to the survey tracklines and are artifacts of data collection and environmental conditions. The pink, green, and yellow boxes outline areas shown in Figure 6 to illustrate changes in backscatter intensity between 1996, 1998, and 2000. See text for a description of this image and major features. Bathymetric contour interval is 5 m (red lines).

about 30 cm) as well as the intensity of sound reflected from the sea floor, which is referred to as backscatter intensity. High backscatter intensity generally indicates the presence of rocks and coarse-grained sediments, while low backscatter intensity indicates the presence of finer grained sediments. Direct observations using bottom photographs, video, and grab samples are needed to verify interpretations of the sea floor geology based on backscatter intensity.

IMAGES OF THE HARS SEA FLOOR

In this fact sheet, the topography and backscatter intensity data measured by the multibeam mapping system are presented in three types of images. Each of these images highlights different features and characteristics of the sea floor.

(1) A *shaded relief image* (Fig. 2) visually shows small topographic features (with relief of a few meters) that could not be effectively shown by contours alone at this scale. The image was created by vertically exaggerating the topography four times and then artificially illuminating the relief by a light source positioned 45° above the horizon from the north. In this image, topographic features are enhanced by strong illumination on the north-facing slopes and by shadows cast on the south-facing slopes.

(2) A *shaded relief image, colored by backscatter intensity*, combines the high-resolution view of topography with a measure of sediment characteristics (Figs. 4, 5B, and 6). In these images, the backscatter intensity is represented by a suite of eight colors ranging from blue, which represents low intensity (fine-grained sediments), to red, which represents high intensity (rock outcrops and coarse-grained sediments). These data are draped over the shaded relief image. The resultant image displays light and dark intensities within each color band that result from a feature's position with respect to the light source. For example, north-facing slopes, receiving strong illumination, show as a light intensity within a color band, whereas south-facing slopes, being in shadow, show as a dark intensity within a color band.

(3) A *shaded relief image, colored by bathymetry*, combines the high-resolution view of topography with color to show water depth (Fig. 5A).

THE SEA FLOOR OF THE HARS

Within the HARS, one of the most striking aspects of the sea floor is the variability in backscatter intensity and bottom morphology over distances of a few kilometers or less, caused by both natural and anthropogenic processes. This fact sheet presents companion images showing the sea floor of the HARS as mapped in April 2000 in plan view (Figs. 2, 4, and 6) and in perspective view (Fig. 5). Images of selected areas in 1996, 1998, and 2000 illustrate changes over time (Fig. 6). Major features of the sea floor of the HARS shown in these images include two

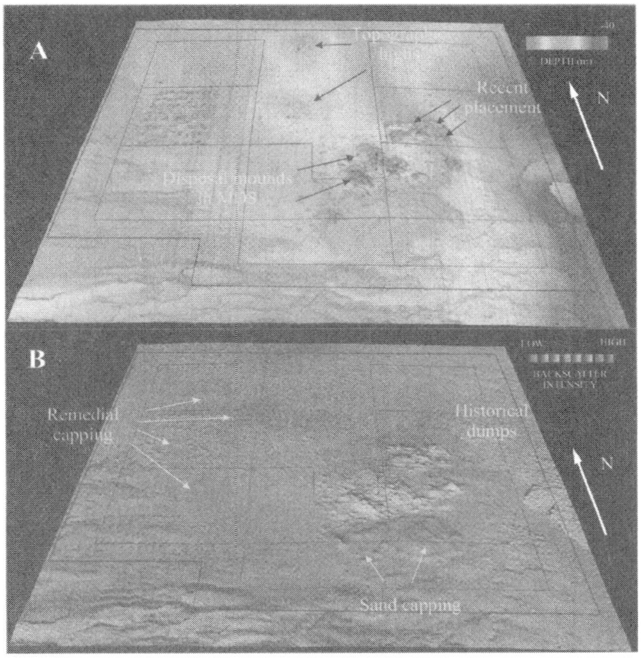

Figure 5. Perspective view of the Historic Area Remediation Site, looking from south to north, based on the multibeam survey carried out in April 2000. A, Shaded relief map with color-coded bathymetry. B, Backscatter intensity draped over shaded relief (see text for a description of the color scheme). The north-trending stripes, running parallel to the survey tracklines, are artifacts of data collection and environmental conditions. The topography, surface features, and the surficial sediments of the HARS have been heavily influenced by the disposal of dredged and other material in this region over the last century, and by recent remedial capping. See text for a description of these images and major features.

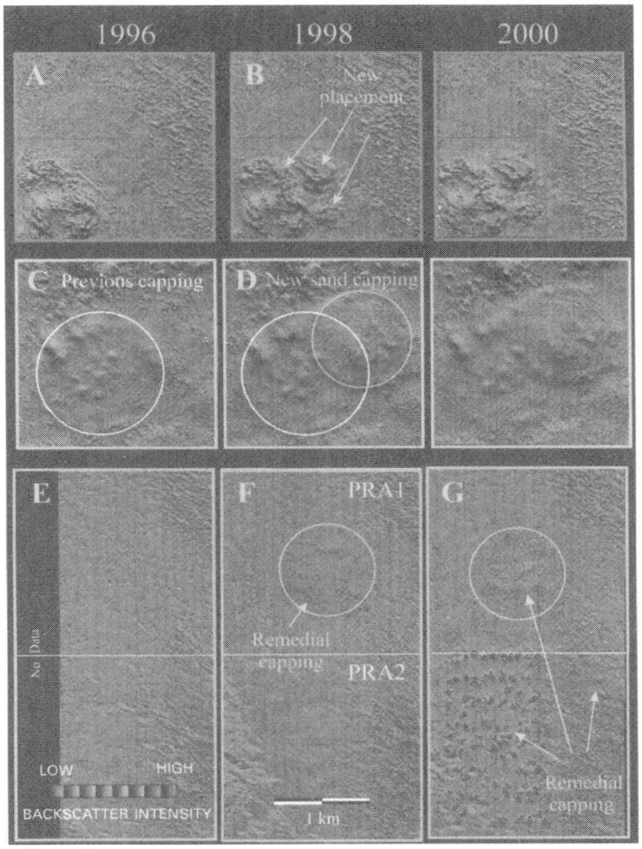

Figure 6. Pseudo-colored backscatter intensity and shaded relief map of parts of the HARS in 1996, 1998, and 2000 (see Fig. 4 for location). These images illustrate changes in the sea floor topography and backscatter intensity that occurred between 1996 and 1998 and between 1998 and 2000 caused by placement of dredged material and by remedial capping. See text for a description of these images and major features.

relatively smooth topographic highs composed of material dumped in the late 1800's and early 1900's ("Topographic highs" in Fig. 5A); mounds of material in the Mud Dump Site ("Disposal mounds in MDS" in Fig. 5A); two circular features where contaminated sediments were placed and then capped with sand, one in the late 1980's, and the other in 1997 ("Sand capping" in Figs. 4 and 5B and "Previous capping" and "New sand capping" in Fig. 6); material deposited between the November 1996 and November 1998 survey ("Recent placement" in Figs. 4 and 5A); many features about 50 m in size interpreted to be individual dumps of material ("Historical dumps" in Figs. 4 and 5B); and material placed as part of remediation activities ("Remedial capping" in Figs. 4, 5B, and 6).

CHANGES IN SURFICIAL PROPERTIES BETWEEN 1996, 1998, AND 2000

Comparison of the topography and backscatter intensity from the three multibeam surveys show how the area changed as a result of dredged material placed before the Mud Dump Site was closed and ongoing remediation of the HARS (see Fig. 2 for locations of placed material from USACE records).

Between 1996 and 1998, changes include (1) mounds of medium backscatter intensity dredged material in the northeastern corner of the MDS, some as high as 6 m, placed between November 1996 and September 1997 (compare panels A and B, Fig. 6); (2) a circular area of low-backscatter intensity material about 1 km in diameter and 2 m thick in the southern part of the MDS associated with sand capping (compare panels C and D, Fig. 6); and (3) a circular area of low backscatter intensity material in PRA1 associated with remedial capping (compare panels E and F, Fig. 6).

Between 1998 and 2000, changes include (compare panels F and G, Fig. 6) (1) increased backscatter intensity in PRA1 due to additional placement of material and consolidation, de-watering, and possible winnowing of the previous cover; (2) a series of crater-like features in the western part of PRA2, 30 to 70 m long and on the order of 20 m wide with elevated rims and central depressions, that were apparently formed as remedial material impacted the soft sediments on the sea floor; and (3) an area of reduced backscatter intensity in the northeastern corner of PRA2 caused by the placement of remedial material.

Resolution limits of the multibeam system, and the amount of material placed over a relatively large area, preclude using the repeated topographic surveys for determining the amount of material placed on the sea floor. However these multibeam data clearly show the overall regional geology and, through comparison of topography and backscatter intensity, document the location of placed material and changes in sediment properties over time.

NOAA AND UNIVERSITY SCIENTISTS STUDY METHYL BROMIDE CYCLING IN THE NORTH PACIFIC

Shari Yvon-Lewis
Atlantic Oceanographic and Meteorological Laboratory

Kelly Goodwin
Sara Cotton
University of Miami

James Butler
Climate Monitoring and Diagnostics Laboratory

Daniel King
University of Colorado

Eric Saltzman
Ryszard Tokarczyk
University of Miami

Patricia Matrai
Brian Yocis
Eileen Loiseau
Bigelow Laboratory for Ocean Sciences

Georgina Sturrock
Commonwealth Scientific and Industrial Research Organization—Australia

As part of a study supported by both NASA and NOAA, scientists from two NOAA laboratories, three universities and CSIRO participated in a research cruise aboard the R/V Ronald H. Brown. The cruise departed Kwajalein, Republic of the Marshall Islands on 14 September 1999 and arrived in Seattle, Washington on 23 October 1999 with stops in Honolulu, Hawaii, Dutch Harbor, Alaska, and Kodiak, Alaska. The objective of this research effort was to obtain reliable measurements of the uptake and emission of methyl bromide and other climatically important halocarbons in tropical to temperate regions of the North Pacific Ocean.

Atmospheric methyl bromide (CH_3Br), which is of both natural and anthropogenic origin, has been identified as a Class I ozone-depleting substance in the amended and adjusted Montreal Protocol on Substances that Deplete Stratospheric Ozone. The role of the ocean in regulating the atmospheric burden of this gas is still somewhat uncertain. Methyl bromide is both produced and destroyed in the ocean through chemical and biological processes. The organisms or reactions that produce CH_3Br at rates sufficient to explain its observed concentrations are not known. Degradation has been shown to occur at rates that are faster than can be explained by known chemical degradation reactions, and evidence suggests that this additional degradation is bacterial consumption of CH_3Br. While recent measurements have shown that, on the whole, the ocean is a net sink for CH_3Br, measurement coverage to date has been limited and sporadic, which restricts our ability to map the spatial and temporal variations that are necessary for understanding how the system will respond to perturbations (e.g., Global Warming).

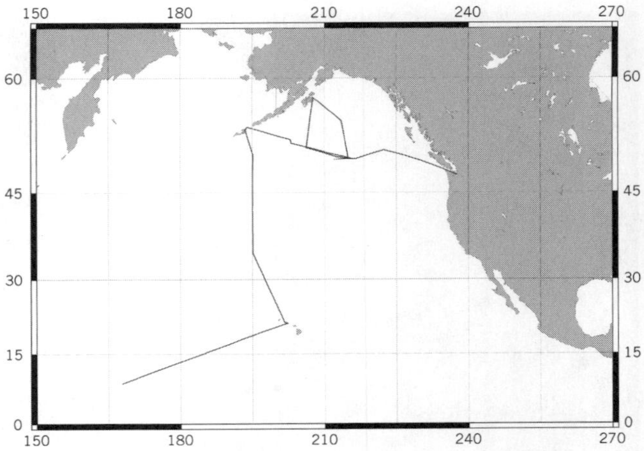

Figure 1. Cruise Track.

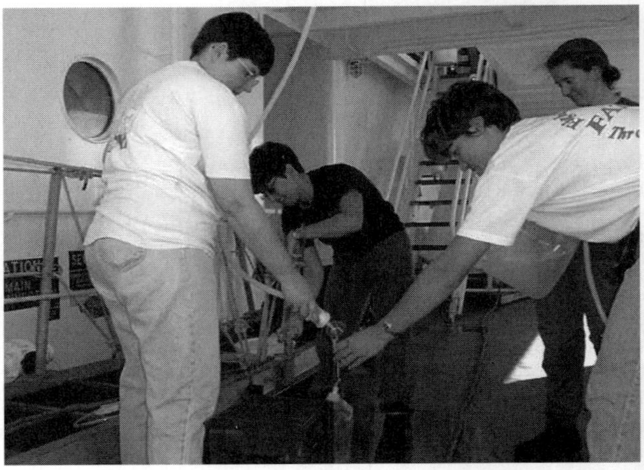

Figure 2. Scientists collecting water samples for production and degradation incubations.

The measurements made during this cruise are designed to help improve our understanding of the role that the oceans play in the cycling of CH_3Br. The program involved instrumentation from two NOAA laboratories and two universities. Measurements were made of the concentrations of CH_3Br and a suite of natural and anthropogenic halocarbons in the air and surface water, degradation rates of CH_3Br in the surface water, production rates of CH_3Br and other natural halocarbons in the surface water, and depth profiles of CH_3Br and other halocarbons. The combined results from these measurements will be used to constrain the budget of CH_3Br in these waters at this time of year. The relative importance of the biological and chemical processes will be examined for tropical and high latitudes. Attempts will also be made to extract relationships between the rates and concentrations measured and satellite measurements in order to develop proxies that can provide global coverage on shorter time scales. At this time, there is insufficient data to examine seasonal and long-term trends in net flux, production, or degradation. Until satellite measurable proxies can be found, additional research cruises are needed to reduce the uncertainty in the global net flux estimate and to map the spatial and temporal variations in the net fluxes, production rates, and degradation rates of CH_3Br and other climatically important halocarbons.

TIDALLY MEDIATED CHANGES IN NUTRIENT CONCENTRATIONS

Paolo Magni
IMC—International Marine Centre
Torregrande-Oristano, Italy

Shigeru Montani
Hokkaido University
Hakodate, Japan

Freshwater runoff during ebb flow and salt water intrusion during the flood may have a major effect on short-term changes in nutrient (ammonium, nitrate+nitrite, phosphate, and silicate) concentrations along an estuary. Time series hourly measurements conducted in a mixed-semidiurnal type estuary (i.e., characterized by two major lower and higher tidal levels) show that these changes are a strong function of both tidal state (e.g., low vs. high tide) and amplitude (e.g., neap vs. spring tide). In particular, the changes in nutrient concentrations are higher during ebb than during flood tide and largest between the lower low tide and the higher high tide of a spring tide. Finally, the importance of investigating simultaneously different stations along the estuarine spine is highlighted, in addition to studying the nutrient distribution based on selected salinity intervals which may reflect only the conditions at a particular tidal state.

BACKGROUND

An important aspect of the high variability of tidal estuaries is related to the effect of the tidal cycle on the physical and chemical characteristics of the water. In particular, on a timescale of hours, freshwater runoff during ebb flow and salt water intrusion during the flood may determine strong changes in salinity and dissolved and particulate compounds. Several studies in riverine and estuarine waters have investigated the distribution of nutrient (e.g., ammonium, nitrate, phosphate, and silicate) concentrations, based on the salinity gradient in the system at a particular tidal state (1–3). Accordingly, plots of nutrients versus salinity are often used to assess the source of different nutrient species, whether from inland, outside the estuary, or within it (3–7).

An evaluation of the distributional pattern of nutrients along an estuary has important ecological implications in relation to the cycling of biophilic elements, such as N (nitrogen), P (phosphorous), and Si (silicon). It is known that numerous processes influence the behavior of nutrients, whether they show conservative mixing or reflect removal or addition along an estuary. Ammonium consumption and ammonium oxidation, for instance, are predominant in the water column, whereas denitrification in sediments is responsible for nitrate removal from the water column (8–10). By contrast, bioturbation and excretion by abundant benthic animals may greatly contribute to the upward flux of regenerated nutrients, such as ammonium and phosphate, which in turn enhance primary production (11,12). Accordingly, it has been shown that regeneration processes within an estuary are consistent with often encountered nonconservative mixing of ammonium (4,8,13,14). This corresponds to the tendency of ammonium concentration to be high at midsalinity ranges, resulting in a poor correlation with salinity. In contrast, nitrate tends to show conservative behavior, as evidence of its riverine origin (2,14,15), although addition (15) or removal (7) is also found. Moreover, it must be considered that in some cases, nitrate versus salinity plots may fail to unravel active nitrate turnover, leading to an approximate balance of sources and sinks (16). As for silicate, a general pattern indicates that estuarine mixing of this nutrient species tends to be conservative (2,7,13). Yet, either silicate removal (7,17) or addition (4) occurs in relation to the development of an algal population in rivers or to a closer interaction with estuarine sediments, respectively, and varies with season (14). A major upward flux of silicate from sediments might also be related to the biological activity and excretory processes of abundant macrofaunal assemblages (18,19).

In addition to these general considerations, the distribution and cycling of nutrients depend strongly on the specific characteristics of each estuary, including water residence time and water depth, nutrient levels, and the extent of salt-water intrusion. Uncles and Stephens (3) showed that saline intrusion was a strong function of the tidal state and a weaker function of freshwater inflow. Accordingly, Balls (2) indicated that conservative mixing of phosphate, is a function of estuarine flushing time, as related to particle–water interaction and chemical speciation (20). In particular, phosphate removal at low salinities may be due to adsorption to iron and aluminium colloidal oxyhydroxides that aggregate and undergo sedimentation (7).

Therefore, it is important that investigations of the distribution of nutrients as a function of salinity are conducted on proper spatial and temporal scales that take into account the extent and variability of salt-water intrusion. Time series surveys represent a valuable approach to quantifying such variability in estuarine waters (21,22). In particular, short-term monitoring surveys during a complete tidal cycle and simultaneous information at different stations can be a powerful tool for evaluating the extent of salt-water intrusion along the estuary and its direct impact on the water chemistry of an estuary. This approach may overcome the limitations of evaluating the distribution and behavior of nutrients based on a particular tidal state at arbitrarily selected salinity intervals. Major drawbacks when working across long distances to cover different salinity ranges include work time to travel between samplings, tidal-water displacements, variations in tidal velocities, and a possible wide range of river discharge. Simultaneous observations at different stations are expected to track changes in nutrient concentrations that occur along the estuarine spine in relation to the extent of salt-water intrusion over time.

Case study

This article reviews the results of 24-hour surveys in a tidal estuary of the Seto Inland Sea, Japan, during a spring tide of May 1995 (13,23). It is aimed to give an example of the effect of freshwater runoff and salt-water intrusion on the spatial and temporal variability of nutrient concentrations during a tidal cycle. The effect of a tide is also evaluated in relation to the *mixed-semidiurnal* behavior (i.e., with pronounced differences between two successive low and high tides) of the estuary. The fieldwork was conducted along a transect line of approximately 1.4 km linking the river to the rear to the subtidal zone. Multiprobe casts were used for hydrologic measurements. Nutrient concentrations were determined every hour in surface waters, simultaneously at a riverine, an intertidal, and a subtidal station, and every two hours at additional depths through the water column at the subtidal station. Details of the sampling scheme and analysis are given in Montani et al. (13) and Magni et al. (23).

At the beginning of the survey, there was a marked salinity gradient along the estuary; surface water salinity was 2.0, 18.5, and 31.3 psu at the riverine, intertidal, and subtidal stations, respectively (Fig. 1). As the lower low tide approached, salinity remained low at the riverine station (Fig. 1a), sharply decreased also at the intertidal station (Fig. 1b) and, subsequently, at the subtidal station (Fig. 1c). By contrast, soon after the lower low tide, salt-water intrusion rapidly caused an abrupt increase in salinity at both the intertidal and the riverine stations, up to >30 psu (Figs. 1a,b). During the flood, there was a homogeneous distribution of high-salinity water along the transect line. At the subtidal station, a major change in salinity as a function of the tidal cycle was also apparent, but restricted to the surface layer, whereas salinity remained constantly >31 psu below the surface (Figs. 1c,d).

The nutrient concentrations were also markedly affected by the tidal cycle. At the beginning of the survey, silicate and nitrate+nitrite concentrations were markedly higher at the riverine station than at the intertidal and subtidal stations, whereas the ammonium concentration was relatively higher at the intertidal station (Figs. 1a, b, c). Approaching the lower low tide, the nutrient concentrations in surface water increased rapidly, especially at the intertidal and subtidal stations. Differently at the riverine station, the ammonium concentrations remained low, suggesting no significant import of this nutrient species through freshwater inflow (Fig. 1a). At the subtidal station, the nutrient concentrations also showed a relatively consistent increase below the surface, yet progressively less noticeable with depth (Fig. 1d).

By contrast, during the flood, as high-salinity water flushed backward into the estuary, the nutrient concentrations dropped to the lowest values at all stations and depths (Figs. 1a, b, c, d); a 7.5-fold and 8.8-fold decrease of silicate and nitrate+nitrite concentrations, respectively, occurred at the riverine station. During the second part of the survey, after the higher high tide, both salinity and nutrient changes were less marked.

Figure 2 summarizes the relationships between salinity and nutrient concentrations. Salinity versus nutrient plots demonstrate that the distributional pattern of nutrients largely varied with station, depth, and the different nutrient species. In particular at the riverine station, silicate and nitrate+nitrite were negatively correlated with salinity; r^2 explained a large portion of total variance (i.e., $r^2 = 0.879$ and 0.796, respectively). By contrast, at this station, ammonium showed a positive correlation with salinity, whereas phosphate did not significantly correlate with salinity. Differently at the intertidal and subtidal stations, the concentrations of all nutrient species in surface waters were negatively and significantly correlated with salinity; levels of confidence varied from $p < .05$ (ammonium) to $p < .001$ (phosphate and silicate).

The variability of both salinity and nutrient concentrations was lowest at the subtidal station below the surface. This test data set (Fig. 2d) indicated that all nutrient species were correlated positively with salinity at a high level of significance ($p < .001$). Relevant plots also highlighted that, within such a restricted variability of salinity, silicate and nitrate+nitrite concentrations comprised narrower values than those of phosphate and ammonium (Fig. 2d). Accordingly, the model equation for the former two nutrient species explained a higher portion of the total variance ($r^2 = 0.423$, and $r^2 = 0.457$, respectively) than that explained by the latter ($r^2 = 0.221$, and $r^2 = 0.245$, respectively).

These results showed that the riverine input was a major source of silicate and, partially, nitrate+nitrite and phosphate. It was also apparent that the increase in nutrient concentrations at the intertidal station and subsequently at the subtidal station, was largest during the first part of the survey (Figs. 1b, c). Companion papers demonstrated that the intertidal zone also plays a major role in nutrient cycling, as a major site of nutrient regeneration within the estuary (12,14,24).

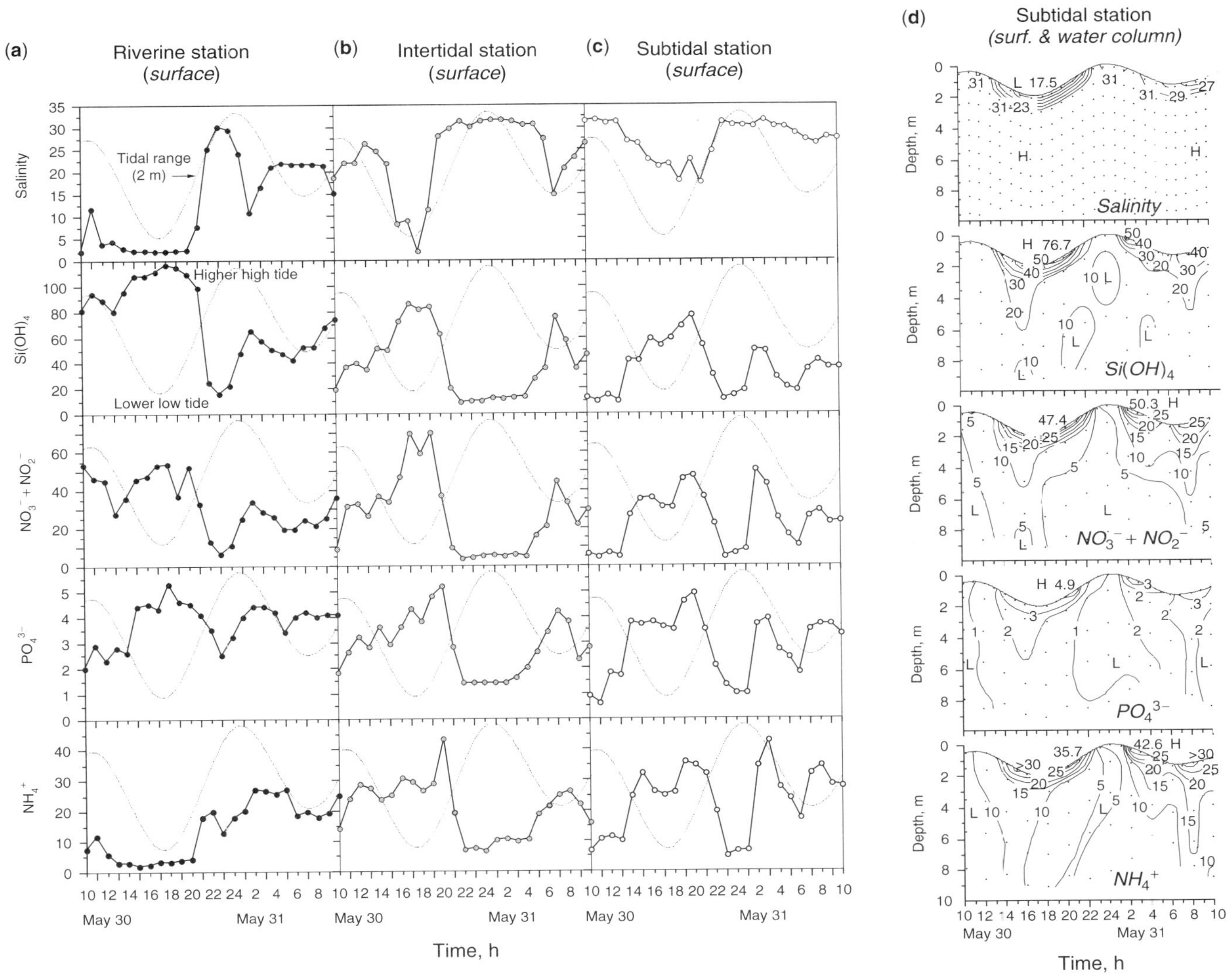

Figure 1. Time series of salinity (psu, practical salinity unit) and nutrient concentration (μM) in surface water at a riverine, intertidal, and subtidal station (Fig. 1a, b, c) and through the water column at the subtidal station (Fig. 1d) during a tidal cycle of a spring tide in a *mixed-semidiurnal* type estuary (Seto Inland Sea, Japan). Data sources: Fig. 1a, b, c from Montani et al. (13) (redrawn from Figs. 4 & 5); Fig. 1d: from Magni et al. (23) (adapted from Figs. 2 & 3).

CONCLUDING REMARKS

This article showed that the effect of salt-water intrusion on the dynamics of nutrients varies strongly both spatially (on relatively short distances) and temporally (on an hour timescale), and that this is much dependent on the tidal state. In particular, it was shown that the effect of tidal amplitude is important in determining the extent of the variations in nutrient concentrations, which were stronger between the lower low tide and the higher high tide. It also indicated that nutrient concentrations were higher during the ebb than during the flood and highest at the surface layer, as strongly correlated inversely with salinity. Finally, this work highlighted the importance of considering simultaneous investigations at different stations along the estuarine spine during a tidal cycle, especially on short distances, besides studying the nutrient dynamics based on selected salinity intervals that may reflect only the situation at a particular tidal state.

BIBLIOGRAPHY

1. Hernández-Ayón, J.M., Galindo-Bect, M.S., Flores-Baez, B.P., and Alvarez-Borrego, S. (1993). Nutrient concentrations are high in the turbid waters of the Colorado River delta. *Estuar. Coast. Shelf Sci.* **37**: 593–602.
2. Balls, P.W. (1994). Nutrient inputs to estuaries from nine Scottish east coast rivers; influence of estuarine processes on inputs to the North Sea. *Estuar. Coast. Shelf Sci.* **39**: 329–352.
3. Uncles, R.J. and Stephens, J.A. (1996). Salt intrusion in the Tweed Estuary. *Estuar. Coast. Shelf Sci.* **43**: 271–293.
4. Balls, P.W. (1992). Nutrient behaviour in two contrasting Scottish estuaries, the Forth and the Tay. *Oceanol. Acta* **15**: 261–277.

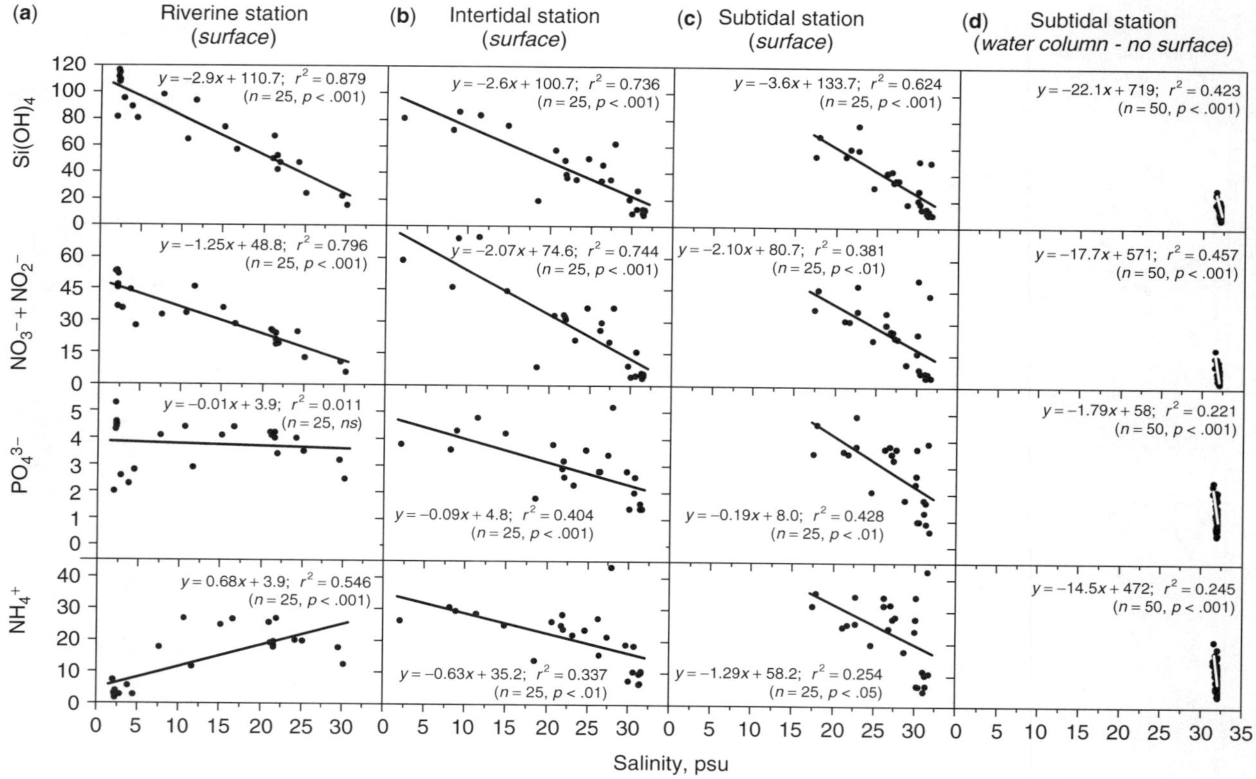

Figure 2. Plots of salinity (psu) versus nutrient concentration (μM) originated from the time series in Fig. 1. Data sources: Fig. 1a, b, c from Montani et al. (13) (redrawn from Figs. 8 to 11); Fig. 1d: after Magni et al. (23).

5. Clark, J.F., Simpson, H.J., Bopp, R.F., and Deck, B. (1992). Geochemistry and loading history of phosphate and silicate in the Hudson estuary. *Estuar. Coast. Shelf Sci.* **34**: 213–233.
6. Page, H.M., Petty, R.L., and Meade, D.E. (1995). Influence of watershed runoff on nutrient dynamics in a southern California salt marsh. *Estuar. Coast. Shelf Sci.* **41**: 163–180.
7. Eyre, B. and Twigg, C. (1997). Nutrient behaviour during post-flood recovery of the Richmond river estuary northern NSW, Australia. *Estuar. Coast. Shelf Sci.* **44**: 311–326.
8. Soetart, K. and Herman, P.M.J. (1995). Nitrogen dynamics in the westerschelde (the Netherlands) using a box model with fixed dispersion coefficients. *Hydrobiologia* **311**: 215–224.
9. Nixon, S.W. et al. (1996). The fate of nitrogen and phosphorus at the land–sea margin of the North Atlantic Ocean. *Biogeochem.* **35**: 141–180.
10. Nedwell, D.B., Jickells, T.D., Trimmer, M., and Sanders, R. (1999). Nutrients in estuaries. *Adv. Ecol. Res.* **29**: 43–92.
11. Herman, P.M.J., Middelburg, J.J., van de Koppel, J., and Heip, C.H.R. (1999). Ecology of estuarine macrobenthos. *Adv. Ecol. Res.* **29**: 195–240.
12. Magni, P., Montani, S., Takada, C., and Tsutsumi, H. (2000). Temporal scaling and relevance of bivalve nutrient excretion on a tidal flat of the Seto Inland Sea, Japan. *Mar. Ecol. Prog. Ser.* **198**: 139–155.
13. Montani, S. et al. (1998). The effect of a tidal cycle on the dynamics of nutrients in a tidal estuary in the Seto Inland Sea, Japan. *J. Oceanogr.* **54**: 65–76.
14. Magni, P. and Montani, S. (2000). Water chemistry variability in the lower intertidal zone of an estuary in the Seto Inland Sea, Japan: Seasonal patterns of nutrients and particulate compounds. *Hydrobiologia* **432**: 9–23.
15. Middelburg, J.J. and Nieuwenhuize, J. (2000). Uptake of dissolved inorganic nitrogen in turbid, tidal estuaries. *Mar. Ecol. Prog. Ser.* **192**: 79–88.
16. Middelburg, J.J. and Nieuwenhuize, J. (2001). Nitrogen isotope tracing of dissolved inorganic nitrogen behaviour in tidal estuaries. *Estuar. Coast. Shelf Sci.* **53**: 385–391.
17. Liss, P.S. and Pointon, M.J. (1973). Removal of dissolved boron and silicon during estuarine mixing of sea and river waters. *Geochim. Cosmochim. Acta* **37**: 1493–1498.
18. Bartoli, M. et al. (2001). Impact of *Tapes philippinarum* on nutrient dynamics and benthic respiration in the Sacca di Goro. *Hydrobiologia* **455**: 203–212.
19. Magni, P. and Montani, S. (2004). Magnitude and temporal scaling of biogenic nutrient regeneration in a bivalve-dominated tidal flat. In: *Encyclopedia of Water*. John Wiley & Sons, New York.
20. Froelich, P.N. (1988). Kinetic control of dissolved phosphate in natural rivers and estuaries: A primer on the phosphate buffer mechanism. *Limnol. Oceanogr.* **33**: 649–668.
21. Yin, K., Harrison, P.J., Pond, S., and Beamish, R.J. (1995). Entrainment of nitrate in the Fraser river estuary and its biological implications. II. Effects of spring *vs.* neap tide and river discharge. *Estuar. Coast. Shelf Sci.* **40**: 529–544.
22. Yin, K., Harrison, P.J., Pond, S., and Beamish, R.J. (1995). Entrainment of nitrate in the Fraser River estuary and its biological implications. III. Effects of winds. *Estuar. Coast. Shelf Sci.* **40**: 545–558.

23. Magni, P., Montani, S., and Tada, K. (2002). Semidiurnal dynamics of salinity, nutrients and suspended particulate matter in an estuary in the Seto Inland Sea, Japan, during a spring tide cycle. *J. Oceanogr.* **58**: 389–402.
24. Montani, S., Magni, P., and Abe, N. (2003). Seasonal and interannual patterns of intertidal microphytobenthos in combination with laboratory and areal production estimates. *Mar. Ecol. Prog. Ser.* **249**: 79–91.

THE ROLE OF OCEANS IN THE GLOBAL CYCLES OF CLIMATICALLY-ACTIVE TRACE-GASES

Robert C. Upstill-Goddard
University of Newcastle upon Tyne
Newcastle upon Tyne, United Kingdom

The oceans are important in the global cycles of a range of trace gases that influence atmospheric chemistry and climate. For some of these, the oceans are a net source to the atmosphere, whereas for others, they are a net sink. Major gases of interest are summarized in Table 1, along with their net flux directions and their principal roles in the troposphere and stratosphere. Carbon dioxide (CO_2), methane (CH_4), and nitrous oxide (N_2O) are major greenhouse gases: CO_2 currently accounts for more than one-half of enhanced global warming, whereas CH_4 and N_2O, respectively, account for about 15% and 6% (1). Volatile sulphurs are also implicated in climate forcing and they play important roles in atmospheric chemistry. As a result of length restrictions, our discussion focuses only on these gases.

GLOBAL PARTITIONING OF ANTHROPOGENIC CO_2

Several mechanisms contribute to natural CO_2 cycling between the atmosphere and the Earth's surface. The largest natural exchanges occur through respiration and photosynthesis on land and in the oceans, and by solubility-driven uptake in the oceans, and the net result of these exchanges is a natural carbon cycle in overall balance. By comparison with these natural fluxes, the flux of anthropogenic CO_2, which is derived primarily from the burning of fossil fuels, cement production, deforestation, and other land-use changes, is rather small, $\sim 7 \pm 1$ Gt C per year (1) (1 Gt = 1 gigatonne = 10^{15} g), which is nevertheless large enough to significantly disturb the natural CO_2 cycle.

As a result of its unreactive nature, the residence time of CO_2 in the troposphere is of the order of 50–200 years, hence anthropogenic CO_2 tends to accumulate there. In fact, tropospheric CO_2 has risen from about 280 ppmv (parts per million by volume) preindustrially, as determined from ice core studies (2), to approaching 380 ppmv by mid-2004 (http://www.cmdl.noaa.gov/). However, this corresponds to somewhat less than one-half of the known anthropogenic release (1). During the 1980s (the latest decade for which estimates of all carbon sources and sinks are available), the mean rate of tropospheric CO_2 growth was only about 3.3 ± 0.1 Gt C per year, the remainder, about 3.7 ± 1.0 Gt C per year, having been absorbed by "sinks" located in the oceans and within the terrestrial biosphere (1). The fraction of anthropogenic CO_2 absorbed by these sinks is, however, not constant. Large fluctuations are evident in the continuous tropospheric records that date back to the late 1950s (1), and these fluctuations are believed to relate directly to short-term variations in global climate. For example, increased atmospheric growth rates correlate with El Niño climate warming events, whereas cooling periods, such as that which followed the eruption of Mount Pinatubo in the early 1990s, seem to be associated with reduced atmospheric growth. These variations are thought principally to reflect changes in the balance of terrestrial primary production (photosynthesis) versus respiration, decomposition, and the combustion of organic material (3).

High background variability has precluded directly measuring the relative magnitudes of the oceanic and terrestrial CO_2 sinks, hence they have hitherto been estimated using models, often with conflicting results (4). However, recent techniques based on measuring carbon and oxygen isotopes in air (5) now enable the partitioning of anthropogenic CO_2 between these reservoirs to be determined with greater certainty, and it is now generally agreed that the ocean and land sinks are of about the same magnitude (although the uncertainties are large), i.e., about 1.9 ± 0.6 Gt C per year (6).

Table 1. Some Atmospherically Active Trace Gases with Important Marine Sources or Sinks

Gas		Net Flux	Effect in the Atmosphere	Troposphere	Stratosphere
Carbon Dioxide	(CO_2)	Into Ocean	Infra-red activity	√	
Methane	(CH_4)	Out of Ocean	Infra-red activity Atmospheric redox Ozone regulation	√	
Nitrous Oxide	(N_2O)	Out of Ocean	Infra-red activity Ozone regulation	√	
Carbon Monoxide	(CO)	Out of Ocean	Atmospheric redox	√	
Dimethyl sulphide	(DMS)	Out of Ocean	Acidity, Cloud formation	√	
Carbonyl Sulphide	(COS)	Out of Ocean	Sulphate aerosol (cooling)	√	√
Carbon Disulphide	(CS_2)	Out of Ocean	Source of COS	√	√
Organohalogens (Natural)		Out of Ocean	Atmospheric redox	√	√
Nonmethane hydrocarbons		Out of Ocean	Atmospheric redox		

MECHANISMS OF OCEAN CO_2 UPTAKE

Compared with many other atmospheric trace gases, CO_2 is rather soluble in seawater, where it occurs as three principal dissolved species that together comprise seawater dissolved inorganic carbon (DIC), which are bicarbonate (HCO_3^-, 91% of DIC), carbonate (CO_3^{2-}, 8% of DIC), and dissolved CO_2 (1% of DIC).

The capacity of the oceans to absorb atmospheric CO_2 is ultimately buffered by the DIC system. CO_2 uptake results in an increase both in the partial pressure of CO_2 (pCO_2) and in the concentration of HCO_3^-, which is produced through the reaction of CO_2 with CO_3^{2-}. With further CO_2 uptake, its conversion to HCO_3^- becomes limited by the decreasing CO_3^{2-} availability, the result being that more CO_2 remains in solution, further decreasing the capacity for CO_2 uptake at increasing tropospheric levels. As an illustration, for atmospheric CO_2 to rise by about 100 ppmv relative to today, the increase in seawater DIC would only be about 60% of that which has accompanied the approximately 100 ppmv rise in tropospheric CO_2 since the industrial revolution (1,6).

Studies based on the downward penetration of chemical tracers, such as ^{14}C, coupled with simple box models of ocean mixing (7), or more sophisticated General Circulation Models (8), show that tropospheric CO_2 equilibrates with the surface ocean mixed layer rather rapidly, on the order of a few years. In contrast, CO_2 penetration into the deep ocean interior is constrained by relatively slow vertical transport; the whole ocean equilibrates with the atmosphere on a timescale of more than 1000 years. As a result of this slow rate of downward mixing, with few exceptions, anthropogenic CO_2 has yet to penetrate below about 1000 m depth. Transport of CO_2 into the ocean interior occurs via the thermohaline circulation, in which cool surface waters in the high latitude North Atlantic (Greenland Sea) and Southern Oceans sequester CO_2 from the troposphere and sink because of their higher density. This downward transport is balanced in regions where deep water with a high capacity for CO_2 uptake wells up to the surface, such as occurs in parts of the tropics and the Southern Ocean. The sinking waters eventually spread laterally toward the subtropics, giving rise to a relatively uniform distribution of anthropogenic CO_2. Eventually, on the 1000-plus year timescale of deep ocean mixing, the deep CO_2-rich waters formed in this way will again well up to the surface, the principal site for this being the Equatorial Pacific. As this water warms during upwelling, its pCO2 will increase, resulting in CO_2 loss to the atmosphere by out gassing.

One other potential removal process for tropospheric CO_2 is through reaction with $CaCO_3$ contained in deep sea sediments; however, their response time to changes in tropospheric CO_2 is several thousand years (9).

According to some coupled ocean-atmosphere models, one possible consequence of global warming is an increase in the intensity of vertical stratification in the oceans (1), which would reduce the rate of surface to deep water mixing and, consequently, the uptake rate of tropospheric CO_2.

In addition to chemically and physically driven uptake, tropospheric CO_2 is also processed through the so-called "biological pump," in which organic matter produced via photosynthesis and cycled through the upper ocean food-web is ultimately transported downward via sinking organic particles or through vertical biomass migrations. This sinking flux of organic carbon is remineralized or respired back to inorganic carbon at depth with an accompanying release of dissolved inorganic nutrients (principally nitrate and phosphate). Model simulations suggest that without the biological pump, tropospheric CO_2 could be about 150 ppmv higher than at present (1). However, the likely response of the biological pump to increasing tropospheric CO_2 is uncertain. Most current evidence tends to discount increased productivity on the grounds of limitation by the supply of nutrients, which are seasonally depleted in most surface waters. However, extensive regions exist of the subarctic Pacific, equatorial Pacific, and Southern Ocean with abundant nitrate and phosphate throughout the year but with very low phytoplankton productivity, the so-called high-nutrient-low-chlorophyll (HNLC) regions. Their low productivity likely reflects a deficiency in the supply of a minor nutrient such as iron (10). Although predicting climate-induced changes in the supply rate of iron to the oceans is far from straightforward, these regions could conceivably play a significant role in the future ocean uptake of anthropogenic CO_2. It is also conceivable that global warming-induced changes in stratification described earlier could also modify the ocean's biological carbon cycle; however, the consequences of this are difficult to predict given the intrinsic complexity of the system (11).

THE GLOBAL ROLE OF CH_4 AND N_2O

As well as being important greenhouse gases, CH_4 and N_2O play important roles in atmospheric chemistry; N_2O is involved in the stratospheric cycling of NO_x (reactive nitrogen oxides) and ozone (12), and CH_4 takes part in reactions that govern levels of tropospheric ozone and hydroxyl radical ($\cdot$ OH) and stratospheric H_2O (13). Like CO_2, N_2O and CH_4 are currently increasing in the troposphere, both by about 0.3% per year (1). However, the role of the oceans in their global budgets differs from that for anthropogenic CO_2 in two fundamental ways. First, the oceans are a net source of tropospheric N_2O and CH_4, and second, for both, the marine source is one of several global sources whose relative magnitudes remain rather uncertain.

GLOBAL SOURCE UNCERTAINTIES

The marine system is one of two major N_2O sources, the other being terrestrial soils. Combustion, biomass burning, and fertilizers make additional minor contributions. Although the uncertainties are large, the oceans are thought to contribute around 20% of the natural global source (1); however, the contribution is larger with anthropogenic sources included (see below). For CH_4, many more global sources exist, in descending order of magnitude, these are: agriculture, wetlands, fossil fuels, biomass burning, termites, oceans, CH_4 hydrates, and landfills. However, the range of uncertainties is no better than it is for N_2O, ranging from ±100% (e.g., wetlands), to in excess of ±2000% (e.g., CH_4 hydrates) (1). According to these data, marine waters contribute only about 3% of the total CH_4 source to the troposphere, but this may be an underestimate (see below).

SOURCES OF SEAWATER N_2O

In seawater, N_2O develops as a byproduct during microbial nitrification (the conversion of dissolved ammonium, NH_4^+, to nitrate, NO_3^-) and as a reactive intermediate during microbial denitrification (the reduction of NO_3^- to gaseous nitrogen). Nitrification principally occurs in oxygenated waters, although it is inhibited by light, whereas denitrification is restricted to anoxic sediments and O_2-deficient waters. For these reasons, N_2O production is insignificant in most open ocean surface waters. In extremely O_2-deficient waters, N_2O can be consumed during denitrification because of its use as an electron acceptor by denitrifying bacteria (14). Coupling of the two processes, in which the NO_3^- developing from nitrification is consumed by denitrification, occurs both in marine sediments (15) and around the fringes of O_2-depleted waters in the open ocean (16).

The net rate of N_2O production by nitrification and denitrification is influenced by several factors in addition to dissolved O_2 availability, including the supply rates of NO_3^- and NH_4^+, the composition of the microbial ecosystem, and in sediments, physicochemical aspects such as porosity and grain size. Consequently, both processes show pronounced seasonal variability. These aspects, coupled with a nonuniform distribution of N_2O source regions, makes the total marine source of N_2O difficult to quantify.

Current data indicate around two-thirds of the marine source of tropospheric N_2O to derive from the open ocean, with the remainder coming from coastal waters. However, these estimates contain uncertainties resulting from incomplete spatial and seasonal sampling and difficulties related to estimating sea-to-air fluxes.

Open ocean emissions are approximately equally distributed between the northern and southern hemispheres. Most of the N_2O is located below the surface-mixed layer and highest concentrations occur at around 500–1000 m depth, where a high O_2 demand results from the bacterial decomposition of sinking organic particles. Exchange of this water with the atmosphere is usually slow, except during wintertime when the surface-mixed layer deepens because of cooling and wind-driven mixing, entraining waters from below. Regions experiencing strong seasonal upwelling are especially strong sources of tropospheric N_2O, which include the Tropical North Pacific, the Arabian Sea/northwestern Indian Ocean, the equatorial upwelling, and along the coasts of northwest Africa and western central and South America. In these areas, the upwellings bring N_2O-rich waters to the surface along with a plentiful supply of nutrients that fuels high primary productivity. The resultant large downward flux of organic particles gives rise to strongly O_2-deficient waters that replenish the deep water inventory of N_2O through nitrification-denitrification coupling (17). Some areas of the subtropical gyres and the North Atlantic seem to be weak sinks for tropospheric N_2O in winter and weak sources in summer.

The major coastal N_2O source regions are estuaries, open coastal shelf waters being generally at or close to equilibrium with tropospheric N_2O levels. Both water column nitrification associated with high suspended particle populations (18) and sediment denitrification (19) have been identified in estuaries.

Whereas oceanic N_2O emissions are considered to be wholly natural, significant N_2O in coastal waters appears to derive indirectly from anthropogenic activity. In particular, the use of fertilizers is reflected in enhanced transport of nitrogen to coastal waters and a consequent increase in N_2O production and tropospheric flux. As much as 90% of current estuarine N_2O emissions and 25% of continental shelf emissions may be anthropogenic and consistent with the geographic distribution of fertilizer use, human population, and atmospheric nitrogen deposition, more than 80% of these anthropogenic sources are located in the Northern Hemisphere mid-latitudes between 20°N and 66°N (20).

SOURCES OF SEAWATER CH_4

Like N_2O, seawater CH_4 has a microbial source (microbial methanogenesis). Methanogenesis is inhibited by dissolved O_2 and therefore usually occurs in anoxic sediments or in waters that are strongly O_2-depleted. Even so, CH_4 concentrations in the oxygenated surface ocean are on average about 30% above the tropospheric equilibrium value, most likely reflecting methanogenesis by O_2-tolerant methanogens inside bacterially maintained "anoxic microniches" in the guts of zooplankton and/or in particles (21). Consequently, the open ocean represents a small CH_4 source. In addition, regions of much higher CH_4 concentration occur in upwelling areas, associated with enhanced primary productivity as for N_2O (16,22). Based on the available open ocean data, the marine contribution to tropospheric CH_4 is about 10 Tg CH_4 per year, which is equivalent to about 3% of the total global source (1).

Coastal waters have much higher CH_4 concentrations than the open ocean but have, until recently, been excluded from global CH_4 source estimates because of a lack of data. For estuaries, values of 100 to 200 times the background equilibrium value are common. Such CH_4 levels reflect direct inputs from rivers, coastal seawater, underlying sediments, and *in situ* production/consumption from water column methanogenesis and microbial CH_4 oxidation. Recent work suggests that correctly accounting for these regions could increase the estimated marine CH_4 source by around 50% (23).

A potentially even larger CH_4 source may be geologically sourced CH_4 from natural marine seeps, which are most common on shallow continental shelves (24). Seeps are episodic in nature, and the CH_4 fluxes developing from them are predominantly by bubbles, which complicates making accurate measurements. As a result of CH_4 losses because of bubble dissolution and subsequent CH_4 oxidation in the water column, shallow water seeps are much more effective than deep water seeps at contributing CH_4 to the troposphere. Revised source estimates that take account of seep occurrences suggest that the total marine source of tropospheric CH_4 could be as much as 40 Tg yr^{-1} (25). If so, the oceans could be a much more important source of tropospheric CH_4 than previously thought.

VOLATILE SULPHUR COMPOUNDS

The marine system is an important source of sulphur globally; in particular, it is the principal source of biogenic atmospheric sulphate aerosol, which plays important roles in atmospheric chemistry and climate.

The predominant volatile sulphur in surface seawater is dimethyl sulphide (DMS) [$(CH_3)_2S$], which is a byproduct of algal metabolism and accounts for about 95% of marine sulphur emissions and 20% of total global sulphur emissions (1). DMS is rapidly oxidized by free radicals in the lower troposphere. Sulphur dioxide (SO_2) is the major reaction product and is subsequently transformed to sulphate aerosol through gas-to-particle reactions. This process is hypothesized to impact directly the radiative forcing of global climate, primarily through changes to cloud albedo (26). Carbonyl sulphide (COS) has also been implicated in global climate forcing. Compared with other volatile sulphurs, COS has a long tropospheric residence time of around 2–6 years. Consequently, it is transported into the stratosphere, where its photo-oxidation is believed to be an important source of sulphate aerosol, which is thought to impact Earth's radiation balance (27) and stratospheric ozone levels (28).

DIMETHYL SULPHIDE

A global database of more than 15,000 measurements of surface seawater (29) revealed distinct annual DMS cycles in the open ocean at mid to high latitudes. In the northern hemisphere, open ocean DMS increases during spring–summer, whereas in the southern hemisphere, concentrations peak six months later. These patterns relate to the timing of phytoplankton blooms and seasonal changes in mixed layer depth. In contrast, tropical regions show weak seasonality; DMS is elevated in the upwelling regions off western Africa and South America, but these concentrations are lower than those at high latitudes during summer. On coastal shelves, DMS is spatially and temporally variable, broadly correlating with seasonal primary productivity and the presence of algal blooms associated with upwelling at water mass boundaries (hydrographic fronts) (29). Similar concentrations are found in many estuaries, notable exceptions being those with high concentrations of suspended particles. Seasonal patterns are, however, rather different, with maximum concentrations occurring during late winter/early spring. The available data indicate that coastal regions may be larger emitters of DMS per unit area than the global mean, however, because of its much larger surface area, the open ocean is the most important source of tropospheric DMS (29).

CARBONYL SULPHIDE

Direct marine emissions are thought to account for about 20% of the global COS source (30), although the tropospheric oxidation of marine-derived DMS and carbon disulphide (CS_2) may increase this to as much as 55% (31). COS in seawater primarily develops from the photodecomposition of humic-like colored dissolved organic matter (CDOM) (32), although a small nonphotochemical source has also been inferred. The distribution of sea surface COS, therefore, corresponds closely to the concentration and reactivity of CDOM, which primarily derives from terrestrial sources. Consequently, COS concentrations are highest in estuaries, which are about an order of magnitude higher than those of adjacent coastal waters (32).

COS undergoes hydrolysis removal in seawater at a similar rate to that for its photo-production, hence COS shows a pronounced diel cycle in surface seawater with concentrations peaking in the early afternoon and declining to a minimum just before sunrise (33). Strong seasonal variation also exists. In the mid to high latitude open ocean the balance of photo-chemical production versus hydrolysis removal can lead to these regions becoming a seasonal sink for tropospheric COS (34). Net ocean uptake of COS has also been found in the subtropical ocean gyres (35). Taking account of these findings, the marine COS source is most likely dominated by the contribution from coastal and shelf areas.

BIBLIOGRAPHY

1. Houghton, J.T., Ding, Y., Griggs, D.J., Noguer, M., van der Linden, P.J., Dai, X., Maskell, K., and Johnson, C.A. (Eds.). (2001). *Climate Change 2001: The Scientific Basis*. Cambridge University Press.
2. Barnola, J.M. et al. (1995). CO_2 evolution during the last millennium as recorded by Antarctic and Greenland ice. *Tellus* **47B**: 264–272.
3. Le Quéré, C., Orr, J., Monfray, P., and Aumont, O. (2000). Interannual variability in the oceanic sink of CO_2 from 1979 through 1997. *Glob Biogeochem Cyc.* **14**: 1247–1265.
4. Siengenthaler, U. and Sarmiento, J.L. (1993). Atmospheric carbon dioxide and the ocean. *Nature* **365**: 119–125.
5. Keeling, R.F. and Shertz, S.R. (1992). Seasonal and interannual variations in atmospheric oxygen and implications for the global carbon cycle. *Nature* **358**: 723.
6. Sarmiento, J.L. and Gruber, N. (2002). Sinks for anthropogenic carbon. *Physics Today* **55**: 30–36.
7. Broecker, W.S., Takahashi, T., Simpson, H.J., and Peng, T-H. (1979). Fate of fossil fuel carbon dioxide and the global carbon budget. *Science* **206**: 409–418.
8. Cox, P.M., Betts, R.A., Jones, C.D., Spall, S.A., and Totterdell, I.J. (2000). Acceleration of global warming due to carbon-cycle feedbacks in a coupled climate model. *Nature* **408**: 184.
9. Archer, D.E., Kheshgi, H., and Maier-Reimer, E. (1997). Multiple timescales for neutralization of fossil fuel CO_2. *Geophys Res. Lett.* **24**: 405–408.
10. Martin, J.H. (1992). Iron as a limiting factor. In: *Primary Productivity and Biogeochemical Cycles in the Sea*. P.G. Falkowski and A. Woodhead (Eds.). Plenum Press, New York, pp. 123–137.

11. Sarmiento, J.L., Hughes, T.M.C., Stouffer, R.J., and Manabe, S. (1998). Simulated response of the ocean carbon cycle to anthropogenic climate warming. *Nature* **393**: 245–249.
12. Nevison, C. and Holland, E. (1997). A re-examination of the role of anthropogenically fixed nitrogen on atmospheric N_2O and the stratospheric O_3 layer. *J. Geophys Res.* **10**: 15809–15820.
13. Crutzen, P.J. (1990). Methane's sinks and sources. *Nature* **350**: 390–381.
14. Cohen, Y. and Gordon, L.I. (1979). Nitrous oxide in the oxygen minimum of the of the eastern tropical north Pacific: Evidence for its consumption during denitrification and possible mechanisms for its production. *Deep Sea Res.* **25**: 509–524.
15. Vanraaphorst, W. et al. (1992). Nitrogen cycling in 2 types of sediments of the southern North-Sea (frisian front, broad fourteens)—field data and mesocosm results. *Neth. J. Sea Res.* **28**: 293–316.
16. Naqvi, S.W.A. et al. (1998). Budgetary and biogeochemical implications of N_2O isotope signatures in the Arabian Sea. *Nature* **394**: 462–464.
17. Suntharalingam, P. and Sarmiento, J.L. (2000). Factors governing the oceanic nitrous oxide distribution: Simulations with an ocean general circulation model. *Glob. Biogeochem. Cyc.* **14**: 429–454.
18. Barnes, J. and Owens, N.J.P. (1998). Denitrification and nitrous oxide concentrations in the Humber estuary, UK, and adjacent coastal zones. *Mar. Poll. Bul.* **37**: 247–260.
19. Dong, L.F., Nedwell, D.B., Underwood, G.J.C., Thornton, D.C.O., and Rusmana, I. (2002). Nitrous oxide formation in the Colne estuary, England: The central role of nitrite. *Appl. Environ. Microbiol.* **68**: 1240–1249.
20. Seitzinger, S.P. and Kroeze, C. (2000). Global distribution of N_2O emissions from aquatic systems: natural emissions and anthropogenic effects. *Chenmosphere* **2**: 267–279.
21. Lamontagne, R.A., Swinnerton, J.W., Linnenbom, V.J., and Smith, W.D. (1973). Methane concentrations in various marine environments. *J. Geophys. Res.* **78**: 5317–5324.
22. Upstill-Goddard, R.C., Barnes, J., and Owens, N.J.P. (1994). Nitrous oxide and methane during the 1994 SW monsoon in the Arabian Sea/noerthwestern Indian Ocean. *J. Geophys. Res.* **104**: 30,067–30,084.
23. Upstill-Goddard, R.C., Barnes, J., Frost, T., Punshon, S., and Owens, N.J.P. (2000). Methane in the southern North Sea: Low-salinity inputs, estuarine removal, and atmospheric flux. *Glob. Biogeochem. Cyc.* **14**: 1205–1217.
24. Judd, A.G. (2003). The global importance and context of methane escape from the seabed. *Geo-Marine Lett.* **23**: 147–154.
25. Kvenvolden, K.A., Lorenson, T.D., and Reeburgh, W.S. (2001). Attention turns to naturally occurring methane seepage. *EOS* **82**: 457.
26. Charlson, R.J., Lovelock, J.E., Andreae, M.O., and Warren, S.G. (1987). Oceanic phytoplankton, atmospheric sulphur, cloud albedo and climate. *Nature* **326**: 655–661.
27. Turco, R.P., Whitten, R.C., Toon, O.B., Pollack, J.B., and Hamill, P. (1980). OCS, stratospheric aerosols and climate. *Nature* **283**: 283–286.
28. Chin, M. and Davies, D.D. (1995). A reanalysis of carbonyl sulphide as asource of stratospheric background sulphur aerosol. *J. Geophys. Res.* **100**: 8993–9005.
29. Kettle, A.J. et al. (1999). A global database of sea surface dimethylsulfide (DMS) measurements and a procedure to predict sea surface DMS as a function of latitude, longitude, and month. *Glob. Biogeochem. Cyc.* **13**: 399–444.
30. Watts, S.F. (2000). The mass budgets of carbonyl sulfide, dimethyl sulfide, carbon disulfide and hydrogen sulfide. *Atmos. Environ.* **34**: 761–779.
31. Kettle, A.J. et al. (2001). Assessing the flux of different volatile sulfur gases from the ocean to the atmosphere. *J. Geophys. Res.* **106**: 12193–12209.
32. Uher, G. and Andreae, M.O. (1997). Photochemical production of carbonyl sulfide in North Sea water: A process study. *Limnol. Oceanogr.* **42**: 432–442.
33. Ulshöfer, V.S., Flöck, O.R., Uher, G., and Andreae, M.O. (1996). Photochemical production and air-sea exchange of carbonyl sulfide in the eastern Mediterranean Sea. *Mar. Chem.* **53**: 25–39.
34. Ulshofer, V.S., Uher, G., and Andreae, M.O. (1995). Evidence for a winter sink of atmospheric carbonyl sulfide in the northeast Atlantic Ocean. *Geophys. Res. Lett.* **22**: 2601–2604.
35. Weiss, P.S., Johnson, J.E., Gammon, R.H., and Bates, T.S. (1995). A reevaluation of the open ocean source of carbonyl sulfide to the atmosphere. *J. Geophys. Res.* **100**: 23083–23092.

PACIFIC MARINE ENVIRONMENTAL LABORATORY—30 YEARS OF OBSERVING THE OCEAN

NOAA—Pacific Marine
Environmental Laboratory

Sept. 29, 2003—Although the NOAA Pacific Marine Environmental Laboratory in Seattle, Wash., celebrates its 30th anniversary this year, its staff has spent 43 years at sea. The figure of 15,654 days at sea was one of the many facts presented during the lab's anniversary celebration in August. That, along with 1,290 published journal articles and 352,000,000 hits on the PMEL Web page indicate that there's a lot going on out on Sand Point.

For two-thirds of its life, the lab has been under the direction of Eddie Bernard. An oceanographer by training, Bernard became director in 1983, a decade after the former Pacific Oceanographic Laboratory became PMEL. "We have dedicated people at PMEL who devote a lot of energy and creativity to the work we do," he said.

NATIONAL TSUNAMI MITIGATION PROGRAM

Some of that creativity and energy became evident when in 1994 the U.S. Senate asked NOAA to come up with a plan to reduce the risk of tsunamis to coastal residents. What resulted was the National Tsunami Hazard Mitigation Program, chaired by Bernard and composed of representatives from federal, state and local agencies from West Coast states, Alaska and Hawaii, working to save lives and property. "The National Tsunami Mitigation Program initiated by PMEL is a unique and effective partnership," said Rich Eisner of the California Governor's

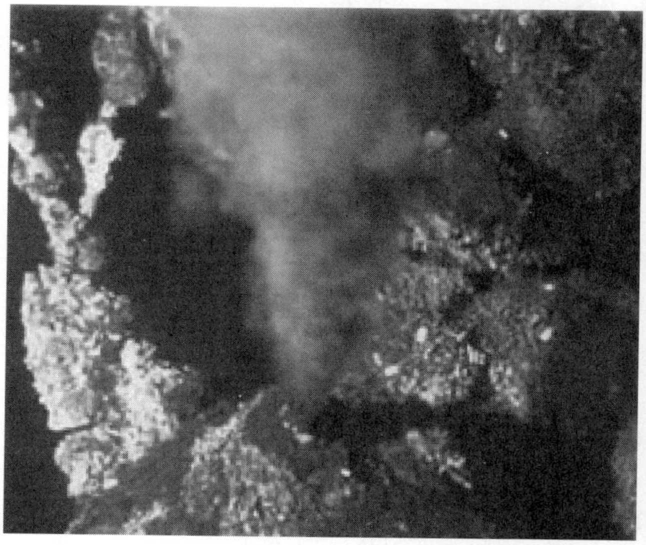

Office of Emergency Services, one of the program members. "The integration of science and mitigation policy, the warning centers, and local emergency management, and the application of new technologies fostered by PMEL have been successful beyond expectations." Among the technological innovations is the system of buoys along the West Coast that serve as warning devices, or, as Bernard calls them, "tsunameters." The tsunami program also includes a public education component that teaches coastal residents what to do in case of a possible tsunami. Signs now indicate evacuation routes and some coastal communities have been designated "TsunamiReady" for their efforts to educate and protect their residents.

UNDERWATER VOLCANOES AND VENTS

As one of NOAA's "wet" labs, PMEL focuses on a variety of ocean issues. When underwater volcanoes or vents, were first discovered in the Galapagos Islands 26 years ago, PMEL was among the first to start investigating these unusual underwater communities, where unique marine life thrive on the chemical soup spewed from the sea floor. "We may be taking drugs in the future made of enzymes that are more compatible with our bodies than synthetic compounds, which may have side effects," Bernard said. "What's spewing from the ocean floor could someday give us resistance to some new strains of infection."

FISHERIES OCEANOGRAPHY COORDINATED INVESTIGATIONS PROGRAM

PMEL began as a "small research laboratory with emphasis on water quality and environmental impact issues" in the waters off the West Coast extending to the equatorial Pacific Ocean. It now has an international reputation in many areas, especially its ability to collect ocean data and to work collaboratively in projects that cover many disciplines. One example is the Fisheries Oceanography Coordinated Investigations program that assists in forecasting fish stocks to help ensure a reliable supply and lower costs to consumers. "In 1985, Eddie Bernard took a big risk," said Doug DeMaster of the NOAA Marine Fisheries Service. That risk was offering to establish with the Alaska Fisheries Science Center and his counterpart, William Aron, a cutting-edge, applied science program across NOAA line offices. Eighteen years after its inception, NOAA's FOCI has published more than 450 scientific articles and was awarded the Department of Commerce Bronze Medal in 2002 for "scientific achievements that have advanced fisheries oceanography and marine ecology and have contributed to building sustainable fisheries in the North Pacific." DeMaster noted, "Today, if you attend a FOCI meeting, you cannot tell which scientists are from NOAA Research and which are from NOAA Fisheries. In 1985, it took vision and courage to blur the lines between line offices. Today, it seems only natural."

PACIFIC TROPICAL ATMOSPHERE OCEAN ARRAY

Understanding the natural systems is a key element of the lab. "The ocean is dynamic, it moves all of the time," Bernard said. "We're now in the third generation of observing systems. In the equatorial Pacific, we have the world's longest continuous time series of open ocean data—25 years." The equatorial Pacific also proved to be the place to be if humans wanted early warning of El Niño events. El Niño is a disruption of the ocean-atmosphere system in the tropical Pacific having important consequences for weather around the globe. After the 1982–83 El Niño, considered the most intense in the 20th century, the challenge was given to develop some sort of early warning device so people could prepare for the devastating and beneficial aspects of an El Niño and its counterpart, La Niña. Once again, in 1994, PMEL harnessed the creativity and talent of dedicated scientists and came up with the Pacific Tropical Atmosphere Ocean (TAO) array, the world's largest ocean observing system. Bobbing in the Pacific are 70 buoys measuring and relaying surface wind, sea surface temperature, upper ocean temperatures and currents, air temperature and relative humidity in real-time via satellite. "We knew we were onto something when we linked data from the TAO buoys to the Internet and attracted millions of hits from all over the world," Bernard said. "All it takes is time, money and commitment."

UNDERWATER ACOUSTICAL MONITORING

Always eager to hear what the Earth has to say, PMEL scientists also listen to the planet via underwater acoustical monitoring. Using a variety of methods, including underwater hydrophones, PMEL listens for seismic activity, marine mammals and ship traffic. The systems also have picked up some so-far unidentified sounds. "People tend to think the ocean is quiet beneath the surface," said Christopher Fox, director of the ocean acoustics project. "But it's pretty noisy down there." Some things are easy to identify, Fox said, such as whales and ship traffic. But visitors to the ocean acoustics Web site can listen to such unidentified sounds that the lab has dubbed "Train," "Upsweep," "Whistle" and "Bloop."

After 30 years, PMEL knows that the Earth still holds countless tantalizing secrets. And PMEL scientists and staff are eager to unlock those secrets. "As the planet aspirates, it provides new opportunities," Bernard

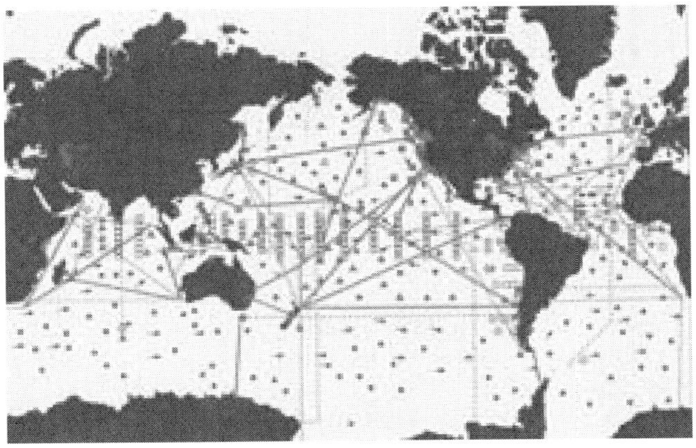

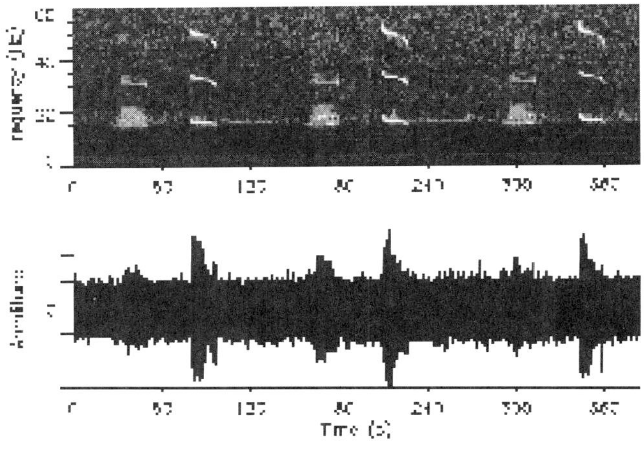

said. "It's an ongoing science experiment with enormous challenges and rewards."

RELEVANT WEB SITES

NOAA Pacific Marine Environmental Laboratory

Dr. Eddie Bernard, Director, Pacific Marine Environmental Laboratory

NOAA National Tsunami Hazard Mitigation Program

NOAA Vents Program

Fisheries Oceanography Coordinated Investigations program

NOAA Marine Fisheries Service

NOAA Alaska Fisheries Science Center

NOAA El Niño Theme Page

NOAA La Niña Theme Page

NOAA Pacific Tropical Atmosphere Ocean (TAO) Project

NOAA Vents Program: Underwater Acoustical Monitoring

NOAA's ALASKA FISHERIES SCIENCE CENTER

Ocean Explorer: Sounds in the Sea

MEDIA CONTACT

Jana Goldman, NOAA Research, (301) 713–2483

SEAWATER TEMPERATURE ESTIMATES IN PALEOCEANOGRAPHY

Giuseppe Cortese
Alfred Wegener Institute for Polar and Marine Research (AWI)
Bremerhaven, Germany

BACKGROUND

Climatic issues, such as global warming, greenhouse gases release into the atmosphere, and the role mankind plays in affecting the climate of our planet have recently gained international interest.

As a result, political panels and decision-makers are starting to look with growing interest at the results of committees established with the purpose of analyzing past, present, and future climate change.

One of these panels [Intergovernmental Panel on Climate Change (IPCC)], established by the World Meteorological Organization (WMO) and the United Nations Environment Programme (UNEP), regularly publishes reports (available also online at http://www.grida.no/climate/ipcc_tar/) intended for scientists, politicians, and anybody interested in and/or concerned by climate change.

The most powerful tools currently available to the scientific community for the analysis and prediction of climate are computer models, which simulate the functioning of Earth as a complex system. These models differ based on which components of the climate system (atmosphere, ocean, biosphere, cryosphere, etc.) they include and account for, their spatial scale and degree of detail, and on the different sets of physicochemical equations they use to describe and parameterize phenomena taking place at a subgrid spatial scale.

Regardless of setup and characteristics, though, models should be benchmarked against an observational base (e.g., to test how well the model reproduces a known distribution of a climatic variable). Models are moreover usually initialized by a set of observations (to let the incorporated physics modify the starting distribution).

Some of these datasets are readily available in the form of measurements performed over historic timescales: One of the best known examples is the atmospheric CO_2 concentration curve measured at Mauna Loa, Hawaii, since the late 1950s (1).

Other important climatic variables cannot be observed or measured directly, and indirect methods, so-called "proxies," have been developed to obtain information on them. Some of the methods that estimate seawater temperature on geological timescales are described in the following.

INTRODUCTION

The heat balance of Earth is strongly affected, to a first approximation, by the amount of solar radiation reaching the top of the atmosphere. The motion of currents in the atmosphere and in the ocean redistributes this heat between low and high latitudes, as the low latitudes receive a higher amount of heat compared with the higher latitudes. The ocean can be regarded as the "thermostat" of such a heat machine, because water has a much higher heat capacity compared with air, and because the turnover time for the ocean (thousands of years) is several orders of magnitude higher than for the atmosphere (days). One of the main features of the large-scale circulation in the ocean is the formation of deep waters at high latitudes, in the Weddell and Ross Seas (around Antarctica), and in the Labrador and Greenland/Iceland Seas (in the northern North Atlantic), because of the cooling of enormous amounts of warm surface waters advected to these locations by ocean currents (e.g., the Gulf Stream in the North Atlantic). When these waters move to higher latitudes, they cool, become denser, and sink to the ocean depths, where they start a reversed journey, from high to low latitudes. During this cooling, they also release latent heat and moisture, which help to mitigate the climate of the coastal areas they fringe, including the whole of western Europe.

Indications exist, from geological records and model simulations (2), that during glacial times, this oceanic overturning did not operate with the same intensity as today, and/or that the places where deep waters formed were displaced compared with today. As seawater temperature, together with salinity, determines the density of seawater (and therefore it strongly influences its circulation and sinking characteristics), a variety of methods have been developed to determine this important climatic variable in the distant, geological past.

METHODS

Sediment cores collected from the bottom of the World Ocean represent an ideal archive to trace the history of the ocean. Some of these cores can be accurately dated, and often they contain the fossilized remains of organisms that lived in past oceans, and reacted to changes in their characteristics.

The most useful of these organisms, because of their high fossilization potential, species diversity, and sensitivity to a range of climatic variables, are micro- and nannofossils. As their name implies, these organisms range from nanometer to millimeter size and leave fossilizable body parts, with the most commonly used groups belonging to algae (diatoms, nannoplankton) and protozoans (radiolarians, foraminifera). Most species in these groups live close to the sea surface, and each of them is adapted to a particular set of sea surface water variables, including temperature, salinity, and macro- and micro-nutrient content. Several techniques have been developed to exploit this close link between species occurrences/abundances and past environmental variables. Our first global view of surface water temperatures in the ocean during the Last Glacial Maximum (ca. 18,000 years Before Present) came from one such technique, namely, Q-Mode Principal Component Analysis applied to foraminifer and radiolarian abundance data in sediment cores (3).

Semiquantitative Floral/Faunal Estimates

Some of the first attempts to obtain seawater temperature information from microfossil assemblages tried to derive this value directly from a formula. The latter was based on the presence/absence, in the fossil samples, of species having a well-known distribution in the modern ocean, and being representative of different climatic zones (e.g., equatorial, subtropical, temperate, etc.). This approach, initially developed for diatoms (4), has also been applied to radiolarians (5). Information on species diversities of modern planktonic foraminifera in the Indian Ocean have also been used, in the late 1970s, to estimate ocean paleotemperatures (6).

Transfer Functions

With the expansion of the knowledge about the biogeography of most plankton groups (and the ecological and environmental significance of many species), and the development of computers (allowing the implementation of more sophisticated algorithms and techniques), transfer functions made their breakthrough in paleoceanography. They were first described and applied to planktonic foraminifera by Imbrie and Kipp (7), who illustrated their utilization in paleoclimatology. The species assemblage present in a collection of modern sediment samples (calibration dataset), containing up to several hundreds of samples, is "simplified" into a limited number of faunal factors/assemblages. A multivariate regression is then used to calibrate these simplified assemblages to the desired environmental variable (e.g., surface seawater temperature) measured, in the water column, at the same locations where the surface sediment samples were recovered. The value of the environmental variable at a certain time in the past is estimated by calculating "pseudo-factors" for each of the past samples (i.e., the value the calibrated assemblages have in these samples) and replacing them in the calibration equation. One main assumption of this method is that the considered environmental variable (in our case, surface temperature) plays a major role in influencing the biogeographic distribution of the considered microfossil species, and that it is therefore responsible for most of the variance observed in the calibration dataset. Transfer functions have been successfully applied to past ocean surface temperature reconstructions by using planktonic foraminifera (8–14), radiolarians (15–22), and diatoms (23,24).

Geochemical Methods

Several methods have been recently developed to estimate past ocean temperatures by measuring chemical elements or organic compounds, which are either incorporated in the organism shell or produced during its life cycle. The general approach is to develop a calibration equation by growing the desired species in culture under a variety of environmental conditions, and measuring how its geochemical composition changes. Additionally, a calibration equation can also be obtained by analyzing extensive collections of recent sediments, covering areas where the environmental variables display a wide range. The resulting equation is then applied to fossil samples to derive the past value of the desired variable.

Mg/Ca

The relative amount of magnesium compared with calcium a certain planktonic foraminifera species incorporates in its test depends on temperature, and the Mg/Ca ratio can therefore also be used as a paleo-thermometer. The calibration equation is derived from laboratory culture studies, based on different species, including *Globigerinoides sacculifer* (25), *Globigerina bulloides*, and *Orbulina universa* (26). Revised calibrations and applications to geological records are being continuously developed (27–31). Another paleo-thermometer derived from elemental ratios in foraminifera tests is the Sr/Ca ratio (32).

Alkenones

Several species of nannoplankton, living in the shallower layer of the ocean (less than ca. 50 m depth), produce organic compounds named long-chain (C_{37}-C_{39}) unsaturated ketones, also known as alkenones. The two species being responsible for most of the alkenone production are *Emiliania huxleyi* and *Gephyrocapsa oceanica*. Although it is still not clear what the function of such compounds is, they have become a very useful tool for the estimation of past seawater temperature. It has been demonstrated (33,34) that the three different varieties of the C_{37} unsaturated compounds ($C_{37:2}$, $C_{37:3}$, $C_{37:4}$), characterized by 2, 3, or 4 double bonds, display abundance variations related to seawater temperature. The alkenone unsaturation ratio (35), $U^{k'}_{37} = [C_{37:2}]/([C_{37:2}] + [C_{37:3}])$, was calibrated to temperature (36) by growing *E. huxleyi* in culture under a range of temperatures. Since then, alkenones have been widely applied in paleoclimatology (37–45).

Other Methods

Other methods commonly used to estimate past ocean temperatures are as follows the modern analogue technique and its variations (46–51); artificial neural networks (52,53); and the $\delta^{18}O$ isotopic composition of planktonic foraminifera (54–58). The study of the isotopic signal stored in corals, although covering shorter time intervals than marine sediment cores, provides excellent temporal resolution archives (down to a few weeks), which allow to analyze seasonal climate variability (59,60).

BIBLIOGRAPHY

1. Keeling, C.D., Whorf, T.P., Wahlen, M., and van der Plicht, J. (1995). Interannual extremes in the rate of rise of atmospheric carbon dioxide since 1980. *Nature* **375**: 666–670.
2. Ganopolski, A. and Rahmstorf, S. (2001). Rapid changes of glacial climate simulated in a coupled climate model. *Nature* **409**: 153–158.
3. CLIMAP. (1976). The surface of the ice-age Earth. *Science* **191**: 1131–1137.
4. Kanaya, T. and Koizumi, I. (1966). Interpretation of diatom thanatocoenoses from the North Pacific applied to a study of core V20-130 (studies of a deep-sea core V20-120, part IV). *Sci. Rep. Tohoku Univ. Ser. 2.* **37**: 89–130.
5. Nigrini, C. (1970). Radiolarian assemblages in the North Pacific and their application to a study of Quaternary sediments in core V20-130. *Memoir of the Geological Society of America* **126**: 139–183.
6. Williams, D.F. and Johnson, W. (1975). Diversity of recent planktonic foraminifera in the southern Indian Ocean and late Pleistocene paleotemperatures. *Quaternary Res.* **5**: 237–250.
7. Imbrie, J. and Kipp, N.G. (1971). A new micropaleontological method for quantitative paleoclimatology: Application to a late Pleistocene Caribbean core. In: *Late Cenozoic Glacial Ages*. K. Turekian (Ed.). Yale University Press, New Haven, CT, pp. 71–181.
8. CLIMAP. (1981). *Seasonal Reconstructions of the Earth's Surface at the Last Glacial Maximum*. Geological Society of America, Map and Chart Series MC-36.
9. CLIMAP. (1984). The Last Interglacial ocean. *Quaternary Research* **21**: 123–224.
10. Mix, A.C. and Morey, A.E. (1996). Climate feedback and pleistocene variations in the atlantic south equatorial current. In: *The South Atlantic: Present and Past Circulation*. G. Wefer, W.H. Berger, G. Siedler, and D.J. Webb (Eds.). Springer-Verlag, Berlin, pp. 503–525.
11. Ortiz, J.D. and Mix, A.C. (1997). Comparison of Imbrie-Kipp transfer function and modern analog temperature estimates using sediment trap and core top foraminiferal faunas. *Paleoceanography* **12**(2): 175–190.
12. Mix, A.C., Morey, A.E., Pisias, N.G., and Hostetler, S.W. (1999). Foraminiferal faunal estimates of paleotemperature: Circumventing the no-analog problem yields cool ice age tropics. *Paleoceanography* **14**: 350–359.
13. Feldberg, M.J. and Mix, A.C. (2002). Sea-surface temperature estimates in the Southeast Pacific based on planktonic foraminiferal species; modern calibration and Last Glacial Maximum. *Marine Micropaleontol.* **44**: 1–29.
14. Niebler, H.S. et al. (2003). Sea-surface temperatures in the Equatorial and South Atlantic Ocean during the Last Glacial Maximum (23-19 ka). *Paleoceanography* **18**: 1069.
15. Moore, T.C., Jr. (1973). Late Pleistocene-Holocene oceanographic changes in the northeastern Pacific. *Quaternary Res.* **3**(1): 99–109.
16. Morley, J.J. (1979). A transfer function for estimating paleoceanographic conditions, based on deep-sea surface distribution of radiolarian assemblages in the South Atlantic. *Quaternary Res.* **12**: 381–395.
17. Pisias, N.G., Roelofs, A., and Weber, M. (1997). Radiolarian-based transfer functions for estimating mean surface ocean temperatures and seasonal range. *Paleoceanography* **12**(3): 365–379.
18. Bjørklund, K.R., Cortese, G., Swanberg, N., and Schrader, H.J. (1998). Radiolarian faunal provinces in surface sediments of the Greenland, Iceland and Norwegian (GIN) Seas. *Marine Micropaleontology* **35**: 105–140.
19. Abelmann, A., Brathauer, U., Gersonde, R., Sieger, R., and Zielinski, U. (1999). Radiolarian-based transfer function for the estimation of sea surface temperatures in the Southern Ocean (Atlantic sector). *Paleoceanography* **14**(3): 410–421.
20. Brathauer, U. and Abelmann, A. (1999). Late quaternary variations in sea surface temperatures and their relationship to orbital forcing recorded in the Southern Ocean (Atlantic sector). *Paleoceanography* **14**(2): 135–148.
21. Cortese, G. and Abelmann, A. (2002). Radiolarian-based paleotemperatures during the last 160 kyrs at ODP Site 1089 (Southern Ocean, Atlantic Sector). *Palaeogeography, Palaeoclimatology, Palaeoecology* **182**(3–4): 259–286.
22. Dolven, J.K., Cortese, G., and Bjørklund, K.R. (2002). A high-resolution radiolarian-derived paleotemperature record for the Late Pleistocene-Holocene in the Norwegian Sea. *Paleoceanography* **17**(4): 1072.
23. Koc-Karpuz, N. and Schrader, H.J. (1990). Surface sediment diatom distribution and Holocene paleotemperature variations in the Greenland, Iceland and Norwegian Sea. *Paleoceanography* **5**(4): 557–580.
24. Zielinski, U., Gersonde, R., Sieger, R., and Fütterer, D. (1998). Quaternary surface water temperature estimations: Calibration of a diatom transfer function for the Southern Ocean. *Paleoceanography* **13**: 365–383.
25. Nürnberg, D., Bijma, J., and Hembelen, C. (1996). Assessing the reliability of magnesium in foraminiferal calcite as a proxy for water mass temperatures. *Geochim. Cosmochim. Acta* **60**: 803–814.
26. Lea, D.W., Mashiotta, T.A., and Spero, H.J. (1999). Controls on magnesium and strontium uptake in planktonic foraminifera determined by live culturing. *Geochim. Cosmochim. Acta* **63**: 2369–2379.
27. Hastings, D.W., Russell, A.D., and Emerson, S.R. (1998). Foraminiferal magnesium in *G. sacculifer* as a paleotemperature proxy in the equatorial Atlantic and Caribbean surface oceans. *Paleoceanography* **13**(2): 161–169.
28. Mashiotta, T.A., Lea, D.W., and Spero, H.J. (1999). Glacial–interglacial changes in Subantarctic sea surface temperature and d18O-water using foraminiferal Mg. *Earth Planetary Sci. Lett.* **170**: 417–432.
29. Nürnberg, D., Müller, A., and Schneider, R.R. (2000). Paleo-sea surface temperature calculations in the equatorial east Atlantic from Mg/Ca ratios in planktic foraminifera: A comparison to sea surface temperature estimates from $U^{k'}{}_{37}$, oxygen isotopes, and foraminiferal transfer function. *Paleoceanography* **15**(1): 124–134.
30. Dekens, P.S., Lea, D.W., Pak, D.K., and Spero, H.J. (2002). Core top calibration of Mg/Ca in tropical foraminifera: Refining paleotemperature estimation. *Geochem. Geophys. Geosyst.* **3**(4): 10.1029/2001GC000200.
31. Lea, D.W., Pak, D.K., Peterson, L.C., and Hughen, K.A. (2003). Synchroneity of tropical and high-latitude Atlantic temperatures over the last glacial termination. *Science* **301**: 1361–1364.

32. Cronblad, H.G. and Malmgren, B.A. (1981). Climatically controlled variation of Sr and Mg in Quaternary planktonic foraminifera. *Nature* **291**: 61–64.

33. Brassell, S.C., Eglinton, G., Marlowe, I.T., Pflaumann, U., and Sarnthein, M. (1986). Molecular stratigraphy: A new tool for climatic assessment. *Nature* **320**: 129–133.

34. Prahl, F.G., Collier, R.B., Dymond, J., Lyle, M., and Sparrow, M.A. (1993). A biomarker perspective on prymnesiophyte productivity in the northeast Pacific Ocean. *Deep-Sea Res. I* **40**: 2061–2076.

35. Prahl, F.G., Mühlhausen, L.A., and Zahnle, D.L. (1988). Further evaluation of long-chain alkenones as indicators of paleoceanographic conditions. *Geochim. Cosmochim. Acta* **52**: 2303–2310.

36. Prahl, F.G. and Wakeham, S.G. (1987). Calibration of unsaturation patterns in long-chain ketone compositions for palaeotemperature assessment. *Nature* **330**: 367–369.

37. Lyle, M.W., Prahl, F.G., and Sparrow, M.A. (1992). Upwelling and productivity changes inferred from a temperature record in the central equatorial Pacific. *Nature* **355**: 812–815.

38. Rostek, F.G. et al. (1993). Reconstructing sea surface temperature and salinity using $\delta^{18}O$ and alkenone records. *Nature* **364**: 319–321.

39. Zhao, M., Rosell, A., and Eglinton, G. (1993). Comparison of two $U^{k}{}_{37}$-sea surface temperature records for the last climatic cycle at ODP Site 658 from the sub-tropical Northeast Atlantic. *Palaeogeography, Palaeoclimatology, Palaeoecology* **103**: 57–65.

40. Ohkouchi, N., Kawamura, K., Nakamura, T., and Taira, A. (1994). Small changes in the sea surface temperature during the last 20,000 years: molecular evidence from the western tropical Pacific. *Geophys. Res. Lett.* **21**: 2207–2210.

41. Sikes, E.L. and Keigwin, L.D. (1994). Equatorial Atlantic sea surface temperature for the last 30 kyr: A comparison of $U^{k'}{}_{37}$, $\delta^{18}O$, and foraminiferal assemblage temperature estimates. *Paleoceanography* **9**(1): 31–45.

42. Schneider, R.R., Müller, P.J., and Ruhland, G. (1995). Late quaternary surface circulation in the east equatorial South Atlantic: evidence from alkenone sea surface temperatures. *Paleoceanography* **10**(2): 197–219.

43. Schneider, R.R., Müller, P.J., Ruhland, G., Meinecke, G., Schmidt, H., and Wefer, G. (1996). Late quaternary surface temperatures and productivity in the East-Equatorial South Atlantic: Response to changes in Trade/Monsoon wind forcing and surface water advection. In: *The South Atlantic: Present and Past Circulation*. G. Wefer, W.H. Berger, G. Siedler, and D.J. Webb (Eds.). Springer-Verlag, Berlin, pp. 527–551.

44. Müller, P.J., Cepek, K., Ruhland, G., and Schneider, R.R. (1997). Alkenone and coccolithophorid species changes in Late Quaternary sediments from Walvis Ridge: Implications for the alkenone paleotemperature method. *Palaeogeography, Palaeoclimatology, Palaeoecology* **135**: 71–96.

45. Rühlemann, C., Mulitza, S., Müller, P.J., Wefer, G., and Zahn, R. (1999). Warming of the tropical Atlantic Ocean and slowdown of thermohaline circulation during the last deglaciation. *Nature* **402**: 511–514.

46. Hutson, W.H. (1980). The Agulhas Current during the late Pleistocene: analysis of modern faunal analogs. *Science* **207**: 64–66.

47. Prell, W.L. (1985). *The Stability of Low Latitude Sea Surface Temperatures: An Evaluation of the CLIMAP Reconstruction with Emphasis on the Positive SST Anomalies*. Rep. TR 025 U.S. Dept. of Energy, Washington, DC, pp. 1–60.

48. Pflaumann, U., Duprat, J., Pujol, C., and Labeyrie, L. (1996). SIMMAX: A modern analog technique to deduce Atlantic sea surface temperatures from planktonic foraminifera in deep-sea sediments. *Paleoceanography* **11**: 15–35.

49. Waelbroeck, C. et al. (1998). Improving past sea surface temperature estimates based on planktonic fossil faunas. *Paleoceanography* **13**(3): 272–283.

50. Hale, W. and Pflaumann, U. (1999). Sea-surface temperature estimations using a Modern Analog Technique with foraminiferal assemblages from Western Atlantic Quaternary sediments. In: *Use of Proxies in Paleoceanography—Examples from the South Atlantic*. G. Fischer and G. Wefer (Eds.). Springer, Berlin, pp. 69–90.

51. Hendy, I.L. and Kennett, J.P. (2000). Dansgaard-Oeschger cycles and the California Current System: Planktonic foraminiferal response to rapid climate change in Santa Barbara Basin, Ocean Drilling Program Hole 893A. *Paleoceanography* **15**: 30–42.

52. Malmgren, B.A. and Nordlund, U. (1997). Application of artificial neural networks to paleoceanographic data. *Palaeogeography, Palaeoclimatology, Palaeoecology* **136**: 359–373.

53. Malmgren, B.A., Kucera, M., Waelbroeck, C., and Nyberg, J. (2001). Comparison of statistical and artificial neural network techniques for estimating past sea-surface temperatures from planktonic foraminifer census data. *Paleoceanography* **16**(5): 520–530.

54. Emiliani, C. (1955). Pleistocene temperatures. *J. Geol.* **63**: 538–578.

55. Shackleton, N.J. and Opdyke, N.D. (1973). Oxygen isotope and palaeomagnetic stratigraphy of equatorial Pacific Core V28-238: Oxygen isotope temperatures and ice volumes on a 10^5 year and 10^6 year scale. *Quaternary Res.* **3**: 39–55.

56. Broecker, W.S. (1986). Oxygen isotope constraints on surface ocean temperatures. *Quaternary Research* **26**: 121–134.

57. Mulitza, S., Dürkoop, A., Hale, W., Wefer, G., and Niebler, H.S. (1997). Planktonic foraminifera as recorders of past surface-water stratification. *Geology* **25**: 335–338.

58. Bemis, B.E., Spero, H.J., Bijma, J., and Lea, D.W. (1998). Reevaluation of the oxygen isotopic composition of planktonic foraminifera: Experimental results and revised paleotemperature equations. *Paleoceanography* **13**(2): 150–160.

59. Felis, T. et al. (2000). A coral oxygen isotope record from the northern Red Sea documenting NAO, ENSO, and North Pacific teleconnections on Middle East climate variability since the year 1750. *Paleoceanography* **15**: 679–694.

60. Nyberg, J., Winter, A., Malmgren, B., and Christy, J. (2001). Surface temperatures in the eastern Caribbean during the 7th century AD average up to 4 °C cooler than present. *Eos Trans. AGU* **82**(47).

PHYSICAL OCEANOGRAPHY

Pacific Fisheries Environmental
Laboratory, NOAA

A great deal of the patterns and fluctuations observed in our living marine resources are attributable to the impact of physical processes in the environment on marine ecosystems and their components. For this reason, PFEL places a strong emphasis on research that examines the role of the physical environmental variability on marine ecosystems in general and commercially important fish stocks specifically. The objectives of the physical oceanography task are:

- perform research on the temporal and spatial scales of environmental variability in eastern boundary current systems in relation to other marine ecosystems
- provide environmental input to SWFSC research programs, particularly the coastal groundfish program
- provide high quality marine information to the research community.

Research is performed at PFEL which integrates environmental and biological data sets, investigating the linkages between environmental variability and fluctuations in the abundance and distribution of marine populations on a continuum of scales (from global, basin-wide spatial scales to the scale of local upwelling centersand on time scales from decades down to days). Physical oceanography research is directed to:

- large-scale climatic variability
- environment/recruitment relationships in eastern boundary current ecosystems
- mesoscale (smaller scale) processes affecting coastal circulation and fisheries recruitment

Examples of research on the large-scale variability include studies of recurring temperature changes off the west coast of the U. S. and their effects on groundfish recruitment, and the investigation of environmental changes in the California Current region associated with recent ENSO events. Much of the mesoscale research focuses on relating environmental variability on day-to-year and 5–100 nm scales to patterns and events in the life history of groundfish (e.g., recruitment success). The physical oceanography research program is linked closely to those of the other tasks at PFEG and to research programs at the other SWFSC labs. PFEG scientists also are involved in numerous cooperative studies with oceanographers and

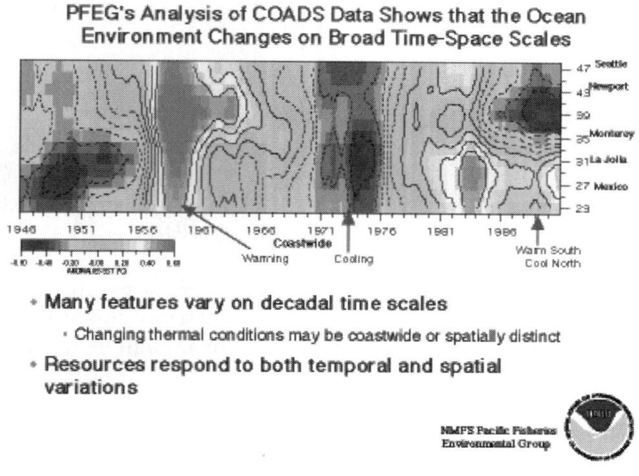

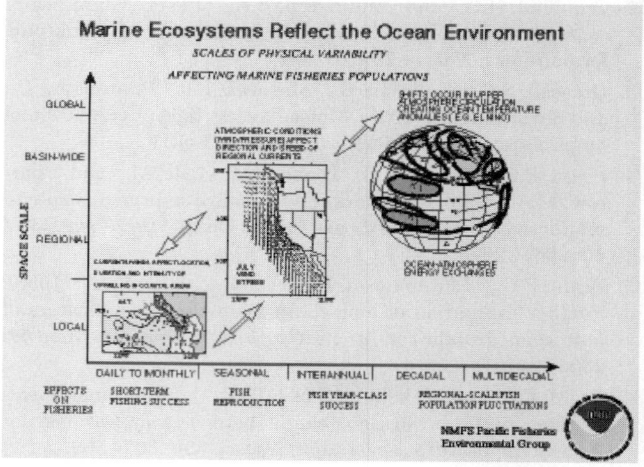

fisheries scientists at many federal and state government, academic and privately supported research institutions.

Expertise in physical oceanography at PFEG and the linkages to the Navy's Fleet Numerical Meteorology and Oceanography Center (FNMOC) and Naval Postgraduate School, as well as numerous other government, academic, and private research facilities, has historically meant that this task serves regionally, nationally and internationally as a resource to other ocean scientists. Within the SWFSC, many cooperative research programs have been developed and planned. As an example, the task provides physical oceanographic expertise to the Tiburon Laboratory Rockfish Recruitment surveys each spring, to relate ocean variability off central California to rockfish recruitment. PFEG physical oceanographers are asked frequently to attend workshop and present seminars as experts on environmental-fishery linkages, and represent SWFSC, NMFS and NOAA on numerous committees and working groups.

COASTAL WATER POLLUTANTS

Upadhyayula V. K. Kumar
Choa Chu kang Ave-4
Singapore

POLLUTANT INPUT INTO THE MARINE ENVIRONMENT

Among all the diversity of human activities and sources of pollution, we can distinguish three main ways that pollutants enter the marine environment:

- direct discharge of effluents and solid wastes into the seas and oceans (industrial discharge, municipal waste discharge, coastal sewage, and others);
- land runoff into the coastal zone, mainly from rivers;
- atmospheric fallout of pollutants transferred by the air mass onto the sea surface.

Certainly, the relative contribution of each of these channels to the combined pollution input into the sea will be different for different substances and in different situations. At present, the signs and consequences of human activity can be found everywhere on Earth. One of the typical features of marine pollution is global spreading of a number of contaminants. Numerous data undoubtedly indicate the existence of a large-scale (global) field of background contamination of the hydrosphere.

Another important feature of marine pollution is the existence of increased pollution levels in the enclosed seas and coastal waters, compared with the open ocean.

Contamination levels also increase during the transition from the southern parts of all oceans to the north, where the main industrial centers and main pollution sources are concentrated. Besides the general distribution pattern of pollution sources, there are two other factors explaining the relative stability of global pollutant distribution in the world ocean: the relative confinement of large-scale water circulation within the limits of each hemisphere and the predominance of the zonal transport of the traces in the atmosphere.

Another distinctive and repeatedly registered feature of the general picture of contaminant distribution in the marine environment is their localization at the water–atmosphere and water–bottom sediment boundaries. Practically everywhere and for all trace components (primarily for oil hydrocarbons), their concentrations are considerably (usually hundreds and thousands of times) higher in the surface microlayer of water and in the upper layer of bottom sediments. These boundaries provide the biotopes for the communities of hyponeuston and benthos, respectively.

The existence of elevated levels of contaminants in zones of high bioproductivity is extremely alarming ecologically. These zones include the water layer up to 100 m from the water surface (photic layer) and boundaries of natural environments (water–atmosphere and water–bottom sediment, as previously mentioned), as well as enclosed seas, estuaries, and coastal and shelf waters. In particular, in shelf and coastal zones, which occupy 10% of the world ocean surface and less than 3% of its volume, the most intense processes of bioproduction, including the self-reproduction of the main living resources of the sea, take place. The main press of anthropogenic impact is also concentrated here.

The number and diversity of pollution components is growing as well. Contaminants that are globally distributed are combined here with hundreds and thousands of ingredients of local and regional distribution. Most of these substances are wastes and discharges from different local industries and activities.

Based on the extreme diversity of marine pollution components, the variety of their sources, the scales of distribution, and the degree of hazards, these pollutants can be classified in different ways, depending on their composition, toxicity, persistence, sources, volumes, and so on.

TYPES AND FORMS OF WATER POLLUTION

Water pollution is attributed to various sources and can be broadly divided into three categories: domestic, industrial, and radioactive wastes, which can be categorized in the following forms: (1) thermal pollution; (2) the addition of pathogenic organisms, leading to public health hazard; (3) oil pollution; (4) the addition of inert, insoluble mineral material; (5) the addition of biodegradable organic material that will result in the depletion or complete removal of dissolved oxygen; (6) toxicity due to the presence of synthetic organic compounds and salts of heavy metals; (7) enhanced eutrophication; (8) acid deposition or discharges; and (9) radioactivity.

Domestic Sewage

The objectionable features of domestic sewage are its population of pathogenic bacteria; its high content of organic matter, which gives it a high oxygen demand; its nutrient content, which gives it the potential of supporting large populations of algae and other plants, which in turn may be as objectionable as the sewage itself; and the obvious aesthetics. The first and last of these can be overcome by proper treatment in biological sewage treatment plants, but the effluent from these plants is generally rich in nutrients. So, the effluent discharged into nearby inshore waters may result in large objectionable weed crops and other plants. An additional difficulty of disposing of domestic sewage and even the effluents from sewage treatment plants in the sea is that the density of the sewage is invariably less than that of seawater. Thus, the sewage tends to float on the surface unless introduced in a region of strong currents, will mix with seawater, and is diluted only slowly.

Industrial Wastes

These can be highly varied in composition and present a variety of special problems. The wastes may be toxic to plants or animals. They may be highly acidic or basic. If they contain large quantities of organic matter, their BOD may be objectionably high. Surface-active ingredients such as detergents may cause objectionable foaming or disrupt normal bacterial populations. The settable chemicals tend to settle to the seabed, react with mineral content there, and change the entire infrastructure of the seabed itself, which is causing accumulation of heavy metals in the sediments of several seas.

Pathogens

A variety of pathogenic organisms, including viruses, bacteria, protozoa, and parasitic worms, exist in seawater and can cause diseases in plants, animals, and people. Impacts include human illness, seafood contamination, and recreational beach closures. Pathogens are discharged to coastal waters through both point and nonpoint sources, especially from insufficiently treated sewage that is released from septic systems on land and on ships and from agriculture and storm water runoff. Higher concentrations tend to occur after storms and related overflow of sewer systems, making it difficult to interpret trends and temporal fluctuations.

Nutrients

Important parameters for monitoring nutrient pollution in coastal waters include the following: nitrogen and phosphorus concentrations, maximum bottom dissolved oxygen levels, extent and duration of anoxic and hypoxic conditions, extent of SAV, chlorophyll-*a* concentrations, turbidity, and duration and extent of algal blooms (by type). Some parameters are important in assessing the vulnerability of an area to pollutants, such as nitrogen and phosphorus, or in determining baseline conditions of the area.

Oil

Petroleum residues can contaminate marine and coastal waters through various routes: accidental oil spills from tankers, pipelines, and exploration sites; regular shipping and exploration operations, such as exchange of ballast water; runoff from land; and municipal and industrial wastes. Although the main global impact is due to tar balls that interfere with recreational activities at beaches, the impact of petroleum hydrocarbon concentrations on marine organisms in the neuston zone in the ocean—particularly fish eggs and larvae—requires more attention. Large-scale oil spills from tankers often make the headlines; yet, nonpoint sources, such as regular maritime transportation operations and runoff from land, are actually considered the main contributors to the total oil discharge into the ocean, although conclusive statistics are lacking. Runoff and routine maintenance of the oil infrastructure, it is estimated, account for more than 70% of the total annual oil discharge into the ocean. Both the number and amount of *accidental* oil spills have been monitored and seem to have declined in the past decade. A single catastrophic event can, however, influence the statistics significantly *and* have a localized, yet tremendous impact on the ecosystem.

Tables 1 and 2 show the variety of oil pollution sources and give expert estimates of the scales of distribution and impact of each of these sources on the marine environment. Even though these estimates can vary up to one to two orders of magnitude (especially for natural oil sources, atmospheric input, and river runoff), the main anthropogenic flows of oil pollution into the marine environment come from land-based sources (refineries, municipal wastes, and river runoff) and transportation activity (tanker oil transportation and shipping). Polycyclic aromatic hydrocarbons (PAHs),

Table 1. Sources and Scale of Oil Pollution Input Into the Marine Environment

	Environment		Scale of Distribution and Impact		
Types and Source of Input	Hydrosphere	Atmosphere	Local	Regional	Global
Natural:					
Natural seeps and erosion of bottom sediments	+	–	+	?	–
Biosynthesis by marine organisms	+	–	+	+	+
Anthropogenic:					
Marine oil transportation (accidents, operational discharges from tankers, etc.)	+	–	+	+	?
Marine nontanker shipping (operational, accidental, and illegal discharges)	+	–	+	?	–
Offshore oil production (drilling discharges, accidents, etc.)	+	+	+	?	–
Onland sources: sewage waters	+	–	+	+	?
Onland sources: oil terminals	+	–	+	–	–
Onland sources: rivers, land runoff	+	–	+	+	?
Incomplete fuel combustion	–	+	+	+	?

Note: +, –, and ? mean, respectively, presence, absence, and uncertainty of corresponding parameters.

Table 2. Estimates of Global Inputs of Oil Pollution Into the Marine Environment (Thousands Tons/Year of Oil Hydrocarbon)

Source	1973*	1979**	1981*	1985***	1990***
Land-based sources:					
Urban runoff and discharges	2,500	2,100	1,080 (500–1,250)	34%	1,175 (50%)
Coastal refineries	200	60	100 (60–600)	–	–
Other coastal effluents	–	150	50 (50–200)	–	–
Oil transportation and shipping:					
Operational discharges from tankers	1,080	600	700 (400-1,500)	45%	564 (24%)
Tanker accidents	300	300	400 (300–400)	–	–
Losses from nontanker shipping	750	200	320 (200–600)	–	–
Offshore production discharges	80	60	50 (40–60)	2%	47 (2%)
Atmospheric fallout	600	600	300 (50–500)	10%	306 (13%)
Natural seeps	600	600	200 (20–2,000)	8%	259 (11%)
Total discharges	**6,110**	**4,670**	**3,200**	**100%**	**2,351**

especially benzo(a)pyrene, enter the marine environment mostly from atmospheric deposition.

Table 2 illustrates the general trend of declining total input of oil pollution into the world ocean over the years. The global situation reflected in this table certainly may differ at the regional level, which depends on natural conditions, the degree of coastal urbanization, the population density, industrial development, navigation, oil and gas production, and other activities. For example, in the North Sea, offshore production input reached 28% of the total input of oil pollution in last decade, instead of the "modest" 2% on the world scale shown in Table 2, which equaled the annual input of more than 23,000 tons of oil products at the background of their general changeable flow of 120,000–200,000 tons a year in the North Sea. One can expect similar situations in other regions of intensive offshore oil and gas development, for example, in the Gulf of Mexico, Red Sea, Persian Gulf, and Caspian Sea. The persistent pollution in oil production areas in the Caspian Sea or the amounts of annual discharges (about 40 million tons of produced waters polluted by oil products) during offshore drilling in the Gulf of Mexico. At the same time, no reliable balance estimates exist for these regions.

The continental shelf of the Gulf of Mexico is also distinctive for intense seepage of natural liquid and gaseous hydrocarbons, which can lead to formation of oil slicks and tar balls on the sea surface, which makes assessing and identifying anthropogenic oil pollution more difficult. In any case, the input of oil hydrocarbons from natural sources into the Gulf of Mexico is larger than in many other areas.

In the Baltic Sea, the Sea of Azov, and the Black Sea, the leading role in oil input most likely belongs to land-based sources, which are dominated by river inflow. The Danube River alone annually brings about 50,000 tons of oil to the Black Sea, half of the total oil input of about 100,000 tons.

Observations in the Caribbean basin, where annually up to 1 million tons of oil enter the marine environment, showed that about 50% of this amount came from tankers and other ships. In the Bay of Bengal and the Arabian Sea, oil pollution inputs from tanker and other ship discharges equal, respectively, 400,000 tons and 5 million tons of oil a year. The most intense tanker traffic exists in the Atlantic Ocean and its seas, which accounts for 38% of international maritime oil transportation. In the Indian and Pacific Oceans, this portion is, respectively, 34% and 28%.

Enforcing stricter requirements on activities accompanied by oil discharges led to a global decline of oil pollution inputs in the marine environment mentioned before. A number of dramatic events show the vulnerability of making an optimistic prognosis about decreasing oil pollution at the regional and global levels. For instance, catastrophic large-scale events took place in the Persian Gulf during and after the 1991 Gulf War. Between 0.5 and 1 million tons of oil were released into coastal waters. Besides, products of combustion of more than 70 million tons of oil and oil products were emitted into the atmosphere. Another large-scale accident occurred in Russia in September–November 1994. About 100,000 tons of oil were spilled on the territory of the Komi Republic, which threatened to cause severe oil pollution for the Pechora River basin and, possibly, Pechora Bay.

It must be remembered that catastrophes, in spite of the obvious consequences and all the attention they attract, are inferior to other sources of oil pollution in their scales and degree of environmental hazard. Land-based, oil-containing discharges and atmospheric deposition of products of incomplete combustion can accordingly give 50% and 13% of the total volume of oil pollution input into the world ocean (see Table 2). These diffuse sources continuously create relatively low but persistent chronic contamination across huge areas. Many aspects of the chemical composition and biological impacts of these contaminants remain unknown.

Persistent Organic Pollutants

Persistent organic pollutants (POPs) consist of a number of synthetic compounds, including industrial polychlorinated biphenyl's (PCBs); polychlorinated dioxins and furans; and pesticides, such as DDT, chlordane, and heptachlor, that do not exist naturally in the environment. A number of POPs often persist in the environment and accumulate through the food chain or in the sediment to a toxic level that is directly harmful to aquatic organisms and humans.

The marine environment collects contaminants from the air, but also from ocean currents, rivers discharging into the ocean, and sea ice that transports POP-laden particles.

Figure 1 presents some examples of contaminant levels in seawater. Hexachlorocyclohexane dominates the picture, except for Russian waters where PCB levels are high, up to 15 nanograms per liter in the Kara Sea. These high levels seem to mirror the high input of PCBs from Russian rivers.

Levels in seawater can also be used to shed light on the mechanisms that transport contaminants to the Arctic. Detailed measurements in the Bering and Chukchi Seas show that hexachlorocyclohexane levels in the water increase along a south–north gradient, which has been suggested as evidence for a cold-condensation theory; those semivolatile contaminants condense out of the atmosphere as temperatures drop. Less volatile contaminants, such as PCBs, DDT, and chlordane, were present at lower levels in the Bering and Chukchi Seas than in more temperate latitudes.

Concentrations of organic contaminants in Arctic marine sediments are, in general, extremely low compared with freshwater sediments, and are ten to a hundred times lower than in the Baltic Sea. The most apparent geographic trends are that concentrations of PCBs, hexachlorocyclohexane, and hexachlorobenzene are higher closer to the shore along the Norwegian coast than in the open sea. They are also higher in gulfs and river mouths along the Russian coast and around Svalbard.

Heavy Metals

Heavy metals exist naturally in the environment, and it is sometimes difficult to distinguish variations developing from anthropogenic inputs and those from the

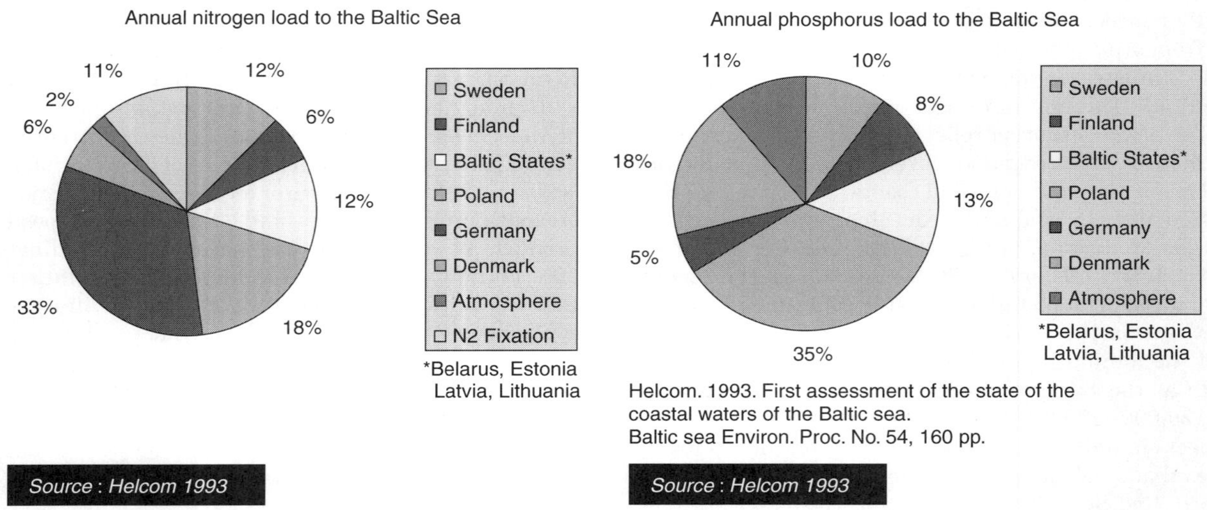

Figure 1. Nutrient release to coastal waters taking the Baltic Sea as an example.

natural hydrologic cycle and the atmosphere. Among the trace metals commonly monitored are cadmium, copper, mercury, lead, nickel, and zinc. When they accumulate through the food chain at moderate to high concentrations, some of these metals can affect the human nervous system.

The marine environment receives heavy metals from atmospheric deposition, river runoff, and local pollution. The relative importance of these sources will differ between regions. For example, rivers carrying metal-laden sediments deposit almost their entire load in the shelf seas, and only a minor portion reaches the deep ocean. Natural sources of metals are important and, in many cases, it is found that they are the main source to the marine environment.

Mining has contaminated ocean waters with several heavy metals. One documented example is in the fjord outside the Black Angel zinc mine in Greenland, where the levels of lead in the bottom water are up to 200 micrograms of lead per kilogram of water. These high lead levels are also reflected in seaweed, blue mussels, prawns, and in some fish; in capelin, lead levels are up to 5 micrograms per gram in the bone. However, no one has been able to document any biological effects in the fish. Cadmium levels in the water are also high, up to 2.5 micrograms per kilogram of water; but in contrast to lead, the animals in the fjord have cadmium levels close to background.

The cryolite mine in Ivittuut in southern Greenland has also contaminated the nearby water. Lead levels of 18 micrograms per kilogram of water have been measured. At Strathcona Sound in northern Baffin Bay, a lead-zinc mine has released lead, making concentrations in the fjord water one to two orders of magnitude higher than background concentrations in the open ocean. Some of the lead has also been taken up by seaweed and crustaceans. Outside a lead-zinc mine in east Greenland, shorthorn sculpins also have elevated levels of lead, whereas the fish outside the cryolite mine on southern Greenland have not been affected.

The mines at Ivittuut and Strathcona Sound have also contaminated their respective fjords with cadmium, but the levels are much lower than those outside the Black Angel mine. At these sites, the cadmium is not affecting the local sediment, nor are elevated levels found in nearby plants and animals.

Metal levels in Arctic Ocean water away from local sources are generally similar to global background levels. Today's global lead concentrations in oceans are generally more than ten times higher than those in prehistoric times. The levels are consistently higher in surface waters than in deeper layers. One might expect the lead levels in the upper Arctic sediments to mirror this increased long-distance transport, but this does not seem to be the case.

Recent seawater analyses from Pechora Bay and Kara and Laptev Seas show very high lead levels, ranging from 0.16 to 0.5 micrograms per kilogram of water. However, these data require confirmation before any conclusions are drawn.

Filter-feeders such as mussels take up lead from sediment particles. The concentration increases slightly with increasing shell length, indicating a moderate accumulation as the mussel ages. However, lead levels are low in crustaceans as well as in fish. The highest levels, 0.05 micrograms per gram of liver, have been recorded in Orkdalsfjorden in Norway.

Lead does not seem to accumulate in fish-eating birds or in marine mammals. In general, levels in marine mammals are low.

Cadmium Levels are High in Marine Biota. Cadmium levels in seawater fall within what could be considered natural background levels. Moreover, there is no indication from sediments that the levels have increased from preindustrial times, nor have temporal trends been detected.

An interesting phenomenon relating to cadmium is that its concentration increases farther away from the coast,

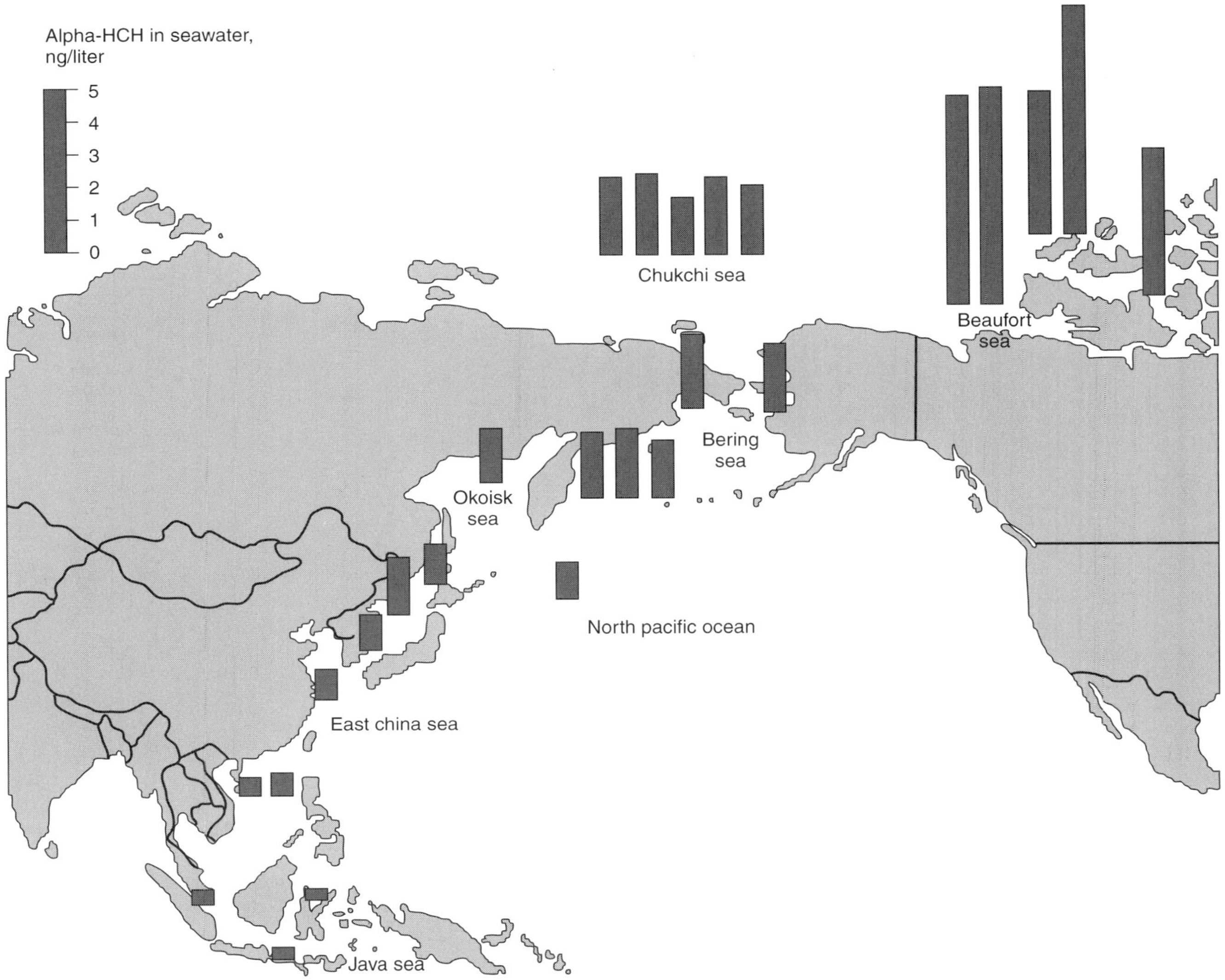

Figure 2. Alpha-hexachlorocyclohexane concentration in seawater increases from south to north, illustrating the cold-condensation effect (1).

which is probably connected with the change in salinity of the water. The result is that cadmium levels in both plants and animals are higher in the open ocean than in the inner region of large fjords, even when there are local sources contaminating the water. Cadmium accumulates with age in mussels and crustaceans. In general, the levels in crustaceans are higher than global background levels but show large variations.

Mercury levels are high and may be increasing. Several sets of data indicate that mercury levels are higher in the upper layers of Arctic marine sediments than in the layers representing preindustrial inputs see the upper right diagram of Fig. 3. Mercury is enriched even in the marine sediments taken at the North Pole. Natural processes may have caused these profiles, but they could indicate that human activities have increased the environmental mobility of existing stores of mercury.

Radioactive Wastes

These can somewhat be divided into high and low level wastes, depending on their activity. Radioactive wastes are characterized by losing their radioactivity with time. Some nuclides lose it quickly; others very slowly. A second consideration is that radioactive elements will enter the biological cycle and therefore the food web. High level radioactive wastes pose a complex problem in their disposal; the low level has been and is being disposed of directly in the sea. Reprocessing plants have added radionuclides to the sea.

Spent nuclear fuel is often processed to recover plutonium. Water used in reprocessing contains a mix of different radionuclides, and some of this contaminated water has been released routinely into the sea. In Europe, three reprocessing plants are relevant to the Arctic because of transport of radionuclides by ocean

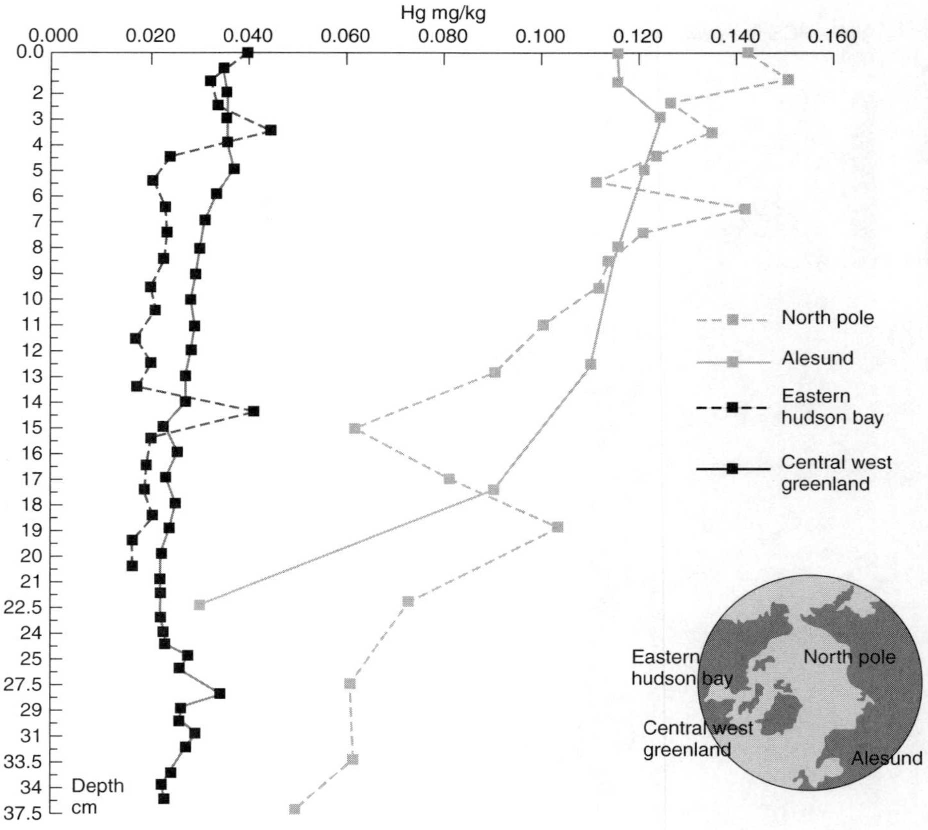

Figure 3. Mercury concentration at different depths in marine sediment cores (1).

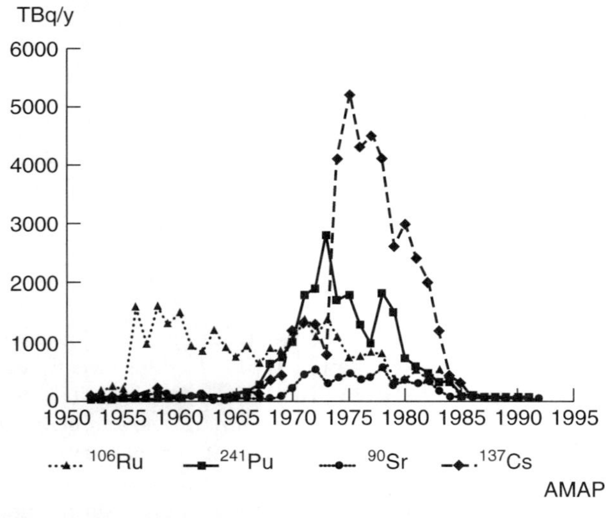

Figure 4. Discharges of beta-emitters from the Sellafield nuclear reprocessing plant, 10^{12} bequerels per year.

currents: Sellafield (formerly Windscale) in Cumbria on the northwest coast of England; La Hague near Cherbourg, France; and Dounreay in northeast Scotland. Sellafield has been the most important source of radionuclides to the Arctic marine environment because of the scale of its discharge. The effluent has been released into the sea and carried north by ocean currents. The releases, which started in 1952, are well documented. Figure 4 indicates the rates of liquid discharges from 1952 to 1992. Cesium-137 dominates. The peak of the release for most radionuclides was in the mid-to-late 1970s.

Underwater Weapon Tests Have Contaminated Chernaya Bay. Chernaya Bay is a fjord inlet, connected to the Barents Sea, on the southwestern coast of Novaya Zemlya. The former Soviet Union used the bay to conduct underwater tests of nuclear bombs in 1955 and 1957 and in the vicinity of the bay in 1961. As a result of these detonations, the bottom sediments of the bay are contaminated by elevated levels of radioactive plutonium and cesium, as well as other radioactive isotopes. However, the mobility of radionuclides in sediment is low and may at present cause only insignificant exposure of people. Exposure of biota is unknown.

Today, the inventory of plutonium in Chernaya Bay is similar to other sites of major plutonium contamination, such as the most contaminated areas of Bylot Sound (where a B-52 bomber crashed) and the Irish Sea in the vicinity of the Sellafield reprocessing plant.

Three underground nuclear detonations were carried out by the United States on Amchitka Island in the Aleutian Islands in 1965, 1969, and 1971. These detonations caused radioactive contamination of deep groundwater and rock around the shot cavities. Long-term monitoring activity is planned for this site to 2025. In 1996, aboveground radioactive contamination was detected at the site.

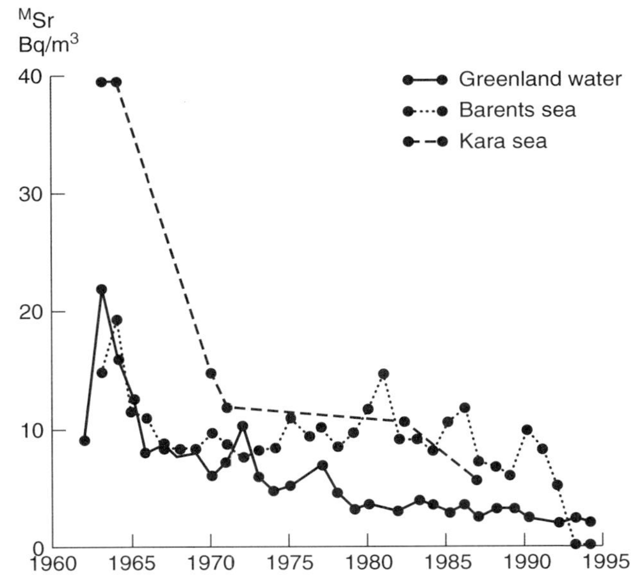

Figure 5. Time trends of activity concentrations of strontium-90 in seawater.

A major direct input of radionuclides into the marine environment has been from European nuclear reprocessing plants, particularly Sellafield on the shore of the Irish Sea. Currents transport the material along the Norwegian coast and into the Arctic Ocean. After 6 to 8 years, some of the contamination leaves the Arctic by way of the East Greenland Current, but much of it stays in the Arctic Basin much longer.

Environmental radiocesium levels have been measured since the early 1970s. As can be seen in Fig. 6, the releases of cesium-137 from Sellafield are virtually mirrored in the levels found in the Barents Sea after a transport time of 4 to 5 years. The peak in concentration in the early 1980s is probably the highest level that has ever occurred in that area of the ocean. The Chernobyl accident in 1986 added cesium to the Arctic Ocean and continues to do so via outflow from the Baltic Sea. Figure 6 shows the recent levels of cesium-137 in seawater around the Arctic. Strontium-90 has been measured in surface seawater collected around Greenland and the Barents Sea; see Fig. 5. During the past 35 years, levels in the waters around Greenland have decreased, approximately half was removed or decayed every 13.5 years. This value is probably representative of the Arctic Ocean as a whole. The highest levels of cesium-137 in people were recorded in the mid-1960s; see the Fig. 6. For the following 20 years, the human body burden decreased by a factor of 3 to 7. However, in 1986, the Chernobyl fallout changed the trend in areas directly affected by the accident.

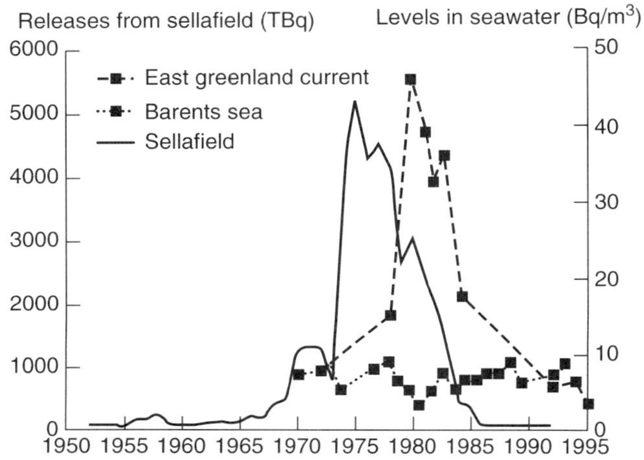

Figure 6. Releases of cesium-137 from Sellafield nuclear reprocessing plant (10^{12} bequerels per cubic meter).

Assessment of Receiving Waters

The effects of the ocean waste disposal are the result of a complex relation between two variables, concentration and time. The effect of the oceanic environment on the effluent is of critical significance. After discharge to the ocean, the effluent experiences changes in its physical and microbiological properties, which vary, each point as a function of time. An accurate prediction of pollution conditions or environmental impact depends on knowledge of the oceanographic conditions in the area. These conditions vary in time and space. A significant and representative quantitative judgement requires refined statistical analysis. A statistical description of receiving coastal waters should be based on adequate observation of an entire area for a sufficiently long period of time (at least 1 year). The factors operating should be recorded simultaneously to provide a comprehensive picture of the physical and microbiological properties of the area. The data collected should enable probability distributions of the variables to be derived to select a coherent and suitable set of design parameters. Continuous and periodic records should be taken to cover all typical oceanographic conditions at stations strategically located at different sites and should record the following:

(1) Currents (direction and speed) distributed in time/space to permit a comprehensive study of coastal water circulation patterns.
(2) A continuous record of tides and winds in the area.
(3) The density field obtained from time and space distribution of salinity and temperature enabling the stratification conditions of the coastal waters to be determined.
(4) The waves recorded periodically, covering the most critical conditions and enabling a probability distribution of the wave characteristics (height, period, and direction) to be derived for structural design.
(5) Periodic dispersion and diffusion experiments using dye tracers.
(6) Simultaneously with the oceanographic survey a program on bacterial concentration decrease should be carried out, covering typical oceanographic conditions. The experiments should be made *in situ*,

preferably in the existing continuous sewage field. Simultaneously sampling for bacteria and dye tracer direction will enable bacterial die-off rates and dilution to be investigated and the derivation of bacterial concentration decrease from both these effects. During concentration decrease experiments, the water temperature, solar radiation, and other climatic data should be collected.

Mainly tides and winds, to a lesser controlled extent by the density field, govern coastal currents near the shore. A comprehensive study of coastal circulation requires a continuous record of winds and tides and simultaneous measurements of vertical profiles of temperature, salinity, and currents. A detailed vector statistical analysis of the current field is an important requirement. The seasonal or climatological associations between oceanographic and meteorological data and *in situ* bacterial assays are of paramount importance; their interactions determine major design parameters. Statistical analysis of the sewage field is required to predict the seasonal variations in initial dilution and final height of the rise. This analysis should be based on simulations of alternative diffuser discharges at various density stratification conditions.

Oceanic Processes

To assess the oceanic equivalent dilution factors, the following formulations may be used.

Initial Dilution. The initial dilution, RI, and terminal rise height, Y_m, were estimated, assuming that the diffuser is a finite line source of buoyancy flux only, by the equation

$$\mathrm{RI} = \mathrm{FI} \cdot Y_m$$

where $\mathrm{FI} = f(F, \theta)U/q$; in which $f(f, \theta)$ is an experimental factor, which depends on θ and a type of Froude number $F = u3/B$. The current and density stratification are the oceanographic variables that directly interface in the initial dilution. We may consider a critical stratification condition in which this interface extends to a constant depth and the dilution layer increases linearly with the bottom slope according to a relation of the type $Y_m = a + b(x - x_0)$, where x is the distance from a horizontal plume perpendicular to the shore. The subsequent dilutions RM, represent a minor part of the overall receiving water reductions. Thus the total physical dilution can be evaluated as $\mathrm{RF} = \mathrm{RI} \cdot \mathrm{RM}$,

$$\mathrm{RF} = \mathrm{FI}(1 + \mathrm{FM}x)[a + b(x - x_0)]$$

All soluble pollutants experience this composite dilution factor. Small particulate matter will also disperse proportionately. Floatable matter may disperse to a lesser extent, remaining visually detectable and liable to be carried to shore by currents and surface winds. The removal of this material is required, and an initial dilution of 100 is required to provide sufficient emulsification for these materials.

Dilution Equivalent to Bacterial Concentration Decreases. Most experiments on bacterial concentration decrease have been shown to fit Chick's law very satisfactorily. $C(t) = C_0 \cdot 10 - t/T90$, where C_0 = coliform concentration at the origin, and T_{90} = time required for 90% reduction. Thus the dilution equivalent to the concentration decrease may be computed directly from

$$\mathrm{RB} = C_0/C(x) = 10'x/X_{90}$$

where $X_{90} = u \cdot T_{90}$ = distance required for the coliform concentration reduction of 90%.

Dilution Equivalent to Sedimentation. The disappearance rate of coliforms due to sedimentation of solids depends on the degree of removal in the treatment plant. Data collected by experts revealed the fact that T_{90} values decrease as the effluent TSS increases. Therefore, sedimentation effects are already incorporated in the field experimental results for the concentration decrease.

Dilution Equivalent to Treatment. The relation between removal rates and corresponding factor RT is $\mathrm{RT} = 100/[100 - T\ (\%)]$.

Dilution Equivalent to Disinfection. The dilution equivalent to disinfection is evaluated by the equation $\mathrm{RD} = 1/\mathrm{rd}$, where rd is the bacteria reduction factor. Chlorination of less treated sewage produces organic chlorine compounds, which are toxic and deleterious to the environment. Therefore, chlorination may compromise the already recognized ecologically beneficial effects of enrichment of coastal waters by the supply of nutrients and organic matter from those effluents.

Overall Equivalent Dilution. Assuming that $\mathrm{RS} = 1$, the overall equivalent dilution can be evaluated by the equation

$$\mathrm{R_{total}} = \mathrm{RT} \cdot \mathrm{RD} \cdot \mathrm{RI} \cdot \mathrm{RM}(x) \cdot \mathrm{RB(x)}$$

$$\mathrm{R_{total}} = \mathrm{RR} \cdot f(F, \theta)(bu/q)[a + b(x - x_0)] \cdot [1 + (KL/b_o)x] \cdot 10'x/u \cdot T_{90}$$

For a given set of parameters, $R_{\mathrm{total}} = \mathrm{CE/CP}$, Q and the oceanic parameters are $u(u, \theta)$, T_{90} and KL are the total equivalent approximation, a linear function of the diffuser length and an exponential quadratic function of the outfall length.

The Effect of Current. Separating the current factors including the current speed u from the above equation, a dilution function may be defined as $Fu = u \cdot 10'x/uT_{90}$, which represents the effect of current on overall oceanic dilution. The figure given shows curves of Fu as a function of u for the various values of the parameter $u_{90} = x/T_{90}$, that is, the velocity for a 90% concentration decrease. It may be seen that the curves show "inflections" that have been connected by the line. The inflection increases as u_{90} decreases or as T_{90} increases. This inflection divides the graph into two domains:

The initial dilution domain (linear)

The concentration decrease dilution factor dilution factor domain (exponential).

For strong currents, the initial dilution is the dominant factor, and for a given T_{90}, the only way to increase Fu is by initial dilution mixing depth, which can be attained by increasing the outfall length. For light currents, the dilution factor equivalent to die-off dominates.

Thus, the current data is collected, as it is useful in the design in

1. analysis of coastal circulation hydrodynamics;
2. prediction of initial dilution;
3. prediction of far field dilution and transport;
4. Prediction of waste field impaction probability.

Thus, a comprehensive oceanographic study is necessary to describe the characteristics of the receiving waters. Therefore, an adequate program of oceanographic investigation *in situ*, including bacterial concentration decrease, is obligatory.

The existence of elevated levels of contaminants in zones of high bioproductivity is extremely alarming ecologically. These zones include the water layer up to 100 m from the water surface (photic layer) and boundaries of natural environments (water–atmosphere and water–bottom sediment, as previously mentioned) as well as enclosed seas, estuaries, and coastal and shelf waters. In particular, in shelf and coastal zones, which occurs in 10% of the world ocean surface and less than 3% of its volume, the most intense processes of bioproduction, including self-reproduction of the main living resources of the sea, take place. The main press of anthropogenic impact is also concentrated here.

EFFECTS OF POLLUTANTS ON MARINE HABITAT

To estimate the hazard of different pollutants, we should take into account not only their hazardous properties, but other factors, too. These include the volumes of their input into the environment, the ways and scale of their distribution, the patterns of their behavior in water ecosystems, their ability to accumulate in living organisms, the stability of their composition, and other properties, such as the extreme diversity of marine pollution components, the variety of their sources, the scales of distribution, and the degree of hazards. Pollutants can be classified in different ways, depending on their composition, toxicity, persistence, sources, volumes, and so on.

To analyze large-scale pollution and its global effects, it is common to distinguish groups of the most widespread pollutants, which include chlorinated hydrocarbons, heavy metals, nutrients, oil hydrocarbons, surface-active substances, and artificial radionuclides. These substances form the so-called background contamination that exists now any place in the hydrosphere.

Depending on the type of impact on water organisms, communities, and ecosystems, the pollutants can be grouped in the following order of increasing hazard:

- substances that cause mechanical impacts (suspensions, films, solid wastes) that damage the respiratory organs, digestive system, and receptive ability;
- substances that provoke eutrophic effects (e.g., mineral compounds of nitrogen and phosphorus, and organic substances) that cause mass rapid growth of phytoplankton and disturb the balance, structure, and functions of water ecosystems;
- substances that have saprogenic properties (sewage with a high content of easily decomposing organic matter) that cause oxygen deficiency followed by mass mortality of water organisms and appearance of specific microflora;
- substances causing toxic effects (e.g., heavy metals, chlorinated hydrocarbons, dioxins, and furans) that damage the physiological processes and functions of reproduction, feeding, and respiration;
- substances with mutagenic properties (e.g., benzo(a)-pyrene and other polycyclic aromatic compounds, biphenyls, radionuclides) that cause carcinogenic, mutagenic, and teratogenic effects.

Some of these pollutants (especially chlorinated hydrocarbons) cause toxic and mutagenic effects. Others (decomposing organic substances) lead to eutrophic and saprogenic effects. Oil and oil products are a group of pollutants that have complex and diverse composition and various impacts on living organisms—from physical and physicochemical damage to carcinogenic effects.

Discharge of heated waters can change the structure and function of coastal marine communities. Impacts of fly ash from coal-fired power plants, hot salty water, and residual chlorine are also important. Dumping of fly ash in coastal waters and into the atmosphere has caused severe impacts on spinner dolphins and mangroves in an area of the south coast of India, and has reportedly changed the number of species of plankton.

Effects of Marine Oil Spills

Oil spills can have a serious economic impact on coastal activities and on those who exploit the resources of the sea. In most cases, such damage is temporary and is caused primarily by the physical properties of oil creating nuisance and hazardous conditions. The impact on marine life is compounded by toxicity and tainting effects resulting from the chemical composition of oil, as well as by the diversity and variability of biological systems and their sensitivity to oil pollution.

Biological Effects of Oil. Simply, the effects of oil on marine life are caused by either the physical nature of the oil (physical contamination and smothering) or by its chemical components (toxic effects and accumulation leading to tainting). Marine life may also be affected by cleanup operations or indirectly through physical damage to the habitats in which plants and animals live.

The main threat posed to living resources by the persistent residues of spilled oils and water-in-oil emulsions ("mousse") is physical smothering. The animals and plants most at risk are those that could come into contact with a contaminated sea surface: marine mammals and reptiles; birds that feed by diving or form flocks on the sea; marine life on shorelines; and animals and plants in mariculture facilities.

The most toxic components in oil tend to be those lost rapidly through evaporation when oil is spilled. Because of this, lethal concentrations of toxic components leading to large-scale mortality of marine life are relatively rare, localized, and short-lived. Sublethal effects that impair the ability of individual marine organisms to reproduce, grow, feed, or perform other functions can be caused by prolonged exposure to a concentration of oil or oil components far lower than will cause death. Sedentary animals in shallow waters such as oysters, mussels, and clams that routinely filter large volumes of seawater to extract food are especially likely to accumulate oil components. Although these components may not cause any immediate harm, their presence may render such animals unpalatable if they are consumed by humans, due to the presence of an oily taste or smell, which is a temporary problem as the components that cause the taint are lost when normal conditions are restored.

The ability of plants and animals to survive contamination by oil varies. The effects of an oil spill on a population or habitat must be viewed in relation to the stresses caused by other pollutants or by any exploitation of the resource. In view of the natural variability of animal and plant populations, it is usually extremely difficult to assess the effects of an oil spill and to determine when a habitat has recovered to its prespill state. In recognition of this problem, detailed prespill studies are sometimes undertaken to define the physical, chemical, and biological characteristics of a habitat and the pattern of natural variability. A more fruitful approach is to identify which specific resources of value might be affected by an oil spill and to restrict the study to meeting defined and realistic aims related to such resources.

Impact of Oil on Specific Marine Habitats. Within each habitat, a wide range of environmental conditions prevails, and often there is no clear division between one habitat and another.

Plankton is a term applied to floating plants and animals carried passively by water currents in the upper layers of the sea. Their sensitivity to oil pollution has been demonstrated experimentally. In the open sea, the rapid dilution of naturally dispersed oil and its soluble components, as well as the high natural mortality and patchy, irregular distribution of plankton, make significant effects unlikely.

In coastal areas, some marine mammals and reptiles, such as turtles, may be particularly vulnerable to adverse effects from oil contamination because of their need to surface to breathe and to leave the water to breed. Adult fish that live in nearshore waters and juveniles in shallow water nursery grounds may be at greater risk to exposure from dispersed or dissolved oil.

The risk of surface oil slicks affecting the sea bed in offshore waters is minimal. However, restrictions on the use of dispersants may be necessary near spawning grounds or in some sheltered, nearshore waters where the dilution capacity is poor.

The impact of oil on shorelines may be particularly great where large areas of rocks, sand, and mud are uncovered at low tide. The amenity value of beaches and rocky shores may require the use of rapid and effective cleanup techniques, which may not be compatible with the survival of plants and animals.

Marsh vegetation shows greater sensitivity to fresh light crude or light refined products, although weathered oils cause relatively little damage. Oiling of the lower portion of plants and their root systems can be lethal, whereas even a severe coating on leaves may be of little consequence especially if it occurs outside the growing season. In tropical regions, mangrove forests are widely distributed and replace salt marshes on sheltered coasts and in estuaries. Mangrove trees have complex breathing roots above the surface of the organically rich and oxygen-depleted muds in which they live. Oil may block the openings of the air breathing roots of mangroves or interfere with the trees' salt balance, causing leaves to drop and the trees to die. The root systems can be damaged by fresh oil that enters nearby animal burrows; the effect may persist for some time and inhibit recolonization by mangrove seedlings. Protection of wetlands, by responding to an oil spill at sea, should be a high priority because physical removal of oil from a marsh or from within a mangrove forest is extremely difficult.

Living coral grows on the calcified remains of dead coral colonies, which form overhangs, crevices, and other irregularities, inhabited by a rich variety of fish and other animals. If the living coral is destroyed, the reef itself may be subject to wave erosion. The proportion of toxic components, the duration of oil exposure, as well as the degree of other stresses, largely determine the effects of oil on corals and their associated fauna. The waters over most reefs are shallow and turbulent, and few cleanup techniques can be recommended.

Birds that congregate in large numbers on the sea or shorelines to breed, feed, or moult are particularly vulnerable to oil pollution. Although oil ingested by birds during preening may be lethal, the most common cause of death is from drowning, starvation, and loss of body heat following damage to plumage by oil.

Impact of Oil on Fisheries and Mariculture. An oil spill can directly damage the boats and gear used for catching or cultivating marine species. Floating equipment and fixed traps extending above the sea surface are more likely to become contaminated by floating oil, whereas submerged nets, pots, lines, and bottom trawls are usually well protected, provided they are not lifted through an oily sea surface. Experience from major spills has shown that the possibility of long-term effects on wild fish stocks is remote because the normal overproduction of eggs provides a reservoir to compensate for any localized losses.

Cultivated stocks are more at risk from an oil spill: Natural avoidance mechanisms may be prevented in the case of captive species, and oiling of cultivation equipment may provide a source for prolonged input of oil components and contamination of the organisms. The use of dispersants very close to mariculture facilities is ill advised because tainting by the chemical or by the dispersed oil droplets may result.

An oil spill can cause loss of market confidence because the public may be unwilling to purchase marine products from the region, irrespective of whether the seafood is actually tainted. Bans on fishing and harvesting marine products may be imposed following a spill to maintain market confidence and to protect fishing gear and catches from contamination.

Mercury levels in marine animals, including bivalves and crustaceans, are generally low, whereas mercury seems to accumulate in fish. The highest values in fish are from northern Canada.

For seals and whales, concentrations often exceed 0.5 micrograms per gram of muscle, especially in older individuals. Livers from ringed seals in the western Canadian Arctic have very high levels of mercury; up to 205 micrograms per gram of liver have been measured. Levels in livers of bearded seals from the Amundsen Gulf are higher than those of both global background and other Arctic areas, as are mercury levels in toothed whales and polar bears. Some of the highest levels, 280 microgram per gram liver (wet weight), have been recorded in pilot whales from the Faroe Islands.

The effects of these mercury levels on the animals are difficult to assess, because some of the mercury may be inactivated by high selenium levels. Moreover, the scientific focus so far has been on tissues relevant for human consumption, and very little information is available on the target organs for mercury, such as the brain. There are no effect studies from the Arctic. However, even for the most exposed animal populations in the western Canadian Arctic and in Greenland, selenium should be abundant enough to protect against mercury poisoning.

Mercury is a major concern because the levels in some animals high in the food chain indicate that the environmental load may have increased in recent years. For example, mercury levels in ringed seals from western Canada show that they accumulated mercury about three times faster during the late 1980s and early 1990s than in the early 1970s. Similar increases have been seen in ringed seals from northwest Greenland taken in 1984 and 1994 and in beluga livers from the western Canadian Arctic. Interpreting these findings is difficult because natural variations that may affect the trends are unknown. Moreover, other data, such as those from Atlantic walrus and ringed seal from central-east Greenland, have not indicated any temporal trends. Very little information is available on temporal trends in Arctic marine fish, but measurements from the Baltic Sea from 1980 to 1993 seem to confirm observations that mercury levels are increasing.

The seas are being polluted by organic and inorganic wastes from sewage, from agricultural and industrial wastes, and from runoff containing oil, hydrocarbons, and heavy metals. All of these contribute to sediment runoff and increased turbidity. Siltation of coral regions is also caused by excessive deforestation and land clearing for commercial crops. Construction and land reclamation has caused changes in water circulation and has increased sedimentation. On the coral reefs, there has been extensive overexploitation of resources by heavy fishing pressure, including very destructive methods such as blasting, coral mining, and cyanide poisoning for live fish collection.

Industrial Pollution

Industrial pollutants that affect coral reefs include nutrients from sewage and organic matter, fertilizer runoff, detergents containing phosphorus, and thermal discharge—the heated water from the cooling systems of power plants and other industries. These all cause nutrient overload, the growth of aquatic plant life, and depletion of dissolved oxygen, or eutrophication, which retards coral growth by decreasing light penetration and changing the dynamics of fish assemblages. Other industrial pollutants include heavy metals and other toxic substances.

The coral reefs bordering major cities throughout Southeast Asia have been largely destroyed. Pollution from oil refineries and drilling platforms, it has been shown, kill reef fish and have negative effects on growth rates, recruitment, and feeding of corals. Thermal pollution from hot water discharge from industrial areas is an additional threat to reef species, many of them cannot withstand sudden and drastic increases in temperature.

Sedimentation. Land-based human activities often cause sedimentation, a major source of reef degradation. As more people move to coastal cities on the South China Sea, there has been a big increase in construction and land reclamation. Land reclamation and sedimentation have been particularly intensive in Singapore. Land was reclaimed by dumping sand and dirt directly onto coral reef flats and shallow water.

These add to the erosion of beaches and sediment runoff that smothers corals and leads to the degradation of a reef. Increased sedimentation also leads to a change in the composition of marine fauna, favoring more resilient species. Sedimentation also comes from soil erosion from unsound agricultural practices, mismanagement of watersheds, exploitation of mangroves, land reclamation and construction, oil drilling, and dumping of terrestrial and marine mine tailings.

Overfishing. Overfishing is a force extremely destructive to corals in the South China Sea. Densities of fish are greatly decreased by overfishing. Coral is damaged by destructive fishing techniques and by removal for trade. It is estimated that 10–15% of the total fish yield in the Philippines comes from coral reef fisheries.

Fishing degrades a reef in several ways. Destructive and illegal fishing methods are common, especially in the Philippines, Indonesia, and Malaysia. These methods include dynamite blasting and cyanide fishing. Overfishing not only depletes fish stocks of target species but also changes the dynamics of the entire reef. Decreases in herbivores can lead to algal blooms that overtop coral growth and can cause mass mortality.

Blooms of noxious algae have increased in the past 20 years worldwide and are being blamed on inputs of excess nutrients due to human activities. Some of these noxious algae produce powerful nerve toxins that can cause massive fish kills or even kill a person who unsuspectingly eats shellfish that was harvested from waters tainted with toxic algae. The case of the "Cell from Hell" now blooming in East Coast waters (North Carolina, Virginia, Maryland) is especially noteworthy. Until recently, *Pfiesteria* was only a curiosity of academic specialists. In the past few years, this organism has been blamed for fish kills unprecedented in their size and has been linked to neurological damage in people who worked or swam in these waters (memory loss, learning difficulties, and decreases in white blood cell content upward of 20% have been recorded in people who were exposed to *Pfiesteria*). Blooms of *Pfiesteria* have been linked to nutrient enrichment of coastal waters due to non-point pollution from agriculture. Nutrients in waters allow huge population increases of toxic organisms in water that were unknown or rare. The U.S. EPA has pledged to adopt new standards for nutrient inputs to waters.

It is hard to imagine an organism more bizarre than *Pfiesteria*. When no fish prey are present, it goes into a cyst form and settles to the bottom, lying dormant in the sediments. It can also emerge to form an amoeba that feeds on algae in the water column, and even can become a photosynthetic plankton-like organism, except that it "steals" the chloroplasts of algae from its algal prey and uses photosynthesis only to supplement its nutrient supply in the water column. In the presence of certain species of fish, however, it becomes a "monster" predator capable of mass fish kills. As a "predatory" dinoflagellate, it produces different types of toxins that do an incredible array of damage to fish. Some toxins attack internal organs. Another works on the fish immune system. And one toxin actually strips the skin off of the fish. Those who have witnessed the power of *Pfiesteria* report thousands of fish flopping and thrashing on the water surface, and fish actually beaching themselves, fleeing the water as if on fire.

SUMMARY

The introduction by man, directly or indirectly, of substances or energy into the marine environment (including estuaries) results in such deleterious effects as harm to living resources, hazards to human health, and hindrance to marine activities, including fishing, impairment of quality for use of seawater, and reduction of amenities. As the uses of coastal waters and the ocean have increased, pollution of the ocean waters has increased in turn. River pollution has also had an impact on the ocean as the rivers transport material to the ocean and, as a result, make it the ultimate sink for the world's waste.

The following chart summarizes the sources of wastes and their effects.

Sources and Effects of Marine Pollution.

Type	Primary Source/Cause	Effect
Nutrients	Runoff approximately 50% sewage, 50% from forestry, farming, and other land use. Also airborne nitrogen oxides from power plants, cars, etc.	Feed algal blooms in coastal waters. Decomposing algae depletes water of oxygen, killing other marine life. Can spur algal blooms (red tides), releasing toxins that can kill fish and poison people.
Sediments	Erosion from mining, forestry, farming, and other land-use; coastal dredging and mining.	Cloud water; impede photosynthesis below surface waters. Clog gills of fish. Smother and bury coastal ecosystems. Carry toxins and excess nutrients.
Pathogens	Sewage, livestock.	Contaminate coastal swimming areas and seafood, spreading cholera, typhoid, and other diseases.
Alien species	Several thousand per day transported in ballast water; also spread through canals linking bodies of water and fishery enhancement projects.	Outcompete native species and reduce biological diversity. Introduce new marine diseases. Associated with increased incidence of red tides and other algal blooms. Problem in major ports.
Persistent toxins (PCBs, heavy metals, DDT, etc.)	Industrial discharge; wastewater discharge from cities; pesticides from farms, forests, home use etc.; seepage from landfills.	Poison or cause disease in coastal marine life, especially near major cities or industry. Contaminate seafood. Fat-soluble toxins that bioaccumulate in predators can cause disease and reproductive failure.

Oil	46% from cars, heavy machinery, industry, other land-based sources; 32% from oil tanker operations and other shipping; 13% from accidents at sea; also offshore oil drilling and natural seepage.	Low level contamination can kill larvae and cause disease in marine life. Oil slicks kill marine life, especially in coastal habitats. Tar balls from coagulated oil litter beaches and coastal habitat. Oil pollution is down 60% from 1981.
Plastics	Fishing nets; cargo and cruise ships; beach litter; wastes from plastics industry and landfills.	Discarded fishing gear continues to catch fish. Other plastic debris entangles marine life or is mistaken for food. Plastics litter beaches and coasts and may persist for 200 to 400 years.
Radioactive substances	Discarded nuclear submarine and military waste; atmospheric fallout; also industrial wastes.	Hot spots of radioactivity. Can enter food chain and cause disease in marine life. Concentrate in top predators and shellfish, which are eaten by people.
Thermal	Cooling water from power plants and industrial sites.	Kills off corals and other temperature-sensitive sedentary species. Displaces other marine life.

READING LIST

Brown, L. (1997). Can we raise grain yields fast enough? *Worldwatch* **10**(4): 9–17.

Engleman and LeRoy. (1996). *Sustaining Water: Population and the Future of Renewable Water Supplies*. Population Action International.

EPA. (1994). *National Water Quality Inventory, 1992: Report to Congress*. Office of Wetlands, Oceans and Watersheds, US EPA, Washington, DC.

Kane, H. (1997). Eco-farming in Fiji. *Worldwatch* **10**(4): 29–34.

Nixon, S. et al. The fate of nitrogen and phosphorus at the land-sea margin of the North Atlantic Ocean. *Biogeochemistry* **35**: 141–180.

NRC (National Research Council). (1993). *Managing Wastewater in Coastal Urban Areas*. NRC, Washington, DC.

NRC (National Research Council). (1994). *Priorities for Coastal Science*. NRC, Washington, DC.

Hotta, K. and Dutton, I.M. (Eds.). (1995). *Coastal Management in the Asia-Pacific Region: Issues and Approaches*. Japan International Marine Science and Technology Federation, Tokyo.

Web Sites

http://www.nos.noaa.gov/icri/state.html

http://www.cutter.com/osir/

http://www.igc.apc.org/wri/indictrs/reefrisk.htm

http://www.enn.com

http://www.env.gov.sg

Journals and Newspapers

Environet

The Straits Times

Science and Technology-The Hindu

TRACE ELEMENT POLLUTION

Bobby J. Presley
Texas A&M University
College Station, Texas

INTRODUCTION

All three words in the term "trace element pollution" need to be defined. Defining an element, that is a chemical element, is relatively straightforward. It is a substance consisting entirely of atoms having the same number of protons. All such substances are listed by name and symbol on the periodic charts of the elements found in many science textbooks. Definitions for "trace" and "pollution" are not so straightforward.

It is sometimes convenient to classify the chemical elements making up a complex substance or matrix into major, minor, and trace, based on their relative amounts (concentrations). Most authors designate those elements present in a matrix at one part per million or less by weight as trace elements. Although trace elements are in low concentration in the environment, they can be either essential or harmful to organisms, depending on the element and the circumstances.

A definition of pollution can be inferred from the widely accepted definition of marine pollution given by the United Nations Group of Experts on the Scientific Aspects of Marine Pollution (GESAMP). They say pollution is "the introduction by man, directly or indirectly, of substances or energy to the marine environment resulting in such deleterious effects as harm to living resources; hazards to human health; hindrance of marine activities including fishing; impairment of the quality for use of seawater; and reduction of amenities." This definition can be applied to all environments, not just to the marine environment.

By the GESAMP definition, pollution must be harmful and must be caused by human activity. In some cases, it is easy to show both a human cause and harm to the environment. In other cases, one or both parts of the definition can be hard to prove. For example, it might be possible to show that the concentration of a trace element, such as mercury, is elevated above normal values in, for example, fish. It might be harder to show harm to the fish or to the consumer of the fish and it might be hard to show that the enrichment is because of human activity. In such a case, it might be better to say contamination has been documented, but not pollution.

ABUNDANCE, DISTRIBUTION, AND BEHAVIOR OF TRACE ELEMENTS

Trace elements occur naturally in the environment. Unlike pesticides, plastics, organic solvents, and other manufactured products, the mere presence of trace elements does not imply pollution, contamination, or even human activity. Sufficiently sensitive analytical techniques can detect some amount of all elements in almost any substance. To determine the amount accurately and precisely is, however, a challenge, and to decide what portion of the element is natural and what portion is because of human activity is an even bigger challenge.

In order to evaluate the significance of trace element occurrences in the environment, information is needed on:

1. Amounts (concentrations) in various compartments (air, soil, water, organisms, etc.).
2. Sources to the environment (both natural and human).
3. Transport mechanism and pathways between compartments (continents to oceans, water to organisms, etc.).
4. Transfer mechanism within compartments (shallow to deep water, fish gill surface to other organs, etc.).
5. Ultimate fate of the element (burial in sediments, mixing throughout the ocean, etc.).
6. Effect of the element on organisms (both short and long term).

As a first step in determining the significance of a given amount of a trace element in a particular environment, it is essential to know the ranges in concentration expected to occur naturally in various media (soil, water, organisms, etc.). Most trace elements were tied up in igneous rocks on the primitive earth. Table 1 gives average concentrations of selected elements in the uppermost part of the earth's crust along with data for organisms, soil, sediment, and water. These values can be considered to be "background" concentrations or at least concentrations not greatly influenced by human activity.

As igneous rocks "weather" to give soil and sediment some fraction of each element in the rock, it becomes dissolved in rivers, lakes, and the ocean. For most trace elements, however, a large fraction is retained in the solids during weathering. For this reason, the natural (background) trace element concentration in soils and sediments varies, depending on the igneous rocks they came from.

Many trace elements behave similarly during weathering, soil formation, erosion, and deposition. For example, fine-grained clay minerals become enriched in most trace elements and quartz sand and carbonate minerals become depleted. It is, therefore, important to have information on the grain size and mineralogy of soils and sediments when evaluating their trace element content. Thus, a given concentration, say 20 ppm Cu, might be background for a clay sediment but be contamination in a sand.

Reliable data on trace elements dissolved in rivers, lakes, and seawater is more difficult to produce than is data on rocks, soil, and sediment. Concentrations are generally a thousand or more times lower (Table 1), and this causes sensitivity and matrix interference problems for even the newest analytical instruments. It is also difficult to avoid contaminating samples with trace elements during

Table 1. Average Concentration of Selected Elements in Various Materials. Values for Sediment, Soil, Crust, Oysters and Fish are μg/g Dry Weight (ppm) Except Ca and Fe in Percent Dry Weight. Values for Water are μg/l (ppb)

Material	Ref.	As	Ba	Ca	Cr	Cu	Fe	Mn	Ni	Pb	V	Hg	Zn
Continental Crust	1	1.5	550	3.00	35	25	3.5	600	20.	20.	60	–	71
Gulf Coast Seds	2	6	–	–	44	10	1.8	330	16	15	–	0.050	60
Gulf Slope Seds	3	8.6	660	11.3	54	27	2.76	300	38	17	100	0.028	81
San Joaquin Soil	4	17.7		1.9	130	34.6		538	88	18.9	112	1.4	106
Sediment Criteria	5	55	–	–	145	390	–	–	50	110	–	1.3	270
Average Seawater	6	1.7			0.2	0.25	0.05	0.2	0.5	0.002	1.5	0.001	0.4
Miss. River Water	7		65			1.5	2.2	1.4	1.5	0.008	1.0		0.27
Seawater Criteria	8	69			1100	4.8			74	210		1.8	90
Fresh Water Criteria	9	340			16	13			470	65		1.4	120
Gulf Oysters	10	10.3			0.57	146	294	14.4	1.77	0.64		0.13	2150
Tuna Fish	11	3.8		100	0.75	2.8	72	0.6	0.5	0.5		4.1	17.4

1. Average metal levels in upper continental crust (95% igneous rock). Taylor and McClennon (1).
2. Median estuarine (inshore) surface sediment metal levels from the U.S. Environmental Protection Agency's northern Gulf of Mexico (GOM) Environmental Mapping and Assessment Program 1991–1993.
3. Average surface sediment metal levels observed among 43 stations ion the Gulf of Mexico Slope. BJ Presley, unpublished
4. Agricultural soil from the San Joaquin Valley, CA. US NIST Standard Reference Material #2709.
5. ER-M values from Long and Morgan (2), indicating sediment metal levels at which biological effects are often seen.
6. Average seawater values from Bruland, (3).
7. Average Mississippi River dissolved values from Shiller, (4).
8. Maximum Contaminant Concentration for seawater, US EPA, (5).
9. Maximum Contaminant Concentration for fresh water, US EPA, (5).
10. Average concentration in 485 pooled samples of 20 oysters each from the Gulf of Mexico. Presley (6).
11. Fillets from 16 Large Mediterranean Sea Tuna. International Atomic Energy Agency reference material # 350.

collection, storage, and analysis. For these reasons, nearly all of the dissolved trace element data published before 1970, and much of the recent data, is unreliable. Much of the published data for both fresh and seawater is too high by factors of up to 100. Only data that has been produced by a lab using a well-documented quality assurance program should be accepted.

Trace element concentrations in organisms are generally intermediate between those in sediments and those in water. However, concentrations vary widely with the specie of organism and with the specific organ within organisms. Livers, for example are enriched in some elements and kidneys in others and some species of organisms are highly enriched in one or another trace element. Trace element concentrations in organisms can also change with season, life stage, health, food supply, etc. Thus, identification of abnormal trace element concentrations is difficult unless reliable data is available for the same organism over a wide area and/or over some time period.

Good general compilations of reliable trace element data for organisms are not as available as are those for soil and sediment. General guidance can be acquired from Furness and Rainbow (7) or similar publications. Many recent journal articles give trace element data for specific organisms from specific locations, but it is not always clear which are background levels.

In addition to natural sources, many different human activities can add trace elements to the environment. Mining and metal processing are classic sources of contamination, but other manufacturing, transportation, and waste disposal practices can also be important. In the United States, the EPA's "Superfund" program has spent many millions of dollars to clean up trace element contamination at dozens of sites around the country. The kinds of practices that led to this gross contamination are very rare today. Environmental regulations and public pressure have caused industry to greatly reduce trace metal releases to the environment. However, as the world population grows, the Earth's surface is increasingly disturbed by agriculture, petroleum production, forestry, urban development, civil conflicts, and war, all of which make trace elements more available for uptake by humans and other organisms.

As discussed above, numerous possible sources exist of trace elements to the environment. It is almost always difficult to determine which possible source is most important for any given element at any given location. Although it is important to identify the source of trace elements, their environmental impact depends not on source but on concentration and behavior. Behavior, including mobility, transport, transfer, and biological uptake, depends strongly on the chemical and physical form of the trace element. In this respect, the size of the trace element specie or the particle with which it is associated is critical, as this will control its transport and settling behavior in air and water. A given trace element will behave differently physically, chemically, and biologically in each of its different forms, and it will partition itself among the various possible forms in response to environmental conditions. It is important to note that many trace elements are "particle-reactive" and will quickly associate with particles if added to the environment in a dissolved form. Trace element concentrations are almost always much higher in particles than in dissolved forms.

Although the behavior of a trace element, including its biological behavior, depends on its form, there is some form of most trace elements that will affect the health of organisms, including humans. At least 20 trace elements have long been known to be essential to health [e.g., (8)]. Diseases because of trace element deficiencies are well known among both humans and other organisms. A number of trace elements such as Cu, Ni, Zn, and Se are essential to life at very low concentrations but toxic at slightly higher concentrations. Good data on trace element concentrations in the environment are needed in order to know whether too little or too much of a giver element is present.

The toxic effects of both essential and nonessential trace elements are well known, in the case of As and Pb, human toxicity has been known for more than 2000 years. For other trace elements, toxic effects are less well recognized. In general, however, for all trace elements, an optimal concentration in the environment and in the organism gives optimal function (growth, reproduction, etc.) and higher or lower concentrations result in less than optimal function and possibly death.

In order for a trace element in the environment to have an effect on an organism, the element must, of course, be taken up by the organism. For plankton and other aquatic plants, this uptake is directly from solution, but for animals, some, or most, of the uptake might be from food or from ingestion of nonfood particles. In any case, at some point the trace element must be in a soluble form and be transferred across cell membranes and possibly transferred to some vital organ within the organism. The form of the trace element is very important in controlling these transfers and the resulting effects, but both environmental conditions (pH, temperature, etc.) and the type of organism and its condition (age, health, etc.) also play a role. Factors that influence the toxicity of trace elements have been discussed by Bryan (9), Luoma (10), and many others, and the large differences in sensitivity to trace elements exhibited by different organisms are well known [e.g., (11)].

ASSESSING BIOLOGICAL IMPACTS

The effect of a trace element on organisms depends on the abundance, distribution, and behavior of the trace element. As discussed above, these are difficult to determine and are subject to complex and incompletely understood processes. The environmental impact of waste disposal or other human activity is, therefore, often controversial. Environmental groups and industry often engage in public fingerpointing and lawsuits over specific activities. Often, more money is spent on lawyers than on attempts to scientifically document impacts. One reason for this is the difficulty in clearly documenting harmful effects in the field, especially at the population or ecosystem level. Laboratory toxicity testing is not easy, but it can usually show dose-response relationships that allow establishment of trace element concentrations above which harmful effects to a given organism are likely to result. Such results are, however, usually difficult to apply to the complex conditions in the field, as is discussed below.

The simplest laboratory toxicity tests are those that use death of the organism as the only indicator of effect. This crude measurement has been much criticized, but it does establish the rough relative toxicity for various trace elements to various organisms. This test will show, for example, that Cu is much more toxic to most plankton than is As. More subtle effects can also be sought in laboratory cultures of various organisms, for example, changes in metabolism, ability to reproduce, find food, grow, etc. A vast amount of literature exists on methods for detecting sublethal effects of toxins on organisms [e.g., (2,12)]. Different sublethal tests often give different results in rating the relative toxicities of different trace elements, but they have the potential for indicating possible long-term effects on organisms that might not show up in short-term acute tests.

Most laboratory toxicity tests use water as free from trace elements, complexing ligands, organic matter, etc. as is possible, so the response of the test organisms can be more clearly related to the trace element added in the test procedure. Consequently, a trace element is almost always less toxic in the environment than it is in the laboratory, because of complexing, adsorption, and other interactions in the environment.

Laboratory toxicity tests, even when they try to imitate the environment by using ambient water, multiple trace elements, varying salinities, temperatures, different life stages of organisms, etc., can never truly duplicate natural conditions. It is useful, then, to look for effects of pollutants in the environment, especially at the population level. This is, however, a difficult task, because of the natural temporal and spatial variability in abundance and health of organisms.

As a result of the relative expense and time involved in toxicity tests, and their sometimes ambiguous results, many environmental assessment programs seek only to determine concentrations of trace elements in the environment and to look for enrichments caused by human activity. If a trace element enrichment is detected, its significance can then be resolved by toxicity testing or detailed ecological field analysis. In any case, trace element enrichments could be sought in air, water, sediment, or organisms.

Water analysis might seem a logical way to detect trace element enrichments in the environment. Furthermore, the significance of trace element concentrations in water can be judged because the US EPA has published values for each element above which harm to organisms is likely (Table 1). However, ambient water, be it river water, groundwater, rainwater, or seawater, is notoriously hard to collect, store, and analyze for trace element content, as was discussed above. Water concentrations can also change over short time scales in some circumstances, which further complicates their use.

Soil or sediment can usually be more easily analyzed accurately for trace element content than can water. Soil and sediment also integrate trace element input over some time scale, so they don't need to be sampled as often as does water. Another advantage of sediment analysis is that it gives a historical view of pollutant input at sites where sediment is laid down layer by layer, year after year. Dates can be assigned to the different layers by use of radio-isotopes, pollen identification, or other means. Furthermore, sediment layers from prehistoric times give a background value for each trace element that can be compared with values in near-surface layers in order to quantify human-induced enrichments.

Recognizing gross sediment contamination is easy. Any sediment that is several-fold higher in a given trace element than the average crustal abundance of that element is contaminated unless some unusual mineralogy exists. However, it is harder to recognize subtle contamination because of difficulties in establishing an exact background concentration for a given location. Values from prehistoric depths in the sediment column are a possible background, as noted above. Another background is sediment well away from any point source of pollutant input. In using either of these methods, care should be taken to compare similar sediment types or to compare element to element concentration ratios rather than absolute concentrations (13).

Another problem with using sediment data is that only some unknown fraction of the trace element in the sediment is likely to be available to organisms, which has been much discussed in the literature [e.g., (14)], especially in conjunction with disposal of dredge spoil [e.g., (15)]. Many authors have suggested leaching sediments with dilute acids or other solutions [e.g., (14)] in order to remove only the trace element that could potentially be removed by an organism. Another suggestion that has been much debated is the ability of sulfide in the sediment to limit availability of trace elements to organisms [e.g., (16)]. Thus, although it generally agreed that only a fraction of the total trace element in sediments is available to organisms, no consensus exists on how to measure that fraction.

Long and Morgan [(2) and elsewhere] suggest another way to identify sediment that is potentially harmful to organisms because of chemical contamination. They compiled published matching biological health and chemical data from numerous field, laboratory, and modeling studies. The data was then ranked from the lowest to the highest contaminant concentration where any adverse biological effect was reported. From the ranking they derived two guideline concentrations for each contaminant. These two values separate the data into values that (1) rarely, (2) occasionally, or (3) frequently cause adverse biological effects. These derived values have been widely used in monitoring programs. See Table 1 for some of the actual values.

If both water and sediment offer analytical and data interpretation challenges, would it not be better to analyze organisms in order to assess trace element contamination? Certainly, advantages to this approach exist. For one thing, there is no question as to whether the element is available to organisms. For another, concentrations are often high enough to make analyses relatively easy, at least for common trace elements such as Cu and Zn. There are, however, problems, for example, deciding what organisms to analyze. It is not practical to analyze every organism at a given location, or even to analyze a representative specie from each major taxonomic group. What, then, should be analyzed?

Farrington (17) summarized the rationale for using common mussels (*Mytilus* sp.), various oyster species (*Crassostrea* and *Ostrea*), and other bivalves as "sentinel" organisms in monitoring studies in the marine environment. This approach has resulted in a very large worldwide data set for trace elements in bivalves. In the United States, the National Oceanic and Atmospheric Administration's "National Status and Trends Program" (NS&T) has been analyzing bivalves from the entire U.S. coastline since 1986 and has produced an especially useful and high-quality data set. As a result of the NS&T and similar data, bivalves should be the first choice for organisms to analyze in marine environmental monitoring programs. Many different kinds of plants and animals have been used in nonmarine environmental monitoring studies, everything from plankton and moss to polar bears. It all depends on what is available and the purpose of the monitoring. In general, organisms that have low natural variability in trace element concentration and are geographically widespread and easy to collect should be selected.

Data from good, long-term environmental monitoring programs can help answer the question "are things getting better or worse." Since strict environmental laws took effect in the United States in the 1970s, billions of dollars have been spent on pollution-control devices and cleanup of polluted sites. Have the efforts worked? (O'Connor (18,19) looked for temporal trends in the NS&T data discussed earlier. He found that for 2744 combinations of 14 chemicals and 196 collection sites over a 10-year time period, only 88 increases and 348 decreases in concentration are significant at the 95% confidence level. Chance alone predicts 69 increases and 69 decreases, so by this analysis, it is quite possible that no real increases occurred over that 10-year period and environmental quality along the U.S. coastline may have improved. This finding is consistent with observations of other environmental scientists. At least for chemical contaminants, environmental laws have worked and the U.S. environment is cleaner now than it was in the 1970s.

BIBLIOGRAPHY

1. Taylor, S.R. and McClennon, S.M. (1985). *The Continental Crust: Its Composition and Evolution*. Blackwell Scientific Publications, Oxford, p. 312.
2. Long, E.R. and Morgan, L.G. (1990). *The Potential for Biological Effects of Sediment-sorbed Contaminants Tested in the National Status and Trends Program*. NOAA Tech. Memo. NOS OMA 52, Rockville, MD, p. 167.
3. Bruland, K.W. (1983). Trace elements in sea-water. In: *Chemical Oceanography*. Vol. 8, J.P. Riley and R. Chester (Eds.). Academic Press, London, pp. 158–220.
4. Shiller, A.M. (1997). Dissolved trace elements in the Mississippi River: Seasonal Interannual and decadal variability. *Geochemica et Cosmochemica Acta* **61**: 4321–4330.
5. U.S. Environmental Protection Agency. (2002). *National Recommended Water Quality Criteria: 2002*. EPA-822-R-02-047, US EPA Office of Water, Washington, DC.
6. Presley, B.J. (1990). Trace metals in Gulf of Mexico oysters. *The Science of the total Environment* **97/98**: 551–593.
7. Furness, R.W. and Rainbow, P.S. (1990). *Heavy Metals in the Marine Environment*. Lewis Publishers, p. 264.
8. Mertz, W. (1981). The essential trace elements. *Science, V.* **213**: 1332–1338.
9. Bryan, G.W. (1976). Factors influenceing toxicity of heavy metals. In: *Marine Pollution*. R. Johnson (Ed.). Academic Press, London, pp. 185–302.
10. Luoma, S.N. (1983). Bioavailability of trace metals to aquatic organisms-A review. *Science of the Total Environment* **28**: 1–29
11. U.S. Environmental Protection Agency. (1987). *Biomonitoring to Achieve Control of Toxic Effluents*. EPA/625/8-87/013, US EPA Office of Water, Washington, DC.
12. U.S. Environmental Protection Agency. (1988). *Short-Term Methods For Estimating Chronic Toxicity of Effluents and Receiving Waters to Marine and Estuarine Organisms*. Environmental Protection Agency, EPA 60014-87/028 Cincinnati, OH.
13. Trefry, J.H. and Presley, B.J. (1976). Heavy Metals in Sediments from San Antonio Bay and the Northwest Gulf of Mexico. *Environmental Geology* **1**: 283–294.
14. Campbell, P.G.C. (1988). *Biologically Available Metals in Sediments*. National Research Council of Canada, ISSN 0316-0114, Ottawa, Canada, p. 298.
15. Lake, J., Hoffman, G.L., and Schimmel, S.C. (1985). *Bioaccumulation of Contaminants from Black Rock Harbor Dredged Material by Mussels and Polychaetes*. U.S. Army, Corp of Engineers, Tech. Report D-85-2, Vicksburg, MS.
16. Allen, H.E., Fu, G., and Deng, B. (1993). Analysis of acid-volatile sulfide (AVS) and simultaneously extracted metals (SEM) for estimation of potential toxicity in aquatic sediments. Environ. *Toxicol. and Chem.* **12**: 1441–1453.
17. Farrington, J.W. (1983). Bivalves as sentinels of coastal chemical pollution; The mussel (oyster) watch. *Oceanus* **26**: 18–29.
18. O'Connor, T.P. (1998). Mussel Watch results from 1986 to 1996. *Mar. Poll. Bull.* **37**: 14–19
19. O'Connor, T.P. (1990). *Recent Trends in Environmental Quality: Results of the First Five Years of the NOAA Mussel Watch Project*. NOAA/ORCA, Silver Spring, MD.

CORAL REEFS AND YOUR COASTAL WATERSHED

U.S. Environmental Protection Agency—Oceans and Coastal Protection Division

Coral reefs are among the world's richest ecosystems, second only to tropical rain forests in plant and animal diversity. However, they are extremely sensitive environments that have special temperature, salinity, light, oxygen, and nutrient requirements. If environmental conditions fall outside the acceptable range of these requirements, the health and dynamics of a coral reef community can be severely disrupted. That's why coral communities are sensitive indicators of water quality and the ecological health of the coastal watershed. They respond to alterations within the entire coastal watershed, such as changes in freshwater flows and nutrient inputs. Consequently, pollution from the destruction and alteration of surrounding coastal watersheds can directly affect the health and productivity of a coral reef.

WHAT ARE CORAL REEFS AND WHAT DO THEY DO?

Coral reef ecosystems are unique, biologically diverse systems recognized as valuable economic and environmental resources. Many people think coral reefs are made of plants or rocks, but they are actually made of animals! A coral polyp is a delicate, limestone-secreting animal. The limestone serves as a skeleton that either is embedded within the living tissue of the coral or encloses the animal. A coral reef is made up of colonies of these coral polyps. There are several benefits of coral reefs.

- Coral reefs are an important recreational and aesthetic resource for people visiting or living in coastal areas. People use coral reefs for fishing, underwater photography, scuba diving, and snorkeling.
- Coral reefs provide protection for harbors and beaches, which are often found behind reefs because the reefs provide natural protection from heavy wave action caused by coastal storms.
- Coral reefs are home to a number of species of fish and other marine species, including many that we rely on for food and economic purposes.
- Coral reefs also serve as a laboratory for students and scientists to study and learn about complex ecological and biological processes. In addition, the reefs yield many biological treasures that are increasingly being recognized as natural sources of biomedical chemicals.

SOME IMPACTS ON CORAL REEFS

Coral reef habitats are extremely sensitive to disturbances, such as various forms of pollution and physical contact. Pollution of coastal watersheds poses a threat to the existence of coral reefs. Impacts can result from activities occurring near the reef itself or from areas within the coastal watershed that drain to the reef. Disturbances and pollution can lead to diseases in coral such as bleaching (when the algae that give corals their color die). Natural occurrences, such as hurricanes, can adversely impact coral reefs through high-energy storm surges and the resulting resuspension of sediment. However, reefs are usually able to recover from natural disturbances.

People using the reef can have an adverse impact on reef resources. Portions of a coral reef can be broken by the impact of boat anchors and boat groundings. Divers and snorkelers can harm the reef by simply touching it or by removing the corals. Suntan oil from swimmers and snorkelers can harm or even kill sensitive corals. Dragging hooks, fishing line, and nets across the coral reef, as well as placing and recovering lobster traps on reefs, can be damaging. Overfishing also harms coral reefs by removing important species that eat the algae growing on corals. When these fish species are removed, the algae overgrow the corals, smothering them.

Marine debris, trash floating on the ocean or resting on the ocean floor, comes from many sources, including boaters, divers, improper disposal of trash on land, storm water runoff to rivers and streams, ships and other vessels, and offshore oil platforms. Marine debris can harm fish species and other aquatic organisms that use the reef. Trash that lands on the reef can kill corals by continually rubbing against it or smothering it.

An excessive amount of *nutrients* from improperly treated sewage, atmospheric deposition, agricultural and urban runoff, and cleaning products high in phosphates can harm coral reef habitats. In excess levels, nutrients overstimulate the growth of aquatic plants and algae. When nutrient levels increase, the delicate balance that exists between corals and algae is destroyed and the algae can overgrow the corals. When this situation is prolonged, the corals are smothered and die beneath the algal carpet. This, in turn, affects the fish and other aquatic organisms using the area, leading to a decrease in animal and plant diversity and affecting use of the water for fishing and swimming.

Some of the leading causes of nearshore coral decline can be related to *land development* and nearshore

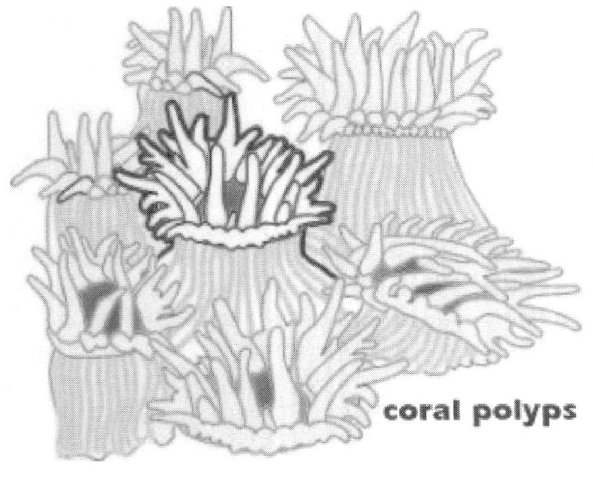

construction that are not environmentally sensitive. Sediment, silt, and other suspended solids wash off plowed fields, construction and logging sites, urban areas, strip-mined land, and eroded stream banks when it rains. Increases in coastal sediment are also caused by construction of seawalls, docks, and marinas; land-clearing; boats running through shallow waters, disturbing and suspending silts with their propellers; and snorkelers and divers kicking up sediment. Sediment can block sunlight that is essential for the survival of some corals, which live in a very close relationship with microscopic plants (algae) that require sunlight to survive. In addition, heavy sedimentation can bury corals, inhibiting their growth or killing them.

Pathogens are disease-causing microorganisms such as viruses, bacteria, and parasites. Pathogens are harmful to corals, causing disease and scarring in many species. These microorganisms enter water bodies from sources such as: inadequately treated sewage, storm water drains, septic systems, runoff from livestock pens, and boats that discharge sewage.

Coral reefs are vulnerable to the introduction of a wide variety of *toxic substances*, including metals (such as mercury and lead), toxic organic chemicals (such as PCBs and dioxin), pesticides, and herbicides found in sources like industrial discharges, urban and agricultural runoff, mining activities, and runoff from landfills. Some toxic substances bind to sediment and are transported to coastal waters through sedimentation. These toxic substances can cause scarring, death, or reproductive failure in fish, shellfish, and other marine organisms. In addition, they can accumulate in fish tissue, leading to fish consumption advisories. The sensitivity of corals makes them especially vulnerable to the introduction of toxic substances.

WHAT IS EPA DOING TO PROTECT CORAL REEFS?

International Coral Reef Initiative

In 1994, EPA, along with the State Department, the National Oceanic and Atmospheric Administration, and the Department of the Interior, formed an international coalition to coordinate information and bring higher visibility to the need for coral reef ecosystem preservation. The coalition became the *International Coral Reef Initiative* (ICRI), which now includes a membership of more than 90 countries.

EPA's Watershed Approach. EPA has joined with others to promote the Watershed Approach nationally as a means to further restore and maintain the physical, chemical, and biological quality of our nation's waters, including coral reefs. By addressing issues on a watershed scale, those areas that pose the greatest risk to human and ecological health can be targeted, several pollutants can be addressed at one time, the public can be involved in cleaning up the environment and protecting coral habitats, and integrated solutions for environmental protection can be considered. This is particularly important given the contribution of activities and sources of pollution within the larger watershed to the decline of coral reefs. Through the Watershed Approach, integrated coastal zone management tools and watershed concepts can be applied in the development of comprehensive management and conservation plans. The Watershed Approach aims

Coral reefs are sensitive indicators of water quality and the ecological health of the coastal watershed.

to determine protective approaches for controlling identified stressors to coral reef ecosystems. EPA's *Coastal Watershed Protection Strategy* specifically provides technical assistance and support to priority coastal watersheds, such as *National Estuary Programs* (NEPs) and other coastal waters identified by states.

Other EPA Programs. In assisting coastal states with the development of their *Coastal Nonpoint Pollution Control Programs*, EPA and other federal agencies developed guidance specifying management measures for sources of nonpoint pollution (diffuse runoff of pollutants) in coastal waters. In its program, a state or territory describes how it will implement nonpoint source pollution controls. EPA also works with other federal agencies to protect human health and coral reefs by reducing marine debris. The efforts include the establishment of the *National Marine Debris Monitoring Program*, which looks at the origins and amounts of marine debris deposited along U.S. coasts. EPA and the Coast Guard work together to regulate the transportation of municipal and commercial waste on vessels and to issue regulations for the manufacture, maintenance, and efficiency of marine sanitation devices (boat toilets), as well as the establishment of "no discharge zones" for vessel sewage. EPA also regulates the discharge of pollutants from facilities into sensitive marine waters. EPA assists states in the development of *water quality standards* designed to protect human health and aquatic life. This assistance includes the development of criteria for water quality that accurately reflects the most up-to-date scientific knowledge about the effects of pollutants on aquatic life, such as corals, and human health.

What Can You Do to Help Protect Coral Reefs?

You can do several things to help protect coral reefs and your coastal watershed:

- *Be Informed and Involved*. Learn about coral reefs and their importance to your coastal watershed. Participate in training or educational programs that focus on reef ecology. Be an informed consumer; ask the store owner or manager from what country the coral was taken and whether that country has a management plan to ensure that the harvest was legal and sustainable over time. Support the creation and maintenance of marine parks and reserves. Become a citizen volunteer. As a volunteer you might be involved in taking water quality measurements, tracking the progress of protection and restoration projects, or reporting special events like fish kills and storm damage. Volunteer for a reef cleanup or a beach cleanup. If you don't live near a coast, get involved in your local watershed program. Report dumping or other illegal activities.
- *Take Responsibility for Your Own Backyard*. Determine whether additional nutrients or pesticides are needed before you apply them, and look for alternatives to fertilizers and pesticides where the chance of runoff into surface waters might occur. Even if you live far from a coral reef ecosystem, these products might ultimately affect the waters that support coral. Consider selecting plants and grasses with low maintenance requirements. Water your lawn conservatively; the less water you use, the less runoff will eventually find its way into the oceans.
- *Practice Good Housekeeping*. Learn about procedures for disposing of harmful household wastes so they do not end up in sewage treatment plants that can't treat them or in landfills not designed to receive hazardous materials. Around the house, keep litter, pet waste, leaves, and grass clippings out of street gutters and storm drains to prevent their entrance into streams that might flow to reefs. Use the minimum amount of water needed when you wash your car to prevent waste and runoff. Never dump any household, automotive, or gardening wastes into a storm drain. They might end up on the reef. Take used motor oil, paints, and other hazardous household

materials to proper collection sites such as approved service stations or designated landfills. Always follow label directions for the use and disposal of household chemicals. Keep your septic tank in good working order. The improper disposal of wastes and hazardous materials can lead to water quality problems and harm to the sensitive coral reef habitats.

- *Respect the Reef.* Help keep the reef healthy by following local guidelines, recommendations, regulations, and customs. If you dive, don't touch the coral. Keep your fins, gear, and hands away from the coral since this contact can hurt you and will damage the delicate coral animals. Stay off the bottom because stirred-up sediment can settle on corals and smother them. Avoid entering sensitive habitat areas with your boat or other motorized watercraft. Maintain your boat engine to prevent oil and gas leaks. Keep all waste produced during your excursions in a safe place to be disposed of properly when you're back on land. If you go boating near a coral reef, don't anchor your boat on the reef. Use mooring buoy systems if they are available. Maintain and use your marine sanitation devices properly. Conserve energy and keep your auto in good running condition. By conserving energy, harmful air emissions leading to air deposition are minimized.

SEA LEVEL AND CLIMATE

Richard Z. Poore
Christopher Tracey
U.S. Geological Survey
Reston, Virginia

Richard S. Williams, Jr.
U.S. Geological Survey
Woods Hole, Massachusetts

INTRODUCTION

Global sea level and the Earth's climate are closely linked. The Earth's climate has warmed about 1 °C (1.8 °F) during the last 100 years. As the climate has warmed following the end of a recent cold period known as the "Little Ice Age" in the 19th century, sea level has been rising about 1 to 2 millimeters per year due to the reduction in volume of ice caps, ice fields, and mountain glaciers in addition to the thermal expansion of ocean water. If present trends continue, including an increase in global temperatures caused by increased greenhouse-gas emissions, many of the world's mountain glaciers will disappear. For example, at the current rate of melting, all glaciers will be gone from Glacier National Park, Montana, by the middle of the next century (Fig. 1). In Iceland, about 11 percent of the island is covered by glaciers (mostly ice caps). If warming continues, Iceland's glaciers will decrease by 40 percent by 2100 and virtually disappear by 2200.

Figure 1. Grinnell Glacier in Glacier National Park, Montana; photograph by Carl H. Key, USGS, in 1981. The glacier has been retreating rapidly since the early 1900's. The arrows point to the former extent of the glacier in 1850, 1937, and 1968. Mountain glaciers are excellent monitors of climate change; the worldwide shrinkage of mountain glaciers is thought to be caused by a combination of a temperature increase from the Little Ice Age, which ended in the latter half of the 19th century, and increased greenhouse-gas emissions.

Most of the current global land ice mass is located in the Antarctic and Greenland ice sheets (Table 1). Complete melting of these ice sheets could lead to a sea-level rise of about 80 meters, whereas melting of all other glaciers could lead to a sea-level rise of only one-half meter.

GLACIAL-INTERGLACIAL CYCLES

Climate-related sea-level changes of the last century are very minor compared with the large changes in sea level that occur as climate oscillates between the cold and warm intervals that are part of the Earth's natural cycle of long-term climate change.

Table 1. Estimated Potential Maximum Sea-Level Rise from the Total Melting of Present-Day Glaciers. [Modified from Williams and Hall (1993). See also http://pubs.usgs.gov/factsheet/fs50-98/]

Location	Volume (km^3)	Potential Sea-Level Rise (m)
East Antarctic ice sheet	26,039,200	64.80
West Antarctic ice sheet	3,262,000	8.06
Antarctic Peninsula	227,100	.46
Greeland	2,620,000	6.55
All other ice caps, ice fields, and valley glaciers	180,000	.45
Total	32,328,300	80.32

During cold-climate intervals, known as glacial epochs or ice ages, sea level falls because of a shift in the global hydrologic cycle: water is evaporated from the oceans and stored on the continents as large ice sheets and expanded ice caps, ice fields, and mountain glaciers. Global sea level was about 125 meters below today's sea level at the last glacial maximum about 20,000 years ago (Fairbanks, 1989). As the climate warmed, sea level rose because the melting North American, Eurasian, South American, Greenland, and Antarctic ice sheets returned their stored water to the world's oceans. During the warmest intervals, called interglacial epochs, sea level is at its highest. Today we are living in the most recent interglacial, an interval that started about 10,000 years ago and is called the Holocene Epoch by geologists.

Sea levels during several previous interglacials were about 3 to as much as 20 meters higher than current sea level. The evidence comes from two different but complementary types of studies. One line of evidence is provided by old shoreline features (Fig. 2). Wave-cut terraces and beach deposits from regions as separate as the Caribbean and the North Slope of Alaska suggest higher sea levels during past interglacial times. A second line of evidence comes from sediments cored from below the existing Greenland and West Antarctic ice sheets. The fossils and chemical signals in the sediment cores indicate that both major ice sheets were greatly reduced from their current size or even completely melted one or more times in the recent geologic past. The precise timing and details of past sea-level history are still being debated, but there is clear evidence for past sea levels significantly higher than current sea level.

POTENTIAL SEA-LEVEL CHANGES

If Earth's climate continues to warm, then the volume of present-day ice sheets will decrease. Melting of the current Greenland ice sheet would result in a sea-level rise of about 6.5 meters; melting of the West Antarctic ice sheet would result in a sea-level rise of about 8 meters (Table 1). The West Antarctic ice sheet is especially vulnerable, because much of it is grounded below sea level. Small changes in global sea level or a rise in ocean temperatures could cause a breakup of the two buttressing ice shelves (Ronne/Filchner and Ross). The resulting surge of the West Antarctic ice sheet would lead to a rapid rise in global sea level.

Figure 2. Wave-cut terraces on San Clemente Island, California. Nearly horizontal surfaces, separated by step-like cliffs, were created during former intervals of high sea level; the highest terrace represents the oldest sea-level high stand. Because San Clemente Island is slowly rising, terraces cut during an interglacial continue to rise with the island during the following glacial interval. When sea level rises during the next interglacial, a new wave-cut terrace is eroded below the previous interglacial terrace. Geologists can calculate the height of the former high sea levels by knowing the tectonic uplift rate of the island. Photograph by Dan Muhs, USGS.

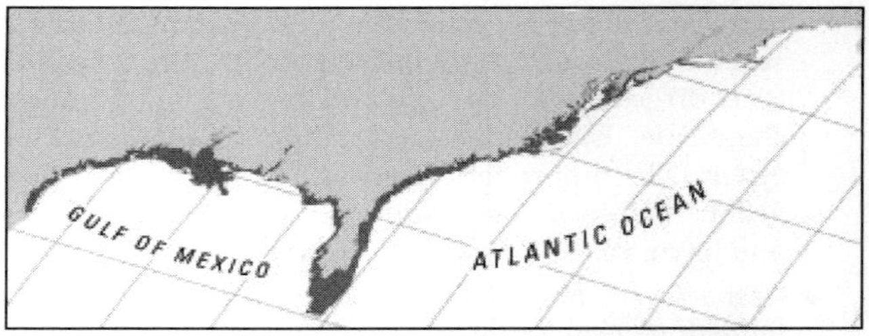

Figure 3. Red shows areas along the Gulf Coast and East Coast of the United States that would be flooded by a 10-meter rise in sea level. Population figures for 1996 (U.S. Bureau of the Census, unpublished data, 1998) indicate that a 10-meter rise in sea level would flood approximately 25 percent of the Nation's population.

Reduction of the West Antarctic and Greenland ice sheets similar to past reductions would cause sea level to rise 10 or more meters. A sea-level rise of 10 meters would flood about 25 percent of the U.S. population, with the major impact being mostly on the people and infrastructures in the Gulf and East Coast States (Fig. 3).

Researchers at the U.S. Geological Survey and elsewhere are investigating the magnitude and timing of sea-level changes during previous interglacial intervals. Better documentation and understanding of these past changes will improve our ability to estimate the potential for future large-scale changes in sea level.

READING LIST

Fairbanks, R.G. (1989). A 17,000-year glacio-eustatic sea level record; influence of glacial melting rates on the Younger Dryas event and deep-ocean circulation. *Nature* **342**(6250): 637–642.

Williams, R.S., and Hall, D.K. (1993). Glaciers, in Chapter on the cryo-sphere, in Gurney, R.J., Foster, J.L., and Parkinson, C.L., eds., *Atlas of Earth Observations Related to Global Change*. Cambridge University Press, Cambridge, UK, pp. 401–422.

THE PERMANENT SERVICE FOR MEAN SEA LEVEL

S. Jevrejeva
S. Holgate
P.L. Woodworth
Proudman Oceanographic Laboratory
Birkenhead, United Kingdom

Mean sea level (MSL) is the average level of the sea, relative to the level of the land on which the measurements are being made, recorded over an extended period such as a month, year, or the lunar nodal period of 18.6 years by an instrument called a tide gauge (or coastal sea level recorder). MSL data are used in a wide range of scientific applications including studies into climate change, ocean circulation variability and geology, as well as in practical applications such as surveying and the establishment of national leveling datums.

The Permanent Service for Mean Sea Level (PSMSL) is the global data bank for such MSL information, and it has since 1933 been responsible for the collection, publication, analysis, and interpretation of sea level data from the global network of tide gauges. It is based at the Proudman Oceanographic Laboratory (POL) in Liverpool, U.K. and is a member of the Federation of Astronomical and Geophysical Data Analysis Services (FAGS) established by the International Council for Science (ICSU). The PSMSL is supported by FAGS, the Intergovernmental Oceanographic Commission of the United Nations Educational, Scientific and Cultural Organisation (IOC/UNESCO), and NERC.

The database of the PSMSL contains almost 53,000 station-years of monthly and annual mean values of sea level from nearly 2000 tide gauge stations around the world received from almost 200 national authorities (see Fig. 1). On average, approximately 2000 station-years of data are entered into the database each year. All data are readily available from the PSMSL website: www.pol.ac.uk/psmsl.

Data for all stations are included in the PSMSL METRIC (or total) dataset. The METRIC monthly and annual means for any one station-year are necessarily required to be measured to a common datum, although, at this stage, datum continuity between years is not essential. The year-to-year datum checks become essential, however, if the data are subsequently to be included in the PSMSL 'Revised Local Reference (RLR)' component of the dataset. The RLR dataset contains records for which time series analysis of sea level changes can be performed. Long records from this dataset have been the basis of all analyses of secular changes in global sea level during the last century. The geographical distribution of longer RLR records contains significant geographical bias toward the Northern Hemisphere, a situation that is being rectified by means of international collaboration.

Aside from its central role of operation of the global sea level data bank, the PSMSL has a responsibility as a member of FAGS to provide the sea level community with as full a Service as possible with regard to the acquisition, analysis, and interpretation of sea level data. Consequently, the PSMSL provides a range of advice to tide gauge operators and data analysts. It has occupied a central planning and management role in the development of the Global Sea Level Observing System (GLOSS) of the IOC. Through GLOSS and via other routes, the PSMSL provides advice and training to national sea level authorities and individual sea level scientists and technologists. In addition to the provision of training materials (e.g., tide gauge operation manuals), the PSMSL supplies software packages suitable for tidal data analysis and quality control purposes.

In addition to the training courses associated with GLOSS, the PSMSL has every few years hosted important study groups and international conferences on sea level science. The study groups have concerned themselves with topics such as the use of global positioning system (GPS) receivers at tide gauge sites to determine the local rates of vertical land movement and have been held under the auspices of the International Association for the Physical Sciences of the Ocean (IAPSO) Commission on Mean Sea Level and Tides (CMSLT), the scientific body to which

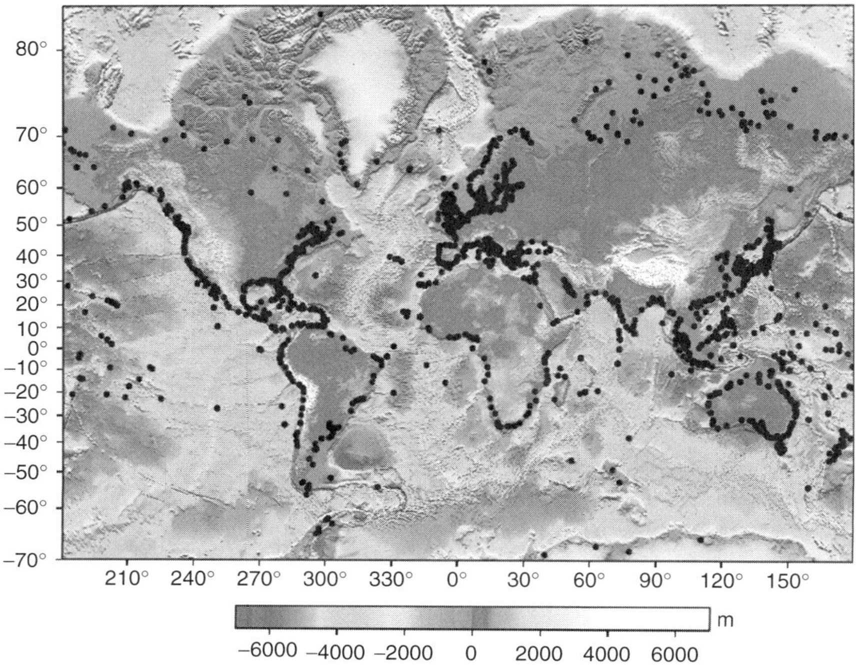

Figure 1. Distribution of PSMSL stations.

the PSMSL reports formally. The PSMSL hosted a major Symposium in Vienna in 1991 as part of the International Union of Geodesy and Geophysics (IUGG) Congress, an international conference at the Linnean Society in London in 1993 as part of its 60th anniversary celebrations, co-organized "Tidal Science 96" at the Royal Society in London in 1996, and took a major part in "A Celebration of UK Sea Level Science" at the Royal Society in 2004. The proceedings of each of these conferences have since been published. A further major conference is planned in 2008 for the PSMSL 75th anniversary.

Probably the most important recent scientific publications with which the PSMSL is associated are those of the First (1990), Second (1995), and Third (2000) Scientific Assessments of Intergovernmental Panel on Climate Change (IPCC). The PSMSL Director has been a lead author for the sea level chapters in each of the IPCC studies. Major conclusions have been that global sea level has indeed risen by approximately 10–20 cm during the past century and may rise by amounts several times larger during the next 100 years.

The PSMSL is conscious that developments in technology have expanded the field of sea and land level studies. During the 1990s, satellite radar altimetry and GPS recording become established techniques, whereas space gravity offers the potential for being an effective source of sea-level-related information in the future. Therefore, the PSMSL maintains full participation with altimeter and space gravity working groups in view of the importance of those techniques to sea level research. PSMSL personnel have Principal Investigator status for the TOPEX/Poseidon, Jason, ERS, and Envisat altimeter missions, in addition to the GRACE and GOCE space gravity projects. The major challenge for the future, to which the PSMSL is committed, is to see the established tide gauge and new space-based techniques closely linked within one coherent global sea level monitoring system.

READING LIST

Woodworth, P.L. and Player, R. (2003). The Permanent Service for Mean Sea Level: an update to the 21st century. *J. Coastal Res.* **19**: 287–295.

Woodworth, P.L., Aarup, T., Merrifield, M., Mitchum, G.T., and Le Provost, C. (2003). Measuring progress of the Global Sea Level Observing System. *EOS, Trans. Am. Geophys. Union.* **84**(50): 565.

MARINE AND ESTUARINE MICROALGAL SEDIMENT TOXICITY TESTS

Ignatio Moreno-Garrido
Luís M. Lubián
Julián Blasco
Institute of Marine Sciences of Analucia
Cadiz, Spain

The term microphytobentos refers to the microscopic algae that live on the submerged (temporary or not) floor in fresh water, estuarine, or marine environments. Microphytobenthos are mainly composed of mobile, pennate diatoms and cyanophytes. In marine and estuarine environments, those organisms can be found in habitats such as salt marshes, submerged vegetation beds, intertidal (sand or mud) flats, or subtidal sediments where light permits microalgal growth. As the presence of these photosynthetic organisms is not always evident, MacIntyre et al. (1) borrowed the term "secret garden" from the homonymous book published in 1888 and written by Frances Hodgson Burnett (1849–1874), in order to make a literary allusion to microphytobenthos. In fact, habitats where microphytobentos are the only primary producers are recognized as "unvegetated" areas, but the concentration of chlorophyll *a* in the upper 0.5 cm of the sediments where those organisms live generally exceed the depth-integrated chlorophyll in the entire overlying water column (2). In some cases, chlorophyll from microphytobenthos can be up to six orders of magnitude higher than that for the overlying water (3). Nevertheless, some authors estimate that primary production in the bottom would be lower than in the free plankton in spite of that exceptional data (4). But in some habitats, biomass from benthic microalgae can match or even exceed biomass of bacteria present in the same space (5). Thus, microphytobenthos necessarily play a key role in the benthic trophic webs (6–8). In certain biocenosis, the organisms of the microphytobenthos are the main—sometimes the only—source of carbon for grazers or bacteria (7,9,10). The microphytobenthos are also very important in relation to the stability of coastal and estuarine sediments. Although some cyanobacteria can show hydrophobicity as a mechanism to attach sediment particles or other cells (11), the main strategy of microphytobenthos to keep attached to the sediment is the production of agglutinant molecules [carbohydrates (CH) or exopolysaccharides (EPS)] (12–15). The presence of these molecules has important trophic implications, but their role in the maintenance of the structure of the upper part of the sediment is also important (16–19), because the film of adherent substances produced by microphytobenthos increases the sediment stability. Toxicants such as herbicides can alter microphytobenthic growth and the production of CH and EPS can be diminished. A loss of sediment stability can induce an increase in the turbidity of the water and a higher rate of deposition of fine particles on submerged higher plants or macroalgae, thus reducing even more the primary production of the whole and adjacent systems (20).

The importance of sediments in accumulation of xenobiotics in coastal and estuarine environments has been pointed out. Most chemical contaminants entering marine or estuarine environments eventually accumulate in sediments due to different reasons, including higher salinity values that diminish solubility of such substances in water. Sediment can act as a sink for these substances but also as a subsequent source for the same (25). On the other hand, and except in cases of extreme contamination, chemical data by themselves do not predict hazard (21,22). Thus, bioassays are needed to assess the potential toxicity of sediments.

In spite of the importance of microphytobenthos, few efforts have been made to develop standard toxicity test on these organisms. Guidelines from the SETAC (23) and recent revisions (24–27) about toxicity testing on benthic organisms offer good information on macro and meiofauna but completely ignore microphytobenthos. The reason for this "exclusion" cannot be found in the lack of importance of benthic microalgae, because of all the reasons expressed above. Probably the most important reasons for the scarcity of work on microphytobenthos ecotoxicology are the difficulties that this biological material cause.

First, it is not easy to efficiently remove microphytobenthos from sediments. Size and weight of microalgae on the sediment match part of the sediment particles mixed with them. There are descriptions of techniques that use the migration capacity of the microphytobenthic organisms in order to remove them from sediment. It is well known that microphytobenthos vertically migrate through the sediment as a function of the light and tide conditions. During low tide and light conditions, cells unbury and remain at the surface of the sediment, but during high tide or night conditions, microphytobenthic cells bury themselves again, in order to avoid being removed by the current or waves or grazed during nonphotosynthetic conditions (1,28,29). This vertical movement can be exploited to make cells migrate through a plankton net separated from the sediment by one or more lens tissue papers (8). Other authors improved on this method by covering the plankton net with a few millimeters of silica powder, where living mobile cells accumulate (30). Less effective seems to be the technique that uses the adherent capacity of microphytobenthos for removing cells by the disposition of a cover glass on the sediment (19). It is supposed that cells will attach to the cover glass, but it is not easy to ensure the efficiency and repeatability of the technique. Another approach is to take cores (made of Plexiglas, PVC, or other materials) (31) from the upper sediment and resuspend subsamples of the previously sliced sediment in order to directly count (and taxonomically identify) cells by light microscopy (4). This latter method ensures the integration of all species (motile or not). The use of fluorescence microscopy can help in distinguishing photosynthetic cells from debris: with a blue filter and a barrier filter of 530 nm, chlorophyll emits a bright red fluorescence that clearly reveals cells and facilitates their localization and count, something that is difficult if this technique is not used (32,33) (Fig. 1). Other techniques use density differences to separate diatoms from debris by centrifugation in a Percoll gradient (34). Although this technique seems to be good for isolating cells, the percentage of cell recovery is low (near 5% of total population in natural locations).

Due to the difficulties of handling cells among the sediment, several works were limited to analyzing photosynthetic pigments or photosynthesis in the upper sediments (35–37) as biomarkers for the microphytobenthic biomass. The analysis of degraded pigment molecules in sediments can be useful as a biomarker for grazing pressure (38).

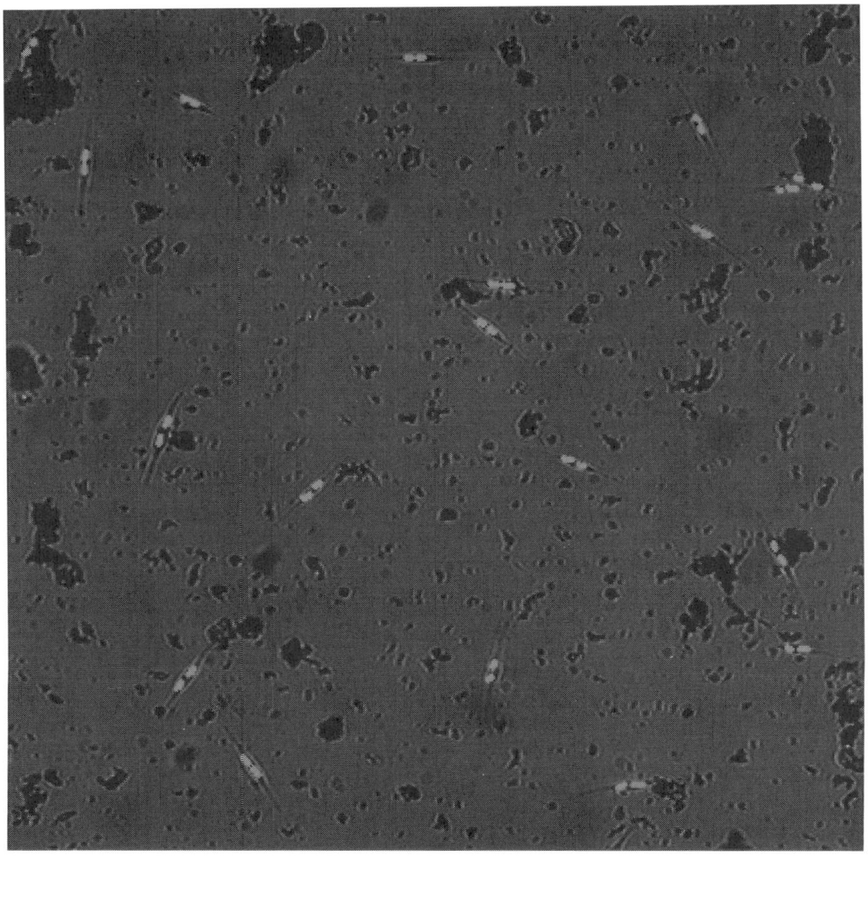

Figure 1. Cells of the diatom *Cylindrotheca closterium* among sediment particles, observed under fluorescence microscopy. Using a barrier filter of 530 nm, the two chloroplasts of each cell are bright red, facilitating the counting and location of the algal cells.

But difficulties in handling microphytobenthos do not end there. The disposition in "patches" of microphytobenthos in field locations is evident, providing a spatial heterogeneous distribution of the cells (39,40). This spatial distribution is sometimes conditioned by the presence of ridges and runnels (41) in the sediment surface, provoked by the natural dynamics of the ecosystem. This must be taken into account when estimations of microphytobenthic organism density are intended to be developed in actual locations (15). Delgado (4) described spatial heterogeneity in the delta of the Ebro River as being insignificant over distances between 10 cm and 10 m. Another consideration is that in emerged (low tide) conditions, there is a process of gradual compaction of sediment due to dewatering, which implies a higher density but a lower content of total pigments, exopolysaccharides, or individuals (42).

In spite of all this, *in situ* or *in vitro* bioassays involving microphytobenthos should be considered as powerful tools to determine potential toxicity of sediments (24). Some efforts have been made in this direction. Wong et al. (43) described a microalgal toxicity test on sediments from the coast of Hong Kong, but they used a planktonic (and not benthic) species (the chlorophyte *Dunaliella salina*) that, additionally, belongs to a genus that demonstrates strong resistance to toxicants. Abalde et al. (44) did not find growth inhibition of populations of *D. tertiolecta* at levels of 8 mg $\cdot$ L^{-1} Cu, and Moreno-Garrido et al. (45) found that *D. salina* was the most resistant species to Cu and Cd among the assayed microalgal strains. Each microalgal species shows different sensitivity for toxicants, but there are references showing similar (on the same order of magnitude) sensitivities for very different taxons (46). In this respect, *Isochrysis galbana*, *Cylindrotheca (Nitzschia) closterium*, and *Nannochloropsis species* showed similar responses to water soluble fraction of petroleum. Phytoplankton (free swimming or floating microalgae), periphyton (microalgae growing on solid substrate), and epipsammon (microalgae growing more or less attached to sand) showed comparable toxicity sensitivity responses to paraquat and simazine (47). Other works describe sediment toxicity tests on elutriates or extracts (48). Tolun et al. (49) described experiments where natural sediment toxicity tests on *Phaeodactylum tricornutum*, based on elutriates and bulk exposure, are compared: in the case of elutriates, the authors found different degrees of growth inhibition, although for direct exposure to sediments all organisms died, supporting the idea that direct exposure to sediment will give more realistic (or more sensitive) responses because part of the toxicants could not be extracted in elutriates or extracts (50). A very interesting approach to sediment toxicity tests on benthic algae was that made by Dahl and Blank (31), in which epipsammic communities were transported to a laboratory where they were kept and used in subsequent measurements of metabolic activities and short-term toxicity tests. Cairns et al. (51) defended the use of microorganisms in toxicity tests, because they can show very high sensitivities to toxicants and thus should be included in regulatory-proposed guidelines. But incorporation in those guidelines of multispecies tests that could be more "environmentally realistic" are very slow, fundamentally because of methodological questions (regarding replication and reproducibility). On the other hand, predictions based on multispecies tests are no more accurate than those based on monospecific bioassays, which are cheaper and more reproducible than a multispecies bioassay (51).

As far as we know, the first attempts to develop a standardized, repeatable protocol for sediment toxicity testing involving a microphytobenthic strain and direct exposure of microalgal cells to sediment have only recently taken place (32,33). In those works, populations of the benthic diatom *Cylindrotheca closterium* were exposed to sediment spiked with heavy metals or tensides. The test also considers the effect of particle size distribution on growth of the tested microalgal strain, which could mask actual responses of algae to present toxicants in other experiments such as those described by Tolun et al. (49). The test is simple, repeatable, and cheap, and it does not require special facilities other than those found in any laboratory. It is based on the 72-h algal growth inhibition test from OECD (52), adapted to sediments and marine or estuarine habitats. *Cylindrotheca closterium*, formerly known as *Nitzschia closterium*, demonstrated to be a good subject in other toxicity bioassays and there is a good pool of previous information about this species (53–58). This species is cosmopolitan for temperate coastal waters, ubiquitous, easy to handle in the laboratory, fast growing, sensitive to toxicants, and presents very low nutrient requirements. Nevertheless, other species have been assayed and compared with *C. closterium* in order to detect toxicity in natural sediments. EC_{50} values were calculated for three benthic diatoms exposed to sediment obtained from six different locations from Aveiro Lagoon (Portugal). Those values are shown in Table 1. Locations I, II, and VI were less toxic to microalgae than locations III, IV, and V, since *C. closterium* is slightly more sensitive than the other two species assayed for the majority of the samples. When a chemical analysis of the samples was performed and a similarity analysis carried out crossing toxicity values with possible substances involved (heavy metals, C, N, PCBs), it was found that some heavy metals (Sn, Zn, Hg, Cu, and Cr) had a great effect (Sn the greatest), as shown by the more than 50% of similarity between samples that showed significant growth inhibition for the benthic diatoms assayed (unpublished data).

Other recent work from Adams and Stauber (59) also describes a whole-sediment toxicity test on a benthic microalgal diatom (*Entomoneis* cf. *punctulata*). In this case, a flow cytometer is used to detect viability of living algae by the use of a fluorochrome (fluorescein diacetate,

Table 1. EC_{50} Values for Sediments from Six Locations at the Aveiro Lagoon (Portugal) and Three Benthic Diatoms

	Sample Sites					
Algal Species	I	II	III	IV	V	VI
P. tricornutum	N.I.[a]	N.I.	27	51	62	N.I
C. closterium	92	90	62	31	61	N.I.
Navicula sp.	N.I.	N.I.	64	71	65	N.I.

[a] If 100% of the sediment does not exhibit a growth inhibition value of 50%, EC_{50} is denoted as N.I. (no inhibition).

FDA). In this technique, used previously (60) to determine the toxic response in planktonic microalgal cells, FDA absorbed by living or dead cells is only hydrolyzed by nonspecific esterases inside the living cells. Hydrolyzed FDA is converted into fluorescein and can be detected by flow cytometry techniques. Although the FDA technique is less sensitive than growth inhibition measurements (60), flow cytometry opens a wide field of possibilities for microalgal toxicity testing on sediments, because several different fluorochromes can be used to measure quite different cellular parameters used as biomarkers.

BIBLIOGRAPHY

1. MacIntyre, H.L., Geider, R.J., and Miller, D.C. (1996). Microphytobenthos: the ecological role of the "secret garden" of unvegetated, shallow-water marine habitats. I. Distribution, abundance and primary production. *Estuaries* **19**(2A): 186–201.
2. Nelson, J.R. et al. (1999). Benthic microalgal biomass and irradiance at the sea floor on the continental shelf of the South Atlantic Bight: spatial and temporal variability and storm effects. *Continental Shelf Res.* **19**: 477–505.
3. Conde, D. et al. (1999). Comparison of the areal amount of chlorophyll *a* of planktonic and attached microalgae in shallow coastal lagoon. *Hydrobiologia* **408/409**: 285–291.
4. Delgado, M. (1989). Abundance and distribution of microphytobentos in the bays of Ebro delta (Spain). *Estuarine Coastal Shelf Sci.* **29**: 183–194.
5. La Rosa, T., Mirto, S., Mazzola, A., and Danovaro, R. (2001). Differential responses of benhtic microbes and meiofauna to fish-farm disturbance in coastal sediments. *Environ. Pollut.* **112**: 427–434.
6. Miller, D.C., Geider, R.J., and MacIntyre, H.L. (1996). Microphytobenthos: the ecological role of the "secret garden" of unvegetated, shallow-water marine habitats. II. Role in sediment stability and shallow-water food webs. *Estuaries* **19**(2A): 202–212.
7. Middelburg, J.J. et al. (2000). The fate of intertidal microphytobenthos carbon: an *in situ* ^{13}C labelling study. *Limnol. Oceanogr.* **45**(6): 1224–1234.
8. Absil, M.C.P. and Van Scheppingen, Y. (1996). Concentration of selected heavy metals in benthic diatoms and sediment in the Westerchelde Estuary. *Bull. Environ. Contam. Toxicol.* **56**: 1008–1015.
9. Puddu, A. et al. (2003). Bacterial uptake of DOM released from P-limited phytoplankton. *FEMS Microbiol. Ecol.* **46**(3): 257–268.
10. Dalsgaard, T. (2003). Benthic primary production and nutrient cycling in sediments with benthic microalgae and transient accumulation of macroalgae. *Limnol. Oceanogr.* **48**(6): 2138–2150.
11. Fattom, A. and Shilo, M. (1984). Hydrophobicity as an adhesion mechanism of benthic cyanobacteria. *Appl. Environ. Microbiol.* 135–143.
12. de Winder, B., Staats, N., Stal, L.J., and Paterson, D.M. (1999). Carbohydrate secretion by phototrophic communities in tidal sediments. *J. Sea Res.* **42**: 131–146.
13. Goto, N., Mitamura, O., and Terai, H. (2001). Biodegradation of photosynthetically produced extracellular organic carbon from intertidal benthic algae. *J. Exp. Mar. Biol. Ecol.* **257**: 73–86.
14. Staats, N., Stal, L.J., and Mur, L.R. (2000). Exopolysaccharide production by the epipelic diatom *Cylindrotheca closterium*: effects of nutrient conditions. *J. Exp. Mar. Biol. Ecol.* **249**: 13–27.
15. Taylor, I.S. and Paterson, D.M. (1998). Microspatial variation in carbohydrate concentrations with depth in the upper millimetres of intertidal cohesive sediments. *Estuarine Coastal Shelf Sci.* **46**: 359–370.
16. Austen, I., Andersen, T.J., and Edelvang, K. (1999). The influence of benthic diatoms and invertebrates on the erodability of an intertidal mudflat, the Danish Wadden Sea. *Estuarine Coastal Shelf Sci.* **49**: 99–111.
17. Blanchard, G.F. et al. (2000). The effect of geomorphological structures on potential biostabilisation by microphytobenthos on intertidal mudflats. *Continental Shelf Res.* **20**: 1243–1256.
18. de Brower, J.F.C., Bjelic, S., Deckere, E.M.G.T., and Stal, L.J. (2000). Interplay between biology and sedimentology in a mudflat (Biezelingse Ham, Westerchelde, The Netherlands). *Continental Shelf Res.* **20**: 1159–1177.
19. Madsen, K., Nilsson, P., and Sundbäck, K. (1993). The influence of benthic microalgae on the stability of a subtidal sediment. *J. Exp. Mar. Biol. Ecol.* **170**: 159–177.
20. Mason, C.F. et al. (2003). The role of herbicides in the erosion of salt marshes in eastern England. *Environ. Pollut.* **122**: 41–49.
21. Munawar, M. and Munawar, I.F. (1987). Phytoplankton bioassays for evaluating toxicity of *in situ* sediment contaminants. *Hydrobiologia* **149**: 87–105.
22. O'Connor, T.P. and Paul, J.F. (2000). Misfit between sediment toxicity and chemistry. *Mar. Pollut. Bull.* **40**(1): 59–64.
23. SETAC (Society of Environmental Toxicology and Chemistry—Europe). (1993). Guidance document on sediment toxicity tests and bioassays for freshwater and marine environments. In: *Workshop on Sediment Toxicity Assessment*. I.R. Hill, P. Matthiessen, and F. Heimbach (Eds.). November 1993, Renesse, The Netherlands.
24. Chapman, P.M. and Wang, F. (2001). Assessing contamination in estuaries. *Environ. Toxicol. Chem.* **20**(1): 3–22.
25. Lamberson, J.O., DeWitt, T.H., and Swartz, R.C. (1992). Assessment of sediment toxicity to marine benthos. In: *Sediment Toxicity Assessment*. G.A. Burton (Ed.). Lewis Publishers, Boca Raton, FL.
26. Traunspurger, W. and Drews, C. (1996). Toxicity analysis of freshwater and marine sediments with meio- and macrobenthic organisms: a review. *Hydrobiologia* **328**: 215–261.
27. Shin, P.K.S. and Lam, W.K.C. (2001). Development of a marine sediment pollution index. *Environ. Pollut.* **113**: 281–291.
28. Guarini, J.M., Blanchard, G.F., and Gros, P. (2000). Quantification of the microphytobenthic primary production in European intertidal mudflats—a modelling approach. *Continental Shelf Res.* **20**: 1771–1788.
29. Brunnet, C. and Lizon, F. (2003). Tidal and diel periodicities of size-fractionated phytoplankton pigment signatures at an offshore station in the southeastern English Channel. *Estuarine Coastal Shelf Sci.* **56**: 833–843.
30. Blanchard, G.F. et al. (2000). Measurement of ingestion rate of *Hydrobia ulvae* (Pennant) on intertidal epipelic microalgae: the effect of mud snail density. *J. Exp. Mar. Biol. Ecol.* **255**: 247–260.
31. Dahl, B. and Blank, H. (1996). Use of sand-living microalgal communities (epipsammon) in ecotoxicological testing. *Mar. Ecol. Prog. Ser.* **144**: 163–173.

32. Moreno-Garrido, I., Hampel, M., Lubián, L.M., and Blasco, J. (2003). Sediment toxicity tests using benthic marine microalgae *Cylindrotheca closterium* (Ehremberg) Lewin & Reimann (Bacillariophyceae). *Ecotoxicol. Environ. Safety* **54**: 290–295.
33. Moreno-Garrido, I., Hampel, M., Lubián, L.M., and Blasco, J. (2003). Marine benthic microalgae *Cylindrotheca closterium* (Ehremberg) Lewin & Reimann (Bacillariophyceae) as a tool for measuring toxicity of linear alkylbenzene sulfonate (LAS) in sediments. *Bull. Environ. Contam. Tox.* **70**(2): 242–247.
34. Sevecke, S., Morawietz, T., Krumbein, W.E., and Rhiel, E. (1997). Use of percoll gradient centrifugation for the isolation of diatoms from Wadden Sea sediments; diatom yields, species recoveries and population diversity. *Microbiologia* **13**(2): 173–184.
35. Bennet, A., Bianchi, T.S., Means, J.C., and Carman, K.R. (1999). The effects of polycyclic aromatic hydrocarbon and grazing on the abundance and composition of microphytobentos in salt marsh sediments (Pass Fourchon, LA). I. A microcosm experiment. *J. Exp. Mar. Biol. Ecol.* **242**: 1–20.
36. Bennet, A., Bianchi, T.S., and Means, J.C. (2000). The effects of PAH contamination and grazing on the abundance and composition of microphytobenthos in salt marsh sediments (Pass Fourchon, LA, USA): II: The use of plant pigments as biomarkers. *Estuarine Coastal Shelf Sci.* **50**: 425–439.
37. Light, B.R. and Beardall, J. (2001). Photosynthetic characteristics of sub-tidal benthic microalgal populations from a temperate, shallow water marine ecosystem. *Aquat. Bot.* **70**: 9–27.
38. Cartaxana, P., Jesus, B., and Brotas, V. (2003). Pheophorbide and pheophytin a-like pigments as useful markers for intertidal microphytobentos grazing by *Hydrobia ulvae*. *Estuarine Coastal Shelf Sci.* **58**: 293–297.
39. Seuront, L. and Spilmont, N. (2002). Self-organized criticality in intertidal microphytobentos path patterns. *Physica A* **313**: 513–539.
40. Rainey, M.P. et al. (2003). Mapping intertidal estuarine sediment grain size distributions through airborne remote sensing. *Remote Sensing Environ.* **86**: 480–490.
41. Taylor, I.S. and Paterson, D.M. (1998). Microspatial variation in carbohydrate concentrations with depth in the upper millimetres of intertidal cohesive sediments. *Estuarine Coastal Shelf Sci.* **46**: 359–370.
42. Perkins, R.G. et al. (2003). Changes in microphytobenthic chlorophyll *a* and EPS resulting from sediment compaction due to de-watering: opposing patterns in concentration and content. *Continental Shelf Res.* **23**: 575–586.
43. Wong, C.K.C., Cheung, R.Y.H., and Wong, M.H. (1999). Toxicological assessment of coastal sediment in Hong Kong using a flagellate, *Dunaliella tertiolecta*. *Environ. Pollut.* **105**: 175–183.
44. Abalde, J. et al. (1995). Response of the marine microalga *Dunaliella tertiolecta* (Chlorophyceae) to copper toxicity in short time experiments. *Bull. Environ. Contam. Toxicol.* **54**: 317–324.
45. Moreno-Garrido, I., Lubián, L.M., and Soares, A.M.V.M. (1999). Growth differences in cultured populations of *Brachionus plicatilis* Müller caused by heavy metal stress as function of microalgal diet. *Bull. Environ. Contamin. Tox.* **63**: 392–398.
46. Tsvetnenko, Y. and Evans, L. (2002). Improved approaches to ecotoxicity testing of petroleum products. *Mar. Pollut. Bull.* **45**: 148–156.
47. Bonilla, S., Conde, D., and Blank, H. (1998). The photosynthetic responses of marine phytoplankton, periphyton and epipsammon to the herbicides paraquat and simazine. *Ecotoxicology* **7**: 99–105.
48. Baun, A., Justensen, K.B., and Nyholm, N. (2002). Algal test with soil suspensions and elutriates: a comparative evaluation for PAH-contaminated soils. *Chemosphere* **46**: 251–258.
49. Tolun, L.G. et al. (2001). The pollution status and the toxicity of the surface sediments in Izmit Bay (Marmara Sea), Turkey. *Environ. Int.* 163–168.
50. Liß, W. and Ahlf, W. (1997). Evidence of whole-sediment, porewater and elutriate testing in toxicity assessment of contaminated sediments. *Ecotoxicol. Environ. Safety* **36**: 140–147.
51. Cairns, J., Jr., McCormick, P.V., and Belanger, S.E. (1993). Prospects for the continued development of environmentally-realistic toxicity tests using microorganisms. *J. Environ. Sci.* **5**(3): 253–268.
52. OECD (Organization for the Economic Cooperation and Development). (1998). *OECD Guideline for Testing of Chemicals: Alga, Growth Inhibition Test.*
53. Florence, T.M., Lumsden, B.G., and Fardy, J.J. (1983). Evaluation of some physicochemical techniques for the determination of the fraction of dissolved copper toxic to the marine diatom *Nitzschia closterium*. *Anal. Chim. Acta* **151**: 281–295.
54. Morris, E.P. and Kromkamp, J.C. (2003). Influence of temperature on the relationship between oxygen- and fluorescence-based esimates of photosynthetic parameters in a marine benthic diatom (*Cylindrotheca closterium*). *Eur. J. Phycol.* **38**: 133–142.
55. Rachlin, J.W., Warkentine, B., and Jensen, T.E. (1983). The growth response of the diatom *Chlorella saccarophyla, Navicula incerta* and *Nitzschia closterium* to selected concentrations of cadmium. *Bull. Torrey Bot. Club* **110**(2): 217–223.
56. Pablo, F., Stauber, J.L., and Buckney, R.T. (1997). Toxicity of cyanide and cyanide complexes to the marine diatom *Nitzschia closterium*. *Water Res.* **31**(10): 2435–2442.
57. Stauber, J.L. and Florence, T.M. (1985). The influence of iron on copper toxicity to the marine diatom, *Nitzschia closterium* (Ehrenberg) W. Smith. *Aquat. Toxicol.* **6**: 297–305.
58. Stauber, J.L. and Florence, T.M. (1990). Mechanism of toxicity of zinc to the marine diatom *Nitzschia closterium*. *Mar. Biol.* **105**: 519–524.
59. Adams, M.S. and Stauber, J.L. (2004). Development of a whole-sediment toxicity test using a benthic marine microalga. *Environ. Toxicol. Chem.* **23**(8): 1957–1968.
60. Hampel, M. et al. (2001). Acute toxicity of LAS homologues in marine microalgae. Esterase activity and inhibition growth as endpoints of toxicity. *Ecotoxicol. Environ. Safety* **48**: 287–292.

MARINE STOCK ENHANCEMENT TECHNIQUES

Jana Davis
Smithsonian Environmental Research Center
Edgewater, Maryland

The United Nations estimates that 28% of all ocean fishery stocks are overexploited or severely depleted, and another 47% is fully exploited (1). Included in this statistic are species that are long-lived and short-lived, open-ocean and coastal, migratory and sedentary. About 75% of fishery stocks are unable to withstand further exploitation, so only 25% of stocks remain to satisfy the increasing human demand, as human population grows, for fishery products.

As a result of the pressure on fishery populations and the degradation of many of these resources, a suite of fishery management techniques have been developed. One such technique, stock enhancement, is broadly defined as the (human) activity by which the population of a species is increased. In contrast, more traditional management techniques generally focus on limiting human activities that reduce the population of a species.

Traditional measures may focus on reducing fishing effort, for example, by limiting the number of days fishermen can fish, restricting the amount of gear a fishermen can use, regulating the type of gear a fisherman can deploy, or setting restrictions on the types of individuals caught by size or gender. Not only does this latter method serve to reduce catch, it strives to shift the pressure away from more reproductively important individuals toward those with less likely contributions to the next generation over the course of their lifetimes.

GOALS OF STOCK ENHANCEMENT: A NONTRADITIONAL MANAGEMENT TECHNIQUE

The technique of stock enhancement, while often used in concert with more traditional catch-reducing techniques, does not describe limitations to human activities but rather the active improvement of a fishery stock. Enhancement efforts can encompass many approaches. The population can be bolstered both by increasing carrying capacity (the number of individuals that can be supported by their habitat) or by increasing the number of individuals themselves (2,3). The former method is effective only when the factor limiting population is related to habitat, food, or some other resource. The latter method, a special case of aquaculture, is effective only when the habitat can hold more individuals than it does at present.

The key to success in stock enhancement is to identify the factor(s) that limit the population in question and then increase the level of that factor. Incorrect identification of this factor could lead to wasted time, money, and energy. Consider the hypothetical case of a coral reef fish. If the fish population is not habitat-limited, but is recruitment-limited, adding more coral reef habitat would likely serve only to spread out the population spatially. No additional recruits would be available to settle on the new habitat substrate. On the other hand, if the fish is habitat-limited, adding additional individuals will not increase the ultimate population, as there would be no space for them to occupy.

Enhancement directed toward habitat-limited populations has included efforts to enhance stocks of both fish, especially salmonids, and invertebrates (4–7). Often these cases are referred to as "habitat restoration," stressing the action rather than the consequence. Most often when people refer to "stock enhancement," they are referring to hatchery-raised individuals, usually juveniles, that are released into the wild to bolster recruitment-limited populations directly. These added juveniles are raised outside of the system, for example, in, an aquaculture facility or a field pen. The juveniles may be offspring of wild parents collected from the field (8) or offspring from generations of parents held in captivity (9).

The goal of stock enhancement is to raise the individuals beyond the initial phase of high early life-history mortality and then release them into the wild. The stage at which they are released is determined through economic optimization, survivorship maximization, or a combination of both. Maintaining individuals, especially heterotrophs, in a hatchery or a pen is expensive. The expense tends to increase as organisms grow. One might calculate the optimum release size or age as a trade-off between the cost of maintaining an individual in the hatchery and the survivorship advantage that the hatchery offers. Often, as individuals grow, the survivorship advantage of the hatchery environment over the natural environment begins to reverse. Many organisms are cannibalistic when held in extremely high densities (10), a trend that increases as organisms grow. Disease transmission also becomes a problem at high densities. Lowering densities requires an expensive solution of creating more tank or pen space. Therefore, at a certain point, mortality rates are actually lower if the organisms are released into the wild than if held in the hatchery.

The optimum release point is different for each species. For example, Kemp's ridley sea turtles are released at the age of 1 year (11). Blue crabs are released at the age of 2 to 3 months (12). Individuals that are never released but harvested in the hatchery are part of a complete aquaculture program such as, for example, farm-raised salmon, catfish, or shrimp (13,14).

In populations truly at risk, the goal of stock enhancement is not simply to provide more fish to catch. Ideally, after release, these hatchery-reared individuals contribute to the spawning stock and have a potential exponential impact on overall population generation after generation, depending on the degree of recruitment limitation. In these cases, a purely economic model to predict optimum size at release is not appropriate. Stock enhancement therefore has a potential role beyond fishery application and into the realm of threatened or endangered species protection.

STOCK ENHANCEMENT EXAMPLES

Stock enhancement techniques have been applied to many types of organisms worldwide. Finfish populations have been the most common recipients of hatchery-raised juveniles, including many salmon from both the Atlantic and Pacific (15,16), Japanese flounder (17), Hawaiian mullet (18), Nassau grouper (19), and Chesapeake Bay striped bass (20), just to name a few. Holothurians have been hatchery-raised for stock enhancement (21). Bivalves, such as quahogs (22) and soft-shell clams (23), and gastropods, such as abalones (24) and queen conch (25), have been the subject of stock enhancement efforts. Finally, crustaceans are beginning to receive more attention as possible stock enhancement targets; programs have been initiated for decades with American and

European lobsters (26,27), more recent programs for prawn (28) and the Japanese swimming crab (3), and a new exploratory program established for the Chesapeake Bay blue crab (8).

Stock enhancement efforts extend worldwide. In Japan, stock enhancement programs are in place for at least 34 finfish and 12 crustaceans (3). Programs relying on various specific techniques exist in South America, North America, Africa, Europe, Asia, Australia/New Zealand, and in the developed world as well as the developing world (29–31).

Specific techniques to optimize stock enhancement depend on the life-history traits of the species. Sessile organisms may be enhanced by seeding areas with young individuals (32). Depending on how closely the success of the enhancement effort is to be monitored, more mobile species may be released into isolated areas where they can be followed, similar to the idea of stocking freshwater lakes (8,28). For extremely mobile open-ocean species, juveniles are released into the open ocean (3), and even if tagged, large numbers cannot be followed over time.

PROBLEMS OF STOCK ENHANCEMENT

Enhancement has been a controversial management method for several reasons (15,29,33). First, hatchery-raised animals may not survive well in the wild. The hatchery environment can offer conditions very different from the natural environment: food is likely to be different, the method of foraging is different, and therefore the hatchery-raised organisms may have mortality-threatening inexperience with natural prey after their introduction into the wild. They may similarly be inexperienced in avoiding predators. Holding tanks may restrict movement, may have unnatural substrates, may have unnatural light regimes, or may have different flow regimes. As a result of all of these differences, stock enhancement efforts, often funded by the taxpayer, may not be successful (34). Many studies have noted differences in hatchery-raised and wild individuals in factors such as behavior, morphology, growth rates, and therefore survivorship (17,35,36).

A second concern is that hatchery-raised individuals may be too successful. Hatchery animals may compete with and displace wild animals (2,37). If at any point the population becomes habitat-limited, rather than recruitment-limited (or if the stock enhancement effort pushes the population over the carrying capacity), the survivorship of wild individuals may actually decrease. The ultimate result would then be a decline in the population of wild individuals, even if the overall population has increased.

A third concern is that hatchery-raised organisms may carry diseases into the wild, affecting the survivorship of wild individuals. Fourth, increases in stock size due to hatchery successes may provoke a rise in fishing effort and therefore greater pressure on the remaining wild individuals (34,38). Stock enhancement efforts are generally performed in concert with traditional management techniques, such as lowering catch and/or restricting catch to certain segments of the population. However, if the human population increases, pressure may be put on managers to reduce fishing regulations. Again, the ultimate result would be a decline in the survivorship rates of wild individuals, even if overall population has increased.

Finally, and perhaps the reason for the controversy generating the most attention in recent years is genetics, or "gene dilution." Even if efforts in the hatchery are expended to obtain brood stock from many different parents, genetic variability in the brood stock is unlikely to approach that of the wild parental stock. The offspring produced will be more similar to each other than the offspring of wild stock. Opponents of stock enhancement fear that releasing genetically relatively homogenous hatchery individuals will reduce genetic variability in wild populations as hatchery and wild individuals interbreed. The ultimate result after several to many generations might be a decline in population because lack of genetic diversity can limit a population's response to environmental change.

The controversy over stock enhancement has been fueled by the fact that most enhancement efforts have not been studied quantitatively (37,41,42). Quantitative, hypothesis-driven study did not begin until the late 1980s (43). However, now that the necessity to understand better the outcomes of enhancement programs has been recognized, in many cases the enhancement process has been refined. Methods to select better candidate species have been developed (44). Advances in tagging techniques have allowed better assessment and comparison of hatchery and wild animal survivorship (44). Most importantly, calls have been made for quantitative study of small-scale enhancement efforts before investment in large-scale programs begins (42,43).

OPTIMIZING STOCK ENHANCEMENT SUCCESS

Stock enhancement is becoming a more often commonly used method for addressing declines in fished species. Many of the problems identified above can be addressed to a comfortable degree before the program is initiated. For example, several programs are considering conditioning individuals to limit the differences between hatchery-raised and wild organisms before release into the wild (12,45,46). In this way, survivorship rates of hatchery organisms are increased, along with the output per enhancement program dollar spent.

In addition, carrying capacities of targeted release areas can be determined before programs are initiated to (1) determine whether the case is appropriate for stock enhancement and (2) determine optimum release densities of hatchery-raised organisms. Certain microhabitats within a targeted release region may have higher carrying capacity and greater food or refuge resources than others, and therefore distribution of hatchery-raised organisms can be optimized.

Methods of determining whether stock enhancement is successful are more difficult. Determining success requires knowledge about how well hatchery-raised organisms survive, how well wild organisms survive to allow comparison, and the contribution of hatchery organisms to the total population. As programs are developed, a wide range of survivorship and contribution values have been reported. For example, survivorship to fishery size was 3–4% for stocked panaeid shrimp (28), 21% for red drum in Texas (47), and up to 30% for Japanese flounder (3). Even among programs for the same species, values range widely. For example, in some European lobster programs, no hatchery-raised lobsters were recaptured in the fishery, and in others, 10–35% of landed lobsters were of hatchery origin (37).

Often the steps to quantify the success of a stock enhancement program take years of scientific study and require laborious study efforts, such as tagging and resampling individuals over time. Such efforts have been deemed mandatory by critics of stock enhancement programs before public monies are used to support these efforts to bolster fishery stocks, one of the world's most important natural resources.

BIBLIOGRAPHY

1. Food and Agriculture Organization of the United Nations. (2002). *The State of World Fisheries and Aquaculture*. United Nations, Rome, Italy.
2. Castro, K.M., Cobb, J.S., Wahle, R.A., and Catena, J. (2001). Habitat addition and stock enhancement for American lobsters, *Homarus americanus*. *Mar. Freshwater Res.* **52**: 1253–1261.
3. Masuda, R. and Tsukamoto, K. (1998). Stock enhancement in Japan: Review and perspective. *Bull. Mar. Sci.* **62**: 337–358.
4. Beamesderfer, R.C.P. and Farr, R.A. (1997). Alternatives for the protection and restoration of sturgeons and their habitat. *Environ. Biol. Fish.* **48**: 407–417.
5. Hendry, K., Cragge-Hine, D., O'Grady, M., Sambrook, H., and Stephen, A. (2003). Management of habitat for rehabilitation and enhancement of salmonid stocks. *Fish. Res.* **62**: 171–192.
6. Peterson, C.H., Grabowsi, J.H., and Powers, S.P. (2003). Estimated enhancement of fish production resulting from restoring oyster reef habitat: Quantitative valuation. *Mar. Ecol. Prog. Ser.* **264**: 149–264.
7. Rodwell, L.D., Barbier, E.B., Roberts, C.M., and McClanahan, T.R. (2003). The importance of habitat quality for marine reserve—fishery linkages. *Can. J. Fish. Aquat. Sci.* **60**: 171–181.
8. Davis, J.L.D., Young-Williams, A.C., Hines, A.H., and Zohar, Y. (in press). Assessing the feasibility of stock enhancement in the case of the Chesapeake Bay blue crab. *Can. J. Fish. Aquat. Sci.*
9. Iguchi, K., Watanabe, K., and Nishida, M. (1999). Reduced mitochondrial DNA variation in hatchery populations of ayu (*Plecoglossus altivelis*) cultured for multiple generations. *Aquaculture* **178**: 235–243.
10. Kestemont, P. et al. (2003). Size heterogeneity, cannibalism and competition in cultured predatory fish larvae: Biotic and abiotic influences. *Aquaculture* **227**: 333–356.
11. Caillouet, C.W., Fontaine, C.T., Manzella-Tirpak, S.A., and Shaver, D.J. (1995). Survival of head-started Kemp's ridley sea turtles (Lepidochelys kempii) released into the Gulf of Mexico or adjacent bays. *Chelate Conserv. Biol.* **1**: 285–292.
12. Davis, J.L.D., Eckert-Mills, M.G., Young-Williams, A.C., Hines, A.H., and Zohar, Y. (in press). Morphological conditioning of a hatchery-raised invertebrate, *Callinectes sapidus*, to improve field survivorship after release. *Aquaculture* **133**: 1–14.
13. Fleming, I.A. et al. (2000). Lifetime success and interactions of farm salmon invading a native population. *Proc. R. Soc. London Ser. B* **267**: 1517–1523.
14. Wirth, F.F. and Davis, K.J. (2003). Seafood dealers' shrimp-purchasing behavior and preferences with implications for United States shrimp farmers. *J. Shellfish Res.* **22**: 581–588.
15. Washington, P.M. and Koziol, A.M. (1993). Overview of the interactions and environmental impacts of hatchery practices on natural and artificial stocks of salmonids. *Fish. Res.* **18**: 105–122.
16. Thorpe, J.E. (1998). Salmonid life-history evolution as a constraint on marine stock enhancement. *Bull. Mar. Sci.* **62**: 465–475.
17. Kellison, G.T., Eggleston, D.B., and Burke, J.S. (2000). Comparative behaviour and survival of hatchery-reared versus wild summer flounder (*Paralichthys dentatus*). *Can. J. Fish. Aquat. Sci.* **57**: 1870–1877.
18. Leber, K.M., Blankenship, H.L., Arce, S.M., and Brennan, N.P. (1996). Influence of release season on size-dependent survival of cultured striped mullet, *Mugil cephalus*, in a Hawaiian estuary. *Fish. Bull.* **95**: 267–279.
19. Roberts, C.M., Quinn, N., Tucker, J.W., and Woodward, P.N. (1995). Introduction of hatchery-reared Nassau grouper to a coral reef environment. *N. Am. J. Fish. Manage.* **15**: 159–164.
20. Secor, D.H. and Houde, E.D. (1998). Use of larval stocking in restoration of Chesapeake Bay striped bass. *ICES J. Mar. Sci.* **55**: 228–239.
21. Battaglene, S.C., Seymour, J.E., and Ramofafia, C. (1999). Survival and growth of cultured juvenile sea cucumbers, *Holothuria scabra*. *Aquaculture* **178**: 293–322.
22. Walton, W.C. and Walton, W.C. (2001). Problems, predators, and perception: Management of quahog (hardclam), *Mercenaria mercenaria*, stock enhancement programs in southern New England. *J. Shellfish Res.* **20**: 127–134.
23. Beal, B.F., Kraus, F., and Gayle, M. (2002). Interactive effects of initial size, stocking density, and type of predator deterrent netting on survival and growth of cultured juveniles of the soft-shell clam, *Mya arenaria* L., in eastern Maine. *Aquaculture* **208**: 81–111.
24. Tegner, M.J. and Butler, R.A. (1985). The survival and mortality of seeded and native red abalones, *Haliotis rufescens*, on the Palos Verdes Penninsula. *Calif. Fish Game* **71**: 150–163.
25. Stoner, A.W. and Glazer, R.A. (1998). Variation in natural mortality: Implications for queen conch stock enhancement. *Bull. Mar. Sci.* **62**: 427–442.
26. Bannister, R.C.A. and Addison, J.T. (1998). Enhancing lobster stocks: A review of recent European methods, results, and future prospects. *Bull. Mar. Sci.* **62**: 369–387.
27. Beal, B.F., Chapman, S.R., Irvine, C., and Bayer, R.C. (1998). Lobster (Homarus americanus) culture in Maine: A community-based, fishermen-sponsored, public stock enhancement program. *Can. Ind. Rep. Fish. Aquat. Sci.* **244**: 47–54.
28. Davenport, J., Ekaratne, S.U.K., Walgama, S.A., Lee, D., and Hills, J.M. (1999). Successful stock enhancement of a lagoon prawn fishery at Rekawa, Sri Lanka, using cultured post-larvae of penaeid shrimp. *Aquaculture* **180**: 65–78.
29. Bell, J.D. and Gervis, M. (1999). New species for coastal aquaculture in the tropical Pacific: Constraints, prospects and considerations. *Aqua Int.* **7**: 207–223.
30. Cowx, I.G. (1999). An appraisal of stocking strategies in the light of developing country constraints. *Fish. Manage. Ecol.* **6**: 21–34.

31. Phillips, B.F. and Liddy, G.C. (2003). Recent developments in spiny lobster aquaculture. In: Phillips, B., Megrey, B.A., and Zhou, Y. (Eds). *Proc. 3rd World Fish. Congr.*: Feeding the world with fish in the next millenium—the balance between production and environment. American Fisheries Society Symposium Vol. 38, pp. 43–57.
32. Barbeau, M.A. et al. (1996). Dynamics of juvenile sea scallop (*Placopecten magellameus*) and their predators in bottom seeding trials in Lanenburg Bay, Nova Scotia. *Can J. Fish. Aquat. Sci.* **53**: 2494–2512.
33. Li, J. (1999). An appraisal of factors constraining the success of fish stock enhancement programmes. *Fish. Manage. Ecol.* **6**: 161–169.
34. Hilborn, R. (1998). The economic performance of marine stock enhancement projects. *Bull. Mar. Sci.* **62**: 661–674.
35. Mana, R.R. and Kawamura, G. (2002). A comparative study on morphological differences in the olfactory system of red sea bream (*Pagrus major*) and black sea bream (*Acanthopagrus schlegeli*) from wild and cultured stocks. *Aquaculture* **209**: 285–306.
36. Davis, J.L.D. et al. (2004). Differences between hatchery-raised and wild blue crabs (*Callinectes sapidus*): Implications for stock enhancement potential. *Trans. Am. Fish. Soc.*
37. Bannister, R.C.A., Addison, J.T., and Lovewell, S.R.J. (1994). Growth, movement, recapture rate and survival of hatchery-reared lobsters (*Homarus gammarus* (Linnaeus, 1758)) released into the wild on the English East Coast. *Crustaceana* **67**: 156–172.
38. Bannister, R.C.A. (2000). Lobster *Homarus gammarus* stock enhancement in the U.K.: Hatchery-reared juveniles do survive in the wild, but can they contribute significantly to ranching, enhancement, and management of lobster stocks? *Can. Spec. Publ. Fish. Aquat. Sci.* **130**: 23–32.
39. Tringali, M.D. and Bert, T.M. (1998). Risk to genetic effective population size should be an important consideration in fish stock-enhancement programs. *Bull. Mar. Sci.* **62**: 641–659.
40. Utter, F. (1998). Genetic problems of hatchery-reared progeny released into the wild, and how to deal with them. *Bull. Mar. Sci.* **62**: 623–640.
41. Heppell, S.S. and Crowder, L.B. (1998). Prognostic evaluation of enhancement programs using population models and life history analysis. *Bull. Mar. Sci.* **62**: 495–507.
42. Crowe, T.P. et al. (2002). Experimental evaluation of the use of hatchery-reared juveniles to enhance stocks of the topshell *Trochus niloticus* in Australia, Indonesia and Vanuatu. *Aquaculture* **206**: 175–197.
43. Leber, K.M. (1999). Rational for an experimental approach to stock enhancement. In: *Stock Enhancement and Sea Ranching*. B.R. Howell, E. Moksness, and T. Sväsand (Eds.). Blackwell Science, Oxford, pp. 63–75.
44. Blackenship, H.L. and Leber, K.M. (1995). A responsible approach to marine stock enhancement. *Am. Fish. Soc. Symp.* **15**: 167–175.
45. Olla, B.L. and Davis, M.W. (1989). The role of learning and stress in predator avoidance of hatchery-reared coho salmon (*Oncorhynchus kisutch*) juveniles. *Aquaculture* **76**: 209–214.
46. Berejikian, B.A. (1995). The effects of hatchery and wild ancestry and experience on the relative ability of steelhead trout fry (*Oncorhynchus mykiss*) to avoid a benthic predator. *Can. J. Fish. Aquat. Sci.* **52**: 2476–2482.
47. McEachron, L.W., Colura, R.L., Bumgaurdner, B.W., and Ward, R. (1998). Survival of stocked red drum in Texas. *Bull. Mar. Sci.* **62**: 359–368.

PHYSICAL AND CHEMICAL VARIABILITY OF TIDAL STREAMS

Paolo Magni
IMC—International Marine Centre
Torregrande-Oristano, Italy

Shigeru Montani
Hokkaido University
Hakodate, Japan

This article focuses on the variability of major hydrologic parameters of tidal streams that occur on different temporal and spatial scales within an individual estuary. Rapid changes in water temperature and salinity take place on a timescale of <1 hour, as a strong function of the tidal state (e.g., low vs. high tide) and amplitude (e.g., spring vs. neap tide). Large day–night fluctuations of dissolved oxygen (DO) concentration are also demonstrated and discussed in relation to the extent of *in situ* biological processes. In addition, the variability of these parameters is evaluated seasonally at controlled low tide. Finally, the year-round distribution of nutrient concentrations (i.e., ammonium, phosphate, silicate, and nitrate+nitrite) in low tide ebbing water is presented and compared with the variability in nutrient concentrations in the sediment porewater of an adjacent exposed sandflat.

SHORT-TERM (HOUR) VARIABILITY

A tidal stream crossing the interidal zone of an estuary in the Seto Inland Sea, Japan, was selected to describe daily fluctuations in water temperature, salinity, and DO concentration (1). Continuous measurements lasting 24 to 28 hours were made during two different tidal cycles, a spring tide of May 1995 and a neap tide of May 1996, using a CTD recorder (Alec AST-1000) placed about 10 cm from the bottom sediment (2,3).

Rapid and drastic changes in the selected parameters occurred during both surveys in 1995 (Fig. 1a) and 1996 (Fig. 1b). Major differences between the two surveys were also found, consistent with spring–neap tidal contrasts (4). Fluctuations in water depth were stronger in the 1995 survey than in the 1996 survey due to a larger amplitude in the tidal cycle during a spring tide (Fig. 1a). In particular, two major lower and higher tidal levels, typical of *mixed-semidiurnal* estuaries, occurred on May 30–31, 1995, as predicted by the local Maritime Safety Agency (2); the maximum tidal range was ca. 2 m (Fig. 1a).

Accordingly, the temporal changes in salinity were largest during the 1995 survey; salinity dropped from >31 to 2.0 psu, coincident with the lower low tide, and rapidly increased up to >31 psu again before the following higher high tide (Fig. 1a). At the subsequent higher low tide, the decrease in salinity was less marked, and minimum values were about 25 psu. An overall more restricted fluctuation of salinity occurred in the 1996

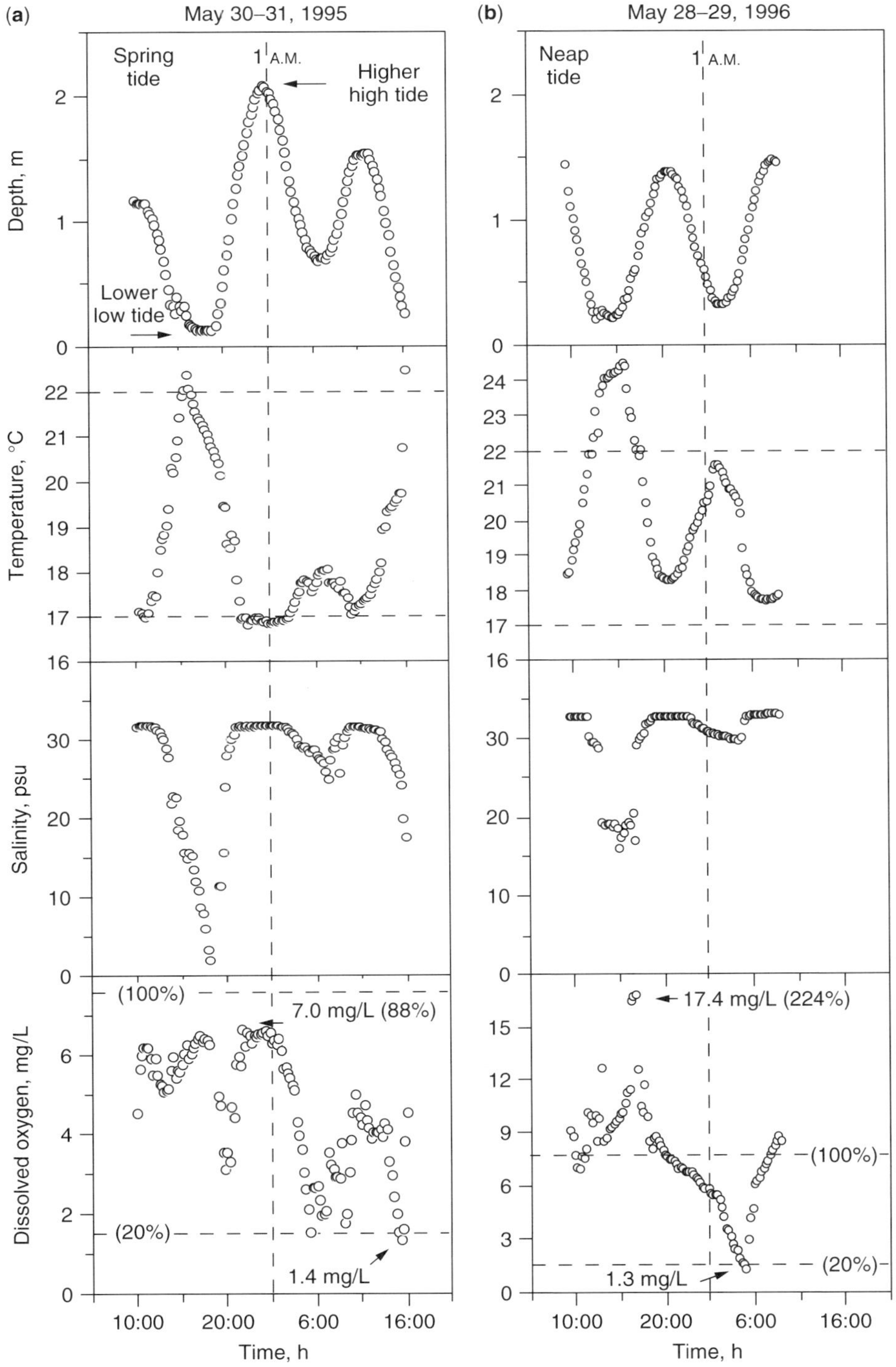

Figure 1. Daily fluctuations of water depth, temperature, salinity and dissolved oxygen (DO) concentration of a tidal stream in an estuarine intertidal zone (Seto Inland Sea, Japan). Measurements were made using a CTD cast placed ca. 10 cm from the bottom sediment at the end of May of 2 consecutive years (a: 1995; b: 1996). Relevant sensors were logged every 15 min. Notes: horizontal dashed lines in temperature and DO boxes individuate major differences within the same range of values (17–22 °C and 20–100% of air saturation, respectively) recorded in the 2 years. In DO boxes, absolute minimum and maximum values are also indicated. The vertical dashed line is arbitrarily depicted at 1 A.M. to highlight the temporal differences in tidal state and amplitude between the two surveys.

survey, consistent with the reduced amplitude of the tide (Fig. 1b) and a rainfall regime more limited in spring 1996 than in spring 1995 (1). These results show that saline intrusion is a strong function of tidal state (e.g., low vs. high tide) and amplitude (e.g., spring vs. neap tide). They also demonstrate that tidal streams may experience, within a few hours, strong changes in salinity that cause very different habitat conditions at the same location in an estuary. These may vary from oligohaline (0.51–5 psu) to euhaline (30.1–40 psu) conditions, in agreement with salinity ranges given by the Venice Conference (5).

The daily fluctuation in water temperature varied according to the tidal cycle and showed a temporal pattern opposite to that of salinity during both surveys (Fig. 1a, b). The range of temperature was relatively larger in May 1996 (17.7–24.5 °C) than in 1995 (16.9 °C–22.3 °C). In both cases, it was apparent that warmer waters were brought into the estuary by the freshwater runoff during ebb flow. These results are consistent with the period and site of measurements; physical processes of heat transfer in spring–summer (and water cooling in winter) are most effective in the upper and shallower riverine zone of the estuary (1).

Dissolved oxygen (DO) concentration was subjected to strong daily fluctuations, partly as a function of the tidal state (and water depth). The two surveys also showed major differences (Figs. 1a and b). In the 1995 survey, DO concentration was mostly within normoxic values, ranging from 1.4 mg/L to 7.0 mg/L. Differently in the 1996 survey, DO concentration showed a larger variation, in much higher values up to 17.4 mg/L.

Elevated DO concentrations are related to ecosystem processes of primary production (6). It is known that shallow lagoons and coastal areas dominated by seagrass or macroalgae are subjected to oversaturation of DO (i.e., >100% of air saturation), especially during the warm period (7,8). Viaroli et al. (7) reported DO oversaturation up to 150% in the near-bottom water of a coastal lagoon dominated by the macroalga *Ulva*. This was followed by the outbreak of a dystrophic crisis, complete anoxia through the water column at some stations of the lagoon. The development of large amounts of macroalgae (*Ulva* sp.) also tends to occur in the estuarine sand flat of this study coincident with increasing temperature and solar radiation during the spring. Extended beds of macroalgal biomass were present during the field measurements of the 1996 survey (personal observations). Accordingly, during the daytime measurements of May 28, DO concentration rapidly rose to oversaturation; a major increase of >200% of air saturation occurred between 16:00 and 16:30 (Fig. 1b), indicating a period of major oxygen production by macroalgae. Similar to the extremely high DO values found in this study, Piriou & Ménesguen (9) reported that the *in vitro* growth of *Ulva* under light saturation and nutrient enrichment raised the DO concentration to 22 mg/L, 4 hours after the experiment started.

By contrast, during nighttime, a progressive decrease in DO concentration occurred, down to hypoxic values of <1.4 mg/L or <14% of air saturation (10). Subsequently, at the next daylight of May 29, the DO concentration was back to normoxic values, reaching oversaturation (Fig. 1b). Differently during the 1995 survey, a much reduced macroalgal biomass was found that tended to be rapidly flushed away by heavy rainfall in spring 1995 (1). This was associated with a more a restricted range of DO concentration that never reached oversaturation during the daytime; whereas, similar to the 1996 survey, hypoxic values occurred during nighttime (Fig. 1a, b). Thus, the different extent of macroalgal coverage indicated how a large DO fluctuation might strongly depend on the extent of biological processes, such as photosynthesis and primary production in the presence of light and respiration at night.

SEASONAL VARIABILITY

This section presents the results of monthly measurements of water temperature, salinity, and DO concentration at controlled low tide on the same tidal stream where the short-term measurements of May 1995 and May 1996 were conducted (see above section and Fig. 1). Between January 1995 and April 1996, the water temperature varied from 3.6 °C ± 0.3 (December) to 29.0 °C ± 1.8 (July) (Fig. 2). Statistically significant differences were demonstrated for the specific geographical area (latitude 34°21′ N, longitude 43°21′E) between a warm period (May–October, water temperature 23.5 ± 4.2 °C) and a cold period (November–April, water temperature 10.0 ± 4.4 °C) (1).

The seasonal variability in low-tide salinity (Fig. 2) was consistent with the results obtained from the short-term measurements of May 1995 and May 1996 during a complete tidal cycle (Fig. 1). In particular, the salinity recorded during the seasonal survey was lowest on May 16, 1995 at 4.6 ± 1.5 psu (Fig. 2). Accordingly, the short-term measurements of May 1995 showed a salinity decrease during low tide to <5 psu (Fig. 1a). By contrast, reduced rainfall during the second half of 1995 and the first half of 1996 (1) corresponded to a progressive increase in salinity in the low-tide stream water during that period. In particular, salinity was >25 psu between the end of September 1995 and April 1996 and reached the highest values in November 1995 at 31.4 ± 1.1 psu (Fig. 2). Consistently, also during the short-term measurements of May 1996, the decrease in salinity during low tide was more limited than that found in May 1995 (Fig. 1a, b).

The DO concentration also varied greatly from 5.4 ± 1.7 mg/L (September 1995) to 15.1 ± 1.1 mg/L (April 1996)

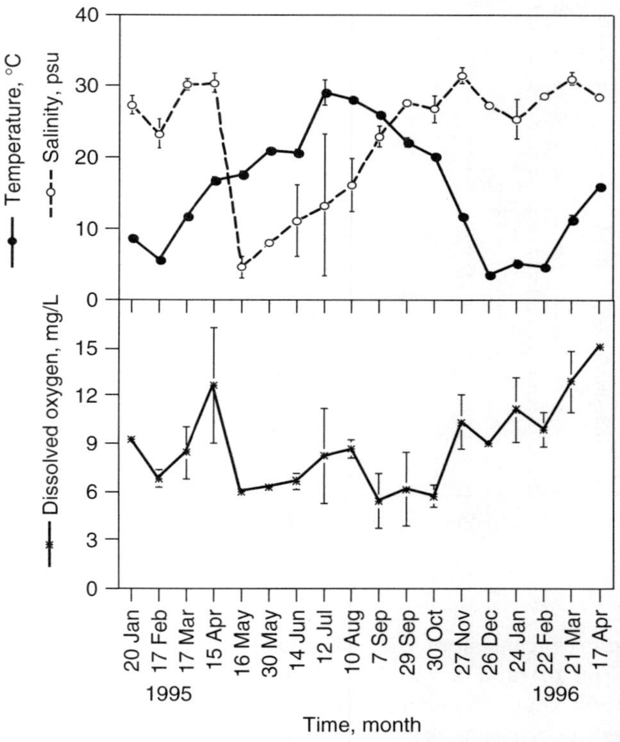

Figure 2. Seasonal variability in temperature, salinity, and dissolved oxygen (DO) concentrations in estuarine ebbing waters during a daytime low tide (Seto Inland Sea, Japan). Data source: Reference 1 (partially redrawn from Fig. 2, as mean values ± standard deviation of Stns. H1 and H2) (with kind permission of Kluwer Academic Publishers).

(Fig. 2). The temporal distribution of the DO concentration highlighted that, after a progressive increase from February 1995 (6.8 ± 0.5 mg/L) to April 1995 (12.6 ± 3.6 mg/L), a noticeable drop in DO concentration to 6–7 mg/L occurred in late spring, May 1995. This tended to persist throughout the following months (Fig. 2). In the second part of the seasonal survey, the DO concentration progressively increased again from relatively low values of 5.7 ± 0.7 mg/L^{-1} in October 1995 to 15.1 mg/L in April 1996 (Fig. 2). The temporal variability of the DO concentration was also consistent with the differences found during the short-term measurements. In particular, the DO level hardly reached oversaturation during the short-term measurements of spring 1995 (Fig. 1a), but it was up to >200% of air saturation in spring 1996 (Fig. 1b).

Nutrients

Nutrients, such as ammonium (NH_4^+-N), nitrate+nitrite ($NO_3^- + NO_2^-$-N), phosphate (PO_4^{3-}-P) and silicate [$Si(OH)_4$-Si], play a major role in controlling the growth, abundance, and structure of primary producers in the ocean (11,12). A qualitative and quantitative evaluation of their distribution represents therefore a fundamental step in assessing the dynamics of biophylic elements (e.g., carbon, nitrogen, phosphorus, and silicon) and the system processes of production and respiration (6). The variability in nutrient concentrations in transitional aquatic systems, such as estuaries and tidal streams, is related to the close interaction between water chemistry, tidal hydrology, and sedimentary processes (1,13,14).

Several approaches have been undertaken on different temporal and spatial scales to evaluate such variability. Time-series surveys of nutrient distribution have been carried out in relation to the ebb–flood tidal state (15–18) and the spring–neap tidal amplitude (4,19,20) or by controlled sampling regimes, such as around high tide (16,21). Other studies in coastal marine systems have used either intact cores (22,23) or *in situ* chambers (24) to quantify the nutrient flux at the sediment–water interface. However, a satisfactory evaluation of the actual physicochemical and biological processes from field enclosed experiments may be controversial because of the nonnatural state of the system, which is shielded from large water movements (25,26). This is particularly true in intertidal environments that are subjected to strong hydrodynamic tidal energy and are more rapidly and directly influenced by "external" variables (e.g., rainfall) and *in situ* benthic processes (e.g., nitrogen uptake by primary producers and release by macrozoobenthos) (27–33).

This work provides an example of the variability of nutrient concentrations in tidal stream water, consistent with large seasonal changes in temperature, salinity, and DO concentration (Fig. 2). A comparison with the nutrient concentrations in the sediment porewater from an adjacent sand flat exposed at low tide is given; it shows how major differences may occur between two closely interacting compartments of an estuary.

The seasonal distribution of nutrient concentrations in ebbing water and in sediment porewater is depicted in Fig. 3, and their annual range of variability is given in Table 1. Ammonium and phosphate concentrations

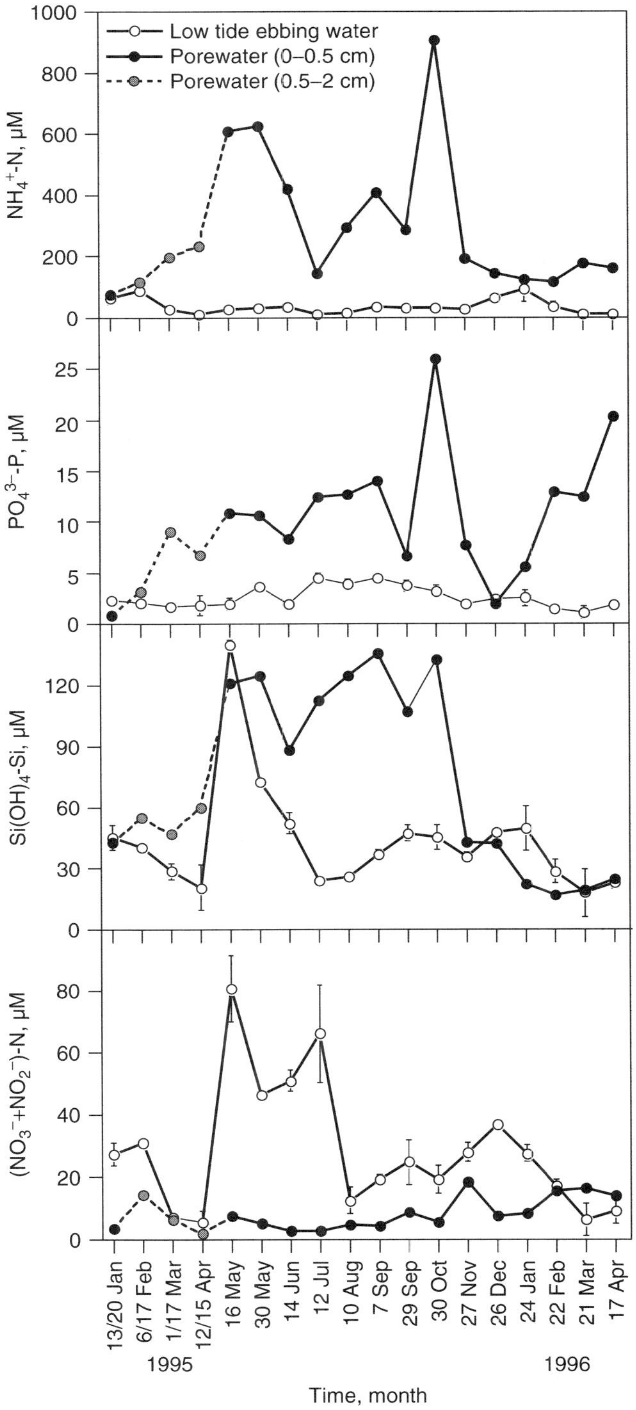

Figure 3. Seasonal variability in nutrient concentrations in estuarine ebbing waters and surface (0–0.5 cm) sediment porewater on an adjacent sand flat measured during a daytime low tide (Seto Inland Sea, Japan). Grey circles refer to porewater nutrient concentrations in subsurface (0.5–2 cm) sediments. On the horizontal axis, two different days from January 1995 to April 1995 refer to the dates of porewater (first day) and ebbing water (second day, see also Fig. 2) sampling. Data sources: ebbing water, data from Reference 1 (partially redrawn from Fig. 3, as mean values ± standard deviation of Stns. H1 and H2) (with kind permission of Kluwer academic Publishers).

Table 1. Annual Minimum and Maximum Values (± Standard Deviation and Period of The Year) of Nutrient Concentrations (μM) in Low Tide Ebbing Water and in the Porewater of Surface (0–0.5 cm) Sediments from an Adjacent Emerged Sand Flat (Seasonal Variation in Fig. 3)

	Ebbing Water		Sediment Porewater	
	Minimum	Maximum	Minimum	Maximum
NH_4^+	10.2 ± 6.8 (Apr '95)	90.8 ± 39.2 (Jan '96)	75.1 (Jan '95)	908 (Oct 30, '95)
PO_4^{3-}	1.1 ± 0.6 (Mar '96)	4.5 ± 0.4 (Jul '95)	0.9 (Jan '95)	26.0 (Oct 30, '95)
$Si(OH)_4$	17.9 ± 11.7 (Mar '96)	139.6 ± 2.5 (May 16, '95)	17.1 (Feb '96)	135.7 (Sept 7, '95)
$NO_3^- + NO_2^-$	5.7 ± 3.5 (Apr '95)	80.7 ± 11.0 (May 16, '95)	2.8 (Jul '95)	18.6 (Nov '95)

in ebbing water are much lower than those found in porewater in all seasons and on most occasions. However, a noticeable peak of ammonium concentration in ebbing water was observed during the colder months (December–February), coincident with a general decrease in ammonium concentration in the porewater. To some extent, this was also true for phosphate, which showed relatively lower values in porewater than in ebbing water in January 1995 and December 1995 (Fig. 3). Silicate concentration tended to follow a pattern similar to that of ammonium and phosphate, especially between spring and late summer. However, the extent of the differences in silicate concentration between porewater and ebbing water was more limited in autumn and spring. For the study area examined, it is suggested that a diffusive upward flux of nutrients from sediments, if solely based on the water–sediment concentration gradients (29,34) would be relatively limited, yet highly variable with season.

Different from the spatial distribution of ammonium, phosphate, and silicate, nitrate+nitrite concentration was higher in ebbing water than in porewater; major peaks occurred during periods of enhanced rainfall, as indicated by the sharp decrease in salinity on May 16 (Fig. 2). This is consistent with previous companion studies that demonstrated that freshwater input is a major contributor of nitrate+nitrite to the estuary of this work (1,2).

Finally, it can be highlighted that low nitrate concentration in porewater, further decreasing through the sediment column, commonly occurs due to numerous processes within sediment, such as nitrate ammonification and/or denitrification processes (35,36).

BIBLIOGRAPHY

1. Magni, P. and Montani, S. (2000). Water chemistry variability in the lower intertidal zone of an estuary in the Seto Inland Sea, Japan: Seasonal patterns of nutrients and particulate compounds. *Hydrobiologia* **432**: 9–23.
2. Montani, S. et al. (1998). The effect of a tidal cycle on the dynamics of nutrients in a tidal estuary in the Seto Inland Sea, Japan. *J. Oceanogr.* **54**: 65–76.
3. Magni, P., Montani, S., and Tada, K. (2002). Semidiurnal dynamics of salinity, nutrients and suspended particulate matter in an estuary in the Seto Inland Sea, Japan, during a spring tide cycle. *J. Oceanogr.* **58**: 389–402.
4. Vörösmarty, C.J. and Loder, T.C. III. (1994). Spring-neap tidal contrasts and nutrient dynamics in a marsh-dominated estuary. *Estuaries* **17**: 537–551.
5. Anonymous. (1959). Symposium on the classification of brackish waters, Venice. *Arch. Oceanogr. Limnol.* **2** (suppl.): 1.
6. D'Avanzo, C., Kremer, J.N., and Wainright, S.C. (1996). Ecosystem production and respiration in response to eutrophication in shallow temperate estuaries. *Mar. Ecol. Prog. Ser.* **141**: 263–274.
7. Viaroli, P., Pugnetti, A., and Ferrari, I. (1992). *Ulva rigida* and decomposition processes and related effects on nitrogen and phosphorus cycles in a coastal lagoon (Sacca di Goro, Po River Delta). In: Marine eutrophication and population dynamics. *Proc. 25th Eur. Mar. Biol. Symp.* G. Colombo et al. (Eds.). Ferrara, Italy, pp. 77–84.
8. Ziegler, S. and Benner, R. (1998). Ecosystem metabolism in a subtropical, seagrass-dominated lagoon. *Mar. Ecol. Prog. Ser.* **173**: 1–12.
9. Piriou, J-Y. and Ménesguen, A. (1992). Environmental factors controlling the *Ulva* sp. bloom in Brittany (France). In: Marine eutrophication and population dynamics. *Proc. 25th Eur. Mar. Biol. Symp.* G. Colombo et al. (Eds.). Ferrara, Italy, pp. 111–115.
10. Diaz, R.J. and Rosenberg, R. (1995). Marine benthic hypoxia: A review of its ecological effects and the behavioural responses of benthic macrofauna. *Oceanogr. Mar. Biol. Annu. Rev.* **33**: 245–303.
11. Justic, D., Rabalais, N.N., and Turner, T. (1995). Stoichiometric nutrient balance and origin of coastal eutrophication. *Mar. Poll. Bull.* **30**: 41–46.
12. Valiela, I., McClelland, J., Hauxwell, J., Beher, P.J., Hersh, D., and Foreman, K. (1997). Macroalgal blooms in shallow estuaries: Controls and ecophysiological and ecosystem consequences. *Limnol. Oceanogr.* **5**: 1105–1118.
13. Schubel, R.J. and Pritchard, D.W. (1986). Responses of upper Chesapeake Bay to variations in discharge of the Susquehanna River. *Estuaries* **9**: 236–249.
14. Page, H.M., Petty, R.L., and Meade, D.E. (1995). Influence of watershed runoff on nutrient dynamics in a southern California salt marsh. *Estuar. Coast. Shelf Sci.* **41**: 163–180.
15. Caffrey, J.N. and Day, J.W., Jr. (1986). Control of the variability of nutrients and suspended sediments in a Gulf

Coast estuary by climatic forcing and spring discharge of the Atchafalaya River. *Estuaries* **9**: 295–300.

16. Balls, P.W. (1994). Nutrient inputs to estuaries from nine Scottish east coast rivers; influence of estuarine processes on inputs to the North Sea. *Estuar. Coast. Shelf Sci.* **39**: 329–352.
17. Uncles, R.J. and Stephens, J.A. (1996). Salt intrusion in the Tweed Estuary. *Estuar. Coast. Shelf Sci.* **43**: 271–293.
18. Magni, P. and Montani, S. (2004). Tidally-mediated changes of nutrient concentrations. *Encyclopedia of Water (OC-136)*. John Wiley & Sons, New York.
19. Yin, K., Harrison, P.J., Pond, S., and Beamish, R.J. (1995). Entrainment of nitrate in the Fraser River estuary and its biological implications. II. Effects of spring *vs.* neap tide and river discharge. *Estuar. Coast. Shelf Sci.* **40**: 529–544.
20. Yin, K., Harrison, P.J., Pond, S., and Beamish, R.J. (1995). Entrainment of nitrate in the Fraser River estuary and its biological implications. III. Effects of winds. *Estuar. Coast. Shelf Sci.* **40**: 545–558.
21. Balls, P.W. (1992). Nutrient behaviour in two contrasting Scottish estuaries, the Forth and the Tay. *Oceanol. Acta* **15**: 261–277.
22. Cowan, J.L.W., Pennock, J.R., and Boyton, W.R. (1996). Seasonal and interannual patterns of sediment–water nutrient fluxes in Mobile Bay, Alabama (USA): Regulating factors and ecological significance. *Mar. Ecol. Progr. Ser.* **141**: 229–245.
23. MacIntyre, H.L. and Cullen, J.J. (1996). Primary production by suspended and benthic microalgae in a turbid estuary: Time-scale of variability in San Antonio Bay, Texas. *Mar. Ecol. Progr. Ser.* **145**: 245–268.
24. Yamada, H. and Kayama, M. (1987). Liberation of nitrogenous compounds from bottom sediments and effect of bioturbation by small bivalve, *Theora lata* (Hinds). *Estuar. Coast. Shelf Sci.* **24**: 539–555.
25. Nakamura, Y. (1994). Effect of flow velocity on phosphate release from sediment. *Water Sci. Technol.* **30**: 263–272.
26. Miller-Way, T. and Twilley, R.R. (1996). Theory and operation of continuous flow systems for the study of benthic–pelagic coupling. *Mar. Ecol. Progr. Ser.* **140**: 257–269.
27. Matsukawa, Y., Sato, Y., and Sasaki, K. (1987). Benthic flux of nutrient salts on an intertidal flat. *Nippon Suisan Gakkaishi* **53**: 985–989.
28. Sundbäck, K., Enoksson, V., Granéli, W., and Pettersson, K. (1991). Influence of sublittoral microphytobenthos on the oxygen and nutrient flux between sediment and water: A laboratory continuous-flow study. *Mar. Ecol. Prog. Ser.* **74**: 263–279.
29. Feuillet-Girard, M., Gouleau, D., Blanchard, G., and Joassard, L. (1997). Nutrient fluxes on an intertidal mudflat in Marennes-Oléron Bay, and influence of the emersion period. *Aquat. Living Resour.* **10**: 49–58.
30. Trimmer, M., Nedwell, D.B., Sivyer, D.B., and Malcolm, S.J. (1998). Nitrogen fluxes through the lower estuary of the river Great Ouse, England: The role of the bottom sediments. *Mar. Ecol. Prog. Ser.* **163**: 109–124.
31. Usui, T., Koike, I., and Ogura, N. (1998). Tidal effect on dynamics of pore water nitrate in intertidal sediment of a eutrophic estuary. *J. Oceanogr.* **54**: 205–216.
32. Oldham, C.E. and Lavery, P.S. (1999). Porewater nutrient fluxes in a shallow fetch-limited estuary. *Mar. Ecol. Prog. Ser.* **183**: 39–47.
33. Takayanagi, K. and Yamada, H. (1999). Effects of benthic flux on short term variations of nutrients in Aburatsubo Bay. *J. Oceanogr.* **55**: 463–469.
34. Barbanti, A., Ceccherelli, V.U., Frascari, F., Reggiani, G., and Rosso, G. (1992). Nutrient regeneration processes in bottom sediments in a Po delta lagoon (Italy) and the role of bioturbation in determining the fluxes at the sediment–water interface. *Hydrobiologia* **228**: 1–21.
35. Rysgaard, S., Christensen, P.B., and Nielsen, L.P. (1995). Seasonal variation in nitrification and denitrification in estuarine sediment colonized by benthic microalgae and bioturbating infauna. *Mar. Ecol. Prog. Ser.* **126**: 111–121.
36. Sundbäck, K., Miles, A., and Göransson, E. (2000). Nitrogen fluxes, denitrification and the role of microphytobenthos in microtidal shallow-water sediments: An annual study. *Mar. Ecol. Prog. Ser.* **200**: 59–76.

BLACK WATER TURNS THE TIDE ON FLORIDA CORAL

NASA—Goddard Space Flight Center

In early 2002, a patch of "black water" spanning over 60 miles in diameter formed off southwestern Florida and contributed to severe coral reef stress and death in the Florida Keys, according to results published from research funded by NASA, the U.S. Environmental Protection Agency and the National Oceanic and Atmospheric Administration (NOAA). The "black water" contained a high abundance of toxic and non-toxic microscopic plants.

Chuanmin Hu and other colleagues at the Institute for Marine Remote Sensing of the University of South Florida (USF), St. Petersburg, Fla., and colleagues from the Florida Fish and Wildlife Conservation Commission (FFWCC) and the University of Georgia, co-authored an article on this phenomenon that appeared as the cover

Figure 1. SeaWiFS Image of Blackwater, True color date Jan. 9, 2002. Courtesy The SeaWiFs Project and ORBIMAGE, Scientific Visualization Studio.

story of a recent issue of the American Geophysical Union's Geophysical Research Letters.

"The water appeared black in satellite imagery because the concentration of the microscopic plants and other dissolved matters were high," Hu said. Because plants and dissolved matter absorb sunlight, they reduce the amount of light normally reflected from the ocean. When a red-tide bloom occurs the water takes on various hues of red or brown. While not all microscopic plants contribute to red tides, the darker hue created by both the plankton and the harmful algal blooms made the water appear black when seen from the satellite.

When Hu and his colleagues examined the data collected by divers from the dark water area in the Florida Keys, they discovered a 70 percent decrease in stony coral cover, a 40 percent reduction of coral species, and a near-elimination of sponge colonies at two reef sites after the dark water passed. By examining satellite images and field survey data, the authors concluded that the coral reef ecosystem was stressed by microscopic organisms and toxins contained in the dark water.

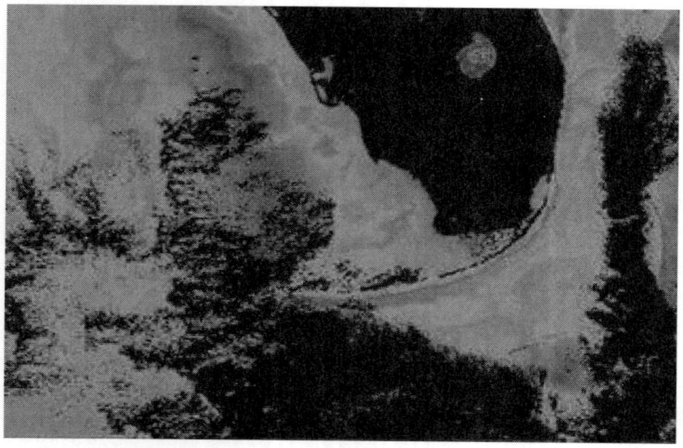

Figure 2. SeaWiFS Image of Blackwater, date Jan. 9, 2002 False color visualization showing chlorophyll. Reds and oranges represent high concentrations of chlorophyll. Courtesy The SeaWiFs Project and ORBIMAGE, Scientific Visualization Studio.

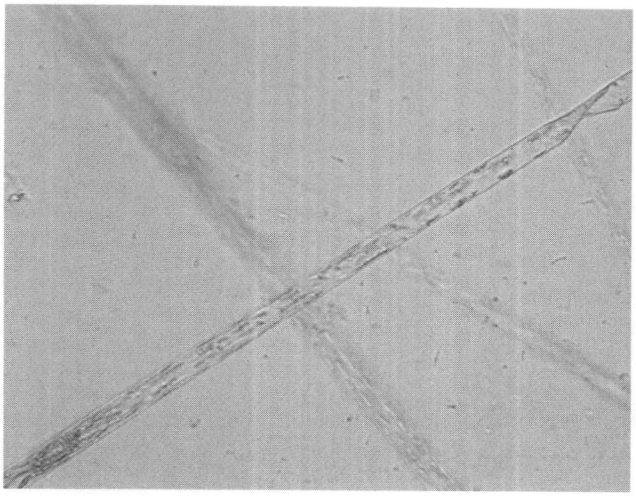

Figure 3. Microscopic image of a dinoflagellate, called Rhizosoleniacea, a cause for Red Tide. Rhizosoleniaceae blooms occur seasonally off Florida's Gulf coast due to freshwater outflow from the Everglades. Courtesy Florida Marine Research Institute.

The "black water" event caused alarm among local fishermen, divers, and the public, as the color of the water was unusual and fish seemed to avoid this large area of dark water. Satellite instruments such as the Sea-viewing Wide Field-of-view Sensor (SeaWiFS) aboard Orbimage's SeaStar satellite and the Moderate Resolution Imaging Spectroradiometer (MODIS) aboard NASA's Terra and Aqua satellites provide information on ocean color that allows scientists to monitor the health of the water and the shallow benthic (ocean bottom) environment. The SeaWiFS and MODIS measurements of the dark water led to a number of investigations to help clarify the issues and to provide answers to the public's concerns.

During January 2002, SeaWiFS detected the dark-colored water in the Florida Bight, just southwest of the Everglades. In fall 2001, the SeaWiFS images showed an extensive red tide off Florida's central west coast, near Charlotte Harbor.

Red tides occur every year off Florida and are known to cause fish kills, coral stress and mortality, and skin and respiratory problems in humans. They are caused by high concentration of microscopic plants called dinoflagellates. Other microorganisms called cyanobacteria can also cause harmful algal blooms. The waters containing this red tide migrated to the south along the coast.

Winter storms caused large amounts of fresh water to drain from the Everglades into Florida Bight (the curve in

Figure 4. NASA's Terra satellite acquired this image of a red tide bloom along Florida's west coast December 22, 2001. Notice the dark reddish color of the ocean. Image courtesy Jacques Descloitres, MODIS Land Rapid Response Team at NASA GSFC.

the shoreline from the Keys north to Everglades National Park on the mainland), carrying high levels of nutrients such as silicate, phosphorus, and nitrogen to the sea. These caused a bloom of the microscopic marine plants known as diatoms in the same patch. The bloom turned the water dark and the "black water" patch re-circulated for several months in a slow clockwise motion off southwest Florida in the Florida Bight. Slowly, the dark water drifted farther south and toward the Florida Keys. By May 2002, the "black water" had moved through passages in the Florida Keys, dispersing into the Atlantic and the Gulf Stream.

Co-authors on this research article included Serge Andrefouet and Frank E. Muller-Karger of USF; Keith E. Hackett, Michael K. Callahan, and Jennifer L. Wheaton of FFWCC, St. Petersburg, Fla.; and James W. Porter of the University of Georgia, Athens, Ga.

NASA funded part of this research as part of its Earth Science mission to understand and protect our home planet. NASA's Earth Science Enterprise is dedicated to understanding the Earth as an integrated system and applying Earth System Science to improve prediction of climate, weather, and natural hazards using the unique vantage point of space.

SHALLOW WATER WAVES

Stefano Pagliara
Pietro Chiavaccini
Universitá di Pisa
Pisa, Italy

INTRODUCTION

Shallow water conditions happens when the ratio of water depth d to the wave length L is less than 1/25, i.e., $d/L < 0.04$. In this condition, there is a combination of high wave steepness, and ratios of wave height to water depth are close to unity. Shallow water waves approaching the coast are subjected to refraction, shoaling, and breaking.

THE SHALLOW WATER THEORIES

The Linear Theory

The linear wave theory (1), in which the nonlinear terms of the equation of motion are omitted, describes water motion by two nondimensional parameters, the relative depth d/L and the wave steepness H/L, which represent respectively the influence of the depth on wave characteristics (celerity, length, and height) and the measure of the wave dimension in respect of height. For large values of relative depth, the linear theory cannot be applied.

Wave profile in linear theory is defined by

$$\eta = a\cos(kx - \omega t) = a\cos\left(\frac{2\pi x}{L} - \frac{2\pi x}{T}\right) \tag{1}$$

The celerity is

$$C = \frac{L}{T} = \left(\frac{gT}{2\pi}\right)\tan h\left(\frac{2\pi d}{L}\right) = \sqrt{\left(\frac{gL}{2\pi}\right)\tan h\left(\frac{2\pi d}{L}\right)} \tag{2}$$

In shallow water, as $\frac{d}{L} \to 0 \tan h\left(\frac{2\pi d}{L}\right) \to \frac{2\pi d}{L}$ and celerity becomes dependent only on depth

$$C = \sqrt{gd} \tag{3}$$

Wave train propagates with a speed different from the celerity, called group velocity, defined by

$$C_G = nC = \frac{1}{2}\left[1 + \frac{2\pi d}{L}\frac{1}{\text{senh}\left(\frac{2\pi d}{L}\right)}\right]C \tag{4}$$

In shallow water, the group velocity becomes $C_G = C$ because $n \to 1$, so wave train propagates with the same speed of the single waves of the train.

In reality, a wave approaching the coast is modified by the bottom: its height grows and loses characteristics of linearity, which implies a more detailed formulation.

The nonlinear theories, used more to describe shallow water waves, are the cnoidal wave theory and the solitary wave theory.

The Cnoidal Wave Theory

The theory of cnoidal waves was first developed by Korteweg and de Vries (2) and can be used to described unidirectional waves in shallow water, where $d/L < 1/8$ and $Ur > 20$. If the water height is small, the cnoidal wave profile becomes the sinusoidal one.

The cnoidal wave is periodic with the profile given by

$$\eta(x,t) = hCn^2\left[2K(\kappa)\left(\frac{x}{L} - \frac{t}{T}\right), \kappa\right] \tag{5}$$

where $K(k)$ is the complete elliptic integral of the first kind of modulus k, $Cn(r)$ is the Jacobian elliptic function, and h is the vertical coordinate of the water surface about the trough level at the horizontal coordinate. To simplify the use of *cnoidal* theory, Wiegel (3) developed graphs and table of easy appliance.

The Solitary Wave Theory

The solitary wave is a progressive wave consisting of a single crest and is not oscillatory as the other examined. The wave form is (4)

$$\eta(x,t) = H\sec h^2\left(\sqrt{\frac{3H}{4d^3}}(x - Ct)\right) \tag{6}$$

where C (wave celerity) is defined by

$$C = \sqrt{gd}\left(1 + \frac{H}{2d}\right) \tag{7}$$

According to the dispersion relation of the linear wave theory, the celerity of water waves in shallow water (Eq. 3) is smaller than solitary wave phase velocity because of the inclusion of terms that depend on H/d.

The solitary wave describes enough well waves approaching shallow water, even if wave period or wave length are not associated with the theory. When the wave pass from deep to shallow water their crests peak up and are separated by flat troughs appearing like a series of solitary waves. Solitary theory appears reasonable even if the periodicity is neglected, because in shallow water the period is not particularly significant, but rather the water depth becomes important.

As the solitary wave advances into shallow water, its height increases, the crests becomes greater and sharper, the trough becomes longer and flatter: in this condition, a wave is well represented by solitary waves. The solitary wave is not an oscillatory wave, as those obtained with other theories, but a translation one. Water particles, as a wave passes, are subjected to a translation in wave direction, whereas in the oscillatory waves it moves forward and backward, remaining after a period in its original position.

Wave profiles, according to the theories illustrated above, are presented in Fig. 1.

WAVE REFRACTION AND SHOALING

Waves passing from deep to shallow water are subjected to refraction in which the direction of their travel changes in such a way that approaching the coast the crests tends to become more parallel to the depth contours (Fig. 2). To determine the variation of the wave direction, Snell's law can be applied

$$\frac{\sin\theta_1}{C_1} = \frac{\sin\theta_2}{C_2} = \text{cons} \tag{8}$$

where q_1 and q_2 are the angles between adjacent wave crests and the respective bottom contours, whereas C_1 and C_2 are the wave celerity at the two depths.

With a regular bottom (straight and parallel contours), the relation can be applied directly between the angle at any depth and the deep water angle approach.

With an irregular bottom, wave refraction may cause a spreading or a convergence of the wave energy. This effect can be easily illustrated taking in account the wave rays (Fig. 2), defined as the lines drawn normal to the wave crests and directed in the wave advance. If the wave rays spread, the wave crests become longer and the energy flux, assumed constant between two rays, must be extended over a greater length. The opposite (energy concentration) happens if the waves rays converge.

Actually, the calculations of the rays are made by software based on models such as a mild slope, parabolic, or Boussinesq wave model.

Another effect of the change in the wave length in shallow water is that the wave height increases. This effect is the consequence of the energy conservation in

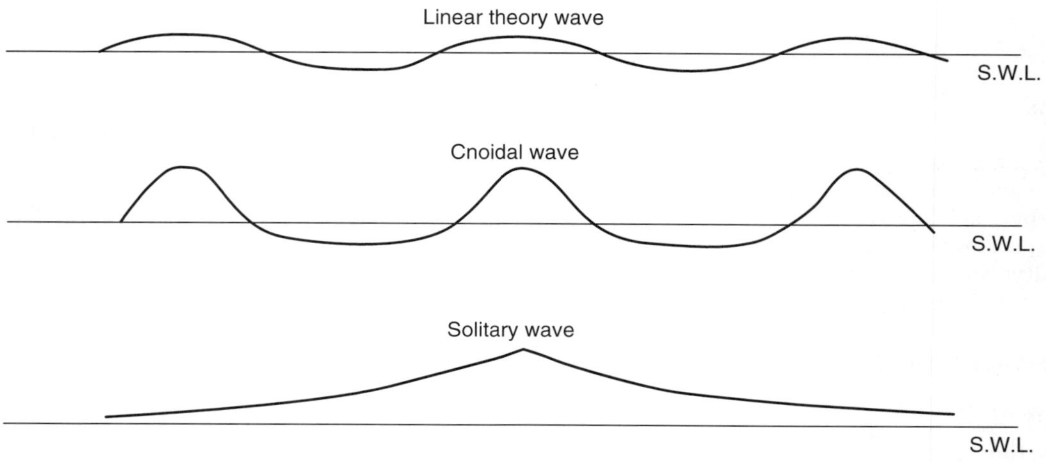

Figure 1. Wave profiles.

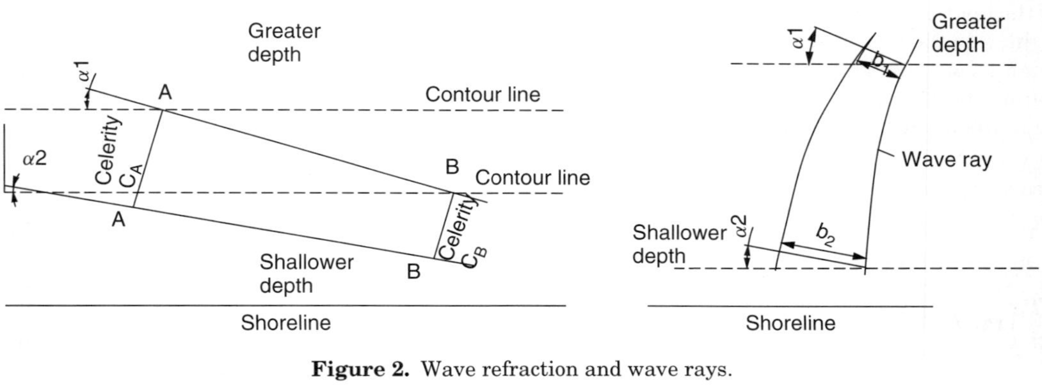

Figure 2. Wave refraction and wave rays.

concert with the decrease of the celerity approaching the shallow water. This phenomenon is referred to as shoaling.

The effects of shoaling and refraction in water can be expressed by the following formula:

$$H = H_0 K_s K_r \tag{9}$$

where H_0 is wave height in deep water, K_s is the shoaling coefficient

$$K_s = \sqrt{\frac{C_{G0}}{C_G}} = \sqrt{\frac{1}{2n \tan h \frac{2\pi d}{L}}} \tag{10}$$

with $n = \frac{1}{2}\left(1 + \frac{2\pi d}{L} \frac{1}{\text{senh}\left(\frac{2\pi d}{L}\right)}\right)$ and K_r is the refraction coefficient

$$K_r = \sqrt{\frac{b_0}{b}} \tag{11}$$

where b_0 and b are the distance between two adjacent rays, respectively, in deep water and at a generic depth. For straight and parallel contours lines, the refraction coefficient becomes

$$K_r = \sqrt{\frac{\cos\theta_0}{\cos\theta}} \tag{12}$$

WAVE BREAKING

Waves shoaling causes the increasing of wave height until its physical limit because of steepness of wave H/L. When this limit is reached, the wave breaks and dissipates energy. Battjes (5) has shown that the breaking wave characteristics can be correlated to a parameter, called surf similarity x, which is defined as

$$\xi = \frac{\tan\beta}{\sqrt{\frac{H_0}{L_0}}} \tag{13}$$

where $\tan b$ is the beach slope and H_0 and L_0 are the wave height and length in deep water, respectively.

Three common type of breakers are recognized (Fig. 3):

— spilling ($x < 0.5$), in which each wave gradually peaks until the wave becomes unstable and cascades down as white water, bubbles, and foam;

— plunging ($0.5 < x < 3.3$), in which the shoreward face of the wave becomes vertical, curls over, and plunges forward and downward as an intact mass of water;

— surging ($3.3 < x < 5$), in which the base of the wave, while it is peaking up, finds the shore, and then the crest collapses and disappears.

When the surf similarity x is >5, reflection happens and no breaking occurs.

According to Galvin (6), a fourth breaker, called collapsing, intermediate between plunging and surging types, exists.

It is difficult to identify which type of breaker can verify, because it depends on the individual heights and interactions of the waves. Plunging and surging breakers can be seen on a beach during the same storm. However, spilling breakers are typical of very low sloping beaches with waves of high steepness values; plunging waves occur in steeper beaches and waves of intermediate steepness; surging, instead, is associated with high gradient beaches with waves of low steepness.

The breaking happens when the water particle velocity is greater than wave celerity. According to solitary wave theory, this condition is described by (7)

$$\left(\frac{H_b}{d_b}\right)_{\max} = 0.78 \tag{14}$$

where the subscript b denotes the breaking.

Laboratory tests pointed out that Equation (14) is verified more for oscillatory waves than solitary wave. However, it is considered fundamental to express the relation between the relative depth and the breaking condition.

Other parameters playing a role in wave breaking exist, such as beach slope and bottom roughness. An empirical relation considers the beach slope m (8):

$$\left(\frac{H_b}{d_b}\right)_{\max} = 0.75 + 25m - 112m^2 + 3870m^3 \tag{15}$$

in which as the slope increases, the breaking happens more and more nearshore.

Another expression was developed by Goda (9)

$$\left(\frac{H_b}{d_b}\right)_{\max} = 0.17\frac{L_0}{d_b}\left[1 - \exp\left(-\frac{15\pi d_b}{L_0}(1 + 15m^{4/3})\right)\right] \tag{16}$$

where L_0 is the deep water wave lenght.

For irregular waves (waves with different height and period), Kamphuis (10) proposed two criteria based on

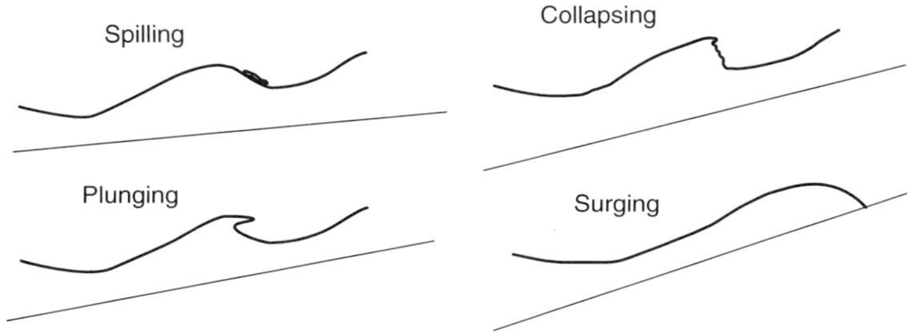

Figure 3. Breaking types.

extensive model tests

$$H_{sb} = 0.095e^{4m} L_{bp} \tan h \left(\frac{2\pi d_b}{L_{bp}} \right) \quad (17)$$

$$\frac{H_{sb}}{d_b} = 0.56e^{3.5m} \quad (18)$$

where H_{sb} is the significant wave breaking, L_{bp} the breaking wave length, and d_{b} is the breaking depth.

BIBLIOGRAPHY

1. Airy, G.B. (1845). Tides and waves. *Encyclopedia Metropolitana*. **192**: 241–396.
2. Korteweg, D.J. and de Vries, G. (1895). On the change of form of long waves advancing in a rectangular canal and on a type of long stationary wave. *Phil. Mag.* **5**: 442–443.
3. Wiegel. (1960).
4. Russell, J.S. (1884). *Report on waves. 14th Meeting of the British Association for the Advancement of Science.* pp. 311–390.
5. Battjes, J.A. (1974). Surf similarity. *Proceedings of the 14th Coastal Engineering Conference, ASCE.* pp. 466–497.
6. Galvin, C.J. (1968). Breaker type classification on three laboratory beaches. *J. Geophys. Res.* **73**: 3651–3659.
7. Munk, W.H. (1949). The solitary wave theory and its application to surfs problems. *Ann. New York Acad. Sci.* **51**: 376–424.
8. SPM. (1984).
9. Goda, Y. (1970). A synthesis of breaker indices. *Trans. Japan Soc. Civil Engrs.* **2**: 227–230.
10. Kamphuis, J.W. (1991). Wave transformation. *Co. Eng.* **15**: 173–184.

FURTHER READING

CERC U.S. Corps of Eng. (1984). *Shore Protection Manual.* Vicksburg, VA.

WATER WAVES

Nai Kuang Liang
National Taiwan University
Taipei, Taiwan

INTRODUCTION

As one stands at the coast, an endlessly moving succession of irregular humps and hollows can be seen reaching to the shore. This is the water wave generated by wind. There are different kinds of water waves, which are driven by different forces. Beside wind waves, the tide and the tsunami are other well-known water waves. Different Water waves are distinguished by their wavelength, which is defined as the length between successive humps. The wavelength of a wind wave is shorter and is easily recognized.

A water wave is an important physical phenomenon in an ocean, sea, or lake. It influences beach morphology, maritime structures, and human activities very much. Because the wind varies tremendously, a typhoon or hurricane may generate huge waves. This storm usually causes disasters. Tidal and wind waves contain a lot of energy. Electric power can be generated by them.

BASIC CHARACTERISTICS OF A WATER WAVE

The water wave was studied mathematically in the nineteenth century as a form of oscillatory wave. In 1847, Stokes (1) published his famous paper entitled "On the theory of oscillatory waves." Stokes' wave theory has been widely used till now. The elements of an oscillatory wave are wave height, wave period, and wavelength. The wave height H is the distance between the wave crest and trough. The wavelength L is the distance between successive crests, and the wave period T the time difference between successive crests. In deep water, the water depth H is larger than the half of the wavelength L. After Stokes, L is a function of the wave period T in the following equation:

$$L = \frac{gT^2}{2\pi} \quad (1)$$

The wave celerity C is then

$$C = \frac{L}{T} = \frac{gT}{2\pi} \quad (2)$$

in which g is the earth's gravitational acceleration. Obviously, the celerity C is proportional to the wave period T. The water depth is less than $L/2$, so then the wave particle movement touches the bottom. Tides and tsunamis usually are shallow water waves or long waves. Equations 1 and 2 become the following:

$$L = \frac{gT^2}{2\pi} \tanh\left(\frac{2\pi h}{L}\right) \quad (3)$$

$$C = \frac{gT}{2\pi} \tanh\left(\frac{2\pi h}{L}\right) \quad (4)$$

A wave contains kinetic and potential energy. The average energy E per unit sea surface area is the following:

$$E = \frac{1}{8}\rho g H^2 \quad (5)$$

The wave energy propagates in a group velocity, Cg, which is given by the following equation (2):

$$Cg = \frac{C}{2}\left(1 + \frac{4\pi h/L}{\sinh 4\pi h/L}\right) \quad (6)$$

The wave energy flux P is as follows:

$$P = E \cdot Cg \quad (7)$$

A real ocean wave is not oscillatory, as Stokes' theory described, but irregular in a stochastic process. Pierson (3,4) introduced the technique of communication engineering to the ocean wave and proposed the random wave theory. The ocean wave would be a superposition of sinusoidal wave components of different directions, amplitudes, frequencies, and angular phases, in which the phase is a random variable of equal probability density between—π and π. Then the ocean wave can be represented by a power spectrum. However, for convenience, engineers usually use the significant wave to represent the ocean wind wave. For a group of N wave heights measured at a point, waves are ordered from the largest to the smallest and assigned a number from 1 to N to them. The significant wave height $H_{1/3}$ is defined as the average of the first (highest) $N/3$ wave heights (5). The order of the wave period is accompanied by its wave height as a pair. The significant wave period $T_{1/3}$ is defined as the average of the first $N/3$ wave periods.

WIND WAVE FORECASTING

During the Second World War, two famous American oceanographers, Dr. H.U. Sverdrup and Dr. W.H. Munk, were assigned by the U.S. Navy to develop a wind wave forecasting scheme for the Normandy landing operations. The work was originally completed in 1943 and classified and published in 1947(6). Later, the scheme was extensively patched and amended by Bretschneider(7). Therefore, the scheme has been named the SMB method. In the SMB method, the wind speed, wind duration, and fetch are the main parameters, where fetch is defined as the wind blowing distance in the water area.

Modern ocean wave modeling was initiated in 1956 and extensively developed and revised (8). So far it is still developing. The basic concept is the evolution of energy spectrum F governed by the energy balance equation:

$$\frac{df}{dt} = S_{\text{in}} + S_{\text{nl}} + S_{\text{ds}} \quad (8)$$

in which S_{in} is the energy input flux from wind to wave spectrum components, S_{nl} the energy flux exchange due to nonlinear wave–wave interaction, and S_{ds} the energy flux output due to dissipation. The present operating ocean models are WOM, NWW3, etc.

TYPHOON WAVE

The typhoon or hurricane is an atmospheric eddy that originates in tropical or subtropical ocean regions. The typhoon wind speed is usually very high so that the typhoon wave plays a significant role in the design of coastal and offshore structures. Because the wind velocity changes rapidly, the generation process within typhoons is complicated. Parametric typhoon or hurricane wave prediction models were developed (9,10), which pointed out that the maximum wave exists at the right side of the typhoon center as one faces the forward direction.

The radius of a typhoon is about 200–400 km. When a typhoon is still far away from a location, the swell may arrive because the swell energy propagates usually faster than the typhoon moves. The typhoon can be regarded as a point source of wave generation. If Person A, who does not move, throws balls to Person B at a fixed time interval and the speed of the ball relative to the ground is constant, then Person B receives balls at the same time interval. If Person A moves toward Person B, Person B receives balls at a shorter time interval. If the ball is like energy, Person A throws more energy flux to Person B, as Person A moves toward Person B. This is the same as the well-known Doppler effect. As a whistling train moves toward an observer, the sound heard by the observer is higher in frequency than the actual whistle at the source. The observer receives more sound energy flux. As shown in Equation 7, wave energy flux P is wave energy $E \times$ group velocity Cg, which is proportional to the wave period. The wave energy is proportional to H^2 (Equation 5). Because the typhoon wave moving speed has little effect on the swell period, then the swell height is enhanced, as the typhoon approaches a station (11). There were cases where swell heights were larger than those inside the typhoon because the moving velocity was close to the group velocity of the swell.

BIBLIOGRAPHY

1. Stokes, G.G. (1847). On the theory of oscillatory waves. *Trans. Cambridge Philos. Soc.* **8**: 441.
2. Kinsman, B. (1965). *Wind Waves: The Generation and Propagation on the Ocean Surface*. Prentice-Hall, Englewood Cliffs, NJ.
3. Pierson, W.J., Jr. (1952). *A Unified Mathematical Theory for the Analysis, Propagation and Refraction of Storm Generated Ocean Surface Waves, Parts I and II*. New York University, College of Engineering, Research Division, Department of Meteorology and Oceanography prepared for the Beach Erosion Board, Department of the Army, and Office of Naval Research, Department of the Navy.
4. Pierson, W.J., Jr. (1955). Wind-generated gravity waves. In: *Advances in Geophysics*. Vol. 2. Academic Press, New York, pp. 93–178.
5. Dean, R.G. and Dalrymple, R.A. (1984). *Water Wave Mechanics for Engineers and Scientists*. Prentice-Hall, Englewood Cliffs, NJ.
6. Sverdrup, H.U. and Munk, W.H. (1947). *Wind, Sea and Swell: Theory of Relations for Forecasting*. U. S. Navy Hydrographic Office Pub. No. 601.
7. Bretschneider, C.L. (1952). The generation and decay of wind waves in deep water. *Trans. Am. Geophys. University* **33**(3): 381–389.
8. Komen, G.J., Cavaleri, L., Donelan, M., Hasselmann, K., Hasselmann, S., and Janssen, P.A.E.M. (1994). *Dynamics and Modeling of Ocean Waves*. Cambridge University Press.
9. Bretschneider, C.L. and Tamaye, E.E. (1976). Hurricane wind and wave forecasting techniques. *Proc. 15th Int. Coastal Eng. Conf.*, Honolulu, HI, pp. 202–237.
10. Young, I.R. (1988). Parametric hurricane wave prediction model. *J. Waterway Port Coastal Ocean Eng.* **114**(5): 637–652.
11. Liang, N.K. (2003). The typhoon swell Doppler effect. *Ocean Eng.* **30**: 1107–1115.

WOODS HOLE: THE EARLY YEARS

Northeast Fisheries Science
Center—NOAA

The beginning of Woods Hole dates back to the early 17th century. Five years before the settlement of Jamestown,

Virginia, and 18 years before the Pilgrims landed at Provincetown and Plymouth, Bartholomew Gosnold coasted along Cape Cod and Marthas Vineyard, and about May 31, 1602, he is believed to have landed at what is now known as Woods Hole. The Town of Falmouth, of which Woods Hole is presently a part, was first settled in 1659–61 when several persons were granted permission to purchase land. The date of the settlement of Woods Hole took place 17 years later. The town (Falmouth) was incorporated on June 4, 1686, and called Succonessett, the name which later, probably in 1694, was changed to Falmouth. On July 23, 1677, the land around Little Harbor of Woods Hole was divided among the 13 settlers in "lots of 60 acres upland to a share" and an "Indian deed" confirming the land title was signed by Job Notantico on July 15, 1679. Fishing, hunting, and sheep breeding were the principal occupations of the early settlers and their descendents. Later on a grist mill was built and salt was made by solar evaporation of sea water in pans built along the banks of Little Harbor.

These quiet, rural conditions, devoid of adventure, persisted until about 1815, when Woods Hole became an important whaling station from which ships operated on the high seas. The whaling industry in the United States became a very profitable business, and Woods Hole was a part of it. In 1854, the total receipts for the American whaling fleet amounted to $10.8 million, the largest part of this amount resulted from whaling carried out by Massachus tts captains. Woods Hole participated in these activities and prospered. It is known that between 1815 and 1860, not less than nine whaling ships were making port at the Bar Neck wharf, which was located where the U.S. Navy building of the Woods Hole Oceanographic Institution now stands. The place was busy processing oil and whalebone and outfitting ships. A bake house for making sea biscuits for long voyages stood next to the present "Old Stone Building" built in 1829 as a candle factory. This conspicuous old landmark on Water Street of Woods Hole, identified by an appropriate bronze plaque, now serves as a warehouse for the Marine Biological Laboratory for storing preserved zoological specimens. About 1860, whaling became less profitable and Woods Hole entered into the second phase of its economic life which was dominated by the establishment and operation of a new commercial venture known as the Pacific Guano Works.

During the years from 1863 to 1889, when the *Pacific Guano Works* was in operation the life of Woods Hole centered around the plant which was built at Long Neck near the entrance to what is known now as Penzance Point. Many large sailing vessels carrying sulphur from Italy, nitrate of soda from Chile, potash from Germany, and many schooners under the American flag loaded with guano and phosphorus from the Pacific Coast of South America were anchored in Great Harbor waiting for their turn to unload their cargoes. The number of laborers regularly employed by the Guano Company varied from 150 to 200 men, mostly Irishmen brought in under contract. Several local fishermen found additional employment as pilots for guano ships. The company maintained a store where various goods such as leather, lead pipe, tin, coal, wood, and other items were bought and sold. The store acted also as a labor housing agency. Through efforts of the business manager of the Guano Company, the Old Colony Railroad was persuaded to extend its branch from Monument Beach to Woods Hoie. The establishment of well-organized and reliable transportation to Boston was an important factor in the future life of the community.

The Pacific Guano Works was established by the shipping merchants of Boston who were seeking cargo for the return voyage of their ships. The guano deposits of one of the Pacific islands seemed to furnish this opportunity. As soon as the joint stock company was organized in 1859 with the capital of $1 million, arrangements were made almost immediately by which the newly formed concern came into possession and control of Howland Island. This island is located in the middle of the Pacific Ocean at longitude 177 deg. W., a short distance north of the Equator, about 1, 500 miles true south from Midway Island of the Hawaiian archipelago. At the same time appropriate plant and docking facilities were built at Woods Hole and 33 large sailing ships became available for hauling guano. Unlike the well-known guano islands off the coast of Peru, Howland Island is located in the zone of abundant rainfall. Consequently, the guano deposits of the island were leached of organic components and consisted of highly concentrated phosphate of lime.

Fertilizer produced by the company was made by restoring the lost organic matter of the phosphate rock

1887 Map of Woods Holl (Fisheries is bottom left)

by adding the right proportion of organic constituents which were obtained from menhaden, pogy, and other industrial fish which abound in Cape Cod waters. The rock was pulverized and purified by washing; fish brought in by local fishermen were first pressed to extract oil, and the residue digested with sulphuric acid, washed, and dried. Acid was produced locally from sulphur imported from Sicily, and the digestion of fish flesh was carried out in large lead-lined vats. The plant was well equipped with machinery needed for the process and even had a chemical laboratory where chemists made the necessary analyses. Various sheds for storage and drying, barracks for laborers, and a business office completed the facilities.

When the deposits of phosphate rock on Howland Island were exhausted, the company acquired title to the Greater and Lesser Swan Islands from the U.S. Government. These islands are located in the Caribbean Sea at latitude 17 deg. N. and longitude 83 deg. W. off the coast of Honduras. The islands are only 400 miles from Key West, Florida, and 500 miles from New Orleans. They contained good-quality phosphate rock and being much closer to Woods Hole greatly reduced the voyage time and cost of delivery. Further expansion of the company consisted in the acquisition of Chisolm's Island near the coast of South Carolina, construction of a plant for cracking and washing phosphate rock on the Ball River side of the island, and establishment of a processing plant in Charleston, S.C. From the initial production (in 1865) of 7, 540 sacks of fertilizer weighing 200 pounds each, the output reached 11, 420 tons in 1871 and continued to grow until the combined annual production in 1879 of the works at Woods Hole and Charleston reached from 40, 000 to 45, 000 tons of guano fertilizer.

Spencer Baird, Secretary of the Smithsonian Instution and first commissioner of the U.S. Commission of Fish and Fisheries arrived in Woods Hole in 1871. Baird was greatly impressed by the idea of utilizing menhaden and other fishes for the production of guano fertilizer and considered it a worthwhile project. In a letter dated October 18, 1875, to John M. Glidden, treasurer of the Pacific Guano Works Company, Baird urged him "to make a display of your wares at the centennial (in Philadelphia), as this is one of the most important interests in the United States." He writes further that "there is no species (of fish) worked up elsewhere comparable to the movement with the menhaden, or pogy, as to numbers and the percentage of oil. The combination, too, of the pogy scrap with the South Carolina phosphates and the guanos of the West Indies and of the PacificA are also quite novel, and as being especially an American industry, are eminently worthy of full appreciation."

While the scientists, agriculturalists, and stockholders of the company thought very highly of the guano works, the existence of a malodorous plant was not appreciated by the residents of Woods Hole who suffered from a strongly offensive odor whenever the wind was from the west. Woods Hole might have continued to grow as one of the factory towns of Massachusetts but, fortunately for the progress of science and good fortune of its residents (except those who invested their savings in the shares of Pacific Guano Works), the company began to decline and became bankrupt in 1889.

Cessation of business and heavy monetary losses brought financial disaster to many residents of Woods Hole. The gloom prevailing in the village after the closing of the guano works began to dissipate, however, with the development of Woods Hole as a place of scientific research and with the increasing tourist trade. The factory buildings were torn down, the chimney which dominated the Woods Hole landscape was dynamited, and over 100,000 pounds of lead lining the acid chambers were salvaged. Large cement vats and the remnants of the old wharf remained; in the following years the latter became a favored place for summer biologists to collect interesting marine animals and plants.

The years from 1871 to the death of Baird in 1887 were the formative period of the new era of *Woods Hole as a scientific center*. In historical documents and in old books the present name Woods Hole is spelled in a different way. The old name "Woods Holl" is considered by some historians of Cape Cod to be a relic of the times prior to the 17th century when the Norsemen visited the coast. The "Holl", supposed to be the Norse word for "hill", is found in the old records. The early settlers gave the name "Hole" to inlets or to passages between the islands, such as "Robinson's Hole" between Naushon and Pasque Islands, or "Quick's Hole" between Pasque and Nashawena Islands, and Woods' Hole between the mainland and Nonamesset Island. In 1877 the Postmaster General ordered the restoration of the original spelling "Wood's Holl", which remained in force until 1896 when the United States Post Office changed it back to Woods

Hole and eliminated the apostrophe in Wood's. The change was regretted by the old timers and by C. O. Whitman who had given the specific name "hollensis" to some local animals he described.

At the time of his arrival at Woods Hole in 1871, Baird was well known to the scientific circles of this country and abroad as a naturalist, student of classification and distribution of mammals and birds, and as a tireless collector of zoological specimens. He maintained voluminous correspondence with the scientists in the United States and Europe, and was Permanent Secretary of the recently organized American Association for the Advancement of Science. To the general public he was known as a contributor to a science column in the New York Herald and author of many popular magazine articles. His newly acquired responsibilities as Commissioner of Fisheries greatly added to his primary duties as Assistant Secretary of the Smithsonian Institution which was primarily responsible for the establishment of the National Museum in Washington. As a scientist, Baird belonged to the time of Louis Agassiz, Th. H. Huxley, and Charles Darwin. Like Agassiz he attended medical college but never completed his studies, although the degree of M. D. honoris causa was later conferred upon him by the Philadelphia Medical College.

In the words of Charles F. Holder, "he was a typical American of the heroic type. A man of many parts, virtues, and intellectual graces, and of all the zoologists science has given the world.... he was most prolific in works of practical value to man and humanity."

Commissioner Baird attended many Congressional hearings and conferences with state officials and fishermen at which the probable causes of the decline of fisheries were discussed and various corrective measures suggested. From the lengthy and frequently heated discussions and evidence presented by the fishermen and other persons familiar with the fisheries problems, he became convinced that an alarmingly rapid decrease in the catches of fish had continued for the last 15 or 20 years. Such a decline was particularly noticeable in the case of scup, tautog, and sea bass in the waters of Vineyard Sound. It was logical, therefore, that the new Commissioner of Fisheries would select for his initial activities the New England coastal area where the fishing industry was of greatest importance as a politico-economical factor. Woods Hole, however, was not a significant fishing center.

In the "Fisheries and Fishlng Industry of the United States" prepared and edited by Goode (1884–87) for the 1880 Census, the fishing activity at Woods Hole is described in the following words: "Of the male inhabitants only seven are regularly engaged in fishing, the remainder being employed in the guano factory, in farming and other minor pursuits.... There is one ship carpenter in Wood's Holl, but he finds employment in his legitimate business only at long intervals. Of sailmakers, riggers, caulkers, and other artisans there are none. Four men are employed by Mr. Spindel, during the height of the fishing season, in icing and boxing fish. The boat fishery is carried on by seven men from April until September, inclusive. Only three species of fish are usually taken, namely, scup, tautog, and sea bass. The total catch of each fisherman is about 15 barrels, or about 2400 pounds. In addition about 6,720 lobsters are annually taken."

Before selecting a location for permanent headquarters for the work on fishery management and conservation, Baird undertook extensive explorations of the fishing grounds off the entire New England Coast. Section 2 of the Joint Resolution Number 8 of Congress gave the Commissioner full authority to carry out the necessary research. In part it reads as follows "and further resolved, That it shall be the duty of the said Commissioner to prosecute investigations and inquiries on the subject, with the view of ascertaining whether any and what diminution in the number of the food-fishes of the coast and the lakes of the United States has taken place; and, if so, to what causes the same is due; and also, whether any and what protective, prohibitory, or precautionary measures should be adopted in the premises; and to report upon the same

to Congress." Section 4 of the same Resolution contains an important clause which authorizes the Commissioner of Fisheries "to take or cause to be taken, at all times, in the waters of the seacoast of the United States, where the tide ebbs and flows, and also in the waters of the lakes, such fish or specimens thereof as many in his judgement, from time to time, be needful or proper for the conduct of his duties as aforesaid, any law, custom, or useage of any State to the contrary notwithstanding."

The significant words "where the tide ebbs and flows" were interpreted by Baird in a very broad scientific sense which extended the authority for his investigations to the offshore areas of the open ocean.

Pounds and weirs were most frequently accused by the public as destructive methods of fishing responsible for the decline in the abundance of food fishes along the coast. Although Baird gave very serious consideration to the possible destructiveness of fixed nets, traps, pounds, pots, fish weirs, and other stationary apparatus, he was fully aware of the complexity of the factors which may cause the decline in fish populations. He discusses this difficult problem in a paper entitled "Report on the condition of the sea fisheries of the south coast of New England" and published as the first section of the voluminous First Report of the Commissioner of Fish and Fisheries for 1871. Of the causes which may have contributed to the decrease of summer shore fisheries of the south side of Massachusetts and Rhode Island, a fact which he considered as well established by the testimonies of competent persons, he lists the following: (1) decrease or disappearance of the food of commercial fishes; (2) migration of fishes to other localities; (3) epidemic diseases and "peculiar atmospheric agencies, such as heat, cold, etc."; (4) destruction by other fishes; (5) man's activities resulting in the pollution of water, in overfishing, and the use of improper apparatus.

The biologist of today will recognize in this statement Baird's broad philosophical approach to the major problem of fishery biology. The outlined program combined oceanographical and meteorological investigations with the studies of biology, ecology, parasitology, and population dynamics of various fish species. Baird's program of research is as comprehensive and valid today as it was 90 years ago.

No time was lost in initiating this program. Woods Hole was selected as the base of the sea coast operations during the first summer and Vinal N. Edwards became the first permanent federal employee of the fisheries service. In spite of the insignificance of local fisheries, this locality offered a number of advantages which were recognized by Baird. Communication with Boston, New York, and Washington was good and promised to be better with the expected opening of the railroad branch in 1872. Being centrally located in relation to principal fishing grounds of New England and having good dock facilities and water of sufficient depth for sea going vessels, Woods Hole was a suitable base for visiting the offshore grounds. Furthermore, it was believed that the alleged decrease in food fishes was most clearly manifested in the region around Vineyard Sound. The small yacht Mazeppa of the New Bedford Custom House and the revenue-cutter

Moccasin attached to the custom-house at Newport, R.I., were placed at the disposal of Baird; and the Light-House Board granted permission to occupy some vacant buildings and the wharf at the buoy-station on the west bank of Little Harbor. The Secretary of the Navy came to Baird's assistance by placing at his command a small steam launch which belonged to the Boston Navy Yard and by giving many condemned powder tanks which could be used for the preservation of specimens. Nets, dredges, tanks, and other gear were provided by the Smithsonian Institution. Cooperation of the various governmental agencies was authorized by Congress which in Section 3 of the Resolution specified that "the heads of the Executive Departments be, and they are hereby directed to cause to be rendered all necessary and practicable aid to the said Commissioner in the prosecution of the investigations and inquiries aforesaid."

This provision of the law was of great value. It is apparent, however, that the success in obtaining cooperation authorized by law depended a great deal on the personal characteristics of Baird, his great ability of getting along with people, and his remarkable power of persuasion, These qualifications played the major role in his success in organizing the Commission's work and also in obtaining the cooperation of scientists as well as that of fishermen and businessmen.

The investigation during the first summer consisted primarily in collecting large numbers of fishes and studying their spawning, rate of growth, distribution, and food. In the course of this work nearly all the fish pounds and traps, some 30 in number, in the vicinity of Woods Hole, were visited and their location recorded. There was no difficulty in obtaining the owners' permission to examine these installations and to collect the needed specimens. Altogether 106 species of fish were secured, photographed, and preserved for the National Museum. Of this number 20 or more species had not previously been known from Massachusetts waters. Information gained in this manner was supplemented by the testimonies of various fishermen who presented their ideas either for or against the use of traps and pounds. Among them was Isaiah Spindel, who at the request of Baird, prepared a description of a pound net used at Woods Hole and explained its operation. In the following years Spindel became an influential member of the group of local citizens who supported Baird's plan of establishing a permanent marine station at Woods Hole.

The ship Moccasin under the command of J. G. Baker was engaged in taking samples of plankton animals, in determining the extent of beds of mussels, starfish, and other bottom invertebrates, and in making temperature observations.

One of the principal collaborators in the studies conducted at Woods Hole in 1871 was A. E. Verrill of Yale University, a professor whom Baird appointed as his assistant and placed in charge of the investigations of marine invertebrates. Dredging for bottom animals during the first summer was carried out on a relatively small scale from a chartered sailing yacht Mollie and a smaller vessel used in the immediate vicinity of Woods Hole. Extensive collections were made by wading on tidal flats exposed at low water.

Zoological work attracted considerable interest among the biologists of this country. Many of them stopped at Woods Hole for greater or lesser periods and were encouraged by Baird to use the facilities of the Fish Commission. The group included such well known men as *L. Agassiz*, A. Hyatt, W. G. Farlow, Theodore Gill, Gruyure Jeffries of England, and many others. The first year's work extended until the early part of October. Before returning to Washington, Baird commissioned Vinal N. Edwards of Woods Hole to continue the investigation as far as possible. By the end of the first year a general plan of study of the natural histories of the fishes and the effect of fishing on fish populations was prepared with the assistance of the well-known ichthyologist, Theodore N. Gill. His old "Catalogue of the fishes of the Eastern Coast of North America from Greenland to Georgia", was revised and the next text including the recently collected data concerning the Massachusetts fishes, appeared in the First Report of the U.S. Commissioner of Fish and Fisheries. The plan of investigation suggested by Gill was adopted by Baird as a guide for the work of his associates for the purpose of "securing greater precision in the inquiries." The plan is composed of 15 sections, such as Geographical distribution, Abundiance, Reproduction, etc., with detailed subdivisions under each one. A questionnaire containing 88 different items was included in order to facilitate the inquiries conducted among the fishermen. The scope of the highly comprehensive program is complete enough to be useful today; marine biologists of today would probably only rephrase it, using modern terminology. During the first year of operations conducted at Woods Hole, Baird and his associates laid down the foundation of the new branch of science which we now call fishery biology or fishery science.

AN ANALYSIS OF THE INTERNATIONAL MARITIME ORGANIZATION–LONDON CONVENTION ANNUAL OCEAN DUMPING REPORTS

Christine Dickenson
Iver W. Duedall
Florida Institute of Technology
Melbourne, Florida

Trends are analyzed for types and quantities of permitted wastes, primarily dredged material, sewage sludge, and industrial waste, to be dumped at sea by member countries to the London Convention (LC) from 1992 and 1995 through 1998. In 1972, the Inter-Governmental Conference on the Dumping of Wastes at Sea led to the creation of the London Dumping Convention (now called the London Convention) to help regulate the dumping of wastes at sea. The act of dumping, as defined by the LC, is international disposal at sea of any material and in any form, from vessels, aircraft, platforms, or other artificial structures. The first consultative meeting of the LC contracting parties was held in 1976 by the Inter-Governmental Maritime Consultative Organization [now called the International Maritime Organization (IMO)]. During this meeting, the procedure for the reporting of permits issued, on an annual basis, for dumping at sea by contracting parties, was determined. The IMO-LC annual reports on permitted wastes list the number of permits issued by member countries, the types and quantities of wastes permitted for dumping at sea, and the location and designation of dump sites. Now with nearly 25 years of dumping records available, we are able to see trends in permitted dumping activity. In 1976, the first year of permitted dumping records, the combined amount of permitted wastes was nearly 150 million tons. In the last four years (1995 through 1998), where data are available, the total amount of wastes permitted to be dumped by the LC contracting parties as between 300 and 350 million tons. Currently, a majority of these wastes are being disposed of in the East Asian Seas and the North Sea and the largest quantity of waste being dumped is dredged material. Although the disposal of wastes at sea is considered to be a major issue, it is only responsible for about 10% of the total anthropogenic contaminants entering the ocean. Unfortunately, the long-term impacts of this dumping in the ocean are still largely unknown.

INTRODUCTION

"Historically, most coastal countries used the sea for waste disposal. It was generally the most economic way to manage waste, since land usually had, and still has, a high price tag while the sea has no private owner in the normal sense. In addition, dilution processes served the illusion that dumping at sea does not cause any permanent damage. So why risk contaminating land or drinking water with wastes if the sea was close by?" (1). "Accurate worldwide records on the amounts of wastes disposed at sea prior to 1976 are virtually impossible to obtain" (1). However, as a result of the international activities leading to the creation of the London Dumping Convention (LC) in 1972, information is now available on the number of permits issued by many countries for disposal at sea, their dumpsite locations, and the kinds and quantities of wastes that have been dumped (1).

According to the LC, the most common form of ocean dumping today is disposal from ships or barges (1,2). Wastes are loaded on these vessels and then taken to the dumpsites. Dumpsites are chosen based on the kind of waste and the ocean's properties at each site (2). Liquid wastes are generally disposed of in more dispersive environments, where mixing will rapidly dilute the dumped material. Solid wastes, on the other hand, are usually disposed of in less dispersive near-shore sites to keep the solids confined.

Here we specifically analyze trends in types and quantities of permitted wastes to be dumped at sea by member countries to the LC from 1992 and 1995 through 1998. The types of wastes that have been dumped in the ocean include industrial waste, sewage sludge, dredged material, incineration at sea, and radioactive wastes. Industrial waste may include both liquid and solid wastes and it may contain such items as acid-iron waste, fish-processing liquids, metal refinery wastes, and gas pipeline flushing wastes (2). An overall reduction in the dumping of these types of wastes has been achieved over the years by switching to alternative disposal methods, reusing wastes, and creating cleaner production technologies. The dumping of industrial wastes at sea has been prohibited since 1996 (3). "Sewage sludge is an anaerobic waste product from treatment of municipal wastewater. The sludge is in aqueous form containing about 3% suspended particles by weight" (2). Alternatives for the disposal of sewage sludge at sea include incineration, deposit on land, and agriculture use (3). "Dredged materials range from clean sands to heavily contaminated fine grained materials" (2). "Physical properties of dredged materials, including grain size, bulk density, water content, and geotechnical characteristics, are especially variable due to the kind or type of sediment being dredged, which is itself dependent on geological and watershed characteristics, as well as to the operational procedures used in dredging and disposal" (2). Incineration at sea is defined as the burning of liquid chlorinated hydrocarbons as well as other halogenated compounds in which all ash is dumped into the sea (3). This type of dumping was phased out early in 1991. The dumping of radioactive wastes, however, might be the most harmful practice of all. Although the dumping of high level radioactive wastes has never been allowed under the London Convention, it has still occurred in some cases, and even though a moratorium was placed on the dumping of low level radioactive wastes in 1983, this type of dumping still occurred (3,4). It was not until 1994 that this act became legally binding. Finally, the other waste category includes such wastes as inert geological materials, decommissioned vessels, scrap metals, and fish wastes (3).

METHODS

Annual reports on permitted dumping at sea by contracting parties were obtained from the IMO-LC (5). There is approximately a three-year delay before the most recent year's records are put into report format and published. Older records that only existed in hard copy form had to be entered into Excel spreadsheets to create a new database. Early on, it was decided that all wastes would be reported in tons to make comparisons from year to year easier. Many of the older reports had wastes recorded in cubic meters. These records were converted into tons by multiplying the cubic meters by the density of the waste (1.3 g/cm^3 for dredged material and other varying factors as indicated by each permit). Newer records, however, were received in Excel format and, for the most part, they were already converted into tons, making them very easy to add to the database.

The primary categories of wastes of importance to this study include dredged material, sewage sludge, industrial waste, and other matter. From the 1992 and 1995 through 1998 reports (5), it was desired to find out which waste was being dumped in the largest quantity, as well as which countries were dumping the most wastes and which water bodies were receiving the most wastes. The more recent data is compared to earlier reports from 1976 to 1985. Comparisons of the different sources of pollutants in the oceans and comparisons between land dumping and sea dumping are also made. Then the past, as well as the future, of ocean dumping can be assessed.

RESULTS

Tables in the IMO-LC annual reports (5) on permitted wastes (IMO, 1992, 1995–1998) list the number of permits issued to member countries, the types and quantities of wastes permitted for disposal at sea, and the location and designation of dumping sites. It is important to know that these reports are reflective of what has been permitted to be dumped and not of what has actually been dumped. The accuracy of these records, therefore, is somewhat questionable (1). It is also important to remember that not all of the contracting parties report their activities every year and some reports from contracting parties may be incomplete (2). For example, in 1995, thirty-eight of the seventy-five contracting parties registered did not report their activities. There is also no way of knowing how much noncontracting parties are dumping (2).

In 1992, 473 million tons were reported as permitted for disposal at sea. In the last four years (1995 through 1998) where data are available, the total amount of wastes permitted to be dumped by the LC contracting parties was 351 million tons in 1995, 312 million tons in 1996, 309 million tons in 1997, and 348 million tons in 1998. Figure 1 illustrates these data.

As seen in Fig. 2, a majority of these wastes are being disposed of in the East Asian Seas (Fig. 3) and the North Sea (Fig. 4). The "other areas" category is represented mostly by the United States and its disposal in the Gulf of Mexico (Fig. 5).

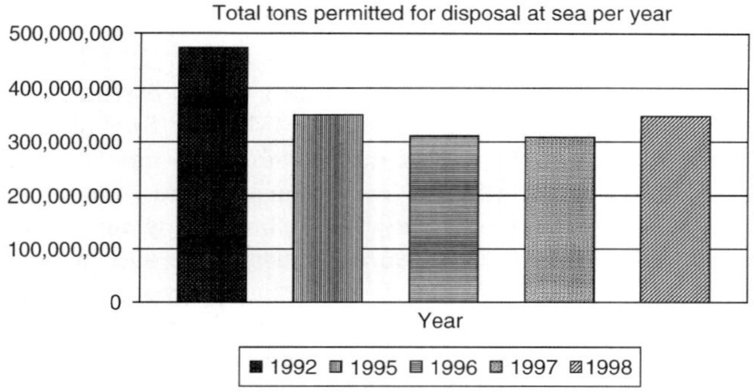

Figure 1. London Convention permitted dumping from 1992 and 1995 through 1998.

The countries responsible for dumping the most wastes, in descending order, are China, the United States, Hong Kong, the United Kingdom, Germany, and Belgium (Table 1).

As shown in Fig. 6, the largest quantity of waste being dumped is dredged material. At 327 million tons in 1998, it is the largest quantity by far, with sewage sludge coming in as the next highest quantity with only 15 million tons.

DISCUSSION

In 1976, the first year of permitted dumping records, the combined amount of permitted wastes was nearly 150 million tons (1). By 1985, this amount had grown to approximately 300 million tons (1). This apparent doubling, however, partly corresponds to an increase in the number of contracting parties to the LC. It is also important to remember that not all of the contracting parties report their activities every year and some reports from contracting parties may be incomplete. Therefore, it is not possible to provide a highly accurate interpretation of these data and the reader should be cautioned that any analysis must be considered approximate.

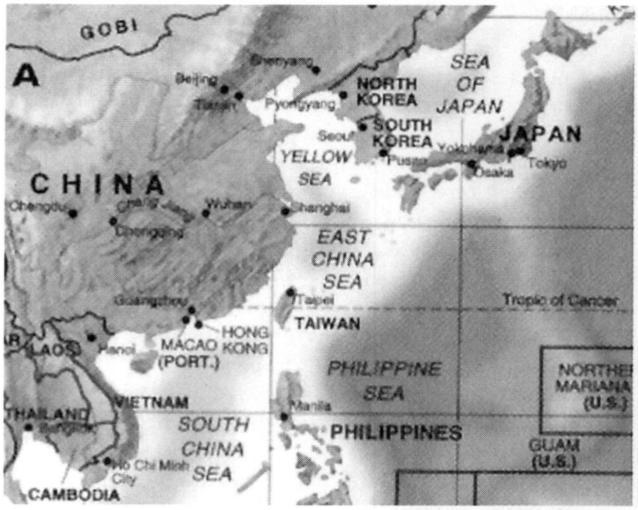

Figure 3. China and Hong Kong dumping in the East Asian Seas. (From Ref. 6).

Looking back to the data in Fig. 1, waste disposal in the ocean seems to continue to increase into the early 1990s with 473 million tons in 1992. However, by 1995 through 1998 this amount had leveled off to between 300 and 350 million tons. This decrease could be representative of the many changes in policy occurring within the LC in the 1990s to intentionally decrease the amount of waste disposed of in the ocean (3). It is also known that a major dredging project took place in Hong Kong in the early 1990s in order to expand the airport there (3). This may explain why the amount of waste dumped in 1992 was so high. The East Asian Seas section of Fig. 2 supports this theory, with data from 1992 being much higher than in the following years. Yearly fluctuations like this occur due to the variation in maintenance dredging and new works associated with shipping activities (3).

According to the IMO-LC web page, "in early 1991, incineration at sea operations came to a halt, ahead of

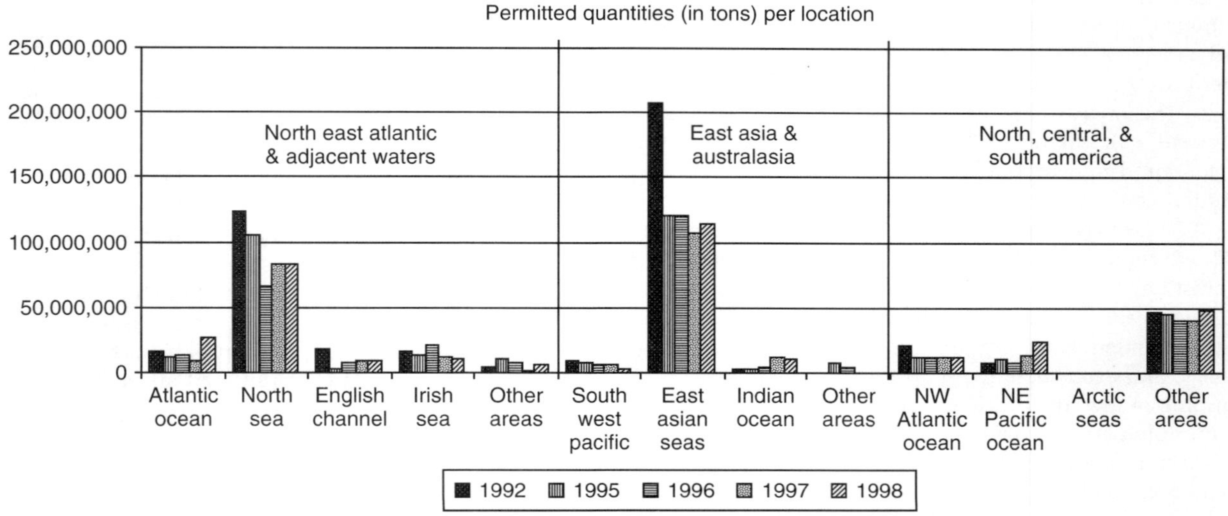

Figure 2. Locations of dumping sites with their respective amounts of wastes dumped.

Figure 4. The United Kingdom, Germany, and Belgium dumping in the North Sea. (From Ref. 6).

the agreed global deadline of December 31, 1992.... In 1990, contracting parties to the LC agreed to phase-out sea disposal of industrial waste effective by January 1, 1996.... In 1993, contracting parties started a detailed

Table 1. Highest Combined Tons for 1992 and 1995–1998

Rank	Country	Tons
1	China	295,482,060
2	United States	293,272,190
3	Hong Kong	285,437,056
4	United Kingdom	246,759,318
5	Germany	132,149,000
6	Belgium	127,507,436

review of the LC, leading to the adoption of a few crucial amendments.... These amendments consolidated in a legally binding manner the prohibition to dump all radioactive wastes or other radioactive matter and of industrial wastes, the latter as per January 1, 1996, as well as the prohibition of incineration at sea of industrial wastes and of sewage sludge" (3).

According to reports by contracting parties, no permits for the dumping of industrial waste have been issued since 1996. Before this phase-out, Japan and the Republic of Korea were responsible for most of the industrial waste being dumped (3). The amount of sewage sludge being dumped at sea decreased in the early 1990s, reflecting the phase-out of this practice by several countries, Ireland and the United Kingdom being the most recent. Currently, only Japan, the Philippines, and the Republic of Korea dispose of sewage sludge at sea (3).

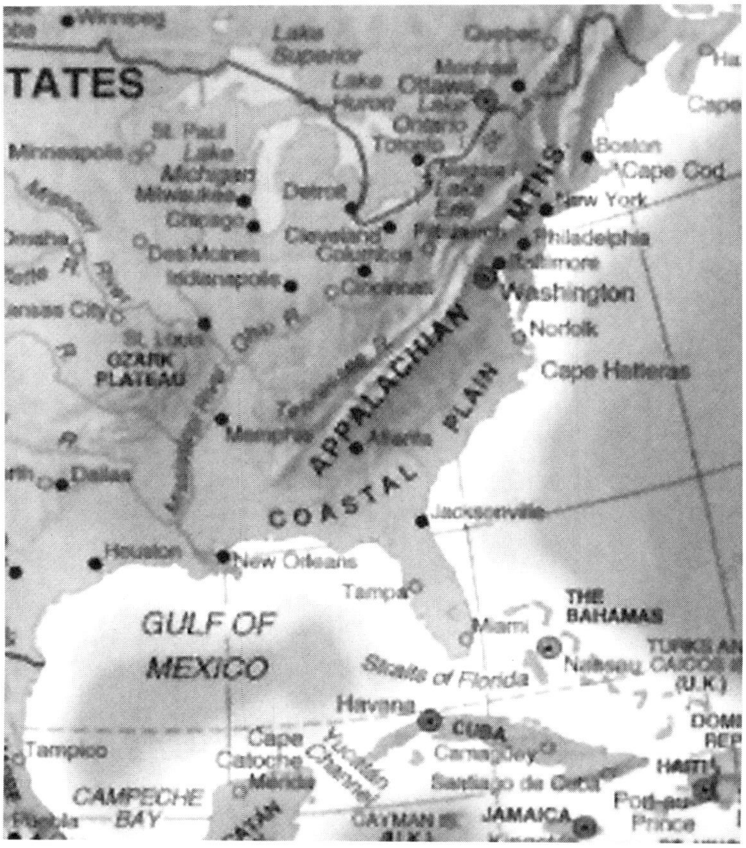

Figure 5. The United States dumping in the Gulf of Mexico and the northwest Atlantic. (From Ref. 6).

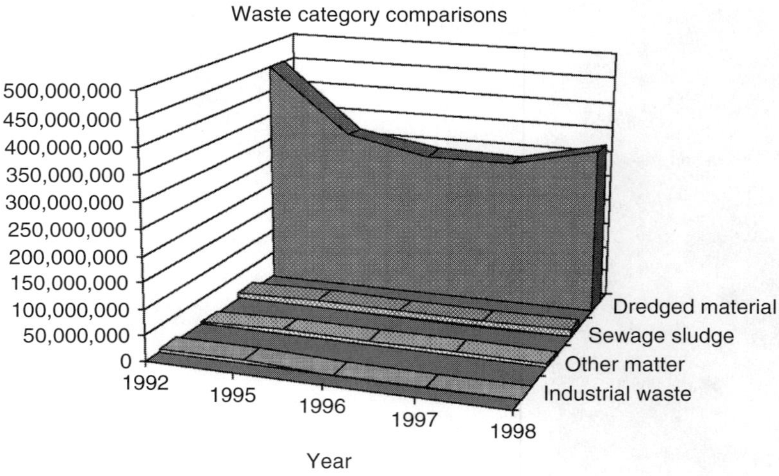

Figure 6. Dredged material, by far, makes up the largest quantity of waste being dumped at sea.

Dredged material makes up the majority of what is being dumped at sea (see Fig. 6). Unlike other wastes, the amount of dredged material dumped each year tends to fluctuate greatly due to variable maintenance dredging and new works associated with shipping activities (3). For example, there were huge dredging efforts going on in Hong Kong with the extension of the airport there in 1992. One hundred forty-four million tons of dredged material were permitted to be disposed of at sea that year by Hong Kong. Arguably, this could be responsible for the 1994 anomaly of the combined total of 473 million tons, meaning that the average amount of wastes dumped annually actually leveled out to between 300 and 350 million tons much earlier than the later 1990s.

The interactions of the wastes dumped at sea with seawater and the toxicity of these wastes to organisms, including humans, are of great importance to the scientific community. Although the long-term effects of disposal at sea are unknown, many studies have looked at some of the short-term effects, particularly in relation to radioactive wastes that have been dumped. Studies conducted in the Arctic by the IAEA and the U.S. Congress Office of Technology Assessment found that (1) releases from dumped objects were small and confined to the immediate vicinity of the dumped objects, (2) projected future doses to the general public were small, (3) doses to marine organisms were insignificant in the context of populations, and (4) remediation on radiological grounds was not warranted (7). A study by Hill et al. (8) assesses the possible effects of 40 years of dredging, and how it might have affected benthos in the Irish Sea. Zooplankton and micronekton communities, off eastern Tasmania, as studied by Bradford et al. (9), were found to be affected by the presence of a warm-core eddy instead of jarosite wastes dumped within the vicinity. As de La Fayette (7) said, it is difficult to draw conclusions from data that has been collected because more research, monitoring, and prevention projects are still needed for us to understand all of the factors that are involved in knowing how, when, and where ocean stored wastes might affect us and the environment.

Although such concern is felt about disposal at sea, it is not the largest contributor of pollutants to the oceans. As seen in Fig. 7, the IMO-LC puts it in fourth place, making up only 10% of the total pollutants in the oceans. Runoff and land-based discharges are the largest source, making up 44%. Even maritime transportation, at an estimated 12%, pollutes the oceans more than dumping does.

In an attempt to compare ocean dumping to disposal in landfills, we look at the United States. According to Zero Waste America (10), in 2001, 409 million tons of municipal waste were generated in the United States. Of that, 278 million tons were disposed of in landfills. Table 2 illustrates dumping on land and in the ocean by the United States during 1992 and 1995 through 1998.

Those five years of ocean dumping, totaling 293 million tons, is only 15 million tons more than what was disposed of in landfills in 2001 alone. In conclusion, we note that much more waste is being disposed of in landfills than in the ocean. Therefore, shouldn't we be more concerned about the land?

CONCLUSION

Although the environmental impacts of dumping at sea are still largely unknown, it is comforting that the LC is moving in a positive direction toward more regulations for the better protection of our oceans. With industrial

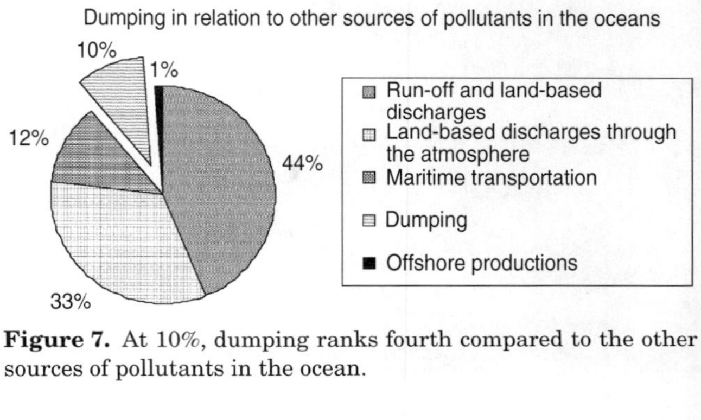

Figure 7. At 10%, dumping ranks fourth compared to the other sources of pollutants in the ocean.

Table 2. Ocean Dumping versus Land Dumping in the United States

	Tons in Millions	
Year	Land	Ocean
1992	241	67
1995	249	58
1996	238	46
1997	236	53
1998	238	69

Data taken from Reference 10.

waste already phased out and sewage sludge on its way to becoming completely phased out, that leaves dredged material and "other matter" as the future of ocean dumping. Other matter mostly makes up inert geological material from mining and excavations; bulky wastes such as steel equipment, scrap metal, and concrete; fish wastes; obsolete ammunition and explosives; discontinued oil platforms; spoiled food; and other random wastes. This category may also eventually be phased out, but dredged material will most likely continue to be dumped since this form of waste, in the normal sense, either came "from the ocean floor" or somewhere close to it being that two-thirds of dredged material is connected with maintenance operations to keep harbors, rivers, and other waterways from being blocked up. Unfortunately, according to the IMO-LC, about 10% of dredged material is moderately to heavily contaminated from a variety of sources including shipping, industrial and municipal discharges, and land runoff. Whether we find somewhere to dispose of dredged material on land, or we continue to dispose of it in the sea, it will continue to affect our environment.

Acknowledgments
We are grateful to Rene Coenen, Office of the London Convention, International Maritime Organization, for kindly providing the LC dumping records. We are also thankful to Ruth Caulk and Melissa Sheffer for assisting in the preparation of the ocean dumping spreadsheets.

BIBLIOGRAPHY

1. Duedall, I.W. (1990). A brief history of ocean disposal. *Oceanus* **33**(2): 29.
2. Connell, D.W. and Hawker, D.W. (1992). *Pollution in Tropical Aquatic Systems*. CRC Press, Boca Raton, FL.
3. London Convention. (2003). Available: http://www.london-convention.org.
4. Ahnert, A. and Borowski, C. (2000). Environmental risk assessment of anthropogenic activity in the deep sea. *J. Aquat. Ecosys. Stress Recov.* **7**: 299–315.
5. IMO. (1972). *Convention on the Prevention of Marine Pollution by Dumping of Wastes and Other Matter, London Convention 1972*. Final Reports on Permits Issued in 1992, 1995, 1996, 1997, and 1998.
6. Karls, F. (1999). *World History: The Human Experience—The Modern Era*. Glencoe/McGraw-Hill, New York.
7. de La Fayette, L. (1998). The London Convention 1972: preparing for the future. *Int. J. Mar. Coastal Law* **13**(4): 515–536.
8. Hill, A.S. et al. (1999). Changes in Irish Sea benthos: possible effects of 40 years of dredging. *Estuarine Coastal Shelf Sci.* **48**: 739–750.
9. Bradford, X. et al. (1999).
10. Zero Waste America. (2001). Available: http://www.zerowasteamerica.org/Statistics.htm.

MARINE SOURCES OF HALOCARBONS

Robert M. Moore
Dalhousie University
Halifax, Nova Scotia, Canada

Halogenated gases are important in the atmosphere by virtue of their ability to carry chlorine, bromine, and iodine to high altitude, where they can act as efficient catalysts of ozone destruction. The effect on ozone of chlorine release in the stratosphere by certain manufactured chlorofluorocarbons is well known, but naturally occurring halogen carriers account for a part of the background cycle of ozone breakdown. The oceans constitute an enormous reservoir of dissolved halogens in the form of halide ions, but the transport of this material to the upper atmosphere is inefficient because sea salt particles, introduced to the atmosphere as sea spray from breaking waves and bubble-bursting, have short lifetimes, being readily washed out by rain. For this reason the relatively minute concentrations of dissolved halogenated gases, such as methyl chloride occurring in surface seawater at concentrations around 10^{-10} M, have the potential to drive significant transport of chlorine into the upper atmosphere. Since the role of anthropogenic chlorine and bromine gases in stratospheric ozone depletion was recognized, there has been renewed interest in the ocean as a source of halogenated trace gases. The emission of iodine compounds from the ocean has more recently been shown to be potentially important in aerosol production as well as in affecting ozone concentrations over the ocean (1).

The fluxes of trace gases such as those between the ocean and atmosphere is most commonly calculated from concentration measurements in the surface ocean and the overlying atmosphere together with empirical relationships between gas exchange coefficients and wind velocity (2). Supersaturation of a gas at the sea surface with respect to the atmosphere will support an outward flux, the magnitude of which is strongly dependent on wind speed, and to a lesser extent on temperature. Gas fluxes can be highly variable spatially and seasonally, and, in general, estimated global fluxes will have a substantial degree of uncertainty due to the sparseness of the concentration measurements combined with a large uncertainty in the exchange velocity. Ideally, gas fluxes would be calculable based on a firm understanding of the processes governing both the production and loss of the gas in the upper ocean and the processes controlling gas exchange. This remains a distant goal, and current studies are directed at developing an understanding of these processes.

Some of the first measurements made of halogenated methanes in seawater suggested that the ocean could be the predominant source of methyl chloride accounting for most of the 4 million tonnes that must be added annually to the atmosphere to support the observed concentrations (3). More recently, it has been demonstrated that, in the case of methyl chloride, the ocean is just one of many sources (including biomass burning, woodrot fungi, and vegetation) and can probably account for only about 15% of the total flux to the atmosphere (4). In contrast, the ocean is still recognized to be the most important source of atmospheric methyl iodide.

Impetus for the study of sources of methyl bromide came with the discovery of the efficiency with which bromine could catalyze stratospheric ozone loss, working either directly or in concert with chlorine. This led to concern about the possible role of methyl bromide, manufactured for use as a fumigant, in observed ozone loss. The existence of natural sources and sinks for methyl bromide made it more difficult to quantify the deleterious effect of the anthropogenic component. Extensive studies of the distribution of methyl bromide in the oceans and of its flux across the air–sea interface in different regions demonstrated the existence of both production and loss processes within the ocean (5). The balance between these determines whether a particular region will emit CH_3Br to the atmosphere or take it up. A major oceanic sink results from the reaction between CH_3Br and chloride ion, the rate being strongly dependent on temperature. Other losses include chemical hydrolysis and also biological uptake. Bromine is carried to the atmosphere by many other gases produced in the ocean, notably bromoform and dibromomethane (6).

Early studies pointed to an association between certain halogenated compounds in seawater, such as methyl iodide, and coastal beds of macrophytes. The more general phenomenon of organohalogen production by seaweeds has long been known, and a wide range of halogenated organic compounds of varying molecular weight and complexity have been identified (7). The focus here is on halogenated compounds that play a role in atmospheric chemistry, so these are typically volatile compounds. Two such compounds that have been shown to have major sources in macrophytes are di- and tribromomethane (CH_2Br_2 and $CHBr_3$). The latter has been found to occur within certain seaweeds at concentrations up to the percent level. It is apparent that haloperoxidase enzymes are involved in the biosynthesis of organohalogens by macrophytes.

In spite of the variety and abundance of halogenated methanes in seaweeds and the flux of halogen to the atmosphere that can thus be supported in the coastal zone, the overall importance of these algae on an oceanwide scale is small on account of their limited distribution (8), although, in the case of bromoform and dibromomethane, Carpenter and Liss, (9) estimated that macrophytes are a major source. For this reason there has been interest in the possible production of the same compounds by marine microalgae, which are ubiquitous in the sunlit waters of the world's oceans. An approach that has been adopted with some success to study this question has been the use of laboratory cultures of specific algae. This has demonstrated the capacity of a range of phytoplankton of different algal classes to produce compounds such as CH_3I, CH_2Br_2, $CHBr_3$, and CH_2I_2 (10,11), although species vary in the spectrum of compounds that they produce as well as in the production rates. There has been some progress in elucidating the production mechanism of these compounds with the identification of haloperoxidase enzymes in a few phytoplankton species (11). However, much uncertainty remains regarding the oceanic source of methyl chloride, bromide, and iodide. For although there are published studies that point to the capacity of some phytoplankton to produce methyl halides (12), the measured rates when normalized to biomass concentrations have typically been found to be quite inadequate to account for the observed fluxes from the ocean into the atmosphere (10). This may be due in part to the limitations of laboratory studies of phytoplankton cultures: relatively few species can be grown successfully in the laboratory and the conditions are very dissimilar to those in the ocean itself. Alternative possible explanations include the involvement of microbial processes (13), zooplankton, and photochemistry, or a combination of sources. One laboratory study has provided some evidence for the production of methyl iodide in seawater irradiated with simulated sunlight (14), but it has yet to be demonstrated that this process is significant in the ocean.

Open ocean studies of halocarbon distributions show that there is no simple correlation with phytoplankton biomass, quantified as chlorophyll *a* concentration. This means that the production of a particular compound cannot be attributed uniformly to all species of phytoplankton. More success has been obtained in studies that measured a series of photosynthetic pigments and evidence is forthcoming from one such study that CH_3Br has a source in Prymnesiophytes (15).

Measurements of dissolved methyl halides in the ocean typically show relatively high concentrations in the surface mixed layer, frequently with a maximum directly beneath, declining to levels at or near detection limits in deep waters (16). Such profiles are broadly consistent with a source at or near the surface and net consumption at greater depths. The maximum may be explained as occurring in a zone of production beneath the surface mixed layer that is poorly ventilated, so with reduced emission to the atmosphere. However, certain halocarbons have quite different distributions with depth. The chlorofluorocarbons such as CFC11 and CFC12 may have higher concentrations below the surface and at intermediate depths, with levels diminishing to low values in some of the deepest waters of the ocean. These distributions are well understood to be the result of transport of these anthropogenic gases from the atmosphere either by direct exchange at the surface or via downward mixing and advection of cold, dense, ventilated waters to intermediate and abyssal depths. There is now evidence that some halocarbons that have relatively short atmospheric lifetimes (e.g., dichloromethane and tri- and tetrachloroethylene) are also introduced to the ocean by the same processes, and that these gases have much longer lifetimes in the ocean than in the atmosphere. This apparently explains the presence of dichloromethane in the near bottom waters of the Labrador Sea (17).

Chloroform is likely to undergo the same processes but may have an additional *in situ* source.

Loss processes occurring in the ocean can be inferred for most organohalogens by their lower concentrations at depth than at the surface. They have been studied most thoroughly for methyl bromide and, for this compound, are known to include bacterial removal and chemical loss through hydrolysis and reaction with chloride ion (18). Other loss processes affecting a range of halogenated compounds apparently exist in waters that are depleted in dissolved oxygen (19). It should be noted a loss process for one halocarbon might represent a source for another: thus, reaction of both CH_3Br and CH_3I with chloride ion yields CH_3Cl. The photocatalyzed loss of CH_2I_2 is a source of CH_2ICl.

Continued measurement of halocarbons both in the ocean and atmosphere will improve our knowledge of the magnitude, geographic distribution, and seasonality of their fluxes, but a greater challenge is likely to be identifying more accurately their sources. That knowledge is essential if we hope to be able to predict how the fluxes of these atmospherically reactive gases may change in the future.

BIBLIOGRAPHY

1. O'Dowd, C.D. et al. (2002). Marine aerosol formation from biogenic iodine emissions. *Nature* **417**: 632–636.
2. Wanninkhof, R. (1992). Relationship between wind speed and gas exchange over the ocean. *J. Geophys. Res.* **97**: 7373–7382.
3. Harper, D.B. (2000). The global chloromethane cycle: biosynthesis, biodegradation and metabolic role. *Nat. Prod. Rep.* **17**: 337–348.
4. Moore, R.M. (2000). The solubility of a suite of low molecular weight organochlorine compounds in seawater and implications for estimating the marine source of methyl chloride to the atmosphere. *Chemosphere: Global Change Sci.* **2**: 95–99.
5. Lobert, J.M. et al. (1995). *Science* **267**: 1002–1005.
6. Quack, B. and Wallace, D.W. (2003). Air–sea flux of bromoform: controls, rates, and implications. *Global Biogeochem. Cycles* **17**(1): 1023–1029.
7. Gribble, G.W. (2003). The diversity of naturally produced organohalogens. In: *The Handbook of Environmental Chemistry*. Vol. 3, Part P, G.W. Gribble (Ed.). Springer, New York, pp. 1–15.
8. Baker, J.M. et al. (2001). Emissions of CH_3Br, organochlorines and organoiodines from temperate macroalgae. *Chemosphere Global Change Sci.* **3**: 93–106.
9. Carpenter, L.J. and Liss, P.S. (2000). On temperate sources of bromoform and other reactive organic bromine gases. *J. Geophys. Res.* **105**: 20,539–20,547.
10. Manley, S.L. and de la Cuesta, J. (1997). Methyl iodide production from marine phytoplankton cultures. *Limnol. Oceanogr.* **42**: 142–147.
11. Moore, R.M., Webb, M., Tokarczyk, R., and Wever, R. (1996). Bromoperoxidase and iodoperoxidase enzymes and production of halogenated methanes in marine diatom cultures. *J. Geophys. Res.* **101**: 20,899–20,908.
12. Scarratt, M.G. and Moore, R.M. (1996). Production of methyl chloride and methyl bromide in laboratory cultures of marine phytoplankton. *Mar. Chem.* **54**: 263–272.
13. Amachi, S., Kamagata, Y., Kanagawa, T., and Muramatsu, Y. (2001). Bacteria mediate methylation of iodine in marine and terrestrial environments. *Appl. Environ. Microbiol.* **67**: 2718–1722.
14. Moore, R.M. and Zafiriou, O.C. (1994). Photochemical production of methyl iodide in seawater. *J. Geophys. Res.* **99**: 16,415–16,420.
15. Baker, J.M. et al. (1999). Biological production of methyl bromide in the coastal waters of the North Sea and open ocean of the northeast Atlantic. *Mar. Chem.* **64**: 267–285.
16. Moore, R.M. and Groszko, W. (1999). Methyl iodide distribution in the ocean and fluxes to the atmosphere. *J. Geophys. Res.* **104**: 11,163–11,171.
17. Moore, R.M. (2004). Dichloromethane in N. Atlantic Ocean waters. *J. Geophys. Res.* **109**: 9004.
18. King, D.B. and Saltzman, E.S. (1997). Removal of methyl bromide in coastal seawater: chemical and biological rates. *J. Geophys. Res.* **102**: 18,715–18,721.
19. Tanhua, T., Fogelqvist, E., and Bastürk, Ö. (1996). Reduction of volatile halocarbons in anoxic seawater, results from a study in the Black Sea. *Mar. Chem.* **54**: 159–170.

FOOD CHAIN/FOODWEB/FOOD CYCLE

Andrew Juhl
Lamont–Doherty Earth Observatory of Columbia University
Palisades, New York

A representation of the feeding relationships of the organisms, or groups of organisms, within an ecological community will be shown here. By showing which organisms feed on which other organisms, the pathways of energy flow through the ecosystem can be followed. "Food chain" is an older term and is currently less used than "foodweb" or "food cycle." The newer terms were coined to acknowledge the complexity of feeding relationships within most ecosystems. Examples of marine planktonic and intertidal foodwebs will be used below to illustrate specific points, but the concepts are applicable to any ecosystem.

The simplest food chain is organized in a strict hierarchy with primary producers (organisms that generate organic matter by fixing inorganic carbon, usually through photosynthesis) eaten by herbivores (primary consumers), the herbivores eaten by carnivores (secondary consumers), on to tertiary consumers, quaternary consumers, and soon. Each level within the food chain is termed a trophic level. Figure 1 shows an example of a very simple food chain for a planktonic ecosystem where phytoplankton are the primary producers, eaten by zooplankton herbivores. Zooplankton, in turn, are eaten by fish. Each group of organisms is represented by a shape, and the feeding connection and direction of energy flow are shown with an arrow. Simple food chains have heuristic value and are the basis of many quantitative ecosystem models. However, they can be criticized for oversimplification and for missing major groups of organisms.

One step toward a more realistic picture of an ecosystem is to disaggregate the trophic levels. Ideally, each level

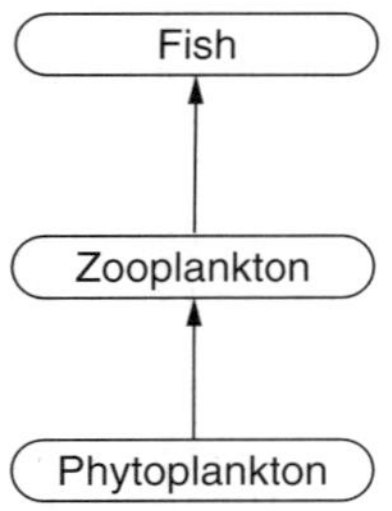

Figure 1. A simple food chain for a planktonic ecosystem.

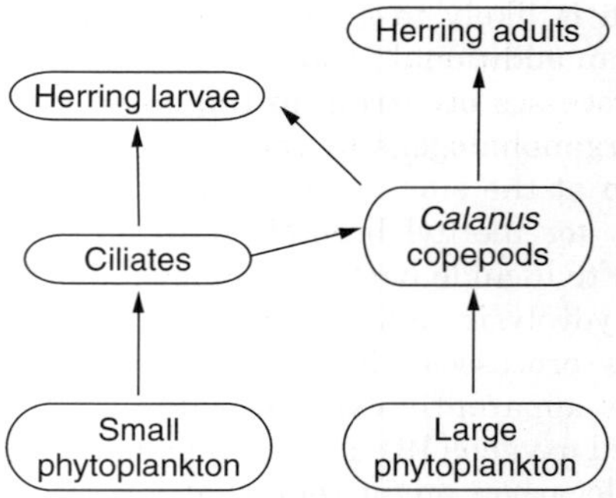

Figure 2. A more complex foodweb for the planktonic ecosystem in Fig. 1. Some groups are taxonomically identified, others are separated by size. The copepods are omnivorous, feeding on both phytoplankton and ciliates. Different feeding roles are shown for adult and larval herring.

could be broken down into its constituent species. One problem to consider is whether rare species should be depicted. A food chain diagram could include any feeding relationship that occurs within a system, whether common or unusual. This could lead to so many boxes and connecting arrows that any heuristic value would be jeopardized. In addition, the required information to fully disaggregate species groups may not be available. For example, microorganisms can be difficult to distinguish to the species level. Practical considerations often lead researchers to lump microorganisms into size classes and functional groups (e.g., primary producers >2 μm, 2–20 μm, <20 μm). In practice, the degree of aggregation of species is dictated by the objectives that motivate construction of the food chain diagram.

It is also more realistic to deemphasize the hierarchy implied by trophic levels. Deviations from strict trophic levels are common in aquatic ecosystems. For example, omnivorous species can prey on different trophic levels. A planktonic copepod may sometimes feed on small zooplankton, acting as a secondary consumer. The same copepod may also feed on phytoplankton, thereby acting as a primary consumer (1). Mixotrophy is another complication, common in planktonic ecosystems. Mixotrophic organisms can operate as primary producers, yet they may also consume other primary producers, or even consumers (2). Some aquatic organisms can occupy different trophic levels during different developmental stages. For example, eggs and early larval stages of the copepod *Acartia tonsa* are fed upon by the heterotrophic dinoflagellate *Protoperidinium* cf. *divergens* (3). However, adult *A. tonsa* feed voraciously on the dinoflagellate (4). This is an example of an "ontogenetic niche," or a niche that changes as the organism develops (5). Accommodating such complexities is one reason why the term "foodweb" is generally preferred over "food chain" (see Fig. 2). The many arrows necessary to accurately depict feeding relationships within most ecosystems resemble a web more than a linear chain.

Decomposers are frequently neglected when foodweb diagrams are constructed (6,7), and yet organic production from all trophic levels eventually cycles through decomposers. As the importance of heterotrophic bacteria and microzooplankton to planktonic biogeochemical cycles began to be realized during the 1970s and 1980s, depictions of plankton foodwebs began to include bacterial decomposers and their consumers in what has become known as the "microbial loop" (8,9) (see Fig. 3). Including decomposers in foodwebs of other ecosystems would certainly be more realistic, although it may not be necessary for every research goal.

QUANTITATIVE FOODWEBS

Even a complicated foodweb diagram is a qualitative description of an ecosystem and may only be representative of a single point in time. One approach to making the foodweb quantitative is to include an estimate of the energy flux along each pathway (10). Energy produced at the foodweb base will be dissipated at successive trophic levels. Knowing or assuming the energy dissipation at each step allows an estimate of the maximum production possible at higher trophic levels. For example, maximum potential fisheries production of the world's oceans has been derived from global estimates of planktonic primary production (11,12). In these analyses, different areas of the world's oceans were assumed to have simple food chains of varying length, resulting in variable efficiency of energy transfer until a trophic level with harvestable biomass was reached. Ocean regions with short food chains had greater fishery production per unit primary production than areas with long food chains.

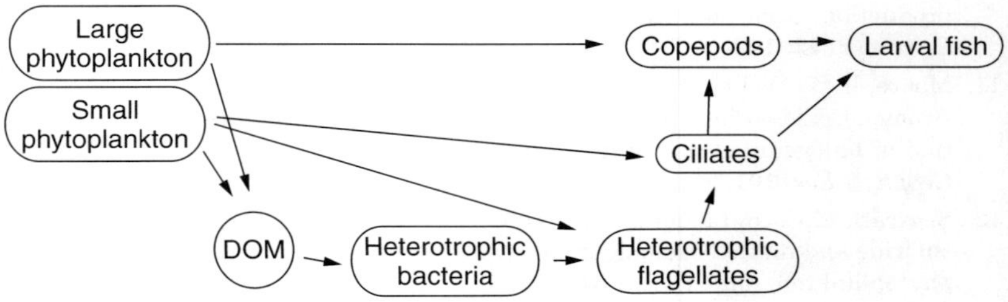

Figure 3. A planktonic foodweb that includes a representation of the "microbial loop" as an alternate pathway from phytoplankton to larger organisms. The loop, shown with thicker arrows, begins with phytoplankton producing dissolved organic matter (DOM). DOM feeds heterotrophic bacteria that are preyed upon by small, unicellular predators (heterotrophic flagellates and ciliates). These protists are food for copepods and larval fishes.

The biological components of many ecosystem models can also be viewed as attempts to make quantitative foodwebs. Models of plankton ecosystems (13,14) are usually not based on energy transfer. Instead, fluxes to and from organisms, and the standing biomass of the organisms, are expressed in units of an important element, often nitrogen, phosphorus, or carbon. Nevertheless, the flux pathways between organisms describe a food chain or foodweb.

Assigning interaction strengths to particular pathways is another approach to making foodwebs less qualitative. One can define strong or weak interactions within a foodweb that are not necessarily related to the flux of energy or material mediated by those interactions. A strong interaction is indicated if removal of one component species leads to pronounced change in the overall community (15). Alternately, removal of a species may cause little change, indicating a weak interaction. The definition of pronounced change is subjective; however, local extinctions and other large shifts in species diversity of a community would likely be acceptable demonstration that a pronounced change had occurred. For example, Paine (16) described two foodwebs for Pacific rocky intertidal communities, one for a coastal region of Washington State, the other for an Alaskan bay. The same 44 species or groups were present in both areas but the interaction strengths were quite different. In Washington, experimental removal of the starfish *Pisaster ochraceus* led to competitive elimination of multiple species of sessile invertebrates and benthic algae. The excluded species were replaced by the blue mussel, *Mytilus californicus*. When the starfish was present, it prevented the mussel from excluding the other species. Thus, the strong interaction between *Pisaster* and *Mytilus* was critical to maintaining a diverse intertidal community, although *Pisaster* preyed on many species and the bulk of its energy came through paths without *Mytilus*. Additional strong interactions within the foodweb were found, but removal of some species had little or no effect on the remaining community. By contrast, in Alaska, *Pisaster* only weakly interacted with other species.

A predator like *Pisaster* (in Washington), whose presence maintains community diversity by preventing competitive exclusion, is known as a "keystone" predator (17). Many other aquatic predators have since been described as keystone predators, including sea otters (18), mosquito larvae (19), and largemouth bass (20).

GENERAL PROPERTIES OF FOODWEBS

In the search for general rules of ecology, it is not surprising that compilations of documented foodwebs have been studied to determine whether foodwebs of different ecosystems share common properties. Certainly, the basic trophic structure with primary producers, one or more levels of consumers, and decomposers is common to vastly different ecosystems. Examples of other properties that might follow general rules include the ratio of predatory to prey species, the number of links per species, the fraction of total species in each trophic level, and the fraction of omnivorous species (21–23). Empirical foodwebs have also been investigated with respect to more abstract metrics, such as connectance—the ratio of actual links within a foodweb to the number of topologically possible links (21,24,25). However, the search for general properties within compilations of foodwebs has been criticized because, as mentioned above, the individual foodwebs used as raw data in such studies are frequently incomplete and overaggregated (6,7,26). The problems associated with documented foodwebs cannot be denied; however, it has been argued that some rules are sufficiently robust to overcome the biases (27).

The search for general rules relating to foodwebs is also approached using theoretical foodweb models. A theoretical foodweb uses a system of equations to represent the interactions of different organisms within an idealized ecosystem. Properties such as the number of links per species and connectance can be modified in different versions of the model. The stability of each resultant system of equations can be examined as various properties of the foodweb are modified (24,27,28). Theoretical foodwebs rapidly provide results for experiments that are difficult or impossible to conduct with real organisms. However, if the system of equations used does not adequately represent a natural condition, extrapolating results to the real world may be problematic.

Whether qualitative or quantitative, foodwebs are models of how the organisms within a community interact. They focus on feeding relationships, neglecting other modes of interaction, such as competition. Like any model, they are open to a variety of criticisms. It is important to recognize that it is not desirable for a model to contain all the complexities of nature; there would otherwise be no advantage to creating the model. Rather, a model should capture those aspects of nature that are necessary for understanding the question at hand. So long as that context is maintained, foodwebs will continue to be one of the most useful tools for teaching, studying, and understanding ecological communities.

BIBLIOGRAPHY

1. Landry, M.R. (1981). Switching between herbivory and carnivory by the planktonic marine copepod *Calanus pacificus*. *Mar. Biol.* **65**: 77–82.
2. Porter, X. (1988). Phagotrophic phytoflagellates in microbial food webs. *Hydrobiologia* **159**: 89–97.
3. Jeong, H.J. (1994). Predation by the heterotrophic dinoflagellate *Protoperidinium cf. divergens* on copepod eggs and early naupliar stages. *Mar. Ecol. Prog. Ser.* **114**: 203–208.
4. Jeong, H.J. (1994). Predation effects of the calanoid copepod *Acartia tonsa* on a population of the heterotrophic dinoflagellate *Protoperidinium cf. divergens* in the presence of co-occurring red-tide dinoflagellate prey. *Mar. Ecol. Prog. Ser.* **111**: 87–97.
5. Werner, E.E. (1984). The ontogenetic niche and species interactions in size-structured populations. *Annu. Rev. Ecol. System.* 393–425.
6. Tobin, J.E. (1991). Scientific correspondence. *Nature* **352**: 482.
7. Wardle, D.A. (1991). Scientific correspondence. *Nature* **352**: 482.
8. Pomeroy, L.R. (1974). The ocean's food web, a changing paradigm. *Bioscience* **24**: 499–503.

9. Azam, F., Fenchel, T., Meyer-Reil, L., and Thingstad, F. (1983). The ecological role of water-column microbes in the sea. *Mar. Ecol. Prog. Ser.* **10**: 257–263.
10. Paine, R.T. (1966). Food web complexity and species diversity. *Am. Nat.* **100**: 65–75.
11. Ryther, J.H. (1969). Photosynthesis and fish production in the sea. *Science* **166**: 72–76.
12. Pauly, D. and Christensen, V. (1995). Primary production required to sustain global fisheries. *Nature* **374**: 255–257.
13. Biktashev, V.N., Brindley, J., and Horwood, J.W. (2003). Phytoplankton blooms and fish recruitment rate. *J. Plankton Res.* **25**: 21–33.
14. Hakanson, L., Boulion, V.V., and Ostapenia, A.P. (2003). The influence of biomanipulations (fish removal) on the structure of lake food webs, case studies using the LakeWeb-model. *Aquat. Ecol.* **37**: 87–99.
15. MacArthur, R.H. (1972). Strong, or weak, interactions? *Trans. Conn. Acad. Arts Sci.* **44**: 177–188.
16. Paine, R.T. (1980). Food webs: linkage, interaction strength and community infrastructure. *J. Anim. Ecol.* **49**: 667–685.
17. Paine, R.T. (1969). A note on trophic complexity and community stability. *Am. Nat.* **103**: 91–93.
18. Estes, J.A. and Palmisano, J.F. (1975). Sea otters: their role in structuring nearshore communities. *Science* **185**: 1058–1060.
19. Cochran-Stafira, D.L. and von Ende, C.N. (1998). Integrating bacteria into food webs: studies with *Sarracenia purpurea inquilines*. *Ecology* **79**: 880–898.
20. Schindler, D.E., Hodgson, J.R., and Kitchell, J.F. (1997). Density-dependent changes in individual foraging specialization of largemouth bass. *Oecologia* **110**: 592–600.
21. Pimm, S.L., Lawton, J.H., and Cohen, J.E. (1991). Food web patterns and their consequences. *Nature* **350**: 669–674.
22. Havens, K. (1992). Scale and structure in natural food webs. *Science* **257**: 1107–1109.
23. Link, J. (2002). Does food web theory work for marine ecosystems? *Mar. Ecol. Prog. Ser.* **230**: 1–9.
24. May, R.M. (1972). Will a large complex system be stable? *Nature* **238**: 413–414.
25. Rejmanek, M. and Stary, P. (1979). Connectance in real biotic communities and critical values for stability of model ecosystems. *Nature* **280**: 311–313.
26. Paine, R.T. (1988). Food webs: road maps of interactions or grist for theoretical development? *Ecology* **69**: 1648–1654.
27. Pimm, S.L. and Lawton, J.H. (1977). Number of trophic levels in ecological communities. *Nature* **268**: 329–331.
28. Sugihara, G., Schoenly, K., and Trombla, A. (1989). Scale invariance in food web properties. *Science* **245**: 48–52.
29. Pimm, S.L. and Lawton, J.H. (1978). On feeding on more than one trophic level. *Nature* **275**: 542–544.

PLANKTON

Andrew Juhl
Lamont–Doherty Earth Observatory of Columbia University
Palisades, New York

INTRODUCTION

Plankton are aquatic organisms living in the water column that drift, float, or swim only weakly. The term derives from the Greek word *planktos*, meaning "to drift or "to wander." It is a descriptive classification, implying a certain type of lifestyle, rather than characteristics associated with taxonomy, size, or ecological role. The following information focuses on marine plankton, but much of the information is directly applicable to freshwater plankton.

Plankton can be contrasted with nekton, aquatic organisms capable of vigorous swimming (such as large fishes, squids, and marine mammals), and with benthos, those aquatic organisms that live on or at the bottom. It is generally thought that organisms classified as plankton do not have much ability to control their horizontal position relative to large-scale currents. However, there are no hard rules defining how strongly an organism may swim and still be considered planktonic. The boundaries between plankton, nekton, and benthos are nebulous. In fact, some organisms can be placed in each of the groups during different developmental stages. For example, the blue crab, *Callinectes sapidus*, has a planktonic larval stage (1). After 30+ days as plankton, these larvae settle to the bottom, becoming benthic juveniles (2). Although they spend most of their adult life on the bottom, blue crab adults swim surprisingly well (*Callinectes* translates to "beautiful swimmer") and females will traverse long distances to spawn (3). Despite a loose definition, plankton is a meaningful term that describes one of the most common aquatic lifestyles.

CATEGORIES OF PLANKTON

The organisms that comprise plankton are exceedingly diverse. Many Archaea and Eubacteria (the two major taxonomic groups of bacteria), most algal divisions, and most animal phyla have some representatives that are planktonic. Plankton also range in size from less than 1 μm (bacterioplankton) to several meters across (some jellyfish and colonies of planktonic tunicates). The range of taxa and sizes results in very different ecological roles for different plankton. Several categorization schemes are commonly used to organize the diversity of plankton into groups that share particular properties.

Phytoplankton, Zooplankton, and Bacterioplankton

Phytoplankton are photosynthetic plankton, mostly single-celled, microscopic algae that use light energy to fix carbon dioxide dissolved in water. This process makes solar energy available to them for their own biochemical needs. When they are eaten by other organisms, and when they release dissolved organic matter into the water, the energy becomes available to nonphotosynthetic organisms. Phytoplankton are therefore considered the base of most aquatic foodwebs (primary producers). Phytoplankton commonly include photosynthetic prokaryotes (cyanobacteria) and photosynthetic eukaryotes, such as diatoms, dinoflagellates, cryptophytes, and chlorophytes.

Zooplankton are eukaryotic, planktonic heterotrophs. They are usually phagotrophic, meaning that they ingest particles. However, there are examples of osmotrophic zooplankton (feeding on dissolved organic matter) (4).

Also, many eggs and larvae that are temporarily planktonic do not feed, but only use stored energy reserves. Both examples would still be considered zooplankton. Common zooplankton include single-celled, heterotrophic protists, such as ciliates and choanoflagellates, and multicellular animals, such as copepods and euphausiids.

Gelatinous zooplankton are a subcategory of zooplankton. Although taxonomically diverse, they share a body composition that is jellylike, mostly lacking hard, structural components. Their bodies have very high water content and are often transparent. Cnidarian medusae known as "jellyfish" are the most commonly known examples. Additional examples of gelatinous zooplankton are found in the ctenophores, mollusks (heteropods), and urochordates (planktonic tunicates such as salps, doliolids, and larvaceans). Although there are small gelatinous forms, the largest zooplankton are all gelatinous. Because of their fragile body plan, gelatinous zooplankton are difficult to collect quantitatively. Their importance to marine ecosystems may therefore be underestimated. Gelatinous zooplankton are rare in fresh water.

Bacterioplankton, or planktonic bacteria, includes representatives of both Eubacteria and Archaea. Photosynthetic bacteria are sometimes included within this category, creating an overlap with phytoplankton. Generally, however, the category implies heterotrophic bacteria. Heterotrophic bacterioplankton are osmotrophs. They either take up dissolved organic matter freely available in the water, or they use external enzymes to solubilize organic matter from particles, making it available for transport into their cells.

Holoplankton Versus Meroplankton

Organisms that are planktonic their entire lives are holoplankton. The term derives from the Greek word *holos*, meaning "entirely." Meroplankton are organisms that are planktonic during only a portion of their lives. The Greek word *meros* means "mixed." Most meroplankton are larval forms that are benthos or nekton as adults. For example, many fish larvae are planktonic. As they grow and their swimming ability improves, they eventually become nekton. Similarly, many mollusks have planktonic larvae that eventually settle to the bottom and become benthos. Because ecological conditions for the plankton are often very different from those facing the adults, meroplanktonic larvae often do not resemble the adults. In numerous cases, meroplanktonic larval stages were first scientifically described as distinct species; only later was the connection to the adult stage realized. Organisms that are entirely planktonic except for a benthic resting stage, such as some cladocerans or dinoflagellates, are still generally considered holoplanktonic.

Plankton are also commonly separated based on size. Viruses are abundant in aquatic environments and may be the smallest plankton, for those who consider viruses to be living. Bacterioplankton are the smallest cellular plankton and may be less than 1 μm in length. The largest plankton are cniderian medusae (jellyfish) that may be more than a meter across the bell, with tentacles that extend over many meters. Planktonic tunicate colonies (especially Pyrosomes) can also be many meters across, although the individuals within the colony are only millimeters to centimeters in length.

Size categories are arbitrary, but Sieburth (5) described the following categories currently used by many researchers: picoplankton (organisms 0.2–2 μm in length), nanoplankton (2–20 μm), microplankton (20–200 μm), mesoplankton (0.2–20 mm), macroplankton (2–20 cm), and megaplankton (20–200 cm). Viruses would be considered femtoplankton. These size categories are often combined with other categories. For example, picophytoplankton and mesozooplankton are common terms in oceanographic literature.

COLLECTING AND STUDYING PLANKTON

In general, the number of individuals per unit volume of water increases as planktonic organisms decrease in size. Nevertheless, natural waters are typically a very dilute suspension of even the smallest organisms and concentrating the organisms is often necessary for analyses. High numbers of small organisms (pico-, nano-, microplankton) can be collected simply by taking a water sample with a bucket or a hydrographic bottle. Concentrating such small organisms by filtration is often done after water collection. The filter pore size used determines the smallest organisms collected on the filter surface. Use of different filter pore sizes allows separation of the plankton sample into several size classes. Larger organisms (microplankton and up) may be too rare for quantitative collection in a whole-water sample. They are often collected by dragging plankton nets through the water. Plankton nets typically have a defined mesh size so that only organisms larger than the mesh will be collected. Plankton nets therefore size fractionate and concentrate before collection. A larger mesh net will filter a greater volume of water before it becomes clogged and ineffective. Larger mesh nets can also be dragged through the water more rapidly. It is therefore important to match the net mesh with the size, concentration, and swimming ability of the organisms one desires to collect. Because of their small size, taxonomic identification of most plankton requires microscopic examination after collection.

Buckets, bottles, nets, and microscopes are still used to study plankton, but these traditional approaches are now complemented with a myriad of advanced techniques. For example, on the smallest scales, methods of molecular biology can be used to study gene expression in individual cells of picoplankton (6). On the largest scales, the global distribution of phytoplankton can be studied in satellite images (7).

SPATIAL PATTERNS OF PLANKTON

The dominant large-scale horizontal pattern in plankton abundance is a decline as one goes away from the coast. One may have to travel tens to hundreds of miles offshore for this pattern to become clear. Nevertheless, the central areas of the oceans that are furthest from land generally have the lowest abundance of plankton, while coastal areas have the highest. This is true for all types of plankton (8–10). However, it should be kept in mind that many species of plankton are adapted to live in open-ocean environments. Therefore, certain planktonic species may actually be more abundant in the open ocean than nearer the coast. Exceptions to the general offshore pattern in plankton abundance can be caused by mesoscale eddies (11,12), oceanographic fronts (13,14), upwelling zones (15), and other phenomena (16). Although areas far from shore tend to have low plankton abundance, the diversity of species can be very high by comparison to coastal locations (17).

By definition, plankton have a limited ability to control their horizontal movement. Horizontal patterns of plankton abundance are therefore regulated by environmental factors external to the organism: currents, light, temperature, nutrient availability, and so on. These factors also contribute to vertical patterns in plankton abundance. However, individual behavior is also important in forming vertical patterns because many planktonic organisms do have the ability to control their vertical position within the water column.

Because light from the sun is rapidly absorbed by water, and because phytoplankton need light to photosynthesize, actively growing phytoplankton are restricted to the upper portion of the water column. In coastal waters, sufficient light for photosynthesis may only penetrate the top few meters (or less). In these conditions, vertical phytoplankton abundance often directly correlates with light; the highest phytoplankton abundance is at the surface, with a sharp decline as depth increases (9). This pattern is mostly caused by higher phytoplankton growth where more light is available. In addition, some phytoplankton can swim enough to change their vertical position in the water column (by as much as 10 m/d) (18). Similarly, other phytoplankton species are positively buoyant due to gas or lipid inclusions. Therefore, high phytoplankton abundance near the surface can sometimes be due to upward movement of phytoplankton (19,20). By contrast, in the open ocean, while sufficient light for photosynthesis is generally found within the upper 100–200 m, phytoplankton abundance is usually low near the surface. Highest phytoplankton abundance in open-ocean waters often occurs in a subsurface layer (9,21). Although light levels are always highest near the surface, phytoplankton also require inorganic nutrients to grow, just like other plants. In offshore regions, low phytoplankton abundance near the surface is generally attributable to low nutrient availability. Greater phytoplankton stocks can be supported at deeper depths where both nutrients and light are available.

Different phytoplankton species are often found at different depths (22,23) and may be specialized to live under the conditions associated with certain depth strata (24,25). However, vertical mixing from tides, winds, surface cooling, internal waves, and other physical processes can homogenize the vertical patterns of phytoplankton abundance and species composition.

Because zooplankton do not require light, they can be found at all ocean depths. However, greater zooplankton abundance is generally found in the surface waters where phytoplankton occur. This should not imply that zooplankton and phytoplankton abundances are necessarily directly related. While many zooplankton feed on phytoplankton, many feed on bacteria or other zooplankton. Correlations between zooplankton and phytoplankton abundance patterns can be direct and causal for some groups of zooplankton, that is, for zooplankton that feed on phytoplankton. However, other zooplankton are only indirectly, or coincidentally, associated with phytoplankton.

Many common mesozooplankton, especially crustacean mesozooplankton such as copepods and euphausiids, vertically migrate between the deeper and upper water layers on a daily basis. Such diel vertical migrations of zooplankton may cover several hundred meters each way (26). The most compelling explanation for this behavior is that the zooplankton migrate down during the day to avoid predation, then rise to the upper layers at night to feed (27,28). At night, migrating zooplankton are less likely to be detected by predators in the surface layers that depend on sight (such as fishes) to catch prey. Darkness, whether due to depth or time of day, provides one of the few opportunities to hide in the relatively featureless, planktonic habitat. Although daily vertical migrations are common in both fresh and marine waters throughout the world, it should be pointed out that most zooplankton species probably do not vertically migrate. Many species live out their lives associated with a particular depth strata (29). For example, Marlow and Miller (30) found that only around 10% of zooplankton species in the oceanic Gulf of Alaska vertically migrated. However, those vertically migrating species made up a large fraction of the total zooplankton biomass. Even species that do vertically migrate can change their migration pattern in response to local conditions (31) or over the course of development (32).

TEMPORAL PATTERNS OF PLANKTON

The abundance of plankton varies seasonally in many places, especially in higher latitude waters where seasonality is most pronounced. A pattern observed in many temperate lakes, estuaries, and the open North Atlantic Ocean is the spring phytoplankton bloom (33–35). The cycle begins with low phytoplankton abundance in winter. In spring or early summer, phytoplankton abundance increases dramatically (a bloom). The zooplankton biomass often also increases, although there may be a time lag between the increase in phytoplankton and the later increase in zooplankton. Eventually, the stocks of both phyto- and zooplankton decline to moderate levels that are sustained through the summer. A second bloom can occur in autumn, before plankton levels decline again for the winter.

Much research effort has focused on describing and understanding seasonal patterns of plankton exemplified by the spring-bloom cycle. It provides the best known temporal pattern in plankton stocks. However, in many lower latitude waters and the open North Pacific Ocean, seasonal cycles are either absent (36), subtle (37,38), or complicated by the impact of episodic events such as upwelling, mesoscale eddies, or storms (11,15,39).

Temporal patterns of plankton in polar seas are often dictated by the timing of sea ice melt (40,41).

GLOBAL IMPORTANCE OF PLANKTON

Although most planktonic organisms are individually small and seemingly ineffectual, plankton are important in aggregate because of their ubiquitous occurrence on this mostly water-covered planet. Outstanding examples of the global importance of plankton include their roles in the production of aquatic resources that humans value, in the Earth's climate, in oxygen production, and in sediment formation. Aquatic ecology and global biogeochemical processes cannot be understood without considering the role of plankton.

As the base of the aquatic foodweb, plankton are a critical food resource for aquatic organisms that are valuable to humans. The connection may be direct, as when valuable species such as oysters and anchovies feed on phytoplankton. The connection may also be indirect, for instance, if a commercially valuable predatory fish, such as a tuna, eats the anchovies. Either way, the occurrence of valued species is often coincident with the distribution of plankton. Large, motile species such as large fishes, seabirds, and whales concentrate their activities in locations where plankton are most abundant (42–44). This dependence of larger organisms on plankton means that commercial fisheries yield can be extrapolated from production of the plankton (45,46). Areas with higher plankton production have much greater fisheries yield.

Plankton influence the Earth's climate by removing carbon dioxide from the atmosphere. Carbon dioxide is the main atmospheric gas responsible for the "greenhouse" effect that regulates the Earth's temperature. Photosynthesis by phytoplankton fixes carbon dioxide from the atmosphere, transforming the carbon into organic compounds. Most of the carbon fixed by phytoplankton during photosynthesis is grazed by zooplankton and respired, releasing carbon dioxide back to the atmosphere (47,48). Nevertheless, some of the fixed carbon sinks to the deep ocean as the cells and bodies of plankton, zooplankton fecal matter, and other organic aggregates. Although the standing biomass of phytoplankton is less than 1% of the biomass of terrestrial plants (49), the amount of carbon fixed by phytoplankton and land plants is roughly equal (49,50). Because carbon fixation is coupled to the release of oxygen, phytoplankton are also responsible for approximately half of the Earth's oxygen production. It is estimated that one-third of the carbon fixed by phytoplankton eventually sinks to the deep ocean, effectively removing the carbon from the atmosphere for thousands of years (51). Determining the amount of carbon removed by plankton from the atmosphere and transported to the deep ocean, and predicting how that process may change in the future, are major questions in climate research.

Phytoplankton also influence climate by releasing sulfur compounds to the atmosphere. They are thought to be the main natural source of volatile sulfur compounds to the troposphere (52), mostly in the form of dimethylsulfide. The pathways leading to dimethylsulfide release from the oceans are incompletely understood, but phytoplankton primarily release the precursors to dimethylsulfide when their cells are ruptured during autolysis, viral attack, or grazing by zooplankton. Enzymes of phyto- and bacterioplankton then transform the released precursors to dimethylsulfide, which is released to the atmosphere. The sulfur aerosols stemming from dimethylsulfide of phytoplankton scatter incoming solar radiation. These aerosols also act as cloud condensation nuclei, further increasing atmospheric backscatter of solar radiation. Therefore, phytoplankton (and their interactions with other planktonic organisms) alter cloud formation and the amount of solar radiation reaching the Earth's surface (53).

A fraction of the carbon fixed by phytoplankton becomes a component of deep-sea sediments, removing it from the atmosphere for millions of years (54). Other inorganic structural components of plankton, such as the silicate tests of diatoms and radiolarians, and the calcareous shells of coccolithophores, pteropods, and foraminifera, also contribute to ocean bottom sediments. Over 60% of the ocean bottom is covered with sediments that are primarily planktonic in origin (55). Over geological time, such sea-bottom sediments, formed in previous ages, have been uplifted. Thus, even the peaks of mountains can be composed of material that originated from plankton.

BIBLIOGRAPHY

1. Epifanio, C.E. (1995). Transport of blue crab (*Callinectes sapidus*) larvae in waters off mid-Atlantic states. *Bull. Mar. Sci.* **57**: 713–725.
2. von Montfrans, J. et al. (1995). Settlement of blue crab postlarvae in Western North Atlantic estuaries. *Bull. Mar. Sci.* **57**: 834–854.
3. Tankersley, R.A., Wieber, M.G., Sigala, M.A., and Kachurak, K.A. (1998). Migratory behavior of ovigerous blue crabs *Callinectes sapidus*: evidence for selective tidal-stream transport. *Biol. Bull.* **195**: 168–173.
4. Manahan, D.T. (1990). Adaptations by invertebrate larvae for nutrient acquisition from seawater. *Am. Zool.* **30**: 147–160.
5. Sieburth, J.M. (1979). *Sea Microbes*. Oxford University Press, New York.
6. Bird, C. and Wyman, M. (2003). Nitrate/nitrite assimilation system of the marine picoplanktonic cyanobacterium *Synechococcus* sp. strain WH 8103: effect of nitrogen source and availability on gene expression. *Appl. Environ. Microbiol.* **69**: 7009–7018.
7. Yoder, J.A. and Kenelly, M.A. (2003). Seasonal and ENSO variability in global ocean phytoplankton chlorophyll derived from 4 years of SeaWiFS measurements. *Global Biogeochem. Cycles* **17**: 1112.
8. Reid, J.L., Brinton, E., Fleminger, A., Venrick, E.L., and McGowan, J.A. (1978). Ocean circulation and marine life. In: *Advances in Oceanography*. H. Charnock and G. Deacon (Eds.). Plenum Publishing, New York, pp. 65–130.
9. Mullin, M.M. (1986). Spatial and temporal scales and patterns. In: *Plankton Dynamics of the Southern California Bight*. R.W. Eppley (Ed.). Springer, New York, pp. 216–273.
10. Cavender-Bares, K.K., Karl, D.M., and Chisholm, S.W. (2001). Nutrient gradients in the western North Atlantic Ocean: relationship to microbial community structure and

comparison to patterns in the Pacific Ocean. *Deep-Sea Res. I* **48**: 2373–2395.

11. The Ring Group. (1981). Gulf Stream cold-core rings: their physics, chemistry, and biology. *Science* **212**: 1091–1100.
12. Falkowski, P.G., Ziemann, D., Kolber, Z., and Bienfang, P.K. (1991). Role of eddy pumping in enhancing primary production in the ocean. *Nature* **352**: 55–58.
13. Marra, J., Houghton, R.W., and Garside, C. (1990). Phytoplankton growth at the shelf-break front in the Middle-Atlantic Bight. *J. Mar. Res.* **48**: 851–868.
14. Franks, P.J.S. (1992). Sink or swim: accumulation of biomass at fronts. *Mar. Ecol. Prog. Ser.* **82**: 1–12.
15. Small, L.F. and Menzies, D.W. (1981). Patterns of primary productivity and biomass in a coastal upwelling region. *Deep-Sea Res.* **28A**: 123–149.
16. Ryan, J.P., Polito, P.S., Strutton, P.G., and Chavez, F.P. (2002). Unusual large-scale phytoplankton blooms in the equatorial Pacific. *Prog. Oceanogr.* **55**: 263–285.
17. McGowan, J.A. and Walker, P.W. (1993). Pelagic diversity patterns. In: *Species Diversity Patterns in Ecological Communities*. R.E. Ricklefs and D. Schluter (Eds.). University of Chicago Press, Chicago, IL, pp. 203–214.
18. Kamykowski, D. (1995). Trajectories of autotrophic marine dinoflagellates. *J. Phycol.* **31**: 200–208.
19. Eppley, R.W., Holm-Hansen, O., and Strickland, J.D.H. (1968). Some observations on the vertical migration of dinoflagellates. *J. Phycol.* **4**: 333–340.
20. Ault, T.R. (2000). Vertical migration by the marine dinoflagellate *Prorocentrum triestinium* maximises photosynthetic yield. *Oecologia* **125**: 466–475.
21. Cullen, J.J. and Eppley, R.W. (1981). Chlorophyll maximum layers of the Southern California Bight and possible mechanisms of their formation and maintenance. *Oceanol. Acta* **4**: 23–32.
22. Venrick, E.L. (1982). Phytoplankton in an oligotrophic ocean: observations and questions. *Ecol. Monogr.* **52**: 129–154.
23. Chisholm, S.W., Olsen, R.J., Zettler, E.R., Goericke, R., Waterbury, J.B., and Welschmeyer, N.A. (1988). A novel free-living prochlorophyte abundant in the oceanic euphotic zone. *Nature* **334**: 340–343.
24. Takahashi, M., Ichimura, S., Kishino, M., and Okami, N. (1989). Shade and chromatic adaptation of phytoplankton photosynthesis in a thermally stratified sea. *Mar. Biol.* **100**: 401–409.
25. Ferris, M.J. and Palenik, B. (1998). Niche adaptation in ocean cyanobacteria. *Nature* **396**: 226–228.
26. Enright, J.T. (1977). Copepods in a hurry: sustained high-speed upward migration. *Limnol. Oceanogr.* **22**: 118–125.
27. Bollens, S. and Frost, B.W. (1989). Predator-induced diel vertical migration in a planktonic copepod. *J. Plankton Res.* **11**: 1047–1065.
28. DeRobertis, A., Jaffe, J.S., and Ohman, M.D. (2000). Size-dependent visual predation risk and the timing of vertical migration in zooplankton. *Limnol. Oceanogr.* **45**: 1838–1844.
29. McGowan, J.A. and Walker, P.W. (1979). Structure in the copepod community of the North Pacific Central gyre. *Ecol. Monogr.* **49**: 195–226.
30. Marlow, C.J. and Miller, C.B. (1975). Patterns of vertical distribution and migration of zooplankton at Ocean Station 'P'. *Limnol. Oceanogr.* **20**: 824–844.
31. Ohman, M.D. (1990). The demographic benefits of diel vertical migration by zooplankton. *Ecol. Monogr.* **60**: 257–281.
32. Osgood, K.E. and Frost, B.W. (1994). Ontogenetic diel vertical migration behaviors of the marine planktonic copepods *Calanus pacificus* and *Metridia lucens*. *Mar. Ecol. Prog. Ser.* **104**: 13–25.
33. Sverdrup, H.U. (1953). On the conditions for the vernal blooming of phytoplankton. *J. Cons. Int. Explor. Mer.* **18**: 287–295.
34. Heinrich, A.K. (1962). The life histories of plankton animals and seasonal cycles of plankton communities in the oceans. *J. Cons. Int. Explor. Mer.* **27**: 15–24.
35. Lund, J.W.G. (1965). The ecology of the freshwater phytoplankton. *Biol. Rev.* **40**: 231–293.
36. Hayward, T.L., Venrick, E.L., and McGowan, J.A. (1983). Environmental heterogeneity and plankton community structure in the central North Pacific. *J. Mar. Res.* **41**: 711–729.
37. Frost, B.W. (1993). A modelling study of processes regulating plankton standing stock and production in the open subarctic Pacific Ocean. *Prog. Oceanogr.* **32**: 17–56.
38. Venrick, E.L. (1993). Phytoplankton seasonality in the central North Pacific: the endless summer reconsidered. *Limnol. Oceanogr.* **38**: 1135–1149.
39. Chang, J., Chung, C., and Gong, G. (1996). Influences of cyclones on chlorophyll *a* concentration and *Synechococcus* abundance in a subtropical western Pacific coastal ecosystem. *Mar. Ecol. Prog. Ser.* **140**: 199–205.
40. Park, G.M., Yang, S.R., Kang, S-H., Chung, K.H., and Shim, J.H. (1999). Phytoplankton biomass and primary production in the marginal ice zone of the northwestern Weddell Sea during austral summer. *Polar Biol.* **21**: 251–261.
41. Rat'kova, T.N. and Wassman, P. (2002). Seasonal variation and spatial distributions of phyto- and protozooplankton in the central Barents Sea. *J. Mar. Sys.* **38**: 47–75.
42. Hyrenbach, K.D., Fernandez, P., and Anderson, D.J. (2002). Oceanographic habitats of two sympatric North Pacific albatrosses during the breeding season. *Mar. Ecol. Prog. Ser.* **233**: 283–301.
43. Baumgartner, M.F. and Mate, B.R. (2003). Summertime foraging ecology of North Atlantic right whales. *Mar. Ecol. Prog. Ser.* **264**: 123–135.
44. Beaugrand, G. and Reid, P.C. (2003). Long-term changes in phytoplankton, zooplankton and salmon related to climate. *Global Change Biol.* **9**: 801–817.
45. Ryther, J.H. (1969). Photosynthesis and fish production in the sea. *Science* **166**: 72–76.
46. Pauly, D. and Christensen, V. (1995). Primary production required to sustain global fisheries. *Nature* **374**: 255–257.
47. Calbet, A. (2001). Mesozooplankton grazing effect on primary production: a global comparative analysis in marine ecosystems. *Limnol. Oceanogr.* **46**: 1824–1830.
48. Calbet, A. and Landry, M.R. (2004). Phytoplankton growth, microzooplankton grazing, and carbon cycling in marine systems. *Limnol. Oceanogr.* **49**: 51–57.
49. Field, C.B., Behrenfeld, M.J., Randerson, J.T., and Falkowski, P.G. (1998). Primary production of the biosphere: integrating terrestrial and oceanic components. *Science* **281**: 237–240.
50. Longhurst, A., Sathyendranath, S., Platt, T., and Caverhill, C. (1995). An estimate of global primary production in the ocean from satellite radiometer data. *J. Plankton Res.* **17**: 1245–1271.
51. Falkowski, P.G., Barber, R.T., and Smetacek, V. (1998). Biogeochemical controls and feedbacks on ocean primary production. *Science* **281**: 200–206.

52. Simo, R. (2001). Production of atmospheric sulfur by oceanic plankton: biogeochemical, ecological and evolutionary links. *Trends Ecol. Evol.* **16**: 287–294.

53. Steinke, M., Malin, G., and Liss, P.S. (2002). Trophic interactions in the sea: an ecological role for climate relevant volatiles? *J. Phycol.* **38**: 630–638.

54. Raven, J.A. and Falkowski, P.G. (1999). Oceanic sinks for atmospheric CO_2. *Plant Cell Environ.* **22**: 741–755.

55. Kennett, J.P. (1982). *Marine Geology*. Prentice-Hall, Englewood Cliffs, NJ.

MAJOR IONS IN SEAWATER

MARIE DE ANGELIS
Humboldt State University
Arcata, California

Seawater contains all of the natural elements. However, its composition is dominated by relatively few ionic species. These species, which comprise >99.8% of the mass of the total dissolved salts in the ocean, are known as the major ions and are defined as dissolved species having a concentration in seawater of >1 mg kg^{-1} (1 ppm) (Table 1). Na^+ and Cl^- alone account for 86% of all dissolved chemical species in seawater.

The source of Cl^-, and most anions in seawater, is *outgassing* of the Earth's interior, or *volcanic emissions*. The majority of the major cations are derived from the effects of continental weathering and are delivered to the ocean via rivers. The gross composition of seawater is the result of the partitioning of elements between continental rock and seawater over geologic time. The major ions represent the more soluble elements, which have preferentially partitioned into seawater.

The major ions are found to occur in nearly constant ratios to each other throughout most of the world's oceans. That is, although the salinity or the total amount of salt dissolved in seawater varies from location to location in the ocean, the ratios of the major ions to each other remains nearly constant, which is true from ocean to ocean as well as from surface to deep waters (Table 2).

Table 1. Major Ions in Seawater

Major Ion	g kg^{-1} Seawater at Salinity of 35.000 ppt	% by wt	g ion/g Cl^-
Chloride, Cl^-	19.353	55.30	1
Sodium, Na^+	10.781	30.77	0.5561
Sulfate, $SO_4{}^{2-}$	2.712	7.75	0.1400
Magnesium, Mg^{2+}	1.284	3.69	0.0668
Calcium, Ca^{2+}	0.4119	1.18	0.0213
Potassium, K^+	0.399	1.14	0.0206
Bicarbonate, $HCO_3{}^{2-}$	0.126	0.41	0.0075
Bromide, Br^-	0.0673	0.19	0.0035
Boron, B (as $B(OH)_3$	0.0257	0.013	0.00024
Strontium, Sr^{2+}	0.00794	0.023	0.00041
Fluoride, F^-	0.00130	0.0037	0.00006

Concentrations are presented in terms of g of ion per kg of seawater, % by weight in seawater, and the weight ratio g ion per g Cl^-.

Table 2. Ion to Chloride Ratios in Various Oceans and Selected Seas for Na^+ and K^+

Ocean or Depth Interval	Na^+/Cl^-	K^+/Cl^-
Atlantic	0.5552	0.0206
Pacific	0.5555	0.0206
Indian	0.5554	0.0207
Red Sea	0.5563	0.0206
Mediterranean	0.5557	0.0206
0–100 m	0.5554	0.0206
700–1500 m	0.5557	0.0206
>1500 m	0.5555	0.0206
MEAN	0.5555+/−0.0007	0.0206+/−0.0002

Also shown are ion to chloride ratios for selected depth intervals in the world ocean. Mean values are for the world ocean over all depth intervals.

This observation is known as "The Rule of *Constant Proportions*" or conservative behavior. As Cl^- is the single most abundant ionic species in seawater, the Rule of Constant Proportion is usually expressed as the ratio of major ions to chloride ion (e.g., Na^+/Cl^-). Exceptions to this rule exist for calcium (Ca^{2+}), strontium (Sr^{2+}), and bicarbonate ($HCO_3{}^{2-}$) as a small fraction of the total concentration of these species participate in biological reactions resulting in slight variations of ion/Cl^- ratios between surface and deep water.

The *residence time* (τ) of a chemical species in the ocean can be defined as the average time an individual atom for a given element remains in seawater before being permanently removed. Assuming that the major source of dissolved salts to the ocean is from riverine input, residence time of a given ion can be calculated as the total mass (in g) of ion in the ocean (M_{ocean}) divided by the mass of ion delivered by rivers annually ($Flux_{in}$) (in g yr^{-1}).

$$\tau(\text{years}) = \frac{M_{ocean}}{Flux_{in}}$$

$Flux_{in}$ can be estimated from the average concentration of the ion in the world's rivers (in g L^{-1}) multiplied by the total volume of river water entering the ocean annually (in L yr^{-1}). M_{ocean} can be estimated from mean seawater concentrations (in g L^{-1}) multiplied by the volume of the ocean (in L). The residence times for the major ions calculated by this method are extremely large, on the order of millions to hundreds of millions of years (Table 3).

The long residence times of major ions reflects the relatively low reactivity of these chemical species. Except

Table 3. Mean Seawater and River Water Concentrations and Residence Times for Several Major Ions

Species	Mean SW (mg L^{-1})	Mean RW (mg L^{-1})	$\frac{SW}{RW}$	τ (10^6 years)
Cl^-	19, 350	5.75	3, 365	123
Na^+	10, 760	5.25	2, 090	75
Mg^{2+}	1, 294	3.35	386	14
$SO_4{}^{2-}$	2, 712	8.25	338	12
K^+	399	1.3	307	11
Ca^{2+}	412	13.4	31	1.1
Br^-	67	0.02	3, 350	123

for Ca^{2+}, Sr^{2+}, and HCO_3^{2-}, no biological processes exist that accelerate the removal of these species to the oceanic sediment. The main removal mechanisms for major ions are kinetically slow reactions such as precipitation or incorporation into clay minerals in the sediment or oceanic crust. Thus, major ions are conservative or remain in constant proportion to each other because the major ions have very long residence times relative to either the residence time of water in the ocean (40,000 years) or the oceanic circulation time (1000 years). Even if the salt content (salinity) of seawater changes from location to location in the ocean, the ratios of the major ions do not change because they react on much longer time scales than the removal or addition of water by precipitation or evaporation. As the residence time of major ions greatly exceeds the average oceanic circulation time of approximately 1000 years, no change is seen in major ion concentrations from ocean to ocean or between surface water and deep water

The major ion composition of river water and seawater are different (Table 3). The composition of present-day river water reflects the weathering of continental rock, whose composition reflects the long-term partitioning of chemical species between the continents and seawater. The composition of seawater with respect to its major ion composition is not believed to have changed over the past several million years despite the constant input of river water of different composition. The oceans are believed to be at steady-state equilibrium with respect to the major ions. That is, the composition of seawater remains constant because the flux of major ions into the ocean equals the flux of major ions out of the ocean.

Although the major source of major ions is river input, hydrothermal circulation of seawater through oceanic basalt, particularly at relatively low temperatures, may be a significant source of some major cations (Na^+, K^+). To achieve steady-state equilibrium, the sum of the sinks or fluxes out of major ions must balance the input from rivers. Several processes, most occurring at extremely slow rates over large areas of the oceanic seafloor, have been identified as sinks for major ions. These processes include: (a) Cation exchange in which clay minerals within the oceanic sediment as well as clays delivered to the oceans by rivers exchange cations in seawater to form new clay minerals relatively enriched in Na, K, and Mg at the expense of Ca. (b) Trapping and eventual burial of seawater within interstitial water of marine sediments. This process removes the more concentrated ions in seawater (Na, Cl). (c) Evaporite formation during some periods of Earth's history, in which large deposits of minerals derived from seawater have formed when seawater was trapped and evaporated from shallow, closed basins. Although the areal extent of such basins is limited at the present time, formation of evaporites is an important removal mechanism for some major ions, including Na, Cl, and SO_4^{2-}, over geologic time. (d) *Sea spray* transported to land can result in net removal of Na and Cl. (e) *Reverse weathering* involving reactions between ions in seawater and cation-poor aluminosilicates derived from continental weathering. This process results in the formation of new cation-rich clay minerals and CO_2, resulting in the net removal of Na, K, and Mg from seawater. (f) *Hydrothermal circulation* involving the reaction of major ions between seawater and oceanic basalt is a net sink for Mg^{2+} and SO_4^{2-}.

READING LIST

Mackenzie, F.T. and Garrels, R.M. (1966). Chemical balance between rivers and oceans. *Amer. J. Sci.* **264**: 507–525.

Holland, H.D. (1984). *The Chemical Evolution of the Atmosphere and Oceans*. Princeton University Press, Princeton, NJ, p. 584.

Pilson, M.E.Q. (1998). *An Introduction to the Chemistry of the Sea*. Prentice-Hall, Englewood Cliffs, NJ, p. 431.

TSUNAMI

Tsunami (also called Seismic Sea Wave, and popularly, Tidal Wave), an ocean wave produced by a submarine earthquake, landslide, or volcanic eruption. These waves may reach enormous dimensions and have sufficient energy to travel across entire oceans. From the area of the disturbance, the waves will travel outward in all directions, much like the ripples caused by throwing a rock into a pond. The time between wave crests may be from 5 to 90 minutes, and the wave speed in the open ocean will average 450 miles per hour.

Tsunamis reaching heights of more than 100 feet have been recorded. As the waves approach the shallow coastal waters, they appear normal and the speed decreases. Then, as the tsunami nears the coastline, it may grow to great height and smash into the shore, causing much destruction.

1. Tsunamis are caused by an underwater disturbance—usually an undersea earthquake. Landslides, volcanic eruptions, and even meteorites can also generate a tsunami.
2. Tsunamis can originate hundreds or even thousands of miles away from coastal areas. Local geography may intensify the effect of a tsunami. Areas at greatest risk are less than 50 feet above sea level and within one mile of the shoreline.
3. People who are near the seashore during a strong earthquake should listen to a radio for a tsunami warning and be ready to evacuate at once to higher ground.
4. Rapid changes in the water level are an indication of an approaching tsunami.
5. Tsunamis arrive as a series of successive "crests" (high water levels) and "troughs" (low water levels). These successive crests and troughs can occur anywhere from 5 to 90 minutes apart. They usually occur 10 to 45 minutes apart.

The Tsunami Warning System, a cooperative international organization and operated by the United States Weather Service, has been in operation since the 1940s. The headquarters of the center is located in Hawaii. An associated Alaska Regional Tsunami Warning System is located in Alaska. Tsunami prediction essentially

Figure 1. Network of tide and seismograph stations that are part of the Tsunami Warning System, headquartered in Hawaii. (*National Oceanic and Atmospheric Administration.*).

commences with earthquake monitoring and prediction information. Inputs from these systems are linked with information from a series of tide monitoring installations. Locations of tide stations and of seismograph stations in the Tsunami Warning System are shown in Fig. 1.

When inputs indicate conditions are favorable for a tsunami, a watch is issued for the probable affected area. Warnings are issued when readings from various tidal stations appear to match the seismographic information. Because of the complexity of the factors involved and a large degree of uncertainty nearly always present, there is a tendency to issue watches and warnings as a safety precaution even though a tidal wave of significance may not develop. Unfortunately, after awhile, persons in likely areas to be affected grow callous to watches and tend to ignore them. A tsunami that hit Hilo, Hawaii in 1960 killed 60 residents even though they had been warned of the coming event an ample 6 hours in advance of the strike.

This, however, was not the reason for the most devastating Tsunami in history which occurred on December 26, 2004 in the Indian Ocean as a result of a 9.0 earthquake off the shore of Banda Aceh, Indonesia. The resulting Tsunami quickly hit the Indonesian coastline, but hours later struck Sri Lanka, India, The Maldives, and later yet, Keyna and Somalia on the East coast of Africa. More than 150,000 lives were lost because no network existed to communicate the likely result of the 9.0 quake in the Indian Ocean. No such event had occurred in this area of the world in over 120 years since Kakatoa erupted in 1883.

Twenty three earthquake monitoring stations picked up the seismic shocks in Indonesia itself, and the U.S. Geological Survey's worldwide monitoring system, with 120 stations, pinpointed the quake immediately. However, on that fateful Sunday, the few warning telephone calls that were made went unanswered. Thus, lack of communication, more than lack of technology, caused the extreme loss of life.

The material destruction, however, touching so many countries, was a result of the movement of the Earth's crust beneath the Indian Ocean extending along a fracture believed to have been 600 miles in length, ultimately creating a tidal wave nearly 1000 miles long.

Such horror can never again touch the Indian Ocean nations as they continue to build a Tsunami Watch System containing all the elements of the early warning capability that has existed in the Pacific Ocean since the middle of the 20th century.

In the late 1970s, scientists suggested an improved method for making tsunami predictions. For a number of years, specialists have suggested that better analysis and interpretation of seismic waves produced by earthquakes may improve the prediction of tsunamis. Seismic waves range from very short-period waves that result from the sharp snap of rocks under high stress to very long-period waves, due mainly to the slower movements of large sections of the ocean floor. Many researchers believe that tsunamis result mainly from the vertical movement of these large blocks, leading to a tentative conclusion that the strength of seismic waves of very long period may be the best criterion for an earthquake's ability to generate a tsunami.

Part of the problem is that most seismographs installed in the system are not very sensitive to very long-period waves, and thus a given earthquake cannot be analyzed effectively in terms of its potential for producing a tsunami. Equipment has been refined so that, today, shorter period waves are used to locate an earthquake and 20-second waves are used to calculate the magnitude. However, some scientists feel that the true magnitude of some earthquakes can only be determined by measuring the characteristics of longer waves, such as 100-second waves. Brune and Kanamori (University of California at San Diego) have observed that the Chilean earthquake of 1960 had a magnitude of 8.3 when calculated on the basis of 20-second waves, but its magnitude was 9.5, or more than 10 times larger in wave amplitude and more than 60 times larger in energy released, when calculated by Kanamori's method, which attempts to include the energy release represented by very long-period seismic waves. Other scientists are coming to the viewpoint that many warnings could be omitted if predictions were based on longer waves. Two of the first long-period seismographs incorporated in the Tsunami Warning System were installed in Hawaii and on the Russian island of Yuzhno-Sakhalinsk, which is northeast of Vladivostok. Later, an installation was made at the Alaska Warning Center in Palmer. Whether the latest reasoning proves successful must await a number of years of experience with the earthquakes of the future and the resulting tsunamis.

CONTINUING TSUNAMI RESEARCH

During the 2000s, research pertaining to the fundamentals of tsunamis and the development of mathematical models of the phenomenon continues. Considerable attention is being directed to specific regions, including the west coasts of Mexico and Chile, the southwestern shelf of Kamchatka (Russia) and, in the United States, the generation of tsunamis in the Alaskan bight and in the Cascadia Subduction Zone off the west coasts of Washington (Puget Sound) and Oregon.

Research also is being directed toward the development of simple and more economic warning systems, particularly in the interest of the developing countries, the coasts of which border on the Pacific Ocean.

The Tsunami Warning System previously described requires millions of dollars for equipment, maintenance, and operation, well beyond the means of some countries. Also, some scientists believe that more localized equipment installations could possibly serve local shore communities better while costing less. These observations, however, do not challenge the need and validity for the larger tsunami network. The National Oceanic and Atmospheric Administration (NOAA) has developed a system costing in the range of $20,000 that can be installed and operated by non experts. The system has been undergoing trials at Valparaiso, Chile, a city that has been struck by nearly 20 tsunamis within the past two centuries. A sensor (accelerometer) is installed in bedrock under the city and can measure tectonic activity in excess of 7.0 on the Richter scale. These measurements are interlocked with level sensors.

Researchers B.F. Atwater and A.L. Moore (University of Washington), in their attempts to model earthquake and tsunami activity in the area over the last thousand years, have reported what they believe to have been a large earthquake on the Seattle fault some time between 1000 and 1100 years ago. The researchers report, "Water surged from Puget Sound, overrunning tidal marshes and mantling them with centimeters of sand. One overrun site is 10 km northwest of downtown Seattle; another is on Whidbey Island, some 30 km farther north. Neither site has been widely mantled with sand at any other time in the past 2000 years. Deposition of the sand coincided—to the year or less—with abrupt, probably tectonic subsidence at the Seattle site and with landsliding into nearby Lake Washington. These findings show that a tsunami was generated in Puget Sound, and they tend to confirm that a large shallow earthquake occurred in the Seattle area about 1000 years ago."

Simulations of tsunamis from great earthquakes on the Cascadia subduction zone have been carried out by M. Ng, P.H. Leblond, and T.S. Murty (University of British Columbia). A numerical model has been used to simulate and assess the hazards of a tsunami generated by a hypothetical earthquake of magnitude 8.5 associated with rupture of the northern sections of the subduction zone. The model indicates that wave amplitudes on the outer coast are closely related to the magnitude of sea-bottom displacement (5 meters). The researchers observe, "Some amplification, up to a factor of 3, may occur in some coastal embayments. Wave amplitudes in the protected waters of Puget Sound and the Straits of Georgia are predicted to be only about one-fifth of those estimated on the outer coast."

READING LIST

Atwater, B.F. and Moore, A.L (December 4, 1991). A tsunami about 1000 years ago in Puget Sound, Washington. *Science*: 1614.

Bernard, E.N. (Ed.). (July 1989). *Tsunami Hazard: A Practical Guide for Tsunami Hazard Reduction*, (Papers from Symposium at Novosibirsk, Russia). International Union of Geodesy and Geophysics, New York.

Bryant, E. (2001). *Tsunami: The Underrated Hazard*. Cambridge University Press, New York.

Collins, E. (February 1988). Wave Watch. *Sci. Amer*: 28.

Kubota, I. (Spring 1987). Japan's Weather Service and the Sea. *Oceanus*: 71.

Lander, J.F., Lockridge, P.A., and Kozuch, M.J. (1997). *Tsunamis Affecting the West Coast of the United States, 1806–1992*. DIANE Publishing Company, Collingdale, PA.

Lander, J.F. (1997). *Tsunamis Affecting Alaska, 1737–1996*. DIANE Publishing Company, Collingdale, PA.

Ng, M.K-F., Leblond, P.H., and Murty, T.S. (November 30, 1990). Simulation of Tsunamis from Great Earthquakes on the Cascade Subduction Zone. *Science*: 1248.

Prager, E.J. (1999). *Furious Earth: The Science and Nature of Earthquakes, Volcanoes, and Tsunamis*. McGraw-Hill Professional Book Group, New York.

Soloviev, S.L., Kim, K.S., Solovieva, O.N. et al. (2000). *Tsunamis in the Mediterranean Sea, 2000 B.C.–2000 A.D.* Kluwer Academic Publishers, Norwell, MA.

Staff. (1997). *Long-Wave Runup Models*. World Scientific Publishing Company, Inc., River Edge, NJ.

Tsuchiya, Y. and Shuto, N. (1995). *Tsunami: Progress in Prediction, Disaster Prevention and Warning*. Kluwer Academic Publishers, Norwell, MA.

WEB REFERENCES

International Tsunami Information Center (ITIC): http://www.shoa.cl/oceano/itic/frontpage.html.

National Tsunami Hazard Mitigation Program: http://www.pmel.noaa.gov/tsunami/.

Tsunami Links: http://www.pmel.noaa.gov/tsunami-hazard/links.html.

BALLOONING AND METEOROLOGY IN THE TWENTIETH CENTURY

Linda Voss
U.S. Centennial of Flight Commission

Balloons are ideal for gathering meteorological information and have been used for that purpose throughout their history. *Meteorological* measurements of wind and air pressure have gone hand in hand with the earliest balloon launches and continue today. Balloons can climb through the denser air close to the Earth to the thinner air in the upper atmosphere and collect data about wind, the different layers of the atmosphere, and weather conditions as they travel.

The first meteorological balloon sondes, or "registering balloons," were flown in France in 1892. These balloons were relatively large, several thousand cubic feet, and carried instruments to record barometric pressure (barometers), temperature (thermometers), and humidity (hygrometers) data from the upper atmosphere. They were open at the base of the balloon and were inflated with a lifting gas, which could be hydrogen, helium, ammonia, or methane. The lifting gas in the balloon exited through the opening as the balloon expanded during its ascent and the air became thinner and the pressure dropped. At the end of the day, as the lifting gas cooled and took up less space, the balloon descended very slowly. The meteorologists had to wait until the balloon descended all the way to Earth to

A balloon equipped for meteorological observations. A German balloon ascent in the late 1800s. 17 Balloon Equipped for Meteorological Observations.

Two men performing balloon tests for the U.S. Weather Bureau.

A zero-pressure balloon being inflated at Alice Springs, Australia.

Weather balloons are used daily to carry meteorological instruments to 20 miles (30 kilometers) and above into the atmosphere to measure temperature, pressure, humidity, and winds. The balloons are made of rubber and weigh up to 2.2 pounds (one kilogram). More than 200,000 weather soundings are made with such balloons worldwide each year.

FIG. 2. FIRST ASCENSION IN AMERICA, ST. LOUIS, SEPTEMBER 15, 1904.

Preparing to launch America"s first "ballon-sonde." Since this first launch on September 15, 1904, in St. Louis, Missouri, literally millions of weather balloons have been launched by the National Weather Service and its predecessor organization. From: The Principles of Aerography, by Alexander McAdle, 1917.

retrieve their instruments, which often had drifted up to 700 miles (1,126 kilometers) from their launch point.

The German meteorologist Assmann solved the problem of drifting balloons and retrieval of instruments in 1892 by introducing closed rubber balloons that burst when they reached a high altitude, dropping the instruments to Earth by parachute much closer to the launch site. These balloons also had fairly constant rates of ascent and descent for more accurate temperature readings. Assmann also invented a psychrometer, a type of hygrometer used to measure humidity in the air that laboratories generally use.

In the 1930s, meteorologists were able to get continuous atmospheric data from balloons when the radiosonde was developed. A radiosonde is a small, radio transmitter that broadcasts or radios measurements from a group of instruments. Balloons, usually unmanned, carry the transmitter and instruments into the upper atmosphere. The radiosonde transmits data to Earth while measuring humidity, temperature, and pressure conditions.

Today, three types of balloons are commonly used for meteorologic research.

Assmann's rubber, or neoprene, balloon is used for measuring vertical columns in the atmosphere, called vertical soundings. The balloon, inflated with a gas that causes the balloon to rise, stretches as it climbs into thin air, usually to around 90,000 feet (27,400 meters). Data is taken as the balloon rises. When the balloon has expanded from three to six times its original length (its volume will have increased 30 to 200 times its original amount), it bursts. The instruments float to Earth under a small parachute. The neoprene balloon can either carry radiosondes that transmit meteorological information or be tracked as a pilot balloon, a small balloon sent aloft to show wind speed and direction. Around the world, balloons equipped with radiosondes make thousands of soundings of the winds, temperature, pressure, and humidity in the upper atmosphere each day. But these balloons are launched and tracked from land, which limits what the radiosondes can measure to less than one-third of the Earth's surface.

Zero-pressure plastic (usually polyethylene) balloons were first launched in 1958. They carry scientific instruments to a predetermined atmospheric density level. Zero-pressure balloons are the best for extremely high altitudes because the balloons can be lighter and stress on them can be distributed over the surface of the balloon.

About the same time, the Air Force Cambridge Research Laboratories (AFCRL) started working on super-pressure balloons, which were made from mylar. The development of mylar plastic films and advances in electronic miniaturization made constant-altitude balloons possible.

Mylar is a plastic that can withstand great internal pressure. The mylar super-pressure balloon does not expand as it rises, and it is sealed to prevent the release of gas as the balloon rises. By the time the balloon reaches the altitude where its density equals that of the atmosphere, the gas has become pressurized because the heat of the sun increases the internal gas pressure. However, because mylar can withstand great internal pressure, the volume of the balloon remains the same. By carefully calculating the weight of the balloon and whatever it is carrying, the altitude at which the balloon will achieve equilibrium and float can be calculated. As long as the pressure inside the balloon remains the same, it will remain at that altitude.

These balloons could be launched to remain aloft at specified altitudes for weeks or months at a time.

Moreover, satellites could be used to track and request data from many balloons in the atmosphere to obtain a simultaneous picture of atmospheric conditions all over the globe. Another advantage of super-pressure balloons is that, since they transmit their data to satellites, they can gather data from over oceans as well as land, which is a limitation of balloons equipped with radiosondes.

The AFCRL program resulted in the Global Horizontal Sounding Technique (GHOST) balloon system. With GHOST, meteorologists at last achieved their goal of semi-permanent platforms floating high in the atmosphere.

Eighty-eight GHOST balloons were launched starting in March of 1966. The GHOST balloons and their French counterpart, EOLE, (the name Clement Ader used for one of his aircraft—named after the Greek god of the wind) used strong, plastic super-pressure balloons to trace air circulation patterns by drifting with the wind at constant density altitudes. Many super-pressure balloons were aloft at a time, grouped at constant density levels. Each balloon had a sensing device and transmitting system for gathering information on its position and weather data and transmitted atmospheric and weather data to weather satellites. They first transmitted their data to the NASA *Nimbus-4* meteorological satellite in 1970.

In 1966, a GHOST balloon circled the Earth in 10 days at 42,000 feet (12,801 meters). By 1973, NASA had orbited scientific instrument packages aboard sealed balloons at altitudes up to 78,000 feet (23,774 meters). Other GHOST balloons remained aloft for up to a year. The program lasted for 10 years.

The ultimate of the super-pressure balloons was the balloon satellite *Echo I.* Launched into space in 1960, the balloon inflated to a sealed volume by residual air, benzoic acid, and a chemical called anthraquinone.

Constant-altitude, super-pressure balloons continue to fly over the oceans and land surfaces. These balloons have been relied on for decades to provide extensive knowledge of global meteorology and improve worldwide weather forecasting.

READING LIST

Crouch, Tom D. *The Eagle Aloft: Two Centuries of the Balloon in America*. Washington, DC.: Smithsonian Institution Press, 1983.

Tannenbaum, Beulah and Harold E. *Making and Using Your Own Weather Station*. New York. Venture Books, 1989.

Vaeth, J. Gordon. "When the Race for Space Began." *U.S. Naval Institute Proceedings*. August 1961.

ONLINE REFERENCES

Cowens, John. "Building a Psychrometer," Greenwood Elementary, La Grande, OR 97850. September 28, 1993. http://explorer.scrtec.org/explorer/explorer-db/html/783750680-447DED81.html

"Hygrometer," Microsoft® Encarta® Online Encyclopedia 2000. 1997–2000 Microsoft Corporation. http://encarta.msn.com.

Lally, Vincent E. "Balloon: Modern Scientific Ballooning," Microsoft® Encarta® Online Encyclopedia 2000. 1997–2000 Microsoft Corporation. http://www.encarta.msn.com//find/concise.asp?ti=761576250&sid=3#s3. http://www.encarta.msn.com.

"Measuring the Weather," USAToday.com. http://www.usatoday.com/weather/wmeasur0.htm.

"Meteorology," Microsoft® Encarta® Online Encyclopedia 2000. 1997–2000 Microsoft Corporation. http://encarta.msn.com.

Warner, Lucy, ed. "Forecasts: Observing and Modeling the Global Atmosphere," *UCAR at 25*. University Center for Atmospheric Research. Boulder, Colorado. Oct. 17, 2000. http://www.ucar.edu/communications/ucar25/forecasts.html.

"Weather Instruments to Make." http://asd-www.larc.nasa.gov/SCOOL/psychrometer.html

To find out how you can make your own psychrometer, link to The CERES S'COOL Project at http://asd-www.larc.nasa.gov/SCOOL. The link has lots of information about making weather observations. Making a Psychrometer is in the Table of Contents or go to http://explorer.scrtec.org/explorer/explorer-db/html/783750680-447DED81.html.

To learn more about weather instruments and even set up your own weather station to report to the U.S. National Weather Service, go to http://www.usatoday.com/weather/wmeasur0.htm.

Educational Organization	Standard Designation (where applicable)	Content of Standard
National Science Education Standards	Content Standard D	Students should develop an understanding of energy in the earth system.
International Technology Education Association	Standard 9	Students will develop an understanding of engineering design.
International Technology Education Association	Standard 10	Students will develop an understanding of the role of trouble shooting, research and development, innovation, and experimentation in problem solving.
International Technology Education Association	Standard 7	Students will develop an understanding of the influence of technology on history.

BAROMETRIC EFFICIENCY

Todd Rasmussen
The University of Georgia
Athens, Georgia

Water levels in wells are often observed to fluctuate as the air pressure changes. Blaise Pascal described this effect in 1663 (1) and was the first to attribute the water level changes in wells to changes in atmospheric pressure. He noted that water levels declined as the barometric pressure increased, and vice versa.

The barometric efficiency, BE, is used to relate changes in water levels, ΔW, to changes in barometric (air) pressure, ΔB:

$$BE = -\frac{\Delta W}{\Delta B} \quad (1)$$

where consistent units (e.g., feet of water, mm of Hg, hPa) are used for both water levels and barometric pressure (2,3).

To understand why this relationship exists, we can conceive of an aquifer that is entirely isolated from the atmosphere. Such an aquifer maintains a constant total head, $H = H_o$, within the aquifer and is entirely unaffected by atmospheric pressure changes. Water level elevations, W, in the aquifer are measured in an open borehole.

The total head within the well is the sum of the water level elevation plus the barometric pressure exerted on the water surface, $H = W + B$. If the total head within the well is equal to the total head within the aquifer, $H_o = W + B$, and the total head within the aquifer is constant, then the water level varies inversely with barometric pressure, $W = H_o - B$.

The change in water level is just the negative of the change in barometric pressure, $\Delta W = \Delta(H_o - B) = -\Delta B$, so that the barometric efficiency is 100 percent, $BE = -\Delta W/\Delta B = -(-\Delta B)/\Delta B = 1$.

A second example assumes that the total head in the hypothetical aquifer increases with barometric pressure, $H = H_o + B$, so that the head in the aquifer goes up and down over time. In this case, the water levels in the well remain unchanged, and the barometric efficiency of the aquifer is zero.

These two extremes are shown in Fig. 1. Note that the barometric pressure is identical in both cases. In the left figure, the barometric efficiency is 100%, so that the water level varies inversely with barometric pressure. The total head is the sum of the barometric pressure and water levels, so it remains unchanged. In the right figure, the barometric efficiency is zero, so that the water level is unaffected by barometric pressure changes and the total head varies directly with barometric pressure.

Most aquifers lie between these two extremes, however, and the actual response depends, in large part, on whether the aquifer is confined or unconfined. Both of these cases are described below.

CONFINED AQUIFERS

For confined aquifers, Jacob (4) showed that increasing the load on the ground surface increases the load on the aquifer. This additional weight is either carried by the mineral grains or by the water within the aquifer pores. If the entire weight is borne by the mineral grains, and these grains do not deform with the increase in load, then the total head within the aquifer remains unchanged, and the barometric efficiency is 100%.

The barometric efficiency is smaller if some of the weight is carried by the fluid. A very low barometric efficiency in confined aquifers occurs when the fluid within

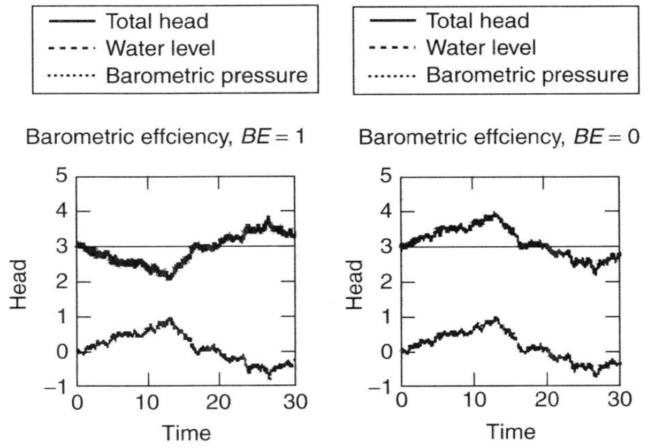

Figure 1. Effect of barometric pressure on water levels in wells. Both figures show barometric pressure (lower, dashed line), total head (solid line), and water levels (dotted line). Left figure shows effects when the barometric efficiency is 100%, $BE = 1$, and right figure shows effects when the barometric efficiency is zero, $BE = 0$.

the aquifer bears most of the weight. Examples of such aquifers include poorly consolidated sedimentary aquifers or horizontal fractures that extend great distances.

Figure 2 shows the extreme conditions. In the left figure, the entire increase in load caused by an increase in barometric pressure is carried by the mineral grains. In this case, the pore fluids are not affected by the increase in load, the total head remains unchanged, and water levels drop in an amount equal to the increase in barometric pressure. In the right figure, the mineral grains do not carry the load and the fluid carries the increased load, causing an increase in total head.

In confined aquifers, the barometric response is virtually instantaneous. A change in barometric pressure should cause an immediate change in water levels in wells. Water levels in large-diameter wells may not respond immediately, however, because of the time required for water levels to adjust to the new level. Instead of a rapid

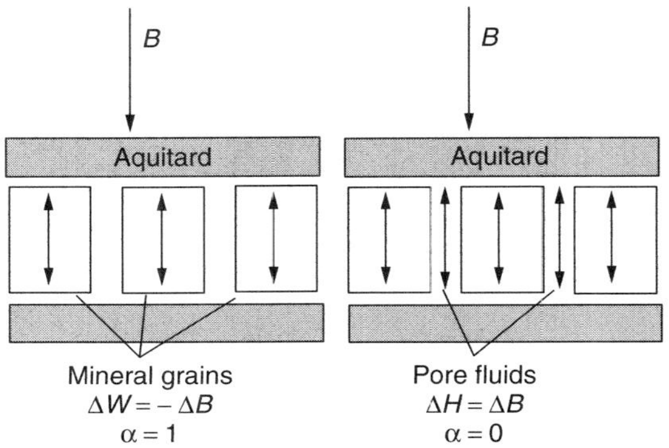

Figure 2. Effect of barometric pressure on total head in confined aquifers. The mineral grains carry the load in the left figure, whereas the pore fluids carry the load in the right figure. Most confined aquifers, however, are intermediate between these two extremes.

response to pressure changes, a slow response may be found in these wells.

This phenomena is called *borehole storage* and is a function of the diameter of the borehole, the length of the screened interval relative to the aquifer thickness, the aquifer hydraulic properties (e.g., transmissivity and storativity), and the rate at which the barometric pressure is changing.

The effects of borehole storage can be eliminated by placing a packer in the well below the water surface, but above the screened interval, which eliminates the need for the water level to change in response to changes in barometric pressure. Some change in volume may result from the slight compressibility of water, but this effect is very small. A gauge-type pressure sensor (i.e., internally vented to the atmosphere) is then placed below the packer to measure the water pressure.

Several techniques are available for estimating the confined aquifer barometric efficiency, including linear regression and Clark's Method (5). Clark's method is an unbiased technique for estimating the barometric efficiency that performs well when a water level trend is present in the data (6). Regression provides a better estimate, but only when the trend function can be accurately specified (7).

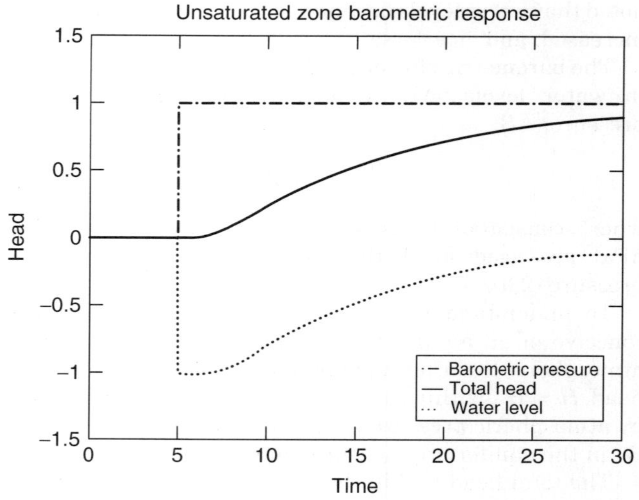

Figure 3. Effect of barometric pressure on total head and water levels in confined aquifers. The barometric pressure in this example takes a step from zero to one at $t = 10$. The total head slowly rises as the barometric pressure makes its way through the unsaturated zone. The water levels initially fall to maintain the constant total head, but then return to normal once the barometric pressure signal reaches the water table.

UNCONFINED AQUIFERS

Barometric pressure changes commonly do not affect water levels in wells located in shallow, unconfined aquifers, which is because the air pressure moves rapidly through the unsaturated zone and causes an immediate increase in total head within the aquifer.

In deeper unconfined aquifers, however, the typical response to barometric pressure changes is to see an immediate inverse response (i.e., an increase in barometric pressure causes an equivalent and opposite water level response), followed by a slow decay back to the original water level (5,8,9).

To understand this complex response, we focus on how total head responds to barometric pressure changes. Initially, no increase in total head within the unconfined aquifer occurs because the water held within pores is not confined by an overlying confining unit. The total head within the aquifer can rise over time, however, as air diffuses downward through the unsaturated zone.

The total head in the unconfined aquifer increases once the air pressure change reaches the water table. The time required for the total head to respond to barometric pressure changes is a function of the depth of the water table and the air diffusivity within the unsaturated zone. The air diffusivity is higher in coarse, dry soils and is lower in wet soils or in soils with a low air permeability.

As the water level response in observation wells is the total head minus the barometric pressure, the fact that the total head does not change initially means that a rapid water-level response occurs to barometric pressure changes. In fact, the barometric efficiency is one in wells with a deep unsaturated zone—or where the air permeability of the unsaturated zone is low—followed by a gradual decrease as air diffuses downward to the water table.

This concept is illustrated graphically in Fig. 3. Note that the barometric pressure is simplified to a simple step change from zero to one. The total head does not initially respond, but slowly rises over time as the barometric pressure moves through the unsaturated zone and eventually reaches the water table. The observed water level in a well placed in this aquifer responds immediately (as long as borehole storage can be neglected), but eventually decays back to its original level

The barometric efficiency of unconfined aquifers is not single valued like the confined aquifer barometric efficiency. Instead, the response includes a delay that can be estimated using regression deconvolution (5,9). The resulting response can be used to estimate the unsaturated air diffusivity of the unsaturated zone.

BIBLIOGRAPHY

1. Pascal, B. (1973). *The Physical Treatises of Pascal.* Octagon Books, New York.
2. Bear, J. (1972). *Dynamics of Fluids in Porous Media.* Dover Publications, New York.
3. Walton, W.C. (1970). *Groundwater Resource Evaluation.* McGraw-Hill, New York.
4. Jacob, C.E. (1940). On the flow of water in an elastic artesian aquifer. *Amer. Geophys. Union Trans.* **21**: 574–586.
5. Rasmussen, T.C. and Crawford, L.A. (1997). Identifying and removing barometric pressure effects in confined and unconfined aquifers. *Ground Water.* **35**(3): 502–511.
6. Clark, W.E. (1967). Computing the barometric efficiency of a well. *J. Hydraul. Divi. Amer. Soc. Civil Engin.* **93**(HY4): 93–98.
7. Davis, D.R. and Rasmussen, T.C. (1993). A comparison of linear regression with Clark's method for estimating barometric

efficiency of confined aquifers. *Water Resources Res.* **29**(6): 1849–1854.

8. Weeks, E.P. (1979). Barometric fluctuations in wells tapping deep unconfined aquifers. *Water Resources Res.* **15**(5): 1167–1176.
9. Spane, F.A., Jr. (2002). Considering barometric pressure in groundwater flow investigations. *Water Resources Res.* **38**(6): 1078.

READING LIST

Spane, F.A., Jr. and Mercer, R.B. (1985). *HEADCO: A Program for Converting Observed Water Levels and Pressure Measurements to Formation Pressure and Standard Hydraulic Head*. Pacific Northwest National Laboratory, Rockwell Hanford Operations, Richland, WA. Report PNL-10835.

CERES: UNDERSTANDING THE EARTH'S CLOUDS AND CLIMATE

NASA—Langley
Research Center

The Clouds and the Earth's Radiant Energy System (CERES) instrument is one of several launched aboard the Earth Observing System's (EOS) Aqua spacecraft in 2002. Scientists use observations from the CERES instrument to study the energy exchanged between the Sun; the Earth's atmosphere, surface and clouds; and outer space.

The CERES Aqua instruments are the fourth and fifth CERES instruments in orbit. NASA launched the first CERES instrument aboard the Tropical Rainfall Measuring Mission satellite or TRMM in November 1997. Results of the TRMM mission show that the first CERES provided better measurement capabilities than any previous satellite instrument of its kind. Two other CERES instruments are currently orbiting the Earth on the EOS Terra spacecraft, launched in late 1999. Early CERES Terra results give new insights into the effects of clouds on climate and how the climate system changes from decade to decade. Two CERES instruments on each of the Terra and Aqua spacecraft will provide global coverage of energy radiated and reflected from the Earth. Scientists use measurements from both satellites' orbits to improve observations of the daily cycle of radiated energy.

NASA Langley Research Center manages the CERES mission. Langley's highly successful Earth Radiation Budget Experiment (ERBE) provided the foundation for the design of the CERES instrument. ERBE used three satellites to provide global energy measurements from 1984 through the 1990s. The TRW Space & Electronics Group in Redondo Beach, Calif., built all six CERES instruments.

WHAT CERES WILL MEASURE

CERES measures the energy at the top of the atmosphere, as well as estimate energy levels in the atmosphere and at the Earth's surface. Using information from very high resolution cloud imaging instruments on the same spacecraft, CERES will determines cloud properties, including altitude, thickness, and the size of the cloud particles. All of these measurements are critical for advancing the understanding of the Earth's total climate system and the accuracy of climate prediction models.

BALANCING THE EARTH'S ENERGY BUDGET

The balance between Earth's incoming and outgoing energy controls daily weather and climate (long-term weather patterns). Sunlight or solar energy is the planet's only incoming energy source. Heat emitted from the sunlight reflected by the Earth's surface, atmosphere and clouds make up the planet's outgoing energy. Scientists have been working for decades to understand this critical energy balance, called the Earth's "energy budget."

The energy received from the Sun is at short wavelengths, while the energy emitted by the surface of the Earth, the atmosphere and clouds is at long wavelengths. Greenhouse gases in the atmosphere absorb the long wavelength energy or heat emitted by the Earth. Increases in the amounts of greenhouse gases produced by both natural processes or human activities can lead to a warming of the Earth's surface. Such changes may, in turn, alter the planet's daily weather and climate.

Clouds and small particles in the atmosphere called aerosols also reflect some sunlight back into space. Major sources of aerosols include windblown dust, emissions from the burning of fossil fuels, such as gasoline, and the burning of forests and agricultural fields.

CLOUD EFFECTS

One of the most intriguing questions facing climate modelers today is how clouds affect the Earth's climate

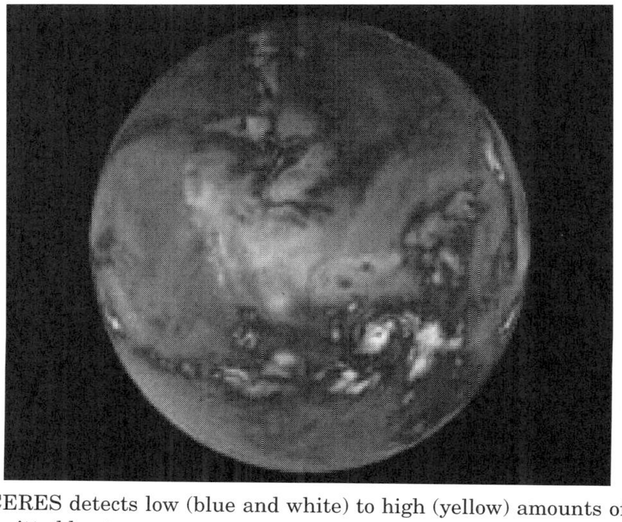

CERES detects low (blue and white) to high (yellow) amounts of emitted heat

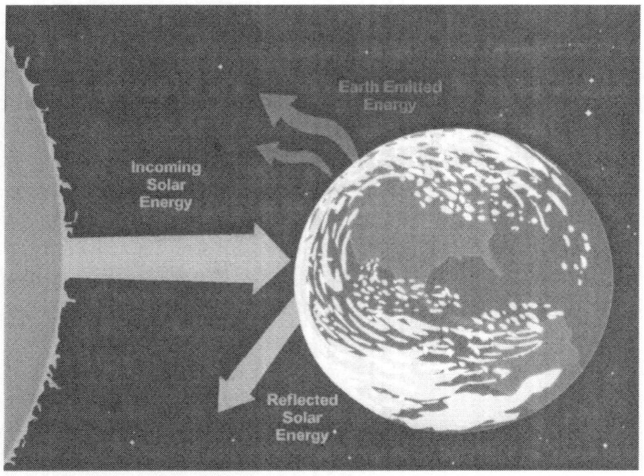

Earth's radiation budget is the balance between incoming and outgoing energy

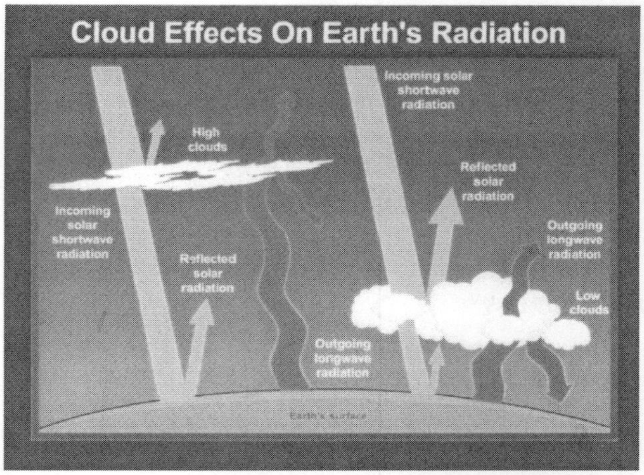

and vice versa. The U.S. Global Change Research Program classifies understanding the role of clouds and the Earth's energy budget as one of its highest scientific priorities. Understanding cloud effects requires a detailed knowledge of how clouds absorb and reflect sunlight, as well as how they absorb and re-emit outgoing heat emitted by the planet. For example, low, thick clouds primarily reflect incoming solar energy back to space causing cooling. Thin, high clouds, however, primarily trap outgoing heat and produce warming. To date, satellite studies have found that clouds have an overall cooling effect on the Earth.

Analyses of satellite data also indicate that clouds which form over water are very different from clouds which form over land. These differences affect the way clouds reflect sunlight back into space and how much heat emitted from the Earth the clouds absorb and re-emit. For example, over the equator in the eastern Pacific Ocean during El Niño events, there is a significant decrease in the amount of energy emitted by the Earth due to increased cloudiness. El Niño events occur when portions of the eastern Pacific Ocean become considerably warmer than normal, causing an increase in cloudiness over the region. These changes can affect weather patterns around the world.

WATER VAPOR EFFECTS

Water vapor in the atmosphere also impacts our daily weather and climate, though scientists are only beginning to understand how this complex mechanism works. Water vapor acts like a greenhouse gas and absorbs outgoing heat to warm the Earth. Because water vapor also condenses to make clouds, additional water vapor in the atmosphere also may increase the amount of clouds.

FUTURE MISSIONS

One additional CERES instrument is available to fill the gap between Aqua and the next generation of highly accurate Earth radiation budget measurements. These observations are expected to be made on the National Polar-orbiting Operational Environmental Satellite System (NPOESS) starting around 2010. To continue the 22-year record of global energy measurements, the next CERES mission should launch in 2007.

EDUCATIONAL OUTREACH

As a CERES instrument passes overhead, students worldwide are observing clouds and then sending their observations to NASA Langley's Atmospheric Sciences Data Center (ASDC). At the ASDC, scientists store data for further analysis by the CERES science team. The student observations are part of a global educational outreach program called the Students' Cloud Observations On-Line (S'COOL) project. Since the project began five years ago, S'COOL has reached over 1,000 schools in all 50 states and 57 other countries on five continents.

COMMERCIAL APPLICATIONS

CERES supports commercial applications by providing data about weather and sunlight at the Earth's surface for the renewable energy industry via an innovative Web site (http://eosweb.larc.nasa.gov/sse/). The Surface Meteorology and Solar Energy Project maintains the site. In the first three years of operation, the number of registered users of the Web site, including major energy companies, financial institutions and federal agencies, has grown to over 2,000 from nearly 100 countries. With 35,000 hits per month since January 2001, SSE is the most accessed Web site at the ASDC.

CHINOOK

ARTHUR M. HOLST
Philadelphia Water Department
Philadelphia, Pennsylvania

Wind is defined as the movement of air. Although we commonly define wind as a gentle breeze or a harsh gust of cold air, there is a wind phenomenon that can increase the temperature instead of lowering it. It is called a Chinook wind. Chinooks are most commonly associated with the Rocky Mountain range in North American but can also be found in the Swiss Alps and the Andes. They can increase temperatures high enough to melt the snow in their path as they travel down the mountainside. The Chinook wind falls under the classification of katabatic wind, which means that it moves downhill.

The name "Chinook" was taken from the Chinook Indians that lived along the Rocky Mountains until the early 1800s when the tribe became extinct due to disease. Although the tribe died off, their legends and tales live on. One such legend is that of the Chinook wind, which in its literal translation means "snow-eater."

It is possible for a Chinook to take place anytime during the year, but its effects are much more dramatic during the winter months. Chinook winds cause dramatic increases in the temperature on the eastern side of the Rockies and can send temperatures into the fifties and sixties. A temperature change of this magnitude can take anywhere from an hour to a day. The heat produced is a reaction formed from the change of gas to liquid in the atmosphere. These warm gusts of air then cause the evaporation of any snow on the ground, hence the name snow-eater. There is no definite length of time that a Chinook will last. On average, it can last hours or days.

Chinooks are the end result of the warm moist air of the Pacific Ocean moving up and down the Rocky Mountains. A westerly wind collects the warm moist air in the Pacific Ocean and carries it to the coast where it meets the western side of the Rockies. As the air makes its way up the mountain, the air becomes cooler, and the precipitation is squeezed out of it. Through this process warm, dry air is produced as it makes its way down the eastern side of the mountain. This process is repeated over and over again on each mountain that is in the way of the wind and each time produces warmer, dryer air. The air that makes its way down the last mountain is extremely dry and warm. This wind is called the Chinook wind. These warm gusts of air can reach speeds up to 100 mph.

The results from Chinook winds are both positive and negative. Chinook winds cause evaporation of the snow covering the ground, so the length of the grazing season is extended. A longer grazing season decreases the need to stock up on feed for animals. A negative result is the decreased amount of precipitation due to the quick evaporation of snow. Less precipitation causes hardships in planting.

READING LIST

Alberta Agriculture, Food, and Rural Development. (2002). *Weather in Alberta*. Available: http://www.agric.gov.ab.ca/agdex/000/7100001b.html. (March 11).

The Alberta Traveller. (2002). *Chinook Winds*. Available: http://www.traveller.babelfish.com/weather_guides/Chinook.html. (March 11).

Black Hills Weather. (2002). *The Snow Eating Wind*. Available: http://www.blackhillsweather.com/chinook.html. (March 11).

Ho, I. (2002). *Snow-Eater. The Catalyst*. Available: http://www.brown.edu/Students/Catalyst/archive/spring99/ch.html. (March 11).

Under the Chinook Arch. http://www.nucleus.com/~cowboy/Misc/Chinooks.html. (11 March 2002).

GLOBAL CLIMATE CHANGE

Geophysical Fluid Dynamics Laboratory—NOAA

In 1967, two GFDL scientists, Syukuro Manabe and Richard Wetherald, published what is now regarded as the first credible calculation of the effect of increased carbon dioxide on the climate. They calculated that a doubling of atmospheric carbon dioxide would warm the earth's surface by about 2 °C. This result laid the foundation for what has become an international, multi-disciplinary research effort on global warming.

Manabe, in collaboration with oceanographer Kirk Bryan and other scientists at GFDL, has continued to lead the international effort to develop the coupled ocean-atmosphere climate models that are crucial to understanding and predicting the impact of greenhouse gases.

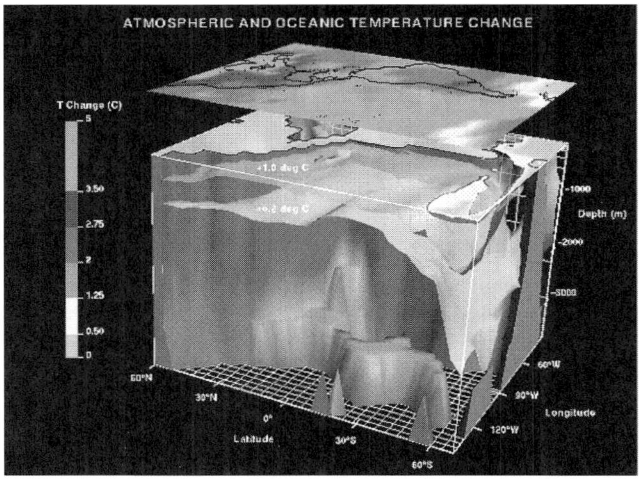

Three-dimensional view of projected surface air temperature and ocean warming (°C) due to greenhouse gases as calculated by a low-resolution GFDL coupled ocean-atmosphere climate model. The top panel shows the surface air temperature change over North and South America and surrounding regions. The three-dimensional box illustrates the depth to which a 1 °C and 0.2 °C warming has penetrated in the model's Pacific Ocean. The gray surface depicts the model's ocean floor. Note the deep mixing of the ocean warming signal in the southern hemisphere ocean near Antarctica. The temperature changes are projections of the warming due to greenhouse gases by the latter half of the twenty-first century in the absence of further increases in sulfate aerosol forcing. Results shown are based on years 61–80 of a transient CO_2 increase experiment (+1% per year compounded). [*Source*: adapted from Syukuro Manabe and Ronald Stouffer, Nature, 15 July 1993.]

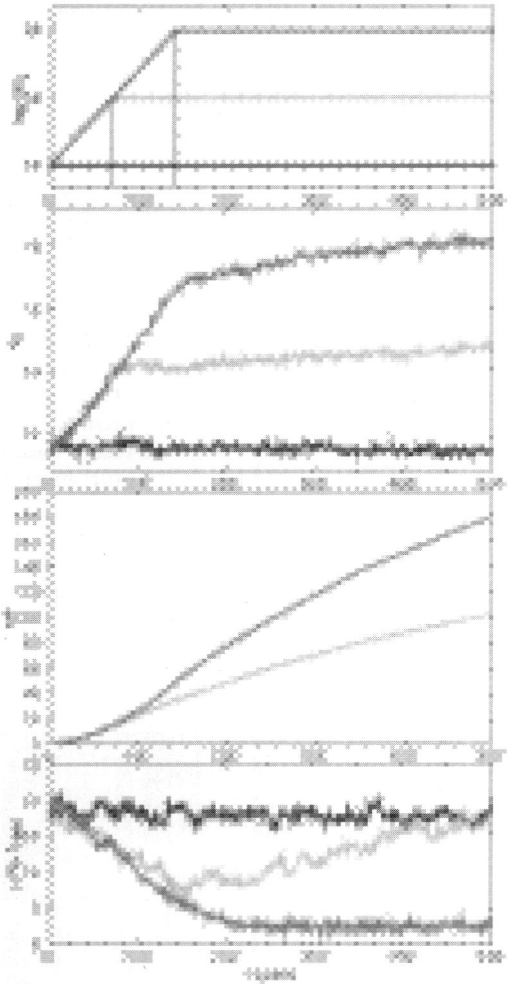

Impact of increasing CO_2 on the earth's climate as simulated in a GFDL coupled ocean-atmosphere climate model. Shown are timeseries of: a) prescribed CO_2 concentration on a logarithmic scale in comparison to present levels; b) global mean surface air temperature (°C); c) global mean increase of sea level (cm) due to thermal expansion; and d) intensity of the North Atlantic Ocean's meridional overturning circulation (10**6 m**3/sec). The labels "Control", "2xCO_2", and "4xCO_2" refer to separate experiments where CO_2 either remains constant (Control), or increases at 1% per year (compounded) to double (2xCO_2) or quadruple (4xCO_2) the current concentration. Note that the sea level rise estimates do not include the effect of melted continental ice sheets. With this effect included, the total rise could be larger by a substantial factor. [*Source*: Syukuro Manabe and Ronald Stouffer, Nature, 15 July 1993.]

A PROBLEM FOR CENTURIES TO COME?

In a recent paper, published 26 years after Manabe's pioneering one-dimensional CO_2 sensitivity study, he and Ron Stouffer of GFDL used a three-dimensional coupled ocean-atmosphere model to examine possible CO_2-induced climate changes over several centuries. Earlier studies had focused on shorter time horizons.

In their scenario, CO_2 quadruples over a period of 140 years, then no longer increases. This perturbation is enough to cause the ocean's global thermohaline circulation to almost disappear in the model. The global thermohaline circulation is important because it is responsible for a large portion of the heat transport from the tropics to higher latitudes in the present climate. In addition, sea level continues rising steadily for centuries after the CO_2 increase is halted. From this perspective, global warming can no longer be viewed as just as a problem of our own lifetimes, but as a legacy—with uncertain consequences—now being passed forward to many future generations.

EVALUATING CLIMATE MODELS

GFDL scientists, including Tony Broccoli and Tom Delworth, are searching for innovative ways to evaluate climate models and to distinguish between human-induced climate change and natural climate variability. Measurements of the current climate, historical observations, and glimpses of earth's climate during the ice ages and other past climates all provide opportunities to test climate models. Through research on climate models and observations, scientists at GFDL will continue to evaluate and refine the climate models that will be needed to help answer critical policy-relevant questions about global warming and its consequences.

OBSERVATIONS OF CLIMATE AND GLOBAL CHANGE FROM REAL-TIME MEASUREMENTS

David R. Easterling
Thomas R. Karl
(from *Handbook of Weather, Climate, and Water: Dynamics, Climate, Physical Meteorology, Weather Systems, and Measurements*, Wiley 2003)

INTRODUCTION

Is the planet getting warmer?
Is the hydrologic cycle changing?
Is the atmospheric/oceanic circulation changing?
Is the weather and climate becoming more extreme or variable?
Is the radiative forcing of the climate changing?

These are the fundamental questions that must be answered to determine if climate change is occurring. However, providing answers is difficult due to an inadequate or nonexistent worldwide climate observing system. Each of these apparently simple questions are quite complex because of the multivariate aspects of each question and because the spatial and temporally sampling required to address adequately each question must be considered on a global scale. A brief review of our ability to answer these questions reveals many successes, but points to some glaring inadequacies that must be addressed in any attempt to understand, predict, or assess issues related to climate and global change.

IS THE PLANET GETTING WARMER?

There is no doubt that measurements show that near-surface air temperatures are increasing. Best estimates suggest that the warming is around 0.6 °C (+−0.2 °C) since the late nineteenth century (1). Furthermore, it appears that the decade of the 1990s was the warmest decade since the 1860s, and possibly for the last 1000 years. Although there remain questions regarding the adequacy of this estimate, confidence in the robustness of this warming trend is increasing (1). Some of the problems that must be accounted for include changes in the method of measuring land and marine surface air temperatures from ships, buoys, land surface stations as well as changes in instrumentation, instrument exposures and sampling times, and urbanization effects. However, recent work evaluating the effectiveness of corrections of sea surface temperatures for time-dependent biases, and further evaluation of urban warming effects on the global temperature record have increased confidence in these results. Furthermore, by consideration of other temperature-sensitive variables, e.g., snow cover, glaciers, sea level and even some proxy non-real-time measurements such as ground temperatures from boreholes, increases our confidence in the conclusion that the planet has indeed warmed. However, one problem that must be addressed is that the measurements we rely upon to calculate global changes of temperature have never been collected for that purpose, but rather to aid in navigation, agriculture, commerce, and in recent decades for weather forecasting. For this reason there remain uncertainties about important details of the past temperature increase and our capabilities for future monitoring of the climate. The IPCC (1) has summarized latest known changes in the temperature record, which are summarized in Fig. 1.

Global-scale measurements of layer averaged atmospheric temperatures and sea surface temperatures from instruments aboard satellites have greatly aided our ability to monitor global temperature change (2–4), but the situation is far from satisfactory (Hurrell and Trenberth, 1996). Changes in satellite temporal sampling (e.g., orbital drift), changes in atmospheric composition (e.g., volcanic emissions), and technical difficulties related to overcoming surface emissivity variability are some of the problems that must be accounted for, and reduce the confidence that can be placed on these measurements (5). Nonetheless, the space-based measurements have shown, with high confidence, that stratospheric temperatures have decreased over the past two decades. Although perhaps not as much as suggested by the measurements from weather balloons, since it is now known that the data from these balloons high in the atmosphere have an inadvertent temporal bias due to improvements in shielding from direct and reflected solar radiation (6).

IS THE HYDROLOGIC CYCLE CHANGING?

The source term for the hydrologic water balance, precipitation, has been measured for over two centuries in some locations, but even today it is acknowledged that in many parts of the world we still cannot reliably measure true precipitation (7). For example, annual biases of more than 50% due to rain gauge undercatch are not uncommon in cold climates (8), and, even for more moderate climates, precipitation is believed to be underestimated by 10 to 15% (9). Progressive improvements in instrumentation, such as the introduction of wind shields on rain gauges, have also introduced time-varying biases (8). Satellite-derived measurements of precipitation have provided the only large-scale ocean coverage of precipitation. Although they are comprehensive estimates of large-scale spatial precipitation variability over the oceans where few measurements exist, problems inherent in developing precipitation estimates hinder our ability to have much confidence in global-scale decadal changes. For example, even the landmark work of Spencer (10) in estimating worldwide ocean precipitation using the microwave sounding unit aboard the National Oceanic and Atmospheric Administration (NOAA) polar orbiting satellites has several limitations. The observations are limited to ocean coverage and hindered by the requirement of an unfrozen ocean. They do not adequately measure solid precipitation, have low spatial resolution, and are affected by the diurnal sampling inadequacies associated with polar orbiters, e.g., limited overflight capability. Blended satellite/*in situ* estimates also show promise (11); however, there are still limitations, including a lack of long-term measurements necessary for climate change studies.

Information about past changes in land surface precipitation, similar to temperature, has been compared with other hydrologic data, such as changes in stream flow, to ascertain the robustness of the documented changes of precipitation. Figure 1 summarizes some of the more important changes of precipitation, such as the increase in the mid to high latitude precipitation and the decrease in subtropical precipitation. Evidence also suggests that much of the increase of precipitation in mid to high latitudes arises from increased autumn and early winter precipitation in much of North America and Europe. Figure 2 depicts the spatial aspects of this change, reflecting rather large-scale coherent patterns of change during the twentieth century.

Other changes related to the hydrologic cycle are summarized in Fig. 1. The confidence is low for many of the changes, and it is particularly disconcerting relative to the role of clouds and water vapor in climate feedback effects.* Observations of cloud amount long have been made by surface-based human observations and more recently by satellite. In the United States, however, human observers have been replaced by automated measurements, and neither surface-based or spaced-based data sets have proven to be entirely satisfactory for detecting changes in clouds. Polar orbiting satellites have an enormous difficulty to overcome related to sampling aliasing and satellite drift (12). For human observers changes in observer schedules, observing biases, and incomplete sampling have created major problems in data

*An enhancement or diminution of global temperature increases or decreases due to other causes.

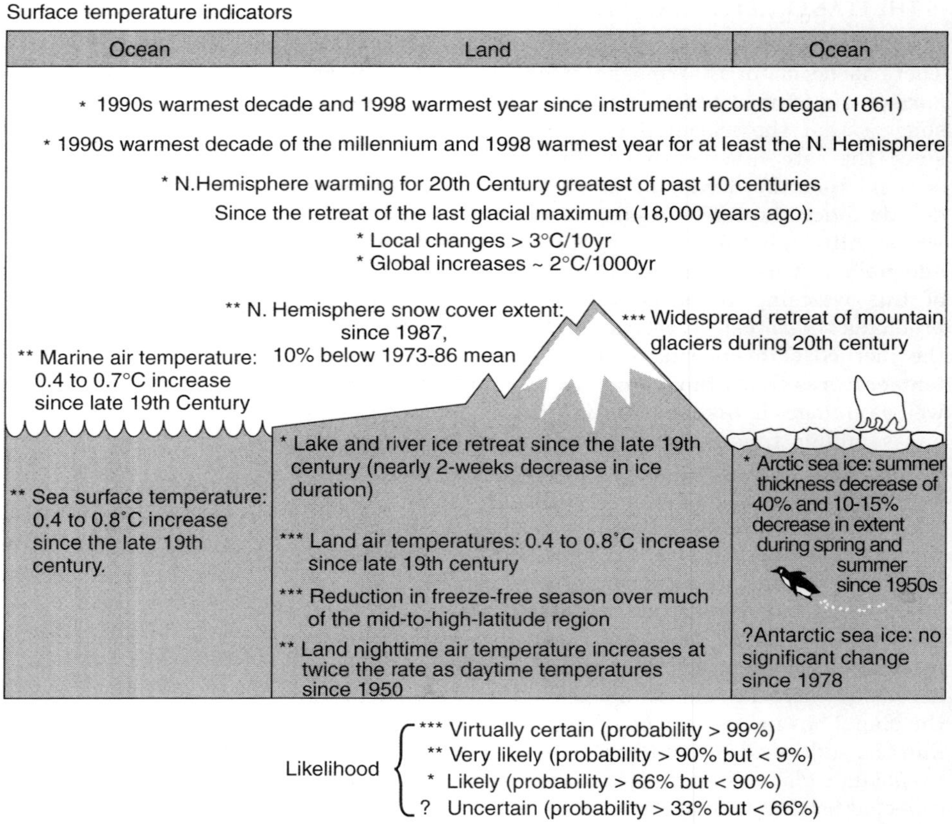

Figure 1. Schematic of observed variations of selected indictors regarding (**a**) temperature and (**b**) the hydrologic cycle (based on Ref. 1).

interpretations, now compounded by a change to new automated measurements at many stations. Nonetheless, there is still some confidence (but low) that global cloud amounts have tended to increase. On a regional basis this is supported by reductions in evaporation as measured by pan evaporimeters over the past several decades in Russia and the United States, and a worldwide reduction in the land surface diurnal temperature range. Moreover, an increase in water vapor has been documented over much of North America and in the tropics (1).

Changes in water vapor are very important for understanding climate change since water vapor is the most important greenhouse gas in the atmosphere. The measurement of changes in atmospheric water vapor is hampered by both data processing and instrumental difficulties for both weather balloon and satellite retrievals. The latter also suffers from discontinuities among successive satellites and errors introduced by changes in orbits and calibrations. Upper tropospheric water vapor is particularly important for climate feedbacks, but, as yet, little can be said about how it has varied over the course of the past few decades.

IS THE ATMOSPHERIC/OCEANIC CIRCULATION CHANGING?

Surprisingly, there is a considerable lack of reliable information about changes in atmospheric circulation, even though it is of daily concern to much of the world since it relates to day-to-day weather changes. Analyses of circulation are performed every day on a routine basis, but the analysis schemes have changed over time, making them of limited use for monitoring climate change. Moreover, even the recent reanalysis efforts by the world's major numerical weather prediction centers, whereby the analysis scheme is fixed over the historical record, contains time-varying biases because of the introduction of data with time-dependent biases and a changing mix of data (e.g., introducing satellite data) over the course of the reanalysis (13). Even less information is available on measured changes and variations in ocean circulation.

A few major atmospheric circulation features have been reasonably well measured because they can be represented by rather simple indices. This includes the El Niño–Southern Oscillation (ENSO) index, the North Atlantic Oscillation (NAO) index, and the Pacific–North American (PNA) circulation pattern index. There are interesting decadal and multidecadal variation, but it is too early to detect any long-term trends. Evidence exists that ENSO has varied in period, recurrence interval, and strength of impact. A rather abrupt change in ENSO and other aspects of atmospheric circulation seems to have occurred around 1976–1977. More frequent ENSOs with rare excursions into its other extreme (La Niña) became much more prevalent. Anomalous circulation regimes associated with ENSO and large-amplitude PNA patterns persisted in the North Pacific

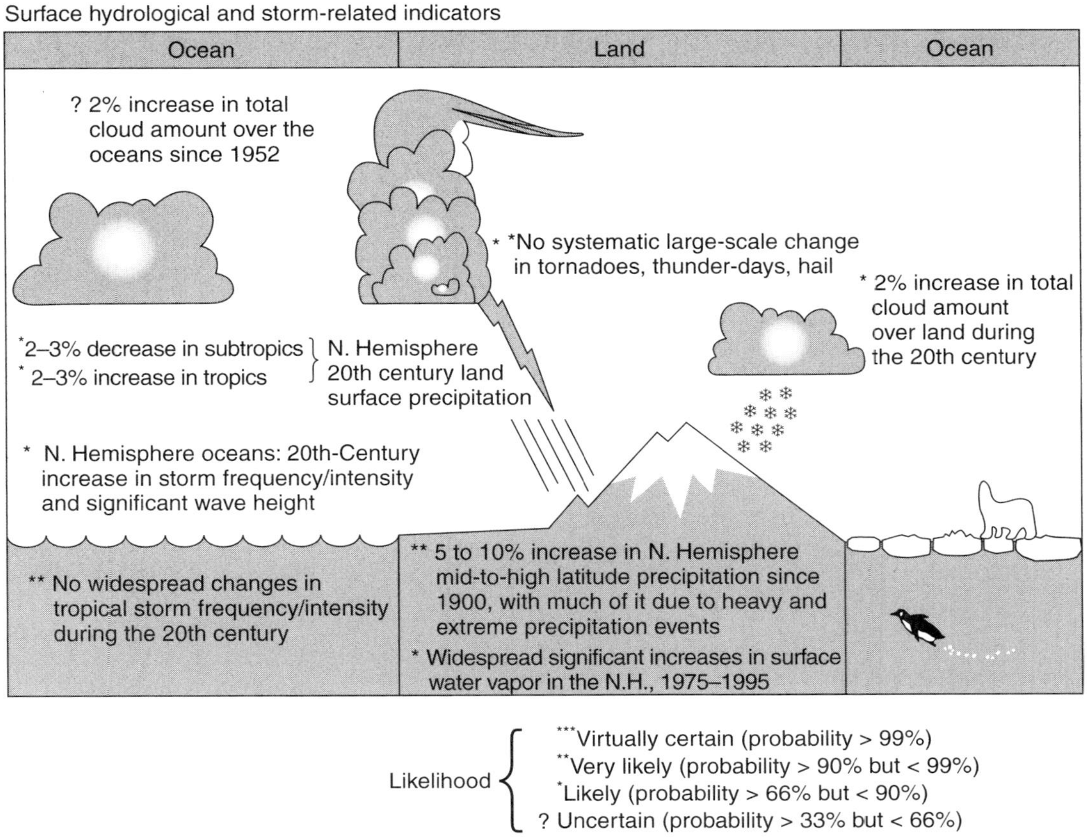

Figure 2. Precipitation trends over land 1900–1999. Trend is expressed in percent per century (relative to the mean precipitation from 1961–1990) and magnitude of trend is represented by area of circle with green reflecting increases and brown decreases of precipitation.

from the late 1970s into the late 1980s, affecting temperature anomalies. Moreover, the NAO has been persistent in its association with strong westerlies into the European continent from the late 1980s until very recently when it abruptly shifted. As a result, temperature anomalies and storminess in Europe have abruptly changed over the past 2 years compared to the past 7 or 8 years.

Increases in the strength of the Southern Hemisphere circumpolar vortex during the 1980s have been documented (14,15) using station sea level pressure data. This increase was associated with a delayed breakdown in the stratospheric polar vortex and ozone deficit in the Antarctic spring. A near-global sea level pressure data set has been used to identify changes in circulation patterns in the Indian Ocean. Allan et al. (16) and Salinger et al. (17) find that circulation patterns in the periods 1870–1900 and 1950–1990 were more meridional than those in the 1900–1950 period, indicating intensified circulation around anticyclones. These changes may be related to changes in the amplitude of longwave troughs to the south and west of Australia and the Tasman Sea/New Zealand area and a subsequent decrease in precipitation in Southwest Australia (18,19).

IS THE WEATHER AND CLIMATE BECOMING MORE EXTREME OR VARIABLE?

Climate and weather extremes are of great interest. Due to inadequate monitoring as well as prohibitively expensive access to weather and climate data held by the world's national weather and environmental agencies, only limited reliable information is available about large-scale changes in extreme weather or climate variability. The time-dependent biases that affect climate means are even more difficult to effectively eliminate from the extremes of the distributions of various weather and climate elements. There are a few areas, however, where regional and global changes in weather and climate extremes have been reasonably well documented (20).

Interannual temperature variability has not changed significantly over the past century. On shorter time scales and higher frequencies, e.g., days to a week, temperature variability may have decreased across much of the Northern Hemisphere (8). Related to the decrease in high-frequency temperature variability there has been a tendency for fewer low-temperature extremes, but widespread changes in extreme high temperatures have not been noted.

Trends in intense rainfall have been examined for a variety of countries. Some evidence suggests an increase in intense rainfalls (United States, tropical Australia, Japan, and Mexico), but analyses are far from complete and subject to many discontinuities in the record. The strongest increases in extreme precipitation are documented in the United States and Australia (21).

Intense tropical cyclone activity may have decreased in the North Atlantic, the one basin with reasonably consistent tropical cyclone data over the twentieth century, but even here data prior to World War II is difficult to assess regarding tropical cyclone strength. Elsewhere, tropical cyclone data do not reveal any long-term trends, or if they do they are most likely a result of inconsistent analyses. Changes in meteorological assimilation schemes have complicated the interpretations of changes in extratropical cyclone frequency. In some regions, such as the North Atlantic, a clear trend in activity has been noted, as also in significant wave heights in the northern half of the North Atlantic. In contrast, decreases in storm frequency and wave heights have been noted in the south half of the North Atlantic over the past few decades. These changes are also reflected in the prolonged positive excursions of the NAO since the 1970s.

IS THE RADIATIVE FORCING OF THE PLANET CHANGING?

Understanding requires a time history of forcing global change. The atmospheric concentration of CO_2, an important greenhouse gas because of its long atmospheric residence time and relatively high atmospheric concentration, has increased substantively over the past few decades. This is quite certain as revealed by precise measurements made at the South Pole and at Mauna Loa Observatory since the late 1950s, and from a number of stations distributed globally that began operating in subsequent decades. Since atmospheric carbon dioxide is a long-lived atmospheric constituent and it is well mixed in the atmosphere, a moderate number of well-placed stations operating for the primary purpose of monitoring seasonal to decadal changes provides a very robust estimate of global changes in carbon dioxide.

To understand the causes of the increase of atmospheric carbon dioxide, the carbon cycle and the anthropogenic carbon budget must be balanced. Balancing the carbon budget requires estimates of the sources of carbon from anthropogenic emissions from fossil fuel and cement production, as well as the net emission from changes in land use (e.g., deforestation). These estimates are derived from a combination of modeling, sample measurements, and high-resolution satellite imagery. It also requires measurements for the storage in the atmosphere, the ocean uptake, and uptake by forest regrowth, the CO_2 and nitrogen fertilization effect on vegetation, as well as any operating climate feedback effects (e.g., the increase in vegetation due to increased temperatures). Many of these factors are still uncertain because of a paucity of ecosystem measurements over a sustained period of time. Anthropogenic emissions from the burning of fossil fuel and cement production are the primary cause of the atmospheric increase.

Several other radiatively important anthropogenic atmospheric trace constituents have been measured for the past few decades. These measurements have confirmed significant increases in atmospheric concentrations of CH_4, N_2O, and the halocarbons including the stratospheric ozone destructive agent of the chloroflourocarbons and the bromocarbons. Because of their long lifetimes, a few well-placed high-quality *in situ* stations have provided a good estimate of global change. Stratospheric ozone depletion has been monitored both by satellite and ozonesondes. Both observing systems have been crucial in ascertaining changes of stratospheric ozone that was originally of interest, not because of its role as a radiative forcing agent, but its ability to absorb ultraviolet (UV) radiation prior to reaching Earth's surface. The combination of the surface- and space-based observing systems has enabled much more precise measurements than either system could provide by itself. Over the past few years much of the ozonesonde data and satellite data has been improved using information about past calibration methods, in part because of differences in trends between the two observing systems.

Figure 3 depicts the IPCC (9) best estimate of the radiative forcing associated with various atmospheric constituents. Unfortunately, measurements of most of the forcings other than those already discussed have low or very low confidence, not only because of our uncertainty about their role in the physical climate system, but because we have not adequately monitored their change. For example, estimates of changes in sulfate aerosol concentrations are derived from model estimates of source emissions, not actual atmospheric concentrations. The problem is complicated because of the spatially varying concentrations of sulfate due to its short atmospheric lifetime. Another example is measurements of solar irradiance, which have been taken by balloons and rockets for several decades, but continuous measurements of top-of-the-atmosphere solar irradiance did not begin until the late 1970s with the *Nimbus*

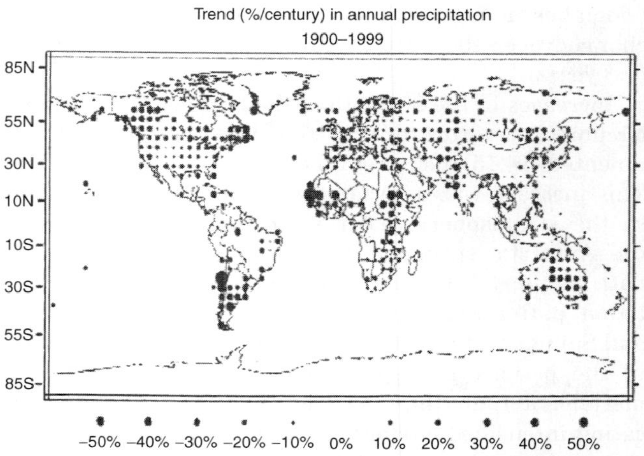

Figure 3. Estimates of globally and annually averaged radiative forcing (in W/m_{-2}) for a number of agents due to changes in concentrations of greenhouse gases and aerosols and natural changes in solar output from 1750 to the present day. Error bars are depicted for all forcings (from Ref. 1).

7 and the Solar Maximum Mission satellites. There are significant absolute differences in total irradiance between satellites, emphasizing the critical need for overlaps between satellites and absolute calibration of the irradiance measurements to determine decadal changes. Spectrally resolved measurements will be a key element in our ability to model the effects of solar variability, but at the present time no long-term commitment has been made to take these measurements. Another important forcing that is estimated through measured, modeled, and estimated changes in optical depth relates to the aerosols injected high into the atmosphere by major volcanic eruptions. The aerosols from these volcanoes are sporadic and usually persist in the atmosphere for at most a few years. Improved measurements of aerosol size distribution and composition will help better understand this agent of climate change.

WHAT CAN WE DO TO IMPROVE OUR ABILITY TO DETECT CLIMATE AND GLOBAL CHANGE?

Even after extensive reworking of past data, in many instances we are incapable of resolving important aspects concerning climate and global change. Virtually every monitoring system and data set requires better data quality, continuity, and fewer time-varying biases if we expect to conclusively answer questions about how the planet has changed, because of the need to rely on observations that were never intended to be used to monitor the physical characteristics of the planet of the course of decades. Long-term monitoring, capable of resolving decade-to-century-scale changes, requires different strategies of operation.

In situ measurements are currently in a state of decay, decline, or rapid poorly documented change due to the introduction of automated measurements without adequate precaution to understand the difference between old and new observing systems. Satellite-based systems alone will not and cannot provide all the necessary measurements. Much wiser implementation and monitoring practices must be adopted for both space-based and surface-based observing systems in order to adequately understand global change. The establishment of the Global Climate Observing System (GCOS) is a high priority (22), and continued encouragement by the World Meteorological Organization (WMO) of a full implementation of this system in all countries is critical. Furthermore, in the context of the GCOS, a number of steps can be taken to improve our ability to monitor climate and global change.

These include:

1. Prior to implementing changes to existing environmental monitoring systems or introducing new observing systems, standard practice should include an assessment of the impact of these changes on our ability to monitor environmental variations and changes.
2. Overlapping measurements in time and space of both the old and new observing systems should be standard practice for critical environmental variables.
3. Calibration, validation, and knowledge of instrument, station, and/or platform history are essential for data interpretation and use. Changes in instrument sampling time, local environmental conditions, and any other factors pertinent to the interpretation of the observations and measurements should be recorded as a mandatory part of the observing routine and be archived with the original data. The algorithms used to process observations must be well documented and available to the scientific community. Documentation of changes and improvements in the algorithms should be carried along with the data throughout the data archiving process.
4. The capability must be established to routinely assess the quality and homogeneity of the historical

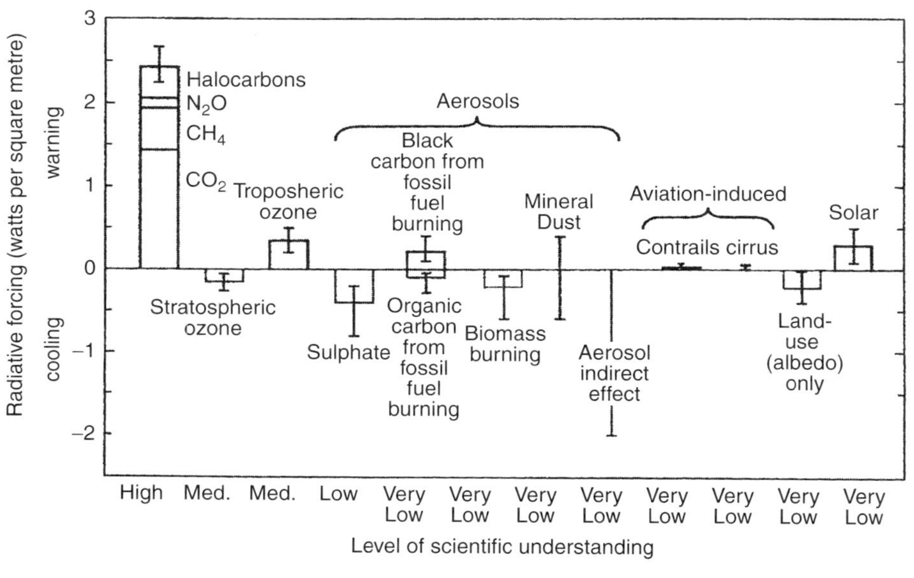

Figure 4. Global, annual-mean radiative forcings (Wm^{-2}) due to a number of agents for the period from pre-industrial (1750) to the present. The height of the vertical bar denotes the central or "best" estimate, no bar indicates that it is not possible to provide a "best" estimate. The vertical line indicates an estimate of the uncertainty range and the level of scientific understanding is a subjective judgement about the reliability of the forcing estimate based on such factors as assumptions, degree of knowledge of the physical/chemical mechanisms, etc. (From Ref. 1).

database for monitoring environmental variations and change, including long-term high-resolution data capable of resolving important extreme environmental events.

5. Environmental assessments that require knowledge of environmental variations and change should be well integrated into a global observing system strategy.
6. Observations with a long uninterrupted record should be maintained. Every effort should be made to protect the data sets that document long-term homogeneous observations. Long term may be a century or more. A list of prioritized sites or observations based on their contribution to long-term environmental monitoring should be developed for each element.
7. Data-poor regions, variables, regions sensitive to change, and key measurements with inadequate temporal resolution should be given the highest priority in the design and implementation of new environmental observing systems.
8. Network designers, operators, and instrument engineers must be provided environmental monitoring requirements at the outset of network design. This is particularly important because most observing systems have been designed for purposes other than long-term monitoring. Instruments must have adequate accuracy with biases small enough to resolve environmental variations and changes of primary interest.
9. Much of the development of new observation capabilities and much of the evidence supporting the value of these observations stem from research-oriented needs or programs. Stable, long-term commitments to these observations, and a clear transition plan from research to operations, are two requirements in the development of adequate environmental monitoring capabilities.
10. Data management systems that facilitate access, use, and interpretation are essential. Freedom of access, low cost, mechanisms that facilitate use (directories, catalogs, browse capabilities, availability of metadata on station histories, algorithm accessibility and documentation, on-line accessibility to data, etc.), and quality control should guide data management. International cooperation is critical for successful management of data used to monitor long-term environmental change and variability.

BIBLIOGRAPHY

1. IPCC (2001). *Climate Change, 2001: The Scientific Basis. Contribution of Working Group 1 to the Third Assessment Report of the Intergovernmental Panel on Climate Change*. J.T. Houghton, Y. Ding, D.J. Griggs, M. Noguer, P.J. van der Linden, X. Dai, K. Maskell, and C.A. Johnson (Eds.). Cambridge University Press, New York.
2. Spencer, R.W. and Christy, J.R. (1992). Precision and radiosonde validation of satellite gridpoint temperature anomalies, Part I: MSU channel 2. *J. Climate* **5**: 847–857.
3. Spencer, R.W. and Christy, J.R. (1992). Precision and radiosonde validation of satellite gridpoint temperature anomalies, Part II: A tropospheric retrieval and trends during 1979–90. *J. Climate* **5**: 858–866.
4. Reynolds, R.W. (1988). A real-time global sea surface temperature analysis. *J. Climate* **1**: 75–86.
5. National Research Council (NRC) (2000). *Reconciling Observations of Global Temperature Change, Report of the Panel on Reconciling Temperature Observations*. National Academy Press, Washington, DC.
6. Luers, J.K. and Eskridge, R.E. (1995). Temperature corrections for the VIZ and Vaisala radiosondes. *Appl. Meteor.* **34**: 1241–1253.
7. Sevruk, B. (1982). Methods of correcting for systematic error in point precipitation measurements for operational use. *Hydrology Rep.* **21**. World Meteorological Organization, Geneva, 589.
8. Karl, T.R., Knight, R.W., and Plummer, N. (1995). Trends in high-frequency climate variability in the twentieth century. *Nature* **377**: 217–220.
9. IPCC (1992). *Climate Change, 1992, Supplementary Report*, WMO/UNEP. J.T. Houghton, B.A. Callander, and S.K. Varney (Eds.). Cambridge University Press, New York, pp. 62–64.
10. Spencer, R.W. (1993). Global oceanic precipitation from the MSU during 1979–1992 and comparison to other climatologies. *J. Climate* **6**: 1301–1326.
11. Huffman, G.J., Adler, R.F., Arkin, P., Chang, A., Ferraro, R., Gruber, A., Janowiak, J., McNab, A., Rudolf, B., and Schneider, U. (1997). The global precipitation climatology project (GPCP) combined precipitation dataset. *Bull. Am. Meteor. Soc.* **78**: 5–20.
12. Rossow, W.B. and Cairns, B. (1995). Monitoring changes in clouds. *Climatic Change* **31**: 175–217.
13. Trenberth, K.E. and Guillemot, C.J. (1997). Evaluation of the atmospheric moisture and hydrologic cycle in the NCEP reanalysis. *Clim. Dyn.* **14**: 213–231.
14. van Loon, H., Kidson, J.W., and Mullan, A.B. (1993). Decadal variation of the annual cycle in the Australian dataset. *J. Climate* **6**: 1227–1231.
15. Hurrell, J.W. and van Loon, H. (1994). A modulation of the atmospheric annual cycle in the Southern Hemisphere. *Tellus* **46A**: 325–338.
16. Allan, R.J., Lindesay, J.A., and Reason, C.J.C. (1995). Multidecadal variability in the climate system over the Indian Ocean region during the austral summer. *J. Climate* **8**: 1853–1873.
17. Salinger, M.J., Allan, R., Bindoff, N., Hannah, J., Lavery, B., Leleu, L., Lin, Z., Lindesay, J., MacVeigh, J.P., Nicholls, N., Plummer, N., and Torok, S. (1995). Observed variability and change in climate and sea level in Oceania. In: *Greenhouse: Coping with Climate Change*. W.J. Bouma, G.I. Pearman, and M.R. Manning (Eds.). CSIRO, Melbourne, Australia, 100–126.
18. Nicholls, N. and Lavery, B. (1992). Australian rainfall trends during the twentieth century. *Int. J. Climatology* **12**: 153–163.
19. Allan, R.J. and Haylock, M.R. (1993). Circulation features associated with the winter rainfall decrease in southwestern Australia. *J. Climate* **6**: 1356–1367.
20. Easterling, D.R., Meehl, G., Parmesan, C., Changnon, S., Karl, T., and Mearns, L. (2000). Climate extremes: observations, modeling, and impacts. *Science* **289**: 2068–2074.

21. Easterling, D.R., Evans, J.L., Groisman, P.Ya., Karl, T.R., Kunkel, K.E., and Ambenje, P. (2000). Observed variability and trends in extreme climate events: A brief review. *Bull. Am. Meteor. Soc.* **81**: 417–426.

22. Spence, T. and Townsend, J. (1995). The Global Climate Observing System (GCOS). *Climatic Change* **31**: 131–134.

READING LIST

Elliott, W.P. (1995). On detecting long-term changes in atmospheric moisture. *Climatic Change* **31**: 219–237.

Groisman, P.Y., and Legates, D.R. (1995). Documenting and detecting long-term precipitation trends: Where we are and what should be done. *Climatic Change* **31**: 471–492.

Hurrell, J.W. and Trenberth, K.E. (1996). Satellite versus surface estimates of air temperature since 1979. *J. Climate* **9**: 2222–2232.

OVERVIEW: THE CLIMATE SYSTEM

ROBERT E. DICKINSON

(from *The Handbook of Weather, Climate, and Water: Dynamics, Climate, Physical Meteorology, Weather Systems, and Measurements*, Wiley 2003)

The climate system consists of the atmosphere, cryosphere, oceans, and land interacting through physical, chemical, and biological processes. Key ingredients are the hydrological and energy exchanges between subsystems through radiative, convective, and fluid dynamical mechanisms. Climate involves changes on seasonal, year-to-year, and decadal or longer periods in contrast to day-to-day weather changes. However, extreme events and other statistical measures are as, or more, important than simple averages. Climate is seen to impact human activities most directly through the occurrence of extremes. The frequency of particular threshold extremes, as, for example, the number of days with maximum temperatures above 100 °F, can change substantially with shifts in climate averages.

THE ATMOSPHERE

The atmosphere is described by winds, pressures, temperatures, and the distribution of various substances in gaseous, liquid, and solid forms. Water is the most important of these substances. Also important are the various other radiatively active ("greenhouse") gases, including carbon dioxide and liquid or solid aerosol particulates. Most of the mass of the atmosphere is in the troposphere, which is comprised of the layers from the surface to about 12 km (8 km in high latitudes to 16 km at the equator) where the temperature decreases with altitude. The top of the troposphere is called the tropopause. Overlying this is the stratosphere, where temperatures increase with altitude to about 50 km or so (Fig. 1). The tropospheric temperature decreases with altitude are maintained by vertical mixing driven by moist and dry convection. The temperature increases with altitude in the stratosphere in response to increasing heating per the unit mass by ozone absorption of ultraviolet radiation. The variation of temperature structure with latitude is indicated in Fig. 2. The troposphere is deepest in the tropics because most thunderstorms occur there. Because of this depth and stirring by thunderstorms, the coldest part of the atmosphere is the tropical tropopause. In the lower troposphere temperatures generally decrease from the equator to pole, but warmest temperatures shift toward the summer hemisphere, especially in July. Longitudinally averaged winds are shown in Fig. 3. Because of the geostrophic balance between wind and pressures, winds increase with altitude where temperature decreases with latitude. Conversely, above about 8 km, where temperatures decrease toward the tropical tropopause, the zonal winds decrease with altitude. The core of maximum winds is referred to as the jet stream. The jet stream undergoes large wavelike oscillations in longitude and so is usually stronger at a given latitude than in its longitudinal average. These waves are especially noticeable in the winter hemisphere as illustrated in Fig. 4.

GLOBAL AVERAGE ENERGY BALANCE

Solar radiation of about 342 W/m^{-2} entering Earth's atmosphere is absorbed and scattered by molecules. The major gaseous absorbers of solar radiation are water vapor in the troposphere and ozone in the stratosphere. Clouds and aerosols likewise scatter and absorb. Clouds are the dominant scatterer and so substantially enhance the overall planetary reflected radiation, whose ratio to incident solar radiation, about 0.31, is referred to as *albedo*. Thermal infrared radiation, referred to as *longwave*, is controlled by clouds, water vapor, and other greenhouse gases. Figure 5 (4) illustrates a recent estimate of the various terms contributing to the global

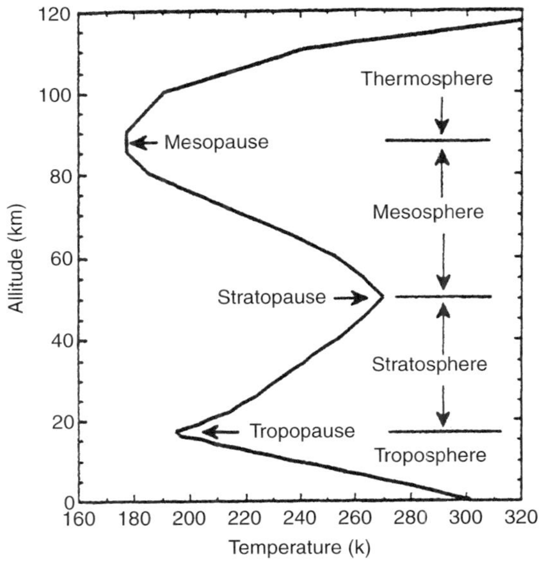

Figure 1. Main zones of the atmosphere defined according to the temperature profile of the standard atmosphere profile at 15°N for annual-mean conditions (1).

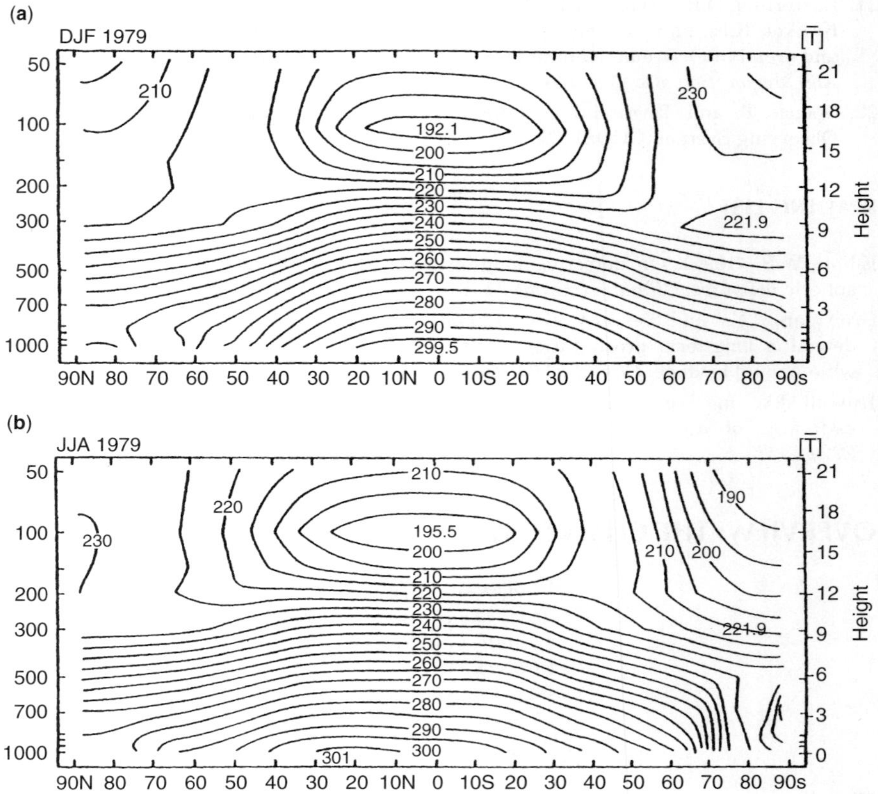

Figure 2. Zonal mean meridional cross sections of temperature during two seasons. (*a*) December 1978–February 1979 and (*b*) June–August 1979. Height scale is approximate and the contour interval is 5 K (2).

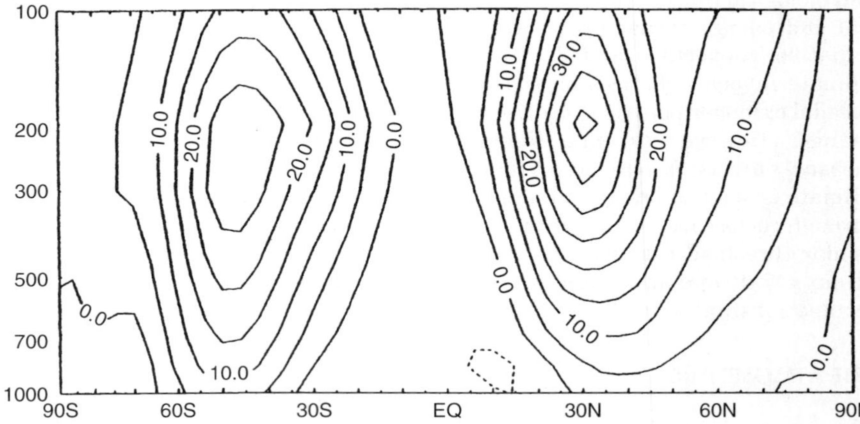

Figure 3. Meridional cross sections of longitudinally averaged zonal wind (top panels, m/s) for DJF (3).

energy balance. The latent heat from global average precipitation of about 1.0 m per year is the dominant nonradiative heating term in the atmosphere.

Because of the seasonally varying geometry of Earth relative to the sun, and the differences in cloudiness and surface albedos, there are substantial variations in the distribution of absorbed solar radiation at the surface and in the atmosphere, as likewise in the transfer of latent heat from the surface to the atmosphere. This heterogeneous distribution of atmospheric heating drives atmospheric wind systems, either directly or through the creation of available potential energy, which is utilized to maintain random occurrences of various kinds of instabilities, such as thunder-storms and wintertime cyclonic storm systems. These dynamical systems hence act to redistribute energy within the atmosphere and so determine the distributions of temperature and water vapor. Likewise, the balances at the surface between fluxes of radiative, latent, and thermal energies determine surface temperatures and soil moistures. The properties of the near-surface air we live in are determined by a combination of surface and atmospheric properties, according to processes of the atmospheric boundary layer. Thus climatic anomalies in surface air may occur either because of some shift in atmospheric circulation patterns or through some modification of surface properties such as those accompanying deforestation or the development of an urban area.

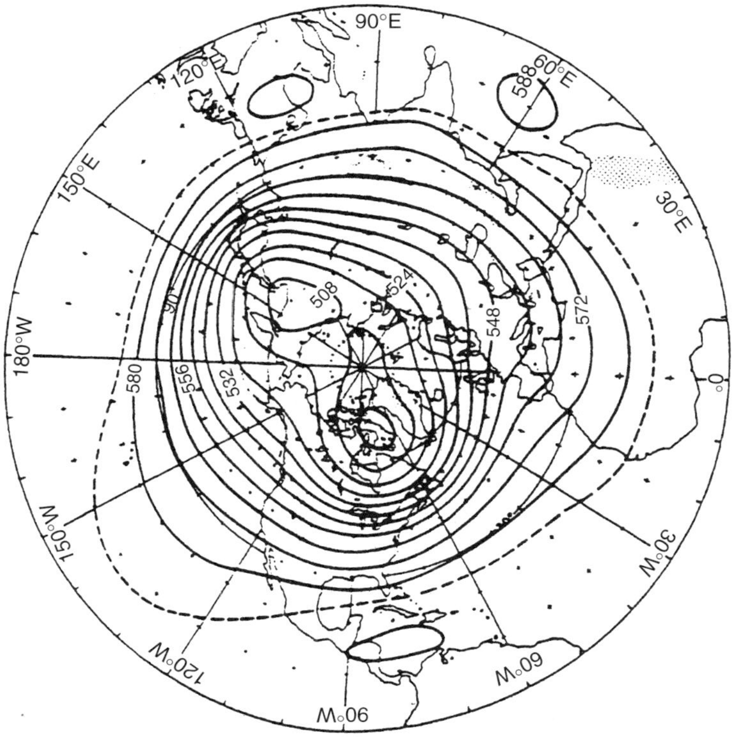

Figure 4. Mean 500-mb contours in January, Northern Hemisphere. Heights shown in tens of meters (3).

THE ATMOSPHERIC BOUNDARY LAYER

The term *boundary layer* is applied in fluid dynamics to layers of fluid or gas, usually relatively thin, determining the transition between some boundary and the rest of the fluid. The atmospheric boundary layer is the extent of atmosphere that is mixed by convective and mechanical stirring originating at Earth's surface. Such stirring is commonly experienced by airplane travelers as the bumps that occur during takeoff or landing, especially in the afternoon, or as waves at higher levels in flying over mountainous regions. The daytime continental boundary layer, extending up to several kilometers in height, is most developed and vigorously mixed, being the extent to which the daytime heating of the surface drives convective overturning of the atmosphere. The land cools radiatively at night, strongly stabilizing the atmosphere against convection, but a residual boundary layer extends up to about 100 m stirred by the flow of air over the underlying rough surface. This diurnal variation of fluxes over the ocean is much weaker and the boundary layer is of intermediate height. The temperature of the atmosphere, when stirred by dry mixing, decreases at a rate of 9.8 K/km. Above the boundary layer, temperatures decrease less rapidly with height, so that the atmosphere is stable to dry convection. A layer of clouds commonly forms at the top of the daytime and oceanic boundary layers and contributes to the convection creating the boundary layer through its radiative cooling (convection results from either heating at the bottom of a fluid or cooling at its top). Also, at times the clouds forming near the top of the boundary layer can be unstable to moist convection, and so convect upward through a deep column such as in a thunderstorm.

ATMOSPHERIC HYDROLOGICAL CYCLE

The storage, transport, and phase changes of water at the surface and in the atmosphere are referred to as the hydrological cycle. As already alluded to, the hydrological cycle is closely linked to and driven by various energy exchange processes at the surface and in the atmosphere. On the scale of continents, water is moved from oceans to land by atmospheric winds, to be carried back to the oceans by streams and rivers as elements of the land hydrological cycle. Most of the water in the atmosphere is in its vapor phase, but water that is near saturation vapor pressure (relative humidity of 100%) converts to droplets or ice crystals depending on temperature and details of cloud physics. These droplets and crystals fall out of the atmosphere as precipitation. The water lost is replenished by evaporation of water at the surface and by vertical and horizontal transport within the atmosphere. Consequently, much of the troposphere has humidities not much below saturation. Saturation vapor pressure increases rapidly with temperature (about 10% per kelvin of change). Hence, as illustrated in Fig. 6, the climatological concentrations of water vapor vary from several percent or more when going from near-surface air to a few parts per million near the tropical tropopause. Water vapor concentrations in the stratosphere are close to that of the tropical tropopause, probably because much of the air in the lower stratosphere has been pumped through the tropical tropopause by moist convection.

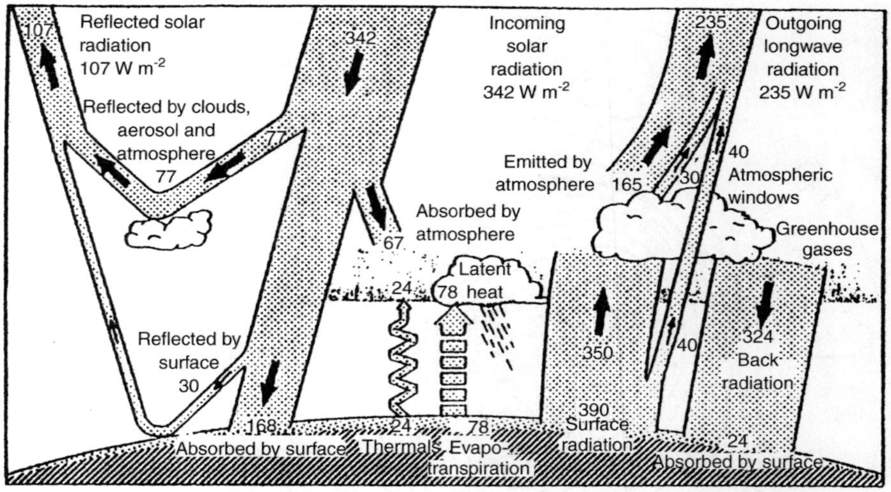

Figure 5. Earth's annual global mean energy budget based on the present study. Units are W/m^2 (4).

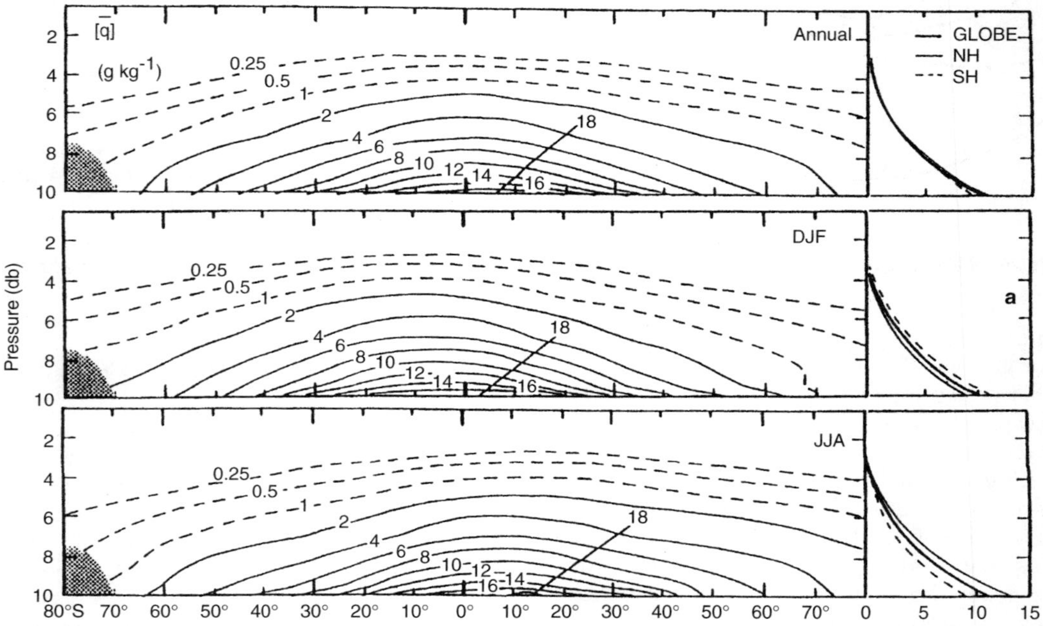

Figure 6. Zonal mean cross sections of the specific humidity in g/kg for annual, DJF, and JJA mean conditions. Vertical profiles of hemispheric and global mean values are shown on the right (5).

CLIMATE OF THE STRATOSPHERE

The dominant radiative processes in the stratosphere are the heating by absorption of solar ultra violet (UV) radiation and cooling by thermal infrared emission from carbon dioxide and, to a lesser extent, ozone molecules. The stratospheric absorption of UV largely determines how much harmful UV reaches the surface. Ozone in the upper troposphere and lower stratosphere additionally adds heat by absorption of thermal emission from the warmer surface and lower layers. The stratosphere, furthermore, enhances the greenhouse warming of CO_2 in the troposphere through substantial downward thermal emissions to the troposphere.

How changes of ozone change stratospheric and tropospheric radiative heating depends on the amounts of overlying ozone and, for thermal effect, on pressure and radiative upwelling depending on underlying temperatures.

Besides radiative processes, stratospheric climate is characterized by its temperature and wind patterns and by the chemical composition of its trace gases. At mid-stratosphere, temperature increases from winter pole to summer pole with an accompanying eastward jet stream in the winter hemisphere extending upward from the tropospheric jet steam. This wind configuration allows planetary wave disturbances to propagate into the stratosphere, contributing significant temporal and longitudinal variabilities. Conversely, the westward jet, found in the summer stratosphere attenuates wave disturbances from below, and so is largely zonally symmetric, changing only with the seasonal heating patterns.

THE CRYOSPHERE

The term *cryosphere* refers to the components of the climate system dominated by water in its frozen phase, that is, in high latitudes and extratropical winter conditions. Elements include snow, its distribution and depths, sea ice, its distribution and properties, high-latitude ice caps, and temperate glaciers. The largest volume of frozen water is stored in ice caps, and glaciers. This storage acts to remove water from the oceans. How it changes with climate change is, hence, of interest for determining changing sea levels.

Ice is highly reflective of sunlight, especially in crystal form. The loss of solar heating because of this high albedo acts to substantially reduce high-latitude temperatures especially in spring and early summer where near-maximum solar radiation sees white snow-covered surfaces. This high albedo can be substantially masked by cloud cover and, over land, tall vegetation such as conifer forests.

THE OCEAN

Oceans are a major factor in determining surface temperatures and fluxes of water into the atmosphere. They store, release, and transport thermal energy, in particular, warming the atmosphere under wintertime and high-latitude conditions, and cooling it under summer and tropical conditions.

How the oceans carry out these services depends on processes coupling them to the atmosphere. Atmospheric winds push the oceans into wind-driven circulation systems. Net surface heating or cooling, evaporation, and precipitation determine oceanic densities through controlling temperature and salinity, hence oceanic buoyancy. This net distribution of buoyancy forcing drives "thermohaline" overturning of the ocean, which acts to transport heat. Climate of the surface layers of the ocean includes the depth to which waters are stirred by waves and net heating or cooling. Heating acts to generate shallow warm stable layers, while cooling deepens the surface mixed layers. Under some conditions, convective overturning of cold and/or high-salinity water can penetrate to near the ocean bottom.

BIBLIOGRAPHY

1. Hartmann, D.L. (1994). Atmospheric temperature, Chapter 1.2. In: *Global Physical Climatology*. Academic Press, San Diego.
2. Grotjahn, R. (1993). Zonal average observations, Chapter 3. In: *Global Atmospheric Circulations: Observations and Theories*. Oxford University Press, New York.
3. Holton, J.R. (1992). The observed structure of extratropical circulations, Chapter 6.1. In: *An Introduction to Dynamic Meteorology*. Academic, San Diego.
4. Kiehl, J.T. and Trenberth, K.E. (1997). Earth's annual global mean energy budget. *J. Clim.* **78**: 197–208.
5. Peixoto, J.P. and Oort, A.H. (1992). Observed atmospheric branch of the hydrological cycle, Chapter 12.3. *Physics of Climate*. American Institute of Physics, New York.

CLIMATE AND SOCIETY

Michael H. Glantz
(from *The Handbook of Weather, Climate, and Water: Atmospheric Chemistry, Hydrology, and Societal Impacts*, Wiley 2003)

At the turn of the twentieth century, scholars who wrote about the interplay between climate and society did so based on their perceptions of climate as a boundary constraint for the development prospects of a society. Perceptions of climate were used as an excuse to dominate societies in Africa, Asia, and Latin America. As a result, climate–society studies soon became viewed as a colonial ploy to control populations in developing areas in the tropics. Perhaps the most cited book in this regard was written by Ellsworth Huntington, *Climate and Civilization*, published in 1915 (1). In his view, inhabitants of the tropics were destined to accept lower levels of economic and social development because their climate setting was not conducive to lively (i.e., productive) human activity or an aggressive work ethic. According to Huntington, tropical climate was the main culprit causing people in the tropics to be less productive than people in temperate regions. Huntington argued that the temperate climate has an energizing effect on humans. With the growing belief that such an argument was racist in intent, Huntington's work was challenged, and discussion of the various ways in which climate might influence human behavior was stifled for decades, notwithstanding a few notable exceptions. One such exception is entitled *Climate and the Energy of Nations* (2) in which Markham referred to the "air-conditioning revolution," a revolution based on the development and spread of a new technology into the tropical areas. Markham asserted that technology brings islands of temperate-zone climate into the tropics, thereby generating a more aggressive work ethic.

Following the end of World War II and the onset of the Cold War between Soviet-style communism and Western capitalism and democracy, attention of governments turned to Cold War conflicts, avoidance of nuclear war, searches for allies, and decolonization. The major Cold War nations were in a competition to show that *their* approach to economic development was the only way for the newly independent countries to follow. A main stated objective was their intent to assist these countries to become food secure based on the nation's resources. Consideration of climate was making its way back into the discussions of economic development in developing countries. Once again interest was raised with regard to climatic constraints to economic development in tropical countries.

In the 1950s and 1960s, attention focused on decolonization and political development of the newly independent states (e.g., 3). In the mid-1970s, a World Bank report about the economic prospects for developing countries—*The Tropics and Economic Development: A Provocative Inquiry into the Poverty of Nations*—hinted at the economic, social, and political problems caused by climate variability from one year to the next. Its author (4) noted that recurrent droughts in northeast Brazil are a chronic constraint on the region's economic development prospects. His reference to interannual climate variability was brief and unelaborated. However, climate as a boundary constraint was starting to give way to climate as something that societies might be able to forecast and cope with, at least in its extremes.

In the 1970s, attention focused on how the vagaries of weather exposed hundreds of millions of people to hunger and, depending on the socioeconomic situation in a particular country, to famine as well (e.g., 5,6). Thus, there was a growing number of examples of the notion that climate was not really a boundary constraint to the level of development that a people or culture could expect to attain. This notion began to give way to the belief that variability in climate, from one year to the next or one decade to the next, could be coped with so as to soften the impacts of climate variability and weather extremes on agriculture and livestock and, more generally, on the productivity of the land's surface (e.g., 7,8).

Recall that the 1970s was a disruptive decade with respect to climate: 5 years of drought in the West African Sahel (5); failure of the Soviet harvest and subsequent large-scale, low-cost grain purchases by the Soviet Union in the early 1970s (9); the global food crisis (10); talk of a possible return to an ice age (e.g., 11,12); the Ethiopian famine (13); drought-related coups in sub-Saharan Africa; drought in the wheat-producing Canadian prairie provinces (7); the first drop in global fish catches since the end of World War II (10), and so forth.

A devastating 5-year drought from 1968 to 1973 in the West African Sahel and its associated death and environmental destruction in the region drew attention to the impacts on household and village responses to prolonged, multiyear droughts. Widespread droughts around the globe in 1972–1973, famines in West Africa and Ethiopia, blamed for the most part on an El Niño event, along with the drop in fish landings, prompted the U.N. Secretary General to convene a series of UN-sponsored world conferences on food (1974), population (1974), human settlements (1976), water (1977), desertification (1977), climate (1979) and technology (1979).

Thus, toward the middle of the 1970s, at least five new major climate-related scientific issues emerged: the effect of chlorofluorocarbons (CFCs) on the ozone layer in the stratosphere, talk of an impending Ice Age suddenly shifted to talk of a human-induced global warming, acid rain, desertification, and El Niño. Each of these issues raised interest in climate–society interactions to higher levels among researchers in different disciplines, government agencies, economic sectors, the media, and the public. Societies around the globe responded (and continue to respond) in different ways to each of these climate-related issues. For example, desertification is an environmental issue that is of great concern to African countries.

North Americans, however, refused to accept the view that desertification could occur in the U.S. West as a result of mismanagement of the land's surface, while noting that *desertification* was the plight of poor developing countries in Africa. The term desertification first appeared in a report on the destruction of dry forests in central Africa by a French forester (14). Since then, the concept of desertification has been expanded to include such land degradation processes as soil erosion, wind deflation, soil salinization, water logging, livestock overgrazing, and soil trampling. While many of these processes were exposed during the prolonged drought in the West African Sahel and then labeled as desertification, it is not difficult to show that similar processes also take place in the U.S. West.

The acid rain issue was addressed in the United States with the implementation by the U.S. Congress of a decade-long national assessment called NAPAP (National Acid Precipitation Assessment Program). Stratospheric ozone depletion was addressed globally in the 1980s with the development of international legal instruments culminating in the Montreal Protocol of 1987 and, later, amendments to it (15).

It was in the early 1970s, 1972–1973 to be exact, that an El Niño event (defined briefly as an invasion of warm water from the Western Pacific into the central and eastern equatorial Pacific Ocean) attracted global attention. An event in 1982–1983, the biggest in a century until that time, captured the full attention of the scientific community and various governments as a natural phenomenon that spawned hazards around the globe. Such hazards included, but were not limited to, droughts, floods, frosts, fires and food shortages, famine, and disease. Ever since the mid-1970s, research funding of El Niño–related research has been growing along with international interest in the phenomenon and its societal and environmental impacts. The extraordinary El Niño event of 1997–1998 helped to make El Niño and its cold counterpart, La Niña, household words throughout much of the world. Only at the end of the twentieth century did La Niña events become of serious interest to the El Niño research and forecasting communities (16). This belated interest is even more surprising given the scientific observation that tropical storms and hurricanes in the Atlantic Basin and in the Gulf of Mexico tend to increase in number during La Niña events and drop in number during El Niño events.

Global warming is an environmental issue that arose out of discussions and governmental and scientific concerns about the possibility of a global cooling. It was first suggested in 1896 by Swedish chemist Arrhenius (17,18) that the burning of coal by human societies would add enough extra carbon dioxide into the air to eventually heat up Earth's atmosphere by a few degrees Celsius. This issue was revisited in the 1930s by Callendar (19), who thought that a human-induced global warming of the atmosphere could stave off the imminent recurrence of an ice age. The issue was again revisited in the 1950s when global warming was looked at in neutral terms, as an experiment that societies were performing on the chemistry of the atmosphere, for which the outcome is unknown (20).

It was not until the mid-1970s that human-induced global warming began to be viewed as an adverse event for future generations of human societies and ecosystems that might not be able to adapt to the rate of warming expected to occur. The cause of the warming was attributed to the increasing amounts of greenhouse gases (CO_2, CFCs, CH_4, NO_x, collectively referred to as GHGs) being emitted into the atmosphere as a result of human activities. Carbon dioxide is a product of the burning of fossil fuels, and its amount in the atmosphere has been rising since the onset of the Industrial Revolution in the late 1700s. Tropical deforestation also contributes carbon dioxide to the atmosphere. Tropical forests have served as sinks for carbon dioxide, pulling it out of the air and storing it. When trees are felled, decompose or burned, the stored carbon is emitted into the air.

Chlorofluorocarbons (CFCs), a greenhouse gas as well as a stratospheric "ozone eater," are man-made chemicals first discovered in the 1920s for use as a refrigerant. Methane resulting from livestock rearing (e.g., cattle, pigs) and from the increasing number of landfills is another greenhouse gas. Nitrous oxides are used by farmers in fertilizers and have been widely applied to agricultural lands around the globe in increasing amounts since the end of World War II. Of these major greenhouse gases, carbon dioxide is seen at the main culprit in the global warming debate.

Current scientific research suggests that the level of climate change that might be expected (at current rates of greenhouse gas emissions) is on the order of 1.5 to 4.5 °C by the end of the twenty-first century (21–23). Concerned with the prospects of a changing global climate, many nations have come together to call for a technical assessment of the state of the science through the Intergovernmental Panel on Climate Change (IPCC).

The degree of warming, however, is dependent on numerous factors: the rate at which GHGs continue to be emitted into the atmosphere, the shift by societies to alternative energy sources, the rate of tropical deforestation, the residence time of GHGs in the atmosphere (several of these gases will remain in the atmosphere for decades to centuries), the development of methods to sequester carbon (i.e., taking it from the atmosphere and binding it in some way in Earth's land surface, vegetation, or oceans), and so forth. Some degree of global warming is inevitable, given the residence time of the GHGs already emitted into the atmosphere. This means that societies around the globe, from local to national, must attempt to ascertain how a warmer global climate regime might affect regional and local climates. Will there be more extreme climate events (such as droughts, floods, frosts, fires) or fewer? These societies must also seriously consider nationally, as well as collectively in cooperation with other countries, the most effective way(s) to cope with the potentially adverse impacts of some degree of human-induced global warming.

Coping mechanisms for climate change likely to occur decades in the future can be divided into three categories: preventive, mitigative and adaptive measures. *Preventive* measures are designed to prevent the increased buildup of GHGs in human-induced global warming, acid rain, desertification, and El Niño. Each of these issues raised interest in climate–society interactions to higher levels among researchers in different disciplines, government agencies, economic sectors, the media, and the public. Societies around the globe responded (and continue to respond) in different ways to each of these climate-related issues. For example, desertification is an environmental issue that is of great concern to African countries.

Climate-related surprise is not a black-and-white condition. People are hardly ever either totally surprised or never surprised. There are shades of surprise with regard to human responses to the same climate-related event. They can be hardly surprised, mildly surprised, somewhat surprised, very surprised, extremely surprised or totally surprised (NB: each of these examples was taken from the scientific literature). Myers (24, p. 358) introduced the interesting notion of "semisurprised." Thus, surprise may best be described in "fuzzy" terms with the degree of surprise dependent on several intervening variables such as personal experience, core beliefs, expectations, or knowledge about a phenomenon or about a geographic location.

One could argue that there are knowable as well as unknowable surprises (25). Knowable refers to the fact that some climate surprises can be anticipated (24). For example, certain parts of the globe are drought prone. It is known that drought will likely recur. What is not known is exactly when it will take place, how long it will last, how intense it will be, or where its most devastating impacts are likely to occur. El Niño is in this category. While we have now come to expect these events to recur, we do not know when that will happen or what it will be like. The uncertainty then cascades down the "impacts chain," and as we speculate about likely impacts of an El Niño, the degree of uncertainty will increase.

Take, for example, the 1997–1998 El Niño. Even with the best monitoring and observing system in the world focused on minute changes in various aspects of the tropical Pacific Ocean, forecasters and modelers were unable to predict the onset of one of the biggest El Niño events in the past 100 years. Nor were they able to predict the course of development of that event. They were better than in earlier times, however, at predicting some of its impacts on societies in certain parts of the globe, especially those where the influences of changes in the sea surface temperatures in the tropical Pacific are known to be strong.

Societies (and their scientists) are on a learning curve with regard to the various ways that climate variability and climate change might affect climate-related human activities. They must avoid becoming complacent as a result of a belief that they fully understand atmospheric processes or their impacts. They must accept that there will be climate surprises in the future, even if the global climate does not change. They must learn from past experiences on how best to cope with the vagaries of climate (26).

Many countries now realize that climate-related problems do not stop at international boundaries. There are many transboundary issues that demand regional (if not international) cooperation, given that countries share river basins, inland seas, airsheds, the global atmosphere as well as the onslaught and impacts of extreme meteorological events such as droughts, floods, and tropical storms.

While climate-related anomalies cannot be prevented, societal preparation for, and response to, their adverse impacts can be improved through better knowledge of the direct and indirect ways in which atmospheric processes interact with human activities and ecological processes. The enhancement of such knowledge will lead to better forecasts as well as better computer modeling of the interactions among land, sea, and air. A society forewarned of climate-related hazards is forearmed to cope with those hazards more effectively.

BIBLIOGRAPHY

1. Huntington, E. *Civilization and Climate*. Yale University Press, New Haven, Connecticut, 1915, rev, ed. 1924, reprinted by University Press of the Pacific, 2001.
2. Markham, S.F. (1944). *Climate and the Energy of Nations*. Oxford University Press, London.
3. Pye, L. (1966). *Aspects of Political Development*. Little, Brown and Co., Boston, Massachusetts.
4. Kamarck, A.M. (1976). *The Tropics and Economic Development: A Provocative Inquiry into the Poverty of Nations*. The Johns Hopkins University Press, Baltimore, Maryland.
5. Glantz, M.H. (Ed.). (1976). *The Politics of Natural Disaster: The Case of the Sahel Drought*. Praeger Press, New York.
6. Sen, A. (1981). *Poverty and Famines: An Essay on Entitlement and Deprivation*. Oxford University Press, Oxford, UK.
7. Glantz, M.H. (Ed.). (1977). *Desertification: Environmental Degradation in and around Arid Lands*. Westview Press, Boulder, Colorado.
8. Hare, K. (1977). Connections between climate and desertification. *Environ. Conserv.* **4**(2): 82.
9. Trager. (1975).
10. Brown, L., and Eckholm, E.P. (1974). *By Bread Alone*. Praeger Press, New York.
11. Ponte, L. (1976). *The Cooling*. Prentice-Hall, Inc., Englewood Cliffs, New Jersey.
12. *Weather Conspiracy: The Coming of the New Ice Age*, Ballentine Books, New York, compiled by The Impact Team, 1977.
13. Wolde Mariam, M. (1984). *Rural Vulnerability to Famine in Ethiopia, 195777*, Vikas Publishing House, New Delhi.
14. Aubreville, A. (1949). *Climate, Forests and Desertification in Tropical Africa*, Sociéte d'Editions Géographiques, Maritimes et Coloniales.
15. Benedick, R.E. (1998). *Ozone Diplomacy: New Directions in Safeguarding the Planet*. Harvard University Press, Cambridge, Massachusetts, Enlarged Edition.
16. Glantz, M.H. (Ed.). (2002). *La Niña and Its Impacts: Facts and Speculation*. United Nations University Press, Tokyo, Japan.
17. Arrhenius, S. (1896). On the influence of carbonic acid in the air upon the temperature of the ground. *Philos. Mag.* **41**: 237–276.
18. Arrhenius, S. (1908). *Worlds in the Making*. Harper & Brothers, New York.
19. Callendar, G.S. (1938). The artificial production of carbon dioxide and its influence on temperatures. *Q. J. Roy. Meteor. Soc.* **64**: 223–237.
20. Revelle, R.R., and Suess, H.E. (1957). Carbon dioxide exchange between atmosphere and ocean and the question of an increase of atmospheric CO_2 during the past decades. *Tellus* **9**: 18–27.
21. IPCC (1990). *Climate Change: The IPCC Scientific Assessment*. Cambridge University Press, Cambridge, UK.
22. IPCC (1996). *Climate Change 1995: The Science of Climate Change*, Contribution of Working Group I to Second Assessment Report, Cambridge University Press, Cambridge, UK.
23. IPCC (Intergovernmental Panel on Climate Change). (2001). *Climate Change 2001: Impacts, Adaptation, and Vulnerability*, Contribution of Working Group II to Third Assessment Report, Cambridge University Press, Cambridge, UK.
24. Myers, N. (1995). Environmental unknowns. *Science* **269**: 358–360.
25. Streets, D.G., and Glantz, M.H. (2000). Exploring the concept of climate surprise. *Global Environ. Chang.* **10**: 97–107.
26. Glantz, M.H. (Ed.). (1988). *Societal Responses to Regional Climatic Change: Forecasting by Analogy*. Westview Special Study, Boulder, Colorado.

READING LIST

Smith, J.B., Bhatti, N., Menzhulin, G., Benioff, R., Bodyko, M.I., Campos, M., Jallow, B., and Rijsberman, F. (Eds.). (1996). *Adapting to Climate Change: An International Perspective*. Springer-Verlag, New York.

WHAT IS CLIMATOLOGY?

National Drought Mitigation Center

"CLIMATE IS WHAT YOU EXPECT. WEATHER IS WHAT YOU GET."

Weather is the condition of the atmosphere over a brief period of time. For example, we speak of today's weather or the weather this week. Climate represents the composite of day-to-day weather over a longer period of time.

People in Minneapolis–St. Paul expect a white Christmas, and people in New Orleans expect very warm, humid summers. And a traveler going to Orlando, Florida, in March will not pack the same kind of clothing as a traveler going to Vail, Colorado, in March. These examples show how climate influences our daily lives. Additionally:

- Our houses are designed based on the climate where we live.
- Farmers make plans based on the length of the growing season from the last killing freeze in the spring to the first freeze in the fall.
- Utility companies base power supplies on what they expect to be the maximum need for heating in the winter and the maximum need for cooling in the summer.

A climatologist attempts to discover and explain the impacts of climate so that society can plan its activities, design its buildings and infrastructure, and anticipate the effects of adverse conditions. Although climate is not weather, it is defined by the same terms, such as temperature, precipitation, wind, and solar radiation.

Climate is usually defined by what is expected or "normal", which climatologists traditionally interpret as the 30-year mean. By itself, "normal" can be misleading unless we also understand the concept of variability. For example, many people consider sunny, idyllic days normal in southern California. History and climatology tell us that this is not the full story. Although sunny weather is frequently associated with southern California, severe floods have had a significant impact there, including major floods in 1862 and 1868, shortly after California became a state. When you also factor in severe droughts, most recently those of 1987–94, a more correct statement would be that precipitation in southern California is highly variable, and that rain is most likely between October and April.

The misconception that weather is usually normal becomes a serious problem when you consider that weather, in one form or another, is the source of water for irrigation, drinking, power supply, industry, wildlife habitat, and other uses. To ensure that our water supply, livelihoods, and lives are secure, it is essential that planners anticipate variation in weather, and that they recognize that drought and flood are both inevitable parts of the normal range of weather.

HOW DOES CLIMATOLOGY HELP US PREPARE FOR DROUGHT?

Climatology provides benchmarks, such as the drought of record. The drought of record is the drought remembered as having the greatest impact on a region. Most of us are not consciously aware of how much the climate fluctuates from one decade or century to the next. One way for reservoir managers, municipal water suppliers, and other planners to check reality is to compare expectations of water supply against a region's drought of record. But caution is necessary here: although the weather conditions could recur, the impacts would likely be very different. For most of the country, the drought of record was 30 to 60 years ago, and population concentrations and water use patterns have shifted substantially since then. Planners need to consider and watch out for a variety of problems and misconceptions.

Specifically, climatology answers crucial questions such as:

- How often does drought occur in this region?
- How severe have the droughts been?
- How widespread have the droughts been?
- How long have the droughts lasted?

Examining water supplies and understanding the impact of past droughts help planners anticipate the effects of drought:

- What would happen if the drought of record occurred here now?
- Who are the major water users in the community, state, or region?
- Where does our water supply come from and how would the supplies be affected by a drought of record?
- What hydrological, agricultural, and socioeconomic impacts have been associated with the various droughts?
- How can we prepare for the next drought of record?

CLOUD SEEDING

Arthur M. Holst
Philadelphia Water Department
Philadelphia, Pennsylvania

Cloud seeding is the deliberate treatment of certain clouds or cloud systems to affect the precipitation process within those clouds. The treating of clouds can alter weather conditions, thus cloud seeding is also called weather modification. The technology that led to cloud seeding has been developed only during the past 60 years. Yet as this technology becomes more efficient, increased worldwide application of cloud seeding for practical use is likely to continue into the future.

THE CLOUD SEEDING PROCESS

To understand how cloud seeding works, one must understand some facts about the weather. All air in the atmosphere contains moisture. Warm air rises from the earth's surface and begins to cool. As the air cools, the moisture condenses into tiny droplets that make up clouds. A cloud is almost 100% air. The tiny droplets composing clouds are not heavy enough to fall to the ground until they merge with millions of other droplets at temperatures below 32 °F and interact with dust, salt, or sand particles.

One type of cloud seeding, known as cold cloud seeding, introduces silver iodide and other agents to enhance ice crystallization in clouds colder than 32 °F. This is often called as the static seeding effect. Once the droplets within the clouds freeze, the resulting ice crystals grow at the expense of the water droplets surrounding them, a process called sublimation. Other crystals grow through contact with neighboring droplets; this is known as riming. Through these two processes, the tiny crystals form snowflakes. If these snowflakes are heavy enough, they fall from the clouds and, depending on the temperatures below, will come to the earth as snow, raindrops, or a mixture.

The other type of cloud seeding, warm cloud seeding, produces rainfall from clouds that are above 32 °F. This process involves introducing additional condensation nuclei (salt particles), which cause additional water droplets to condense within the clouds. If the collision of these particles makes the droplets heavy enough, precipitation can fall from the clouds. In each case, it will usually take the clouds 20 to 30 minutes to produce rain, making it crucial to monitor cloud movement.

USES OF WEATHER MODIFICATION

The primary goal of most weather modification projects is to increase levels of precipitation or to suppress fog or hail. Water agencies, local municipalities, farmers, ranchers, hydroelectric power facility operators, ski resort owners, and others sponsor cloud seeding activities. To date, cloud seeding has successfully stimulated precipitation in more than 50 countries. More than half of the United States now has some type of regulation concerning cloud seeding programs.

More recently, cloud seeding has been used to suppress certain damaging effects of weather. Airports have employed programs to disperse fog levels and increase flight visibility. And in areas damaged by hail, programs have been undertaken to decrease hailstorms.

As technology continues to enhance the weather modification process, it can be assumed that more entities will sponsor cloud seeding programs. Cloud seeding technology is highly portable and very flexible to changing weather conditions. For those who use it, cost/benefit ratios are typically very favorable.

EFFECTIVENESS OF CLOUD SEEDING

From the earliest experiment, which produced just a few droplets of rain, cloud seeding has progressed to make some significant and valuable impacts on weather. Various studies have documented the effects of different programs. For the most part, these programs are overwhelmingly successful. To augment precipitation, well-designed and well-conducted projects yield an average winter precipitation of 5 to 20% more in continental regions and 5 to 30% more in coastal areas; they have yielded as much as 100% more in warm weather. Increases depend on a variety of factors, including spatial coverage of suitable cloud systems and the frequency of different systems.

ENVIRONMENTAL HEALTH CONCERNS OF WEATHER MODIFICATION

To date, no significant environmental problems have been attributed to cloud seeding programs, though government agencies, private firms, and research institutions have conducted many studies. Researchers believe that any negative effects are minimal because relatively little seeding material is used compared to the large surface area that is targeted. For example, the most common seeding material, silver iodide, will yield a concentration of less than 0.1 microgram per liter in rainwater or snow. The U.S. Public Health Service states that the acceptable amount of silver iodide in water is 50 micrograms per liter.

One other common question about cloud seeding is whether the stimulation of rainfall in one area results in decreased rainfall in other areas. It does not. Clouds are inefficient in the way they gather and distribute moisture. They never gather or release all the moisture that is available; rather, clouds gather only about 1% of the moisture in the atmosphere. Even if a seeding program tripled the efficiency of cloud formation, the cloud would still contain only about 3% of the moisture, leaving 97% available. Furthermore, some analyses of precipitation data downwind from seeding projects have indicated small increases in precipitation. To date, no scientific studies have shown that some areas receive precipitation at the expense of their neighbors.

FUTURE OF CLOUD SEEDING

Despite all of the advances made during the past 60 years, weather modification continues to receive relatively little support. Many states and the federal government acknowledge and regulate cloud seeding, but few provide significant funding for weather modification studies or projects. Most scientists insist that more research must be done and more data must be gathered before weather-changing programs garner public trust on a larger scale. However, most researchers are optimistic that new studies and technological advances will continue to advance the science of weather modification.

READING LIST

American Society of Civil Engineers. (1995). *Guidelines for Cloud Seeding to Augment Precipitation.* ASCE Manuals and Reports on Engineering Practice No. 81.

Stauffer, N.E. and Williams, K. (2000). *Utah Cloud Seeding Program, Increased Runoff/Cost Analysis.* Technical Report, Utah Department of Natural Resources, Division of Water Resources.

Weather Modification Advisory Board. (1978). *The Management of Weather Resources. Report to the Secretary of Commerce,* 2 volumes.

Klien, D.A. (1978). *Environmental Impacts of Artificial Ice Nucleating Agents.* Dowden, Hutchinson & Ross, Inc., Stroudsburg, PA.

Brown, K.J., Elliot, R.D., and Edelstein, M.W. (1978). *Transactions of Total-Area Effects of Weather Modification.* Report to the National Science Foundation on a workshop held August 8–12, 1977, Fort Collins, CO.

Cloud Seeding (Weather Modification) Frequently Asked Questions—http://www.xmission.com/~nawc/wmfaq.html.

The Science of Cloud Seeding—http://twri.tamu.edu/twripubs/WtrResrc/v20n2/text-1.htm.

Cloud Seeding Circus.com—www.cloudseedingcircus.com.

Al Weather Modification Page—www.atmos-inc.com/weamod.html.

CONDENSATION

ALDO CONTI
Frascati (RM), Italy

Condensation is the physical process by which a vapor changes to its liquid state. Condensation happens when the temperature of the vapor decreases below the so-called dew point. In physics, this sort of process is called a phase change because the matter involved changes its state (i.e., from gaseous to liquid, in this case). Water condensation is evident in our atmosphere; it produces fog, mist, dew, clouds, and rain, depending on the conditions. It can be seen on a cold sheet of glass, where condensed water forms a maze of droplets.

Condensation depends mainly on temperature, and it is a process that happens at the molecular level. When the temperature is high, the molecules in a vapor have plenty of energy and collide at high speed, which means that the molecules bounce immediately and do not stay together long enough to establish a bond. In this situation, the vapor remains. But when the temperature, and therefore the speed, decreases, then the molecules can stick to each other. The result is a droplet of liquid. Condensation is ruled not only by temperature, but by pressure as well, which is the reason why one speaks about dew point and not dew temperature. As a general rule, the dew point temperature increases with pressure.

Condensation, with evaporation, is very important in the water cycle of the earth. By condensation, water falls back as rain, hail, or snow and becomes available again for human use.

COSMIC WATER

D.L. Marrin
Hanalei, Hawaii

Once believed by science to be the substance that distinguished the earth from the rest of the universe, it is now understood that water is ubiquitous in the cosmos—not only as ice and vapor, but also as liquid. Sophisticated scientific instruments can detect cosmic water on the basis of the light, or other electromagnetic, waves that it absorbs and emits (1). Unlike the legendary "waters of chaos" that gave rise to the material world, water's component hydrogen and oxygen atoms owe their existence to the Big Bang and to the stars, respectively.

HYDROGEN AND OXYGEN

Hydrogen is both the simplest and the most abundant atom in the universe. It represents about 75% of the atomic mass in the cosmos. The word *hydrogen* literally means "water-forming" according to the Greek language from which it is derived. Hydrogen atoms are generally traced back to the so-called Big Bang, when a tremendous amount of energy was released and subsequently expanded into what we call our universe. As the newborn universe began to cool, subatomic and atomic particles (e.g., quarks, protons, electrons) were initially created and later drawn together by a number of fundamental forces to form atoms. Possessing one proton and one electron, hydrogen is believed to have been the first atom created. As more hydrogen atoms were created in the early universe, they coalesced into dense gas clouds that contained much of the conventional matter.

Oxygen is the third most abundant atom in the cosmos—behind hydrogen and helium. Because helium is a very inert (nonreactive) atom, water is sometimes described as an interaction between the two most abundant "reactive" atoms in the cosmos. Unlike hydrogen, the origins of the oxygen atom are rooted in dying stars rather than in the Big Bang. As stars near the end of their stellar life, they begin to cool and to switch from hydrogen to helium as a source for nuclear fusion. The cooling stars then enter a phase where they become increasingly dense as intense gravitational energy compresses them into an extremely unstable state that may explode during the final stages of compression (2). This explosion or *supernova* releases the outer layer of the star, which contains many common heavier atoms (e.g., oxygen, carbon), into space as interstellar clouds. Dust grains comprising these clouds are composed of both silicate (oxygen-rich) and carbonaceous (carbon-rich) minerals that are available to react with hydrogen (1).

INTERSTELLAR SPACE

In the interstellar realms of galaxies, water exists predominantly as ice—adhering to the tiny particles that comprise the ubiquitous dust and gas clouds. Water is the primary molecular ice attached to these particles, although methane, carbon monoxide, and water–ammonia mixtures may also be present depending on physical conditions in the gas clouds (3). Water ice in the 10 K temperatures and vacuum conditions of interstellar space is what physical chemists refer to as *amorphous ice* or *glassy water*; it is relatively unstructured compared to the highly crystalline ices that are formed at higher temperatures (e.g., those characteristic of the earth's surface and atmosphere). Amorphous ice is so unstructured that it can flow, not unlike a viscous version of liquid water (4). Some astrophysicists posit that the simple organic molecules responsible for biological life may have been created in this strange ice (4). As interstellar temperatures rise above 150 K (as often occurs near stars), amorphous ice irreversibly transits to more familiar crystalline ice. The various phases and molecular structures of water as a function of temperature are shown in Table 1.

Although water's component atoms are plentiful in interstellar dust and gas clouds, creation of molecular water requires either converting O and OH species directly to water ice on the surfaces of dust grains or producing water vapor via heat energy—usually in the form of stellar radiation (5). The latter process requires that water vapor adhere to dust grains, where the newly formed water molecule is protected from the same ionizing radiation that created it. Scientists currently believe that stars facilitate the creation of water vapor and also that water vapor assists in the birthing of stars. Stars are being born and are dying on an ongoing basis, such that star birthing regions (e.g., the Orion Cloud Complex of the Milky Way Galaxy) generate up to 20% of a galaxy's luminosity as gas and dust clouds are gravitationally compressed into newborn stars. Recent data indicate that these cloud complexes contain an extremely high concentration of water vapor, which has been estimated of the order of 1 part in 2000 (6). The superabundance of water in stellar nurseries (about 20 times greater than that in similar interstellar clouds)

Table 1. Representative Temperature Ranges and Cosmic Locations for the Three Phases of Water

Location	Temperature, K[a]	Phase	Comments
Stellar surfaces	4000 to 50,000	None	Water molecules do not exist. Only hydrogen and oxygen atoms or plasma.
Stellar/planetoid surfaces	<3500	Vapor	Water exists at the surface of cool stars and in cooler regions of some hot stars.
Some planetoids	273 to 373 (at 1 atmosphere)	Liquid	Few places in the cosmos possess the requisite temperatures and pressures.
Interplanetary/interstellar space	>100	Solid (crystalline)	Crystalline ices possess both a cubic and a more common hexagonal structure.
Interstellar space	10 to 150	Solid (amorphous)	The depths of interstellar space are cold enough to produce glassy water.

[a]These temperatures, which are measured on the Kelvin scale, are only approximate and vary depending on environmental conditions (e.g., pressure, rate of cooling or heating).

may permit the gas and dust to cool sufficiently so that condensation can proceed and stars are eventually formed.

As hot winds (in the form of shock waves) are sent out during stellar birthing, the cloud must be cooled—initially by molecular hydrogen and subsequently by water and other simple gases (7). Water vapor is created during the interstellar cloud shocks as oxygen reacts explosively with hydrogen, causing the water vapor concentration to increase substantially during star birthing. Scientists have theorized one of two eventual fates for water created in star birthing. One is that the intense heat of the fledgling star rapidly dissociates water into its component atoms. The other is that the water is deposited on dust grains that later form the star's planetoids. The origin of earthly water is usually attributed to either this second process or the impact of large comets, which are believed to have been more prevalent during the earth's early history.

STARS

Two of the brightest supergiants in our galaxy, Betelgeuse and Antares, have water in their *photospheres*, which constitutes the visible portion of a star (8). A star's photosphere is where its gases transit from opaque to transparent, permitting us to see the stars that are located closest to the earth. This stellar water is actually present within the star itself, not simply as a component of the surrounding dust and gas cloud from which the star was birthed. Aging supergiant stars release massive amounts of water as they die; however, the exact source and role of this water are not yet known.

In addition to cool stars, water has been discovered in the photosphere of at least one hot or *main sequence* star—the Sun. Water cannot exist on the surface of the Sun, where temperatures of 6000 K dissociate the water molecule into its component hydrogen and oxygen atoms. Water can exist in the dark centers of sunspots, where temperatures are less than 3500 K (9). Sunspots are relatively calm solar regions where strong magnetic fields filter out the energy emanating from the intense interior, rendering them both the coolest and darkest regions of the Sun. Water is a major player in determining a star's *radiative opacity*, which describes the extent to which light escapes from stars into interstellar space (10). In this role, water impedes the outward flow of radiation from stars by absorbing energy within certain wavelengths and, thus, renders the star more opaque than it would otherwise appear.

COMETS AND METEORS

Comets are one of the few interstellar objects that are commonly associated with water, predominantly as ice. Comets are composed primarily of water ice that incorporates many of the other simple molecules in interstellar dust and gas clouds (e.g., carbon monoxide, methane, ammonia). Comets are most easily recognized by their unmistakable tails, which can extend millions to hundreds of millions of kilometers behind the icy body of the comet. The tail consists of dust and ionized particles (mostly water ice) that are always transported away from the Sun by the *solar wind*, which is an ionized stream of particles consisting predominantly of protons and electrons. The ionization of water ices is the primary mechanism influencing the properties of a comet's tail, including the steam jets that release tons of water vapor per second from the comet. It is now believed that these steam jets result from solar-induced changes in ice's molecular geometry (e.g., a transition from the amorphous to crystalline structure). Large comets are generally accepted as a source of planetary water, but controversy surrounds the hypothesis that many small comets hitting the planet's upper atmosphere also contribute significantly to the volume of water on the earth.

The first liquid water in the solar system, it was projected, made its appearance on meteors just 20 million years after our Sun and its debris emerged from the interstellar dust and gas cloud (11). Although liquid water is rarely present on the surface of meteors today, the chemical interaction of water with primitive rocks produced carbonate minerals, suggesting that the chemical processes of water evaporation and condensation were among the earliest in the solar system. Recently, a small meteorite found in southwestern North America contained actual liquid water within its salt crystals, which were believed to have been created from the original interstellar cloud that gave rise to the solar system.

PLANETOIDS

Most planet-sized bodies in our solar system (and probably in others) are now known or suspected to contain water in some form. A number of recent missions have revealed a Martian landscape that almost certainly indicates the large-scale flow of liquid water. The surface features of Mars (e.g., flood plains, river beds, mud deposits) suggest the recent presence of liquid water and also the mineralogy of Martian rocks could have resulted only from aqueous processes. Moreover, it has been hypothesized that Mars may have also once possessed surface oceans. The Jovian moon, Europa, is another of the solar system planetoids that probably contains liquid water located tens of kilometers beneath its icy surface. The liquid water underlying Europa's surface ice is believed to be an ocean containing saltwater that may be similar in composition to the seawater of the earth's oceans. Unlike the earth, the heat required to maintain water in a liquid phase on Europa is believed to originate from an internal source such as volcanic activity rather than from the heat of the Sun.

Data collected from the Infrared Space Observatory indicate the presence of water in the upper atmospheres of our solar system's gas giant planets and on one of Saturn's moons (1). The source of water in these planet's atmospheres is attributed to comets or to water-containing interplanetary dust. Based on recently developed techniques for measuring a suite of stellar characteristics (e.g., orbital velocity, position, brightness), the search for planets has been extended beyond our solar system to other star systems in the galaxy (12). Various planets have been identified orbiting stars in the constellations of Leo, Pegasus, Virgo, and Ursa Major that probably possess surface temperatures ranging from slightly less than 100 °C down to −100 °C. Planets or moons characterized by this temperature range could possess water in solid, gaseous, and liquid phases.

BIBLIOGRAPHY

1. Salama, A. and Kessler, M. (2000). *ISO and Cosmic Water*. European Space Agency Bulletin 104, pp. 31–38.
2. Lewis, J. (1995). *Physics and Chemistry of the Solar System*. Academic, San Diego, pp. 30–31.
3. Tielens, A.G.G.M., Hagen, W., and Greenberg, J.M. (1983). Interstellar ice. *J. Phys. Chem.* **87**: 4220–4229.
4. Blake, D. and Jenniskens, P. (2001). The ice of life. *Sci. Am.* **August**: 44–51.
5. O'Neill, P.T. and Williams, D.A. (1999). Interstellar water and interstellar ice. *Astrophys. Space Sci.* **266**: 539–548.
6. Cowen, R. (1998). Unveiling the hidden universe. *Sci. News* **153**: 328–330.
7. Harwit, M., Neufield, D.A., Melnick, G.J., and Kaufman, M.J. (1998). Thermal water vapor emission from shocked regions in Orion. *Astrophys. J. Lett.* **497**: 105–108.
8. Jennings, D.E. and Sada, P.V. (1998). Water in Betelgeuse and Antares. *Science* **279**: 844–847.
9. Oka, T. (1997). Water on the sun: Molecules everywhere. *Science* **277**: 328–329.
10. Tsuji, T. (1986). Molecules in stars. *Annu. Rev. Astron. Astrophys.* **24**: 89–125.
11. Endress, M., Zinner, E., and Bischoff, A. (1996). Early aqueous activity on primitive meteorite parent bodies. *Nature* **379**: 701–703.
12. Boss, A. (1996). Extrasolar planets. *Phys. Today* **September**: 32–38.

THE WATER CYCLE

NASA—Goddard Space Flight Center

INTRODUCTION

As seen from space, one of the most unique features of our home planet is the water, in both liquid and frozen forms, that covers approximately 75% of the Earth's surface. Believed to have initially arrived on the surface through the emissions of ancient volcanoes, geologic evidence suggests that large amounts of water have likely flowed on Earth for the past 3.8 billion years, most of its existence. As a vital substance that sets the Earth apart from the rest of the planets in our solar system, water is a necessary ingredient for the development and nourishment of life.

HYDROLOGIC HISTORY

The notion that water is continually circulating from the ocean to the atmosphere to the land and back again to the ocean has interested scholars through most of recorded history. In Book 21 of the lliad, Homer (ca. 810 B.C.) wrote of "the deep-flowing Oceanus, from which flow all rivers and every sea and all springs and deep wells." Thales (ca. 640 B.C.–ca. 546 B.C.) and Plato (ca. 427 B.C.–347 B.C.) also alluded to the water cycle when they wrote that all waters returned by various routes to the sea. But it wasn't until many centuries later that scientific measurements confirmed the existence of a water (or hydrologic) cycle. Seventeenth century French physicists Pierre Perrault (1608–1680) and Edmond Mariotte (1620–1684) separately made crude precipitation measurements in the Seine River basin in France and correlated these measurements with the discharge of the river to demonstrate that quantities of rainfall and snow were adequate to support the river's flow.

WATER, WATER, EVERYWHERE

Water is everywhere on Earth and is the only known substance that can naturally exist as a gas, liquid, and solid within the relatively small range of air temperatures and pressures found at the Earth's surface. In all, the Earth's water content is about 1.39 billion cubic kilometers (331 million cubic miles) and the vast bulk of it, about 96.5%, is in the global oceans. Approximately 1.7% is

Table 1. One Estimate of Global Water Distribution. Estimates of Groundwater are Particularly Difficult and Vary Widely Amongst Sources, with the Value in this Table Being Near the High End of the Range. Using the Values in This Table, Groundwater Constitutes Approximately 30% of Fresh Water, Whereas Ice (Including Ice Caps, Glaciers, Permanent Snow, Ground Ice, and Permafrost) Constitute Approximately 70% of Fresh Water. With Other Estimates, Groundwater is Sometimes Listed as 22% and Ice as 78% of Fresh Water

	Volume (1000 km^3)	Percent of Total Water	Percent of Fresh Water
Oceans, Seas, & Bays	1,338,000	96.5	—
Ice caps, Glaciers, & Permanent Snow	24,064	1.74	68.7
Groundwater	23,400	1.7	—
Fresh	(10,530)	(0.76)	30.1
Saline	(12,870)	(0.94)	—
Soil Moisture	16.5	0.001	0.05
Ground Ice & Permafrost	300	0.022	0.86
Lakes	176.4	0.013	—
Fresh	(91.0)	(0.007)	0.26
Saline	(85.4)	(0.006)	—
Atmosphere	12.9	0.001	0.04
Swamp Water	11.47	0.0008	0.03
Rivers	2.12	0.0002	0.006
Biological Water	1.12	0.0001	0.003
Total	**1,385,984**	**100.0**	**100.0**

Source: Gleick, P.H., 1996: Water resources. In *Encyclopedia of Climate and Weather*, ed. by S. H. Schneider, Oxford University Press, New York, vol. 2, pp. 817–823.

stored in the polar icecaps, glaciers, and permanent snow, and another 1.7% is stored in groundwater, lakes, rivers, streams, and soil. Finally, a thousandth of 1% exists as water vapor in the Earth's atmosphere (Table 1).

A MULTI-PHASED JOURNEY

The hydrologic cycle describes the pilgrimage of water as water molecules make their way from the Earth's surface to the atmosphere, and back again. This gigantic system, powered by energy from the sun, is a continuous exchange of moisture between the oceans, the atmosphere, and the land (Fig. 1).

Studies have revealed that the oceans, seas, and other bodies of water (lakes, rivers, streams) provide nearly 90% of the moisture in our atmosphere. Liquid water leaves these sources as a result of evaporation, the process by which water changes from a liquid to a gas. In addition, a very small portion of water vapor enters the atmosphere through sublimation, the process by which water changes from a solid (ice or snow) to a gas. (The gradual shrinking of snow banks, even though the temperature remains below the freezing point, results from sublimation.) The remaining 10% of the moisture found in the atmosphere is released by plants through transpiration (Fig. 2). Plants take in water through their root systems to deliver nutrients to their leaves, then release it through small pores, called stomates, found on the undersides

Figure 1. In the hydrologic cycle, individual water molecules travel between the oceans, water vapor in the atmosphere, water and ice on the land, and underground water.

of their leaves. Together, evaporation, sublimation, and transpiration, plus volcanic emissions, account for all the water vapor in the atmosphere. While evaporation from the oceans is the primary vehicle for driving the surface-to-atmosphere portion of the hydrologic cycle, transpiration is also significant. For example, a cornfield 1 acre in size can transpire as much as 4000 gallons of water every day.

After the water enters the lower atmosphere, rising air currents carry it upward, often high into the atmosphere,

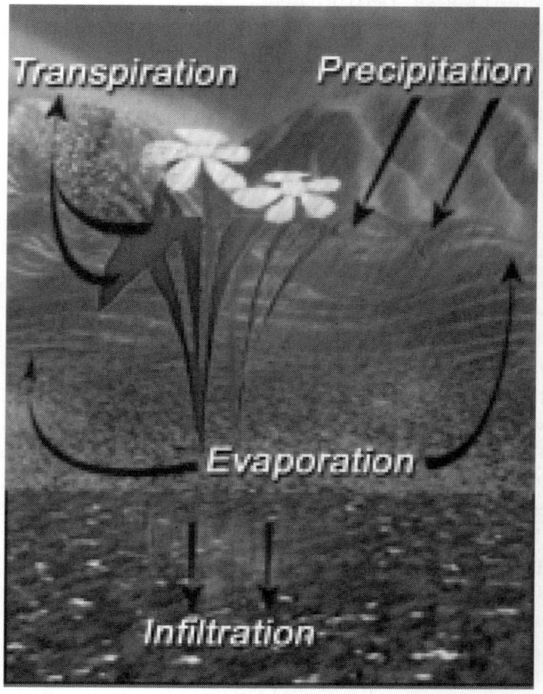

Figure 2. Plants return water to the atmosphere through transpiration. In this process, water evaporates from pores in the plant's leaves, after being drawn, along with nutrients, from the root system through the plant.

where the air cools and loses its capacity to support water vapor. As a result, the excess water vapor condenses (i.e., changes from a gas to a liquid) to form cloud droplets, which can eventually grow and produce precipitation (including rain, snow, sleet, freezing rain, and hail), the primary mechanism for transporting water from the atmosphere back to the Earth's surface.

When precipitation falls over the land surface, it follows various routes. Some of it evaporates, returning to the atmosphere, and some seeps into the ground (as soil moisture or groundwater). Groundwater is found in two layers of the soil, the "zone of aeration," where gaps in the soil are filled with both air and water, and, further down, the "zone of saturation," where the gaps are completely filled with water. The boundary between the two zones is known as the water table, which rises or falls as the amount of groundwater increases or decreases (Fig. 3). The rest of the water runs off into rivers and streams, and almost all of this water eventually flows into the oceans or other bodies of water, where the cycle begins anew (or, more accurately, continues). At different stages of the cycle, some of the water is intercepted by humans or other life forms.

Even though the amount of water in the atmosphere is only 12,900 cubic kilometers (a minute fraction of Earth's total water supply that, if completely rained out, would cover the Earth's surface to a depth of only 2.5 centimeters), some 495,000 cubic kilometers of water are cycled through the atmosphere every year, enough to uniformly cover the Earth's surface to a depth of 97 centimeters. Because water continually evaporates, condenses, and precipitates, with evaporation on a global basis approximately equaling global precipitation, the total amount of water vapor in the atmosphere remains approximately the same over time. However, over the continents, precipitation routinely exceeds evaporation, and conversely, over the oceans, evaporation exceeds precipitation. In the case of the oceans, the routine excess of evaporation over precipitation would eventually leave the oceans empty if they were not being replenished by additional means. Not only are they being replenished, largely through runoff from the land areas, but, over the past 100 years, they have been over-replenished, with sea level around the globe rising by a small amount. Sea level rises both because of warming of the oceans, causing water expansion and thereby a volume increase, and because of a greater mass of water entering the ocean than the amount leaving it through evaporation or other means. A primary cause for increased mass of water entering the ocean is the calving or melting of land ice (ice sheets and glaciers).

Figure 3. The water table is the top of the zone of saturation and intersects the land surface at lakes and streams. Above the water table lies the zone of aeration and soil moisture belt, which supplies much of the water needed by plants.

Throughout the hydrologic cycle, there are an endless number of paths that a water molecule might follow. Water at the bottom of Lake Superior may eventually fall as rain in Massachusetts. Runoff from the Massachusetts rain may drain into the Atlantic Ocean and circulate northeastward toward Iceland, destined to become part of a floe of sea ice, or, after evaporation to the atmosphere and precipitation as snow, part of a glacier. Water molecules can take an immense variety of routes and branching trails that lead them again and again through the three phases of ice, liquid water, and water vapor. For instance, the water molecules that once fell 100 years ago as rain on your great grandparents' farmhouse in Iowa might now be falling as snow on your driveway in California.

THE WATER CYCLE AND CLIMATE CHANGE

Amongst the highest priorities in Earth science and environmental policy issues confronting society are the potential changes in the Earth's water cycle due to climate change. The science community now generally agrees that the Earth's climate will undergo changes in response to natural variability, including solar variability, and to increasing concentrations of greenhouse gases and aerosols. Furthermore, agreement is widespread that these changes may profoundly affect atmospheric water vapor concentrations, clouds, and precipitation patterns. For example, a warmer climate, directly leading to increased evaporation, may well accelerate the hydrologic cycle, resulting in an increase in the amount of moisture circulating through the atmosphere. Many uncertainties remain, however, as illustrated by the inconsistent results given by current climate models regarding the future distribution of precipitation.

THE AQUA MISSION AND THE WATER CYCLE

As mentioned earlier, the hydrologic cycle involves evaporation, transpiration, condensation, precipitation, and runoff. NASA's Aqua satellite monitors many aspects of the role of water in the Earth's systems, and will do so at spatial and temporal scales appropriate to foster a more detailed understanding of each of the processes that contribute to the hydrologic cycle. These data and the analyses of them nurture the development and refinement of hydrologic process models and a corresponding improvement in regional and global climate models, with a direct anticipated benefit of more-accurate weather and climate forecasts.

Aqua's contributions to monitoring water in the Earth's environment involves all six of Aqua's instruments: the Atmospheric Infrared Sounder (AIRS), the Advanced Microwave Sounding Unit (AMSU), the Humidity Sounder for Brazil (HSB), the Advanced Microwave Scanning Radiometer-Earth Observing System (AMSR-E), the Moderate Resolution Imaging Spectroradiometer (MODIS), and Clouds and the Earth's Radiant Energy System (CERES). The AIRS/AMSU/HSB combination provides more-accurate space-based measurements of atmospheric temperature and water vapor than have ever been obtained before, with the highest vertical resolution to date as well. Since water vapor is the Earth's primary greenhouse gas and contributes significantly to uncertainties in projections of future global warming, it is critical to understand how it varies in the Earth system.

The water in clouds is examined with MODIS, CERES, and AIRS data; and global precipitation is monitored with AMSR-E. The cloud data includes the height and areal coverages of clouds, the liquid water content, and the sizes of cloud droplets and ice particles, the latter sizes being important to the understanding of the optical properties of clouds and their contribution to the Earth's albedo (reflectivity). HSB and AMSR-E, both making measurements at microwave wavelengths, have the ability to see through clouds and detect the rainfall under them, furthering the understanding of how water is cycled through the atmosphere.

Frozen water in the oceans, in the form of sea ice, is examined with both AMSR-E and MODIS data, the former allowing routine monitoring of sea ice at a coarse resolution and the latter providing greater spatial resolution but only under cloud-free conditions. Sea ice can insulate the underlying liquid water against heat loss to the often frigid overlying polar atmosphere and also reflects sunlight that would otherwise be available to warm the ocean. AMSR-E measurements allow the routine derivation of sea ice concentrations in both polar regions, through taking advantage of the marked contrast in microwave emissions of sea ice and liquid water. This continues with improved resolution and accuracy, a 22-year satellite record of changes in the extent of polar ice. MODIS, with its finer resolution, permits the identification of individual ice floes, when unobscured by clouds.

AMSR-E and MODIS also provide monitoring of snow coverage over land, another key indicator of climate change. Here too, the AMSR-E allows routine monitoring of the snow, irrespective of cloud cover, but at a coarse spatial resolution, while MODIS obtains data with much greater spatial detail under cloud-free conditions.

As for liquid water on land, AMSR-E provides an indication of soil moisture, which is crucial for the maintenance of land vegetation, including agricultural crops. AMSR-E's monitoring of soil moisture globally should permit, for example, the early identification of signs of drought episodes.

THE AQUA SPACECRAFT

Aqua is a major mission of the Earth Observing System (EOS), an international program centered in NASA's Earth Science Enterprise to study the Earth in detail from the unique vantage point of space. Focused on key measurements identified by a consensus of U.S. and international scientists, EOS is further enabling studies of the complex interactions amongst the Earth's land, ocean, air, ice and biological systems.

The Aqua spacecraft circles the Earth in an orbit that ascends across the equator each day at 1:30 p.m. local time and passes very close to the poles, complementing the 10:30 a.m. measurements being made by Terra, the first of the EOS spacecraft, launched in December 1999. The instrument complement on Aqua is designed to provide information on a great many processes and components of the Earth system, including cloud formation, precipitation, water vapor, air temperature, cloud radiative properties, sea surface temperature, surface wind speeds, sea ice concentration and temperature, snow cover, soil moisture, and land and ocean vegetation. The individual swaths of measurements will be compiled into global images, with global coverage of many variables being obtained as frequently as every two days or, with the help of numerical models, combined every 6 or 12 hours into comprehensive representations of the Earth's atmospheric circulation and surface properties. In combination with measurements from other polar orbiting satellites, Aqua measurements also provide accurate monthly-mean climate assessments that can be compared with and assimilated into computer model simulations of the Earth's climate.

The Earth Observing System has three major components: the EOS spacecraft, an advanced ground-based computer network for processing, storing, and distributing the collected data (the EOS Data and Information System); and teams of scientists and applications specialists who study the data and help users in universities, industry, and the public apply it to issues ranging from weather forecasting and climate prediction to agriculture and urban planning.

CYCLONES

Arthur M. Holst
Philadelphia Water Department
Philadelphia, Pennsylvania

Cyclones are hazardous weather conditions characterized by extreme gusts of wind moving in a circular pattern, low pressure, and intense rain. They form over tropical or subtropical waters; however, some can reach land, where they have in the past inflicted great amounts of damage on buildings and communities.

Based on wind strength, cyclones are given various names. Tropical depressions have maximum surface winds of less than 39 mph. Those cyclones whose maximum winds are between 39 and 74 mph are known as tropical storms. Upon attaining 74-mph surface winds, a cyclone is known as one of a number of names that are all regional equivalents of the same type of storm.

The names of cyclones are typhoon (in the NW Pacific Ocean), hurricane (in the North Atlantic, NE Pacific, and South Pacific Ocean), tropical cyclone (in the SW Indian

Ocean), severe cyclonic storm (in the North Indian Ocean), and severe tropical cyclone (in the SW Pacific and SE Indian Oceans). Cyclones always build over tropical seas. There are seven basins in which the majority of tropical cyclones form: the Atlantic Basin, the North Indian Basin, the Southwest Indian Basin, the Southeast Indian-Australian Basin, the Australian-Southwest Pacific Basin, the Northwest Pacific Basin, and the Northeast Pacific Basin. Heat gives cyclones their energy, and thus the water over which a cyclone forms must be at least as warm as 80°F (i.e., tropical). Other conditions necessary for a cyclone include a fast cooling atmosphere, moisture in the middle troposphere area, a distance of more than 300 miles from the equator to be influenced by the Coriolis force, and minimal vertical wind shear. This vertical wind shear is the result of differences between winds in the lower and upper portions of the atmosphere. However, the main contributor to the formation of a cyclone is a disturbance in the form of a thunderstorm or group of showers. When all of these factors come together, conditions are right for a tropical cyclone, although the presence of each of these factors does not guarantee the formation of a cyclone. Cyclones are spontaneous; a minute variation in one variable can be the difference between a hurricane and a thunderstorm.

The circular area in the center of a cyclone, known as the "eye," has conditions quite different from those in the region surrounding it. Calmness and a light breeze characterize the eye. Temperatures are generally higher than those in the surrounding area, and the sky is usually very clear.

Since the 1970s, cyclones have been regularly named by various people and agencies, depending on where the cyclones originated. They are named to ensure ease of description among newscasters, weatherpeople, and the general public. Names are often taken from rotating lists, one name for each letter of the alphabet. When an exceptional cyclone occurs, its name is taken out of use ("retired") to avoid confusion.

Strong cyclones often cause extensive damage to whatever they encounter. Damage can range from crop destruction to total devastation of building structures, depending on the severity of the cyclone. They become most dangerous as they hit land and spawn tornadoes. These are formed when tropical cyclones begin to lose their power. The diameter of a tropical cyclone is measured in kilometers; the diameter of a tornado is measured in meters. In 1980, one of the most destructive cyclone-spawned tornadoes in the United States caused nearly $100,000,000 worth of damage to the Austin, Texas, area. Along with destruction, cyclones (and the subsequent tornadoes) cause death. Large objects lifted from the path of the extreme wind are tossed about like lethal weapons. In 1964, twenty-two people were killed by a tornado that hit the Los Angeles area in the United States. The most damaging cyclone in history was Hurricane Andrew, which caused over $26 billion in damage to the Southeastern United States. The most deadly cyclone ever may have been the Bangladesh Cyclone, which killed at least 300,000 people in 1970. One of the ways in which a hurricane will cause damage is through the storm surge. This is the name given to the phenomenon whereby sea level in the area of a cyclone rises due to cyclonic winds.

Once a cyclone reaches 74 mph maximum winds and is considered a hurricane, it's intensity is rated by the Saffir–Simpson Scale, used since the 1970s by the National Oceanic and Atmospheric Administration (NOAA). Hurricanes are referred to as category one, category two, category three, category four, or category five. Category one hurricanes have winds between 74 and 95 mph. They cause negligible damage to buildings and other structures, although they can damage mobile homes and road signs. They carry a storm surge of about 5 feet. Category two hurricanes have maximum winds of 96–110 mph. They can cause minor damage to buildings and sizable damage to mobile homes and trees. Their storm surge is usually around 7 feet. Category three hurricanes have winds from 111–130 mph. They cause noticeable damage to buildings and trees; buildings near the shoreline are often destroyed by flooding. Evacuation of certain low-lying areas can be necessary for category three hurricanes, and their storm surge ranges from 9–12 feet. Category four hurricanes have winds between 131 and 155 mph. These hurricanes can cause massive damage to smaller structures. They destroy trees and mobile homes utterly. Evacuation of large areas can be necessary, as the storm surge can reach up to 18 feet. Category five hurricanes are the most powerful; they have maximum sustained winds of more than 155 mph. Residential as well as industrial structures are often destroyed. Damage is catastrophic, and major evacuations usually take place. The storm surge exceeds 18 feet above normal.

Cyclones can range from relatively simple tropical storms to devastating hurricanes whose winds swirl at furious speeds. Their effects can be disastrous and long-lasting.

READING LIST

Bureau of Meteorology. (2002). *Surviving Cyclones.* Available: http://www.bom.gov.au/info/cyclone/. (March 13).

Department of Public Safety. (2002). "*General Hurricane Information.*" Available: http://www.escambia-emergency.com/geninfo.asp. (March 13).

Landsea, Christopher W. *FAQ: Hurricanes, Typhoons, and Tropical Cyclones.* Available: http://www.aoml.noaa.gov/hrd/tcfaq/tcfaqA.html. (March 13).

National Weather Service Houston/Galveston. (2002). *Tropical Cyclone FAQ.* Available: http://www.srh.noaa.gov/hgx/index3/tropical3/geninfo3.htm. (March 13).

Hurricane Research Division: FAQ Available: http://www.aoml.noaa.gov/hrd/tcfaq/F1.html.

Hurricane Research Division: FAQ Available: http://www.aoml.noaa.gov/hrd/tcfaq/E7.html.

Hurricane Research Division: FAQ Available: http://www.aoml.noaa.gov/hrd/tcfaq/B1.html.

Hurricane Research Division: FAQ Available: http://www.aoml.noaa.gov/hrd/tcfaq/A10.html.

Hurricane Research Division: FAQ Available: http://www.aoml.noaa.gov/hrd/tcfaq/A1.html.

Wonders of Weather. Available: http://school.discovery.com/lessonplans/programs/wondersofweather/vocab.html.

The Saffir–Simpson Hurricane Scale. Available: http://www.nhc.noaa.gov/aboutsshs.shtml.

WATER CYCLE

Arthur M. Holst
Philadelphia Water Department
Philadelphia, Pennsylvania

The water cycle—a series of steps through which virtually all water on earth is constantly cycling. The names of these steps vary, but they generally follow this order: storage, evaporation, transpiration, condensation, precipitation, infiltration, and runoff.

Due to the nature of the water cycle, there is really no "beginning" or "end", but for simplicity, the cycle can be said to begin at the storage stage. Storage is the storing of water on, in, and above the earth. There are about 340 million cubic miles of water on the earth. This water is stored in many places: glaciers, seas, rivers, lakes, polar ice caps, and all living things on the planet. The largest percentage of water, approximately 97.25%, exists as salt water in the ocean. Water is also stored in the earth's atmosphere and as groundwater. Groundwater in the top few miles of the earth's crust is easily obtainable, whereas groundwater further down is chemically attached to rocks and usually cannot be accessed.

Evaporation is an essential way in which water is transferred from the earth to the atmosphere. During evaporation, water molecules change from a liquid state directly into a gaseous state, known as water vapor. Changes in temperature and air pressure on the earth cause millions of gallons of water to evaporate into the atmosphere each day. The majority of the water that is transferred into the atmosphere is primarily in a liquid state before the change; however, a small percentage of new water vapor comes from ice. The process of water changing state from a solid directly to a gas is called sublimation. Sublimation occurs on glaciers and polar ice sheets. When the sun strikes these massive ice structures, the temperature and pressure allow no room for a liquid, only a solid and a gas.

Transpiration is another way that water is transferred into the atmosphere. Although similar in process to evaporation, transpiration differs because the medium from which the water comes is not earth, it is plants. Transpiration is the process through which water absorbed from the ground by plant roots evaporates into the atmosphere. After the plants absorb the water from the soil, the water moves up through the veins of the plant, eventually reaching the leaves (or the plant's equivalent). These leaves have pores, or stomata, which allow water to evaporate. Large forests can release massive amounts of water into the atmosphere through transpiration.

Once the water has been transferred into the atmosphere by either evaporation or transpiration, the molecules will eventually condense. This process is called condensation. Condensation is the reciprocal process of evaporation. During condensation, water changes state from a gas to a liquid. Once again, a change in either temperature or pressure initiates the change. Condensation is most visible in the formation of dew, fog, and clouds. Dew usually appears in the morning and is water that has condensed on any solid object during the rapid temperature change that occurs from day to night. If the water vapor is dense enough around objects when the temperature change happens, dew is formed. Fog is really a low level cloud formation. When the conditions are right, fog is formed on or near the ground. The only difference between fog and cloud formations is their locations with reference to the earth. Clouds form when water vapor in the atmosphere encounters a critical temperature or air pressure, causing the water vapor to change state into a liquid. Water molecules then "stick together" causing physical cloud structures to emerge. When the molecules in the cloud that are stuck together reach a certain critical mass, they release the moisture causing precipitation to begin.

Precipitation is the process by which water is transferred from the medium of the atmosphere to the medium of the earth. Precipitation literally comes in all shapes and sizes, including rain, snow, and ice. The differentiating factor that determines the form water takes when precipitating is air temperature. The lower the temperature, the more likely it is that the precipitation will be either snow or ice (the closer the temperature is to the freezing point of water, 32 °F or 0 °C the more likely). Droplets of rain form around small particles of dust or dirt to form a cohesive raindrop as a vehicle for the trip to the earth.

Regardless of the form it takes on its descent, water will take one of two paths once it arrives on the earth's crust. It will either seep into the ground or flow into larger bodies of water such as streams, seas, oceans, and rivers. The former process is infiltration, and the latter is runoff. Infiltration restores groundwater to the water table that has been lost from wells. During the infiltration process, the water is purified by cascading amongst rocks and minerals that draw impurities out of the water.

Water that does not infiltrate the earth's crust runs off into streams, rivers, seas, and oceans. Runoff replenishes bodies of water that have lost water due to evaporation. Runoff can occur either above the land or below it. Large amounts of water can run across the land immediately after precipitation. Smaller amounts of water run through the ground and along the water table until they reach a larger body of water.

So thusly does the water complete its cycle throughout the different media of the earth. This process is essential for the survival of life on the earth because it allows water to permeate every accessible region of the crust and atmosphere, thereby allowing life to flourish wherever water exists.

READING LIST

Hydrologic Cycle. (2002). Available: http://www.und.nodak.edu/instruct/eng/fkarner/pages/cycle.htm. (April 1).

The Water Cycle: Scientific Concepts. (1995–1998). Available: http://mbgnet.mobot.org/fresh/cycle/concepts.htm. (March 28, 2002).

Gedzelman, S.D. (2001). The Water Cycle. In: *Microsoft Encarta Encyclopedia Deluxe 2001.*

DEGREE DAY METHOD

Carlos D. Messina
University of Florida
Gainesville, Florida

An adequate description and prediction of plant development is critical for understanding plant growth and its responses to the environment. Only then we can assess the impact of a changing environment on plant productivity and survival, understand plant adaptation, design adequate agricultural production systems, and optimize natural resources management. Of particular importance for water resource management are the effects of plant ontogeny on plant water use and productivity throughout the regulation of leaf area. This determines the patterns of plant water demand and the partitioning between transpiration and evaporation components. Therefore, plant development dictates the selection of cultivars that best fit the water availability patterns under dryland conditions, and the amount, frequency and timing for irrigation otherwise.

Plant development is responsive to environmental cues such as photoperiod, water availability, and temperature. Figure 1a shows an example of the close association between ambient temperature and plant development for boll growth duration in cotton. Rameur first suggested in 1735 that the duration of any developmental stage was related to temperature and that this duration can be characterized by a thermometric constant, which can be predicted by the sum of daily air temperatures (2). Since then, several methods based on the concept of normalizing time by temperature to predict developmental rates were developed (2,3) and applied in life sciences. Although we can formulate some hypothesis to explain the relationship between plant development and temperature based on the effects of temperature on circadian oscillations (4), lipid composition and membranes fluidity, and cell division (5), the scientific basis of these empirical methods remains elusive. It was shown, however, that degree day based methods significantly improved the description and prediction of phenological events relative to other approaches based on the time of the year or the number of days (3).

Because of the close coupling between the timescale of plants and temperature, Ritchie and NeSmith (6) proposed that the most appropriate term to describe plant development would be "thermal time" or degree days (DD), whose units are °C-day. The degree day method is based on the effects of temperature on developmental rates, rather than on the duration of a phase. Figure 1b shows the linear relationship between temperature and rate of development, the reciprocal of the duration of the phase in days. The base temperature (T_b), temperature at which the rate of development is zero, is obtained by extrapolating this linear relationship to the intersection with the x-axis:

$$T_b = -\frac{y_0}{\alpha}$$

where y_0 is the intercept and α is the slope of the linear regression equation (Fig. 1b). Then, the rate of development (RD) can be calculated as

$$\mathrm{RD}(t) = \alpha\,(T_a - T_b)$$

where T_a is the mean temperature. Integrating the rate of development in time,

$$\int \mathrm{RD}\,dt = \int \alpha\,(T_a - T_b)\,dt$$

and considering a daily time step ($\Delta t = 1$) for integration, we can estimate the cumulative development (CD) at time n (t_n), provided that $t_0 = 0$, as

$$\mathrm{CD} = \alpha \sum_n (T_a - T_b)$$

For constant temperatures, given that development is complete when CD equals one, the degree days above a certain base temperature for a given phase are

$$\mathrm{DD} = (T_a - T_b)\,n$$

or

$$\mathrm{DD} = \frac{1}{\alpha}$$

For the example of cotton shown in Fig. 1, boll growth requires 1000 °C-days above a base temperature

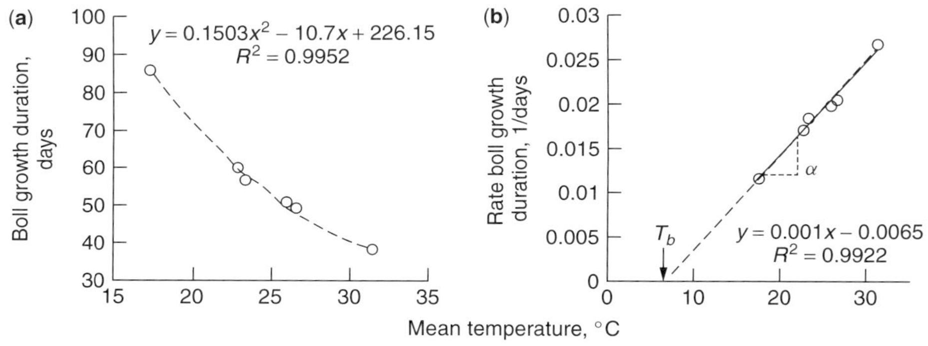

Figure 1. Responses of boll growth duration in cotton (*Gossypium hirsutum*, L.) development to air temperature. (**a**) Boll growth duration, (**b**) rate of boll growth duration (1/d). T_b denotes base temperature (see below), and α is the slope of the regression line. Data is from (1).

of 6.5 °C to reach maturity. Although this formulation for calculating DD is useful for estimating the requirements of the plant to complete a given developmental phase, in natural environments, temperature fluctuates and makes this procedure inadequate. The following canonical form to calculate DD is frequently used:

$$\mathrm{DD} = \sum_{n}\left[\left(\frac{T_{\max}+T_{\min}}{2}\right) - T_{\mathrm{b}}\right]$$

where $T_{\max}$ and $T_{\min}$ are maximum and minimum daily temperatures, n is the number of days of temperature observations, and $[(T_{\max}+T_{\min})/2] = T_{\mathrm{b}}$ when $[(T_{\max}+T_{\min})/2] < T_{\mathrm{b}}$. An alternative method, frequently used in calculating DD for corn, compares $T_{\max}$ and $T_{\min}$ with T_{b} individually before calculating their average. Significant discrepancies in calculated DD between these procedures may arise under some circumstances (3).

These procedures for calculating *DD* are adequate, provided that (1) the daily temperature does not fall below T_{b} and does not exceed an upper threshold for a significant part of the day, (2) the temperature of the growing plant tissue is the same as the air temperature, and (3) the response of the rate of development to temperature is linear over the range of temperature that the crop experiences (5). When these assumptions are not satisfied, alternative approaches are required. For example, McMaster et al. (5) and Vinocur and Ritichie (7) used soil or apex temperatures to calculate DD because the growing points were close to the soil rather than at the height where the temperature is normally measured. In simulation models such as CERES (8), hourly temperatures are approximated from $T_{\max}$ and $T_{\min}$, and if the temperature is below T_{b}, DD is set to zero for that part of the day. For a thorough discussion of the limitations of the degree day method, see Ritchie and NeSmith (5).

The degree day method can be found in the literature in a different mathematical form under the concept of physiological day. This model is generally used to simulate development in legumes and is particularly adequate for incorporating the effects of photoperiod on development (9). The rate of development is calculated as the product of two functions:

$$\mathrm{RD}(t) = f(P) \times f(T)$$

One function accounts for the effects of photoperiod (P) and the other for the effects of temperature (T) on development. The model predicts relative development; the maximum rate of development is standardized to 1.0. At optimum temperature (T_{opt}) and photoperiod ($P < P_{\min}$ in short day species and $P > P_{\max}$ in long day species), the rate of progress in calendar days equals the rate of progress in physiological days. When conditions deviate from the optimum, the rate of progress per day decreases and becomes a fraction of a physiological day. Then, cumulative development is measured in photothermal time units. The effects of the photoperiod on the rate of development are nonlinear:

$$\begin{aligned}
f(P) &= 0, && \text{if } P > P_{\max}\\
f(P) &= 1 - \frac{P - P_{\min}}{P_{\max} - P_{\min}}, && \text{if } P_{\min} < P \le P_{\max}\\
f(P) &= 1, && \text{if } P \le P_{\min}
\end{aligned}$$

In short day species, the relative rate of development is maximum below a threshold $P_{\min}$, above which the development decreases to zero at $P_{\max}$. Both $P_{\min}$ and $P_{\max}$ are characteristic of the species and cultivar (Table 1). A similar function is used to describe the effects of temperature:

$$\begin{aligned}
f(T) &= 0, && \text{if } T \le T_{\mathrm{b}}\\
f(T) &= \frac{T - T_{\mathrm{b}}}{T_{\mathrm{opt1}} - T_{\mathrm{b}}}, && \text{if } T_{\mathrm{b}} < T < T_{\mathrm{opt1}}\\
f(T) &= 1, && \text{if } T_{\mathrm{opt2}} \ge T \ge T_{\mathrm{opt1}}\\
f(T) &= \mathbf{1} - \frac{T - T_{\mathrm{opt2}}}{T_{\mathrm{upper}} - T_{\mathrm{opt2}}}, && \text{if } T_{\mathrm{upper}} > T > T_{\mathrm{opt2}}\\
f(T) &= 0, && \text{if } T \ge T_{\mathrm{upper}}
\end{aligned}$$

where T_{opt1} and T_{opt2} define a plateau whose rate of development is maximum. For temperatures above T_{opt2} and below T_{opt1}, the rate of development decreases linearly to zero at T_{upper} and T_{b}, respectively. Table 1 shows characteristic values for selected cereals and legumes.

In this chapter, a derivation of the degree day method is shown along with a mathematical variant that incorporates the effects of the photoperiod. The literature about models that uses degree days is vast, it is not

Table 1. Cardinal Temperatures and Photoperiods for Selected Models of Cereals and Legumes[a]

	Temperature, °C				Photoperiod, h	
	T_{b}	T_{opt1}	T_{opt2}	T_{upper}	$P_{\min}$	$P_{\max}$
Legumes						
Soybean						
Vegetative	7	28	35	45		
Early reproductive	6	26	30	45	11.7–14.6	15.5–21.0
Late reproductive	−48	26	34	45		
Bean						
Vegetative	4	27	35	45		—
Early reproductive	5	22	35	45	12.2–13.2	—
Late reproductive	0	18	35	45		
Peanut[b]						
Vegetative	11	28	30	55		
Early reproductive	11	28	28	55	—	—
Late reproductive	5	26	26	55		
Cereals						
Wheat & Barley	0	15	—	—	20	—
Millet	10	36	—	—	12	125–150[c]
Rice	9	—	—	—	11.7–12.8	35–189[c]
Maize	8	34	—	—	12.5	0.3–0.8[d]
Sorghum	8	34	—	—	12.5–13.6	30–90[c]

[a] Adapted from Reference 10.
[b] insensitive to photoperiod.
[c] DD (°C-day) per hour increase in photoperiod.
[d] Units are in days per hour increase in photoperiod (11).

the intent to review it here. However, the rationale and concept presented here covers most of the models currently used. Summerfield et al. (12) discuss additive photothermal models and Jones et al. (13) provide further details on multiplicative ones. Ritchie and NeSmith (6) and McMaster and Wilhelm (3) discuss some limitations of the degree day method. As emphasized in these previous papers, to obtain reliable and accurate predictions or descriptions of plant development, the parameters and the method used for their estimation must be consistent.

BIBLIOGRAPHY

1. Reddy, K.R., Davidonis, G.H., Johnson, A.S., and Vinyard, B.T. (1999). Temperature regime and carbon dioxide enrichment alter cotton roll development and fiber properties. *Agron. J.* **91**: 851–858.
2. Wang, J.Y. (1960). A critique of the heat unit approach to plant-response studies. *Ecology* **41**: 785–790.
3. McMaster, G.S. and Wilhelm, W.W. (1997). Growing degree-days: one equation, two interpretations. *Agric. Forest Meteorol.* **87**: 291–300.
4. Michael, T.P., Salome', P.A., and McClung, C. (2003). Two *arabidopsis* circadian oscillators can be distinguished by differential temperature sensitivity. *Proc. Nat. Acad. Sci. USA* **100**: 6878–6883.
5. McMaster, G.S. et al. (2003). Spring wheat leaf appearance and temperature: extending the paradigm? *Ann. Bot.* **91**: 697–705.
6. Ritchie, J.T. and NeSmith, D.S. (1991). Temperature and crop development. In: *Modeling Plant and Soil Systems*. R.J. Hanks and J.T. Ritchie (Eds.). *Agronomy* 31. American Society of Agronomy, Madison, WI, pp. 5–29.
7. Vinocur, M.G. and Ritchie, J.T. (2001). Maize leaf development biases caused by air-apex temperature differences. *Agron. J.* **93**: 767–772.
8. Ritchie, J.T., Singh, U., Godwin, D.C., and Bowen, W.T. (1998). Cereal growth, development and yield. In: *Understanding Options for Agricultural Production*. G.Y. Tsuji, G. Hoogenboom, and P.K. Thornton (Eds.). Kluwer Academic, Dordrecht, the Netherlands, pp. 79–98.
9. Boote, K.J., Jones, J.W., and Hoogenboom, G. (1998). Simulation of crop growth: CROPGRO model. In: *Agricultural Systems Modeling and Simulation*. R.M. Peart and R.B. Curry (Eds.). Marcel Dekker, New York, pp. 651–693.
10. Jones, J.W., Tsuji, G.Y., Hoogenboom, G., Hunt, L.A., Thornton, P.K., Wilkens, P.W., Imamura, D.T., Bowen, W.T., and Singh, U. (1998). Decision support system for agrotechnology transfer. In: *Understanding Options for Agricultural Production*. G.Y. Tsuji, G. Hoogenboom, and P.K. Thornton (Eds.). Kluwer Academic, Dordrecht, the Netherlands, pp. 157–177.
11. Kiniry, J.R. (1991). Maize phasic development. In: *Modeling Plant and Soil Systems*. R.J. Hanks and J.T. Ritchie (Eds.). *Agronomy* 31. American Society of Agronomy, Madison, WI, pp. 55–69.
12. Summerfield, R.J. et al. (1993). Towards the reliable prediction of time to flowering in 6 annual crops .2. *soybean* (*Glycine max*). *Exp. Agric.* **29**: 253–289.
13. Jones, J.W., Boote, K.J., Jagtap, S.S., and Mishoe, J.W. (1991). Soybean development. In: *Modeling Plant and Soil Systems*. R.J. Hanks and J.T. Ritchie (Eds.). *Agronomy* 31. American Society of Agronomy, Madison, WI, pp. 71–90.

DESERTIFICATION

Aldo Conti
Frascati (RM), Italy

Desertification is a degradation of the top layer of the soil that reduces its ability to support plant life and to produce food. As a result, the soil becomes dusty and dry, and it is easily carried away by erosion, which affects wild and domestic animals, wild plants and crops and, finally, humans. Desertification can be the result of human activities or of natural climate changes. In the latter case, the process is very slow and can take several thousand years to produce its effects. A common misconception about desertification is that it spreads from a desert core. The truth is that land degradation can occur where land abuse has become excessive. If it is not stopped in time, desertification spreads from that spot. Eventually, many of these spots merge and form a large homogeneous area. Another misconception is that desertification is the result of droughts. In fact, well-managed land can recover from even a long period of drought with very little adverse effect as soon as rains return. But it is true that droughts can increase the pace of desertification already taking place.

Desertification is a term that has been in use since at least 1949, when Aubreville, a perceptive and well-informed botanist and ecologist, published a book on *Climate, Forets, et Desertification de l'Afrique Tropicale*. Aubreville observed desertification in tropical Africa and understood immediately that the culprit was not the Sahara desert gaining land. He noticed instead that the main reasons behind desertification were tree cutting, indiscriminate use of fire to clear the land, and cultivation.

Many processes can lead to desertification. Logging, for instance, makes the soil unstable on mountain slopes. Eventually, all the soil runs down as a dangerous landslide or mud river and leaves behind exposed rocks, unable to support any life. What is left behind are barren mountains, particularly evident in China. In Europe, one of the leading causes of desertification is overgrazing. In the past, wild animals used to move following the rainfall, always grazing the richest areas. In modern times, the use of fences prevented these movements, and the result was heavy overgrazing that left the soil exposed to erosion. In many areas, desertification is the result of agriculture. A typical example of this is salinization of soil that happens normally when the soil is overirrigated. The water that is not used by plants evaporates, leaving behind salts that concentrate in the soil. Eventually, the concentration of salts in the soil becomes so high that plants cannot survive, again exposing the land to erosion. In same cases, desertification is the more direct result of urbanization, mining, and recreational activities. In any case, the adverse effects are still the same.

Nowadays, desertification is a serious problem that affects, according to some estimates, up to 30% of dry lands. Worldwide, desertification is making approximately 12 million hectares useless for cultivation every year. But land degradation and desertification are by no means new problems, despite the attention focused on them in recent years. During the first conquest of Africa, it was normal to clear patches of land with fire and then use them to grow crops. After three or four seasons, the land was depleted of nutrients and unable to support any plant life. Moreover, there is some historical evidence that serious and extensive land deterioration was already occurring several centuries before. It started in arid regions, and it had three epicenters: the Mediterranean Sea and other places where destructive changes in soil and plant cover had occurred, but were small in extent or not well known.

Luckily, desertification in many areas has been stopped, but very little effort has been made to restore the land to its original productivity. Today, desertification can be defeated using techniques already known, if financial resources are available and the political will to act is present. For instance, only in the last few decades, satellite images have allowed a better understanding and monitoring of the problem on a large scale.

There are several possible remedies available, even at the local level. The first is to avoid cutting trees or, at least, to replace them with new ones. Plants, a major soil stabilizer, can alone stop erosion. Moreover, the use of available local water and ways to control the salinity of the soil can be very effective. On this topic, genetic engineering is trying to help. Scientists are working on the development of crops that can survive higher salinity, both as a way to use the land and produce food and as a way to save lost soils. Curiously, one of the remedies until now used against desertification is to pollute the soil. In Iran, oil is sprayed over semiarid land with crops. The oil covers the seedlings, retains the moisture, and prevents them from being blown away by the wind until they grow large enough.

As stated at the beginning, desertification is sometimes a slow natural process. As an example, it should be enough to say that 20,000 years ago, the Sahara Desert was a lush forest. This is proved by the fossils of the animals that used to live in this forest. Moreover, pictures taken using radar from the Space Shuttle allow us to identify numerous dry riverbeds under a few meters of sand. This is already happening again due to the global warming on the earth. Many deserts are expanding, even though it is not possible to find a human reason for that. Desertification seems to threaten, in particular, all countries in the Mediterranean region. The coasts of North Africa are already disappearing into the sand, but the same might start happening in Europe. The heat wave of 2003 caused lots of problems. In Italy, many crops were heavily affected, and in the whole Mediterranean region, but in particular in Spain and Portugal, large areas of forest were destroyed by fires.

DEW

Aldo Conti
Frascati (RM), Italy

Dew forms when water condenses on objects on the surface near the ground and forms a thin layer or many droplets. Dew forms normally during the night when the air temperature decreases and approaches the dew point. Objects on the surface cool down, too, by radiative cooling, facilitating the condensation of small droplets of water. Dew formation is helped by the high humidity of the bottom layer of air, close to the surface. This layer can supply the needed water and prevent the evaporation of the dew already deposited. Strong winds can inhibit the formation of dew. Turbulence mixes a larger layer of air and creates a more homogeneous distribution of humidity and heat, thus preventing the formation of the right conditions near the ground. Dew forms more easily on surfaces that cool efficiently, like metals, which is the reason that cars are often covered with dew in the morning. But dew is often seen on grass or plants because their transpiration creates a thin layer of very humid air.

Dew is very important in the ecology of many deserts, especially those along the western coasts of Africa and South America. In some of them, water from condensation of dew caused by night cooling often exceeds that of rainfall. But dew can be an interesting source of water for human consumption, too. For instance, data recorded in the Negev desert of Israel, a country where water is really scarce, have shown that dew falls for 200 days each year. Moreover, dew is plentiful on many little islands that, surrounded by water, have high humidity but where no water dwells. Some islands of the Mediterranean Sea suffer from a chronic water shortage, which is the reason that one of the first dew-collecting plants has been built in Ajaccio (Corsica Island in France). It is also interesting to describe an old project studied in India and unfortunately never actually realized. The idea was to pump water at 4.5 °C from the sea at a depth of 500 meters. The pumping scheme called for the use of four 1.2-meter pipes and wind-powered pumps. A heat exchanger of 130,000 square meters could then condense, every day, more than 600 cubic meters of dew.

But many animals and plants have already learned how to survive on dew. A place particularly interesting for this kind of adaptation is the Namib desert in Namibia (Africa). Here, many insects in the early morning sit on top of sand dunes trying to catch some dew (or fog) on their wings and legs. In particular, the beetle *Stenocara* has even modified the surface of its hard shell surface to trap mist more efficiently. This beetle shell is covered with bumps whose peaks are smooth like glass and attract water. The troughs around the bumps are covered with a wax that repels water. The water is therefore collected by the peaks, and when a droplet is big enough to touch the water repelling valley it rolls down to the animal's mouth. Some researchers think that the easy trick of the animal can be used in dew collecting devices. A similar surface can be prepared in many ways, even with an ink-jet printer that sprays hydrophilic ink onto an acetate sheet. Dew-collecting devices have the problem of making the water run off the collecting surface.

Another Namibian creature well adapted to survive on dew is the plant *Welwitschia mirabilis*. Despite its look, it is a close relative of pine trees. It has a short trunk split in two and from each side grows a single leaf that can be several meters long. Welwitschias are among the oldest plants on the earth, and some of them are 1500 years old. The trick that allows this plant to survive in one of the driest deserts of the planet is a two-part root system. Welwitschias have a long root that goes deep in the soil trying to reach the water table and extends for several meters away from the plant. These roots collect all the dew that forms on the top layer of soil. Moreover, all the water that condenses on the leaves forms big drops that eventually run down, so that the plant can water its own roots.

DEW DESERTS

Giora J. Kidron
The Hebrew University of Jerusalem
Jerusalem, Israel

The term dew deserts denotes deserts that receive dew precipitation throughout the year. Dew is defined as condensation of atmospheric vapor and does not include vapor condensation that stems from the wet ground, i.e., distillation. We term dew desert as a desert that receives at least 10 mm of dew in >0.1 mm of daily precipitation. As a result of dew precipitation, primary production in dew deserts is much higher, affecting the entire food chain.

Like the Atacama Desert in South America, the Namib Desert in South Africa, and the Sonora Desert in Baja California (North America), dew only affects those portions of the desert close to a sea or ocean, i.e., to a large body of water that can serve as an adequate vapor source. As most deserts occupy mainly the interior of the continents, the term applies only to a small part of near sea deserts.

At present, information regarding the extent of dew precipitation within deserts is scarce. Most of the available information pertains to the heart of the Negev Desert where annual precipitation is 90–100 mm (1). There, a total of 33 mm was annually measured (2). Nevertheless, based on rain isohyets and the proximity to a sea or ocean (3), one may hypothesize that parts of the Sahara and the Arabian Peninsula are in fact dew deserts. Certain areas within the fog deserts may also be dew deserts (4–6).

Although daily dew amounts may usually range between 0.1 and 0.3 mm and the total annual sum may be much smaller than that of rain precipitation, the occurrence of a steady and constant source may be of great importance in arid and semiarid zones (7–9). As the threshold for organism activity was found to be 0.03 mm of dew (10), amounts of 0.1–0.3 mm are still sufficient to allow growth activity.

Although small, the dew amounts received may allow the establishment and distribution of many species, otherwise not capable of inhabiting the site. Consequently, higher biomass and higher species diversity may characterize the dew deserts. Biomass increase per millimeter of dew may be much higher than that calculated for rain (7) and that extrapolated, by adding the addition of water from dew to that of rain. The dew may thus be an important source of moisture for the primary food chain in arid and semiarid zones (11).

Many groups of micro-organisms were hypothesized to use dew. Among them, cyanobacteria, hypolithic algae and cyanobacteria that occupy the underside of stones (12,13), and lichens (14,15). Among the lichens are the endolithic lichens, embedded within calcite or dolomite crystals at the upper 1–5 mm of the rock and stone surface (16); epilithic and epedaphic lichens that dwell on the rock and soil surface, respectively (17); foliose (18); and fruticose lichens (18,19). In addition, bugs, ants, and beetles were reported to use dew directly by drinking it (20), whereas isopods and beetles were reported to use it indirectly by consuming wet food (20). When wet, snails in the Negev Desert were reported to successfully graze on endolithic lichens. By doing so, the snails disintegrate the limestone, a weathering phenomenon that can take place only upon rock moistening (21).

Although cyanobacteria, hypolithic algae, and cyanobacteria, epedaphic, endolithic, epilithic, foliose, and fruticose lichens were all hypothesized to use dew, positive evidence for dew use under field conditions was obtained only from endolithic, epilithic, foliose, and fruticose lichens. These lichens were shown to photosynthesize for 2–4 h in the morning following dew in the Negev Desert (Fig. 1) (10,18,22). One should note that whereas lithobiontic lichens (inhabiting rocks and stones) were shown to use dew (22), dew use by micro-organisms inhabiting the soil is still controversial. Whereas Lange et al. (22) and Veste et al. (23) showed the use of dew by epedaphic lichens inhabiting soil and sand, respectively, and indirect evidence, expressed by the development of sexual and vegetative reproduction organs, were also monitored in mosses in the Negev Desert (24), the use of dew by endedaphic cyanobacteria inhabiting sand is not certain. According to Jacobs et al. (25), dew may moisten the surface. However, other reports indicate that dew moistening of the soil is rare (24,26). According to Bunnenberg and Kühn (26), an amount of 0.13 mm of dew at 9 cm above ground amounted to 0.03 mm only at the surface. Similarly, Kidron et al. (24) showed that average dew precipitation of 0.1 mm measured on glass plates at 0.7 cm above ground amounted to only 0.034 mm at the surface. As cyanobacteria need liquid water for growth (27), and the necessary threshold for net photosynthesis was 0.1 mm (28), only rarely sufficient dew will moisten the surface to reach >0.1 mm, and thus the contribution of dew to the growth and development of endedaphic cyanobacteria may be negligible.

The uncertainty regarding the use of dew by certain micro-organisms and at certain habitats is mainly linked to two factors: (a) Does dew condense at all habitats? (b) Is the amount supplied sufficient for use and for net carbon gain? As for the second question, rapid dew evaporation during the morning may result in a net loss of carbon, as was shown for the lichen *Ramalina maciformis* following mornings of low dew precipitation (10).

Whereas the efficiency to which different micro-organisms may use dew may be species-dependent, no controversy exists as to the perquisite conditions necessary, i.e., the capability of the dew to condense at their habitat. This ability depends, of course, on the existence of sufficient moisture in the air. It also depends

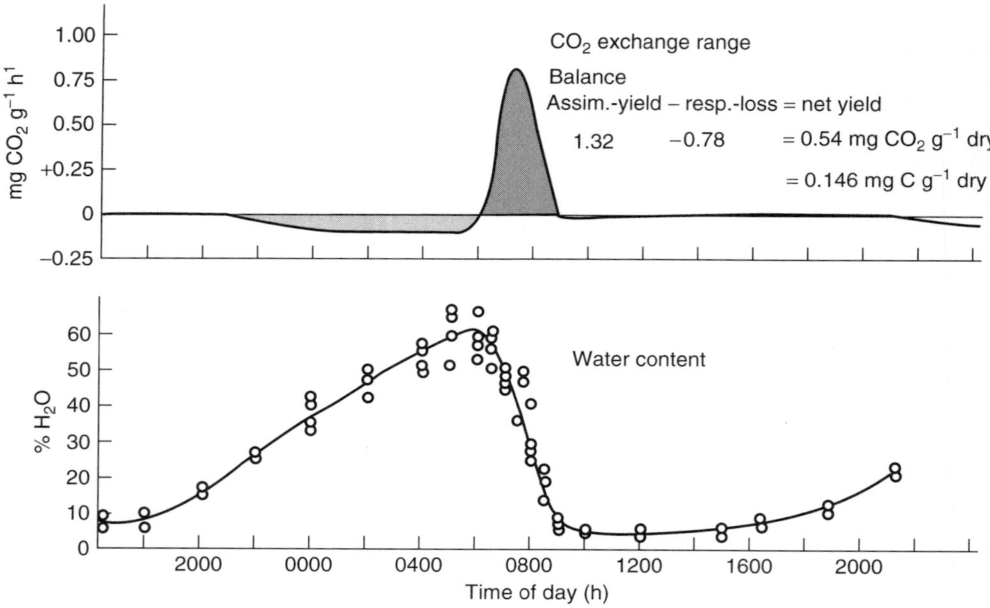

Figure 1. Apparent CO_2 exchange rate (above) and the thalli water content (below) of *Ramalina maciformis* during 24 h in September 1967. Net carbon yield resulted in 0.146 mg C per gram of dry weight a day (18).

on substrate temperatures that will dictate whether the dew point temperature, necessary for vapor condensation, will be reached.

Once reached, dew condensation will take place. As dew was found to condense at a more or less constant rate (29), the time during which the dew point is reached may dictate, to a large extent, the overall amount of dew that will be condensed. Furthermore, as phototrophic microorganisms necessitate light hours for photosynthesis, the length of time during which dew is available may dictate, to a large extent, the overall net gain of organic carbon owing to dew use. Thus, variables that may affect dew condensation and dew duration should be considered once dew availability in different habitats is examined. Factors such as topographical elevation, height above ground, aspect, location along the slope, and angle may all affect dew condensation.

Recent dew measurements carried in the Negev Desert aimed to examine the role of the above-mentioned factors. Thus, in order to obtain continuous dew amounts and duration, a simple and inexpensive method was adapted. The Cloth-Plate Method (CPM) consists of $10 \times 10 \times 0.2$ glass plates glued at their bottom to $10 \times 10 \times 0.5$ cm plywood plates, thus creating an identical substratum (30). Velvet-like cloth ($6 \times 6 \times 0.15$ cm) are attached each afternoon to the center of the glass plate and collected throughout the following morning into glass jars that are immediately sealed and then weighed in a nearby lab, oven dried (in 70 °C) and then weighed again, and their moisture content is calculated. By placing plates next to each other within a certain habitat, the CPM facilitates inexpensive large-scale continuous measurements (Fig. 2). The readings are also not affected by wind, as might be the case with some other devices (31).

Figure 2. Dew measurements by the CPM in the Negev Highlands.

The CPM was used to assess the possible role of altitude and distance from the sea. When simultaneous measurements at three locations at 250, 550, and 1000 m above sea level being 37, 55, and 98 km from the Mediterranean Sea were carried out in the Negev Desert, positive correlations between dew and fog amounts and altitude were found (Fig. 3). Dew precipitation, as well as fog precipitation, increased with altitude, with the most elevated location receiving more than twice the amount obtained at the topographically lowest location, which was the case, although the most elevated location was also the farthest away from the Mediterranean Sea, i.e., from a vapor source. The data thus indicate that, within 100 km from the sea, topographical height may compensate for the increase in distance from the vapor source. In this regard, one should note that elevation plays a major role in controlling dew and fog precipitation also in the Atacama Desert (4).

The CPM facilitates careful analysis of dew precipitation. When continuous dew measurements were carried out, a typical condensation and evaporation pattern was obtained during the morning hours (Fig. 4). Dew condensation was found to continue also after dawn and sunrise, explained by radiation-induced air turbulence (30). As for the relationship between dew amounts and height above ground, dew values showed high variability (33,34). When dew was measured on a hilltop in the Negev Highlands at 0.7, 10, 20, 30, 40, and 50 cm above ground, maximal values were recorded at 10 cm above ground (Fig. 5). The findings reflected two effects: the warming effect of the soil as a result of a nocturnal heat flux that rises from the deeper horizons of the soil, responsible for the decrease in dew values near the ground (31,34,35), and the air turbulence at height, resulting from higher wind velocities, responsible for lower dew values farther away from the ground (35,36).

High variability in dew amounts and duration was also found in the Negev Desert when dew was measured at 0.7 and 40 cm above ground along limestone slopes (approximately 50 m long) of four aspects (north, south, east, and west) within a second-order drainage basin (Fig. 6). Whereas at 0.7 cm above ground, the hilltops and the bottom parts of the northern and western aspects received the highest amounts, the wadi beds received approximately half these values (Fig. 6a). The lowest values were received at the south-facing midslope (located at the lee side of the prevailing north north-west winds), being approximately a quarter of the maximal values recorded. At 40 cm above ground, a reverse trend was obtained with the south-facing midslope and the wadi beds receiving the highest values (Fig. 7). And thus, whereas the 0.7 cm above-ground results were not in accordance with the classical model that predicts high dew values

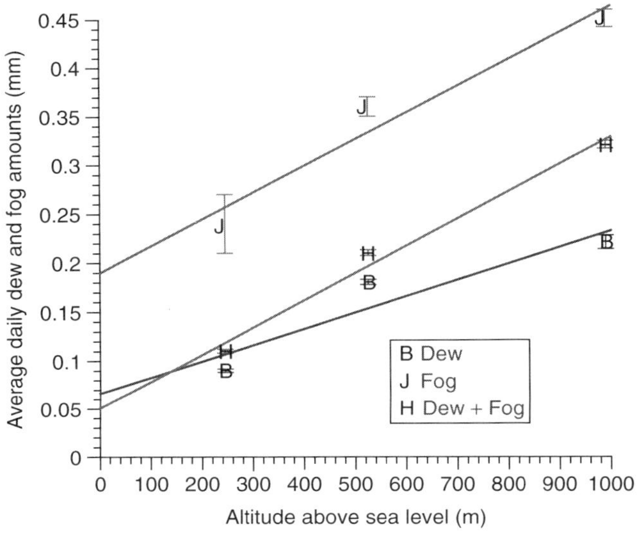

Figure 3. The relationships between dew and fog precipitation with altitude in the Negev Desert (32).

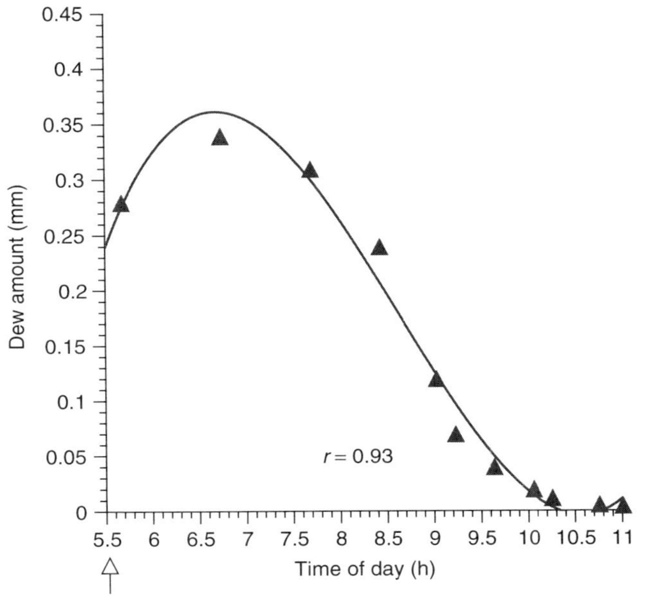

Figure 4. A typical condensation and evaporation pattern during the morning hours (30).

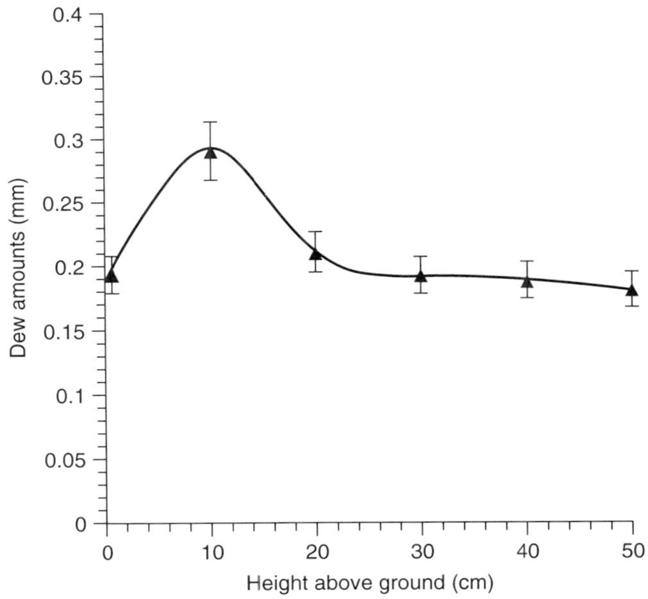

Figure 5. Dew amounts with height above ground (30).

at the wadi beds (because of nocturnal katabatic wind) and higher amounts at the lee side of the wind, i.e., at the south-facing slope (because of undisturbed inversion), the 40 cm above ground measurements corresponded to the classical model. The discrepancy was explained by the overwhelming impact of the rock surface temperatures on the dew values at 0.7 cm above ground, with south-facing rock surfaces being 3–8 degrees higher than the north-facing rock surfaces throughout the night. As for the higher amounts of dew received at 0.7 cm above ground at the hilltops in comparison with the wadi

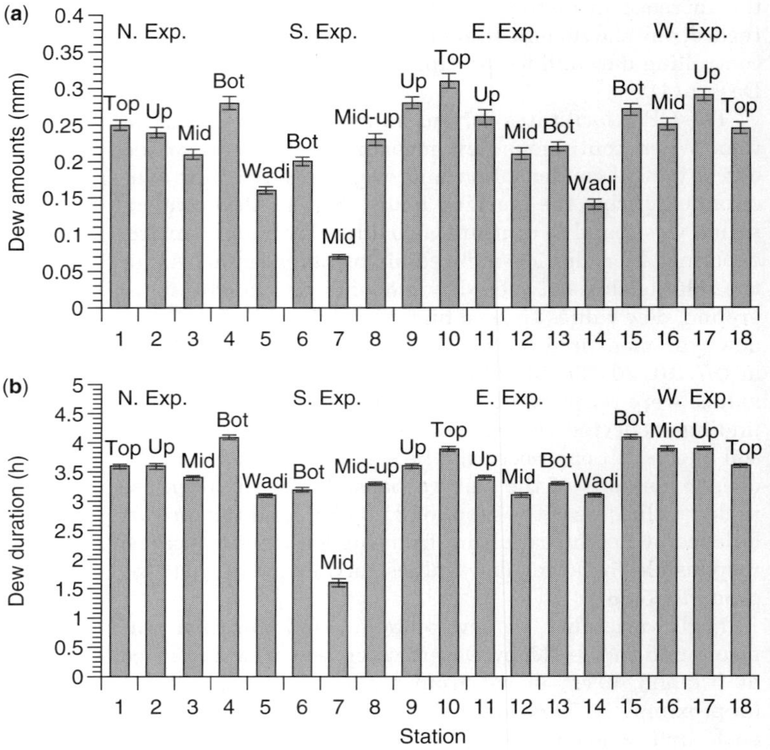

Figure 6. Dew amounts (**a**) and duration (**b**) within 18 stations located at 4 aspects of a second-order drainage basin in the Negev Desert. Top = hilltop, Up = Upper slope, Mid = Midslope, Bot = Bottom slope (37).

beds, it was hypothesized that the afternoon winds act as a cooling agent, facilitating a much earlier drop in surface temperatures (and thus reaching much faster the necessary dew point temperature) at the wind-exposed hilltops (37).

Aspect and slope location also dictate dew duration, seen as being of even greater ecological importance to organisms than the maximal amount (38–40). Generally, a positive correlation between dew amounts and duration was found (34,37), and thus, by affecting dew amounts, aspect and slope position also affect dew duration (Fig. 6b). Nevertheless, dew duration was also affected by slope location and aspect beyond these relations. Although condensation was found to take place in all habitats also after sunrise (for usually 0.5–1.0 h after sunrise), condensation at the sun-sheltered habitats of the bottom north- and west-facing slopes continued for up to 1.5 h following sunrise (41). As a result of the higher maximal values and the limited desiccation effect of the sun in these habitats, these habitats were characterized by longer dew duration.

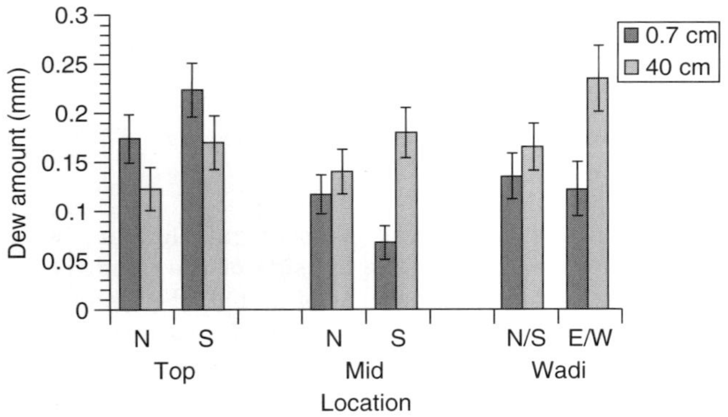

Figure 7. Dew measurements at 0.7 and 40 cm above ground at three habitats within a second-order drainage basin in the Negev Desert. N = northern exposure, S = southern exposure, Wadi N/S = wadi between the northern and southern exposure, Wadi E/W = wadi between the eastern and western exposures (37).

The substrate angle is another factor found to affect dew amounts and duration. When cloths were attached to $50 \times 50 \times 10$ cm wooden boxes, having sides of different angles (30, 45, 60, 75, and 90°) placed at different aspects (facing north, south, west, and east) on a hilltop, a decrease in dew amounts with an increase in angle, from 30° to 90°, was found and dew amounts were positively correlated with $\cos(\theta)$ (Fig. 8). Thus, dew amounts obtained at an angle of 90° was approximately a quarter of the values obtained at a horizontal surface and at an angle of 30°, both of which received similar values. This difference was explained by the slower rate of nocturnal cooling that takes place with an increase in angle in accordance with the lower proportion of sky seen by the substrate (43). No preferential condensation in accordance with aspect was found. Nevertheless, dew duration was aspect-dependent with a decrease in duration following the order west>north>south>east (Fig. 9). Thus, daylight dew duration was approximately double at the west-facing sun-sheltered angle than at the sun-exposed east-facing aspect.

The variability in dew amounts and duration may have important consequences for micro-organisms. For

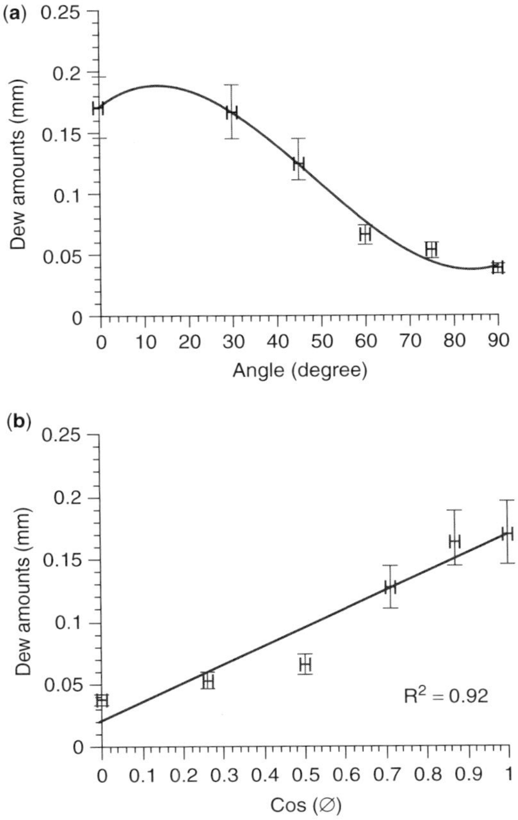

Figure 8. The relationships between dew amounts with (**a**) angle (in degrees) and (**b**) cos(θ) as measured in the Negev Desert (42).

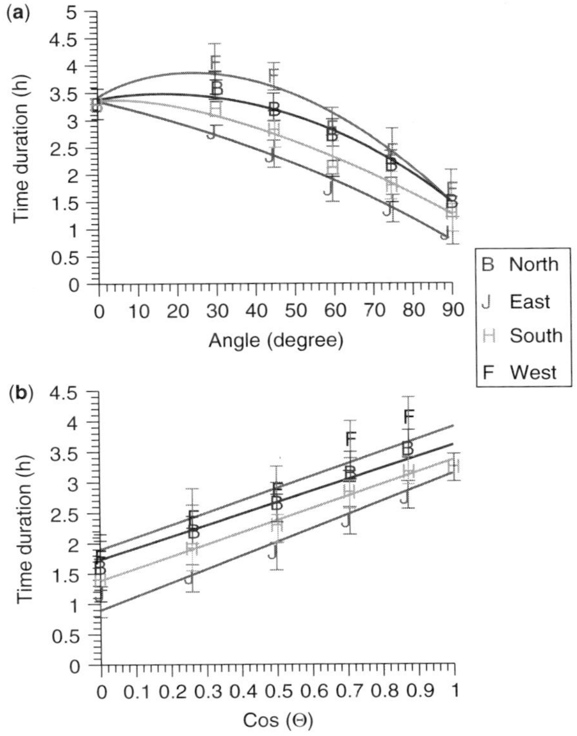

Figure 9. The relationships between dew duration with (**a**) angle (in degrees) and (**b**) cos(θ) in the Negev Desert (42).

most micro-organisms, the ability to use dew is primarily a function of the amount condensed and the duration during which dew is available, both of which are habitat-dependent. Thus, when cloths were attached to different substrates, high variability in dew amounts was received on loose cobbles, partially embedded cobbles, and rock surfaces (Fig. 10). Loose cobbles that rapidly cool at night were found to receive approximately twice the amounts of embedded cobbles and more than 4 times the amounts condensed on a nearby rock (44). Whereas the cobbles were primarily inhabited by endolithic lichens, the rock surfaces were mainly inhabited by epilithic lichens. Similarly, lichen growth on loose cobbles was controlled by angle-induced dew. Thus, the top of loose cobbles, inhabited by lichens, received approximately twice the amount condensed on the uninhabited side of the cobble (45). Consequently, dew contribution to the ecosystem biomass may be highly important. For instance, in research conducted by Kappen et al. (22,46) in the Negev Desert, the biomass of lichens (with *Ramalina maciformis* predominating) in the northern aspect was over 200 g m^{-2}, three to five times as much than at the other aspects. Although some of the differences can certainly be attributed also to use of rain-induced moisture, the fact that the main water source for *Ramalina maciformis* and other endolithic and epilithic lichens in the Negev Desert was dew pointed to the importance of dew on lichen biomass. The difference in lichen cover of the sun-sheltered and the sun-exposed slopes points also to the role of dew in lichen distribution (47). As lichens may serve as food for snails, the whole food chain may be affected by dew precipitation.

Dew precipitation may also affect vascular plants. The effect may not always be positive, as dew may hasten plant fungal infection (7,48). Dew may preferentially accumulate on leaves because of their lower temperature (0.5–2 °C lower in comparison to the ambient air) (35,36) and thus, wet the leaves for several hours during the morning (34).

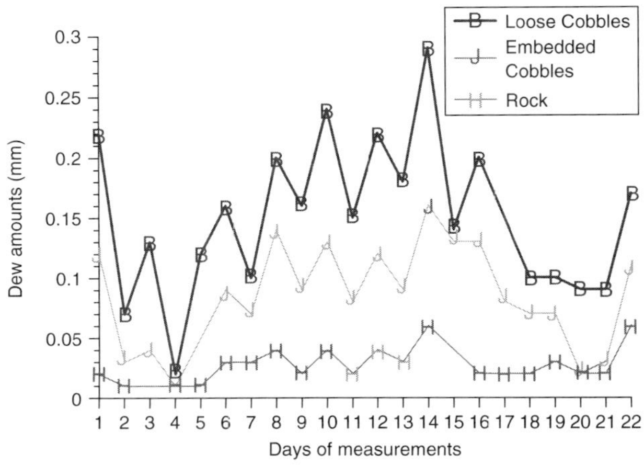

Figure 10. Simultaneous dew amounts as obtained on loose and embedded cobbles inhabited by endolithic lichens, and rock surface inhabited by epilithic lichens (44).

However, no conclusive evidence of direct dew use by vascular plants is yet available, and the research conducted on this topic is controversial (7,8,31,49,50). Dew, however, may have an indirect effect, facilitating stomata opening (51), longer hours of photosynthesis (39), reducing transpiration (7), and affecting flowering (7). Dew may facilitate seedling survival (52), recovery from wilting (34), plant growth (34), and yield (34,53). By affecting micro-organisms and plant growth, and by facilitating insect survival (20), dew also indirectly affects the nutrient cycling (54) and soil-forming processes (55), which in turn has an important impact on the ecosystem and may thus call for a specific awareness of the dew deserts.

BIBLIOGRAPHY

1. Rosenan, N. and Gilad, M. (1985). Meteorological data. In: *Atlas of Israel*. Carta, Jerusalem.
2. Evenari, M. (1981). Ecology of the Negev Desert, a critical review of our knowledge. In: *Developments in Arid Zone Ecology and Environmental Quality*. H. Shuval (Ed.). Balaban ISS, Philadelphia, PA, pp. 1–33.
3. Reitan, C.H. and Green, C.R. (1968). Appraisal of research on weather and climate of desert environments. In: *Deserts of the World, An Appraisal of Research Into Their Physical and Biological Environments*. W.G. McGinnies, B.J. Goldman, and P. Paylore (Eds.). The University of Arizona Press, Tucson, AZ, pp. 19–92.
4. Redon, J. and Lange, O.L. (1983). Epiphytic lichens in the region of a Chilean “fog oasis” (Fray Jorge). 1. Distributional patterns and habitat conditions. *Flora*. **174**: 213–243.
5. Lange, O.L., Meyer, A., Ullmann, I., and Zellner, H. (1991). Microclimate conditions, water content and photosynthesis of lichens in the coastal fog zone of the Namib Desert: measurements in the fall. *Flora*. **185**: 233–266.
6. Lange, O.L., Meyer, A., Zellner, H., and Heber, U. (1994). Photosynthesis and water relations of lichen soil crusts: field measurements in the coastal fog zone of the Namib Desert. *Functional Ecol*. **8**: 253–264.
7. Wallin, G.R. (1967). Agrometeorological aspects of dew. *Agric. Meteorol*. **4**: 85–102.
8. Evenari, M., Shanan, L., and Tadmor, N. (1971). *The Negev, the Challenge of a Desert*. Harvard Univ. Press, Cambridge, MA, p. 345.
9. Noy-Meir, I. (1973). Desert ecosystems: environment and producers. *Ann. Rev. Ecol. Syst*. **4**: 25–51.
10. Kappen, L. etal. (1979). Ecophysiological investigations on lichens of the Negev Desert, IV: Annual course of the photosynthetic production of *Ramalina maciformis* (Del.) Bory. *Flora*. **168**: 85–108.
11. Shachak, M. and Steinberger, Y. (1980). An algae-desert snail food chain: energy flow and soil turnover. *Oecologia*. **46**: 402–411.
12. Berner, T. (1974). *The Ecophysiology of the Hypolithic Algae in the Negev Highlands of Israel*. Ph.D. Thesis, The Hebrew University of Jerusalem.
13. Berner, T. and Evenari, E. (1978). The influence of temperature and light penetration on the abundance of the hypolithic algae in the Negev Desert of Israel. *Oecologia*. **33**: 255–260.
14. Friedmann, E.I., Lipkin, Y., and Ocampo-Paus, R. (1967). Desert algae of the Negev. *Phycologia*. **6**: 185–195.
15. Friedmann, E.I. and Galun, M. (1974). Desert algae, lichens and fungi. In: *Desert Biology II*. G.W. Brown (Ed.). Academic Press, New York, pp. 165–212.
16. Fry, E.I. (1922). Some types of endolithic lichens. *Ann. Bot*. **35**: 541–562.
17. Golubic, S., Freiedmann, I., and Schneider, J. (1981). The lithobiontic ecological niche, with special reference to microorganisms. *J. Sedimentary Petrol*. **51**: 475–478.
18. Lange, O.L., Schulze, E-D., and Koch, W. (1970). Ecophysiological investigations on lichens of the Negev Desert, II: CO_2 gas exchange and water conservation of *Ramalina maciformis* (Del.) Bory in its natural habitat during the summer dry period (Technical translation 1655. National Research Council of Canada). *Flora*. **159**: 38–62.
19. Lange, O.L. (1969). Ecophysiological investigations on lichens of the Negev Desert. I. CO_2 gas exchange of *Ramalina maciformis* (Del.) Bory under controlled conditions in the laboratory (Technical translation 1654. National Research Council of Canada). *Flora*. **158**: 324–359.
20. Broza, M. (1979). Dew, fog and hygroscopic food as a source of water for desert arthropods. *J. Arid. Environ*. **2**: 43–49.
21. Shachak, M., Jones, C.G., and Granot, Y. (1987). Herbivory in rocks and the weathering of a desert. *Science* **236**: 1098–1099.
22. Lange, O.L., Schulze, E-D., and Koch, W. (1970). Ecophysiological investigations on lichens of the Negev Desert, III: CO_2 gas exchange and water metabolism of crustose and foliose lichens in their natural habitat during the summer dry period. *Flora*. **159**: 525–538.
23. Veste, M., Littmann, T., and Friedrich, H. (2001). Microclimate boundary conditions for activity of soil lichen crusts in sand dunes of the north-western Negev Desert, Israel. *Flora*. **196**: 465–474.
24. Kidron, G.J., Hernstadt, I., and Barzilay, E. (2002). The role of dew as a moisture source for sand microbiotic crusts in the Negev Desert, Israel. *J. Arid. Environ*. **52**: 517–533.
25. Jacobs, A.F.G., Heusinkveld, B.G., and Berkowicz, S.M. (2000). Dew measurements along a longitudinal sand dune transect, Negev Desert, Israel. *Int. J. Biometeorol*. **43**: 184–190.
26. Bunnenberg, C. and Kühn, W. (1977). Application of the β-absorption method to measure dew on soil and plant surfaces. *Int. J. Appl. Radiation Isotopes* **28**: 751–754.
27. Lange, O.L., Kilian, E., and Ziegler, H. (1986). Water vapor uptake and photosynthesis of lichens: performance differences in species with green and blue-green algae as phycobionts. *Oecologia* **71**: 104–110.
28. Lange, O.L. et al. (1992). Taxonomic composition and photosynthetic characteristics of the “biological soil crusts” covering sand dunes in the Western Negev Desert. *Func. Ecol*. **6**: 519–527.
29. Zangvil, A. (1996). Six years of dew observation in the Negev Desert, Israel. *J. Arid. Environ*. **32**: 361–372.
30. Kidron, G.J. (1998). A simple weighing method for dew and fog measurements. *Weather* **53**: 428–433.
31. Noffsinger, T.L. (Ed.). (1965). Survey of techniques for measuring dew. In: *Humidity and Moisture, Measurement and Control in Science and Industry*. A. Wexler (Ed.). Reinhold Publishing Corporation, New York, pp. 523–531.
32. Kidron, G.J. (1999). Altitude dependent dew and fog in the Negev desert, Israel. *Agric. Forest Meteorol*. **96**: 1–8.
33. Lloyd, M.G. (1961). The contribution of dew to the summer water budget of Northern Idaho. *Bull. Am. Meteorol. Soc*. **42**: 572–580.
34. Duvdevani, S. (1964). Dew in Israel and its effect on plants. *Soil Sci*. **98**: 14–21.
35. Long, I.F. (1958). Some observations on dew. *Meteorol. Mag*. **87**: 161–168.

36. Monteith, J.L. (1957). Dew. *Quart. J. Royal Meteorol. Soc.* **83**: 322–341.
37. Kidron, G.J., Yair, A., and Danin, A. (2000). Dew variability within a small arid drainage basin in the Negev highlands, Israel. *Quart. J. Royal Meteorol. Soc.* **126**: 63–80.
38. Lange, O.L., Geiger, I.L., and Schulze, E-D. (1977). Ecophysiological investigations on lichens in the Negev Desert. V. A model to simulate net photosynthesis and respiration of *Ramalina macifomis*. *Oecologia.* **28**: 247–259.
39. Kappen, L. et al. (1980). Ecophysiological investigations on lichens of the Negev Desert, VII: The influence of the habitat exposure on dew imbibition and photosynthetic productivity. *Flora.* **169**: 216–229.
40. Lange, O.L. and Tenhunen, J.D. (1982). Water relations and photosynthesis of desert lichens. *J. Hattori. Bot. Lab.* **53**: 309–313.
41. Kidron, G.J. (2000). Analysis of dew precipitation in three habitats within a small arid drainage basin, Negev Highlands, Israel. *Atmos. Res.* **55**: 257–270.
42. Kidron, G.J. (in press). Angle and aspect dependent dew precipitation in the Negev Desert, Israel. *J. Hydrol.*
43. Oke, T.R. (1978). *Boundary Layer Climates*. John Wiley & Sons, New York, p. 372.
44. Kidron, G.J. (2000). Dew moisture regime of endolithic and epilithic lichens inhabiting calcareous cobbles and rock outcrops, Negev Desert, Israel. *Flora.* **195**: 145–153.
45. Kidron, G.J. (2002). Causes of two patterns of lichen colonization on cobbles in the Negev Desert, Israel. *Lichenologist.* **34**: 71–80.
46. Kappen, L., Lange, O.L., Schulze, E-D., Evenari, M., and Buschbom, U. (1975). Primary production of lower plants (lichens) in the desert and its physiological basis. In: *Photosynthesis and Productivity in Different Environments.* J.P. Cooper (Ed.). Cambridge University Press, Cambridge, UK, pp. 133–143.
47. Danin, A. and Garty, J. (1983). Distribution of cyanobacteria and lichens on hillsides of the Negev Highlands and their impact on biogenic weathering. *Z. Geomorph.* **27**: 423–444.
48. Duvdevani, S., Reichert, I., and Palti, J. (1946). The development of downy and powdery mildew of Cucumbers as related to dew and other environmental factors. *Palestine J. Bot.* (Rehovot Series). **2**: 127–151.
49. Stone, E.C. (1957). Dew as an ecological factor. I. A review of the literature. *Ecology* **38**: 407–413.
50. Waisel, Y. (1958). Dew absorption by plants of arid zones. *Bull. Res. Counc. Israel* **6D**: 180–186.
51. Schulze, E-D. et al. (1972). Stomatal responses to changes in humidity in plants growing in the desert. *Planta* **108**: 259–270.
52. Stone, E.C. (1957). Dew as an ecological factor. II. The effect of artificial dew on the survival of *Pinus ponderosa* and associated species. *Ecology* **38**: 414–422.
53. Rotem, J. and Reichert, J. (1964). Dew - a principal moisture factor enabling early blight epidemics in a semi-arid region of Israel. *Plant Disease Reptr.* **48**: 211–215.
54. Jones, C.G. and Shachak, M. (1990). Fertilization of the desert soil by rock-eating snails. *Nature* **345**: 839–841.
55. Shachak, M., Jones, C.G., and Brand, S. (1995). The role of animals in an arid ecosystem: Snails and isopods as controllers of soil formation, erosion and desalinization. *Adv. GeoEcol.* **28**: 37–50.

DEW POINT

Aldo Conti
Frascati (RM), Italy

The dew point is the temperature to which air must be cooled at constant pressure until it reaches saturation and the water, or any other vapor, begins condensing. The result is the formation of fog. When the dew point falls below freezing, it is called the frost point. In this case, water forms ice crystals directly. The dew point is also a good indicator of the amount of water contained in the air. The more humid the air, the higher the dew temperature. The dew point is often reached in the evening, when the air cools down, which is why fog is more common in the evening, or why one normally finds that dew or frost are common in the morning but disappear as soon as the temperature increases.

Dew point and relative humidity can be measured with an instrument called a wet- and dry-bulb psychrometer. The dry bulb is a normal thermometer that measures the actual air temperature. The wet bulb is another thermometer, but has a bulb wrapped in a piece of cloth dampened by a string that dips into a bottle of distilled water. Unless the air is so humid that it is very close to saturation, the wet bulb thermometer measures a lower temperature, because it is cooled by the evaporation of water. The difference between the dry-bulb and wet-bulb temperatures is called the wet-bulb depression. It is then possible to calculate the dew point as a function of the wet-bulb depression and of the dry-bulb temperature.

To calculate the dew point starting from the psychrometer readings, the first step is to calculate the saturation vapor pressure in millibars, corresponding to the dry- and wet-bulb temperature:

$$E_s = 6.11 \times 10.0[7.5T/(237.7 + T)]$$

$$E_{swb} = 6.11 \times 10.0[7.5T_{wb}/(237.7 + T_{wb})]$$

where T and T_{wb} are the readings of the dry- and wet-bulb temperatures.

Now we are ready to calculate the actual mixing ratio of the air:

$$W = [(T - T_{wb})(C_p) - L_v(E_{swb}/P)]/[-(T - T_{wb})(C_{pv}) - L_v]$$

where
C_p = specific heat of dry air at constant pressure (J/g)~1.005 J/g
C_{pv} = specific heat of water vapor at constant pressure (J/g)~4.186 J/g
L_v = Latent heat of vaporization (J/g)~2500 J/g
T = air temperature in °C
T_{wb} = wet bulb temperature in °C
E_{swb} = saturation vapor pressure at the wet-bulb temperature (mb)
P = atmospheric pressure at the surface ~1013 mb at sea level

We can now use the following formula to obtain the saturation mixing ratio for the air:

$$W_s = E_s/P$$

$$\mathrm{RH} = W/W_s$$

Now we can use the relative humidity to calculate the actual vapor pressure (E) of the air as follows:

$$E = \mathrm{RH} \times E_s$$

The dew point temperature is then

$$T_d = [-430.22 + 237.7 \times \ln(E)]/[-\ln(E) + 19.08]$$

Let us now assume a psychrometer that gives the following readings at a pressure of 1013 mbar:

$$T = 30\,^{\circ}\mathrm{C}$$

$$T_{wb} = 20\,^{\circ}\mathrm{C}$$

Then,

$$E_{swb} = 23.34 \text{ mbar}$$

$$E_s = 42.31 \text{ mbar}$$

The mixing ratio becomes

$$W = 0.019$$

$$W_s = 0.042$$

The relative humidity is then,

$$\mathrm{RH} = 0.45 \text{ (or 45\%)}$$

The actual vapor pressure is then,

$$\mathrm{E} = 18.95 \text{ mbar}$$

and the dew point is

$$T_d = 16.7\,^{\circ}\mathrm{C}$$

Instead of going through all these formulas, it is possible to use tables that give the dew point temperature as a function of the readings for different pressures.

DROUGHTS

Arthur M. Holst
Philadelphia Water Department
Philadelphia, Pennsylvania

A drought is a period of time during which weather is in one way or another excessively dry. A drought's severity is determined by the degree of moisture deficiency in the affected area, as measured by various methods. Droughts are a standard, recurring, and normal feature of long-term climate. There are four generally accepted categories of drought: meteorological, socioeconomic, hydrologic, and agricultural.

Meteorological droughts involve a lack of precipitation during a given time period. Their specific definition depends mostly on the region in question; a lack of precipitation for some areas would be excessive precipitation in others. Some regions receive rain year-round, and some receive virtually none. Likewise, some regions receive precipitation consistently throughout the year, and others have seasonal precipitation patterns. A departure from any of these climatological characteristics as abnormally low precipitation can cause a meteorological drought. Meteorological droughts are usually measured in terms of their duration and in contrast to established precipitation averages for specified time periods for a specified area. This type of drought generally precedes the other three types.

Agricultural droughts need not involve a severe lack of precipitation. They are marked by damage to crops and plants due to insufficient water supplies. Because water from rain and snow is more likely to run into creeks and rivers instead of filtering into the soil due to human intervention, agricultural droughts can sometimes occur in spite of noticeable rainfall. Depending on a plant's species and stage of growth, it needs varying amounts of water from the soil; when the majority of crops cannot obtain the amount of water they need, an area is said to be undergoing an agricultural drought. This type of drought is most damaging to farmers and those in poor countries, and it generally comes after the onset of a meteorological drought.

A hydrologic drought involves water shortages in reservoirs, lakes, and streams due to lack of precipitation. These droughts are measured by stream flow and water levels in local reservoirs. Local climate is not the only factor that causes a hydrologic drought; altered land usage, dams, and degraded soil quality can all have effects as well. Reduced precipitation in one area can cause a hydrologic drought in another because various areas are hydrologically connected by their rivers, lakes, and other bodies of water. The primary cause of this type of drought is generally a meteorological drought; hydrologic droughts usually come some time after agricultural droughts.

The final category of drought is socioeconomic. This type of drought, like the other three, involves a lack of precipitation. It occurs when reduced precipitation causes a shortage of water as related to the demands of the local populace. The socioeconomic drought is the only type of drought that has a major effect on the general population, and it is generally the last of the four types to occur.

Aside from lack of rain, the three most important variables to look at during a drought are air temperature, humidity, and circulation patterns in the atmosphere. Atmospheric circulation patterns that produce little or no precipitation are often associated with droughts. Climate abnormalities are another component of a drought. Precipitation levels that would be considered a drought in one area are normal in another that has a different climate. To understand the cause of a drought, scientists study the circulation patterns of the atmosphere across global distances, using computer-generated atmospheric simulator models.

Droughts also have direct effects on communities that experience this phenomenon. Both water quality and cost are negatively affected by droughts. Changes in the taste and odor of drinking water can also occur. Those who draw their water from wells may have to drill more deeply to reach the lowered level of the water table.

During a drought, water conservation is important. Restrictions on watering gardens, washing cars, and using sprinkler systems are applied when a drought emergency or warning is in effect. There are generally three stages of drought preparedness: drought watches, drought warnings, and drought emergencies. During a drought watch, no restrictions are enforced, but citizens are encouraged to curb water usage because the possibility of drought has been established for an area. During a drought warning, restrictions are still not enforced, but citizens are instructed to reduce water usage further in response to impending drought conditions. The final stage is a drought emergency. This is the most serious stage, often involving mandatory water usage guidelines to ensure that sufficient water is available for critical needs. At this stage, officials try to avoid local shortages by evenly distributing the water available.

Droughts can have a great impact in many ways. There are three main types of impact they can have: economic, social, and environmental. A lack of water in the soil for crops can cause farmers great financial difficulty. The resulting decrease in crop production has a ripple effect on the economy, causing short-term and long-term problems. One of the biggest short-term problems is unemployment. Certain industries can also be affected by long-term problems. For instance, loss of tax revenue can hinder tourism. Brush and trees can become very dry, and an outbreak of destructive fires can result. Drought can harm the logging industry, fisheries, and hydroelectric power generation, all economic effects. It can also cause insect problems in agriculture, as well as erosion and disease. Environmental effects include reduced biodiversity, degraded air quality, and perhaps most importantly, devastating forest fires. Social impacts include induced emigration, rampant famine, and, indirectly, greater poverty. These are merely a few of the multitude of impacts, direct and indirect, that drought can have.

READING LIST

All About Droughts. (2002). Available: http://205.156.54.206/om/drought.htm. (March 13).

Hanson, Ronald L. Evaporation and Droughts, *U.S. Geological Survey*. (2002). Available: http://geochange.er.usgs.gov/sw/changes/natural/er/. (March 13).

Impacts of Drought. (2002). Available: http://enso.unl.edu/ndmc/enigma/impacts.htm. (March 13).

McNab, A. and Karl, T. Climate and Droughts, *National Oceanic and Atmospheric Administration*. (2002). Available: http://geochange.er.usgs.gov/sw/changes/natural/drought/. (March 13).

Stormfax Guide to Droughts, *Stormfax*. Available: http://www.stormfax.com/drought.htm.

Understanding Your Risk: Impacts of Drought. Available: http://www.drought.unl.edu/risk/impacts.htm.

All About Droughts. Available: http://www.nws.noaa.gov/om/drought.htm.

Understanding and Defining Drought. Available: http://www.drought.unl.edu/whatis/concept.htm.

Drought Information Center. Available: http://www.dep.state.pa.us/dep/subject/hotopics/drought/drtterm.htm.

DROUGHT INDICES

Michael J. Hayes
Climate Impacts
Specialist—National Drought
Mitigation Center

INTRODUCTION

Drought indices assimilate thousands of bits of data on rainfall, snowpack, streamflow, and other water supply indicators into a comprehensible big picture. A drought index value is typically a single number, far more useful than raw data for decision making.

There are several indices that measure how much precipitation for a given period of time has deviated from historically established norms. Although none of the major indices is inherently superior to the rest in all circumstances, some indices are better suited than others for certain uses. For example, the Palmer Drought Severity Index has been widely used by the U.S. Department of Agriculture to determine when to grant emergency drought assistance, but the Palmer is better when working with large areas of uniform topography. Western states, with mountainous terrain and the resulting complex regional microclimates, find it useful to supplement Palmer values with other indices such as the Surface Water Supply Index, which takes snowpack and other unique conditions into account.

The National Drought Mitigation Center is using a newer index, the Standardized Precipitation Index, to monitor moisture supply conditions. Distinguishing traits of this index are that it identifies emerging droughts months sooner than the Palmer Index and that it is computed on various time scales.

Most water supply planners find it useful to consult one or more indices before making a decision. What follows is an introduction to each of the major drought indices in use in the United States and in Australia.

PERCENT OF NORMAL

Overview: The percent of normal is a simple calculation well suited to the needs of TV weathercasters and general audiences.

Pros: Quite effective for comparing a single region or season.

Cons: Easily misunderstood, as normal is a mathematical construct that does not necessarily correspond with what we expect the weather to be.

The percent of normal precipitation is one of the simplest measurements of rainfall for a location. Analyses using the percent of normal are very effective when used for a single region or a single season. Percent of normal is also easily misunderstood and gives different indications of conditions, depending on the location and season. It is calculated by dividing actual precipitation by normal precipitation—typically considered to be a 30-year mean—and multiplying by 100%. This can be calculated for a variety of time scales. Usually these time scales range from a single month to a group of months representing a particular season, to an annual or water year. Normal precipitation for a specific location is considered to be 100%.

One of the disadvantages of using the percent of normal precipitation is that the mean, or average, precipitation is often not the same as the median precipitation, which is the value exceeded by 50% of the precipitation occurrences in a long-term climate record. The reason for this is that precipitation on monthly or seasonal scales does not have a normal distribution. Use of the percent of normal comparison implies a normal distribution where the mean and median are considered to be the same. An example of the confusion this could create can be illustrated by the long-term precipitation record in Melbourne, Australia, for the month of January. The median January precipitation is 36.0 mm (1.4 in.), meaning that in half the years less than 36.0 mm is recorded, and in half the years more than 36.0 mm is recorded. However, a monthly January total of 36.0 mm would be only 75% of normal when compared to the mean, which is often considered to be quite dry. Because of the variety in the precipitation records over time and location, there is no way to determine the frequency of the departures from normal or compare different locations. This makes it difficult to link a value of a departure with a specific impact occurring as a result of the departure, inhibiting attempts to mitigate the risks of drought based on the departures from normal and form a plan of response (1).

STANDARDIZED PRECIPITATION INDEX (SPI)

Overview: The SPI is an index based on the probability of precipitation for any time scale.

Who uses it: Many drought planners appreciate the SPI's versatility.

Pros: The SPI can be computed for different time scales, can provide early warning of drought and help assess drought severity, and is less complex than the Palmer.

Cons: Values based on preliminary data may change.

Developed by: T.B. McKee, N.J. Doesken, and J. Kleist, Colorado State University, 1993.

SPI Values	
2.0+	extremely wet
1.5 to 1.99	very wet
1.0 to 1.49	moderately wet
−.99 to .99	near normal
−1.0 to −1.49	moderately dry
−1.5 to −1.99	severely dry
−2 and less	extremely dry

Monthly maps: http://www.drought.unl.edu/monitor/spi.htm; http://www.wrcc.dri.edu/spi/spi.html.

The understanding that a deficit of precipitation has different impacts on groundwater, reservoir storage, soil moisture, snowpack, and streamflow led McKee, Doesken, and Kleist to develop the Standardized Precipitation Index (SPI) in 1993. The SPI was designed to quantify the precipitation deficit for multiple time scales. These time scales reflect the impact of drought on the availability of the different water resources. Soil moisture conditions respond to precipitation anomalies on a relatively short scale. Groundwater, streamflow, and reservoir storage reflect the longer-term precipitation anomalies. For these reasons, McKee et al. (2) originally calculated the SPI for 3-, 6-, 12-, 24-, and 48-month time scales.

The SPI calculation for any location is based on the long-term precipitation record for a desired period. This long-term record is fitted to a probability distribution, which is then transformed into a normal distribution so that the mean SPI for the location and desired period is zero (3). Positive SPI values indicate greater than median precipitation, and negative values indicate less than median precipitation. Because the SPI is normalized, wetter and drier climates can be represented in the same way, and wet periods can also be monitored using the SPI.

McKee et al. (2) used the classification system shown in the SPI values table to define drought intensities resulting from the SPI. McKee et al. (2) also defined the criteria for a drought event for any of the time scales. A drought event occurs any time the SPI is continuously negative and reaches an intensity of −1.0 or less. The event ends when the SPI becomes positive. Each drought event, therefore, has a duration defined by its beginning and end, and an intensity for each month that the event continues. The positive sum of the SPI for all the months within a drought event can be termed the drought's "magnitude".

Based on an analysis of stations across Colorado, McKee determined that the SPI is in mild drought 24% of the time, in moderate drought 9.2% of the time, in severe drought 4.4% of the time, and in extreme drought 2.3% of the time (2). Because the SPI is standardized, these percentages are expected from a normal distribution of the SPI. The 2.3% of SPI values within the "Extreme Drought" category is a percentage that is typically expected for an "extreme" event (Wilhite 1995). In contrast, the Palmer Index reaches its "extreme" category more than 10% of the time across portions of the central Great Plains. This standardization allows the SPI to determine the rarity of a current drought, as well as the probability of the precipitation necessary to end the current drought (2).

The SPI has been used operationally to monitor conditions across Colorado since 1994 (4). Monthly maps of the SPI for Colorado can be found on the Colorado State University website (http://ulysses.atmos.colostate.edu/SPI.html). It is also being monitored at the climate division level for the contiguous United States by the National Drought

Mitigation Center and the Western Regional Climate Center (WRCC). You can download the SPI program and sample files here.

PALMER DROUGHT SEVERITY INDEX (THE PALMER; PDSI)

Overview: The Palmer is a soil moisture algorithm calibrated for relatively homogeneous regions.
Who uses it: Many U.S. government agencies and states rely on the Palmer to trigger drought relief programs.
Pros: The first comprehensive drought index developed in the United States.
Cons: Palmer values may lag emerging droughts by several months; less well suited for mountainous land or areas of frequent climatic extremes; complex—has an unspecified, built-in time scale that can be misleading.
Developed by: W.C. Palmer, 1965.
Weekly maps: http://www.cpc.ncep.noaa.gov/products/analysis_monitoring/regional_monitoring/palmer.gif.

Palmer Classifications	
4.0 or more	extremely wet
3.0 to 3.99	very wet
2.0 to 2.99	moderately wet
1.0 to 1.99	slightly wet
0.5 to 0.99	incipient wet spell
0.49 to −0.49	near normal
−0.5 to −0.99	incipient dry spell
−1.0 to −1.99	mild drought
−2.0 to −2.99	moderate drought
−3.0 to −3.99	severe drought
−4.0 or less	extreme drought

In 1965, W.C. Palmer developed an index to measure the departure of the moisture supply (5). Palmer based his index on the supply-and-demand concept of the water balance equation, taking into account more than just the precipitation deficit at specific locations. The objective of the Palmer Drought Severity Index (PDSI), as this index is now called, was to provide measurements of moisture conditions that were standardized so that comparisons using the index could be made between locations and between months (5).

The PDSI is a meteorological drought index, and it responds to weather conditions that have been abnormally dry or abnormally wet. When conditions change from dry to normal or wet, for example, the drought measured by the PDSI ends without taking into account streamflow, lake and reservoir levels, and other longer-term hydrologic impacts (6). The PDSI is calculated based on precipitation and temperature data, as well as the local Available Water Content (AWC) of the soil. From the inputs, all the basic terms of the water balance equation can be determined, including evapotranspiration, soil recharge, runoff, and moisture loss from the surface layer. Human impacts on the water balance, such as irrigation, are not considered. Complete descriptions of the equations can be found in the original study by Palmer (5) and in the more recent analysis by Alley (7).

Palmer developed the PDSI to include the duration of a drought (or wet spell). His motivation was as follows: an abnormally wet month in the middle of a long-term drought should not have a major impact on the index, or a series of months with near-normal precipitation following a serious drought does not mean that the drought is over. Therefore, Palmer developed criteria for determining when a drought or a wet spell begins and ends, which adjust the PDSI accordingly. Palmer (5) described this effort and gave examples, and it is also described in detail by Alley (7). In near-real time, Palmer's index is no longer a meteorological index but becomes a hydrological index referred to as the Palmer Hydrological Drought Index (PHDI) because it is based on moisture inflow (precipitation), outflow, and storage, and does not take into account the long-term trend (6).

In 1989, a modified method to compute the PDSI was begun operationally (8). This modified PDSI differs from the PDSI during transition periods between dry and wet spells. Because of the similarities between these Palmer indices, the terms *Palmer Index* and *Palmer Drought Index* have been used to describe general characteristics of the indices.

The Palmer Index varies roughly between −6.0 and +6.0. Palmer arbitrarily selected the classification scale of moisture conditions based on his original study areas in central Iowa and western Kansas (5). Ideally, the Palmer Index is designed so that a −4.0 in South Carolina has the same meaning in terms of the moisture departure from a climatological normal as a −4.0 in Idaho (7). The Palmer Index has typically been calculated on a monthly basis, and a long-term archive of the monthly PDSI values for every climate division in the United States exists with the National Climatic Data Center from 1895 through the present. In addition, weekly Palmer Index values (actually modified PDSI values) are calculated for the climate divisions during every growing season and are available in the *Weekly Weather and Crop Bulletin*. These weekly Palmer Index maps are also available on the World Wide Web from the Climate Prediction Center at http://www.cpc.ncep.noaa.gov/products/analysis_monitoring/regional_monitoring/palmer.gif.

The Palmer Index is popular and has been widely used for a variety of applications across the United States. It is most effective measuring impacts sensitive to soil moisture conditions, such as agriculture (1). It has also been useful as a drought monitoring tool and has been used to trigger actions associated with drought contingency plans (1). Alley (7) identified three positive characteristics of the Palmer Index that contribute to its popularity: (1) it provides decision makers with a measurement of the abnormality of recent weather for a region; (2) it provides an opportunity to place current conditions in historical perspective; and (3) it provides spatial and temporal representations of historical droughts. Several states, including New York, Colorado, Idaho, and Utah, use the Palmer Index as one part of their drought monitoring systems.

There are considerable limitations when using the Palmer Index, and these are described in detail by Alley (7) and Karl and Knight (6). Drawbacks of the Palmer Index include:

- The values quantifying the intensity of drought and signaling the beginning and end of a drought or wet spell were arbitrarily selected based on Palmer's study of central Iowa and western Kansas and have little scientific meaning.
- The Palmer Index is sensitive to the AWC of a soil type. Thus, applying the index for a climate division may be too general.
- The two soil layers within the water balance computations are simplified and may not be accurately representative of a location.
- Snowfall, snow cover, and frozen ground are not included in the index. All precipitation is treated as rain, so that the timing of PDSI or PHDI values may be inaccurate in the winter and spring months in regions where snow occurs.
- The natural lag between when precipitation falls and the resulting runoff is not considered. In addition, no runoff is allowed to take place in the model until the water capacity of the surface and subsurface soil layers is full, leading to an underestimation of runoff.
- Potential evapotranspiration is estimated using the Thornthwaite method. This technique has wide acceptance, but it is still only an approximation.

Several other researchers have presented additional limitations of the Palmer Index. McKee et al. (4) suggested that the PDSI is designed for agriculture but does not accurately represent the hydrological impacts resulting from longer droughts. Also, the Palmer Index is applied within the United States but has little acceptance elsewhere (9). One explanation for this is provided by Smith et al. (10), who suggested that it does not do well in regions where there are extremes in the variability of rainfall or runoff. Examples in Australia and South Africa were given. Another weakness in the Palmer Index is that the "extreme" and "severe" classifications of drought occur with a greater frequency in some parts of the country than in others (1). "Extreme" droughts in the Great Plains occur with a frequency greater than 10%. This limits the accuracy of comparing the intensity of droughts between two regions and makes planning response actions based on a certain intensity more difficult.

CROP MOISTURE INDEX (CMI)

Description: A Palmer derivative, the CMI reflects moisture supply in the short term across major crop-producing regions and is not intended to assess long-term droughts.
Pros: Identifies potential agricultural droughts.
Developed by: W.C. Palmer, 1968.
Weekly maps: http://www.cpc.ncep.noaa.gov/products/_analysis_monitoring/regional_monitoring/cmi.gif.

The Crop Moisture Index (CMI) uses a meteorological approach to monitor week-to-week crop conditions. It was developed by Palmer (11) from procedures within the calculation of the PDSI. Whereas the PDSI monitors long-term meteorological wet and dry spells, the CMI was designed to evaluate short-term moisture conditions across major crop-producing regions. It is based on the mean temperature and total precipitation for each week within a climate division, as well as the CMI value from the previous week. The CMI responds rapidly to changing conditions, and it is weighted by location and time so that maps, which commonly display the weekly CMI across the United States, can be used to compare moisture conditions at different locations. Weekly maps of the CMI are available as part of the USDA/JAWF Weekly Weather and Crop Bulletin (http://www.usda.gov/oce/waob/jawf/wwcb.html).

Because it is designed to monitor short-term moisture conditions affecting a developing crop, the CMI is not a good long-term drought monitoring tool. The CMI's rapid response to changing short-term conditions may provide misleading information about long-term conditions. For example, a beneficial rainfall during a drought may allow the CMI value to indicate adequate moisture conditions, while the long-term drought at that location persists. Another characteristic of the CMI that limits its use as a long-term drought monitoring tool is that the CMI typically begins and ends each growing season near zero. This limitation prevents the CMI from being used to monitor moisture conditions outside the general growing season, especially in droughts that extend over several years. The CMI also may not be applicable during seed germination at the beginning of a specific crop's growing season.

SURFACE WATER SUPPLY INDEX (SWSI; PRONOUNCED "SWAZEE")

Description: The SWSI is designed to complement the Palmer in the state of Colorado, where mountain snowpack is a key element of water supply; calculated by river basin, based on snowpack, streamflow, precipitation, and reservoir storage.
Pros: Represents water supply conditions unique to each basin.
Cons: Changing a data collection station or water management requires that new algorithms be calculated, and the index is unique to each basin, which limits interbasin comparisons.
Developed by: Shafer and Dezman, 1982.

The Surface Water Supply Index (SWSI) was developed by Shafer and Dezman (12) to complement the Palmer Index for moisture conditions across the state of Colorado. The Palmer Index is basically a soil moisture algorithm calibrated for relatively homogeneous regions, but it is not designed for large topographic variations across a region and it does not account for snow accumulation and subsequent runoff. Shafer and Dezman designed the SWSI to be an indicator of surface water conditions and described the index as "mountain water dependent", in which mountain snowpack is a major component.

The objective of the SWSI was to incorporate both hydrological and climatological features into a single index value resembling the Palmer Index for each major river basin in the state of Colorado (Shafer and Dezman 1982). These values would be standardized to allow comparisons between basins. Four inputs are required within the SWSI: snowpack, streamflow, precipitation, and reservoir storage. Because it is dependent on the season, the SWSI is computed with only snowpack, precipitation, and reservoir storage in the winter. During the summer months, streamflow replaces snowpack as a component within the SWSI equation.

The procedure to determine the SWSI for a particular basin is as follows: monthly data are collected and summed for all the precipitation stations, reservoirs, and snowpack/streamflow measuring stations over the basin. Each summed component is normalized using a frequency analysis gathered from a long-term data set. The probability of non-exceedence—the probability that subsequent sums of that component will not be greater than the current sum—is determined for each component based on the frequency analysis. This allows comparisons of the probabilities to be made between the components. Each component has a weight assigned to it depending on its typical contribution to the surface water within that basin, and these weighted components are summed to determine a SWSI value representing the entire basin. Like the Palmer Index, the SWSI is centered on zero and has a range between −4.2 and +4.2.

The SWSI has been used, along with the Palmer Index, to trigger the activation and deactivation of the Colorado Drought Plan. One of its advantages is that it is simple to calculate and gives a representative measurement of surface water supplies across the state. It has been modified and applied in other western states as well. These states include Oregon, Montana, Idaho, and Utah. Monthly SWSI maps for Montana are available from the Montana Natural Resource Information System (http://nris.state.mt.us/wis/SWSInteractive/).

Several characteristics of the SWSI limit its application. Because the SWSI calculation is unique to each basin or region, it is difficult to compare SWSI values between basins or regions (13). Within a particular basin or region, discontinuing any station means that new stations need to be added to the system and new frequency distributions need to be determined for that component. Additional changes in the water management within a basin, such as flow diversions or new reservoirs, mean that the entire SWSI algorithm for that basin needs to be redeveloped to account for changes in the weight of each component. Thus, it is difficult to maintain a homogeneous time series of the index (8). Extreme events also cause a problem if the events are beyond the historical time series, and the index will need to be reevaluated to include these events within the frequency distribution of a basin component.

RECLAMATION DROUGHT INDEX

Description: Like the SWSI, the RDI is calculated at the river basin level, incorporating temperature as well as precipitation, snowpack, streamflow, and reservoir levels as input.

Who uses it: The Bureau of Reclamation; the State of Oklahoma as part of their drought plan.

Pros: By including a temperature component, it also accounts for evaporation.

Cons: Because the index is unique to each river basin, interbasin comparisons are limited.

Developed by: The Bureau of Reclamation, as a trigger to release drought emergency relief funds.

RDI Classifications	
4.0 or more	extremely wet
1.5 to 4.0	moderately wet
1 to 1.5	normal to mild wetness
0 to −1.5	normal to mild drought
−1.5 to −4.0	moderate drought
−4.0 or less	extreme drought

The Reclamation Drought Index (RDI) was recently developed as a tool for defining drought severity and duration, and for predicting the onset and end of periods of drought. The impetus to devise the RDI came from the Reclamation States Drought Assistance Act of 1988, which allows states to seek assistance from the Bureau of Reclamation to mitigate the effects of drought.

Like the SWSI, the RDI is calculated at a river basin level, and it incorporates the supply components of precipitation, snowpack, streamflow, and reservoir levels. The RDI differs from the SWSI in that it builds a temperature-based demand component and a duration into the index. The RDI is adaptable to each particular region and its main strength is its ability to account for both climate and water supply factors.

Oklahoma has developed its own version of the RDI and plans to use the index as one tool within the monitoring system designated in the state's drought plan. The RDI values and severity designations are similar to the SPI, PDSI, and SWSI.

DECILES

Description: Groups monthly precipitation occurrences into deciles so that, by definition, "much lower than normal" weather cannot occur more often than 20% of the time.

Who Uses It: Australia.

Decile Classifications	
Deciles 1–2: lowest 20%	much below normal
Deciles 3–4: next lowest 20%	below normal
deciles 5–6: middle 20%	near normal
deciles 7–8: next highest 20%	above normal
deciles 9–10: highest 20%	much above normal

Pros: Provides an accurate statistical measurement of precipitation.
Cons: Accurate calculations require a long climatic data record.
Developed by: Gibbs and Maher, 1967.

Arranging monthly precipitation data into deciles is another drought-monitoring technique. It was developed by Gibbs and Maher (14) to avoid some of the weaknesses within the "percent of normal" approach. The technique they developed divided the distribution of occurrences over a long-term precipitation record into tenths of the distribution. They called each of these categories a *decile*. The first decile is the rainfall amount not exceeded by the lowest 10% of the precipitation occurrences. The second decile is the precipitation amount not exceeded by the lowest 20% of occurrences. These deciles continue until the rainfall amount identified by the tenth decile is the largest precipitation amount within the long-term record. By definition, the fifth decile is the median, and it is the precipitation amount not exceeded by 50% of the occurrences over the period of record. The deciles are grouped into five classifications.

The decile method was selected as the meteorological measurement of drought within the Australian Drought Watch System because it is relatively simple to calculate and requires less data and fewer assumptions than the Palmer Drought Severity Index (10). In this system, farmers and ranchers can only request government assistance if the drought is shown to be an event that occurs only once in 20–25 years (deciles 1 and 2 over a 100-year record) and has lasted longer than 12 months (15). This uniformity in drought classifications, unlike a system based on the percent of normal precipitation, has assisted Australian authorities in determining appropriate drought responses. One disadvantage of the decile system is that a long climatological record is needed to calculate the deciles accurately.

BIBLIOGRAPHY

1. Willeke, G., Hosking, J.R.M., Wallis, J.R., and Guttman, N.B. (1994). *The National Drought Atlas*. Institute for Water Resources Report 94–NDS–4, U.S. Army Corps of Engineers.
2. McKee, T.B., Doesken, N.J., and Kleist, J. (1993). The relationship of drought frequency and duration to time scales. Preprints, 8th Conference on Applied Climatology, pp. 179–184. January 17–22, Anaheim, CA.
3. Edwards, D.C. and McKee, T.B. (1997). Characteristics of 20th century drought in the United States at multiple time scales. Climatology Report Number 97–2, Colorado State University, Fort Collins, CO.
4. McKee, T.B., Doesken, N.J., and Kleist, J. (1995). Drought monitoring with multiple time scales. Preprints, 9th Conference on Applied Climatology, pp. 233–236. January 15–20, Dallas, TX.
5. Palmer, W.C. (1965). Meteorological drought. Research Paper No. 45, U.S. Department of Commerce Weather Bureau, Washington, DC.
6. Karl, T.R. and Knight, R.W. (1985). Atlas of Monthly Palmer Hydrological Drought Indices (1931–1983) for the Contiguous United States. Historical Climatology Series 3–7, National Climatic Data Center, Asheville, North Carolina.
7. Alley, W.M. (1984). The palmer drought severity index: limitations and assumptions. *Journal of Climate and Applied Meteorology* **23**: 1100–1109.
8. Heddinghaus, T.R. and Sabol, P. (1991). A review of the palmer drought severity index and where do we go from here? In: *Proc. 7th Conf. on Applied Climatology*, pp. 242–246. American Meteorological Society, Boston.
9. Kogan, F.N. (1995). Droughts of the late 1980s in the United States as derived from NOAA polar-orbiting satellite data. *Bulletin of the American Meteorological Society* **76**(5): 655–668.
10. Smith, D.I., Hutchinson, M.F., and McArthur, R.J. (1993). Australian climatic and agricultural drought: Payments and policy. *Drought Network News* **5**(3): 11–12.
11. Palmer, W.C. (1968). Keeping track of crop moisture conditions, nationwide: The new Crop Moisture Index. *Weatherwise* **21**: 156–161.
12. Shafer, B.A. and Dezman, L.E. (1982). Development of a Surface Water Supply Index (SWSI) to assess the severity of drought conditions in snowpack runoff areas. In: *Proceedings of the Western Snow Conference*. Colorado State University, Fort Collins, CO, pp. 164–175.
13. Doesken, N.J., McKee, T.B., and Kleist, J. (1991). Development of a surface water supply index for the western United States. Climatology Report Number 91–3, Colorado State University, Fort Collins, CO.
14. Gibbs, W.J. and Maher, J.V. (1967). Rainfall deciles as drought indicators. *Bureau of Meteorology Bulletin* No. 48, Commonwealth of Australia, Melbourne.
15. White, D.H. and O'Meagher, B. (1995). Coping with exceptional droughts in Australia. *Drought Network News* **7**(2): 13–17.

READING LIST

Gommes, R. and Petrassi, F. (1994). Rainfall variability and drought in Sub-Saharan Africa since 1960. Agrometeorology Series Working Paper No. 9, Food and Agriculture Organization, Rome, Italy.

Le Houérou, H.N., Popov, G.F., and See, L. (1993). Agrobioclimatic classification of Africa. Agrometeorology Series Working Paper No. 6, Food and Agriculture Organization, Rome, Italy.

Wilhite, D.A. (1995). Developing a precipitation-based index to assess climatic conditions across Nebraska. Final report submitted to the Natural Resources Commission, Lincoln, NE.

Wilhite, D.A. and Glantz, M.H. (1985). Understanding the drought phenomenon: The role of definitions. *Water International* **10**(3): 111–120.

THE EARTH OBSERVING SYSTEM: AQUA

NASA—Goddard Space
Flight Center

EARTH SYSTEM SCIENCE

Beginning in the 1960s, NASA pioneered the study of the atmosphere from the unique perspective of space with the launch of its Television Infrared Observation Satellite (TIROS-1). Thanks to new satellite and computer technologies, it is now possible to study the Earth as a global system. Through their research, scientists are better understanding and improving their forecasting of short-term weather phenomena.

Long-term weather and climate prediction is a greater challenge that requires the collection of better data over longer periods. Since climate changes occur over vast ranges of space and time, their causes and effects are often difficult to measure and understand. Scientists must obtain long-term data if they are to reach a full understanding of the interactions among the Earth's physical and biological systems. NASA's Earth Observing System (EOS) will help us to understand the complex links among air, land, water and life within the Earth system.

WHAT IS AQUA?

NASA's commitment to studying the Earth as a global system continues with the Aqua spacecraft (originally called EOS PM-1), representing a key contribution by NASA to the U.S. Global Change Research Program. Aqua carries six state-of-the-art instruments to observe the Earth's oceans, atmosphere, land, ice and snow covers, and vegetation, providing high measurement accuracy, spatial detail, and temporal frequency. This comprehensive approach to data collection enables scientists to study the interactions among the four spheres of the Earth system—the oceans, land, atmosphere, and biosphere.

Aqua, Latin for "water," is named for the large amount of information that the Aqua spacecraft will collect about the Earth's water cycle. In particular, the Aqua data will include information on water vapor and clouds in the atmosphere, precipitation from the atmosphere, soil wetness on the land, glacial ice on the land, sea ice in the oceans, snow cover on both land and sea ice, and surface waters throughout the world's oceans, bays, and lakes. Such information will help scientists improve the quantification of the global water cycle and examine such issues as whether or not the cycling of water might be accelerating.

In addition to information about the water cycle, Aqua also provides information on many additional elements of the Earth system. For instance, Aqua enables studies of the fluxes of radiation from the Sun and from the Earth that combine to constitute the Earth's radiation balance. It also enables studies of small particles in the atmosphere termed "aerosols" and such trace gases in the atmosphere as ozone, carbon monoxide, and methane. The trace gases each have a potential contribution to global warming, whereas the aerosols are more likely to have a cooling effect. Aqua also provides observations on vegetation cover on the land, phytoplankton and dissolved organic matter in the oceans, and the temperatures of the air, land, and water. All of these measurements have the potential to contribute to improved understanding of the changes occurring in the global climate and the role of the interactions among the various elements of the climate system.

One of the most exciting of the potential practical benefits likely to derive from the Aqua data is improved weather forecasting. Aqua carries a sophisticated sounding system that allows determination of atmospheric temperatures around the world to an accuracy of 1° Celsius in 1-km-thick layers throughout the troposphere, the lowest portion of the atmosphere. The troposphere extends to an altitude of about 10–15 km, depending on location, and contains most of the global cloud cover. The anticipated 1° Celsius accuracy far exceeds current accuracies from satellite observations and, in conjunction with the moisture profiles also obtainable from the Aqua sounding system, offers the potential of improved weather fore-casting. NASA is working with the U.S. National Oceanic and Atmospheric Administration and the European Centre for Medium-Range Weather Forecasts to facilitate the incorporation of the Aqua data in their weather forecasting efforts.

INTERNATIONAL COLLABORATION

Aqua is a joint project of the United States, Japan, and Brazil.

THE SPACECRAFT

The spacecraft was designed and built by TRW in Redondo Beach, California. Aqua is based on TRW's modular, standardized AB1200 common spacecraft bus. This design features common subsystems scalable to the mission-specific needs of Aqua as well as future missions. Instrument payloads can be attached on a "mix and match" basis without changing the overall design or subsystem support requirements.

THE INSTRUMENTS

The Atmospheric Infrared Sounder (AIRS), built by BAE Systems, was provided by NASA's Jet Propulsion Laboratory in Pasadena, California. AIRS is the highlighted instrument in the AIRS/AMSU-A/HSB triplet centered on measuring humidity, temperature, cloud properties, and the amounts of greenhouse gases throughout the atmosphere. AIRS/AMSU-A/HSB will improve weather forecasting, establish the connection between severe weather and climate change, examine whether the global water cycle is accelerating, and detect the effects of greenhouse gases.

The Advanced Microwave Scanning Radiometer for EOS (AMSR-E), built by Mitsubishi Electronics Corporation, was provided by Japan's National Space Development Agency. AMSR-E measures precipitation rate, cloud water, water vapor, sea surface winds, sea surface temperature, ice, snow, and soil moisture.

The Advanced Microwave Sounding Unit (AMSU-A), built by Aerojet and provided by NASA's Goddard Space Flight Center (GSFC) in Greenbelt, Maryland, obtains temperature profiles in the upper atmosphere (especially the stratosphere) and will provide a cloud-filtering capability for tropospheric temperature observations. The EOS AMSU-A is part of the closely coupled AIRS/AMSU-A/HSB triplet.

The Clouds and the Earth's Radiant Energy System (CERES), built by TRW, was provided by NASA's Langley Research Center in Hampton, Virginia. This instrument measures the Earth's total thermal radiation budget, and, in combination with Moderate Resolution Imaging Spectro-radiometer (MODIS) data, provides detailed information about clouds. The first CERES instrument was launched on the Tropical Rainfall Measuring Mission (TRMM) satellite in November 1997; the second and third CERES instruments were launched on the Terra satellite in December 1999; and the fourth and fifth CERES instruments was on board the Aqua satellite launched in May 2002. The pairs of CERES on both Terra and Aqua satellites allow coincident measurements by one CERES scanning in lines perpendicular to the path of the satellite and by the other CERES scanning in lines at various angles with respect to the satellite's path.

The Humidity Sounder for Brazil (HSB), built by Matra-Marconi, was provided by Brazil's Instituto Nacional de Pesquisas Espaciais, the Brazilian Institute for Space Research. The HSB obtains humidity profiles throughout the atmosphere. The HSB is the instrument in the AIRS/AMSU-A/HSB suite that allows humidity measurements even under conditions of heavy cloudiness and haze.

MODIS, built by Raytheon Santa Barbara Remote Sensing, was provided by GSFC. MODIS is a 36-band spectroradiometer measuring visible and infrared radiation and obtaining data that is used to derive products ranging from vegetation, land surface cover, and ocean chlorophyll fluorescence, to cloud and aerosol properties, fire occurrence, snow cover on the land, and sea ice cover on the oceans. The first MODIS instrument was launched on board the Terra satellite.

Aqua was launched in May 2002 aboard a Delta 7920-10L launch vehicle from Vandenberg Air Force Base, California. The stowed spacecraft is 8.8 ft (2.68 m) × 8.2 ft (2.49 m) × 21.3 ft (6.49 m). Deployed, Aqua is 15.8 ft (4.81 m) × 54.8 ft (16.70 m) × 26.4 ft (8.04 m). The spacecraft, at launch, weighed 6784 lbs with a full propellant load of 508 lbs and is powered by 4.6 kilowatts of electric power from its solar array.

Aqua was launched into a circular 680-km orbit. Over a period of days after separation from the launch vehicle, it was commanded by the ground to raise its orbit to the prescribed 705-km (438-mile) orbit. This was necessary in order to allow for proper phasing of Aqua with other spacecraft in orbit and the polar ground stations used for communications. The spacecraft was ultimately be positioned in a near-polar (98°) orbit around the Earth in synchronization with the Sun, with its path over the ground ascending across the equator at the same local time every day, approximately 1:30 p.m. The early afternoon observation time contrasts with the 10:30–10:45 a.m. equatorial crossing time (descending in this case) of the Terra satellite. The two daytime crossing times account for why the Terra and Aqua satellites were originally named "EOS AM" and "EOS PM," respectively. The combination of morning and afternoon observations allows studies concerning the diurnal variability of many of the parameters discussed above.

MANAGEMENT

Overall management of the Aqua mission is located at GSFC, which is managing the integration and testing of the spacecraft. The Aqua data is processed, archived, and distributed using distributed components of the Earth Observing System Data and Information System (EOSDIS). EOSDIS also provides the mission operations systems that perform the functions of command and control of the spacecraft and the instruments. NASA's Kennedy Space Center is responsible for the launch operations, including Boeing's Delta launch vehicle and the prelaunch integrated processing facility. The U.S. Air Force is responsible for all range-related matters. GSFC manages EOS for NASA's Earth Science Enterprise (ESE), headquartered in Washington, DC.

DATA PROCESSING AND DISTRIBUTION

Aqua provides a major part of a 15-year environmental dataset focusing on global change. The Aqua instruments produce more than 750 gigabytes of data per day, which is equivalent to 75 personal computer hard disks at 10 gigabytes each per day. This massive amount of information is handled using EOSDIS, in addition to its present accumulation of nearly 3000 gigabytes per day.

EOSDIS provides the high-performance computing resources needed to process, store, and rapidly transmit terabytes (thousands of gigabytes) of the incoming data every day. EOSDIS has several distributed sites that perform these functions: Distributed Active Archive Centers (DAACs) that process, store and distribute the data, and Science Investigator-led Processing Systems that process the data and send them to the DAACs for storage and distribution. EOSDIS uses an "open" architecture to allow insertion of new technology while enabling the system to support the changing mission and science needs throughout the EOS Program.

GOALS AND OBJECTIVES

NASA's ESE identified several high-priority measurements that EOS should make to facilitate a better understanding of the components of the Earth system—the atmosphere, the land, the oceans, the polar ice caps, and the global energy budget. The specific objectives of Aqua include:

- producing high-spectral resolution obtaining 1 K/1 km global root-mean-square temperature profile accuracy in the troposphere by 1 year after launch;
- extending the improved TRMM rainfall characterization to the extra tropics;
- producing global sea surface temperature daily maps under nearly all sky conditions for a minimum of 1 year;
- producing large-scale global soil moisture distribution for regions with low vegetation;
- producing calibrated global observations of the Earth's continents and ocean surfaces 150 days after the mission is declared operational;
- capturing and documenting three seasonal cycles of terrestrial and marine ecosystems and atmospheric and cloud properties;
- producing three sets of seasonal/annual Earth radiation budget records;
- producing improved measurements of the diurnal cycle of radiation by combining Aqua measurements with Terra measurements for months of overlap;
- producing combined cloud property and radiation balance data to allow improved studies of the role of clouds in the climate system; and,
- capturing, processing, archiving, and distributing Aqua data products, by 150 days after the mission is declared operational.

A NEW PERSPECTIVE

Complemented by Terra, aircraft and ground-based measurements, Aqua data enable scientists to distinguish between natural and human-induced changes. The EOS series of spacecraft are the cornerstone of NASA's ESE, a long-term research effort to study the Earth as a global environment.

More information on EOS and the science related to it can be found at the EOS Project Science Office website at http://eospso.gsfc.nasa.gov and at the Earth Observatory website at http://earthobservatory.nasa.gov. Further information on Aqua can be found at http://aqua.nasa.gov

ENTROPY THEORY FOR HYDROLOGIC MODELING

Vijay P. Singh
Louisiana State University
Baton Rouge, Louisiana

INTRODUCTION

Entropy theory has recently been employed in a broad range of applications in hydrology, and new applications continue to unfold. This paper revisits entropy theory and its application to hydrologic modeling.

Hydrologic systems are inherently spatial and complex, and our understanding of these systems is less than complete. Many of the systems are either fully stochastic or part stochastic and part deterministic. Their stochastic nature can be attributed to randomness in one or more of the following components that constitute them: (1) system structure (geometry), (2) system dynamics, (3) forcing functions (sources and sinks), and (4) initial and boundary conditions. As a result, a stochastic description of these systems is needed, and entropy theory enables the development of such a description.

Engineering decisions concerning hydrologic systems are frequently made with less than adequate information. Such decisions may often be based on experience, professional judgment, rules of thumb, crude analyses, safety factors, or probabilistic methods. Usually, decision-making under uncertainty tends to be relatively conservative. Quite often, sufficient data are not available to describe the random behavior of such systems. Although probabilistic methods allow for a more explicit and quantitative accounting of uncertainty, their major difficulty occurs due to the lack of sufficient or complete data. Small sample sizes and limited information render estimation of probability distributions of system variables by conventional methods difficult. This problem can be alleviated by using entropy theory that enables determining the least biased probability distributions based on limited knowledge and data. Where the shortage of data is widely rampant, as is normally the case in many countries, entropy theory is particularly appealing.

Since the development of entropy theory by Shannon in the late 1940s and of the principle of maximum entropy (POME) by Jaynes in the late 1950s, there has been a proliferation in applications of entropy. The real impetus to entropy-based modeling in hydrology was, however, provided in the early 1970s, a great variety of entropy-based applications have since been reported, and new applications continue to unfold. This article aims to revisit entropy theory and to underscore its usefulness for both modeling and decision-making in hydrology.

ENTROPY THEORY

Entropy theory is comprised of three main parts: (1) Shannon entropy, (2) the principle of maximum entropy, and (3) the principle of minimum cross entropy. Before discussing these parts, it will be instructive to discuss briefly the meaning of entropy.

Meaning of Entropy

Entropy originated in physics. It is an extensive property like mass, energy, volume, momentum, charge, or number of atoms of chemical species, but unlike these quantities, it does not obey a conservation law. The entropy of a system is an extensive property, so the total entropy of the system equals the sum of the entropies of individual parts. The most probable distribution of energy in a system is the one that corresponds to the maximum entropy of the system. This occurs under the condition of dynamic equilibrium. During evolution toward a stationary state, the rate of entropy production per unit mass should be minimum, compatible with external constraints. This is the Prigogin principle. In thermodynamics, entropy is decomposed into two parts: (1) entropy exchanged between the system and its surroundings and (2) entropy produced in the system. According to the second law of thermodynamics, the entropy of a closed and isolated system always tends to increase. In hydraulics, entropy is a measure of the amount of irrecoverable flow energy that the hydraulic system expands to overcome friction. The system converts a portion of its mechanical energy to heat energy which then is dissipated to the external environment. Thus, the process equation in hydraulics expressing the energy (or head) loss, it can be argued, originates in the entropy concept.

Entropy has been employed in thermodynamics as a measure of the degree of ignorance about the true state of a system. If there were no energy loss in a hydraulic system, the system would be orderly and organized. The energy loss and its causes make the system disorderly and chaotic. Thus, entropy can be interpreted as a measure of the amount of chaos within a system. Algebraically, it is proportional to the logarithm of the probability of the state of the system. The constant of proportionality is the Boltzmann constant, and this defines Boltzmann entropy.

Shannon Entropy

Shannon (1) developed the entropy theory for expressing information or uncertainty. To understand the informational aspect of entropy, we perform an experiment on a random variable X. There may be n possible outcomes $x_1, x_2, \ldots, x_n$, whose probabilities are $p_1, p_2, \ldots, p_n$; $P(X = x_1) = p_1, P(X = x_2) = p_2, \ldots, P(X = x_n) = p_n$. These outcomes can be described by

$$P(X) = (p_1, p_2, \ldots, p_n); \ \sum_{i}^{n} p_i = 1; p_i \geq 0, i = 1, 2, \ldots, n \tag{1}$$

If this experiment is repeated, the same outcome is not likely, implying that there is uncertainty as to the outcome of the experiment. Based on one's knowledge about the outcomes, the uncertainty can be more or less. For example, the total number of outcomes is a piece of information, and the number of those outcomes of nonzero probability is another piece of information. The probability distribution of the outcomes, if known, provides a certain amount of information. Shannon (1) defined a quantitative measure of uncertainty associated with a probability distribution or the information content of the distribution in terms of entropy, $H(P)$ or $H(X)$, called Shannon entropy or informational entropy as

$$H(X) = H(P) = -\sum_{i=1}^{n} p_i \ln p_i = E[-\ln p] \tag{2}$$

If the random variable X is continuous, then Shannon entropy is expressed as

$$\begin{aligned} H(X) &= -\int_0^\infty f(x) \ln[f(x)]\, dx \\ &= -\int \ln[f(x)]\, dF(x) = E[-\ln f(x)] \end{aligned} \tag{3}$$

where $f(x)$ is the probability density function (PDF) of X, $F(x)$ is the cumulative probability distribution function of X, and E [.] is the expectation of [.].

Thus, entropy is a measure of the amount of uncertainty represented by the probability distribution and is a measure of the amount of chaos or of the lack of information about a system. If complete information is available, entropy = 0; otherwise, it is greater than zero. The uncertainty can be quantified using entropy by taking into account all different kinds of available information. Shannon entropy is the weighted Boltzmann entropy.

Principle of Maximum Entropy

In search of an appropriate probability distribution for a given random variable, entropy should be maximized. In practice, however, it is common that some information about the random variable is available. The chosen probability distribution should then be consistent with the given information. There can be more than one distribution consistent with the given information. From all such distributions, we should choose the distribution that has the highest entropy. To that end, Jaynes (2) formulated the principle of maximum entropy (POME), a full account of it is presented in a treatise by Levine and Tribus (3). According to POME, the minimally prejudiced assignment of probabilities is that which maximizes entropy subject to the given information, that is, POME takes into account all of the given information and at the same time avoids considering of any information that is not given.

If no information about the random variable is available, then all outcomes are equally likely, that is, $p_i = 1/n$, $i = 1, 2, 3, \ldots, n$. It can be shown that Shannon entropy is maximum in this case and may serve as an upper bound of entropy for all cases involving some information. In a more general case, let the information available about P or X be

$$p_i \geq 0, \sum_{i=1}^{n} p_i = 1 \quad (4)$$

and

$$\sum_{i=1}^{n} g_r(x_i) p_i = a_r \quad r = 1, 2, \ldots, m \quad (5)$$

where m is the number of constraints, $m + 1 \leq n$ and g_r is the rth constraint. Equations 4 and 5 are not sufficient to determine P uniquely. Therefore, there can be many distributions that satisfy Eqs. 4 and 5. According to POME, there will be only one distribution that will correspond to the maximum value of entropy, and this distribution can be determined using the method of Lagrange multipliers which will have the following form:

$$p_i = \exp[-\lambda_0 - \lambda_1 g_1(x_i) - \lambda_2 g_2(x_i) \ldots - \lambda_m g_m(x_i)]$$
$$i = 1, 2, \ldots, n \quad (6)$$

where λ_i, $i = 0, 1, 2, \ldots, m$, are Lagrange multipliers that are determined by using the information specified by Eqs. 4 and 5.

Because the POME-based distribution is favored over those with less entropy among those that satisfy the given constraints, according to Shannon entropy as an information measure, entropy defines a kind of measure on the space of probability distributions. Intuitively, distributions of higher entropy represent more disorder, are smoother, are more probable, are less predictable, or assume less. The POME-based distribution is maximally noncommittal with regard to missing information and does not require invoking ergodic hypotheses.

Principle of Minimum Cross Entropy

According to Laplace's principle of insufficient reason, all outcomes of an experiment should be considered equally likely, unless there is information to the contrary. On the basis of intuition, experience, or theory, a random variable may have an a priori probability distribution. Then, Shannon entropy is maximum when the probability distribution of the random variable is one which is as close to the a priori distribution as possible. This is called the principle of minimum cross entropy (POMCE) which minimizes Bayesian entropy (4). This is equivalent to maximizing Shannon entropy.

To explain POMCE, let us suppose that we guess a probability distribution for a random variable x as $Q = \{q_1, q_2, \ldots, q_n\}$ based on intuition, experience, or theory. This constitutes the prior information in the form of a prior distribution. To verify our guess, we take some observations $X = (x_1, x_2, \ldots, x_n)$ and compute some moments of the distribution. To derive the distribution $P = \{p_1, p_2, \ldots, p_n\}$ of X, we take all the given information and make the distribution as near to our intuition and experience as possible. Thus, POMCE is expressed as

$$D(P, Q) = \sum_{i=1}^{n} p_i \ln \frac{p_i}{q_i} \quad (7)$$

where the cross entropy D is minimized. If no a priori distribution is available and if, according to Laplace's principle of insufficient reason, Q is chosen as a uniform distribution U, then Eq. 7 takes the form

$$D(P, U) = \sum_{i=1}^{n} p_i \ln\left[\frac{p_i}{1/n}\right] = \ln n \left(\sum_{i=1}^{n} p_i \ln p_i\right) \quad (8)$$

Hence, minimizing $D(P, U)$ is equivalent to maximizing Shannon entropy. Because D is a convex function, its local minimum is its global minimum. Thus, a posterior distribution P is obtained by combining a prior Q with the specified constraints. The distribution P minimizes the cross (or relative) entropy with respect to Q, defined by Eq. 7, where the entropy of Q is defined as in Eq. 2. Cross-entropy minimization results asymptotically from Bayes' theorem.

JOINT ENTROPY, CONDITIONAL ENTROPY, AND TRANSINFORMATION

If there are two random variables X and Y whose probability distributions are $P(x) = \{p_1, p_2, \ldots, p_n\}$ and $Q(y) = \{q_1, q_2, \ldots, q_n\}$, which are independent, then Shannon entropy of the joint distribution of X and Y is the sum of the entropies of the marginal distributions expressed as

$$H(P, Q) = H(X, Y) = H(P) + H(Q) = H(X) + H(Y) \quad (9)$$

If the two random variables are dependent, then Shannon entropy of the joint distribution is the sum of the marginal entropy of one variable and the conditional entropy of the other variable conditioned on the realization of the first. Expressed algebraically,

$$H(X, Y) = H(X) + H(Y \mid X) \quad (10)$$

where $H(Y|X)$ is the conditional entropy of Y conditioned on X. The conditional entropy can be defined as

$$H(X|Y) = -\sum_{i=1}^{n} \sum_{j=1}^{m} p(x_i, y_j) \ln(p(x_i \mid y_j)) \quad (11)$$

It is seen that if X and Y are independent, then Eq. 10 reduces to Eq. 9. Furthermore, the joint entropy of dependent X and Y will be less than or equal to the joint entropy of independent X and Y, that is, $H(X, Y) \leq H(X) + H(Y)$. The difference between these two entropies defines transinformation $T(X, Y)$ or $T(P, Q)$ expressed as

$$T(X, Y) = H(X) + H(Y) - H(X, Y) \quad (12)$$

Transinformation represents the amount of information common to both X and Y. If X and Y are independent, $T(X, Y) = 0$. Substitution of Eq. 10 in Eq. 12 yields

$$T(X, Y) = H(Y) - H(Y|X) \quad (13)$$

Equation 13 states that stochastic dependence reduces the entropy of Y.

ENTROPY AS A MODELING TOOL

Although entropy theory has been applied in recent years to a variety of problems in hydrology, its potential as a decision-making tool has not been fully exploited. A brief discussion follows highlighting this potential. Fundamental to the concepts presented below is the need for probability distributions that can be derived by using entropy theory.

Information Content of Data

One frequently encounters a situation in which one exercises freedom of choice, evaluates uncertainty, or measures information gain or loss. The freedom of choice, uncertainty, disorder, information content, or information gain or loss has been variously measured by relative entropy, redundancy, and conditional and joint entropies employing conditional and joint probabilities. As an example, in the analysis of empirical data, the variance has often been interpreted as a measure of uncertainty and as revealing gain or loss in information. However, entropy is another measure of dispersion—an alternative to variance. This suggests that it is possible to determine the variance whenever it is possible to determine entropy measures, but the reverse is not necessarily true. However, variance is not the appropriate measure if the sample size is small.

To measure correlation or dependence between any two variables, an informational coefficient of correlation r_0 is defined as a function of transinformation, T_0, as

$$r_0 = [1 - \exp(-2T_0)]^{0.5} \tag{14}$$

The transinformation, given by Eq. 14, expresses the upper limit of common information between two variables and represents the level of dependence (or association) between the variables. It represents the upper limit of transferable information between the variables, and its measure is given by r_0. The ordinary correlation coefficient r measures the amount of information transferred between variables under specified assumptions, such as linearity and normality. An inference similar to that of the ordinary correlation coefficient, r, can be drawn by defining the amount (in percent) of transferred information by the ratio T/T_0, where T can be computed in terms of ordinary r.

Criteria for Model Selection

Usually, there are more models than one needs, and so a model has to be chosen. Akaike (5) formulated a criterion, called the Akaike information criterion (AIC), for selecting the best model from amongst several models as

$$\text{AIC} = 2\log(\text{maximized likelihood}) + 2k \tag{15}$$

AIC provides a method of model identification and can be expressed as minus twice the logarithm of the maximum likelihood plus twice the number of parameters used to find the best model. The maximum likelihood and entropy are uniquely related. When there are several models, the model that gives the minimum value of AIC should be selected. When the maximum likelihood is identical for two models, the model that has the smaller number of parameters should be selected, for that will lead to a smaller AIC and comply with the principle of parsimony.

Hypothesis Testing

Another important application of entropy theory is testing of hypotheses (6). By using Bayes' theorem in logarithmic form, an evidence function is defined for comparing two hypotheses. The evidence in favor of a hypothesis over its competitor is the difference between the respective entropies of the competition and the hypothesis under test. Defining surprisal as the negative of the logarithm of the probability, the mean surprisal for a set of observations is expressed. Therefore, the evidence function for two hypotheses is obtained as the difference between the two values of the mean surprisal multiplied by the number of observations.

Risk Assessment

In common language, risk is the possibility of loss or injury and the degree of probability of such loss. Rational decision-making requires a clear and quantitative way of expressing risk. In general, risk cannot be avoided, and a choice has to be made between risks. There are different types of risk, such as business risk, social risk, economic risk, safety risk, investment risk, and occupational risk. To put risk in proper perspective, it is useful to clarify the distinction between risk, uncertainty, and hazard.

The notion of risk involves both uncertainty and some kind of loss or damage. Uncertainty reflects the variability of our state of knowledge or state of confidence in a prior evaluation. Thus, risk is the sum of uncertainty plus damage. Hazard is commonly defined as a source of danger that involves a scenario identification (e.g., failure of a dam) and a measure of the consequence of that scenario or a measure of the ensuing damage. Risk encompasses the likelihood of converting that source into the actual delivery of loss, injury, or some form of damage. Thus, risk is the ratio of hazard to safeguards. By increasing safeguards, risk can be reduced, but it is never zero. Awareness of risk reduces risk, so awareness is part of safeguards. Qualitatively, risk is subjective and is relative to the observer. Risk involves the probability of a scenario and its consequence resulting from the occurrence of the scenario. Thus, one can say that risk is probability and consequence. Kaplan and Garrick (7) analyzed risk using entropy.

HYDROLOGIC MODELING USING ENTROPY THEORY

A historical perspective on entropy applications in environmental and water resources is given in Singh and Fiorentino (8) and Singh (9). Harmancioglu and Singh (10) discussed the use of entropy in water resources. A brief synopsis of entropy-based applications follows.

Derivation of Probability Distributions

Frequency distributions that satisfy the given information are often needed. Entropy theory is ideally suited to that end. POME has been employed to derive a variety of distributions; some have found wide applications in environmental and water resources. Many of these distributions have been summarized in Singh and Fiorentino (8) and by Singh (9). Let $p(x)$ be the probability distribution of X that is to be determined. The information on X is available in terms of constraints given by Eq. 2. Then, the entropy-based distribution is given by Eq. 6. Substitution of Eq. 5 in Eq. 2 yields

$$\exp(\lambda_0) = Z = \sum_{i=1}^{n} \exp\left[-\sum_{j=1}^{m} \lambda_j g_j(x_i)\right] \quad (16)$$

where Z is called the partition function and λ_0 is the zeroth Lagrange multiplier. The Lagrange parameters are obtained by differentiating Eq. 16 with respect to the Lagrange multipliers:

$$\begin{aligned} \frac{\partial \lambda_0}{\partial \lambda_j} &= -a_j = E[g_j], \quad j = 1, 2, 3, \ldots, m \\ \frac{\partial^2 \lambda_0}{\partial \lambda_j^2} &= \mathrm{Var}[g_j] \\ \frac{\partial^2 \lambda_0}{\partial \lambda_j \partial \lambda_k} &= \mathrm{Cov}[g_j, g_k] \\ \frac{\partial^3 \lambda_0}{\partial \lambda_j^3} &= -\mu_3[g_j] \end{aligned} \quad (17)$$

where E[.] is the expectation, Var [.] is the variance, Cov [.] is the covariance, and is the third moment about the centroid, all for g_j.

When there are no constraints, then POME yields a uniform distribution. As more constraints are introduced, the distribution becomes more peaked and possibly skewed. In this way, entropy reduces from a maximum for the uniform distribution to zero when the system is fully deterministic.

Parameter Estimation

It is desirable to estimate parameters of a distribution in terms of the given constraints. Entropy theory accomplishes precisely that. Singh (11) described POME-based estimation for a number of probability distributions used in hydrology and environmental and water resources. He also discussed a comparison of the POME-based method with the methods of moments, maximum likelihood estimation, and some others. The comparison shows that the POME-based method is comparable to some methods and is better than others.

Entropy-Spectral Analysis for Flow Forecasting

Maximum entropy spectral analysis (MESA) was introduced by Burg (12). It has several advantages over conventional spectral analysis methods. It has short and smooth spectra that have high-degree resolutions (13). The statistical characteristics that are used in stochastic model identification can also be estimated using MESA, thus permitting integration of spectral analysis and computations related to stochastic model development. Ulrych and Clayton (14) reviewed the principles of MESA and the closely related problem of autoregressive time series modeling. Shore (15) presented a comprehensive discussion of minimum cross-entropy spectral analysis.

The relationship between spectrum $W(f)$ with frequency f of a stationary process $x(t)$ and entropy $H(f)$ can be expressed as

$$H(f) = \frac{1}{2}\ln(2\,w) + \frac{1}{4\,w}\int_{w}^{+w} \ln[W(f)]\,df \quad (18)$$

where w is the frequency band. Equation 18 is maximized subject to the constraint equations given as autocorrelations until log m:

$$\rho(n) = \int_{w}^{+w} W(f)\exp(i_2\pi f n\Delta t)\,df, \qquad m \le n \le +m \quad (19)$$

where Δt is the sampling time interval and $i = (-1)^{1/2}$. Maximization of Eq. 19 is equivalent to maximizing

$$H(f) = \int_{w}^{+w} \ln[W(f)]\,df \quad (20)$$

which is known as Burg entropy. The spectrum $W(f)$ can be expressed in terms of the Fourier series as

$$W(f) = \frac{1}{2\,w}\sum_{n=\infty}^{\infty} \rho(n)\exp(i2\pi n f\Delta t) \quad (21)$$

Substituting Eq. 21 in Eq. 20 and maximizing lead to MESA.

Jaynes (16) has shown that MESA and other methods of spectral analysis, such as Schuster, Blackman–Tukey, maximum likelihood, Bayesian, and autoregressive (AR, ARMA, or ARIMA) models are not in conflict and that AR models are a special case of MESA. Krstanovic and Singh (17,18) employed MESA for long-term stream flow forecasting. Krstanovic and Singh (19,20) extended the MESA method to develop a real-time flood forecasting model. Padmanabhan and Rao (21,22) applied MESA to analyze rainfall and river flow time series. Rao et al. (23) compared a number of spectral analysis methods with MESA and found that MESA is superior. Eilbert and Christensen (24) analyzed annual hydrologic forecasts for central California and found that dry years might be more predictable than wet years. Dalezios and Tyraskis (25) employed MESA to analyze multiple precipitation time series.

Regional Precipitation Analysis and Forecasting

The Burg algorithm or MESA can be applied to identify and interpret multistation precipitation data sets and to explore spectral features that lead to a better understanding of rainfall structure in space and time (25).

Then, multistation rainfall time series can be extrapolated to develop regional forecasting capabilities.

Grouping of River Flow Regimes

An objective grouping of flow regimes into regime types can be employed as a diagnostic tool for interpreting the results of climate models and flow sensitivity analyses. By minimizing an entropy-based objective function (such as minimum cross entropy), a hierarchical aggregation of monthly flow series into flow regime types can, therefore, be effectively performed, which will satisfy chosen discriminating criteria. Such an approach was developed by Krasovskaia (26) who applied it to a regional river flow sample for Scandinavia for two different formulations of discriminating criteria.

Basin Geomorphology

Entropy plays a fundamental role in characterizing landscape. Using entropy theory for the morphological analysis of river basin networks, Fiorentino et al. (27) found that the connection between entropy and the mean basin elevation is linearly related to basin entropy. Similarly, the relation between the fall in elevation from the source to the outlet of the main channel and the entropy of its drainage basin, it was found, is linear and so also was the case between the elevation of a node and the logarithm of its distance from the source. When a basin was ordered following the Horton–Strahler ordering scheme, a linear relation was found between drainage basin entropy and basin order. This relation can be characterized as a measure of basin network complexity. Basin entropy, it was also found, is linearly related to the logarithm of the magnitude of the basin network. This relation led to a nonlinear relation between the network diameter and magnitude where the exponent, it was found, is related to the fractal dimension of the drainage network.

Design of Hydrologic Networks

The purpose of measurement networks is to gather information in terms of data. Fundamental to evaluating these networks is the ability to determine if the networks are gathering the needed information optimally. Entropy theory is a natural tool for that determination. Krstanovic and Singh (28,29) employed the theory for space and time evaluation of rainfall networks in Louisiana. The decision whether to keep or to eliminate a rain-gauge was based entirely on the reduction or gain of information at that gauge. Yang and Burn (30) employed a measure of information flow, called directional information transfer index (DIT), between gauging stations in the network. The value of DIT varies from zero, where no information is transmitted and the stations are independent, to one where no information is lost and the stations are fully dependent. Between two stations of one pair, the station that has the higher DIT value should be retained because of its greater capability of inferring information at the other side.

Rating Curve

Moramarco and Singh (31) employed entropy theory to develop a method for reconstructing the discharge hydrograph at a river section where only water level is monitored and discharge is recorded at another upstream section. The method, which is based on the assumption that lateral inflows are negligible, has two parameters linked to remotely observed discharge and permits, without using a flood routing procedure and without the need of a rating curve at a local site, relating the local river stage to the hydraulic condition at a remote upstream section.

CONCLUDING REMARKS

Entropy theory permits determining of the least biased probability distribution of a random variable, subject to the available information. It suggests whether or not the available information is adequate and, if not, then additional information should be sought. In this way, it brings the model, the modeler, and the decision-maker closer. As an objective measure of information or uncertainty, entropy theory allows communicating with nature, as illustrated by its application to the design of data acquisition systems, the design of environmental and hydrologic networks, and the assessment of the reliability of these systems or networks. In a similar vein, it helps better understand the physics or science of natural systems, such as landscape evolution, geomorphology, and hydrodynamics. A wide variety of seemingly disparate or dissimilar problems can be meaningfully solved by using entropy.

BIBLIOGRAPHY

1. Shannon, C.E. (1948). A mathematical theory of communications, I and II. *Bell System Technical Journal* **27**: 379–443.
2. Jaynes, E.T. (1957). Information theory and statistical mechanics, I. *Physical Review* **106**: 620–630.
3. Levine, R.D. and Tribus, M. (Eds.). (1978). *The Maximum Entropy Formalism*. The MIT Press, Cambridge, MA, p. 498.
4. Kullback, S. and Leibler, R.A. (1951). On information and sufficiency. *Annals of Mathematical Statistics* **22**: 79–86.
5. Akaike, H. (1973). Information theory and an extension of the maximum likelihood principle. *Proceedings, 2nd International Symposium on Information Theory*. B.N. Petrov and F. Csaki. (Eds.). Publishing House of the Hungarian Academy of Sciences, Budapest, Hungary.
6. Tribus, M. (1969). *Rational Description: Decision and Designs*. Pergamon Press, New York.
7. Kaplan, S. and Garrick, B.J. (1981). On the quantitative definition of risk. *Risk Analysis* **1**(1): 11–27.
8. Singh, V.P. and Fiorentino, M. (Eds.). (1992). *Entropy and Energy Dissipation in Water Resources*. Kluwer Academic Publishers, Dordrecht, the Netherlands.
9. Singh, V.P. (1998). *Entropy-Based Parameter Estimation in Hydrology*. Kluwer Academic Publishers, Boston, MA.
10. Harmancioglu, N.B. and Singh, V.P. (1998). Entropy in environmental and water resources. In: *Encyclopedia of Hydrology and Water Resources*. R.W. Hershey and R.W. Fairbridge (Eds.). Kluwer Academic Publishers, Boston, MA, pp. 225–241.
11. Singh, V.P. (1998). The use of entropy in hydrology and water resources. *Hydrological Processes* **11**: 587–626.

12. Burg, J.P. (1975). *Maximum entropy spectral analysis*. Unpublished Ph.D. thesis, Stanford University, Palo Alto, CA, p. 123.
13. Fougere, P.F., Zawalick, E.J., and Radoski, H.R. (1976). Spontaneous life splitting in maximum entropy power spectrum analysis. *Physics of the Earth and Planetary Interiors* **12**: 201–207.
14. Ulrych, T. and Clayton, R.W. (1976). Time series modeling and maximum entropy. *Physics of the Earth and Planetary Interiors* **12**: 188–199.
15. Shore, J.E. (1979). *Minimum cross-entropy spectral analysis*. NRL Memorandum Report 3921, Naval Research Laboratory, Washington, DC.
16. Jaynes, E.T. (1982). On the rationale of maximum entropy methods. *Proceedings of the IEEE* **70**: 939–952.
17. Krstanovic, P.F. and Singh, V.P. (1991). A univariate model for long-term streamflow forecasting: 1. Development. *Stochastic Hydrology and Hydraulics* **5**: 173–188.
18. Krstanovic, P.F. and Singh, V.P. (1991). A univariate model for long-term streamflow forecasting: 2. Application. *Stochastic Hydrology and Hydraulics* **5**: 189–205.
19. Krstanovic, P.F. and Singh, V.P. (1993). A real-time flood forecasting model based on maximum entropy spectral analysis: 1. Development. *Water Resources Management* **7**: 109–129.
20. Krstanovic, P.F. and Singh, V.P. (1993). A real-time flood forecasting model based on maximum entropy spectral analysis: 2. Application. *Water Resources Management* **7**: 131–151.
21. Padmanabhan, G. and Rao, A.R. (1986). Maximum entropy spectra of some rainfall and river flow time series from southern and central India. *Theoretical and Applied Climatology* **37**: 63–73.
22. Padmanabhan, G. and Rao, A.R. (1988). Maximum entropy spectral analysis of hydrologic data. *Water Resources Research* **24**(9): 1591–1533.
23. Rao, A.R., Padmanabhan, G., and Kashyap, R.L. (1980). Comparison of recently developed methods of spectral analysis. *Proceedings, Third International Symposium on Stochastic Hydraulics*. Tokyo, Japan, pp. 165–175.
24. Eilbert, R.F. and Christensen, R.O. (1983). Performance of the entropy minimax hydrological forecasts for California water years 1948–1977. *Journal of Climate & Applied Meteorology* **22**: 1654–1657.
25. Dalezios, N.R. and Tyraskis, P.A. (1989). Maximum entropy spectra for regional precipitation analysis and forecasting. *Journal of Hydrology* **109**: 25–42.
26. Krasovskaia, R. (1997). Entropy-based grouping of river flow regimes. *Journal of Hydrology* **202**: 173–1191.
27. Fiorentino, M., Claps, P., and Singh, V.P. (1993). An entropy-based morphological analysis of river basin networks. *Water Resources Research* **29**(4): 1215–1224.
28. Krstanovic, P.F. and Singh, V.P. (1992). Evaluation of rainfall networks using entropy: 1. Theoretical development. *Water Resources Management* **6**: 279–293.
29. Krstanovic, P.F. and Singh, V.P. (1992). Evaluation of rainfall networks using entropy: II. Application. *Water Resources Management* **6**: 295–314.
30. Yang, Y. and Burn, D.H. (1984). An entropy approach to data collection network design. *Journal of Hydrology* **157**: 307–324.
31. Moramarco, T. and Singh, V.P. (2001). Simple method for relating local stage and remote discharge. *Journal of Hydrologic Engineering, ASCE* **6**(1): 78–81.

EVAPORATION

THEODORE A. ENDRENY
SUNY-ESF
Syracuse, New York

Evaporation of water is a solar energy driven phase change from liquid to vapor that maintains the hydrologic cycle by transferring liquid water at the earth's surface to water vapor in the atmosphere, where it may lift, condense, and precipitate to earth as liquid water. In this discussion, evaporation is defined to include the closely associated transpiration process as a subcategory. The coupled processes are often called as evapotranspiration, or ET, where transpiration focuses on the transport of liquid water through the plant roots, stem, and leaf prior to evaporation through the leave's stomata. Further, when the discussion does not make a clear distinction between potential and actual evaporation, potential should be assumed. Potential evaporation refers to the amount of water that available energy and diffusion processes can transfer into atmospheric vapor, which is typically greater then the amount actually transferred due to limits on soil water volumes and resistances in the path.

PHYSICAL CONTROLS ON EVAPORATION

Evaporation is effectively a two-step process that first, requires that water changes phase to a vapor state and second, requires that the vapor is transported by advection and/or diffusion into unsaturated air. The phase change alone does not completely satisfy the requirement for evaporation when dynamic equilibrium exists at the boundary between liquid and vapor and condensation of the vapor saturated air returns liquid water to the surface to maintain no net water loss. Transport, therefore, ensures removal of the water vapor and a net loss of heat and mass from the liquid surface. In the vapor state, evaporated water is invisible to the human eye, which detects wavelengths between 0.4 and 0.7 μm, but is detectable in other areas of the electromagnetic spectrum. As such, though clouds are derived from and contain evaporated water, they are not vapor and instead reveal condensed water droplets that geometrically scatter light.

The first step in changing the phase of water from liquid to vapor requires an input of solar energy, which is stored at the surface and supports nighttime evaporation. Nearly 52% of solar energy absorbed at the earth's surface is used for vaporization. This energy is called latent heat of vaporization, or λ, and is a function of water temperature. When the phase changes directly from frozen water to vapor, known as sublimation, a greater amount of energy is required. Phase changes require energy to separate the hydrogen bond based, attractive intermolecular forces holding water molecules in an organized pattern and close proximity. Latent heat is a name that suggests dormant, or invisible, and is used to indicate that, unlike the measurable effect of sensible heat on air temperature, the solar energy used in evaporation remains 'hidden' from thermometer measurement.

As an example, at the standard pressure of 1013.3 milibar (mb), heating liquid water from 0 to 100 °C requires 4186.8 joules (J) of heat per kilogram of water per degree C temperature change and is detectable with a thermometer. In contrast, the latent heat of vaporization converting 100 °C water into 100 °C vapor does not have a measurable impact on water temperature. The latent heat, in megajoules per kilogram (MJ kg^{-1}), required per °C of water temperature, T, is given as

$$\lambda = 2.501 - 0.002361 \cdot T$$

It is apparent that the majority of heat input to evaporate water, whether at 0 °C or 99 °C, is that for the phase change. A relatively constant 1350 $Js^{-1}\,m^{-2}$ stream of solar energy entering the earth's upper atmosphere provides energy to evaporate water just like a stove can boil and evaporate water. When 50% of this solar constant strikes the earth's surface after atmospheric attenuation (e.g., scattering and reflection), it requires approximately 1 hour and 10 minutes to generate the latent heat needed to evaporate 1 kg of water at 20 °C.

Latent heat, though not detectable as affecting air temperature, is stored with the vapor in greater vibrational, rotational, and translational movement of vapor molecules. The removal of this vapor from the liquid, therefore, causes a measurable loss of energy and temperature from the remaining liquid, producing a cooling effect. A familiar example of this process is that wet skin, from sweat or a shower, cools faster in moving rather than still air because the wind speeds evaporation that takes heat from the body. Knowing that heat is used for evaporation, it is now clear that a covered pot will boil faster than a counterpart uncovered pot do to a lower loss of heat to the net evaporation of water. Heat stored in vapor is later released back into the environment when the vapor vibrational speeds slow, and it condenses into water, called latent heat of condensation. When vapor passes directly to solid frozen water, called deposition, a greater amount of heat stored in rotational, vibrational, and translational molecular movement is released into the environment.

The second physical step in evaporation is transport by advection and/or diffusion, which provides the net movement of water molecules from the liquid water surface of soil, plants, or lakes to atmospheric vapor. Vapor and wind gradients exert the principal controls on removal of water vapor beyond the saturated layer of air that maintains condensation–evaporation equilibrium dynamics. Fundamental barriers to transport beyond the layer of dynamic equilibrium include stagnant air and saturated air above the evaporating surface, conditions readily created within and at the surface of soil and plant systems. Net evaporation therefore increases with steeper wind and saturation gradients, which are defined as the change in wind or saturation with distance above the evaporating surface. Work on fluid velocity and turbulence gradients in the mid-1900s by Prandtl and von Karman has been used to estimate momentum, sensible heat, and vapor transport from wind speed measurements.

Estimates of atmospheric wind and vapor conditions above the evaporating surface provide important data for estimating wind and saturation gradients and predicting barriers to vapor transport and net evaporation. Meteorological stations are frequently equipped with anemometers and thermometers at 2 meters (m) above the ground surface to help establish the wind and vapor gradients controlling evaporation. Wind profiles, it is assumed, begin with stagnant air at the no-slip boundary, or zero-plane displacement height, and increase logarithmically. In a landscape broken by tree canopies, a 2 m wind measurement may be inadequate to represent observations. Research has shown that dynamic turbulence and eddies created by such forested heterogeneity result in increased wind and evaporation rates that exceed the estimated atmospheric potential.

Based on Dalton's work on individual pressures of multiple atmospheric gases summing to the observed atmospheric pressure, vapor is often reported as a partial pressure and can be derived from measurements of temperature. In the following equation, vapor pressure is reported in kilopascals (kPa) and temperature in °C. Initially dry warm air can absorb more water than initially dry cold air before reaching saturation.

$$e = 0.6108 \exp\left(\frac{17.27\,T}{237.3 + T}\right)$$

The dry-bulb temperature is used to estimate the total amount of vapor the air could absorb prior to saturation; the dew point temperature represents the temperature to which the air must cool for total saturation. When the dry-bulb and dew point temperatures are equal, the air is fully saturated. The ratio of actual to saturated vapor pressures is the relative humidity. The dew point temperature is estimated by using a psychrometer that measures the difference between dry-bulb and wet-bulb thermometers, called the wet-bulb depression, together with lookup tables relating wet-bulb depression to dew point temperature. The dry-bulb is a normal thermometer measuring air temperature, but a moist piece of cloth typically covers the wet-bulb, and evaporation of the water from the cloth causes the temperature to drop. Chilled mirrors hygrometers, hair hygrometers, and vapor pressure sensors are also used to detect the vapor content in the air.

EVAPORATION MEASUREMENT AND ESTIMATION

Evaporation is fundamental to both energy and water balances, yet despite the importance of evaporation to hydrologic assessment of the paths, quantities, and quality of water in the lithosphere, biosphere, and atmosphere, the complexity of the process has prevented easy or exact techniques for measuring and estimating it. The relative accuracy of yearly river basin evaporation estimates is high, as the estimate time frame and spatial area become smaller, but the simple application of energy and water balance models to solve for evaporation becomes less tractable. A variety of measurement and estimation techniques have been developed for these smaller scales, such as hourly, daily, and monthly evaporation from reservoirs, farm fields, and single plants.

In general, evaporative fluxes from the land surface are more difficult to measure than from open water, given that an immeasurable number of irregular, tiny and unique soil and leaf surfaces are involved in this phase change and that suction gradients draw water to this evaporating interface from unobserved reserves of unknown volume. Fluxes from open water, though relatively homogeneous, still provide challenges when subsurface inflows and outflows are poorly understood and significant and wind and water advected energy influencing evaporation is heterogeneously distributed. Hence, the numerous methods developed for estimating evaporation are categorized based on the type of surface, availability of water, and the importance of stored energy, water-advected energy, and air-advected energy.

Actual evaporation can be measured by using a water balance approach, a turbulent-transfer approach, a potential evaporation approach, or a water quality approach. The water balance approach can function with measurements of a mass balance being kept for a water pan, such as a the Class-A Pan of the National Weather Service, a soil and plant system, such as in a weighing lysimeter, or of a small, enclosed atmosphere. Turbulent-transfer methods, which derive evaporation from estimates of momentum or heat flux, can provide estimates for larger heterogeneous areas, but assume that the air sampled by the field instrumentation used for the Bowen ratio or eddy-correlation method is well mixed to represent the upwind land area by a given fetch. Water quality methods include techniques that track concentrations of dissolved solids, which enrich when evaporation removes the water solvent, and isotope tracer studies that show heavier isotopes are enriched by preferred evaporation of lighter isotopes. Sap flow monitoring in trees provides another technique to measure the flux of water from the ground to the atmosphere.

Mathematical estimates of evaporation rates have been approached by using an equally wide variety of techniques and include temperature-based, aerodynamic-based, radiation-based, and combination-based methods. Temperature-based methods, such as the monthly time-step Thornthwaite equation, use air temperature and length of day, as well as an assumed humidity, to compute the potential evaporation, and they have been adapted to suit several different climates and regions. Aerodynamic methods assume that solar radiation is not limiting and consider only wind speed and turbulence as controls on the transport of water vapor away from the surface. Radiation-based methods likewise assume that wind turbulence and eddies are not limiting and use measured incoming radiation and the latent heat of vaporization to compute evaporation flux. A popular form of this equation is the Priestly–Taylor, which increased it by a factor of 30% to account for added aerodynamic transfer. The combination method, known most extensively for the Penman–Monteith equation, uses air temperature, net radiation, wind speed, and relative humidity vapor gradients to derive minute by minute and daily evaporation rates.

HYDROLOGIC IMPACTS OF EVAPORATION

Observations of terrestrial river discharge reveal that more water precipitates on land than evaporates from land and that more water evaporates from oceans than precipitates on oceans. Precipitation totals may vary from year to year, but evaporative demands are rather steady, which creates a greater relative fluctuation in river discharge than in precipitation. This is illustrated by considering a 20% decrease in annual precipitation from 100 to 80 cm, where 50 cm went to evaporation in both years, and discharge dropped by 40% from 50 to 30 cm. Evaporation and its impact on liquid and vapor water volumes and the partitioning of solar energy into latent and sensible heat create and maintain a range of climatic conditions, from microclimates on the scale of a tree canopy to macroclimates that describe the global distribution of plants.

The volume of water evaporated from the ocean and land surface is greatest at the meteorological equator, or intertropical convergence zone (ITCZ), and smallest at the poles, which is the result of a similar longitudinal distribution of insolation intensity. Sinking Hadley cell air at the 30° latitude belts, which warms to absorb greater amounts of water vapor, is the cause of a belt of deserts in this region. The distribution of incoming solar radiation, which is greatest at the equator and smallest at the poles, is the driving force explaining the meridional (across lines of latitude) distribution of evaporation. Wind transport of this evaporated vapor from the equatorial region to the midlatitudes and poles, where latent energy is released to the atmosphere as sensible heat during condensation, is one of only a few processes that help to maintain the earth's energy balance.

Global water balance numbers reveal that a relatively small volume of evaporated vapor resides in the atmosphere. The earth's atmosphere has a volume of 12,900 km^3, and contains just 0.001% of all global water. As evaporated vapor, it receives 71,000 $km^3\,yr^{-1}$ from land, 1000 $km^3\,yr^{-1}$ from lakes, and 505,000 $km^3\,yr^{-1}$ from oceans, and this flux rate into its total volume equals a residence time of 8.2 days, or just over a week before evaporated water precipitates. The atmospheric vapor precipitates at a volumetric rate of 577 $km^3\,yr^{-1}$, of which 119,000 $km^3\,yr^{-1}$ falls on land. Observation and estimation of river discharge at 47,000 $km^3\,yr^{-1}$ was used to deduce the amount evaporated from land, which is 61%

Table 1. Continental Average Estimated Evaporation

Continent	Area, km^2	Evaporation, mm yr^{-1}	Evaporation, %
Antarctica	14,100,000	28	17
Europe	10,000,000	375	57
Asia	44,100,000	420	60
South America	17,900,000	946	60
North America	24,100,000	403	62
Africa	29,800,000	582	84
Australia	7,600,000	420	94
Total Land	148,900,000	480	64

Table 2. Average Estimated Evaporation Based on Studies of Precipitation and Runoff

Watershed by River Name	Continent [Nation(s)]	Basin Size, km^2	Evaporation[a], %
Brahmaputra	Asia (Tibet/Bangladesh)	589,000	35
Irrawaddy	Asia (Burma)	431,000	40
Yangtzekiang	Asia (China)	1,970,000	50
Amazon	South America (six nations)	7,180,000	53
Orinoco	South America (Venezuela)	1,086,000	54
Lena	Asia (Russia)	2,430,000	54
Mekong	Asia (China)	795,000	57
Yenesei	Asia (Russia)	2,599,000	58
Ganges	Asia (China)	1,073,000	58
Saint Lawrence	North America (Canada, U.S.)	1,030,000	67
Amur	Asia (Russia)	1,843,000	68
Congo	Africa (7 nations)	3,822,000	75
Ob	Asia (Russia)	2,950,000	76
Mississippi	North America (U.S.)	3,224,000	79
La Plata	South America (five nations)	2,650,000	80
Average	–	2,224,800	60

[a]Evaporation may include basin transfers because it was not directly measured but derived from precipitation and runoff measurements.

evaporating, and this differs from the continental average of 64% reported in Table 1.

For the continental United States, which contains deserts in Arizona and rain forests in Washington, the average annual percentage of precipitation converted to evaporation is approximately 62%. This value is similar to global patterns but varies considerably from that measured on other continents (see Table 1) and larger watersheds (see Table 2).

The agricultural impact of evaporation is both the cause of nutrient uptake and growth in plants, as well as the loss of soil water and plant stress. Maintenance of optimal water levels in the soil, called field capacity, when gravitational water has drained, is the goal of many irrigation projects. If irrigation causes evaporation to exceed local precipitation, then salts will be drawn to the soil surface, which often creates osmotic gradients at the root interface that kill the agricultural crop. Agricultural irrigation to satisfy the high evaporation demand of sunny agricultural land, such as California's Central Valley, has become a direct competitor for use as a public water supply.

READING LIST

Shuttleworth, W.J. (1992). Evaporation. In: *Handbook of Hydrology*. D.R. Maidment (Ed.). McGraw Hill, New York.

Dingman, S.L. (1994). *Physical Hydrology*. Prentice-Hall, Englewood Cliffs, NJ.

EVAPOTRANSPIRATION

Jose O. Payero
University of Nebraska-Lincoln
North Platte, Nebraska

Liquid water from a surface can be transformed into water vapor by either evaporation or by transpiration. *Evaporation* is the process for converting liquid water to water vapor and removing it from the evaporating surface. *Transpiration* is the vaporization of water contained in plant tissues and the removal of vapor to the atmosphere through leaf stomata (1,2). Evaporation and transpiration occur simultaneously in cropped surfaces, and there is no easy way of quantifying the magnitude of each component. For practical applications, such as irrigation scheduling and irrigation system design, it has been more useful to consider both processes combined. The combination of these two processes is called *evapotranspiration*. The proportion of each component in a cropped surface is affected by factors such as vegetative cover, available water in the soil, and surface wetness. For an annual crop like corn, evaporation is the dominant component of evapotranspiration at the beginning of the season when the soil surface is exposed to solar radiation. As the crop grows and the canopy covers the surface, evaporation is minimized, and transpiration becomes the dominant component.

IMPORTANCE OF EVAPOTRANSPIRATION

Evapotranspiration is an important process in agriculture and other natural sciences, such as hydrology, because it is an important component of the hydrologic cycle. It represents the water that is effectively lost from the earth's surface, and can no longer be controlled by humans. This type of water loss is often called consumptive use. Other types of processes that usually cause water losses from a given area on the earth's surface, such as runoff and deep percolation, do not involve a change in the physical state of water and therefore, water can still be controlled to some degree by humans.

Plants use water as a solvent and transport mechanism for nutrients and other chemicals, as a reagent for the chemical reactions involved in their physiological processes (such as photosynthesis), and as a component of cell cytoplasm, which allows plant tissues to stay turgid. Most of the water consumed by plants, however, is used in evapotranspiration. Evapotranspiration has the important function of regulating the temperature of plants, keeping them cool within a temperature range that favors growth. When the water supply in the soil is limited, for instance, plants respond by closing their stomata. This restricts the rate of evapotranspiration, and the temperature of the canopy tends to increase (3). This increase in canopy temperature has been used to estimate the rate of evapotranspiration of crops and as a way to detect crop water stress for irrigation scheduling (4). Because most of the water consumed by plants is lost in evapotranspiration, it takes a considerable amount of water for a crop to produce one unit weight of dry matter, as shown for different crops in (Fig. 1).

Researchers have shown that crop yield is often linearly related to crop evapotranspiration, up to the point where yield is limited by factors other than water (6). Therefore, if evapotranspiration is limited, yield is usually reduced. For this reason, in regions where rainfall is not sufficient to provide enough water for crops to keep evapotranspirating at a nonlimited rate, irrigation is required to obtain adequate crop yields. The nonlimited rate of evapotranspiration, however, can also be maintained by applying excess water. Application of excess water, however, has been linked to undesired side effects such as drainage problems, salinization of soils, soil erosion, and pollution of surface and groundwaters. In places where irrigation water needs to be pumped, pumping excess water also represents higher production cost. Therefore, it is considered ideal to schedule irrigation according to crop water needs. This requires, among other things, good knowledge of crop evapotranspiration rates.

MEASURING EVAPOTRANSPIRATION

Measuring the rate of evapotranspiration of crops and other surfaces is complex and is a subject that has attracted considerable research. Many methods have been devised to measure and estimate evapotranspiration. *Methods for measuring* evapotranspiration include the use of lysimeters, scintillometers, micrometeorological techniques such as the Bowen ratio and eddy covariance

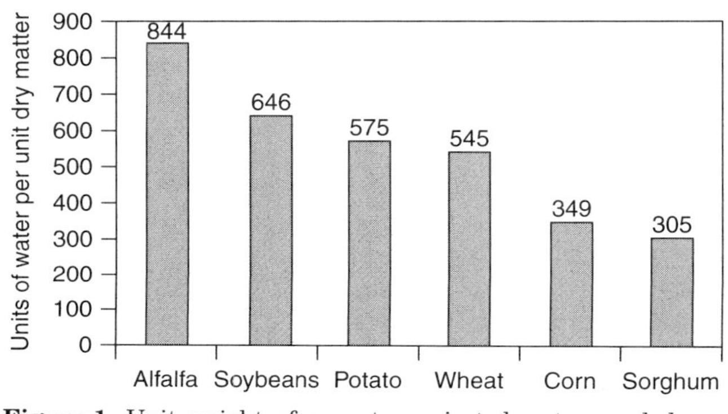

Figure 1. Unit weight of evapotranspirated water needed per unit weight of dry matter produced by different crops (adapted from Ref. 5).

methods, and measurement of sap flow. Methods to measure evapotranspiration usually require expensive instrumentation and specialized training, and therefore have been confined mostly to research applications. Because of this, considerable effort has been devoted to develop methods for estimating evapotranspiration.

ESTIMATING EVAPOTRANSPIRATION

Methods for estimating evapotranspiration include the use of different kinds of devices and a variety of equations. Devices such as evaporation pans, atmometers, and soil moisture monitoring devices have long been used to estimate evapotranspiration. The use of equations, however, is by far the most common method to estimate evapotranspiration. A great variety of equations have been proposed though the years, ranging from very simple to very complex models (1,7).

Simple models usually try to estimate evapotranspiration based on empirical relationships involving one or several meteorological variables. Complex models consider the physics governing the evapotranspiration process and try to include all factors that significantly contribute to the process (1,8). Complex models can be further divided into single-layer models and multiple-layer models. Single-layer models, such as the Penman–Monteith model, consider the crop canopy as a "big leaf," taking into account only processes that occur between the top of the canopy and the atmosphere. Multiple-layer models, on the other hand, also take into account those processes that take place below the crop canopy. Multiple-layer models are more theoretically sound than single-layer models, but their complexity makes them impractical for widespread use. Because a single-layer model is sufficiently accurate for most practical applications and relatively simple to apply, it has been proposed as the recommended method for estimating evapotranspiration (9).

Considerable effort has been made to estimate evapotranspiration using inputs obtained from remote sensing platforms, such as satellites or airplanes (1,10,11), and others have even tried to estimate evapotranspiration by measuring the flux of stable isotopes (12). Evapotranspiration, however, is more often estimated from equations that use meteorological data as input, as well as inputs that describe the characteristics of the evaporating surface. This is a convenient method because meteorological data are readily available in most places. Most nations and states support a network of meteorological stations and offer this information to the public in various ways. Meteorological data commonly used to estimate evapotranspiration include solar radiation, air temperature, relative humidity, and wind speed.

A detailed procedure for calculating evapotranspiration has been described by Allen et al. (1). The method involves a two-step process. One step consists of calculating the evapotranspiration rate of a reference crop, either clipped grass or alfalfa, which is usually known as reference evapotranspiration. In older literature, this was also called potential evapotranspiration. It represents a measure of the evaporating demand of the atmosphere for a short, cropped surface that effectively covers the ground, is growing healthily, and is not short of water, that is, a condition in which transpiration is not limited by stress, and the evapotranspiration demand of the atmosphere is met. The second step involves adjusting the reference evapotranspiration to match the conditions of the specific surface or crop being considered. This is done by calculating an adjustment factor, usually known as the crop coefficient. Depending on the accuracy required, calculating crop coefficients can also be a simple or complex process (7,13). Multiplying the reference evapotranspiration by the crop coefficient then results in the evapotranspiration rate for the crop or surface in question.

REQUIREMENTS FOR EVAPOTRANSPIRATION

Procedures used to estimate evapotranspiration try to simplify the complexities of the physical and physiological processes that affect evapotranspiration to a manageable number of quantifiable variables. They try to recognize that for the evapotranspiration process to take place, it is necessary to have

1. energy
2. water
3. space in the atmosphere to hold the water vapor
4. a transport mechanism for the water vapor to move from the surface to the atmosphere.

Evapotranspiration is an energy-driven process. It takes approximately 2.45 megajoules of energy to evaporate 1 kilogram of water at 20 °C. The sun supplies the energy needed for evapotranspiration from the earth's surface. Part of the solar energy that reaches the evaporating surface, however, is reflected back to the atmosphere and cannot be used for evapotranspiration. Of the energy that stays on the evaporating surface, known as net radiation, not all is used in evapotranspiration. The energy balance of a surface also includes energy that is absorbed or released by the soil (soil heat flux), by the air (latent heat flux), and that used in evapotranspiration (latent heat flux). All of these types of heat fluxes take place simultaneously from a given surface, and their proportions depend on the characteristics of the surface and weather conditions.

The amount of energy used for evapotranspiration includes short-wave radiation that comes directly from the sun, long-wave radiation or heat that comes from the surface, and advective heat that is transported horizontally by hot wind to the evaporating surface. The amount of energy available for evapotranspiration varies with latitude, day of the year, time of day, atmospheric conditions, and the characteristics of the surface itself. Figure 2 shows how the theoretical clear-sky solar radiation for different latitudes varies throughout the year. Figure 3 shows how measured solar radiation varies during the day and the effect of cloudiness in reducing the amount of solar energy that reaches the surface. As a general rule, the higher the amount of energy available at the evaporating surface, the higher the evapotranspiration rate, assuming that all other conditions required for evapotranspiration are met.

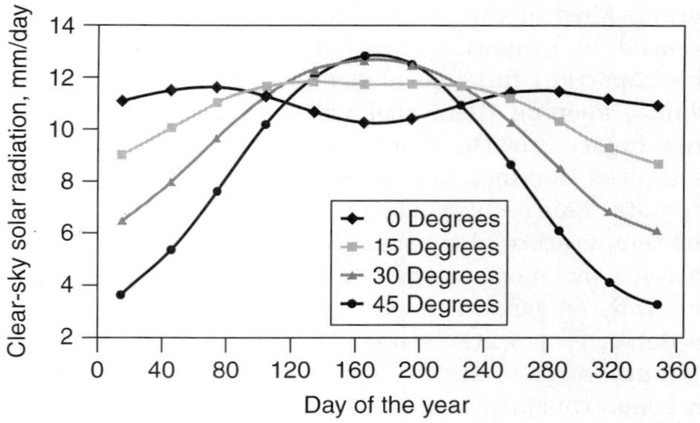

Figure 2. Theoretical clear-sky solar radiation values for different northern latitudes and day of the year at sea level.

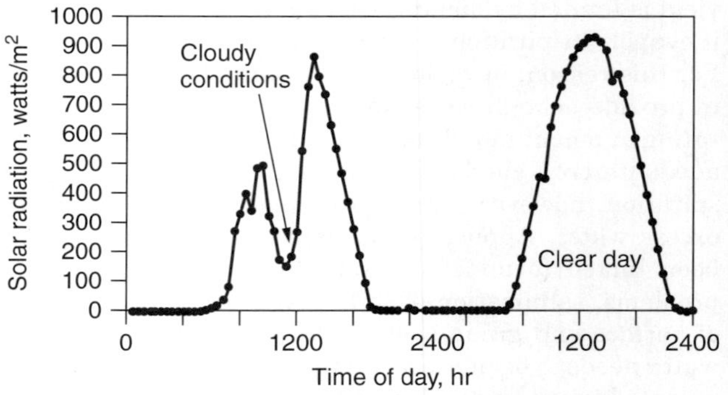

Figure 3. Measured solar radiation in North Platte, NE, on July 15 and 16, 2001.

Water is the next essential requirement for evapotranspiration. If water is not available, the energy available to evaporate water is then used to heat the air and the soil. Under desert conditions, for instance, where energy is usually plentiful, but water is normally very limited, the evapotranspiration rate can be very small. Most of the energy is converted to sensible heat flux, soil heat flux, and a very small portion or none of the energy is converted to latent heat flux. Similarly, when crops are under water stress, the evapotranspiration rate is reduced.

Space in the atmosphere to hold the water vapor is also needed for evapotranspiration to proceed. When water is evaporated from a surface, the water vapor will travel to the atmosphere where it will be stored, provided that the atmosphere is not saturated. If the atmosphere is already saturated, it will not be able to store the additional water vapor, and evapotranspiration will be restricted. The drier the air in contact with the evaporating surface, the more evapotranspiration is enhanced. For this reason, the humidity of the air, usually expressed as relative humidity, is an important factor to consider when estimating evapotranspiration.

The last requirement for evapotranspiration is a transport mechanism for moving water vapor from the surface to the atmosphere. There are two basic transport mechanisms for water vapor. The first is turbulence.

Turbulence is created when air moves horizontally over a rough surface. The friction created by the contact of the air with the rough elements of the surface creates a vertical component of the wind speed, which creates eddies of different sizes. The size of these eddies depends on the roughness of the surface and on the magnitude of the wind speed. These eddies carry the water vapor to the atmosphere. The other transport mechanism is buoyancy. Hot air is less dense that cold air; as the air close to the surface becomes hotter than the air above, it tends to ascend, carrying with it the water vapor. Turbulence is the most important transport mechanism in most instances, so wind speed is a very important factor in determining evapotranspiration. The higher the wind speed and the rougher the surface, the more turbulence, and the higher the evapotranspiration rate will be.

MAGNITUDE OF EVAPOTRANSPIRATION

Evapotranspiration rates are most commonly expressed in units of water depth per unit time, such as millimeters per day (mm day^{-1}) or inches per month (in $months^{-1}$). Since it takes energy to evaporate water, water depths can also be expressed in terms of energy received per unit area. Therefore, evaportranspiration is often expressed in units of energy per unit area per unit time, such as watts per squared meter (w m^{-2}), or megajoules per squared meter per day (MJ m^{-2} day^{-1}). It can, however, also be expressed in units of energy per unit area, such as watts m^{-2}, or MJ m^{-2} day^{-1}. Many of the factors affecting evapotranspiration are so dynamic that the magnitude of the evapotranspiration rates for a given surface will vary from day to day, from place to place, and throughout the day. Figure 4 shows the calculated daily evapotranspiration rate for corn in North Platte, Nebraska, during the 2000 growing season. It shows the typical large variations in evapotranspiration rate that can be expected from day to day, as a result of normal daily changes in weather conditions. It also shows a seasonal pattern, as a response to the seasonal changes in available energy and to the changing water demand of the crop during its growing cycle.

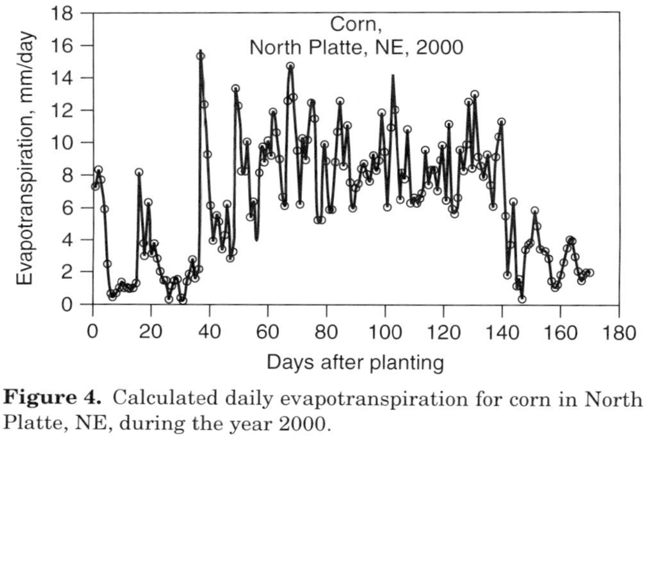

Figure 4. Calculated daily evapotranspiration for corn in North Platte, NE, during the year 2000.

BIBLIOGRAPHY

1. Allen, R.G., Pereira, L.S., Raes, D., and Smith, M. (1998). *Crop evapotranspiration—guidelines for computing crop water requirements*. Irrigation and Drainage Paper No. 56, FAO, Rome, Italy.
2. Allen, R.G. et al. (2002). Evapotranspiration from a satellite-based surface energy balance for the Snake Plain Aquifer in Idaho. *Proceedings of the 2002 USCID/EWRI Conference*. July 9–12, San Luis Obispo, CA, pp. 167–178.
3. Gates, D.M. (1964). Leaf temperature and transpiration. *Agronomy Journal* **56**(3): 273–278.
4. Yazar, A., Howell, T.A., Dusek, D.A., and Copeland, K.S. (1999). Evaluation of Crop Water Stress Index for LEPA irrigated corn. *Irrigation Science* **18**: 171–180.
5. Hoeft, R.G., Nafziger, E.D., Johnson, R.R., and Aldrich, S.R. (2000). *Modern Corn and Soybean Production*, 1st Edn. MCSP Publications, Champaign, IL.
6. Schneekloth, J.P. et al. (1991) Crop rotations with full and limited irrigation and dryland management. *Transactions of the ASAE* **34**(6): 2372–2380.
7. Doorenbos, J. and Pruitt, W.O. (Eds.). (1977). *Crop Water Requirements*. Irrigation and Drainage Paper No. 24, FAO, Rome, Italy.
8. Suttleworth, W.J. and Wallace, J.S. (1985). Evaporation from sparse crops—an energy combination theory. *Quarterly Journal of the Royal Meteorological Society* **111**: 839–853.
9. EWRI. (2001). *The ASCE Standardized Reference Evapotranspiration Equation*. Environmental and Water Resource Institute of the American Society of Civil Engineers. Standardization of Reference Evapotranspiration Task Committee Report. Available: http://www.kimberly.uidaho.edu/water/asceewri/.
10. Moran, M.S. et al. (1989). Mapping surface energy balance components by combining Landsat Thematic Mapper and ground-based meteorological data. *Remote Sensing of Environment* **30**: 77–87.
11. Payero, J.O. (1997). *Estimating Evapotranspiration of Reference Crops Using the Remote Sensing Approach*. Ph.D. Dissertation, Utah State University, Logan, UT.
12. Yakir, D. and Sternberg, L.daS.L. (2000). The use of stable isotopes to study ecosystem gas exchange. *Oecologia* **123**: 297–311.
13. Wright, J.L. (1982). New evapotranspiration crop coefficients. *Proceedings of the ASCE* **108**: 57–75.

FOG

Arthur M. Holst
Philadelphia Water Department
Philadelphia, Pennsylvania

Fog is a cloud that materializes near the ground or over water. It is a reaction that occurs when the temperature near the ground cools to the temperature required to produce dew and causes the water vapor in the air to become visible in the form of a cloud of precipitation. There are different types of fog, which occur when different variables are involved. The two conditions necessary for fog formation are mild or no winds and air temperatures that are equivalent to the temperature at the dew point. Mild or no winds are important because obstructions in the way of a strong wind can cause it to blow in an up and down motion, which brings warmer air down and pushes the colder air up. Cool air is a requirement for fog formation so wind prevents it from happening.

Fog, like most weather hazards, can cause serious problems for society. The only time this condition is not applicable is when fog forms over water due to the absence of obstructions. In this case, the reverse reaction occurs, and the fog worsens as the wind grows faster. Fall nights are often said to contain the perfect conditions for producing fog. At night, the ground cools which also cools the air directly above it. This reaction causes droplets of water to become suspended in the air resulting in the creation of fog. However, these perfect conditions don't last forever. By morning, the heat from the sun begins to warm the ground, and the fog evaporates.

There are many types of fog, but the four most common are evaporation fog, upslope fog, precipitation fog, and radiation fog which is commonly called ground fog. Evaporation fog is found mostly out at sea, which is why it is also known as "sea fog." It is the reaction of moist air moving over colder water. The fog conditions worsen as the wind speed increases. Upslope fog is produced by moist air that is sent by a strong wind up into mountainous regions. Precipitation fog is created when precipitation in the form of rain or snow hits drier air and causes water vapor to materialize instantly. Radiation fog is seen at night when the ground is cool. The air that comes in contact with the ground also becomes cool and creates water vapor which soaks the ground and causes the formation of dew.

Like most weather hazards, fog can have severe impacts on society. Fog can cause health problems in polluted areas because the water vapor produced can become acidic. Driving through a thick cloud severely reduces a person's visibility and is a major contributor to car accidents. Even worse than the occasional car accident due to fog are the aircraft and boating accidents that can occur. Throughout history, fog has been known as the silent murderer that has taken hundreds of lives at sea in catastrophic boating accidents.

READING LIST

Joseph, P. *Weather Wisdom.* (2002). Available: http://www.touch-tmj4.com/4weather/wxwisdom/fog/fogtypes.asp. (March 19).

Malan, J. *Weather Wisdom.* (2002). Available: http://www.touch-tmj4.com/4weather/wxwisdom/fog/fog.asp. (March 19).

Malan, J. *Weather Wisdom.* (2002). Available: http://www.touch-tmj4.com/4weather/wxwisdom/fog/fogdisasters.asp. (March 19).

Parsons, L. *Fog.* (2002). Available: http://www.activeangler.com/articles/safety/articles/lee_parsons/fog.asp. (March 19).

Understanding Clouds and Fog. USA Today. (2002). Available: http://www.usatoday.com/weather/wcloud0.htm. (March 19).

Williams, J. *Fall nights often perfect for forming fog. USA Today.* (2002). Available: http://www.usatoday.com/weather/wrfog.htm. (March 19).

COASTAL FOG ALONG THE NORTHERN GULF OF MEXICO

Timothy Erickson
National Weather Service New Orleans/Baton Rouge Forecast Office
Slidell, Louisiana

INTRODUCTION

Coastal fog is a major problem for all traffic along the United States' coastlines. Lives and large monetary losses have occurred because of coastal fog. These losses have been realized over land, in the air, and on the water. Huge strides have been taken to understand and combat the coastal fog forecasting problem by the National Weather Service. Studies and research continue to improve these fog forecasting techniques. Tens of thousands of people use the Gulf of Mexico as their home or as part of their occupation in many different ways, and billions of dollars in products and property are carried and moved through the channels and river systems to and from the Gulf of Mexico each year. All are affected by coastal fog many times throughout the fall, winter, and spring. This research project took place along the northern Gulf of Mexico from the upper Texas coast to the Mississippi coastline.

METHODOLOGY

The synoptic and mesoscale patterns used in this research were from the fall, winter, and spring of 1998, 1999, 2000, and 2001. Fog was defined as water droplets suspended in the air reducing visibility to one half-mile or less. Shower and thunderstorm activity reducing visibility to these levels was not used. No records exist for fog contributions to the hydrologic cycle. Water contributions caused by fog are dismissed as false tips in rain gauges. The reference to "boundary layer" in this article will be the layer of atmosphere from the surface to the base of the lowest inversion.

Variables used in this research were surface pressure, rainfall, moisture advection, wind direction, wind speed, water temperatures, ambient temperature, and ambient dew point temperature from land observations and production platforms in the northern Gulf of Mexico. Parameters used from previous studies by Johnson and Graschel (1) were air temperature, dew point temperature, wind direction, wind speed, ceiling heights, and visibilities from oil and gas production platforms at an average altitude of 35 ms over the northern Gulf of Mexico. Gulf of Mexico sea surface temperatures were provided by the Tropical Prediction Center oceanographer in Miami, Florida.

SCALES OCCURRING WITH COASTAL FOG

Mesoscale—Widespread

Horizontal range would normally be from 50 to several hundred miles.

Microscale—Locally

Horizontal range could be at a point close to 50 miles.

CONDITIONS THAT HELP PRODUCE COASTAL FOG

Radiative

Heat is transmitted via long wave radiation (infrared) away from Earth. This radiation causes nocturnal cooling of the ground and subsequent atmospheric boundary layer. Radiational cooling is responsible for developing a stable layer near the surface by cooling the layer of air in contact with the ground while leaving a relatively warmer layer immediately above this cooler layer. This stability by warmer air above cooler air with increasing altitude is called an inversion. This feature is important in forecasting the depth of coastal fog. As the radiative process continues, the layer will cool further until saturation is achieved and fog can be produced, given the pressure does not change much during the process.

Coastal Interface. Temperature and moisture gradients are strong where the ocean meets land, especially at night during the fall, winter, and spring. During prime radiational cooling processes, this interface can become moist and stable horizontally from land to water when there is no forcing. These conditions were found when a thin ribbon of fog developed at this interface of land and water. The fog stretched linearly for more than 100 miles and vertically for hundreds of feet. Studies on how this type of fog develops are ongoing but horizontal skew-t sounding-like profiles may be a tremendous help in what is occurring at this land–water junction when this fog forms. No horizontal sounding-like profiles exist, and this is the first mention of such a profile. But if they were available, they could give some valuable insight not only into the coastal "ribbon" fog formation but also where thunderstorms are more likely to develop along with many other variables.

Frontal

This is another inversion producing process. Coastal fog can be developed with the assistance of cold, warm, or stationary fronts.

Cold Frontal. An inversion develops by cooler air displacing warmer air at the surface. The interface where these air masses meet in the vertical is known as a frontal inversion.

Coastal fog formation under a frontal inversion is nonexistent with fast-moving cold fronts, because of stronger winds that cause a deeply mixed layer, along with dry, cold air advection behind the front.

When warm air is lifted by a cold front, it causes displacement and mid-altitude air ahead of the front becomes subsident. This process causes compressional warming, which forms another inversion well ahead of the cold front. As the surface cools at night, the boundary layer becomes stable well ahead of the front. If there has been rain ahead of the cold front with a prefrontal trough, there will be plenty of moisture and a strong inversion that may develop some dense coastal fog. Even without a prefrontal trough or rainfall, the subsidence inversion and gulf moisture ahead of a slow-moving cold front will sometimes be enough to provide conditions promoting dense coastal fog.

Warm Frontal. Relatively cool saturated or nearly saturated air ahead of a warm front is gently displaced by warmer air. The cool air is normally not very deep (100–300 ms) and is topped by warmer air at relatively low altitudes, which causes an inversion to form where the two air masses meet in the vertical. Cooler air cannot hold as much moisture as warm air. As the air ahead of the warm front is already moisture laden, cooler, and stable, water vapor condenses to small particles, which causes fog to form. This result happens frequently along the coast during the cooler months of the year and sometimes far inland. A similar scenario along the coast was also described by Hsu (2). The dynamics of warmer, higher dew point air flowing over cooler waters was described by Kotsch (3) and Mullan (4).

Coastal fog can occur behind warm fronts but is not as common. The air is, by definition of a warm front, warmer than the air ahead of it. The warmer air expands and therefore can hold more water vapor, which will normally dissipate any fog behind a warm front if no overwhelming positive moisture flux is occurring in the warm sector. When this flux occurs, it can be far too great for even the warm air to hold and the water vapor will condense, which causes fog to form. During research, this result occurred only when an inversion was still present behind a warm front.

Stationary and Slow Moving Fronts. These types of fronts are aggressive at developing coastal fog, which can occur in two ways ahead of a cold front that becomes stationary or slow moving. One is when moisture is not displaced and wind speeds are very light behind the trough preceding the cold front. The second is when the wind fetch is well over the marine environment and it brings warm moist air back over cooler waters or land. Warm frontal fog conditions were explained above.

Two unusual fog days occurred as post-cold frontal events. A cold front passed through the southern Louisiana coastal region, and dew points and dry bulb temperatures cooled. Coastal fog then began to develop over many sites from east central Texas to south central Louisiana. The air mass change was more negative for the dry bulb temperature than for the dew point temperature, and the cold front slowed from about 10 knots to less than 5 knots as it reached the coastline. This decrease caused the air behind the front to become saturated. Frontal forcing weakened, and consequently northerly winds weakened and created perfect conditions for coastal fog development. Post-cold frontal fog is rare, but when it does occur, visibilities can plummet to less than one quarter of a mile quickly.

Hsu's (2) description on frontal fog production is similar to this research. Johnson and Gracshel (1) called frontal fog "mixing fog."

Maritime (Sea Fog)

When fog develops via any process over any large marine area, it is called marine or sea fog. Marine fog may be stationary or be advected elsewhere. It is during the advection process that coastal fog occurs.

Dormant. As the cooler boundary layer air over land "drains" offshore, it causes stability to decrease, which makes mixing possible. The warm gulf gives off water vapor to the cooler air, which causes the vapor to condense, making water droplets. This process quickly deteriorates when only little "drainage" is occurring. Latent heat released because of condensation warms the marine boundary layer too fast, which makes the stability in this layer neutral. Fog will form in the near shore waters if this latent heating can be exhausted through the top of the maritime boundary layer. Fog development caused by this process does not occur often. When it does, the fog stays near the coast, rarely moving anywhere because of the very slow, almost laminar flow of air moving off land. When this type of fog affects the shoreline, it is also called coastal fog. The other processes mentioned above may produce nonadvective marine fog that stays in the near shore waters.

Advancing. This process in which coastal fog is achieved is the most common. The air in the boundary layer is well mixed but does not entrain air from above the inversion. Fog that forms either onshore or offshore by any process and moves elsewhere is an advective fog. When fog advects from one offshore location to another or to the coast, it is said to be advancing. The most common type of advancing fog is found with post-strong cold front events. Winds from these fronts move the top layer of warm shelf water away from the coast. This water is replaced by cooler upwelled water. A strong temperature gradient exists where the cooler upwelled water meets the warm water offshore. When the next cold front produces return flow from the warmer water offshore, moisture-laden air saturates while moving over the cooler water. Extensive areas of fog or very low cloud ceilings form before moving inland, which causes extremely low visibilities stretching for miles along the coast. When the moisture came in the form of low clouds during research, the cloud deck would often descend to the surface producing very dense coastal fog as well. As warm air overlays cooler water, it produces a stable boundary layer. No vertical mixing occurs by air moving up or down. Mixing was produced solely by wind advancing the fog. The wind did not reach through the inversion layer, and therefore, no entrainment occurred.

A similar scenario was recorded by Binhau (5) and Hsu (2). Their contributions showed this environment to be stable, with surface winds from the southeast through southwest. This scenario holds true for all locations along the northern Gulf of Mexico.

Advective

Anytime fog forms elsewhere and is forced to another location, it is a type of advection fog. A well-mixed boundary layer also exists with advective fog, but again, air from above the inversion is not mixed downward. In some cases, the atmosphere will be capable of supporting fog but it cannot produce it. When this happens, fog may be brought into the area by wind or by the slow movement of an entire layer of air. It is important that this air mass does not mix with air from above the inversion. If this mixture occurs, dry, relatively warmer air will be mixed into the boundary layer, causing the fog to erode.

Conglomerate of Two or More Types from Above

This fog producer is the most common. Two or more of the above processes usually develop fog that forms almost anywhere on Earth. One major variable may occur at the time of fog formation, but most of the time fog is supported by another equal or weaker variable, for example, frontal-induced marine fog. That is, fog develops over the marine environment by a cold front, which brings warm, moist air back over cooler waters. The fog develops mainly because of the marine environment, where it derives its moisture, but it could only do so as a result of forcing by the cold front.

VARIABLES PRESENT DURING COASTAL FOG PRODUCTION

Variables Always Present During Coastal Fog Formation

Negative or Neutral Omega Within or Just Above the Boundary Layer. When lift occurred during the research, fog would dissipate or simply not form. The lift causes mixing through the inversion, which brings dry, relatively warm air into the boundary layer.

Weak or No Positive Vorticity in the Boundary Layer. Coastal fog developed under boundary layer-negative vorticity regimes, but it would not develop under moderate-to-strong positive vorticity in the boundary layer. When weak vorticity occurred within or at the top of the boundary layer, fog would lift and become a low-level cloud ceiling. As the vorticity center moved past, the ceiling would once again descend to the surface, which caused coastal fog to form.

Inversion. An inversion is always present during any and all fog development and duration.

Variables Present During Research for Each Condition

Radiation Fog Variables

Winds 0 to 3 Knots. Radiation fog events are created without mixing through the inversion. Wind greater than 3 knots was found to create too much mixing and dissipated coastal fog during radiative events.

Moisture Advection or Rainfall Within 36 Hours. Moisture is needed for any fog to develop. With no moisture advection, it was found that moisture input from rainfall would be sufficient inside 36 hours. This field was dependant on several variables. These variables included amount of rainfall, ground moisture, insolation, and amount of rainfall coupled with timing. The best results were found with light rain episodes during the early morning with strong insolation during the day.

Neutral or Negative Omega. Neutral or negative lift from some height above down to the top of the boundary layer was always found when coastal fog formed during this research. Positive lift caused mixing through the inversion, which dissipated the fog.

Weak-to-No Boundary Layer Positive Vorticity. Coastal fog formed under negative and neutral vorticity regimes. Coastal fog was either displaced or was not present when moderate-to-strong positive vorticity was found within the boundary layer. Coastal fog was also present with weak positive vorticity, but a few interesting findings occurred. As a weak positive vorticity center moved through the boundary layer, coastal fog would lift, which created a low-level ceiling from 100 to 400 feet. When the vorticity maxima passed, the low-level ceiling would descend to the ground, which caused coastal fog to return. This phenomenon was known for creating "bouncy" fog conditions where visibilities would swing wildly from as little as 0 to as much as 4 miles.

Clear Skies. Clear skies were the overwhelming majority of sky conditions experienced during radiational coastal fog events. The minority sky condition consisted of very high thin cirrus clouds. No radiational coastal fog events occurred during any other cloud conditions.

Outside Downtown Areas. Radiational coastal fog events during this study were found outside the downtown areas of cities. The heat island effect was enough to dissipate any fog trying to form inside these areas.

Moisture Advection or Rain Within 36 Hours. The highest frequency of coastal fog during frontal regimes was found when a prefrontal trough passed. Moisture was input by both the front causing return flow from the marine environment and the prefrontal trough causing rainfall.

Clear Skies or Very High Cloud Ceiling. High cloud ceilings were noted several times throughout the research when frontal-induced coastal fog developed. Winds of 10 to 20 knots were noted above a shallow inversion, which allowed heat from the boundary layer to escape and be carried away.

Surface Winds of 0 to 12 Knots. During radiative conditions, coastal fog only formed when wind speeds were 0 to 3 knots. During advective or advancing conditions, coastal fog was carried to or along the coast when wind speeds were 4 to 12 knots.

No coastal fog formed during this research when wind speeds were greater than 12 knots.

Onshore Winds of At Least 4 Knots and Not More Than 12 Knots. During marine-induced coastal fog, winds were necessary to advance the fog from the marine environment to the coast. But wind speeds greater than 12 knots caused the fog to dissipate by mixing with air above the inversion. When these winds did not reach the top of the boundary layer, no mixing occurred through the inversion and the fog would simply lift, developing a low-level cloud ceiling.

No Moisture Advection. Surprisingly, moisture preconditioning was not necessary when marine fog caused coastal fog. The marine fog was the moisture advection.

Clear or Very High Ceilings. Several cases showed coastal fog production via a low-level cloud deck. When heat was capable of radiating through the inversion at the top of the low-level clouds, the cloud deck lowered or deepened toward the surface through the night, finally reaching the ground, which caused coastal fog to form. It was also noted that the lower the cloud deck, the shorter the time frame coastal fog would develop. A very general rule of thumb was realized when the boundary layer was capable of supporting fog. On average, it took about an hour for the cloud ceiling to descend 100 feet. Cloud bases higher than 1000 feet never reached the surface during marine-induced coastal fog events.

Temperature. Johnson and Graschel (1) found temperature differences of several variables to be important when maritime fog developed. As indicated in their article, "Sea Fog and Stratus: A Major Aviation and Marine Hazard In The Northern Gulf Of Mexico," the differences between water and air temperatures, as well as between water and dew point temperatures, were the most important variables producing marine fog when the right atmospheric conditions were in place. The graphs below show these parameters versus relative humidity (RH) values.

Johnson and Graschel's study (1) as well as this project found RH values of 98% or greater always present with coastal fog.

Figure 1 shows the continental shelf region along the northern Gulf of Mexico, which is shown at the 200-m depth contour. It is also the region where a cool water temperature of 20 °C (68 °F) or less was found to be critical for marine fog development in the northern gulf during the right atmospheric conditions. These findings may not be the same at other locations around the globe because fog development depends on temperature gradients, over water and/or land, which are relative. Water temperature findings close to these were also accomplished by Binhau (5).

Figure 2 shows RH versus (Ta - Tw), where RH is relative humidity and (Ta - Tw) is the difference between the ambient air temperature and the water temperature.

In Fig. 3, in regard to the positive (+) numbers, when the water temperature is cooler than the air temperature, the air must be moving between 4 knots to as high as 12 knots for coastal fog to form. During this process, the moisture-laden warmer air loses its heat to the cooler water. As the air cools to its dew point, condensation takes place. Latent heat is released during the condensation process. The air can lose this added heat to the cooler water below and through the top of the boundary layer

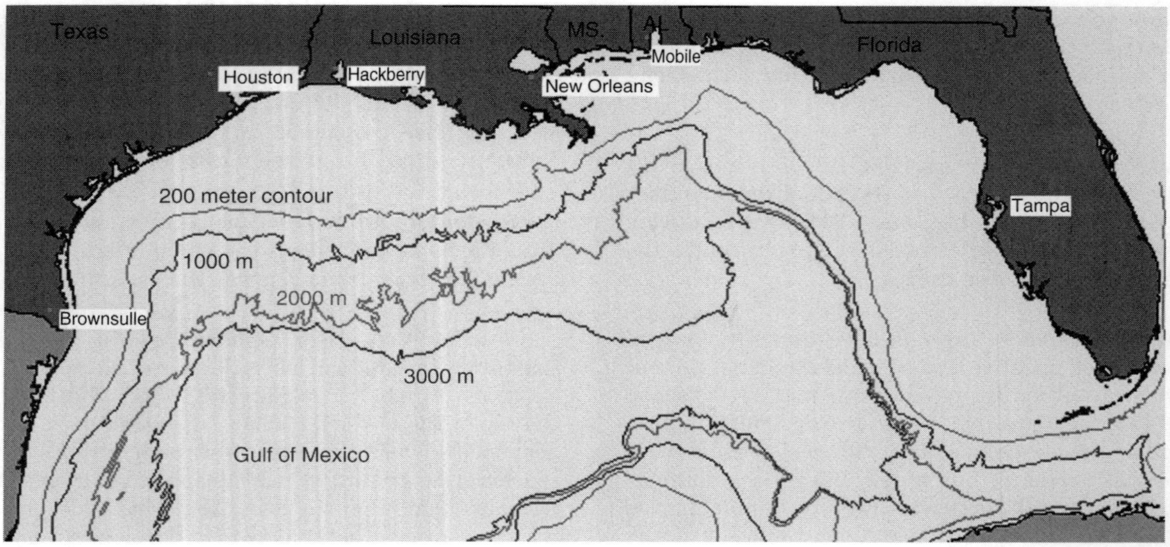

Figure 1. Continental Shelf-Northern Gulf at Mexico.

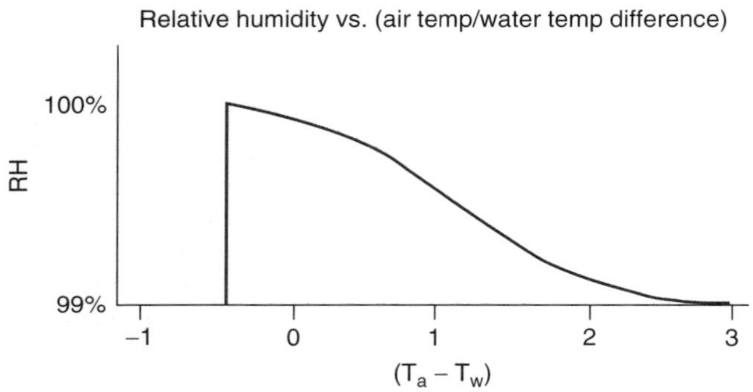

Figure 2. This graph shows the relationship between air temperature and water temperature when coastal fog developed with respect to the lowest relative humidity value found during research.

via long wave radiation, so fog can begin to form. This process is efficient and is the main reason for coastal fog developing from advancing marine fog.

The positive numbers in Fig. 3 show a stable boundary layer also described by Binhau (5) and Hsu (2).

In Fig. 4, in regard to the negative (−) numbers, when the water temperature is warmer than the air temperature, the air must be calm or moving no faster than 3 knots for coastal fog to form. During this process, when coastal tidal marshes exist, a column of cooler dry air will have a net gain of water vapor, which causes the dew point temperature to rise. When the air and dew point temperature are very close, condensation will begin. As a result of condensation, latent heat is released. The column will need to lose this heat through the top of the boundary layer before fog will begin to form. Even though marine fog occurs by this process, it is not an efficient coastal fog producer unless fog initially forms along the coast. Most of the time, fog will develop over the marine environment and be stationary or drift farther offshore.

Negative numbers in Fig. 4 show an unstable boundary layer as described by Binhau (5). Binhau showed that this type of fog may be produced over the marine environment even with northerly winds of 30 knots. But with winds of this magnitude, coastal fog will never develop along the northern gulf coast because the winds would force it out to sea.

Figure 5 shows RH versus (Td − Tw), where RH is relative humidity and (Td − Tw) is the difference between the ambient dew point temperature and the water temperature.

In Fig. 6, in regard to the positive (+) numbers, it is important to remember that the atmosphere as well as the ocean is always trying to reach a state of equilibrium.

When the ambient dew point temperature is higher than the water temperature, water molecules can easily move to and become a part of the ocean surface. Water molecules find it hard to break away from the waters' surface during these conditions, and therefore, a net moisture flux from air to water occurs. The air temperature is always equal to or greater than the dew point. Hence, there will be a transfer of heat from the air to the water as well. This process cools the air temperature, but saturation is difficult to achieve because there is a net loss of water vapor to the water surface. The air continues to cool and dry until temperatures of the water, the air, and the dew point equal or become very close. This process eventually causes saturation and can, but rarely does, cause coastal fog to form when atmospheric conditions are right and (Td—Tw) is zero or very close. Normal occurrences of fog on the positive side of Fig. 6 are when fog develops elsewhere and advances into the area.

In Fig. 7, in regard to the negative(−) numbers, when the water temperature is warmer than the dew point, water molecules can easily break away from the water surface to the air. Regardless of the air temperature,

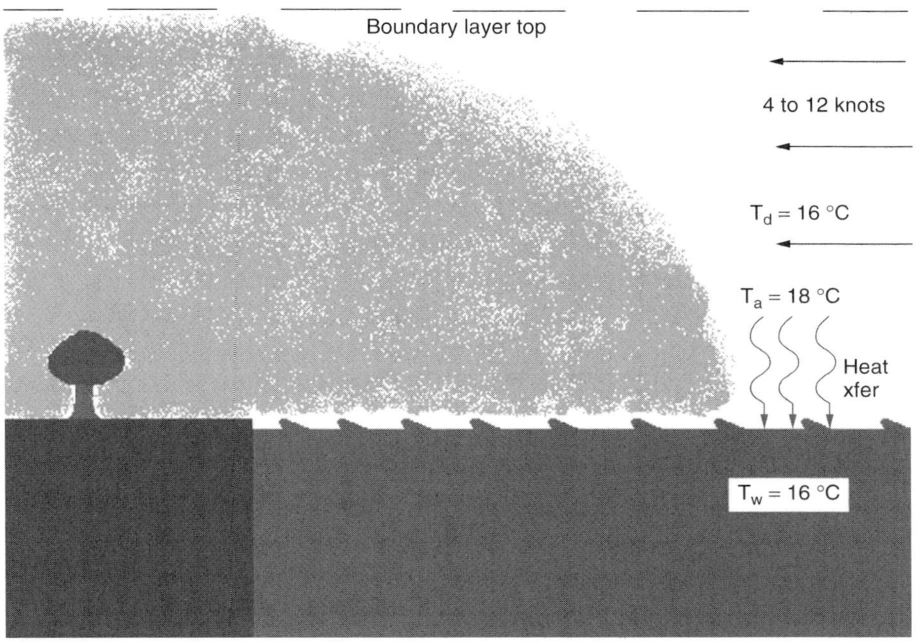

Figure 3. Stable Boundary Layer.

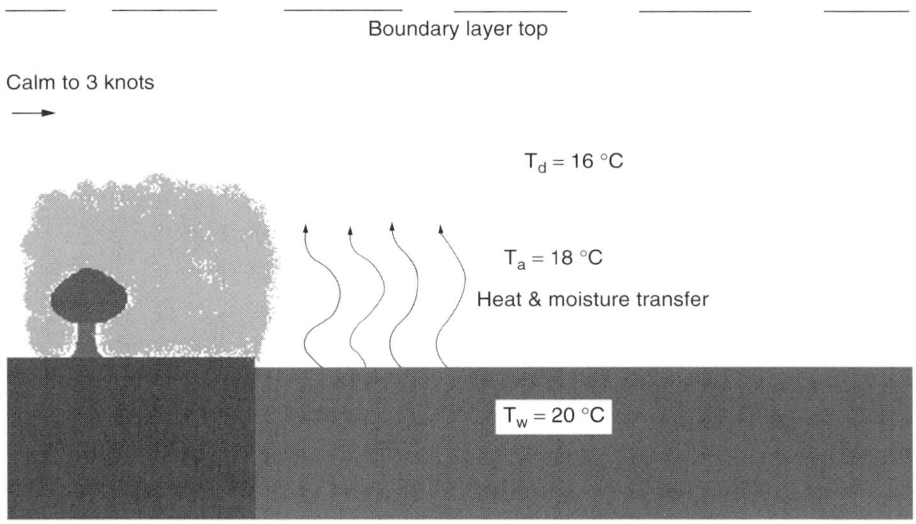

Figure 4. This picture shows a light offshore wind, dew point temperature (Td), air temperature (Ta), and water temperature (Tw).

the water will try to modify it until the water and air temperatures are equal. Even if saturation is achieved, moisture will continue to be added to the air until the dew point is equal to the water temperature. As moisture is added to the air at saturation, condensation begins and latent heat is released to the air until the temperatures of the water, dew point, and air are equal. This saturation can easily develop fog near or offshore. Normally, marine fog produced by this process occurs with little airflow (0–3 knots). After marine fog forms, it can be easily forced to the coast as winds increase ahead of the next cold front. This coastal fog producer is aggressive.

As Fig. 7 depicts, this process occurs with a minimal separation of air, dew point, and water temperatures. When separations are too large, there may be too much

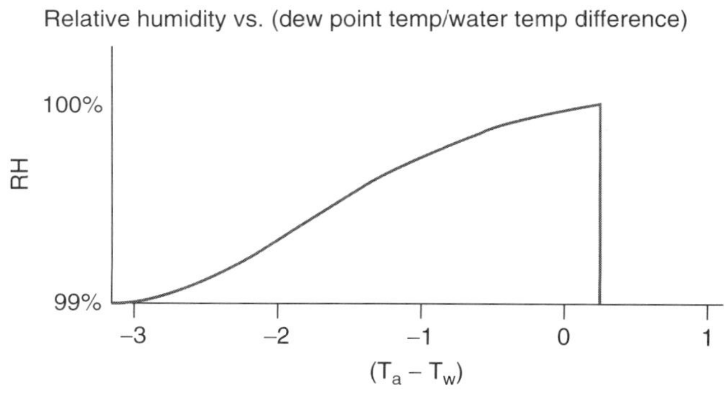

Figure 5. This graph shows the relationship between dew point temperature and water temperature when coastal fog developed with respect to the lowest RH value found during research.

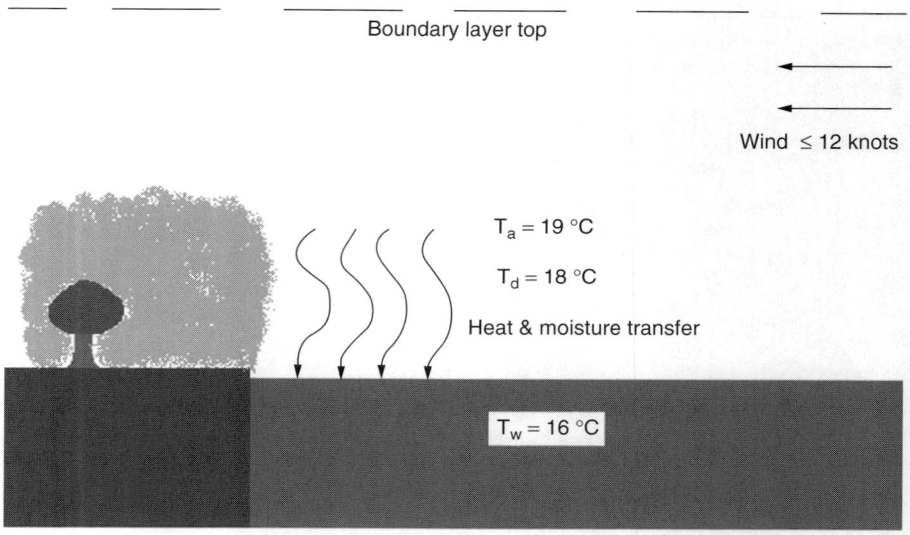

Figure 6. This picture shows winds of 12 knots or less, air temperature (Ta), dew point temperature (Td), and water temperature (Tw).

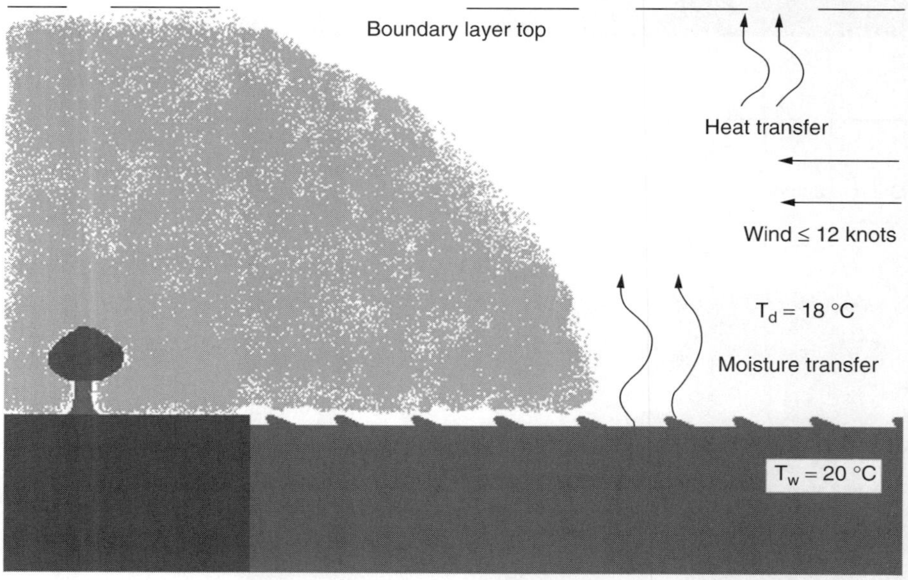

Figure 7. This picture shows wind speeds less than or equal to 12 knots, dew point temperature (Td), and water temperature (Tw).

dry air to modify before the next system brings this air mass back to shore. If the air is moving more than 3 knots, there will be a continual replacement of dry air and/or air from above the inversion may be mixed into the boundary layer and saturation will not be achieved.

Advective Fog Variables

Winds of At Least 4 Knots and Not More Than 12 Knots. Advective fog, which causes coastal fog, is different from advancing marine fog in a couple of ways, i.e., where the fog developed and the wind fetch. Fog that develops onshore and moves to the coast is only known as advective fog. Whenever fog moves to the coast, the wind direction will always be in a direction from the fog to the coastal location. All other variables that are needed for marine fog to advance to the coast are also needed for advective fog.

No Moisture Advection. Even though moisture may be in place, advective fog does not need to have a premoistened atmosphere for coastal fog to be produced. Moisture advection can be induced by the fog moving to the coast.

Clear or Very High Cloud Ceilings. Cloud conditions were found to be the same as for marine fog.

Conglomerate Fog Variables. The variables for each condition associated with coastal fog development have to be present when coastal fog forms under more than one condition.

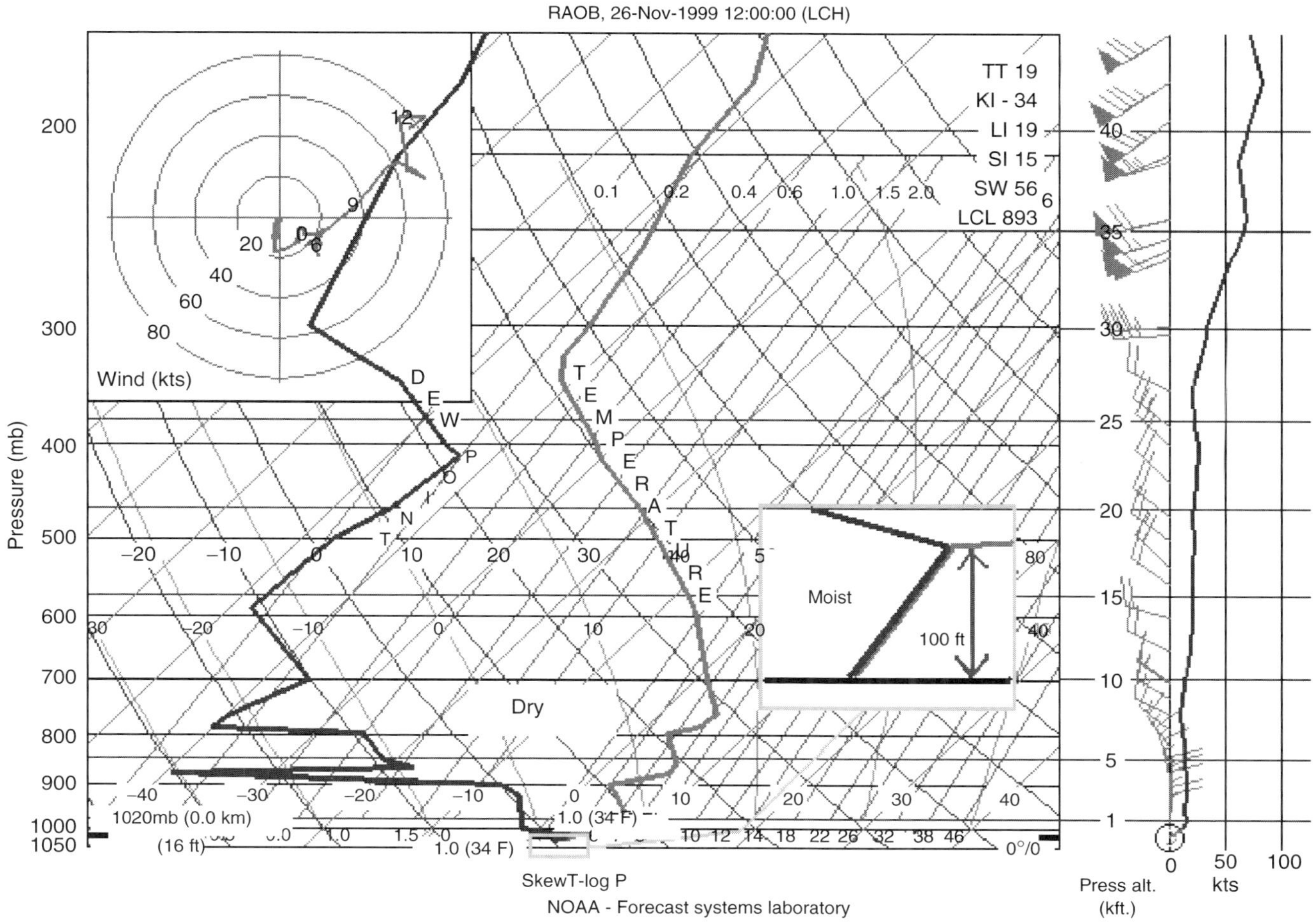

Figure 8. Sounding Profile-Radiative.

SOUNDING PROFILES PRESENT DURING COASTAL FOG

A sounding profile similar to Fig. 8 was always present shortly before and during radiative coastal fog events. This actual sounding shows a very shallow (~100 feet) moist layer capped by three strong inversions.

A sounding profile similar to Fig. 9 was always present shortly before and during advective and maritime coastal fog events. This actual sounding shows a very deep (~1000 feet) moist layer capped by a moderate inversion.

Fog rarely develops beneath the type of inversion in Fig. 10. This actual sounding shows a deep moist layer capped by a frontal inversion; coastal fog occurred before the passage of this cold front, but it quickly dissipated by the time this sounding was taken.

Frontal-induced and conglomerate fog was found with all three types of sounding profiles. Frontal-induced coastal fog occurred more often under the first and second profile types.

Sounding profiles are important for forecasting the depth of coastal fog or any type of fog. Maritime, advective, and frontal-induced coastal fog can only be as deep as the height from the surface to the base of the inversion. These types of fog produced the deepest fog along the coast as well as inland, and consequently they took longer to dissipate.

During research, most radiation-induced coastal fog formed only under a low-level inversion. The inversion could be as high as 100 feet or as low as a few feet from the ground. The inversion was frequently strong. Air parcels were not able to penetrate the inversion, but it would not stop radiative heat transfer from the surface via long-wave radiation. Radiative fog did not always follow the depth rule. It could be from the surface to the base of the inversion deep or as shallow as a few inches, even when the inversion was much higher.

An interesting find during the project was horizontal stability and moisture profiles during radiative fog formation conditions along the coast. The horizontal profile shown below from the coast along with the radiative vertical profile from inland locations were both present when a thin "ribbon" of fog formed along the coast (see Fig. 11). This fog was found to run linearly along the coast and stretched upward for hundreds of feet. The inland vertical sounding did not support such a high

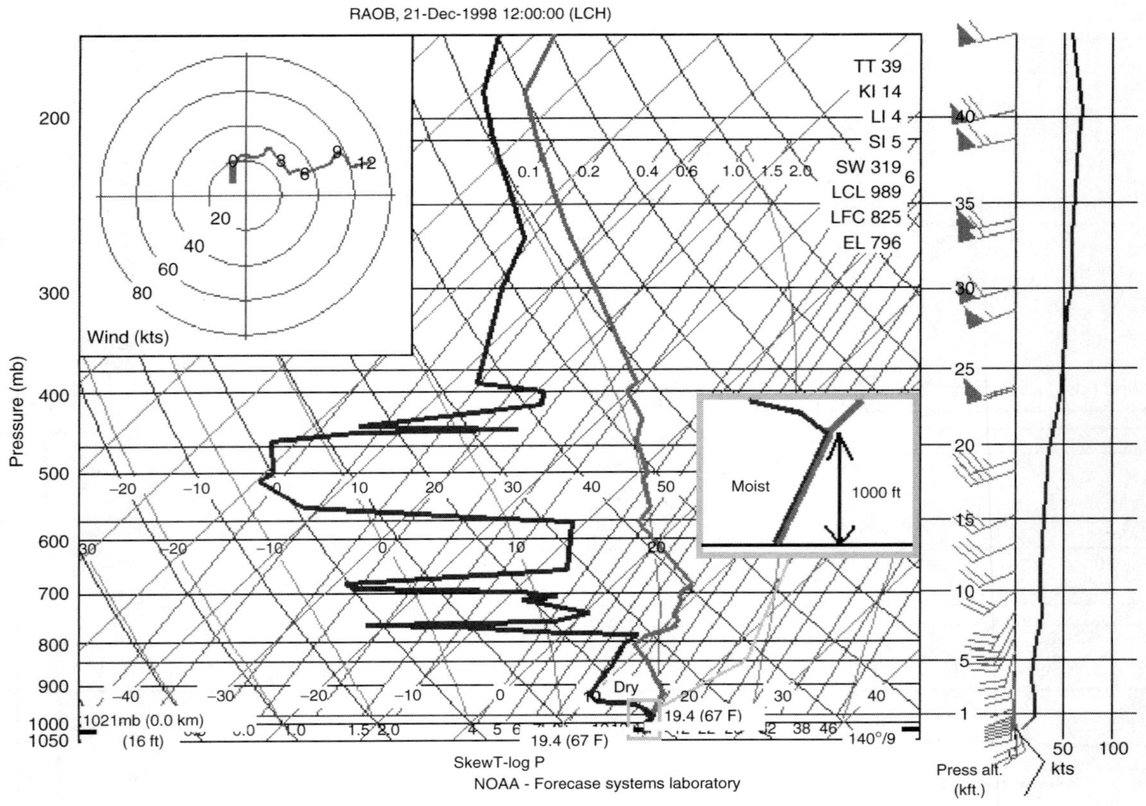

Figure 9. Sounding Profile-Advective.

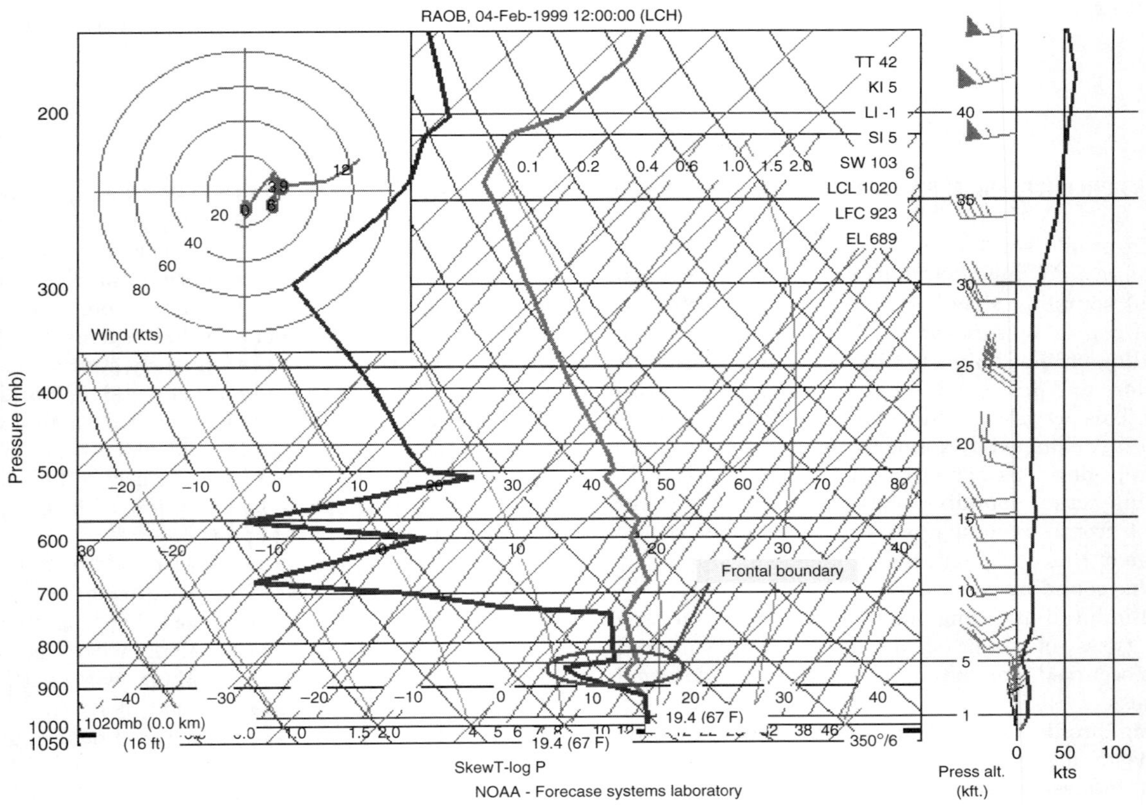

Figure 10. Sounding Profile-Inversion.

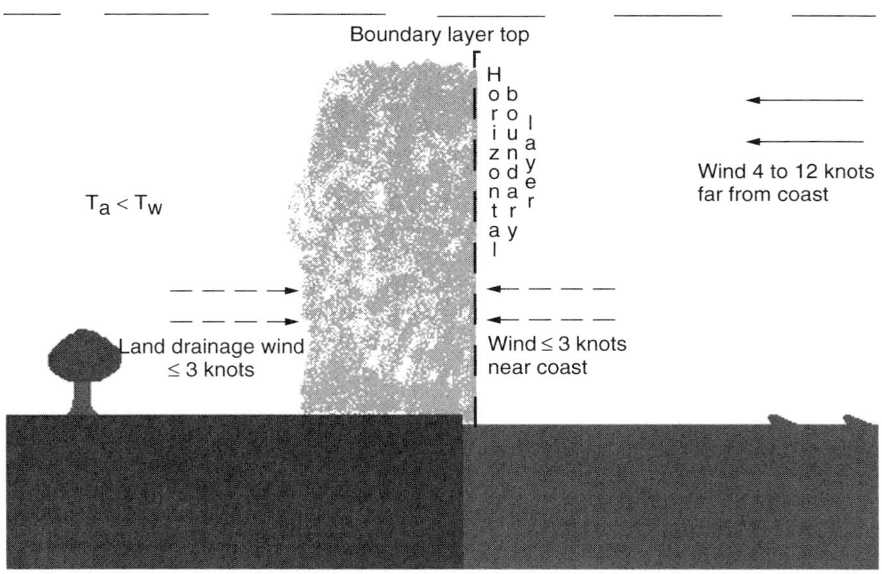

Figure 11. This illustration shows a hypothetical representation on how a very thin ribbon of fog develops at the coast.

depth of coastal fog, but this vertical profile may have been different along the coast because strong moisture and temperature gradients existed. Research into this phenomenon is ongoing.

CONCLUSION

Fog has never received the attention it deserves because it does not make for an explosive story like hurricanes and tornadoes. But fog is blamed for large monetary losses as well as property losses. Each year, fog is blamed for indirectly taking more lives than hurricanes in the United States. Fog also donates a small but significant amount of water to the hydrologic cycle. In some places, such as the western high coast region of South America, fog is the only way insects and grasses receive water.

When fog develops, there is always an inversion in place, which means the boundary layer is disconnected from the remaining atmosphere above with respect to mixing. This process is called decoupling. When the inversion erodes and mixing resumes through this layer, it is said to be coupled. As a result, pollutants released to the environment will remain in the boundary layer during decoupled conditions and will mix out during coupled conditions. Petrochemical plants and other facilities-producing pollutants that are dispersed to the environment can use fog as an indicator for when not to release waste products.

Results during research show there are numerous variables and observations from the microscale environment to consider when forecasting coastal fog conditions. Current National Weather Service numerical models do not solve for microscale conditions, and therefore, forecasters must rely on pattern recognition to resolve these issues.

BIBLIOGRAPHY

1. Johnson, G.A. and Graschel, J. (1990). *Sea Fog and Stratus: A Major Aviation and Marine Hazard in the Northern Gulf of Mexico.*
2. Hsu, S.A. (1988). *Coastal Meteorology*. Academic Press, New York, p. 52.
3. Kotsch, W.J. (1983). *Weather for the Mariners*. Naval Institute Press, Annapolis, MD.
4. Mullan, D.S. (1984). Low stratus and sea fog in the North Irish Sea. *Weather (GB)* **39**: 235–239.
5. Binhau, W. (1985). *Sea Fog*. Springer-Verlag, New York.

RAIN FORESTS

Aldo Conti
Frascati (RM), Italy

Rain forests are, by definition, those forests that receive more than 2500 mm of rain each year. Rain forests are characterized by very dense vegetation dominated by tall trees and huge biodiversity. Rain forests exist in many parts of the planet, but most of them are along the equator, where the weather is stable throughout the year and there is never a dry season. Rain forests do not have seasons at all. The amount of rain is almost constant during the year, and the temperature seldom dips below 16 °C. Rain forests cover 7% of the earth's land surface and 2% of its total surface, but are home to more than half of all animal and plant species. Despite the fact that rain forests cover less than 10% of the earth, they support a third of its plant matter. The largest tropical rain forest in the world is the Amazon Rain Forest, which lies in the countries of Brazil, Bolivia, Peru, Ecuador, and Colombia. There are rain forests in Africa, mainly in the Congo, and in Oceania.

The large amount of rain also creates some of the biggest rivers and flood plains of the world.

Unfortunately, rain forests are in danger. They lie mainly in poor countries, where the economic situation forces people to use all resources, which is why most rain forests have shrunk dramatically in size over the last few decades. What is in danger is their huge biodiversity. Some researchers estimate that the Amazon Forest alone might host up to 10 million species of animals, mainly insects. At the rate the forest is disappearing, most of the estimates say that we are losing something like 100 of these species every day, before even being described by science.

Most of the animals in rain forests have adapted to live in the upper level of the canopy, where food is really plentiful. The constant weather means that there are flowers and fruits at any time of the year. The largest group of animals is that of insects that can easily climb trees and often have a highly symbiotic relationship with the plant life of the forest. Ants and termites are the most abundant animal of every rain forest. Animals are very important for the survival of the forest. Underneath the canopy, the wind is not strong enough to disperse efficiently the seeds produced by plants. Plants rely therefore on insects, that very often pick up the seeds and drop them some distance away. But the most important animals for seed dispersal are birds. Birds in rain forests eat mainly fruits, and the seeds can pass through their digestive system unscathed. By the time a bird excretes its load of seeds, it has normally flown a long distance from the plant, which ensures a high level of genetic mixing, healthy and helpful for plants. There are even some seeds that do not germinate unless they have gone through the digestive system of birds. More than one-fourth of all bird species in the world today live in tropical rain forests.

To have an idea of the importance of rain forests, some figures might be interesting. A single square meter of rain forest supports between 45 and 80 kg of biomass, far more than any other biome. This biomass gains more importance considering that most of it is made of carbon removed from our atmosphere. One hectare of rain forest can contain 200 species of trees and more than 40,000 species of insects. In Panama, scientists discovered fully 80% of the world's currently known beetle species on only 19 trees. Once, researchers discovered over 600 new species of beetle by studying a single species of tree.

Although not as rich in species as their Asian or Amazonian counterparts, African rain forests contain more than half of that continent's animal and plant species, even though they cover less than 7% of its total land area.

Because of the huge biodiversity, most species have evolved to occupy very specialized niches of the environment, which means that many species depend on each other and cannot survive without each other. Deforestation and many other human activities disrupt these complicated relationships.

Rain forests could play a crucial role in feeding the whole world's population. Many vegetables and fruits such as bananas and peppers that we consume come from rain forests. Peanuts come from rain forests, as do many drinks (coffee, tea, cola), oils (palm, coconut), flavorings (cocoa, vanilla), and other foods (beans, grains, fish). And many more vegetables are still there ready to be discovered. Moreover, researchers have identified over 200 plants that produce potential cancer fighting substances. And this considering that only 1% of plants have been intensively screened for such properties. Tropical rain forests do not offer only goods. They are a vital part of the hydrogeologic cycle of the planet and act as a global air purifier, absorbing huge amounts of carbon dioxide and releasing oxygen.

Despite their importance, rain forests everywhere are exposed to huge threats. Often, forests are cleared with fire to make room for cultivation. One plot is used for a few years until the soil is exhausted, and then farmers move on to clear another patch, putting the lushest forests in danger of desertification. But other industrial interests, including timber and mining, are taking advantage of rain forests. Part of the danger comes also from animal and plant species introduced from other environments. All these activities result, every year, in a rather large loss of rain forest. It is difficult to estimate the extent of the damage, as data are not plentiful or reliable. Nevertheless, it is true that in some countries, like Madagascar, the whole forest has almost disappeared in a few decades. As a result, human activities might be inducing the most important mass extinction since the fall of dinosaurs 65 million years ago. According to some research, up to 10% of the world's species might disappear in the next 25 years. But the truth is that over 50% of the earth's plants and animal are in danger. Nearly 20% of known endangered vertebrates are threatened by introduced species. Cultures are going extinct, too. Since the turn of the century, 90 tribes of indigenous peoples have been wiped out in Brazil alone. The pace of annihilation is increasing; 26 of those tribes were killed or scattered in the past decade. Everything should be done to halt this loss that many scientists think might affect the earth's climate, too, on a global scale.

FROST

Arthur M. Holst
Philadelphia Water Department
Philadelphia, Pennsylvania

Essentially, a type of dew, frost is ice formed by the condensation of atmospheric water vapor on a surface. It generally occurs at night, and when frost forms, it can sometimes be seen in patterns of ice crystals. Frost can be extremely damaging to outdoor crops and plants. Researchers and weather services throughout the world monitor the effects of damaging frosts. Frost forms on any surface, including cars, grass, and buildings. Frost forms through one of two processes: the formation of dew that subsequently freezes, or deposition, which is the process wherein a gas changes to a solid. In frost formation, this gas is water vapor, and the solid state is ice crystals. These two processes will occur when the air is saturated. Saturation occurs when the air holds as much water vapor as is possible at its temperature and pressure. The temperature at which frost is deposited is known as the frost point temperature.

Condensing water vapor must have something upon which to condense. If the temperature of the ground is below 0 °C, then the deposit will initially be as dew. Over time, this dew will eventually freeze, forming frost. However, if the dew point of the air is below 0 °C, the deposit will be hoarfrost, which is ice that forms directly through deposition, without initially forming as dew. Hoarfrost is also known as black frost, because unlike regular frost, it is not visible as white crystals (normal frost is called white frost).

It is possible for frost to form even if the air temperature is above freezing. Frost formation depends solely on the air's dew point. However, the temperature of a surface affects whether or not dew, frost, or hoarfrost form because colder objects radiate less heat into the air surrounding them, keeping that air's dew point down. The formation of frost is also governed by a process known as radiational cooling. Frost formation requires a surface temperature below 0 °C, so cold surfaces are necessary. At night, certain surfaces will cool much faster than the surrounding air and other surfaces because all objects radiate heat at all times. During the day, objects generally recoup any lost energy from energy received from the Sun. However, at night, objects no longer receive heat radiated by the Sun, and so less energy is generally received, resulting in a rapidly decreasing nighttime temperature. Frost is more likely to form on clear nights because surfaces cool faster when no clouds radiate heat. On cloudy nights, the clouds may radiate enough heat to surfaces to prevent frost from forming.

Frost typically forms under conditions of light or no wind and sufficiently cold temperature. Winds cause air turbulence, and this turbulence mixes the air, which inhibits frost formation. Typically, frost will form more easily overnight because temperatures tend to be lower and the air moves more slowly than it does during the day. Due to radiational cooling, frost forms less often in areas where many buildings, trees, and other objects are; it also forms less often near bodies of water.

Multiple factors are used by scientists and meteorologists to determine whether or not frost will occur on a given night. One is the general weather of that day/night. The situation most favorable to frost formation is a cloudy day followed by a clear night because clouds prevent the Sun from adequately heating the soil. Humidity is also used. If the dew point is over 5.5 °C at night, frosts are unlikely. If it is below 2.2 °C, a frost is highly probable.

In areas where frost forms, local weather services will designate the type of frost that might be deposited in the region. These designations are light, heavy, and killing. A light frost will have no destructive effects on vegetation. A heavy frost is a significant deposit but is not likely to affect the staple vegetation of a region. A killing frost is severely destructive to vegetation and can decimate an entire crop. In the United States, frost warnings will be issued only until October 15th west of a line from Frederick, Maryland, to Charlottesville, Virginia. East of this line, warnings are not issued past November 1st. In the spring, frost warnings are issued only if there is a possibility that crops and other plants could be damaged.

Killing frosts are monitored by weather services, as regions attempt to predict possible arrival dates, so that crop producers can better prepare for their arrival. Various methods can be employed to diffuse the effects of a harmful frost, such as placing small heating systems throughout a crop area, continuously sprinkling water on crops throughout the night, or, on a smaller scale, placing simple screening cloths around a home garden. Farmers and scientists are continuing to experiment with new methods to avoid the harmful effects of black frost on vegetation. Frost harms plants by forming ice in and around cells. The water around the cells is purer than that inside the cell, so it will usually freeze first. Those plants that have greater quantities of solutes within their cells are more resistant to frost and can withstand cold temperatures more easily than those plants with little solute in their water. Plant damage from frost is determined by the type of plant, the stage of growth of the plant, and the length of time the temperature is low enough for frost formation.

READING LIST

The Encyclopedia of the Atmospheric Environment on-line. Available: http://www.doc.mmu.ac.uk/aric/eae/Weather/Older/Frost.html.

The Weather Doctor. Available: http://www.islandnet.com/~see/weather/whys/frost.htm.

About Frost by Steve Horstmeyer. Available: http://www.shorstmeyer.com/wxfaqs/frost/frost.html.

UCLA Department of Atmospheric Science on-line. Available: http://www.atmos.ucla.edu/web/.

Iowa State University—On-line Case Studies. Available: http://www.iitap.iastate.edu/jhodson/idot/frost/jan1/frost1.html.

Dew and Frost. Available: http://www.nssl.noaa.gov/~cortinas/1014/112_2.html.

ASK—A—SCIENTIST Archive, Argonne National Laboratory. Available: http://newton.dep.anl.gov/askasci/wea00.htm.

When Does The Last Frost Usually Occur? Available: http://www.crh.noaa.gov/gjt/frost.htm.

How Does Frost Injury Occur? Available: http://www.gov.on.ca/OMAFRA/english/crops/field/news/croppest/cp0798_w.htm.

Sterling Watch/Warning Definitions. Available: http://www.erh.noaa.gov/er/lwx/Defined/.

Cooling Air. Available: http://www.doc.mmu.ac.uk/aric/eae/Weather/Older/Cooling_Air.html.

Frosty Cars. Available: http://www.weatherwise.org/qr/qry.frostycar.html.

Frost Damage, Control, and Prevention—Fruit and Vines. Available: http://www.pir.sa.gov.au/pages/agriculture/horticulture/frostdamagefs.pdf.

Frost Risk Assessment and Damage. Available: http://www.pir.sa.gov.au/pages/agriculture/horticulture/frost_risk.htm:sectID=445&tempID =11.

FROST DAMAGE

Arthur M. Holst
Philadelphia Water Department
Philadelphia, Pennsylvania

Frost is ice formed by the condensation of atmospheric water vapor on a surface. When low temperatures are present in a certain region, the potential for frost damage to plant life exists in that region. Frost damage can injure plants permanently or slow plant development. Several factors dictate the extent and severity of frost damage to plants in any given region.

Frost damage can have profound effects on agriculture and ecosystems. Frost is most likely to form on a cold, clear night that has been preceded by a cold and cloudy day. Lack of wind is vital to a developing frost. The little heat that was stored in the ground during the day will dissipate more quickly overnight. If the dew point has dropped below freezing, the result, is a heavy frost. As conditions change from day to night, an air temperature drop of 1 °C per hour is a signal that a frost is likely to develop.

Even if the conditions of freezing temperatures and a calm night are present, some areas will be more susceptible to frost damage than others. Fields that have lighter soils, which dry out faster, fail to insulate the soil below. This will prevent the natural warming by soil radiation of the air directly above the soil surface and increase the likelihood that a heavy frost will form. Recently cultivated fields will suffer the same fate. Low-lying valleys, where cold dense air cannot be affected by winds, can be heavily damaged. Areas of a field that are close to the edge of a crop formation are also susceptible. In these areas, grass near a crop formation acts as a blanket or insulator, preventing warm soil from heating cold air directly above it. Finally, areas that have recently been treated with herbicides are more susceptible to frost damage. Herbicide stress on a plant can be compounded by cold stress from the weather, increasing the possibility of frost damage.

Frost damage is evidenced by a variety of symptoms on plants. Plant leaves are the best indicators of frost damage. Frost damage on the youngest leaves of a plant (the top leaves) is often seen as a burn on the tips of the leaves. More severe damage is a darkening of the entire leaf. In the most extreme form, the entire plant takes on a black appearance. This darkening is evidence that the frost penetrated and destroyed cell membranes of the plant. Severe darkening of this kind is sometimes referred to as a "killing frost." Whether a killing frost has set in on plant life may not be noticeable until a day or two after the frost. If the plant has turned almost completely brown, chances of recovery are not good. However, a closer look may reveal that the lower part of the plant, or the pseudostem, is still green, a good sign that some recovery from the damage may be possible. The three most important factors in determining the ability of plant life to withstand damaging frost are the plant's maturity, its health prior to a frost, and the weather immediately following the frost. Susceptibility to frost damage increases as plant development increases. As the growing season progresses, the chances of weather conditions conducive to a damaging frost decrease. Young plants are less susceptible to frost damage that will lead to a plant's death because their growing points are still below ground, insulated from freezing temperatures. However, should frost injury occur at this young stage, it could severely delay growth as the season progresses and affect the overall harvest. More mature plants present more opportunities for frost damage. Mature plants have more exposed leaves and may have growing points above the earth's surface. Damage to the outside of a mature plant can constrict future growth. Plant health prior to a frost also determines the ability of a plant to recover. If plants have been continually exposed to cold stress, herbicides, excessive moisture, or disease, even the most minimal frost can be debilitating. Finally, the weather following frost damage plays an important role in the plant's recovery. If warm temperatures follow frost damage, a plant's ability to recover increases.

Worldwide, 5–15% of all agricultural production is lost to frost damage each year. Frost can also cause a loss of food supplies for an animal species by killing leaves, seeds, and fruits.

READING LIST

www.epa.gov—*United States Environmental Protection Agency*—Office of Pesticide Programs.

www.sciencedaily.com/releases/2000/10/001017073120.htm—*Science Daily Magazine*—Climate Change Shifts Frost Seasons & Plant Growth.

www.agric.gov.ab.ca/crops—*Alberta Agriculture, Food, And Rural Development*—Frost Damage to Cereals.

www.pioneer.com/usa/crop—*Pioneer Hi-Bred International, Inc.*—Microclimatic Effects on Frost Damage, Early Season Frost Damage to Corn.

www.pir.sa.gov.au/pages/agriculture—*PIRSA Agriculture*—Frost Risk Assessment and Damage.

Carter, P.R. (1995). Late spring frost and post-frost clipping effect on corn growth and yield. *J. Prod. Agric.*.

Kunkel, K.E. and Hollinger, S.E. (1995). Late spring freezes in the central USA: Climatological aspects. *J. Prod. Agric.*

Nielsen, R.L. (1999). Assessing frost damage to young corn. Purdue pest management and crop production newsletter. *Purdue Univ.*, 27 May 1999.

THE GLOBAL WATER CYCLE

U.S. Global Change Research Program

USGCRP-supported research on the global water cycle focuses on: (1) the effects of large-scale changes in land use and climate on the capacity of societies to provide adequate supplies of clean water; and (2) how natural processes and human activities influence the distribution and quality of water within the Earth system and to what extent the resultant changes are predictable. Specific areas include:

identifying trends in the intensity of the water cycle and determining the causes of these changes (including feedback effects of clouds on the global water and energy budgets as well as the global climate system); predicting precipitation and evaporation on timescales of months to years and longer; and modeling physical/biological processes and human use of water, to facilitate efficient water resources management.

The USGCRP budget includes $311 million in FY 2003 for research and observations related primarily to the Global Water Cycle. The Global Water Cycle program studies the movements and transformations of water, energy, and water-borne materials through the Earth system and their interactions with ecosystems. The movements and transformations of water are important because they appear to control the variability of the Earth's climate and they provide an essential resource for the development of civilization and the Earth's environment. Figure 1 schematically illustrates the movements and transformations. This cycling involves water in all three of its phases—solid, liquid, and gaseous—and exchanges large amounts of energy as water moves and undergoes phase changes. Therefore, the water cycle operates necessarily on a broad continuum of time and spatial scales.

Water vapor is a greenhouse gas that maintains temperatures in a range required by life on Earth. Many of the uncertainties in the current projections of the effects of the atmospheric buildup of carbon dioxide are related to the feedbacks between the climate and the water cycle. While warmer temperatures enable the atmosphere to hold more water leading to further warming, the complex interactions among changing cloudiness, precipitation patterns, land cover, and decreasing snow and ice cover have limited the quantitative understanding of the links between water and climate warming.

Water is not evenly distributed over the globe, nor is it always accessible for human use. Society is becoming more vulnerable to variations in the water cycle as a result of expanding populations and increasing water use. The increasing demands for water accompanied by the growing economic losses from droughts and floods place pressure on the science community to develop the knowledge and tools needed to manage our limited water resources more effectively. There are large potential paybacks from increased investments in scientific research to improve the monitoring and prediction of the global water cycle variations and in water management applications.

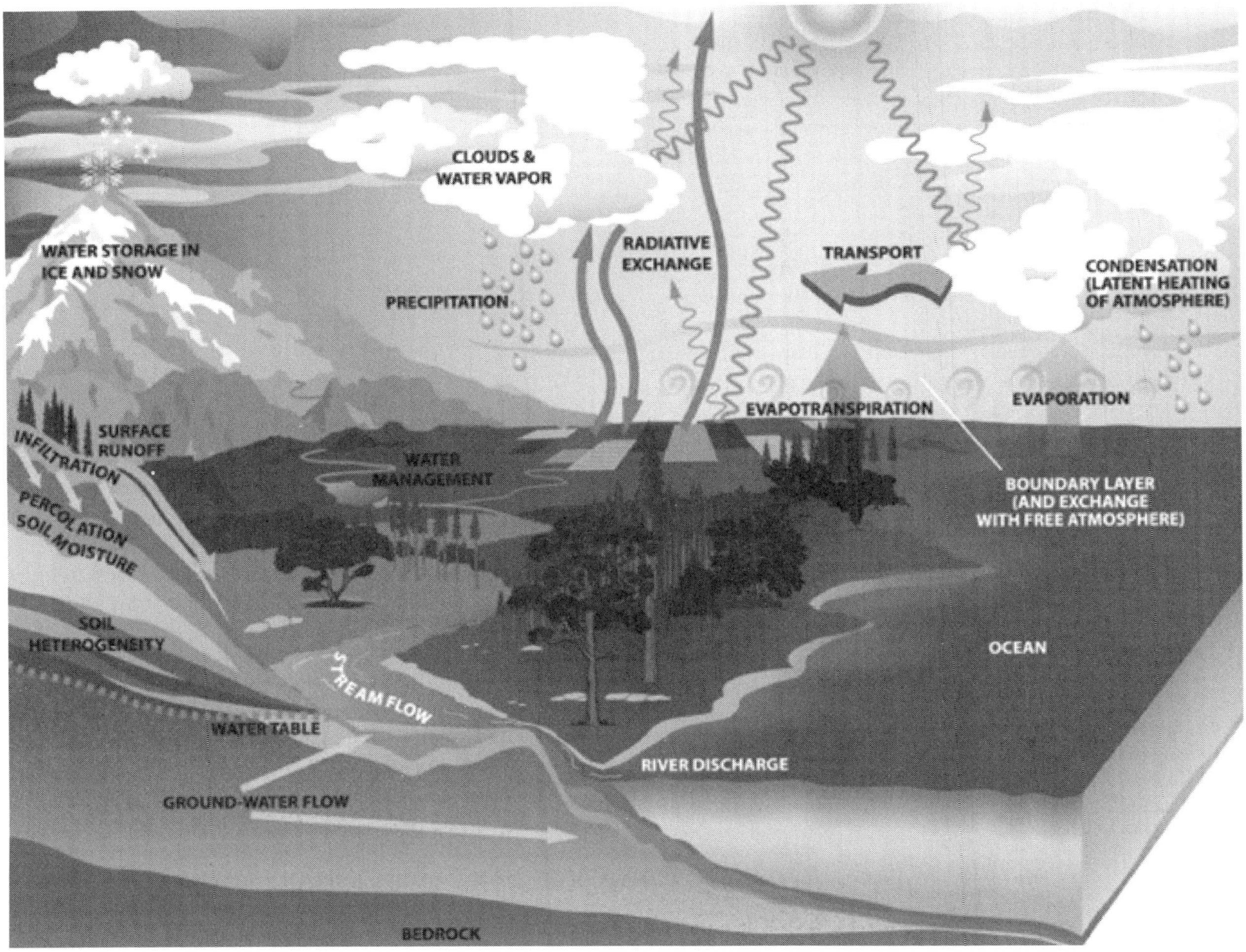

Figure 1. Conceptualization of the water cycle.

On a national basis, near-crisis situations have occurred in several dry southwestern river basins, including the Colorado and Rio Grande, where over-allocation has taken place. Recent drought conditions and rapid development in these basins have exposed the intensity of competition that exists over the available water resources. The development of a capability to predict where water management crises will emerge due to a drought or extended flood conditions is a priority for the Global Water Cycle program. The ability to provide probabilistic forecasts of rainfall and snowfall at various time and space scales is at the center of all potential applications of climate change science and climate information systems. The program has research activities directed at developing experimental predictions that will ultimately benefit society through better protection of human health and assets, and more efficient water system management and infrastructure planning.

Human activity is an integral part of the water cycle. A recent USGCRP-commissioned report, *A Plan for a New Science Initiative on the Global Water Cycle*, issued in 2001, concluded that, among other priorities, there is a pressing need to determine the causes of water cycle variations on both global and regional scales, and to what extent these variations are induced by human activities. In view of this emerging link between water science and water resource issues, the USGCRP global water cycle strategic plan addresses two major questions: (1) What are the effects of large-scale changes in land use and climate on the capacity of societies to provide adequate supplies of clean water, and (2) how do natural processes and human activities influence the distribution and quality of water within the Earth system and to what extent are resultant changes predictable?

Stakeholders are helping to define the Global Water Cycle program at the catchment and larger river basins scales. Users are interested in better forecasts of precipitation, runoff, and soil moisture. Reservoir management decisions require forecast lead times of up to seasons and, in some cases, years. For planning reservoirs, dam recommissioning, and water control infrastructure, and developing new proposals for water law, projections of water variability are required on the decadal to century timescales.

The USGCRP Global Water Cycle program focuses on characterizing, explaining, and predicting variability and long-term changes in the global water cycle and their impacts. To address the issues arising from the intimate role of the water cycle in controlling climate variability on seasonal to multidecadal timescales, the program investigates the pathways of water movement between the biosphere and surface hydrologic systems, the atmosphere, and the oceans, as well as feedback processes between climate, weather, and biogeochemical cycles. Because the biosphere is a substantial regulator of the Earth's carbon cycle, the global water cycle maintains a considerable influence upon the global pathways of carbon. Globally, the cycling of water and its associated energy and nutrient exchanges among the atmosphere, ocean, and land determine the Earth's climate and cause much of climate's natural variability.

A critical contribution of the USGCRP to Federal water activities lies in the benefits that come from drawing together the wide range of programs and expertise from different agencies with the capabilities of the academic community to address these complex issues. The elements of the management structure that the USGCRP has put in place during the past year include: (1) Interagency Global Water Cycle working group, (2) Global Water Cycle scientific steering group, and (3) Global Water Cycle program office. The linkages between the global water cycle, the global carbon cycle, and climate will be explored in the coming year through this strengthened program management structure.

SEE ALSO:

Water Cycle [also available: PDF Version]. Chapter 5 from the *Strategic Plan for the Climate Change Science Program (July 2003)*. See also the draft white paper, *The Global Water Cycle and Its Role in Climate and Global Change* [PDF] (posted 27 Nov 2002).

Water Cycle. Presentation from Breakout Session 8 of the *US Climate Change Science Program: Planning Workshop for Scientists and Stakeholders*, 3–5 December 2002, Washington, DC.

Climate Variability—Atmospheric Composition—Water Cycle. Presentation from Breakout Session 19 of the *US Climate Change Science Program: Planning Workshop for Scientists and Stakeholders*, 3–5 December 2002, Washington, DC.

GROUND-BASED GPS METEOROLOGY AT FSL

Seth I. Gutman
Kirk L. Holub
NOAA Forecast Systems Laboratory

INTRODUCTION

Water vapor is one of the most important components of the Earth's atmosphere. It is the source of precipitation, and its latent heat is a critical ingredient in the dynamics of most major weather events. As a greenhouse gas, water vapor also plays a critical role in the global climate system: it absorbs and radiates energy from the sun and affects the formation of clouds and aerosols and the chemistry of the lower atmosphere. Despite its importance in climate and weather prediction, water vapor has been one of the most poorly measured and least understood components of the Earth's atmosphere. Researchers at FSL and elsewhere are utilizing recent technology to reverse this situation.

The ability to use the Global Positioning System (GPS) to make accurate refractivity measurements under all weather conditions has led to the development of a promising new meteorological observing system for NOAA. The first and most mature application of ground-based GPS meteorology involves the measurement of integrated (total column) precipitable water vapor (IPW) in the atmosphere. The GPS-IPW technique is more advantageous than conventional water vapor observing systems because of its low-cost, high-measurement accuracy, all weather operability, and long-term measurement stability. Further, GPS-IPW requires no external calibration, operates unattended for long periods with high reliability, and is easily maintained. Since GPS-IPW measurements are compatible with satellite data retrievals, they provide an independent method for calibrating and validating global satellite observations.

These positive attributes, however, are accompanied with one major disadvantage: GPS-IPW provides no direct information about the vertical distribution of water vapor in the atmosphere. In an attempt to mitigate this deficiency, researchers at government laboratories and universities around the world are investigating the best ways to use GPS-IPW as a "proxy quantity" for moisture profiles in weather forecasting.

In this article we discuss how IPW is now calculated from GPS signal delays and the potential use of slant-path measurements in numerical weather prediction models. Preliminary results of the effect of GPS-IPW on numerical weather prediction, the demonstration network, data and product availability, and plans for the operational network are also described.

CALCULATING IPW FROM GPS SIGNAL DELAYS

GPS signals are delayed as they pass through the Earth's atmosphere (Fig. 1). The signal delay caused by the presence of free electrons in the ionosphere makes the largest contribution to the total atmospheric delay. Because the ionosphere is a dispersive medium, the velocity of the GPS signals is frequency dependent and its impact can be effectively eliminated by using dual frequency receivers.

Below the ionosphere, in the electrically neutral portion of the atmosphere, refraction (that is, slowing and bending) of the GPS signal is caused by changes in temperature, pressure, and water vapor. Most of this delay occurs in the troposphere, which extends from about 9 km at the poles to about 13 km at the equator. The primarily tropospheric delay consists of a hydrostatic (or dry) component caused by the mass of the atmosphere and a wet component (the wet delay) caused by the dipole moment of the water vapor molecule. The contributions of the wet and dry components of the tropospheric signal delay are in the same proportion as the wet and dry components of the atmosphere.

FSL currently collects GPS observations from a demonstration network of 55 sites (Fig. 2) and processes them to produce IPW measurements every 30 minutes using the scheme shown in Fig. 3.

The first step in obtaining IPW from GPS observations is to determine the zenith-scaled delay caused by the neutral atmosphere. This delay is commonly referred to as the zenith tropospheric delay (ZTD), and is calculated from carrier phase and range observations made by networks of GPS receivers. The calculation is made using GPS analysis software such as GAMIT (GPS At MIT), which in addition to the GPS observations, requires improved satellite orbits and parameters describing the orientation of the Earth in space and time. Next, the ZTD is separated into its wet and dry components using additional observations made by collocated surface meteorological sensors. The zenith- scaled hydrostatic delay (ZHD) is caused by the mass of the atmosphere directly above the site and can be estimated with great accuracy from a surface pressure measurement. The wet signal delay (ZWP) is caused by water vapor along the paths of the radio signals to all satellites in view, about 6 to 8 with the current GPS satellite constellation.

ZWP is calculated simply by subtracting the hydrostatic delay from the tropospheric delay. The resulting wet delay can be mapped into IPW with an error of about 5 degrees using a quantity that is proportional to the mean vapor pressure-weighted temperature of the atmosphere (Tm). Tm may be estimated from a climate model, the surface temperature derived from a numerical weather prediction model, or measured directly using remote sensing techniques. FSL is planning to utilize model-derived Tm estimates operationally.

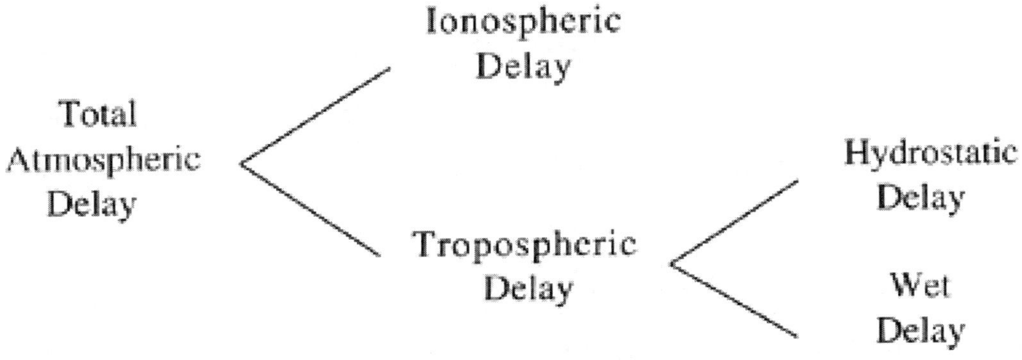

Figure 1. Signal delays caused by the atmosphere.

Figure 2. A map of the NOAA-FSL Global Positioning System Integrated Precipitable Water (GPS-IPW) Demonstration Network (55 sites) as of October 1999.

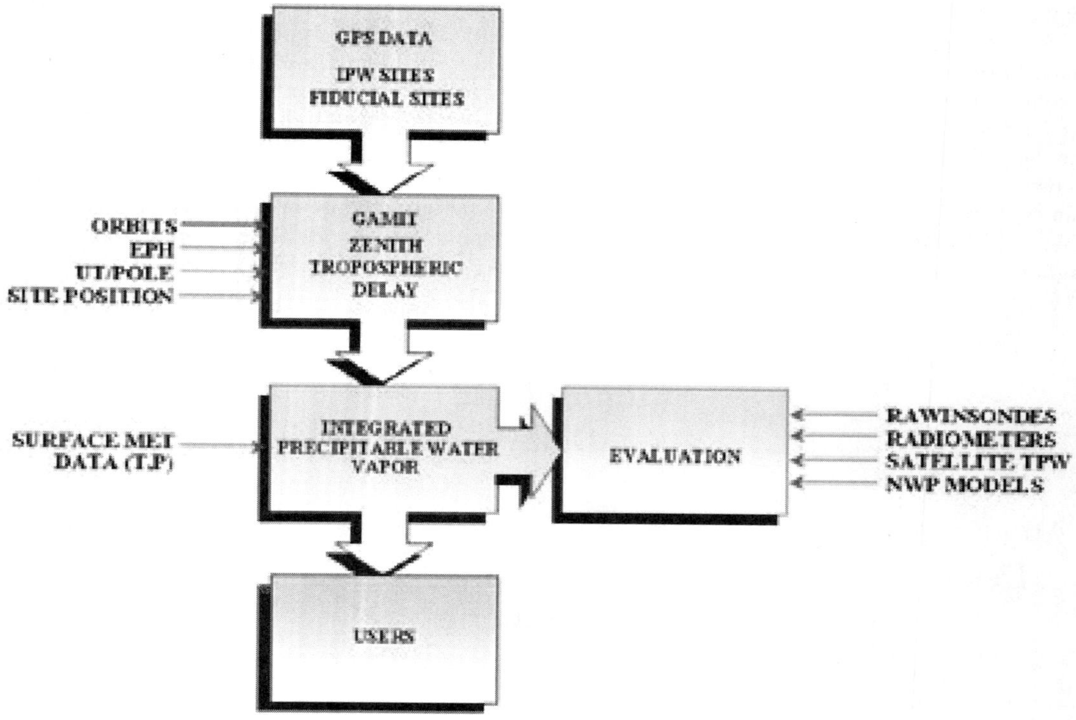

Figure 3. The FSL-developed data processing scheme used to produce IPW measurements.

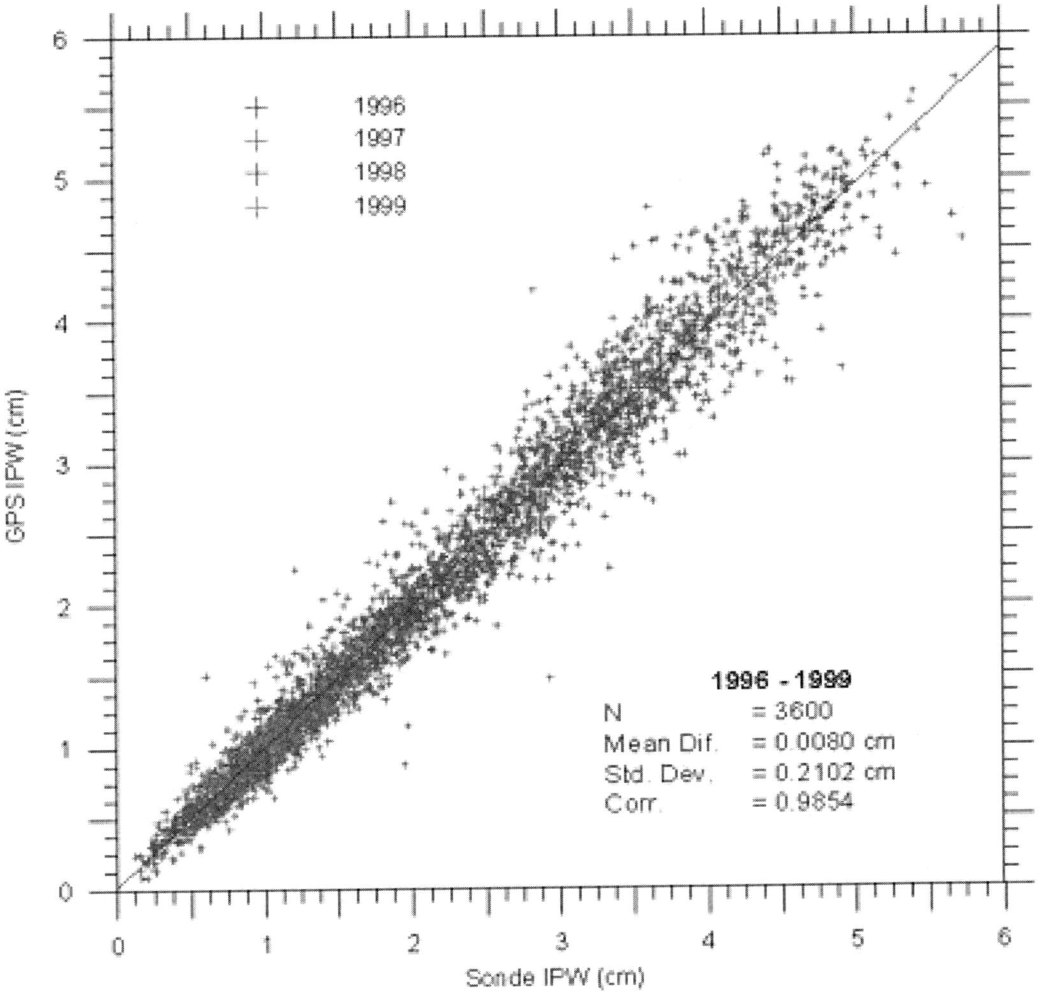

Figure 4. Scatterplot of GPS and rawinsonde observations of integrated precipitable water vapor at the ARM CART site near Lamont, Oklahoma, between January 1996 and September 1999.

Integrated precipitable water calculated from GPS signal delays is physically identical to integrated measurements or retrievals made by other upper-air observing systems including rawinsondes, ground-based microwave water vapor radiometers, or satellite microwave and infrared instruments including sounders and interferometers. Comparisons of GPS and radiosonde-derived total column water vapor have been carried out continuously since 1996 under all weather conditions at the DOE Atmospheric Radiation Measurement site near Lamont, Oklahoma. Results from 3600 comparisons (to September 1999) indicate a mean difference of 0.08 mm and a standard deviation of 2.1 mm (Fig. 4).

SLANT-PATH SIGNAL DELAY MEASUREMENTS

Recent investigations by FSL director A.E. MacDonald and Yuanfu Xie (of the Forecast Research Division) of the potential use of line-of sight estimates of path-integrated water vapor (derived from slant-path GPS signal delay measurements) to retrieve the 3-D moisture field have been very interesting and potentially significant. The experiments involve assimilating simulated slant-path moisture measurements from a wide area network of closely spaced stations into the Quasi-Nonhydrostatic (QNH) model using variational techniques. In recent research, their simulations indicate that it may be possible to recover the three- dimensional structure of the moisture field from a densely spaced network of ground-based GPS receivers making a single line-of- sight, or slant path, measurement of the signal delay to all satellites in view. The configuration of the GPS satellite constellation as seen from Boulder, Colorado, between 1200 and 1300 UTC on 28 September 1999 is shown in Fig. 5. A GPS satellite moves across the sky at the rate of about 30 degrees per hour. Although 10 satellites are visible above the horizon in this example, six to eight would be more typical at any one time.

Making a slant-path signal delay measurement with the same accuracy as a zenith-scaled measurement is

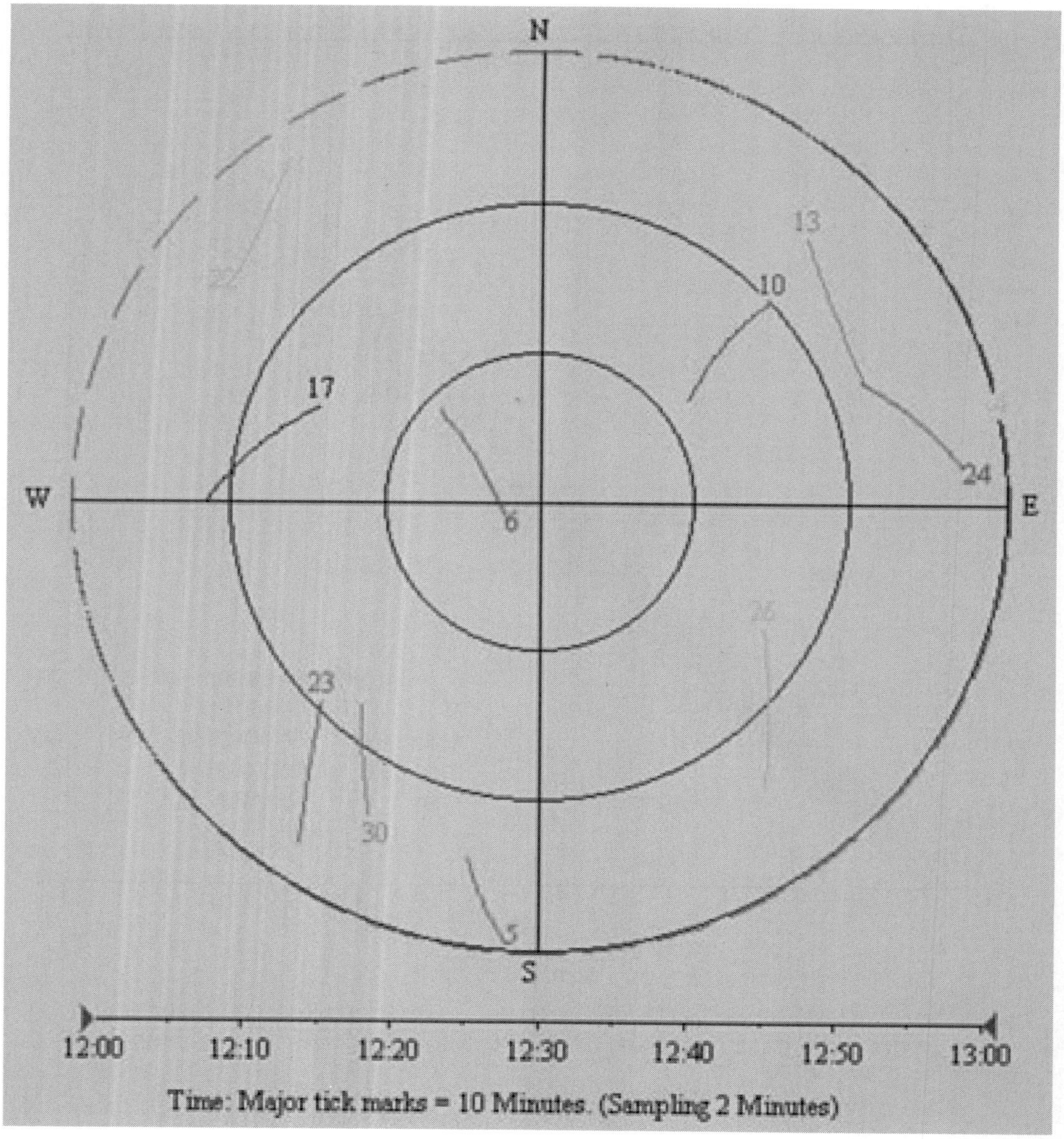

Figure 5. Configuration of the GPS satellite constellation as seen from Boulder between 1200 and 1300 UTC 28 September 1999.

not trivial. The sources of measurement error that are successfully managed through geodetic modeling of the zenith tropospheric signal delay will have to be dealt with in other ways. Although some of the most important information about the structure of the atmosphere can be obtained from low-angle observations, measurement errors increase significantly along with the negative impact of multipath reflections from nearby obstacles as satellites approach the horizon. One way to mitigate these problems is to utilize advanced GPS receivers and antennas that maximize the ability to track satellites under all conditions and reject multipath reflections. Unfortunately, not all problems can be eliminated through the selection of hardware, and advanced data processing techniques will be needed as well.

Research at Scripps Institution and the University of Hawaii into ways to monitor the accuracy of GPS orbit predictions suggests that these techniques can also be used to reduce systematic errors in slant-path signal delay or refractivity measurements to individual satellites.

EFFECT OF GPS-IPW DATA ON THE ACCURACY OF NUMERICAL WEATHER PREDICTION

Since 1997, parallel runs with and without GPS have been carried out using the research version of the Rapid Update Cycle (RUC-2) model to assess how GPS-IPW data affect the accuracy of numerical weather prediction. Results from the first two years using optimal interpolation techniques have been encouraging despite the fact that the observations came from only a limited number of widely spaced sites. Model runs using data acquired from more sites over a larger area through September 1999 confirm improvements

Figure 6. GPW-IPW installation at the NOAA Profiler Network site at Platteville, Colorado.

in forecast accuracy, especially under conditions of active weather. Therefore, NOAA meteorologists expect significant improvements in short-term forecasts of clouds, precipitation, and severe weather when high-resolution numerical weather prediction models routinely use data from a nationwide network of GPS-IPW systems in conjunction with data from other observing systems and advanced data assimilation techniques.

The decision to implement ground-based GPS Meteorology (GPS-Met) as a next-generation upper-air observing system will be supported in part by promising assessments such as this one. In anticipation of a favorable decision, network design and implementation options for a national network of ground-based GPS- IPW systems are being evaluated at FSL.

GPS-IPW DEMONSTRATION NETWORK

The rapid development of the GPS-IPW Demonstration Network for meteorological remote sensing has been made possible by a fortuitous synergy with the positioning and navigational applications of GPS by the U.S. Coast Guard and U.S. Department of Transportation. As of October 1999, the data acquisition component of the demonstration network consisted of 55 GPS-IPW systems operating in the continental United States and Alaska.

Thirty-four systems are currently installed at NOAA Profiler Network (NPN) sites, seven at sites belonging to other NOAA organizations or institutions affiliated with NOAA, 11 belong to the U.S. Coast Guard Maritime Differential GPS (DGPS) system, and three are at the Department of Transportation Nationwide Differential GPS facilities. Typical sites from each organization are illustrated in Figs. 6–9.

In addition to supporting the assessment of GPS as a possible next-generation upper-air observing system, the GPS-IPW Demonstration Network is designed to help NOAA accomplish the following tasks:

- Evaluate the engineering and scientific bases of ground- based GPS-Met, including advanced data acquisition and processing techniques.

Figure 7. GPS-IPW installation at the Scripps Institution of Oceanography at La Jolla, California.

- Develop and test strategies to build, monitor, operate and maintain large networks of GPS reference stations for meteorological remote sensing.
- Develop techniques to acquire, process, and quality control GPS observations and data products.
- Provide observations and derived meteorological products to users (such as forecasters, modelers, researchers) and data archives.
- Transfer ground-based GPS-Met technologies to operational use in weather forecasting and climate monitoring.
- Other possible applications under investigation include calibration and validation of environmental satellite data and improved positioning and navigation services.

All ground-based observing systems in the GPS-IPW Demonstration Network consist of dual-frequency GPS receivers and antennas, and collocated surface meteorological sensors.

These systems are located at sites where shelter, power, and communications are available to operate and collect data from the instruments, and transmit these data in real or near real time to one of two locations. The generalized flow of data and products from the network is illustrated in Fig. 10.

DATA AND PRODUCT AVAILABILITY

GPS and surface meteorological observations from the GPS-IPW Demonstration Network sites are available to the general public in near real time through the NOAA National Geodetic Survey. Information and raw data may be acquired via the Web.

Processed data, including GPS signal delays and integrated precipitable water vapor, are available shortly after improved NAVSTAR GPS satellite orbits and Earth Orientation Parameters are available from one of the International GPS Service (IGS) tracking stations. This usually occurs within 24 hours of the close of the day, but efforts to accelerate the process and make improved orbits available within 1–3 hours are well underway. IPW and other products may be acquired from the FSL Demonstration Division, *GPS-Met Observing Systems Branch*.

Figure 8. GPS-IPW installation at the U.S. Coast Guard Differential GPS site at Cape Canaveral Air Force Station.

PLANS FOR NETWORK EXPANSION AND IMPROVEMENT

Our primary goals in 2000 are to continue to expand the demonstration network, demonstrate distributed data processing using low-cost PCs instead of high-end workstations, and implement real-time data processing.

Expansion of the Demonstration Network. Now that all NOAA Profiler Network sites have been equipped with GPS-IPW systems, expansion of the network will mostly proceed by installing GPS Surface Observing System (GSOS) packages at U.S. Coast Guard Maritime DGPS and Department of Transportation NDGPS sites. Depending on the availability of funds and the status of interagency agreements under review, during the next year we hope to install 21 new systems at DGPS sites (mostly in the Mississippi Valley and Great Lakes regions), 11 NDGPS sites, and one at the Department of Energy ARM facility at Point Barrow, Alaska (Fig. 11).

Implementation of Real-time Data Processing. We define real-time data processing as acquiring and processing GPS and ancillary observations to yield signal delay or IPW calculations within a single numerical weather prediction assimilation cycle. In the case of the Rapid Update Cycle, running operationally at the National Centers for Environmental Prediction, this is approximately 75 minutes.

Real-time data processing techniques are being tested and evaluated in a collaborative effort involving FSL, the Scripps Permanent Orbit Array Center, and the University of Hawaii at Manoa. Techniques involve acquiring data from a subset of the IGS global tracking network and using these observations to produce an improved retrospective orbit with only about 2-hour latency. An orbit prediction that covers the data gap is also made, and it is this short-term prediction that is used to calculate IPW. In theory, the error in a prediction that spans only 2 or 3 hours will be proportionally less than an error made over an interval of 36–48 hours.

Real-time quality control techniques are also under evaluation. The most promising involve continuous monitoring of the relative positions of a number of sites, and using these data to infer changes in orbit

Figure 9. GPS-IPW installation at the DOT National Differential GPS site at Whitney, Nebraska.

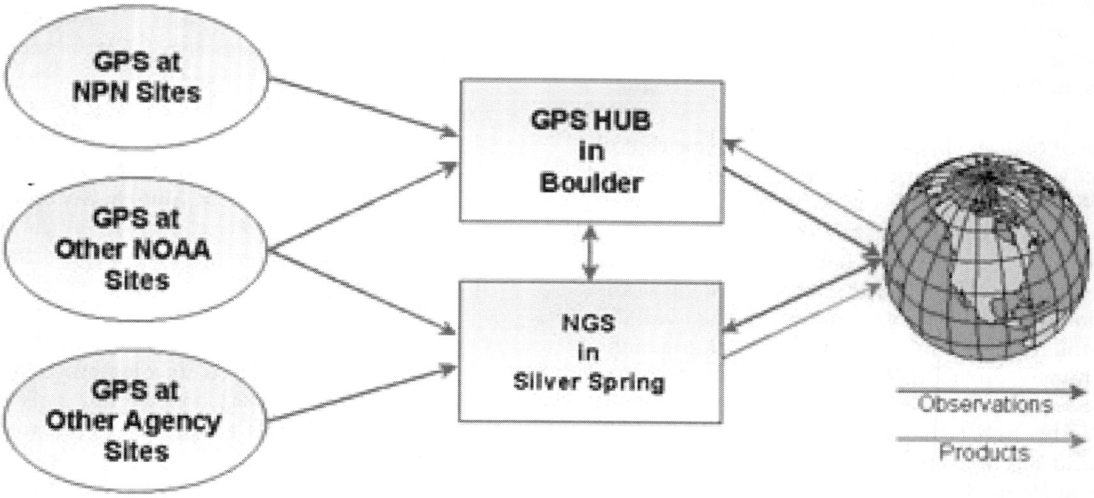

Figure 10. Flow of data and products from the GPS-IPW Demonstration Network.

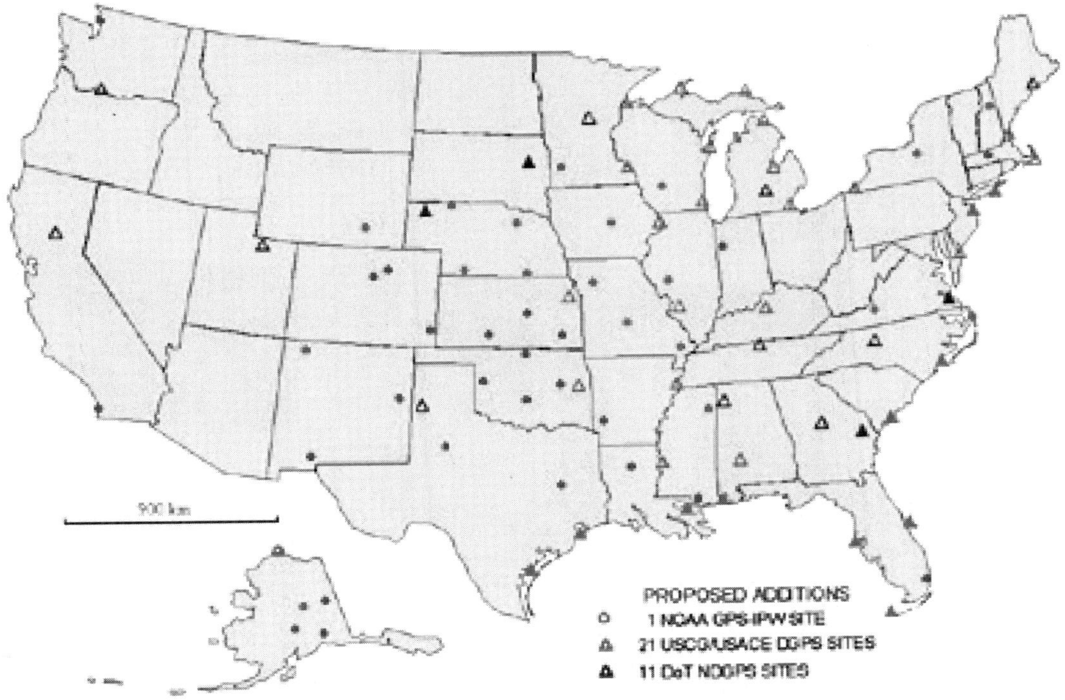

Figure 11. Planned expansion of the Demonstration Network to about 92 sites during 2000.

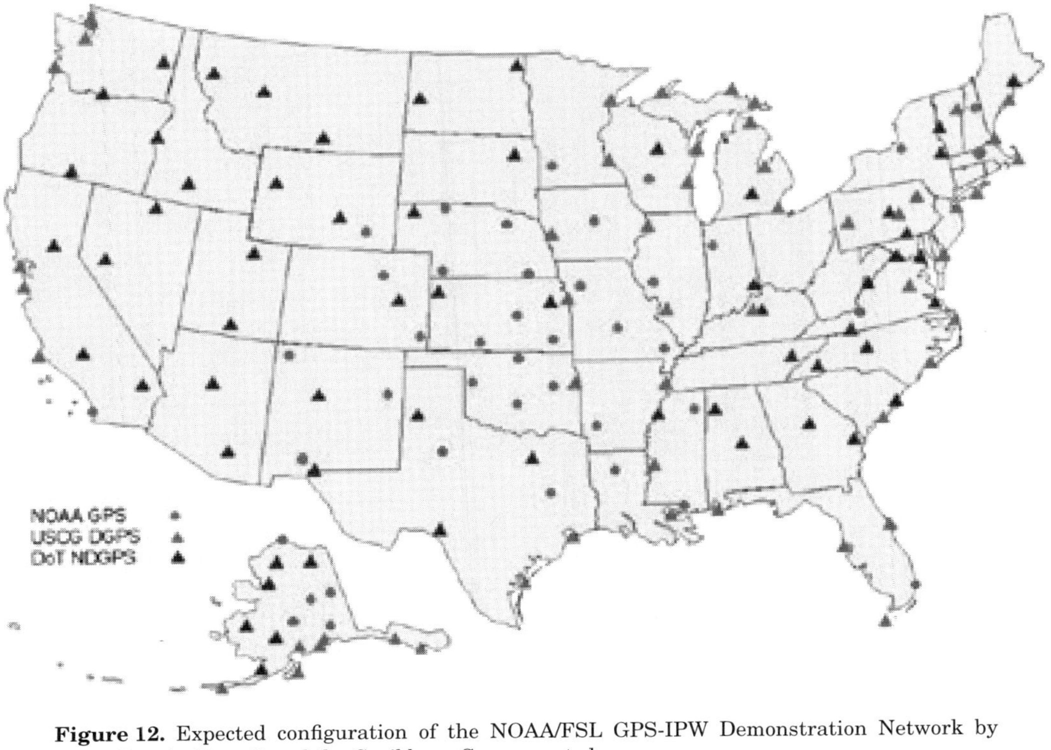

Figure 12. Expected configuration of the NOAA/FSL GPS-IPW Demonstration Network by 2005. Sites in Hawaii and the Caribbean Sea are not shown.

Figure 13. An Automated Surface Observing System (ASOS) installation at Cape Hatteras, North Carolina.

Figure 14. GPS receiver placement on top of FSL's new office building, the David Skaggs Research Center.

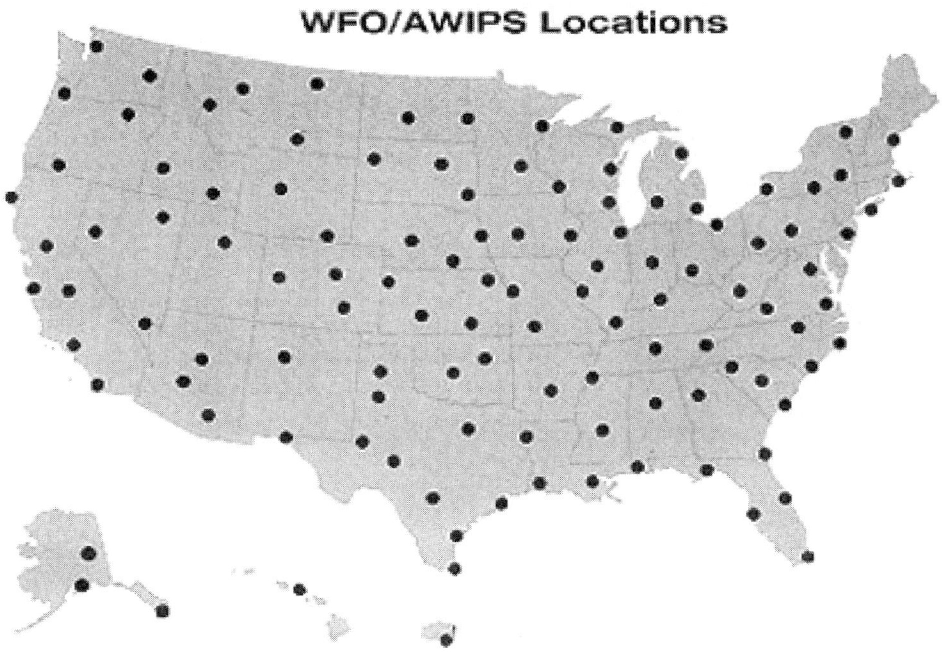

Figure 15. A map of the AWIPS offices.

accuracy for specific satellites in the constellation. When a problem is encountered, the satellites are removed temporarily from the ephemerides until an updated orbit can be produced.

Distributed Data Processing. Recent advances in low-cost PC processor speed and memory will be utilized to perform data processing in a fully distributed environment. During this year we have demonstrated the ability to partition a large network into smaller subnetworks, process each independently in substantially less time with no significant loss of accuracy and precision.

Operational GPS-IPW Network Implementation Strategy. The expansion of the GPS-IPW Demonstration Network to about 200 sites, and the transition from retrospective to real-time data processing will enable us to assess the impact of these data on weather forecast accuracy. Based on the results of these studies, a decision to implement ground-based GPS-IPW as a next-generation upper-air observing system for NOAA is expected. The following plan has been developed to expand the demonstration network to an operational network of about 1000 sites with an average station spacing of somewhat less than 100 km (Fig. 12).

- Once transition of the GPS-IPW Demonstration Network to operational status has become a reality, receivers and antennas will be upgraded. Communications will transfer from FTS-2000 to the AWIPS [Advanced Weather Interactive Processing System] communications systems via the Internet.
- GPS receivers will be added to about 800 Automated Surface Observing (ASOS) sites (sample site shown in Fig. 13). The reason for collocating GPS at ASOS sites is to take advantage of the surface meteorological data and site infrastructure, including shelter, power, data communications, field maintenance, and logistics support. This will minimize implementation time and life-cycle cost. The GPS antenna installation at a typical ASOS site will resemble the one at FSL's new location, the David Skaggs Research Center (Fig. 14).
- Data processing hardware, software, and training will be provided to all AWIPS offices (Fig. 15).

[Editor's Note: More information on the topics covered here is available by contacting Seth Gutman, who can provide copies of published articles which include a list of references.]

CLIMATE AND WATER BALANCE ON THE ISLAND OF HAWAII

James O. Juvik
D.C. Singleton
G.G. Clarke
University of Hawaii
Hilo, Hawaii

INTRODUCTION

The island of Hawaii, with a surface area of only 10,455 km, exhibits a spectacular range of climatic diversity comparable with that found on large continents. Three major factors contribute to this climatic diversity:

1. Topographic relief. The volcanic mountains of Mauna Kea and Mauna Loa reach summit elevations of 4,205 m and 4,168 m, respectively. The attitudinal range provides for a diversity of temperatures, and the mountains themselves are barriers that induce orographic precipitation.
2. Large-scale synoptic wind field. The strong and persistent northeast trade winds interact with the island topography to produce distinctive windward and leeward climates. The associated upper-level trade wind inversion exerts a particularly strong control on mountain precipitation gradients.
3. Local circulation. Differential heating and cooling of the land, water, mountain, and lowland areas on Hawaii give rise to localized wind regimes which add to the island's climatic diversity.

KOPPEN CLIMATIC ZONES

Integrating the attitudinal temperature gradients with the annual, seasonal, and spatially variable rainfall regimes results in a diverse combination of climatic environments. The Koppen climate classification uses monthly temperature and precipitation characteristics in a descriptive system that distinguishes broad regional and global climatic zones. The system has been often criticized for its empirical approach and lack of emphasis on "dynamic processes" (e.g., 1); however, as a "first approximation" the Koppen classification offers useful insights into regional climatic patterns.

Four broad Koppen climatic zones are distinguished on the island of Hawaii. They are organized primarily as concentric attitudinal bands on the mountain slopes. Figure 2 illustrates the spatial distribution of these Koppen climatic types on the island. The map was constructed on the basis of temperature (absolute or extrapolated) and precipitation data from 55 island stations. Discussions of the zones follow.

Humid Tropical Zone (A climates)

Characterized by warm temperatures throughout the year and relatively high annual rainfall, humid tropical climates occupy the lower slopes of the island from sea level to about 450 m (slightly higher in warmer areas of leeward

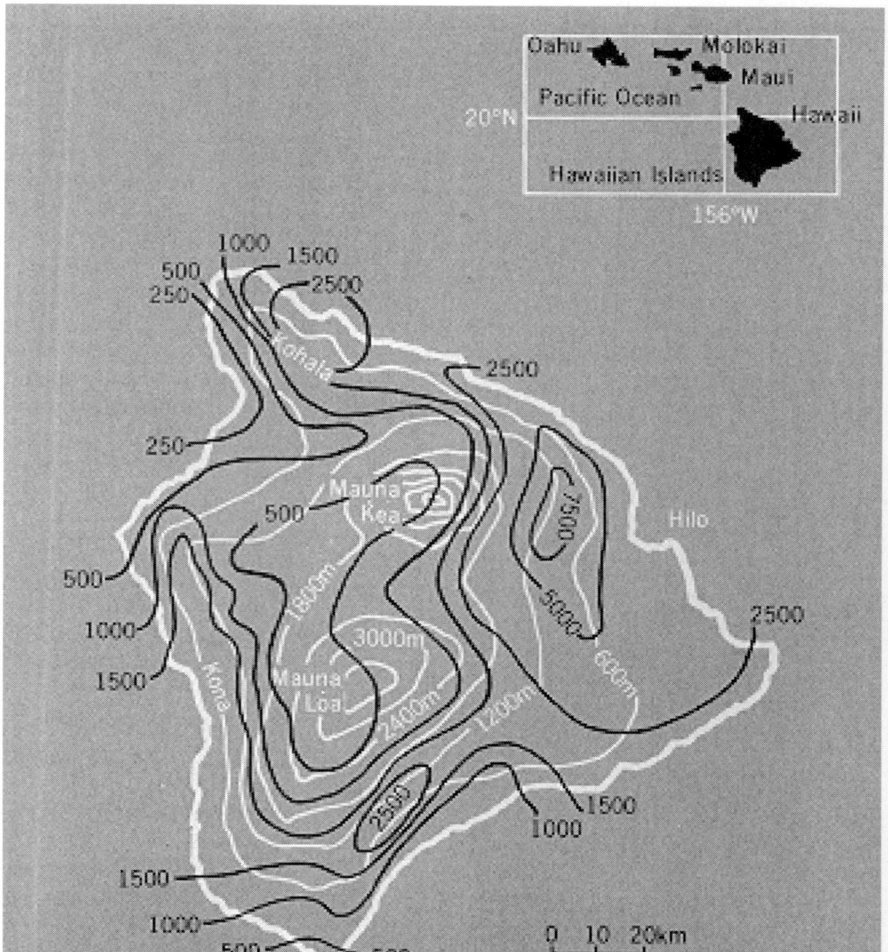

Figure 1. Distribution contours of mean annual rainfall (mm), superimposed on topographic map of the island of Hawaii. (Redrawn from 2,3).

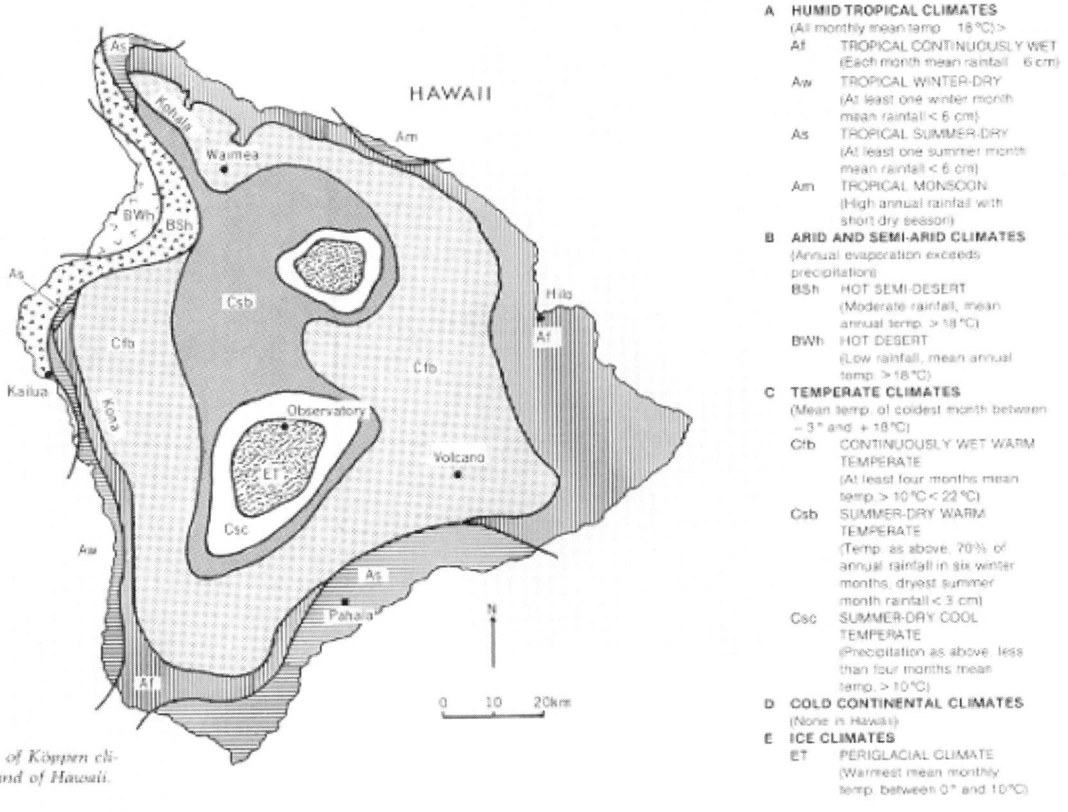

Figure 2. Distribution of Köppen climate types on the island of Hawaii.

Kona). This tropical zone may be further differentiated on the basis of rainfall seasonality. Large-scale synoptic disturbances in winter (mid-latitude cyclonic storms) produce substantial rainfall that is to some extent independent of slope aspect or elevation, and as a result most locations on the island exhibit an absolute winter maximum in rainfall. However, windward areas of Hawaii also receive substantial orographic rainfall throughout the year, with the result that there is no distinct dry season (Af climate). Lowland areas on the island that are transitional in location between windward and leeward receive less orographic rainfall (since they are not oriented normal to trade wind flow) and exhibit a distinctive summer dry season (As climate). Humid summer-dry climates are not common anywhere in the world, since for most tropical locations rainfall is at a maximum in the summer, the result of increased convective instability in the high-sun period. Outside of Hawaii the As climate type occurs only in southern Madras (India) and adjacent northern Sri Lanka.

The leeward or Kona coast of Hawaii contains the only extensive area of summer maximum rainfall in the Hawaiian archipelago (Aw, winter-dry climate). Isolated from the prevailing trade wind flow by intervening high mountains, the Kona coast's dominant circulation pattern is formed by a localized land-sea breeze regime. Increased land surface temperatures in summer strengthen the daily sea breeze regime and increase convective instability, leading to a high frequency of afternoon thundershowers. The vertical structure necessary for thundershower development is further assured by the high mountains, which exclude the trade wind aloft and limit the potential for strong vertical wind shear. The presence of a strong shearing force would otherwise tend to destabilize these leeward convectional cells. Although there is generally a summer rainfall maximum throughout Kona, the Aw climate gives way to Af at elevations above 400 m, where, by virtue of general orographic position, there is adequate precipitation in all months.

Arid and Semi Arid Zones (B climates)

A classic rain shadow desert exists on the leeward side of the Kohala mountains. Smaller and lower (maximum elevation 1,670 m) than Mauna Kea and Mauna Loa, the Kohala mountains are incapable of blocking out trade wind flow to leeward. Having become depleted of moisture during windward ascent, the trades warm adiabatically to leeward, promoting a hot, arid zone. With only 190 mm of annual rainfall, Kawaihae on the leeward Kohala coast is the driest location in the Hawaiian archipelago. The Koppen system distinguishes two climatic subtypes, the true desert (BWh climate) and the semidesert (BSh climate) on the basis of relative aridity. In leeward Kohala the true desert gives way to semi-desert at higher elevations.

[1]The Temperate (C) and Polar (E) climates as originally proposed by Koppen were not applied in high-altitude tropical environments, which, because of their orographic complexity,

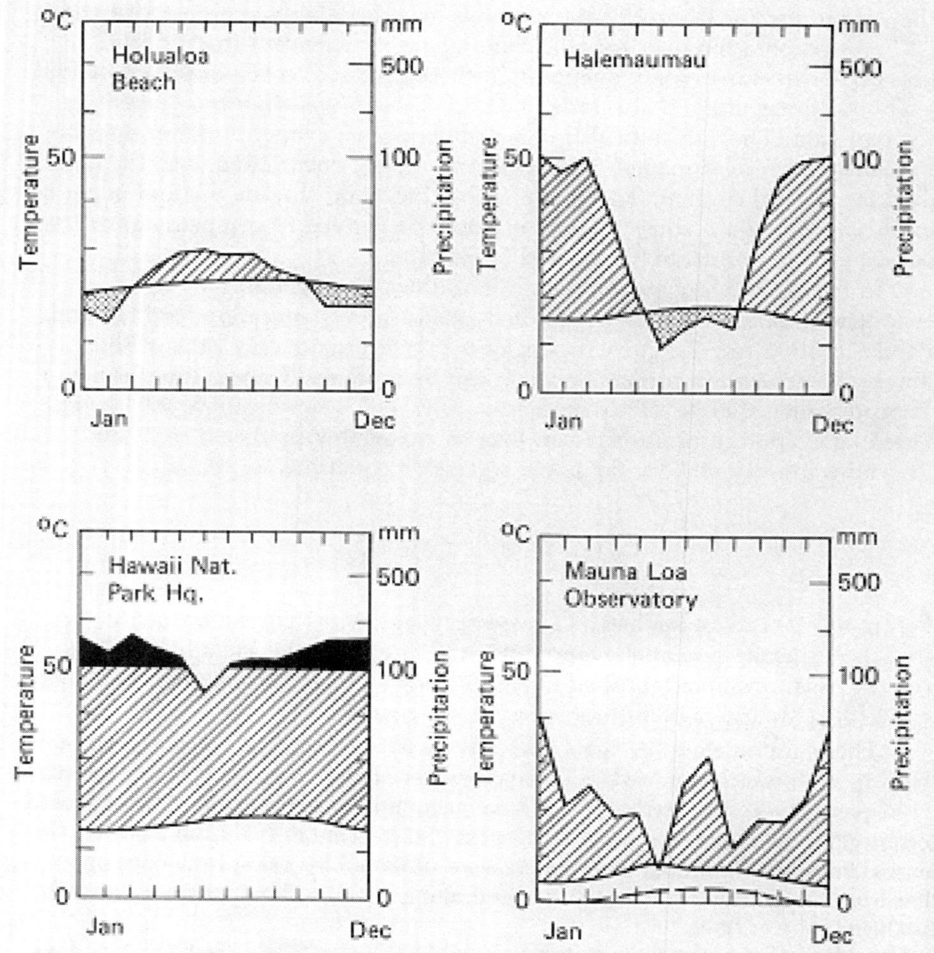

Figure 3. "Walter" climate diagrams for four Hawaii island stations (4).

Temperate Zone (C climates)[1]

Average air temperature in Hawaii decreases with altitude at the rate of about 0.55øC/100 m (Price, 1973). When the criteria of the Koppen classification are used, at elevations above 400–500 m on the mountain slopes, tropical climates grade to temperate as a result of decreasing average temperatures. As a result of the moderating influence of altitude, almost two-thirds of the "tropical" island of Hawaii possesses a temperate upland climate. The majority of this zone is characterized by warm summers and adequate precipitation in all months (Cfb climate). Except for the absence of a stronger season variability the upland Hawaii climates are analogous to those of similar Koppen designation in Pacific coastal areas of North America. Ascending orographic clouds compressed between the rising mountain slope and an upper-air temperature inversion produce frequent ground level mountain fog, an important moisture source for upland vegetation (Juvik and Perreira, 1974). At still higher elevations on Mauna Kea and Mauna Loa (above 2,000 m) there is a tendency toward summer drought. The increased strength and frequency of the trade wind inversion in summer (modal elevation 1,800 m) inhibits the vertical penetration of orographic clouds and precipitation to the higher slopes. This summer-dry zone (Gsb climate) also occurs at a lower elevation in leeward Kohala, Mauna Kea, and Mauna Loa, where summer orographic precipitation is largely absent. Above 2,500 m on both Mauna Loa and Mauna Kea the summer-dry regime changes from warm to cool (Gc climate).

lack of meteorological data, and absence of strong seasonality, were simply designated as highland climates (H). In more recent global and regional climatic maps of the tropics, highland C and E climatic areas are frequently portrayed in order to show approximate attitudinal analogs for these broad latitudinal climatic zones.

Alpine (periglacial) Zone (E climates)

Above 3,200-m level on Mauna Kea and Mauna Loa all months have a mean temperature below 10 °C, and the climates are classified as periglacial (ET). Nighttime freezing is common throughout the year. Although it exhibits a winter maximum, annual rainfall is very low (200–400 mm) and variable. Above the 3,500-m level, winter snowfall accounts for a substantial portion of the seasonal precipitation. Koppen used the 10 °C (warmest month) boundary to separate the C and E climates on the basis that trees will not normally grow where mean temperatures fall below this level. Hence E climates characterized the treeless arctic tundra. The upper tree line of Mauna Kea (3,000 m) corresponds fairly closely to the C/E boundary mapped in Fig. 2. On Mauna Loa the tree line is much lower for edaphic reasons (recent lava).

WATER BALANCE

The preceding discussion of Koppen climatic zones on Hawaii provides a general overview of the dramatic range in regional climatic diversity found on the island. However, this descriptive approach says little about the direct linkage of climate to physical and biological processes at the earth/atmospheric interface.

An integration of seasonal moisture supply (precipitation) with the evaporation and transpiration demands of the environment (determined primarily by solar energy inputs) provides an index of moisture surplus or deficit. Such indices can illuminate direct process/response relationships between climate and the terrestrial ecosystem.

In an initial survey of water balance climatology on the island of Hawaii, Mueller-Dombois (4) constructed a series of climate diagrams 21 stations (see Fig. 3). This type of diagram, popularized by Walter and Lieth (5), portrays seasonal curves of mean monthly temperature and precipitation. According to Muller-Dombois (6) an index of precipitation efficiency is built into the diagrams by making one degree of temperature (Celsius) equal to two millimeters of precipitation in the scaling of the two ordinates. This is based on the assumption that monthly potential evapotranspiration (in millimeters) is roughly equal to twice the mean monthly temperature (7). Wherever the precipitation curve drops below the temperature curve, a drought season is indicated. Thus the graph is transformed into a water balance diagram with the temperature curve interpreted as an index of potential evapotranspiration.

A serious problem inherent in this graphing technique is the tendency to approximate evapotranspiration with a simple linear function of air temperature (i.e., the 2:1 ratio). Chang (8,9) has reviewed the problems of temperature-based estimations of potential evapotranspiration and points out that solar radiation rather than temperature is the primary forcing function in the evaporative process. Temperature-based estimates of evaporation implicitly assume a strong correlation between temperature and solar radiation. In Hawaii, as a result of advection and the buffering effect of the surrounding marine environment, there is generally poor correlation between solar radiation and temperature. In Hilo, for example, the range in mean monthly air temperature is only 1.4 °C between June (24.2 °C) and December (22.8 °C). By contrast, the receipt of solar radiation in June (563 Ly; see solar radiation data for 1965 from Ref. 10) is more than twice that in December (263 Ly).

It is obvious from the above comparison that temperature-based estimates of evapotranspiration cannot be expected to portray realistically the seasonal fluctuations implied in the radiation data. However, in the absence of a dense network of solar radiation monitoring stations on the island, upon which more sophisticated spatial modeling of evapotranspiration might be based, it is necessary to revert to some form of temperature-derived estimation in a "first approximation" of water balance regimes.

Thornthwaite (11) has developed perhaps the most widely adopted method of estimating potential evapotranspiration. His empirical formula is based essentially on air temperature:

$$E = 1.6\ (10T/I)^a. \tag{1}$$

Potential evapotranspiration E is computed from mean monthly temperature T and an empiric "heat index" I, which itself is an exponential function of temperature; a is a constant.

To obtain mean monthly evapotranspiration, the values derived from eq. (1) are corrected for mean daylength and number of days in the month. The Thornthwaite equation, although subject to the general limitations of all temperature-based methods, might be expected to give better results than the Walter method in Hawaii,

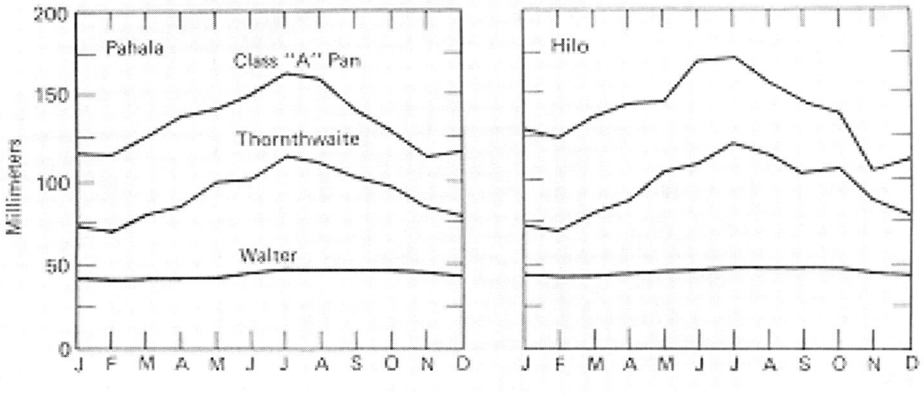

Figure 4. Monthly estimated potential evapotranspiration and measured pan evaporation for Hilo and Pahala data.

since potential evapotranspiration is expressed as an exponential rather than linear function of temperature.

In Fig. 4, monthly values of potential evapotranspiration derived by both the Walter and the Thornthwaite methods have been plotted along with class "A" pan-evaporation data for Hilo and Pahala. It is evident that Walter grossly underestimates pan evaporation (here assumed to be equal to potential evapotranspiration) and also fails to detect the seasonal rhythm apparent in the pan data. Thornthwaite also underestimates pan evaporation but does so in a fairly consistent manner and achieves a strong covariation with the pan data in seasonal rhythm. This suggests that the Thornthwaite method might be useful in Hawaii if a correction factor could be derived to compensate for the consistent underestimation exhibited in Fig. 4.

In Fig. 5, monthly values of the Thornthwaite potential evapotranspiration estimate are plotted against pan evaporation for Hilo and Pahala. With a regression coefficient of 0.844, approximately 71% of the observed variation in pan evaporation can be explained by variation in the Thornthwaite estimate. (The regression coefficient is significant at 0.01 level.) Potential evapotranspiration Y can thus be reasonably predicted from the Thornthwaite values X by the linear regression equation

$$Y = 42.8 + 1.016\,(X) \tag{2}$$

Before eq. (2) can be applied as a general (island-wide) correction factor for the Thornthwaite potential evapotranspiration estimate, it must be verified that the relationship established in Fig. 5 (for two lowland locations) is equally valid for mid- and high-altitude areas of the island.

There are no class "A" pan evaporation data for inland mountain areas of Hawaii with which the lowland-derived correction factor might be compared. However, Juvik and Clarke (12) have accumulated limited experimental data on mountain evaporation gradients in Hawaii Volcanoes National Park on the east flank of Mauna Loa.

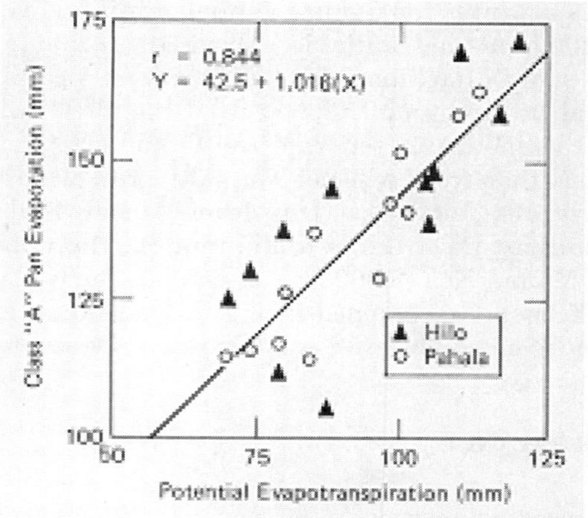

Figure 5. Relationship between monthly class "A" pan evaporation and estimated monthly potential evapotranspiration (Thornthwaite) for Hilo and Pahala data.

These data were obtained by using four constant-level pan evaporimeters (Fig. 6) situated along an attitudinal transect between sea level and 2,000 m.

In Fig. 7, measured mean daily evaporative rates (averages from 133 days of simultaneous readings taken from September 1974 through May 1975) are plotted against elevation. There is a clear linear decrease in evaporation over the attitudinal range surveyed (approximately 0.72 mm/day/1,000 m). Figure 7 also shows the corrected (eq. 2) Thornthwaite potential evapotranspiration values (mean of 9 months, September to May) derived from temperature-recording stations that occur near the evaporimeter transect. There is good agreement between the Thornthwaite and the evaporimeter values (differences range from 1% to 12%), largely because air temperature also decreases linearly with elevation. On the basis of the close agreement in Fig. 7, the corrected Thornthwaite estimate was considered acceptable to use for all areas of the island

Figure 6. Constant-level pan evaporimeter. a) field installation with inner tube reservoir; b) evaporimeter detail. Note foam insulation around evaporation pan.

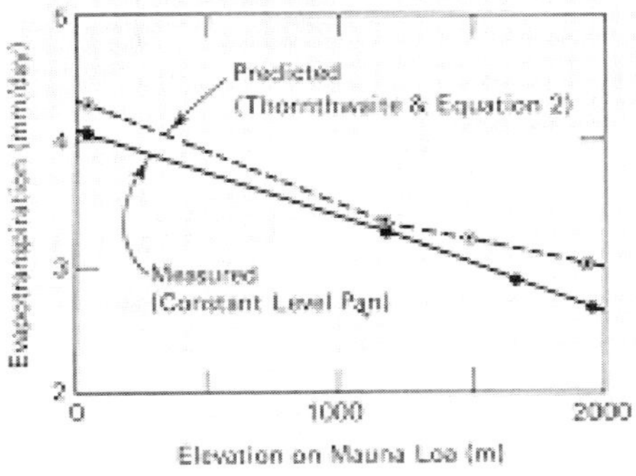

Figure 7. Relationship between measured and predicted evapotranspiration along an attitudinal transect in Hawaii Volcanoes National Park on the east flank of Mauna Loa (pan data from 12).

in the derivation of monthly and annual potential evapotranspiration from temperature data.

Corrected Thornthwaite estimates of monthly and annual potential evapotranspiration were computed from standard tables (13) and eq. (2), for 30 stations on the island of Hawaii. The evapotranspiration data were then integrated with monthly precipitation values to produce seasonal water balance diagrams (Fig. 8). Because some of the water surplus received in the wet season is stored as soil moisture for utilization during dry periods, the computation of seasonal water balance must incorporate a parameter describing the moisture storage capacity of the soil.

On the geologically youthful island of Hawaii, soils are not generally well developed except for limited areas where ash deposits occur in high-rainfall zones. Recent lava flows exhibiting little or no soil development cover substantial portions of the island. The average depth to bedrock for 75 different Hawaii island soil types and subtypes has been calculated at 0.89 m with only moderate variation (14), lending quantitative credence to this stated geological youthfulness of the island. Soil moisture storage capacity has not been well studied for most Hawaiian soil types. A value of 125 mm/m is the average moisture capacity for ten different soil types for which data are available. If this is assumed to be representative, then the average soil moisture storage capacity for all Hawaii island locations would be 111.2 mm (i.e., 0.89×125). In the construction of water balance diagrams for all island stations this value was rounded off to 100 mm so that standard moisture depletion tables could be employed in water balance calculations (13).

The 30 water balance diagrams constructed for the island depict both steep gradients and pronounced regional differences in seasonal moisture surplus and deficit. In Fig. 8 the difference between annual precipitation and potential evapotranspiration has been mapped in four zones:

1. Annual surplus exceeding 1,000 mm. This zone comprises 20% (2,100 km^2) of the island area and is restricted to the high-rainfall regions of windward Mauna Kea, Mauna Loa, and the summit area of Kohala. The annual moisture surplus in this zone

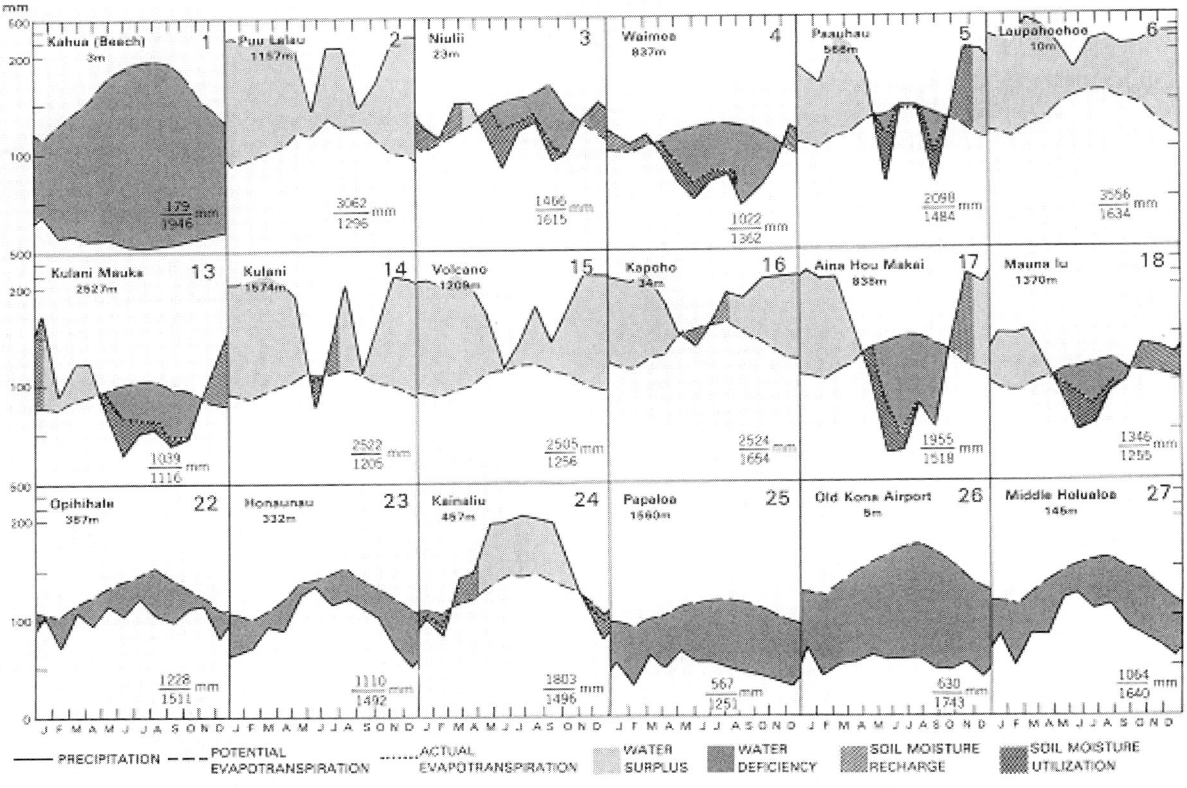

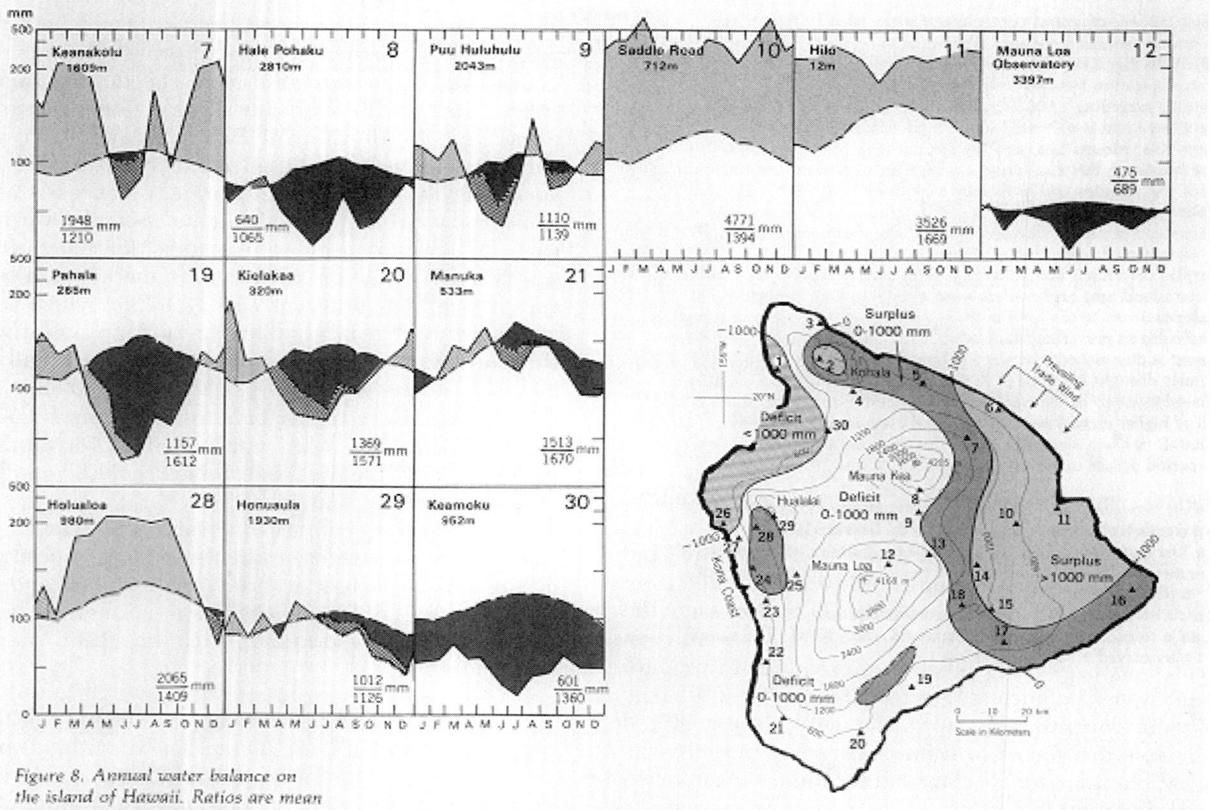

Figure 8. Annual water balance on the island of Hawaii. Ratios are mean annual precipitation over mean annual potential evapotranspiration in millimeters.

ranges as high as 3,377 mm (station 10) at middle elevations. All stations within this zone (stations 2, 6, 10, 11, 14, and 15) exhibit an absolute winter maximum in precipitation, and a secondary summer maximum also occurs at elevated stations where summer orographic precipitation is exaggerated (e.g., stations 2 and 14).

2. Annual surplus between 0 and 1,000 mm. This zone comprises 21% (2,200 km^2) of the island area and extends from middle to high elevations on the windward slopes down to sea level in those areas where slope aspect is not oriented perpendicular to prevailing trade wind flow, and thus the orographic rainfall component is diminished. For the windward stations there is typically a moderate summer drought (stations 5, 7, 9, 13, and 17) from 2 to 5 months long. The increased strength of the trade wind inversion in particular limits summer rainfall at higher elevations. The localized core area of high convectional rainfall in Kona also falls within this moisture zone. However, here the deficit period occurs in winter (stations 24, 28, and 29) and is not severe.
3. Annual deficit between 0 and 1,000 mm. This zone comprises 54% (5,600 km^2) of the island area and occupies a predominantly leeward location on Kohala, Mauna Kea and Mauna Loa. The drought period may be concentrated in either the summer (on the windward side) or the winter (on the Kona side) and is typically 6 to 12 months long.
4. Annual deficit exceeding 1,000 mm. This zone comprising 5% (550 km^2) of the island area is restricted to leeward Kohala and north Kona. The annual moisture deficit may exceed 1,900 mm (station 1).

SUMMARY

The Thornthwaite water balance diagrams and map demonstrate graphically the tremendous climatic diversity on the island of Hawaii. Although the Koppen map (Fig. 2) shows only a relatively small portion of the island to be arid or semi-arid (BWh and BSh), from the water balance analysis it is evident that nearly 60% (zones 3 and 4 above) of the island experiences an annual moisture deficit.

Acknowledgments

This research was supported in part by grants from the Hawaii Natural History Association and the U.S. Department of Interior, Office of Water Resources Research.

BIBLIOGRAPHY

1. Carter, D.B. and Mather, J.R. (1966). Climatic classification and environmental biology. *Publ. Climatol.* **19**(4): 305–390.
2. Taliaferro, W.J. (1959). *Rainfall of the Hawaiian Islands*. Hawaii Water Authority, p. 394.

3. State of Hawaii (1970). *An inventory of basic water resources dataÄisland of Hawaii*. Rep. R34, Dept. of Land and Natural Resources, Honolulu, Hawaii, p. 188.
4. Mueller-Dombois, D. (1966). Climate. Chap. IV. In: *Atlas for Bioecology Studies in Hawaii Volcanoes National Park*. Maxwell S. Doty and D. Mueller-Dombois. (Eds.). U.S. National Park Service, p. 507. (Republished as Hawaii Agr. Exp. Sta. Bull. 89, 1970.)
5. Walter, H. and Lieth, H. (1960). *Klimadiagram-Weltatlas*, Jena.
6. Mueller-Dombois, D. (1976). *The Major Vegetation Types and Ecological Zones in Hawaii Volcanoes National Park and Their Application in Park Management and Research*. In Proceedings of the First Conference in Natural Sciences, Hawaii Volcanoes National Park, edited by S. W. Smith, Department of Botany, University of Hawaii, Honolulu, 149–161.
7. Gaussen. (1954).
8. Chang, J.-H. (1959). An evaluation of the 1948 Thornthwaite classification. *Ann. Assoc. Amer. Geogr.* **49**(1): 24–30.
9. Chang, J.-H. (1968). *Climate and Agriculture*. Aldine Publishing Co., Chicago, p. 304.
10. Lof, G.O.C., Duffie, J.A., and Smith, C.O. (1966). *World Distribution of Solar Radiation*. Rep. 21, Solar Energy Laboratory, Univ. of Wis., p. 59. (plus maps).
11. Thornthwaite, C.W. (1948). An approach toward a rational classification of climate. *Geogr. Rev.* **38**: 55–94.
12. Juvik, J.O. and Clarke, G.G. (1976). *Topoclimatic Gradients in Hawaii Volcanoes National Park (abstract)*. In Proceedings of the First Conference in Natural Sciences, Hawaii Volcanoes National Park, edited by S. W. Smith, Department of Botany, University of Hawaii, Honolulu, p. 113.
13. Thornthwaite, C.W., and J.R. Mather (1957). Instructions and tables for computing potential evapotranspiration and the water balance. *Publ. Climatol.* **10**(3): 1–311.
14. Sato, H.H. et al. (1973). *Soil Survey of Island of Hawaii, State of Hawaii, U.S.D.A.* Soil Conservation Service, p. 115. (plus maps).

READING LIST

Juvik, J.O. and Perreira, D.J. (1974). Fog interception on Mauna Loa, Hawaii. *Proc. Assoc. Amer. Geogr.* **6**: 22–25.

Price, S. (1973). Climate. In: *Atlas of Hawaii*. R.W. Armstrong. (Ed.). University of Hawaii Press, Honolulu, pp. 53–60.

HEAT OF VAPORIZATION

NARAINE PERSAUD
Virginia Polytechnic Institute
and State University
Blacksburg, Virginia

HEAT OF VAPORIZATION

The heat of vaporization (here denoted as L_v) is defined as the heat added (or given off) when unit mass undergoes isobaric phase transformation in any closed two-phase, one-component liquid/vapor system. In engineering and meteorology, L_v is used in a restricted sense to mean the heat of vaporization of the two-phase liquid water/water vapor system. Although much of the subsequent discussion focuses on L_v, for this specific system as an example, the concepts covered are universally applicable to all fluid/vapor systems.

As illustrated below (Fig. 1), for the liquid water/water vapor system, L_v represents the heat gained when unit mass of water in the system evaporates in the isobaric phase transformation H_2O (liquid) → H_2O (vapor). For the reverse phase change H_2O (vapor) → H_2O (liquid), i.e., condensation, L_v is lost from the system. This seemingly simple phase transition H_2O (liquid) ↔ H_2O (vapor) is the fundamental driving process of the earth's hydrological cycle, the working principle of the steam engine that ushered humanity into the industrial revolution along with its (often negative) social and environmental pollution consequences, and the physical mechanism that maintains the body temperature of plants and warm-blooded animals.

In general, the state of any closed two-phase, one-component system is defined by the state variables temperature (T in °K), saturation vapor pressure (P in Pascal), and volume (V in m^3). The behavior of any such system (generally termed as PVT systems) is usually represented as a family of experimental constant temperature curves (isotherms) on a P-V coordinate plane called an Amagat–Andrews diagram. The general shape of these experimental isotherms is illustrated below (Fig. 2).

For the liquid water/water vapor system, the liquid and vapor phases coexist in equilibrium only at P-V coordinates between 2 and 3 along an isotherm, provided

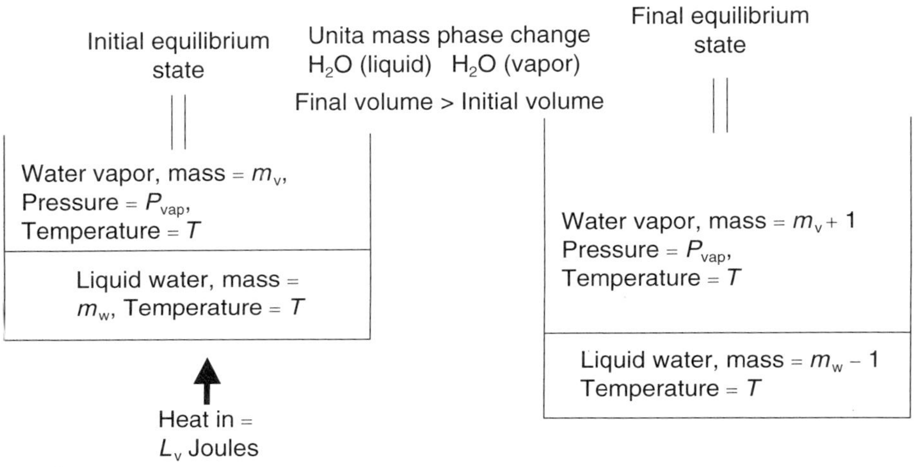

Figure 1. Liquid Water/Water Vapor System.

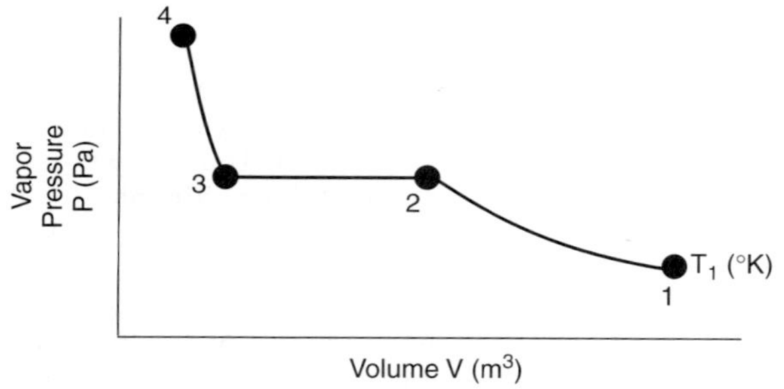

Figure 2. Amagat–Andrews Diagram.

that T_1 is above the triple point temperature of water (0.01 °C, i.e., the temperature at which ice, liquid water, and water vapor can coexist in equilibrium), and below the critical temperature (374 °C, i.e., the temperature above which it is impossible to produce condensation by increasing the pressure). Between 2 and 1, the system can exist as vapor only, and as liquid between 3 and 4. Thus the liquid-vapor phase transition at a given temperature can only take place at constant pressure or vice-versa. Consequently, as shown in Fig. 1, isobaric liquid vaporization and condensation is necessarily an isothermal process, implying that the triple-point saturation vapor pressure is fixed (it is 611 Pa), and so is the saturation vapor pressure at the critical temperature (it is 2.21×10^7 Pa = 218.2 atm). Similar isotherms and parameters exist for all liquid/vapor systems.

This observed behavior of closed two-phase, one-component systems is of course predicted by Gibb's Phase Rule (1), namely $F + N = C + 2$, where F = degrees of freedom, i.e., the smallest number of intensive variables (such as pressure, temperature, concentration of components in each phase) that must be specified to completely describe the state of the system; N = number of phases, i.e., distinct subsystems of uniform chemical composition and physical properties; and C = the number of components, i.e., the number of independent chemical constituents meaning those constituents whose concentration can be varied independently in the different phases. In a liquid/vapor system $P = 2$ and $C = 1$, and therefore $F = 1$, implying only one intensive variable is needed to specify the state of the system. Therefore, temperature and pressure cannot be fixed independently. For the liquid water/water vapor system, this means physically that at a given temperature between the triple point and critical temperature, water vapor will evaporate or condense to achieve the equilibrium saturation vapor pressure as would be evidenced in a complete Amagat–Andrews diagram for water (2,3).

The earth's atmosphere and oceans can be considered as a vast closed two-phase liquid water/moist air system. Consequently, for most practical engineering and meteorological applications, one is interested in the heat of vaporization of the liquid water/moist air system rather than a pure liquid water/water vapor system. Fortunately, the presence of the other gases (collectively called dry air) in the liquid water/moist air system has negligible effect on the saturation vapor pressure. The reason is that the dry air component in the liquid water/moist air system remains unchanged and is always in the gaseous state during phase transition at temperatures and pressures of practical interest. Therefore, it can be considered as a closed subsystem as opposed to the open liquid water and water vapor subsystems. Consequently, results obtained from an analysis of the thermodynamics of the pure system are applicable to the natural liquid water/moist air system.

Energy conservation required under the first law of thermodynamics implies that heat (Q) exchanged reversibly with the surroundings between equilibrium states of any closed two-phase PVT system is consumed by any internal energy change (ΔU) of the liquid and vapor phases associated with the mass change from one phase to the other, and any mechanical work ($\pm P\Delta V$) realized as the volume of the system increases (positive work) or decreases (negative work). Stated mathematically, $Q = \Delta U + P\Delta V$, or in a differential form, $\delta Q = dU + pdV$. Here, P is the saturation vapor pressure (P_{vap} in Fig. 1). As entropy (S) is defined as Q/T, then $\delta Q = TdS$. The first law can therefore be restated in terms of exact differentials as $TdS = dU + PdV$. Dividing by dV at constant T and rearranging gives $dU/dV = T(dS/dV) - P$. Using the Maxwell relation $(\partial S/\partial V)_T = (\partial P/\partial T)_V$ (1), dS/dV can be replaced (for fixed T) by dP/dT. The equation becomes $dU/dV = T(dP/dT) - P$.

At a given pressure and temperature, the internal energy of the system (U in Joules) can be partitioned as $m_w u_w + m_{\text{vap}} u_{\text{vap}}$, where m_w, m_{vap} and u_w, u_{vap} represent the masses and the specific internal energies (internal energy per unit mass in J kg^{-1}) of the water and water vapor in the system. Similarly, the volume (V in m^3) of the system can be partitioned as $m_w v_w + m_{\text{vap}} v_{\text{vap}}$, where v_w, v_{vap} represent the specific volume (volume per unit mass in J kg^{-1}) of the water and water vapor in the system. If, as illustrated in Fig. 1, the system internal energy changes by ΔU from U to $U + \Delta U$ as a result of L_v Joules of heat absorption to convert unit mass of water to water vapor, then $U + \Delta U = (m_w - 1)u_w + (m_w + 1)u_{\text{vap}}$, and therefore $\Delta U = (u_w - u_{\text{vap}})$. Similar reasoning shows that, if the volume changes from V to $V + \Delta V$ in the process, then $\Delta V = (v_{\text{vap}} - v_w)$. The mechanical work because of volume change is $P\Delta V$, where P (the saturation vapor pressure) is a constant at a fixed temperature. Therefore, the heat absorbed (or released) by the system for isobaric phase transition of unit mass in the liquid water/water vapor system ($L_v$by definition) $= \Delta U + P\Delta V$. Dividing by ΔV gives $\Delta U/\Delta V = (L_v/\Delta V) - P$. Substituting $\Delta V = (v_{\text{vap}} - v_w)$ gives $\Delta U/\Delta V = [L_v/(v_{\text{vap}} - v_w)] - P$, or in a differential form, $dU/dV = [L_v/(v_{\text{vap}} - v_w)] - P$. [It should be noted that because enthalpy (H) is defined as $H = U + PV$, then $\Delta U + P\Delta V = \Delta H$, and therefore L_v is the same as the specific enthalpy change ($\Delta h = \Delta H$ per unit mass) for phase transition of unit mass in the liquid water/water vapor system.]

Combining the results for dU/dV from the two previous paragraphs gives $T(dP/dT) - P = [L_v/(v_{\text{vap}} - v_w)] - P$, and therefore $T(dP/dT) = L_v/(v_{\text{vap}} - v_w)$, which can be rearranged to obtain the general forms of the Clapeyron (Emile Clapeyron, 1799–1864) equation $dP/dT = L_v/[T(v_{\text{vap}} - v_w)] = \Delta h/[T(v_{\text{vap}} - v_w)]$ or $L_v = [T(v_{\text{vap}} - v_w)]dP/dT$.

The Clapeyron equation can be used to obtain L_v at a given temperature T for any liquid provided one can obtain values of $(v_{vap} - v_w)$ and an accurate representation of dP/dT. Values of $(v_{vap} - v_w)$ can be obtained from tabulated measurements (3). Alternatively, because $v_{vap} \gg v_w$ at the low pressures, $(v_{vap} - v_w)$ can be taken as equal to v_{vap}. Assuming further that at low pressures water vapor behavior closely approximates that of an ideal gas, then $v_{vap} = RT/P$, where R is the specific gas constant for water $= 8.314/0.018 = 461.9\ \mathrm{J\,kg^{-1}\,°K^{-1}}$. The Clapeyron equation becomes $L_v = (RT^2/P)dP/dT$, and this form is referred to as the Clausius–Clapeyron equation (Rudolph Clausius, 1822–1888).

Unfortunately, no single function has been shown to adequately represent vapor pressure data for various liquid/vapor systems over wide ranges of T. For the liquid water/water vapor system, values of dP/dT for a given value of T can be obtained by finite differencing of tabulated values. As an example, consider using the above equation to estimate L_v for water at human body temperature of 37 °C. Tabulated values are $v_w = 1.007\ \mathrm{cm^3 g^{-1}} = 0.01007\ \mathrm{m^3\,kg^{-1}}$, $v_{vap} = 22,760\ \mathrm{L\,kg^{-1}}, = 22.760\ \mathrm{m^3\,kg^{-1}}$, $P = 5.940$ kPa at 36 °C and 6.624 kPa at 38 °C, which gives $(v_{vap} - v_w) = 22.760 - 0.01007 = 22.75\ \mathrm{m^3\,kg^{-1}}$. By finite differencing, $dP/dT \approx \Delta P/\Delta T = (6.624 - 5.940)/(38 - 36) = 0.342\ \mathrm{kPa\ °C^{-1}} = 0.342\ \mathrm{kPa\ °K^{-1}} = 342\ \mathrm{Pa\ °K^{-1}}$ (a temperature difference of 1 °C is the same as a difference of 1°K). At $T = 37 + 273.16 = 310.16$°K, $L_v = 310.16°\mathrm{K}\ (22.75\ \mathrm{m^3\,kg^{-1}}) \times 342\ \mathrm{Pa\ °}K^{-1} = 2399\ \mathrm{kJ\,kg^{-1}}$. The actual value (tabulated as Δh) is $2414\ \mathrm{kJ\,kg^{-1}}$, an error of <1%.

Sometimes it is more efficient to have semianalytical expressions for $P(T)$ and hence dP/dT. Examples of such expressions for $P(T)$ are the Antoine equation, $\ln P = A + [B/(T + C)]$, and the Rankine equation, $\ln P = A + B/T + C \ln T$, where T is (usually) in °K, and A, B, C are empirical parameters obtained by fitting these equations to experimental data on phase equilibrium for a specific liquid/vapor system for various temperature ranges. Antoine's equation coefficients for the vapor pressure of quite a number of chemical compounds are available online in the NIST Chemistry Webbook (3). This reference lists Antoine coefficients for liquid water based on the vapor pressure data of Bridgeman and Aldrich (4) over several temperature ranges as:

Antoine Equation Coefficient	Temperature Range (°K)			
	273 to 303	304 to 333	334 to 363	344 to 373
A	5.40221	5.20389	5.0768	5.08354
B	1838.675	1733.926	1659.793	1663.125
C	−31.737	−39.485	−45.854	−45.622

These coefficients apply to Antoine's equation in the form $\log_{10} P = A - [B/(T + C)]$ with P in bar and T in °K. At 37 °C, this equation gives $P = 6.280$ kPa compared with the measured tabulated value of 6.274 kPa. The derivative of Antoine equation in this case is $dP/dT = P[2.3026B/(T + C)^2]$, which gives dP/dT at 37 °C $= 6.280 \times 0.0545 = 0.342\ \mathrm{kPa\ °K^{-1}}$, which is the same as obtained by finite differencing of tabulated values.

For the liquid water/water vapor system, the equation for $P(T)$ developed by Richards (5) covers a wide temperature range and gives estimates for dP/dT close to those obtained by finite differencing of tabulated data. This equation is:

$$P(T) = ae^{f(T_r)}$$

where P is in millibar (mb), e is the base of natural logarithms, $T_r = 1 - \dfrac{373.15}{T}$, T is the temperature in °K, and $a = 1013.25$. The function $f(T_r) = 13.3185T_r - 1.9760\ (T_r)^2 - 0.6445\ (T_r)^3 - 0.1299\ (T_r)^4$. The derivative (with respect to T) is:

$$\frac{dP}{dT} = P(\frac{373.15}{T^2})(\frac{df(T_r)}{dT_r})$$

At 37 °C, $T_r = -0.2031$, the calculated value of $P = 62.78$, $mb = 6.278$ kPa compared with the tabulated measured value of 6.274 kPa. Using $P = 62.78$, mb gives $dP/dT = 62.78 \times 0.00388 \times \{13.3185 - (2 \times 1.9760 \times -0.2031) - [3 \times 0.6445 \times (-0.2031)^2] - [4 \times 0.1299 \times (-0.2031)^3]\} = 0.2434 \times \{13.3185 + 0.8026 - 0.0797 + 0.0044\} = 0.2434 \times 14.046 = 3.419\ \mathrm{mb\ °K^{-1}} = 341.9\ \mathrm{Pa\ °K^{-1}}$, practically the same as the value of $342\ \mathrm{Pa\ °K^{-1}}$ obtained above from finite differencing of tabulated data.

Applying the Clausius–Clapeyron equation with $P = 6.280$ kPa at 37 °C by the Antoine equation, gives $L_v = 461.9 \times [(310.16)^2/6.280] \times 0.342 = 2,420\ \mathrm{kJ\,kg^{-1}}$ compared with the actual value of $2414\ \mathrm{kJ\,kg^{-1}}$.

BIBLIOGRAPHY

1. Cengel, Y.A. and Boles, M.A. (2002). *Thermodynamics: An Engineering Approach*, 4th Edn. McGraw-Hill Series in Mechanical Engineering, Boston, MA.
2. Wagner, W. and Pruss, A. (2002). The IAPWS formulation 1995 for the thermodynamic properties of ordinary water substance for general and scientific use. *J. Phys. Chem. Ref. Data* **31**: 387–535.
3. Lemmon, E.W., McLinden, M.O., and Friend, D.G. (2003). Thermophysical properties of fluid systems. In: *NIST Chemistry WebBook, NIST Standard Reference Database Number 69*. P.J. Linstrom and W.G. Mallard (Eds.). National Institute of Standards and Technology, Gaithersburg, MD. Available: http://webbook.nist.gov.
4. Bridgeman, O.C. and Aldrich, E.W. (1964). Vapor pressure tables for water. *J. Heat Transfer* **86**: 279–286.
5. Richards, J.M. (1971). Simple expression for the saturation vapor pressure of water in the range −50 to 140 °C. *Brit. J. Appl. Phys.* **4**: L15–L18.

HYDROLOGIC HISTORY, PROBLEMS, AND PERSPECTIVES

Jason A. Hubbart
University of Idaho
Moscow, Idaho

HYDROLOGIC HISTORY

Since humans first settled along the banks of lakes and rivers, there has been a keen interest in the proper management of fresh water resources both as a necessity for life as well as to avoid potential health hazards. Along the Indus in Pakistan, the Tigris and Euphrates in Mesopotamia, the Hwang Ho in China, and the Nile in Egypt, the first "hydraulic engineers" created canals, levees, dams, subsurface water conduits, and wells as early as 5000–6000 years ago. Hydrologic information became vital to these early civilizations. The flow rates and yields of rivers were monitored by the Egyptians as early as 3800 years ago, and rainfall measuring instruments were first used approximately 2400 years ago by Kautilya of India (1).

The idea that a global hydrologic cycle actually exists dates back at least 3000 years when early Greek philosophers such as Thales, Anaxagoras, Herodotus, Hippocrates, Plato, and Aristotle conceptualized the basic ideas governing the process (2). Many initial ideas established by the Greeks about the hydrologic cycle were quite reasonable. However, many of the initial mechanisms concerning the routes by which water returned from the sea and entered rivers were devoid of as much logic. Despite apparent gaps in hydrologic mechanisms, the Romans developed aqueduct systems reflecting an extensive practical understanding of hydrology and hydraulics and did so using the basic hydrologic ideas established and passed along by the Greeks (3). During the Renaissance, Leonardo da Vinci (1500) in France proclaimed, on the basis of field observations, that the waters in rivers come from precipitation. During that time, any unrealistic mechanisms proclaimed by the Greek philosophers concerning the hydrologic cycle were either refuted or modified (4,5). In the seventeenth century, the modern scientific approach to studying the hydrologic cycle was assumed by the Frenchmen Pierre Perault and Edme Marriotte. By the 1670s and 1680s, they had published data and calculations to support the contention that precipitation was the precursor to streamflow. By 1700, Edmun Halley, an English scientist, added to the work of Perault and Marriotte by estimating the quantity of water involved in the hydrologic cycle of the Mediterranean Sea and surrounding lands.

Substantial progress was made during the eighteenth century in applications of mathematics, fluid mechanics, and hydraulics by scientists like Pitot, Bernoulli, Euler, Chezy, and other European professionals. The term "hydrology" arrived at its current meaning around 1750, and by 1800, the work of English physicist and chemist John Dalton solidified the current understanding of the global hydrologic cycle (2).

Until the 1800s, the physical processes governing groundwater flow had confounded scientists and created barriers to understanding the hydrologic cycle. These barriers were eliminated in 1856, when the French engineer, Henry Darcy, introduced his law describing flow through porous media. Other advances in the hydrologic sciences were made throughout the 1800s. Poiseuille, DuPuit, DuBoys, Stokes, Manning, Reynolds, and others contributed substantially to fluid mechanics, hydraulics, and sediment transport during this period of time (3). Also during the 1800s, literary publications began to surface, increasingly frequently in the last half of the century. Many works examined relationships between precipitation and stream flow out of necessity for engineers designing bridges and other structures. During this time, the close association between hydrology and civil engineering was established. Daniel Mead published the first English-language text in hydrology in 1904, and Adolf Meyer followed with his text in 1919. Both texts were written for civil engineers (1). The association of hydrology and civil engineering established during this time, it has been argued, both enhanced as well as possibly inhibited the development of hydrology as a science (6).

The first half of the twentieth century saw great advancements in the hydrologic sciences starting with the addition of the Section of Scientific Hydrology in the International Union of Geodesy and Geophysics in 1922. This was followed by the addition of the Hydrology Section of the American Geophysical Union in 1930. These were the first formal recognitions of the scientific status of hydrology. Many individuals contributed substantially in their areas of hydrologic expertise during the early and middle decades of the twentieth century, including A. Hazen, E. J. Gumbel, H. E. Hurst, and W. B. Langbein in statistical applications of hydrologic data; O. E. Meinzer, C. V. Theis, C. S. Slichter, and M. K. Hubbert who contributed to the development of theoretical and practical aspects of groundwater hydraulics; L. Prandtl, T. Von Karman, H. Rouse, V. T. Chow, G. K. Gilbert, and H. A. Einstein in sediment transport and stream hydraulics; R. E. Horton and L. B. Leopold who contributed greatly to runoff processes and quantitative geomorphology; W. Thornthwaite and H. E. Penman in furthering the understanding of hydroclimatologic processes and modeling evapotranspiration; and A. Wolman and R. S. Garrels who contributed greatly to understanding and modeling water quality. It was not until the 1960s that detailed field studies attempting to understand the physical processes by which water enters streams began to emerge (3).

At the emergence of the twenty-first century, many new breakthroughs in the hydrologic sciences are imminent. In the forthcoming years, breakthroughs will describe the relationships between hydrologic regimes to current and future climate change and the effects of hydrologic processes on landform development. New findings will also include modeling regional evapotranspiration rates and geomorphologic water transport (1).

Population growth and land use changes have resulted in sizable impacts on surface water and groundwater resources. Increasing demands for fresh water has led to water stress in many parts of the world. Increasing quantities of water are used for agriculture, and increasing water quantities withdrawn from subsurface water bodies and large rivers have had significant influences on the water cycle. Population growth and land use changes have led to polluting of surface waters and groundwater resources as well as soil deterioration through erosion, salinization, and of course pollution. The ability to meet challenges of increased demand and to propose mitigating strategies will require intensive studies looking into how population growth and associated land use change and water demands influence the local, regional, and global water cycles.

HYDROLOGIC PROBLEMS AND PERSPECTIVES

Many specific scientific problems or challenges related to hydrologic study exist. These include assessing the impacts that growing populations have on water demand, land use change on groundwater reserves, and river discharges. Once successfully concluded, these assessments must be used as tools for predicting the extent of expected water stress in the coming decades due to changes in the supply and demand of water resources. Scientists must strive to understand the extent to which increased evaporation and transpiration through increased withdrawal of groundwater and surface water will lead to changes in local and global water cycles. There is a growing need for improved understanding of pollutant related flows in groundwaters and the self-cleaning capacities of aquifers and rivers related to different pollutants. Thus, further work is needed to describe the impacts of population growth and land use change on the quality of groundwater reserves, soils, and rivers and also what the effects may be on the natural ecosystems that depend on them.

Studies will be needed to understand landscape processes better, including hill slope forms, channel networks, and the processes responsible for transporting water, sediments, and pollutants. Hydrologic research in past decades has focused attention mainly on modeling the dynamic response of channel networks. Progress in understanding catchment scale flow will require that relevant hill slope flow also be carefully studied. This will likely require collaborative research approaches among hydrologists, geomorphologists, geologists, pedologists, and others to identify scaling relationships in flow domain structures.

There remains a need for improved understanding of the interactions between climate change and the hydrologic cycle. Further understanding and documentation is required to predict apparent changes that have led to an intensification of the hydrologic cycle due to climate change. Critical questions need to be asked concerning current changes observed in water availability and whether critical regions can be identified where the interactions between water availability and human demand become problematic. Conducting this type of research necessitates better understanding of past climates and paleohydrologic behavior. At present, the ability to predict changes in the fluxes of the water cycle under various scenarios of climate change is poor. Therefore, it is desirable to create improved methods of portraying the interactions between water systems and natural and human induced environments.

Due to the inherently complex nature of hydrologic systems, the characterization and subsequent modeling of these systems is a continual challenge to scientists. There are a number of unresolved issues that continue to impede the ability of scientists to analyze and predict behavioral changes of hydrologic systems. One of the most predominant features of hydrologic systems is the spatial heterogeneities and temporal variabilities that occur perpetually and persistently on multiple scales. Modeling physical processes encumbered by such diversity is a constant challenge. For these reasons, theoretical upscaling and downscaling attempts to develop quantitative links among process descriptions at various hydrologic scales has been so arduous. New or modified algorithms need to be devised, which will govern observed variabilities and define constitutive relations for each system. In the past, many empirical observations made on simple systems have been often erroneously applied to more complicated systems. Hydrologic systems are quite complex, and equations valid at one scale are not necessarily valid on higher scales, so one may question the applicability of theories that are originally developed for a simple system and vice versa.

Moreover, hydrologic processes are nonlinear and there tend to be strong couplings among them. As a result, the effect of one process is influenced by the occurrence of other processes, and a negligible effect of one process can lead to significant impacts on another. Clearly, there is a need to develop rigorous physical theories to describe the synergistic relationships of coupled nonlinear hydrologic systems on various scales. Intense study on a small scale is not leading to reliable working policy for addressing watershed processes critical to broad scale water management. As such, work needs to be undertaken designed specifically to improve our ability to model and predict water flows and movement. Finally, these models must also be appropriate and applicable to modeling and accurately predicting of flows and outputs in ungauged basins.

Hydroelectric dams, important for generating electricity, have transformed large fluvial systems into chained lakes. These large waterworks have generated a number of substantial environmental disruptions, including the erection of what are often unsurpassable obstacles for fish, explosive growth of floating aquatic plants, and eutrophication of dam reservoirs. The large development projects that connect watersheds and improve continental navigational networks also affect wetlands. Timber harvesting can likewise have negative effects on the production and regulation of water flows because increased soil erosion rates may increase the amount of suspended sediments, which can affect the quality of the water resources and the functioning of dams and reservoirs. On this basis, studies are needed to clarify the most appropriate management practices for harvesting timber.

Water resource management faces many other monumental problems. Academia needs to understand the current crisis completely to become innovative in producing new technologies and in advocating arrangements to satisfy demand. The sciences must be able to integrate and upscale. The field of ecohydrology (ecology and hydrology) is a testimony to the potential success of integrating sciences and should serve as a model for future interdisciplinary cooperatives. However, from a historical context, academia is easily drawn apart to the point that each discipline often fails to identify with statements of findings produced by other disciplines. The public is poorly served by divided or polarized science. Hydrologists are not unique integrators amongst scientists, but interdisciplinary research is required in hydrology to solve the current water-related problems. The goal of strengthening the identity of hydrology as a discipline is commensurate with hydrology's desire to serve the needs of society better. For example, the general public is much more aware of water problems (too much, too little, too dirty) than of the discipline of hydrology and how it relates to water problems. However, people agree that objectives intended to strengthen the hydrologic sciences will result in hydrology that is better equipped to serve the needs of society. These are therefore compatible ideals, well suited to serve both coalitions.

Hydrologic resources management has evolved through incremental refinements to its current point by using information on precipitation, streamflows, water demands, and other processes based to a high degree on a measurement technology that is often archaic in design. Now, scientific advances offer technologies that can be used to add much more information and provide much greater detail. The academic community is often the source of new science and of new technology designed to organize and disseminate information better. Academia is also often the source of courses and literature intended to inform people and make them comfortable with information resources. The world has a fixed quantity of water to use to satisfy a widening diversity and deepening intensity of human and environmental needs. Therefore, academic must make a concerted effort to inform the general public of the limitations of hydrologic resources.

A system for water resources management must also be sustainable to succeed. It is a major challenge to preserve options in a dynamic world where changes are caused by anthropogenic activities, environmental evolution, and geologic processes. These are the challenges that await educators, scientists, and water resource managers in the future.

Ultimately, a management system that addresses periodic severe crises may be less sustainable than that designed to mitigate long-term contingencies. The greatest difficulty lies in protecting future options as social preferences change. Regardless of origin, scientists, educators, politicians, and lawmakers must all cooperate to meet these challenges.

Hydrology and water use are obviously considered from a variety of perspectives, including social, economic, legal, scientific, and environmental, to determine how differing viewpoints affect the quality and quantity of water supplies. Researchers study pollution carried by water, which includes oceans, rivers, streams, rain, snow, and ice, and devise methods to clean and control it. Some study weather related problems such as flood forecasting, drought management, acid rain, and global warming. Others manage water resources so that the goals of all water users are achieved efficiently while protecting the environment. To extenuate and manage water quality issues adequately, they must be dealt with from all perspectives. Some water quality problems that can be common to entire regions include toxic contamination from industry, waste disposal, and eutrophication from human sewage. Bacterial pollution of water supplies is a perpetual dilemma that has adverse effects on human health. The major concerns centered on high bacterial and organic loads range from poor quality drinking water, eutrophication, and disappearance of aquatic life to food contamination and the prevalence of waterborne diseases.

Rivers can be polluted from a number of different sources. The largest contributors include industrial and urban wastewaters from large cities, wastewater from mining industries, and agricultural runoff. As a result, many river water resources today are chemically and biologically contaminated. Large quantities of agricultural contaminants are likewise disposed of in streams flowing to the sea, where there has been found clear evidence of elevated levels of phosphorous, nitrates, potassium, pesticides such as dichlorodiphenyltrichloroethane (DDT), and highly organic effluents. These elevated levels of pollutant concentrations caused by the indiscriminate discharge of highly pollutant-loaded effluents into waterbodies are also seen in deep underground aquifers, which are often contaminated by the same negligence.

Experiences during the past century have contained many useful lessons that are now being used to shape water management practices to address effectively the threats of water shortages and pollution that are growing worldwide. A half-century ago, water resources planning focused on building facilities to capture and deliver water for beneficial uses and to contain floods to reduce damage. Planners were certain that full resource development would foster economic growth and serve broad social needs. Governments built projects that proved financially costly, environmentally destructive, and politically divisive. Environmental harm increased as projects drained rivers and wetlands and converted forests, meadows, and deserts into fields and cities. Political differences deepened as jurisdictions sought to develop their own water resources with little regard for the needs of neighbors. Disputes mounted among towns, states, and nations. As water resources were more fully developed, projects became larger and more costly, and their adverse environmental impacts became more severe. Fiscal and environmental concerns caused the political process to halt construction as society looked for alternative approaches to basic human needs. Now, water withdrawals are approaching the upper limit to what nature can supply, and freshwater resources are being reduced by pollution and threatened by climate change.

Throughout history, humans have been concerned with the proper management of hydrologic resources. This task has been less than facile simply because water availability and population induced threats vary greatly among locations and over time. There is a need to think more creatively to gain deeper understanding of the inherent heterogeneities and diversities within hydrologic systems. Proper management of water resources will need to address the problems and challenges of growing regional and global populations with complex societies but will also need to sustain the needs of the earth's natural ecosystems.

BIBLIOGRAPHY

1. Eagleson, P.E. et al. (1991). *Opportunities in the Hydrologic Sciences*. National Academy Press, Washington, DC, pp. 39–45.
2. Nace, R.L. (1974). *General Evolution of the Concept of the Hydrological Cycle. Three Centuries of Scientific Hydrology*. UNESCO-World Meteorological Organization-International Association of Hydrological Sciences, Paris, pp. 40–48.
3. Dingman, S.L. (1994). *Physical Hydrology*, 1st Edn. Prentice-Hall, Upper Saddle River, NJ, pp. 4–5.
4. Adams, F.D. (1938). *The Birth and Development of the Geological Sciences*. Williams and Wilkins, Baltimore, MD, p. 8.
5. Biswas, A.K. (1970). *History of Hydrology*. North-Holland, Amsterdam, pp. 83–84.
6. Klemes, V. (1986). Dilettantism in hydrology: Transition of destiny. *Water Resour. Res.* **22**: 177–188.

HUMIDITY—ABSOLUTE

NARAINE PERSAUD
Virginia Polytechnic Institute and State University
Blacksburg, Virginia

HUMIDITY—ABSOLUTE

Absolute humidity is one of many measures to quantify the water vapor content of the atmosphere. It is defined as the mass (m_v) of water vapor per unit volume of moist air. It can be denoted as ρ_v with SI units of g (vapor) m^{-3} moist air or in corresponding engineering units as lb ft^{-3} or (more commonly) as grain ft^{-3}. Conversion factors are 1 lb = 453.6 g = 7000 grain and 1 m^3 = 35.31 ft^3. Other terms in use for this quantity are vapor concentration and density of water vapor in a moist air mixture. Absolute humidity is quite often incorrectly used to mean specific humidity, which is the mass of water vapor per unit mass of moist air, and sometimes to mean the mixing ratio, which is the mass of water vapor per unit mass of dry air. Absolute humidity, specific humidity, and mixing ratio are important in the study of evaporation from water bodies, crop transpiration, soil-water balances, and in ventilation and air conditioning engineering applications. These interrelated measures are rarely measured directly.

Absolute humidity can be estimated indirectly in terms of psychrometric measures of atmospheric water vapor content and known physical parameters. The key assumption in making such estimates is that the mixture of gases in moist atmospheric air behaves ideally. With this assumption, a volume (Vm^3) of moist air at a given pressure (P in Pa; usually at $P = 1.013 \times 10^5 Pa = 1$ atm) and temperature T°K obeys the relationship PV = nRT, where n is the number of moles of mixture and R is the ideal gas constant = 8.314 J mol^{-1} $°K^{-1}$. The number of moles of mixture can be taken as the mass of mixture/mean molecular weight (M_{wm}) of the mixture, which gives PV = mRT/$M_{wm} = mR * T$, where $R* = R/M_{wm}$ and is called the specific ideal gas constant. Dividing both sides by V gives $P = mR * T/V = \rho_m R * T$, where $\rho_m = m/V$ is the density of the moist air.

The partial pressure P_i of the ith constituent in a volume V of an ideal gas mixture at temperature T can be defined as the pressure it would exert if it alone occupied the volume V, implying that $P_i V = n_i RT$. As for the entire mixture, the total pressure $PV = nRT$, which implies (by summation over i) that $P = \sum P_i$ and (by division) that $P_i/P = n_i/n$. $P = \sum P_i$ is known as Dalton's Law of Partial Pressures. Similarly, the partial volume V_i of the ith constituent T can be defined as the volume it would occupy if exerted and if it alone existed at the total pressure of the mixture. This definition implies $PV_i = n_i RT$, and therefore $V = \sum V_i$ (know as Amagat's Law) and $V_i/V = n_i/n$. Combinations of these results show that $P_i/P = V_i/V = n_i/n$. As the water vapor content by volume in moist air rarely exceeds 0.04, this result implies that its partial pressure rarely exceeds 0.04 atm.

Application of the foregoing ideal relations to estimate absolute humidity from psychrometric meterological observation would be straightforward if the composition of the atmospheric air were constant. However, this is not so, because the water vapor content by volume (i.e., volume water vapor per unit volume of atmospheric air) is the most variable constituent of the atmosphere. It can vary from as little as $2 \times 10^{-6}\%$ (at high altitudes, in extremely cold conditions such as Antartica, and in deserts) to between 4 and 5% (in warm equatorial regions close to large waterbodies). On the other hand, the other constituents of atmospheric air are relatively constant. Therefore, a quantity measured on a volume V of moist air can be regarded as a combination of the quantity measured on the water vapor (V_v) and dry air (V_d) volume components. Thus, the mass m_m of moist air = mass water vapor m_v + mass of dry air m_d. Similarly, the density of moist air $\rho_m = m_m/V = m_v/V + m_d/V = \rho_v + \rho_d$ where ρ_v (the partial density of the vapor component) is by definition the absolute humidity, and ρ_d is the partial density of dry air in the vapor/dry air mixture. The quantity ρ_v/ρ_m represents the mass of water vapor per unit mass of moist air, i.e., the specific humidity. The total moles $n = n_v + n_d$, and therefore, because mass = moles times molecular weight, $\rho_m = [n_v M_{wv}/V] + [n_d M_{wd}/V]$, where M_{wv} and M_{wd} represent molecular weight of water (18.016 g mol^{-1}) and dry air (28.97 g mol^{-1}), respectively. As the molecular weight of water is less than that of dry air, the foregoing indicates that the density of a given volume V of moist air would be less than the density of the same volume of dry air (<1.29 kg m^{-3}). Indeed, if this were not so, instead of our poetry-inspiring cloudy skies, most of the earth would be blanketed by dense surface fogs most of the time.

Pursuing this line of reasoning, the total pressure P of a volume V of moist air at temperature T would be equal to $P_v + P_d$, where P_v and P_d are the partial pressures of the water vapor and dry air. P_v is usually denoted as e, implying $P_d = P - e$. Applying the ideal gas relations above for a volume V of moist air at atmospheric pressure P and temperature T, $e = \rho_v R_v T$, and therefore ρ_v(the absolute humidity) = $e/R_v T = eM_{wv}/RT$, because $R_v = R/M_{wv}$ is the specific gas constant for water vapor. Dividing numerator and denominator by M_{wd} (molecular weight of dry air) gives:

$$\rho_v = \frac{e\left(\frac{M_{wv}}{M_{wd}}\right)}{R_v T} = \frac{0.622e}{R_d T}$$

where R_d is the specific gas constant for dry air (=287.04 J kg^{-1} °K^{-1}).

Thus, at a given temperature T (in °K), the absolute humidity can be obtained if the vapor pressure e were known.

Observations of humidity are widely reported as relative humidity (RH), which is strictly defined as the quotient of the mixing ratio (w) of actual (vapor unsaturated) air at a temperature T and the mixing ratio ($w*$) in vapor-saturated air at the same temperature. The mixing ratio w is defined as the mass of water vapor per unit mass of dry air, and therefore $w = \rho_v/\rho_d$. Therefore, $RH = \mathrm{w}/w*$. As at a given T, $\rho_d = P_d/(R_d T)$ and $P_d = P - e$, therefore from the above equation for ρ_v, $w = 0.622e/(P - e)$. At saturation, the mixing ratio $w* = 0.622e*/(P - e*)$, where $e*$ is the saturation vapor pressure at temperature T. As e and $e*$ are, in general, small compared with P, $w \approx 0.622e/P$, and $w* \approx 0.622e*/P$, implying that $w/w*$ is very nearly equal to $e/e*$, i.e., the ratio of the vapor pressure in actual (vapor unsaturated) air at a temperature T to the vapor pressure in vapor-saturated air at the same temperature. Therefore, e can be obtained from reported RH values if $e*$ were known.

Many equations exist for estimating $e*$ at a given temperature T°K. Based on thermodynamic theory, $e*$ is a function of T and can be obtained by integrating the Clausius–Clapeyron equation $de*/dT = (0.622\,Le*)/(R_d T^2)$, where L is the latent heat of evaporation (J kg^{-1}) at temperature T, R is the specific gas constant for dry air = 287.04 J kg^{-1} °K^{-1}, and T is the absolute temperature (°K). As L is a function of T, the Clausius–Clapeyron equation can be integrated for a given function $L(T)$ to obtain $e*(T)$. One such integral is that of Goff and Gratch (1) and Goff (2) that has become the basis for the published Smithsonian and World Meterorological Organization (WMO) standard tables of $e*(T)$. Gibbins (3) reviewed this and many other equations developed for estimating $e*(T)$. One developed by Richards (4) gives values that differ by no more than 0.01% of the tabulated Goff–Gratch standard. This equation is:

$$e*(T) = ae^{f(T_r)}$$

where $e*$ is in millibar (mb), e is the base of natural logarithms, $T_r = 1 - \frac{373.15}{T + 273.15}$, T is the temperature in °C, and $a = 1013.25$. The function $f(T_r) = 13.3185T_r - 1.9760\ (T_r)^2 - 0.6445\ (T_r)^3 - 0.1299\ (T_r)^4$. For example, at 20°C, $T_r = -0.2729$, $f(T_r) = -3.7694$, and $e*(20°\mathrm{C}) =$ 23.372 mb (=2337.2 Pa) compared with the tabulated standard value of 23.373 mb.

Another computationally less tedious equation for $e*(T)$ commonly used in ventilation and air conditioning engineering calculations is the Magnus equation (also referred to as the Magnus–Teten formula) valid for air temperatures between 0–50°C:

$$e*(T°\mathrm{C}) = 6.112\,\mathrm{e}^{\frac{17.62T}{243.12+T}}$$

where T is the temperature in °C and $e*$ is in mb. This formula gives $e*(20°\mathrm{C}) = 23.325$ mb.

BIBLIOGRAPHY

1. Goff, R.M. and Gratch, S. (1946). Low pressure properties of water from −160 to 212°F. *Trans. Amer. Soc. Heating Ventilation Engineers* **52**: 95–121.
2. Goff, J.A. (1957). Saturation pressure of water on the new Kelvin temperature scale. *Trans. Amer. Soc. Heating Ventilation Engineers* **63**: 347–354.
3. Gibbins, C.J. (1990). A survey and comparison of relationships for the determination of the saturation vapour pressure over plane surfaces of pure water and of pure ice. *Ann. Geophysica* **8**: 859–886.
4. Richards, J.M. (1971). Simple expression for the saturation vapor pressure of water in the range −50 to 140°C. *Brit. J. Appl. Phys.* **4**: L15–L18.

RELATIVE HUMIDITY

CLIMENT RAMIS
ROMUALDO ROMERO
SERGIO ALONSO
University of the Balearic
Palma de Mallorca, Spain

INTRODUCTION

If samples of air collected at different times of the year and over different points of the earth between the surface and 80 km altitude are analyzed, it is observed that the proportions of gases that form air are, except for a few components, nearly constant in all samples.

That fixed composition is nitrogen (N_2), the most abundant gas with 78.08% in volume concentration, oxygen (O_2) with 20.95%, and argon (Ar) with 0.93%. Other less abundant gases, known as trace gases, are helium (He), neon (Ne), hydrogen (H_2), carbon dioxide (CO_2), ozone (O_3), and water vapor (H_2O), accounting for all together less than 0.04% in volume concentration. The atmospheric layer that has this composition is called the homosphere.

Among the trace gases, ozone shows a particular vertical distribution. The greater concentrations are found at altitudes between 20 and 35 km. This layer is known as the ozonosphere. The concentrations near the earth's surface and over 40 km are very low. On the other hand, the ozone concentration is at a maximum in spring and at a minimum in autumn in the Northern Hemisphere.

Water vapor distribution exhibits great spatial and temporal variability. At any given point, the quantity of water vapor varies continuously, and it is always different from one place to another. An image from the Meteosat satellite in the water vapor (WV) channel permits visualizing the spatial variability (Fig. 1). This spatial variability is particularly high in the vertical, such that about 75% of the water vapor in the atmosphere is in the lower 3000 m, and practically all of this trace gas is trapped in the lowest atmospheric layer or troposphere (10–12 km deep). The high spatial and temporal variability of water vapor requires the definition of variables that, having a simple formulation, permit us to quantify its abundance.

Water itself is a very important component because it can exist in three states (solid, liquid, and gas) within the atmospheric temperature range. Many processes in the atmosphere can cause water to change state. Particularly relevant is when water vapor condenses into liquid water because this process is accompanied by a release of heat that can be absorbed by the air. Similarly, when liquid water freezes or water vapor sublimes, latent heat is released. On the contrary, ice melting, ice sublimation, and liquid evaporation imply heat absorption. The quantification of water vapor content is therefore some measurement of the latent heat available to the air, which, if released, can strongly affect the atmospheric stability to vertical motions and dynamic processes such as the development of pressure systems.

In the remainder of this article, some indexes that are used to quantify water vapor content in the atmosphere are described.

VAPOR PRESSURE AND SATURATION VAPOR PRESSURE

The air is a mixture of gases. The pressure exerted by air is the sum of the partial pressures exerted by each of its components (Dalton's law). The partial pressure exerted by water vapor, denoted as vapor pressure (e), is then a measure of the water vapor content in the atmosphere. A state can be represented by a point in a diagram whose coordinates are (temperature, vapor pressure) = (T, e).

Figure 1. WV image from Meteosat satellite on November 13, 2002 at 1500 UTC (from Dundee University, UK).

For each temperature, a maximum vapor pressure exists that is known as the saturation vapor pressure (e_s). That saturation value cannot be surpassed, and if water vapor in excess occurs, condensation or sublimation results. Figure 2 shows the state curve (T, e_s), with respect to liquid water for temperatures above 0 °C and with respect to supercooled water (e_s) and ice (e_i) below 0 °C. The values of saturation vapor pressure are determined experimentally. A list of values as a function of temperature can be found in Smithsonian Meteorological Table (2). Table 1 includes some e_s values at several temperatures.

Efforts have been carried out to adjust the curve (T, e_s) with respect to liquid water (above and below 0 °C) by an analytic formula. Following is the expression of Tetens (3),

$$e_s(t) = 6.11 * 10\left(\frac{7.5t}{t + 237.3}\right) \quad (1)$$

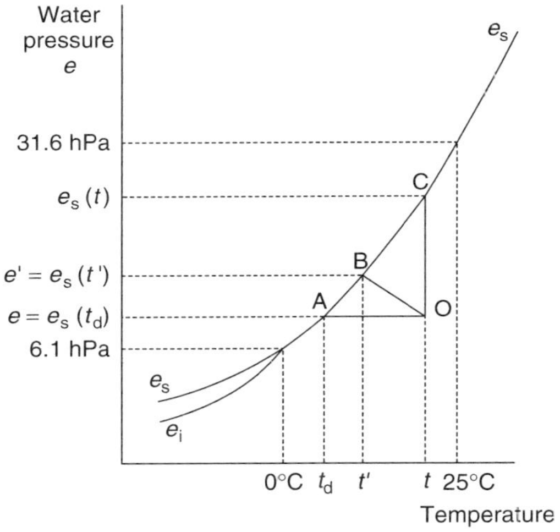

Figure 2. State curve of saturation vapor pressure as a function of temperature. e_s = pressure over a plane surface of liquid water; e_i = pressure over a plane surface of ice. See text for additional information contained in the figure (Diagram is based on Fig. 2 of Reference 1).

Table 1. Values of Saturation Vapor Pressure Over a Plane Surface of Liquid Water at Different Temperatures

t, °C	e_s, hPa
−20	1.2
−10	2.9
0	6.11
10	12.28
20	23.38
30	42.43
40	73.76
50	123.30

Figure 3. Psychrometer. Note the wet thermometer with its bulb covered by wet muslin (photograph courtesy of INM, Spain).

that of Bolton (4),

$$e_s(t) = 6.112 * \exp\left(\frac{17.67t}{t + 243.5}\right) \quad (2)$$

and the modified Clausius–Clapeyron formula,

$$\ln e_s = 53.67957 - \frac{6473.70}{t + 273.16} - 4.8451 \ln(t + 273.16) \quad (3)$$

where t is the temperature in °C and e_s, is the saturation vapor pressure in hPa. The Boltons expression is precise to 0.3% in the range −35 °C to 35 °C and the modified Clausius–Clapeyron method to 0.01% between −40 °C and 40 °C. Both intervals cover a wide spectrum of temperatures in which condensation phenomena are important. Other more complex formulas for saturation vapor pressure on liquid water and on ice can be obtained by polynomial adjustments (5).

Given an air parcel of state (T, e) represented by point O in Fig. 2, two very simple processes permit the air to reach saturation. Suppose that at constant temperature (t), water vapor is added to the parcel so that the vapor pressure is increased. This process eventually leads to saturation, that is, the vapor pressure would reach $e_s(t)$ (point C in the diagram). Analogously, the temperature could be diminished while maintaining the water pressure constant and a lower temperature (t_d) would be reached in which the water pressure would equal the saturation value, that is, $e_s(t_d) = e$ (point A). The temperature at which this happens is known as the "dew point." If the temperature decreased further, the vapor pressure would equally decrease following curve e_s and water vapor would condense.

Another saturation mechanism would consist of adding liquid water to the air parcel. That liquid water would evaporate, getting the necessary heat from the same air. The temperature would diminish at the same time that the vapor pressure would increase. Thus, with enough liquid water available, saturation would eventually be reached at some temperature t' (point B in Fig. 2). That temperature, known as the "wet bulb temperature," is the smallest possible temperature an air mass can acquire by evaporation of water in its body.

DETERMINATION OF VAPOR PRESSURE

The vapor pressure at the surface in the real atmosphere is determined indirectly by the psychrometer. That instrument is composed of two identical thermometers placed in a stationary screen box; one has its bulb dampened in muslin (Fig. 3), called the wet thermometer. The temperature difference between the dry and the wet thermometers depends on the water vapor content of the atmosphere; the smaller the vapor content, the greater the temperature difference. Assuming that the air circulating near the wet thermometer effectively reaches the saturation vapor pressure according to the process explained above, its temperature is simply t′ (Fig. 2), and we refer to the dry bulb temperature as t. The following relation is satisfied:

$$e = e_s(t') - \frac{8c_p}{5Lv}p(t - t') \tag{4}$$

where c_p ($c_p = 1005$ Jkg^{-1}K^{-1}) is the specific heat of dry air at constant pressure, L_v ($L_v = 2501 \times 10^6$ J kg^{-1}) the latent heat of condensation, and p is the pressure, which permits determination of the vapor pressure e from the dry and wet bulb thermometer readings. Commonly, tables with double entries of t' and $t - t'$ are constructed which permit determination of e (2).

RELATIVE HUMIDITY

Once a vapor pressure value is determined, the whole picture is not complete unless that value is compared to the saturation vapor pressure at the temperature of the air.

Relative humidity (h) is thus defined as the ratio (expressed in a percentage) of the vapor pressure e and the saturation vapour pressure $e_s(t)$ at the corresponding temperature (1,6):

$$h = 100\frac{e}{e_s(t)} \tag{5}$$

Keeping in mind the dew point definition introduced earlier,

$$h = 100\frac{e_s(t_d)}{e_s(t)} \tag{6}$$

and $e_s(t_d)$ can never be greater than $e_s(t)$. For an interpretation, see Fig. 2. When $h = 100\%$, saturation has been reached and condensation is produced.

Some surfaces exposed to the environment often acquire a temperature that is different from that of the air. For those surfaces, an equivalent relative humidity can be defined as

$$h = 100\frac{e}{e_s(t_s)} = 100\frac{e_s(t_d)}{e_s(t_s)} \tag{7}$$

where t_s is now the temperature of the surface. A surface cooler than the air ($t_s < t$), for example, would then promote saturation more easily than the surrounding air. This explains why dew sometimes forms on plant leaves although 100% relative humidity has not been reached in the atmosphere. Analogous expressions could be defined considering the saturation pressure with respect to ice.

Hygrometers are instruments that measure relative humidity directly. Digital hygrometers are generally based on some change in the electrical properties of some material as a function of the relative humidity of the air, whereas analogic hygrometers use the variation of length that the human hair (especially that of a blond woman!) experiences with a change in relative humidity.

If the relative humidity is known, the dew point can be determined by inverting some of the formulas for calculating the saturation vapor pressure (Eqs. 1–3). Conversion tables are also available for that purpose (2).

OTHER HUMIDITY INDEXES

There are several other humidity indexes apart from relative humidity. They are often more useful than the latter for adequate formulation of the moist thermodynamic processes that occur in the atmosphere. The following are emphasized:

Mixing Ratio (*r*)

It is defined as the quantity of water vapor per unit mass of dry air. Its units are therefore g g^{-1} but generally it is expressed in g kg^{-1}:

$$r = \frac{m_v}{m_d} \tag{8}$$

Both the water vapor and dry air portions of the sample, which have partial pressures e and $p - e$, respectively, comply with the equation for the ideal gases; then,

$$e = \frac{m_v}{V}R_vT \tag{9}$$

$$p - e = \frac{m_d}{V}R_dT \tag{10}$$

where R_v and R_d are the specific gas constants of water vapor and dry air, respectively, V is the volume, and T the temperature. Combining Eqs. 9 and 10,

$$r = \frac{m_v}{m_d} = \frac{R_d}{R_v}\frac{e}{p - e} = \varepsilon\frac{e}{p - e} \tag{11}$$

and using the constants $R_d = 287.04$ J kg^{-1} K^{-1} and $R_v = 461.50$ J kg^{-1} K^{-1}, Eq. 11 is

$$r = 0.622\frac{e}{p - e} \tag{12}$$

Within the interval of temperatures and atmospheric pressures, $e << p$, and therefore

$$r \cong 0.622\frac{e}{p} \tag{13}$$

If the vapor pressure corresponds to its saturation value, then a saturation mixing ratio is defined as

$$r_s \cong 0.622\frac{e_s}{p} \tag{14}$$

that depends only on pressure and temperature.

Using Eqs. 13 and 14, the relative humidity is

$$h = 100\frac{e}{e_s(t)} \cong 100\frac{r}{r_s(t)} \tag{15}$$

Therefore, the numerical values of h deduced from vapor pressure or from the mixing ratio are practically the same.

The exact relation between the relative humidity and the mixing ratio is

$$h = 100\frac{r}{r_s(t)}\frac{1+\dfrac{r_s(t)}{\varepsilon}}{1+\dfrac{r}{\varepsilon}} \tag{16}$$

Specific Humidity (q)

Specific humidity is defined as the relation between the vapor mass and the mass of moist air that contains this vapor, that is, the quantity of water vapor per unit mass of moist air:

$$q = \frac{m_v}{m_r + m_d} \tag{17}$$

The units are g g^{-1} although it is generally expressed in g kg^{-1}.

Using the same formulation introduced previously for the mixing ratio,

$$q = \varepsilon\frac{e}{p - e(1-\varepsilon)} \tag{18}$$

And according to the same previous arguments,

$$q \cong \varepsilon\frac{e}{p} \tag{19}$$

and therefore

$$h \cong 100\frac{q}{q_s(t)} \tag{20}$$

The same comments noted earlier for the mixing ratio (r) are valid for the specific humidity in relation to relative humidity.

Knowledge of the specific humidity permits us to calculate the specific parameters for humid air, for example, the specific heat

$$c_p' = (1-q)c_p + qc_v \tag{22}$$

where c_p and c_v are the specific heats of dry air and water vapor at constant pressure, respectively ($c_p = 1005$ J kg^{-1}K^{-1}, $c_v = 1.85\ 10^3$ J kg^{-1}K^{-1}); and the specific gas constant

$$R' = (1-q)R_d + qR_v \tag{23}$$

BIBLIOGRAPHY

1. McIntosh, D.H. and Thom, A.S. (1978). *Essentials of Meteorology*. Wykeham.
2. (1980). *Smithsonian Meteorological Tables*, 5th Edn.
3. Murray, F.W. (1967). On the computation of saturation vapour pressure. *J. Appl. Meteorol.* **6**: 203–204.
4. Bolton, D. (1980). The computation of equivalent potential temperature. *Mon. Weather Rev.* **108**: 1046–1053.
5. Flatan, P.J., Walks, R.L., and Cotton, W.R. (1992). Polynomial fits to saturation vapour pressure. *J. Appl. Meteorol.* **31**: 1507–1513.
6. Wallace, J.M. and Hobbs, P.V. (1977). *Atmospheric Science. An Introductory Survey*. Academic Press, San Diego, CA.

HURRICANES: MODELING NATURE'S FURY

Geophysical Fluid Dynamics
Laboratory—NOAA

Few natural phenomena evoke a greater sense of awe and mystery than hurricanes, the powerful and destructive tropical storms that often menace coastal regions. Scientists at GFDL, led by Yoshio Kurihara, have worked steadily for two decades to develop physically based models that can be used to understand the basic mechanisms producing hurricanes and to predict the movement and intensity of individual storms.

A PREDICTION BREAKTHROUGH

Improvements in hurricane prediction have been made possible by improved observations, models, and methods for incorporating observations into the forecast models. Kurihara and collaborators Bob Tuleya and Morris Bender realized that current weather analyses were inherently inadequate as "initial conditions" for such models. They developed a method to circumvent much

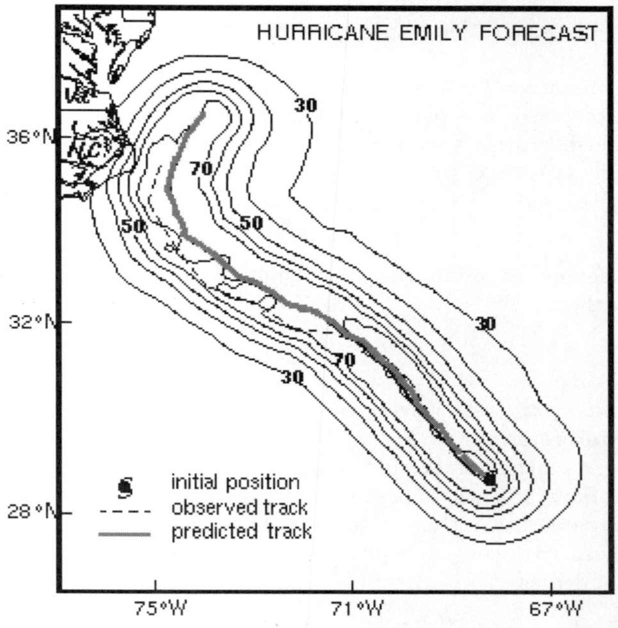

Three-day forecast of Hurricane Emily's track (red line) and intensity produced by the GFDL hurricane model. Thin contours show the maximum predicted wind speed (in knots) during the forecast period. The forecast was available at 8 a.m., Sunday, Aug. 29, 1993. The hurricane caused extensive damage along North Carolina's Outer Banks beginning late Tuesday afternoon (Aug. 31) before recurving to the northeast and out to sea, as predicted by the GFDL model

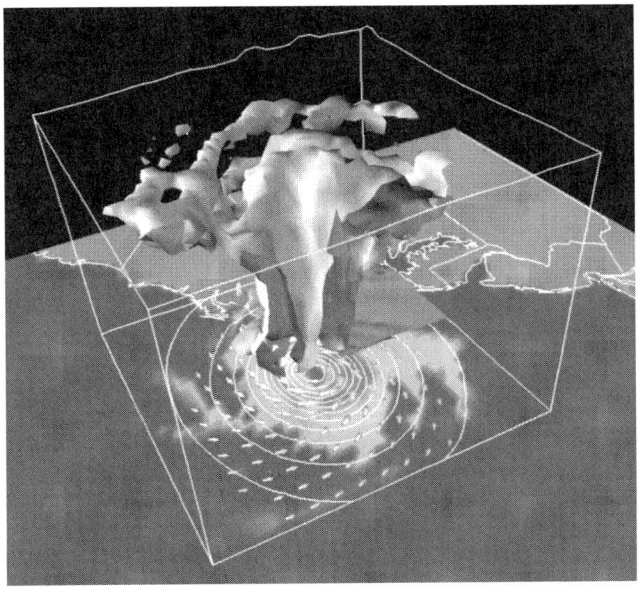

Three-dimensional view of Hurricane Emily near the North Carolina coast as modeled by the dynamical hurricane prediction system developed at GFDL. The view is a 48-hour forecast (valid 8 p.m., August 31, 1993) obtained from the GFDL model. This forecast system provides improved hurricane track predictions roughly two days in advance of hurricane conditions and has the potential to provide useful forecasts of storm intensity and precipitation as well. In the case of Hurricane Emily, the model successfully predicted the occurrence of hurricane-force winds in the Cape Hatteras, N.C. vicinity. In the figure, the three-dimensional "cloud-like" volume (which is sliced vertically through the storm center) depicts the 91% relative humidity surface; the colored shading at the earth's surface shows precipitation (higher values are red); red and white arrows show vertical motions and surface winds, respectively; white contours depict sea-level pressure; the dark blue surface and solid green background represent the Atlantic Ocean and the U.S. East Coast, respectively. Note the "eye-wall" in the relative humidity surface at the storm center and the heavy precipitation north of the storm

of the initial data problem and achieved a breakthrough in hurricane prediction.

The new forecast system achieved a series of remarkably successful forecasts, compared to other available methods for Hurricanes Andrew, Iniki, and Emily. For example, while conventional models predicted that Hurricane Emily (August 1993) would make landfall in the vicinity of Georgia, the GFDL forecast system correctly predicted that the hurricane would deliver a glancing blow to North Carolina's Outer Banks before veering back out to sea. Based on its impressive performance during the 1993 and 1994 hurricane seasons, the model has been adopted as the operational hurricane prediction model at the National Meteorological Center.

HURRICANE EMILY FORECAST

Given the expense and inconvenience of hurricane evacuations and emergency preparations, the public benefits greatly if unnecessary hurricane warnings can be avoided. Furthermore, residents are likely to take hurricane warnings more seriously if the forecasts are shown to be more reliable. With potential damage from a single storm exceeding $25 billion (e.g., Hurricane Andrew in 1992), reliable short-term preparations, while expensive to undertake, can result in large net savings by the public.

A NEW COASTAL FORECAST SYSTEM

In a complementary effort, GFDL is cooperating with scientists at Princeton University and NOAA's National Meteorological Center (NMC) and National Ocean Service (NOS) to develop a coastal forecast system for the East Coast. The system is designed to provide nowcasts and forecasts of currents, storm surges, and other information to coastal residents, the fishing and marine transport industries, and government users. The forecast system uses a high-resolution ocean model–developed by Princeton's George Mellor in collaboration with GFDL–and an atmospheric model from NMC.

HYDROLOGIC CYCLE

Sharad K. Jain
National Institute of Hydrology
Roorkee, India

Vijay P. Singh
Louisiana State University
Baton Rouge, , Louisiana

The hydrologic cycle, the endless circulation of water between the earth, the oceans, and the atmosphere, is fundamental to the study of the waters of the earth. The cycle entails a huge turnover of water driven by the energy of the Sun and gravity. It has a profound influence on the landscape and the earth's climate. The major components of the hydrologic cycle are precipitation, evapotranspiration, interception, infiltration, overland and channel flow, and groundwater flow. Quantitatively, the hydrologic cycle is represented by a mass balance or continuity equation. The relative significance of the terms of this equation depends on the space and timescale. Human activities and the hydrologic cycle are interactive and influence the earth's climate.

INTRODUCTION

The hydrologic cycle is a fundamental concept in hydrology and is amongst a number of cycles known to be operating in nature, such as the carbon cycle, the nitrogen cycle, and other biogeochemical cycles. The National Research Council (1) defines the hydrologic cycle as "the pathway of water as it moves in its various phases to the atmosphere, to the earth, over and through the land, to the ocean and back to the atmosphere." This cycle has no beginning or end, and water is present in all three states, solid, liquid,

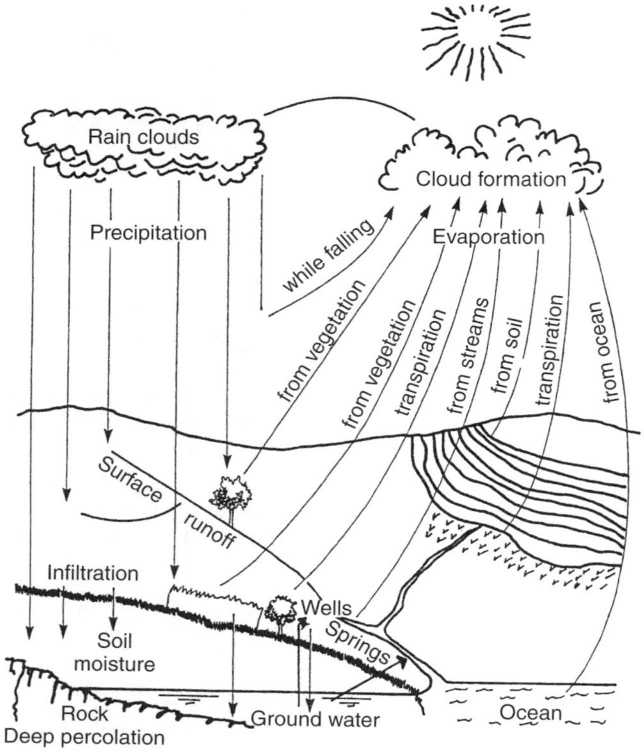

Figure 1. Pictorial representation of the hydrologic cycle (2).

and gas, in the cycle. A pictorial representation of the hydrologic cycle is given in Fig. 1.

The hydrologic cycle, also known as the water cycle, is a concept that considers the processes of motion, loss, and recharge of the earth's waters. It connects the atmosphere and two storages of the earth system: the oceans and the landsphere (lithosphere and pedosphere). The water that is evaporated from the earth and the oceans enters the atmosphere. Water leaves the atmosphere through precipitation. The oceans receive water from the atmosphere by precipitation and from the landsphere through stream flow and groundwater flow. The only way that water leaves oceans is by evaporation. The landsphere receives water by precipitation. The water leaves this sphere by evaporation, transpiration, evapotranspiration, stream flow, and groundwater flow. Evaporation and precipitation are the processes that take place in the vertical plane, whereas stream flow and groundwater flow occur mostly in the horizontal plane.

The exchange of water among the oceans, land, and the atmosphere has been termed 'the turnover' by Shiklomanov (3). This turnover affects the global patterns of the movement of ocean waters and gases in the atmosphere, thereby greatly influencing climate. Besides, water is a very good solvent, and hence, geochemistry is an integral part of the hydrologic cycle. The hydrologic cycle is, thus, the integrating process for the fluxes of water, energy, and the chemical elements (4). Usually, rain and snow are considered the purest form of water, although they may also be mixed with pollutants in the atmosphere. During the journey on the earth, many chemical compounds are mixed with water, and consequently water quality undergoes a change.

The hydrologic cycle can also be visualized as a perpetual distillation and pumping system. In this endless circulation of water, the glaciers and snow packs are replenished, the quantity of river water is replenished, and its quality is restored. From the view point of water use, the land phase of the hydrologic cycle is the most important. The necessity and utility of studying the hydrologic cycle arises because water is essential for the survival of life and is an important input in many economic activities. But water of the desired quality may not be available when needed.

In view of the complexities and extensive coverage, the study of the complete hydrologic cycle is truly interdisciplinary. For instance, the atmospheric part is studied by meteorologists, the pedospheric part by soil scientists, the lithosphere by geologists, and the part pertaining to oceans falls in the domain of oceanographers. The domain of a hydrologist is confined mainly to the land phase, and this in itself is a very large domain. A host of other professionals may be interested in studying the hydrologic cycle, including energy utility managers, chemists, agricultural engineers, public health officers, and inland navigation managers. It is hard to draw lines demarcating various domains in such an interdisciplinary subject, and the boundaries are often blurred. Teams of people from various specialties often work together on a problem dealing with the components of the hydrologic cycle.

COMPONENTS OF THE HYDROLOGIC CYCLE

The hydrologic cycle can be subdivided into three major systems: The oceans are the major reservoir and source of water, the atmosphere functions as the carrier and deliverer of water, and the land is the user of water. The amount of water available at a particular place changes with time because of changes in the supply and delivery. Globally, water movement is a closed system, but locally, it is an open system. A systems representation of the hydrologic cycle is shown in Fig. 2.

The major components of the hydrologic cycle are precipitation (rainfall, snowfall, hale, sleet, fog, dew, and drizzle), interception, depression storage, evaporation, transpiration, infiltration, percolation, moisture storage in the unsaturated zone, and runoff (surface runoff, interflow, and base flow).

Evaporation of water takes place from the oceans and the land surface mainly due to solar energy. The moisture moves in the atmosphere as water vapor which precipitates on land surface or oceans as rain, snow, hail, or sleet. Part of this precipitation is intercepted by vegetation or buildings. Of the amount reaching the land surface, part infiltrates into the soil, and the remaining water runs off the land surface to join streams. These streams finally discharge into the ocean. Some of the infiltrated water percolates deep to join groundwater, and some comes back to the streams or appears on the surface as springs.

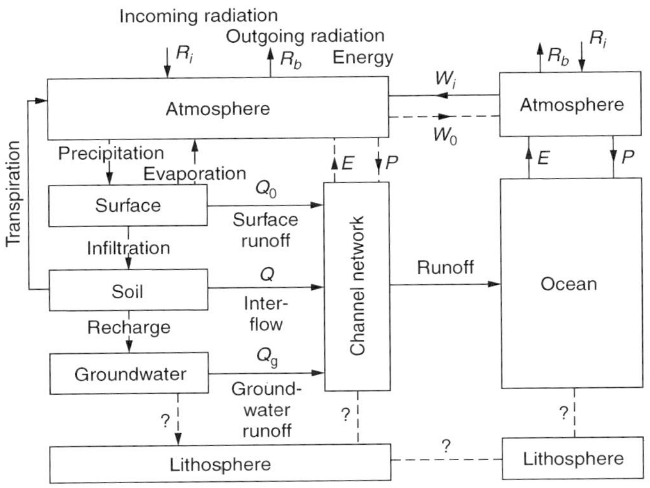

Figure 2. Systems representation of the hydrologic cycle (5).

This immense movement of water is driven mainly by solar energy: the excess of incoming radiation over outgoing radiation. Therefore, the Sun is the prime mover of the hydrologic cycle. The energy for evaporating water from streams, lakes, ponds, oceans, and other open water bodies comes from the Sun. A substantial quantity of moisture is added to the atmosphere by transpiration of water from vegetation. Living beings also supply water vapor to the atmosphere through perspiration. Gravity has an important role in the movement of water on the earth's surface, and anthropogenic activities also have an increasingly important influence on water movement.

An interesting feature of the hydrologic cycle is that at some point in each phase, there usually occur (1) transportation of water, (2) temporary storage, and (3) a change of state. For example, in the atmospheric phase, there is vapor flow, vapor storage in the atmosphere, and condensation or formation of precipitation created by a change from vapor to either the liquid or solid state. Moreover, in the atmosphere, water is in the vapor form, whereas it is mostly (saline) liquid in the oceans.

Scales for Studying the Hydrologic Cycle

Depending on the purpose of study, the hydrologic cycle is studied on many different spatial scales. Horton (6) clearly recognized the diversity of scales when he stated *Any natural, exposed surface may be considered as a unit area on which the hydrologic cycle operates. This includes, for example, an isolated tree, even a single leaf or twig of a growing plant, the roof of a building, the drainage basin of a river system or any of its tributaries, an undrained glacial depression, a swamp, a glacier, a polar ice-cap, a group of sand dunes, a desert playa, a lake, an ocean, or the Earth as a whole* (Quoted in Ref. 4). A qualitative representation of the hydrologic cycle is given in Fig. 3.

From the view point of hydrologic studies, two scales are readily distinct, the global scale and the catchment scale.

Global Scale. From a global perspective, the hydrologic cycle is comprised of three major systems: the oceans, the atmosphere, and the landsphere. Precipitation, runoff, and evaporation are the principal processes that transmit water from one system to the other, as illustrated in Fig. 4. This illustration depicts a global geophysical view of the hydrologic cycle and shows the interactions among the earth (lithosphere), the oceans (hydrosphere), and the atmosphere. Study on the global scale is necessary to understand the global fluxes and global circulation patterns. The results of these studies form important inputs to water resource planning for a national or regional water resource assessment, weather forecasting, and study of climate changes. These results may also form the boundary conditions of small-scale models/applications.

Catchment Scale. While studying the hydrologic cycle on a catchment scale, the spatial coverage can range from a few square km to thousands of square km. The timescale could be a storm lasting for a few hours to a study spanning many years. When the water movement of the earth system is considered, three systems can be recognized: the land (surface) system, the subsurface system, and the aquifer (or geologic) system. The dominant processes of the hydrologic cycle of the land system, are precipitation, evapotranspiration, infiltration, and surface runoff. The land system itself comprises three subsystems: the vegetation subsystem, the structural subsystem, and the soil subsystem. These subsystems subtract water from precipitation through interception, depression, and detention storage. This water is either lost to the atmospheric system or enters the subsurface system. The exchange of water among these subsystems takes place through infiltration, exfiltration, percolation, and capillary rise. Figure 5 shows the schematic of the hydrologic cycle on a global scale in the earth system and a microscale view of the cycle in the land system.

Timescales in the Hydrologic Cycle. The time required for the movement of water through various components of the hydrologic cycle varies considerably. The velocity of stream flow is much higher than the velocity of groundwater. The time-step size for an analysis depends on the purpose of study, the availability of data, and how detailed the study is. The estimated periods of renewal of water resources in water bodies on the earth are given in Table 1. The time step should be sufficiently small so that the variations in the processes can be captured in sufficient detail, but at the same time, it should not put undue burden on data collection and computational efforts.

The range of the spatial and temporal dimensions of many processes related to the hydrologic cycle is shown in Fig. 6.

MATHEMATICAL REPRESENTATION OF THE HYDROLOGIC CYCLE

The quantities of water going through the various components of the hydrologic cycle can be evaluated by the so-called hydrologic equation, which is a simple spatially lumped continuity or water budget equation:

$$I - Q = \Delta S \quad (1)$$

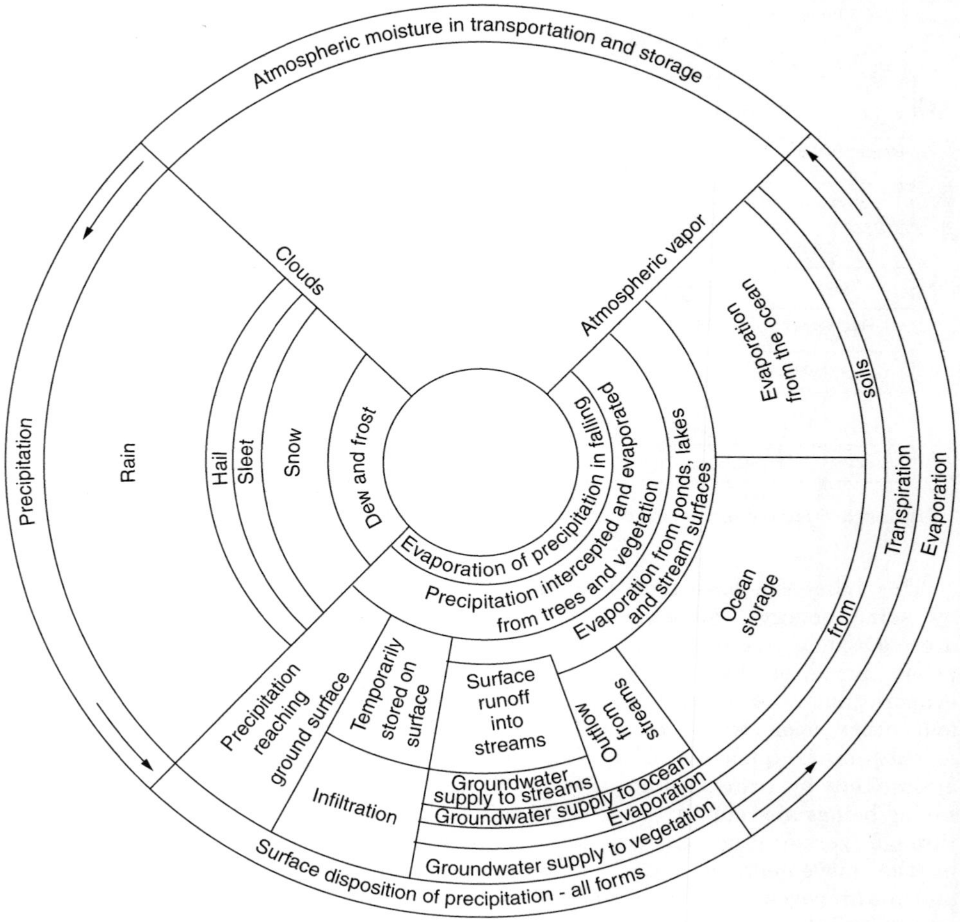

Figure 3. Qualitative representation of the hydrologic cycle (6).

Table 1. Periods of Water Resource Renewal on the Earth[a]

Water of Hydrosphere	Period of Renewal
World ocean	2500 years
Ground water	1400 years
Polar ice	9700 years
Mountain glaciers	1600 years
Ground ice of the permafrost zone	10000 years
Lakes	17 years
Bogs	5 years
Soil moisture	1 year
Channel network	16 days
Atmospheric moisture	8 days
Biological water	Several hours

[a]Reference 3.

where I = inflow of water to a given area during any given time period, Q = outflow of water from the area during the selected time period, and ΔS = change in storage of water in the given area during the time period. If I and Q vary continuously with time, then Eq. 1 can be written as

$$dS(t)/dt = I(t) - Q(t) \tag{2}$$

By integrating, this equation can also be written as

$$\int dS(t) = \int [I(t) - Q(t)]\, dt$$

or

$$S(t) - S(0) = \int_0^t I(t)dt - \int_0^t Q(t)dt = V_I(t) - V_0(t) \tag{3}$$

where $S(0)$ is the initial storage at time $t = 0$, $S(t)$ is the storage at time t, and $V_0(t)$ and $V_I(t)$ are volumes of outflow and inflow at time t. Each of the terms of this lumped equation is the result of a number of other terms. They can be subdivided and even eliminated from the equation, depending on the temporal and spatial scale of the study. The continuity equation is one of the governing equations of almost all hydrologic problems. For a watershed, Eq. 1 may be written as

$$P + Q_{SI} + Q_{GI} - E - Q_{SO} - Q_{GO} - \Delta S - \varepsilon = 0 \tag{4}$$

where P is precipitation, Q_{SI} is surface inflow, Q_{GI} is groundwater inflow, E is evaporation from the watershed, Q_{SO} is surface water outflow, Q_{GO} is groundwater outflow, ΔS is the change in the storage of water in the watershed, and ε is a discrepancy term. For large watersheds, Q_{GI}

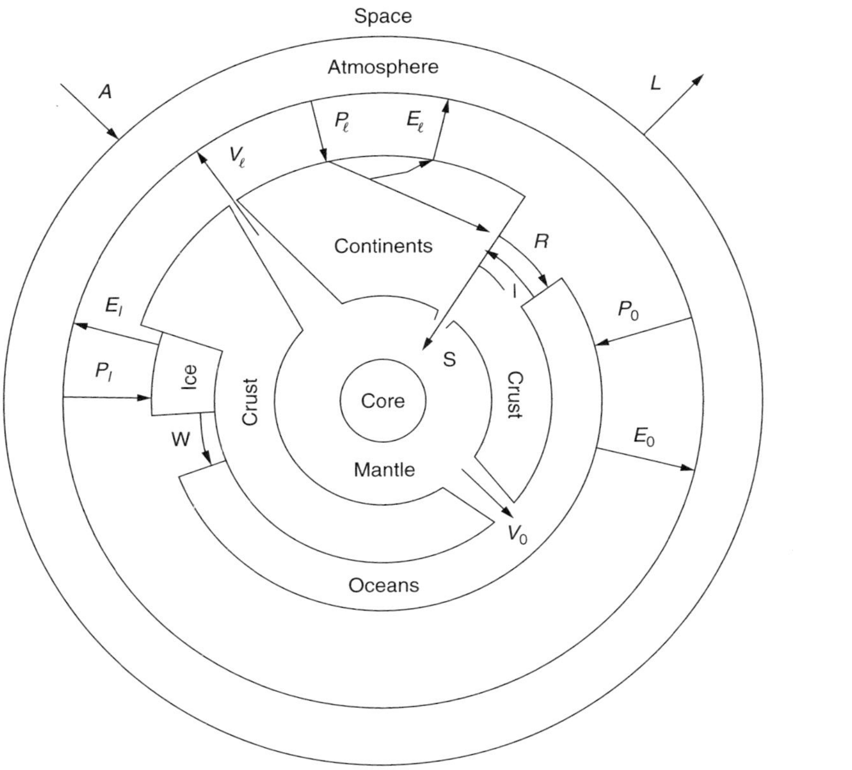

A = Additions of water from space
E_0 = Evaporation from oceans
E_I = Evaporation (i.e., sublimation) from ice
E_ℓ = Evapotranspiration from land
I = Intrusion of seawater into continental aquifers
L = Loss of water to space
P_0 = Precipitation on oceans
P_I = Precipitation on ice
P_ℓ = Precipitation on land
R = Runoff from continents
S = Subduction of water - containing crust
V_0 = Volcanic venting to oceans
V_ℓ = Volcanic venting to atmosphere
W = Wastage of ice sheets to ocean

Figure 4. The hydrologic cycle as a global geophysical process (4).

and Q_{GO} are usually negligible. The discrepancy term is included in Eq. 4 because the sum of all other terms may not be zero due to measurement errors and/or simplifying assumptions. However, a small value of the discrepancy term does not necessarily mean that all other terms have been correctly measured/estimated.

Depending on the specific problem, the terms of Eq. 1 may be further subdivided. For example, when applying the hydrologic equation for short time intervals, the change in the total water storage (ΔS) may be subdivided into several parts: changes of moisture storage in the soil (ΔM), in aquifers (ΔG), in lakes and reservoirs (ΔL), in river channels (ΔS_C), in glaciers (ΔS_G), and in snow cover (ΔS_S). Thus, ΔS can be expressed as

$$\Delta S = \Delta M + \Delta G + \Delta L + \Delta S_C + \Delta S_G + \Delta S_S \qquad (5)$$

The hydrologic equation may be applied to any time interval; computation of the mean annual water balance for a basin is the simplest because it is possible to disregard changes in water storage in the basin (ΔS), which are difficult to measure and compute. In general, the shorter the time interval, the more stringent are the requirements for measuring or computing the components, and the more subdivided are the terms of Eq. 1. This results in a complex equation which is difficult to close with acceptable errors.

The hydrologic equation may be applied to areas of any size, but the complexity of computation depends greatly on the extent of the area under study. The smaller the area, the more complicated is its water balance because it is difficult to estimate components of the equation. Finally, the components of the hydrologic equation may be expressed in terms of the mean depth of water (mm), as a volume of water (m^3), or as flow rates (m^3/s or mm/s).

Global Water Balance

According to estimates (7), the annual average depth of precipitation on the land surface is about 108×10^3 km^3. Out of this, about 61×10^3 km^3 is returned to the atmosphere as evapotranspiration, and the runoff from land to oceans is 47×10^3 km^3. As far as the water balance of oceans is concerned, the depth of precipitation over them is about 410×10^3 km^3, 47×10^3 km^3 of water is received as runoff from the land, and 457×10^3 km^3 is lost as evaporation. If we consider the water balance of the atmosphere, 457×10^3 km^3 water is received as evaporation from oceans and 61×10^3 km^3 from land. The precipitation over oceans is

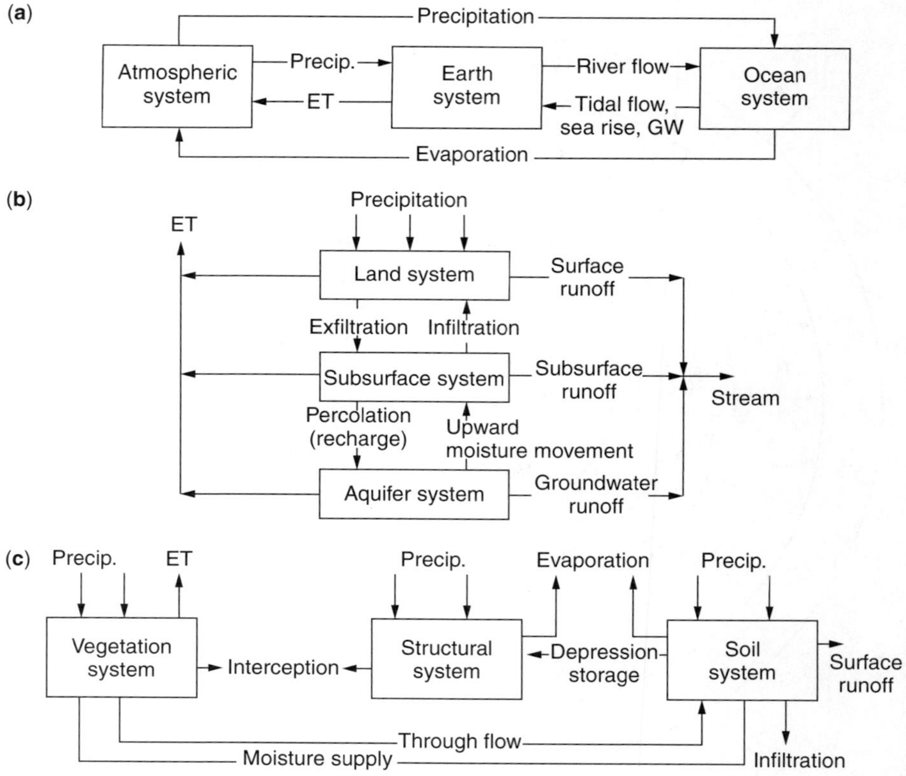

Figure 5. (**a**) A global schematic of the hydrologic cycle (2). (**b**) A schematic of the hydrologic cycle of the earth system (2). (**c**) A detailed schematic of the hydrologic cycle in the land system (2).

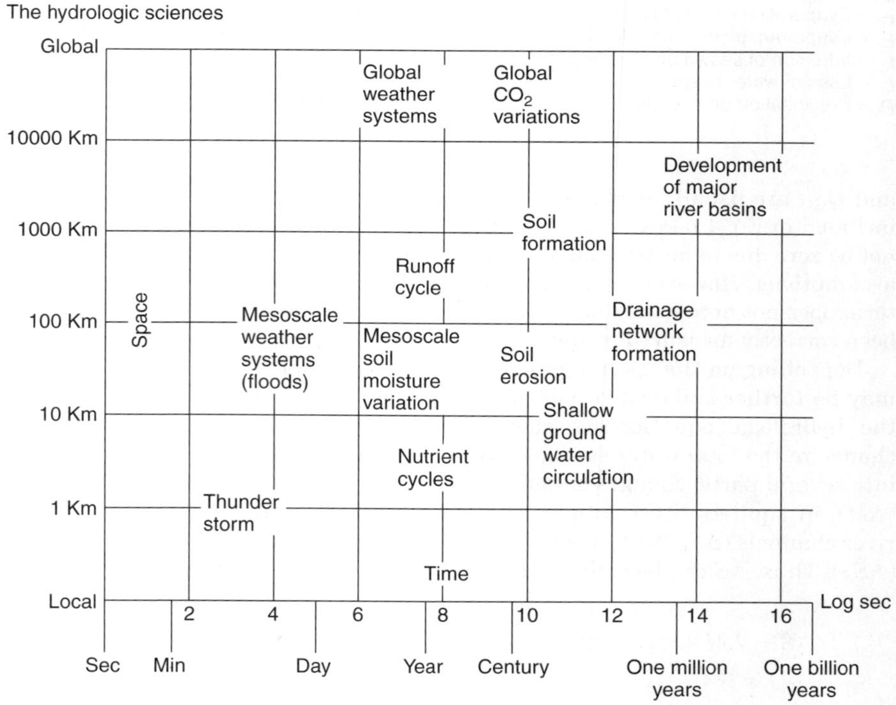

Figure 6. Illustrative range of process scales (4).

410×10^3 km^3, and it is 108×10^3 km^3 over land. Table 2 gives quantitative values of water in various forms on the earth.

The storages and fluxes of the hydrologic cycle are shown in Fig. 7.

INFLUENCE OF HUMAN ACTIVITIES AND LAND-USE CHANGES ON THE HYDROLOGIC CYCLE

Watersheds are subjected to many types of changes, major or minor, for various reasons. Some of them are natural

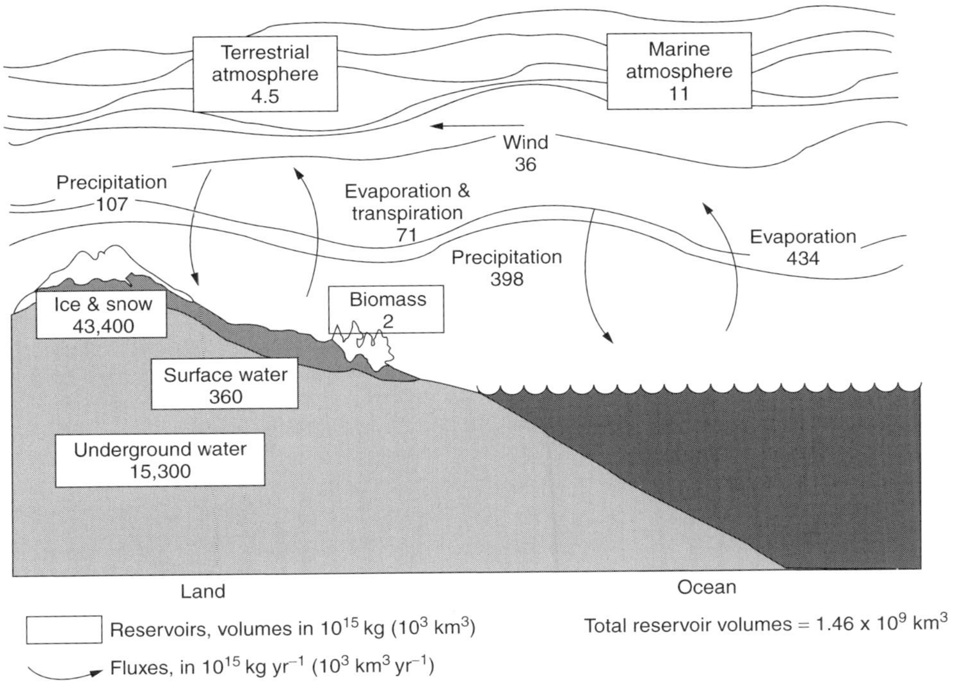

Figure 7. Volumes and fluxes of the global hydrologic cycle (9).

changes, and some are due to human activities. Watershed changes affect virtually all elements of the hydrologic cycle. The quality of water is significantly deteriorating at many places due to industrial and agricultural activities. There has been a growing need to quantify the impact of major human-induced changes on the hydrologic cycle to anticipate and minimize the potential environmental detriment and to satisfy the water resource requirements of society. Even if the water of adequate quantity were present at a place, its use may be limited because of poor quality. The classical and the modern viewpoints of the human role in the hydrologic cycle are shown in Fig. 8.

Most watershed changes can be distinguished as point changes and nonpoint changes. Structural changes, such as dam construction, channel improvement, and detention storage are examples of point changes and affect watershed response in terms of evaporation, seepage, and residence time. Forestry, agriculture, mining, and urbanization are nonpoint land-use changes that affect catchment response. A qualitative discussion of the hydrologic consequences of watershed changes is in order.

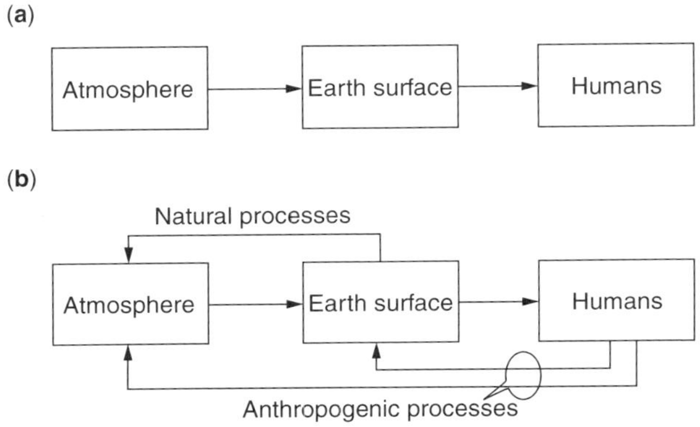

Figure 8. (**a**) Classical viewpoint of the human role in the hydrologic cycle. (**b**) Modern viewpoint of the human role in the hydrologic cycle (1).

Effects of Agricultural Changes

These changes imply that a land area that was forested or barren earlier, is now being cultivated. As a result, the vegetal cover changes; the land slope may be altered a little bit, soil crusting and infiltration characteristics change, and artificial bounds may be placed. The effect of these changes on the hydrologic regime is pronounced and may be multiplicative. Water may be withdrawn from the groundwater zone, or canal irrigation may be introduced leading to noticeable changes in water table behavior. The impacts are also noticed in evapotranspiration, overland flow, channel flow, and infiltration. Fertilizers, pesticides, and insecticides that are applied to crops affect the quality of runoff from agricultural areas.

Effects of Urbanization

A land area that was being used for forestry or agriculture might be transformed into an urban area where houses, roads, parks, parking lots, and sewers are constructed. A large increase in the paved (impervious) surface takes place which considerably reduces infiltration, and the removal of storm water is accelerated. Urban development usually increases the volume and peak of direct runoff for a given rainfall. The time of travel of water is reduced, resulting in lower lag time and a lower time of concentration. However, it is possible to control these effects by providing detention storage or changing the landscape and sizing the storm drains. In brief, the hydrologic effects of urbanization are

Table 2. Water Reserves in Various Phases of the Hydrologic Cycle[a]

	Distribution Area, 10^3 km^2	Volume, 10^3 km^3	Layer, m	Percentage of Global Reserves	
				Of Total Water	Of Freshwater
World ocean	361,300	1,338,000	3,700	96.5	–
Groundwater	134,800	23,400	174	1.7	–
Freshwater	–	10,530	78	0.76	30.1
Soil moisture	–	16.5	0.2	0.001	0.05
Glaciers and permanent snow cover	16,227	24,064	1,463	1.74	68.7
Antarctic	13,980	21,600	1,546	1.56	61.7
Greenland	1,802	2,340	1,298	0.17	6.68
Arctic islands	226	83.5	369	0.006	0.24
Mountainous regions	224	40.6	181	0.003	0.12
Ground ice/permafrost	21,000	300	14	0.022	0.86
Water reserves in lakes	2,058.7	176.4	85.7	0.013	–
Fresh	1,236.4	91	73.6	0.007	0.26
Saline	822.3	85.4	103.8	0.006	–
Swamp water	2,682.6	1.47	4.28	0.0008	0.03
River flows	148,800	2.12	0.014	0.0002	0.006
Biological water	510,000	1.12	0.002	0.0001	0.003
Atmospheric water	510,000	12.9	0.025	0.001	0.04
Total water reserves	510,000	1,385,984	2.718	100	–
Total freshwater reserves	148,800	35,029	235	2.53	100

[a] Reference 8.

1. increased water withdrawals from surface and subsurface water bodies; sometimes demands exceed the available natural resources;
2. increased peak flow and diminishing baseflow of streams;
3. reduced infiltration;
4. increased pollution of rivers and aquifers that endanger the ecology;
5. increased withdrawals from ground water; and
6. changes in local microclimate.

Effects of Forest Activities

These activities may be directed toward planting trees as well as cutting them. The immediate effect of forest activities is change in vegetal cover. When a forested area is deforested and forest litter removed, the interception of precipitation is virtually eliminated. Litter removal changes the infiltration capacity of soil and has a pronounced effect on raindrop impact and resulting soil erosion. When vegetation is lost, evapotranspiration is generally decreased. These changes amount to increased production of direct runoff, reduced surface roughness, and decreased recharge to groundwater. The hydrograph of direct runoff rises more quickly because of the reduced time to peak. However, when additional trees are planted in an area, the effect is reversed, though the impact takes place gradually as the trees grow.

Effects of Structural Changes

Typical structural changes include a dam, a weir, and channel improvement. A dam–reservoir is constructed for many purposes. Regardless of its intended function, it does affect the hydrology of the stream on which it is built. In general, the peak of the outflow from a reservoir is less and the flow may be more even than in the preproject condition. The volume of flow downstream may be considerably less in the afterproject scenario if the reservoir water is diverted elsewhere.

In contrast with reservoirs, channel improvements have an opposite effect. For example, decreasing channel roughness increases flow velocity and hence peak discharge for the same channel size. The removal of vegetation, lining of the channel, and proper maintenance can greatly reduce roughness. Other alterations, such as straightening the channel, maintenance of bands, or increasing slope, significantly affect travel time and flow velocity. Depending on the bed material, infiltration through the bed and banks also modifies flow characteristics.

Impact of Climate Changes on the Hydrologic Cycle

An analysis of measured data series as well as paleoclimatological records suggests that the climate of the earth may be undergoing a significant change. The climatic change may have important impacts on the hydrologic cycle (the converse is also true). The change may be taking place due to natural and human causes. The major changes that could affect climate are changes in vegetation (changes in evapotranspiration, soil moisture, albedo, and radiation balance), increased use of water for day-to-day needs, burning of fossil fuels, and industrial activities. Large-scale water transfers from one basin to another may also cause climate change in the long run. The change manifests itself in changed patterns of spatial and temporal variability in the components of the hydrologic cycle.

The increased emission of greenhouse gases is believed to be the cause of a gradual increase in the earth's temperature. An increase in the temperature of the atmosphere would lead to higher evapotranspiration; changes in precipitation patterns, timing, and distribution; melting of polar ice caps; and recession of glaciers. Higher melting of polar ice and glaciers and thermal expansion of sea water will cause a rise in the seawater level and inundation of islands of low elevations as well as cities adjacent to seas. Another possibility is that an increase in temperature may mean more precipitation, some of which will be snow at the poles, leading to an additional accumulation of ice. The coupled atmosphere–ocean general circulation models are widely used to study the response of climate to various changes. Notwithstanding a large number of studies, what is going to happen and where? is not known with a sufficient degree of certainty and accuracy.

SUMMARY AND CONCLUSIONS

Water is central to the environment and is considered the driver of nature. All studies of water are concerned with some part of the hydrologic cycle. The occurrence and availability of water are closely linked to human development. As a result of many human activities, the features/characteristics of many components of the hydrologic cycle are undergoing changes that are likely to have long-term effects on climate. It is necessary to predict the magnitude and extent of these changes so that the required ameliorative measures can be initiated in time.

BIBLIOGRAPHY

1. NRC. (1982). *Scientific Basis of Water Resource Management*. National Research Council, National Academy Press, Washington, DC.
2. Singh, V.P. (1992). *Elementary Hydrology*. Prentice-Hall, Englewood Cliffs, NJ.
3. Shiklomanov, I.A. (1999). *World Water Resources: Modern Assessment and Outlook for the 21st Century*. (Prepared in the framework of IHP, UNESCO), State Hydrology Institute, St. Petersburg.
4. NRC. (1991). Opportunities in the Hydrologic Sciences. *Committee on 'Opportunities in the Hydrologic Sciences' of National Research Council*. National Academy Press, Washington, DC.
5. Singh, V.P. (1988). *Hydrologic Systems*. Vol. I. Prentice-Hall, Englewood Cliffs, NJ.
6. Horton, R.E. (1931). *The Field, Scope, and Status of the Science of Hydrology*. pp. 189–202. in Trans. AGU, Reports and Papers, Hydrology. National Research Council, Washington, DC.
7. Seckler et al. (1998). *World Water Demand and Supply, 1990 to 2025: Scenarios and Issues*. Research Report 19. International Water Management Institute. Colombo, Sri Lanka.
8. UNESCO. (1978). *World Water Balance and Water Resources of the Earth*. United Nations Educational, Scientific, and Cultural Organization, Paris.
9. NRC. (1986). *Global Change in Geosphere-Biosphere*. National Research Council, National Academy Press, Washington, DC.

HYDROSPHERE

Yoseph Negusse Araya
Open University
Milton Keynes, United Kingdom

INTRODUCTION

The hydrosphere [Greek *hydor* water and *sphera* sphere] refers to the water on or surrounding the surface of the globe, as distinguished from those of the lithosphere (the solid upper crust of the earth) and the atmosphere (the air surrounding the earth). More specifically, the hydrosphere includes the region that includes all the earth's liquid water, frozen and floating ice, water in the upper layer of soil, and the small amounts of water vapor in the earth's atmosphere. The hydrosphere is the major setting for the earth's hydrologic cycle [see Hydrologic Cycle].

ORIGIN OF WATER ON EARTH

The most significant feature of the earth, in contrast to our neighboring planets, is the liquid water that covers more than two-thirds of the planet's surface. This water came about during the early days of the formation of the earth, when the earth's surface cooled down and the oxygen and hydroxides contained in the accreted material, diffused toward the surface. These gases then cooled and condensed to form the earth's oceans. It is believed that since then, there has been little loss or gain in the overall quantity of the hydrosphere, despite minor fluctuations such as gain from continued degassing and infalling comets and loss in the upper layers of the atmosphere caused by ultraviolet light breaking up water molecules.

DISTRIBUTION AND QUANTITY OF WATER ACROSS THE GLOBE

The earth's water has six major reservoirs in which water resides. These include the oceans, the atmosphere (split into two reservoirs, one over the land and one over the oceans), surface water (including water in lakes, streams, and the water held in the soil), groundwater (water held in the pore spaces of rocks below the surface), and snow and ice. The locations of some major reservoirs on earth are shown in Fig. 1.

The approximate contribution of the different components of the reservoirs to the hydrosphere, the annual recycled volumes, and the average replacement periods are shown in Table 1.

Table 1 highlights the enormous disparity between the huge volume of salt water and the tiny fraction of freshwater and, in addition, the long residence time of polar ice and groundwater, as opposed to the brief period for which water remains in the atmosphere. Some 96.5% of the total volume of the world's water, it is estimated, exists in the oceans and only 2.5% as freshwater. Of this freshwater, nearly 70% occurs in the ice sheets and

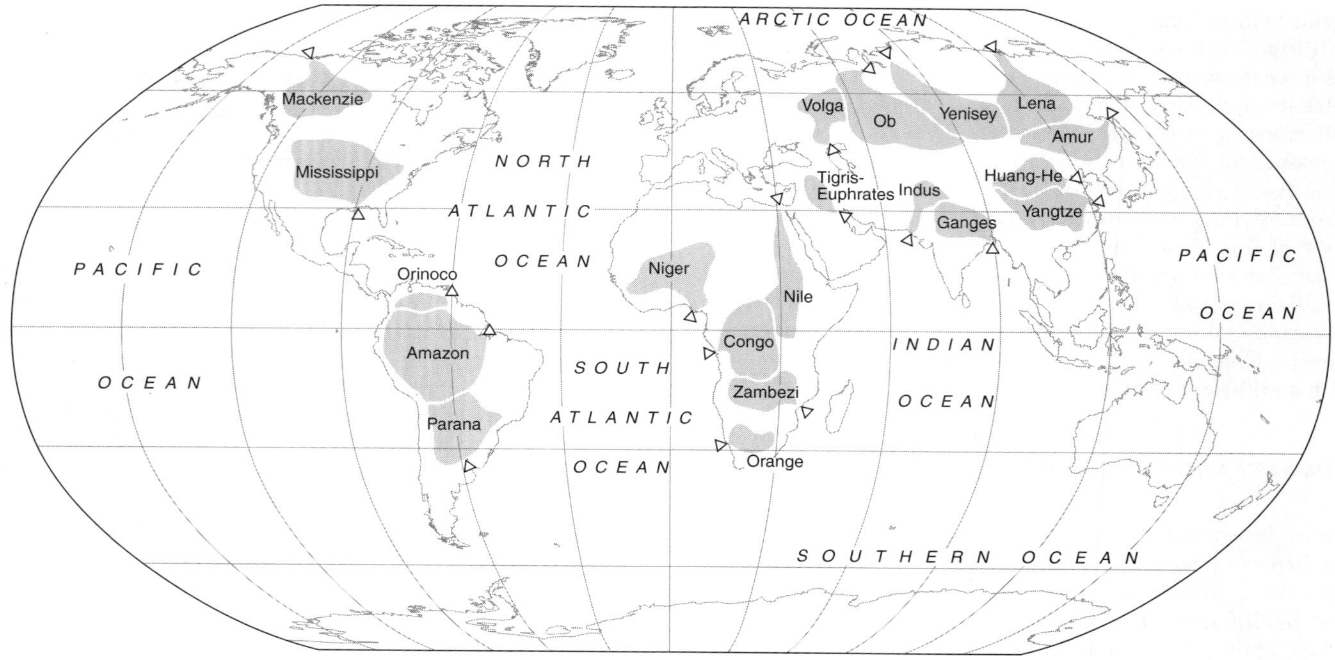

Figure 1. The locations of some major global water reservoirs: oceans and surface water drainage basins (1).

Table 1. The Distribution of Water Across the Globe[a]

Location	Volume 10^3 km^3	% of Total Volume in Hydrosphere	% of Freshwater	Volume Recycled Annually km^3	Renewal Period, Years
Ocean	1,338,000	96.5	—	505,000	2,500
Groundwater (gravity and capillary)	23,400[b]	1.7		16,700	1,400
Predominantly fresh groundwater	10,530	0.76	30.1		
Soil moisture	16.5	0.001	0.05	16,500	1
Glaciers and permanent snow cover	24,064	1.74	68.7		
Antarctica	21,600	1.56	61.7		
Greenland	2,340	0.17	6.68	2,477	9,700
Arctic Islands	83.5	0.006	0.24		
Mountainous regions	40.6	0.003	0.12	25	1,600
Ground ice (permafrost)	300	0.022	0.86	30	10,000
Water in lakes	176.4	0.013	—	10,376	17
Fresh	91.0	0.007	0.26		
Salt	85.4	0.006	—		
Marshes and swamps	11.5	0.0008	0.03	2294	5
River water	2.12	0.0002	0.006	43,000	16 days
Biological water	1.12	0.0001	0.003		—
Water in the atmosphere	12.9	0.001	0.04	600,000	8 days
Total volume in the hydrosphere	1,386,000	100	—		
Total freshwater	35,029.2	2.53	100		

[a] Reference 2.
[b] Excluding groundwater in the Antarctic estimated at 2 million km^3, including predominantly freshwater of about 1 million km^3.

glaciers in the Antarctic, Greenland and in mountainous areas, whereas a little less than 30%, it is calculated, is stored as groundwater in the world's aquifers.

Water moves through the reservoirs by a variety of processes, at different rates, and for unique residence times within any reservoir. This flow of water constitutes the Earth's hydrologic cycle [see HYDROLOGIC CYCLE]. A brief summary of the major processes involved in this movement, along with the amount of water transferred per unit time are shown in Fig. 2.

THE BIOGEOCHEMISTRY OF THE HYDROSPHERE

The quality of natural water in the various reservoirs of the hydrosphere depends on a number of interrelated factors. These factors include geology, climate, topography, biological processes, land use, and the time for which the water has been in residence. Table 2 gives a comparison of major elements in selected reservoirs.

Rainwater has a low concentration of nutrients compared to the other reservoirs because it originates as evaporated water vapor and also has a relatively short residence time in the atmosphere. Even so, it is never pure. The major constituents originate from dissolution of aerosol particles, which are formed from natural processes, such as evaporation of sea spray or human activities, such as burning of fossil fuels. Naturally rain water has a slightly acid pH (about 5.5). This results from the formation of mild carbonic acid, when rainwater reacts with atmospheric carbon dioxide:

$$CO_2 + H_2O \longrightarrow H_2CO_3$$

In areas of high emission of sulfur dioxide or nitrogen oxide gases from industrial activities or fossil fuel burning, hydrolysis of rainwater may result in more acidic rain and a pH as low as 4 [see Acid Rain].

River waters have an intermediate concentration of ions compared to that of rainwater and oceans. The main factor controlling the composition or river water is the weathering reaction between rainfall and rocks through which this water passes. An example is that of calcite in limestone, which reacts with carbonic acid of rainfall, as

$$CaCO_3 + H_2CO_3 \longrightarrow Ca^{2+} + 2HCO_3^{-}$$

Lakes also have an intermediate concentration of ions compared to those of river and seawater. Lake waters constitute a reservoir of freshwater, and their composition depends on four factors: the hydrology (e.g., the relative importance of groundwater or surface water inputs, evaporation), the surrounding geology (e.g., carbonate rocks or granite), temperature-driven circulation patterns, and anthropogenic factors (e.g., acid rain, agricultural fertilizers). In some instances, evaporation of water from lakes formed in closed basins may result in a high concentration of salts, as opposed to areas of high rainfall.

Sea and ocean waters are dominated by sodium and chloride ions, followed by sulfate and magnesium. Surface sea water is alkaline at an average pH of about 8. Seawater tends to have a more or less uniform composition in the major elements. But concentrations of minor constituents, including trace and heavy metals and nutrients, vary with depth and location, resulting in marked differences in biological productivity. Organisms living on the surface of the seawater are also involved in changes in its composition via removal of nutrients and breakdown of organic matter at different depths.

Ground water composition is the result of the rock type in which it is confined (e.g., limy is to calcium as argillaceous is to silica); the chemical processes of dissolution, hydrolysis, oxidation-reduction; and biological processes. Moreover, anthropogenic contaminants such as excess fertilizers and heavy metals may also affect the composition of groundwater.

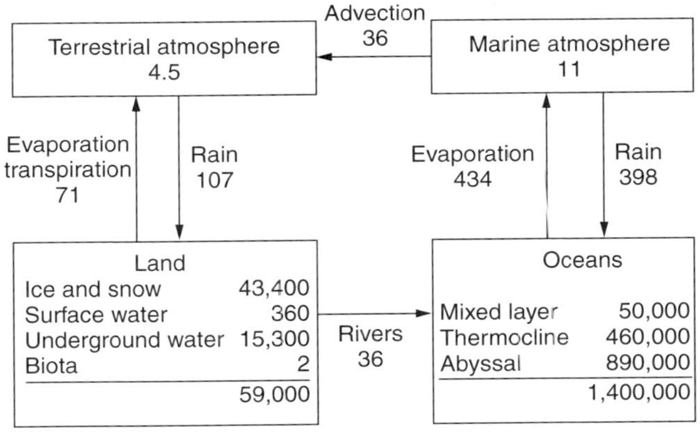

Figure 2. Estimates of global water reservoirs (in 10^{15} kg and 10^{15} kg/yr) global water cycle fluxes (3).

Table 2. Chemistry of Some Hydrospheric Components (in Parts per Million—ppm)[a]

Major Element	Average Seawater	Average Natural River Water	Average Rainwater
Chloride (Cl^-)	19,000	5.75	3.79
Sodium (Na^+)	10,500	5.15	1.98
Sulphate (SO_4^{2-})	2,700	8.25	0.58
Magnesium (Mg^{2+})	1,350	3.35	0.27
Calcium (Ca^{2+})	410	13.4	0.09
Potassium (K^+)	390	1.3	0.3
Bicarbonate (HCO_3^-)	142	52	0.12
Bromide (Br^-)	67	0.02	
Strontium (Sr^{2+})	8	0.03	
Silica (SiO_2)	6.4	10.4	—
Boron (B)	4.5	0.01	
Fluoride (F^-)	1.3	0.1	

[a]References 1 and 4.

Ice is a pure solid and has thus only few impurities in its structure, but particulate matter and gases may be trapped within it. Analysis of successively trapped gases or other anthropogenic substances such as carbon dioxide in polar ice caps, has been used to study consecutive changes in the atmospheric composition of past times.

EFFECT OF HUMAN BEINGS ON THE HYDROSPHERE

During the last 200 or so years, the sharp rise in population, urbanization, industrial development, and intensification of agricultural practices have combined to affect most natural waterbodies of the earth. This is due to the transport of waste products from those activities by surface water, groundwater, and the atmosphere. The scale and intensity of this pollution vary considerably; there are global problems such as the presence of heavy metals, regional problems such as acid rain, and much more localized ones such as groundwater contamination.

Overall, globally, organic material from domestic sewage, municipal waste, and agroindustrial effluent is the most widespread pollutant. The sewage contains

Table 3. The World's Major Water Quality Issues[a]

Issue Scale	Waterbodies Polluted[b,c]	Sector Affected	Time Lag Between Cause and Effect	Effects Extent
Organic pollution	Rivers++ Lakes + Groundwater +	Aquatic environment	< 1 year	Local to district
Pathogens	Rivers ++ Lakes + Groundwater +	Health ++	< 1 year	Local
Salinization	Groundwater ++ Rivers +	Most uses Aquatic environment Health	1–10 years	District to region
Nitrate	Rivers + Lakes + Groundwater ++	Health	> 10 years	District to region
Heavy metals	All bodies	Health Aquatic environment Ocean fluxes	< 1 to > 10 years	Local to global
Organics	All bodies	Health Aquatic environment Ocean fluxes	1–10 years	Local to global
Acidification	Rivers ++ Lakes ++ Groundwater +	Health Aquatic environment	> 10 years	District to region
Eutrophication	Lakes ++ Rivers +	Aquatic environment Most uses Ocean fluxes	> 10 years	Local
Sediment load (increase and decrease)	Rivers + Lakes	Aquatic environment Most uses Ocean fluxes	1–10 years	Regional
Diversion, dams	Rivers + Lakes + Groundwater ++	Aquatic environment Most uses	1–10 years	District to region

[a] References 5 and 6.
[b] +: serious on global scale.
[c] ++: very serious issue on global scale.

pathogens that lead to disease and mortality among the populations using this water.

Moreover, this organic material has also high concentrations of nutrients, particularly nitrogen and phosphorus, which cause eutrophication (i.e., nutrient enrichment) of lakes and reservoirs. This eutrophication promotes abnormal plant growth and oxygen depletion, which destroys aquatic ecosystems. Excess fertilizers from agricultural production areas also have similar consequences.

Acidification of surface waters as a result of acid rain has adverse effects on aquatic life and also human health. Salinization, the high concentration of salts in the soils of irrigated areas, as a result of poor drainage and high evaporative loss also causes water pollution. Sediments in the form of suspended load may affect physical structures, for example, silting up of dams and damage to aquatic life. Table 3 is a summary of these problems by the type of waterbodies polluted and the extent and reach of the effects.

In general, it may be noted that the effect of pollution can reach far beyond the vicinities of its origin. Moreover, the effects may not be noticed until substantial time has elapsed and it is too late. This calls for constant monitoring and control strategies to save the earth's most precious resource, water.

BIBLIOGRAPHY

1. Ernst, W.G. (Ed.). (2000). *Earth Systems, Processes and Issues*. Cambridge University Press, Cambridge, UK.
2. Shiklomanov, I.A. (in press). *World Water Resources at the Beginning of the 21st Century*. Cambridge University Press, Cambridge, UK.
3. Chahine, M.T. (1992). The hydrological cycle and its influence on climate. *Nature* **359**: 373–380.
4. Meybeck, M. (1979). Concentration des Eaux Fluviales en Eléments Majeurs et Apports en Solution aux Océans. *Revue de Géologie Dynamique et de Géographie Physique* **21**: 215–246
5. United Nations Environment Programme. (1991). *Freshwater Pollution*. Nairobi, United Nations Environment Programme/Global Environment Monitoring System Environment.
6. World Health Organization. (1991). *GEMS/Water 1990–2000. The Challenge Ahead*. UNEP/WHO/UNESCO/WMO Programme on Global Water Quality Monitoring and Assessment, Geneva.

READING LIST

World Water Assessment Programme. (2003). *Water for People, Water for Life: The United Nations World Water Development Report*. UNESCO, Berghahn Books, Paris.

HYDROLOGIC CYCLE, WATER RESOURCES, AND SOCIETY

JASON A. HUBBART
University of Idaho
Moscow, Idaho

THE HYDROLOGIC CYCLE

Water is more or less constantly moving and changing from one state to another (solid, liquid, or vapor) while interacting with the physical processes in the atmosphere, lithosphere, and biosphere. These changes and movements of water are linked together into a grand scheme called the hydrologic cycle. Components of the hydrologic cycle include the water vapor and clouds in the atmosphere, and also liquid surface waters (oceans, lakes, and streams) on continents as well as groundwater. Other important components of the hydrologic cycle include glacial ice held on continents and water in biomass. Plants and animals are about 70% water by volume. Water evaporates in enormous quantities from the oceans and then falls as precipitation either on land or ocean. That portion which falls on land evaporates, evapotranspires, runs off, or infiltrates by some measure. Eventually, all water ends up back in the ocean. Between the various stages of the hydrologic cycle, water moves between temporary storage areas often called reservoirs. These movements are controlled by climatic conditions, which include rain, snow, wind, and other meteorologic processes.

Various pathways exist in the hydrologic cycle (Fig. 1). In nonvegetated systems, interception processes usually do not occur; however, within vegetated systems, precipitation can be intercepted and lost to evaporation. Some precipitation may reach the ground by stemflow but most often reaches the soil by a process called throughfall. Precipitation that reaches the soil surface can then infiltrate and recharge soil waters. From there, it can percolate through soil to deeper soil layers or groundwater supplies. The rate of evaporation from soils or surface waters (including oceans) can be influenced by several factors, including temperature, relative humidity, and wind.

Warmer air can hold more water vapor, thereby generating increased evaporation rates. Drier air can accept much more water vapor than air that is nearly saturated, and windy conditions generally accelerate evaporation. Transpiration from plants takes place via water loss from the stomata, the gas exchange organs on the leaves of plants. The stomata also act as an important cooling mechanism for plants by means of latent heat absorbed and transferred from the plants. This process also supplies the tensional forces necessary to help draw water into roots from the surrounding soil.

Transpiration is often lumped with evaporation and is then called evapotranspiration. Besides evapotranspiration, water can be converted to water vapor in the atmosphere by sublimation. When water is heated enough, it becomes water vapor, which results in evaporation. Sublimation is the result of ice going directly from a solid to water vapor without passing through a liquid state.

Air can hold only a certain amount of water vapor. The amount depends largely upon temperature; thus, as mentioned, hotter air can hold more water vapor than colder air. When air is saturated with water vapor, the water vapor condenses into droplets of water, forming clouds. When the droplets gain enough mass, they fall to the ground as rain or snow. Water that falls as snow or rain evaporates, sublimates, runs off the ground surface (runoff), or soaks into the ground (infiltration).

Whether the water runs off or infiltrates is controlled by a number of mechanisms including precipitation rate, soil water content, slope, and vegetation.

As rates of precipitation increase, runoff rates tend to increase. In extreme cases, rapid cloudbursts of precipitation can result in flash floods and potential land slides. Prior to rainfall, the relative amount of soil water content can dictate the amount of water that soil can hold before saturation. Generally, more water will infiltrate into dry soil than into wet soil, and more water will run off steeper slopes than off moderate slopes. Vegetation can counteract this process by holding more water than hill slopes by themselves. Hillsides, which have undergone timber harvesting or have been burned, often suffer severely from mudslides and flooding. Runoff processes within the hydrologic cycle include stream flow and groundwater flow, each of which eventually results in flow to ocean reservoirs. Lakes and streams may occur in low points of the landscape due to streamflow, or where the groundwater table emerges above the surface. Other contributors of the runoff process include melting of glacial ice and calving of icebergs into the sea.

Globally, precipitation exceeds evapotranspiration over continents, evaporation exceeds precipitation over ocean reservoirs, and excess water moves from continents to oceans as runoff. The ratio of precipitation rate to evaporation rate can vary dramatically between regions. For example, in tropical rainforests, precipitation may greatly exceed evapotranspiration (1).

Often, rates of hydrologic activity are measured in terms of "mean residence time," or the average amount of time that water remains in its various states. For example, the mean residence time of a water molecule in the atmosphere is very short, usually from days to a week or two. Water tends to move quickly through plant and animal tissues but can be stored in cells for a much longer period of time. It can take weeks to months for water to move through surface drainage networks, depending on the complexity of the geomorphologic network. Water may be stored for months to years in soil water, and individual water molecules may remain in deep groundwater, glaciers, and ocean basins for decades to 10,000 or more years.

Traditionally, quantities of water transferred within the global hydrologic cycle are described in terms of cubic kilometers (km^3) by virtue of the immense quantities held in respective reservoirs. Each km^3 of water contains 10^{12} liters and weighs 10^{15} grams (1). Often, changes in the water cycle are discussed in terms of depth, for example, centimeters per year (cm/yr) or millimeters per hour (mm/hr). Annually, approximately 100 cm of water are evaporated from the surface area of oceans.

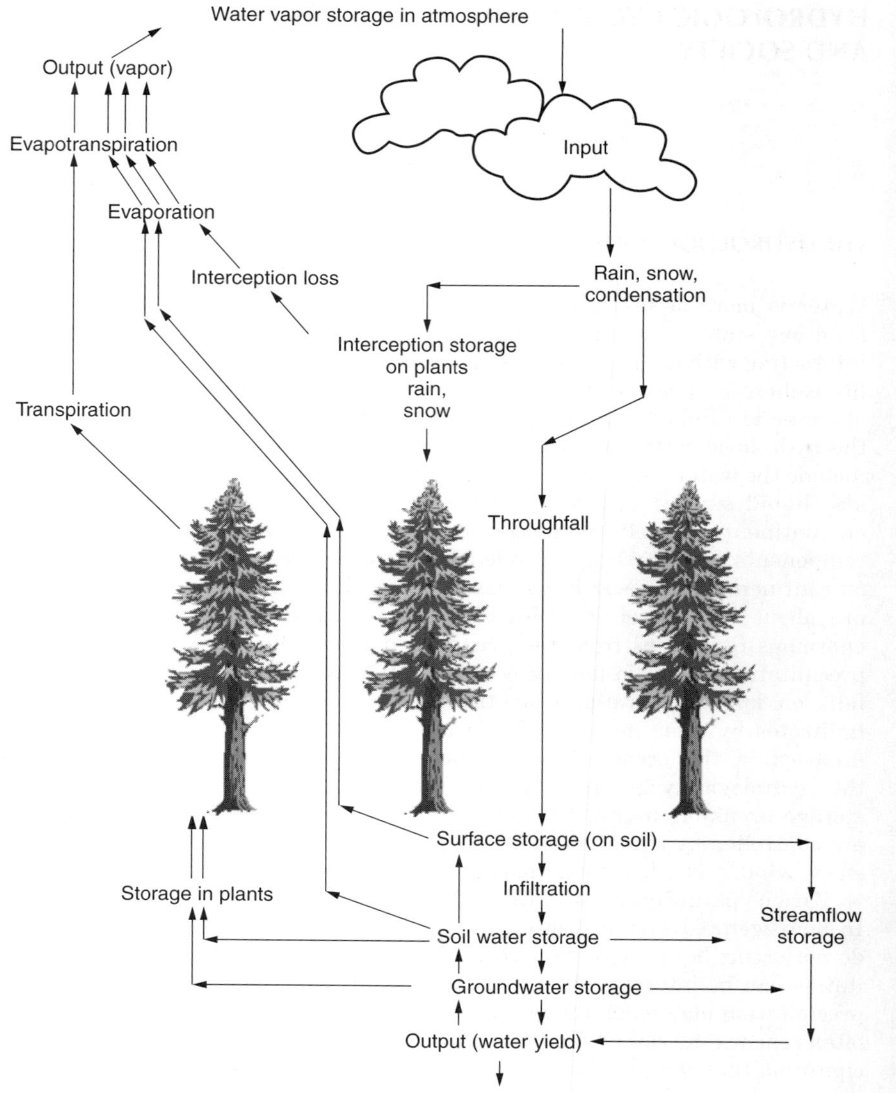

Figure 1. The hydrologic cycle, consisting of water storage compartments and the solid, liquid, and gaseas phases associated within and between each compartment.

Oceans, which also happen to be the dominant pool in the global water cycle, contain approximately 1,350,000 km^3 of water. Of this, approximately 425,000 km^3 are evaporated to the atmosphere each year, 385,000 km^3 are returned directly to the oceans via precipitation, and 40,000 km^3 are delivered to land by rainfall (net transport to land). At any given time, the atmosphere contains about 13,000 km^3 water, or approximately 0.3 cm of rainfall. The large quantities of annual water movement through the atmosphere result in 111,000 km^3 being precipitated upon land and vegetation and 71,000 km^3, either evaporated or transpired (evapotranspiration). Polar and glacial ice reservoirs account for approximately 33,000,000 km^3 water, water held in soils is equal to approximately 122,000 km^3, and groundwaters hold approximately 15,300,000 km^3 (1–3).

WATER RESOURCES

Water resources refer to the supply of groundwater and surface water in a given area. Water resources may also reference the current or potential value of the resource to the community and the environment. The maximum rate at which water is potentially available for human use and management is often considered the best measure of the total water resources of a given region. Approximately 30% of the world's fresh water is in liquid form and therefore potentially accessible for human use and management at any given time (4). The rest is either locked up in polar or glacial ice or water vapor. Of the 30%, almost all is held in groundwaters. Historically, attempts to develop global assessments of available water resources have resulted in limited applicability. The usefulness of resulting aggregated quantities, based on streamflow and population calculations, which lead to measurements in relative abundance and shortages of water regionally, have often been unreliable. The extreme difficulty in preparing a global assessment stems from the general lack of sufficient and reliable information on water availability, quality, and water use in many areas of the world.

Efforts to balance supply and demand and plans for a sustainable future are severely hampered by this lack of reliable information. Studies of water resources leading to meaningful assessments have been found realistic only if conducted regionally or locally. Only then has proper accounting of seasonal and interannual variability of stream flow as well as interactions between groundwater and surface water been appropriately accounted for. Likewise, only then have the potentials for reusing the water as it proceeds downstream and the balance of in-stream and withdrawal uses been appropriately managed. Reduced scale also allows for the importance of water quality in determining suitability for use and, most importantly, the realistic evaluation of social, economic, and political factors that help determine per capita water use (5–8).

Despite these constraints, some attempts to describe global water use have led to reasonable conclusions. For example, estimates indicate that since 1900, global water withdrawal has increased about ninefold and per capita withdrawal has quadrupled. Globally, the largest use of water is for irrigation (70%); industry uses 20%, and 10% is for direct human consumption (9). As a result, humans now withdraw about 35% of the world's reliable runoff. At least another 20% of this runoff is left in streams to transport goods by boats, dilute pollution, and sustain fisheries and wildlife (10). Obviously, these percentages may vary greatly from one region to another, depending on natural precipitation and the degree of development and human population in the region (9). There may be further variance because the distribution of water resources over the landmass of the earth is uneven and unrelated to population demographics or economic development. Because of these complexities and constraints, humans have attempted to increase available water resources by increasing precipitation in various anthropogenic manners, including cloud seeding. Humans have also attempted to decrease evapotranspiration by modifying vegetation, sometimes genetically. The effects of these manipulations are usually minimal and most often temporary but unfortunately often have serious environmental, social, economic, and legal ramifications (4).

SOCIETY

Water is central to survival. Without it, plant and animal life would be impossible. Water is a central component of the earth's ecosystems and provides important controls on weather and climate. Water is likewise central to economic well-being via delivery of direct precipitation and agricultural irrigation, forestry, navigation, waste processing, and hydroelectricity. There are more than 250 international river basins, and in each basin, water moving across diplomatic boundaries can create sensitive and potentially dangerous political challenges. Our present understanding of climate and the consequences of future climate change (whether natural or anthropogenic), together with population growth and economic development, necessitate that water resources are of increasing interest and importance for the foreseeable future.

Reflecting the impacts of population growth alone, water use is increasing everywhere. If per capita consumption of water resources continues to rise at its current rate, humans could be using over 90% of all available freshwater by the year 2025, leaving just 10% for all other living organisms. This forecast does nothing to reflect the high likelihood of climate change. Many studies point to serious consequences for water resources due to climate change, especially in arid regions. For instance, a 1 °C to 2 °C increase in the mean annual air temperature and a 10% decrease in precipitation can create a 40 to 70% reduction in annual runoff in regions that already experience insufficient moisture. The same is true for water resources of large economic regions. American scientists have calculated that a 2 °C climate warming and a 10% reduction in precipitation, can decrease water resources by 1.5 to 2.0-fold in those regions of the country located in arid climate zones alone.

It is well known that the severity of water management problems is determined by the ratio of available water resources to demand. Based on predicted climate changes, plans for further development in irrigated agriculture, reservoir construction, and most water-consuming types of industry will have to be revised. All such revisions will need to result in the greatest positive impacts on hot arid regions. These regions already have difficulties with water supplies and undergo persistent conflicts between different water users and consumers. Regardless, as water use responds to changing demand with some degree of inertia, regional water availability will be determined primarily by the changing water regime within a given territory. The water management systems of river basins will be complex and will have to be extremely flexible and capable of efficient control over water resources under different climate scenarios. In this respect, the regions with large capabilities in river runoff regulation, as is the case in many parts of the United States of America, Canada, and Europe, will have considerable advantages relative to solving water supply and flood regulation, as opposed to those regions with more natural river systems such as regions located in southern and southeastern Asia and South America.

Relatively little account has been taken of the qualitative or quantitative depletion of water resources due to increasing pollution of natural freshwater sources. On this basis alone, all water resource estimates could be construed as optimistic. The major sources of intensive pollution of waterways and waterbodies are contaminated industrial and municipal wastewater as well as water runoff originating from irrigated areas. This problem can be no more acute than in the industrially developed and densely populated regions where relatively little wastewater purification takes place. It is estimated that in 1995, the volume of wastewater was 326 km^3/year in Europe, 431 km^3/year in North America, 590 km^3/year in Asia, and 55 km^3/year in Africa. Many of these countries discharge large portions of their wastewater containing harmful substances into the hydrologic system with no preliminary purification. Prime water resources are thus polluted and their subsequent use becomes unsuitable, especially as potable supplies. Every cubic meter of contaminated wastewater discharged into water bodies and watercourses can potentially spoil between eight and ten cubic meters of pure water. This means that most parts of the world are already facing the threat of catastrophic depletion of water resources due to pollution alone.

Forecasts of future population, industry, and power generation growth are used to predict water withdrawal and water availability. The values obtained for 2025 are ±10–15% for those regions of predominantly developed countries and ±20–25% for regions with predominantly developing countries. To be able to plan for and properly manage these potential future demands for fresh water, it will be vital to achieve close cooperation between scientists and water resource managers. Scientists from different countries and international organizations will have to collaborate and deal with many problems related to hydrology, climatology, and other related disciplines in an cooperative effort to understand better the complex use and protection of water resources.

In human history, water has been central to society and remains so today. Over time, we have grown more aware of the accelerated growth of industry and population correlated to increased uses of water and how that use relates to the earth's fixed supply. In recent times, it has become clearer that human prosperity and prospects for survival vary with the amount and distribution of fresh, unpolluted water. Each year there are millions more humans but no more water than before (11). The same amount of water exists now as when the earth was formed. Ultimately, although finite in quantity, water is a renewable resource. There can be little argument that exponential population growth coupled with an increasing demand for freshwater is resulting in increasing pressures on the resource. The continued degradation, abuse, and pollution of the environment and its resources may cause irremediable damage to our own health and that of the planet on which all life depends. Despite strong feedback mechanisms between hydrologic and meteorologic processes, which have historically led to poor representations of hydrologic processes, increased understanding will lead to improved management strategies for the future and the various processes of the hydrologic cycle.

BIBLIOGRAPHY

1. Schlesinger, W.H. (1997). *Biogeochemistry, An Analysis of Global Change*. Academic Press, San Diego, CA, pp. 346–349.
2. L'vovitch, M.I. (1973). The global water balance. *earth observing soc. (EOS), Trans. Am. Geophys. Union* **54**: 28–42.
3. Chahine, M.T. (1992). The hydrologic cycle and its influence on climate. *Nature* **359**: 373–380.
4. Dingman, S.L. (1994). *Physical Hydrology*, 1st Edn. Prentice-Hall, Upper Saddle River, NJ, pp. 67–68.
5. L'vovitch, M.I. (1974). *World Water Resources and Their Future*. Translated by R.L. Nace. American Geophysical Union, Washington, DC, p. 56.
6. Barney, G.O. (Ed.). (1980). *The Global 2000 Report to the President of the United States*. Pergamon Press, New York, pp. 328–331.
7. Laurenson, E.M. (1987). Assessment of water resources—group report. In: *Resources and World Development*. D.J. McLaren and B.J. Skinner (Eds.). Wiley-Interscience, New York, p. 16.
8. Shiklomanov, I.A. (1989). Climate and water resources. *Hydrol. Sci. J.* **34**: 495–529.
9. Wright, R.T. and Nebel, B.J. (2002). *Environmental Science Toward a Sustainable Future*, 8th Edn. Prentice-Hall, Upper Saddle River, NJ, pp. 214–217.
10. Miller, G.T. (2004). *Living in the Environment. Principles, Connections, and Solutions*, 13th Edn. Brooks/Cole, Pacific Grove, CA, p. 314.
11. Ward, A.D. and Elliot, W.J. (Eds.). (1995). *Environmental Hydrology*. Lewis, CRC Press, Boca Raton, FL, p. 9.

ISOHYETAL METHOD

Sharad K. Jain
National Institute of Hydrology
Roorkee, India

Vijay P. Singh
Louisiana State University
Baton Rouge, Louisiana

The isohyetal method is used to estimate the mean precipitation across an area by drawing lines of equal precipitation. The method uses topographic and other data to yield reliable estimates.

INTRODUCTION

An accurate assessment of the mean areal precipitation is needed in many hydrologic analyses. Precipitation observations from gages are point measurements, and precipitation exhibits appreciable spatial variation over relatively short distances. Numerous methods of computing areal precipitation from point measurements have been proposed, such as the arithmetic average method, the Thiessen polygon method, the Isohyetal method, and kriging. The choice of any method depends on the quality and nature of data, precision required (1), the availability of time and resources, and the preference of the analyst.

Let the precipitation data be available at n stations, spread over an area and P_i be the observed depth of precipitation at the ith station. Using a linear interpolation technique, an estimate of the precipitation across the area can be expressed by

$$P* = \sum_{i=1}^{n} P_i W_i \tag{1}$$

where W_i is the weight at the ith station. Spatial averaging techniques differ in the method of evaluating these weights. The weights of an optimal interpolation technique are chosen so that the variance of error in estimation is a minimum.

The simplest technique for computing the average precipitation depth across a catchment is to take an arithmetic average of the observed depths at gauges within the area for the time period of concern. If the gauges are relatively uniformly distributed across the catchment and the precipitation depths do not vary greatly, this technique yields good results. However, gauges are not always uniformly distributed, there may be large variations in precipitation depths, and topography and orography influence storm movement and precipitation. For these reasons, the arithmetic average is not a preferred method; a technique that can take into account the spatial variability of precipitation is needed.

ISOHYETAL METHOD

Isohyets are contours of equal precipitation analogous to contour lines on a topographic map. In the isohyetal method, precipitation values are plotted at their respective stations on a suitable base map, and isohyets are drawn to create an isohyetal map. Isohyetal lines are based on interpolation between rain gauge stations. While constructing isohyets, it is assumed that rainfall between two stations varies linearly, unless abrupt changes in topography indicate otherwise.

Obviously, this method works best when there are a number of rain gauges. In regions of little or no physiographic influence, isohyetal contours may be drawn taking into account the spacing of stations, the quality, and the variability of the data. In regions of pronounced orography, where precipitation is influenced by topography, analysts take into consideration the orographic effects and storm orientation to interpolate between station values, thus making full use of their knowledge of orographic effects and storm morphology. The areas enclosed by two successive isohyets are multiplied by the average of the isohyets to obtain the volume of precipitation needed to compute the average rainfall depth across the basin. However, a more accurate value of mean areal rainfall is obtained if the lengths of the isohyets are measured as well as the area between the isohyets. If b is the length of the lower value isohyet B, a is the length of the higher value isohyet A, and Δi is the isohyet interval (A–B), then it can be shown that mean rainfall r for the segment is

$$r = B + \frac{\Delta i(2a + b)}{3(a + b)} \quad (2)$$

An isohyetal map resembles a topographical map, except that it contains contours of equal precipitation depths. The topographic parameters that are used to draw isohyets are the elevation of the station, the proximity to mountains and their slope, and the proximity to valleys. If linear interpolation between stations is used, the results will be essentially the same as those obtained by methods, such as the Thiessen polygon method. Clearly, the accuracy of the method depends high on the skill of the analyst; an improper analysis may lead to serious errors. This technique of computing the average precipitation is laborious, if done manually, because the precipitation values need to be interpolated, isohyetal lines are to be drawn, and areas are to be measured for each storm.

Steps of the Isohyetal Method

1. The location of each station is plotted on a map of the catchment.
2. The observed depth of precipitation at each station is marked on the map.
3. The increment of isohyets is chosen, depending on the variation of precipitation values and the scale of the map.
4. Isohyets at selected increments are plotted. In flat areas, linear interpolation between observed precipitation depths is performed. If the area has mountains, valleys, etc., isohyets are drawn keeping in view the influence of these features on precipitation.
5. Starting either from the outermost or the innermost isohyet, the average precipitation between two consecutive isohyets is obtained as the arithmetic average of the value of two isohyets. For example, between the 10 and 20 cm isohyets, the average precipitation would be 15 cm.
6. The area between the two isohyets is measured by using a planimeter or by counting the number of boxes (if the map is drawn on graph paper) and using the scale of the map.
7. Multiplying the area between isohyets by the average precipitation gives the volume of precipitation for that area. The total catchment precipitation is the sum of all such values.
8. The average depth of precipitation is obtained by dividing the sum by the area of the catchment:

$$P = \frac{\sum_{i=1}^{n} P_i A_i}{\sum_{i=1}^{n} A_i} \quad (3)$$

where A_i is the area between successive isohyets and P_i is the average precipitation across this area.

Figure 1 shows an isohyetal map for a catchment that was drawn using point rainfall data. The area enclosed by each isohyet was calculated as given in Table 1. To compute the average catchment rainfall, the average isohyetal value is worked out for each isohyet (the maximum observed rainfall was 108 cm and the minimum 38 cm). The average value multiplied by the area enclosed by the corresponding isohyets gives the volume of precipitation for that area. The volumes for different areas are summed and divided by the area of the catchment to get the average catchment rainfall. The computations are shown in Table 1.

Features of the Isohyetal Method

Important features of the isohyetal method are

- This method is widely regarded as the most accurate method for determining the average precipitation across an area.

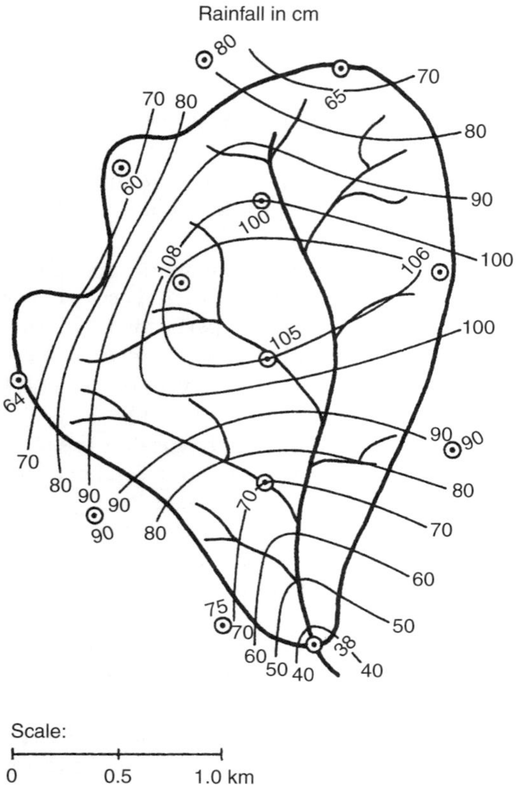

Figure 1. The isohyetal method for computing mean areal rainfall.

Table 1. Estimate of Mean Areal Rainfall by the Isohyetal Method

Isohyet Value, cm	Average Value, cm	Area Enclosed, km^2	Net Area, km^2	Rainfall Volume, km^2-cm
105	106.5	0.79	0.79	84.14
100	102.5	1.52	0.73	74.83
90	95	2.57	1.05	99.75
80	85	3.47	0.90	76.50
70	75	4.50	1.03	77.25
60	65	5.18	0.68	44.20
50	55	5.39	0.21	2.20
<40	39	5.41	0.02	0.78
	Total		5.41	459.65
Average catchment rainfall = 459.65/5.41 = 84.96 cm				

- The isohyetal method permits using available pertinent data, such as orographic features and storm morphology.
- This method is subjective because drawing isohyets depends on the individual analyst. Consequently, there is a potential for mistakes.
- The method is time-consuming.
- The isohyetal method is usually applied in the past tense. It is not generally applied in the present to provide inputs for modeling or forecasting.
- This method can be used for spatial averaging of other variables, such as evapotranspiration and snow depth.

There are many computer softwares (2,3) that can be used to draw an isohyetal map from point data. Most of these use a statistical technique known as kriging. The use of software relieves the hydrologist of the tedious job of drawing a map, interpolating precipitation from point data, and drawing isohyets. But most software follow a mathematical algorithm and do not use topographic or storm features. While manually drawing isohyets, most people use judgment rather than actual calculations in interpolation. An alternative is using a computer to draw the isohyetal map which can be refined by the hydrologist later using additional information, experience, and judgment. This results in a considerable saving of time and provides better results.

BIBLIOGRAPHY

1. Singh, V.P. (1989). *Hydrologic Systems*. Vol. II, *Watershed Modeling*. Prentice-Hall, New Jersey.
2. Diskin, M.H. and Davis, P.R. (1970). *A Computer Program for Interpolation and Plotting of Isohyetal Points*. Agricultural Research Service, ARS 41–177, United States Department of Agriculture, Washington, DC.
3. Shearman, R.J. and Salter, P.M. (1975). An objective rainfall interpolation and mapping technique. *Hydrological Sciences Bulletin* **20**(3): 353–363.

WHAT ABOUT METEOROLOGY?

Pacific Northwest National Laboratory—Shrub-Steppe Ecology Series

BACKGROUND INFORMATION

Meteorology is the science of the atmosphere. The word meteorology comes from the Greek meteors, which means lofty or elevated and logos, meaning discourse or description. The field of meteorology includes both weather and climate and is concerned with the physical, dynamical (a force producing change or motion), and chemical state of the Earth's atmosphere and interactions between the atmosphere and the Earth's surface. Humans' earliest observations of climate and weather date to hunters, gatherers, and farmers, who studied signs of changing weather in the sky. Today, scientists who study weather are called meteorologists.

Weather and climate are closely related, but big differences exist between them. Weather refers to the condition of the atmosphere at a given time and location. It describes daily local changes in temperature, humidity, wind, clouds, and precipitation (rain, snow). Climate, in contrast, describes the average of these conditions over a long period of time–month, year, or more (annual precipitation, monthly mean, high, and low temperatures) and variability.

WHAT ABOUT THE COLUMBIA BASIN'S CLIMATE?

As you know, the climate of the Columbia Basin tends to be warm and dry. The Cascade Mountains to the west of the Basin greatly influence our climate by means of their "rain shadow" effect, which causes moisture (rain or snow) to be released from the clouds before they reach the area. This limits the amount of precipitation we receive each year. For example, the average rainfall in the lower Columbia Basin is 6 to 7 inches per year, which includes about 13 inches of snow (13 inches of snow equals about 1.5 inches of rain). Because the mountains significantly reduce the amount of moisture in the air, fewer clouds exist in the Columbia Basin, which means more sunshine and very warm temperatures from late spring through early autumn. The average annual temperature recorded at the Hanford Meteorology Station–located on the Hanford Site near Richland–is 53.3 degrees Fahrenheit. The warmest month is July, which averages 76.4 degrees Fahrenheit, while the coolest month is January, which averages 30.5 degrees Fahrenheit. The hottest temperature ever recorded at Hanford was 113 degrees Fahrenheit on August 4, 1961; the coldest was −23 degrees Fahrenheit on February 1 and 3, 1950.

HOW IS WIND CREATED?

In the Columbia Basin, we are all familiar with wind, the kind that can whip up to 50 miles per hour, blowing dust from nearby fields, tumbling trees, and tearing roofs off buildings. But how is wind created? Scientists say it has to do with air pressure. If you have seen what happens to a tire if it's punctured or a balloon when you blow it up and then hold the neck and let the air come back out, you get an idea of this pressure. Air inside a tire or balloon is at high pressure, and air outside of the tire or balloon is at lower pressure. Air comes out of the high-pressure tire or balloon into the lower pressure air of its surroundings. A similar occurrence happens in the Earth's atmosphere. Air moves from regions of high pressure to regions of low pressure (though not in a straight line). This movement of air is what we know as wind. The greater the difference between high pressure and low pressure, the stronger the wind.

WHAT ARE CLOUDS, FOG, RAIN, AND SNOW?

Clouds are formed as air rises into the atmosphere. They are created by very tiny water droplets (it takes a million cloud droplets to make one raindrop). Scientists classify clouds by their shape and by their height (low, medium, and high). Stratus and cumulus are two kinds of low-level clouds. Altostratus and altocumulus are medium-level clouds, and cirrus and cirrostratus are high-level clouds.

Fog is a stratus-type cloud that reaches the ground. As cloud droplets grow, they become heavy and eventually may become so heavy they fall to the ground as rain. In parts of the world where air temperatures are cold enough, water droplets in the clouds freeze into ice crystals and join together into snowflakes. When the snowflakes become big and heavy enough they fall from the clouds–and if the air near the ground remains cold enough (usually near or below freezing) they remain as snowflakes. If the air above and near the ground is too warm, they melt and fall as rain.

WHAT DO METEOROLOGISTS DO?

To become a meteorologist you must go to college and take courses in mathematics, physics, chemistry, and many different courses in meteorology or atmospheric sciences. Meteorologists may have a variety of different jobs. Some meteorologists that many people call "weathermen" may study day-to-day weather events and provide forecasts for the public. Other meteorologists may fly in research aircraft to study clouds, rain, or hurricanes. Meteorologists conducting tornado research may use cars and trucks to try to "chase" these violent storms so they can understand them and provide better forecasts to save lives. Some meteorologists also work mainly in offices with computers. They develop computer models to try to better understand the Earth's atmosphere. Others, primarily interested in research, may work in colleges and universities or in government or private laboratories. Those involved in weather forecasting may work for the government (such as the National Weather Service) or for private weather companies or television/radio stations.

> "Science is constructed of facts as a house is of stones. But a collection of facts is no more a science than a heap of stones is a house."-Henri Poincare
>
> Keep in mind this fact sheet is intended to be used only as background information to support your effort to encourage inquiry-based science, which parallels the way scientists uncover knowledge and solve problems.

SUGGESTED ACTIVITIES

1. You can have fun determining the monthly average(s) for your community.
   Were you surprised to read that the average temperature at the Hanford Meteorology Station

in July was 76.4 degrees Fahrenheit and 30 degrees Fahrenheit in January? Did you think the average temperature would be higher for our area? Lower?

Here's How: You will need a calendar, pencil, calculator, and access to the local newspaper. Each day check the paper for high and low temperature from the previous day.

1. Record both the high and low temperatures on your calendar.
2. At the end of the month, add all the high temperatures for your community and divide by the number of days in the month. Do the same for the low temperatures.
3. What is the average high? Low?

If you track the weather for a year then you can compute a 1-year average (which describes the climate of your community). What was the maximum temperature for the month? For the year? The minimum temperature for the month? For the year? How much hotter than the average was the hottest day? The coolest night? Ask a friend or relative in another city to keep same data on the temperatures in their community. How do the temperatures they record differ from yours? What are some of factors that might influence the temperature?

2. Do you know about weather in other areas of the world?
You will need a local newspaper, a world map, pencil, and paper that has a grid on it (see example below). Each member of your class or family should select a city from the map to track (make sure the city is one whose temperature and other aspects of weather are reported in your paper. As a group, decide how often you will check the temperature, precipitation, humidity, cloudiness, and windiness in the cities you have selected. Remember to record your data on the data sheet. If you are not familiar with the symbols used for the weather in the newspaper, check the key or talk to a teacher or your parent(s). Watch for news on weather phenomena in your city. Share what you have learned about your city at a weekly class or family meteorological meeting. How could this information help you plan a trip? Decide where to live? (Adapted from Nick Walker, The "Weather Dude" Seattle, Washington)

City Name __________ County __________

Date	Temperature	Precipitation	Cloud Cover	Windspeed

3. Did you ever wonder how the weather and climate impact plants and animals?
Pick a site in your backyard, or in a park that you can visit on a regular basis. Take along a notebook and sit quietly in the spot you picked. Record in your journal the date, time of day, and the weather (What is the temperature? Is it cloudy, windy, humid?). What animals do you see (insects, birds, fish, mammals)? What are they doing? Where do they go in winter? Can they be found in other habitats? If you can identify the birds, note their names in your notebook. If you don't know the types of birds you see quickly sketch the birds and note any special markings. Later look them up in a bird guide and add the names to your notebook. What plants do you see? Are shoots just starting to emerge from the ground? Are the leaves frostbitten? Visit your favorite spot throughout the year, in different kinds of weather and record your observations in your notebook. Weather and climate influence the life cycle of plants and animals but what other factors influence the plants and animals you observed? How would those plants and animals be affected if the climate changed?

Acknowledgments
Development and printing was funded by the Partnership for Arid Lands Stewardship (http://www.pnl.gov/pals). Written by: Dana Hoitink. Series Editor: Georganne O'Connor; Design: WinSome Design; Printing: Eagle Printing and Graphic Design. First Printing: April 1999.

BIBLIOGRAPHY

1. ***Cloudy with a Chance of Meatballs.*** 1982. Judith Barrett, Judi Barrett. Simon & Schuster. New York.
2. ***Full Option Science System (FOSS),*** Encyclopedia Britannica Educational Corp. Chicago, ***Air and Weather Module,*** Grades K-2.
3. ***How the Weather Works.*** Michael Allaby. 1995. Reader's Digest Association Incorporated. Pleasantville, New York.
4. ***National Audubon Society Field Guide to North American Weather.*** 1991. David M. Ludlum, Alfred A. Knopf. New York.
5. ***Nature Journaling.*** Clare Walker Leslie and Charles Roth. 1998. Storey Books. Pownal, Vermont.
6. ***Puddle Jumpers: Fun Weather Projects for Kids.*** Jennifer Storey Gillis. 1996. Storey Books. Pownal, Vermont.
7. ***Ranger Rick's NatureScope,*** Wild About Weather. National Wildlife Federation. Washington, DC.
8. ***Science and Technology for Children (STC).*** Carolina Biological Supply Company. Burlington, North Carolina, Weather Unit - Grades K-2.
9. ***Weather: Nature Company Discoveries Library.*** Time-Life Books. 1996. San Francisco, California.
10. ***Weather,*** Brian Cosgrove. 1991. Eyewitness Books, Knopf. New York.

WEB SITES

1. **The Hanford Meteorological Station:** http://etd.pnl.gov:2080/HMS/
2. **The "Weather Dude":** http://www.wxdude.com/
3. **The NOAA Weather Page:** http://www.esdim.noaa.gov/weather_page.html

4. **EarthWatch—Weather on Demand:** http://www.earthwatch.com/
5. **How the Weather Works**—http://www.weatherworks.com/
6. **NOAA Weather Education**—http://www.nws.noaa.gov/om/edures.htm
7. **The Weather Network**—http://www.theweathernetwork.com/

BASIC RESEARCH FOR MILITARY APPLICATIONS

W.D. Bach, Jr.
(from *The Handbook of Weather, Climate, and Water: Dynamics, Climate, Physical Meteorology, Weather Systems, and Measurements*, Wiley 2003)

> If you know the enemy and know yourself, your victory will never be endangered; if you know Heaven and know Earth, you may make your victory complete
>
> —Sun Tzu, The Art of War X, 31 circa 500 BC

The ancient Chinese general succinctly summarizes necessary ingredients for success in war. Understanding that "Heaven" represents the atmospheric environment, the need to know the weather is deeply rooted in military preparation and tactics. History is replete with examples of commanders using weather as an ally or suffering its bad effects. Even with today's modern technology, the dream of an all-weather military has not become a reality. Thus the military continues to seek ways to use the atmosphere as a "combat multiplier" in the order of battle.

MILITARY PERSPECTIVE

The military must understand the atmosphere in which it operates. That understanding will ultimately depend on scientific understanding of the atmospheric processes, an ability to use all of the information available, and an ability to synthesize and display that information in a quickly understandable fashion.

The military's needs for understanding the atmosphere's behavior has changed as a result of geopolitics and advancing technology. On the battlefield, future enemies are more apt to resort to chemical or biological attacks as a preferred method of mass destruction. The atmospheric boundary layer is the pathway of the attack. Increased reliance on "smart" weapons means that turbulent and turbid atmospheric effects will influence electromagnetic and acoustic signals propagating through the boundary layer. Intelligence preparation of the battlespace, a key to successful strategies, depends on full knowledge of the enemy, weather, and terrain. It requires an ability to estimate atmospheric details at specific locations and future times to maximize strategic advantages that weather presents a commander. It also avoids hazards and strategic disadvantages.

A convergent theme of basic research in the atmosphere is the interdependence of research progress in acoustic and electromagnetic propagation with the measurement and modeling of the dynamical boundary layer. The scattering of propagating energy occurs because of fine structure changes in the index of refraction in the turbulent atmosphere. The connection between the two—the refractive index structure function parameter, C_n^2—is proportional to the dissipation rate of the turbulent kinetic energy.

Military operations are often in time (seconds to hours) and space (millimeters to tens of kilometers) scales that are not addressed by conventional weather forecasting techniques. Atmospheric information is required for meteorological data-denied areas. Furthermore the real conditions are inhomogeneous at various scales. Significant adverse effects may have short lifetimes at high resolution. Such breadths of scales make accurate quantification of important boundary layer processes very difficult. Carefully planned and executed experiments in the uncontrolled laboratory of the atmosphere are needed for almost every phase of the research. The combined range of atmospheric conditions and potential propagation types and frequencies that are of interest to the military is too broad to condense into a few nomograms. The propagation studies require intensive and appropriate meteorological data to understand the atmospheric effects. Understanding of heterogeneous atmospheric fields arising from inhomogeneous conditions requires measurements achievable only by remote sensors.

Chemical and biological agents constitute major threats to military field units as well as to civil populations. Electromagnetic propagation and scattering are the principal means of remote or in-place detection and identification of agents. Current models of atmospheric transport and diffusion of the agents over stable, neutral, and convective conditions are marginally useful and based on 40-year-old relationships. Significantly improved models of the transport and diffusion are needed to estimate concentrations and fluctuations at various time and space scales for simulated training, actual combat, and for environmental air quality.

RESEARCH ISSUES

Basic research for the military in the atmospheric sciences comes under two broad, interdependent scientific efforts:

Atmospheric wave propagation, which is concerned about the effects of the atmosphere, as a propagating medium, on the transmission of electromagnetic (EM) or acoustic energy from a source to a receptor, and atmospheric dynamics, which is concerned with quantifying the present and future state of the atmosphere.

Atmospheric Wave Propagation

Electromagnetic and acoustic propagation and scattering is a large subject that addresses problems in both characterizing the battlespace and in detecting targets for engagement and situational awareness. Although there has been much previous research in the general area of wave propagation and scattering, there remains a pressing need for research that specifically address issues related to propagation and scattering in realistic atmospheric environments.

Electro-optical and infrared propagation in the atmosphere is limited by turbulence and by molecular and aerosol extinction by both absorption and scattering. The relative importance of turbulence and extinction is dependent on the wavelength of the radiation and the atmospheric conditions. Modern, high-resolution imaging systems are often turbulence limited, rather than diffraction limited, impairing long-range, high-resolution target acquisition, recognition, and identification. Laser-based systems are limited in their ability to retain spatio-temporal coherence and effectively focus at tactical ranges. Like acoustic signals, electromagnetic signals are primarily useful for target detection or ranging and for remote sensing of the atmosphere itself. Although much research has been done in the area of electromagnetic extinction, and current models are generally good, applications of these models for remote-sensing purposes is an area of active research. A large body of research exists in the area of atmospheric turbulence effects for electromagnetic propagation, but with less conclusive results. Hence understanding of atmospheric turbulence for optical and infrared propagation is still an active area of research.

Understanding of the interaction between propagation and turbulence is rather straightforward in the theoretical sense. Practically, the time scales of the EM propagation interactions with the air are much shorter, $O(10^{-9}$ s) along a path, than the capability to locally sample the atmospheric turbulence or density fluctuations, $O(10^{-1}$ s) and to characterize the turbulence spectra, $O(10^{+3}$ s). Acoustic interactions also occur on short time scales. Second, the turbulent structure of the atmosphere between the source and detector is largely unknown, so the scale of the disturbance is undetermined. Furthermore, with models, grid volume samples are realized at $O(10^{0}$ s) in large eddy simulation models and at $O(10^{2}$ s) in fine mesoscale models. To make the connection between propagation effects, turbulence characteristics, and atmospheric models, new measurement techniques are needed. In careful field campaigns to relate propagation to atmospheric conditions, the biggest problem is independent ground truth.

Atmospheric Dynamics

The military emphasis on global mobility requires weather forecasting support at all scales from global to engagement scales. In recent years, the explosion in computer capacity enabled highly complex numerical weather prediction models on most of these scales. Model accuracy has improved even as complexities are added. Model results have become more accepted as a representation of the atmospheric environment. As computation power increases, finer and finer scales of motion are represented in smaller grid volumes. Lately, large eddy simulations represent dynamics at grid sizes of a few meters in domains of 5 km^2 by 1 km deep.

As the models go to finer scales, the variability imposed by large-scale synoptic and mesoscale influences the Atmospheric Boundary Layer (ABL) is modified by both the cyclical nature of solar radiation and metamorphosis due to the stochastic behavior of clouds and other natural processes as well as anthropogenic causes. The resulting (stable and unstable) boundary layers are sufficiently different that current models of one state do not adequately capture the essential physics of the other or the transition from one state to the other. Accurate predictions become more difficult, principally because the atmosphere is poorly represented. Data are lacking at the scales of the model resolutions. Parameterizations required to close the set of equations are inadequate. Observations are lacking to describe the four-dimensional fields of the forecast variables at the model resolution. The models are unlikely to improve until better theories of small-scale behavior are implemented and observations capable of testing the theories are available.

For the military, techniques to represent the inhomogeneous boundary layer in all environments—urban, forest, mountains, marine, desert—is absolutely essential for mission performance. In most cases, this must be done with limited meteorological data. Furthermore, the military is more frequently interested in the effects—visibility, trafficability (following rain), ceiling—than in the "weather" itself. Nevertheless, without high-quality, dependable models that represent the real atmosphere in time and space, the effects will not be representative.

Basic research attempts to improve modeling capability by increasing the knowledge base of the processes of the atmosphere. Until new measurement capabilities are developed and tested, our ability to characterize the turbulent environment—affecting propagation and dispersion of materials—will be severely limited.

PROTOTYPE MEASUREMENT SYSTEMS

Military basic research has participated in several new techniques to measure winds and turbulence effects in the atmospheric boundary layer. These techniques concentrate on sampling a volume of the atmosphere on the time scales of (at least) the significant energy containing eddies of the atmosphere.

The Turbulent Eddy Profiler, developed at the University of Massachusetts, Amherst, is a 90-element phased-array receiving antenna operating with a 915-MHz 25-kW transmitter. It is designed to measure clear air echoes from refractive index fluctuations. The received signal at each element is saved for postprocessing. These signals are combined to form 40 individual beams simultaneously pointing in different directions every 2 s. In each 30-m range gate of each beam, the intensity of the return, the radial wind speed is computed from the Doppler shift, and the spread of the Doppler spectrum is calculated. These data are displayed to show a four-dimensional evolution of refractive index structures in the boundary layer. Further processing gives estimates of the three components of the wind velocity. Combining the wind vectors with refractive index structures has shown significant horizontal convergence occurring with strong refractive index structures. At 500 m above the ground the volume represented is approximately a 30-m cube.

The Volume Imaging Lidar at the University of Wisconsin measures backscatter from atmospheric aerosols in three dimensions at 7.5-m range resolutions. The evolution of boundary layer structures can be seen in a variety of experiments. Capabilities to estimate horizontal winds at selected altitudes have been developed and demonstrated.

An eye-safe, scanning Doppler lidar operating at 2 μm has been jointly developed with NOAA/ERL/ETL to measure radial wind speeds at 30-m resolution. The error of the measurement is < 0.2 m/s. Various scanning approaches have shown several different evolutions of the morning transition and breakdown of a low-level jet.

Another lidar system has been developed at the University of Iowa to measure the horizontal wind in the boundary layer at about 5-m increments every minute. Teamed with existing FMCW radars, having comparable vertical resolution, should add to meteorological understanding of layers of high refractive index.

CURRENT MEASUREMENT CAPABILITIES

Reliable measures of atmospheric temperature and moisture at high resolution in space and time are still lacking. Some progress has been made but does not yet achieve the resolutions of the wind measurements.

Prototype and existing wind/wind field data do not yet measure the turbulence. Although the Doppler spectrum width is an indicator of the eddy dissipation rate, few researchers or equipment developers report the variable or its spatial variability. To date, none have done so operationally.

CONCLUSIONS

Some progress is being made by the military to develop necessary instrumentation to measured atmospheric fields at high time and spatial resolution. This is motivated by the need to improve the theory and subsequently the models of atmospheric motions at small scales to provide the armed services with reliable models on which to gain superiority over the adversary and successfully complete the mission.

UNCERTAINTIES IN RAINFALL–RUNOFF MODELING

Patrick Willems
Hydraulics Laboratory
Leuven, Belgium

WHY DEAL WITH RUNOFF PREDICTION ERRORS OR UNCERTAINTIES?

Errors and uncertainties in the prediction of rainfall–runoff are often substantial. Consideration of these errors is therefore crucially important. When water decision-making is based on the rainfall–runoff prediction, for instance, by means of rainfall–runoff, uncertainties affect model outcomes and, consequently, the decisions. Hence, good modeling practice provides not only the results of model predictions but also the accuracy of these results. It supplies decision makers with important additional information on the uncertainty in the data and information that they use as a basis for their decisions. After quantification of the runoff prediction errors, water policies can be set up for which the efficiency can be guaranteed up to specified acceptable risk levels. Combined with the uncertainty analysis, a sensitivity analysis provides the modeler and decision maker with information about the importance of various types of model limitations and sources of uncertainty.

RAINFALL INPUT ERRORS AND UNCERTAINTIES

Rainfall is the driving force of the rainfall–runoff process. Therefore, both for watersheds and for urban drainage catchments, the rainfall input errors are one of the most important sources of uncertainty in the runoff prediction process. These errors have therefore been studied extensively in the literature. In general, they can be classified as rainfall measurement errors and errors by the simplification in the real spatial variability of the rainfall.

Rainfall Measurement Errors

Both rain gauge and radar data are subject to errors in the registration of the rainfall intensity (1–6). For rain gauges, this might be caused by:

- *Rain Losses by Wind Effects, Position, and Shelter*. Rain gauges installed above the ground level or in the lee of trees or buildings typically record less rainfall because of stronger wind effects. In comparison with rain gauges installed at ground level in a pit, gauges at higher elevations can have wind-induced losses up to 5–15% (7–9). The presence of a wind shield on the gauge can reduce this influence (10–13).
- *Rain Losses by Wetting and Evaporation*. All gauges using collecting funnels have losses of rain water remaining on the funnel surface without passing into the collecting chamber, which subsequently evaporates.
- *For Tipping Bucket Rain Gauges: Loss of Water During the Tipping Movements of the Buckets*. This bias can be corrected by calibration of the rain gauge (14–18).
- *For Tipping Bucket Rain Gauges: Resolution Error*. This is the error because of resolution of the recorded rainfall intensity, where the resolution corresponds to the volume of one tipping-bucket. This type of error is only important for small rainfall volumes.

Simplification in the Real Temporal and Spatial Variability of Rainfall

Rainfall–runoff models require simplification in the real spatial distribution of the rainfall. Lumped conceptual models require a larger averaging of the rainfall over the catchment in comparison with semidistributed and fully distributed models. The accuracy of areal rainfall estimates strongly depends on the simplification of the rainfall variability in both space and time. For the spatial dimension, the degree of simplification depends on the rain gauge network density more than on the interpolation method. The effect of the rain gauge network density was first studied by McGuinness (19) and afterwards by Bras and Rodriguez-Iturbe (20); Dawdy and Bergman (21); Wilson et al. (22); Troutman (23); Bastin et al. (24); Schilling (25); Lebel et al. (26); Fontaine (27); Peters-Lidard and Wood (28); Roux et al. (29); Yoo and Ha (30); and Lau and Sharpe (31). Areal reduction factors (or more generally, areal correction factors) have been developed for bias correction when areal rainfall estimates are based on rain gauge data (32,33). Uncertainty estimates for such areal correction factors can be found in Willems (34). They are based on simulations with a spatial rainfall generator. One of the problems with the accuracy of the areal correction factors is locating the desired point within the given area. If the point is away from the center of the area, the correction factors may be vastly different in comparison with correction factors based on historical storms located centrally in the catchment. Correction factors based on individual storm events also may not necessarily look like the climatological rainfall patterns for the whole catchment region. The climatological patterns indeed have to be considered with a reasonable level of confidence (35).

When radar data is available, this data can be used in addition to the rain gauge data. Radar data mainly provides detailed information on the spatial coverage and movement of the rain storms. The rainfall intensity registrations are less accurate, mainly because of the ambiguity of Z–R relationships for conversion of radar reflectivity to rainfall rate. Radar data, thus, can be used in a complementary way to the rain gauge data to increase the accuracy of the spatial interpolation (36,37). The accuracy and benefits of radar data depend, however, largely on the type of the radar and on the data processing (38–41).

PROPAGATION OF INPUT ERRORS THROUGH THE MODEL

To calculate the propagation of the rainfall input errors and related uncertainties to the uncertainty in the runoff prediction, different techniques can be used (the same techniques are also used for propagation of errors or uncertainties on other input variables and parameters). The most popular technique is the Monte Carlo method [either using Simple Random Sampling, Importance Sampling (e.g., 42), or Latin Hypercube methods]. These techniques also allow performing sensitivity analysis, on which the relative contribution of uncertain inputs to the output uncertainty is obtained. Alternatives to the Monte Carlo technique are based on analytical variance calculation. Only the first two moments of the error distributions are considered ("first order second method" reliability analysis methods; 43–47). These methods usually involve some degree of simplification of the model response behavior.

The significance of rainfall input on runoff hydrograph properties, such as runoff volumes, time to peak runoff, and peak runoff rates, has been studied extensively in the literature (21,22,48–54).

In a limited number of studies, the importance of the rainfall input errors in comparison with other uncertainty sources is investigated. For urban drainage stormwater modeling, Schilling and Fuchs (55), Lei and Schilling (56), Willems and Berlamont (57), and Van Mameren and Clemens (58) have made such assessment, and concluded that the input of rainfall and the paved catchment area (which is not so accurately known in urban drainage modeling) are by far the most important uncertainty sources. For watersheds, Andréassian et al. (59) have shown that rainfall–runoff models are more sensitive to the spatial rainfall distribution than to the spatial distribution of watershed (land-surface) parameters.

UNCERTAINTIES IN INPUT DATA AND MODEL PARAMETERS

Other inputs in rainfall–runoff models consist of evapotranspiration estimates, but also in geometric catchment characteristics (at least input of the catchment area is needed), and for distributed rainfall–runoff models, geological, soil type, and land-use maps.

When runoff predictions are based on modeling, the uncertainty on the model results does not only originate from model input uncertainties, but also from uncertainties on the model parameters and the model structure.

Model parameters are most often derived from calibration on the basis of simultaneous observations for the model input and model output variables. Parameter uncertainties will then develop when erroneous and/or limited time series of measurements are used for calibration or when the calibration procedure is not optimal. When inputs are biased (systematic deviation from their real values), the calibrated model parameters might also be biased. This problem has been shown, for instance, by Troutman (23,60,61) and Chaubey et al. (62), investigating the effect of rainfall input errors on runoff volumes. It is obvious that parameter uncertainties decrease when a longer time series of observations is used for calibration.

Parameter uncertainties strongly depend on the parameter calibration process. Optimization algorithms are most often used for parameter calibration (63–69). These algorithms also provide information on the confidence limits of the calibrated model parameter values, as they are based on statistical considerations (see the linear regression theory in any basic statistical handbook as the most simple example). The shuffled complex evolution algorithm (SCE-UA) is one of the most advanced algorithms (70,71). Other techniques are based on maximum likelihood estimation [in combination with Kalman filter in the recursive parameter calibration techniques of Jakeman et al. (72) and Young (73)]. Parameter uncertainty estimates were calculated and discussed for hydrological models by Johnston and Pilgrim (74), Kuczera (75), and Kuczera and Mroczkowski (76).

Goodness-of-fit statistics used in the optimization are the mean squared error, the coefficient of efficiency (77),

Young's Information Criterion (73), etc.; for an overview, see Legates and McCabe (78).

Parameter optimization may have difficulties because of the different sources of model errors. For complex models, it might be true that different parameter sets give equally acceptable goodness-of-fits. This problem together with related issues are discussed by several authors, and are denoted by the terms "non-uniqueness," "ambiguity," "equifinality," "nonidentifiability," "overparameterization," etc. (79–81). It may, in addition to other parameter optimization problems (such as the presence of many local optima in the objective function), explain underestimations in the parameter uncertainty estimates.

One method for inferring parameter uncertainties (and also for total model uncertainty), at the same time addressing the above-mentioned problem, is the generalized likelihood uncertainty estimator (GLUE) proposed by Beven and Binley (82). After it was used in a number of studies presented in the hydrological literature (e.g., 83–85), it became more and more popular. GLUE rejects the existence of an optimum model and parameter set and assumed that multiple likely models and parameter sets exist [the "equifinality" concept of Beven (81,86,87)]. This approach is opposed to the more classic approach of identifying the true or optimal model in face of uncertainty. Goodness-of-fit statistics in the GLUE approach are used as likelihood measures. A number of parameter sets is generated via sampling, and the performance of each trial assessed through a likelihood measure. When the likelihood measure reaches a lower limit, then the tried parameter set is called nonbehavioral and is rejected. To the behavioral sets, likelihood-based weights can be applied and parameter and model output uncertainties assessed (88). The problem of the GLUE approach is, however, that is does not account for input uncertainties (89), apart from other statistical problems that make the GLUE inappropriate for uncertainty estimation [e.g., subjective likelihood functions are used and no formal representation is made of the model error (66,90)]. It can, at most, be used for identifying the most sensitive parameters of the model.

For all cases where models are calibrated, the uncertainty in individual parameter values strongly depends on the model sensitivity to each specific parameter. Sensitivity analysis was therefore conducted in many studies.

Input and parameters strongly depend on the scale (temporal and spatial). Often, because of averaging effects, the uncertainty is likely to be much smaller for a larger scale average than for a smaller scale average (see also 91).

Model structure uncertainties, as the last type of model uncertainties considered, originate from the modeler's limited understanding of the physical reality. These can be assessed by comparison of different model structures for the hydrological system under study, or based on the analysis of the model residuals (considering model structure uncertainties as rest term after subtraction of the input and parameter uncertainties; see next sections).

MODEL RESIDUALS

The total uncertainty in the model results (the model output) can be quantified by calculation of the standard deviation or variance (squared standard deviation) of the model residuals (differences between modeled and observed runoff values). This variance can be further split up in the mean error (the bias or systematic deviation in the model output) and the random error (representing the uncertainty after bias correction). An illustration is given in the scatterplot of Fig. 1, where residuals are shown for all hydrographs in a runoff series. The mean deviation and the random error standard deviation are represented by the solid and dotted lines. Both may depend on the runoff value; typically, higher errors are found for higher runoff values. Transformation of both axes in the scatterplot can be done to normalize the residuals—to derive a normal distribution for the residuals (Gaussian residuals) independently of the runoff value (homoscedastic residuals). The Box–Cox (BC) transformation (92) is most flexible; as—depending on its parameter value—it can cover the whole range of weak and strong transformations. The strongest transformation in this range is the logarithmic transformation. The logarithmic transformation is useful in case the standard deviation of the residuals is shown proportional to the runoff value. In this case, the model has a constant relative error. In most practical cases, neither the absolute error nor the relative error is constant, and a BC transformation is needed (see applications by 93). Another transformation applied to reach Gaussian residuals is the normal quantile transform (94). In the case of heteroscedastic residuals, dependence between the residuals and the runoff value has to be considered (95–97), which can be done in the easiest way by the use of a meta-Gaussian model where both the residuals and the runoff discharges are transformed into a Gaussian distribution, and the correlation between both is considered (89,94).

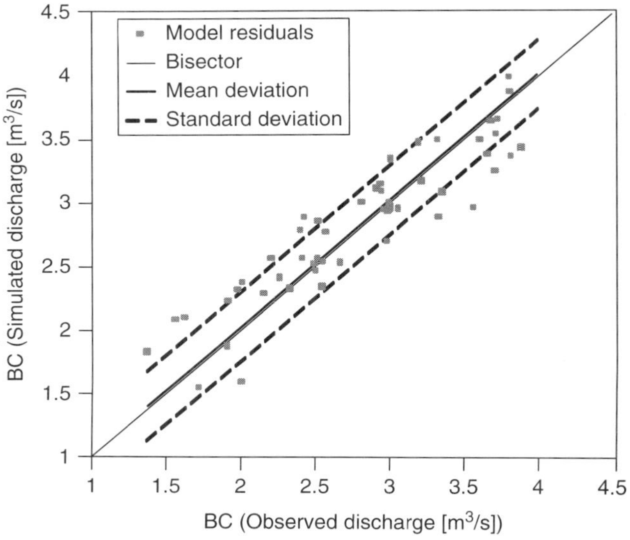

Figure 1. Scatterplot showing model residuals.

SERIAL DEPENDENCE

When runoff series with small timesteps are considered (e.g., hourly or smaller than the recession constants of one of the subflows in the catchment), residuals calculated from runoff values at different timesteps will have a serial dependence. Often, this serial dependence is higher for dry periods in comparison with wet periods, which is caused by the longer recession constant of the baseflow component in comparison with the quick runoff components. To derive independent residuals, a prior selection can be made of nearly independent values from the flow series. These can be peak values during subsequent quick-flow hydrographs, cumulative volumes during these hydrographs, low-flow values during subsequent baseflow recession periods, etc.

Other approaches use the full time series of residuals and describe both the amplitude and the serial correlation by means of stochastic models (98–101). These stochastic models are needed, for instance, on the basis of the recursive parameter estimation techniques.

ERRORS ON RUNOFF OBSERVATIONS

The total residual variance is larger than the total variance of the errors in the model output. Errors on the observed runoff values indeed also take part in the residual variance. These runoff discharges are most often not directly measured, but calculated from river discharge observations. River discharges are most often closely related to the average rainfall–runoff discharge upstream of the river flow monitoring station. Effects of routing along the river and flooding (over bank storage), may explain differences between both. As river flow monitoring stations are only available at a limited number of locations, they are lacking in providing the modeler with a detailed spatial coverage of runoff data. Errors in runoff observations can thus be classified in the following categories:

- errors in the flow data (by uncertainties in the rating curve for limnigraphic stations or in the accuracy of the flow monitoring device);
- differences between the flow at the monitoring station and the upstream runoff because of routing and flooding influences;
- differences between the runoff at a single point along the catchment or river and the average runoff upstream of the flow monitoring station (only relevant for distributed hydrological models).

MODEL ERROR

After compensation for the runoff observation errors, the total variance of the errors in the modeled runoff can be derived from the total residual variance. After calculation of the contribution of the input and parameter uncertainties (by propagation through the model), the remaining uncertainty can then be considered as the model structure uncertainty. When observations are available for some of the submodel output variables, the model structure uncertainties can be decomposed for different submodels.

Such a separate description of the different types of model uncertainties has the advantage that the contributions of the different sources of uncertainty to the total uncertainty in the model output can be quantified and compared. The comparison of the uncertainties resulting from the data, on the one hand, and the uncertainties resulting from the model structure, on the other hand, is especially valuable. Ragas et al. (102) called these two types of uncertainties respectively "operational" and "fundamental" uncertainties. As a general rule, a balance between the two is recommended. Whenever the operational uncertainties dominate, much more attention should be given to data collection than to model structure improvement in an attempt to improve the model results. Whenever the fundamental uncertainties dominate, the reverse is recommended. By comparing the contributions of the different uncertainty sources, efficient measures can thus be determined to reduce the total uncertainty in the model results.

IGNORANCE

Previous paragraphs only consider "quantifiable uncertainty." The real uncertainty may, however, be larger than this quantifiable uncertainty. It indeed might occur that the system structure is depending on specific influences, which were not observed in the past, but which may occur in an unpredictable manner in the future, which is a specific and special type of "lack of knowledge" that we cannot include in our uncertainty predictions because it is not quantifiable. Such additional nonquantifiable uncertainty is usually referred to as "ignorance." Ignorance occurs when we are missing relevant knowledge. Two types of ignorance have to be considered (see also 103):

- *Recognized Ignorance or Accepted Ignorance:* we realize and accept that we are ignorant, and communicate about this;
- *Total Ignorance:* we do not realize this lack of knowledge and are in complete ignorance (or "ignorance about the fact that we are ignorant").

Our confidence in a model may range from being certain (called "determinism" by Harremoës, see Fig. 2) to accepting that we are ignorant (zero confidence,

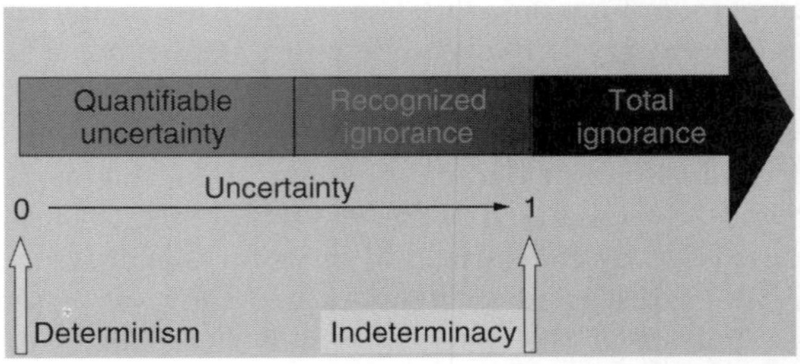

Figure 2. Illustration of the level of uncertainty from the ideal of what we know to total ignorance (103).

"indeterminacy" in Fig. 2). Regardless of our state of confidence, we may be correct or incorrect as a result of ignorance. Nonquantifiable uncertainties, thus, also may contribute to the runoff prediction uncertainty, and may, in some cases, be more serious than the quantifiable uncertainties. Therefore, a need to communicate these uncertainties exists in the decision support (104). When essential processes are not incorporated in the structure of the hydrological model because of insufficient knowledge about the processes, then it would be wrong to forget about this recognizable ignorance and not to communicate this potential source of uncertainty. The communication still can be supported by quantitative information, by running some scenarios on the unknown processes, and to report on the potential results (e.g., epistemic uncertainty, comparing a set of competing models based on different assumptions for the unknown processes).

BIBLIOGRAPHY

1. Larson, L.W. and Peck, E.L. (1974). Accuracy of precipitation measurements for hydrological modelling. *Water Resources Res.* **10**(4): 857–863.
2. Sevruk, B. (1989). Reliability of precipitation measurements. In: *Proceedings of the International Workshop on Precipitation Measurement*. B. Sevruk (Ed.). World Meteorological Organization, Geneva, pp. 13–19.
3. Goodison, B.E., Sevruk, B., and Klemm, S. (1989). WMO solid precipitation measurement intercomparison: Objectives, methodology, analysis, Atmospheric Deposition. *IAHS Publ.* **179**: 57–64.
4. Groisman, P.Y., Koknaeva, V.V., Belokrylova, T.A., and Karl, T.R. (1991). Overcoming biases of precipitation measurement: A history of the USSR experience. *Bull. Am. Meteorol. Soc.* **71**(11): 1725–1732.
5. Groisman, P.Y. and Legates, D.R. (1994). The accuracy of United States precipitation data. *Bull. Am. Meteorol. Soc.* **75**(2): 215–227.
6. Peck, E.L. (1997). Quality of hydrometerological data in cold regimes. *J. Am. Water. Resour. Assoc.* **32**(1): 125–134.
7. Braak, C. (1945). *Wind Influence on rain gauge measurements (in Dutch)*. Royal Dutch Meteorological Institute of the Netherlands, Report 102 '48. Mededeelingen en Verhandelingen', Rijksuitgeverij, 's-Gravenhage, the Netherlands.
8. Neff, E.L. (1977). How much rain does a rain gage gage?. *J. Hydrol.* **35**: 213–220.
9. Sevruk, B. (1996). Adjustment of tipping-bucket precipitation gauge measurements. *Atmospheric Res.* **42**: 237–246.
10. Alter, C.J. (1937). Shielded storage precipitation gauges. *Mon. Weather Rev.* **65**: 262–265.
11. Larkin, H.H. (1947). A comparison of the Alter and Nipher shields for precipitation gages. *Bull. Am. Meteorol. Soc.* **28**(4): 200–201.
12. Yang, D. et al. (1999). Quantification of precipitation measurement discontinuity induced by wind shields on national gauges. *Water Resources Res.* **35**(2): 491–508.
13. Duchon, C.E. and Essenberg, G.R. (2001). Comparative rainfall observations from pit and aboveground rain gauges with and without wind shields. *Water Resources Res.* **37**(12): 3253–3263.
14. Marsalek, J. (1981). Calibration of the tipping bucket raingauge. *J. Hydrol.* **53**: 343–354.
15. Adami, A. and Da Deppo, L. (1985). On the systematic errors of tipping bucket recording raingauges. *Proceedings of the International Workshop on the Correction of Precipitation Measurements*. World Meteorological Organization, Instruments and Observing Methods Report, no. 24, Geneva, pp. 27–30.
16. Niemczynowicz, J. (1986). The dynamic calibration of tipping-bucket raingauges. *Nordic. Hydrol.* **17**: 203–214.
17. Luyckx, G. and Berlamont, J. (2001). Simplified method to correct rainfall measurements from tipping bucket rain gauges. *Proceedings of the World Water and Environmental Resources (EWRI) Congress—Urban Drainage Modeling (UDM) Symposium*. Orlando, May 20–24.
18. La Barbera, P., Lanza, L.G., and Stagi, L. (2002). Tipping bucket mechanical errors and their influence on rainfall statistics and extremes. *Water Sci. Technol.* **45**(2): 1–9.
19. McGuinness, J.L. (1963). Accuracy of estimating watershed mean rainfall. *J. Geophys. Res.* **68**(16): 4763–4767.
20. Bras, R.L. and Rodriguez-Iturbe, I. (1976). Rainfall network design for runoff prediction. *Water Resources Res.* **12**(6): 1197–1208.
21. Dawdy, D.R. and Bergmann, J.M. (1969). Effect of rainfall variability on streamflow simulation. *Water Resources Res.* **5**(5): 958–966.
22. Wilson, C.B., Valdes, J.B., and Rodriguez-Iturbe, I. (1979). On the influence of the spatial distribution of rainfall on storm runoff. *Water Resources Res.* **15**(2): 321–328.
23. Troutman, B.M. (1983). Runoff prediction errors and bias in parameter estimation induced by spatial variability of precipitation. *Water Resources Res.* **19**(3): 791–810.
24. Bastin, G., Lorent, B., Duque, C., and Gevers, M. (1984). Optimal estimation of the average areal rainfall and optimal selection of raingauge locations. *Water Resources Res.* **20**(4): 463–470.
25. Schilling, W. (1984). Effect of spatial rainfall distribution on sewer flows. *Water Sci. Technol.* **16**(8/9): 177–188.
26. Lebel, T., Bastin, G., Obled, C., and Creutin, J.D. (1987). On the accuracy of areal rainfall estimation: a case study. *Water Resources Res.* **23**(11): 2123–2134.
27. Fontaine, T.A. (1991). Predicting measurement error of areal mean precipitation during extreme events. *Water Resour. Bull.* **27**(3): 509–520.
28. Peters-Lidard, C.D. and Wood, E.F. (1994). Estimating storm areal average rainfall intensity in field experiments. *Water Resources Res.* **30**(7): 2119–2131.
29. Roux, C., Guillon, A., and Comblez, A. (1995). Space-time heterogeneities of rainfalls on runoff over urban catchments. *Water Sci. Technol.* **32**(1): 209–215.
30. Yoo, C. and Ha, E. (2002). Basin average rainfall and its sampling error. *Water Resources Res.* **38**(11): 1259, doi:0.1029/2001WR001081.
31. Lau, J. and Sharpe, W. (2003). Comparison between distributed point rainfall measurements and spatially correct rainfall data in urban drainage runoff and quantity predictions. *Proceedings of the 6th International Workshop on Precipitation in Urban Areas*. Pontresina, December 4–7.
32. Rodriquez-Iturbe, I. and Mejia, J.M. (1974). On the transformation of point rainfall to areal rainfall. *Water Resources Res.* **10**(4): 729–735.
33. Lebel, T. and Laborde, J.P. (1988). A geostatistical approach for areal rainfall statistics assessment. *Stochast. Hydrol. Hydraul.* **2**: 245–261.

34. Willems, P. (2001). Stochastic description of the rainfall input errors in lumped hydrological models. *Stochastic Environ. Res. Risk Assess.* **15**: 132–152.
35. Silverman, Rogers, and Dahl. (1981). On the sampling variance of raingauge networks. *J. Appl. Meteorol.* **20**: 1468–1478.
36. Grimes, D.I.F., Pardo-Igúzquiza, E., and Bonifacio, R. (1999). Optimal areal rainfall estimation using raingauges and satellite data. *J. Hydrol.* **222**: 93–108.
37. Sokol, Z. (2003). Utilization of regression models for rainfall estimates using radar-derived rainfall data and rain gauge data. *J. Hydrol.* **278**: 144–152.
38. Joss, J. and Waldvogel, A. (1990). Precipitation measurements and hydrology, In: *Radar in Meteorology*. D. Atlas (Ed.). Amer. Meteor. Society, pp. 577–606.
39. O'Connel, P.E. and Todini, E. (1996). Modelling of rainfall, flow and mass transport in hydrological systems: an overview. *J. Hydrol.* **175**: 3–16.
40. Collier, Ch.C. (1996). *Applications of Weather Radar Systems*. Praxis Publishing Ltd., p. 390.
41. Harrison, D.L., Driscoll, S.J., and Kitchen, M. (2000). Improving precipitation estimates from weather radar using quality control and correction methods. *Meteorol. Appl.* **6**: 135–144.
42. Kuczera, G. and Parent, E. (1998). Monte Carlo assessment of parameter uncertainty in conceptual catchment models: the Metropolis algorithm. *J. Hydrol.* **211**: 69–85.
43. Wood, E.F. (1976). An analysis of the effects of parameter uncertainty in deterministic hydrologic models. *Water Resources Res.* **12**(5): 925–932.
44. Garen, D.C. and Burges, S.J. (1981). Approximate error bounds for simulated hydrographs. *J. Hydraul. Divis ASCE.* **107**(HY11): 1519–1534.
45. Yen, B.C., Chang, S.T., and Melching, C.S. (1986). First-order reliability analysis. In: *Stochastic and Risk Analysis in Hydraulic Engineering*. B.C. Yen (Ed.). Water Resources Publications, Littleton, CO, pp. 1–34.
46. Kuczera, G. (1988). On the validity of first-order prediction limits for conceptual hydrologic models. *J. Hydrol.* **103**: 229–247.
47. Melching, C.S., Yen, B.C., and Wenzel, H.G. (1990). A reliability estimation in modeling watershed runoff with uncertainties. *Water Resources Res.* **26**(10): 2275–2286.
48. Singh, V.P. (1977). Sensitivity of some runoff models to errors in rainfall excess. *J. Hydrol.* **33**: 301–318.
49. Beven, K.J. and Hornberger, G.M. (1982). Assessing the effect of rainfall pattern of precipitation on modelling stream flow hydrographs. *Water Resources Bull.* **18**(5): 823–829.
50. Hamlin, M.J. (1983). The significance of rainfall in the study of hydrological processes at basin scale. *J. Hydrol.* **65**: 73–94.
51. Corradini, C. and Singh, V.P. (1985). Effect of spatial variability of effective rainfall on direct runoff by geomorphologic approach. *J. Hydrol.* **81**: 27–43.
52. Seliga, T.A., Aron, G., Aydin, K., and White, E. (1992). Storm Runoff Simulation Using Radar Rainfall Rates and A Unit Hydrograph Model (SYN-HYD) Applied to GREVE Watershed. In: *American Meteorological Society, 25th International Conference on Radar Hydrology*, pp. 587–590.
53. Faures, J., Goodrich, D.C., Woolhiser, D.A., and Sorooshian, S. (1995). Impact of small-scale spatial variability on runoff modeling. *J. Hydrol.* **173**: 309–326.
54. Shah, S.M.S., O'Connell, P.E., and Hosking, J.R.M. (1996). Modelling the effects of spatial variability in rainfall on catchment response. 1. Formulation and calibration of a stochastic rainfall field model. *J. Hydrol.* **175**: 67–88.
55. Schilling, W. and Fuchs, L. (1986). Errors in stormwater modelling—a quantitative assessment. *J. Hydraul. Engineer.* **112**(2): 111–123.
56. Lei, J.H. and Schilling, W. (1994). Parameter uncertainty propagation analysis for urban rainfall runoff modelling. *Water Sci. Technol.* **29**(1-2): 145–154.
57. Willems, P. and Berlamont, J. (1999). Probabilistic modelling of sewer system overflow emissions. *Water Sci. Technol.* **39**(9): 47–54.
58. Van Mameren, H. and Clemens, F. (1997). Guidelines for hydrodynamic calculations on urban drainage in the Netherlands: overview and principles. *Water Sci. Technol.* **36**(8-9): 247–252.
59. Andréassian, V. et al. (2004). Impact of spatial aggregation of inputs and parameters on the efficiency of rainfall-runoff models: A theoretical study using chimera watersheds. *Water Resources Res.* **40**: W05209, doi:10.1029/2003WR002854.
60. Troutman, B.M. (1982). An analysis of input errors in precipitation-runoff models using regression with errors in the independent variables. *Water Resources Res.* **18**(4): 947–964.
61. Troutman, B.M. (1985). Errors and parameter estimation in precipitation-runoff modelling, 1. Theory. *Water Resources Res.* **21**(8): 1195–1213.
62. Chaubey, I., Haan, C.T., Grunwald, S., and Salisbury, J.M. (1999). Uncertainty in the model parameters due to spatial variability of rainfall. *J. Hydrol.* **220**: 48–61.
63. Duan, Q., Gupta, V.K., and Sorooshian, S. (1992). Effective and efficient global optimization for conceptual rainfall-runoff models. *Water Resources Res.* **28**: 1015–1031.
64. Gan, T.Y. and Biftu, G.F. (1996). Automatic calibration of conceptual rainfall-runoff models: optimisation algorithms, catchment conditions, and model structure. *Water Resources Res.* **32**(12): 3513–3524.
65. Gupta, H.V., Sorooshian, S., and Yapo, P.O. (1998). Towards improved calibration of hydrologic models: multiple and non-commensurable measures of information. *Water Resources Res.* **34**: 751–763.
66. Thiemann, M., Torsset, M., Gupta, H., and Sorooshian, S. (2001). Bayesian recursive parameter estimation for hydrologic models. *Water Resources Res.* **37**: 2521–2535.
67. Beven, K. and Young, P. (2003). Comment on "Bayesian recursive parameter estimation for hydrologic models" by M. Thiemann, M. Torsset, H. Gupta, and S. Sorooshian. *Water Resources Res.* **39**(5): COM 1-1–COM 1-4.
68. Gupta, H.M., Thiemann, M., Torsset, M., and Sorooshian, S. (2003). Reply to comment by K. Beven and P. Young on "Bayesian recursive parameter estimation for hydrologic models," *Water Resources Res.* **39**(5): COM 2-1–COM 2-5–.
69. Yapo, P.O., Gupta, H.V., and Sorooshian, S. (1996). Automatic calibration of conceptual rainfall-runoff models: sensitivity to calibration data. *J. Hydrol.* **181**: 23–48.
70. Duan, Q. (1994).
71. Vrugt, J.A., Gupta, H.V., Bouten, W., and Sorooshian, S. (2003). A shuffled complex evolution Metropolis algorithm for optimization and uncertainty assessment of hydrological model parameters. *Water Resources Res.* **39**(6): 1201, doi:10.1029/2002WR001642.
72. Jakeman, A.J., Littlewood, I.G., and Hornberger, P.G. (1990). Computation of the instantaneous unit hydrograph

and identifiable component of flow with application to two small upland catchments. *J. Hydrol.* **117**: 275–300.

73. Young, P. (1986). Time series methods and recursive estimation in hydrological systems analysis, In: *River Flow Modelling and Forecasting*. Craijenhoff, D.A. and Moll, J.R. (Eds.). Reidel, Dordrecht, the Netherlands, pp. 129–180.
74. Johnston, P.R. and Pilgrim, D. (1976). Parameter optimisation for watershed models. *Water Resources Res.* **12**(3): 477–486.
75. Kuczera, G. (1983). Improved parameter inference in catchment models: 1. evaluating parameter uncertainty. *Water Resources Res.* **19**: 1151–1162.
76. Kuczera, G. and Mroczkowski, M. (1998). Assessment of hydrologic parameter uncertainty and the worth of multiresponse data. *Water Resources Res.* **34**: 1481–1489.
77. Nash, J.E. and Sutcliffe, J.V. (1970). River flow forecasting through conceptual models, 1. A discussion of principles. *J. Hydrol.* **10**: 282–290.
78. Legates, D.R. and McCabe, G.J. (1999). Evaluating the use of "goodness-of-fit" measures in hydrologic and hydroclimate model validation. *Water Resources Res.* **35**(1): 233–241.
79. Gupta, V.K. and Sorooshian, S. (1983). Uniqueness and observability of conceptual rainfall-runoff model parameters: the percolation process examined. *Water Resources Res.* **19**(1): 269–276.
80. Beck, M.B. and Halfon, E. (1991). Uncertainty, identifiability and the propagation of prediction errors: a case study of Lake Ontario. *J. Forecasting*. **10**: 135–162.
81. Beven, K.J. (1993). Prophesy, reality and uncertainty in distributed hydrological modelling. *Adv. Water Resour.* **16**: 41–51.
82. Beven, K.J. and Binley, A. (1992). The future of distributed models: model calibration and uncertainty prediction. *Hydrological Processes*. **6**: 279–298.
83. Cameron, D.S., Beven, K.J., Tawn, J., Blazkova, S., and Naden, P. (1999). Flood frequency estimation by continuous simulation for a gauged upland catchment (with uncertainty). *J. Hydrol.* **219**: 169–187.
84. Cameron, D.S., Beven, K.J., and Naden, P. (2000). Flood frequency estimation by continuous simulation under climate change (with uncertainty). *Hydrol. Earth Syst. Sci.* **4**: 393–405.
85. Blazkova, S. and Beven, K.J. (2002). Flood frequency estimation by continuous simulation for a catchment treated as ungauged (with uncertainty). *Water Resources Res.* **38**: 1–14.
86. Beven, K.J. (2001). How far can we go in distributed hydrological modelling. *Hydrol. Earth Systems Sci.* **5**: 1–12.
87. Beven, K.J. (2002). Towards an alternative blueprint for a physically based digitally simulated hydrologic response modeling system. *Hydrol. Processes* **16**: 189–206.
88. Freer et al. (1996).
89. Montanari, A. and Brath, A. (2004). A stochastic approach for assessing the uncertainty of rainfall-runoff simulations. *Water Resources Res.* **40**: W01106, doi:10.1029/2003WR002540.
90. Clarke. (1994).
91. Heuvelink, G.B.M. (1998). *Error Propagation in Environmental Modelling with GIS*. Taylor & Francis, London, p. 127.
92. Box, G.E.P. and Cox, D.R. (1964). An analysis of transformations. *J. Royal Statistical Soc.* 211–243, discussion 244–252.
93. Thyer, M., Kuczera, G. and Wang, Q.J. (2002). Quantifying parameter uncertainty in stochastic models using the Box-Cox transformation. *J. Hydrol.* **265**(1-4): 246–257.
94. Kelly, K.S. and Krzysztofowicz, R. (1997). A bivariate meta-Gaussian density for use in hydrology. *Stochast Hydrol. Hydraul.* **11**: 17–31.
95. Sorooshian, S. (1980). Parameter estimation of rainfall-runoff models with heteroscedastic streamflow errors: the noninformative case. *J. Hydrol.* **52**: 127–138.
96. Sorooshian, S. and Dracup, J.A. (1980). Stochastic parameter estimation procedures for hydrologic rainfall-runoff models: correlated and heteroscedastic error cases. *Water Resources Res.* **16**: 430–442.
97. Sorooshian, S. and Gupta, V.K. (1983). Evaluation of maximum likelihood parameter estimation techniques for conceptual rainfall-runoff models: influence of calibration data variability and length on model credibility. *Water Resources Res.* **19**(1): 251–259.
98. Whitley, R.J. and Hromadka, T.V. II (1991). Evaluating uncertainty in design storm runoff. *Water Resources Res.* **27**(10): 2779–2784.
99. Jakeman, A.J. and Hornberger, G.M. (1993). How much complexity is warranted in a rainfall-runoff model. *Water Resources Res.* **29**(8): 2637–2649.
100. Young, P. and Beven, K.J. (1994). Data-based mechanistic modelling and the rainfall-flow nonlinearity. *Environmetrics* **5**: 335–363.
101. Lees, M.J., Price, N., Wheater, H.S., and Peach, D. (1998). A rainfall-runoff simulation model for South Galway, Ireland, In: *Hydrology in a Changing Environment*. Vol. 3. H. Wheater and C. Kirby (Eds.). John Wiley & Sons, New York, pp. 93–104.
102. Ragas, A.M.J., Haans, J.L.M., and Leuven, R.S.W.E. (1997). Selecting water quality models for discharge permitting. *Eur. Water Pollut. Control* **7**(5): 59–67.
103. Harremoës, P. (2003). The role of uncertainty in application of integrated urban water modeling. *Proceedings of the IMUG Conference*, Tilburg, April 23–25.
104. van der Sluijs, J.P. et al. (2003). *RIVM/MNP Guidance for Uncertainty Assessment and Communication: Detailed Guidance (RIVM/MNP Guidance for Uncertainty Assessment and Communication Series*. Vol. 3. Utrecht University, Utrecht.

READING LIST

Fankhauser, R. (1998). Influence of systematic errors from tipping bucket raingauges on recorded rainfall data. *Water Sci. Technol.* **37**(11): 121–129.

Melching, C.S., Yen, B.C., and Wenzel, H.G. (1990). A reliability estimation in modeling watershed runoff with uncertainties. *Water Resources Res.* **26**(10): 2275–2286.

Luyckx, G., Willems, P., and Berlamont, J. (1998). Influence of the spatial variability of rainfall on sewer system design, In: *Hydrology in a Changing Environment*. Vol. 3. H. Wheater and C. Kirby (Eds.). John Wiley & Sons, New York, pp. 339–349.

Radwan, M., Willems, P., and Berlamont, J. (2004). Sensitivity and uncertainty analysis for river water quality modelling. *J. Hydroinformat.* **06.2**: 83–99.

Willems, P. (2000). *Probabilistic Immission Modelling of Receiving Surface Waters*. Ph.D thesis, Faculty of Engineering, K. U. Leuven, Belgium.

MONSOON

Arthur M. Holst
Philadelphia Water Department
Philadelphia, Pennsylvania

Cyclic air masses accompanied by either much or little rain, carried by winds that reverse direction semiannually, are called monsoons. The word monsoon comes from the Arabic word "mausim," which is translated as "season." Monsoons occur over almost every continent, but they come in varying degrees of strength, as judged by rainfall production and wind speed. The most powerful and well-known monsoons occur in India and Southeast Asia.

Monsoons are caused by a difference in temperature, and thus air pressure, between landmasses and oceans. They are especially strong in Southeast Asia because the area is geographically conducive to monsoon weather. There is much more land in the Northern Hemisphere, namely, the Asian continent, and there is much more water in the Southern Hemisphere, namely the Indian Ocean. Due to the earth's tilt as it revolves around the Sun, the Northern Hemisphere receives more sunlight during the summer, and the Southern Hemisphere receives more sunlight during the winter. This means that there is always a marked difference between the air temperature over land and the air temperature over the sea in Southern Asia and India. The two variations that result are called wet and dry monsoons. Locally, they are called wet season (summer) and dry season (winter). Wet monsoons occur when the land in the Northern Hemisphere becomes heated by extensive sunlight. Due to convection, the warm air over land rises into the atmosphere, and cooler ocean air slides in to take its place. This cool air carries vast amounts of moisture, which is dropped on the landmass, in this case India and SE Asia.

Dry monsoons occur in the winter when the reverse occurs, and the oceans are heated by sunlight. The air over them rises, and cool air from the land leaves to take its place. This cold air takes moisture with it; thus, there is little rainfall over land.

The precipitation released during the wet season can fall quickly and unpredictably, sometimes resulting in floods, but usually resulting in enormous quantities of much needed rainfall for the farmers and residents of Southeast Asia. Due to this fact, monsoons are integral to the lives and well-being of virtually all of the people who live in India and Southeast Asia, as they have been for millennia. However, despite their importance to the area, they can cause great damage at times, as well. It is not uncommon for abnormal monsoon conditions to result in massive flooding or extended droughts. In fact, a 1987 Indian drought due to variations in the local monsoons was one of the worst of the twentieth century. And parts of India can receive 40 feet of rain or more in a span of less than 4 months.

The monsoon seasons are the distinctive feature of the Southeast Asian climate, but they do exist to lesser degrees elsewhere in the world. Africa, Australia, and the Southwestern United States all undergo minor monsoon seasons during the year. Generally, their monsoon weather is less pronounced than that of Southeast Asia because they do not have its unique geographical properties.

READING LIST

"The Asian Monsoons". (April 16, 2002). Available: http://www.uea.ac.uk/~f821864/monsoon.html.

"The Indian Monsoon And Its Impact On Agriculture". (April 16, 2002). Available: http://www.usda.gov/oce/waob/jawf/profiles/specials/monsoon/monsoon.htm.

"Monsoon". (April 17, 2002). Available: http://www.encyclopedia.com/html/m/monsoon.asp.

"Monsoon". (April 17, 2002). Available: http://www.pbs.org/wnet/nature/monsoon/html/intro.html.

"Monsoons". (April 14, 2002). Available: http://library.thinkquest.org/C003603/english/monsoons/glossary.shtml?tqskip1 = 1& tqtime = 0412.

Rosenberg, M. "Monsoons". (April 14, 2002). Available: http://geography.about.com/library/weekly/aa102599.htm.

AFRICAN MONSOONS

Mike Bettwy
Goddard Space Flight Center, NASA

In the world of weather, we've all heard of the "butterfly effect," that is, if a butterfly flaps its wings in China, it will affect the weather in New York a few weeks later. While this may be an exaggeration, a new study offers more evidence of how seasonal weather patterns can impact conditions thousands of miles away.

One cyclic weather pattern is the monsoon—a seasonal reversal in wind direction—caused by the fact that land and ocean heat up at different rates. Land quickly responds to heating, while oceans, with a much larger heat capacity,

This photograph of heavy rainfall in Ghana was taken August 19, 2001 during an ongoing NASA/USAID project. The West African monsoon season typically lasts from May-October, with the largest totals observed during June-September. Click on image to enlarge. Credit: Distinguished Professor, Dr. James W. Jones, University of Florida take longer to react. The difference in temperature leads to a difference in pressure, ultimately creating a wind shift which changes rainfall patterns. The monsoon system in West Africa is especially important, as it brings vital rainfall to support life and maintain local economies.

Now, new research by NASA and University of Maryland scientists suggests that these naturally occurring annual swings in temperature and pressure create two distinct monsoons in Africa, one in late spring and another in mid-summer. Using data from satellites, including NASA's Tropical Rainfall Measuring Mission (TRMM), researchers were able to detect a distinct northward "jump" in the monsoon from late spring to mid-summer.

In the first episode, large differences in sea surface temperatures between coastal waters and water further out to sea enhances rainfall on the West African Coast. By July, a second monsoon appears. This downpour is related to African Easterly Waves, low-pressure systems that form on the African continent during the summer months. On average, about 60 waves are generated over North Africa each year, and previous research suggests they may kick off small circulations that develop into hurricanes over the North Atlantic Ocean.

Between May and October, about two Easterly Waves per week travel from the African continent onto the cool Eastern Atlantic. These waves generally fall apart, but their remnants do survive and head towards the Western Atlantic and Caribbean, where they regenerate. Only about one in ten Easterly Waves survive to develop into gale-force tropical storms or full-fledged hurricanes. But, almost 85 percent of major hurricanes (wind speeds 111 mph or greater) began as Easterly Waves. Some studies have also suggested that many of the tropical cyclones in the Eastern Pacific Ocean can also be traced back to Africa.

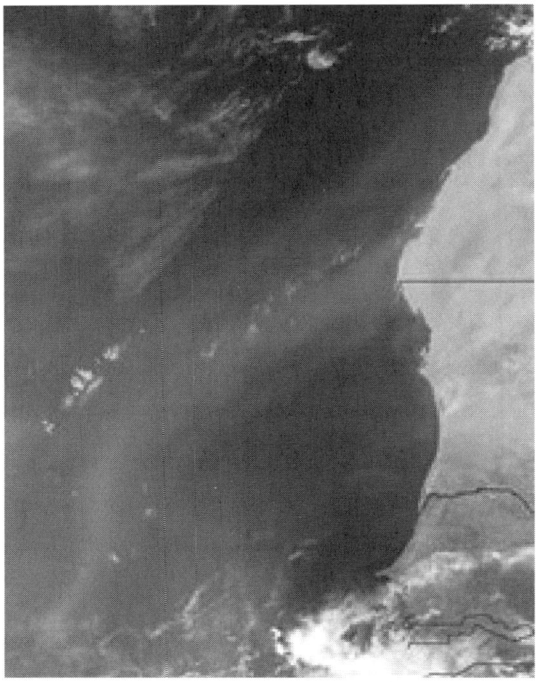

The Moderate Resolution Imaging Spectroradiometer (MODIS), aboard NASA's Aqua satellite, captured this true-color image of Saharan Desert dust blowing southwestward off the coasts of Morocco and Mauritania in West Africa. The light brown plume can be seen wending its way over the Cape Verde Islands in the Eastern Atlantic Ocean. Click on image to enlarge. Credit: NASA/GSFC

While not all Easterly Waves will form into hurricanes, Florida, like Africa, receives a large chunk of its annual rainfall from these storm systems. But other waves that originate to the north of Ghana are dry and pick up debris and dust from Africa's Sahara Desert. A strong high-pressure system over the Atlantic Ocean, known as the Bermuda High, helps steer these waves toward the Florida coast. Easterly Waves tend to be drier in El Niño years, which means they're less likely to produce hurricanes, but more likely to bring pollutants and dust toward the United States.

PERMANENT FROST

Arthur M. Holst
Philadelphia Water Department
Philadelphia, Pennsylvania

INTRODUCTION

The last geographic remnant of the last great Ice Age is permafrost, or permanently frozen ground. Permafrost exists today in many places on the earth, including the northernmost parts of America, Eurasia, and the Antarctic. In the coldest regions, permafrost has developed and existed for millions of years. Today, it covers about 20% of the earth's surface.

Frozen soil or rock that remains at a temperature of 0 °C (32 °F) or below for 2 or more years is permafrost. Permafrost is defined solely by temperature and is not determined by the amount of snow cover, soil moisture content, or its location. Permafrost may contain more than 30% ice or no ice at all, and may be topped by meters of snow cover or no snow at all. The defining characteristic of permafrost is its temperature.

REMAINS OF THE GREAT ICE AGES

What remains from the last Ice Age is nearly one-fifth of the Earth's surface still permanently frozen. During the past 2 million years, the great ice sheets that covered much of the earth's Northern Hemisphere prevented the surface below them from thawing. This freeze seeped deeper and deeper into the earth, forming thick layers of permanently frozen ground, permafrost. As time passed, much of this permafrost thawed due to the end of the Ice Age. As the earth's surface began to warm again, permafrost levels declined. However, across the Northern Hemisphere and in Antarctica, much of the ground remains permanently frozen, even during the summer months. Other permafrost forms when the ground cools enough during the winter to prevent it from thawing in the following summer.

Though the determining characteristic of permafrost is ground temperature, the determining cause is not air temperature but atmospheric climate. Temperature instead affects the distribution and thickness of permafrost.

GEOGRAPHY OF PERMAFROST

Permafrost is located throughout the Northern and Southern Hemispheres. The largest continuous mass of permafrost in the Southern Hemisphere is beneath the thick ice of Antarctica. This colossal mass alone comprises about 9% of the earth's surface. Permafrost is more widespread in the Northern Hemisphere. It is found as far north as Greenland and as far south as the heights of the Himalayas. Most of the Northern Hemisphere's permafrost, which makes up 23% of the hemisphere, occurs in the Eastern Hemisphere in parts of Siberia, eastern Russia, northern parts of Mongolia and China, and in the Tibetan Plateau. Other significant permafrost comprises most of Alaska (85% of the state) and parts of northwestern Canada. More than 50% of the surface of Canada is permafrost. Farther north, permafrost levels decline rapidly as land gives way to the Arctic Ocean; yet, a shelf of permafrost has been discovered even under the sea.

Permafrost is separated into two categories: continuous and discontinuous. Continuous implies permafrost everywhere except underneath bodies of water. Discontinuous permafrost is broken into two subdivisions: widespread and sporadic. In the widespread permafrost zone, 50–90% of the land is permafrost. In the sporadic zone, only 10–50% of the ground is permafrost.

LIFE IN A PERMAFROST REGION

A widely held misconception is that a permafrost zone is barren and inhospitable to any form of plant or animal life. On the contrary, the Arctic permafrost zone is home to a variety of mammals, birds, insects, and evergreen forest plant life that covers much of the Northern Hemisphere's permafrost zone.

In Arctic summers, the terrain is alive with animal life. Insects and other invertebrates inhabit the shallow wetlands. Birds feed from these insects and the plant life; they are provided ideal breeding and nesting grounds during the summer months. Snow geese and tundra swans are among the more common visitors of more than 100 bird species, which also include some hawks, owls, and cranes. Polar bears and caribou are also common in the Arctic's permafrost region. Musk oxen, other species of bears, several rodent species, foxes, squirrels, and hares often accompany these Arctic inhabitants. However, when the brief summer comes to a close and the frigid cold reasserts itself upon the region, most animal life leaves in what is essentially a mass exodus.

The trees and shrubs that grow in permafrost forests are much smaller than those in the south and are far more resistant to freezing conditions. Forests tops are evidenced by species of spruce, birch, and aspen trees, and are carpeted below by shrubs, heath, and moss.

PERMAFROST FEARS

Scientists fear that changes to the permafrost regions pose a serious threat to humans and animals. The Earth's natural warming, or human-influenced global warming, has begun to melt the thick permafrost. Also, removal of vegetation and organic cover, as well as forest fires, can also have temperature-raising effects on permafrost regions, causing further thawing of the Permafrost. If this problem continues, the permafrost could turn into a slush mixture. This would cause permanent damage to buildings, roads, pipelines, and other infrastructure in the Arctic region. In particular, oil pipelines, originally set in solid permafrost, would be disrupted. The results could be disastrous, releasing oil into an already fragile ecosystem. Scientists believe this chain of events has begun, but admit that their investigations still remain at a very early stage. However, in Alaska and other Arctic regions, builders are already taking more precautions before building structures on permafrost.

READING LIST

Detecting Permafrost 'til the Cows Come Home. Alaska Science Forum. (1989). Available: http://www.gi.alaska.edu.

Scientists Fear Cycle of Global Warming. ABC News. (2001). Available: http://www.abcnews.com.

Permafrost at the Geological Survey of Canada. The Geological Survey of Canada. (2001). Available: http://sts.gsc.nrcan.gc.ca/permafrost/general.htm.

The Frozen Underground. The Weather Notebook. (2002). Available: http://www.weathernotebook.org.

New permafrost data may affect climate models study. University of Delaware Update. (2002). Available: http://www.udel.edu.

Heidorn, K.C. (2002). The Weather Doctor Almanac. Ph.D. Available: http://www.islandnet.com.

United States Environmental Protection Agency online database. (2002). Available: http://www.epa.gov.

Available: http://www.geography.uc.edu.

Seifert, R. (1994). *Permafrost: A Building Problem in Alaska*. The University of Alaska Fairbanks, Cooperative Extension Service.

Brown, R.J.E. (1967). *Permafrost Investigation in British Columbia and Yukon Territory*. National Research Council of Canada, Division of Building Research. Technical Paper No. 253, Ottawa, Canada.

RADAR USE IN RAINFALL MEASUREMENTS

Theodore A. Endreny
SUNY-ESF
Syracuse, New York

Meteorologists, hydrologists, and other environmental scientists use microwave radar (radio detection and ranging) techniques to observe remotely the location, type, and intensity of atmospheric precipitation. Doppler radar units provide an additional measurement of storm movement and wind rotational strength. Hence, microwave radar units are used for remotely sensing precipitation and winds.

Radar operates by emitting electromagnetic radiation into the air at various scan heights and then detecting the part of that radiation that is reflected back by hydrometeors (e.g., backscattered) to the radar device and considered reflectivity. Hydrometeors refer to liquid and frozen precipitation, but backscattering can also occur from large flocks of birds, insects, and other large airborne particles. Based on the density, size, and shape of the hydrometeors (e.g., spherical liquid vs. angular frozen crystals) and the polarization of the radar beam, the algorithms processing the radar signal can determine precipitation type and intensity.

RAINFALL AS ENERGY SPECTRUM

The wavelengths of radiation that are most commonly used for measuring precipitation are 3 cm (X-band), 5 cm (C-band), and 10 cm (S-band). National Weather Service (NWS) Weather Surveillance Radar 1988 Doppler (WSR-88D) can detect the radiation at various distances. Beam focusing algorithms are used to ensure that the beam penetrates a limited volume and distance and its signal-to-noise ratio is maximized. When radar beams are backscattered from intense precipitation, detection rises up to a 258-km radius from the radar unit, but reflectivity in weak to moderate precipitation usually limits detection to a 147-km radius.

WSR-88D Doppler radar stations were deployed in the 1990s to replace a network of non-Doppler radar units that were first deployed in the 1950s, after rapid advances in radar during WWII. The WSR-88D (also known as NEXRAD, for next generation weather radar) stations consist of a radar data acquisition unit (RDA) and data post-processing units. The RDA contains the antenna, pedestal, radome (a white spherical radar antenna cover), transmitter, receiver, and signal processor to transmit and then receive the microwave signal. The process of transmitting a signal, receiving and listening for a return signal, and then transmitting the next signal, takes place nearly 1300 times each second. For the 5-minute period to rotate 360° and survey precipitation, the WSR-88D transmits for only 0.6 seconds but receives for 4 minutes and 59.4 seconds to conduct surveillance of beam reflectivity.

Circular scans of 360° about the horizon by the WSR-88D referred to as volume coverage patterns (VCPs), are used to monitor the surrounding volume of the atmosphere. VCPs vary with weather forecasting needs, and there are currently two clear-air mode and two precipitation-mode VCPs. The differences between VCPs affect both the height and length of atmosphere sampled, which are inversely related such that scans that sample at greater heights will be limited to shorter ranges from the radar unit.

In both clear-air mode VCPs, the radar antenna transmits at five tilt angles, which are 0.5°, 1.5°, 2.5°, 3.5° and 4.5° (known as the 1st, 2nd, 3rd, 4th, and 5th tilt) above the horizon, completing a full VCP scan in 10 minutes (see Fig. 1). In both clear-air mode VCPs, two sweeps are taken at the 1st and 2nd tilt, the first sweep to monitor radar reflectivity and the second to monitor Doppler velocity, and a single sweep is taken for the 3rd to 5th tilt when reflectivity and velocity monitoring are combined. Distinguishing the two clear-air mode VCPs is the 3rd tilt, where one mode conducts two sweeps and the other mode one sweep, to monitor reflectivity and velocity. After the radar performs a full 360° revolution at each tilt angle, the radar has completed a full volume scan, and the radar processing algorithms can compute the precipitation rates. Each tilt has a limited range of detection from the radar, which is greatest at the 1st tilt and decreases with higher tilts. Precipitation VCPs begin tilts at 0.5° and end tilts at 19.5° to sample a greater volume of atmosphere. In one precipitation-mode VCP, a total of 14 tilts is sampled in 5 minutes (see Fig. 2); in the other, slower VCP, 9 tilts are scanned in 6 minutes (see Fig. 3).

Ground-based Doppler radar precipitation estimates are often overestimates by a factor of 2 due to their inability to scan the lower volume without ground interference. They often extend a prediction of upper level precipitation readings and do not account for evaporation in the lower 1 km of the atmosphere. Gauge measurements

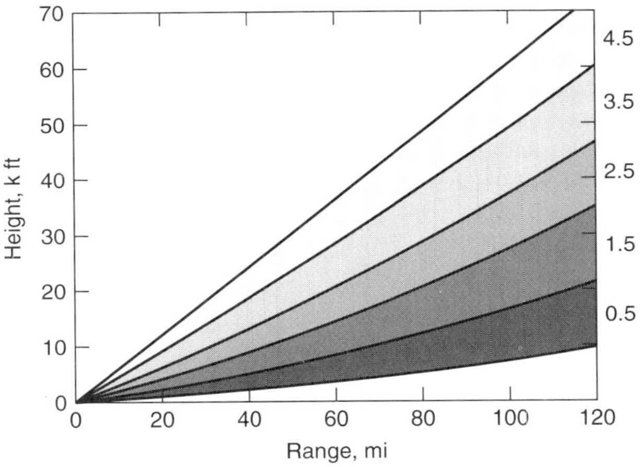

Figure 1. Clear-air mode VCP scan tilt and heights (thousand feet) for 0 to 120 mile ranges.

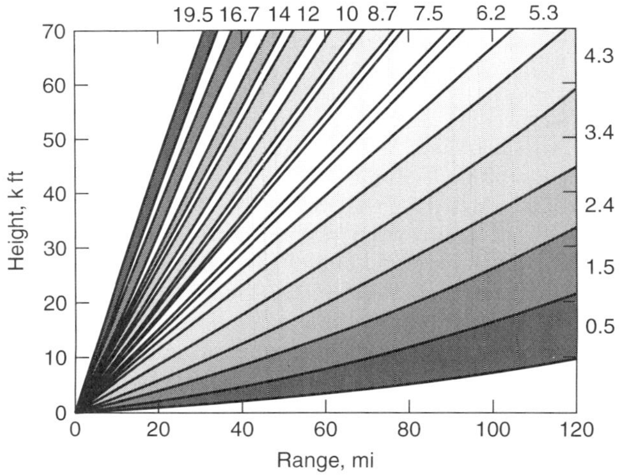

Figure 2. Precipitation-mode VCP scan tilt and heights (thousand feet) for 0 to 120 mile ranges in fast mode.

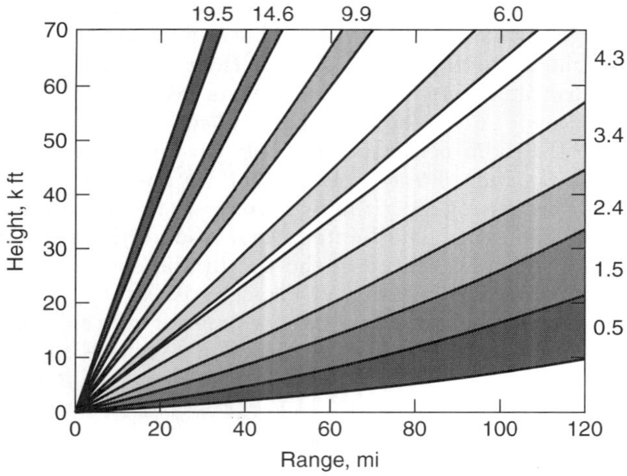

Figure 3. Precipitation-mode VCP scan tilt and heights (thousand feet) for 0 to 120 mile ranges in slow mode.

recorded at the ground may be used to correct some of this bias, and the adjusted radar estimates can provide better spatial estimates than traditional Thiessen polygon and other methods of interpolating point measurements. Extensive research on removing radar bias is ongoing.

Separating transmitting, attenuating, and receiving aspects of the radar is most helpful in understanding Doppler radar. The pulsed-Doppler radar is transmitted by coordinating oscillation and amplification. After the stabilized local oscillator (STALO) generates a nearly perfect sinusoidal continuous wave (cw) signal, which is modulated on and off, it is then amplified to create the powerful microwave signal. These pulses are phase coherent such that their interaction with a moving hydrometeor could displace the phase and allow for Doppler interpretation of hydrometeor movement path and velocity. The microwave beam is manipulated to have a fixed diameter, and an antenna is used to focus the radiation into a narrow angular region, or beam width, containing the peak radiation intensity.

Radar attenuation describes the power loss in the microwave beam due to absorption and scatter during its path through the atmosphere. The total power extracted from the wave is proportional to the combined absorbance and total-scatter cross section encountered and is determined by the highly sensitive radar receiving antennas.

The radar receiver detects the radar reflectivity factor, or Z, which is given units of $mm^6\ m^{-3}$, and is proportional to the sample volume of the radar, the number of hydrometeors within that volume, and the 6th power of the diameter of the hydrometeors. Determining the diameter of the hydrometeors is one of the most sensitive steps in determining rainfall rates, for there is often more than one diameter in the volume scanned. Much of the work in this area stems from observations made by Marshall and Palmer in 1948, determining the raindrop diameter for different raindrop size distributions. The Z factor is most commonly expressed as decibels Z, or dBZ, which is 10 times the $\log_{10}$ of Z.

Rainfall rates, R, are computed from the receiver measured Z factor using an empirical power law expressed as $R = \alpha Z^{\beta}$, where the α and β constants are determined from collecting actual rainfall data. For different climates, different values of α and β have been determined; values for α range from 0.013 to 0.028, and those for β range from 0.62 to 0.76. Currently, the NWS uses four different combinations of α and β to estimate rainfall rates for different weather systems. Methods for determining snowfall rates, which are expressed as S, are not as stable as for rainfall rates. Rainfall rates for WSR-88D radars associated with detected dBZ values are provided in Table 1. Note that most bird, insect, and dust detections are usually less than 30 dBZ.

Rainfall products are initially stored in binary HRAP units, which are part of a spherical coordinate system that records intensities. When HRAP units are transformed to a Cartesian system, typically presented at spatial scales of 1 to 4 km bins of average rainfall rates, for time periods per volumetric scan (about 5 minutely), 15 minutely, 1 hourly, 3 hourly, 6 hourly, and storm total precipitation. These rainfall rate products can be viewed on-line in spherical coordinates. They also can be downloaded and reconfigured or purchased from various vendors in the Cartesian coordinate system.

Products available from the WSR-88D radar units linked to the Internet provide "real-time" reflectivity images. These products are categorized as a base reflectivity image from the 0.5° tilt and a composite reflectivity image showing the greatest values sampled from all tilts. The base reflectivity maximum short-range distance provides a 227-km sample radius, and its maximum long-range distance is a 456-km sample radius; both have a spatial resolution at 1 km × 1 km grids. The composite reflectivity maximum short-range distance provides a 227-km sample radius at 2 km × 2 km grids, and the long-range distance provides a 419-km sample radius, but spatial resolution is at a 4 km × 4 km grid.

Algorithms process the data from the integrated volume scan and search for anomalous propagation (AP), such as backscatter from a nonmoving target such as the ground. One method of detecting AP is to look for an abrupt decrease in backscatter between the 1st and 2nd scan, which would indicate that the 1st scan captured surface terrain. Other methods may compare signal strength between different polarization angles or concurrently analyze strength with knowledge of the zero-degree isotherm, or radiosonde determined elevation above which precipitation should be frozen. Another algorithm

Table 1. Rainfall Rates

Reflectivity dBZ	Rain Rate in h^{-1}	Rain Rate cm h^{-1}
65	16.0	40.6
60	8.0	20.3
55	4.0	10.2
52	2.5	6.4
47	1.3	3.2
41	0.5	1.3
36	0.3	0.6
30	0.1	0.3
20	Trace	Trace

technique is to discriminate hail from rainfall or snow by capping the maximum dBZ at 50 or 53.

Radar measured rainfall is also observed from space by the Precipitation Radar instrument, owned by the Japanese, which rides aboard the Tropical Rainfall Measuring Mission (TRMM), a joint U.S. and Japanese satellite. This satellite, which has a low earth orbit of 350 km, was launched in November 1997 to measure precipitation in tropical areas (approximately 30°N and S of the equator) at a resolution of 4.3 km horizontal at 250 m vertical increments and a swath width of 220 km. It is sensitive to a rain rate of 0.5 mm h^{-1}.

Numerous private companies offer rainfall radar, but WSR-88D units are nationally deployed by the NWS, under the federal Department of Commerce and National Atmospheric and Oceanic Administration (NOAA), as well as the Department of Defense and the Federal Aviation Administration. The NWS-88D radar units are deployed in every state, except Connecticut, New Hampshire, and Rhode Island, and in the territories of Guam and Puerto Rico. Most states contain more than one unit; the NWS placed them so that their observed volumes would overlap and create complete coverage for the United States. Daily archives of NEXRAD data files can be downloaded on the Internet at: http://www.nws.noaa.gov/radfiles.html. Current NEXRAD observations are viewable at: http://weather.noaa.gov/radar/mosaic/DS.p19r0/ar.us.conus.shtml.

READING LIST

Doviak, R.J. and Zrnic, D.S. (1993). *Doppler Radar and Weather Observations*. Academic Press, San Diego, CA.

NOAA National Weather Service. Radar Information. Available: http://weather.noaa.gov/radar/radinfo/radinfo.html.

Smith, J.A. (1996). *Radar Hydro Meteorology (CIV 591)*. Princeton University, Princeton, NJ.

RAINFALL

James A. Smith

(from *The Handbook of Weather, Climate, and Water: Atmospheric Chemistry, Hydrology, and Societal Impacts*, Wiley 2003)

The capability to measure rainfall advanced dramatically in the last quarter of the twentieth century. The advances have been paced by remote-sensing technologies including both ground-based weather radar and satellite-borne instruments. The most dramatic developments have centered around the capability to monitor precipitation globally from satellite sensors. This measurement capability provides a variety of avenues for hydroclimatological analysis and forecasting. Advances in ground-based radar technologies and deployment of dense networks of rain gages has enhanced the ability to measure rainfall at short time scales (less than 1 h) and small spatial scales (less than 1 km). These time and space scales are often most relevant for water management applications. A brief summary of rainfall measurement and analysis capabilities is presented in the following three sections and organized by the three principal measurement technologies: rain gage, radar, and satellite.

RAIN GAGES

Networks of rain gages play a key role in hydrologic applications ranging from flood forecasting to design of high-hazard structures and water supply management. A wide variety of recording and nonrecording rain gages are used for hydrologic applications. Review and discussion of rain gage technologies are presented in the work by Sumner (1).

There exist several inherent sources of error that affect all types of rain gages. All rain gages suffer from errors due to modification of the wind field by the gage [see Robinson and Rodda (2) for detailed discussions]. The magnitude of errors depends on wind speed, siting characteristics, and type of precipitation (3–9). Rain gage measurement of rainfall is especially difficult in a variety of settings, including mountain ridges, forests, and water bodies. Measurement errors for snow are typically much larger than for rain and are generally in the form of catch deficiencies (3).

Rain gage networks serve as the basis for climatological assessments of precipitation that are used for a wide range of applications (see, e.g., 10). Three of the principal types of climatological analyses that are used for water management applications are illustrated in Figs. 1 to 3. Assessments of average rainfall conditions, in a variety of forms, are central to activities involving industrial, municipal, and agricultural water use. Mean annual precipitation is shown in Fig. 1 [see also Groisman and Legates (3) for a discussion of biases in rain gage analyses of mean precipitation]. Global assessments of continental precipitation have been developed from rain gage observations by Legates (11) [see also Legates and Wilmott (12)]. Precipitation frequency analysis plays a central role in engineering design problems, especially in urban areas (13). The 15-min, 100-year rainfall magnitude for the United States (14) is illustrated in Fig. 2. The network of gages that have the temporal resolution to provide short-term precipitation frequency analyses, such as those in Fig. 2, is far less dense than the rain gage network used to produce mean annual precipitation maps. Consequently, it is difficult to assess the true geographic variability of extreme rainfall rates. It is likely that geographic features, such as mountains and land–water boundaries, exert a pronounced influence on the frequency of extreme rainfall rates. The density of the network, however, is not adequate to resolve these geographic variations. Design of high-hazard structures, such as spillways on major dams, is determined through probable maximum precipitation (PMP) analyses (15,16). Rain gage data sets, in the form of storm catalogs, play a central role in PMP analyses. Storm catalogs for PMP analyses consist of gage observations from specific events. Consequently,

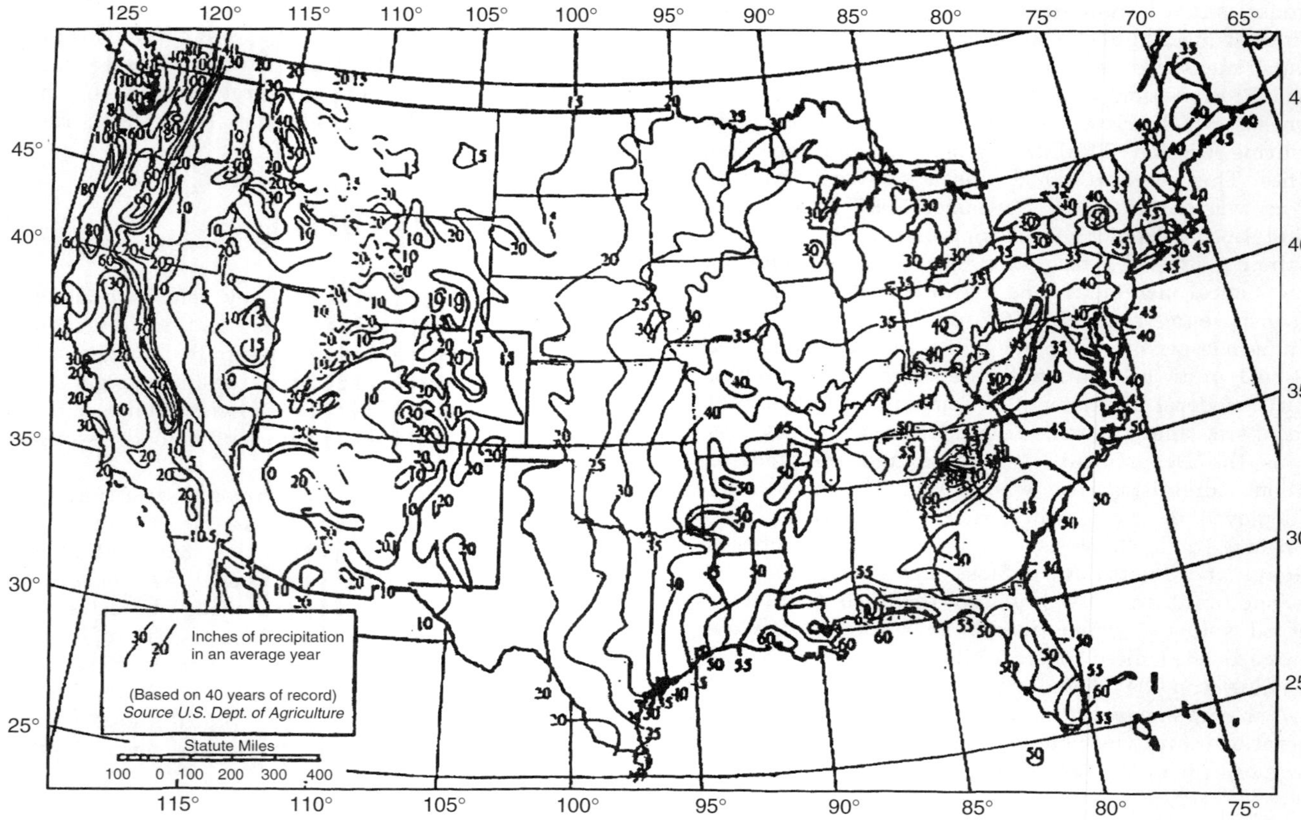

Figure 1. Mean annual precipitation (inches) for the United States from rain gage observations.

the density of gage observations in regions experiencing catastrophic rainfall is critical for PMP analyses. The 6-h, 200 mi^2 PMP for the eastern United States is shown in Fig. 3. The greatest uncertainties in PMP analyses are for small areas (less than 200 mi^2), short time periods (6 h and less), and for regions of complex terrain (17).

RADAR

Implementation of the NEXRAD (next-generation weather radar) system of WSR-88D (weather surveillance radar—1988 Doppler) radars has resulted in dramatic advances in rainfall measurement capabilities for the United States (18). Operational National Weather Service (NWS) rainfall products derived from WSR-88D observations provide rainfall analyses for the United States at 1-h time resolution and spatial resolution of approximately 4 km (19).

The hourly digital product (HDP) rainfall estimates are created by the WSR-88D radar product generator on a 131×131, 4-km grid centered at each radar site. The range over which rainfall products are constructed for each site is approximately 230 km. The algorithm used to construct this product (20) consists of the following steps: (1) quality control, including identification and elimination of anomalous propagation returns, (2) conversion of radar reflectivity factor to rainfall rate through a Z-R relationship, (3) correction for range effects, (4) aggregation of rainfall estimates to hourly, 4-km grid scale, and (5) bias correction using rain gage observations. The HDP product is the base rainfall product from the NEXRAD system. Detailed assessments of HDP algorithm performance are presented in Smith et al. (21) and Baeck and Smith (22) [see also Joss and Waldvogel (23), Wilson and Brandes (24), and Anagnostou and Krajewski (25)].

In a second stage of WSR-88D rainfall processing, multisensor precipitation analyses employ rain gage observations and the 4-km HDP rainfall fields in an optimal estimation framework developed by Krajewski (26) and Seo (27,28). These rainfall fields are subsequently composited into a regional mosaic. The regions that comprise the individual mosaics correspond to the watershed boundaries that delimit the NWS River Forecast Center areas of coverage. Algorithms used for mosaicking of multiple, overlapping radar coverages are described in Seo et al. (29). A national, hourly precipitation analysis is produced at the National Centers for Environmental Prediction (NCEP).

The national 4-km, hourly rainfall mosaic produced by NWS from rain gage and WSR-88D rainfall products will provide an important source of rainfall information for climatological analyses, especially as the observing period

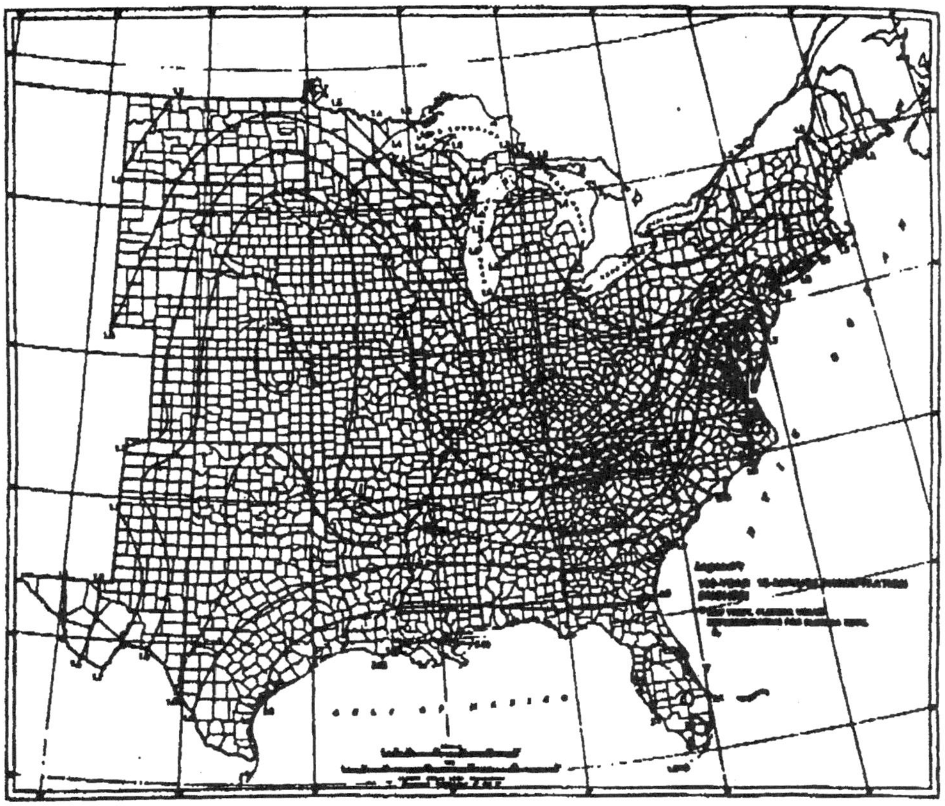

Figure 2. The 100-year, 15-min rainfall magnitudes (inches) for the United States east of the Rocky Mountains.

increases. Radar observations have not generally served as the basis for climatological analyses of rainfall [see, however, Baeck and Smith (30) for an exception]. Issues of bias in radar rainfall estimation must be addressed for radar-based rainfall databases to be most useful for climatological studies (21).

Radar polarimetric measurements (31–34), which utilize the capability of radar to transmit and receive electromagnetic radiation at alternating polarization, hold promise for providing significant improvements in rainfall estimates. Polarization measurements have been shown to be quite useful for quality control algorithms, including detection of bright band, hail, and AP [anomalous propagation (of radar waves, due to sharp gradients of water and air density)], as well as for algorithms for estimating rainfall rate (31,35). The NEXRAD network was designed for eventual implementation of polarization measurements by the WSR-88D.

Radar has provided a significant component of the observational basis for studying storms that produce extreme rainfall. Chappell (36) and Doswell et al. (37) summarize key elements of heavy rainfall producing storms with particular emphasis on storms that produce large point rainfall accumulations through small net storm motion [see also Maddox et al. (38)]. These storms have been termed quasi-stationary convective systems (36). Houze et al. (39) provide a detailed summary of radar-derived storm structure for severe thunderstorms in the central United States [see also Perica and Foufoula-Georgiou (40) and Steiner et al. (41)].

WSR-88D observations, and the rainfall products derived from these observations, have provided a new playing field for hydrologic application and science. Many hydrologic problems that were previously not possible to address due to an absence of information concerning rainfall, have been attacked from an observational perspective. Numerous examples can be drawn from flood hydrology. Figure. 5 illustrates a storm total rainfall analysis constructed for the rapidan storm of June 27, 1995 (42). More than 600 mm of rain fell on the east slope of the Virginia Blue Ridge during a 12-h period resulting in record unit discharge for the United States east of the Mississippi River and catastrophic landslides and debris flows. Fluvial and geomorphic impacts of the rapidan storm rival those described in the classic study by Hack and Goodlett (43) for the June 1949 storm in the Shenandoah Mountains. The chief difference between studies of the 1949 and 1995 storms is rainfall measurement at the 1-km horizontal scale and 6-min time scale for the 1995 storm that allows direct assessment of hydrologic processes.

SATELLITE

Satellite-borne instruments have proven useful for monitoring precipitating cloud system since the 1960s. Steady

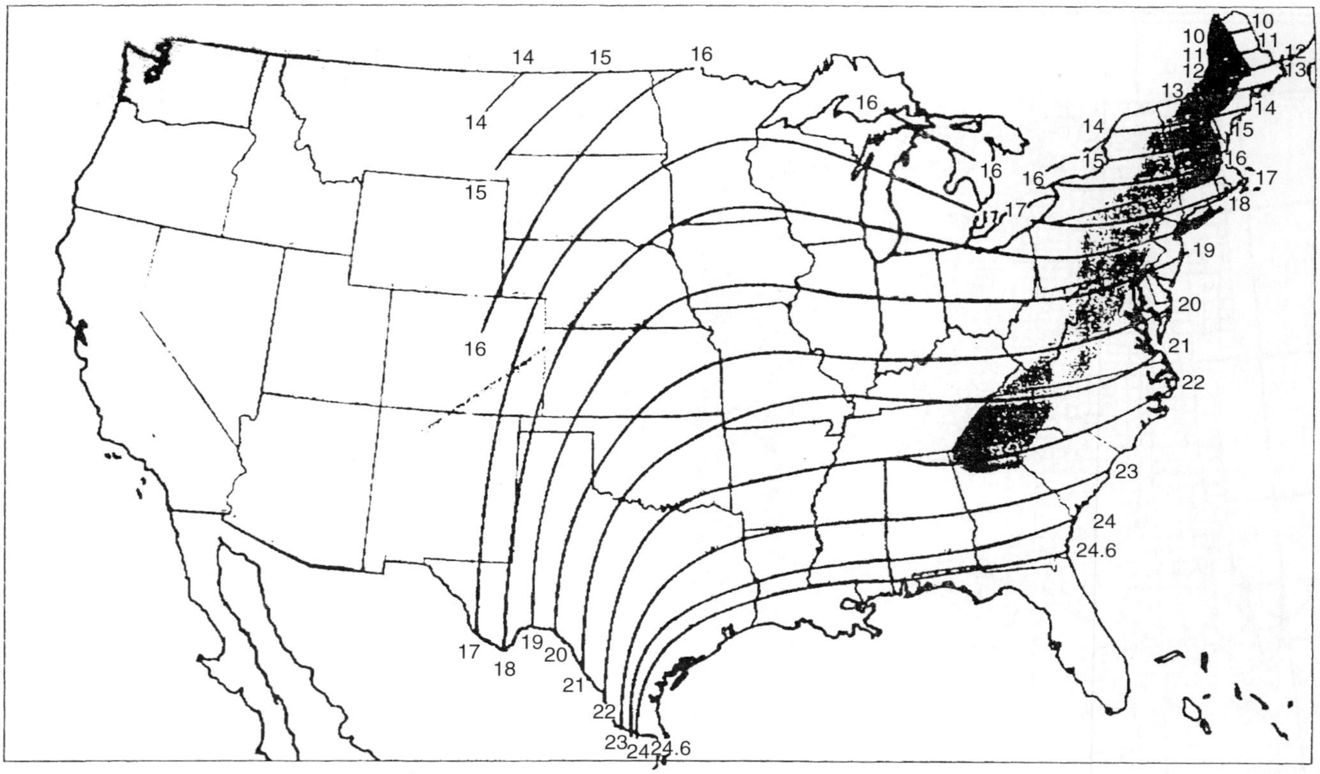

Figure 3. The 6-h, 200 mi^2 PMP magnitudes (inches) for the eastern United States.

progress has been made in developing algorithms for removing rainfall accumulations from passive satellite observations in the microwave (44,45) and infrared (46–48) portions of the electromagnetic spectrum. This progress is reflected in rapidly advancing capabilities for hydroclimatological analysis (49–53). Techniques for quantitative precipitation estimation from satellite sensors are reviewed below.

A geostationary, infrared-based satellite algorithm (46) has been developed and implemented for heavy rainfall measurement. This algorithm is specifically designed for deep, moist convective systems. Estimated precipitation rates are based on the cloud top temperature obtained from the 10.7-μm infrared channel. The empirical equations used to relate cloud top temperature and rainfall rate were calibrated from radar data sets consisting of observations from thunderstorm systems. A moisture correction factor obtained from the precipitable water and mean relative humidity is used to adjust the estimates for different moist environments. The technique of relating rain rate and cloud top temperature tends to overestimate the rain area in some cases and rain rate in others. The technique is also subject to underestimation of rain rates in warm cloud top environments and overestimation of cold top storms in strong wind shear environments.

The *Tropical Rainfall Measuring Mission* (*TRMM*) satellite (54) is designed to measure tropical precipitation and its variation. With the inclusion of a precipitation radar, TRMM provides the first opportunity to estimate the vertical profile of the latent heat that is released through condensation. The TRMM rainfall data will be particularly important for studies of the global hydrological cycle and for testing the ability of climate models to simulate climate accurately on the seasonal time scale.

The TRMM instruments for rainfall observation consist of a precipitation radar, a multifrequency microwave radiometer, and a visible and infrared (VIS/IR) radiometer. The precipitation radar provides measurements of the three-dimensional rainfall distribution over both land and ocean. The precipitation radar will permit the measurement of rain over land where passive microwave channels have difficulty. The horizontal resolution is approximately 4 km, the range resolution is 250 m, and the scanning swath width is 220 km. The multichannel microwave radiometer provides information on vertically integrated precipitation, its areal distribution, and its intensity. Rainfall analyses using the microwave radiometer are best suited for open ocean conditions. The visible infrared (IR) scanner provides high-resolution information on cloud coverage, type, and cloud top temperatures and serves as the link between these data and the long and virtually continuous coverage by the geosynchronous meteorological satellites. The instrument, with a swath width of 720 km, will provide cloud distributions by type and height and rain estimates from brightness temperatures at a horizontal resolution of approximately 2 km.

Satellite IR observations from geostationary satellites have been used extensively for assessing the climatology of extreme rainfall producing storms. An extensive

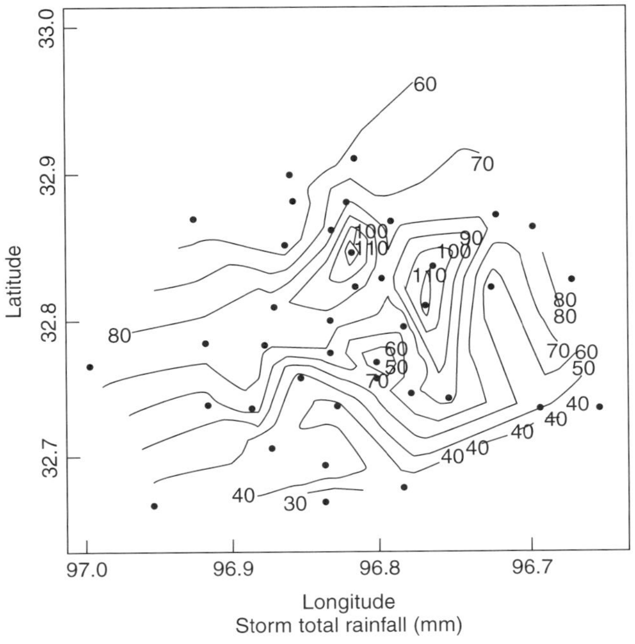

Figure 4. Storm total rainfall (mm) from the Dallas Metropolitan Area mesonet for the Dallas hailstorm of May 5, 1995. The dimensions of the surrounding box are approximately 30 × 30 km.

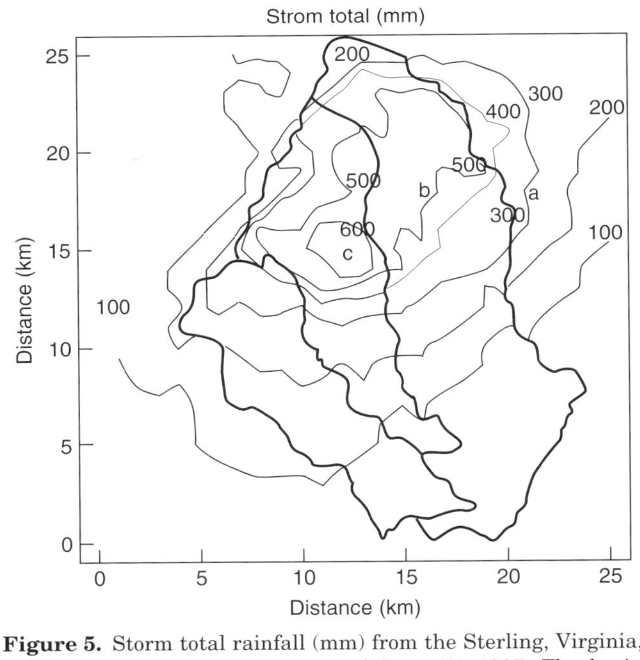

Figure 5. Storm total rainfall (mm) from the Sterling, Virginia, WSR-88D for the rapidan storm of June 27, 1995. The basin boundary for the 295-km^2 watershed and boundaries for 3 subwatersheds are shown in solid lines.

climatology has been developed for mesoscale convective complexes (55) based on IR-based assessments of cloud properties. Numerous studies have examined the links between mesoscale convective complexes (MCCs), and the more general category of mesoscale convective systems, and heavy rainfall [see Houze (56)].

BIBLIOGRAPHY

1. Sumner, G. (1988). Precipitation measurement and observation, in *Precipitation: Process and Analysis*. Wiley, Chichester Chapter 7.
2. Robinson, A.C., and Rodda, J.C. (1969). Rain, wind and the aerodynamic characteristics of raingauges. *Met. Mat.* **98**: 113–120.
3. Groisman, P.Y., and Legates, D.R. (1994). The accuracy of United States precipitation data. *Bull. Am. Meteor. Soc.* **75**(2): 215–227.
4. Sevruk, B. *Methods of correction for systematic error in point precipitation measurement for operational use*, WMO Publication No. 589, OHR No. 21, 1982.
5. Sevruk, B. *Wind-induced measurement error for high intensity rains*, Proceedings, WMO/IAHS International Workshop on Precipitation Measurements, St. Moritz, Switzerland, 1989, pp. 199–204.
6. Nystuen, J.A., Proni, J.R., Black, P.G., and Wilkerson, J.C. (1996). A comparison of automatic rain gauges. *J. Atmos. Oceanic Technol.* **13**: 62–73.
7. Steiner, M., Smith, J.A., Burges, S.J., Alonso, C.V., and R.W. Darden (1999). Effect of bias adjustment and rain gauge data quality control on radar rainfall estimation. *Water Resour. Res.* **35**(8): 2487–2503.
8. McCollum, J.R., and Krajewski, W.F. (1998). Uncertainty of monthly rainfall estimates from rain gauges in the Global Precipitation Climatology Project. *Water Resour. Res.* **34**(10): 2647–2654.
9. Larson, L. and Peck, E. (1974). Accuracy of precipitation measurements for hydrologic modeling. *Water Resour. Res.* **10**(4): 857–863.
10. Frei, C. and Schaer, C. (1998). A precipitation climatology of the Alps from high-resolution raingauge observations. *Int. J. Climatol.* **18**(8): 873–900.
11. Legates, D.R. (1987). A climatology of global precipitation. *Climatology* **40**(1), .
12. Legates, D.R. and Willmott, C.J. (1990). Mean Seasonal and Spatial Variability in Gauge-Corrected, Global Precipitation. *Int. J. Climatol.* **10**(2): 111–127.
13. Urbonas, B.R., and Roesner, L.A. (1992). Hydrologic design for urban drainage and flood control. In: *Handbook of Hydrology*. D. Maidment (Ed.). McGraw-Hill, New York.
14. Frederick, R.H., Myers, V.A., and Auciello, E.P. *Five-to 60-minute Precipitation Frequency for the Eastern and Central United States*. NOAA Technical Memo, NWS Hydro-35, June 1977.
15. Hansen, M. (1987). Probable maximum precipitation for design floods in the United States. *J. Hydrol.* **96**: 267–278.
16. World Meteorological Organization (WMO) (1986). *Manual for estimation of Probable Maximum Precipitation*, 2nd Edn. WMO-No. 332, WMO, Geneva, Switzerland.
17. National Research Council. (1994). *Estimating Bounds on Extreme Precipitation Events*. National Academy Press, Washington, DC.
18. Klazura, G.E., and Imy, D.A. (1993). A description of the initial set of analysis products available from the NEXRAD WSR-88D System. *Bull. Am. Meteor. Soc.* **74**: 1293–1311.
19. Hudlow, M.D., Smith, J.A., Walton, M.L., and Shedd, R.C. (1991). NEXRAD—New era in hydrometeorology. In: *Hydrological Applications of Weather Radar*. I. Cluckie and C. Collier (Eds.). Ellis-Norwood, New York, pp. 602–612.

20. Fulton, R.A., Breidenbach, J.P., Seo, D.-J., and Miller, D.A. (1998). The WSR-88D rainfall algorithm. *Weather Forecast.* **13**(2): 377–395.
21. Smith, J.A., Seo, D.-J., Baeck, M.L., and Hudlow, M.D. (1996). An intercomparison study of NEXRAD precipitation estimates. *Water Resour. Res.* **32**(7): 2035–2045.
22. Baeck, M.L., and Smith, J.A. (1998). Rainfall estimation by the WSR-88D for heavy rainfall events. *Weather Forecast.* **13**: 413–436.
23. Joss, J. and Waldvogel, A. (1989). Precipitation measurement and hydrology: A review. In: *Battan Memorial and Radar Conference*. David Atlas (Ed.). American Meteorological Society, Boston.
24. Wilson, J.W., and Brandes, E.A. (1979). Radar measurement of rainfall—a summary. *Bull. Am. Meteor. Soc.* **60**: 1048–1058.
25. Anagnostou, E., and Krajewski, W.F. (1998). Calibration of the WSR-88D precipitation processing subsystem. *Weather Forecast.* **13**(2): 396–406.
26. Krajewski, W.F. (1987). Co-kriging of radar and rain gage data. *J. Geophys. Res.* **92**(D8): 9571–9580.
27. Seo, D.-J. (1998). Optimal estimation of rainfall fields using radar and rain gage data. *J. Hydrol.* **208**(1,2), pp. 25–36.
28. Seo, D.-J. (1998). Real-time estimation of rainfall fields using rain gage data under fractional coverage conditions. *J. Hydrol.* **208**(1,2), pp. 37–52.
29. Seo, D.J., Fulton, R., and Breidenbach, J. (1998). *Rainfall estimation in the WSR-88D era for operational hydrologic forecasting in the National Weather Service, Symposium on Hydrology, Phoenix.* American Meteorological Society, pp. J60–J62.
30. Baeck, M.L., and Smith, J.A. (1995). Climatological analysis of manually digitized radar data for the United States. *Water Resour. Res.* **31**(12): 3033–3049.
31. Zrnic, D.S. (1996). Weather radar polarimetry—trends toward operational applications. *Bull. Am. Meteorol. Soc.* **77**(7): 1529–1534.
32. Zrnic, D.S., and Ryzhkov, A.V. (1996). Advantages of rain measurements using specific differential phase. *J. Atmos. Ocean. Tech.* **13**(2): 454–464.
33. Ryzhkov, A.V., and Zrnic, D.S. (1996). Assessment of rainfall measurement that uses specific differential phase. *J. Appl. Meteorol.* **35**(11): 2080–2090.
34. Aydin, K., Bringi, V.N., and Liu, L. (1995). Rain rate estimation in the presence of hail using S-band specific differential phase and other radar parameters. *J. Appl. Meteor.* **34**: 404–410.
35. Petersen, W.A., Carey, L.D., Rutledge, S.A., Knievel, J.C., N.J. Doesken, Johnson, R.H., McKee, T.B., Vonder Haar, T., and Weaver, J.F. (1999). Mesoscale and radar observations of the Fort Collins flash flood of 28 July 1997. *Bull. Am. Meteor. Soc.* **80**(2): 191–216.
36. Chappell, C. (1989). Quasistationary convective events. In: *Mesoscale Meteorology*. P. Ray (Ed.). American Meteorological Society, Boston.
37. Doswell III, C.A., Brooks, H.E., and Maddox, R.A. (1996). Flash flood forecasting: An ingredients-based methodology. *Weather Forecast.* **11**(4): 560–581.
38. Maddox, R., Chappell, C., and Hoxit, L. (1979). Synoptic and meso-α scale aspects of flash flood events. *Bull. Am. Meteor. Soc.* **60**(2): 115–123.
39. Houze, R.A., Smull, B.F., and Dodge, P. (1990). Mesoscale organization of springtime rainstorms in Oklahoma. *Monthly Weather Rev.* **118**: 613–654.
40. Perica, S., and Foufoula-Georgiou, E. (1996). Linkage of scaling and thermodynamic parameters of rainfall: Results from midlatitude mesoscale convective systems. *J. Geophys. Res.* **101**(D3): 7431–7448.
41. Steiner, M., Houze, R.A., Jr., and Yuter, S.E. (1995). Climatological characterization of three-dimensional storm structure from operational radar and raingage data. *J. Appl. Meteor.* **34**: 1978–2007.
42. Smith, J.A., Baeck, M.L., Steiner, M., and Miller, A.J. (1996). Catastrophic rainfall from an upslope thunderstorm in the Central Appalachians: the Rapidan Storm of June 27, 1995. *Water Resour. Res.* **32**(10): 3099–3113.
43. Hack, J. and Goodlett, J. *Geomorphology and Forest Ecology of a Mountain Region in the Central Appalachians, U.S. Geological Survey Professional Paper 347*, 1960.
44. Negri, A.J., Adler, R.F., Nelkin, E.J., and Huffman, G.J. (1994). Regional rainfall climatologies derived from special sensor microwave imager (SSM/I) data. *Bull. Am. Meteor. Soc.* **75**: 1165–1182.
45. Adler, R.F., Huffman, G.J., and Keehn, P.R. (1994). Global tropical rain estimates from microwave-adjusted geosynchronous IR data. *Remote Sensing Rev.* **11**: 125–152.
46. Vicente, G.A., and Scofield, R.A. (1997). *Real time rainfall rate estimates derived from the GOES-8/9 satellites for flash flood forecasting, numerical modeling and operational hydrology*, Preprints, 13th Conference on Hydrology, Long Beach, CA, American Meteorological Society, pp. J115–J118.
47. Huffman, G.J., Adler, R.F., Rudolf, B., Schneider, U., and P.R. Keehn (1995). A technique for combining satellite-based estimates raingauge analysis, and NWP model precipitation information into global precipitation estimates. *J. Climate* **8**(5): 1284–1295.
48. Adler, R.F., and Negri, A.J. (1988). A satellite infrared technique to estimate tropical convective and stratiform rainfall. *J. Appl. Meteor.* **27**: 30–51.
49. Krajewski, W.F., Ciach, G.J., and McCollum, J.R. (2000). Initial validation of the global precipitation climatology project monthly rainfall over the United States. *J. Appl. Meteorol.* **39**(7): 1071–1086.
50. Adler, R.F., Huffman, G.J., Bolvin, D.T., Curtis, S., and Nelkin, E.J. (2000). Tropical rainfall distributions determined using TRMM combined with other satellite and rain gauge information. *J. Appl. Meteor.* **39**(12): 2007–2023.
51. Huffman, G.J., Adler, R.F., Arkin, P., Chang, A., Ferraro, R., Gruber, A., Janowiak, J., McNab, A., Rudolf, B., and Schneider, U. (1997). The Global Precipitation Climatology Project (GPCP) combined precipitation dataset. *Bull. Am. Meteor. Soc.* **78**(1): 5–20.
52. Huffman, G.J., Adler, R.F., Morrissey, M.M., Bolvin, D.T., Curtis, S., Joyce, R., McGavock, B., and Susskind, J. (2001). Global precipitation at one-degree daily resolution from multisatellite observations. *J. Hydrometeor.* **2**(1): 36–50.
53. Kummerow, C. (2000). The status of the Tropical Rainfall Measuring Mission (TRMM) after two years in orbit. *J. Appl. Meteor.* **39**(12): 1965–1982.
54. Simpson, J., Adler, R.F., and North, G. (1988). A proposed tropical rainfall measuring mission (TRMM). *Bull. Am. Meteor. Soc.* **69**: 278–295.
55. Maddox, R.A. (1980). Mesoscale convective complexes. *Bull. Am. Meteor. Soc.* **61**(11): 1374–1387.
56. Houze, R.A.(1993). *Cloud Dynamics*. Academic, New York.

READING LIST

Barros, A.P. and Lettenmaier, D.P. (1994). Dynamic modeling of orographically induced precipitation. *Rev. Geophys.* **32**(3): 265–284.

Seo, D.J., Breidenbach, J., and Johnson, E. (1999). Real-time estimation of mean field bias in radar rainfall data. *J. Hydrol.* **223**(3,4): 131–147.

RAINFALL AND RUNOFF

Theodore A. Endreny
SUNY-ESF
Syracuse, New York

Rainfall is generated by multiple weather patterns, summarized into frontal, orographic, and convective, that affect the duration and intensity of storm accumulation. Climate data analyses reveal regional and local trends in the seasonal and annual rainfall patterns of duration and intensity and have been reported in map and tabular form for durations of 5 minutes to 24 hours and a variety of return periods in Technical Papers 40 and 49 by the U.S. Weather Bureau and in HYDRO-35 by the National Weather Service. A large variation in rainfall spatial and temporal duration and intensity has been observed for large and small area, long and short time events, and a combined system of drop-size spectrometers, recording rain gauges, and microwave radar have helped to capture this detail nationally at the resolution of 5 minutely volume scans for 1 km^2 bins, aggregated to rain-gauge-corrected hourly totals at 4 km^2 bins. Internal warm and cold cloud precipitation dynamics, low-level evaporation of falling precipitation, and distinction between rain, sleet, freezing rain, and snow still provide technological challenges. Time series plots of rainfall are referred to as hyetographs, maps of equal rainfall amounts are known as isopleths, and National Weather Service NEXRAD Doppler radar gauge-corrected hourly 4 km^2 grid based rainfall maps are known as Stage III products. The spatial aggregation of rainfall point estimates from rain gauges is performed using surface fitting, weighting, and interpolation schemes, including the method of Thiessen polygons.

RAINFALL OBSERVATION AND ESTIMATION

Estimates of expected precipitation intensities, as a function of duration and return period (frequency), are referred to as intensity-depth-frequency (IDF) curves (see Fig. 1), and are computed using climatic precipitation data or regional coefficients, as found in such methods as HYDRO-35. Engineering analyses of the failure of various structures, such as drainage culverts and retention basins, are examined using the IDF curve for storms of regulated frequency and duration. The time of concentration, or the time for water to travel from its hydraulically most distant watershed point to the outlet, is used to design peak runoff concentration. Because the rainfall intensity diminishes with duration, the watershed will generate the greatest instantaneous volume of runoff for the time of concentration. Probable maximum precipitation (PMP) rates represent known and estimated regional maximum storm volumes for a given duration and are required inputs in the human safety analysis of critical infrastructure, such as nuclear power plants and large reservoirs and dams where significant loss of life is possible in cases of failure.

The temporal persistence or absence of sequential rainfall affects the earth's surface antecedent moisture and response to the current rainfall, as do surface slopes, ground cover, and roughness. Surface geologic, geographic, and biological features and controls on rainfall partition percentages and timing are relatively steady or predictable across time; surface antecedent moisture is ephemeral and has greater variability in partitioning between surface storage, groundwater recharge, and surface and subsurface runoff to rivers, lakes, and oceans. Antecedent moisture is categorized for the growing and dormant seasons, and in a Natural Resources Conservation Service (NRCS) method, antecedent moisture is a function of the previous 5-day precipitation totals. At the annual time step, however, soil moisture is relatively constant, absent periods of drought. As such, the fact that perennial, intermittent, and ephemeral streams drain most coastal areas indicates that rainfall exceeds evaporation and that runoff is a unique land feature. The atmosphere precipitates nearly 116,000 $km^3\ yr^{-1}$ over land surfaces, but only 44,700 $km^3\ yr^{-1}$ drains to the oceans in rivers, and only 2,200 $km^3\ yr^{-1}$ drains in groundwater. The difference is evaporated mostly from the land surface.

RAINFALL PARTITIONING AT THE LAND SURFACE

Landscape rather than plot analysis is the preferred method for predicting the partitioning of rainfall into runoff, which can travel as surface, unsaturated subsurface (through flow and interflow), and saturated subsurface flow toward the receiving water. Several theories on the partitioning and pathways of rainfall runoff exist, some of which are illustrated in Fig. 2, which isolate the interaction of rainfall and soil saturation conditions. In general, Hortonian overland flow expects infiltration to be exceeded uniformly in space and time and predicts an entire hillslope of overland runoff. This theory does not explain observed conditions in most areas, but it works reasonably well in many dry western U.S. states. Betson flow expects infiltration to have some spatial variation where it is exceeded by rainfall, and was derived from mathematical analysis and not located in the hillslope. In separate field studies by Hewlett and Dunne, this phenomenon of a dynamic spatial and temporal zone of runoff was observed and called the concept of variable source area (VSA). Whether this runoff originates as the result of rainfall rates exceeding infiltration, known as infiltration excess overland flow, or as the result of saturation from below, preventing infiltration, known as saturation excess overland flow, remains difficult to distinguish. In contrast, assuming that the system is a gaining stream where the terrestrial water table is above the streambed, then base flow is relatively constant for a storm runoff.

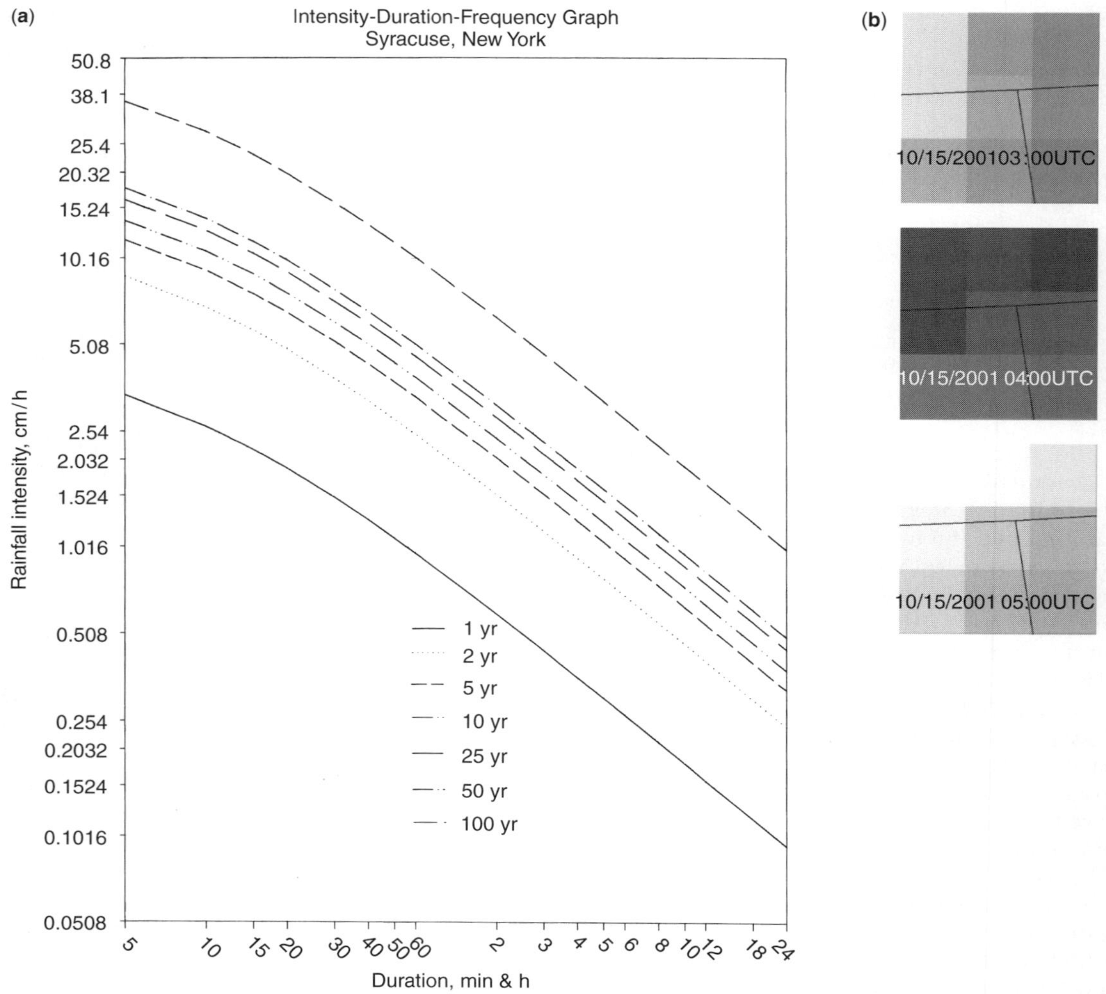

Figure 1. (**a**) The intensity–duration–frequency (IDF) curve for Syracuse, NY, where intensity decreases with duration, but increases with return period, or frequency; (**b**) Illustrations of space and time varying rainfall intensity, represented by 4km x 4km NEXRAD Stage III precipitation grids, where white indicates 0 mm/hour and dark gray indicates 14.5 mm/hour. The total area of 3 x 3 grids is 144 km^2, located at 45 degrees Latitude North by 74 Degrees Longitude West. The dark lines connect at the point where Franklin and Clinton County, New York border Ontario, Canada.

The distinction between infiltration excess and saturation excess has been examined theoretically, comparing observed precipitation rates, infiltration capacity rates, and hydraulic conductivity values, longer period (hourly) averaged precipitation rates seldom exceed infiltration capacity. Shorter period rates (5 minutely) of intense convective storms often do exceed soil infiltration capacity. In arid climates and agricultural areas where soils crust, infiltration is significantly retarded, and precipitation excess often generates runoff. In most other pervious areas, long-duration surface runoff is typically generated by saturation from below, where the local or perched water table has reached the surface. Observations that distinguish between these two events have been advanced using increased spatial and temporal remote sensing monitoring technologies of soil moisture and rainfall dynamics. As watershed rainfall runoff studies increasingly consider urban and suburban dynamics, a greater percentage of the watershed is paved, making the distinction of a runoff-generating mechanism a moot point.

RUNOFF OBSERVATION AND ESTIMATION

Runoff generated from rainfall can be expressed as a volume, a volume per time, as a depth per watershed area, or as a depth per watershed area per time. The ratio of the total runoff and total observed rainfall per time period gives the runoff ratio, which may be presented for climatological or storm periods and has an average U.S. value of 38%. Table 1 provides runoff ratio values for several global regions, and Table 2 provides values for several river basins, which range from 65 to 20%. It is apparent that not all rainfall is converted into storm runoff, and this was first quantified in the 1670s on the upper Seine River in France, but that significant percentages go toward recharging soils and groundwater supplies. The amount of rainfall that is converted into runoff is labeled the effective precipitation, which is an important distinction from observed (e.g., rain gauge measured) precipitation. The depth difference between observed and effective rainfall is explained by the filling by rainfall of a series of storages, including canopy interception, depression storage, detention storage, and soil moisture storage. Areas where observed precipitation is equal to effective, or in stream observed runoff, include rain on water and rain on impervious surfaces (e.g., pavement, rooftops, frozen soil) that are directly connected to the point of measured discharge.

Runoff in channels, or discharge, is measurable using a variety of techniques, including monitoring dilution of injected tracers; monitoring stage height per time for discharge calibrated control structures; monitoring total volumes of discharge per time; and monitoring depth integrated velocities with physical floats, impellers, and Doppler signals together with cross-sectional areas. A runoff time series follows a traditional trend of rising and falling discharge per time, known as a hydrograph. The hydrograph is often separated to distinguish different flow contributions, first delineating a steady volume of base flow or groundwater inflow mostly unaffected by storm dynamics and second, delineating the overlying pulse of event flow. The event flow is potentially separated into separate types of response by using tracers and mixing models, such as slow and fast from the through flow, interflow, pressure wave release, macropore or pipe flow, surface flow, and in-stream rainfall.

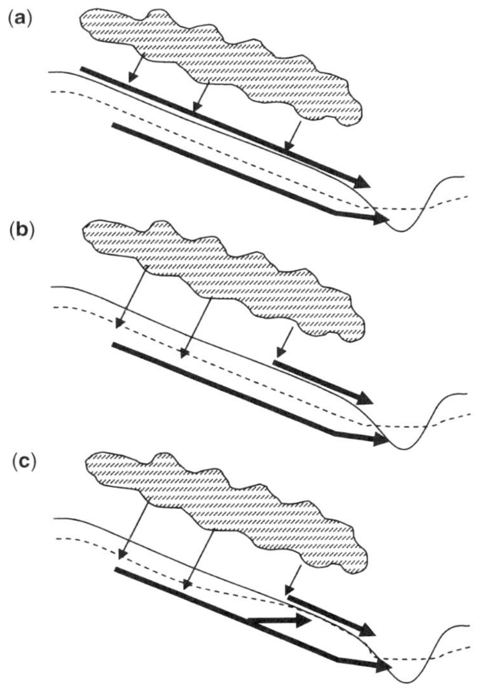

Figure 2. In all figures, the thin solid line is the hillslope surface, the dashed line is the watertable and stream surface, the hashed cloud is rain, the thin arrows are rainfall with or without infiltration, and the thick arrows are surface and subsurface runoff. (**a**) Hortonian concept for overland runoff; the entire hillslope contributes, and base flow is below the water table; (**b**) Betson concept of a with partial hillslope area contributions, but not derived from observations of variable areas; and (**c**) Dunne and Hewlett concept of a variable source area where the water table rises to the ground surface and generates overland runoff.

ENGINEERING VERSUS SCIENTIFIC METHODS

Discharge as total storm runoff, peak storm runoff, or storm time series runoff is estimated by using an equally wide array of methods, and each uses a specific method that first distinguishes the effective precipitation. The most common engineering approach to predicting storm total runoff is the curve number, or SCS, method that partitions observed rainfall into effective rainfall for 6 to 24 hour storms based on antecedent moisture, land cover type, management condition, and hydraulic soil type, which ranges from high infiltration A soils to poor infiltration D soils. The most common engineering approaches to predicting storm peak runoff are either by the curve number based TR-55 method or the runoff coefficient based rational method, which uses a land cover type runoff coefficient to convert observed rainfall intensity into an effective intensity. The most common engineering approach to predicting time series runoff is the unit hydrograph, which requires a user input of the effective rainfall across time, provides a transfer function known as a convolution integral technique to output volume per time discharge, and represents the observed rising and falling limbs of incrementally changing watershed contribution. Unit hydrographs are created for unit depth rainfall (e.g., 1 cm) of distinct duration and based on the principle that runoff response is linear and additive, are weighted by greater or lesser effective precipitation amounts. Empirical, synthetic, and geomorphic techniques exist to create unit hydrographs.

Alternative analytic approaches exist to partition observed rainfall into estimated runoff and, through validation studies on several watersheds, are increasingly demonstrated as accurate. These approaches are based on solving a series of equations for estimating interception volumes, filling detention, depression, and retention storages, partitioning soil infiltration volumes, and possibly accounting for evaporation rates, where initial estimates of storages are either input or acquired from model spin ups where evaporation is a controlling process. As the watershed size increases, effective values for observed precipitation and parameters for soil and plant processes present greater model input challenges. Distinctions between these mechanistic models include their representation of the watershed in space, as either distributed or lumped, and their simulation through time, as either storm or continuous. New modeling techniques are experimenting with parameter optimization through genetic algorithms that mutate and select the most robust parameter strings, and neural networks that seek optimal, but nonphysical, solutions that have best met the objective functions, which typically minimize differences between observed and predicted runoff. Popular rainfall runoff models include those used for operational forecasting

Table 1. Continental Average Estimated Runoff

Continent	Area km^2	Runoff mm yr^{-1}	Runoff Ratio %
Antarctica	14,100,000	28	83
Europe	10,000,000	375	43
Asia	44,100,000	420	40
South America	17,900,000	946	40
North America	24,100,000	403	38
Africa	29,800,000	582	16
Australia	7,600,000	420	6
Total Land	148,900,000	480	36

Table 2. Average Estimated Runoff Ratio for Major River Basins

Watershed by River Name	Continent [Nation(s)]	Basin Size km^2	Runoff Ratio %
Brahmaputra	Asia (Tibet/Bangladesh)	589,000	65
Irrawaddy	Asia (Burma)	431,000	60
Yangtzekiang	Asia (China)	1,970,000	50
Amazon	South America (six nations)	7,180,000	47
Orinoco	South America (Venezuela)	1,086,000	46
Lena	Asia (Russia)	2,430,000	46
Mekong	Asia (China)	795,000	43
Yenesei	Asia (Russia)	2,599,000	42
Ganges	Asia (China)	1,073,000	42
Saint Lawrence	North America (Canada, U.S.)	1,030,000	33
Amur	Asia (Russia)	1,843,000	32
Congo	Africa (seven nations)	3,822,000	25
Ob	Asia (Russia)	2,950,000	24
Mississippi	North America (U.S.)	3,224,000	21
La Plata	South America (five nations)	2,650,000	20
Average		2,224,800	40

by the National Weather Service, such as the SAC-SMR (Sacramento Soil Moisture Accounting) or on those capturing governing topographic controls on generating runoff, such as TOPMODEL. A partial list of agency and research based surface water models is provided by the USGS Surface Water Modeling Clearinghouse website.

RUNOFF ROUTING

Explicit routing of runoff across the landscape, through a retention basin, and along a section of channel upstream of the gauging station may provide additional insighting time-based runoff estimates. In each case, routing has been examined using fully dynamic equations of energy and continuity, written in their complete form (e.g., St. Venant) or neglecting acceleration and pressure gradients (e.g., kinematic), solved using numerical methods of finite element and finite difference. Simpler solutions that still consider the effects of friction, such as the Darcy–Weisbach and Manning equations have been used to route flow across land, as have stand-alone continuity equations that simply push water from cell to cell at a given slope-based velocity. The algorithms used to partition flow in grid-based networks of the watershed terrain map (e.g., digital elevation map files with draped land cover) are varied, ranging from a single path to flow into a maximum of all eight neighboring lower cells. Routing across triangular irregular network grids and polygons is also used. Flow through a reservoir, known as level-pool routing, and along a channel (e.g., the Muskingum method), attenuates the inflow hydrograph based on increasing storage surface area and stage-discharge relationships and wave celerity and propagation rates, respectively. Both methods are presented in most hydraulic engineering handbooks.

SPECIAL TOPICS: WATER YEAR, RUNOFF QUANTITY AND RUNOFF QUALITY

Annual time series analysis of precipitation hyetographs and runoff hydrographs reveals trends that give rise to regional characterization of U.S. hydrologic zones. The hyetograph analysis considers rain and liquid equivalent snowfall. West of the 100th meridian, rainfall is typically less than 20 inches annually. California has increased precipitation during the winter when a Pacific blocking high moves northward; southwestern states have summer monsoon rains; central, midwestern, and southeastern states have increased precipitation during summer convective showers, and eastern states have rather steady annual precipitation. Runoff often follows separate trends, given the land surface seasonal function in partitioning rainfall. Snow accumulation in the winter often keeps winter runoff low; snowmelt in the spring and higher altitudes in the summer release a pulse of the largest stream flow volumes. Spring and summer vegetation growth creates soil moisture deficits that allow for larger soil storage of rainfall and reduced stream flow. The fall dieback of vegetation reduces evaporation rates, and the soil storage fills and remains filled, which increases stream flow. The annual rainfall–runoff cycle that leads to a renewal of soil moisture reserves and increased runoff has led to the creation of the water year, beginning on October 1. Anomalies to this pattern exist, and events such as warm rain on snow, which melts the pack in midwinter but does not infiltrate into the frozen soils, can create huge winter floods. Likewise, the global ocean and atmospheric patterns set in place by systems such as the El Niño southern oscillation can cause southeastern spring and summer droughts when there are usually heavy rains.

Special topics in runoff hydrology include flood and drought analysis as well as water quality, which are useful in human safety, infrastructure design, and environmental management. The Water Resources Council 1972 method continues to provide guidance on fitting annual maximum discharge observations onto a Type III Pearson distribution to forecast return interval floods that exceed observed record lengths. Although 10, 50, and 100 year events have significant discharge rates and therefore erosive power, channel patterns are shaped by flows of the more frequent 1.5 return intervals. Drought analysis is critical to sizing drinking water reservoirs and wastewater treatment discharge dilution plans and are typically based on 7-day, 10-year low flows, or for ungauged rivers, on regional relationships derived from similar streams that were gauged. Water quality has become more directly tied to rainfall runoff analysis after the successful control of many municipal and industrial point source pollution dischargers, leaving the nonpoint source (NPS) pollution from agricultural fields, commercial parking lots, urban streets, and residential lawns. NPS runoff from these sources, it has been observed, explains the large fraction of soil, nutrient, chemical, metal, organic, and bacterial loads entering the 40% of national waterways that are still noncompliant with Clean Water Act requirements for fishing and swimming access.

READING LIST

Dingman, S.L. (1994). *Physical Hydrology*. Prentice-Hall, Englewood Cliffs, NJ.

Mays, L.W. (2001). *Water Resources Engineering*. John Wiley & Sons, New York.

REMOTE SENSING OF APPLICATIONS IN HYDROLOGY

Jessica Black
NOAA/CREST CCNY
New York, New York

Theodore A. Endreny
SUNY-ESF
Syracuse, New York

REMOTE SENSING DEFINED

Remote sensing is the science of observing and analyzing an area, object, or phenomenon based on data collected by sensors physically separated from the target. The sensors are mounted on platforms such as ground-based, airplane, or satellite platforms. Physical radiation transfer is the foundation for remote sensing. Typically, the observation is of radiation reflected or emitted from the object of interest. There are two types of remote sensing: active and passive. In active remote sensing, the object is illuminated by radiation of a known wavelength (or frequency); the backscatter from the object is recorded by the sensor. In passive remote sensing, the sensor records radiation generated from natural sources, such as the sun, or radiation emitted from the earth or atmosphere. Radiation that the sensor receives has a spectral response that reveals properties of the atmosphere and object of interest. By extending remote sensing applications to improve the spatial and temporal resolution of certain key water resource variables, scientists can better understand and quantify water resource processes.

METHODOLOGIES

Qualitative and quantitative approaches for analyzing remotely sensed data provide discrete and continuous information about an area. The qualitative approach separates observed spectral data into different classes and generates map-based images showing the spatial distribution of the classes. For example, if trying to map vegetation in an area that has certain known vegetation classes, the qualitative approach produces a map where every pixel is assigned a vegetation class or a 'no class.'

The quantitative approach provides numerical information about the area or phenomenon of interest. For example, by computing the normalized difference vegetation index (NDVI) using visible and near-infrared data and generating a map where a value is attached to every pixel, a map user has an idea how dense the vegetation is and where the denser areas lie.

For either of these approaches, several different classification methods can be used. All classification methods fall under two general categories: supervised classification and unsupervised classification. Supervised classification requires the scientist to collect data from the area of interest and assign these data to pixels within the image. Examples of supervised classification methods are the parallelepiped method and the minimum distance method. Examples of unsupervised classification are the chain method and the ISODATA method.

RESOLUTION AND SPECTRAL REGIONS

To determine the type of data required for a remote sensing application, scientists need to identify and quantify the key data characteristics of the phenomenon of interest. These characteristics may include the temporal resolution, spatial resolution, radiometric resolution, and spectral resolution. All four characteristics determine the platform, sensor, and capital necessary for the project.

The spatial resolution determines the minimal size of objects that can be seen from the sensor; the smaller the area covered, the higher the resolution. The temporal resolution is the time between data acquisitions across the

Table 1. Resolution Types and Potential Applications

Resolution Type	Resolution	Applications	Sensor Examples
Spatial	10–100 m	Local hydrological processes	ASAR; MSS
Spatial	1–5 km	Local meteorologic processes	AVHRR/2; MODIS; SSM/I; ERBE
Spatial	>25 km	Basin-wide studies; global studies	
Temporal	Every 30 min	Meteorologic studies	VHRR; VISSR
Temporal	2/day	Meteorologic studies	AVHRR
Temporal	Every 2 weeks	Land use	ETM+
Temporal	Once	Maps	

same area. The radiometric resolution is the sensitivity of a sensor to differences in signal strength as it records the incoming radiant flux. The radiometric resolution is often determined by the number of bits (6 bits, 8 bits, etc.), available for storing the data. The spectral resolution is the range of wavelengths detected by the selected sensor (1). Different wavelengths or frequencies can be used for different applications. For examples of different uses for different resolutions and spectral regions, see Tables 1 and 2.

Almost all spectral regions have been used for passive remote sensing, ranging from the visible (VIS), infrared (IR), and microwave (MW) regions (Table 3); however, active remote sensing has been limited to microwave regions. Each of these regions breaks down to smaller regions; some of these regions are more commonly used than others. For example, remote sensors will capture red, blue, and green within the visible region more often than purple/violet or yellow because of the way certain physical environments reflect these particular wavelengths. Within the IR region, near-infrared (NIR) and thermal-infrared (TIR) radiation react to or are emitted from certain components of the physical environment more than other wavelengths in that region. The bands of microwave radiation wavelengths and frequencies are denoted by the letters K-, W-, X-, C-, S-, and L-.

Table 2. Spectral Bands, Bandwidths, and Their Potential Applications[a]

Spectral Band	Bandwidth	Applications	Sensor Examples
Blue	0.45–0.50 μm	Water penetration, land use, vegetation characteristics, sediment	ETM+; ASTER; MISR; TM
Green	0.50–0.60 μm	Green reflectance of healthy vegetation	AATSR; ASTER; MISR; TM
Red	0.60–0.70 μm	Vegetation discrimination because of red chlorophyll absorption	AATSR; ASTER; AVHRR/2; MISR
Panchromatic	0.50–0.75 μm	Mapping, land use, stereo pairs	ASTER;
Near-infrared	0.75–0.90 μm	Biomass, crop identification, soil crop, land–water boundaries	AATSR; ASTER; AVHRR/2
Midinfrared	1.5–1.75 μm	Plant turgidity, droughts, clouds, snow-ice discrimination	AATSR
Thermal infrared	8–12.5 μm	Relative temperature, thermal discharges, vegetation classification, moisture studies, thermal inertia	AATSR; AVHRR/2
Microwave: Shortwave	0.1–5 cm	Snow cover, depth, vegetation water content	ALT Duel-frequency Radar Altimeter; MWR
Microwave: Long wave	5–24 cm	Melting snow, soil moisture, water–land boundaries, penetrate vegetation	SAR; AMI-SAR

[a] Compiled based on Reference 2, p. 11, Table 1.1.

Table 3. Wavelengths and Frequencies for VIS, IR, and MW

Region	Wavelength (λ)	Frequency (υ)
Visible: R,B,G	0.4–0.7 (μm)	750,000–428,571 GHz
Infrared: NIR, TIR	0.7–100 (μm)	428,571–3000 GHz
Microwave shortwave: K-, W-band	0.1–5 (cm)	300–6 GHz
Microwave long wave: X-, C-, S-, L-band	5–24 (cm)	6–1.25 GHz

Basic Terms and Equations

Basic terms and equations fundamental to the theory of remote sensing applications in water resources are given below. Wavelength, λ (m), is the length of the cycle of vibration, given as

$$\lambda = \frac{c}{v} \tag{1}$$

where c is the speed of light, approximated as 3×10^8 m/s, and v is the frequency of vibration (1/s). Note that the speed of light is adjusted for density differences between space and the earth's atmosphere. Calculation of brightness temperature (3), T_B, is based on the measured excitance, or vibration, from a narrow spectral band, as a function of emissivity, ε, and temperature, T:

$$T_B(\lambda) = \varepsilon(\lambda)T \tag{2}$$

The terms backscatter (3), emission (3) and reflectance (3) refer to different types of energy received by a spaceborne sensor from the targeted area. Backscatter is the reflectance of microwave radiation from the target, defined by

$$\frac{P_r}{P_r} = \frac{\lambda^2 G_t G_r}{(4\pi)^3 R_t^2 R_r^2}\sigma \tag{3}$$

where P_r is the received power, P_t is the transmitted power, λ is the wavelength of the radar beam, G is the antenna gain, R is the distance (range) between target and antenna, and σ is the radar scattering cross section. Reflectance occurs when radiation bounces off the target and is redirected. Specular reflectance occurs when ~100% is redirected in the same direction, and diffuse reflectance occurs when 0% is redirected in the same direction. Reflectance is defined by

$$r_{s,\lambda} = \frac{\pi L_{s,\lambda}}{(E_{dir,\lambda} + E_{dif,\lambda})} \tag{4}$$

where $r_{s,\lambda}$ is the spectral surface reflectance, $L_{s,\lambda}$ is the reflected radiance at the surface, $E_{dir,\lambda}$ is the direct irradiance, and $E_{dif,\lambda}$ is the diffuse irradiance. Excitance (emission) is the radiation that comes from the earth's surface, such as heat, and is defined by

$$L_{\infty,\lambda} = L_{0,\lambda} t_\lambda(0,\infty) + \int_{z=0}^{z=\infty} L_{B,\lambda}[T(x)] \frac{\partial t_\lambda(z,\infty)}{\partial z} dz \tag{5}$$

where t_λ is the spectral transmissivity, $L_{B,\lambda}$ is the spectral blackbody radiance, $L_{0,\lambda}$ is the surface emitted radiance, and $L_{\infty,\lambda}$ is the upwelling radiance.

Water resource remote sensing applications covered in this article include evapotranspiration, flood mapping, precipitation, soil moisture, surface snow, waterbodies, and wetland detection. Each topic is discussed in detail, and basic algorithms are provided where possible. These algorithms, adapted from different situations, may have limitations, when applied to new situations. These limitations, noted and explained in the article, should be considered when applying the equations to different environments.

EVAPOTRANSPIRATION

Evaporation is the physical change of liquid water to a vapor. Transpiration is the evaporation of water from plant tissues into the atmosphere. Evapotranspiration (ET) is the combination of these two processes. Remote sensing of evapotranspiration remains an imperfect science in part because scientists are still largely unable to measure evapotranspiration directly. Scientists must rely on creating evapotranspiration algorithms and defining the variables within the algorithm using remote sensing. Even with these algorithms, scientists still require ground data for variables not determined through remote sensing methods (4).

The two spectral regions utilized for ET are VIS and IR (4). Scientists can analyze two types of ET, actual ET and potential ET, using different methods. Actual ET, the ET that occurs in any real situation, becomes more difficult to quantify in remote sensing, compared to potential ET(PET), the ET that occurs in when the soil/vegetated area has 100% soil moisture.

Currently, many algorithms exist to calculate ET. For this article, two equations were selected that calculate some idealized evaporation rates (potential and reference crop), and then different methods are explained for calculating actual ET. The first equation, developed by Priestley and Taylor (5), calculates the reference crop evaporation assuming that approximately 30% of the evaporation is driven by turbulent transfer. This equation depends upon remotely sensed thermal properties.

$$E_{rc} = \alpha \frac{\Delta}{\Delta + \gamma}(R_n - G) \tag{6}$$

where E is the evaporation rate for the reference crop; R_n is the net radiation; G is the ground heat flux; Δ is the slope of the vapor pressure versus temperature curve; and α is the empirical evaporation constant, determined as 1.26 or dependent on soil moisture (6,7).

The second equation, derived by Choudhury (8), calculates global potential evaporation. The data for this equation need to be collected from a variety of satellites that measure visible and thermal bands (4):

$$E_0 = \frac{\Delta R_{ni} + \rho c_p \dfrac{D}{c r_e}}{\Delta + \gamma \dfrac{(r_s + r_h)}{r_e}} (\text{mmd}^{-1}) \tag{7}$$

where Δ (hPa K^{-1}) is the differential of saturated vapor pressure with respect to temperature evaluated at air temperature T_a (K), γ (hPa K^{-1}) is the thermodynamic psychrometric constant, R_{ni} (mm d^{-1}) is the thermal net radiation, ρ is the density of air, c_p is the specific heat of air, D (hPa) is the vapor pressure deficit, r_s(s m^{-1}) is the surface resistance, r_H (s m^{-1}) is the aerodynamic resistance for heat transfer from a virtual source height where the canopy is at a reference level in the atmosphere, r_e(s m^{-1}) is the effective resistance to heat transfer obtained by putting the aerodynamic resistance r_H and the resistance to long-wave radiative transfer r_R in parallel, and c (mm d^{-1}) is the constant equal to the product of latent heat of evaporation and density of water.

The Choudhury (8) equation, based only on PET from grassy areas, needs to be adjusted before application to other types of land cover (9,10):

$$K_c = \frac{E_m}{E_0} \tag{8}$$

where K_c is the crop coefficient; E_m is the basic evaporation equation result with changes in albedo, aerodynamic effective resistance, and the minimum surface resistance relevant to the crop/landcover to be studied; and E_0 is the basic evaporation equation result.

To calculate actual evapotranspiration, scientists employ numerical models because of the complexity of the data requirements. One method, the integration of soil-vegetation-atmosphere-transfer (SVAT) models with planetary boundary layer (PBL) models, have the advantage of calculating ET from a higher altitude, thereby reducing surface spatial heterogeneity (4). Some problems to consider when tackling the calculation of actual evapotranspiration are the difficulty of validation and the inability to derive air humidity from remotely sensed data (4).

FLOOD MAPPING

Flood mapping is the process of delineating areas that may be flooded or have been flooded. Remote sensing flood mapping deals mostly with areas that have been flooded. The spectral regions used for this type of mapping are NIR (only for cloud-free conditions) and MW. NIR produces maps where the waterbodies are dark in contrast to surrounding soil and vegetation. Barber et al. (11) determined that C-HH data (active microwave data) also produced flood delineation maps collecting specular reflection, as long as the incident angles were greater than 45°. Active microwave sensors have an advantage over NIR sensors for flood mapping because they can penetrate clouds (12).

PRECIPITATION: RAINFALL AND SNOWFALL

The detection and estimation of precipitation can be either ground-based or space-based. All three spectral regions are used, VIS, IR, and MW, and each spectral region requires a different approach to calculating precipitation. Common instruments for ground-based rainfall and snowfall estimation are weather radars. These instruments take vertical measurements of snow/rainstorms and receive the backscattering from the individual drops or flakes. Spaceborne instruments include infrared and visible sensors (13).

For ground-based weather radars, the theoretical basis for rainfall estimation consists of the following equations (13,14):

$$Z = \int_0^\infty D^6 N_V(D)dD \tag{9}$$

$$R = 6\pi \times 10^{-4} \int_0^\infty D^3 v(D) N_V(D)dD \tag{10}$$

$$v(D) = cD^\gamma \tag{11}$$

where Z is the radar reflectivity factor, $N_V(D)\ dD$ is the drop size distribution per volume of air, D is the spherical raindrop diameter, R is the precipitation rate, $v(D)$ represents the functional relationship between the raindrop terminal fall speed in still air and the equivalent spherical raindrop diameter, $c = 3.778$ (if v is expressed in m s^{-1} and D in mm), and γ is 0.67. A limitation of Equation (9) is that raindrops are not round, they are oblate. To take this limitation into account, an adjustment is made in the original Z reflectivity equation where dual polarization is used. A limitation of this method is that single point measurements for the differential reflectivity may not properly represent rainfall at the land surface (13).

The relationship between the radar reflectivity factor, the radar properties, and backscatter is defined by the following equation (15):

$$\overline{P_r} = C\frac{|K|^2}{r^2}Z \tag{12}$$

where $\overline{P_r}$ is the mean power (W) received from raindrops at range r (km), C is the radar constant and $|K|^2$ is a coefficient related to the dielectric constant of water (~ 0.93). The limitations of this equation are that it holds for Rayleigh scattering theory, where targets are small compared to radar wavelength, perfect radar calibration is necessary, and attenuation is absent (14). The relationship between Z and R (sometimes denoted as S for snowfall), the precipitation rate, is the following:

$$Z = AR^B \tag{13}$$

where A and B are coefficients that vary with climatic and location characteristics but are independent of R (Table 4). This equation is theoretically based upon Eq. 14 and empirically based on numerical evidence (15). There are two methods for calculating A and B: regression analysis and parameterization using the raindrop size distribution (Eqs. 9–11) (14).

For snowfall, Z_e is considered instead of Z because of the physical differences between snowflakes and water drops (16):

$$Z_e = \frac{|K|_i^2}{|K|_W^2} \cdot Z \tag{14}$$

where Z_e is the radar reflectivity of snow/hail, $|K|_i^2$ is the dielectric constant for ice and $|K|_W^2$ is the dielectric constant for water. The 'correct' values for $|K|_i^2$ depend on the particle size selected (or determined). Some examples of selected particle sizes and their respective dielectrics can be found in Table 5. A typical value for the dielectric constant for water ($|K|_W^2$) is 0.93.

Two major physical differences exist between snow and rain that make Eq. 13 less accurate for snowfall. Upwelling of snow due to the snowflake density/size

Table 4. Empirical Constant Values and Respective Precipitation[a]

A	B	Precipitation Type	Reference
140	1.5	Drizzle	(19)
250	1.5	Widespread rain	(19)
200	1.6	Stratiform rain	(20)
31	1.71	Orographic rain	(21)
500	1.5	Thunderstorm rain	(19)
486	1.37	Thunderstorm rain	(22)
2000	2	Aggregate snowflakes	(23)
1780	2.21	Snowflakes	(24)
1050	2	Snowfall (X-Band)	(25)
229	1.09	Snowfall (X-Band)	(26)
427	1.09	Snowfall (X-Band)	(27)

[a]Adapted from Reference 13 pg 113, Table 6.2; $Z(\text{mm}^6\ \text{m}^{-3})$, $R(\text{mm h}^{-1})$

Table 5. Snowfall Diameter Types and Respective Dielectric Constants $|K|_i^2$

Diameter Size/Type	Dielectric Constant Value
Melted deep diameters	0.208
Ice sphere diameters	0.176

and the inhomogeneity of snowflake type and density do not follow the basic assumptions required for the Z–R relationship and decrease its overall accuracy (17). However, Collier (13) states that with careful quality control and adjustment of radar measurements, snowfall estimations can be as accurate as rainfall up to 50 km from the radar site.

For infrared satellite measurements, two algorithms commonly used are the cloud indexing algorithm and the life history algorithm. There are also several modeling methods that use artificial neural networking (ANN) in combination with remote sensing data. Cloud indexing is based on the idea of a classification type where the brightness temperature of every pixel is calculated and compared to a threshold brightness temperature. The threshold brightness temperature represents the temperature below which a cloud pixel precipitates. The algorithm is adjusted using empirically derived constants. One of the major limitations of this algorithm is that it is relevant only to cumuliform clouds and precipitating clouds. It is not relevant to cold nonprecipitating clouds like cirrus. However, this adjustment can be overcome by using wavelengths of 10.8 and 12 μm to differentiate between the two types of clouds (18). Another major limitation is calibration and validation due to the lack of spatial and temporal correlation between rain gauges and the algorithm. Rain gauges are point measurements whereas the algorithm measures over space. An alternative to rain gauges for calibration/validation is ground-based radar, but it too has inaccuracies (13):

$$\text{PI} = A_0 + \sum_i A_i\, S_i(\text{TBB}_i) \tag{15}$$

where TBB_i is the Pixel brightness temperature, T_0 is the brightness temperature threshold that is determined empirically, A_0 is a constant that is determined empirically, A_i is a constant that is determined empirically, S is the surface area, which is a function of brightness temperature, and PI is the precipitation index.

The life history method takes into account the growth and decay of a particular cloud or storm, and rain rates are adjusted to the cloud's stage of growth. The variables and limitations of this equation are similar to those of the cloud indexing equation (13).

$$\text{PI} = A_0 + A \cdot S(\text{TBB}) + A' \frac{d}{dt} S(\text{TBB}) \tag{16}$$

Modeling programs that use the IR spectral region, such as precipitation estimation from remotely sensed information using artificial neural networks (PERSIANN), are alternative methods of calculating rainfall and snowfall more precisely than the simple algorithms above.

The use of the passive MW spectral region in precipitation detection and estimation is a hot topic in research now and is producing very favorable results. Microwave estimation of precipitation is favorable because it measures precipitation more directly than IR, where the radiation is absorbed or scattered by the precipitation. The distinction between the two methods of estimation, absorption/emission and scattering, is important because both have limitations. Absorption/emission of microwave radiation by liquid water is relatively high and increases as frequency increases. This increase in emission drowns out signals of light precipitation. Scattering occurs when ice particles are in the lower parts of the cloud before the particles become liquid. However, the existence of ice particles must then be connected to the physical precipitation in some manner (13).

SOIL MOISTURE

Soil moisture is defined as the water retained in the soil after infiltration. The spectral regions used for remote sensing are MW and TIR. The spatial resolution for this work is very important, and is one of the reasons that the microwave region is preferentially used by hydrologists. The use of thermal infrared and passive microwave requires more algorithmic work compared to active microwave. Generally, soil moisture is measured by developing an algorithm that contains variables quantifiable through remote sensing analysis (28). For TIR, the following equations can be used:

$$\text{DT}_s = T_s(\text{PM}) - T_s(\text{AM}) = f\left(\frac{1}{D}\right) \tag{17}$$

$$D = w \cdot Q_c \cdot k \tag{18}$$

where DT_s is the diurnal temperature, the difference between the afternoon surface temperature $T_s(\text{PM})$ and the early morning surface temperature $T_s(\text{AM})$; D is the diurnal inertia; w is the day length; Q_c is the volumetric heat capacity, and k is the thermal capacity.

Based on this equation, a scientist would then develop a relationship between the temperature change and the soil moisture. The limitations of this equation are that it can be used only for bare soil. Clouds, surface topography, and local meteorologic conditions can decrease accuracy due to their absorption of thermal radiation. The penetration depth that can be considered is small because the thermal radiation is not penetrating but emanating (28).

Microwave soil moisture equations are based on dielectric constants or the emanation of microwaves from the soil surface. Passive microwave soil moisture calculations require the brightness temperature, ground emissivity, and the dielectric constant. Scientists will then use the final temperature brightness outcome to develop an empirical relationship to known soil moisture values (28,29):

$$T_{\mathrm{B}} = t(H) \cdot \lfloor rT_{\mathrm{sky}} + (1-r)T_{\mathrm{soil}} \rfloor + T_{\mathrm{atm}} \qquad (19)$$

where T_{B} is the brightness temperature, $t(H)$ is the atmospheric transmissivity for a radiometer at height H above the soil, r is the smooth surface reflectivity, T_{soil} is the thermometric temperature of the T_{soil}, T_{atm} is the average thermometric temperature of the atmosphere, and T_{sky} is the contribution from the reflected sky brightness.

Active microwave soil moisture measurement has problems similar to passive, requiring an empirical development of a relationship between backscatter and soil moisture. The spatial resolution of active microwave sensors has an advantage over passive and thermal infrared sensors. Both types of microwave remote sensing can also be used for change detection (28). Some limitations to consider for microwave sensing are surface roughness and measurement depth. Surface roughness is spatially and temporally variable, which affects measurements on the watershed scale (12). Measurement depth is important for calculating the correct dielectric constant (28).

SURFACE SNOW

Surface snow is important as a source of freshwater for many areas. Scientists are interested in many characteristics of surface snow that will help them to estimate quantity and time to runoff. The most important characteristics are identification of extent, snow-covered area (SCA), depth, density, and snow water equivalent (SWE). Each of these characteristics requires different spectral data, and each can use different techniques during analysis.

Surface snow extent can be measured using VIS and MW from satellite platforms. The method employed for snow extent is detection. Each spectral region has advantages and limitations. The visible region usually has good resolution (depending upon the type of satellite) but cannot provide data at night and is not good for distinguishing between clouds and snow.

Passive microwaves (SSM/I) are limited to very large areas due to poor resolution ($\sim$25 km) and detect only dry snow. Passive microwave data are also hard to validate due to large pixel size. Active microwave data, such as SAR, can discriminate between wet and dry snow with a high resolution but have trouble with forested areas (30).

Surface snow water equivalent can be calculated using microwave and gamma radiation. The gamma radiation is detected by using an airplane platform, and microwaves are detected by using an airplane or satellite platform. The following equation was developed for gamma radiation calculation of SWE. When using gamma radiation to calculate SWE, soil moisture can attenuate the radiation, and this affects the accuracy (20,31).

$$\mathrm{SWE} = \frac{1}{A}\left(\ln\frac{C_0}{C} - \ln\frac{100 + 1.11M}{100 + 1.11M_0}\right) \qquad (20)$$

where SWE (g/cm^2) is the snow water equivalent, C and C_0 are uncollided terrestrial gamma counts over snow and bare ground, M and M_0 are the percent soil moisture in snow covered and bare soil areas, and A (cm^2/g) is the radiation alternation coefficient in water.

The SWE based on passive microwave data is a function of brightness temperature and empirical constants. These constants allow the "moving" algorithm from location to location. When calculating SWE, the effect of vegetation on backscattering and emissions must be considered. Forested areas tend to decrease the accuracy of results. A major limitation of passive microwave snow detection and SWE estimation is that passive microwaves sensors cannot detect wet snow. Wet snow increases brightness temperatures much more than dry snow, thereby decreasing the accuracy of the original equation. Even slightly wet snow is hard to detect with passive microwaves.

Active microwaves, such as SAR, can detect wet snow but cannot distinguish dry snow from no-snow areas. This suggests that multiband methods have a future in SWE determination, but the sensor technology may not be available (20,32).

$$\mathrm{SWE} = A + B\left(\frac{T_{\mathrm{B}}(f_1) - T_{\mathrm{B}}(f_2)}{(f_2 - f_1)}\right) \qquad (21)$$

where A is the offset of the regression of the brightness temperature difference, B is the slope of the regression of the brightness temperature difference, f_2 is the high scattering channel, commonly 37 GHz, f_1 is the low scattering channel, commonly 18 or 19 GHz, and SWE (mm) is the snow water equivalent.

WATERBODIES: DETECTION, COLOR, CLARITY, AND QUALITY

Remote sensing can be used to observe several different characteristics of waterbodies such as waterbody detection; waterbody perimeter calculation; and waterbody clarity, color, and quality. Waterbody detection is the process of determining the existence of water bodies within an area, whether they are ephemeral or permanent. The regions used for this type of detection are VIS, IR, and MW. IR is absorbed by water, so waterbodies show up as black areas on false colored IR images, and when using

VIS, radiation is reflected (12). Passive microwave emissions can be used to calculate the brightness temperature, and this allows distinguishing between water and floating ice. Open water has a lower brightness temperature (12).

Examples of satellite sensors and platforms for waterbody detection are SSM/I, Landsat TM, and NOAA AVHRR (LIC). A problem to consider with the spectral regions of VIS and IR is that VIS and IR are constrained by clouds. A general limitation to consider is that the area to be analyzed should exceed the spatial resolution by a factor of 10 (33).

To determine water quality, clarity, or color requires using the same spectral regions, VIS and IR, but with a different method of analysis. Using linear and nonlinear equations based on the radiance, reflectance, or energy of the radiation, scientists relate these to the parameter of interest. There are three approaches for creating this relationship: empirical, semiempirical, and analytical. The empirical methods use measured spectral properties and measured water quality parameters. The semiempirical methods use spectral/optical characteristics to approximate the appropriate wavelength for the water quality parameter. The analytical method uses optical properties of the water quality parameters to develop a model for the spectral quality of the water, and this model is introduced into an equation that relates the remote sensing data, the water quality parameters, and the optical parameters. Some examples of parameters that can be measured are suspended sediments, chlorophylls (algae), DOM (humus), oil, and temperature (Table 6) (34). An example of an empirical/semiempirical equation is

$$Y = A + BX \tag{22}$$

or

$$Y = AB^X \tag{23}$$

where Y is the measured radiance, reflectance, or energy; X is the water equality parameter of interest; A is the empirically derived constant; and B is an empirically derived constant (35). An example of an analytical equation is

$$R = r_{\mathrm{i}}\frac{b}{a+b} = r_i\omega_b \tag{24}$$

where R is the reflectance, r_{i} is the radiance to reflectance conversion, a is the absorption, b is the backscattering, and ω_{b} is the backscattering albedo (34,36).

WETLAND DETECTION

The detection of wetlands has become extremely important for environmental agencies as more and more shoreline areas become developed. The main difficulty in identifying wetlands is the submerged aquatic vegetation. This vegetation can make water detection difficult with VIS and near- or midinfrared. However, thermal infrared allows comparison between aquatic vegetation and terrestrial vegetation because the water/vegetation combination has higher thermal inertia compared to that of land (12). The major limitation of this spectral region is that when the surrounding area is supersaturated with water, the thermal infrared comparison is difficult (12). Another method is to use active microwave radar data, such as SAR, and a qualitative approach to distinguish plants from open water and land.

ADDITIONAL TOPICS

Topics not covered in this article, but remain important in the area of water resources and hydrology, are land cover mapping, area delineation, slope delineation, flood forecasting, bathometry, ice and glaciers, and permafrost. Land cover mapping was discussed briefly

Table 6. Water Quality Parameters and Spectral Bands

Water Quality Parameter	Spectral Region	Note
Suspended sediments	700–800 μm(38)	Other scientists have found that optimum wavelength depends on concentration(37).
Chlorophyll (algae)	0.69 μm (red edge)	Distinction between chlorophyll and suspended sediment is difficult given the spectral data available; however, one should concentrate on the relationship between chlorophyll-a and the red edge of the spectrum(34)
Temperature	10.4–12.5 μm	
Oils	8–14 μm	Most commonly used sensors are those that are placed on airborne platforms, e.g., planes, because immediate and constant coverage is required for an oil spill; however, Landsat TM can also be used for monitoring. Spatial resolution is very important for patches and windrows. Infrared thermal remote sensing is most important for cleanups; ultraviolet is important as well because thin layers of oil emit high ultraviolet radiation patterns; passive microwave is good because oil emits more microwaves than water, but spatial resolution is poor (34).

Satellites useful for Water Resources[a]

Variable	Satellite	Central Wavelength or Frequency Required for Observation	Resolution	Coverage
Snow-covered area (SCA)	NOAA	0.62, 10.80 μm	1 km	2 per day
	SPOT	0.59, 0.69, 0.89 μm	10–25 m	16 day[b]
	GOES	0.64 μm (visible)	2 km	2 per hour
Snow depth	Nimbus 7	37 GHz	30 km	2 per hour
	DMSP			
Snow water equivalent (SWE)	SSM/I			
	MOS-1, MSR	19.3 GHz, 37.0 GHz	25 km	2 per day
		23.0 GHz, 31 GHz	23–32 km	2 per day
	ERS-1,2	C-band(5.3 GHz)V V	30 km	35 days
Wet snow	RADARSAT	SAR C-BAND(5.3 GHz) H H SAR	8–25 m	16 days[c]
Surface temperature	NOAA	12.0 μm (band 4)	1 km	2 per day
	Landsat TM	12.5 μm (band 6)	80 m	16 days
Evapotranspiration	NOAA	0.62, 0.91, 10.80, 12.0 μm	1 km	2 per day
	GOES	0.64,11.5 μm	2–8 km	2 per hour
	Meteosat			
	GOES	0.65 μm (visible)		
Precipitation	TRMM	Infrared	3 km	2 per hour
	Landsat TM	0.52, 0.60, 0.96, 0.90, 1.75, 2.35, 12.5 μm	30 m	16 days
Land cover/land use and vegetation	Landsat	0.55, 0.65, 0.75, 0.9 μm	80 m	16 days
	MSS NOAA	0.62, 0.91 μm (visible)	1 km	2 per day
	SPOT	0.59, 0.69, 0.89 μm	10–25 m	16 day[b]
	ERS-1,2	C-band (5.3 GHz) V V	30 m	35 days
Soil Moisture	RADARSAT	C-band (5.3 GHz) H H	8–25 m	16 days[c]
Groundwater	Landsat	0.95 μm (band 7, near IR)	80 m	16 days
	SPOT			
	ERS-1,2	0.59, 0.69, 0.89, μm	10–25 m	16 days[b]
	Radarsat	C-band (5.3 GHz) V V	30 m	35 days
	Landsat TM	C-band (5.3 GHz) H H	8–25 m	16 days[c]
	Landsat	0.52, 0.60, 0.69, 0.90, 1.75, 2.35, 12.5 μm	30 m	16 days
Surface Water	MSS	0.55, 0.65, 0.75, 0.9 μm	80 m	16 days

[a]Compiled from Reference (12).
[b]Coverage frequency can be up to 2–3 days in northern latitudes using the off-nadir pointing capability of the satellite.
[c]Coverage frequency can be increased depending on incidence angle.

in the opening of this article. Terrain elevation analysis from imagery remotely obtained by stereopairs and interferometry provides the main method for area and slope delineation. Flood forecasting may occur by real-time monitoring of water surface elevations with light detection and ranging (LIDAR) or by simulation of soil moisture dynamics, land surface runoff, and river routing with remotely sensed inputs of vegetation, terrain, and precipitation.

CONCLUSION

There are several limitations of remote sensing of water resources to keep in mind while delineating a water resource application. The first limitation, that parameters are never measured directly, cannot be solved easily but rather calls for an increase in confidence in calibration and validation methods for the equation and the ability to select equations/algorithms or methods proper for the particular study (2). A lack of suitable algorithms for available data can also hinder studies. To a certain extent, remote sensing is new for water resources, so data required for specific algorithms may not be accessible. Calculating parameters such as evaporation may require expensive mapping programs that cannot be afforded by certain research institutes and universities. Finally, much of VIS and IR remote sensing depends on the weather, such as the presence of clouds, and other uncontrollable factors.

APPENDIX

Common Acronyms

AVHRR	Advanced very high resolution radiometer
DEM	Digital elevation model
EOS	Earth Observing System
ERS-1,2	European Resources Satellite 1,2
GIS	Geographic Information Systems
IR	Infrared radiation
NOAA	National Oceanic and Atmosphere Administration
NDVI	Normalized difference vegetation index
PBL	Planetary boundary layer model
RADARSAT	Radar satellite
SAR	Synthetic aperture radar
SPOT	Système Probatoire pour l'Observation de la Terre
SSM/I	Special sensor microwave imager

SWE	Snow water equivalent
TIR	Thermal infrared radiation
VIS	Visible radiation

BIBLIOGRAPHY

1. Jensen, J.R. (1996). *Introductory Digital Image Processing: A Remote Sensing Perspective*, 2nd Edn. Prentice-Hall, Upper Saddle River, NJ.
2. Schultz, G.A. and Engman, E.T. (2000). *Remote Sensing in Hydrology and Water Management*. Springer, New York.
3. Rott, H. (2000). Physical principles and technical aspects of remote sensing. In: *Remote Sensing in Hydrology and Water Management*. G.A. Shultz and E.T. Engman (Eds.). Springer, New York.
4. Meneti, M. (2000). Evaporation. In: *Remote Sensing in Hydrology and Water Management*. G.A. Shultz and E.T. Engman (Eds.). Springer, New York.
5. Priestly, C.H.B. and Taylor, R.J. (1972). On the assessment of surface heat flux and evaporation using large-scale parameters. *Mon. Weather Rev.* **100**: 81–92.
6. Barton, I.J. (1978). A case study comparison of microwave radiometer measurements over bare and vegetated surfaces. *J. Geophys. Res.* **83**: 3513–3517.
7. Davies, J.A. and Allen, C.D. (1973). Equilibrium, potential and actual evaporation from cropped surfaces in southern Ontario. *J. Appl. Meteorol.* **12**: 649–657.
8. Choudhury, B.J. (1997). Global pattern of potential evaporation calculated from the Penman–Monteith equation using satellite and assimilated data. *Remote Sensing Environ.* **61**: 64–81.
9. D'Urso, G. and Meneti, M. (1996). Performance indicators for the statistical evaluation of digital image classifications. *ISPRS J. Photogr. Remote Sensing* **51**(2): 78–90.
10. D'Urso, G. and Meneti, M. (1995). Mapping crop coefficients in irrigated areas from Landsat TM images. *SPIE Int. Soc. Opt. Eng.* **2585**: 41–47.
11. Barber, D.G., Hocheim, K.P., Dixon, R., Mosscrop, D.R., and McMullan, M.J. (1996). The role of Earth observation technologies in flood mapping: A Manitoba case study. *Can. J. Remote Sensing* **22**(1): 137–143.
12. Pietroniro, A. and Leconte, R. (2000). A review of Canadian remote sensing applications in hydrology, 1995–1999. *Hydrol. Processes* **12**: 1641–1666.
13. Collier, C.G. (2000). Precipitation. In: *Remote Sensing in Hydrology and Water Management*. G.A. Shultz and E.T. Engman (Eds.). Springer, New York.
14. Uijlenhoet, R. (2001). Raindrop size distributions and radar reflectivity-rain rate relationships for radar hydrology. *Hydrol. Earth Syst. Sci.* **5**(4): 615–627.
15. Battan, L.J. (1973). *Radar Observation of the Atmosphere*. The University of Chicago Press, Chicago, IL.
16. Smith, P.L. (1984). Equivalent radar reflectivity factors for snow and ice particles. *J. Climate Appl. Meteorol.* **23**(8): 1258–1260.
17. Matrosov, S.Y. (1998). A dual-wavelength radar method to measure snowfall rate. *J. Appl. Meteorol.* **37**(11): 1510–1521.
18. Kidder, S. and Vonder Harr, T. (1995). *Satellite Meterology*. Academic Press, New York, pp. 307–343.
19. Joss, J., Schram, K., Thams, J.C., and Waldvogel, A. (1970). *On the Quantitative Determination of Precipitation by Radar*. Wissenschaftliche Mitteilung Nr 63, Zurich, Eidegenossische Kommision zum Studium der Hagelbildung und der Hagelabwehr.
20. Marshall, J.S. and (Mc)Palmer, W. (1948). The distribution of raindrops with size. *J. Meteorol.* **5**: 165–166.
21. Blanchard, D.C. (1953). Raindrop size distribution in Hawaiian rains. *J. Meteorol.* **10**: 457–473.
22. Jones, D.M.A. (1956) Raindrop size distribution in Hawaiian rains. *J. Meteorol.* **10**: 457–473.
23. Gunn, K. and Marshall, J. (1958). The distribution with size of aggregate snowflakes. *J. Meteorol.* **15**: 452–561.
24. Sekhon, R.S. and Srivastava, R.C. (1970). Snow size spectra and radar reflectivity. *J. Atm. Sci.* **27**: 229–267.
25. Puhakka, T. (1975). *On the Dependence of Z-R Relation on the Temperature in Snowfall. Preprints*. 16th *Radar Meterol. Conf.*, Houston, TX, American Meteorology Society, pp. 504–507.
26. Boucher, R.J. and Wieler, J.G. (1985). Radar determination of snowfall rate and accumulation. *J. Climate Appl. Meteorol.* **15**: 452–461.
27. Fujiyoshi, Y., Endoh, T., Yamada, T., Tsuboki, K., Tachbana, Y., and Wakaham, G. (1990). Determination of a Z–R relationship for snowfall using a radar and high sensitivity snow gauges. *J. Appl. Meteorol.* **29**: 147–152.
28. Engman, E.T. (2000). Soil Moisture. In: *Remote Sensing in Hydrology and Water Management*. G.A. Shultz and E.T. Engman (Eds.). Springer, New York.
29. Schmugge, T. (1990). Measurements of surface soil moisture and temperature. In: *Remote Sensing of Biosphere Functioning*. R.J. Hobbs and H.A. Mooney (Eds.). Springer-Verlag, New York.
30. Rango, A., Walker, A.E., and Goodison, B.E. (2000). Snow and ice. In: *Remote Sensing in Hydrology and Water Management*. G.A. Shultz and E.T. Engman (Eds.). Springer, New York.
31. Carroll, T.R. (1990). Airborne and Satellite Data Used to Map Snow Cover Operationally in the U.S. and Canada. *Proc. Int. Symp. Remote Sensing Water Resour.*, Enschede, The Netherlands, **147**: 155.
32. Rott, H. (1993). Capabilities of Microwave Sensors for Monitoring Areal Extent and Physical Properties of the Snowpack. *Proc. NATO Adv. Res. Workshop Global Environ. Change Land Surf. Processes Hydrol.*, Tuscon, AZ.
33. USEPA. (1999). *Remotely Sensing Water Resources: an EPA/NASA Workshop*. Office of Water (4503F), United States Environmental Protection Agency, Washington, DC.
34. Ritchie, J.C. and Schiebe, F.R. (2000). Water quality. In: *Remote Sensing in Hydrology and Water Management*. G.A. Shultz and E.T. Engman (Eds.). Springer, New York.
35. Morel, A. and Gordon, H.R. (1980). Report of the working group on water color. *Boundary Layer Meteorol.* **18**: 343–355.
36. Dekker, A.G., Malthus, T.J., and Hoogenboom, H.J. (1995). The remote sensing of inland water quality. In: *Advances in Remote Sensing*. F.M. Danson and S.E. Plummer (Eds.). John Wiley, Chichester, pp. 123–142.
37. Curran, P.J. and Novo, E.M.M. (1988). The relationship between suspended sediment concentration and remotely sensed spectral radiance: A review. *J. Coastal Res.* **4**: 351–368.
38. Ritchie, J.C., Schiebe, F.R., McHenry, J.R., Wilson, R.B., and May, J. (1976). Sun angle, reflected solar radiation and suspended sediments in north Missippi reservoirs. In: *Remote Sensing of Earth Resources Vol. IV*. F. Shahrokhi (Ed.). University of Tennessee Space Institute, Tullahoma, TN.

ATMOSPHERIC SCIENTISTS

U.S. Department of Labor,
Bureau of Labor Statistics

SIGNIFICANT POINTS

- Almost 4 in 10 atmospheric scientists work for the Federal Government, which is the largest employer of such workers.
- A bachelor's degree in meteorology, or in a closely related field with courses in meteorology, is the minimum educational requirement; a master's degree is necessary for some positions, and a Ph.D. is required for most research positions.

NATURE OF THE WORK

Atmospheric science is the study of the atmosphere—the blanket of air covering the Earth. *Atmospheric scientists*, commonly called *meteorologists*, study the atmosphere's physical characteristics, motions, and processes, and the way in which it affects the rest of our environment. The best known application of this knowledge is in forecasting the weather. Aside from predicting the weather, scientists also attempt to identify and interpret climate trends, understand past weather, and analyze today's weather. However, weather information and meteorological research also are applied in air-pollution control, agriculture, forestry, air and sea transportation, defense, and the study of possible trends in the Earth's climate, such as global warming, droughts, or ozone depletion.

Atmospheric scientists who forecast the weather, known professionally as *operational meteorologists*, are the largest group of specialists. They study information on air pressure, temperature, humidity, and wind velocity; and apply physical and mathematical relationships to make short-range and long-range weather forecasts. Their data come from weather satellites, weather radars, sensors, and weather stations in many parts of the world. Meteorologists use sophisticated computer models of the world's atmosphere to make long-term, short-term, and local-area forecasts. More accurate instruments for measuring and observing weather conditions, as well as high-speed computers to process and analyze weather data, have revolutionized weather forecasting. Using satellite data, climate theory, and sophisticated computer models of the world's atmosphere, meteorologists can more effectively interpret the results of these models to make local-area weather predictions. These forecasts inform not only the general public, but also those who need accurate weather information for both economic and safety reasons, such as the shipping, air transportation, agriculture, fishing, forestry, and utilities industries.

The use of weather balloons, launched a few times a day to measure wind, temperature, and humidity in the upper atmosphere, is currently supplemented by sophisticated atmospheric monitoring equipment that transmits data as frequently as every few minutes. Doppler radar, for example, can detect airflow patterns in violent storm systems—allowing forecasters to better predict tornadoes and other hazardous winds, and to monitor the storms' direction and intensity. Combined radar and satellite observations allow meteorologists to predict flash floods.

Some atmospheric scientists work in research. *Physical meteorologists*, for example, study the atmosphere's chemical and physical properties; the transmission of light, sound, and radio waves; and the transfer of energy in the atmosphere. They also study factors affecting the formation of clouds, rain, and snow; the dispersal of air pollutants over urban areas; and other weather phenomena, such as the mechanics of severe storms. *Synoptic meteorologists* develop new tools for weather forecasting using computers and sophisticated mathematical models of atmospheric activity. *Climatologists* study climactic variations spanning hundreds or even millions of years. They also may collect, analyze, and interpret past records of wind, rainfall, sunshine, and temperature in specific areas or regions. Their studies are used to design buildings, plan heating and cooling systems, and aid in effective land use and agricultural production. Environmental problems, such as pollution and shortages of fresh water, have widened the scope of the meteorological profession. *Environmental meteorologists* study these problems and may evaluate and report on air quality for environmental impact statements. Other research meteorologists examine the most effective ways to control or diminish air pollution.

WORKING CONDITIONS

Most weather stations operate around the clock, 7 days a week. Jobs in such facilities usually involve night, weekend, and holiday work, often with rotating shifts. During weather emergencies, such as hurricanes, operational meteorologists may work overtime. Operational meteorologists also are often under pressure to meet forecast deadlines. Weather stations are found everywhere—at airports, in or near cities, and in isolated and remote areas. Some atmospheric scientists also spend time observing weather conditions and collecting data from aircraft. Weather forecasters who work for radio or television stations broadcast their reports from station studios, and may work evenings and weekends. Meteorologists in smaller weather offices often work alone; in larger ones, they work as part of a team. Meteorologists not involved in forecasting tasks work regular hours, usually in offices. Those who work for private consulting firms or for companies analyzing and monitoring emissions to improve air quality usually work with other scientists or engineers; fieldwork and travel may be common for these workers.

EMPLOYMENT

Atmospheric scientists held about 7,700 jobs in 2002. The Federal Government was the largest single employer of civilian meteorologists, accounting for about 2,900. The National Oceanic and Atmospheric Administration (NOAA) employed most Federal meteorologists in National Weather Service stations throughout the Nation; the remainder of NOAA's meteorologists worked mainly in research and development or management. The U.S. Department of Defense employed several hundred civilian meteorologists. Others worked for professional, scientific, and technical services firms, including private weather consulting services; radio and television broadcasting; air carriers; and State government.

Although several hundred people teach atmospheric science and related courses in college and university departments of meteorology or atmospheric science, physics, earth science, or geophysics, these individuals are classified as college or university faculty, rather than atmospheric scientists. (See the statement on **postsecondary teachers** elsewhere in the *Handbook*.)

In addition to civilian meteorologists, hundreds of Armed Forces members are involved in forecasting and other meteorological work. (See the statement on **job opportunities in the Armed Forces** elsewhere in the *Handbook*.)

TRAINING, OTHER QUALIFICATIONS, AND ADVANCEMENT

A bachelor's degree in meteorology or atmospheric science, or in a closely related field with courses in meteorology, usually is the minimum educational requirement for an entry-level position as an atmospheric scientist.

The preferred educational requirement for entry-level meteorologists in the Federal Government is a bachelor's degree—not necessarily in meteorology—with at least 24 semester hours of meteorology courses, including 6 hours in the analysis and prediction of weather systems, 6 hours of atmospheric dynamics and thermodynamics, 3 hours of physical meteorology, and 2 hours of remote sensing of the atmosphere or instrumentation. Other required courses include 3 semester hours of ordinary differential equations, 6 hours of college physics, and at least 9 hours of courses appropriate for a physical science major—such as statistics, chemistry, physical oceanography, physical climatology, physical hydrology, radiative transfer, aeronomy, advanced thermodynamics, advanced electricity and magnetism, light and optics, and computer science. Sometimes, a combination of education and appropriate experience may be substituted for a degree.

Although positions in operational meteorology are available for those with only a bachelor's degree, obtaining a second bachelor's degree or a master's degree enhances employment opportunities and advancement potential. A master's degree usually is necessary for conducting applied research and development, and a Ph.D. is required for most basic research positions. Students planning on a career in research and development need not necessarily major in atmospheric science or meteorology as an undergraduate. In fact, a bachelor's degree in mathematics, physics, or engineering provides excellent preparation for graduate study in atmospheric science.

Because atmospheric science is a small field, relatively few colleges and universities offer degrees in meteorology or atmospheric science, although many departments of physics, earth science, geography, and geophysics offer atmospheric science and related courses. Prospective students should make certain that courses required by the National Weather Service and other employers are offered at the college they are considering. Computer science courses, additional meteorology courses, a strong background in mathematics and physics, and good communication skills are important to prospective employers. Many programs combine the study of meteorology with another field, such as agriculture, oceanography, engineering, or physics. For example, hydrometeorology is the blending of hydrology (the science of Earth's water) and meteorology, and is the field concerned with the effect of precipitation on the hydrologic cycle and the environment. Students who wish to become broadcast meteorologists for radio or television stations should develop excellent communication skills through courses in speech, journalism, and related fields. Those interested in air quality work should take courses in chemistry and supplement their technical training with coursework in policy or government affairs. Prospective meteorologists seeking opportunities at weather consulting firms should possess knowledge of business, statistics, and economics, as an increasing emphasis is being placed on long-range seasonal forecasting to assist businesses.

Beginning atmospheric scientists often do routine data collection, computation, or analysis, and some basic forecasting. Entry-level operational meteorologists in the Federal Government usually are placed in intern positions for training and experience. During this period, they learn about the Weather Service's forecasting equipment and procedures, and rotate to different offices to learn about various weather systems. After completing the training period, they are assigned a permanent duty station. Experienced meteorologists may advance to supervisory or administrative jobs, or may handle more complex forecasting jobs. After several years of experience, some meteorologists establish their own weather consulting services.

The American Meteorological Society offers professional certification of consulting meteorologists, administered by a Board of Certified Consulting Meteorologists. Applicants must meet formal education requirements (though not necessarily have a college degree), pass an examination to demonstrate thorough meteorological knowledge, have a minimum of 5 years of experience or a combination of experience plus an advanced degree, and provide character references from fellow professionals.

JOB OUTLOOK

Employment of atmospheric scientists is projected to increase about as fast as the average for all occupations through 2012. The National Weather Service has completed an extensive modernization of its weather forecasting equipment and finished all hiring of meteorologists needed to staff the upgraded stations. The Service has no plans to increase the number of weather stations or the number of meteorologists in existing stations. Employment of meteorologists in other Federal agencies is expected to remain stable.

On the other hand, job opportunities for atmospheric scientists in private industry are expected to be better than for those in the Federal Government over the 2002-12 period. As research leads to continuing improvements in weather forecasting, demand should grow for private weather consulting firms to provide more detailed information than has formerly been available, especially to weather-sensitive industries. Farmers, commodity investors, radio and television stations, and utilities, transportation, and construction firms can greatly benefit from additional weather information more closely targeted to their needs than the general information provided by the National Weather Service. Additionally, research on seasonal and other long-range forecasting is yielding positive results, which should spur demand for more atmospheric scientists to interpret these forecasts and advise weather-sensitive industries. However, because many customers for private weather services are in industries sensitive to fluctuations in the economy, the sales and growth of private weather services depend on the health of the economy.

There will continue to be demand for atmospheric scientists to analyze and monitor the dispersion of pollutants into the air to ensure compliance with Federal environmental regulations outlined in the Clean Air Act of 1990, but related employment increases are expected to be small. Opportunities in broadcasting are rare and highly competitive, making for very few job openings.

EARNINGS

Median annual earnings of atmospheric scientists in 2002 were $60,200. The middle 50 percent earned between $39,970 and $76,880. The lowest 10 percent earned less than $30,220, and the highest 10 percent earned more than $92,430.

The average salary for meteorologists in nonsupervisory, supervisory, and managerial positions employed by the Federal Government was about $74,528 in 2003. Meteorologists in the Federal Government with a bachelor's degree and no experience received a starting salary of $23,442 or $29,037, depending on their college grades. Those with a master's degree could start at $35,519 or $42,976; those with the Ph.D., at $51,508. Beginning salaries for all degree levels are slightly higher in areas of the country where the prevailing local pay level is higher.

RELATED OCCUPATIONS

Workers in other occupations concerned with the physical environment include **environmental scientists and geoscientists, physicists and astronomers, mathematicians**, and **civil, chemical**, and **environmental engineers**.

SOURCES OF ADDITIONAL INFORMATION

Information about careers in meteorology is available on the Internet from:

- American Meteorological Society. Internet: http://www.ametsoc.org/AMS. Phone: (617)-227-2425.

Information on obtaining a meteorologist position with the Federal Government is available from the U.S. Office of Personnel Management through a telephone-based system. Consult your telephone directory under U.S. Government for a local number or call (703) 724–1850; Federal Relay Service: (800) 877–8339. The first number is not toll-free, and charges may result. Information also is available from the Internet site: http://www.usajobs.opm.gov/.

OOH ONET CODES

19–2021.00

Suggested citation: Bureau of Labor Statistics, U.S. Department of Labor, *Occupational Outlook Handbook, 2004-05 Edition*, Atmospheric Scientists, on the Internet at http://www.bls.gov/oco/ocos051.htm (visited August 14, 2004).

RAIN SIMULATORS

Joseph Holden
University of Leeds
Leeds, United Kingdom

Rain simulators have been extensively used in hydrological, pedological, and geomorphological problems (1–13). Simulators allow much greater control of the rainfall variable for experimental purposes and have been used in both field and laboratory applications (13). The amount, intensity, and duration of rain can be controlled, along with other parameters such as drop-size distribution and water chemistry, to varying degrees, depending on the system of application. Unlike a ring infiltrometer, the surface does not pond immediately but will only do so at some later stage, if the input is great enough. The time to ponding will depend on the application rate as well as on the hydraulic properties of the soil. Surface runoff can be collected to determine infiltration rates (by subtracting surface runoff rates from application rates) and erosion, either as a whole sample for smaller plots or in a subsampling strategy for larger plots.

TYPES OF RAINFALL SIMULATORS AND CHOICE OF APPROPRIATE EQUIPMENT

There is a range of simulator designs, and a database catalogue has been developed incorporating design information by the British Geomorphological Research Group (www.geog.le.ac.uk/bgrg). Some other designs are published in Costin and Gilmour (14); Romkens et al. (15); Bowyer-Bower and Burt (16); Esteves et al. (17); and Waddington and Devito (18). There are two rainfall simulator systems: spray nozzles (including rotating sprays) and drip-screens. Spray systems generally supply rainfall in pulses to the ground (14), and rotating sprays deliver rainfall across a large surface area although intensity usually decreases with distance from the rotating nozzle. The spray-type systems often provide rainfall at terminal velocities which approach that of natural rainfall. Drip systems use hypodermic syringes (15) or other drop formers across a fixed grid to produce rainfall across relatively small surface areas. The drop formers are often not raised high enough to allow representative terminal velocities, but they generally allow a constant rainfall rate whose drop sizes are more easily controlled than in the spray systems (16).

Various factors, including money, time available, the purpose of the experiment, and the local climatic conditions whose simulation is desired, influence the choice of simulator. Table 1 lists some of the factors of importance in choosing a type of rainfall simulator and the most suitable type of simulator for each factor. The amount of water used to provide rainfall at any given intensity is generally much less for a drip-screen design than for spray-type simulators (16). This may be crucially important in remote arid or semiarid areas. The use of a drip-type simulator generally means, however, that smaller plot areas will be used for analysis. Using appropriate pressure regulators, rain simulators in the field or laboratory can be attached to a mains water supply. It usual to control the amount and intensity of rain during a simulation by using a pressure regulator, but often a simple tap connected to a calibrated manometer board will suffice (16).

Table 1. Desirable Characteristics of a Rainfall Simulator

Desired Attribute	Best Suited Simulator
Easily transportable	Drip-type
Efficient use of water	Drip-type
Protection from the wind to increase accuracy of rainfall application volumes into selected plot	Drip-type
Reproduction of high intensity rainfall to reflect expected natural conditions	Spray-type
Reproduction of low intensity rainfall to reflect expected natural conditions	Drip-type
Long duration rainfall	Drip-type
Achievement of 'natural' terminal velocity	Spray-type
Control over drop-size distribution	Drip-type (but not very flexible)
Attainment of desired drop size relative to rainfall intensity (i.e., flexibility)	Spray-type
Accurate replication of rainfall parameters	Drip-type
Uniformity of rainfall across the whole plot	Drip-type
Large plot area	Spray-type
Adaptability to difficult terrain and vegetation	Spray-type

Drop-Size Distribution

Drops of a uniform size formed by a rain simulator do not represent those in natural rainfall. If the experiment requires simulating a natural drop-size distribution (19–21), then it is often necessary to have a wire mesh that scatters and breaks up drops into different sizes. For drip-type simulators, Bowyer-Bower and Burt (16) found that suspending a wire mesh below the drop former of 3 mm × 3 mm spacing and 1 mm diameter wire was suitable. Salles et al. (22) demonstrated that it is necessary to sample at least 10,000 drops to estimate the drop-size distribution from a rain simulator at an accuracy of 3% or less. Several methods are available for examining drop-size distribution, including light diffraction and filter paper collection. An economic and simple approach is to allow drops to fall into finely sieved flour and then to bake the pellets and sieve them through meshes of different diameters. Larger intensity rainfall is associated with larger raindrop size (21), but it is often difficult to replicate changing drop sizes as intensity is altered using rain simulators, although spray-type simulators are more flexible.

Achievement of Terminal Velocity and Energy Characteristics of the Rain

Where erosion or soil crusting processes are the focus of study, it is essential to know the kinetic energy of the raindrops produced by a rain simulator (23). Kinetic energy is crucial for soil detachment and transport. Natural rain normally falls from great heights by which time the droplets have achieved terminal velocity (maximum velocity in air—they cannot fall any faster because air resistance prevents any further acceleration due to gravity). Droplets that fall at terminal velocity have more kinetic energy than those falling from small heights. Larger drops require a greater height to reach their terminal velocity. Droplets of 1 mm in diameter require around 2.5 m, and droplets of 2 mm in diameter require around 6 m to reach terminal velocity. Thus, if a rainfall simulator is not positioned high enough, then the kinetic energy of the drops will be significantly less than that of natural rainfall. Knowledge of the drop size distribution is therefore important for determining the rain splash energy provided.

Spatial and Temporal Coverage of a Plot

For experiments that are concerned only with some aggregate or mean effect of simulated rain, then variations in rainfall delivery across a plot may be unimportant. Where rainfall simulation is being used to study small-scale processes that are themselves spatially variable (such as rill initiation), then knowledge of the simulator's inherent variability is vital (24). Rainfall delivery may be measured quite simply by placing a series of collectors across a trial plot and recording rainfall delivery to the ground across a range of rainfall intensities and durations. Many spray systems have greater intensity rains closer to the spray nozzle, and intensity declines as the distance from the nozzle increases.

Typical Measurements

Depending on the nature of the experiment, it is usually necessary to create a bounded plot when simulating rain. This is particularly important if a parameter of interest is infiltration. It is usual to bound the plot on three sides and then to collect the surface runoff from the unbounded fourth (downslope) side of the plot. The bounding prevents leakage out of the plot, so that more accurate estimates of infiltration rates can be made. Measurement of the surface runoff can indicate the infiltration rate within the plot, as long as the application rate is known: At steady state, Infiltration = Rainfall intensity—Surface Runoff. Units of length are used for the three variables in the equation, and so one must remember to divide the surface runoff volumes by the plot area. There will be a time lag in response so that in the early stages of the experiment when the soil surface is wetting and water is stored in detention pools, then the previous equation will not hold. The equation also ignores evaporation and storage on vegetation surfaces. If the interest is in infiltration rates, it is often necessary to run the experiment long enough to allow runoff rates to stabilize. Early in the experiment, runoff rates are likely to increase over time, but in theory should eventually stabilize if the rainfall intensity remains constant. It should then be possible to apply the theoretical infiltration curve of Philip (25) which is an exponential curve fitted through the first and last data point of the order of $I = A + Bt^{-0.5}$, where I = infiltration rate at time t and A and B are constants. This equation may be solved using simultaneous equations (i.e., equations at two different times) to allow constants A and B to be evaluated.

In practice, however, surface runoff from the plot on most slopes of interest will rarely stabilize completely because of surface ponding and episodic cut and fill of microtopographical features on the surface and the burst out of water followed by a period of pool refill and microchannel change (12). It may therefore be better to use a mean value across a sample of data points to estimate infiltration. If the rainfall intensity is lower than the infiltration rate into the surface of investigation, then it may be that no surface runoff is observed. If surface runoff is collected, it is also possible to collect sediment, solutes, and microorganisms moved by the runoff.

Typically, results are scaled up to the hillslope and sometimes the catchment scale. However, this should be done only with great caution. First, most hillslopes have a high natural heterogeneity such that plots may not be representative. Second, coupling of water, sediment, nutrients, or microorganisms within a plot may not be the same as on a hillslope. Sediment moved from the top to bottom of a plot, for example, may not move from the top to bottom of a hillslope during a real storm because there may be natural zones of deposition and erosion and larger hillslopes may not provide the good coupling seen in a plot-scale experiment.

Almost all field rainfall simulator experiments have collected runoff solely from the surface layer and have disregarded any lateral through-flow in deeper soil layers. However, using an appropriate apparatus, it is possible to investigate soil water flow mechanisms using a rain simulator by digging a pit at the lower end of the plot and collecting runoff from different layers of the subsurface (12). However, pit disturbance may change the local hydrologic gradients resulting in unwanted changes of the natural processes, particularly, if the layers of investigation are deep below the surface.

IMPORTANCE OF RAINFALL INTENSITY USED

Some rainfall simulators can reproduce natural rainfall typical of tropical storms across 50-m plots (e.g., Reference 17 reproduced 70 mm h^{-1} storms), whereas others can reproduce gentle rains typical of temperate parts of the world whose intensities are as low as 3 mm h^{-1} (12). The choice of rainfall intensity used in a simulation may be very important, and many hydrologic studies have ignored this issue. Most studies have simulated only one rainfall intensity. Some studies, however, found that as rainfall intensity is increased on a simulation plot, the steady infiltration rate also increases (12,26). This is very important and has major implications for conclusions drawn from experiments. There are likely to be physical reasons for this phenomenon, including increased ponding development on the soil surface as rain intensity is increased that leads to a greater hydraulic head that forces more water into the soil (27–29); increased ponding that results in dispersing and suspending soil crusts and surface particles, allowing pores to unblock; and most soil plots are nonuniform; some parts allow much more infiltration than others (30). Therefore, if rainfall intensity is increased, then proportionally more water may infiltrate into particular parts of a plot. This will result in an increased plot-averaged value of the infiltration rate. Thus, those experiments that have simulated unrealistic rainfall intensities (e.g., 150 mm h^{-1} in the British Isles where rainfall very rarely exceeds 12 mm h^{-1}) may have unrealistic results. The results from an intense rainfall simulation may assume that the infiltration capacity of the soil is 80 mm h^{-1}, for example, but if a more realistic intensity had been used, then the result may be an infiltration capacity of 6 mm h^{-1}. Thus, conclusions drawn about runoff production in soil water processes, irrigation, and erosion may be misleading, and care should be taken in choosing an appropriately representative range of rainfall intensities.

BIBLIOGRAPHY

1. De Ploey, J., Savat, J., and Moeyersons, J. (1976). The differential impact of some soil factors on flow, runoff creep and rainwash. *Earth Surface Processes* **1**: 151–161.
2. Lusby, G.C. (1977). Determination of runoff and sediment yield by rainfall simulation. In: *Research Techniques, Erodibility and Sediment Delivery*. T.J. Toy (Ed.). Geobooks, Norwich, pp. 19–30.
3. Bork, H.R. and Rohdenburg, H. (1981). Rainfall simulation in southeast Spain: analysis of overland flow and infiltration. In: *Soil Conservation, Problems and Prospects*. R.P.C. Morgan (Ed.). John Wiley, Chichester, pp. 293–302.
4. Imeson, A.C. (1983). Studies of erosion threshold in semi-arid lands: field measurements of soil loss and infiltration in northern Morocco. In: *Rainfall Simulation, Runoff and Soil Erosion*. J. De Ploey (Ed.). Catena Supplement 4: pp. 79–89.

5. Pilgrim, D.H. and Huff, D.D. (1983). Suspended sediment in rapid subsurface stormflow on a large field plot. *Earth Surface Processes* **8**: 451–453.
6. Imeson, A.C. and Kwaad, F.J.P.M. (1990). The response of tilled soils to wetting by rainfall and the dynamic character of soil erodibility. In: *Soil Erosion on Agricultural Land*. J. Boardman, I.D.L. Foster, and J.A. Dearing (Eds.). John Wiley, Chichester, pp. 3–14.
7. Morgan, R.P.C. (1995). *Soil Erosion and Conservation*. Longman, London.
8. Bergkamp, G. (1998). A hierarchical view of the interactions of runoff and infiltration with vegetation and microtopography in semiarid shrublands. *Catena* **33**: 201–220.
9. Cerda, A. (1998). Relationships between climate and soil hydrological and erosional characteristics along climatic gradients in Mediterranean limestone areas. *Geomorphology* **25**: 123–134.
10. Foster, I.D.L., Fullen, M.A., Brandsma, R.T., and Chapman, A.S. (2000). Drip-screen rainfall simulators for hydro- and pedo-geomorphological research: The Coventry experience. *Earth Surface Processes and Landforms* **25**: 691–708.
11. Duiker, S.W., Flanagan, D.C., and Lal, R. (2001). Erodibility and infiltration characteristics of five major soils of southwest Spain. *Catena* **45**: 103–121.
12. Holden, J. and Burt, T.P. (2002). Infiltration, runoff and sediment production in blanket peat catchments: implications of field rainfall simulation experiments. *Hydrological Processes*. **16**: 2537–2557.
13. Holden, J. and Burt, T.P. (2002). Laboratory experiments on drought and runoff in blanket peat. *European Journal of Soil Science* **53**: 675–689.
14. Costin, A.B. and Gilmour, D.A. (1970). Portable rainfall simulator and plot unit for use in field studies of infiltration, runoff and erosion. *Journal of Applied Ecology* **7**: 193–200.
15. Romkens, M.J.M., Glenn, L.F., Nelson, D.W., and Roth, C.B. (1975). A laboratory rainfall simulator for infiltration and soil detachment studies. *Soil Science Society of America Proceedings* **39**: 158–160.
16. Bowyer-Bower, T.A.S. and Burt, T.P. (1989). Rainfall simulators for investigating soil response to rainfall. *Soil Technology* **2**: 1–16.
17. Esteves, M. et al. (2000). The 'EMIRE' large rainfall simulator: design and field testing. *Earth Surface Processes and Landforms* **25**: 681–690.
18. Waddington, J.M. and Devito, K.J. (2001). Portable irrigation system for studying hillslope and wetland runoff generation processes. *Hydrological Processes* **15**: 281–287.
19. Laws, J.O. and Parsons, D.A. (1943). The relation of raindrop size to intensity. *Transactions of the American Geophysical Union* **24**: 452–60.
20. Best, A.C. (1950). The size distribution of raindrops. *Quarterly Journal of the Royal Meteorological Society* **76**: 16–36.
21. Hudson, N.W. (1971). *Soil Conservation*. Batsford, London.
22. Salles, C., Poesen, J., and Borselli, L. (1999). Measurement of simulated drop size distribution with an optical spectro pluvimeter: sample size considerations. *Earth Surface Processes and Landforms* **24**: 545–556.
23. Epema, G.F. and Riezebos, H.Th. (1983). Fall velocity of waterdrops at different heights as a factor influencing erosivity of simulated rain. In: *Rainfall Simulation, Runoff and Soil Erosion*. J. DePloey (Ed.). Catena Supplement 4: pp. 2–17.
24. Lascelles, B., Favis-Mortlock, D.T., Parsons, A.J., and Guerra, A.J.T. (2000). Spatial and temporal variation in two rainfall simulators: implications for spatially explicit rainfall simulation experiments. *Earth Surface Processes and Landforms* **25**: 709–721.
25. Philip, J.R. (1957). The theory of infiltration 1. The infiltration equation and its solution. *Soil Science* **83**: 345–357.
26. Bowyer-Bower, T.A.S. (1993). Effects of rainfall intensity and antecedent moisture on the steady-state infiltration rate in a semi-arid region. *Soil Use and Management* **9**: 69–76.
27. Schiff, L. (1953). The effect of surface head on infiltration rates based on the performance of ring infiltrometers and ponds. *Transactions of the American Geophysical Union* **34**: 257–266.
28. Philip, J.R. (1958). The theory of infiltration 6. Effect of water depth over soil. *Soil Science* **85**: 278–286.
29. Schmid, B.H. (1989). On overland flow modelling: can rainfall excess be treated as independent of flow depth? *Journal of Hydrology* **107**: 1–8.
30. Hawkins, R.H. (1982). Interpretations of source area variability in rainfall-runoff relations. In: *Rainfall-runoff relationship. Proceedings of the International Symposium on rainfall runoff modelling held May 18–21, 1981, Mississippi State University*. Water Resources Publications, Littleton, CO, 303–324.

SNOW DENSITY

Pratap Singh
National Institute of Hydrology
Roorkee, , India

Density is one of the fundamental and important characteristics of snow. It is defined as the mass of snow per unit volume. To determine the density of snow, a core sample of snow is extracted from the accumulated snow on the ground using a snow sampler of known volume. The extracted snow sample is weighed and its mass is determined. The density of the sampled snow is determined by dividing the mass by the volume. The most common unit used for representing the density is g/cc or kg/m^3.

Freshly fallen snow is very light and has a very low density. The density of fresh snow varies from about 0.01 to 0.20 g/cc, but a rule of thumb often used for the average density of fresh snowfall is 0.1 g/cc when actual measurements are not made. Once snow accumulates on the ground and stays over time, the density of the snow increases. Metamorphic changes in snow, which involve readjustment of the shape and size of crystals, contribute to changes in the density of snow. Metamorphism also results from the compaction of snow caused by the pressure of overlying layers of snow and can also be accelerated by strong winds, warm temperatures, and intermittent melting of the snow cover. The density of snow changes significantly during the first 2–3 months when transformation of fresh snow into fine granular snow continues and recrystallization processes dominate. During a period of time, snow first transforms into firn, defined as snow that has survived for at least one melt season. When climatic conditions allow snow to remain

Table 1. Typical Densities of Snow in Different Forms[a]

Snow Type	Density g/cc
New snow at low temperature in calm air	0.01–0.03
New snow immediately after falling in calm air	0.05–0.07
Damp new snow	0.10–0.20
Settled snow	0.20–0.30
Depth hoar	0.20–0.30
Windpacked snow	0.35–0.40
Firn	0.40–0.65
Very wet snow and firn	0.70–0.80
Glacier ice	0.85–0.95

[a]Ref. 1.

for several years, it gradually transforms into glacier ice. Glacier ice is impermeable to air and water; some air is present only as bubbles. Glacier ice is the most compacted form of snow. It attains the maximum density at this stage, which is very close to the density of water, 1.0 g/cc. The density of snow at different stages is shown in Table 1.

To estimate the amount of water stored as snow in a basin, information on the density and depth of the snowpack is needed. Usually, such information is collected through snow surveys, which are organized to obtain the density and depth of snow at representative points in the basin. Snow density is determined for the total depth of snowpack because it varies with depth. The average density of a snowpack is determined by taking the average of the snow density obtained at different depth levels of the snowpack. The snow water equivalent at any point can be obtained by multiplying the snow density by the depth. In addition to the texture and wetness of snow, information on the density of snow near the surface is also required to ascertain the suitability of snow for skiing, which is one of the most important recreational activities in Europe and North America.

BIBLIOGRAPHY

1. Singh, P. and Singh, V.P. (2001). *Snow and Glacier Hydrology*. Kluwer Academic Publishers, Dordrecht, The Netherlands, p. 221.

FLATTOP MOUNTAIN SNOTEL SNOWPACK: WATER YEAR 2004

Global Change Research Program—U.S. Geological Survey

These materials prepared by the Global Change Research Program—U.S. Geological Survey, Northern Rocky Mountain, Glacier Field Station. Dan Fagre, Program Coordinator, email: dan_fagre@usgs.gov. For more information,

visit the Glacier Field Station—Global Change Research Program web site.

The Flattop Mountain SNOTEL (SNOw TELemetry) station is one of nearly 600 similar stations operated throughout the western United States by the Natural Resources Conservations Service (NRCS, U.S. Department of Agriculture). These stations measure and record Snow Water Equivalent (SWE), which is the weight of snow water equivalent to inches of water. Most SNOTEL sites also measure temperature and precipitation; many now measure snow depth as well. Various agencies and organizations use the data from the stations to forecast water availability, floods, and avalanche hazard.

The Flattop Mountain SNOTEL station has operated since October 1st, 1969. To help calculate water storage and availability during agricultural growing seasons, the data are compiled by water year; a water year runs from 1 October - 30 September. SWE records from the station extend for 33 water years (1970–2002). Precipitation records from the station start water year 1979, with temperature records starting water year 1983.

The Flattop Mountain SNOTEL station is located at approximately 6300 feet in elevation on Flattop Mountain, a high plateau between the Lewis and Livingston Ranges in Glacier National Park, Montana. The site is three air miles south of the Continental Divide. During the winter, complex combinations of weather and terrain determine snowfall at the site. Westerly weather systems predominate, bringing moisture from the Gulf of Alaska or Pacific Ocean. Less frequent northerly systems spill drier Arctic air through passes on the Continental Divide as they slide south along the Rocky Mountain Front. Flattop Mountain's position between the Livingston and Lewis Ranges amplifies the effects of these large-scale weather systems; the two ranges rise 2–4000 feet above the site and orographically wring moisture from both westerly and upslope easterly storms. Flattop Mountain is a useful indicator of snowfall throughout Glacier National Park because it is subject to the factors that influence conditions elsewhere in the park.

The graph below shows the current water year SWE at Flattop Mountain along with snow depth and average daily temp:

Since the last update of this webpage (May 10th), the snowpack at Flattop Mtn. SNOTEL station continued its annual melt. Conditions in May and early June have generally been wet, with average temperatures. The station recorded. 1" or more of precipitation on 27 of 44 days, for a total of 8.3" of precipitation, 5.2" of that during May. Nearly all of that fell as rain; the station has measured just 1" of SWE gain in May and .5" in June. The highest one-day rainfall was .9" on May 28th, while the longest stretch of days with precipitation was May 25th–May 30th. It was followed by the longest period without precipitation, May 31st–June 4th. The precipitation of the past four weeks is common for this time of year; the station typically shows a secondary peak in precipitation known locally as the "June Monsoon."

The station recorded the warmest temperatures early in the month. After a mid-month winter storm, temperatures remained cool. Maximum and minimum temperatures for May were 16.7 C (60 F) on May 1st and -8 C (17.6 F) on May 12th. The highest mean daily temperature for the month was 8.2 C (47.8 F) on May 7th. The mean monthly temperature was 3.7 C (38.6 F), which is identical to the mean monthly temperature for May for the 21 years (1982–2003) that the station has recorded temperatures.

The average temperatures for the month ended a trend of above-average temperatures that dominated the winter. November has so far been the only month of the 2004 Water Year with mean temperatures below average. From December through April, mean monthly temperatures have remained above average, with December and January warmer than the mean (2.6 C/4.7 F and .5 C/1 F respectively), and February and March substantially warmer than average (3.2 C/5.8 F for February; 5 C/9 F for March). The difference between 2004 and the 21-year average narrowed in April and closed in May. The Mean Monthly Temperature Comparison Graph below shows this pattern. While the graph shows a relatively normal curve to the temperatures, the curve appears shifted a month forward. That is, early winter temperatures are typical mid-winter temperature, and late winter/early spring temperatures are typical of late spring. A. pdf version of this graph is available at Mean Monthly Temperature Graph.

The warm temperatures of the winter had a significant effect on the snowpack at the station. Despite a dry February, SWE gain for February-April was 15.9", which is average. The warm temperatures, however, have resulted in 7" of SWE loss in the same period, leaving the snowpack considerably below average when it peaked on April 20th, a week earlier than the 34-year average of April 27th. The SWE total was 37.4", or 81% of the 34-year average

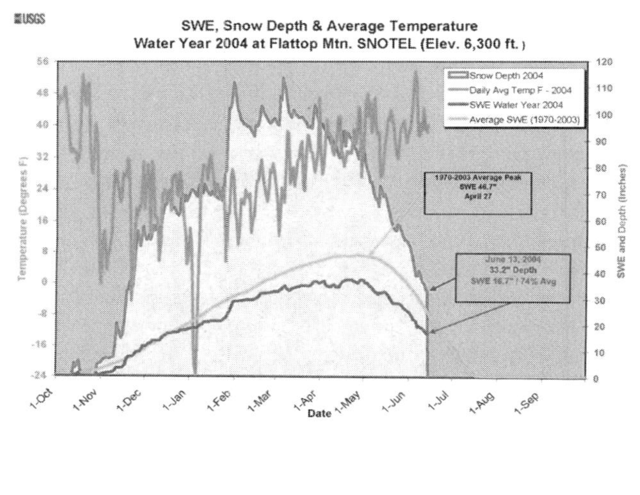

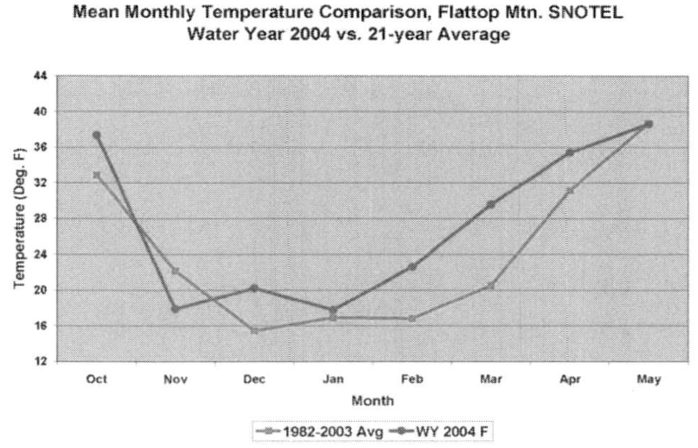

of 46.74". In addition to the 7" of SWE loss in March and April, the station lost 1.2" of SWE in October, for a total loss of 8.2"—nearly enough to make up the 9.3 inch deficit on April 20th. It's thus reasonable to conclude that the below average snowpack at Flattop Mtn. SNOTEL is primarily the result of warm temperatures rather than lack of precipitation. A graph showing the monthly SWE gain at Flattop Mountain SNOTEL is below; a. pdf version of the graph can be found at Monthly SWE 2004.

Since April 30th, the snow pillow at Flattop has recorded 22" of SWE loss and 1.5" of SWE gain. The station measured 13" of SWE loss and 1" of SWE gain in May. The greatest SWE loss occurred May 18th (1.1") and June 3rd (1.4"). As of midnight, June 14th, the SWE total at Flattop was 16.7", which represents 74% of the 34-year average of 25.3". Snow depth was 33". Despite the below-average SWE total, conditions at the station are well within the range of the past 34 water years. The maximum SWE recorded on this day (Julian Day 165) was 48.6" in 1974; the minimum was 0" in 1987, a 48.6" range. All told, there have been 12 years in the 34-year history of the station in which SWE total on this date was lower. The SWE total on this date last year was 22.6", with a snow depth of 44.5". A graph comparing SWE for significant water years is below.

A more detailed. PDF version of this graph is available for printing or viewing at SWE Comparison Graph.

In its June10th El Nino/ENSO Discussion, the National Weather Service Climate Prediction Center (CPC) reported that oceanic and atmospheric conditions In May throughout the Pacific Basin were typical of neutral phase of the El Nino Oscillation. The CPC reported that Sea Surface Temperature (SST) anomalies have been warmer than average in the western and central equatorial Pacific and cooler than average in the eastern equatorial Pacific. These anomalies have led to precipitation and wind speed gradients across the equatorial Pacific, with precipitation and wind speeds generally higher in the western portion of the equatorial Pacific. Forecasts of conditions for the next 3 months are split, with some models predicting neutral conditions and others El Nino conditions. The CPC concluded that ENSO-neutral conditions are more likely through August 2004, while noting that trends after that are uncertain.

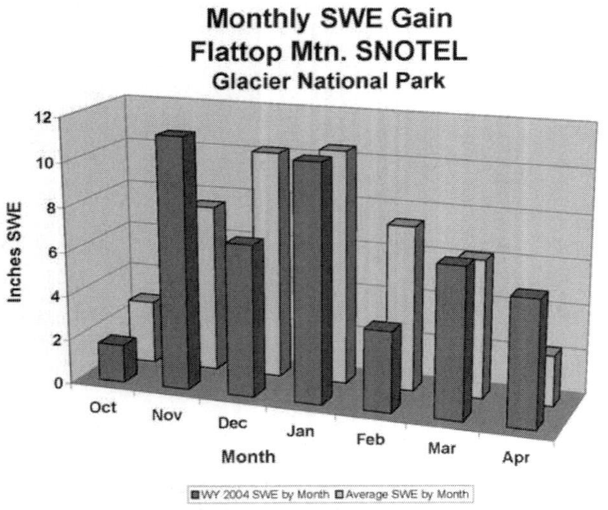

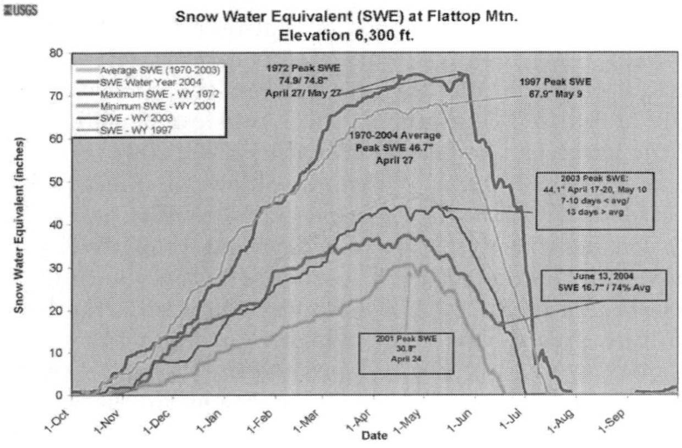

For more information, see the climate prediction center's El Nino/ENSO Discussion.

Also available for viewing are graphs for Water Year 2003, Water Year 2002, Water Year 2001, Water Year 2000, and Water Year 1999.

For details on how snowpack affects the spring opening of Logan Pass, visit our **Spring Opening of the Going-to-the-Sun Road** web page (Logan Pass is located at 6646 feet elevation along the upper edge of the McDonald drainage, approximately 10 miles southeast of the Flattop Mountain SNOTEL station).

- For the most recent SNOTEL data, graphs, and maps of snow distribution visit the Montana NRCS - Snow, Water, and Climate Services Snow Products web page—they provide *snow depth!* data for the Flattop Mountain SNOTEL station under the Daily Products-Reports heading. Also, their detailed map of Montana SNOTEL sites is a great resource.
- The NRCS, National Water & Climate Center provides access to all SNOTEL data as well as in-depth publications describing the SNOTEL program.
- Avalanche forecasts for Flathead National Forest, Kootenai National Forest, & Glacier National Park can be found at the Glacier Country Avalanche Center, which also links to avalanche advisories for areas throughout the western US & Canada.
- For current & historic river level data, visit the USGS - Montana Current Streamflow Conditions web page.

SNOW AND SNOWMELT

Pratap Singh
National Institute of Hydrology
Roorkee, India

Snow is a solid form of precipitation. The formation of snow crystals inside a cloud depends on the presence of ice nuclei, water vapor, motion of the cloud, and the temperature of the cloud. Snow crystals are formed when the cloud temperature is below freezing. These crystals experience diffusional, accretional, and aggregational growth inside the cloud. At temperatures greater than about −5 °C, crystals usually stick together, and aggregation of ice crystals forms snowflakes. Snowflakes obtain their maximum size when the temperature of a cloud is near 0 °C. Aggregation of ice crystals ceases at temperatures below −20 °C. Due to changes in the ambient condition of the cloud, the size and form of the crystals change when they descend through a cloud. In a typical cloud, a snowflake of 1 mm diameter can grow to 10 mm in about 20 minutes. Because of the irregular shapes of snowflakes, measurement of their linear dimensions is difficult. The maximum diameter of snowflakes may range from 0.1 mm to several centimeters.

Snow falls from the clouds mainly in the form of branched hexagonal crystals or stars. Other observed shapes include hexagonal rods, needles, and plates. Snowflakes have a large surface area and fall slowly compared with raindrops. Therefore, snowflakes scavenge more atmospheric aerosols than rain. Precipitation as snow occurs when the atmospheric temperature is near 0 °C or less. Compared to rainfall, snowfall occurs more uniformly, but its distribution on the ground is highly influenced by wind. To obtain the water equivalent of snow, the actual depth of snow unaffected by wind and its density are needed.

Accumulation of snow on the ground leads to the development of a snowpack, which stores water in the form of snow. Depending upon climatic conditions, snow on the ground can stay for a period as long as several months. The global distribution of snow shows that a major portion of the Northern Hemisphere is covered by snow during winter. Snow has great scientific interest because of its practical importance and utility in many fields, especially, in water resources, climate, and winter sports. For many countries, such as the United States, Canada, India, China, Pakistan, Afghanistan, Russia, Nepal, and European countries, snowmelt runoff is a vital source of water for drinking, irrigation, and hydropower generation. Several important rivers, such as the Columbia River in the United States and Canada, the Rhine River in Europe, and the Indus River in Asia, get substantial contributions from snowmelt runoff. Melting snow provides more than 70% of the water supply for the western United States (1).

In general, snow starts melting in the spring when the temperature begins rising. The depth and extent of the snowpack is reduced due to snow melting, and water is released from the snowpack. Snowmelt runoff estimates are needed for forecasting seasonal water yields, river regulation, reservoir operation, determination of design floods, design of hydrologic and hydraulic structures, and planning flood control programs. The snowmelt is estimated either using the energy balance approach or the temperature index method. The energy balance method requires information on radiation energy, sensible and latent heat, energy transferred through rainfall onto the snow, and heat conduction from the ground to the snowpack. When all the components for an energy balance computation are known, the melt rate can be expressed as (2)

$$M = \frac{1000\,Q_m}{\rho_w L \beta} \tag{1}$$

where M is the depth of melt water (mm d^{-1}), Q_m is the net energy flux available for melting (kJm^{-2} d^{-1}), L is the latent heat of fusion (333.5 kJ kg^{-1}), ρ_w is the density of water (1000 kg m^{-3}), and β is the thermal quality of the snow. The thermal quality of snow depends on the amount of free water content (generally 3–5%) and the temperature of the snowpack. Assuming $\beta = 0.97$ for a thermally ripened snowpack, Eq. 1 reduces to

$$M = 0.0031\,Q_m \tag{2}$$

Snowmelt follows an almost diurnal pattern of energy availability, and thus, it is possible to determine the diurnal variation in snowmelt by applying the diurnal distribution of energy received on the snow surface. The albedo, the most important parameter controlling the absorption of solar radiation, is very high for snow. For fresh snow, the albedo is about 0.80–0.90, suggesting that most of the short-wave radiation is reflected back to the atmosphere from the snow surface.

In the absence of detailed energy data for computing snowmelt, the temperature index method is considered the best substitute for the energy balance and is widely used. The air temperature, expressed as degree-days, is used for snowmelt computation. Temperatures are the most readily available data; therefore, the temperature index method is extensively used. The most common expression for estimating snowmelt using temperature is given as

$$M = D(T_i - T_b) \tag{3}$$

where M is the depth of melt water (mm d^{-1}) produced in unit time, D is the degree-day factor (mm °C^{-1} d^{-1}), T_i is the index air temperature (°C), and T_b is the base temperature (usually, 0 °C). Daily mean temperature is most commonly used as an index of temperature for snowmelt. The degree-day factor is used to convert the degree-days to the depth of snowmelt. A wide range of degree-day factors (0.7–9.2 mm °C^{-1} d^{-1}) has been reported in the literature. However, the majority of the reported values of the degree-day factor range from 3–5 mm °C^{-1} d^{-1} (2). The degree-day factor is influenced by the physical properties of the snow and, therefore, changes with time. Usually, temperature data used for melt estimation are available at a few locations in a basin. These available temperatures can be interpolated/extrapolated to different altitudes using the temperature lapse rate (usually, 0.65 °C/100 m).

Snow melts on the upper surface of the snowpack exposed to the atmosphere. In some cases, a little melting is also possible from the lower surface of the snowpack due to ground heat transfer. Snow is a porous medium, and melt water generated on the surface percolates through the snowpack before appearing as runoff. In the beginning of the melt season, the upper part of the snowpack may be warm (0 °C), but the lower part may be cold (subfreezing). Under such conditions, melt water produced on the surface may freeze in the snowpack to raise the temperature to 0°C. Unless the whole snowpack is isothermal at 0°C, no melt runoff is produced.

BIBLIOGRAPHY

1. Chang, A.T.C., Foster, J.L., Gloersen, P., Campbell, W.J., Josberger, E.G., Rango A., and Danes, Z.F. (1987). Estimating snowpack parameters in the Colorado River basin. In: *Proc. Large Scale Effects of Seasonal Snow Cover*. IAHS Publ. No. 166, pp. 343–353.
2. Singh, P. and Singh, V.P. (2001). *Snow and Glacier Hydrology*. Kluwer Academic Publishers, Dordrecht, The Netherlands, p. 221.

SNOW SURVEYS

Pratap Singh
National Institute of Hydrology
Roorkee, India

The basic objective of conducting snow surveys is assessing the spatial distribution of the snow water equivalent accumulated in the basin at a specific time. The distribution of snow is affected by several factors, including topography, orientation, vegetation, and wind speed. During strong winds, the drifting and blowing of snow create a highly uneven spatial distribution of snow depth. Therefore, to account for uneven distribution of snow in a basin, snow surveys are needed. Snow surveys are primarily a field activity that include determining the depth and vertically integrated density. An estimate of the snow water equivalent at a site is computed by multiplying the snow depth by the density of the accumulated snow. Snow surveys are made at regular intervals during both accumulation and ablation periods. Snow surveys conducted just after winter are considered especially important because they provide information about the total snow water equivalent accumulated during the winter. Such information helps in assessing the water resource potential of the deposited snow and is used for forecasting the volume of water expected from snowmelt.

The distribution of the snow water equivalent is needed as input to snowmelt models used for predicting runoff. In addition to hydrologic applications, agricultural, transportation, and recreational activities also require such data (1).

To carry out snow surveys, first, a snow course is established in the basin. The snow course represents a preselected line of marked sampling points in the basin. It is selected by considering representative sites to provide a reliable index of the water stored as snow across the entire basin. Based on the topographical features of the basin, the length of the snow course and a suitable number of observations are chosen for snow surveying. Once the snow course is established in the basin, measurements are repeated every year on the same snow course. Each sampling point of a snow course is located by measuring its distance from a reference point marked on a map. In mountainous regions, snow is distributed much more unevenly and therefore, the length of a snow course is shorter, say, only a few hundred meters, compared to flat regions. Moreover, sampling points are closely located in mountainous areas. The average of the snow water equivalent at all sampling points on the snow course gives information on the availability of water in the basin in the form of snow. Depending on the area covered by surveying, surveys can be made by surveyors using snowshoes, skis, or snow vehicles.

The most common instruments used for snow surveying include a graduated tube that has a snow cutter fixed at its lower end and a balance for determining the weight of the snow sample. The extracted core is retained in the sampler, and the sampler plus core is weighed. The weight of the empty sampler is predetermined. Therefore, the difference in weight with and without the core gives the value of the snow water equivalent of the core. Some instruments, such as the spring balance, give a direct value of the snow water equivalent of a sample when it is weighed. Proper documentation of records is made after sampling the site. In addition to the depth, density, and snow water equivalent for each sampling site, the snow course number and details of site (latitude, longitude, and elevation), weather conditions, vegetation, etc., are also recorded. Such data help in interpreting data collected. The date and timing of the beginning and end of the snow surveys and name of the snow surveyor are also recorded. A complete set of information should be prepared in the desired format just after finishing snow surveys.

An alternative method used for surveying the snow is based on the detection of natural gamma radiation emitted from the ground surface. The gamma radiation flux near the ground originates primarily from natural radioisotopes in the soil. Potassium, uranium, and thallium in the soil are the main sources of radiation. Typically, more than 95% of the gamma radiation is emitted from the top 20 cm of soil. Snow on the ground attenuates the emission of natural gamma radiation from the earth's surface, and the magnitude of attenuation provides an estimate of the snow water equivalent. A greater water equivalent of the snow on the ground will provide a higher attenuation of radiation. Therefore, presnow or no-snow measurements of gamma radiation level are required along the traverse to be followed for surveying. This approach can be used for a ground survey as well as for an aerial survey by aircraft. In a ground survey, a hand carried detector provides a means for measuring the averaged water equivalent along the length of the snow course. The equipment includes a portable gamma-ray spectrometer that uses a small scintillation crystal to measure the rays across a wide spectrum. Recently, radars are also being used for estimating snow water equivalent in basins. Such a radar is installed on a small carrier attached to a snow scooter, that is driven over the snow following the snow course.

In aerial natural gamma radiation surveys, a gamma-ray sensor and a recording device are installed in a low-flying aircraft. Flying of aircraft along the predetermined line is very essential in this type of survey because a relatively small deviation from the flight line may introduce errors due to radiation from other types of rocks. Thus, the flight line must be selected very carefully, so that the aircraft can follow it from year to year. Such aerial surveys may be made at altitudes up to 300 m but are most accurate at the lowest altitude possible for the terrain because of significant atmospheric attenuation of the radiation. This approach is considered the most reliable for level terrain but can also be used for hilly areas where elevation differences up to 400 m can be surveyed. There are some advantages and disadvantages of this approach. The disadvantage is that the aerial survey can be used only to measure the water equivalent of 10 to 100 mm of show (2). The advantage of natural gamma radiation method is that manual observations of the depth, density, or snow water equivalent are not required and a larger area can be surveyed, in a short time.

BIBLIOGRAPHY

1. Singh, P. and Singh, V.P. (2001). *Snow and Glacier Hydrology*. Kluwer Academic Publishers, Dordrecht, The Netherlands.
2. World Meteorological Organisation. (1981). *Guide to Hydrological Practices. Vol. I, Data Acquisition and Processing*. WMO No. 168.

A STATISTICAL APPROACH TO CRITICAL STORM PERIOD ANALYSIS

F. Ruiz
M. Abad
A. Rodríguez-Ramírez
L.M. Cáceres
J. Rodríguez Vidal
Universidad de Huelva. Avda. de las Fuerzas Armadas
Huelva

R. Pino
J.M. Muñoz
Universidad de Sevilla. Avda. Reina Mercedes
Sevilla

INTRODUCTION

Periodicity analysis is a natural and integral part of meteorological studies on different timescales. Numerous studies have focused on the possible cyclicity of several meteorologic data and/or phenomena, such as wind velocities, atmospheric pressures, sunspots, hurricanes, and currents (1–4).

In numerous coastal areas, winter storms are common meteorologic phenomena characterized by wind measuring 10 or higher on the Beaufort scale (89 km per hour, 55 miles per hour). Storm periods are a significant problem for coastal managers; they cause considerable property damage to beach–nearshore profiles with tourist-economic uses, fisheries, harbors, private vacation homes, and public earthworks (5). Their effects may vary over successive sectors of the same coast with different hydrodynamic features (6,7). These events increase coastal erosion and move sands rapidly offshore, whereas lower energetic conditions may cause gradual beach accretion (8). Consequently, managers need tools (i.e., event periodicity, sediment transport, modeling approach) to evaluate the possible consequences (9–11).

In this article, we analyze some mathematical procedures used in time series analysis. These robust methods are applied to calculate an approach to critical storm periodicity on the southwestern Spanish coast.

THE STATISTICAL BASIS FOR CRITICAL STORM PERIOD ANALYSIS

Introduction: The Predicting Nature

Statistical techniques permit us to analyze the structure and behavior of natural phenomena, their relationships, and temporal evolution. In this last case, high variability levels appear in the time series analysis, and consequently some statistical approaches may be useful to study their periodicity. The following step is prediction, but sometimes it is not possible to reach this objective because of great variability in the variables that describe the nature of a reliable prediction or insufficient levels of correlation between both explanatory variables and objective variables.

Even though adequate levels of prediction cannot be reached, statistical methodology allows us to obtain useful conclusions on the natural phenomena under study. The analysis of a "storm period" is a particular case of this type of study. Prediction is difficult, but it is possible to reach conclusions about past behavior. To this end, time series and spectral analysis are especially useful.

Spectral Analysis

Spectral analysis is a set of statistical and mathematical methods for identifying hidden periodicities in a time series. Moreover, spectral analysis is concerned with estimating the spectrum over the whole range of frequencies (or their inverses, the time periods). These techniques are widely used in electrical engineering, physics, computer science, geology, and environmental sciences (12–15).

From a mathematical viewpoint, spectral analysis is a modification of Fourier analysis to suit it for stochastic rather than deterministic functions of time. Fourier analysis is oriented to approximate a function by a sum of sine and cosine terms that is called the Fourier series representation. We distinguish two classes of spectral techniques, univariate and bivariate.

Univariate Spectral Analysis. The Wiener–Khintchine theorem states that the variance of any stationary stochastic process can be expressed as the integral in the interval (0, p) of a function called the spectral density function (or spectrum); so a spike of this function identifies a frequency with an important contribution to the variance, and a periodicity is associated with that frequency. Basically, univariate spectral analysis computes an estimate of the spectrum from the Fourier series representation of the available time series, and a graphical representation, called a periodogram (Fig. 1), is used to identify possible spikes in the underlying spectral density function (16).

This method has been used to calculate the periodicity of the heart rate and locomotor activity of several organisms (17), orbital cyclicities in geologic sequences (18), and seismology (19). In meteorology, it may be applied to the relationships among wind and waves (20) and analysis of the periodicity and persistence of daily rainfalls (21).

Bivariate Spectral Analysis. When we try to identify periodic movements common to a pair of time series, we need bivariate spectral analysis techniques. The definition of a cross-spectrum lets us generalize the univariate spectrum of this situation, although its interpretation is not so clear. Several statistical measures can be computed from this cross-spectrum, but some controversy is associated with the selection of the one most appropriate. We can remark on one of these measures for its easy interpretation: coherence. It measures the square of the linear correlation between two time series at a given frequency. It takes values in [0,1], and the closer it is to one, the more closely related are the two process at that frequency, so identifying a common periodic term.

In meteorology, these techniques have been used in investigating tropospheric ozone formation and decomposition processes (22), correlation of sea levels and atmospheric variables (23), analysis of the interactions between biosphere and atmosphere (24), and monitoring present-day climatic conditions (25).

A CASE STUDY: THE RECENT STORM RECORD ON THE SOUTHWESTERN SPANISH COAST

Study Area

The Huelva coast (SW Spain) is composed of large sandy beaches (145 km long), interrupted only by estuarine mouths (Fig. 2: Guadiana, Piedras, Tinto-Odiel, and Guadalquivir). The littoral morphology of this area is linked mainly to a mesotidal regime (2 m mean range), medium wave energy, and the fluvial sedimentary inputs of the Guadiana (144 m^3/s) and Guadalquivir (185 m^3/s) Rivers, a littoral net sediment flow oriented toward the east (26–29).

This coast is heavily "defended" by modern protective groins, causing either total (Vila Real do Santo Antonio,

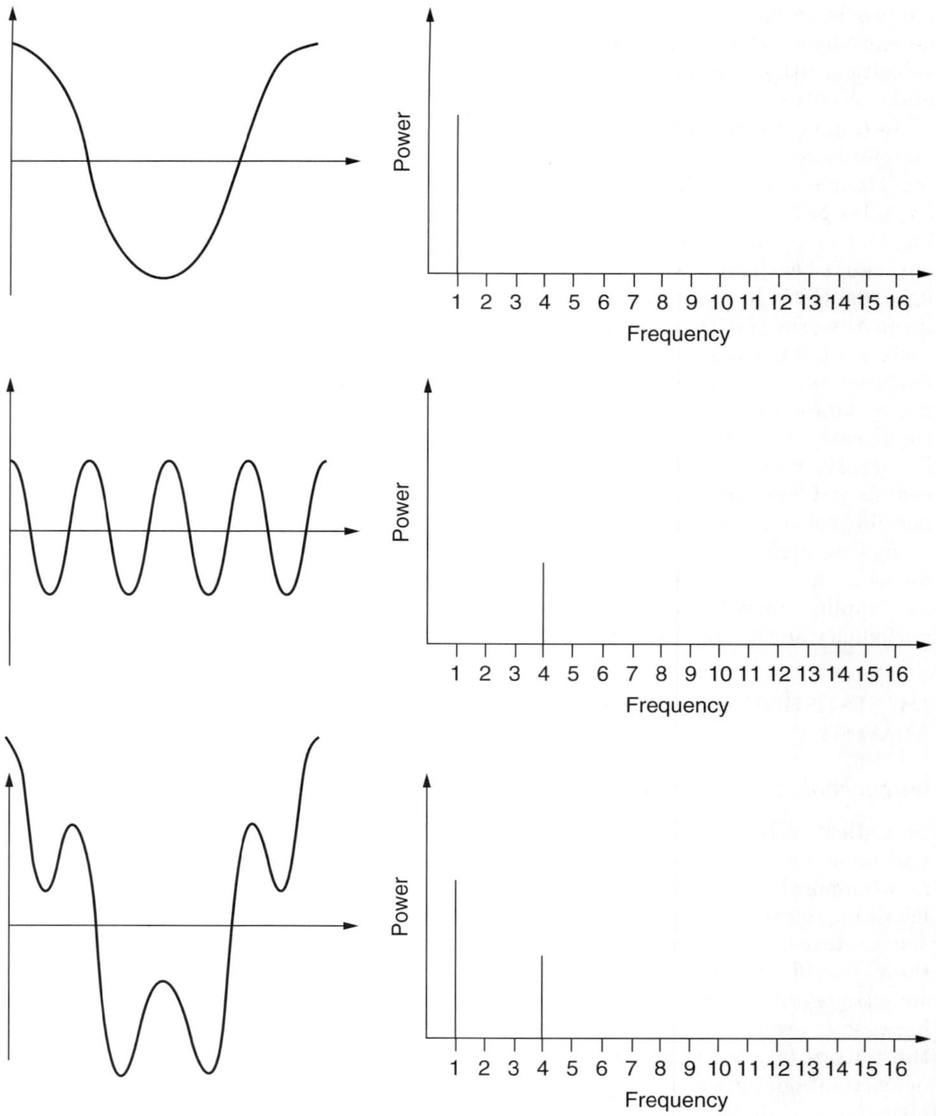

Figure 1. Fourier power spectra (right) resulting from various idealized data sets (left). Modified from Reference 15.

Huelva) or partial (Isla Cristina, Punta Umbria) obstacles to sedimentary transition. Consequently, two sedimentary units (Ayamonte-Punta Umbría and Mazagón-Doñana) may be delimited in this human-altered area, directly related to the effects of the main groins (30).

Each unit includes two different zones: (1) beaches affected by erosive processes (Ayamonte, Isla Cristina, La Antilla, Punta Umbría, Mazagón, Matalascañas) and (2) prograding spits (El Rompido, Doñana). These beaches constitute an important tourist area and are visited by more than 1 million people between May and September. This "sun and beach" tourism needs continuous management during the winter months due to periodic storm damage. The causes of this periodicity have not been studied but may be initially related to climate variables, such as the North Atlantic Oscillation Index (NAO; 31). The NAO is an index that represents the differences in atmospheric pressure at sea level between the Azores and Iceland (32). Positive values are associated with low cyclone activity in southern Europe and conversely (33).

The Winter Storm Record

The recent record (1956–1996) of critical storm periods was obtained from two daily journals (Odiel and ABC), wave heights were recovered from the Huelva harbour archives, and the wind speeds were supplied by the Spanish National Institute of Meteorology at the Huelva station (period 1961–1996). Data were normalized to standard scales (Fig. 3).

Between 1956 and 1996, eight winter storm periods on the southwestern Spanish coast were identified (Fig. 3, C): 1962–1963, 1969–1970, 1972–1973, 1978–1979, 1981–1982, 1987–1988, 1989–1990, and 1995–1996. In most cases, these high-energy periods are concentrated in December and January. An additional periodicity may be inferred in the fair weather conditions, the intervals separating two consecutive storm periods (5). After the

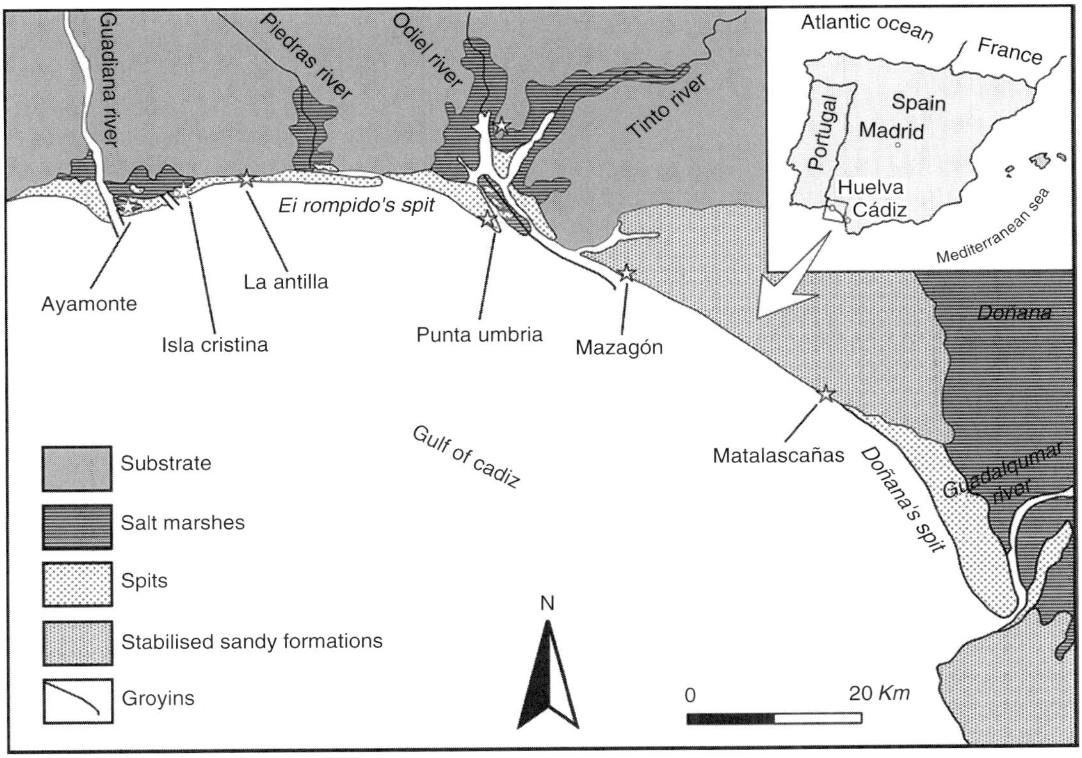

Figure 2. Location map, including the main tourist beaches, spits and groins. 1: Plio-Pleistocene substrate; 2: salt marshes; 3: spits; 4: stabilized sandy formations. Main groins. a: Vila Real do Santo Antonio; b: Huelva.

first storm period (1962–1963), a long interval (6–7 years) was found until the next period, followed by a shorter cycle (2–3 years).

An analysis of the storm features indicates a direct relation between the direction and speed of the prevailing winds and the causes of these events. In all cases, these winds come from the third quadrant at high to very high speed (Fig. 3: 8–12) on the Beaufort scale. Consequently, a more detailed study of the third quadrant velocities is needed to establish the periodicity of the storm wind.

Meteorologic Variables and Storm Periodicity

The application of the Fourier transformation to the time series of autumn–winter wind speeds permits us to distinguish two more probabilistic frequencies (Fig. 4a: 6–7 years and 9–10 years), which are also found in the storm series. In most cases, a 9-year poststorm period contains two storms (period 2 and period 3), situated at 6–7 and 9–10 years after period 1, respectively. This initial coincidence would indicate a possible statistical interrelationship between both variables, but the small record of recent storm periods permits only a first approximation.

This hypothesis was tested in the period 1996–2000 by analyzing the new record of storm periods. A high-energy period (1998–1999) happened in agreement with the hypothetical storm sequence, which predicts new storms in 1995–1996 (period 2) and 1997–1998 (period 3) after 1989 (period 1). In this hypothetical 9-year sequence, the next storm period will be between 2004 and 2005, although a new positive date does not totally confirm the initial hypothesis. We will need 20–30 storm periods (a century, at least) for an adequate, statistical test of this correlation.

This possible periodicity may be related to cyclical changes of the cyclone regime, which are expressed in NAO values (Fig. 4a). In the analysis of the wind–NAO spectral series, the 3-year frequency shows the higher probabilities (Fig. 4b), indicating some degree of common behavior between these two variables.

This frequency agrees with the two main frequencies (6 and 9–10 years) obtained for the wind series. Moreover, an additional coincidence was found by contrasting the evolution of the NAO values and the storm record. Except for the 1989–1990 winter, the remaining storm periods were characterized by negative values of the NAO index, indicating a possible relation between storms and high cyclone phases. In the future, the NAO values collected during storm periods may even be positive due to the increasing tendency of this variable in the last decades.

ADDITIONAL MANAGEMENT IMPLICATIONS

A continuous, yearly study in 20–30 years will be interesting to contrast the first approach now presented to estimate the periodicity of critical storm periods

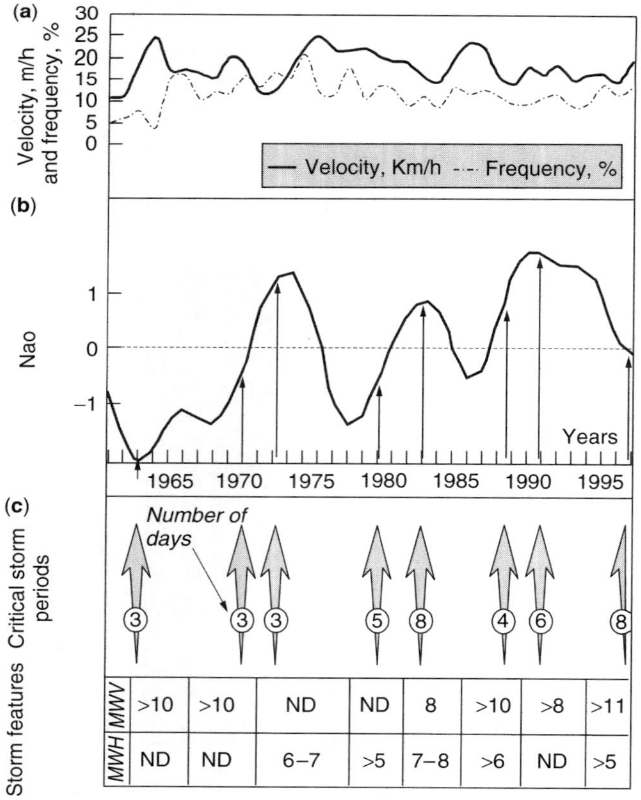

Figure 3. (**a**) Velocity and frequency of the third quadrant winds (period 1961–1996); (**b**) NAO record; (**c**) Critical storm periods between 1960 and 1996, with the maximum wind velocities (MWV; Beaufort scale) and maximum wave heights (MWH, in meters). ND: No data.

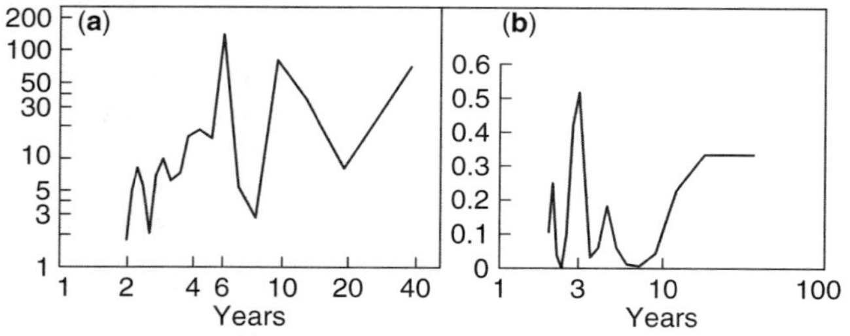

Figure 4. Periodicity diagram of the third quadrant wind speeds; Coherence values between NAO and wind speeds (X axis on log scale).

and its consequence more rigorously. Some additional investigations are necessary:

1. A short-term, periodic analysis of beach–nearshore profiles at the main beaches, with special attention to significant changes during these storms.
2. A first approach to long-term coastal development in examining the roles of waves, tides, storms, and sediments in coastal dynamics, as pointed out by Carter (34).
3. An evaluation of the sedimentary effects produced by the two main groins of the sector (Vila Real do Santo Antonio and Huelva), with indications of possible new emplacements and/or changes in their characteristics (mainly length and alignment).
4. Coastal strategies, including future redistribution of the littoral space (protection zones, demolition of shorefront residences, and relocation of both public infrastructures and private homes).

Taking into account the total damage occasioned by storms in the 1989–1999 period (probably up to $ 20 million), a moderate inversion ($2 million/year) may be very profitable in the next decade.

CONCLUSIONS

In the last 40 years (1956–1996), the analysis of storm periods suffered by the Huelva littoral (SW Spain) indicates a concentration in the winter months (December–January) and statistical behavior similar to third quadrant wind speeds. After the first storm period registered (1962–1963), two new high-energy periods were found in the following 10-year intervals: period 2, at 6 years and period 3, at 9–10 years. The first test of this hypothesis was positive in the 1997–2000 period.

This first approach holds out the prospect of a possible statistical prediction of critical storm periodicity in the next decades, in connection with new investigations into meteorologic changes, sedimentological studies, and coastal management strategies in this area.

Acknowledgments

The financial support of the DGYCIT (projects AMB99-0226-C03-03 and BTE2000-1153) and the Almonte Corporation are gratefully acknowledged. This article is a contribution to IGCP 396 and 437.

BIBLIOGRAPHY

1. Gazis, P.R. (1996). Long-term enhancements in solar wind speed. *J. Geophys. Res.* **101**: 415.
2. Elsner, J.B., Kara, A.B., and Owens, M.A. (1999). Fluctuations in North Atlantic hurricane frequency. *J. Climate* **12**: 427–437.
3. Rodbell, D.T. et al. (1999). An approximately 15,000 year record of El Niño-driven alluviation in southwestern Ecuador. *Science* **283**: 516–520.
4. Artigas et al. (2004).
5. Lee, G., Nicholls, R.J., and Birkemeier, W. (1998). Storm-driven variability of the beach-nearshore profile at Duck, North Carolina, USA, 1981–1991. *Mar. Geol.* **148**: 163–177.
6. Ballesta, M., Morales, J.A., and Acosta, E.A. (1998). Efecto erosivo de los temporales del invierno 1995–1996 sobre la playa de Mazagón (Huelva, SO de España). *Rev. Soc. Geol. España* **11**: 285–296.
7. Rodríguez-Ramírez, A., Cáceres, L.M., Rodríguez Vidal, J., and Cantano, M. (2000). Relación entre clima y génesis de crestas/surcos de playa en los últimos cuarenta años (Huelva, Golfo de Cádiz). *Rev. Cuat. Geomorf.* **14**: 109–113.
8. Komar, P.D. (1976). *Beach Processes and Sedimentation*. Prentice-Hall, Upper Saddle River, NJ.

9. Ahyerre, M., Chebbo, G., Tassin, B., and Gaume, E. (1998). Storm water quality modelling, an ambitious objective?. *Water Sci. Technol.* **37**: 205–213.
10. Gönnert, G. (1999). The analysis of storm surge climate change along the German coast during the 20th century. *Quat. Int.* **56**: 115–121.
11. Williams, J.J. and Rose, C.P. (2001). Measured and predicted rates of sediment transport in storm conditions. *Mar. Geol.* **179**: 121–133.
12. Jenkins, G.M. and Watts, D.G. (1968). *Spectral Analysis and Its Applications*. Holden-Day, London.
13. Cohn, B.P. and Robinson, J.E. (1976). A forecast model for Great Lakes water levels. *J. Geol.* **84**: 455–465.
14. Chatfield, C. (1989). *The Analysis of Time Series: An Introduction*. Chapman and Hall, London.
15. Swan, A.R.H. and Sandilands, M. (1995). *Introduction to Geological Data Analysis*. Blackwell Science, Malden, MA.
16. Kokoszka, P. and Mikosch, T. (2000). The periodogram at Fourier frequencies. *Stochastic Processes Their Appl.* **86**: 49–79.
17. Bojsen, B.H., Witthøfft, H., Styrishave, B., and Andersen, O. (1998). *In situ* studies on heart rate and locomotor activity in the freshwater crayfish, *Astacus astacus* (L.) in relation to natural fluctuations in temperature and light intensity. *Freshwater Biol.* **39**: 455–465.
18. Scarparo Cunha, A.A. and Koutsoukos, E.A.M. (2001). Orbital cyclicity in a Turonian sequence of the Cotinguiba Formation, Sergipe Basin, NE Brazil. *Cretaceous Res.* **22**: 529–548.
19. Baisch, S. and Bokelmann, G.H.R. (1999). Spectral analysis with incomplete time series: An example from seismology. *Comput. Geosciences* **25**: 739–750.
20. Beard, A.G., Williams, P.J.S., Mitchell, N.J., and Muller, H.G. (2001). A spectral climatology of planetary waves and tidal variability. *J. Atmos. Solar-Terrestrial Phys.* **63**: 801–811.
21. Kottegoda, N.T., Natale, L., and Raiteri, E. (2004). Some considerations of periodicity and persistence in daily rainfalls. *J. Hydrol.* **296**: 23–37.
22. Sebald, L., Treffeisen, R., Reimer, E., and Hies, T. (2000). Spectral analysis of air pollutants. Part 2: Ozone time series. *Atmos. Environ.* **34**: 3503–3509.
23. Fenoglio-Marc, L. (2001). Analysis and representation of regional sea-level variability from altimetry and atmospheric-oceanic data. *Geophys. J. Int.* **145**: 1–18.
24. Baldocchi, D., Falge, E., and Wilson, K. (2001). A spectral analysis of biosphere-atmosphere trace gas flux densities and meteorological variables across hour to multi-year time scales. *Agric. Forest Meteorol.* **107**: 1–27.
25. Sondag, F. et al. (2003). Monitoring present day climatic conditions in tropical caves using an Environmental Data Adquisition System (EDAS). *J. Hydrol.* **273**: 103–118.
26. Vanney, J.R. (1970). *L'hydrologie du Bas Guadalquivir. Serv. Publ. Dpto. Geograf. Aplic.*, CSIC, Madrid.
27. CEEPYC. (1979). *Centro de Estudios y Experimentación de Puertos y Costas "Ramón Iribarren"*. Plan de estudio de la dinámica litoral de la Provincia de Huelva. Informe Dirección General de Puertos y Costas, Madrid.
28. Borrego, J. and Pendón, J.G. (1989). Influencia de la actividad humana sobre la evolución sedimentaria de un sector de la ría de Huelva (España). *Geolis* **3**: 125–131.
29. Cuena, G.J. (1991). *Proyecto de regeneración de las playas de Isla Cristina*. Informe del Servicio de Costas, Ministerio de Obras Públicas y Turismo.
30. Ojeda, J. (1989). La dinámica litoral reciente de la Costa Occidental de Andalucía. In: F. Díaz del Olmo and J. Rodríguez Vidal (Eds.). *El Cuaternario en Andalucía Occidental.*; *AEQUA Monografías* **1**: 123–132.
31. Hurrell, J.W. (1996). Influence of variations in extratropical wintertime teleconnections on Northern Hemisphere temperature. *Geophys. Res. Lett.* **23**: 665–668.
32. Kutzbach, J.E. (1970). Large-scale features of monthly mean Northern Hemisphere anomaly maps of sea-level pressure. *Mon. Weather Rev.* **98**: 708–716.
33. Rodwell, M.J., Rowell, D.P., and Folland, C.K. (1999). Ocean forcing of the wintertime North Atlantic Oscillation and European climate. *Nature* **398**: 320–323.
34. Carter, R.W.G. (1988). *Coastal Environments*. Academic Press, London.

SUBLIMATION

Edinara Adelaide Boss
Artur-Nogueira-SP, Brazil

CONCEPT

Sublimation occurs when the matter passes directly from the solid phase to the vapor phase, as is shown in Fig. 1. Examples include dry ice, how a frost-free refrigerator works, and naphthalene.

Sublimation is an endothermic process:

$$\Delta H_{\text{sublimation}} > 0$$

The heat required to change the phase of a pure substance is given by

$$Q = m \cdot L$$

where m is the mass of the pure substance and L is the latent heat.

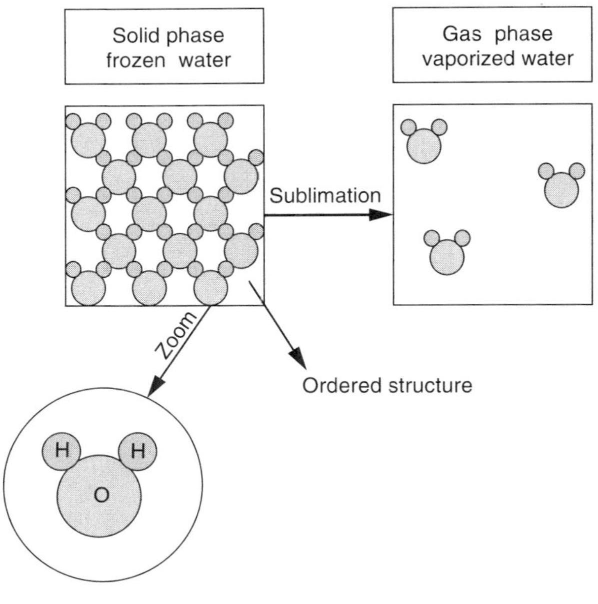

Figure 1. Water sublimation.

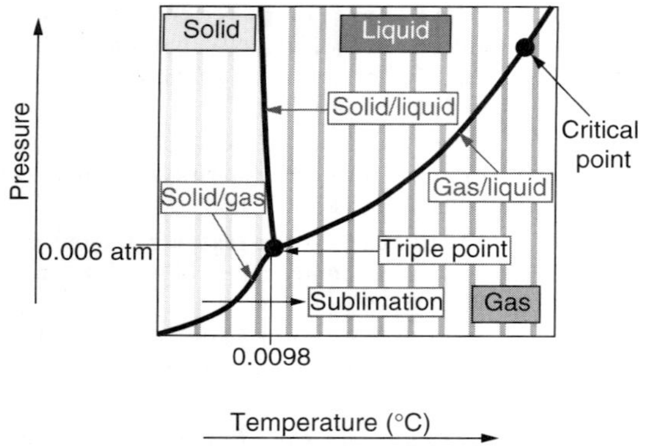

Figure 2. Phase diagram of pure water.

The water latent heat for sublimation is 2.83×10^6 J/Kg.

Figure 2 presents the water phase diagram, which usually shows the pressure versus the temperature.

The diagram is divided into three areas, which represent the solid, liquid, and gaseous phases.

Triple point is a particular condition of temperature and pressure, which the three phases for a substance are in equilibrium.

Above the critical point, the liquid and gas phases are indistinguishable. The critical temperature is the minimum temperature for liquefaction of a gas using pressure. The critical pressure is the pressure required for liquefaction. The critical temperature of water is 374 °C, and the water critical pressure is 218 atm.

Sublimation phenomenon happens below the triple point in the water phase diagram. The sublimation curve is the line between the solid and gas phases.

At the triple point of the water phase diagram, the temperature is 0.0098 °C and the pressure is 0.006 atm.

APPLICATION

Freeze-drying is a process by which a water or aqueous substance is removed from a frozen material or frozen solution by sublimation. Figure 3 shows an example of a frozen solution in a freeze-drying tray.

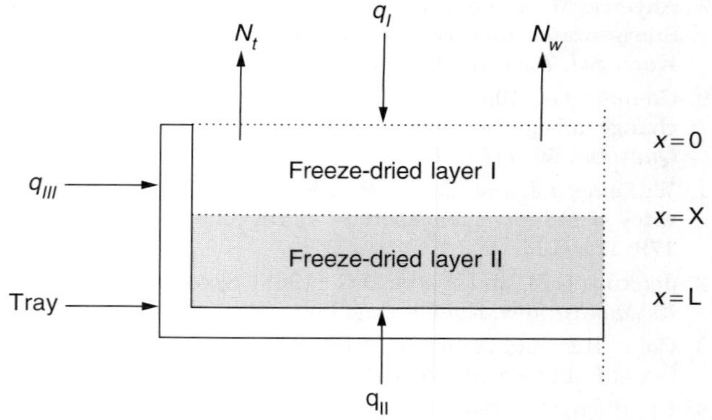

Figure 3. Diagram of a frozen material on a tray during freeze-drying. X is the position of the sublimation interface between the freeze-dried (layer I) and the frozen material (layer II).

Here q_I, q_{II}, and q_{III} are the heat, which can be supplied to the surface by conduction, convection, or radiation. N_w and N_t represent the mass flow of the water vapor and the total mass flow, respectively, in the dried layer.

This process involves the following three periods: freezing, primary drying, and secondary drying periods (1) as shown in Fig. 4.

A batch freeze-dryer is shown in Fig. 5. This equipment is composed of a stainless steel cylinder chamber with eight connectors for connection of the bottles. Above the cylinder chamber, a cover is built by transparent material. At the cylinder, there is a system with three trays. The vacuum pump localizes outside the freeze-dryer.

The freeze-drying process has a wide range of applications, including pharmaceutical products, foodstuffs (the organoleptic properties are important and have to be maintained), and other industrial byproducts (conservation of living microorganism, dehydration, or concentration of heat labile).

Another example of freeze-drying is the soluble coffee process, which produces a porous product by sublimation. This porous material is eager for water.

The performance of the overall freeze-drying process depends significantly on the freezing stage. The material to be processed is cooled to a temperature below the solidification. The shape of the pores, the pore size distribution, and the pore connectivity of the porous network of the dried layer formed by the sublimation of the frozen water during the primary drying stage depend on the ice crystals formed during the freezing stage.

The solvent is removed by sublimation under a vacuum pressure, and heat addition in the primary drying stage is carried out. A significant amount of the latent heat of sublimation is also kept when the water molecules sublime and enter the vapor phase. Because of this result, the temperature of the frozen product is reduced. Then it is necessary to supply heat to the product. The heat could be provided by conduction, convection, and/or radiation.

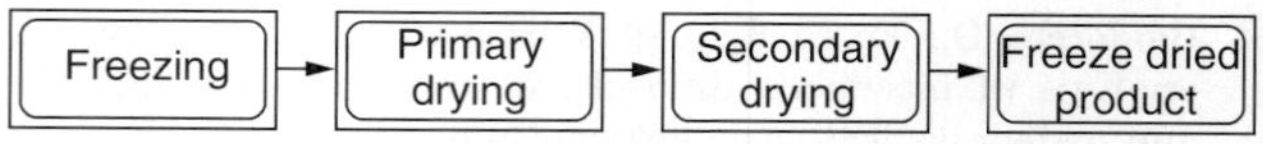

Figure 4. Freeze-drying periods.

1. Glass camera with trays of inoxidable steel.
2. Manifold-exits of silicon and serpentine interns.
3. System of control of refrigeration and pressure.
4. Bomb of vacuous.

Figure 5. Freeze-drying equipment.

During the secondary drying stage the solvent is removed from the chamber and a small amount of sorbed water can be removed by desorption.

A dynamic mathematical model for the freeze-drying stages was proposed by Boss et al. (2). This model represents the process well.

The low temperature of the freeze-drying process and the low water activity resulting from this process minimize some reactions of degradation such as Maillard's reaction, protein denaturation, and enzymatic reaction (3). Color and flavor are important quality attributes in food, and the freeze-drying process preserves more of the pigments and reduces more of the transport rates than does the conventional drying processes. The hot air drying causes irreversible damages to the texture and incomplete rehydration. The freeze-drying process proposes good preservation of the original product characteristics and an increase in the stability of the product during the storage.

BIBLIOGRAPHY

1. Boss, E.A., Maciel Filho, R., and Toledo, E.C.V. (2004). Real time optimization for freeze-drying process. In: *14th European Symposium on Computer Aided Process Engineering—Escape 14*, Portugal, May.
2. Boss, E.A., Costa, N.A., Maciel Filho, R., and Toledo, E.C.V. (2004). Freeze drying process: real time model and optimization. *Chemical Engineering and Processing*. Available: http://www.sciencedirect.com/science/journal/0255270/.
3. Liapis, A.I., Millman, M.J., and Marchello, J.M. (1985). An analysis of the lyophilization process using a sorption-sublimation model and various operational policies. *AIChE J.* **31**(10): 1594–1604.

TRANSPIRATION

KOICHI MIYASHITA
Okayama University
Kurashiki, Japan

Transpiration is the evaporation of water from the plant to the atmosphere. The moved water speed in a plant by the transpiration depends on a vapor pressure deficit that is the difference of a water vapor pressure between the atmosphere and the leaf surface. Two types of the transpiration exist. One is the stomatal transpiration through stomata, and the other is the cuticular transpiration through a cuticular. When the stomata open, the cuticular transpiration is much less than the stomatal transpiration. However, the cuticular transpiration may become large when the stomata are almost closed or when the leaf is cracked and damaged. Figure 1 shows that the water vapor in a stomatal aperture diffuses to the atmosphere through the stomata and the boundary layer of a leaf, in the case of the open stomata.

The transpiration rate from a leaf (Tr) can be generally expressed as

$$Tr = \frac{c_p \rho}{\gamma L} \frac{e_s - e_a}{r_s + r_a}$$

where C_p is the specific heat capacity, ρ is the density of air, γ is the psychrometer constant, L is the latent heat, e_s is the water vapor pressure of the stomatal aperture, e_a is the water vapor pressure of the atmosphere, r_s is the stomatal resistance, and r_a is the leaf boundary layer resistance that is effected by a leaf size and a shape.

The transpiration rate (Tr) is proportional to the difference ($e_s - e_a$) of a water vapor pressure between the stomata and the atmosphere. Usually, in the case of a plant with the water fully supplied, the water vapor pressure of the stomatal aperture is the saturated water vapor pressure at a leaf temperature. The more atmosphere dries, the more ($e_s - e_a$) become large. Then, the transpiration rate increases.

In general, the plant transpiration has the diurnal pattern. It is great in the daytime. On the other hand, the transpiration is small at the nighttime. In the daytime, the stomatal resistance (r_s) becomes small because of the opening of stomata, depending on photosynthesis activities by the solar radiation. Moreover, the exchange of a water vapor becomes active because of the increase of the wind speed on a leaf surface. Then, the leaf boundary layer resistance (r_a) decreases. Therefore, the transpiration rate (Tr) becomes large for the reason that the transpiration is inversely proportional to the sum total ($r_s + r_a$) of a stomatal resistance (r_s) and a leaf boundary layer resistance (r_a).

Meteorological factors significantly affect the transpiration. Several experimental results have been published about the relationships between the meteorological factors and the transpiration. Transpiration completely depends on the physiological activities as long as stomatal aperture does not change. However, the plant may regulate stomatal opening and transpiration. A temporary reduction of stomatal aperture is often caused by a decrease of light intensity, dry air, water deficit, high temperature, and toxic gases. Haseba (1) showed by a simulation analysis between meteorological factors and physiological activity that the transpiration rate increases with increasing a wind speed under a high air temperature, a low relative humidity, a low solar radiation, and a large stomatal conductance and that decreases in an adverse condition. He also showed that the transpiration may not change between two extreme conditions even if a wind speed increases.

The increase in a solar radiation and an air temperature makes the transpiration increase because the saturated water vapor pressure in the stomatal pore (e_s) increases by rising leaf temperature. On the other hand, the transpiration rate is inhibited as the relative humidity of the ambient air is great and the water vapor deficit decreases.

For example, transpiration is also influenced by deficit of the soil moisture. Transpiration does not change if the soil moisture slightly decreases. However, if the soil water content greatly decreases, to keep the water content plant regulate stomatal aperture decreases transpiration.

Transpiration is necessary for the growth of plants. The flow of water through the plant by transpiration increases the absorption of minerals from the soil. Changes in the water balance and the amount of water available in the soil

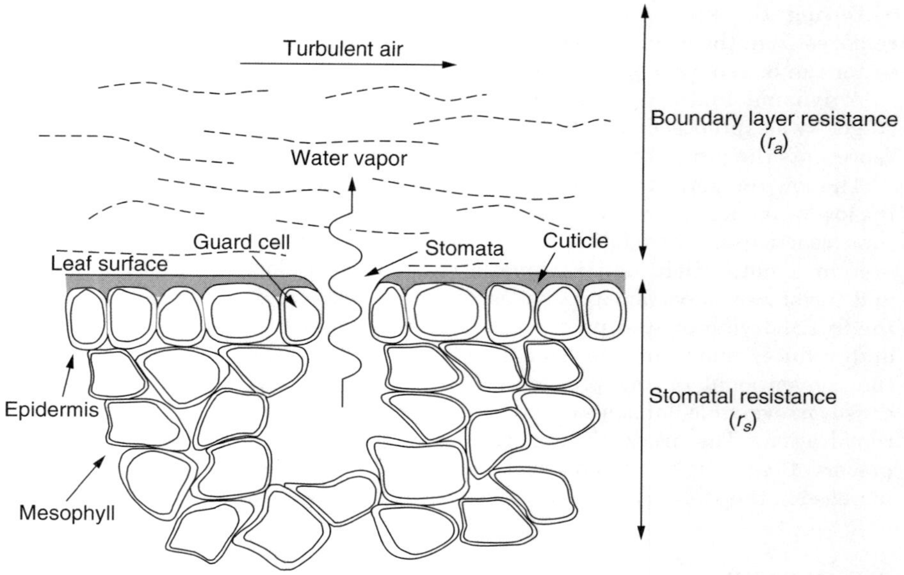

Figure 1. Conceptual diagram of the transpiration from a leaf in case of the open stomata. The curvilinear arrow is the flow of a water vapor from the stomatal aperture to the atmosphere.

can be crucial for plants. There is a rising concern that the worldwide water resources may be exhausted by global warming. Under the drought stresses, a plant does not only decrease the transpiration along with the decrease of stomatal conductance, but also it often may cause damage. Miyashita et al. (2) showed that the recovery level of transpiration in the kidney bean after drought stress gradually decreased as the periods of water stress.

The other effect of transpiration is to decrease the leaf temperature. In particular, the adjustability of a leaf temperature prevents too high of a temperature of a plant from the strong solar radiation. It is well known that the leaf temperatures are different among the plant species even if meteorological conditions are the same, which is mainly the difference of transpiration.

The differences (*Ts-Ta*) between leaf temperature (*Ts*) of worldwide barley cultivars (north America cvs, north African cvs, Chinese cvs, Ethiopian cvs, southwest Asiatic cvs, Japanese cvs, Korean cvs, Nepalese cvs, Turkish cvs, and European cvs) and ambient air temperatures (*Ta*) were measured under the same field (3). The temperature differences (*Ts-Ta*) of barley cultivars in the habitats with the cool meteorological phenomena such as Europe were large under high-temperature conditions. These show the differences of capability in the transpiration among the geographical varieties. Not only the plant species, but also the relationships between transpiration and the meteorological conditions at the places of origin are recognized.

For the plant canopy, such as a forest or an agricultural land, the transport of water vapor from the canopy to the atmosphere includes both the transpiration from plants and the evaporation from the wetted surfaces of plants and soils. Total of transpiration and evaporation is known and described as evapotranspiration. Transpiration of plant canopy is approximately proportional to the green mass, although the transpiration rate of individual leaves decreases with increasing leaf densities because of shielding of radiation and wind. Therefore, transpiration from the canopy is affected by development of the foliar mass. Canopy transpiration has the seasonal and diurnal patterns. Total annual and daily transpiration of stands of plant in various climatic regions, which are obtained from data of numerous authors, are summarized (4). Community transpiration of tropical rainforest is 1500 ~ 2000 mm/year, that of grain fields 400 ~ 500 mm/year and that of open vegetation in arid desert 0.01 ~ 0.4 mm/day. Under similar climatic conditions, forests with greater mass transpire more than grass lands, which in turn transpire more than heath. In addition to accurate observational data, the detailed simulation models, which include all basic hydrological parameters, are now available in calculation of stand transpiration. These quantitative data about transpiration of plant vegetations are important for forestry and landscape management, irrigation management, and prediction of climatic change.

As high CO_2 in the future decreases stomatal aperture (or stomatal conductance), transpiration may decrease if CO_2 increases and leaf temperature is constant. However, high leaf temperature at higher ambient temperature may partly counteract the expected suppression of transpiration. Despite higher leaf temperature, water-use efficiency (the ratio of photosynthesis to transpiration) increases at high CO_2 mainly because of increased photosynthesis (5).

MEASUREMENT OF TRANSPIRATION

Transpiration in the single leaf, the individual plant, and the canopy level measured by many methods is as follows: (1) The weighing method. The total weight of a plant with a pot is precisely measured at a constant time. The amount of a drop of a weight is comparable with the transpiration. As similar methods, there are the leaf cutting method for a single leaf and the lysimeter method for many plants. (2) The chamber method. The plant is covered by a transparent box with an inlet and an outlet port for the air flow. The humidity of an outlet port increases by the transpiration if an air flows. Transpiration rate can be calculated from the difference of an absolute humidity and an air flow rate. This method can be widely used for a single leaf, an individual plant, and a canopy. (3) The sapflow method. There is a transpiration stream within a plant vessel from a root to a leaf. It is in proportion to the increasing amount of a stem temperature when a stem is warmed by a constant quantity of heat. This method is used for plants (tomato, soybean, sugarcane, sweet potato, and etc) or trees. (4) Micrometeorological method. Evapotranspiration from a canopy is measured by a micrometeorological observation. Also, the Bowen ratio energy balance method, the aerodynamic method, the turbulent transport method, and so on are available.

BIBLIOGRAPHY

1. Haseba, T. (1973). Agro-meteorological studies of transpiration in plant leaf in relation to the environment. *Mem. Coll. Agr., Ehime Univ.* **18**: 1–141.
2. Miyashita, K., Tanakamaru, S., Maitani, T., and Kimura, K. (In press). Recovery responses of photosynthesis, transpiration, and stomatal conductance in kidney bean following drought stress. *Environ. Exp. Bot.*
3. Miyashita, K., Maitani, T., and Takeda, K. (1999). Comparison of surface temperature of barley varieties under high temperature conditions. *J. Agric. Meteorol.* **55**(4): 323–328.
4. Larcher, W. (1995). *Physiological Plant Ecology*, 3rd Edn. Springer, Berlin, pp. 264–275.
5. Imai, K. and Okamoto-Sato, M. (1991). Effects of temperature on Co2 dependence of gas exchanges in C3 and C4 crop plants. *Jpn. J. Crop Sci.* **60**: 139–145.

WATERSPOUT

Arthur M. Holst
Philadelphia Water Department
Philadelphia, Pennsylvania

A funnel-shaped weather disturbance which exists over a body of water, a waterspout is often referred to as a tornado on water. This is accurate for one type of waterspout, the aptly named tornadic waterspout. It is somewhat inaccurate with regard to the much more common type of waterspout, the fair weather waterspout.

Tornadic waterspouts are tornadoes over water. These are stronger, larger, and faster than fair weather waterspouts. However, they do tend to form over land as true tornadoes before moving over water. Fair weather waterspouts form differently from tornadoes. Rather than developing downward, they begin at the surface, building up toward the sky.

For waterspouts to form, a series of things must occur. First, a small portion of air over the surface of water must become heated. As this warm air begins to rise, the cool air around it starts to form a small vortex of rotating air. This vortex of swirling air is carried upward by rising water vapor; waterspouts usually form over warm water because there is more water vapor present. Upon rising to an overhead cloud mass, the vortex forms a funnel cloud. At this point, a fair weather waterspout begins its five-stage life cycle.

The first stage of fair weather waterspout development involves the formation of a light spot on the surface of the water, which is surrounded by a darker area. The second stage is the formation of a spiral pattern on the surface around the dark spot. This spiral is colored lightly in some places and darkly in others. The third stage is when the funnel winds begin to pick up speed. The waterspout becomes more visible as sea spray is picked up from below.

This water swirls around the dark spot, known as a cascade. Like a hurricane, it appears to center around an eye. The fourth stage is the stage at which the waterspout is most powerful. Winds can be 60 mph or more. The funnel cloud can be hundreds or even thousands of feet high. Almost all the destruction caused by a waterspout occurs during this stage. The fifth and final stage is the decay of the waterspout. Often, rain begins to fall from the top of the waterspout. As these vortices require warm air to sustain themselves, the resulting downdraft of cool air speeds the process of dissipation of the waterspout. The entire life span of a fair weather waterspout is not often longer than 15 minutes.

Fair weather waterspouts can cause some damage, but it is always over water; they are rarely a threat upon reaching land. Tornadic waterspouts, however, are much more destructive. Their winds can exceed 150 miles per hour, they can last a half-hour or longer, and they are much more likely to appear during severe weather disturbances, such as tropical storms, tropical depressions, and hurricanes. These waterspouts can sink smaller ships and can cause extensive damage to sails and masts on larger ships. However, most waterspouts merely result in large amounts of rain.

READING LIST

The Five Stage Life Cycle of A Waterspout. (8 April 2002). Available: http://www.ziplink.net/users/jfrantis/cycle_ws.htm.

Smith, Bruce B. Waterspouts. (8 April 2002). Available: http://www.crh.noaa.gov/apx/science/spouts/waterspouts.htm.

Tornadoes: Waterspouts. (8 April 2002). Available: http://library.thinkquest.org/C003603/english/tornadoes/waterspouts.shtml.

Waterspouts. (8 April 2002). Available: http://www.srh.noaa.gov/eyw/HTML/spoutweb.htm.

Williams, J. *Waterspouts: Tornadoes Over Water*. USA Today. (8 April 2002). Available: http://www.usatoday.com/weather/wspouts.htm.

UNITED STATES WEATHER BUREAU

Arthur M. Holst
Philadelphia Water Department
Philadelphia, Pennsylvania

From 1890–1970, the United States Weather Bureau provided various weather-related services to the nation. Weather analysis had been considered for hundreds of years, but real progress was not made until the nineteenth century, when techniques and technology began to make accurate and useful weather forecasting possible, the main purpose of the United States Weather Bureau, its predecessors, and its successor, the National Weather Service. The first national American weather service was initiated on February 9, 1870, when President Ulysses S. Grant signed into existence what was then known as The Division of Telegrams and Reports for the Benefit of Commerce. This service began in November 1870. It lasted until 1890, when it was renamed the United States Weather Bureau, and Congress reassigned it from the Department of War to the Department of Agriculture at the behest of President Benjamin Harrison. Then, in 1940, President Franklin Delano Roosevelt reassigned the Weather Bureau again, this time to its current home, the Department of Commerce. The United States Weather Bureau existed until 1970, when its name was again changed, this time by President Richard Nixon, to the National Weather Service, the name it is known by today.

During the 1800s and early 1900s, servicemen for the Weather Bureau often communicated by telegraph, sharing readings of atmospheric pressure, temperature, and the like. They also communicated with seafaring ships in this manner, receiving the first weather report from a ship in 1905. From 1898 to 1933, the Weather Bureau also used kites regularly to ascertain atmospheric conditions. As technology progressed, so did the tools used by the Weather Bureau. Starting in 1909, the Bureau began using weather balloons, a more versatile tool, for weather measurements. And in 1925, it began to use airplanes in Washington, DC., as a more precise tool. The Air Commerce Act of 1926 make the Bureau responsible for providing weather-related services to civil aviators.

Besides forecasting daily weather, the Bureau also made progress in forecasting, tracking, and predicting more specific types of inclement weather. In 1943, an airplane was flown into the eye of a hurricane for the first time, a technique that now provides for hurricane readings that could not otherwise be obtained. In 1948, members of the Air Weather Service issued the first official tornado warning. And in 1952, the Bureau launched a program to forecast and warn of severe locally thunderstorms.

Throughout the Bureau's 70 years of existence, the United States government also tried to incorporate it into an international network of weather forecasting. In 1951, the World Meteorological Organization was formed, and at its head was U.S. Weather Bureau President Francis W. Reichelderfer. Ten years later, President John F. Kennedy called for a worldwide weather forecasting system in his State of the Union address. This all came together in 1963, when the World Weather Watch was established. This program still ensures today the free exchange of data, analysis, and equipment among WMO members (including the NWS), allowing for better weather forecasting and more standardized methods and measurements.

READING LIST

<http://www.randomhouse.com/features/isaacsstorm/science/history.html>

<http://www.nws.mbay.net/history.html>

<http://205.156.54.206/ pa/special/history/125thsk.htm>

<http://www.gfz-potsdam.de/ewc98/abstract/mildner.html>

<http://www.crh.noaa.gov/lmk/history1.htm>

WEATHER FORECASTING THROUGH THE AGES

NASA—Goddard Space Flight Center

INTRODUCTION

Imagine a rotating sphere that is 12,800 kilometers (8000 miles) in diameter, has a bumpy surface, is surrounded by a 40-kilometer-deep mixture of different gases whose concentrations vary both spatially and over time, and is heated, along with its surrounding gases, by a nuclear reactor 150 million kilometers (93 million miles) away. Imagine also that this sphere is revolving around the nuclear reactor and that some locations are heated more during one part of the revolution and others during another part of the revolution. And imagine that this mixture of gases continually receives inputs from the surface below, generally calmly but sometimes through violent and highly localized injections. Then, imagine that after watching the gaseous mixture, you are expected to predict its state at one location on the sphere one, two, or more days into the future. This is essentially the task encountered day by day by a weather forecaster (Ryan, Bulletin of the American Meteorological Society, 1982).

EARLY HISTORY

The art of weather forecasting began with early civilizations using reoccurring astronomical and meteorological events to help them monitor seasonal changes in the weather. Around 650 B.C., the Babylonians tried to predict short-term weather changes based on the appearance of clouds and optical phenomena such as haloes. By 300 B.C., Chinese astronomers had developed a calendar that divided the year into 24 festivals, each festival associated with a different type of weather.

Around 340 B.C., the Greek philosopher Aristotle wrote *Meteorologica*, a philosophical treatise that included theories about the formation of rain, clouds, hail, wind, thunder, lightning, and hurricanes. In addition, topics such as astronomy, geography and chemistry were also addressed. Aristotle made some remarkably acute observations concerning the weather, along with some significant errors, and his four-volume text was considered by many to be the authority on weather theory for almost 2000 years. Although many of Aristotle's claims were

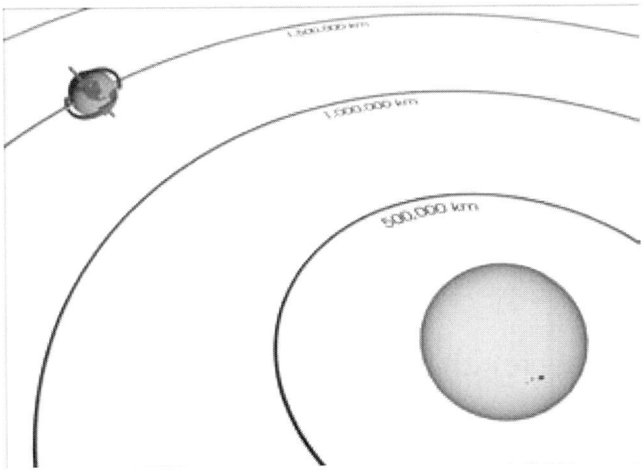

Earth rotates on its axis once every 23 hours, 56 minutes, and completes one revolution around the sun every 365.25 days

Aristotle, as sculpted by the Greek sculptor Lysippos

erroneous, it was not until about the 17th century that many of his ideas were overthrown.

Throughout the centuries, attempts have been made to produce forecasts based on weather lore and personal observations. However, by the end of the Renaissance, it had become increasingly evident that the speculations of the natural philosophers were inadequate and that greater knowledge was necessary to further our understanding of the atmosphere. In order to do this, instruments were needed to measure the properties of the atmosphere, such as moisture, temperature, and pressure. The first known design in western civilization for a hygrometer, an instrument to measure the humidity of air, was described by Nicholas Cusa (c. 1401–1464, German) in the mid-fifteenth century. Galileo Galilei (1564–1642, Italian) invented an early thermometer in 1592 or shortly thereafter; and Evangelista Torricelli (1608–1647, Italian) invented the barometer for measuring atmospheric pressure in 1643.

While these meteorological instruments were being refined during the seventeenth through nineteenth centuries, other related observational, theoretical, and technological developments also contributed to our knowledge of the atmosphere; and individuals at scattered locations began to make and record atmospheric measurements. The invention of the telegraph and the emergence of telegraph networks in the mid-nineteenth century allowed the routine transmission of weather observations to and from observers and compilers. Using these data, crude weather maps were drawn and surface wind patterns and storm systems could be identified and studied. Weather-observing stations began appearing all across the globe, eventually spawning the birth of synoptic weather forecasting, based on the compilation and analysis of many observations taken simultaneously over a wide area, in the 1860s.

With the formation of regional and global meteorological observation networks in the nineteenth and twentieth centuries, more data were becoming available for observation-based weather forecasting. A great stride in monitoring weather at high altitudes was made in the 1920s with the invention of the radiosonde. Small lightweight boxes equipped with weather instruments and a radio transmitter, radiosondes are carried high into the atmosphere by a hydrogen or helium-filled balloon that ascends to an altitude of about 30 kilometers before bursting. During the ascent, these instruments transmit temperature, moisture, and pressure data (called soundings) back to a ground station. There, the data are processed and made available for constructing weather maps or insertion into computer models for weather prediction. Today, radiosondes are launched every 12 hours from hundreds of ground stations all over the world.

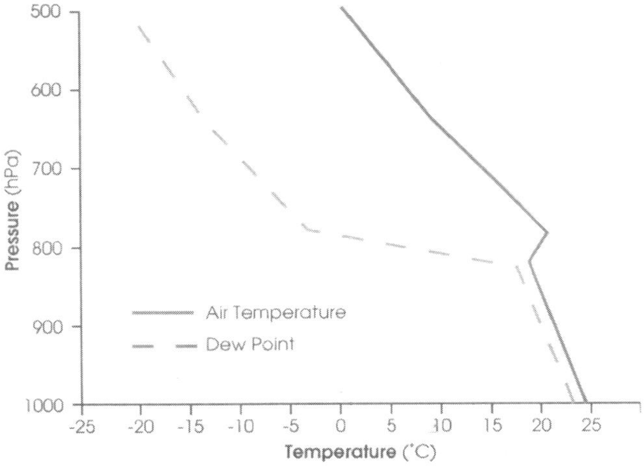

A schematic sounding of air temperature and dewpoint derived from radiosonde data. This sample schematic sounding includes a temperature "inversion" (temperatures increasing with height) at about 800 hPa and reflects atmospheric conditions that frequently precede the development of severe thunderstorms and possibly tornadoes. [1 hectoPascal (hPa) = 1 millibar (mb).]

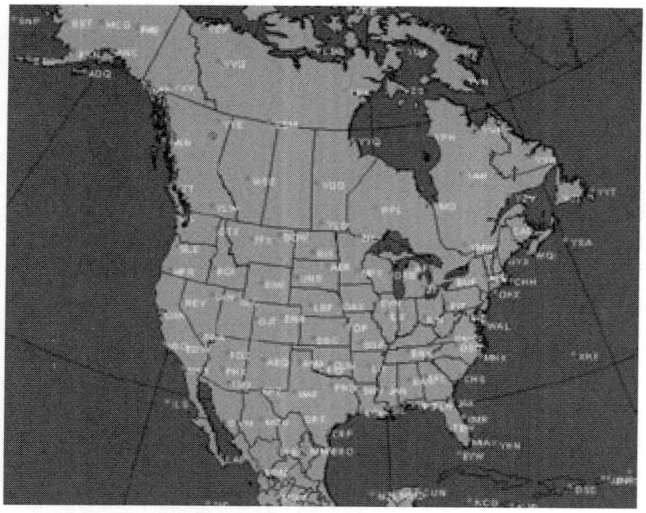

The North American network of upper air ground stations, each indicated by a three-letter station identifier. Radiosondes are launched and tracked from each location every twelve hours

TOWARDS NUMERICAL PREDICTION

Over the past few centuries, physical laws governing aspects of the atmosphere have been expressed and refined through mathematical equations. The idea of numerical weather forecasting—predicting the weather by solving mathematical equations—was formulated in 1904 by Vilhelm Bjerknes (1862–1951, Norwegian) and developed by British mathematician Lewis Fry Richardson (1881–1953, British). Despite the advances made by Richardson, it took him, working alone, several months to produce a wildly inaccurate six-hour forecast for an area near Munich, Germany. In fact, some of the changes predicted in Richardson's forecast could never occur under any known terrestrial conditions. Adding to the failure of this effort, a six-hour forecast is not particularly useful if it takes weeks to produce.

Courageously, Richardson reported his results in his book *Weather Prediction by Numerical Process*, published in 1922. In one of the chapters of this work, Richardson describes a scheme for predicting the weather before it actually happens, a scheme involving a roomful of people, each computing separate sections of the equations, and a system for transmitting the results as needed from one part of the room to another. Unfortunately, Richardson's estimated number of human calculators needed to keep pace with weather developments was 64,000, all located in one very large room.

Richardson's work highlighted the obvious fact that a large number of calculations had to be made very rapidly in order to produce a timely forecast. In the late 1940s, using one of the earliest modern computers, significant progress toward more practical numerical weather forecasts was made by a team of meteorologists and mathematicians at the Institute for Advanced Study (IAS) in Princeton, New Jersey. Mathematician John von Neumann (1903–1957, Hungarian-American) directed the construction of the computer and put together a team of scientists led by Jule Charney (1917–1981, American) to apply the computer to weather forecasting. Charney determined that the impracticality of Richardson's methods could be overcome by using the new computers and a revised set of equations, filtering out sound and gravity waves in order to simplify the calculations and focus on the phenomena of most importance to predicting the evolution of continent-scale weather systems. In April 1950, Charney's group made a series of successful 24-hour forecasts over North America, and by the mid-1950s, numerical forecasts were being made on a regular basis.

MODERN TOOLS OF THE TRADE

Modern technology, particularly computers and weather satellites, and the availability of data provided by coordinated meteorological observing networks, has resulted in enormous improvements in the accuracy of weather forecasting. Satellites, in particular, have given forecasters routine access to observations and data from remote areas of the globe. On April 1, 1960, the polar-orbiting satellite TIROS 1 (the first in the series of Television and Infrared Observation Satellites) was launched. Although the spacecraft operated for only 78 days, meteorologists worldwide were ecstatic over the pictures of the Earth and its cloud cover that TIROS relayed back to the ground.

Over the past 40 years, satellite sensor technology has advanced enormously. In addition to providing visual images, satellites can also provide data that allow calculation of atmospheric temperature and moisture profiles and other environmental variables. This is done using a variety of instruments, among them atmospheric

British mathematician Lewis Fry Richardson (1881–1953). (Photo credit: Bassano and Vandyk Studios.)

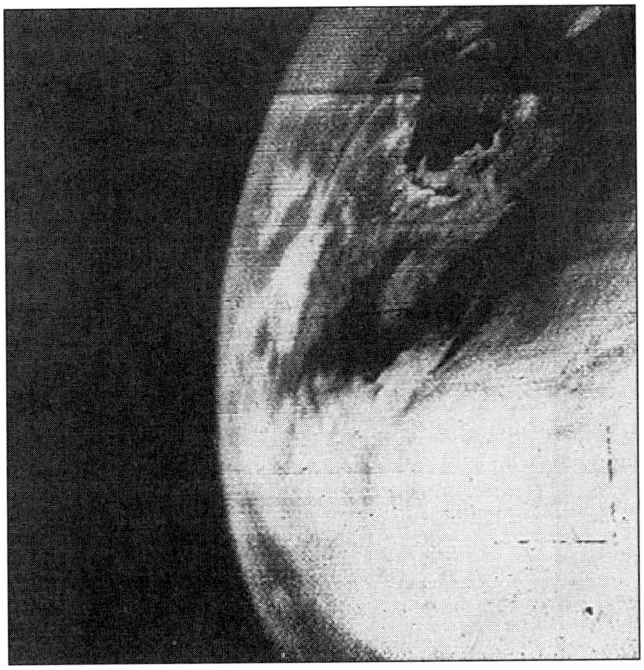

The first picture of Earth from a weather satellite, taken by the TIROS-1 satellite on April 1, 1960. Although primitive in comparison with the images we now receive from satellites, this first picture was a major advance

sounders, which measure quantities at various levels in atmospheric columns. The data retrieved from sounder measurements taken from a satellite can be made similar to radiosonde observations, with the major advantage that the satellite data are more complete spatially, filling in gaps between weather ground stations, which often are hundreds or even thousands of kilometers apart.

THE AQUA SPACECRAFT

In 2002, the Atmospheric Infrared Sounder (AIRS), the Advanced Microwave Sounding Unit (AMSU), and the Humidity Sounder for Brazil (HSB) were launched together on NASA's Earth Observing System (EOS) Aqua satellite, a satellite that also carries a Moderate-Resolution Imaging Spectroradiometer (MODIS), two Clouds and the Earth's Radiant Energy System (CERES) sensors, and an Advanced Microwave Scanning Radiometer for EOS (AMSR-E). The AIRS/AMSU/HSB sounding system provides the capability for determining atmospheric temperature and moisture more accurately than ever before from space-based measurements. These measurements provide to the National Oceanic and Atmospheric Administration (NOAA), the European Centre for Medium-Range Weather Forecasts (ECMWF), and the weather community at large for assimilation into operational numerical weather prediction models. It is expected that assimilation of global AIRS/AMSU/HSB data, complementing other operational observations, should lead to a substantial improvement in the accuracy of mid- and long-range weather forecasts.

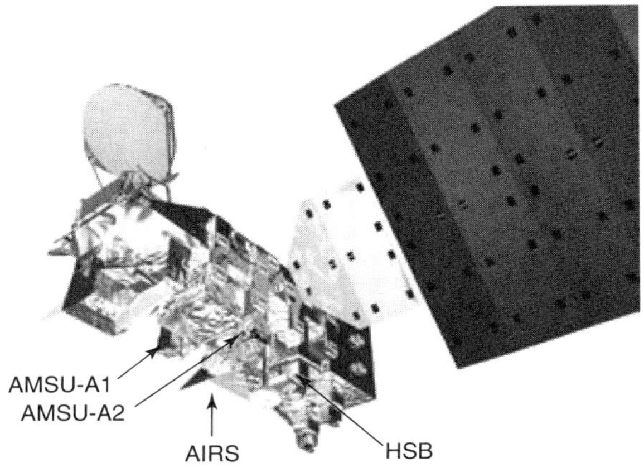

The Aqua spacecraft, with pointers to the AMSU-A1, AMSU-A2, AIRS, and HSB instruments

The AIRS/AMSU/HSB instrument suite builds on the technical heritage of NOAA's High Resolution Infrared Sounder (HIRS) and Microwave Sounding Unit (MSU). The HIRS/MSU combination was the National Weather Service's (NWS's) operational weather sounding system for nearly twenty years, flying on numerous NOAA polar orbiting satellites. This system was enhanced in the late 1990s by the replacement of the four-channel MSU by a 20-channel AMSU, which includes Aqua's AMSU and HSB channels. Looking toward further improvements in weather forecasts, the NWS has set measurement requirements for temperature at an accuracy of 1 °C in layers 1 km thick and humidity at an accuracy of 20% in layers 2 km thick in the troposphere (the lower part of the atmosphere, where weather systems are of most relevance to human life and property). AIRS/AMSU/HSB will meet these requirements, allowing meteorologists to improve and extend their predictions and reduce the number of significant prediction mistakes, like failing to predict a major storm prior to a few hours before its arrival.

BENEFITS TO SOCIETY

The economic benefits of more accurate weather forecasts are immense. For example, a more accurate 24-hour forecast of heavy rain and thunderstorms along a cold front could allow airline dispatchers enough time to reroute their airplanes appropriately and thereby help alleviate costly delays. Being able to pinpoint a wintertime low temperature in Florida could be the deciding factor in whether orange grove farmers make the correct decision regarding deployment or non-deployment of freeze prevention methods to save their crops. Better information on wind patterns could aid the National Hurricane Center in producing a more accurate forecast of a hurricane's track and might enable a reasoned decision to be made regarding evacuating thousands of families out of Miami or Jacksonville. In military operations, there is a considerable historical record of instances when weather conditions have altered the course of battles, including examples when accurate forecasting has been a deciding factor in one side's victory.

Sample crops that could benefit from more accurate weather forecasts aiding farmers in selecting harvesting times and in protecting their crops from freezing temperatures

CLOSING

Only fifty years ago, weather forecasting was an art, derived from the inspired interpretation of data from a loose array of land-based observing stations, balloons, and aircraft. Since then it has evolved substantially, based on an array of satellite and other observations and sophisticated computer models simulating the atmosphere and sometimes additional elements of the Earth's climate system. All this has been made possible by advances in satellite technology, a sweeping acceleration in worldwide communications, and overwhelming increases in computing power. Aqua's AIRS/AMSU/HSB combination should further these advances, enabling more accurate predictions over longer periods.

OVERVIEW OF WEATHER SYSTEMS

John W. Nielsen-Gammon
(from *Handbook of Weather, Climate, and Water: Dynamics, Climate, Physical Meteorology, Weather Systems, and Measurements*, Wiley 2003)

INTRODUCTION

Meteorology is the only science in which it is common for a practitioner to appear every day on the evening news. Weather forecasting has immediate interest to people because it affects their day-to-day lives. The impact is great because weather itself changes from day to day. If weather were all good, or all bad, perhaps no one would notice—when's the last time you thought about the quality of the ground under your feet?

FOUNDATIONS OF WEATHER FORECASTING

The attempt to understand the weather is driven largely by the desire to be able to forecast the weather. However, the atmosphere is a complicated dynamical system, and an understanding of even the basic physical principles came only recently. Before that, scientists either attempted to understand the physical causes of weather events or recorded "weather signs," rules for weather forecasting. Because the physical understanding was incorrect, and most weather signs had no physical basis, neither approach was very successful. True progress in the study of the atmosphere that would lead to success in weather forecasting required a complete understanding of physical principles and a comprehensive set of observations of the atmosphere.

Aristotle's *Meteorologica*, written around 340 BC, was the leading book on meteorological principles in the Western world for two millenia. With his theories on the fundamental physical nature of the universe, Aristotle sought to explain a wide range of meteorological phenomena. For example, he argued that wind is caused by the hot, dry exhalation of the earth when struck by sunlight. A leading book on weather signs, *De Signis Tempestatum*, written by Aristotle's pupil Theophrastus, focused on such maxims as "A dog rolling on the ground is a sign of violent storm" and "Reddish sky at sunrise foretells rain." Since some of these maxims are still in use today, it appears that Theophrastus's work has outlived that of Aristotle.

Astronomers during the Middle Ages often argued that weather could be predicted by careful examination of the sky, clouds, and stars. This method of weather forecasting is sometimes successful, but it more closely resembled astrology than modern-day meteorology. Astrological forecasting schemes grew more advanced and intricate with time, while the scientific study of weather systems made little or no progress.

Scientific progress required basic observational data for describing weather systems. Only with good observations could theories be tested and the correct nature of the atmosphere be determined. The primary tools for observation of weather conditions are the thermometer (for temperature), the barometer (for air pressure), and the hygrometer (for humidity). Prior to and during the Middle Ages, these basic weather instruments did not exist. These instruments were invented in the sixteenth and seventeenth centuries and refined throughout the eighteenth century. The resulting observations of the atmosphere eventually established the basis for meteorological breakthroughs of the nineteenth and twentieth centuries.

Meanwhile, Isaac Newton had laid the foundation for the development of physical laws governing the motion of objects. The great mathematician Leonhard Euler of Germany, in 1755, rewrote those laws in a form that applied to a continuous fluid such as air or water. But Euler's laws were incomplete. They described how air can be neither created nor destroyed and how air accelerates (and wind blows) in response to forces acting on it. Missing from Euler's equations were the crucial relationship between temperature and pressure and the consequences of evaporation and condensation.

With the new instruments, the study of air became an experimental science. Through numerous laboratory experiments, scientists gradually discovered the fundamental physical laws governing the behavior of gases. In particular, the missing key relationship, the first law of thermodynamics, became known from experiments during the first half of the nineteenth century.

Now that the basic physical laws were known, a comprehensive set of observations of the atmosphere was necessary for further progress. Weather observations from a single careful observer were insufficient, but dozens of observations taken simultaneously across Europe or the United States gave a much clearer picture of the distribution of weather elements within a winter storm. Such "synoptic" observations were also used directly in weather forecasting. As rapid long-distance communication became possible through the telegraph, individual weather systems could be tracked and therefore forecasted. The coordinated observations that have taken place since the middle of the nineteenth century form the foundation of modern weather forecasting and our understanding of weather systems of all scales.

DEVELOPING AN UNDERSTANDING OF WEATHER SYSTEMS

Progress in understanding meteorology in the nineteenth century was largely based on thermodynamics. James Pollard Espy of the United States conducted a variety of laboratory experiments involving the condensation of water vapor in air. In 1841, based on his experiments and those of others, Espy correctly described the basic principle of thunderstorm formation: Condensation of water vapor within ascending air causes the air to become warmer than its surroundings and thus to continue rising. Espy went on to assert that the resulting lower pressure near the ground accounts for the low pressure observed in large-scale storm systems.

Espy's theory lay dormant until later in the nineteenth century, when the complete first law of thermodynamics was established. With this development, scientists were able to put Espy's theory of convection on a solid mathematical footing. Karl Theodor Reye of Germany, for example, determined the specific conditions under which such ascending air would be unstable. Their ideas form our understanding of the fundamental nature of individual thunderstorms to this day.

Espy also suggested that upward motion and the release of latent heat by evaporation might be the driving force for large weather systems. The widespread development of synoptic observations finally established that surface winds spiral inward toward a large-scale low-pressure system. The strong temperature contrasts within low-pressure systems suggested that the ascent would be driven, at least initially, by a current of warm air impinging on a current of cold air and rising. The rising air, and the resulting latent heating from condensation, would cause surface pressures to fall. This theory for the cause of low-pressure systems gained rapid acceptance for both theoretical and observational reasons in the 1870s.

A leading proponent of this thermal theory of cyclones was the American William Ferrel. Earlier, Ferrel had completed the application of Euler's equations to the atmosphere by formally including the effect of Earth's rotation as the Coriolis force. Ferrel also saw a clear analog to the role of equatorial convection in the general circulation, which drives ascent in the tropics and descent in the subtropics.

Unfortunately for its proponents, the thermal theory was wrong. Coordinated mountaintop observations in Europe, analyzed by Julius von Hann, showed that above the ground the centers of low-pressure systems were actually cooler than the centers of high-pressure systems. This observation illustrates the self-limiting nature of moist convection: The downward motion outside the convective cell causes warming of that surrounding air, so that eventually the instability is eliminated.

WEATHER AND ENERGY

The above discussion leads to a key question for the understanding of weather systems: How do weather systems grow and maintain themselves? As noted above, the basic dynamics of ordinary moist convection were well established, both conceptually and mathematically, by the end of the nineteenth century. Convection was also understood to be an important driver of the general circulation, or at least the tropical and subtropical circulation known as the Hadley cells.

By the turn of the century, the work of Hann and others showed that convection was not the energy source for extratropical cyclones, the midlatitude migratory low-pressure systems. Attention shifted from the role of latent heat release to the role of the large horizontal temperature gradients that were systematically observed within midlatitude low-pressure systems. In 1903, Max Margules, born in the Ukraine and working in Austria, calculated the amount of kinetic energy that could be obtained from the rising of hot air and the sinking of cold air in a low-pressure system. He found that the amount of energy that could be converted into kinetic energy was comparable to the actual amount of kinetic energy in a mature storm system. Margules had identified the correct energy source for extratropical cyclones. The process by which cyclones form and move was described by researchers working under Vilhelm Bjerknes in Bergen, Norway, shortly after World War I. Finally, a comprehensive theory by Jule Charney in 1947 explained that the structure and intensity of low-pressure systems was a consequence of the growth of unstable eddies on a large-scale horizontal temperature gradient.

Hurricanes did not have the prominent horizontal temperature variations found in midlatitude weather systems; indeed, many hurricanes seemed almost perfectly symmetrical. The prominence of convection throughout the hurricane, particularly within the eyewall, suggested that convection was fundamentally important in the development and maintenance of hurricanes. But how did the hurricane organize itself or maintain itself against large-scale subsidence within its environment? The currently accepted theory, published by Kerry Emanuel and Richard Rotunno in 1986 and 1987, relies on radiative cooling of the subsiding air at large distances from the hurricane. Emanuel also noted that as air spirals inward toward the eye it would become more unstable because it would be gaining heat and moisture from the sea surface at a progressively lower pressure.

Theories satisfactorily describing organized convection such as mesoscale convective systems and supercells also have been slow to develop. One reason for this is that no observing systems could accurately describe the structure of organized convection until the development of radar. Indeed, the term "mesoscale" was coined specifically to refer to in-between sizes (10 to 500 km) that were too large to be adequately observed at single locations and too small to be resolved by existing observing networks. The widespread use of weather radar in the 1950s helped fill the observation gap. The development of Doppler radar (for measuring winds within precipitating systems) for research purposes in the 1970s and as part of a national network in the 1990s helped even more. With comprehensive radar observations, it became clear that the long lifetime of organized convection was due to a storm keeping its updraft close to the leading edge of its cold, low-level outflow. The storm could then take advantage of the ascent caused by the cold air undercutting warm air without having its supply of warm air cut off completely. But even radar was not sufficient. The development of the first dynamical descriptions of supercells and squall lines in the 1980s, by Joseph Klemp, Richard Rotunno, Robert Wilhelmson, and Morris Weisman, relied upon numerical computer-generated simulations of the phenomena to provide an artificial data set unavailable from observations.

While the above discussion has focused on self-contained weather systems, certain "weather producers" may be thought of as byproducts of these weather systems. Many forms of severe weather, such as hail, lightning, and tornadoes, are essentially side effects of convection (organized or otherwise) but have little or no influence on the convection itself. Other mesoscale phenomena, such as sea breezes, upslope precipitation, and downslope windstorms, do not represent instabilities at all. Instead they are called "forced" weather phenomena because they are driven by such external features as solar heating gradients and topographic obstacles to the large-scale flow.

FORECASTING

The massive strides in weather forecasting during the past 50 years are due in large part to our growing understanding of the nature and dynamics of important weather systems. Forecasting of many phenomena has evolved from an exercise in extrapolation to specific predictions of the evolution of weather systems to computer simulations that accurately forecast the evolution of weather systems. Currently, the most skillful weather forecasts involve the prediction of large-scale extratropical weather systems.

One very important advance in our ability to forecast large-scale weather systems was the development of the omega equation, first presented in simplified form by Richard Sutcliffe in 1947. The omega equation is named after the Greek symbol that represents the change in pressure of an air parcel. In its original form, the omega equation is based upon a simplification of the equations of motion that assumes that all motions are large-scale and evolve slowly. When this approximation is made, it is possible to diagnose vertical motion entirely from large-scale pressure and temperature variations. Vertical motion is important for weather prediction because it is fundamentally related to clouds and precipitation, but large-scale vertical motion also can be used to infer the evolution of low-level and upper-level wind patterns. Vertical motion became the key to understanding the evolution of weather. In forecast offices, maps were designed and widely distributed that made it easy to look at the large-scale fields and diagnose vertical motion.

A second important advance was the development of numerical weather prediction, or NWP. NWP, discussed more extensively in the chapter by Kalnay in the dynamics part of this *Handbook*, has completely transformed the forecasting of large-scale weather systems. In the past, forecasters relied on their diagnosis of the current weather patterns to make forecasts. Nowadays, computers can make much more accurate forecasts than humans alone, and the best possible forecast is obtained by humans working with computer forecast output. The forecaster's task has become one of identifying likely errors in the model forecast, based on the forecaster's knowledge of systematic errors within the model, errors in the initial analysis, and computer forecast scenarios that run counter to experience.

In contrast to large-scale weather systems, individual convective storms are not directly simulated, nor accurately forecasted, by the numerical weather prediction models currently in use. Forecasting convective storms still relies heavily on extrapolation and a sound knowledge of what sorts of storms are likely to be produced by certain larger-scale weather conditions. The modern era of forecasting severe weather began in 1948, when two Air Force weather forecasters (Maj. Ernest Fawbush and Capt. Robert Miller) had responsibility for forecasting at Tinker Air Force Base near Oklahoma City. After a damaging tornado struck the base in 1948, the two meteorologists were given the task of identifying the days when tornadoes were likely to strike the base. We now know that forecasting the specific path of a tornado is essentially impossible, but when a similar low-pressure system evolved a few days later, the forecasters were compelled by the base commander to use a newly minted severe weather warning system to issue a forecast of a possible tornado at the base. Amazingly, the forecast came true, and preventive measures taken before the second tornado struck the base prevented considerable casualties to aircraft and personnel. This weather forecast, possibly the most serendipitous in the history of humankind, led directly to modern tools for diagnosing the likelihood of severe weather from large-scale conditions and laid the foundation for modern tornado forecasting.

CONCLUSION

This historical review has emphasized that progress in our understanding of the weather has required four ingredients: the need to understand the science of the atmosphere before one can hope to make accurate forecasts; a complete knowledge of the basic physical underpinnings of atmospheric behavior; observations adequate to test the theories and point the way toward new ones; and numerical models to provide simulations of the atmosphere more complete than any set of observations. The following chapters of this part of the *Handbook* discuss the current understanding of the weather phenomena discussed above, improvements in which have lead to vastly improved weather forecasts.

READING LIST

Frisinger, H.H. (1983). *The History of Meteorology: To 1800*. American Meteorological Society, Boston, MA.

Kutzbach, G. (1979). *The Thermal Theory of Cyclones: A History of Meteorological Thought in the Nineteenth Century*. American Meteorological Society, Boston, MA.

Miller, R.C., and Crisp, C.A. (1999). The first operational Tornado Forecast—twenty million to one. *Weather and Forecasting* **14**(4): 479–483.

UNIT HYDROGRAPH THEORY

Vijay P. Singh
Louisiana State University
Baton Rouge, Louisiana

INTRODUCTION

In 1932, Sherman (1) developed the concept of the unit hydrograph (UH) for determining the surface or direct runoff hydrograph (DRH) from the effective rainfall hyetograph (ERH). The unit hydrograph, or simply unit graph, of a watershed is defined as the DRH resulting from one unit (1 inch or 1 centimeter) of effective rainfall (ER) that occurs uniformly across the watershed at a uniform rate during a unit period of time. This unit period is not necessarily equal to unity; it can be any finite period up to the time of concentration. This unit period of ER is the period for which the UH is determined for a watershed. As soon as this period changes, so does the UH for the watershed. This means that there can be as many UHs for a watershed as periods of ER. Often used are a 1-hour UH, 6-hour UH, 12-hour UH, or a 1-day UH. Here the duration of 1 hour, 6 hours, 12 hours, or 1 day is not the duration for which the UH occurs, but it is the duration of the ER for which the UH is derived. The ER is assumed constant during its duration D for which the UH is derived, so the effective rainfall intensity (ERI) is $1/D$.

Since its inception, the unit hydrograph concept has been widely applied in surface-water hydrology. The principle underlying the UH has considerable appeal and has, therefore, been applied to a wide range of problems in environmental and water resources. For example, Rendon-Herrero (2) employed the UH concept to derive unit sediment graphs. Williams (3) extended the same concept to derive instantaneous unit sediment graphs. Thomann (4) used it to derive water quality graphs, and Dooge (5) employed it for groundwater recharge. It has also been used by Snyder and Asmussen (6) for subsurface flow, by Venetis (7) for groundwater flow, by Singh et al. (8) for irrigation flow, and by Westestrom and Singh (9) for snowmelt runoff.

The objective of this chapter is to describe the unit hydrograph theory and illustrate its application to build hydrologic models in four select areas.

UNIT HYDROGRAPH THEORY

The fundamental premise of unit hydrograph theory is that the hydrologic system is linear and time invariant. Although the assumptions implied or made in the theory are, strictly speaking, seldom true, the theory, nevertheless, does yield reasonable results in practice. The theory is applicable to a wide range of problems in hydrology and water resources.

The unit hydrograph is based on these postulates: (1) For a given drainage basin, the duration of surface runoff is constant for all uniform-intensity storms of the same duration, regardless of differences in the total volume of surface runoff. (2) For a given drainage basin, if two uniform-intensity storms of the same length produce different total volumes of surface runoff, then the rates of surface runoff at corresponding times, after the beginning of the two storms, are in the same proportion to each other as their total volumes of surface runoff. (3) The time distribution of surface runoff from a given storm period is independent of concurrent runoff from antecedent storm periods.

Postulates 1 and 2 together make up the principle of proportionality. In other words, if the duration of the ER is fixed but its volume changes, then the duration of the ERH does not change, but its ordinates do in proportion to the volume of ER. The third postulate is the principle of superposition, which allows the decomposition of a complex ERH into rectangular blocks or pulses and then superimposing, on one another, hydrographs of these rectangular pulses, each of steady intensity, to obtain the total DRH.

As stated in the definition of the unit hydrograph, it is assumed that the ER is uniformly distributed within its duration. This means that the rainfall intensity is uniform throughout the watershed while rain falls. The other assumption is that it is uniformly distributed throughout the watershed. This means that uniform runoff occurs from all parts of the watershed. The ER is defined as that portion of rainfall that produces surface runoff, so all abstractions must be subtracted from rainfall, and for the ER to be uniform, the abstractions must be uniform throughout the watershed.

These postulates, together with the definition and the assumptions, constitute what is now referred to as the unit hydrograph theory. The theory fundamentally assumes that the watershed is linear and time invariant, that is, the DRH is derived from the ERH by a linear operation. Once the assumptions of linearity and time invariance are recognized, the theory can be applied to any linear, time-invariant system (10,11) and is not restricted only to computing the surface runoff hydrograph. For this reason, it has been employed to derive the unit sediment hydrograph, the unit pollutant graph, the unit groundwater hydrograph, and the unit interflow hydrograph, among others. Thus, in a generic sense, the behavior of a linear, time-invariant system that receives input and produces output can be described by the theory. By conservation of mass, the volume of input must be equal to the volume of output.

The dimensions of the UH can be expressed in two ways. First, if the UH is derived from the DRH due to an ER of specified duration whose depth is not unity, then the dimensions of UH ordinates will be equal to the dimensions of runoff divided by the dimension of the depth of ER. Runoff can, however, be expressed either in terms of discharge in volumetric rate units or specific discharge in volumetric rate per unit area or intensity units. Therefore, the dimensions of UH ordinates are either the dimensions of discharge per unit depth or the dimensions of discharge per unit area per unit depth. Let L denote the dimension of length and t the dimension of time. Then, the two forms of the dimensions of UH ordinates are (1) L^2/t and (2) $1/t$.

LIMITATIONS OF THE UNIT HYDROGRAPH THEORY

The unit hydrograph theory is only an approximation, and its postulates are not strictly valid. Nevertheless, it is a useful practical tool and is amenable to linear mathematics whose methods are simple and easily understood.

An ER of a specified duration seldom occurs uniformly across a watershed of reasonable size. Spatial variation generally becomes more pronounced as the watershed size increases. Rainstorms that produce intense rainfall usually do not extend across large areas. The nonuniform distribution of rainfall can cause variation in the hydrograph shape. However, errors in the UH due to nonuniformity in areal distribution can be minimized by decomposing the watershed into subwatersheds, each small enough to ensure approximately uniform areal distribution, by deriving the UH for each subwatershed, and then by appropriately combining these UHs to develop the UH for the entire watershed.

An ER seldom occurs uniformly, even for as short a duration as 5 minutes. However, the effect of temporal variation of rainfall intensity on the UH shape depends principally on the watershed size. For example, rainfall bursts lasting only a few minutes may produce well-defined peaks in the runoff hydrograph from small watersheds but may cause little change in the runoff hydrograph shape from large watersheds. The effect of temporal variation can be minimized by employing the instantaneous unit hydrograph (IUH).

All watersheds in nature are nonlinear; some are more linear, and some are less. They are linear only by assumption. If the hydrographs from storms of the same duration are compared, their ordinates are not in proportion to their corresponding runoff amounts. The peaks of the UHs for small rainfalls are usually lower than those for larger ones.

MATHEMATICAL REPRESENTATION

To avoid ambiguity in interpreting the UH, it must be emphasized that it corresponds to a particular duration of an ER and has a volume of one unit (i.e., 1 inch or 1 cm). Because the ER, it is assumed, occurs uniformly, its duration is sufficient to define its uniform intensity. This duration is referred to as the unit storm duration and may change from one application to another. Thus, for a given watershed, there can be a large number of UHs; each corresponds to a specific duration of ER.

Let an ERH of D-hour duration be denoted as

$$I(t) = I,\ 0 \le t < D; \quad I(t) = 0, \quad t \ge D \tag{1}$$

where I is a constant ER intensity (L/t), the D-hour UH is $h(D, t)$ and the resulting DRH is $Q(t)$. Then, applying the principle of proportionality, one obtains

$$Q(t) = h(D, t)I(t)D = h(D, t)ID \tag{2}$$

The quantity ID denotes the volume of ER. This equation states that the DRH is proportional to the ER volume. When, $ID = 1$, the DRH and UH are numerically the same.

Let an ERH be a complex hyetograph composed of an n number of pulses, each of duration D hours. Let the intensity of pulse j be $I_j, j = 1, 2, 3, \dots, n$, and the DRH of pulse j be $Q_j(t)$. If the D-hour UH is $h(D, t)$, then the DRH due to the complex hyetograph can be expressed as

$$Q(t) = \sum_{j=1}^{n} Q_j[t - (j-1)D] \tag{3}$$

Now, note that Q_1 starts at $t = 0$, Q_2 starts at $t - D$, and Q_j starts at $t - (j-1)D$. The individual DRHs can be expressed as

$$\begin{aligned}
&Q_1(t) = h(D, t)I_1 D \\
&Q_2(t - D) = h(D, t - D)I_2 D, \qquad h(D, t - D) = 0, t \le D \\
&Q_3(t - 2D) = h(D, t - 2D)I_2 D, \quad h(D, t - 2D) = 0, t \le 2D \\
&\vdots \\
&Q_n[t - (n-1)D] = h[D, t - (n-1)D]I_n D \\
&\qquad\qquad h[D, t - (n-1)D] = 0, t \le (n-1)D
\end{aligned} \tag{4}$$

Thus, the DRH can be expressed as

$$Q(t) = \sum_{j=1}^{n} h[D, t - (j-1)D]I_j D \tag{5}$$

The DRH at any time t is the result of the ERH pulses up to that time but not subsequent to it, so Equation 5 must be modified as

$$Q(t) = \sum_{j=1}^{t} h[D, t - (j-1)D]DI_j D \tag{6}$$

For simplicity, time t can be taken as a discrete variable that takes on values at D time intervals.

The difficulty that develops from the dependence of the UH on the duration of D of the ER is circumvented by letting D tend to be instantaneous. The UH so obtained is called the instantaneous unit hydrograph (IUH). Thus, the IUH, $h(0, t) = h(t)$ is a hypothetical UH due to the ER whose duration tends to zero as a limit, but whose volume remains unity (say, 1 cm). Clearly, the IUH is independent of the duration of ER. Mathematically,

$$h(t) = h(0, t) = \lim_{D \to 0} h(D, t) \tag{7}$$

and

$$\delta(t) = \lim_{D \to 0} I(t, D)D \tag{8}$$

where $\delta(t)$ is the Dirac delta function at $t = 0$. Physically, this function can be thought of as a spike of infinitesimally small thickness and infinitely large height such that the area under the spike is 1. Therefore, Eq. 6 can be expressed as

$$Q(t) = \int_0^t h(t - \tau)I(\tau)d\tau, \quad h(\tau) = 0, \quad \tau \leq 0 \tag{9}$$

Because mathematical operations are linear, Eq. 9 can also be written as

$$Q(t) = \int_0^t h(\tau)I(t - \tau)d\tau \tag{10}$$

where

$$h(t) \geq 0 \quad \text{for any } t \geq 0$$

$$\lim_{t \to \infty} h(t) = 0$$

and

$$\int_0^\infty h(s)ds = 1$$

Experimental evidence shows that

$$\int (t - \bar{t})h(t)dt \succ 0 \tag{11}$$

where $\bar{t}$ is the centroid of the IUH located at the t axis:

$$\bar{t} = \int_0^\infty th(t)dt \tag{12}$$

Equation 12 states that the IUH must be skewed to the right, that is, its peak must be to the left of the centroid, and a long tail extends to the right.

If the duration of the ER is indefinitely long and its intensity is one unit per unit of time (say, 1 cm/h), then the hydrograph so obtained is termed the summation hydrograph or S-hydrograph. This hydrograph assumes a deformed S-shape, and its ordinates ultimately approach the rate of ER in the limit or at the time of equilibrium. If the ERH is broken down into pulses each of duration D hours, then the SH will be obtained by superimposing the DRHs due to individual pulses. For this ERH, the SH, $U(t)$, is obtained from Eq. 6 as

$$U(t) = \sum_{j=1}^{t} h[D, t - (j - 1)D]D = \sum_{j=1}^{t} Q[D, t - (j - 1)D] \tag{13}$$

Clearly, if D tends to 0, $h[D, t - (j - 1)D]$ tends to $h(t)$, the lower limit for summation will approach the origin, and

$$U(t) = \int_0^t h(s)ds \tag{14}$$

Now, the relations between IUH, UH, and SH can be derived. From the definitions of the UH and IUH, one can write

$$h(D, t) = \frac{1}{D}\int_{t-D}^{t} h(s)ds \tag{15}$$

Similarly, the relation between UH and SH can be expressed as

$$h(D, t) = \frac{1}{D}[U(t) - U(t - D)] \tag{16}$$

As D tends to 0,

$$h(t) = \frac{dU(t)}{dt} \tag{17}$$

It may be instructive to summarize that the three hydrographs—IUH, UH, and SH—are ascribed to three distinct characteristics of the effective rainfall—volume, duration, and intensity—as shown in Table 1:

Also note that the time to the peak of $h(D, t)$ can be determined by differentiating Eq. 16 with respect to t and equating the derivative to zero:

$$h(t) - h(t - D) = 0 \tag{18}$$

Equation 18 states that the peak of $h(D, t)$ occurs when the ordinate of the IUH is equal to the ordinate at D time units earlier. The SH attains its maximum value at a time equal to D hours less than the time base of the initial $h(D, t)$.

UNIT HYDROGRAPH MODELS

The unit hydrograph theory has been extensively applied to develop a range of hydrologic models in surface water hydrology, groundwater hydrology, and water quality hydrology. Four illustrative examples are presented.

Rainfall–Runoff Models

The unit hydrograph theory was developed for rainfall–runoff modeling and has been most extensively applied therein. There is a large range of UH-based models, including empirical, conceptual, and physically based (10). However, for economy of space, only a simple conceptual model will be presented to illustrate the power of the UH theory. To develop a conceptual model,

Table 1. Effective Rainfall Characteristics

Hydrograph	Volume	Duration	Intensity
IUH	Unity	Zero	Indefinite
UH	Unity	Finite	1/duration
SH	Indefinite	Indefinite	Unity

a watershed is represented by a network of linear elements, such as reservoirs and channels. This network can be defined using the laws of geomorphology or arbitrarily to mimic best the flow pattern in a watershed. Rainfall is reduced to effective rainfall by subtracting infiltration. Then, an instantaneous ER is routed through the network of reservoirs and/or channels to obtain the IUH of the watershed. By convoluting this IUH with the ERH, the DRH is obtained.

A linear reservoir can be defined by a linear relation between storage S and surface runoff Q as

$$S = KQ \tag{19}$$

where K is the residence time. The water balance of the reservoir can be expressed as

$$\frac{dS}{dt} = I(t) - Q(t) \tag{20}$$

where I is ER and t is time. Substituting Eq. 20 in Eq. 19 yields

$$\frac{dQ}{dt} + \frac{1}{K}Q = \frac{I}{K} \tag{21}$$

For an instantaneous $I(t)$, represented by $\delta(t)$, $Q(t)$ will reduce to $h(t)$. Therefore, Eq. 21 can be expressed as

$$\frac{dh}{dt} + \frac{1}{K}h = \frac{\delta(t)}{K} \tag{22}$$

For $h(0) = 0$, the solution of Eq. 22 is

$$h(t) = \frac{1}{K}\exp(-t/K) \tag{23}$$

If n linear reservoirs are arranged in series, then the IUH of this cascade is

$$h(t) = \frac{1}{K\Gamma(n)}\left(\frac{t}{k}\right)\exp\left(-\frac{t}{K}\right) \tag{24}$$

Equation 24 is a gamma function and is known as the Nash model (12).

A linear channel is defined by a linear rating curve or a linear relation between flow cross section $A(t)$ and discharge Q:

$$A = C(x)Q \tag{25}$$

where $C(x)$ is a constant at a channel section, varies with distance, and has the connotation of the inverse of the velocity; and x is the distance along the channel. The continuity equation for a channel in one-dimensional form can be written as

$$\frac{\partial Q}{\partial x} + \frac{\partial A}{\partial t} = 0 \tag{26}$$

Substituting Eq. 25 in Eq. 26 produces

$$\frac{\partial Q}{\partial x} + C(x)\frac{\partial Q}{\partial t} = 0 \tag{27}$$

The inflow to the channel is upstream and can be denoted as $Q(x_0, t)$. A linear channel simply translates the flow by the time equal to the travel time, T. Thus, the solution of Eq. 27 is

$$Q(x_1, t) = Q(x_0, t - T), \quad T = C(x_1 - x_0) \tag{28}$$

where x_1 denotes the downstream end of the channel reach. If the inflow is given by the delta function $\delta(t)$ at the upstream end of the channel, the IUH of the channel is

$$h(t) = \delta(t - T) \tag{29}$$

The linear reservoirs and linear channels can be arranged in a way that best represents the watershed. Dooge (13) employed a network of linear channels and reservoirs to represent the watershed. All operations are linear, so the IUH of the network can be derived with little difficulty. For any $I(t)$, the DRH can be determined by convolving the IUH with $I(t)$ as

$$Q(t) = \int_0^t h(t-\tau)I(\tau)d\tau \tag{30}$$

For practical purposes, Eq. 30 can also be expressed in discrete form.

Channel Flow Routing

Channel flow is routed by using either a systems-based approach or a hydraulic approach. The unit hydrograph theory is applicable to both approaches if they are linear. The systems-based approach based on the Muskingum method is the most popular. This method is comprised of Eq. 20 and a linear storage–discharge relation expressed as

$$S = k[wI + (1 - w)Q] \tag{31}$$

where S is the storage of water within the channel reach, I is the inflow rate, Q is the outflow, and w and k are routing parameters. For a delta function inflow, $\delta(t)$, the IUH of the Muskingum method is obtained by substituting Eq. 31 in Eq. 20 and solving the combined equation:

$$h(t) = -\frac{w}{1-w}\delta(t) + \frac{1}{k(1-w)^2}\exp\left[-\frac{t}{k(1-w)}\right] \tag{32}$$

where $I(0)$ is $I(t)$ at $t = 0$. By convoluting Eq. 32 with $I(t)$, the outflow discharge $Q(t)$ can be obtained.

On the other hand, in a hydraulic approach, St. Venant equations are employed. The unit hydrograph theory has been applied to several linearized forms of these equations. One simple approximation is the kinematic diffusion approximation (14,15) expressed as

$$\left(gh_0 - \frac{u_0^2}{4}\right)\frac{\partial^2 Q}{\partial x^2} = 3gS_0\frac{\partial Q}{\partial x} + \frac{2gS_0}{u_0}\frac{\partial Q}{\partial t} \tag{33}$$

where h_0 is the flow depth at the reference discharge, u_0 is the velocity at the reference discharge, Q is the perturbation discharge around the reference discharge, g is the acceleration due to gravity, S_0 is the bed slope, x is the distance along the channel, and t is time. The reference discharge may correspond to steady uniform flow. For instantaneous flow at the upstream end of the channel reach, $\delta(t)$, the IUH of the channel reach described by Eq. 33 is

$$h(t) = \frac{x}{\sqrt{4\pi D t^2}} \exp\left[-\frac{(x-at)^2}{4Dt}\right] \quad (34)$$

where $a = 1.5\,u_0$, and

$$D = \frac{Q_0}{2S_0}\left(1 - \frac{F^2}{4}\right), \quad F = \frac{u_0}{\sqrt{gh_0}} \quad (35)$$

where F is the Froude number corresponding to the reference discharge and D is hydraulic diffusivity. By convoluting Eq. 34 with the upstream discharge, the outflow (or routed) discharge can be obtained.

Groundwater Flow

The unit hydrograph theory has been applied to model groundwater recharge, stream–aquifer interaction, aquifer response due to pumping, and a whole host of groundwater problems. Only the problem of stream–aquifer interaction will be discussed to illustrate an application of the unit hydrograph theory in groundwater hydrology. To that end, suppose that a seminifinite aquifer interacts with a stream. The aquifer is confined, homogeneous, and isotropic. The governing equation for one-dimensional flow can be written as

$$\frac{\partial^2 H(x,t)}{\partial x^2} = \frac{1}{D}\frac{\partial H(x,t)}{\partial t} \quad (36)$$

where H is the hydraulic head and D is the aquifer diffusivity equal to the ratio of transmissivity (T) to storativity S_c. The initial and boundary conditions can be written as

$$h(x,0) = 0 \quad (37a)$$

$$h(0,t) = I(0,t) \quad (37b)$$

$$h(0,t) \to 0, \text{ as } x \to \infty \quad (37c)$$

The IUH of Eq. 36 can be obtained using the Laplace transform (16) as

$$h(x,t) = \frac{x}{2t}\left[\frac{S_c}{\pi t T}\right]^5 \exp\left[-\frac{S_c}{T}\frac{x^2}{4t}\right] \quad (38)$$

Equation 38, when convoluted with $I(0,t)$, will yield the aquifer response.

SEDIMENT TRANSPORT

Analogous to the IUH, in the instantaneous unit sediment graph (IUSG), h_S is defined as the distribution of sediment from an instantaneous burst of rainfall that produces one unit of runoff and is considered the product of the IUH and the sediment concentration distribution $C(t)$:

$$h_S(t) = h(t)C(t) \quad (39)$$

In the IUSG, $C(t)$, it is assumed, varies linearly with V_Q, the amount of effective rainfall or direct runoff. Analogous to the ERH, Chen and Kuo (17) defined the effective sediment erosion intensity per unit depth (e.g., foot) (ESEI), S_e. Then, similar to the surface runoff hydrograph given by Equation 6, the sediment discharge, Q_S, can be expressed as

$$Q_S(t) = \sum_{j=1}^{n} h_S[D, t-(j-1)D]D)S_e(t)D \quad (40)$$

where D is the time interval of each pulse. The dimensions of Q_S and h_S are FT^{-1} and T^{-1}, respectively. A value of 1 hour or less may be appropriate for D. Singh and Chen (18) established that the sediment yield, Y, and the direct runoff volume are related as

$$Y = aV_Q^b \quad (41)$$

where a and b are parameters. A similar relation exists between $Q(t)$ and $Q_S(t)$. As $Q(t)$ varies with ERI, so does $Q_S(t)$ with ESEI. In other words, the effective rainfall intensity or depth produces direct runoff; in turn, the direct runoff produces the direct sediment yield. Conceptually, therefore the ERI generates its counterpart ESEI which when combined with IUSG produces a sediment graph. The relation between $S_e(t)$ and the effective rainfall is of the same form as that between $Q_S(t)$ and $Q(t)$. The sum of S_e ordinates must equal Y as that of ER ordinates must equal V_Q.

CONCLUSIONS

The unit hydrograph theory is a powerful tool that is applicable to a wide range of problems in hydrology and environmental engineering. In practice, this theory is still routinely applied. Although most hydrologic and environmental systems are nonlinear, their linear part dominates in most cases. That is why this theory works so well.

BIBLIOGRAPHY

1. Sherman, L.K. (1932). Stream flow from rainfall by the unit graph method. *Engineering News Record* **108**: 501–505.
2. Rendon-Herrero, O. (1978). Unit sediment graph. *Water Resources Research* **14**(5): 889–901.
3. Williams, J.R. (1978). A sediment graph model based on an instantaneous unit sediment graph. *Water Resources Research* **14**(4): 659–664.
4. Thomann, R.V. (1972). *Systems Analysis and Water Quality Management*. McGraw-Hill Book Company, New York.
5. Dooge, J.C.I. (1960). The routing of groundwater recharge through typical elements of linear storage. *IASH Bulletin* **52**: 286–300.
6. Snyder, W.M. and Asmussen, L.E. (1973). Subsurface hydrograph analysis by convolution. *Journal of Irrigation and Drainage Division* ASCE **98**(IR3): 405–418.
7. Venetis, C. (1968). On the impulse response of an aquifer. *IAHS Bulletin* **13**(3): 136–139.

8. Singh, V.P., Scarlatos, P.D., and Prasad, S.N. (1990). An improved Lewis-Milne equation for the advance phase of border irrigation. *Irrigation Science* **11**: 1–6.
9. Westerstrom, G. and Singh, V.P. (2000). An investigation of snowmelt runoff on experimental plots in Lulea, Sweden. *Hydrological Processes* **14**: 1869–1885.
10. Singh, V.P. (1988). *Hydrologic Systems: Vol. 1. Rainfall-Runoff Modeling*. Prentice Hall, Englewood Cliffs, NJ.
11. Singh, V.P. (1989). *Hydrologic Systems: Vol. 2. Watershed Modeling*. Prentice Hall, Englewood Cliffs, NJ.
12. Nash, J.E. (1957). *The Form of the Instantaneous Unit Hydrograph* IASH Publication, **45**(3): 114–121.
13. Dooge, J.C.I. (1959). A general theory of the unit hydrograph. *Journal of Geophysical Research* **64**(2): 241–256.
14. Dooge, J.C.I. (1973). *Linear Theory of Hydrologic Systems*. Technical Bulletin No. 1468, p. 327, U.S. Department of Agriculture, Agricultural Service, Beltsville, MD.
15. Singh, V.P. (1996). *Kinematic Wave Modeling in Water Resources: Surface Water Hydrology*. John Wiley & Sons, New York.
16. Hall, F.R. and Moench, A.F. (1972). Application of the convolution equation to stream-aquifer relationships. *Water Resources Research* **8**(2): 487–493.
17. Chen, V.J. and Kuo, C.Y. (1986). A study of synthetic sediment graphs for ungaged watershed. *Journal of Hydrology* **84**: 35–54.
18. Singh, V.P. and Chen, V.J. (1982). On the relation between sediment yield and runoff volume. In: *Modeling Components of Hydrologic Cycle*. V.P. Singh (Ed.). Water Resources Publications, Littleton, CO, pp. 555–570.

WEATHER AND THE ATMOSPHERE

David P. Stern
Goddard Space Flight Center, NASA

An optional extension of section (S-1) "Sunlight and the Earth." A more detailed (but qualitative) discussion of heat flow in the atmosphere and related weather processes, including the roles of buoyancy, convection and humidity.

Around and around goes the wind, and from its circuits returns the wind.

Ecclesiastes, ch. 1, v.6

PRESSURE AND CONVECTION

Let us start with the **flow of air.** Suppose a "parcel of air" is heated near the ground (by **conduction** of heat, the flow of heat due to direct contact). Heat makes it expand, it becomes less dense than the surrounding air and **buoyant,** and it rises like a hot-air balloon (or like a drop of oil in a bottle of water).

At the higher levels of the atmosphere, this warm bubble again gives up its heat (to other flows or perhaps to cold space), cools down, and other bubbles coming from below push it to the side, where it descends again. (diagram on the board). Such a circulating flow is called **convection.**

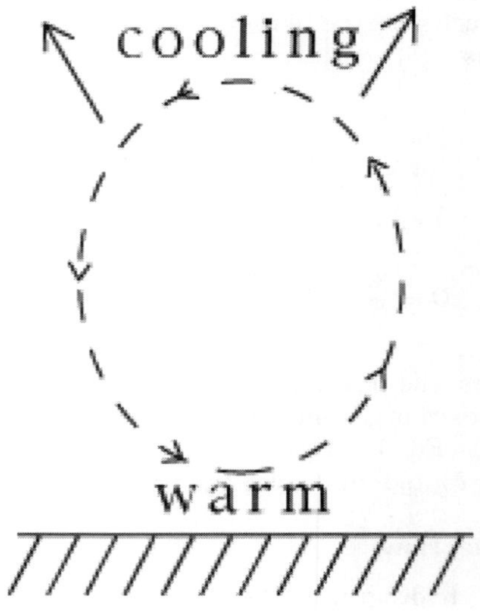

More generally, convection is any flow which

1. picks up heat at one place,
2. drops it at another, and
3. is driven by this transport of heat.

The important thing to remember when dealing with convective flows, is that the **higher** one is in the atmosphere, the lower are the **pressure and density** of the air. What compresses it is the **weight of air above it** which it must support. On top of Mt. Everest, less air is piled up on top and therefore the pressure is lower.

At ground level, the compressing weight of the atmosphere amounts to about 1 kilogram on each square centimeter. That pressure does not bother our bodies, because the air inside us is at the same pressure, and the fluids of the body (like blood) do not compress easily. For the same reason, fish have no problems with depth—even at a depth of 100 meters, with a pressure 11 times larger (10 kilogram water above each square centimeter, plus the weight of the atmosphere) they feel no discomfort.

(Divers too can stand such pressure, provided the air they breathe is similarly compressed—except that the mixture must be changed, otherwise they may take in too much oxygen, and too much nitrogen is dissolved in their bloodstream.)

At an altitude of about 5 kilometers, only **half** the atmosphere is above us, the other half is below, so only the weight of half the atmosphere must be supported, and the pressure is reduced to one half.

By "Boyle's law" (named for Robert Boyle, 1627–91), the **density** is also reduced to one half (ignoring any variation in temperature). Rising an **additional** 5 kilometers, the pressure again falls by half, to 1/4 of what it was on the ground, and at 15 kilometers, it is halved again to about 1/8. All this is approximate and depends on temperature, but the trend should be clear.

The cabin of a jetliner flying at 10 km must be sealed and pressurized, because passengers breathing air at 1/4 the sea-level density would be starved for oxygen and might lose consciousness. On the very rare occasions when a jetliner loses its pressure, masks connected to oxygen canisters drop down automatically, allowing the passengers to breathe normally while the pilot quickly descends to a lower altitude.

WEATHER

First, **a word of caution:** what follows below is **a very crude and simplified** picture of a much more complicated process. While the circulation in the drawing below resembles the one proposed by Hadley in 1735, the actual situation is much more involved.

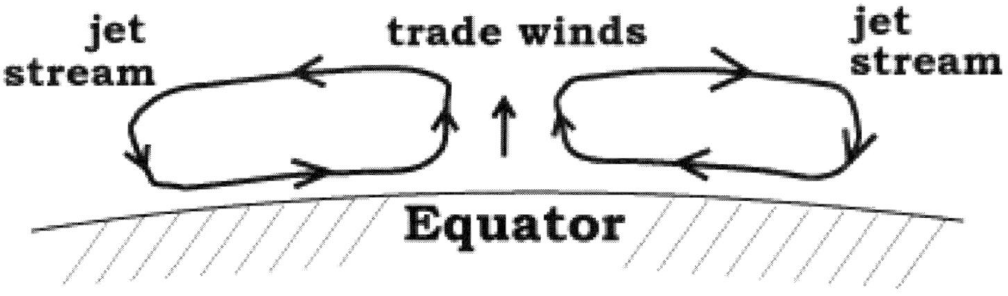

(a) Local Weather

When the atmosphere is stable, the higher we go, the **cooler** the air is.

Air is **warmest** near the ground, which absorbs receives heat from sunlight. It is **coldest** above the level where jetliners fly, at 10–15 kilometers, the region from where it **radiates** most of its heat into space. That is why mountaintops are cold and the highest mountains have snow on their tops.

(Still higher layers may get quite hot again, by absorbing UV and "extreme UV," but they have little effect on what goes on below them).

How Exactly Does This Happen?

Suppose some "parcel of air" (dry air, for now—humidity is an additional factor, considered later) is heated by the ground and rises. Higher up the pressure is lower, so the air expands: but expansion **cools** it.

Similarly, if for some reason the parcel is blown down, is compressed again and **heated** by the compression. Such up-and-down motions happen all the time, and the net result is that when conditions are stable, the temperature drops at a **steady rate** as we go higher.

The motion of the rising parcel of air depends on its surroundings. It always cools by expansion—but is it still warmer than the still air around it? **If it is, it continues to rise; if not, it stops.** As will be seen, this is where the **humidity** of air has an important effect.

[On an ordinary day, direct heating by the ground only carries the air a few hundred meters, perhaps a kilometer, creating above the ground a "boundary layer" with many small convective flows. Large scale motions like thunderstorms usually occur higher up (see below).]

(b) Global Weather

Convection also operates on a global scale. The greatest heating occurs near the equator, and air heated there **rises and flows poleward,** to cooler regions of the Earth.

The rotation of the Earth greatly modifies this flow, by the *Coriolis effect*, as explained below.

- At the equator the east-west motion of the air **matches** that of the equator.
- **Away** from the equator, however, the Earth's surface is **closer** to the rotation axis, the distance it covers in 24 hours is **smaller,** and so its west-to-east speed is **slower**. If the air moving **away** from the equator (right side of the drawing above) **persists** at its original west-to-east speed, it will **overtake** the local surface, making the predominant winds blow from **west to east** ("westerlies").
- The cooler air returns equatorward (middle part of the drawing above) at **lower** altitudes, completing the loop. If it **still** kept its original west-to-east speed, it would **again** match the local rotation of the equator.
- Actually, the air loses velocity to friction with the ground. Therefore, the time it returns to the equator, it **lags behind** the rotating ground, and the average wind is **easterly**.

In the age of sailing ships, sea-captains took advantage of this system. Sailing from Spain to America, they would

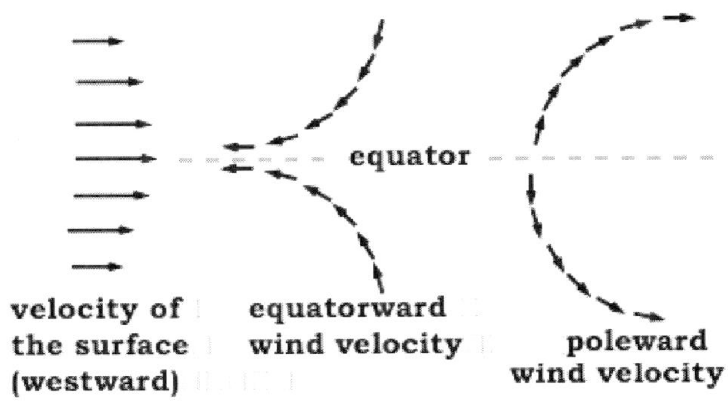

go closer to the equator, a more southern route that took advantage of the easterly "trade winds." Sailing back home they would go further north and use the westerlies. The many Spanish wrecks off Florida, some containing quite rich cargo, were lost on this home voyage back to Spain, loaded with gold and silver from Mexico and South America.

Jetliners flying across the US cannot swing quite as far. However, when flying west to east, their pilots often exploit the fast core of the westerlies, known as the **jet stream** and flowing at high altitudes. Flying westward they try to avoid the jet stream.

WATER VAPOR

Instead of heating the Earth, sunlight can **evaporate water** from it—especially from the oceans, which cover most of the Earth's surface. Humid air may be viewed as containing **additional energy**, invested by the Sun when its heat evaporated the water. While heat drives convection, humidity may **amplify** it.

Hot humid air is what drives thunderstorms, and a warm ocean surface is also the traditional birthplace of violent tropical storms, known as hurricanes in America and typhoons in Asia.

We look at two examples of humidity in action.

1. In a **thunderstorm**, hot humid air rises, as in ordinary convection. As it rises to regions of lower pressure, it cools by expanding. However, cold air cannot hold as much humidity as warm air, and the extra water is therefore squeezed out. In moderate convection it forms **clouds** (as in the second example below), but in a vigorous thunderstorm, there is too much of it and it turns into **rain**.

Giving up water **heats** the air, or rather, slows down its cooling, because the heat invested by the Sun in evaporating the water is now passed back to the air. As a result, the rising air is still warmer than the air layers around it, and it **continues to rise vigorously**. It squeezes out still more rain and forms the tall thunderstorm clouds, which pilots know well to avoid.

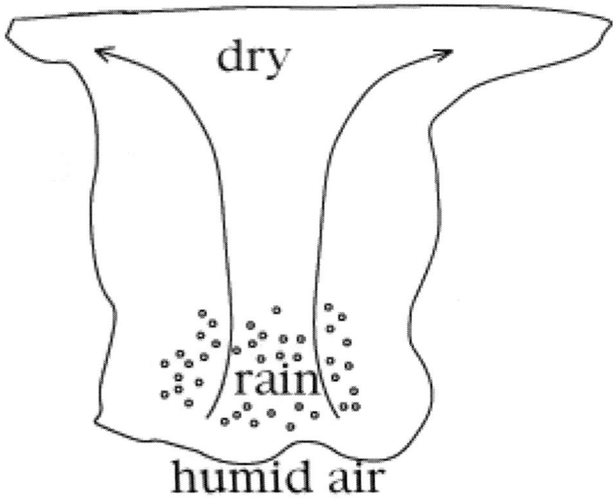

Thunderstorm convection

(In very vigorous thunderstorms, the "updrafts" of rising air may rise so quickly that they blow raindrops into the higher and colder sections of the cloud, where they **freeze,** producing **hail.** Some hailstones are picked up again and again, adding layers of ice on each upwards journey. That is how unusually big hailstones can form.)

2. On a hot clear day, many fluffy small **clouds** may form. A light plane flies across the land, and every time it comes under a cloud, the pilot feels that it is **bodily lifted.** What is happening?

—Here is the reason. The heating of the ground by the Sun has created many small convection currents rising upwards. Their air contains humidity, not enough for a serious thunderstorm, but enough to produce small clouds when droplets of water condense as the rising air cools.

The small clouds mark the top of the "boundary layer" near the ground, with many small circulating flows. Each cloud sits on top of a rising convection current, which lifts the airplane as it flies across it. Since "what goes up must come down" the pilot can also expect downdrafts between the clouds, where the air goes down again, as part of the circulating convection. Such up-and-down motions may make passengers on low-flying airplanes quite airsick!

Further Exploring. Strictly for those interested in history: the original 1896 article in which Svante Arrhenius proposed the greenhouse effect. Note that carbon dioxide was then called "carbonic acid."

Tidbit. Who first wrote "**Everybody talks about the weather, but nobody does anything about it**"? Most would claim it was Mark Twain (just type the first 5 words into a search engine and see!), but it ain't necessarily so. It looks like Twain's style, but actually the words first appeared in an editorial in the Hartford (Conn.) Courant on 24 August 1897, written by **Charles Dudley Warner.**

Warner was a good friend of Twain, who himself had lived in Hartford for many years (he left before 1897). He was a newspaperman in Hartford and the two had collaborated on an 1873 book "The Gilded Age." Warner is also remembered for other quotes, e.g., "Politics make strange bedfellows."

Concerning the quote, see on the web http://www.m-w.com/wftw/99aug/082499.htm. and more about Charles Dudley Warner (and Twain) at http://courant.ctnow.com/projects/twain/warner.htm.

VAPOR PRESSURE

Naraine Persaud
Virginia Polytechnic Institute and State University
Blacksburg, Virginia

Vapor pressure is the partial pressure exerted by the vapor of a substance in the gaseous phase of a closed heterogeneous system at equilibrium with condensed phases (i.e., solid or liquid) of the substance present in the system. It is the macroscopic measure of the tendency of the molecules of a given species in the condensed phases of the system to escape into the gaseous phase. A heterogeneous system means a mixture of two or more distinct, physically or chemically homogeneous phases. Mixture implies the phases are mechanically separable; distinct implies presence of unequivocally defined physical boundaries (interfaces) between phases; and homogeneous implies each phase is uniform throughout chemically and physically. The simplest possible conceptualization for a heterogeneous system is any closed two-phase system consisting of the solid or liquid of a pure substance in equilibrium with its vapor at a given temperature. In more complex heterogeneous systems, the condensed phases may be completely miscible or immiscible liquids, aqueous solutions, discontinuous liquid aerosols, or solid/liquid emulsions, and the vapors may be mixed with other vapors and gases present in the gaseous phase. Although the entire system is closed (permitting only energy transfer) with respect to its surroundings, the subsystems (phases) are open, permitting both energy and mass transfer between phases.

The equilibrium vapor pressure is of interest and importance in many applications encountered in water/atmospheric science and engineering such as quantifying water vapor pressure in the air above ice or liquid water, above an aqueous solution, and in the presence of curved air–water interfaces (menisci, bubbles, and droplets). Covering all these facets of vapor pressure would necessitate a discussion that far exceeds the limits of an encyclopedia article on vapor pressure. Our subsequent discussion will therefore focus on the fundamental relationships that can be universally applied to quantitatively describe and predict phase equilibrium in any closed heterogeneous system—in particular, a discussion of the Gibbs chemical potential and phase rule. We will then demonstrate applications of these principles to the simplest (and perhaps the most common) case of the vapor pressure in a two-phase system of a pure substance with specific reference to the solid/liquid/vapor system for pure water.

In a quintessential contribution to classical thermodynamics (entitled "On the equilibrium of heterogeneous substances") published between 1875 and 1876 in the *Transactions* of *the Connecticut Academy of Arts and Sciences*, Josiah Willard Gibbs (1839–1903) formulated seminal concepts governing equilibrium in *any* closed heterogeneous system (1,2). Specifically, Gibbs addressed reversible changes in the equilibrium state of closed heterogeneous systems at uniform temperature and pressure due to changes in the chemical composition or physical state of any of the phases through chemical reactions or mass transfer between phases.

From the first law of thermodynamics, for a reversible change in equilibrium state of a closed system, the differential change (gain) in the internal energy $dU = \delta q + \delta w$, where δq and δw represent the small amounts of the (positive) total heat or total work transferred into the system. Heat or work transferred out of the system is negative, implying a loss in U. The total work can be divided into any PV work (δw_{PV}) done on the system ($\delta w_{PV} = -P\,dV$ since dV is negative), and any (reversible) non-PV work done on the system ($\delta w_{\mathrm{n}PV}$). Since dS (entropy change) is defined as $\delta q/T$, then $\delta q = T\,dS$ and $dU = T\,dS - P\,dV + \delta w_{\mathrm{n}PV}$. In the absence of any external force fields and surface effects, Gibbs identified the non-PV work with changes in composition of each phase of the system. Gibbs therefore extended the internal energy function U for pure substances to include composition variables as $U(S, V, n_1, n_2, \ldots, n_c)$, where $n_1, n_2, \ldots, n_C$ represent the minimum number (C) of variables (termed as components) required to characterize the composition of a given phase j (of J phases), specified either as number of moles (n_i), mole fractions ($x_i = n_i/n_{\mathrm{total}}$), or mass fractions. Mathematically, the C variables represent the basis vectors of the compositional space of a phase, and $(n_1, n_2, \ldots, n_C)$ can be written as a compositional vector $\mathbf{n}_i$. Since dU is an exact differential, for the entire multiphase system, $dU = (\partial U/\partial S)_{V,ni}dS + (\partial U/\partial V)_{S,ni}dV + \sum_j \sum_i (\partial U/\partial n_i)_{S,V,\mathbf{n}_j} dn_i$, where the summation is over $j = 1, 2, \ldots, J$ phases and $i = 1, 2, \ldots, C$ components. Comparison to $dU = T\,dS - P\,dV + \delta w_{\mathrm{n}PV}$ implies $(\partial U/\partial S)_{V,ni} = T$, $(\partial U/\partial V)_{S,ni} = -P$, and $\sum_j \sum_i (\partial U/\partial n_i)_{S,V,\mathbf{n}_j} dn_i = \delta w_{\mathrm{n}PV}$.

The intensive partial molar internal energies $\partial U/\partial n_i$ are a measure of how much the internal energy of the system changes if one changed the ith species in any phase by a small amount (dn_i mole) while keeping the amounts of the other species (denoted as vector $\mathbf{n}_j$) together with S, V in the phase constant. For a given phase, Gibbs termed the intensive partial molar quantity ($\partial U/\partial n_i$) the chemical potential (function) of the ith component in the phase and denoted it as μ_i. For a given phase $\mu_i = f(S, V, \mathbf{n}_j)$ represents the slope of the $U(S, V, \mathbf{n}_j)$ space for that phase. Physically, the chemical potential represents the generalized force for defining energy (work) due to changes (displacements) in composition ($\mathbf{n}_j$) of any of the phases due to reaction or mass transfer. Since they represent partial molar quantities, μ_i values depend on the composition of the phase. Chemical potential gradients driving such non-PV work transfer between phases or between system and surroundings are analogous to temperature and pressure gradients that drive energy transfer as heat and PV work.

In order to transform the system coordinates from (S, V) to the more convenient and practical (P, T), one can apply a Legendre transformation on the basic $U(S, V)$ state function in order to eliminate S and V in favor of P and T. Gibbs applied the Legendre transformation to develop the state potential function $G = U + PV - TS$ defined on the five primitive thermodynamic state functions (internal energy U, pressure P, volume V, temperature T, and entropy S). The total differential of this function is $dG = d(U + PV - TS) = dU + P\,dV + V\,dP - T\,dS - S\,dT$. Substituting $dU = T\,dS - P\,dV + \delta w_{\mathrm{n}PV}$ gives $dG = -S\,dT + V\,dP + \delta w_{\mathrm{n}PV}$. Therefore, if some (reversible) process changes the system equilibrium state at constant temperature and pressure, $dG = \delta w_{\mathrm{n}PV}$ and $\int dG = \Delta G = w_{\mathrm{n}PV}$ represents the work exclusive of PV work that must be done on the system (or by the system if ΔG is negative) to achieve the change in equilibrium state. Gibbs termed the state function G as the free energy of the system. Since the state function H (enthalpy) was defined as $H = U + PV$, the free energy function is formally defined as $G = H - TS$, and is therefore alternatively called the free enthalpy. Since for any phase $dG\ (P, T, \mathbf{n}_i) = (\partial G/\partial P)_{T,\mathbf{n}_i} dP + (\partial G/\partial T)_{P,\mathbf{n}_i} dT + \sum_i (\partial G/\partial n_i)_{P,T\mathbf{n}_j} dn_i$, then by comparison with the above total differentials dG and dU, $(\partial G/\partial P)_{T,\mathbf{n}_i} = V$, $(\partial G/\partial T)_{P,\mathbf{n}_i} = -S$, and $\mu_i = (\partial U/\partial n_i)_{S,V,\mathbf{n}_j} = (\partial G/\partial n_i)_{P,T,\mathbf{n}_j}$ for a given phase.

As the four extensive thermodynamic energy functions (i.e., U, G, H, and A) present the same physical information in different coordinates, Legendre transformations made on $U(S,V,\mathbf{n}_i)$ can be used to develop the other two energy functions, enthalpy $H(S, P, \mathbf{n}_i)$ and Helmholtz free energy $A(T, V, \mathbf{n}_i)$. The partial molar values of these functions give equivalent definitions for $\mu_i = (\partial H/\partial n_i)_{S,P,\mathbf{n}_j} = (\partial A/\partial n_i)_{T,V,\mathbf{n}_j}$. In the case of a single pure substance $\mathbf{n}_j$ has only one element, and the value of μ for a given phase is simply the value of the state energy function divided by the moles of the pure substance comprising the phase.

Extensive thermodynamic equilibrium state functions (the four energy functions included) of the form $f(x_1, x_2, \ldots)$ are first-order homogeneous functions of degree one, meaning that $f(\lambda x_1, \lambda x_2, \ldots) = \lambda f(x_1, x_2, \ldots)$, where λ is an arbitrary constant scaling all *extensive* variables in the basis vector set $\mathbf{x}_i$. Applying Euler's theorem for first-order homogeneous functions in this case would give $f(x_1, x_2, \ldots) = \sum_i (\partial f/\partial x_i)\ x_i$. As an example, applying these concepts for the Gibbs energy function $G(T, P, \mathbf{n}_i)$ gives (since T and P are intensive variables), $G(T, P, \lambda\mathbf{n}_i) = \lambda G(T, P, \mathbf{n}_i)$ and by Euler's theorem $G(T, P, \mathbf{n}_i) = \sum_i n_i (\partial G/\partial n_i)_{P,T,\mathbf{n}_j} = \sum_i n_i \mu_i$. Therefore, $dG = \sum_i n_i d\mu_i + \sum_i \mu_i dn_i = V\,dP - S\,dT + \sum_i \mu_i dn_i$. Subtraction of these two expressions for dG gives $V\,dP - S\,dT + \sum_i n_i d\mu_i = 0$. This result is called the Gibbs–Duhem equation. Similar "Gibbs–Duhem" equations can be derived for any extensive state function of the form $f(x_1, x_2, \mathbf{n}_i)$ using Euler's theorem (although strictly this name refers to the Gibbs function). Such results are essential starting points in the analysis of solution/vapor systems.

For internal equilibrium (meaning nothing changes macroscopically) in any heterogeneous system, it is intuitive (albeit there are formal proofs) that the intensive state functions P, T, and μ_i (the chemical potential of each component) must be uniform throughout the system. Achieving uniformity of μ_i implies that mass transfers between phases occur such that the phase with the smaller chemical potential will grow at the expense of the phase with the larger chemical potential until equilibrium is achieved. At equilibrium since T, P, μ_i are fixed, the system cannot deliver non-PV work and therefore $dG = 0 = \sum_i \sum_j \mu_{ij} dn_{ij}$, where the summation is over $j = 1, 2, \ldots, J$ phases and $i = 1, 2, \ldots, C$ components.

Among thermodynamic intensive state functions, the chemical potential function $\mu(P, T, \mathbf{n}_i)$ serves as a powerful tool in analyzing processes (i.e., changes in equilibrium state) in heterogeneous systems of variable composition, especially those involving phase changes, chemical reactions, or electrochemical and surface phenomena. At equilibrium, $\mu(P, T, \mathbf{n}_i)$ defines a surface (sometimes called the Gibbs surface) representing all possible equilibrium states of the system. Each point on the Gibbs surface represents the minimum value of $G(P, T, \mathbf{n}_i)$ at a given T, P. From the relationships for $G(T, P, \mathbf{n}_i)$ in the preceding paragraphs, and Euler's condition for exactness, $(\partial \mu/\partial T)_{P,\mathbf{n}_i} = \partial/\partial n_i (\partial G/\partial T)_{P,\mathbf{n}_j} = -(\partial S/\partial n_i)_{P,\mathbf{n}_j} = -s$ (partial molar entropy) and similarly $(\partial \mu/\partial P)_{T,\mathbf{n}_i} = (\partial^2 G/\partial n_i\ \partial P)_{T,\mathbf{n}_j} = v$ (partial molar volume) for a given phase.

Gibbs also quantified the minimum number (D) of unconstrained intensive variables (such as T, P, μ_i) required to unambiguously define the equilibrium state of a closed heterogeneous system. Each such variable constituted one of D "internal degrees of freedom" of the system. As discussed above, components (C) are distinct chemical species that can be independently varied and whose total quantities are conserved; phases (J) are the coexisting mechanically separable, distinct, homogeneous subsystems.

Gibbs considered a system containing C components present in J phases designated $1, 2, 3, \ldots, J$. In each phase, in addition to P and T, there are $C - 1$ possible independent composition variables since the sum of the moles in that phase is fixed. The total *possible* number of degrees of freedom describing the system is therefore $2J + J(C - 1) = JC + J$. If the system is in a state of internal equilibrium, P, T, and μ_i are equal in all phases. These constitute restraints that reduce the total possible degrees of freedom. Equality of P and T in each phase provides $2(J - 1)$ restraints. Equality of μ_i in all phases imposes $J - 1$ restraints for each component (meaning that for the ith component one can write $J - 1$ equalities for that μ_i), and therefore $C(J - 1)$ restraints for C components. The difference between the potential degrees of freedom and the restraints gives the actual degrees of freedom of the system. This gives $D = JC + J - (2J - 2 + JC - C) = JC + J - 2J + 2 - JC + C = C - J + 2$. This result is the well-known Gibbs phase equilibrium rule and is universally applicable to all heterogeneous systems at equilibrium. D (sometimes called the variance) is the number of unconstrained variables, that is, intensive quantities that can be independently varied without changing the number of phases. It may be interesting to note that it is no coincidence that this result resembles Descartes's topological formula ($F = 2 + E - V$) relating the number of faces (F), edges (E), and vertices (V) of simple polyhedra. In this context, the phase rule characterizes the topology of the Gibbs surface for the phases at equilibrium.

The foregoing concepts and relationships are *sine qua non* to the understanding and quantifying of all heterogeneous phase equilibrium systems. As an example, let us consider the experimental phase equilibrium for a pure ice/liquid water/water vapor system. It can be represented as the locus of points in a vapor pressure (P in pascal) versus temperature (T in kelvin) coordinate plane representing states for which these phases are in equilibrium. Such a plot, termed a phase diagram, for water is shown in Fig. 1. There are three regions in the phase diagram corresponding to the three phases of water. The phase diagram shows three possible phase transitions liquid ↔ vapor (vaporization), solid (ice) ↔ liquid (fusion or melting), and solid (ice) ↔vapor (sublimation). Applying the phase rule to Fig. 1, all three phases can coexist at only one combination of P and T known as the triple point. At the triple point the system has no degree of freedom, implying that the triple point is unique for all pure substances. The two lines separating the vapor from the liquid and solid phases specify the pressure at which the vapor of the substance is in equilibrium with its liquid or solid phases at a given temperature. For example, along the sublimation line, ice and vapor can coexist in equilibrium. Along the phase equilibrium lines the system has one degree of freedom, and hence the vapor pressure is fixed if T is specified and vice versa. Since the equilibrium vapor pressure for a given system may be established at any temperature, and because vapor formation is endothermic, an increase in temperature will tend to shift the equilibrium state toward higher vapor pressure and vice versa. Therefore, vapor pressure is necessarily an increasing function of the temperature of the system. The temperature at which the vapor pressure exactly equals the atmospheric pressure ($1\ \text{atm} = 1.103 \times 10^5$ Pa) is called the normal boiling point of the substance (100 °C for water). Above the critical temperature (374 °C), it is not possible to produce condensation by increasing the pressure, and the system has two degrees of freedom (i.e., P and T). Similar phase diagrams exist for all pure substances except that the fusion line usually tilts to the right of the triple point. Water is one of few substances (another is bismuth) whose melting temperature is lowered as P increases so that its fusion line tilts to the left of the triple point.

For a single-component phase diagram (as in Fig. 1), application of the foregoing general equilibrium relationships are greatly simplified. A question of practical importance is: What is the relation $P(T)$ between vapor pressure (P) and T along the equilibrium sublimation and vaporization curves? Using subscript α to denote any of the condensed phases (solid, liquid) and the vapor phase as β, then at any point along the phase transition curves $\mu_\alpha = \mu_\beta$. Applying Taylor's theorem to expand $\mu_\alpha\ (T, P)$ and $\mu_\beta\ (T, P)$ in the neighborhood of the same arbitrary point P_0, T_0 [i.e., about $\mu_\alpha\ (T_0, P_0) = \mu_\beta\ (T_0, P_0)$] on the

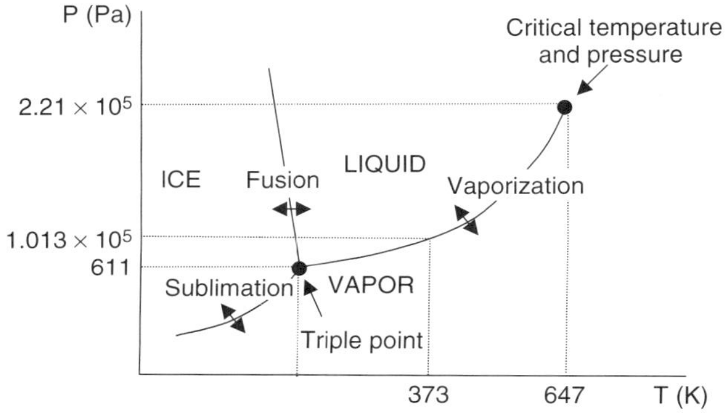

Figure 1.

phase transition curve by one gets $\mu_\alpha\ (T_0 + dT, P_0 + dP) = \mu_\alpha\ (T_0, P_0) + (\partial\mu_\alpha/\partial T)_P dT + (\partial\mu_\alpha/\partial P)_T dP \ldots$ equal to the similar expansion for $\mu_\beta\ (T, P)$ about $\mu_\beta\ (T_0, P_0)$. Equating the terms involving dP and dT in these expansions gives $dP/dT = -(\partial\mu_\alpha/\partial T - \partial\mu_\beta/\partial T)/(\partial\mu_\beta/\partial P - \partial\mu_\alpha/\partial P)$. As shown in a previous paragraph $(\partial\mu_\alpha/\partial T)_{P,n\beta} = -s_\alpha$ and $(\partial\mu_\alpha/\partial P)_{T,n\beta} = v_\alpha$, where s and v represent partial molar entropy and volume.

However, in the special case of only one component, partial molar values of extensive state functions are the same as their molar values, and therefore $-s_\alpha = -S_\alpha$ and $v_\alpha = V_\alpha$, where S_α and V_α denote molar entropy and volume. $-S_\beta$ and V_β are similarly defined, and therefore $dP/dT = (S_\beta - S_\alpha)/(V_\beta - V_\alpha)$. This important general equation for the slope of the phase equilibrium line in phase diagrams for single-component systems was first proposed by French physicist and engineer Emile Clapeyron in 1834. As, by definition, $(S_\beta - S_\alpha) = (H_\beta - H_\alpha)/T$, then $dP/dT = (H_\beta - H_\alpha)/[T(V_\beta - V_\alpha)]$. By convention, extensive property changes (such as ΔS, ΔH, ΔG) for phase transition are taken such that their values are positive. Along the vaporization curve in Fig. 1, $\Delta H_{\text{vap}} = H_{\text{vap}} - H_{\text{liquid}}$ and $\Delta V_{\text{vap}} = V_{\text{vap}} - V_{\text{liquid}}$. Similarly, along the sublimation curve, $\Delta H_{\text{sub}} = H_{\text{vap}} - H_{\text{solid}}$ and $\Delta V_{\text{sub}} = V_{\text{vap}} - V_{\text{solid}}$.

The foregoing reasoning assumes that the gaseous phase consists only of the vapor. If, however, other gases are present (e.g., water vapor in air), this would increase the external pressure on the condensed phase and tend to increase mass transfer to the vapor phase. For this system, $\mu_\alpha\ (P)|_T = \mu_\beta\ (P)|_T$ at equilibrium (for fixed T). Taylor's series expansion, described above, reduces to $dP_\alpha\ (\partial\mu_\alpha/\partial P)_T = dP_\beta\ (\partial\mu_\beta/\partial P)_T$ and therefore $dP_\alpha/dP_\beta = V_\beta/V_\alpha$, where V is the molar volume, P_β is the vapor pressure, and P_α is the total pressure exerted on the condensed phase. This assumes that the only effect of the other gases is increasing the total pressure on the condensed phase, meaning they behave as a closed subsystem with respect to the vapor phase. V_β (molar volume of the vapor phase) is usually $>>$ V_α (molar volume of the condensed phase) and therefore V_α can be assumed constant. Assuming ideal behavior of the vapor phase, $V_\beta = RT/P_\beta$. Substitution and rearranging gives $dP_\beta/P_\beta = (V_\alpha dP_\alpha)/(RT)$. Integration between $(P_\beta)_1$ at $(P_\alpha)_1$ and $(P_\beta)_2$ at $(P_\alpha)_2$ gives $\ln[(P_\beta)_2/(P_\beta)_1] = [V_\alpha/RT][(P_\alpha)_2 - (P_\alpha)_1]$. For water, using $V_\alpha = 18\ \text{cm}^{-3} \cdot \text{mol}^{-1} = 1.8 \times 10^{-7}\ \text{m}^3 \cdot \text{mol}^{-1}$ at 277.16 K (4 °C), $R = 8.314\ \text{J}{\cdot}\text{mol}^{-1}{\cdot}\text{K}^{-1}$, and 1 atm ($1.013 \times 10^5$ Pa) change in external pressure, we find the right-hand side of the solution = 0.000791. This implies the ratio of the vapor pressures $= e^{0.000791} = 1.00079$ and therefore an increase in vapor pressure of 0.079%. This shows that the effect of the presence of other "inert" gases that increase the external pressure on the condensed phase is generally small and can be ignored for most cases of practical interest.

The Clapeyron equation can be integrated exactly if functions $(\Delta H)_{\text{tr}}(T, P)$ and $(\Delta V)_{\text{tr}}(T, P)$ can be defined. Here, the subscript tr denotes either solid/vapor or liquid/vapor transition. Such functions are usually semianalytical at best. They are often applicable over short ranges of T and P and therefore values for pure substances are usually presented as tables (3). These tabulations usually list H, S, and V for the pure phases as values per unit mass obtained by dividing the molar values by the molecular weight in appropriate mass units.

Reasonable approximations can be made if one assumes (1) $V_{\text{vap}} >> V_{\text{liquid}}$ or V_{solid}, (2) ΔH_{tr} is constant over the range of temperature of interest, and (3) the vapor behaves as an ideal gas. In this case, $v_{\text{vap}} = RT/P$, where v_{vap} is the specific volume of the vapor ($\text{m}^3 \cdot \text{kg}^{-1}$) and R ($\text{J} \cdot \text{kg}^{-1} \cdot \text{K}^{-1}$) is the specific ideal gas constant ($R = 8.314\ \text{J} \cdot \text{mol}^{-1} \cdot \text{K}^{-1}$/molecular weight of the vapor in $\text{kg} \cdot \text{mol}^{-1}$). After some rearrangement, the Clapeyron equation becomes $dP/dT = d(\ln P)/d(1/T) = -(\Delta h)_{\text{tr}}/R$, where $(\Delta h)_{\text{tr}}$ is the specific enthalpy change ($\text{J} \cdot \text{kg}^{-1}$), that is, the latent heat for the phase transition. Integration of this approximation of the Clapeyron equation (called the Clausius–Clapeyron equation) between $P = P_1$ at $T = T_1$ and $P = P_2$ at $T = T_2$ gives $\ln(P_2/P_1) = (\Delta H_{\text{tr}}/R)\,(1/T_1 - 1/T_2)$. Indefinite integration gives $\ln P = [-(\Delta h)_{\text{tr}}/(RT)] + C$ (an integration constant). This implies that if observations of $\ln P$ are plotted versus $1/T$, the (negative) slope of the line multiplied by $(-R)$ gives an estimate of $(\Delta h)_{\text{tr}}$ over a short temperature range of interest. Discontinuities exist as the upper limit of the temperature range approaches the triple point or the critical temperature.

In practical applications over wider ranges of temperature, empirical equations are preferred (4). For water, the Goff–Gratch (5–7) and Magnus–Teten (8) equations for $P(T)$ are commonly used in the atmospheric sciences. The Goff–Gratch function for water is regarded as the standard (9) and is given by $P(T) = ke^{f(\theta)}$, where $\theta = 373.16/T$ and

$$f(\theta) = a\theta + b\ln(\theta) + ce^{d/\theta} + fe^{g\theta}$$

where P is in mb, T is in K, $k = 7.95357242 \times 10^{10}$, $a = -18.1973$, $b = 5.0281$, $c = -70242.1852$, $d = -26.1205$, $f = 58.0692$, and $g = -8.0395$.

The corresponding function for ice is $P(T) = ke^{f(\theta)}$, where $\theta = 273.16/T$ and

$$f(\theta) = a\theta + b\ln(\theta) + c/\theta$$

where P is in mb, T is in K, $k = 5.75185606 \times 10^{10}$, $a = -20.9470$, $b = -3.5665$, and $c = -2.0189$.

The Magnus–Teten function is $P(T) = ae^{f(T)}$ with $a = 6.1078$ (P at triple point) and $f(T) = c_1\ (T - 273.16)/(T - c_2)$, where P is in mb and T is in K. For water $c_1 = 17.2694$ and $c_2 = 35.86$. For ice $c_1 = 21.8746$ and $c_2 = 7.66$

Values of P at $10\,^\circ\text{C} = 283.16$ K for water are 12.272 mb by the Goff–Gratch equation and 12.279 mb by the Magnus–Teten equation. Values of P at $-10\,^\circ\text{C} = 263.16$ K for ice are 2.597 by the Goff–Gratch equation. The corresponding value by the Magnus–Teten formula is 2.595 mb.

BIBLIOGRAPHY

1. Gibbs, J.W. (1928). *The Collected Works of J. Willard Gibbs*. Longmans, Green and Co., New York.
2. Wheeler, L.P. (1951). *Josiah Willard Gibbs, the History of a Great Mind*. Yale University Press, New Haven, CT.
3. Lemmon, E.W., McLinden, M.O., and Friend, D.G. (2003). Thermophysical properties of fluid systems. In: *NIST Chemistry WebBook*. P.J. Linstrom and W.G. Mallard (Eds.). *NIST Standard Reference Database*, Number 69, National Institute of Standards and Technology, Gaithersburg, MD. Available: http://webbook.nist.gov.
4. Gibbins, C.J. (1990). A survey and comparison of relationships for the determination of the saturation vapour pressure over plane surfaces of pure water and of pure ice. *Ann. Geophys.* **8**: 859–886.
5. Goff, R.M. and Gratch, S. (1946). Low pressure properties of water from −160 to 212 °F. *Trans. Am. Soc. Heating Ventilation Eng.* **52**: 95–121
6. Goff, J.A. (1957). Saturation pressure of water on the new Kelvin temperature scale. *Trans. Am. Soc. Heating Ventilation Eng.* **63**: 347–354.
7. Fukuta, N. and Gramada, C.M. (2003). Vapor pressure measurement of supercooled water. *J. Atmos. Sci.* **60**: 1871–1875.
8. Murray, F.W. (1967). On the computation of saturation vapor pressure. *J. Appl. Meteorol.* **6**: 203–204.
9. World Meteorological Organization. (1988). *WMO Technical Regulations: General Meteorological Standards and Recommended Practices*, *WMO* No. 49. WMO, Geneva.

ADIABATIC COOLING

NARAINE PERSAUD
Virginia Polytechnic Institute
and State University
Blacksburg, Virginia

When any closed gaseous system expands against an external pressure with no compensatory input of heat from its surroundings to offset the pressure–volume work done by the system, the expansion results in adiabatic cooling of the gas. The vice versa process results in adiabatic heating. The term adiabatic (from the Greek word meaning "not passing through") implies an idealized process and a perfectly insulated system. Real, imperfectly insulated systems would permit some compensatory heat flux. However, if the effect of such fluxes on the system temperature is small relative to the cooling effect, the expansion can be considered as adiabatic for practical purposes.

Adiabatic cooling and heating is the principal mechanism underlying the continual turbulent vertical mixing (autoconvection) in the earth's troposphere, that is, the lowest 9 km (polar regions) to 18 km (tropics) thick atmospheric layer comprising about 80% of the atmospheric mass where all weather processes occur. It is especially important in the planetary boundary layer (i.e., the lowest 1–2 km thick layer) where surface effects on vertical motions are more manifest. Cloud formation and precipitation, observed temperature decrease with height, and energy and mass transfers in the troposphere all depend on adiabatic cooling. This thermophysical process is therefore sine qua non to the continuity of the earth's hydrological cycle. The focus here is on the physics of adiabatic cooling in the troposphere and the resulting temperature decreases with elevation.

A global average of close to 30% of incoming solar short-wave radiation is absorbed at the earth's surface layers. The resulting increase in the temperature of these layers and heat transfer to the air is the most important driver for turbulent vertical convection and associated adiabatic cooling in the troposphere. The earth's atmosphere is a quasiuniform thermophysical system, meaning that pressure gradients rather than the pressure of the atmosphere as a whole can be defined. To facilitate the analysis of tropospheric adiabatic cooling, the concept of an air parcel is introduced. A parcel (or less commonly an eddy) is a discrete closed volume of tropospheric air with uniform properties that is small enough so that the pressure at any point is the same. This implies that, although each parcel has its own individual structure and evolution in time, the observed state and dynamics of tropospheric air is the net result of the motion of these parcels. Because of the many forces involved in such motions, applying Newton's second law to describe the hydrodynamics of these parcels in the earth's atmosphere would result in a complex set of three-dimensional vector differential equations in spherical coordinates. However, only the relatively simpler vertical component of the vector equations is required for analysis of adiabatic cooling. This component is dominated by the vertical pressure and gravitational forces (1).

In order to understand the interplay of these forces, consider a prizm-shaped, initially motionless parcel in a *dry* atmosphere air above the earth's surface as shown in Fig. 1.

It is assumed that tropospheric air is continuously stratified meaning pressure (P) and density (ρ_d) are continuous and decreasing functions of vertical height (z) above the surface. The subscript d refers to dry air. The volume of the prizm (V) is ($\Delta x\ \Delta y\ \Delta z$). The net force on the prizm is $-V\rho_d g$ (downward weight) + $(\Delta x\ \Delta y)(dP/dz)\ \Delta x$ (upward net pressure buoyancy force). Since the prizm is assumed static, the net force per unit mass (acceleration) = 0 and therefore $(1/\rho_d)\ dP/dz + g = 0$. This equation (termed the hydrostatic or aerostatic equation) states the necessary condition for vertical hydrostatic equilibrium in the troposphere, namely, that gravity must be balanced by the vertical component of the atmospheric pressure gradient force.

Any disturbance of this hydrostatic equilibrium would result in vertical motion of the parcel. In particular, if the density of the parcel (ρ_p) is different from that of the surrounding air (ρ_0), the parcel would accelerate. The buoyancy acceleration is $g[(\rho_0/\rho_p) - 1]$ and is positive (upward) if $\rho < \rho_0$ and vice versa (assuming the vertical pressure gradient force is fixed). Substituting $g = 9.81\ \text{m·s}^{-2}$, a 10% decrease in the density of the parcel would result in an acceleration of close to $1\ \text{m·s}^{-2}$. Such buoyancy accelerations are the direct result of temperature changes caused by heating (from surface or otherwise) of air parcels. Assuming dry air behaves as an ideal gas $\rho_d = P/R_dT$ [where R_d is the specific gas constant for dry air = $8.314\ \text{J·mol}^{-1}\text{·K}^{-1}/0.289$ kg (mol dry air)$^{-1}$ and T is the absolute temperature (K)], the buoyancy acceleration can be written as $g[(T/T_0) - 1]$, where T and T_0 are the temperatures of the parcel and surrounding air at a given elevation.

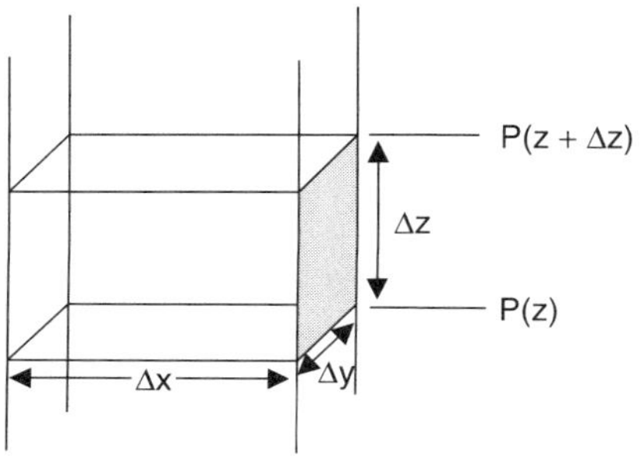

Figure 1. Schematic representation of a retarding basin.

For purposes of analysis of adiabatic cooling, an air parcel is considered as a perfectly insulated bubble of air submersed in the atmosphere, although in reality it could exchange heat with its surroundings. The thermal conductivity of dry air (k_{da} in $\text{W·m}^{-1}\text{·K}^{-1}$ or $\text{cal·s}^{-1}\text{·cm}^{-1}\text{·K}^{-1}$) at 1-atm pressure is linearly related to temperature over the range 263–473 K. It increases from 0.023823 $\text{W·m}^{-1}\text{·K}^{-1}$ (0.0000568 $\text{cal·s}^{-1}\text{·cm}^{-1}\text{·K}^{-1}$) at 273.16 K by 0.0000712 $\text{W·m}^{-1}\text{·K}^{-1}$ (0.0019 $\text{cal·s}^{-1}\text{·cm}^{-1}\text{·K}^{-1}$) for each kelvin unit (or °C) change in temperature. The conductivity of moist air is always less than or equal to k_{da}, the difference widening in nonlinear manner with increasing temperature and increasing humidity. Upward acceleration and accompanying expansion of the parcel occur relatively quickly, and because of the low thermal conductivity values of moist air, any heat transfer to the parcel is negligible. This assumption makes possible analysis of the thermal physics of an ascending air parcel as an adiabatic cooling process.

From the integral form of the first law of thermodynamics, a change in internal energy (ΔU) of a parcel equals absorption (gain) or release (loss) of heat (q) minus mechanical work done *by* the parcel (loss) or work done *on* the parcel (gain). Assuming expansion/compression takes place reversibly (i.e., ideally balanced quasistatic equilibrium is always maintained during the process), the mechanical pressure–volume (PV) work done by the parcel against the external pressure (P) is given as $P(\Delta V)$, where ΔV is the volume change (positive for expansion). Thus, in general, $q = \Delta U + P(\Delta V)$. The ideal gas equation of state for unit mass of dry air is $PV = R_dT$. Assuming reversible adiabatic expansion (i.e., process is isentropic), $q = 0$ and $\Delta U = -P(\Delta V) = -R(\Delta T)$ for reversible adiabatic cooling of an ideal gas.

For dry air (a homogeneous gaseous mixture) U in general is a function of absolute temperature (T) and P. However, for an ideal gas, U is a function of T only as demonstrated by James Prescott Joule (1818–1889). Also, U is a state function implying that ΔU is the same for a given temperature change (ΔT) of the parcel, regardless of how ΔT was achieved. Two possibilities are discussed below, namely, changing T at constant volume or at constant pressure.

The specific heat of any substance is the heat input required to cause unit temperature change in unit mass of the substance. In general, specific heat is a function of temperature. The specific heat for gases at constant volume and constant pressure are defined as $C_v(T) = (dq/dT)|_v$ and $C_p(T) = dq/dT|_p$. From the first law, for dry air $C_v < C_p$ since "extra" heat is required at constant pressure for expansion work done against the surrounding pressure. Over tropospheric temperature ranges of practical importance, the specific heats for dry air (denoted as C_{vd} and C_{pd}) can be considered as constants.

Several important results follow from applying the first law to the analysis of adiabatic reversible cooling of a dry air parcel in a quasiuniform atmosphere assuming ideal behavior and constant C_{vd} and C_{pd}. Applying $\Delta U = q - P(\Delta V)$ to the heating of unit mass dry air through ΔT, at constant P, implies that $C_{pd}(\Delta T) - \Delta U = P(\Delta V) = R_d(\Delta T)$. But ideal behavior implies that ΔU can be taken as $C_{vd}(\Delta T)$, and therefore $C_{pd}(\Delta T) - C_{vd}(\Delta T) = R(\Delta T)$, which simplifies to $C_{pd} - C_{vd} = R_d$.

Taking the total differential of the equation of state for unit mass of dry air assuming ideal behavior (i.e., $PV = R_dT$) gives $V\,dP + P\,dV = R_d dT$ or $P\,dV = R_d dT - V\,dP$. Substituting this, together with $R_d = C_{pd} - C_{vd}$, and $dU = C_{vd}dT$ into the differential form of the first law (i.e., $\delta q = dU - P\,dV$) gives $\delta q = C_{vd}dT - V\,dP + (C_{pd} - C_{vd})dT = C_{pd}dT - V\,dP$. The result $\delta q = C_{pd}dT - V\,dP$ is in effect a restatement of the differential form of the first law for dry air.

For an adiabatic ideal gas process $\delta q = 0$ and $C_{pd}dT = V\,dP = [(R_dT)/P]dP$. Separating variables gives $d(\ln T) = (R_d/C_{pd})d(\ln P)$. Integrating between $P = P_0$ at $T = T_0$ and $P = P$ at $T = T$, and rearranging gives $T/T_0 = (P/P_0)^{R_dC_{pd}}$. This result is one of the three adiabatic ideal gas process P–V–T relationships known as Poisson equations (Siméon Denis Poisson, 1781–1840). It describes the quasistatic reversible adiabatic expansion for an ideal gas with constant specific heats from an initial temperature T_0 and pressure P_0 to a new temperature T and pressure P. Usually the ratio $\gamma_d = C_{pd}/C_{vd}$ is defined. Dividing $C_{pd} - C_{vd} = R_d$ by C_{pd} and rearranging gives $C_{pd} = (R_d\gamma_d)/(\gamma_d - 1) = R_d/\kappa_d$ with $\kappa_d = (\gamma_d - 1)/\gamma_d$. For dry air $R_d = 8.314\ \mathrm{J{\cdot}mol^{-1}{\cdot}K^{-1}}/0.029\ \mathrm{kg\ (mol\ dry\ air)^{-1}} = 286.7\ \mathrm{J{\cdot}kg^{-1}{\cdot}K^{-1}}$, $C_{pd} = 1004\ \mathrm{J{\cdot}kg^{-1}{\cdot}K^{-1}}$, and $\kappa_d = 0.286$.

The result $T/T_0 = (P/P_0)^{\kappa_d}$ can be solved for $T_0 = T(P/P_0)^{\kappa_d} = \theta_d$. This definition of θ_d with $P_0 = 1$ bar (1000 mb) is termed the potential temperature. It is the temperature that the parcel initially at T, P would assume when changed adiabatically to 1000 mb. From its definition, $\ln\theta = \ln T + \kappa_d\ [\ln P_0 - \ln P]$ and $d(\ln\theta) = d(\ln T) - \kappa_d\ \mathrm{d}(\ln P)$. Dividing $\delta q = C_{pd}dT - [(RT)/P]dP$ by $C_{pd}T$ gives $\delta q/C_{pd}T = d(\ln T) - \kappa_d d(\ln P)$ and therefore $d(\ln\theta_d) = \delta q/C_{pd}T$. As $\delta q = 0$ for an adiabatic process, $d(\ln\theta_d) = 0$, implying the potential temperature is a parameter that remains invariant during the adiabatic ascent of a parcel. As $T = (\theta_d/P_0)P^{0.286}$, plots of T versus $P^{0.286}$ for different θ_d result in a family of straight lines.

Using $\delta q = C_{pd}dT - V\,dP$ with $\delta q = 0$ for adiabatic cooling, gives $dT/dP = V/C_{pd} = 1/(\rho_dC_{pd})$ because V per unit mass (specific volume) is the reciprocal of the density (ρ_d). As shown above, for hydrostatic equilibrium $dP = -\rho_d g\,dz$. Substituting this result for dP in the previous equation gives $-dT/(\rho_d g\,dz) = 1/(\rho_dC_{pd})$ or $dT/dz = -g/C_{pd} = -(g\kappa_d)/R_d$. Writing this result as $C_{pd}dT = -g\,dz$ shows that adiabatic cooling can be interpreted physically as the dry air parcel trading its internal energy ($C_{pd}dT$) to bootstrap its ascent in the troposphere to acquire gravitational potential energy ($-g\,dz$). This interpretation is appropriate since as already discussed the atmosphere is a quasiuniform thermophysical system.

The temperature of an adiabatically cooled parcel of dry (ideally behaved) air ascending in a tropospheric air layer that is in vertical hydrostatic equilibrium therefore decreases linearly with altitude with slope $= -(g\kappa_d)/R_d$. This important theoretical result is known as the dry adiabatic lapse rate (DALR). For $\gamma_d = 1.4$, $\kappa_d = 0.286$, and together with $R_d = 286.7\ \mathrm{J{\cdot}kg^{-1}{\cdot}K^{-1}}$, $g = 9.81\ \mathrm{m{\cdot}s^{-2}}$, gives $dT/dz = 0.0098\ \mathrm{K{\cdot}m^{-1}}$. Assuming a layer of dry air in the troposphere is rapidly and perfectly mixed by vertical convection of a large number of air parcels rising and falling adiabatically, the temperature would be expected to decrease linearly by approximately 1 K (same as 1 °C) per 100 m change in elevation. The layer is isentropic (constant entropy with elevation) and therefore the potential temperature (θ_d) is constant.

The DALR is a theoretical value achievable only in a turbulent well-mixed layer of dry tropospheric air. However, real tropospheric air is always moist because a varying amount of water vapor is always present. To facilitate including this factor in analysis of the behavior of rising moist parcels, the water vapor and dry air are treated as distinct components of a (one-phase) ideal gas mixture. In this case, three independent variables are needed to define an equation of state for moist air, namely, two (usually P,T) of the P–V–T variables together with a (humidity) variable that measures the mass ratio of water vapor in the air. Among several possible measures of this ratio the specific humidity (r_q) is the one traditionally used. It is defined as the mass of water vapor per unit mass of moist air.

Consider a volume (V) of moist air of mass (m_m) = mass water vapor (m_v) + mass of dry air (m_d). The subscripts m, v, d denote properties associated with the moist mixture, vapor, and dry air, respectively. The density $\rho_m = m_m/V = m_v/V + m_d/V = \rho_v + \rho_d$ (i.e., the sum of the partial densities), and therefore specific humidity $(r_q) = \rho_v/\rho_m$. The total pressure P of the volume V of moist air at temperature T (K) would be equal to $P_v + P_d$ (i.e., the sum of the partial pressures). P_v is usually denoted as e, implying $P_d = P - e$. Using ideal gas relations, $e = \rho_vR_vT$ and $P - e = \rho_dR_dT$, where $R_v = R/M_v$, $R_d = R/M_d$ denote the specific gas constants for water vapor and dry air, $R = 8.314\ \mathrm{J{\cdot}mol^{-1}{\cdot}K^{-1}}$ is the universal gas constant, and M is molecular weight ($M_v = 0.018\ \mathrm{kg{\cdot}mol^{-1}}$ and $M_d = 0.029\ \mathrm{kg{\cdot}mol^{-1}}$). This gives $\rho_v = e/(R_vT) = (eM_{wv})/(RT)$ and $\rho_d = (P - e)/(R_dT)$ Dividing numerator and denominator of the right-hand side for ρ_v by M_d gives $\rho_v = (0.622\ e)/(R_dT)$, where $0.622 = M_v/M_d$ (usually denoted as ε), and R_d is the specific gas constant for dry air ($= 286.7\ \mathrm{J{\cdot}kg^{-1}{\cdot}K^{-1}}$). Summing ρ_v and ρ_d, rearranging algebraically, and simplifying gives $\rho_m = [P/(R_dT)]\{1 - [e\ (1 - \varepsilon)/P]\}$. This shows that density of moist air is slightly less than the density of dry air [$= P/(R_dT)$] at a given pressure and temperature, a fact first noted by Sir Isaac Newton in 1717 in his book entitled *Optiks*. The value of the expression in curly brackets increases with decreasing P for a given e. Taking a maximum possible value of $e = 4500$ Pa at mean sea level with $P = 1$ atm $= 1.013 \times 10^5$ Pa, and $\varepsilon = 0.622$ gives $1 - [e\ (1-\varepsilon)/P] = 0.96$ and implies $\rho_m \approx \rho_d$. Using these results gives, after substituting and simplifying, $r_q = \rho_v/\rho_m = (\varepsilon e)/\{P - [(1-\varepsilon)e]\} \approx \varepsilon e/P$. The mixing ratio r_w defined as mass vapor per unit mass dry air in the mixture $= \rho_v/\rho_d = (\varepsilon e)/(P - e) \approx r_q$.

Eliminating e between ρ_v and ρ_d, rearranging algebraically, and simplifying gives the equation of state for moist air as $P(T, r_q) = \rho_m R_d T(1 + 0.61\ r_q)$. This is equivalent to the ideal gas equation for dry air with $T(r_q) = T(1 + 0.61\ r_q)$ or with $R_m = R_d(1 + 0.61\ r_q)$. In the former case, $T(r_q)$ is termed the virtual temperature (denoted as T_v) and the moist air equation of state is written as $P = \rho_m R_d T_v$. T_v is the temperature dry air would assume in order to have the same density as moist air with specified values of P, T, and r_q.

The specific heats C_{pm}, C_{vm} for moist air would also be functions of r_q. Partitioning the heat for a temperature change (ΔT) in mass (m_m) of moist air at a given r_q expanding at constant pressure gives $m_m C_{pm}(\Delta T) = m_v C_{pv}(\Delta T) + m_d C_{pd}(\Delta T)$. Dividing by $m_m(\Delta T)$ gives $C_{pm} = r_q C_{pv} + (1 - r_q)C_{pd}$ since r_q is by definition m_v/m_m and $m_d = m_m - m_v$. Similar reasoning gives $C_{vm} = r_q C_{vv} + (1 - r_q)C_{vd}$. Values given in the Smithsonian Meteorological Tables (2) are (in J·kg^{-1}·K^{-1}): $C_{pv} = 1846$, $C_{pd} = 1005$, $C_{vv} = 1386$, and $C_{vd} = 716$. Substituting these values and simplifying gives $C_{pm} = C_{pd}(1 + 0.84\ r_q)$ and $C_{vm} = C_{vd}(1 + 0.94\ r_q)$. From these $\gamma_m = C_{pm}/C_{vm} = \gamma[(1 + 0.84\ r_q)/(1 + 0.94\ r_q)]$, where $\gamma = C_{pd}/C_{vd} = 1.4$ is the ratio of the specific heats for dry air as described previously. Multiplying numerator and denominator of the term in square brackets by $(1 - 0.94\ r_q)$ and neglecting r_q raised to powers >1 gives $\gamma_m \approx \gamma(1 - 0.1\ r_q)$. Similarly, $\kappa_m = R_m/C_{pm} = [R_d(1 + 0.61 r_q)]/[C_{pd}(1 + 0.84\ r_q)]$. Multiplying numerator and denominator of the term in square brackets by $(1 - 0.84\ r_q)$ and neglecting r_q raised to power >1 gives $\kappa_m \approx \kappa_d(1 - 0.23\ r_q)$, where $\kappa_d = R_d/C_{pd}$.

γ_m is lower for moist air than dry air but does not affect the DALR appreciably as long as the air does not become saturated. For example, using the foregoing results gives the unsaturated moist adiabatic lapse rate (MALR) as $dT/dz = -(g\kappa_m)/R_m$, where $\kappa_m = (\gamma_m - 1)/\gamma_m$ and $R_m = R_d(1 + 0.61\ r_q)$. At 20 °C and 50% saturation the vapor pressure ≈ 1170 Pa. Using the equations for ρ_v and ρ_m with $R_d = 286.7$ J·kg^{-1}·K^{-1} and $P = 1.013 \times 10^5$, Pa gives $r_q = \rho_v/\rho_m = 0.0072$ kg vapor (kg moist air)$^{-1}$, $\gamma_m = 1.4 \times 0.999 = 1.399$, $\kappa_m = 0.285$, $R_m = 288.0$ J·kg^{-1}·K^{-1}, and MALR $= -0.0097$ K·m^{-1} compared with -0.0098 K·m^{-1} for the DALR. As was done for dry air, the potential temperature invariant for moist unsaturated air (θ_m) can be defined as $\theta_m = T[P/P_0]^{\kappa_m}$, where as discussed above, $\kappa_m \approx \kappa_d(1 - 0.23\ r_q)$ with $\kappa_d = R_d/C_{pd}$.

If a rising moist air parcel becomes cooled to the dew point, condensation occurs (if condensation nuclei are present). The elevation at which this occurs is called the lifting condensation level (LCL). The dew point temperature at the LCL (T_{dL}) is different than at the surface because, although $r_q(\approx \varepsilon e/P)$ remains the same, the vapor pressure decreases as the ascending parcel expands. The dew point (T_{d0}) at the surface depends on the observed ambient temperature (T) and relative humidity $= (e/e_s)|_T$ of the air. The relationship can be obtained using the Clausius–Clapeyron equation $de_s/e_s = (L\ dT)/(R_v T^2)$ for the slope of the equilibrium saturation vapor pressure (e_s) versus temperature curve, where L is the latent heat of vaporization and $R_v(= 461.8\ \mathrm{J \cdot kg^{-1} \cdot K^{-1}})$ is the specific gas constant for water vapor (3). L (J·kg^{-1}) is usually a weak function of temperature approximated as $L = 3.071 \times 10^6 - 2134\ T$ K. L is usually taken as a constant value $= 2.5 \times 10^6$ J·kg^{-1} corresponding to $T = 273$ K. Integrating the Clausius–Clapeyron equation between $e_s = e$ at T to $e_s = e_s$ at T_d gives (after substituting for L and R_v), $T - T_{d0} = -(1.845 \times 10^{-4} T)T_{d0} \ln[(e/e_s)_T]$. When required, reasonable estimates of $e_s(T)$ can be obtained by integrating the Clausius–Clapeyron equation on the half-open interval using the triple point of water ($e_s = 611$ Pa at $T = 273.13$ K) as the lower bound. After substituting numerical values for L and R_v, rearranging and simplifying the result is $\ln e_s$(in Pa) $= \ln 611 + [19.83 - (5417/T)]$.

The foregoing implies T_{dL} depends on the surface T_{d0} and on the dew point lapse rate (DPLR). Logarithmic differentiation of $r_{qs} = \varepsilon e_s/P$ gives $dr_{qs}/r_{qs} = de_s/e_s - dP/P = 0$ and therefore $de_s/e_s = dP/P$. Substituting the Clausius–Clapeyron equation for de_s/e_s and the aerostatic equation dP/P gives $L/[R_v(T_d)^2]\ dT_d = (-g/R_d T)dz$, where T_d is the dew point of the ascending parcel at elevation z and T is the corresponding air temperature, and $R_d \approx R_m$. Substituting numerical values $L = 2.5 \times 10^6$ K·kg^{-1}, $R_v = 461.8$ J·kg^{-1}·K^{-1}, $R_d = 286.7$ J·kg^{-1}·K^{-1}, and $g = 9.81$ m·s^{-1} gives dT_d/dz (in K·m^{-1}) $= -(T_d)^2/(0.158 \times 10^6)T$. The LCL occurs when T by cooling at the DALR $= T_d$ by cooling at the DPLR. Therefore, the LCL is obtained by solving $T_0 - 0.0098\,\mathrm{LCL} = T_{d0} - [(T_d)^2/(0.158 \times 10^6 T)]$ LCL, where T_0 is the surface air temperature, which gives $\mathrm{LCL} = (T_0 - T_{d0})/\{0.0098 - [(T_d)^2/(0.158 \times 10^6 T)]\}$. T_d and T are functions of elevation but in practice are considered equal to T_{d0} and T_0, respectively, in estimating the LCL. For example, given $T_0 = 20$ °C (293 K) and $T_{d0} = 1$ °C (274 K), the $\mathrm{LCL} = 19/\{0.0098 - [(274)^2/(0.158 \times 10^6 \times 293)]\} = 2.323$ km. Alternatively, it has been shown that $[(T_d)^2/(0.158 \times 10^6 T)]$ ranges from about 1.7 to 1.9 K·km^{-1} (4). Using an average value of 1.8 K·km^{-1}, the LCL can be estimated as LCL (km) $= (T_0 - T_{d0})/8$. Using this alternate formula in the above example gives $\mathrm{LCL} = 19/8 = 2.38$ km.

For ascent above the LCL, latent heat of condensation is released into the parcel, but the parcel remains saturated at the given temperature. A change $d(r_{qs})$ in r_{qs} (where the subscript s denotes saturation) would result in latent heat released into the parcel $= -L\,d(r_{qs})$, where L denotes the latent heat of vaporization. As was done previously, substitution of the hydrostatic equation $dP = -\rho_s g\ dz$ into the first law as $\delta q = C_{ps}dT - V\ dP$ gives (for unit mass) $\delta q = C_{ps}dT + g\ dz$. Since no water is lost from the parcel $\rho_s = \rho_m = P/R_m T$. Substituting $-Ld(r_{qs})$ for δq gives

$C_{ps}dT + g\,dz = -L\,d(r_q)$. Assuming almost all of the heat released by condensation goes to heating the dry air component in the parcel would imply $C_{ps} \approx C_{pd}$. Making the substitution, dividing throughout by $C_{pd}dz$, and rearranging gives $dT/dz = -(L/C_{pd})(dr_{qs}/dz) - g/C_{pd}$. This is the saturated adiabatic lapse rate (SALR). Since r_{qs} is a function of T and P, dividing its total differential (dr_{qs}) by dz gives $dr_{qs}/dz = [(\partial r_{qs}/\partial P)]_T(dP/dz) + [(\partial r_{qs}/\partial T)]_P(dT/dz)$. Substituting this result with $dP/dz = -\rho_m g$ into the SALR expression, rearranging, and simplifying gives

$$\text{SALR} = \frac{-g}{C_{pd}}\left(\frac{[1 - \rho_m L(\partial r_{qs}/\partial P)_T]}{[1 + (L/C_{pd})(\partial r_{qs}/\partial T)_P]}\right)$$

Differentiating $r_{qs} \approx \varepsilon e_s/P$, where $e_s(T)$ is the saturation vapor pressure gives $[(\partial r_{qs}/\partial P)]_T = -(\varepsilon e_s)/P^2$ and $[(\partial r_{qs}/\partial T)]_P = (\varepsilon/P)(de_s/dT)$. Assuming constant L and ideal gas behavior, the Clausius–Clapeyron equation (3) for the slope of the saturation vapor pressure versus temperature curve is $de_s/dT = Le_s/(R_v T^2)$, where R_v is the specific gas constant for water vapor. Substitution of these results together with $\rho_m \approx \rho_d = P/R_d T$ into the equation for the SALR and simplifying gives a more practical form as (4):

$$\text{SALR} = \frac{-g}{C_{pd}}\left(\frac{\left(1 + \dfrac{Lr_{qs}}{R_d T}\right)}{\left(1 + \dfrac{L^2 r_{qs}}{C_{pd} R_v T^2}\right)}\right)$$

The value of $(Lr_{qs}/R_d T)$ is usually small. In many applications this term is neglected, resulting in the simpler expression usually encountered in atmospheric science literature. The SALR represents the saturated lapse rate above the LCL. It is assumed that the system remains closed, that is, r_q remains the same (all the water and heat from condensation remains within the ascending parcel), and the ascent continues as a reversible saturated-adiabatic process. However, the system would no longer remain closed if (as may be expected) some or all of the water and heat from condensation at a given elevation is immediately released from the parcel. This would imply an irreversible and nonadiabatic process. However, the heat lost is small relative to the heat content of the parcel and the ascent in this case is termed as pseudoadiabatic (meaning "as if it were adiabatic"). The SALR given above is considered equal to the pseudoadiabatic rate for all practical purposes.

However, the SALR is considerably lower than the DALR $\approx$ MALR and is a function of temperature. The SALR is about 4 K·km^{-1} for humid air masses near the surface. It increases to about 9 K·km^{-1} as the humidity decreases with increasing elevation. An average value is about 5.4 K·km^{-1}. Like its counterparts θ_d and θ_m for dry and unsaturated air, the potential temperature invariant for saturated adiabatic ascent (θ_e termed the equivalent potential temperature) is defined as $d(\ln\theta_e) = \delta q/(C_{pd}T) = (-L\,dr_{qs})/(C_{pd}T) = -d[(Lr_{qs})/(C_{pd}T)]$. Integrating (on the half-open interval with lower bound $\theta_e = \theta_d$ at $r_{qs} = 0$) and exponentiating gives $\theta_e = \theta_d e^{[Lr_{qs}/(C_{pd}T)]}$. Physically, θ_e is the temperature attained if moist air at temperature T is expanded pseudoadiabatically, until all the water has condensed out and is removed from the parcel, and then is compressed adiabatically to the standard pressure of 1000 mb.

The theoretical lapse rates developed in the foregoing discussion serve as reference values for inferring tropospheric conditions from the observed environmental lapse rate (ELR) at a given location. Departures of the ELR from these reference lapse rates provide a measure of the stability (i.e., the autoconvective propensity or the tendency for vertical motions induced by density stratification) in a given layer of tropospheric air.

Below the LCL, when the ELR = DALR $\approx$ MALR (lapse rate is adiabatic) a rising parcel will always be at the same temperature as its surroundings. Therefore, it would have the same density and would neither tend to return nor continue its displacement. The stability of a tropospheric layer with these conditions prevailing is termed neutral. If the ELR $<$ DALR (lapse rate is subadiabatic) an ascending parcel will quickly become colder (more dense) than the surrounding air and will return to its original position. The layer is termed absolutely stable. Stable air resists vertical movement. If the reverse is true and the ELR $>$ DALR (lapse rate is superadiabatic) an ascending parcel will become warmer (less dense) than the surrounding air and will continue to rise. Except in the heated layer next to the surface, such instabilities do not persist since they are rapidly counteracted by vigorous autoconvection and vertical mixing as fast as they develop. In terms of θ_d, these stability conditions correspond to stable ($\theta_d < 0$), neutral ($\theta_d = 0$), and unstable ($\theta_d > 0$). In a similar manner, saturated air parcels will be stable, neutral, or unstable with respect to vertical displacements depending on whether the ELR is $<, =,$ or $>$ the SALR or $\theta_e <, =,$ or > 0.

The tropospheric lapse rate for the standard atmosphere (5) is $-6.5\,^\circ$C·km^{-1}. Parameters for the standard atmosphere have been specified by the International Civil Aeronautical Organization (ICAO) and are needed for standardization in aviation, aeronautics, and meterology. Using subscript *sa* to denote the standard atmosphere, solving $-0.0065\ \text{K}\cdot\text{m}^{-1} = -[g(\gamma_{sa} - 1)/(R_d\gamma_{sa})]$ for γ_{sa}, with $g = 9.81$ m·s^{-2} and $R_d = 286.7$ J·kg^{-1}·K^{-1}, gives $\gamma_{sa} = 1.23$ compared to the value of $\gamma_d = 1.4$ for dry air. Other parameters for the standard atmosphere are surface temperature = 288.15 K, pressure = 101, 325 Pa, density = 1.225 kg·m^{-3}, and mean molecular weight = 28.964 g·mol^{-1}.

The standard lapse rate implies $T(z) = T_0 - \beta z$, where $T(z)$ is the temperature (K) at elevation z (m), T_0 is the surface temperature, and $\beta = 0.0065$ K·m^{-1}. Assuming ideal gas behavior, substituting $T = T_0 - \beta z$ into the aerostatic equation $dP/dz = (-Pg/RT)dz$, separating variables, and integrating on a half-open interval with lower bound $P = P_0$ at $T = T_0$ gives, after some rearranging, $\beta z/T_0 = 1 - (P/P_0)^{R\beta/g}$. These relationships can be used to calculate P, T of tropospheric air for the standard atmosphere at any given elevation. As an example, for $z = 10$ km and $T_0 = 288.15$ K, $T = 288.15\ \text{K} - (0.0065\ \text{K}\cdot\text{m}^{-1} \times 10^4\text{m}) = 223.2$ K. Substituting $T_0 = 288.15$ K, $P_0 = 1.01325 \times 10^5$ Pa, $R = 286.7$ J·kg^{-1}·K^{-1}, and $\beta = 0.0065$ K·m^{-1} gives P at 10 km and $P = 0.26$ atm with $T = 223$ K.

BIBLIOGRAPHY

1. Brown, R.A. (1990). *Fluid Mechanics of the Atmosphere*. Academic Press, San Diego, CA.
2. List, R.J. (1971). *Smithsonian Meterological Tables*, 6th Edn. Smithsonian Institution Press, Washington, DC.
3. Salby, M.L. (1996). *Fundamentals of Atmospheric Physics*. Academic Press, San Diego, CA.
4. Tsonis, A.A. (2002). *An Introduction to Atmospheric Thermodynamics*. Cambridge University Press, Cambridge, UK.
5. NOAA. (1976). *The U.S. Standard Atmosphere*. U.S. Government Printing Office, Washington, DC (available from National Technical Information Office, Springfield, VA, product number ADA-035-6000).

RAIN AND ROCKS: THE RECIPE FOR RIVER WATER CHEMISTRY

Mohammed Riajul Islam
University of Idaho
Moscow, Idaho

INTRODUCTION

The dissolved and suspended loads of most of the world's major rivers are dominated by weathering processes. Particularly, chemical weathering of rocks and soils is one of the essential processes in the geochemical cycling of elements in rivers (1,2) as well as in the continent–river–ocean system (3). The climatic factors such as moisture and temperature mainly affect the weathering processes (2,4). The most soluble ions that will end up in the ocean through rivers are released because of interactions between water and rocks at the earth's surface (5–7). Weathering's importance has led to a huge number of detailed geochemical studies on the world's major river systems, including the Ganges-Brahmaputra (3,8), other major Himalayan rivers (9,10), the Amazon (11), the Mackenzie (12), rivers of the Canadian Shield (5,12), the Congo (13–15), some large Chinese rivers (16,17), the Mississippi, Indus, and some other U.S. rivers (18,19), the Loire River in France (20), and the upper Niger river (21).

Aerosols are the primary source of chemical elements transported by rainwater (22), which are removed from the atmosphere by rainout or by washout. Atmospheric aerosols (sea salt, crustal dust, and biogenic aerosols) have been found as the major sources of dissolved species in river water (14,23,24). Dissolved components from rainwater come from those originating from sea salt aerosols; from those originating from terrestrial aerosols (soil dust, biological emissions); and from those originating from anthropogenic sources (industry, agriculture, burning of fossil fuels, etc.) (25).

Along with the geological characterization of the watershed, meteorological parameters are critically important in evaluating rainfall effects, as demonstrated by Grosbois et al. (20). These investigators have demonstrated, in accordance with Negrel and Roy (25), that the Loire River basin was characterized by its diversity of air-mass trajectories divided into four major origins: (1) westerly, which originates from the Atlantic (26); (2) northeasterly to northwesterly, which originates from the North Atlantic and North Sea and passes over Great Britain and the industrialized countries from France to Eastern Europe; (3) easterly, with a continental origin and a geochemical signature influenced by polluted countries and a large forested area; and (4) southeasterly to southwesterly, which originates in the Mediterranean Sea and is influenced by natural Saharan aerosols and pollution from Spain.

The radiogenic isotope ^{87}Sr is produced because of the radioactive decay of ^{87}Rb isotope. The existence of ^{87}Sr—which is often expressed as the $^{87}Sr/^{86}Sr$ ratio in a given geological material—is a function of its age and Rb/Sr ratio. As a result, Sr released by aerosols having different origins results in different $^{87}Sr/^{86}Sr$ ratios in rainwater. The Sr variations are caused mainly because of the mixture of different sources with different Sr isotope compositions (25). Sr isotope ratios in precipitation have been reported by a wide number of investigators from various parts of the world (27–30).

METHODOLOGY OF STUDYING ROCKS–RECHARGE INTERACTIONS AND RAINWATER CHEMISTRY

Geology of the Study Areas

Consideration of geology is crucial in assessing the link between river water chemistry and local bedrock geochemistry, as demonstrated by English et al. (31) in their Seti river studies in the Himalayan area, which receives high rainfall (32). Over 70% of the TDS in the Seti River is derived from carbonate weathering in the Tethyan sedimentary series/Greater Himalayan series and upper Nawakhot Group of the Lower Himalayan as bedrock and sediment control riverine TDS and Sr. The investigators also found certain rock types in the river bedload that have been derived from related bedrock because of weathering. Furthermore, tributary waters display unique characteristics directly related to the local geology (i.e., bedrocks, climate, hydrology, etc.). The river watersheds, tributaries, local rainfall, humidity, and temperature data must be considered for the study of rocks–recharge interactions as well as river water chemistry. Table 1 shows the effects of various local bedrocks on river water chemistry in different regions.

Sampling

Systematic sampling of precipitation (25), river and tributary waters, bedrock, and river bed sediments would identify geochemical controls on stream chemistry (3,9,10,20,31).

The dissolved elements in stream waters provide information on the source of the atmospheric contribution (36). The marine contribution decreases with increasing distance from the ocean, whereas terrestrial contributions become dominant. However, it is very difficult to identify different kinds of sources (e.g., terrestrial aerosols, anthropogenic aerosols), unless isotopes (e.g., $^{87}Sr/^{86}Sr$)

Table 1. Chemical Compositions of Surface Water in Various Rivers, μmol/L

Rivers (Ref.)	Location	pH	TDS, mg/L	Ca^{2+}	Mg^{2+}	Na^+	K^+	Cl^-	SO_4^{2-}	NO_3^-
Yaou catchment	Precipitation—dry season	6.1	3.8	5	4.2	33	5.1	23.1	6.4	6.4
(French Guiana) (33)	Precipitation—wet season	6.1	1.7	3	2.9	9.1	3.1	8.7	3.7	0.6
Ganges, Bangladesh (34)	Rajshahi—August	7.6	125	466	201	175	69	94	76	0.3
	Rajshahi—March	8.5	374	1418	612	619	110	234	64	2.7
Brahmaputra, Bangladesh (34)	Chilmari—August	7.6	105	393	168	104	52	25	55	bdl[a]
	Chilmari—March	8.5	145	590	214	196	75	53	149	1.6
Trisuli, India (34)	Betrawati—April	—	70	324	49	67	39	19	46	bdl
	Betrawati—August	—	107	418	136	200	43	49	133	1.9
Nyong basin river, Cameroon (35)	Ayos	5.5	—	1314	975	758	1809	829	966	bdl
	Mefou	6.5	—	3856	1097	3677	4833	3295	2994	1195
Seti River, Nepal (31)	Seti above Liyangwan	—	87	531	222	82	44	40	127	—
	Seti below Taru Gad	—	91	501	300	80	38	29	122	—

[a] bdl, below detection limit.

are obtained from atmospheric contributions. The major and trace element data of sediments does provide clues to source area (37).

For rainwater collection, an automatic precipitation sampler is used (25). The basic requirements for the collector are automatic detection and collection of rainfall, dry fallout elimination, collection of frozen precipitation, and avoidance of sample contamination, particularly for trace element analysis.

Chemical Analysis: Rainwater

Rainwater is analyzed for conductivity, temperature, and pH—all determined on site. Electrical conductivity and water temperature are determined with a microprocessor conductivity meter. pH is measured with any standard pH meter. For major and trace elemental analyses, AAS, ICP-MS, and IC (ion chromatography) are used. Bicarbonate is analyzed by HCl titration and Gran's method (25).

Chemical separation and mass spectrometry procedures for Sr followed the standard method used at BRGM. Strontium is separated by using a cation exchange column (i.e., DOWEX AG50X8) with 2 *N* HCl as eluant. After chemical separation, one-fifth of each sample is analyzed using a Finnigan MAT 262 multiple collector mass spectrometer. The $^{87}Sr/^{86}Sr$ ratios are normalized to $^{86}Sr/^{88}Sr = 0.1194$.

Bedrock and river sediments can be analyzed for major and trace elements using XRF, ICP-MS, and AAS.

RESULTS AND INTERPRETATIONS

Meteorological data are taken generally from local meteorological offices. Average air mass trajectories and weather patterns enable one to identify air mass origins. For example, Negrel and Roy (25) demonstrated the limits of the four air-mass trajectories and four directions of the air-mass movements in the Massif Central area in France. Rainwater chemistry and Sr isotope values are determined for the evaluation of rainwater. The mean pH value of rainwater is 5.22 ± 0.65 (ranges 4.29–6.2) as determined in Massif Central, France (25), whereas the pH of pristine rainwater is moderately acidic with a theoretical value of 5.7 (4). Any higher values reported in France could be the result of dissolution of windblown dust with high $CaCO_3$ content. Rainwater often has a pH value below 5.7 because of the presence of H_2SO_4, weak organic acids, anthropogenic emission of H_2SO_4, and/or HNO_3. The concentrations of elements in rainwater vary greatly with the variation of sampling period (25). In France, the ion imbalance between cations and anions in rainwater is well studied (26,38,39). Significant temporal variation in $^{87}Sr/^{86}Sr$ ratios is also present (709198 to 0.713141) and is particularly associated with changes in seasons. Table 2 indicates the influence of meteorological and anthropogenic effects on rainwater chemistry.

Quantification of Marine and Terrestrial Contributions Using Concentration Data

A sodium (Na) reference (hereafter referred to as Y_{ref}) is used to determine sea salt (ss) and terrestrial contributions in rainwaters (rw) (46–48). To calculate the contribution of the sea salt component (ss) with seawater (sw) characteristics, the following equation is used:

$$X_{ss} = Y_{ref} x (X/Y)_{sw} \tag{1}$$

The contribution of the terrestrial component (t) is the difference between the total composition of rainwater (rw) and the sea salt (ss) contribution:

$$X_t = X_{rw} - X_{ss} \tag{2}$$

In calculating the contribution for a given element *X*, the main obstacle to resolve is the selection of the reference species (marine or terrestrial). An element to be used as a marine or terrestrial reference must be free from fractionation during aerosol formation, transportation, or deposition by rain impaction. In addition, the element must be absolutely of marine or terrestrial origin.

Table 2. Chemical Compositions of Representative Rainwater in Various Regions, μequiv/L

Regions (Ref.)	pH	Cl^-	SO_4^{2-}	NO_3^-	NH_4^+	Ca^{2+}	Mg^{2+}	Na^+	K^+
Buyukcekmece, Istanbul (mean) (40)	5.36	215.11	231.89	78.41	16.41	473.6	259.84	113.13	73.6
Patras, Greece (41)	5.16	114.3	46.1	19.4	16.3	98.5	30.4	90.2	6.6
Montseny, Spain (42)	6.4	28.4	46.1	20.1	22.9	57.5	4	9.8	22.3
Sardinia, Italy (43)	5.18	322	90	29	25	70	77	252	17
Silent Valley, India (mean) (44)	—	43	20	21	3	43	14	46	4
Morvan, France (45)	5.25	14.7	29.4	18.8	57	12.4	6.4	13.3	—

Generally, Na is used as a reference element for sea salt and Mg for coastal or marine rains (4). However, both Na and Mg contents in rainwater can be influenced by nonmarine inputs (48–50). Chlorine could be used as a reference if the Cl/Na ratios of rainwater do not differ from those of seawater (1.16 mol/mol) (4).

Calcium comes mainly from a natural terrestrial source, and a marine source may not be significant for K and Sr. Nitrate (NO_3) and SO_4 come mainly as a result of human activities (e.g., fuel combustion) (4). However, major element chemistry cannot characterize perfectly the various terrestrial end members. For this reason, Sr isotope compositions are used to get additional information regarding the different possible terrestrial sources and their respective proportions.

To determine the possible terrestrial inputs, Losno et al. (38) used the Al content in rainwater to determine Na from terrestrial input (t), according to the following equation:

$$Na_t = Al_{rw}(Na/Al)_t \quad (3)$$

Strontium Isotope Systematics

From the mixture of sea salt and terrestrial dust, the terrestrial isotope ratio $(^{87}Sr/^{86}Sr)_t$ in each rainwater sample can be calculated using Sr isotope systematics (30). The binary mixing equation for these two end members is demonstrated by

$$(^{87}Sr/^{86}Sr)_{rw} = \alpha(^{87}Sr/^{86}Sr)_{sw} + (1-\alpha)(^{87}Sr/^{86}Sr)_t \quad (4)$$

where $(^{87}Sr/^{86}Sr)_{rw}$ is the measured isotope ratio in rainwater, $(^{87}Sr/^{86}Sr)_{sw}$ is the isotope ratio of seawater (0.70917 ± 0.00001), $(^{87}Sr/^{86}Sr)_t$ is the crustal isotope ratio, and α is $Sr_{marine}/Sr_{terrestrial} + Sr_{marine}$ (determined in Eqs. 1–3).

Hence, we obtain

$$(^{87}Sr/^{86}Sr)_t = \{(^{87}Sr/^{86}Sr)_{rw} - \alpha(^{87}Sr/^{86}Sr)_{sw}\}/(1-\alpha) \quad (5)$$

The terrestrial ratios in the rain samples significantly vary, which is indicative of terrestrial aerosol heterogeneities because isotopic variations in the crustal source demonstrate the different terrestrial components with different isotope ratios in the mixture. The isotopic composition of Sr in dusts can be calculated by studying $^{87}Sr/^{86}Sr$ ratios of river water. For example, on a carbonate watershed, the river water's $^{87}Sr/^{86}Sr$ ratio is the same as that of the rocks (25).

Identification and Quantification of Natural and Anthropogenic Inputs in River Water

Attention must be given to the different contributions of the different end-member inputs to the total dissolved load, as any calculation of the export rate of chemical species (natural or anthropogenic origin) requires the quantification of each end member. This problem can be solved using a mass balance calculation.

In general, a mass balance equation for the budget can be expressed as follows (51,52):

$$\text{rock weathering} + \text{atmospheric input} + \text{human activities input} = \text{dissolved load} + \text{solid load}$$

Determination of an output–input budget (53,54) and determination of element origins are required for the mass balance approach. The mass balance approach may evaluate the contribution of atmospheric inputs, anthropogenic impacts, and rock weathering processes.

Atmospheric Inputs Correction

Dissolved species in surface water may contain a significant amount of solutes from rainfall (51). The target of the atmospheric input correction is to quantify and subtract the portion of the elements carried by rainwater in the chemical composition of the river water. Knowledge of the chemical composition of the rainfall on the total watershed is required for the quantification of atmospheric input because of rainfall (24,51).

Classically (51), for any element Z in the river $[(Z)_r]$, correction of the atmospheric contribution to a river (r) is estimated by reference to the Cl concentration, called $(Cl)_{ref}$, multiplied by the Z/Cl ratios of rainwater (rw):

$$(Z)_{rw} = (Cl)_{ref} \cdot (Z/Cl)_{rw} \quad (6)$$

$$(Z)_{r,rw\ corrected} = (Z)_{initial\ r} - (Z)_{rw} \quad (7)$$

Atmospheric chloride ions originate from sea salt and human activities. Except for salt rock, Cl release because of rock weathering has not been demonstrated clearly (55). Moreover, through the hydrological cycle, Cl behaves conservatively (55), so it is used as a reference of atmospheric inputs in many unpolluted hydrosystems (3,56). However, chloride ions can also have an anthropogenic origin (57). For the mass balance equations, the number of Cl ions issued from rainwater, known as $(Cl)_{ref}$, has to be determined. The $(Cl)_{ref}$ is calculated with each mean weighted chloride content for the sampling spots multiplied by the concentration factor F of each region. The concentration effect of evapotranspiration is represented by the factor F and is related to total amount of rainwater P (in mm) and the evapotranspiration process E (in mm) by the equation

$$F = P/(P - E) \quad (8)$$

The general equation to calculate $(Cl)_{ref}$ is

$$(Cl)_{ref} = (Cl)_{SA} \cdot 0.25 \cdot F_{SA} + (Cl)_{MC} \cdot 0.75 \cdot F_{MC} \quad (9)$$

where $(Cl)_{SA}$ and $(Cl)_{MC}$ correspond to the weighted concentration of chloride in study areas.

The residual chloride $(Cl)_{res}$ in the river is attributed only to human activities:

$$(Cl)_{res} = (Cl)_{measured} - (Cl)_{ref} \quad (10)$$

Budgets of Human Activities

When the influence of rainwater is subtracted from the river water composition, the general mass balance equation indicates that the dissolved load carried out by the river results from rock weathering and human activities. It is needed to quantify the influence of anthropogenic inputs in order to evaluate the chemical weathering on the river watershed (20).

It is widely known that phosphorus and nitrogen forms and potassium and chloride ions are mainly derived from agricultural fertilizers, animal waste, and municipal and industrial sewage (58–60). These species are indicative of both population and agricultural impacts, whereas sulfates are bacterial oxidation products of some industrial products (61,62).

The contributions of the different end members to the dissolved load can be calculated using the estimation of chemical signature of the pure end members. For each chemical species, the contributions for the silicate (SIL), carbonate (CAR), and anthropogenic (ANT1, ANT2) end members are calculated for each sample of river water using the following equation, as demonstrated by Grosbois et al. (20):

$$[X]_{riv} = a[X]_{SIL} + b[X]_{CAR} + c[X]_{ANT1} + d[X]_{ANT2} \quad (11)$$

Conditionally, $a + b + c + d = 1$, where X represents the amounts (in μmol/L) of X in the SIL, CAR, ANT1, and ANT2 end members, and a, b, c, and d represent the proportions (%) of each end-member contribution.

Natural and Anthropogenic Export in the River Watershed

Using the mixing equation (Eq. 11), it is possible to determine natural and anthropogenic exports from the river watershed. The export rates ω (in t/d) because of rock weathering or anthropogenic additions can be calculated for each end member (SIL, CAR, ANT1, and ANT2) and for each hydrological period as follows:

$$\omega = \Sigma(\alpha[Ca^{2+}] + \alpha[Mg^{2+}] + \alpha[K^{+}] + \alpha[Na^{+}] + \alpha[HCO_3^{-}] + \alpha[Si]) \cdot W \quad (12)$$

where [X] is the concentration (g/L), W represents the discharge (m^3/s) measured at the sampling points during the considered hydrological period, and α represents the proportion of an end member for the sample (20). The rock weathering and anthropogenic rates are calculated and ω results behave differently in accordance with the hydrological period.

Contribution of the Lithology

Comparing $^{87}Sr/^{86}Sr$ and elemental concentration ratios (e.g., Ca/Na, Mg/Na, and HCO_3/Na) for the rivers draining different monolithological watersheds (i.e., silicate, carbonate, etc.), the respective contribution of each lithology can be deciphered (14,63). The following methodology can be used to correct all ratios for atmospheric input:

$$X_{atm} = Cl_r(X/Cl)_{rw}$$

where X_{atm} is the percentage of element coming from the atmosphere, Cl_r is the chloride concentration in the river, and $(X/Cl)_{rw}$ is the ratio measured in the local rainwater (64).

CONCLUSIONS

Chemical weathering of rocks and soils is one of the essential processes in the geochemical cycling of elements in rivers. Although climatic factors such as moisture and temperature play a major role in releasing those elements, atmospheric aerosols have been identified as a major source of dissolved species in river water. Rainwater is significantly impacted by sea salt aerosols, terrestrial aerosols, and anthropogenic sources. For the study of rocks–recharge interactions as well as river water chemistry, the river watersheds, tributaries, local rainfall, humidity, and temperature must be considered. Furthermore, attention must be given to the different contributions of the different end-member inputs, either natural or anthropogenic, to the total dissolved load, as any calculation of the export rate of chemical species requires the quantification of each end member.

BIBLIOGRAPHY

1. Garrels, R.M. and McEnzie, F.T. (1971). *Evolution of Sedimentary Rocks*. W.W. Norton, New York.
2. Drever, J.I. (1988). *The Geochemistry of Natural Waters*, 2nd Edn. Prentice-Hall, Englewood Cliffs, NJ.
3. Sarin, M.M. et al. (1989). Major ion chemistry of the Ganga-Brahmaputra River system, weathering processes and fluxes to the Bay of Bengal. *Geochim. Cosmochim. Acta* **53**: 997–1009.
4. Berner-Kay, E. and Berner, R.A. (1987). *The Global Water Cycle. Geochemistry and Environment*. Prentice-Hall, Englewood Cliffs, NJ.
5. Millot, R., Gaillardet, J., Dupre, B., and Allegre, C.J. (2002). The global control of silicate weathering rates and the coupling with physical erosion: new insights from rivers of the Canadian Shield. *Earth Planet. Sci. Lett.* **196**: 83–98.
6. Krishnaswami, S. et al. (1992). Strontium isotopes and rubidium in the Ganga-Brahmaputra River system: weathering in

the Himalaya, fluxes to the Bay of Bengal and contributions to the evolution of oceanic $^{87}Sr/^{86}Sr$. *Earth Planet. Sci. Lett.* **109**: 243–253.

7. Derry, L.A. and France-Lanord, C. (1996). Neogene Himalayan weathering history and river $^{87}Sr/^{86}Sr$: impact on the marine Sr record. *Earth Planet. Sci. Lett.* **142**: 59–72.
8. Galy, A., France-Lanord, C., and Derry, L.A. (1999). Th strontium isotopic budget of Himalayan rivers in Nepal and Bangladesh. *Geochim. Cosmochim. Acta* **63**: 1905–1925.
9. Dalai, T.K., Singh, S.K., Trivedi, J.R., and Krishnaswami, S. (2002). Dissolved rhenium in the Yamuna River system and the Ganga in the Himalaya: role of black shale weathering on the budgets of Re, Os, and U in rivers and CO_2 in the atmosphere. *Geochim. Cosmochim. Acta* **66**: 29–43.
10. Quade, J., English, N., and DeCelles, P.G. (2003). Silicate versus carbonate weathering in the Himalaya: a comparison of the Arun and Seti River watersheds. *Chem. Geol.* **202**: 275–296.
11. Gibbs, R.J. (1967). The geochemistry of the Amazon River system: Part I. The factors that control the salinity and the composition and concentration of the suspended solids. *Geol. Soc. Am. Bull.* **78**: 1203–1232.
12. Millot, R., Gaillardet, J., Dupre, B., and Allegre, C.J. (2003). Northern latitude chemical weathering rates: clues from the Mackenzie River basin, Canadian. *Geochim. Cosmochim. Acta* **67**: 1305–1329.
13. Probst, J-L. et al. (1992). Dissolved major elements exported by the Congo and the Ubangi Rivers during the period 1987–1989. *J. Hydrol.* **135**: 237–257.
14. Negrel, P., Allegre, C.J., Dupre, B., and Lewin, E. (1993). Erosion sources determined by inversion of major and trace element ratios and Sr isotopic ratios in river water. The Congo basin case. *Earth Planet. Sci. Lett.* **120**: 59–76.
15. Gaillardet, J., Dupre, B., and Allegre, C.J. (1995). A global geochemical mass budget applied to the Congo Basin rivers: erosion rate and continental crust composition. *Geochim. Cosmochim. Acta* **59**: 3469–3485.
16. Ming-hui, H., Stallard, R.F., and Edmond, J.M. (1982). Major ion chemistry of some large Chinese rivers. *Nature* **298**: 550–553.
17. Han, G. and Liu, C-Q. (2004). Water geochemistry controlled by carbonate dissolution: a study of the river waters draining karst-dominated terrain, Guizhou Province, China. *Chem. Geol.* **204**: 1–21.
18. Keasler, K.M. and Loveland, W.D. (1982). Rare earth elemental concentrations in some Pacific Northwest rivers. *Earth Planet. Sci. Res.* **61**: 68–72.
19. Goldstein, S.J. and Jacobsen, S.B. (1988). Rare earth elements in river waters. *Earth Planet. Sci. Res.* **89**: 35–47.
20. Grosbois, C., Negrel, Ph., Fouillac, C., and Grimaud, D. (2000). Dissolved load of the Loire River: chemical and isotopic characterization. *Chem. Geol.* **170**: 179–201.
21. Picouet, C., Dupre, B., Orange, D., and Valladon, M. (2002). Major and trace element geochemistry in the upper Niger River (Mali): physical and chemical weathering rates and CO_2 consumption. *Chem. Geol.* **185**: 93–124.
22. Junge, C.E. (1963). *Air Chemistry and Radioactivity*. Academic Press, New York.
23. Gibbs, R.J. (1970). Mechanisms controlling world water chemistry. *Science* **170**: 1088–1090.
24. Likens, E.G., Boremann, F.H., Pierce, R.S., Eaton, J.S., and Johnson, N.M. (1977). *Biogeochemistry of Forested Ecosystem*. Springer-Verlag, Berlin.
25. Negrel, Ph. and Roy, S. (1998). Rain chemistry in the Massif Central (France): a strontium isotopic and major elements study. *Appl. Geochem.* **13**: 941–952.
26. Jaffrezo, J.L. (1987). Étude du lessivage des aerosols atmospheriques per les precipitations en milieu urbain. Ph.D. thesis. Université de Paris VII.
27. Graustein, W.C. and Armstrong, R.L. (1983). The use of $^{87}Sr/^{86}Sr$ ratios to measure atmospheric transport into forested watershed. *Science* **219**: 298–392.
28. Gosz, J.R. and Moore, D.I. (1989). Strontium isotope studies of atmospheric inputs to forested watersheds in New Mexico. *Biogeochemistry* **8**: 115–134.
29. Dupre, B., Negrel, Ph., Seimbille, F., and Allegre, C.J. (1994). $^{87}Sr/^{86}Sr$ ratios variation during a rain event. *Atmos. Environ.* **28**: 617–620.
30. Herut, B., Starinsky, A., and Kartz, A. (1993). Strontium in rainwater from Israel, sources, isotopes and chemistry. *Earth Planet. Sci. Lett.* **120**: 77–84.
31. English, N.B., Quade, J., DeCelles, P.G., and Garzione, C.N. (2000). Geologic control of Sr and major element chemistry in Himalayan rivers, Nepal. *Geochim. Cosmochim. Acta* **64**: 2549–2566.
32. Central Bureau of Statistics. (1998). *Statistical Pocket Book, Nepal*. His Majesty's Government National Planning Commission Secretariat, Kathmandu, Nepal.
33. Freyssinet, P. and Farah, A.S. (2000). Geochemical mass balance and weathering rates of ultramafic schists in Amazonia. *Chem. Geol.* **170**: 133–151.
34. Galy, A. and France-Lanord, C. (1999). Weathering processes in the Ganges-Brahmaputra basin and the riverine alkalinity budget. *Chem. Geol.* **159**: 31–60.
35. Viers, J. et al. (2000). Major and trace element abundances, and strontium isotopes in the Nyong basin rivers (Cameroon): constraints on chemical weathering processes and elements transport mechanisms in humid tropical environments. *Chem. Geol.* **169**: 211–241.
36. Elbaz-Poulichet, F. et al. (1999). Trace element geochemistry in the upper Amazon drainage basin (Bolivia). *Chem. Geol.* **157**: 319–334.
37. Vital, H. and Stattegger, K. (2000). Major and trace elements of stream sediments from the lowermost Amazon River. *Chem. Geol.* **168**: 151–168.
38. Losno, R., Bergametti, G., Carlier, P., and Mouvier, G. (1991). Major ions in marine rain water with attention to sources of alkaline and acidic species. *Atmos. Environ.* **25A**: 763–770.
39. Galloway, J.N., Likens, G.E., Keene, W.C., and Miller, J.N. (1982). The composition of precipitation in remote areas of the world. *J. Geophys. Res.* **87**: 8771–8786.
40. Basak, B. and Alagha, O. (2004). The chemical composition of rain water over Buyukcekmece Lake, Istanbul. *Atmos. Res.* **71**: 275–288.
41. Glavas, S. and Moschonas, N. (2002). Origin of observed acidic–alkaline rains in a wet-only precipitation study in a Mediterranean coastal site, Patras, Greece. *Atmos. Environ.* **36**: 3089–3099.
42. Aliva, A. and Alarcon, M. (1999). Relationship between precipitation chemistry and meteorological situations at a rural site in NE Spain. *Atmos. Environ.* **33**: 1663–1667.
43. Le Bolloch, O. and Guerzoni, S. (1995). Acid and alkaline deposition in precipitation of the western coast of Sardinia, Central Mediterranean (40 °N, 8 °E). *Water Air Soil Pollut.* **85**: 2155–2160.

44. Rao, P.S.P. et al. (1995). Rain water and throughfall chemistry in the Silent Valley forest in south India. *Atmos. Environ.* **29**: 2025–2029.
45. Plaisance, H., Coddeville, P., Roussel, I., and Guillermo, R. (1996). A qualitative determination of the source locations of precipitation constituents in Morvan, France. *Environ. Technol.* **17**: 977–986.
46. Hofmann, E.J., Hofmann, G.L., and Duce, R.A. (1974). Chemical fractionation of alkaline earth metals in atmospheric particulate matter over the North Atlantic. *J. Geophys. Res.* **8**: 675–688.
47. Bonsang, B., Nguyen, B.C., Gaudry, A., and Lambert, G. (1980). Sulfate enrichment in marine aerosols owing to biogenic gaseous sulfur compounds. *J. Geophys. Res.* **85**: 7410–7416.
48. Suzuki, T. and Tsunogai, S. (1988). Origin of calcium in aerosols over the western North Pacific. *J. Atmos. Chem.* **6**: 363–374.
49. Ichikuni, M. (1978). Calcite as a source of access Ca in rain water. *J. Geophys. Res.* **83**: 67249–67688.
50. Keene, W.C.A.A., Pszenny, P., Galloway, J., and Hawley, M.E. (1986). Sea-salt corrections and interpretations of constituent ratios in marine precipitation. *J. Geophys. Res.* **91**: 6647–6658.
51. Meybeck, M. (1983). Atmospheric inputs and river transport of dissolved substances. *IAHS Publ.* **141**: 173–192.
52. Drever, J.I. and Zobrist, J. (1992). Chemical weathering of silicate rocks as a function of elevation in the Southern Swiss Alps. *Geochim. Cosmochim. Acta* **56**: 3209–3216.
53. Velbel, M.A. (1985). Geochemical mass balances and weathering rates in forested watersheds of the Sothern Blue Ridge. *J. Sci.* **285**: 901–930.
54. Mast, M.A. and Drever, J.I. (1990). Chemical weathering in the Loch Vale watershed, Rocky Mountain National Park, Colorado. *Water Resour. Res.* **26**: 2971–2978.
55. Meybeck, M. (1979). Concentrations des eaux fluviales en elements majeurs et apports en solution aux oceans. *Rev. Geol. Dyn. Geogr. Phys.* **21**: 215–246.
56. Gaillardet, J., Dupre, B., Allegre, C.J., and Negrel, P. (1997). Chemical and physical denudation in the Amazon River basin. *Chem. Geol.* **142**: 141–173.
57. Sherwood, W.C. (1989). Chloride loading in the South Fork of the Shenandoah River, Virginia, U.S.A. *Environ. Geol. Water Sci.* **14**(2): 99–106.
58. Meybeck, M. (1986). Composition chimique des ruisseaux non pollues en France. *Sci. Geol. Bull.* **39**(1): 3–77.
59. Dojlido, J.R. and Best, G.A. (1996). Chemistry of water and pollution. *J. Hydrol.* **180**: 1–4.
60. Ramos, C. (1996). Effect of agricultural practices on the nitrogen losses to the environment. *Fert. Environ.* 355–361.
61. Mackenzie, R.T. and Garrels, R.M. (1966). Chemical mass balance between rivers and the oceans. *Am. J. Sci.* **264**: 507–525.
62. Probst, J-L. (1986). Dissolved and suspended matter transported by the Girou River (France): mechanical and chemical erosion rates in calcareous molasses basin. *Hydrol. Sci. J.* **31**: 1–3.
63. Gaillardet, J., Dupre, B., Louvat, P., and Allegre, C.J. (1999). Global silicate weathering and CO_2 consumption rates deduced from the chemistry of large rivers. *Chem. Geol.* **159**: 3–30.
64. Freydier, R. (1997). Chimie des precipitations en Afrique intertropicale: cations et anions majeurs, elements traces et acides organiques. These de doctorat. Université Paul Sabatier, Talouse, France.

READING LIST

Berner-Kay, E. and Berner, R.A. (1996). *Global Environment, Water, Air and Geochemical Cycles.* Prentice-Hall, Englewood Cliffs, NJ.

PHYSICS AND CHEMISTRY OF WATER

ACID RAIN AND SOCIETY

PAULETTE MIDDLETON
(from *The Handbook of Weather, Climate, and Water: Atmospheric Chemistry, Hydrology, and Societal Impacts*, Wiley 2003)

INTRODUCTION

Acid rain is one of many manifestations of how actions of society can have adverse effects on human health and welfare. Now more than ever before, the breadth of socioeconomic as well as environmental impacts associated with air pollutants, connections among pollutant contributions to these many impacts, and the implications of these connections are being recognized for policy making and development of management strategies.

It can no longer be argued that it is very costly to mediate acid rain and related air quality concerns. Assessments are beginning to suggest that multiple benefits associated with addressing acid rain in combination with other issues outweigh the costs of control of key responsible pollutants. In addition, when innovative strategies, which include market trading and incentives for conservation and use of clean fuels, are initiated, the costs of pollutant management become even lower. As factors that are not easily quantified monetarily are considered more directly in assessments, the benefits become even greater.

Recent analyses show that the implementation of the acid-rain-related part of the 1990 Clean Air Act Amendments has resulted in reductions in acidity in the north-eastern United States. Improvements in acid-related impacts have also been suggested. However, projections of future conditions over the next 20 to 50 years suggest that, unless more dramatic steps are taken, the overall burden of harmful pollutants could continue to rise in general in different parts of the country. Dramatic reductions in sulfur oxides, the current main contributors to acidity, alone are probably not enough. Planned reductions in nitrogen oxides may or may not be adequate. Similarly, continued close monitoring, if not increased management, of volatile organic compounds and fine particulates, and reassessment of their importance to acid rain and related concerns will be important over the next few years.

Of course, all of these issues are part of the bigger international picture of energy, environment, and economy. While the United States may be more aggressively and wisely addressing acid rain and related issues here at home, the projections for future fossil fuel use worldwide must be considered for the sake of regional air quality as well as the global climate condition. Capital investment in cleaner technologies such as renewable energy and the promotion of conservation strategies worldwide could bring long-term environmental and economic benefits that far surpass the initial costs. The alternative, continued growth in fossil fuel usage in developing countries, could exacerbate air quality problems that already exist in many of these areas and, in the long term, could cause the same acid rain damages experienced in many parts of North America and western Europe. In addition, it would contribute to adverse long-term carbon-dioxide-induced climate change.

As a contribution to our understanding of the atmospheric pollution problem and our role in the solution, this chapter summarizes:

- Acid rain and its relationship with other major issues
- U.S. response to the acid rain issue
- Current assessment of progress on reducing the effects of acid rain
- Projections and speculation on the future of acid rain

ACID RAIN: THE PHENOMENON

The relationships between chemical emissions into the atmosphere and the effects of the chemicals on various ecosystems, human health, and materials are highly complex. Many harmful chemicals (i.e., air pollutants) chemically interact to form other pollutants, which are perhaps even more harmful than the originally emitted chemicals. The most prominent examples of these dangerous chemical products are acid rain, ozone, and aerosols. In addition, many air pollutants are thought to interact in a synergistic fashion to cause even more harm as a group rather than individually. An example of the possible synergism is the hypothesized impact of acid rain and ozone on forest ecosystems. Understanding the causes and effects of acid rain and related air quality issues has become an important mission of atmospheric scientists around the globe.

DEFINITION OF ACID RAIN

Acid rain is the general term used to describe the removal, by rainfall, of acidic pollutants from the atmosphere. Acids also can be removed by other forms of precipitation, such as snow or fog. Acid pollutants may also fall as dry particles or gases that form acids when later combined with moisture. The term *acid deposition* is used to include all the possible forms of acid pollutant removal from the atmosphere, but *acid rain* remains the popular term.

The majority of the deposited acids are nitric acid and sulfuric acid. In some of the more rural regions of the world, organic acids also are important. In very remote areas where the level of acid is low, the "natural background" acid is carbonic acid, which is associated with carbon dioxide in the air. The overall acidity of precipitation also depends on the basic (or alkaline) constituents of the precipitation. Major bases include ammonia and geologic materials, such as dust and fly ash.

Acidity is measured in terms of a pH scale, which is a measure of the log of the hydrogen ion concentration in the precipitation. The scale runs from 0 to 14 with 0 being very acidic and 14 being very alkaline. A midscale value of 7 is considered neutral. A change in 1 pH unit indicates 10-fold increase or decrease in acidity. Unpolluted rain water is considered to have a pH of about 5.6. This acidity is assumed to contain only carbonic acid. In the highly polluted eastern states of the United States., the average acidity of water has a pH between 4 and 5. Even in some remote areas of the world, pH values of 5.2 have been found. These acidity levels suggest that there is a long-range transport of nitrogen and sulfur chemicals.

SOURCES OF ACIDITY

Atmospheric acids mainly are produced in the air as a result of complex chemical reactions of the acid precursor gases. Direct emissions of acids such as sulfuric acid, hydrogen chloride, and hydrogen fluoride have been estimated but are not thought to play a significant role in the acidic deposition processes (1). Sources of the harmful chemical precursors affecting the acidity of deposited materials can be natural or human caused. The three pollutants of most concern in the acidic deposition process are sulfur dioxide (SO_2), nitrogen oxides (NO_x) and volatile organic compounds (VOCs). Fossil-fuel-based power plants and motor vehicles are the major sources of all of these acid precursor pollutants in industrialized areas of the world (2,3).

Sulfur gases are primarily emitted from point sources, involving the combustion of coal in particular. Natural sources of sulfur gases include sea spray, volcanoes, and biologic activity. These sources, however, are at least a factor of 10 *less* than the human-caused emission for major industrial areas such as the United States. Nitrogen gases also result primarily from human activities involving fossil-fuel-derived energy use related to transportation and utilities. Major natural sources include soils and lightning and are thought to make up a significant portion of the overall emission totals, more than has been estimated for the sulfur gases. Estimates of these levels, however, are uncertain (1). The VOCs that produce the organic acids and influence the chemistry producing sulfuric and nitric acids also come mainly from automobile use. However, for the VOCs, natural production from vegetation can be quite significant. In highly vegetated, low industrialized regions natural sources become the dominant producers of VOCs.

Estimates of alkaline particulate and ammonia emissions indicate that there is a high potential for acid neutralization in some parts of the United States. The estimates, however, are subject to a high degree of uncertainty (1). On a global scale, emissions of these important acid neutralizers are among the least well-known chemical emissions (2).

EFFECTS OF ACID RAIN

The effects of acid deposition are the subject of continuing controversy. The northeastern United States has experienced the worst reported impacts in the United States (1). Severe damages attributed to acid rain also have been documented for parts of western Europe (4).

The most sensitive systems to acid deposition are poorly buffered lakes and streams. Buffering capacity refers to the availability of alkaline minerals from soil or rocks to neutralize the acids. When minerals are dissolved in a lake, buffering is able to diminish acid effects. However, this buffering ability or alkalinity can be used up with additions of acidic pollutants. Low alkalinity lakes have the greatest potential for damage, since their neutralizing minerals can be quickly depleted.

Vegetation is exposed to wet acidic deposition through rain, snow, and by direct contact with low, acid-laden clouds. There is currently no widespread forest or crop damage in the United States related to these possible pathways. However, cloud acidity, together with a complex combination of other factors (e.g., ozone, soil acidification, climate) contribute to reduced cold tolerance in high-elevation spruce in the eastern United States and in Europe. This can contribute to damage to trees above cloud level during winters with particularly low temperatures.

Adverse effects on forests in other regions of the world are associated with ozone, as is the case with high-elevation pines in California, or they are closely related to localized soil nutrient deficiencies, as is the case with sugar maples in eastern Canada. Acidic deposition may increase leaching rates of important base cations, principally magnesium and calcium, in forest solids and may be a contributing factor in sugar maple decline in some areas.

Generally, controlled experiments on trees and crops have indicated that ozone, at concentrations near ambient levels, adversely affects forests and crops primarily by growth reduction. Other controlled experiments have demonstrated that normal levels of atmospheric sulfur and nitrogen deposition cause no negative direct effects. Some areas actually may benefit through nutrient enrichment by nitrogen and sulfur deposition.

Computer models project that continued acidic deposition could result in long-term deficiencies of nutrients in some soils. However, currently, there is no evidence to indicate that forest health in general is currently affected by nutrient deficiency or will be affected in the next half century.

Air pollution and acidic deposition contribute to the corrosion of metals and deterioration of stone in buildings, statues, and other cultural resources. Although air pollution is an important concern for cultural objects, the magnitude of its effect on construction materials has been difficult to assess. Many construction materials have protective coatings such as paints; therefore, maintenance and service of protective coatings have an important role in determining the ultimate impact of air pollutants. Paints may also be affected by ambient levels of air pollution.

Another related side effect of acid rain is visibility degradation. Fine particles in the atmosphere containing sulfate, nitrate, and other chemical constituents, which when deposited are associated with acid deposition, cause visibility degradation while in the air. These fine particles have been the major factor in the reduction of visibility in rural and urban areas in the eastern United States since the beginning of the century. In the U.S. West, visibility degradation is being reported in major urban areas and in national parks and wilderness areas.

Direct adverse effects of these pollutants on humans occur largely through the respiratory system. Sensitive populations with existing respiratory or cardiovascular problems, such as those with asthma, are especially susceptible. These effects have been mainly associated with the acid precursor gases and ozone. Studies of the effects of acidic aerosols, composed primarily of nitric acid, ammonium bisulfate, and sulfuric acid, are still relatively new. It has been found that on rare occasions acid levels approach 10 times the long-term mean levels for sulfuric acid. Although substantial uncertainty exists, the body of data raises concern that acidic aerosols alone, or in concert with other pollutants, may be contributing to health effects in exposed populations at current concentration levels.

Human health also can be affected indirectly by pollutants related to acid deposition. People who eat large amounts of fish from acidic lakes or streams may experience exposure to methylmercury in some regions of the country. Drinking water from acidic sources may contain significantly elevated levels of lead. It is unlikely, however, that exposure to humans by this pathway is important, except in isolated cases.

SOCIAL RESPONSE TO ACID RAIN

Historically, toxic effects have been observed in populations acutely exposed to high concentrations of air pollutants. As early as the Middle Ages in London prohibitions on coal burning were instituted in response to perceived health effects of mixtures of dust, soot, and fog. The industrial revolution brought the air pollution issue to the United States, where air quality management continued to be considered local in nature into the twentieth century.

The severity of air pollution impacts became very obvious during the London "killer fog" of 1952, when a mixture of particulates, sulfur dioxide, and acidic fog was associated with severe respiratory effects and approximately 4000 deaths. Emergence of air pollution as a public health issue in the 1950s, as a result of this and other deadly episodes, led to the development of federally funded research programs, culminating in the Clean Air Act and in the establishment of the Environmental Protection Agency (EPA) in 1970. These were major stimuli for the establishment of the U.S. National Ambient Air Quality Standards (NAAQS) that today restrict the atmospheric concentration of pollutants such as sulfur dioxide, nitrogen oxides, and ozone.

Other countries around the world have been developing institutional responses to the threat to human health of air pollution. Air pollution effects on the environment, however, had been slower to be recognized as a serious issue. Acid rain and its ecological effects were first documented in England at the end of the nineteenth century and became regional issues in northwestern Europe and the northeastern United States and eastern Canada only recently—in the late 1960s. During this period and into the 1970s, the mounting anecdotal evidence of the effects of acid rain on aquatic and terrestrial ecosystems launched acid rain as perhaps the first pollution threat to the environment to receive international attention.

The origins of the pressures to regulate acid rain in the United States were primarily twofold. First, Canada protested, lobbied, and publicized its contention that major environmental damage was occurring in its eastern provinces because of acid deposition, and that the major sources of acid precursors were in the United States Second, elected officials and citizens in the eastern and New England states echoed the same concerns, elevating the acid rain controversy to the level of a growing interregional conflict between receptor states and polluting states (5).

The U.S. responses to these concerns took the form of federal research and eventually control programs. The first step was the Acid Precipitation Act of 1980, which created the National Acid Precipitation Assessment Program (NAPAP). During its first 10 years, the research and periodic assessments conducted by NAPAP improved the understanding of the scientific processes and effects of acid deposition. The monitoring and research conducted in the 1980s and the subsequent integrated assessment completed in 1990 provided the scientific knowledge base for Title IV, the Acid Deposition Control Program, of the 1990 Clean Air Act Amendments.

Title IV is designed to reduce the adverse effects of acid deposition through the reductions in annual emissions of sulfur dioxide (SO_2) and nitrogen oxides (NO_x), the precursors to acid rain. Recognizing that the principal sources of acid rain precursors in the atmosphere are emissions from the combustion of fossil fuels, control measures were initiated to reduce emissions from electric utilities. However, rather than the traditional command-and-control approach to regulation, alternative methods of compliance were allowed. These methods included technological adaptation (e.g., scrubbers, higher-efficiency boilers), fuel-switching, and an innovative SO_2 emissions allowance trading program. This represented the first national effort to use market-based incentives to achieve environmental goals.

Due to the innovative nature of using market-based incentives for environmental regulation, Congress set up a mechanism for checking how well trading was working. As part of this activity, Congress asked NAPAP to assess the costs and economic impacts of the acid deposition control program as well as the effectiveness and benefits associated with the various human health and welfare effects. The effects included visibility, materials, and cultural resources damages and ecosystem effects. NAPAP was also asked to consider the deposition levels needed to protect sensitive ecosystems. The results of the assessment of Title IV are to be reported to Congress quadrennially, beginning with the 1996 Report to Congress (6).

CURRENT CONDITIONS

As of the completion of the first report to Congress, several observations have been made regarding the success of Title IV. It appears that the market-based approach has lowered compliance costs. Costs are lower than expected, probably due to a number of factors such as railroad deregulation, technological innovation, and lower operating costs for scrubbers. In addition, all affected utilities have fulfilled the compliance requirements of Title IV. In the first annual reconciliation of allowances and emissions, SO_2 allowances matched or exceeded SO_2 emissions. NO_x reductions have not been as dramatic. This is expected since mandates on NO_x reductions are not in place yet. However, NO_x emissions from all sources in 1995 were 1.5 million tons below 1980 levels. Utilities were responsible for 53% of that reduction.

Statistically significant reductions in acidity and sulfate in precipitation were reported at monitoring sites in the Midwest, mid-Atlantic, and northeastern United States. There is no real evidence of statistically significant decreases in nitrate concentration. Changes in aquatic ecosystems have not yet been detected. However, over the last 15 years, lakes and streams throughout many areas of the United States have experienced decreases in sulfate concentration in response to decreased emissions. While there is some evidence of recovery from acidification in New England, Adirondack lakes continue to acidify, suggesting that additional reductions may be needed in these areas (6).

Sulfur and nitrogen deposition has caused adverse impacts on certain sensitive forest ecosystems in the United States, with high-elevation spruce in the eastern United States being most sensitive. Other sensitive forests are apparently not experiencing the same effects in mortality and growth, at least for now, but some of the same processes appear to be slowly occurring.

The leaching of soil nutrients by continued acidic deposition is a gradual process that will eventually impact forest nutrition and growth in many areas. The recent reductions in sulfur should result in some small immediate improvements in sensitive forests, but large improvements will be slow to occur.

Reduced emissions of sulfur oxides are expected to reduce sulfate concentration and its contribution to haze. It is difficult to assess the extent to which recent reductions have contributed to changes in visibility over the past few years since meteorological and other factors determine the overall changes in visibility. Information is needed over the long term.

The recent reductions in SO_x and NO_x emissions are expected to reduce fine particulates and, as a result, lead to improved human health. It is suggested that reduced emissions will lead to a reduction in premature mortality from cardiovascular and respiratory causes and to a dramatic reduction in the number of asthma symptom days.

One difficulty in determining effects at this time is that many impacts have a response times that are longer than the few years since the passage of Title IV. Visibility and acute health effects can be detected on the order of hours to days. Episodic aquatic effects and soil and plant processes in the forest ecosystem respond on the order of days and weeks to months. Chronic human health, chronic aquatic effects, and forest health, on the other hand, indicate response times on the order of years to decades. Effects on forest solid nutrient reserves and effects on materials begin to show up on the order of decades to centuries. These latter effects are more on the order of climate change impacts response times.

The difference in response times, of course, makes an evaluation of actions taken in the early 1990s difficult to quantify. Improvements in health and visibility can serve as indicators of positive change. However, as already noted, even changes in visibility cannot be directly attributed to sulfate reductions alone, since other factors such as meteorological variability play a role in determining visibility changes especially in the humid part of the eastern United States.

KEEPING A BROAD BASIS OF ASSESSMENT AND ACTION

To review, acid deposition is an end product of a complex series of interactions among atmospheric chemical species emitted by both natural and human sources. For policy assessment purposes, the most important groups of chemical species are compounds containing sulfur and nitrogen compounds that are emitted from factories, power plants, and automobiles based on fossil-fuel combustion. In addition, volatile organic carbon compounds and fine particles play a role in modulating chemical processes and acidity. Some key compounds remain unchanged in the atmosphere and some are neutralized, but others are oxidized into more acidic forms through a complicated series of chemical, meteorological, physical, and biological interactions.

Decisions about the control of acid deposition must deal with the environmental impacts of estimated future emission levels as well as present levels. Projections depend on many complex and interacting socioeconomic factors. The predictability of how rapidly and to what extent fossil fuels will be replaced by clearer and safer fuels (society may change transportation and other energy use habits) and the inter-relationships among countries of the world becoming driving influences for these changes is highly uncertain. Given the demonstrated value of examining multiple causes and effects together, it will continue to be important to keep the base of assessment broad in spite of the uncertainties. The elements of such assessments are illustrated in Fig. 1.

On issues related to acid rain, other policy discussions going on throughout North America also illustrate the growing awareness of the interconnections, as well as the need to capitalize on the relationship in developing strategies for the future. For example, EPA, through the Federal Advisory Committee Act (FACA), is leading the development of combined ozone, particulate matter (aerosols), and regional haze implementation program rules and guidance. Other activities at the regional level, such as the Western Governors' Association (WGA) Air Quality Initiative, the Ozone Transport and Analysis Group (OTAG), the Southern Appalachian Mountain

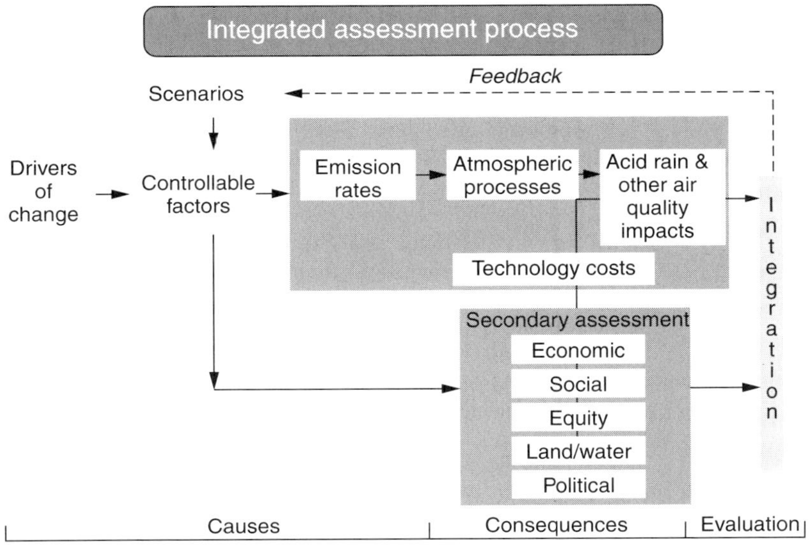

Figure 1. Important elements to consider in an assessment of acid rain in the context of other important environmental, energy, and economic concerns.

Initiative (SAMI), the Southern Oxidant Study (SOS), and the North American Research Strategy for Tropospheric Ozone (NARSTO), are addressing various science and policy issues associated with ozone, particulate matter, and regional haze.

On a continental scale, the Commission on Environmental Cooperation (CEC) is working on North American strategies for addressing transboundary concerns, which include ozone, particulate matter, regional haze, and acid rain along with other hazardous pollutants. Finally, on a much broader scale, the U.S. global climate change program is addressing regional climate assessment for areas throughout the United States. The regional air quality concerns addressed by FACA and the climate concerns to be considered in these discussions are closely linked through the development of aerosols that can influence climate on regional scales as well as produce other problems. On a larger policy implementation level, the two are linked through the development of energy strategies aimed at reducing greenhouse gas emissions and the emission of other more traditionally harmful air pollutants.

All of these programs and approaches share the same fundamental concerns. The role of natural or background processes and the role of chemical interactions in determining the levels of impacts in different regions continue to be fundamental overarching scientific questions. The implementation issues of emissions trading versus pollution prevention versus technological controls are also part of each aspect of the various debates. Assessments of trade-offs for any decisions made with respect to any of these issues must consider the less quantifiable, and sometimes more uncertain, impacts associated with heath, social impacts and values, equity, and related environmental concerns about water and soil quality as well as air. Given these strong interconnections, it is important to make the best use of research and policy-making resources across organizations addressing acid rain and related issues, where energy and the environment are key factors.

BIBLIOGRAPHY

1. U.S. National Acid Precipitation Assessment Program (NAPAP). (1991). *Acidic Deposition: State of Science and Technology, Summary Report of the U.S. National Acid Precipitation Assessment Program*. Washington, DC.
2. Graedel, T.E. et al. (1993). A compilation of inventories of emissions to the atmosphere. *Global Biogeochem. Cycles* **7**: 1–26.
3. Middleton, P. (1995). Sources of air pollutants. In: *Composition, Chemistry, and Climate of the Atmosphere*. Van Nostrand Reinhold, New York.
4. Graedel, T.E. and Crutzen, P.J. (1989). The changing atmosphere. *Sci. Am.* **261**(Sept.): 58.
5. Rhodes, S.L., and Middleton, P. (1983). The complex challenge of controlling acid rain. *Environment* **25**: 6–9, 31–38.
6. U.S. National Acid Precipitation Assessment Program (NAPAP). (1998). *NAPAP 1996 Report to Congress*, http://www.nnic.noaa.gov/CENR/NAPAP/NAPAP.96.htm.

ADSORPTION CAPACITY OF ACTIVATED CARBON FOR WATER PURIFICATION

Carl L. Yaws
Prasad K. Narasimhan
Lamar University
Beaumont, Texas

Adsorption on activated carbon is an effective method for water purification. In this article, results for the adsorption capacity of activated carbon for removing organic chemicals from water are presented for a wide variety of substances—alcohols, ketones, aldehydes, acetates, benzenes, chloroalkanes, chloro-olefins, and bromoalkanes. The results use the adsorption capacity as a function of the adsorption potential of the chemical in the aqueous liquid. The correlation constants are displayed in an easy to use tabular format that is especially applicable to rapid engineering usage.

Correlation results and experimental data are in favorable agreement.

INTRODUCTION

Physical and thermodynamic data for organic chemicals are especially helpful to engineers and scientists in industry. In particular, capacity data for adsorption of organic chemicals on activated carbon is becoming increasingly important in engineering and environmental studies because of more and more stringent governmental regulations involving substances in water.

In this article, the objective is to provide results for the adsorption capacity of activated carbon for removing organic chemicals from water. Such results are usable in engineering and environmental studies. As an example of such usage, capacity data from the correlation are useful in the engineering design of carbon adsorption systems for water purification.

CORRELATION FOR ADSORPTION CAPACITY

Adsorption capacity was correlated as a function of adsorption potential (Polyani theory) of a chemical in the aqueous phase using the following equation:

$$\log Q = A + B\, T/V \log\,(x_s/x) \tag{1}$$

Q = adsorption capacity at equilibrium, g compound/100 g carbon
A and B = regression coefficients
T = temperature, K
V = molar volume of liquid, cm^3/mol
x_s = solubility in water, parts per million by weight, ppm(wt)
x = concentration in water, parts per million by weight, ppm(wt)

The coefficients (A and B) are given in Table 1 for a wide variety of organic chemicals in water—alcohols, ketones, aldehydes, acetates, benzenes, chloroalkanes, chloro-olefins, and bromoalkanes. The table also provides the representative formula and an example compound for the chemical families. The correlation should not be used for other chemical families.

The coefficients were determined from regression of available data. In preparing the correlation, a literature search was conducted to identify data source publications (1–10). The excellent compilations by Arbuckle (1); Sonthier, Crittenden, and Summer (2); and Speth (3,4) were used for adsorption capacity. The compilations by Howard and Meylan (5); Mackay, Shiu, and Ma (6); Verschueren (7); Yalkowsky (8); and Yaws (9) were used for water solubility. Liquid molar volume was ascertained from data compilations by Yaws (9,10). The publications were screened and copies of appropriate data were made. These data were then keyed into the computer to provide a database of adsorption capacity at different adsorption potentials in the aqueous phase. The database also served as a basis to check the accuracy of the correlation.

Graphs of adsorption capacity as a function of the adsorption potential are shown in Figs. 1–5, for the substances. The graphs discloses favorable agreement of correlation results and experimental data.

In general, experimental data available in the literature are very limited for the adsorption capacity of activated carbon for organic compounds in aqueous liquids at low concentrations (parts per million range). In view of the very limited experimental data, estimates in the tabulation should be considered rough approximations. Adsorption capacity is also complicated by differences in carbon, lots of carbon, source of water, and concentration ranges (4). If initial feasibility studies using the correlation are favorable, follow-up experimental determination of equilibrium adsorption capacity is recommended.

Examples

The correlations for water solubility are useful in engineering applications involving process wastewater. Examples are shown below.

Table 1. Adsorption Capacity of Activated Carbon—Aqueous System[a]

$\log_{10} Q = A + B\, T/V \log_{10}(x_s/x)$ ($Q - g$ of Compound/100 g of carbon, x – ppm(wt))

Chemical Family	A	B	Formula	Example Compound
Alcohols	1.7250	−0.1210	$C_nH_{2n+2}O$	1-Butanol
Ketones	1.7250	−0.1210	$C_nH_{2n}O$	Diethyl ketone
Aldehydes	1.7250	−0.1210	$C_nH_{2n}O$	Butyraldehyde
Acetates	1.7250	−0.1210	$C_nH_{2n}O_2$	Butyl acetate
Benzenes	1.8893	−0.1199	$C_{n+6}H_{2n+6}$	Ethyl benzene
Chloroalkanes	2.6021	−0.1931	$C_nH_{2n+2-x}Cl_x$	Chloroform
Chloro-olefins	2.2175	−0.1599	$C_nH_{2n-x}Cl_x$	Trichloroethylene
Bromoalkanes	3.0607	−0.1991	$C_nH_{2n+2-x}Br_x$	Bromoform

[a] F 400 Activated carbon, Calgon.
Q—adsorption capacity at equilibrium, g compound/100 g carbon.
x_s—solubility in water at 25 °C, parts per million by weight, ppm(wt).
x—concentration in water at 25 °C, parts per million by weight, ppm(wt).
T—temperature, K.
V—molar volume of liquid, cm^3/mole.
A and B—regression coefficients.

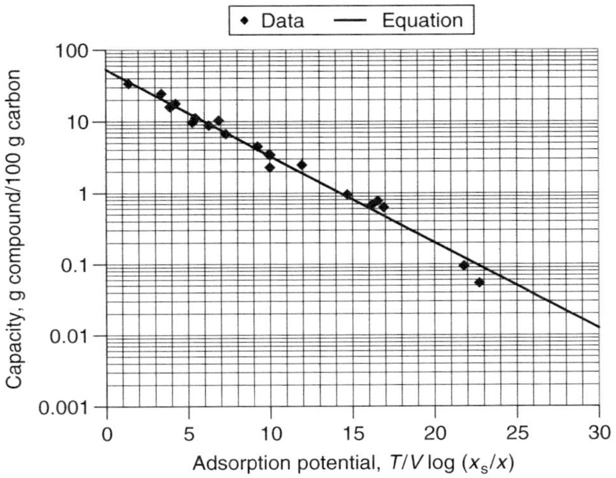

Figure 1. Adsorption capacity of activated carbon for organic oxygen chemicals (alcohols, ketones, aldehydes, and acetates) in water.

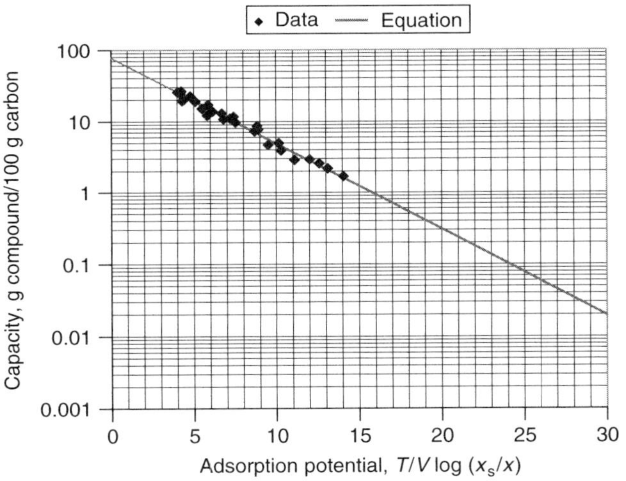

Figure 2. Adsorption capacity of activated carbon for benzenes in water.

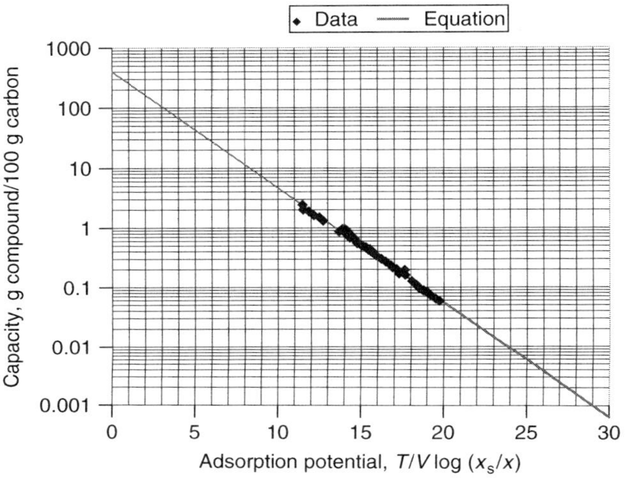

Figure 3. Adsorption capacity of activated carbon for chloroalkanes in water.

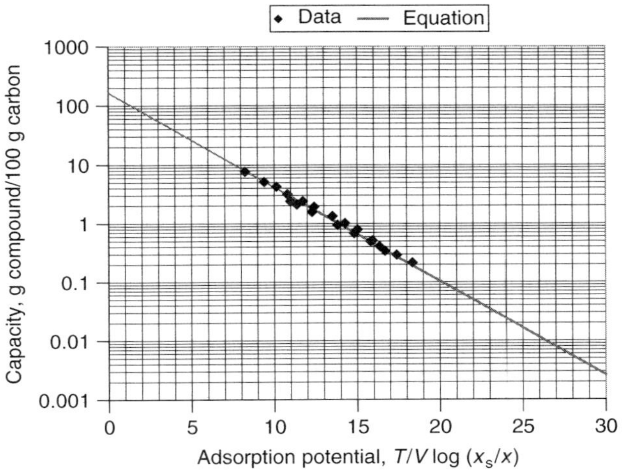

Figure 4. Adsorption capacity of activated carbon for chloro Olefins in water.

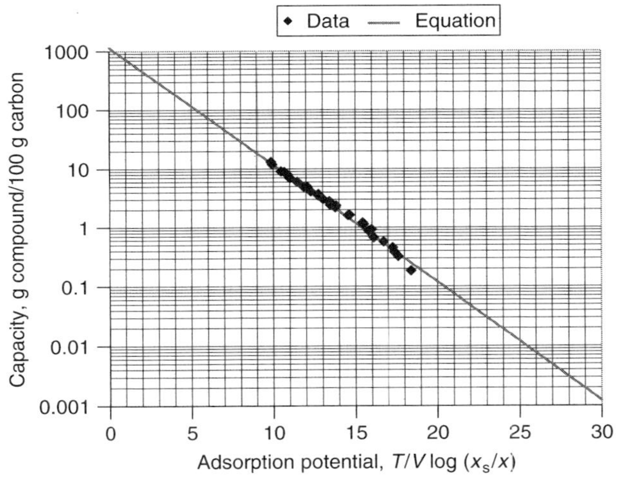

Figure 5. Adsorption capacity of activated carbon for bromoalkanes in water.

Example 1. A chemical spill of n-hexanol occurs in a body of water at ambient conditions. Additional data are

Concentration in water, ppm(wt)	96.31
Temperature, K	298.15
Molecular weight, g/mol	102.177
Liquid density, g/cm^3	0.816
Liquid molar volume, cm^3/mol	125.216
Solubility in water, ppm(wt)	5875

Estimate the adsorption capacity of activated carbon to purify the water.

The correlation for adsorption capacity may be used. Substitution of the coefficients for alcohols and above data in the correlation equation yields

$$\log_{10} Q = 1.7250 - 0.1210 \times [298.15/125.216 \log_{10}(5{,}875/96.31)] = 1.21063$$

$$Q = 16.24 \text{ g } n\text{-hexanol}/100 \text{ g carbon}.$$

The calculated value and data compare favorably (16.24 vs. 17.79; deviation = 1.55/17.79 = 8.7%).

Example 2. A chemical spill of toluene occurs in a body of water at ambient conditions. Additional data are

Concentration in water, ppm(wt)	2.211
Temperature, K	298.15
Molecular weight, g/mol	92.141
Liquid density, g/cm^3	0.865
Liquid molar volume, cm^3/mol	106.521
Solubility in water, ppm(wt)	542.4

Estimate the adsorption capacity of activated carbon to purify the water.

The correlation for adsorption capacity may be used. Substitution of the coefficients for benzenes and the above data in the correlation equation yields

$$\begin{aligned}\log_{10} Q &= 1.8893 - 0.1199 \\ &\quad \times [298.15/106.521 \log_{10}(542.4/2.211)] \\ &= 1.0874 \\ Q &= 12.23 \text{ g toluene}/100 \text{ g carbon}\end{aligned}$$

The calculated value and data compare favorably (16.23 vs. 12.96; deviation = 0.73/12.96 = 5.6%).

Acknowledgments
The Texas Hazardous Waste Research Center (Lamar University, Beaumont, Texas) provided partial support for this work.

BIBLIOGRAPHY

1. Arbuckle, W.B. (1981). *Environ. Sci. Technol.* **15**(7): 812.
2. Sontheimer, H., Crittenden, J.C., and Summers, R.S. (1988). *Activated Carbon for Water Treatment*, 2nd Edn. DVGW—Forschungsstelle, FRG.
3. Speth, T.F. (1986). Masters Thesis, Michigan Technological University, Houghton, MI.
4. Speth, T.F. and Miller, J.R. (1990). *J. AWWA* **72**(February).
5. Howard, P.H. and Meylan, W.M. (Eds.). (1997). *Handbook of Physical Properties of Organic Chemicals*. CRC Press, Boca Raton, FL.
6. Mackay, D., Shiu, W.Y., and Ma, K.C. (1992, 1992, 1993, 1995, 1997). *Illustrated Handbook of Physical-Chemical Properties and Environmental Fate for Organic Chemicals*. Vols. 1–56. Lewis, New York.
7. Verschueren, K. (1996, 2001). *Handbook of Environmental Data on Organic Chemicals*, 3rd and 4th Edn. Van Nostrand Reinhold, New York.
8. Yalkowsky, S.H. (1990–2004). *Aquasol Database*. University of Arizona, Tucson, AZ.
9. Yaws, C.L. (1999). *Chemical Properties Handbook*. McGraw-Hill, New York.
10. Yaws, C.L. (2003). Yaws Handbook of Thermodynamic and Physical Properties of Chemical Compounds. electronic ed. www.knovel.com, Knovel Corporation, Norwich, NY.

ADSORPTION OF ORGANIC COMPOUNDS

Suparna Mukherji
IIT Bombay
Powai, Mumbai, India

INTRODUCTION

Organic compounds in the aqueous phase have a strong tendency to interact with surfaces, which results in their immobilization on the surfaces and consequent removal from the aqueous phase. This tendency is particularly high for nonpolar hydrophobic organic compounds (HOCs) characterized by low aqueous solubility. In water treatment applications, low levels of HOC contamination of both natural and anthropogenic origin are effectively removed by passage through a bed packed with activated carbon. In the subsurface environment, organic compounds in groundwater have a strong tendency to associate with soil. Thus, in contaminated zones soils and sediments are heavily loaded with organic contaminants, and the loading on these natural geosorbents per unit mass is eventually significantly higher than the aqueous phase concentration. The desorption of sorbed contaminants from soil acts as a long-term source, contaminating groundwater over decades. The association of HOCs with soil is typically termed sorption, which is inclusive of both absorption (partitioning) and adsorption (interface accumulation) (1).

Although sorption is primarily considered to be a reversible process, it significantly affects the fate and transport of organic compounds (2). It affects transformation processes such as photolysis and biodegradation. Thus, sorbed compounds are not readily available to microorganisms even though they may be readily biodegradable from aqueous solution. Aside from the fraction of the organics that are irreversibly bound to the surfaces, through reactions such as polymerization reactions, the reversibly bound fraction can desorb back into aqueous solution. However, the rate of desorption varies with the degree of penetration into the surface sites on the solids. A part of the reversibly bound fraction desorbs rapidly, while the remaining desorbs at a very slow rate over an extended period of time. This is responsible for the hysteresis and aging phenomena often reported in the literature. Hysteresis implies that isotherms relating the loading of a compound on a sorbent to aqueous phase concentration at equilibrium obtained in adsorption studies are not valid for desorption. Aging is the phenomenon of slow desorption observed for sorbents that have been exposed to pollutants over a long period of time. Thus, desorption experiments in laboratory contaminated soil may not be representative of field contaminated soil.

EQUILIBRIUM CONSIDERATIONS IN SORPTION OF HYDROPHOBIC ORGANIC COMPOUNDS ON GEOSORBENTS

Adsorption Isotherms

The maximum loading of an organic compound on a sorbent (q_e, mg/g) is a function of the solution phase concentration (C_e, mg/L) at equilibrium. This relationship is defined for a constant temperature and is known as an isotherm. Since the characteristics of the geosorbents are highly variable, each geosorbent yields a unique isotherm that needs to be obtained experimentally. Isotherms are obtained by setting up multiple batch reactors (sealed bottles) in which a fixed volume of the aqueous phase (V, L) containing varying concentrations of the organic compound (C_o, mg/L) is exposed to a fixed mass of sorbent (M, g). Parallel control reactors without sorbent are also included to determine the system losses. The reactors are tumbled end over end until equilibrium is achieved, and the loading on the sorbent (q_e) is determined by liquid phase mass balance as shown in Eq. 1 (1):

$$q_e = \frac{(C_o - C_e)}{(M/V)} \quad (1)$$

Equation 2 indicates how q_e can be determined after loss correction; it assumes system losses are linearly related to C_e and yields a constant slope S, which is estimated using the controls (3):

$$q_e = \frac{[C_o - C_e(1 + S/V)]}{(M/V)} \quad (2)$$

The experimentally obtained isotherms are fitted to models, such that the model and its parameters indicate the nature of the relationship between q_e and C_e. Three models most commonly used for fitting sorption data are the linear model, the Langmuir model, and the Freundlich model. The model expression, parameters, and the linear transformation typically used for parameter estimation are provided in Table 1 (1).

The constant K_D in the linear model can be determined by one-parameter linear regression. The constants Q_a^o and b in the Langmuir model and the constants K_F and n in the Freundlich model can be determined by two-parameter linear regression by using appropriate data transformations. However, a direct estimation of model parameters by nonlinear regression, which can readily be performed using statistical packages, is more desirable since the transformations performed distort the error structure, thus violating a basic assumption inherent in least squares linear regression (4). The Langmuir model assumes monolayer sorption, reversible sorption, and equal energy sorption sites. It is linear at low concentrations and levels off at a high concentration and shows the characteristic shape depicted in Fig. 1. The Freundlich model is empirically derived and typically provides the best fit to data since it allows for multilayer sorption and heterogeneous sites with variable energies (1). The characteristic shape of this isotherm for various n values is demonstrated in Fig. 1. Low n values, significantly lower than unity, are indicative of greater heterogeneity in site energy distribution and more favorable sorption for low C_e values. The K_F value reflects the sorption capacity.

Sorption in Various Geosorbent Domains

Sorption of nonpolar organic compounds on geosorbents can occur in various domains that are distributed within

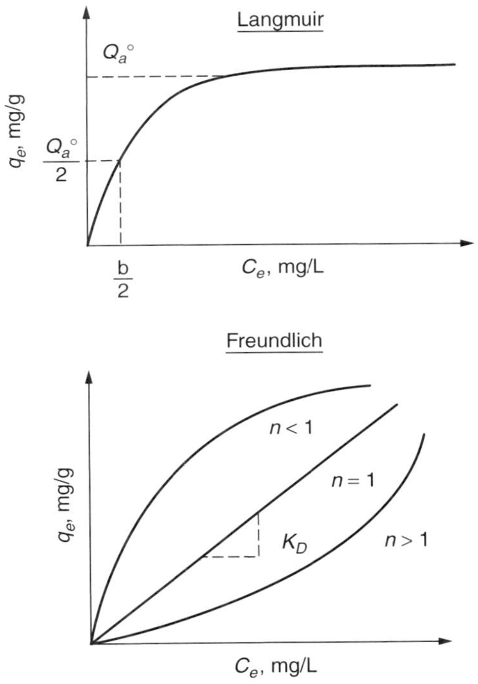

Figure 1. Characteristic shape of the Langmuir and Freundlich isotherm.

Table 1. Isotherm Models and Linear Transformations for Parameter Estimation

Model	Equation	Parameters	Linearization[a] ($y = a + mx$) and Parameter Estimation
Linear	$q_e = K_D C_e$ $K_D = f_{OC} K_{OC}$	K_D, distribution coefficient; f_{OC}, fraction organic carbon in sorbent; K_{OC}, organic carbon normalized coefficient	$a = 0, K_D = m$
Langmuir	$q_e = \frac{Q_a^o b C_e}{1 + bC_e}$	Q_a^o, monolayer sorption capacity; b, relates to sorption energetics	Reciprocal transform: $\frac{1}{q_e}$ versus $\frac{1}{C_e}$ $Q_a^o = 1/a; b = a/m$
Freundlich	$q_e = K_F C_e^n$	K_F, relates to sorption capacity; n, relates to sorption energetics	Log transform: $\log q_e$ versus $\log C_e$ $K_F = 10^a; n = m$

[a] Here a is the intercept and m is slope.

the geosorbent. Thus, the overall q_e is the sum total of q_e in the various domains. The domains have been conceptualized as the amorphous organic matter domain, the condensed organic matter domain, and the mineral domain. While organic matter is typically considered to be of natural origin, organic matter of anthropogenic origin can also exert a significant influence when present (5). Such organic matter includes nonaqueous phase liquids and combustion residues such as soot.

Sorption in Amorphous and Condensed Organic Matter Domains. It is well established that soil organic matter has a strong influence on the sorption of organic compounds and a linear correlation exists over a wide range of organic carbon content from about 0.1% to 20%. The distribution coefficient K_D for sorption of an organic compound on soil (assuming linear isotherm) can thus be normalized using the fraction organic carbon on a soil to yield the organic carbon normalized coefficient (K_{OC}). The K_{OC} value is strongly dependent on the organic compound of interest and correlates strongly with the octanol–water partition coefficient, K_{OW}, and the solubility S of the organic compounds. Linear correlations between log K_{OC} and log K_{OW} have been developed for specific groups of compounds as summarized in various reviews and handbooks (6,7). However, these values should be used as preliminary estimates.

It is well established that the linear model is often inadequate and the Freundlich model provides better fits to data, particularly when data is obtained over a wide C_e range. For phenanthrene sorption on various sorbents, Weber et al. (8) have demonstrated how the individually measured (single point) K_D and corresponding K_{OC} values vary as a function of C_e, when sorption is well characterized by the Freundlich isotherm. This has been attributed to the heterogeneity of sorption sites. The organic matter in the sorbent includes proteins, lignin, cellulose, and macromolecules that are altered over geologic time scales. This organic matter originates from organisms, such as microorganisms (algae and bacteria), zooplanktons and phytoplanktons, and vascular plants, depending on the depositional environment of the sorbents.

The linear model was originally proposed for sorption of organic compounds on various soils. Soil organic matter is proposed to be amorphous in nature and is often termed “soft” carbon, which is easily oxidized (e.g., humic and fulvic acids). Sorption of organics on soft carbon is like a partitioning phenomenon; that is, organic contaminants are distributed over the entire volume of soil organic matter. This is essentially a physisorption process characterized by low sorption energies. The K_{OC} values that are determined from correlations based on K_{OW} and aqueous solubility widely reported in the literature are only representative of sorption in the soft organic matter fraction.

As the organic matter gets aged by weathering processes over geologic time scales, indicated by a lowering in the O/C and H/C ratios, it gets more condensed, the degree of cross-linking increases, its aromaticity increases, and it becomes “hard” and insoluble (difficult to oxidize), for example, kerogen present in shales. Sorption on condensed organic matter is a true adsorption phenomenon, characterized by accumulation of organic compounds at the interface. The condensed organic matter has a significantly higher sorption capacity for organic compounds compared to the amorphous organic matter, and the sorption nonlinearity increases (characterized by decrease in Freundlich exponent n) with increase in the degree of condensation as diagenesis progresses. The sorption energies are also higher as in chemisorption processes. The distributed reactivity model proposed by Weber et al. (9) indicates that sorption in geosorbents is the sum total of sorption in the individual components; however, shale particles in a subsurface sand, rich in condensed organic carbon, were found to dominate the sorptive interactions. The failure of subsurface remediation technologies designed using linear isotherms have been attributed to increased nonlinearity.

Physisorption involving weak intermolecular forces is a reversible process. Thus, sorption on soft amorphous carbon typically present in soil is reversible. The adsorption isotherm and desorption isotherm are coincident and both are linear, as shown in Fig. 2. For sorption on hard condensed carbon such as kerogen, the adsorption and desorption isotherms are not coincident and significant hysteresis exists as demonstrated in Fig. 2. Moreover, both sorption and desorption are characterized by significant nonlinearity.

Sorption in Mineral Domain. The mineral surfaces exert greater influence on the sorption of polar organic compounds and the sorption is strongly dependent on the solution pH, pK_a values of the polar organic compound, and pH_{PZC} (point of zero charge of the sorbent). For nonpolar organic compounds, although the influence of organic matter content and nature dominates the sorptive interaction, mineral surfaces can also play a role, particularly when the organic matter content is less than

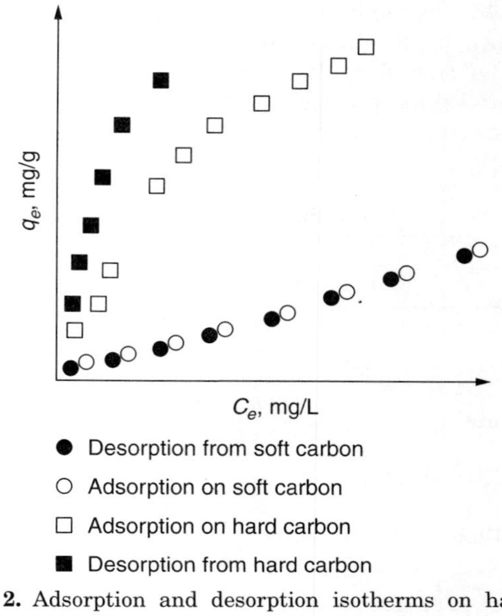

Figure 2. Adsorption and desorption isotherms on hard and soft carbon.

0.1–0.2%. Aquatic systems and subsurface environments are often characterized by low organic matter content. Moreover, the clay liners used in landfills can play a role in the retardation of organic compounds by virtue of sorptive interactions.

Sorption on minerals is dependent on the area of the mineral surface per unit mass of the sorbent. Area of the exposed mineral surface is higher when the organic carbon content is lower, since the natural organic matter competes with the organic compound for attaching on to the mineral surface. Thus, porous silica and the finer clay fraction exert a greater influence. Overall sorption also depends on the type of mineral surfaces that are present—clays (montmorillonite, kaolinite), silica, alumina, and iron oxide. Competition between sorption on natural organic matter and on mineral surfaces may depend on the proportion of swelling clays, the determining parameter being the ratio of clay surface area to the fraction of organic carbon. The interactions on mineral surfaces are typically due to van der Waals forces, that is, dipole–dipole, dipole–induced dipole, and hydrogen bonding interactions. Sometimes the interactions are stronger; chemisorption and clay colloid induced polymerization dominate sorption. Such specific binding of organics on the mineral surfaces may cause sorption to appear irreversible (7).

Factors Affecting Sorption Equilibrium

Factors other than geosorbent characteristics that affect sorption have been discussed elaborately by Schwarzenbach et al. (7). These include the following:

- *Temperature*: Adsorption is typically an exothermic process and K_D decreases with increasing temperature. Heat of adsorption associated with physisorption is typically a few hundred calories per degree per mole.
- *Nature of the Compound*: Within a group of organic compounds, K_D and K_{OC} increase as the aqueous solubility decreases and as the octanol–water partition coefficient K_{OW} increases. These linear free energy relationships have been widely utilized for obtaining preliminary estimates of the linear sorption coefficient.
- *Solution pH*: The effect of pH on the sorption of nonpolar organic compounds is not very significant. For ionizable organics the effect is significant.
- *Salinity*: As salinity increases, the aqueous solubility of organic compounds decreases; thus, K_D increases. However, typical levels of dissolved salts present in aquatic environments do not cause much impact.
- *Organic Colloids*: Colloids are macromolecules that remain in aqueous solution. Thus, in aqueous solution the organic compound is present as truly dissolved species and also as species associated with colloidal macromolecules. This causes an apparent lowering in the K_D values.

RATE CONSIDERATIONS IN SORPTION OF HYDROPHOBIC ORGANIC COMPOUNDS

The instantaneous local equilibrium assumption with respect to sorption processes is often made for modeling the transport of organic compounds through subsurface systems. These transport models assume that the sorption kinetics is so rapid that rate considerations can be ignored and that the sorption isotherm is linear (constant K_D). With these assumptions the net effect of sorption is manifested by a reduction in the advective velocity and dispersion coefficient by a constant factor, referred as the retardation factor (2). The retardation factor (R) can be estimated if K_D and porous media characteristics, such as the porosity (ε) and solid density (ρ_s) or bulk density (ρ_B), are known, as shown in Eq. 3:

$$R = 1 + \frac{\rho_B K_D}{\varepsilon} = 1 + \left(\frac{1-\varepsilon}{\varepsilon}\right)\rho_s K_D \qquad (3)$$

However, the assumption of instantaneous equilibrium is typically a poor assumption and this often leads to failure of the models in predicting the experimentally observed fate and transport phenomenon. Nonequilibrium sorption can affect the transport of contaminants by causing effects such as early breakthrough, decrease in peak breakthrough concentration, and tailing/asymmetry in the concentration profile.

Sorption rates are typically slow at the intraparticle level. The various mechanisms implicated in slow sorption include (1,10) low film transfer coefficient (k_f) for mass transfer across the boundary layer separating the sorbent from the aqueous phase; slow diffusion through natural organic matter matrices along the sorbent surfaces due to low surface diffusion coefficient (D_S); and slow diffusion through tortuous nanopores due to low pore diffusion coefficient (D_p). However, the relative contribution of the various mechanisms is difficult to predict. The adsorption–desorption bond energies may also play a role in determining the process kinetics. High energy adsorption sites within the organic matter matrix and in the nanopores may be responsible for slow sorption kinetics. A conceptualization of factors affecting adsorption rates is illustrated in Fig. 3. Film transfer and intraparticle diffusion constitute resistances in series, and since intraparticle diffusion is typically slower, it has a dominant impact on the slow kinetics. The two intraparticle diffusion mechanisms—surface and pore diffusion—represent parallel mechanisms, and the faster of the two is thought to control overall mass transfer.

Typically, the sorption and desorption of organics on geosorbents is characterized by a rapid phase, followed by a slow phase—that is, it is bimodal. The magnitude of the slow fraction is quite significant. Pignatello and Xing (10) have effectively illustrated how slow sorption can affect the experimentally determined K_D values. With increasing contact time in equilibrium studies from short (1–3 days) to long (several months), the K_D was demonstrated to increase from 30% to tenfold for a range of organic compounds. The time to reach true equilibrium in batch systems may vary from 1 wk to several months.

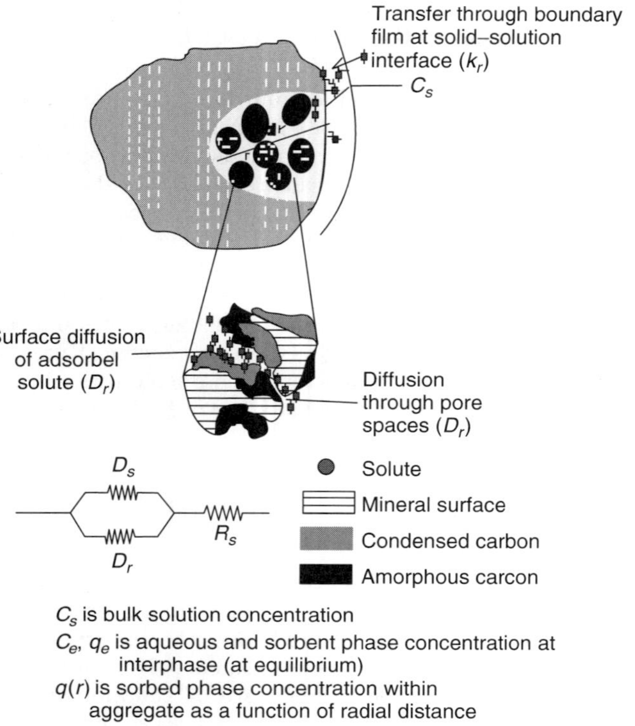

Figure 3. Rate considerations in sorption processes on a geosorbent.

Slow sorption leads to a chemical fraction that resists desorption. If sorption is allowed to occur over a long time period (aging), the contribution of slow sorption and a desorption resistant fraction increases; that is, the slow fraction following a rapid release may vary from 10% to 96% depending on aging time and the type of geosorbent. The desorption resistant fraction is persistent in natural environments and is also difficult to extract. Thus, for historically contaminated field samples, it is difficult to ensure complete recovery by the extraction techniques employed. Hatzinger and Alexander (11) effectively demonstrated how the biodegradation of phenanthrene and 4-nitrophenol was significantly reduced in aged soils.

Sorption–desorption kinetics is significantly affected by diffusion processes; however, Fickian diffusion alone cannot explain experimental and field-scale observations. It has been reported that the slow fraction is inversely dependent on the initial concentration. This is partly explained by isotherm nonlinearity (Freundlich exponent significantly lower than unity); however, it also appears that intraparticle retardation increases as the concentration within a particle declines. Moreover, sorption is reported to be kinetically hysteretic—the rate of filling the slow sites appears to be faster than the rate at which they are emptied, thus indicating an apparent irreversibility. Several researchers have used stochastic modeling for describing the slow sorption phenomenon on geosorbents; that is, the rate constants and diffusion coefficients are assumed to be distributed according to a statistical density function (e.g., two-parameter gamma distribution).

BIBLIOGRAPHY

1. Weber, W.J., Jr. and DiGiano, F.A. (1996). *Process Dynamics in Environmental Systems*. John Wiley & Sons, Hoboken, NJ.
2. Weber, J.J., Jr., McGinley, P.M., and Katz, L.E. (1991). Sorption phenomena in subsurface systems: concepts, models and effects on contaminant fate and transport. *Water Res.* **25**: 499–528.
3. Lion, W.L., Sauffer, T.B., and MacIntyre, W.G. (1990). Sorption of hydrophobic compounds on aquifer materials: analysis methods and the effect of organic carbon. *J. Contam. Hydrol.* **5**: 215–234.
4. Berthouex, P.M. and Brown, L.C. (1994). *Statistics for Environmental Engineers*. Lewis Publishers CRC Press, Boca Raton, FL, pp. 273–278.
5. Luthy, R.G. et al. (1997). Sequestration of hydrophobic organic contaminants by geosorbents. *Environ. Sci. Technol.* **31**: 3341–3347.
6. Lyman, W.J. (1990). Adsorption coefficient for soils and sediments. In: *Handbook of Chemical Property Estimation Methods*. W.J. Lyman, W.F. Reehl, and D.H. Rosenblatt (Eds.). ACS, Washington, DC, Chap. 4.
7. Schwarzenbach, R.P., Gschwend, P.M., and Imboden, D.M. (1993). *Environmental Organic Chemistry*. John Wiley & Sons, Hoboken, NJ, Chap. 11.
8. Weber, W.J., Jr., Huang, W., and LeBoeuf, E.J. (1999). Geosorbent organic matter and its relationship to the binding and sequestration of organic contaminants. *Colloids Surf. A* **151**: 167–179.
9. Weber, W.J., Jr., McGinley, P.M., and Katz, L.E. (1992). A distributed reactivity model for sorption by soils and sediments. 1. Conceptual basis and equilibrium assessments. *Environ. Sci. Technol.* **26**: 1955–1962.
10. Pignatello, J.J. and Xing, B. (1996). Mechanisms of slow sorption of organic chemicals to natural particles. *Environ. Sci. Technol.* **30**: 1–11.
11. Hatzinger, P.B. and Alexander, M. (1995). Effect of aging of chemicals in soil on their biodegradability and extractability. *Environ. Sci. Technol.* **29**: 537–547.

AGE DATING OLD GROUNDWATER

Gholam A. Kazemi
Shahrood University of Technology
Shahrood, Iran

INTRODUCTION

First, two closely interrelated terms, groundwater age and groundwater residence time, need to be explained and differentiated. Groundwater age is the amount of time that has lapsed since the water molecule, whose age is of concern, was recharged into the system till this molecule is sampled for age-dating. Groundwater residence time is the time it takes for a single molecule of water to travel from the recharge area to the discharge area of the aquifer, the time interval between infiltrating into, and exfiltrating out of, the subsurface media. Groundwater age and residence time are comparable to the age and lifetime of a person, respectively.

Groundwater age and residence time are illustrated in Figs. 1a,1b. Note that some researchers mix up these two

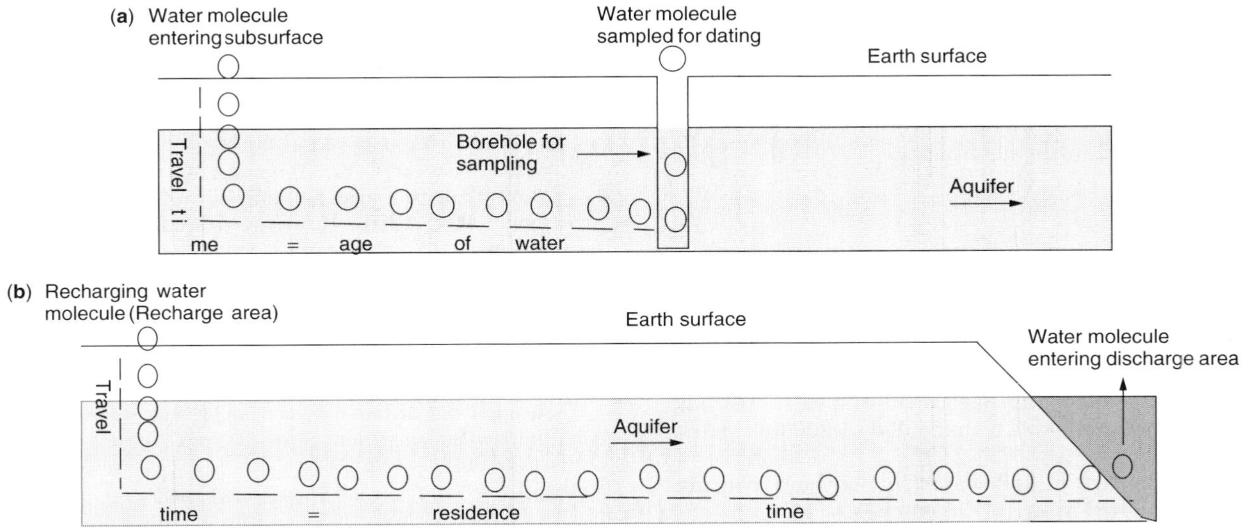

Figure 1. (**a**) Illustration of groundwater age (**b**) Illustration of groundwater residence time.

concepts and do not clearly differentiate between them. Mazor and Nativ (1, p. 219), for instance, refer to the age of groundwater as hydraulic age: "Hydraulic ages relate to travel time, the time needed for a volume of groundwater to move across a flow path Δl." Similarly, their definition of residence time "or isotopic age" is no different from the definition of "groundwater age" given by many other researchers. A number of workers use slightly different terms for groundwater age such as mean age, transit time, turnover time (2) or mean travel time (3). A recent useful discussion on groundwater age is that by Bethke and Johnson (4).

DISCUSSION

There is neither clear definition nor many citations of the term "old groundwaters" in the literature (in contrast to "young groundwaters"). Examples of limited available references include Coplen (5), Zhao et al. (6), Le Gal La Salle et al. (7), and Park et al. (8).

Old groundwaters may be defined as those that are too old to be dated using techniques like ^3H, CFCs, ^{85}Kr, $^{36}Cl/Cl$ and $^3H-^3He$. These techniques are specifically applied to age-date young groundwater. Four methods have been in use for dating old groundwaters. In the order of the age range they can measure, these include (1) ^{39}Ar, (2) ^{14}C, (3) ^{36}Cl, and (4) ^4He. A brief description of each dating technique and example(s) from each method follow. Note that other less known/used methods such as ^2H, ^{18}O, ^{32}Si, and ^{81}Kr for dating old groundwater have received very little attention (9).

^{39}Ar

Argon-39 whose half-life is 239 years can be used to date groundwater ages that range from 200 to 1000 years (10). This method has not been used widely.

^{14}C

The ^{14}C method is the most extensively used method for dating old groundwater. ^{14}C, also known as radiocarbon, has a half-life of 5730 years; it can enter groundwater systems by various geochemical and biological processes. The concentration of ^{14}C in soil CO_2 is 100 pmc (percent modern carbon) in prenuclear bomb times (pre-1952) and up to 200 pmc (about 130 pmc, on average) in postbomb years (11). Once water reaches the saturated zone of an aquifer, it is isolated from the atmosphere, and its ^{14}C decays. ^{14}C measurement is useful for dating groundwater ranging in age from 200 to 50,000 years. Values of 90 pmc or more, indicate post-1952 water, and values less than 50 pmc indicates ages up to 50,000 years.

Examples

1. The age of palaeochannel groundwaters of Western Australia is estimated at 20,000 years (12).
2. The ages of groundwater in the upper part of the Mojave River Basin, California, USA, and in the regional aquifer close to the Mojave River were estimated at less than 2400 and 20,000 years, respectively (13).
3. Le Gal La Salle et al. (7) computed that the age of groundwater in the Iullemeden Basin (Niger) is older than 25,000 years.
4. The age of groundwater brines in the Dead Sea area, determined by Yechieli et al., was 9000 years (14).
5. Kazemi (15) calculated that the age of groundwater in various parts of the Buckinbah Creek Catchment, Eastern Australia, is between 970 and 13,000 years.

^{36}Cl

^{36}Cl, whose half-life is 301,000 years, is useful for dating very old groundwater. It is produced naturally in the deep subsurface through thermal neutron capture on the stable nuclide chlorine-35; at and near the land surface by spallation and neutron capture related to secondary cosmic particle bombardment; by cosmic particle interaction with atmospheric gases, particularly argon; and artificially through the activation of ^{35}Cl in sea water by the high flux of neutrons from testing of fusion devices in the western Pacific starting in late 1952 and extending through mid-1958 (16).

Rainfall dissolves atmospheric ^{36}Cl, and carries it to the subsurface where decays, and its initial value is reduced according to the following equation:

$$C = C_0 e^{\lambda t}$$

where C is the concentration of ^{36}Cl at any time t, C_0 is the initial concentration of ^{36}Cl, and λ is the decay constant of $2.25\times10^{-6}\alpha^{-1}$. The longer ^{36}Cl resides in the aquifer, the less is its concentration, and hence the older the ambient groundwater. The subsurface production of ^{36}Cl is not large enough to affect the age calculations (8). The age range of the ^{36}Cl method is from 50,000–1,200,000 years.

Example. The age of soil water in Southern Nevada, USA, was estimated at up to 120,000 years (17).

Limitation. If chlorinity of the water sample exceeds ~75–150 mg/kg, the ^{36}Cl method cannot be used for groundwater dating (8). Note that the ratio of $^{36}Cl/Cl$ is used for dating young groundwater (18).

^4He

Helium accumulates in basin groundwater from radioactive decay of the uranium and thorium in the basin strata, as well as from any flux into the basin from the underlying crust and mantle (6). If the rate at which the *in situ* production and basal flux have supplied helium to flowing groundwater is known, then it is possible to calculate the length of time that the groundwater of known helium content has resided in the subsurface. The age range of this method is 10^4 to 10^8 years and due to semiquantitative nature of the method, it is referred to by some researchers as an age indicator rather than a dating method (11).

Example. Using this method, the age of the Baden region of Switzerland was estimated at several hundred thousand years (10).

BIBLIOGRAPHY

1. Mazor, E. and Nativ, R. (1992). Hydraulic calculation of groundwater flow velocity and age: Examination of the basic premises. *J. Hydrol.* **138**: 211–222.
2. Maloszewski, P. and Zuber, A. (1982). Determining the turnover time of groundwater systems with the aid of environmental tracers. 1. Models and their applicability. *J. Hydrol.* **57**: 207–231.
3. Portniaguine, O. and Solomon, D.K. (1998). Parameter estimation using groundwater age and head data, Cape Cod, Massachusetts. *Water Resour. Res.* **34**(4): 637–645.
4. Bethke, C.M. and Johnson, T.M. (2002). Paradox of groundwater age: Correction. *Geology* **30**(4): 385–388.
5. Coplen, T.B. (1993). Uses of environmental isotopes. In: *Regional Ground-water Quality*. W.M. Alley (Ed.). ITP, New York, pp. 255–293.
6. Zhao, X. et al. (1998). Controls on the distribution and isotopic composition of helium in deep ground-water flows. *Geology* **26**(4): 291–294.
7. Le Gal La Salle, C., Fontes, J.C., Andrews, J.N., Schroeter, P., Karbo, A., and Fifield, K.L. (1995). Old groundwater circulation in the Iullemeden basin (Niger): Preliminary results of an isotopic study. In: *Application of Tracers in Arid-Zone Hydrology*. E.M. Adar and C. Leibundgut (Eds.). *IAHS publications* no. 232, IAEA, Vienna, pp. 129–139.
8. Park, J., Bethke, C.M., Torgerson, T., and Johnson, T.M. (2002). Transport modeling applied to the interpretation of groundwater ^{36}Cl age. *Water Resour. Res.* **38**(5): 1–15.
9. Plummer, L.N., Michel, R.L., Thurman, E.M., and Glynn, P.D. (1993). Environmental tracers for age dating young groundwater. In: *Regional Ground-Water Quality*. W.M. Alley (Ed.). ITP, New York, pp. 227–254.
10. Mazor, E., Jaffe, F.C., Fluck, J., and Dubois, J.D. (1986). Tritium corrected ^{14}C and atmospheric noble gas corrected ^4He applied to deduce ages of mixed groundwaters: Examples from the Baden region, Switzerland. *Geochim. Cosmochim. Acta* **50**: 1611–1618.
11. Mazor, E. (1991). *Applied Chemical and Isotopic Groundwater Hydrology*. Open University Press, Milton Keynes.
12. Turner, J.V., Rosen, M.R., Fifield, L.K., and Allan, G.L. (1995). Chloride-36 in hypersaline palaeochannel groundwaters of Western Australia. In: *Application of Tracers in Arid-Zone Hydrology*. E.M. Adar and C. Leibundgut (Eds.). *IAHS publications* no. 232, IAEA, Vienna, pp. 15–33.
13. Izbicki, J.A., Martin, P., and Michel, R.L. (1995). Source, movement and age of groundwater in the upper part of the Mojave River Basin, California, USA. In: *Application of Tracers in Arid-Zone Hydrology*. E.M. Adar and C. Leibundgut (Eds.). *IAHS publications* no. 232, IAEA, Vienna, pp. 43–56.
14. Yechieli, Y., Ronen, D., and Kaufman, A. (1996). The source and age of groundwater brines in the Dead Sea area, as deduced from ^{36}Cl and ^{14}C. *Geochimica Cosmochim. Acta* **60**(11): 1909–1916.
15. Kazemi, G.A. (1999). *Groundwater Factors in the Management of Dryland and Stream Salinity in the Upper Macquarie Valley, New South Wales, Australia*. Unpublished Ph.D. Thesis, University of Technology, Sydney, Australia.
16. Davis, S.N., Moysey, S., Cecil, L.D., and Zreda, M. (2003). Chlorine-36 in groundwater of the United States: Empirical data. *Hydrogeol. J.* **11**: 217–227.
17. Tyler, S.W., Chapman, J.B., Conrad, S.H., and Hammermeister, D. (1995). Palaeoclimate response of a deep vadose zone in southern Nevada, USA, as inferred from soil water tracers. In: *Application of Tracers in Arid-Zone Hydrology*. E.M. Adar and C. Leibundgut (Eds.). *IAHS publications* no. 232, IAEA, Vienna, pp. 351–361.
18. Cook, P.G. and Solomon, D.K. (1997). Recent advances in dating young groundwater: Chlorofluorocarbons, $^3H/^3He$ and ^{85}Kr. *J. Hydrol.* **191**: 245–265.

AMMONIA

Robert M. Burgess
U.S. Environmental Protection Agency
Narragansett, Rhode Island

Ammonia is a primary breakdown product of nitrogen-containing organic matter in aquatic environments. Because a great deal of organic matter and waste is introduced into the hydrologic cycle by natural and anthropogenic processes, ammonia can be found in aquatic environments, especially those exposed to large amounts of human or domestic animal wastes. Ammonia is also of great importance in understanding the global cycling of several key elements, including nitrogen and carbon. Ammonia is one of the compound uniquely critical to life which when not present in adequate amounts in aquatic environments can be associated with negative biological effects and when in excess can result in adverse ecological and toxic effects. When insufficient ammonia is available to plants, which assimilate ammonia to grow, they may experience metabolic problems due to nutrient limitations. Conversely, excess ammonia is commonly associated with eutrophic conditions in water bodies and toxicity to some organisms. Further, nitrous oxide (N_2O) a molecule involved in the environmental cycling of ammonia is considered a potential 'greenhouse gas' which can also destroy stratospheric ozone. Because of these adverse effects, ammonia is frequently considered a pollutant in aquatic systems.

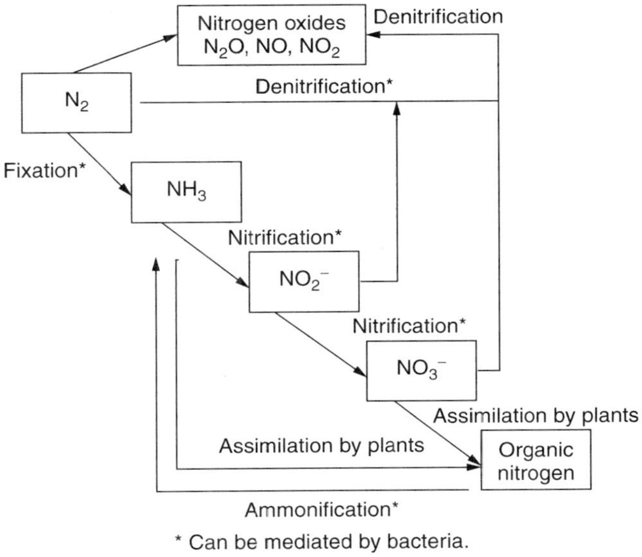

Figure 1. Nitrogen cycle.

BACKGROUND

The word 'ammonia' first came into chemical-related use in approximately 1799 and was derived from the Egyptian deity, Ammon. Ammonia referred to the gas emitted by camel excreta in the vicinity of temples honoring Ammon. The form of ammonia known as ammonium (NH_4^+) was named by the eminent Swedish chemist Berzelius in 1808.

Today, it is known that ammonia consists of one nitrogen atom and three hydrogen atoms (NH_3) and has a molecular weight of 17. Ammonia is formed from the bacterial decomposition of organic matter containing nitrogen. For example, the degradation of the simple organic molecule urea [$(NH_2)_2CO$] in water:

$$(NH_2)_2CO(l) + H_2O(l) \longrightarrow 2\ NH_3(g) + CO_2(g) \quad (1)$$

results in the formation of the gases, un-ionized ammonia (NH_3) and carbon dioxide (CO_2). In the same way, complex organic molecules in industrial and municipal wastes and plant and animal matter also form NH_3 when degraded. In aqueous solution, ammonia exists in two forms as a function of pH, ionic strength, and temperature:

$$NH_4^+ + OH^- \longleftrightarrow NH_3 + H_2O \quad (2)$$

Ionic strength and temperature influence ammonia speciation, but pH is the controlling variable. Consequently, with a base dissociation constant (pK_b) of 9.5, at a pH less than 9.5, most of the ammonia in solution is ionic ammonium (NH_4^+); at a pH greater than 9.5, the majority is NH_3. Consequently, in seawater and freshwater, the majority of ammonia is in the NH_4^+ form.

CYCLING

As noted above, ammonia is a key component of the greater global nitrogen cycle. As shown in Fig. 1, the fixation of nitrogen gas (N_2) forms NH_3, and the ammonification of organic nitrogen also results in forming ammonia. Ammonia is assimilated by plants and used to form organic nitrogen molecules, including amino acids, and then proteins. Conversely, ammonia can be formed by the ammonification of amino acids and proteins. Through nitrification, ammonia can also be oxidized to form nitrite (NO_2^-) which is further oxidized to form nitrate (NO_3^-). Nitrate is another nitrogen molecule that is assimilated by plants. Anaerobic conditions retard nitrification and cause an elevation of NH_3 concentrations. Denitrification of nitrite and nitrate results in forming N_2 and several nitrogen oxides, including N_2O, which completes the nitrogen cycle. Nitrification, denitrification, and ammonification are all reactions that occur abiotically but can also be moderated by bacteria. In nature, fixation requires microbial mediation. Synthetic fixation can be performed using the process developed by Haber and Bosch in 1910–1914 to produce fertilizer. Energetically, the path leading to the formation of organic nitrogen compounds is most favored compared to the reverse path resulting in nitrogen gas (Fig.1).

Equation 1 gives an example of NH_3 formed by ammonification. The formation of NH_3 by fixation of N_2, as shown in Equation 3, is more complicated and involves linking the nitrogen cycle to the carbon cycle. In this reaction, a simple carbon molecule (CH_2O) is oxidized to form CO_2; the inclusion of N_2 and an exchange of hydrogen ions (H^+) and electrons (e^-) results in ammonium formation:

$$CH_2O(s) + H_2O(l) + N_2(g) + 8H^+ + 6e^- \longrightarrow CO_2(g) + 4H^+ + 4e^- + 2\ NH_4^+(l) \quad (3)$$

Without the involvement of the carbon cycle, this reaction would not be thermodynamically favored, and N_2 fixation would not take place. Because Equation 3 is thermodynamically favored, microbes mediate the reaction and acquire energy.

SOURCES AND CONCENTRATIONS

Several studies have attempted to quantify ammonia sources and concentrations around the world. In these studies, the atmosphere is considered the ultimate sink for global ammonia, although, it should be noted, other studies speculate that the oceans are the ultimate sink. Primary sources of ammonia to the global environment from four different studies are listed in Table 1. Despite the different information used to assemble the table, the sources show remarkable similarity. The range of total ammonia released to the environment is 45 to 75 TgN yr^{-1}. Clearly, domesticated animals are the largest sources of ammonia globally, followed by soil, the oceans, and fertilizers. Lesser amounts of ammonia originate from biomass burning and human waste, and even lesser from fuel combustion and automobiles. The primary sources of ammonia are all systems containing bacteria that significantly enhance the nitrogen fixation and ammonification reactions discussed before.

As may be expected, concentrations of ammonia in aquatic systems vary based on sources and environmental conditions (Table 2).

Overlying fresh and marine waters that are well mixed, aerobic, and lack a direct source of ammonia maintain relatively low concentrations in the high parts per billion range ($\mu g\ L^{-1}$ or ppb); sediment interstitial waters, especially those that are anoxic, support elevated levels in the lower parts per million range ($mg\ L^{-1}$ or ppm). Sewage effluents and sludges can have extremely high concentrations of ammonia from human and animal waste. As a concentrated form of nitrogen-containing organic matter, sludge is potentially very elevated in concentrations of ammonia. Because of the elevated concentrations of ammonia in sewage effluents, waste treatment facilities frequently aerate the liquid to enhance nitrification and counter anaerobic conditions. Finally, rainwater can have elevated ammonia concentrations as a result of the diffusion of NH_3 into precipitation from the atmosphere.

Table 1. Sources of Ammonia in the Environment

Source	Amount, TgN yr^{-1}[a]			
	Study 1	Study 2	Study 3	Study 4
Domestic animals	32	26	24.5	22
Ocean surface	13	—	7	8
Undisturbed soils	10	15	5.1	6
Fertilizers	9	3	6.4	9
Biomass burning	5	5	2	6
Human waste	4	3	—	3
Coal combustion	2	< 2	—	0.1
Automobiles	0.2	0.2	—	0.2
Total	75	54	45	54

[a]Teragrams nitrogen per year (tera = 10^{12}).

Table 2. Ammonia Concentrations in Aquatic Samples

Aquatic sample	Total Ammonia Concentration ($mg\ L^{-1}$)
Coastal marine waters	0.24–0.58
Freshwaters	<1.0
Marine sediment interstitial waters	1–11
Freshwater sediment interstitial waters	1–53
Sewage effluent	16–33
Sewage sludge	1000–2030[a] ($mg\ kg^{-1}$)
Rainwater	0.05–2.60[a]

[a]Expressed as NH_4^+.

EUTROPHICATION

Eutrophication is the process of enriching a water body with nutrients (see *Eutrophication*). In marine systems, nitrogen-containing compounds (e.g., NO_3^-, NH_3) most often cause eutrophic conditions, in freshwater systems, phosphorus-containing compounds are currently considered most important. In most instances of eutrophication, anthropogenic sources of nutrients, including fertilizers and domestic animals, are suspected as the causes. Consequently, eutrophication is considered a major environmental problem in areas where human populations are present. Eutrophication causes several adverse environmental effects. The first effect is the occurrence of large algal blooms. These blooms can reduce light penetration into the water throughout the system which may adversely affect other vegetative life. Next, the accumulation of decaying organic matter enhances the growth of bacteria which causes a decline in water column dissolved oxygen concentrations. The resulting anaerobic conditions can kill aquatic animals and foster an increase in ammonia concentrations in the water and adjacent sediments. Because anaerobic environments support the persistence of ammonia, it is frequently associated with eutrophic conditions. Thus, the presence of ammonia perpetuates the potential for eutrophic conditions to arise. Another adverse effect of eutrophication and elevated concentrations of ammonia is volatilization of NH_3 to the atmosphere followed by transport and deposition in other regions. The occurrence of eutrophication has many implications, including severe disruptions of ecological processes and potential ammonia-related toxicity to aquatic organisms.

TOXICITY

Ammonia toxicity to aquatic organisms is extremely well documented. Table 3 has a small collection of selected ammonia toxicity values. The toxicity of ammonia is dominated by the NH_3 form; NH_4^+ demonstrates little contribution to overall toxicity. Because the concentration of NH_3 is a direct function of solution pH, very frequently, the toxicity of total ammonia to organisms is observed to increase as pH increases. This trend is explained by the commensurate increase in NH_3 concentration as pH increases. Interestingly, the toxicity of NH_3 actually decreases as pH increases, indicating that NH_3 is less toxic at higher pHs.

Ammonia toxicity, as measured by effects on survival, tends to occur in the low parts per million concentration range, although this varies by species. Sublethal effects to

Table 3. Ammonia toxicity to aquatic organisms

Species	Medium	End Point	Toxicity Value[a] (mg/L)
Mysid shrimp (*Americamysis bahia*)	Marine	Mortality	2.82
Mysid shrimp (*Americamysis bahia*)	Marine	Sublethal	0.23
American lobster (*Homarus americanus*)	Marine	Mortality	2.21
Inland silverside (*Menidia beryllina*)	Marine	Survival	0.88
Inland silverside (*Menidia beryllina*)	Marine	Sublethal	0.06
Striped bass (*Morone saxatilus*)	Marine	Mortality	0.91
Winter flounder (*Pseudpleuronectes americanus*)	Marine	Mortality	0.53
Cladoceran (*Daphnia magna*)	Freshwater	Mortality	2.69
Cladoceran (*Daphnia magna*)	Freshwater	Sublethal	0.63
Rainbow trout (*Oncorhynchus mykiss*)	Freshwater	Mortality	0.39
Rainbow trout (*Oncorhynchus mykiss*)	Freshwater	Sublethal	0.02
Smallmouth bass (*Micropterus dolomieui*)	Freshwater	Mortality	1.01
Smallmouth bass (*Micropterus dolomieui*)	Freshwater	Sublethal	0.60

[a]Mortality values are LC_{50}s; sublethal values are EC_{50}s.

aquatic organisms, including effects on growth and reproduction, occur in the high parts per billion concentration range but also vary by species. In general, fish species in both fresh and marine waters are the most sensitive to ammonia, as compared to invertebrate species. For example, Table 3 shows that in marine waters, the inland silverside (a minnow) is approximately three times more sensitive to ammonia than the mysid shrimp, based on a survival end point. Similarly, in freshwater, the rainbow trout is about seven times more sensitive than a cladoceran (Table 3). These differences in sensitivity are also apparent based on sublethal effects; for example, sublethal effects in trout occur at concentrations 30 times lower than sublethal effects in a cladoceran (Table 3)! The extreme sensitivity of fish species to ammonia has significant implications for the aquaculture and fisheries industries with regard to ensuring that their wastes do not adversely affect their products.

Because nitrogen-containing organic matter accumulates in sedimentary aquatic environments, elevated concentrations of ammonia are often found in sediments. This is especially true in eutrophic settings enriched with anthropogenically derived nutrients. Consequently, concerns about ammonia toxicity are frequently focused on sediment-dwelling organisms and not the water column species in aquatic systems. Of course, if there is evidence that ammonia from point (e.g., sewage treatment plant) or nonpoint sources is being released into aquatic systems, ammonia toxicity may be expected.

ANALYSIS

The two most common ways to measure ammonia in aqueous solution are by spectrometry and ion-selective electrodes. Both methods function well in freshwater and seawater, including clean and polluted samples. The spectroscopic method involves a reaction between ammonia and phenol and sodium hypochlorite with sodium nitroprusside to form indophenol blue which is measured by spectrometry. A variation on this method substitutes sodium salicylate for the phenol, eliminating the production of *o*-chlorophenol, a toxic substance, and resulting in comparable sensitivity. In the ion-selective method, changes in the measured electrode potential are compared to a reference potential. These potentials are proportional to the amount of NH_3 in the sample solution allowing for a determination of ammonia concentration. Both methods have advantages and disadvantages. The spectrometric method is the more sensitive technique, detection limits are in the mid- to high parts per billion range, but it suffers from being a relatively slow method which uses somewhat hazardous reagents. The ion-selective method is relatively fast allowing analyses to be completed in minutes but is relatively insensitive; detection limits are in the upper parts per billion to lower parts per million range. Both methods also differ in terms of the form of ammonia measured. Spectrometric methods measure NH_4^+ whereas the ion-selective method measures NH_3. Given that the two forms of ammonia behave differently in the environment, it is important to understand which is being measured.

READING LIST

Historical

Lewis, R.J. (1993). *Hawley's Condensed Chemical Dictionary*. Van Nostrand Reinhold Company, New York, p. 1275.

Partington, J.R. (1989). *A Short History of Chemistry*. Dover, New York, p. 415.

General Chemistry

Hampson, B.L. (1977). Relationship between total ammonia and free ammonia in terrestrial and ocean waters. *Journal of Cons. in Exploration of the Mer.* **37**: 117–122.

Khoo, K.H., Culberson, C.H., and Bates, R.G. (1977). Thermodynamics of the dissociation of ammonium ion in seawater from 5 to 40 °C. *Journal of Solution Chemistry* **6**: 281–290.

O'Neill, P. (1985). *Environmental Chemistry*. George Allen and Unwin, Boston, MA, p. 232.

Sources and Concentrations

Bini, C. and Bresolin, F. (1998). Soil acidification by acid rain in forest ecosystems: a case study in northern Italy. *The Science of the Total Environment* **222**: 1–15.

Friedl, G., Dinkel, C., and Wehrli, B. (1998). Benthic fluxes of nutrients in the northwestern Black Sea. *Marine Chemistry* **62**: 77–88.

Park, G.S. and Park, S.Y. (2000). Long-term trends and temporal heterogeneity of water quality in tidally mixed estuarine waters. *Marine Pollution Bulletin* **40**: 1201–1209.

Phillips, B.M., Anderson, B.S., and Hunt, J.W. (1997). Measurement and distribution of interstitial and overlying water ammonia and hydrogen sulfide in sediment toxicity tests. *Marine Environmental Research* **44**: 117–126.

Rosenfeld, J.K. (1979). Ammonium adsorption in nearshore anoxic sediments. *Limnology and Oceanography* **24**: 356–364.

Sarda, N. and Burton, G.A. (1995). Ammonia variation in sediments: spatial, temporal and method-related effects. *Environmental Toxicology and Chemistry* **14**: 1499–1506.

Schlesinger, W.H. and Hartley, A.E. (1992). A global budget for atmospheric NH_3. *Biogeochemistry* **15**: 191–211.

Valiela, I. (1984). *Marine Ecological Processes*. Springer-Verlag, New York, p. 546.

Wetzel, R.G. (1983). *Limnology*. W.B. CBS College Publishing, Saunders Company, New York, p. 767.

Wong, J.W.C., Li, K., Fang, M., and Su, D.C. (2001). Toxicity evaluation of sewage sludges in Hong Kong. *Environment International* **27**: 373–380.

Eutrophication and Toxicity

Ankley, G.T., Katko, A., and Arthur, J.W. (1990). Identification of ammonia as an important sediment-associated toxicant in the Lower Fox River and Green Bay, Wisconsin. *Environmental Toxicology and Chemistry* **9**: 313–322.

Burgess, R.M. et al. (2003). Removal of ammonia toxicity in marine sediment TIEs: a comparison of ulva lactuca, zeolite and aeration methods. *Marine Pollution Bulletin* **46**: 607–618.

Erickson, R.J. (1985). An evaluation of mathematical models for the effects of pH and temperature on ammonia toxicity to aquatic organisms. *Water Research* **19**: 1047–1058.

Kohn, N.P. et al. (1994). Acute toxicity of ammonia to four species of marine amphipod. *Marine Environmental Research* **38**: 1–15.

Morell, J.M. et al. (2001). Nitrous oxide fluxes in Caribbean and tropical Atlantic waters: evidence for near surface production. *Marine Chemistry* **74**: 131–143.

National Research Council. (2000). *Clean Coastal Waters: Understanding and Reducing the Effects of Nutrient Pollution*. National Academy of Sciences, National Academy Press, Washington, DC, p. 377.

Russo, R.C. (1985). Ammonia, Nitrite, and Nitrate. In: *Fundamentals of Aquatic Toxicology*. G.M. Rand and S.R. Petrocelli (Eds.). Hemisphere Publishing Corporation, New York, p. 666.

U.S. Environmental Protection Agency. (1989). *Ambient Water Quality Criteria for Ammonia (Saltwater)—1989*. EPA 440-5-88-004. Office of Regulation and Standards, Washington, DC.

U.S. Environmental Protection Agency. (1998). *1998 Update of Ambient Water Quality Criteria for Ammonia*. EPA 822-R-98-008. Office of Water, Washington, DC.

Viessman, W. and Hammer, M.J. (1985). *Water Supply and Pollution Control*. Harper & Row Publishers, New York, p. 797.

Analytical Chemistry

Bower, C.E. and Holm-Hansen, T. (1980). A salicylate-hypochlorite method for determining ammonia in seawater. *Canadian Journal of Fisheries and Aquatic Science* **37**: 794–798.

Gilbert, T.R. and Clay, A.M. (1973). Determination of ammonia in aquaria and in sea water using the ammonia electrode. *Analytical Chemistry* **45**: 1757–1759.

BERYLLIUM IN WATER

JAY CLAUSEN
AMEC Earth and
Environmental, Inc.
Westford, Massachusetts

INTRODUCTION

Beryllium was discovered by Vauquelin in 1798 as an oxide in beryl and emeralds. The metal was isolated in 1828 by Bussy and Wohler but its use was limited until it was discovered that a mixture with copper produced an alloy six times stronger than steel (1). Beryllium is a Group 2 (IIA) element found naturally as a rare alkaline earth metal in the earth's crust and is the lightest metal. As a pure metal, beryllium has a steel gray color and is present in a variety of materials, such as coal, oil, rock, soil, and volcanic dust. It can be found concentrated in igneous intrusive rocks such as granites, specifically beryllium-rich pegmatite. Beryllium ions are small enough to substitute for silicon in igneous rocks. Beryllium is found in some 50 mineral species; the most important are beryl [$Be_3Al_2(SiO_3)_6$], bertrandite [$Be_4Si_2O_7(OH)_2$], chrysoberyl [$BeAl_2O_4$], and phenacite [Be_2SiO_4]. These minerals can be present at elevated concentrations in water due to the weathering of the host rock. Beryllium, however, does not dissolve readily under natural conditions. For this reason, precious forms of beryl, better known as either aquamarine (blue or blue-green) or emerald (green), can be found in nature.

In addition to its presence in its natural state, beryllium can also be redistributed into the environment through anthropogenic activities such as burning coal (2–4) and oil and disposal of metal manufacturing waste (5). Prior to the 1940s, there were few applications for beryllium. Specialized metal components requiring lightweight and strength have resulted in its increased usage during the last 60 years. Most of the beryllium ore that is mined is converted into alloys (mixtures of metals). Beryllium can be found in computer, aircraft, spacecraft, and automotive parts; gyroscopes; sports equipment; nuclear reactor reflectors; X-ray machines; mirrors; and other specialized engineering components.

Beryllium metal and its salts are extremely toxic. Inhalation of fine beryllium dust can result in the lung disease, berylliosis, also known as chronic beryllium disease, and possibly cancer (6–9). The U.S. Department of Health and Human Services lists beryllium as a Class A carcinogen (2). As a consequence, beryllium is typically included as one of several metals analyzed as part of environmental investigations.

Numerous analytical methods are available for analyzing beryllium: inductively coupled plasma-mass spectrometry (ICP-MS), optical emission spectrometry (ICP-OES), atomic absorption spectrometry (AAS), colorimetric, and fluorimetric methods (10,11). A detailed listing of the colorimetric methods can be found in Taylor et al. (12). The preferred method for soil and water analysis in the environmental business is ICP-MS following United

States Environmental Protection Agency (USEPA) methods 200.7 and 6010. The ICP-MS instrumentation has greater sensitivity and reduced interference from other metals, compared to other analytical methods. A typical reporting limit of beryllium in environmental investigations using the ICP-MS method is 5 ppb; typical method of detection limits for soil and water are 0.1 μg/kg and 0.7 μg/L, respectively. AAS is also an USEPA approved process following method 7090 or 7091. Due to beryllium's low concentration and subsequent interference with other metals in environmental samples, a number of separation and preconcentration methods have been developed using chelating agents, ion-exchange resins, sorbants, and surfactants.

CHEMISTRY

Most beryllium oxides and hydroxides have very low solubilities, although the chloride and sulfate oxides are very soluble. However, the chloride and sulfate oxides hydrolyze to relatively insoluble hydroxide species. Baes and Mesmer (13) found that beryllium hydroxide solubilities range from 0.9 to 100 μg/L in the pH range of 6 to 8.5. Iron, manganese, and zinc beryllium silicates are believed more soluble because they contain divalent sulfur, which can be easily oxidized (14).

The dominant dissolved species found under environmental conditions are Be^{2+}, $BeOH^{+}$, $Be(OH)_2$, and $Be(OH)_3^{-}$. Thermodynamic data for the dominant beryllium species are presented in Table 1.

Figure 1, a phase diagram for beryllium under normal temperatures and atmospheric conditions at a concentration of 90 μg/L, suggests the solid species of BeO dissolves to form Be^{2+} under acidic conditions. Increasing the concentration of beryllium in solution results in a shift of the BeO to the Be^{2+} transition zone at higher pH. However, this is contrary to known environmental conditions, whereby beryllium can form insoluble mineral species such as beryl, emerald, and bertrandite. Brookins (18) suggested that the discrepancy between thermodynamic calculations and reality may be attributable to the presence of metastable species, which affect the solubility of the various beryllium species.

Table 1. Thermodynamic Data for Beryllium

Species[a,b]	$\Delta G^0{}_f$, kcal/gfw[c]	Reference
Be (metal)	0	15
Be^{2+} (aq)	− 90.76	16
BeO (s)	−138.70	16
$BeO_2{}^{2-}$ (aq)	−152.99	16
$Be_2O_3{}^{2-}$ (aq)	−298	17
Be_2O^{2+} (aq)	−218	17
$Be(OH)_2$ (s)	−194.79	16
BeO $\cdot$ $Be(OH)_{2(s)}$	−338	17
BeS (s)	−55.9	17
$BeCl_2$ (s)	−111.8	17
$BeSO_4$ (s)	−260.2	17

[a] Aq: aqueous.
[b] s: solid.
[c] Kcal/gfw–kilo calories per gram formula weight.

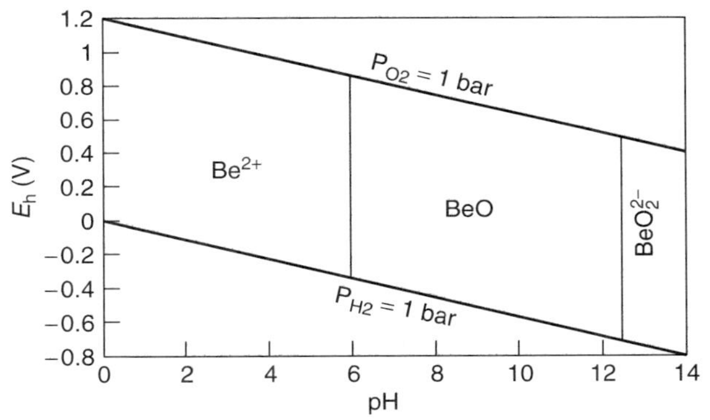

Figure 1. E_h–pH diagram for the Be–O–H system at a concentration of 90 μg/L under standard temperature and pressure.

SOURCE MATERIAL

Beryllium is a widely distributed metal that has a crustal average of 3 ppm, and its average concentration is 5 ppm in granitic rock, according to Krauskopf (19). The weathering product of rock is typically a kaolinite or hydrolysate, which accumulates in the B-soil horizon. Hydrobiotite, a feldspar mineral, weathers to montmorillonite, which sorbs beryllium (20). Beryllium concentrations in soil often correlate with aluminum and magnesium levels because both metals can be involved in cation exchange reactions with montmorillonite minerals. Berg et al. (21) found a correlation between aluminum and beryllium in a study near an aluminum smelter. Similarly, a study by Asami and Kubota (22) found a weak correlation, 0.505 at a level of significance of 0.1% for n=113 samples, between beryllium and aluminum for soils collected from several different countries. An earlier study by Asami and Fukazawa (23) found a similar relationship for soils collected in Japan near a beryllium manufacturing facility.

A number of studies have been conducted to determine the range of beryllium concentrations in surficial soils (0.3 to 6 ppm) throughout the world (22,24–34). A study by Anderson et al. (35) of four eastern U.S. Piedmont soils suggested higher beryllium concentration ranges for surface soil of 6.3 to 13.3 and 7 to 30.5 mg/kg for deeper soils. However, the number of samples collected was very small and insufficient to develop statistically meaningful values. In a study in Germany, Grote et al. (25) found beryllium levels range from 0.05 to 1 mg/kg for rural soils and 1 to 2.5 mg/kg for urban soils. This study suggests that elevated levels of beryllium can be found as a result of anthropogenic activities.

Fishbein (36) reports beryllium can be found in coal at concentrations ranging from 0.1 to 1000 ppm and in oil at a concentration of 0.8 ppb. Data of beryllium concentration in different types of U.S. coal by weight are presented in USEPA (37) and summarized in Taylor et al. (12). Weathering of soil and rock and combustion of oil and coal can account for the presence of beryllium in water.

Surface Water

A number of studies have been conducted on surface waters such as lakes, rivers, and seawater (Table 2).

Table 2. Concentration of Beryllium in Various Waters

Source of Water	Location[a]	n	Mean, μg/L	Range, μg/L[b]	Reference
Seawater	NA	NA	0.0006	NA	38
Seawater	NA	NA	NA	0.00036–0.009	39
River water	NA	NA	NA	ND–0.01	40
River water	US/Canada	59	NA	ND–0.22	41
River water	US	NA	<0.05	ND–0.28	42
River water	US, downgradient of manufacturing plants	NA	NA	18–21	36
River water	Czechoslovakia—Be rich bedrock, acid rain	39	11.97	2–24	43
Surface water	US	1577	0.19	0.01–1.22	44
Surface water	US	NA	NA	ND–1.2	42
Surface water	Eastern US/Siberia	NA	NA	0.1–0.9	45
Surface water	USSR	NA	NA	ND–0.1	20
Surface water	Czechoslovakia	3028	0.0054	0.0022–65	46
Finished water	US	380	0.1	0.02–0.17	44
Finished water	US	NA	NA	0.01–0.7	47
Finished water	US	NA	NA	ND–0.17	42
Finished water	US	NA	0.2	0.01–1.2	36
Bottle water	Europe	NA	NA	0–2.5	11
Groundwater	USSR, granitic area	NA	NA	0.01–30	20
Groundwater	India, downgradient of Be manufacturing plant	NA	NA	0.001–0.002	24

[a]NA: not available.
[b]ND: nondetectable.

In some cases, it appears that "finished water," treated drinking water, was analyzed. However, the studies in many cases do not indicate whether the original source of water was a river, lake, or groundwater. A number of studies also state that surface water was sampled but do not elaborate if this consisted of river or lake water.

A study of 59 surface water bodies and 15 rivers in the United States and Canada found a maximum beryllium concentration of 0.22 μg/L (41). However, the study did not specify which waterbodies were studied and whether raw or finished water was analyzed. Similarly, Kopp (44) analyzed 1577 surface water samples from the United States and found a concentration range of 0.01 to 1.22 μg/L and a mean of 0.19 μg/L. Griffitts et al. (45) reported a range of beryllium of 0.1 to 0.9 μg/L in eastern U.S. and Siberian surface waters. Soviet surface water rarely exceeded 0.1 μg/L, according to Grigor'yev (20).

In Hem's (48) review of several studies, it was concluded "that river waters normally can be expected to carry no more than a few tenths of a microgram per liter of beryllium in solution." However, elevated beryllium concentrations were found in streams where the beryllium-rich source rock was being attacked by acid rain (46). Skirivan et al. (43) suggest that the concentration of fluoride ions and pH significantly control beryllium concentrations in surface waters.

Studies by Skirivan et al. (46) and Burba et al. (49) suggest beryllium can be sorbed onto suspended organic matter. The work of Grigor'yev (20) found between 50 and 73% of the total beryllium in river water samples is associated with suspended solids, which supports the Burba et al. (49) findings. However, some streams at low pH contain dissolved beryllium. Upon discharge of low pH stream water into a higher pH pond, the mobile beryllium was scavenged from the water by solid species containing hydrated metal oxides (aluminum, iron, and manganese) residing on organic matter (43,46,50,51). At higher pH, the beryllium complexes of $BeOH^+$ and $Be(OH)_2$ form, which then can be adsorbed onto colloidal particles (52). Skirivan et al. (43) and Veseley et al. (46) found the highest beryllium concentrations in surface water were associated with low pH waters. At a pH less than 4, beryllium is principally found as the Be^{2+} cation (43,52).

Fishbein (36) found beryllium levels of 18 to 21 μg/L from effluent samples collected downgradient of power plants and metal manufacturing facilities, although the locations were not specified. This study suggests that elevated beryllium levels in surface water are a possibility near metal manufacturing facilities.

Groundwater

Very limited information is available on the concentration of beryllium in groundwater in the general literature. Data is likely to be available from environmental studies of potential contaminated sites; however, this information is not readily accessible.

No general regional baseline studies of beryllium in groundwater have been published in the peer-reviewed literature. However, Grigor'yev (20) studied groundwater from a granitic intrusive area and found levels range from 0.01 to 30 μg/L. Additionally, a study by Korkish et al. (11) was conducted with European bottled water, which likely originated as ground-fed spring water. The Korkish et al. (11) study found beryllium levels range from nondetectable to 2.25 μg/L. Burba et al. (49) present data on two mineral water samples collected in Germany, which

Table 3. Beryllium Groundwater Concentrations for Paducah, KY[a]

Formation	Depth	n	Mean, μg/L	Range, μg/L
Upper continental recharge system	shallow	83	0.92	0.2–3
Regional gravel aquifer	mid	237	1.0	0.04–4.4
McNairy	deep	14	0.76	0.04–2.2

[a]Reference 53.

again presumably have a groundwater origin. A sample from Essen had a beryllium concentration of 9 μg/L, and a sample from Bad Pyrmont had a level of 55.8 μg/L.

To place the previous discussion in context, the US EPA maximum contaminant level (MCL) for beryllium is 4 μg/L. The MCL is the highest level of a contaminant that is allowed in drinking water and is an enforceable standard. The drinking water equivalent level (DWEL) for beryllium is 70 μg/L. A DWEL is a lifetime exposure concentration protective of adverse, noncancer health effects, which assumes that all exposure to a contaminant is from drinking water.

Several studies of potential contamination sites have been conducted where beryllium was one of several metals analyzed. One specific study focused on a beryllium manufacturing plant in India (34). This study found groundwater concentrations ranged from 0.001 to 0.002 μg/L, which suggests the beryllium levels represent background conditions.

Another detailed groundwater study (Table 3) was conducted at an uranium gaseous diffusion facility and included beryllium data for groundwater in a shallow unconfined system, a mid-depth confined aquifer, and a deep confined aquifer (53). The pH of this low ionic strength groundwater ranged from 5.3 to 7.9 and was oxidizing. No known source of beryllium existed at this facility, so the beryllium concentrations were likely representative of ambient background conditions.

CONCLUSION

The available information on the presence of beryllium in water samples is spotty, many of the studies lack specific information on type of water sample (lake, river, or groundwater), where, when, and why samples were collected. However, taken as a whole, the studies indicate that the presence of beryllium in water samples should be low at concentrations in the tenths to single ppb range. Beryllium water concentrations near landfills, metal manufacturing sites, coal regions, or locations of beryllium-enriched pegmatitic rock can have elevated levels, typically at the tens of ppb level. Higher beryllium water concentrations are limited by the low solubility of most mineral species.

BIBLIOGRAPHY

1. Walsh, K.A. (1973). Beryllium. In: *The Encyclopedia of Chemistry*, 3rd Edn. Van Nostrand Reinhold, New York.
2. USDHHS. (2002). 10th Report on Carcinogens. U.S. Department of Health and Human Services (USDHHS), Public Health Service, National Toxicology Program. Available: http://ehis.niehs.nih.gov (accessed July 2003).
3. Nriagu, J.O. (1990). Global metal pollution. *Environment* **32**: 7–33.
4. Nriagu, J.O. and Pacyna, J.M. (1988). Quantitative assessment of worldwide contamination of air, water, and soils by trace metals. *Nature* **333**: 135–139.
5. Bhat, P.N. and Pillai, K.C. (1997). Leachability and immobilization of beryllium from solid waste (red-mud) generated in processing beryl. *Water Air Soil Pollut.* **94**: 297–306.
6. Eisenbud, M. (1998). The standard for control of chronic beryllium disease. *Appl. Occup. Environ. Hyg.* **12**: 25–31.
7. Saltini, C. et al. (1998). Immunogenetic basis of environmental lung disease: Lessons from the berylliosis model. *Eur. Respir. J.* **12**: 1463–1475.
8. Reeves, A.L. (1995). Toxicity of detrimental metal ions, Beryllium. In: *Handbook of Metal–Ligand Interactions, Biological Fluids: Bioorganic Medicine*. Vol. II. Dekker, New York, pp. 772–779.
9. Hofele, J., Sietz, M., Grote, M., and Janssen, E. (1994). Untersuchungen zu berylliumgehalten von kartoffeln und boden. *Agribiological Res.* **47**: 3–4.
10. Sato, A. (1981). *Practice On Furnace Atomic Absorption Spectrometry*. Kodansha, Tokyo, Japan.
11. Korkisch, J., Sorio, A., and Steffan, I. (1976). Atomic-absorption determination of beryllium in liquid environmental samples. *Talanta* **28**: 289–294.
12. Taylor, T.P. et al. (2003). Beryllium in the environment: A review. *J. Environ. Sci. Health* **A38**: 439–469.
13. Baes, C.F., Jr. and Mesmer, R.E. (1976). *The Hydrolysis of Cations*. John Wiley & Sons, New York.
14. Hormann, P.K. (1969). Beryllium. In: *Handbook of Geochemistry*. K.H. Edepohl (Ed.). Springer-Verlag, New York.
15. Rossini, F.D., Wagman, D.D., Evans, W.H., Levine, S., and Jaffe, I. (1952). *Selected Values of Chemical Thermodynamic Properties*. National Bureau Standards Circular 500. Washington, DC.
16. Wagman, D.D. et al. (1982). The NBS tables of chemical thermodynamic properties, selected value for inorganic C1 and C2 organic substances in SI units. *J. Phys. Chem. Reference Data.* **2**: 392.
17. Latimer, W.M. (1952). *Oxidation Potentials*, 2nd Edn. Prentice-Hall, New York.
18. Brookins, D.G. (1988). *E_h-pH Diagrams for Geochemistry*. Spinger-Verlag, New York.
19. Krauskopf, K.B. (1979). *Introduction to Geochemistry*, 2nd Edn. McGraw-Hill, New York.
20. Grigor'yev, N.A. (1986). Distribution of beryllium at the surface of the earth. *Int. Geol. Rev.* **28**: 127–179 and **28**: 326–371.
21. Berg, T., Royset, O., Steines, E., and Vadset, M. (1995). Atmospheric trace element deposition: principal component analysis of ICP-MS data from moss samples. *Environ. Pollut.* **88**: 67–77.
22. Asami, T. and Kubota, M. (1995). Background levels of beryllium in several countries. *Environ. Geochem. Health* **17**: 32–38.
23. Asami, T. and Fukazawa, F. (1981). Beryllium pollution of soils surrounding beryllium factories. *J. Sci. Soil Manure* **52**: 1–7.

24. Bhat, P.N. and Pillai, K.C. (1996). Beryllium in environmental air, water, and soil. *Water Air Soil Pollut.* **95**: 133–146.

25. Grote, M., Sietz, M., Hofele, J., and Kowalsik, J. (1997). Investigations on the phytoavailability of beryllium and vanadium. *Fresenius Environ. Bull.* **6**: 037–042.

26. Kabata-Pendia, A. and Pendias, H. (1992). *Trace Elements in Soils and Plants.* 2nd Edn. CRC Press, Boca Raton, FL.

27. Schacklette, H.T. and Boerngen, J.G. (1984). *Element Concentrations in Soils and Other Surficial Materials of the Conterminous United States*. USGS Professional Paper 1270. Washington, DC.

28. Bowen, H.J.M. (1979). *Environmental Chemistry of the Elements*. Academic Press, New York, p. 60.

29. Ure, A.M. and Bacon, J.R. (1978). Comprehensive analysis of soils and rocks by spark-source mass spectrometry. *Analyst* **103**: 807.

30. Conner, J.J. and Shacklette, H.T. (1975). *Background Geochemistry of Some Rocks, Soils, Plants, and Vegetables in the Conterminous United States*. USGS Professional Paper 574-F. Washington, DC, pp. 33–35.

31. Meehan, W.R. and Smythe, L.E. (1967). Occurrence of beryllium as a trace element in environmental materials. *Environ. Sci. Technol.* **1**: 839–844.

32. Gorham, F. and Swaine, D.J. (1965). The influence of oxidizing and reducing conditions upon the distribution of some elements in lake sediments. *Limnol. Oceanogr.* **10**: 268–279.

33. Vinogradov, A.P. (1959). *The Geochemistry of Rare and Dispersed Chemical Elements in Soils.* 2nd Edn. Consultants Bureau, New York.

34. Swaine, D.J. (1955). *The Trace Element Content of Soils.* Commonwealth Agricultural Bureau, Farnham, UK.

35. Anderson, M.A., Bertsch, P.M., and Miller, W.P. (1990). Beryllium in selected southeastern soils. *J. Environ. Qual.* **19**: 347–348.

36. Fishbein, L. (1984). Overview of analysis of carcinogenic and/or mutagenic metals in biological and environmental samples, I. Arsenic, beryllium, cadmium, chromium, and selenium. *Int. J. Environ. Anal. Chem.* **50**: 874–880.

37. USEPA. (1998). *Study of Hazardous Air Pollutant Emissions from Electric Utility Steam Generating Units—Final Report to Congress*. U.S. Environmental Protection Agency, Research Triangle Park, NC.

38. Goldberg, E.D., Broecker, W.S., Gross, M.G., and Turekian, K.K. (1971). *Radioactivity in the Marine Environment*. National Academy of Science, Washington, DC, pp. 137–146.

39. Measures, C.I. and Edmond, J.M. (1986). Determination of beryllium in natural waters in real time using electron capture detection gas chromatography. *Anal. Chem.* **58**: 2065–2069.

40. Merrill, J.R., Lyden, E.F.X., Honda, M., and Arnold, J.R. (1960). The sedimentary geochemistry of beryllium isotopes. *Geochim. Cosmochim Acta.* **18**: 108–129.

41. Duram, W.H. and Haffty, J. (1961). *Occurrence of Minor Elements in Water*. USGS Circular 445. Washington, DC.

42. SDWC. (1977). *Drinking Water and Health*. Safe Drinking Water Committee of the National Academy of Sciences, Washington, DC.

43. Skirivan, P., Stastny, M., Kotkova, P., and Burain, M. (1996). Partition of beryllium and several other trace elements in surface waters. *Scientia Agriculturae Bohemica* **27**: 131–145.

44. Kopp, J.F. (1969). *The occurrence of trace elements in water. Proc. 3rd Annu. Conf. Trace Substances Environ. Health.* University of Missouri, Columbia, MD, pp. 59–73.

45. Griffits, W.R., Allaway, W.H., and Groh, D.H. (1977). The relation of other substances and trace elements to health and disease. In: *Beryllium In Geochemistry and the Environment*. Vol. II. National Academy of Science, Washington, DC.

46. Veseley, J., Benes, P., and Sevcik, K. (1989). Occurrence and speciation of beryllium in acidified freshwaters. *Water Res.* **23**: 711–717.

47. APHA. (1976). *Standard Methods for the Examination of Water and Waste Water*, 13th Edn. American Public Health Association, Washington, DC.

48. Hem, J.D. (1989). *Study and Interpretation of the Chemical Characteristics of Natural Water*, 3rd Edn. USGS Water-Supply Paper 2254. Washington, DC.

49. Burba, F., Willmer, P.G., Betz, M., and Fuchs, F. (1983). Atomic absorption determination of beryllium traces in natural waters. *Int. J. Environ. Anal. Chem.* **13**: 177–191.

50. Borovec, Z., Tolar, V., and Mraz, L. (1994). The state of accumulation and source relations of Co, Ni, Cr, Be, and Zn in silt-clay sediments of the Labe river. *Geology Carp. Ser. Clays.* **45**: 3–9.

51. Borovec, Z. (1993). Partitioning of silver, beryllium, and molybdenum among chemical fractions in the sediments from the lake (Elbe) river in central Bohemia, Czech Republic. *GeoJournal* **29**: 359–364.

52. Benes, P. and Jiranek, V. (1974). Adsorption and colloidal behavior of carrier-free ^{7}Be in aqueous solutions. *Radiochimica Acta* **21**: 49–53.

53. Moore, G.K. (1995). *Inorganic Soil and Groundwater Chemistry Near Paducah Gaseous Diffusion Plant, Paducah, Kentucky*. Oak Ridge National Laboratory, Oak Ridge, TN.

READING LIST

Bashir, W. and Paull, B. (2001). Sensitive and selective ion chromatographic method for the determination of trace beryllium in water samples. *J. Chromatogr. Anal.* **910**: 301–309.

Burguera, J.L. et al. (2000). Determination of beryllium in natural and waste waters using on-line flow-injection preconcentration by precipitation/dissolution for electrothermal atomic absorption spectrometry. *Talanta* **52**: 27–37.

Capitan, F. et al. (1989). Determination of beryllium in water by ion-exchange spectrofluorimetry. *Analyst* **114**: 969–973.

Hoshino, H., Monura, T., Nakano, K., and Yotsuyanagi, T. (1996). Complexation of Be(II) ion with 1-(2,4-dihydroxy-1-phenylazo)-8-hydroxy-3,6-napthalenedisulfonate as the chemical basis of the selective detection of ultratrace Be by reversed-phase high-performance liquid chromatography. *Anal. Chem.* **68**: 1960–1965.

Kubova, J., Nevoral, V., and Sresko, V. (1994). Determination of beryllium trace contents in mineral waters after preconcentration on a chelating ion-exchanger. *Fresenius J. Anal. Chem.* **348**: 287–290.

Majee, S. and Das, J. (1998). Use of chelating ion exchanger containing acetoacetanilide group selective separation of beryllium. *Indian J. Chem.* **27A**: 983–986.

Narayana, R. and Khopkar, S.M. (1985). Separation of beryllium(II) from other metallic ions with aliquot 336S by extraction chromatography of its malonato complex. *Analyst* **110**: 1295–1298.

Rose, A.W., Hawkes, H.E., and Webb, J.S. (1979). *Geochemistry in Mineral Exploration*. Academic Press, London.

Singh, H.B., Agnihotri, N.K., and Singh, V.K. (1998). Determination of trace amounts of beryllium using derivative spectrophotometry in non-ionic micellar medium. *Talanta* **47**: 1287–1296.

Voloschick, I.N., Litvina, M.L., and Rudenko, B.A. (1995). Ion chromatographic determination of beryllium in rock and waste waters with a chelating sorbent and conductimetric detection. *J. Chromatogr. Anal.* **706**: 315–319.

DISSOLVED ORGANIC CARBON

Jamie R. Lead
University of Birmingham
Birmingham, United Kingdom

DISSOLVED ORGANIC CARBON

Of the three words in the title of this entry, little explanation is needed for the last two. The first, however, needs some clarification. The term has traditionally been used to denote material that has been filtered through a certain type of standard membrane (0.45 μm nominal pore size). Material passing through the filter (filtrate) was termed the dissolved phase, and material retained by the filter (retentate) was termed the solid phase. This filtration step was performed for a number of reasons such as reduction in the complexity of the sample and partial sterilization of the filtrate due to the removal of microorganisms. The procedure is still widely used in the scientific and regulatory community. However, it has long been recognized as an inadequate definition because of the presence of colloidal carbon (1–1000 nm in size) and the artifacts inherent in filtration (1,2). The use of a single filtration step thus leads to results that are neither thermodynamically accurate nor biogeochemically relevant. There have been a number of theoretical and practical attempts to delimit the colloidal and dissolved phases. Now, the main practical delimiter is based on ultrafiltration using a nominal pore size of 1,000–10,000 daltons (ca 1 nm) (5). This is a more useful definition for understanding biogeochemical processes but remains entirely operational. In this article, the term 'dissolved' will be used in the wider sense to incorporate a fraction of the colloidal material.

The importance of dissolved organic carbon (DOC) lies in its role in the biogeochemical carbon cycle, as a necessary basis for aquatic microbial processes and in aquatic chemistry (6). Although the global burden of carbon is dominated by sediments (7), dissolved organic carbon is of obvious importance as carbon is the basis of life, an energy and nutrient source for microorganisms. DOC is also of significance in trace pollutant binding (and consequently in pollutant transport and bioavailability), in the charge and aggregation behavior of colloids, in weathering, in pH buffering, and in acid–base and redox chemistry of natural waters. It thus has a large and essential role in determining the chemical composition of natural waters. Although microbiological and chemical aspects are frequently taught and researched in isolation, it may be impossible to understand the composition and regulation of natural waters without simultaneously considering the microbiological and chemical processes because they are so closely interlinked (8). Any distinction between the two processes is therefore likely to be arbitrary and misleading.

The chemical structure of much of dissolved organic carbon is poorly understood because of its complexity, its variability over both space and time, and the ease with which its structure is changed by sampling and analytical procedures (1,9). As DOC is physicochemically diverse, a number of partially successful subclassifications have been used based on size, chemistry, and source. However, for most DOC, it is impossible to define either an exclusive structural or functional definition or classification system, as has been done for biochemical polymers. Here, we will discuss two types of DOC, humic substances and biologically produced chemicals. In addition, anthropogenic pollutants may also be important, for instance, in human and ecological health, although not in terms of total environmental loads of DOC.

Humic substances (HS) are often the dominant type of DOC, particularly in 'organic-rich' waters such as marshes and in dystrophic waters where microbial activity is limited. HS are a general class of complex, biogenic, macromolecular species. They are heterogeneous and polydisperse and do not have one single structure but exhibit a range of characteristic properties such as molar mass and charging behavior. It has been suggested that heterogeneity is their essential, defining characteristic (10). They are the products of biological decomposition. However, they differ from the biologically produced chemicals of the next section because they have undergone a substantial degree of chemical and biological degradation (humification), resulting in a kinetically stable product resistant to further degradation. They are defined operationally by the extraction system used to separate them from other components such as clays, oxides, and nonhumic organic material. It is important to realize that they are not defined either by structure or function because their structure is too complex and they have no clearly defined 'function' in the biochemical sense. HS are usually divided into three main groups based on their relative solubilities. It is hoped that these reflect to some extent the solubility of HS in the environment and properties such as solid:solution partitioning and transport. In any case, the division results in a simplification of the HS which has analytical advantages. The three fractions are

- fulvic acid—soluble at all pHs.
- humic acid—soluble at high pH, precipitated in acid (pH 2).
- humin—insoluble in acid and base.

The extraction and fractionation scheme have been standardized by the International Humic Substances Society (IHSS). In general terms and considering their heterogeneity, fulvic acids and aquatic HS are more soluble, have a greater density of functional groups, carry a greater charge, and have a lower C:O ratio than humic acid or terrestrial HS, respectively. HS are important in a number of processes and have been much studied. Information has been collected on properties such as their acid–base, metal binding and redox chemistry, photochemistry, structure, functional group content, molar mass, and role as a nutrient. However, a fundamental, unanswered question relates to the extraction and fractionation scheme: To what extent are their properties significantly modified by the extraction system and therefore how are extracted HS different from 'natural' HS?

Biologically produced chemicals consist of a number of well-defined chemicals produced directly from microbial activity. These can further divided into (1) simple monomers and (2) more complex polymers. The monomers include amino acids and monosaccharides that are relatively simple compounds whose structure can be ascertained precisely. In the main, they are produced and consumed rapidly and thus have a short residence time (a few days), low concentration, and are of relatively minor importance. Their concentrations depend on the relative rate of supply and the rate of demand, both controlled primarily by microbial activity. The compounds may be 'free' or bound to humic substances. As for HS, their concentration and fraction of total DOC are variable, depending on the water type and season. More complex, biologically produced polymers, sometimes denoted as biopolymers, are less well studied than HS. However, they can have concentrations similar to HS and can make up 30–40% of total DOC, particularly in eutrophic waters. They consist primarily of proteins and polysaccharides from sources such as plant and microbial cell walls, exudates secreted by microbes, and internal cellular proteins and polysaccharides from microbes after cell lysis. It is possible that HS and biopolymers have differing roles in aquatic systems. For instance, it is well known that HS tends to stabilize colloids and particles and reduce aggregation. However, it is possible that the biopolymers tend to increase aggregation by bridging mechanisms (11) that have important consequences for the chemical and biological composition of natural waters.

In summary, DOC, including some colloidal OC, is clearly extremely important in biological and chemical activity and interactions in the environment. However, despite the progress made and the vast amount of research performed, many fundamental questions await answers.

BIBLIOGRAPHY

1. Lead, J.R., Davison, W., Hamilton-Taylor, J., and Buffle, J. (1997). **3**: 219–232.
2. Horowitz, A.J. et al. (1996). *Environ. Sci. Technol.* **30**: 954–963.
3. Gustafsson, O. and Gschwend, P.M. (1997). *Limnology and Oceanography* **42**: 519–528.
4. Stumm, W. (1993). *Colloids and Surfaces A: Physicochemical and Engineering Aspects* **73**: 1–19.
5. Martin, J-M., Dai, M-H., and Cauwet, G. (1995). *Limnology and Oceanography* **40**: 119–131.
6. Thurman, E.M. (1985). *Organic Geochemistry in Natural Waters*. Martinus Nijhoff/Dr W. Junk, The Netherlands.
7. Butcher, S.S., Charlson, R.J., Orians, G.H., and Wolfe, G.V. (Eds.). (1992). *Global Biogeochemical Cycles*. Academic Press, London.
8. Buffle, J. and de Vitre, R.R. (1994). *Chemical and Biological Regulation in Aquatic Systems*. Lewis Publishers, New York.
9. Wilkinson, K.J., Balnois, E., Leppard, G.G., and Buffle, J. (1999). *Colloids and Surfaces A: Physicochemical and Engineering Aspects*. Vol. 155, pp. 287–310.
10. MacCarthy and Suffet. (1989). *Chemical Modeling in Aqueous Species*. American Chemical Society Symposium.
11. Buffle, J. et al. (1998). *Environmental Science and Technology* **32**: 2887–2899.

MECHANISMS OF WATER ADSORPTION ON CARBONS

Piotr A. Gauden
Artur P. Terzyk
Gerhard Rychlicki
Sylwester Furmaniak
Radosław P. Wesołowski
Marek Wiśniewski
N. Copernicus University
Toruń, Poland

Piotr Kowalczyk
Faculty of Science, Chiba University
Chiba, Japan

The properties of water are of importance in many scientific disciplines such as chemistry, biology, geology, nanotechnology, and materials technology. Moreover, the adsorption of water on activated carbon (AC) is an important topic in many different areas of science and technology because water is the most common solvent in nature. The adsorption and phase behavior of polar fluids in carbon pores has been studied extensively, but our understanding of the adsorption of water on carbonaceous materials is still incomplete (1–4).

In recent years, a number of experimental and simulation studies of adsorption of water in pores have appeared in the literature. Some studies have assumed that the adsorption behavior of water in graphite pores is hydrophobic. Although principally hydrophobic adsorbents may contain significant numbers of adsorption centers that can interact with water, it is generally believed that the combination of weak carbon–water dispersive attractions and strong water–water associative interactions are responsible for the complex behavior of water confined in carbonaceous pores.

On the other hand, surface heterogeneity can be introduced by incorporating some polar groups on the surface and those active sites lead to enhanced water adsorption. The loss of water–water H-bonds in graphitic pores is compensated for by the ability of water molecules to form H-bonds with the surface sites in activated carbon pores; the result is a change in the molecular level adsorption phenomenon from capillary condensation in graphite pores to continuous pore filling in activated carbon pores. This leads to an important practical conclusion that water adsorption measurements can be used to determine the number of surface-active centers.

The next important problem is the structure of water adsorbed in pores. Some experimental X-ray diffraction (XRD) (5), differential scanning calorimetry (DSC) (6), dielectric relaxation spectroscopy (DRS) (6), and theoretical studies (6–8) showed the presence of a well-ordered structure of water adsorbed in the graphite micropores at ambient temperature. The adsorbed water has a more ordered structure than bulk liquid, but less than ice, and the structure of water molecules depends on the width of pores (5,8).

One of the most spectacular methods is calorimetry (2,9–22). Careful measurements of the energetic effects accompanying water adsorption on carbon blacks and carbons were studied by scientists from the so-called "Russian school of adsorption" in the 1950s (15,16) and the 1980s (17,18) and by others (9–11,13,14,19–22). Heats of adsorption of water indicate that adsorption is a strong function of surface chemistry. The primary high-energy sites have a significant influence on adsorption and the enthalpy of adsorption at low relative pressures. Three components can be delineated according to measured differential heats of adsorption: (1) chemical adsorption, (2) physical adsorption, and (3) condensation and/or micropore filling. From the point of view of the thermodynamic verification of theoretical models, the first is the most interesting and important. Chemisorption of water on carbons has been previously reported (2,9–11,17,20–22). Summing up, analysis of the experimental data points out that significant discrepancies exist between the role of specific surface groups (e.g., various types, location, density) and their energy of interaction with water molecules and a quantitative description of the isotherm over the entire relative pressure range that is universal for all carbons with low, medium, and high densities of surface groups.

Recently, knowledge of the water–carbon intermolecular interaction and adsorption of water in the microporous region of ACs has been greatly increased by quantum methods (QM) and molecular simulations. A natural starting point for computer simulations is the development of a model potential for the interaction between water and benzene (23,24). Although model potentials already exist for the water/benzene and water/graphite systems (3,23,25), the reliability of these models has not been widely tested. The chemistry and reaction mechanisms of adsorption, as well as of intercalation, are less known, and ab initio QM methods provide a way to examine these phenomena on the atomic level.

The existing water/graphite model potentials give binding energies for a single water monomer interacting with the surface ranging from −6.91 to −18.00 kJ/mol (25). There also exists an experimental value of −15.07 kJ/mol for the binding energy (26). However, these values seem somewhat low in the light of the most recent high-level ab initio results for the water–benzene system, for which the electronic binding energy has been estimated at −19.32 kJ/mol (27). A large basis set at MP2 level (second-order Mřller–Plesset perturbation theory) calculations on a series of fragment models up to $C_{96}H_{24}$ in size was used by Feller and Jordan (28) to estimate the electronic binding energy of an individual water molecule to single-layer graphite (-24.28 ± 1.67 kJ/mol). The interaction between water molecules and graphite structures was clearly stronger with the $C_{96}H_{24}$ model than for the smaller ones. The opposite result was observed for a smaller basis and the Hartree–Fock method (28). These calculations highlight the difficulty of approaching the complete basis set limit in extended systems, such as graphite, using traditional atom-centered basis sets. Feller and Jordan shown (29) that the estimated water–graphite binding energy is slightly larger than that for the interaction between two water molecules (-20.92 ± 0.42 kcal/mol) (30). At the same time, this interaction is significantly smaller than the average electronic binding energy of water monomers in bulk water, where the hydrogen bonding network involves each water molecule in (up to) four hydrogen bonds (30). Nonetheless, the estimated water–graphite interaction is sufficiently large to lead to the conclusion that in carbon, the interaction of water with the graphite walls plays a significant role in determining the structure of the confined water. On the other hand, in the water–benzene complex, the important attractive interactions are the dipole-quadrupole, dipole-induced dipole (induction), and dispersion contributions (23,24,27).

Ruuska and Pakkanen (28) investigated interactions between water molecules and one- and two-layer graphite using the ab initio Hartree–Fock (HF) method. Three one-layer models were used as a reference to investigate the influence of the adsorption site (on top, bridge, and hollow), the orientation of the water molecule, and the size of the model. Only minor differences were found in the three adsorption sites. The behavior of water molecules between two graphite planes was examined using the two-layer model. The coronene planes were moved from a distance of 9 to 3 Å, and the water molecule was placed in the middle of the layers. It is clear from results published by Ruuska and Pakkanen (28) that a water molecule requires relatively much space because it doubles the interlayer distance. Additionally, the features of the two-layer graphite with a water molecule between are similar to those of graphite intercalates.

A more complex problem considered by QM is the interaction of water with a graphite-like surface containing oxygenated hydrophilic groups. It is well known that adsorption is strongly dependent on the presence and arrangement of hydrophilic groups. Tarasevich and Aksenenko (31) used the semiempirical PM3 method to study the interaction of very few water molecules with a partially oxidized graphite surface. The main conclusion of this work is that the water molecules are adsorbed at the hydrophilic center (i.e., the carboxyl) with the formation of microclusters consisting (at initial coverages) of two to three molecules. In the light of these results, Tarasevich and Aksenenko (31) suggested that water adsorption measurements cannot be used to determine the number of active centers even at hydrophobic, and, far less, at hydrophilic surfaces, and methanol was recommended for this purpose.

Hamad and co-workers (32) reported the quantum chemical characterization of a hydrophilic site modeled by a carboxyl group attached to one of the carbon atoms in the graphite layer. They supply values of point charges at the

–COOH sites that have been used to derive a pair potential for the water-active site interaction. Picaud et al. (33) presented molecular dynamics simulations based on a classical potential issued from QM calculations (30) to provide a view of the dynamics and the structure of a water layer on a model surface that consists of a planar graphitic layer plus different numbers of active carboxylic sites. The results of these simulations show that the –COOH groups act as strong trapping sites for a few water molecules that then become nucleation centers for other water molecules that form larger aggregates tied to the –COOH.

The strong effect of PACs on water adsorption was indicated on the basis of the computer simulations (3,8,33–45). The adsorption of water molecules on a hydrophobic surface has been investigated (8,34,35,38–41,44). On the other hand, several types of polar oxygen-containing sites [e.g., carboxyl (42) carbonyl (34,42,43), and hydroxyl (37,42) groups or H-sites (40)] were placed on the surface of carbon with different densities and local distributions to determine the influence of each of these factors on the adsorption of water. Computer simulation models are still being developed and more and more complicated systems have been analyzed. To provide a realistic pore representation and a description of water adsorption, recently, Thompson and Gubbins (3,43,45) suggested the application of a model based on reverse Monte Carlo. In this model, surface sites have been added at random points on the edges of graphene microcrystals possessing random sizes and structures. The major drawback to advanced numerical calculations is the computing power and time needed to simulate isotherms. Therefore, uncomplicated analytic approaches are often favored to simplify consideration and cut down significantly the computation time, for instance, so-called local isotherms obtained from computer simulation are described by a theoretical model [for example, proposed by Talu and Meunier (46)] (37).

Muller and Gubbins (36) showed, that for nonactivated carbon slit pores almost no adsorption occurs until a sharp vertical rise in the adsorption curve is seen and the pore fills suddenly (i.e., capillary condensation occurs). If the surface is doped by active sites, the adsorption isotherm changes drastically, strong bonds are formed between these sites and water molecules, and these adsorbed molecules become nucleation sites for other water molecules to adhere. Thus, these results confirm the role of primary surface sites in the enhancement of water adsorption. Furthermore, Jorge et al. (42) showed that the local distribution of primary sites (carboxyl, hydroxyl, and carbonyl) has a strong effect on the low-pressure part of the adsorption isotherm, whereas the overall site density affects mainly the vapor–liquid phase transition. The type of oxygen-containing group was not of critical importance because more complex groups can effectively be represented by simpler sites. On the other hand, McCallum et al. (37) proposed two alternative mechanisms for water adsorption onto activated carbon walls due to surface sites regularly arranged or a random array (the same site density in both cases was assumed). Five stages are observed from the analysis of both types of simulated adsorption isotherms. Moreover, their behavior is similar to experimental ones.

The theory describing the mechanism of water adsorption on carbonaceous materials proposed by Dubinin and Serpinsky (DS) (1,3,16,47) is probably the most famous. Although the fundamentals of this model were developed 50 years ago, it is still widely used mainly due to its simplicity. Dubinin et al. (16) introduced a phenomenological model of adsorption of water molecules on energy privileged sites which act as primary adsorption centers for water on a predominantly hydrophobic carbon surface. On the other hand, water molecules adsorbed on these sites produce new centers (called elsewhere secondary ones) for further adsorption and cluster formation. Note that the hyperbolic behavior of this equation leads to an infinite adsorbed water concentration at higher pressures. It is well known that the DS model describes well only the data measured on nonporous surfaces and on strongly hydrophobic carbons. The next main limitation is the assumption of unlimited adsorption space. Therefore, the original DS equation has been improved by Dubinin et al. (47), Barton et al. (48–50), and others (10,11,51–55) by taking into consideration new functions describing a decrease in the number of secondary adsorption sites.

On the other hand, it is well known that the simplest method leading to an increase in the affinity of a carbon surface for water vapor is, for example, oxidation of this surface. In such a case, a more hydrophilic surface chemical composition leads to a rise in adsorption, especially in the initial range of the water isotherm (an increase in the enthalpy of adsorption and enthalpy of immersion in water are observed simultaneously (2,9–11,15,17,19–21,56)). For hydrophilic surfaces, DS-like equations (1,3,10,11,51–55) are insufficient and, for example, the D'Arcy and Watt (DW) adsorption isotherm can be applied (57) (however, some attempts to improve the DS model to correct the description of this type of experimental data were recently proposed by Gauden and Terzyk (52,53)). In the original DW model, it is assumed that sorption of water takes place independently on strongly (high-energy) binding primary sites (which are hydrophilic groups) and to a less extent on weakly (low-energy) binding primary sites (located elsewhere on the carbon surface). Moreover, the term describing multilayer adsorption on the so-called secondary centers (which need not be completely occupied before monolayer formation) is assumed in the D'Arcy and Watt isotherm. As shown by Evans et al. (58,59) and others (1,10,11,60,61), the DW equation satisfactorily describes water adsorption data on different adsorbents (not only on carbons). Note that in the opinion of Barton and co-workers (59), as well as in the opinion of the authors of the current study (54,55), it appears that D'Arcy and Watt's isotherm contains some inconsistencies. A serious flaw in this simple treatment is the assumption that Langmuir and the original DS isotherms describe the adsorption of water molecules on the adsorbent independently (i.e., they are connected with various centers). Moreover, the further disadvantage of this model is an unsatisfactory description of experimental adsorption data at higher pressures and enthalpy of adsorption data (10,11,51,54,55) due to the assumption of one kind of surface center on the carbon surface. Therefore, we recently generalized the original D'Arcy and Watt equation taking into account the influence of the concentration of the surface primary sites on the amount of the secondary ones (54,55). Next, we assumed that the adsorbent has regions of heterogeneity on the molecular level (i.e., the various functional groups and/or defects on the surface of the adsorbent). Moreover, water molecules can adsorb on all types of high-energy primary sites obeying the Langmuir isotherm. Summing up, the multisite generalized D'Arcy and Watt (MSGDW) model was proposed (54,55). It can be noticed that finite adsorption space is taken into account by the term $(1 - ka^n)$. From the analysis of the isotherm generated as well as the enthalpy plots, it is seen that these curves are similar to those measured using adsorption calorimetry (10,11,18,20,54,55,63,65), especially at low coverages.

Note that the theories proposed by Dubinin, Serpinsky and D'Arcy and Watt do not give information on the dimensions of adsorbed water clusters (however, it is obvious that their sizes are confined to the available adsorption space). On the other hand, it is well known that water molecules in the bulk state form dimers (64); however, the number of dimerized molecules is not significant at room temperature and standard pressure. In a hydrophobic micropore, water molecules are expected to be more dimerized due to the compressed condition there. In narrow pores, the dimers grow to a greater cluster, which has been used to model the elemental structure of water (1,8,44,65). On the other hand, the pore width is limited, and so the size of the clusters is restricted by the finite pore dimension. Iiyama et al. (66) stated that water forms clusters containing at least five water molecules per cluster.

Different authors, for example, Do and Do (67) and Neitsch et al. (68), applied the above mentioned fundamental information about the size of water clusters to develop new water adsorption models. As a result, the model proposed by Do and Do (67) is based on the growth of a water cluster around surface functional groups, the creation of a pentamer, and penetration of these clusters into the micropore (they remain in the fine pores if they have sufficiently high dispersion energy). Note that the Do and Do model can be considered a so called "hybrid" (similarly to the DW and MSGDW equations) in which the first term describes adsorption on primary sites (it is a BET-type) and the second term refers to desorption from the secondary sites and the adsorption of the pentamer into the micropores (so the equation describes a micropore filling mechanism). Neitsch et al. (68) modified this model. They assumed that the size of the water clusters (pentameter in the original Do and Do model) is the best fitting parameter.

Another well-known model describing water adsorption was proposed by Talu and Meunier (46). Their theoretical isotherm equation assumed the following mechanism of adsorption of water molecules: at low pressures, the behavior of the system is controlled by vertical interactions of a molecule to the surface and the molecules are adsorbed on active sites. At intermediate coverages where the inflection point occurs, the adsorbed molecules form clusters via hydrogen bonding. At high surface coverages where the plateau is observed, the adsorption is limited by micropore filling.

Malakhov and Volkov (69) proposed a cooperative multimolecular adsorption theory describing adsorption of alcohols on glassy polymers. This model was successfully adopted by Rutherford to water adsorption on activated carbons (70). In this theory, the movement of adsorbate molecules to a single adsorption site is taken into account. The cooperative influence is induced by the adsorbed molecule (primary stage) which promotes the entry of other molecules to adjacent centers. In the secondary stage, the model considers a unit triad where a central adsorption site promotes adsorption on either side of the central site. Additionally, secondary centers are defined as side associates forming from the primary surface sites and hence allowing the formation of dimers, trimers, and more complicated clusters. Summing up, the cooperative multimolecular adsorption theory is based on the assuming the existence of one kind of primary and three kinds of secondary centers. However, this theory is not valid to describe adsorption at high relative pressures (70).

Several predictive models have been proposed (in our opinion the most important are mentioned above). Unfortunately, none of them satisfactorily describe the isotherms and the corresponding differential heat of adsorption over the whole pressure range. Moreover, some fundamental aspects are not taken into account in the theoretical considerations: the role of specific surface groups (e.g., various type, location, density), their interaction with water molecules, and the structure of water confined in pores.

BIBLIOGRAPHY

1. Mowla, D., Do, D.D., and Kaneko, K. (2003). *Chem. Phys. Carbon* **28**: 229.
2. Vartapetyan, R.Sh. and Voloshchuk, A.M. (1995). *Usp. Khim.* **64**: 1055. (in Russian).
3. Brennan, J.K., Bandosz, T.J., Thomson, K.T., and Gubbins, K.E. (2001). *Colloids Surf. A* **187–188**: 539.
4. Terzyk, A.P. (2003). *J. Colloid Interf. Sci.* **268**: 301.
5. Iiyama, T., Nishikawa, K., Suzuki, T., and Kaneko, K. (1997). *Chem. Phys. Lett.* **274**: 152.
6. Sliwinska-Bartkowiak, M. et al. (2001). *Phys. Chem. Chem. Phys.* **3**: 1179.
7. Sanfelix, P.C., Holloway, S., Kolasinki, K.W., and Darling, G.R. (2003). *Surf. Sci.* **532**: 166.
8. Ohba, T., Kanoh, H., and Kaneko, K. (2004). *J. Am. Chem. Soc.* **126**: 1560.
9. Tsutsumi, K. and Mizoe, K. (1989). *Colloids Surf. A* **37**: 29.
10. Terzyk, A.P., Gauden, P.A., and Rychlicki, G. (1999). *Colloid Surf. A* **148**: 271.
11. Terzyk, A.P., Rychlicki, G., and Gauden, P.A. (2001). *Colloid Surf. A* **179**: 39.
12. Vartapetyan, R.Sh., Voloshchuk, A.M., Isirikyan, A.A., Olyakov, N.S., and Tarasevich, Yu.I. (1995). *Colloids Surf. A* **101**: 227.
13. Stoeckli, F., Jakubov, T., and Lavanchy, A. (1994). *J. Chem. Soc., Faraday Trans.* **90**: 783.

14. Carrasco-Marin, F., Meuden, A., Ceneteno, T.A., Stoeckli, F., and Moreno-Castilla, C. (1997). *J. Chem. Soc. Faraday Trans.* **93**: 2211.
15. Avgul, N.N., Dzigit, O.M., Kiselev, A.V., and Sherbakov, K.D. (1953). *Izv. AN SSSR* **2**: 105. (in Russian).
16. Dubinin, M.M., Zaverina, E.D., and Serpinsky, V.V. (1955). *Trans. Faraday Soc.* 1760. (in Russian).
17. Dubinin, M.M. and Isrikikyan, A.A. (1989). *Izv. AN SSSR* **10**: 2183. (in Russian).
18. Vartapetyan, R.Sh., Voloshchuk, A.M., Dubinin, M.M., and Yakubov, T.S. (1987). *Izv. AN SSSR* **9**: 1934. (in Russian).
19. Miura, K. and Morimoto, T. (1988). *Langmuir* **4**: 1283.
20. Rychlicki, G. (1989). *Polish J. Chem.* **63**: 225.
21. Philips, J., Kelly, D., Radovic, L., and Xie, F. (2000). *J. Phys. Chem. B* **104**: 8170.
22. Groszek, A.J. (2001). *Carbon* **39**: 1857.
23. Dang, L.X. and Feller, D. (2000). *J. Phys. Chem. B* **104**: 4403.
24. Tarakeshwar, P. et al. (1999). *J. Chem. Phys.* **111**: 5838.
25. Vernov, A. and Steele, W.A. (1992). *Langmuir* **8**: 155.
26. Avgul, N.N. and Kiselev, A.V. (1970). *Chem. Phys. Carbon* **6**: 1.
27. Feller, D. (1999). *J. Phys. Chem.* **103**: 7558.
28. Ruuska, H. and Pakkanen, T.A. (2003). *Carbon* **41**: 699.
29. Feller, D. and Jordan, K.D. (2000). *J. Phys. Chem. A* **104**: 9971.
30. Feyereisen, M.W., Feller, D., and Dixon, D.A. (1996). *J. Phys. Chem.* **100**: 2993.
31. Tarasevich, Y.I. and Aksenenko, E.V. (2003). *Colloids Surf. A* **215**: 285.
32. Hamad, S., Mejias, J.A., Lago, S., Picaud, S., and Hoang, P.N.M. (2004). *J. Phys. Chem. B* **108**: 5405.
33. Picaud, S., Hoang, P.N.M., Hamad, S., Mejias, J.A., and Lago, S. (2004). *J. Phys. Chem. B* **108**: 5410.
34. Malenkov, G.G. and Dubinin, M.M. (1984). *Izv. AN SSSR* **6**: 1217. (in Russian).
35. Slasli, A.M., Jorge, M., Stoeckli, F., and Seaton, N.A. (2003). *Carbon* **41**: 479.
36. Muller, E.A. and Gubbins, K.E. (1998). *Carbon* **36**: 1433.
37. McCallum, C.L., Bandosz, T.J., McGrother, S.C., Muller, E.A., and Gubbins, K.E. (1999). *Langmuir* **15**: 533.
38. Striolo, A., Chialvo, A.A., Cummings, P.T., and Gubbins, K.E. (2003). *Langmuir* **19**: 8583.
39. Brodskaya, E.N., Zakharov, V.V., and Laaksonen, A. (2001). *Langmuir* **17**: 4443.
40. Galle, J. and Vortler, H.L. (2001). *Surface Sci.* **481**: 39–53.
41. Spohr, E., Trokhymchuk, A., and Henderson, D. (1998). *J. Electroanal. Chem.* **450**: 281.
42. Jorge, M., Schumacher, C., and Seaton, N.A. (2002). *Langmuir* **18**: 9296.
43. Brennan, J.K., Thomson, K.T., and Gubbins, K.E. (2002). *Langmuir* **18**: 5438.
44. Kimura, T. et al. (2004). *J. Phys. Chem. B* **108**: 14043.
45. Thompson, K.T. and Gubbins, K.E. (2000). *Langmuir* **16**: 5761.
46. Talu, O. and Meunier, F. (1996). *AIChE J.* **42**: 809.
47. Dubinin, M.M. and Serpinsky, V.V. (1981). *Carbon* **19**: 402.
48. Barton, S.S., Evans, M.J.B., and MacDonald, J.A.F. (1991). *Carbon* **29**: 1099.
49. Barton, S.S., Evans, M.J.B., and MacDonald, J.A.F. (1992). *Carbon* **30**: 123.
50. Barton, S.S., Evans, M.J.B., Liang, S., and MacDonald, J.A.F. (1996). *Carbon* **34**: 975.
51. Wesołowski, R.P., Gauden, P.A., Terzyk, A.P., and Furmaniak, S. (2004). *Carbon* **53**: 42.
52. Terzyk, A.P. and Gauden, P.A. (2002). *4th Int. Symp. Coal Struc.—Struc. Reactivity Carbonaceous Mater.*, Gliwice, Poland.
53. Gauden, P.A. *J. Colloid Interface Sci.*, in press.
54. Gauden, P.A., Furmaniak, S., Terzyk, A.P., Rychlicki, G., and Wesołowski, R. (2004). *ISSHAC—5 Proc.* Gdańsk, p. 138.
55. Furmaniak, S., Gauden, P.A., Terzyk, A.P., Wesołowski, R.P., and Rychlicki, G. *Carbon* submitted.
56. Barton, S.S. and Koresh, J.E. (1983). *J. Chem. Soc., Faraday Trans.* **79**: 1147.
57. D'Arcy, R.L. and Watt, I.C. (1970). *Trans. Faraday Soc.* **66**: 1236.
58. Evans, M.J.B. (1987). *Carbon* **25**: 81.
59. Barton, S.S., Evans, M.J.B., and MacDonald, J.A.F. (1994). *Langmuir* **10**: 4250.
60. Venkataramani, B. (1993). *Langmuir* **9**: 3026.
61. Sychev, M. et al. (2001). *Microporous Mesoporous Mater.* **47**: 311.
62. Vartapetyan, R.Sh., Voloshchuk, A.M., Dubinin, M.M., Chekulina, L.A., and Chepurnoi, S.G. (1986). *Izv. AN SSSR* **11**: 2407. (in Russian).
63. Dubinin, M.M. (1981). *Izv. AN SSSR* **1**: 9. (in Russian).
64. Curtis, L.A., Frurip, D.J., and Blender, M. (1979). *J. Chem. Phys.* **71**: 2703.
65. Kaneko, K., Hanzawa, Y., Iiyama, T., Kanda, T., and Suzuki, T. (1999). *Adsorption* **5**: 7.
66. Iiyama, T., Ruike, M., and Kaneko, K. (2000). *Chem. Phys. Lett.* **331**: 359.
67. Do, D.D. and Do, H.D. (2000). *Carbon* **38**: 767.
68. Neitsch, M., Heschel, W., and Suckow, M. (2001). *Carbon* **39**: 1421.
69. Malakhov, A.O. and Volkov, V.V. (2000). *Polymer Sci. Ser. A* **42**: 1120.
70. Rutherford, S.W. (2003). *Carbon* **41**: 622.

THE EFFECT OF CARBON SURFACE CHEMICAL COMPOSITION ON THE MECHANISM OF PHENOL ADSORPTION FROM AQUEOUS SOLUTIONS

Artur P. Terzyk
Gerhard Rychlicki
Piotr A. Gauden
N. Copernicus University
Toruń, Poland

Piotr Kowalczyk
Faculty of Science,
Chiba University
Chiba, Japan

Experimental studies of solute adsorption from dilute solutions, especially from aqueous dilute solutions, are carried out in many laboratories, yet the theoretical foundations of this phenomenon are still insufficiently solved (1). Adsorption of phenols on carbons is an important subject especially due to ecological aspects. Therefore, an enormous number of papers has been published in this field. The most important reviews dealing with this problem were recently published by Radovic and co-workers (2) and by Moreno - Castilla (3). The authors consider the changes in the surface chemistry of carbons with changes in pH, and discuss different mechanisms of adsorption proposed in the literature.

The main regularities associated with adsorption of phenol on carbons are as follows (2):

(usually observed) a decrease in adsorption with a rise in surface acidity (as measured by Boehm titration) and with surface oxidation, problems in determining real equilibrium;

chemisorption is inhibited by the presence of oxygen surface complexes; therefore, the sites responsible for phenol chemisorption are carbon sites of the active surface area, that is, oxygen free sites located at the edges of graphene layers whereas physisorption occurs all over the surface;

irreversible adsorption is caused by oxidative coupling in the presence of molecular oxygen.

The authors mentioned pointed out the importance of the papers by Coughlin and Ezra (4) who found that adsorption of phenol decreases after oxidation of carbon, and they suggested that surface oxygen complexes increase the affinity of water to carbon and, therefore, solvent molecules can effectively block some micropores. Magne and Walker (5) suggested that phenol adsorbs by physisorption and chemisorption, but they did not define the mechanism of adsorption. Mattson and co-workers (6), Kipling (7), and Puri (8) emphasized the important role of surface carbonyls (the interaction of the π electrons of the benzene ring with the partial positive charge of the carbonyl carbon atom) in the adsorption of phenol and benzene. Coughlin and Ezra (4) suggested that oxygen chemisorbed on carbon appears to influence phenol adsorption only at low concentrations, when the molecules are thought to be adsorbed parallel to the surface due to nonpolar forces operating over the entire phenol molecules. At higher concentrations, phenol molecules are probably adsorbed in the vertical or "end-on" position due to intramolecular interaction with the hydroxyl groups directed away from the carbon surface. Under these conditions, surface oxygen has little influence on phenol adsorption. They also suggested that the decrease in phenol adsorption after oxidation can be caused by the removal of electrons from the π electron system of the basal planes as a result of oxygen chemisorption. They pointed out that adsorption of phenol takes place between the phenol aromatic ring and the π electron system of carbon basal planes.

In their paper, Radovic et al. (2) pointed out the two possible factors influencing the mechanisms of phenol adsorption on carbons. Citing the paper of Coughlin and Ezra (4) and Mattson and co-workers (9), they concluded that the mechanism of phenol adsorption is determined by so-called "$\pi-\pi$ interactions" and "donor–acceptor complex" formation. The first mechanism assumes that oxygen atoms bound to the carbon reduce the π electron density and weaken the dispersion forces between phenol π electrons ring and the π electrons of carbons. Second, one postulates that the adsorption mechanism is based on the formation of donor–acceptor complexes between the surface carbonyl groups (electron donors) and the aromatic rings of phenol acting as the acceptor and the decrease in phenol adsorption after carbon oxidation can be due to the oxidation of surface carbonyls (9).

The energetics of phenol adsorption was studied by Kiselev and co-workers (10) who considered the effect of the substitution of the benzene ring on the enthalpy of phenol adsorption on graphitized carbon black at "zero" coverage. They showed that the substitution of hydrogen for an OH group in the benzene ring increases the potential energy of interaction of the phenol molecule with the surface of graphitized carbon by about 15 kJ/mole and the substitution for an NH_2 group by about 14 kJ/mole.

The study by Jankowska and co-workers (11) paid attention to the role of basic groups in phenol adsorption, showing that they play a dominant role in this process. They also pointed out that the rise in the ratio of the immersion heat of carbon in benzene to the immersion heat in water leads to a rise in phenol adsorption. For that reason, their study is of extreme importance from the point of view of the adsorption mechanism of phenol on carbons. The authors concluded that phenol adsorption is limited by the micropore volume of the carbons, and the actual position beneath this limit is determined by their acid–base characteristics. The study by these authors performed with Koganovski (12) also pointed out that in the range of low equilibrium concentrations, the pore volume of carbon and also the ratio of acidic/basic group concentration determine phenol adsorption. The larger this ratio, the lower the adsorption. The results of Koganovski and co-workers were also confirmed in the paper by Tarasevich and Rak (13) showing that the modification of coal leading to the introduction of basic surface sites favors the adsorption of weak acid in micropores. There was also Speculation about the role of surface functionalities in phenol adsorption on carbons at infinite dilution. Biniak and co-workers (14) studied the isotherms of phenol adsorption on different carbons, and the main attention was paid to the correlation between the value of adsorbability and acidic properties of carbon surface; however, a linear relationship between these two values was possible only if it was assumed that the acidic groups diminish phenol adsorbability with different contributions. They also pointed out that moderately and slightly acidic groups are the most important ones influencing the mechanism of phenol adsorption on carbons.

Zawadzki (15), applying FTIR spectroscopy, showed that oxidation of the carbon surface does not always lead to a decrease in phenol adsorption. Contrary to others, he showed that phenol uptake increases after oxidation. He explained his results by showing that surface oxidation increases the binding energy between phenol and the carbon surface. The interaction with the surface is partially a specific one and occurs with the participation of π electrons of the adsorbed molecules. The main role is played by hydrogen bonds between phenol and surface oxides. Therefore, he concluded that the effect of the competition between water and phenol in the adsorption on surface-active sites is negligible and cannot result in a decrease in phenol adsorption after oxidation by nitric acid.

Calleja and co-workers (16) compared the different models of diffusion showing that the pore diffusion and linear models describe the kinetic adsorption curves of phenol on carbons satisfactorily. The study by Halhouli and co-workers (17) is one rarely met in the literature, where the effect of temperature on phenol adsorption was studied. They showed that the decrease in pH has no effect on the adsorption capacity of charcoal (unspecified by them) and the rise in the temperature increases the adsorption of phenol.

Singh and co-workers (18) pointed out the role of hydrogen bonding between phenolic hydrogen and chemisorbed oxygen present as oxygen surface groups. They also calculated the diameter of the phenol molecule as 0.62 nm. It is interesting that the computer simulation studies of liquid phenol (19) led to the diameter of 0.6 nm.

Loughlin and co-workers reported differences in the kinetics of phenol adsorption on carbons in the presence or absence of oxygen and proposed a model (20). They found that phenol diffusion is pore and surface diffusion and the reaction between phenol and oxygen is observed to be first order. They explained the results obtained as follows: both the molecules of oxygen and phenol diffuse to the surface of carbon and are adsorbed. Reaction occurs on the carbon surface and is catalyzed by it. If the order is assumed to be first, the dissociation of the oxygen molecules occurs before their reaction with phenol. On the contrary, if the order is second, the adsorbed oxygen molecules react directly with phenol.

Abuzaid and Nakhla (21) continued the study on this subject showing that temperature increases the rates of adsorption as well as phenol polymerization. They found values of diffusion energy around 15 kJ/mole, of the same magnitude as determined by others (22,23). The detailed study of phenol adsorption in oxic and anoxic conditions together with the results published by other authors led Terzyk and co-workers to the free-radical mechanism of phenol polymerization on carbons (22–24). The authors mentioned also concluded that the incorporation of surface oxygen groups, called "lactones" in Boehm's method, leads to a decrease in the amount of chemically bound phenol.

Barton and co-workers (25) studied the effect of chemical heterogeneity on phenol adsorption. They showed that the fit of adsorption data described by the Dubinin–Radushkevich (DR) equation is good, and the characteristic adsorption energy values varied from 19–22 kJ/mole. Stoeckli et al. (26,27) showed that the Dubinin–Astakhov (DA) adsorption isotherm equation is applicable for describing phenol adsorption data on a series of microporous carbons and that the parameter n of this equation can be successfully approximated as 4. Following this idea, Terzyk developed a relationship between the characteristic energy of phenol adsorption ($n = 4$) and the average micropore diameter of carbon slit-like micropores (22,23).

The results of Barton et al. (25) supported the previous study by Singh (28) showing that phenol and aniline are more strongly adsorbed (from water) by basic than by acidic carbon. Their explanation is as follows: the process of phenol adsorption via hydrogen bonding with acidic surface sites should be accompanied by an energetically unfavored process of water–phenol bond breaking. Therefore, adsorption takes place mainly by the formation of electron donor–acceptor complexes between basic sites (principally π electron-rich regions on the basal planes of the carbon and the aromatic ring of the adsorbate). It appears, however, that further adsorption takes place by pore filling. The presence of water adsorbed by oxygen groups decreases the adsorption of phenol. The rise in phenol adsorption after oxidation as well as alkylation of coal was found by Jaoui et al. (29).

An interesting study of phenol and benzene adsorption on zeolites was recently done by Klemm and co-workers (30). They showed that the most important interactions in phenol adsorption are coulombic with Na^+ cations. An interesting study of the adsorption of phenols, stressing the role of solute–solvent interactions, was done by Dargaville and co-workers (31). Laszlo and co-workers (32) pointed out that phenol molecules and also water are present on the surface; therefore, the competition between them should be taken into account in calculating the cross-sectional area of adsorbed phenol; moreover, some pores can be inaccessible to adsorbed phenol–solvent aggregates.

Interesting results have been recently published by Nevskaia and co-workers (33). They showed that for nonoxidized carbon, the mechanism of adsorption is based mainly on the interaction between π electrons in phenol and π electrons of carbon; after oxidation, donor–acceptor complexes arise between basic surface oxygen groups and phenol aromatic rings. The study of the kinetics of phenol adsorption performed by the same group (34) showed that for nonmodified carbon, diffusion within the porous structure controls the adsorption rate; for oxidized carbon, the controlling step is the adsorbate–surface interaction, and a shorter time is required to reach the adsorption equilibrium in comparison with the original carbon.

Mahajan et al. (35) pointed out that both water and phenol can form hydrogen bonds with surface groups. Salvador and Merchan (36), studying the thermal desorption of phenolic compounds adsorbed on carbons, concluded that the values of adsorption energies are in the range of the energy of hydrogen bonds. Moreover, they concluded that this energy does not increase with ring substitution but rather with an increase in the electrophilic nature of the substituents.

A very important paper confirming this point of view was also published recently by Franz and co-workers (37). The authors emphasized the importance of the hydrogen bonding created by the molecules of solute and surface groups and suggested a mixed mechanism between the π–π dispersion interactions and bonding of water molecules to surface groups (mainly "carboxylic"). They also postulated that this effect is different for different solute molecules and it is absent in phenol adsorption from water solutions (i.e., water adsorption on active surface groups predominates). The final conclusion was that Mattson's mechanism is not a driving force for adsorption of aromatics on carbons.

Terzyk (38), studying the adsorption of a series of organic compounds on carbons, concluded that the effect of carbon surface chemistry on the adsorption of organics decreases with a rise in temperature. Thus it seems obvious that, differences in the mechanisms of adsorption (more accurately the domination of one of them), observed by different authors are caused by the temperature dependence of the energy of intermolecular interactions. The dispersion forces are temperature independent, whereas the electrostatic interactions between dipoles of adsorbed molecules (and solvent) and surface polar groups generally decrease with a rise in temperature. This implies that the measurements of the temperature dependence of adsorption can be a very useful tool for elaborating the mechanisms of this process. The study published recently by Garcia - Araya and co-workers (39) pointed out the importance of papers showing the temperature dependence of adsorption from solutions on carbons. Costa et al. (40) attributed the rise in phenol adsorption with temperature to an activated adsorption effect, Chiou and Manes (41) concluded that this results from packing effects, and Mattson and co-workers (6) that it is chemisorption. In the opinion of the authors of this study, chemisorption is usually small, so the temperature dependence of adsorption is mainly due to the decrease in the amount of electrostatically interacting molecules with surface groups located at the entrances to small micropores where the so-called "primary micropore filling" process occurs.

A more general mechanism of adsorption was proposed by Radovic and co-workers (2). They pointed out the importance of the pH of the solution and the surface chemistry of carbon, as well as the relative importance of electrostatic and dispersive adsorbate/adsorbent interactions. They also concluded that even though different surface functional groups are responsible for the amphoteric character of carbon, their specific identification is less important than the determination of the isoelectric point and the point of zero charge. The authors concluded that adsorption of aromatic anions occurs primarily on the basal plane of carbon and the electron-withdrawing effects of surface groups suppress the interaction of basal planes with the adsorbate's aromatic rings. Adsorption of aliphatic anions also occurs predominantly on the basal plane, and the same electron-withdrawing effects enhance the interaction with carboxyl anions. Although electrostatic interactions are important, the $\pi-\pi$ dispersion interactions appear to be dominant in the adsorption of aromatic solutes. They also pointed out that if dispersion forces dominate the functionalization (of the carbon adsorbent or the aromatic adsorbate), which decreases the π electron density, this leads to suppressed or weaker adsorption.

In his review, Moreno-Castilla (3) also pointed out the importance of the irreversible adsorption of phenols on carbons. This process was studied extensively in the pioneering paper of Grant and King (42) and by Tamon et al. (43,44) and others (22,24). The occurrence of this effect complicates the observation of regularities during the adsorption process.

It can be seen that despite an enormous number of papers, there are still unsolved problems:

The role of porosity; some authors implied that the adsorption of phenol is mainly controlled by the porous structure, and others thought that it is not. There is no correlation between phenol uptake and total surface area of carbons, and the details of the role of surface functional groups in phenol adsorption are still unknown; although different general mechanisms of phenol adsorption have been proposed, we still lack the general adsorption mechanism taking into account the "micropore filling" process and also answering the question, exactly what types of surface functional groups influence the value of phenol adsorption beneath the previously mentioned limit of micropore filling. The irreversibility of phenol adsorption on carbons is fairly well established, however, the detailed mechanism of this process as well as the well documented answer to the question, what types of surface functional groups (and why) are responsible for this process are still unknown. The influence of surface functional groups on the kinetics of phenol adsorption has not yet been determined, that is it is not known what types of surface functionalities determine the kinetics of adsorption of that compound.

Recently, Terzyk (22) presented the results of phenol adsorption–desorption (and kinetic) measurements at three temperatures and at two pH levels on a series of microporous activated carbons. Comparing the results of phenol adsorption with those determined for paracetamol, acetanilide, and aniline, it was shown that at a low equilibrium concentration, phenol adsorption decreases with the basic and carboxylic group content of the carbon surface. A linear correlation was observed between the average adsorbability of phenol and the total concentration of surface bases and carboxyls [the same author showed that the results for two Polish adsorbents follow the same line (23)]. The plot of the results tabulated in the paper by Biniak and co-workers (14) shows a similar tendency Therefore, carboxyls, that is, the most acidic and some of the basic groups, play the most important role in phenol adsorption at low concentrations. For adsorption at the acidic pH level, the value of adsorbability was correlated with the enthalpy of carbon immersion in water (therefore, with the total content of surface acidic groups and the pH of the zero point charge). Consequently, at an acidic pH, where all the carbons studied are below the pH of the point of zero charge, the more polar the surface of carbon, the lower the adsorbability of phenol. Therefore, it can be presumed that at infinite dilution and neutral pH, the positive surface charge resulting from adsorption of protons by the strongest surface bases leads to an increase in water adsorption and blocks some more active sites. The same effect is caused by undissociated surface acidic groups, and both of them decrease phenol adsorbability, compared to that observed for nonmodified carbon. The same author showed that in the range of micropore filling the pore blocking effect by phenol molecules dominates, and this effect weakens with a rise in temperature. Thus, the mechanism of phenol adsorption process is mixed, and

$\pi - \pi$ interactions between the phenol aromatic ring and pore walls are "screened" by the blocking of pore entrances by water and phenol molecules (interactions *via* hydrogen bonds) (38).

BIBLIOGRAPHY

1. Jaroniec, M. and Madey, R. (1988). *Physical Adsorption on Heterogeneous Solids*. Elsevier, Amsterdam, p. 251.
2. Radovic, L.R., Moreno-Castilla, C., and Rivera-Utrilla, J. (2001). *Chem. Phys. Carbon* **27**: 227.
3. Moreno-Castilla, C. (2004). *Carbon* **42**: 83.
4. Coughlin, R.W. and Ezra, F.S. (1968). *Environ. Sci. Technol.* **2**: 291.
5. Magne, P. and Walker, P.L. (1986). *Carbon* **24**: 101.
6. Mattson, J.S., Mark, H.B., Jr., Malbin, M.D., Weber, W.J., Jr., and Crittenden, J.C. (1969). *J. Colloid Interface Sci.* **31**: 116.
7. Kipling, J.J. (1967). *Porous Carbon Solids*. R.L. Bond (Ed.). Academic Press, London, p. 65.
8. Puri, B.R. (1970). *Chem. Phys. Carbon* **6**: 191.
9. Mattson, J.S. and Mark, H.B. (1971). *Activated Carbon. Surface Chemistry and Adsorption from Solution*. Marcel Dekker, New York, p. 209.
10. Beljakova, L.D., Kiselev, A.V., and Kovaleva, N.V. (1966). *Zhurn. Fiz. Khim.* **40**: 1494.
11. Jankowska, H., Świątkowski, A., Gelbin, D., Radeke, K.H., and Seidel, A. (1986). *Ext. Abstr. Int. Carbon Conf. Carbon 86*, Baden–Baden, Germany, p. 397.
12. Koganovski, A.M., Zaidel, A., and Radeke, K.H. (1987). *Khim. Technol. Vody* **6**: 500.
13. Tarasevich, Y.I. and Rak, V.S. (1988). *Kolloid. Zhurn* **60**: 84.
14. Biniak, S., Kaźmierczak, J., and Świątkowski, A. (1989). *Adsorption Sci. Technol.* **6**: 182.
15. Zawadzki, J. (1989). *Chem. Phys. Carbon* **21**: 147.
16. Calleja, G., Serna, J., and Rodriguez, J. (1993). *Carbon* **31**: 691.
17. Halhouli, K.A., Darwish, N.A., and Al-Dhoon, N.M. (1995). *Sep. Sci. Technol.* **30**: 3313.
18. Singh, B., Madhusudhanan, S., Dubey, V., Nath, R., and Rao, N.B.S.N. (1996). *Carbon* **34**: 327.
19. Mooney, D.A., Muller-Plathe, F., and Kremer, K. (1998). *Chem. Phys. Lett.* **294**: 135.
20. Loughlin, K.F., Hassan, M.M., Ekhator, E.O., and Nakhla, G.F. (1996). In *Fundamentals of Adsorption*. M.D. LeVan (Ed.). Kluwer, Boston, MA, p. 537.
21. Abuzaid, N.S. and Nakhla, G.F. (1997). *Sep. Sci. Technol.* **32**: 1255.
22. Terzyk, A.P. (2003). *J. Colloid Interface Sci.* **268**: 301.
23. Terzyk, A.P. (2003). *Adsorption Sci. Technol.* **21**: 539.
24. Terzyk, A.P., Rychlicki, G., and Gauden, P.A. (2003). *Carbon 03*. Extended Abstracts, Oviedo, Spain.
25. Barton, S.S., Evans, M.J.B., and MacDonald, J.A.F. (1997). *Polish J. Chem.* **71**: 651.
26. Stoeckli, F., Lopez–Ramon, M.V., and Moreno–Castilla, C. (2001). *Langmuir* **17**: 3301.
27. Stoeckli, F. and Hugi-Cleary, D. (2001). *Russ. Chem. Bull.* **50**: 2060.
28. Singh, D.H. (1971). *Indian J. Chem.* **9**: 1369.
29. Jaoui, M., Gruber, R., Cagniant, D., Nosyrev., I.E., Rogalski, M., and Achard, C. (1997). *Proc. ICCS'97*. P&W Druck and Verlag, Essen, p. 1831.
30. Klemm, E., Wang, J., and Emig, G. (1998). *Microporous Mesoporous Mater.* **26**: 11.
31. Dargaville, T.R., Guerzoni, F.N., Looney, M.G., and Solomon, D.H. (1996). *J. Colloid Interface Sci.* **182**: 17.
32. Laszlo, K., Botta, A., Nagy, L.G., and Cabasso, I. (1999). *Colloids Surf. A.* **151**: 311.
33. Nevskaia, D.M. and Guerro-Ruiz, A. (2001). *J. Colloid Interface Sci.* **234**: 316.
34. Nevskaia, D.M., Santianes, A., Munoz, V., and Guerro-Ruiz, A. (1999). *Carbon* **37**: 1065.
35. Mahajan, O.P., Moreno-Castilla, C., and Walker, P.L. (1980). *Sep. Sci. Technol.* **15**: 1733.
36. Salvador, F. and Merchan, M.D. (1996). *Carbon* **12**: 1543.
37. Franz, M., Arafat, H.A., and Pinto, N.G. (2000). *Carbon* **38**: 1807.
38. Terzyk, A.P. (2004). *J. Colloid Interface Sci.* **275**: 9.
39. Garcia-Araya, J.F., Beltran, F.J., Alvares, P., and Masa, F.J. (2003). *Adsorption* **9**: 107.
40. Costa, E., Calleja, G., and Marijuan, L. (1989). *Adsorption Sci. Technol.* **5**: 213.
41. Chiou, C.C.T. and Manes, M. (1974). *J. Phys. Chem.* **78**: 622.
42. Grant, T.M. and King, C.J. (1990). *Ind. Eng. Chem. Res.* **29**: 264.
43. Tamon, H. and Okazaki, M. (1996). *J. Colloid Interface Sci.* **179**: 181.
44. Tamon, H., Atsushi, M., and Okazaki, M. (1996). *J. Colloid Interface Sci.* **177**: 384.

CARBONATE GEOCHEMISTRY

KENNETH F. STEELE
University of Arkansas
Fayetteville, Arkansas

Calcite ($CaCO_3$) is one of the most common minerals at or near the surface of the earth and thus is one of the main contributors to carbonate geochemistry. Less common carbonate minerals include dolomite [$MgCa\ (CO_3)_2$]; aragonite, a calcite polymorph (mineral of the same composition as calcite but having a different atomic structure); azurite and malachite (copper hydroxycarbonate minerals); siderite ($FeCO_3$); and rhodochrosite ($MnCO_3$) which can be important spatially and economically.

Although calcite does occur in igneous and metamorphic rocks, the mineral is present dominantly in the sedimentary environment which is the focus of this article. Calcite may be precipitated to form a relatively pure monomineralic rock (limestone), be present as a cement binding sediments into indurated rock, occurs simply as a trace mineral in rocks, and composes many shells and fossils. Weathering of calcite is relatively easy and depends largely on the amount of carbonic acid present. Other carbonate minerals, except dolomite, generally are dissolved and precipitated in a manner geochemically similar to calcite.

CALCITE SOLUBILITY PRODUCTS

The solubility of $CaCO_3$ depends on which polymorph is considered and the reference for the equilibrium constant.

For example, at 25 °C, the ion concentration products (K_{sp}) for calcite and aragonite (the most common polymorphs of $CaCO_3$) are $10^{-8.48}$ and $10^{-8.34}$, respectively (1). Krauskopf and Bird (2) provide calcite and aragonite K_{sp} values of $10^{-8.35}$ and $10^{-8.22}$, respectively. The K_{sp} values for calcite solubility from Krauskopf and Bird (2) are 35% greater than those from APHA (1). A similar relationship exists for aragonite; Krauskopf and Bird (2) present K_{sp} values for aragonite that are 31% higher. Variation in solubility experiments can yield significant differences in the K_{sp}.

Temperature

Counter to the typical increase of mineral solubility with increasing temperature observed for most minerals, many carbonate minerals are more soluble in cold water. For example, the K_{sp} for calcite at 0 °C and at 50 °C are $10^{-8.02}$ and $10^{-8.63}$, respectively (3). These values represent about a fourfold difference in calcite solubility caused by temperature alone. A thorough discussion of the thermodynamic reactions of carbonate minerals solubilities is beyond the scope of this article. A simplified explanation why carbonate minerals are more soluble in cold water is that the dissolution reaction for carbonate minerals is exothermic, which results in higher temperatures favoring the solid phase over dissolved ions. A more detailed discussion for this unusual behavior can be found in Langmuir (4).

In addition to an increase in solubility products with decreasing temperature, carbon dioxide is more soluble at lower temperatures, further favoring carbonate mineral dissolution in cooler environments. More carbonic acid is present in cold water at a given partial pressure of carbon dioxide (p_{CO_2}), and carbonic acid concentration is the controlling factor for the solubility of carbonate minerals under natural conditions. Carbon dioxide is obtained from the air or from decomposition of organic matter that releases carbon dioxide which reacts with water to form carbonic acid. These factors combine to form the calcite compensation depth (about 4500 m) below which calcite dissolves (2).

When geothermal waters reach the surface of the earth and precipitate tufa, a calcium carbonate rock, some reaction must overcome the temperature decrease, or these waters would dissolve calcite rather than deposit tufa as the water cools. The required reaction is the loss of CO_2 due to lower p_{CO_2} pressure at the earth's surface and subsequent decrease in the amount of carbonic acid, which controls the amount of dissolved calcite (see Calcite Solubility below).

Pressure

The increased partial pressure of CO_2 near the surface of the earth increases the amount of CO_2 that dissolves in water, therefore, increasing the solubility of carbonate minerals such as calcite. Pressure alone does not affect the solubility of calcite as much as the effect of temperature. Nonetheless, where pressure is large, for example, deep within the ocean, its effect alone can increase calcite solubility about twofold (2).

Ionic Strength

Other ions in solution shield Ca^{2+} and CO_3^{2-} ions from interacting and precipitating. Faure (5) describes this situation as "... the activity of the ions in electrolyte solutions is less than their concentration. It is plausible to expect that the interference by the other ions increases with their concentrations and charges." Activity can be defined as the concentration of an ion at zero ionic strength. The effect of the concentration and charge of ions is represented by the ionic strength (I) of the solution:

$$I = 0.5\ \Sigma\ m_i z_i^2 \tag{1}$$

where m_i is the molar concentration of an ion (i) and z is the charge of the ion.

Ionic strength can be use to calculate the activity coefficient γ which relates molar concentration to activity; γ is 1 or less.

$$m_i(\gamma) = a \tag{2}$$

where a is the activity of the ion.

γ can be calculated from the Debye–Hückel equation (Eq. 3) or other similar equations that use I and some constants that relate to the dielectric constant of water and temperature (A and B) and effective size (a) of ions. Values of the constants can be obtained from Faure (5).

$$-\log \gamma = \frac{Az^2 I^{1/2}}{I + aBI^{1/2}} \tag{3}$$

SATURATION INDEX

Waters tend to precipitate $CaCO_3$ when oversaturated with $CaCO_3$, and tend to dissolve $CaCO_3$ if undersaturated with $CaCO_3$. The Saturation Index (SI) is the most widely used method to determine the amount of calcite that will be precipitated or dissolved. Many water chemistry computer programs calculate RS (relative saturation) which is related to SI as

$$SI = \log RS \tag{4}$$

where RS = ratio of $CaCO_3$ activity product (K_a) to $CaCO_3$ solubility product (K_{sp}).

SI values of 1 indicate saturation, negative values indicate undersaturation, and positive values indicate oversaturation.

MEASURED ALKALINITY

Alkalinity is the measure of the acid-consuming ability (power) of a solution expressed as mg/L $CaCO_3$. In nature, the main ions that neutralize H^+ ions are HCO_3^- and CO_3^{2-} (6). Although OH^-, PO_4^{3-} and other negative ions can neutralize H^+ ions, generally the concentrations of these ions is so small in the natural environment that HCO_3^- and CO_3^{2-} are the only significant acid-neutralizing ions. In some situations, especially in anthropogenically influenced situations, OH^- can be important (e.g., the impact of cement on water chemistry); thus, OH^- is included in the following calculations. Alkalinity is usually determined by titrating a 50 or 100 mL water sample with 0.02 N H_2SO_4 to a pH of 8.3 for CO_3^{2-} and/or OH^-. This 8.3 pH end point is often referred to as the phenolphthalein (or *P*) end point because this indicator changes colors at pH 8.3. Titration continues past 8.3 to a pH of about 4.5 as any HCO_3^- (originally present or formed from partial neutralization of CO_3^{2-}) is neutralized. The pH of 4.5 can vary from 4.3–4.9 depending on the presence and amounts $CaCO_3$ and certain anions (1). This final end point is referred to as total alkalinity. These pH titration end points are based on inflection points on the titration curve of pH versus milliliters of acid added to the water sample titration curve for neutralization of the bases. The amount of acid necessary to reach the phenolphthalein end point is equivalent to the amount of base ($CaCO_3$) neutralized and is referred to as *P* alkalinity. The amount of $CaCO_3$ to neutralize "completely" or "totally" the H^+ ions added is referred to as *T* alkalinity. Thus, *T* must be larger than or equal to *P*. There are five alkalinity conditions possible: (1) only HCO_3^-, (2) only CO_3^{2-}, (3) only OH^-, (4) HCO_3^- plus CO_3^{2-}, and (5) CO_3^{2-} plus OH^-. Bicarbonate and OH^- ions cannot coexist under typical natural conditions because these ions react to form CO_3^{2-} ions and H_2O. These five situations are discussed briefly below.

Bicarbonate Alkalinity. If the initial pH of a sample is less than 8.3, there is only bicarbonate alkalinity and no *P* alkalinity. Or one can state that there is only *T* alkalinity (the amount of acid necessary to lower the pH below 4.5).

Carbonate Alkalinity. There are two abrupt changes in pH during alkalinity titration, one at 8.3 and the other at 4.5. If only CO_3^{2-} ions are present, the amount of acid necessary to lower the pH below 8.3 is one-half the total acid (i.e., *T*) necessary to lower the pH below 4.5. The first abrupt pH change at 8.3 represents the conversion of carbonate to HCO_3^- (i.e., half of the carbonate has been neutralized), and the second abrupt pH change at 4.5 represents neutralization of the HCO_3^- created from carbonate. If only CO_3^{2-} ions are present, $T = 2P$.

Hydroxide Alkalinity. If the initial pH is greater than 8.3 and the addition of acid rapidly lowers the pH below 4.5, there is only hydroxide alkalinity and thus $T = P$.

Bicarbonate Plus Carbonate Alkalinity. Because P represents the conversion of CO_3^{2-} to HCO_3^-, the same amount of acid (*P*) is required to neutralize the HCO_3^- ions formed from the CO_3^{2-} ions; 2*P* represents complete neutralization of CO_3^{2-} ions present. In addition, HCO_3^- originally present is included in *T*, which means that $T > 2P$. Because CO_3^{2-} ions are present, there will be two abrupt changes in pH (at 8.3 and 4.5).

Carbonate Plus Hydroxide Alkalinity. When both CO_3^{2-} and OH^- ions are present, the amount of acid required to lower the pH below 8.3 is that necessary to convert CO_3^{2-} to HCO_3^- plus that necessary to neutralize OH^- ions (i.e., the OH^- milliequivalent/L (meq/L) concentration is equal to T minus CO_3^{2-} meq/L). In addition, T-P is equal to 1/2 the CO_3^{2-} (the amount of CO_3^{2-} converted to HCO_3^{2-} concentration) or CO_3^{2-} meq/L equals 2(T-P). These two relationships can be combined to yield OH^- meq/L equals to 2P-T. These conditions constrain the *P* and *T* titration relationship to $T < 2P$.

Table 1. Alkalinity Determined by Titration

Result of Titration	Bicarbonate Alkalinity	Carbonate Alkalinity	Hydroxide Alkalinity
$P = 0$	T	0	0
$2P = T$	0	T	0
$P = T$	0	0	T
$P > 0.5T$ or $T > 2P$	$T - 2P$	$2P$	0
$P < 0.5T$ or $T < 2P$	0	$2(T - P)$	$2P - T$

Table 1 summarizes the equations necessary to determine the portion of alkalinity due to HCO_3^-, CO_3^{2-} ions, and OH^-. At pH 9, the OH^- concentration is only 1.0 mg/L as $CaCO_3$ and can generally be neglected in alkalinity determinations for water whose pH is less than 9.0.

CALCULATION OF BICARBONATE, CARBONATE, AND HYDROXIDE ALKALINITY

Bicarbonate, carbonate, and hydroxide alkalinity can be calculated from total alkalinity (TA) if the pH is known. These concentrations are often calculated, especially for carbonate and hydroxide ion concentrations that are generally low and difficult to measure. The ion concentrations must be given in molar (M) concentrations, which are indicated by brackets in this article. For most natural situations, Eq. 5, which takes into account charge differences, is adequate for determining the various sources of alkalinity:

$$2\text{TA} = [\text{HCO}_3^-] + 2[\text{CO}_3^{2-}] + [\text{OH}^-] \quad (5)$$

TA is commonly expressed as mg/L $CaCO_3$ and must be converted to moles/L (M) to make the following calculations. The conversion of mg/L to M $CaCO_3$ is accomplished by using Eq. 6

$$(\text{TA moles/L CaCO}_3) = \frac{\text{mg/L CaCO}_3}{100{,}000\ \text{mg/mole}} \quad (6)$$

Rearranging the second disassociation constant for carbonic acid (Eq. 7) in terms of $[CO_3^{2-}]$ yields Eq. 8:

$$K_2 = \frac{[\text{H}^+][\text{CO}_3^{2-}]}{[\text{HCO}_3^-]} = 10^{-10.3} \quad (7)$$

$$[\text{CO}_3^{2-}] = \frac{K_2[\text{HCO}_3^-]}{[\text{H}^+]} \quad (8)$$

Substitution of Eq. 8 for $[CO_3^{2-}]$ in Eq. 5 and expressing $[OH^-]$ in terms of $\frac{K_w}{[H^+]}$, based on Eq. 9, yields Eq. 10.

$$[\text{OH}^-] = \frac{K_w}{[\text{H}^+]} = \frac{10^{-14}}{[\text{H}^+]} \quad (9)$$

$$2\text{TA} = [\text{HCO}_3^-] + 2\left[\frac{K_2[\text{HCO}_3^-]}{[\text{H}^+]}\right] + \frac{K_w}{[\text{H}^+]} \quad (10)$$

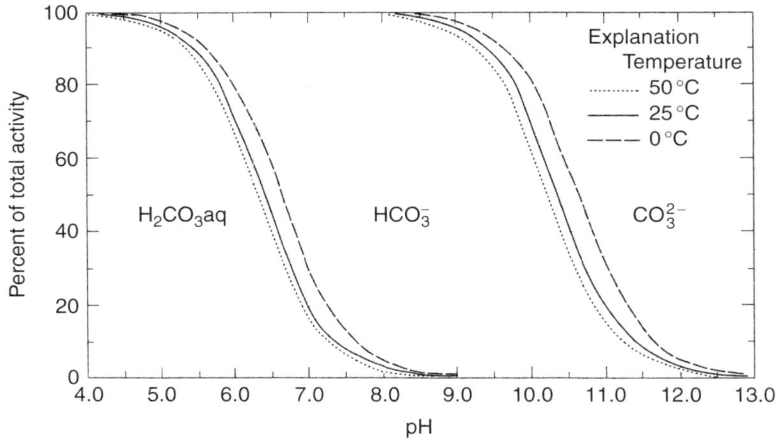

Figure 1. Plot of percent of total activity of carbonate species versus pH at 0, 25 and 50 °C. From Hem (7).

Rearranging Equation 10 to solve for HCO_3^- (Eq. 12) is carried out by the following step (Eq. 11) and expressing K_w as exponent of 10 and negative pH as exponent of 10.

$$2\text{TA} - \frac{K_w}{[H^+]} = [HCO_3^-]\left(1 + \frac{2K_2}{[H^+]}\right) \tag{11}$$

$$[HCO_3^-] = \frac{2\text{TA} - 10^{Kw+pH}}{1 + 2K_2 10^{pH}} \tag{12}$$

Once $[HCO_3^-]$ has been calculated (or determined by titration), this concentration can be substituted in Equation 8 and $[CO_3^{2-}]$ can be calculated. To determine the hydroxide alkalinity, only the pH must be known (Eq. 9).

To convert these molar alkalinity concentrations (Eqs. 8, 9, and 12) to mg/L, the molar concentrations must be multiplied by the molecular weight of $CaCO_3$ (100,000 mg/mole) and for $[HCO_3^-]$ and $[OH^-]$, a factor of 1/2 is required because 1 mole of $CaCO_3$ is equivalent to 2 moles of HCO_3^- and 2 moles of OH^-, whereas 1 mole of CO_3^{2-} is equivalent to 1 mole of $CaCO_3$.

These are the equations necessary for calculating bicarbonate, carbonate, and hydroxide alkalinities as $CaCO_3$ mg/L:

Bicarbonate alkalinity ($CaCO_3$ mg/L)

$$= 50{,}000\ \left(\frac{2\text{TA} - 10^{Kw+pH}}{1 + 2K_2 10^{pH}}\right) \tag{13}$$

Carbonate alkalinity ($CaCO_3$ mg/L)

$$= 100{,}000\ (K_2[HCO_3^-](10^{pH}) \tag{14}$$

Hydroxide alkalinity ($CaCO_3$ mg/L)

$$= 50{,}000\ (10^{Kw+pH}) \tag{15}$$

CARBONATE SPECIES

Concentrations of the carbonate species $[H_2CO_3]$, $[HCO_3^-]$ and $[CO_3^{2-}]$ can be calculated if the total dissolved inorganic carbonate (DIC) concentration and pH are known. [DIC] is often expressed as HCO_3^- because it is the dominant ion at typical pH values (Fig. 1). Figure 1 shows the concentrations of the carbonate species for an aqueous solution with respect to the pH and temperature of the solution. Note that activities (concentrations) are higher at lower temperatures. The pertinent equilibrium reactions are the two disassociation constants for H_2CO_3 (Eqs. 8 and 16).

$$K_1 = \frac{[H^+][HCO_3^-]}{[H_2CO_3]} = 10^{-6.4} \tag{16}$$

$$[\text{DIC}] = [H_2CO_3] + [HCO_3^-] + [CO_3^{2-}] \tag{17}$$

By rearranging K_1 (Eq. 16) to express $[H_2CO_3]$ in terms of $[HCO_3^-]$, by rearranging K_2 (Eq. 7) to express $[CO_3^{2-}]$ in terms of $[HCO_3^-]$, and then substituting these values in Equation 17, the following equation is obtained:

$$[\text{DIC}] = [[H^+]\ [HCO_3^-](K_1^{-1})] + [HCO_3^-] + [[H^+]^{-1}\ [HCO_3^-](K_2)] \tag{18}$$

Factoring out $[HCO_3^-]$, rearranging Eq. 16 to solve for $[HCO_3^-]$, and expressing $[H^+]$ as pH (negative exponent of 10) yields the following:

$$[HCO_3^-] = \left[\frac{[\text{DIC}]}{(10^{-pH-K_1}) + (10^{pH+K_2}) + 1}\right] \tag{19}$$

Because K_1, K_2, and K_w are temperature dependent, the concentrations of the carbonate species must be determined by using the appropriate values for a specific temperature.

CALCITE SOLUBILITY

If only calcite were present in an aqueous solution with no carbon dioxide, one could determine the solubility of calcite by obtaining the square root of the solubility product for calcite (Eq. 20). Reported K_{sp} values for calcite vary with the reference and range from $K_{sp} = 10^{-8.3}$ to K_{sp} $10^{-8.5}$ at 25 °C (1,2). The value of $10^{-8.3}$ will be used throughout this article.

$$[CaCO_3] = [Ca^{2+}][CO_3^{2-}] = K_{sp} = 10^{-8.3} \tag{20}$$

Hydrolysis

Although one can calculate the solubility of Ca^{2+} as $10^{-4.15}$ M or 6.8×10^{-5} from Equation 20, the CO_3^{2-} ion will undergo hydrolysis (Eq. 21):

$$CaCO_3 + H_2O = Ca^{2+} + HCO_3^- + OH^- \quad (21)$$

$$K = [Ca^{2+}][HCO_3^-][OH^-] = 10^{-12} \quad (22)$$

Inspection of Eq. 21 indicates that all three product ions are equal. The cube root of 10^{-12}, yields a concentration of 10^{-4} M for all three ions; thus, the pH of the solution is 10. At this high pH, there is appreciable disassociation of HCO_3^- to CO_3^{2-} and H^+ (which is quickly neutralized). Then these CO_3^{2-} ions undergo hydrolysis producing equal amounts of HCO_3^- and OH^- (Eq. 23).

$$CO_3^{2-} + H_2O = HCO_3^- + OH^- \quad (23)$$

Furthermore, water with a pH of 10 has a $[CO_3^{2-}]/[HCO_3^-]$ ratio of 1/2 (from Eq. 7). The total carbonate in this situation is the sum of HCO_3^- and CO_3^{2-} (H_2CO_3 is negligible at this high pH—see Fig. 1), and total carbonate is equal to Ca^{2+} (Eq. 24).

$$\Sigma[CO_3^{2-}] = [HCO_3^-] + [CO_3^{2-}] = [Ca^{2+}] \quad (24)$$

Because $\dfrac{[CO_3^{2-}]}{[HCO_3^-]} = 0.5$, Eq. 7 can be modified to

$$[Ca^{2+}] = [HCO_3^-] + 0.5\,[HCO_3^-] = 1.5\,[HCO_3^-] \quad (25)$$

Instead of Ca^{2+} having the same concentration as HCO_3^- and OH^-, Ca^{2+} will be equal to 1.5 $[HCO_3^-]$ and 1.5 $[OH^-]$ because $[HCO_3^-] = [OH^-]$ (Eq. 23).

Equation 22 is then modified to

$$K = [Ca^{2+}]\ 1.5\,[HCO_3^-]\ 1.5\,[OH^-] \qquad K = 2.25 \times 10^{-12} \quad (26)$$

Because $[Ca^{2+}] = 1.5[HCO_3^-] = 1.5[OH^-]$, the Ca^{2+} concentration can be estimated by the cube root of 2.25×10^{-12} (Eq. 26) or 1.3×10^{-4} (Eq. 26). M (equivalent to 5.2 mg/L). This value is about twice that obtained in Equation 20 and represents the lowest equilibrium concentration of Ca^{2+} for pure water *not* in contact with CO_2. Using this method results in a pH change from 10.00 to 9.94. Recalculation using this method based on the new pH of 9.94 yields a 2% change in $[Ca^{2+}]$ and a new pH of 9.93. Another iteration of these calculations yields a $[Ca^{2+}]$ difference of 1% and the pH remains 9.93.

Carbonic Acid

The main factor controlling carbonate solubility in nature is the amount (partial pressure) of CO_2 that forms H_2CO_3 (Eq. 27). The partial pressure of CO_2 in the atmosphere is 0.0003 or 0.03% of the atmospheric pressure at the surface of the earth (i.e., 1 bar or atmosphere). Partial pressure is equal to moles per liter (5); thus, there is $10^{-5.0}$ M H_2CO_3 in water in contact with the atmosphere (Eqs. 26 and 27).

$$CO_2 + H_2O = H_2CO_3 \quad (27)$$

$$K = \frac{[H_2CO_3]}{[CO_2)(H_2O]} = \frac{[H_2CO_3]}{(0.0003)(1)} = 10^{-1.5} \quad (28)$$

$$[H_2CO_3] = 3 \times 10^{-4}(10^{-1.5}) = 10^{-5.0}M \quad (29)$$

In most natural environments, the appropriate equation for calcite solubility is Equation 30 where CO_2 is present:

$$CaCO_3 + H_2CO_3 = Ca^{2+} + 2HCO_3^- \quad (30)$$

Equation 29 (2) takes into account the presence of CO_2 at the earth's surface which forms carbonic acid (Eq. 27) that readily dissolves $CaCO_3$:

$$K = \frac{[Ca^{2+}][HCO_3^-]^2}{[H_2CO_3]} = 10^{-4.4} \quad (31)$$

$[HCO_3^-] = 2[Ca^{2+}]$ from inspection of Eq. 30; therefore,

$$K = 10^{-4.4} = \frac{[Ca^{2+}][2Ca^{2+}]^2}{[H_2CO_3]} = \frac{4[Ca^{2+}]^3}{10^{-5.0}} \quad (32)$$

The Ca^{2+} concentration calculated from Equation 32 is 4.7×10^{-4} M (equivalent to 18.8 mg/L). This is the solubility of calcite at surface conditions, 25 °C and 1 atmosphere pressure, which provides a "minimum" concentration of atmospheric carbonic acid. More CO_2 can be dissolved in cold water, and the CO_2 partial pressure in soil can be rather high, leading to increased calcite solubility. Decaying vegetation in soil can have a CO_2 partial pressure in excess of 0.1 atmospheres (2) which can significantly raise H_2CO_3 concentrations of infiltrating meteoric water. The solubility of calcite in water in contact with 0.1 atmosphere of CO_2 can be calculated as in Eqs. 28 and 32, producing a calcite solubility of 3.2×10^{-3} M (equivalent to 128 mg/L Ca^{2+}). This value is not an absolute maximum but does give an approximate upper limit for Ca^{2+} concentrations from limestone near the surface of the earth. Other acids, for example, sulfuric from the oxidation of sulfide minerals such as pyrite (FeS_2), can also have a significant effect on calcite solubility locally.

CAVE FORMATION

Carbonic acid-rich water forms limestone caves, which is the most common type of cave. When the water table is high, carbonic acid-rich water dissolves the limestone (calcite). Later when the water table drops, a "void" filled with air is formed and a large void is called a cave. Smaller amounts of water rich in Ca^{2+} and HCO_3^- may continue to flow through the void. When these waters enter the void, CO_2 partial pressure decreases and CO_2 is released. This degassing of CO_2 drives the reactions in Equations 27 and 30 to the left, leading to the precipitation of calcite and formation of stalagmites, stalactites, and other cave features. Although evaporation may play a small role in the deposition of calcite, the loss of CO_2 is the most important factor.

HARDNESS

Water hardness is a measure of how difficult it is for soap to lather; most of the divalent ions in water must be precipitated by soap before soap can produce suds. The most abundant divalent ions in nature are Ca^{2+} and Mg^{2+}, although Fe^{2+} and other divalent ions can be important in some waters. The major source of Ca^{2+} and Mg^{2+} is carbonate minerals; therefore, hardness is expressed as mg/L $CaCO_3$. The range in hardness values with descriptive terms is shown in Table 2.

In the past, hardness was determined by a titration analysis of divalent ions (7). Today, hardness is calculated by converting Ca^{2+} and Mg^{2+} molar concentrations to mg/L $CaCO_3$ by multiplying the mg/L concentrations of the two ions by the inverse of the equivalent weights of the ions and summing the two concentrations (Eq. 31):

$$\text{Hardness as mg/L } CaCO_3 = 2.497\ (Ca^{2+}\ \text{mg/L}) + 4.118\ (Mg^{2+}\ \text{mg/L}) \qquad (33)$$

Some scientists consider hardness more of a water quality (use) factor that relates to scaling in hot water heaters and in industrial setting, rather than an important geochemical factor. However, comparison of hardness and alkalinity (both expressed as mg/L $CaCO_3$) can yield helpful water chemistry information. If hardness exceeds alkalinity, the excess is termed "noncarbonate" hardness, meaning that there are noncarbonate mineral sources of Ca^{2+} and Mg^{2+} ions. The remainder of the hardness is considered "carbonate" hardness derived from carbonate minerals. In this discussion, it is assumed that there is "no" hydroxide alkalinity. If alkalinity exceeds hardness, there must be carbonate mineral sources that do not contain Ca^{2+} and/or Mg^{2+}, for example, $NaHCO_3$, or cation exchange has occurred with clay minerals (e.g., Ca^{2+} replacing Na^+).

DOLOMITE

Dolomite has a very small K_{sp}; values range from 10^{-17} to 10^{-19} (2). The K_{sp} is too small to produce large thicknesses of dolomite and furthermore there is no true dolomite being precipitated today. Some "protodolomite" is precipitated in restricted environments, such as evaporite lagoons. Dolomite should not exist, and yet there are thick sequences of dolostone throughout the world and throughout the geologic record—hence, the "dolomite problem."

Dolostone characteristically preserves fossils poorly, is coarse grained, and commonly has cavities/pore spaces, which all indicate replacement of limestone as the origin of the rock. In addition, it is known that saline water, high pH, a low Ca^{2+}/Mg^{2+} ratio, and elevated temperatures all favor dolostone formation. Based on these factors, various marine and deep water reflux models have been proposed, as well as mixing models for fresh water and sea water, but no "perfect" model has been developed for the formation of dolomite. The dolomite atomic structure is highly ordered which may affect the amount of time necessary to develop the structure. Dissolution of dolomite also is complicated because dolomite undergoes incongruent dissolution as Mg^{2+} dissolves before Ca^{2+} (2). At any rate, the formation and dissolution of dolomite is complex and not well understood in comparison with other carbonate minerals.

Table 2. Hardness Ranges. Units are in mg/L of Calcium Carbonate[a]

0–60	Soft
51–120	Moderately hard
121–180	Hard
>180	Very hard

[a]Reference 7.

BIBLIOGRAPHY

1. Eaton, A.D., Clesceri, L.S., and Greenberg, A.E. (Eds.). (1995). *Standard Methods for the Examination of Water and Wastewater*. American Public Health Association, Washington DC, pp. 2-25–2-38 and 4-17–4-18.
2. Krauskopf, K.B. and Bird, D.K. (1995). *Introduction to Geochemistry*, 3rd Edn. McGraw-Hill, New York.
3. Garrels, R.M. and Christ, C.L. (1965). *Solutions, Minerals and Equilibria*. Harper & Row, New York.
4. Langmuir, D. (1997). *Aqueous Environmental Geochemistry*. Prentice-Hall, Upper Sanddle River, NJ.
5. Faure, G. (1998). *Principles and Applications of Geochemistry*, 2nd Edn. Prentice-Hall, Upper Saddle River, NJ.
6. Drever, J.I. (1997). *The Geochemistry of Natural Waters-Surface and Groundwater Environments*, 3rd Edn. Prentice-Hall, Upper Saddle River, NJ.
7. Hem, J.D. (1985). *Study and Interpretation of the Chemical Characteristics of Natural Water*. 3rd Edn. U.S. Geological Survey Water Supply Paper 2254, U.S. Geological Survey, Alexandria, VA, p. 263.

CARBONATE IN NATURAL WATERS

Robert Y. Ning
King Lee Technologies
San Diego, California

INTRODUCTION

Carbonate is the divalent anion resulting from complete neutralization of carbonic acid. In natural waters, it is in dynamic equilibrium with carbon dioxide gas, carbonic acid, bicarbonate anion, and solid forms of insoluble carbonate minerals—mainly calcium carbonate:

$$CO_2(g) + H_2O \Longleftrightarrow H_2CO_3 \Longleftrightarrow HCO_3^- \Longleftrightarrow CO_3^{2-} \Longleftrightarrow \text{Carbonate(s)}$$

Various aspects of the CO_2–carbonate systems in natural waters have been extensively reviewed (1–6). Practical interests lie in the areas of geochemistry, biochemistry, water treatment and usage, and environment.

GEOCHEMISTRY

Carbonate Dissolution

Typically, limestone [$CaCO_3$] and dolomite [$CaMg(CO_3)_2$] rocks are dissolved by infiltrating rainwater which has taken up biogenic CO_2 in the soil zone. As this solution percolates downward, continuously dissolving solid carbonates, the hydrogeochemical system will change from "open" with respect to a CO_2 gas phase in the unsaturated zone to "closed" with respect to that phase in the saturated zone. Thus, the solution passes through various stages of a system, which are intermediate between the limiting cases, in general. In modern times, the dissolution of carbonate rocks is enhanced by the action of acids of sulfur and nitrogen oxides from atmospheric pollution.

Carbonate Deposition

Carbonate rocks, mainly aragonite and calcite, constitute about 11% of the total sediments of the earth's crust. Bicarbonate salts are relatively soluble in water, but calcium carbonate is not. Freshwater does not sustain extreme supersaturation of calcium carbonate. In most cases, the phase precipitated is calcite—the polymorph of calcium carbonate that has the lowest solubility product. Carbonate solubility in seawater is governed essentially by the same parameters, but kinetically the precipitation of calcium carbonate is complicated by high concentrations of Mg^{2+} and the actions of marine organisms.

pH Buffer

The pH of natural waters on land is controlled mainly by the bicarbonate–carbonate system. Most of the CO_2 absorbed into the ocean is in the form of HCO_3^- (93%) and CO_3^{2-} (6%). The pH of the upper regions of the sea is controlled by the carbonate as well as the borate systems. In the lower regions of the sea, pH appears to be controlled by the precipitation and solution of various minerals. The bicarbonate alkalinity of natural waters has the following typical ranges: rainwater 0 to 2.5 ppm (mg/L), river and groundwater 6 to 600 ppm, and seawater 140 to 159 ppm.

Geochemical Carbon Cycle

The atmospheric CO_2 concentration of about 0.04% on a timescale of million of years is controlled mainly by a combination of degassing from the earth's interior balanced by the uptake of carbon dioxide via the weathering of calcium and magnesium silicate minerals (7,8). Carbon dioxide in water as carbonic acid is converted to dissolved calcium and magnesium bicarbonate on land during weathering. This carbon form is stored in lakes and aquifers or carried to the sea by rivers and precipitated there as calcium and magnesium carbonate minerals. The overall reactions shown here were first formulated by Urey (8) in 1952.

$$CO_2 + CaSiO_3 \Longleftrightarrow CaCO_3 + SiO_2$$

$$CO_2 + MgSiO_3 \Longleftrightarrow MgCO_3 + SiO_2$$

The reactions from left to right represent carbon dioxide absorption via weathering. The reactions from right to left represent degassing due to the thermal decarbonation of carbonates after burial at sufficient depth in the earth's crust. Added to the Urey reactions should also be the organic carbon subcycle. It exerts an important additional control on the atmospheric carbon dioxide level. This involves the removal of carbon dioxide from the atmosphere through photosynthesis by organisms, the burial of organic matter in sediments, and the weathering of organic matter (or its thermal decomposition) which returns carbon dioxide. The overall reaction is represented as

$$6\,CO_2 + 6\,H_2O \Longleftrightarrow (CH_2O)_6 + 6\,O_2$$

Together, these reactions regulate carbon dioxide on a geologic timescale.

BIOCHEMICAL ASPECTS

Biological Pump

Life has developed on the earth in the water phase and in general requires buffering of the pH at around 7.0. Here carbonate–bicarbonate plays a major role. The process of photosynthetic fixation of CO_2 in the photic zone of the upper ocean in the presence of sunlight and nutrients is balanced by the export of organic carbon (soft tissues, secretions) and carbonate (bones, shells) out of the mixed layer into the deep ocean by gravity. Chemically, the "biological pump" is composed of the organic carbon pump and carbonate pump (2):

$$CO_2 + H_2O + \text{energy} \Longleftrightarrow (CH_2O)_X \text{ (organic carbon)} + O_2$$

$$Ca^{2+} + 2\,HCO_3^- \Longleftrightarrow CaCO_3 \text{ (bones, shells)} + H_2O + CO_2$$

The organic carbon pump sequesters CO_2 from the mixed surface layer and dumps it as particulate organic carbon into the deeper ocean, sometimes seen as white flakes of "marine snow." The carbonate pump, on the other hand, acts in the direction opposite to the organic carbon pump by producing CO_2.

Oceanic Chalk

Much of the calcium carbonate is in the form of hard parts in calcareous organisms (e.g., foraminifera, cocolithophores). Other marine carbonates (aragonite, magnesian, calcite) are metastable phases whose origin is in the reluctance of the Mg^{2+} to be dehydrated and converted to dolomite, [$MgCa(CO_3)_2$] and magnesite,

[$MgCO_3$]. Carbonate sediments in sunlit waters of tropical seas originate as organically secreted solids. They occur either as skeletal debris, as organically secreted carbonate mud, or in various combinations of the two. More than 70% of the total $CaCO_3$ contained in large reef systems is in the form of fine, unconsolidated sand deposited in thick beds across large areas adjacent to a living reef frame (4). In addition to skeletal particles, nonskeletal carbonate particles are important, for example, ooids, pellets, pelloids, intraclasts, and grapestones that produce thick carbonate deposits in the rock record. Enormous petroleum reservoirs discovered in carbonate rocks (Devonian of Canada, Pennsylvanian and Permian strata of New Mexico and Texas, Mesozoic and Cenozoic deposits in Middle East) point to the intimate association of organic carbon with carbonate deposits.

Biological Carbonates

Calcite and aragonite, two polymorphs of $CaCO_3$ represent the bulk of the carbonates formed in biological systems. The solubility products of these two forms are 4.7×10^{-9} (calcite) and 6.9×10^{-9} (aragonite), which shows that aragonite is less stable than calcite. Surface seawater to a depth of 3,000 m is commonly supersaturated with respect to aragonite and calcite (6). This disequilibrium state is attributed to a series of inhibitors, most notably magnesium and dissolved organic matter, that interfere with the nucleation of a $CaCO_3$ seed crystal. By the same token, it appears to offer an optimal condition for the biocontrolled crystallization of $CaCO_3$ to sustain an abundance of marine life.

WATER TREATMENT AND UTILIZATION

Carbonate Scaling

Relevant to water treatment is the fact that natural waters contain high concentrations of soluble bicarbonate salts in dynamic equilibrium with carbonic acid (dissolved carbon dioxide) and insoluble carbonate salts. To estimate the amount of the three forms in water, it is important to know that carbonic acid (H_2CO_3) is a divalent acid whose $pK_1 = 6.35$ and $pK_2 = 10.33$. At pH = 6.35, the unionized acid and the bicarbonate (HCO_3^-) ion exist in equal concentrations. pH is a logarithmic scale, so at pH = 7.35 (one unit higher), the solution would contain 90% bicarbonate and 10% carbonic acid, and at pH = 8.35, the solution would contain 99% bicarbonate and 1% carbonic acid. The pH of 8.2–8.4 (phenolphthalein end point in alkalinity titrations) is also significant in that above this pH, carbonate (CO_3^{2-}) ions begin to appear. At pH = 9.3, about 10% of the bicarbonate ions would have converted to carbonate ion, and at pH = 10.33, carbonate and bicarbonate ion concentrations would be equal. Calcium carbonate is quite insoluble (14 mg/L water at 25°C), so natural waters are usually found in the pH range of 6–8; bicarbonate ions range between 6 to 600 mg/L, and calcium ranges from 2 to 200 mg/L. Seawater representing 98% of blended natural waters on the surface of the earth typically contains 410 mg/L of Ca^{2+} ions and 143 mg/L of HCO_3^- ions. Magnesium carbonate is about 100 times more soluble than calcium carbonate, so the magnesium ion concentration in seawater is much higher (about 1310 mg/L). Carbonate scaling potentials in surface waters are focused mainly on calcium carbonate crystallization because it is the first carbonate scale to form under treatment or use (9,10).

Calcium Carbonate Scaling Indexes

The Langelier saturation index (LSI) and the form modified for high salinity water called the Stiff and Davis stability index (S&DSI) are widely used to predict calcium carbonate scaling potentials in cooling systems and boiler waters (9,10). These indexes are widely applied for predicting the same in reverse osmosis systems (11). The formulas for calculating the indexes (9–11) are given here:

$$\mathrm{LSI} = \mathrm{pH} - \mathrm{pH_s}$$

where $\mathrm{pH_s}$ is the pH at which the water is saturated in calcium carbonate and is calculated by the following:

$$\mathrm{pH_s} = (9.3 + A + B) - (C + D)$$

where

$A = (\mathrm{Log_{10}[TDS]} - 1)/10$
$B = -13.12 \times \mathrm{Log_{10}}(^\circ\mathrm{C} + 273) + 34.55$
$C = \mathrm{Log_{10}}[\mathrm{Ca^{2+}\,as\,CaCO_3}] - 0.4$
$D = \mathrm{Log_{10}}[\mathrm{alkalinity\ as\ CaCO_3}]$

In this model, we see that calcium carbonate saturation is affected by salinity (total dissolved solids, TDS) and temperature along with calcium and bicarbonate ion concentrations. The Langelier index loses its accuracy in predicting the solubility of calcium carbonate in water when salt concentrations exceed 4,000 mg/L (11). Stiff and Davis modified Langelier's formula for highly saline waters by using empirically derived values. The Stiff and Davis stability index (S&DSI) is calculated as follows:

$$\mathrm{S\&DSI} = \mathrm{pH} - p_{\mathrm{Ca}} - p_{\mathrm{ALK}} - K$$

where

p_{Ca} is the negative log of the calcium concentration expressed as molarity;
p_{ALK} is the negative log of the total alkalinity concentration expressed as molarity;
K is a constant whose value depends on the water temperature and ionic strength.

At an index of less than 0, water will have a very limited scaling potential. At increasingly positive index values, the scaling potential increases. By projecting the pH, TDS, calcium, and alkalinity concentrations, calcium carbonate scaling indices can be calculated.

Carbonate Scale Control with Acid versus Antiscalant

Carbonate scale control in the purification or use of water and wastewater typically involve using either acid to reduce pH and carbonate concentration or antiscalant to inhibit the growth of calcium carbonate crystals, or a combination of the two. Sodium hexametaphosphate

(SHMP) is a generic antiscalant widely used due to its low cost. It acts, as do the newer organic antiscalants, by inhibiting the growth of calcium carbonate seed crystals (11,12). The limitations of SHMP are that it is stable only as an anhydrous solid. It is difficult to dissolve in the plant, and once dissolved in water, it slowly hydrolyzes to phosphate ions which precipitate calcium, magnesium, aluminum, iron, and other ions as insoluble phosphate salts; in addition, it is a stimulant for microbial growth. As an antiscalant for simultaneous control of other types of scales, it is much less effective than the newer types of organic antiscalants. Great advances have been made in the development and use of organic antiscalants. In a line of six distinctly differentiated organic antiscalants, all classified under ANSI-NSF Standard 60 for drinking water production by reverse osmosis (13), the maximum calcium carbonate scaling control possible ranges from LSI = 2.8 to 3.2 and S&DSI = 3.9 to 4.5. Under these high limits for calcium carbonate control, the differentiated antiscalants are selected more on the basis of simultaneous control of calcium, strontium, and barium sulfates, calcium fluoride, silica, iron, aluminum, and heavy metal scales. In each case when used, acidification of feedwater is not necessary.

ENVIRONMENTAL IMPACT

Concern for the environmental impact of the natural CO_2 balance in the biosphere is raised by the modern trend of deforestation, industrialization, burning of fossil fuels, and population explosion. Understanding the natural processes is the starting point for monitoring environmental changes and modulating human activities.

Natural Balance

The earth's atmosphere contains approximately $2,000\times10^9$ tons of CO_2, the oceans $130,000\times10^9$ tons, and the rocks of the earth together with fossil fuels approximately $40,000\times10^9$ tons. The concentration of CO_2 in the atmosphere is maintained at a relatively constant value of about 0.03% by weight by an annual exchange between the oceans and the air of approximately 100×10^9 tons of CO_2 gas. Fluctuations of CO_2 in the atmosphere occur due to photosynthesis in daylight hours, which fixes approximately 60×10^9 tons of CO_2 per year, and to biological respiration and decay, which release about the same amount.

Ocean Storage of CO_2

It is known that CO_2 will form solid hydrates with water at temperatures below 10°C and pressures higher than 4.4 MPa (450 m depth); at 0°C, the hydrate is stable at pressures higher than 1.2 MPa (125 m depth). The cold deep waters of the world's oceans are unsaturated with respect to CO_2 and have an enormous and under used capacity to dissolve CO_2. The high dissolved inorganic carbon level of the oceans (38,000 Gt C) would be little changed, even if they were to gain all the carbon in the known fossil fuel reserves (4,000 Gt C). The idea of injecting human-produced CO_2 directly into the deep ocean has attracted much interest (2).

BIBLIOGRAPHY

1. Stumm, W. and Morgan, J.J. (1996). *Aquatic Chemistry*, 3rd Edn. John Wiley & Sons, New York.
2. Wong, C.S. and Hirai, S. (1997). *Ocean Storage of Carbon Dioxide*. International Energy Assoc., Greenhouse Gas R&D Programme, Cheltenham, UK. Available: http://www.ieagreen.org.uk.
3. Matthess, G., Frimmel, F., Hirsch, P., Schulz, H.D., and Usdowski, H-E. (Eds.). (1992). *Progress in Hydrogeochemistry*. Springer-Verlag, New York.
4. Bhattacharyya, A. and Friedman, G.M. (Eds.). (1983). *Modern Carbonate Environments*. Hutchinson Ross Publishing Co., Stroudsburg, PA.
5. Loewenthal, R.E. and Marais, G.V.R. (1976). *Carbonate Chemistry of Aquatic Systems: Theory & Application*. Ann Arbor Science, Ann Arbor, MI.
6. Degens, E.T. (1976). *Inorganic Biochemistry. Topics in Current Chemistry*. Vol. 64, Springer-Verlag, New York.
7. Berner, R.A. (1995). A.G. Hogbom and the development of the concept of the geochemical carbon cycle. *American Journal of Science* **295**: 491–495.
8. Urey, H.C. (1952). *The Planets: Their Origin and Development*. Yale University Press, New Haven, CT, p. 242.
9. Kemmer, F.N. (Ed.). (1988). *The Nalco Handbook*. 2nd Edn. McGraw Hill, New York, pp. 4.10–4.18 and 6.4–6.6.
10. Betz Handbook of Industrial Water Conditioning. (1991), 9th Edn. Betz Laboratories, Trevose, PA, pp. 183, 184 and 223–225.
11. Byrne, W. (1995). *Reverse Osmosis: A Practical Guide for Industrial Users*. Tall Oaks Publishing Inc., Littleton, CO, pp. 111–117.
12. Ning, R.Y. (1999). Operational Characteristics of Reverse Osmosis Process Chemistry. *Ultrapure Water* **16**(4): 39–42.
13. *The six antiscalants are available from King Lee Technologies (San Diego, CA) under the Pretreat Plus™ name.* General information can be found on the web site: http://www.kingleetech.com.

CHLORINE-36 AND VERY OLD GROUNDWATERS

GHOLAM A. KAZEMI
Shahrood University of Technology
Shahrood, Iran

INTRODUCTION

Chlorine has nine isotopes whose mass numbers range from 32 to 40, but only three of these occur naturally. The remaining six have half-lives shorter than 1 hour, (1), and this is why they do not occur in nature. The three abundant isotopes of chlorine include two stable isotopes, chlorine-35 and chlorine-37, with 75.53% and 24.47% abundances, respectively, (2) and one radioactive isotope, chlorine-36, with a half-life of 301,000 ± 2000 years (some references ± 4000 years). The natural mixture of these three isotopes makes up environmental chlorine.

PRODUCTION OF CHLORINE-36

Chlorine-36 is produced by the following processes:

1. cosmic ray splitting of ^{40}Ar in the atmosphere:

$$^{40}\text{Ar} + \text{x} \longrightarrow {}^{36}\text{Cl} + \text{x}' + \alpha$$

   The amount of ^{36}Cl produced by this process depends mainly on the geographic latitude and is greater in middle latitudes. For example, the fallout rate of ^{36}Cl at Stripa in Sweden at 57 °C latitude is 14±3 atoms m^{-2} s^{-1} (3), the mean atmospheric flux of ^{36}Cl over the United States is 19.6 ± 4.5 atoms m^{-2} s^{-1}, and the mean global ^{36}Cl flux is 30.5 ± 7 atoms m^{-2} s^{-1} (4).
2. neutron activation of stable ^{35}Cl in the subsurface:

$$^{35}\text{Cl} + {}_0{}^1\text{n} \rightarrow {}^{36}\text{Cl}$$

   The rate of production of ^{36}Cl by this process varies depending on the type of rocks, minerals, and solutions, and the availability of neutron sources (neutrons are produced by bombardment of rocks with alpha particles as well as through spontaneous fission of ^{238}U). For instance, the ^{36}Cl production rate at the top of Murray Group Limestone was calculated by Davie et al. (1) (using the following equation) as 0.09 atoms/cm^3/year.

$$P_{36} = {}^{36}\text{Cl}/{}^{40}\text{Ca} \times \rho/\text{MW} \times N_A \times \log_e{}^2/t_{1/2}$$

   where P_{36} is production rate of ^{36}Cl, $^{36}Cl/^{40}Ca$ is the equilibrium ratio of ^{36}Cl to ^{40}Ca, ρ is the density of the limestone, MW is the molecular weight of limestone, N_A is Avogadro's number, and $t_{1/2}$ is the half-life of ^{36}Cl.
3. Weapon testing of fusion devices starting in late 1952 and extending through mid-1958 has led to the production of up to 70,000 atoms m^{-2} s^{-1} (5). Chlorine-36 produced by this phenomenon is used as a marker to age-date young groundwaters that are less than 50 years old (6,7).

It has been recently suggested that the operation of two nuclear reactors in the United States has led to the production of ^{36}Cl (8).

MEASUREMENT OF ^{36}CL CONTENT OF WATER SAMPLES AND REPORTING THE RESULTS

To analyze water samples for ^{36}Cl content, the following steps are undertaken (4,8): A specified volume of a water sample, depending on its chlorine concentration, is selected (the selected volume should contain a minimum of 2 mg chlorine). In some studies, sample volumes of up to 4 liters have been collected. The sample is then prepared for target loading. This means preparing about 20–350 mg pure AgCl and pressing it into copper sample holders. The samples are prepared for target loading by adding $AgNO_3$ solution to the samples to precipitate AgCl. The precipitate (AgCl) is then purified by dissolving it with NH_4OH. Next, $Ba(NO_3)_2$ is added to precipitate unwanted sulfur as $BaSO_4$. HNO_3 is added next to neutralize the solution and in the last stage, the AgCl is reprecipitated with $AgNO_3$ for target loading. If the Cl^- concentration of the sample is very low, a carrier must be added to it. The carrier may be AgBr of very low sulfur and chlorine content. The prepared water samples are then analyzed for ^{36}Cl content using an accelerator mass spectrometer (AMS). Usually, standard and blank samples are tested to identify the reproducibility and accuracy of the measurements. Davie et al. (1) reported that ^{36}Cl measurements by AMS have a long reproducibility of about 6% and a sensitivity for $^{36}Cl/Cl$ of a few parts in 10^{15}.

The measured chlorine-36 is reported in two ways:

1. as a ratio of $^{36}Cl/Cl \times 10^{15}$ (the number of ^{36}Cl atoms to total number of chlorine atoms). The multiplication factor (10^{15}) is used because the actual concentration of ^{36}Cl is very small.
2. as atoms of chlorine-36 per liter of water using the following equation (8):

$$^{36}\text{Cl (in atoms per liter)} = 1.699 \times 10^4 \times B \times R$$

where B is the concentration of chlorine in mg/l and R is the (^{36}Cl/total Cl) × 10^{15} ratio. Therefore, to determine the concentration of ^{36}Cl, the concentration of Cl^- in water samples must be measured by conventional methods such as titration or ion chromatography.

Example

The $^{36}Cl/Cl \times 10^{15}$ of a water sample is 550 and its chlorine concentration is 200 mg/l. What is the ^{36}Cl concentration in this sample?

$$^{36}\text{Cl (in atoms per liter)} = 1.699 \times 10^4 \times B \times R$$

^{36}Cl (in atoms per liter) = $1.699 \times 10^4 \times 200$ mg/l $\times 550 = 186.89 \times 10^7$ atoms/liter (see Table 1 for more examples).

APPLICATIONS OF CHLORINE-36 AND PRINCIPLES OF THE ^{36}CL DATING METHOD

Chlorine-36, in this article, is treated as a tool to age-date very old groundwaters. However, it has many other applications such as the geochronological study of young volcanic rocks (9), estimation of groundwater recharge (10,11), calculation of dispersivity (12) and a mixing study of two waterbodies (13).

The principle of this method is simple. It is based on the radioactive decay of ^{36}Cl in the subsurface groundwater system. Above the earth's surface, chlorine-36 produced (atmospheric ^{36}Cl) with an initial value, C_0, enters groundwater by rainwater infiltration. After time t, it decays and reaches a new concentration, C, according to the following equation:

$$C = C_0 e^{-\lambda t}$$

Table 1. ^{36}Cl and Cl^- Measurements of Murray Mallee Samples[a]

Well	$^{36}Cl/Cl \times 10^{-15}$[b]	Cl^- mg/l	$^{36}Cl^-$ Atoms/l $\times 10^{6}$[b]
1	19 + 4	264	85 + 18
2	22 + 4	551	205 + 37
3	35 + 6	361	215 + 37
4	32 + 4	576	315 + 39
5	28 + 6	1067	510 + 110
6	3p17	314	195 + 37
7	31 + 7	475	250 + 56
8	31 + 7	468	245 + 55
9	43 + 7	804	585 + 95
10	22 + 8	747	280 + 100
11	32 + 15	752	410 + 190
12	42 + 7	2023	1450 + 240
13	21 + 6	3191	1140 + 330
14	39 + 7	2573	1710 + 310
15	33 + 6	1607	900 + 160
16	22 + 4	5327	1990 + 360
17	13 + 4	32,600	7200 + 2200
18	22 + 6	5566	2080 + 570

[a]From Reference 1, with permission from Geoscience Australia.
[b]The number after + is the error range.

From the initial concentration, C_0, and the present concentration, C, one can calculate the length of time that the ^{36}Cl has resided in the subsurface groundwater system. For example, if the concentration of ^{36}Cl in a groundwater sample is 2000 atoms/liter and its initial concentration (in the infiltrating rainwater) was 3000 atoms/liter, then the time "t" would be 176,136 years. This means that the water molecule, which contain the ^{36}Cl in question, entered the groundwater system 176,136 years ago.

However, many inaccuracies and problems surround C_0 and C values, the initial and the present concentration of ^{36}Cl, respectively. We do not accurately know the concentration of ^{36}Cl in the rainwater when it entered the subsurface system. Davis et al. (8) present six ways to find the C_0 value and argue that all contain weaknesses and inaccuracies. These approaches include:

1. calculation of theoretical cosmogenic production and fallout,
2. measuring ^{36}Cl in present-day atmospheric precipitation and using it as C_0,
3. assuming that shallow groundwater contains a record of C_0,
4. extracting ^{36}Cl from vertical depth profiles in desert soils,
5. recovering ^{36}Cl from cores of glacial age, and
6. calculating subsurface production of ^{36}Cl for water that has been isolated from the atmosphere for more than 1 million years.

The situation is no better with regard to C. The assumption in using the decay equation is that the ^{36}Cl atoms which enter subsurface groundwater system behave as an isolated packet when they migrate through the flow system (piston flow theory). As explained above, ^{36}Cl is produced in the atmosphere, and it is also produced in the subsurface. There are other subsurface sources of ^{36}Cl such as what is called "dead chlorine," chlorine present in fluid inclusions; ancient formation waters; saline water from compacting clays; and chlorine from salt beds and evaporites. In addition, subsurface mixing (mixing of low and high ^{36}Cl content waters), cross formational flow, diffusion between aquitard and aquifer, and dilution and evaporation processes all unite to complicate the job of finding which C should be used in the decay equation, the C that is measured in the laboratory or the C that is obtained when the contributions/effects of all the above factors have been eliminated.

Apart from ^{4}He which is considered an age indicator not a dating method (14), ^{36}Cl is the only method to age-date very old groundwaters quantitatively. However, in only a very few cases were groundwater ages satisfactorily determined. In many instances, the chlorine-36 dating study was not successful due to many factors such as the uncertainty in estimating C_0, the effect of dilution/evaporation processes on the value of C, and the much longer half life of ^{36}Cl compared with the age of the groundwater (1,15,16). Other counterproductive factors include relatively recent introduction of the method, less interest in very old groundwaters, a limited number of well-defined regional aquifers worldwide for which this method is appropriate, many analytical considerations and the unavailability of equipment in a large number of countries, difficulty in interpreting the results (initial value problem, local interferences, many sources of ^{36}Cl, etc,), and inherent inadequacies [for example, Park et al. (19) show that if the chlorinity of the water sample exceeds ~75–150 mg/kg, the $^{Cl}36$ method cannot be used for groundwater dating]. These may be the reasons that an important text book such as that by Mazor (14) does not include a section on ^{36}Cl to describe it as an age-dating tool. A valuable recent paper by Park et al. (17) describes the principles and limitations of age-dating very old groundwater by ^{36}Cl.

CASE STUDIES

The work by Davis and Schaeffer (18) was the first attempt to apply ^{36}Cl isotopes to environmental problems. However, Davie et al. (1) argue that extensive application of ^{36}Cl in environmental studies started only after the introduction of the accelerator mass spectrometry (AMS) technique for detecting ^{36}Cl by Elmore and his colleagues in 1979 (19). Because earlier methods of analysis such as decay counting required 13 g of chlorine, a difficult task, and it had much less sensitivity than AMS.

Examples of aquifers subjected to chlorine-36 dating include the Great Artesian Basin (GAB) of Australia (20,21); the fractured metamorphic rock aquifer in the Clare Valley, South Australia (22); the Milk River aquifer of the western Canada sedimentary basin (23,24); the Aquia aquifer in the Maryland Coastal Plain (25); the palaeochannel groundwater of Kalgoorlie in Western Australia (15); and the groundwater brines of the Dead Sea area (16). Of these aquifers, GAB is perhaps the most studied as far as chlorine-36 is concerned. A list of directly related publications dealing with chlorine-36 in various aquifers around the globe is included at the end of this article for further consultation. It is interesting to note that all publications dealing with ^{36}Cl have been produced by groups of authors; there is no single author publication, which may be an indication of the complexity of the matter.

One of the early studies dealing with ^{36}Cl in groundwater is that by Davie et al. (1), in which the ^{36}Cl content of limestone groundwater in the Murray Basin of Australia was measured. The results of their analysis, as an example, are presented in Table 1. This publication was selected for detailed explanation because it has not been cited elsewhere and it is a valuable study. Table 1 shows that the $^{36}Cl/Cl$ ratios of the Murray Mallee samples are low, generally lower than those reported by Davis et al. (8) for groundwaters of the United States (It may be inappropriate to compare the two study areas because of the large difference in their sizes). This study showed that the ratio of $^{36}Cl/Cl$ in the aquifer increases along the flow path most probably because of significant localized recharge.

The age of the deep Jurassic confined aquifer in the GAB was estimated by Bentley et al. (20) at $(1200 \pm 500) \times 10^3$ years (much older than the hydraulic age). Notle et al. (23) measured the ^{36}Cl concentration of 12 samples from the Milk River aquifer, Alberta, Canada, and found that the ages of the groundwater ranged from 0.6–2 Ma at a distance of 80 km from the recharge area.

By analyzing 17 samples from water wells and seeps, Turner et al. (15) showed that the $^{36}Cl/Cl$ ratios and ^{36}Cl content of palaeochannel groundwater in Kalgoorlie, Western Australia, ranged from $32–129 \times 10^{-15}$ and 1460 to 79,000 atoms/l, respectively. Similarly, the $^{36}Cl/Cl$ ratios of 17 samples from the Continental Terminal aquifer in Niger calculated by Le Gal La Salle et al. (26) ranged from $3–27 \times 10^{-15}$.

Davis et al. (8) measured the concentration of ^{36}Cl in the groundwater of the United States to determine what they call the "preanthropogenic levels" (natural = pristine = background level) of ^{36}Cl in groundwater. This study which was subsequently completed by Moysey et al. (4) is the most comprehensive study worldwide of ^{36}Cl in groundwater so far; a total of 183 water samples from unpolluted wells and springs were analyzed. Davis et al. (8) showed that natural ratios of $^{36}Cl/Cl$ are lowest near the coast and increase to a maximum in the central Rocky Mountains of the United States (less than 50×10^{-15} in Florida to more than 1200×10^{-15} in the Rocky Mountains). They attribute this pattern to an inland decrease in the stable chlorine content of atmospheric precipitation; the increase in the ratio of $^{36}Cl/Cl$ inland is due to an inland decrease in stable chlorine, not an increase in ^{36}Cl.

BIBLIOGRAPHY

1. Davie, R.F. et al. (1989). Chlorine-36 measurements in the Murray Basin- preliminary results from the Victorian and South Australian Mallee region. *BMR J. Aust. Geol. Geophys.* **11**(2/3): 261–272.
2. Marques, J.M., Andrade, M., Aires-Barros, L., Graca, R.C., Eggenkamp, H.G.M., and Antunes da Silva, M. (2001). 87Sr/86Sr and 37Cl/35Cl signatures of CO2-rich mineral waters (N-Portugal): Preliminary results. In: *New Approaches Characterizing Groundwater Flow*. K.P. Seiler and S. Wohnlich (Eds.). Swets and Zeitlinger Lisse, Lisse, Vol. 2, pp. 1025–1029.
3. Bowen, R. (1988). *Isotopes in the Earth Sciences*. Elsevier, London.
4. Moysey, S., Davis, S.N., and Cecil, L.D. (2003). The distribution of meteoric $^{36}Cl/Cl$ in the United States: A comparison of models. *Hydrogeol. J.* Online Publication, 10.1007/s10040-003-0287-z.
5. Bentley, H.W., Phillips, F.M., and Davis, S.N. (1986). Chlorine-36 in the terrestrial environment. In: *Handbook of Environmental Isotope Geochemistry*. P. Fritz and J.C. Fontes (Eds.). Elsevier, New York.
6. Coplen, T.B. (1993). Uses of environmental isotopes. In: *Regional Ground-Water Quality*. W.M. Alley (Ed.). ITP, New York, pp. 255–293.
7. Cook, P.G. and Solomon, D.K. (1997). Recent advances in dating young groundwater: chlorofluorocarbons, $^{3}H/^{3}He$ and ^{85}Kr. *J. Hydrol.* **191**: 245–265.
8. Davis, S.N., Moysey, S., Cecil, L.D., and Zreda, M. (2003). Chlorine-36 in groundwater of the United States: Empirical data. *Hydrogeol. J.* **11**: 217–227.
9. Leavy, B.D., Philips, F.M., Elmore, D., Kubic, P.W., and Gladney, E. (1987). Measurement of $^{36}Cl/Cl$ in young volcanic rocks: An application of accelerator mass spectrometry to geochronology. *Nucl. Instrum. Methods Phys. Res.* **B29**: 246–250.
10. Cook, P.G. and Robinson, N.I. (2002). Estimating groundwater recharge in fractured rocks from environmental ^{3}H and ^{36}Cl, Clare Valley, South Australia. *Water Resour. Res.* **38**(8): 10.1029/2001WR000772.
11. Cecil, L.D., Beasley, T.M., Pittman, J.R., Michel, R.L., Kubik, P.W., Sharma, P., Fehn, U., and Gove, H.E. (1992). Water infiltration rate in the unsaturated zone at the Idaho National Engineering Laboratory estimated from chlorine-36 and tritium profiles and neutron logging. In: *Water Rock Interaction*. Vol. 1, Y.F. Kharaka and A.S. Maest (Eds.). Balkema, Rotterdam, pp. 709–714.
12. Cecil, L.D., Welhan, J.A., Green, J.R., Frape, S.K., and Sudicky, E.R. (2000). Use of chlorine-36 to determine regional scale aquifer dispersivity, eastern Snake River Plain aquifer, Idaho/USA. *Nucl. Instrum. Methods Phys. Res.* **B172**: 679–687.
13. Calf, G.E., Bird, J.R., Kellet, J.R., and Evans, W.R. (1998). *Origins of Chlorine Variation in the Murray Basin- Using Environmental Chlorine-36*. Australian Bureau of Mineral Resources, Geology and Geophysics Record 1988/43, Groundwater series No. 16, pp. 28–30.
14. Mazor, E. (1991). *Applied Chemical and Isotopic Groundwater Hydrology*. Open University Press, Melton Keynes.
15. Turner, J.V., Rosen, M.R., Fifield, L.K., and Allan, G.L. (1995). Chlorine-36 in Hypersaline Palaeochannel Groundwaters of Western Australia. In: *Application of Tracers in Arid-zone Hydrology*. E.M. Adar and C. Leibundgut (Eds.). *IAHS publication* no. 232, Vienna, pp. 15–33.
16. Yechieli, Y., Ronen, D., and Kaufman, A. (1996). The source and age of groundwater brines in the Dead Sea area, as deduced from ^{36}Cl and ^{14}C. *Geochim. Cosmochim. Acta* **60**(11): 1909–1916.
17. Park, J., Bethke, C.M., Torgerson, T., and Johnson, T.M. (2002). Transport modeling applied to the interpretation of groundwater ^{36}Cl age. *Water Resour. Res.* **38**(5): 10.1029/2001WR000399.

18. Davis, R.F. and Schaeffer, O.A. (1955). Chlorine-36 in nature. *Ann. NY Acad. Sci.* **62**: 105–122.
19. Elmore, D. et al. (1979). Analysis of ^{36}Cl in environmental water samples using an electrostatic accelerator. *Nature* **272**: 22–25; errata 246.
20. Bentley, H.W. et al. (1986). Chlorine-36 dating of very old groundwater 1. The Great Artesian Basin, Australia. *Water Resour. Res.* **22**: 1991–2001.
21. Love, A.J., Herczeg, A.L., Sampson, L., Cresswell, R.G., and Fifield, K.L. (2000). Sources of chlorine and implications for ^{36}Cl dating of old groundwater, southwestern Great Artesian Basin. *Water Resour. Res.* **36**: 1561–1574.
22. Cook, P.G. (2003). Groundwater ages in fractured rock aquifers. *Proc. Int. Conf. Groundwater Fractured Rocks*, Prague, IHP-VI, series on groundwater no. 7, pp. 139–140.
23. Notle, E. et al. (1991). Measurements and interpretations of ^{36}Cl in groundwater, Milk River aquifer, Alberta, Canada. *Appl. Geochem.* **6**: 435–445.
24. Phillips, F.M., Bentley, H.W., Davis, S.N., Elmore, D., and Swanick, G.B. (1986). Chlorine-36 dating of very old groundwater, 2, Milk River Aquifer, Alberta, Canada. *Water Resour. Res.* **22**: 2003–2016.
25. Purdy, C.B. et al. (1996). Aquia Aquifer dissolved Cl$^-$ and ^{36}Cl/Cl: Implications for flow velocities. *Water Resour. Res.* **32**: 1163–1171.
26. Le Gal La Salle et al. (1995). Old groundwater circulation in the Iullemeden basin (Niger): Preliminary results of an isotopic study. USA. In: *Application of Tracers in Arid-zone Hydrology*. E.M. Adar and C. Leibundgut (Eds.). *IAHS publication* no. 232, IAEA, Vienna, pp. 129–139.

READING LIST

Andrews, J.N., Fontes, J.C., Michelot, J.L., and Elmore, D. (1986). Insitu neutron flux, ^{36}Cl production and groundwater evolution in crystalline rocks at Stripa, Sweden. *Earth Planet Sci. Lett.* **77**: 49–58.

Fritz, P. and Fontes, J.C. (Eds.). (1986). *Handbook of Environmental Isotope Geochemistry*. Elsevier, New York.

Davis, S.N. et al. (2000). Chlorine-36 in groundwater containing low chlorine concentrations. *Groundwater* **38**: 912–921.

Hendry, M.J., Wassennar, L.I., and Kotzer, T. (2000). Chlorine and chlorine isotopes (^{36}Cl and δ^{37}Cl) as tracers of solute migration in a thick, clay rich aquitard system. *Water Resour. Res.* **36**: 285–296.

Torgersen, T. and Phillips, F.M. (1993). Reply to "Comment on chlorine 36 dating of very old groundwater, 3. Further results on the Great Artesian Basin, Australia by T. Torgersen et al., " by J.N. Andrews and J.C. Fontes. *Water Resour. Res.* **29**: 1875–1877.

CHLOROFLUOROCARBONS (CFCs)

Gholam A. Kazemi
Shahrood University of Technology
Shahrood, Iran

INTRODUCTION

The following paragraph, extracted mostly from a comprehensive paper by Badr et al. (4), is a brief yet valuable introduction to chlorofluorocarbons (CFCs). Thereafter, an overview of CFC applications in hydrologic studies, especially for age-dating young groundwater, is presented.

Before 1930, the refrigerants commonly used were ammonia (R-717, the most popular refrigerant in the industrial and heavy commercial sectors), carbon dioxide (R-747), ethyl chloride (R-160), isobutane (R-600a), methyl chloride (R-40), methylene chloride (R-30), and sulfur dioxide (R-764). Because of either, toxicity or fire hazard, the need for a safer refrigerant became clear. The new refrigerant had to be nonflammable, have a boiling point between −40 and 0 °C, be highly stable, and low in toxicity. In response to these requirements, the CFC family of refrigerants was developed. Dichloromonofluoromethane (R-21) was the first CFC developed that had attractive thermodynamic properties for use as a refrigerant, but due to its low toxicity, stability, and inertness, it had never been used commercially as a refrigerant. The first CFC suitable for commercial applications was dichlorodifluoromethane (R-12), whose discovery was announced in 1930 at the American Chemical Society meeting in Atlanta, Georgia. CFCs have been among the most useful chemical compounds ever developed. There is no known natural source of CFCs (5).

In addition to their uses as working fluids in refrigerators and air-conditioning systems, CFCs have also been employed, since 1950, as blowing agents for foams and plastics, aerosol propellants, solvents for cleaning precision and delicate electronic equipment, as well as for dry cleaning, in sterilizers for surgical instruments and catheters, and as working media in Rankine-cycle engines for waste-heat recovery systems. By 1974, approximately one million tonnes per year of CFCs were being produced worldwide. Within the European Countries (EC), the main uses for CFCs (according to 1988 consumption figures) are as aerosol propellants (38%), foam-blowing agents (35%), solvents (17%), and refrigerants in cooling systems (10%). In the USA, CFCs are used for the following purposes: foam-blowing (34%), refrigeration and air conditioning (35.5%), solvents (18%), sterilizers (5.4%), and other applications (7.0%). CFCs along with some other gases harm the earth's atmosphere by two phenomena: (1) depletion of the ozone layer and (2) the greenhouse effect (or global warming). A major cause of environmental concern for CFCs is their long atmospheric lifetime (55 years for CFC-11 and 140 years for CFC-12). Hydrochlorofluorocarbons (HCFCs) are now replacing CFCs and recent data suggests that the growth rate of CFCs concentration in the atmosphere has begun to decrease (6).

The concentration of CFCs in the atmosphere is measured regularly at a few stations worldwide. The atmospheric concentrations for CFC-11, CFC-12, and CFC-113, those species that are used to age-date young groundwaters (less than 50 years old), are illustrated in Fig. 1. Industrial use of CFC-12 started in the 1930s, CFC-11 in the 1950s, and CFC-113 in the 1970s (7). Atmospheric concentrations of these species show little spatial variation; only 10% variation is observed among average concentrations in Ireland, Oregon, Barbados,

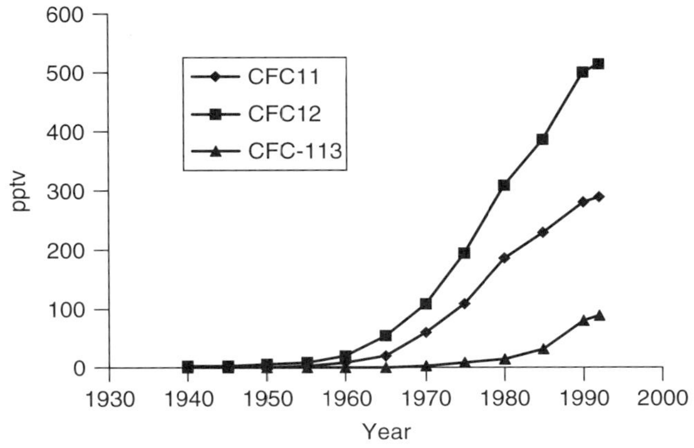

Figure 1. Atmospheric concentrations of CFC-11, CFC-12, and CFC-113 (pptv) from 1940 to 1992. (*Source*: Smethie et al., 1992, cited in Ref. 8. redrawn)

Figure 2. Apparatus for sampling groundwater for CFC analysis.

Samoa, and Tasmania (source: Cunnold et al., 1994, cited in Ref. 8).

CFCs in hydrological studies—Age dating young groundwaters

The use of CFCs in hydrology began when Thompson et al. (1) discussed the usefulness of a special type of fluorocarbon known as Freon 11, as a tracer in both groundwater and surface water hydrology. They found good agreement between CCl_3F data and hydrologic and tritium age data. Brown (2) also examined the use of fluorocarbons as groundwater tracers through soil column studies. There was no significant use of or development in CFC dating techniques during the 1980s, possibly due to the availability of other methods such as the tritium method and the low resolution of the available CFC analytical techniques. From 1992, the application of CFCs in groundwater studies, especially for age dating and determination of rainfall recharge became important (3). In groundwater textbooks, not many contain a section on the CFC age-dating technique. One text that deals briefly with the topic is by Domenico and Schwartz (9). A book chapter by Plummer et al. (10) is a valuable resource that covers most aspects of the CFC age-dating technique.

SAMPLING GROUNDWATER FOR CFCS

In sampling groundwater for CFC analysis, care must be exercised to prevent the sample from contacting air. The average CFC concentration in the air is about 300 times that in groundwater. Therefore, any small contamination from air will have a very large impact on the concentration of CFCs in the groundwater sample. A sophisticated sampling apparatus like that developed by CSIRO (Commonwealth Scientific and Industrial Research Organization), Australia, is necessary to meet this condition (Fig. 2). Prior to sampling, about three volumes of well water are purged. A 6-mm nylon tube and a stainless steel bailer are sent downhole to obtain the sample. Both nylon tube and bailer are initially flushed with ultrahigh purity (UHP) nitrogen to remove any air from the system. The sample containers are 62 cm^3 borosilicate glass ampules that are flushed, prior to filling, with UHP nitrogen for two minutes. After filling, the ampules are flame sealed using a propane–oxygen torch. The CFC content of the samples is measured using a purge-and-trap gas chromatograph with an electron capture detector (ECD). An illustrative and practical text that deals with sampling for CFCs is that by Kazemi (11).

In another method of sampling for CFCs, the ground water is pumped via a copper tube for a considerable length of time. While the water is flowing through the tube, a small length of the tube is cut and flame sealed quickly to avoid contamination from air. One other method of sampling involves pumping a large volume of groundwater into a big container and then sampling groundwater from the water inside the container.

HOW TO CALCULATE CFC AGE

Apparent CFC ages are obtained by converting measured CFC concentrations in groundwater to equivalent air concentrations using known solubility relationships (12) and the recharge temperature (13). Therefore, the first step is to measure the CFC concentration in groundwater (usually less than 400 pg/kg). The second step is to determine the air temperature at which the rainfall recharge occurs. This is usually average summer temperature (for regions of summer dominated rainfall), average winter temperature (for regions of winter dominated rainfall), and mean annual air temperature (for regions of all year round rainfall). The third step is to calculate the solubility of the CFCs at the temperature considered (summer, winter, or annual) by using the Warner and Weiss (12) relationship. The final step is to use all three above and the following formula to obtain the equivalent atmospheric concentration (EAC) of CFCs:

$$\mathrm{EAC} = \frac{\mathrm{CFCs_{gw}}}{S \times 121} \text{(adapted from Ref. 13)}$$

where EAC is the equivalent atmospheric concentration, $\mathrm{CFCs_{gw}}$ is the concentration of CFCs in the groundwater, S is solubility, and 121 is the molecular weight of CFCs in units of g/mol. The EAC is then compared to the graphs of CFC concentrations in the atmosphere (Fig.1), and the year when the recharge occurred is determined.

LIMITATIONS AND POSSIBLE SOURCES OF ERROR IN THE CFC DATING TECHNIQUE

Error in Estimating Recharge Temperature and Excess Air

Temperature plays a major role in the solubility of CFCs in rainwater. If the air temperature at the time of recharge is overestimated, then the solubility of CFCs that are calculated will be greater than real values. This leads to overestimation of groundwater ages. Note that the error in estimating the recharge temperature is minimal for waters recharged before 1985 (3).

Excess air is the air bubbles trapped during recharge. The CFC content of these air bubbles is slowly dissolved in the groundwater, leading to a rise in the CFC concentration in groundwater. This results in underestimating groundwater ages.

Thick Unsaturated Zone (More Than 5 Meters) and the Associated Error

It is assumed in the CFC technique that the concentration of CFCs immediately above the water table is the same as the CFC concentration in air. If there is a thick unsaturated zone and the soil water content is high, then CFCs may be dissolved preferentially into the soil water. Consequently, the CFC concentration above the water table differs from the CFC concentration in air. Furthermore, there will be a lag between the time that rainwater, containing CFCs, leaves the earth surface and the time that it enters the groundwater. This phenomena leads to overestimating the groundwater age.

Contamination During Sampling

There is always a possibility that groundwater could be contaminated during sampling. Apart from air contamination, CFCs may enter the sample through sampling equipment.

Microbial Degradation of CFCs in the Aquifer

Experiments by Oremland et al. (14) have shown that CFCs are degraded by microbial activity especially in the soil zone. However, the magnitude of degradation is low. If CFCs are decomposed to a large extent, then it is difficult to obtain the true age of the groundwater.

Sorption and Desorption of CFCs from the Aquifer Matrix, Diffusion, and dispersion

Russell and Thompson (15) showed that CFCs are adsorbed to and released from soil particles. This has two implications for the CFC dating technique:

1. Rainwater infiltrating through the soil zone adsorbs some CFCs from soil particles (in addition to the CFCs gained from the atmosphere), and hence it is enriched in CFCs.
2. Some of the CFC content of infiltrating rainwater may be adsorbed to soil particles, and hence it is depleted in CFCs.

In the saturated zone, CFCs are adsorbed to the aquifer matrix. Some time later, the adsorbed CFCs may be released to the groundwater system. Diffusion and dispersion of CFCs in groundwater leads to underestimation of CFC concentration in groundwater and hence overestimation of groundwater age.

Mixing of Water of Different Ages

The problem of mixing of groundwaters from different sources and different aquifers that surrounds most of the hydrological calculations affects the CFCs technique, too. Groundwater in fractured rock aquifers is the most problematic because of the large difference between the age of fracture water and the age of matrix water (15). This problem may present the biggest challenge to the CFC dating technique but could be alleviated, to some extent, by comparing CFC ages with hydraulic ages or ages obtained by other techniques.

Uncertainty in Estimating the Atmospheric Input Function

There is always some uncertainty in estimating the historical concentration of CFCs in the atmosphere at a particular site, especially if the site is far from CFC measurement stations. This could lead to underestimation or overestimation of groundwater ages. However, the nearly homogeneous concentration of CFCs in the atmosphere across long distances reduces the uncertainty.

EXAMPLES OF THE APPLICATION OF CFCS IN GROUNDWATER STUDIES

Reilly et al. used CFC and tritium dated ages as well as a numerical simulation technique to quantify the groundwater flow rate, recharge ages of shallow groundwater, and mixing properties of the groundwater system in Locust Grove, Maryland, U.S.A. They showed that the two methods together (the age-dating method and the simulation technique) enabled a coherent explanation of the flow paths and rates of movement; at the same time, they indicated weaknesses in understanding the system. They also showed that the numerical simulations in the absence of the environmental tracer information provided nonunique and therefore, uncertain estimates of the quantities of water and flow paths within the groundwater system.

Bohlke and Denver (18) used CFC age dating along with chemical and isotopic techniques to resolve the history and fate of nitrate contamination in two agricultural watersheds of the Atlantic Coastal Plain, Maryland, USA.

Cook and Solomon (19) showed that only the thick unsaturated zone would have a major influence on the result of CFC dating. Cook et al. (20) showed that the problem of excess air and degradation of CFCs are two factors that may negatively affect CFC dating results. They also concluded that

1. CFC-11 appears to be degraded both in the highly organic unsaturated zone and below the 3.5-m depth in the aquifer.
2. Degradation of CFC-11 resulted in apparent ages that greatly overestimate groundwater travel times at all depths.
3. There is no evidence for degradation of either CFC-12 or CFC-113, whereas CFC-113 shows evidence of retardation. It was, therefore, decided that the best age tracer that matches hydraulic ages is CFC-12.

Cook et al. (21) used CFCs age data in County Musgrave on Eyre Peninsula and the Clare Valley, both in South Australia, to determine the recharge rate to the shallow groundwater. The results showed that the recharge rate to the aquifer was about 20–50 mm per year.

Szabo et al. (22) dated the shallow groundwater in the southern New Jersey coastal plain using CFCs and $^{3}H/^{3}He$. They found an excellent correlation between ages obtained from the two methods and also with those obtained by a groundwater flow model.

Cook and Solomon (8) compared the results of CFC, $^{3}H/^{3}He$, and ^{85}Kr measurements to date young groundwater, to estimate the groundwater flow path, and to determine vertical and horizontal flow velocities. They noted that in most cases the results from the three methods are in good agreement. However, where there is a difference between the ages, the difference could be used to

1. obtain other information on flow processes;
2. determine the mixing of water of different ages; and
3. identify a fracture flow system because of the different rates at which these tracers diffuse between fractures and matrix.

The residence times of groundwater seepage to streams in the New Jersey coastal plain were determined by Modica et al. (23) using CFC data. They found a negative correlation between the age of groundwater and its nitrate concentration.

CFC groundwater ages proved to be a helpful tool for managing salinized catchments in Eastern Australia (24).

Having measured the CFC ages of groundwater from a number of aquifers in Eastern Australia, Kazemi (11) showed that the groundwater does not become older along the flow line. He also showed a negative correlation between the age of groundwater and the maximum fluctuation in the water table.

MacDonald et al. (25) used CFC concentration in the groundwater of the Permian Aquifer in Dumfries, Scotland, to identify trends in groundwater quality. They clearly demonstrated a positive correlation between the percentage of modern water and the concentration of nitrate in the aquifer; the higher the percentage of modern water in the aquifer, the higher its nitrate concentration.

WORLDWIDE CFC MEASUREMENT LABORATORIES AND COSTS

As of 1999, the following organizations were able to measure CFC concentration in groundwater (11):

1. The Center for Groundwater Studies (CGS), Adelaide, Australia
2. United States Geological Survey, Reston, Virginia
3. Institute fur Umveltphysik, University of Heidelberg, Germany
4. Geological Survey, Switzerland
5. Institute of Geological and Nuclear Studies, New Zealand
6. University of Waterloo, Canada.

The cost of analyzing groundwater for CFCs was about US $100 per sample as of 1999. The total cost of a single analysis is about double considering the expenses of (1) the labor, (2) the ultra high purity (UHP) C size nitrogen gas cylinder, (3) the oxygen and fuel (for sealing ampules), and (4) the shipment of samples to laboratories and hiring of sampling equipment (11).

BIBLIOGRAPHY

1. Thompson, G.M., Hayes, J.M., and Davis, S.N. (1974). Fluorocarbon tracers in hydrology. *Geophys. Res. Lett.* **1** (4): 177–180.
2. Brown, J.D. (1980). *Evaluation of Fluorocarbon Compounds as Groundwater Tracers: Soil Column Studies*. M.Sc Thesis, University of Arizona, Tucson, AZ.
3. Busenberg, E. and Plummer, L.N. (1992). Use of chlorofluorocarbons (CCl_3F and CCl_2F2 as hydrologic tracers and age-dating tools: The alluvium and terrace system of central Oklahoma. *Water Resour. Res.* **28**(9): 2257–2283.
4. Badr, O., Probert, D.S., and O'Callaghan, P.W. (1990). Chlorofluorocarbons and the environment: Scientific, economic, social, and political issues. *Appl. Energy* **37**: 247–327.
5. Lovelock, J.E. (1971). Atmospheric fluorine compounds as indicators of air movements. *Nature* **230**: 379.
6. Schoeberl, M.R. (1999). Ozone and stratospheric chemistry. In: *EOS Science Plan*. M.D. King (Ed.). NASA, Greenbelt, MD, pp. 309–337.
7. Plummer, L.N. and Busenberg, E. (1999). Chlorofluorocarbons. In: *Environmental Tracers in Subsurface Hydrology*. P.G. Cook and A.L. Herczeg (Eds.). Kluwer, Dordrecht, The Netherlands, pp. 441–478.
8. Cook, P.G. and Solomon, D.K. (1997). Recent advances in dating young groundwater: Chlorofluorocarbons, $^{3}H/^{3}He$ and ^{85}Kr. *J. Hydrol.* **191**: 245–265.
9. Domenico, P.A. and Schwartz, F.W. (1998). *Physical and Chemical Hydrogeology*, 2nd Edn. John Wiley & Sons, New York.
10. Plummer, L.N., Michel, R.L., Thurman, E.M., and Glynn, P.D. (1993). Environmental tracers for age dating young groundwater. In: *Regional Ground-Water Quality*. W.M. Alley (Ed.). ITP, New York, pp. 227–254.
11. Kazemi, G.A. (1999). *Groundwater Factors in the Management of Dryland and Stream Salinity in the Upper Macquarie Valley, New South Wales, Australia*. Unpublished Ph.D. Thesis, University of Technology, Sydney, Australia.
12. Warner, M.J. and Weiss, R.F. (1985). Solubilities of chlorofluorocarbons 11 and 12 in water and seawater. *Deep Sea Res.* **32**: 1485–1497.)
13. Cook, P.G. and Herczeg, A.L. (1998). Groundwater chemical methods for recharge studies. Part 2 of the *Basics of Recharge and Discharge*. L. Zhang (Ed.). CSIRO Publishing, Australia.

14. Oremland, R.S., Lonergan, D.J., Culbertson, C.W., and Lovely, D.R. (1996). Microbial degradation of hydrochlorofluorocarbons ($CHCl_2F$ and $CHCl_2CF_3$) in soils and sediments. *Appl. Environ. Microbiol.* 1818–1821.
15. Russell, A.D. and Thompson, G.M. (1983). Mechanisms leading to enrichment of the atmospheric fluorocarbons CCl_3F and CCl_2F_2 in groundwater. *Water Resour. Res.* **19**: 57–60.
16. Cook, P.G. (2003). Groundwater Ages in Fractured Rock Aquifers. In: *Proc. Int. Conf. Groundwater Fractured Rocks*. Prague, Krasny-Hrkal-Bruthans (Eds.). IHP-VI, series on groundwater no. 7, pp. 139–140.
17. Reilly, T.E., Plummer, L.N., Phillips, P.J., and Busenberg, E. (1994) The use of simulation and multiple environmental tracers to quantify groundwater flow in a shallow aquifer. *Water Resour. Res.* **30** (2): 421–433.
18. Bohlke, J.K. and Denvor, J.M. (1995). Combined use of groundwater dating, chemical and isotopic analyses to resolve the history and fate of nitrate contamination in two agricultural watersheds, Atlantic Coastal Plain, Maryland. *Water Resour. Res.* **31** (9): 2319–2339.
19. Cook, P.G. and Solomon, D.K. (1995). Transport of atmospheric trace gases to the water table: Implications for groundwater dating with chlorofluorocarbons and krypton-85. *Water Resour. Res.* **31** (2): 263–270.
20. Cook, P.G. et al. (1995). Chlorofluorocarbons as tracers of groundwater transport processes in a shallow, silty sand aquifer. *Water Resour. Res.* **31** (3): 425–434.
21. Cook, P.G., Love, A., and Dowie, J. (1996). Recharge estimation of shallow groundwater using CFC age dating. *Miner. Energy South Aust. J.* **3**: 32–33.
22. Szabo, Z. et al. (1996). Age dating of shallow groundwater with chlorofluorocarbons, tritium, helium and flow path analysis, southern New Jersey coastal plain. *Water Resour. Res.* **32** (4): 1023–1038.
23. Modica, E., Buxton, H.T., and Plummer, L.N. (1998). Evaluating the source and residence times of groundwater seepage to streams, New Jersey coastal plain. *Water Resour. Res.* **34**(11): 2797–2810.
24. Kazemi, G.A., Milne-Home, W.A., and Keshwan, M. (1998). Groundwater age dating with CFC in the Macquarie River Basin, NSW, Australia; Implications for dryland salinity management. In: *Proc. IAH 28th Congr. 'Gambling Groundwater'*, Las Vegas, NV, Sep–Oct. 1998, pp. 549–554.
25. MacDonald, A.M., Darling, W.G., Ball, D.F., and Oster, H. (2003). Identifying trends in groundwater quality using residence time indicators: An example from the Permian Aquifer of Dumfries, Scotland. *Hydrogeol. J.* **11** (4): 504–517.

COAGULATION AND FLOCCULATION

Kirsten Exall
National Water Research Institute
Burlington, Ontario, Canada

INTRODUCTION

Removal of suspended or colloidal solids is an integral component of both drinking water and wastewater treatment. Natural waterbodies from which drinking water is obtained contain a variety of particulates, including clays, silt, organic detritus, and microorganisms. Municipal and industrial wastewaters contain high concentrations of solids, as well as nutrients, organics, and pathogens. Many contaminants of concern to human and environmental health are commonly associated with particles (e.g., certain heavy metals and polynuclear aromatic hydrocarbons, PAHs) or are of colloidal size themselves (e.g., pathogenic microorganisms such as *Cryptosporidia*) (2). Removal of suspended or colloidal particles is therefore necessary to reduce a variety of health hazards.

The addition of chemical coagulants to aid in removing solids and organic matter is a fundamental step in conventional drinking water treatment. In wastewater treatment, coagulants may be added to enhance primary settling (known as chemically enhanced primary treatment or advanced primary treatment) or to improve separation of sludge in the secondary clarifier. Additionally, flocculants may be applied during sludge thickening and dewatering; this application has recently been discussed by Bache (1). This article focuses on chemical coagulant addition and subsequent flocculation processes during drinking water and wastewater treatment; various factors that influence the effectiveness of these processes are discussed.

COAGULATION AND FLOCCULATION

Colloidal Stability

Colloidal particles, as broadly defined, exist in the size range of nanometers to micrometers. Such particles do not readily settle out of suspension in natural water or wastewater due to very low settling velocities and do not aggregate quickly due to electrostatic or steric stabilizing influences. Many colloidal materials in natural water and wastewater have negatively charged surfaces, which lead to the formation of an electrical double layer (the charged surface with adsorbed ions and a more diffuse, surrounding ion cloud). Repulsion between like-charged surfaces of neighboring colloids imparts electrostatic stability to the particles. The degree of stability depends on the magnitude of the surface charge, which can be indirectly measured by the zeta potential of the particle. Further discussion of the origins, impact, and modeling of the electrical double layer and electrostatic stabilization (as they relate to water treatment) can be found in Letterman et al. (2).

Steric stabilization may occur when polymer segments adsorb on a particle's surface, leaving loops and tails extending into solution. Interactions between adsorbed polymers on nearby colloids then lead to repulsion, as mixing or compression of the extended polymer segments increases local concentration of those segments (which may directly repel one another) and reduces the volume available to each segment (a reduction in entropy). Steric stability in natural waters can arise from the adsorption of materials such as humic substances on colloid surfaces (2).

MECHANISMS OF COAGULATION AND FLOCCULATION

Most colloids are not indefinitely stable and aggregate after some time. Coagulation in water treatment uses physicochemical processes to accelerate the kinetics of destabilization (coagulation) and aggregation (flocculation). Coagulation involves adding a chemical agent to destabilize colloids in the water and requires intensive mixing to disperse the coagulant rapidly. Flocculation, on the other hand, the formation of aggregates, or flocs, of the destabilized colloids, requires gentle mixing to allow interparticle collisions to form flocs without shearing the aggregates (2). Separation of the flocs and other particulate matter from the water may then occur by gravity settling (sedimentation), dissolved air flotation, or filtration (3).

Coagulation processes include double layer compression, charge neutralization, sweep coagulation, and complex formation; these processes are described below:

- *Double layer compression* occurs on addition of simple salts (e.g., Na^+) to a colloidal suspension; this mechanism of colloid destabilization is impractical for water and wastewater treatment because the salt concentrations required are high and particle aggregation is slow.
- *Charge neutralization* is a common mechanism of contaminant removal employed in water treatment, whereby cationic (positively charged) coagulant species adsorb to the negatively charged particle surface, neutralizing the surface charge and thereby reducing electrostatic stability. Addition of excess coagulant, or overdosing, can result in the reversal of surface charge and restabilization of the dispersion (2). In the case of polymeric coagulants, neutralization is described by the electrostatic charge patch model, where polymer species adsorb to portions of the colloid surface, resulting in the formation of charge-neutral or positively charged "patches," which cause electrostatic attraction and aggregation of colloidal species (4,5).
- In *sweep coagulation*, voluminous metal hydroxide flocs bring about removal of colloidal contaminants through enmeshment as they move through the suspension. The relationship between floc structure and operating problems during sweep coagulation has been reviewed by Bache et al. (6).
- Dissolved contaminants, such as organic species or phosphate, may also be removed by *formation of insoluble complexes* or by adsorption to flocs (2).

Flocs are aggregates of primary particles bound together by intermolecular forces. Flocculation, or floc growth, occurs through interparticle collisions. Both particle collision frequency and collision efficiency (sticking) influence the formation of flocs; floc strength and mixing intensity, or shear, influence the final floc size (7). Polymer flocculants (described below) can improve flocculation through adsorption and *interparticle bridging*. In this mechanism of flocculation, segments of a polyelectrolyte are bound to one colloid particle, but loops and dangling chains of the polymer extend into solution and can attach to nearby particles, leading to an increase in the size of the agglomerate (4,5). It may be difficult to distinguish between charge neutralization and bridging when using polymers for flocculation, because both processes can take place simultaneously (8).

COAGULANTS AND FLOCCULANTS

Coagulation has long been used as an aid to sedimentation in water and wastewater treatment. Common coagulants include both metal salts and charged polymers (also termed polyelectrolytes); flocculants are typically composed of high molecular mass polymers.

Inorganic Coagulants

When simple inorganic coagulants, such as aluminum sulfate (alum) or ferric chloride, are added to water, the salts dissociate, and the metal ions undergo hydrolysis to form a variety of hydroxoaluminum or hydroxoiron species. In these reactions, the acidity of the metal cation causes deprotonation of waters of hydration. The resultant species may be cationic, anionic, or neutral and can include an insoluble metal hydroxide; the distribution of species depends on the concentration of the metal ion, pH, and other materials in the water.

Successive reactions in a solution of an aluminum salt lead to the formation of the species Al^{3+}, $Al(OH)^{2+}$, $Al(OH)_2{}^+$, $Al(OH)_3$, and $Al(OH)_4{}^-$. Note that for the sake of simplicity, the waters of hydration (e.g., $Al(H_2O)_6{}^{3+}$) are conventionally not shown. The cationic species can destabilize colloids through surface charge neutralization. At higher concentrations or pH, both Al^{3+} and Fe^{3+} are converted to insoluble hydroxides, which form voluminous, amorphous flocs. These flocs are responsible for sweep coagulation and can enmesh colloidal material as they pass through the suspension. Depending on the conditions of hydrolysis, polynuclear aluminum species (those containing more than one aluminum ion) may also be formed. These include dimeric [e.g., $Al_2(OH)_2{}^{4+}$], trimeric [$Al_3(OH)_4{}^{5+}$], and tridecameric ($Al_{13}O_4(OH)_{24}{}^{7+}$) forms (9). Concerns exist over the potential health effects of aluminum in drinking water, so coagulation conditions should be chosen to minimize residual levels of the metal in the finished water (2). Similar reactions occur when using iron salts, and equivalent mononuclear and polynuclear iron species exist (2,9).

Prehydrolyzed Coagulants

Various prehydrolyzed coagulants have been prepared and investigated, including polyaluminum chloride (PACl) (e.g., 10,11), polyaluminum silicate sulfate (PASS) (12), polyaluminum sulfate (PAS, also known as polyaluminum hydroxysulfate, PAHS) (e.g., 13,14), and various iron-based and mixed Fe-Al coagulants (15,16). PACl is the most commonly used of these (9). Prehydrolyzed coagulants are formed by partial neutralization of a simple salt solution, such as aluminum chloride, by a base. The degree of prehydrolysis is expressed as a hydroxide-to-metal ion ratio and can be expressed as a percent basicity (10), where

$$\text{Basicity (\%)} = 100 * [OH^-]/(3 * [M_T]) \quad (1)$$

Prehydrolysis favors the formation of soluble polynuclear species, such as the tridecamer (9,10). The rate of particle destabilization is faster with PACl than with alum, and the prehydrolyzed product, PACl, is less dependent on pH for efficient coagulation. The performance of prehydrolyzed aluminum coagulants is also less affected by low temperatures than that of alum [9, and references therein]. Though effective in organic matter removal, prehydrolyzed coagulants are not superior to alum in this regard (9) and may even display higher sensitivity to organic matter concentration (19). Other reported benefits of prehydrolyzed coagulants have included lower aluminum residuals, reduced sludge volume, less alkalinity consumption, and improved settling efficiency (9,18).

Organic Polymers

Synthetic polymers are commonly used as coagulants or flocculants in water and wastewater treatment (4,20,21). Typical polymers include polyacrylamide (PAM), polydiallyldimethyl ammonium chloride (PolyDADMAC), and epichlorohydrindimethylamine (Epi-DMA). Many polymers are derivatized (chemically altered) to contain charged functional groups, such as quaternary ammonium groups, to improve coagulating characteristics (4). The behavior of a polymer is governed by the nature of its charge, whether cationic (positively charged), anionic (negatively charged), zwitterionic (containing both positively and negatively charged groups), or nonionic. Although anionic and nonionic polymers are effective in removing colloidal contaminants under some conditions and may be used as flocculants in conjunction with inorganic coagulants, the polymers used as primary coagulants in water and wastewater treatment are most often cationic (2,4,20).

Synthetic polymers may be used as either coagulants or flocculants, depending on the molecular character and mode of action on suspended solids. Polymer coagulants typically have a low molecular mass and high charge density and commonly act on colloids to bring about coagulation through charge neutralization. In this case, removal is considered to proceed by the electrostatic charge patch model (discussed above). As with inorganic coagulants, charge reversal and restabilization is possible with polymer coagulants. Polymer flocculants, on the other hand, generally have very high molecular mass and low charge density and typically act through a bridging mechanism. Flocculants are also commonly known as coagulant aids because they are often used in conjunction with inorganic or polymeric coagulants (4,5).

The effectiveness of polymers (particularly the large molecular mass flocculants) can depend strongly on their conformation, which in turn depends on the preparation of polymer solution. Many flocculants are supplied in concentrated, highly entangled forms, often as dispersions or emulsions in mineral oils. The polymer must be dissolved in water and at least partially disentangled, or "activated," for optimal flocculation behavior. The intensity of mixing during dispersion can impact the dissolution process; a balance must be achieved between rapid dispersion of the polymer and rupture of polymer chains under high shear conditions (8). Other factors that may influence the effectiveness of a polymer solution include other ions in the water (8) and the age of a polymer solution (22).

Concerns over the use of synthetic polymers in water and wastewater treatment have been raised with regard to the toxicity of residual trace levels in the effluent and formation of toxic by-products on reaction of residues with disinfectants (e.g., 23,24). Polymers are expected to be largely partitioned into the solid phase, but the amount of residual polymer remaining in the liquid stream depends on the dosage. Careful dosage selection and control is therefore needed. Other concerns exist with respect to contaminant monomer concentrations (e.g., acrylamide in polyacrylamide formulations); many countries regulate the composition and allowed dosages of polymers used in drinking water treatment (2,25,26).

Natural Coagulants

Natural products, including chitosan (a derivative of chitin, found in the shells of crustaceans), starch derivatives, seed extracts of such plants as *Moringa oleifera*, tannin- and lignin-based materials, and alginate have also been studied for use in municipal water or wastewater treatment (4,27). These products have been examined primarily for use in developing countries, but often display many of the advantages of synthetic polymeric coagulants and flocculants, while providing the additional benefits of biodegradability and reduced toxicity (27).

Ballasted Flocculation

Under conditions that would lead to the formation of poorly settling flocs, such as low solids and high organics concentrations, ballasted flocculation may be used. In this technique, additional solids are added with (or shortly after) the coagulant to improve floc formation and settling. Polymers may also be dosed to improve adhesion of the forming flocs to the solids, and the solids may be recovered and recycled through the process (4,28). Various solids have been used as ballast, including bentonite, fly ash, powdered magnetite, and microsand (3,29).

ORGANIC MATTER REMOVAL

Organic species exist in many forms in natural waters and wastewaters and, even when not directly harmful to human health, can be carriers of contaminants such as pesticides and can interfere with treatment processes. Reaction of organic matter (OM) with disinfectants may also lead to the formation of harmful disinfection by-products, including trihalomethanes (2). In drinking water treatment, the term "enhanced coagulation" refers to the practice of optimizing coagulation conditions for removing natural organic matter (NOM), which is typically accomplished through increased coagulant dosage and adjustment of pH (30).

In general, the presence of organic matter increases the required dosage of a metal salt coagulant. Once the coagulant demand is met, turbidity and NOM removal occur simultaneously (31–33). A linear relationship often exists between OM concentration and the coagulant dose required for coagulation. The presence of such a "stoichiometric" relationship between the concentration of humic substances and alum dosage was first described by Hall and Packham (31) and has since been observed with various coagulants and organic species (e.g., 33,34).

The nature of the organic matter present affects its removal by coagulation. Many researchers (e.g., 32,33) have observed preferential removals of hydrophobic, dissolved organic carbon compounds over hydrophilic organic fractions and of higher molecular mass organic matter over lower molecular mass compounds.

Metal ion concentration and pH are important factors governing the mechanism of OM removal by coagulation. Two distinct regions of organic matter or color removal have been observed (32,34,35), one near pH 5.5 and one centered near pH 7. The two domains have been explained in terms of removal mechanisms. The lower pH and aluminum concentration region is thought to represent removal by charge neutralization/precipitation, where cationic aluminum species interact directly with OM to bring about removal (2,32). The higher pH/higher coagulant dosage domain is thought to indicate sweep coagulation, where organic species adsorb directly to the surface of formed $Al(OH)_3$ flocs (32,35). Removal of OM may not occur by a single, simple mechanism, however. It has been suggested that even the OM that is ultimately removed by adsorption to $Al(OH)_3$ surfaces may first be complexed by cationic coagulant species (17,33).

The presence of organic matter also increases the required dosage of polymer coagulants and flocculants, thereby greatly increasing the cost of treatment (4,36). As with metal salts, the more hydrophobic OM fractions are more easily removed by polymer, and increasing polymer charge density improves OM removal (37). Polymer molecular mass also influences organic matter removal, although the effect is inconsistent (e.g., 37,38).

FACTORS AFFECTING EFFICIENCY OF COAGULATION AND FLOCCULATION

Many variables may influence the progression of coagulation and flocculation processes, but the primary factors include pH, temperature, mixing conditions, and other ions.

pH Dependence

The speciation of aluminum and iron in aqueous solution are highly dependent on pH. Under acidic conditions, cationic coagulant species (e.g., $Al(OH)^{2+}$) are dominant; in highly basic solution, the anionic hydroxoaluminum or iron form ($Al(OH)_4^-$ or $Fe(OH)_4^-$, respectively) predominates. At intermediate pH, both the pH and metal ion concentration affect the speciation of aluminum or iron, in turn influencing the interactions between the coagulant species and colloidal solids or organic matter (2). The optimum pH for solids removal and organic matter removal may differ slightly and depends on the mechanism of coagulation (32,34,35).

To minimize residual aluminum or iron in treated water, it is preferable to conduct coagulation under conditions of minimum metal hydroxide solubility. Aluminum hydroxide exhibits a minimum in solubility at 25 °C at approximately pH 6.3; the minimum solubility of iron at the same temperature is at approximately pH 8 (2). In general, synthetic polymers are much less sensitive to pH than metal salt coagulants (4,20).

Temperature

It has long been known that cold temperatures increase the coagulant dose required and change the nature of the flocs produced during coagulation by simple salts (39). Many reasons for the temperature effects have been suggested, including changes in water viscosity (which could affect mixing efficiency), change in the rate and extent of hydrolysis, and changes in the size and strength of flocs formed at lower temperatures (40,41).

The effect of pH varies in cold waters; the optimum pH value for coagulation and flocculation changes with both coagulant concentration and temperature (39). Kang and Cleasby (40) noted that the value of pK_w is temperature-dependent, and adjusting the pH of the solution so that a constant pOH value was held in cold and warm waters was found partially effective in reducing the impact of low temperature on flocculation kinetics.

Prehydrolyzed coagulants are generally less susceptible to temperature changes than are simple aluminum or iron salts (11,18). Synthetic polymers are also not as sensitive to water temperature as metal salt coagulants, although storage of polymer solutions at elevated temperatures may increase the rate of degradation and lowers the effective polymer concentration (22).

Mixing Effects

Coagulation and flocculation processes require both a rapid mix phase to achieve complete and uniform distribution of the coagulant in the water to be treated and a slow mix phase, which promotes interparticle collisions so that flocs may form and grow. Both mixing phases influence the efficiency of contaminant removal.

Although the intensity of rapid mixing is not particularly important for sweep coagulation processes (where contaminant removal occurs by physical interaction between amorphous metal hydroxide flocs and colloids), charge neutralization with metal salt coagulants requires contact between the rapidly changing cationic aluminum or iron species and the colloids in the water before hydrolysis is complete and the metal hydroxide solid forms. For effective charge neutralization, the coagulant species must therefore be distributed throughout the water sample as quickly as possible. Optimization of the rapid mix parameters improves turbidity removal considerably for a given water sample and lowers the required coagulant dosage (42). To optimize the rapid mixing phase of treatment, the intensity of mixing (often expressed as the velocity gradient, G), duration of mixing, and container geometry should all be considered (44).

Increased agitation accelerates floc formation and aggregation through interparticle collisions; if mixing is too vigorous, however, turbulent shear forces cause the breakup of flocs. Interparticle collisions occur through transport by fluid shear, which is controlled by characteristics of the slow mixing phase of treatment, such as vessel geometry, flow patterns, and the turbulence and intensity of mixing (2). Floc breakup may occur through surface erosion of primary particles or fracture of the relatively fragile floc aggregate and is dependent on the intensity of the slow mix phase. The competing processes of aggregation and floc breakup result in a stable floc size for a given set of formation conditions (7).

As discussed above, mixing intensity is significant in the activation of polymer flocculants (8). It is also an important parameter in the application of polymer coagulants and flocculants. In general, higher intensity mixing (higher energy input) is required for the dispersion of polymers than for metal salts, but care must be taken not to shear the polymer (20,43).

Other ions

A number of aspects of water quality impact coagulation processes, including the alkalinity of the water. At very low alkalinity, small additions of an acidic metal coagulant (such as alum) can depress the pH of the water below that of efficient coagulation; at very high alkalinity, coagulant demand may be extremely high (32).

Some anions, such as fluoride, phosphate and certain organic acids, can strongly complex cationic metal species, inhibiting coagulation by reducing free cations in solution (2). The presence of divalent cations, such as Ca^{2+}, typically broadens the pH range for effective organic matter removal and lowers the coagulant dosage required. This is likely due to the complexation of such cations with anionic organic functional groups, lowering the stability of the organic species and thereby decreasing demand on the coagulant (32,45). Sulfate may enhance coagulation by suppressing charge reversal or by catalyzing the kinetics of precipitation through the formation of outer sphere complexes with aluminum ions (45). The presence of silicate or sulfate during prehydrolysis of aluminum salts may also influence aluminum speciation, thereby altering the coagulant composition and activity (12).

The selection and effectiveness of polymers may also be affected by the ionic strength of the system. The addition of simple salts (such as NaCl) causes a contraction in polymer conformation, reducing the effectiveness of polymer flocculants (5).

The conflicting effects of many variables make prediction of coagulant demand for a given water type difficult, so the required dosage of coagulant or flocculant is typically determined empirically by laboratory jar tests.

SUMMARY

Coagulants and flocculants are commonly used in the destabilization and aggregation of suspended solids during water and wastewater treatment. Both inorganic salts and organic polymers may be used as coagulants, and various synthetic and natural organic polymers are applied as flocculants. Organic matter removal is another common objective of coagulation during drinking water treatment, although it generally requires a higher coagulant dosage. Other factors that influence the effectiveness of coagulants and flocculants include pH, temperature, mixing conditions, and other ions in the water. To maximize the efficiency of contaminant removal while minimizing residuals, careful coagulant or flocculant dose selection and control are essential.

BIBLIOGRAPHY

1. Bache, D.H. (2002). Polymers in alum sludge dewatering: developments and issues. In: *Chemical Water and Wastewater Treatment VII*. H.H. Hahn, E. Hoffmann, and H. Odegaard (Eds.). IWA, Cornwall, pp. 287–296.
2. Letterman, R.D., Amirtharajah, A., and O'Melia, C.R. (1999). Coagulation and flocculation. In: *Water Quality and Treatment: A Handbook of Community Water Supplies*, 5th Edn. R.D. Letterman (Ed.). McGraw-Hill, Toronto, pp. 6.1–6.66.
3. Gregory, R., Zabel, T.F., and Edzwald, J.K. (1999). Sedimentation and flotation. In: *Water Quality and Treatment: A Handbook of Community Water Supplies*, 5th Edn. R.D. Letterman (Ed.). McGraw-Hill, Toronto, pp. 7.1–7.87.
4. Bolto, B.A. (1995). Soluble polymers in water purification. *Prog. Polym. Sci.* **20**: 987–1041.
5. Mortimer, D.A. (1990). Synthetic polyelectrolytes—a review. *Polym. Int.* **25**(1): 29–41.
6. Bache, D.H. et al. (1999). Sweep coagulation: structures, mechanisms and practice. *J. Water SRT: Aqua* **48**(5): 201–210.
7. Gregory, J. and Yukselen, M.A. (2002). Break-up and reformation of flocs formed by hydrolyzing coagulants and polymeric flocculants. In: *Chemical Water and Wastewater Treatment VII*. H.H. Hahn, E. Hoffmann, and H. Odegaard (Eds.). IWA, Cornwall, pp. 29–38.
8. Kim, Y.H. (1990). Importance of flocculant preparation for use in solid/liquid separation. In: *Solid/Liquid Separation: Waste Management and Productivity Enhancement*. H.S. Muralidhara (Ed.). Battelle Press, Columbus, OH, pp. 289–304.
9. Gregory, J. and Duan, J. (2001). Hydrolyzing metal salts as coagulants. *Pure Appl. Chem.* **73**(12): 2017–2026.
10. Pernitsky, D.J. and Edzwald, J.K. (2003). Solubility of polyaluminum coagulants. *J. Water SRT: Aqua* **52**(6): 395–406.
11. Van Benschoten, J.E. and Edzwald, J.K. (1990). Chemical aspects of coagulation using aluminum salts I. Hydrolytic reactions of alum and polyaluminum chloride. *Water Res.* **24**(12): 1519–1526.
12. Boisvert, C. and Jolicoeur, J.P. (1999). Influences of sulfate and/or silicate present in partially prehydrolyzed Al(III) flocculants on Al(III) speciation in diluted solutions. *Colloid Surf. A: Physicochem. Eng. Aspects* **155**(2–3): 161–170.
13. Shen, Y-H. and Dempsey, B.A. (1998). Fractions of aluminum coagulants as a function of pH. *Environ. Technol.* **19**(8): 845–850.
14. Koether, M.C., Deutschman, J.E., and vanLoon, G.W. (1997). Low-cost polymeric aluminum coagulant. *J. Environ. Eng.* **123**(9): 859–864.
15. Gray, K.A., Yao, C., and O'Melia, C.R. (1995). Inorganic metal polymers: Preparation and characterization. *J. AWWA* **87**(4): 136–146.
16. Jiang, J. and Graham, N.J.D. (1998). Preliminary evaluation of the performance of new pre-polymerized inorganic coagulants for lowland surface water treatment. *Water Sci. Technol.* **37**(2): 121–128.

17. Edzwald, J.K. (1993). Coagulation in drinking water treatment: Particles, organics and coagulants. *Water Sci. Technol.* **27**(11): 21–35.
18. Lind, C. (1994). Reduce residuals with PACl coagulants. *Public Works* **Dec. 1994**: 43–47.
19. Exall, K. and vanLoon, G.W. (2000). Using coagulants to remove organic matter. *J. AWWA* **92**(11): 93–102.
20. Fettig, J., Ratnaweera, H., and Odegaard, H. (1991). Synthetic organic polymers as primary coagulants in wastewater treatment. *Water Supply* **9**: 19–26.
21. Bolto, B.A. et al. (1996). The use of soluble organic polymers in waste treatment. *Water Sci. Technol.* **34**(9): 117–124.
22. Gehr, R. and Kalluri, R. (1983). The effects of short-term storage and elevated temperatures on the flocculation activity of aqueous polymer solutions. *Water Pollut. Res. J. Can.* **18**: 23–43.
23. Rowland, C.D., Burton, G.A., and Morrison, S.M. (2000). Implication of polymer toxicity in a municipal wastewater effluent. *Environ. Toxicol. Chem.* **19**(8): 2136–2139.
24. Lee, J-F., Liao, P-M., Tseng, D-H., and Wen, P-T. (1998). Behavior of organic polymers in drinking water purification. *Chemosphere* **37**(6): 1045–1061.
25. Dentel, S.K., Chang, L-L., Raudenbush, D.L., Junnier, R.W., and Abu-Orf, M.M. (2000). Analysis and fate of polymers in wastewater treatment. Water Environment Research Federation, WERF Report 94-REM-2.
26. Letterman, R.D. and Pero, R.W. (1990). Contaminants in polyelectrolytes used in water treatment. *J. AWWA* **82**(11): 87–97.
27. Kawamura, S. (1991). Effectiveness of natural polyelectrolytes in water treatment. *J. AWWA* **83**(10): 88–91.
28. Desjardins, C., Koudjonou, B., and Desjardins, R. (1999). Laboratory study of ballasted flocculation. *Water Res.* **36**: 744–754.
29. Ding, Y., Dresnack, R., and Chan, P.C. (2001). Mechanism and feasibility of colloidal particle removal by a microcarrier process. *Int. J. Environ. Stud.* **58**: 535–548.
30. Edzwald, J.K. and Tobiason, J.E. (1999). Enhanced coagulation: US requirements and a broader view. *Water Sci. Technol.* **40**(9): 63–70.
31. Hall, E.S. and Packham, R.F. (1965). Coagulation of organic color with hydrolyzing coagulants. *J. AWWA* **58**(9): 1149–1166.
32. Randtke, S.R. (1988). Organic contaminant removal by coagulation and related process combinations. *J. AWWA* **80**(5): 40–56.
33. Kuo, C-J. and Amy, G.L. (1988). Factors affecting coagulation with aluminum sulfate II. Dissolved organic matter removal. *Water Res.* **22**(7): 863–872.
34. Van Benschoten, J.E. and Edzwald, J.K. (1990). Chemical aspects of coagulation using aluminum salts II. Coagulation of fulvic acid using alum and polyaluminum chloride. *Water Res.* **24**(12): 1527–1535.
35. Dempsey, B.A. (1984). Removal of naturally occurring compounds by coagulation and sedimentation. *CRC Crit. Rev. Environ. Control* **14**(4): 311–331.
36. Narkis, N. and Rebhun, M. (1997). Flocculation in presence of organic macromolecules of natural water and secondary effluents. *Water Sci. Technol.* **36**(4): 85–91.
37. Bolto, B.A. et al. (1999). Experimental evaluation of cationic polyelectrolytes for removing natural organic matter from water. *Water Sci. Technol.* **40**(9): 71–79.
38. Kam, S-K. and Gregory, J. (2001). The interaction of humic substances with cationic polyelectrolytes. *Water Res.* **35**(15): 3557–3566.
39. Camp, T.R., Root, D.A., and Bhoota, B.V. (1940). Effects of temperature on rate of floc formation. *J. AWWA* **32**(11): 1913–1927.
40. Kang, L-S. and Cleasby, J.L. (1995). Temperature effects on flocculation kinetics using Fe(III) coagulant. *J. Environ. Eng.* **121**(12): 893–901.
41. Hanson, A.T. and Cleasby, J.L. (1990). The effects of temperature on turbulent flocculation: Fluid dynamics and chemistry. *J. AWWA* **82**(11): 56–73.
42. Rossini, M., Garrido, J.G., and Galluzzo, M. (1999). Optimization of the coagulation—flocculation treatment: Influence of rapid mix parameters. *Water Res.* **33**(8): 1817–1826.
43. Young, S., Stanley, S.J., and Smith, D.W. (2000). Effect of mixing on the kinetics of polymer-aided flocculation. *J. Water SRT: Aqua* **49**(1): 1–8.
44. Mhaisalkar, V.A., Paramasivam, R., and Bhole, A.G. (1991). Optimizing physical parameters of rapid mix design for coagulation—flocculation of turbid waters. *Water Res.* **25**(1): 43–52.
45. Hundt, T.R. and O'Melia, C.R. (1988). Aluminum—fulvic acid interactions: mechanisms and applications. *J. AWWA* **80**(4): 176–186.

CONDUCTIVITY-ELECTRIC

David M. Gray
Mettler-Toledo Thornton Inc.
Bedford, Massachusetts

Electrical conductivity is the ability of a material to carry electrical current. In water, it is generally used as a measure of mineral or other ionic concentration. Conductivity is a measure of the purity of water or the concentration of ionized chemicals in water. However, conductivity is only a quantitative measurement: it responds to all ionic content and cannot distinguish particular conductive materials in the presence of others. Only ionizable materials will contribute to conductivity; materials such as sugars or oils are not conductive.

In a metal conductor, electrical current is the flow of electrons and is called *electronic* conductance. In water, electrical current is carried by ions because electrons do not pass through water by themselves. This is *electrolytic* conductance.

When a voltage is applied between two inert electrodes immersed in a solution, any ions between them will be attracted by the electrode that has an opposite charge. Ions will move between electrodes and produce a current depending on the electrical resistance of the solution. This is the basis of conductivity measurement—an application of Ohm's law. For high purity waters, it is common to express conductivity as its reciprocal, resistivity.

To prevent altering a sample by major ionic movement and electrochemical reaction at the electrodes, alternating current is always used for measurement. Using ac, the polarity changes frequently enough that ions do not move or react significantly. Measuring systems must control the voltage, frequency, and current density to minimize errors due to electrode polarization and capacitance. Modern instrumentation may change one or more of these variables automatically, depending on the conductivity range being measured.

UNITS OF MEASURE

Conductivity is the conductance of a standardized volume of water of 1 cm^2 cross-sectional area and electrodes spaced 1 cm apart, as shown in Fig. 1. The same results will be obtained with various geometries as long as the ratio of length/area is equal to 1 cm^{-1}, the cell constant. Units of measure are related as listed in Table 1. Most conductivity values are quite small, so unit multiplier prefixes of milli (m, 10^{-3}) and micro (μ, 10^{-6}) are commonly used.

The standardized geometry of the cell constant ensures that the conductivity measurement is a property of the sample and not of the sensor. The cell constant is determined with precision by calibration in standard solutions of known conductivity. Accepted standard potassium chloride solutions are established in an ASTM standard method. Standard reference materials are also available from NIST.

Cell constants other than 1 cm^{-1} may be used as long as the measuring instrument readout is normalized accordingly. A lower cell constant sensor is needed to enable the measuring instrument to measure accurately in low conductivity (high resistivity) samples. Higher cell constants are needed to measure in high conductivity samples. The exact requirements depend on the measuring instrument.

Most practical cells do not use the parallel plate electrode arrangement of Fig. 1. They have greater durability and allow more convenient installation in other arrangements. For example, typical pure water sensors for on-line measurement use concentric electrodes that maintain the spacing and geometry for 0.01 to 0.1/cm constant. A variety of two-electrode process conductivity sensors is illustrated in Fig. 2.

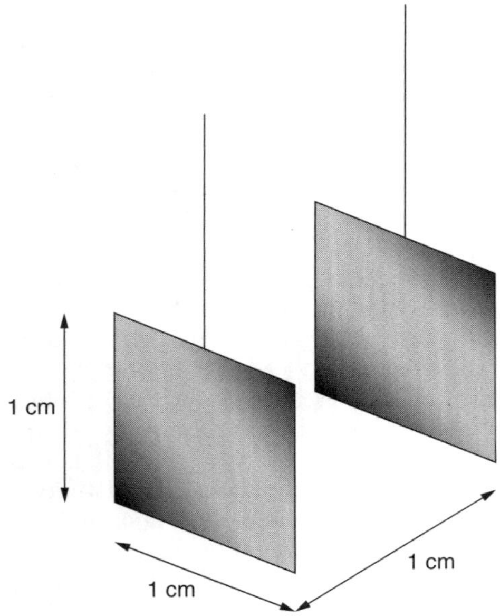

Figure 1. Cell constant = 1 cm electrode spacing divided by 1 cm^2 cross-sectional area of sample.

Table 1. Units Related to Conductivity Measurement

Measurement	Application	Units
Resistance	Electrical circuit	ohm (Ω)
Conductance	Electrical circuit	ohm^{-1} (Ω^{-1}) = siemens (S) = mho (now obsolete)
Resistivity	High purity water	ohm · cm (Ω· cm)
Conductivity	Most water samples	siemens/cm (S/cm) = mhos/cm (now obsolete), siemens/m (S/m)[a]

[a]Most users employ units of S/cm. However, SI conductivity units used in some parts of the world are S/m which can easily be confused. 1 S/cm = 100 S/m.

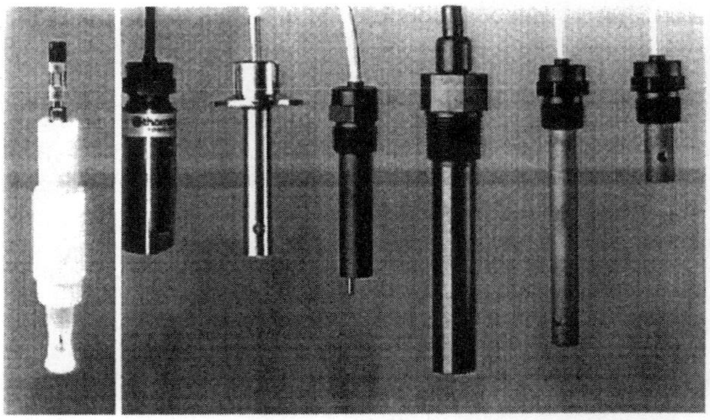

Figure 2. Conventional two-electrode conductivity sensors: 0.1 cm^{-1} in retractable housing, 0.1 cm^{-1} in flow chamber, sanitary 0.1 cm^{-1}, 10 cm^{-1} insertion, 50 cm^{-1} insertion, long 0.1 cm^{-1} insertion, short 0.1 cm^{-1} insertion.

The conductivity of very dilute solutions can be calculated from physical chemistry data based on Equation 1 which sums the conductivity contribution of all ions in the solution:

$$\Lambda = \rho\Sigma(\lambda_i c_i) \tag{1}$$

Λ = conductivity
ρ = density of water
λ_i = equivalent ionic conductance of ion 'i'
c_i = concentration of ion 'i'

TEMPERATURE EFFECTS

Conductivity is affected by temperature because water becomes less viscous and ions can move more easily at higher temperatures. Conventionally, conductivity measurements are referenced to 25 °C, though occasionally a 20 °C reference is used. The variation with temperature is apparent in Equation 1 because λ_i and, to a lesser degree, ρ are temperature dependent. The conductivity of most ions increases by about 2.2% of their value per °C which allows for simple temperature compensation. This is suitable for most midrange conductivity measurements. Very low and very high conductivity samples require special handling of temperature effects.

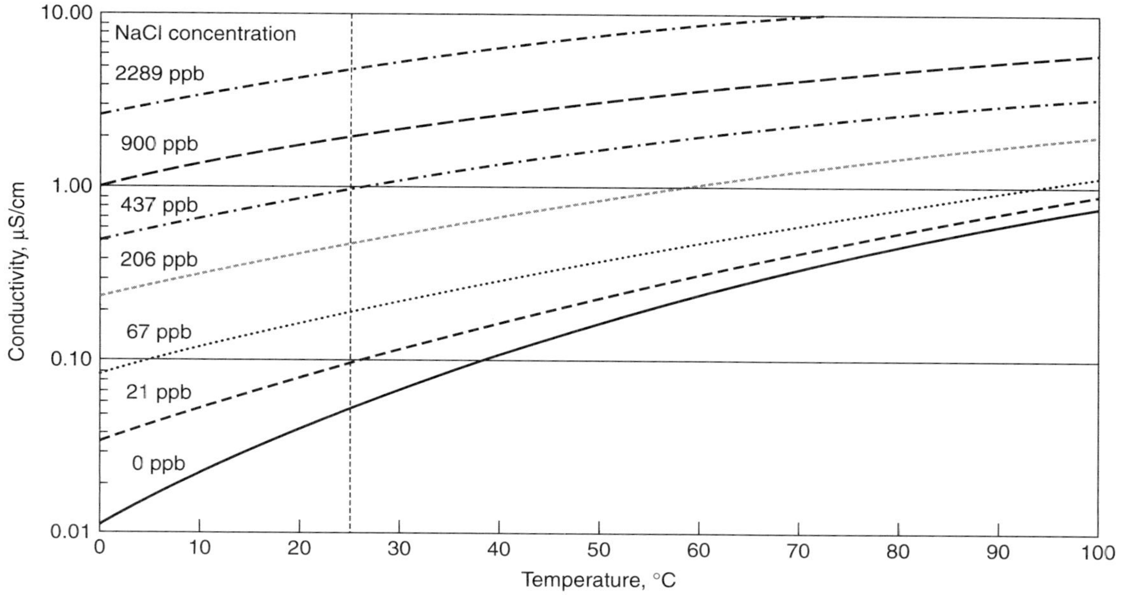

Figure 3. High purity water conductivity vs. temperature.

As water is purified below 5 μS/cm (>0.5 MΩ·cm), the small amount of hydrogen (H^+) and hydroxide (OH^-) ions from water itself become a significant part of total conductivity, that is, the concentrations, c_i, in Eq. 1, of H^+ and OH^- must be included. Water self-ionizes to a much greater degree at higher temperatures, producing a higher concentration of ions. Those ions are also more conductive. The temperature coefficient increases to 4–7%/°C, depending both on purity level and temperature; both effects are nonlinear. These effects are evident in Fig. 3. Specialized high purity temperature compensation must be employed for accurate measurements in this low conductivity range.

Highly concentrated solutions also exhibit deviant behavior. In modest concentration ranges, conductivity is roughly proportional to concentration. At higher concentrations, Eq. 1 does not apply, and ionic interference restricts the mobility of the ions. Conductivity levels off and in many cases decreases as concentration increases, as shown in Fig. 4. Measurement near the peaks of these curves is ambiguous and cannot be relied on for inference of concentration because there are two possible values.

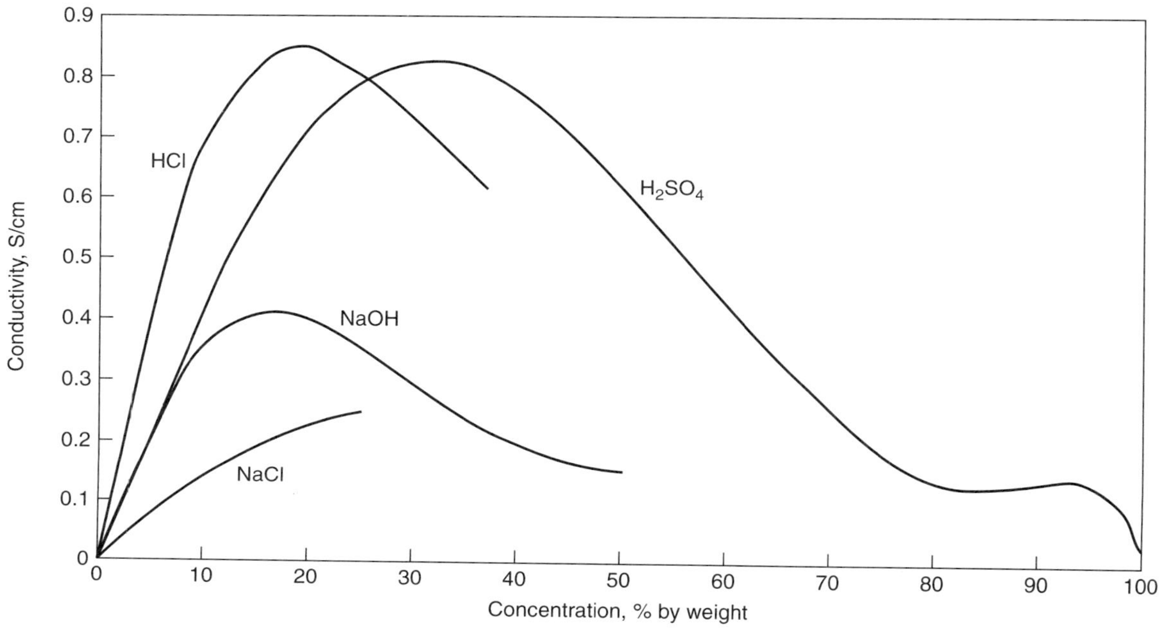

Figure 4. Conductivity vs. concentration at 25°C.

Highly conductive solutions also have a lower temperature influence, typically less than 2%/°C.

ALTERNATIVE MEASUREMENT TECHNOLOGIES

Figure 1 and its description refer to the conventional two-electrode sensor and measurement technique. Four-electrode conductivity measurement uses a sensor incorporating four electrodes. It is useful for highly conductive and/or dirty water samples which would foul the surfaces or plug the narrow passages of conventional high constant two-electrode sensors.

Four-electrode measurement applies ac through the sample via two outer drive electrodes, as shown in Fig. 5. These electrodes may become fouled, and the circuit will compensate to maintain the ac current level constant. Two inner measuring electrodes are used to sense the voltage drop through the portion of solution between them. The circuit makes a high impedance ac voltage measurement, drawing negligible current and making it much less affected by additional resistance due to fouling of the measuring electrode surfaces. Sensors for four-electrode conductivity measurement are shown in Fig. 6.

Inductive (also known as noncontact, electrodeless, or toroidal) conductivity measurement is made without any direct electrical contact with the sample. The sensor consists of two parallel coils sealed within a doughnut-shaped insulated probe, as shown in Figs. 7 and 8. The instrument energizes one coil with ac. A weak ac current is induced in the surrounding sample, depending on its conductivity. That current, in turn, induces a signal in the measured coil which provides the measurement signal. The sample acts like the core of a transformer. A temperature sensor is incorporated into the probe body to enable compensation in the instrument.

Because there are no electrodes in contact with the sample, extreme fouling conditions can be tolerated. Coatings can cause errors only if there is so much accumulation that they reduce the diameter of the hole.

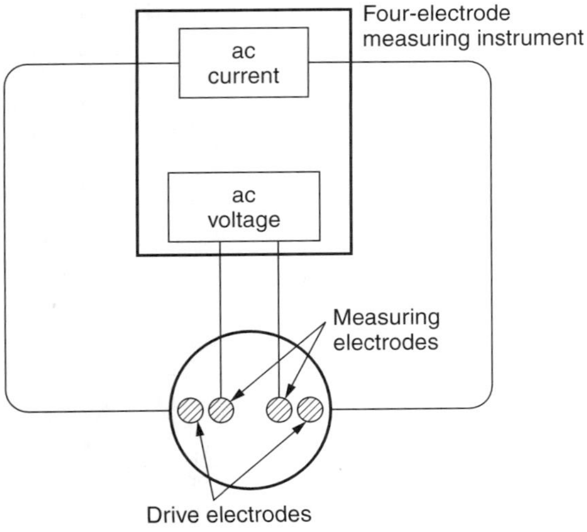

Figure 5. Four-electrode conductivity measurement.

Figure 6. Four-electrode conductivity sensors.

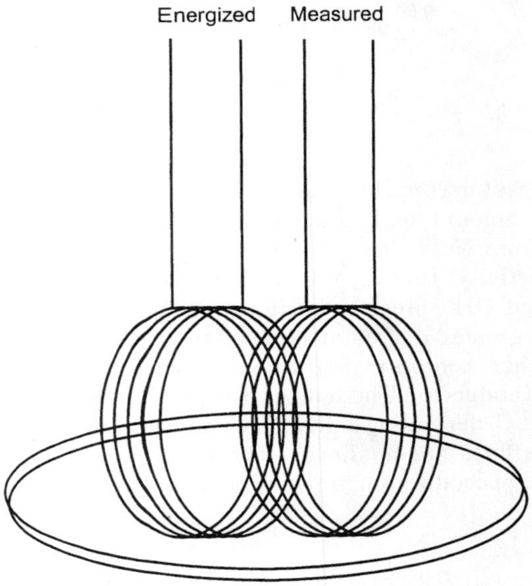

Figure 7. Inductive conductivity measurement.

Care is needed in installating an inductive sensor to allow for the specified spacing around it. Otherwise, the cross section of sample immediately around the probe (cell constant) will be affected, and recalibration may be required.

TOTAL DISSOLVED SOLIDS (TDS)

TDS is sometimes inferred from conductivity and is reported in units of parts per million. However, the relationship of conductivity and concentration is not standardized and to be meaningful, should be specified whenever TDS units are used. Typical conversions are based on sodium chloride (which may also be called salinity) at approximately 0.5 ppm TDS per μS/cm. Alternatively, a "natural water" mineral composition including bicarbonates would have a conversion of 0.6–0.7 ppm TDS per μS/cm. Conversions may also be slightly nonlinear with concentration.

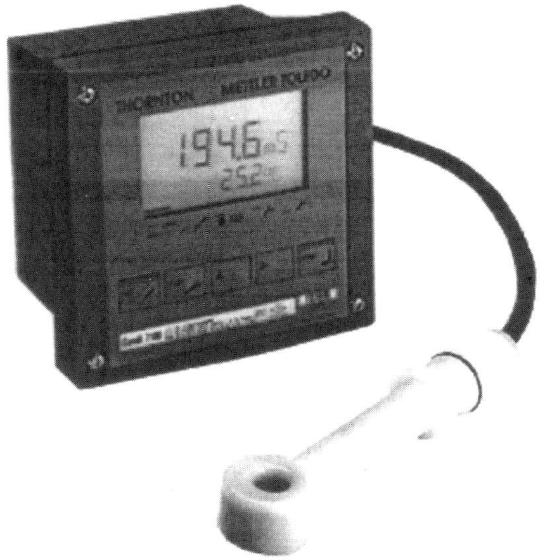

Figure 8. Inductive conductivity measurement equipment.

READING LIST

Standard Test Methods for Electrical Conductivity and Resistivity of Water. D1125, American Society for Testing and Materials, W. Conshohocken, PA.

Certificate of Analysis, Aqueous Electrolytic Conductance Standard Reference Materials 3190-3193, 3198-3199. National Institute of Standards and Technology, U.S. Department of Commerce, Gaithersburg, MD.

Standard Test Method for Electrical Conductivity and Resistivity of a Flowing High Purity Water Sample. D5391, American Society for Testing and Materials, W. Conshohocken, PA.

Weast, R.C. (Ed.). (1978). *CRC Handbook of Chemistry and Physics*, 59th Edn. CRC Press, West Palm Beach, FL, pp. D265–D314.

CONSERVATION AND THE WATER CYCLE

Natural Resources
Conservation Service

Water is probably the natural resource we all know best. All of us have had firsthand experience with it in its many forms—rain, hail, snow, ice, steam, fog, dew.

Yet, in spite of our daily use of it, water is probably the natural resource we least understand. How does water get into the clouds, and what happens to it when it reaches the earth? Why is there sometimes too much and other times too little? And, most important, is there enough for all the plants, and all the animals, and all the people?

Water covers nearly three fourths of the earth; most is sea water. But sea water contains minerals and other substances, including those that make it salty, that are harmful to most land plants and animals. Still it is from the vast salty reservoirs, the seas and oceans, that most of our precipitation comes—no longer salty or mineral laden. Water moves from clouds to land and back to the ocean in a never ending cycle. This is the water cycle, or the hydrologic cycle.

Ocean water evaporates into the atmosphere, leaving impurities behind, and moves across the earth as water vapor. Water in lakes, ponds, rivers, and streams also evaporates and joins the moisture in the atmosphere. Soil, plants, people, and animals, and even factories, automobiles, tractors, and planes, contribute moisture. A small part of this moisture, or water vapor, is visible to us as fog, mist, or clouds. Water vapor condenses and falls to earth as rain, snow, sleet, or hail, depending on region, climate, season, and topography.

Every year about 80,000 cubic miles of water evaporates from oceans and about 15,000 cubic miles from land sources. Since the amounts of water evaporated and precipitated are almost the same, about 95,000 cubic miles of water are moving between earth and sky at all times.

Storms at sea return to the oceans much of the water evaporated from the oceans, so land areas get only about 24,000 cubic miles of water as precipitation. Precipitation on the land averages 26 inches a year, but it is not evenly distributed. Some places get less than 1 inch and others more than 400 inches.

The United States gets about 30 inches a year, or about 4300 billion gallons a day. Total streamflow from surface and underground sources is about 8.5 inches a year, or about 1200 billion gallons a day. This is the amount available for human use—homes, industry, irrigation, recreation.

The difference between precipitation and streamflow -21.5 inches a year, or 3100 billion gallons a day—is the amount returned to the atmosphere as vapor. It is roughly 70 percent of the total water supply. It includes the water used by plants.

People can exist on a gallon or so of water a day for drinking, cooking, and washing though we seldom do or have to. In medieval times people probably used no more than 3 to 5 gallons a day. In the 18th century, especially in Western nations, people were using about 95 gallons a day. At present in the United States, people use about 1500 gallons a day for their needs and comforts including recreation, cooling, food production, and industrial supply.

When water hits the ground some soaks into the soil, and the rest runs off over the surface. The water that soaks into the soil sustains plant and animal life in the soil. Some seeps to underground reservoirs. Almost all of this water eventually enters the cycle once more.

People can alter the water cycle but little, so their primary supply of water is firmly fixed. But we can manage and conserve water as it becomes available—when it falls on the land. If we fail to do so we lose the values that water has when used wisely.

Water management begins with soil management. Because our water supply comes to us as precipitation falling on the land, the fate of each drop of rain, each snowflake, each hailstone depends largely on where it falls—on the kind of soil and its cover.

A rainstorm or a heavy shower on bare soil loosens soil particles, and runoff—the water that does not soak into the soil—carries these particles away. This action, soil erosion by water, repeated many times ruins land for most uses. Erosion, furthermore, is the source of sediment that fills streams, pollutes water, kills aquatic life, and shortens the useful life of dams and reservoirs.

Falling rain erodes any raw earth surface. Bare, plowed farmland, cleared areas going into housing developments, and highway fills and banks are especially vulnerable.

In cities and suburbs, where much of the land is paved or covered—streets, buildings, shopping centers, airport runways—rainwater runs off as much as 10 times faster than on unpaved land. Since this water cannot soak into the soil, it flows rapidly down storm drains or through sewer systems, contributing to floods and often carrying debris and other pollutants to streams.

Grass, trees, bushes, shrubs, and even weeds help break the force of raindrops and hold the soil in place. Where cultivated crops are grown, plowing and planting on the contour, terraces, and grassed waterways to carry surplus water from the fields are some of the conservation measures that slow running water. Stubble mulching protects the soil when it has no growing cover. Small dams on upper tributaries in a watershed help control runoff and help solve problems of too much water one time and not enough another time.

Throughout the world the need for water continues to increase. Population growth brings demands for more water. Per capita use of water, especially in industrialized countries, is increasing rapidly.

It is our management of the precipitation available to us that determines whether or not we have both the quantity and the quality of water to meet our needs.

It is our obligation to return water to streams, lakes, and oceans as clean as possible and with the least waste.

THE WATER CYCLE is an endless process of water circulation going on throughout the world.

To trace the movement of water through the cycle, begin at the far right of the diagram. There the sun's energy is transferring water from the sea and earth to the atmosphere in the form of water vapor. The soil and inland water bodies through *evaporation* and plants through *transpiration* add large amounts of water vapor to the atmosphere, but most of it comes from the oceans. Man, animals, and machines add small amounts by means of *respiration* and *combustion*.

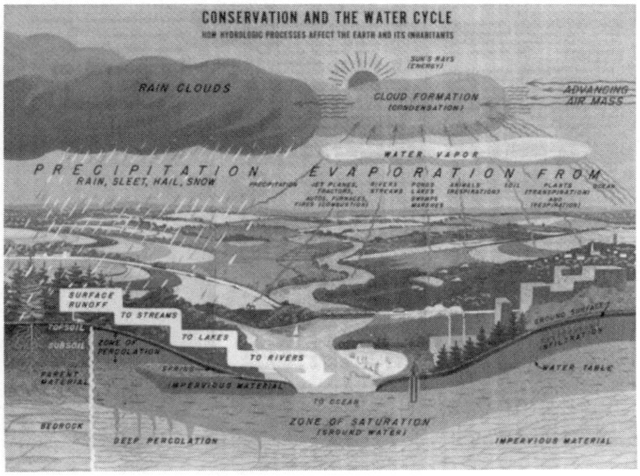

Air masses (top of diagram) carry the water vapor across the earth, and the water vapor condenses into *precipitation*.

At the left, precipitation falls as rain, snow, sleet, hail. Some evaporates while falling and returns to the atmosphere. A small amount is intercepted and held by plants or by buildings, automobiles, and other structures and machines until it evaporates back into the atmosphere.

Most of the precipitation soaks into the soil; the part that doesn't run to the sea by way of streams and rivers. Ground water gets there more slowly.

Misuse and poor management of the soil will decrease the amount of water that soaks into the soil and increase the amount that runs off over the surface. Runoff on bare land leads to erosion. Grass, trees, and other plants hold the soil in place and slow the runoff, allowing more water to soak into the soil.

Some of the water that soaks into the soil is used by plants. Part of it percolates beyond the reach of plant roots to the water table, to underground reservoirs, and to springs and artesian wells.

Runoff on its way to the sea can be intercepted and stored for industrial or household use, and it can be diverted for irrigation.

Little water has been added or lost through the ages. The water cycle prevails in all places and at all times with neither beginning nor end.

DEFLUROIDATION

Eli Dahi
Environmental Development Corporation
Søborg, Denmark

DEFINITION

Defluoridation refers to methods of water treatment that reduce the concentration of fluoride in the water, normally in order to make it safe for human consumption. Some water treatments that have the capacity of reducing the fluoride concentration along with most other anions, or anions and cations, in the water are not considered defluoridation methods. Thus, general demineralizing methods like distillation, reverse osmosis, electrodialysis, and resin deanionization, which are able to remove fluoride fully or partly from the water, are not considered defluoridation methods. On the other hand, methods that only remove fluoride without any addition or reduction of other parameters are not yet discovered, which is why the expression "fluoride removal" lacks precision. Defluoridation is used to characterize methods that reduce the fluoride ion specifically, without other ***major*** changes to the quality of the treated water.

OBJECTIVES AND EFFECTS

The objective of defluoridation is to prevent or to mitigate ***endemic fluorosis***. The term endemic fluorosis covers a wide variety of clinical manifestations of accumulative fluoride poisoning, mainly because of long-term consumption of relatively high concentration fluoride water. In mild cases, fluorosis is observed as a more or less repulsive mottling of teeth, ***dental fluorosis*** (Fig. 1). More severe cases are known as ***skeletal fluorosis*** (Fig. 2), including deformities like knock knees (genu varum), bowleg (genu valgum), antero-posterior bowing of tibia, humpback (kyphosis), exostosis, and others.

Also, ***nonskeletal fluorosis*** includes swollen joints, muscular tenderness, rheumatism-like pain especially in the joints, systematic rigidity, stiffness, and impaired movement, all symptoms that give a feeling of rapid aging. Also, gastric problems, burning sensation during urination, mental retardation, and paraplegia are reported as clinical manifestation of fluorosis.

SAFE CONCENTRATIONS

According to WHO, the "optimum" and the maximum water concentrations are **0.8 and 1.5 mgF/L**, respectively. These guidelines are adopted on the background of studies carried out mainly in low fluoride areas and in the decades before the widespread fluoridation of toothpaste. Presently, in a fluorotic area, most professionals consider **0.5 and 1.0 mg/L** as the desired and maximum allowable concentrations, respectively. In a defluoridation process, the aim is to treat down to 0.5 mg/L, whereas 1 or 1.5 mg/L can be tolerated as the saturation level. Experience, as well as mass balances, show that once the drinking and cooking water is within these limits, the users do not need to worry about the daily intake of fluoride from other sources like food and normal use of toothpaste, even when residing in a fluorotic area.

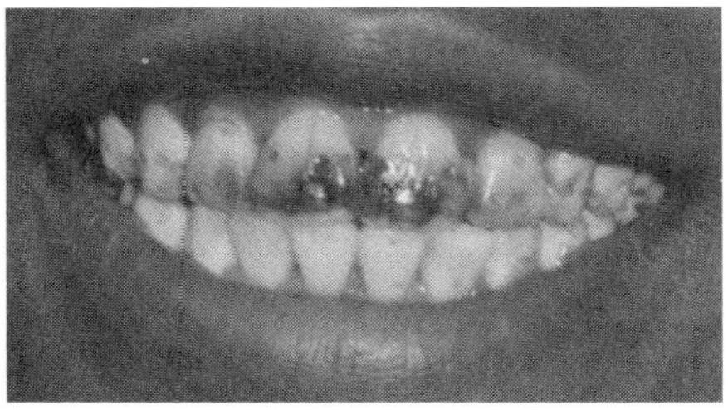

Figure 1. Dental fluorosis.

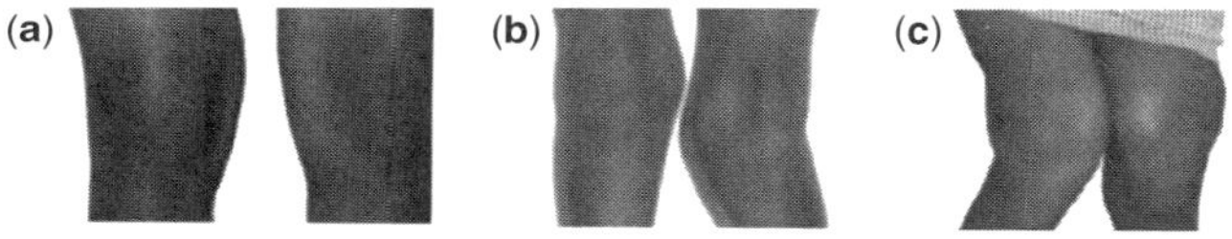

Figure 2. Skeletal fluorosis. (**a**) Stage I: swollen joints, pain, and rigidity. (**b**) Stage II: deformed but still moving. (**c**) crippled, only moving short and with difficulties and pain.

FLUORIDE OCCURRENCE

Fluoride occurs in all natural waters. Ocean waters contain about 1 mg/L, whereas most fresh surface and subsurface waters contain less than 0.5 mg/L. Groundwater is, in general, subject to more variation at higher range. In fluorotic areas, groundwater and/or surface waters may contain up to 20–30 mg/L. As an extreme case, the concentration of 90 mg/L has been recorded in one Rift Valley Lake, where the water is soft but alkaline and the evaporation is high. Fluoride-rich waters that are subject to defluoridation for human consumption are in the range 1.5–20 mg/L.

APPROPRIATE TECHNOLOGY

Defluoridation differs from most other water treatments in four main aspects:

First, because fluoride does not deteriorate the piped water quality technically or oganoleptically, it is normally only required for the small part of the water that is used for drinking and cooking.

Second, defluoridation of large quantities of water, out of which only a small part is used for human consumption, is environmentally unsound, which is because of accumulation of correspondingly large quantities of toxic sludge that would create a new problem, more severe to deal with.

Third, most large cities and towns do not need defluoridation, because they are historically built in sites privileged to have good water resources.

Fourth, defluoridation is easily avoided in industrialized countries, as they have no difficulties in prospecting and transporting low fluoride water.

APPROPRIATE CRITERIA

As a result of the above-mentioned four factors, defluoridation is often required in small scale and in rural areas in developing countries. As such, many defluoridation approaches have been launched, proclaiming success without unbiased field proof and thorough optimization. Further confusion comes with the fact that the process or the technical setup that may work in one context of socio-economic and environmental conditions may fail in another. Local availability and acceptability of the required materials, fluoride contamination level, and water quality are major factors to be considered when selecting the process and the design that minimizes the capital and running costs.

THREE OPERATIONAL LEVELS

Defluoridation technology has to be simple, affordable, reliable, and operational at least at three different levels:

1. Domestic level, e.g., in the kitchen where piped water may not be available, see Fig. 3.
2. Institutional level, e.g., in schools or a working place where piped water may be available and the defluoridator is protected, see Fig. 4.
3. Community level, e.g., in a village or a market where the defluoridator is supplied with water often

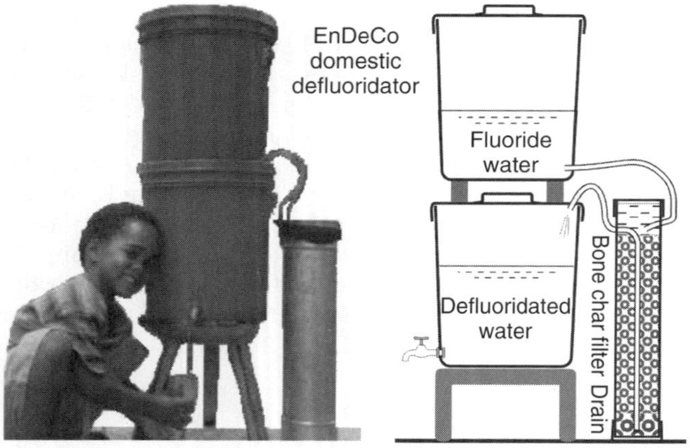

Figure 3. An example of field-tested and optimized domestic defluoridator. It can treat 3000 L of water containing 5 mgF/L, suitable for one family and one-half yearly recharge.

intermittently through a pipe or a hand or power pump, see Fig. 5.

TREATMENT PRINCIPLES

Defluoridation is carried out in one of three different processes:

1. A sorption process in a plug flow filter column with a medium that has a certain capacity of absorption, adsorption, or ion exchange of the fluoride. This process requires recharge or regeneration of the medium upon saturation. Numerous media are known to have defluoridation properties, see Table 1. However, because of capacity, availability, and subsequent water quality limitations, only bone char and activated alumina are worth mentioning. The plants shown in Figs. 1–3 are based on bone char sorption, which is the process of choice if acceptable

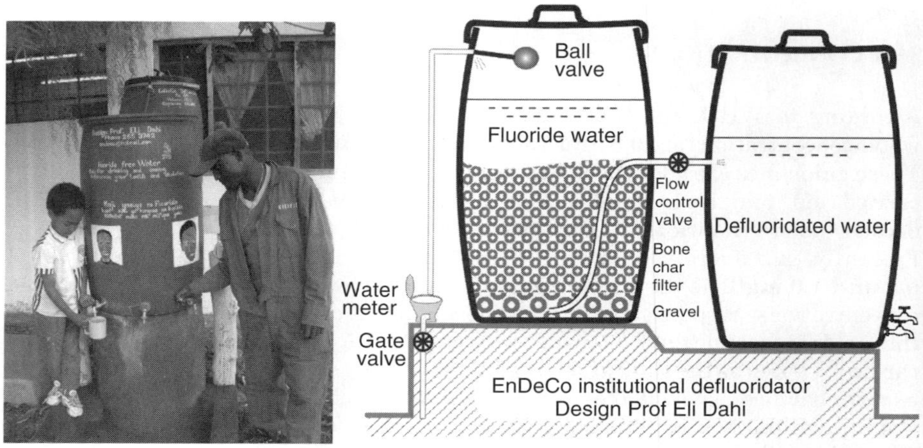

Figure 4. The EnDeCo Institutional Defluoridator. It can be made at any desired capacity, normally in the range 20–160 m^3 for a raw water fluoride of 5 mg/L.

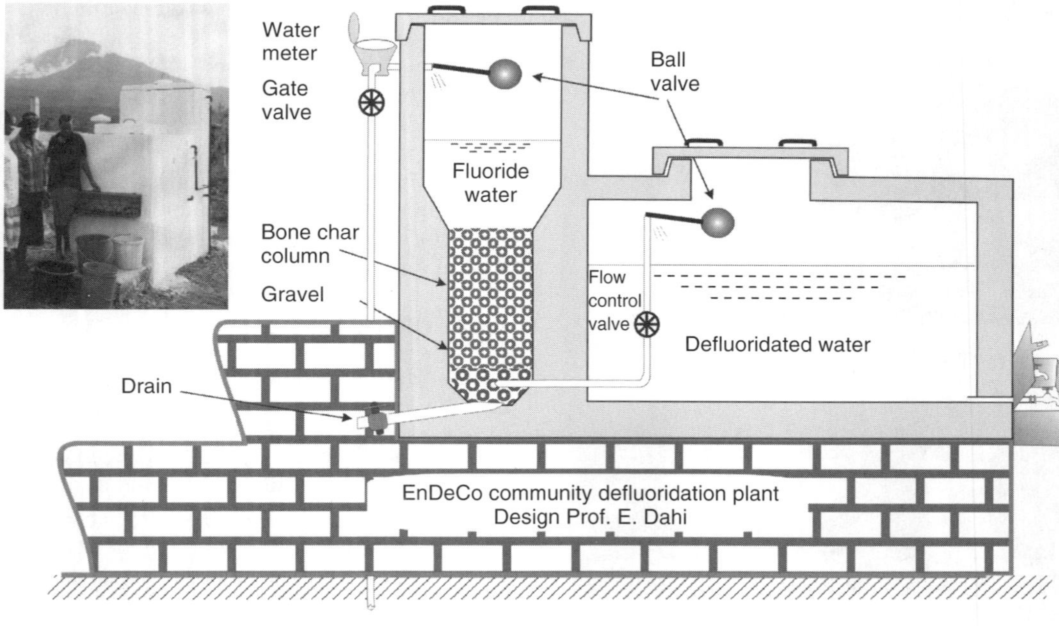

Figure 5. An example of a community defluoridator that has been operating continuously since 1994.

Table 1. Some Natural Media Capable to Sorb Fluoride

Magnesite	Apophyllite	Natrolite
Stilbite	Clinoptilolite	Gibbsite
Goethite	Kaolinite	Halloysite
Bentonite	Verimiculite	Zeolite(s)
Serpentine	Alkalkine soil	Aiken soil
Acidic clay	Kaolinitic clay	China clay
Fuller's earth	Diatomaceous earth	Ando soil

to the community. Alternatively, the same setups can be used in the activated alumina process, which is often preferred, e.g., in strictly vegetarian societies that consider the use of (cow) bone char as unethical.

2. A coprecipitation process in batch containers, where the precipitating chemicals are totally mixed with the raw water. This process requires filling of water and chemicals, mixing and settling, and subsequent withdrawal of the treated water and the produced sludge. In the so-called Nalgonda technique, a high dosage of alum and lime are used in a coagulation/sedimentation process, where the precipitation of aluminium hydroxide instantly binds a part of the water fluoride. A domestic approach for the Nalgonda process is given in Fig. 6.
3. Contact precipitation is a process where the water is mixed with the precipitating chemicals, calcium and phosphate ions, flowing into a catalytic filter column. So far, this process is only known to operate in a fluoride-saturated bone char, being the catalyst, dosed with calcium chloride and sodium dihydrogen phosphate. Contact precipitation is considered to be the process of choice in the future, when the local availability of bone/bone char reaches its limit.

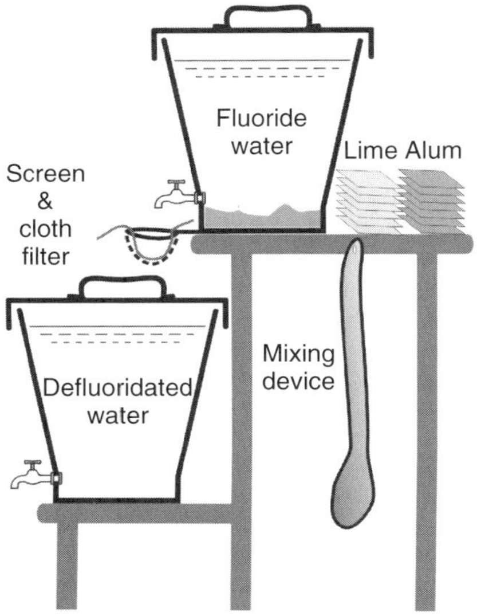

Figure 6. The Nalgonda process as optimized for use in the Rift Valley.

SORPTION DESIGN

The sorption process may be designed on the basis of daily water demand, capacity of the medium, the raw water contamination level, proposed length of operation period, and the bulk density of the medium. The required column treatment capacity and the amount of medium required in a filter column are calculated (Eqs. 1 and 2). From the bulk volume of the medium (Eq. 3), the dimensions of the column are derived (Eqs. 4 and 5):

$$\kappa = D \cdot P \tag{1}$$

$$M = \frac{D \cdot P \cdot (F_r - F_t)}{1000 \cdot C} \tag{2}$$

$$V_m = \frac{M}{\sigma} \tag{3}$$

$$d = \sqrt{\frac{4000 \cdot V_m}{\pi \cdot h}} \tag{4}$$

$$H = 1.2 \cdot h \tag{5}$$

where D is daily water demand, L/day.
C is capacity of the medium, g/kg.
F_r is raw water contamination level, mgF/L.
F_t is treated water concentration, (0.5) mgF/L.
P is proposed length of operation period, days
σ is bulk density of the medium, kg/L.
κ is required column capacity, L.
M is amount of medium required in a filter column, kg.
V_m is bulk volume of the medium, L.
d is inner diameter of the filter column, cm.
h is height of medium in the column, cm.
H is height of the column, cm.

Unfortunately, the capacity of the medium, and its purity, is subject to batch variation. Further, it depends on the fluoride concentration in the raw water. In practice, however, the capacities of 4, 1, and 0.03 g/kg are for good qualities of bone char, activated alumina, and activated clay, respectively.

NALGONDA PROCESS DESIGN

In the Nalgonda and similar processes, the container used has to have a volume well over the daily usage, which will allow for comfortable mixing in the container as well as include provision for the sludge water loss. A plastic bucket of 20 L is often sufficient for the daily use of a family. About 18 L can be treated at a time, out of which about 16 can be used.

The Nalgonda process design is complicated and must be confirmed empirically, which is mainly because of the coprecipitation that lacks stoichiometry, the great variation in media quality and solubility, and the efficiency being dependent on the raw water quality, in particular the fluoride concentration, pH, and the alkalinity. The best possible estimation is based on the Freundlich equation:

$$m = \frac{V_b \cdot (F_i - F_o)}{\alpha \cdot F_o^{1/\beta}} \tag{6}$$

where m is amount of alum required for a daily treatment of a batch, in g.
F_r is fluoride concentration in the raw water, in mg/L.
F_t is residual fluoride concentration in the treated water, in mg/L.
V_b is volume of water to be treated in batch, L.
α is sorption capacity constant, $L^{(1-1/\beta)}$ $mg^{2/\beta}g^{-1}$.
β is sorption intensity constant.

Any resulting pH between 6.2 and 7.6 is adequate. For pH $= 6.7$ and required residual fluoride between 1 and 1.5 mg/L, $\alpha = 6$ and $\beta = 1.33$. The amount of lime required may be 20–50% of the alum dosage.

Alum and lime are added simultaneously to the raw water bucket where it is dissolved/suspended by stirring. The operator should stir fast while counting to 60, i.e., about 1 minute, and then slowly while counting to 300, i.e., about 5 minutes. The mixture is left for settling for about one hour. The treated water is then tapped through the cloth into the treated water bucket from where it is stored for daily drinking and cooking. This separation of the water from the sludge is essential in order to avoid escape of the toxic aluminium to the treated water and in order to avoid the detachment of removed fluoride from the aluminium hydroxide flocs during storage.

TESTING OR METERING

The coprecipitation technique, although arbitrary in nature, is reliable in the sense that the same removal is obtained if the procedure followed exactly.

On the other hand, the process results in poor removal efficiency, and it demands burdensome daily preparation.

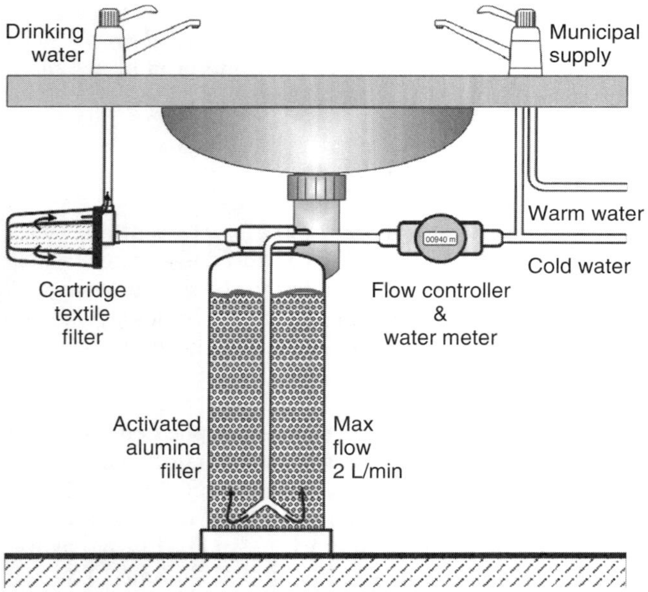

Figure 7. A safe and user-friendly setup of defluoridation based on sorption, in case of activated alumin. The system, containing 8 kg AA, is commercially available in Scandinavia and declared to be able to treat 2000 L of water containing 5 mg/L.

The sorption process is less burdensome. It is also able to provide high removal efficiencies, even at high fluoride levels in the raw water. However, the operator has to watch for the break through the column at saturation. As fluoride is oganoleptically neutral, this cannot be done without either regular testing of the treated water fluoride or checking the accumulated water flow.

Testing the fluoride concentration in the treated water was impossible without expensive laboratory equipment. Now, simple, reliable, and cheap test kits are available.

Figure 7 shows a user-friendly setup for household defluoridation. The secondary cartridge filter ensures that no toxic aluminium can escape with the treated water, and the water meter ensures that accumulated water flow can be compared with declared capacity.

READING LIST

E. Dahi et al. (Eds.). (1995, 1997, 2000, 2004). *Proceedings of 1st–4th International Workshops on Fluorosis Prevention and Defluoridation of Water*. International Society of Fluoride Research.

DEUTERIUM

Gholam A. Kazemi
Shahrood University of Technology
Shahrood, Iran

INTRODUCTION

The nucleus, the chief ingredient of the atom, consists of two main types of particles: protons (positive charge) and neutrons (neutral). The number of neutrons is either equal or slightly greater than the number of protons. Some elements have only one type of atom, but most have several types. Isotopes of an element have the same number of protons in the atomic nucleus but a different number of neutrons; this results in the same atomic number but different atomic weight. The average atomic weights of the elements have been determined from two sets of data: the atomic weight of each isotope and the relative abundances of the isotopes (1). Isotopes with a higher number of neutrons are commonly referred to as heavy isotopes of that element.

HYDROGEN ISOTOPES

The hydrogen atom is believed to have been created during the "Big Bang" by proton–electron combinations (2). Together with an oxygen atom, it forms the water molecule and has three isotopes: ^1H (common hydrogen), ^2H (deuterium, D), and ^3H (tritium, T). Deuterium, discovered by Urey et al. (3) in 1932, is much heavier than ^1H. Common hydrogen has one proton, deuterium has one proton and one neutron, and tritium has one proton and two neutrons. Deuterium (also referred to as the heavy stable isotope of hydrogen) and common hydrogen are stable, but tritium is radioactive. In seawater, the relative abundances of hydrogen isotopes are H (99.984%), D (0.016%), and T (5×10^{-5}%) (1). In the hydrosphere, there is only one deuterium atom per 6700 H atoms (4). Note that the arena of this article is limited to deuterium as it is a part of the water molecule. Hence, deuterium that may exist in other materials such as methane will not be dealt with here.

ANALYZING WATER SAMPLES FOR DEUTERIUM AND REPORTING THE RESULTS

To analyze water samples for deuterium, first the sample is reduced to H_2. There are a variety of techniques to achieve this; most have been reviewed by Vaughn et al. (5). A common method is the use of zinc in quartz ampoules at 700 °C. The analysis is then carried out using a mass spectrometer such as a Finnigan MAT 250 triple collector mass spectrometer (6). The precision attainable in many laboratories in sample preparation and analysis of water samples for deuterium is 1‰ (7). The subsidized cost of analysis for deuterium as of 1999 is about A$35 per sample (approximately US$25) (8).

An automated system for high precision deuterium analysis on large numbers of water samples was introduced by Vaughn et al. (5) at the Institute of Arctic and Alpine Research at the University of Colorado. This system does not require sample preparation and is capable of extended unattended operation, an attractive option when analyzing thousands of samples during a weekend.

The results of deuterium analysis are expressed in delta (δ) units per mil (parts per thousand or ‰) differences relative to an arbitrary standard known as standard mean ocean water (SMOW):

$$\delta‰ = [(R - R_{\text{standard}})/R_{\text{standard}})] \times 1000$$

where R is the isotopic ratio ($^2H/^1H$) of the sample and R_{standard} is the isotopic ratio ($^2H/^1H$) of the standard (9). Negative δD values represent water that has less than SMOW δD (i.e., they are isotopically depleted), and positive values represent water that has more than SMOW δD (i.e., they are isotopically enriched). In this system, therefore, δD of SMOW is zero. However, a more recent unit to express these isotopes has been proposed (10). In this system, the hydrogen isotopic ratios are reported relative to VSMOW (Vienna standard mean ocean water) on a normalized scale, so that the (δ^2H) of SLAP (standard light Antarctic precipitation) relative to VSMOW is −428‰ (10). Therefore, the δ^2H of VSMOW is 0‰.

DEUTERIUM IN THE GLOBAL WATER CYCLE

At the first onset of the water cycle, the deuterium content of the ocean water is increased when water evaporates to form clouds (see ISOTOPE FRACTIONATION article for more information). Thus, water vapor in the atmosphere (in the form of clouds or in any other form) has less deuterium compared to ocean water. The evaporated water (clouds) is then condensed into precipitation to form rivers, surface waters, lakes, and groundwater, all of which is isotopically lighter than ocean water, unless subjected to considerable evaporation. The final destination of surface and ground water that has less deuterium than ocean water is the ocean that was originally enriched in deuterium due to evaporation. Therefore, the deuterium content of ocean water remains constant over time. [However, the early ocean was plausibly deuterium-depleted in comparison with the present-day ocean, see Lecuyer et al. (11)]. Therefore, it is evident that the deuterium content of water in any part of the hydrologic cycle is influenced by a variety of parameters; the most important is evaporation. The factors that influence the deuterium content of atmospheric precipitation are described here (7):

Altitude Effect. Generally the higher the altitude, the more depleted the precipitation (−0.15 to −0.50‰ per 100-m elevation);

Latitude Effect. The deuterium content decreases as latitude increases.

Continental Effect. The deuterium content decreases inland from the coast.

Seasonal Variation. The deuterium content of winter rainfall is less than that of summer rains.

Amount Effect. The greater the rainfall, the more depleted the deuterium.

Apparent Temperature Effect. This effect is similar to the effect of latitude because the higher the latitude, the lower the temperature. Therefore, it is not easy to find the cause of the deuterium difference, whether deuterium depletion is due to the decrease in temperature or due to the increase in latitude. Because of this complexity, Mazor (1) ignored the latitude effect and did not mention it as a factor that influences deuterium content.

Having considered the many factors above, it is now easy to realize that atmospheric precipitation in each part of the world is unique in its deuterium content. Harmon Craig (12) studied the deuterium and ^{18}O content of a large number of precipitation samples worldwide. He drew a line to express the relationship between these two isotopes in a global perspective which is now known as the Craig Line, the World Meteoric Water Line (WMWL), or sometimes, Global Meteoric Water Line (GMWL). The Craig line equation is $\delta D = 8\delta^{18}O + 10$. In this equation, 8 (coefficient) is the slope of the line, and 10 (the intercept) is called the deuterium excess ($d = \delta D - 8\delta^{18}O$). The lower the slope, the higher the effect of evaporation on the water samples studied. A typical evaporation line has a slope of five. The higher the deuterium excess, the higher the effect of evaporative enrichment on the water samples in question. Generally speaking, samples plotted to the right of the Craig line are isotopically enriched, and those plotted to the left are depleted. However, it should be stressed that each region has its own meteoric water line, a local meteoric water line (LMWL), that should be prepared by analyzing the deuterium and ^{18}O content of a number of local atmospheric precipitation samples. The equation of a number of LMWLs for various parts of the world is given by Mazor (1).

APPLICATIONS OF DEUTERIUM AS A TRACER—ISOTOPIC HYDROLOGY

In almost all water related studies, both deuterium and ^{18}O are measured simultaneously. The applications of stable isotopes of oxygen ($\delta^{18}O$) and hydrogen (^{2}H) in water studies, especially for analysis of flow path and fractionation processes dates back to the early 1950s (9). Undoubtedly, measurement of deuterium in water, whether it is river, lake, ground, rain, or porewater is the fundamental of an important field of knowledge "isotopic hydrology," a technique or science whose many applications have been summarized by Payne (13). The textbook by Mazor (1) and the recent works by Clarke and Fritz (14) and Cook and Herczeg (15) deal extensively with the topic.

To interpret the deuterium data, the most common procedure is to plot δD and $\delta^{18}O$ on a diagram on which the Craig line and, if possible, LMWL are also illustrated. It is then possible to interpret the data, for example drawing an evaporation line, calculating the slope of the line, and determining the deuterium excess. As an example, the results of an isotopic hydrology study by Kazemi (8) on the Buckinbah Creek Watershed in Eastern Australia are presented in Table 1 and Fig. 1. As can be seen from Fig. 1, only two samples show the effect of evaporative enrichment. From Fig. 1, it can also be inferred that the deep groundwater, compared to shallow groundwater, has been subjected generally to less enrichment by evaporation. Note that the approximate positions of summer and winter precipitation samples are also shown in Fig. 1.

The deuterium content of water samples is measured to (1) find the cause of groundwater salinization (16,17); (2) estimate groundwater inflow into wetlands (18,19); (3) separate the streamflow hydrograph into baseflow and an event water component (20,21); (4) calculate surface evaporation via a stable isotope evaporation model (22); (5) study the mechanism, timing, and area of recharge to groundwater (23–26); (6) study the pattern of groundwater circulation and identify the hydrogeologic characteristics of karst groundwater flow (27); (7) identify the origin of groundwater, whether groundwater was recharged in modern time or in the Holocene (28); (8) study the infiltration processes in clayey soils (29); (9) determine groundwater discharge from forests (30); and (10) be used as a tracer in groundwater artificial recharge schemes (31,32).

Case Studies

A large number of case studies are regularly published through the International Atomic Energy Agency

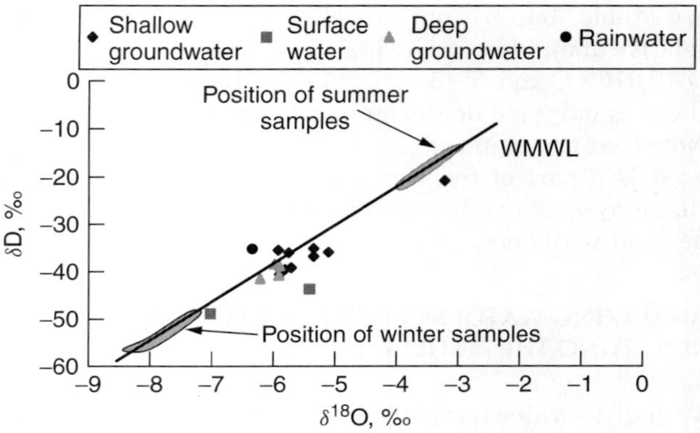

Figure 1. Deuterium versus oxygen-18 plot of water samples from the Buckinbah Creek Watershed, NSW, Australia (8). The Craig (World Meteoric Water Line) is also illustrated.

(IAEA) and International Association of Hydrological Sciences publications, IAHS (33) and IAEA (34). Examples of the deuterium content of groundwater from the Amman–Zarkan Basin in Jordan are −32 to −26‰ and from the Aladag karstic aquifer in Turkey −100.1 to −58.4‰ (33).

BIBLIOGRAPHY

1. Mazor, E. (1991). *Applied Chemical and Isotopic Groundwater Hydrology*. Open University Press, Milton Keynes.
2. Ottonello, G. (1997). *Principles of Geochemistry*. Columbia University Press, New York.
3. Urey, H.C., Brickwedde, F.G., and Murphy, G.M. (1932). An isotope of hydrogen of mass 2 and its concentration (abstract). *Phys. Rev.* **39**: 864.
4. Drever, J.I. (1997). *The Geochemistry of Natural Waters: Surface and Groundwater Environments*. Prentice-Hall, Englewood Cliffs, NJ.
5. Vaughn, B.H. et al. (1998). An automated system for hydrogen isotope analysis of water. *Chem. Geol.* **152**: 309–319.
6. Jorgensen, N.O. and Heinemeier, J. (2003). Isotope Studies of Saline Groundwater in a Sandstone Aquifer on Bornholm, Denmark. *Proc. Int. Conf. Groundwater Fractured Rocks*, Prague, IHP-VI, series on groundwater no. 7, pp. 205–206.
7. Coplen, T.B. (1993). Uses of environmental isotopes. In: *Regional Ground-Water Quality*. W.M. Alley (Ed.). ITP, New York, pp. 255–293.
8. Kazemi, G.A. (1999). *Groundwater Factors in the Management of Dryland and Stream Salinity in the Upper Macquarie Valley, New South Wales, Australia*. Unpublished Ph.D. Thesis, University of Technology, Sydney, Australia.

Table 1. The Results of Stable Isotopes Analysis of Surface, Ground, and Rainwater Samples in the Buckinbah Creek Watershed, NSW, Australia[a]

Site[b]	Sgw	Sgw	Sgw	Sgw	Sgw	Sgw	Sgw	Sgw	Sgw	Sgw	Sw	Sw	Dgw	Dgw	Dgw	Dgw	Rw
$\delta^{18}O$‰	−5.8	−5.7	−5.8	−5.4	−5.9	−5.4	−5.9	−5.1	−5.9	−3.2	−7	−5.4	−5.9	−6	−5.9	−6.21	−6.34
δD ‰	−40	−39	−36	−37		−35	−36	−36	−41	−21	−49	−44	−41	−38	−38.7	−41.5	−35.2

[a]Reference 7.
[b]Sgw: Shallow groundwater; Dgw: Deep groundwater; Sw: Surface water; Rw: Rainwater.

9. Freeze, R.A. and Cherry, J.A. (1979). *Groundwater*. Prentice-Hall, Englewood Cliffs, NJ.
10. Hornberger, G.M. (1995). New manuscript guidelines for the reporting of stable hydrogen, carbon, and oxygen isotopic ratio data. *Water Resour. Res.* **31**(12): 2895.
11. Lecuyer, C., Gillet, P., and Robert, F. (1998). The hydrogen isotope composition of seawater and the global water cycle. *Chem. Geol.* **145**: 249–261.
12. Craig, H. (1961). Isotopic variations in meteoric water. *Science* **133**: 1702–1703.
13. Payne, B.R. (1988). The status of isotope hydrology today. *J Hydrol.* **100**: 207–237
14. Clarke, I. and Fritz, P. (1997). *Environmental Isotopes in Hydrogeology*. Lewis Publishers, Boca Raton, FL.
15. Cook, P.G. and Herczeg, A.L. (Eds.). (1999). *Environmental Tracers in Subsurface Hydrology*. Kluwer, Dordrecht, The Netherlands.
16. Stimson, J. et al. (1993). Causes of groundwater salinisation in a low lying area of Cochabamba Valley, Bolivia. In: *Applications of Isotope Techniques in Studying Past and Current Environmental Changes in the Hydrosphere and Atmosphere*. IAEA publication, Vienna, pp. 185–199.
17. Kazemi, G.A. and Milne-Home, W.A. (1999). The application of strontium, oxygen-18 and deuterium isotopes in the estimation of the solute budget and hydrology of a small watershed in Eastern Australia. *Eos Trans. AGU* **80**(17): Spring Meet. Suppl. S139.
18. Hunt, R.J., Krabbenhoft, D.P., and Anderson, M.P. (1996). Groundwater inflow measurements in wetland systems. *Water Resour. Res.* **32**(3): 495–507.
19. Krabbenhoft, D.P., Bowser, C.J., Kendall, C., and Gat, J.R. (1994). Use of oxygen-18 and deuterium to assess the hydrology of groundwater-lake systems. In: *Environmental Chemistry of Lake and Reservoirs*. American Chemical Society, Washington, DC, pp. 66–90.
20. Hooper, R.P. and Shoemaker, C.A. (1986) A comparison of chemical and isotopic hydrograph separation. *Water Resour. Res.* **22**(10): 1444–1454.
21. Gunyakti, A., Altinbilek, H.D., and Qasem, H. (1995). Study of small basin hydrology in Anatolia using environmental isotope analysis. *Hydrolog. Sci. J.* **40**(1): 43–53.
22. Meyers, J.B., Swart, P.K., and Meyers, J.L. (1993). Geochemical evidence for groundwater behavior in an unconfined aquifer, South Florida. *J. Hydrol.* **148**: 249–272.
23. Sami, K. (1992). Recharge mechanisms and geochemical processes in a semi-arid sedimentary basin, Eastern Cape, South Africa. *J. Hydrol.* **139**: 27–48.
24. Leaney, F.W. and Herczeg, A.L. (1995). Regional recharge to a karst aquifer estimated from chemical and isotopic composition of diffuse and localized recharge, South Australia. *J. Hydrol.* **164**: 363–387.
25. Feurdean, L., Feurdean, V., Zamfirescu, F., and Sarbu, S. (2003). Evidence and constraints revealed by environmental stable isotopes in the chemoautotrophic system of the Movile Cave, Romania. *Proc. Int. Conf. Groundwater Fractured Rocks*, Prague, IHP-VI, series on groundwater no. 7, pp. 201–202.
26. Feurdean, L., Feurdean, V., and Oraseanu, I. (2001). The use of deuterium as natural tracer to point out groundwater flow from karst area. Case study. In: *New Approaches Characterizing Groundwater Flow*. K.P. Seiler and S. Wohnlich (Eds.). Swets and Zeitlinger, Lisse, pp. 59–63.
27. Bayari, C.S. and Gunay, G. (1995). Combined use of environmental isotopic and hydrochemical data in differentiation of groundwater flow patterns through the Aladag karstic aquifer, Turkey. In: *Application of Tracers in Arid-Zone Hydrology*. IAHS publication no. 232, pp. 99–117.
28. Guendouz, A. et al. (2003). Hydrogeochemical and isotopic evolution of water in the Complex Terminal aquifer in the Algerian Sahara. *Hydrogeol. J.* **11**: 483–495.
29. Mathieu, R. and Bariac, T. (1996). An isotopic study (^{2}H and ^{18}O) of water movements in clayey soils under a semiarid climate. *Water Resour. Res.* **32**(4): 779–789.
30. Thorburn, P.J., Hatton, T.J., and Walker, G.R. (1993). Combining measurements of transpiration and stable isotopes of water to determine groundwater discharge from forests. *J. Hydrol.* **150**: 563–581.
31. Jorgensen, N.O and Helleberg, B.B. (2002). Stable isotopes (^{2}H and ^{18}O) and chloride as environmental tracers in a study of artificial recharge in Denmark. In: *Management of Aquifer Recharge for Sustainability*. P. Dillon (Ed.). Balkema, Lisse, pp. 245–250.
32. Le Gal La Salle, C. et al. (2002). Isotope contribution to geochemical investigations for aquifer storage and recovery. In: *Management of Aquifer Recharge for Sustainability*. P. Dillon (Ed.). Balkema, Lisse, pp. 265–268.
33. IAHS. (1995). *Application of Tracers in Arid Zone Hydrology*. IAHS Pub No. 232, p. 542.
34. IAEA. (1979). *Isotopes in Lake Studies*. International Atomic Energy Agency, Vienna, Austria.

DISTILLED WATER

Aldo Conti
Frascati (RM), Italy

Water is never pure in nature. It normally contains some salts and minerals in solution and even bacteria and pollutants. Most of these substances, salts and minerals, in particular come from rocks and soils with which the water has been in contact. All these impurities make water useless for some purposes. It must be cleaned of pollutants and bacteria for human consumption, for instance. But for some other uses, even salts and minerals are dangerous. At home, in our vapor iron, water should be free of calcium carbonate that forms a layer on the heating resistor and eventually breaks it. The easiest method of purification is distillation, during which water is heated until it turns into vapor. The vapor is then collected on a cold surface, where it condenses. Because salts and other impurities do not vaporize, they normally remain in the boiling solution. Distilled water is never perfectly pure, because there are substances with nearly the same boiling point as that of water and some liquid droplets can be carried by the steam. Nevertheless, distilled water can be 99.9% pure and, if needed, can be distilled a second time. Distillation is very efficient in removing heavy metals, radionuclides, and solid particles, and it kills bacteria and viruses.

From a chemical point of view, even the purest distilled water is always a solution containing a very small amount of hydroxide (OH^-) and hydronium (H_3O^+) ions, which happens because water, like many other solvents, reacts with itself. At room temperature, one can expect one molecule in every ten million to dissociate and donate a hydrogen atom to another molecule.

Distilled water does not remain pure for long. If left exposed to air, it absorbs carbon dioxide very efficiently and reacts with it producing carbonic acid (H_2CO_3), which is why distilled water often becomes acidic. If one leaves distilled water exposed to air for 24 hours, a pH between 5.5 and 6.0 can be expected. To neutralize this acidity, it is sufficient to boil the water for 5 to 10 minutes, so that it releases the dissolved carbon dioxide, bringing the pH very close to 7.

The acidity induced by carbon dioxide is enough to make distilled water unhealthy for human consumption. Moreover, water contains electrolytes (sodium, potassium, chloride) and many other elements, such as magnesium, that are essential for our body. Most, if not all, of these substances can be found in food, too, but nevertheless, drinking large amounts of distilled water is not advisable.

ELECTRICITY AS A FLUID

David P. Stern
Mauricio Peredo
Goddard Space Flight Center, NASA

The preceding sections did not follow the conventional order:

First **electric currents** were brought up, in connection with Oersted's work.

Then **electric charges,** in connection with electrons.

That may be OK, because most of us are **familiar** with currents and charges. **Currents** run electric lights, radios, TVs, clocks and appliances in the home, and **"static" electric charge** causes papers and clothes to cling together in dry weather. An electric current is essentially the continuous flow of electric charge. **This section** goes a little further, to the concepts of **voltage** and of the **electric field**.

OHM'S LAW

It is easier to understand electricity if we regard electric charge as a sort of **fluid,** like water, as scientists did for the first 200 years. Yes, it does consist of individual electrons, but those are so small that any large charge behaves like a continuous fluid. In the same way sand pours like a fluid, and water in a glass is usually regarded as a fluid, even though it consists of individual molecules.

Using a pump we can push water through a pipe around a closed circuit (top drawing). The rate at which it flows past any point in the pipe–measured in **gallons** (or liters) **per second**–depends on the **pressure** produced by the

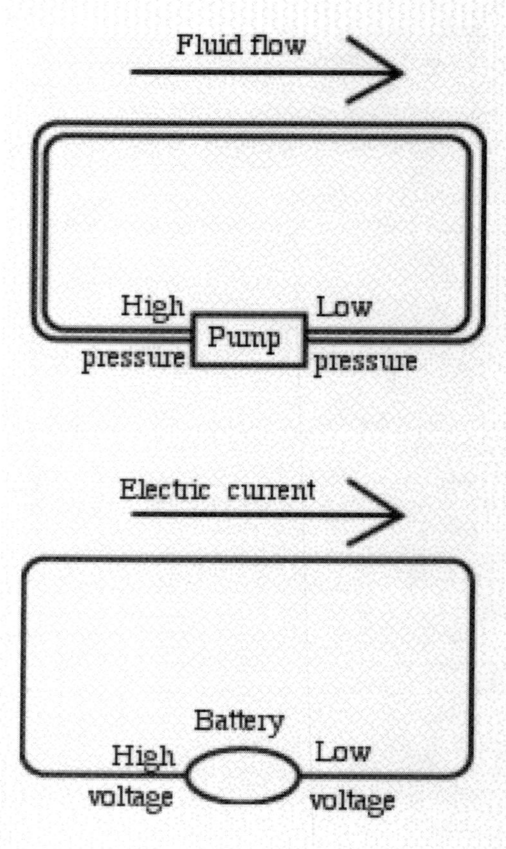

pump (measured in pounds per square inch, or kilograms per square centimeter). More accurately, it depends on the **pressure difference** between the entrance to the pipe (left) and the exit from the pipe (right).

The greater the pressure difference, the greater the flow. In addition, given a certain pressure difference, a **fatter** pipe will carry more water, and a **longer** one will resist the flow and carry less.

ALL THIS MIRRORS EXACTLY THE BEHAVIOR OF THE ELECTRIC FLUID

No pipes here, though: electricity in our homes and appliances usually flows in metal wires, most often of copper. Electrons in a metal can jump from atom to atom, and that way carry negative charge around the circuit.

Like a fluid, they are driven by a kind of electric pressure, known as **voltage,** because it is measured in units known as **volts,** named after the Italian scientist Alessandro Volta. An electric battery produces (by a chemical process) a **voltage difference V** between its two ends, and therefore acts like a pump (bottom drawing).

The **electric current I** flows from high voltage to low voltage and is measured in units known as **Amperes,** named for André-Marie Ampere whom we met in section #2. And as with water, we expect that if we increase the driving voltage V, the driven current I will also increase. In fact, the two are pretty much proportional: double the voltage, and you get double the current. That relation is known as **Ohm's Law,** after Georg Ohm who first formulated it.

Ohm's law (with some extra details concerning the length and thickness of the wire) is usually among the first things taught in electricity classes, and many students therefore view it as one of the fundamental laws of electricity. **It isn't.** It holds quite well for metal wires, but as will be shown in section #7a, it fails badly in fluorescent tubes–while in space currents exist which flow without any voltage driving them (section #10a).

ELECTRIC FIELDS

One small caution here. By long tradition, the direction of the electric current is defined as the direction in which **positive** charges move. We may blame Ben Franklin for deciding–by pure guess–what kind of electricity is called "positive" and which "negative." A century later it was found that most electric currents were carried by negative electrons, which move in the opposite direction.

One may therefore argue that the flow direction which should be assigned to electric fluid is really the **opposite** of what we say it is. But it's much too late to change the old convention.

In our homes, electric currents and effects of electricity are usually channeled along insulated wires. **In 3-dimensional space,** on the other hand, electric phenomena tend to spread out. If the way electric current flows in the home resembles water flow in pipes, then in space the flow is often like ocean currents or air motion in the atmosphere, spread out in 2 and even 3 dimensions.

In a wire, voltage depends only on one dimension–on the distance along the wire. **In space,** every point can have its own voltage. Currents that flow may depend on such voltages, but don't look to Ohm's law for guidance, because the flow of current is primarily dictated by magnetic fields, which makes a difference.

The **3-dimensional voltage distribution** is often called the **electric field.**

Compare:

In a **magnetic field** the **direction of the field** is that of the force. If **isolated N magnetic poles** existed, they too could do so, by moving in the direction of the field, while S poles would move in the opposite direction. The poles at the ends of a compass needle move this way, and thus line up the needle in the direction of the field.

Similarly

An **electric field** is a region where electric forces can be felt by charged objects, and the **direction of the field** is the one in which **positive** charges would move. A positive ion moves in the direction of the field, a negative electron moves in the opposite direction.

If **both** magnetic and electric fields are present, the motion of ions and electrons gets complicated. **That** however is left for a later section.

ANALYSIS OF AQUEOUS SOLUTIONS USING ELECTROSPRAY IONIZATION MASS SPECTROMETRY (ESI MS)

William Henderson
University of Waikato
Hamilton, New Zealand

INTRODUCTION

Electrospray ionization mass spectrometry was developed primarily by Fenn and co-workers in the late 1980s for the analysis of macromolecules such as polyethers and proteins (1). Since then, the technique has undergone an exponential growth in applications, in many areas of chemical and biochemical sciences. A considerable number of textbooks (2,3) and major reviews (4) describing the experimental basis of the technique (5–8) and applications have appeared (8–10) and ESI MS has been widely used for the analysis of aqueous solutions.

THE ESI TECHNIQUE

The key features of the technique are:

1. The ability to analyze materials directly from solution. Almost any solvent, including water, can be used, depending on the solubility of the analyte under study.
2. Ionization is very gentle; this is of great importance for thermally labile analytes such as proteins or weakly bound metal complexes.
3. The technique provides information on mixtures of compounds, which may be in equilibrium in solution, and is applicable to paramagnetic materials (such as many transition metal and lanthanide ions and complexes thereof).
4. The technique is well suited for the analysis of both positive- and negative-ion species.
5. As with all mass spectrometry techniques, miniscule amounts of sample are required, and detection limits are usually very low.

In electrospray ionization, a solution of the compound is nebulized, at atmospheric pressure, from a metal capillary held at a high potential. Nebulization is assisted by an inert gas, typically nitrogen. This produces a spray of charged liquid droplets that, through processes of droplet fission, solvent evaporation, and ion evaporation, produce gas phase ions. The ions are then detected using one of a number of mass analyzers (such as quadrupole or ion trap). Fragmentation of the initially formed ions can be achieved by the technique of collision-induced dissociation (CID), whereby ions undergo collisions with gas and/or solvent molecules. The application of a *cone voltage*—a potential difference that accelerates the ions—is one such method that can effect fragmentation of the original ions.

For analyte species that are already charged, the electrospray process simply transfers existing solution ions into the gas phase. For analytes that are uncharged, ionization must occur for the material to be detected by the mass spectrometer; a number of processes may operate, depending on the actual system, but can include protonation, deprotonation, metallation (e.g., attachment of adventitious alkali metal cations present in aqueous solutions stored in glass), and oxidation.

APPLICATIONS OF ESI MS TO THE ANALYSIS OF AQUEOUS SOLUTIONS

Aqueous solutions are highly amenable to analysis by ESI MS. In many cases, an organic cosolvent is added, to reduce the surface tension of the resulting solvent mixture, which aids in spray formation (7). The surface activity of an analyte ion can affect its ESI response; a high surface activity will result in a high response in the ESI mass spectrum. As an example, an equimolar solution of Cs^+ and $C_{10}H_{21}NMe_3{}^+$ in 1:1 methanol–water gives a substantially more intense signal for the high surface activity ammonium cation (11).

ESI MS can be used for both qualitative and quantitative analysis. As an example, the negative-ion electrospray mass spectrum of an aqueous solution of $NaBrO_3$ (containing 5% methanol) is shown in Fig. 1. Under gentle ionization conditions (cone voltage 30 V), the BrO_3^- ion is exclusively observed at *m/z* (mass-to-charge) 127, but at a high cone voltage (120 V), fragmentation occurs, giving BrO_2^- (*m/z* 111), BrO^- (*m/z* 95), and Br^- (*m/z* 79). The distinctive two-line pattern for each ion arises due to the isotopic signature of bromine (^{79}Br 100%, ^{81}Br 97.3% relative abundances) and *m/z* values are given for the ^{79}Br isotopomer. Electrospray mass spectrometry has been used to quantify levels of oxyhalide anions in aqueous solutions; for example, bromate (12) and other oxyhalide anions (13) can be measured at sub-ppb levels.

For a metal salt M^+ X^- dissolved in water or an organic solvent–water mixture, the ions that are typically observed over a wide mass range are ion clusters of the type $[(MX)_n + M]^+$ in positive-ion mode and $[(MX)_n + X]^-$ in negative-ion mode; salts containing doubly charged ions behave in a similar manner (14).

The tendency for ions to retain a hydration (or more generally a solvation) sphere in the gas phase can easily be investigated using ESI MS. As an example, ESI MS of the sodium salts of the doubly charged sulfur oxyanions SO_4^{2-} (sulfate), $S_2O_6^{2-}$ (dithionate), and $S_2O_8{}^{2-}$ (peroxydisulfate) reveals differences in the gas phase stabilities of these ions. The unsolvated $SO_4{}^{2-}$ ion cannot be detected, and instead a series of hydrated ions $[SO_4 + nH_2O]^{2-}$ ions are observed with $n = 5–16$. Ions with $n < 3$ are unstable and undergo charge separation to give $HSO_4{}^-$ and OH^-. However, for $S_2O_6{}^{2-}$ and $S_2O_8{}^{2-}$, the unsolvated ions can be observed in the gas phase, because of the greater charge delocalization, and hence lower charge density of these ions, which results in higher gas phase stabilities (15).

Metal cations also have a strong tendency to retain a hydration sphere, with divalent cations being more

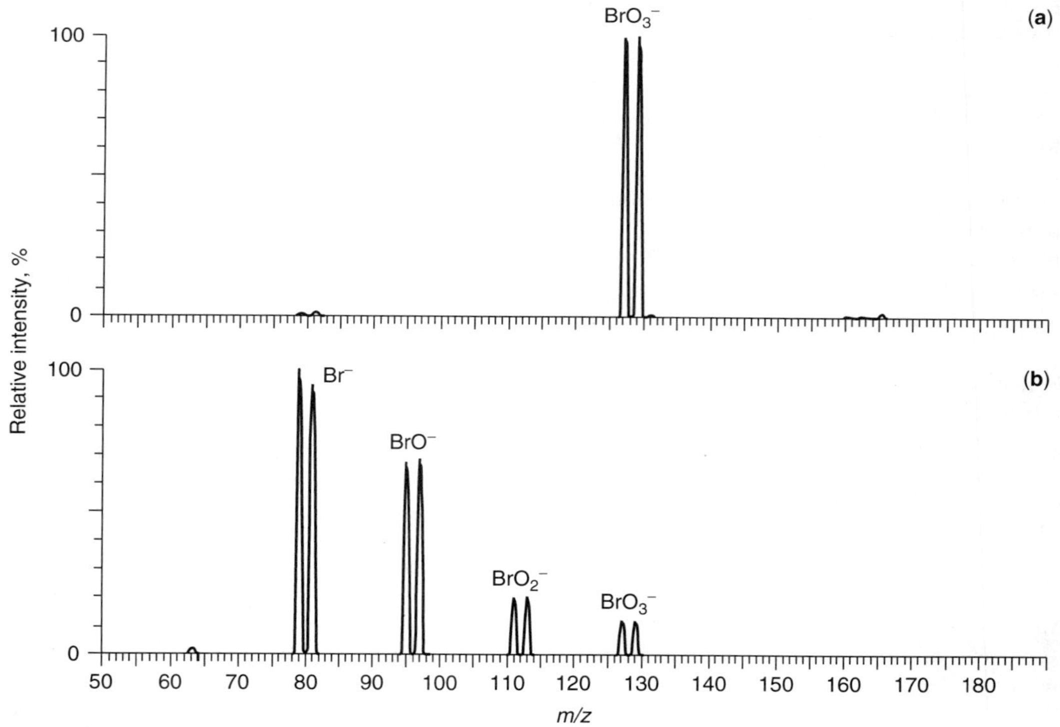

Figure 1. Negative-ion electrospray mass spectra of an aqueous solution of potassium bromate, $KBrO_3$, containing 5% of methanol to aid in spray stabilization. Spectrum (**a**) is recorded at a cone voltage of 30 V and shows exclusively the $BrO_3{}^-$ ion, while spectrum (**b**) is recorded at 120 V and additionally shows the series of fragment ions $BrO_2{}^-$, BrO^-, and Br^-.

strongly solvated than monocations, as expected. For example, $[M(H_2O)_n]^{2+}$ ions have been generated for a range of divalent metal ions, and ion–water binding energies were determined (16). For trivalent cations, and to a lesser extent for divalent cations, charge reduction will occur in the gas phase if the hydration sphere is unable to stabilize the charge on the metal cation. For example, for a solvated trivalent cation

$$[M(H_2O)_n]^{3+} \rightarrow [M(OH)(H_2O)_{n-2}]^{2+} + H_3O^+$$

In the case of copper(II) systems, reduction to copper(I) is often seen. For other metals, under forcing CID conditions, further charge reduction to low valent metal ion species can occur (17): for example, the lanthanide ions give $[Ln(OH)(H_2O)_n]^{2+}$, which give ions LnO^+, $LnOH^{2+}$, and ultimately Ln^+ (18). This experimental regime is often referred to as *bare metal* or *elemental* mode, and mass analysis of the resulting ions can be used to provide isotopic information on the elemental ion concerned.

In addition to the applications described above, the list below gives a selection of some applications of ESI MS in the analysis of aqueous (or mixed aqueous–organic) solutions:

- Biochemical materials of many types (19), including proteins (20)
- Polyoxometallate anions (21)
- Aqueous polysilicate anions (22)
- Polyhalide (23) and polyselenide (24) anions
- Polyphosphates (25)
- Sulfur oxyanions (15,26)
- Chromium(III) and chromium(VI) species (27)
- Oxyhalide anions EO_n^- ($E = Cl$, Br, I; $n = 0–4$) (12,13,28)

BIBLIOGRAPHY

1. Whitehouse, C.M., Dreyer, R.N., Yamashita, M., and Fenn, J.B. (1985). *Anal. Chem.* **57**: 675–679.
2. Gross, J.H. (2004). *Mass Spectrometry—A Textbook*. Springer, New York.
3. Cole, R.B. (Ed.). (1997). *Electrospray Ionization Mass Spectrometry: Fundamentals, Instrumentation and Applications*. Wiley-Interscience, Hoboken, NJ.
4. Gross, M.L. and Caprioli, R.M. (2003–2006). *The Encyclopedia of Mass Spectrometry*. Elsevier, New York.
5. Kebarle, P. (2000). *J. Mass Spectrom.* **35**: 804–817.
6. Kebarle, P. and Tang, L. (1993). *Anal. Chem.* **65**: 972A–986A.
7. Cech, N.B. and Enke, C.G. (2001). *Mass Spectrom. Rev.* **20**: 362–387.
8. Stewart, I.I. (1999). *Spectrochim. Acta* **54B**: 1649–1695.
9. Fenn, J.B. et al. (1990). *Mass Spectrom. Rev.* **9**: 37–70.
10. Colton, R., D'Agostino, A., and Traeger, J.C. (1995). *Mass Spectrom. Rev.* **14**: 79–106.
11. Zhou, S. and Cook, K.D. (2001). *J. Am. Soc. Mass Spectrom.* **12**: 206–214.
12. Charles, L., Pépin, D., and Casetta, B. (1996). *Anal. Chem.* **68**: 2554–2558.
13. Charles, L. and Pépin, D. (1998). *Anal. Chem.* **70**: 353–359; Barnett, D.A. and Horlick, G. (1997). *J. Analyt. Atom. Spectrom.* **12**: 497–501.
14. Hao, C. et al. (2001). *J. Mass Spectrom.* **36**: 79–96; Hao, C. and March, R.E. (2001). *J. Mass Spectrom.* **36**: 509–521.
15. Blades, A.T. and Kebarle, P. (1994). *J. Am. Chem. Soc.* **116**: 10761–10766.
16. Blades, A.T., Jayaweera, P., Ikonomou, M.G., and Kebarle, P. (1990). *J. Chem. Phys.* **92**: 5900–5906; Jayaweera, P., Blades, A.T., Ikonomou, M.G., and Kebarle, P. (1990). *J. Am. Chem. Soc.* **112**: 2452–2454; Peschke, M., Blades, A.T., and Kebarle, P. (1999). *Int. J. Mass Spectrom.* **185–187**: 685–699.
17. Cheng, Z.L., Siu, K.W.M., Guevremont, R., and Berman, S.S. (1992). *Org. Mass Spectrom.* **27**: 1370–1376.
18. Stewart, I.I. and Horlick, G. (1994). *Anal. Chem.* **66**: 3983–3993.
19. Gaskell, S.J. (1997). *J. Mass Spectrom.* **32**: 677–688.
20. Mann, M. (1990). *Org. Mass Spectrom.* **25**: 575–587.
21. Deery, M.J., Howarth, O.W., and Jennings, K.R. (1997). *J. Chem. Soc. Dalton Trans.*: 4783–4788; Lau, T-C., Wang, J., Guevremont, R., and Siu, K.W.M. (1995). *J. Chem. Soc. Chem. Commun*: 877–878.
22. Bussian, P. et al. (2000). *Angew. Chem. Int. Ed. Engl.* **39**: 3901–3905.
23. McIndoe, J.S. and Tuck, D.G. (2003). *Dalton Trans.*: 244–248.
24. Raymond, C.C., Dick, D.L., and Dorhout, P.K. (1997). *Inorg. Chem.* **36**: 2678–2681.
25. Choi, B.K., Hercules, D.M., and Houalla, M. (2000). *Anal. Chem.* **72**: 5087–5091.
26. Stewart, I.I., Barnett, D.A., and Horlick, G. (1996). *J. Analyt. Atom. Spectrom.* **11**: 877–886.
27. Gwizdala, A.B. III, Johnson, S.K., Mollah, S., and Houk, R.S. (1997). *J. Analyt. Atom. Spectrom.* **12**: 503–506.
28. Charles, L. and Pépin, D. (1998). *Anal. Chem.* **70**: 353–359.

FENTON'S REACTION AND GROUNDWATER REMEDIATION

James A. Jacobs
Environmental Bio-Systems, Inc.
Mill Valley, California

Hydrogen peroxide, an *in situ* chemical oxidizer in the form of Fenton's Reaction, can rapidly treat waste waters, sludges, and contaminated soils and groundwater. The effect of the Fenton's Reaction is organic contaminant destruction, toxicity reduction, and biodegradability improvement. In addition, removal of biological oxygen demand (BOD), chemical oxygen demand (COD), odor, and color occur. If Fenton's Reaction is allowed to be carried to completion, then all organic molecules will ultimately be broken into carbon dioxide and water. Fenton's Reaction is a nonselective oxidizer. Fenton's Reaction also called Fenton's Reagent or Fenton's Chemistry in the literature.

Soils contaminated with toxic and recalcitrant organic wastes such as petroleum hydrocarbons and chlorinated solvents can be effectively treated with Fenton's Reaction (1–3). *In situ* chemical oxidation (ISCO) uses contact chemistry of the oxidizing agent to react with petroleum hydrocarbons, volatile organic compounds, munitions, certain pesticides, and wood preservatives in the vadose zone or in the groundwater. The gasoline additive, methyl tertiary butyl ether (MtBE), has been shown to break down with Fenton's Reaction (4–6).

HISTORY

Hydrogen peroxide is one of the earliest chemical oxidants to be discovered and used in industry. It was discovered in the late 1700s and was commercialized in the early 1800s. Hydrogen peroxide works as a remedial treatment chemical in two ways: free radical production and direct chemical oxidation using hydrogen peroxide.

A British Professor H. J. H. Fenton (7,8) described the exothermic and somewhat violent reaction of hydrogen peroxide with iron salts (ferrous sulfate). A very thick and heavy brown foam forms early in the reaction with soil. Fenton's chemistry or Fenton's Reaction uses a transition metal catalyst or an acid to enhance the oxidation chemical reaction of hydrogen peroxide by producing the hydroxyl radical. The hydroxyl radical is a powerful oxidation agent, second behind fluorine.

During the 1930s, the reaction mechanisms were described, although even today, the details are not completely understood because of the large number of side reactions. Commercial reactors have been available for the water treatment industry over the past 20 years. Within the past decade, *in situ* groundwater treatment with Fenton's Reaction have been developed. Within the last several years, commercialization of these methods has become more commonplace.

REACTIONS

Fenton's Reaction produces the hydroxyl free radical, which cleaves and oxidizes organic compounds such as gasoline resulting in producing successively smaller organic compounds. The intermediate compounds of the Fenton's Reaction are generally mono-carboxylic and di-carboxylic acids. These products are nontoxic, naturally occurring, and can be oxidized further to carbon dioxide and water (9).

The breakdown products of the Fenton's Reaction as well as the hydrogen peroxide and iron do not produce additional organic compounds or inorganic solids such as permanganate and dichromate, because there are no carbon atoms in the treatment chemicals. Although many organic compounds can be oxidized with Fenton's Reaction, compounds resulting from the oxidation products of larger molecules, such as small chlorinated alkanes, n-paraffins, and short-chain carboxylic acids, appear to resist further destruction by the Fenton's Reaction. Laboratory bench tests are highly recommended to define feasibility and optimal chemistry on a site-by-site basis.

For *in situ* chemical oxidation, the metal catalyst is usually provided by iron oxides within the soil or fill material, or added separately as a solubilized iron salt, such as iron sulfate. In addition, pH adjustment using an acid such as sulfuric (H_2SO_4) is common because the chemical oxidation is more rapid and efficient under lower pH conditions (pH 2–4 is optimal). The supportive chemical processes that essentially result in the destruction of petroleum hydrocarbons, and other volatile organic compounds are well documented (1–3).

EQUATIONS

When chemical oxidant H_2O_2 is injected at concentrations of 10% to 35% into the subsurface, it decomposes readily into reactive hydroxyl radicals ($OH^{\bullet}$) and water. The hydroxyl radical ($OH^{\bullet}$) in the subsurface can rapidly mineralize hydrocarbon, solvent, and other contaminants to water and carbon dioxide. This reaction is enhanced in the presence of iron. Iron is naturally occurring in soil and groundwater or can be added during the injection process, if needed. The reaction is based on the principle of Fenton's Reaction where iron and hydrogen peroxide react to form hydroxyl radicals and other byproducts as shown in (1) (1–3).

$$\begin{aligned}
Fe^{2+} + H_2O_2 &\rightarrow OH^{\bullet} + OH^- + Fe^{3+}\\
Fe^{3+} + H_2O_2 &\rightarrow Fe^{2+} + HO_2/O_2{}^- + H^+\\
Fe^{3+} + HO_2/O_2 &\rightarrow Fe^{2+} + O_2 + H^+\\
&\quad \text{(iron catalyst gets regenerated)}\\
OH^{\bullet} + H_2O_2 &\rightarrow HO_2/O_2{}^- + H_2O\\
OH^{\bullet} + Fe^{2+} &\rightarrow Fe^{3+} + OH^-\\
HO_2/O_2{}^- + Fe^{2+} + H^+ &\rightarrow Fe^{3+} + H_2O_2
\end{aligned} \tag{1}$$

$OH^{\bullet}$ = hydroxyl radical

The process is self-replicating because the reaction of ferric iron with hydrogen peroxide to generate the perhydroxyl radical also occurs (10). Equation (2) is another way to view the reaction:

$$Fe^{3+} + H_2O_2 \rightarrow H^+ + {\bullet}OOH^- + Fe^{2+} \tag{2}$$

The perhydroxyl radical is a less powerful oxidant than is the hydroxyl radical, but it does generate further ferrous ions that in turn stimulate further reaction with the hydrogen peroxide to produce more hydroxyl radicals (10).

Chlorinated solvents such as trichloroethylene (TCE) and tetrachloroethylene (PCE) have been the focus of *in situ* applications of Fenton's Reaction. The double bonds, C=C, that characterize chlorinated ethenes are more reactive than are the single C–C bonds of chlorinated ethanes. Therefore, PCE and TCE are more susceptible to chemical oxidation than is TCA. Although these chemicals are susceptible to chemical oxidation, relative resistance to oxidation goes from highest to lowest: PCE, TCE, vinyl chloride, phenanthrene, benzene, and hexane. The oxidation reaction for a common solvent, TCE, forms several unstable daughter products (epoxides), then breaks down to ketones and aldehydes, finally yielding carbon dioxide, water, and chloride ions (11). The oxidation of TCE is shown in (3):

$$4OH^{\bullet} + C_2HCl_3 \rightarrow 2CO_2{}^- + 3Cl + 5H^+ \tag{3}$$

The hydroxyl radical that attacks the carbon–hydrogen bonds is capable of degrading many chlorinated solvents, chloroalkenes, esters, aromatics, pesticides, and other recalcitrant compounds such as MTBE, PCP, and PCB (2). The Fenton's chemistry reaction is highly complex. The iron cycles between the Fe(II) and Fe(III) oxidation states yield the hydroxyl radical and other byproducts (11).

Residual H_2O_2 not used in the oxidation process breaks down to water and oxygen in a matter of hours. In addition to the reaction described in (2), there are a large number of competing reactions including the free radical scavengers, most importantly, carbonate and bicarbonate alkalinity, that will greatly affect the overall reaction scheme. In addition, H_2O_2 can serve as an oxygen source for microbes in the subsurface to enhance biodegradation of contaminants.

IN SITU ISSUES

Fenton's Reaction has been effective in the relatively low concentrations in waste water treatment plants, but not in *in situ* applications. In this scenario, solid oxygen demand is limited in the treatment water and competing reactions are small by comparison with those in soils. Soil contains competing reactions that will react with the hydrogen peroxide, including organic carbon, unoxidized minerals, and biologically produced chemicals. Therefore Fenton's Reaction treatment plans in the field may take significantly more applications than the bench testing might indicate because of the competing reactions that occur in the soil and groundwater. More information about hydrogen peroxide decomposition in the presence of aquifer material is described in Miller and Valentine (12).

Although Fenton's chemistry has been documented for well over 100 years, it has been employee safety and handling issues that have kept large numbers of environmental contractors from using *in situ* chemical oxidation technology. Although this technology can be used safely, significant safety planning, worker training, personal protective equipment, on-site supervision, and monitoring must be an integral part of all oxidation projects. The main challenges for *in situ* applications are delivery of the oxidants, which are difficult to handle, and pH control of the subsurface environment where the reaction must occur. Since most soil has adequate concentrations of naturally occurring iron, addition of iron is not needed at many sites. The typical iron to hydrogen peroxide ratios is 1:5 to 10 wt/wt. Iron levels less than 25 to 50 mg/L can require excessive reaction times (10 to 24 hours) (13). Improper control of the Fenton's reaction will only generate oxygen and water, not the high intensity Fenton's oxidation chemistry.

DELIVERY OPTIONS

Delivery of the Fenton's Reaction chemistry requires close spacing of the injection ports/wells because of the rapid nature of the reactions. The liquids are generally pumped using pressures of 100 to 600 psi for probe rods or 1000 to 3000 psi using lances. Pouring hydrogen peroxide into wells or trenches may work for treating water in those areas, but because of the rapid reaction rates, pressure is needed to move the peroxide away from the injection location. The high pressure is also needed to ensure mixing in the subsurface and a reasonable radius of influence. Fenton's Reaction, although powerful, is not generally recommended or cost effective for low-concentration groundwater plumes where the concentrations of organics is less than 5 mg/L. Variations in the effective porosity, permeability, lithologic heterogeneities, BOD, COD, pH, and contaminant mass affect the outcome of *in situ* applications of Fenton's Reaction.

Hydrogen peroxide reacts in an optimal manner in lower pH settings, with lower alkalinity readings. In some cases, strong acids, most notably sulfuric acid, lower the pH. Trace chloride from chlorinated compounds will likely combine with sodium or calcium ions to form salts or with hydrogen to form weak acids. Careful evaluation of soil and water chemistry with a bench test with soil and water samples is recommended before the start of any injection process. Because of the rapid reaction time of the Fenton's Reaction, subsurface spacing of injection ports must be relatively close. The lower the effective porosity of the treatment zone, the closer the injection port spacing. Based on field experience, clays and silts, which are problematic to remediate *in situ*, typically require 0.6 to 1.0 m spacing, whereas injection ports for clean sand and gravels can be placed at 2.0 to 3.0 m spacing.

Chemical compatibility of the injection equipment components and safety procedures become critical with the injection of strong acids, bases, oxidants, and other chemicals. Although handling hydrogen peroxide and other oxidants requires significant safety training and planning, the oxidant is effective at remediation of a variety of organic contaminants and is relatively inexpensive. The reaction time for hydrogen peroxide in the subsurface is usually within seconds to minutes rising under ideal conditions to hours. A rise in temperature in the subsurface illustrates the exothermic nature of Fenton's Reaction process.

Based on field research, the optimum reaction temperature is relatively low, as measured on the reaction foam in the field using infrared thermal meters in 35 °C to 41 °C. With increases in temperature beyond about 57 °C, the peroxide becomes more volatile as it starts to produce a wispy white vapor. The gas is water vapor and carbon dioxide, the products of chemical oxidation. Subsurface reaction temperatures are measured with an infrared thermal meter. Reaction temperatures in the range of 82 °C to 93 °C are explosive and unsafe. Temperatures of subsurface chemical reactions can be monitored and lowered by adding water, lowering concentrations of the catalyst or oxidant, and reducing injection pressures. Effective porosity of the treatment zone may be reduced by the formation of metal oxide precipitates.

CASE STUDIES

Fenton's Reaction has been successful on a variety of sites throughout the United States since the early to mid-1990s. Reductions of organic contaminants such as DNAPLs in full-scale treatment projects are as high as 90% to 99%, depending on the contaminant, the subsurface variables, and the starting concentrations. Specific Fenton's Reaction case examples are included in a variety of publications, including U.S. EPA, (14), U.S. Navy (9), U.S. Navy (15), and Jacobs and Testa (16).

BIBLIOGRAPHY

1. Watts, R.J., Smith, B.R., and Miller, G.C. (1991). Treatment of octachlorodibenzo-p-dioxin (OCDD) in surface soils using catalyzed hydrogen peroxide. *Chemosphere* **23**: 949–956.
2. Watts, R.J. et al. (1992). Hydrogen peroxide for physicochemically degrading petroleum-contaminated soils. *Remediation* **2**: 413.
3. Watts, R.J. and Stanton, P.C. (1994) *Process Conditions for the Total Oxidation of Hydrocarbons in the Catalyzed Hydrogen Peroxide Treatment of Contaminated Soils*. WA-RD, Vol. 337.1, Washington State Department of Transportation, Olympia, WA.
4. Jacobs, J., Guertin, J., and Herron, C. (Eds.). (2001). *MTBE and its Effect on Soil and Groundwater Resources*. Lewis Publishers/CRC Press, Boca Raton, FL.
5. Leethem, J.T. (2002). In-situ chemical oxidation of MTBE: A case study of the successful remediation of a large gasoline release. *Contaminated Soil Sediment and Water* **July/August**: 70–75.
6. Kelly, K.L., Marley, M.C., and Sperry, K.L. (2002). In-situ chemical oxidation of MTBE. *Contaminated Soil Sediment and Water*. **July/August**: 36–40.
7. Fenton, H.J.H. (1893). *Proc. Chem. Soc.* **9**: 113.
8. Fenton, H.J.H. (1894). *J. Chem. Soc.* **65**: 899.
9. U.S. Navy. (2001). In Situ Chemical Oxidation Process using Fenton's Reagent, Navy Pollution Abatement Ashore Technology Demonstration/Validation Program. Available: http://p2ashore.nfsec.navy.mil/cgi-bin/project_descriptions/cfpr_results.cfm?PROJECT.
10. Vance, D.B. (2002). New Applications in Redox Reactions for In-Situ Groundwater Remediation. Available: http://2the4net/ThreeRedox.htm.
11. Suthersan, S.S. (2002). *Natural and Enhanced Remediation Systems*. CRC Press/Lewis Publishers, Boca Raton, FL.
12. Miller, C.M. and Valentine, R.L. (1995). Hydrogen peroxide decomposition and quinoline degradation in the presence of aquifer material. *Water Resources* **29**(10): 2353–2359.
13. U.S. Peroxide. (2002). Reference Library Peroxide Applications: Fenton's Reagent, Iron-Catalyzed Hydrogen Peroxide. Available: http://www.h2o2.com/applications/industrialwastewater/fentonsreagent.html.
14. U.S. EPA. (2000). Abstracts of Remediation Case Studies, Vol. 4, EPA 542-R-00-006. Available: http://www.epa.gov/tio/download/frtr/abstractsvol4.pdf.
15. U.S. Navy. (2003). Available: http://enviro.nfesc.navy.mil/erb/restoration/technologies/remed/phys_chem/phc-43.asp.
16. Jacobs, J. and Testa, S.M. (2003). Design considerations for in-situ chemical oxidation using high pressure jetting technology. *AEHS Soil, Sediment and Water* **March/April**: 51–60.

READING LIST

Watts, R.J., Udell, M.D., Rauch, P.A., and Leung, S.W. (1990). Treatment of Pentachlorophenol Contaminated Soils Using Fenton's Reaction. *Hazardous Waste Hazardous Materials* **7**: 335–345.

WHERE WATER FLOATS

NASA Marshall Space Flight Center

The **Fluid Physics Research Program** is implemented at the Lewis Research Center in Cleveland, Ohio and is managed by the Microgravity Research Program Office located at Marshall Space Flight Center in Huntsville, Alabama.

A fluid is any material that flows in response to an applied force, and assumes the shape of its container. Examples include liquids, gases, and granular systems such as soil and sand.

Fluid physics is vital to understanding, controlling, and improving all of our industrial and natural processes. The engines used to propel a car or an airplane, the shape of the wings of an airplane that allow it to fly, the operations of boilers that generate steam used to produce over 90% of the world's electric power, the understanding of how cholesterol is transported in our bloodstream and effects heart disease, and how pollutants are transported and dispersed in air and water are just a few examples of how fluid physics affects our everyday life and forms the very basis for an industrial society. Areas of NASA's research in fluid physics include the following:

Complex Fluids are fluids containing structures or aggregates on supermolecular scales; examples include colloids, emulsions, liquid crystals, granular solids, and polymer solutions.

Multiphase Flow and Heat Transfer includes the study of processes such as boiling, condensation, flows and heat transfer in fluid-fluid and fluid-solid mixtures.

Interfacial Phenomena investigates the shape and dynamics of solid-liquid-gas interfaces such as during a coating process, the liquid-gas interface as in partially filled containers, and the effect of energy and mass transfer on these interfaces.

Dynamics and Instabilities deal with fluid flows and stability of thin and thick films, geological and biological flows, liquid bridges, drops and bubbles, and electrochemical and chemical transport. Pattern formation and turbulence are other interesting areas investigated under this topic.

FRESHWATER

Gholam A. Kazemi
Shahrood University of Technology
Shahrood, Iran

INTRODUCTION

Water is an important solvent and is sometimes called the universal solvent (1). It dissolves, to the extent possible, the material with which it comes in contact. Salinity is a measure of various constituents dissolved in water, including solids, liquids, and gases. Total dissolved solids (TDS), another form of expressing the amount of dissolved ingredients in water, is the sum of total solids that have been dissolved in water [note that TDS at 180 °C = Sum of ions + SiO_2 − 0.5082 × mg/lHCO_3; (2)]. However, many references do not differentiate between TDS and salinity because the absolute majority of the dissolved constituents of water are usually the solids. Both parameters have units of mg/l or ppm. These two units, ppm and mg/l, are almost the same except for high salinity waters for which 1 mg/l = 1 ppm × density of water. The salinity of groundwater ranges from 25 mg/l in quartzite springs to more than 300,000 mg/l in brine (3), and the salinities of distilled water, River Murray in South Australia, seawater, and the Dead Sea are 0, 553, 35,000, and 385,000 mg/l, respectively (4).

Waters are classified into various types based on their salinity or TDS content. Two widely used classification systems are illustrated in Table 1. The Australian Water Resources Council (AWRC) (1976) (5) has devised a considerably different classification, shown in Table 2, which can be regarded as one of limited local use.

DEFINITION

Based on the two widely used classification systems above, freshwaters are those whose salinity or TDS is less than 1000 mg/l. Interestingly, freshwater has identical definitions in both systems; other water types, for example, brines, have slightly to considerably different meanings. Despite this equality, Manning (8) believes that criteria for terms such as fresh, when applied to water quality, are not universal. He writes (p. 221) "It depends on the individual tastes and on the purpose for the water. One person's unacceptable saline water might be another's only drinking water."

In a generalized sense, freshwaters may be regarded as all waters in the world, excluding saline waters of ocean, seas, and saline lakes. Based on this, global freshwater resources are estimated at 36.02 × 10^6 km^3 accounting for about 2.6% of the total world water of 1.348 × 10^9 km^3 (9). If one adopts this generalization, then, rivers, freshwater lakes, most groundwaters, and soil moisture all are eligible to be categorized as freshwater resources.

The term "freshwater," may be misleading. It may suggest that freshwaters are those that are suitable for drinking, but this is not the case. In other words, freshwater is an arbitrary chemical denomination and does not strictly imply suitability of this type of water for drinking. Various guidelines, such as those established by the United States, European Community, and World Health Organization, suggest a TDS of less than 500 mg/l to 1500 mg/l for drinking water, which does not completely conform with the TDS of freshwater. For instance, a particular type of freshwater may not be suitable for drinking because one or more of its constituents exceeds the desirable criteria. Water from an irrigation field,

Table 2. AWRC Classification of Water

Water Type	TDS
Fresh	<500 mg/l
Marginal	500–1,000
Brackish	1000–3000
Saline	>3000

Table 1. Water Types Based on TDS Value—(Different Classification Systems)

Water Type[a]	TDS mg/l[a]	Water Type[b]	TDS mg/l[b]
Fresh	0–1000	Fresh	0–1000
Brackish	1000–10,000	Slightly saline	1,000–3,000
Saline	10,000–100,000	Moderately saline	3,000–10,000
Brine	>100,000	Very saline	10,000–35,000
		Briny	>35,000

[a] *Source*: Reference 6, in Reference 3.
[b] *Source*: Reference 7—proposed by USGS.

for example, may be chemically classified as fresh, but due to its high nitrate, phosphate, or toxic components is more likely to be rejected as a source of drinking water. Similarly, water that flows through, or emerges from, geologic media containing sulfur deposits will not be suitable for drinking, though chemically fresh, because of its low pH value. Hundreds or thousands of similar cases can be described. However, in normal conditions and in the majority of cases, so-called freshwater meets drinking water standards and can be consumed.

SOME GENERAL POINTS ABOUT FRESHWATER

1. All types of wet atmospheric precipitations such as rain, snow, and hail turn into freshwater.
2. Surface waters are generally lower in TDS than groundwaters and, thus, are more likely to be classified as freshwater compared with groundwater. It is not incorrect to say that river waters are generally fresh.
3. Various groundwaters (excluding connate water, intruded sea water, magmatic and geothermal water, and water affected by salt leaching) generally have a TDS of less than 10,000 mg/l (10). Thus, they may or may not be fresh.
4. Generally the electrical conductivity (EC) of groundwater in hard rock regions is 120 μS/cm or less (11) and therefore, is fresh.
5. Water emanating from upland catchments and water of lowland rivers that have typical ECs less than 700 μS/cm(12) are fresh.
6. Igneous, volcanic, carbonate, and metamorphic rocks generally yield freshwater.
7. Freshwater may turn into other types of water (brackish, saline, etc) by various means, including human-induced pollution, sea/saltwater intrusions, overexploitation, and other natural causes.

BIBLIOGRAPHY

1. Kemmer, F.N. (1988). *The NALCO Water Handbook*, 2nd Edn. McGraw-Hill, New York.
2. Hounslow, A.W. (1995). *Water Quality Data: Analysis and Interpretation*. Lewis, New York.
3. Todd, D.K. (1980). *Groundwater Hydrology*, 2nd Edn. John Wiley & Sons, New York.
4. Taylor, S. (1993). *Dryland Salinity: Introductory Extension Notes*. Department of Conservation and Land Management of New South Wales, Australia.
5. Australian Water Resources Council. (1976). *Review of Australia's Water Resources 1975*. AGPS Canberra, for the Department of National Resources.
6. Carroll, D. (1962). *Rainwater as a Chemical Agent of Geologic Processes—A Review*. US Geological Survey Water-Supply Paper 1535.
7. Hem, J.D. (1985). *Study and Interpretation of the Chemical Characteristics of Natural Waters*, 3rd Edn. U.S. Geological Survey Water-Supply Paper 2254, Denver, CO.
8. Manning, J.C. (1997). *Ápplied Principles of Hydrology*. Prentice-Hall, Upper Saddle River, NJ.
9. McDonald, A. and Kay, D. (1988). *Water Resources: Issues and Strategies*. Longman Scientific and Technical, New York.
10. Rail, C.D. (2000). *Groundwater Contamination*. Vol. 1: *Contamination, Sources and Hydrology*. Technomic, Lancaster, PA.
11. Bender, S. (2003). *Low Mineralized Water: Challenge for Hydrochemical Investigations in Hard-rock Area. Proc. Int. Conf. Groundwater Fractured Rocks*, Prague, IHP-VI, series on groundwater no. 7, pp. 195–196.
12. Tebbutt, T.H.Y. (1995). *Principles of Water Quality Control*. Reprint of 4th Edn. Butterworth Heinemann, Oxford.

DISSOLVED GASES

M. Boström
Linköping University
Linköping, Sweden

B.W. Ninham
University of Regensburg
Regensburg, Germany
and
Australian National University
Canberra, Australia

Gases dissolved in water play an important role in solution chemistry and in biological processes in solutions. The story that we will present of dissolved gases in water is in some sense twofold. We begin with the more conventional description and a general discussion of gases dissolved in groundwater, in seawater, and in blood, and decompression sickness. We then give a short introduction to a new and rapidly developing scientific field within colloid and solution chemistry that has demonstrated a profound role for dissolved gas in, for example, colloidal interactions, hydrophobic interactions, conductance, surface tension, and bubble fusion. We argue strongly that different scientific disciplines that consider dissolved gases in very different ways can benefit tremendously from better knowledge of work in other disciplines.

DISSOLVED GASES IN GROUNDWATER

Dissolved gases and dissolved organic matter (hydrophobic acids, bases, and neutrals) are ubiquitous in water. A large number of dissolved organic materials are composed of humic acids that are heterogeneous organic macromolecules. These have an important role in binding of metals and pesticides. Humic substance concentrations range from 20 μg/L in groundwaters to 30 mg/L in surface waters. Groundwater also contains dissolved gases due to exposure to the atmosphere, gas production reactions with minerals, and biological activity. These gases are certainly at least as important as dissolved matter. The most common gases are those that constitute the atmosphere; O_2, N_2, and CO_2. Methane, H_2S, N_2O, and other gases are also present in groundwater due to biogeochemical processes. Dissolved gases can have a tremendous impact. The presence of dissolved H_2S in water can, for example, give rise to the smell of rotten eggs, and above an H_2S concentration of 1 mg/L, the water is no longer suitable for consumption. Dissolved oxygen, on the other hand, is vital for biology. The concentration of dissolved oxygen (at 20 °C) at saturation is 9.1 mg/L. This is not that far above the minimum concentration of dissolved oxygen that can maintain aquatic life (which is around 5 mg/L). Aerobic bacteria can break down oxygen and leave an oxygen concentration close to this lower limit (it may even become too low to sustain aquatic life) (1). Another major dissolved gas is carbon dioxide. The total concentration of carbon oxides ($[CO_2(aq)] + [H_2CO_3] + [HCO_3^-] + [CO_3^{2-}]$) is around 2.3 mM in seawater, 0.1–5 mM in river water, 0.5–8 mM in groundwater, and 10–50 μM in rainwater (2).

DISSOLVED GASES IN THE OCEANS

A large number of measurements of the fluxes of gases between the atmosphere and the ocean have been carried out to determine gases dissolved in seawater (3,4). The first determination of N_2 dissolved in Atlantic seawater was performed in 1938 (5). A large number of methods have been used to measure dissolved gases in water (e.g., Winkler method, gasometric methods, mass spectroscopic methods (6), and gas chromatographic methods (4,7)). The mechanisms of gas exchange in the ocean are not entirely understood. Studies of gas exchange between air and water are often based on the simple theoretical surface film model developed by Liss and Slater (3). However, it is a complicated system and more recent attempts have been made to include different complications. Gas exchange depends, for example, on the partial pressure of a gas in air, its solubility coefficient, concentration in bulk surface water, wind velocity (8), temperature, salinity, molecular diffusivity, bubbles, and bubble size (9). Another complicating factor for oxygen and carbon dioxide dissolved in the ocean is that biological processes affect them. Oxygen profiles as a function of depth under the ocean surface are very different compared to the profile of N_2 (4) because O_2 is produced in the surface euphotic zone by photosynthesis and consumed by decomposition of organic matter in the aphotic zone.

The composition of gases in seawater (or in any other aqueous solution) under solution equilibria differ from their atmospheric composition. Dissolved gases that have been studied include carbon dioxide (CO_2), carbon monoxide (CO), nitrous oxide (N_2O), nitrogen (N_2), oxygen (O_2), methane (CH_4), sulfur dioxide (SO_2), dimethyl sulphide (DMS), and argon (Ar). Many of these gases are related to global warming, and there has been some discussion of the ocean as a source of methane to the atmosphere (4,10). Some of these dissolved gases are also contained in air bubbles. For example, Nakayama et al. (4) report values around 0.6 mM/kg for N_2, 0.35 mM/kg for O_2, and 16.5 μM/kg for Ar in the surface (0–50 m depth) region of the north Pacific Ocean and a mean value of 18.4 μmol/kg of dissolved air bubbles in the upper 30 m of water (4).

DISSOLVED GASES IN BIOLOGICAL SYSTEMS

Dissolved gases are very important in biological systems, and all forms of life visible to the human eye require oxygen. Hemoglobin is an important transport protein embedded in the cell membrane of the red blood cell. Its main function is to collect oxygen in the lungs, where it is abundant, and deliver it to the tissues around the body. O_2 that is used by most organisms is converted to CO_2. The carbon dioxide that diffuses in the blood is mainly in the form of dissolved CO_2 because conversion to carbonic acid is a comparably slow process. Long before this conversion takes place, it has usually diffused into the red cell. Once inside, conversion to carbonic acid takes place well buffered by the hemoglobin in the cell wall. While O_2 is delivered to the tissues, CO_2 is absorbed and transported out of the system (11).

Large volumes of gas dissolve in the tissues of divers. Gases such as nitrogen and helium are not metabolized by our bodies and need to be eliminated during slow decompression. Gradual decompression for several days is needed following extended human dives to pressures above 20 atm to avoid excessive gas supersaturation and bubble formation that may lead to decompression sickness (12,13). An interesting biochemical method to tackle decompression sickness has been investigated in experiments on rats. Suspensions of the human colonic microbe *Methanobrevibacter smithii* were introduced into the rats. This microbe that converts 4 mol of H_2 and 1 mol carbon dioxide (which is a gas that is abundant in tissues) to 1 mol methane and water, significantly reduces the risk of decompression sickness in rats (12). Decompression sickness begins at a much lower gas pressure than traditionally anticipated (less than 2 atm). The most likely explanation for this has only recently emerged and is discussed in the following section.

THE ROLE OF DISSOLVED GASES IN COLLOID SCIENCE AND BIOCHEMISTRY

Literature has accumulated during the last decade in physical chemistry journals focusing on the effects of dissolved gases on, for example, bubble fusion (14), emulsion stability (15,16), colloid interactions (17), hydrophobic forces (18,19), and optical cavitation (20). Bubble fusion depends strongly on salt concentration and ionic species but obviously also on the amount of dissolved gas (14). A mechanism has been proposed for hydrophobic forces that are based on surface adsorption and nucleation of dissolved gases in water. Suggestions have been made that dissolved gas molecules can be involved in propagating perturbations induced by hydrophobic surfaces. This could possibly give rise to the extraordinarily long-range hydrophobic interaction observed. There is clear evidence that degassing a system can significantly reduce both the magnitude and range of the hydrophobic interaction. Bubble formation occurs much more readily near hydrophobic surfaces. We have plenty of hydrophobic surfaces in our bodies, which is highly relevant for the understanding of decompression sickness. It is a plausible explanation why decompression sickness begins at a much lower pressure than expected and why physical exertion before decompression can increase the severity. Exertion is likely to damage tissue mildly and increase the number of hydrophobic sites where such cavitation can occur (14). The interested reader is referred to the references suggested, but we would like to mention one very interesting proposal made by Karaman et al. (15). They argued, quite convincingly, that there might be an unanticipated role for dissolved gases in biochemistry. Chemical reactivity can depend on dissolved gas through enhanced radical production in microbubbles formed from gas aggregates. Experiments in molecular biology have revealed that negatively charged restriction enzymes lock onto negatively charged DNA and cut it at particular sequences. Enzymes and DNA have the same charge, so this cannot be explained in terms of electrostatics, and the source of the energy to drive the catalysis is not explained. If cavitation occurs on contact, free radical $OH^{\bullet}$ production follows, and the transfer of an electron to $Cl^{\bullet}$ does provide sufficient energy. Experiments reveal that above 0.2 M NaCl, where cavitation is inhibited, the enzyme loses its catalytic property entirely and reversibly. Below 0.2 M NaCl, the addition of vitamin C, a well-known free radical scavenger, totally inhibits the enzyme. Dissolved gas can play an important role in harnessing weak molecular forces cooperatively to influence vital biochemical reactions. Interestingly, the cutting efficiency of DNA by enzymes also depends on the choice of background salt following a so-called Hofmeister series (see article on the "Hofmeister Effect" in this volume), in particular on the supposedly irrelevant choice of negative co-ions (21). One can show that this experimental finding further strengthens the proposed mechanism for the activation energy.

THE ROLE OF DISSOLVED GASES IN SURFACE TENSION OF ELECTROLYTES

In the past, there has been a discussion of the ion specificity found for the surface tension of salt solution. In the air–water interfacial region (which is roughly the first 6 Å closest to the air–water interface), there is a gradual transition of the relative amount of water and dissolved gases. This means that an ion near the interface experiences a force that originates from the change in solvation energy when the ion moves into the surface region (with its increasing concentration of dissolved gas and decreasing concentration of water molecules). This is besides previously discussed image potentials and dispersion potentials (originating from the polarizability of ions). Inclusion of solvation energy effects acting on ions near the air–water interface gives exactly the right order of magnitude and the right ion specific sequence for the surface tension of different salt solutions (22). If one furthermore includes dispersion potentials acting on ions, one improves the result further. There are clearly other effects that influence the surface tension of salt solutions, but it is clear that solvation energy effects, related to the interfacial profiles of dissolved gases and water molecules, together with ionic dispersion potentials between ions and the interface, play vital parts. References and a discussion of ion specific effects can be found in the article "Hofmeister Effects" in this volume.

Some attempts have been made to model theoretically the influence of dissolved gases, but there is in most cases still a long way to go until it is firmly understood. But it is very clear from experiments that dissolved gases play a very important, sometimes vital, role in a large number of different chemical and biological systems in water.

BIBLIOGRAPHY

1. Lester, J.N. and Birkett, J.W. (1999). *Microbiology and Chemistry for Environmental Scientists and Engineers*. E & F.N. Spon, London.
2. Stumm, W. and Morgan, J.J. (1981). *Aquatic Chemistry: An Introduction Emphasizing Chemical Equilibria in Natural Waters*. John Wiley & Sons, New York.
3. Liss, P.S. and Slater, P.G. (1974). *Nature* **247**.
4. Nakayama, N., Watanabe, S., and Tsunogai, S.J. (2002). *Oceanography* **58**: 775.
5. Rakestraw, N.W. and Emmel, V.M.J. (1938). *Mar. Res.* **1**: 207.
6. Emerson, S., Quay, P.D., Stump, C., Wilber, D., and Knox, M. (1991). *Biogeochem. Cycles* **5**: 49.
7. Gamo, T. and Horibe, Y. (1980). *Bull. Chem. Soc. Jpn.* **53**: 2839.
8. Liss, P.S. and Merlivat, L. (1986). In: *The Role of Air-Sea Exchange in Geochemical Cycling*. P. Buat-Menard (Ed.). Kluwer, Dordrecht, The Netherlands.
9. Merlivat, L. and Memery, L. (1983). *J. Geophys. Res.* **88**: 707.
10. Bates, T.S., Kelly, K.C., Johnson, J.E., and Gammon, R.H. (1996). *J. Geophys. Res.* **101**: 6953.
11. Zubay, G.L. (1998). *Biochemistry*, 4th Edn. Wm. C. Brown, Dubuque, IA.
12. Weathersby, P.K., Homer, L.D., and Flynn, E.T. (1982). *J. Appl. Physiol.* **53**: 940.
13. Kayar, S., Miller, T., Wolin, M., Aukhert, E., Axley, M., and Kiesow, L. (1998). *Am. J. Physiol.* **275**: R677.
14. Craig, V.S.J., Ninham, B.W., and Pashley, R.M. (1993). *J. Phys. Chem.* **97**: 10192.
15. Karaman, M.E., Ninham, B.W., and Pashley, R.M. (1996). *J. Phys. Chem.* **100**: 15503.
16. Maeda, N., Rosenberg, K.J., Israelachvili, J.N., and Pashley, R.M. (2004). *Langmuir* **20**: 3129.
17. Alfridsson, M., Ninham, B.W., and Wall, S. (2000). *Langmuir* **16**: 10087.
18. Craig, V.S.J., Ninham, B.W., and Pashley, R.M. (1999). *Langmuir* **15**: 1562.
19. Wennerström, H. (2003). *J. Phys. Chem. B* **107**: 13772.
20. Bunkin, N.F. et al. (1997). *Langmuir* **13**: 3024.
21. Kim, H-K., Tuite, E., Norden, B., and Ninham, B.W. (2001). *Eur. Phys. J. E* **4**: 411.
22. Boström et al. (2004). Submitted.

HARD WATER

GHOLAM A. KAZEMI
Shahrood University of Technology
Shahrood, Iran

INTRODUCTION

In defining hard water, first it is necessary to describe hardness, a closely related chemical property of water.

Hardness is defined by some as a property of water that prevents lather formation with soap and produces scale in hot water systems (1); by others, in simpler terms, as the soap consuming capacity of water (2); and in broader terms, by Freeze and Cherry (3), as the "metallic ion content of water which reacts with sodium soaps to produce solid soaps or scummy residue and which reacts with negative ions, when the water is evaporated in boilers, to produce solid boiler scale."

The ions which result in, or produce, hardness include Mg^{+2}, Ca^{+2}, Sr^{+2}, Fe^{+2}, and Mn^{+2}, and to a lesser extent, Ba^{+2} and Zn^{+2} and other divalent ions. The first two are the most influential. There are different forms of hardness (4):

1. *Total hardness*: Ca and Mg expressed as $CaCO_3$
2. *Calcium hardness*: Ca expressed as $CaCO_3$
3. *Magnesium hardness*: Mg expressed as $CaCO_3$
4. *Temporary hardness*: Equals alkalinity, if alkalinity is less than total hardness
5. *Permanent hardness*: The amount of hardness greater than temporary hardness
6. *Noncarbonate hardness*: Permanent hardness

Of the six different forms of hardness, the most important and the one that is commonly known as hardness, is total hardness. It is calculated from the following simplified equation:

$$\text{Hardness} = 2.5\ \text{Ca} + 4.1\ \text{Mg} \quad (\text{units in mg/l})$$

Note that there are equations to calculate each of the six different forms of hardness.

Example: The Ca and Mg concentrations in a water sample are 20 and 15 mg/l, respectively. What is its hardness?

$$\text{Hardness} = 2.5 \times 20 + 4.1 \times 15 = 111.5\ \text{mg/l}$$

As stated, each of the six forms of hardness can be calculated by analyzing a water sample for the relevant ion(s), followed by using the relevant equation. However, there are techniques for measuring total hardness directly such as (a) a standard solution of soap, that was in use for many years (2); and (b) Schwarzenbach et al.'s method of colorimetric titration (5), which is sometime referred to as the EDTA method.

The author believes that earlier, the measurement of hardness was a common and an inseparable practice in water analysis. To agree with this statement, one needs to look at a relatively old handbook and read, "The common denominator of the majority of water problems is hardness" (Ref. 2, p. 4.1, first published in 1979). Somehow, in contrast, recent hydrochemical analysis of water samples does not usually include a measurement of hardness. For instance, there is not a single report of hardness in the hundreds of hydrochemistry papers presented at the International Association of Hydrogeolgists (IAH) conferences in Prague (2003) and in Munich (2001) and in the recent past (1997–2001) annual symposia of the Geological Society of Iran. Other aspects of water chemistry and water biology, such as nutrients, heavy metals, bacteriological species, chlorinated hydrocarbons and many more contaminants have become so widespread and are considered more important and more dangerous than hardness. Rejection of water for drinking or other uses, these days, is many times more likely to be due to its harmful DDT level than because of its high hardness value. One other possible reason is that hardness and hard water topics are only important in a narrow field of water, e.g., hot water boilers.

The second point is that industrialization, urbanization, agricultural practices, and most environmentally unfriendly activities are not likely to result in an increase in the hardness of water resources. These activities do not usually generate hardness producing ions such as Ca^{+2} or Mg^{+2}, though, they may produce other trivial (in terms of hardness) ions such as Mn^{+2} and Fe^{+2}.

DEFINITION AND SPECIFICATIONS OF HARD WATER

Todd (6, p. 282) reports that Hippocrates (460–354 B.C.) was the first to use the term hard and soft for water when he, in his treatises on public hygiene, wrote "consider the waters which the inhabitants use, whether they be marshy and soft, or hard and running from elevated and rocky situations, and then if saltish and unfit for cooking... for water contributes much to health." However, it is probably safe to say that hard and soft water in Hippocrates' writings are likely to mean different things than what we today call hard and soft water.

The definition of hard water varies slightly among researchers. The most frequently used classification system for hardness is that of Sawyer and McCarty (7), presented in Table 1. There is another slightly different classification system by Hem (8), which is referred to in some publications as the classification of the U.S. Geological Survey (9).

Based on Table 1, hard water is defined as water whose hardness is about 120 to 300 mg $CaCO_3$/l. However, some researchers argue that there is no exact definition for determining hard or soft water, because it depends, at least partially, on what one is accustomed to (9). What is hard water for one person, may be moderately hard for another and vice versa. The very hard water category in Table 1 can be grouped into the hard water category as it has the specifications of hard water. It is, however, unclear to which category one can assign moderately hard water, should hard water and soft water be the only categories available.

Table 1. Different Types of Water Based on Hardness

Water Type	Hardness, mg $CaCO_3$/l	
Soft	below 75[a]	Below 60[b]
Moderately hard	75–150[a]	60–120[b]
Hard	150–300[a]	120–180[b]
Very hard	Over 300[a]	Over 180[b]

[a]Reference 7.
[b]Reference 8.

John C. Manning (9, p. 228), a very well known authority in the field of hydrology for his Manning's roughness coefficient, succinctly yet nicely describes the characteristics of hard water; "Really hard waters, are unmistakable. If you have ever stayed in a motel supplied with hard water and tried to take a shower with a tiny bar of soap many motels supply their guests, you know how frustrating it is to see the soap disappear before you are probably lathered and bathed. On the other hand most people like water with a little hardness. It usually tastes better, and it washes the soap off when you bathe. Very soft water feels slick on your body, as though the soap were still there. Water with a little hardness, which is to say with calcium and magnesium ions, is favored for irrigation, too, as it tends to maintain higher soil permeability and allow infiltration to take place."

The explanations given above can be summarized as "those waters that require considerable amounts of soap to produce a foam or lather and that also produce scale in hot water pipes, heaters, boilers, and other high water temperature units, are called hard water." By the end of this article, the author confesses that the reader may become confused as what is hard water and what is hardness.

The economic disadvantages of hard water include increased soap consumption and higher fuel costs for boilers due to poor heat conduction resulting from scale formation. The term scale describes a continuous, adherent layer of foreign material formed on the waterside of a surface through which heat is exchanged (5). The thermal conductivity of $CaCO_3$ scale is as low as 0.0022 cal/s.cm. °C, 418 times lower than that of copper which is 0.92 cal/s.cm. °C. Hence, precipitation of $CaCO_3$ due to use of hard water significantly reduces the thermal conductivity of boilers. As a result, guidelines have been produced to set the limit for hardness of waters used in various industries; a few are presented as examples in Table 2. In a similar fashion, 80 mg/l of hardness is set as the limit for domestic water (10).

It is generally accepted that there is no health hazard associated with hard water (1). However, a recent study in Taiwan (11) shows that there is a significant negative relationship between drinking water hardness and colon cancer mortality, an important finding for the Taiwanese water industry and human health. This is in line with Tebbutt's (1) report that there is some statistical evidence to suggest that artificially softened waters may increase the incidence of some forms of heart disease. All these confirm Muss's (12) observation that in the United States, death rates from heart and circulatory diseases are lower in states where public water supplies are harder.

Table 2. Recommended Hardness Limit for Industrial Process Water[a]

Use	Hardness Limit, mg $CaCO_3$/l
Carbonated beverages	200–250
Confectionery	75
Dairy	180
Food canning and freezing	25–400 (depends on the type of food)
Food equipment washing	10
Food processing, general	10–250
Laundering	0–50
Paper and pulp, fine	100 (Ca hardness, 50)
Paper, groundwood	200
Paper, kraft, bleached	100
Paper, kraft, unbleached	200
Paper, soda, and sulfate pulps	100

[a]Reference 6.

WHAT TO DO WITH HARD WATERS

When needed, there are a few methods to change a hard water to a soft water to suit the purpose, an exercise called water softening (1):

1. *Lime softening*: This method involves adding lime and is suitable for calcium hardness of the carbonate form. In a typical experiment, total hardness and carbonate hardness of 150 and 125 mg/l are reduced, after softening, to 65 and 40 mg/l, respectively.
2. *Lime—soda softening*: This technique involves adding soda ash (Na_2CO_3), and is suited to all forms of calcium hardness. In a typical experiment, total hardness and carbonate hardness of 200 and 100 mg/l are reduced, after softening, to 40 and 40 mg/l, respectively.
3. *Excess—lime softening*: This is suitable for magnesium carbonate hardness. Chlorination may be unnecessary after such softening, but carbonation is necessary.
4. *Excess—lime soda softening*: This method involves adding lime and soda ash. It is a complicated process which is suited to all forms of magnesium hardness and produces a large volume of sludge.
5. *Ion exchange*: This method takes place through three approaches: (a) zeolite, (b) hydrogen cycle cation exchangers, and (c) strong ion exchanger. The Na^+ ions of zeolites replace Ca^{+2} and Mg^{+2} ions in the water, a reaction like ion exchange between Na_2clay and water. A natural sodium cycle zeolite has an exchange capacity of about 200 gram equivalents/m^3. Hydrogen cycle cation exchangers exchange all cations for hydrogen and so, produce zero-hardness water. A strong anion exchanger removes all ions.

HARDNESS OF WATER—WORLDWIDE CASES

Limestone and dolostone aquifers that contain gypsum and anhydrite in their matrix may yield hard water, whose hardness is from 200–300 mg/l or more (10). Groundwaters are generally higher in hardness than surface waters (4). The hardness of groundwater in Akwa Ibom State, Nigeria, ranges from 1.2–65 mg/l (13). In the Jornada del Muetro Basin of New Mexico, the groundwater hardness reaches 260 mg/l (14). The hardness of shallow groundwater in Balatina and Arpineni in the Republic of Moldovia reaches 55.5 mg-ecv/l (15), and in the Damman

Aquifer of Bahrain, reaches 3115 mg/l and an average of 1296 mg/l (16). The Mississippi River at Vicksburg has a hardness of 172 mg/l (2).

BIBLIOGRAPHY

1. Tebbutt, T.H.Y. (1995). *Principles of Water Quality Control*. Reprint of 4th Edn. Butterworth Heinemann, Oxford, UK.
2. Kemmer, F.N. (1988). *The NALCO Water Handbook*, 2nd Edn. McGraw-Hill, New York.
3. Freeze, R.A. and Cherry, J.A. (1979). *Groundwater*. Prentice-Hall, Englewood Cliffs, NJ.
4. Hounslow, A.W. (1995). *Water Quality Data: Analysis and Interpretation*. Lewis, New York.
5. McCoy, J.W. (1981). *The Chemical Treatment of Boiler Water*. Chemical Publishing, New York.
6. Todd, D.K. (1980). *Groundwater Hydrology*, 2nd Edn. Jonn Wiley & Sons, New York.
7. Sawyer, C.N. and McCarthy, P.L. (1967). *Chemistry for Sanitary Engineers*, 2nd Edn. McGraw-Hill, New York.
8. Hem, J.D. (1985). *Study and Interpretation of the Chemical Characteristics of Natural Waters*, 3rd Edn. U.S. Geological Survey Water-Supply Paper 2254, Denver, CO.
9. Manning, J.C. (1997). *Applied Principles of Hydrology*. Prentice-Hall, Upper Saddle River, NJ.
10. Bell, F.G. (1998). *Environmental Geology, Principles and Practice*. Blackwell Science, Cambridge, UK.
11. Yang, C.Y. and Hung, C.F. (1998). Colon cancer mortality and total hardness levels in Taiwan's drinking water. *Arch. Environ. Contamination Toxicol.* **35**(1): 148–151.
12. Muss, D.L. (1962). Relationship between water quality and deaths from cardiovascular disease. *J. Am. Water Works Assoc.* **54**: 1371–1378.
13. Akujieze, C.N., Coker, S.J.L., and Oteze, G.E. (2003). Groundwater in Nigeria—a millennium experience—distribution, practice, problems and solutions. *Hydrogeol. J.* **11**(2): 259–274.
14. Schulze-Makuch, D., Goodell, P., Kretzschmar, T., and Kennedy, J.F. (2003). Microbial and chemical characterization of a groundwater flow system in an intermountain basin of southern New Mexico, USA. *Hydrogeol. J.* **11**(3): 401–412.
15. Melian, R., Myrlian, N., Gouriev, A., Moraru, C., and Radstake, F. (1999). Groundwater quality and rural drinking-water supplies in the Republic of Moldova. *Hydrogeol. J.* **7**(2): 188–196.
16. Zuber, W.K. (1999). The Damman aquifer in Bahrain - Hydrochemical characterization and alternatives for management of groundwater quality. *Hydrogeol. J.* **7**(2): 197–208.

AN ANALYSIS OF THE IMPACT OF WATER ON HEALTH AND AGING: IS ALL WATER THE SAME?

HOWARD REMINICK
Ohno Institute on Water and Health
Willoughby, Ohio

The human body was created to be in balance or return to balance. We call this homeostasis. Our body relies on its chemical processing plant to keep a metabolic balance. The body knows when it is being depleted of essential minerals and nutrients, which creates imbalance. When this happens, the body has to draw from its reserves. But this can only be temporary, until these reserves are replenished.

Our body knows when its energy level is low. If nothing is done, if we continue to ignore the body's call for help, we let our body begin on the path of its destruction. When the body is neglected, it becomes vulnerable to attack from invaders from the outside. It also is open to attack from the inside by free radicals. Bacteria, which are essential for our survival, suddenly turn on us and destroy us.

The body reacts to regain its balance by calling on the immune system. Sometimes it is too late if the body has been in a contaminated state too long, which can result from the poor quality of water we put in our body every day.

Water has the power to transfer information from cell to cell and bring back balance from imbalance. Water is vital to maintain life and organization of our cells. Organized cells are healthy, productive cells. To create harmony and balance, whether in our bodies or in the environment, water must also be balanced.

The body knows when it is drying out. The body can keep its ability to remain balanced only if it is given enough good water. Not just clean water, but water that can provide a good internal environment for the body to function and carry on its purpose to remain balanced or return to being balanced.

Water holds some of the most basic and, therefore, significant secrets of life. This is not a new idea. All ancient civilizations described this in their attempt to explain the structure of life. Even then as today, many believe that water within the cell has a "vital force." Cells are the fundamental units of life, so then they are responsible for all life processes and are the force that permeates all nature. If that force is available in any one substance, then it must be in water.

The earth is 70% water. Our body is 70% water. A cell is 70% water. DNA is 70% water. Is this just a coincidence? About 3 billion years ago, life was born in water. The first life form existed in pure, clean water. There is a direct connection between the quality, content and even the structure of the water in our body that can make the difference between sickness and health.

One of the most important books written, which explains the necessity of water for health is *Your Body's Many Cries For Water*, by F. Batmanghelidj, a medical doctor, who presents the most compete description, and further warns us that this relationship is not being addressed by general medical practice (1). He attributes the basis of all disease to dehydration and overly prescribed medications. He also warns that dehydration is a world epidemic, which can be controlled by attention to the body's hydration needs.

Understanding disease and aging begins with knowing what our body needs throughout its lifetime to maintain health and vitality. Premature aging can result from loss of water volume without adequate replacement or from contaminated water in our bodies. Because of environmental toxins and chemicals, food and water sources have been depleted of essential minerals. As a result, the body's chemistry becomes unbalanced, which creates a breakdown in the body's internal environment, making it susceptible to disease and rapid aging.

The body must receive at least 2.5 liters of water daily to replace loss. Although beverages such as coffee, tea, soda, and beer contain water, they do not count as hydration replacement. These beverages actually use water to be processed in our body, which increases dehydration. Only water can replace water in our body.

Symptoms of the body's deterioration begin to appear when the body loses 5% of its total water requirements. In a healthy, normal adult, this appears as fatigue and general discomfort. Although an adult can survive a 5% water loss, this can be fatal to an infant. In elderly persons, a 5% water loss may not be fatal, but the body chemistry becomes abnormal, especially the percent of electrolytes being overbalanced by sodium. Continuous water loss over time will speed up aging. Lost water must be replaced (Fig. 1).

Because water is an excellent solvent, vital nutrients are absorbed in the cells. The body must receive the best quality of water available to ensure that its internal environment is in the best operational state. This allows the body to be at its best to respond to distress and imbalance by making corrections, returning the body to homeostasis.

Every function such as digestion, absorption, elimination of waste products, and production of glucose to provide energy is maintained if the body has water that is not only pure, but bio-active. This bioactivity is created by magnetic influence throughout the body. Minerals absorbed in the body's water are carried throughout the body to become ionizing agents for all cells. The effect of biomagnetism on hemoglobin in the blood is critical. Hemoglobin consists of iron, which is the catalyst in its reaction to the body's biomagnetism in its role of stimulating nerve and muscle impulses.

It should now be evident that there are essential issues about the water in our body that must be considered if we are to remain as healthy as possible. First and foremost, all water is not the same. Although all water is H_2O, that is where it ends. And there is no way to fool the body. Water must be able to prevent toxins and chemical substances from accumulating and creating destructive influences

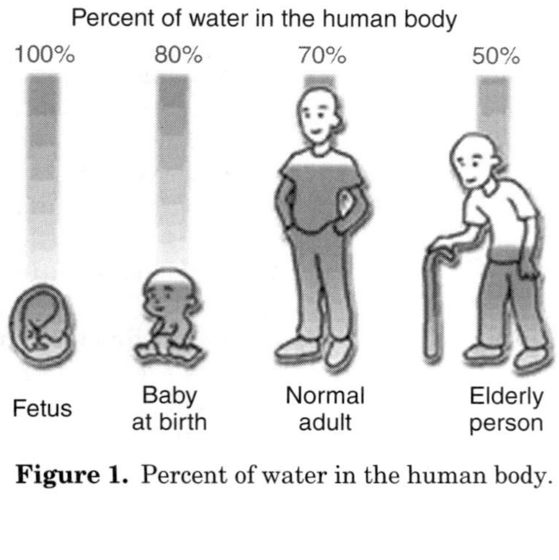

Figure 1. Percent of water in the human body.

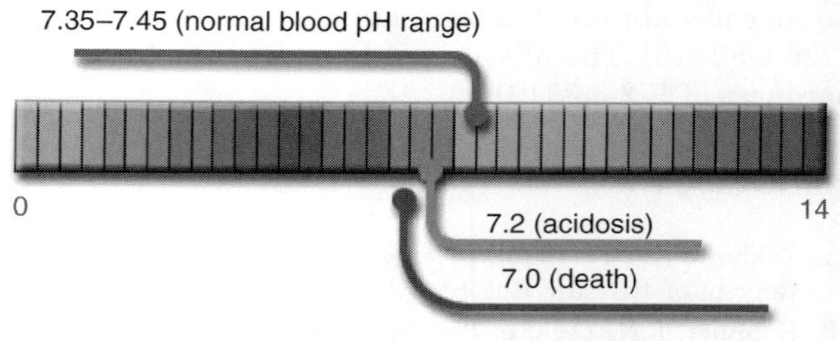

Figure 2. 7.35–7.45 (normal blood pH range).

in cells. Water must bring all minerals and nutrients required for metabolism into each cell and remove any substance that can damage the cell.

Another critical need in the body is the pH balance of blood and body fluids. Normal blood pH has a very small window of normalcy (Fig. 2). It must range between 7.35 and 7.45. There must be an adequate amount of oxygen in the blood. Too little will diminish all organic function. Too much can cause oxidation, which is what happens when an apple is left exposed. Because of the ever-increasing levels of pollution and toxicity in our environment, which includes food and water supplies, our blood tends to be acidic.

THE INTERACTIVE EFFECT OF MAGNETIZATION AND WATER ON CELL BIOLOGY

Everything that happens in our bodies every day of our lives depends on the presence of magnetism. Magnetism in our body is created by the magnetic field, which dominates and controls the stability of our planet Earth. It also controls the stability of all life on Earth.

Does magnetization of the water do anything to change its basic physical properties? The answer is yes. The changes are very small but measurable. Magnetized water exhibits a change in the strength with which water molecules are able to bind with one another. Magnetized water can disrupt hydrogen bonding in water. Water, charged with magnetic energy, allows minerals and other organic substances to provide intermediate pathways of flow for magnetic energy to be directed to where it is needed in the body. The fact that magnetized water possesses altered binding forces between water molecules and various proteins, salts, and mineral substances appears to endow it with the ability to dissolve buildups of various types of salts and minerals.

There is a fascinating fact about the level of energy in moving water in our body and its effect on ionization. Because all living things are known to function largely by the influence of bio-magnetism, it seems reasonable to provide the body with water that is magnetically activated. Normally, water forms large molecular clusters, which slows movement and absorption into cells. This can be seen by microscopic analysis [most water molecules will aggregate around the rim of a slide (Figs. 3, 4, and 5). Naturally magnetized water will disperse throughout the slide, indicating free flow]. Because water molecules, which are not adequately magnetized, will lose order when they are not adequately magnetized, this can affect intracellular hydration.

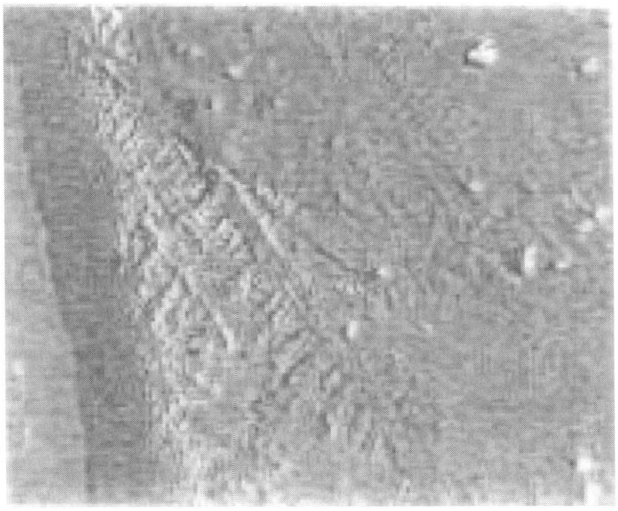

Figure 3. Microscopy photo of an unmagnetized water molecular structure. Note sparse formation of crystals and thick molecular clustering appearing as a mass and gathered at rim of slide.

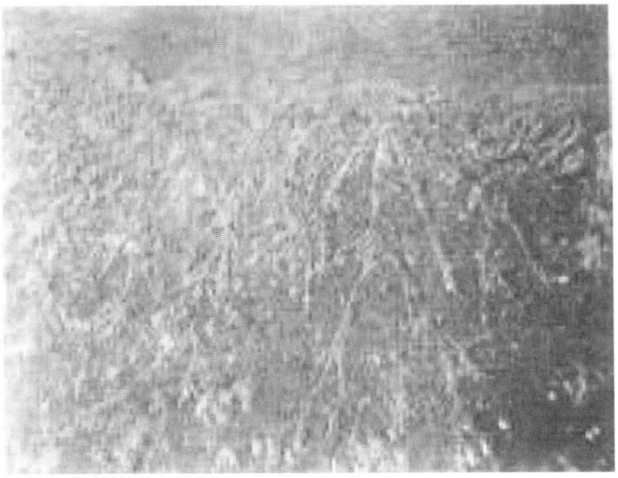

Figure 4. Microscopy photo of naturally magnetized water. Note large number and density of crystals and individually fragmented molecules, remaining evenly dispersed on slide.

Water has long been known to be affected by magnetism. Magnetization can significantly reduce the surface tension of water and increase the pH, making it more alkaline. Water molecules are known to have a small interaction with magnetism, creating chains of organized molecules, which can influence the biochemistry of the water in the body. The intricate biological requirements needed to maintain electrochemical balance have been reported in books and journals about human physiology. Magnetically activated water may increase the percentage of intracellular hydration, which is a major factor in controlling the aging process and aging-related diseases.

RELATIONSHIP OF MAGNETISM TO CELL STABILITY

What is a living cell? It appears as one large molecule tied together by a matrix of water molecules. Like the atoms that compose them, they are tied together by ionic bonds,

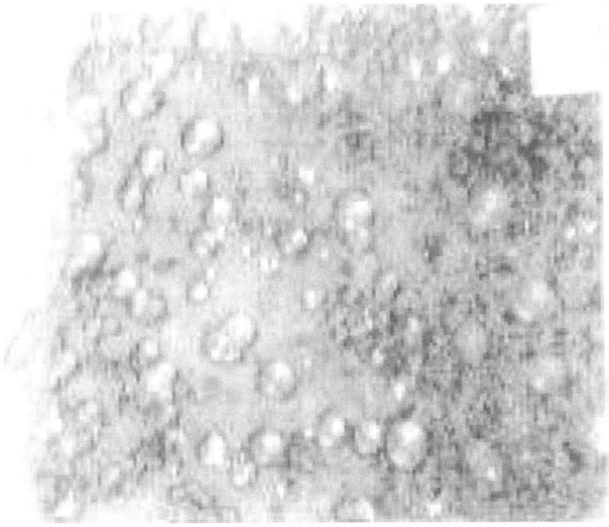

Figure 5. Microscopy photo showing detailed individual molecules of naturally magnetized water. They will not re-form in large clusters over time.

consisting of positive and negative charges on their outer surfaces. What creates these bonds?

If salts like sodium chloride are added to water, the sodium and chloride ions (Na^+ and Cl^-) immediately become surrounded by water molecules to dissipate their charges. In doing so, adjacent water molecules take on the charges to form H_2ONa^+, which possesses the unique property of being able to transfer the small positively charged nucleus of one of its hydrogen atoms to a neighboring water molecule. Then the neighboring molecule becomes positively charged as H_3O^+, due to the extra hydrogen proton bonded to its oxygen atom. This charged molecule transfers one of its protons to another neighbor, cascading in a linear fashion to water molecules surrounding negatively charged ions. Thus, the stability of a particular molecule is the electromagnetic bond of energies between atoms.

Magnetism affects the electron stability of cells. Electrons of a cell are established in pairs. Think of an unstable element moving around the body, looking for an electron to make itself stable. It will steal an electron from wherever it can, at the expense of a stable cell, which needs paired electrons to remain stable. When a cell loses an electron from a pair, it will lose energy and die, sacrificing itself to the "thief," which is how bacteria and cancer strive. This is believed to be the activity creating free radicals, that is, strong, unstable magnetic activity as a result of a cell with electron imbalance aggressively seeking another electron, thereby creating imbalance in another cell.

One of the first things we learn about the complex molecules of living cells is the way in which their atoms are bonded together. Magnetism takes place when electrons orbit around their atoms. Electrons have a property called spin, which makes them act like miniature magnets with two opposite poles. If neighboring electrons spin so their poles are aligned in the same direction, they become magnetic. The electrons circling around atoms resist attempts to disturb their usual orbits, so they can create their own magnetic field.

The molecular parts which compose the cells of our bodies are much the same as those that composed the earliest forms of life on Earth. If we wish to comprehend the composition of life and its molecular roots and determine how these molecules move and flow in a spontaneous and miraculous fashion to sustain and perpetuate themselves, then, like nature, we must begin with water. Genetics defines which molecules are produced; water defines how they move and associate with each other.

Modern scientific technology has revealed the basis of molecular biology in the cell, its DNA, and how critical water is to its order. If DNA is dehydrated, like other bonding components of the cell, it loses its orderly, repeating structures and takes on more random shapes and forms. In other words, DNA loses its normal, functional properties.

It is now clear that all information needed to direct molecules into their functional shapes is contained in the molecules themselves in association with each other and the surrounding water medium. DNA, operating in harmony with water, determines shapes that are adopted as they are produced and move within the cell. It appears that DNA molecules themselves are the "code" and water the medium where the code is expressed.

As long as environment and temperature are maintained within the cell and there is adequate magnetization to keep cell biology functioning normally, then cellular reproduction, along with normal DNA coding, will continue. If, however, the content and effectiveness of the water inside and outside the cell contain destructive toxins, the degradation and deterioration of cells will increase quickly, resulting in rapid aging and disease.

THE EFFECT OF WATER ON CELLS

Fluid fluctuations affect blood volume and cellular function, which can be life threatening. When the amount of fluid within various cellular compartments is relatively constant, there is an exchange of solutes and water between compartments to maintain unique compositions. Individuals with more body fat have proportionately less total body water and are more susceptible to fluid imbalances that cause hydration.

The total body water percentage decreases with age, resulting in inadequate cellular hydration. Most critical is the decrease in intracellular fluid, which is due, in part, to increased fat and decreased muscle, and decreased ability to regulate the sodium and water balance. With age, kidney function becomes less efficient in producing urine, and responses for conserving sodium weaken.

There are two major issues about the water in our body that must be resolved if the body is to remain in a state of harmony and stability. First, the water must be able to prevent toxins and chemical substances from accumulating and creating destructive influences in the cells. Water must bring all minerals and nutrients required for metabolism into the cell and remove any substances that can damage the cell. Also, the cell wall must be protected from damage and invasion. Second, water is involved in every function of the body, so it must act as a conductor of electrochemical activity, such as neurotransmission, by moving water from one nerve cell to another smoothly and effectively.

When cells are studied under magnetic resonance imaging (MRI), water is observed to act very differently from when it is not exposed to a magnetic field. Molecules are re-formed in smaller clusters in a linear arrangement. The molecules are lined up and move in and out of the cells easily. In contrast, when water is photographed after removal from MRI exposure, molecules are randomly ordered and cluster with neighboring water molecules to form large, molecular clusters. (Interestingly, when cancer cells are studied under MRI influence, their water molecules are also organized and calm, in contrast to their aggressive, violent movement without MRI exposure.) However, naturally magnetized water remains organized in small clusters, whether exposed to MRI or not.

THE PASSAGE OF WATER IN AND OUT OF CELLS

Movement of water occurs because of hydrostatic pressure and osmosis in capillaries. As water flows from artery to vein, changes in pressure create movement across membranes. Any changes in pressure allow proteins, minerals, and other nutrients carried by the blood to escape into spaces between vessels and deprive the cells of vital needs to sustain life. In addition, when the water in the blood is acidic and contaminated with chemicals, it enters the cells and changes their structure, including changes in DNA, which is how disease begins and attacks structures in the body.

Osmosis is the diffusion of water across a membrane that permits the free passage of water but not solutes. Molecules of solutes (sugars, amino acids, and ions) dissolved in water are in constant motion, which creates diffusion, the random movement toward zones where the concentration of the molecules is lower. This tends to distribute molecules uniformly. Because of diffusion, both solvent (water) and solute molecules in a cell will move from areas where concentration is greater to an area where their concentration is less.

As water molecules continue to diffuse inward toward the area of lower concentration of unbound water molecules, pressure within the cell increases, which is called hydrostatic pressure. This is the force that must be applied to stop osmotic movement of water across a membrane. When two areas are separated by a membrane, what happens depends on whether the molecule can pass freely through the membrane; most solutes in cells cannot do so.

Solutes are water soluble and are kept within the cell. Water molecules, however, can diffuse across the membrane into the cell, thus diluting the high concentration of solutes within the cell so that it matches more closely the lower concentration in the solution outside the cell (extracellular matrix). Eventually, however, the cell will reach an equilibrium, and the osmotic force driving water inward is counterbalanced exactly by the hydrostatic pressure driving water out. As a result, the water and solute concentrations will be the same in both compartments.

The nature and proportion of solutes in the body's water determine its effectiveness in hydrating and detoxifying cells so that a homeostatic, internal environment can be maintained to ensure proper cell metabolism, which requires bioenergy. A key component in dispensing bioenergy is the interactive effect of water and biomagnetism. This interactive bioenergy is seen in the effectiveness of osmosis and the osmolarity of the water in the extracellular matrix, as well as in the cells. Through the influence of the body's bio-magnetism, greater penetration of ions through the cell membrane, transported by water, will increase the health and stability of cells.

Osmosis is created by a magnetic force in our body, which keeps the movement in balance. Osmotic activity occurs as a result of osmolarity, the total concentration of all solute particles in a solution. The extent to which water's concentration is decreased by solutes depends on the number, not the type, of solute particles in a solution, because one molecule or one ion of solute displaces one water molecule. Thus, a solution's osmolarity is a count of the number of solute particles (including ions) dissolved in one liter of water and is seen in the solution's ability to cause osmosis. The measurement used to describe the solute concentration of body fluids such as water is the milliosmol. The normal osmolality of body fluids should average around 270 milliosmoles.

Osmolarity controls the distribution and movement of water between body compartments. The osmolality of intracellular and extracellular fluid tends to equalize and, therefore, provides a measure of body fluid concentration and the body's hydration status. Osmolarity measures the effective change in pressure for water, assuming that all the solutes are impenetrable. When a solute is added to water, the volume is expanded and includes the original water plus the volume occupied by the solute particles. In measuring osmolarity, the volume of water is, therefore, reduced by an amount equal to the volume of added solute.

When equal volumes of solutions of different osmolarity are separated by a membrane that is permeable to all molecules in the system, net diffusion (tendency of the molecules or ions to scatter evenly) of both solute and water occurs, each moving down its own concentration gradient. A solution that changes the tone or shape of a cell by altering its internal water volume is described by its osmotic activity, or tonicity. Tonicity is the tendency of a solution to resist expansion of intracellular fluid volume.

Solutions with concentrations of nonpenetrating solutes equal to those found in cells are isotonic, or have the same tonicity. An isotonic solution has the same osmolality as the intracellular fluid (ICF) or the extracellular fluid (ECF), which is required to maintain normal cell metabolism. An example of an isotonic solution is 5% dextrose in water and a normal saline solution of 0.9% NaCl. This solution is isotonic because there is no net movement of the active particles in the solution into cells and the particles are not metabolized. Extracellular fluids and most intravenous solutions are isotonic.

Solutions that contain fewer nonpenetrating solutes are called hypotonic. Cells placed in hypotonic solutions plump up rapidly as water rushes into them. Distilled water represents the most extreme example of hypotonicity. Because it contains no solutes, distilled water pulls water into cells without proper resistance or osmotic pressure and causes them to swell or burst.

A hypertonic solution has a concentration greater than 285 to 294 milliosmoles. An example is a 3% saline solution. Water is pulled out of the cells so that the cells shrink. Tonicity is critical to cell metabolism when correcting water and solute imbalances by administering different types of replacement solutions, such as those used in IV therapy solutions.

Balanced osmolarity can be the difference between the life and death of an organism. An example is seawater, overbalanced with sodium ions, or distilled water, which can be deadly if infused because it has no osmolarity. If distilled water is present on both sides of a permeable membrane, no net osmosis occurs, even though water molecules continue to move in both directions through the membrane. If, however, the solute concentration on both sides of the membrane differs, water concentration differs as well, because as the solute concentration increases, water concentration decreases.

There is a consideration when determining osmotic activity. In living plant cells, as water diffuses into the cell, the point is reached where the hydrostatic pressure (the back pressure exerted by water against the membrane) within the cell is equal to its osmotic pressure (tendency to resist further net water entry). Generally, the higher the amount of nonpenetrable solutes within the cell, the higher the osmotic pressure and the greater the hydrostatic pressure that must be present to resist further net water entry. Osmotic pressure is also necessary to control water loss.

However, these pressure changes do not occur in living animal cells, which lack rigid walls and are surrounded only by flexible plasma membranes. Osmotic imbalances cause animal cells to swell or shrink, due to net water gain or loss, until the solute concentration on both sides of the plasma membrane is the same or the membrane is stretched to its limit. Many molecules, particularly intercellular proteins and selected ions, are prevented from diffusing through the plasma membrane. Any changes in their concentration produce changes in water concentration on the two sides of the membrane and result in net loss or gain of water by the cell.

THE EXTRACELL MATRIX AND WATER

Our cells are continuously bathed in an extracellular fluid called interstitial, found outside the cell. Because it contains such a large variety of ingredients, such as amino acids, sugars, fatty acids, vitamins, hormones, neurotransmitters, and salts, it is sometimes referred to as the "soup" that feeds the cells. It is also the dumping ground for waste products removed from the cells. To remain healthy, each cell must absorb substances from this "soup" in the exact amounts it needs at specific times and reject the rest.

The human body can maintain homeostasis only if the extracellular fluid (ECF) volume is closely regulated to maintain blood pressure and salt balance. Its osmolarity must be regulated to prevent swelling or shrinking of the cells. The maintenance of water balance is important in regulating ECF osmolarity. To maintain stability, the ECF must respond quickly to changes in volume and osmolarity (2).

In humans, kidneys can spare water by eliminating wastes in concentrated urine. This urinary concentration relies on water reabsorption driven by an osmotic gradient formed along the collecting ducts, consisting of NaCl and urea at similar osmolarities. Survival of renal cells is induced by NaCl, but not urea, because urea is a membrane-permeant solute. With excess saline in the tissues, there is a decrease in water excretion to bring the salt concentration back to normal, which allows the cell concentration to remain constant and prevent shrinking by fluid moving out of the cells to reach equilibrium within the cells.

The environment of a cell is directly affected by the environment outside the cell. Cells have a reciprocal relationship with their environment. The environment surrounding a cell consists of a structured, basic substance, called the extracellular matrix (ECM). This enormous space of connective tissues has a significant effect on determining genetic outcomes of the cell. Water is its primary regulation system. The ion composition of the elements in the extracellular space is activated in water.

Understanding how a cell maintains its integrity requires understanding the role of the extracellular matrix in the body in the health and longevity of its internal environment. Most body cells are in contact with a jelly-like substance composed of proteins and polysaccharides. These molecules are secreted by the cells into an organized mesh in the extracellular matrix, where they serve as a sort of glue that helps hold cells together. In reference to the cells, this can be viewed as the mortar between bricks of a wall. Depending on the time of day, this "mortar" can be gelatin-like or liquid because of the impact of the sympathetic and parasympathetic nervous systems, which perform different functions at different times of the day.

During the 12-hour "dumping cycle" (3:00 A.M. to 3:00 P.M.), controlled by the sympathetic nervous system, the "mortar" is gelatose to remove and retain wastes from cells and is acidic. During the 12-hour "feeding cycle" (3:00 P.M. to 3:00 A.M.), controlled by the parasympathetic nervous system, the "mortar" is liquid to transport nutrients to cells and is alkaline. It is the effect of magnetically activated water in the extracell matrix that assures an optimal delivery system and that these cycles function properly with pH balance (Fig. 6).

The ECM is the largest system in the body. It is virtually responsible for all vital functions that take place within the body. All cells depend on an intact ECM for their existence. It regulates the cell's environment and is part of every aspect of the immune system because the lymphatic system is connected with it. It takes care of cell nutrition through internal circulation of water and

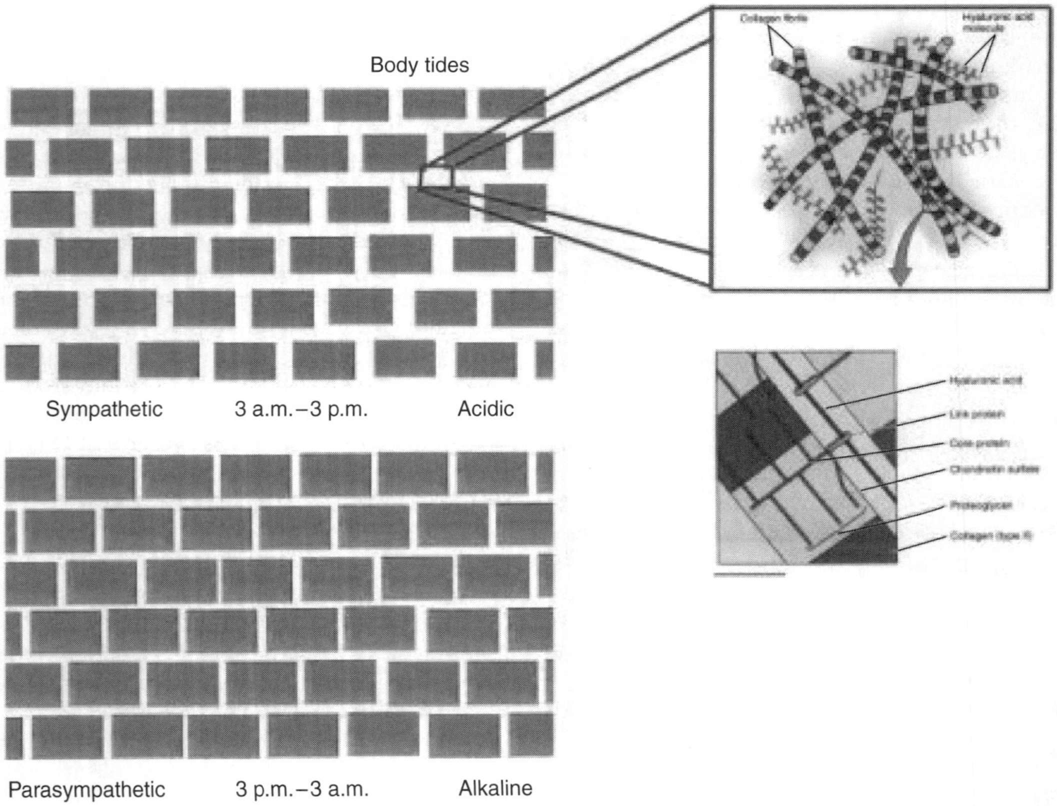

Figure 6. ECM Cycle.

removal of wastes. Disease originates in dysfunctions in this space, as it affects organs and other systems throughout the body.

More than having a supporting and space-filling function, the ECM provides regeneration for specific organ functions, as well as mediation of nerve and vascular functions. Everything that comes out of the blood takes a complicated route through the connective tissue to the cell and then into the lymphatic system. It is even more vital to the health of the human body because the nerves and blood vessels do not come into direct contact with the cells at any point; therefore, the ECM becomes the mediator for nervous and nutritional flow. As connective tissue is really a living part of the body, everything attributable to the life and health of the body, and all its basic functions, has to be attributed to the extracellular matrix.

The extracellular matrix provides the communication of connective tissue over the entire organism. The entire field of activity and information related to communication is the extracellular fluid. Connective tissue is found everywhere between the bloodstream and the principal structural parts. Everything that comes out of the blood takes a complicated route through the connective tissue to the primary, functioning cell and then into the lymphatic system, which is connected with it. It takes care of cell nutrition through internal circulation of water and the removal of wastes. It regulates the cell's environment and is part of every defense process.

The extracellular matrix is connected to the endocrine gland system via capillaries and the central nervous system via the peripheral nerve endings. Both systems are connected to each other in the brain stem, so superior regulatory centers can be influenced by the ECM. The advantage of the interconnection of these systems results in a significant increase in the performance capacity of the ECM, which can be attained only by highly functional, interconnected sets of systems, functioning as one.

When looking for the source of disease and aging, we must first look at the water in our bodies. I would like to repeat the statement that led me to search for the causes of aging diseases: If the body is composed of 70% water, then the condition of the water in our body can determine how well our body maintains its state of health. The structure of all living tissue is predominately water, a plain and simple substance. Knowing the importance of water in the body can help us learn how to keep the body from breaking down. Further, understanding that a living organism is composed mostly of water should help us better appreciate the need to supply our body with the best water possible.

The signature of life is found in the water we drink every day. A body can remain in a state of health only if it is given enough water, not just clean water, but water that has the structure, properties, and activity to provide a good internal environment for a body to function at its best. We become sick, not just from dehydration, but from the inadequacy of the water we drink. The question becomes, "is all water the same?" The answer is a definite no. Only water that is bioactive with magnetic energy can provide ideal hydration.

After years of investigation on the biological effects of magnetized water on human health, it became apparent that all magnetized water is not the same, nor does each behave the same in the body. Water that is magnetized with devices displays structural changes initially but will return to its original formation because every water source is unique; each has its own frequency, its own signal, due to its unique memory. Artificially altering water, therefore, will not create a permanent structural change. In all of our studies, only water that is naturally magnetized, which creates the water's memory, will perform as the optimal hydration source consistently and aid the body in maintaining a better internal environment, resulting in better health and more natural aging.

To learn more about the phenomenon of naturally magnetized water, go to http://www.ohno.org.

BIBLIOGRAPHY

1. Batmanghelidj, F. (1997). *Your Body's Many Cries for Water*. Global Health Solutions, Vienna, VA.
2. Sherwood, L. (1997). *Human Physiology: From Cells to Systems*, 3rd Edn. Wadsworth, Belmont, CA.

READING LIST

Abrams, A. (1913). *Human Energy*. Borderland Sciences, San Francisco, CA.

Ackerman, J.M. (2001). The Biophysics of the Vascular Autonomic Signal and Healing, *New Perspectives, Journal of the Center for Frontier Sciences of Temple University* **10**(2).

Collins, J.C. (2000). *Water: The Vital Force of Life*. Molecular Presentations, Kinderhook, NY.

Collins, J.C. (1991). *The Matrix of Life*. Molecular Presentations, East Greenbush, NY.

Davis, R.D. and Rawls, W.C., Jr. (1996). *Magnetism and Its Effects on the Living System*. Acres, Metairie, LA.

Dubrov, A.P. (2001). Russian Advances in the Study of the Effects of Environmental Factors and Electromagnetic Medicine. *New Perspectives, Journal of the Center for Frontier Sciences of Temple University* **10**(2).

Ganong, W.F. (1993). *Review of Medical Physiology*, 16th Edn. Appleton & Lange, Norwalk, CT.

Gerber, R. (1996). *Vibrational Medicine*. Bear, Santa Fe, NM.

Marieb, E.N. (1995). *Human Anatomy and Physiology*, 3rd Edn. Benjamin/Cummings, Redwood City, CA.

McCance, K.L. and Huether, S.E. (1994). *Pathophysiology: The Biologic Basis for Disease in Adults and Children*. Mosby, St. Louis, MD.

Oschman, J.L. (2001). Energy Medicine: The Scientific Basis. *New Perspectives, Journal of the Center for Frontier Sciences of Temple University* **10**(2).

Oschman, J.L. (2000). *Energy Medicine: The Scientific Basis*. Churchill Livingstone, Edinburgh, UK.

Osyka, F. and Edmonds. (1991). *Critical Necessity of Negative/Positive Ion Balance: The Ion Effect*. Alpine Industries, Greeneville, TN.

Philpott, W.H. and Kalita, D.K. (2000). *Magnet Therapy*. Alternative Medicine.com Books, Tiburon, CA.

Pischinger, A. (1975). *Matrix and Matrix Regulation*. Haug International, Brussels.

HEAVY WATER

GHOLAM A. KAZEMI
Shahrood University of Technology
Shahrood, Iran

Heavy, heavy water
Won't wash away the sins of the father
Unholy, holy water
Leading us like lambs to the (slaughter)

James Young
(URL Ref. http://www.styxnet.com/styxlyrics/hw2.htm)

INTRODUCTION

This article should be studied in conjunction with the closely related articles on Deuterium and Isotope Fractionation. In some ways, it might have been slightly more appropriate if the Deuterium and Heavy Water articles were combined into one. For example, the whole introduction and some parts of the text of both articles overlap, making it more or less difficult to draw a sharp boundary between the two.

Each of the atoms of hydrogen and oxygen, which form the water molecule, has three isotopes; ^1H, ^2H (D), ^3H (T) for hydrogen and ^{16}O, ^{17}O, and ^{18}O for oxygen. The average terrestrial abundances of these isotopes are given in Table 1 (1).

The abundances of these isotopes in the hydrosphere, biosphere, and lithosphere and in different stages of the water cycle vary slightly from these figures.

Considering the six isotopes of hydrogen and oxygen, 18 combinations of H–O–H or "18 varieties of water" are possible (2,3). These and their respective molecular weights are shown in Table 2.

Table 1. Average (%) Terrestrial Abundances of Hydrogen and Oxygen Isotopes[a]

^1H (Proteum - stable)	99.985	^{16}O (stable)	99.762
^2H (Deuterium - stable)	0.015	^{17}O (stable)	0.038
^3H (Tritium -radioactive[b])	$< 10^{-14c}$	^{18}O (stable)	0.200

[a]Reference 1.
[b]$t_{1/2} = 12.3$ years
[c]Some references give 10^{-14}–10^{-16}.

Of the above eighteen varieties of waters, $^1H_2{}^{16}O$ has a mass of 18 and is the most common; it is generally known as "water" or sometime called ordinary or light water. Deuterium oxide ($D_2{}^{16}O$) has a mass of 20 and is known as "heavy water." Some references use the term "semiheavy water" for $HD^{16}O$. Washburn and Urey (4) were the first to enrich deuterium oxide in water in 1932. Some of the 18 varieties of water are extremely rare. An interesting, though trivial, point is that the molecular weight of the most common form of water $^1H_2{}^{16}O$ (18), equals the number of different varieties of water, 18.

CHARACTERISTICS OF HEAVY WATER

The physicochemical properties of heavy water and deuterium are compared with those of water and hydrogen in Table 3.

The extent of dissociation of heavy water is about half that of ordinary water:

$$D_2O \rightarrow D^+ + OD^-$$

and

$$H_2O \rightarrow H^+ + OH^-$$

Table 2. Eighteen Varieties of Water and Their Respective Masses

$H_2{}^{16}O$	18	$D_2{}^{16}O$	20	$T_2{}^{16}O$	22	$HD^{16}O$	19	$HT^{16}O$	20	$DT^{16}O$	21
$H_2{}^{17}O$	19	$D_2{}^{17}O$	21	$T_2{}^{17}O$	23	$HD^{17}O$	20	$HT^{17}O$	21	$DT^{17}O$	22
$H_2{}^{18}O$	20	$D_2{}^{18}O$	22	$T_2{}^{18}O$	24	$HD^{18}O$	21	$HT^{18}O$	22	$DT^{18}O$	23

Table 3. Physicochemical Properties of Heavy Water, Water, Deuterium, and Hydrogen[a]

Characteristics	Heavy water	Water	Deuterium	Hydrogen
Density at 25 °C (g/ml)	1.1045	0.99707		0.0899
Boiling point, °C	101.42	100	−249.55	−252.77
Freezing point, °C	3.8	0	−254.5	−259.2
Molecular volume at 20 °C	18.092	18.016		
Viscosity, poise at 20 °C	12.6	10.09		
Solubility of NaCl, g/100 g	30.5	35.9		
Critical pressure, atm			16.432	12.807
Evaporational enthalpy, kJ/mol	46.5	40.7		
Dielectric constant	78.25[b]	78.54[b]	1.277[c]	1.228[c]
Ionization potential, eV	12.6	12.6	15.46	15.427
Bond length at 298 K, Å	104.45	0.958 angle	0.7416	0.7461
Bond strength kJ/mol	—	Diatomic	443.546	436.002

[a]References 5–7. Note that in many references, the boiling and freezing points of H_2 and D_2 are given in K.
[b]At 20°K.
[c]At 25 °C.

This means that a bond to deuterium breaks 18 times more slowly than the same bond to hydrogen (6). This is the reason that the vapor pressure of heavy water is lower than that of ordinary water.

The amazing behavior of heavy water is its inertness in normal metabolic processes. For example, rats will die of thirst if given heavy water exclusively, and plants will not grow in pure heavy water (8). Experiments with mice showed that heavy water, due to a slightly different reaction rate compared to ordinary water, inhibits mitosis and causes progressive damage of tissues that need rapid regeneration. Human drinking of up to 25% heavy water is not harmful, but the main problem is consequent lack of enough ordinary water in the body. After several days of ingesting only heavy water, the body fluids contain about 50% heavy water, and at this point symptoms begin due to the decrease in cell division rates of rapidly dividing tissues, such as hair roots and stomach linings. Drinking heavy water may lead to the remission of aggressive cancer cell, but the, effect predicted is not great enough to make this a useful therapy (URL Ref. http://en2.wikipedia.org/wiki/Heavy_water#Toxicity).

In slight contrast to these publications, Kushner et al. (9) report that, so long as it has not been used in nuclear reactors, D_2O itself poses little risk to the environment because:

1. D_2O itself is not toxic to animals, except under exceptional and very improbable circumstances;
2. D_2O is nonradioactive; and
3. The amount of the ^3H, found in heavy water is very low, <6 kBq/kg D_2O (<0.15 μCi/kg D_2O), barely above background levels.

More information on the biological aspects of heavy water may be found in Katz (10). Dahl (11) provides good coverage of many aspects of heavy water, including discovery, uses, production techniques, its relationship to developments in nuclear physics and chemistry, social history, and a chronology of events relating to heavy water.

APPLICATIONS OF HEAVY WATER

Heavy water is used (1) as a neutron moderator, (2) to detect neutrinos, and (3) in the biology/biotechnology field. Biotechnological applications of heavy water are in the early stages of research and have not yet received widespread attention.

Neutron Moderator

The poem from James Young, prefacing this article, is directed toward this application of heavy water, its use in nuclear reactors. In nuclear reactors, an important factor in producing a chain reaction in uranium is the energy distribution of neutrons, which peaks near 1 MeV. To reduce the neutrons energies from 1 MeV to thermal energies (so they can react with the uranium in the reactor) would require at least 2000 collisions (with something else). The most common materials used to slow down neutron energy are heavy water and graphite (12). A huge nuclear reactor that uses heavy water as a moderator is CANDU in Canada (Fig. 1). The advantage of heavy water is that it has a low probability of absorbing neutrons. Ordinary water also acts as a moderator but because light water absorbs neutrons, reactors using light water must use enriched uranium rather than natural uranium; otherwise criticality is impossible. A nation that has a sufficiently powerful heavy water reactor can use it to turn uranium into bomb-usable plutonium, entirely bypassing uranium enrichment and all of the related technological infrastructure. Heavy water reactors have been used for this purpose by India, Israel, Pakistan, and North Korea. Due to its usefulness in nuclear

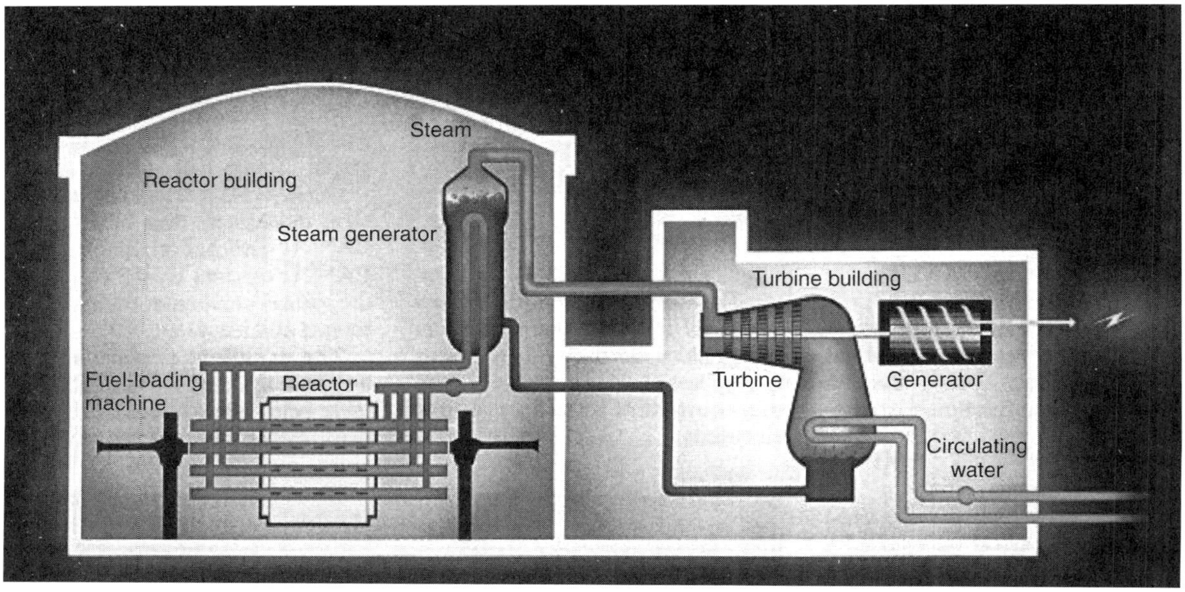

Figure 1. The layout of the CANDU heavy water reactor in Canada (URL Ref http://www.nucleartourist.com/type/candu.htm).

weapons programs, heavy water is subject to government control in several countries. In addition, heavy-water-moderated reactors can be used to make tritium (URL Ref. http://www.fas.org/nuke/intro/nuke/heavy.htm).

Neutrino Detector

The neutrino, discovered in 1956, is an elementary particle that has no charge, zero rest mass and travels at the speed of light (13). The detection of this particle is a difficult and complicated task. The Sudbury Neutrino Observatory (SNO), near Sudbury, Ontario, is a heavy-water Cherenkov detector that is designed to detect neutrinos produced by fusion reactions in the Sun. It is located 6800 feet underground in an old mine (to shield it from cosmic rays) and uses 1000 tonnes of heavy water on loan from Atomic Energy of Canada Limited (AECL), contained in a 12-meter diameter acrylic vessel. Neutrinos react with heavy water (D_2O) to produce flashes of light called Cherenkov radiation. This light is then detected by an array of 9600 photomultiplier tubes mounted on a snoball surrounding the heavy water vessel. The detector laboratory is extremely clean to reduce background signals from radioactive elements present in the mine dust which would otherwise hide the very weak signal from neutrinos (http://sps.la.asu.edu/~steve/paper/sno/node6.html).

Biological and Biotechnological Applications of Heavy Water

Kushner et al. (9) describe the potentially extensive use of heavy water and deuterated compounds in biotechnological applications. They report as follows:

1. Many microorganisms can grow in "pure" (99.6–99.8%) D_2O; usually, after a period of adaptation, they can produce a very large number of deuterated compounds that have many applications in optical technology, in deuterated drugs and fibers, etc.
2. Deuterated glycerol has interesting potential uses as a precursor of transparent plastics and other compounds and can be formed by *Dunaliella* spp, a salt-tolerant alga that grows in D_2O.
3. Heavy water has some direct therapeutic uses in both animals and humans. Nontoxic levels of heavy water, for instance, reduce induced hypertension and associated changes in rats.
4. Heavy water has been widely used in measuring body water space in children and in adults at various stages of nutrition and exercise.
5. Applications of deuterium in environmental studies are numerous. (These are dealt with in the Deuterium article.)

PRODUCTION OF HEAVY WATER

Heavy water occurs naturally in limited, yet precisely unknown, quantities. Production of heavy water includes separating these limited quantities (not actually making them) from other ordinary water molecules. In the hydrosphere, one in every 6700 molecules of water contains deuterium in its composition. If ^{16}O is the other atom of all of these "one in 6700" molecules, then the ratio of heavy water/water in the hydrosphere is 1/6700. This ratio, though most likely realistic, may be smaller because some of the oxygen atoms in water molecules that contain deuterium are ^{17}O and ^{18}O. In other words, not all deuterium isotopes join with ^{16}O to form heavy water; some are combined with ^{17}O and ^{18}O to form other varieties of water depicted in Table 2.

The discovery of deuterium in 1932 by Urey and his colleagues (14), may be considered the onset of the artificial production of heavy water. However, Gilreath (15, p. 61) in this regards, writes, "In 1933, G. N. Lewis and his collaborators at the University of California succeeded, by the continuous electrolysis of a very large amount of water obtained from an old commercial electrolytic cell, in preparing about 0.1 cc of water which contained practically only deuterium atoms of mass 2. The specific gravity of this water was about 10% greater than that of ordinary water which was in accordance with the molecular weight of D_2O." This paragraph may indicate that the *first drop* of heavy water was produced in 1933.

There are a few methods for artificially enriching the deuterium content of ordinary water and, thus, producing heavy water. These include distillation, electrolysis, and chemical exchange.

Distillation

The foundation of this method is based on the difference between the vapor pressures and boiling points of H and D. This was the principle used by Urey to discover deuterium. The most efficient process for enriching heavy water by this method uses a fractionating column in a continuous process, in which an upward-directed stream of vapor is kept in contact with a downward-directed stream of liquid; the two streams constantly exchange molecules. The molecules of H_2O have a relatively greater tendency to enter the vapor stream, and the molecules of D_2O enter the liquid (15). In 1943, the first plant based on distillation was built to produce heavy water for use as a moderator in nuclear reactors. However, the plant operated for less than a year only because (1) it was found that the chemical exchange method is more efficient and (2) it was learned that graphite is superior to D_2O as a moderator (15).

Electrolysis

The principle of this method, a common method for laboratory scale production of heavy water and described in many chemistry textbooks, is simple. Passage of an electric current through water leads to splitting water into hydrogen and oxygen. The deuterium atoms move more slowly because they are heavier and, so, it is possible to separate them. Finally, the resulting deuterium is reacted with oxygen to form heavy water. The electrolytic technique consists of up to seven stages, depending on the purity of heavy water required; the higher the purity of the required heavy water, the greater the number of stages of the process. This technique takes a vast amount of electrical energy. In a typical experiment, 2400 liters of ordinary water produce 83 ml of D_2O that is 99% pure (6).

Chemical Exchange

Minute differences between chemical properties and the reaction rates of hydrogen and deuterium form the foundation of this method. In this process, water flows from the top of a tower that contains a catalyst, downward to meet a mixture of steam and hydrogen moving upward. The catalyst promotes a rapid isotopic exchange reaction between the hydrogen and the water molecules in the steam; hence, there is a concentration of deuterium in the steam in the form of heavy water molecules. The flow of water from the top of tower causes condensation of the steam and, consequently, enrichment of deuterium in the outflow from the bottom of the tower. The outflow is then electrolyzed to increase its deuterium content and it is, then, repeatedly reprocessed in the system to achieve a high concentration of D_2O (15). The refined/advanced version of this method is now known as the Girdler Sulfide process (9—see the following section about the Bruce Heavy Water Plant for more information about this method).

WORLDWIDE PRODUCTION OF HEAVY WATER (THE FOLLOWING INFORMATION HAS BEEN GATHERED FROM WEB SITES)

As can be inferred from the previous explanations, a huge technical infrastructure is required to produce heavy water, and, thus, only a few countries such as Argentina, Canada, India, and Norway have the capacity to produce significant amounts.

The first commercial heavy water plant was the Norsk Hydro facility in Norway (built in 1934, capacity 12 metric tons per year); this is the plant which was attacked by the Allies to deny heavy water to Germany during World War II.

India is the world's second largest producer of heavy water through its Heavy Water Board of the Department of Energy (URL Ref: http://www.heavywaterboard.org), but accidents and shutdowns may have led to limitations on its production rate.

The Bruce Heavy Water Plant in Ontario, Canada (Fig. 2), built in 1979, has a capacity of 700 metric tons/year; it is the world's largest producer of D^2O (1979; 700 metric tons/year). It uses the Girdler Sulfide (GS) process which incorporates a double cascade in each step. In the upper ("cold," 30–40 °C) section, deuterium from hydrogen sulfide preferentially migrates into water. In the lower ("hot," 120–140 °C) section, deuterium preferentially migrates from water into hydrogen sulfide. An appropriate cascade arrangement actually accomplishes enrichment. In the first stage, the gas is enriched from 0.015% deuterium to 0.07%. The second column enriches this to 0.35%, and the third column achieves an enrichment between 10% and 30% deuterium. This product is sent to a distillation unit for finishing to 99.75% "reactor-grade" heavy water. Only about one-fifth of the deuterium in the plant feed water becomes heavy water product. The production of a single pound of heavy water requires 340,000 pounds of feed water. Heavy water produced by Bruce is used in CANDU reactors.

Figure 2. Bruce Heavy Water Plant in Ontario, Canada (Toronto Star, Online Edition, Dec. 2, 2003).

The Bruce Power Plant, which presently operates in the shutdown mode, was going to be decommissioned a few years ago, but its license was renewed in 2002 for 2 more years (Nuclear Energy Overview, 4 November 2002 cited in World Nuclear Association, WNA, News Briefing 02.45, 30 Oct to 5 Nov. 2002). The latest news headline about Bruce Power Plant, appeared in Toronto Star online edition, Dec. 2, 2003:

"Safety commission finds 'serious event' at Bruce

A defect in the backup shutdown system of a reactor at the Bruce B nuclear generating station went undetected for more than 3 weeks in December, according to a newly released report by the Canadian Nuclear Safety Commission."

There are a few website publications which claim that some countries, Iran, for example, are pursuing plans for heavy water reactors. One such report is that by J. Boureston and C. Mahaffey from Janes International Security which can be found at: http://www.janes.com/security/international_security/news/jir/jir031114_1_n.shtml. There are also locality maps of various heavy water reactors and production plants, photographs, statistics, and many other heavy-water-related publications available at various websites. A useful site is that of the Heavy Water Board of India at http://www.heavywaterboard.org.

BIBLIOGRAPHY

1. IUPAC. (1992). Isotope compositions of the elements-1989. *Pure Appl. Chem.* **63**: 991–1002.
2. Freeze, R.A. and Cherry, J.A. (1979). *Groundwater*. Prentice-Hall, Englewood Cliffs, NJ.
3. Hem, J.D. (1985). *Study and Interpretation of the Chemical Characteristics of Natural Waters*, 3rd Edn. U.S. Geological Survey Water-Supply Paper 2254, Denver, CO.
4. Washburn, E.W. and Urey, H.C. (1932). *Proc. Natl. Acad. Sci.* **18**: 496.
5. Lagowski, J.J. (1973). *Modern Inorganic Chemistry*. Marcel Dekker, New York.

6. Sienko, M.J. and Plane, R.A. (1979). *Chemistry: Principles and Applications*. McGraw-Hill, New York.
7. Attendorn, H.G. and Bowen, R.N.C. (1997). *Radioactive and Stable Isotope Geology*. Chapman & Hall, London.
8. Bowen, R. (1988). *Isotopes in the Earth Sciences*. Elsevier, London.
9. Kushner, D.J., Baker, A., and Dunstall, T.G. (1996). Biotechnological Potential of Heavy Water and Deuterated Compounds. *Proc. Papers 1996 Risk Assessment Res. Symp.*, Chapter 2.
10. Katz, J.J. (1960). The biology of heavy water. *Sci. Am.* **July**: 106–115.
11. Dahl, P.F. (1999). *Heavy Water and the Wartime Race for Nuclear Energy*. Institute of Physics, London.
12. Yang, F. and Hamilton, G.H. (1996). *Modern Atomic and Nuclear Physics*, International Editions. McGraw-Hill, New York.
13. Goldstein, A.M. (1986). *A Dictionary of Physics*. CBS Publishers and distributors, Delhi, India.
14. Urey, H.C., Brickwedde, F.G., and Murphy, G.M. (1932). An isotope of hydrogen of mass 2 and its concentration (abstract). *Phys. Rev.* **39**: 864.
15. Gilreath, E.S. (1985). *Fundamental Concepts of Inorganic Chemistry*. 19th printing, International Student Edition. McGraw-Hill, Singapore.

HENRY'S LAW

Roger C. Viadero Jr.
West Virginia University
Morgantown, West Virginia

OVERVIEW AND THEORY

William Henry (1774–1836) established the fundamental relationship to define the partitioning of a soluble (or partially soluble) chemical species between gas and liquid phases, at equilibrium, in a dilute aqueous solution. For the transfer of a chemical species, C, from the gas phase to the aqueous phase, presented schematically in Fig. 1, the following chemical reaction and dimensionless equilibrium constant, k, can be written for a dilute solution:

$$C_g \xleftrightarrow{k} C_{aq} \tag{1}$$

$$k = \frac{[C_{aq}]}{[C_g]} \tag{2}$$

where $[C_g]$ = gas phase concentration (mol/L) of C and $[C_{aq}]$ = aqueous phase concentration (mol/L) of C.

Upon recognizing that the molar concentration of a gas is related to the partial pressure of that gas via the ideal gas law presented in Eq. 3, the expression for the dimensionless equilibrium constant can be rewritten as in Eq. 4:

$$C_g = \frac{n}{V} = \frac{p_g}{RT} \tag{3}$$

$$k = \frac{[C_{aq}]}{[C_g]} = \frac{[C_{aq}],}{\frac{p_g}{RT}} \tag{4}$$

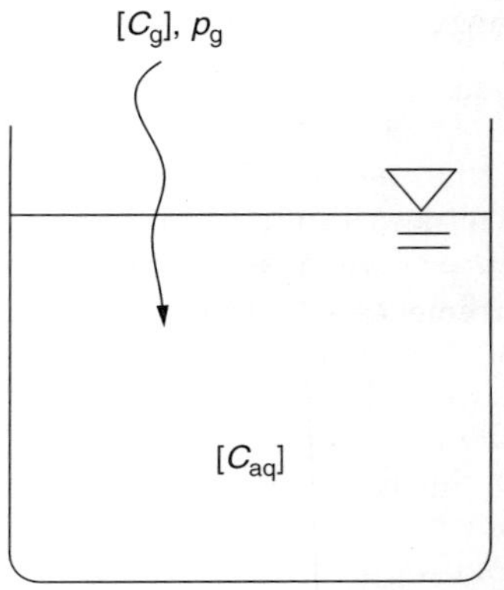

Figure 1. Transfer of chemical species C from the gas phase to the aqueous phase.

where n = number of moles of C, V = volume occupied by the gas (L), p_g = partial pressure of the gas (atm), R = ideal gas constant (0.082057 atm/L/K/mol) and T = temperature of the gas (K). By rearranging the relationship presented in Eq. 4, a familiar form of Henry's law is realized:

$$C_{aq} = k_H p_g \tag{5}$$

where k_H = Henry's law constant (mol/L/atm). The dimensionless equilibrium constants, k, and k_H, are related according to Eq. 6:

$$k_H = \frac{k}{RT} \tag{6}$$

Values of k_H for representative inorganic chemical species are presented in Table 1 (1,2). There are such a large number of volatile or semivolatile organic chemicals of environmental importance; readers are referred to the following representative supporting literature for detailed enumerations of Henry's law constants for organic species (3–10).

When using tabulated values of Henry's law constants or dimensionless equilibrium constants, care must be taken to examine the direction of transport being considered because the direction of reaction affects the units of k_H. For example, if the development of k_H presented above were reformulated for volatilization of C (from the aqueous to the gas phase), values of k_H would have units of atm L/mol and would thus be the inverse of the case developed above. The inconsistency encountered in the units of k_H is due partially to the

Table 1. k_H Values for Representative Inorganic Chemical Species[a]

Reaction	k_H (mol/L/atm)[b]
$CO(g) \leftrightarrow CO(aq)$	9.55×10^{-4}
$SO_2(g) \leftrightarrow SO_2\ (aq)$	1.23
$NH_3(g) \leftrightarrow NH_3\ (aq)$	57.5
$NO_2(g) \leftrightarrow NO_2\ (aq)$	1.00×10^{-2}
$O_3(g) \leftrightarrow O_3(aq)$	9.33×10^{-3}

[a] References 1 and 2.
[b] At 25 °C and 1 atm.

difference in conventions used by engineers and chemists and general lack of consensus with regard to a standard for the direction of transport (1,11).

LIMITATIONS OF HENRY's LAW

Three main limitations must be considered when applying Henry's law: (1) the state of equilibrium, (2) the solute concentration, and (3) the veracity of using standard data for calculations in contaminated water systems. The application of Henry's law is limited to cases where chemical equilibrium has been reached in a dilute solution (<1% solute concentration). However, the concentrations of most water constituents are normally less than 1%, so the application of Henry's law is typically valid. Tabulated values of k_H are generally based on experiments using pure water. As a result, the application of such data to contaminated water systems can yield unpredictable results (1). Under such circumstances, experiments may be needed to determine the appropriate value of k_H for the contaminated solution. Similar care should be taken when working with solvent systems other than water (e.g., when determining the solubility of a gas in an organic solvent).

APPLICATION OF HENRY'S LAW

By evaluating the relative magnitude of k_H for particular species, it is possible to learn about the solubility and general tendency of those species to partition between water and air. For instance, in the transport of a soluble chemical species from the gas phase to the aqueous phase, a compound that has a high k_H would be more soluble than one that has a lower k_H and would partition into the water phase where it may be transported downgradient or adsorb onto solid surfaces. In contrast, a species that has a low k_H would be less water soluble and tend to remain in the gas phase.

The surface of the earth is dominated by an air/water interface, so the interaction between gas and aqueous phase chemicals is extremely important in understanding the origins, fate, and transport of water and dissolved contaminants through the hydrogeochemical cycle.

Consider a sample of pure water which is completely isolated from the atmosphere. At standard conditions, the pH of the water would be neutral (pH = 7). However, when the pure water is allowed to reach equilibrium with an atmosphere containing $CO_2(g)$, the following series of reactions occurs:

$$CO_2(g) \overset{k_H}{\longleftrightarrow} CO_2(aq) \tag{7a}$$

$$CO_2(g) + H_2O \overset{k_1'}{\longleftrightarrow} \Leftrightarrow H_2CO_3{}^* \tag{7b}$$

$$H_2CO_3{}^* \overset{k_1}{\longleftrightarrow} H^+ + HCO_3{}^- \tag{7c}$$

$$HCO_3{}^- \overset{k_2}{\longleftrightarrow} H^+ + CO_3{}^{2-} \tag{7d}$$

At standard conditions, the partial pressure of $CO_2(g)$ is 3.162×10^{-4} atm (12). $CO_2(aq)$ and carbonic acid, H_2CO_3, act as weak acids and the equilibrium pH of the water is ~5.6. In this case, the atmosphere serves as a constant source of CO_2, and the total concentration of carbonate species increases with pH as presented in Fig. 2, where

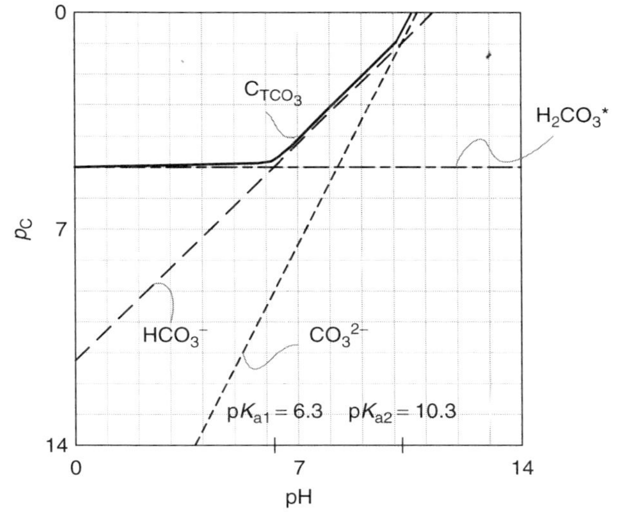

Figure 2. Plot of p_C versus pH for an aqueous system in equilibrium with atmospheric $CO_2(g)$ at standard conditions.

$$C_{T_{CO_3}} = [H_2CO_3{}^*] + [HCO_3^-] + [CO_3^{2-}] \tag{8}$$

Further, any change in the partial pressure of CO_2 would result in a corresponding shift in the equilibrium pH. Failure to account for the presence and effects of $CO_2(g)$ on the pH would result in an entirely incorrect interpretation of acid–base chemistry which would then adversely impact conclusions drawn relative to environmentally important chemical reactions (e.g., precipitation/dissolution reactions, complexation reactions, and oxidation/reduction reactions).

BIBLIOGRAPHY

1. Stumm, W. and Morgan J. (1996). *Aquatic Chemistry*, 3rd Edn. John Wiley & Sons, New York.
2. Mackay, D., Shiu, W., and Ma, K. (1995). *Illustrated Handbook of Physical-Chemical Properties and Environmental Fate for Organic Chemicals*, Vol. IV: *Oxygen, Nitrogen, and Sulfur Containing Compounds*. Lewis, Boca Raton, FL.
3. Burkhard, L., Armstrong, D., and Andren, A. (1985). Henry's law constants for the polychlorinated biphenyls. *Environ. Sci. Technol.* **19**: 590–596.
4. Mackay, D. and Shiu, W. (1981). A critical review of Henry's law constants for chemicals of environmental interest. *J. Phys. Chem. Ref. Data* **10**: 1175–1199.
5. Mackay, D., Shiu, W., and Ma, K. (1992). *Illustrated Handbook of Physical-Chemical Properties and Environmental Fate for Organic Chemicals*, Vol. I: *Monoaromatic Hydrocarbons, Chlorobenzenes, and PCBs*. Lewis, Boca Raton, FL.
6. Mackay, D., Shiu, W., and Ma, K. (1992). *Illustrated Handbook of Physical-Chemical Properties and Environmental Fate for Organic Chemicals*, Vol. II: *Polynuclear Aromatic Hydrocarbons, Polychlorinated Dioxins, and Dibenzofurans*. Lewis, Boca Raton, FL.
7. Mackay, D., Shiu, W., and Ma, K. (1993). *Illustrated Handbook of Physical-Chemical Properties and Environmental Fate*

for Organic Chemicals, Vol. III: *Volatile Organic Chemicals*. Lewis, Boca Raton, FL.
8. Murphy, T., Mullin, M., and Meyer, J. (1987). Equilibration of polychlorinated biphenyls and toxaphene with air and water. *Environ. Sci. Technol.* **21**: 155–162.
9. Rice, C., Chernyak, S., and McConnell, L. (1997). Henry's law constants for pesticides measured as a function of temperature and salinity. *J. Agric. Food Chem.* **45**: 2291–2298.
10. U.S. EPA. (1998). *How to determine Henry's law constants, F_m Values, F_r Values, and F_e values for organic compounds, Appendix J to part 60*. U.S. Environmental Protection Agency, Washington, DC.
11. Benjamin, M. (2002). *Water Chemistry*. McGraw-Hill. New York.
12. Snoeyink, V. and Jenkins, D. (1980). *Water Chemistry*. John Wiley & Sons. New York.

HOFMEISTER EFFECTS

M. Boström
Linköping University
Linköping, Sweden

W. Kunz
University of Regensburg
Regensburg, Germany

B.W. Ninham
Australian National University
Canberra, Australia
and
University of Regensburg
Regensburg, Germany

During the 1880s and 1890s, Franz Hofmeister (Professor of Pharmacology at the Charles University in Prague) and some of his coworkers demonstrated that salts with common cation (positive ion) but differing in anion (negative ion) have different effectiveness in stabilizing protein suspensions (1,2). They can be arranged in a sequence that seemed later to be universal. Reference (2) includes an English translation of two of Hofmeister's papers. As can be seen in Fig. 1, depending on the choice of salt, different concentrations are required to precipitate a certain concentration of proteins from whole egg white (3). Hofmeister effects, or Hofmeister sequences, refer to the relative effectiveness of anions or cations on a wide range of phenomena. They have remained unexplained by theories of physical chemistry until recently. But, as we will discuss later, some encouraging steps have recently been taken toward resolving the problem. Understanding Hofmeister effects is essential in understanding how different salt ions in water influence solution chemistry and are critical to biological processes that are ubiquitous.

Hofmeister effects have been observed in an enormous number of systems: solubility of salts; solubility of oxygen; surface tension of salt solutions (4); pH measurements (5); buffers; ion binding to micelles, proteins (6), and membranes; conformational changes of rhodopsin (membrane

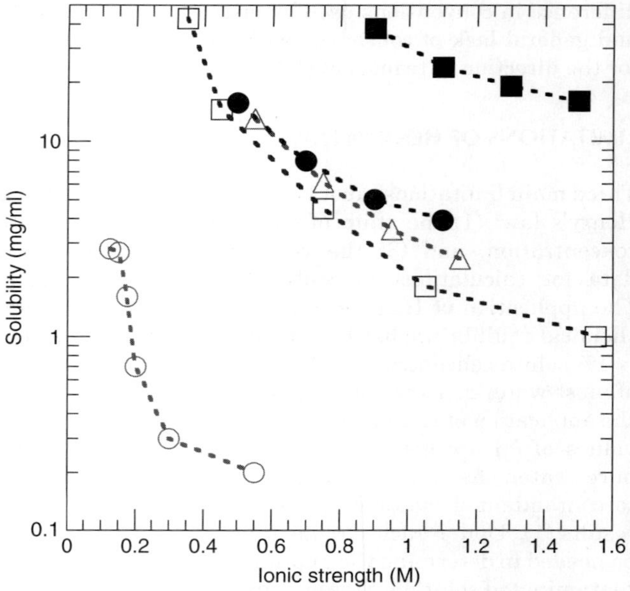

Figure 1. Variation, as a function of salt type, of hen-egg-white protein (lysozyme) solubility. Depending on the choice of salt, different concentrations are required to precipitate a certain concentration of proteins from whole egg white. Data taken from Ref. 3 are shown for NH_4OAC (filled squares), NH_4Cl (filled circles), KCl (open triangle), NaCl (open square), and KSCN (open circle).

bound protein acting as light receptor) (7); transport across human red blood cells (8); molecular forces (9,10); and colloid stability. Another example is the cutting efficiency of DNA in buffered salt solutions that not only follows a Hofmeister series: A change in supposedly irrelevant choice of buffer has the ability to reverse the Hofmeister series (11). A corresponding change in buffer has recently been demonstrated to reverse, in the same way, the Hofmeister series in pH measurements. Even bacterial cell growth follows the Hofmeister series (12).

Understanding Hofmeister effects is especially important for biological and medical applications. Although many biochemical functions require specific ions at certain concentrations, too high concentrations of, for example, NaCl can have deleterious effects on enzymatic activities (13,14). One important biological example is the human red blood cell that is sensitive to salt. During gas exchange, involved in buffering of blood and in O_2 and CO_2 transport in blood, the chloride ion moves between the plasma and the red blood cell intracellular fluid (15). It is well known experimentally that transport of Cl^-, Na^+, K^+, and water molecules through the cell membrane, and the corresponding cell volume changes, depend strongly on the choice of background anions (8). The rate constant for chloride exchange is, e.g., lowered to 6.5% of its original value when 120 mM of the chloride ions are replaced with thiocyanate ions. The interactions of ions with cell membranes and with membrane bound proteins (hemoglobin) determine key regulatory mechanisms. Taboy et al. (16), investigating hemoglobin, found a larger redox potential in the presence of NO_3^- than in a Cl^- salt solution (with a maximum redox potential around a salt concentration of 0.2 M). This finding is interesting because it demonstrates that different anions have a different affinity for hemoglobin and may provide useful insights as to why the affinity of chloride ions is larger for deoxyhemoglobin than for oxyhemoglobin (17). This finding has important biological consequences, but the reasons have not been clear. Even though it is a different story, it is worth noticing that the difference between two typical Hofmeister anions, thiocyanate and chloride, has also been exploited in detoxification of anticancer agents (18).

These systematic Hofmeister series demonstrate that common underlying interactions are involved, which is encouraging and strongly suggests that general and truly predictive theories may not be far away. Clearly experiments and theory that explain these effects are valuable and can provide insights into the underlying mechanisms that control vital biological systems. Having explained why it is important to understand Hofmeister effects, we now proceed to discuss some recent theoretical attempts to understand it.

The traditional hand waving "explanation" for the Hofmeister effects has been that the ions exert their effects indirectly by changing the water structure. People have used words such as: "hard and soft ions," "hydrophobic and hydrophilic ions," "lyotropic series," "chaotropic and cosmotropic ions," "salting-in," and "salting-out." When words like these and simple theories have proven insufficient to accommodate experimental results, a large number of additional effects and interactions have been invoked (accompanied by numerous adjustable fitting parameters). Although "understanding" of experiments has occasionally been claimed, there has been no predictability.

Until recently, the theories used were mainly based on electrostatic interactions between charged particles. These theories failed to explain why a change from one monovalent salt to another (i.e., salts with the same charges) can have huge effects. The main attempts to explain Hofmeister phenomena involved extension of theory to allow different "hydrated ion sizes," parameters that vary from one situation to another. However, such attempts do not work and Hofmeister series do not correlate directly with any additional physical or chemical parameter. The problem is that there must be a coupling, or cooperative effects, which are not so easy to characterize. An understanding has only recently emerged that the polarizability of ions, neglected in previous theories, plays an essential role. Polarizability is a quantity that measures the response of a particle to a perturbing electromagnetic field. Although two charged particles can interact via electrostatic forces, a quantum mechanical attraction also exists, the van der Waals force, between two charged neutral particles that is related to the polarizabilities of the two particles. This van der Waals force is also present for the interaction between charged particles but has usually been assumed to be negligibly small compared with the electrostatic force. Although electrostatic forces often dominate at low salt concentrations, van der Waal forces become important at high salt concentrations where the electrostatic forces are screened. Ninham et al. (19) demonstrated that van der Waals forces between ion and interface often dominate over electrostatic forces above and around the salt concentration we have in our blood (i.e., 0.15 molar salt), which is exactly the concentration regime where Hofmeister effects are observed. As can be seen in Fig. 2, ion-specific bubble fusion is, for instance, inhibited, following an abrupt transition with increased salt concentration, in sodium chloride solutions above the salt concentration we have in our blood (20). This is why ocean water is foamy, but fresh water is not. Today the salt concentration of the ocean is around 0.6 M, but the concentration in the Permian Sea was close to what we have in our blood, which means that even a small decrease in salt concentration would have had dramatic effects on bubble fusion. To have a sufficiently high salt concentration in the Permian Sea must have been a critical condition when early life emerged. A related topic is that the local salt concentration, e.g., in the Baltic Sea, may have dropped suddenly below the critical 0.15 M concentration after ice melted at the end of ice ages. A speculation on how this may have caused mass extinction of plankton can be found in the references (1).

At present, two parallel and complementary lines of work demonstrate that both surface effects and bulk effects are important for the understanding of Hofmeister effects. We will first briefly discuss surface related effects. The theoretical papers by Boström et al. have focused on surface phenomena, i.e., the van der Waals interaction between ions and interface. Boström et al. (21) demonstrated, e.g., that this is one important reason why the charge of lysozyme (hen-egg-white protein) at pH = 4.5 is higher in 0.1 M NaSCN than in 0.1 M NaCl. The negatively charged thiocyanate anions (SCN^-) are more attracted toward the protein, by larger van der Waals forces, than are the chloride ions. A higher local

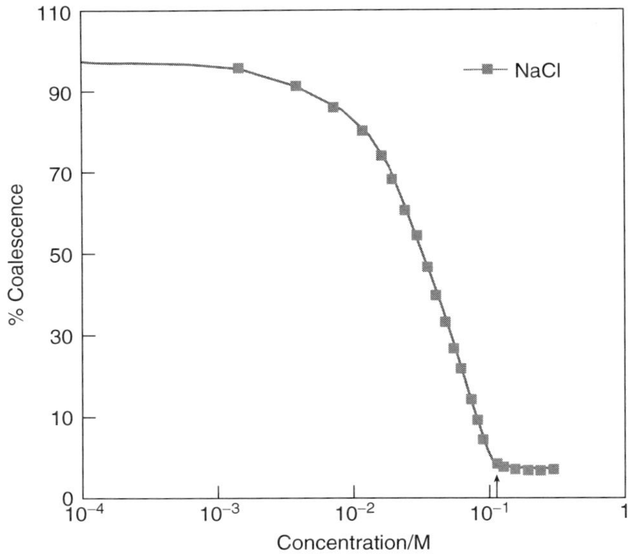

Figure 2. Effect of NaCl on bubble coalescence. The arrow indicates the concentration of NaCl in the extracellular fluid of humans. Bubble fusion is ion specific, and remarkably, with some salts, such as NaOAc, essentially no transition occurs as more salt is added to the solution. Data taken from Craig et al. (20).

anion concentration with SCN^- ions near the protein surface gives rise to a larger local concentration of hydronium ions (H_3O^+) at the protein surface, which in turn leads to more hydrogen ions being bound to the protein charge groups (i.e., a larger protein charge) in a thiocyanate (or iodide) salt compared with a chloride salt. Recent extension by Boström et al. of this theoretical work (investigating theoretically solutions with different buffers and proteins) has demonstrated that this also influences the ion-specific buffer capacity of a lysozyme solution at typical protein concentrations. Remarkably, in as far as both pH in solution and charge of hen-egg-white protein is concerned, one can often to a good approximation replace a certain concentration of NaCl with a smaller concentration of NaSCN and find the same final result. A similar result has been observed in experiments on enzymatic activity by P. Baudin et al. and in pH measurements in protein (cytochrome c) solutions by P. Baglioni et al. This kind of surface related effects has also offered a partial explanation (22) for the experimental observation by Bonneté et al. (6) that the interaction between two lysozyme proteins in salt solutions follows a Hofmeister series (the interaction is, e.g., more attractive in $NaNO_3$ solutions than in NaCl solutions).

The other line of work has focused on interactions in the bulk solution between ions and between ion and water molecules. It is important to stress that van der Waals forces are only one consequence of the ion polarizability. Other consequences are dipole-induced dipole and induced dipole-induced dipole forces between ions and between ions and water molecules. These forces have been in focus in molecular simulations. It has been repeatedly demonstrated that the static polarizability of ions must be included in computer molecular simulations to get reasonable results for the solvation energy of ions in water (23,24). This, however, is not the entire story. To get correct results for the activity coefficients (25) and solvation energies (26) of ions in water, strong evidence exists that one has to include the entire dynamic ionic polarizability. In several important contributions (27–29), attempts have been made to include forces that originate from the static polarizability correctly in computer simulations that include water molecules. These ion–ion and ion–water interactions are also important for the air–water interface. The air–water interface, and the surface tension of salt solutions, is more complicated than one might initially have guessed and not yet fully understood. Promising simulations (28) seem to give the right trend in Hofmeister sequence for the surface tension change with added salt. But these simulations give results that differ greatly from experimental surface tension changes with added salt. This is also true for pure water where simulations without fitting parameters give incorrect results for the surface tension and freezing temperature. One reason for these problems must be that simulations so far have not included many body dispersion forces properly. But simulations together with experiments are still useful tools for further understanding. As demonstrated by Jungwirth and Tobias (28), some highly polarizable anions, such as bromide, may, contrary to conventional wisdom (which is based on electrostatic theories), go to the air–water interface. It has also been demonstrated that one important effect is that the water molecules near an air–water interface are not isotropically ordered (28), which effectively gives an anisotropic film near the interface that strongly influences the interaction between ions and the air–water interface. These interactions (ion–ion, ion–water molecule, ion interface) influence each other in a self-consistent manner, which makes this a complicated but solvable problem. It is clearly important to take into account all of the complementary aspects discussed here for further progress. Our most recent results suggest a key role for interfacial dissolved gases, and the corresponding change in ion solvation energy, behind the observed surface tension changes (see article on Dissolved Gases). Theories, accompanied by refined simulations and experiments, are on the way that may substantially improve the understanding of the Hofmeister effect in the near future. It is also worth remarking that until now no theoretical work has included the effect of dissolved gasses properly. These gases are known to influence colloid interactions (30), bubble–bubble interaction (20), conductance, pH, and most likely the surface tension of salt solutions. It is not only CO_2 that is important but also other dissolved gasses. The interested reader can find more details about Hofmeister effects in different reviews (1,31–34).

BIBLIOGRAPHY

1. Kunz, W., Lo Nostro, P., and Ninham, B.W. (2004). *Curr. Op. Colloid Int. Sci.* **9**: 1.
2. Kunz, W., Henle, J., and Ninham, B.W. (2004). *Curr. Op. Colloid Int. Sci.* **9**: 19.
3. Ries-Kautt, M.M. and Ducruix, A.F. (1989). *J. Biol. Chem.* **264**: 745.
4. Boström, M., Williams, D., and Ninham, B.W. (2001). *Langmuir* **17**: 4475.
5. Boström, M. et al. (2003). *J. Phys. Chem. B* **207**: 2875.
6. Bonneté, F., Finet, S., and Tardieu, A. (1999). *J. Crystal Growth* **196**: 403.
7. Vogel, R. (2004). *Curr. Op. Colloid Int. Sci.* **9**: 133.
8. Payne, J.A., Lytle, C., and McManus, T. (1990). *J. Am. J. Physiol.* **259**: C819.
9. Pashley, R. et al. (1986). *J. Phys. Chem.* **90**: 1637.
10. Boström, M., Williams, D., and Ninham, B.W. (2001). *Phys. Rev. Lett.* **87**: 168103.
11. Kim, H-K., Tuite, E., Norden, B., and Ninham, B.W. (2001). *Eur. Phys. J. E* **4**: 411.
12. Lo Nostro, P. et al. (2004). *Curr. Op. Colloid Int. Sci.* **9**: 97.
13. Yancey, P. et al. (1982). *Science* **217**: 1214.
14. Baudin, P. et al. (2004). *Curr. Op. Colloid Int. Sci.* **9**: 43.
15. Prange, H. et al. (2001). *J. Appl. Pysiol.* **91**: 33.
16. Taboy, C. et al. (2000). *J. Biol. Chem.* **275**: 39048.
17. Rollema, H., de Bruin, S., Janssen, L., and van Os, G. (1975). *J. Biol. Chem.* **250**: 1333.
18. Hovinen, J., Petterson-Fernholm, T., Lahti, M., and Vilpo. (1998). *J. Chem. Res. Toxicol.* **11**: 1377.
19. Ninham, B.W. and Yaminsky, V. (1997). *Langmuir* **13**: 2097.
20. Craig, V., Ninham, B.W., and Pashley, R. (1993). *Nature* **364**: 317.

21. Boström, M., Williams, D., and Ninham, B.W. (2003). *Biophys. J.* **85**: 686.
22. Tavares, F.W., Bratko, D., Blanch, H., and Prausnitz, J.M. (2004). *J. Phys. Chem. B* **108**: 9228.
23. Stuart, S.J. and Berne, B.J. (1996). *J. Phys. Chem.* **100**: 11934.
24. Grossfield, A., Ren, P., and Ponder, J.W. (2003). *J. Am. Chem. Soc.* **125**: 15671.
25. Kunz, W., Belloni, L., Bernard, O., and Ninham, B.W. (2004). *J. Phys. Chem. B* **108**: 2398.
26. Boström, M. and Ninham, B.W. (2004). *J. Phys. Chem. B* **108**: 12593.
27. Ahn-Ercan, G., Krienke, H., and Kunz, W. (2004). *Curr. Op. Colloid Int. Sci.* **9**: 92.
28. Jungwirth, P. and Tobias, D. (2002). *J. Phys. Chem. B.* **106**: 6361.
29. Netz, R.R. (2004). *Curr. Op. Colloid Int. Sci.* **9**: 192.
30. Alfridsson, M., Ninham, B.W., and Wall, S. (2000). *Langmuir* **16**: 10087.
31. Cacace, M.G., Landau, E.M., and Ramsden, J.J. (1997). *Q. Rev. Biophys.* **30**: 241.
32. Collins, K.D. and Washabaugh, M.W. (1985). *Q. Rev. Biophys.* **18**: 323.
33. Leontidis, E. (2002). *Curr. Op. Colloid Int. Sci.* **7**: 81.
34. Abernethy, J.L. (1967). *Proc. California Association of Chem. Teachers* **44**: 177.

CLATHRATE HYDRATES

Carolyn Ann Koh
E. Dendy Sloan
Colorado School of Mines
Golden, Colorado

DEFINITION OF CLATHRATE HYDRATES

Clathrate hydrates are crystalline inclusion compounds consisting of a network of polyhedral water cages (host), which enclathrate small hydrophobic molecules (guests; typically <0.9 nm). Clathrate hydrates can be described as solid solutions (1), such that the empty hydrate lattice is unstable with respect to ice. When guest molecules occupy the water cages, the hydrate lattice is stabilized by van der Waals interaction forces. Typically, a water cage has a maximum occupancy of one guest molecule.

Simple clathrate hydrates contain only one type of guest molecule within the hydrate lattice, while *binary clathrate hydrates* contain two different types of guest molecules. *Mixed clathrate hydrates* refer to hydrates containing two or more different guest species. When the guest species are gas molecules, such as methane, ethane, or carbon dioxide, clathrate hydrates are also called *gas hydrates*. It is not necessary for all the cages in these clathrate hydrate structures to be occupied; for example, methane hydrate can be prepared with 90% of the cages occupied by methane (2). The cage filling depends on pressure, temperature, and the nature of the guest species. Hence, these structures are often referred to as "nonstoichiometric compounds."

COMMON CLATHRATE HYDRATE STRUCTURES

The three most common types of clathrate hydrate structures are sI, sII, and sH hydrate. sI hydrate predominates in natural environments, while sII hydrate is found mainly in artificial environments (3). sH hydrate may occur in either environment but is less common than either sI or sII hydrate. These three structures have distinct crystallographic properties (see Table 1 and References 2 and 4 for further details). Both sI and sII hydrates have cubic crystal structures (Fig. 1), while sH hydrate has a hexagonal crystal structure. sI and sII hydrates contain two different types of water cages (Fig. 2). The water cages are described by the general notation X^n, where X is the number of sides of a cage face and n is the number of cage faces having these X sides. For example, the pentagonal dodecahedral 5^{12} water cage is composed of 12 *five*-membered water rings. A unit cell of sI hydrate contains two small (5^{12}) + six large tetracaidecahedral ($5^{12}6^2$) cages. A unit cell of sII hydrate contains six small (5^{12}) + eight large hexacaidecahedral ($5^{12}6^4$) cages (sII). A unit cell of sH hydrate is comprised of three types of water cages: three small (5^{12}) cages, two irregular dodecahedral ($4^35^66^3$) cages, and a single large icosahedral ($5^{12}6^8$) cage (Fig. 2).

The type of clathrate hydrate structure that forms depends on the size of the guest molecule; for example, CH_4 and CO_2 both form sI hydrate; C_3H_8 and C_4H_8O (tetrahydrofuran, THF) form sII hydrate; while a larger guest molecule such as methylcyclohexane in the presence of methane forms sH hydrate (2,4). Table 2 illustrates the range of different guest molecules that form sI, sII,

Table 1. Crystallographic Properties of the Common Clathrate Hydrate Structures

Property	sI	sII	sH
Space group	*Pm3n*	*Fd3m*	*P6/mmm*
Unit cell parameters (nm)	$a = 1.20$	$a = 1.70$	$a = 1.21, c = 1.01$
Average cavity radius (nm)	0.395[2(5^{12})](S)	0.391[16(5^{12})](S)	0.391[3(5^{12})](S)
[number of cavities per unit cell (cavity type)]	0.433[6($5^{12}6^2$)](L)	0.473[8($5^{12}6^4$)](L)	0.406[2($4^35^66^3$)](S)
			0.571[1($5^{12}6^8$)](L)
Symmetry of cages	M$\underline{3}$(S), $\underline{4}$2m(L)	3m (S), $\underline{4}$3m (L)	Mmm(S), $\underline{6}$m2(S), 6/mmm(L)
Ratio of numbers of small/large cavities	0.33	2	5
Number of water molecules per unit cell	46	136	34
General unit cell formula	X·5.75H_2O	X·17H_2O	5(X,Y)·34H_2O

L, large cage; S, small cage; X,Y, guest molecules.

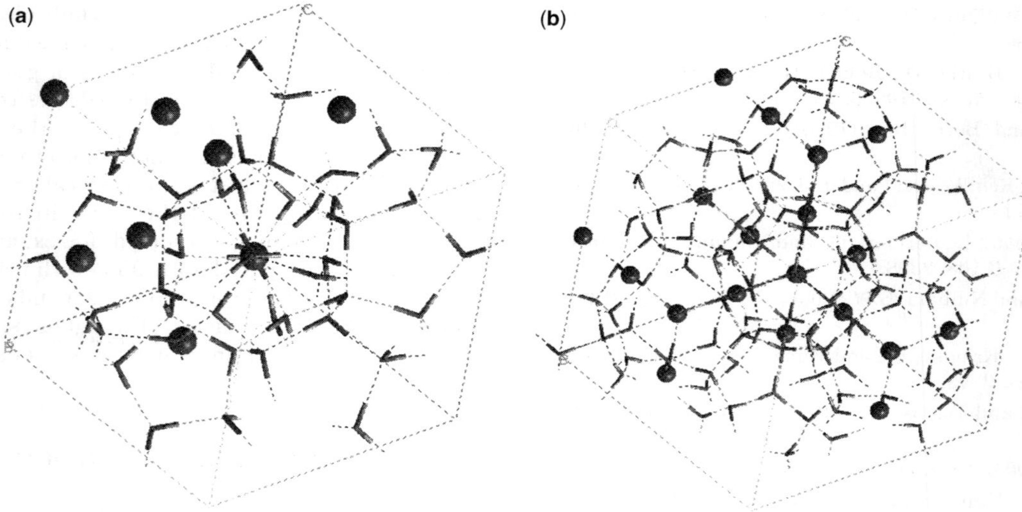

Figure 1. The partial unit cell structures of (**a**) sI hydrate and (**b**) sII hydrate. Dotted lines represent hydrogen bonds between water molecules. Spheres represent guest molecules.

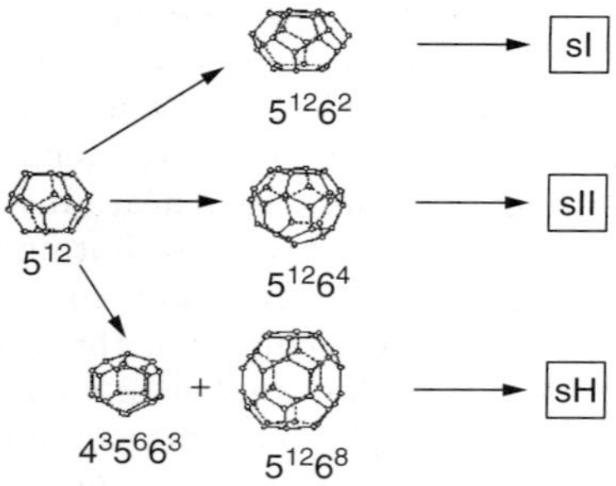

Figure 2. Water cages of sI, sII, and sH hydrates.

and sH hydrates. Typically, high pressure (3–11 MPa) and low temperature (275–285°K) conditions are required to stabilize gas hydrate systems. Table 2 gives the dissociation pressures of different clathrate hydrates and the occupancy of the corresponding hydrate guest molecules at around 277°K (the temperature at which most natural and industrial hydrates are found). Clathrate hydrates will be stable at 277°K and at a pressure above the dissociation pressure. Although CH_4 and C_2H_6 are both sI hydrate formers, a binary CH_4/C_2H_6 mixture can exhibit sI/sII transitions with varying pressure and/or composition. In contrast, a binary CH_4/CO_2 mixture (where both pure components are sI hydrate formers) forms only sI hydrate (Table 2).

NOVEL CLATHRATE HYDRATE STRUCTURES

Between the time that clathrate hydrates were first discovered in 1810 (6) and upto 1996, only three main structures were identified. Yet, within the last eight years (1997–2004), there have been a number of exciting reports of new clathrate hydrate structures. In addition to the discovery of multiple occupancies of the hydrate water cages, at least three new clathrate hydrate structures have been discovered. Some of these new structures are discussed next.

As mentioned earlier, typically only one guest molecule is accommodated within each clathrate hydrate cage. However, recent measurements and computer calculations have demonstrated that clathrate hydrate structures can consist of water cages containing more than one guest molecule. Specifically, neutron diffraction measurements show that double occupancy of the large cage of sII hydrate can occur for both nitrogen hydrate and oxygen hydrate at higher pressures (7). Similar conclusions have been derived from molecular dynamics computer calculations (8). Multiple cage occupancy has also been proposed in sI, sII, and sH argon hydrates from first principle and lattice dynamics calculations (9).

Multiple occupancy of the water cages in H_2 hydrate was discovered using neutron and X-ray diffraction and Raman spectroscopy by Mao et al. (10,11). H_2 molecules were reported to exhibit double and quadruple occupancy in the small and large cages of sII hydrate, respectively (10,11). However, recently Lokshin and co-workers in collaboration with the Mao group (10) reported that D_2 molecules only singly occupy the small cages of D_2 hydrate (12). This discovery that H_2 can form a clathrate hydrate is especially noteworthy since it was previously assumed that small molecules such as H_2 and He are too small to stabilize a clathrate hydrate structure. The pure H_2 hydrate structure is typically formed at very high pressures (near 1 GPa). However, the addition of a promoter molecule, for example, THF, to form a binary H_2/THF hydrate will stabilize H_2 within the sII framework at pressures two orders of magnitude lower than that in pure hydrogen hydrates (13).

A complex clathrate hydrate structure, 1.67 choline hydroxide-tetra-n-propylammonium fluoride·$30.33H_2O$, was identified in 1999 using single-crystal X-ray diffraction (14). A space group of R-3 was assigned to this structure. This new clathrate hydrate consists of alternating stacks of sH and sII hydrates. Unlike

Table 2. Different Hydrate Guest Molecules and Their Clathrate Hydrate Structures, Dissociation Pressures, and Water Cage Occupancy

Guest Molecule	Structure Formed	Occupancy of Water Cages[a]	P,kPa; T, K[b]
CH_4	sI	5^{12} (86%) and $5^{12}6^2$ (95%)	4170, 277.6
C_2H_6	sI	5^{12} (0.3%) and $5^{12}6^2$ (97%)	814, 277.6
CO_2	sI	5^{12} (70%) and $5^{12}6^2$ (98%)	3847, 277.6
C_3H_8	sII	$5^{12}6^4$ (~ 100%)	455, 277.8
C_2H_4O (EO)[c]	sI	$5^{12}6^2$ (~ 100%)	101.3, 284.2
C_4H_8O (THF)[c]	sII	$5^{12}6^4$ (~ 100%)	101.3, 277.5
CH_4+ C_2H_6	Binary sI/sII	sII–C_1: 5^{12} (79%), $5^{12}6^2$ (18%)	90.4% C_1: 2096, 277.6
		sII–C_2: $5^{12}6^2$ (80%)	
		sI–C_1: 5^{12} (63%), $5^{12}6^2$ (3%)	56.4% C_1: 1564,[d] 277.6
		sI–C_2: 5^{12} (0.2%), $5^{12}6^2$ (97%)	
CH_4+ CO_2	Binary sI	CH_4: 5^{12} (66%), $5^{12}6^2$ (41%)	66% C1: 2840, 277.0
		CO_2: 5^{12} (14%), $5^{12}6^2$ (55%)	
Methylcyclohexane + CH_4	sH	5^{12} (CH_4), $4^35^66^3$ (CH_4), $5^{12}6^8$ (mch)	2137, 277.6
t-Butylmethyl ether + CH_4	sH	5^{12} (CH_4), $4^35^66^3$ (CH_4), $5^{12}6^8$ (*t*-btme)	2550, 277.5

[a]A Gibbs energy minimization program, CSMGem, has been used to estimate the occupancy of each water cage (5).
[b]Values are from experimental data extracted from Reference 4.
[c]Dissociation temperatures are given for these compounds since they are stable at atmospheric pressure.
[d]Dissociation pressure calculated from CSMGem (5).

the common (sI, sII, and sH) hydrates, which contain hydrophobic guest molecules, the choline hydroxide guest in this complex structure exhibits both hydrophobic and hydrophilic modes of hydration. Choline hydroxide occupies the large cages ($5^{12}6^4$ and $5^{12}6^8$), displacing one of the host water molecules of the $5^{12}6^8$ cages. The tetra-*n*-propylammonium guest was reported to occupy fused supercages ($4^35^{30}6^3$ and 5^{36}). It was suggested that the closed-cage version of this structure could occur in natural and industrial environments with natural gas molecules as the guest species.

Contrary to previous reports of bromine forming a number of different hydrate structures, single-crystal X-ray diffraction measurements by Udachin et al. (15) showed that bromine hydrate only forms a tetragonal structure of space group P4$_2$/*mmm* (cf. Ref. 16). The composition of bromine hydrate was found to be $Br_2 \cdot nH_2O$, where $n = 8.62–10.68$. The tetragonal framework is comprised of small 5^{12} cages and large $5^{12}6^2$ and $5^{12}6^3$ cages. The large cages are occupied by varying amounts of bromine molecules. The small cages were either vacant or partially occupied by O_2 or N_2 molecules, which were incorporated during crystallization of the bromine/water solution in air.

sT clathrate hydrate, a trigonal structure of space group *P*321, was discovered in 2001 (17). The structure was determined using single-crystal X-ray diffraction and NMR spectroscopy. All three types of large cages ($5^{12}6^3$, $5^{12}6^2$, and $4^15^{10}6^3$) in the structure are occupied by dimethyl ether guest molecules. The ratio of small to large cages is 0.25, which is significantly smaller than other known structures. As a consequence, this structure is most efficient at minimizing void space when the small cages are empty and has a higher density (1.074 g·cm^{-3}) than sII or sI hydrate. The authors suggest that this higher density may result in sT being favored over sI and sII for intermediate sized guests at high pressure. However, they also acknowledge that other factors need to be accounted for, such as the degree of ring strain of the four-membered water rings.

Methane hydrate has been shown from neutron and X-ray diffraction to undergo structural transformations to unconventional structures at elevated pressures (2,18,19). sI hydrate transforms to MH-II, a hexagonal structure, at around 1 GPa, which further transforms at around 2 GPa to an orthorhombic ice-filled structure MH-III. The ice-filled structure of methane hydrate was found to be stable up to 42 GPa (19).

Similarly, THF hydrate has been reported to undergo a structural transition from sII hydrate to an unusual structure at high pressure (0.8 GPa) (20). From neutron diffraction, the new high pressure phase of THF hydrate was assigned to an orthorhombic structure of space group *Pnma*. The water framework is reported to be formed exclusively from one type of 14-hedron cage, $4^45^46^6$. The unit cell stoichiometry of this new structure is 4(THF-d_8)·24D_2O.

CLATHRATE HYDRATES IN NATURAL AND INDUSTRIAL ENVIRONMENTS

Clathrate hydrates, containing predominantly biogenic methane, occur naturally in ocean sediments and permafrost regions (2–4). The amount of carbon in these natural gas hydrate deposits is estimated to be twice that of combined fossil fuel reserves [1 m^3 of methane hydrate in which 90% of the cages are occupied contains the equivalent of 156 m^3 of methane under standard conditions (STP)]. Significant efforts have been made over recent years to evaluate the feasibility of recovering these natural gas hydrate deposits. Oceanic gas hydrate deposits are currently considered to be too disperse and expensive to warrant near-term exploration (3). Conversely, permafrost deposits have significantly higher hydrate concentrations than hydrates found in ocean sediments. As a consequence, pilot drilling and characterization are being performed in Alaska and Canada in the permafrost, and in Japan in the ocean (3). However, extensive research and development is still required before these natural gas hydrate deposits can be produced safely and economically.

Natural gas hydrates also occur in gas and oil subsea pipelines, when the fluid in the pipeline reaches the temperature and pressure conditions that are favorable for hydrate formation. This can result in blockage of the pipelines and, as a consequence, catastrophic economic loss and ecological and safety risks (2,4). The typical method used to prevent hydrate formation within subsea pipelines is to add a thermodynamic inhibitor, which shifts the hydrate formation conditions to lower temperatures and/or higher pressures (2,4).

However, in many deepwater production scenarios, thermodynamic inhibition can become uneconomical and even prohibitive due to the high concentrations required under these conditions. Therefore, new technologies are currently being developed to control hydrate formation within these deepwater pipelines. These new methods include:

1. The addition of low dosage hydrate inhibitors (LDHIs) that are effective at concentrations below about 1 vol%. There are two broad classes of LDHIs: kinetic hydrate inhibitors (KHIs) and antiagglomerant inhibitors (AAs). KHIs operate by delaying nucleation and/or crystal growth. AAs prevent hydrate crystals from agglomerating and forming a blockage by maintaining the hydrates in the form of a suspended slurry, which allows fluid flow to occur unimpeded.
2. "Cold flow," whereby hydrates could be pumped through the pipeline without the need of inhibitors (21). Sintef-BP researchers have reported that the addition of water to a flow of dry hydrate results in the formation of further dry hydrate. It is suggested that this implies that if there is a cold flow of oil/gas containing dry hydrates, warm well streams containing water can then be added, with the water from the well being converted to dry hydrates.

Numerous technological applications have been proposed for clathrate hydrates. Highlights of some of these applications are given as follows. The discovery of the ability to store H_2 in a pure H_2 hydrate at high pressures by Mao et al. (10,11) and more recently within a binary H_2/THF hydrate at lower pressures (13) presents an innovative technology for hydrogen storage. This technology seems particularly feasible since the successful proof-of-concept that hydrogen can be stored at mild conditions (of T and P) within a binary clathrate hydrate lattice using promoter molecules (13). The advantage of using clathrate hydrate materials for hydrogen storage is that the hydrogen is stored in molecular form, without the need of a chemical reaction for its release (13). This increases the overall energy efficiency of the system. Furthermore, all stored H_2 should be completely released by decomposition of the hydrate at near-ambient temperature and pressure conditions. The storage of methane within a sI hydrate framework is also currently being evaluated as an alternative method to fuel ships or to transport stranded gas (3).

Methane hydrate has also been considered a potential source for climate change (3). The clathrate gun hypothesis (also known as the "late Quaternary climate change") suggests that methane released from methane hydrate around 15,000 years ago caused immense global warming (22). However, this hypothesis remains controversial.

The use of gas hydrates for cool storage has been explored since the early 1980s (23). The R141b gas hydrate (sII hydrate) was selected as a cool-storage system, since the R141b refrigerant offers significant advantages over Freon hydrate formers, including a lower saturation pressure and lower cost. R141b hydrate can be formed under mild conditions of 0.1 MPa and around 278°K. A further potential application of clathrate hydrates is CO_2 sequestration. This would involve injecting liquid CO_2 (droplets) into the ocean at depths of greater than 500 m where a solid interface of CO_2 hydrate will form between CO_2 and water (24,25). Injecting CO_2 into deep marine sediments would not be expected to result in long-term effects to biological systems and could circumvent the problem of eventually releasing the disposed CO_2 into the atmosphere.

Other proposed applications of gas hydrates include a range of separation processes, such as the treatment of aqueous and gaseous pollutant streams [e.g., recovery of chlorinated hydrocarbons from contaminated groundwater (26) or carbon dioxide from flue gases (27), respectively], isotopic fractionation of oxygen and hydrogen (28), and marine desalination (29).

CONCLUSION

The increasing utilization of molecular tools, such as X-ray and neutron diffraction, Raman and NMR spectroscopy, and computer simulation calculations, is producing a wealth of new and interesting findings on the structural and dynamic properties of clathrate hydrates. Further understanding of the hydrate structural transitions occurring over a range of temperature, pressure, and composition and the kinetics of these transitions will be important in advancing the range of new technologies being developed based on clathrate hydrates. This new molecular-level insight of hydrate structure and dynamics could have implications in the broader fields of inclusion compounds and crystal growth processes.

Acknowledgment

Thanks to K. Hester for his help with the CSMGem calculations.

BIBLIOGRAPHY

1. van der Waals, J.H. and Platteeau, J.C. (1959). Clathrate solutions. *Adv. Chem. Phys.* **2**: 1.
2. Koh, C.A. (2002). Towards a fundamental understanding of natural gas hydrates. *Chem. Soc. Rev.* **31**: 157–167.
3. Sloan, E.D. (2003). Fundamental principles and applications of natural gas hydrates. *Nature* **426**: 353–359.
4. Sloan, E.D. (1998). *Clathrate Hydrates of Natural Gases*, 2nd Edn. Marcel Dekker, New York.
5. CSMGem Program, Colorado School of Mines.
6. Davy, H. (1811). *Philos. Trans. R. Soc. London* **101**: 1–35.

7. Chazallon, B. and Kuhs, W.F. (2002). *In situ* structural properties of N_2-, O_2-, and air-clathrates by neutron diffraction. *J. Chem. Phys.* **117**: 308–320.
8. van Klaveren, E.P. et al. (2001). Molecular dynamics simulation study of the properties of doubly occupied N_2 clathrate hydrates. *J. Chem. Phys.* **115**: 10500–10508.
9. Inerbaev, T.M. et al. (2004). Theoretical study of clathrate hydrates with multiple occupation. *J. Incl. Phen. Macrocyclic Chem.* **48**: 55–60.
10. Mao, W.L. et al. (2002). Hydrogen clusters in clathrate hydrates. *Science* **297**: 2247–2249.
11. Mao, W.L. and Mao, H.K. (2004). Hydrogen storage in molecular compounds. *Proc. Natl. Acad. Sci.* **101**: 708.
12. Lokshin, K.A. et al. (2004). Structure and dynamics of hydrogen molecules in the novel clathrate hydrate by high pressure neutron diffraction. *Phys. Rev. Lett.* **93**: 125503–125504.
13. Floruuse, L.J. et al. (2004). Stable low pressure hydrogen clusters stored in a binary clathrate hydrate. *Science*.
14. Udachin, K.U. and Ripmeester, J.A. (1999). A complex clathrate hydrate structure showing bimodal guest hydration. *Nature* **397**: 420–423.
15. Udachin, K.A., Enright, G.D., Ratcliffe, C.I., and Ripmeester, J.A. (1997). Structure, stoichiometry, and morphology of bromine hydrate. *J. Am. Chem. Soc.* **119**: 11481–11486.
16. Allen, K.W. and Jeffrey, G.A. (1963). *J. Chem. Phys.* **38**, 2304.
17. Udachin, K.A., Ratcliffe, C.I., and Ripmeester, J.A. (2001). A dense and efficient clathrate hydrate structure and unusual cages. *Angew. Chem. Int. Ed.* **40**: 1303–1305.
18. Loveday, J.S. et al. (2001). Transition from cage clathrate to filled ice: the structure of methane hydrate III. *Phys. Rev. Lett.* **87**: 215501–215504.
19. Hirai, H. et al. (2003). Retention of filled ice structure of methane hydrate up to 42 GPa. *Phys. Rev. B* **68**: 172102–172104.
20. Kurnosov, A.V. et al. (2004). New clathrate hydrate structure: high-pressure tetrahydrofuran hydrate with one type of cavity. *Angew. Chem. Int. Ed.* **43**: 2922–2924.
21. Evensen, K. (2004). Cold flow techniques set to thaw out hope for tie backs. *Upstream Technol.* **76**.
22. Kennett, J.P., Cannariato, G., Hendy, I.L., and Behl, R.J. (2003). *Methane Hydrates in Quaternary Climate Change: The Clathrate Gun Hypothesis*. American Geophysical Union, Washington, DC.
23. Bi, Y. et al. (2004). Influence of volumetric-flow rate in the crystallizer on the gas hydrate cool-storage process in a new gas hydrate cool-storage system. *Appl. Energy* **78**: 111–121.
24. Rehder, G. et al. (2004). Dissolution rates of pure CH_4 hydrate and CO_2 hydrate in undersaturated seawater at 1000-m depth. *Geochim. Cosmochim. Acta* **68**: 285.
25. Radhakrishnan R. and Trout, B.L. (2002). A new approach for studying nucleation phenomena using molecular simulations: application to CO_2 hydrate clathrates. *J. Chem. Phys.* **117**: 1786.
26. Seo, Y. and Lee, H. (2001). A new hydrate-based recovery process for removing chlorinated hydrocarbons from aqueous solutions. *Environ. Sci. Technol.* **35**: 3386–3390.
27. Kang, S-P. and Lee, H. (2000). *Environ. Sci. Technol.* **34**: 4397–4400.
28. Maekawa, T. (2004). Experimental study on isotopic fractionation in water during gas hydrate formation. *Geochem. J.* **38**: 129–138.
29. Max, M.D. (2002). Desalination using positively buoyant or negatively buoyant/assisted buoyancy hydrate. US Patent 6,497,794.

HYDRATION

Carolyn Ann Koh
Colorado School of Mines
Golden, Colorado

DEFINITION OF HYDRATION

Hydration can be defined as the process by which water is added to produce a *hydrate*. The process is reversible, that is, water can be reextracted from the hydrate. A second meaning of *hydration* involves a *hydration reaction*. This is where water is permanently and chemically combined with a reactant in a way that it can no longer be reextracted. The second meaning of hydration will not be considered here.

Hydration to form a hydrate can occur when water (*solvent*) interacts with *solute* molecules to form a solution, which involves *hydrophobic hydration*. Hydrophobic hydration is the main focus here.

HYDRATION OF HYDROPHOBIC MOLECULES

Hydration of nonpolar molecules is usually associated with hydrophobic effects in which water cavities or hydration shells are formed. This is accompanied by a loss in entropy and aggregation of the nonpolar (hydrophobic) molecules. Hence, ordering of the water hydration shell around hydrophobic molecules has been attributed to “clathrate” behavior in which the water hydration shell is dominated by pentagons compared to bulk liquid water (1). However, thermodynamic studies (2) suggest that although there are larger numbers of pentagons in the solvation shell compared to the bulk, the hydration shell also contains significant numbers of hexagons and larger polygons. It was suggested that the existence of larger polygons in the hydration shell indicates that the clathrate analogy is too simple to explain water organization along hydrophobic surfaces (2).

The water hydration shell around nonpolar molecules, such as methane, has been studied before, during and after gas clathrate hydrate formation using neutron diffraction with H/D isotopic substitution coupled with empirical potential structure refinement (EPSR) computer simulations (3–5). Gas clathrate hydrates are crystalline inclusion compounds that are formed when water and nonpolar molecules, such as methane, come into contact at high pressure and low temperature (6). Figure 1 illustrates the expansion of the water hydration shell and increased ordering of the second hydration shell around methane after clathrate hydrate formation.

Hydrophobic behavior has been observed for a concentrated solution of 0.86 mole fraction of tertiary butanol in water (7). This behavior is generally seen in more dilute solutions. A solution of tertiary butanol in water presents

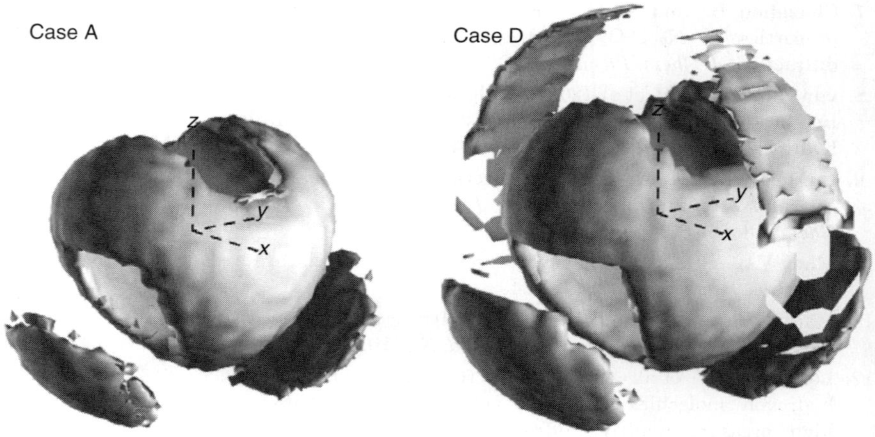

Figure 1. The hydration shell structure around methane: Case A, before hydrate formation (180 bar, 18 °C); case D, after hydrate formation (180 bar, 4 °C).

a classic example of an ampliphile–water solution. Neutron diffraction with H/D isotopic substitution coupled with EPSR computer simulations was used to examine the intermolecular correlations between alcohol and water molecules. The results revealed that "water pockets" are created within the concentrated alcohol–water solution, which reduce the degree of alcohol–alcohol direct hydrogen bonding. The pockets were found, on average, to contain between two and three water molecules. These water molecules exhibit a strong tendency to interact with the polar hydroxyl groups on the alcohol. As a result, the nonpolar groups became more tightly packed (i.e., exhibiting hydrophobic association) compared to that in the pure liquid alcohol. In the case of more dilute solutions of tertiary butanol in water (0.06 mole fraction of alcohol), the structural data do not support the conventional view that hydrophobic processes are dominated by perturbations of the first hydration shell. Conversely, significant changes are observed in the second hydration shell of water, with a "compression and tightening" of water molecules in the second shell (8).

A two-moment information theory was developed to describe the hydrophobic effect (9). This theory was shown to have a clear connection with the molecular principles of statistical thermodynamics. It was suggested that this theory could also be applied to describe hydrophobic effects on biopolymer structure in aqueous solution.

HYDRATION OF IONS

Concentrated solutions of sodium hydroxide in water have been studied using neutron diffraction with H/D isotopic substitution and EPSR computer simulations. The solute was found to affect the tetrahedral network of hydrogen-bonded water molecules (cf. when high pressure is applied to pure water). The results also indicated that there was a competition between the hydrogen-bonding interactions and Coulomb forces in determining the orientation of water molecules within the cation solvation shell (10).

The effect of adding sodium chloride as a salting-out agent to a dilute solution of tertiary butanol in water (0.02 mole fraction of alcohol) has also been studied using neutron diffraction (11). Contrary to previous understanding on salting-out (12), these measurements show that an anion bridge is formed between the polar ends of nearby alcohol molecules. Hence, there is significant enhancement of polar–polar interactions of the alcohol hydroxyl groups. As a result of the anion bridge formation, further hydrophobic interactions occur between the nonpolar/nonpolar groups.

CONCLUSION

Hydration of nonpolar molecules and salts and hydrophobic hydration have been studied in detail over the last couple of decades using computer simulation and theory. Recently, microscopic tools such as neutron diffraction with H/D isotopic substitution have been applied to study hydration of nonpolar molecules and salts. Neutron diffraction measurements coupled with molecular simulation provide a powerful tool to verify existing models of hydration of nonpolar molecules and salts, as well as combinations of these species.

Acknowledgment

The author acknowledges funding support from Chevron Texaco Energy Technology Company, EPSRC, the ISIS facility, and the CSM Hydrate Consortium. Thanks also to P. Buchanan, N. Aldiwan, and Y. Lui.

BIBLIOGRAPHY

1. Franks, F. (1973). *Water. A Comprehensive Treatise*. Vol. 2, Plenum, New York, pp. 1–48.
2. Head-Gordon, T. (1995). Is water structure around hydrophobic groups clathrate-like? *Proc. Natl. Acad. Sci.* **92**: 8308–8312.
3. Koh, C.A. et al. (2000). Water ordering around methane during hydrate formation. *J. Chem. Phys.* **113**: 6390–6397.
4. Buchanan, P. et al. (2003). *In situ* neutron diffraction studies of methane hydrate formation and decomposition. *J. Chem. Eng. Data* **48**: 778–782.
5. Koh, C.A. (2002). Towards a fundamental understanding of natural gas hydrates. *Chem. Soc. Rev.* **31**: 157–167.
6. Koh, C.A. and Sloan, E.D. (2005). Clathrate hydrates. In: *Encyclopedia of Water*. John Wiley& Sons, Hoboken, NJ.
7. Bowron, D.T. and Moreno, S.D. (2002). The structure of a concentrated aqueous solution of tertiary butanol: water

pockets and resulting perturbations. *J. Chem. Phys.* **117**: 3753–3761.

8. Bowron, D.T., Soper, A.K., and Finney, J.L. (2001). Temperature dependence of the structure of a 0.06 mole fraction tertiary butanol–water solution. *J. Chem. Phys.* **114**: 6203–6219.
9. Hummer, G. et al. (1996). An information theory model of hydrophobic interactions. *Proc. Natl. Acad. Sci.* **93**: 8951–8955.
10. Botti, A. et al. (2004). The microscopic structure of concentrated NaOH solutions. *J. Chem. Phys.* **120**: 10154–10162.
11. Bowron, D.T. and Finney, J.L. (2003). Structure of a salt–amphiphile–water solution and the mechanism of salting out. *J. Chem. Phys.* **118**: 8357–8372.
12. Sacco, A., De Cillis, F.M., and Holz, M.J. (1998). *Chem. Soc. Faraday Trans.* **94**: 2089.

THE MIRAGE OF THE H_2 ECONOMY

Reuel Shinnar
The City College of the CUNY
New York

In a recent paper (1), I analyzed the new "hydrogen economy" and explained why, for most uses, it makes no sense to convert fuels, electricity, or any other form of energy to hydrogen, only to use the hydrogen as a fuel for fuel cells. As H_2 is not a resource but an inefficient energy carrier, it would appear that the H_2 economy is just a mirage. If alternative energy sources are ever developed, an economy based on electricity would be cheaper (by a factor of three) and much easier to adopt than H_2. Oddly, I had covered the same topic twenty years ago in a paper that analyzed the hydrogen economy of the 1970s (2). Then, the investment in H_2 research was about 10 billion wasted dollars before it was phased out. In this column, I will try to explain why the scientific establishment engages in such obviously futile behavior and why this mirage attracts so many followers.

In the mid 1970s, I became aware of a strange phenomenon. Many leading scientific journals were suddenly giving enormous coverage to hydrogen, the fuel of the future; a new term, hydrogen economy, was born. International meetings attracted thousands of participants and the United States, Germany, and Japan funded the necessary development with vast resources. The central idea, put forward by the International Atomic Energy Research Center in Vienna, used a high-temperature nuclear reactor as the heat source. In this reactor, pressurized helium was used as the coolant. To generate electricity, the hot helium was first expanded through a turbine and then used for steam generation before being fed back to the reactor.

The "thermochemical cycle," the name applied to this new process, does not include a turbine, and the heat from cooling the hot helium is used to drive a set of closed-cycle chemical reactions; the net result is that water is split into hydrogen and oxygen.

Following is a hypothetical example of such a reaction:

$$\text{MeO} + \text{Heat} \longrightarrow \text{Me} + 1/2\text{O}_2$$
$$\text{Me} + \text{H}_2\text{O} \longrightarrow \text{H}_2 + \text{MeO} + \text{Heat}$$

where Me is any metal. No such reaction was ever discovered, and all cycles proposed were quite complex. One does not need to be an experienced cost estimator to realize that such a cycle would be very expensive and have a low thermal efficiency.

The H_2 was supposed to be used to replace natural gas in existing pipelines. Several obvious questions were not asked:

1. A pipeline for methane cannot be used to transport H_2. To avoid leakage, one needs totally different valves and connectors. Furthermore, as the volume of H_2 is triple that of methane, one needs larger pipes and different compressors. Of course, the same factors would apply to the distribution of the gases to customers. None of the valves or controls would be tight enough, and all burners would have to be changed.
2. As nuclear reactors are available, why not use electricity instead of methane for most applications? Electricity is far more efficient than H_2 or methane (by a factor of two) for most uses. Let me illustrate. Today, in the United States and the world, a large amount of electricity is generated from natural gas. This electricity, on a BTU basis, costs by a factor of about 2.2 more than the natural gas feed. The fact that we do it obviously means that electricity is more valuable than natural gas. At the same price, it would have driven out natural gas. H_2 as a fuel has an even lower value than natural gas. If we now convert electricity to H_2, we again lose at least one-third of the energy. Thus, H_2 made from electricity has a value about four times lower than the electricity. This simple calculation illustrates the stupidity of the idea to convert nuclear or solar energy to hydrogen. At that time, no one was foolhardy enough to suggest using H_2 generated by a nuclear reactor as fuel for a fuel cell. Furthermore, an electricity distribution system is already in place and can grow with demand.
3. Why not make H_2 by electrolysis, a well-known process? Nobody asked why—in contradiction to our cumulative technical experience—a set of chemical reactions should generate a large increase in ΔT more cheaply or efficiently than electrolysis. The main item in the cost of electricity is the nuclear reactor, and if a thermochemical cycle is less efficient, it is inherently more expensive.
4. The cost of switching to hydrogen is astronomical. To produce the equivalent of one million BTU of hydrogen per day by electrolysis would require a 25-kilowatt nuclear reactor at an investment cost of \$150,000, which means that a standard nuclear reactor of one gigawatt costing \$6 billion will be able to supply 40,000 million BTU/day of H_2 the

equivalent of 6700 barrels of oils. Building the capacity for supplying a new pipeline that could carry 12% of the current U.S. capacity of natural gas would have to be built from the start at a cost of close to $1 trillion. For thermochemical cycles, the cost would be significantly higher. All of this must be invested before a pipeline can be started. The same energy needs could be supplied by electrical energy at only one-quarter the cost because electrical energy has a higher efficiency for almost all purposes. The investment for electricity can be spread over many years because a grid is already in existence. Furthermore, just the energy from the nuclear reactions would cost $50 per million BTU H_2.

5. The most dangerous of all fuels, pressurized H_2 ignites with an invisible flame. Its explosive limits are very wide, and its minimum ignition energy is ten times smaller than that of any other gaseous fuel. This problem occurs not only when H_2 is being distributed or used, but even more so when it is generated in a thermochemical cycle. When H_2 is produced by electrolysis, one can locate the plant far away from the nuclear reactor generating the electricity, which, however, is not true for helium, which must be pumped in such large quantities that it cannot be sent over long distances at reasonable cost. Society has accepted the risks of explosion in conventional fuels but tries to limit exposure. Thus, even a small propane storage tank may not be transported through a tunnel, but what level of risk from a potential explosion at a nuclear power plant complex is acceptable? None of these critical questions were asked then—nor are they being asked today.

These questions have been on my mind for a long time. In the "age of technology," how could the scientific community have been misled by such obvious nonsense? And why had so many engineering firms and large companies come up with totally unrealistic cost estimates, at least by a magnitude too low, and thermal efficiency estimates that are unrealistically high?

The intellectual climate of the time can be illustrated by a small story. *Science* carried a six-page article, featured on the cover, about a new cycle developed by General Electric that promised to produce H_2 at $2.50/million BTU H_2 (at that time it would have cost $7.00/million BTU to produce H_2 from oil). For fun, I flow-sheeted the process and published the mass and heat balance in a letter to *Science* (3) suggesting that if GE can reduce costs to $2.50, they should forget about hydrogen and change the economy. By comparison, us poor chemical engineers could not achieve such results at a cost of $250/million BTU H_2. Three months later, I was invited to GE's central laboratory where the Director of Chemical Research offered me a consulting job. The head of the lab was a chemist who had read about the need for a thermochemical cycle, and he actually developed the first one that worked. His engineers had given him this cost estimate. When my letter was published, he asked Corporate Engineering to make their own estimate and they confirmed my numbers. This story illustrates a key aspect of the hydrogen economy. Many excellent scientists lack the skills needed to deal with economics and costs. Later, I published another paper demonstrating, from elementary design principles, that any thermochemical process must, by necessity, cost much more than electrolysis (2).

It is true that many scientists could have erred because they were not knowledgeable about process economics. However, I also came to realize that other forces were driving the program. Although they are inherently safer, high-temperature nuclear reactors of the type required by thermochemical cycles were deemed less competitive and more costly than boiling water reactors, which created an opportunity for their developers to seek more government support. A large nuclear research community was hungry for grants and contracts.

Interestingly and unfortunately, the episode of the 1970s has been ignored or forgotten. However, before we place large national resources behind an effort to again deal with such major problems as renewable energy, global warming, and energy independence, we have to understand what went wrong the first time a so-called "hydrogen economy" was attempted.

The paper I just published is about the reincarnation of the hydrogen economy. This time, all the proposed processes are feasible, but the program itself seems to make even less sense. As before, the main emphasis is on the replacement of conventional fuels with a national distribution system of H_2 at a tremendous loss in efficiency in the production and transmission steps. As before, the same intractable safety and cost problems still persist.

Similar driving forces remain at work. Large research budgets are projected to find a use for a technology in which DOE has invested huge sums of money to little or no avail. This time, it is fuel cells powered by hydrogen, a technology that is good for space applications, where H_2 is used as a rocket fuel, and therefore available in small quantities, and other specialty applications, but useless for applications that couple H_2 generation with the generation of electricity. Such systems have half the efficiency of combined cycle turbines, cost over five times more (1,4), and create at least double the greenhouse emission, probably much more.

Today, there even is talk about switching cars from gasoline to H_2, which makes the transition problem even more difficult as the new supplying system cannot start locally, and this would also require the use of pressurized H_2 at 6000 psi, a pressure very seldom used in industry. Nobody mentions that in industry an H_2 tank with the capacity proposed for a family car requires storage in a special room with a blowout wall. Any technician approaching such a pressurized tank to check the valves swings a wooden broom to prevent being fried by an invisible flame which may come from a leaking valve, as above 1500 psi, an H_2 leak is self-igniting (1). For a gas station, a minimum-sized storage tank would have the explosive power of two thousand tons of TNT or two of the largest bombs used by the Air Force. A terrorist could just open a valve and detonate a small bomb after 20 seconds.

The most obvious fallacy of the new H_2 economy, however, is that it solves all of our problems (5,6). For those of us who accept the first and second laws of thermodynamics, it is obvious that the use of H_2 would increase the use of other energy resources. Arguments are made that, once we have a hydrogen distribution system, we could capture the CO_2 produced and sequester it. However, we have no idea how to sequester such vast amounts of CO_2 safely. A successful policy can only be formulated after all the elements are solved and a good deal is known about costs. At present, the United States releases 100 million tons of sequestered CO_2 back into the atmosphere each year.

In a system in which the basic science is known, costs are reduced by innovative ideas that develop during the implementation stage of a project rather than by more research. The cost of liquid natural gas plants and combined cycle power plants decreased in the last ten years by a factor of two even though the basic technology did not change because many were built to compete in the market. So, research, to be effective, needs to be coupled with a clear plan for implementation.

In order to promote the use of fuel cells or other new technologies, which are by one to two orders of magnitude more expensive than available technologies, the research community has invented the myth of "learning curves." In the last 40 years, computers became, through miniaturization, cheaper by several magnitudes, but this option does not apply to energy systems. In the last 20 years, the cost of solar cells was reduced by less than a factor of two despite a tremendous research budget. Better engineering and mass production can reduce the cost of most new technologies by a factor of two to three. However, this would require the building of large plants based on competitive bidding, not a new research breakthrough. Large research budgets over the past 40 years have not made the cost of fuel cells more competitive.

The alternative sources that really reduce CO_2 emissions include solar and nuclear energy and biomass. The use of H_2 just makes every one of these options several times more expensive. Solar and nuclear energy produce electricity, and, for most purposes, electricity is far more efficient to use than H_2. To provide electricity to the motor in a H_2 car requires three to four times the solar or nuclear capacity required for providing the electricity directly via a battery. Although we don't have good electric cars yet, Toyota has just come out with a hybrid car with a plug-in battery large enough for a 40-mile drive. Such a car could reduce total gasoline consumption by 80%, but electric cars would only reduce greenhouse emissions if the electricity were generated by solar energy. As for biomass, which is a valuable but limited future resource, conversion by fermentation to alcohol followed by conversion to gasoline or diesel is far cheaper and thermally more efficient than conversion to hydrogen.

In the early 1990s, an Israeli company, Luz (7), built solar electric power plants using a high-temperature heat transfer fluid, which included sufficient storage to supply Los Angeles with 300 megawatts of electricity on a reliable basis. The Luz plant had another major advantage; the design allowed for conventional fuel to be used as a backup because, even in the desert, 10% of the days may be rainy. To be profitable, all Luz needed was a 50% tax break on the investment. When Luz became large, the tax breaks were taken away. If Luz had been allowed to build a few gigawatts, the price would have come down and other companies would have come in; we might have had affordable solar energy today.

We have today the resources and the technology to do something useful about global warming if we really had the resolve. For example, for \$50 billion a year over twenty years, we could replace 40% of coal power plants with solar power plants using proven technology; we just have to engineer them better and provide them with sufficient storage capacity. We are the only western nation who has enough sun and enough area for solar power. Alternatively, we could eliminate about 80% of our gasoline used for cars. We could easily afford to do both, but we have to realize that no research can make solar, nuclear, or other alternative energy competitive with cheap oil or gas.

Our society has solved far more difficult large-scale technical problems successfully, e.g., space travel and the most advanced defense weapons in the world. Applying our capabilities to global warming or to resource depletion to preserve a livable world to our grandchildren would require not only conviction but a strong political will to do so.

To give a recent example, thirty years ago, we faced the problem of increasing pollution from our coal power plants. We had the technology (scrubbers) to reduce the pollution substantially. It would have cost about \$20–30 billion, but politics and the influence of senators from coal-producing states prevented the adoption of such a solution. Instead, we spent \$20 billion on "clean coal" research programs that failed to produce results that could help solve this problem. However, the "clean coal" effort did provide large research budgets along with the illusion that we were accomplishing something. Thirty years later, the same plants are producing more electricity and consuming more coal—still poisoning the fish we eat with mercury, causing smog that blocks us from seeing the sky, destroying our forests, and, most importantly, impairing the health of all of us more than any other form of pollution.

The private car is another example of the obstacles that a serious national policy faces. In the 1970s, in response to the Arab oil embargo, oil price increases, and an alleged likelihood that the oil resource base would become depleted within the next several decades, Presidents Nixon, Ford, and Carter proposed new energy policy initiatives that included the goal of improving auto fuel economy. These initiatives resulted in the Corporate Average Fuel Economy standards (CAFÉ), which required that the fleet fuel economy of new light-duty vehicles (5500 lbs or less) reach 27.5 mpg by a certain, date a 30% reduction in gasoline demand. This standard limited the ability of the companies to produce large cars and stationwagons. The same provision set lower standards in mileage and emissions for light-duty trucks weighing less than 5500 lbs.

Nine years ago, the car industry realized that one could produce large cars (SUVs) that fit all the definitions of a light truck, such that one should be able to fold all seats to the floor to provide a major space occupying the car and several others. It was a great success. As light trucks convertible to luxury passenger cars do not count as passenger cars for the CAFE laws and are not considered in the fleet average, this allowed car companies again to produce gas guzzlers. Eight years later, 30% of the passenger cars sold in the United States are SUVs, and all the efficiency gains of 30 years have been wiped out.

As we face the potential of global warming and resource depletion that might endanger the future of our grandchildren, we escape our responsibilities by pretending that all problems can be solved better and at lower cost through the mirage of more research. Thus, the H_2 mirage allows car companies to continue to produce gas-guzzling, highly polluting SUVs while touting the glorious future, an "H_2" car that will miraculously be energy efficient and pollution-free. And, as in the 70s, thousands in the scientific and political communities are promoting this mirage to the public.

Thus, mirages like the "hydrogen economy" play an important psychological function in our society, and many in the technical community play along. The role engineers and scientists should be playing today is to educate the public and the politicians about the real options. We should be reminded of an old cartoon in the Pogo series that bears the caption "We have met the enemy and he is us."

BIBLIOGRAPHY

1. Shinnar, R. (2003). The hydrogen economy, fuel cells, and electric cars. *Technology in Society* **25**(4): 453–576.
2. Shinnar, R., Shapira, D., and Zakai, S. (1981). A thermochemical and hybrid cycles for hydrogen production. A differential economic comparison with electrolytes. *IEC Process Design & Development* **20**: 581.
3. Shinnar, R. (1975). Thermochemical hydrogen generation heat requirements and costs. *Science* **188**: 1036.
4. Shinnar, R. (2003). Fuel-cell faux pas. *Chemical & Engineering News* **81**(14): 4.
5. Rifkin, J. (2003). The hydrogen economy. *Environmental Magazine*. January.
6. Bush, G.W. (2003). *State of the Union Address*. January 28.
7. Shinnar, R. (1993). The rise and fall of luz. *ChemTech* **23**: 50–53.

READING LIST

Energy Research of the Department of Energy. (2001). *Was It Worth It?* National Research Council.

Effectiveness and Impact of Corporate Average Fuel Economy (CAFE) Standards. (2002). National Research Council.

HYDROGEN ION

Gholam A. Kazemi
Shahrood University of Technology
Shahrood, Iran

INTRODUCTION

This article includes an introduction to the ion and its properties followed by an overview of the characteristics of the hydrogen atom. The latter has been extracted mostly from Sienko and Plane (1). This article should be studied in conjunction with the pH and Hydronium Ion articles.

An ion in general terms can be any grouping of one or more atoms that carries a charge (1). The characteristics of an ion include:

1. the sign (type) of charge: positive (cation) or negative (anion);
2. the number of charges or valence (z): monovalent divalent, etc;
3. the ionic radius (r): the distance from the center of the nucleus to the outer shell of the atomic particle;
4. the ionic potential: z/r, the ratio of the charge of the ion to the ionic radius;
5. the electronegativity: the power of the ion to attract electrons;
6. the molar volume: the volume of one mole (cm^3/mol); and
7. the ionic conductance: the ability to carry electrical current.

When an ion is surrounded by many other ions/molecules such as in a solution, its other properties such as ionic strength and ion activity coefficient come to light. Ottonello (2) is a major reference where many properties of various ions are tabulated.

Hydrogen, the lightest of all gases, is colorless, odorless, and tasteless. Together with helium, it occupies the first period of the periodic chart of elements and is the most abundant element in the solar system. After helium, it has the second lowest boiling temperature of all elements (2). Because of its low boiling point, liquid hydrogen is used as a cryogenic fluid (to produce low temperatures), and because of its high heat of combustion (120 kJ/g), it is a valuable rocket fuel. Molecular hydrogen, H_2, can combine directly with most elements. Mixtures of H_2 and O_2 can be explosive. Much of the industrial consumption of hydrogen is in hydrogenation reactions, the addition of H_2 to other molecules. Water and power are required to produce hydrogen. In its compounds, hydrogen is found in three oxidation states, +1, −1, and 0. Hydrogen is such an important element that it has its own association. The National Hydrogen Association (WWW. Hydrogenus.org), for example, was formed in 1989 (in the U.S.) to act as a focal point in hydrogen related matters. Anyone interested in water related matters should know the important characteristics of hydrogen because of its place in the water molecule.

DEFINITION OF A HYDROGEN ION

A hydrogen ion, H^+, is a monovalent cation that occurs when a hydrogen atom loses its only electron:

$$H \longrightarrow H^+ + e^-$$

Therefore, the hydrogen ion has only one proton and no electron. This is why "proton" is another name for hydrogen ion. Some references state that there is no such ion as a hydrogen ion and what occurs, is actually a hydronium ion, H_3O^+, rather than a hydrogen ion (3). The hydride ion, H^-, occurs when a hydride such as NaH dissociates:

$$NaH \longrightarrow Na^+ + H^-$$

In this reaction, the hydride ion radius is 1.46 Å. Molecular hydrogen is formed from the reaction of a hydride ion and a hydrogen ion (4):

$$H^+ + H^- \longrightarrow H_2$$

${H_3}^+$ is an ion that is artificially generated by the interaction of H^+ and H_2. For instance, generation of ${H_3}^+$ is the most critical specification of the ion source in a mass spectrometer (5).

CHARACTERISTICS OF THE HYDROGEN ION

Some major characteristics of the hydrogen ion are presented in Table 1.

In chemistry and geochemistry, it is assumed that the standard heat capacity, C_{p}^0; standard partial molal Gibbs free energy from the elements, G_{f}^0; standard partial molal enthalpy of formation from the elements, H_{f}^0; standard partial molal entropy, S_0^{p}; and standard partial molal volume, V_{p}^0, of the hydrogen ion equal zero at all temperatures and pressures (2,7).

MEASUREMENT OF HYDROGEN ION CONCENTRATION

More than 100 years ago, Arrhenius suggested that acid properties are attributable to the presence of an easily detachable hydrogen ion, H^+ (1). Now, everyone knows that the concentration of hydrogen ion in any solution is a measure of the solution's acidity or pH ($\mathrm{pH} = -\log H^+$). Thus, the measurement of H^+ concentration is better described in the pH article of the encyclopedia and is only briefly dealt with here.

The H^+ ion is the easiest ion concentration to measure. This is done using a hydrogen electrode, which is formed by bubbling H_2 over a platinum electrode surface (7). A variety of these electrodes are commercially available; Metrohm pH meter Model 744 and Orion pH meter Model 420A are examples. Measurement of pH cannot be more accurate than 0.02 pH units because the standard solutions or buffers which have been calibrated by the U.S. Bureau of Standards (NBS) have an absolute accuracy in the range of ±0.01 to 0.02 pH units (7). In field conditions, where disturbing factors are ample, the accuracy is usually ±0.05 pH units. Colorimetry is one older method of measuring H^+ concentration.

Table 1. Notable Characteristics of the Hydrogen Ion, H^{+a}

Type	Cation
Valence	1
Radius[b]	—
Ionic potential (IP)[c]	—
Electronegativity	2.2
Molar volume	0 (This is based on convention)
Ionic conductance	$349.8\ \mathrm{cm}^2 \times (\mathrm{equiv}\Omega)^{-1}$
Charge to mass ratio (q/m)	9.6×10^7 C/kg

[a]References 4, 6, 7.
[b]It is difficult to define because the electron of the hydrogen atom has gone, and only the proton remains. So, there is no outer shell left to measure its distance to the proton.
[c]It cannot be calculated because there is no specified ionic radius.

Some Examples

Pure neutral water at 25 °C has a pH of 7, a H^+ concentration of 10^{-7} mg/l. A 1-molal solution of HCl will have an H^+ concentration of 1 and a 0.01-molal solution will have an H^+ concentration of 10^{-2} mg/l. The concentration of hydrogen ion in a 0.01-molal solution of a weak such as H_2CO_3 is only $10^{-4.2}$ mg/l, and it is $10^{-10.8}$ mg/l in an aqueous solution of pH 3.2.

Many authors prefer to use the term "hydrogen ion activity" rather than hydrogen ion concentration. The reason that the "pH" concept has been developed is the very low concentration of hydrogen ion and the difficulty in comprehending such low, sometime ambiguous, numbers. In solutions, the concentration of H^+ never reaches zero; there is no solution that is 100% basic or 100% acidic.

PRODUCTION OF HYDROGEN ION IN WATER

Hydrogen ion is produced in natural water from the three following main sources as well as from humic and fluvic acids, volcanic gases, acid rain, and short-chain organic acids present in some oil field brines (8).

1. Hydrolysis

$$H_2O \longrightarrow H^+ + OH^-$$

As stated above, however, it might be more appropriate to write the previous equation in the form:

$$2H_2O \longrightarrow H_3O^+ + OH^-$$

2. Dissociation

$$H_2CO_3 \longrightarrow H^+ + {HCO_3}^-$$

3. Oxidation

$$2FeS_2 + 7.5O_2 + 7H_2O \longrightarrow 2Fe(OH)_3 + 8H^+ + 4{SO_4}^{2-}$$

IMPORTANCE OF THE HYDROGEN ION

Some of the important functions and behaviors of the hydrogen ion are described as following. Considering the huge and universal importance of H^+, these should be regarded only as examples.

1. The radius of the hydrogen ion is very close to the size of the neutron. This similarity is the principle used to design and develop compensated neutron

logs (CNL) and neutron probes. The former is used as an important tool for calculating the porosity of petroleum reservoirs and aquifers (9,10), and the latter is of extensive use in agricultural practices for measuring soil moisture (11).

2. The concentration of hydrogen ion, pH, is a master variable of the groundwater system because the hydrogen ion participates in many important chemical reactions such as (12):

   a: mineral dissolution—precipitation

   $$CaCO_3 + H^+ = Ca^{2+} + HCO_3^-$$

   b: aqueous complexation

   $$Fe^{3+} + H_2O = Fe\,(OH)^{2+} + H^+$$

   c: adsorption –desorption

   $$\equiv FeOH + Cu^+ = \equiv FeOCu + H^+$$

3. The hydrogen ion plays a major role in denitrification through which various forms of nitrogen in the soil revert to N_2 (6):

   $$4NO_3^- + 4H^+ \longrightarrow 2N_2 + 5O_2 + 2H_2O$$

   It also plays a major role in redox reactions and acid–base reactions.

4. In sanitary engineering and water treatment, pH is an important factor in chemical coagulation, disinfection, water softening, corrosion control, and biological processes (13).
5. The most remarkable function of hydrogen ion is its control of the acidity or alkalinity of solutions.

BIBLIOGRAPHY

1. Sienko, M.J. and Plane, R.A. (1979). *Chemistry: Principles and Applications*. McGraw-Hill, New York.
2. Ottonello, G. (1997). *Principles of Geochemistry*. Columbia University Press, New York.
3. Fetter, C.W. (1988). *Applied Hydrogeology*. 2nd Edn. CBS Publishers and Distributors, Delhi.
4. Lagowski, J.J. (1973). *Modern Inorganic Chemistry*. Marcel Dekker, New York.
5. Attendorn, H.G. and Bowen, R.N.C. (1997). *Radioactive and Stable Isotope Geology*. Chapman & Hall, London.
6. Evangelou, V.P. (1998). *Environmental Soil and Water Chemistry: Principles and Applications*. John Wiley & Sons, New York.
7. Langmuir, D. (1997). *Aqueous Environmental Geochemistry*. Prentice-Hall, Upper Saddle River, NJ.
8. Hounslow, A.W. (1995). *Water Quality Data: Analysis and Interpretation*. Lewis, New York.
9. Hearst, J.R., Nelson, P.H., and Paillet, F.L. (Eds.). (2001). *Well Logging for Physical Properties: A Handbook for Geophysicists, Geologists, and Engineers*, 2nd Edn. John Wiley & Sons, New York.
10. Todd, D.K. (1980). *Groundwater Hydrology*, 2nd Edn. John Wiley & Sons, New York.
11. Wilson, E.M. (1984). *Engineering Hydrology*, 3rd Edn. Macmillan, London.
12. Deutsch, W.J. (1997). *Groundwater Geochemistry: Fundamentals and Applications to Contamination*. Lewis, Boca Raton, FL.
13. Tebbutt, T.H.Y. (1995). *Principles of Water Quality Control*. Reprint of 4th Edn. Butterworth Heinemann, Oxford.

THE HYDRONIUM ION

Carl W. David
University of Connecticut
Storrs, Connecticut

Naked protons cannot exist in aqueous solution. Even when one writes

$$HCl(g) \rightarrow H^+(aq) + Cl^-(aq) \quad (1)$$

one knows that the $H^+(aq)$ is a representation of a significantly more complicated moiety (or, to be more precise, set of moieties). The simplest representation of $H^+(aq)$ is the hydronium ion (sometimes also referred to as the oxonium ion or the hydroxonium ion),

$H_3O^+(aq)$, but this is just the first in a series the next few members of which are $H_5O_2^+$, $H_7O_3^+$, $H_9O_4^+$, and in summary $(H_2O)_nH_3O^+$, the hydrated hydronium ions (Fig. 1).

The hydronium ion is isoelectronic with ammonia (NH_3) and therefore, ought to be pyramidal with the oxygen at the apex of the pyramid and the three protons disposed below, with symmetry C_{3v}. Like ammonia, it should invert, with the O[or N] passing through the plane of the three protons from one side to the other. In the planar transition state, when all four nuclei are coplanar, the symmetry should be D_{3h}. Also, like ammonia, there should be a barrier to this umbrella inversion; appropriate tunneling has been observed spectroscopically (ν_2). It is thought that the O–H distance is about 0.96 Å, and the H–O–H angle about 115°. The barrier to inversion, a saddle point on the energy hypersurface, is thought to be between 1 and 5 kcal/mol (Fig. 2).

The simplest occurrence of the hydronium ion can be seen in the gas phase water dimer, $(H_2O)_2$. The proton transfer reaction inside the dimer would be

$$H_2O\text{–}H_2O(g) \rightarrow H_3O^+(g) : OH^-(g) \quad (2)$$

forming an ion pair (Fig. 3).

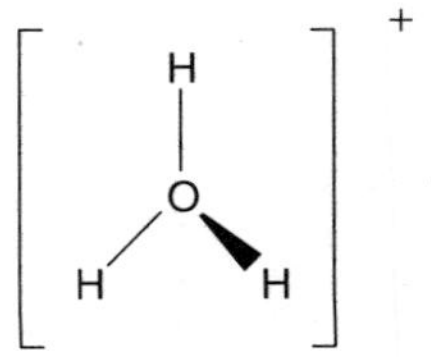

Figure 1. The hydronium cation.

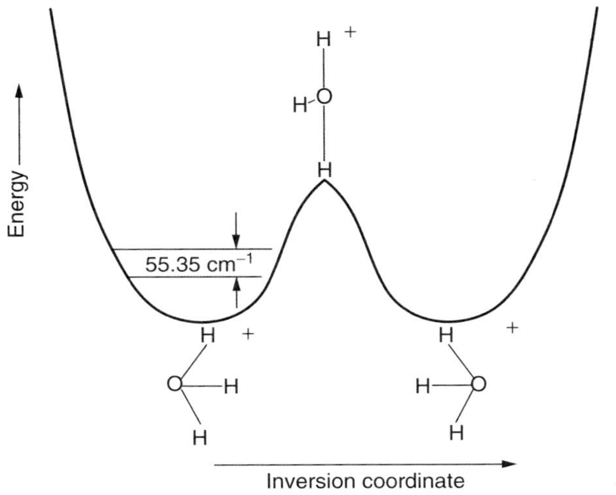

Figure 2. Potential energy surface for the hydronium ion's inversion coordinate.

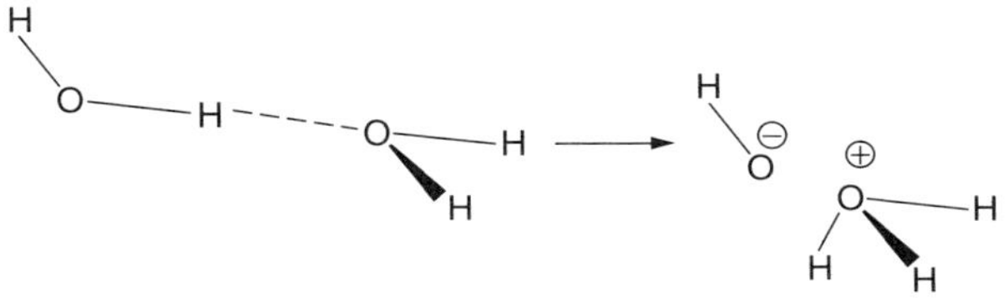

Figure 3. Dimer water forming an ion pair through proton transfer.

In the dimer, one molecule of water is a proton donor to the other (which is a proton acceptor). The former water is a Bronsted acid, the latter a Bronsted base. On the line of the hydrogen bond, $HO–H_{donor}::::OH_2$, the movement of the donor proton from one oxygen to another is equivalent to the proton transfer (above). The O–O distance in the dimer is about 3 Å, and the OH distances are about 1 Å, so the shift of 1 Å in a proton's position (between two oxygens) creates an $OH^-:::H_3O^+$ ion pair.

Hydrated versions of this ionization process illustrate that the actual process in liquid water is not cleanly definable. Figure 4 shows the results of three water molecules involved in ionization, whereas Fig. 5 shows the results with four water molecules.

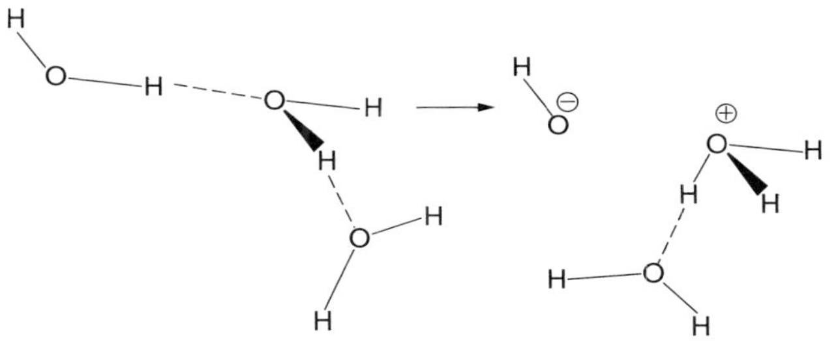

Figure 4. With three molecules of water, one can form a hydrated hydronium cation and a hydroxide anion.

Clearly, in liquid water, the hydroxide and hydronium ions are solvated but not statically enough to allow definition of a moiety.

In aqueous media, for the analog of the proton transfer reaction, one could write

$$2H_2O(l) \rightarrow H_3O^+(aq) + OH^-(aq) \tag{3}$$

the prototype of all acid–base chemistry in an aqueous medium.

In aqueous acidic media, the hydronium ion is rarely written. Instead, the older notation is employed. For instance, for acetic acid, CH_3COOH, one would write

$$CH_3COOH(aq) \Leftrightarrow H^+(aq) + CH_3COO^-(aq) \tag{4}$$

partial dissociation into a proton and a acetate anion, rather than the more accurate

$$H_2O(l) + CH_3COOH(aq) \Leftrightarrow H_3O^+(aq) + CH_3COO^-(aq) \tag{5}$$

The conductance properties of water are attributable in part to the hydronium cation and its peculiar properties. The Grotthuss mechanism for proton transfer in ice employs hydronium ions hydrogen bound to adjacent water molecules, in which, like the dimer, an excess proton migrates (or tunnels) from one oxygen to an adjacent one, moving only a tiny distance, whereupon the charge has shifted enormously (one O–O distance).

A similar mechanism for $OH^-(aq)$ charge migration most likely also obtains. Here a water donor adjacent to the hydroxide anion shifts its donating proton to the hydroxide, thereby itself becoming a hydroxide anion (Fig. 6).

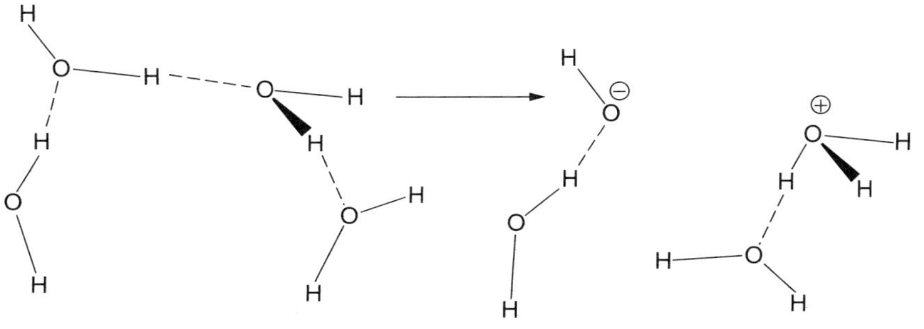

Figure 5. With four waters, hydrated hydroxide and hydrated hydronium ions are possible.

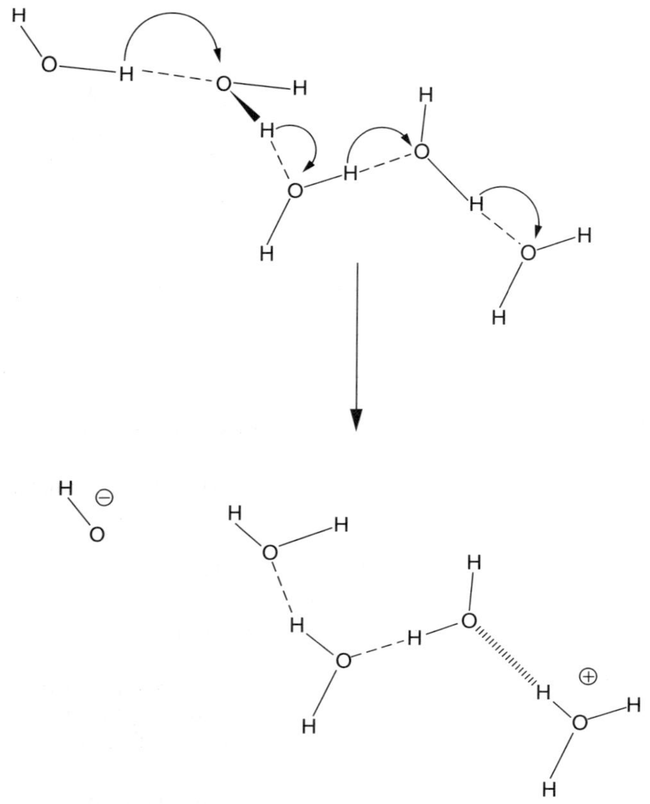

Figure 6. The Grotthuss mechanism for positive and negative charge creation and migration in aqueous solution.

INFILTRATION AND SOIL MOISTURE PROCESSES

Paul R. Houser

(from *The Handbook of Weather, Climate, and Water: Atmospheric Chemistry, Hydrology, and Societal Impacts*, Wiley 2003)

Infiltration is the process of water entry from surface sources such as rainfall, snowmelt, or irrigation into the soil. The infiltration process is a component in the overall unsaturated *redistribution* process (Fig. 1) (1) that results in *soil moisture* availability for use by vegetation transpiration, exfiltration (or evaporation) processes, chemical transport, and groundwater recharge. Soil moisture, in turn, controls the partitioning of subsequent precipitation into infiltration and runoff, and the partitioning of available energy between sensible and latent heat flux.

Because of the importance of soil moisture on multiple processes, its definition can be elusive (2); however, it is most often described as moisture in the unsaturated surface layers (first 1 to 2 m) of soil that can interact with the atmosphere through evapotranspiration and precipitation (3).

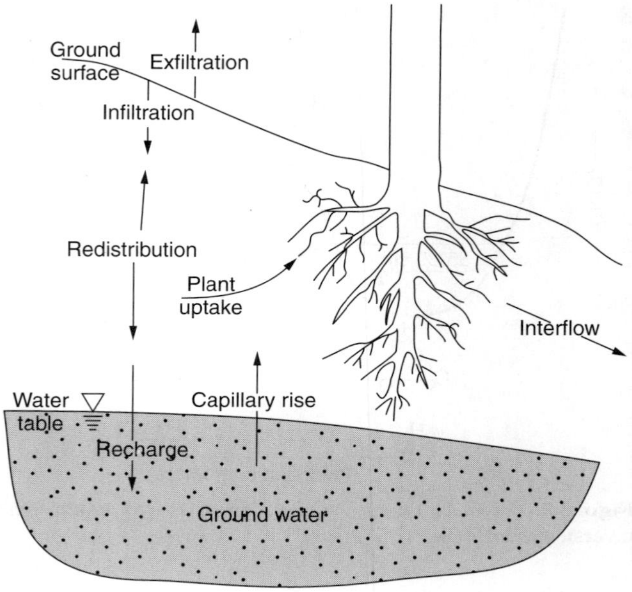

Figure 1. Unsaturated zone definition and active processes (1).

CONTROLS ON INFILTRATION AND SOIL MOISTURE

To characterize soil moisture and infiltration, the physical controls on these processes must be considered. The primary soil controls will be considered in this chapter; however, other factors such as soil chemistry, thickness, soil layering or horizons, and preferential flow paths, as well as vegetation cover, tillage, roughness, topography, temperature, and rainfall intensity also exert important controls (4).

A soil's particle size distribution has a large impact on its hydraulic properties. Soil particles less than 2 mm in diameter are divided into three texture groups (sand, silt, and clay) that help to classify broad soil types and soil water responses (Fig. 2) (5). The type of clay and the coarse material over 2 mm in diameter can also have a significant impact on soil water properties. An overview of methods for determining particle size properties is given by Gee and Bauder (6).

Bulk density, ρ_b (M/L^3) is the ratio of the weight of dry solids to the bulk volume of the soil, and *porosity*, ϕ (M^3/M^3), is the total volume occupied by pores per unit volume of soil:

$$\phi = \frac{V_a + V_w}{V_s} = 1 - \frac{\rho_b}{\rho_m} \tag{1}$$

where V_s (L^3) is the total volume of soil, V_a (L^3) is the volume of air, V_w (L^3) is the volume of water, and ρ_m (ML^{-3}) is the particle density (normally about 2.65 g/cm^3).

The volumetric water content, or soil moisture, θ (L^3 L^{-3}) is the ratio of water volume to soil volume:

$$\theta = \frac{V_w}{V_s} = \frac{W_w}{W_d}\frac{\rho_b}{\rho_w} \tag{2}$$

where W_w (M) is the weight of water, W_d (M) is the weight of dry soil, and ρ_w (M/L^3) is the density of water.

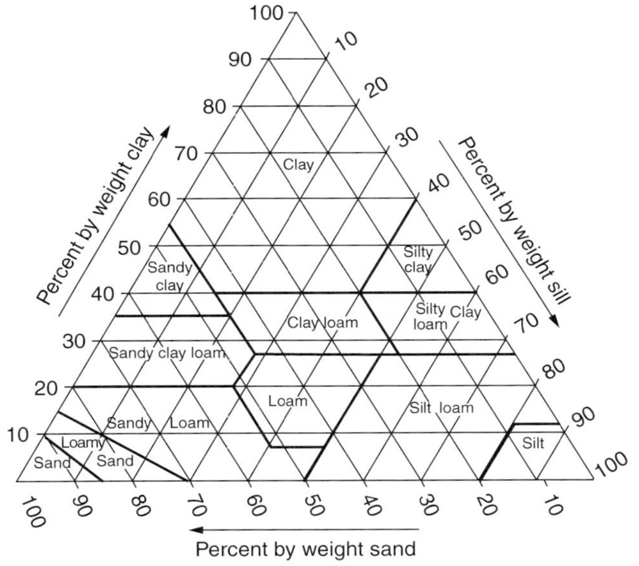

Figure 2. Soil textural triangle describing the relationship between texture and particle size distribution (5).

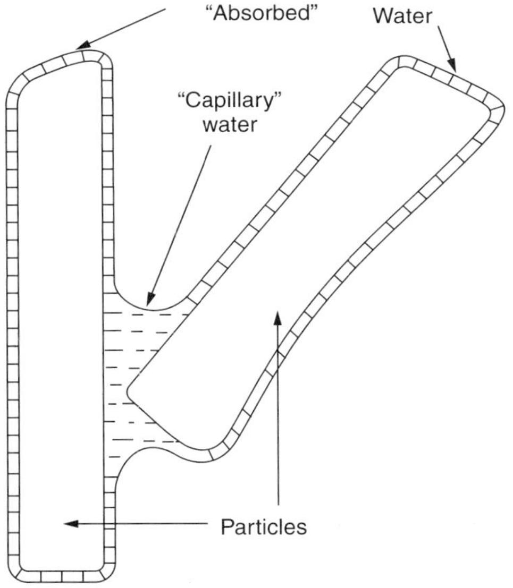

Figure 3. Capillarity and adsorption combine to produce suction (7).

Soil moisture can vary in both time and space, with a theoretical range from 0 to φ, but for natural soils the range is significantly reduced due to isolated pore space and tightly held or "adsorbed" water (Fig. 3) (7). If a soil is saturated, then allowed to drain until the remaining water held by surface tension is in equilibrium with gravitational forces, it is at *field capacity*, θ_f. Vegetation can remove water from the soil until the *permanent wilting point*, θ_w, is reached. Therefore, the *available water content* for plant use, $\theta_a = \theta_f - \theta_w$. Typical ranges of porosity, field capacity, and wilting point for different soils are given in Fig. 4 (8).

In unsaturated soils, water is held in the soil against gravity by surface tension (Fig. 3). This tension, suction, or

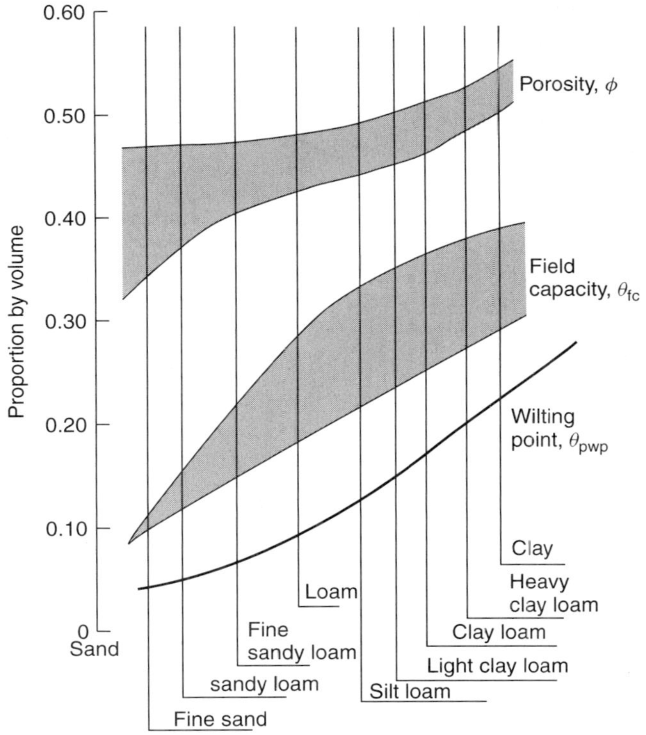

Figure 4. Water holding properties of various soils (8).

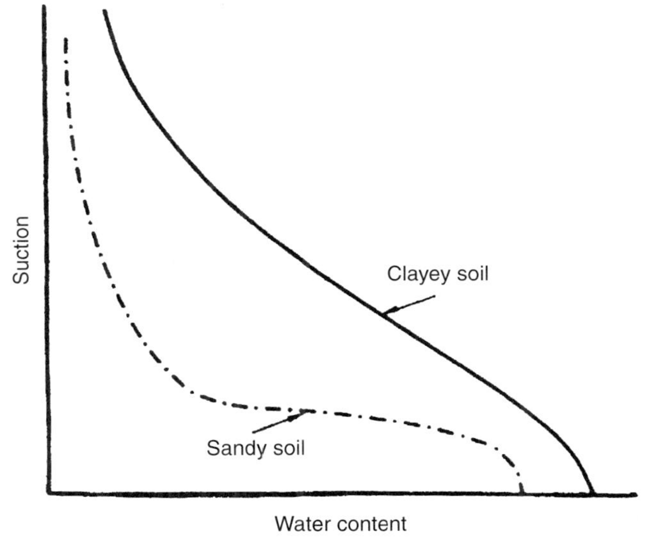

Figure 5. Effect of texture on water retention characteristics (9).

matric potential, ψ (L), increases as the radii of curvature of the meniscus or water content decreases (Fig. 5) (9). Matric potential is expressed in reference to atmospheric pressure, so for saturated soil $\psi = 0$ and for unsaturated soil $\psi < 0$.

The *hydraulic conductivity*, K (L/T), is a measure of the ability of the soil to transmit water that varies nonlinearly over a large range depending on both soil properties and water content (Fig. 6) (10). Many laboratory and field hydraulic conductivity measurement methods exist for

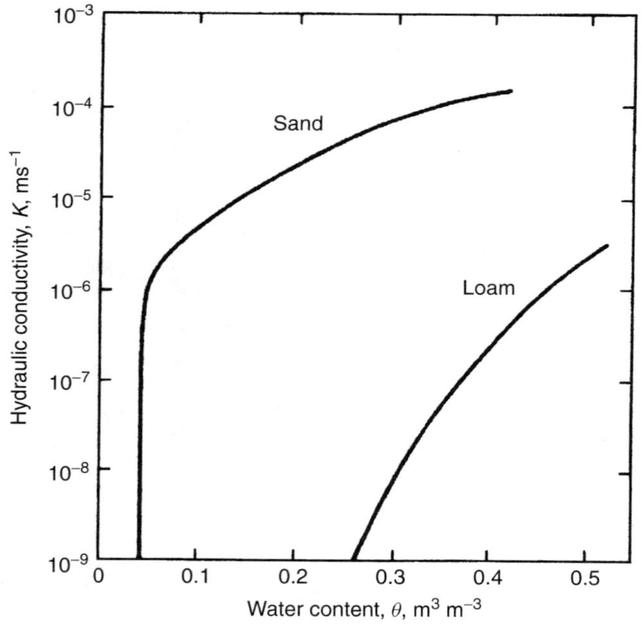

Figure 6. Effect of texture and soil moisture on hydraulic conductivity (10).

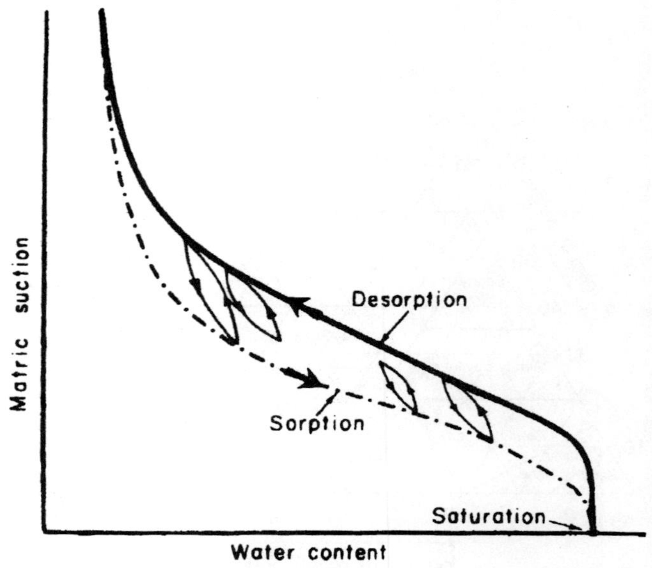

Figure 7. Changes in water retention characteristics between sorption and desorption (16).

use with various soils; see Bouwer and Jackson (11) or Green et al. (12) for details.

Soil water content can significantly impact infiltration by (1) increasing the hydraulic conductivity, which increases infiltration, and (2) reducing the surface tension that draws moisture into the soil, which reduces infiltration. The net effect of these impacts depends on the water content itself, the water input rate, and duration and the distribution of hydraulic conductivity.

The *water retention characteristic* describes a soil's ability to store and release water and is defined by the relationship between soil moisture and the matric potential (Fig. 5). This is a power function relationship that has been described by Brooks and Corey (13) and Van Genuchten, (14) among others. The water tension characteristic is usually measured in air pressure chambers where the water content of a soil sample can be monitored over a wide pressure range (15).

The water retention relationship may actually change between drying and wetting due to the entrapment of air in soil pores (Fig. 7) (16). For practical applications, this effect, called *hysteresis*, is usually neglected (17).

PRINCIPLES OF SOIL WATER MOVEMENT

Through experiments on saturated water flow through sand beds, Darcy (18) found that the rate of flow, Q (L^3/T), through a cross-sectional area A (L^2), is directly proportional to head loss (e.g., water elevation difference), ΔH (L), and inversely to the flow path length, Δl (L):

$$Q = KA\frac{\Delta H}{\Delta l} \tag{3}$$

Combining *Darcy's law* with the law of conservation of mass results in a description of unsaturated flow called *Richards equation* (19):

$$\frac{\partial \theta}{\partial t} = \frac{\partial}{\partial z}\left(\frac{K}{C}\frac{\partial \theta}{\partial z}\right) - \frac{\partial K}{\partial z} \tag{4}$$

where $C = -\partial\theta/\partial\psi$ is the water content change in a unit soil volume per unit matric potential, ψ change. The Richards equation is the basis for most simulations of infiltration and redistribution of water in unsaturated soil. Using some approximations, analytical solutions of the Richards equation are available (20,21) that show good agreement with observations (22). The Richards equation is based on saturated flow theory, and does not account for all of the processes active in natural systems, so it may not always perform well (23).

INFILTRATION ESTIMATION

Some basic principles that govern the movement of water into the soil can be used to predict infiltration. The *infiltration capacity*, f(L), is the maximum rate that a soil in a given condition can absorb water and generally decreases as soil moisture increases. If the *rainfall rate* is less than the infiltration capacity, then infiltration proceeds at the capacity rate. However, if the rainfall rate exceeds the infiltration capacity, then infiltration proceeds at the capacity rate, and the excess rainfall ponds on the surface or runs off. As the time from the onset of rainfall increases, infiltration rates decrease due to soil moisture increases, raindrop impact, and the clogging of soil pores, until a steady-state infiltration rate is reached (Fig. 8) (24). Existing infiltration models use empirical, approximate, or physical approaches to predict infiltration.(25)

Empirical. Empirical infiltration models generally utilize a mathematical function whose shape as a function of

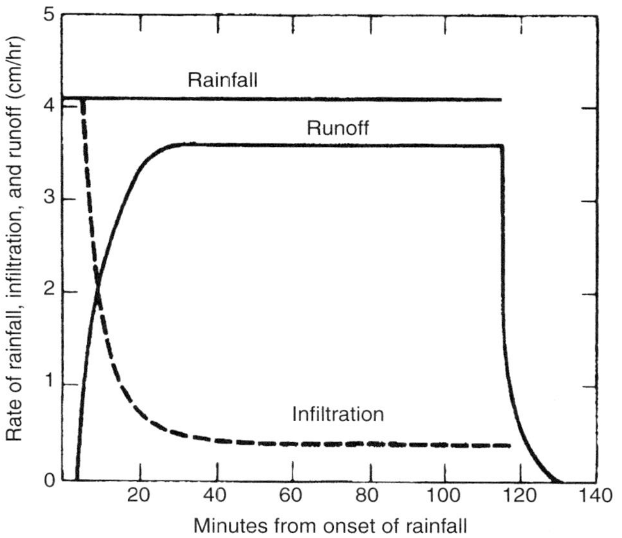

Figure 8. Idealized relationship between rainfall, infiltration, and runoff rates (24).

time, t, matches observations and then attempts a physical explanation of the process.

Kostiakov (26) proposed the simple infiltration rate, f(L/T) model:

$$f = \alpha\gamma t^{\alpha-1} \tag{5}$$

where α and γ are constants that have no particular meaning and must be evaluated by fitting the model to experimental data.

Horton's (27) infiltration model has been widely used in hydrologic simulation. It relates infiltration capacity to initial infiltration rate, and f_0, the constant infiltration rate at large times, f_c:

$$f = f_c + (f_0 - f_c)e^{-\beta t} \tag{6}$$

where β is a soil parameter describing the rate of decrease of infiltration.

Approximate. Analysis approximations to the Richards equation are possible if several simplifying assumptions are made. Most approximate infiltration models treat the soil as a semi-infinite medium, with the soil saturating above a wetting front.

Green and Ampt (28) assumed in a soil with constant hydraulic properties, the matric potential at the moving wetting front is constant, leading to a discontinuous change in soil moisture at the wetting front:

$$f = K\left[1 + \frac{(\varphi - \theta_i)S_f}{F}\right] \tag{7}$$

where S_f (L) is the effective suction at the wetting front, θ_i is the initial water content, and F (L) is the accumulated infiltration.

Phillip (29) proposed that the first two terms in a series of powers of $t^{1/2}$ could be used to approximate infiltration:

$$f = \tfrac{1}{2}St^{1/2} + A \tag{8}$$

where S is a parameter called sorptivity, t is time from ponding, and A is a constant that depends on soil properties. In this model, the infiltration rate approaches a constant equal to the hydraulic conductivity at the surface water content, and the wetting front advances without changing its shape and approaches a constant velocity.

Physical. Recent advances in numerical methods and computing has facilitated the practical application of the Richards equation to realistic flow problems. Such packages can simulate water infiltration and redistribution using the Richards equation and including precipitation, runoff, drainage, evaporation, and transpiration processes (30).

INFILTRATION MEASUREMENT

Infiltration rates can be measured at a point using a variety of methods described here, each appropriate for certain conditions. However, because of the large temporal and spatial variability of infiltration processes, catchment average infiltration rates may be desired, which can be obtained through the water balance analysis of rainfall–runoff observations (31).

Ring Infiltrometer. This simple method is most appropriate for flood irrigation or pond seepage infiltration. A cylindrical metal ring is sealed at the surface and flooded. Intake measurements are recorded until steady-state conditions are reached (32). If the effects of lateral flow are significant, then a double-ring infiltrometer can be used. Due to ponding conditions within the ring, observed infiltration rates are often higher than under natural conditions (33).

Sprinkler Infiltrometer. This method is appropriate for quantifying infiltration from rainfall. Artificial rainfall simulators are used to deliver a specified rainfall rate to a well-defined plot. Runoff from the plot is measured, allowing computation of the infiltration rate (34,35).

Tension Infiltrometer. The tension or disk infiltrometer employs a soil contact plate and a water column that is used to control the matric potential of the infiltrating water. By varying the tension, the effect of different size macropores can be determined (36,37).

Furrow Infiltrometer. This method is useful if information on infiltration of flowing water in irrigation furrows is desired. Either the water added to a small section of blocked off furrow to maintain a constant depth or the inflow—outflow of a furrow segment can be monitored to determine the infiltration characteristics of the system (38).

SOIL MOISTURE MEASUREMENT

Soil water content can be determined directly using gravimetric techniques or indirectly by inferring it from a property of the soil (39,40).

Gravimetric. The oven-drying soil moisture measurement technique is the standard for calibration of all other methods but is time consuming and destructive. The method involves obtaining a wet soil sample weight, drying the sample at 105 °C for 24 h, then obtaining the dry sample weight [see Eq. (2)].

Neutron Thermalization. High-energy neutrons are emitted by a radioactive source into the soil and are preferentially slowed by hydrogen atoms. The number of slow neutrons returning to the detector are a measure of soil moisture.

Gamma Attenuation. The attenuation in soil of gamma rays emitted from caesium-137 is directly related to soil density. If the soil's bulk density is assumed to be constant, then changes in attenuation reflect changes in soil moisture (41).

Time-Domain Reflectometry (TDR). TDR measures the soil's dielectric constant, which is directly related to soil moisture, by measuring the transmit time of a voltage pulse applied to a soil probe.

Tensiometric Techniques. This method measures the capillary or moisture potential through a liquid-filled porous cup connected to a vacuum gage. Conversion to soil moisture requires knowledge of the water retention characteristic.

Resistance. The electrical resistance or conductivity of a porous block (nylon, fiberglass, or gypsum) imbedded in the soil depends primarily on the water content of the block. However, because of salinity and temperature sensitivity, measurements of these sensors are of limited accuracy (42).

Heat Dissipation. Changes in the thermal conductivity of a porous block imbedded in the soil depend primarily on the water content of the block. The dissipation of a heat pulse applied to the block can be monitored using thermistors, then the soil water content can be determined from calibration information.

Remote Sensing. Soil moisture can be remotely sensed with just about any frequency where there is little atmospheric absorption (43). But, it is generally accepted that long wavelength, passive microwave sensors have the best chance of obtaining soil moisture measurements that contain little error introduced by vegetation and roughness and offer great potential to remotely sense soil moisture content with depth due to differential microwave absorption with varying dielectric constant (44).

SPATIAL AND TEMPORAL VARIABILITY

Natural soils exhibit considerable spatial heterogeneity in both the horizontal and vertical directions, and at all distance scales from the pore to the continent to a degree that it is difficult to capture this variability in routine measurements [45,46]. This large variations in soil properties, infiltration, and soil moisture over relatively small areas makes it difficult to transfer the understanding of processes developed at a point to catchment scales. Many hydrological models assume that a single spatially representative average soil property can be used to characterize catchment (or even larger) scale processes. It is clear from the nonlinear character of soil water processes that catchment average infiltration cannot be computed based on catchment average soil properties. It is also clear that the physical meaning of a soil property, say porosity, is relative to the volume over which it is averaged [47]. However, there is a need to understand and reduce this complexity for the purposes of prediction and management. Several approaches, including dividing the catchment into hydrologically similar subareas [48], various statistical approaches [49], and scaling and similarity theory [50,51] have made headway toward an understanding of infiltration and soil moisture spatial variability, but are not being widely used in practical applications.

One of the most important recent findings in this regard is the scale invariance of soil water movement(i.e., the Richards equation) remain uniform across spatial scales [52]. This new understanding of the underlying symmetry of the Richards equation may help to facilitate a workable scale invariant analytical soil water dynamical model.

Finally, there is acontinuing need for the observation of soil properties, soil moisture, and infiltration processes at multiple scales to facilitate understanding and prediction of these complex and socially significant processes. It is likely that remote sensing of soil moisture and other land surface factors will be instrumental in this respect.

BIBLIOGRAPHY

1. Dingman, S.L. (1994). *Physical Hydrology*. Prentice-Hall, Englewood Cliffs, NJ, p. 211.
2. Lawford, R.G. (1992). An Overview of soil moisture and its role in the climate system. In: F.J. Eley, R. Granger and L. Martin (Eds.). *Soil Moisture; Modeling and Monitoring for Regional Planning*, National Hydrology research Centre Symposium No.9 Poceedings, pp. 1–12.
3. Schmugge, T., Jackson, T.J., and McKim, H.L. (1979). Survey of in-situ and remote sensing methods for soil moisture determination. In: *Satellite Hydrology*. American Water Resources Association.
4. Rawls, W.J., Ahuja, L.R., Brakensiek, D.L., and Shirmohammadi, A. (1993). Infiltration and soil water movement. In: *Handbook of Hydrology*. D.R. Maidment (Ed.). McGraw-Hill, New York, pp. 5.1–5.51.
5. Hillel, D. (1982). *Introduction to Soil Physics*. Academic, New York, p. 29.
6. Gee, G.W., and Bauder, J.W. (1986). Particle size analysis. In: *Methods of Soil Analysis, Part I—Physical and Mineralogical Methods*, 2nd Edn. A. Klute (Ed.). *American Society of Agronomy Monograph* 9, pp. 383–411.
7. Hillel, D. (1980). *Fundamentals of Soil Physics*. Academic, New York, p. 143.
8. Dunne, T. and Leopold, L. (1978). *Water in Environmental Planning*. W.H. Freeman, New York, p. 175.
9. Hillel, D. (1980). *Fundamentals of Soil Physics*. Academic, New York, p. 150.
10. Marshall, T.J., and Holmes, J.W. (1988). *Soil Physics*, 2nd Edn. Cambridge University Press, p. 87.
11. Bouwer, H. and Jackson, R.D. (1974). Determining soil properties. In: *Drainage for Agriculture*. J. van Schilfgaard (Ed.). American Society of Agronomy, Madison, WI, pp. 611–672.

12. Green, R.E., Ahuja, L.R., and Chong, S.K. (1986). Hydraulic conductivity, diffusivity, and sorptivity of unsaturated soils—field methods. In: *Methods of Soil Analysis, Part I—Physical and Mineralogical Methods*, 2nd Edn. A. Klute (Ed.). American Society of Agronomy Monograph 9, pp. 771–798.
13. Brooks, R.H. and Corey, A.T. (1964). *Hydraulic Properties of Porous Media*, Hydrology Paper 3, Colorado State University, Fort Collins, CO.
14. Van Genuchten, M.Th. (1980). A closed-form equation for predicting the hydraulic conductivity of unsaturated soils. *Soil Sci. Soc. Am. J.* **44**: 892–898.
15. Cassel, D.K. and Klute, A. (1986). Water potential: Tensiometry. In: *Methods of Soil Analysis, Part I—Physical and Mineralogical Methods*, 2nd Edn. A. Klute (Ed.). American Society of Agronomy Monograph 9, pp. 563–596.
16. Hillel, D. (1980). *Fundamentals of Soil Physics*. Academic, New York, p. 153.
17. Rawls, W.J., Ahuja, L.R., Brakensiek, D.L., and Shirmohammadi, A. (1993). Infiltration and Soil Water Movement. In: *Handbook of Hydrology*. D.R. Maidment (Ed.). McGraw-Hill, New York, pp. 5.5–5.6.
18. Darcy, H. (1856). *Les fontaines publiques de la ville de Dijon*. Dalmont, Paris.
19. Richards, L.A. (1931). Capillary conduction of liquids in porous mediums. *Physics* **1**: 318–333.
20. Kuhnel, V., Dooge, V.C.I., Sander, G.C., and O'Kane, J.P.J. (1990). Duration of atmosphere-controlled and soil-controlled phases of infiltration for constant rainfall at a soil surface. *Ann. Geophys.* **8**: 11–20.
21. Sposito, G. *Recent advances associated with soil water in the unsaturated zone*. In: *Reviews of Geophysics*, Supplement, U.S.National Report to International Union of Geodesy and Geophysics 1991–1994, pp. 1059–1065.
22. Whisler, F.D. and Bouwer, H. (1970). A comparison of methods for calculating vertical drainage and infiltration in soils. *J. Hydrol.* **10**: 1–19.
23. Nielsen, D.R., Biggar, J.W., and Davidson, J.M. (1962). Experimental consideration of diffusion analysis in unsaturated flow problems. *Soil Soc. Am. Proc.* **26**: 107–111.
24. Dunne, T. and Leopold, L. (1978). *Water in Environmental Planning*, W.H. Freeman, New York, p. 169.
25. Rawls, W.J., Ahuja, L.R., Brakensiek, D.L., and Shirmohammadi, A. (1993). Infiltration and soil water movement. In: *Handbook of Hydrology*. D.R. Maidment (Ed.). McGraw-Hill, New York, pp. 5.21–5.23.
26. Kostiakov, A.N. (1932). On the dynamics of the coefficient of water-percolation in soils and on the necessity for studying it from a dynamic point of view for purposes of amelioration. *Trans. Sixth Comm. Intern. Soil. Sci. Soc. Russian, part A*, 17–21.
27. Horton, R.E. (1940). An approach toward a physical interpretation of infiltration-capacity. *Soil Sci. Soc. Am. J.* **5**: 399–417.
28. Green, W.H., and Ampt, G.A. (1911). Studies on soil physics: 1. Flow of air and water through soils. *J. Agric. Sci.* **4**: 1–24.
29. Phillip, J.R. (1957). The theory of infiltration: 1. The infiltration equation and its solution. *Soil Sci.* **83**: 345–357.
30. Ross, O.J. (1990). Efficient numerical methods for infiltration using Richards equation. *Water Resour. Res.* **26**: 279–290.
31. Soil Conservation Service, Hydrology. (1972). In: *SCS National Engineering Handbook*. U.S. Department of Agriculture, Washington, DC, Sec. 4.
32. Bouwer, H. (1986). Intake rate: Cylinder infiltrometer. In: *Methods of Soil Analysis, Part I—Physical and Mineralogical Methods*, 2nd Edn. A. Klute (Ed.). *American Society of Agronomy Monograph* 9, pp. 825–844.
33. Lukens, R.P. (Ed.). (1981). *Annual Book of ASTM Standards*, Part 19: *Soil and Rock, Building Stones*, pp. 509–514.
34. Agassi, M., Shainberg, I., and Morin, J. (1988). Effects on seal properties of changes on drop energy and water salinity during a continuous rainstorm. *Aust. J. Soil Res.* **26**: 1–10.
35. Peterson, A. and Bubenzer, G. (1986). Intake rate sprinkler infiltrometer. In: *Methods of Soil Analysis, Part I—Physical and Mineralogical Methods*, 2nd Edn. A. Klute (Ed.). *American Society of Agronomy Monograph* 9, pp. 845–870.
36. Ankeny, M.D., Kaspar, T.C., and Horton, R. (1988). Design for an automated tension infiltrometer. *Soil Sci. Soc. Am. J.* **52**: 893–896.
37. Perrouix, K.M. and White, I. (1988). Designs of disc permeameters. *Soil Sci. Soc. Am. J.* **52**: 1205–1215.
38. Kincaid, D.C. (1986). Intake rate: Border and furrow. In: *Methods of Soil Analysis, Part I—Physical and mineralogical Methods*, 2nd Edn. A. Klute (Ed.). *American Society of Agronomy Monograph* 9, pp. 871–887.
39. Gardner, W.H. (1986). Water content. In: *Methods of Soil Analysis, Part I—Physical and Mineralogical Methods*, 2nd Edn. A. Klute (Ed.). *American Society of Agronomy Monograph* 9, pp. 493–544.
40. Schmugge, T.J., Jackson, T.J., and McKim, H.L. (1980). Survey of methods for soil moisture determination. *Water Resour. Res.* **16**(6): 961–979.
41. Jury, W.A., Gardner, W.R., and Gardner, W.H. (1991). *Soil Physics*, 5th Edn. Wiley, pp. 45–47.
42. Hillel, D. (1982). *Introduction to Soil Physics*. Academic, New York, 1982, pp. 57–89.
43. Schmugge, T.J. (1985). Remote sensing of soil moisture. In: *Hydrological Forecasting*. John Wiley & Sons, New York.
44. Ulaby, F.T., Bradley, G.A., and Dobson, M.C. (1979). Potential application of satellite radar to monitor soil moisture. In: *Satellite Hydrology, American Water Resources Association*, pp. 363–370.
45. Jury, W.A., Gardner, W.R., and Gardner, W.H. (1991). *Soil Physics*, 5th Edn. John Wiley & Sons, New York, pp. 268–293.
46. Dingman, S.L. (1994). *Physical Hydrology*. Prentice-Hall, Englewood Cliffs, NJ, pp. 247–250.
47. Bear, J. (1972). *Dynamics of Fluids in Porous Media*. Elsevier, New York.
48. Springer, E.P. and Cundy, T.W. (1987). Field-scale evaluation of infiltration parameters from soil texture for hydrologic analysis. *Water Resour. Res.* **23**: 325–334.
49. Berndtsson, R., and Larson, M. (1987). Spatial variability of infiltration in a semi-arid environment. *J. Hydrol.* **90**: 117–133.
50. Sharma, M.L., Gander, G.A., and Hunt, G.C. (1980). Spatial variability of infiltration in a watershed. *J. Hydrol.* **45**: 101–122.
51. Wood, E.F., Sivaplan, M., and Beven, K. (1990). Similarity and scale in catchment storm response. *Rev. Geophys.* **28**: 1–18.
52. Sposito, G. (1995). *Recent advances associated with soil water in the unsaturated zone*. In: *Reviews of Geophysics*, Supplement, U.S. National Report to International Union of Geodesy and Geophysics 1991–1994, pp. 1059–1065.

ION EXCHANGE AND INORGANIC ADSORPTION

K. Kadirvelu
Jyotsna Goel
Centre for Fire, Explosives, and Environmental Safety
Timarpur, India

The presence of inorganic ions in water is of major concern for water quality because they can interfere with legitimate use of water for various applications. Inorganic ions, particularly heavy metals, are seriously questionable because of their toxicity and threat to human life and the environment.

Ion exchange and adsorption share so many common features in regard to removal of inorganic ions from water that they can be grouped together as unified processes for inorganic ion treatment. These processes involve the increase in concentration of a particular component at the surface or interface between two phases.

The objective of this article is to enumerate application of adsorption and ion exchange processes for inorganic ion removal from water. This article presents a review of potential adsorbents and ion exchangers for inorganic ion removal. It introduces important topics regarding inorganic ion removal by ion exchange and adsorption processes. Factors effecting metal removal are also presented and discussed in sections of the article. Research results from batch studies are presented to illustrate basic concepts. Accordingly, emphasis is placed on providing a review of adsorbents and ion exchangers available for inorganic ion removal.

ADSORPTION AND ION EXCHANGE: IN GENERAL

Adsorption involves, in general, the accumulation (or depletion) of solute molecules at an interface (including gas-liquid interfaces, as in foam fractionation, and liquid-liquid interfaces, as in detergency). Adsorbent surfaces are often physically and/or chemically heterogeneous, and bonding energies may very widely from one site to another. In discussing the fundamentals of adsorption, it is useful to distinguish between physical adsorption, involving only relatively weak intermolecular forces, and chemisorption, which involves, essentially, the formation of a chemical bond between the sorbate molecule and the surface of the adsorbent. The general features, which distinguish physical adsorption and chemisorption, are presented in Table 1.

Table 1. Distinction Between Physical Adsorption and Chemisorption (1)

Physical Adsorption	Chemical Adsorption
Low heat of adsorption (<2 or 3 times latent heat of evaporation.)	High heat of adsorption (> or 3 times latent heat of evaporation.)
Nonspecific.	Highly specific.
Monolayer or multilayer.	Monolayer only.
No dissociation of adsorbed species. Only significant at relatively low temperatures.	May involve dissociation. Possible over a wide range of temperature.
Rapid, nonactivated, reversible.	Activated, may be slow and irreversible.
No electron transfers, although polarization of sorbate may occur.	Electron transfer leading to bond formation between sorbate and surface.

Ion exchange is an effective means of removing heavy metals from wastewaters. It is a reversible chemical reaction, where the removal of heavy metals is accomplished by the exchange of ions on the resin for those in wastewater. In ion exchange, ions of positive charges (cations) and negative charges (anions) from the fluid (from aqueous solution) replace dissimilar ions of the same charge in solids. The ion exchanger contains permanently found functional groups of opposite charge-type. Cation exchangers generally contain bound sulfonic acid groups; less commonly, groups are carboxylic, phosphonic, and phosphinic, and so some resins involve quaternary ammonium groups (strongly basic) and amino groups (weakly basic).

Most ion exchangers in large-scale use are based on synthetic resins—either preformed and then chemically reacted as styrene or formed from active monomers. Natural zeolites were the first ion exchangers, and natural and synthetic zeolites are in use today.

Ion exchange may be thought of as a reversible reaction and of chemically equivalent quantities. A common example of ion exchange is the familiar water-softening reaction

$$Ca^{++} + 2NaR \Leftrightarrow CaR_2 + 2Na^{+} \quad (1)$$

where R represents a stationary univalent anionic site in the electrolyte network of the exchanger phase.

FACTORS AFFECTING INORGANIC IONS REMOVAL BY ADSORPTION

Factors affecting inorganic ion removal by adsorbent include (a) nature of the adsorbent (b) solution pH, (c) presence of foreign ions, and (d) temperature.

Nature of the Adsorbent

Surface Area and Pore Structure. Surface area with active functional groups is one of the principle characteristics affecting the adsorption capacity of an adsorbent. The adsorption capacity of a solid adsorbent is generally proportional to the specific surface area of the mass adsorbent; that is, the adsorption of certain solutes increases with an increasing surface area. Table 2 shows the physical properties of a few adsorbents that have been already employed for inorganic removal.

However, the specific surface area alone is frequently inadequate to explain the adsorption capacity of porous solids, such as activated carbons, for different solutes. In view of this, pore size distribution in activated carbon indicates that microspores contribute a major portion of the specific surface area.

Table 2. Properties of Adsorbents Employed for Inorganic Ion Removal (2–6)

Property	Calgon CAL	Coir-Pith Carbon	AC from *Arudo Donax*	AC Cloth	Carbon Aerogel
Surface area (m^2/g)	1100–1300	575	1142	1689	700
Porosity (%)	65.0	93.11	76.0	66.5	36.2
Micropore Volume (g/ml)	0.12	0.0913	0.47	0.964	0.10
Bulk density (g/ml)	0.56	0.12	0.5	0.13	0.644
Moisture content (%)	2.0	5.88	9.3	2.0	18.0
Ash content (%)	7.0	7.87	21.56	13.0	3.5

Depending on their preparation, active carbons also contain larger pores, known as macro and mesopores in the classification proposed by Dubinin (7) and now adopted by the International Union of Pure and Applied Chemistry (IUPAC) (8). The definition of the different types of pores is based on their width w, which represents the distance between the walls of a slit-shaped pore or the radius of a cylindrical pore. One distinguishes micropores, for which w does not exceed 2 nm; macropores, for which w exceeds about 50 nm; and mesopores of intermediate size (2.0–50 nm).

A linear relationship has also been demonstrated between the specific surface area and sorbate molecular size (9).

Chemistry of Adsorbent Surface. The nature and the presence of specific functional groups on the surface of the adsorbent impart certain characteristics that affect the adsorption process. It is now known that surface functional groups are formed during the carbon (adsorbent) activation process.

Attempts have been made to identify and estimate the surface oxygen chemical structures (functional groups) using several physical and chemical techniques, which include neutralization of acid and base (known as Boehm's method); desorption of the oxide layer; potentiometer, thermometric, and radiometric titrations; direct analysis of oxide layers by specific chemical reaction; Infrared (IR) spectroscopy; and X-ray photoelectron spectroscopy (XPS). The surface morphology and metal ion distribution adsorbent samples can be visualized via scanning electron microscopy (SEM). Figure 1 shows SEM micrograph for carbon aerogel equilibrated with Pb (II). SEM of activated carbon surfaces before and after the adsorbent was equilibrated with the metal ion solution clearly shows the presence of Pb (II) ions.

As a result of these investigations, various researchers have postulated the existence of functional groups, such as carboxyic, phenolic, lactonic, aldehydes, ketones, quinones, hydroquinone, anhydrides, and ethereal structures. Also, analysis of physicochemical properties of adsorbent ions provides a deep insight into adsorption mechanism involved during the adsorption process.

Solution pH

Extensive research investigations have been reported for adsorption of inorganic ions at different solution pH values (9–11). The dependence of adsorbent's adsorption capacity for inorganic ion removal from aqueous solution pH has been attributed to changes in surface chemical characteristics of adsorbent and species distribution of the inorganic ions varying with solution pH. In general, the cationic species removal will increase with increasing solution pH (Fig. 2), whereas in the case of inorganic ions present as anionic species, removal generally increases with decreasing pH, which is mainly because of electrostatic repulsion, which can be understood by removal of Cr (VI) by various adsorbents at lower pH range.

Specifically, pH affects the status of the outer hydration sheaths of the metal ion, metal speciation of inorganic ion in aqueous medium, complexation and solubility, and the electrochemical behavior of the carbon surface.

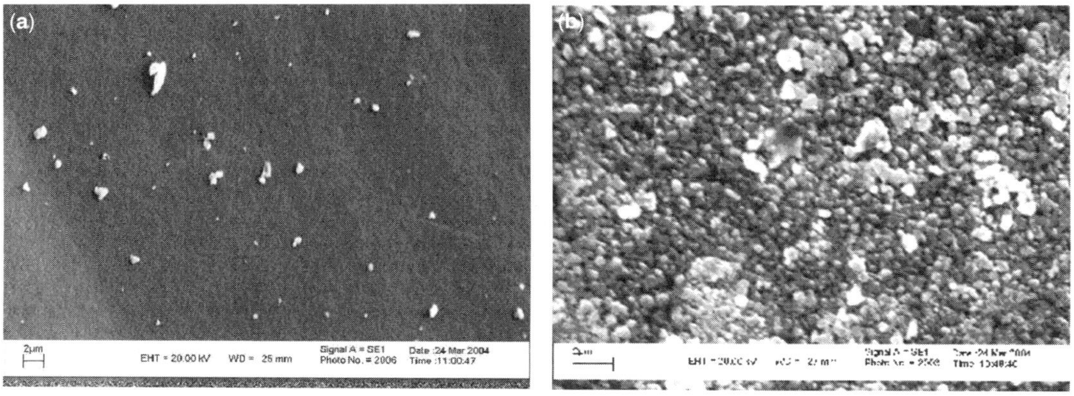

Figure 1. (**a**) SEM of carbon aerogel; (**b**) SEM of carbon aerogel loaded with lead.

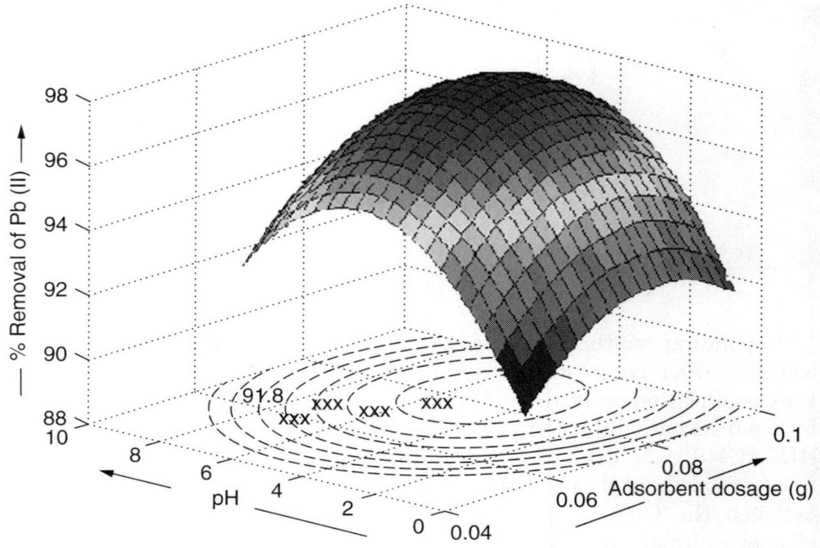

Figure 2. Plot for percent removal of Pb (II) versus pH and adsorbent dosage using carbon aerogel as adsorbent.

Figure 3 depicts possible inorganic ion removal mechanisms for carbon aerogel in relationship to pH, at which solution precipitation begins for Pb (II), Hg (II), and Ni (II). Figure 3 shows clearly that at any pH, the removal of metal ions is greater by adsorption in the presence of carbon aerogel than by precipitation in the absence of carbon. A rapid increase in percentage removal of metal ions is observed after pH 5.4 and 7.5 for Pb (II) and Ni (II), respectively, because of precipitation in the absence of carbon aerogel.

The adsorption of metal ions depends on both the nature of the adsorbent surface and the species distribution of the metal ions in the aqueous solution, which is, in turn, governed by solution pH. Removal noticed in a pH range of 3–5 for Pb (II) and Hg (II), because of their partial hydrolysis resulting in the formation of hydrolyzed metal ion species $[PbOH]^+$, $[HgOH]^+$ Pb $(OH)_2$, and Hg $(OH)_2$. In the case of Ni (II), the maximum adsorption range was shifted to 4–7, which may be because of formation of $[NiOH]^+$ and Ni $(OH)_2$ at these pH (12). Low solubilities of hydrolysis metal ion species may be another reason for maximum adsorption. At pH values below pH_{prec},

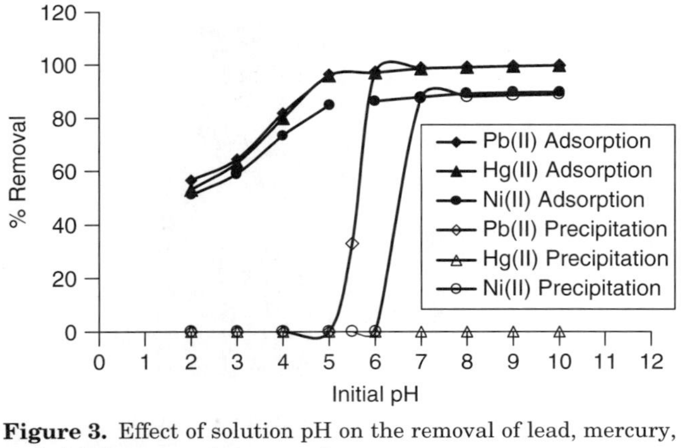

Figure 3. Effect of solution pH on the removal of lead, mercury, and nickel using carbon aerogel.

adsorption was hypothesized to be the primary removal mechanism. Also, it has been reported that metal removal by activated carbon was found to be inversely proportional to the solubility of the metal (13).

Namasivayam and Kadirvelu (14) reported an ion exchange phenomenon for Hg (II) removal by the formation of the following surface complexes when Hg (II) is present in the solution with the functional groups present on GAC (ACG-50).

$$2(C_XOH^+) + Hg^{2+} \longrightarrow (C_XO)_2Hg^{2+} + 2H^+ \quad (2)$$

According to the surface complexation theory, Hg (II) removal increases as pH function increases can be explained on the basis of a decrease in adsorption competition between proton and metal ion species for the adsorbent surface sites by the decrease in positive surface charge, which results in a lesser columbic repulsion forward of the Hg (II) ion.

Presence of Foreign Ions

The presence of complexing ligands and competing adsorbates can alter metal removal from that observed in the mono-component system. The degree of competition is dependent on the type and concentration of competing ions, number of surface sites, and affinity of the surface for adsorbate and the ionic property of adsorbate. Presence of other cocations can influence the adsorption capacity of adsorbent with respect to one adsorbate. This phenomenon is illustrated in Fig. 4, where the adsorption equilibrium curve for Hg (II) removal by GAC is considerably effected by the presence of other metal ions like Cu (II) and Pb (II) (15,16). Similar results have been reported for the effect of EDTA on lead removal using GAC column studies (17).

Temperature Effect

As the process of adsorption is spontaneous, it is accompanied by a decrease in the system's free energy. A

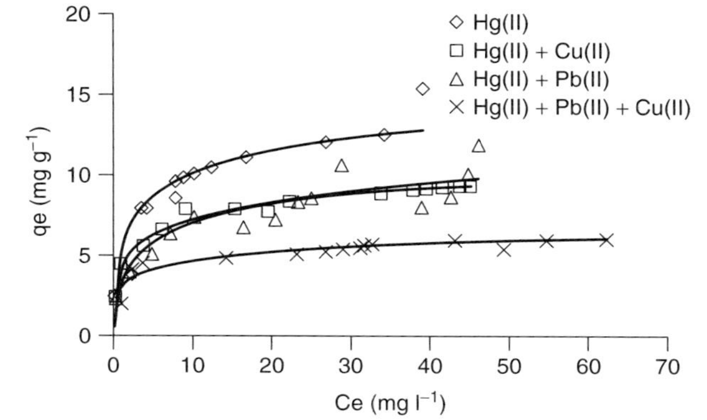

Figure 4. Equilibrium curve for Hg (II) ion in binary and tertiary system with Cu (II) and Pb (II) ions using GAC as adsorbent.

decrease in entropy always exists because of loss of degrees of freedom of the solute in passing from the dissolved state to the adsorbate state. It follows from the relationship

$$\Delta G = \Delta H - T\Delta S \quad (3)$$

that the adsorption process must always be exothermic; that is ΔH must have negative value regardless of the nature of interaction forces. An increase in temperature, therefore, will result in a reduction of the equilibrium adsorption capacity, whereas lower temperature will favor an increased capacity.

As reported by J. Goel et al. (6), according to the adsorption isotherm, the amount of Hg (II) by carbon aerogel increases with increasing temperature. Values of the thermodynamic parameters, such as ΔG^0, ΔH^0, and ΔS^0, describing mercury uptake by carbon aerogel were calculated using the following thermodynamic equations:

$$\Delta G^0 = -RT.\ln K \quad (4)$$

$$\Delta H^0 = [R \times T_1 \times T_2/(T_2 - T_1)] \times \text{Ln}\,(K_2/K_1) \quad (5)$$

$$\Delta S^0 = (\Delta H^0 - \Delta G^0)/T \quad (6)$$

where R is the gas constant and K, K_1, and K_2 are the equilibrium constant at temperature T, T_1, and T_2, respectively. The equilibrium constants was calculated from:

$$K = C_{\text{eq.s}}/C_{\text{eq.1}} \quad (7)$$

where $C_{\text{eq.s}}$ and $C_{\text{eq.1}}$ were the equilibrium concentration of Hg (II) in solution and on the adsorbent, respectively. Values found for ΔG^0, ΔH^0, and ΔS^0, as presented in Table 3, are indicative of the spontaneous nature of the uptake process.

The positive ΔH^0 value confirms the endothermic nature of the sorption process. As diffusion is an endothermic process, it would be expected that increased solution temperature would result in increased uptake of Hg (II) ions from aqueous solution.

Anoop Krishnan et al. (18) has reported that the endothermic nature of the adsorption process, where uptake of mercury on the bagasse piths activated carbon, is favored by increase in temperature. Panayotova (19) and Raji et al. (20) have also presented alike thermodynamics for the uptake of copper and mercury, respectively.

Table 3. Thermodynamic Parameters for the Mercury Adsorption on Carbon Aerogel

Temperature (K)	% Removal	ΔG^0 (J mol^{-1})	ΔH^0 (Jmol^{-1})	ΔS^0 (Jmol$^{-1}K^{-1}$)
293	60	−987.714	+25708.82	+91.11444
303	68	−1898.86	+18990.88	+68.94302
313	73	−2588.29	+32921.47	+113.4497
323	80	−3722.79	+46023.63	+154.0137
333	87	−5262.92	+28133.9	+100.2908
343	90	−6265.83	—	—

ADSORPTION MECHANISM—ION EXCHANGE PROCESS

With some of the adsorbents, like coir pith carbon, adsorption mechanisms have been explained based on the ion exchange model (12). A pure carbon surface is considered to be nonpolar, but in actual practice, some carbon–oxygen complexes C_x O, CO_x and C_xO_2 are usually present (21–23). The surface oxygen complexes hydrolyze water molecules as shown below:

$$C_xO + 2H_2O \longrightarrow C_xOH_2{}^{2+} + 2OH^- \quad (8)$$

$$CO_x + xH_2O \longrightarrow C(OH)^x + xOH^- \quad (9)$$

$$C_xO^2 + H_2O \longrightarrow C_xO^{2+} + 2OH^- \quad (10)$$

As the carbon is prepared with H_2SO_4, $(NH_4)_2S_2O_8$, and $NaHCO_3$, groups such as C_xONa^+, $C_xONa_2{}^{2+}$, C_xSO_3Na, and C_xONH_4 are also present. Na^+ in the above groups also exchange with H^+ in the medium as follows:

$$C_xONa^+ + H^+ \longrightarrow C_xOH^+ + Na^+ \quad (11)$$

$$C_xONa^+ + 2\,H^+ \longrightarrow C_xOH_2{}^{+2} + Na^+ \quad (12)$$

$$C_xONa_2{}^{2+} + H^+ \longrightarrow C_xOH_2{}^{+2} + 2\,Na^+ \quad (13)$$

$$CxSO_3Na + H^+ \longrightarrow C_xSO_3H + Na^+ \quad (14)$$

Excess of Na^+ was introduced into carbon where it was washed with $NaHCO_3$ to neutralize any free sulfuric acid (H_2SO_4) during the preparation of carbon. When metal ion is present in solution, its adsorption will free some H^+ and the pH increase will be lower than in the blank. At the same time, Na^+ will also be released according to reactions as given below:

$$2C_xOH^+ + M^{2+} \longrightarrow (C_xO)_2M^{2+} + 2H^+ \quad (15)$$

$$C_XOH_2{}^{2+} + M^{2+} \longrightarrow C_XOM^2 + 2H^+ \quad (16)$$

$$2C_xONa^+ + M^{2+} \longrightarrow (C_xO)_2M^{2+} + 2Na^+ \quad (17)$$

$$C_XONa_2{}^{2+} + M^{2+} \longrightarrow C_XOM^2 + 2Na^+ \quad (18)$$

$$2C_xSO_3H^+ + M^{2+} \longrightarrow (C_xSO_3)_2M^+ + 2H^+ \quad (19)$$

$$2C_xSO_3Na + M^{2+} \longrightarrow (C_xSO_3)M + 2Na^+ \quad (20)$$

which indicates that apart from exchange of H^+ ions on the adsorbent with M^{2+} ions, significant amount of Na^+ ions, which were present in the adsorbent, were also exchanged with M^{2+} ions.

ION EXCHANGERS FOR INORGANIC ION REMOVAL

By ion exchange, undesirable ions are replaced by others, which do not contribute to contamination of the environment. The method is technologically simple and enables efficient removal of even traces of impurities from solutions. Examples of selective removal of heavy metal ions by ion exchange are presented in Table 4. They include removal of Pb (II), Hg (II), Cd (II) Ni (II), V (IV, V), Cr (III, VI), Cu (II), and Zn (II) from water and industrial wastewaters by various means of modern type of ion exchangers.

By ion exchange, either all ions can be removed from a solution or substances are separated. Therefore, selective removal of ionic contamination and complete deionization can be distinguished. The choice between selective removal and complete deionization depends mainly on the composition of solution and on the extent of decontamination required.

Hg (II) ions sorption on the ion exchanger Imac TMR (Styrenedivinylbenzene Copolymer) can be described by means of the following reactions:

$$2R\text{–}SH + Hg^{2+} \longrightarrow R\text{–}S\text{–}Hg\text{–}S\text{–}R + 2H^{+} \quad (21)$$

$$RSH + HgCl^{+} \longrightarrow R\text{–}S\text{–}HgCl + H^{+} \quad (22)$$

where R is the styrenedivinylbenzene copolymer. After Hg (II) ions sorption, the ion exchanger Imac TMR can be regenerated by means of the concentrated hydrochloric acids.

Metal ions and other cationic impurities are removed on the cation exchanger, whereas chromates are removed on the anion exchangers:

$$nRSO_3H + M^{+} \longrightarrow nRM^{n+} + nH^{+} \quad (23)$$

$$2ROM + CrO_4^{2-} \longrightarrow R_2CrO_4 + 2OH^{-} \quad (24)$$

$$R_2CrO_4 + CrO_4^{2-} + H^{+} \longrightarrow R_2Cr_2O_7 + OH^{-} \quad (25)$$

The NaOH solutions are applied for anion exchanger regeneration. Further developments of conventional ion exchangers, like polystyrenesulphonate cation exchangers and strongly basic anion exchangers, were aimed at large exchange capacity and better chemical, thermal, and mechanical stability. Kinetic capability of ion exchangers could be improved by means of macroporous ion exchangers developed in the 1970s. The range of application of the ion exchange method was remarkably extended, first by invention and application of new organic and inorganic ion exchangers.

ADSORBENTS FOR INORGANIC ION REMOVAL

Large numbers of adsorbents have been developed so far for the removal of inorganic ions, which primarily

Table 4. Selective Removal of Heavy Metal Ions by Ion Exchangers

Metal Ion	Ion Exchangers	Composition and Functional Groups
Pb (II)	Leawtit TP 207 (24)	Chelating ion exchanger with functional iminodiacetate group
	Ethanotricarboxylate, 1,1-ddicarboxylate-2-etanophosphonate (25)	Gel type Copolymer Of Vinyl benzene Chloride, Styrene and Divinyl Benzene
	Polyacrylamidoglycol Resin (26)	Amide, OH and carboxylic groups
	Clinoptilolite (27)	Aluminosilicate structure
Hg (II)	Imac TMR	Styrenedivinylbenzene Copolymer
	Varian BTAM (Nike), Varian BTKM, Varian BSM, Ionac SR-3 (28–31)	Containing nitrogen and sulphur functional group (22), Thiocarbamate group (23), Mercapathile group (24), Thiourea functional group (25)
	Dowex A-1 (32)	Iminodiacetate group
Cd (II)	Dowex 50 W-X4	Polystyrenesulphonic cation exchanger
	Diphonix (33)	Sulphonic groups and diphosphonic groups
	KB-2 and KB-2-12P (34)	Polyacrylate carboxylic cation exchangers
	Phillipsite and Chabazite (35)	Natural zeolite from volcanic rocks of Central Italy
Ni (II) and V (IV, V)	Wofatit CA-20 (36)	Macro porous carboxylic cation exchanger
	Chelex −100 (37)	Iminodiacaetae ion exchanger with Na/H/Ni in exchange system
	Sodium titnate, $Na_4Ti_9O_{20}$ (38)	Layered material with exchangeable Na ions in the titanium oxide layer.
Cr (III, VI)	Lewatit MP-500A (39)	Basic anion exchanger
	Amberlite IR-120 (40)	Polystyrenesulphone cation exchanger

Table 5. Summary Table of Maximum Reported Adsorption Capacities (mg/g)

Sorbent	Source	Cd (II)	Cr (III)	Cr (VI)	Hg (II)	Pb (II)
Bark	(41–44)	32	19.45		400	182
CPEI cotton	(45)				1000	
Chitin	Cr, Hg, Pb (46)	558	92		100	796
Chitosan	(47, 48)	16.50		57	1123	58
Clay	(49, 50)	28				116
Dead biomass	(51)			43.0		
Leaf mould	(46–52)				150	1587
Lignin	Cd, Cr, Hg and Pb (46)	87	17		632	1587
Modified wool	(53)	46.65				
Moss	(54, 55)	5.058	76	43.9	16.2	135
Peat	(54, 55)					
Sawdust	(56)			16.05		230
Seaweed	(57, 58)	215				344
Xanthate	(59, 60)	33.27	19.67		1.149	18
Coir pith carbon	(3, 61)	94.0			154.0	264.0
Fly ash	(62)	198			.073	444.0
Red mud	(63)	67.0			7.1	165.0
Peanut hull carbon	(64)	89.0			220.0	210.0
Furnace Sludge	(65)	7.4				68.0

include activated carbon prepared from a variety of materials of biological origin, industrial wastes, dead biomass, organic wastes, etc. Besides, several low-cost adsorbents such as chitin, chitosin, clay, fly ash, peat, moss, zeolites, xanthates, seaweed/algae/alginate, lignin, and bark/tannin-rich material have also been explored for heavy metal containing wastewaters (37–56). Table 5 presents some of the highest adsorption capacity reported for various low-cost adsorbents.

Although a significant number of low-cost adsorbent materials from sources, the commercial activated carbon (CAC) is still being used intensively to date. Based on its size and shape, activated carbon is classified as follows: Powder (PAC), granular (GAC), Fibre (ACF), and cloth (ACC). As a result of the different sources of raw materials, the extent of chemical activation, and the physicochemical characteristics, each type of activated carbon has its specific application with respect to physicochemical properties, as well as depending on the inherent advantage and disadvantages in wastewater treatment.

Acknowledgment

Authors are grateful to Sh. A. K. Kapoor, Director, Dr. Chitra Rajagopal, Scientist "F", CFEES, DRDO, for granting permission to write this article and their constant encouragement and support.

BIBLIOGRAPHY

1. Ruthven, D.M. (1984). John Wiley & Sons, New York.
2. Netzer, A. and Hughes, D.E. (1984). *Water Res.* **18**: 927.
3. Kadirvelu, K. and Namasivayam, C. (2000). *Environ. Technol.* **21**: 1091–1097.
4. Basso, M. C., Cerella, E.G., and Cukierman, A.L. (2002). *Ind. Eng. Chem. Res.* **41**: 180–189.
5. Kadirvelu, K., Faur-Brasquet, C., and Le Cloirec, P. (2000). *Langmuir* **16**: 8404–8409.
6. Goel, J., Kadirvelu, K., Rajagopal, C., and Garg, V. K. (2004). *Carbon*. Elsevier Press, Amsterdam, In press.
7. Dubinin, M. M. (1983). *Carbon* **19**: 321.
8. IUPAC. (1972). Manual of symbols and terminology, Appendix 2, Pt.1, colloids and surface Chemistry. *Pure and Appl. Chem.* **3**: 578.
9. Calgon Corporation. *Basic Concept of Adsorption on Activated Carbon*. Calgon, Pittsburgh, PA.
10. Kadirvelu, K., Thamaraiselvi, K., and Namasivayam, C. (2001). *Sep. and Purifi. Technol.* **24**: 497–505.
11. Ricou, P., Lecuyer, I., and Le Cloirec, P. (1998). *Environ. Technol.* **19**: 1005–1062.
12. Kadirvelu, K. (1998). Characterization of coir pith carbon and its utilization in the treatment of metal bearing wastewater. Ph.D. Thesis, Bharthiar University, Coimbatore, India, pp. 59–60.
13. Sigworth, E.A. and Smith, S.B. (1972). *J. AWWA* **64**: 386–391.
14. Namasivayam, C. and Kadirvelu, K. (1999). *Carbon* **37**: 243–256.
15. Goel, J., Kadirvelu, K., and Rajagopal, C. (2004). *Ads. Sci. Tech.* **22**: 257–274.
16. Goel, J., Kadirvelu, K., and Rajagopal, C. (2004). *Environ. Technol.*
17. Reed, B.E. and Arunachalam, S. (1995). *ASCE J. Env. Eng.* **121**(9): 653–663.
18. Panayotova, M.I. (2001). *Waste Mgmt.* **21**: 671–676.
19. Anoop Krishnan, K. and Anirudhan, T.S. (2002). *J. Haz. Mater.* **B92**: 161–183.
20. Raji, C. and Anirudhan, T.S. (1996). Removal of Hg (II) from aqueous solution by sorption on polymerized sawdust. *Ind. J. Chem. Tech.* **3**: 49–54.
21. Huang, C.P. and Ostovic, F.B. (1978). *J. Env. Eng. Div. ASCI.* **104**: 863–878.
22. Sharma, D.C. and Forster, C.F. (1994). *Bioresource Technol.* **47**: 257–264.
23. Yenki, M. K.N. and Natarajan, G. S. (1991). *Separation Sci. Technol.* **26**: 661–674.

24. Bayer, A.G. (2000). Leverkusen.
25. Trochimczuk, W.A. (1999). *Europ. Polym. J.* **35**: 1457–1464.
26. Rivas, B.L., Maturana, H.A., Villegas, S., and Pereira, E. (1998). *Polym. Bull.* **40**: 721–728.
27. Petruzzelli, D., Pagano, M., and Tiravanti, G. (1995). *Ind. Eng. Chem. Res.* **34**: 2612–2617.
28. Bolto, B.A., Pawlowski, L.E., and Spoon, F.N. Ltd. (1987). London.
29. Warshawsky, A. and Erlich-Rogozinski, S. (1977). *Microm. J.* **22**: 362–368.
30. Dorfne, K. (1991). Walter de Gruyter, Berlin.
31. Inoue, K. et al. (1993). *Bunseki Kagaku.* **42**: 725–731.
32. Rengan, K. (1997). *J. Radioanal. Nucl. Chem.* **219**: 211–215.
33. Hubicki, Z., Jakowicz, A., and Lodyga, A. Application of the ions from waters and sewages. In: *Adsorption and Environmental Protection, Studies in Surface Science and Catalysis*. A. Dabrowski (Ed.). Vol. 120. Elsevier, New York.
34. Kauspedience, D., Snukiskis, J., and Gefeniene, A. (1998). *J. Radioanal. Nucl. Chem.* **229**: 129–132.
35. Colella, C., De Gennaro, M., Langella, A., and Pansini, M. (1993). Cadmium removal from wastewater using chabazide and phillipsite. In: *Natural Zeolites 93*. D.W. Ming and F.A. Mupton (Eds.). Brockport, NY, p. 377.
36. Halle, K., Fischwasser, K., and Fenz, B. (1982). *Technol. Umweltschutz.* **25**: 1201032.
37. Leinonen, H. and Lehto, J. (2000). *Rect. Funct. Polym.* **43**: 1–6.
38. Lehto, J. et al. (1996). *J. Radio Nucl. Chem.* **208**: 435–443.
39. Untea, I., Tudorache, E., and Florea, C. (2000). *Revista de Chimie.* **51**: 123–126.
40. Drela, I. (1996). *Rudy Met. Niezelaz.* **41**: 321–323.
41. Randell, J.M., Bermann, R.L., Garrett, V., and Waiss, A.C. (1974). *J. Foorest Prod. J.* **24**(9): 80–84.
42. Teles de Vasconcelos, L.A. and Gonzalez Beca, C.G. (1994). *Eur. Water Poll. Control* **4**(1): 41–51.
43. Masri, M.S. and Friedman, M. (1974). *J. Appl. Polymer Sci.* **18**: 2367–2377.
44. Randall, J.M., Hautala, E., and Waiss, A.C., Jr. (1974). Removal and recycling of heavy metals ions from mining and industrial waste water streams with agricultural byproducts. *Proceedings of the Fourth Mineral Waste Utilization Symposium*, Chicago, IL, pp. 329–334.
45. Roberts, E.J. and Rowland, S.P. (1973). *Env. Sc. Technol.* **7**(6): 552–555.
46. Masri, M.S., Reuter, F.W., and Friedman, M. (1974). *J. Appl. Poly. Sc.* **18**: 675–681.
47. Prades, E.G. et al. (1994). *J. Chem. Tech. Biotechnol.* **59**: 289–295.
48. Cadena, F., Rizvi, R., and Peters, R.W. (1990). Feasibility studies for the removal of heavy metals from solution using tailored bentonite. In hazardous and industrial wastes. *Proceedings of the 22nd Mid-Atlantic Industrial Waste Conference*. Drexel University, Philadelphia, PA, pp. 77–94.
49. Matis, K.A. and Zoublious, A.I. (1994). *Biotech. Bioeng.* **44**: 354–360.
50. Niu, H., Xu, X.S., and Wang, J.H. (1993). *Biotech. Bioeng.* **42**: 785–787.
51. Sharma, D.C. and Forster, C.F. (1994). *Biores. Technol.* **49**: 31–40.
52. Srivastava, S.K., Singh, A.K., and Sharma, A. (1994). *Environ. Technol.* **15**: 353–361.
53. Low, K.S., Lee, C.K., Cheong, F.Y., and Lim, G.S. (1993). *J. Environ. Sci. Health A* **28**(8): 1689–1704.
54. Tong, P., Baba, Y., Adachi, Y., and Kawaza, K. (1991). *Chem Lett.* **9**: 1529–1532.
55. Kertman, S.V., Kertman, G.M., and Chibrikova, Z.H. (1993). *J. Appl. Chem. USSR* **66**(2): 465–466.
56. Dikshit, V.P. (1989). *Nat. Acad. Sci. Lett.* **12**(12): 419–421.
57. Holan, Z.R., Volesky, B.N., and Prasetyo, I. (1993). *Biotech. Bioeng.* **41**(8): 819–825.
58. Leush, A., Holan, Z.R., and Volesky, B. (1995). *J. Chem. Tech. Biotechnol.* **62**: 279–288.
59. Bricka, R.M. and Hill, D.O. (1989). Metal immobilization by solidification of hydroxide and xanthate sludges. In: *Environmental Aspects of Stabilisation and Solidifications of Hazardous and Radioactive Wastes*. P.L. Cote and T.M. Gilliam (Eds.). ASTM, Philadelphia, PA, pp. 257–272.
60. Flynn, C.M., Jr., Carnahan, T.G., and Lindastrom, R.E. (1980). Adsorption of heavy metal ions by xanthated sawdust. Report of Investigation #8427. United States. Bureau of Mines.
61. Kadirvelu, K. and Namasivayam, C. (2003). *Advances in Environmental Research*.
62. Apak, R., Tutem, E., Hugul, M., and Hizal, J. (1998). *Water Res.* **32**: 430–440.
63. Suemitsu, R., Venishi, R., Akashi, I., and Nakana, M. (1986). *J. Appl. Poly. Sci.* **31**: 75–83.
64. Periasamy, K. (1993). Adsorptive removal of heavy metals from synthetic indisustrial wastewater using peanut hull carbon. Ph.D. Thesis, Bharthiar University, Coimbatore, India.
65. Lopez-Delgado, C.P. and Girgis, B.S. (1998). *Water Res.* **32**: 1169–1177.

IRON

William R. Roy
Illinois State Geological Survey
Champaign, Illinois

CHEMISTRY OF IRON

The average concentration of iron in the earth's crust is about 5%, and it occurs in a wide variety of oxide, carbonate, silicate, and sulfide compounds either as metallic ores or in rock-forming minerals. Iron also occurs naturally in water from the dissolution of iron-containing minerals. Iron in water can occur in a dissolved form, a colloidal form that is combined with organic matter, and as suspended, iron-containing particles. Dissolved iron occurs in water in two oxidation states: ferrous iron, Fe (II), and ferric iron, Fe (III). In solution, iron can occur as a cation (Fe^{2+}) and in combination with other ions to form complexes and neutral ion pairs. Iron does not generally occur as the trivalent cation (Fe^{3+}) in significant concentrations in water. Trivalent iron generally exists in combination with other constituents. Depending on the occurrence of anions, such as sulfate, carbonate, and phosphate, and the pH and reduction-oxidation potential (E_h), the major iron species found in natural waters include a sulfate complex ($FeSO_4{}^0$),

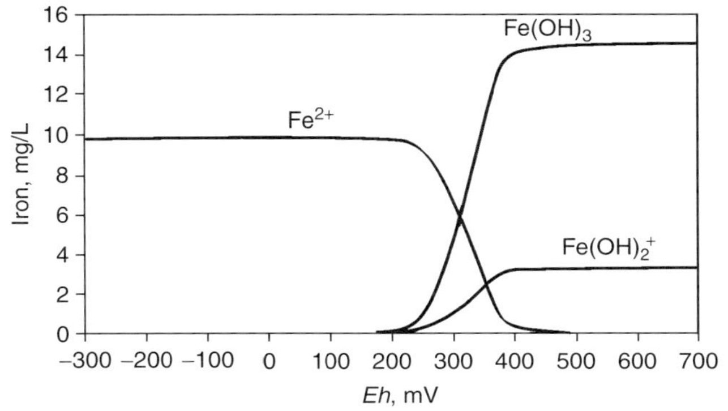

Figure 1. The distribution of 10 mg/L of iron in water at 15 °C and pH 7.0 as a function of oxidation-reduction potential (*Eh*).

a carbonate form [$Fe(HCO_3)_2$], phosphate complexes ($FeHPO_4{}^0, FeH_2PO_4{}^+$), and various hydroxide species such as $FeOH^+$, $Fe(OH)_2{}^+$, $Fe(OH)_3{}^0$, $Fe(OH)_4{}^-$, and $FeOH^{2+}$. The relative proportion of ferrous iron and ferric iron depends on the E_h of the water; ferrous iron is the dominant form in oxygen-poor water (reducing conditions). In oxygen-rich water (oxidizing conditions), ferric forms of iron dominate. For example, in a solution containing 10 mg/L of iron [Fig. 1], about 98% of the iron would occur as Fe^{2+} in a relatively reducing system. At an *Eh* greater (i.e., more oxidizing) than about 200 mV, ferrous iron is converted to the ferric forms $Fe(OH)_3{}^0$ and $Fe(OH)_2{}^+$. At an *Eh* of 700 mV, about 96% of the iron occurs in ferric forms. When a groundwater sample is collected, it may initially appear clear. However, if the sample is allowed to react with air and oxidize, the ferrous iron in the initially clear, reduced sample will be converted to ferric forms:

$$4Fe^{2+} + O_2 + 8OH^- + 2H_2O \rightleftharpoons 4Fe(OH)_3 \quad (1)$$

As a consequence, the sample may develop a reddish-brown color, and iron-containing particles may settle at the bottom of the container. The rate at which ferrous iron is oxidized depends on temperature, pH, and the amount of dissolved oxygen and iron.

Microorganisms can influence the oxidation state and distribution of iron in water. Iron can be used as an energy source via oxidation by microbial activity or reduced because iron can act as an electron acceptor. For example, *Crenothrix* and *Leptothrix* are diverse groups of iron-precipitating bacteria that are widely distributed in freshwater and soil. After ferrous iron is oxidized to the sparingly soluble ferric form, it may precipitate as ferric hydroxide. The bacterial cell sheaths can become coated with the precipitate, and the resulting gelatinous mass can clog water pipes and drains. There are bacteria and fungi that can reduce ferric iron. However, much less is known about iron-reducing microorganisms (1). The *Geobacter* are a group of bacteria that can reduce ferrous iron.

HEALTH AND REGULATORY ISSUES

Iron is not toxic in human consumption. Iron is an essential element for the formation of hemoglobin. The recommended dietary allowance for iron in healthy adults is 10 mg per day for men and 15 mg per day for premenopausal women (2). Dissolved iron in potable water is more of a nuisance than a potential heath issue. It can impart an unpleasant metallic taste to drinking water and alter the taste of food cooked in the water. The taste threshold for ferrous iron is about 0.1 mg/L. Concentrations exceeding 0.3 mg/L of iron can stain plumbing and clothing (3). The secondary maximum contaminant level for public water supplies is 0.3 mg/L, and 1.0 mg/L for freshwater aquatic life (4). However, bioassays by Roy et al. (5) suggested that concentrations less than 1.0 mg/L may be toxic to freshwater aquatic life.

LEVELS OF IRON IN WATER

The amount of dissolved iron in water depends on pH and *Eh*, but concentrations of less than 0.01 to 10 mg/L are typical for natural waters (6). Well-aerated surface water usually contains less than 0.05 mg iron/L. Iron concentrations in aquifers in Illinois have ranged from <0.05–21 mg/L (7). Iron in acid-mine drainage from coal mines has ranged from 10–5,625 mg/L (5,6).

SOLUBILITY OF IRON IN WATER

The amount of dissolved iron in water can be limited by the solubility of iron-containing solid phases. Schwab and Lindsay (8) concluded that when $pH + pE$ (pE is related to the redox potential *Eh* by $pE = (EhF)/(2.3RT)$ where R = gas constant, T = absolute temperature, and F = Faraday constant) is less than 8.5, the concentration of ferrous iron in groundwater is limited by the solubility of siderite, $FeCO_3$:

$$Fe^{2+} + CO_3{}^{2-} \rightleftharpoons FeCO_3 \quad (2)$$

When $pE + pH$ is greater than 8.5, the solubility of ferrosic hydroxide, $Fe_3(OH)_8$, limits the amount of ferrous iron:

$$2Fe^{3+} + Fe^{2+} + 8H_2O \rightleftharpoons Fe_3(OH)_8 + 8H^+ \quad (3)$$

When the concentration of iron is greater than the solubility product of either siderite or ferrosic hydroxide, the iron precipitates, from solution. Ferrosic hydroxide is thermodynamically unstable, and other minerals may form, causing ferrosic hydroxide to dissolve. For example, amorphous ferric hydroxide, [$Fe(OH)_{3,a}$] and goethite (α-FeOOH) are less soluble than ferrosic hydroxide and given time, may control iron solubility. Moreover, amorphous ferric hydroxide can dehydrate (9) to form the more stable mineral hematite (Fe_2O_3):

$$2\,Fe(OH)_{3,a} \rightleftharpoons Fe_2O_3 + 3H_2O \quad (4)$$

The prevalent form of iron in surface water in the iron–carbon dioxide–water system can be illustrated by constructing a pH–pE stability diagram [Fig. 2]. This type of diagram illustrates which forms of iron are most thermodynamically stable at various values of pH

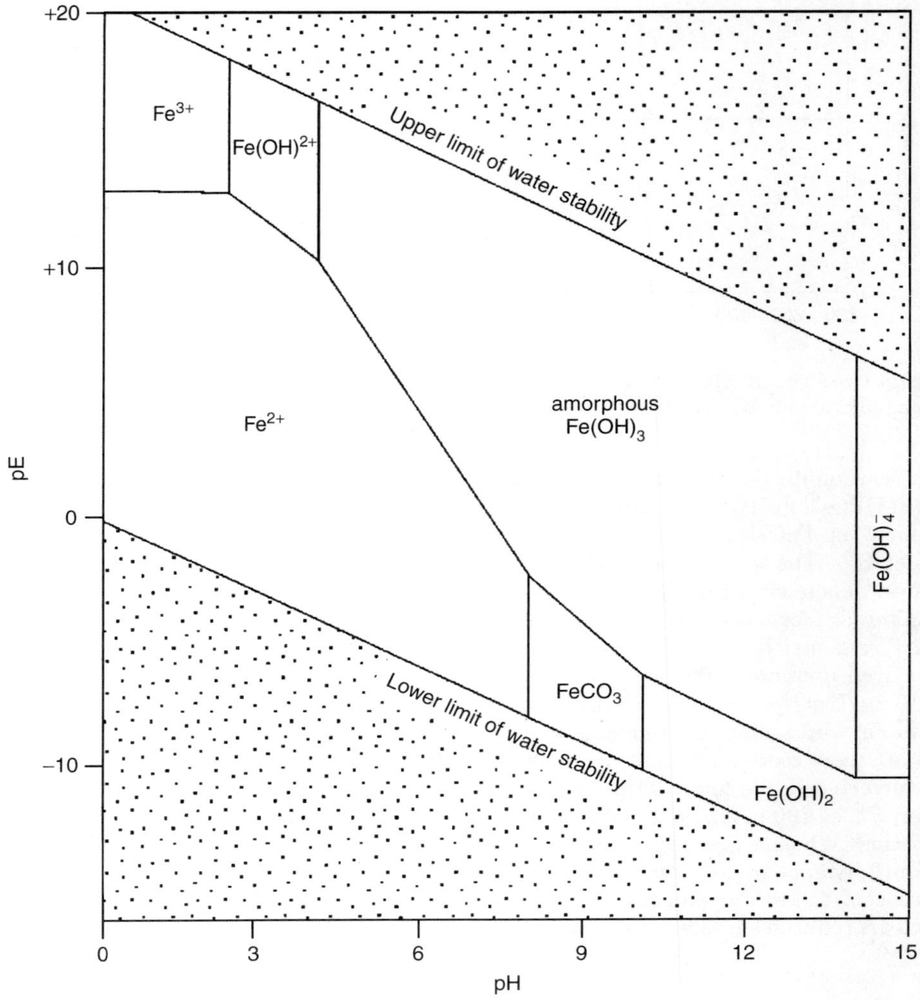

Figure 2. p*E*–pH diagram for the iron–carbon dioxide–water system at 25 °C and 1 atm (11).

and p*E* rather than at specific concentrations. Ferrous iron is the stable form in equilibrium with relatively acidic and reducing waters, whereas ferric iron, as both Fe^{3+} and $FeOH^{2+}$, is in equilibrium with acidic and oxidizing waters. In relatively less oxidizing and alkaline waters, the most thermodynamically stable forms may be amorphous iron hydroxide, siderite ($FeCO_3$), or solid ferrous hydroxide.

The concentration of iron in water may also be influenced by sorption–desorption interactions with mineral surfaces. Iron oxides and hydroxides have been studied as sorbents for other metals, but ionic iron is subject to the same electrostatic interactions that attract cations to clays and organic matter. Ferric iron can be sorbed by silica, clay minerals, and iron and manganese oxides (10). The sorption of ferrous iron has not been studied in detail.

METHODS FOR DETERMINING IRON

The amount of iron dissolved in water can be determined quantitatively by atomic absorption spectroscopy (AAS), atomic emission spectroscopy, and by colorimetric methods (12). In the AAS method, a water sample is aspirated into an air-acetylene flame and atomized. A monochromatic beam of light at a wavelength corresponding to iron is passed through the flame. The amount of light energy sorbed is proportional to the amount of iron present. When analyzed at a wavelength of 248.3 nm, the optimal concentration range for iron is about 0.3 to 10 mg/L.

Dissolved iron can be determined by argon-supported inductively coupled plasma (ICP) spectroscopy. In ICP, the liquid sample is aspirated into a high-temperature (6000–8000 K) flame, and atomic emission is measured to determine the concentration of the iron present. The concentration of iron is proportional to the intensity of the emissions. Iron has many such emission lines in its spectrum, but the wavelength of 259.94 nm is commonly used for concentrations ranging from about 0.007 to 100 mg/L.

Dissolved iron also can be determined by the phenanthroline method. In this method, iron in solution is reduced to the ferrous form, and then allowed to react with 1, 10-phenanthroline to form a reddish-orange product. The absorbance of the complex is measured in the range of 460–520 nm. Both AAS and ICP methods determine total iron, but the amount of ferric and ferrous iron can be determined separately by the phenanthroline method (13).

METHODS FOR REMOVING IRON FROM WATER

The presence of large concentrations of dissolved iron is a common water-quality problem, especially for groundwater (1). Methods for removing iron are summarized in Lehr et al. (3) and Vigneswaran and Visvanathan (2). The oxidation of soluble, ferrous iron into relatively insoluble ferric iron, followed by filtration of the suspended mass is common to many of the methods. Depending on its concentration in the untreated water, the oxidation of ferric iron can be promoted by aeration, catalytic oxidation with manganese oxides, and chlorination. Household water softeners are designed to remove calcium and magnesium from hard water using an ion exchange resin, but they can also remove dissolved iron. Lime softening of municipal water supplies can also promote the precipitation of ferric iron. The presence of iron-fixing bacteria may necessitate the use of relatively large amounts of bleach ("shot chlorination") to reduce the amount of iron in the final product.

BIBLIOGRAPHY

1. Chapelle, F.H. (2001). *Ground-Water Microbiology and Geochemistry*. John Wiley & Sons, New York.
2. Vigneswaran, S. and Visvanathan, C. (1995). *Water Treatment Processes. Simple Options*. CRC Press, Boca Raton, FL.
3. Lehr, J.H., Gass, T.E., Pettyjohn, W.A., and DeMarre, J. (1980). *Domestic Water Treatment*. McGraw-Hill, New York, pp. 172–178.
4. U.S. EPA. (1999). *National Recommended Water Quality Criteria–Correction*. EPA 822-Z-99-001.
5. Roy, W.R., Skelly, T.M., Krapac, I.G., Griffin, R.A., Dickerson, D.R., Schuller, R.M., Suloway, J.J., and Shimp, N.F. (1984). *Geochemical Properties and Toxicological Effects on Aquatic Life of Coal Wastes*. Environmental Geology 107, Illinois State Geological Survey, Champaign, IL, p. 124.
6. Patterson, J.W. (1975). *Wastewater Treatment Technology*. Ann Arbor Science, Ann Arbor, MI.
7. Warner, K.L. (2000). *Analysis of Nutrients, Selected Inorganic Constituents, and Trace Elements in Water from Illinois Community-Supply Wells, 1984–91*. Water-Resources Investigations Report 99-4152, U.S. Geological Survey.
8. Schwab, A.P. and Lindsay, W.L. (1983). Effect of redox and pH on the solubility and availability of iron. *Soil Sci. Soc. Am. J.* **47**:201–205.
9. Garrels, R.M. and Christ, C.L. (1965). *Solutions, Minerals, and Equilibria*. Freeman Cooper, San Francisco, CA.
10. Anderson, M.A. and Rubin, A.J. (1981). *Adsorption of Inorganics at Solid–Liquid Interfaces*. Ann Arbor Science, Ann Arbor, MI.
11. Stumm, W. and Morgan, J.J. (1981). *Aquatic Chemistry*. John Wiley & Sons, New York.
12. American Public Health Association. (1992). *Standard Methods for the Examination of Water and Wastewater*, 18th Edn. Washington, DC, Section 3500. Iron, 3-65–3-68.
13. Stucki, J.W. (1981). The quantitative assay of minerals for Fe^{2+} and Fe^{3+} using 1, 10-phenanthroline: II. A photochemical method. *Soil Sci. Soc. Am. J.* **45**: 638–641.

ISOTOPES

Aldo Conti
Frascati (RM), Italy

Isotopes are atoms with the same atomic number, **Z**, but different atomic weight, **A**. Different isotopes, therefore, have the same chemical but different physical properties, such as melting and boiling point. The name isotope means "at the same place" and comes from the periodic table, where different isotopes of the same element are at the same place. The scientific way of labeling isotopes reflects this. They have the symbol of the corresponding element, but followed by a hyphen and the number of nucleons (neutrons and protons) of the isotope. So, C-12 is the normal carbon (six protons and six neutrons) and C-14 is a heavy unstable isotope (six protons and eight neutrons).

Isotopes of the same element share the same chemical properties because they depend only on the number of protons, *Z*. But different isotopes can have very different weights, which is particularly evident in hydrogen. Normal hydrogen has only one proton and no neutrons. Deuterium, or H-2, has one neutron and its weight is nearly twice that of hydrogen. Then, tritium, H-3, has two neutrons and is three times as heavy, which is the reason why, when deuterium combines with oxygen, the resulting molecule is called heavy water. Due to the extra weight of deuterium, heavy water boils at 101.42 °C (214.56 °F) and freezes at 3.81 °C (38.86 °F), at standard pressure. The difference in weight affects the speed of chemical reactions involving water. For this reason, heavy water is not really safe to drink. Experiments performed with mice have shown that large amounts of heavy water in the body can inhibit mitosis, which affects rapidly dividing tissues, such as the stomach lining.

Isotopes do not differ only in physical properties. The extra neutrons make some of the nuclei highly unstable, which is the reason that some isotopes are radioactive. A typical example is uranium. Its isotopes, such as U-235, decay radioactively following a long chain, which leads to lead. Some of this decay involves a neutron becoming a proton by emitting an electron. In this case, the nucleus becomes a new element, with the atomic number increased by one. The energy emitted during the decay can be harvested in nuclear plants or used destructively in atomic bombs.

The instability of their nuclei makes isotopes useful for many applications. A famous one is carbon dating. Carbon comes mainly in two isotopes, the normal C-12 and the heavy, unstable, C-14. When an organism grows, it uses them in the same proportion in which they are found in the environment. After the death of the organism, the ratio changes continuously, as the heavy C-14 decays. C-14 has a half life of 5730 years, which means that after this period half of it has gone. By this method, it is possible to date any biological remains or artifact up to about 50,000 years old. For older objects, the amount of C-14 becomes too small to be measured reliably. Isotopes are often used as labels. By replacing a stable atom in a molecule with an unstable radioactive isotope, it becomes possible to follow the chemical reactions the molecule undergoes. Our ability to detect radioactive isotopes makes them very useful in the fields of medicine and biology, where they are used as tracers.

ISOTOPE FRACTIONATION

Gholam A. Kazemi
Shahrood University of Technology
Shahrood, Iran

INTRODUCTION

The following paragraph extracted from the major textbook by Bowen (1) describes fractionation as it is dealt with in general chemistry. The subsequent paragraph defines fractionation in general hydrochemistry. Then, isotope fractionation is discussed.

In general chemistry, fractionation is the separation of a mixture, usually of chemically related components, into fractions that possess different properties. A mixture of liquids having different boiling points, may be separated by collecting fractions that boil at different temperatures in a fractionating column. This is a long, vertical column containing plates or bubble caps attached to a still. As a consequence of internal reflux, a gradual separation occurs between high and low boiling point fractions of the liquid mixture.

In hydrochemistry, the fractionation factor is a ratio which shows the depletion or addition of ions in rainwater compared to seawater (2). In other words, the fractionation factor is the ratio of two ions in rainwater to the ratio of the same ions in seawater. For example, for Cl^- relative to Na^+, the fractionation factor F_{Na} is (2):

$$F_{Na} = \frac{(Cl/Na)_{rainwater}}{(Cl/Na)_{seawater}}$$

In this equation, the units of concentration can be mg/l, mmol/l, meq/l, etc.

Example: What is the fractionation factor of Na for a rainwater sample whose chloride and sodium concentrations are 1 and 0.5 mg/l, respectively?

$$Cl/Na_{rainwater\ sample} = 1/0.5 = 2$$

$$Cl/Na_{seawater} = 1.81$$

$$F_{Na} = \frac{(2)}{(1.81)} = 1.11$$

A fractionation factor greater than unity indicates that the rainwater sample is depleted (in comparison to seawater) in the ion in question. If it is less than unity, it shows the enrichment (in comparison to seawater) of the sample in the ion. Appelo and Postma (2) also propose the fractionation factor as a quantitative approach to find the activity of a single species that forms part of a given total concentration, for example, for HCO_3, $m_{HCO_3^-} = \alpha \times \Sigma CO_2$, where α is the fractionation factor. Some researchers use "enrichment factor (Ef)" in an equation slightly different from the fractionation factor to deduce source/s of solutes in atmospheric precipitation (3,4). If the Ef is greater than unity, the ion in question (in the sample) is derived partly or wholly from a source other than oceanic origin. If it is less than unity, the ion could have been derived wholly from oceanic sources:

$$Ef = \| \frac{(Ion/Cl)_{sample}}{(Ion/Cl)_{seawater}}$$

In this example, the enrichment factor equals 0.91.

$$Ef = \frac{(0.5)}{(0.55)} = 0.91$$

DEFINITION

Isotope fractionation is defined by Domenico and Shwartz (5) as a change in an isotopic ratio (for example, $^{18}O/^{16}O$) because of a chemical reaction (for example, water changes to water vapor). It is also defined by Coplen (6) as the fractionation or partitioning of isotopes by physical or chemical processes which is proportional to the differences in their masses. Coplen also classifies isotopic fractionation processes into two categories, chemical and physical isotopic fractionation processes: physical processes such as ultrafiltration or gaseous diffusion of ions or molecules. Chemical processes are those that lead to redistribution of the isotopes of an element among phases or chemical species and are divided into two broad categories equilibrium isotopic reactions and kinetic isotopic reactions (6).

Gilreath (7) in a simple yet valuable text describes the fundamentals of isotope fractionation in detail and refers to it as "isotope separation." Anyone intending to deal with isotope fractionation is advised to consult this text first. The point of primary importance is that fractionation of isotopes is important in the geologic environment *only* for elements of low atomic number. Bowen (1) describes three main processes which cause fractionation: (1) isotope exchange reactions; (2) unidirectional reactions; and (3) physical processes such as evaporation, condensation, crystallization, melting, adsorption, desorption, and diffusion. Isotopes may also be fractionated in the laboratory by a variety of means (URL Ref. http://www.c14dating.com/frac.html). One method that may result in considerable fractionation during isotopic analysis is the recently introduced technique of secondary-ion mass spectrometry (SIMS) (8).

The following are the factors that control the rate of isotope fractionation:

1. Fractionation is minimized when reactions are fast (9).
2. The higher the mass difference between various isotopes of the element, the more effective the fractionation process. Deuterium, for example, has the largest mass difference from common hydrogen compared to isotopes of other elements. Hence, it is subject to greater fractionation.

3. Isotope fractionation decreases as temperature increases—at finite temperatures, all isotopic species are well mixed (6).
4. Isotopic exchange between minerals and water in deep, basinal flow systems or in geothermal systems induced by high temperature helps the fractionation process to proceed (5).
5. In evaporation from a pond, fractionation is more efficient if the vapor produced is constantly removed, for example, by wind (10).

Fractionation Factor

The fractionation factor is the ratio of isotopes (for example, $^{18}O/^{16}O$) in a liquid phase to the same ratio in the vapor phase;

$$\alpha = \frac{(^{18}O/^{16}O)_{\text{liquid}}}{(^{18}O/^{16}O)_{\text{vapor}}} \tag{11}$$

or more clearly/generally defined by Haldorsen et al. (11) "as the increase or decrease of a given isotope in a substance that undergoes a phase change or simply in terms of delta notation:

$$\alpha = \frac{1000 + \delta^{18}O_{\text{liquid}}}{1000 + \delta^{18}O_{\text{vapor}}}$$

Example: In the condensation of water vapor to liquid, typical values of $\delta^{18}O$ for the liquid and vapor are, -5% and -14%, respectively. Therefore, the fractionation factor is,

$$\alpha = \frac{1000 + (-5)}{1000 + (-14)} = \frac{995}{986} = 1.0092$$

Each fractionating process results in a specific fractionation factor. For example, the oxygen fractionation factor in freezing differs from that from condensation. The former and the latter are 1.0111 and 1.003, respectively. This means that freezing is more effective than condensation for isotope fractionation; freezing leads to stronger fractionation than condensation.

Fractionation factors approach unity as temperature increases; when $\alpha = 1$, the isotope ratios become equal in both phases being considered, and hence no further isotopic fractionation can occur (1). Higher fractionation factors mean more effective fractionation processes.

Isotopic Fractionation Curve

A curve defines the equilibrium isotope fractionation between coexisting phases as a function of temperature (1). It usually takes the form of an exponentially decreasing curve; there is a negative correlation between isotope fractionation and temperature. The shape of this curve is different for different isotopes.

Fractionation of Various Environmental Isotopes

As can be inferred from the information provided so far, isotope fractionation is an extensive topic in chemistry/geochemistry and has been dealt with in detail in many references such as Bowen (1), Gat (12), Gilreath (7), and Ottonello (13); the last is a thorough mathematical treatment of the topic. Here, in the *Encyclopedia of Water*, the discussion of isotope fractionation is limited to the isotopes of hydrogen and oxygen, specifically, the atoms that form the water molecule, as well as other isotopes that are of frequent use in water related/hydrologic studies. The latter group includes isotopes of the elements C, N, Cl, S, Sr, and less used U, Th, and Si. These are, often, referred to as environmental isotopes (6). Some workers such as Haldorsen et al. (11) classify the environmental isotopes into two groups: hydrologic isotopes (O, D, T) and biochemical isotopes (C, N, S). Although this classification covers only a limited number of environmental isotopes, it is sound for fractionation. For the first group, the main fractionating processes are physical or chemical reactions, biological/bacteriological/microbial processes are dominant in the fractionation of the second group.

Before discussing all environmental isotopes one by one, it should be pointed out that in some cases, considerable isotope fractionation has led to the usefulness of those isotopes in hydrologic studies, for example, H and O isotopes (enrichment and depletion processes are the fundamentals for studying lake–groundwater interaction, recharge temperature, and many other applications). In contrast, in some other cases, lack of isotope fractionation has been helpful. Strontium isotopes, for example, are used to study the origin of solutes in the streams because they are not subjected to substantial fractionation.

H and O

Hydrogen and oxygen atoms have three isotopes each (see also Heavy Water and Deuterium articles): ^{1}H, ^{2}H, ^{3}H, and ^{16}O, ^{17}O, and ^{18}O. Water molecules containing the heavy isotopes are less likely to evaporate and more likely to condense than those containing lighter isotopes. Hence, when a body of water evaporates, the vapor phase is depleted and the residue, the liquid phase, is enriched in heavy isotopes, ^{2}H, ^{3}H, ^{17}O, and ^{18}O. During melting/freezing of water, heavy isotopes are slightly more inclined to freeze and less inclined to melt than lighter isotopes, making the melt water depleted in ^{18}O and deuterium compared to the snowpack. Ultrafiltration (passage of water through shale micropores) enriches the residue (not passed through water) in O and H isotopes (6), but transpiration is not a fractionating process. The principles of hydrogen isotope fractionation are the scientific foundation for producing heavy water, which is used in the nuclear industry.

The fractionation factor for deuterium is greater than that for oxygen-18 because of a larger difference between deuterium and hydrogen masses compared to the difference between ^{16}O and ^{18}O masses. The fractionation factors of these two elements at various temperatures are calculated via the following equations (1):
For oxygen,

$$1000 \ln (\alpha\ ^{18}O) = 2.644 - 3.026 \times 10^3/T + 1.534 \times 10^6/T^2$$

For hydrogen,

$$\ln (\alpha\ D) = 0.0771 + 13.436 \times 10^3/T^2$$

(note that the temperature is in K).

Example:
The fractionation factor for ^{18}O at 293.15 K (20 °C) is

$$\ln(\alpha\ ^{18}O) = (2.644 - 3.026 \times 10^3/293.15 + 1.534 \times 10^6/293.15^2)/1000 = 1.01000$$

The fractionation factor for ^{18}O during calcite precipitation from water is 1.0349, which means that calcite is 35% more enriched in ^{18}O relative to the water from which it precipitates. However, not all precipitated calcite is richer in ^{18}O compared to "parent water." For instance, if calcite precipitation is due to freezing, the ^{18}O content of calcite is lower (11).

C

Oxidation–reduction processes in which the heavier isotope is concentrated in the more oxidized forms are important in carbon fractionation. The most influential factor in carbon isotope fractionation is selective uptake of lighter CO_2 by plants. During photosynthesis, plants discriminate against the heavier isotopes of carbon; the isotope ^{13}C in organic matter (soil) is depleted by 1.8% compared to its natural ratios in the atmosphere. It is believed that all organisms discriminate against ^{14}C about twice as much as against ^{13}C and the ratio between the stable ^{12}C and ^{13}C atoms can be used to correct for the initial depletion of ^{14}C. The amount of isotope fractionation depends on the photosynthetic pathway used by the plant. Plants are divided into three categories in this regard (11). At high temperatures, equilibrium fractionation of isotopes may occur between CH_4 and CO_2, and this is sensitive to changes in pH, oxygen fugacity, and ionic strength as well as to changes in temperature (13). As an example, the fractionation factor for ^{13}C in $CaCO_3$ in equilibrium with CO_2 is 1.01074.

Cl

The author has not come across a remark in the literature about the occurrence of fractionation in chlorine isotopes. This may have happened because chloride is a highly mobile, soluble, and geochemically conservative (is not removed from solution by mineral interaction or secondary mineral formation) element and has a relatively high mass.

N

Fractionation of nitrogen isotopes occurs mostly through biological processes and some isotope exchange reactions. Kinetic isotope fractionation often dominates nitrogen reactions and because of this, during denitrification, $\delta^{15}N$ values of residual nitrate increase exponentially as concentrations decrease (6). Bowen (1) reports that conversion of ammonia to nitrate (nitrification) by nitrifying organisms produces isotope fractionation up to 20% and even greater values (up to 30%) are associated with denitrification.

S

Sulfur in near-surface environments frequently changes between the oxidized and the reduced forms, so it acquires a wide variety of $^{34}S/^{32}S$ ratios. Isotope fractionation occurs in many processes of the sulfur cycle. The principal reaction leading to the fractionation of sulfur isotopes in nature is the reduction of sulfate ion by anaerobic bacteria (13). Bacterial reactions are important because they are kinetic, fast, favor the lighter isotope, and result in depleting ^{34}S in natural biogenic H_2S and hence enrich the remaining sulfate (11). Brucher et al. (14) studied isotope fractionation during bacterial sulfate reduction and found that in addition to the variations in the rates of bacterial sulfate reduction, genetic and physiological differences between various genera of sulfate-reduction bacteria strongly influence isotope fractionation.

In the precipitation of sulfide minerals from solution, the fractionation and enrichment of ^{34}S depend on the relative bond strengths and the temperature of solution (13). Sulfide minerals in newly deposited sediment can be enriched in ^{32}S by about 50% compared with associated marine sulfate (1). The equilibrium isotope fractionation factor between SO_4^{2-} and H_2S is 1.074 and is 1.020 between water and dissolved sulfate at 88 °C (6). During dissolution of evaporites, the largest reservoir of sulfate, no sulfur fractionation takes place.

Sr

Because of its relatively high mass, no detectable isotope fractionation accompanies precipitation of Sr within minerals (6).

ARTIFICIAL ISOTOPE FRACTIONATION FOR "SEPARATION OF ISOTOPES"

The most interesting and applied part of the isotope fractionation topic is artificial separation ("fractionation") of isotopes, a subject that is scientifically simple, but, in engineering terms very complicated. Historically, the first successful attempt to separate isotopes was that by Aston (the inventor of the mass spectrograph and the Nobel prize winner in 1922) who separated neon into two fractions of different atomic weights by repeated diffusion. The topic gained importance with the discovery, in 1939, that ^{235}U is a fissionable material and in succeeding years that uranium-235 is the key material for producing nuclear weapons and power (7).

VARIOUS TECHNIQUES FOR SEPARATING ISOTOPES ARTIFICIALLY

As stated before, separating isotopes is an intensive complicated engineering task, and only a few countries have the capacity. Four principles that constitute the foundation of isotope separation should be considered when devising a separation process (7):

1. At a given temperature, gaseous molecules, regardless of mass, have the same average kinetic energy.

2. The gravitational or centrifugal forces acting on a particle of matter are directly proportional to its mass.
3. The electric or magnetic forces affecting ionized atoms or molecules follow Coulomb's laws for these forces.
4. Among the lighter elements, isotopic species may vary slightly in their chemical reactivities.

Considering these four principles, a variety of techniques have been developed for isotopic enrichment (7). Each of these techniques has advantages and disadvantages and is suitable for specific types of conditions and isotope(s).

1. The gaseous-diffusion method is used for separating and enriching uranium isotopes. The production plant which uses this technique is huge; the huge diffusion plant at Oak Ridge, USA, in operation from 1945, represents one of the greatest technical achievements in the history of humankind.
2. The electromagnetic method is suitable only where cheap electricity is available.
3. The thermal diffusion method is good for laboratory-scale production.
4. The centrifuge method was not developed into a large-scale production plant.
5. The electrolytic method is used only for separating hydrogen and deuterium.
6. The chemical exchange method is used mostly for separating C and N.
7. The distillation method is only for enriching deuterium.

A sideline and an interesting coincidence is that while this article was being written, the Board of Governors of International Atomic Energy Agency (IAEA) postponed its decision on Iran's (the home country of the author) uranium enrichment and nuclear program activities (CNN Online edition, Nov. 22, 2003). The uranium enrichment and nuclear activities program of Iran has been in the news headlines for quite a long time.

BIBLIOGRAPHY

1. Bowen, R. (1988). *Isotopes in the Earth Sciences*. Elsevier, London.
2. Appelo, C.A.J. and Postma, D. (1999). *Geochemistry, Groundwater and Pollution*. Balkema, Rotterdam.
3. Kazemi, G.A. and Mehdizadeh, H. (2001). Solute content and chemical composition of atmospheric precipitation of Shahrood Region, *Proc. of 5th Iranian Geol. Convention*, Tehran, September 2001, pp. 401–405 [in Persian].
4. Panettiere, P., Cortecci, G., Dinelli, E, Bencini, A., and Guidi, M. (2000). Chemistry and sulfur isotopic composition of precipitation at Bologna, Italy. *Appl. Geochem.* **15**: 1455–1467.
5. Domenico, A.P. and Shwartz, F.W. (1991). *Physical and Chemical Hydrogeology*. John Wiley & Sons, New York.
6. Coplen, T.B. (1993). Uses of environmental isotopes New York. In: *Regional Ground-Water Quality*. W.M. Alley (Ed.). ITP, pp. 255–293.
7. Gilreath, E.S. (1985). *Fundamental Concepts of Inorganic Chemistry*. 19th printing, International Student Edition. McGraw-Hill, Singapore.
8. MacRae, N.D. (1995). Secondary-ion mass spectrometry and geology. *Can. Mineralogist* **33**: 219–236.
9. Langmuir, D. (1997). *Aqueous Environmental Geochemistry*. Prentice-Hall, Upper Saddle River, NJ.
10. Mazor, E. (1991). *Applied Chemical and Isotopic Groundwater Hydrology*. Open University Press, Milton Keynes.
11. Haldorsen, S., Riise, G., Swanson, B., and Sletten, R.S. (1997) Environmental isotopes as tracers in catchments. In: *Geochemical Processes, Weathering and Groundwater Recharge in Catchments*. O.M. Saether and P.D. Caritat (Eds.). Balkema, Rotterdam, pp. 185–210.
12. Gat, J.R. (1981). Isotopic fractionation. In: *Stable Isotope Hydrology, Deuterium and Oxygen-18 in the Water Cycle*. J.R. Gat and R. Gonfiantini (Eds.). IAEA, technical report series, No. 210, pp. 21–34.
13. Krauskopf, K.B. (1979). *Introduction to Geochemistry*, 2nd Edn. McGraw-Hill, New York.
14. Bruchert, V., Knoblauch, C., and Jorgensen, B.B. (1999). Sulfur isotope fractionation during bacterial sulfate reduction. In: *Geochemistry of Earth's Surface*. H. Armannsson (Ed.). Balkema, Rotterdam, pp. 303–305.

MARIOTTE BOTTLE—USE IN HYDROLOGY

Joseph Holden
University of Leeds
Leeds, United Kingdom

A Mariotte bottle is a device that allows constant flow of a fluid from a container, even when the fluid level in the container changes. It is named after Edme Mariotte (1620–1684), a French physicist, who, independently from Boyle, discovered the law that states that the volume of a gas varies inversely with its pressure. Mariotte bottles have a wide range of applications in engineering and science (such as fuel supply) and are used in a variety of hydrologic applications ranging from river discharge gauging to infiltration measurement.

Figure 1 shows the general design of a Mariotte bottle. It consists of a bottle that has an escape pipe or tap to allow fluid to exit and a vent tube to allow air to enter the bottle. The bottle is sealed so that the only location where air can enter is at that bottom of the vent tube, b. The fluid (such as water) exits through a tap (or any sort of escape tube) at c, at a rate that is controlled by the size of the escape tube and the head of water. The head of water, h will always be equal to the distance between the outlet tap and the bottom of the vent tube. Therefore, regardless of the water level inside the bottle, there will be a constant head of water.

If the vent tube is kept at a constant level, then as a result of the reduction of the fluid in the bottle, the air above it expands and its pressure drops. When the pressure of the air has dropped so that it plus the pressure of the water column from c to b (the vent tube opening) is less than the atmospheric pressure of the air outside the Mariotte bottle (and therefore also inside the vent tube), then air enters the Mariotte bottle at b and rises through the fluid. Therefore, the pressure at c will always be equal

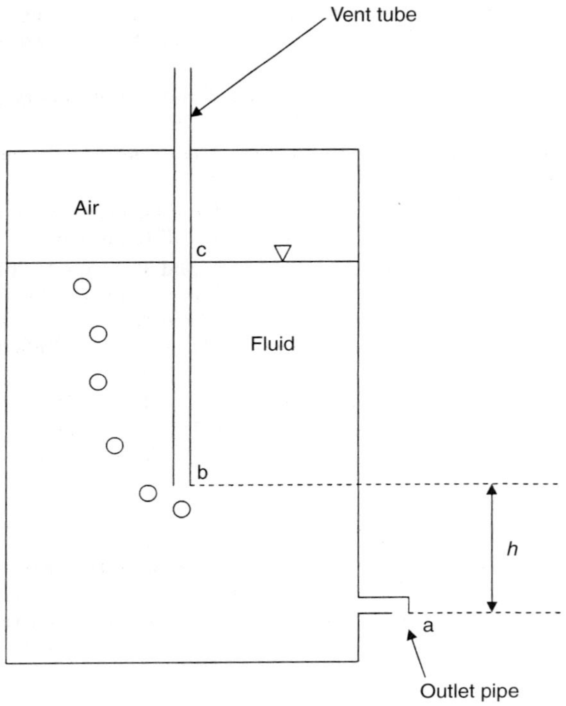

Figure 1. A Mariotte bottle.

to atmospheric which is also equal to the pressure at b. The excess pressure, under which the fluid is ejected, is the pressure of the fluid between the levels b and a. It remains constant until the fluid drops to the level of b, and the discharge from the escape pipe will remain constant until then. Thus, Mariotte bottles can be used wherever there is an application for a constant flow of water from a supply reservoir.

If the vent tube is raised to a greater height, the head will increase, and therefore, the rate of water flowing out of the bottle will increase. It is therefore possible to control the rate of water release from the bottle by controlling the head of water, and this is, in turn, controlled by the height of the base of the vent tube above the outlet at a. If the vent tube is moved down to the level of the escape opening, a, the flow stops because then the height of the column becomes zero. The discharge also ceases if the vent tube is sealed preventing air from entering the bottle.

There is a wide range of common hydrologic applications for Mariotte bottles. They are used frequently for adding a constant rate of solution (such as a dye or tracer) to watercourses. Most commonly this is for discharge gauging where, if a constant rate of solution (such as sodium chloride solution) is added to river, then the discharge of the river can be calculated based on the equation,

$$Q = q(C_i - C_d)/(C_d - C_b)$$

where Q is the river discharge, q is the rate of injection of tracer solution, C_i is the concentration of the tracer, C_b is the background concentration in the river, and C_d is the downstream sampled concentration of the tracer.

It is often useful to use a Mariotte bottle to supply a constant rate of water and also, if the bottle is graduated,

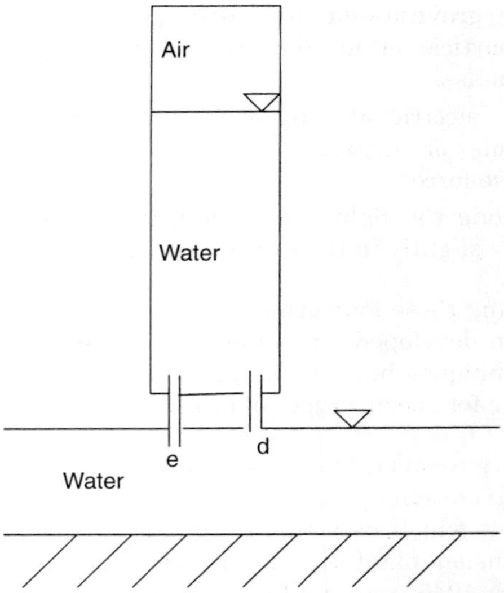

Figure 2. Mariotte bottle used to keep a constant water level above the ground surface.

then it can be used to determine how much water has been supplied during any given period. Frequently, Mariotte bottles are used in instruments designed to measure soil infiltration rates such as rain simulators or other types of infiltrometers where a constant rate of water supply to the soil surface might be required.

In constant head infiltrometers (see entry for INFILTROMETERS), the Mariotte principle is used to ensure that there is a constant depth of ponded water above the soil surface. This is done by placing the vent tube for the Mariotte bottle at the same end of the bottle as the water exit pipe as in Fig. 2. Whenever the water level on the ground drops below the height of the base of the vent tube, d, air is allowed to enter the water supply bottle, and water can be released from e until the water level on the ground has risen back to the base of the vent tube. In this way, a constant height or head of water can be maintained on a soil surface. This principle can also be used for irrigation to maintain a constant surface water level or a constant groundwater level.

MARS EXPLORATION ROVER MISSION

GUY WEBSTER
Jet Propulsion Laboratory
Pasadena, California

DONALD SAVAGE
NASA Headquarters
Washington

OPPORTUNITY ROVER FINDS STRONG EVIDENCE MERIDIANI PLANUM WAS WET

Scientists have concluded the part of Mars that NASA's Opportunity rover is exploring was soaking wet in the past.

This image, taken by Opportunity's microscopic imager, shows a portion of the rock outcrop at Meridiani Planum, Mars, dubbed "Guadalupe." View Opportunity images from this press release.

Evidence the rover found in a rock outcrop led scientists to the conclusion. Clues from the rocks' composition, such as the presence of sulfates, and the rocks' physical appearance, such as niches where crystals grew, helped make the case for a watery history.

"Liquid water once flowed through these rocks. It changed their texture, and it changed their chemistry," said Dr. Steve Squyres of Cornell University, Ithaca, N.Y., principal investigator for the science instruments on Opportunity and its twin, Spirit. "We've been able to read the tell-tale clues the water left behind, giving us confidence in that conclusion."

Dr. James Garvin, lead scientist for Mars and lunar exploration at NASA Headquarters, Washington, said, "NASA launched the Mars Exploration Rover mission specifically to check whether at least one part of Mars ever had a persistently wet environment that could possibly have been hospitable to life. Today we have strong evidence for an exciting answer: Yes."

Opportunity has more work ahead. It will try to determine whether, besides being exposed to water after they formed, the rocks may have originally been laid down by minerals precipitating out of solution at the bottom of a salty lake or sea.

The first views Opportunity sent of its landing site in Mars' Meridiani Planum region five weeks ago delighted researchers at NASA's Jet Propulsion Laboratory, Pasadena, Calif., because of the good fortune to have the spacecraft arrive next to an exposed slice of bedrock on the inner slope of a small crater.

The robotic field geologist has spent most of the past three weeks surveying the whole outcrop, and then turning back for close-up inspection of selected portions. The rover found a very high concentration of sulfur in the outcrop with its alpha particle X-ray spectrometer, which identifies chemical elements in a sample.

"The chemical form of this sulfur appears to be in magnesium, iron or other sulfate salts," said Dr. Benton Clark of Lockheed Martin Space Systems, Denver. "Elements that can form chloride or even bromide salts have also been detected."

At the same location, the rover's Mössbauer spectrometer, which identifies iron-bearing minerals, detected a hydrated iron sulfate mineral called jarosite. Germany provided both the alpha particle X-ray spectrometer and the Mössbauer spectrometer. Opportunity's miniature thermal emission spectrometer has also provided evidence for sulfates.

On Earth, rocks with as much salt as this Mars rock either have formed in water or, after formation, have been highly altered by long exposures to water. Jarosite may point to the rock's wet history having been in an acidic lake or an acidic hot springs environment.

The water evidence from the rocks' physical appearance comes in at least three categories, said Dr. John Grotzinger, sedimentary geologist from the Massachusetts Institute of Technology, Cambridge: indentations called "vugs," spherules and crossbedding.

Pictures from the rover's panoramic camera and microscopic imager reveal the target rock, dubbed "El Capitan," is thoroughly pocked with indentations about a centimeter (0.4 inch) long and one-fourth or less that wide, with apparently random orientations. This distinctive texture is familiar to geologists as the sites where crystals of salt minerals form within rocks that sit in briny water. When the crystals later disappear, either by erosion or by dissolving in less-salty water, the voids left behind are called vugs, and in this case they conform to the geometry of possible former evaporite minerals.

Round particles the size of BBs are embedded in the outcrop. From shape alone, these spherules might be formed from volcanic eruptions, from lofting of molten droplets by a meteor impact, or from accumulation of minerals coming out of solution inside a porous, water-soaked rock. Opportunity's observations that the spherules are not concentrated at particular layers in the outcrop weigh against a volcanic or impact origin, but do not completely rule out those origins.

Layers in the rock that lie at an angle to the main layers, a pattern called crossbedding, can result from the action of wind or water. Preliminary views by Opportunity hint the crossbedding bears hallmarks of water action, such as the small scale of the crossbedding and possible concave patterns formed by sinuous crestlines of underwater ridges.

The images obtained to date are not adequate for a definitive answer. So scientists plan to maneuver Opportunity closer to the features for a better look. "We have tantalizing clues, and we're planning to evaluate this possibility in the near future," Grotzinger said.

JPL, a division of the California Institute of Technology in Pasadena, manages the Mars Exploration Rover project for NASA's Office of Space Science, Washington.

For information about NASA and the Mars mission on the Internet, visit http://www.nasa.gov/.

Images and additional information about the project are also available at http://marsrovers.jpl.nasa.gov/ and http://marsrovers.jpl.nasa.gov/relocate.html?relocate= http://athena.cornell.edu/.
http://marsrovers.jpl.nasa.gov and http://athena.cornell .edu.

REMOVAL OF ORGANIC MICROPOLLUTANTS AND METAL IONS FROM AQUEOUS SOLUTIONS BY ACTIVATED CARBONS

Basu Saha
Loughborough University
Loughborough, United Kingdom

INTRODUCTION

World population growth has triggered agricultural and industrial growth for the sustenance of the day-to-day demands of the population. However, the increase in agricultural and industrial growth has also led to a corresponding growth in pollution related to the two activities. The use of pesticides to boost food production has caused increased contamination of drinking water, thus putting pressure on governments to invest in the provision of clean water. The same can also be said about the increase in industrial activities, which has seen, among many of the pollutants, the increase of heavy metals in the water bodies. Since the installation of water pollution control schemes are not profit-generating activities, there is a lot of ongoing research into finding efficient but inexpensive ways of separating pollutants from drinking water.

The ban on the use of atrazine for nonagricultural purposes in the United Kingdom in 1993 has resulted in the introduction of alternative herbicides. Among them is benazolin, which can be a potential groundwater pollutant owing to its high water solubility when compared with that of atrazine. A report compiled by the Environment Agency (1) revealed that, for the first time in England and Wales, benazolin has become one of the 10 most frequently occurring herbicides in surface water. The potential harmful effects of benazolin as a herbicide necessitates that it must be removed from water bodies if present at concentrations higher than that required by environmental regulations. However, conventional water treatment processes, such as filtration, disinfection, coagulation, and sedimentation, have been reported to be more effective in removing less water-soluble and easily degradable pesticides (2). Ozone technology is a potentially attractive technique in wastewater treatment. However, in some cases, the ozone degradation products can be equally harmful. Lambert et al. (3) have investigated degradation efficiency of benazolin and other herbicides using ozone and a combination of ozone and hydrogen peroxide. They realized that for benazolin, doses of 1.0 mg O_3/l, 2.0 mg O_3/l, and 3.1 mg O_3/l, degradation efficiencies of 55%, 68%, and 88% could be achieved, respectively. The combination of ozone and hydrogen peroxide further increased the degradation efficiency, but they recommended complementary techniques for other herbicides. Adsorption on activated carbon is one of the well-established and effective techniques for the removal of herbicides from water. The adsorptive efficiency of activated carbon is because of its well-developed internal pore structure, surface area, and surface reactivity. Typically, activated carbon consists of interwoven micropores, mesopores, and macropores, and these are responsible for the high surface area of the carbons. The existence of surface functional groups in an activated carbon matrix means that they can be manipulated by thermal or chemical treatments to produce adsorbents that are tailored for particular functions.

As wastewater streams consist of a mixture of pollutants that can generally be classified as organic and inorganic pollutants, tailoring an adsorbent to remove both classes of pollutants would give a significant process advantage. Among the inorganic pollutants, heavy metals form a major part in the classification. Heavy metals constitute a class of environmental pollutants that are of particular concern in the treatment of industrial wastewater because of their potential danger to humans and the environment. Therefore, they must be treated if present at concentrations higher than those specified by the environmental regulations.

Activated carbons have been proven to be an effective adsorbent for the removal of organic compounds from water (4,5). Research has also shown that they have a potential to remove heavy metals, especially in its modified state. For example, a number of authors have oxidized activated carbons to enhance sorption of heavy metals (6–9). The oxidation processes have mainly been by use of nitric acid, ozone, and air. The three methods have always produced the same results, that of producing weakly acidic functionality on the surface of the carbon. Saha et al. (10) compared the metal sorption capacity of a conventional carbon and its oxidized form. The adsorption of herbicides and trace metal ions onto activated carbon is discussed in this chapter. In particular, we discuss the effect of modification of activated carbon on the uptake of copper and benazolin. The effect of post-treatment of the oxidized carbons on the uptake of the target pollutants is also highlighted. The adsorbents were characterized in a bid to understand their chemical and physical properties, which gives a better understanding of the nature of the association of the target pollutants and the adsorbents.

BACKGROUND

Herbicides have undoubtedly contributed to the growth of agricultural productivity. For example, it has been reported that a \$4 billion U.S. investment in herbicide control saved \$16 billion U.S. in crops every year (11). However, the benefit realized from the use of herbicides does not factor in the negative effects they have on human health.

As a result of their negative impact on human health and the environment, stringent legislation has been imposed on the purity of drinking water. The European Union (E.U.) agreed to have the drinking water Directive Admissible Concentration for single herbicide or pesticide at a maximum of 0.1 μg/L. In Great Britain, the National Rivers Authority (12) showed that eulan and permethrin (mainly in effluents from the textile industry) had exceeded their Environmental Quality Standard (EQS) limits. Also, diazinon, a sheep dip insecticide, had also exceeded its EQS limits. One major concern was that the most commonly used, herbicides atrazine, diuron, bentazone, isoproturon, and mecoprop exceeded the standard 0.1 μg/L in surface waters, which has been and always will be a major concern simply because most groundwater sources used for drinking water have no treatment facilities designed to remove herbicides. Also, the most common water treatment processes, such as sedimentation, disinfection, coagulation, and filtration, can effectively remove only selected herbicides, especially those that have low solubility in water. On the other hand, advanced water treatment processes (mainly using adsorption onto activated carbon) have proved to be the most efficient and reliable method for the removal of aqueous dissolved organic herbicides. Hence, a lot of studies have gone into trying to improve the removal efficiency of the existing adsorbents. In improving the adsorption capacity of adsorbents, attention has been turned on to the number of variables that affect the adsorption capacity of the activated carbons.

As pollution in wastewater is because of a cocktail of pollutants, the use of activated carbon could be a viable way of simultaneously adsorbing pollutants of a different nature, which has a further advantage of cutting down on the number of wastewater processing units. Hence, the objective of this research is to investigate the effect of simultaneous adsorption of a heavy metal (copper) and a selected herbicide (benazolin) on as-received and surface-modified activated carbons. The adsorbents chosen for this work are as-received F400 supplied by Chemviron (USA) and various surface-modified samples of F400 that were prepared in the author's laboratory. Benazolin is highly soluble in water (solubility 600 mg/l) when compared with atrazine (solubility 33 mg/l). Properties, structural formula, and three-dimensional representation of benazolin are given in Table 1 and Fig. 1, respectively. Copper has been chosen as a representative of heavy metals. In order to understand the sorption mechanism of the target pollutants, physical and chemical characterization were performed for all the adsorbents. The other objective of the study is to investigate the mechanism of adsorption of the chosen pollutants. If the mechanism is fully understood, it is possible to produce tailor-made engineered adsorbents for various purposes in water treatment.

EXPERIMENTAL

Adsorbent Materials

A coal-based commercial granular activated carbon, F400 was used for surface modification with nitric acid. The

Table 1. Properties and Structural Formula of Benazolin

Herbicide	Chemical Formula	Structural Formula	Solubility in Water (mg/L)	pKa
Benazolin	$C_9H_6ClNO_3S$	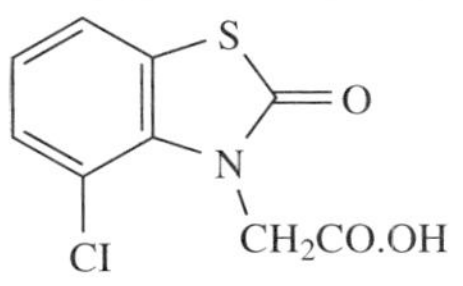	600	3.04

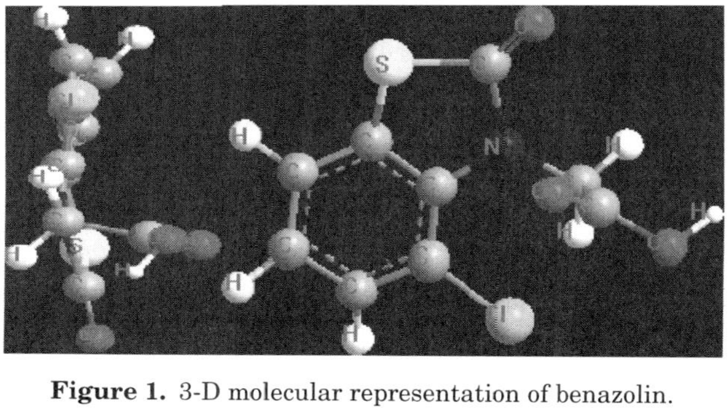

Figure 1. 3-D molecular representation of benazolin.

carbon sample was obtained from Chemviron (USA). Before the oxidation process, the carbon was washed in distilled water and dried at 383°K for 24 hours. The dried sample was then subjected to different modes of modification using part of a method described by Saha et al. (7) and Chingombe et al. (13).

Sample Preparation

A known amount of the washed and dried sample was introduced into the reaction vessel that contained nitric acid and distilled water at a ratio of 1:1 (v/v). The reaction mixture was heated to 363°K, and the reaction was allowed to continue for 6 hours with continued stirring. The reaction solution was removed and a fresh reaction mixture of nitric acid and water was introduced. The reaction was allowed to continue for an additional 3 hours under the same reaction conditions. The oxidized carbon was then washed with distilled water until no further change in pH could be detected.

The resultant water-washed samples (after oxidation) were divided into three portions. The first portion was denoted AC1, and this is a sample that was washed with water after the oxidation reaction. AC2 was a sample that was washed with 0.1 M sodium hydroxide to remove humic acids that were formed during the oxidation process. The washing was continued until no further coloration from the humic acids could be detected. Reconditioning of the sample to the hydrogen form was performed using 0.1 M hydrochloric acid, and further washing with distilled water was done until the pH of the supernatant solution stabilized to a pH of about 4.0. The conversion was deemed complete when the inlet and the outlet pH of the liquid flowing in the column were the same. The third sample, denoted AC3, was a result of heating the third portion of the oxidized sample at a temperature of 580°K under vacuum of 2 mmHg for 12 hours to remove humic acids.

ANALYSIS OF METAL ION AND HERBICIDE

Copper

The copper ion content was analyzed using a Varian SpectraAA-200 atomic absorption spectrophotometer (AAS). The AAS was set on the flame mode using air-acetylene mixture.

HPLC Analysis Method

As preconcentration of benazolin before analyzing on the HPLC was required, a trace analysis method had to be developed for the quantification of the herbicide. A Hewlette Packard 100 series HPLC consisting of a diode array detector, column thermostat, auto sampler, and a binary pump was used for analysis. Jones Chromatograph supplied a Genesis C18 silica column (4 μm, 150 × 3.0 mm) combined with a 1 cm Genesis guard column. The detector, flow rate, and injection volume were set at 222 nm, 0.88 ml/min, and 100 μl, respectively, and this gave the retention time of benazolin at 2.01 minutes. The column temperature was also set at 311°K. The eluent consisted of a solution of 40% acetonitrile and 60% buffer (10 mmol KH_2PO_4 at pH 3). The calibration curve for benazolin showed that correlations were linear over the range (0–8 ng/μl).

Solid phase extraction (SPE) technique was employed for the preconcentration of benazolin before analysis. The complete procedure for SPE is given in Table 2.

Table 2. Procedures for Solid Phase Extraction (SPE)

Conditioning	500 mg, s-triazine, 6 ml SPE columns were conditioned by passing 10 ml of HPLC-grade methanol followed by 10 ml of 2% MeOH in water under gravity.
Extraction	500 ml sample solution were passed through the column at a flow rate of 1 ml/min.
Elution	6 ml of acetonitrile was allowed to soak in the columns for 20 minutes before being passed through at approximately 0.5 ml/min and the eluent collected in 7 ml vials. The columns were then blown dry with nitrogen and the solvent was evaporated to dryness using a stream of nitrogen.
Reconstitution	500 μl of 40% acetonitrile, 60% 10 mmol KH_2PO_4 buffer at pH 3.0 was added to the vials to redissolve the analytes. The vials were shaken vigorously and their contents transferred to HPLC auto-sample vials.

CHARACTERIZATION OF ADSORBENTS

All samples were characterized to investigate the effect of the treatment and the chemical as well as the physical characteristics of these materials. The characterization included scanning electron micrographs (SEM), Brunauer-Emmett-Teller (BET) and Langmuir surface area measurements, pore size distribution, Fourier Transform Infra Red spectroscopy (FTIR) analysis, sodium capacity measurement, pH titration, zeta potential measurements, and X-ray photoelectron spectroscopy (XPS) analysis and elemental analysis. The detailed characterization procedure and the results have been reported by Chingombe et al. (13). The results indicated that a significant change occurred in the structure of the precursor by oxidation and subsequent treatments. Characterization of the adsorbents revealed that modification of the carbon by oxidation resulted in the introduction of weakly acidic functional groups, and the presence of such groups was confirmed by FTIR, pH titration, zeta potential measurements, and sodium capacity results. Sodium capacity was enhanced by a factor of up to 8 for the oxidized carbons compared with as-received samples.

Batch Adsorption Experiments

The sorption of copper on the F400 carbon range is shown in Fig. 2. The results clearly show that AC1 and AC2 have the same adsorptive capacity for copper when compared with AC3 and unoxidized F400. However, among the adsorbents employed, as-received F400 showed the least adsorptive capacity. The enhanced capacity observed on AC1, AC2, and AC3 could be as a result of the oxidation process, which generated carboxylic functional groups as evidenced in the FTIR spectra and the high sodium capacity of the above sorbents. These groups

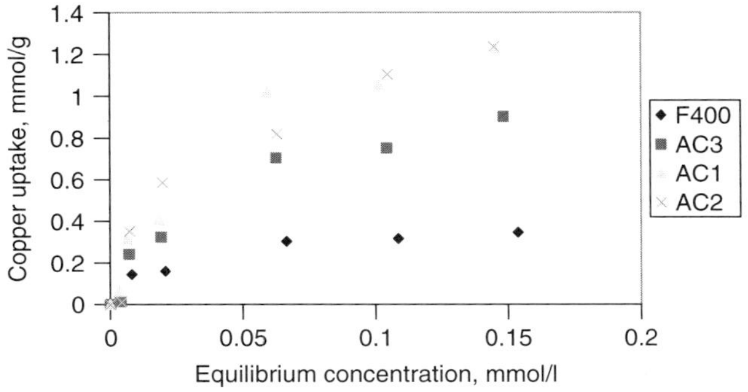

Figure 2. Equilibrium sorption results of copper on F400 and the oxidized carbons.

would be dissociated in aqueous solution, and they are responsible for copper sorption. The oxidized and heat-treated sample AC3 showed reduced capacity compared with AC1 and AC2, which could be because of the decomposition of the carboxylic acid groups during the heat treatment. Toledo-Bautista et al. (14) made similar observations when they adsorbed Cr(III) on a range of conventional and modified carbons.

The formation of metal surface complexes on the oxidized carbon involving cooperative action seems quite likely. Approximate calculation of oxygenated functional group density per unit area for F400(ox) yields a value of 0.02 functional groups per square angstrom. Given that the oxidized carbons evaluated in the present study possess a large proportion of pores in the region of 10–20 angstrom and the diameter of the hydrated metal ions is approximately 8 angstrom (15), it is reasonable to assume a cooperative binding mechanism (see Fig. 3).

Figure 4 shows the adsorption of benazolin on all the adsorbents under study. The results clearly show that benazolin adsorption on F400 is slightly higher than AC2 and reasonably higher than AC1 and AC3. However, AC1 and AC3 seem to have similar sorptive performance, whereas AC2 showed a higher performance than the other oxidized carbons, which can be expected because there was a significant loss in micropores for samples AC1 and AC3 as obtained by the pore size distribution analysis (13). It is generally accepted that sorption of an organic molecule takes place mainly in the micropore region of the adsorbents, because in this region, the contact points between the adsorbent and the molecules is higher than in mesopores and macropores. Also, it is reported that an overlap of potential forces exists in the micropores, which results in increased adsorption forces. Therefore, naturally, the capacity of AC1 and AC3 is expected to be low. Another factor in the difference between the adsorption capability of the adsorbents is the electrostatic interaction between the adsorbent and the adsorbate.

The zeta potential results show that the isoelectric point (IEP) of the oxidized carbons are significantly lower (1.5) than that of the conventional carbon (6.0) because of the introduction of weakly carboxylic acid groups with dissociation between pH 2 and 6. As the experiments are performed at pH 5, it would imply that the surface of the oxidized carbons is predominantly negative because of the dissociation of the carboxylic groups. Also, the pK_a value of benazolin (3.04) is lower than the pH of the solution experiment, and this condition would render mutual repulsion between the adsorbent and the adsorbate. Adsorption of benazolin is more favorable on F400 because of its high IEP. Streat and Horner (4), when comparing adsorption of some herbicides (including benazolin) on F400 and hypercross-linked polymers, realized that F400 outperformed the polymeric sorbents. They postulated

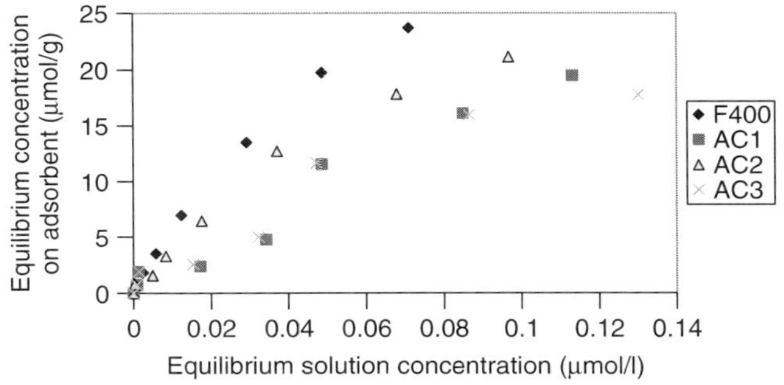

Figure 4. Adsorption of benazolin on F400 and the oxidized carbons.

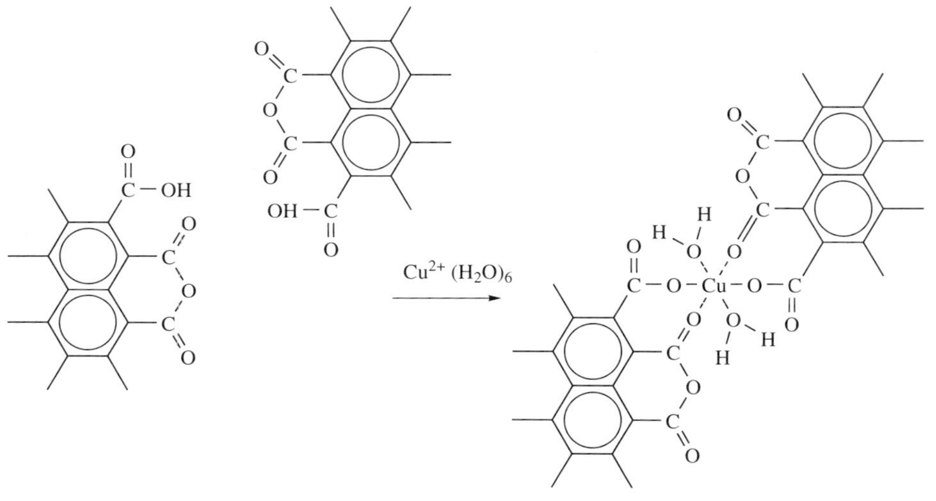

Figure 3. Postulated complexation reaction between copper (II) and oxidized carbon surface.

that adsorption of benazolin on F400 was through the dispersion interaction of the π electrons in the respective aromatic systems by a donor-acceptor mechanism. They further noted that the chloro (–Cl) group on benazolin is an electron-withdrawing group, therefore implying that a reduction of electron density occurs in the π ring system of the herbicide. Hence, benazolin would act as an electron acceptor in the donor-acceptor mechanism involving F400. This observation is consistent with investigations carried out by Radovic et al. (16) when they considered the adsorption of aromatics possessing electron-withdrawing and electron-donating functional groups on conventional and chemically modified activated carbons. Chemical modification involved samples that were subjected to nitration and others that were oxidized. Heat-treated carbons were also considered. They used nitrobenzene and aniline as the target adsorbates. Nitrobenzene contains nitro ($-NO_2$) group, which is known to be electron withdrawing, whereas aniline possesses an amino ($-NH_2$) group, which is electron donating. As nitration and oxidation reduces the π-electron density in the graphene layers of the carbon surface, which would in turn favor adsorption of aromatics with electron donor functional groups, it was realized that maximum adsorption uptake of aniline (with electron donor group) were attained on oxidized carbon surfaces. On the other hand, maximum adsorption uptake for nitrobenzene were found on heat-treated carbon surfaces (which possess higher π-electron density than the oxidized carbons). Hence, from the view point of Radovic et al. (16), it would be reasonable to postulate that benazolin adsorption is favored by adsorbents with high π-electron density. However, in the oxidized carbons employed in this work, such a mechanism would be highly unlikely because of the introduction of oxygen groups in the form of carboxylic acids. The carboxylic acid functional groups are electron withdrawing, and this condition reduces the π-electron density in the ring systems of the carbon. Therefore, the dispersion interaction between the herbicide and the oxidized carbon is reduced. Hence, unoxidized F400 would be expected to have a higher benazolin capacity than the oxidized carbons.

Figures 5 and 6 show the adsorption isotherms of benazolin in the presence of copper at 5 ppm and 10 ppm for unoxidized F400 and AC1 and AC3 samples, respectively. For both cases, it is apparent that benazolin

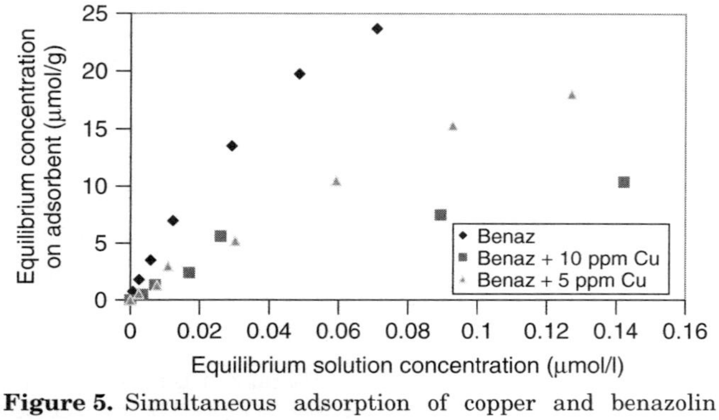

Figure 5. Simultaneous adsorption of copper and benazolin on F400.

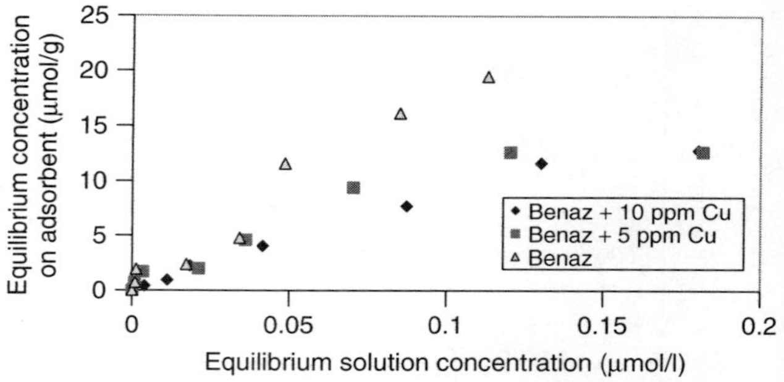

Figure 6. Simultaneous adsorption of copper and benazolin on AC1.

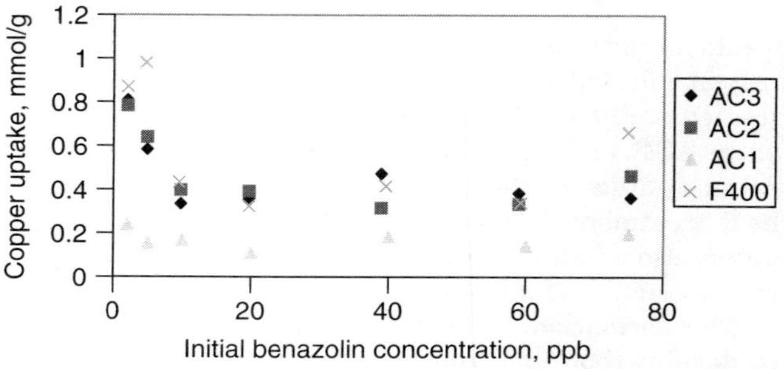

Figure 7. Effect of the initial concentration of benazolin on copper uptake.

uptake is distinctly affected in the presence of copper ions. Also, Fig. 7 shows the effect of the initial concentration of benazolin on the uptake of copper. The results clearly show that as the initial concentration of benazolin increases, a corresponding decrease occurs in the uptake of copper up to the point when the initial concentration of benazolin is at approximately 20 ppb, that is when the uptake of copper normalizes. The immediate conclusion that could be drawn from the two sets of results is that a competition exists for the same adsorption sites by copper and benazolin. As mentioned previously, the adsorption of benazolin is favored through electron donor-acceptor type of a mechanism, which involves π electron in the ring systems of the carbon. Biniak et al. (6) showed that copper ions had the capacity to interact with the π electron in the ring systems of the carbon. In their work, they employed XPS and FTIR to study the state of the copper adsorbed on different carbon surfaces. They investigated three types of modified activated carbons. One of the modified carbons was obtained by annealing a conventional carbon in a vacuum at 1000°K. This carbon had a low oxygen content of 2.72% compared with oxidized and ammonia-treated carbons, which had 10.10% and 6.20%, respectively. The annealed carbon was virtually devoid of acidic functional groups. XPS and FTIR results indicated that the dominant mechanisms of copper adsorption on heat-treated active carbons could be dipole-dipole (π-d) interactions between graphene layers and metal ionic species as well as spontaneous electrochemical reduction of copper ions. Hence, it is expected that benazolin and copper would compete for the same sites derived from the π electrons in the graphene layers of the modified adsorbents. However, conclusive evidence of such competition would only be proved by employing an annealed carbon adsorbent.

CONCLUSIONS

Oxidation of the conventional carbon and post-treatment of the resultant carbons had an influence on their uptake for copper and benazolin. Oxidation resulted in the introduction of weakly acidic carboxylic functionality, which markedly enhanced the uptake of copper ions. The isoelectric point (IEP) was also reduced as a result of oxidation. Oxidation of the carbon resulted in altered pore size distribution. Reduction in microporosity was observed, and it also explained the low uptake of benazolin by oxidized sample compared with as-received F400. However, post-treatment by washing with sodium hydroxide seemed to open up the pores for AC2 sample. Simultaneous adsorption of benazolin and copper showed that direct competition existed for the active sites, and possibly the two adsorbates targeted the π electron cloud on the graphene layers of the adsorbent. Copper uptake by the conventional and modified samples decreased in the presence of benazolin and vice versa. The decrease in metal uptake was dependent on the concentrations of metals in multicomponent aqueous solutions. The effect of copper on benazolin uptake was more pronounced than the effect of benazolin on copper uptake from water. Further work is necessary to fully understand the precise mechanisms of adsorption of complex aromatic molecules onto carbon, and this forms the basis of our continued work. Research is in progress in the author's laboratory to provide conclusive evidence on the interaction of these adsorbates with the modified carbon surface.

The environment is challenged by other micropollutants, namely polychlorinated hydrocarbons, organic oxygen compounds, surfactants, pesticides, aliphatic intermediates developing from the chemical industry, and, of course, endocrine disrupting compounds (EDCs). Further study of activated carbon continues to explore for potential solutions to these important problems.

Acknowledgments

I would like to express my sincere gratitude to Professor Michael Streat for his support, continuous encouragement, and advice for conducting the research in the area of adsorption and ion exchange and development of tailored sorbent materials for environmental remediation. I am indebted to Dr. Hadi Tai and Purazen Chingombe for their meticulous experimental work. Funding from European Community, EPSRC, Severn Trent Water Ltd., and Commonwealth Scholarship Commission is gratefully acknowledged.

BIBLIOGRAPHY

1. Environment Agency. (2002). *The Annual Report of the Environment Agency pesticide monitoring programme*. UK.
2. Kouras, A., Zouboulis, A., Samara, C., and Kouimtzis, T. (1998). *Environ. Pollut.* **103**: 193–202.
3. Lambert, S.D., Graham, N.J.D., and Croll, B.T. (1996). *Ozone Sci. Engineer.* **18**: 251–269.
4. Streat, M. and Horner, D.J. (2000). *Trans. IChemE.* **78**(B2): 363–382.
5. Streat, M., Malik, D.J., and Saha, B. (2004). In: *Ion Exchange and Solvent Extraction*. Vol. 16. A.K. SenGupta and Y. Marcus (Eds.). Marcel Dekker, New York, pp. 1–84.
6. Biniak, S., Pakula, M., Szymanski, G.S., and Swiatkowski, A. (1999). *Langmuir* **15**: 6117–6122.
7. Saha, B., Tai, M.H., and Streat, M. (2001). *Trans. IChemE.* **79**(Part B): 211–217.
8. Jia, Y.F., Xiao, B., and Thomas, K.M. (2002). *Langmuir* **18**: 470–478.
9. Saha, B., Tai, M.H., and Streat, M. (2003). *Chem. Eng. Res. Des.* **81**: 1343–1353.
10. Saha, B., Tai, M.H., and Streat, M. (2001). *Trans. IChemE.* **79**(Part B): 345–351.
11. Department of the Environment. (1996). *Pesticides in Water*. Department of the Environment Report, The Working Party on the Incidence of Pesticides in Water.
12. National Rivers Authority. (1998). Pesticides in the aquatic environment. *Water Quality Series* **26**: 12.
13. Chingombe, P., Saha, B., and Wakeman, R.J. (2004). *16th International Congress of Chemical and Process Engineering*, Prague, Czech Republic.
14. Toledo-Bautista, I., Utrilla-Rivera, J., Garcia-Ferro, M.A., and Castillo-Moreno, C. (1994). *Carbon* **32**(1): 93–100.
15. Nightingale, E.R.J. (1959). *Phys. Chem.* **63**: 1381–1386.
16. Radovic, L.R. et al. (1997). *Carbon*. **35**(9): 1339–1348.

MOLECULAR NETWORK DYNAMICS

D.L. Marrin
Hanalei, Hawaii

Perhaps the twentieth century's most startling discovery about liquid water is that it functions as a vast interconnected network, rather than as a random collection of independent molecules. Individual water molecules serve as network *elements*, which constitute the building blocks for the fundamental water network. Elements in the water network are connected to one another via magnetic-type linkages known as *hydrogen bonds*. In solid water (ice), each water molecule hydrogen bonds with all four of its nearest neighbors in forming a perfect *tetrahedron*, which is a three-sided pyramid comprising the most basic molecular geometry of water (see Fig. 1). As ice melts into liquid water, some of the hydrogen bonds between water molecules are broken (causing the tetrahedron to distort), whereas the remaining bonds transition to an ultra-dynamic state whereby they are switched as rapidly as a trillion times per second. The dynamics of this unique molecular network are believed to underlie many of water's anomalous properties.

NETWORK BEHAVIORS

This frantic switching of hydrogen bonds, which constitutes the underlying dynamism of water's molecular network, permits water to flow and to behave as a liquid even though it retains much of the molecular geometry of a solid. Physical chemists have long modeled liquid water as a space-filling network of individual water molecules, within

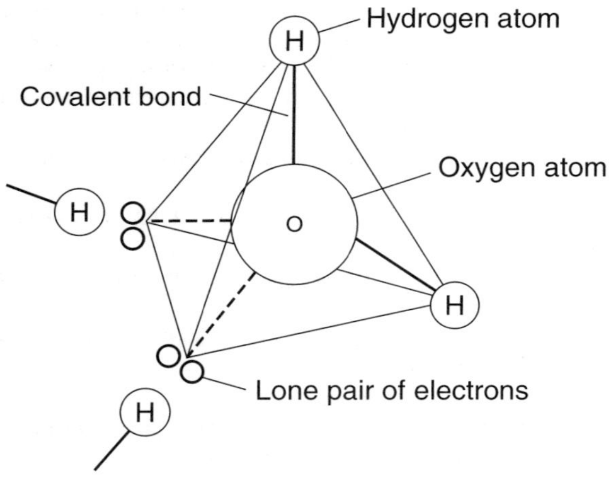

Figure 1. This diagram represents the simplest tetrahedral unit of water's H-bonded network (1). Two of the four hydrogen atoms comprising the tetrahedron are covalently bonded to the central oxygen atom (forming an individual H_2O molecule), whereas the remaining two hydrogen atoms (donated by each of two neighboring water molecules) are hydrogen bonded to the oxygen's lone pairs of electrons. Every water molecule in the network can act as both a hydrogen donor and a hydrogen acceptor in bonding with its nearest neighbors.

which all potential hydrogen bonds were described as either unbroken (ON) or broken (OFF) (2). More recently, scientists have characterized the shuffling (i.e., breaking and forming) of hydrogen bonds comprising liquid water's network as a nonlinear process and, consequently, have been unable to decipher the rules that govern the changing of molecular interconnections. This inability does not imply that the switching rules are haphazard—only that science's observational and mathematical techniques are currently overwhelmed by water's dynamism. Even scientists who study much simpler computer-generated systems have found it difficult to identify switching rules solely by analyzing network patterns and behaviors. Instead, researchers arbitrarily set the switching rules for these artificial systems and then observe corresponding network behaviors.

Whereas network hierarchies within liquid water have only recently been investigated, structural hierarchies are well described. Water exhibits structural self-organization on at least three distinct levels: molecules, clusters of molecules (e.g., hydration shells), and clusters of clusters (e.g., micelles). Self-organization implies that water molecules or molecular clusters arrange themselves into increasingly complex states. It is this self-organization that reportedly permits a network of simple interconnected elements, along with their switching rules, to display coherent behaviors (3). Although the primary water network is held together by short-range hydrogen bonds between neighboring molecules, it may either behave as an extensive system with global coherence or as individual smaller regions with local order (4). These two behaviors have been classically linked to network patterns displaying chaotic or disordered (i.e., bulk liquid water) and more ordered (i.e., water clusters) properties, respectively.

BIOLOGICAL STRUCTURE AND FUNCTION

Molecular biologists originally considered water to be merely an inert solvent within which biomolecules (e.g., proteins and nucleic acids) structured and activated themselves. More recently, water has been perceived as a key structural and functional component of many biomolecules, whose interaction with the surrounding environment is often mediated through *hydration shells*, which are immense water networks surrounding the biomolecules (5). Water releases its hydrogen bonds with biomolecular components so that protein or DNA assembly may begin. Conformational changes in the biomolecule then restructure both the internal (bound or integral) and external (hydrating) water that, in turn, facilitates more changes in the biomolecule. This iterative process appears to repeat until assembly and folding mechanisms yield a bioactive crystal.

In addition to the structural importance of water in biomolecules, the insertion and removal of water also has profound effects on their functioning. One of the most remarkable functions that proteins perform is that of a biochemical catalyst or *enzyme*, which permits chemical reactions to proceed at rates that are millions of times faster than would occur under ambient conditions. Water appears to play several roles in the functioning of enzymes. First, water molecules located at the interface of the protein-substrate complex often provide the free energy that actually binds the enzyme to its substrate. Second, water may be excluded from the active site until the moment of catalysis; at which time, it is allowed to enter the active site and reconfigure hydrogen bonds (permitting the reaction to proceed). Finally, the enzymes actually structure the water network or networks around them, thus enhancing the rate of catalysis (6).

In addition to its role as a structural component of biomolecule, water hydrates DNA as it would any other solute in an aqueous solution. It appears that water is most highly ordered adjacent to the DNA molecule and is less ordered with each successive hydration shell. This highly ordered water is sometimes referred to as *vicinal* or biological water (6). As was the case for proteins, each subunit of the DNA seems to have its own water network that interacts with other networks and, through successive hydration shells, with bulk water. If the geometry of the DNA molecule is contorted, the hydration shells and associated geometric clusters seem to be likewise affected. Similarly, the DNA molecule may contort in response to changes initiated by the water that hydrates it. The reasons for twisting and folding DNA molecules may include stuffing it into chromosomes or unraveling it during transcription and replication.

COMPLEX DYNAMICS

Scientists working with the pentagonal water cluster, which serves as a basic building block for many large hydration shells, observed that water's vibrational states are caused by a combination of bending and stretching motions of its bonds. Molecular vibrations are expressed in the units of cycles per second, or *hertz*. Although water's covalent bonds vibrate at about 10^{14} hertz (one-hundredth of a picosecond per cycle), its hydrogen bonds vibrate within the lower frequency range of 10^{12} to 10^{13} hertz (one-tenth to one picosecond per cycle). The time frames for one cycle of these bond vibrations are shorter than those measured for the exchange hydrogen bonds in water's bulk liquid network (more than one picosecond). One *picosecond* equals a time period of one-trillionth of a second (10^{-12} seconds). In addition to the dynamics associated with vibrating and exchanging bonds, water's molecular network performs several concurrent activities that have been clocked at both faster and slower rates.

As an example, *ultrafast vibrational energy* is transferred in water on timescales less than one-tenth of a picosecond (7). This ultrafast transfer could play a role in liquid water's transporting vibrational energy among certain solutes (e.g., proteins, DNA) and, perhaps, in transferring information within water's local network. Recent spectroscopy studies have suggested that liquid water's network may also function on timescales of less than a *femtosecond* (one-thousandth of a picosecond), which is at least 100 times faster than the ultrafast vibrational energy. Moreover, some of the studies indicate that most molecules in the liquid network form only two (rather than three or four) hydrogen bonds with their neighbors, perhaps creating long chains of molecules within a disordered cluster network (8). In contrast to these sub-picosecond dynamics, the reorientation of water molecules in a spatially restricted environment (e.g., the interface of a micelle) occurs on the much slower timescale of hundreds to thousands of picoseconds. Although further research is required to substantiate the recent observations, models describing both the structure and dynamics of water's molecular network are rapidly evolving.

SEAWATER

Thus far, the discussion of water's network has dealt with either pure water or dilute aqueous solutions in which solutes comprise only a small fraction (e.g., less than 0.1%) of the total solution. By contrast, the salts (ions) in seawater account for about 3.5% of the total solution. Unlike the structure of pure water's bulk phases and clusters, the molecular structure of seawater is poorly described as a result of difficulties associated with employing the investigative techniques that routinely probe the structure of pure water networks. Moreover, the high concentration of ions in seawater is hypothesized to significantly affect the water network. As the percentage of water molecules comprising hydration shells increases, the network structure of bulk water is compromised (see Fig. 2). A salting-out effect is observed at very high ion concentrations, when the interaction between water molecules and ions is sufficiently weakened to permit the formation of salt crystals.

The most pertinent changes in the molecular water network that have been observed or hypothesized in transitioning from dilute aqueous solutions to seawater include (1) the aforementioned ion–ion interactions, (2) overlapping hydration shells in which a single water molecule simultaneously hydrates more than one ion,

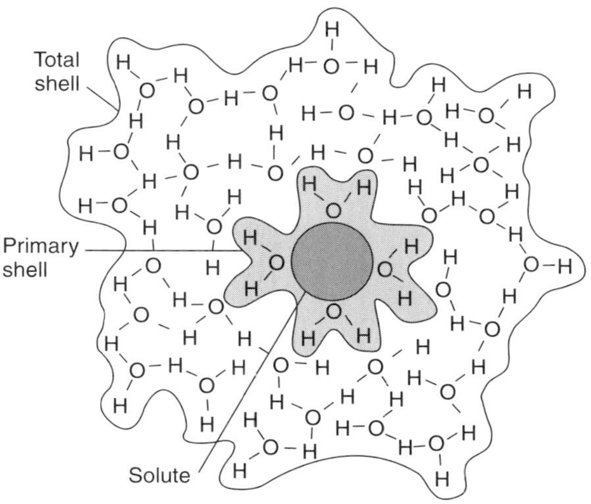

Figure 2. This diagram illustrates the primary and total hydration shells surrounding an ionic, or electrically charged, solute (1). In this case, a positively charged solute (e.g., a sodium ion in seawater) attracts the negatively charged oxygen atoms of surrounding water molecules. A solute's effect on the orientation of water molecules in the liquid network is greatest within the primary shell and diminishes as a function of distance from the hydrated ion. Theoretically, the total hydration shell includes all water molecules in the network that are affected by the solute.

and (3) varying orientations of the water molecules that comprise ion hydration shells (9). These factors are believed to profoundly affect the structure and dynamics of the water network comprising the world's oceans, which contain 97% of the planetary water. Perhaps scientists will better understand the oceans when the secrets of seawater's molecular network dynamics are more fully revealed.

BIBLIOGRAPHY

1. Marrin, D.L. (2002). West. In: *Universal Water*. Inner Ocean Publ., Maui, HI, pp. 100–114.
2. Rice, S. (1975). Conjectures on the structure of amorphous solid and liquid water. *Topics Curr. Chem.* **60**: 109–200.
3. Varela, F., Thompson, E., and Rosch, E. (1991). *The Embodied Mind*. MIT Press, Cambridge, UK, pp. 87–90.
4. Blumberg, R.L., Stanley, H.E., Geiger, A., and Mausbach, P. (1984). Connectivity of hydrogen bonds in liquid water. *J. Chem. Phys.* **80**: 5230–524.
5. Westhof, E. (1993). Structural water bridges in nucleic acids. In: *Water and Biological Macromolecules*. E. Westhof (Ed.). CRC Press, Boca Raton, FL, Chapter 7.
6. Clegg, J. (1979). Metabolism and the intracellular environment: the vicinal-water network model. In: *Cell-Associated Water*. W. Drost-Hansen and J. Clegg (Eds.). Academic Press, New York, pp. 363–413.
7. Nitzan, A. (1999). Ultrafast relaxation in water. *Nature* **402**: 472–475.
8. Zubavicus, Y. and Grunze, M. (2004). New insights into the structure of water with ultrafast probes. *Science* **304**: 974–976.
9. Robinson, G., Wilse, S.B.Z., Singh, S., and Evans, M. (1996). *Water in Biology, Chemistry and Physics*. World Scientific, Singapore, pp. 245–259.

IN SITU CHEMICAL MONITORING

Melissa A. Singer Pressman
Joseph H. Aldstadt, III
Genetic Technologies, Inc.
Testing Institute
Waukesha, Wisconsin

INTRODUCTION

For routine monitoring of environmental systems, *in situ* measurements offer several distinct advantages over laboratory analysis (1). *In situ* analysis allows much higher rates of sample throughput. As a result of the rapid dynamics of most environmental systems, it is important to collect frequent samples to prevent undersampling of the phenomena under study, which is essential to allow for a more accurate characterization of the true dynamics of the system and, therefore, a better understanding of the biogeochemical processes involved (2). Additionally, *in situ* analysis eliminates sample contamination that inevitably develops during the transport and storage of samples to the laboratory. Furthermore, current research efforts in several laboratories have demonstrated that relatively inexpensive monitors can be fielded that can self-calibrate, can be remotely controlled, and can provide reliable measurements in realistic environments (3). In particular, it is the inability to provide high data quality at trace levels in complex matrices that has prevented *in situ* measurements from gaining more favor.

In situ monitors have been classified broadly as either "chemical sensors" or "chemical analyzers." Chemical sensors depend on diffusion to bring the analyte to the transducer surface. This type of sensor is mechanically simple, sensitive, and practical from the standpoint of low power requirements and small footprint, but often suffers from poor selectivity and the inability to self-calibrate. More importantly, sensors are based on reversible chemical reactions, i.e., the interaction of the analyte with the transducer surface, which makes it difficult for these sensors to detect environmentally relevant levels; that is, to detect trace levels of a chemical species, the equilibrium binding (formation) constants must be very large. For the reaction of analyte A with ligand B to form a complex AB, the rate equations are (3):

$$aA + bB \rightarrow AB rate = -(d/dt)\,[A] = -(d/dt)\,[B] = k_1[A]^a[B]^b$$

$$xAB \rightarrow A + B rate = -(d/dt)\,[AB] = k_{-1}[AB]^x$$

The equilibrium constant for the dissociation of the complex is equal to the ratio of the rate constants for the dissociation of the analyte-ligand complex and the formation of the complex, as shown in the following equation (3):

$$K_D = k_{-1}/k_1$$

If the complex dissociates slowly, the time required to return the signal to a steady baseline ("clear down" time) is introduced. One way to avoid this problem is to continuously renew the reagent at the sensor surface. This strategy depends on molecular diffusion to transport the analyte to and from the sensor surface; this type of sensor is called a "renewable reagent sensor" (4–6).

Chemical analyzers, on the other hand, depend on *active* transport of the analyte into and away from the monitor. Although analyzers lack the mechanical simplicity of sensors, the addition of valves and pumps in the design can more importantly provide a means by which to accommodate self-calibration of the monitor. The addition of these mechanical capabilities also allows for multistep, complex chemistries to be used as a way to increase sensitivity and selectivity of the monitor.

Continuous flow analysis (CFA) is a more general term than so-called "chemical analyzers," referring to any measurement process in which the concentration of an analyte is determined based on a steady flow of sample through the instrument (7). Nonchromatographic CFA-based measurement systems are widely used for industrial process control and environmental monitoring. As the instrument is often placed in a hostile environment in these types of applications, the relative simplicity of the CFA approach is attractive (8). Initial attempts to use CFA-based instruments were direct adaptations of benchtop (i.e., laboratory) CFA systems that were ruggedized (e.g., water-proofed) and taken to the sampling site (9). This approach was burdensome, expensive, and far from robust.

An example of an early effort of this type is illustrated by the work of Elliott et al. in using a segmented continuous flow analyzer, often called an "autoanalyzer," for shipboard chemical analyses (10). These researchers monitored nitrate concentrations in the Everglades using a commercially available autoanalyzer that they modified for the nitrate derivatization chemistry. In their design, air bubbles were created in-line to segment the liquid sample stream as a means to prevent uncontrolled dispersion

More recently, *in situ* CFA monitors have been developed that are not merely ruggedized bench-top instruments. One of the more interesting instruments, and the model which inspired this research, was described in work published in 1994 by Jannasch et al. at the Moss Landing Marine Laboratory in Monterey, CA (11). This group of chemical oceanographers was interested in accurately measuring nitrate ion in seawater with high sampling frequency. To accomplish this goal, they designed a CFA monitor "from the ground up" and incorporated several novel features. In doing so, they created a new paradigm in the science of chemical measurements.

In the system developed by Jannasch et al., a well-characterized, irreversible colorimetric method was used to measure nitrate. The dissolved nitrate was determined by reduction to nitrite in a neutrally buffered (imidazole) solution by an in-line copperized cadmium surface. The nitrite then merged with an incoming reagent stream containing sulfanilamide (*p*-aminobenzenesulfonamide or

PABSA) in acid to undergo reaction to form a diazonium ion. This species was then reacted with *N*-(1-naphthyl)ethylenediamine dihydrochloride (NED) to form an azo dye of high molar absorptivity that was detected by molecular absorbance spectroscopy.

The novel *in situ* monitor that the Moss Landing group developed used nonmechanical osmotic pumps to both deliver reagents to the system and to propel the continuous flow sample stream through the microscale manifold (~1 mm diameter conduits) (11). For the two reagent osmotic pumps (i.e., for PABSA and NED), the flow was driven by the osmotic pressure differential developed across the semipermeable membrane that separated a saturated brine solution from a solution with deionized water (12). The pressure created from the diffusion of deionized water across the semipermeable membranes resulted in compression of inner bags containing the reagents and their pulse-free delivery to the manifold (at ~4 μL per hr). For the sample pump, conventional osmotic pumps were modified by removing the inner bags and mounting the rigid permeable membranes in an enclosed container. Four of these modified osmotic pumps were then coupled to draw sample through the manifold at 12 μL per hr.

Microscale flow manifolds can be fabricated by "cold-forging" PVC: A wire bent in the desired flow microconduit configuration was pressed onto the surface of a flat-milled sheet of PVC. Following removal of the wire, ~0.5-mm diameter holes were drilled into the plate for connection of components. The optical system in the Moss Landing nitrate analyzer used a simple light-emitting diode (LED) source with photodiode detector, with data collected by a low-power programmable datalogger. The monitor exhibited a detection limit of 0.11 μM and a linear response to 20 μM nitrate. Measurements were taken every 2 min and the response time was ~30 min. Quality control samples were measured automatically every 48 hours. Jannasch et al. demonstrated the *in situ* monitor's stability by operating it continuously for month-long periods in large tanks at the Monterey Bay Aquarium.

The Moss Landing group's research incorporated several novel features—mainly the microscale flow design using nonmechanical osmotic pumps for reagent delivery and sample intake—into an *in situ* monitor capable of frequent and reliable measurements in environmentally complex samples. In addition, through the periodic addition of nitrate standards, the monitor automatically self-calibrated by correcting for baseline drift. Another advantage to this system was its flexibility; using different irreversible chemistries, the *in situ* monitor could be readily adapted to measure other analytes for optical detection and quantitation. A drawback of the system was that it detected relatively high levels of analyte (i.e., micromolar levels), albeit levels that are relevant to studying this important nutrient. Concerns over particles in the water eventually clogging the microconduit channels are less serious, because small particles (approx. <50 μm) cannot be drawn into the manifold at the low flow rates used.

In 1999, Byrne and coworkers at the University of South Florida developed a monitor based on long path-length absorbance spectroscopy (LPAS), which could be used for the determination of nitrate and several other biologically important chemicals in seawater at much lower levels. These workers used the same well-characterized chemistry as Jannasch et al., but they benefited from a 1-m liquid core waveguide as the absorbance flow cell. Another difference with the Byrne group's approach was in the use of mechanical (peristaltic) pumps instead of osmotic pumps—resulting in a five-fold increase in response time. This monitor was field tested recently in the Gulf of Mexico, and detection limits for nitrate were observed to be approximately 2 nM (13,14).

The next key development was reported recently by Worsfold and coworkers at the University of Plymouth. A portable *in situ* monitor for the determination of trace phosphate in surface waters was developed (15,16). This work demonstrated a monitor that was built with a forged plate similar in design to the work done by Jannasch et al. The phosphate monitor had a throughput of ~225 determination per hr and a detection limit of 0.15 μM with 2.0% RSD ($n = 9$) reported. This monitor was evaluated by mapping the phosphate concentration in Port Phillip Bay in Australia for a brief (2 hr) period. The agreement between determinations obtained using the *in situ* monitor and data obtained using manual sampling aboard the ship followed by laboratory analysis was within ±10%.

INSTRUMENT DEVELOPMENT FOR *IN SITU* MONITORING OF CR(VI)

For the *in situ* measurement of relatively low levels of important biological nutrients such as nitrate or phosphate, the work of the Jannasch, Byrne, and Worsfold groups (and others) was pioneering. However, clearly the next challenge in developing *in situ* monitoring instruments was the ability to measure relevant levels of trace metal ions in surface waters by the *in situ* approach. Hexavalent Cr was chosen as the target analyte because it is regulated at nanomolar levels (low parts per million) in surface waters. In addition, the metropolitan-industrial areas of southern Lake Michigan, specifically the Milwaukee Harbor area, contain industrial sources that generate a concentration of total Cr in surface water that ranges from 0.1 to greater than 100 μg/L (17). These reasons, coupled with the well-characterized spectrometric chemistry (18–21), made Cr a logical choice as a representative analyte to study for the development of an *in situ* monitor that measures trace metals in the environment.

A benchtop method for determining hexavalent chromium in high particulate-containing surface waters by sequential injection (SI) was developed (22). For utility in particulate-rich waters, the relative performance of two membrane-based methods for sampling was compared. The first membrane approach was based on a commercial design known as the "supported capillary membrane sampler (SCMS)" that uses tubular membranes; the second approach was based on a conventional parallel-plate dialyzer (PPD) design that uses planar membranes. The membranes were evaluated using the colorimetric method for the determination of hexavalent chromium by complexation with 1,5-diphenylcarbazide (DPC). Thin-walled (~200 μm) microporous (pore size ~0.2 μm) polypropylene membranes were equilibrated with DPC during each sampling period. Formation of the DPC-Cr(VI) complex allowed for efficient membrane transport; without the membrane, Cr(VI) transport decreased by ~90%. Factors optimized included reagent concentration, sampling time, flow rate, and spectrophotometric conditions. Optimal conditions for the reagents were 2.00 mM DPC and 0.100 M nitric acid, and sampling times of 600 s and 900 s for the planar and tubular designs, respectively. The planar (PPD) design increased the sensitivity relative to the tubular (SCMS) design by ~225%. The PPD-SI method was applied to the determination of dissolved Cr(VI) in high particulate-containing surface water samples. Figures-of-merit included a detection limit of <20 μg/L, precision of 1.1% relative standard deviation at 100 μg/L ($n = 4$), and selectivity for dissolved Cr(VI) in several surface water samples with high levels of particulate matter and potential interferences.

The optimized method conditions from the SI studies was then applied to the design, fabrication, and testing of an *in situ* remote monitor for the measurement of sub-part per million levels in surface waters (23,24). The monitor was based on a continuous flow analysis (CFA) design in which a narrow conduit (1.0 mm) manifold was hot-forged between Plexiglas plates. A miniature low-power peristaltic pump was used to draw sample (1.5 mL/hr) into the manifold where chromium (VI) in the sample was derivatized to its diphenylcarbazide complex. The absorbance of the complex was then measured in a light-emitting diode/photodiode detection cell. Solenoid-actuated osmotic pumps were used to introduce quality control standards at user-determined periodicities to achieve acceptable data quality. Figures-of-merit included a detection limit of 0.25 mg/L (7 min cycle time), precision of 2.5% relative standard deviation at 1 mg/L ($n = 3$), and selectivity for dissolved Cr(VI) in several surface water samples with moderate levels of particulate matter. Continuous testing of the monitor over a three-week period using Milwaukee Harbor water fortified with Cr(VI) yielded results that were in close agreement (±5%) with an ICP-MS reference method.

BIBLIOGRAPHY

1. Hanrahan, G., Ussher, S., Gledhill, M., Achterberg, E.P., and Worsfold, P.J. (2002). High temporal and spatial resolution environmental monitoring using flow injection with spectroscopic detection. *TrAC, Trends Analytic Chem.* **21**: 233.
2. Rawls, R. (1998). Wanted: instruments to view ocean's dynamics. *Chem. Engineer. News.* **31**.
3. Johnson, K.S. and Jannasch, H.W. (1994). Analytical chemistry under the sea surface: monitoring ocean chemistry *in situ*. *Naval Research Rev.* **3**: 4.
4. DeGrandpre, M.D., Baehr, M.M., and Hammar, T.R. (1999). Calibration-free optical chemical sensors. *Analytical Chem.* **71**: 1152.
5. Inman, S.M., Stromvall, E.J., and Lieberman, S.H. (1989). Pressurized membrane indicator system for fluorogenic-based fiber-optic chemical sensors. *Analytica. Chimica. Acta.* **217**: 249.
6. Degrandpre, M.D. (1993). Measurement of seawater pCO2 using a renewable-reagent fiber optic sensor with colorimetric detection. *Analytical Chem.* **65**: 331.
7. Ruzicka, J. and Hansen, E.H. (1988). *Flow Injection Analysis*, 2nd Edn. John Wiley & Sons, New York.
8. Johnson, K.S., Coale, K.H., and Jannasch, H.W. (1992). Analytical chemistry in oceanography. *Anal. Chem.* **64**: 1065A.
9. Valcarcel, M., Cardenas, S., and Gallego, M. (2002). Continuous flow systems for rapid sample screening. *Trac-Trends Analytic Chem.* **21**: 251.
10. Elliott, C.L., Snyder, G.H., and Cisar, J.L. (1989). A modified Autoanalyzer-II method for the determination of NO_3^- in water using a hollow-Cd reduction coil. *Communications in Soil Science and Plant Analysis* **20**: 1873.
11. Jannasch, H.W., Johnson, K.S., and Sakamoto, C.M. (1994). Submersible, osmotically pumped analyzers for continuous determination of nitrate in-situ. *Analytic Chem.* **66**: 3352.
12. Pump, A.O. (2000). *Technical Note*. Alzet, Inc.
13. Kaltenbacher, E.A., Byrne, R.H., and Steimle, E.T. (2001). Design and applications of a chemical sensor compatible with autonomous ocean-sampling networks. *IEEE J. Oceanic Engineer.* **26**: 667.
14. Bryne, R.K. (1999). Autonomous *in situ* analysis of the upper ocean. *Sea Technol.* **71**.
15. Hanrahan, G., Gledhill, M., Fletcher, P.J., and Worsfold, P.J. (2001). High temporal resolution field monitoring of phosphate in the River Frome using flow injection with diode array detection. *Analytica. Chimica. Acta.* **440**: 55.
16. Lyddy-Meaney, A.J., Ellis, P.S., Worsfold, P.J., Butler, E.C.V., and McKelvie, I.D. (2002). A compact flow injection analysis system for surface mapping of phosphate in marine waters. *Talanta.* **58**: 1043.
17. Environmental Defense Fund. (2004). Available: http://www.scorecard.org/.
18. Miller-Ihli, N.J. (1992). Chromium. *Technol. Instrum. Anal. Chem.* **12**: 373.
19. Lynch, T.K. and Wilson, J. (1984). Speciation of metals in solution by flow injection analysis. Part 1: Sequential spectrophotometric and atomic absorption detectors. *Analyst.* **109**: 839.
20. Losi, M.E., Amrhein, C., and Frankenberger, W.T. (1994). *Rev. Environ. Contamin. Toxicol.* **136**: 91.
21. Kotas, J. and Stasicka, Z. (2000). Chromium occurrence in the environment and methods of its speciation. *Environ. Pollut.* **107**: 263.
22. Pressman, M.A. and Aldstadt, J.H. (2003). A comparison of diffusion samplers for the determination of hexavalent chromium by sequential injection spectrophotometry. *Microchem. J.* **74**: 47–57.
23. Pressman, M.A. (2003). *Development of Novel Analytical Techniques and Methods for the Determination of Trace Chromium in the Environment*, Ph.D. dissertation, University of Wisconsin-Milwaukee.
24. Pressman, M.A. and Aldstadt, J.H. (2005). *Development and Testing of an in situ Chemical Monitor for Quantitation of Hexavalent Chromium in Natural Waters*, in preparation.

NITROGEN

Kenneth F. Steele
Kristofor R. Brye
University of Arkansas
Fayetteville, Arkansas

INTRODUCTION

Nitrogen (N) is ubiquitous in the environment and in the correct form is an essential nutrient for plants and animals. About 26.5% of Earth's nitrogen is in rocks and essentially all of the remainder (73%) is in the atmosphere as an inert monotomic gas (N_2) that comprises 78% of the atmosphere (1). Only a small amount of nitrogen is present in the hydrosphere and biosphere, but nitrogen is a crucial nutrient there. For nitrogen to be used by organisms, it must be in a usable form; it must be combined or "fixed" with oxygen or hydrogen. The most common fixed form of nitrogen in ground water is the anionic oxide—nitrate (NO_3^-), and the second most common form is the cationic form–ammonium (NH_4^+). Approximately 85% of the nitrogen in stream water is in the form of organic N and most of the rest (NO_3^- and NH_4^+) is derived from the decomposition of organic matter (2). N_2 is converted to organic nitrogen predominantly by microorganisms and certain plants. About 96% of the nitrogen in soil is bound in organic matter, and nitrate is a product of organic matter degradation (3). Nitrogen occurs in five oxidation states, −3, 0, +1, +3 and +5, that are primarily the result of bacteria-induced oxidation. The nitrogen species are listed in Table 1.

NITROGEN CYCLE

Fixation by plants can occur; leguminous plants convert N_2 to organic nitrogen compounds, but most plants cannot use N_2 directly. Most plants require atmospheric nitrogen converted to ammonium or nitrate ions (4). Organic nitrogen is not directly available to plant roots and has to be mineralized in the soil by a chain of microbial processes (Fig. 1). The release of ammonium and ammonia (total ammonia) from organic matter is referred to as mineralization (3) or ammonification. Total ammonia is readily oxidized to nitrate. The conversion of nitrogen gas to fixed form and from fixed form to nitrogen gas includes four main conversion reactions that form the essential parts of the nitrogen cycle (Fig. 1):

1. fixation
2. ammonification
3. nitrification
4. denitrification

Table 1. Common Forms of Nitrogen in the Environment and Typical Concentrations of Nitrogen

Species	Oxidation State	Typical Source	Typical Concentration in Water as N
N_2	0	Atmosphere	14 mg/L
NO_3^-	+5	Oxidation of organic matter/ atmosphere	<0.6 mg/L background to 250 mg/L for contaminated sources
NO_2^-	+3	Intermediate form in nitrification, denitrification, and reduction of NO_3^-	<0.1 mg/L
NH_3	−3	Fixation of N_2	<0.01 mg/L
NH_4^+	−3	Ionization of NH_3	<0.01 mg/L
N_2O	+1	Reduction of NO_2^-	Negligible

Fixation

For N_2 gas to be used by plants, it must first be fixed with hydrogen to form ammonium ions. Fixation is a microbial process, and even the few types of plants that can use N_2 directly are ultimately dependent on microbes also. For example, legumes can use N_2 because of the symbiotic relationship between the plant and *Rhizobium* bacteria that live in nodules in the plant roots. Other means of fixation include conversion of atmospheric N_2 to nitrate by lightning and industrial processes, such as the Haber–Bosch process for producing ammonia (NH_3 gas) (4).

Ammonification

Ammonification (also referred to as mineralization), the process of converting organic nitrogen (e.g., urea) to ammonium (Eq. 1), generally occurs during decomposition of animal and plant matter (4). Dissolution of ammonium carbonate releases ammonium ions:

$$\underset{\text{Urea}}{H_2NCONH_2} + 2H_2O \longrightarrow \underset{\text{Ammonium carbonate}}{(NH_4)_2CO_3} \quad (1)$$

As indicated by Eqs. 2 and 3, the relationship between NH_3 and NH_4^+ is pH dependent. Under alkaline conditions (pH > 9), formation of NH_3 is favored due to neutralization of NH_4^+ by hydroxide (OH^-) (Eq. 2). In acid conditions (Eq. 3), ammonia gas is transformed into

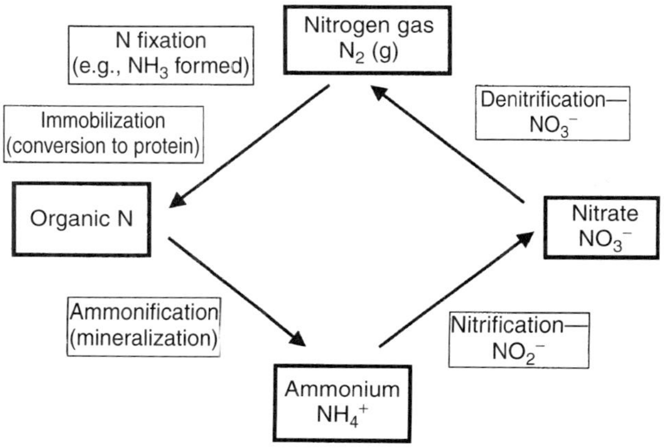

Figure 1. Simplified nitrogen cycle.

ammonium ions. For example, at 15 °C, ammonia gas makes up 0.027% of total ammonia ($NH_3 + NH_4^+$) at a pH of 6.0, but makes up 2.7% of total ammonia at a pH of 8.0. Although 2.7% is a small amount, it is a hundredfold increase of ammonia (5).

$$NH_3 + H_2O = NH_4^+ + OH^- \quad (2)$$

$$NH_3 + H^+ = NH_4^+ \quad (3)$$

Other factors that affect the concentration of NH_3 in water are temperature and ionic strength. The concentration of NH_3 increases as temperature increases and decreases as ionic strength increases. At 15 °C and pH 8.0, the percent NH_3 in solution is 2.7%, whereas, at 30 °C and pH 8.0, the concentration increases to 7.5% (5). The effect of ionic strength is minor for fresh to slightly saline waters.

Nitrification

Nitrification is the biological oxidation of ammonium ions to nitrite followed by conversion of nitrite to nitrate. Biological oxidation is carried out by *Nitrosomonas* and *Nitrobacter* bacteria (6,7) as shown in Eqs. 4 and 5. In addition to these two bacteria, (8) indicate that several different autotrophic organisms mediate nitrification:

$$2NH_4^+ + 3O_2 \xrightarrow[\text{bacteria}]{\textit{Nitrosomonas}} 2NO_2^-(\text{nitrite}) + 4H^+ + 2H_2O \quad (4)$$

$$2NO_2^- + O_2 \xrightarrow[\text{bacteria}]{\textit{Nitrobacter}} 2NO_3^-(\text{nitrate}) \quad (5)$$

Because oxidation of nitrite to nitrate is a rapid reaction and nitrite is metastable, nitrite is not abundant (Table 1) (9). Note that if both oxygen and nitrate are present in soil, bacteria preferentially use oxygen because it yields more energy (1). Therefore, nitrification typically occurs in oxic conditions, whereas denitrification occurs in anoxic conditions.

Denitrification

Denitrification is the biological reduction of nitrate to nitrogen gas by heterotrophic bacteria, such as *Pseudomonas denitrificans* and *Thiobacillus denitrificans* (8). These bacteria require organic carbon for energy (Equations 6 and 7).

$$NO_3^- + \text{organic carbon} \xrightarrow{\text{Heterotrophic bacteria}} NO_2^- \quad (6)$$

$$NO_2^- + \text{organic carbon} \xrightarrow{\text{Heterotrophic bacteria}} N_2 + CO_2 + H_2O \quad (7)$$

Because oxidation of organic carbon by oxygen yields more energy, generally the oxygen concentration must be very low (anoxic conditions) for nitrate to be used (4,6).

Analytical Terminology

Analytically, organic nitrogen (proteins, peptides, nucleic acid, urea, and other compounds) and ammonia are determined together and are referred to as "Kjeldjahl nitrogen," a term that refers to the analytical method used. By expressing all nitrogen species as N, the concentrations of the various species can be added and subtracted. Organic N (−3 charge) is determined by analyzing ammonia by other methods and subtracting this value from the Kjeldhal nitrogen concentration (10).

Inorganic N is the sum of nitrate, nitrite, and total ammonia (ammonia plus ammonium). Some analytical methods determine the sum of nitrate and nitrite, but the results are sometimes incorrectly reported as nitrate because nitrite concentrations are considered very low. Total N is the sum of organic N and inorganic N.

AQUATIC NITROGEN

Despite the large proportion of nitrogen in the atmosphere, its abundance in water is low-measured as a few mg/L N, mostly in the inert form of nitrogen. Typically, the most environmentally important form of nitrogen is nitrate; however, dissolved organic nitrogen makes up the majority of total dissolved nitrogen in watersheds in New England (11). Note that for primary producers ammonium may be more available to organisms than nitrate (12). The various common species of nitrogen in the environment and the estimated concentration of each are presented in Table 1. Mueller and others (13) have demonstrated that background concentrations of nitrate as N are less that 0.6 mg/L.

Nitrogen species are subject to oxidation/reduction (redox) and reactions with hydrogen ions (pH). In Fig. 2, the NO_3^-/NO_2^- pe–pH boundary is dotted to indicate that NO_2^- is metastable. Nitrate, at a pe above the NO_3^-/NO_2^- boundary, is reduced at the NO_3^-/N_2 boundary to form nitrogen. Thus, there is little NO_3^- available to form NO_2^-. Because nitrite is a metastable species, it occurs only in low concentrations in natural waters where denitrification is occurring (9). Nitrate is present in the N_2 field, but under natural conditions, it is present in concentrations less than those of N_2. Ammonium ions are the dominant nitrogen species at low pe and pH. Ammonia is the dominant species above pH 9 (Fig. 2) and thus is not present in significant concentrations in natural waters; most natural waters have a pH range of 5–9. Dissolved organic N can also be relatively mobile.

Mobility

Nitrate is very soluble in water and because of its negative charge, is not adsorbed by humus or clay. Nitrate not taken up by plants can be transported from the upper soil zone to lower soil zones and eventually to groundwater. Ammonium ions are not nearly as mobile as nitrate ions, in part, because their positive charge allows their transport to be retarded by adsorption on humus and clay. In some situations, ammonium ions can be transported to deep soil horizons (3).

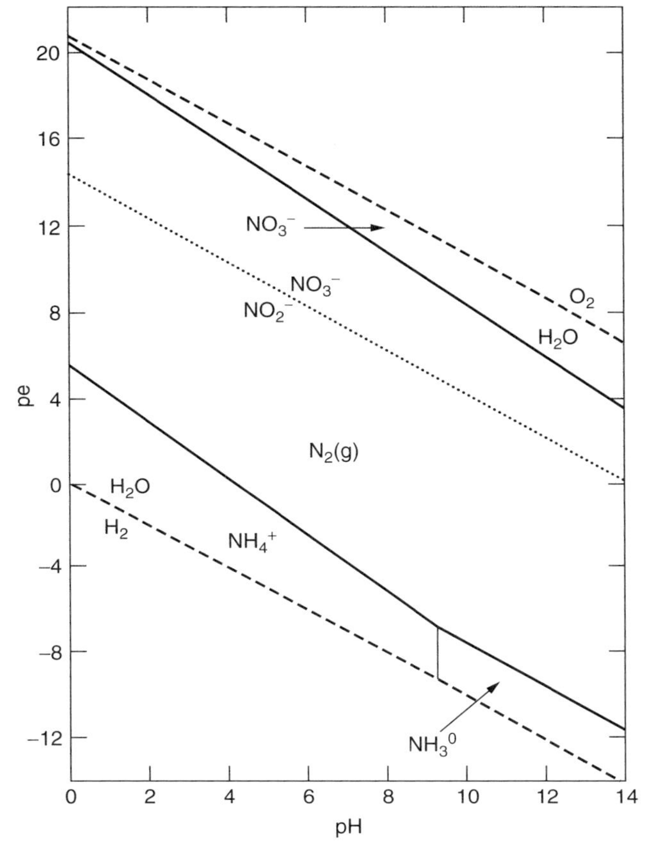

Figure 2. Stability fields for common nitrogen species in natural waters. The diagram is based on the following conditions: temperature of 25 °C, partial pressures of H_2 and O_2 are each 1 atmosphere, partial pressure of N_2 is 0.77 atmosphere, and total aqueous N concentration is 0.001 moles per liter.

Toxicity

Ammonia is a pungent, colorless gas that is very soluble in water and is toxic to fish at concentrations greater than 0.2 mg/L. Dissolved ammonia reacts with water to form ammonium ions, which are not nearly as toxic (1/50th) as the gas (3). The ammonia–ammonium equilibrium is controlled to a large extent by temperature and pH; only highly alkaline waters favor the formation of ammonia. The pH range of most natural waters is such that ammonium ions are dominant compared to ammonia gas (Fig. 2). The U.S. Environmental Protection Agency bases its ammonia criteria for water on a complex set of equations that include the effect of temperature and pH on ammonia concentration. The equations were developed to account for acute and chronic toxicity to particular types of fish. In addition, the criteria are based on specific periods of time and the frequency of exceeding these criteria (14).

Typically nitrate is toxic to warm-blooded animals only where large amounts of nitrite are formed by the reduction of nitrate. Nitrate can be reduced in the gastrointestinal tract to nitrite, especially in infants. Nitrite enters the bloodstream and reacts with hemoglobin to produce methemoglobin that reduces the number of hemoglobin cells available for oxygen exchange. These reactions can be life threatening for children less than 3 years old. Babies who consume water with nitrate concentrations as low as 1 mg/L as N risk methemoglobinia or "blue baby disease" because of the impairment of oxygen transport. There are also indications that nitrate (as related to nitroso compounds) may be associated with gastric cancer and perhaps liver and kidney cancer. The relative consumption of nitrate from water compared to consumption of dietary nitrate is a factor that concerns researchers attempting to determine the risk of cancer from water (15,16). The maximum concentration limit (MCL) of 10 mg/L N for nitrate in drinking water is based on minimizing methemoglobinia and is considered safe concerning cancer.

CONTAMINATION

The causes of pollution of water are divided into point sources and nonpoint sources. Nitrogen contamination of water is generally related to nitrate from sewage and septic systems (point sources) and from agricultural sources. The agricultural sources are commercial and animal manure fertilizers (nonpoint sources) (17,18) and from confined animal operations (point sources). Agriculture is generally considered the main source of nitrogen pollution of water (19). During rainstorms, streams often exhibit a dilution of nitrate concentration that indicates that most of the nitrate is transported to these streams by groundwater. Excessive nutrients in runoff can result in eutrophication of surface waters, for example, the Neuse River in North Carolina, U.S.A. (20). Best management practices have been developed to minimize the negative impacts of fertilizers (17).

ISOTOPES

There are two stable isotopes of nitrogen, ^{14}N and ^{15}N. Although the whole earth abundance of ^{14}N is 99.63‰ of total nitrogen, due to fractionation, the ^{15}N:^{14}N ratio is 0.003677 (21) in the atmosphere. Differences in the ratio of the two isotopes in the various nitrogen species are a result of the faster chemical reaction of the lighter isotope than the heavier isotope. Isotopic ratios of nitrogen are expressed as

$$\delta^{15}N = [(R_{sample}/R_{air}) - 1] \times 1000$$

where R is the abundance ratio of ^{15}N to ^{14}N. The international standard for $\delta^{15}N$ is air, which by definition has a $\delta^{15}N$ of 0‰. Values greater than 0‰ indicate enrichment in ^{15}N, and values less than 0‰ indicate depletion of ^{15}N. Sources of nitrate have been determined using $\delta^{15}N$. The premise is that agricultural activity and sewage (including septic tank effluent) may have distinct $\delta^{15}N$ values. A definitive determination of source is sometimes thwarted by the possibility of overlapping $\delta^{15}N$ values between possible sources and changes in δ^{15} N due to volatilization of NH_3 from the soil and denitrification of nitrate in water. The origin of the nitrate in water can be found by using $\delta^{18}O$ to recognize different sources of nitrate and the influence of denitrification on $\delta^{15}N$ (21,22). A number of studies have had success using N and often O isotopes to determine the source of nitrate contamination (23,24).

BIBLIOGRAPHY

1. Hitchon, B., Perkins, E.H., and Gunter, W.D. (1999). *Introduction to Ground Water Geochemistry*. Geoscience, Alberta, Canada, pp. 195–200.
2. Berner, E.K. and Berner, R.A. (1987). *The Global Water Cycle*. Prentice-Hall, Inc., Englewood Cliffs, NJ, p. 397.
3. Canter, L.W. (1996). *Nitrates in Groundwater*. CRC Lewis, Boca Raton, FL.
4. Jeurgens-Gschwind, S. (1989). Ground water nitrates in other developed countries (Europe)–relationships to land use patterns. In: *Nitrogen Management and Ground Water Protection*. R.F. Follett (Ed.). Elsevier, New York, pp. 75–138.
5. U.S. EPA. (1976). *Quality Criteria for Water*. U.S. Environmental Protection Agency, Washington, DC, pp. 10–13.
6. Manahan, S. (1999). *Environmental Chemistry*. CRC Press, Boca Raton, FL.
7. Stumm, W. and Morgan, J.J. (1996). *Aquatic Chemistry*, 3rd Edn. John Wiley & Sons, New York, pp. 927–934.
8. Kendall, C. and Aravena, R. (1999). Nitrate isotopes in groundwater systems. In: *Environmental Tracers in Subsurface Hydrology*. P.G. Cook and A.L. Herczeg (Eds.). Kluwer Academic, Boston, pp. 261–298.
9. Kehew, A.E. (2001). *Applied Chemical Hydrogeology*. Prentice-Hall, Upper Saddle River, NJ.
10. Eaton, A.D., Lenore, C.S., and Greenberg, A.E. (Eds.). (1995). *Standard Methods for the Examination of Water and Wastewater*. American Public Health Association, Washington, DC, pp. 4-75–4-96.
11. Campbell, J.L. et al. (2000). Dissolved organic nitrogen budgets for upland, forested ecosystems in New England. *Biogeochemistry* **49**: 123–142.
12. Peterson, B.J. et al. (2001). Control of nitrogen export from watersheds by headwater streams. *Science* **292**: 86–90.
13. Mueller, D.K., Hamilton, P.A., Helsel, D.R., Hitt, K.J., and Ruddy, B.C. (1995). Nutrients in ground water and surface water of the United States—an analysis of data through 1992. U.S. Geological Survey Water-Resources Investigations Report 95-4031, p. 74.
14. U.S. EPA. (1999). *1999 Update of Ambient Water Quality Criteria for Ammonia. U.S. Environmental Protection Agency*. National Service Center for Environmental Publications, Cincinnati, OH.
15. Bogardi, I. and Kuzelka, R.D. (Eds.). (1991). *Nitrate Contamination: Exposure, Consequence and Control*. NATO ASI Series. Springer-Verlag, New York.
16. Follett, R.F. and Walker, D.J. (1989). Ground water quality concerns about nitrogen. In: *Nitrogen Management and Ground Water Protection*. R.F. Follett (Ed.). Elsevier, New York, pp. 1–22.
17. Steele, K.F. (Ed.). (1995). *Animal Waste and the Land–Water Interface*. CRC Press, Lewis, Boca Raton, FL.
18. Steele, K.F. and McCalister, W.K. (1991). Potential nitrate pollution of ground water in limestone terrain by poultry litter, Ozark Region, USA. In: *Nitrate Contamination: Exposure, Consequence and Control*. I. Bogardi and R.D. Kuzelka (Eds.). NATO ASI Series, Springer-Verlag, New York, pp. 209–218.
19. Spalding, R. and Exner, M. (1993). Occurrence of nitrate in groundwater—A review. *J. Environ. Qual.* **22**: 392–402.
20. Paerl, H.W., Pinckney, J.L., Fear, J.M., and Peierls, B.L. (1998). Ecosystem responses to internal and watershed organic matter loading: Consequences for hypoxia in the eutrophying Neuse River estuary, North Carolina, USA. *Mar. Ecol. Prog. Ser.* **166**: 17–25.
21. Clark, I.D. and Fritz, P. (1997). *Environmental Isotopes in Hydrogeology*. Lewis, CRC Press, Boca Raton, FL, pp. 150–152.
22. Ostrom, N.E., Peggy, H., and Ostrom. (1999). Nitrogen isotopes. In: *Encyclopedia of Geochemistry*. C.P. Marshall and R.W. Fairbridge (Eds.). Kluwer Academic, Dordrecht, The Netherlands, pp. 431–433.
23. Karr, J.D., Showers, Hinson, W.J., and Thomas, H. (2002). Nitrate source identification using d ^{15}N in a ground water plume near an intensive swine operation. *Ground Water Monitoring and Remediation* 68–75.
24. Aravena, R., Evans, M.L., and Cherry, J.A. (1996). Stable isotopes of oxygen and nitrogen in source identification of nitrate from septic systems. *Ground Water* **31**: 180–186.

OSMOSIS-DIFFUSION OF SOLVENT OR CAUSED BY DIFFUSION OF SOLUTES?

KARL ERIK ZACHARIASSEN
Norwegian University of Science and Technology
Trondheim, Norway

Osmosis is the movement of water through a semipermeable membrane separating two solutions to the solution with the higher solute concentration. The membrane is semipermeable in the sense that it is permeable to water, but not to the solutes. The osmotic movement of water can be blocked by the application on the high concentration side of a hydrostatic pressure, which counteracts the so-called osmotic pressure that moves the water through the membrane.

The prevailing theory ascribes osmosis to a diffusion of water molecules through the membrane down a water concentration gradient across the membrane. The solutes are supposed to dilute the water, and the dilution will be strongest at the side with the highest solute concentration.

However, several investigators claim that this explanation fails to explain important characteristics of osmotic phenomena. Osmotic movement of water is considerably faster than should be expected from a diffusive movement of water. In addition, the water concentration of solutions of different solutes are not correlated with the osmotic activities of the solutions. Furthermore, the osmotic pressure is related to the total concentrations of dissolved particles, and van t'Hoff pointed out that the number of solute particles and the solution volume are related to the osmotic pressure in the same way as the number of molecules, pressure, and volume of a gas, i.e., osmotic phenomena, follow the universal gas laws. No theory has been proposed to explain the agreement between osmotic phenomena and gas laws on the basis of diffusion of water down a water concentration gradient.

In 1903, George Hulett (1) published a theory that explains the agreement between osmotic phenomena and gas laws and that is in agreement with the other observations on osmotic phenomena. Hulett emphasized that the solute particles diffuse inside solutions in the same way as gas molecules diffuse within the space in which the gas is distributed. In the same manner, as gas molecules create a pressure when they collide with the gas boundary, solute particles create a pressure when they collide with the boundary of the solution. Both for gases and solutions this pressure will tend to cause the volume to expand, which is easy for a gas if the boundaries can yield to the pressure and allow expansion of the volume. However, the water in which the solutes of a solution are distributed is essentially incompressible, and the expansive pressure created by the diffusing solutes cannot lead to an expansion unless more water can enter the solution and increase the solution volume. The semipermeable membrane of osmotic processes allows such an influx of water from a solution where a lower solute concentration creates a lower expanding pressure, whereas the solutes that create the pressure are prevented from passing. Water will subsequently move across the membrane until the pressures on both sides are equal.

According to the gas laws, an ideal 1 molar solution (1 mole of particles dissolved in a volume of 1 L), which is separated from pure water by a semipermeable, will create a pressure of 22.1 atm, which corresponds to a water column of impressive 224 m. It will have a melting point of $-1.86\,^{\circ}$C, and a solution with this osmotic concentration is said to be a 1 osmolar solution.

Proportionality exists among the osmolality, the osmotic pressure, and the melting point depression. This proportional relationship with osmotic pressure is also displayed by vapor pressure and boiling point elevation. Because of the exact proportionality that exists among these parameters, they are frequently referred to as the colligative parameters (from the Latin *colligare* = bind together).

A closer inspection of the relationship between mole-based concentration measures and the colligative properties reveals that it is the molal concentration that is most closely related to the colligative parameters. For this reason, some investigators prefer the term *osmolality*.

UNITS FOR OSMOTIC CONCENTRATIONS

Some confusion exists about the units for osmotic concentration. The officially approved unit for osmolality is Osmoles/kg water. Some investigators find that this unit is misleading, because it invites us to consider an Osmole as a defined amount of solute, which when diluted in 1 kg of water, gives an osmolality of 1 Osmole/kg water. Accordingly, when diluted in 2 kg of water, the same osmole of solutes should give an osmolality of 0.5 Osmole/kg water, but this is never the case. An Osmole is not a defined amount of solute, and the concept of osmolality has a meaning only for solutions. For this reason, some investigators use the unit Osm, where the letter *m* emphasizes the close relationship between osmotic and molal concentrations, which have *m* as the official unit.

To add to the confusion, the official unit for osmolality has led some authors to introduce Osmoles/L solution as a unit for osmolarity. However, the melting point is claimed to be $-1.86\,^{\circ}$C both for a 1 osmolal and for a 1 osmolar solution. Hence, the distinction between units for osmolality and osmolarity is merely semantic and of no practical consequence. Osmolality and osmolarity are different terms for the same thing, and furthermore, they are just quasi-measures for concentration. Solutions of different solutes with the same molar or molal concentrations all have different osmolalities, and for a given type of solute, no exact proportionality exists between molal concentrations on one side and colligative parameters on the other, which is at least in part because the amount of hydration water associated with each solute particle differs substantially from substance to substance, and for dissolved salts, the degree of dissociation varies from substance to substance and with their concentration.

Although osmolality is not a precise chemical concentration measure, it is a useful tool for comparing total solute concentrations of fluids in physiology, and it is a useful measure to control the chemical solute concentrations of body fluids and solutions.

BIBLIOGRAPHY

1. Hulett, G. (1903). Beziehung zwischen negativem Druck und osmootischen Druck. *Z. Phys. Chem.* **42**: 353–368.

Hammel, H.T. and Scholander, P.F. (1976). *Osmosis and Tensile Solvent*. Springer Verlag, Berlin.

PARTITIONING AND BIOAVAILABILITY

Jim Philp
Fiona Stainsby
Sandra Dunbar
Napier University
Edinburgh, Scotland, United Kingdom

INTRODUCTION

As soon as a chemical is spilled it becomes a pollutant, but the fate of that pollutant depends to a large degree on its chemistry, which determines how it will partition to various "compartments," or phases, of the environment. This idea is enshrined in risk-based strategies for assessing contaminated land and water. In many cases a contaminant spill on land quickly becomes a water pollution issue, be it surface or groundwater. As many countries are heavily reliant on groundwater as a source of potable water, investigating the fate of pollutants is an essential prerequisite to any remedial action plan for contaminant cleanup.

Partitioning is governed by some fundamental properties of chemicals, but partitioning should be viewed in relation to the toxicity of the chemical in question and its degradability in the various environmental compartments. In this article, the important properties of chemicals that relate to partitioning are examined, and the theory is reinforced by example. The examples chosen are benzene, phenol, and some chlorinated phenols. All of these compounds are important industrial chemicals.

Bioavailability is a simple concept, but practically it is a complex issue that remains to be resolved. The bioavailability concept is also enshrined in the risk assessment approach to contaminated land and water. If a pollutant is present in soil or water in high concentration, but is not available to the biota, then surely it presents minimal risk. Regardless of the apparent simplicity of the concept, there is great urgency to find methods for quantifying bioavailability. Some approaches to solving this quandary are discussed.

PROPERTIES THAT DETERMINE PARTITIONING

The compartments of the environment relevant to partitioning are air, water, soil, and sediment and, through bioaccumulation/magnification through food chains, biota. Partitioning into the biota is beyond the scope of this article, and it will deal principally with partitioning to nonliving compartments. Figure 1 represents a common scenario in environmental remediation, contamination from old gas stations.

Vapor Pressure

This is the partial pressure of a chemical in a gas phase that is in equilibrium with the pure liquid or solid phase of the chemical. Vapor pressure is greatly influenced by temperature, and figures are usually quoted at a constant temperature (usually 20 °C or 25 °C). If the volume of the containing vessel is also known, the vapor pressure can be used to calculate the concentration of the chemical in the headspace of the vessel from the ideal gas law:

$$\frac{n}{V} = \frac{P}{RT} \quad (1)$$

where P = vapor pressure (atm)
V = volume
n = number of moles of the chemical
R = gas constant
T = absolute temperature

Another way to think of vapor pressure is that it is the pressure that a vapor exerts on the atmosphere: thus it will be clear that it is the driving force for volatilization. Vapor pressures of organic compounds are often compared to the vapor pressure of mercury (atmospheric pressure is 760 mmHg). Thus the vapor pressure of benzene is 87.2 mmHg at 25 °C. Volatilization potential can be misleading in the absence of information on the water solubility of the compound, however. A compound that has a low vapor pressure may still have a high tendency to "escape" if the water solubility is low. For example in Table 1, 2,4,6-trichlorophenol has a much lower vapor pressure than 3-chlorophenol, and yet its tendency to escape to air is higher. This can be explained when the water solubilities are compared: the much lower value for 2,4,6-trichlorophenol results in it having a Henry's constant about a magnitude order higher than 3-chlorophenol, even though its vapor pressure is much lower.

Henry's Law Constant

To study partition phenomena, then, Henry's law constant is more valuable than vapor pressure. Henry's law constant is the ratio of the equilibrium concentration of a compound in air and its equilibrium concentration in water. It can therefore be regarded as the partition coefficient between air and water, and is the best indicator of the tendency of a chemical to volatilize from water. Henry's law constant can be written as:

$$H = \frac{P}{S} \quad (2)$$

where H = Henry's law constant (atm-m^3/mol)
P = vapor pressure (atm)
S = water solubility (mol/m^3)

Water Solubility

This is of central importance to understanding partition phenomena in the environment. It is the maximum, or

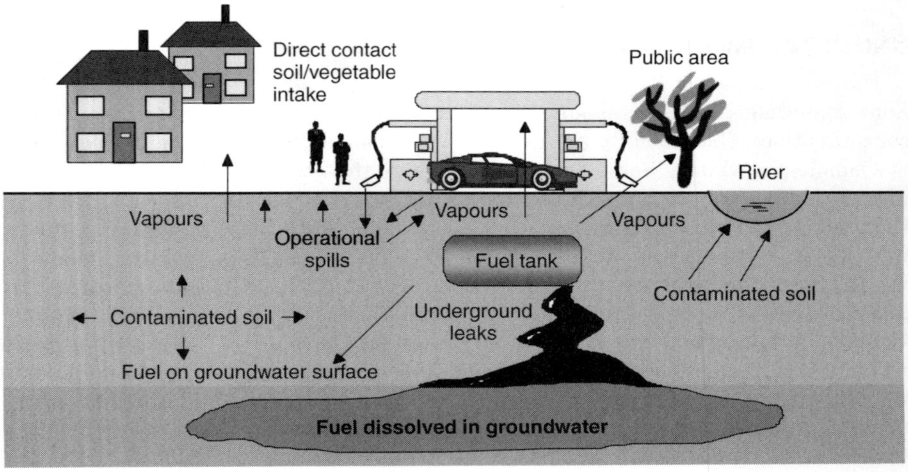

Figure 1. Partitioning of fuel as a contaminant from a gas station that has a leaking underground storage tank. Because fuels are multicomponent liquids, different components partition to different compartments of the environment.

Table 1. Important Partition Variables for a Range of Compounds. The Figures were Computed from U.S. EPA Software (EPI Suite v.3.11 and ECOSAR)

	Vapor Pressure, mmHg, 25 °C	Henry's Constant	Water Solubility, mg/L	Log K_{ow}	Log K_{oc}	Level III Fugacity Mass Amount, %	Toxicity Fish 96-h LC_{50}, mg/L	Biodeg. Probability, MITI Non-Linear
Benzene	87.2	5.55×10^{-3}	1790	2.13	2.22	Air: 37.6 Water: 48.4 Soil: 13.8 Sediment: 0.20	59.2	.73
Phenol	0.323	3.33×10^{-7}	77, 900	1.46	2.43	Air: 1.33 Water: 47.3 Soil: 51.3 Sediment: 0.098	27.7	.71
3-chloro-phenol	0.119	3.45×10^{-7}	26, 000	2.50	2.64	Air: 1.35 Water: 39.4 Soil: 59.1 Sediment: 0.21	15	0.39
2,4,6-trichloro-phenol	0.00536	2.6×10^{-6}	800	3.69	3.07	Air: 2.55 Water: 22.9 Soil: 72.4 Sediment: 2.17	3.7	0.043
Pentachloro-phenol	1.08×10^{-5}	2.45×10^{-8}	14	5.12	3.53	Air: 0.0665 Water: 5.59 Soil: 70.9 Sediment: 23.4	0.8	.0031

saturation, concentration of a compound that dissolves in water. It also is significantly affected by temperature. As the water solubility of compounds is defined by a host of factors, comparisons should always be made at a constant temperature, usually 25 °C. In natural water systems, the concentration of pollutant chemicals is usually rather low, making this especially important.

Polarity refers to the extent to which charge is unevenly distributed within a molecule. Water itself is highly polar. Therefore charged or highly polar organic and inorganic molecules, hydrophilic molecules, are readily soluble in water. Conversely, nonpolar, hydrophobic molecules find solution in water a less energetically favorable state, and they have low water solubility. In the examples in Table 1, benzene has moderate water solubility, but the addition of the polar hydroxyl group makes phenol much more water soluble. The solubility of any solute in any solvent is mediated by forces of attraction; the main ones are:

- van der Waals forces;
- hydrogen bonding;
- dipole–dipole interactions.

The size and shape of a molecule is also critical in determining its overall water solubility. Large molecules tend to have lower water solubility because they have a higher molar volume. The addition of halogens to an organic molecule might be expected to increase the water solubility of the new molecule as halogens are electronegative (electron-withdrawing) species and would thus tend to increase the polarity of the molecule. However, in the increasingly chlorinated series of phenols in Table 1, this is not the case: increasing chlorination very markedly decreases the solubility of chlorophenols because the halogens increase the molecular volume of organic compounds (1).

Other chemical functional groups affect water solubility in rather specific ways. The presence of polar functional groups tends to increase the water solubility of molecules compared to unsubstituted hydrocarbons: alcohols, amines, ethers, ketones, and organic acids are good examples. Even with these examples, though, a general rule of increasing molecular size and decreasing water solubility still exists. The short-chain fatty acid acetic acid is much more water soluble than a long-chain fatty acid. The additions to the molecule are not more polar functional groups but simply more nonpolar $C–H_2$ groups that make the molecule larger but less polar.

The situation is further complicated by more specific chemical reactions. The chlorophenols in Table 1 illustrate this point. Chlorophenols behave as weak acids when dissolved in water due to the loss of a proton from the hydroxyl group to leave a phenoxide ion. The more chlorines that are added to the ring, the more electrons are withdrawn from the electron-rich phenoxide ion. This lowers the phenoxide negative charge and therefore lowers its ability to hold a proton. Thus the more chlorines on the ring, the more easily a proton is released, and the greater the acidity. The ionized form is usually several orders of magnitude more water soluble than the neutral un-ionized form. The relative proportion of each species is highly dependent on pH.

Log K_{ow}, a Measure of Hydrophobicity

Polarity and water solubility do not give direct measures of the hydrophobicity of a compound. A hydrophobic compound might be expected to dissolve in a lipophilic phase, and this is the basis for the octanol–water partition coefficient, K_{ow}. This is the ratio of a chemical's concentration, at equilibrium, when dissolved in octanol and in water, when both phases are saturated with each other.

This test was devised by the pharmaceutical industry to study partitioning of drugs as it roughly equates to the partitioning between water and body fat. It has proven very useful in predicting how chemicals partition in the environment. Large molecules of low polarity have a high K_{ow} and are more likely to partition to solids in the environment. Small, polar molecules dissolve in water rather than octanol, have a low K_{ow}, are much less likely to partition to solids in the environment, and so are more likely to remain in an aqueous phase:

$$K_{ow} = \frac{\text{concentration of pollutant in water-saturated octanol (mg/L)}}{\text{concentration of pollutant in octanol-saturated water (mg/L)}} \quad (3)$$

The range of values for K_{ow} is very wide, and tabulated values are more often expressed as log K_{ow}. In the examples in Table 1, it can be seen that, as the molecule becomes larger through increasing chlorination, the water solubility decreases and log K_{ow} increases.

Log K_{oc}, Partitioning to Soil

The most basic statement of the composition of soil is that it contains sand, clay, organic matter, water, and air. Although typically the organic matter content of soil is in the range of 0.1 to 5%, this is where organic pollutants sorb, not to the soil as a whole. The soil adsorption coefficient, K_{oc}, takes account of this. For a liquid organic pollutant,

$$K_{oc}\ (\text{mL/g}) = \frac{\text{mass of pollutant sorbed to soil organic matter (mg/g)}}{\text{mass of pollutant in the aqueous phase (mg/mL)}} \quad (4)$$

The range of values for K_{oc} is very wide, and therefore the use of log K_{oc} is preferred. The soil partition coefficient is greater, the more hydrophobic the pollutant. However, it has been measured for far fewer chemicals than K_{ow}. It might be expected to be a more realistic measure than K_{ow}, but as data are relatively rare for K_{oc}, a series of correlation equations has been developed that relate K_{oc} to K_{ow}. The increase in log K_{ow} seen in Table 1 as chlorination increases is mirrored by increasing log K_{oc}.

Fugacity

The term fugacity literally means the "tendency to flee," and refers to a molecule's driving force to escape from the compartment in which it currently exists. With knowledge of all of a chemical's partition coefficients, it is possible to predict its relative concentration in air, water, and soil at equilibrium. The gradient of fugacity between two compartments (e.g., soil and soil porewater) determines the potential for the chemical to move from one compartment to the other. Fugacity therefore has units of pressure, but this can be related to the chemical concentration through modeling, and thus the distribution through various compartments predicted (Table 1).

THE RELATIONSHIP TO TOXICITY AND BIODEGRADATION

Hydrophobicity and Toxicity

Particularly K_{ow} and log K_{ow} have become increasingly used as predictors of the behavior of pollutants in the environment. It has been correlated with toxicity. The most common mode of action regarding the toxicity of industrial pollutants is probably narcosis. Nonpolar narcosis results from the perturbation of cellular membranes (2) due to the entry of hydrophobic (lipid-soluble) molecules into the phospholipid bilayer of the membrane. Narcosis has been shown to be a nonspecific physiological effect which is independent of chemical structure, and log K_{ow} is highly correlated with nonpolar narcosis toxicity (3).

Phenols have greater toxicity than predicted by nonpolar narcosis (4). They act by polar narcosis, which probably results from a strong hydrogen-bonding group that makes the compound more polar. Nevertheless, there is still a strong correlation between the toxicity of these compounds and log K_{ow}. Pentachlorophenol has greater toxicity as a result of a further mechanism: its acidity makes it a weak acid respiratory uncoupler (WARU). Terada (5) noted that WARUs possess an acid-dissociable moiety (e.g., a hydroxyl group), a bulky hydrophobic moiety, such as an aromatic ring, and multiple electronegative groups, such as halogens. During weak acid uncoupling, ATP synthesis is inhibited with no effect on the respiratory chain (6) because electron transport and phosphorylation reactions are uncoupled from each other.

Hydrophobicity and Biodegradability

It is believed that only molecules of hydrocarbons that are dissolved in the aqueous phase are available for intracellular metabolism (7). The rate at which a particular organic compound dissolves in water is critical to both the toxicity and biodegradability of the compound as this governs the rate of transfer to the organism (in biodegradability, microorganisms are important). The rate of transfer is determined by the equilibrium and actual concentration in the bulk phase and the aqueous phase. This is central to the concept of bioavailability.

From the compounds in Table 1, the increasing degree of chlorination decreases the rate of biodegradation. Increasing chlorination also leads to a marked drop in water solubility and a corresponding increase in hydrophobicity, which influence biodegradability. The toxicity of the compound also adversely affects biodegradation. The interaction between toxicity and biodegradation is not well understood, but they are clearly linked and competing processes: at low concentrations, a compound may be biodegradable, but this is arrested at higher concentrations by toxicity. If the concentration is very low for biodegradation, then this may not be enough to support microbial growth and reproduction, the concept of threshold elaborated in detail by Alexander (8).

BIOAVAILABILITY

Bioavailability is a concept recently borrowed from nutrition to explain the behavior of chemical pollutants in the environment (9). There are several definitions in circulation. In its simplest form (10), bioavailability is defined as "the amount of contaminant present that can be readily taken up by microorganisms and degraded." But this ignores bioavailability to higher organisms. Another definition that takes account of this and alludes to the risk assessment practice (11) is "a measure of the potential of a chemical for entry into ecological or human receptors."

The bioavailability quandary mentioned previously has been accurately articulated by the International Union of Pure and Applied Chemistry (IUPAC).

> "In the IUPAC Conference of Pesticide Chemistry in London in 1998, there was a controversial discussion on the topic whether agrochemicals in the environment are really hazardous. The conventional wisdom is that chemicals in soil are available to microorganisms, plant roots, and soil fauna like earthworms and animals via dermal exposure. Then bioaccumulation through the food web may induce indirect exposure to higher organisms. National governments are simply reducing the threshold levels of chemicals in the environments for larger safety margins in their guidelines. However, the question raised at the Congress was that chemical residues in the environment are not always bioavailable, so that *the actual exposure of biota to chemicals is rather different from the amount (concentration) present in the environment* (authors' italics). In addition, the persistence and efficacy of agrochemicals are affected by their bioavailability in soil. There is a lack of comprehensive understanding of the bioavailability of chemicals with different chemical characteristics, such as non-polar organochlorines, polar agrochemicals and etc. Thus, there is a strong need to clarify the scientific basis for bioavailability: definition, estimation methods and affecting factors."

Measuring Bioavailability

Measuring or even estimating bioavailability is problematic as it is affected by many individual and interacting conditions relating to soil and water chemistry, pollutant chemistry and partitioning, and biological transformation and concentration factors (12). Moreover, the "aging" process brings about a time-dependent decrease in bioavailability as compounds become sequestered in soil over time by ill-understood mechanisms (13). The techniques currently being researched can be categorized broadly as direct or indirect and biological or chemical (14), although, as pointed out by these authors, only organisms can determine whether a chemical is bioavailable, so direct chemical methods are not possible.

Direct Biological Methods. These are techniques that measure the actual amount of a chemical taken up by a target organism, and this may ultimately be the most accurate measure of bioavailability, although it is by no means routine. For example, the determination of chemical levels in earthworms is relatively expensive and time-consuming. The great strength of these direct techniques, however, is that they integrate all the biotic and abiotic modifying factors of chemical bioavailability (14).

Indirect Biological Methods. It is possible to observe a response to a chemical in an organism without actually measuring the concentration of the chemical. A range of responses is possible, such as lethality, enzyme induction or inhibition, and reproductive effects. These are regarded as indirect biological methods because the effect may be quantifiable in the organism, but the chemical concentration remains unknown.

A promising, novel approach is the use of genetically modified microorganisms that can detect and quantify specific pollutants. This type of biosensor is based on the highly specific genetic control mechanisms used by microorganisms to ensure that specific proteins are expressed only when needed; for example, to detoxify a particular toxic substance. This control is exerted by inducible promoters, consisting of a specific DNA sequence upstream of the genes to be controlled and a DNA-binding protein that either activates or prevents transcription in response to the presence or absence of the target compound. Biosensors of this type are easily generated by fusing such a controllable promoter to a reporter gene that generates a detectable signal when the promoter is activated (15). The most popular reporter is the bacterial *lux* system that produces light (Fig. 2), because light can be easily quantified and, due to its rarity in biology, there is no interference from the background biochemistry of the host organism.

Such biosensors can be both highly specific and responsive to very low concentrations, and the microbial cell membrane makes this approach truly a measure of bioavailability. There are limitations to developing such biosensors for detecting organic pollutants as often the

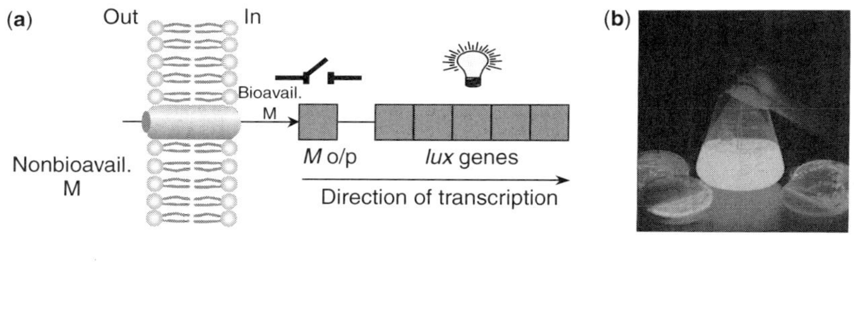

Figure 2. (**a**) Schematic of a genetically modified whole bacterial cell biosensor for detecting bioavailable metal. The metal may be expected to be transported into the bacterial cell via a specific transporter protein mediating uptake and efflux. If the metal is in a nonbioavailable form, for example, as a precipitate, it will not be transported. The interaction of the metal with the specific metal regulatory element initiates transcription resulting in synthesis of the luciferase enzyme and the production of light. (**b**) The naturally bioluminescent bacterium *Vibrio fischeri* producing light from colonies and liquid culture.

required specificity is lacking. For heavy metals, specificity may not be a problem. Nevertheless, careful selection of the host strain is required for a variety of reasons (16). Such biosensors offer the prospect of a much more routine measure of bioavailability.

Indirect Chemical Methods. Indirect chemical methods usually involve extracting a fraction of the chemical (metals or organics) from a soil; extractability is defined by the chemical itself, the nature of the extractant(s), and the experimental conditions applied. The origins of this approach are in sequential extraction procedures (17) that attempt to quantify the speciation of toxicants into those weakly bound and those strongly bound to the soil matrix. These tests often assume that the weakly bound toxicants are those that are more bioavailable, although evidence for this is incomplete. A recent advance in this technique is sequential accelerated solvent extraction (18) that reduces sample preparation time and maintains relatively constant extraction conditions. A technique under development is the use of selective supercritical fluid extraction (19).

A number of methods are based around the human gastrointestinal tract (the physiologically based extraction test, PBET; 20). To mimic human conditions, they incorporate gastric juices and enzymes, mix at 37 °C, and use soil residence times similar to those found in children after ingestion of food. Analysis of all solutions produced in the PBET extractions is completed by analytical chemistry techniques under matrix-matched conditions.

However sophisticated, any such technique is ultimately limited by the lack of an organism. There will be no universal chemical test for bioavailability as the phenomenon is organism- and even species-dependent (21).

BIBLIOGRAPHY

1. Yalkowsky, S.H. and Valvani, S.C. (1979). Solubilities and partitioning. 2. Relationships between aqueous solubilities, partition coefficients and molecular surface areas of rigid aromatic hydrocarbons. *J. Chem. Eng. Data* **24**: 127–129.
2. Cronin, M.T.D. and Schultz, T.W. (1998). Structure-toxicity relationships for three mechanisms of action of toxicity to *Vibrio fischeri*. *Ecotoxicol. Environ. Saf.* **39**: 65–69.
3. Cronin, M.T.D. and Dearden, J.C. (1995). Review: QSAR in toxicology. 1. Prediction of aquatic toxicity. *Quant. Struct. Act. Relationships* **14**: 1–7.
4. Cronin, M.T.D. and Schultz, T.W. (1997). Validation of *Vibrio fischeri* acute toxicity data: mechanism of action-based QSARs for non-polar narcotics and polar narcotic phenols. *Sci. Total Environ.* **204**: 75–88.
5. Terada, H. (1990). Uncouplers of oxidative phosphorylation. *Environ. Health Perspect.* **87**: 213–218.
6. Schultz, T.W. and Cronin, M.T.D. (1997). Quantitative structure-activity relationships for weak acid respiratory uncouplers to *Vibrio fischeri*. *Environ. Toxicol. Chem.* **16**: 357–360.
7. Sikkema, J., de Bont, J.A.M., and Poolman, B. (1995). Mechanisms of membrane toxicity of hydrocarbons. *Microbiolo. Rev.* **59**: 201–222.
8. Alexander, M. (1999). Threshold. In: *Biodegradation and Bioremediation*. Academic Press, San Diego, pp. 105–116.
9. Madsen, E.L. (2003). *Report on Bioavailability of Chemical Wastes with Respect to the Potential for Soil Bioremediation*. U.S. Environmental Protection Agency report *EPA/600/R-03/076*.
10. Maier, R. (2000). Bioavailability and its importance to bioremediation. In: *International Society for Environmental Biotechnology: Environmental Monitoring and Biodiagnostics*. J.J. Valdes (Ed.). Kluwer, Boston, pp. 59–78.
11. Loehr, R.C. and Webster, M.T. (1997). Effects of treatment on contaminant availability, mobility, and toxicity. In: *Environmentally Acceptable Endpoints in Soil*. D.G. Linz and D.V. Nakles (Eds.). American Academy of Environmental Engineers, Annapolis, MD, pp. 137–386.
12. Hund-Rinke, K. and Kördel, W. (2003). Underlying issues in bioaccessibility and bioavailability: experimental methods. *Ecotoxicol. Environ. Saf.* **56**: 52–62.
13. Chung, N. and Alexander, M. (2002). Effect of soil properties on bioavailability and extractability of phenanthrene and atrazine sequestered in soil. *Chemosphere* **48**: 109–115.
14. Lanno, R., Wells, J., Conder, J., Bradham, K., and Basta, N. (2004). The bioavailability of chemicals in soil for earthworms. *Ecotoxicol. Environ. Saf.* **57**: 39–47.
15. Philp, J.C., French, C., Wiles, S., Bell, J.M.L., Whiteley, A.S., and Bailey, M.J. (in Press). Wastewater toxicity assessment by whole cell biosensor. In: *Handbook of Environmental Chemistry*. D. Barceló (Ed.). Vol. 5. Water Pollution, Emerging Organic Pollutants in Wastewaters, Springer-Verlag, Berlin.
16. Rensing, C. and Maier, R.M. (2003). Issues underlying use of biosensors to measure metal bioavailability. *Ecotoxicol. Environ. Saf.* **56**: 140–147.
17. Tessier, A., Campbell, P.G.C., and Bisson, M. (1979). Sequential extraction procedure for the speciation of particulate trace metals. *Anal. Chem.* **51**: 844–851.
18. Tao, S., Guo, L.Q., Wang, X.J., Liu, W.X., Ju, T.Z., Dawson, R., Cao, J., Xu, F.L., and Li, B.G. (2004). Use of sequential ASE extraction to evaluate the bioavailability of DDT and its metabolites to wheat roots in soils with various organic carbon contents. *Sci. Total Environ.* **320**: 1–9.
19. Björklund, N.T., Bøward, S., Pilorz, K., Mathiasson, L., and Hawthorne, S.B. (2000). Introducing selective supercritical fluid extraction as a new tool for determining sorption/desorption behavior and bioavailability of persistent organic pollutants in sediment. *J. Biochem. Biophys. Methods* **43**: 295–311.
20. Ruby, M.V., Davis, A., Schoof, R., Eberle, S., and Sellstone, C.M. (1996). Estimation of lead and arsenic bioavailability using a physiologically based extraction test. *Environ. Sci. Technol.* **30**: 422–430.
21. Reid, B.J., Jones, K.C., and Semple, K.T. (2000). Bioavailability of persistent organic pollutants in soils and sediments—a perspective on mechanisms, consequences and assessment. *Environ. Pollut.* **108**: 103–112.

READING LIST

Hemond, H.F. and Fechner-Levy, E.J. (2000). *Chemical Fate and Transport in the Environment*. Academic Press, San Diego.

Landis, W.G. and Yu, M-H. (2003). *Introduction to Environmental Toxicology*, 3rd Edn. Lewis, Boca Raton, FL.

Nirmalakhandan, N.N. and Spreece, R.E. (1988). Prediction of aqueous solubility of organic compounds based on molecular structure. *Environ. Sci. Technol.* **22**: 328–338.

Watts, R.J. (1997). *Hazardous Wastes: Sources, Pathways, Receptors*. John Wiley & Sons, New York.

PHYSICAL PROPERTIES

Kevin S. McLeary

STRUCTURE

Water is a molecule composed of two atoms of hydrogen bonded to an atom of oxygen. It has a volume of 2.97 × 10^{-11} μ^3 (29.7 Å^3) (1). As a result of the hybridization of the oxygen atom, the molecule is bent at an angle of 104.45°. Due to the nonlinear structure and to the difference in electronegativity between oxygen (3.44) and hydrogen (2.20), the molecule has a dipole moment of 1.85 Debye (2).

Water molecules are held together by intermolecular attractions. In water, the intermolecular attractions are hydrogen bonds. The negatively charged oxygen atom in one molecule of water is electrically attracted to the positively charged hydrogen atom of another molecule. Hydrogen bonding governs the structure of water up to the boiling point (2).

SPECIFIC WEIGHT (γ)

The specific weight of a substance is its weight per unit of volume. In U.S. Customary Units, it is often expressed as pounds-force per cubic foot (3). In SI units, it is often expressed in kN/m^3. Although a value of 62.4 pounds-force per cubic foot (9.81 kN/m^3) is commonly used for the majority of engineering purposes, the specific weight of water varies with temperature, from 62.42 pounds-force per cubic foot (9.805 kN/m^3) at 32 °F (0 °C) to 59.8 pounds-force per cubic foot (9.399 kN/m^3) at 212 °F (100 °C) (4). Table 1 shows the variability in the specific weight of water with temperature in U.S. Customary Units. Table 2 shows the variability in the specific weight of water with temperature in SI Units.

Specific weight is also a function of gravity, and it varies slightly with elevation. One cubic foot of water at 32 °F at sea level has a specific weight of 62.42 pounds-force per cubic foot (9.805 kN/m^3); the same volume of water at 10,000 feet above sea level has a specific weight of 62.36 pounds-force per cubic foot (9.796 kN/m^3). Because the earth is not a perfect sphere, but bulges at the equator, specific weight is also a weak function of latitude. However, these variations are small enough to warrant the use of a constant value of 62.4 pounds-force per cubic foot (9.81 kN/m^3) at all elevations and latitudes (4).

DENSITY (ρ)

The density of a fluid is defined either as its mass per unit of volume or as the ratio of its specific weight to the acceleration due to gravity. Density ranks among the most important hydraulic properties of water (1). It is a factor in the settling velocity of particles and in the filtration of water through a porous medium. A fluid's Reynolds number, the most important dimensionless number in hydraulics, is a direct function of density. The Reynolds number is used in calculating friction losses, pipe sizing, and pumping requirements

In U.S. Customary Units, density is expressed in slugs/ft^3, and in SI units, it is expressed in kg/m^3. Like specific weight, density varies with temperature. The density of pure water varies from 1.940 slugs/ft^3 (999.8 kg/m^3) at 32 °F (0 °C) to 1.860 slugs/ft^3 (958.4 kg/m^3) at 212 °F (100 °C). Tables 1 and 2 show the variation of density in U.S. Customary Units and SI units, respectively.

SPECIFIC GRAVITY

Specific gravity is defined as the ratio of the density of one fluid at a specified temperature to that of a reference fluid at a reference temperature. For liquids, the reference liquid is usually water, although varying reference temperatures have been used, for example, 39° and 68 °F (4° and 20 °C). Although specification of the reference temperature is required to ensure accuracy, it is often not given because the density of water varies by only 0.2% over the usual temperature range (5). The specific gravity of water is usually taken as 1.0.

ABSOLUTE (DYNAMIC) VISCOSITY (μ)

The viscosity of a fluid is expressed as a measure of the resistance to flow. Mathematically, it is defined as the ratio of the shear stress of a fluid to its velocity gradient or as the proportionality constant between the shear stress and the velocity gradient:

$$\frac{F}{A} = \mu \frac{dv}{dy} \tag{1}$$

Viscosity can be measured by taking two plates of area A, which are separated by a fluid whose thickness is y. The bottom plate is held stationary. The top plate is moved at a constant velocity v. The force required to maintain the constant velocity is measured and used to calculate the viscosity μ of the fluid.

The chief cause of viscosity in fluids is molecular cohesion. Molecular cohesion decreases as temperature is increased, therefore, viscosity also decreases as temperature increases. Changes in pressure do not affect viscosity (5). Impurities also exert little influence on viscosity (1). Tables 1 and 2 show the effect of temperature on viscosity.

KINEMATIC VISCOSITY (ν)

There are many times in fluid mechanics and hydraulics when the fluid's absolute viscosity is divided by its density. To ease calculations, this term has been defined as kinematic viscosity because it has kinematic units of length and time, ft^2/sec or m^2/s.

Both absolute viscosity and density are a function of temperature, so kinematic viscosity is also a function of temperature. Tables 1 and 2 show how kinematic viscosity varies with temperature.

Table 1. Physical Properties of Water in U.S. Customary Units[a]

Temp., °F	Specific Weight, γ, lb_f/ft^3	Density, ρ, $slug/ft^3$	Modulus of Elasticity[b] $E/10^3$, $lb_f/in.^2$	Dynamic Viscosity, $\mu \times 10^5$, $lb_f \cdot sec/ft^2$	Kinematic Viscosity, $\nu \times 10^5$, ft^2/sec	Surface Tension,[c] σ, lb_f/ft	Vapor Pressure, p_v, $lb_f/in.^2$
32	62.42	1.940	287	3.746	1.931	0.00518	0.09
40	62.43	1.940	296	3.229	1.664	0.00614	0.12
50	62.41	1.940	305	2.735	1.410	0.00509	0.18
60	62.37	1.938	313	2.359	1.217	0.00504	0.26
70	62.30	1.936	319	2.050	1.059	0.00498	0.36
80	62.22	1.934	324	1.799	0.930	0.00492	0.51
90	62.11	1.931	328	1.595	0.826	0.00486	0.70
100	62.00	1.927	331	1.424	0.739	0.00480	0.95
110	61.86	1.923	332	1.284	0.667	0.00473	1.27
120	61.71	1.918	332	1.168	0.609	0.00467	1.69
130	61.55	1.913	331	1.069	0.558	0.00460	2.22
140	61.38	1.908	330	0.981	0.514	0.00454	2.89
150	61.20	1.902	328	0.905	0.476	0.00447	3.72
160	61.00	1.896	326	0.838	0.442	0.00441	4.74
170	60.80	1.890	322	0.780	0.413	0.00434	5.99
180	60.58	1.883	318	0.726	0.385	0.00427	7.51
190	60.36	1.876	313	0.678	0.362	0.00420	9.34
200	60.12	1.868	308	0.637	0.341	0.00413	11.52
212	59.83	1.860	300	0.593	0.319	0.00404	14.70

[a]References 3 and 7.
[b]The modulus of elasticity is given at atmospheric pressure.
[c]Surface tension is given for water in contact with air.

Table 2. Physical Properties of Water in SI Units[a]

Temp., °C	Specific Weight, γ, kN/m^3	Density, ρ, kg/m^3	Modulus of Elasticity,[b] $E/10^6$, kN/m^2	Dynamic Viscosity, $\mu \times 10^3$, $N{\cdot}s/m^2$	Kinematic Viscosity, $\nu \times 10^6$, m^2/s	Surface Tension,[c] σ, N/m	Vapor Pressure, p_v, kN/m^2
0	9.805	999.8	1.98	1.781	1.785	0.0765	0.61
5	9.807	1000.0	2.05	1.518	1.519	0.0749	0.87
10	9.804	999.7	2.10	1.307	1.306	0.0742	1.23
15	9.798	999.1	2.15	1.139	1.139	0.0735	1.70
20	9.789	998.2	2.17	1.002	1.003	0.0728	2.34
25	9.777	997.0	2.22	0.890	0.893	0.0720	3.17
30	9.764	995.7	2.25	0.798	0.800	0.0712	4.24
40	9.730	992.2	2.28	0.653	0.658	0.0696	7.38
50	9.689	988.0	2.29	0.547	0.553	0.0679	12.33
60	9.642	983.2	2.28	0.466	0.474	0.0662	19.92
70	9.589	977.8	2.25	0.404	0.413	0.0644	31.16
80	9.530	971.8	2.20	0.354	0.364	0.0626	47.34
90	9.466	965.3	2.14	0.315	0.326	0.0608	70.10
100	9.399	958.4	2.07	0.282	0.294	0.0589	101.33

[a]References 3 and 7.
[b]The modulus of elasticity is given at atmospheric pressure.
[c]Surface tension is given for water in contact with air.

SURFACE TENSION (σ)

Surface tension is caused by the attractive forces between molecules in a liquid. In a bulk liquid, each molecule is attracted to its neighboring molecules, and the total force on the molecule is zero. However, a molecule at the surface experiences an attraction to other molecules deeper in the liquid, but there are no molecules outside to balance these attractions. The molecule at the surface experiences a net attractive force toward the bulk liquid.

Surface tension is the amount of work necessary to increase the surface area of the liquid by one unit. It has units of force per unit area ($ft\text{-}lb_f/ft^2$, or lb_f/ft) and decreases as the temperature of the liquid increases. It is also a function of the gas contacting the liquid (5). Table 1 gives the variation of surface tension of water in contact with air as a function of temperature in U.S. Customary Units, and Table 2 shows the variation in SI units.

VAPOR PRESSURE (P_V)

Water is a collection of a large number of molecules in a liquid phase that move about at random. The molecules collide with each other and rebound in different directions. Each molecule moves with a certain velocity (energy). Molecules on the surface of a liquid moving away from the bulk liquid have no other molecule to collide with. If such a molecule has sufficient energy to escape the intermolecular attractions, it may escape the liquid into the vapor space above the liquid. It evaporates. Evaporated water molecules also move about at random. Some of the water vapor condenses into the liquid water phase.

If the water were placed in a closed container at a constant temperature and the air evacuated from the vapor space above the liquid, one could measure the pressure exerted by the water vapor. The pressure would increase until equilibrium is attained—when the rate of molecules evaporating and the rate of molecules condensing are equal. The pressure of the water vapor then is constant. This is the vapor pressure of the water for the given temperature. Table 1 gives the vapor pressure of water in psi. Table 2 gives the vapor pressure in kN/m^2.

MODULUS OF ELASTICITY (*E*)

Although water is usually considered incompressible, it is slightly compressible. The modulus of elasticity is defined as the increase in pressure that results in a specific percentage change in volume:

$$E = \frac{\Delta p}{\Delta V/V} \tag{2}$$

where Δp is the increase in pressure, ΔV is the resulting change in volume, and V is the original volume of water.

The modulus of elasticity has units of pressure and is a function of temperature. Tables 1 and 2 show the variation of the modulus of elasticity with temperature.

SPEED OF SOUND (*C*)

Sound is the propagation of a pressure wave through a material and is dependent on the material's modulus of elasticity and density. The speed of sound, c, in a liquid is given by

$$c = \sqrt{\frac{Eg_c}{\rho}} \tag{3}$$

The speed of sound in water is important in calculating the pressure increase due to water hammer. When the velocity in a pipe is decreased suddenly, the pipe experiences an increase in pressure. The increase in pressure, Δp, caused by a decrease in velocity of Δv, is

$$\Delta p = \frac{\rho c \Delta v}{g_c} \tag{4}$$

HEAT CAPACITY (C_P)

Heat capacity is defined as the amount of heat needed to raise the temperature of a given amount of substance by 1°. It is usually expressed in units of heat per unit of substance per degree. The term "heat capacity" is often used for this quantity of heat on a molar basis. If a mass basis is used, the term "specific heat" is frequently used.

The heat capacity of water is the basis of the unit of heat in both the English and SI systems. Therefore, the heat capacity (specific heat) of water is 1 Btu/pound-mass/°F in the English system and 1 cal/g/°C in the SI system.

Heat capacities of liquids are not functions of temperature, except at high reduced temperatures. (A reduced temperature is a thermodynamic quantity relating the temperature of a material to its critical temperature. The critical temperature for water is 647.1 K.) (6). At reduced temperatures exceeding 0.7–0.8, liquid heat capacities are large and are strong functions of temperature.

BIBLIOGRAPHY

1. Fair, G.M., Geyer, J.C., and Okun, D.A. *Water and Wastewater Engineering*. John Wiley & Sons, New York.
2. Atkins, P.W. *Physical Chemistry*, 2nd Edn. W.H. Freeman, San Francisco.
3. Metcalf & Eddy, Inc. *Wastewater Engineering: Treatment, Disposal, and Reuse*, 3rd Edn. McGraw-Hill, New York.
4. F.S. Merritt (Ed.). *Standard Handbook for Civil Engineers*, 3rd Edn. McGraw-Hill, New York.
5. Lindeburg. *Civil Engineering Reference Manual*.
6. Smith, J.M. and van Ness, H.C. *Introduction to Chemical Engineering Thermodynamics*, 3rd Edn. McGraw-Hill, New York.
7. Culp/Wesner/Culp. *Handbook of Public Water Systems*. Van Nostrand-Reinhold, New York.

ENVIRONMENTAL PHOTOCHEMISTRY IN SURFACE WATERS

Mark A. Young
University of Iowa
Iowa City, Iowa

INTRODUCTION

Surface waters make up a large fraction of the available terrestrial surface area, accounting for over 70% of the total area with most of that consisting of ocean waters. A correspondingly large fraction of the incoming solar flux will impinge on natural waters, penetrating to a depth determined by the scattering and absorption properties of the water system. While most of the short wavelength solar radiation is attenuated in the atmosphere, mainly by absorbing species such as O_2 and O_3, the surface flux of energetic photons in the near-UV and visible remains appreciable. While water itself is not photoactive, the irradiated volume of typical surface waters contains a plethora of potential chromophores. The chromophoric material consists of dissolved species and suspended particulate matter that can have strong electronic absorptions that significantly overlap with the available solar flux. Absorption of solar photons can initiate a rich variety of photochemical processes with sufficient kinetic rates to manifest a substantial impact on the chemistry taking place in the aqueous environment. Photochemically driven chemistry, in turn, can have enormous implications for global geochemical cycles, climate, and biogenic processes, making photochemistry a critical consideration in the study of environmental processes.

BACKGROUND

Natural surface waters can be either saltwater (marine, or ocean, waters) or freshwater (continental waters). Within each of these reservoirs, widely varying environments are found: remote and coastal areas in the oceans, and rivers, lakes, streams, bogs/marshes on the continents, as well as interfacial environments, such as estuaries. In addition to differences in salinity and pH, there will be variations in the types and concentrations of chromophoric species residing in each reservoir. Potential chromophores include dissolved organic material (DOM), inorganic NO_x^- ions, metal complexes, anthropogenic pollutants, and colloidal matter consisting of crustal minerals and particulate organic material (POM). Organic chromophores may also adsorb on, or complex with, the surface of mineral colloids and participate in charge transfer processes. Likewise, metals can become incorporated into organic material. The distinction between dissolved and colloidal material is not sharp and often depends on the analytical procedure used to assay concentrations. The relative importance of the various chromophores will depend on the details of the aqueous environment. Chromophores can absorb solar photons and undergo photochemical transformation while yielding a variety of primary and secondary reactive species, as depicted schematically in Fig. 1. The photogenerated intermediates can, in turn, react with other species present in the water at rates that are often diffusion controlled to form final products. In some instances, the chromophore may be regenerated, resulting in a photocatalytic cycle that can be highly efficient (Fig. 1).

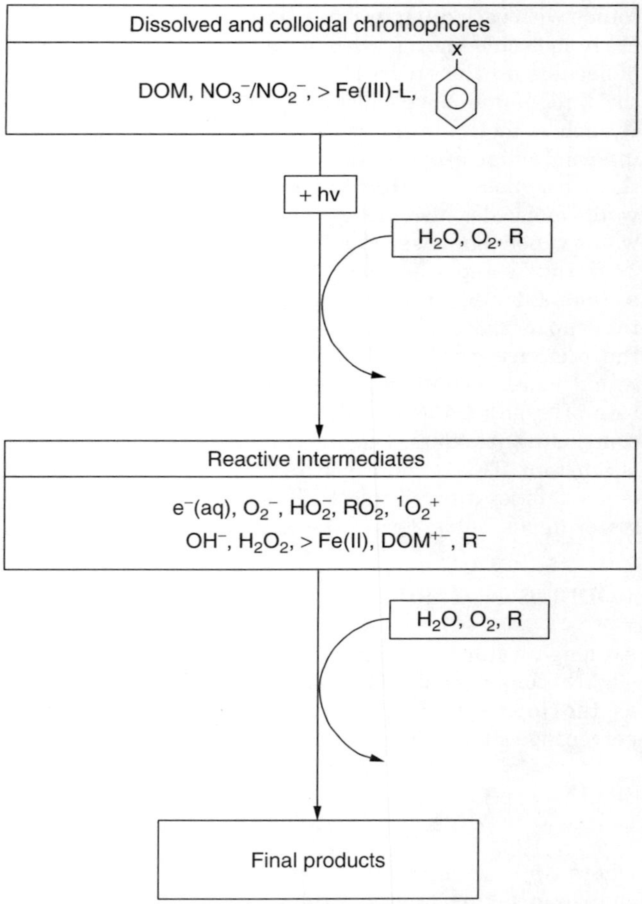

Figure 1. Schematic representation of photochemical processes in surface waters. Chromophores consist of DOM, nitrate/nitrite ions, dissolved and colloidal metal complexes [represented by Fe(III)], and anthropogenic pollutants (such as substituted aromatics). Reactive intermediates consist of solvated electrons, various radical species, both cationic and anionic, peroxides, and reduced metal centers.

Excitation of a chromophore can initiate a number of photophysical and photochemical processes that are outlined in the following equations:

$$S + h\nu \xrightarrow{J} S^* \quad \text{excitation;} \quad (1)$$

$$S^* \xrightarrow{k_L, k_I} S\ (+h\nu) \quad \text{luminescence, internal conversion, and intersystem crossing;} \quad (2)$$

$$S^* + Q \xrightarrow{k_Q} S + Q \quad \text{quenching;} \quad (3)$$

$$S^* \longrightarrow S^+ + e^- \quad \text{photoionization;} \quad (4)$$

$$S^* \underset{[R]}{\xrightarrow{k_R}} P \quad \text{reaction;} \quad (5)$$

$$S^* + R \xrightarrow{k_{ET}} S + R^* \quad \text{energy transfer;} \quad (6)$$

$$S^* + R \xrightarrow{k_{CT}} S^{+/-} + R^{-/+} \quad \text{charge transfer.} \quad (7)$$

Photophysical processes, luminescence (fluorescence and phosphorescence), internal conversion, intersystem crossing—Eq. 2—and collisional quenching—Eq. 3—will return the chromophore to the electronic ground state and convert the photon energy to heat. The excited state may also ionize, Eq. 4, or react, Eq. 5, converting the photon energy into chemical potential. The reaction could be unimolecular, such as dissociation, or bimolecular in nature, involving other components of the water system. Products might include the reactive intermediates in Fig. 1, which can then participate in secondary chemistry. Energy transfer, or sensitization, Eq. 6, could also induce a concurrent change in the electronic spin state, yielding a reactive triplet excited state.

The photochemical sequence is initiated by absorption of a photon, Eq. 1, with a photolytic rate constant, J. From a simple consideration of Beer's Law, J is seen to depend on the incident solar flux, I, and the (molar, decadic) extinction coefficient of the chromophore, ε. The relative efficiency, or yield, of a reaction can be specified by the quantum yield, ϕ, which is the number of product species formed divided by the number of photons absorbed. The absorption coefficient and quantum yield are generally wavelength dependent. The flux depends on wavelength as well as the solar zenith angle and other factors, such as latitude, longitude, and meteorological conditions, such as cloud cover. The rate constant for a photoinduced process can then be written as an integral over the wavelength region of interest:

$$J = 2.303 \int \varepsilon(\lambda)\phi(\lambda)I(\lambda)d\lambda = 2.303 \sum_i \overline{\varepsilon}(\lambda_i)\overline{\phi}(\lambda_i)\overline{I}(\lambda_i)\Delta\lambda_i \quad (8)$$

The flux in this case is termed the spectral actinic flux and is a spherically averaged irradiance that takes into account radiation incident on the chromophore over 4π steradians. For practical purposes, the integral is often evaluated as a discrete sum over small wavelength intervals, as expressed in the last term of Eq. 8, where $\overline{\varepsilon}$, $\overline{\phi}$, and $\overline{I}$ represent average quantities over the interval $\Delta\lambda_i$ centered at λ_i.

The solar actinic flux at noon (zenith angle = 0°) for a northern latitude (47.4°N) is shown in Fig. 2. The flux is expressed in terms of the photon flux per unit wavelength interval with units of einstein $\cdot$ $m^{-2} \cdot s^{-1} \cdot nm^{-1}$, where an einstein corresponds to 1 mole of photons (Avogadros's number, $N_A = 6.022 \times 10^{23}$ mol^{-1}). The UV-B radiation (280–320 nm) is strongly attenuated by the ozone layer, effectively limiting the useful actinic flux to about 300 nm. Shorter wavelength radiation does reach the surface but at 293 nm, the flux has decreased by more than two orders of magnitude relative to 300 nm. UV-A (320–400 nm) is not strongly absorbed by ozone and the flux increases up to the maximum in the visible at approximately 575 nm. A photon has a molar energy of

$$\overline{E} = N_A \frac{hc}{\lambda} \text{ J} \cdot \text{einstein}^{-1} \quad (9)$$

A 300-nm photon has an energy of 399 kJ $\cdot$ $einstein^{-1}$, which is on the order of typical bond enthalpies (e.g., C–C, 348 kJ; O–O, 146 kJ; C–OH, 322 kJ; benzylic C–H, 368 kJ; phenolic O–H, 362 kJ; C–Cl, 342 kJ). Over the wavelength range of photochemical interest, 300–500 nm, Fig. 2 indicates that the noon-time flux to surface waters is about 4 einstein $\cdot$ $m^{-2} \cdot h^{-1}$. Typical penetration depths range from 1 to 20 m and so the molar rate of photon absorption is $0.2–4.2 \times 10^{-3}$ einstein$\cdot L^{-1} \cdot h^{-1}$. Dissolved organic carbon (DOC) is the major absorbing species in most natural waters (see Fig. 2) and has an approximate average concentration of 5 mg C$\cdot L^{-1}$. While highly diverse in nature, an average chromophore molecular weight of 120 g C$\cdot mol^{-1}$ (10 carbon atoms) can be assumed and so the absorber concentration is approximately 0.04 mM. The estimated photon absorption rate for a well mixed volume would correspond to an average excitation rate of 5–100 h^{-1}, meaning that a near-surface chromophore is excited approximately once each minute (2,3). Furthermore, it is estimated that the accumulated annual radiation dose is about 1300 times larger (2). Given the time scales for transport and chemistry in surface waters, it is clear that photochemically driven reactions will be an important component of chemical cycles in the hydrosphere (Fig. 2).

REACTIVE INTERMEDIATES

A variety of reactive intermediates have been identified from the irradiation of natural waters at relevant solar wavelengths (see Fig. 1). Selected properties for some of the important species are listed in Table 1, along with estimated steady-state concentrations in typical surface waters (generally corresponding to a noon-time solar flux). Under conditions of constant solar illumination, pseudo-first-order kinetics result and the steady-state concentration for an intermediate, $[I]_{SS}$, can be used in conjunction with the second-order rate constant, k, for a particular reaction to calculate the reaction half-life as

$$\tau = \frac{\ln 2}{k[I]_{SS}} \quad (10)$$

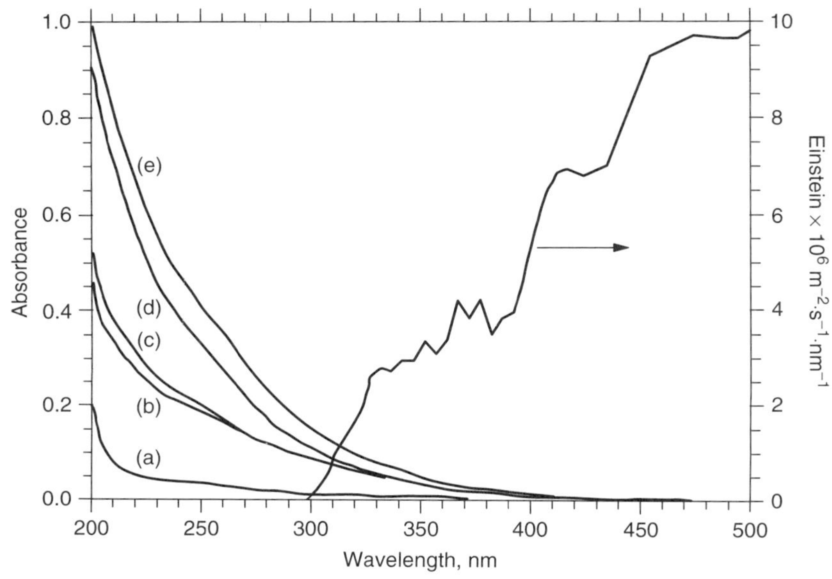

Figure 2. Left ordinate: Absorption spectra of surface water samples with varying concentrations and sources of DOC in mg C$\cdot L^{-1}$: (**a**) = 3.0, (**b**) = 7.8, (**c**) = 13.4, (**d**) = 13.4, (**e**) = 15.4. Path length is 1.0 cm. Data are adapted from Reference 7. Right ordinate: Solar actinic flux for noon-time sun (zenith angle = 0°) for a northern, midlatitude. Data is from Reference 1.

and, thus, estimate the importance of a given reaction pathway. For example, half-lives for reactions with phenol, a representative organic, aromatic compound, have been included in Table 1.

Primary transient species are produced rapidly, on the order of <1 ns. These transients can then react with other components in the aquatic environment to produce secondary products. Besides H_2O, one of the most important interactions involves dissolved oxygen, which, in well aerated surface waters, will be present in concentrations determined by Henry's Law,

$$[O_2] = K_H P_{O_2} \quad (11)$$

where $K_H = 1.3 \times 10^{-3}\ M^{-1} \cdot atm^{-1}$ is the Henry's Law constant (8). The mixing ratio of O_2 in the atmosphere is 21%, yielding a concentration of $[O_2] = 0.27$ mM at 298 K. Primary radicals will react with dissolved O_2 at diffusion controlled rates (diffusion controlled rate constant, $k_D \approx 7 \times 10^9\ M^{-1} \cdot s^{-1}$ at 298 K) corresponding to a time scale of ≈ 0.4 μs, which will limit the lifetime of many photogenerated transients (the "oxygen wall") (9).

e^-(aq)

Electrons generated from photoionization of DOM are rapidly solvated in aqueous environment to produce, e^-(aq). The solvated electron reacts with dissolved oxygen to form the superoxide,

$$e^-(aq) + O_2 \xrightarrow{k} O_2^- \quad (12)$$

where $k \approx 2 \times 10^{10}\ M^{-1} \cdot s^{-1}$ [which is larger than the diffusion limit due to a tunneling mechanism for e^-(aq) transport] (4,5). The time scale for Eq. 12 is on the order of 200 ns. The short lifetime coupled with measured typical production rates leads to a very small steady-state concentration of $[e^-(aq)]_{SS} \approx 10^{-17}$–$10^{-15}$ M (2,9). While reaction with e^-(aq) has been shown to degrade organic pollutants (2,10), it is unlikely to play an important role in photochemical processes in most surface waters.

O_2^-

Superoxide, O_2^-, can be formed from reaction with e^-(aq), as above, in charge transfer reactions with sufficiently reducing species, and via electron trapping at the surface of semiconductor colloids. Photolysis of DOM is a major source of superoxide but photolysis of trace metal complexes may also contribute to O_2^- formation (11). Superoxide is a mild reductant/oxidant (Table 1) and can participate in both one-electron oxidation and reduction reactions. The steady-state concentration is uncertain but from limited formation rate measurements, the concentration is $[O_2^-]_{SS} \approx 10^{-9}$–$10^{-7}$ *M* (9,10). The relatively high concentration suggests that superoxide may be important in photochemical degradation of various species (7). However, O_2^- is unstable in water with respect to disproportionation to form hydroperoxyl radical and hydrogen peroxide (3,12)

$$O_2^- + H^+ \overset{K}{\rightleftharpoons} HO_2 \quad \log K = -4.8 \quad (13)$$

$$HO_2/O_2^- + H^+ \longrightarrow H_2O_2 + O_2 \quad (14)$$

$RO_2^\bullet$

Peroxyl radicals can be formed from the photolysis of DOM and metal complexes. Formation rates for RO_2 species are thought to linearly scale with the DOM concentration in typical waters (2), but the reactivity of the radical is highly dependent on the type of humic substance involved, which is itself not well characterized. These uncertainties make it difficult to estimate a steady-state concentration for peroxyl radicals. HO_2 is a mild oxidant/reductant (Table 1) but DOM-derived peroxyl radicals exhibit significant

Table 1. Selected Properties of Some Important Reactive Intermediates in Aqueous Photochemistry[a]

	Radical				
Property	e^-(aq)	O_2^-	HO_2	OH	$^1O_2^*$
ΔG_f^0 (kJ·mol^{-1})	276[b]			13[b]	
ΔH_f^0 (kJ·mol^{-1})	277[b]	80[c]	138[c]	−7[b]	94.3[d]
E^0 (V)[e]	−2.9[f]	−0.33[g]	0.79[h]	1.9[i]	0.65[j]
$[I]_{SS}$ (*M*)	10^{-17}–10^{-15}	10^{-9}–10^{-7}		10^{-19}–10^{-17}	10^{-14}–10^{-13}
k (M^{-1}·s^{-1})[k]	3.0(7)	5.8(2)	8.5(3)	1.0(10)	0.2(7)
τ (days × 10^3)	0.3–30	0.0001–0.01		0.08–8	0.04–0.4

[a]The rate constant data is for reaction with C_6H_5OH (phenol) except for the HO_2 radical, which is for reaction with 1,4-$C_6H_4(OH)_2$.
[b]Data from Reference 4.
[c]Data from Reference 5.
[d]Gas phase value.
[e]Data from Reference 6.
[f]$H_2O + e^- \rightarrow e^-(aq)$
[g]$O_2 + e^- \rightarrow O_2^-$
[h]$HO_2 + e^- \rightarrow HO_2^-$
[i]$OH + e^- \rightarrow OH^-$
[j]$^1O_2^* + e^- \rightarrow O_2^-$
[k]Number in parentheses indicates the exponent value. Data from Reference 7.

reactivity with alkylphenols (13) and so may be an important reaction channel for some pollutants.

$^1O_2$*

Singlet oxygen, $^1O_2$*, is an electronically excited state of oxygen with 94.3 kJ of excess energy. Triplet excited ^{3}DOM*, generated by intersystem crossing from the initially excited singlet state, can sensitize the formation of $^1O_2^*$ through energy transfer, Eq. 6. Most of the singlet oxygen will be rapidly quenched back to the triplet ground state by H_2O ($\tau \approx 4$ μs) and only a fraction of the $^1O_2^*$ will be available for subsequent reaction (2). The production of $^1O_2^*$ generally scales with the amount of DOC and steady-state concentrations of $0.3–3.0 \times 10^{-14}$ M per mg·L^{-1} of DOC have been measured for typical surface waters (14). Singlet oxygen is a reactive electrophile, particularly toward cyclic 1,3 dienes, phenolate anions, and some polynuclear aromatics (2,14,15) and can play an important role in aqueous photochemistry.

OH

The hydroxyl radical is a powerful oxidant (Table 1) that exhibits high reactivity toward a range of organic species. It is mainly formed from photodissociation of NO_3^-/NO_2^- but also via photolysis of metal-hydroxides, decomposition of H_2O_2, and DOM photochemistry. The hydroxyl radical is rapidly scavenged by DOM and the steady-state concentration of OH in surface waters is generally small, $[OH]_{SS} \approx 10^{-19}–10^{-17}$ M (9,10). The OH radical readily reacts with many organic species, via abstraction and addition to double bonds, but the low steady-state concentration suggests that hydroxyl will not be the dominant photooxidant for most surface waters. For example, a "typical" OH rate constant (4) is 5×10^9 M$^{-1} \cdot$ s^{-1}, corresponding to a reaction half-life of $\tau \approx 160$ d at the higher OH concentrations. However, for environments with a larger concentrations of NO_x^- relative to DOC, hydroxyl radical reactions can be important pathways due to near diffusion-controlled OH rate constants (2).

H_2O_2

Hydrogen peroxide is a relatively stable, nonradical species produced from photochemical processes in natural surface waters. It is not a primary product but, rather, is generated in secondary reactions such as the disproportionation of superoxide, Eqs. 13 and 14. Hydrogen peroxide absorbs only weakly in the UV ($\varepsilon = 1$ M$^{-1} \cdot$ cm^{-1} at 310 nm) (16), and decomposition to yield hydroxyl radicals may be dominated by biological mediation and redox cycles involving trace metals

$$H_2O_2 \longrightarrow 2OH \qquad (15)$$

The slow decomposition of H_2O_2 leads to relatively high steady-state concentrations in surface waters, $[H_2O_2]_{SS} = 10^{-8}–10^{-7}$ M (2,9).

CHROMOPHORES

NO_3^-/NO_2^-

Inorganic nitrate and nitrite ions are perhaps the most well defined and easily understood photoactive component of natural waters. Nitrate concentration in surface waters is on the order of 1 mg·L^{-1} although it may be substantially higher in waters subject to an influx of agricultural drainage due to the heavy use of nitrogen fertilizers. The absorption band maxima for nitrate and nitrite are at 302 nm ($\varepsilon = 3.1$ M^{-1}·cm^{-1}) and 352 nm ($\varepsilon = 9.6$ M^{-1}·cm^{-1}), respectively (17,18). Ultraviolet excitation of nitrite leads to photodissociation (18):

$$NO_2^- + h\nu \longrightarrow NO + {}^\bullet O^- \qquad (16)$$

The $^\bullet O^-$ species, the conjugate base of hydroxyl radical, is rapidly protonated at pH < 12 [pK_a($^\bullet$OH) = 11.9] (4):

$$^\bullet O^- + H^+ \longrightarrow {}^\bullet OH \qquad (17)$$

The quantum yield for $^\bullet$OH production from nitrite varies from 0.015 to 0.08 over the wavelength range 298–371 nm (18,19). Photolysis of the nitrate ion yields either nitrite ion or nitrogen dioxide, which can then reform nitrate and nitrite:

$$NO_3^- + h\nu \longrightarrow NO_2^- + O \qquad (18)$$

$$NO_3^- + h\nu \longrightarrow {}^\bullet NO_2 + {}^\bullet O^- \qquad (19)$$

$$2{}^\bullet NO_2 + H_2O \longrightarrow NO_2^- + NO_3^- + 2H^+ \qquad (20)$$

For wavelengths ≥290 nm, the quantum yield for formation of $^\bullet$OH from NO_3^-, via Eq. 19, is in the range of $9.2–17 \times 10^{-3}$ (17,18). Hydroxyl is a highly reactive radical and a strong oxidizing agent (see above). Various laboratory studies have shown that trace organic molecules, such as biologically refractory pollutants, are degraded by irradiation in the presence of nitrate/nitrite (17–19).

DOM

Dissolved organic matter (DOM) is a major and ubiquitous component of natural surface waters, having numerous sources such as soils, sediments, and aqueous biological activity (20). DOM concentrations are often expressed in terms of the dissolved organic carbon (DOC) fraction, which comprises about 60% of the total DOM. Continental waters have typical levels of 1–10 mg C·L^{-1} DOC, although some waters, bogs, marshes, and shallow streams, for instance, can have concentrations exceeding 50 mg C·L^{-1}. Marine DOC is about 1 mg C·L^{-1} and represents an enormous carbon reservoir, approximately equal in size to the CO_2 reservoir in the atmosphere (2). DOM is nominally defined as the organic content of water that passes through a filter of some pore size, 0.1–0.5 μm, and so will have a colloidal component. The DOC in surface waters can be divided into two fractions; a lower molecular weight nonhumic fraction and higher molecular weight humic component. The nonhumic fraction is about 10–60% of the DOC and is made up of small compounds of mainly biogenic origin, such as peptides, carbohydrates, fats, and chlorophyll, which are subject to rapid biological degradation. In addition, anthropogenically derived species, such as biocides (often halogenated compounds) and polycyclic aromatic hydrocarbons (PAHs, from combustion processes) are present. The higher molecular weight fraction (>300 Da), comprising 40–90% of the DOC, is termed humic matter (HM). Humic matter is a complex material composed of condensed biopolymers and is much more biologically refractory. HM is mildly acidic, has a significant aromatic content, and is highly functionalized with carboxyl, phenol, hydroxyl, carbonyl, ether, and ester functional groups present. HM can be further divided based on solubility properties into humic acids (HAs), which precipitate under acidic conditions, and fulvic acids (FAs), which are soluble at all pH values. FAs are more highly substituted and have a higher oxygen content than HAs and also have a lower average molecular weight.

DOM is a highly colored material with an absorption spectrum that overlaps significantly with the solar actinic flux (see Fig. 2). The UV absorption of surface waters has been found to be highly correlated with the DOC content (10,21,22). Studies have shown that photoexcited DOM can yield many of the radical species outlined in Fig. 1 and Table 1 through primary and secondary processes (9). The radical production rate varies with the character of the DOM and the concentration, as well as the location and type of surface water environment. One study found that coastal marine waters yielded radical production rates of $0.18–1.8\ \mu M{\cdot}d^{-1}$ in the summer due to irradiation of humic material in the surface layer (to a depth of 1.5 m) (23). The active spectral region was determined to be 300–385 nm, suggesting a quantum yield for radical production of $1–2 \times 10^{-3}$. Measurement of light absorption rates in highly colored, organic-rich waters that exhibit significant absorption at longer, visible wavelengths suggest the radical production rate will be orders of magnitude higher (9). These radicals dominate the photochemistry of most natural water samples, particularly in marine environments.

The DOM will be altered by photolysis, yielding a cationic radical product, for instance, but it can also be attacked by the photogenerated radicals. DOM decomposition can lead to photobleaching, allowing solar radiation to penetrate deeper into the water and impacting microbial photosynthesis. DOM is generally biologically refractory, resulting in long carbon reservoir residence times, estimated at 6000 years for deep ocean DOM (24). Photodecomposition of DOM may be an important removal pathway with significant implications for the global carbon cycle (9,15,24,25). The lower molecular weight degradation products are biologically labile and can provide a major fraction of the biological carbon uptake in certain surface waters (21,25).

Fe(III/II)

Iron is an abundant crustal metal and is one of the most important trace metals in natural water systems. The Fe-(hydrous)oxides (e.g., α-FeOOH, goethite; α-Fe_2O_3, hematite; γ-FeOOH, lepidocrocite; $Fe(OH)_3$, ferrihydrite) may be present as dissolved species but are predominantly in colloidal form. Iron has two common oxidation states, Fe(III) and Fe(II), and participates in redox reactions that are an important component of aqueous chemical cycles. Fe(III) can form strong complexes with naturally occurring organic ligands, such as polycarboxylate anions (e.g., oxalate, malonate, citrate), and humic and fulvic acids (11,26). Photoexcitation of a Fe(III) complex promotes a ligand to metal charge transfer process yielding a free, oxidized ligand and a reduced, Fe(II), center. Subsequent reactions generate reactive intermediates (11,12), as schematically depicted below for the oxalato ($C_2O_4^{2-}$) ligand:

$$>\mathrm{Fe(III)C_2O_4^-} + h\nu \longrightarrow\ >\mathrm{Fe(II)} + \mathrm{C_2O_4^-} \quad (21)$$

$$\mathrm{C_2O_4^-} \longrightarrow \mathrm{CO_2} + \mathrm{CO_2^-} \quad (22)$$

$$\mathrm{CO_2^-} + \mathrm{O_2} \longrightarrow \mathrm{CO_2} + \mathrm{O_2^-} \quad (23)$$

$$>\mathrm{Fe(II)} + \mathrm{O_2} \longrightarrow\ >\mathrm{Fe(III)} + \mathrm{O_2^-} \quad (24)$$

$$>\mathrm{Fe(II)} + \mathrm{H_2O_2} \longrightarrow\ >\mathrm{Fe(III)} + \mathrm{OH} \quad (25)$$

where ">Fe(III/II)" represents either an iron colloid surface site or a dissolved species. The superoxide will form hydrogen peroxide and hydroperoxyl radicals, as discussed above. In colloidal material, the surface Fe(II) is generally oxidized back to Fe(III) by the ligand or O_2. However, due to weakened binding with neighbor atoms, the reduced species may also be released to the solution (3,26). The result is photodissolution of the mineral colloid and an increase in soluble iron species, facilitating redox reactions and promoting microbial activity.

BIBLIOGRAPHY

1. Finlayson-Pitts, B.J. and Pitts, J.N., Jr. (1986). *Atmospheric Chemistry: Fundamentals and Experimental Techniques*. Wiley-Interscience, Hoboken, NJ.
2. Hoigne, J., Faust, B.C., Haag, W.R., Scully, J.F.E., and Zepp, R.G. (1989). Aquatic humic substances as sources and sinks of photochemically produced transient reactants. In: *Aquatic Humic Substances*. Vol. 219, I.H. Suffet and P. MacCarthy (Eds.). American Chemical Society, Washington, DC, pp. 363–381.
3. Stumm, W. and Morgan, J.J. (1996), 3rd Edn. John Wiley & Sons, New York.
4. Buxton, G.V., Greenstock, C.L., Helman, W.P., and Ross, A.B. (1988). Critical review of rate constants for reactions of hydrated electrons, hydrogen atoms and hydroxyl radicals (OH/O^-) in aqueous solution. *J. Phys. Chem. Ref. Data* **17**: 513–886.
5. Spinks, J.W.T. and Woods, R.J. (1990). *An Introduction to Radiation Chemistry*, 3rd Edn. John Wiley & Sons, Hoboken, NJ.
6. Wardman, P. (1989). Reduction potentials of one-electron couples involving free radicals in aqeous solution. *J. Phys. Chem. Ref. Data* **18**: 1637–1755.
7. NDRL Radiation Chemistry Data Center. Available: http://www.rcdc.nd.edu/.
8. Seinfeld, J.H. and Pandis, S.N. (1998). *Atmospheric Chemistry and Physics: From Air Pollution to Climate Change*. John Wiley & Sons, Hoboken, NJ.

9. Mopper, K. and Kieber, D.J. (2000). Marine photochemistry and its impact on carbon cycling. In: *The Effects of UV Radiation in the Marine Environment*. Vol. 10. S.D. Mora, S. Demers, and M. Vernet (Eds.). Cambridge University Press, Cambridge, UK, p. 101–129.
10. Cooper, W.J., Zika, R.G., Petasne, R.G., and Fischer, A.M. (1989). Sunlight-induced photochemistry of humic substances in natural waters: major reactive species. In: *Aquatic Humic Substances*. Vol. 219. I.H. Suffet and P. MacCarthy (Eds.). American Chemical Society, Washington, DC, pp. 333–362.
11. Faust, B.C. (1994). A review of the photochemical redox reactions of iron(III) species in atmospheric, oceanic, and surface waters: influences on geochemical cycles and oxidant formation. In: *Aquatic and Surface Photochemistry*. G.R. Helz, R.G. Zepp, and D.G. Crosby (Eds.). Lewis Publishers, Boca Raton, FL, pp. 3–37.
12. Zuo, Y. and Hoigne, J. (1992). Formation of hydrogen peroxide and depletion of oxalic acid in atmospheric water by photolysis of iron(III)-oxalato compounds. *Environ. Sci. Technol.* **26**: 1014–1022.
13. Faust, B.C. and Hoigne, J. (1987). Sensitized photooxidation of phenols by fulvic acid and in natural waters. *Environ. Sci. Technol.* **21**: 957–964.
14. Haag, W.R. and Hoigne, J. (1986). Singlet oxygen in surface waters. 3. Photochemical formation and steady-state concentration in various types of waters. *Environ. Sci. Technol.* **20**: 341–348.
15. Aguer, J-P., Richard, C., and Andreux, F. (1997). Comparison of the photoinductive properties of commercial, synthetic and soil-extracted humic substances. *J. Photochem. Photobiol. A* **103**: 163–168.
16. Bolton, J.R. and Cater, S.R. (1994). Homogeneous photodegradation of pollutants in contaminated water: an introduction. In: *Aquatic and Surface Photochemistry*. G.R. Helz, R.G. Zepp, and D.G. Crosby (Eds.). Lewis Publishers, Boca Raton, FL, pp. 467–490.
17. Zepp, R.G., Hoigne, J., and Bader, H. (1987). Nitrate-induced photooxidation of trace organic chemicals in water. *Environ. Sci. Technol.* **21**: 443–450.
18. Alif, A. and Boule, P. (1991). Photochemistry and the environment part XIV. Phototransformation of nitrophenols induced by excitation of nitrite and nitrate ions. *J. Photochem. Photobiol. A* **59**: 357–367.
19. Marbury, S.A. and Crosby, D.G. (1994). The relationship of hydroxyl reactivity to pesticide persistence. In: *Aquatic and Surface Photochemistry*. G.R. Helz, R.G. Zepp, and D.G. Crosby (Eds.). Lewis Publishers, Boca Raton, FL, pp. 149–161.
20. Beckett, R. (1990). The surface chemistry of humic substances in aquatic systems. In: *Surface and Colloid Chemistry in Natural Waters and Water Treatment*. R. Beckett (Ed.). Plenum Press, New York, pp. 3–20.
21. Kieber, R.J., Zhou, X., and Mopper, K. (1990). Formation of carbonyl compounds from UV-induced photodegradation of humic substances in natural waters: fate of riverine carbon in the sea. *Limnol. Oceanogr.* **35**: 1503–1515.
22. Miller, W.L. (1994). Recent advances in the photochemistry of natural dissolved organic matter. In: *Aquatic and Surface Photochemistry*. G.R. Helz, R.G. Zepp, and D.G. Crosby (Eds.). Lewis Publishers, Boca Raton, FL, pp. 111–127.
23. Zafiriou, O.C. and Dister, B. (1991). Photochemical free radical production rates: Gulf of Maine and Woods Hole-Miami transect. *J. Geophys. Res.* **96**: 4939–4945.
24. Mopper, K. et al. (1991). Photochemical degradation of dissolved organic carbon and its impact on the oceanic carbon cycle. *Nature* **353**: 60–62.
25. Kieber, D.J., McDaniel, J., and Mopper, K. (1989). Photochemical source of biological substrates in sea water: implications for carbon cycling. *Nature* **341**: 637–639.
26. Sulzberger, B., Laubscher, H., and Karametaxas, G. (1994). Photoredox reactions at the surface of iron(III) (hydr)oxides. In: *Aquatic and Surface Photochemistry*. R. Helz, R.G. Zepp, and D.G. Crosby (Eds.). Lewis Publishers, Boca Raton, FL, pp. 53–73.

ISOTOPE EXCHANGE IN GAS-WATER REACTIONS

Tom Wydrzynski
Warwick Hillier
The Australian National University
Canberra, Australia

INTRODUCTION

Molecular oxygen began to accumulate in Earth's atmosphere about 2.2–2.7 billion years ago, and it is widely accepted that it originated from the photosynthetic oxidation of water (1). The accumulation of O_2 in the atmosphere marked the most important event in the evolution of life on Earth as the atmosphere changed from a highly reducing to an oxidizing condition. All preexisting life forms had to either find an anaerobic ecological niche or adapt to the ever-increasing O_2 levels of the atmosphere.

Today on Earth, the photosynthetic O_2 generators are widely distributed on land and in the oceans in the form of plants, algae, and cyanobacteria. The current atmospheric pool of O_2 contains about 4×10^{19} mols and takes about 3.7 million years to turnover. The unique enzyme complex that produces the O_2 by photosynthetic organisms is called photosystem II (PSII), which acts as a water/plastoquinone oxido-reductase (2). PSII contains a special photochemical reaction center, which at excitation can generate strong oxidants that can be coupled to the splitting of two water molecules into O_2, four protons, and four electrons. The protein motif within PSII that catalyzes the water oxidation reaction contains a cluster of four Mn ions, a Ca ion, and a redox-active tyrosine and is often referred to as the O_2 evolving complex (OEC).

The oxidation of water by PSII is driven photochemically by the capture of light quanta (3,4). The chlorophyll binding proteins associated with peripheral regions of the PSII complex serve as light-harvesting antenna and direct spectrally diverse quanta to the special reaction center chlorophyll cluster called P680. At absorption of a light quantum, P680 generates an excited state with its pheophytin partner (P680-Pheo)*, which in turn is in equilibrium with the exciton in the antenna and the radical pair $P680^+$-$Pheo^-$. The time course for this reaction has been under some debate, but it is in the order of 3–20 ps. The primary charge separation is rapidly followed by a charge stabilization event, in which the electron is rapidly transferred to a tightly bound plastoquinone molecule, Q_A, in 300–500 ps, which results in the formation of $P680^+$-Pheo Q_A^-. The electron hole in $P680^+$ is subsequently filled with an electron derived from a redox-active tyrosine, Y_Z, located in the D1 protein subunit within some 40–380 ns, which results in the formation of Y_Z^{ox} P680-Pheo Q_A^-. The electron at Q_A^- is then transferred to a second bound plastoquinone molecule, Q_B, in 100–200 μs, whereas Y_Z^{ox} is reduced by the Mn_4Ca cluster with S-state-dependent kinetics (see below) in the range of 30–1300 μs.

One key finding for understanding the water oxidation reaction is the observation that for dark-adapted, PSII-containing samples illuminated by a series of brief, saturating light flashes, O_2 is released with a damped periodicity of four. The peak O_2 yields occur on the third, seventh, eleventh, and so on flashes and eventually damp out after the third or fourth cycle (5, see Figs. 2B and 2C). The phenomenological model to explain this observation was developed by Kok et al. (6,7) and entails a cyclic reaction sequence in which the OEC traverses through five intermediate states called the S states. Beginning in S_0 and progressing to S_4, each S state is advanced by a single quantum event at P680. When it reaches S_4, O_2 is released within 1–2 ms, S_0 is regenerated, and the cycle begins anew (see Fig. 1).

The damping in the O_2 flash yield oscillations (which is caused by the desynchronization of the forward S state transitions within a given population of O_2 evolving centers) is explained by a certain probability for a "miss" (α) to occur during a flash in which an electron hole fails to accumulate in the OEC and, depending on the flash duration, by a certain probability for a "double hit" (β) to occur in which a double excitation event takes place. The maximum O_2 yield on the third flash is explained by the condition that the cycle initially starts in the S_1 state after complete dark-adaptation. The higher S states (S_2 and S_3) exhibit lifetimes of tens of seconds to several minutes, depending on the temperature, and deactivate to S_1 if not advanced by another incoming quantum, whereas the S_0 state slowly relaxes in the dark to the S_1 state (8). The S_4 state being metastable reacts immediately to produce O_2 and the S_0 state.

DURING WHICH STEP IN THE S STATE CYCLE DOES THE SUBSTRATE WATER BIND?

In the original Kok et al. hypothesis, it was implied that the two substrate water molecules entered the reaction sequence during the last step, just before O_2 release. Although this notion has been considered throughout the literature, the most recent models for the OEC generally predict that the substrate water binds to the catalytic site at the beginning of the S state cycle and that a concerted four-electron event occurs during the S_3 to S_4 to S_0 transition to create the O–O bond (9,10). However, definitive experimental evidence to show at which step in the S state cycle the substrate water binds and the O–O bond forms has until recently remained elusive.

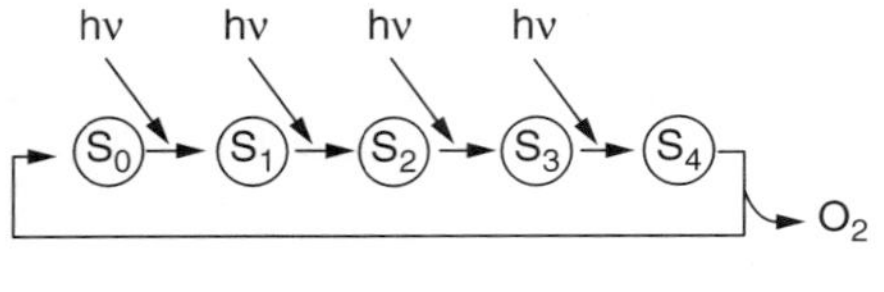

Figure 1. OEC Cycle.

The binding of the substrate water to the catalytic site can be probed by oxygen isotope exchange measurements between the solvent water and the photogenerated O_2 using mass spectrometry. Basically, the measurement involves the transfer of a photosynthetic sample into labeled water of known oxygen isotopic composition (i.e., ^{18}O-enriched water) and then determining the extent of isotopic incorporation into the photogenerated O_2.

In the first oxygen isotope exchange measurements (11–13), the fastest detectable rate of exchange was limited to ~30 s, mainly because of the long stabilization time of the "open" chamber system that introduced isotopically labeled water to the sample. However, in the more recent measurements from our group, the stabilization time was greatly improved, by a factor of nearly 5000, through the introduction of a "closed" chamber system (14,15). The results from this experimental approach revealed for the first time that one substrate water molecule is bound to the catalytic site throughout the S-state cycle, whereas the second substrate water molecule is bound in at least the S_2 and S_3 states, and that the O–O bond indeed forms during the last step of the reaction sequence. This article documents the experimental approach and the arguments that come to these conclusions.

RAPID ^{18}O EXCHANGE MEASUREMENTS DURING PHOTOSYNTHETIC O_2 EVOLUTION

To probe the substrate water bound at the OEC, excess ^{18}O isotope (added as ^{18}O-labeled water) is introduced into the sample to set up a new isotopic distribution, because natural abundance ^{18}O in the solvent water is low at about 0.2%. The effect of the new isotopic distribution on the photogenerated O_2 can then be measured at m/e = 32, m/e = 34, and m/e = 36 with an in-line mass spectrometer (e.g., the VG Instruments IsoPrime stable isotope spectrometer Manchester UK) for the $^{16}O^{16}O$, $^{16}O^{18}O$, and $^{18}O^{18}O$ di-oxygen products, respectively. From these measurements, the extent of ^{18}O incorporation during the O_2 evolving process can be determined. Figure 2(a) shows the O_2 flash yields at these three masses for spinach thylakoid membrane samples after the injection of an aliquot of $H_2{}^{18}O$ (e.g., 98.5% enriched water from ISOTECH, Miamisberg, Ohio) that gives a final ^{18}O isotopic enrichment in the sample of $\varepsilon \sim 12\%$. In this case, after complete equilibration, the absolute yields (as determined by the extrapolated lines shown in the figure for the third flash) at the three masses follow closely the theoretical isotope equilibrium distribution: $32{:}34{:}36 = (1-\varepsilon)^2 : 2\varepsilon(1-\varepsilon) : \varepsilon^2 = 100\%$, which indicates that no major isotopic discrimination occurs during the O_2 evolving process, as shown earlier from steady-state measurements (16).

Figure 2(b) shows a plot of the normalized O_2 yields at m/e = 32 as a function of flash number. The oscillatory

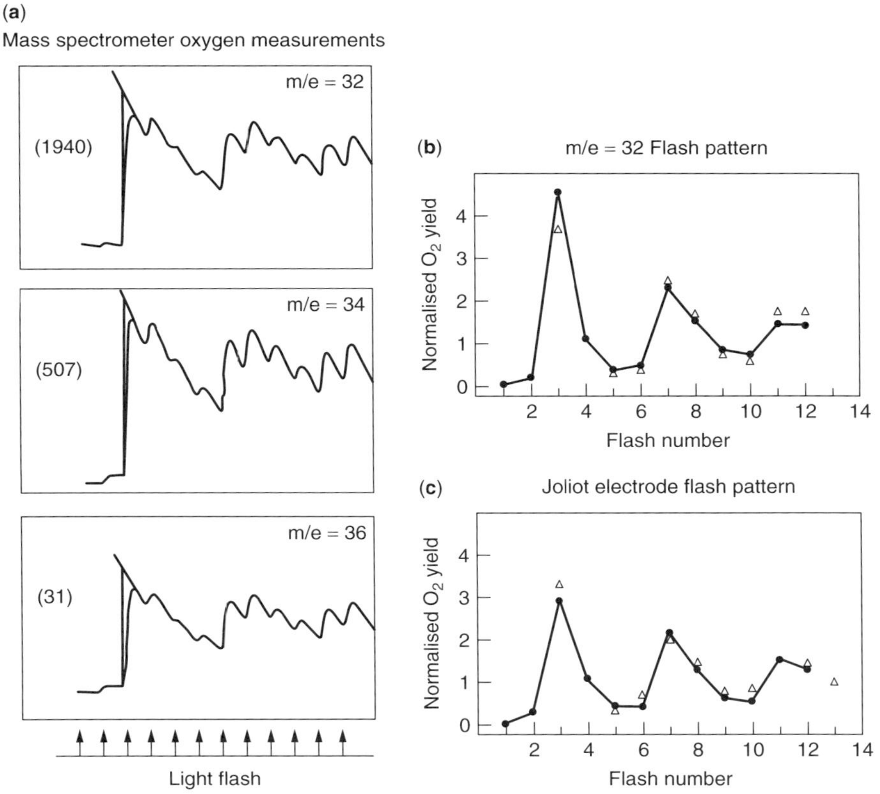

Figure 2.

pattern in the mass spectrometric measurements follows very closely the flash pattern in Fig. 2(c), which was obtained by the conventional amperometric method using a so-called Joliot-type oxygen electrode. Both types of measurements give rise to similar Kok et al. parameters, which indicates that the two techniques are measuring the same phenomenon.

To get information on the binding affinities of the substrate water in the OEC, the rate of isotope exchange between ^{18}O-labeled water and the photogenerated O_2 needs to be determined. To achieve this, our group designed a "closed" chamber system in which the sample volume is considerably reduced (160 μl) and a stirrer is incorporated to facilitate rapid mixing. By employing a solenoid-actuated Hamilton syringe, ^{18}O-labeled water can be rapidly injected and equilibrated with the sample within 6–8 ms. Thus, by measuring the extent of isotope incorporation as a function of the time after injection of the ^{18}O-labeled water, the rate of isotope exchange can be determined.

To evaluate the S-state dependence on the rate of isotope exchange, the "closed" chamber is optically coupled to a bank of three xenon flash lamps (e.g., FX-193 lamp, 4 μF @ 1 kV capacitor from EG & G, Salem, MA) via a 3-to-1 fiber optic to variously illuminate the sample with brief (FWHH ~8 μs), saturating light flashes. The samples are first given a series of activating flashes to preset the sample into a particular S state (see below). Then, after the rapid addition and equilibration of the ^{18}O-labeled water, a subsequent turnover flash sequence (spaced 5–10 ms apart, depending on S state, see below) is applied to photogenerate O_2. As the actual instrumental response time is relatively slow (being about 10 s because of the diffusion of the O_2 gas across a semipermeable membrane into the inlet line from the sample chamber), the flash spacings of a subsequent flash sequence that normalizes the oxygen signals are increased, typically to 20 s. As such, to retard the deactivation reactions of the higher S states during these flash spacings (see above), the temperature of the sample is reduced (usually to 10 °C), which is accomplished by enclosing the sample chamber in a cooling block. Details of the setup have been published earlier (63,64) and are given in Fig. 3.

S-STATE DEPENDENCE ON THE RATE OF ISOTOPE EXCHANGE BETWEEN ^{18}O-LABELED WATER AND THE PHOTOGENERATED O_2

To determine S-state dependence on the rate of isotope exchange, the mass measurements are made as a function of time between the addition and equilibration of ^{18}O-labeled water with the sample in a particular S-state and the first peak yield of the photogenerated O_2. Thus, for example, for the S_3 state, an S_1-enriched sample suspended in $H_2{}^{16}O$ is given two preflashes to advance the OEC into the S_3 state. $H_2{}^{18}O$ is then rapidly injected and followed by a variable delay time (Δt), which allows for the time-dependent isotopic exchange to occur between the solvent water and the substrate water. A single flash is then given, and the oxygen signals at m/e = 34 and m/e = 36 are measured. The oxygen signals are then collected on subsequent flashes (in which complete

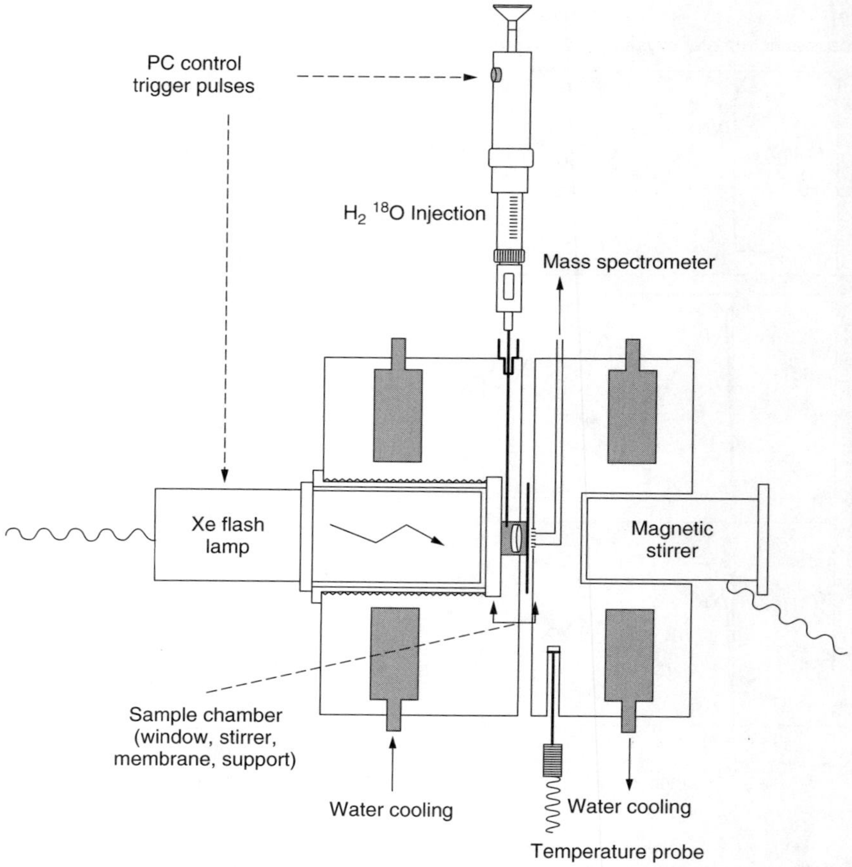

Figure 3.

exchange has occurred because of the 20 s flash spacings) to normalize the data among the different sample aliquots measured at different Δt values.

The complete ^{18}O exchange kinetics were determined for the different S states using the injection/flash protocols depicted in Fig. 4.

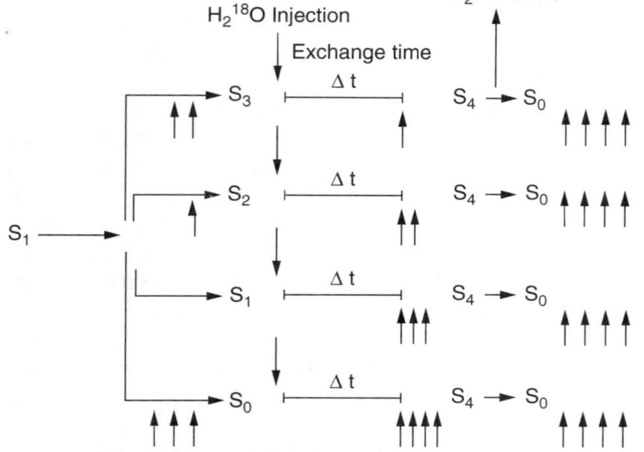

Figure 4.

The basic protocol involves the following procedure. Before loading, the samples are thawed in the dark at room temperature and given one saturating preflash followed by a 10-min dark period at room temperature to enrich the S_1 population. The S_1-enriched sample is then loaded into the sample chamber in the dark, equilibrated to 10 °C for 10–12 min, and then given several activating flashes (spaced 1 s apart). At a given S state, $H_2{}^{18}O$ is rapidly injected into the chamber ($t_{1/2\ (mixing)} \sim 4$ ms) and a variable delay time (4 ms $< \Delta t <$200 s) is provided to allow the water bound in the OEC to exchange with the ^{18}O-enriched solvent water. After the delay time, several rapid turnover flashes (spaced at 5, 10, and 10 ms apart for S_2, S_1, and S_0 states, respectively) are then applied to advance the sample through to the S_3 to S_4 to S_0 transition, where the mass ratio of the photogenerated O_2 (Y_M) is determined. After this process, several normalization flashes (spaced 20 s apart) are given and their separate O_2 yields ($Y_{n(1\rightarrow 4)}$) are measured. The Y_M is corrected for background O_2 contributions, double hits, and the injection response as described earlier (17,18) to give Y_C.

The resolvable ^{18}O exchange kinetics for spinach thylakoid samples as a function of the S-states are shown in Fig. 5. Here, the corrected, normalized yields of O_2 (Y_C) measured at m/e = 34 for the mixed labeled $^{16}O^{18}O$ product (left side of the figure) and at m/e = 36 for the double labeled $^{18}O^{18}O$ product (right side of the figure) are presented. Each point is a plot of Yc at a particular delay time (Δt) between the injection of $H_2{}^{18}O$ into the sample and the turnover flash sequences that photogenerated O_2. The m/e = 36 data exhibit only a single kinetic phase and are fit with a simple exponential function:

$$^{36}Y = (1 - \exp(-^{36}k\ t)) \quad (1)$$

In contrast, the m/e = 34 data on the left side in Fig. 5 reveals two distinct kinetic phases. The insets show expanded time ordinates for the fast phases. The two phases, however, are unequal in amplitude, with the fast phase constituting slightly more than half of the total signal. The basis for this difference in amplitude is well explained by the enrichment condition for two independent, exchanging sites. As the apparent kinetics of the two phases differ by at least a factor of 10, the fast phase of exchange is virtually complete before the slow phase begins. Thus, at short Δt only one substrate water molecule is exchanging at the catalytic site, which means at a typical ^{18}O enrichment of $\varepsilon = 12\%$, the mass distribution at 32:34:36 for the two oxygen isotopes will be 88:12:0. On the other hand, at longer Δt, when the second substrate water molecule is also exchanging, the mass distribution will be 77.44:21.12:1.44. Therefore, the relative contributions of the fast and slow phases will be unequal, with the fast phase representing ~57% (i.e., 12/21.12) of the total amplitude and the slow phase ~43%. The 0.57:0.43 distribution between the two phases is found consistently in the m/e = 34 data. Thus, the exchange kinetics in this case is fit exactly by the sum of two exponentials:

$$^{34}Y = 0.43(1 - \exp(-^{34}k_1\ t)) + 0.57(1 - \exp(-^{34}k_2\ t) \quad (2)$$

The solid lines in Fig. 5 represent the best fits to the data points based on Eqs 1 and 2 for the m/e = 36 and m/e = 34 data, respectively. The exchange rates from the corresponding kinetic fits at 10 °C are given in Table 1. The entire S-state results for the thylakoids have been presented earlier (17,18).

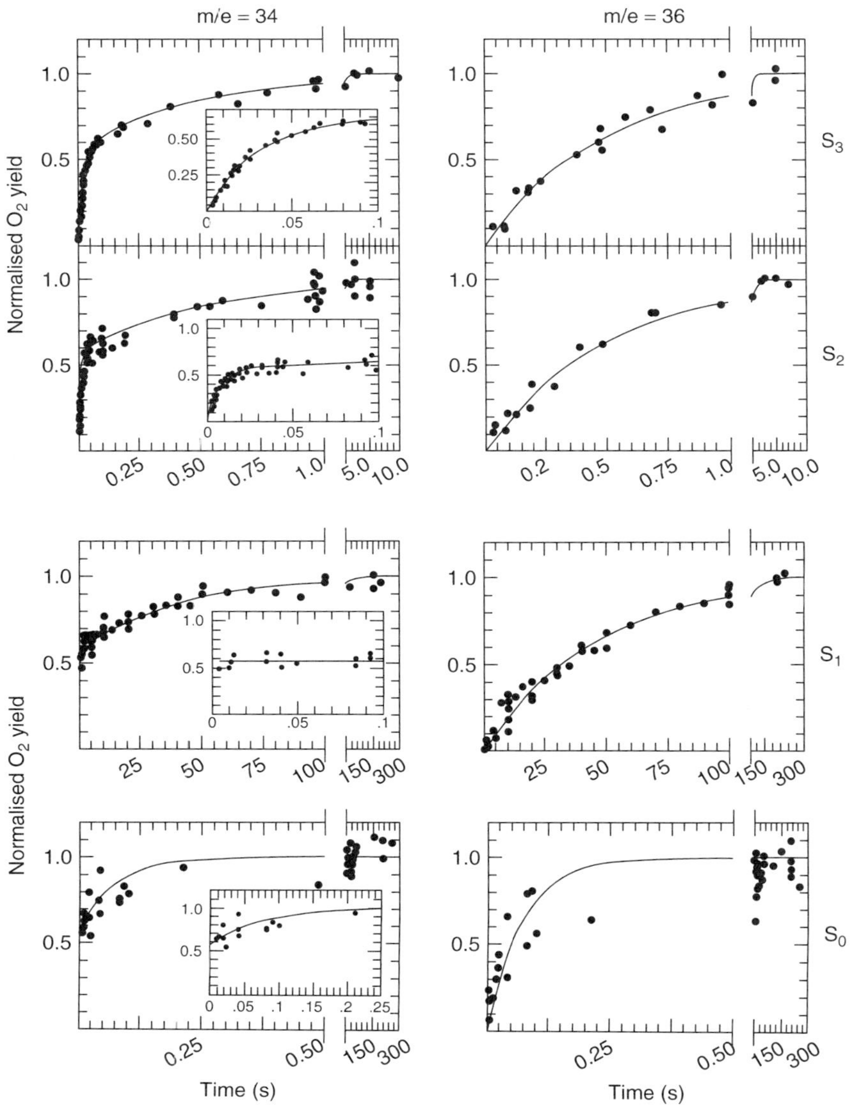

Figure 5.

Table 1. Rate Constants for the ^{18}O Exchange in Spinach Thylakoid Samples as a Function of the S States at 10 °C

S-State	^{36}k (s^{-1})	$^{34}k_1$ (s^{-1})	$^{34}k_2$ (s^{-1})
S_3	2.1 ± 0.2	1.9 ± 0.2	36.8 ± 1.9
S_2	2.2 ± 0.1	1.9 ± 0.3	120 ± 6
S_1	0.022 ± 0.002	0.021 ± 0.002	>175
S_0	18 ± 3	8 ± 2	>175

Adapted from Ref. 18.

IMPLICATIONS OF SUBSTRATE WATER BINDING FOR THE MECHANISM OF O_2 EVOLUTION

Several striking features are found in the rate constants listed in Table 1. In each S state, the slow phase of exchange was resolvable, whereas the fast phase of exchange could be resolved only in the S_2 and S_3 states. For the S_0 and S_1 states, the fast phase of exchange is equal to or faster than the response time of the currently available experimental setup ($\sim 100 - 175\ s^{-1}$). Most importantly, the rate constants for the slow phase kinetics in the m/e = 34 data are virtually identical to the rate constants for the monophasic kinetics in the m/e = 36 data. This last result clearly indicates that the extent of ^{18}O incorporation into the photogenerated O_2 is limited throughout the S-state cycle by the substrate water molecule undergoing the slow isotope exchange process and that the biphasic behavior in the m/e = 34 data does not originate from PSII heterogeneity. Rather the two kinetic components in the m/e = 34 data reflect the binding of the substrate water at two separate sites.

The most relevant question to ask is at which S state does the substrate water bind. Based on the kinetic resolution of our current mass spectrometric setup, a measurable ^{18}O exchange kinetic will conclusively show the existence of bound water in a particular S state, which is justified by the observations that the O_2 release on the S_3 to S_4 to S_0 transition takes place in ~2 ms, whereas the rest of the S-state advancement from S_0 to S_3 takes at most another 1 ms (8). The catalytic site thus can make one complete turnover in ~3 ms. Therefore, any ^{18}O exchange that is slower than 3 ms would be indicative of bound water. From the rate constants for the slow phase of ^{18}O exchange listed in Table 1 for the m/e = 34 data, it is clear that one substrate water molecule is bound to the catalytic site through the S-state cycle. Interestingly, the results for the fast phase of exchange show that the second substrate water molecule is bound in at least the S_2 and S_3 states, but they are inconclusive for the S_0 and S_1 states because the fast exchange kinetics in these S states are not resolvable by our current setup (i.e., faster than 6–8 ms).

Even though the measurable exchange rates listed in Table 1 are slower than the total turnover time of the catalytic site, the magnitudes of these exchange rates in general are all remarkably fast, where the slowest exchange is measured in the S_1 state with $t_{1/2}$ ~50 s. Exchange rates in this magnitude range indicate relatively labile binding sites and can be interpreted for two overall mechanisms: (1) exchange of water bound to a metal site or (2) diffusional or isotope equilibration processes, where the solvent water must be translocated to the binding site across a physical barrier.

Rates of whole water exchange at a metal site can span a tremendous range (as much as 18 orders of magnitude), and examples for various transition metals have been summarized earlier (19–22). The most significant factors governing water exchange at a metal site are the charge and ionic radius of the metal center and any electronic occupancy of d orbitals. As a general rule, water exchange rates will decrease (by several orders of magnitude) as the metal center (M^n) is oxidized and charge is increased; i.e., k_{ex} (M^n-OH_2) > k_{ex} (M^{n+1}-OH_2). Furthermore, metal-bound water can undergo deprotonation in response to an increase in charge (and acidity) at the metal center. Thus, it may also be expected for water exchange to decrease at deprotonation for a given oxidation state; i.e., k_{ex} (M^n-OH_2) > k_{ex} (M^n-OH). In the case of the deprotonation of a neighboring ligand (L), the opposite effect may occur because the effective charge of the metal center is reduced, As such the rate of water exchange will tend to increase; i.e., k_{ex} (HL-M^n-OH_2) < (L-M^n-OH_2). The protonation state will thus be defined by the pK of the aquo metal complex.

Any number of possible intermediates in the water oxidation reaction can exist, and it is important to appreciate that the ^{18}O exchange measurements could reflect not only whole water and hydroxyl exchange but also the exchange of μ-oxo (–O–), peroxo (–O–O–), and oxo (=O) intermediates. Indeed various models in the literature on O_2 evolution invoke one or more of these intermediates (9,10). Although a dearth still exists in studies of ^{18}O exchange in model compounds, it is expected that the isotope exchange in μ-oxo (–O–), perxo (–O–O–), and oxo (=O) intermediates is much slower (in the order of hours to days) than whole water exchange because of tighter bonding and lack of protonation.

The other type of mechanism that could limit the ^{18}O exchange could be diffusion or isotope equilibration across a physical barrier. The catalytic site for O_2 evolution in PSII is located within the protein domain away from the solvent interface. In this situation, the structural barriers provided by the protein matrix may be mediated through specific water chains or water channels (23), something akin to what has been proposed for the cytochrome c oxidase (24,25). Such a water channel may provide an important function in optimizing O_2 formation by controlling the accessibility of the water substrates and O–O bond formation at the catalytic site, thereby minimizing side reactions that lead to the formation of reactive oxygen intermediates (23). The inability to resolve the fast phase of exchange in the S_0 and S_1, states (Table 1) could indicate that the second substrate water molecule only enters the reaction sequence at the S_2 state. However, such a possibility can only be clarified by resolution of the fast phase kinetics or by the progressive improvement in the resolution of crystal structure of PSII (26–28).

The S-state dependence of the ^{18}O exchange kinetics for the substrate water are clearly complex, but these results place limits on the types of mechanisms that may be involved in the oxidation of water during photosynthesis. Although the S-state-dependent changes in the kinetics may in the first instance be interpreted for metal-centered oxidations and/or substrate or other ligand protonation/deprotonation events, second-order effects mediated by the protein environment, such as H-bonding networks and structural conformations, will also have to be considered. To sort out the fine details, additional measurements will have to be made under different biochemical conditions and in mutants of the PSII. This work is currently in progress.

Acknowledgments
The authors gratefully acknowledge the various contributions to this research by Johannes Messinger, Murray Badger, and Garth Hendry. This research was supported in part by a Human Frontier Science Program grant (RGP 29/2002).

BIBLIOGRAPHY

1. Blankenship, R.E. and Hartman, H. (1998). *T.I.B.S.* **23**: 94–97.
2. Wydrzynski, T. and Satoh, K. (Eds.). (In press). *Photosystem II: The Water/Plastoquinone Oxido-Reductase in Photosynthesis*. Springer-Kluwer Academic Publishers, Dordrecht, The Netherlands.
3. Diner, B.A. and Rappaport, F. (2002). *Annu. Rev. Plant. Biol.* **53**: 551–580.
4. Barter, L.M.C., Durrant, J.R., and Klug, D.R. (2003). *Proc. Natl. Acad. Sci. USA.* **100**: 946–951.
5. Joliot, P., Barbieri, G., and Chabaud, R. (1969). *Photochem. Photobiol.* **10**: 309–329.
6. Kok, B., Forbush, B., and McGloin, M. (1970). *Photochem. Photobiol.* **11**: 457–475.
7. Forbush, B., Kok, B., and McGloin, M. (1971). *Photochem. Photobiol.* **14**: 307–321.
8. Joliot, P. and Kok, B. (1978). *Bioenergtics of Photosynthesis*. Govindjee (Ed.). Academic Press, New York.
9. Nugent, J.H.A. (Ed.). (2001). Special Issue: Photosynthetic Water Oxidation. *Biochim. Biophys. Acta* **1503**.
10. Messinger, J. and Lubitz, W. (Eds.). Special Issue: Photosynthetic water oxidation. *Phys. Chem. Chem. Phys.* (In press).
11. Radmer, R. and Ollinger, O. (1980). *F.E.B.S. Lett.* **110**: 57–61.
12. Radmer, R. and Ollinger, O. (1985). *F.E.B.S. Lett.* **195**: 285–289.
13. Bader, K.P., Thibault, P., and Schmid, G.H. (1987). *Biochim. Biophys. Acta* **893**: 564–571.
14. Messinger, J., Badger, M., and Wydrzynski, T. (1995). *Proc. Natl. Acad. Sci. USA* **92**: 3209–3213.
15. Hillier, W., Messinger, J., and Wydrzynski, T. (1998). *Biochemistry* **37**: 16908–16914.
16. Guy, R.D., Fogel, M.L., and Berry, J.A. (1993). *Plant. Phys.* **101**: 37–47.
17. Hillier, W. and Wydrzynski, T. (2000). *Biochemistry* **39**: 4399–4405.
18. Hillier, W. and Wydrzynski, T. (In press). *Phys.Chem. Chem. Phys.*
19. Hillier, W. and Wydrzynski, T. (2001). *Biochim. Biophys. Acta* **1503**: 197–209.
20. Hendry, G. and Wydrzynski, T. (2003). *Biochemistry* **42**: 6209–6217.
21. Lincoln, S.F. and Merbach, A.E. (1995). *Adv. Inorg. Chem.* **42**: 1–88.
22. Richens, D.T. (1997). *The Chemistry of Aqua Ions*. John Wiley, West Sussex, UK.
23. Wydrzynski, T., Hillier, W., and Messinger, J. (1996). *Physiol. Plant.* **96**: 342–350.
24. Iwata, S., Ostermeier, C., Ludwig, B., and Michel, H. (1995). *Nature* **376**: 660–669.
25. Tsukihara, T. et al. (1996). *Science* **272**: 1136–1144.
26. Zouni, A. et al. (2001). *Nature* **409**: 739–743.
27. Kamiya, N. and Shen, J-R. (2003). *Proc. Natl. Acad. Sci. USA.* **100**: 98–103.
28. Ferreira, K.N. et al. (2004). *Science* **303**: 1831–1838.

RADON IN WATER

Babs A. Makinde-Odusola
Riverside, California

Some groundwater contains the naturally occurring radioactive gas, radon-222 (radon). Radon dissolved in groundwater is released into indoor air when plumbed in water is agitated indoors. Radon released from domestic water typically contributes less than 2% of total indoor radon. This article describes the health risks from radon in domestic water and the efforts within the United States to regulate radon in drinking water. The U.S. National Research Council (NRC) found that the primary health risks from radon in domestic water are lung cancer and stomach cancer, primarily to ever-smokers. Radon concentration in air could reach levels considered unsafe for underground miners at some water treatment facilities that aerate water as part of the treatment process. Inhalation of radon gas and its progeny may result in lung cancer (inhalation risk). An estimated 11% of the health risk is from direct ingestion of radon dissolved in drinking water (ingestion risk). The NRC found that radon is a greater source of cancer risk than any other regulated chemical in drinking water. Some ecological studies contradict using the linear nonthreshold (LNT) model to assess the health risks from exposure to radon at low levels typical of indoor air. The U.S. Environmental Protection Agency (EPA) relied on the LNT model to assess health risks and propose standards and goal for radon in drinking water. Some water utilities could opt for treatment to comply with the radon in drinking water standard. Some other water utilities might avoid treatment by adopting a multimedia approach to radon mitigation. The multimedia approach could be more cost-effective and achieve greater health risk reduction. Noneconomic considerations may influence that decision. Health risk reduction and cost analyses (HRRCA) suggest that most of the benefits would accrue to ever-smokers and smaller water systems could incur most of the costs.

INTRODUCTION

Radon, in this article, refers to the specific isotope, radon-222, the daughter product of radium-226 and a member of the uranium-238 decay series. The decay scheme for uranium-238 is illustrated in Drago (1) and

NRC (2). Uranium is found in trace amount in soils everywhere. Radon escapes as a gas from soils into groundwater, indoor air, and outdoor (ambient) air. The U.S. Geological Survey (3) reported that soil air within the United States contains between 200 and 2,000 pCi/L (1 pCi/L = 37 $Bq\,m^{-3}$) of radon. Radon is a chemically inert, colorless, and odorless gas whose half-life is about 3.82 days. Radon emits alpha (α) particles as it decays to a series of solid short-lived radionucleotides, such as polonium-218 and lead-214. Those physical properties are important in understanding the health risks from exposure to radon and the mitigation of such risks.

The United Nations Scientific Committee on the Effects of Atomic Radiation (4) estimated that about 50% of the average annual effective dose from natural sources of ionizing radiation worldwide is from exposure to the alpha-particles produced by radon and its progeny. Potential contributors of radon to indoor air include soil gas, natural gas, building materials, and to a lesser extent—domestic water. Radon emanating from soils under homes is the primary source of radon in indoor air, which can range from <1 to 3,000 pCi/L (3).

Radon in drinking water is typically of concern to sources that rely on groundwater, where there had not been sufficient opportunity for significant decay nor release into the atmosphere before reaching indoor plumbing. A portion of the radon in plumbed water used indoors is released into the air, especially during agitation (e.g., laundering, dishwashing, humidifying, toilet flushing, etc.), which increments the level of radon in indoor air. The NRC (2) estimated that the average transfer coefficient is about 0.01% (1 in 10,000) of radon in water; that is 10,000 pCi/L of radon in water would increase the level of radon in indoor air by 1 pCi/L. The U.S. EPA (5) estimated that less than 2% of radon in indoor air emanates from domestic water. This author prefers to emphasize "domestic water" because of the widespread use of bottled water for drinking and because the NRC (2) found that the major part of the health risk is from inhalation of radon and its progeny.

Some countries have policies to reduce public exposure to radon, especially in indoor environments. Cole (6) provides the historical development of radon risk policies in the United States. The U.S. Indoor Radon Abatement Act of 1988 established the goal of reducing indoor radon concentrations to natural ambient level. The NRC (2) estimated that the arithmetic national ambient average level of radon was 0.4 pCi/L. The U.S. EPA established a voluntary action level of 4 pCi/L for radon in indoor air. The indoor radon action level in other countries is as high as 27 pCi/L (7). Indoor radon guidelines are 22 pCi/L and 11 pCi/L in Canada and the European Union, respectively (7). The U.S. EPA (5) reported that more than 2500 lives were "saved" through the year 2000 since the inception of its voluntary indoor radon abatement program in 1986.

Many countries are developing policies to reduce exposure to radon in domestic water. The World Health Organization (WHO) is creating a database of national radon regulations [http://www.who.int/ionizing_radiation/env/radon/en/]. The rest of this article focuses on efforts within the United States. The NRC (2) found that radon is a greater source of cancer risk than any other regulated chemical in drinking water. Evaluating, and where necessary, mitigating the level of indoor radon is common during real estate transactions in the northeastern United States, even though compliance with the action level is voluntary. That practice could extend to testing for radon in privately owned wells after the EPA finalizes the maximum contaminant level (MCL) for radon in drinking water. Regulating radon in domestic water had been very controversial in the United States and dates back to 1986. In 1986, Congress directed the EPA to establish a mandatory standard for radon in drinking water as part of the amendments to the Safe Drinking Water Act (SDWA). The EPA (8) proposed the radon in drinking water rule using the risk management approach typically used for drinking water contaminants. Amendments to the SDWA in 1996 directed the EPA to withdraw the proposed rule of 1991 and repropose after an independent risk assessment by the NRC. The amendments include provisions for a multimedia approach to achieving an equal or greater level of health risk reduction. In 1999, the EPA (5) proposed 300 pCi/L and 4,000 pCi/L as the MCL and alternate MCL, respectively, for radon in drinking water. The EPA is expected to finalize the radon in drinking water rule after 2004. Key inputs and major milestones in developing the radon rule are illustrated in Fig. 1 of Article MW-66 (9).

OCCURRENCE OF RADON IN DOMESTIC WATER

Radon levels in water, like those in ambient air and indoor air, vary temporally and spatially. The EPA (5) analyzed data from the Safe Drinking Water Information System (SWDIS) to estimate the nationwide occurrence of radon in community water systems (CWS). The EPA (10) found geographic disparity in its occurrence. New England CWS had the highest arithmetic and geometric mean concentrations of 2933 pCi/L and 1214 pCi/L, respectively. Some privately owned wells, not monitored nor regulated by the EPA, might contain radon at much higher levels. The USGS (3) reported that the radon level in groundwater ranges from about 100 pCi/L to nearly 3,000,000 pCi/L. The EPA (5) estimated the population-weighted national average concentration of radon in CWS at 213 pCi/L, and a credible range of 213 to 240 pCi/L. In general, smaller CWS that rely on groundwater have the highest levels of radon. Radon levels exceeded 300 pCi/L at more than 51% and 16% of the CWS serving fewer than 100 people and 10,000 people, respectively (5). Radon levels exceeded 300 pCi/L and 4000 pCi/L in the domestic water of more than 16 million people and 77,000 people, respectively (5). The EPA (5) and NRC (2) have additional information on the occurrence and spatial distribution of radon in domestic water, indoor air, and ambient air. SWDIS would contain additional radon occurrence data when the radon rule is finalized and CWS begin compliance monitoring.

Radon levels at a particular well could fluctuate significantly. The EPA (5) proposed a monitoring regime that would require CWS using groundwater to monitor for radon initially at each entry point to the distribution system, quarterly for 1 year. Compliance would be based

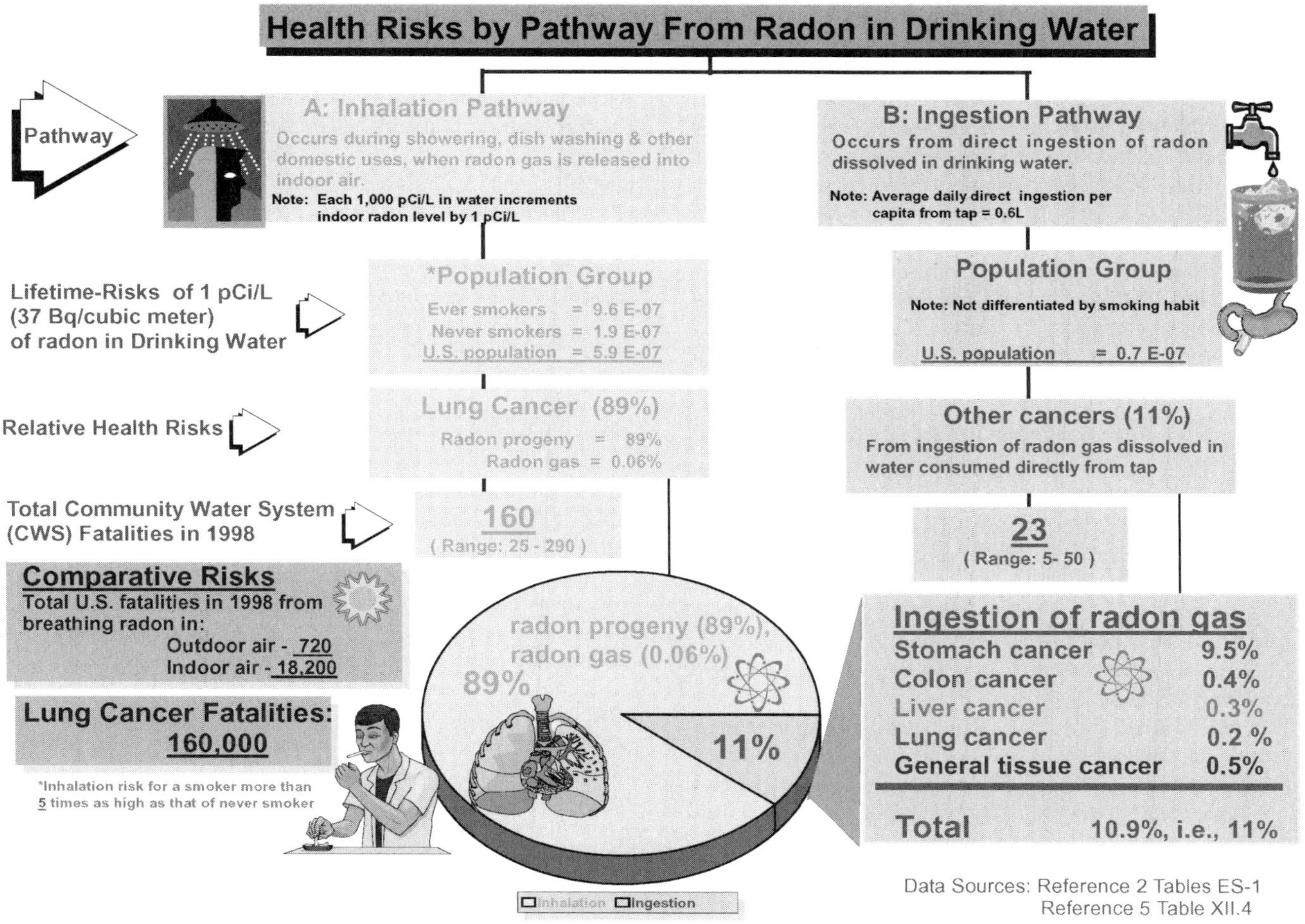

Figure 1. Estimated fatalities in the U.S. by pathway from radon in drinking water provided by CWS.

on the average of four consecutive quarterly samples. Zikovsky and Chah (11) found that concentrations of radon are lognormal in distribution and that the arithmetic mean of concentrations might create a false impression about exposure levels. The EPA (5) estimated that 26.8% of sources whose radon concentrations averaged between 200 and 300 pCi/L during the first four quarterly measurements would have long-term exceedances of 300 pCi/L. Field et al. (12) found higher levels of radon in domestic water at the point of use than prior to distribution because of increments from decay of the radium-226 deposits within the older portions of water distribution systems. The EPA (5) identified man-made sources of radon in groundwater to include releases from wastes from radium or thorium processing and from phosphate and uranium mining and milling operations.

RISK ASSESSMENT OF RADON IN DOMESTIC WATER

The health risks from radon in domestic water are primarily lung cancer and stomach cancer. Radon is a human carcinogen based on epidemiological studies of underground miners, cellular mutagenesis studies, and experimental research in animals (13). A linear nonthreshold (LNT) model is often used to assess the health risks of exposure to radon. The use of the LNT model to assess the health effects from exposure to radon at typical indoor concentrations is controversial. The National Council on Radiation Protection and Measurements (14) evaluated the evidence for and against the LNT dose–response hypothesis and found that the LNT model is more plausible. The NCRP (14) did not exclude other dose–response relationships because of growing evidence that adaptive responses, bystander effects, and other variables may modify the dose–response relationship. The NRC Committee on the Biological Effects of Ionizing Radiation (BEIR) VI did not exclude the possibility of a threshold relationship at the very low levels of radon exposure typically found in indoor air. The NRC (2) supported using the LNT relationship to derive quantitative estimates of cancer risks in the absence of credible evidence to the contrary. The NRC (2) recommended reconsideration of the LNT assumption when the "probability of an exposed cell becoming transformed" and the "latent period before malignant transformation is complete" are known. Lorimore et al. (15) observed the potential for interaction between irradiated and nonirradiated cells in the production of genetic damage and its significant implications for risk estimation. Brenner and Sachs (16) suggested that bystander effects from alpha-particle damage could cause downwardly curving dose–response relations and inverse dose-rate effects.

Parsons (17) presented a biological model of radiation hormesis that invalidated the LNT. Cohen (18) tested the LNT model by analyzing lung cancer mortality rates and average radon concentration in homes in 1601 U.S. counties while addressing problems typical of ecological studies. Cohen (18) found a "strong tendency for lung cancer rates to decrease with increasing radon exposure in sharp contrast to the increase" expected from the LNT model. The NRC (2) recommended investigating the significant negative trend of county lung cancer mortality with increasing home radon concentration established in the ecological study by Cohen (18). Puskin (19) found similar "strongly negative correlations for cancers strongly linked to cigarette smoking" but did not find "such correlations" for cancers not linked to smoking. Puskin (19) explained that the ecological correlation (18) "provides no substantial evidence for a protective effect of low level radon exposure." Van Pelt (2003) found that altitude (elevation) explains some of the inverse association between radon and lung cancer. Van Pelt (2003) postulated that the decrease in lung cancer is "caused by the carcinogenic effect of higher absolute oxygen concentration in the inspired air at lower."

The most direct way to assess the effects of radon in homes is through case-control studies (20). Krewski et al. (13) summarizes case-control studies of residential radon and lung cancer in North America. Field and Duport (21) presented points and counterpoints regarding lung cancer risks from exposure to residential radon. The NRC (2) reviewed available epidemiological studies of indoor radon and concluded "it is low and difficult, if not impossible to detect given the high background lung-cancer mortality in the populations studied."

Inhalation of radon gas, especially its progeny, causes lung cancer primarily among ever-smokers. BEIR VI (20) evaluated the risks of domestic radon to ever-smokers and never-smokers separately. CDC (22) defines ever-smokers as people who had smoked at least 100 cigarettes in their entire lifetimes. Alavanja (23) found many similarities between the pathologic lung-cancer mechanisms observed among those exposed to tobacco smoke and those exposed to radon gas.

NRC (2) conducted the most comprehensive review of the health risks of radon in domestic water. That review included epidemiological studies, BEIR studies, and earlier EPA (8,24) risk assessments. Both BEIR VI (20) and the NRC (2) reviewed the mechanistic information on alpha-particle-induced carcinogenesis. The NRC (2) postulated that the biological effects of exposure to radon at the low level of typical indoor air are initiated by the passage of single alpha-particles that have very high linear energy transfer (LET). Hei et al. (25) described the mutagenesis of a single alpha-particle in mammalian cells. Genes involved at low dose are different from those involved at higher doses and adaptive responses in cells exposed to low-level ionizing radiation allow the cells to better withstand a much higher dose later (priming effect) [http://www.llnl.gov/str/JulAug03/Wyrobek.html].

The NRC (2) developed a new model of ingested radon diffusion because its review of published ingestion risks found a wide range in the estimates. The uncertainties were mainly due to the assumed model of radon absorption in the gastrointestinal tract. The NRC (2) noted the lack of data on the extent to which radon diffuses through the stomach wall and recommended research to reduce the uncertainty regarding ingestion risks. Details of the mathematical models of radon diffusion in the stomach and the physiologically based pharmacokinetic (PBPK) model of radon behavior in the body can be found in the NRC (2). The NRC (2) estimated the ingestion risks using estimates of the amount of water ingested, the fraction of radon lost from water before ingestion, and calculations of the dose absorbed by tissues exposed. The NRC (2) estimated the confidence limits on the ingestion risk but noted, "the actual risk from ingested radon could be as low as zero depending on the validity of the LNT dose–response hypothesis." Yamaoka el al. (2004) found inhalation of radon gas from radon enriched water induced a small amount of active oxygen in the blood, decreased the levels of lipid peroxide and total cholesterol in the blood of osteoarthritis patients alleviating pain related to peroxidation reactions and immune depression.

Table 1 is a summary of the risks estimated by the NRC (2). Table 1 shows that males have higher lifetime risks than females and that the inhalation risk to ever-smokers is five times as high as those for never-smokers of either sex. Notice that the aggregate health risk to the general population is thrice as high as the risk for never-smokers due to the inclusion of ever-smokers. Figure 1 shows the relative distribution of cancer fatalities by exposure pathways. Notice in Fig.1 that 89% of the health risks are from inhalation of the radioactive by-products of radon released from water into indoor air. About 11% of the health risk is from direct ingestion of tap water (0.6 L/day directly from a tap). An earlier risk assessment by the EPA (5) estimated that the proportion of ingestion and inhalation risks were 53% and 47%, respectively.

Both the NRC (2) and the EPA (5) found that the hazards of exposure to radon during showering are likely to be small because the buildup of radon progeny is less than 4% of the maximum possible value during a typical shower that lasts 10 minutes. The NRC (2) found it highly unlikely that radon in drinking water has substantial teratologic or reproductive effects. The NRC (2) found that information was insufficient to determine radon risks to "sensitive" sub-populations, such as children. However, about 30% of fatal stomach cancers are attributed to exposure before age 10 (2). The NRC (2) suggested that there might be an opportunity interval for lung-cancer risk reduction from exposure during childhood. Bilban and Vaupotic (26) found higher chromosome aberrations in pupils at elementary schools that had high indoor radon levels. Fewer than 34% of lung cancer patients are diagnosed before age 65 (27).

Many researchers have estimated fatalities from radon in domestic water. Mills (28) estimated 60, 400, and 5000 annual fatalities from radon in surface water, public groundwater, and private wells sources, respectively. The NRC (2) used the risk estimates (Table 1) and information regarding the occurrence of radon in CWS, daily water

Table 1. Lifetime and Unit Risks Posed by Exposure to Radon in Drinking Water at 1 pCi/L (37 Bqm^{-3})[a]

Exposure Pathway	Lifetime Risk[b]			Unit Risk
	Male	Female	U.S. Population	U.S. Population
Inhalation (ever-smokers)	1.15×10^{-6}	7.40×10^{-7}	9.62×10^{-7}	1.31×10^{-8}
Inhalation (never-smokers)	2.18×10^{-7}	1.48×10^{-7}	1.85×10^{-7}	2.43×10^{-9}
Inhalation (population)	7.77×10^{-7}	4.44×10^{-7}	5.92×10^{-7}	7.90×10^{-9}
Ingestion* (population)	5.55×10^{-8}	8.51×10^{-8}	7×10^{-8}	9×10^{-10}
Total risk (inhalation and ingestion)	8.14×10^{-7}	5.18×10^{-7}	6.66×10^{-7}	8.89×10^{-9}
*Breakdown of Ingestion Risks				
Stomach			5.92×10^{-8}	
Colon			2.18×10^{-9}	
Liver			2.15×10^{-9}	
Lung			1.26×10^{-9}	
General tissue			2.92×10^{-9}	
Total ingestion			7×10^{-8}	

[a]Data sources: Reference 2, pp. 16, 17; Reference 5, Table X11.3.
[b]Annual unit risk can be obtained by dividing lifetime unit risks by the expected lifetime for each population category. Expected lifetime in years: Ever-smokers = 73.7; never-smokers = 76.1; combined population = 74.9.

Table 2. Health Risks from Radon in Drinking Water

a. Estimated Annual U.S. Fatalities Due to Radon in *Drinking Water

Health Risk	Source of Estimate	Central Estimate	Range
Lung cancer	NRC (2)	160	25–290
Lung cancer	EPA (5)	148	70–263
Stomach cancer	NRC (2)	23	5–50
Stomach cancer	EPA (5)	18	3–38

b. Occurrence of Lung Cancer due to Radon by Exposure Pathway

Pathway	Central Estimate	Range
Outdoor air (2)	720	120–300
Indoor air (5)	18200	3,000–33,000
*Drinking Water (2)	160	25–290
*Drinking Water (5)	148	70–263

c. Comparative Health Risks of Radon in *Drinking Water

Health Risk	Annual Fatalities (Central Estimates)		% of All Causes Due to *Drinking Water
	All Causes	*Drinking Water	
Lung cancer (2)	160,000	160	0.10%
Stomach cancer (2)	14,000	18	0.13%

Data sources: NRC(2)page 18; EPA(5).
Fatalities based on 1998 data.
*Potable water delivered by a community water system (CWS).

intake, and smoking patterns among several factors to derive estimates of cancer fatalities in the United States. Table 2 summarizes the estimated U.S. fatalities due to radon in domestic water relative to other sources. The NRC estimated that in 1998 radon from soil gas caused about 18,200 lung cancer fatalities in the United States. The NRC (2) also estimated that 160 lung cancer fatalities were due to radon in domestic water. Notice that an estimated 720 fatalities were from an unavoidable background level of radon in outdoor air. The NRC (2) estimated that 23 stomach cancer fatalities were from ingesting radon in drinking water. The EPA (5) estimated that radon in domestic water caused 148 lung cancer and 18 stomach cancer fatalities annually. The EPA (5) also estimated (screening level) that an additional 36 fatalities annually were from radon in water delivered outside residences, at schools, offices, etc.

RISK MANAGEMENT

The EPA (8) first proposed 300 pCi/L and 0 pCi/L as the maximum contaminant level (MCL) and MCL goal (MCLG), respectively, for radon in drinking water. Many criticized the proposed rule because it was not considered cost-effective. In addition, a multimedia approach could be more cost-effective and achieve greater health risk reduction. In 1996, as part of the amendments to the SDWA, Congress directed the EPA to withdraw the proposed rule and re-propose using an updated risk assessment provided by the NRC. Congress also directed the EPA to develop an alternate MCL (AMCL) and guidelines for multimedia mitigation (MMM) programs (MMMP). The NRC (2) defined the AMCL as the concentration of radon in water that would cause an increase of radon in indoor air that is no greater than the level of radon naturally present in outdoor air. The NRC (2) determined that the AMCL should be about 4000 pCi/L based on the national average outdoor air radon concentration of 0.4 pCi/L and the water to air transfer factor of 10,000. EPA is only to establish an AMCL if the MCL is more stringent than the water equivalent of the national average of radon in outdoor air.

The EPA (5) used NRC risk estimates (Table 1) to repropose the radon rule (MCL = 300 pCi/L; MCLG = 0 pCi/L, AMCL = 4000 pCi/L). CWS whose radon levels are within the range of the MCL and the AMCL have the option of complying with either the MCL or the less stringent AMCL. CWS that choose to comply with the AMCL must participate in an approved and verifiable MMMP that achieves at least an equivalent level of risk reduction as treatment to the MCL. CWS may opt to develop individual MMMP in accordance with EPA guidelines if the primacy state chooses not to develop an MMMP. The EPA (5) proposed four criteria required for approval of MMMP: involve the public in developing MMMP; include quantitative goals to measure the effectiveness of MMMP; include implementation plans, strategies and specific activities for achieving quantitative goals; and include plans for measuring and reporting results. Examples of MMMP include mitigation of radon levels in existing construction and use of radon-resistant techniques for new construction. CWS that choose the AMCL approach are not required to treat, but to participate in approved MMMP. The implemented MMMP might not reduce the ingestion risk, but reduce the inhalation risks, though not necessarily in the same community, nor in the same homes. Some CWS systems may opt to purchase water from other sources or consolidate with other CWS in lieu of treating or implementing the MMMP.

HEALTH RISK REDUCTION AND COST–BENEFIT ANALYSES (HRRCA)

Radon is the first drinking water contaminant for which the EPA (27) used the new cost–benefit analyses framework. The HRCCA (27) included detailed cost–benefit analyses by water system size for each considered MCL and MMMP adoption. In estimating benefits, the EPA (27) used the value of \$5.8 million (1997 dollars) as the value of statistical life (VSL) for each avoided cancer fatality and willingness to pay (WTP) of \$536,000 for nonfatal cancers. The EPA (5) estimated that the cost for MMMP is \$0.7 million per avoided lung cancer (the estimated average costs of indoor radon screening for all homes and mitigation cost for each home at or above the 4-pCi/L indoor air action level). The EPA (5) estimated annualized national costs at \$408 million and monetized annual benefits at \$362 million if 100% of impacted CWS choose to comply with the proposed MCL and assuming costs of capital of 7%. The estimated national annualized costs and monetized benefits are \$60 million and \$301 million, respectively, under a scenario that assumes that 95% of states implement MMMP and 90% of the CWS in each state comply with the AMCL. The EPA (5) prefers that states adopt the AMCL approach and develop MMMP but the NRC (2) noted that noneconomic considerations, such as equity, could be deciding factors in whether to undertake MMMP. The EPA (5) estimated that annual MCL compliance costs per household are about \$270 and \$10 for CWS serving 25 to 100 people and more than 100,000 respectively. The HRRCA shows that small CWS would incur most of the treatment costs but that the majority benefits would go to larger systems under the MCL approach. Some have argued that the EPA underestimated the costs of compliance and overstated the benefits. Costs could be higher because of unique local conditions and environmental permitting. Hahn and Burnett (29) found that the proposed radon rule was unjustifiable based on cost–benefit analyses.

TREATMENT TECHNOLOGIES

CWS may choose to comply with either the MCL or, under certain circumstances, with the AMCL. Treatment of radon in domestic water will reduce the health risks from both direct ingestion and inhalation through a decrement. Deb (30) reported on the decrements in indoor radon level after water treatment in three New Hampshire communities. Deb (30) found that significant reduction in the indoor radon level, occurred only in homes served by the CWS where water (radon concentration >20,000 pCi/L in water) was the major source of radon in indoor air.

The EPA (5) identified aeration as the best available technology (BAT) for removing radon from domestic water in large CWS. Article DW-32 (31) contains additional information regarding radon removal from drinking water. Aeration treatment methods such as a packed tower, diffused bubble, spray, slat tray, and cascade aeration are very efficient in removing radon from water because of the volatility of radon gas. Liquid-phase granular activated carbon (GAC) removes radon by adsorption but requires longer empty bed contact time (EBCT) ranging from 20 to 130 minutes, and its performance depends on water quality (1). The EPA (5) identified point-of-entry (POE) GAC as one of the affordable small systems compliance technologies (SSCTs). Drago (1) reported removal efficiencies of 20% to greater than 99% at 34 water treatment facilities and found that construction costs for either BAT are site specific. Either treatment technology requires disinfection after treatment, which EPA believes, will reduce the microbial risk and provide the additional benefit of complying with the proposed groundwater rule (32). The EPA (5,27) found that high levels of dissolved iron and manganese co-occurring with radon in water may complicate and increase radon treatment costs.

The EPA (5) reviewed the permitting and other environmental issues of either technology. Beta-gamma radiation from radon progeny might pose concern to GAC operators (1). A major concern of aeration technologies is off-gas emissions. Fisher et al. (33) found that radon concentration in air could reach levels considered unsafe for underground miners at some water treatment facilities that aerate water as part of the treatment process. The NRC (2) and EPA (5) recommended monitoring for radon in air around facilities treating groundwater for any contaminant and mitigation, if necessary, to reduce exposure. The EPA (5) modeled risks from potential air emissions of radon off-gas from aeration facilities based on data from 20 CWS whose radon levels ranged from 1000 pCi/L to 16,000 pCi/L. The EPA (5) estimated a maximum lifetime risk of 2×10^{-5} and a crude national estimate of 0.09 cancer fatalities per year if all CWS comply with the MCL. An air dispersion modeling study for the East Valley Water District, San Bernardino, California, suggests that the PTA (Packed Tower Aeration) downwind risk estimates could exceed the local regulatory risk threshold (34). The CDM (34) model predicted that a maximum individual cancer risk at some locations near the PTA could exceed 10^{-5} and 0.2 as the mean cancer burden for the service area. Preliminary design parameters for PTA to mitigate risks from radon emissions were not feasible from design/construct perspectives, for example, stack height up to 200 feet (34).

CONCLUSIONS

The preferred risk management model for radon in domestic water differs from those of other regulated drinking water contaminants because

- Amendments to the SDWA in 1996 require the EPA Administrator to take into account the costs and benefits of control programs for radon from other sources in setting the MCL.
- The major source of radon in indoor air is not regulated. Domestic water, which contributes on average less than 2% of radon in indoor air, would be regulated. Areas that have the highest indoor radon level may not coincide with service areas of CWS that have the highest level of radon in domestic water.
- The majority of health risks is not from direct ingestion of water, but rather from radon released during domestic use. Drinking water standards are usually set based on risks primarily due to ingestion.
- The majority of health risks are due to synergy with voluntary lifestyle choice—smoking. An ongoing public health antismoking campaign may result in significant reductions in the population of ever-smokers, thus reducing overall population unit risks. Though the EPA did not specifically categorize ever-smokers as a "sensitive subpopulation," their inclusion in risk analyses and benefit analyses significantly increased the benefits of adopting the rule.
- CWS may choose to comply with the AMCL rather than the MCL. In some scenarios, health benefits from risk reduction might accrue to communities outside the service area of the CWS. Health benefits might accrue to selected individuals rather than uniform risk reduction to all consumers.

Ayotte et al. (35) estimated that a 0.05% reduction in the prevalence of smoking would prevent as many deaths from lung cancer as mitigation of indoor radon. The NRC (2) noted "except in situations where concentrations of radon in water are very high, the reduction of radon in water will generally not make a substantial reduction in the total radon-related health risks to occupants of dwellings served by the water supply." The NRC (2) expressed similar concerns about the lack of data regarding the "overall effects of radon-resistant construction methods on reducing concentrations of radon in indoor air radon concentrations."

BIBLIOGRAPHY

1. Drago, J.A. (1998). *Critical Assessment of Radon Removal Systems for Drinking Water Supplies*. American Water Works Association Research Foundation, Denver, CO.
2. National Research Council. (1999). *Risk Assessment of Radon in Drinking Water*. National Academy Press, Washington, DC.
3. U.S. Geological Survey. (2000). *The Geology of Radon*. U.S. Geological Survey, Denver, CO.
4. United Nations Scientific Committee on the Effects of Atomic Radiation. (1993). *Report to the General Assembly*. United Nations, New York. Available: http://www.unscear.org/reports/2000_1.html
5. U.S. Environmental Protection Agency. (1999). *National Primary Drinking Water Regulations; Radon-222: Proposed Rule*. 40 CFR Parts 141 and 142. Federal Register (November 2, 1999) **64**(211): 59246–59378. [64 FR 59246]. U.S. Govt. Printing Office, Washington, DC.
6. Cole, L.A. (1993). *Element of Risk: The Politics of Radon*. AAAS Press, Washington, DC.
7. World Health Organization. (2002). *Radon and Health*. Final Information Sheet. October 2002. Available: http://www.who.int/ionizing_radiation/env/Radon_Info_sheet.pdf
8. U.S. Environmental Protection Agency. (1991). *National Primary Drinking Water Regulations; Radionuclides: Proposed Rule*. 40 CFR Parts 141 and 142. Federal Register (July 18, 1991) **56**(138): 33050–33127. [56 FR 33050]. Govt. Printing Office, Washington, DC.
9. Makinde-Odusola, B. (2004). MW-66 drinking water quality standards—United States. *Encyclopedia of Water*. John Wiley & Sons, New York.
10. U.S. Environmental Protection Agency. (1999). *Methods Occurrence, and Monitoring Document for Radon*, Draft. U.S. Environmental Protection Agency, Office of Ground Water and Drinking Water.
11. Zikovsky, L. and Chah, B. (1990). The lognormal distribution of radon concentration in ground water. *Ground Water* **28**(5): 673–676.
12. Field, R.W., Fisher, E.L., Valentine, R.L., and Kross, B.C. (1995). Radium-bearing pipe scale deposits: implications for national waterborne radon sampling methods. *Am. J. Public Health* **85**: 567–570.

13. Krewski, D. et al. (2002). A combined analysis of North-American case control studies of residential radon and lung cancer: An update. *Radiat. Res.* **158**(6): 785–790.
14. National Council on Radiation Protection and Measurements. (2001). *Evaluation of the Linear -Nonthreshold Dose-Response. Model for Ionizing Radiation NCRP Report No. 136*. NCRP, Bethesda, MD. Available: http://www.ncrp.com/rpt136.html.
15. Lorimore, S.A. et al. (1998). Chromosomal instability in the descendants of unirradiated surviving cells after a-particle irradiation. *Proc. Natl. Acad. Sci. USA* **95**: 5730–5733.
16. Brenner, D.J. and Sachs, R.K. (2003). Domestic radon risks may be dominated by bystander effects—but the risks are unlikely to be greater than we thought. *Health Phys.* **85**(1): 103.
17. Parsons, P.A. (2002). Radiation hormesis: Challenging LNT theory via ecological and evolutionary considerations. *Health Phys.* **82**(4): 513.
18. Cohen, B.L. (1995). Test of the linear-no threshold theory of radiation carcinogenesis for inhaled radon decay products. *Health Phys.* **68**(2): 157–174.
19. Puskin, J.S. (2003). Smoking as a confounder in ecologic correlations of cancer mortality rates with average county radon levels. *Health Phys.* **84**(4): 526–532.
20. National Research Council. (1999). *Health Effects of Exposure to Radon (BEIR VI)*. National Academy Press, Washington, DC.
21. Field, R.W. and Duport, P.J. (2003). Exposure to residential radon causes lung cancer. W.R. Hendee, Moderator, Point/Counterpoint. *Med. Phys.* **30**(4): 485–488.
22. Centers for Disease Control. (1995). *Morbidity Mortality Wkly. Rep.*, Cigarette smoking among adults—United States 1993.
23. Alavanja, M.C. (2002). Biologic damage resulting from exposure to tobacco smoke and from radon: Implication for preventive interventions. *Oncogene* **21**(48): 7365–7375.
24. U.S. Environmental Protection Agency. (1994). *Report to the United States Congress on Radon in Drinking Water, Multimedia Risk and Cost Assessment of Radon*. Office of Drinking Water, Washington, DC.
25. Hei, T.K. et al. (1997). Mutagenic effects of a single and an exact number of α particles in mammalian cells. *Proc. Natl. Acad. Sci. USA* **94**: 3765–3770.
26. Bilban, M. and Vaupotic, J. (2001). Chromosome aberrations study of pupils in high radon level elementary school. *Health Phys.* **80**(2): 157–163.
27. U.S. Environmental Protection Agency. (1999). *Health Risk Reduction and Cost Analysis for Radon in Drinking Water Regulations: Notice and Request for Comments and Announcement of Stakeholder Meeting*. 40 CFR Parts 141 and 142. Federal Register (February 26, 1999) **64**(38): 9559–9599. [64 FR 9559], U.S. Govt. Printing Office, Washington, DC.
28. Mills, W.A. (1990). Risk assessment and control management of radon in drinking water. In: *Radon, Radium and Uranium in Drinking Water*. C.R. Cothern and P.A. Rebers (Eds.). Lewis, Chelsea, MI, pp. 27–37.
29. Hahn, R.W. and Burnett, J.K. (2001). *The EPA's Radon Rule: A Case Study in How Not to Regulate Risks*. Regulatory Analysis 01-01. AEI-Brookings Joint Center for Regulatory Studies. Available: http://www.aei.brookings.org
30. Deb, A.K. (1992). *Contribution of Waterborne Radon to Home Air Quality*. American Water Works Association Research Foundation, Denver, CO.
31. DW-32. (2004). Methods of Reducing Radon in Drinking Water. *Encyclopedia of Water*. John Wiley & Sons, New York.
32. U.S. Environmental Protection Agency. (2000). *National Primary Drinking Water Regulations: Groundwater Rule; Proposed Rule 40 CFR Parts 141 and 142*. Federal Register (FR) **65**(91): 30194–30274). Available: http://www.access.gpo.gov/su_docs/aces/aces140.html (accessed Jan 2001).
33. Fisher, E.L., Fuortes, L.J., and Field, R.W. (1996). Occupational exposure of water-plant operators to high concentrations of radon-222 Gas. *J. Occup. Environ. Med.* **38**: 759–764.
34. CDM. (1999). *Radon Removal Preliminary Cost Analysis and Risk Assessment*. Prepared for the East Valley Water District, San Bernardino, CA.
35. Ayotte, P. et al. (1998). Indoor exposure to ^{222}Rn: A public health perspective. *Health Phys.* **75**(3): 297–302.
36. Yamaoka et al. (2004). Study on biologic effects of radon and thermal therapy on osteoarthritis. *Journal of Pain 2004* **5**: 20–25.
37. Van Pelt, W.R. (2003). Epidemiological associations among lung cancer, radon exposure and elevation above sea level–a reassessment of Cohen's county level radon study. *Health Physics* **85**(4): 397–403.

Note: All referenced URLs were valid as of July 2003.

SILICA IN NATURAL WATERS

ROBERT Y. NING
King Lee Technologies
San Diego, California

INTRODUCTION

Silica (1–4) and silicates (5–7) are generic names given to the family of silicon dioxide compounds derived from the dehydration–polymerization of silicic acid [$Si(OH)_4$]. Silicic acid, variously called orthosilicic acid, monosilicic acid, or silicic acid monomer, is a weak tetravalent acid that has pK_a values of 9.9, 11.8, and 12. Although silicic acid is prevalent in natural waters and is readily available from acidification of commercial crystalline sodium silicate or metasilicate (8), it has a great propensity to polymerize by dehydration to form Si–O–Si anhydride bonds:

$$\text{n } Si(OH)_4 \longrightarrow (OH)_3Si{-}O{-}Si(OH)_3 \text{ dimer} \longrightarrow \text{Oligomers} \longrightarrow \text{Colloidal polymers} \longrightarrow (SiO_2)_n$$

Thus silicon dioxide (SiO_2, silica) is the final and most stable product in normal conditions of silico-oxygen acid polymerization. The molecule of silica represented by the formula SiO_2 is polymeric and is more accurately represented by the formula $(SiO_2)_n$, where n is a large number.

The rate of silicic acid polymerization is strongly pH-dependent and is influenced by silicic acid concentration. It is very fast in neutral and slightly alkaline solutions, and extremely slow at low pH values of 2 to 3. Dilute polysilicate solutions that are undersaturated with respect to amorphous silica depolymerize to monosilicic acid.

Silicates imply more complex forms of silica in which silicic acid co-polymerizes with hydroxides of other metal elements covering almost the entire periodic table. This gives rise to the immense variety of rocks found on earth, which are copolymers of silicic acid in ultimate dehydrated forms. In solution, silicic acid readily reacts with alcohols and other hydroxyl-group containing organic compounds to form anhydrides. This provides for a great diversity of products and interactions in the biosphere.

CLASSIFICATION

Silica in natural waters can generally be classified in three categories: reactive soluble, nonreactive soluble (colloidal), and nonreactive insoluble (particulate). Reactivity refers to staining by molybdate reagent in the colorimetric assay of dissolved silica. The rate of this staining reaction is quantitatively reproducible only with low molecular weight polysilicic acids and varies inversely with the degree of polymerization of silicic acid (9). Monosilicic acid reacts with molybdic acid under assay conditions within 2 minutes, and disilicic acid reacts completely in about 10 minutes. Higher polysilicic acids require much longer times for substantial but incomplete reactions. After removal of suspended insoluble silica by filtration using typically a 0.45 micron filter, total dissolved silica (reactive + colloidal) can be measured by atomic emission or absorption spectroscopy. The colloidal silica value is the difference between the total and reactive silica. Normal detection limits of reactive silica in 25 mL samples are 0.1 mg/L. A greater sensitivity of 0.001 mg/L is possible in 100 mL sample sizes (9).

Agglomerated silica and silicates can reach a size, in solution, where they are no longer ions, but are actually colloidal particles that can remain in suspension indefinitely and are very difficult to remove by filtration.

GEOCHEMISTRY

Silica (SiO_2) in nature occurs in the crystalline forms of quartz, cristobalite, and tridymite, and also in the form of amorphous silica and glass. In number of atoms as well as weight, silicon is exceeded only by oxygen in the earth's crust. More than 95% of the volume of the earth's crust is composed of quartz and a small number of other rock-forming silicates.

Second only to carbon, silicon forms the largest number of compounds with other elements. The large number of carbon compounds is due to the fact that the bond energies of the C–C, C–O, and C–H bonds are of about equal magnitude, and they will therefore be formed with about the same probability. In contrast, the bond energy of the Si–O bond is considerably higher than that of the Si–H bond and more than twice that of the Si–Si bond. As a consequence, instead of the common C–C–C chains of carbon chemistry, chains of the type Si–O–Si–O are the skeletons of silicon chemistry (5).

Typical concentrations of dissolved silica in natural waters are:

Lakes	1–3 ppm
Major rivers	3–15 ppm
Seawater	1–10 ppm
Wells	2–60 ppm
Wells in volcanic and oil fields	50–300 ppm

In river and seawater, silica is not found exclusively in the colloidal state, but also in molecular solution, because its saturation concentrations are not attained. Despite the propensity for monosilicic acid to polymerize, it has been shown that silica in seawater is nearly exclusively in the monosilicic acid form in concentrations in the range of <0.006 to 10.1 ppm (11, 12). The concentration is low at the surface and increases with depth. Samples of lake and river water contain only monosilicic acid; there was no evidence of the presence of dimeric silicic acid (13). Silica in ground waters also appears to be in low molecular weight forms rather than the colloidal form (14). The concentration of particulate silica in seawaters of several oceans has been determined as 0.0006 to 0.43 ppm SiO_2 (12). It is thought that it is mainly of biogenic origin. Colloidal silica is depolymerized by seawater, which is considered at least one order of magnitude undersaturated with respect to amorphous silica.

DISSOLUTION OF SILICATE ROCKS

Pure silica such as quartz is completely dehydrated silicic acid. Depolymerization by rehydration as represented below represents dissolution to form soluble silicic acid:

$$SiO_2(s) + 2H_2O \longrightarrow Si(OH)_4$$

The dissolution rate depends on temperature and pH, it is minimum at pH = 2 to 3 and rises rising about 100-fold at pH = 11. Heat-accelerated hydrolysis would be expected in volcanic regions.

Silicic acid is also generated from the weathering of silicate rocks. Silicates are basically dehydrated forms of metal ion salts of silicic acid, which can be decomposed by carbonic acid. The following two reactions represent important dissolution processes of silicates (1):

$$2NaAlSi_3O_8 + 2CO_2 + 6H_2O \longrightarrow Al_2Si_4O_{10}(OH)_2 + 2Na^+ + 2HCO_3^- + 2Si(OH)_4$$

$$KMgFe_2AlSi_3O_{10}(OH)_2 + 1/2O_2 + 3CO_2 + 11H_2O \longrightarrow Al(OH)_3 + 2Fe(OH)_3 + K^+ + Mg^{2+} + 3HCO_3^- + 3Si(OH)_4$$

Field measurements of the weathering rate on a per square meter catchment area basis from different geographic areas are remarkably similar: 10^{-2} to 10^{-1} moles $m^{-2}yr^{-1}$. It is estimated that 10^5 m^2 surface area of mineral grains active in weathering is available per m^2 of geographic area and that only a few monolayers of rock silicates are dissolved per year (1). The silica concentrations on porewater increase continuously from the earth's surface to a depth of 3 meters. The silica concentration decreases slightly with depth either by precipitation or—under the influence of groundwater—by dilution. The total annual chemical load delivered from the continents to the oceans is estimated at 39.3×10^{14} g of which SiO_2 amounts to 4.26×10^{14} g (15).

DEPOSITION OF SILICA AND SILICATES

Most natural waters are undersaturated in amorphous silica and silicates, but unusual conditions found in geothermal waters (16) and industrial concentration or uses of waters do lead to deposition of silica and silicates. Biochemical extraction and deposition of silica by living organisms (see below) constitute a major deposition pathway.

Up to a pH of approximately 9, the solubility of amorphous silica (SiO_2) is relatively constant. It has been reported by various investigators as 100 to 150 ppm at 25 °C; the dissolved species is $Si(OH)_4$. Amorphous silica is metastable relative to quartz. The rate of crystallization of quartz is so slow at ordinary temperatures that the solubility of amorphous silica represents the upper limit of dissolved aqueous silica.

The precipitation of amorphous silica and silicates apparently proceeds in a series of steps. Polymerization by dehydration accompanied by cross-linking reactions and aggregation by van der Waal forces leads to negatively charged silica colloids. Further aggregation eventually leads to soft then hard gels.

At suitably high pHs, if $Fe(OH)_3$, $Al(OH)_3$, $Ca(OH)_2$, and $Mg(OH)_2$ are present, these hydroxides readily participate in the copolymerization resulting in the incorporation of Fe, Al, Ca, and Mg into complex anhydrous silicate structures. The strong tendency of magnesium hydroxide to react with silica, even in the monomeric state, is shown by the fact that the addition of 300 ppm of $Mg(OH)_2$ to water will reduce the soluble silica content from 42 to 0.1 ppm. The order of precipitability starting with the most readily precipitated metal is copper, zinc, manganese, cadmium, lead, nickel, silver, magnesium, and calcium. Metals that react in acidic solutions with silicates are Fe^{3+}, Co^{2+}, Zn^{2+}, and Cu^{2+} (4). Aluminum seems to occupy a unique position in combining with silica. Aluminosilicate gels (zeolites) can be prepared that have ion-exchange properties. A gel composition of $5SiO_2{:}Al_2O_3{:}Na_2O$ has an exchange capacity of 6% of its weight of CaO, and almost all of the sodium ions are exchangeable. Colloidal silica and silicates are active as flocculants and precipitate with other natural colloidal matter in rivers, lakes and the sea.

GEOCHEMICAL SILICA CYCLE

Geologically, the silica cycle is intimately associated with the carbonate cycle (see Carbonate in Natural Waters). The atmospheric CO_2 concentration of about 0.04% on a timescale of millions of years is controlled mainly by a combination of degassing from the earth's interior balanced by the uptake of carbon dioxide via the weathering of calcium and magnesium silicate minerals (17,18). The carbon dioxide in water as carbonic acid is converted to dissolved calcium and magnesium bicarbonate on land during weathering. This carbon form is stored in lakes and aquifers or is carried to the sea by rivers and precipitated there as calcium and magnesium carbonate minerals. The overall reactions shown here were first formulated by Urey (18) in 1952.

$$CO_2 + CaSiO_3 \Longleftrightarrow CaCO_3 + SiO_2$$

$$CO_2 + MgSiO_3 \Longleftrightarrow MgCO_3 + SiO_2$$

The reactions from left to right represent carbon dioxide absorption via weathering. The reactions from right to left represent degassing due to thermal decarbonation of carbonates after burial at sufficient depths in the earth's crust. Added to the Urey reactions should also be the organic carbon subcycle. It exerts an important additional control on the atmospheric carbon dioxide level. This involves the removal of carbon dioxide from the atmosphere through photosynthesis by organisms, the burial of organic matter in sediments, and the weathering of organic matter (or its thermal decomposition) which returns the carbon dioxide. The overall reaction is represented as:

$$6CO_2 + 6H_2O \Longleftrightarrow (CH_2O)_6 + 6O_2$$

Together, these reactions regulate carbon dioxide and silica on a geologic timescale.

BIOCHEMICAL ASPECTS

The importance of silicon for life on the earth results primarily from the fact that the fertility of soil relies in large part on the ability of clay minerals to absorb and release water and several cations that are indispensable for plant nutrition. This process is fundamental to the life of plants and based on them, to the life of those animals which in turn feed upon plants. Data on the physiology of silicon is limited, except that in a number of species, its presence is needed for normal growth.

Dissolved silica compounds play a substantial role in the cells of living organisms. Large amounts of silica are found in horsetail, rice, feather grass, reed and bamboo, where it contributes to the strength of the leaves and stems, and in the skeletons of diatoms which consist of very pure SiO_2. Silicon compounds are vital components in the metabolism of many bacteria, especially those that live in hot springs. The bacterium *Proteus mirabilis* even substitutes silicon for phosphorous in phospholipids. Silicon is also present as a trace element in higher animals and in humans, where it constitutes about 10 mg per kg live weight. It is essential in the cells of connective tissue and is involved in the biosynthesis of collagen, the substance that forms hair and nails, and in the formation of bony tissue.

The enrichment of silicon in mineralized tissues of unicellular organisms, such as diatoms and radiolaria, and a few metaozoa, such as sponges and gastropods is spectacular (19). The mineral phase is always amorphous silica in various stages of hydration as $SiO_2{\cdot}nH_2O$. Individual diatom populations require a certain level of monosilicic acid, and declining ambient silica concentrations in natural environments may influence the sequence of seasonal successions. The present day ocean contains a few ppm (parts per million) of dissolved silica; diatoms are often cited as the primary agent of silica removal from the world's ocean and keeping a low silicic acid level. Others favor inorganic extraction as the main regulating mechanism because common silicates take up dissolved silica from silica-rich seawater. The data suggest that the world ocean was low in silicon for most of its history.

BIBLIOGRAPHY

1. Stumm, W. and Morgan, J.J. (1996). *Aquatic Chemistry*. 3rd Edn. John Wiley & Sons, New York.
2. Iler, R.K. (1979). *The Chemistry of Silica*. John Wiley & Sons, New York.
3. Sosman, R.B. (1965). *The Phases of Silica*. University Microfilms International, Ann Arbor, MI (originally from Rutgers University Press, New Brunswick, NJ).
4. Iler, R.K. (1955). *The Colloidal Chemistry of Silica and Silicates*. Cornell University Press, Ithaca, NY.
5. Liebau, F. (1985). *Structural Chemistry of Silicates*. Springer-Verlag, New York.
6. Babushkin, V.I., Matveyev, G.M., and Mchedlov-Petrossyan, O.P. (1985). *Thermodynamics of Silicates*. Springer-Verlag, New York.
7. Eitel, W. (1954). *The Physical Chemistry of the Silicates*. University of Chicago Press, Chicago, IL.
8. Alexander, G.B. (1953). *J. Amer. Chem. Soc.* **75**: 2887.
9. Alexander, G.B. (1953). *J. Amer. Chem. Soc.* **75**: 5655.
10. Kemmer, F.N. (Ed.). (1988). *The Nalco Water Handbook*. 2nd Edn. McGraw-Hill, New York.
11. Isshiki, K., Sohrin, Y., and Nakayama, E. (1991). *Marine Chemistry* **32**: 1.
12. Kido, K. and Nishimura. (1975). *Deep-Sea Research* **22**: 323.
13. Okada, T. and Kuwamoto, T. (1985). *Anal. Chem.* **57**: 258.
14. George, S., Steinberg, S.M., and Hodge, V. (2000). *Chemosphere* **40**: 57.
15. Matthess, G., Frimmel, F., Hirsch, P., Schulz, H.D., and Usdowski, H-E. (Eds.). (1992). *Progress in Hydrogeochemistry*. Springer-Verlag, New York.
16. Henley, R.W. (1983). *Geothermics* **12**: 307.
17. Berner, R.A. (1995). *Am. J. Science* **295**: 491.
18. Urey, H.C. (1952). *The Planets: Their Origin and Development*. Yale Univ. Press, New Haven, CT, p. 242.
19. Degens, E.T. (1976). *Topics in Current Chemistry*. Vol. 64, *Inorganic Biochemistry*. Springer-Verlag, New York.

SODIUM IN NATURAL WATERS

NITISH PRIYADARSHI
Ranchi University
Ranchi, Jharkhand, India

INTRODUCTION

Sodium is a light alkali metal which floats on water in its pure state. However, it is very chemically active and is rarely found in its pure state. All natural waters contain sodium; nearly all sodium compounds readily dissolve in water, and it naturally leaches from rocks and soils.

The sodium ion, the sixth most abundant metallic ion in the earth's crust, is a natural constituent of both food and water from several sources. Most water supplies contain less than 20 mg of sodium per liter, but in some countries, levels can exceed 250 mg/liter. Saline intrusion, mineral deposits, seawater spray, sewage effluents, industrial brines, chemicals used in paper processing, soap, glass, and glass-wool industries, food processing, slaughtering and meat packing industries and salt used in road deicing all contribute significant quantities of sodium to water. In addition, water treatment chemicals, such as sodium fluoride, sodium bicarbonate, and sodium hypochlorite can together result in sodium levels as high as 30 mg/liter. Domestic water softeners can give levels of over 300 mg/liter. A much more important source of sodium in fluvial waters is the input from leaching of halite from sedimentary or evaporite deposits. More than 40% of the sodium in fluvial waters comes from this source. Sodium is also derived from the weathering of silicate minerals, principally sodic plagioclase feldspar. This source accounts for more than 20% of the sodium in river water. It is also an important element in mica, amphibole, and pyroxene.

A Na^+ restricted diet is recommended to patients suffering from hypertension or congenital heart disease; in such cases, the intake of high Na^+ through drinking water may prove critical (1). The current drinking water standard for sodium is 20 milligrams per liter.

SODIUM IN NATURAL WATER

Groundwaters

The Na content of groundwaters is a function of (1) weathering of Na plagioclase from bedrock, followed by (2) exchange of Ca^{2+} for Na^+ on the surfaces of newly formed clay minerals. Feth et al. (2) considered the weathering reactions which supplies Na^+ to groundwater as:

$$3Na_{0.66}Ca_{0.33}Al_{1.33}Si_{2.66}O_8 + 14H_2O + 4CO_2$$
$$\rightarrow 2H_4Al_2Si_2O_9 + 2Na^+ + Ca^{2+} + 4H_4SiO_4 + 4HCO_3^-$$
$$\text{(kaolinite)}$$

and

$$6Na_{0.66}Ca_{0.33}Al_{1.33}Si_{2.66}O_8 + 6CO_2 + 18H_2O$$
$$\rightarrow 3Na_{0.66}Al_{2.66}Si_{3.33}O_{10}(OH)_2 + 2Ca^{2+} + 2Na^+$$
$$+ 6H_4SiO_4 + 6HCO_3^- \text{(montmorillonite)}$$

These theoretical calculations apply well to the actual minerals in the rocks and soils of the area studied (western U.S.)

Garrels (3) suggested that the Na^+ content of groundwater would rise due to weathering of plagioclase feldspars. If the system is closed, the waters eventually proceed through kaolinite stability. If calcite is precipitating, this montmorillonite will be sodic and hold the Na^+

content of the water relatively steady. If calcite is not precipitating, Ca^{2+} will be used by the montmorillonite, and the Na^+ content of the groundwater will continue to rise. Low concentrations of Na^+ ions are associated with the minerals, kaolinite and quartz (4).

Sathyanarayanan and Periakali (5), in their study of Na concentration in groundwater in Ultrabasic and Peninsular Gneissic Rocks of Tamil Nadu State of India, suggested that weathering of plagioclase feldspar and biotite in the Peninsular gneiss could be the source of Na in groundwater.

In short, the Na^+ content of groundwater is a function of weatherable minerals, pH, reaction time versus drainage time, the supply of Ca^{2+} to form Ca montmorillonite, sluggish drainage conditions, intensive irrigation practices, extensive use of chemical fertilizers, inadequate sanitary facilities, and precipitation dominance controlling water chemistry (6–9). In addition, groundwater can contain large amounts of Na^+ from the dissolution of evaporites, from saltwater intrusion, and smaller amounts from cyclic sea salts in original precipitation.

Rivers

The concentration of elements in river water varies with time at any single sampling site as a function of discharge, tributary supply, and groundwater discharge. In areas where rivers are draining evaporite deposits, the Na content can be very high. For example, streams in the Brazos River basin (Texas) have as much as 99,800 ppm Na and an average of 35,900 ppm (10).

The unweighted average Na content of 639 rivers and streams is 38.6 ± 5.9 ppm ($s/\sqrt{n}$). The range is 0.1 to 1.540 ppm, and the median is 6.1 ppm Na (11,12). The mean content may be used as a worldwide estimate, but the population is actually polymodal with a mean for humid and temperate weathering zones, a mean for arid climates, and an extremely high mean for rivers draining evaporite deposits (13). Comparison shows that river water and seawater are opposite in character for Na concentration. In seawater, Na>Mg>Ca; in average river water, Ca>Na>Mg (14). Livingstone's (11) weighted mean Na^+ content of world rivers was 6.3 ppm.

Lake Waters

The sodium content of lake water is a function of river and groundwater supply followed by increase due to evaporation where this process exceeds input. The evaporation may reach a stage of Na mineral precipitation, which occurs at a high level of Na^+.

Seawater

The sodium released into solution during weathering is transported by rivers to seas/oceans and represents a primary cyclic reservoir. Early knowledge of the average composition of seawater was provided by Dittmar in 1884; he carefully analyzed 77 water samples, representative of all oceans and taken both from the surface and from the depths. These samples had been collected on the voyage around the world of H.M.S. Challenger (1872–1876). He determined halides, sulfate, magnesium, calcium, and potassium. Sodium was calculated by difference by subtracting the sums of the chemical equivalents of the negative and positive ions. This procedure was followed because Dittmar was unable to achieve satisfactory direct determinations for sodium. He estimated 30.61% sodium in sea water.

In the valuable treatise, The Data of Geochemistry, (1924), F. W. Clarke calculated that the 300,000,000 cubic miles of oceanic waters owed their salinity to 4,800,000 cubic miles of dissolved salts which had been derived from the igneous rocks of the earth's crust. He confined his computations to the sodium (Na) percentages, reckoning an average of 1.14% in seawater and 2.90% of this element in igneous rocks. The average concentration of sodium in seawater is 1.1×10^7 g/L (14). Earlier, Goldberg (15) estimated 10,500 ppm Na^+ in seawater. Calculations of Garrels and Thompson (16) indicated that 99% of the Na in seawater exists as uncomplexed Na^+.

TREATMENT

Sodium may be removed from a water supply by ion exchange. The ion exchange resin in the water softener is enriched with calcium and removes the sodium from the water. Other demineralization techniques, such as reverse osmosis, distillation, electrodialysis and freeze–thaw processes, should be effective for removing sodium from water. If a large supply of water is required, it may be necessary, if feasible, to blend it with another water source low in sodium.

In addition to water treatment, another method of removing sodium from drinking water would require the construction or reconstruction of a water well. This frequently involves installing additional casing beyond the depth normally required by regulations.

BIBLIOGRAPHY

1. Holden, W.S. (1970). *Water Treatment and Examination*. J & Churchill, London.
2. Feth, J.H., Roberson, C.E., and Polzer, W.L. (1964). *Sources of Mineral Constituents in Water from Granite Rocks of the Sierra Nevada*. U.S. Geological Survey Water Supply Papers, 1535-I, pp. 1–70.
3. Garrels, R.M. (1967). Genesis of some groundwaters from igneous rocks. In: *Researches in Geochemistry*, 2nd Edn. P.H. Abelson (Ed.). John Wiley & Sons, New York, pp. 405–420.
4. Garrels, R.M. (1976). A survey of low temperature water-mineral relations. In: *Interpretation of Environmental Isotope and Hydrochemical Data in Groundwater Hydrology*. International Atomic Energy Agency, Vienna, Austria.
5. Sathyanarayanan, M. and Periakali, P. (2003). Geochemistry of groundwater in Ultrabasic and Peninsular gneissic rocks, Salem District, Tamil Nadu. *J. Geol. Soc. India* **62**(1): 63–73.
6. Gibbs, R.J. (1970). Mechanisms controlling world water chemistry. *Science*. **170**: 1088–1090.
7. Prakasa Rao, J. (1997). *Environmental Evaluation of Groundwater Quality of a Developing Urban Area of Andhra Pradesh, India*. Unpublished. Ph.D. Thesis, Andhra University, Visakhapatnam, p. 104.

8. Subba Rao, N. (1998). Groundwater quality in crystalline terrain of Guntur District, Andhra Pradesh. *Visakha Sci. J.* **2**: 51–54.
9. Ballukrya, P.N. and Ravi, R. (1999) Characterization of groundwater in the unconfined aquifers of Chennai city, India: Part I- Hydrogeochemistry. *J. Geol. Soc. India* **54**: 1–11.
10. Baker, R.C., Hughes, L.S., and Yeast, I.D. (1964). *Natural Sources of Salinity in the Brazos River, Texas*. U.S. Geological Survey Water Supply Papers. 1669-cc, pp. 1–81.
11. Livingstone, D.A. (1963) *Chemical Composition of Rivers and Lakes*. U.S. Geological Survey Professor. Papers 440-G, pp. 1–64.
12. Anderson, P.W. and George, J.R. (1966). *Water-Quality Characteristics of New Jersey Streams*. U.S. Geological Survey Water Supply Papers 1819-G, pp. 1–48.
13. Wedepohl, K.H. (1970). *Handbook of Geochemistry*. Vol. II/2, Springer-Verlag, Berlin.
14. Mason, B. and Moore, C.B. (1985). *Principles of Geochemistry*, 4th Edn. Wiley Eastern Limited, New Delhi, p. 242.
15. Goldberg, E.D. (1957). Biogeochemistry of trace metals. In: *Ecology*, vol.1. *Treatise on Marine Ecology and Paleocology*. J.W. Hedgpeth (Ed.). Geological Society America Member **67**, pp. 345–358.
16. Garrels, R.M. and Thompson, M.E. (1962). A chemical model for sea water at 25 °C and one atmosphere total pressure. *Am. J. Sci.* **260**: 57–66.

SOFT WATER

Gholam A. Kazemi
Shahrood University of Technology
Shahrood, Iran

INTRODUCTION

At the outset, it should be pointed out that soft and hard waters are two sides of a coin. Hence, the respective articles, this one and the Hard Water article are alike in many aspects. In defining soft water, it is necessary first to describe hardness, a closely related chemical property of water. Hardness is defined by some as a property of water that prevents lather formation with soap and produces scale in hot water systems, (1); by others, as the soap consuming capacity of water (2); and in broader terms by Freeze and Cherry (3), as the "metallic ion content of water which reacts with sodium soaps to produce solid soaps or scummy residue and which reacts with negative ions, when the water is evaporated in boilers, to produce solid boiler scale."

The ions which result in, or produce, hardness include Mg^{+2}, Ca^{+2}, Sr^{+2}, Fe^{+2}, and Mn^{+2}, and to a lesser extent, Ba^{+2} and Zn^{+2} and other divalent ions. The first two of these, are the most influential. There are different forms of hardness (4):

1. Total hardness: Ca and Mg expressed as $CaCO_3$
2. Calcium hardness: Ca expressed as $CaCO_3$
3. Magnesium hardness: Mg expressed as $CaCO_3$
4. Temporary hardness: Equals alkalinity, if alkalinity is less than total hardness
5. Permanent hardness: The amount of hardness greater than temporary hardness
6. Noncarbonate hardness: Permanent hardness

Of the six different forms of hardness, the most important and the one that is commonly known as hardness, is total hardness. It is calculated by the following simplified equation:

$$\text{Hardness} = 2.5\ \text{Ca} + 4.1\ \text{Mg}\ (\text{units in mg/l})$$

Note that there are equations to calculate each of the six different forms of hardness.
Example: The Ca and Mg concentrations in a water sample are 20 and 15 mg/l, respectively. What is its hardness?

$$\text{Hardness} = 2.5 \times 20 + 4.1 \times 15 = 111.5\ \text{mg/l}$$

As stated, each of the six forms of hardness can be calculated by analyzing a water sample for the relevant ion(s), followed by using the relevant equation. However, there are techniques to measure total hardness directly such as (1) a standard solution of soap, that was in use for many years (2); and (2) Schwarzenbach et al's. method of colorimetric titration (5), which is sometime referred to as the EDTA method.

The author believes that earlier, measuring hardness was a common and an inseparable practice in water analysis. To agree with this statement, one needs to look at a relatively old handbook and read, "*The common denominator of the majority of water problems is hardness*" (2, p. 4.1, first published in 1979). Somehow, in contrast, recent hydrochemical analysis of water samples does not usually include a measurement of hardness. For instance, there is not a single report of hardness in the hundreds of hydrochemistry papers presented at the International Association of Hydrogeolgists (IAH) conferences in Prague (2003) and in Munich (2001) and in the recent past (1997–2001) annual symposia of the Geological Society of Iran. This occurred, perhaps, because other aspects of water chemistry and water biology, such as nutrients, heavy metals, bacteriological species, chlorinated hydrocarbons, and many contaminants have become so widespread that they are considered more important and more hazardous than hardness. This is to say that rejecting water for drinking or other uses, these days, is many times more likely to be due to its harmful DDT level than because of its high hardness value. One other possible reason is that hardness and soft water topics are important only in some narrow fields of water, e.g., hot water boilers.

The second point is that industrialization, urbanization, agricultural practices, and most environmentally unfriendly activities are not likely to result in an increase in the hardness of water resources because these activities do not usually generate hardness-producing ions such as Ca^{2+} or Mg^{2+}, though, they may produce other trivial (in terms of hardness) ions such as Mn^{2+} and Fe^{2+}.

The author, here, wishes to draw the reader's attention to one other point and pose a question as well. He believes that "softness" would have had an identical application and a similar meaning to hardness if it would have been there. Therefore, it is interesting to speculate as to what was the rationale for use of the term "hardness", not softness. One reason might be that hard water is generally easier to notice because of its unwanted and disappointing characteristics and our predecessors looked for something to describe it. In other words, the term hardness was created to explain easily noticeable hard water, not to describe soft water. This is why there is no such practice as water hardening, changing soft water to hard water, in contrast to the well-established exercise of water softening which changes hard waters to soft waters.

DEFINITION AND SPECIFICATIONS OF SOFT WATER

Todd (6, p. 282) reports that Hippocrates (460–354 B.C.) was the first to use the terms hard and soft for water, when in his treatises on public hygiene, he wrote, "consider the waters which the inhabitants use, whether they be marshy and soft, or hard and running from elevated and rocky situations, and then if saltish and unfit for cooking... for water contributes much to health." However, it is probably safe to say that hard and soft water in Hippocrates' writings, are likely to mean different things from what we today call hard and soft water.

The most frequently used classification system for water based on hardness is that of Sawyer and McCarty (7), presented in Table 1. There is another slightly different classification system by Hem (8), which is referred to in some publications as the classification of the U.S. Geological Survey (9).

The definition of soft water varies slightly among researchers. Based on Table 1, soft water is defined as water whose hardness is less than 60 or 75 mg $CaCO_3$/l. This means that waters with less than 15 mg Ca^{2+}/l and 10 mg Mg^{2+}/l are soft; higher concentrations in waters belong to the other categories such as hard water or very hard water. However, some researchers argue that there is no exact definition for determining soft or hard water because it depends, at least partially, on what one is accustomed to (9). What is soft water for one person, may be moderately hard for another and vice versa. The very hard water category in Table 1 can be grouped into the hard water category as it has the specifications of hard water. It is, however, unclear to which category one can assign moderately hard water, if hard water and soft water are the only categories available.

Table 1. Different Types of Water Based on Hardness

Water Type	Hardness mg $CaCO_3$/l	
Soft	below 75[a]	Below 60[b]
Moderately hard	75–150[a]	60–120[b]
Hard	150–300[a]	120–180[b]
Very hard	Over 300[a]	Over 180[b]

[a] Reference 7.
[b] Reference 8.

Manning (9), discredits soft water and states that very soft water when used for bathing feels slick on the body, as though the soap is still there. He also explains that most people prefer slightly hard water because of its taste and its ability to wash the soap off the body. Soft waters are not very good for irrigation either because they may lead to a lower infiltration rate compared with slightly hard water (9). In contrast to hard water, soft water produces a foam or lather easily but does not produce scale in hot water pipes, heaters, boilers, and other high water temperature units. Therefore, using soft water is an economically worthwhile practice because it leads to a decrease in soap consumption and lowers fuel costs for boilers due to good heat conduction.

The recommended water for domestic use should have a hardness of 80 mg/l; i.e., it should be soft (10). This is why water softening is a common practice where the water supply has a hardness greater than 80–100 mg/l (3). Soft water is also required for some industrial processes such as food equipment washing, confectionery, food canning and freezing, food processing, and laundering (6).

SOFT WATER AND HEALTH

There is a general perception among the public that soft water is healthier to drink. Studies, though not satisfactorily conclusive, suggest the opposite, especially with regard to artificially softened water. For instance, Tebbutt (1995) (1) reports that there is some statistical evidence to suggest that artificially softened waters may increase the incidence of some forms of heart disease. Also, a recent study in Taiwan (11) shows that there is a significant negative relationship between drinking water hardness and colon cancer mortality, a finding considered important for the Taiwanese water industry and human health. All these confirm Muss's (12) observation that in the United States, death rates from heart and circulatory diseases are lower in states where public water supplies are harder. Furthermore, artificially softened water may taste salty because its sodium content increases in the softening process.

AVAILABILITY OF NATURALLY SOFT WATER

Soft water may exist naturally or may be produced artificially, as stated before. Atmospheric precipitations (snow and rain) are soft waters. Similar to these are waters that emanate from upland catchments with an average total hardness of 10 mg/l (1). In contrast, most groundwaters, excluding those that occur in fractured igneous rocks, are hard. All ground water and surface water that pass through, or are in contact with, carbonate and other Ca–Mg bearing rocks are hard. Lowland rivers with a hardness of 200 mg/l (1) are also hard. Therefore, only a limited percentage of freshwater resources of the world, groundwaters and river waters, is soft.

Examples of naturally occurring soft waters include the Vilnius well fields groundwater (intermorainic and alluvial deposit aquifer), Vilnius District, Lithuania, that have unbelievably low hardness values of 3.9–8.5 mg/l (13), groundwater in Akwa Ibom State of Nigeria (fluvio-volcanic rocks) with a hardness of 1.2–65 mg/l (Akujieze, 2003) (14), and shallow basement groundwater of Lithembwe dambo, Malawi, with 1–14 mg/l of calcium and 1–7 mg/l of magnesium (15).

BIBLIOGRAPHY

1. Tebbutt, T.H.Y. (1995). *Principles of Water Quality Control*. Reprint of 4th Edn. Butterworth Heinemann, Oxford, UK.
2. Kemmer, F.N. (1988). *The NALCO Water Handbook*, 2nd Edn. McGraw-Hill, New York.
3. Freeze, R.A. and Cherry, J.A. (1979). *Groundwater*. Prentice-Hall, Englewood Cliffs, NJ.
4. Hounslow, A.W. (1995). *Water Quality Data: Analysis and Interpretation*. Lewis, New York.
5. McCoy, J.W. (1981). *The Chemical Treatment of Boiler*. Water Chemical Publishing, New York.
6. Todd, D.K. (1980). *Groundwater Hydrology*, 2nd Edn. John Wiley & Sons, New York.
7. Sawyer, C.N. and McCarthy, P.L. (1967). *Chemistry for Sanitary Engineers*, 2nd Edn. McGraw-Hill, New York.
8. Hem, J.D. (1985). *Study and Interpretation of the Chemical Characteristics of Natural Waters*, 3rd Edn. U.S. Geological Survey Water Supply Paper 2254, Denver, CO.
9. Manning, J.C. (1997). *Applied Principles of Hydrology*. Prentice Hall, Upper Saddle River, NJ.
10. Bell, F.G. (1998). *Environmental Geology, Principles and Practice*. Blackwell Science, Cambridge, UK.
11. Yang, C.Y. and Hung, C.F. (1998). Colon cancer mortality and total hardness levels in Taiwan's drinking water. *Arch. Environ. Contamination Toxicol.* **35**(1): 148–151.
12. Muss, D.L. (1962). Relationship between water quality and deaths from cardiovascular disease. *J. Am. Water Works Assoc.* **54**: 1371–1378.
13. Klimas, A.A. (1995). Impacts of urbanization and protection of water resources in the Vilnius District, Lithuania. *Hydrogeol. J.* **3**(1): 24–35.
14. Akujieze, C.N., Coker, S.J.L., and Oteze, G.E. (2003). Groundwater in Nigeria—a millennium experience—distribution, practice, problems and solutions. *Hydrogeol. J.* **11**(2): 259–274.
15. Chilton, P.J. and Foster, S.D. (1995). Hydrogeological characterization and water-supply potential of basement aquifers in tropical Africa. *Hydrogeol. J.* **3**(1): 36–49.

SOLUBILITY OF CHEMICALS IN WATER

Carl L. Yaws
Prasad K. Narasimhan
Helen H. Lou
Lamar University
Beaumont, Texas

Ralph W. Pike
Louisiana State University
Baton Rouge, Louisiana

Correlations are presented for the solubility of chemicals in water. A wide variety of compounds are covered—alcohols, ketones, ethers, aldehydes, acids, acetates, amines, monochloroalkanes, and monobromoalkanes. The new correlation provides reliable solubility values down to very low concentrations (parts per million range). The correlation is based on the boiling point of the compound. The correlation and experimental data are in favorable agreement. The results are useful in process engineering for wastewater minimization.

IMPORTANCE OF WATER SOLUBILITY

The solubility of compounds in water is very important. This importance will increase in the future because of heath, safety, and environmental considerations. The following brief discussion illustrates the importance even at very low concentrations.

For health involving human exposure to substances in air, the threshold limit value (TLV) for *sec*-butanol in air is given as 100 ppm (parts per million, vol) by the Occupational Safety and Health Act (1). A concentration of only 0.0005 mol fraction of *sec*-butanol in water provides about 236 ppm of *sec*-butanol in air at the air–water interface. This concentration of 236 ppm in air exceeds the threshold limit value of 100 ppm.

For safety, the lower explosion limit (LEL) for *sec*-butanol in air is given as 1.7% by Yaws (2). A concentration of only 0.04 mol fraction of *sec*-butanol in water provides about 1.9% of *sec*-butanol in air at the air–water interface. This concentration of 1.9% in air exceeds the lower explosion limit of 1.7%.

For environmental, consider a spill of *sec*-butanol in contact with water. The water will become saturated with *sec*-butanol. At saturation, the solubility of *sec*-butanol in water is about 0.181 weight fraction or 0.0508 mol fraction (3). This concentration of only 0.0508 mol fraction at saturation will provide about 24,000 ppm or 2.5% of *sec*-butanol in air at the air–water interface. This concentration of 24,000 ppm or 2.4% greatly exceeds both the threshold limit value of 100 ppm and the lower explosion limit of 1.7%.

This brief discussion (using *sec*-butanol as an example) indicates that very low concentrations of a chemical in water can provide concentrations in air at the air–water interface that exceed the threshold limit value for human exposure and the lower explosion limit for flammability.

CORRELATION FOR WATER SOLUBILITY

In earlier work by Yaws and co-workers (2), the water solubility of hydrocarbons was correlated as a function of the boiling point of the compound. In this article, it was determined that the boiling point method is also applicable to the correlation of the water solubility of chemicals:

$$\log_{10}(S) = A + B\,T_B + C\,T_B{}^2 + D\,T_B{}^3 \qquad (1)$$

where S = solubility in water at 25 °C, ppm by weight,
T_B = boiling point of compound, K
A, B, C, and D = regression coefficients

Table 1. Solubility in Water[a]

	$\log_{10} S = A + B\,T_B + C\,T_B^2 + D T_B^3$				(S, ppm(wt), T_B, K)			
Chemical Family	A	B	C	D	T_{BMIN}	T_{BMAX}	Formula	Example Compound
Alcohols	45.6398	−2.3859E−01	4.8739E−04	−3.7160E−07	370	620	$C_nH_{2n+2}O$	*n*-Pentanol
Ketones	45.2000	−2.3859E−01	4.8739E−04	−3.7160E−07	350	590	$C_nH_{2n}O$	Diethyl ketone
Ethers	11.4000	−2.2000E−02	0.0000E+00	0.0000E+00	300	560	$C_nH_{2n+2}O$	Dipropyl ether
Aldehydes	12.4200	−2.2000E−02	0.0000E+00	0.0000E+00	320	560	$C_nH_{2n}O$	1-Hexanal
Acids	14.3000	−2.2000E−02	0.0000E+00	0.0000E+00	440	560	$C_nH_{2n}O_2$	*n*-Hexanoic acid
Acetates	13.0000	−2.3382E−02	0.0000E+00	0.0000E+00	330	600	$C_nH_{2n}O_2$	Ethyl acetate
Primary amines	11.4300	−1.8000E−02	0.0000E+00	0.0000E+00	400	560	$C_nH_{2n+3}N$	*n*-Hexyl amine
Secondary amines	11.3600	−1.8000E−02	0.0000E+00	0.0000E+00	360	560	$C_nH_{2n+3}N$	Dipropyl amine
Tertiary amines	11.8200	−2.0000E−02	0.0000E+00	0.0000E+00	360	560	$C_nH_{2n+3}N$	Triethyl amine
Monochloroalkanes	−7.4500	1.0050E−01	−2.7288E−04	1.9987E−07	300	590	$C_nH_{2n+1}Cl$	2-Chlorobutane
Monobromoalkanes	−7.1700	1.0050E−01	−2.7288E−04	1.9987E−07	300	590	$C_nH_{2n+1}Br$	1-Bromobutane

[a] S-solubility in water at 25 C, parts per million by weight, ppm(wt).
A, B, C, and D-regression coefficients.
T_B-boiling point temperature of compound, K.
T_{BMAX}-maximum temperature, K.
T_{BMIN}-minimum temperature, K.

The range of boiling point depends on the chemical family.

The coefficients (A, B, C, and D) are given in Table 1 for a wide variety of chemicals—alcohols, ketones, ethers, aldehydes, acids, acetates, amines, monochloroalkanes, and monobromoalkanes. The table also provides the boiling point range (T_{BMIN}, T_{BMAX}) for each chemical family. The correlation is applicable to compounds whose boiling points are in this range. The correlation should not be used for compounds whose boiling points are outside the boiling point range.

The coefficients were determined from regression of available data. In preparing the correlation, a literature search was conducted to identify data source publications (2–13). The excellent compilations by Horvath (5–6); Howard and Meylan (7); Mackay, Shiu, and Ma (9); Verschueren (11); Yalkowsky (3); and Yaws (2) were used extensively. The publications were screened and copies of appropriate data were made. These data were then keyed in to the computer to provide a database for which experimental data are available. The database also served as a basis to check the accuracy of the correlation.

Graphs of water solubility versus boiling point are shown in Figs. 1–11 for the chemical families. The graphs indicate favorable agreement of correlation values and experimental data.

For ketones, the graphs disclose that more data are available for alcohols. The graphs, also disclose that the data for ketones appear to be below and approximately parallel to the more extensive data for alcohols. This suggests that the solubility of ketones is lower and approximately parallel to the water solubility of alcohols.

For the other oxygenated compounds (ethers, aldehydes, acids, and acetates), the graphs disclose that the data for oxygenated compounds appear to be above and approximately parallel to the data for benzenes. This suggests that the solubility of oxygenated compounds is higher and approximately parallel to the water solubility of benzenes.

For the nitrogen compounds (primary amines, secondary amines, and tertiary amines), the graphs disclose

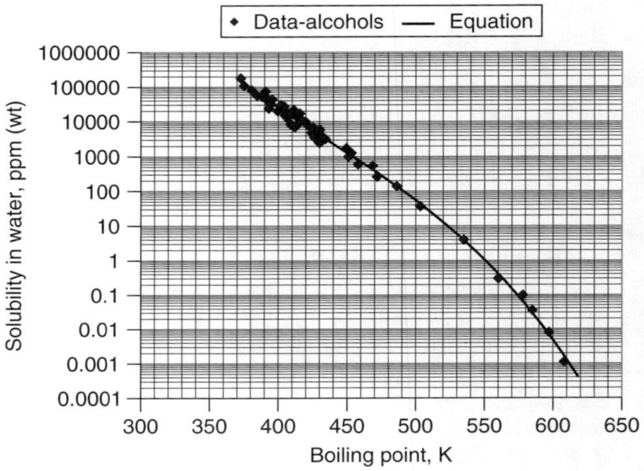

Figure 1. Solubility of alcohols in water.

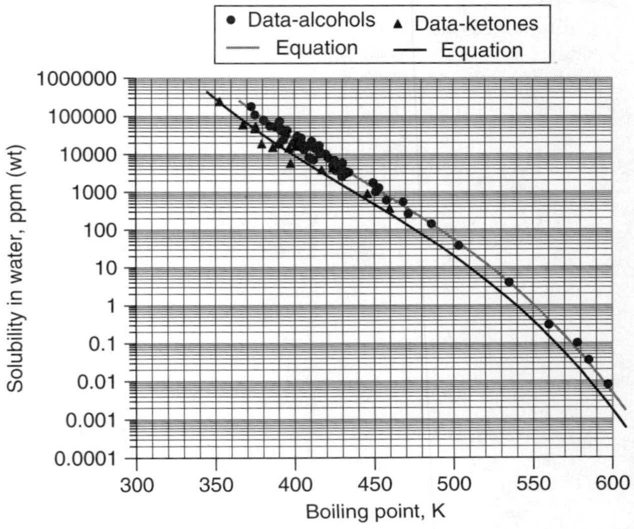

Figure 2. Solubility of ketones in water.

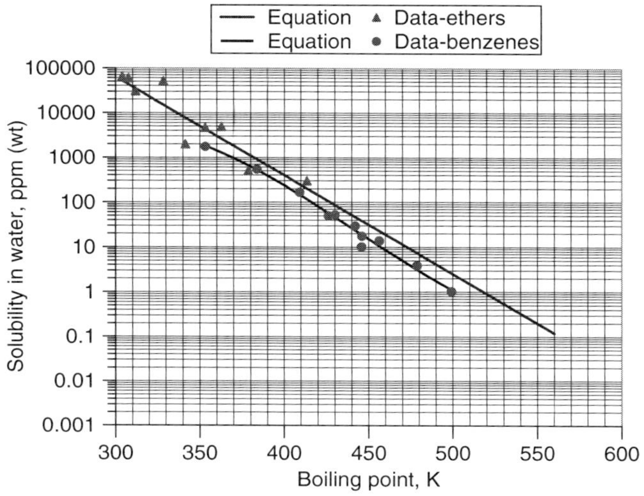

Figure 3. Solubility of ethers in water.

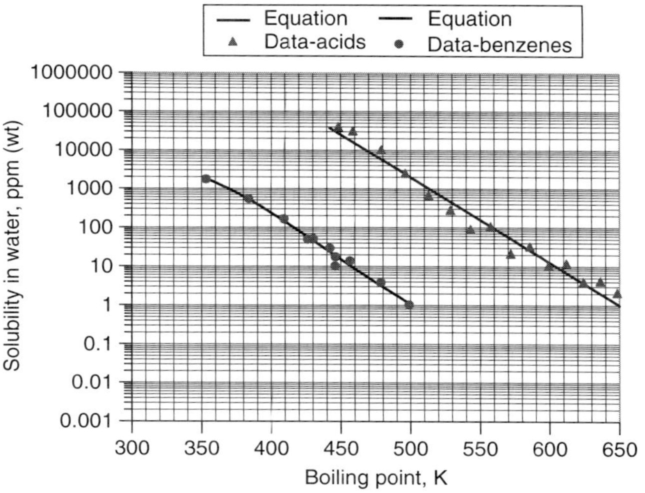

Figure 5. Solubility of acids in water.

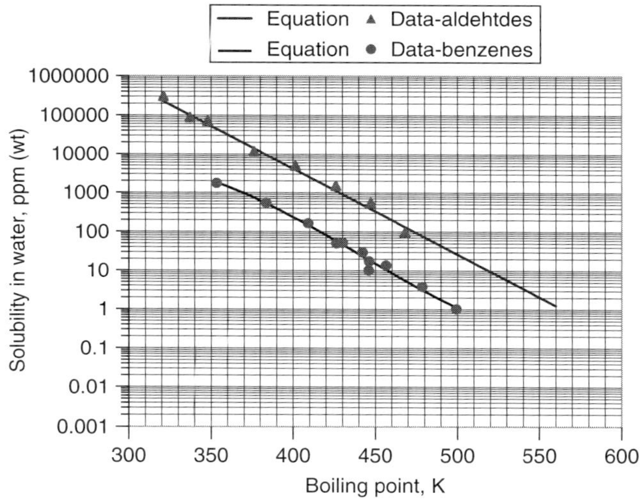

Figure 4. Solubility of aldehydes in water.

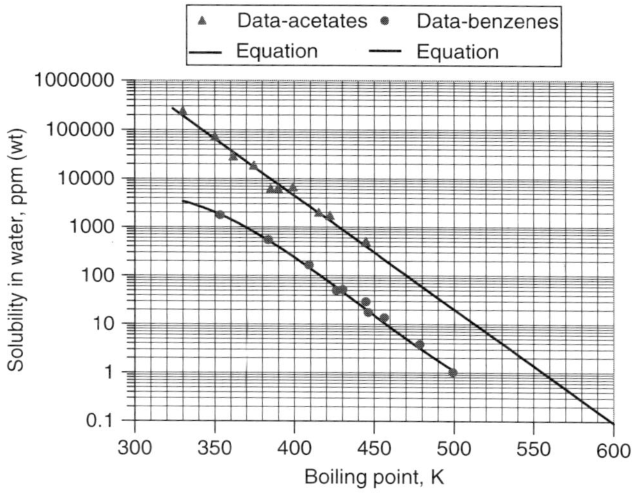

Figure 6. Solubility of acetates in water.

that the data for amines appear to be above and approximately parallel to the data for benzenes. This suggests that the solubility of amines is higher and approximately parallel to the water solubility of benzenes.

For monochloro and monobromoalkanes, the graphs disclose that the data appear to be lower and approximately parallel to the data for benzenes. This suggests that the solubility of monochloro- and monobromoalkanes is lower and approximately parallel to the water solubility of benzenes.

EXAMPLES

The correlations for water solubility are useful in engineering applications involving process wastewater. Examples are shown below.

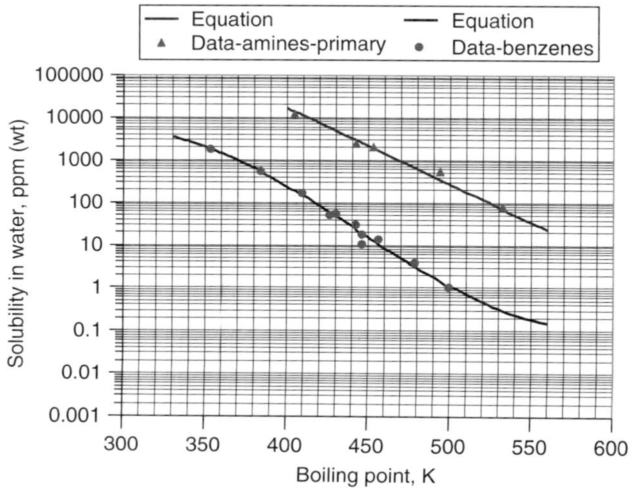

Figure 7. Solubility of primary amines in water.

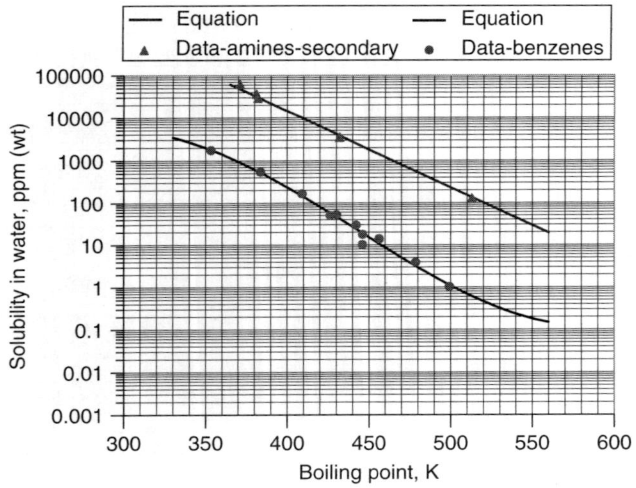

Figure 8. Solubility of secondary amines in water.

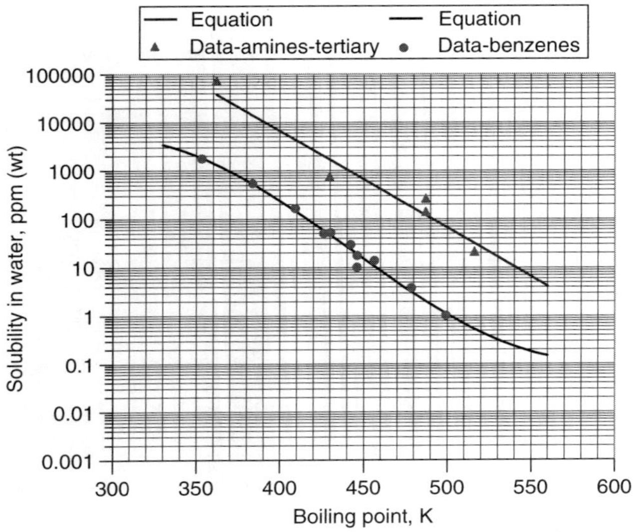

Figure 9. Solubility of tertiary amines in water.

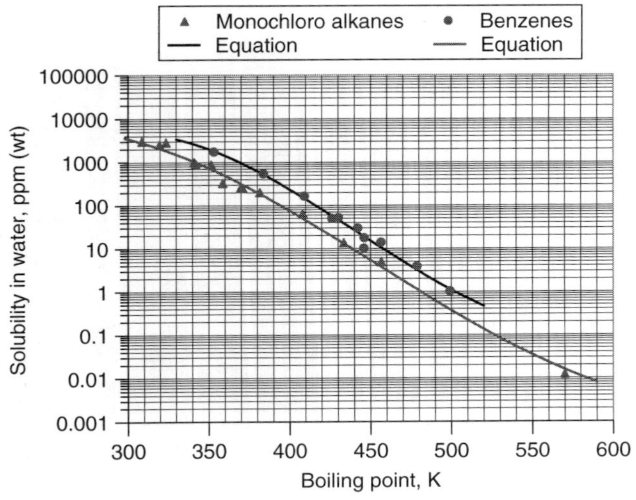

Figure 10. Solubility of monochloroalkanes in water.

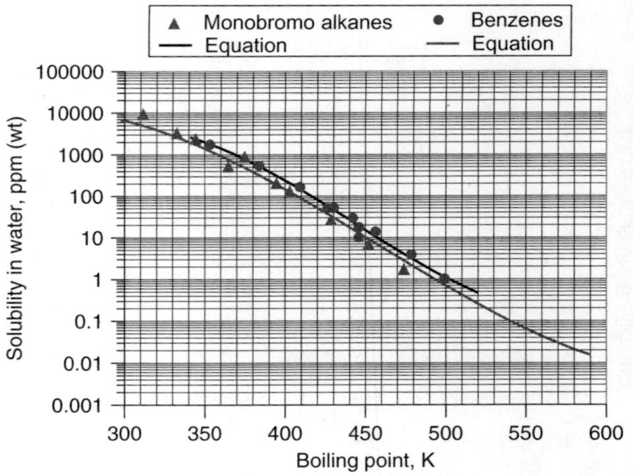

Figure 11. Solubility of monobromoalkanes in water.

Example. A chemical spill of dodecanol occurs into a body of water at ambient conditions. Estimate the concentration in the water at saturation.

The correlation for alcohols maybe used to determine the solubility in water. Substitution of the coefficients and boiling point of dodecanol in the correlation equation yields

$$\log_{10}(S) = A + B\,T_B + C\,T_B{}^2 + D\,T_B{}^3$$
$$= 45.6398 - 2.3859\text{E}{-01}(535.00)$$
$$+ 4.8739\text{E}{-04}(535.00)^2 - 3.7160\text{E}{-07}(535.00)^3$$
$$S = 3.93 \text{ ppm (wt)}.$$

The calculated value and data compare favorably (3.93 vs. 4.00; deviation = 0.07/5 = 1.8%).

Example. A chemical spill of diethyl ketone occurs into a body of water at ambient conditions. Estimate the concentration in the water at saturation.

The correlation for ketones maybe used to determine the solubility in water. Substitution of the coefficients and boiling point of diethyl ketone in the correlation equation yields

$$\log_{10}(S) = A + B\,T_B + C\,T_B{}^2 + D\,T_B{}^3$$
$$= 45.2000 - 2.3859\text{E}{-01}\ (375.14)$$
$$+ 4.8739\text{E}{-04}(375.14)^2 - 3.7160\text{E}{-07}\ (375.14)^3$$
$$S = 45,257 \text{ ppm (wt)}.$$

The calculated value and data compare favorably (45,257 vs. 48,235; deviation = 1,758/48,235 = 3.6%).

Acknowledgment

The Texas Hazardous Waste Research Center (Lamar University, Beaumont, Texas) provided partial support for this work.

BIBLIOGRAPHY

1. *Guide to Occupational Exposure Values.* (1997). American Conference of Governmental Industrial Hygienists, Cincinnati, OH.

2. Yaws, C.L. (1999). *Chemical Properties Handbook*. McGraw-Hill, New York.
3. Yalkowsky, S.H. (1990–2004). *Aquasol Database*. University of Arizona, Tucson, AZ.
4. Hine, J. and Mukherjee, P.K. (1975). *J. Org. Chem.* **40**: 292.
5. Horvath, A.L. and Getzen, F.W. (1999). *J. Phys. Chem. Ref. Data* **28**(3): 649–777.
6. Horvath, A.L. (1982). *Halogenated Hydrocarbons: Solubility-Miscibility with Water*. Marcel Dekker, New York.
7. Howard, P.H. and Meylan, W.M. (Eds.). (1997). *Handbook of Physical Properties of Organic Chemicals*. CRC Press, Boca Raton, FL.
8. *Liquid–Liquid Equilibrium Data Collection*. (1979). Dechema Chemistry Data Series, Frankfurt/Main, Germany, Vol. V, part 1.
9. Mackay, D., Shiu, W.Y., and Ma, K.C. (1992, 1992, 1993, 1995, 1997). *Illustrated Handbook of Physical-Chemical Properties and Environmental Fate for Organic Chemicals*. Vols. 1–5, Lewis, New York.
10. Suzuki, T. (1991). *J. Computer-Aided Molecular Design* **5**: 149.
11. Verschueren, K. (1996, 2001). *Handbook of Environmental Data on Organic Chemicals*, 3rd and 4th Edn. Van Nostrand Reinhold, New York.
12. Donald, M., Shiu, W-Y., and Ma, K-C. (1995). *Illustrated Handbook of Physical-Chemical Properties and Environmental Fate for Organic Chemicals*. Vol. 4, Lewis, New York.
13. (2001). *Handbook of Environmental Data on Organic Chemicals*, 4th Edn. John Wiley & Sons, New York.

SOLUBILITY OF HYDROCARBONS IN SALT WATER

Carl L. Yaws
Prasad K. Narasimhan
Lamar University
Beaumont, Texas

INTRODUCTION

Physical and thermodynamic property data are required to design and operate industrial processes. In particular, the water solubility of hydrocarbons in the sea and ocean is becoming increasingly important because of more and more stringent regulations regarding the environment.

In this article, results are presented for the solubility of hydrocarbons in salt water. The results are applicable to the complete range of salt concentrations, including water without salt to water saturated with salt. The results are intended for use in engineering applications. As an example of such usage, solubility values from the correlation are useful in determining the distribution of a hydrocarbon spill upon contact with seawater. Solubility values at other salt concentrations may also be ascertained.

CORRELATION OF SALT WATER SOLUBILITY

The correlation of the solubility of hydrocarbons in salt water containing salt is based on a series expansion in salt concentration:

$$\log_{10} S = A + B\,X + C\,X^2 \qquad (1)$$

where S = solubility in salt water at 25 °C, parts per million by weight, ppm(wt)
X = concentration of salt (NaCl) in water, parts per million by weight, ppm(wt)
$A, B,$ and C = regression coefficients

The coefficients (A, B, and C) are given in Table 1. The values in the table were determined from regression of the data sources (1–10) for water solubility. The compilations by Price (7) and Sutton and Calder (8) were especially helpful. The values presented are applicable to a wide variety of hydrocarbons (alkanes, cyclopentanes, cyclohexanes, benzenes (no and single substitution), and benzenes (double and triple substitution). The range of application for the correlations is about C_5 to C_{16} compounds.

Graphs of water solubility are shown in Figs. 1–3 for representative hydrocarbons (pentane, methylcyclopentane, and benzene). In the figures, solubility values are plotted at salt concentrations ranging from water without salt to water saturated with salt. The graphs discloses favorable agreement of correlation values and experimental data.

Examples

The correlations for solubility are useful in engineering applications involving hydrocarbons in contact with water containing salt. Examples are shown below.

Example 1. A chemical spill of *n*-pentane occurs in a body of saltwater [salt = 10,000 ppm(wt)] at ambient conditions. Estimate the concentration in the saltwater.

The values in Table 1 are used for the coefficients. Introducing the boiling point of *n*-pentane to calculate A and using the tabulated values for B and C provides

$$\begin{aligned} A &= -17.7916 + 1.7865\text{E}{-01}\, T_B - 5.0255\text{E}{-04}\, T_B^{\,2} + 4.1218\text{E}{-07}\, T_B^{\,3} \\ &= -17.7916 + 1.7865\text{E}{-01}(309.22) - 5.0255\text{E}{-04}(309.22)^2 + 4.1218\text{E}{-07}(309.22)^3 \\ &= 1.5850 \\ B &= -4.5956\text{E}{-06} \\ C &= 2.2978\text{E}{-12} \end{aligned}$$

Substitution of these coefficients and the salt concentration in the correlation equation for solubility in saltwater yields

$$\begin{aligned} \log_{10} S &= A + B\,X + C\,X^2 \\ &= 1.5850 - 4.5956\text{E}{-06}(10{,}000) + 2.2978\text{E}{-12}(10{,}000)^2 \\ &= 1.5393 \\ S &= 34.62 \text{ ppm (wt)} \end{aligned}$$

Table 1. Solubility in Salt Water[a]

	$\log_{10} S = A + BX + CX^2$	S-solubility, ppm (wt), X(salt) – ppm(wt)	
Chemical Family	A, B, and C	Formula	Example Compound
Alkanes	$A = -17.7916 + 1.7865\text{E}{-01}T_B - 5.0255\text{E}{-04}\ T_B{}^2 + 4.1218\text{E}{-07}\ T_B{}^3$ $B = -4.5956\text{E}{-06}$ $C = 2.2978\text{E}{-12}$	C_nH_{2n+2}	n-pentane
Cyclopentanes	$A = 9.55 - 2.2987\text{E}{-02}T_B$ $B = -4.5987\text{E}{-06}$ $C = 2.2993\text{E}{-12}$	C_nH_{2n}	Methyl cyclopentane
Cyclohexanes	$A = 9.75 - 2.2987\text{E}{-02}T_B$ $B = -4.5987\text{E}{-06}$ $C = 2.2993\text{E}{-12}$	C_nH_{2n}	Cyclohexane
Benzenes (no and single substitution)	$A = 5.725 + 3.6778\text{E}{-03}\ T_B - 3.0346\text{E}{-05}\ T_B{}^2$ $B = -3.2714\text{E}{-06}$ $C = 1.6357\text{E}{-16}$	$C_{n+6}H_{2n+6}$	Benzene
Benzenes (double and triple substitution)	$A = 5.375 + 3.6778\text{E}{-03}\ T_B - 3.0346\text{E}{-05}\ T_B{}^2$ $B = -3.2714\text{E}{-06}$ $C = 1.6357\text{E}{-16}$	$C_{n+6}H_{2n+6}$	Benzene

[a] S-solubility in salt water at 25 C, parts per million by weight, ppm(wt).
X-salt (NaCl) concentration in water, ppm(wt).
Salt concentration range is $X = 0$ to 358,700 (saturated).
Salt in seawater is 34,472 ppm(wt).
A, B, and C—regression coefficients.
T_B—boiling point of compound, K.
T_{BMAX}—maximum temperature, $K = 560$.
T_{BMIN}—minimum temperature, $K = 300$.
Correlation should not be used for compounds having boiling points outside T_BMAX and T_BMIN.
Range of application for correlation is about C_5 to C_{16}.

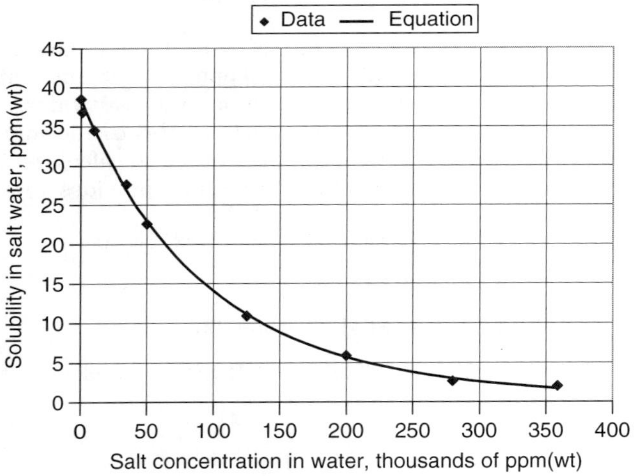

Figure 1. Solubility of n-pentane in salt water.

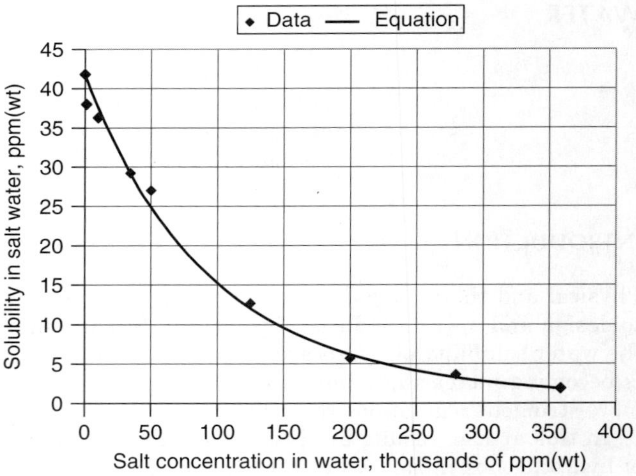

Figure 2. Solubility of methyl cyclopentane in salt water.

The calculated value and data compare favorably (34.62 vs. 34.50; deviation = 0.12/34.50 = 0.3%).

Example 2. A chemical spill of benzene occurs in seawater [salt = 34,472 ppm(wt)] at ambient conditions. Estimate the concentration in the seawater.

The values in Table 1 are used for the coefficients. Introducing the boiling point of benzene to calculate A and using the tabulated values for B and C provides

$$A = 5.725 + 3.6778\text{E}{-03}\ T_B - 3.0346\text{E}{-05}\ T_B{}^2$$
$$= 5.725 + 3.6778\text{E}{-03}\ (353.31)$$
$$- 3.0346\text{E}{-05}\ (353.31)^2$$
$$= 3.2364$$
$$B = -3.2714\text{E}{-06}$$

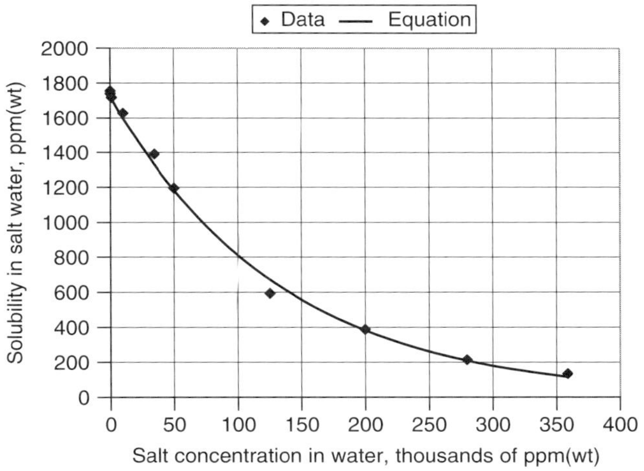

Figure 3. Solubility of benzene in salt water.

$C = 1.6357\text{E}{-16}.$

Substitution of these coefficients and the salt concentration in the correlation equation for solubility in saltwater yields

$$\begin{aligned}\log_{10} S &= A + B\,X + C\,X^2\\ &= 3.2364 - 3.2714\text{E}{-06}(34{,}472)\\ &\quad + 1.6357\text{E}{-16}(34{,}472)^2\\ &= 3.1236\\ S &= 1329.24 \text{ ppm (wt)}\end{aligned}$$

The calculated value and data compare favorably (1329.24 vs. 1391.00; deviation = 61.76/1391.00 = 4.4%).

Acknowledgment
The Texas Hazardous Waste Research Center (Lamar University, Beaumont, Texas) provided partial support for this work.

BIBLIOGRAPHY

1. Donald, M., Shiu, W-Y., and Ma, K-C. (1995). *Illustrated Handbook of Physical-Chemical Properties and Environmental Fate for Organic Chemicals*. Vol. 4, Lewis, New York.
2. (2001). *Handbook of Environmental Data on Organic Chemicals*, 4th Edn. John Wiley & Sons, New York.
3. Hine, J. and Mukherjee, P.K. (1975). *J. Org. Chem.* **40**: 292.
4. Howard, P.H. and Meylan, W.M. (Eds.). (1997). *Handbook of Physical Properties of Organic Chemicals*. CRC Press, Boca Raton, FL.
5. Mackay, D., Shiu, W-Y., and Ma, K-C. (1992, 1992, 1993, 1995, 1997). *Illustrated Handbook of Physical-Chemical Properties and Environmental Fate for Organic Chemicals*. Vols. 1–5, Lewis, New York.
6. Price, L.C. (1976). *Bull. Am. Assoc. Pet. Geologists* **60**(2): 213.
7. Sutton, C. and Calder, J.A. (1974). *Environ. Sci. Technol.* **8**(7): 654.
8. Suzuki, T. (1991). *J. Comput.-Aided Molecular Design* **5**: 149.
9. Verschueren, K. (1996, 2001). *Handbook of Environmental Data on Organic Chemicals*, 3rd and 4th Edn. Van Nostrand Reinhold, New York.
10. Yalkowsky, S.H. (1990–2004). *Aquasol Database*. University of Arizona, Tucson, AZ.
11. Yaws, C.L. (1999). *Chemical Properties Handbook*. McGraw-Hill, New York.

SOLUBILITY OF HYDROCARBONS AND SULFUR COMPOUNDS IN WATER

Carl L. Yaws
Prasad K. Narasimhan
Helen H. Lou
Lamar University
Beaumont, Texas

Ralph W. Pike
Louisiana State University
Baton Rouge, Louisiana

Correlations are presented for the solubility of hydrocarbons and organic sulfur compounds in water. A wide variety of compounds are covered—alkanes, olefins, diolefins, acetylenes, cyclopentanes, cyclohexanes, benzenes, mercaptans, thiophenes, and sulfides. The new correlation provides reliable solubility values down to very low concentrations (parts per million range). The correlation is based on the boiling point of the compound. Correlation and experimental data are in favorable agreement. The results are useful in process engineering for wastewater minimization.

IMPORTANCE OF WATER SOLUBILITY

The solubility of compounds in water is very important. This importance will increase in the future because of heath, safety, and environmental considerations. The following brief discussion illustrates the importance even at very low concentrations.

For health involving human exposure to substances in air, the threshold limit value (TLV) for *n*-pentane in air is given as 600 ppm (parts per million, vol) by the Occupational Safety and Health Act (1). A concentration of only 0.0000001 mol fraction of *n*-pentane in water provides about 7000 ppm of *n*-pentane in air at the air–water interface. This concentration of 7000 ppm in air exceeds the threshold limit value of 600 ppm.

For safety, the lower explosion limit (LEL) for *n*-pentane in air is given as 1.4% by Yaws (2). A concentration of only 0.000001 mol fraction of *n*-pentane in water provides about 7% of *n*-pentane in air at the air–water interface. This concentration of 7% in air exceeds the lower explosion limit of 1.7%.

For environmental, consider a spill of *n*-pentane in contact with water. The water will become saturated. At saturation, the solubility of *n*-pentane in water is about 0.0000385 weight fraction or 0.00000916 mol fraction, as given by Yalkowsky (3). This concentration of only

0.00000916 mol fraction at saturation will provide about 644,000 ppm or 64.4% of n-pentane in air at the air–water interface. This concentration of 644,000 ppm or 64.46% greatly exceeds both the threshold limit value of 600 ppm and the lower explosion limit of 1.4%.

This brief discussion (using n-pentane as an example) indicates that very low concentrations (ppm or less) of compounds in water can provide concentrations in air at the air–water interface that exceed the threshold limit value for human exposure and the lower explosion limit for flammability.

CORRELATION FOR WATER SOLUBILITY

In earlier work by Yaws and coworkers (2), the water solubility of various chemical types was correlated as a function of the boiling point of the compound. In this article it was determined that the boiling point method is also applicable to correlation of the water solubility of hydrocarbons and organic sulfur compounds:

$$\log_{10}(S) = A + BT_B + CT_B^2 + DT_B^3 \quad (1)$$

where S = solubility in water at 25 °C, ppm by weight,
T_B = boiling point of compound, K
$A, B, C,$ and D = regression coefficients

The range for boiling point is 298 K to about 560 K.

The coefficients (A, B, C, and D) are given in Table 1 for a wide variety of hydrocarbons and sulfur compounds—alkanes, olefins, diolefins, acetylenes, cyclopentanes, cyclohexanes, benzenes, mercaptans, thiophenes, and sulfides. The table also provides the boiling point range (T_{BMIN}, T_{BMAX}) for which the correlation is applicable. The correlation should not be used for compounds whose boiling points are outside the boiling point range.

The coefficients were determined from regression of available data. In preparing the correlation, a literature search was conducted to identify data source publications (2–10). The excellent compilations by Howard and Meylan (5); Mackay, Shiu, and Ma (6); Verschueren (8); Yalkowsky (3); and Yaws (2) were used extensively. The publications were screened and copies of appropriate data were made. These data were then keyed in to the computer to provide a database for which experimental data are available. The database also served as a basis to check the accuracy of the correlation.

Graphs of water solubility versus boiling point are shown in Figs. 1–8 for the compound families. The graphs indicate favorable agreement of correlation values and experimental data.

For olefins and diolefins, the graphs disclose that more data are available for alkanes. The graphs also disclose that the data for olefins and diolefins appear to be above and approximately parallel to the more extensive data for alkanes. This suggests that the solubility of olefins and diolefins is higher and approximately parallel to the water solubility of alkanes.

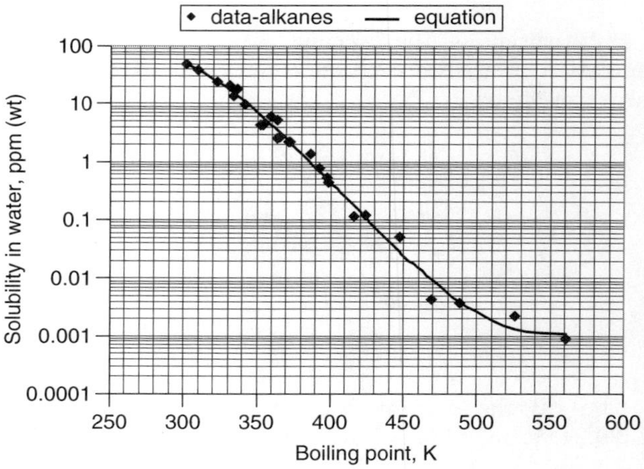

Figure 1. Solubility of alkanes in water.

Table 1. Solubility in Water[a]

$\log_{10} S = A + BT_B + CT_B^2 + DT_B^3$					(S - ppm(wt), $T_B - K$)			
Chemical family	A	B	C	D	T_{BMIN}	T_{BMAX}	Formula	Example Compound
Alkanes	−17.6520	1.77811E−01	−5.00907E−04	4.11124E−07	298	560	C_nH_{2n+2}	n-Hexane
Olefins	−17.0300	1.77811E−01	−5.00907E−04	4.11124E−07	298	560	C_nH_{2n}	1-Hexene
Diolefins	−16.5610	1.77811E−01	−5.00907E−04	4.11124E−07	298	560	C_nH_{2n-2}	1,5-Hexadiene
Acetylenes	−15.8350	1.77811E−01	−5.00907E−04	4.11124E−07	298	560	$C_{n+2}H_{2n+2}$	1-Hexyne
Cyclohexanes	−16.7000	1.77811E−01	−5.00907E−04	4.11124E−07	298	560	C_nH_{2n}	Methyl cyclohexane
Cyclopentanes	−16.9000	1.77811E−01	−5.00907E−04	4.11124E−07	298	560	C_nH_{2n}	Methyl cyclopentane
Benzenes (single substitution)	−24.0080	2.21196E−01	−5.55632E−04	4.18830E−07	298	560	$C_{n+6}H_{2n+6}$	Toluene
Benzenes (multiple substitution)	−23.6500	2.21196E−01	−5.55632E−04	4.18830E−07	298	560	$C_{n+6}H_{2n+6}$	p-Xylene
Mercaptans	−6.9000	1.00500E−01	−2.72880E−04	1.99870E−07	298	560	$C_nH_{2n+2}S$	Propyl mercaptan
Thiophenes	−6.8500	1.00500E−01	−2.72880E−04	1.99870E−07	298	560	$C_nH_{2n-4}S$	2-Methyl thiophene
Sulfides	−6.5390	1.00500E−01	−2.72880E−04	1.99870E−07	298	560	$C_nH_{2n+2}S$	Ethyl methyl sulfide

[a] S = solubility in water at 25 C, parts per million by weight, ppm(wt).
$A, B, C,$ and D = regression coefficients.
T_B = boiling point of compound, K.
T_{BMAX}-maximum temperature, K.
T_{BMIN}-minimum temperature, K.

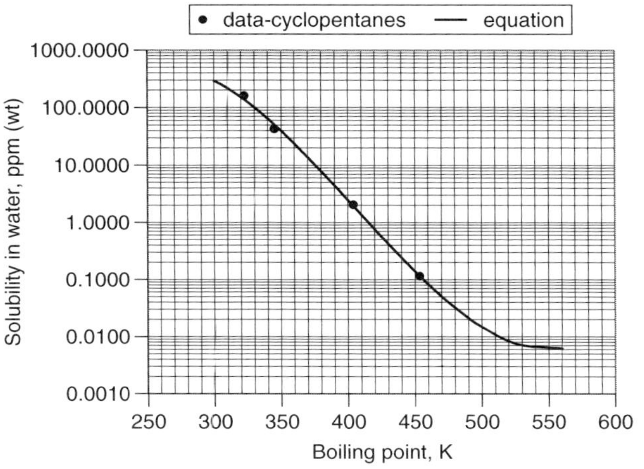

Figure 2. Solubility of cyclopentanes in water.

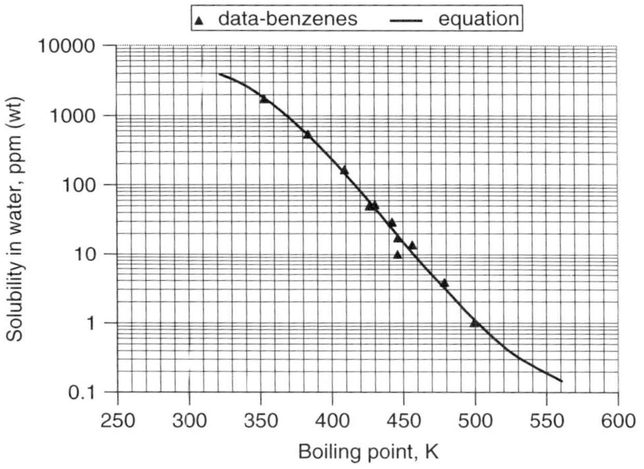

Figure 3. Solubility of benzenes in water.

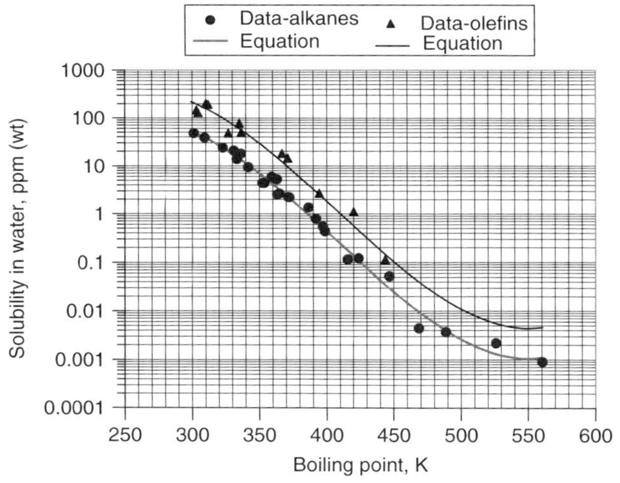

Figure 4. Solubility of olefins in water.

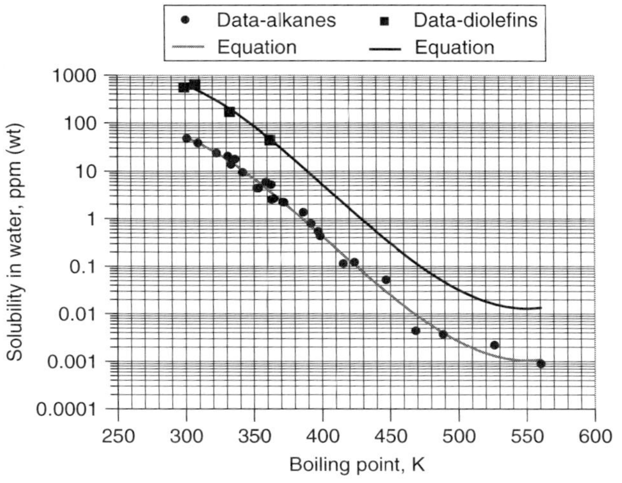

Figure 5. Solubility of diolefins in water.

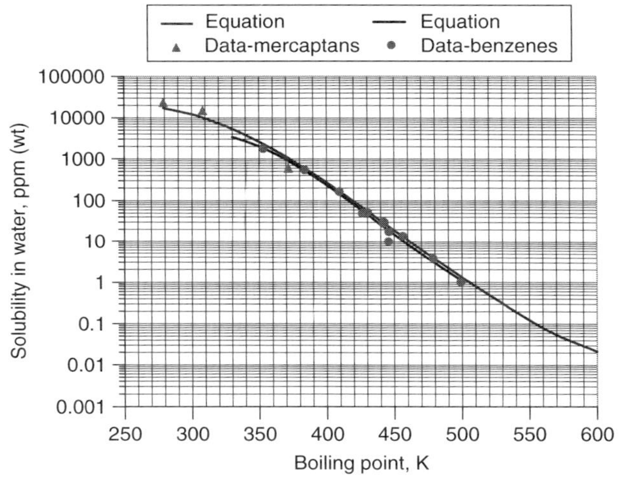

Figure 6. Solubility of mercaptans in water.

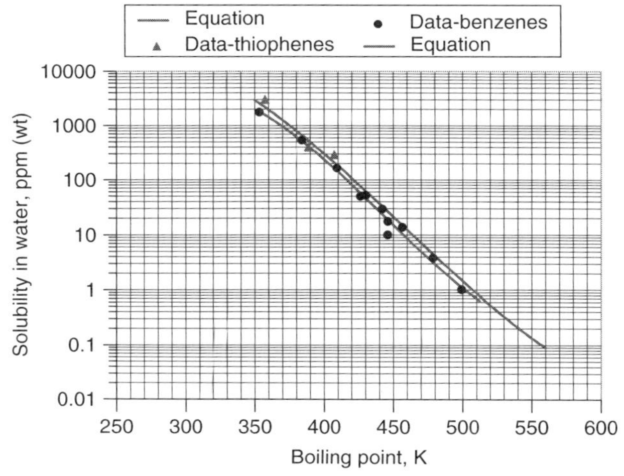

Figure 7. Solubility of thiophenes in water.

For sulfur compounds (mercaptans, thiopheolefins, and diolefins), the graphs disclose that more data are available for benzenes. The graphs also, disclose that the data for sulfur compounds appear to be slightly above and approximately parallel to the more extensive data for benzenes. This suggests that the solubility of sulfur

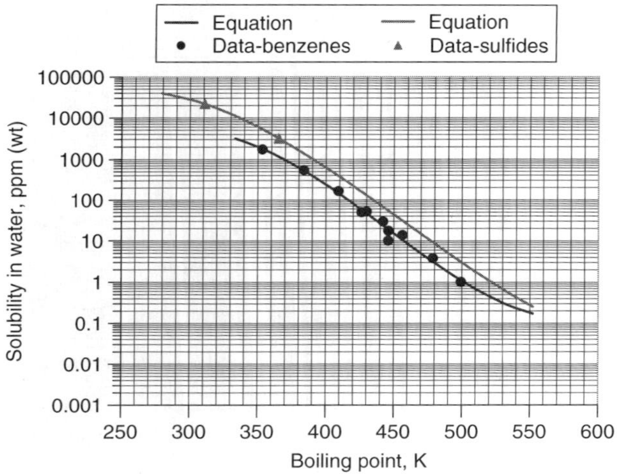

Figure 8. Solubility of sulfides in water.

compounds is higher and approximately parallel to the water solubility of benzenes.

Examples

The correlations for water solubility are useful in engineering applications involving process wastewater. Examples are shown below.

Example 1. A chemical spill of n-hexane occurs into a body of water at ambient conditions. Estimate the concentration in the water at saturation.

The correlation for alkanes maybe used to determine the solubility in water. Substitution of the coefficients and boiling point of n-hexane in the correlation equation yields:

$$\begin{aligned}\log_{10}(S) &= A + BT_B + CT_B{}^2 + DT_B{}^3\\ &= -17.6520 + 1.77811\text{E}{-01}(341.88)\\ &\quad - 5.00907\text{E}{-04}(341.88)^2\\ &\quad + 4.11124\text{E}{-07}(341.88)^3\\ S &= 10.46 \text{ ppm (wt)}\end{aligned}$$

The calculated value and data compare favorably (10.46 vs 9.47; deviation = 0.99/9.47 = 10.4%).

Example 2. A chemical spill of toluene occurs into a body of water at ambient conditions. Estimate the concentration in the water at saturation.

The correlation for benzenes (single substitution) maybe used to determine the solubility in water. Substitution of the coefficients and boiling point of toluene in the correlation equation yields

$$\begin{aligned}\log_{10}(S) &= A + BT_B + CT_B{}^2 + DT_B{}^3\\ &= -24.0080 + 2.21196\text{E}{-01} * (383.78)\\ &\quad - 5.55632\text{E}{-04} * (383.78)^2\\ &\quad + 4.18830\text{E}{-07} * (383.78)^3\\ S &= 524.68 \text{ ppm (wt)}\end{aligned}$$

The calculated value and data compare favorably (524.68 vs 542.4; deviation = 17.72/542.4 = 3.3%).

Acknowledgments
The Texas Hazardous Waste Research Center (Lamar University, Beaumont, Texas) provided partial support for this work. Portions of this material appeared in *Oil & Gas Journal* **89**; 79 (April 8, 1991), *Chem. Eng.*, **100**; 122 (Oct., 1993), *Oil & Gas Journal* **89**: 86 (Sept. 16, 1991), *Oil & Gas Journal* **95** (35): 80 (Aug. 29, 1994), and *Chem. Eng.* **110**: 60 (Aug., 2003). These portions are reprinted by special permission.

BIBLIOGRAPHY

1. *Guide to Occupational Exposure Values*. (1997). American Conference of Governmental Industrial Hygienists, Cincinnati, OH.
2. Yaws, C.L. (1999). *Chemical Properties Handbook*. McGraw-Hill, New York.
3. Yalkowsky, S.H. (1990–2004). *Aquasol Database*. University of Arizona, Tucson, AZ.
4. Hine, J. and Mukherjee, P.K. (1975). *J. Org. Chem.* **40**: 292.
5. Howard, P.H. and Meylan, W.M. (Eds.). (1997). *Handbook of Physical Properties of Organic Chemicals*. CRC Press, Boca Raton, FL.
6. Mackay, D., Shiu, W-Y., and Ma, K-C. (1992, 1992, 1993, 1995, 1997). *Illustrated Handbook of Physical-Chemical Properties and Environmental Fate for Organic Chemicals*. Vols. 1–5, Lewis, New York.
7. Suzuki, T. (1991). *J. Comput.-Aided Molecular Design* **5**: 149.
8. Verschueren, K. (1996, 2001). *Handbook of Environmental Data on Organic Chemicals*, 3rd and 4th Edn. Van Nostrand Reinhold, New York.
9. Donald, M., Shiu, W-Y., and Ma, K-C. (1995). *Illustrated Handbook of Physical–Chemical Properties and Environmental Fate for Organic Chemicals*. Vol. 4, Lewis Publishers, New York.
10. (2001). *Handbook of Environmental Data on Organic Chemicals*, 4th Edn. John Wiley & Sons, New York.

SORPTION KINETICS

TH.D. KARAPANTSIOS
M.X. LOUKIDOU
K.A. MATIS
Aristotle University
Thessaloniki, Greece

INTRODUCTION

Sorption is, by definition, a general term describing the attachment of charged species (such as toxic metal ions) from a solution to a coexisting solid surface. Sorption kinetics may be controlled by several independent processes that can act in series or in parallel (see Fig. 1). These processes fall in one of the following general categories:

1. bulk diffusion,
2. external mass transfer (film diffusion),

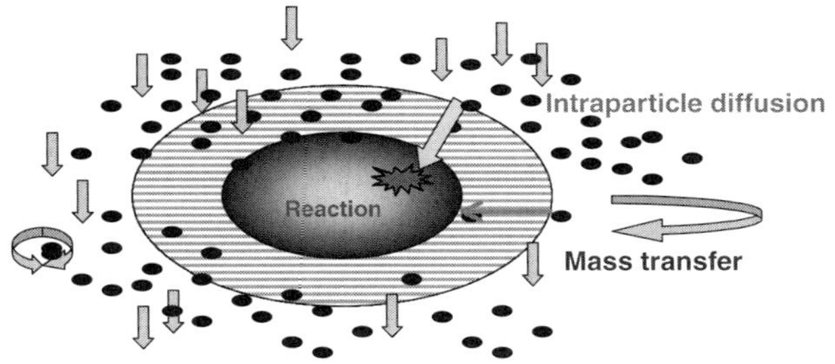

Figure 1. Illustration of the mechanism during sorption to describe suitably and model the respective process kinetics.

3. chemical reaction (chemisorption), and
4. intraparticle diffusion.

Kinetic analyses allow estimating sorption rates and also lead to suitable rate expressions characteristic of possible reaction mechanisms (1–3).

At sufficiently high agitation speed in a reaction vessel, the bulk diffusion step can be safely ignored because then sorption onto sorbent particles is decoupled from mass transfer in the bulk mixture. Apart from that, it is quite common that more than one process can contribute to system performance at the same time. In this case, the extensive interrelationships between the various equations make the overall kinetic model exceedingly complicated to evaluate. A rather simplifying approach to circumvent this problem is to assume that each one of the co-current processes dominate over the others (i.e., the rate-controlling step) at specific time regimes of the process and then study them independently (4); this approach was also adopted in this article.

Many studies engaged so far to examine sorption phenomena involved analysis of batch experiments where data were sampled at even time intervals over the entire course of the process. As a result, fast changing kinetic data characteristic of the phenomena just after the onset of sorption could not be accurately depicted on an adequately short timescale. Thus, a primary objective of this article is to investigate the kinetic mechanism of metal sorption on sorbent particles, putting more emphasis on samples collected at short times after the initiation of the process, where the major part of the adsorption occurs.

To identify the most appropriate mechanism for a process, several models must be checked for suitability and consistency in a broad range of system parameters. The model selection criteria proposed by Ho et al. (5) for sorption of pollutants in aqueous systems were used herein as a general guideline. According to this, several reaction-based and diffusion-based models were tested in simulating our data. The finally chosen kinetic models are those that fit the data closely and also represent reasonable sorption mechanisms.

EQUILIBRIUM AND KINETIC MODELING

An overall mass balance of the sorbate across the sorbent surface can be written as

$$X\,q_t + C_t^s = C_o \Leftrightarrow q_t = \frac{(C_o - C_t^s)}{X} \tag{1}$$

where q is the specific metal uptake (mg of metal per g of sorbent), C is the metal bulk concentration (mg/L) and X is the sorbent feed per unit volume of solution (g/L). Subscripts o and t denote conditions at the beginning and any other instant of the process, respectively. The superscript s denotes conditions at the sorbent interface. For adequately long adsorption times: $q_t \approx q_e$ and $C_t^s \approx C_e^s \approx C_e$, where the subscript e denotes equilibrium conditions. The known Langmuir and Freundlich adsorption models have usually been employed to describe equilibrium metal uptake:

$$q_e = \frac{q_{max} b C_e}{1 + bC_e} \tag{2a}$$

$$q_e = K_F C_e^{1/n} \tag{2b}$$

q_{max} (mg/g), b (L/mg), and K_F [mg [g (mg/L)$^{-1/n}$]$^{-1}$] and n (dimensionless) are constants.

To compare measurements from various experiments for a kinetics investigation, it is necessary to introduce a dimensionless degree of conversion. Thus, by normalizing the remaining ion concentration, C_t, with respect to some reference value, an index of sorption is defined. Taking advantage of the values of C_t before the onset of sorption, C_o, and for completed sorption, $C_\infty = C_e$, the following degree of conversion is proposed:

$$\alpha = \frac{C_o - C_t}{C_o - C_e} \tag{3}$$

Kinetic studies customarily use the basic conversion rate equation:

$$\frac{d\alpha}{dt} = k(T)f(\alpha) \tag{4}$$

where $f(\alpha)$ is a conversion-dependent function and k(T) is the reaction rate constant. Sorption in heterogeneous liquid–solid systems has usually been considered to follow Arrhenius kinetics, which means that the temperature dependence of the rate constant k may be described by the well-known expression:

$$k(T) = A\exp(-E/RT) \tag{5}$$

However, to apply the model in such systems, the morphology of the reactive surface and the final sorption capacity must not vary with temperature, conditions commonly overlooked by many workers (6).

The example used in this article deals with an investigation of the removal and depletion of cadmium and chromates from aqueous solutions by biosorption on *Aeromonas caviae* (the sorbent). This microorganism is often present in groundwater and generally in aquatic environments. Biosorption, the uptake of heavy metals by dead biomass, has gained credibility during recent years as it offers a technically feasible and economical approach (7–10); the latter author illustrated that adsorption is different from ion exchange. Below, evidence is also provided that the system examined is a complex process. More than one sorption model were often reported to describe a case study correctly (11). Attention should be paid to any possible experimental error.

The Langmuir isotherm equation was often found to fit quite satisfactorily the experimental data (12,13). Experimental adsorption isotherms of metal cations or oxyanions, obtained with different sorbent concentrations, temperatures, etc. are usually presented as figures and then, tables display the results of fitting the Langmuir and/or the Freundlich model to the data obtained; the values of the respective correlation coefficient (r^2), if high, are often used to demonstrate a satisfactory description. The assumptions involved in these models were well described, among others, by Volesky (9). The Langmuir isotherms obtained during the equilibrium study lend further support to the notion of a monolayer (14).

On the other hand, the Freundlich isotherm is heavily used in industry. This model is often accurate for higher concentrations, but is inaccurate at low concentrations due to its power-law nature. The equilibrium properties do not change proportionally with the sorbent addition, possibly due to an aggregation of solids at higher loads and hence reducing the effective adsorption area (15).

From the chemical reaction category (chemisorption), the best fit for the data sets of this study is achieved by second-order chemical reactions (16). The solution of the standard second-order reaction, based on constant stoichiometry of one metal ion per binding site, is (e.g., 4):

$$C_t = \frac{C_o}{1 - \dfrac{C_o}{C_e}\exp(-k_1 C_e t)} \tag{6}$$

where k_1 is the reaction rate constant [L*(mg^{-1} of metal)*min^{-1}]. This adsorption model has been very effective in describing the kinetics of adsorption of gases on solids. Nevertheless, when the data were plotted, it was shown that Eq. 6 clearly fails to capture the steep concentration gradient of the early removal stage (16). This was a direct indication that adsorption on solids from a liquid phase is a process different from adsorption from a gas phase, where traditionally the remaining bulk concentration dictates the kinetics.

If the rate of sorption depends not on bulk concentration but on uptake by the sorbent, this can be described by the so-called Ritchie second-order equation according to which one metal ion occupies two binding sites (17):

$$q_t = q_e\left[1 - \left(\frac{1}{1 + k_2 t}\right)\right] \tag{7}$$

where k_2 is the reaction rate constant (min^{-1}). When in the above treatment, q_e does not dictate sorbate uptake, then a pseudo-second-order rate expression is more appropriate:

$$\frac{t}{q_t} = \frac{1}{k_m q_m^2} + \frac{1}{q_m} t \tag{8}$$

where k_m is the reaction rate constant [g of sorbent*(mg^{-1} of metal)*min^{-1}] and q_m is a numerically determined parameter which, under ideal second-order rate control, corresponds to q_e. In the literature (5), various other kinetic equations have been attempted: zero, first (forward or reversible)-order, Langmuir–Hinshelwood, and the Elovich-type.

Equations 7 and 8 provided quite a suitable description of data for advancing time (see Fig. 2). It is noteworthy that both models adequately capture the rapid rate of adsorption during the first minutes of the experiments. This implies that the metal uptake by the sorbent is a satisfactory rate-controlling parameter under a second-order reaction mechanism. The numerically best fit values of the rate parameters of these equations were displayed in a table (16). The predicted equilibrium sorption capacities were quite close to the experimental values for both models. Nevertheless, the rate constant of the pseudo-second-order model, k_m, is monotonically correlated with changes in the biomass load and in the bulk concentration, features that have been encountered in the past in biosorption (18).

In contrast, the rate constant of the Ritchie second-order equation, k_2, fluctuated, beyond any physical reasoning. In addition, Equation 8 exhibited better fitting statistics. Despite the goodness of fit for sorption at 40 and 60 °C, the reaction rate constant of both models varied randomly with temperature. Preliminary calculations,

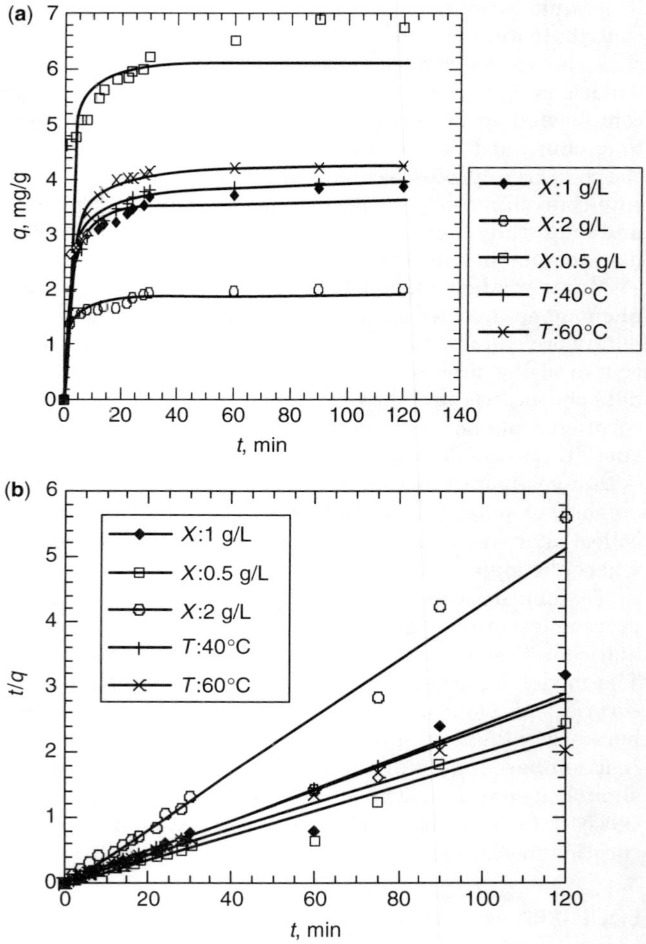

Figure 2. Comparison of experimental removal curves vs. theoretical predictions based on (**a**) the Ritchie second-order equation (at an initial cadmium concentration of 5 mg/L) and (**b**) the pseudo-second-order equation (at an initial cadmium concentration of 50 mg/L). Reprinted from Reference 16.

using the Arrhenius model between two temperatures every time, always gave activation energies below 10 kJ/mol, which is far less than expected for reaction-controlled sorption processes. The morphological changes in the biomass (sorbent) surface at different temperatures and the dependence of sorption capacity on temperature may be blamed for this irregularity.

DIFFUSION KINETICS AND MODELING

For intraparticle diffusion, Crank (19) proposed a model that takes into account the continuously decreasing bulk concentration due to sorbate removal. This gives rise to a time-dependent boundary condition for the concentration at the surface of the sorbent particle. The solution of the diffusion equation for such a boundary condition and a concentration independent diffusivity is

$$\alpha = 1 - 6\sum_{n=1}^{\infty} \frac{\exp(-\xi p_n^2 t)}{9\Lambda/(1-\Lambda) + (1-\Lambda)p_n^2} \tag{9}$$

where p_n is given by the nonzero roots of

$$\tan(p_n) = \frac{3p_n}{3 + p_n^2/(1-\Lambda)} \tag{10}$$

and $\Lambda \equiv (C_o - C_\infty)/C_o$ is the fraction of metal ultimately adsorbed by the sorbent.

Equation 9 can be solved numerically to determine ξ, the effective diffusional time constant, which for particle (also called micropore) diffusion control equals D_c/R_c^2; D_c and R_c are the intraparticle diffusion coefficient (m^2/s) and mean particle radius (m), respectively. The same expression is the solution of the diffusion equation for (macro)pore diffusion control but only where the equilibrium isotherm is linear ($q_e \approx K^*C_e$) for the concentration range under investigation. Then ξ equals $(D_p/R_p^2)/[1 + (1-\varepsilon_p)K/\varepsilon_p]$, where $\varepsilon_p D_p/[1 + (1-\varepsilon_p)K]$ is the effective macropore diffusivity, K is the equilibrium constant, ε_p is the void volume fraction, and R_p is the radius of the particle (14).

The experimental procedure in this article is based on measuring the remaining metal concentration in the bulk, so Λ is always above 0.5. For Λ greater than about 0.1, the effect of diminishing bulk concentration becomes significant, and under these conditions, the assumption of constant metal concentration at the surface of the sorbent leads to an erroneously high diffusivity. Provided that the concentration step employed in a sorption experiment corresponds to a linear section of the equilibrium isotherm, the constant K can be replaced by the local slope of the isotherm, dq_e/dC_e. This is approximately valid for the concentrations employed in this work, as can be seen from the experimental isotherms. Our interest is chiefly focused on the short time region ($\sqrt{D_{c,p}t/R_{c,p}^2} < 0.2$) where Eq. 9 converges slowly, so at least 200 terms are used in the summation to achieve satisfactory accuracy.

External mass transfer has been customarily analyzed in literature by adopting a pseudo-first-order reaction model (5,20). This approach assumes that the sorbate concentration at the sorbent surface is zero at all times. However, this is not true, particularly where a significant quantity of sorbate is adsorbed rapidly at the beginning of the process. A more realistic model should consider instead that a rapid equilibrium is established between the sorbate at the interface and that present on the sorbent surface (21), and this concept is adopted in this article.

If one combines the mass balance across the sorbent surface, Equation 1, the Langmuir adsorption isotherm, Equation 2a, and the rate of change equation in the bulk concentration, one ends up with (21)

$$\frac{dC_t}{dt} = -K_m S(C_t - C_t^s) \tag{11}$$

$$\frac{dC_t^s}{dt} = \left(\frac{K_m S}{X q_{max} b}\right)[(C_t - C_t^s)(1 + bC_t^s)^2] \tag{12}$$

where K_m is the external mass transfer coefficient (m/s) and S is the specific surface area of the sorbent particles per unit volume of the reactor (m^2/m^3). Using the following dimensionless variables: $C^* = C_t/C_o$, $C_s^* = C_t^s/C_o^s$, and $t^* = t/\tau$, where C_o^s is determined from Eqs. 1 and 2a and τ is the total adsorption time, Eqs. 11 and 12 convert to

$$\frac{dC^*}{dt^*} = -K_m S\tau(C^* - C_s^*) \tag{13}$$

$$\frac{dC_s^*}{dt^*} = \left(\frac{K_m S\tau}{X q_{max} b}\right)[(C^* - C_s^*)(1 + bC_o C_s^*)^2] \tag{14}$$

which is a system of two first-order ordinary differential equations that must be solved simultaneously. The initial conditions are $C^* = 1$ and $C_s^* = 0$ at $t^* = 0$. Note that normalizing C_t^s with respect to C_o^s —and not C_o, as Puranik et al. (21) did —and also t with respect to τ, improved the stability and convergence characteristics of the solution markedly due to comparable spreading of all variables over the computational domain.

Equations 9–10 and 13–14 are solved numerically to determine ξ and $K_m S$, respectively. The nonlinear numerical regression to fit experimental data to those equations is performed by the Levenberg–Marquardt method, which gradually shifts the search for the minimum of the sum of the errors squared (SSE), from the steepest descent to quadratic minimization—Gauss–Newton (22):

$$\text{SSE} = \sum_i \left[\frac{(q_{exp,i} - q_{cal,i})^2}{q_{exp,i}^2}\right] \tag{15}$$

Figure 3a presents the results of fitting Eq. 9 to biosorption data obtained with different initial concentrations, solids loads, and temperatures. It is apparent that, despite some scatter in measurements, the finite volume diffusion model can fairly well describe the entire range of data, also including the steep concentration gradient at short times. The values of Λ, ξ, and the computed values of D_c were displayed in a table (14). Such behavior has been customary as a consequence of the decreasing slope of a nonlinear equilibrium curve, example, a Langmuir isotherm, which

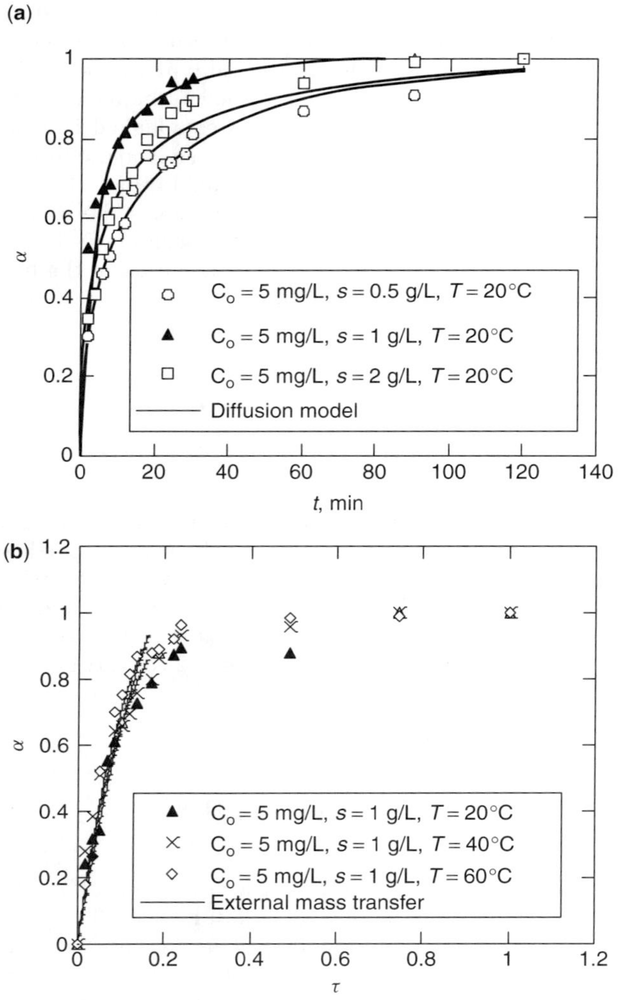

Figure 3. Experimental degree of conversion, α, vs. predictions based on the solution of (**a**) the diffusion equation, for various adsorbent loads and (**b**) the mass transfer equation, for various temperatures (both at an initial chromium concentration of 5 mg/L). Reprinted with permission from Reference 14; copyright (2004) American Chemical Society.

causes the diffusivity to increase rapidly as concentration increases (11).

A way to check further on the possibility of a pore diffusion-controlled mechanism is to perform desorption kinetic tests with adsorbent previously used for sorption. For a nonporous sorbent particle, transport of solute inside the particle may be neglected, and it can be assumed that biosorption occurs mainly at the particle surface. This idea can be effectively extended to cases of relatively large macropores, where the metal ions may have ready access to react with internal surface sites.

It was argued that, at the beginning of the process, chromium is sorbed according to quite a fast and highly favorable chemical mechanism, such as ion exchange, but soon external film diffusion comes into play (14). On this account, if one ignores the very first minute of sorption, the remaining curves were fitted pretty well by the model Eqs. 13 and 14. Figure 3b displays these results. Near the end of sorption (for α higher than 0.9), a much slower process, intraparticle diffusion, gradually becomes the rate-controlling step. Yet, this is a regime of no practical significance.

The computed mass transfer rate constant, K_mS, was presented in a table (14). As regards the effect of changing the solids load, it appears that K_mS is virtually not affected. To assess the statistical significance of the determination, a derivative time series analysis has been performed where $d\alpha/dt$ is plotted versus t and then, a Hanning low-pass filter was applied to flatten out the signal undulations until $d\alpha/dt$ versus α became a reasonably smooth curve (23). Next, the integrated smoothed signal $\alpha(t)$ was processed as before and resulted in K_mS values closely comparable to the values obtained from the unfiltered data, with no preferential trend in biomass load. This is an additional positive sign that external mass transfer may be the predominant mechanism of the sorption process after the initial fast chromium removal.

As a conclusion in the example thoroughly studied, we may say that the analysis conducted has not been capable of providing strong evidence in favor of any of the examined mechanisms because several diverse kinetic models were successful in fitting the experimental data. Perhaps, the questioning of surface kinetics by Levenspiel (24) is known, concluding that it is good enough to use the simplest available correlating rate expression.

NOMENCLATURE

- q: specific metal uptake (mg/g)
- C: metal bulk concentration (mg/L)
- X: sorbent loading per unit volume of solution (g/L)
- q_{max}: Langmuir constant (mg/g)
- b: Langmuir constant (L/mg)
- K_F: Freundlich constant {mg [g (mg/L)$^{-1/n}$]$^{-1}$}
- n: Freundlich constant (dimensionless)
- E: apparent activation energy (J/mol)
- A: preexponential factor (min^{-1})
- n: apparent reaction order
- R: gas constant (8.3136 J/mol/K)
- T: temperature (K)
- t: time (min)
- k: rate constant
- k_1: second-order chemical reaction rate constant [L*(mg^{-1} of metal)*min^{-1}]
- k_2: second-order Ritchie reaction rate constant (min^{-1})
- k_m: pseudo-second-order reaction rate constant [g of sorbent*(mg^{-1} of metal)*min^{-1}]
- D_c: intraparticle diffusion coefficient (m^2/s)
- R_c: mean particle radius (m)
- R_p: radius of a particle
- K_m: external mass transfer coefficient (m/s)
- S: specific surface area of sorbent particles per unit volume of the reactor (m^2/m^3)

Greek Letters

- α: degree of conversion
- Λ: fraction of metal ultimately adsorbed by the sorbent

ξ: numerically determined parameter
ε_p: void volume fraction

BIBLIOGRAPHY

1. Tien, C. (1994). *Adsorption Calculations and Modelling*. Butterworth-Heinemann, Boston.
2. Yiacoumi, S. and Tien, C. (1995). *Kinetics of Metal Ion Adsorption from Aqueous Solutions*. Kluwer Academic, Boston.
3. Xu, Y. and Schwartz, F.W. (1994). Sorption of Zn^{2+} and Cd^{2+} on hydroxyapatite surfaces. *Environ. Sci. Technol.* **28**: 1472–1480.
4. Hill, C.G. (1977). *An Introduction to Chemical Engineering Kinetics and Reactor Design*. John Wiley & Sons, New York, pp. 24–166.
5. Ho, Y.S., Ng, J.C.Y., and McKay, G. (2000). Kinetics of pollutant sorption by biosorbents: review. *Sep. Purific. Methods* **29**: 189–232.
6. Smith, J.M. (1981). *Chemical Engineering Kinetics*, 3rd Edn. McGraw-Hill, Singapore, pp. 37–92, 636–645.
7. Aksu, Z. (2001). Equilibrium and kinetic modeling of cadmium(II) biosorption by *C. vulgaris* in a batch system: Effect of temperature. *Sep. Purific. Tech.* **21**: 285–294.
8. Butter, T.J., Evison, L.M., Hancock, I.C., Holland, F.C., Matis, K.A., Philipson, A., Sheikh, A.I., and Zouboulis, A.I. (1998). The removal and recovery of cadmium from dilute aqueous solutions by biosorption and electrolysis at laboratory scale. *Water Res.* **32**: 400–406.
9. Volesky, B. (1999). Biosorption of Heavy Metals. CRC Press, Boca Raton, FL, Chap. 6. Available: http://ww2.mcgill.ca/biosorption/publication/index.htm.
10. Zouboulis, A.I., Matis, K.A., and Hancock, I.C. (1997). Biosorption of metals from dilute aqueous solutions. *Sep. Purif. Methods* **26**: 255–295.
11. Smith, E.H. (1996). Uptake of heavy metals in batch systems by a recycled iron-bearing material. *Water Res.* **30**: 2424–2434.
12. Mandjiny, S., Zouboulis, A.I., and Matis, K.A. (1995). Removal of cadmium from dilute solutions by hydroxyapatite. Part I. Sorption studies. *Sep. Sci. Tech.* **30**: 2963–2978.
13. Kefala, M.I., Zouboulis, A.I., and Matis, K.A. (1999). Biosorption of cadmium ions by *Actinomycetes* and separation by flotation. *Environ. Pollut.* **104**: 283–293.
14. Loukidou, M.X., Karapantsios, Th.D., Zouboulis, A.I., and Matis, K.A. (2004). Diffusion kinetic study of chromium(VI) biosorption by *Aeromonas caviae*. *Ind. Eng. Chem. Res.* **43**: 1748–1755.
15. Esposito, A., Pagnanelli, F., Lodi, A., Solisio, C., and Vegliò, F. (2001). Biosorption of heavy metals by *Sphaerotilus natans*: An equilibrium study at different pH and biomass concentration. *Hydrometallurgy* **60**: 129–141.
16. Loukidou, M.X., Karapantsios, Th.D., Zouboulis, A.I., and Matis, K.A. (2003). *Cadmium(II) biosorption by Aeromonas caviae: Kinetic modelling. 15th Intl. Biohydrometallurgy Symp.*, Athens.
17. Ritchie, A.G. (1977). Alternative to the Elovich equation for the kinetics of adsorption of gases on solids. *J. Chem. Soc.-Farad. Trans.* **73**: 1650–1653.
18. Cheung, C.W., Porter, J.F., and McKay, G. (2001). Sorption kinetic analysis for the removal of cadmium ions from effluents using bone char. *Water Res.* **35**: 605–612.
19. Crank, J. (1975). *Mathematics of Diffusion*. Oxford University Press, London.
20. Dzul Erosa, M.S., Saucedo, T.I., Navarro Mendoza, R., Avila Rodriquez, M., and Guibal, E. (2001). Cadmium sorption on chitosan sorbents: Kinetic and equilibrium studies. *Hydrometallurgy* **61**: 157–167.
21. Puranik, P.R., Modak, J.M., and Paknikar, K.M. (1999). A comparative study of the mass transfer kinetics of metal biosorption by microbial biomass. *Hydrometallurgy* **52**: 189–197.
22. Bates, D.M. and Watts, D.G. (1988). *Nonlinear Regression Analysis and its Applications*. John Wiley & Sons, New York.
23. Bendar, J.S. and Piersol, A.G. (1986). *Random Data: Analysis and Measurement Procedures*. John Wiley & Sons, New York.
24. Levenspiel, O. (1999). *Chemical Reaction Engineering*, 3rd Edn. John Wiley & Sons, New York, p. 380.

SOUND IN WATER

D.L. Marrin
Hanalei, Hawaii

As sound travels through an elastic medium such as water, a wave is created that imparts energy to the individual molecules—causing them to compress and then to relax along the path of the wave. Because the medium is elastic, the distance separating individual water molecules is permitted to vary just slightly, so that the propagation of sound may be defined as a periodic variation in pressure that is transmitted via adjacent molecules. Sound emanating from a point source in water usually (but not always) results in a three-dimensional waveform that spreads spherically through the medium (1). Similarly, a planar source of sound in water can be considered multiple point sources that create a wave consisting of all the individual waves.

Sound waves are *longitudinal*, meaning that molecules (e.g., water or air) comprising the elastic medium move in a direction parallel to the propagating wave. Because water has a significantly higher density than air, water transmits sound more rapidly and efficiently than air. The water molecules oscillate, or move back and forth over a very short longitudinal distance; hence, sound waves are classified as *mechanical*. By contrast, light waves and microwaves are classified as *electromagnetic* and do not involve the movement of water molecules for propagation. The speed of sound in water is about 1500 meters per second or 3300 miles per hour (i.e., approximately five times greater than that in air), which varies slightly with temperature, pressure, and salinity.

FREQUENCY AND AMPLITUDE

The frequency of a sound wave is the number of waves that pass by a given point during a specified period of time. A single wave consists of a cycle that includes both compression (squeezing together) and rarefaction (rebounding apart); the distance over which a cycle is continually repeated is known as the wavelength. The period of time over which waves are normally counted is one second; therefore, sound frequencies are expressed as cycles per second or *hertz*. Whereas high-frequency sounds possess short wavelengths, low-frequency sounds possess long wavelengths. Frequency ranges and estimated intensity levels for common underwater sounds are listed in Table 1. Frequency-dependent differences in underwater sound attenuation result in low-pitched sounds that travel farther than high-pitched sounds generated at similar intensities.

The amplitude of a sound wave is the maximum longitudinal displacement (i.e., during one cycle of compression and rarefaction) of water molecules relative to their resting, or equilibrium, position. The intensity of a sound wave is proportional to its amplitude and is often quantified in units of *decibels*. The decibel scale is based on a ratio between the intensities (expressed as a power) of a measured sound and a reference sound that corresponds to a standard pressure. Many of the man-made sounds listed in Table 1 are actually more intense than naturally produced sounds, contributing, at least in part, to the controversy regarding the possible effects of anthropogenic noise on marine organisms. Sound intensities are commonly presented for a distance of only 1 meter from the source and, as such, are rarely indicative of ambient underwater noise (e.g., the cumulative intensity of sounds measured at greater distances from their respective sources).

FATE OF OCEANIC SOUND

The propagation of sounds in the ocean is a rather complex matter owing to differences in water density as a function of depth, latitude, seafloor topography, and many other factors. Changes in seawater temperature, salinity, and pressure contribute to changes in density that, in turn, affect sound waves traveling through the medium. Generally, sound is attenuated by *spreading* (proportional to distance), *reflection* (due to solid structures or boundaries between water of different densities), *scattering* (due to rough surfaces), *absorption* (conversion of acoustic to thermal energy), and *refraction* (deflection of sound waves from a straight path) (5). Ambient noise in a shipping channel can exceed that outside a channel by as much as 45 decibels (re 1 μPa), but noise in a channel is attenuated about 100 decibels from that measured directly beneath ships (2).

Due to the vertical stratification of oceans, sound behaves differently in turbulent shallow waters from the way it does in either more isothermal middepth waters or waters immediately overlying the bottom (see Table 2). Sound transmission at the surface is highly dependent on local conditions (e.g., wind, precipitation, swells, bubbles) and is usually restricted to low-frequency sounds. However, most of the ocean lies between shallow and bottom waters, creating a relatively thick layer that transmits a spectrum of sound waves that is far less attenuated than that in either shallower or deeper waters. Both humans and whales have taken advantage of this *sofar channel* to broadcast signals and songs, respectively, over vast oceanic distances (6). Sounds generated within the sofar channel itself or at shallower depths alongside seamounts are entrained and transmitted within this deep isothermal layer as a result of many contributing physical processes.

UNDERWATER DETECTION

Detecting and identifying an underwater sound from a particular source is often difficult due to the aforementioned effects of the environment and to the myriad of background noises from which any single sound must be distinguished. Sea sounds are detected by hydrophones (i.e., the underwater equivalent of a microphone) that are often positioned in specific arrays, thus permitting the relative position of a sound source to be determined. This type of *passive* detection may be used to locate oceanic precipitation, breaking surface waves, underwater explosions, undersea volcanoes, shipping traffic, and marine organisms that emit audible sounds (e.g., crustaceans, fishes, mammals) (7).

Active detection techniques, such as sonar, are routinely used to locate underwater objects because the objects themselves need not emit sound and because the results

Table 1. Approximate Ranges for the Frequencies and Intensity Levels of Common Underwater Sounds. Reported Ranges were Complied from Data Presented in Various Sources[a]

Sound Source	Frequency, Hertz	Intensity Level, Decibels re 1 μPa at 1 m[b]
Ship engines and propellers	10–5000	160–190
Navigation and profiling sonars	100–3000	180–230
Explosive devices/air guns	1000–17,000	190–260
Military surveillance sonars	1000–10,000	190–235
Icebreaking/drilling operations	20–1000	100–150
Whale songs and moans	10–8000	120–190
Dolphin clicks and whistles	500–25,000	100–180
Cetacean echolocation	10,000–150,000	130–230
Snapping shrimp colony	2000–15,000	180–190
Lightning strikes/undersea volcanoes/earthquakes	0.1–20,000	Up to 260

[a]Reference 2–4.
[b]Intensity levels are presented in units of decibels relative to a reference pressure of 1 micropascal at a distance of 1 meter from the sound source. Sound level attenuation is a function of distance from the source, frequency range, and various environmental factors.

Table 2. Approximate Depths and Representative Sound Speeds for Various Water Layers in the Ocean. Reported Ranges were Complied from Data Presented in Various Sources[a]

Vertical Zone	Depth, Meters	Speed, m/s	Comments
Overlying air	NA[b]	330	Significantly lower acoustic impedance than water; minimal transfer of sound energy
Surface water layer	0 to 50	1515	Influenced by weather (e.g., wind, rain, temp.) and turbulence that affect sound waves
Seasonal thermal layer	50 to 800	1485–1505	Sound channel (mostly for low frequencies); limited vertical thickness; refracts sound waves originating shallower and deeper
Deep isothermal layer	800 to 4000+[c]	1485–1525	Optimal sound channel (for all frequencies); extensive vertical thickness and horizontal extent; minimal sound attenuation
Bottom water layer	4000+ to seafloor[c]	1525	Influenced by seafloor topographic features that reflect and scatter sound waves
Underlying rock	NA[b]	2000–6000	Higher acoustic impedance than water; moderate transfer of sound energy

[a]References 2, 6, and 7.
[b]Not applicable.
[c]Indicates that the transition between deep isothermal layers and bottom water layers may exceed 4000 meters, depending on the exact location within an ocean basin.

are generally more interpretable. The most basic type of sonar is echolocation, whereby a sound pulse (acoustic energy) is introduced into the water. Depending on the sonic wavelength and the size of the objects, a portion of that acoustic energy is reflected back to the source (e.g., a ship) where it is detected, transduced, and analyzed. The distance, movement, size, and even shape of some objects may be determined using sophisticated sonars (e.g., Doppler, multibeam) and data analysis techniques (8). Several marine organisms—most notably cetaceans—are believed to use echolocation as a means of finding or assessing prey, predators, conspecifics, seabed features, and environmental conditions (2).

ULTRASOUND AND INFRASOUND

Most of the underwater sounds discussed thus far lie within the *sonic* range of humans (i.e., frequencies of 20 to 20,000 hertz), but water also transmits mechanical waves of *ultrasonic* (>20,000 hertz) and *infrasonic* (<20 hertz) frequencies. Dolphins emit whistles at frequencies as high as 30,000 hertz and complex echolocation clicks at frequencies exceeding 300,000 hertz (8). Obviously, the upper limit of hearing of dolphins far exceeds that of humans. In addition to marine organisms, ambient oceanic noise associated with bubbles (i.e., trapped gases), surface winds, waves, and sea spray can reach ultrasonic frequencies. Man-made ultrasounds include everything from complex signals for communication to guided waves for inspecting underwater pipelines (9).

The moans of baleen whales are the most common source of biologically produced infrasound, which has been documented down to about 10 hertz (8). By contrast, a number of geologic and meteorologic events (e.g., volcanoes, earthquakes, sea ice cracking, hurricanes) produce infrasonic waves that propagate through the ocean at frequencies of approximately 0.1 to 10 hertz. Even the breaking of large surf, either onto the shore or on the water itself, creates infrasound at frequencies of 1 to 5 hertz (10). Anthropogenic infrasound in the ocean is created primarily via shipping traffic, low-frequency active sonar (LFAS), and acoustic thermography. A thermography technique known as ATOC generates tones down to about 1 hertz, whereas LFAS sweeps multiple tones of 100 to 500 hertz across each other and produces resulting tones as low as 0.1 hertz (2). Both ATOC and LFAS generate peak intensity sounds that are greater than those of ships.

BIBLIOGRAPHY

1. Leighton, T.G. (1994). *The Acoustic Bubble*. Academic, San Diego, Chap. 1.
2. Stocker, M. (2003). Ocean bio-acoustics and noise pollution. *J. Acoust. Ecol.* **4**: 16–29.
3. Science of Sound in the Sea. (2004). *Office of Marine Programs*. University of Rhode Island, HTML format.
4. Greene, C.R. and Moore, S.E. (1995). Man-made noise. In: *Marine Mammals and Noise*. W.J. Richardson et al. (Eds.). Academic, San Diego, Chap. 6.
5. Haines, G. (1974). *Sound Underwater*. Crane Russak, New York, pp. 22–25.
6. Payne, R. (1995). *Among Whales*. Scribner, New York, Appendix.
7. Medwin, H. and Clay, C.S. (1998). *Fundamentals of Acoustical Oceanography*. Academic, San Diego, Chap. 1 and 10.
8. Thompson, D.H. and Richardson, W.J. (1995). Marine mammal sounds. In: *Marine Mammals and Noise*. W.J. Richardson et al. (Eds.). Academic, San Diego, Chap. 7.
9. Na, W.B. and Kundu, T. (2002). Underwater pipeline inspection using guided waves. *J. Pressure Vessel Technol.* **124**: 196–200.
10. Garcés, M., Hetzer, C., Marrifield, M., Willis, M., and Aucan, J. (2003). Observations of surf infrasound in Hawai'i. *Geophys. Res. Lett.* **30**: 2264–2266.

WATER ON THE SPACE STATION

Patrick L. Barry
Tony Phillips
NASA

Rationing and recycling will be an essential part of life on the International Space Station. In this article, Science@NASA explores where the crew will get their water and how they will (re)use it.

November 2, 2000—Future astronauts poised to blast off for an extended stay on the International Space Station (ISS) might first consider dashing to the restroom for a quick splash at the lavatory, or better yet, a luxurious hot shower. Once on board the ISS, spacefarers are in for a steady diet of sponge baths using water distilled from—among other places—their crewmates breath!

If you're squeamish, read no farther, because the crew will eventually include lab rodents—and they'll be breathing, too. All of the denizens of the space station lose water when they exhale or sweat. Such vapors add to the ambient cabin humidity, which is eventually condensed and returned to the general water supply.

Sometimes it's better *not* to think about where your next glass of water is coming from!

Rationing and recycling will be an essential part of daily life on the ISS. In orbit, where Earth's natural life support system is missing, the Space Station itself has to provide abundant power, clean water, and breathable air at the right temperature and humidity—24 hours a day, 7 days a week, indefinitely. Nothing can go to waste.

In this article, the first of a series about the practical challenges of living in space, *Science@NASA* will examine how the Space Station's **E**nvironmental **C**ontrol and **L**ife **S**upport **S**ystem (ECLSS), under continuing development at the Marshall Space Flight Center, will help astronauts use *and re-use* their precious supplies of water. Future installments will explore air management, thermal control and fire suppression—in short, all of the things that will make the Space Station comfortable and safe.

MAKING A SPLASH IN SPACE

Before recycling can begin, there has to be some water to start with.

"We have plenty of water on the Space Station now," says Jim Reuter, leader of the ECLSS group at the Marshall Space Flight Center. "The Russian module Zarya is packed with contingency water containers (CWCs) that were carried over from the Space Shuttle during assembly missions earlier this year. They look like duffle bags and each one holds about 90 lbs."

"But it's expensive to ferry water from Earth," he added. "We have to recycle. There's already a Russian-built water processor in orbit that collects humidity from the air. Here at Marshall we're building a regenerative system that will be able to recycle almost every drop of water on the station and support a crew of seven with minimal resupplies."

The ECLSS Water Recycling System (WRS), developed at the MSFC, will reclaim waste waters from the Space Shuttle's fuel cells, from urine, from oral hygiene and hand washing, and by condensing humidity from the air. Without such careful recycling 40,000 pounds per year of water from Earth would be required to resupply a minimum of four crewmembers for the life of the station.

Not even research animals are excused from the program.

"Lab animals on the ISS breath and urinate, too, and we plan to reclaim their waste products along with the

Shuttle pilot Terry Wilcutt with 7 contingency water containers destined for the space station Mir

crew's. A full complement of 72 rats would equal about one human in terms of water reclamation," says Layne Carter, a water-processing specialist at the MSFC.

It might sound disgusting, but water leaving the space station's purification machines will be cleaner than what most of us drink on Earth.

"The water that we generate is much cleaner than anything you'll ever get out of any tap in the United States," says Carter. "We certainly do a much more aggressive treatment process (than municipal waste water treatment plants). We have practically ultra-pure water by the time our water's finished."

MIMICKING MOTHER EARTH

On Earth, water that passes through animals' bodies is made fresh again by natural processes. Microbes in the soil break down urea and convert it to a form that plants can absorb and use to build new plant tissue. The granular soil also acts as a physical filter. Bits of clay cling to nutrients in urine electrostatically, purifying the water and providing nutrients for plants.

Water excreted by animals also evaporates into the atmosphere and rains back down to the Earth as fresh water—a natural form of distillation.

Water purification machines on the ISS partly mimic these processes, but they do not rely on microbes or any other living things.

"While you try to mimic what's happening on Earth—which is so complicated if you really think about it—we have to use systems that we can control 100 percent," said Monsi Roman, chief microbiologist for the ECLSS project at MSFC. ECLSS depends on machines—not microbes—because, "if a machine breaks, you can fix it."

The water purification machines on the ISS will cleanse wastewater in a three-step process.

The first step is a filter that removes particles and debris. Then the water passes through the "multi-filtration beds," which contain substances that remove organic and inorganic impurities. And finally, the "catalytic oxidation reactor" removes volatile organic compounds and kills bacteria and viruses.

EVERY DROP COUNTS

Once the water is purified, astronauts will do everything possible to use it efficiently. "On the ground, people flick on the faucet and they probably waste a couple of liters of water just because it's free and the water pressure is high," notes Carter.

"On the ISS, the water pressure will be about half what you might experience in a typical household," Carter said. "We don't use faucets on the ISS, we use a wash cloth. It's much more efficient. If you're an astronaut, you'll wet the wash cloth with a spray nozzle and then use the cloth to wash your hands."

On the space station, people will wash their hands with less than one-tenth the water that people typically use on Earth. Instead of consuming 50 liters to take a shower, which is typical on Earth, denizens of the ISS will use less than 4 liters to bathe.

Even with intense conservation and recycling efforts, the Space Station will gradually lose water because of inefficiencies in the life support system.

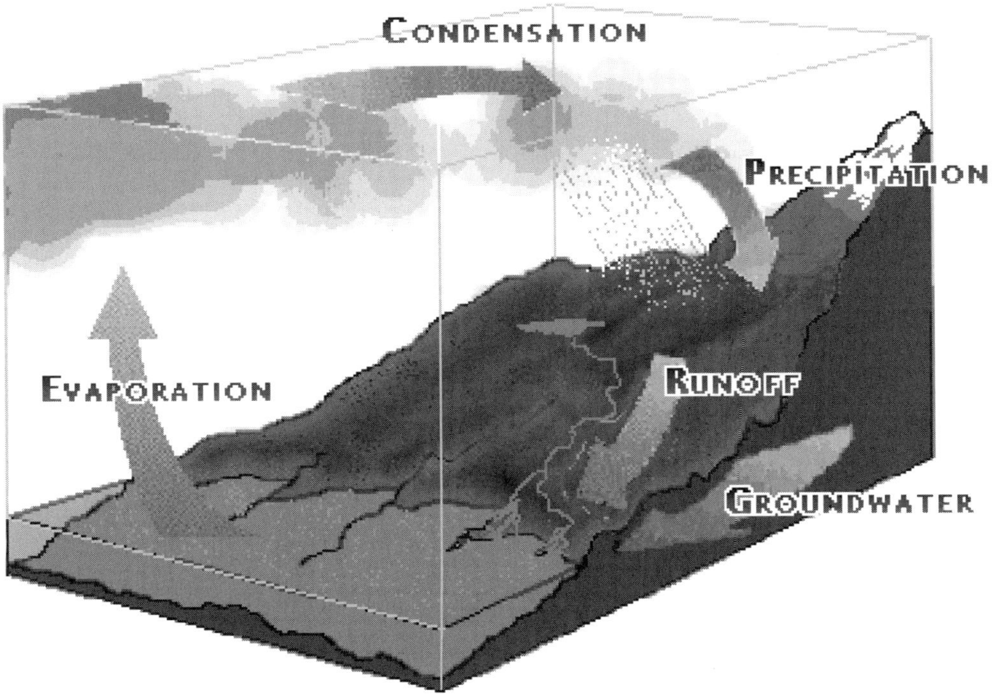

When water evaporates from the ocean and surface waters, it leaves behind impurities. In the absence of air pollution, nearly pure water falls back to the ground as precipitation

"We will always need resupply, because none of the water reprocessing technology that is available right now for space flight ... is 100 percent efficient. So there's always some minimal loss," said Marybeth Edeen, deputy assistant manager of environmental control and life support at NASA's Johnson Space Center.

Water is lost by the Space Station in several ways: the water recycling systems produce a small amount of unusable brine; the oxygen-generating system consumes water; air that's lost in the air locks takes humidity with it; and the CO2 removal systems leach some water out of the air, to name a few.

Lost water will be replaced by carrying it over from the Shuttle or from the Russian Progress rocket. The Shuttle produces water as its fuel cells combine hydrogen and oxygen to create electricity, and the Progress rocket can be outfitted to carry large containers of water.

NASA scientists will continue to look for ways to improve the life support systems of the Space Station, reducing water losses and finding ways to reuse other waste products. If the water recycling systems can be improved to an efficiency of greater than about 95 percent, then the water contained in the Station's food supply would be enough to replace the lost water, Edeen said.

One of the "nodes" that will become a part of the Space Station. The ECLSS life support equipment will be housed in Node 3, which is scheduled to be attached to the station in October 2005

Sign up for our EXPRESS SCIENCE NEWS delivery

"It takes processes that are slightly more efficient than we have developed for the space station to do that," Edeen said. "Those are the next generation water processing systems. Those are being developed now, but they're not ready for space flight yet."

The ECLSS life support system will join the Space Station as part of Node 3, which is scheduled to launch in October 2005. Until then, the environment inside the ISS will be maintained primarily by life support systems on the Russian Zvezda Service Module.

WEB LINKS

International Space Station—NASA's Web page for the International Space Station

Wheels in the Sky—Science@NASA article about humanity's dreams of a space station from the science fiction fantasies of the Nineteenth Century to Wernher von Braun's catalytic vision in the 1950s

Advanced Life Support Web Page—from the Johnson Space Flight Center

Environmental Control and Life Support Systems—describes the life support systems being developed at Marshall Space Flight Center

STRONTIUM ISOTOPES IN WATER AND ROCK

Gholam A. Kazemi
Shahrood University of Technology
Shahrood, Iran

INTRODUCTION

A brief summary of strontium and strontium isotopes, their concentrations in natural materials, and their evolution in the geologic system is given in the following paragraph. These introductory notes have been taken from a benchmark paper by Capo et al. (1). Anyone intending to deal with strontium isotopes should consult that paper as well as the paper by Clow et al., (2). Textbooks such as those by Faure (3) and Faure (4) provide excellent coverage of strontium isotopes in rocks. Being an encyclopedia of water, the emphasis here is more on strontium isotopes in water.

Strontium has four naturally occurring isotopes, whose approximate abundances are ^{84}Sr, 0.56%; ^{86}Sr, 9.87%; ^{87}Sr, 7.04%; and ^{88}Sr, 82.53%. All four isotopes are stable, although the fraction of ^{87}Sr varies due to radioactive decay of ^{87}Rb. The ionic radius of strontium (1.18 Å) is similar to that of calcium (1.00 Å), and, hence, Sr substitutes for Ca in minerals, including plagioclase feldspar, apatite, sulfates such as gypsum and anhydrite, and carbonates (calcite, dolomite, and especially aragonite). The average concentration of Sr in soils is 240 ppm but may fall below 10 ppm or exceed 1000 ppm in some cases. Weathering of bed rock or sediments can be a significant source of strontium in soil. Soil parent materials often have distinct Sr isotopic signatures. The average concentration of Sr in crustal rocks is 370 ppm but can vary from 1 ppm in some ultramafic rocks to several percent in some aragonitic corals. If Rb and Sr are incorporated into a mineral or rock at its formation and the system remains closed with respect to those elements, then the amount of ^{87}Sr increases over time as radioactive ^{87}Rb decays; the amounts of ^{84}Sr, ^{86}Sr, and ^{88}Sr remain constant. Therefore, older rocks in general have higher $^{87}Sr/^{86}Sr$ ratios than younger ones that have the same initial Rb/Sr ratio. Over geologic time, rocks of a given age, composed of minerals that have a high Rb/Sr ratio (e.g., granites in the continental crust), develop a higher $^{87}Sr/^{86}Sr$ than rocks that have a lower Rb/Sr ratio (e.g., oceanic basalts). Thus, $^{87}Sr/^{86}Sr$ ratios in geologic materials are indicators of both age and geochemical origin. The strontium in rocks is released by weathering, cycled through vegetation and animals, and eventually enters the oceans, primarily via rivers. Minerals within a crystalline rock or sediment generally have variable Rb/Sr ratios and therefore a range of $^{87}Sr/^{86}Sr$ ratios. River water has a Sr concentration that varies from ~6 to 800 ppb and averages ~60 ppb. The Sr concentration in seawater is about 8 ppm. Precipitation near ocean basins generally has a strontium isotopic composition similar to that of seawater (0.709). However, the concentration of Sr in rainfall is generally lower than that in seawater by several orders of magnitude (usually <1 ppb). Strontium leaves the oceans, the largest reservoir of dissolved Sr, primarily by deposition in marine carbonate. A small amount of Sr is also transferred directly from the oceans to the atmosphere and hence transferred to the continents in precipitation.

SR ISOTOPES IN WATER AND ROCK: SOURCES AND IMPORTANCE

The Sr isotopic ratio in natural water reflects the contribution made by various minerals as they dissolve in water. Rivers are the major provider of strontium to the oceans and define, to a great extent, the marine $^{87}Sr/^{86}Sr$ ratio. Strontium in rivers is mainly derived from two end members: limestone and evaporites as one end member (high Sr and low $^{87}Sr/^{86}Sr$ ratio of 0.706–0.709) and silicate rocks as the other end member that are low in Sr and have a higher $^{87}Sr/^{86}Sr$ ratio. Processes such as atmospheric deposition, cation exchange, and biological cycling, too, influence the Sr isotope ratio in water. However, unless there have been recent anthropogenic or natural disturbances, annual gains or losses of Sr from cation exchange and biological pool should be minimal.

Sr isotopes in water samples are used (1) as a chemical tracer to determine the sources of atmospheric contributions to soil cation reservoirs over time (1); (2) to determine the contribution of various components to the water balance of a lake (5); (3) to distinguish weathering reactions from cation exchange processes (6); (4) to investigate the origin of brine, brine migration, and the relationship between brine and diagenetic phases (7); (5) to characterize weathering reactions and identify sources of dissolved Ca in stream water (2); (6) as a tracer of cation sources in stream water (8); (7) to correlate and date marine sequences of late Neogene age (9); (8) to identify the origin of salts in groundwater (10); (9) to investigate the effect of carbonate dissolution on the chemical composition of groundwater (11); and (10) to indicate the influence of seawater on fresh groundwater (12).

Analysis of Sr Isotopes in a water sample

The strontium isotopes of a water sample are measured under clean laboratory conditions using a mass spectrometer, such as a Sector 54 type (13), a Finnigan Mat 261 multicollector model (2), or a VG354 thermal ionization brand (9). To analyze water for a Sr isotope ratio using a Sector 54 Type Mass Spectrometer, an initial aliquot of the sample is placed in a microwave oven to be evaporated to dryness. For water that has a total Sr content of more than 0.5 ppm, the aliquot is 2 mL, and evaporation time is about 1 hour; for water that has low concentrations such as rainwater, 500 mL is required which takes much longer to evaporate to dryness. Then 3 mL of 6 molar HCl is added to the residue to dissolve it and left to stabilize for a few hours. The following steps are then taken one after another (13):

1. Add 1 mL of 2.5 molar HCl, and evaporate to dryness by heating slowly.
2. Add 1 mL of 6 molar HCl, and leave to dissolve.
3. Transfer to centrifuge vessels, and centrifuge for around 5 minutes.
4. Pour 1 mL of the sample into the column (the column detail is AG W50-X 8).
5. Add 1 mL of 2.5 molar HCl to the column.
6. Add 1 mL of 2.5 molar HCl to the column.
7. add 1 +30 mL of 2.5 molar HCl to the column in two stages, respectively.
8. Add 10 mL of 2.5 molar HCl to the column.
9. Collect Sr by placing a container beneath the column and letting it drain.
10. Add 1 mL of HNO_3, and slowly heat to dryness.
11. Add 1 mL of HNO_3, and slowly heat to dryness.
12. Add 1 mL of Milli-Q water to the sample.
13. Place a small volume of the sample in the holding filament.
14. Heat sample slowly.
15. Add a small volume of phosphoric acid to the sample to facilitate ionization.
16. Heat the sample rapidly to a very high temperature (to burn) to remove any Rb.
17. Place the sample in the sample holder.
18. Transfer sample/s to the mass spectrometer.
19. Start running sample with mass spectrometer.

Analysis of Sr Isotopes in a rock sample

Rocks Other Than Limestone. To analyze a rock sample for Sr isotopes ratio using a Sector 54 Type Mass Spectrometer under clean laboratory condition, first it has to be ground to a 1 micron size. After grinding, for all rock types but limestone, the following steps should be undertaken one after another:

1. 0.1 gram of sample to accuracy of 0.0001 gram is selected.
2. 2 mL of HF acid and 10 drops of HNO_3 are added to each sample.
3. Samples are heated to dryness.
4. 2 mL of HF is again added to each sample.
5. 10 drops of $HClO_4$ are added to each sample, and sample is placed on a hot plate at 150 °C.
6. 5 drops of $HClO_4$ are added to each sample.
7. 5 more drops of $HClO_4$ are added to each sample.
8. Samples are placed on a hot plate at 180 °C.
9. Samples are then placed on a hot plate at 130 °C on the next day, and 1 mL 2.5 M HCl is added to each sample.
10. Samples are transferred to centrifuge tube to be centrifuged for 5 minutes.
11. From this step forward, all the steps are exactly the same as those for water samples from Step 5 onward.

Limestone. Due to its high solubility compared to other rock types, limestone is prepared for analysis using different techniques and the following steps:

1. 0.1 mg of sample is selected.
2. 1 mL of 1 M acetic acid is added to the sample.
3. 2 mL of 1 M acetic acid is added.
4. After reaction subsides, sample is centrifuged, and the residue is decanted to a clean beaker and heated to dryness at 150 °C.
5. 3 mL of 6.25 M HCl is added and evaporated to dryness, and 1 mL more is added again and heated to dryness.
6. 1 mL of 2.5 M HCl is added.
7. The sample is then loaded into columns (like other rock types).
8. The Sr is then collected and dried.
9. The columns are flushed and cleaned.
10. The dry sample (Sr) is again dissolved with 1 mL of 2.5 M HCl and loaded into the columns again. In contrast to the other samples, carbonates have a higher calcium content. Therefore, they need to be cleaned up twice before they are loaded into the mass spectrometer.
11. The rest of the procedure is like that for other rock types.

REPORTING THE RESULTS OF SR ISOTOPIC ANALYSIS

Before reporting the results of the strontium isotopic analysis, they should first be corrected for mass fractionation to $^{88}Sr/^{86}Sr = 8.37521$. Also, the value of isotopic ratio ($^{88}Sr/^{86}Sr$) of the NBS (National Bureau of Standards) 987 Sr metal used as a reference by the spectrometer should be included in the report. Measured strontium isotopic values are expressed as both ratios (ratio of ^{87}Sr to ^{86}Sr) and standard delta notation ($\delta^{87}Sr$) relative to NBS standard 987. To calculate $\delta^{87}Sr$, the following formula is used:

$$\delta^{87}\mathrm{Sr} = \left[\frac{(^{87}\mathrm{Sr}/^{86}\mathrm{Sr})\mathrm{sample}}{(^{87}\mathrm{Sr}/^{86}\mathrm{Sr})\mathrm{seawater}} - 1\right] \times 10^5$$

where $(^{87}Sr/^{86}Sr)_{sample}$ is the measured ratio of $^{87}Sr/^{86}Sr$ in a sample that has been normalized to $^{86}Sr/^{88}Sr = 0.1194$ and $(^{87}Sr/^{86}Sr)_{seawater}$ is the ratio of $^{87}Sr/^{86}Sr$ in modern seawater which is about 0.709172 (9). In most analysis, the precision of the $^{87}Sr/^{86}Sr$ ratios is ±0.0001 or better, and higher precision is of little value because seasonal variations in $^{87}Sr/^{86}Sr$ ratios in rivers are commonly in the third figure (14).

In a majority of cases, the Sr isotopic composition of a water or rock sample is a mixture of two or more end members. To calculate the relative proportion of each end member of a two component mixture, the following equation applies:

$$(^{87}\mathrm{Sr}/^{\mathrm{total}}\mathrm{Sr})_{\mathrm{m}} = x(^{87}\mathrm{Sr}/^{\mathrm{total}}\mathrm{Sr})_{\mathrm{a}} + (1 - x)(^{87}\mathrm{Sr}/^{\mathrm{total}}\mathrm{Sr})_{\mathrm{b}} \quad (1)$$

where the subscript "m" is mixture, "a" is the first end member, "b" is the second end member, x is the proportion of end member a, and $(1-x)$ is the proportion of end member b. To solve this equation, $^{87}Sr/^{total}Sr$ can be calculated from the equation given by (Reference 2) as follows:

$$^{87}\mathrm{Sr}/^{\mathrm{total}}\mathrm{Sr} = (^{87}\mathrm{Sr}/^{86}\mathrm{Sr})/(3 +^{87}\mathrm{Sr}/^{86}\mathrm{Sr}) \quad (2)$$

Example (from Reference 13). In Table 1, the isotopic composition of groundwater (GW) is considered a mixture of soil and rainwater isotopic compositions. The relative proportion of Sr from rainfall and soil to groundwater in Piezometer 40, P40, is computed here. In Equation 1, "m" = groundwater; "a" = rainfall; and "b" = soil; x = proportion of end member a, rainfall; and $1-x$ = proportion of end member b, soil. To calculate the proportion of each end member, first the $^{87}Sr/^{total}Sr$ ratio (Eq. 2) must be computed:

$$\begin{aligned}
^{87}\mathrm{Sr}/^{\mathrm{total}}\mathrm{Sr}\text{ for rainfall} &= (0.7089)/(3 + 0.7089)\\
&= 0.191135\\
^{87}\mathrm{Sr}/^{\mathrm{total}}\mathrm{Sr}\text{ for soil} &= (0.72068)/(3 + 0.72068)\\
&= 0.19368\\
^{87}\mathrm{Sr}/^{\mathrm{total}}\mathrm{Sr}\text{ for groundwater} &= (0.7105)/(3 + 0.7105)\\
&= 0.191091
\end{aligned}$$

Substituting these values in Eq. 1, the relative contribution of rainfall to the Sr budget of groundwater in P40 is calculated as 86%, and only 14% of the Sr is from soils.

Case Studies

Kazemi (13) and Kazemi and Milne-Home (15), through a strontium isotope study of rainwater, surface water, groundwater (shallow and deep), soil, and rocks of a salinized watershed in Eastern Australia (Buckinbah Creek watershed) found that the ratio of $^{87}Sr/^{86}Sr$ of shallow groundwater was similar to that of soil and

Table 1. The $^{87}Sr/^{86}Sr$ of Various Materials in the Buckinbah Creek Watershed in Eastern Australia[a]

GW (Shallow)		GW (Deep)		Surface Water		Rocks[b]		Soils	
B7	0.71197	B4	0.7084	S6	0.7099	Kabadah Fm[c]	0.70555	P45-Top	0.709
P19	0.7058	B3	0.708	S2	0.7087	Wansey Fm[d]	0.70813	P41	0.7373
P37	0.71	B2	0.7083	S5	0.7076	CV	0.72547	P42-Top	0.7135
P45	0.7078	B1	0.7084			Kabadah Fm[e]	0.7051	P40	0.72068
P40	0.7105					Burrawong Ls	0.70812	P40B	0.72298
B5	0.7099						0.71047		0.720692
P41	0.7169			Rain	0.7089				
P39	0.7092			Leaf	0.7115				
P42	0.7125								
P15	0.71023								

[a] *Source*: References 13.
[b] CV: Canowindra Volcanics, Burrawong Ls: Burrawong limestone, Fm: Formation.
[c] Rock type- shale.
[d] Rock type- volcanoclastic sandstone.
[e] Rock type- volcanoclastic.

rainwater (Table 1). They showed that between 20 to 90% of the Sr budget of shallow groundwater and surface waters is derived from rainwater. They further concluded that the deep groundwater is the source of baseflow at the watershed outlet (due to identical $^{87}Sr/^{86}Sr$ ratios of deep groundwater and baseflow).

Hunt et al. (6) used water and strontium isotopes to investigate the hydrology of a natural and a constructed wetland. They showed that the porewater from the natural wetland had a ^{87}Sr content ranging from 0.00 to 0.86 per mil ($^{87}Sr/^{86}Sr = 0.71026$ to 0.71087) and the ^{87}Sr of pore water from the constructed wetland ranged from -2.35 to $+1.24$ per mil ($^{87}Sr/^{86}Sr = 0.70859$ to 0.71114). They attributed this difference to the presence of a thick peat layer in the natural wetland.

The $^{87}Sr/^{86}Sr$ ratio of 32 brine samples from the Upper Jurassic Smackover Formation in southern Arkansas was measured at 0.7071–0.7101, more radiogenic than the Sr in Late Jurassic seawater because of the contribution of Sr from detrital sources (7).

The linear negative correlation between the Sr and $^{87}Sr/^{86}Sr$ ratio (when plotted against each other) has been used by Faure (3), Dogramaci et al. (16), and Singh et al. (17) as proof of the mixing of two waterbodies. Similarly, the negative correlation between the molar ratio of Mg/Ca and Sr isotopes is an indication of the incongruent dissolution of calcitic minerals in the aquifer (11). Incongruent dissolution of carbonates occurs progressively through two mechanisms:

1. Dissolution of calcitic minerals, causing the addition of Sr to groundwater and modification of the $^{87}Sr/^{86}Sr$ ratio; and
2. Reprecipitation of calcitic minerals, thereby removing some of the Sr from the groundwater without changing the Sr isotope ratio, approaching eventually the ratio in calcitic minerals (11).

A large data set of Sr isotopic ratios of the rivers of the Ganges, Orinoco, and Amazon basins is found in Palmer and Edmond (14).

The study by Lyons et al. (18) focuses on the strontium isotopic signature of some streams and lakes in Antarctica. They found that the ratios of $^{87}Sr/^{86}Sr$ of Lake Fryxell, Lake Hoare, and Lake Bonney were 0.70895, 0.71057, and 0.71187–0.71204, respectively. They also found that in Lake Fryxell and Lake Hoare, the $^{87}Sr/^{86}Sr$ became less radiogenic with depth, whereas in Lake Bonney, the ratios increase slightly with depth.

Sr isotopes together with isotopes of Nd and Pb of Cenozoic basalts in eastern Australia were studied by Zhang et al. (19) to show that the isotopic signatures of the Pacific Ocean midoceanic ridge basalts (MORB) type characterize the lava-field basalts in southeastern Australia, whereas the mantle isotopic signatures of Indian Ocean MORB type characterize younger basalts from northeastern Australia.

BIBLIOGRAPHY

1. Capo, R.C., Stewart, B.W., and Chadwick, O.A. (1998). Strontium isotopes as tracers of ecosystem processes: Theory and methods. *Geoderma* **82**: 197–225.
2. Clow, D.W., Alisa Mast, M., Bullen, T.D., and Turk, J.T. (1997). Strontium 87/strontium 86 as a tracer of mineral weathering reactions and calcium sources in an alpine/subalpine watershed, Loch Vale, Colorado. *Water Resour. Res.* **33**(6): 1335–1351.
3. Faure, G. (1986). *Principles of Isotope Geology*. John Wiley & Sons, New York.
4. Faure, G. (2001). *Origin of Igneous Rocks: The Isotopic Evidence*. Springer, Heidelberg.
5. Clauer, N., Chauduhuri, S., Toulkeridis, T., and Blanc, G. (2000). Fluctuations of caspian sea level: Beyond climatic variations. *Geology* **28**(11): 1015–1018.
6. Hunt, R.J., Bullen, T.D., Krabbenhoft, D.P., and Kendall, C. (1998). Using stable isotopes of water and strontium to investigate the hydrology of a natural and a constructed wetland. *Groundwater* **36**(3): 434–443.
7. Stueber, A.M., Pushkar, P., and Hetherington, E.A. (1984). A strontium isotopic study of Smackover brines and associated solids, southern Arkansas. *Geochim. Cosmochim. Acta* **48**: 1637–1649.
8. Miller, E.K., Blum, J.D., and Friedland, A.J. (1993). Determining of soil exchangeable-cation loss and weathering rates using Sr isotopes. *Nature* **362**: 438–441.
9. Hodell, D.A., Mead, G.A., and Mueller, P.A. (1990). Variation in the strontium isotopic composition of seawater (8 Ma to

present): Implications for chemical weathering rates and dissolved fluxes to the ocean. *Chem. Geol.* **80**: 291–307.

10. Morgan, K. and Jankowski, J. (2003). Origin of salts in a fractured bedrock aquifer in the Central West Region, NSW, Australia. *Proc. Int. Conf. Groundwater Fractured Rocks*, Prague, Krasny-Hrkal-Bruthans (Eds.). IHP-VI, series on groundwater no. 7, pp. 221–222.
11. Dogramaci, S.S., Herczeg, A.L., and Bone, Y. (1998). $^{87}Sr/^{86}Sr$ in Groundwater as Indicators of Carbonate Dissolution. *Proc. 9th Int. Symp.* on *water–rock Interaction*, Taupo, New Zealand, March–April, 1998, G.B. Arehart and J.R. Hulston (Eds.). A.A. Balkema, Rotterdam, pp. 211–214.
12. Jorgensen, N.O. and Heinemeier, J. (2003). Isotope studies of saline groundwater in a sandstone aquifer on Bornholm, Denmark. *Proc. Int. Conf. Groundwater Fractured Rocks*, Prague, Krasny-Hrkal-Bruthans (Eds.). IHP-VI, series on groundwater no. 7, pp. 205–206.
13. Kazemi, G.A. (1999). *Groundwater factors in the management of dryland and stream salinity in the upper Macquarie Valley, New South Wales, Australia*. Unpublished Ph.D. Thesis, University of Technology, Sydney.
14. Palmer, M.R. and Edmond, J.M. (1992). Controls over the strontium isotope composition of river water. *Geochim. Cosmochim. Acta* **56**: 2099–2111.
15. Kazemi, G.A. and Milne-Home, W.A. (1999). The application of strontium, oxygen 18 and deuterium isotopes in the estimation of the solute budget and hydrology of a small watershed in Eastern Australia. *Eos Trans. AGU* **80**(17) *Spring Meet.* Suppl. S139, 1999.
16. Dogramaci, S.S., Herczeg, A.L., and Bone, Y. (1998). *Chemical evolution of groundwater in the Murray Basin: Geochemistry and mass transfer modelling*. Abstract No. 49, Geol. Soc. Aust., 14th Aust. Geol. Convention.
17. Singh, K.S. et al. (1998). Chemical and strontium, oxygen, and carbon isotopic compositions of carbonates from the Lesser Himalaya: Implications to the strontium isotope composition of the source waters of the Ganga, Ghaghara, and the Indus rivers. *Geochim. Cosmochim. Acta* **62**: 743–755.
18. Lyons, W.B. et al. (2002). Strontium isotopic signatures of the streams and lakes of Taylor Valley, Southern Victoria Land, Antarctica: Chemical weathering in a polar climate. *Aquatic Geochem.* **8**(2): 75–95.
19. Zhang, M., O'Reilly, S.Y., and Chen, D. (1999). Location of Pacific and Indian mid-ocean ridge-type mantle in two time slices: Evidence from Pb, Sr, and Nd isotopes for Cenozoic Australian basalts. *Geology* **27**(1): 39–42.

TECHNETIUM IN WATER

Jay Clausen
AMEC Earth and Environmental, Inc.
Westford, Massachusetts

INTRODUCTION

Technetium, element 43, was predicted on the basis of its location in the periodic table. Technetium was erroneously reported as having been discovered in 1925 by Noddack and Tacke (1), who called it masurium. However, Element 43 was actually discovered in Italy (2). It was found in a sample of molybdenum, which had been bombarded with deuterons in the University of California—Berkeley cyclotron by E. Lawrence and then sent to Perrier and Segre. Technetium was the first element to be produced artificially.

Twenty-two isotopes of technetium whose masses range from 90 to 111 are known, and all are radioactive (Table 1). Technetium has three long-lived radioactive isotopes: technetium-97 [^{97}Tc] (half-life $\{T_{1/2}\} = 2.6 \times 10^6$ years), ^{98}Tc ($T_{1/2} = 4.2 \times 10^6$ years), and ^{99}Tc ($T_{1/2} = 2.13 \times 10^5$ years). An important isotope is ^{95m}Tc ("m" stands for metastable state) [$T_{1/2} = 61$ days], which is used in tracer work. However, the most useful isotope of technetium is ^{99m}Tc ($T_{1/2} = 6.01$ hours), which is used in many medical radioisotope tests because of its short half-life, easy detection of the gamma ray it emits, and the ability of technetium to bind chemically to many biologically active molecules.

Technetium-99 is a weak, beta-emitting radionuclide, 0.292 million-electron volts (MeV), produced during nuclear fission, which decays to the stable isotope ruthenium-99. It has a specific activity of 3.78×10^{10} transmutations per minute per gram, which corresponds to 17 microcuries per milligram (μCi/mg) (4). Technetium metal is a silver-gray color, tarnishes slowly in the air, and has properties similar to those of rhenium. However, as mentioned earlier, technetium does not occur naturally.

However, ^{99}Tc in terrestrial material has been evaluated since its discovery. In 1961, ^{99}Tc was isolated and identified in African pitchblende (a uranium-rich ore) in extremely minute quantities (5). It was theorized that ^{99}Tc was produced as a result of the spontaneous fission of uranium-238 (^{238}U).

Technetium-99 is produced primarily through anthropogenic processes such as detonation of nuclear weapons, reprocessing of spent nuclear reactor fuel rods containing uranium, and disposal of nuclear waste. Reprocessing involves dissolving the spent nuclear fuel in acid to separate the potentially reusable components of the spent fuel

Table 1. Isotopes, Half-Lives, and Decay Processes for Technetium[a]

Nuclide	Half-Life[b]	Type of Decay	Nuclide	Half-Life	Type of Decay
^{91}Tc	3.2 m	β^+, γ	^{99}Tc	2.13×10^5 y	B^-
^{92}Tc	4.4 m	β^+, γ	^{100}Tc	15.8 s	B^-, γ
^{93m}Tc	43.5 m	ε, γ	^{101}Tc	14 m	B^-, γ
^{93}Tc	2.7 h	$\varepsilon, \beta^+, \gamma$	^{102m}Tc	4.3 m	B^-, γ
^{94}Tc	53 m	β^+, γ	^{102}Tc	6.3 m	B^-, γ
^{94}Tc	4.9 h	$\varepsilon, \beta^+, \gamma$	^{103}Tc	50 s	B^-, γ
^{95m}Tc	60 d	$\varepsilon, \beta^+, \gamma$	^{104}Tc	18.0 m	B^-, γ
^{95}Tc	20 h	ε, γ	^{105}Tc	7.6 m	B^-, γ
^{96m}Tc	52 m	ε, γ	^{106}Tc	36 s	B^-, γ
^{96}Tc	4.3 d	ε, γ	^{107}Tc	21 s	B^-, γ
^{97m}Tc	91 d	γ	^{108}Tc	5 s	B^-, γ
^{97}Tc	2.6×10^6 y	ε	^{109}Tc	1 s	B^-
^{98}Tc	4.2×10^6 y	B^-, γ	^{110}Tc	0.83 s	B^-, γ
^{99m}Tc	6 h	γ, B^-	^{111}Tc	?	?

[a]Reference 3.
[b]d: days; h: hours; m: minutes; s: seconds; y: years.
[c]β^+: positron; β^-: beta particle; ε: internal electron conversion; γ: gamma ray.

(the uranium and plutonium) from the waste component (the fission products). The fission products, which are of no use, must be disposed of properly. In a 1000-megawatt nuclear reactor (thermal fission yield of approximately 6%), approximately 28 g of ^{99}Tc is produced daily from the slow neutron-induced fission of ^{235}U or plutonium-239 [^{239} Pu] (6). The long half-life and mobility of ^{99}Tc makes it an important element to the nuclear fuel industry in terms of its potential environmental impact. This is a particular concern for designing underground nuclear waste storage facilities because ^{99}Tc is the most mobile of all radionuclides present.

There are no known uses of technetium other than ^{99m}Tc for medical purposes. Technetium has superconducting properties and is a good corrosion preventer (7). However, its unstable nature and production of radioactive particles negate its practical use.

ANALYSIS

Analysis for ^{99}Tc typically consists of methyl ethyl ketone (MEK) extraction followed by liquid scintillation counting using United States Environmental Protection Agency (USEPA) Method 900.0. The typical quantification limit is 1 becquerel per liter (Bq/L), which is equivalent to 25 picocuries per liter (pCi/L).

Chemistry

The oxidation states most commonly found for ^{99}Tc are Tc^0, Tc^{+4}, and Tc^{+7}. The most stable chemical species of ^{99}Tc in aqueous solutions is the pertechnetate anion, TcO_4^-, in the Tc^{+7} valence state, which is highly mobile and nonreactive (8–12). However, ^{99}Tc is very sensitive to changes in redox conditions (13). Under reducing conditions, ^{99}Tc can be precipitated by various cations to form anhydrous technetium dioxide, TcO_2, (as Tc^{+4}) or technetium heptoxide, Tc_2O_7, or other insoluble oxides and hydroxides such as Tc_2S_7 or TcS_2 or incorporated into minerals (14–17):

$$TcO_4^- + 4H^+ + 3e^- = TcO_2 + 2H_2O;\ E_0 = +0.738 \text{ volts} \tag{1}$$

In the presence of oxygen, the Tc^{+4} oxide or oxyhydroxide can be reoxidized (14,18):

$$[TcO_{2\,(s)} \text{ or } TcO(OH_{2\,(s)}] \rightarrow TcO_4^- \tag{2}$$

The pertechnetate anion also forms complexes with organic compounds containing –C –O and –COO functional groups, especially under reducing conditions. The thermodynamic data for technetium (Table 2) can be used to develop a phase diagram to identify which species are likely to be present at a given set of E_h and pH conditions (Fig. 1).

The pertechnetate ion is mobile under oxidizing conditions (Fig. 1). Bidoglio et al. (20,21) found in column experiments that the oxidized species persists, even under slightly reducing or anoxic conditions. Technetium can undergo a number of surface reactions on various minerals under reduced conditions.

Table 2. Thermodynamic Data for Technetium[a]

Species[b]	$\Delta G^0{}_f$ kcal/gfw[c]
$TcOH_{(s)}$	−56.1
$Tc(OH)_{2\,(s)}$	−110.2
$Tc_3O_{4\,(s)}$	−206.3
$TcO_{4\,(s)}^-$	−149.1
$TcS_{2\,(s)}$	−51.0
$TcO_{2\,(s)}$	−88.3

[a]Reference 19.
[b]s: solid.
[c]kcal/gfw: kilocalories per gram formula weight

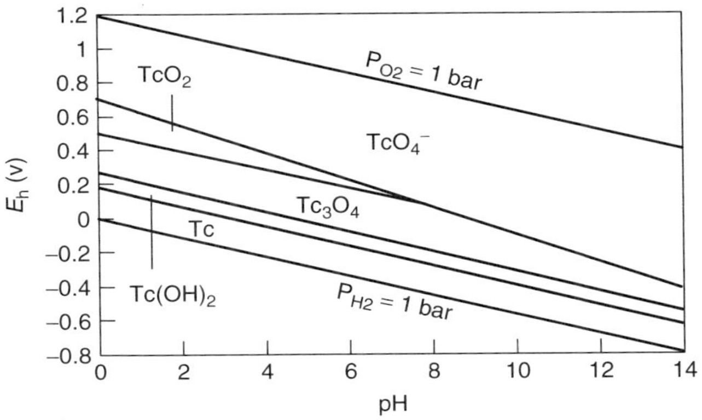

Figure 1. E_h-pH diagram for the Tc–O–H system at a concentration of 90 μg/L under standard temperature and pressure.

The solubility of sodium pertechnetate is reportedly 11 moles per liter [mol/L] (13). The solubility of ^{99}Tc in a reducing environment varies from 1.32×10^{-8} to 2.56×10^{-9} mol/L and is controlled by the Tc species present (14). Redox thus is a key variable affecting solubility (8). The solubility of technetium under reducing conditions in carbonate groundwater is pH dependent. At pH levels between 5 and 7.5, the solubility of Tc^{+4} is less than 10^{-8} mol/L. Pilkington (22) suggests that the solubility of TcO_2 is closer to 10^{-7} mol/L under reducing conditions. At higher pHs of 8 to 11, solubility increases, and the dominant ionic complex is $TcO(OH)_3CO^-{}_3$.

Equally important are the fate-and-transport properties of ^{99}Tc. Under aerobic soil conditions, ^{99}Tc behaves as an inert tracer and readily passes through soil unless there is a high content of organic matter or metal oxide (10,23–30). As Gu and Dowlen (31) report, this is not surprising because soil mineral surfaces are negatively charged as is TcO_4^- and therefore the two repel each other. Once introduced to surface water or groundwater, it can be transported significant distances due to its recalcitrant nature and lack of sorption. Given the concern over ^{99}Tc mobility, extensive studies have been conducted in various geologic media to assess suitable nuclear waste disposal storage facilities. Adsorption coefficients for various minerals are presented in Sheppard et al. (23), Wildung et al. (27), Palmer and Meyer (32), and

Strickert et al. (33). Sheppard et al. (23) indicate that soil redox conditions and organic matter content are the key variables influencing ^{99}Tc mobility in soil.

REGULATORY STANDARDS

The USEPA has set an effective dose equivalent of 4 millirem (mrem)/year for exposure, which they incorrectly calculated as 900 pCi/L (36 Bq/L). The correct activity equivalent to 4 mrem/year is 3790 pCi/L [152 Bq/L] (34). The U. S. Department of Energy (DOE) Order 5400.5 sets a permissible ingested-water limit of 100 mrem/year or 100,000 pCi/L (4000 Bq/L), whereas the U. S. Nuclear Regulatory Agency (NRC) has set a value of 50 mrem/year or 60,000 pCi/L (2400 Bq/L). The NRC has set the maximum activity of ^{99}Tc that can be released to a sanitary sewer at 500 mrem/year or 600,000 pCi/L (24,000 Bq/L). The proposed action level for groundwater cleanup at the Paducah Gaseous Diffusion Plant (PGDP) site is 25 pCi/L (1 Bq/L), which is the quantification limit.

Rainwater

Technetium has been released to the atmosphere in the form of Tc_2O_7 and $HTcO_4$ as a consequence of aboveground detonation of nuclear weapons through the fission of ^{235}U and ^{239}Pu. The nuclear industry also releases TcF_6 in addition to Tc_2O_7 and $HTcO_4$ during uranium enrichment. During a nuclear detonation, elemental ^{99}Tc burns to form Tc_2O_7, which upon cooling and contact with water vapor forms $HTcO_4$. The activity of ^{99}Tc measured in rainwater reportedly ranges from 8.2×10^{-8} to 1.0×10^{-6} μg/L (35,36). Work by Leon (37) and Attrep et al. (36) suggest the concentration of ^{99}Tc in rainwater has increased from the early 1960s through the 1980s. Both papers suggest that the source of this increase in ^{99}Tc in rainwater is from the reprocessing of spent nuclear fuel rather than nuclear weapons testing.

Surface Water

Technetium-99 is a surface water and groundwater contaminant at a number of DOE sites such as the PGDP in Paducah, Kentucky; the Portsmouth Gaseous Diffusion Plant in Piketon, Ohio; Oak Ridge National Laboratory Y-12 Plant, Savannah River Laboratory in Aiken, South Carolina; the Pacific Northwest National Laboratory 200 west area; and the Nevada Test Site (30,38–40).

Currently, the largest annual release of ^{99}Tc to the environment occurs at the two nuclear fuel-reprocessing facilities in Sellafield, United Kingdom. Technetium-99 laden effluent water from these facilities is discharged directly into the Irish Sea.

Contamination has been detected as far west as Svalbard and along the Norwegian coastline, distances of several hundred miles. The total discharge from these two plants amounts to between 70 and 90 tera (TBq)/year of ^{99}Tc (41).

Techentium-99 is reported in the Savannah River as a result of discharges from the U. S. Department of Energy Savannah River Plant located near Aiken, South Carolina (42,43). Technetium-99 is present at Oak Ridge National Laboratory where radioactive liquids generated from routine operations were disposed of in shallow waste pits (44). Nearby seeps have ^{99}Tc activity levels of up to 1185 pCi/L [3,200 Bq/L].

Blaylock and Frank (45) conducted a study in which TcO_4^- was added to a freshwater pond. Their study indicated that 84% of the TcO_4^- added remained in solution; very little was taken up by biota or by settling out into sediments. The vast majority of the ^{99}Tc was removed through drainage, indicating that the ultimate reservoir for technetium released into the environment is the ocean. Similarly, in a study of the Rhone River, it was found that the ^{99}Tc activity varied from 2.5 to 53×10^{-3} pCi/L (0.1 to 2.1×10^{-3} Bq/L) in seven samples collected in 1985 and 1986 (46).

Finally, significant quantities of ^{99}Tc laden effluent water were released to a settling lagoon as well as to several small streams surrounding the PGDP. Activity levels as high as 100,000 pCi/L (4000 Bq/L) were routinely released to surface water. Ferrous iron sulfate was used to precipitate metals in the lagoon before discharge into a nearby stream. The addition of this material shifted the redox state of the lagoon such that TcO_4^- was precipitated from solution. The likely form of the technetium precipitate was Tc_2S_7. In addition, the iron-rich clay and organic-rich stream sediments retained much of the ^{99}Tc, limiting its downgradient migration from the lagoon. Therefore, the ^{99}Tc migrated only several thousand feet from the point of release. However, downgradient of the site, groundwater discharges to the surface water streams and ^{99}Tc-rich water has been measured in several seeps (47).

Ground Water

The most extensive groundwater distribution of ^{99}Tc is at the PGDP where a 2-$\frac{1}{2}$ mile long plume of ^{99}Tc contaminated groundwater is commingled with the solvent trichloroethene. The ^{99}Tc was introduced to the site during the 1970s when fissioned uranium was reprocessed. ^{99}Tc became entrained in processing equipment at the PGDP, which was later cleaned with organic solvents and released to the environment. Gu et al. (48) determined conclusively that the species of ^{99}Tc mobile in groundwater at the PGDP was the pertechnetate anion, TcO_4^-. This later was supported through the development of phase diagrams using site-specific E_h and pH groundwater data (38,39). In addition, Gu and Dowlen (31) and Gu et al. (48) conclusively ruled out colloidal and natural organic matter facilitated transport as viable mechanisms of ^{99}Tc mobilization. Previous studies of PGDP, other sites, and laboratory studies assumed that TcO_4^- was the form of ^{99}Tc in most environmental settings. The estimated partitioning coefficient of ^{99}Tc exceeded 27,000 milliliter per gram [mL/g] (48). Technetium-99 activities as high as 179 Bq/L (4800 pCi/L) were observed in the groundwater plume (38,39). The estimated mass of ^{99}Tc dissolved in the groundwater plume was 4 to 6 lb, which equates to an activity of 1.0 to 1.6×10^{12} Bq/L (28 to 42×10^{12} pCi) of ^{99}Tc.

At the Hanford Site in Washington, ^{99}Tc has been found in groundwater in the Hanford 200 West Area where spent nuclear fuel was reprocessed and then the waste material was stored in underground tanks (30). Technetium-99 levels between 900 and 27,000 pCi/L (36 and 1080 Bq/L) have been found in groundwater.

Technetium-99 activity in shallow groundwater at the Portsmouth Gaseous Diffusion Plant has been observed at the 1000 pCi/L (40 Bq/L) level (49). The causes of contamination are similar to those discussed for the Paducah site. However, because the Portsmouth site received feed material from the Paducah site, the quantity of ^{99}Tc introduced into the site was much lower.

A unique study was conducted by the Lawrence Livermore National Laboratory at the Nevada Test site to assess the distribution of radionuclides in caverns produced by underground nuclear testing. The work by Buddemeier et al. (50) and Schroeder et al. (40) found that ^{99}Tc was present in groundwater associated with two separate nuclear weapons tests. The ^{99}Tc appeared to be associated and transported with colloidal material, but its mobility was reduced upon contact with zones of high levels of iron oxyhydroxides.

There are several known nongovernmental sites containing ^{99}Tc. One such facility is the Fields Brook National Priority Listed (NPL) site in Ashtabula, Ohio. The site was formerly a storage site for liquid waste, which contained ^{99}Tc as well as other radionuclides and solvents. Technetium-99 activities as high as 100,000 pCi/L (4000 Bq/L) have been detected in groundwater. A second site is the uranium fuel plant in Hematite, Montana. Recycled uranium, containing ^{99}Tc, was used to fabricate metallic fuel rods. Low-levels of ^{99}Tc have been observed in groundwater.

REMEDIATION

The technologies explored for remediating ^{99}Tc in water include ion exchange, liquid–liquid extraction, precipitation with various forms of iron, redox manipulation, the FORAGER sponge, granular activated carbon (GAC), and natural attenuation (10,16,18,31–33,48,51–65). The report by Bostick et al. (53) provides an excellent reference list of the many studies conducted on ^{99}Tc removal from aqueous waste streams. At the Portsmouth and Paducah Gaseous Diffusion plants, ion exchange is being used for treating ^{99}Tc-contaminated groundwater. Although, Gu et al. (31) found that GAC was superior to ion exchange due to its low cost, efficiency, and sorption capacity, GAC has not been employed at these sites.

CONCLUSION

Technetium is an element not typically found in nature and has environmental consequence at facilities that enrich or reprocess uranium fissioned material. These facilities are typically government operated. The high solubility of TcO_4^- and its stability across a wide spectrum of redox conditions enables the ion to be extremely mobile once introduced into water. As a consequence, at the few sites where ^{99}Tc has been introduced into the natural environment, it has produced very large plumes of contaminated surface and groundwater.

BIBLIOGRAPHY

1. Noddack, W. and Tacke, I. (1925). Die Ekamangane. *Naturwissenschaften* **13**: 567.
2. Perrier, C. and Segre, E. (1937). *Nature* (London). **140**: 193.
3. Seelmann-Eggebert, W., Pfennig, G., and Munzel, H. (1974). *Chart of the Nuclides*. Gersbach Verlag, Munich, Germany.
4. Schwochau, K. (1981). The analytical chemistry of technetium. *Top. Curr. Chem.* **96**: 109–147.
5. Kenna, B.T. and Kuroda, P.K. (1961). Technetium 99. *J. Inorg. Nucl. Chem. Lett.* **23**: 142.
6. De Alleluia, T.B. (1983). Tc: Technetium, System No. 69. In: *Gmelin Handbook of Chemistry*. Supplement Vol. 1, 8th Edn. Springer-Verlag, Berlin.
7. Cartledge, G.H. (1973). Twenty-year inhibition of corrosion by the pertechnetate ion. *Corrosion* **29**: 361–362.
8. Eriksen, T.E., Ndalamba, P., Bruno, J., and Caceci, M. (1992). The solubility of $TcO_2{\cdot}nH_2O$ in neutral to alkaline solutions under constant p_{CO_2}. *Radiochimica Acta* **58/59**: 67–70.
9. Bruhl, H., Bock, W., Trapp, C., and Winkler, A. (1989). *Ausbreitungsverhalten von Technetium (Selen) in geochemisch unterschiedlichen Grundwasserleitern*. Institut fur Angeandte Geologie Freie Unversitat Berlin. Final report 02 U 5695 A9.
10. Vandergraaf, T.T., Ticknor, K.V., and George, I.M. (1984). *Reactions Between Technetium in Solution and Iron-Containing Minerals Under Oxic and Anoxic Conditions*. Atomic Energy of Canada. Limited Report AECL-1957. Pinawa, Manitoba, Canada, pp. 25–43.
11. Paquette, J., Reid, J.A.K., and Rosinger, E.L.J. (1980). *Review of Technetium Behavior in Relation to Nuclear Waste Disposal*. Whiteshell Nuclear Research Establishment. Technical Record TR-25. Pinawa, Manitoba, Canada.
12. Allard, B., Kigatsi, H., and Torstenfelt, B. (1979). Technetium: Reduction and sorption in granitic bedrock. *Radiochem. Radioanalytical Lett.* **37**: 223–230.
13. Gu, B. and Schulz, R.K. (1991). *Anion Retention in Soil: Possible Application to Reduce Migration of Buried Technetium and Iodine*. U.S. Nuclear Regulatory Commission, Washington, DC.
14. Meyer, R.E., Arnold, W.D., and Case, F.I. (1987). *The Solubility of Electrodeposited Tc(IV) Oxides*. Oak Ridge National Laboratory, Oak Ridge, TN.
15. Lieser, K.H. and Bauscher, C.H. (1987). Technetium in the hydrosphere and in the geosphere 1. Chemistry of technetium and iron in natural waters and influence of the redox potential on the sorption of technetium. *Radiochimica Acta* **42**: 205–213.
16. Lee, S.K. and Bondietti, E.A. (1983). Technetium behavior in sulfide and ferrous iron solutions. *Mater. Res. Soc. Symp. Proc.* **15**: 315–321.
17. Wildung, R.E., Routson, R.C., Serne, R.J., and Garland, T.R. (1974). *Pertechnetate, Iodide, and Methyl Iodide Retention by Surface Soils. In Pacific Northwest Laboratory Annual Report for 1974 to the USAEC Division of Biomedical and Environmental Research*. B.E. Vaughan (Ed.). Pacific Northwest National Laboratory, Richland, WA, pp. 37–40.
18. Meyer, R.E., Arnold, W.D., and Case, F.I. (1984). *Valence Effects on the Sorption of Nuclide on Rocks and Minerals*. Oak Ridge National Laboratory, Oak Ridge, TN.
19. Rard, J.A. (1983). *Critical Review of the Chemistry and Thermodynamics of Technetium and Some of its Inorganic*

Compounds and Aqueous Species. Lawrence Livermore National Laboratory, Livermore, CA.

20. Bidoglio, G., Avogadro, A., and De Plano, A. (1985). Influence of redox environments on the geochemical behavior of radionuclides. *Mater. Res. Symp. Proc.* **50**: 709–716.
21. Bidoglio, G., Chatt, A., De Plano, A., and Zorn, F. (1983). Investigation on technetium species in ground water and migration through clay columns. *J. Radioanalytical Chem.* **79**: 153–164.
22. Pilkington, N.J. (1990). The solubility of technetium in the near-field environment of a radioactive waste repository. *J. Less-Common Met.* **161**: 203–212.
23. Sheppard, S.C., Sheppard, M.I., and Evenden, W.G. (1990). A novel method used to examine variation in Tc sorption among 34 soils, aerated and anoxic. *J. Environ. Radioactivity* **11**: 215–233.
24. Henrot, J., Garten, C.T., Jr., and Blaylock, B.G. (1989). *Behavior of Technetium in Soil: Sorption-Desorption Processes*. Oak Ridge National Laboratory, Oak Ridge, TN.
25. Schulte, E.H. and Scoppa, P. (1987). Sources and behavior of technetium in the environment. *Sci. Total Environ.* **64**: 163–179.
26. Wolfrum, C. and Bunzl, K. (1986). Sorption and desorption of technetium by humic substances under oxic and anoxic conditions. *J. Radioanal. Nucl. Chem.* **99**: 315–323.
27. Wildung, R.E., Garland, T.R., McFadden, K.M., and Cowan, C.E. (1986). Technetium sorption in surface soils. In: *Technetium in the Environment*. G. Desmet and C. Myttenaere (Eds.). Elsevier Applied Science, London, pp. 115–129.
28. Wildung, R.E., Garland, T.R., and Cataldo, D.A. (1977). Accumulation of technetium by plants. *Health Phys.* **32**: 314–317.
29. Bondietti, E.A. and Francis, C.W. (1979). Geologic migration potentials of technetium-99 and neptunium-237. *Science* **203**: 1337–1340.
30. Brown, D.J. (1967). Migration Characteristics of Radionuclides Through Sediments Underlying the Hanford Reservation. In: *Disposal of Radioactive Wastes in to the Ground*. International Atomic Energy Agency Symposium. Vienna, Austria.
31. Gu, B., Dowlen, K.E., and Clausen, J.L. (1996). Efficient separation and recovery of technetium-99 from contaminated groundwater. *Efficient Sep.* **6**: 123–132.
32. Palmer, D.A. and Meyer, R.E. (1981). Adsorption of technetium on selected inorganic ion-exchange materials and on a range of naturally occurring minerals under oxic conditions. *J. Inorg. Nucl. Chem.* **43**: 2979–2984.
33. Strickert, R.G., Friedman, A.M., and Fried, S. (1980). The sorption of technetium and iodine radioisotopes by various minerals. *Nucl. Technol.* **49**: 253–266.
34. Davenport, D. (1994). *Interoffice Memorandum to L. Stinton*. Martine Marietta Energy Systems, Piketon, OH.
35. Erhardt, K.C. and Attrep, M., Jr. (1978). Technetium-99 in the atmosphere. *Environ. Sci. Technol.* **12**: 55–57.
36. Attrep, M., Jr., Enochs, J.A., and Broz, L.D. (1971). Atmospheric technetium-99. *Environ. Sci. Technol.* **5**: 344–345.
37. Leon, M. (1989). Determination and levels of ^{99}Tc in environmental and biological samples. In: *Technetium in the Environment*. G. Desmet and C. Myttenaere (Eds.). Elsevier Applied Science, London, pp. 171–178.
38. Clausen, J.L., Zutman, J.L., Pickering, D.A., and Farrow, N.D. (1995). *Final Report on Drive-Point Profiling of the Northwest Plume and Analysis of Related Data*. Lockheed Martin Energy Systems, Inc., Kevil, KY.
39. Clausen, J.L., Richards, W.L., Korte, N.E., and Liang, L. (1995). ORNL/MMES research into remedial applications of zero-valence metals, 3: Removal of TCE, cis-1,2-DCE, vinyl chloride and technetium. *Am. Chem. Soc. Symp. Reductive Dehalogenation Proc.*, April 2–7, Anaheim, CA, pp. 755–758.
40. Schroeder, N.S., Morgan, D., Rokop, D.J., and Fabryka-Martin, J. (1992). Migration of Technetium-99 in the Alluvial Aquifer at the Nevada Test Site. Los Alamos National Laboratory, LA-UR-92-463, Los Alamos, NM.
41. Brown, J., Kolstad, A.K., Lind, B., Rudjord, A.L., and Strand, P. (1998). *Technetium-99 contamination in the North Sea and in Norwegian Coastal areas 1996 and 1997*. Stralevern Report. 1998:3. Oslo, Norway.
42. Turcotte, M.S. (1982). *Environmental Behavior of Technetium-99*. Savannah River Plant and Laboratory Report DP-1644. Aiken, SC.
43. Anderson, T.J. (1979). Development of chemical isolation and concentration techniques for Tc-99 analysis by resin—bead mass spectrometry. In: *Radioelement Analysis*. Savannah River Laboratory, Aiken, SC.
44. Olsen, C.R. et al. (1986). Geochemical and environmental processes affecting radionuclide migration from a formerly used seepage trench. *Geochim. Cosmochim. Acta* **50**: 593–607.
45. Blaylock, B.G. and Frank, M.L. (1981). Bioaccumulation and distribution of Tc-95m in an experimental freshwater pond. In: *Int. Symp. Migration Terrestrial Environ. Long-Lived Radionuclides Nucl. Fuel Cycle*. Savannah River Laboratory, Aiken, SC.
46. Ballestra, S. et al. (1987). *J. Radiochem.* **115**: 51.
47. Fryar, A. and Lasage, D.M. (2001). Technetium-99 as an in-situ tracer of ground-water/stream interactions and organic co-contaminant fate. *Geol. Soc. Am. Annu. Meet.*, November 5–8, Boston, MA.
48. Gu, B., Clausen, J.L., McDonald, J.A., and McCarthy, J.F. (1994). *Assessment of the Influences of Groundwater Colloids on the Migration of ^{99}Tc at the PGDP Site, Paducah, KY*. Oak Ridge National Laboratory, Oak Ridge, TN.
49. Muck, M.T., Korte, N.E., Liang, L., Marsh, D., Zutman, J.L., and Clausen, J.L. (1995). *Technology Screening Report for Tc Removal from Groundwater Portsmouth Uranium Enrichment Plant, Piketon, Ohio*. Martin Marietta Energy Systems, Inc., Piketon, OH.
50. Buddemeier, R.W., Finkel, R.C., Marsh, K.V., Ruggieri, M.R., Rego, J.H., and Silva, R.J. (1992). Radionuclide migration at the Cheshire and Cambric event sites. In: *Hydrology and Radionuclide Migration Program 1989 Progress Report*. K.V. Marsh (Ed.). Lawrence Livermore National Laboratory, Albuquerque, NM, pp. 5–20.
51. Clausen, J.L., Sturchio, N.C., Heraty, L., Abrajano, T., and Huang, L. (1997). *Evaluation of Natural Attenuation Processes for Trichloroethene and Technetium-99 in the Northwest and Northeast Plumes at the Paducah Gaseous Diffusion Plant Paducah, Kentucky*. Lockheed Martin Energy Systems, Inc., Kevil, KY.
52. Clausen, J.L. and Richards, W.L. (1994). *Treatability Studies Using Iron Filings to Remediate Trichloroethene and Technetium*. Martin Marietta Energy Systems. KY/ER-51. Paducah, KY.
53. Bostick, W.D., Osberne, P.E., Del Cul, G.D., and Simmons, D.W. (1995). *Treatment of Aqueous Solutions Contaminated with Technetium-99 Originating from Uranium Enrichment Activities: Final Report*. Martin Marietta Energy Systems, Inc., Oak Ridge, TN.

54. Ishanullah. (1994). Optimization of various factors for the separation of technetium using anion-exchange resins. *Sep. Sci. Technol.* **29**: 239–247.
55. Del Cul, G.D., Bostick, W.D., Trotter, D.R., and Osberne, P.E. (1993). Technetium-99 removal from process solutions and contaminated groundwater. *Sep. Sci. Technol.* **28**: 551–564.
56. Kawasaki, M., Amori, T., and Hasegawa, K. (1993). Adsorption of pertechnetate on an anion exchange resin. *Radiochimica Acta* **63**: 53–56.
57. Takeuchi, M., Tanaka, S., and Yamawaki, M. (1993). Solvent extraction of Tc(vii) and Re(vii) from nitric acid solution with the mixture of CMPO and TBP. *Radiochimica Acta* **63**: 97–100.
58. Yamagishi, I. and Kubota, M. (1993). Recovery of technetium with active carbon column in partitioning process of high-level liquid waste. *J. Nucl. Sci. Technol.* **30**: 717–719.
59. Karalova, Z.K., Lavrinovich, E.A., and Myasoedov, B.F. (1991). Separation of technetium from accompanying elements in basic solutions by extraction with triphenyltetrazolium chloride. *Sov. Radiochem.* **33**: 266–269.
60. Chen, Q.J., Dahlgaard, H., Hansen, H.J.M., and Aarkrog, A. (1990). Determination of Tc-99 in environmental samples by anion exchange and liquid extraction at controlled valency. *Anal. Chim. Acta* **228**: 163–167.
61. Emery, M.F. and Lim, C.K. (1989). Separation of cationic technetium-99m amine complexes on porous graphiti carbon. *J. Chromatogr.* **479**: 212–215.
62. Bostick, W.D. and Evans-Brown, B.S. (1988). *Sorptive Removal of Technetium from Heavy Metals Sludge Filtrate Containing Nitrate Ion*. Oak Ridge National Laboratory, Oak Ridge, TN.
63. Yamagishi, I. and Kubota, M. (1989). Separation of technetium with active carbon. *J. Nucl. Sci. Technol.* **26**: 1038–1044.
64. Walton, F.B., Paquette, J., Ross, J.P.M., and Lawrence, W.E. (1986). Tc(IV) and Tc(VII) interactions with oxyhydroxides. *Nucl. Chem. Waste Manage.* **6**: 121–126.
65. Fried, S., Friedman, A.M., Cohen, D., Hines, J.J., and Strickert, R.G. (1978). *The Migration of Radioactive Processing Wastes in Selected Rocks*. Argonne National Laboratory, Argonne, IL.

WATER—NATURE'S MAGICIAN

Environment Canada

Water is the solvent, the medium and the participant in most of the chemical reactions occurring in our environment.

Water is at once simple and complex. A water molecule itself is simple, made up of three atoms: two hydrogen and one oxygen, H_20. The configuration of these building blocks produces a molecule with almost magical properties.

WATER—VISIBLE AND INVISIBLE

On earth, water is found as a liquid, as a solid (ice) or as a gas (water vapour).

This article is produced by Environment Canada and is reproduced with permission.

- Canada has about 7% of the world's renewable freshwater supply, compared with 18% for Brazil, 9% for China, and 8% for the United States.
- The ice we skate on in winter is water in its *solid* form. Unlike most substances, which are densest in their solid state, ice is less dense than water and thus floats. If this were not the case, lakes and rivers would freeze from the bottom up. Fish could not survive, and it is unlikely that rivers and lakes in northern countries would ever completely thaw.
- Water vapour forms a kind of global "blanket" which helps to keep the earth warm. Heat radiated from the sun-warmed surface of the earth is absorbed and held by the vapour.

WATER'S MAGICAL PROPERTIES

Water molecules are attracted to each other, creating *hydrogen bonds*. These strong bonds determine almost every physical property of water and many of its chemical properties too.

Boiling and Freezing

Pure water at sea level boils at 100 °C and freezes at 0 °C. At higher elevations (lower atmospheric pressures) water's boiling temperature decreases. This is why it takes longer to boil an egg at higher altitudes. The temperature does not get high enough to cook the egg properly. If a substance is dissolved in water, then the freezing point is lowered. That is why we spread salt on streets in winter to prevent ice formation.

Thermal Properties

Water absorbs or releases more heat than many substances for each degree of temperature increase or decrease. Because of this, it is widely used for cooling and for transferring heat in thermal and chemical processes.

Differences in temperature between lakes and rivers and the surrounding air may have a variety of effects. For example, local fog or mist is likely to occur if a lake cools the surrounding air enough to cause saturation; consequently small water droplets are suspended in the air

Large bodies of water, such as the oceans or the Great Lakes, have a profound influence on climate. They are the world's great heat reservoirs and heat exchangers and the source of much of the moisture that falls as rain and snow over adjacent land masses.

When water is colder than the air, precipitation is curbed, winds are reduced, and fog banks are formed.

Surface Tension

Surface tension is a measure of the strength of the water's surface film. The attraction between the water molecules creates a strong film, which among other common liquids is only surpassed by that of mercury.

This surface tension permits water to hold up substances heavier and denser than itself. A steel needle carefully placed on the surface of a glass of water will float. Some aquatic insects such as the water strider rely on surface tension to walk on water.

Surface tension is essential for the transfer of energy from wind to water to create waves. Waves are necessary for rapid oxygen diffusion in lakes and seas.

Molecules in Motion

Water molecules as well as binding to each other, bind to many other substances such as glass, cotton, plant tissues, and soils. This is called adhesion. For example, in a thin glass tube, when the molecules at the edge reach for and adhere to the molecules of glass just above them, they at the same time tow other water molecules along with them. The water surface, in turn, pulls the entire body of water to a new level until the downward force of gravity is too great to be overcome. This process is called *capillary action*.

Thus water readily wets many materials. Capillary action allows a paper towel or a sponge to be used to soak up spilled water. Without this property, the nutrients needed by plants and trees would remain in the soil.

The Universal Solvent

An extraordinary property of water is its ability to dissolve other substances. There is hardly a substance known which has not been identified in solution in the earth's waters. Were it not for the solvent property of water, life could not exist because water transfers nutrients vital to life in animals and plants.

A drop of rain water falling through the air dissolves atmospheric gases. When rain reaches the earth, it affects the quality of the land, lakes and rivers.

Did You Know

- Raindrops are not tear-shaped. Scientists, using high-speed cameras, have discovered that raindrops resemble the shape of a small hamburger bun.
- About 70% of the human body is water.
- Life on earth probably originated in water.
- More than half of the world's animal and plant species live in the water.
- Almost 75% of the earth is covered in water.
- The human body needs 2 liters of water a day in our climate; we can last only a few days without water.
- Most of our food is water: tomatoes (95%), spinach (91%), milk (90%), apples (85%), potatoes (80%), beef (61%), hot dogs (56%).

THE SUN-POWERED CYCLE

The endless circulation of water from the atmosphere to the earth and its return to the atmosphere through condensation, precipitation, evaporation and transpiration is called the hydrologic cycle.

Heating of the ocean water by the sun is the key process that keeps the hydrologic cycle in motion. Water evaporates, then falls as precipitation in the form of rain, hail, snow, sleet, drizzle or fog. On its way to earth some precipitation may evaporate or, when it falls over land, be intercepted by vegetation before reaching the ground. The cycle continues in three different ways:

- *Evaporation / Transpiration*— On average, as much as 40% of precipitation in Canada is evaporated or transpired.
- *Percolation Into the Ground*— Water moves downward through cracks and pores in soil and rocks to the water table. Water can move back up by capillary action or it can move vertically or horizontally under the earth's surface until it re-enters a surface water system.
- *Surface Runoff*— Water runs overland into nearby streams and lakes; the steeper the land and the less porous the soil, the greater the runoff. Overland flow is particularly visible in urban areas. Rivers join each other and eventually form one major river that carries all of the subbasins' runoff into the ocean.

Although the hydrologic cycle balances what goes up with what comes down, one phase of the cycle is "frozen" in the colder regions during the winter season. During the Canadian winter, for example, most of the precipitation is simply stored as snow or ice on the ground. Later, during the spring melt, huge quantities of water are released quickly, which results in heavy spring runoff and flooding.

FRESHWATER SERIES A-1

Note: A resource guide, entitled Let's Not Take Water For Granted, is available to help classroom teachers of grades 5–7 use the information from the Water Fact Sheets.

FREEZING AND SUPERCOOLING OF WATER

Karl Erik Zachariassen
Norwegian University of Science and Technology
Trondheim, Norway

Water in the crystalline solid state is known as ice, and for pure water at normal atmospheric pressure, ice is the physically stable form at temperatures below 0 °C.

Ice has a specific gravity of 0.92 g/ml, which is about 8% below the corresponding value of water, reflecting the fact that water expands by about 8% when it enters the solid state. The specific heat capacity of ice is 0.502 cal/g.deg, which is about half the corresponding value of water. The amount of heat liberated during the solidification (heat of fusion) is 79.71 cal/g.

The transition of water from the liquid to the solid state is known as freezing, and pure water has a so-called freezing point of 0 °C, which does not imply that pure water will freeze when cooled to this temperature. A small sample of pure water can be cooled to as much as −40 °C before it freezes spontaneously. For this reason, the term freezing point is misleading, and it is often replaced by the more precise terms *melting point* or *equilibrium freezing point*.

A substance that exists in the liquid state at a temperature below the equilibrium freezing point is said to be *supercooled* or *undercooled*, and the temperature where freezing is initiated is referred to as the *supercooling point* or the *temperature of crystallization*.

When a sample of supercooled water freezes, the released heat of fusion will immediately heat the system to the melting point. As a result of the high heat of fusion of water, the melting point temperature will be reached following the freezing of only a small fraction of the sample. From this point, the freezing can proceed only as fast as the released heat of fusion can be removed from the system. When pure water freezes, the temperature will remain at 0 °C until all water is frozen.

Initiation of freezing requires that the water sample contains an ice-like structure, which can act as a nucleus for ice formation, from which ice can grow. Liquid water is assumed to contain aggregates of water molecules organized in an ice-like pattern, but which usually are too unstable to act as nucleators. As the temperature drops, these aggregates become more stable and increase in size. On sufficient cooling, one of these aggregates will eventually reach the size critical for ice nucleation, and freezing will be initiated from this aggregate. When nucleation is initiated in this manner from the water molecules themselves, it is known as *homogeneous* nucleation.

Water samples often contain other substances that organize water molecules in an ice-like pattern and that cause freezing to be initiated at a higher temperature than would otherwise be the case. Such substances are known as *ice nucleators,* and freezing catalyzed by an ice nucleator is referred to as *heterogeneous freezing*. Ice nucleator activity is seen in various types of crystals and dust particles, and biological organisms may synthesize nucleating lipoproteins, which are physiologically important as ice nucleators, and which may have a high nucleating activity. Certain bacteria produce ice nucleators that can nucleate ice formations at above −1 °C.

FACTORS AFFECTING FREEZING

The equilibrium freezing point is influenced by factors such as solute concentration (Fig. 1) and hydrostatic pressure (Fig. 2). The osmolal melting point depression is 1.86 °C/Osm, whereas pressure depresses melting points by about 1 °C/100 bar.

The effect of these parameters on supercooling points depends on whether nucleation is homogeneous or heterogeneous. Homogeneous nucleation temperature is depressed 1.5–3 times more than melting points by increased osmolality and pressure, whereas, at least for the more active nucleators, nucleation temperature depression seems to be strictly equivalent to the corresponding melting point depression.

Nucleation temperature of purified water samples (no potent ice nucleators) also depends on sample volume as shown by Bigg in 1953 (1) (Fig. 3). When potent ice nucleators are present, the nucleation temperature depends only on the activity of the most active nucleator, and it does not matter if this nucleator occurs in a large or small water volume.

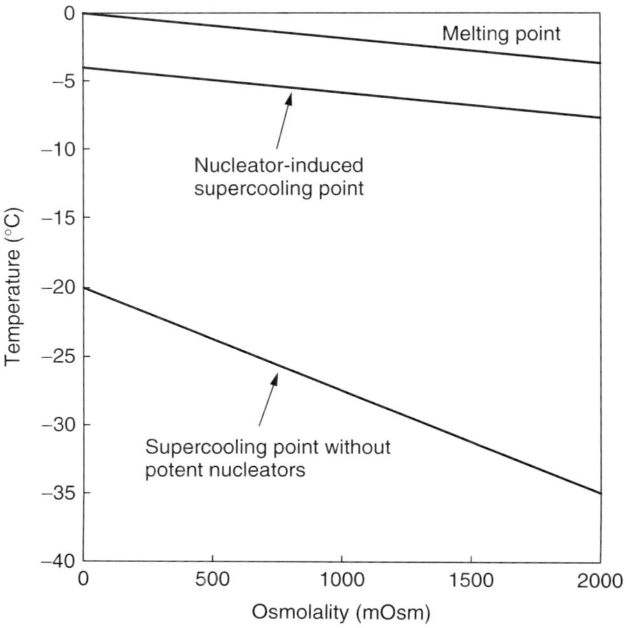

Figure 1.

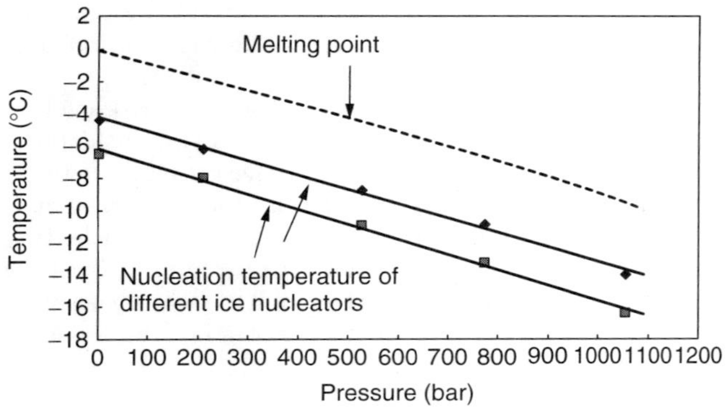

Figure 2.

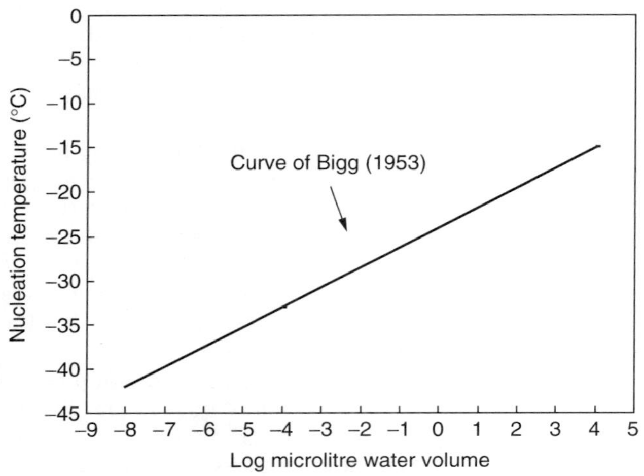

Figure 3.

FREEZING OF SOLUTIONS

When a solution freezes, the solutes are not incorporated in the ice structure, but remain in the liquid fraction surrounding the ice. Here, they become concentrated precisely until the solution reaches a melting point equal to the actual temperature. In this situation, vapor pressure equilibrium will exist between the solution and the ice, and no further freezing will occur until the released heat of fusion is removed from the system. As the heat is removed and the freezing proceeds, a gradual reduction of the temperature of the system will occur as the fluid fraction becomes more concentrated and the melting temperature of the fluid fraction drops. The solute concentration will increase until the eutectic point is reached. At the eutectic point, water and solutes freeze together.

BIBLIOGRAPHY

1. Bigg, E.K. (1953). The supercooling of water. *Proc. Roy. Soc. London. B* **66**: 688–694.

READING LIST

Zachariassen, K.E. and Kristiansen, E. (2000). Ice nucleation and antinucleation in nature. A review. *Cryobiology* **41**: 257–279.

CHEMICAL PRECIPITATION

Bradley A. Striebig
Gonzaga University
Spokane, Washington

Chemical precipitation is the removal of dissolved and suspended solids from aqueous solutions through chemical means. Chemical precipitation includes the removal of dissolved solids by altering the aqueous chemistry to make the dissolved solids less soluble. It also includes removing suspended solids by increasing their size in a process referred to as chemical flocculation. The larger the particle size of the suspended solids, the easier it is to remove them, usually through gravimetric settling in a process called clarification.

USES OF CHEMICAL PRECIPITATION

Chemical precipitation processes are utilized in water and wastewater treatment due to their effectiveness for pollutant removal with relatively low capital investment. The chemical reaction rates are typically very fast compared to biological reactions, so residence times in treatment processes or the reactor size may be significantly smaller than in other treatment techniques. Chemical precipitation processes can also process very toxic wastewaters. However, chemical precipitation processes often require the on-going addition of chemicals to control the pH and polymers to increase the size of the particles. The residuals collected from chemical precipitation processes may also be quite toxic and expensive to dispose of as sludge. Chemical precipitation is an effective tool for water treatment, and the most common uses are briefly described.

Industrial wastewaters often require pretreatment before being released to the sewer or receiving waters. High strength industrial wastewaters, such as those used in electroplating, contain high concentrations of dissolved metals. These metals can be removed from the wastewater by altering the pH of the water to reduce the solubility of the metal ions. The insoluble fractions of the metal ions form solid particles or precipitates, which are removed through filtration or gravimetric clarification. This allows industrial wastewaters to significantly reduce their metals concentrations in order to meet RCRA standards and also reduce the toxicity of the wastewater.

Municipal wastewaters may contain dissolved ions and a large amount of cell mass from biological treatment processes. The focus of chemical precipitation on municipal wastewaters is to remove the cell mass or suspended solids prior to and after biological treatment. The suspended solids are removed by adding chemical polymers, which attach to the existing suspended solids and help combine the individual particles into larger particles called flocs. This process, called coagulation, is accomplished by the chemical destabilization of particles and the subsequent aggregation. The process of gentle mixing and increasing the size of the suspended solids particles is called flocculation. The large floc particles are removed through gravimetric settling in a process called clarification. Thus a large fraction of the suspended solids and also the remaining BOD (in the form of cell mass) can be removed with the aid of chemical precipitation. Chemical precipitation is also commonly used when phosphorus removal is required at municipal wastewater treatment plants.

Industrial process water and recycled wastewater use chemical precipitation to remove hardness and dissolved salts from water. This allows a higher quality of water needed for specific industrial processes. The limits of hardness and salt removal are governed by the solubility of the dissolved ions in solution. Therefore, chemical precipitation has limited uses in industries that require highly purified water. However, chemical precipitation can be used as an inexpensive pretreatment step prior to ion exchange or reverse osmosis processes, which produce very high-quality water.

DISSOLVED SOLIDS

Dissolved solids are defined as the solid residuals in water that pass through a filter with a nominal pore size of 2.0 μm. These solids include dissolved ions and small colloidal solids. The dissolved ions in solution can be predicted based on the solubility product constant for the dissolved species. The solubility product constant, or solubility product, is an equilibrium constant that describes the reaction by which ions dissolve and solids precipitate:

$$A_zB_{y(s)} \rightleftharpoons zA^{y+} + yB^{z-}$$

For ions A and B, the generic equilibrium constant is

$$K_s = \frac{\{A^{y+}\}^z\{B^{z-}\}^y}{\{A_zB_y\}_{(s)}}$$

Many salts will form and precipitate as either hydroxides, with the hydroxide ion (OH^-), or carbonates, with the carbonate ion ($CO_3{}^{2-}$). The concentration of these ions can often be predicted through an equilibrium analysis with the aid of electron or pH diagrams. Since the activity of the solid substance is defined as 1, and the solubility product is a constant value, the activity of the anion of interest (A) can be predicted in the aqueous solution of interest. Solubility constants for a wide array of dissolved ions of interest in water and wastewater are found in the referenced literature (1–3).

In a dilute solution under ideal conditions the activity of the ions of interest can be closely approximated by their concentration:

$$\{A\} \approx [A]$$

However, many wastewaters have high concentrations of dissolved ions and do not behave ideally. An activity coefficient, γ_A, must be used to establish an accurate relationship between the activity of a compound in solution and the concentration:

$$\{A\} \approx \gamma_A[A]$$

The activity coefficient can be estimated from the ionic strength of the solution. The ionic strength is also commonly estimated from measurements of the total dissolved solids (TDS) and the electrical conductivity:

Langelier approximation	$I = 2.5 \times 10^{-5}$ (TDS)[mg/L]
Russell approximation	$I = 1.6 \times 10^{-5}$(conductivity) [μ mho]

One estimate of the activity coefficient is the Debye–Hückel limiting law for solutions with ionic strength less than 0.005 mol/L ($I < 0.005\ M$):

$$\gamma_A \cong -Az^2I^{1/2}$$

where A is a function of the dielectric constant of the solution. For water at 25 °C, $A = 0.51$. The term z is the ionic charge of the ion of interest. For more concentrated solutions ($I > 0.005\ M$), other estimates for the activity coefficient, such as the extended Debye–Hückel model and the Davies approximation can be used.

The potential to precipitate multiple ions from solution can be approximated from the activity of the ions in solution and the solubility product. Often, the pH of the solution is used as a "master variable." Solids lines on a graph of this type indicate the solids equilibrium concentration. For most hydroxide species, a minimum soluble concentration appears between a pH of 8 and 12. Similar graphical representations can be made for sulfide and carbonate species. This minimum solubility can serve as a starting point for estimating and optimizing chemical precipitation of various ions in solution.

The removal of hexavalent chromium (Cr^{6+}) from solution is one illustration of the potential of chemical precipitation processes to reduce the toxicity of wastewater. Hexavalent chromium can be converted to the less toxic trivalent form by reaction at low pH with a reducing agent such as ferrous bisulfite. Figure 1 shows a solution containing hexavalent chromium, which when dissolved in water results in a yellow color. Figure 2 shows the change in color associated with the change in the oxidation state of the chromium from the 6+ state to the 3+ state. When the pH of the solution is raised to 8 or 9, the trivalent chrome becomes much less soluble in solution than the hexavalent form of chromium as shown in Fig. 3. The resulting solids precipitate out of solution to form a concentrated sludge as shown in Fig. 4.

SUSPENDED SOLIDS

Suspended solids are composed of floating matter, settleable matter, and large colloidal matter. This matter

Figure 1. Hexavalent chromium dissolved in water.

Figure 2. Reduction of hexavalent chromium to trivalent chromium.

Figure 3. Formation of insoluble chromium oxide salts.

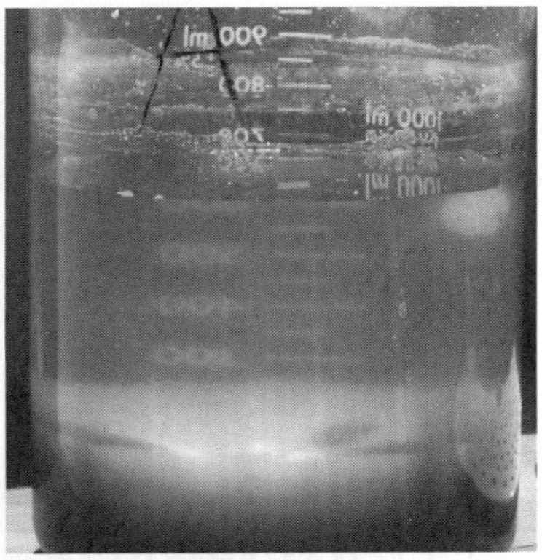

Figure 4. Precipitation of solid insoluble chromium into a sludge blanket.

may be organic or inorganic in nature. In municipal wastewater treatment plants, a significant fraction of the suspended solids consist of biological cell mass. The amount of suspended solids in water is an important parameter in determining water quality. Suspended solids are typically removed through a process of coagulation, flocculation, and gravimetric settling (or clarification). Chemical precipitation or coagulation of the suspended particles involves the destabilization of colloidal particles. Coagulation is usually followed by flocculation, which is the addition of a chemical, typically an organic polymer, to increase particle aggregation.

Suspended solids in water often have a charge that acts at the surface of the particle. Oil droplets, gas bubbles, and other inert substances may have a negative charge due to the preferential absorption of anions. Cellular materials such as microorganisms and proteins develop a charge through the ionization of carboxyl and amino acids functional groups. These charges create a force or zeta potential around the particles, which prevent them from increasing in size through aggregation as illustrated in Fig. 5. Coagulants are added to overcome these surface forces by the addition of potential determining ions or electrolytes. Once these charges have been neutralized, flocculants can be added to increase the speed of aggregation (Fig. 6).

Flocculant polymers that are anionic and nonionic become attached to the particles. These long particles increase the rate of aggregation and become intertwined

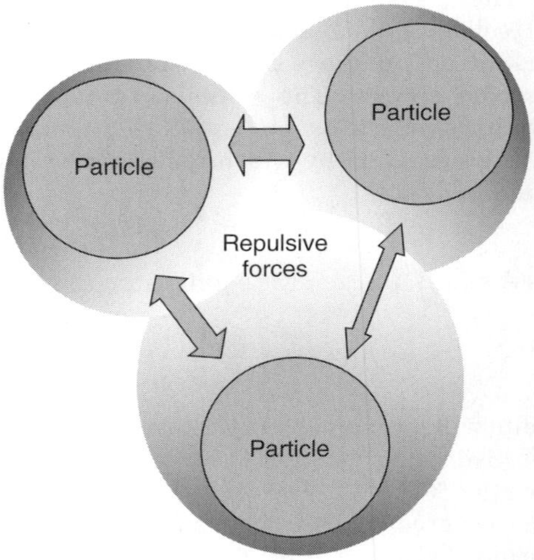

Figure 5. Repulsive forces acting on the surface of suspended particles.

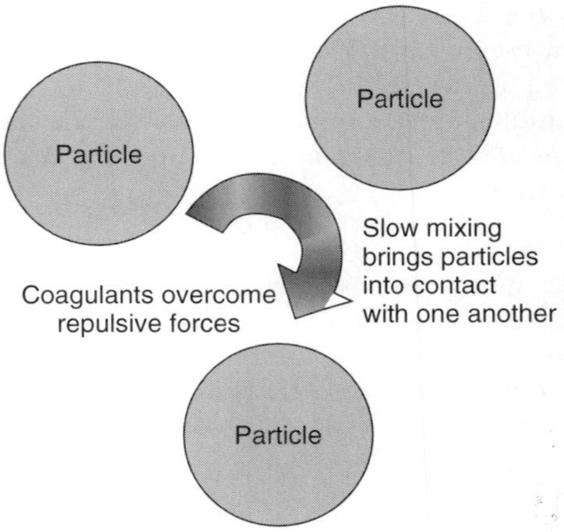

Figure 6. Chemical coagulants overcome repulsive forces and the solution is mixed.

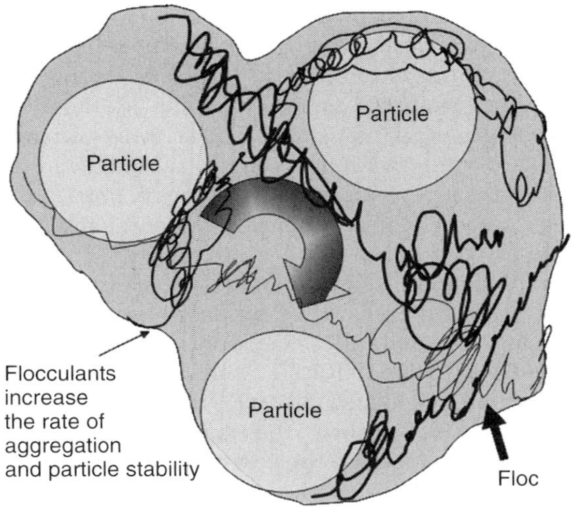

Figure 7. Flocculants increase the rate of aggregation to form large flocs.

during flocculation. This results in a significant increase in the particle size and mass (Fig. 7). The large particles called flocs are stabilized by the polymers, which prevent them from breaking up during the mixing process. The large flocs can then be removed with gravimetric settling in the clarification process. The resultant solids collected in the clarifier are the wastewater solids, or sludge.

Chemical precipitation by coagulation and flocculation has been used in the past to enhance removal of suspended solids and BOD. These processes are widely used to improve the performance of primary clarifiers, in basic physical–chemical processing of industrial wastewaters, for phosphorus removal, and for heavy metal removal.

BIBLIOGRAPHY

1. Tchobanoglous, G., Burton, F.L., and Stensel, H.D. (2003). *Wastewater Engineering Treatment and Reuse*, 4th Edn. Metcalf and Eddy, Inc/McGraw Hill, Boston, pp. 42–52, 475–517.
2. Stumm, W. and Morgan, J.J. (1996). *Aquatic Chemistry Chemical Equilibria and Rates in Natural Waters*, 3rd Edn. John Wiley & Sons, Hoboken, NJ, pp. 252–398.
3. Jensen, J.N. (2003). *A Problem-Solving Approach to Aquatic Chemistry*. John Wiley & Sons, Hoboken, NJ, pp. 418–450.

ANTIMONY IN AQUATIC SYSTEMS

Montserrat Filella
University of Geneva
Geneva, Switzerland

INTRODUCTION

Antimony is a naturally occurring element. It belongs to group 15 of the periodic table of the elements. Antimony can exist in a variety of oxidation states (−III, 0, III, V) but it is mainly found in two oxidation states (III and V) in environmental, biological, and geochemical samples. Although antimony was already known to the ancients, it is still often overlooked, both as an element of environmental concern and as a subject for study, probably because of its lower abundance and the relative insolubility of most of its compounds. This is reflected in the poor standard of existing data on the behavior of antimony in natural waters. However, interest in the study of this element seems to be growing and an increasing number of papers on the subject are being published. A useful series of comprehensive reviews on antimony in the environment has recently been published (1,2).

Antimony has no known biological function and, like arsenic, is toxic. Trivalent species are reported to be more toxic than pentavalent forms (3). The U.S. Environmental Protection Agency (EPA) (4) and the Commission of the European Communities (5) both consider antimony and its compounds to be pollutants of prime interest. The U.S. EPA drinking water standards are: MCLG (maximum contaminant level goal) and MCL (maximum contaminant level), both 6 μg/L (6). The European Union's maximum admissible concentration of antimony in drinking water is 5 μg/L (7). Antimony is on the list of hazardous substances under the Basel convention on the restriction of the transfer of hazardous waste across borders (8).

ABUNDANCE

The relative abundance of antimony in different systems is shown in Table 1. According to Goldschmidt's classification, antimony is a strong chalcophile element of which the most common natural form is the sulfide mineral stibnite, Sb_2S_3. Antimony is occasionally found as a native metal or as an intermediate compound with arsenic. More than 113 different sulfide, antimonide, arsenide, and telluride minerals have been identified. Sulfosalts represent the most common antimony-bearing minerals, usually in some combination with silver, copper, arsenic, mercury, and lead. More than 35 oxide, hydroxide, and oxyhalide mineral species are known. Antimony is also found in seven silicate minerals, all of which are rare (9). Antimony is also a common component of coal and petroleum.

Antimony is present in the aquatic environment as a result of rock weathering, soil runoff, and anthropogenic activities. Typically, concentrations of dissolved antimony in unpolluted waters are less than 1 μg/L. The mean antimony concentration in surface marine waters is 184 ± 45 ng/L (1). However, in the proximity of anthropogenic sources, concentrations can reach up to 100 times natural levels.

Airborne supply to aquatic and terrestrial systems is important for the environmental fate of antimony in systems located far from direct sources of pollution. Atmospheric emission values for antimony are shown in Table 2. The values in this table should be viewed with some caution because the data used comes from relatively old studies. For instance, antimony production was considered to be only 80,000 metric tons/yr (142,000 metric tons in 2003) and, more importantly, the uses of antimony have changed significantly over the past few years. Pacyna and Pacyna (15) estimated natural and anthropogenic emissions into the atmosphere to be 2400 tons/yr and

Table 1. Antimony Geochemical Abundance

Material	Sb Abundance[a]
Cosmic abundance	0.246[b]
Chondrites	0.1
Mean crustal average	0.2
Upper continental crust	0.31
Lower continental crust	0.30
Terrestrial abundance	0.7
Basic rocks (basalt)	0.15
Granitic rocks	0.2
Sedimentary rocks	
Shales	1–2
Carbonates	0.2
Deep sea clays	1

[a]All values in ppm (mg/g), except for the cosmic abundance.
[b]Atomic abundance relative to Si (=1,000,000) deduced from the following mean composition of meteorites: silicate = 100, sulfide = 7, metal = 10.6 parts per weight.
Source: Adapted from Reference 1.

1600 tons/yr, respectively. Very recently, Shotyk and co-workers (16) dismissed all previous values of natural antimony releases as being grossly overestimated. Based on antimony concentrations in ombrotrophic peat from Switzerland and Scotland, these authors suggested that 90–154 metric tons/yr would be more appropriate values.

USES

The first recorded use of antimony as a pigment to make mascara is found in an Egyptian papyrus dating from 1600 B.C. According to Emsley (17), the first uses of antimony date back even further; part of a 5000-year-old vase now in the Louvre is made of almost pure antimony. A peat core from an ombrotrophic Swiss bog revealed significant enrichments of antimony dating back to Roman times, which indicates that the anthropogenic fluxes of this element have exceeded natural fluxes for more than 2000 years. The present day enrichment factor (relative to the element/Sc ratios of typical crust rocks) is on the order of 70 times (as compared to 20 for As and 130 for Pb) (18). The chronology and intensity of the antimony enrichments in peat are similar in many ways to those of lead, which indicates that the extent of the human impact on the geochemical cycle of antimony is comparable to that of lead (16).

The production and use of antimony have steadily increased. World reserves of antimony are in excess of 2 million tons and are located principally in China, South Africa, Bolivia, Russia, and Tajikistan (19). The world's mines produced 142,000 tons of antimony in 2003 (19). Table 3 shows the main uses of primary antimony in the United States in 2003, which can be seen as representative of the use of antimony across the developed world. Uses of antimony have changed over the years. Traditionally, it was used in lead–antimony alloys. Bulk-secondary antimony could be recovered as antimonial lead, most of which was regenerated and then consumed by the battery industry. Today, the main use of antimony is as Sb_2O_3 in flame-retardants, which have applications in, among others, children's clothing and toys and aircraft and automobile seat covers. Antimony trioxide cannot be recycled and is released into the environment.

Table 2. Global Emissions[a] of Antimony into the Atmosphere, 10^3 metric tons/yr

Natural Sources	Sb Emission	Anthropogenic Sources	Sb Emission
Windborne dust (includes industrial sources of dust)	0.8	Energy production (coal, oil, gas)	1.3
Sea salt spray	0.6	Mining	0.1
Volcanic activity	0.7	Smelting and refining	1.4
Forest fires	0.2	Waste incineration	0.7
Biogenic sources	0.3		
Total	*2.6*	*Total*	*3.5*

[a]Adapted from Reference 1. Original estimations come from References 10–13. A completely different estimation, which is often cited in the literature, was published by Lantzy and Mackenzie in 1979 (14). They estimated natural and anthropogenic emissions of antimony into the atmosphere to be 0.98×10^3 metric tons/yr and 38×10^3 metric tons/yr, respectively.

Table 3. Reported Industrial Consumption (20) of Primary Antimony in the United States in 2002 (Metric Tons of Antimony Content)[a]

Uses	Sb Amount[b]
Metal products	
Antimonial lead	910
Bearing metal and bearings	43
Solder	85
Ammunition, cable covering, castings, sheet and pipe, and type metal	1,370
Total	*2,410*
Nonmetal products	
Ammunition primers	W
Ceramics and glass	487
Pigments	597
Plastics	532
Fireworks and rubber products	488
Total	*2,100*
Flameretardants	
Adhesives	W
Plastics	3,680
Rubber	126
Textiles	150
Paper and pigments	767
Total	*4,720*
Grand Total	*9,230*

[a]Secondary antimonial lead was 5600 metric tons in 2003.
[b]W = Withheld to avoid disclosing company proprietary data.

ANTIMONY SOLUTION CHEMISTRY

Both Sb(III) and Sb(V) ions hydrolyze easily in aqueous solution, thus making it difficult to keep antimony ions stable in solution except in highly acidic media. Since most of the techniques used for the study of solution equilibria (solubility measurements, potentiometry, voltammetry, spectroscopy) necessitate the use of macroamounts of reactants, relatively little information is available on the chemical species of antimony that exist in aqueous solution, and on their stability constants in particular. Moreover, most of the published studies are relatively old. Only the system antimony–sulfur has been the object of recent studies, mainly in relation to the understanding of hydrothermal systems.

The following main points can be made regarding the antimony inorganic chemistry in solution (2): (1) According to thermodynamic data, antimony should exist as Sb(V) in oxic systems and as Sb(III) in anoxic ones. (2) At environmental concentration levels, at the pH range commonly found in natural waters and in the absence of sulfur, antimony exists as a soluble species regardless of its oxidation state. Antimony(V) is present as a negative species in solution [$Sb(OH)_6^-$, also represented as SbO_3^-] and Sb(III) as a neutral species [$Sb(OH)_3$, also often quoted as $HSbO_2$]. (3) It is not clear whether Sb(V) and Sb(III) form any chloride-containing species in seawater, but hydroxychloro species would be the most probable entities. (4) Antimony oxides exhibit low water solubility, which may affect the fate of antimony in soils and water, particularly since today antimony is mostly used as Sb_2O_3(s) in flame-retardant applications. Moreover, it is thought that, when released into the atmosphere as an aerosol, antimony is oxidized to antimony trioxide through reactions with atmospheric oxidants. (5) In anoxic systems, and in the presence of sulfur, antimony forms insoluble stibnite, Sb_2S_3(s), and soluble SbS_2^{2-}, depending on the pH. The formation of polymerized species at natural concentration levels (e.g., $Sb_2S_4^{2-}$, $Sb_2S_5^{4-}$, $Sb_4Sb_7^{2-}$) is not proved. The formation of an Sb(V)-containing species (SbS_4^{3-}) has been observed by EXAFS.

Eh–pH diagrams can illustrate the redox behavior of elements like antimony. Figure 1 shows one such diagram for the system Sb–S–H_2O at environmentally realistic concentrations of antimony and dissolved sulfur.

A comprehensive, thermodynamically consistent antimony database of equilibrium constants of antimony with inorganic ligands has recently been published (21).

Barely any data have been reported on Sb(V) binding by low molecular weight organic ligands. More extensive studies have been carried out on the binding of Sb(III), which behaves like a borderline metal. The extent of antimony binding by natural organic matter (NOM) is a point of some debate. A few studies report significant Sb–NOM interactions, but most do not, and observed environmental behavior seems to suggest that Sb–NOM interactions are not relevant in most aquatic systems.

Hydrous oxides of manganese and iron have been reported to significantly sorb Sb(III) but very few systematic studies have been carried out on antimony sorption by pure metal oxides. These oxides will also oxidize the trivalent antimony to pentavalent antimony. Several studies have found a correlation between the adsorption of antimony in soil and oxic sediments and the presence of manganese and iron oxides (22,23).

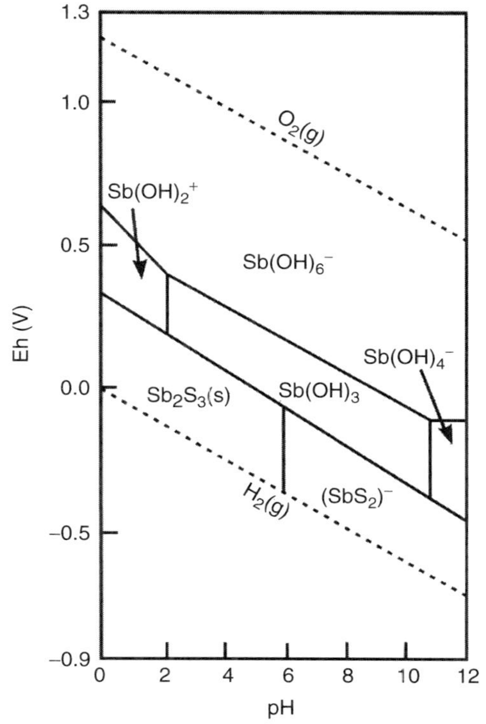

Figure 1. Eh–pH diagram of antimony in the Sb–S–H_2O system at a dissolved antimony concentration of 10^{-8} mol/L and a dissolved sulfur concentration of 10^{-3} mol/L. Adapted from Reference 2.

ANALYTICAL METHODS

The total concentrations of antimony in aqueous and solid environmental samples can readily be determined using current analytical techniques, primarily the result of the development of microwave digestion techniques for solid samples and NAA and plasma-based analyte detection systems. Both ICP-AES and ICP-MS techniques are in widespread use, although ICP-MS has less spectral interference and a lower detection limit (ng/L). An excellent critical review of the methodologies currently available for the environmental analysis of antimony has recently been published (24). However, determining antimony speciation still presents a significant analytical challenge (25). Recent data are almost exclusively based on hydride generation methods and often only oxidation state data are obtained. A lack of stability in the antimony species being investigated, particularly in trivalent forms, throughout the entire analytical process may result in an overestimation of pentavalent species concentrations.

Determining methylated antimony species is hampered by severe molecular rearrangements originating from the demethylation of trimethylantimony during the analysis. This problem has been the subject of extensive literary discussion but a solution has yet to be found (26).

ANTIMONY CONCENTRATIONS IN THE DIFFERENT AQUATIC COMPARTMENTS

The distribution and behavior of antimony in freshwater systems have not been studied extensively. Published concentrations range from a few nanograms/liter to a few micrograms/liter depending on location. They reflect the wide range of physical and chemical conditions existing in freshwater bodies and are very sensitive to the proximity of sources of pollution.

The concentration of antimony in ocean surface waters is about 200 ng/L. The behavior of antimony in the open seas is not considered to be highly reactive. According to some authors, it behaves as a conservative element and to others as a mildly scavenged element with surface (atmospheric) input. The comparison between antimony contents in deep waters of different oceans indicates that no antimony accumulation is produced during deep-water oceanic circulation (27). Antimony concentrations seem to reflect regional geochemical characteristics in coastal environments. Antimony behavior in estuaries varies somewhat according to the characteristics of the estuary. In some cases, it behaves conservatively and in others it shows a midestuary maximum. Behavior has even been seen to vary in the same estuary from one year to another.

Antimony concentrations in sediments are on the order of a few micrograms/gram of sediment and rarely exceed 1 mg/g. Higher concentrations are directly related to anthropogenic sources, mainly proximity to smelting plants and outfalls of sewage and fertilizer facilities.

ANTIMONY SPECIATION IN NATURAL WATERS

Today it is widely accepted that an understanding of biogeochemical processes requires knowledge of the chemical forms, or species, that are present in the natural environment. Despite this, the speciation of antimony in natural aquatic systems has yet to form the subject of many in-depth studies.

Dissolved Versus Particulate Fractions

Weak interaction between anionic species and suspended particles negatively charged in natural waters is to be expected and it is generally accepted that antimony is present as "dissolved" in aquatic systems. Experimental data (2) exist to support this. Fractionation of the so-called dissolved phase into several classes by ultrafiltration showed that most of the dissolved antimony is present in the smallest molecular weight region (2,28).

Redox

Antimony occurs in two oxidation states in natural waters and, thus, its behavior and toxicity can be affected by changes in the redox status of the aquatic environment. Antimony speciation studies have focused mainly on the separation and identification of Sb(III) and Sb(V) species.

Antimony (V) is the predominant species found in oxygenated systems but thermodynamically unstable Sb(III) has also been detected in different oxic marine water, fresh waters, groundwaters, and rain waters. This contrasts with thermodynamic equilibrium predictions, which suggest that Sb(V) concentration should exceed Sb(III) by several orders of magnitude. For thermodynamically unstable species to be present in water, mechanisms for their production and slow rates of interconversion must exist. The presence of Sb(III) has often been attributed to phytoplankton activity. However, although possible, a detailed analysis of published data has so far failed to produce conclusive proof to validate this hypothesis. Recently, the photochemical reduction of Sb(V) to Sb(III) has been suggested as an alternative to the phytoplankton source.

The true speciation of antimony under anoxic conditions remains unclear. According to thermodynamic calculations, in the absence of oxygen, antimony should only be present in the trivalent form. However, the occurrence of pentavalent forms of antimony has been reported in different anoxic systems. A variety of mechanisms have been suggested to explain these results, including delivery of Sb(V) on sinking detritus from oxic waters, formation of thiocomplexes by the pentavalent element, and advection of surface waters containing high concentrations of antimonate. All these mechanisms must be accompanied by relatively slow rates of reduction.

Methylated Species

Methylated species have been detected in surface seawaters, in estuaries, and in some rivers. They usually account for 10% or less of the total dissolved antimony. The monomethyl species is more abundant than the dimethyl form and close to the surface values tend to be higher. Monomethyl, dimethyl, trimethyl, and triethyl antimony derivatives have been detected in sediments from rivers and harbors. However, the identification of methylated antimony species has been questioned because some of these compounds undergo rearrangement reactions when subjected to the hydride generation procedures typically used for aquatic speciation.

The organoantimony compounds detected by Dodd et al. (26) in a freshwater plant (*Potamogetan pectinatus*) and a stibnolipid isolated from a marine diatom by Benson (29) are, as yet, the only organoantimony compounds with unequivocal biological origin to be detected in aquatic environments.

KINETIC ISSUES

As mentioned, the biogeochemical cycle of antimony is more complicated than equilibrium thermodynamic calculations predict (1). Several authors have cited the kinetic stabilization of thermodynamically unstable species to explain their presence in natural systems. Several processes may influence oxidation and reduction rates, including chemical and photochemical transformations and biological mediated reactions. It is well known that, when a redox process is accompanied by hydrolysis reactions, as is likely to be the case for antimony, the overall process can be much slower than expected. At present, kinetic information for redox reactions of antimony in natural waters is limited and rate constants are, for the most part, unknown. Cutter estimated an overall Sb(III) oxidation rate of 0.008 d^{-1} in the 65–85 m layer of the Black Sea from depth profiles (30). However, antimony residence times in seawater are probably longer because this rate included all forms of removal; that is, Sb(III) oxidation and also Sb(III) scavenging by sedimenting particles. Moreover, it was calculated at the oxic–suboxic interface of the Black Sea, where the presence of manganese and iron oxides is likely to increase the Sb(III) oxidation rate (31). Indeed, Belzile and co-workers (32), in a study on the oxidation of Sb(III) in the presence of natural and synthetic iron and manganese oxyhydroxides, showed that Sb(III) can be rapidly oxidized to Sb(V) by both compounds. Iron oxides also proved to play an important role in antimony redox chemistry in lake sediments (23). However, iron and manganese oxides are not likely to be present in oxic waters in significant quantities and thus are unlikely to influence antimony chemistry in oxic systems. It has recently been shown that Sb(III) oxidation by hydrogen peroxide and iodate is strongly pH dependent: the neutral $Sb(OH)_3$ species is unreactive to hydrogen peroxide and the presence of $Sb(OH)_4^-$ is needed for the oxidation to occur (33,34).

The addition of organic substances such as tartaric, lactic, citric, and ascorbic acids to natural or synthetic solutions has been shown to have a stabilizing effect on Sb(III). Solutions of Sb(III) prepared from potassium antimonyl tartrate have been reported to remain stable for long periods. The organic matter present in natural waters probably has a similar stabilizing effect. Ligand stabilization has been used to explain the apparent presence of Fe(II) in oxic solutions that contain relatively high levels of organic matter.

Little information is available on Sb(V) reduction kinetics in anoxic media. The reduction rate for Sb(V) in water at the bottom of the Black Sea was estimated to be 1.1×10^{-6} d^{-1} (35).

ANTIMONY AND AQUATIC BIOTA

Reported concentrations for antimony in fresh water and marine algae range from 0.02 to 1 μg/g dry weight (1). Although algae bioaccumulation and detoxifying mechanisms may play a key role in antimony redox speciation in natural waters, few studies have been carried out. Kantin (36) collected three marine algae near San Diego Bay in the United States and Sb(V) was the dominant species in all of them. Only *Sargassum* sp. contained up to 30% of Sb(III), thus demonstrating an ability to form reduced antimony species. Andreae and Froelich (37) analyzed the chemical speciation of antimony in phytoplankton (largely diatoms) from the eastern North Pacific and found that the trivalent species formed a significant part of the inorganic antimony. Both studies have since been used to justify the frequently cited biological origin of the Sb(III) found in oxic waters. Although no detailed studies exist, antimonate is expected to behave differently in bioavailability than arsenate and phosphate because of the weaker Lewis acidity and larger ionic radius of the antimonate ion in solution. More recently, the algae *Chlorella vulgaris*, isolated from an arsenic-polluted environment, excreted 40% Sb(V) and 60% Sb(III) on exposure to Sb(III), suggesting that a change in oxidation state is used by this alga as a detoxifying mechanism (38).

BIBLIOGRAPHY

1. Filella, M., Belzile, N., and Chen, Y-W. (2002). Antimony in the environment: a review focused on natural waters. I. Occurrence. *Earth Sci. Rev.* **57**: 125–168.
2. Filella, M., Belzile, N., and Chen, Y-W. (2002). Antimony in the environment: a review focused on natural waters. II. Relevant solution chemistry. *Earth Sci. Rev.* **59**: 265–285.
3. Bencze, K. (1994). Antimony. In: *Handbook on Metals in Clinical and Analytical Chemistry*. H.G. Seiler, A. Sigel, and H. Sigel (Eds.). Marcel Dekker, New York, pp. 227–236.
4. United States Environmental Protection Agency. (1979). *Water Related Fate of the 129 Priority Pollutants*. Vol. 1, EP-440/4-79-029A, U.S. EPA, Washington, DC.
5. Council of the European Communities. (1976). Council Directive 76/464/EEC of 4 May 1976 on pollution caused by certain dangerous substances discharged into the aquatic environment of the Community. Official Journal L 129, 18/05/1976, pp. 23–29.
6. United States Environmental Protection Agency. (1999). *National Primary Drinking Water Standards*. Doc. 810-F-94-001, U.S. EPA, Washington, DC.
7. Council of the European Union. (1998). Council Directive 98/83/EC of 3 November 1998 on the quality of water intended for human consumption. Official Journal L 330, 05/12/1998, pp. 32–54.
8. United Nations Environmental Program. (1999). *Basel Convention on the Control of Transboundary Movements of Hazardous Wastes and Their Disposal* (with amended Annex I and two additional Annexes VIII and IX, adopted at the fourth meeting of the Conference of the Parties in 1998), SBC No. 99/001. UNEP, Geneva.
9. Lueth, V.W. (1999). Antimony: element and geochemistry. In: *Encyclopedia of Geochemistry*. C.P. Marshall and R.W. Fairbridge (Eds.). Kluwer Academic Publishers, Dordrecht, The Netherlands, pp. 15–16.
10. Nriagu, J.O. and Pacyna, J.M. (1988). Quantitative assessment of worldwide contamination of air, water and soils by trace metals. *Nature* **333**: 134–139.
11. Nriagu, J.O. (1989). A global assessment of natural sources of atmospheric trace metals. *Nature* **338**: 47–49.
12. Nriagu, J.O. (1990). Global metal pollution. Poisoning the biosphere? *Environment* **32**: 7–11, 28–33.
13. Maeda, S. (1994). Safety and environmental effects. In: *The Chemistry of Organic Arsenic, Antimony and Bismuth Compounds*. S. Patai (Ed.). Wiley, Hoboken, NJ, pp. 725–759.
14. Lantzy, R.J. and Mackenzie, F.T. (1979). Atmospheric trace metals: global cycles and assessment of man's impact. *Geochim. Cosmochim. Acta* **43**: 511–525.
15. Pacyna, J.M. and Pacyna, E.G. (2001). An assessment of global and regional emissions of trace metals to the atmosphere from anthropogenic sources worldwide. *Environ. Rev.* **9**: 169–298.
16. Shotyk, W. et al. (1996). Two thousand years of atmospheric arsenic, antimony, and lead deposition recorded in an ombrotrophic peat bog profile, Jura Mountains, Switzerland. *Earth Planet. Sci. Lett.* **145**: E1–E7.
17. Emsley, J. (2001). *Nature's Building Blocks. An A–Z Guide to the Elements*. Oxford University Press, New York.

18. Shotyk, W., Krachler, M., and Chen, B. (2004). Antimony in recent, ombrotrophic peat from Switzerland and Scotland: comparison with natural background values (5,320 to 8,020^{14}C yr BP) and implications for the global Sb cycle. *Global Biogeochem. Cycles* **18**: GB1016.

19. Carlin, J.F., Jr. (2004). Antimony. In: *U.S. Geological Survey Mineral Commodity Summaries*.

20. Carlin, J.F. (2003). *U.S. Geological Survey Minerals Yearbook*.

21. Filella, M. and May, P.M. (2003). Computer simulation of the low-molecular-weight inorganic species distribution of antimony(III) and antimony(V) in natural waters. *Geochim. Cosmochim. Acta* **67**: 4013–4031.

22. Müller, B. et al. (2002). As, Sb, Mo, and other elements in sedimentary Fe/Mn layers of Lake Baikal. *Environ. Sci. Technol.* **36**: 411–420.

23. Chen, Y-W., Deng, T-L., Filella, M., and Belzile, N. (2003). Distribution and early diagenesis of antimony species in sediments and porewaters of freshwater lakes. *Environ. Sci. Technol.* **37**: 1163–1168.

24. Nash, M.J., Maskall, J.E., and Hill, S.J. (2000). Methodologies for determination of antimony in terrestrial environmental samples. *J. Environ. Monit.* **2**: 97–109.

25. Krachler, M., Emons, H., and Zheng, J. (2001). Speciation of antimony for the 21st century: promises and pitfalls. *Trends Anal. Chem.* **20**: 79–90.

26. Dodd, X. et al. (1996). Antimony speciation in freshwater plant extracts by using hydride generation–gas chromatography–mass spectrometry. *Analyst.* **121**: 223–228.

27. Cutter, G.A. and Cutter, L.S. (1998). Metalloids in the high latitude North Atlantic Ocean: sources and internal cycling. *Mar. Chem.* **61**: 25–36.

28. Stolpe, X., Hassellöv, M., Andersson, K., and Turner, D. (2004). High resolution ICPMS as an on-line detector for flow-field-flow fractionation; multi-element determination of colloidal size distributions in a natural water sample. *Anal. Chim. Acta.* in press.

29. Benson, A. (1988). Antimony metabolites in marine algae. In: *The Biological Alkylation of Heavy Metals*. P.J. Craig and F. Glocking (Eds.). *Special Publication*, Vol. 66, Royal Society of Chemistry, London, pp. 135–137.

30. Cutter, G.A. (1992). Kinetic controls on metalloid speciation in seawater. *Mar. Chem.* **40**: 65–80.

31. Cutter, L.S., Cutter, G.A., and San Diego-McGlone, M.L.C. (1991). Simultaneous determination of inorganic arsenic and antimony species in natural waters using selective hydride generation with gas chromatography/photoionization detection. *Anal. Chem.* **63**: 1138–1142.

32. Belzile, N., Chen, Y-W., and Wang, Z. (2001). Oxidation of antimony(III) by amorphous iron and manganese oxyhydroxides. *Chem. Geol.* **174**: 379–387.

33. Quentel, F., Filella, M., Elleouet, C., and Madec, C-L. (2004). Kinetic studies on Sb(III) oxidation by hydrogen peroxide in natural waters. *Environ. Sci. Technol.* **38**: 2843–2848.

34. Quentel, F., Filella, M., Elleouet, C., and Madec, C-L. (2004). Sb(III) oxidation by iodate in seawater: a cautionary tale. (Submitted).

35. Cutter, G.A. (1991). Dissolved arsenic and antimony in the Black Sea. *Deep-Sea Res.* **38**: S825–S843.

36. Kantin, R. (1983). Chemical speciation of antimony in marine algae. *Limnol. Oceanogr.* **28**: 165–168.

37. Andreae, M.O. and Froelich, P.N. (1984). Arsenic, antimony, and germanium biogeochemistry in the Baltic Sea. *Tellus* **36**: 101–117.

38. Maeda, S. et al. (1997). Bioaccumulation of antimony by *Chlorella vulgaris* and the association mode of antimony in the cell. *Appl. Organomet. Chem.* **11**: 393–396.

WATER LAW AND ECONOMICS

THE CLEAN WATER ACT

Timothy J. Ryan
Ohio University
Athens, Ohio

U.S. water legislation dates back to the late nineteenth century, when the 49th Congress enacted the Rivers and Harbors Act of 1886, which ultimately became the Rivers and Harbors Act of 1899. This law was designed to minimize obstructions in any navigable waterway in the United States, and it prohibited the deposit of any material on banks of navigable waters that would lead to the impeding of navigation. The law basically allowed for fines of $500 to $2,500 as well as imprisonment. Recognizing the continuing public health threat posed by water contamination, in 1948 Congress enacted the Federal Water Pollution Control Act (FWPCA) in order to "enhance the quality and value of our water resources and to establish a national policy for the prevention, control and abatement of water pollution." The original FWPCA aimed to establish programs that would improve sanitary conditions of both surface and groundwater while minimizing or eliminating the pollution of interstate waterways and tributaries. These goals thus differed from the original Rivers and Harbors Act's goals by adding pollution regulation in hopes of improving the public's water supply, habitats of aquatic life, and water for recreational, agricultural, and industrial uses. FWPCA and several subsequent amendments set out the basic legal authority for federal regulation of water quality to the present day.

The original FWPCA, or simply the "Act," was subjected to incremental but limited improvements that eventually resulted in a somewhat disparate collection of water-related requirements. Eleven reorganizations and restructurings of federal agency responsibilities compounded the difficulty of effectively implementing the law. For this reason, the 1972 amendments to the FWPCA restructured the authority for water pollution control and consolidated authority in the Administrator of the then newly formed United States Environmental Protection Agency (EPA). Formed in 1970 by President Nixon in response to the growing public demand for an overall healthier environment, the EPA was the first federal authority to implement pollution control programs for industry. Growing public awareness and concern for controlling water pollution throughout the mid-1970s led to amendments of the Act in 1977. As amended, this law became commonly known as the Clean Water Act (CWA).

The CWA, and it's associated rules found in Title 33 of the Code of Federal Regulations, established the basic priorities and structure for regulating pollution releases into the waters of the United States. The Act's major subchapters include six Titles: (I) Research and Related Programs, (II) Grants for Construction of Treatment Works, (III) Standards and Enforcement, (IV) Permits and Licenses, (V) General Provisions, and (VI) State Water Pollution Control Revolving Funds. The Act continued requirements to set water quality standards for all contaminants in surface waters and made it unlawful to discharge any pollutant from a single, or point, source into navigable waters, unless a permit was obtained under the Act. It also funded the construction of sewage treatment plants under the construction grants program and recognized the need for planning to address the critical problems posed by diffuse, or nonpoint, sources of pollution.

NATIONAL POLLUTANT DISCHARGE ELIMINATION SYSTEM

One of the most significant features of the Clean Water Act is the creation of a National Pollutant Discharge Elimination System (NPDES) under Section 402 of Title IV, Permits and Licensing. The NPDES prohibited industrial sources and publicly owned treatment works from discharging pollutants into navigable waters without a process review and permit. Under NPDES, all point source discharges of pollution into the waters of the United States are required to obtain a permit. Point source discharges come from "any discernible, confined, and discrete conveyance including but not limited to any pipe, ditch, channel, tunnel, conduit, well, discrete fissure, container, rolling stock, concentrated animal feeding operation, landfill leachate collection system, vessel or other floating craft from which pollutants are or may be discharged." The EPA Administrator may issue a permit for discharge on the condition that the discharge meets applicable requirements, which are detailed in the Act. Permitted discharges are only allowed insofar as federal effluent limitations and state water quality standards can continue to be met. The Act also provides that, given EPA approval, a state may administer its own permit program in lieu of the federal program so long as the state effort is deemed at least as effective as the federal controls.

NONPOINT SOURCE POLLUTION CONTROL

Since its use in 1972, the term "nonpoint source pollution" has been used in a variety of sometimes confusing contexts. Most simply, a nonpoint source is any source of pollution that is not otherwise legally considered as a point source. In 1987, Congress enacted Section 319 of Title III, Standards and Enforcement, of the Clean Water Act to establish a national program to control nonpoint sources of pollution. In its guidance for the program, the EPA rigorously defines nonpoint source pollution as that "... caused by rainfall or snowmelt moving over and through the ground and carrying natural

Table 1. Toxic Pollutants Designated Pursuant to Section 307(a)(1) of the Clean Water Act

1. Acenaphthene
2. Acrolein
3. Acrylonitrile
4. Aldrin/Dieldrin
5. Antimony and compounds
6. Arsenic and compounds
7. Asbestos
8. Benzene
9. Benzidine
10. Beryllium and compounds
11. Cadmium and compounds
12. Carbon tetrachloride
13. Chlordane (technical mixture and metabolites)
14. Chlorinated benzenes (other than di-chlorobenzenes)
15. Chlorinated ethanes (including 1,2-di-chloroethane, 1,1,1 trichloroethane, and hexachloroethane)-
16. Chloroalkyl ethers (chloroethyl and mixed ethers)
17. Chlorinated naphthalene
18. Chlorinated phenols (other than those listed elsewhere; includes trichlorophenols and chlorinated cresols)
19. Chloroform
20. 2-chlorophenol
21. Chromium and compounds
22. Copper and compounds
23. Cyanides
24. DDT and metabolites
25. Dichlorobenzenes (1,2-, 1,3-, and 1,4-di-chlorobenzenes)
26. Dichlorobenzidine
27. Dichloroethylenes (1,1-, and 1,2-dichloroethylene)
28. 2,4-dichlorophenol
29. Dichloropropane and dichloropropene
30. 2,4-dimethylphenol
31. Dinitrotoluene
32. Diphenylhydrazine
33. Endosulfan and metabolites
34. Endrin and metabolites
35. Ethylbenzene
36. Fluoranthene
37. Haloethers (other than those listed elsewhere; includes chlorophenylphenyl ethers, bromophenylphenyl ether, bis(dichloroisopropyl) ether, bis-(chloroethoxy) methane and polychlorinated diphenyl ethers)
38. Halomethanes (other than those listed elsewhere; includes methylene chloride, methylchloride, methylbromide, bromoform, dichlorobromomethane
39. Heptachlor and metabolites
40. Hexachlorobutadiene
41. Hexachlorocyclohexane
42. Hexachlorocyclopentadiene
43. Isophorone
44. Lead and compounds
45. Mercury and compounds
46. Naphthalene
47. Nickel and compounds
48. Nitrobenzene
49. Nitrophenols (including 2,4-dinitrophenol, dinitrocresol)
50. Nitrosamines
51. Pentachlorophenol
52. Phenol
53. Phthalate esters
54. Polychlorinated biphenyls (PCBs)
55. Polynuclear aromatic hydrocarbons (including benzanthracenes, benzopyrenes, benzofluoranthene, chrysenes, dibenz-anthracenes, and indenopyrenes)
56. Selenium and compounds
57. Silver and compounds
58. 2,3,7,8-tetrachlorodibenzo-p-dioxin (TCDD)
59. Tetrachloroethylene
60. Thallium and compounds
61. Toluene
62. Toxaphene
63. Trichloroethylene
64. Vinyl chloride
65. Zinc and compounds

and human-made pollutants into lakes, rivers, streams, wetlands, estuaries, other coastal waters, and ground water. Atmospheric deposition and hydrologic modification are also sources of nonpoint pollution." A major source of typically nonpoint source pollution was recognized in urban stormwater. In response, and to address the specific concerns of contaminated stormwater runoff, the National Stormwater Program was established as part of the 1987 Clean Water Act amendments. Although stormwater is considered nonpoint source pollution, the hazardous constituents in the water may come from various points sources such as industry, sewers, or construction. Thus, the National Stormwater Program is controlled through NPDES, previously discussed.

EFFLUENT LIMITATIONS AND WATER QUALITY STANDARDS

In Section 401 of Title IV, Permits and Licensing, the Act prohibits the discharge of pollutants in excess of effluent limitations, and makes it unlawful to discharge radiological, chemical, or biological warfare agents; high-level radioactive waste; or any medical waste into navigable waters. Effluent limitations from point sources other than publicly owned treatment works (POTWs) must be treated using best practicable control technology, or BACT. The effluent limitations provide standards for a variety of pollutants and pollution sources, including but not limited to sources such as coal mining, landfills, construction, iron and steel manufacturing, meat products, pharmaceutical, paper products, and leather and tanning industry. Each source type has a different set of EPA effluent limitation standards to meet. Toxic pollutants defined in the Act (Table 1) require treatment using the best available technology that is economically achievable. If the source discharges into a POTW (e.g., via a connection to a sanitary sewer), it must comply with applicable pretreatment requirements as well as POTW expectations.

WATERSHED RESTORATION PLANS

Section 303(d) of Title III, Standards and Enforcement, of the Act requires that states and the EPA identify rivers, lakes, and coastal waters threatened or already impaired as a result of one or more specific pollutants. The states or EPA must develop "Total Maximum Daily Loads" (TDML) for every water body that has been placed on the EPA-designated threatened or impaired waters list. A TDML is thus a pollution cap or ceiling for the water body, and it has the functional effect of a cleanup plan. An associated plan for the affected water body then determines exactly how to attain that cap. The TMDL must specify problems, determine pollution reductions needed to solve the problems, identify pollution sources, and assign responsibilities for needed action. Also under Title III, various other specific and significant sources of pollution are regulated. Thermal discharges, marine sanitation devices, and oil and hazardous substance discharge are a few aspects of water pollution so regulated. As all of these sources have the potential to be hazardous polluters, the Act sets standards to which each must abide in different manners to ensure the quality of the water.

PLANS, GRANTS, AND FUNDING

The CWA both requires and assists in the development and implementation of waste treatment management plans and practices to achieve the water quality goals of the Act. Plans and practices must provide for treatment of waste prior to any discharge of pollutants into receiving waters, as well as the secure disposal of other pollution sources to avoid migration that could ultimately cause water or other environmental pollution. To accomplish these goals, Title II of the Act outlines a program of grants to states, municipalities, or other agencies for the construction of governmental water treatment facilities, or POTWs. Revisions of the Act in 1981 streamlined the municipal construction grants process, improving the capabilities of treatment plants built under the program. Additional changes in 1987 phased out the construction grants program but replaced it with the State Water Pollution Control Revolving Fund (found in Title VI of the Act), known within professional circles as the Clean Water State Revolving Fund. That funding strategy addressed water quality needs by building on EPA-State partnerships to identify and meet state and regional water quality issues.

KEY AMENDMENTS AND ASSOCIATED WATER PROTECTION LAWS

Safe Drinking Water Act

Over the years, many other environmentally related laws have changed or otherwise impacted parts of the CWA. The Safe Drinking Water Act (SDWA) was originally passed by Congress in 1974 to specifically protect the nation's public drinking water supply. The SDWA authorized the EPA to set national health-based standards for drinking water to protect against both naturally occurring as well as man-made contaminants that may be found in drinking water. These standards are divided into primary and secondary standards. Primary standards are set to protect the public's health, are legally enforceable, and apply to public water systems. Secondary standards are nonenforceable guidelines set to regulate contaminants in the drinking water that may cause negative cosmetic or aesthetic effects, such as tooth discoloration or bad odor. States may choose to adopt these secondary standards as enforceable regulations. The law was amended in 1986 and 1996 and requires various actions to protect drinking water sources, including rivers, lakes, reservoirs, springs, and groundwater.

Oil Pollution Act

As a response to the Exxon Valdez disaster of 1989, when a crude oil tanker ran into a reef on the coast of Alaska and spilled 10.8 million gallons of the oil into the Prince William Sound, Congress passed the Oil Pollution Act in 1990. This act amended the CWA to significantly raise the limits of liability to would-be polluters, and the amount of compensation that must me made to injured parties resulting from an oil spill.

Great Lakes Critical Programs Act

As a result of their immense size and value as a freshwater supply, state, federal, and international steps have been taken to protect the water quality of the Great Lakes. Title I of the Great Lakes Critical Programs Act of 1990, for example, put into place parts of the U.S.-Canadian Great Lakes Water Quality Agreement of 1978, wherein the two nations agreed to jointly reduce certain toxic pollutants entering these resources. That law required the EPA to establish water quality criteria for the Great Lakes for 29 toxic pollutants with a particular focus on persistent bioaccumulative toxics (such agents tend to occur in higher concentration in aquatic biota than in open waters). The criteria are intended to protect human health, aquatic life, and wildlife and include the first-ever EPA wildlife criteria to protect birds and mammals from long-term exposure to mercury, DDT, PCBs, and dioxin. It also required the EPA to help the states implement the criteria on a specific schedule.

With a growing world population and intense industrial demands on a finite water supply, regulations concerning U.S. water resources will continue to change and evolve with time. As the most prominent water quality law presently in effect, the CWA will continue to play a significant roll in the maintenance of water quality for future generations in the face of technological changes and demands.

READING LIST

EPA Laws and Regulations—Clean Water Act History. Available: http://www.epa.gov/region5/water/cwa.htm.

EPA Water History Website. Available: http://www.epa.gov/history/topics/fwpca/05.htm.

Digest of Federal Resource Laws of Interest to the U.S. Fish and Wildlife Service. Federal Water Pollution Control Act (Clean Water Act). Available: http://laws.fws.gov/lawsdigest/fwatrpo.html.

River Network Website—Understanding the Clean Water Act. Available: http://cleanwateract.org/pages/c1.htm.

Knudson, S. Quick Reference: Clean Water Act (CWA). Center for Sustainable Systems, Ann Arbor, MI. Available: http://css.snre.umich.edu/css_doc/CWA.pdf.

Copeland, C. (2002, 2004). CRS Report for Congress—Clean Water Act: Summary of the Law.

Clean Water Act. Available: http://www.boozman.house.gov/UploadedFiles/H20%20%20Clean%20Water%20Act%20A%20 Summary%20 of%20 the%20Law.pdf.

Environmental Literacy Council—Clean Water Act. Available: http://www.enviroliteracy.org/article.php/372.html.

EPA Electronic "Snapshot" of the Clean Water Act (the Act in PDF format). Available: http://www.epa.gov/region5/water/cwa.htm#ECWA.

NOAA Legislative Summaries—Federal Water Pollution Control Act. Available: http://www.csc.noaa.gov/opis/html/summary/cwa.htm.

CLEAN WATER ACT, WATER QUALITY CRITERIA/STANDARDS, TMDLS, AND WEIGHT-OF-EVIDENCE APPROACH FOR REGULATING WATER QUALITY

G. Fred Lee
Anne Jones-Lee
G. Fred Lee & Associates
El Macero, California

INTRODUCTION

The 1972 amendments to the Federal Water Pollution Control Act (Clean Water Act, CWA) established a regulatory framework to maintain and enhance water quality in waterbodies of the United States. The CWA specified that waterbodies were to be classified with respect to their beneficial uses, such as domestic water supply, propagation of fish and aquatic life, and recreation, and that the U.S. EPA was to develop national water quality criteria for chemical and other agents that would protect the designated beneficial uses of waterbodies. The numeric chemical-concentration-based national water quality criteria that it eventually developed were designed to be protective in any water under worst-case exposure conditions. The U.S. EPA criteria, in turn, were to be used by the states as the basis for water quality standards, the benchmark for control of beneficial use-impairment of waterbodies.

The way in which these CWA provisions have been implemented has been to require that a waterbody in which a water quality standard was exceeded be listed as a Clean Water Act section 303(d)-"impaired" waterbody. Such a listing requires that a program be initiated to control the sources of the constituents responsible for the water quality standard violation, which is done through a total maximum daily load (TMDL) assessment through which is determined a load that would not result in exceeding the standard; that load is then apportioned among those discharging the constituent.

Although this TMDL approach was originally adopted in 1972, it was not until the late 1990s that it began to be implemented to any significant extent. This implementation developed from environmental groups filing suit against the U.S. EPA for having failed to implement the TMDL provisions of the Clean Water Act. As part of settling the lawsuit, the U.S. EPA agreed to see to the implementation of TMDLs by the states. Considerable controversy, exists however, about the appropriateness of the TMDL process as it is typically being implemented for regulating water quality caused largely by the use of numeric U.S. EPA national water quality criteria and state standards equivalent to them as the benchmark for evaluation of water quality. Following is a synopsis of how this approach came to be established and issues that remain inadequately addressed in the pursuit of the goals of the CWA. Also described is a "best professional judgment" triad weight-of-evidence approach that provides a technically valid, cost-effective framework for assessing water quality/beneficial-use impairments and their remediation.

EARLY REGULATORY APPROACH—ISSUES STILL NEEDING ATTENTION

Lee (1) discussed several significant technical shortcomings inherent in the TMDL approach adopted by the U.S. EPA for regulating chemical constituents for the protection of water quality. For example, in the 1960s, when toxicity tests were starting to evaluate the toxicity of wastewater discharges, it was often found that a portion of the heavy metals in some discharges was in forms that did not cause toxicity. This finding was in accord with what would be expected based on the aquatic chemistry of heavy metals. It was recognized that heavy metals exist in a variety of chemical forms; given the chemical nature of the forms, it could be expected that not all of them would be equally available/toxic to aquatic life, although this distinction could not be made with chemical analytical techniques. This situation was sufficiently well known that by the early 1970s, the National Academies of Science and Engineering concluded, as part of their development of the *Blue Book of Water Quality Criteria* (2), that heavy metals in wastewater discharges could not be reliably regulated based on chemical concentrations. Rather, because of the numerous unquantifiable factors that control the manifestation of a chemical's toxicity, they recommended a toxicity test approach to determine the availability of the heavy metals, either alone or in combination with other metals or other substances, in a particular water. The National Academies of Science and Engineering Blue Book Criteria were adopted by the U.S. EPA (3) in its 1976 *Red Book Criteria*, which were the first official water quality criteria developed pursuant to the Clean Water Act.

In the early 1980s, however, the U.S. EPA abandoned the toxicity testing approach recommended by the National Academies of Science and Engineering. In its place, it adopted a policy of applying the numeric, worst-case, national water quality criteria (generally, chronic exposure, safe concentrations) to the concentrations of total recoverable metals (i.e., those forms that are measurable after strong acid digestion of the sample), rather than to the available forms, in water quality evaluation and regulation. That approach lead to overregulation of heavy metals because in some situations, substantial parts of the "total recoverable" heavy metals are nontoxic/unavailable.

Some relief from the overregulation of heavy metals was provided when application of the criteria was shifted from "total recoverable" forms to "ambient-water dissolved" forms of metals (4). The shift in focus at that time was not based on the finding of any new information because it had been well-established in the 1960s and 1970s that particulate forms of heavy metals in the water column were nontoxic. Focusing on dissolved forms of heavy metals corrected a long-standing problem in the implementation of the national water quality criteria into state water quality standards. That notwithstanding, the Agency has still not addressed its inappropriate application of water quality criteria for many other constituents, such as organics, to total concentrations rather than properly addressing contaminant availability.

Although deficiencies in the conventional application of the water quality criteria were generally recognized, they were not addressed by the U.S. EPA primarily because the regulations that were developed were not being enforced by either the U.S. EPA or many of the states. As a result, the promulgation took place of the National Toxics Rule in subsequent revisions of the Clean Water Act, through which Congress mandated that states either adopt the U.S. EPA criteria for toxics or have them imposed on them by the U.S. EPA. By the early 1990s, all states had adopted U.S. EPA criteria for "toxics." California's regulations adopting the U.S. EPA criteria as state standards, however, were soon judged invalid through court action because California state law also requires consideration of economic impact of water pollution control regulations. As the California State Water Resources Control Board did not comply with those state requirements, the courts determined that the regulations must be voided. Thus, for many years, California did not have water quality criteria/objectives for "toxics." Finally, in 2000, the U.S. EPA Region 9 imposed what became known as the California Toxics Rule criteria (5). They are the U.S. EPA criteria for "toxics" that were originally adopted in the mid-1980s, or subsequent updates, such as the U.S. EPA (4,6–8). The most recent update of national water quality criteria occurred in 2002, when the U.S. EPA (9) developed its currently recommended national water quality criteria. The U.S. EPA requires that as states update their criteria, they incorporate the 2002 criteria as the state standards.

Lee et al. (10), Lee and Jones (11), and Lee and Jones-Lee (12) discussed alternative approaches for assessing and controlling the impact of contaminants on water quality that took better account of contaminant availability to affect water quality, thus directing the financial resources available first toward defining those constituents that adversely impact the beneficial uses of a waterbody and then toward controlling those constituents to the extent necessary to protect those uses.

PRIORITY POLLUTANT LIST

The 1972 Clean Water Act also mandated that the U.S. EPA develop a list of "priority pollutants" that was to include those chemicals found in water that should receive the highest attention for regulatory action. National water quality criteria were then to be developed for each of those chemicals to protect fish and aquatic life in all waters. Congress, however, did not fund the U.S. EPA adequately to carry out this mandate. Finally, when the U.S. EPA could not develop the list within the timeframe allowed, an environmental group filed suit to force the U.S. EPA to do so. In response to that suit, in the mid-1970s, the U.S. EPA's attorneys and environmental group attorneys, with limited technical input and without public peer review, promulgated what is now known as the "Priority Pollutant" list.

Despite the intention to focus on water quality problems, in actuality, the "Priority Pollutant" list has proven to be detrimental to properly defining the constituents that are significantly adverse to the beneficial uses of waterbodies. The list was not properly peer-reviewed for its inclusion and prioritization of pollutants of real water quality significance. Instead, the list focused primarily on what are known as rodent carcinogens, i.e., those constituents that, at high concentrations over extended periods of exposure, cause cancer in rats, but do not necessarily have a great significance to water quality. Large amounts of public resources have been devoted to analyzing for and then developing control programs for many of the rodent carcinogens, especially the chlorinated solvents, while neglecting many pollutants of greater water quality significance.

In looking at this situation, Dr. Christian Daughton, Chief, Environmental Chemistry Branch, U.S. EPA National Exposure Research Laboratory, indicated that there are more than 22 million organic and inorganic substances, with nearly 6 million commercially available. The current water quality regulatory approach addresses fewer than 200 of those chemicals as potential water pollutants, i.e., the "priority pollutants." He stated, "Regulated pollutants compose but a very small piece of the universe of chemical stressors to which organisms can be exposed on a continual basis" (13).

IMPLEMENTATION OF TMDLS

The establishment of TMDLs focuses on achieving water quality standards in receiving water. This focus led to a review of the TMDL-based water pollution control program by the National Research Council (NRC). The NRC's review (14) discussed technical deficiencies in the U.S. EPA's TMDL program. Waterbodies have been placed inappropriately on the Clean Water Act section 303(d) list of impaired waterbodies on the basis of unreliable evaluation. TMDL goals of achieving worst-case water quality standards for total concentrations are often inappropriate goals for solving real, significant, water quality/use-impairment problems in a technically valid, cost-effective manner. Most importantly, inadequate time and inadequate funding are available to support the development of TMDLs as they are being administered through the U.S. EPA and state regulatory agencies.

The first step toward establishing a more appropriate TMDL process should be an assessment of the appropriateness of the water quality standards that established the 303(d) listing and the standards that serve as TMDL goals to correct water quality impairment. As a considerable part of the TMDL program is directed toward sources such as agricultural runoff and urban runoff that frequently contain constituents that are, in substantial proportion, nontoxic, and unavailable, it is important that the U.S. EPA and the states refocus TMDL programs on controlling toxic, available forms, as opposed to total concentrations of constituents.

BPJ TRIAD WEIGHT-OF-EVIDENCE

Because of the technical inappropriateness and unreliability of using overly simplistic, even though administratively expedient, indicators of impact, increasing attention is being given to the use of a triad "weight-of-evidence" approach as a regulatory tool for water quality impact assessment and management. Although this approach has been configured in several ways with varying degrees of technical validity, a well-accepted, technically appropriate format is a "best professional judgment" (BPJ) evaluation of a triad of key parameters: aquatic life toxicity/bioaccumulation of the contaminant(s); aquatic organism assemblage information in the aquatic system of concern; and chemical kinetic/thermodynamic information pertaining to the contaminant(s) of interest and the aquatic system of concern. These components are described below.

Toxicity/Bioaccumulation

The availability of most chemical contaminants to aquatic life cannot be determined through chemical analysis; the availability can be affected by the character of the ambient water, the nature of the exposure conditions, and the sensitivity of the organisms of concern. Furthermore, routine chemical analysis does not cover all potential toxicants that may be present. Therefore, a key component of a BPJ weight-of-evidence evaluation for water quality impact assessment and management is aquatic life toxicity testing and/or bioaccumulation testing of aquatic organism tissue for potentially hazardous chemicals that are a threat to human health or higher trophic-level organisms that use aquatic life as food. The importance of using the BJP triad framework for the interpretation of the results of this testing is illustrated by the fact that, as discussed by Lee and Jones-Lee (15), finding aquatic life toxicity in laboratory tests of ambient water or sediment cannot be assumed to mean that that toxicity represents a significant impairment of the beneficial uses of the waterbody that is of concern to the public. It is not necessarily possible to equate laboratory-based water column or sediment toxicity with water quality impairment. For example, many sediments have natural toxicity because of low dissolved oxygen, ammonia, and hydrogen sulfide; yet the waterbodies associated with them have excellent fisheries and high water quality. The other aspects of the triad must be used with this information to make a best professional judgment regarding the significance of the toxicity and bioaccumulation information.

Organism Assemblage

Organism assemblage information includes description of the numbers, types, and characteristics of aquatic life and, as appropriate, terrestrial organisms such as fish-eating birds, present at a potentially impacted site. Insight into whether water quality has been adversely affected can be gained from understanding how the organism assemblage at the location compares with that which might be expected to be supported there. However, a variety of physical (flow, temperature, sunlight, sediment, and habitat alteration, etc.), non-potential-pollutant chemical (TDS, nutrients, organic constituents, hardness, alkalinity, etc.), and biological (reproductive cycles, disease, predation, etc.) factors other than chemical potential pollutants can affect the numbers, types, and characteristics of aquatic life in a waterbody's water column or sediments. Therefore, it is critical that this information be considered in the context of a BPJ evaluation in assessing the water quality impacts of chemicals on the beneficial uses of a waterbody.

Chemical Information

Chemical concentration has been one of the most convenient, yet most misunderstood and misused, pieces of information in water quality evaluation and management. Unfortunately, chemical information and "chemistry" have often been considered to be a laundry list of total concentrations of a few regulated constituents having water quality standards, which are mechanically applied to discharges, ambient water, or sediment. This result occurs despite that for half a century, it has been known that the total concentration of a potentially toxic constituent in the water column and/or sediments is an unreliable basis for estimating the impacts of that contaminant on the beneficial uses of a waterbody. Thus, the administratively expedient application of such values directly can distort the significance of that chemical contaminant to water quality.

The reason the total concentration of a selected chemical is unreliable for assessing water quality/use-impairments is that many chemical constituents exist in aquatic systems in a variety of chemical forms, only some of which are toxic or available to affect aquatic life. The aquatic "chemistry," i.e., the kinetics (rates) and thermodynamics (positions of equilibrium) of reactions that a chemical can undergo in a natural water system, is illustrated in the aquatic chemistry "wheel" presented in Fig. 1.

As illustrated, chemicals can undergo oxidation/reduction, volatilization, photochemical, complexation, sorption/desorption, precipitation, hydrolysis, and acid/base reactions in natural water systems. Many reactions alter the availability of the chemical to affect aquatic life. Although measurement of the total concentration of a chemical includes essentially all of these forms, the amount of the "available" forms of the chemical present depends in part on the nature of the chemical and the types and amounts of materials in the water and/or sediment that act to "detoxify" the chemical, i.e., render it nontoxic or unavailable to affect aquatic life through these reactions. These types of materials include organic carbon, sulfides, carbonates, hydrous oxides, clay minerals, and others. The amount of available forms also depends on the rates of reaction, the extent to which these reactions occur, and the comparative availability among the forms.

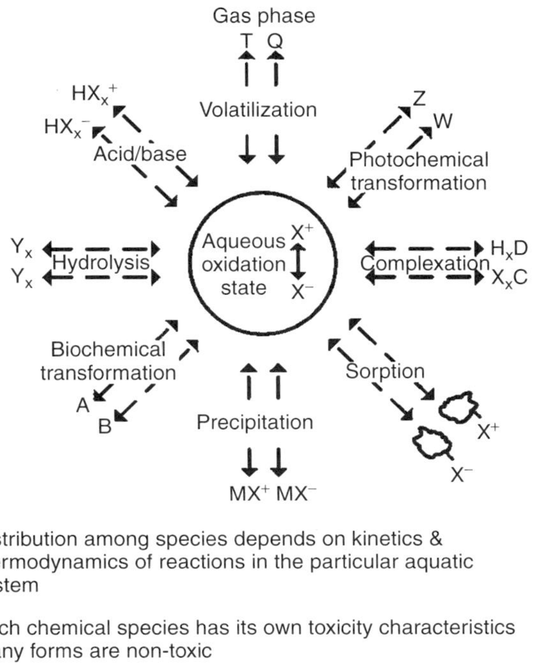

Figure 1. Aquatic chemistry of chemical constituents.

No simplistic method exists by which to reliably quantitatively account for these interactions by mathematical manipulation or chemical analysis. It is for this reason that chemical concentration information, especially total concentrations of chemicals, should not be used alone in assessing water quality or impairment of water quality. The disregard for the aquatic chemistry of contaminants and for the nature and duration of organism exposure in aquatic systems makes the application of worst-case, numeric water quality criteria typically overly restrictive for the protection of beneficial uses of waterbodies. Although chemical concentration information can raise issues to consider and sources that may be worthy of further investigation, it is unreliable for reaching a conclusion about "water quality" or beneficial use-impairment. It is for this reason that the aquatic toxicity/bioaccumulation and organism assemblage information are also critical parts of the triad evaluation, and the need for objective, technically informed, best professional judgment is clear.

CO-OCCURRENCE-BASED SEDIMENT QUALITY GUIDELINES: CAUTION

The issue of sediment quality criteria and guidelines warrants special attention in this discussion. Reliable evaluation and regulation of the impacts of contaminants in sediment systems is substantially more complex and more ill-defined than in ambient waters. Although it has been long understood that no relationship exists between the total concentrations of chemical contaminants in sediments and toxicity, bioaccumulation, or adverse impact on beneficial uses of a water, chemical concentration is simple and numeric, and lends itself to easy decision-making. In an attempt to make simple chemical analysis relevant, some have developed and advanced the use of co-occurrence-based "sediment quality guidelines" (16–18). A group of sediments was evaluated for having some type of biological impact. Then the concentrations of a few selected contaminants in those sediments that exhibit an "impact" were examined. Basically, then, for each contaminant, the lowest concentration associated with an "impacted" sediment was said to be adverse, a cause for concern in any sediment, and the basis for a "sediment quality guideline." No consideration was given to the actual cause of the "impact" reported, to the fact that total concentration is not related to impact, to several chemical and conditions that are well-understood to cause sediment toxicity, or to sediments having that or higher concentrations of the contaminant without exhibiting adverse impacts. The only basis for the so-called "guideline" concentration was the "co-occurrence" of the contaminant in the sediment with some biological impact attributed to that sediment. It is entirely expected that chemical constituents derived from urban industrial areas, although having biological effects, often occur in the presence of other constituents, which, although in nontoxic, unavailable forms, are present in elevated concentrations.

Co-occurrence-based approaches exemplify inappropriate use of chemical information in a water quality assessment but nonetheless are popularized because of their ease of application. As discussed by Lee and Jones-Lee (19–21), no cause-and-effect relationship is established in the co-occurrence-based values. Although so-called "correlations" exist between toxicity and a sediment quality guideline is exceeded, this is coincidental and unreliable for an assessment of the cause of the biological impacts. The fact that a chemical constituent exceeds a particular "sediment quality guideline" does not mean that that constituent is in any way related to biological effects, such as toxicity, bioaccumulation, and/or changes in organism assemblages. The actual cause of the biological response can readily be either a constituent(s) that is not measured or not considered in the scheme or a combination of constituents that, although measured, do not, individually or summed, exceed the "sediment quality guidelines." Thus, no expectation can exist that funds spent to achieve "sediment quality guideline" values will result in any improvement in sediment/water quality or that sediments targeted by the exceedance of guideline values are, in fact, of the greatest concern.

Some try to skirt the fundamental technical flaws in the approach by limiting its use to "screening" sediments. However, using a patently unreliable "screening" tool can do nothing but provide patently unreliable results, which will serve to misdirect concern, responsibility, and funds for remediation, and leave real problem areas unaddressed. Although sediments that exceed one or more "sediment quality guidelines" may, in fact, merit further investigation or remediation, the guideline values are meaningless for making that assessment; under no circumstances should anyone assume that the exceedance of a guideline value represents a cause-and-effect relationship that can determine the likely cause of a biological response. As discussed by O'Connor (22–24), O'Connor and Paul (25), O'Connor et al. (26), Engler (27),

DiToro (28), Chapman (29), Burton (30), and Lee and Jones-Lee (15,19–21), the co-occurrence approaches are obviously technically invalid and unreliable for assessing cause-and-effect. A reliable evaluation and regulatory program must be based on reliable assessments of the cause of the adverse effect.

RECOMMENDED APPROACH FOR INCORPORATION OF CHEMICAL INFORMATION INTO A BPJ TRIAD WEIGHT-OF-EVIDENCE WATER QUALITY EVALUATION

The recommended approach for the use of chemical concentration information in a BPJ triad weight-of-evidence evaluation of the cause of a water quality impairment begins with the reliable definition of the water quality/use-impairment that is of concern. The existence of a contaminant in total concentrations greater than a worst-case-based standard/guideline is not, in itself, an adverse impact unless that contaminant is causing adverse impacts to the beneficial uses of the waterbody. Thus, for example, before measuring the concentrations of copper, lead, zinc, and cadmium that typically occur in street and highway stormwater runoff at concentrations above U.S. EPA worst-case-based water quality criteria and state water quality standards based on those criteria, the chemical impact evaluation approach determines whether the water or sediment of concern is toxic. Jones-Lee and Lee (31) describe an evaluation monitoring approach that has been developed to focus on chemical impacts rather than on chemical concentrations.

If toxicity is found in laboratory tests of a water or sediment, then an assessment should be made as to whether that toxicity translates to an adverse impact on the waterbody's beneficial uses, such as fisheries, survival or reproduction of desired aquatic life, and so on. It should not be assumed that toxicity measured in a standard toxicity test necessarily translates to a toxicity that is significantly altering the numbers, types, and characteristics of desirable forms of aquatic life in a waterbody. This fact is especially true for situations such as urban area and highway stormwater runoff, where there can be short-term pulses of toxicity associated with a runoff event that are not of sufficient magnitude or duration to exceed the critical magnitude/duration coupling needed to be adverse to important forms of aquatic life in a waterbody.

Figure 2 shows a typical relationship between duration of exposure and toxicity; as illustrated, the manifestation of a toxicity response to available forms of a contaminant depends on the duration of exposure. Elevated concentrations of a toxicant can be tolerated by aquatic life, provided that the duration of exposure of this toxicity is shorter than the critical, or threshold, toxicity/duration of exposure coupling for that contaminant and organism in the ambient water. Exposure durations of aquatic life can be affected by the intermittent nature of the introduction of the contaminant, the rate and nature of dilution in the ambient water, avoidance or attraction behavior of the organism, and so on.

If measured toxicity is determined to potentially affect the beneficial uses of the waterbody, then a toxicity identification evaluation (TIE) should be conducted to evaluate the cause of the toxicity (32–34). The TIE process defines the availability of contaminants present and can elucidate the potential availability of the toxicity-causing contaminant(s) from its various sources. This process, then, can direct control toward those sources that are contributing significantly to the toxicity. Not only is this approach more technically sound, but it also addresses the issue of potential impacts of unrecognized, unmeasured, and/or unregulated pollutants that has been the subject of increasing importance in water quality management.

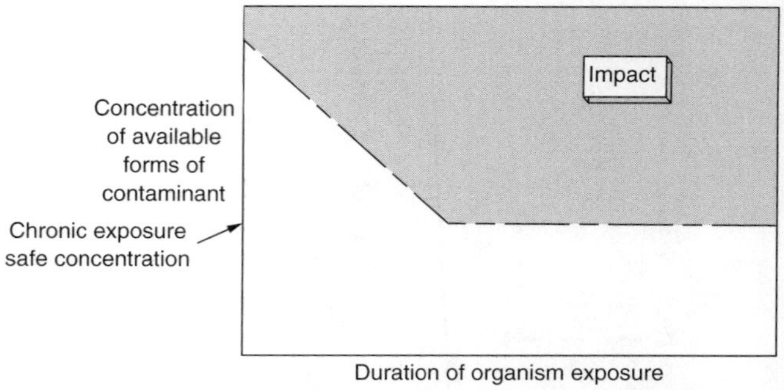

Figure 2. Critical concentration—duration of exposure relationship.

SUMMARY AND CONCLUSION

The federal Clean Water Act and amendments have directed the development of water quality criteria and standards and a framework for identifying and addressing beneficial use-impairments in the waters of the country. In the quest for administratively simple means by which to evaluate and regulate water quality and sediment quality, worst-case-based criteria and standards for a select group of chemicals have been developed. These are applied to the concentrations of contaminants in ambient waters as a means of judging water quality. This approach is known to be unreliable for evaluating or regulating water quality because it does not properly account for the beneficial uses of the water; the nature, behavior, and forms of chemical contaminants; or the unrecognized and unregulated pollutants.

The BPJ triad weight-of-evidence approach described herein provides a technically sound foundation and approach for assessing water quality impacts, causes of the impacts, and directions for controlling the impact. It is not as administratively expedient as simple comparison of chemical concentration data; however, that simplistic approach is not reliable for determining impairments or effecting control of real water quality problems. The BPJ triad weight-of-evidence approach requires the allocation of sufficient funds to determine the characteristics of the constituents/conditions of concern, with particular emphasis on properly defining toxicity and water quality cause-and-effect relationships. It also requires that individuals knowledgeable in aquatic chemistry, aquatic toxicology, and water quality provide guidance on, and appropriate interpretation of the results of, the kinds of chemical and toxicity studies that are needed to appropriately incorporate chemical information into assessing the water quality significance of chemical constituents in impacting the beneficial uses of a waterbody.

Numeric weight-of-evidence approaches in which arbitrary scale factors are assigned to each of the three components of the triad are technically invalid, because the scaling factors do not represent the relationship between a chemical constituent in a water or sediment and its impact on the water quality-beneficial uses of a waterbody.

Significant problems can occur with the use of the BPJ approach in incorporating chemical information into the triad. The use of total concentrations of constituents and/or the exceedance of a co-occurrence-based so-called sediment quality guideline is technically invalid. Such an approach can distort the triad water/sediment quality evaluation because it incorporates information into the triad that is not related to the impact of the chemicals on aquatic-life-related beneficial uses.

The BPJ weight-of-evidence approach should be based on the consensus of a panel of experts who, in a public, interactive, peer-review process, consider the information available, define what additional information is needed, and then render an opinion as to the integrated assessment of the information available on the significance of a particular chemical constituent in impacting the beneficial uses of a waterbody. The characteristics of the components of a BPJ weight-of-evidence approach that focuses on the appropriate use of chemical information are discussed further by Lee and Jones-Lee (35).

BIBLIOGRAPHY

1. Lee, G.F. (2001). Comments on US EPA TMDL Program. U.S. Environmental Protection Agency, Office of Water, Washington DC, December. Available: http://www.gfredlee.com/TMDLcomments.pdf.
2. NAS/NAE. (1973). *Water Quality Criteria of 1972*. National Academies of Science and Engineering, EPA/R3-73-033, U.S. Environmental Protection Agency, Washington, DC.
3. U.S. EPA. (1976). *Quality Criteria for Water*. U.S. Environmental Protection Agency, Washington, DC.
4. U.S. EPA. (1995). *Stay of Federal Water Quality Criteria for Metals; Water Quality Standards; Establishment of Numeric Criteria for Priority Toxic Pollutants; States' Compliance Revision of Metals Criteria; Final Rules. Federal Register* **60**(86): 22228–22237, U.S. Environmental Protection Agency, Washington, DC.
5. U.S. EPA. (2000). *Water Quality Standards; Establishment of Numeric Criteria for Priority Toxic Pollutants for the State of California; Rule*. U.S. Environmental Protection Agency, Region 9, Federal Register 40 CFR Part 131, Vol. 65, No. 97, [FRL-6587-9], RIN 2040-AC44, San Francisco, CA.
6. U.S. EPA. (1987). *Quality Criteria for Water 1986*. EPA 440/5086-001, U.S. Environmental Protection Agency, Office of Water Regulations and Standards, Washington, DC.
7. U.S. EPA. (1996). *1995 Updates: Water Quality Criteria Documents for the Protection of Aquatic Life in Ambient Water*. EPA-820-B-96-001, U.S. Environmental Protection Agency, Office of Water, Washington, DC.
8. U.S. EPA. (1999). *National Recommended Water Quality Criteria—Correction*. US Environmental Protection Agency, Office of Water, EPA 822-Z-99-001, Washington, DC.
9. U.S. EPA. (2002). *National Recommended Water Quality Criteria: 2002*. EPA-822-R-02-047, U.S. Environmental Protection Agency, Washington, DC.
10. Lee, G.F., Jones, R.A., and Newbry, B.W. (1982). Alternative approach to assessing water quality impact of wastewater effluents. *J. Water Pollut. Control Fed.* **54**: 165–174.
11. Lee, G.F. and Jones, R.A. (1987). Assessment of the Degree of Treatment Required for Toxic Wastewater Effluents. *Proc. Int. Conf. on Innovative Biological Treatment of Toxic Wastewaters*, U.S. Army Construction Engineering Research Laboratory, Champaign, IL, pp. 652–677.
12. Lee, G.F. and Jones-Lee, A. (1995). Appropriate use of numeric chemical water quality criteria. *Health and Ecological Risk Assessment* **1**: 5–11; Letter to the Editor, Supplemental Discussion, **2**: 233–234 (1996). Available: http://www.gfredlee.com/chemcri.htm.
13. Daughton, C.G. (2004). Pharmaceuticals and Personal Care Products (PPCPs) as Environmental Pollutants: Pollution from Personal Actions. Presentation at the California Bay-Delta Authority Contaminant Stressor Workshop, February. Additional information on PPCPs and other unregulated chemicals is available at www.epa.gov/nerlesd1/chemistry/pharma/index.htm.
14. NRC. (2001). *Assessing the TMDL Approach to Water Quality Management*. National Research Council, National Academy Press, Washington, DC.
15. Lee, G.F. and Jones-Lee, A. (1996). Evaluation of the Water Quality Significance of the Chemical Constituents in Aquatic Sediments: Coupling Sediment Quality Evaluation Results to Significant Water Quality Impacts. WEFTEC '96, Surface Water Quality and Ecology I & II, Proc. Water Environ. Fed. Annual Conference **4**: 317–328. Available: http://www.gfredlee.com/wef_seda.htm.
16. Long, E.R. and Morgan L.G. (1991). The potential for biological effects of sediment-sorbed contaminants tested in the national status and trends program. Tech. Memo. NOS OMA52, National Oceanic and Atmospheric Administration, Seattle, WA.
17. Long, E.R., MacDonald, D.D., Smith, S.L., and Calder, F.D. (1995). Incidence of adverse biological effects within ranges of chemical concentrations in marine and estuarine sediments. *Environ. Manage.* **19**: 81–97.
18. Long, E.R. and MacDonald, D.D. (1998). Recommended uses of empirically derived, sediment quality guidelines for marine and estuarine ecosystems. *Human and Ecological Risk Assessment* **4**(5): 1019–1039.
19. Lee, G.F. and Jones-Lee, A. (1993). Sediment Quality Criteria: Numeric Chemical vs. Biological Effects-Based Approaches. Proc. Water Environment Federation National Conference, Surface Water Quality & Ecology, pp. 389–400. Available: http://www.gfredlee.com/sedqualcri.html.
20. Lee, G.F. and Jones-Lee, A. (1996). 'Co-Occurrence' in Sediment Quality Assessment. Report of G. Fred Lee & Associates, El Macero, CA, February. Available: http://www.members.aol.com/apple27298/COOCCUR2PAP.pdf.
21. Lee, G.F. and Jones-Lee, A. (2004). Unreliability of Co-Occurrence-Based Sediment Quality 'Guidelines' for Evaluating the Water Quality Impacts of Sediment-Associated Chemicals. Report of G. Fred Lee & Associates, El Macero, CA, September. Available: http://www.gfredlee.com/psedqual2.htm.

22. O'Connor, T.R. (1999). Sediment quality guidelines do not guide. *Environ. Toxicol. Chem. SETAC News* **19**: 28–29.
23. O'Connor, T.R. (1999). Sediment Quality Guidelines Reply-to-Reply. NOAA National Status and Trends Program, May.
24. O'Connor, T.P. (2002). Empirical and theoretical short-comings of sediment-quality guidelines. In: *Handbook on Sediment Quality*. Whittemore et al. (Eds.). Special Publication, Water Environment Federation, Alexandria, VA, pp. 317–325.
25. O'Connor, T.P. and Paul, J.F. (2000). Misfit between Sediment Toxicity and Chemistry. *Mar. Poll. Bull.* **40**: 59–64.
26. O'Connor, T.P. et al. (1998). Comparisons of measured sediment toxicity with predictions based on chemical guidelines. *Environ. Toxicol. Chem.* **17**: 468–471.
27. Engler, R. (2004). U.S. Army Corps of Engineer Waterways Experiment Station, Vicksburg, MS, Personal Communication.
28. DiToro, D. (2002). Sediment Toxicity Prediction—What is Known and Unknown. Presented at Sediment Quality Assessment (SQA5), Aquatic Ecosystem Health and Management Society, Chicago, IL.
29. Chapman, P.M. (2004). 'Paracelsus' Dictum for Sediment Quality (and other) Assessments. Presented at the 2002 Fifth International Symposium on Sediment Quality Assessment (SQA5), In: *Aquatic Ecosystem Health and Management*. M. Munawar, (Ed.). Vol. 7, pp. 369–374.
30. Burton, G.A., Jr. (2002). Summary of Pellston Workshop on Use of Sediment Quality Guidelines (SQGs) and Related Tools for the Assessment of Contaminated Sediments. Presented at the 2002 Fifth International Symposium on Sediment Quality Assessment (SQA5), Chicago, IL.
31. Jones-Lee, A. and Lee, G.F. (1998). Evaluation Monitoring as an Alternative to Conventional Water Quality Monitoring for Water Quality Characterization/Management. Proc. of the NWQMC National Conference "Monitoring: Critical Foundations to Protect Our Waters." U.S. Environmental Protection Agency, Washington, DC, pp. 499–512. Available: http://www.gfredlee.com/wqchar_man.html.
32. U.S. EPA. (1991). *Methods for Aquatic Toxicity Identification Evaluations, Phase I Toxicity Characterization Procedures*, 2nd Edn. EPA/600/6-91/003, U.S. EPA Office of Research & Development, Washington, DC.
33. U.S. EPA. (1993). *Methods for Aquatic Toxicity Identification Evaluations, Phase II Toxicity Identification Procedures for Samples Exhibiting Acute and Chronic Toxicity*. EPA/600/R-92/080, U.S. EPA Office of Research & Development, Washington, DC.
34. Lee, G.F. and Jones-Lee, A. (2002). Issues in Developing a Water Quality Monitoring Program for Evaluation of the Water Quality/Beneficial-Use Impacts of Stormwater Runoff and Irrigation Water Discharges from Irrigated Agriculture in the Central Valley, CA. California Water Institute Report TP 02–07 to the California Water Resources Control Board/Central Valley Regional Water Quality Control Board, California State University Fresno, Fresno, CA, December. Available: http://www.gfredlee.com/Agwaivermonitoring-dec.pdf.
35. Lee, G.F. and Jones-Lee, A. (2004). Appropriate Incorporation of Chemical Information in a Best Professional Judgment 'Triad' Weight of Evidence Evaluation of Sediment Quality. Presented at the 2002 Fifth International Symposium on Sediment Quality Assessment (SQA5), In: *Aquatic Ecosystem Health and Management*. M. Munawar, (Ed.). Vol. 7, pp. 351–356. Available: http://www.gfredlee.com/BPJWOEpaper-pdf.

THE CONSTITUTION AND EARLY ATTEMPTS AT RATIONAL WATER PLANNING

Martin Reuss, Ph.D
Office of History Headquarters
U.S. Army Corps of Engineers

In the first decades of the American Republic, river and harbor bills met overwhelming resistance from the executive branch, mainly because of concerns over constitutionality, one more issue that reverberates through much of America's history. The specific points usually centered on presumed lack of constitutional authority to construct works of mainly local and even private benefit or works on rivers that were not clearly navigable. Pre-Civil War Presidents either practiced an inconsistent policy towards public works or became adamant opponents. Thomas Jefferson objected to federal involvement because it would empower the government at the expense of the common man, burden the taxpayer, lead to projects benefiting one location at the cost of another, and would enrich men at public expense. James Madison, who had been an early internal improvements advocate, on the evening before his departure from the White House vetoed the so-called Bonus Bill that would have provided funds for public works, declaring that the Constitution did not give Congress the power to appropriate money for or to construct public works without an "inadmissible latitude of construction." James Monroe at first thought that Congress could appropriate funds for public works but agreed with Madison that the federal government had no authority to construct the projects. Later, he determined that Congress might construct public works after all but only for those projects that were "national not state, general not local," a clarification that left the proverbial barn door open for defining local and general.

Andrew Jackson saw himself as a friend of internal improvements, but he feared the extension of federal power, sought a clarifying amendment on the appropriate national role in internal improvements, and admitted, like Monroe, that, while the federal government could appropriate money for truly national projects, it could not actually construct the projects itself. Strict constitutional constructionist James K. Polk vetoed every rivers and harbors bill sent to him. He even went to his office on the last day of his administration armed with a veto should Congress try to pass an internal improvements bill at the last moment. Abraham Lincoln, a young Whig congressman from Illinois, succinctly captured the problem in his denunciation of Polk's veto of the 1848 rivers and harbors act. "The just conclusion from all this is that, if the nation refuses to make improvements of the more general kind because their benefits might be somewhat local, a state may, for the same reason, refuse to make an improvement of a local kind because its benefits may be somewhat general. A state may well say to the nation, 'If you will do nothing for me I will do nothing for you.'" In the following decade, Franklin Pierce vetoed five rivers and harbors bills on the grounds of unconstitutionality. As Civil War erupted in the land, the political and philosophical jousting over water projects remained as short of resolution as ever.

Given the fears, hopes, and questions facing the early American republic, it is little wonder that it saw no successful implementation of coordinated public works administration. Perhaps more surprising is that this became a permanent condition in the United States. Funding issues, sectional friction, and constitutional questions invariably posed insurmountable barriers. Prior to the Civil War, the federal government attempted twice to develop and implement a national program of public works. The first was the well-known Gallatin Plan. At the request of Congress, Secretary of Treasury Albert Gallatin proposed in 1808, an ambitious network of roads and canals connecting the Eastern seaboard with the interior and a coastwise water route to shorten distances between major Atlantic seaboard cities. Gallatin argued that the federal government should construct internal improvements that provide "annual additional income to the nation" but are beyond the capacity of private entrepreneurs to build. His formulation harkens back to Adam Smith's The Wealth of Nations and anticipates the term "National Economic Development " that appeared in twentieth century economic jargon. However, his effort fell victim to lack of funds (both private and public), New England opposition to the Jefferson Administration, and, finally, growing preoccupation with real and apparent British threats to the United States, which eventually resulted in war. It is also of more than passing interest that Gallatin himself agreed with President Jefferson that his plan could never be efficiently realized without a constitutional amendment.

The next great attempt occurred in 1824. President's Monroe's vacillation, the growing clout of new states interested in waterborne commerce, and a favorable Supreme Court ruling (Gibbons v. Ogden) that sanctioned federal control over interstate commerce, including rivers, based on the Commerce Clause of the Constitution, allowed passage of the General Survey Act in 1824, after weeks of acrimonious debate. The act carried largely because of support from the Middle Atlantic states (except Delaware) and the new states west of the Appalachians. It authorized the President to use the Army Engineers to survey (not build) roads and canals (not rivers) that may be deemed "of national importance in a commercial or military point of view, or necessary to the transportation of the public mail."

Once the bill passed, Secretary of War John C. Calhoun organized a Board of Engineers for Internal Improvements to determine which routes should be surveyed among the scores suggested. Like Gallatin's plan, this program could have become the beginning of a great nationwide, coordinated system of internal improvements. Instead, once projects were surveyed, they became subject to the same parochialism in Congress that had doomed earlier, similar ideas, and congressmen continued to introduce pet projects for funding despite a contrary recommendation from the army engineers or the absence of a survey altogether. Its planning role severely diminished, the Board of Engineers languished, and a reorganization of the Corps of Engineers in 1830 provided an excuse for its abolishment. Six years later, Congress repealed the General Survey Act, partially a response to the legislature's own abuse of the act, using Army Engineers to survey potential projects of clear local and even private interest. Thus began a contest between rational administration and congressional politics that has remained unresolved and contentious to the present.

This article is based on Martin Reuss, "The Development of American Water Resources, Planners, Politicians, and Constitutional Interpretation," in Paul Slack and Julie Trottier, eds., Managing Water Resources, Past and Present: The Twelfth Annual Linacre Lectures. Oxford University Press, 2004. Forthcoming. Used by permission of Oxford University Press.

The Development of American Water Resources: Planners, Politicians, and Constitutional Interpretation.

The Constitution and Early Attempts at Rational Water Planning.

The Expansion of Federal Water Projects.

ECONOMIC VALUE OF WATER: ESTIMATION

Jorge Ramírez-Vallejo
Harvard University
Cambridge, Massachusetts

"Water has an economic value in all its competing uses and should be recognized as an economic good." The Dublin Statement, 1993.

MOTIVATION FOR VALUING WATER

The global awareness of the vital role water plays in sustainable national development is increasing. Water is vital for sustaining life on the earth. It is crucial for economic and social development, including energy production, agriculture, and domestic and industrial water supplies. Therefore, each unit of water should be used efficiently, equitably, and soundly. The economic value that water generates should be given due attention, when apportioning scarce water resources among competing uses.

Water is often regarded as a naturally occurring, freely available commodity. However, its supply, especially in an unpolluted state, is often limited compared to the demands placed upon it. Hence, water is both a constraint on and an essential input of economic activity, and decisions are required to allocate it among competing uses. Considerable resources are committed to delivering required standards of water services in adequate quantities and qualities. In addition, water-related environmental qualities, such as those associated with wetland regimes, are recognized as valuable to society. These characteristics of water imply that it can be regarded as an economic commodity—an economic good—whose use can be guided by economic principles (1).

Water, as other resources, has value whenever users are willing to pay a price for it rather than do without. The economic value and economic benefit of water can be considered synonymous. The benefits from this resource are derived from various uses. First, we find benefits of water from personal drinking, cooking, and sanitation, and from that those contribute to productive activities on farms and in businesses and industries. These uses are called *consumptive* uses. Second, there are *in-stream* water benefits from other uses such as hydroelectric power generation and waterways transportation. These uses are called *nonconsumptive* uses. The third type of value of water comes from the economic benefits that water has in recreation, aesthetics, and fish and wildlife habitat. As income and leisure time grow, these benefits are becoming more and more important. Fourth, there are the economic benefits of waste disposal using water as a sink for carrying away a wide range of residuals from the processes of human production and consumption. Water is used then to dilute them and in some instances to transform them into less undesirable form. Fifth, we find the *dis-values* of water that come from floods and negative water quality impacts. Finally, we have the *nonuse* value of water that comes from people who are willing to pay for environmental services that might neither be used nor experienced in the future.

Economic valuation is a process of expressing preferences in a money metric. But, why bother valuing water anyway? This is the obvious question that comes to mind when approaching this subject. The key answer to this question is simple but fundamental: for the future well-being of humanity. The longer we ignore or distort water's value, the more overused and misallocated the resource becomes. Without price signals or other indicators of value to help guide policy, we tend to manage the resource blindly.

Estimates of the economic value are key for management decisions on allocation. Economic valuation is also important because it aids in determining if people want proposed projects and helps in estimating the degree to which they are willing to pay for benefits. When public budgets are constrained, this information becomes important for making sound investment decisions. For policy decisions, water values are also important to set the levels that the various users should pay for this resource.

Estimating the economic value of water helps in decisions for optimal allocation of the resource among various uses. Limited supply and increasing demand for water are common around the world. Under these circumstances, the economic value of water for each of its competing uses helps the decision maker decide how much water to allocate to each sector or use, and also helps determine the use of water in time, how much to use now and how much to save for the future.

In 1992, experts from 100 countries, intergovernmental organizations, and NGOs gathered in Dublin for the International Conference on Water and the Environment. The conference influenced preparations for the UN Conference on Environment and Development, held later that year in Rio de Janeiro.[1] Its brief concluding statement remains an authoritative affirmation of the water policy agenda; it contained four "guiding principles." The most important principle from this statement that has generated a great deal of controversy has to do with the consideration that water has an economic value in all its competing uses and should be recognized as an economic good. However, the statement recognizes the basic right of all human beings to have access to clean water and sanitation at an affordable price (see Appendix 1).

ECONOMIC CONCEPTS

To understand better the methodologies for estimating the economic value of water described below, it is important to review two key economic concepts (2).

The Demand Curve and the Price Elasticity of Water Demand

When we talk about willingness to pay for value, we mean the maximum amount that a individual is willing to pay for a product. This is the sum of the product price (P) and the consumer surplus. If we wish to estimate the maximum that the individual is willing to pay for a proposed change, we should estimate the demand curve.

A consumer's willingness to pay for an increment of supply is the corresponding area under the demand curve (price vs. quantity), although the amount the consumer actually pays for the increment is the water price times the quantity. The difference between these two values is what it is called the consumer surplus (the benefit for the consumer that he/she did not pay for it).

Assuming that the price elasticity of demand (ε) is known, is constant over the relevant range, and is not equal to one, and second, that the relevant initial price P_1 and quantity Q_1 can be specified, then the area under the curve (i.e., the desired measure of value—denoted V) for a change in quantity Q_1 to Q_2 is given by

$$V = [(P_1 {Q_1}^{1/\varepsilon})/(1 - 1/\varepsilon)][(Q_1/Q_1^{1/\varepsilon}) - (Q_2/Q_2^{1/\varepsilon})]$$

This expression represents the value of an increment of quantity Q to the final user. This incremental value will provide a measure of benefits to compare with costs in an investment project evaluation.

[1]Awareness of the seriousness of water problems has evolved gradually over the years, and important recommendations have been formulated since the United Nations Water Conference in Mar del Plata (Argentina) in 1977. Other events that have focused on water issues include the 1992 International Conference on Water and the Environment, convened by the World Meteorological Organization (WMO) in Dublin, Ireland, and the 1992 United Nations Conference on Environment and Development (UNCED) in Rio de Janeiro. The Ministerial Conference on Drinking Water and Environmental Sanitation held in Noordwijk in March 1994 adopted an action program. In 1994, the United Nations Commission on Sustainable Development requested a comprehensive assessment of the world's freshwater resources, including projections of future needs the results of the study were considered by the 1997 special session of the UN General Assembly reviewing the implementation of Agenda 21.

Estimates of price elasticities of water demand have shown that water use is consistently price-inelastic when there is no substitute, whereas residual use (i.e., watering the garden) is more elastic and differs significantly across geographical regions and between climate seasons (i.e., winter–summer, rainy–dry) (3).

The price responsiveness of water demand reflects the heterogeneity of demand components. A small portion of the demand is for uses for which there are no substitutes and that are of great necessity to the consumer; other uses are of lesser value and have ready substitutes.

Treatment of Time

The rate of conversion of value between time periods is called the rate of time of preference. Defining the best long-term management of the resource requires balancing the needs of the present and those of the future. Using a discount rate that adequately considers the value of goods or services for future generations creates this balance. A high discount rate implicitly places a low value on water's value to future generations. A low discount rate implies the opposite.

Another issue related to time is the difference between long-run and short-run values of water. The distinction between these two relates to the fixity of certain inputs necessary to make water available for consumption, such as supply infrastructure. Other things remaining constant, we would expect that values estimated for short-run contexts would be larger than values for the long run. Price elasticity of demand is less in the short run than in the long run when adjustments to shortages and infrastructure can be made. Accordingly, willingness to pay in the short run is usually higher than in the long run.

ESTIMATION METHODOLOGIES

The principles for valuing natural assets such as energy and mineral deposits, forests, and aquifers were set forth more than 60 years ago, but the principles for measuring consumptive value of water for household use were set forth a century ago.

Water services in many situations are difficult to value because much of the information needed for valuation is not readily available. When markets exist, they automatically generate useful information, but unfortunately, most of the services provided by water are not traded on markets. Therefore, one has to rely on techniques to value nonmarket goods.

Economists have developed indirect methods (e.g., hedonic price models, travel cost method, averting behavior), which infer values from other behaviors associated with the good. The strengths of indirect methods come from the fact that they rely on observed behaviors of producers and consumers. Examples are the market price as result of human behavior (demand and supply) as a function of the quality of the water nearby. Recent summaries of indirect approaches can be found in Braden and Kolstad (4), Mendelsohn and Markstrom (5), Smith (6,7), and Freeman (8).

The valuation method must be well matched to the context and the water function or service of interest. It is hard to generalize in the abstract about the validity and reliability of specific valuation approaches. The validity of the methodology used depends on the valuation context and the type of service that are of interest. Different approaches are needed to value different services. A valuation framework takes into account how time, institutions, water quality and quantity, hydrologic factors, and services interact to affect the resource's value.

One of the major challenges in valuing water resources is how to integrate the hydrologic and physical components into a valuation scheme. Valuation of the extractive and onsite services of water requires understanding the hydrology and ecology of the water source. Hydrologic information includes variables such as rainfall, runoff, infiltration, water balance, depth to groundwater, aquifer characteristics, and water quality.

DIRECT VALUATION APPROACHES

Market Prices

When market prices exist, the equilibrium price represents the willingness of potential buyers to pay for the good or service of water. In other words, when water has a market and it can be bought and sold, then the economic value of water is its price. However, in some cases due to government intervention such as subsidies or market failures, these prices are distorted and need to be adjusted to obtain the real economic value of the resource.

Transactions between a buyer and a seller are increasingly observed when water rights are defined and other conditions exist that make the water market function. Easter and Hearne (9) and Rosegrant and Binswanger (10) describe examples of water rights markets. If there is a market for perpetual water rights, the value of water can be estimated by using the perpetuity formula $V = A/r$ where r is the interest rate (discount rate), V the value of the perpetual water right, and A the value of water.

Contingent Valuation Method

Economic value is not a fixed, inherent attribute of a good or service but rather depends on time, circumstances, and individual preferences. One technique that attempts to measure total value is the contingent valuation method (CVM). CVM values are elicited directly from individuals via interviews or questionnaires in the form of statements of maximum willingness to pay (WTP) or minimum willingness to accept compensation (WTA) for hypothetical changes in environmental goods, such as water quantity or quality. However, there are numerous methodological controversies associated with the application of CVM, including the way the hypothetical water change (in quantity or quality) is to be specified, the elicitation format for asking valuation questions, the appropriate measure to be elicited, and various types of response biases. CVM provides reliable estimates of value when an individual has a close connection to the resource being valued.

This methodology is used when no value measures can be derived from observing individual choices through a market, such as water quality issues. This approach can potentially measure the economic benefits (or damages) of a wide assortment of beneficial (or adverse) effects in a way that is consistent with economic theory (11).

The CVM method is an ex-ante technique that can be used to measure the value of water without the need for observable behavior (data). In addition, this is the only method (direct or indirect) that measures existent or bequest values (use and nonuse values). It is an expensive method because of the need to carry out a survey, and there are concerns about its use for nonuse value applications.

Using this approach, respondents of surveys are offered conditions simulating a hypothetical market in which they are asked to express willingness to pay for existing or potential environmental conditions not registered on any market. The contingent valuation method (CVM) derives estimates for social benefits attributable to water, when considered a public or nonmarket good. This method is used frequently for projects or programs that affect water quality. The procedural steps for applying the CVM are as follows (12):

The first component describes the setting under which the respondent is to imagine herself/himself. Second are the choice questions, which will be used to infer values of the amenity or policy change. The third element asks questions about the respondent to identify demand shift variables such as age, education, income, and gender. For the forms of questions used and the way to design the survey instrument in CVM studies, see Mitchell and Carson (11) and Cummings et al. (12).

The CVM approach, given its hypothetical attributes, is susceptible to various sources of errors or biases. Some of them are (1) the questionnaire scenario can encourage strategic behavior, (2) the compliance bias or a tendency to fit the responses to the perceived preferences of the interviewer, (3) the possibility that the scenario provides implied values cues to the respondent, and (4) scenario misspecification when the respondent fails to understand the scenario intended by the researcher. These biases demand a careful questionnaire design.

The willingness to pay is estimated by using the following expression:

$$\mathrm{WTP} = f(\Delta Q_j, \Delta P_j, \Delta \mathrm{SP}_j/I_j, H_j)$$

where ΔQ_j defines a specific change in water quality, ΔP_j represents the change in price of water from primary supplies, ΔSP_j represents the change in price of substitute sources of water, I_j represents household income for individual j net of taxes, and Hj represents various nonincome characteristics of individual j's household.

INDIRECT VALUATION APPROACHES

Derived Demand

Water demand (households, farms, etc.) tends to be very site-specific and is influenced by a range of natural and socioeconomic factors. The demand relationship is represented algebraically by the familiar function,

$$Q_w = Q_w(P_w, P_a, P; Y; \mathbf{Z})$$

where Q_w refers to the individual's level of water consumption in a specified time period, P_w refers to the price of water, P_a denotes the price of an alternative water source, P refers to an average price index representing all other goods and services, Y is the consumer's income, and $\mathbf{Z}$ is a vector representing other factors such as climate and consumer preferences (13). This demand curve is usually estimated by using cross-sectional observations (data taken for the same time period) from a sample of water service agencies which exhibit a range of water rates and often variation in other factors influencing demand, such as income, rainfall, and temperature. The general approach is to apply regression analysis to estimate the parameters of the demand equation, parameters that allow us to estimate the price elasticity of demand.

The gross value of water is derived by integrating (area under the curve) the inverse of the econometrically derived demand function, $P(Q_w)$, over the appropriate increment in quantity. If the demand function was estimated for tap water and one is looking for the consumer demand function for raw water for intersectional allocation, then an additional step is necessary. The costs of capturing, transporting, treating, and storing water must be deducted from the estimated willingness to pay for tap water. When tap water (domestic) is priced to recover fully the costs of supplying it, the average revenue can be subtracted from the value of water estimated before. See Booker and Young (14) for an application of this point.

The derived demand approach is usually used to estimate the value of water in the municipal environment. Its main advantages are that it is based on observable data from firms/households using water and is relatively inexpensive. However, using this approach, it is not possible to measure the nonuse value of water and, sometimes, understates the willingness to pay for the resource (2).

The approach described above does not capture the benefit of loss in value due to the reliability dimension (the probability of having an interrupted service). Some other methodologies such as the contingent valuation method have been used in the past to estimate the reliability attribute of water supply. See, for example, Carson (15), Altaf et al. (16), and Howe and Smith (17).

Valuing Water as an Intermediate Good

Water is commonly used for producing goods; it is not a final product itself. Agricultural irrigation, industry, and hydroelectric or thermal power generation are examples of these uses. For intermediate goods, the theory of producer's demand for inputs provides the conceptual basis for valuing the economic value of water (18). The producer's demand for an input is its value as a marginal product (VMP); therefore measuring the economic value of water requires an approximation of the VMP. Three methods of valuing water as an intermediate good are in common use: the residual approach, the hedonic approach, and the alternative cost approach.

The Residual Approach. The "residual" method is the most frequently used approach for estimating the price of intermediate goods, particularly of irrigation water. Under this method, the incremental contribution of each input in the production process is determined. If appropriate prices can be assigned by market forces to all inputs but one, water, for example, the remainder of total value of product is imputed to the remaining residual input. In algebraic terms,

$$P_w = [\text{TVPy} - [(P_k Q_k) + (P_l Q_l) + (P_R Q_R)]]/Q_w$$

where P_w and Q_w are the price and quantity of water, respectively, TVP_y is the total value product, and the parenthesis terms are the price multiplied by the quantity of each of the other inputs in the production process (K, L, R).

This technique is extremely sensitive to small variations in the assumptions about the nature of the production function and about prices. Biases can be introduced by any of several types of errors and omissions. Therefore, this approach is most suitable when water contributes significantly to the value of the output (agricultural production, hydropower generation).

Sometimes, mathematical programming (optimization) models are used to estimate the value of water using the residual approach. These models are built to obtain the optimum allocation of water and other resources (inputs) so as to maximize profits, subject to constraints on resource availability and institutional capacity (19). Mathematical programming is advantageous when a wide range of technological options is to be studied.

The Alternative Cost Approach. This method is also appropriate for measuring the value of water when water is an intermediate good. The technique is attractive under the assumption that, if a given project of specified output costs less than the next best public or private project which can achieve the same output, then the cost of the next best project can be assigned as the benefit to the public project in consideration. This approach has been employed for valuing of water in municipal, industrial, hydroelectric, and thermal electric power projects. Steiner (20) remains the primary authority on this approach. See also Herfindahl and Kneese (21).

Averting Behavior Model

Using this method, water value is estimated by the actions taken to avoid or reduce damage from exposure to water contaminants, such as changing daily routines to avoid exposure. However, averting expenditures and true reduction of pollution differ because such expenditures do not measure all the costs related to pollution that affect the consumer's utility. Therefore, the ability of this valuation approach to provide a lower bound to willingness to pay depends on various assumptions. A list of these conditions is explained in detailed in the literature (22).

The averting behavior technique is based on observable data from actual behavior and choices and is relatively inexpensive. It provides a lower bound of the willingness to pay if certain assumptions are met. However, its estimates do not capture full losses from environmental degradation, and several key assumptions must be met to obtain reliable estimates. It is usually used for expost analysis and is limited to current situations; as in the derived-demand approach, it does not estimate nonuse values.

Hedonic Price Model

This pricing method is based on the premise that people value a good because of the attributes of that good rather than the good itself. These models encompass both land (housing) price models and wage models that account for variations due to environmental attributes. Therefore, by regressing housing prices to the attributes of a house and site, including environmental characteristics, one obtains the marginal value of the characteristic q. In the estimating equation,

$$P = F(\mathbf{S}, \mathbf{N}, \mathbf{Q}) + \varepsilon$$

P is the sale price of the asset, $\mathbf{S}$ is a vector of structural characteristics, $\mathbf{N}$ is a vector of neighborhood characteristics, $\mathbf{Q}$ represents the environmental attribute(s) of interest, F is a function to be estimated by multiple-regression methods, and ε is an error term. $\partial P/\partial Q$ is the marginal value of the characteristic q or the amount of value gained or lost due to a defined change in the environment.

Estimating the economic value of environmental resources by hedonic methods is quite difficult in practice, and the technique is subject to serious limitations (8). First, a large enough sample might be difficult to obtain. Second, if water resources are already in public ownership, market transactions may not be available. Third, sometimes the effect of an environmental attribute or characteristic of price may be small and hard to detect statistically or to disentangle from the effects of all other variables. See a detailed description and examples in Palmquist (23) and McConnell (24).

Travel Cost Method

The travel cost is the most widely used example of the observed indirect methods. It involves two steps: the first is to estimate the individual recreationist's demand for the resource, and the second is to derive statistically the relevant aggregate resource demand curve. Travel cost methods encompass a variety of models, ranging from the simple single-site travel cost model to regional and generalized models that incorporate quality indexes and account for substitution across sites. The basic premise behind these models is that the travel cost incurred in traveling to a site can be regarded as the price of access to the site. Therefore, under some assumptions, it is possible to derive the demand for visits to a site as a function of the price of admission, which is set equivalent to the cost of travel to the site.

This method is based on observable data from actual behavior and choices and is relatively inexpensive. However, sometimes is not easy to observe consumer behavior regarding the choice of recreation sites, and it is limited to the water uses that include travel such as recreation. It does not measure nonuse values.

The major drawback is that sometimes it is difficult to isolate the contribution of water to the attractiveness of a site. In other words, there are some other attributes that contribute to the total site value and are difficult to isolate. Another problem in using this approach is to decide what to include when estimating the cost of travel (cost of gas, oil, tires, etc), and whether or not the opportunity cost of the recreationist's time should be included in the estimate. A detailed explanation of this approach and real world examples are in Walsh (25), McConnell (24), Freeman (8), and Fletcher et al. (26).

Benefit Transfer and Meta-Analysis

The process of benefit transfer consists of adapting estimates relating to sites that have been studied to sites lacking such studies. It is an approach that consists of pooling data from already completed studies and then applying multiple regression analysis using explanatory variables that are site specific and methodology specific. For a detailed description, see Walsh et al. (27). A similar approach but using research findings is called meta-analysis; it has been used for estimating the benefits of natural resources. See, for example, Smith and Kaoru (28).

AN APPLICATION OF VALUATION

Quality of Water

Valuing the quality of water usually requires establishing its relationship to the health conditions of its users. In the past, various methodologies have been used to measure the economic value of changes in health benefits or risks from a change in water quality levels. For this purpose, they have used various tools or techniques such as water demand functions, averting or defensive expenditures, changes in production costs, the hedonic price method, or stated preference methods such as the contingent valuation method or conjoint analysis (29).

Another service supported by water quality is the provision of drinking water for livestock and irrigation. A change in the quality of water used for these purposes may result in a change in the value of livestock or crops, a change in agricultural production costs, or a change in human health or health risks by increasing or decreasing the trace chemicals found on plants ingested by people. The change in the value of crops can be estimated using a market demand function. The change in production costs can be estimated using cost functions. For changes in human health and health risks, one could use the damage avoided approach and stated preference methods. Similar methods should be used for the eventual changes in the value of crops from a commercial fish harvest, commercial hunting harvest, or commercial plant harvest.

A change in surface or groundwater quality can also have an impact on the quantity or quality of recreational services such as fishing, hunting, or just enjoying the scenery. Potential techniques for valuing these effects include the travel cost method and stated preference methods.

On the other hand, the effects of changes in nonuse or passive use services are some of the most difficult effect to value monetarily. A change in nonuse or passive use services results in a direct change in personal utility or satisfaction that may be reflected in the decision on consumption of water. The most viable techniques for measuring these effects are stated preference methods (11,15).

A comprehensive, rigorous and useful discussion of the theory and practice of valuing environmental resources is given in Freeman (8). See also Anderson and Bishop (30), Bentkover et al (31), Johansson (32), Braden and Kolstad (4), and Pearce and Warford (33).

BIBLIOGRAPHY

1. Young, R.A. and Gray, S.L. (1972). *Economic Value of Water: Concepts and Empirical Estimates*. National Water Commission, Washington, DC.
2. National Research Council. (1997). *Committee on Valuing Ground Water, Valuing Ground Water: Economic Concepts and Approaches*. National Academy Press, Washington, DC.
3. Gibbons, D. (1986). *The Economic Value of Water*. Resources for the Future, Washington, DC.
4. Braden, J.B. and Kolstad, C.D. (1991). *Measuring the Demand for Environmental Quality*. Elsevier Science, Amsterdam.
5. Mendelsohn, R. and Markstrom, D.C. (1988). The use of travel cost and hedonic methods in assessing environmental benefits. In: *Amenity Resource Valuation: Integrating Economics with Other Disciplines*. G.L. Peterson, B.L. Driver, and R. Gregory (Eds.). Venture Publishing, State College, PA.
6. Smith, V.K. (1989). Taking stock of progress with travel cost recreation demand methods: theory and implementation. *Mar. Resour. Econ.* **6**: 279–310.
7. Smith, V.K. (1993). Nonmarket valuation of environmental resources: an interpretive appraisal. *Land Econ.* **69**: 1–26.
8. Freeman, A.M., III. (1993). *The Measurement of Environmental and Resource Values: Theory and Methods*. Resources for the Future Press, Washington, DC.
9. Easter, K.W. and Hearne, R. (1995). Water markets and decentralized water resource management: international problems and opportunities. *Water Resour. Bull.* **31**(1): 9–20.
10. Rosegrant, M.W. and Binswanger, H.P. (1995). Markets in tradeable water rights: potential for efficiency gains in developing countries. *World Dev.* **22**(11): 1613–1621.
11. Mitchell, R.C. and Carson, R.T. (1989). *Using Surveys to Value Public Goods: The Contingent Valuation Method*. Johns Hopkins University Press for Resources for the Future, Baltimore, MD.
12. Cummings, R.G., Brookshire, D.S., and Schulze, W.D. (1986). *Valuing Environmental Goods: A State of the Art Assessment of the Contingent Valuation Method*. Rowman and Allanheld, Totawa, NJ.
13. Munasinghe, M. (1992). *Water Supply and Environmental Management: Developing World Applications*. Westview Press, Boulder, CO.
14. Booker, J.F. and Young, R.A. (1985). Modelling Intrastate and Interestate Markets for Colorado River Water Resources. *Environ. Econ. Manage.* **26**(1): 66–87.
15. Carson, R.T. (1991). Constructed markets. In: *Measuring the Demand for Environmental Quality*. J.B. Braden and C.D. Kolstad (Eds.). Elsevier Science, Amsterdam.
16. Altaf, M.A., Whittington, D., Jamal, H., Liu, J.L., and Smith, V.K. (1993). Rethinking rural water supply policy in the Punjab Pakistan. *Water Resour. Res.* **29**(7): 1943–1954.

17. Howe, C.W. and Smith, M.G. (1994). The value of water supply reliability in urban water systems. *Environ. Econ. Manage.* **26**(1): 19–30.
18. Just, R.E., Hueth, D.L., and Schmitz, A. (1982). *Applied Welfare Economics and Public Policy*. Prentice-Hall, Englewood Cliffs, NJ.
19. Williams, H.P. (1990). *Model Building in Mathematical Programming*, 3rd Edn. John Wiley & Sons, New York.
20. Steiner, P.O. (1965). The role of alternative cost in project design and selection. *Q. J. Econ.* **79**(3) August.
21. Herfindahl, O.C. and Kneese, A.V. (1974). *Economic Theory of Natural Resources*. Merrill, Columbus, OH.
22. Abdala, C.W., Roack, B.A., and Epp, D.J. (1992). Valuing environmental quality changes using averting expenditures: An application to ground water contamination. *Land Econ.* **68**(2): 163–169.
23. Palmquist, R.B. (1991). Hedonic methods. In: *Measuring the Demand for Environmental Quality*. J.B. Braden and C.D. Kolstad (Eds.). Elsevier Science Publishers, Amsterdam.
24. McConnell, K.E. (1993). Indirect Methods of Natural Resource Damage Assessment. In: *Valuing Natural Assets: The Economics of Natural Resource Damage Assessment*. R.J. Kopp and V.K. Smith (Eds.). Resources for the Future, Washington, DC.
25. Walsh, R.G. (1986). *Recreation Economic Decisions: Comparing Benefits and Costs*. Vbenture Publishing, State College, PA.
26. Fletcher, J.J., Adamovicz, V.L., and Graham-Tomasi, T. (1990). The travel cost model of recreational demand: theoretical and empirical issues. *Leisure Sci.* **12**: 119–147.
27. Walsh, R.G., Johnson, D.M., and McKean, J.R. (1992). *Benefit Transfer of Outdoor Recreation Demand Studies*. 1968–1988. Water Resources Research Vol. 28.
28. Smith, V.K. and Kaoru, Y. (1990). Signals or noise? Explaining the variation in recreational benefit estimates. *Am. J. Agric. Econ.* **71**(2): 419–433.
29. Bergwstrom, J.C., Boyle, K.J., and Poe, G.L. (2001). *The Economic Value of Water Quality, New Horizons in Environmental Economics*. Edward Elgar, Northampton, MA.
30. Anderson, G.D. and Bishop, R.C. (1986). The valuation problem. In: *Natural Resources Economics: Policy Problems and Contemporary Analysis*. D.W. Bromley (Ed.). Kluwer Nijhoff, Boston, pp. 89–137.
31. Bentkover, J.D., Covello, V.T., and Munpower, J. (Eds.). (1999). *Benefit Assessment: The State of the Art*. Reidel, Dordrecht, the Netherlands.
32. Johansson, P-O. (1987). *The Economic Theory and Measurement of Environmental Benefits*. Cambridge University Press, Cambridge, UK.
33. Pearce, D.W. and Warford, J.J. (1993). *World Without End: Economics, Environment and Sustainable Development*. Oxford University Press (for the World Bank), New York.
34. Young, R.A. (1996). *Measuring Economic Benefits for Water Investment and Policies*. World Bank Technical Report No 33B, Washington, DC.

APPENDIX 1. THE DUBLIN STATEMENT

In commending this Dublin Statement to the world leaders assembled at the United Nations Conference on Environment and Development (UNCED) in Rio de Janeiro in June 1992, the Conference participants urge all governments to study carefully the specific activities and means of implementation recommended in the Conference Report, and to translate those recommendations into urgent action programmes for water and sustainable development.

Guiding Principles

Concerted action is needed to reverse the present trends of overconsumption, pollution, and rising threats from drought and floods. The Conference Report sets out recommendations for action at local, national and international levels, based on four guiding principles.

Principle No. 1. Fresh water is a finite and vulnerable resource, essential to sustain life, development and the environment

Since water sustains life, effective management of water resources demands a holistic approach, linking social and economic development with protection of natural ecosystems. Effective management links land and water uses across the whole of a catchment area or groundwater aquifer.

Principle No. 2. Water development and management should be based on a participatory approach, involving users, planners and policy-makers at all levels

The participatory approach involves raising awareness of the importance of water among policy-makers and the general public. It means that decisions are taken at the lowest appropriate level, with full public consultation and involvement of users in the planning and implementation of water projects.

Principle No. 3. Women play a central part in the provision, management and safeguarding of water

This pivotal role of women as providers and users of water and guardians of the living environment has seldom been reflected in institutional arrangements for the development and management of water resources. Acceptance and implementation of this principle requires positive policies to address women's specific needs and to equip and empower women to participate at all levels in water resources programmes, including decision-making and implementation, in ways defined by them.

Principle No. 4. Water has an economic value in all its competing uses and should be recognized as an economic good

Within this principle, it is vital to recognize first the basic right of all human beings to have access to clean water and sanitation at an affordable price. Past failure to recognize the economic value of water has led to wasteful and environmentally damaging uses of the resource. Managing water as an economic good is an important way of achieving efficient and equitable use, and of encouraging conservation and protection of water resources.

Acknowledgments
It must be acknowledged that this chapter owes a great deal in both substance and form to the paper written by Robert A Young

in 1996 for the World Bank based on the pioneering work by him and S. Lee Gray in 1972 (1,34).

WATER SUPPLY PLANNING—FEDERAL

David L. Feldman
University of Tennessee
Knoxville, Tennessee

The U.S. government has been engaged in water supply planning for domestic, agricultural, commercial, and industrial needs from its inception. Planning was initially undertaken for inland navigation and flood control. Later, primacy was given to irrigation, municipal needs, and hydroelectricity. The Army Corps of Engineers, the largest and most important federal water supply planning entity, was initially given responsibility for all these efforts. In 1902, the Bureau of Reclamation (BOR) was established and given principal responsibility for water supply projects in the West.

GENERAL PLANNING ISSUES

Federal water supply planning has been characterized by recurring policy debates on four issues:

- the value of *structural* measures (e.g., dams, impoundments) versus *nonstructural* measures such as conservation and demand-side management;
- the primacy to be accorded the *philosophies and missions* of competing agencies (e.g., the Corps, BOR) in the planning process;
- the advantages of *comprehensive, conjoint coordination* by a single entity (e.g., Interior Department, Water Resources Council), facilitated by state cost-sharing; and
- the appropriate role of *public participation*, particularly inclusion of environmental NGOs in plan formulation.

These issues have frequently arisen in discussions over specific plans. In the 1940s, for example, the Pick–Sloan Missouri River project, one of the largest federal water supply planning efforts, was conceived as a compromise between the Corps and BOR, which shared responsibility for water supply in the upper Missouri basin though they had different priorities. The BOR envisioned irrigation as the primary objective of main-stem dam construction, and the Corps gave primacy to flood control and hydropower. When these agencies failed to reconcile priorities, Congress opted to allow each to develop separate plans which would later be integrated by a series of review boards charged with managing and overseeing reservoir operations (1).

HISTORY OF FEDERAL WATER SUPPLY PLANNING EFFORTS

For much of the nineteenth century, water supply planning meant the development by states and communities of engineering-based solutions to meet specific stakeholder objectives such as flood control, municipal supply, or water treatment (2). During the twentieth century planning embraced "multiple objectives" undertaken by *all* levels of government. It is now generally conceived as a comprehensive effort to harmonize environmental, economic, financial, and other concerns (3).

Five distinct periods in the evolution of federal water supply planning are discernible: (1) a formative period consisting of federal intervention in water supply in response to specific regional needs; (2) a federal–state cooperative planning era, following the Civil War, whose focus was mostly navigational improvements; (3) a "multiprogram" era, beginning in the 1930s, characterized by establishment of national priorities; (4) an effort to "federalize" water supply planning through greater presidential-level coordination; and (5) a new federal–state partnership coupled with greater sensitivity to environmental protection and restoration needs.

Formative Developments

The earliest federal water supply planning efforts coincided with exploratory surveys of the intermountain West. Led in many cases by Army engineers, one of these expeditions was intended to identify and trace the origins and suitability for development of water supply sources. The journeys of Lewis and Clark (1804–05) and Zebulon Pike (1806–07) in the upper Missouri and Arkansas River basins, respectively, are the best known of these; Stephen H. Long's Central Rocky Mountain-Yellowstone expedition (1819–20) also surveyed regional water resources.

In 1824, the Corps was officially designated as the steward of the nation's water resources. Between this time and the Civil War, the Corps identified mill sites and needs for navigational improvements and engaged in channel widening and other flood hazard mitigation activities (4,5). These efforts generally produced site-specific solutions to local problems. Despite fierce engineering debates, particularly over proposals to harness the Mississippi River for navigation and flood prevention (6), they failed to set broad, consistent development priorities.

The Advent of Federal–State Cooperative Planning

The origins of a more comprehensive approach to water supply planning may be traced to 1879, when explorer and geographer John Wesley Powell decried the system of Western water law, charging that by encouraging the first users of a river to lay claim to its entire flow (the law of prior appropriation), harm is imposed on other users and the in-stream environment. As first director of the U.S. Geological Survey, Powell proposed dividing the West's arid lands into irrigation districts where each farmer was allotted 80 acres. These districts would be drawn to conform to watershed topography, and drainage basins would serve as state boundaries (7,8).

Powell's proposals discouraged Western migration and drew the ire of Congress and Western land development interests. Ironically, this adverse reaction partly hastened federal efforts to initiate large-scale reclamation projects designed, principally, to support Western agriculture.

These projects would be justified (and subsidized) by their flood control and hydroelectric power benefits—a pattern which would characterize federal water supply planning until the late twentieth century. In exchange for broad political support, a reciprocal system of large-scale, publically financed water projects was eventually provided to Eastern states, principally for navigation, power, and flood abatement (9–13).

In 1879 the Mississippi River Commission (MRC) was also established, one of the first efforts to coordinate federal–state water supply planning efforts in a single basin. MRC was formed to deepen and widen the river's navigable channel and provide flood abatement. Limited efforts to accelerate intergovernmental coordination efforts elsewhere occurred under the presidency of Theodore Roosevelt who "was interested in securing state cooperation and assistance in his program" to develop comprehensive management of the nation's rivers (14). However, these efforts were largely unidirectional in their scope: federal construction agencies that had greater fiscal and technical resources continued to make most decisions. Another goal of Teddy Roosevelt—reducing the power of private interests to exploit the development of water resources for their own gain—was not seriously tackled until the 1930s.

Multiprogram Planning and the Great Depression

The first multipurpose federal water supply planning efforts began when the Tennessee Valley Authority (TVA) was established in 1933. The TVA was granted broad power over a seven-state area to engage in regional water supply planning, flood control, hydropower production, navigation, and economic development. In President Franklin Roosevelt's words, it was to be a "... corporation clothed with the power of government but possessed with the flexibility and initiative of private enterprise" (15). Local entities were incorporated into its economic development programs, and long-term political alliances were built to ensure programmatic and funding stability. Most importantly, TVA "supplanted" other federal water resource agencies in its operating region, and the absence of large private power companies also helped it achieve comprehensive breadth (16).

The political conditions that permitted TVA to do all these things were uniquely characteristic of American politics during the 1930s. Some contend that these conditions may no longer exist, exemplified by the fact that later federal efforts to develop TVA-like entities in the Missouri and Columbia basins failed to gain support in Congress and among local elites (13,15). In short, no TVA-like entity has been replicated anywhere else in the United States.

In other regions, FDR's Interior Secretary Harold Ickes formed basin study groups with formal staffs, planning funds, and some powers of program review during this period. These groups had little formal access to Congress and lacked the ability to initiate new projects or overrule decisions of established water agencies regarding what they often deemed as uneconomical or environmentally damaging policies. They also had no control over planning objectives, which continued to be left to the Corps and other "construction agencies" (17).

A parallel, noteworthy development in this period was the federal interstate compact. The Colorado River Compact, one of the oldest of these, allocates both claimed and unused waters in a seven-state basin based on future needs and water-use priorities. The compact entered into force in 1922, but Arizona, due to resentment over what it considered to be an inequitable division of its share of the lower Colorado, refused to ratify it until 1944. When it did so, it tried to use its ratification to leverage federal funds to build the long-coveted Central Arizona Project, designed to divert Colorado River water to Phoenix, Tucson, and surrounding rural communities via aqueduct (18–20).

Although interstate compacts vary in intent, scope of activity, and structure, all derive their legal authority from the compact clause of the U.S. Constitution (Art. I, Sec. 10, clause 3) which mandates Congress's consent when states enter into binding agreements with one another. This endows them with rights to embrace public as well as private facilities and projects within their jurisdiction. It also binds the actions of signatory states *and* federal agencies and authorizes allocation of interstate waters in accordance with the doctrine of equitable apportionment (i.e., in ways that do not disturb or impair rights awarded to parties under Supreme Court decree without their consent). In effect, their decisions have the force of national law.

During the 1960s, in response to deteriorating water quality and droughts in the Northeast, compact commissions were established for the Delaware (DRBC), Susquehanna (SRBC), and Potomac River (PRBC) basins. These were empowered to allocate interstate waters (including groundwater and interbasin diversions), regulate water quality, and manage interstate bridges and ports. The DRBC, as a forerunner of this model, included numerous federal partners such as Interior Department and Corps of Engineers officials (21,22).

Since 1997, most compacts have continued to function without federal appropriations as a result of Congressional adoption of Heritage Foundation recommendations which urged that Congress "defund" programs whose benefits were viewed as more regional than national in scope. The DRBC and SRBC have sustained themselves without federal appropriations, and Congress has not withdrawn from any compact. Also in 1997, the DRBC's Executive Director advised officials of Alabama, Florida, and Georgia during formulation of two new compacts in that region (the ACF-ACT compacts), and the DRBC and SRBC were explicitly adopted as models by these compacts' framers (21).

"Federalizing" Water Supply Planning: The Water Resources Council and Other Efforts

In the early 1960s, Congress embarked on an effort to provide greater coordination of water supply planning. The Water Resources Council (WRC) was established in 1961 to hasten a more holistic approach to planning by encouraging interagency cooperation and integrating water quality and quantity concerns. It was initially composed of representatives from the Departments of Agriculture, Army, Commerce, and Interior.

In 1965, WRC was empowered by the Water Resources Planning Act (PL 89-80) to engage in comprehensive water supply planning by "maintain(ing) a continuing study of the nation's water and related land resources and to prepare periodic assessments to determine the adequacy of these resources to meet present and future requirements" (23). The Act encouraged states to establish "Title II" river basin commissions charged with integrating and harmonizing state assessments of water supply problems and encouraging regional supply planning. These were staffed by state *and* federal representatives.

The WRC employed competitive planning grants to states and river basin commissions for gathering basic data on water; identifying problems; and developing long-term ("level B") plans for "comprehensive, coordinated management of water resources." The commissions also tried to coordinate the efforts of the Corps of Engineers, BOR, and other federal agencies to provide assistance to states and communities in drawing up these plans. The WRC conducted two national assessments in 1968 and 1978. Both comprehensively depicted the nation's current and projected water needs; analyzed water-related problems from national, regional, and major river basin vantage points; and provided a wealth of data on water quality, quantity, and land use trends and problems (23,24).

As long as the WRC's efforts remained confined to preparing broad sets of recommendations for others to weigh and adopt, they generated little controversy and even less public attention. In the mid-1970s, however, WRC was charged by the Office of Management and Budget with developing "principles and standards" to be applied to evaluation of water projects, as well as cost-share guidance. The WRC was also granted power to discourage projects of dubious economic value or potentially adverse environmental impact. Controversy came to a head during the presidency of Jimmy Carter (1977–1981) who actively sought to diminish Congress's tradition of "pork barrel" water projects, charging that they were wasteful of public expenditures, economically unjustifiable, and in many cases, ecologically harmful. Criticism grew in Congress that WRC was exercising an "anti-dam" agenda through budget-cutting and impoundments to eliminate locally popular, necessary projects (25).

As a result of these criticisms and its eventual demise in 1981 under President Reagan, WRC is regarded as a failure. Critics charge that WRC did not ensure meaningful citizen participation at the inception of federal agency planning; overcome disagreements among states, communities, and other traditional beneficiaries of "pork-barrel" projects; or overcome the perception that its planning efforts were perceived as imposed from above. Moreover, it failed to find "an appropriate ideological basis for policy" (26,27).

Mention should also be made of the U.S. National Water Commission (NWC, 1967–1973), which grew out of controversies in the 1960s surrounding proposals to dam the lower Colorado River. Its chief legacy was the attempt to place fiscal discipline upon federal water supply efforts and to identify ways to better coordinate federal agency water supply planning and pollution control activities (28,29). NWC did succeed in placing species diversity, riverine protection, and other environmental issues on a policy footing equal to traditional economic development drivers of federal water supply planning. Moreover, the NWC's final report provided a "penetrating critique of water resources decision making" (29, pp. 4–24). Its recommendations to improve groundwater management, liberalize in-stream flow protection, urge greater wastewater reuse, and increase agricultural water efficiency remain part of many recent water planning efforts.

The 'New' Paradigm: Federal–State Partnership and Environmental Sensitivity

Since the 1980s, a "new" paradigm for federal water supply planning has evolved that is characterized by greater stakeholder participation in formulating policy alternatives; environmentally sound, socially just water resources management; reliance upon drainage basins as planning units; and greater state–federal cost sharing (30). This paradigm, advocates believe, promotes a management approach more accountable to *varied* interests, amenable to "regionally and locally tailored solutions" (29), and protective of fish and wildlife. Eroding federal interest in water supply, exemplified by Congress' 1997 'zeroing-out' of funding for river basin commissions and a general decline in support for new water projects, has helped legitimize this new paradigm (31–33).

In 1980, Congress passed the *Pacific Northwest Electric Power Planning and Conservation Act* [16 U.S.C. § 839b(h)]. The Act's purpose was to restore salmon spawning runs on the Columbia and Snake Rivers by making salmon a "coequal partner" with hydroelectric power in the operation of the Columbia Basin's 150 + dams (34). The Act also established the Northwest Power Planning Council (NPPC), a multistate, multiagency partnership comprised of the Bonneville Power Authority (BPA), the Pacific Northwest Electric Power and Conservation Council, and governor-appointed representatives from Washington, Oregon, Idaho, and Montana. NPPC has encouraged cooperation among utility systems, conservation and efficient power use, greater involvement of state and local governments in regional power planning, and oversight of electric power system planning and regional fish and wildlife recovery. NPPC is developing a long-range plan to ensure low-cost power for the region while protecting and rebuilding fishery and wildlife populations depleted by hydropower development.

Because NPPC represents a bold step toward adopting this "new" paradigm, its efforts have been heralded as an innovative means of bringing stakeholders together. Its activities have even led to creation of a regional "water budget" designed both to protect fisheries and provide for other services (more than 4 million acre-feet for salmon, seasonally allocated on the Columbia and Snake Rivers), and the set aside of 44,000 river miles of prime fish and wildlife habitat from new hydropower development.

Although it had an enormous effect on regional electric policies and regional approaches to fish and wildlife protection and restoration, NPPC has not prevented several species of Snake River salmon from being listed under the *Endangered Species Act*. By 1990, annual fish runs had been reduced from an estimated 10–16 million to fewer than 3 million, despite more than $100 million in expenditures on environmental improvements at dams and decreased hydropower production. By 1995, many West Coast salmonid species had been driven to extinction, and others remain at risk of extinction. Adding to its problems, in 1994, the Ninth Circuit Court of Appeals rejected NPPC's initial Salmon Strategy and criticized the council's "sacrificing the Act's fish and wildlife goals" (35, p. 190). NPPC has promulgated a new plan to provide for endangered salmon recovery as well as to ensure protection of other fish species.

Although not formally a "compact," NPPC does have compact-like powers to develop plans and meld the operations of state and federal water supply entities. The various water budgets and operational and engineering changes developed on the riverine system are products, in part, of data management tools refined in the light of experience. Eleven state and federal agencies, 13 Native American tribes, eight utilities, and numerous fish, forest, and environmental groups are among the stakeholders involved in restoring the Columbia River system.

Despite its mixed record, NPPC has forced environmental issues onto the agenda of water and power agencies in the Pacific Northwest. As one observer has stated, ecological imperatives, aesthetics, and demands for equity "have given rise to a greater consciousness of the ethical implications of water use" (36). Moreover, unlike previous federal efforts, NPPC has encouraged extensive public involvement and planning for long-term threats by incorporating local community and tribal concerns (35,36). Clearly, NPPC is a major departure from previous federal water supply planning efforts. It may also be a window into the future.

CONCLUSIONS: WHITHER FEDERAL WATER SUPPLY PLANNING?

Federal water supply planning efforts began as attempts to promote support for regional power, reclamation, or flood control projects; or to thwart the influence of private interests who wanted to develop such projects. For most of their history, these efforts limited public involvement to a "consultative" role; failed to provide adequate interagency coordination; and, sometimes fell short of being truly comprehensive.

We now know that effective federal water supply planning requires the ability to overcome agency "turf wars" by pursuing objectives defined by regional stakeholders; supplant authority vested in several agencies (e.g., TVA), and comprehensively manage several water supply needs simultaneously (e.g., DRBC and SRBC have mandates over water quality *and* supply). The record of federal planning further suggests that effective efforts are those perceived as economical and efficient, and thus able to generate public support for projects and programs (37). Finally, they must encourage collaboration among those stakeholders most directly affected by water supply problems. Collaboration overcomes resistance to change, facilitates new opportunities for funding, and stimulates resilient policy ideas (38). Today, public involvement is an essential part of all water supply planning efforts in the United States. As we enter a new millennium, questions that remain to be answered include, how do we determine the objectives of future plans, and what should the federal role be in their implementation?

BIBLIOGRAPHY

1. Feldman, D.L. (1995). *Water Resources Management: In Search of an Environmental Ethic*. Johns Hopkins University Press, Baltimore, MD.
2. Kundell, J.E., DeMeo, T.A., and Myszewski, M. (2001). *Developing a Comprehensive State Water Management Plan—A Framework for Managing Georgia's Water Resources*. Sponsored by Home Depot and Georgia Power. Research Atlanta, Inc./Georgia State University, Andrew Young School of Policy Studies, Atlanta, GA.
3. Grigg, N.S. (1996). *Water Resources Management: Principles, Regulations, and Cases*. McGraw-Hill, New York.
4. Merritt, R.H. (1979). *Creativity, Conflict, and Controversy: A History of the St. Paul District, U.S. Army Corps of Engineers*. Government Printing Office, Washington, DC.
5. Holmes, B.H. (1979). *A History of Federal Water Resources Programs and Policies, 1961–1970*. Economics, Statistics, and Cooperatives Service, U.S. Department of Agriculture, Washington, DC.
6. Barry, J.M. (1998). *A Rising Tide: The Great Mississippi Flood of 1927 and How it Changed America*. Touchstone Books, New York.
7. Powell, J.W. (1879). *Lands of the Arid Region of the United States*. GPO, Washington, DC.
8. McNamee, G. (1994). *Gila: The Life and Death of an American River*. University of New Mexico Press, Albuquerque, NM.
9. Foster, C.H. and Rogers, P. (1988). *Federal Water Policy: Toward an Agenda for Action*. Energy and Environmental Policy Center, JFK School of Government, Harvard University, Cambridge, MA.
10. Wilkinson, C.F. (1992). *Crossing the Next Meridian: Water, Power, and the Future of the American West*. Island Press, Washington, DC.
11. Mann, D.E. (1993). Political science: the past and future of water resources policy and management. In: *Water Resources Administration in the United States: Policy, Practice, and Emerging Issues*. M. Reuss (Ed.). Michigan State University Press, E. Lansing, MI, pp. 55–65.
12. Harden, B. (1994). *A River Lost: The Life and Death of the Columbia*. Macmillan, New York.
13. Pitzer, P.A. (1995). *Grand Coulee: Harnessing a Dream*. Washington State Univ. Press, Pullman, WA.
14. Caldwell, L.K. (1947). Interstate cooperation in river basin development. *Iowa Law Rev.* **32**: 232–243.
15. Conkin, P. (1986). Intellectual and political roots. In: *TVA: Fifty Years of Grass-roots Bureaucracy*. E.C. Hargrove and P.C. Conkin (Eds.). University of Tennessee Press, Knoxville, TN, pp. 3–34.
16. Selznick, P.W. (1966). *TVA and the Grassroots*. Harper and Row, New York.

17. Derthick, M. (1974). *Between State and Nation*. Brookings Institution Press, Washington, DC.
18. Sheridan, T. (1998). The big canal: the political ecology of the central arizona project. In: *Water, Culture, and Power: Local Struggles in a Global Context*. J.M. Donahue and B. R. Johnston (Eds.). Island Press, Washington, DC, pp. 163–186.
19. Grant, D.L. (1998). Introduction to interstate allocation problems. In: *Waters and Water Rights*. R.E. Beck (Editor-in-Chief). 1997 Replacement volumes and Supplemental Pocket Parts. Michie, Charlottesville, VA.
20. *Arizona v. California*. (1963). 373 U.S. 546, 565.
21. *Delaware River Basin Commission Annual Report*. (1997). 1998. DRBC, West Trenton, NJ.
22. Weston, T.R. (1999). *Interstate Water Rights and River Basin Compacts in the East*. Unpublished manuscript, Kirkpatrick and Lockhart, LLP, Harrisburg, PA.
23. (U.S. WRC) U.S. Water Resources Council. (1968). *First National Assessment: The Nation's Water Resources*. U.S. Government Printing Office, Washington, DC.
24. (U.S. WRC) U.S. Water Resources Council. (1978). *The Nation's Water Resources, 1975–2000, Volume 1: Summary, 2nd National Water Assessment*. Water Resources Council, December, Washington, DC.
25. Fradkin, P.L. (1996). *A River No More: The Colorado River and the West*. University of California Press, Berkeley, CA.
26. Robie, R.B. (1980). The impact of federal policy on state planning: a cautionary example. *Ame. Water Works Assoc. J.* **72**(February): 70–73.
27. Goslin, I.V. (1978). Colorado river development. In: *Values and Choices in the Development of the Colorado River Basin*. D.F. Peterson and A.B. Crawford (Eds.). University of Arizona Press, Tucson, AZ.
28. U.S. National Water Commission. (1972). *Water Resources Planning. Consulting Panel on Water Resources Planning*. NWC, Washington, DC.
29. *Water in the West: Challenge for the Next Century*. (1998). Report of the Western Water Policy Review Advisory Commission, Springfield, VA, NTIS, June.
30. Cortner, H.A. and Moote, M.A. (1994). Setting the political agenda: paradigmatic shifts in land and water policy. In: *Environmental Policy and Biodiversity*. R.E. Grumbine (Ed.). Island Press, Washington, DC, pp. 365–377.
31. McGinnis, M.V. (1995). On the verge of collapse: the Columbia River system, wild salmon, and the northwest power planning council. *Natural Resources Journal* **35**: 63–92.
32. Miller, K., Rhodes, S.L., and MacDonnell, L.J. (1996). Global change in microcosm: the case of U.S. Water Institutions. *Policy Sci.* **29**: 271–290.
33. Cody, B.A. (1999). *Western Water Resource Issues*. A Congressional Research Service Brief for Congress. Congressional Research Service, Washington, DC, March 18.
34. Blumm, M. (1998). Columbia river basin. In: *Waters and Water Rights*. Vol. 6. R.E. Beck (Ed.). Michie Publishing, Charlottesville, VA.
35. Getches, D.H. (1998). Water in the Columbia river basin: from a source of performence to an instrument for economic growth. In: *Traditional and Modern Approaches to the Environment on the Pacific Rim: Tensions and Values*. H. Coward (Ed.). State University of New York Press, Albany, NY, pp. 177–193.
36. Volkman, J.M. (1996). *A River in Common: The Columbia River, the Salmon Ecosystem, and Water Quality*. Western Water Policy Review Advisory Commission, Denver, CO.
37. American Society of Civil Engineers. (1984). *Environmental Objectives in Water Resources Planning and Management*. Water Resources Planning and Management Division. ASCE, New York.
38. Chrislip, D.D. and Larson, C.E. (1994). *Collaborative Leadership: How Citizens and Civic Leaders Can Make A Difference*. Jossey-Bass, San Francisco, CA.

FLOOD CONTROL ACT OF 1944

Arthur M. Holst
Philadelphia Water Department
Philadelphia, Pennsylvania

The Flood Control Act of 1944, enacted on December 22 of that year, established federal policy regarding jurisdiction over rivers and other bodies of water, specifically watersheds. The act states that all water-related construction and navigation projects will be approved only if they substantially improve navigational capacity and do not excessively interfere with standard water usage. The plan was proposed to prevent flooding and to allow more irrigation, easier navigation, increased river-related recreation, and the production of hydroelectric power. The Army Corp of Engineers was designated to oversee these tasks.

A major component of the Flood Control Act of 1944 is known as the Pick–Sloan plan. Named after its two creators, Glenn Sloan and Lewis Pick, members of the Army Corps of Engineers, this plan authorized the construction of six dams in three states on the Missouri River. The plan created the largest reservoir in North America. Prior to the 1940s, national water navigation and flood control projects were run by four different agencies that had little or no communication and planning among them: the Army Corps of Engineers, the Federal Power Commission, the Reclamation Bureau, and the Soil Conservation Service. As a result of massive and catastrophic flooding that occurred in 1936, all this changed. In March of that year, flooding killed 107 people and caused $270,000,000 of damage. The United States had never before had very many flood control projects; after these floods, it became clear that this needed to change. The federal government responded by passing the 1936 Flood Control Act and 8 years later, the Flood Control Act of 1944.

The Flood Control Act has 15 sections. The first section is composed of rules for water-related planning submissions to Congress, specifically, that they must not interfere with established water uses. These include domestic and municipal use, as well as irrigation and industrial uses. It also asserts the importance of state involvement in projects and plans that will affect them, as determined by the Chief Engineer of the Army Corps.

The second section explains that flood control includes channel and drainage improvements. This section also states that river improvement for flood control is under the War Department (now the Department of Defense) as directed by the Secretary of War (now the Secretary of Defense), and supervised by the Chief Engineer of the Army Corps. Watershed alteration, erosion prevention, and water flow control projects, on the other hand, are stated to be under the control of the Department of Agriculture, as directed by the Secretary of Agriculture. These jurisdictions are subject to change only by Acts of Congress.

The third section deals with the 1936 Flood Control Act. It says that section 3 of that Act still applies to all projects. It also allows a 5-year period in which local interests working on water-related projects can secure whatever local cooperation the Secretary of War deems necessary.

Section 4 authorizes the Chief Engineer of the Army Corps to operate and maintain reservoir facilities in the United States. The Secretary of War is authorized to lease reservoir land but must give preference to federal, state, and local government. The Secretary must also make reservoir areas available to the general public for recreational purposes, as adjudged consistent with the public interest.

Section 5 instructs the Secretary of War to deliver to the Secretary of the Interior all energy produced at reservoirs that is not required for their operation. The Secretary of the Interior is instructed to dispense this energy to the public at low rates and is allowed to build or acquire transmission lines and other facilities with money from Congress only if necessary.

The sixth section of the Act authorizes the Secretary of War to sell surplus reservoir water for reasonable prices, so long as this does not negatively affect standard water uses.

Section 7 instructs the Secretary of War to regulate the usage of reservoir storage intended for navigation and flood control. This section does not generally apply to the Tennessee Valley Authority.

Section 8 states that the Secretary of War may, on advisement of the Secretary of the Interior, allow the use of water for irrigation. The Secretary of the Interior is then authorized to construct and maintain whatever structures are necessary for irrigation. However, the Secretary is allowed to undertake such projects only after receiving Congressional authorization. This section does not apply to dams and reservoirs built for irrigation prior to 1944 by the Army Corps of Engineers.

Section 9 of the Flood Control Act of 1944 refers to specific flood control projects and allocates $200,000,000 for those projects; specifically, the Missouri River project.

The tenth section details many other projects which are authorized by the Act and affect the following areas: the Lake Champlain Basin, the Blackstone River Basin, the Connecticut River Basin, the Thames River Basin, the Housatonic River Basin, the Susquehanna River Basin, the Roanoke River Basin, the Edisto River Basin, the Savannah River Basin, the Mobile River Basin, the Alabama-Coosa River Basin, the Lower Mississippi River, the Red-Ouachita River Basin, the Arkansas River Basin, the White River Basin, the Upper Mississippi River Basin, the Red River of the North Basin, the Missouri River Basin, the Ohio River Basin, the Great Lakes Basin, the Colorado River Basin (Texas), the Brazos River Basin, the Rio Grande Basin, the Great Salt Basin, the Colorado River Basin, the San Diego River Basin, the Ventura River Basin, the Santa Ana River Basin, the Los Angeles-San Gabriel Basin and Ballona Creek, the Pajaro River Basin, the Sacramento-San Joaquin River Basin, the San Joaquin River, the Napa River Basin, the Coquille River Basin, the Nehalem River Basin, the Willamette River Basin, the Columbia River Basin, the Willapa River Basin, the Chehalis River Basin, and the Territory of Hawaii.

Section 11 lists many more areas for which the Secretary of War is instructed to oversee preliminary studies of flood control and other plans. These plans are not to be undertaken until authorized by law.

Section 12 appropriates $950,000,000 for all of these projects. It also appropriates $10,000,000 to the Departments of Agriculture and War for any preliminary studies of flood control improvements, $1,500,000 to the Federal Power Commission for these studies, and a $500,000 emergency fund for emergency flood prevention.

Section 13 authorized various projects to reduce soil erosion and to improve water flow and runoff in these areas: the Los Angeles River Basin, the Santa Ynez River Basin, the Trinity River Basin (Texas), the Little Tallahatchie River Watershed, the Yazoo River Watershed, the Coosa River Watershed, the Little Sioux River Watershed, the Potomac River Watershed, the Buffalo Creek Watershed, the Colorado River Watershed (Texas), and the Washita River Watershed.

Section 14 of the Act deals with the balance of some $10,000,000 appropriated by an earlier act. It is to be used for the work detailed in section 13 of the Flood Control Act of 1944, and no more than 20% is to be spent on any one project.

The 15th and final section of the Act amends an earlier act approved in 1938. It allows the Secretary of War to undertake emergency measures to avert dangerous, destructive, and/or fatal flooding, based on codicil that he may not spend more than $100,000 in any fiscal year.

This Act permanently altered the manner in which water issues are handled in the United States and has saved billions of dollars in property damage by averting flooding.

READING LIST

Flood Control Act of 1944.

http://lwas.fws.gov/lawsdigest/flood.html.

http://www.usace.army.mil/inet/usace-docs/eng-pamphlets/ep870-1-29/c-6.pdf.

GREAT LAKES GOVERNORS' AGREEMENT

Frederic Lasserre
Laval University
Quebec City, Quebec, Canada

Several institutions govern the use of the Great Lakes water and try to coordinate the actions undertaken by national governments (the United States and Canada) as well as state and provincial actions. Some of these institutions are old; some are much more recent and stem from political concerns regarding the quality of the water in the Great Lakes, as well as potential diversions that would have taken large quantities out of the basin.

THE GREAT LAKES GOVERNANCE

The International Joint Commission (IJC) prevents and resolves disputes between the United States of America and Canada under the 1909 Boundary Waters Treaty and "pursues the common good" of both countries as an "independent and objective advisor" to the two governments (1). Transportation was one of the main concerns of water management at the time: there was no water quantity problem, nor was there much talk of large variations in Great Lakes levels that could hamper human or industrial consumption. The negotiators wanted to ensure that water would always be sufficient to allow for the safe passage of ships through locks. Article III of the Treaty states that "no further or other uses or obstructions or diversions, whether temporary or permanent, of boundary waters on either side of the line, affecting the natural level or flow of boundary waters on the other side of the line shall be made, except by authority of the United States or the Dominion of Canada within their respective jurisdictions, and with the approval, as hereinafter provided, of a joint commission, to be known as the International Joint Commission." The IJC is, therefore, as its name underlines, an international institution between two sovereign States.

In particular, the Commission rules upon applications for approval of projects affecting boundary or transboundary waters and may regulate the operation of these projects; it assists the two countries in protecting the transboundary environment, including the implementation of the Great Lakes Water Quality Agreement (1972, renewed in 1978) and improving transboundary air quality, and it alerts the governments to emerging issues along the boundary that may give rise to bilateral disputes. The IJC, in particular, in its *Final Report on Protection of the Waters of the Great Lakes* (2000), recommended that Canadian and U.S. federal, provincial, and state governments should not permit diversion of water from the Great Lakes Basin unless the proponent can demonstrate that the removal would not endanger the integrity of the Great Lakes ecosystem (2).

The Great Lakes Commission (GLC) is the only regional organization whose statutory mandate is to represent the eight Great Lakes states on a variety of environmental and economic issues. It was created in 1955 and was granted Congressional consent in 1968. The Great Lakes Commission is now a binational agency: its members include the eight Great Lakes states—Illinois, Indiana, Michigan, Minnesota, New York, Ohio, Pennsylvania, and Wisconsin; the Canadian provinces of Ontario and Québec have had associate member status since 1999.

The Commission is dedicated to managing the waterbody of the Great Lakes: it promotes the orderly, integrated and comprehensive development, use, and conservation of the water and related natural resources of the Great Lakes Basin and St. Lawrence River. Each jurisdiction appoints a delegation of three to five members comprised of senior agency officials, legislators, or appointees of the governor or premier.

The Commission works actively at the member states' civil servant level. It is responsible for sharing information among members, coordinating state and provincial positions on issues of regional concern, and advocating those positions on which the members agree (3).

The GLC was set up by the member states as the St. Lawrence Seaway construction was well under way. The concern of the states was, at the time, to coordinate their views on ways to manage the economic, ecological, and social impacts of the increasing commercial traffic that would result from opening the Seaway in 1959.

The Council of Great Lakes Governors (CGLG), created in 1983, is a partnership of the Governors of the eight Great Lakes states. It has no legal basis. It is a pressure group that coordinates actions among the member states. It is, therefore, an institution separate from the Great Lakes Commission. Besides, the GLC is more technical, the CGLG more political in its aims and functioning. In recent years, the Canadian premiers of Ontario and Quebec have joined with the Council of Governors in sharing their views on the Great Lakes region.

The CGLG is not limited to Great Lakes water management. For instance, the CGLG coordinated negotiations among member states. The resulting 1988 Economic Development Agreement provides for coordinating the commercial promotion and trade missions of the member states abroad.

The central issue of common concern to all of the Governors was protecting and managing the waters of the Great Lakes. As the primary source of freshwater for the region as well as a trade route that connects the region to the world through the St. Lawrence Seaway, proper management of the lakes is crucial to the health of the residents of the Great Lakes region as well as to the continued growth of the region's economy. Consequently, coordinating Great Lakes regional water policy was the original impetus for creating the Council (4).

A MAJOR IMPETUS: PREVENTING DIVERSION FROM THE GREAT LAKES

The Advent of the Great Lakes Charter

Water pollution was one concern, and the Governors tackled the issue of improving Great Lakes water quality. In 1986, the Governors signed the Toxic Substances Control Agreement (TSCA) that provides for developing programs for companies to adopt pollution-prevention devices.

Diversion control and basin water integrity were another major concern. Governmental archives from the early 1980s attest to western United States lobbying for diversion of Great Lakes water to quench their lack of water. Great Lakes states wanted to resist these projects, both for environmental reasons as well as for political reasons: why would the Great Lakes states give to California an added value at a time when so many firms were leaving the area and moved away to the West Coast (5). The Great Lakes Charter stemmed from this growing concern that Great Lakes water could be diverted to water-scarce regions of the United States, the Midwest, and the West especially. States and provinces from the Great Lakes Basin had discussed their fears of diversion schemes at Mackinac Island (Michigan) in 1982 and at Indianapolis (Indiana) in 1983. The Great Lakes Charter, signed in 1985 by the eight Great Lakes Governors and the premiers of Ontario and Quebec, is the final resolution of a process set up after the Indianapolis meeting. The Charter created a notice and consultation process for Great Lakes diversions. The signatories agreed that no Great Lakes state or province would proceed with any new or increased diversion or consumptive use of Great Lakes water of more than five million gallons per day without notifying, consulting, and seeking the consent of all affected Great Lakes states and provinces. It also calls for coordinating water policies among member states and exchanging hydrologic data.

The Great Lakes states also lobbied the federal government that was in the process of designing a new law governing water use in the United States. The subsequent Water Resources Development Act of 1986 (WRDA) prohibits "any diversion of Great Lakes water by any State, federal agency, or private entity for use outside the Great Lakes basin unless such diversion is approved by the Governor of each of the Great Lakes States." It also prohibits any Federal agency from studying the transfer of Great Lakes water for use outside the Great Lakes Basin, unless done under the auspices of the IJC. To ensure compliance with a 1967 Supreme Court (modified in 1980) consent decree, the Act also appropriates federal resources to monitor and measure Lake Michigan's water flow into the Chicago River.

The Council coordinates the authority granted to the Governors under the Water Resources Development Act of 1986 (WRDA). The Act requires the Governors' unanimous approval on any proposed out-of-basin diversion of water from the Great Lakes Basin.

In 1997, the Great Lakes states entered into a Memorandum of Understanding on the Lake Michigan diversion, concluding a lengthy mediation process on the matter of Illinois' diversion of Lake Michigan water at Chicago. Under the U.S. Supreme Court decrees (1967 and 1980), Illinois was limited to 3,200 cfs each year. Illinois had exceeded that limit by nearly 15%. Illinois will further reduce its annual diversion during the following 14 years to restore to Lake Michigan the excess amount of water it has withdrawn since 1980 and construct new lakefront structures that do not allow leakage. All eight Great Lakes states and the U.S. federal government participated in the discussions. The Province of Ontario, as well as the City of Chicago and the Army Corps of Engineers, were also involved as observers or by providing technical support.

Diversions Are Still Feared: Banning Water Exports?

Diversion projects are far from being outmoded, despite the conclusion set forth by many observers that bulk water is, for now, too expensive to move over large distances. Large and small diversion schemes are still pushed by pressure groups. For instance, in the spring of 1998, the Province of Ontario approved a permit for the Nova Group of Sault Saint Marie Ontario that would have allowed the Nova Group to remove 60 million gallons per year from Lake Superior that would then be sold in Asia. In direct response to concerns expressed by the Great Lakes Governors via the Council, the Ontario government rescinded the permit. The accompanying public uproar led to the previously mentioned review of the issue of bulk removals by the International Joint Commission (IJC) in 2000. The Governors, through the Council, were represented on the study team that advised the IJC on these issues. Again in November 2002, Milwaukee mayor, John Norquist, denounced a diversion project to bring water to far western suburbs, outside the Great Lakes Basin.

So as to address a growing public concern that large water diversion schemes could someday be implemented, despite the Charter and the WRDA, the Council of Great Lakes Governors convened in 1999 and agreed there was a political need to reinforce the provisions of the Charter by

- forging a new, more binding agreement to manage the Great Lakes waters;
- creating a new standard requiring an improvement to the water and water-dependent natural resources of the Great Lakes before allowing new water uses;
- pledging to obtain better information so that the water is managed rationally; and
- lowering the trigger level for diversions that need to be approved under the Charter.

The Great Lakes Governors and the premiers of Ontario and Quebec signed the Great Lakes Charter Annex on June 18, 2001 in Niagara Falls, New York. The Annex, an amendment to the Great Lakes Charter of 1985, provides a blueprint for creating a new set of binding agreements among the states and provinces on the effective management of Great Lakes water resources. It also outlines a series of principles for reviewing water withdrawals from the Great Lakes Basin that is grounded in protecting, conserving, restoring, and improving the Great Lakes ecosystem. The new agreement must be set up by the member States before 2004 and then ratified at the two national levels of Canada and the United States to prevent any attempt to divert water from the Great Lakes Basin. A particular emphasis was placed in the Annex on the need to close loopholes that legal experts claim have arisen for water exports under the North American Free Trade Agreement and the General Agreement on Trade and Tariffs.

The history of the Great Lakes agreements attests to the evolution of the concerns regarding the largest body of freshwater. From a very transportation-oriented perspective, as embodied in the 1909 Boundary Waters Treaty, the focus evolved to pondering the possible impacts of the opening of the St. Lawrence Seaway in 1955, by the Great Lakes Commission, to emphasis placed on water quality and quantity, by the Great Lakes Charter of 1985. The advocacy of a major water diversion by western states pressure groups and recent strong variations in the Great Lakes levels stirred an important public controversy that led the Great Lakes Basin state and province leaders to sign the Annex in 2001 to ban all large water diversion out of the basin, thus implementing recommendations already made by the IJC in 2000.

Acknowledgment
This research was made possible thanks to a grant from the FQRSC.

BIBLIOGRAPHY

1. IJC Status. Available: http://www.ijc.org.
2. Lasserre, F. (2002). La guerre de l'eau aura-t-elle lieu? In: *Espaces et enjeux: Méthodes d'une Géopolitique Critique*. F. Lasserre and E. Gonon (Eds.). Collection Chaire Raoul Dandurand/UQÀM, L'Harmattan, Montréal and Paris, p. 251; Lasserre, F. (2002). L'Amérique a soif. Les États-Unis obligeront-ils Ottawa à céder l'eau du Canada? In: *Eaux et Territoires: Tensions, Coopérations et Géopolitique de l'eau*. F. Lasserre and L. Descroix (Eds.). Presses de l'Université du Québec/L'Harmattan, Quebec City/Paris, p. 315.
3. Great Lakes Commission. Available: http://www.glc.org/about/.
4. Council of Great Lakes Governors. Available: www.cglg.com.
5. Lasserre, F. (2001). L'Amérique a soif. Les besoins en eau de l'Ouest des États-Unis conduiront-ils Ottawa à céder l'eau du Canada? *Revue internationale d'Études canadiennes* **24**: 196–214.

U.S./CANADIAN BOUNDARY WATERS TREATY AND THE GREAT LAKES WATER QUALITY AGREEMENT

David W. Yoskowitz
Texas A&M University—
Corpus Christi
Laredo, Texas

The United States and Canada established the International Joint Commission (IJC) to take on certain responsibilities for transboundary bodies of water. The United States also has a similar arrangement with Mexico known as the International Boundary Water Commission (IBWC). Note that the task of the IBWC by comparison is not as involved because it has only two rivers in its jurisdiction, whereas the IJC has more than 200 transborder rivers (1), plus lakes and common aquifers.

FORM

The IJC was created as a result of the Boundary Waters Treaty of 1909 between the United States and Great Britain because it still controlled Canadian foreign policy. The pressure for a treaty came from within as populations of communities on both sides of the border began to grow and demands on water increased along with it. Initially, obstacles had to be overcome in the legal wording of the document because the two countries look at boundary waters differently. Americans tend to look at boundary water resources from the point of view of equity and equitable utilization, in which rights and obligations of the individual water user are recognized. Canadians, on the other hand, see water issues in terms of equality and the rights and responsibilities of the two states.

Institutionally, it was the early irrigation congresses and conferences of the 1890s that ultimately promoted movement toward considering bilateral agreements and treaties and, as a result, the creation of the IJC (2).

The treaty begins with assurance that navigation along common bodies of water shall remain open, and no country may impose excessive tools on the other for use of canals, locks, etc. Until power production and irrigation became important uses of water near the turn of the century, international water law had focused on navigation rights. A strong impetus for a treaty was to secure navigation rights, although it was less important during negotiation than other boundary water issues (3).

The first time the IJC is mentioned in the Treaty Relating to Boundary Water Between the United States and Canada is in Article III with regard to the authority to approve/disapprove obstructions or diversions of boundary water that affect the natural level or flow of the water on the other side of the line, but it is not until Article VII that establishes the IJC with six commissioners, three from each side, appointed by their respective governments. Its jurisdiction covers all cases involving use, obstruction, or diversion for which approval is required as mentioned in the treaty (3). A majority is required to make a decision, and, if evenly split, separate reports will be made to the respective governments and settled through diplomatic channels (4).

Each country was assured equal and similar rights in the use of the water, but there was an order of precedence:

1. uses for domestic and sanitary purposes; then
2. uses for navigation, including the service of canals for navigation; then
3. uses for power and irrigation.

These reflect the interests of the time and thus are dated because they do not reflect more current interests in fisheries, recreation, wildlife, and the environment in general (3). What makes the IJC important are the investigative powers that were accorded to them in Article IX, which states that, "The IJC is authorized in each case so referred to examine into and report upon the facts and circumstances of the particular questions referred" (4). Reports issued by the IJC are not to be considered decisions, as was made very clear in the treaty, and therefore the governments could ultimately ignore the IJC as they saw fit.

FUNCTION

The Boundary Water Treaty put the IJC into action using some broad guidelines but has changed since its inception at the beginning of the century. Along the northern border, the questions have moved from quantity to quality, and the IJC has a leading role in the answers. The governments have delegated new responsibilities to the IJC based on its track record and reputation for impartiality (3). An existing problem is that the commission has no follow-up implementation authority, which gives it largely an advisory role.

Several categories of management issues/problems dominate the work that the IJC becomes involved in, including upstream–downstream conflicts, common pool resources, and integrated river basin developments (5). Work on these issues can take the form of either an approval or reference. Approval is required if an application is made that falls under the scope of either Article III or IV of the Treaty (3), and usually consists of constructing dams, hydropower projects, or diverting flows. The IJC appoints a technical advisory board of members from each country to conduct the investigative work on whether to approve/disapprove a project. References, on the other hand, take much more time and involve constant monitoring of whatever issue has been referred to the commission. Advisory boards are set up in much the same way as they are in the approval category (3).

The strength of the commission lies in several areas where it could become an instrument for boundary environmental protection and management: (a) The IJC performs administrative and quasi-judicial tasks handling a range of issues; (b) There is experience and legitimacy in dealing with boundary issues on an impartial basis; (c) The IJC is an arbitrator of fact; (d) It is a mediator of policy; (e) Top water managers in each country meet regularly under the IJC umbrella, which has created an informal network of contact among government officials and experts (6).

GREAT LAKES WATER QUALITY AGREEMENT AND THE IJC

The Agreement, first signed in 1972 and renewed in 1978, expresses the commitment of each country to restore and maintain the chemical, physical, and biological integrity of the Great Lakes Basin ecosystem and includes a number of objectives and guidelines to achieve these goals. It reaffirms the rights and obligation of Canada and the United States under the Boundary Waters Treaty.

In 1987, a Protocol was signed amending the 1978 Agreement. The amendments aim to strengthen the programs, practices, and technology described in the 1978 Agreement and to increase accountability for implementing them. Timetables are set for implementing specific programs.

The parties will meet biennially to discuss progress and report periodically to the Commission. New annexes address atmospheric deposition of toxic pollutants, contaminated sediments, groundwater, and nonpoint sources of pollution. Annexes are also added to incorporate the development and implementation of remedial action plans for areas of concern and lakewide management plans to control critical pollutants.

The IJC monitors and assesses progress under the Agreement and advises each government on matters related to the quality of the boundary waters of the Great Lakes system. The Agreement also calls on the IJC to assist the United States and Canada with joint programs under the Agreement and provides for two binational boards—the Great Lakes Water Quality Board and the Great Lakes Science Advisory Board—to advise the Commission.

The general objectives of the agreement, expressed in Article III of the Agreement (7), are to keep the waters:

(a) free from substances that directly or indirectly enter the waters as a result of human activity and that will settle to form putrescent or otherwise objectionable sludge deposits, or that will adversely affect aquatic life or waterfowl;

(b) free from floating materials such as debris, oil, scum, and other immiscible substances resulting from human activities in amounts that are unsightly or deleterious;

(c) free from materials and heat directly or indirectly entering the water as a result of human activity that alone, or in combination with other materials, will produce color, odor, taste, or other conditions in such a degree as to interfere with beneficial uses;

(d) free from materials and heat directly or indirectly entering the water as a result of human activity that alone, or in combination with other materials, will produce conditions that are toxic or harmful to human, animal, or aquatic life; and

(e) free from nutrients directly or indirectly entering the waters as a result of human activity in amounts that create growths of aquatic life that interfere with beneficial uses.

BIBLIOGRAPHY

1. Allee, D.J. (1993). Subnational governance and the international joint—commission: local management of United States and Canadian boundary water. *Natl. Resour. J.* **33**: 133–151.
2. Carroll, J.E. (1986). Water resources management as an issues in environmental diplomacy. *Natl. Resour. J.* **26**: 207–242.
3. Lemarquand, D. (1993). The international joint commission and changing Canada—United States boundary relations. *Natl. Resour. J.* **33**: 59–91.
4. *Treaty Relating to Boundary Water Between the United States and Canada, January 11, 1909*. U.S.—Great Britain, 36 Stat, 2449.
5. Sadler, B. (1993). Shared resources, common future, sustainable management of Canada–United States border waters. *Natl. Resour. J.* **33**: 375–396.
6. Lemarquand, D. (1986). Preconditions to cooperation in Canada–United States boundary waters. *Natl. Resour. J.* **26**: 361–380.
7. *Great Lakes Water Quality Agreement of 1978 Agreement, with Annexes and Terms of Reference, between the United States and Canada signed at Ottawa November 22, 1978 and Phosphorus Load Reduction Supplement signed October 16*, 1983 as amended by Protocol signed November 18, 1987.

GREAT LAKES WATER QUALITY INITIATIVE

Elsie F. Millano
ERM, Inc.
Vernon Hills, Illinois

GENERAL DESCRIPTION

The Great Lakes Water Quality Initiative (the "Initiative") is a program to improve the quality of the waters in the

Great Lakes system. The Great Lakes consist of (1) Lake Ontario, Lake Erie, Lake Huron (including Lake Saint Clair), Lake Michigan, and Lake Superior; and (2) their connecting channels, including the Saint Mary's River, the Saint Clair River, the Detroit River, the Niagara River, and the Saint Lawrence River to the Canadian border. The Great Lakes system includes all streams, rivers, lakes, and other bodies of water within the drainage basin of the Great Lakes within the United States. The effort was undertaken by the U.S. Environmental Protection Agency (U.S. EPA) and the eight Great Lakes U.S. states (Illinois, Indiana, Michigan, Minnesota, Ohio, Pennsylvania, New York, and Wisconsin), supported by the U.S. Fish and Wildlife Service, the National Park Service, environmental groups, municipalities, industry, and academia.

The main objective of the Initiative is to reduce the discharge of bioaccumulative compounds of concern (BCCs) into the Great Lakes system. The first phase of the Initiative consisted of the development of a consistent set of water quality criteria to be applied throughout the Great Lakes system by the Great Lakes States and by any Indian Tribes that have jurisdiction over portions of the Great Lakes system.

The Initiative participants collected available scientific information on persistent toxic chemicals in the Great Lakes system, reviewed and evaluated the data and the different regulations and procedures to establish water quality criteria of the participant states and developed the water quality criteria and procedures issued in 1995 as the Water Quality Guidance for the Great Lakes System (the "Guidance").

The second phase of the Initiative, called the Great Lakes Toxic Reduction Effort (GLTRE), is under development. The GLTRE focuses on reducing BCCs contributed to the Great Lakes system by nonpoint sources.

HISTORY

In the spring of 1989, the Council of Great Lakes Governors unanimously agreed to participate in the Initiative with the U.S. EPA, to establish consistent water quality criteria throughout the Great Lakes system. The Initiative was a response by the United States to continuous contamination of the Great Lakes system by persistent and bioaccumulative toxic (PBT) pollutants (such as polychlorinated biphenyls, dioxins, DDT, DDE, and mercury), despite efforts to bring the Great Lakes system back into health that date back to 1972, when Canada and the United States signed the Great Lakes Water Quality Agreement (GLWQA).

At the time of the 1972 GLWQA, algae growth and reduced dissolved oxygen concentration were reducing the fish population in the Great Lakes system. The 1972 GLWQA specified (1) objectives to reduce the discharge of pollutants that caused nuisance conditions (color, odor, floating oil and scum, aquatic weeds, and algae) and of substances toxic to human, animal, or aquatic life; and (2) numerical targets for reducing loadings of phosphorus into the lower Great Lakes (i.e., Lake Erie and Lake Ontario). The 1972 GLWQA also contained measures to reduce the loading of phosphorus, including effluent limits on all major municipal sewage treatment facilities, limitations on the phosphorus content in household detergents, and reduction in nonpoint source runoff loadings.

During the 1970s, the passage of the Clean Water Act in the United States and similar efforts in Canada reduced the discharge of untreated municipal and industrial wastewaters and provided funding for further evaluating the effects of pollutants on the Great Lakes system. The additional research indicated that PBT substances were responsible for the declining population and accumulation of pollutants in several species of animals, including bald eagles, minks, herring gulls, double-crested cormorants, common and Caspian terns, ospreys, and lake trout among others.

As a result, the 1978 revisions to the GLWQA established a policy calling for virtual elimination of the discharge of PBT substances. In addition, the 1978 GLWQA specified modified phosphorus target loadings for the lower Great Lakes and new phosphorus target loadings for the upper Great Lakes.

In the 1980s, scientists detected effects of PBT substances on humans, consisting of neurological effects on infants whose mothers had eaten Lake Michigan fish contaminated with PBT substances during their pregnancies. In 1987, Canada and the United States amended the GLWQA by Protocol to require the development of (1) Lakewide Management Plans (LaMPs) to control and reduce PBT substances and (2) Remedial Action Plans (RAPs) for 43 recognized local areas of concern in the lakes (now 42, one of them was delisted as such an area in 1994) affected by PBT substances. The 1987 amendments called for greater coordination among different jurisdictional authorities to allow the success of the LaMPs and RAPs, plus more focused attention on PBT substance pollution from land runoff, contaminated sediment, airborne toxic substances, and contaminated groundwater.

As previously indicated, one of the objectives of the Initiative was to develop water quality guidance for BCCs in the Great Lakes system to ensure consistency in requirements among the Great Lakes states. In 1990, the U.S. Congress approved an amendment to the Clean Water Act that required the U.S. EPA to publish water quality guidance for the Great Lakes system in accordance with the 1987 GLWQA and to perform other key activities in the 1987 GLWQA, imposed statutory deadlines for implementing the water quality guidance once issued, and provided funding for programs to meet the 1987 GLWQA requirements. The Water Quality Guidance for the Great Lakes System (the "Guidance") was issued by the U.S. EPA in March 1995 and supplemented in November 2000. The Guidance uses the term BCC to refer to PBT substances. The GLTRE is currently focusing on the other aspects of the 1987 GLWQA.

WATER QUALITY GUIDANCE FOR THE GREAT LAKES SYSTEM

Under United States regulations, each state or Indian tribe has the right to set water quality criteria for the waters under their jurisdiction. These criteria are, in turn, used by each state or Indian tribe to develop water quality standard-based discharge limits for wastewater effluents under the National Pollutant Discharge Elimination System (NPDES). The issuance of the Guidance ensures that all point source discharges to the Great Lakes system will meet a set of consistently developed criteria.

General Description

The proposed regulation was issued in 1993. The final Guidance, which was issued in March 1995 and then supplemented in November 2000, contains

- water quality criteria for a set of 29 chemical constituents to protect aquatic life, human health, and/or wildlife;
- procedures to develop water quality criteria for additional pollutants;
- an antidegradation policy to be adopted, at a minimum, for BCCs;
- a ban on mixing zones for BCCs; and
- procedures to implement the water quality criteria guidance.

Mixing zones, which are established by each state on a point-source basis, allow discharge of compounds at concentrations higher than the surface waterbody standard as long as the concentration in the water near or outside the boundary of the mixing zone meets the water quality criteria.

The March 1995 Guidance contained all five elements listed above, but the ban on mixing zones was vacated in June 1997 by the U.S. Court of Appeals in the District of Columbia and reinstated in November 2000 with modifications. The final Guidance is codified at Part 132 of Title 40 of the Code of Federal Regulations, and includes the six tables and the six appendixes shown in Table 1. The implementation procedures included in Appendix F are listed in Table 2.

The Guidance required the Great Lakes states to adopt and submit to the U.S. EPA for review, within two years of publication of the final Guidance (i.e., March 23, 1997), minimum water quality standards, antidegradation policies, and implementation procedure regulations in compliance with the corresponding Guidance procedures. To date, all states have submitted their proposed regulations and obtained either approval or partial approval of their submittals. When an element of a state's submittal was inconsistent with the Guidance, the U.S. EPA disapproved that element of the submittal and established that the specific state has to meet the Guidance's requirements for that specific element.

Water Quality Criteria

Table 3 presents the 29 chemical constituents for which water quality criteria were issued, along with an indication of whether each compound is a BCC and what population is covered by the water quality criteria. The Guidance includes water quality criteria for

- protecting aquatic life on an acute and chronic basis;
- protecting human health from drinking water and eating fish, or from recreationally using the Great Lakes waters; and
- protecting wildlife against ingestion of aquatic life containing the indicated chemical constituents.

Tables 4 through 7 reproduce the Guidance's water quality criteria, and Table 8 presents the pollutants of initial focus in the Guidance. Only chemical constituents for which sufficient information was available to determine a water quality standard based on requirements included in the Guidance appear in Tables 4 through 7; those chemical constituents include only nine of the 22 BCCs shown in Section A of Table 8.

Table 1. Attachments to the Water Quality Initiative Regulation

Attachment	Description
Table 1	Acute Water Quality Criteria for Protection of Aquatic Life in Ambient Water
Table 2	Chronic Water Quality Criteria for Protection of Aquatic Life in Ambient Water
Table 3	Water Quality Criteria for Protection of Human Health
Table 4	Water Quality Criteria for Protection of Wildlife
Table 5	Pollutants Subject to Federal, State, and Tribal Requirements
Table 6	Pollutants of Initial Focus in the Great Lakes Water Quality Initiative
Appendix A	Methodologies for Development of Aquatic Life Criteria and Values
Appendix B	Methodology for Deriving Bioaccumulation Factors
Appendix C	Methodologies for Development of Human Health Criteria and Values
Appendix D	Methodology for the Development of Wildlife Criteria
Appendix E	Antidegradation Policy
Appendix F	Implementation Procedures

Table 2. List of Implementation Procedures

Procedure #	Description
1	Site-Specific Modification to Criteria and Values
2	Variances from Water Quality Standards for Point Sources
3	Total Maximum Daily Loads, Wasteload Allocations for Point Sources, Load Allocations for Nonpoint Sources, Wasteload Allocations in the Absence of a TMDL, and Preliminary Wasteload Allocations for Purposes of Determining the Need for Water Quality Based Effluent Limits
4	Additivity
5	Reasonable Potential to Exceed Water Quality Standards
6	Whole Effluent Toxicity Requirements
7	Loading Limits
8	Water Quality-Based Effluent Limitations Below the Quantification Level
9	Compliance Schedules

Table 3. Summary of the Parameters for which a Water Quality Criterion/Standard is Presented in Part 132

Parameter	BCC?	Acute Aquatic Life	Chronic Aquatic Life	Human Health	Wildlife
Arsenic (III)		XX	XX		
Benzene				XX	
Cadmium		XX	XX		
Chlordane	Yes			XX	
Chlorobenzene				XX	
Chromium (III)		XX	XX		
Chromium (VI)		XX	XX		
Copper		XX	XX		
Cyanide		XX	XX		
Cyanides				XX	
DDT				XX	
DDT and metabolites	Yes				XX
Dieldrin	Yes	XX	XX	XX	
2,4-Dimethylphenol				XX	
2,4-Dinitrophenol				XX	
Endrin		XX	XX		
Hexachlorobenzene	Yes			XX	
Hexachloroethane				XX	
Lindane	Yes	XX		XX	
Mercury (II)	Mercury	XX	XX		
Mercury (including methylmercury)	Mercury			XX	XX
Methylene chloride				XX	
Nickel		XX	XX		
Parathion		XX	XX		
PCBs (class)	Yes			XX	XX
Pentachlorophenol		XX	XX		
Selenium		XX	XX		
2,3,7,8-TCDD	Yes			XX	XX
Toluene				XX	
Toxaphene	Yes			XX	
Trichloroethylene				XX	
Zinc		XX	XX		

Key:
BCC = Bioaccumulative chemical of concern
DDD = 1, 1-Dichloro-2,2-bis(p-chlorophenyl)ethane
DDE = 1, 1-Dichloro-2,2-bis(p-chlorophenyl)ethylene
DDT = 1, 1, 1-Trichloro-2,2-bis(p-chlorophenyl)ethane
TCDD = Tetrachlorodibenzo-p-dioxin
Notes:
The concentration of 2,3,7,8-TCDD in the wastewater discharge to be compared with the effluent standards for the protection of human health must be calculated by determining the equivalent toxicity of all dioxins and furans.
DDT's metabolites include DDD and DDE.

The Guidance establishes that water quality criteria for protecting aquatic life, human health, and wildlife for pollutants not included in Tables 4 through 7 must be developed by each state using the procedures in Appendixes A, C, and D of the Guidance. In addition, water quality criteria for the chemical constituents or parameters listed in Table 9 can be established on a state-specific basis, which may or may not be in accordance with the procedures in the Guidance.

Antidegradation Policy

The antidegradation policy consists of the following provisions:

- applies, at a minimum, to BCCs;
- includes an antidegradation standard that requires maintaining the water quality necessary to protect existing uses;
- allows the identification of high-quality waters for individual pollutants;
- provides procedures for implementing the antidegradation standard;
- describes the information required for demonstrations to lower water quality; and
- presents the procedures to be followed to evaluate demonstrations to lower water quality and issue decisions on lowering water quality; these demonstrations are subject to public participation and intergovernmental coordination.

Mixing Zones Ban

With respect to the ban on mixing zones for BCCs, the Guidance establishes the following requirements:

- elimination of the use of mixing zones for the BCCs shown in Section A of Table 8 for new dischargers;

- phasing out of mixing zones for BCCs discharged by existing facilities by November 15, 2010; and
- allowing an exception to the mixing zone ban for dischargers that (1) have discharges at higher concentrations of BCCs than specified in the Guidance as a result of water conservation measures that produce an overall reduction of the mass of BCCs discharged, or (2) treat their discharge using the best possible technology and would suffer unreasonable economic effects if further control measures were taken.

GREAT LAKES TOXIC REDUCTION EFFORT

The GLTRE is evaluating nonpoint sources, wet-weather point sources, and atmospheric deposition to determine the need for additional controls on the discharge of BCCs from these sources. Some of the issues being evaluated or to be evaluated include

- additional regulation of BCCs under the Clean Air Act;
- spill prevention planning;
- reduction of BCC discharge from hazardous waste sites being remediated under the Resource Conservation and Recovery Act and the Comprehensive Environmental Response, Compensation, and Liability Act;
- improved pesticide registration procedures;
- additional regulation of combined sewer overflows and stormwater outfalls (wet-weather sources);
- public education on the dangers of BCCs;
- improved reporting of BCCs under the Toxic Release Inventory; and
- procedures to apply the National Contaminated Sediment Management Strategy in the Great Lakes Basin.

Table 4. Acute Water Quality Criteria for Protection of Aquatic Life in Ambient Water[1]

Part	Chemical	CMC (μg/L)	Conversion Factor (CF)
(a)	Arsenic (III)	339.8[a,b]	1.000
	Chromium (VI)	16.02[a,b]	0.982
	Cyanide	22[c]	n/a
	Dieldrin	0.24[d]	n/a
	Endrin	0.086[d]	n/a
	Lindane	0.95[d]	n/a
	Mercury (II)	1.694[a,b]	0.85
	Parathion	0.065[d]	n/a
	Selenium	19.34[a,b]	0.922

Part	Chemical	m(A)	b(A)	Conversion Factor (CF)
(b)	Cadmium[e,f]	1.128	−3.6867	0.85
	Chromium (III)[e,f]	0.819	+3.7256	0.316
	Copper[e,f]	0.9422	−1.700	0.960
	Nickel[e,f]	0.846	+2.255	0.998
	Pentachlorophenol[g]	1.005	−4.869	n/a
	Zinc[e,f]	0.8473	+0.884	0.978

Key:
The term "n/a" means not applicable.
The term "ln" represents the natural logarithm function.[2]
The term "exp" represents the base e exponential function.
CMC is Criterion Maximum Concentration.
CMC(tr) is the CMC expressed as total recoverable.
CMC(d) is the CMC expressed as a dissolved concentration.
CMC(t) is the CMC expressed as a total concentration.
Notes:
[a]CMC = CMC(tr)
[b]CMC(d) = [CMC(tr)]*CF. The CMC(d) shall be rounded to two significant digits.
[c]CMC should be considered free cyanide as CN.
[d]CMC = CMC(t)
[e]CMC(tr) = exp{m(A)*[ln (hardness)] + b(A)}
[f]CMC(d) = [CMC(tr)]*CF. The CMC(d) shall be rounded to two significant digits.
[g]CMC(t) = exp m(A)*[(pH) + b(A)]. The CMC(t) shall be rounded to two significant digits.

Table 5. Chronic Water Quality Criteria for Protection of Aquatic Life in Ambient Water[3]

Part	Chemical	CCC (μg/L)	Conversion Factor (CF)
(a)	Arsenic (III)	147.9[a,b]	1.000
	Chromium (VI)	10.98[a,b]	0.962
	Cyanide	5.2[c]	n/a
	Dieldrin	0.056[d]	n/a
	Endrin	0.036[d]	n/a
	Mercury (II)	0.9081[a,b]	0.85
	Parathion	0.013[d]	n/a
	Selenium	5[a,b]	0.922

Part	Chemical	m(C)	b(C)	Conversion Factor (CF)
(b)	Cadmium[e,f]	0.7852	−2.715	0.85
	Chromium (III)[e,f]	0.819	+0.6848	0.860
	Copper[e,f]	0.8545	−1.702	0.960
	Nickel[e,f]	0.846	+0.0584	0.997
	Pentachlorophenol[g]	1.005	−5.134	n/a
	Zinc[e,f]	0.8473	+0.884	0.986

Key:
The term "n/a"' means not applicable.
The term "exp" represents the base e exponential function.
The term "ln" represents the natural logarithm function.[4]
CCC is Criterion Continuous Concentration.
CCC(tr) is the CCC expressed as total recoverable.
CCC(d) is the CCC expressed as a dissolved concentration.
CCC(t) is the CCC expressed as a total concentration.
Notes:
[a]CCC = CCC(tr).
[b]CCC(d) = [CCC(tr)]*CF. The CCC(d) shall be rounded to two significant digits.
[c]CCC should be considered free cyanide as CN.
[d]CCC = CCC(t).
[e]CCC(tr) = exp{m(C)*[ln (hardness)] + b(C)}.
[f]CCC(d) = [CCC(tr)]*(CF). The CCC(d) shall be rounded to two significant digits.
[g]CMC(t) = exp{m(A)*(pH) + b(A)}. The CMC(t) shall be rounded to two significant digits.

[1]EPA recommends that metals criteria be expressed as dissolved concentrations.
[2]Not included in the regulation.
[3]EPA recommends that metals criteria be expressed as dissolved concentrations.
[4]Not included in the regulation.

Table 6. Water Quality Criteria for Protection of Human Health

Chemical	HNV (μg/L) Drinking	HNV (μg/L) Non-Drinking	HCV (μg/L) Drinking	HCV (μg/L) Non-Drinking
Benzene	1.9E1	5.1E2	1.2E1	3.1E2
Chlordane	1.4E−3	1.4E−3	2.5E−4	2.5E−4
Chlorobenzene	4.7E2	3.2E3		
Cyanides	6.0E2	4.8E4		
DDT	2.0E−3	2.0E−3	1.5E−4	1.5E−4
Dieldrin	4.1E−4	4.1E−4	6.5E−6	6.5E−6
2,4-Dimethylphenol	4.5E2	8.7E3		
2,4-Dinitrophenol	5.5E1	2.8E3		
Hexachlorobenzene	4.6E−2	4.6E−2	4.5E−4	4.5E−4
Hexachloroethane	6.0	7.6	5.3	6.7
Lindane	4.7E−1	5.0E−1		
Mercury (including methylmercury)	1.8E−3	1.8E−3		
Methylene chloride	1.6E3	9.0E4	4.7E1	2.6E3
PCBs (class)			3.9E−6	3.9E−6
2,3,7,8-TCDD	6.7E−8	6.7E−8	8.6E−9	8.6E−9
Toluene	5.6E3	5.1E4		
Toxaphene			6.8E−5	6.8E−5
Trichloroethylene			2.9E1	3.7E2

Key:
DDT = 1, 1, 1-Trichloro-2,2-bis(p-chlorophenyl)ethane
HNV = Human non-cancer value
HCV = Human cancer value
PCBs = Polychlorinated biphenyls
TCDD = Tetrachlorodibenzo-p-dioxin.

Table 7. Water Quality Criteria for Protection of Wildlife

Chemical	Criteria (μg/L)
DDT and metabolites	1.1E−5
Mercury (including methylmercury)	1.3E−3
PCBs (class)	7.4E−5
2,3,7,8-TCDD	3.1E−9

Key:
DDD = 1, 1-Dichloro-2,2-bis(p-chlorophenyl)ethane
DDE = 1, 1-Dichloro-2,2-bis(p-chlorophenyl)ethylene
DDT = 1, 1, 1-Trichloro-2,2-bis(p-chlorophenyl)ethane
PCBs = Polychlorinated biphenyls
TCDD = Tetrachlorodibenzo-p-dioxin
Note:
DDT and metabolites include DDD and DDE.

In addition to the GLTRE efforts to control BCCs, each of the Great Lakes states has prepared a Lake Management Plan (LaMP) under the binational GLWQA, which will evaluate current sources of BCCs and recommend enhancements to any media-specific program that would result in improved quality of the Great Lakes.

READING LIST

Additional information can be found in the web pages listed below. Information presented in this entry is current as of December 2001.

Great Lakes Initiative: http://www.epa.gov/waterscience/GLI/.

Table 8. Pollutants of Initial Focus in the Great Lakes Water Quality Initiative

Section A. Pollutants that Are Bioaccumulative Chemicals of Concern	
Chlordane	4,4′-DDD; p,p′-DDD; 4,4′-TDE; p,p′-TDE
4,4′-DDE; p,p′-DDE	4,4′-DDT; p,p′-DDT
Dieldrin	Hexachlorobenzene
Hexachlorobutadiene; hexachloro-1, 3-butadiene	Hexachlorocyclohexanes; BHCs
alpha-Hexachlorocyclohexane; alpha-BHC	beta-Hexachlorocyclohexane; beta-BHC
delta-Hexachlorocyclohexane; delta-BHC	Lindane; gamma-hexachlorocyclohexane; gamma-BHC
Mercury	Mirex
Octachlorostyrene	PCBs; polychlorinated biphenyls
Pentachlorobenzene	Photomirex
2,3,7,8-TCDD; dioxin	1,2,3,4-Tetrachlorobenzene
1,2,4,5-Tetrachlorobenzene	Toxaphene
Section B. Pollutants that Are Not Bioaccumulative Chemicals of Concern	
Acenaphthene	Acenaphthylene
Acrolein; 2-propenal	Acrylonitrile
Aldrin	Aluminum
Anthracene	Antimony
Arsenic	Asbestos
1,2-Benzanthracene; benz[a]anthracene	Benzene
Benzidine	Benzo[a]pyrene; 3,4-benzopyrene
3,4-Benzofluoranthene; benzo[b]fluoranthene	11,12-Benzofluoranthene; benzo[k]fluoranthene
1,12-Benzoperylene; benzo[ghi]perylene	Beryllium
bis(2-Chloroethoxy) methane	Bis(2-chloroethyl) ether
bis(2-Chloroisopropyl) ether	Bromoform; tribromomethane
4-Bromophenyl phenyl ether	Butyl benzyl phthalate
Cadmium	Carbon tetrachloride; tetrachloromethane
Chlorobenzene	p-Chloro-m-cresol; 4-chloro-3-methylphenol
Chlorodibromomethane	Chloroethane
2-Chloroethyl vinyl ether	Chloroform; trichloromethane
2-Chloronaphthalene	2-Chlorophenol
4-Chlorophenyl phenyl ether	Chlorpyrifos
Chromium	Chrysene

Table 8. *(continued)*

Copper	Cyanide
2,4-D; 2,4-Dichlorophenoxyacetic acid	DEHP; di(2-ethylhexyl) phthalate
Diazinon	1,2:5,6-Dibenzanthracene; dibenz[a,h]anthracene
Dibutyl phthalate; di-n-butyl phthalate	1,2-Dichlorobenzene
1,3-Dichlorobenzene	1,4-Dichlorobenzene
3,3′-Dichlorobenzidine	Dichlorobromomethane; bromodichloromethane
1,1-Dichloroethane	1,2-Dichloroethane
1,1-Dichloroethylene; vinylidene chloride	1,2-trans-Dichloroethylene
2,4-Dichlorophenol	1,2-Dichloropropane
1,3-Dichloropropene; 1,3-dichloropropylene	Diethyl phthalate
2,4-Dimethylphenol; 2,4-xylenol	Dimethyl phthalate
4,6-Dinitro-o-cresol; 2-methyl-4,6-dinitrophenol	2,4-Dinitrophenol
2,4-Dinitrotoluene	2,6-Dinitrotoluene
Dioctyl phthalate; di-n-octyl phthalate	1,2-Diphenylhydrazine
Endosulfan; thiodan	alpha-Endosulfan
beta-Endosulfan	Endosulfan sulfate
Endrin	Endrin aldehyde
Ethylbenzene	Fluoranthene
Fluorene; 9H-fluorene	Fluoride
Guthion	Heptachlor
Heptachlor epoxide	Hexachlorocyclopentadiene
Hexachloroethane	Indeno[1,2,3-cd]pyrene; 2,3-o-phenylene pyrene
Isophorone	Lead
Malathion	Methoxychlor
Methyl bromide; bromomethane	Methyl chloride; chloromethane
Methylene chloride; dichloromethane	Naphthalene
Nickel	Nitrobenzene
2-Nitrophenol	4-Nitrophenol
n-Nitrosodimethylamine	n-Nitrosodiphenylamine
n-Nitrosodipropylamine; n-nitrosodi-n-propylamine	Parathion
Pentachlorophenol	Phenanthrene
Phenol	Iron
Pyrene	Selenium
Silver	1,1,2,2-Tetrachloroethane
Tetrachloroethylene	Thallium
Toluene; methylbenzene	1,2,4-Trichlorobenzene
1,1,1-Trichloroethane	1,1,2-Trichloroethane
Trichloroethylene; trichloroethene	2,4,6-Trichlorophenol
Vinyl chloride; chloroethylene; chloroethene	Zinc

Key:
DDD = 1, 1-Dichloro-2,2-bis(p-chlorophenyl)ethane
DDE = 1, 1-Dichloro-2,2-bis(p-chlorophenyl)ethylene
DDT = 1, 1, 1-Trichloro-2,2-bis(p-chlorophenyl)ethane
TCDD = Tetrachlorodibenzo-p-dioxin
TDE = DDD

Table 9. Pollutants Subject to Federal, State, and Tribal Requirements

Alkalinity	Ammonia
Bacteria	Biochemical oxygen demand (BOD)
Chlorine	Color
Dissolved oxygen	Dissolved solids
pH	Phosphorus
Salinity	Temperature
Total and suspended solids	Turbidity

Federal Register—Environmental Documents: http://www.epa.gov/fedrgstr/.

Code of Federal Regulations: http://www.access.gpo.gov/nara/cfr/index.html.

Great Lakes Water Quality Agreement: http://www.on.ec.gc.ca/glwqa.

Great Lakes National Program Office: http://www.epa.gov/glnpo/.

QUANTITATIVE GROUNDWATER LAW

Marshall Lawson
Land Conservation Legal Services, LLC
Columbia, South Carolina

INTRODUCTION

From ancient times, the law looked upon surface water and groundwater as separate resources, requiring separate rules for ownership and use. Whereas surface water was considered a public good to be shared among competing users, groundwater was equated with private property to be appropriated by the landowner at will.[1]

[1]Compare Dig. 43.12.3 (Paulus, Ad Sabinum 16) ("Rivers which flow [perennially] are public, and their banks are also public.");

The separate classification and treatment of groundwater can be attributed, in part, to a primitive understanding of hydrogeology; a mathematical description of groundwater storage and movement was not formulated until the midnineteenth century and not widely disseminated until the late twentieth century.[2] Without an understanding of aquifer characteristics, legal authorities lacked a basis for developing sophisticated groundwater allocation rules. The occurrence and nature of groundwater was, to them, too "... uncertain ...[to] subject it to the regulations of law...".[3]

Hand in glove with a nascent understanding of hydrogeology was the propensity of nineteenth century English jurists and commentators to follow the perceived equities of Roman water law as "indirect authority" for English water law.[4] The English Reasonable Use Rule, for example, embodies the Roman concept of limited property rights in public surface waters.[5] Likewise, the Roman classification of groundwater as private property free from any equitable rules of sharing appeared well over a millennium latter in English and American common law as Absolute Dominion or the Rule of Capture.[6]

The most important reason, however, for the development of separate legal rules for surface and groundwater centered not upon science or legal history but technology. Prior to the invention of the deep-well turbine pump in the last decade of the nineteenth century, groundwater could not commonly be exploited at rates sufficient to encroach upon the property rights of neighboring landowners. Consequently, there was little need to impose legal restraints on groundwater use.[7] The advent of submersible pumps and power drilling equipment, however, made it possible to develop aquifers at sufficient pumping rates to cause widespread injury to the environment and other property owners in the form of diminished stream flow, saltwater intrusion, land subsidence, well interference, and groundwater mining.[8] Technology had effectively made the once functional Rule of Capture obsolete in theory, if not in practice.

As groundwater conflicts became more common, American courts attempted to mitigate the harsh Rule of Capture by adopting various equitable rules of allocation, starting with the Reasonable Use Doctrine in 1862 followed by the Correlative Rights Doctrine in 1903 and The Restatement (Second) of Torts rule in 1979.[9] In reality, the rough justice dispensed by courts under these doctrines did little to protect groundwater supplies from being diminished or extinguished by competing users, and aquifers remained theoretically susceptible to complete dewatering by overlying property owners.

In response to the failure of the common law to manage groundwater use effectively, many Eastern states enacted legislation modifying common law rights in the use of groundwater. The legislation varies widely in scope and criteria but centers upon the requirement of an administrative permit to pump groundwater above certain predetermined thresholds.[10]

In contrast to the various quasi-riparian rules governing groundwater use in states west of the Mississippi River, a distinctly American system of water use, called the Prior Appropriation Doctrine, arose in the arid West. The system allocates water rights based upon priority of beneficial use and irrespective of contiguous (or in the case of groundwater, overlying) land ownership.

Appropriation proved more suited to the climate and geology of the arid West than riparianism. The central tenets of the riparian doctrine, that water rights accrue

Dig. 43.20.3.1 (Pomponius, Ad Sabinum 34) ("Many may take away water from a river, but in such manner only that their neighbors are not injured, or if the stream is not large, those on the other side.") with Dig. 43.24.11 (Ulpian, Ad Edictum 71) (Labeo stating the well settled rule that groundwater was considered part of the land); Dig. 39.3.1.12 (Ulpian, Ad Edictum 53) (Marcellus holding "that to him who, digging in his own land, turns away the sources of his neighbor's spring, nothing can be done, not even an action for fraud, and surely there should be none if that which he did was for the purpose of benefiting his own field, and not with the intention of injuring his neighbor.") (The Digests or Pandects of Justinian, commenced in A.D. 530 and completed in A.D. 533, comprise part of the Corpus Juris Civilis, the great codification of the Roman Emperor Justinian. The sections of the Digest pertaining to water reveal a remarkable understanding of the equities of water use. Although written almost 1500 years ago and based upon legal principles formed over the preceding 1000 years, the Pandects provide a foundation and, in some cases, a blueprint for many of the principles of Western water law in use today); See also, Earl Finbar Murphy, Waters and Water Rights § 20.02 (Robert Beck ed., 2d ed. 1991).

[2]French Engineer Henry Darcy pioneered the basic equations of modern groundwater hydrogeology in the mid-nineteenth century by quantifying the rate and direction of groundwater flow; see C.W. Fetter, Applied Hydrogeology 76 (1988); Henri-Philbert-Gaspard Darcy, Encyclopedia Britannica, III Micropaedia 883 (1991).

[3]Murphy, supra note 1, at §§ 20.03 [citing Chatfield v. Wilson, 28 Vt. 49, 54 (1856)].

[4]See C.E. Busby, *American Water Rights Law: 5* S.C.L.Q. 106, 116 n. 57 (1952) (The language "no direct authority" in Acton v. Blundell 152 Eng. Rep. 1228 (Ex. Cham. 1843) raises the presumption the court relied on Roman law in adopting a Rule of Capture for groundwater); See also Murphy, supra note 1, at § 20.04 [citing Chasemore v. Richards 11 Eng. Rep. 140, 155 (1857)] (Lord Wensleydale quoting from the Roman sources for the proposition that groundwater lay within the domain of private property); Embrey v. Owen, 155 Eng. Rep. 579, 585 (1851) (Baron Parke adopting the Roman concept of usufructuary rights in public surface waters).

[5]Murphy, English Water Law Doctrines Before 1400, 1 Am J. Leg. Hist. 103, 105–106 (1957) (tracing the Roman concept of surface water as a public good); Embrey v. Owen, 159 Eng. Rep. 579, 585 (1851) ("[Water that] is *publici juris* [is not so] in the sense that it is a *bonum vacans,* to which the first occupant may acquire an exclusive right, but that it is public and common in this sense only: that all may reasonably use it who have a right of access to it; that none can have any property in the water itself, except in particular portion which he may choose to abstract from the [source] and take into his possession, and that during the time of possession only.").

[6]See Greenleaf v. Francis, 35 Mass. 117 (1836); Acton v. Blundell, 152 Eng. Rep. 1223 (Ex. Cham. 1843).

[7]Raphael G. Kazman, Modern Hydrogeology 272 (1988).

[8]Sax, J.L. et al. (2000). *Legal Constraints on Water Resources*, 3rd Edn. p. 345.

[9]See generally, Basset v. Salisbury Mfg. Co., 43 N. H. 569, 82 Am Dec. 179 (1862); Katz v. Walkinshaw, 74 P. 766 (Cal. 1903); Restatement (Second) of Torts § 858 (1979).

[10]Murphy, supra note 1, at § 24.05.

only to contiguous or overlying landowners and are primarily limited to on-tract use, proved impracticable in the West where the federal government owned much of the land, water resources were scarce, and mining and agriculture practices required off-stream water use. Furthermore, the subjective allocation criteria of the various riparian doctrines proved ill suited for managing scarcity.

ORIGIN AND NATURE OF PROPERTY RIGHTS IN GROUNDWATER

Property rights in water, including groundwater, have a long pedigree in European jurisprudence. Roman law is the oldest legal system to which the roots of European water law can be traced with any certainty and, as discussed above, Roman jurists considered groundwater an integral part of the land.[11] As such, the landowner enjoyed a property right in the groundwater itself (res) as well as an unfettered right of use (usufruct).[12] English and American courts firmly embraced this concept of groundwater as an absolute property right in the nineteenth century.[13] Although many U.S. courts and legislatures now define groundwater as a public resource, most, if not all, jurisdictions still acknowledge (at least tacitly) a usufructuary right of the landowner to use the groundwater underlying his property.[14]

As a form of property, regulation of groundwater use is theoretically subject to challenge as a "taking" of private property without compensation in violation of the Takings Clause of the United States Constitution.[15] At least one court has stricken modification of water rights as a violation of the Takings Clause,[16] but the decision is very much an anomaly. Other courts addressing the issue have upheld groundwater regulation in the face of takings clause challenges on one of the following grounds: (1) Permitting has not deprived a landowner of all economic use of his property right and is, therefore, not a compensable taking. (2) Landowners enjoy no constitutionally protected right in groundwater until they actually begin pumping. (3) The regulation is a valid exercise of the sovereign's police power.[17] States have broad police power under the Constitution to regulate for the public health, safety, morals, and general welfare. The sovereign's exercise of this power may not require compensation for diminished property rights resulting from regulation, as long as such regulation is not arbitrary or irrational.[18]

At least one authority would add another defense to takings clause challenges to groundwater regulation. Because the overlying landowner's right to use water under various riparian doctrines is "subject to complete destruction by a more powerful pumper, [the] ... landowner's expectation of exclusive use of groundwater [is] close to illusory."[19] Stated another way, riparian law (or quasi-riparian law for groundwater) holds that groundwater is private property but effectively treats it as common property by failing to provide a mechanism for quantifying and protecting the water right.

On the other hand, it could be argued that permitting systems for groundwater use, depending upon the terms of the permit, might actually strengthen the claim of a landowner to a property right in the use of groundwater. Consistent with this argument, a permit holder could defend his right to pump against future permit holders if such pumping reduced his own allotted yield.[20]

Regardless of the court's willingness to uphold groundwater-use regulation, it should not be assumed that courts or groundwater-use statutes have abolished private property rights in the use of groundwater. Even the Regulated Riparian Model Water Code, which proposes comprehensive regulation of water resources, makes clear that the administrative permit process does not usurp

[11] Ludwig A. Teclaff, Water Law in Historical Perspective 6 (1985); Dig. 43.24.11 (Ulpian, Ad Edictum 71).

[12] Dig. 39.3.1.12 (Ulpian, Ad Edictum 53); Maricopa County Mun. Water Conservation Dist. Number One v. Southwest Cotton Co., 4 P.2d 369, 372 (Ariz. 1931).

[13] See Greenleaf v. Francis, 35 Mass. 117, 121 (1836); Acton v. Blundell, 152 Eng. Rep. 1223, 1228 (Ex. Cham. 1843); Chatfield v. Wilson, 28 Vt. 49, 54 Chasemore v. Richards, 11 Eng. Rep. 140, 155 (1857); see also, John M. Gould, Law of Waters 538 (2d Ed. 1891) ("Water percolating through the ground beneath the surface, either without a definite channel, or in the courses which are unknown and unascertainable, belongs to the realty in which it is found").

[14] Murphy, supra note 1, at § 20.04 ("Notwithstanding the importance of public rights in water, both the common law and the legislatively created appropriation doctrine have made rights to water primarily the domain of private property, and this includes groundwater"); see generally, Bamford v. Upper Republican Natural Resources District 512 N.W. 2d 643 (Neb. 1994); Tequesta v. Jupiter Inlet Co., 371 So.2d 663 (Fla. 1979).

[15] U.S. Const. amend. V (The Takings Clause of the U.S. Constitution bars the taking of property by the federal or state government without compensation); see, e.g., Lucas v. South Carolina Coastal Counsel, 505 U.S. 11003 (1992) (finding, on remand, a compensable taking based upon the South Carolina Coastal Council's refusal to grant a permit to construct two homes on the petitioner's oceanfront property; see generally, Robert Meltz et al., The Takings Issue: Constitutional Limits on Land Use Control and Environmental Regulation 463-70 (1999).

[16] See American Charolaise, Ltd. V. Oklahoma Water Resources Board, 855 P.2d 568 (Okla. 1990) (The Oklahoma Supreme Court struck down legislation limiting riparian water rights, holding that riparian rights constitute private property rights and legislation permitting competing appropriators to divert the water constituted a taking of that property for public use).

[17] See Meltz, supra note 15, at 126 (citing Department of Agric. & Consumer Serv. V. Polk, 276 U.S. 272 (1928); German v. City of Philadelphia, 700 A.2d 444 (Pa. 1997); Hoeck v. City of Portland 57 F.3rd 781, 787 (9th Cir. 1995)). Compare City of Houston v. Crabb, 905 S.W.2d 669, 674-75 (Tex. Ct. App. 1995) (finding a taking where the "[c]ity failed to prove a valid exercise of its police power").

[18] SAX, supra note 8, at 396 (citing Peterson v. Dept. of Ecology, 596 P.2d 285 (1979)); see also Bamford v. Upper Republican Natural Resources Dist., 512 N.W.2d 642 (Neb. 1994); Town of Chino Valley v. City of Prescott, 638 P.2d 1324 (1981); Crookston Cattle Co. v. Minnesota Dept. of Nat'l Resources, 300 N.W.2d 769 (Minn. 1980); Village of Tequesta v. Jupiter Inlet Corp., 371 So.2d 663 (Fla. 1979); Omernik v. State, 218 N.W.2d 734 (Wis. 1974); Knight v. Grimes, 127 N.W. 2d 708 (S.D. 1964); Williams v. City of Wichita, 279 F.2d 375 (10th Cir. 1960).

[19] A. Dan Tarlock et al., Law of Water Rights and Resources § 4.09[1] (1986).

[20] Sax, supra note 8, at 396.

common law rights in water and that "the new system is a regulation rather than a taking of the older riparian rights."[21]

LEGAL CLASSIFICATIONS OF GROUNDWATER

The common law historically distinguished not only between groundwater and surface water, but also distinguished between two types of groundwater: (1) that found flowing in underground streams; and (2) so-called "percolating" groundwater, or that which did not "flow" but rather "oozed" through the void spaces within the aquifer skeleton. Courts generally assign the law of surface streams to the former, and groundwater law to the latter.[22]

As mentioned above, this dual classification arose initially because "percolating" groundwater was envisioned as too "occult in its occurrence and movement" for courts to attach formalistic rules to its use.[23] In addition, courts labored under the false assumption that a large portion of groundwater occurs as an underground analog to surface streams, and, therefore, both flow regimes should be governed by the same principles. In reality, "percolating" groundwater, like stream water, flows in an ascertainable direction and rate, albeit through an aquifer matrix and at orders of magnitude slower than surface water. Furthermore, the occurrence of groundwater as underground streams is extremely rare and confined to karstic terrains or lava flows.[24]

In acknowledgment of the hydraulic interconnection between groundwater flow and stream flow, some modern courts recognize a third legal category of groundwater, the "subflow" of surface streams. The subflow of surface streams, like water flowing in underground streams, was generally made subject to the same law that applied to surface water.[25]

Finally, statutes in a number of jurisdictions exempt certain kinds of groundwater from ordinary groundwater law. For example, groundwater pumped from a mine to allow continued mineral extraction may be exempt from regulation applied to other groundwater. Another category of groundwater often exempted from regulation as groundwater is water hot enough to have value as a geothermal resource. Water extracted from the ground as a by-product of oil and gas development may also be separately treated by statute.[26]

[21]American Society of Civil Engineers, Water Resources Planning & Management Div., the Regulated Riparian Model Water Code; Final Report of the Water Laws Committee, vii (Joseph W. Dellapenna ed,, 1997).
[22]Murphy, supra note 1, at § 20.07 (citing the American & English Encyclopedia of Law 311 (1905); Dickinson v. The Grand Junction Canal Co., 155 Eng. Rep. 953 (Ex. 1852)).
[23]See, e.g., Frazier v. Brown, 12 Ohio St. 294, 311 (1861) ("[T]he existence, origin, movement, and course of such water, and the causes which govern and direct their movements, are so secret, occult and concealed, than an attempt to administer any set of legal rules in respect to them would be involved in hopeless uncertainty, and would be, therefore, practically impossible.").
[24]Fletcher G. Driscoll, Groundwater and Wells, 672 (2d ed. 1986).
[25]Sax, supra note 8, at 360.
[26]Id. at 361.

LEGAL RULES FOR GROUNDWATER USE

Federal Legislation

The United States has no single rule governing groundwater use. Although the federal government regulates a wide range of groundwater pollution sources, it has, thus far, left formulation of quantitative groundwater law to the states, which have in turn adopted their own unique rules and statutes governing groundwater use.[27]

State Groundwater Law

Common Law (Quasi-Riparian Doctrines). Four distinct allocation doctrines for groundwater use apply east of the Mississippi and in a handful of Western states. These doctrines are best described as quasi-riparian because the property right in the groundwater vests with ownership of overlying land. They are (1) the "English rule" of Capture or Absolute Ownership, (2) the "American rule" of Reasonable Use, (3) the "Correlative Rights Rule" of shared access, and (4) the Restatement of the Law of Torts (2d, Tent. Draft 17, § 858A).[28]

A definitive and accurate appraisal of states adhering to a particular groundwater doctrine is problematic due to the blurring of the various rules in case law; an absence of decisional law in some states; and administrative overlays on the common law that modify, in various degrees, common law rights in groundwater.

In general terms, Absolute Ownership survives in only a few jurisdictions.[29] Most Eastern states employ some variation of the Reasonable Use Rule.[30] Several states follow some variation of the Correlative Rights Rule originating in California.[31] The Restatement approach has found little favor and has been adopted in only a handful of states.[32]

Absolute Ownership (Capture). Under this doctrine, also known as the English Rule of Capture, landowners are allowed to withdraw all the water they wish for whatever

[27]Murphy, supra note 1, at § *20.04*.
[28]SAX, supra note 8, at 364.
[29]Id. at 366; see generally, Wiggins v. Brazil Coal & Clay Corp., 452 N.E.2d 958 (Ind. 1983); City of Sherman v. Public Util. Comm'n, 643 S.W.2d 683 (Tex. 1983); St. Amand v. Lehman, 47 S.E. 949 (Ga. 1904) (but putting a restriction on the rule of capture by holding a malicious diversion or wasting of percolating underground waters would be actionable).
[30]See generally, Clinchfield Coal Corp. v. Compton, 139 S.E. 308 (1927); Rouse v. City of Kinston, 123 S.E. 482 (N.C. 1924); Nashville C. & St. L. Ry. V. Rickert, 89 S.W.2d 889 (1935), *cert. Denied* by Tennessee Supreme Court (1936); Sloss-Sheffield Steel & Iron Co. v. Wilkes, 165 So. 764 (Ala. 1938); Barclay v. Abraham, 96 N.W. 1080 (Iowa 1903); Cincinnati, N.O., & Tex. Pac. R.R. v. Gillespie, 113 S.W. 89 (Ky. 1908); Schenck v. City of Ann Arbor, 163 N.W. 109 (Mich. 1917); Pence v. Carney, 52 S.E. 702 (W. Va. 1905); Cline v. American Aggregates Corp.; 474 N.E.2d 324 (Ohio 1984); Rothrauf v. Sinking Spring Water Co., 14 A.2d 87(Pa. 1940); Wood v. Picillo, 443 A.2d 1244 (R.I. 1982); State v. Michels Pipeline Const., Inc., 217 N.W.2d 339 (Wis. 1974)
[31]Katz v. Walkinshaw, 74 P. 766 (Cal. 1903); see also Jones v. Oz-Ark-Val Poultry Co., 306 S.W.2d 111 (Ark. 1957); Prather v. Eisenmann, 261 N.W.2d 766 (Neb. 1978); Vt. Stat. Ann. Tit. 10, § 1410(a)(3) (1984).
[32]SAX, supra note 8, at 366.

purpose on or off the owner's land. The owner could even waste the water, thereby injuring a neighbor, but still face no liability.[33]

American Reasonable Use. Reasonable Use, also known as the American Rule, found favor with American courts as a modification of Absolute Dominion. In effect, Reasonable Use is equivalent to an equitable servitude upon the property right in groundwater requiring that (1) groundwater use be reasonable and (2) such use must be confined to the overlying tract of land.[34]

The term "Reasonable" varied among jurisdictions but generally meant "A waste of water or a wasteful use of water was unreasonable only if it caused harm, and any nonwasteful use of water that caused harm was, nevertheless, reasonable if it was made on or in connection with the overlying land."[35] The doctrine does not articulate a rule of allocation, but some courts imply a rule of sharing that makes the practical application of the doctrine difficult to differentiate from the Correlative Rights Doctrine.[36]

Correlative Rights ("California Rule"). This doctrine, first articulated by the California Supreme Court, requires a sharing of available water equitably among overlying landowners. They share the resource in times of scarcity proportionate either to their aerial land ownership or actual historic use.[37] Even if the water is used beneficially on the overlying land, landowners cannot infringe upon the rights of others by exceeding their proportionate shares.[38]

Off-tract uses are subordinate and are legally protected only in the event "surplus" water is available, that is, if recharge exceeds withdrawals.

If no competition for water exists, then Correlative Rights are synonymous with Reasonable Use.[39]

Restatement (Second) of Torts (Reasonable Use). Although the Restatement rule is similar to reasonable use, it draws no distinction between on-tract and off-tract use. It also differs from Reasonable Use by establishing criteria for liability. A landowner who pumps groundwater from his land and uses it for a beneficial purpose is liable for interference with other's groundwater use if

1. the withdrawal causes unreasonable harm through lowering of the water table;
2. the groundwater occurs in a distinct underground stream;
3. the withdrawal of water has a substantial effect upon a stream, river, or lake.[40]

Priority System (Prior Appropriation). In the West, a uniquely American water use rule called the Prior Appropriation Doctrine gradually supplanted riparian principles for both surface and groundwater. In contrast to riparian doctrines, the acquisition of water rights does not depend upon the ownership of contiguous or overlying land. Water rights vest by appropriating the water and putting it to beneficial use and accrue on a first in time, first in right, basis. Furthermore, water rights are quantified, can be lost through nonuse, and are transferable, as long as other water rights holders suffer no harm.

Prior appropriation developed in the mining camps of the Sierra Nevada during the California gold rush of the 1840s. As discussed before, the nature of mining operations and federal ownership of western lands were incompatible with the riparian doctrine for three reasons: (1) Under the riparian system, water rights accrued only to the owners of land. The miners, however, did not own the land they mined; it belonged to the federal government; (2) Mining operations were often conducted remotely from waterbodies requiring off-stream diversions that violated riparian prohibitions of off-tract transfers of water.[41] (3) In contrast to riparian doctrines, Appropriation provided quantifiable water rights that were enforceable and transferable. These attributes fostered predictability of planning and investment.[42]

Appropriation arose first as a common law doctrine, effectively ratifying water use practices in the mining camps. Over time, the doctrine was codified and even incorporated into some state's constitutions.[43] Most Western states have extended some version of the Prior Appropriation Doctrine, to groundwater, by statute, making it the primary doctrine for groundwater use in the West.[44]

The Appropriation rules for groundwater are essentially the same as those for surface water. In case of a dispute, priority of use is the determining factor in allocating water among competing users; junior claimants may have their supply of groundwater diminished or eliminated by senior claimants in times of scarcity.[45] The realities of hydrogeology, including the time lag between the start of pumping and deleterious effects on water levels, can make

[33]See Greenleaf v. Francis, 35 Mass. 117 (1836) ("Everyone has the liberty of doing in his own ground whatsoever he pleases, even although it should occasion his neighbor some other sort of inconvenience"); Acton v. Blundell, 152 Eng. Rep. 1223, 1228 (Ex. Cham. 1843) (In this case, a landowner's well was dewatered by a neighbor's excavation of a coal mine. The court held the damage to be *damnum absque injuria* [an injury without a remedy] based upon its determination that groundwater flow was inherently unpredictable: "... the person who owns the surface may dig therein; and apply all that is there found to his own purposes at his free will and pleasure; and that if, in the exercise of such right, he intercepts or drains off the water collected from underground springs in his neighbor's well, this inconvenience to his neighbor ... cannot become the ground of an action ...").

[34]Henderson v. Wade Sand & Gravel Co., 388 So.2d 900 (Ala. 1980)

[35]Id.

[36]Murphy, supra note 1, at § 22.01

[37]Joseph W. Dellapenna, Groundwater Law for Mineral Lawyers, Energy & Mineral Law Foundation (1992).

[38]Murphy, supra note 1, at § 20.07.

[39]Sax, supra note 8, at 364; Fletcher G. Driscoll, Groundwater and Wells 673 (2d ed. 1986).

[40]Restatement (Second) of Torts § 858 (1979).

[41]Sax, supra note 8, at 282

[42]Id. at 286

[43]Id. at 382

[44]Id. at 282; Murphy, supra note 1, at §§ 20.04; 20.07.

[45]Clesson Kinney, The Law of Irrigation and Water Rights and Arid Region Doctrine of Appropriation of Waters 598 (1912); Atchison v. Petersen, 87 U.S. 507 (1874); SAX, supra note 8, at 382.

application of the doctrine difficult to enforce in the case of groundwater.[46]

Active Management and Regulation of Groundwater

Permit Statutes ("Regulated Riparianism")[47]. Although some Eastern states enacted groundwater statutes more than a century ago, such legislation did not become common until the second half of the twentieth century. As of 1996, eighteen states had enacted regulated riparian systems for surface water sources, generally including underground water resources as well. Two other states apply a regulated riparian system only to underground water sources.[48]

The common characteristic of such legislation is the requirement of a permit to withdraw groundwater based upon reasonableness of use.[49] An administrative agency is responsible for both the decision to grant or deny a permit and, at least initially, dispute resolution.[50]

These statutes were not originally introduced as a radical revision of the water law of a particular state but emerged gradually through a process of discrete initiatives in the form of well-spacing regulations, prohibitions of wasteful pumping, or licensing of major withdrawals. In several states, it remains unclear whether the law would be better described as basically riparian rights with limited legislative alterations or as comprehensive regulatory schemes referred to as "regulated riparianism."[51]

Supply Side (Conjunctive Use) Management. The Orange County (California) water district (OCWD) takes a novel approach to groundwater allocation. Instead of imposing restrictions on pumping, the district focuses on maintaining groundwater supplies by recharging aquifers with imported surface water. No attempt is made to limit groundwater pumping directly or prevent waste. In fact, OCWD is forbidden from directly limiting the amount of groundwater an individual user may withdraw, although it may discourage use by levying a "basin equity assessment" against pumpers who withdraw more than a percentage of groundwater set by the district. In addition, the OCWD does not attempt to manage groundwater withdrawals to prevent or minimize well interference.[52]

The drawbacks to this type of management model include dependence upon imported recharge water and the cost of transmission and recharge facilities. These costs, however, are largely offset by maintaining a stock of relatively cheap local water in the form of groundwater storage.[53]

[46]Kazman, supra note 7, at 272; DELLAPENNA, *supra* note 37, at 7.

[47]Dellapenna, supra note 37, at 7

[48]Model Water Code, supra note 21, at vi.

[49]Id. at viii.

[50]Dellapenna, supra note 37.

[51]Model Water Code, supra note 21, at vi.

[52]Sax, supra note 8, at 454 (citing William Bloomquist, Dividing the Waters: Governing Groundwater in Southern California, ICS Press, 1992).

[53]Id. at 455.

NATURE OF GROUNDWATER DISPUTES

Most groundwater disputes are the product of a race-to-pump mentality referred to as "the tragedy of the commons."[54] When many users tap the same aquifer, it becomes a common property resource. Without social or legal limits on water use from the common aquifer, there is no incentive to conserve because the conserver does not enjoy an exclusive right to the water that is saved.[55] For example, under the Reasonable Use Doctrine, two neighboring landowners are free to withdraw all the groundwater they can put to beneficial use upon their own land. The reasonableness of the use is judged solely as it relates to the purpose of such use and not its effect upon others or the sustainability of the resource.[56]

The race-to-pump has produced widespread declines in potentiometric surfaces and in severe cases, groundwater mining[57] that, in turn, led to well interference, land subsidence, saltwater intrusion,[58] and diminished stream flow. Unwise groundwater use may even threaten endangered species by desiccating springs.[59] In addition to environmental damage, these phenomena impose additional costs upon water users, ranging from a few thousand dollars incurred by individual well owners for well deepening and increased electrical costs to millions of dollars to provide alternative water supplies for municipal systems threatened by saltwater intrusion.[60]

[54]William H. Rodgers, Jr., Environmental Law 39 (West 1994) [citing Garrett Hardin, *The Tragedy of the Commons*, 162 Science 1243 (1968)].

[55]Tom Tietenberg, Environmental and Natural Resource Economics 207 (1988)

[56]Hornsberger, Oeltjen, & Fischer, *Groundwater: From Windmills to Comprehensive Public Management*, 52 Neb. L. Rev. 179 at 205 (1973).

[57]Driscoll, supra note 24, at 64 (The potentiometric surface is an imaginary line representing the hydrostatic head throughout all or part of a confined aquifer and is the level to which water will rise in a tightly cased well. Unwise groundwater use practices, such as improper well spacing, may draw down the potentiometric surface without actually dewatering the aquifer); Kazman, supra note 7 (Groundwater mining results when the rate of withdrawal exceeds the rate of recharge to an aquifer, decreasing the volume of groundwater held in storage. This phenomenon is especially acute in arid regions where aquifers have little to no natural recharge. Without artificial recharge, it may take thousands of years for such an aquifer to regain its prepumping storage volume).

[58]J. Marshall Lawson, *Transboundary Groundwater Pollution: The Impact of Evolving Groundwater Use Laws on Salt Water Intrusion of the Floridian Aquifer along the South Carolina-Georgia Border,* 9 S. C. Env. L. J. 85 (2000).

[59]Ben F. Vaughn IV & Peter M. Emerson, Protecting the Edwards Aquifer; An Efficient and Ecological Alternative, in Water Marketing: The Next Generation 167 (Terry L. Anderson & Peter J. Hill eds., 1997) (Explosive population growth and unrestricted pumping of groundwater from the Edwards aquifer in central Texas is depleting the flow of water to the aquifer's discharge areas threatening several endangered species with extinction and threatening economic interests dependent on plentiful water supplies. This realization has prompted legislators to rethink the state's longstanding adherence to the Rule of Capture).

[60]Lawson, supra note 58, at 95 [Saltwater intrusion of the Floridian aquifer in the vicinity of Savannah, Georgia, forced

Groundwater disputes may involve adjudication of the denial for new permits to pump groundwater or limitations and conditions upon existing permits.[61] In the case of property damages from groundwater use, courts may apply groundwater law (common law or a statute on point) or even tort law (e.g., nuisance or negligence theories) to the facts of a case to determine pumping rights or fix remedies in the form of an injunction or money damages.[62]

Plaintiffs may be private citizens, corporations, or public entities such as municipalities seeking to define, expand, or defend rights to groundwater. Depending on the jurisdiction, groundwater disputes, not resolved by administrative processes, may be appealed to higher state courts.

In addition to state common law and administrative law, federal law may come into play if groundwater use affects federal water rights claims,[63] impacts a federal statute, such as The Endangered Species Act,[64] or when groundwater use has an interstate dimension such as transportation of groundwater across state lines.[65]

LEGAL REFORM

Until recently, American groundwater law has been a hodgepodge of outdated common law doctrines that lacked an effective mechanism for allocating supplies on a sustainable basis and dealt only with disputes after the fact, during litigation. The need for reform prompted proposals for statutory models that can be roughly divided between emerging market-based approaches based upon the movement of water to higher value uses,[66] on the one hand, and command and control regulatory schemes restricting water use, on the other.[67]

Regulatory proponents believe that agency experts are the best suited to allocate the limited supplies in a sustainable fashion. In their eyes, market-based approaches to natural resource management inevitably lead to competitors vying to monopolize water rights and, thus, deprive others of access to water supplies.

Market proponents, on the other hand, argue that administrative control does not end competition for water but only shifts competition from the market to the political arena thereby promoting uncertainty that tends to work against meaningful conservation.[68] They also decry the assumption long held by legal authorities that control of the production of water, like alcohol, must somehow be tied to the end use of the product and not related to the valuation of the marketplace.[69] The more economical and less environmentally harmful choice to them is a system of tradable property rights in groundwater based upon either a share system or a seniority system.[70] This type of system would allow water to seek its highest value use, benefiting both buyers and sellers and minimizing government regulation with its attendant costs.

Others would like to see a combination of the two models, market incentives with government oversight to protect outside parties and the general public from harmful side effects.[71] Such oversight could include zoning restrictions, assignment of quota rights in common pools, and the right to receive money damages for economic harm to individual water users.[72]

CONCLUSION

Although groundwater law has progressed beyond the ancient Rule of Capture in most jurisdictions, it remains far from the nimble and pristine law required for the wise management of groundwater resources. Effective groundwater law of the future must focus on sustainable use of the resource within the context of both local and regional hydrogeologic realities, a formidable task, given the complexities of hydrogeology and political impediments to meaningful legal reform.

The law should encourage efficient water use by providing quantifiable and transferable property rights in water, and government should play the role of gatekeeper in maintaining environmental integrity and access to water for newcomers. Above all, water should no longer be regarded as so plentiful that it is regarded as a free good or so precious that it can be assigned no value. Where demand exceeds thresholds the environment can safely bear, as in mining or saltwater intrusion, consumers should pay for the realistic cost of alternative water supplies. Realistic water pricing combined with sound laws incorporating the appropriate mix of market incentives and regulation would foster both conservation and the beneficial transfer of water from areas of abundance to areas of scarcity.

Regardless of the mix of regulation and economic incentives, wise use of groundwater in the future will

Public Service Districts on Hilton Head Island, South Carolina, to develop alternative water supplies (AWS) for the island in the form of transmission lines to the Savannah River and the construction of a deep well and desalination facility to treat heavily mineralized water from deeper aquifers. The total cost of AWS was estimated at $24,950,000.00).

[61]See, e.g., Baker v. Ore-Ida Foods, Inc., 513 P.2d 627 (Idaho 1973).

[62]See generally, Adams v. Lang, 533 So.2d 89 (Ala. 1989); Friendswood Development Co. v. Smith Southwest Industries, Inc., 576 S.W.2d 21(Tex. 1978); Finley v. Teeter Stone, Inc. 248 A.2d 106 (Md. 1968); Restatement (Second) of Torts § 858(1)(a).

[63]See, e.g., Cappaert v. United States, 426 U.S. 128 (1976).

[64]16 U.S.C.A. §§ 1531–1534 (1982).

[65]Vaughn & Emerson, supra note 59, at 177; Sporhase v. Nebraska, ex rel. Douglas, 458 U.S. 941 (1982) (The U.S. Supreme Court holding that a Nebraska statutory restriction on the interstate transportation of its groundwater violated the Commerce Clause of the U.S. Constitution); See also U.S. Const. art. I, § 10, cl. 3; art. III, § 2. cl. 1,2.

[66]Vaughn & Emerson, supra note 59, at 167; Terry L. Anderson & Pamela Snyder, Water Markets: Priming the Invisible Pump 161; Tom Tietenberg, Environmental and Natural Resource Economics 203 (1988); James E. Kundell & Diana Tetens, Carl Vinson Inst. of Gov't, Whose Water Is It?: Major Water Allocation Issues Facing Georgia 9 (1988).

[67]See generally, Model Water Code, supra note 21.

[68]Kazman, supra note 7, at 278 (Citing J. Hirshleifer et al., Water Supply Economics, Technology, and Policy 35 (1960)).

[69]Id. at 283.

[70]Vaughn & Emerson, supra note 59, at 180.

[71]Kazman, supra note 7, at 367 (Citing J. Hirshleifer et al., Water Supply Economics, Technology, and Policy 364 (1960)).

[72]Id. at 337.

require the education and cooperation of stakeholders, and also integrating the law with sound scientific, economic, and technological principles.

ISLAMIC WATER LAW

A. Zaharuddin
S. Sabri
University of Malaya
Petaling Jaya Selangor,
Malaysia

ISLAMIC WATER LAWS

Islam covers all aspects of human life. It regulates the relationships among God (Allah), man, and nature. It is based on the recognition of the Creator and of human submission to His will. Muslims believe that everything originates from the One God and everyone is responsible to Him. According to Islam, Allah created nature for the benefit of humans. The relationship between humans and nature is based on harmony; humans are urged to explore and use natural resources in a sustainable manner.

Water is considered one of the most profound elements in Islam. Water is also necessary for cleaning one's home and personal effects, as well as for general hygiene. It is indispensable to agriculture and industry: *He it is who sends down water from the sky. With it, we bring forth plants of every kind* (6:99). It cannot be denied that all human beings rely on water for life and good health, but for Muslims, water is first and foremost, a social good in Islam. It is also regarded as a blessing from Allah that gives and sustains all life in this world. Furthermore, the word "water" appears sixty-three times in the holy book of Muslims, Al-Quran. This indicator shows the importance of water in our daily life.

Before the coming of the Prophet Muhammad, there were no standard water regulations in the Arab world. The only guide concerning water laws was through customs practiced by tribes. There are four fundamentals in Islamic law, the first three can be considered the roots of Muslim dogma. The first is the Holy *Quran*, the divine revelation from Allah to humans through the Prophet Muhammad. It consists of 114 suras (chapters) and 6666 words. It also considered the basis for all legislation.

Second are the *Sunna* and the *Hadith*, which represent the sum total of habits or general pattern of life of the Prophet Mohammad, inclusive of his word, acts, and tacit approval. Third, is the *Ijma*, which means the universal consensus of opinion of the nations or Muslim community. And finally, *Qiyas* or deduction by legal analogy. It consists of the following precedents in deciding new cases, which have not been discussed or found before.

WATER RIGHTS IN ISLAM

In Islam, water is considered a gift from God, so no individual literally owns it. Humans act as the stewards of water as well as other resources that belong to the community. The environment is protected from humans by specifics injunctions against upsetting its natural order through pollution or other activities. This is to ensure that Allah's gifts are available to all living things in the present and future.

Although humans are the most favored creation, we are also responsible for ensuring that water is available to all living things. Allah commands believers to "*make not mischief on earth.*" (2:11); this means that we cannot spoil or degrade natural resources. The prophet Mohammad, very sensibly, forbade urination into stagnant water and advised guarding against this practice. This shows how Islam has underlined that water sources should be guarded from any contempt that can pollute it. This opens the door for penalties or punishment in modern legislation for those who disobey. Apart from that, waterside co-owners who enjoy the right to use water should care for and maintain water sources. They have to bear the burden of maintenance and cleaning operations. The expenses can be taken from taxes or special funds for that purpose.

Muslims believe that ensuring social justice is the cornerstone of Islam. The recognition of water as a vital resource, to which everyone has the right to a fair share, is emphasized by the following hadith, which effectively makes water a community resource to which all, rich or poor, have a right: "*Muslims have a common share in three things, pasture, water, and fire*", as narrated by Abu Daud.

Apart from that, animals cannot be allowed to die of thirst, and the water that remains after humans quench their thirst must be given to them. The immense value of giving water to any creature is reflected by the following hadith: "*A prostitute was forgiven by Allah because she took off her shoe and tying it with her head cover and drew out some water from a well to be given to a panting dog passing by that was about to die of thirst. So Allah forgave her because of that,*" as narrated by Al-Bukhari.

The Al-Quran notes that the gift of water is for flora as well: "*vegetation of all kinds*" (6:99) is nourished by rainwater that Allah sends down. This verse supports the statement that water is made available by Allah so that all life should receive support according to its needs, including humans, animals, and plants. As in Islam, humankind has the first right to resources that Allah has provided for its creation. It is well accepted by Islamic scholars (1) that the priority rights of the use of water are first the right of humans to quench their thirst; second, the right of cattle and household animals; and third the right of irrigation.

Nevertheless most Islamic scholars have concluded that individuals or groups have the right to use, sell, and recover the value added costs of water. These judgments are based totally on two hadiths: the first, about Othman, the third caliph, who purchased a well at Ruma (a place near Medina), proves that wells can be owned and traded (2) and secondly, as Prophet Muhammad said, "*It is better to go to the wood and cut and sell lumber to feed himself than to beg people for help,*" narrated by Muslims. This implies that common property resources such as wood and water can be sold and traded (3).

In his time, the Prophet Muhammad discouraged the selling of water and even forbade the sale of excess water. As noted, he also encouraged Othman to buy the well at Ruma and give away its water free. These examples reflect the prophet's desire for the poor and weak to have access to water. Apart from that, he encouraged people to work for a living, even by taking wood (a free natural resource) and selling it; a number of Islamic scholars believe it should also be applied to water, where we have the right to use, trade, and sell.

As narrated by Abu Hurairah, Allah's Apostle, said, *"No doubt, you had better gather a bundle of wood and carry it on your back (and earn your living thereby) rather than ask somebody who may give you or not."* Obviously, water in its natural state, such as a river or a lake, cannot be bought or sold. However, most prominent Muslim scholars agreed that if infrastructure and knowledge have been invested to withdraw it, for instance, a treatment and distribution system to convey it to people's homes, then the water utility has the right to recover its costs.

To prevent anyone new from diminishing the water flow, Islam has underlined a principle that entitles canals or wells or and any other sources of water that entail the ownership of a certain amount of adjacent land. This is known as *harim* or forbidden area. In this area, there should be no development such as land or tree cutting; it should remain as it is. Most schools of thought in Islam proclaim that no uniform size for a *harim* can be fixed and that the rules sanctioned by local customs must be adhered to, even in a region of 3 meters for a small well up to 100 meters for a spring.

Irrigation Regulations

The general principles of irrigation are that no one may alter an established irrigation system. Water may be drawn in any way, provided this does not entail any derogation of the lawful owner's rights and privileges. The assent of the waterside dwellers must be obtained, and the irrigation rights of other parties must be considered. Lower lying plots of ground are supposed to receive water from natural flows from upstream plots. The owner of the lower plot may not erect any barrier likely to obstruct the natural flow of water. Likewise, the owner of the upper plot may not do anything likely to harm the lower plot (5).

As narrated by Abdullah bin Az-Zubair,

> *An Ansari man quarreled with Az-Zubair in the presence of the Prophet about the Harra Canals which were used for irrigating date palms. The Ansari man said to Az-Zubair, "Let the water pass," but Az-Zubair refused to do so. So, the case was brought before the Prophet who said to Az-Zubair, "O Zubair! Irrigate (your land) and then let the water pass to your neighbor." On that the Ansari got angry and said to the Prophet, "Is it because he (i.e., Zubair) is your aunt's son?" On that the color of the face of Allah's Apostle changed (because of anger) and he said, "O Zubair! Irrigate (your land) and then withhold the water till it reaches the walls between the pits round the trees." Zubair said, "By Allah, I think that the following verse was revealed on this occasion": "But no, by your Lord they can have No faith until they make you judge in all disputes between them."* (4.65)

In addition Islam proclaims water conservation, and it is considered a fixed concept of Islamic teaching. A Muslim is ordered to be economical with water even if he is taking his water from a fast flowing river. The Prophet Muhammad teaches us not to overdo even in worshipping Allah; for instance, he uses only a handful of water in taking his ablution. Further more, as narrated by Anas, The Prophet used to take a bath with one Saor up to five Mudds (1 Sa' is equal to 4 Mudds) of water and used to perform an ablution with one Mudd of water. As in bathing, he also used a single pot of water. As narrated by 'Aisha (one of his wives), The Prophet and I used to take a bath from a single pot called a 'Faraq.

CONCLUSION

Islam is the religion of all times and places. It is a perfect system of temporal values. By practicing its laws, Muslims all over the world attain spirituality and also the pinnacle of all other glories. It was proven that the religious precepts of Islam do not represent an obstacle for efficient management of water resources; on the contrary, proper use of Islamic principles could significantly enhance water management practices in the countries concerned.

Before it came to mean simply law, the Arabic word *sharia* denoted the law of water (1). Therefore, it is not surprising that a detailed examination of the Quran and the Hadith shows that Islam makes a remarkable number of specific statements about water management. It is because Islam rose and developed in a desert area where water resources were extremely important and always considered sacred. It is also connected with the nature of Islam as a monotheistic religion that sought to regulate the behavior of humans according to the commands of Allah.

Actually, there is no major contradiction between what Islam says about the laws of water and the emerging international consensus on the issue. Islamic water laws are not that unique; some of the same principles could be derived by studying other faiths, their holy books, and the lives of their prophets. The major differences existed and were practiced more than 1400 years ago; they have proven adequate, applicable, and practical in our daily life. It was noted that Islam is very flexible and can accommodate all concerns regarding natural resource management, including water. Furthermore, Islamic principles for water management are completely compatible with the Dublin International Conference on Water and Environment and other accepted principles. In addition, Islam considers the protection of biological biodiversity, as well as water quantity and quality.

BIBLIOGRAPHY

1. Mallat, C. (1995). *Water in the Middle East*. I.B Tauris, New York.
2. Kitab Al-muslim (1967). Darul Nada, Kuala Lumpur.
3. Yusof Ali, A. (1977). *The Holy Quran: Text, Translation and Commentary*. American Trust Publication for The Muslim Student Association of the United States and Canada, Plainfield, IN.

4. Zouhaili, O. (1992). *Al-Fiqh wa-Dalalatuh (Islamic Jurisprudence and its proof)*. Dar El-Machariq, Damascus.
5. Querry, A. (1872). *Recueil Des Lois Concernant Les Musulmans Schytes*. Vol. 2, Imprimerie Nationale, Paris.

READING LIST

Agwan, A.R. (1997). *Islam and the Environment*. Synergy Books International, Kuala Lumpur.

Caponera, D.A. (1973). *Water Laws In Moslem Countries*. FAO Publications 20, no. 1, Organisation. Food and Agriculture Organisation, Rome

Faruqui, N.I. (2001). *Water Management in Islam*. United Nation University Press, Tokyo.

http://www.usc.edu/dept/MSA/fundamentals/hadithsunnah.

TRANSBOUNDARY WATERS IN LATIN AMERICA: CONFLICTS AND COLLABORATION

Scott Whiteford
Alicia Jimenez
Michigan State University
East Lansing, Michigan

Tensions over freshwater resources may become more frequent as pressures on water resources grow due to increasing water scarcity, degrading water quality, rapid population growth, unilateral water development, and uneven levels of economic development. As water resources become increasingly overallocated and used, the potential for conflict between nations with competing priorities and management strategies grows. Despite many obstacles, several international organizations are proposing that instead of being zones of conflict, shared water resources can provide a basis for cooperation and benefit-sharing, provided that threats to international waters are recognized and collaborative structures created. In the largest study of transnational water conflict and cooperation, it was found that cooperative interactions between riparian states over the past fifty years have outnumbered conflictive interactions by more than two to one (1).

Acknowledging the benefits of cooperative water management frameworks, policy makers have developed guiding principles and laws for international freshwater management. On a smaller scale, regional bodies and individual governments have developed protocols and treaties governing the management and protection of specific international water bodies (2).

In the Americas, there are approximately 70 international basins shared by Latin American countries (32 in North America and 38 in South America). Of those, 15 of the North American international basins and six of the South American international basins are governed by international treaties (2). These treaties refer to agreements where there is a concern with water as a scarce or consumable resource, a quantity to be managed, or an ecosystem to be improved or maintained (see Table 1).

Although treaties and agreements of cooperation exist, countries of the Americas face a great challenge in collaborating to achieve sustainable development on both sides of their borders. This is particularly important where the actions of one nation can directly or indirectly affect the well-being of its neighbors. Frequently, projects that threaten water quality are located near international borders and create winner–loser confrontations. These situations generate serious international conflicts, sometimes violent, which may extend for years. To resolve these types of prolonged conflicts, collaborative agreements are negotiated which facilitate shared water and land management. This article presents examples of conflicting situations and cooperative efforts among Latin American countries in North America, Central America, and South America. This article examines key examples of binational development that has threatened the quality of water used by two nations and created binational conflict and negotiation.

NORTH AMERICA: MEXICO–U.S. BORDER

In the U.S.–Mexico border region, water scarcity is a growing problem heightened by drought, rapid population growth, tourism, urbanization, and agriculture. The border extends almost 2000 miles and includes multiple watersheds, diverse ecological zones, political units, and a range of economic activities. The two nations have developed a set of institutions to manage water conflict which we do not examine in this article. At the end of this section, we present a list of significant binational disputes and negotiations.

The following is a brief overview of one of major conflicts between the United States and Mexico, the conflict over water quality at the lower end of the Colorado River. Because Mexico is down stream on the Colorado River, which provides much of the water for the western United States, the quality and quality of water flowing from the United States to Mexico has been the subject of dispute and international litigation. See Table 1 for the many treaties written to govern the border water resources between these countries. Although both countries have developed institutions to negotiate the disputes over water rights and responsibilities, the challenge is almost overwhelming, given that the two countries have very different water laws, national institutions, and constituencies (3).

Water conflicts over water quality in the border region arose because of the salinity from drainage water coming into the Colorado from irrigated fields in the United States and then taken from the Colorado for irrigation in the Mexicali Valley (Mexico). This valley is located on the western area of the United States- Mexico border and is one of the richest agricultural zones in North America. The valley draws water from the Colorado River and also from the Mexicali-Mesa de Andrade aquifer that transcends both regions and which is also subject to international dispute. Until recently, the Mexican government has had the dominant role in managing water for both drinking and irrigation in Mexicali Valley; nonetheless, people in Mexicali mobilized for their water rights and put pressure on their local and national governments to negotiate with the United States to protect surface water. The negotiations led to the introduction of water quality control that resolved the crisis. Nevertheless, Mexican farmers are periodically plagued by discharges of poor quality water (3).

Water of the Colorado River is governed by international agreements, but this is not the case for groundwater which was not a source of conflict until the end of the twentieth century. It was assumed to be in great supply and of high quality, but conflicts have arisen with the paving of the irrigation canals on the American side of the border that had been recharging the aquifer pumped by farms on the Mexican side of the border. The quality of the water in the aquifer is also being questioned after recent studies that showed high levels of pesticides and salts. Binational efforts to resolve the groundwater conflicts are beginning, but cooperation between the two countries over groundwater is still being negotiated. Groundwater is not the only transnational water problem being negotiated between the United States and Mexico. Other issues include conflicts over water of the Rio Bravo/Rio Grande; pollution of the Rio Bravo, Tijuana, and Nogales by urban and industrial water; and the construction of radioactive waste dumps

Table 1. Transboundary Water Treaties in the Americas

NORTH AMERICA[b]

Basin	Total Area	Date	Treaty Basin	Signatories	Treaty Name
Artibonite	8,800 km^2	02-20-1979	Frontier or shared waters	Dominican Republic; Haiti	Treaty of peace, friendship, and arbitration between both countries.
Candelaria	12,800 km^2	04-10-1987	Frontier or shared waters	Guatemala; Mexico	Agreement between both countries on the protection and improvement of the environment in the border area
Coatan achute	2,000 km^2	04-10-1987	Frontier or shared waters	Guatemala; Mexico	Agreement between both countries on the protection and improvement of the environment in the border area
Colorado	655,000 km^2	07-16-1994	Colorado	United States of America and Mexico	Minute No. 291 of the International Boundary and Water Commission (IBWC) concerning improvements to the conveying capacity of the international boundary segment of the Colorado River.
		07-18-1985	Frontier or shared waters		Agreement of cooperation regarding pollution along the inland international boundary by discharges of hazardous substances
		08-14-1983	Frontier or shared waters		Agreement on cooperation for the protection and improvement of the environment in the border area
		08-30-1973	Colorado		Agreement on the permanent and definitive solution to the salinity of the Colorado River Basin (IBWC Minute No. 242)
		04-30-1973	Colorado		Agreement extending Minute No. 241
		07-14-1972	Colorado		Agreement Minute No. 241
		08-24-1966	Colorado		Agreement concerning the loan of waters of the Colorado river for irrigation of lands in the Mexicali Valley
		11-14-1944	Colorado, Rio Bravo del Norte, Rio Grande, Tijuana		Treaty between both countries regarding the waters of those four rivers
		11-21-1900	Colorado, Rio Grande		Boundary waters, extension of convention 03-1-1889
		12-22-1899	Colorado, Rio Grande		Boundary waters, extension of convention 03-1-1889
		12-2-1898	Colorado, Rio Grande		Boundary waters, extension of convention 03-1-1889

(continued overleaf)

Table 1. (*continued*)

Basin	Total Area	Date	Treaty Basin	Signatories	Treaty Name
		10-29-1897	Colorado, Rio Grande		Boundary waters, extension of convention 03-1-1889
		11-6-1896	Colorado, Rio Grande		Boundary waters, extension of convention 03-1-1889
		10-1-1895	Colorado, Rio Grande		Boundary waters, extension of convention 03-1-1889
		03-1-1889	Colorado, Rio Grande		Convention on boundary waters
Grijalva	126,800 km^2	04-10-1987	Frontier or shared waters	Guatemala;Mexico	Agreement on the protection and improvement of the environment in the border area
Hondo	14,600 km^2	04-10-1987	Frontier or shared waters	Guatemala; Mexico	Agreement on the protection and improvement of the environment in the border area
Lempa	18,000 km^2	04-9-1938	Lempa, Paz	El Salvador; Guatemala	Treaty for the delimitation of the boundary between both countries
Massacre	800 km^2	02-20-1929	Frontier or shared waters	Dominican Republic; Haiti	Treaty of peace, friendship, and arbitration
Paz	2,200 km^2	04-9-1938	Lempa, Paz	El Salvador, Guatemala	Treaty for the delimitation of the boundary between both countries
Pedernales	400 km^2	02-20-1929	Frontier or shared waters	Dominican Republic, Haiti	Treaty of peace, friendship, and arbitration
Rio Bravo/Rio Grande	656,100 km^2	11-13-1992	Rio Grande	Mexico; United States of America	IBWC Minute No. 289—observation of the quality of the waters along the U.S. and Mexico border
		11-10-1987	Rio Bravo-Rio Grande		Boundary waters agreement
		07-18-1985	Frontier or shared waters		Agreement of cooperation regarding pollution along the inland international boundary by discharges of hazardous substances
		08-14-1983	Frontier or shared waters		Agreement on cooperation for the protection and improvement of the environment in the border area
		10-24-1960	Rio Grande		Agreement to proceed with the construction of Amistad Dam on the Rio Grande
		11-14-1944	Colorado, Rio Bravo del Norte, Rio Grande, Tijuana		Treaty between both countries regarding the waters of those four rivers
		02-1-1933	Rio Bravo/Rio Grande		Convention on the rectification of the Rio Grande (Rio Bravo del Norte)
		05-21-1906	Rio Grande		Convention on the distribution of waters of Rio Grande
		11-21-1900	Colorado, Rio Grande		Boundary waters, extension of convention 03-1-1889
		12-22-1899	Colorado, Rio Grande		Boundary waters, extension of convention 03-1-1889
		12-2-1898	Colorado, Rio Grande		Boundary waters, extension of convention 03-1-1889
		10-29-1897	Colorado, Rio Grande		Boundary waters, extension of convention 03-1-1889
		11-6-1896	Colorado, Rio Grande		Boundary waters, extension of convention 03-1-1889
		10-1-1895	Colorado, Rio Grande		Boundary waters, extension of convention 03-1-1889

Table 1. (*continued*)

Basin	Total Area	Date	Treaty Basin	Signatories	Treaty Name
		03-1-1889	Colorado, Rio Grande		Convention on boundary waters
San Juan	42,200 km^2	03-22-1888	Frontier or shared waters	Costa Rica; Nicaragua	Award of the President of U.S. on the validity of the treaty of limits of 15 April 1858
Suchiate	1,600 km^2	04-10-1987	Frontier or shared waters	Guatemala; Mexico	Agreement on the protection and improvement of the environment on the border area
Tijuana	4,400 km^2	07-2-1990	Rio el Alamar, Tijuana	Mexico, United States of America	Boundary waters: San Diego/Tijuana. IBWC Minute No. 238
		07-18-1985	Frontier or shared waters		Agreement of cooperation regarding pollution of the environment along the inland international boundary by discharges of hazardous substances.
		08-14-1983	Frontier or shared waters		Agreement on cooperation for the improvement of the environment in the border area
		11-14-1944	Colorado, Rio Bravo del Norte, Rio Grande, Tijuana		Treaty between both countries regarding the waters of those four rivers
Yaqui	74,700 km^2	07-18-1985	Frontier or shared waters	Mexico; United States of America	Agreement of cooperation regarding pollution of the environment along the inland international boundary by discharges of hazardous substances
		08-14-1983	Frontier or shared waters		Agreement of cooperation for the protection and improvement of the environment in the border area

SOUTH AMERICA[a]

Basin	Total Area	Date	Treaty Basin	Signatories	Treaty Name
Amazon	5,866,100 km^2	08-2-1988	Beni, Madeira, Mamoré	Bolivia; Brazil	Exchange of notes constituting an agreement for the construction of a hydroelectric plant in Cachuela Esperanza, supplementary to the agreement on economic and technical cooperation
		02-8-1984	Beni	Bolivia, Brazil	Agreement concerning the Cachuela Esperanza hydroelectric plant
		07-3-1978	Amazon	Bolivia, Brazil, Colombia, Ecuador, Guyana, Peru, Surinam, Venezuela	Treaty for Amazonian cooperation
		05-22-1944	Amazon, Chira, Tumbes, Zarumilla	Ecuador, Peru	Declaration and exchange of notes concerning the termination of the process of demarcation of the Peruvian-Ecuadorean frontier
Chuy	200 km^2	03-11-1991	Not specified	Brazil, Uruguay	Complementary agreement to the basic scientific and technical cooperation agreement between both countries on the area of water resources

(*continued overleaf*)

Table 1. (*continued*)

Basin	Total Area	Date	Treaty Basin	Signatories	Treaty Name
		12-20-1933	Frontier or shared waters	Brazil, Uruguay	Convention regarding the determination of the legal status of Brazil and Uruguay frontier
La Plata	2,954,500 km^2	05-6-1997	Cuareim	Brazil, Uruguay	Complementary settlement to the agreement of cooperation for the use of natural resources and development of this basin
		09-2-1995	Pilcomayo	Argentina, Bolivia, Paraguay	Agreement constituting the trilateral commission for the development of this basin
		06-9-1995	Bermejo, Grande de Tarija	Argentina, Bolivia	Agreement for the multiple uses of the resources of the upper basin of the Bermejo River and Grande de Tarija River
		03-11-1991	Cuareim	Brazil, Uruguay	Agreement of cooperation for the use of natural resources and development of this basin
		03-11-1991	Not specified	Brazil, Uruguay	Complementary agreement to the basic scientific and technical cooperation agreement on water resources
		06-29-1983	Pepiri-Guazu, Uruguai	Argentina, Brazil	Decree No. 88.441, agreement for water resources exploitation within the Uruguay and Pepiri-Guaco Rivers
		05-17-1980	Pepiri-Guazu, Uruguai	Argentina, Brazil	Treaty for the development of the water resources contained in the border reaches of Uruguai and Pepiri-Guazu Rivers
		10-19-1979	Paraná	Argentina, Brazil, Paraguay	Agreement on Paraná River projects
		02-26-1975	Uruguay	Argentina, Uruguay	Statute of the Uruguay River
		04-26-1973	Iguassu, Paraná	Brazil, Paraguay	Treaty concerning the hydroelectric utilization of the Paraná River owned in condominium by the two countries
		04-7-1961	Uruguay	Argentina, Uruguay	Treaty on the boundary constituted by the Uruguay River
		01-23-1958	Paraná	Argentina, Paraguay	Agreement concerning the study of the utilization of the water power of the Apipe Falls
		01-20-1956	Acaray, Monday	Brazil, Paraguay	Agreement concerning a study on the utilization of the water power of the Acaray and Monday Rivers
		12-30-1946	Uruguay	Argentina, Uruguay	Agreement concerning the utilization of the rapids of the Uruguay River in the Salto Grande area
		06-1-1945/07-5-1939	Pilcomayo	Argentina, Paraguay	Supplementary boundary treaties
		12-20-1933	Frontier	Brazil, Uruguay	Convention for determination of the legal status of the frontier of both countries
		01-5-1910	Plate	Argentina, Uruguay	Protocol dealing with question of the jurisdiction of the River Plate

Table 1. (*continued*)

Basin	Total Area	Date	Treaty Basin	Signatories	Treaty Name
Lagoon Mirim	55,000 km^2	03-11-1991	Not specified	Brazil, Uruguay	Complementary agreement to the basic scientific and technical cooperation agreement on water resources
		07-7-1977	Juguarao, Mirim Lagoon		Treaty on cooperation for the utilization of the natural resources and the development of the Mirim Lagoon basin
		04-26-1963	Mirim Lagoon		Agreement on establishing an joint commission for the development of the Mirim Lagoon
		12-20-1933	Frontier		Convention on the determination of the legal status of the frontier between Brazil and Uruguay
Lake Titicaca-Poopo system	111,800 km^2	06-21-1993	Desaguadero, Poopó, Salar de Coipasa	Bolivia; Peru	Notas related to the creation of the Autonomous Binational Authority of the basin
		02-19-1957	Lake Titicaca		Agreement concerning a preliminary economic study of the joint utilization of Lake Titicaca waters
		07-30-1955	Lake Titicaca		Preliminary convention concerning a study of the joint utilization of Lake Titicaca
		04-20-1955	Lake Titicaca		Exchange of notes establishing a joint commission for study of the Puno–Guaqui railway line and joint use of Lake Titicaca
		07-17-1935	Lake Titicaca		Preliminary convention for exploitation of fisheries in Lake Titicaca
		06-3-1929	Arica, Tacna	Chile, Peru	Treaty between both countries for the settlement of the dispute regarding Tacna and Arica
Maroni	65,000 km^2	09-30-1915	Maroni, Marowinjine	France, Netherlands (French Guiana, Suriname, and Brazil)	Convention between France and the Netherlands to fix the boundary between Suriname and French Guiana

[a] *Source*: Reference 20.
[b] Includes Central America and the Caribbean.

in Sierra Blanca, Texas; Ruidoso, New Mexico, a toxic waste incinerator in Ciudad Juarez; and several toxic waste dumps in the border region of south Texas. All of these projects threaten the border environment. Meanwhile treatment plants have been built along the border. Major sources of pollution such as the Rio Nuevo, that flows from Mexicali to the Salton Sea in California, are now being treated and are returning a much better quality of water to the stream. Conflicts, negotiations, and cooperation will continue to be the pattern along this border for years.

CENTRAL AMERICA

Despite the importance of transnational rivers in Central America, they have not been the subject of extensive diplomacy. Dam construction projects on the Usumancinta, Salinas, and Pasión Rivers on the Mexico-Guatemala border have generated both conflicts and cooperation. The Mexican and Guatemala governments have signed a memorandum of energy cooperation which has created great concern among people who live in communities along the rivers and are afraid of being displaced by their governments by the construction of new and bigger dams (4). Community leaders have protested the lack of information about future construction plans, especially in the context of the Plan Puebla-Panama. The Plan Puebla-Panama implementation, begun in 2001, framed economic integration and sustainable development for Central America and the southern states of Mexico (5). Funding from the World Bank and the Inter-American Development Bank for the projects includes building dams on the border rivers. In response to such controversial plans, 98 nongovernmental organizations (NGOs) from 21 countries of the Americas and Europe and affected communities have created a regional movement against dams (6). The concern is focused on forced relocation of communities, the destruction of valuable archaeological sites, and loss of biodiversity (4).

Other major transboundary rivers in Central America include the Rio Hondo between Mexico and Belize and the San Juan River that divides Costa Rica and the San Juan. Heavy use of pesticides in Belizean agriculture that are killing fish and impacting coral reefs are creating concern on the Mexican side of the river. Another type of transboundary river problem, navigation rights, is reflected in a dispute between Costa Rica and Nicaragua over the San Juan River. The conflict started in 1998, when Arnoldo Aleman, a former Nicaraguan president used police to impose navigation fees on Costa Rican citizens using the river on the Nicaraguan side. It was only by the intervention of the Costa Rican government that the fees were dropped, much to the relief of people living on both sides of the river (7).

SOUTH AMERICA

Transboundary water management in South America is very challenging. The region includes one of the largest river basins in the world, the Amazon River Basin. The region, known as Amazonia, includes more than 8000 km of borders, including eight South American countries. Its enormous natural, cultural, biological, hydrological, mineral, forest, and medical resource richness has made it an object of intensive exploitation that has endangered the natural balance and health of the basin.

Another important international basin is the Orinoco, located in the south part of this basin where Venezuela and Guyana are located. The Orinoco is affected by gold mining. In 1995, more than one million liters of cyanide waste were spilled in the Omai and Esequibo Rivers in Guyana. The impact of these activities and the contamination of the zone's hydrologic network affect at least 25% of the basin's indigenous population, including the following ethnic groups: Pemon, Yanomami, Piaroa, Guahibo, Yekwana, and 17 other indigenous peoples. There are no treaties or agreements of cooperation on water quality between the countries that share this basin.

In addition to the Amazon, other rivers in South America play an important role in linking as well as dividing countries in northern South America. The Meta River that divides Colombia and Venezuela has supported transportation of produce for centuries (8). Sediments, caused by watershed degradation, create serious difficulties for transportation on the Meta River reducing its capacity for transportation. With funding from the Inter-American Development Bank, a binational dredging project of the Meta River has begun to increase trade by reducing transportation cost, but the project has been opposed by many stakeholders and conservation groups on the grounds that no impact studies were carried out and local populations had little input to the project.

Sovereignty and control over who uses rivers (sailing rights) may be a source of conflict and/or cooperation. Sailing rights in waters controlled by multiple countries are a challenge for integration. An example of this is the Orinoco-Amazonas fluvial interconnection project, which has been promoted by Venezuela's and Brazil's presidents, but integrating the Colombia has been difficult. Colombia wants free sailing rights on the Orinoco, without the approval of Venezuela and Brazil, to facilitate transportation and trade. The Orinoco–Meta integration is important for Colombia because it would provide a corridor between the Pacific and Atlantic Oceans. Both Venezuela and Brazil are reluctant to give Colombia free access and have recommended doing the fluvial interconnection gradually and in specific phases. To open the waterway to commercial vessels will require considerable dredging and construction in Colombia and cooperation from its neighbors (9).

La Plata Basin, which is shared by five countries, Brazil, Argentina, Paraguay, Bolivia, and Uruguay, is also under threat. The Pilcomayo river, part of this basin shared by Argentina, Bolivia, and Paraguay, has been contaminated periodically with arsenic, cadmium, and lead caused by spills of toxic waste containers from Bolivian mining. In the trinational zone, the Pilcomayo is an important source of water for indigenous peoples, agriculture, and its unique wildlife. There are several cooperative agreements for the water management of this river (see Table 1).

The process to build a hydro-via, a water highway, using the Parana and Paraguay Rivers is in its final discussion phase (10). The Paraguay-Parana River Hydro-Via is a plan of the five governments of the La Plata Basin to build 2100 miles of river and canal systems navigable by large barges. The project would significantly reduce the transportation costs of commodities within and beyond the region. For example, it would reduce the cost of transportation of soybeans exported to Europe from South America by half. However, the project was halted in 1997 due to the Rios Vivos Coalition's efforts to demonstrate the irreversible effects on this unique ecosystem, the Pantanal. The Pantanal is the world's largest tropical wetland ecosystem formed by the Parana-Paraguay rivers (11).

The Pantanal is an ecological wonder shared by Bolivia, Brazil, and Paraguay. The ecosystem of the region is based on seasonal flooding of the major rivers, including the transborder rivers; yet there is no trinational agreement to protect the ecosystem and the people who depend upon it. This unique region has serious problems caused by deforestation, expansion of agriculture, contamination of the water by pesticides and derivatives of combustible alcohol production, and illegal hunting of endangered species. There has been an increase in natural habitat loss and a decline in the quality of life

South America has major several shared hydroelectrical dams; over the Parana River are two: Yacyretá (Argentina and Paraguay) and the biggest dam in the world, Itaipú (Brazil and Paraguay). Another is Salto Grande, shared by Argentina and Uruguay over the Uruguay River (10). Five more international dams on the Uruguay and Parana border rivers have been under study.

The Yacyreta binational dam has had great impact on native and rural populations, forcing them to move away from their homes and farms. This mobilization has affected the health of these populations, increasing their risk of contracting malaria and other tropical diseases. The Itapu dam has also had massive social and environmental impacts. One of the most significant impacts has changed the weather of the region periodically creating semihurricane conditions. The building of dams and sharing of electricity reflect a high level of cooperation between the countries that share borders.

South America economic integration and the new infrastructure, including highways and hydroelectric dams in the southern region of South America, have also generated conflicts over transboundary waters. One criticism of the integration process is the lack of wide stakeholder participation and the lack of information to the civil society about these efforts. The projects mentioned before, the Meta River canalization, the Orinoco–Amazonas fluvial interconnection, the Venezuelan–Brazilian electrical interconnection, and the South Cone projects, among others, are usually perceived as local or isolated national efforts, without taking into account their continental scope. In addition, according to Arocha (12), most of the new infrastructural projects will produce considerable disturbances in the natural and social environment, and most likely, the indigenous population will be affected the most.

Several organizations, such as the United Nations Global Environment Facility (GEF) propose a comprehensive, ecosystem-based approach to managing transboundary waters and their drainage basins; these are often called "international water projects"(1). The goal of these organizations is to build transboundary collaboration projects to help each of the countries use the full range of technical, economic, financial, regulatory, and institutional measures needed to operationalize sustainable use strategies for transboundary waterbodies and contributing basins.

One example of a GEF international water project in Latin America is the Bermejo River Basin of Bolivia and Argentina. The GEF became interested in this basin because it has excessive levels of sediment, important transboundary biodiversity, and a binational commission for developing of the basin. One of the objectives of GEF was the involvement of stakeholder groups in the basin to determine their sustainable development future. They facilitated participatory transboundary diagnostic analysis (TDA), a scientific-based assessment of the key transboundary problems and their root causes. Originally, 16 dams were proposed for hydropower production in the basin. This was quite controversial. However, through public participation in the project, the binational commission that was created to develop hydropower transformed itself into a force to consider sustainable development options. Three dams will still be pursued, but the focus on the transboundary environment, strategic processes for participation, and practical measures consistent with sustainable development helped overcome initial transboundary concerns and NGO opposition (1).

The future of transnational rivers in Latin America and their role in enhancing sustainable development depends on binational collaboration. The watershed and basin agreements in place have generally facilitated negotiation and found solutions of mutual benefit. In other transborder water situations, mutually agreed upon conflict resolution mechanisms and management plans have yet to be developed. Until these agreements are in place, binational management will continue to generate conflict. At stake is the development of some of the most important yet vulnerable areas on the continents. Fortunately, progress is being made on almost every border.

BIBLIOGRAPHY

1. Uitto, J.I.D. and Alfred, M. (2002). Management of transboundary water resources: Lessons from international cooperation for conflict prevention. *Geogr. J.* **168**(4): 365–378.
2. UNEP, FAO, Oregon State University. (2002). *Atlas of International Freshwater Agreements*. Kenya; USA.
3. Whiteford, S. and Cortez, A. (in press). Good to the last drop: the political ecology of water and health on the border. In: *Globalization, Water and Health*. L. Whiteford and S. Whiteford (Eds.). School of American Research Press, Santa Fe, NM.
4. Mesoweb. (2002). Maya sites to be flooded by Dam, Mesoweb Reports. Available: http://www.mesoweb.com/reports.dam.html.
5. Delgado, G.C. (2002). El Corredor Biologico Mesoamericano, Tres Ecorregiones y COINBIO: esquemas del Banco Mundial integrados a la politica nacional., Biodiversidadla.org.
6. Bellinghausen, H. (2002). During the Mesoamerican forum, they demanded water, light and land for rural villages. *La Jornada*. Mexico D.F.
7. Gonzalez, A.H. and Mauricio. (2002). Acuerdo evito choque bilateral. *La Nacion*. San Jose.
8. Colmenares, G. (1997). *Tratamiento binacional de asuntos ambientales en regiones fronterizas*. La experiencia Colombo-Venezolana. CEFIR.
9. Nacional, E. (2000). Recomiendan realizar por etapas la integracion de Orinoco-Amazonas. *El Nacional*.
10. Gudynas, E. (1998). *La dimension ambiental de la integracion fisica y energetica en el Cono Sur*. Uruguay. CLAES.
11. IRN. (2001). *Stop the Hidrovia in the Pantanal*. International Rivers Network (IRN).
12. Leon-Oliveros, R. (1997). *Situacion actual y perspectivas de las relaciones fronterizas de los paises miembros del grupo Andino*. CEFIR.
13. Arocha, M. (2001). Ambiente, proyectos de integracion fisica en America Latina y desarrollo sustentable en Venezuela. Analitica.com.

UNITED STATES-MEXICO BORDER WATERS: CONVENTIONS, TREATIES, AND INSTITUTIONS

NANCY A. LOWERY
San Diego, California

SELECTED TREATIES AND CONVENTIONS

Cooperation over shared United States–Mexico waters dates back to the *Treaty of Peace, Friendship, Limits, and Settlements*, know as the Treaty of Guadalupe Hidalgo, signed in 1848. This treaty established the international boundary from the Gulf of Mexico along the Rio Grande to El Paso, Texas, then west to the Pacific Ocean just south of San Diego Harbor, a stretch of 3141 km. For 2001 km of that border, the Rio Grande serves as the boundary line. This is a long and complex boundary with topography ranging from coastal plains to mountain ranges and high desert. The climate is predominately semiarid to arid, has two major watersheds, the Rio Grande (called the Rio Bravo in Mexico) and the Colorado River, many smaller watersheds, and numerous groundwater basins. This was a treaty of intent, not of specifics. The treaty did not deal with details of the boundary line, management of water between the two countries, fluctuation of the river courses, water quality, or groundwater resources. Subsequent agreements would have to deal with these issues.

The introduction of the railroad and irrigated agriculture in the late 1800s spurred settlement, which required that questions left unanswered by the 1848 Treaty be resolved or at least addressed. Therefore several conventions were enacted to solve disputes over the boundary line, establish surveys, and fix boundary monuments to officially demarcate the border (1). These include the *Gadsden Treaty of December 30, 1853* and the *Convention of July 29, 1882* (2).

Changes in the course of the Rio Grande and Colorado rivers and the resulting fluctuations in the international boundary were dealt with by the *Convention of November 12, 1884*. This agreement established rules for maintaining the boundary line, given changes in the riverbed. The *Convention to Avoid the Difficulties Occasioned by the Reason of the Changes which Take Place in the Beds of the Rio Grande and Colorado River*, March 1, 1889, established a binational commission, the International Boundary Commission (IBC). The IBC's role was to apply the 1884 Convention rules and handle disputes related to boundary waters. The IBC had both a United States and a Mexican Section; each had a separate commissioner appointed by the federal governments of the respective countries.

None of these previously mentioned agreements dealt with the apportionment of waters between the two countries. By the end of the nineteenth century, however, allocation issues could no longer be ignored. The growth of irrigated agriculture on the upstream reaches of the Rio Grande meant that in low flow years, farmers on the Mexican side of the border did not receive sufficient water to irrigate crops. By the 1890s, Mexico was pressing for a solution to this problem, especially where agricultural pressures were at their greatest, which was in the section of the Rio Grande that forms the border between New Mexico, Texas, and Chihuahua.

To alleviate stress in that region, the May 12, 1906 *Convention Providing for the Equitable Distribution of the Waters of the Rio Grande for Irrigation Purposes* allocated the waters of the Rio Grande above Ft. Quitman, Texas, between the United States and Mexico. It guaranteed Mexico 74 Mm^3 (60,000 acre-feet) to be delivered at a point just above Ciudad Juárez, Chihuahua, in accordance with a monthly schedule. In exchange for that delivery, Mexico waived any additional claims to the waters from the upper stretch of the river to Ft. Quitman, Texas.

All water allocated under the 1906 Convention is for irrigation; none is set aside for municipal or industrial uses or for instream flows. As water demand has changed, this allocation structure has had ramifications for the region. The Convention also did not deal with water quality issues or apportion groundwater supplies. It did, however, provide for drought or catastrophic conditions. Stating that should it be impossible to deliver the quantity of water promised due to severe drought or destruction of the delivery system, Mexico's supply would be diminished in the same proportion as the supply delivered to U.S. irrigators.

In 1933, the two governments enacted the *Convention for the Rectification of the Rio Grande (Rio Bravo del Norte) in the El Paso–Juárez Valley*. This convention established the Rio Grande Rectification Project to manage the river in the increasingly developed Mesilla–El Paso–Juárez Valleys for flood control and delivery of treaty waters. It provided for joint construction, operation, and maintenance, by the IBC, of the facilities needed for these purposes, which entailed straightening a 155-mile stretch of the river and was the first joint construction treaty between the United States and Mexico.

Continued border growth in the early 1900s caused the lower Rio Grande basin and the Colorado River to experience water demand pressures similar to those that had been felt in the upper Rio Grande basin 50 years earlier, which led to the 1944 *Treaty Relating to the Utilization of Waters of the Colorado and Tijuana Rivers and of the Rio Grande*, considered the most important United States–Mexico water treaty. This treaty expanded the scope of cooperation between the two countries to include issues related to border sanitation and, more importantly, provided the legal framework for resolving water disputes between the two countries. It replaced the IBC with the International Boundary and Water Commission (IBWC), called the Comision Internacional de Limites y Aguas (CILA) in Mexico.

The new IBWC had expanded scope and authority in line with the expanded scope of the 1944 Treaty. It has the status of an international body, has two Commissioners, appointed by the presidents of each country, and permanent offices in El Paso, Texas, and Juárez, Chihuahua. It operates under the Departments of State of each country and has the authority and responsibility to negotiate and carry out agreements between the two countries regarding border water and sanitation, as outlined in the treaties. These agreements are called Minutes. Once approved by the United States and Mexican governments, Minutes are binding on those governments. As of March 2004, there are 311 Minutes (3). The Minute process has been an effective means to address serious border water issues, such as salinity, untreated effluent discharges, and drought mitigation, without having to renegotiate existing treaties or negotiate new ones.

Additionally, the 1944 Treaty (4) sets out, in Article 3, priorities for the joint use of the international waters, which would be used if the Commission were asked to make a determination of priorities. In order of preferential use, they are

- domestic and municipal uses,
- agriculture and stock raising,
- electric power,

- other industrial uses,
- navigation,
- fishing and hunting, and
- any other beneficial uses that may be determined by the Commission.

The 1944 Treaty also apportioned the shared surface waters, not already divided, between the two countries. Waters of the Rio Grande–Rio Bravo from Fort Quitman to the Gulf of Mexico are apportioned in the following manner:

Mexico has the right to:

- all of the waters reaching the main channel of the Rio Grande from the San Juan and Alamo Rivers, including the return flows from the lands irrigated from those two rivers;
- two-thirds of the flow of the main channel of the Rio Grande measured from the Conchos, San Diego, San Rodrigo, Escondido, and Salados Rivers and the Las Vacas Arroyo, subject to the U.S. right to an average 431,721,000 cubic meters (350,000 af/yr) in cycles of 5 consecutive years; and
- one-half of all other flows in the main channel of the Rio Grande downstream from Ft. Quitman not specifically allocated under this Treaty to either of the two countries.

The United States has the right to:

- all of the waters reaching the main channel of the Rio Grande from the Pecos and Devils Rivers; Goodenough Spring; and Alamito, Terlingua, San Felipe, and Pinto Creeks;
- one-third of the flow reaching the main channel of the river from the Conchos, San Diego, San Rodrigo, Escondido, and Salados Rivers and the Las Vacas Arroyo, providing that this third shall not be less than 431,721,000 cubic meters annually (350,000 af/yr) as an average amount in cycles of 5 consecutive years; and
- one-half of all other flows in the main channel of the Rio Grande downstream from Ft. Quitman not specifically allocated under this Treaty to either of the two countries.

In the case of the Colorado River, under the 1944 Treaty, Mexico is to receive 1,850,234,000 cubic meters annually (1.5 million af/yr), to be delivered in accordance with schedules provided in advance by Mexico. Mexico is also assigned any surplus waters not necessary for uses in the United States arriving at the Mexican point of diversion. The total annual amount of water delivered to Mexico is not to exceed 2,096,931,000 cubic meters (1.7 million af/yr).

The amount of Colorado River water allocated to the United States is not specifically mentioned in the 1944 Treaty. Those allocations are set out in the *Colorado River Compact of 1922*. The compact allocates 18,495,000,000 m^3 year (15 million acre-feet) to the compact States of Arizona, California, Colorado, Nevada, New Mexico, Utah, and Wyoming. The Compact, which came into effect before the 1944 Treaty, maintains in Article III that if the need arises for waters of the Colorado to be delivered to Mexico, than Mexico's share would come from excess flows. If surplus waters are unavailable, then the deficit in flows is to be shared equally by the states. Because of water variability and increasing urban and agricultural reliance on the Colorado River, supplying the allocated waters has become a problem (5). The 18.5 Bm^3 allocated under the Compact and the 1.8 Bm^3 assigned under the Treaty are higher than the annual average flow of the river, which is approximately 16.6 Bm^3 or 13.5 million acre feet (6).

In allocating Tijuana River waters, provisions for distribution are not laid out. The 1944 Treaty recommends, however, that the Commission investigate all means for equitable distribution of the waters between the two countries. This investigation and all other exploration of border surface water issues would be carried out by the newly formed IBWC.

Finally, the 1944 Treaty allows for the construction of dams, delivery structures, and flood control structures that are necessary to fulfill treaty obligations. It determined that these structures would be jointly constructed, maintained, and operated, and the costs to each country would be equal to the benefits provided. Since 1944, the IBWC has been responsible for constructing, maintaining, and operating two large dams and storage reservoirs, Falcon and Amistad, numerous diversion and delivery structures, international wastewater treatment plants, and a surface water salinity treatment plant.

On August 29, 1963 the two governments signed the *Convention for the Solution of the Problem of the Chamizal* to resolve a 100-year-old boundary and land dispute between the United States and Mexico. The Rio Grande channel in the area of El Paso, Texas, and Juárez, Chihuahua, had shifted periodically over the centuries, causing approximately 600 acres to move back and forth between the north and the south banks of the river. The Convention settled this dispute by repositioning and stabilizing 4.4 miles of the river, which transferred 437.18 acres from the United States to Mexico.

In 1970, the *Treaty to Resolve Pending Boundary Difference and Maintain the Rio Grande and the Colorado River as the International Boundary between the United States and Mexico* was enacted. This Treaty provides for the maintenance of the boundary line due to shifting river courses, erosion, or other changes in the channel. It also resolved all pending boundary disputes between the two countries. The IBWC is responsible for carrying out the provisions of the Treaty.

SELECTED BINATIONAL AGREEMENTS

There are several other agreements that were signed by the United States and Mexico that are relevant to managing water in the border region. These include the La Paz Agreement and the side agreements of the *North American Free Trade Agreement*.

The *Agreement between the United States of America and the United Mexican States on Cooperation for the*

Protection and Improvement of the Environment in the Border Area (La Paz Agreement) was signed on August 14, 1983. The La Paz Agreement is the framework for cooperation between the United States and Mexico on protecting, improving, and conserving the environment in a 100-kilometer area on either side of the border. This Agreement deals with issues of water, soil, and air pollution and provides for a system of emergency notification.

Five technical annexes were established to deal with specific projects of environmental concern, two of which are water related. Annex I, *Solution of the Border Sanitation Problem at San Diego, California and Tijuana, Baja California*, and Annex II, *Regarding Pollution of the Environment along the Inland International Boundary by Discharges of Hazardous Substances*, were both enacted in 1985. The framework set up by the La Paz Agreement paved the way for future environmental cooperation between the two countries. The 1992–1994 Integrated Border Environment Program (IBEP), the Border XXI Program, and the Border 2012 Program were developed in conformance with the *La Paz Agreement*.

The *North American Free Trade Agreement* (NAFTA) among Canada, United States, and Mexico became effective on January 1, 2004. NAFTA uniquely included side agreements on the environment. These agreements established a Commission on Environmental Cooperation (CEC) among the three signatories. CEC's role is to protect the environmental regulations of each country and to allow individuals and governments to file complaints with the Commission if they believe environmental laws are not being enforced.

The NAFTA side agreements also established the Border Environment Cooperation Commission (BECC) and the North American Development Bank (NADBank). BECC and NADBank are international organizations whose goal is to protect and improve the environment of the United States–Mexico border. BECC's mandate is to certify water, wastewater, and solid waste projects within the border zone. This area is defined using the La Paz Agreement boundaries of 100 kilometers on either side of the border. Certification is based on a set of engineering standards and economic and environmental sustainability concepts. NADBank's role as a development bank is to fund projects certified by BECC. Both of these institutions have played an integral role in water and sanitation infrastructure and management in the border region; due to the certification process, projects cannot be built without BECC approval.

CURRENT SITUATION

The complex set of Treaties, Convention, Compacts, and side agreements that laid the foundation for water allocation and management in the United States–Mexico borderlands have served the region well. Rapid growth, however, has placed stress on both the resource and the legal and institutional structures used to govern it. Most, if not all, border water resources are plagued by increased demand and diminishing supply. In 2000, the border population was approximately 11.8 million (7), primarily living in 14 sets of twin cities stretched out along the border. Based on annual growth rates of between 3% and 5%, that population is expected to reach 19.4 million by 2020 (8).

Current water management on the U.S.–Mexico border, therefore, is defined by three realities: first, seasonal and annual variability of water supply; second, overallocation of that supply; and third, increased competition for the resource due to burgeoning growth and changing priorities. Demand in the early and mid-1900s was sufficiently low that, except in times of severe drought, supply was adequate to meet demand. Growth and changing demand patterns have altered the balance and strained the institutional mechanisms that regulate and manage shared waters. Issues of how to deal with persistent drought, the allocation and joint management of groundwater basins, and environmental concerns surrounding degraded and endangered habitat and species have stretched the limits of the treaties and their institutions. The current debate centers around the question of whether the treaties in force have the flexibility to deal with emerging issues, whether they should be significantly altered, or if new treaties should be enacted (9–12).

BIBLIOGRAPHY

1. Leon Metz provides a detailed description of boundary disputes and surveys in Metz, L. C. (1989). *Border: The U.S.-Mexico Line*. Mangan Books, El Paso, TX.
2. For a complete list of treaties, see United States Department of State, O. o. t. L. A. (2003). *Treaties in Force: A List of Treaties and Other International Agreements of the United States in Force on January 1*, 2003. Washington, DC.
3. A description of the Minutes approved since 1946 can be accessed at www.ibwc.state.gov.
4. corpauUnited States Department of State. (1944). *Treaty Relating to the Utilization of Waters of the Colorado and Tijuana Rivers and of the Rio Grande*. 59 Stat. 1219; TS 994; 9 Bevans 116; 3 UNTS 313.
5. The limited ability of the 1944 Treaty to deal with drought is expertly outlined by Stephen Mumme and Ismael Aguilar Barajas in "Managing border water to the year 2020" in *The U.S.-Mexican Border Environment: Binational Water Management Planning*.
6. Dettinger, M.D. (1995). Coping with severe and sustained drought in the Southwest. *Water Resour. Bull.* October.
7. For an explanation of population and economic growth trends, see Peach, J. and Williams, J. (2000). Population and economic dynamics on the U.S.-Mexican border: Past, present and future. *The U.S.-Mexican Border Environment: A Road Map to a Sustainable 2020*. P. Ganster, San Diego State University Press, San Diego.
8. United States Environmental Protection Agency and Secretaría del Medio Ambiente y Recursos Naturales. (2003). Border 2012: U.S.-Mexico Environmental Program.
9. Lopez, M. (1997). Border tensions and the need for water: An application of equitable principles to determine water allocation from the Rio Grande to the United States and Mexico. *Georgetown Int. Environ. Law Rev.* **9**(2): 489.
10. Mumme, S. (2000). Minute 242 and beyond: Challenges and opportunities for managing transboundary groundwater on the Mexico–U.S. border. *Nat. Resour. J.* **40**(2): 341–378.

11. Brown, C., Castro, J.L. et al. (2003). Comparative analysis of transborder water management strategies: Case studies on the United States–Mexico border. *The U.S.–Mexican Border Environment: Binational Water Management Planning*. S. Michel (Ed.). San Diego State University, San Diego.
12. Mumme, S. and Aguilar Bajaras, I. (2003). Managing border water to the year 2020: The challenge of sustainable development. *The U.S.-Mexican Border Environment: Binational Water Management Planning*. S. Michel (Ed.). San Diego State University, San Diego, pp. 51–94.

NEGOTIATING BETWEEN AUTHORITY AND POLLUTERS: AN APPROACH TO MANAGING WATER QUALITY

Petr Šauer
Antonín Dvořák
University of Economics
Prague, Czech Republic

INTRODUCTION

Water pollution reduction is one of the most complicated challenges of environmental policy. Water quality engineers are able to offer many technical solutions. Some are based on "end-of-pipe" technologies; some represent "cleaner production" solutions. Solutions exist that are already economically efficient for the polluters. Many cleaner production projects are of this type. However, many projects exist that bring net control costs for the polluters, so the polluters do not have an economic interest in them. As these projects can still be efficient from the social point of view (if negative externalities would be internalized), it is important to change the institutional framework of environmental policy. The best institutions are those that not only satisfy ecological efficiency but also support economically efficient solutions, or at least cost-effective solutions, to the given problem. Our contribution is to present a relatively new approach that could lead to cost-effective solutions of water pollution reduction problems. It is based on voluntary negotiations between an authority and polluters where economic tools of environmental policy (environmental fees and subsidies) are used at the same time. The principle of the approach is explained in the and an illustration of an economic laboratory experiment follows.

PRINCIPLE OF THE APPROACH DESCRIBED

The suggested approach is based on (mostly multiple rounds) negotiations between an authority of environmental policy and economic subjects—polluters. The binding agreements are the results of the negotiations. The agreements contain: (final) commitments of the polluters to reduce pollution, commitments of the authority to contribute negotiated financial resources to pollution reduction projects, and a set of consequences in the case of noncompliance with the agreement conditions. If the water pollution charges are introduced, then there must be a commitment of the polluters to pay it, unless the charge has not been already introduced by a law, i.e., the charge already exists. NGO's and various independent experts (like public health experts, ecologists, environmental engineers) can also take part in this procedure.

The positions and performance of the parties taking part in the negotiations are as follows [for the formalized procedure see (1)]:

The polluters have technical solutions (projects) that can reduce their pollution. They know the true information about their net costs of environmental protection but keep it confidential, which creates a so-called information asymmetry between the polluters and the authority. The fact that the polluters can keep the private information confidential and still achieve a cost-effective outcome seems to be one of the important privileges of the approach. During the negotiation process, they apply their own strategies to their benefits maximization when requesting (financial) contributions from the authority and offering water pollution reduction in the single rounds of the negotiations.

The authority has some available information about contributions of individual polluters to the pollution in the region, but it is unfamiliar with the true net costs of environmental protection of polluters and about their criteria of economic efficiency. The authority uses traditional economic tools in the process of the negotiation. The rate of payments for pollution is set. The rate of this payment can also be negotiable in more advanced models of the concept. The authority collects revenues from these payments in a fund and negotiates financial contributions from the fund with the polluters. During the process of negotiations, it maximizes various environmental effects per one unit of the financial contribution, which leads to the cost-effective solutions. Various methods and approaches for a quantification of indicators of these environmental effects can be used, such as risk assessment, multicriteria analyses, etc. Three different sets of environmental indicators could be used in the process: first, charging for the pollution; second, setting priorities for the financial support from the fund, and third, the final control of the concluded agreements. The authority can use various strategies from a spectrum defined by marginal strategies as follows: the first strategy leads to a uniform contribution from the fund per one unit of the positive environmental effect; the second strategy tries to reflect different net costs of environmental protection of the individual polluters.

LABORATORY ECONOMIC EXPERIMENT ILLUSTRATION

Let us illustrate the approach to a laboratory economic experiment. An ideal case of (politically) required water pollution reduction in a river basin was based on both environmental and economic data taken from various projects realized in practice in the Czech Republic. We considered pollution based on chemical oxygen demand (3,700 t/year), dissolved substances (790 t/year), insolved substances (1,320 t/year), phosphorous (18 t/year), and nitrogen (140 t/year). To put it more simply, the pollution was recalculated by using coefficients of harmfulness of the single pollutant. The resulted pollution was 1,454,000 units/year.

15 major polluters exist in the region. Their contribution to the pollution was transferred to the recalculated

units as follows: textile company (A; 30,760), stone manufacturing (B; 5,760), heating station (C; 7,750), pigments manufacturing (D; 4,500), heating station (E; 14), photo-developing (F; 50), metal surface remanufacturing (G; 10,800), textile (H; 102,560), food production (I; 365,200), food production (J; 157,890), waterworks (K; 176,750), municipality I (L; 54,605), municipality II (M; 99,500), municipality III (N; 105,250), and municipality IV (O; 43,790). These polluters contribute, in total, 1,165,108 units, which is approximately 80% of total water pollution in the basin.

The environmental goal for the region was to reduce pollution from all sources at least by about one-third and to reduce pollution from the major sources by about one-half.

The laboratory experiment was conducted by Czech students of Environmental Economics and Policy Specialization at the University of Economics in Prague. The students played roles of managers of polluting entities. The teacher played the role of the authority. Charges that are paid according to Czech law (2) on pollution of surface water were used as a rate of payment for single pollutants to create revenues in the fund.

The negotiating in this experiment took three rounds. Six hypothetical agreements between the authority and the polluters were concluded at the end of the last round: I (pollution reduction 280,700 units), B (5,760), J (151,975), N (89,448), K (88,750), G (7,200). The agreed pollution reduction was 623,833 units in total, which represents about 54% of total pollution from major sources and about 43% of total pollution in the region. When repeating the experiment with teachers of economics, the agreements were concluded with a similar set of polluters C, G, I, J, B, and K. The pollution reduction was, in this case, 535,825 units [for the other experiment see (1)]. The theoretical (computed) pollution reduction satisfying the cost-effectiveness rule was 536,523 units, i.e., very close to the results of both experiments. Also, the subsidies agreed at the end of the experiments were very close to the theoretical (computed) ones: Computed subsidies were 55,799 monetary units. Students agreed with 52,650 and teachers concluded the agreements with the total subsidies at 54,780 monetary units.

Other experiments testing the described approach were conducted using ideal data. Also, in these cases, the experimental outcomes were very close to the computed ones.

In the next step, we used the same data for experimenting with the tradable pollution permit system. We were interested what the result would be if the same (water) pollution reduction target were achieved via trading the permits. In all laboratory experiments, the set of polluters who concluded the agreements with the authority and the set of polluters who sold the permits were about the same. Moreover, the price per one unit of permit was very close to the subsidy per one unit of reduced pollution.

CONCLUSIONS

It would be possible to discuss several practical applications of the approach. It could be used as a model for negotiations between a state environmental protection authority (regulator) and multiple polluters, as described above.

An environmental protection fund could be established and privately owned by the polluters that voluntarily want to solve a pollution reduction problem in a region and want to use the negotiation approach to avoid negative state authority action. The concept could be applied within a single corporation having multiple sources of pollution for solving the "principal-agents" problem where manager's and owner's utility functions are different.

A special application might be to organize a voluntary, experimental workshop of selected polluters and/or experts. Some information that might be useful for environmental policy decision-making, like a rough estimation of the environmental pollution charges, could be generated at the workshop (to receive information that is important).

The approach can also help with an efficient initial distribution of tradable water pollution rights. Moreover, it could contribute to resolving complicated situations (characterized by many pollutants, synergic effects, etc.) before the rights are allocated for limited number of the pollutants.

BIBLIOGRAPHY

1. Sauer, P., Dvorak, A., Lisa, A., and Fiala, P. (2003). A procedure for negotiating pollution reduction under information asymmetry. *Environ Resource Econom.* **24**: 103–119.
2. Czech National Council Act No. 58/1998 Coll. on the Charges for Pollution of Surface Water.

READING LIST

Brink, P. (Ed.). (2002). Voluntary environmental agreements. *Process, Practice and Future Use*. Greenleaf Publishing, Sheffield.

Dinar, A. (Ed.). (2000). *The Political Economy of Water Pricing Reforms*. Oxford University Press, New York.

May, C. (1998). Environmental agreements. *The Role and Effect of Environmental Agreements in Environmental Policies*.

Fiala, P. (1997). Models of cooperative decision-making, In: *Multiple Criteria Decision-Making*. G. Fandel and T. Gal (Eds.). Springer, Berlin, pp. 128–136.

Glasbergen, P. (Ed.). (1998). *Co-operative Environmental Governance: Public-Private Agreements as a Policy Strategy*. Kluwer Academic Publishers, Dordrecht, the Netherlands.

On Environmental Agreements. (1996). *Commission of the European Communities (COM 96–561 final)*. Brussels, Belgium.

OECD. (1999). *Voluntary Approaches for Environmental Policy: An Assessment*. OECD, Paris.

WATER RESOURCE ORGANIZATIONS

Peter E. Black
SUNY ESF
Syracuse, New York

Water management organizations evolved at local, regional, state, and federal levels. They were created, in large part, during the twentieth century, exhibit different degrees of formality and authority, and exist for different purposes, serving different groups consisting of individual qualified members and/or of the public at large. The organizations fill specific missions, roles, and meet demands of members and official segments of society. Knowing their missions and responsibilities makes it easier to seek information or how to influence their responsibilities.

INTRODUCTION

Organizations may be classified in a variety of ways. The water and related land resources agencies lend themselves to classification by mission (or responsibility), level of government, and legal status. Other water organizations may be national in scope but voluntary in membership, responsible for licensed professional activity, or open to all who subscribe to the group's philosophy. Any nongovernmental organization may be chartered under the laws of a state and may be incorporated for protection from liability actions. Some organizations may be tax-exempt, depending on their political nature and/or activity. Organizations may be temporary if their missions are limited; others may experience a shift in mission as their goals are achieved, or the membership finds them no longer relevant.

GOVERNMENTAL ORGANIZATIONS

Federal Agencies

Federal agencies derive their authority from the U.S. Constitution. State agencies, of course, derive theirs from their constitutions, which were approved by the U.S. Congress at the time of statehood, or were subsequently created by state legislatures. All the organizations' chartering legislation is subject to judicial review to ensure constitutionality. For example, the 1911 Weeks Forest Purchase Act (36 Stat 961) asserts in its opening statement of purpose of "the purchase of forest lands necessary to protection of the flow of navigable streams," the federal government may buy abandoned forest lands for inclusion in the National Forest system. Those fourteen words (in quotes) are sufficient to establish the constitutionality of the Act because the Constitution confers the right to regulate navigation on the federal government [Article I, Section 8 (3)]. The principal federal agencies charged with water and related land management are arranged by primary function(s) in Tables 1a to 1e (abstracted from Table 3.1 in Reference 1). Many agencies have other functions as well.

State Agencies

The states usually have a series of departments involved with forests, water, lands, fish and wildlife, and parks and recreation. The specific names and combinations of responsibilities are quite variable but tend to follow some typical patterns and in several different ways administer those permits and programs that federal agencies may have delegated to lower levels either because the states refused to be regulated by the federal government, they wished to have greater control over the regulations, or the federal government had neither the time nor funding to execute the mandated responsibility ("unfunded mandate").

Typically, the states create state-wide agencies such as departments of natural resources or environmental conservation, fish and wildlife, parks and recreation, agriculture and markets, and so forth. These deal with *intrastate* policies and management, regulation, granting of permits, and occasionally research and inventory. For point source pollution control permits under the control of the (federal) Environmental Protection Agency, the latter may have delegated responsibility for issuing National Pollution Discharge Elimination System (NPDES) to the state agency, in which case they are called SPDES permits. Continuing coordination between the state issuing authority and the EPA is part of these cooperative efforts. Some of the agencies may also be involved as partners with similar agencies in other states.

States may create *interstate* organizations to work with neighboring states by authority of the United States Constitution [Commerce Clause, Article I, Section 10 (3)]. These are interstate compacts that may exist for any of several reasons. Most interstate compacts administer interstate rivers and harbors. The managing organization is known as a *commission* (as opposed to a *committee*, which is usually informal. The Commissioner of the appropriate state department may serve—by appointment of the governor—as the representative of the state to an interstate commission.

Local Agencies

States may extend the constitutional right to local government to create agencies such as districts. These are extensions of state government that have the power to set boundaries, hold referenda, tax those holding real property within the district boundaries, enter into contracts, and incur indebtedness. There is a wide variety of districts; many exist for water management, including soil and water conservation districts, flood control districts, conservancy districts, drainage districts, and so on (in addition to the typical neighborhood highway districts, lighting, fire fighting, and school districts).

Typically, local organizations are formal or informal. The former include the *districts* (described above), and the latter are *associations*. Associations are not legal extensions of state government. Without formal boundaries, they may have members who agree to pay dues and may live anywhere. Usually, some common long- or short-term interest is the glue that holds the association together. Currently, there are thousands of watershed initiatives, associations that have a wide variety of collective names. They are watershed, river basin, or lake associations, councils, committees, land trusts, and so forth. They may, of course, be incorporated. Districts are often identified by other descriptive names, such as Resource Conservation Districts, Soil Conservation Districts, Conservation Districts, Natural Resource Districts, and Land Conservation Districts (as is pointed out on the NACD website http://www.nacdnet.org/). In some of the states that have strong county governments, the districts are organized on the basis of and coincident with the boundaries of the county; in others, the district is based on natural drainage.

Table 1a. Federal Land and Water Management Agencies

Agency	Date	Department	Mission, Responsibility
Bureau of Indian Affairs	1824	War, Interior	Manage forest, range and water resources; education, advocacy
Forest Service	1905	Interior, Agriculture	National forests, research, state and private forestry, and international forestry
National Park Service	1916	Interior	Preservation for science and recreation
Fish and Wildlife Service	1940	Interior	Wildlife refuge administration
Bureau of Land Management	1946	Interior	Renewable and mineral resources on public lands

Table 1b. Federal Construction and Operating Agencies

Agency	Date	Department	Mission, Responsibility
Corps of Engineers	1774	Continental Congress, War, Defense	Survey, rivers and harbors, navigation, flood control, water supply, water supply and recreation, dam safety, wetland permits, and aquatic ecosystem restoration
Bureau of Reclamation	1902	Interior	Irrigation and dams therefore
Natural Resources Conservation Service (nee Soil Conservation Service)	1933	Agriculture (Interior for a few years)	Soil and water conservation, upstream flood control, wetlands, education and information; administration of principal agricultural conservation programs; disaster response
Tennessee Valley Authority	1933	Independent	Hydropower development, flood control, economic improvement

Table 1c. Federal Regulation and Enforcement Agencies

Agency	Date	Department	Mission, Responsibility
Interstate Commerce Commission	1887	Independent	Carrier rules and regulation; sets interstate freight rates and routes
Federal Power Commission	1920	Independent	Licenses hydropower sites on navigable streams, public lands
Flood Insurance Administration	1968	Housing and Urban Development	National Flood Insurance Program
Environmental Protection Agency	1970	Independent	Pollution control permits, abatement, research/grant programs

Table 1d. Federal Research and Inventory Agencies

Agency	Date	Department	Mission, Responsibility
Geological Survey	1879	Interior	Mapping, hydrologic data collection, publication, and research; benchmark watershed program
National Weather Service (nee Weather Bureau)	1890	Signal Corps, Agriculture, Commerce, NOAA	Weather data collection, research, and publication; flood and fire weather forecasting
Agricultural Research Service	1953	Agriculture	Soil–water–vegetation research
Office of Water Research and Technology	1964	Interior, now abolished	Promotion, coordination, and funding for water resources research
National Oceanic and Atmospheric Administration	1970	Commerce	Explores, maps, and charts oceans, atmosphere, and space

Table 1e. Federal Coordination and Study Agencies

Agency	Date	Department	Mission, Responsibility
Senate Select Committee on Water Resources	1959–1961	Congress	Federal investigation of all water resource issues
Water Resources Council	1965–1981	Independent (now nonfunded)	Coordination of federal activities, flood control regulations, principles and standards for evaluation of water and related land resources projects and programs
National Water Commission	1968–1973	Independent	Nonfederal study of all water resource programs issues
Council on Environmental Quality	1970	President	Policy, research, advice to President and reports to Congress and the Public
National Study Commission	1972–1977	Independent	Charged to report on necessary adjustments to Clean Water Act halfway to 1983 and 1985 goals

PROFESSIONAL ORGANIZATIONS

There are numerous professional societies (sometimes confusingly referred to as associations) that serve the particular profession, such as architects, engineers, fishery biologists, foresters, geologists, geophysicists, hydrologists, landscape architects, limnologists, oceanographers, public administrators, range managers, soil scientists, surveyors, water works managers, well drillers, wetland managers, wildlife managers, and so forth. Many of these organizations serve, wholly or in part, important professions in water resource activities. Many have both technical and public-oriented publications that might be of interest to the lay public. Annual national and regional or state meetings are also often open to the public in whole or in part. Usually, putting the organization's initials (or full name) into the website URL line will provide the home webpage. Many water resources organizations may be found listed under links at the American Water Resources Association (http://www.awra.org), the principal interdisciplinary professional water resources organization in the United States, at the Environmental Protection Agency (http://www.epa.gov/), or at the website of the Conservation Technical Information Center (http://www.ctic.purdue.edu/CTIC/CTIC.html).

NONGOVERNMENTAL ORGANIZATIONS

Nongovernmental organizations are most often referred to collectively as "NGOs." The number continues to grow. Many of these may also be found at the websites of organizations cited above, by online search, or in the publications of the organizations.

During the past two decades (approximately), water resources management decisions have been effected by partnerships made up of mandated national, state, and local agencies in concert with business, interested individuals, and NGOs. Successful partnerships operate at all levels, with large or small numbers of participants, with and without government sanction, and over time with a shifting membership reflecting past accomplishments and initial goal achievement. Partnerships represent the finest in problem solving challenges and are truly democracy in action.

Partnerships require that all stakeholders[1] have an opportunity to take part in the process; that the participants respond to questions such as "What are our common goals?", "What can we do together better than separately?", "On what do we disagree?", and "Can we agree on an initial objective?"; that no one participant exercises fiscal, organizational, political, or social control over the process or the outcome; that participants treat one another with cordiality and respect; and that there be some identifiable constraint on the challenge, usually for water resources challenges, geographical, but also possibly economic, legal, or disaster or other crisis.

[1]This interesting term is currently used to identify those with a vested interest. However, that is the precise opposite of the original use of the term to identify the person in a poker game who held the bet funds—the stake—while the game was played out, in other words, it was the person who had *no* vested interest!)

SUMMARY

Historical, legal, and cultural events during the last 200 years combine with the individual personalities of leaders and public citizens to leave a massive array of legislation, government agencies, and nongovernmental organizations all charged with or assuming responsibilities for some part of the water resources of the United States. Knowing the authority, extent of influence, nature, and relationships among the organizations is essential for sound water resources management. But, it does not guarantee it: only when people take on the responsibility of participating in the important water resources decision-making process will the vital water resources be effectively managed for a sustainable future. Partnerships are important players in the continuing policy and management decisions over water and related land resources.

BIBLIOGRAPHY

1. Black, P.E. and Fisher, B.L. (2000). *Conservation of Water and Related Land Resources*, 3rd Edn. CRC Press/Lewis, Boca Raton, FL.

READING LIST

Holmes, B.H. (1979). *History of Federal Water Resources Programs and Policies, 1961–70*. USDA, Miscellaneous Publication No. 1379, Superintendent of Documents, USGPO, Washington, DC.

Holmes, B.H. (1972). *History of Federal Water Resources Programs*. Miscellaneous Publication 12233, Department of Agriculture, Washington, DC.

Office of the Federal Register. (1998). *Government Organization Manual, 1998–1999*. National Archives and Records Administration, Washington, DC.

A BRIEF HISTORY OF THE WATER POLLUTION CONTROL ACT IN THE U.S.

David L. Russell
Global Environmental
Operations, Inc.
Lilburn, Georgia

The Rivers and Harbors Act of 1899 was the earliest piece of federal water pollution control legislation in the United States. It regulated the dumping of garbage into rivers and harbors from ships. The act was only moderately successful in its application, but it became important in the 1970s when the EPA used it to regulate discharges into navigable waters.

ORSANCO

The earliest significant start for water pollution control legislation in the United States began with the formation of the Ohio River Sanitary Commission (ORSANCO) in the 1940s. It was formed as a result of efforts by individuals from the U.S. Public Health Service (USPHS) and many of the state Departments of Health. The ORSANCO compact culminated more than 10 years of work to promote public health and reduce the incidence of waterborne disease in communities along the Ohio River. ORSANCO started as an initial effort among four states, and by 1940, the compact had six states. In 1948, the seventh and eighth states signed the compact on the same day that President Harry Truman signed the first comprehensive water pollution control legislation that directed the U.S. Public Health Service to control water pollution to promote public health.

The Public Health Service already had authority, but was reluctant to use it for water pollution control unless there was a specific prominent incident. The real power to control pollution still lay with the states.

By contrast, ORSANCO was effective because the control came from within the states as a cooperative agreement among the Departments of Health. ORSANCO was instrumental in issuing the first set of coordinated standards—the first water treatment design code manual—in 1954. ORSANCO was also effective in coordinating the development of antipollution lawsuits and still is an effective voice in the Ohio River region when it speaks for the state governments. ORSANCO has legal authority to compel individual polluters in any of the compact states to reduce their discharge of pollutants, or even to cease the discharge completely and can, if it chooses, override individual state governments.

The ORSANCO compact is still in force and has remained unchanged for 50 years. The fundamental tenet of the compact is that the discharge from one state shall not be injurious to the public health and well-being of the residents of another state.

It is important to point out that developments in water chemistry, analytical technology, and fundamental research, along with political developments, accompanied the growth and expansion of ORSANCO. Many of these developments were independent, but it was the driving interest in water pollution control and industrial wastes that created the climate for research and development to bloom. Many of the earliest names in the history of water pollution control were involved with ORSANCO as well. Harold Streeter, who is one of the formulators of the Streeter–Phelps equations (which underlie much of river modeling) was an early Commissioner of ORSANCO. State Commissioners on the ORSANCO Board of Governors (from their respective State Health Departments) include Clarence Klassen (Illinois), Blutcher Poole (Indiana), and Fred Waring (Ohio). One of many early workers in the State of Indiania was Dr. Don Bloodgood at Purdue University, who created the Purdue Industrial Waste Conference in 1946. That conference was made possible through the encouragement of some of the ORSANCO Commissioners and was instrumental as a showcase in promoting nationally prominent water pollution control research until its incorporation into the Water Environment Federation Industrial Wastes Program in 2000.

WATER POLLUTION CONTROL ACT OF 1948

In 1948, the Water Pollution Control Act (Ch. 758; P.L. 845) authorized the Public Health Service to prepare comprehensive programs for eliminating or reducing the pollution of interstate waters and tributaries and improving the sanitary condition of surface and underground waters. Under the plan, water conservation and improvement was to be emphasized. The original statute also authorized the Federal Works Administrator under the U.S. Public Health Service to assist states, municipalities, and interstate agencies in constructing treatment plants to prevent discharges of inadequately treated sewage and other wastes into interstate waters or tributaries. More information is available at http://laws.fws.gov/lawsdigest/fwatrpo.html.

Controlling pollution and industrial waste gradually became more important, and the original act was amended to authorize additional water quality programs, standards, and procedures to govern allowable discharges as well as funding for construction grants or general program funding. The authority of the original act was extended several times, through appropriations up to 1956 by Ch. 927; P.L. 579.

Progress to the Mid-1960s

The Federal Water Pollution Control Act Amendments of 1961 (P.L. 87-88) stipulated that, during the planning for any reservoir, federal agencies consider storage to regulate stream flow for water quality control, but the act was placed under section 33 U.S.C. 1252 (transportation) of the U.S. Code. This later provided impetus for the U.S. Army Corps of Engineers and the U.S. Coast Guard to have authority for "transportation related" oil spills and what later became National Pollution Control Discharge Permit issuance. Under the 1961 amendments, the Department of Health, Education, and Welfare was directed to undertake research and determine the effect of pollutants and treatment methods and assess water quality in the Great Lakes. The Secretary of HEW was authorized to take "appropriate measures to ward off pollution of interstate navigable waterways if requested by a state (33 U.S.C. 1254(f)). Few requests were issued under the legislation.

The Department of Heath, Education, and Welfare was also ordered to report to Congress by 1964 on the effects of pollution on U.S. estuaries and the economic impacts of pollution control. HEW was also told to develop recommendations for further actions that would lead to a comprehensive national policy and define the roles of federal, state, and local governments. The interests of Mexico and Canada were initially considered but only as they related to damage from pollution of boundary waters (the Great Lakes and St. Lawrence Seaway, and the Rio Grande River). Congress prohibited the discharge of oil into U.S. waterways under 33USC 466 (transportation legislation).

Congress decided to provide some form of funding to states and cities to construct sewage treatment plants. This Construction Grants Program was initially a relatively modest effort through most of the 1960s, and it was capped until late in the decade at a maximum grant of 33% of the cost of interceptor sewers and treatment works.

PL-660

Public Law 660 was the beginning of a comprehensive pollution control effort. Prior to PL-660, the Public Health Service (Commission Corps) officers were largely responsible for water pollution control. The law that was passed in the early 1960s made several important changes in the funding of water pollution control programs. A new organization, called the Federal Water Pollution Control Agency, was created under the administration of the Department of Interior. The FWPCA also had responsibility for pollution control in Alaska (P.L. 86-70) and an expanded role in construction grants (P.L. 86-624).

The Silent Spring, the Age of Aquarius, and the Vietnam War

The publication of Rachel Carson's *The Silent Spring* in 1967 and the dawning of the "Age of Aquarius,"—the coming of age of the children born after World War II—led to a major departure from the once quiet role of the federal government in pollution control. In the United States, the Vietnam War was a time of domestic turbulence and significant social and political protest. The statements of young radicals ("Don't trust anyone over the age of 30") and a growing dissatisfaction with the progress in solving the war led to social protest and dissatisfaction with other areas of government as well. "Make Love not War" and the countercultural exploration of new religions and ideas, especially those from India, quickly led to the incorporation of the concept of "Mother Earth" and a "Gaia" movement. Among the tenets of this movement were the concepts that the earth was a living sentient being, that all industry was bad and full of polluters, and that all pollution was bad and should be stopped at any cost. During this era, the first ecoterrorists became active in driving spikes in trees to be logged so that they would ruin the timber cutting and debarking machines. Other actions and rhetoric were also significant in defining and shaping the environmental movement.

The Silent Spring presented a graphic description and made a persuasive case for the ecological damage from the indiscriminate use of pesticides, principally DDT. When the book reached bestseller status and was condensed in Reader's Digest, it struck a resonant chord among disaffected youth. When combined with the societal revolution, it quickly led to the inauguration of the first "Earth Day" in 1970. Later, researchers accused Carlson of sloppy and inaccurate research and presentation of a poor case for action, but her literary talents and public opinion silenced the critics.

The Clean Water Restoration Act of 1966, P.L. 89-753, authorized the Departments of Interior and Agriculture and the Water Resources Council to study the effects of pollution, including sedimentation, on fish and wildlife, sport and commercial fishing, recreation, water supply and power, and other specified uses in the estuaries and estuarine zones of the United States.

By 1969, the United Nations entered the political fray about the environment, when the Secretary General (U Thant) predicted Armageddon from environmental causes. He was asking whether mankind could survive another 10 years without immediate and decisive actions to control pollution. Predictably, he condemned the United States as the principal culprit in creating this forthcoming environmental disaster (which it may have been) and demanded that the United States take unilateral action to control pollution. He conveniently ignored the pollution arising from the rest of the world and the benefits from their industrial activity.

In the United States, Congress created the National Environmental Policy Act in 1969 and submitted it to President Nixon for signature. The NEPA was hailed as a new direction for the federal government because, for the first time, certain specific federal actions were to become subject to environmental review and the issuance of an "environmental impact statement." According to President Nixon, this heralded, "a new decade of environmental awareness."

On December 2, President Nixon signed Reorganization Plan Number 3, which created a new cabinet-level federal agency, the Environmental Protection Agency. The plan disbanded the Federal Water Quality Administration and stripped the Department of Interior of environmental enforcement responsibilities and programs. It also demoted the role of the Public Health Service and the Corps of Engineers in managing and controlling air pollution and water pollution, respectively.

It should be noted that at that time, many of the industries located on major waterways, such as the Ohio River, were required to obtain operating discharge permits issued by the Corps of Engineers. Municipalities were, depending on size, operating with state permission and permits. The federal program was much better administered than state programs. Funding for construction of wastewater treatment facilities by municipalities was coordinated through the states, and federal monies for construction grants to municipalities were rapidly increasing to 66% of the cost of those facilities that the states had prioritized.

Other actions in 1970 under PL 91-224 included

1. the development of a National Contingency Plan for preventing discharge of oil to the seas;
2. the development of regulations for determining harmful quantities of oil discharged to U.S. waters;
3. state certification that such discharge will not violate applicable water quality standards;
4. development of performance standards for marine sanitation devices;
5. demonstration projects to control acid or other mine water pollution;
6. control of water pollution within the watersheds of the Great Lakes; and
7. required federal facilities to operate in compliance with applicable water quality standards.

PL 92-500

The passage of PL 92-500 is often considered a major landmark in defining national policy for environmental protection. The reason often cited is the stated goals of the law: to make the waters of the United States *fishable* and *swimmable* by 1985. The law also had several other important provisions in addition to dramatic increases in funding for sewage treatment works.

The financial incentive for municipalities was increased to between 85% and 95% of the applicable construction costs and was supported by a very large pool of money. The consequence of this funding was that the qualified and unqualified began designing and building wastewater treatment works at an enormous rate—so fast that the personnel required for planning, project review, and construction oversight were overwhelmed by applications. Moreover, the incentive for oversight was reduced because the municipalities had only 5% to 15% of the actual cost at risk and the administration of the federal Construction Grants Program was primarily financial, not technical. As a result, a large number of new facilities were built, many of them far in excess of the needs of the communities.

By 1982, the EPA had issued a manual citing the most common mistakes in designing facilities. This manual detailed the most common failures in designing facilities, and, by its very existence, was a ringing condemnation of the many failures of oversight of the Construction Grants Program. Many of the mistakes cited by the EPA were elementary and illustrated a degree of incompetence of many of the design professionals and failure of the regulatory community to exercise adequate planning and oversight. Planning and oversight takes qualified manpower and time and cannot be rushed.

The Construction Grants Program was a well-intentioned failure in a number of other aspects as well. First, it added a significant level of complexity and paperwork requirements to the construction program that increased engineering, administrative, and total project costs. The requirement for Davis–Bacon Act compliance, which mandates that the highest prevailing wage rate would be paid and that union labor would be used whenever and wherever possible, effectively eliminated any potential savings from the use of nonunion labor and discouraged many cost-efficient improvements.

The second item that caused the greatest percentage of failures was directed at the engineering specifications. The EPA required that the project engineer must provide three specifications for each major piece of equipment in a project and that those items would be competitively bid without regard to the quality, only the price. The result was often a direct and dramatic deterioration in the quality of equipment purchased. The equipment was often selected by the contractor who had the option to appeal to the EPA on the basis of potential savings for the project. Many projects became contractor's auctions where the final project bore only a slight resemblance to the engineer's original specifications. The result was that equipment compatibility and reliability issues often arose during construction but were often sidestepped or overlooked because of the insistence on obtaining the lowest costs.

One final impact of the Construction Grants Program under PL 92-500 was the restructuring of the U.S. wastewater treatment industry. The emphasis on equipment price caused severe competition among wastewater treatment equipment manufacturers and reduced profit margins. As a result, companies cut back on research, and that had a 10 to 20-year impact on the wastewater treatment industry. From about 1975 to 1995, new products were being developed and introduced by non-U.S. sources.

Other requirements of PL-92-500 required the Federal Power Commission not to license hydroelectric power projects unless stream flow and water quality objectives could be met; proposed a range of national objectives for water quality designed to restore and maintain the chemical, physical, and biological integrity of the national waterways (33 USC 1252); and required reissuance of permits for pollution control discharges.

This last item was the item that created the National Pollution Discharge Elimination System (NPDES) permit system. The legislation also established minimum standards and requirements and provided funding for states to assume control over their own water quality management programs, once they had demonstrated their compliance with minimum federal standards set by the EPA.

In addition, PL 92-500 significantly expanded provisions related to pollutant discharges by requiring point-source effluent limitations consistent with state water quality standards, enabled the State issuance of water quality standards, and provided initial methodology for identifying and evaluating nonpoint-source pollution. Authority was also provided for developing water quality inventories and toxic and pretreatment effluent standards.

Other provisions clarified and established liability for discharges of oil and hazardous substances, provided for federal oversight and backup responsibility for oil spill cleanup operations, and authorized the Corps of Engineers to issue permits for discharging dredged or fill material into navigable waters at specified disposal sites (Section 404-33 U.S.C. 1344). However, the EPA could block the use of a disposal site based on water quality issues related to the discharge from the site and potential deleterious effects on municipal water supplies, fisheries, wildlife, or other water quality uses.

1977 Amendments

The Clean Water Act was further amended in 1977 (P.L. 95-217) to provide additional requirements for the following: "Best Management Practices" Program as part of the statewide planning program (33 U.S.C. 1288); provision for the Fish and Wildlife Service to provide technical assistance for developing "best management practices;" authorization for the Department of the Interior to complete the National Wetlands Inventory by December 31, 1981; and additional regulations and de minimis standards for the Corps of Engineers to issue general permits for dredging and filling waters and wetlands. These exemptions included normal farming, silviculture, and ranching (33 U.S.C. 1344(f)), and general activities below certain sizes.

1987 Amendments

The Water Quality Act of 1987 (P.L. 100-4) amended the Clean Water Act to provide (1) continuation of the Chesapeake Bay Program and an office dealing with pollution of the bay; (2) establishment of a Great Lakes

National Program Office within EPA and a Great Lakes Research Office within NOAA; (3) a requirement that EPA, the Fish and Wildlife Service, and NOAA conduct research on the harmful effects of pollutants in the Great Lakes, emphasizing the bioaccumulation of these pollutants in aquatic species; (4) require states to develop strategies for cleanup of toxics in waters where the application of "best available technology" (BAT) discharge standards will not meet state Water Quality Standards; (5) an increase in the penalties for violations of Section 404 (Dredge and Fill) permits; and (6) additional state reporting on lake water quality and methods to reduce the adverse impact of excess acidity.

A $15 million funding package has been provided to demonstrate methods to reduce the acidity of lakes (often found in coal and surface mining areas), and a $400 million program has been authorized for states to implement a nonpoint pollution source (agricultural and other areas) control program with EPA oversight. Estuaries of national significance are also to be nominated under this program, and plans must be developed to protect and restore those estuaries and to protect their chemical and biological status.

THE NATIONAL POLLUTION DISCHARGE ELIMINATION SYSTEM

Rolf R. von Oppenfeld
The TESTLaw Practice Group
Phoenix, Arizona

INTRODUCTION

The Clean Water Act (CWA) authorizes the United States Environmental Protection Agency (EPA) to regulate water pollution discharges to federally regulated surface waters via a permitting process. All facilities that discharge pollutants from a point source into waters of the United States must obtain a National Pollution Discharge Elimination System (NPDES) permit from the EPA or an authorized state [42 U.S.C. §§ 1311(a), 1342; 40 C.F.R. § 122.1(b)]. Permits typically allow the permittee to discharge specified pollutants from designated outfalls. See the NPDES program homepage at http://www.epa.gov/npdes/. The NPDES program generally imposes technology-based standards and water-quality-based effluent limits on point source discharges of pollutants to surface waters, although less stringent requirements presently apply to stormwater runoff discharges.

Currently, 46 states and territories are authorized (have primacy) under Section §r402(b) of the CWA to administer the NPDES program. The EPA regions are the lead permitting authority in the other states and territories. State NPDES programs must use federal standards as a baseline; however, states may impose stricter requirements [40 C.F.R. § 123.1(I)]. State-issued permits are also subject to review by the EPA [33 U.S.C. § 1342(d)]. The EPA may make comments, objections, or recommendations to the state. If the state does not change the permit to address the EPA's concerns, the EPA may assume control of the permitting process for that facility. [33 U.S.C. § 1342(d)(4)]. The EPA may also retract program approval from states not administering the program in accordance with federal law [33 U.S.C. § 1342(c)(2)]. When the EPA region is the permitting authority in the state, the state must certify that an authorized discharge will comply with the state water quality standards [33 U.S.C. § 1341(a)].

TYPES OF PERMITS

A NPDES permit is a license to discharge a specified amount of a pollutant or pollutants into a specified receiving water under certain conditions. The NPDES permits establish effluent limitations [33 U.S.C. § 1362(11)]. However, NPDES permits also contain monitoring and reporting requirements, a duty to properly operate and maintain systems, upset and bypass provisions, recordkeeping, and inspection and entry requirements [40 C.F.R. § 122.41]. The source of much information concerning NPDES permits is contained in EPA's *Permit Writers' Manual* located online at http://cfpub1.epa.gov/npdes/writermanual.cfm.

There are two basic types of permits: the individual and general permit. NPDES permits are issued for fixed terms, not to exceed 5 years [33 U.S.C. § 1342(b)(1)(B); 40 C.F.R. § 122.46]. Individual permittees generally must apply for renewal 180 days before the permit expires. EPA has made electronic copies of NPDES permits for major facilities available online at http://www.epa.gov/NPDES/permitdocuments/.

General Permits

The permitting agency develops and issues a general permit. A general permit typically covers multiple facilities within a specified category, industry, or area. General permits are cost-effective and efficient because the agency can authorize discharges from a large number of facilities with numerous common criteria under one blanket permit. Facilities covered by a general permit "self-regulate," although the permitting agency will periodically inspect or investigate potential noncompliances. The procedures for using general permits and the general conditions that apply to all general permits can be found at 40 C.F.R. § 122.28. Most stormwater dischargers may apply for a general permit by filing a Notice of Intent (NOI) pursuant to a federal or state stormwater general permit. Many states also issue general NPDES permits for certain process wastewater discharges generally considered of minimal impact.

Individual Permits

A discharging facility must be covered by an individual permit if it does not qualify for coverage under a general permit. The authorized permitting agency prepares an individual permit for a specific facility. A permittee first provides the required information on an individual permit application.

Once the permit writer determines that the application is complete, the technical review starts. Effluent limitations are determined, and a preliminary permit containing these limits is prepared. (Id). The agency publishes public notice of the draft permit. Typically, the public notice allows for 30 days of public comment. Based on this information, the agency will issue or deny a permit (40 C.F.R. § 122.5).

DISCHARGE LIMITATIONS

Effluent limitations are established in a two-step approach consisting of technology-based and quality-based standards. First, dischargers are required to use treatments that are technologically and economically feasible for the particular industry. Additionally, dischargers may be required to use more stringent treatments to achieve certain water quality goals for the particular waterbody. Limits can be expressed as mass limits, concentration limits, visual observations, monitoring requirements, testing requirements, parameter limits, flow, pH range, or temperature limits. The limits can be imposed as maximum limitations and monthly average limitations.

Technology-Based Effluent Limitations (TBEL)

The NPDES permit system sets technology-based levels of allowable discharges, determined by a series of minimum technology-based national effluent standards for various industrial discharge categories. If no categorical technology-based standards apply, the standards are determined by the permit writer's best professional judgment (BPJ) of the appropriate technology-based reduction. See 40 C.F.R. Part 400 et seq.

TBELs are national effluent guidelines that establish limitations for all types of discharges within a specific industry, such as direct and indirect dischargers, existing and new sources, and specific discharges, including process water, cooling water, and sanitary wastewater. EPA has issued approximately 50 industrial category effluent guidelines (40 C.F.R. §§ 405–471). Guidelines are available online at http://www.epa.gov/docs/epacfr40/chapt-I.info/subch.N.htm.

The CWA created a two-step approach for dealing with the reduction of water pollution through TBELs. The first phase required industrial dischargers to meet pollution control through the best practicable control technology currently available (BPT) [33 U.S.C. § 1311(b)(1)(A)]. The second level of pollution control required using the best available technology economically achievable (BAT) [33 U.S.C. § 1311(b)(2)(A)]. BAT applies to nonconventional pollutants, whereas best conventional pollutant control technology (BCT), which is more lenient, applies to conventional pollutants, as explained below. However, BPT/BAT/BCT standards do not apply to new sources, only to existing sources. New Source dischargers must meet New Source Performance Standards (NSPS), based on the best available demonstrated control technology (BADCT) (33 U.S.C. § 1316). Also, industrial indirect dischargers that discharge to a publicly owned treatment works (POTWs) are not regulated by TBELs but are instead subject to the Pretreatment Program under 33 U.S.C. § 1317(b).

BPT Standard. The BPT standard collectively examines an industry category to determine the types of treatment facilities used in that industry. The EPA looks at the better run facilities and estimates the level of pollution control achieved through the typical technology. The EPA compares the cost of applying the technology to the entire industry with the effluent reduction benefits. See 52 Fed. Reg. 42522 (Nov. 5, 1987). This analysis yields a pollution control level demonstrating the average of the best achieved pollution reduction.

BAT Standard. BAT controls represent the maximum feasible pollutant reduction. EPA considers the technology currently used in that industry and also pilot data and technology used in other comparable industries. BAT does not require the use of a cost–benefit analysis, but the standard must be economically achievable. The BAT standard was designed to apply to toxic and nonconventional pollutants.

BCT Standard. BCT standards require industries to meet a cost-reasonableness test for the discharge of conventional pollutants, such as BOD, TSS, fecal coliform, pH, and oil and grease [51 Fed. Reg. 24973 (July 9, 1986)]. The cost-reasonableness test compares the reasonableness of the cost of the reduction in discharge levels of conventional pollutants to the cost of achieving that reduction. Conventional pollutants are subject to BCT, which are no more stringent than the BPT standard, rather than the more stringent BAT standard [33 U.S.C. § 1311(b)(2)(e)].

NSPS. New sources must comply with the stricter New Source Performance Standards, thus making the definition of a new source highly disputed. The EPA employs the substantial independence test to determine whether construction is leading to a new source or a modification of an existing facility [40 C.F.R. § 122.29(b)(1)(iii)]. This test looks at the degree to which the new unit functions independently of the existing facility, the degree of integration, and the extent to which the new unit is engaged in the same general type of activity as the existing source [49 Fed. Reg. 37998, 38043 (Sept. 26, 1984)].

NSPS are generally more stringent than those for existing sources; ideally, new sources should be built with state-of-the-art treatment technology. Requiring existing sources to retrofit under NSPS would be unreasonable. The major benefit to classification as a new source is that any new source constructed to meet NSPS may not be subjected to more stringent standards for 10 years after the date construction is complete or for the period of depreciation under the Internal Revenue Code, whichever is shorter (33 U.S.C. § 1316). However, the 10-year protection applies only to TBELs, not to WQBELs. At the end of the 10-year period, immediate compliance with the standards in effect at that time is required [33 U.S.C. § 1316(d)].

Best Professional Judgment. When TBELs have not been established for a particular industry sector or where there is a concern over a particular parameter (e.g., the permit writer thinks a receiving water is too polluted), permit writers will apply best professional judgment (BPJ) to establish permit limitations [40 C.F.R. § 125.3(e)]. The permit writer may consider data from the facility itself, from similar industries having similar waste treatment, or potentially applicable technology from an industry outside of the immediate industrial group. See 40 C.F.R. §§ 122.44(a), 125.3(c)(2).

Variances From TBELs. A qualified permittee may obtain a variance or modification from its permit's TBELs. One type of variance is a fundamentally different factors (FDF) variance [33 U.S.C. § 1311(n); 40 C.F.R. §§ 125.30–32]. The discharger must demonstrate that the factors applicable to its facility are fundamentally different from those considered for developing the effluent guidelines for its industrial category (40 C.F.R. Part 125). A variance is not issued merely because the cost of compliance would cause the facility to close. (Id). The applicant must also demonstrate that an objection was raised based on the FDF during effluent guidelines development or show why it was not. (Id).

The EPA may also modify BAT requirements or pretreatment requirements affecting nonconventional and nontoxic pollutants if the economic capability of the discharger requires less stringent limits [33 U.S.C. § 1311(g)]. The permitting agency may grant a variance from BAT for several nonconventional, nontoxic pollutants, such as ammonia, chlorine, color, iron and total phenols, where applicable BATs are unnecessarily stringent. (Id). Both the variance and the modification are granted on a permit specific basis.

Water-Quality-Based Effluent Limitations (WQBEL)

WQBELs are imposed where TBELs are insufficient to ensure water quality. WQBELs rely on the state water quality standards (WQS), and consist of two elements: (1) designated use classification and (2) criteria that are used to protect the designated uses. Surface water quality standards have become increasingly stringent during the last decade. Consequently, the WQBELs required in a NPDES permit for a particular pollutant are often more stringent than the corresponding technology-based effluent limitations.

Water Quality Standards. Water quality standards (WQS) are numeric standards developed for each stretch of a waterbody based on the designated uses. All states must classify their waters according to intended uses, for example, public drinking water, propagation of fish and wildlife, recreation, industrial, agricultural, and other uses [33 U.S.C. § 1313(c)(2)]. The WQS must also attain the CWA's goal of fishable or swimmable, if possible, and must maintain both the current and designated uses under the antidegradation policy, unless the state can demonstrate that it is unattainable or infeasible [40 C.F.R. §§ 131.10(g), 131.12].

Water quality criteria describe the physical, chemical, and biological characteristics of waters necessary to attain the designated uses. The EPA has developed numeric surface water quality criteria for over 150 pollutants [33 U.S.C. § 1314(a)(1); 63 Fed. Reg. 67547 (Dec. 7, 1998)]. Each navigable water is assigned such criteria to ensure that the water can support its designated use.

When the permitting agency determines that a discharge causes, has the reasonable potential to cause, or contributes to an in-stream excursion above the allowable ambient concentration of state numeric criteria within a state water quality standard for an individual pollutant, the NPDES permit must contain effluent limits for that pollutant [40 C.F.R. § 122.44(d)].

Toxicity-Based Limitations. Effluent limitations are sometimes based upon Whole Effluent Toxicity (WET), when it becomes difficult to set individual WQBELs [40 C.F.R. § 122.44(d)(1)(ii), (iii) and (iv)]. When the permitting agency determines that a discharge causes, has the reasonable potential to cause, or contributes to an in-stream excursion of pollutants in amounts or combinations that are *toxic* to humans, animals, plants, or other organisms, the permit must contain effluent limits for WET. WET limitations are incorporated into the NPDES permit and require the permittee to perform toxicity testing on the effluent. Toxicity tests expose various aquatic species to one or more concentrations of effluent in the laboratory to determine the short-term and long-term effects of exposure. Most permits issued since the release of EPA's WET Guidance Document require conducting toxicity tests regularly, such as monthly or quarterly [40 C.F.R. § 136, 65 Fed. Reg. 46457 (July 28, 2000)].

STORMWATER DISCHARGES

Regulation of stormwater runoff has presented challenges because stormwater runoff comes from diverse sources. EPA estimates that about 30% of known pollution to the nation's surface waters results from stormwater runoff. In 1987, Congress required EPA to establish regulations and issue permits for stormwater discharges. "Stormwater" includes stormwater runoff, snowmelt runoff, and surface runoff and drainage [40 C.F.R. § 122.26(b)(13)]. Stormwater does not include street wash water or infiltration, water that enters a sewer system below ground through defective pipes, pipe joints, connections or manholes. Discharges to retention basins that do not flow into waters of the United States or flow to POTWs also do not require stormwater permits.

The stormwater program has been implemented in two parts: Phase I and Phase II. The federal NPDES stormwater program relies on three types of permits: (1) The Construction General Permit, (2) The Multi -Sector General Permit, and (3) The Individual Permit. State agencies with primacy generally have analogous permits. More information on the federal stormwater program is available at http://cfpub.epa.gov/npdes/home.cfm?programid=6.

The Phase I Program

Phase I regulations cover separate municipal storm sewer systems (MS4s) serving populations over 100,000. Discharges into MS4s are treated as discharges to waters of the United States and require a NPDES permit (40 C.F.R. § 122.26). A MS4 is a conveyance or system of gutters, ditches, man-made channels or storm drains, owned by a state, county, municipality, or other public entity, designed or used for conveying stormwater, and is not a combined sewer or part of a POTW [40 C.F.R. § 122.26(b)(8)].

Phase I also regulates discharges from "areas associated with industrial activities" in 11 categories listed at 40 C.F.R. § 122.26(b)(14). The term "areas associated with industrial activities" covers discharges from any point source used for collecting and conveying stormwater, which is directly related to manufacturing, processing, or material storage at industrial facilities [55 Fed. Reg. 48007-15 (Nov. 16, 1990)]. The regulations break this definition down into two parts: what types of facilities have "industrial activity" based on their facility's Standard Industrial Classification (SIC) and what portions of the facilities include industrial activity. The portions of the facilities that include industrial activity include industrial plant yards, material handling sites, refuse sites, shipping and receiving areas, manufacturing buildings, raw material storage areas, etc. [40 C.F.R. § 122.26(b)]. Discharges from areas separate from the industrial activities, such as office buildings and parking lots and discharges from facilities engaged in wholesale, retail, service, or commercial activities are not included. [55 Fed. Reg. 48007 (Nov 16, 1990)]. Facilities covered by Phase I must apply for an NPDES permit, unless otherwise exempt.

Exemptions are provided until March 10, 2003 for industrial activities operated by municipalities with populations less than 100,000. See 64 Fed. Reg. 68722 (Dec. 8, 1999). Light industrial facilities (SIC codes 20–39 and part of 42) with "no exposure" of industrial activities or materials to stormwater are exempt under Phase I.

Permits for MS4s can be issued either for a particular system or on a jurisdiction-wide basis. MS4 permits must control pollution to the maximum extent practicable (MEP) [33 U.S.C. § 1342(p)(3)(B)(iii)]. The MS4 permit must also contain a requirement that prohibits nonstormwater discharges into the system's storm sewers.

Entities regulated under Phase I must obtain coverage under either an Individual Permit or a General Permit. Regulated entities must also implement stormwater pollution prevention plans (SWPPPs) or stormwater management programs; both require BMPs that effectively reduce or prevent the discharge of pollutants to receiving waters.

The Phase II Program

On December 8, 1999, the EPA published the NPDES stormwater management program Phase II Final Rule [64 Fed. Reg. 68722 (Dec. 8, 1999)]. Facility operators covered under Phase II must apply for permit coverage by March 10, 2003, except for oil and gas small construction activities which must apply by March 10, 2005 [67 Fed. Reg. 79828 (Dec. 30, 2002)].

The Phase II rule subjects two new groups of permittees to stormwater management: operators of small MS4s in urbanized areas and small construction activities that disturb 1–5 acres of land. The nonexposure exemption has been extended to include all industrial activities other than construction activities. The Phase II rule impacts about 5000 municipalities and federal systems such as military bases and prisons. Regulated operators must implement stormwater discharge best management practices (BMPs) that reduce or prevent the discharge of pollutants to receiving waters.

Stormwater Discharges Associated with Industrial Activities

Phase I contains a permitting component for industrial stormwater. The operations of industrial facilities that are within the 11 categories of "stormwater discharges associated with industrial activity" [located at 40 C.F.R. § 122.26(b)(14)(i)-(xi)] that either discharge to a MS4 or directly into waters of the United States must receive a NPDES industrial stormwater permit. Operators of industrial facilities can generally obtain either a general or individual permit to satisfy the NPDES stormwater permit requirement.

Conditional No Exposure Exclusion

"No exposure" means that all industrial materials or activities are protected by storms resistant sheltering to prevent exposure to rain, snow, snowmelt, or runoff. Certified facilities are not required to obtain a stormwater permit. Applicants must submit the "No Exposure Certification Form" every 5 years.

The no-exposure certification form is nontransferable. All new owners or operators must complete, sign, and submit a new form to claim the no-exposure exclusion. Upon submission of the new owners' form, the permitting agency may inspect the facility to verify the claim. A facility owner or operator who fails to maintain nonexposure status could be liable for an unauthorized discharge of stormwater. Under Phase II regulations, all industrial activities, except for construction, are eligible for exclusion.

The Multi-Sector General Permit

The Multi-Sector General Permit (MSGP) is designed to regulate certain stormwater discharges from nonconstruction industrial activities [63 Fed. Reg. 52430 (Sept. 30, 1998)]. The MSGP is one large permit subdivided into 29 separate sectors based on the SIC system and/or narrative descriptions. If a facility has operations that can be described by more than one sector, then it must comply with all pertinent requirements.

Facilities eligible for MSGP coverage must file a notice of intent (NOI). The NOI is essentially a promise that the applicant will comply with all MSGP conditions. The applicant must also prepare and implement a Storm Water Pollution Prevention Plan (SWPPP) prior to the submission of the NOI. The applicant can look at the MSGP for guidance on what to include in the SWPPP.

The SWPPP must create a Pollution Prevention Team comprised of individuals who develop and implement the SWPPP. The SWPPP must include a map of the facility, an assessment of the potential sources of stormwater pollution at the facility, and a list of BMPs. Generic BMPs include "good housekeeping," preventive maintenance, spill prevention and response, employee training, record keeping, nonstormwater discharge evaluation, erosion control measures, and stormwater management measures, as appropriate. Comprehensive site inspection/compliance evaluations are required.

Stormwater discharges subject to an existing NPDES permit cannot be covered under the MSGP until the existing permit expires. If the existing permit contains numeric standards that are more stringent than those required in the MSGP, the MSGP will not be available when the existing NPDES permit expires.

Stormwater Discharges Associated with Construction

Operators of both large and small construction activities must obtain a NPDES construction permit if the activity disturbs more than 1 acre or is part of a larger common plan of development or sale that will disturb more than 1 acre and the discharge is to a water of the United States or a MS4. The EPA requires that construction activities obtain a general permit. Effective on July 1, 2003, a new federal construction general permit is applicable to both large and small construction sites [68 Fed. Reg. 39087 (July 1, 2003)]. States with primacy have developed their own construction general permits.

There are two waivers available for small construction sites. The first waiver is the rainfall erosivity factor (R). The construction site operator must determine that the R in the reused universal soil loss equation is less than five. The second waiver is available operators in nonimpaired waters or who certify that stormwater controls are not needed, based on a total daily maximum allowance.

Separate Municipal Storm Sewer Systems (MS4)

Operators of large, medium, and regulated small MS4s must obtain NPDES permits to discharge pollutants and must develop stormwater management programs to prevent harmful pollutants from entering MS4s by stormwater runoff. Medium and large operators must submit permit applications under Phase I. Regulated small MS4s can choose among an individual permit, a general permit, or a modification of an existing Phase I MS4 individual permit. See Phase II Guidance, EPA 833-R-00-002 p. 4-1.

A "large MS4" is any MS4 located in an incorporated place or county with a population of 250,000 or greater, as of the 1990 Census. A "medium MS4" is any MS4 located in an incorporated place or county with a population from 100,000–249,999, as of the 1990 Census (40 C.F.R. Part 122 App. F, G, H). Many MS4s in areas of less than 100,000 population were individually brought into the Phase I program by NPDES. These regulated MS4s are not required to develop a Phase II program. See EPA Phase II Guidance, EPA 833-R-00-002, p. 4-2.

The Phase II program requires that certain small MS4s obtain a NPDES permit. A "small MS4" is any MS4 not already regulated under the Phase I definition of "small," including federally owned systems, such as military bases and veterans hospitals. See Phase II Guidance at p. 4-3. Small MS4s can be regulated if the small MS4 is wholly or partially located within an urbanized area, as defined by the 2000 U.S. Census, designated by NPDES, or designated by the permitting agency [40 C.F.R. § 122.32(a)(1), (2)]. Phase II requires that the permitting agency develop and apply designation criteria to all small MS4s located outside of an urbanized area that serves a population of at least 10,000 and a population density of at least 1,000 people per square mile.

Some MS4s receive a "physically interconnected" designation. Physically interconnected means that one MS4 is connected to a second in a way that facilitates direct discharges into the second system (Phase II Guidance, EPA 833-R-00-002 at p. 4-3-5). The permitting agency must designate most small MS4s located outside of an urbanized area that contributes substantially to pollutant loadings of a physically interconnected MS4 permitted by the permitting agency. The NPDES permitting authority may waive the coverage requirement in limited circumstances. See Phase II Guidance at p. 4-10, 11. Phase II allows operators of regulated small MS4s to choose from three permit options, the general permit, an individual permit, or a modification of an existing individual permit. The permitting agency reserves the right to determine which options are available to a regulated small MS4.

LEGAL PROTECTION FOR IN-STREAM FLOW

Melinda R. Kassen
Trout Unlimited, Colorado
Water Project
Boulder, Colorado

Water is precious. Human society needs water for drinking, other municipal uses, crops, and stock watering, as well as for industrial, aesthetic, and recreational purposes. But water left flowing through the landscape also has value, as it sustains every component of an ecosystem.

Water in rivers and lakes creates aesthetic value, as well as habitat for thousands of water-dependent species. Such species include aquatic species—fish, insects, amphibians, and plant life—as well as the land-based animals that eat them. Water flowing in a natural hydrograph helps maintain water quality. In addition, flowing rivers and streams can provide significant recreational amenities that feed local economies from fishing and boating or secondary activities such as hiking and bird-watching.

For most of recorded history, there has been enough water to supply humanity's needs without worrying about how removing water from streams for human use might affect other water-dependent species. However, because of significant human population growth, enhanced appreciation for the importance of protecting natural ecosystems, and increased concern about sustainability, societies have begun to provide legal protection to water left in a stream for the benefit of both water-dependent species and the people who use and enjoy the natural environment. These concerns have surfaced more in the western part of the United States than in the east, but concerns now arise even in less water-short areas.

Moreover, in view of a recent United Nations statement that one out of three people globally will face a water shortage by the year 2025, making water the commodity likely to cause controversy during the twenty-first century as oil did in the twentieth, understanding competing water uses and values is critical throughout the world.[1]

STATE PROTECTION OF IN-STREAM VALUES—WESTERN STATES[2]

In the United States, states, rather than the federal government, have primary authority over water allocation. Those states west of the 100th meridian are, for the most part, arid or semiarid. Since early in their settlement, they have found a need to reconcile disputes between water users; each of these states follows the doctrine of prior appropriation to allocate water among competing users.[3] Each state's system is unique, but their systems share many features. Most importantly, these systems define the value of water in terms of its use, and until late in the twentieth century, this meant use out of the stream. The prior appropriation system did not recognize any value for leaving water in a stream. Because the prior appropriation system functions as a seniority system, where the earliest user of a river retains always the first right of use, the system has created a challenge today for states and others seeking to protect remaining in-stream values, where water is "unused" except in sustaining the environment or by recreationists in a non-consumptive manner. Because water allocation is primarily a function of state law, the means to protect in-stream values must also be found primarily in the systems states use to allocate quantities of water to various uses.

All but three western states (New Mexico, North Dakota, and Oklahoma) provide some in-stream flow protection.[4] States use three different types of programs: legislatively established in-stream flows, agency actions to create or enforce in-stream flow rights, and private or market mechanisms to protect in-stream flows.

States vary in the methodology they use to quantify in-stream flow protection levels. Some, like the Tennant and R2Cross methods, look at a few simple variables, such as maximum flows, velocity, depth and wetted perimeter, and input them into a model that calculates minimum flows. States whose programs protect only minimum flows, such as Colorado and Wyoming, use these methods.[5] Others, like instream flow incremental methodology (IFIM) or physical habitat simulation system (PHABSIM), model the habitat requirements for aquatic species. Finally, there is the indicators of hydrologic alterations model developed by The Nature Conservancy; this model yields results that more closely mimic a natural hydrograph.

State Legislative Actions

In Montana, Kansas, and Idaho, the legislatures establish or approve agency recommendations for in-stream rights. Montana created a dozen such rights for blue-ribbon trout streams in a one-time effort in 1969.[6] Idaho has legislated nine stream flows or lake levels,[7] and Kansas has adopted statutory protection for 32 watercourses.[8] In addition, in 2000, Montana closed the Bitterroot entirely to new appropriations to protect the recreation and farm based economy of the area, both of which need water flowing downstream.[9]

State Agency Actions

Many Western states allow only state agencies to hold in-stream flows. Even so, neither Kansas nor Texas has recognized any in-stream flow rights; Nebraska has fewer than five. Pursuant to his statutory authority,[10] the Wyoming state engineer has approved 13 applications for new water rights from the state Game and Fish Commission to protect in-stream flows, but has only adjudicated two; adjudication is a prerequisite for enforcement. Sixty-nine applications await decision. Montana gives any state or federal agency the ability to apply for an in-stream reservation of up to 50% of a stream's average annual flow, and Montana's Department of Fish, Wildlife, and Parks now holds in-stream flow reservations on the main-stem Missouri and Yellowstone Rivers.[11]

Colorado and Oregon have the most extensive programs for in-stream protection.[12] In Oregon's program, the responsible state agency limits other diversions based on in-stream levels established for more than 1400 stream reaches.[13] On the streams where these levels exist, it is virtually impossible to obtain a new diversionary right; moreover, Oregon actively enforces its rights. In Colorado, where all water rights are a matter of court decree, the state, which has the exclusive authority to hold in-stream protection flows, has applied for and obtained more than 1350 water rights since 1973.[14] Unlike the situation in Oregon, however, Colorado's in-stream flow decrees do not automatically stop additional future development. Because Colorado also has no field personnel to monitor its rights and few gauges to measure their flows, Colorado has only once made one "call" on the river to demand that

[1]Statement of the UN-sponsored Third World Water Forum, Stockholm, Sweden, released August 13, 2001.
[2]Statistics in this section date from December 2001
[3]The prior appropriation system is the subject of another entry in this Encyclopedia.
[4]Gillilan and Brown, Instream Flow Protection, Island Press 1997, pp. 121-22. A 1998 New Mexico Attorney General opinion expands the definition of beneficial use there to include in-stream protection, but no court has affirmed the opinion and no entity has yet applied for an in-stream flow right.
[5]See, Wyo. Stat. Ann. §43.3.1001(c), (d); Colo. Rev. Stat. 37-92-102(3).
[6]Act of March 13, 1969, ch. 345, Mont. Laws 879-81.
[7]Idaho Code, §67-4301
[8]Kan. Stat. Ann. §82-703(c).
[9]Mont. Stat. §85-2-344.
[10]Wyo. Stat. Ann. §41-2-1001, et seq.
[11]Mont. Code Ann. §85-2-316.
[12]Colorado holds these rights pursuant to Colo. Rev. Stat. §37-92-102(3).
[13]Oregon holds these rights pursuant to Or. Rev. Stat. §§537.332 to. 360.
[14]These rights protect 8100 miles of more than 107,000 miles of state streams.

junior appropriators' diversions be curtailed to satisfy the Board's in-stream right. Instead, Colorado objects to applications for changes of water rights that might adversely affect the state's in-stream flow and negotiates court settlements that rely on others to monitor the state's rights. In both states, only the state agency owner of the right may enforce the rights with a call; members of the public must rely on the agencies.

Idaho has had a minimum stream flow statute on the books since 1937.[15] However, the program does not allow for new in-stream flow protective rights on streams that are otherwise fully appropriated—and most Idaho streams have long been overappropriated, so the statute provides little protection. Instead of obtaining water rights, the Idaho's Department of Water Resources has protected about 100 streams by adopting recommendations in its state water resources plan for minimum stream flow reservations for recreation or natural resources.[16]

The State of Washington follows a similar strategy. In 19 of its 62 water resources inventory areas, as well as the Columbia River, the state has adopted in-stream resource protection programs that establish minimum stream flows.[17] The state conditions later-issued water rights on meeting these flows. In addition, the state agency responsible for water rights has closed portions of the Yakima River Basin to further groundwater appropriations (at least pending further study) to keep some of this basin's water in-stream.[18]

Most states whose agencies have the power to protect in-stream flows acted only after 1970—in a region where traditional water rights date back to the mid-nineteenth century. Given that the prior appropriation doctrine values water rights based on their seniority and given that a senior user may dry up a stream entirely before junior users are entitled to any water, the junior nature of states' in-stream flows still leave agencies unable to protect—let alone restore—healthy stream flows. For example, Colorado has fewer than 15 rights whose priorities date earlier than 1973; in a state where most rivers were overappropriated at the beginning of the twentieth century, these rights can protect against additional future depletions but do little to restore already damaged fisheries to health. The state acquired its few, highly valuable senior rights by donation from cities seeking to protect streamfront amenities, and in one case, from a private conservation group. Wyoming also has the authority to acquire traditional existing rights by transfer or gift and change these rights to instream flow purposes,[19] but the state has never done so.

Some state agencies allowed to hold in-stream rights may not make new appropriations for such rights but may only change existing rights or acquire previously consumptive rights by donation. This is true in Utah, where the Division of Wildlife Resources holds five in-stream flow rights.[20] California and Arizona agencies may also accept such transfers.

Finally, some states can protect in-stream flows by denying other new diversionary applications. A statute directs the Utah state engineer to ensure that no new diversion will "unreasonably affect public recreation or the natural stream environment."[21] However, the state engineer has never relied exclusively, or even primarily, on this provision to deny a new or changed application for a water right. Oregon also has a "public interest" test, and though the state has never denied a new permit based on this test, Oregon has adopted rules pursuant to this provision that make it difficult to obtain a new water development permit where there are sensitive stocks of fish.[22]

Public Trust Doctrine

The public trust doctrine also plays a role in protecting in-stream water resources. Evolved from the laws of the Roman Empire, this doctrine holds that a state holds certain resources in trust for the benefit of society at large because of their public values. The ancient Romans applied the doctrine to the ownership of navigable rivers and harbors, which were held in trust by the state to protect the public's interest in commerce, navigation, and fishing. This concept was subsequently adopted by the English courts and applied to lands beneath tidal waterways.

In this country, states assumed this trust responsibility, including the ability to forbid private ownership of public trust resources. For example, the U.S. Supreme Court read the public trust to require automatic state ownership of submerged lands beneath navigable waters.[23] The Court then added that the states could not sell trust lands if the transfer would unduly burden public fishing, navigation, and commercial rights.[24] Because the public trust doctrine is primarily an aspect of state law, its application varies widely between states. Whereas the Colorado courts have never recognized the public trust doctrine as applied to water resources, California has used the doctrine aggressively to protect both lands and waters. The California Supreme Court found that its state water agency, the State Water Resources Control Board, had to consider public trust values in permitting new water withdrawals and also in reviewing existing withdrawals.[25] Thus, Los Angeles had to curtail existing diversions by one-third from tributaries of Mono Lake until the lake level recovered to a certain elevation.

Private and Market-Based Efforts

In some states, private entities can augment state programs that protect in-stream flows. The states of

[15]Idaho Code §42-1501 et seq.

[16]Idaho Code, sec42-1734A.

[17]Washington State Depts. of Ecology and Fish and Wildlife, "A Guide to Instream Flow Setting in Washington State." Pub. No. 03-11-007, p. 23 (March 2003).

[18]Ground Water/Surface Water Interactions to be Studied in Yakima River Basin, Department of Ecology Focus No. 99–1817, August, 1999.

[19]Wyo. Stat. Ann. §41-3-1007.

[20]Utah Code Ann. §73-3-3(11)(g)(1989 & Supp. 2001).

[21]Utah Code Ann. §73-3-8(1).

[22]OAR 690, Div. 33.

[23]*Pollard v. Hagan*, 44 U.S. 212 (1845).

[24]*Illinois Central Railroad Co. v. Illinois,* 146 U.S. 387 (1892).

[25]*National Audubon Society v. Superior Court of Alpine County*, 658 P.2d 709 (Cal. 1983).

Oregon and Washington have nonprofit water trusts that purchase mostly small quantities of water in the tributaries of targeted river basins dedicated to in-stream flow protection. In most cases, these purchases are for short periods of time. For example, although the Oregon Water Trust, has made more than one dozen permanent acquisitions (where it transfers the rights to the state), most are short-term leases, from 1–10 years. Still, the Trust has brokered more than 50 deals protecting 450 river miles. In addition, the Trust is helping irrigators take advantage of Oregon's conserved water program, which allows a water user who conserves water to use 75% of that water, provided that the user turns 25% back to the stream.[26]

Montana's program allows government agencies or nonprofit entities to lease water that is currently put to beneficial use through a diversion for in-stream flow protection for up to 30 years.[27] The state agency for fish, wildlife, and parks has entered into 10 leases under this provision. Trout Unlimited, a private conservation organization, has entered into fewer than 10, but at least one of TU's leases put 200 cfs into the Madison River, a quantity of substantial and significant magnitude. Montana also allows private water users to change a traditional diversionary right temporarily for in-stream purposes.[28] The change can last 10 years for a direct flow right and up to 30 years for water coming out of storage. Only one water diverter has sought (and been granted) a temporary change, as of 2001.

Alaska, Arizona, Nevada, and California also allow individuals to hold in-stream rights permanently. Anyone may transfer existing rights to in-stream reservations in Alaska,[29] although these reservations must undergo review every 10 years, a hurdle no other water right must jump. Two government agencies have obtained such rights, and several private entities have pending applications.

Under its statute,[30] Arizona has received 85 applications for new in-stream flow rights from individuals, environmental groups, and several federal agencies. Of these, the state has permitted 25 and perfected another 23 (after the applicants demonstrated actual use for 3 years). In addition, in 1997, the state used its Water Protection Fund to purchase unused Central Arizona Project water to sustain riparian habitat.

In California, an individual may not make a new appropriation for an in-stream flow right but may convert an existing diversionary right to in-stream protection; however, in 2001, 8 years after passage of this statutory provision, the Water Resources Control Board had yet to grant a single in-stream permit.

At least one federal agency has obtained a new in-stream flow water right in Nevada.[31] In addition, since 1988, Nevada has allowed the transfer of existing water rights to in-stream use. Both government agencies and private organizations have done so; in one case off-stream rights were transferred to protect streams on the Stillwater Wildlife Refuge.

South Dakota has no express statutory authority for in-stream flows, but the state Supreme Court has held that diversion of water was not necessary to achieve a beneficial use of water under state law.[32] As a result, the state has awarded several in-stream water rights, two new rights to a state agency, one to a federal agency and a change to a private company. In 2001, there were two additional applications pending.

Conclusion

Most states have some power to maintain, if not restore, healthy stream flows, but few have made such protection a priority. Western states water law systems simply do not make in-stream flow protection easy, and at the same time the region must accommodate significant growth. Given the limitations described above (little enforcement, seniority, or measurement), many state programs provide more of an illusion of protection than real water in-stream.

STATE PROTECTION OF IN-STREAM VALUES—EASTERN STATES

The wetter United States east of the Mississippi River follows a system of water allocation different from that of Western states. In the East, the riparian doctrine prevails, under which riparian landowners, those adjacent to a waterbody, have a right to use a portion of the flow of the stream or lake.[33] Although originally, a riparian water user could not diminish the natural flow of the stream,[34] today, holders of riparian rights can consumptively use water—even outside the watershed—if the use is "reasonable."[35] Today, urban, industrial, and agricultural demands have increased to the point that water resources are sometimes scarce in riparian jurisdictions.

To protect the public's interest in a reliable water supply, many riparian states require even landowners to obtain permits before using any water. Essentially, permits serve as a means for a state to quantify existing water uses and give the permit holder a right to use a quantity of water permanently or for a limited time. The permit requirement limits the number and size of water rights that come with riparian property ownership.

Most riparian states recognize water as a public resource and have taken steps to protect public recreational, environmental, and aesthetic values. Thus, these states have passed statutes to protect stream flows. In addition, many of these states have established minimum stream flows, lake levels, and groundwater levels or delegated the authority to set these levels to a state agency.

[26]Or. Rev. Stat. §537.455.

[27]Mont. Code Ann. §85-2-408(2).

[28]*Id.*

[29]Alaska Code Ann. §46.15.145.

[30]Ariz. Rev. Stat. §45–151.

[31]*State v. Morros*, 766 P.2d 263 (1988).

[32]*DeLay v. US Fish and Wildlife Service*, 524 NW2d 855 (S.D. 1994).

[33]*See* National Water Commission, A Summary-Digest of State Water Laws 32 (Richard Dewsnup & Dallin Jensen eds., 1973).

[34]*Tyler v. Wilkinson*, 24 F.Cas. 472 (C.C.R.I. 1827)

[35]*Stratton v. Mt. Hermon Boy's School*, 103 N.E. 87 (Ma. 1913).

For example, Florida has authorized its Water Management Districts to set seasonal minimum flow levels for both surface and groundwater.[36] "Minimum flow" is "...the limit at which further withdrawals would be significantly harmful to the water resources or [area] ecology."[37] Further diversions may not occur below the established minimum. When existing flows or levels do fall below or are projected to do so within 20 years, the state must implement a "recovery or prevention strategy."[38]

New Hampshire's Rivers Management and Protection Program takes a different approach.[39] There, the state develops "river corridor management plans" to maintain in-stream flows along designated rivers after someone nominates the river for protection. Once accepted, the nomination must undergo court review and legislative approval. The rules to implement this program establish a general standard of in-stream flow protection for all designated rivers.[40]

FEDERAL LAW PROTECTION OF IN-STREAM VALUES

State law is the primary vehicle for water allocation decisions, but the federal government also plays a role. This is true in part because the federal government is a major landowner, particularly in the West. In addition, the federal Bureau of Reclamation owns and operates a massive system of water projects that controls the flows on many Western rivers. Finally, the nation's web of environment and natural resources law intersects with the goal of river protection.

Federal lands reserved for specific purposes, such as national parks, monuments, forests, and refuges, almost always need water. At the beginning of the twentieth century, the U.S. Supreme Court held that in reserving lands, Congress had implicitly reserved as well the water to fulfill the purpose of the reservation and also that the water right's priority date would be the date of the initial Congressional reservation, not the later year in which the federal government sought to quantify its right.[41] In 1954, Congress passed the McCarran Amendment, a statute directing the federal agencies to obtain these reserved water rights through the individual state systems of water allocation.[42]

Only in the 1970s did federal agencies begin in earnest to enter state systems to obtain and quantify their rights. Unfortunately, given the Western states' system of prior appropriation, the time lag between the date the federal land was reserved and the year when the federal government sought to confirm its water rights has proved devastating to this federal effort. Much of the West was unsettled in the early twentieth century; since that time, large cities have grown, farming has continued to divert 80–90% of the water used, and new industrial needs for water have been created—from heap leach mining to oil shale extraction to ski area snowmaking. These intervening water users are unwilling to remain passive when the federal government seeks to jump the seniority line and gain water rights for its lands.

As a result, almost all adjudications of reserved rights have been controversial. In most cases involving national forest lands, the Forest Service has failed to obtain meaningful protection for the rivers crossing its lands, even though its mandate includes the command to provide favorable flows.[43] The National Park Service and U.S. Fish and Wildlife Service on behalf, respectively, of national parks and the wildlife refuge system have fared marginally better; yet even these agencies have failed frequently to obtain protective flows. In Colorado, the Park Service failed to obtain a right to protect the rivers within Dinosaur National Monument.[44] More recently, the Idaho Supreme Court denied the U.S. Fish and Wildlife Service a reserved right for the Deer Flat National Wildlife Refuge made up a series of islands in the Snake River.[45]

Some federal agencies also use administrative mechanisms to protect stream flows when making permitting or licensing decisions. Thus, wholly outside state water allocation systems, federal agencies impose bypass flows to fulfill requirements that derive from federal laws, such as the Federal Power Act, the Clean Water Act, and the Federal Land Management and Policy Act, as well as the agencies' authorizing statutes, for example, the Forest Service Organic Act. In part because the use of this authority has sometimes limited validly issued state water rights, agency exercise of this authority has proved highly controversial and is thus used sparingly, if at all.

The Clean Water Act provides two mechanisms for protecting in-stream values to any state willing to use them. First is the state water quality certification program.[46] Under its provisions, a state must certify that a federally permitted or licensed project will meet state water quality standards. Pursuant to this authority, the State of Washington imposed substantial flow restrictions on a hydropower facility on the Olympic peninsula; in 1994, the U.S. Supreme Court upheld these restrictions.[47] Second is the total maximum daily load (TMDL) program" which arguably provides states a mechanism to protect flows through water quality regulation.[48] Here, states must list streams whose water quality is impaired.[49] Then, states must establish TMDLs that will bring the streams back into compliance. Recently, the EPA determined that the federal government cannot impose flow-based conditions in TMDLs, yet the states, which have primary

[36]Fla. Stat. §373.042 (2001).
[37]*Id.* at §373.042 (1)(a).
[38]*Id.* at §373.042 2(a), 2(b).
[39]N.H. Rev. Stat §§ 483:1–483:15.
[40]Rules for the Protection of Instream Flow on Designated Rivers, Chap.Env-ws 1900(2003). This document is available online at: http://www.des.state.nh.us/rules/env-ws1900.pdf.
[41]*Winters v. US, 207 U.S. 564 (1908).*
[42]43 U.S.C.A. §666.

[43]See, e.g., *US v. New Mexico*, 438 U.S. 696 (1978); *Potlatch Corp. v. US*, 12 P.3d 1260 (2000), *US v. City & County of Denver*, 656 P.2d 1 (Colo. 1982) *(US v. Denver)*.
[44]*US v. Denver*.
[45]*US v. State*, 23 P.3d 117 (Idaho 2001).
[46]33 USC §1341.
[47]*PUD No. 1 of Jefferson County v. Washington Department of Ecology*, ## US 700 (1994).
[48]The total maximum daily load (TMDL) program is the subject of a separate entry in this Encyclopedia.
[49]33 USC §1333(d).

responsibility for this program, can. Montana, for example, has adopted a flow-based TMDL for an impaired stream, a tributary of the Yellowstone River.[50]

The Endangered Species Act (ESA) provides federal agencies another handle for restoring stream flows when such flows are necessary to avoid jeopardizing an endangered or threatened species' existence. Again, the use of this authority is highly controversial. In one recent case, a federal court upheld the U.S. Forest Service's imposition of a bypass flow to protect an endangered species as consistent with both the ESA and the Forest Service's own Organic Act.[51] Some water users who have received less water from federal projects as a result of actions taken under the ESA have begun to challenge these federal actions as "takings" of their property rights without compensation, in violation of the 5th amendment to the U.S. Constitution.[52] In the summer of 2001, such a curtailment occurred to benefit both endangered salmon and lake fish sacred to the Klamath tribe that has resulted in a firestorm of litigation and proposed legislation.

On the other hand, ESA mandates have forced some creative responses. The Middle Rio Grande Conservancy District, embroiled in litigation over recovery of the silvery minnow, created a water bank to allow users to lease surplus water, at least some of which may go to keep critical Rio Grande reaches wet. New Mexico and the federal government have also reached an agreement to operate federal reservoirs in the Rio Grande basin to release stored water for in-stream flows to maintain minnow habitat. An even larger effort is beginning in California. There, parties trying to address endangered species concerns in the San Francisco Bay Delta created the Environmental Mitigation Bank to trade water to benefit the ecosystem.[53]

Finally, due to the existence of major federal water storage projects throughout the West, interests have pressured the federal agencies—owners of these projects—to change their operation. Primarily to benefit endangered species, but also to enhance recreation and other values, reoperation of federally owned or licensed storage and hydropower facilities can enhance river flows in a way that restores, or at least maintains, aquatic species. This has occurred to benefit endangered species in the Rio Grande, the Klamath, and the Columbia Rivers, along with many lesser known systems where endangered species live. Through the "CalFed" project, California reoperations have benefited people as well as endangered species.

[50]Montana Department of Environmental Quality, Big Creek Flow Restoration Plan (TMDL), December 2000.

[51]*County of Okanogan v. Nat'l Marine Fisheries Service*, 57ERC 1321, 2003 WL 22455482 (9th Cir. 2003) (not selected for publication in Federal Register).

[52]*Tulare Lake Basin Water Storage District v. United States*, 49 Fed. Cl. 313(2001).

[53]For information about this venture, *see*, CalFed Bay-Delta Program, Programmatic Record of Decision (2000), pp 54–59. [Note: this document is available online at http:// calwater.ca.gov/Programs/Environmental Water account/adobe_pdf/ ROD8-28-00.pdf.]

CONCLUSION

State water law systems grew out of a nineteenth century worldview that believed that all resources are endless. Therefore, it is not a surprise that governments struggle today to protect in-stream values. But, because of the pressures exerted on the system as a result of modern environmental values, decision-makers are seeking ways to ensure that at least some of the country's waterways maintain healthy flows.

WATER QUALITY

U.S. Geological Survey

APPLICABLE STATUTES

Clean Water Act (PL 92–500; 33 U.S.C. 1251 et seq.)

Safe Drinking Water Act (PL 93–523; 42 U.S.C. 300f-j-26)

CLEAN WATER ACT, AS AMENDED (PL 92–500; 33 U.S.C. 1251 ET SEQ)

Purpose

Also known as the Federal Water Pollution Control Act, it is the objective of the Clean Water Act (CWA) "to restore and maintain the chemical, physical, and biological integrity of the Nation's waters". Seven goals and policies are declared in order to meet this objective. One goal was to reach zero discharge of pollutants by 1985. Others include providing funding for the construction of publicly owned treatment works (POTWs), creating a nonpoint source pollution program, and generally, and making the waters of the U.S. "fishable and swimmable". Although not specifically stated in this declaration, the CWA contains the nation's most noteworthy wetlands legislation.

MAJOR PROVISIONS BY SECTION

201–209 (33 U.S.C. 1281–1289)—Grants for Construction of Treatment Works.

These sections originally provided federal grants for the construction of wastewater treatment plants. The program has been phased out by the 1987 amendments in favor of a revolving loan fund.

301 (33 U.S.C. 1311)—Effluent Limitations

The discharge of any pollutant into the nation's waters except for discharges in compliance with the CWA is prohibited, according to this section. Limitations are placed on existing sources which vary according to the nature of the pollutant discharged and to where the outfall is directed.

302 (33 U.S.C. 1312)—Water Quality Related Effluent Limitations

Point sources which interfere with the attainment or maintenance of desired water quality are subject to the imposition of more stringent effluent limitations.

303 (33 U.S.C. 1313)—Water Quality Standards and Implementation Plans

Water quality-based regulatory controls on dischargers known as Water Quality Standards (WQS) are required by states to protect designated uses of water bodies. Technological capability is not a consideration in setting WQS.

304 (33 U.S.C. 1314)—Information and Guidelines

This section requires the EPA to develop water quality criteria and guidelines for effluent limitations, pretreatment programs, and administration of the National Pollutant Discharge Elimination System (NPDES) program.

306 (33 U.S.C. 1316)—National Standards of Performance

A list of categories of effluent sources is presented which specifies that each of the industries listed must conform to technology-based new source performance standards. The standards are to demonstrate the best demonstrated control technology.

307 (33 U.S.C. 1317)—Toxic and Pretreatment Effluent Standards

This requires that industries discharging toxic pollutants meet effluent limits that employ the best available technology economically achievable. Part (b) mandates the establishment of pretreatment standards while part {c} looks at new sources of pollutants into publicly owned treatment works (POTWs).

309 (33 U.S.C. 1319)—Enforcement

Enforcement by the states, as well as compliance orders, and administrative, civil, and criminal penalties are authorized.

311 (33 U.S.C. 1321)—Oil and Hazardous Substance Liability

The Congressional declaration of policy against discharges of oil or hazardous substances in harmful quantities into waters and adjoining shorelines is presented. A spill prevention, control, and countermeasure plan is to be developed by all facilities that handle, transport, and store oil. Any spill or discharge of a harmful quantity of oil must be reported to the National Response Center. Strict liability is assigned to owners or operators of facilities from which oil discharges occur.

319 (33 U.S.C. 1329)—Nonpoint Source Management Programs

This section requires that states identify waters that are not able to meet WQS because of nonpoint sources. The activities responsible for the pollution are to be identified and a management plan is to be created to help correct the nonpoint source problem.

401 (33 U.S.C. 1341)—Certification

Any applicant for a Federal license or permit to conduct any activity which may result in any discharge into the navigable waters, shall provide the licensing or permitting agency a certification from the State in which the discharge will originate.

402 (33 U.S.C. 1342)—National Pollutant Discharge Elimination System

One of the most critical parts of the CWA is the establishment of the National Pollutant Discharge Elimination System which translates standards into enforceable limitations. This program may be administered by the EPA or states under EPA-delegated authority. After an opportunity for public hearing, a permit may be issued for the point source discharge of any pollutant, or combination of pollutants.

404 (33 U.S.C. 1344)—Permits for Dredged or Fill Material

This is the major wetlands provision of the CWA, and largely, in environmental law. The basic gist of this section is that a permit is required from the U.S. Army Corps of Engineers for the disposal of dredged or filled materials into navigable waters, notably wetlands, with EPA concurrence, and notice and opportunity for public hearings.

505 (33 U.S.C. 1365)—Citizen Suits

Any citizen is given the right to file suit against any person in violation of an effluent standard or against EPA for failure to perform nondiscretionary duties.

PERTINENT REGULATIONS

- **33 CFR Part 320—General Regulatory Policies, Army Corps of Engineers**
  The Corps issues permits under several different acts, the CWA being only one of them. The Corps must issue permits for discharges to insure that they comply with the applicable limitations and WQS. Section 320.4 describes the general policies that the Corps will follow in reviewing all Department of the Army permits. Such permits are to consider public interest, effect on wetlands, fish and wildlife, water quality, property ownership, energy conservation, navigation, environmental benefits, and economics, among other things. The pertains particularly to dredge and fill permits covered under 404 of the CWA.
- **33 CFR Part 323, Permits for Discharges of Dredged or Fill Material into Waters of the United States, Army Corps of Engineers**
  Contains definitions and special policies, practices and procedures to be followed by the Corps in connection with review of permit applications for the discharge of dredged or fill materials under 404 of CWA. Permits issued under 404 are required

for the discharge of dredged or fill materials not exempted by 323.4 of this part or permitted by 33 CFR Part 330. Please refer to 323.4 for a detailed listing of exemptions.

- **33 CFR Part 325—Processing of Department of the Army Permits, Army Corps of Engineers**
  Contains general processing procedures for all Department of the Army permits. Attention is given to dredge and fill activities in 325.1(d)(3) & (4). Federal agencies that initiate or authorize proposed actions that include dredge or fill discharge operations must ensure that the appropriate permits are obtained. In states with approved programs, permit application is done through the appropriate State agency.
- **40 CFR Part 110—Discharge of Oil, EPA**
  This section reiterates the mandate of 311(b)(3) of the CWA. It generally prohibits the discharge of oil into navigable waters in such quantities as may be harmful.
- **40 CFR Part 112—Oil Pollution Prevention, EPA**
  This requires that owners or operators of non-transportation related onshore and offshore facilities engaged in any type of oil and gas operation prepare a Spill Prevention Control and Countermeasure Plan, and provide guidelines for preparation of the plan. Such a plan may be required of an applicant for USGS approval of any oil and gas activities.
- **40 CFR Part 122—EPA Administered Permit Programs: The National Discharge Elimination System, EPA**
  Contains definitions and basic permitting requirements for EPA-administered NPDES programs under 318, 402, and 405 of the CWA. Permit applications and special NPDES program requirements are discussed in Subpart B. This is followed by permit conditions in subpart C. Subpart D covers the transfer, modification, revocation and reissuance, and termination of permits. Federal agencies that initiate or authorize proposed actions that include point source operations must ensure that the appropriate permits are obtained. In states with approved programs, permit application is done through the appropriate State agency.
- **40 CFR Part 123—State Program Requirements, EPA**
  Describes the general requirements and additional requirements for states and the EPA to obtain and give approval, revision, and withdrawal of state NPDES programs. State program information shall be made available to EPA upon request. EPA also has the right to review proposed general permits for 90 days.
- **40 CFR Part 125—Criteria and Standards for the National Pollutant Discharge Elimination System, EPA**
  This section prescribes criteria and standards for various requirements imposed as conditions for NPDES permit approval. Some of the criteria expanded upon are those presented for the imposition of technology-based treatment requirements in permits as given under 301(b) and 402(a)(1), those for modifying secondary treatment requirements under 301(h), those for Best Management Practices authorized under 304(e), and those applying to ocean dumping, in Subparts A, G, K, & M, respectively.
- **40 CFR Part 129—Toxic Pollutant Effluent Standards, EPA**
  This section designates toxic pollutant effluent standards and applies to owners or operators of specified facilities discharging into navigable waters. Section 129.4 listed the pollutants to be regulated. Each owner/operator is given 60 days to notify the Regional Administrator of any listed pollutant that is discharged. Much of the regulation gives specific information on each of the toxic pollutants.
- **40 CFR Part 130—Water Quality Planning and Management, EPA**
  Section 303 of the CWA gives the authority for promulgation of water quality standards (WQS) by states. Here, policies are established for water quality planning, management, and implementations of 303. The Water Quality Management process from the CWA provides the authority for a "consistent national approach for maintaining, improving and protecting water quality while allowing States to implement the most effective individual programs". After WQS are set by each state, implementation of the standards may be achieved by issuing permits, building publicly-owned treatment works, or instituting Best Management Practices (BMP) through a water quality management plan. Total maximum daily loads (TMDL) and individual water quality-based effluent limitations are discussed in 130.7. States are required to submit water quality reports to the Regional Administrator in accordance with 305(b) of the CWA (130.8). Final points of the regulation cover state submittals to EPA and program management.
- **40 CFR Part 131—Water Quality Standards, EPA**
  While Part 130 of this title works with water quality management, this section deals with water quality standards (WQS). It is stated that a WQS defines the water quality goals of a water body by designating the uses to be made of the water and by setting criteria necessary to protect the uses. WQS should be set to provide for the protection of fish, shellfish, wildlife, water recreation, and the use and value of public water supplies, agricultural, industrial, and other purposes. The procedures are presented for developing, reviewing, revising and approving WQS by the states and EPA. In 131.12, states are ordered to develop statewide antidegradation policies. WQS shall be reviews at least every three years by the states.
- **40 CFR Part 230—Section 404 (b)(1) Guidelines for Specification of Disposal Sites for Dredged or Fill Material, EPA**

These guidelines were written to restore and maintain the chemical, physical, and biological integrity of the U.S. through the control of discharges of dredged or fill material, with the guiding principle that degradation of sites may represent an irreversible loss of valuable aquatic resources. In Subpart B, it is stated that "no discharge of dredged or fill material shall be permitted if there is a practicable alternative to the proposed discharge which would have less adverse impact on the aquatic ecosystem". The permitting authority must determine the potential effects of dredging or filling activities on the components of the aquatic environment.

- **40 CFR Part 231—Section 404(c)Procedures, EPA**
  Contains procedures for EPA in exercising its authority to veto the specification by the Army Corps of Engineers or by a State of a disposal site for a 404 permit.
- **40 CFR Parts 401-471—Effluent Guidelines and Standards, EPA**
  Prescribes effluent limitations and pretreatment and performance standards, categorized by industries, that must be complied with as conditions for NPDES permit approval. Part 401 gives general provisions and Part 403 covers general pretreatment regulations for existing and new pollution sources. The remaining parts are industry-specific. Of particular interest to the USGS are:

  40 CFR Part 434—Coal Mining Point Source Category
  40 CFR Part 435—Offshore Segment of the Oil and Gas Extraction Point Source Category
  40 CFR Part 436—Mineral Mining and Processing Point Source Category
  40 CFR Part 440—Ore Mining and Dressing Point Source Category

- **518 DM 1—Comprehensive Waste Management, Department of the Interior**
  This chapter defines waste to include solid and hazardous waste, hazardous materials, and hazardous substances. Departmental policies, responsibilities, and functions regarding waste management are presented, with an aim towards prevention of hazardous waste generation. The use of sound waste management practices is mandated.

SAFE DRINKING WATER ACT (PL 93–523; 42 U.S.C. 300F—J-10)

Purpose

In 1974, the Safe Drinking Water Act (SDWA) was enacted with the general intent to protect the quality of drinking water the public receives from public water systems. To accomplish this, the SDWA focuses on two approaches. The first is to assure the quality of drinking water coming from the tap. The other approach is to prevent the contamination of groundwater that may be a source for drinking water.

MAJOR PROVISIONS BY SECTION

§1412 (42 U.S.C. 300g-1)—National Drinking Water Regulations

The EPA is required to promulgate national drinking water regulations, known as Maximum Contaminant Level Goals (MCLGs) and Maximum Contaminant Levels (MCLs) for public water systems. These standards specify minimum levels of drinking water quality and are to be issued for any contaminant that is known or anticipated to have any adverse effect on health.

§1413 (42 U.S.C. 300g-2)—State Primary Enforcement Responsibility; Regulations; Notice of Hearing; Publication in Federal Register; Applications

States are given the authorization to assume primary enforcement of the Act.

§1415–1416 (42 U.S.C. 300g-4 & 5)—Variances—Exemptions

The conditions are offered in which states may issue variances and exemptions.

§1417 (42 U.S.C. 300g-6)—Prohibition on Use of Lead Pipes, Solder, or Flux

Any use of lead in pipes, solder, or flux in public water systems where the water is intended for human consumption is prohibited.

§1422 (42 U.S.C. 300h-1)—State Primary Enforcement Responsibility

This section requires the establishment of state underground injection control programs so as to protect current and potential underground sources of drinking water from contamination.

§1424 (42 U.S.C. 300h-3)—Interim Regulation of Underground Injections

This declares that Federal agencies shall make no commitment of Federal assistance to actions that will contaminate an aquifer designated as the sole or principal drinking-water source for an area which, if contaminated, would create a significant hazard to the public health.

§1428 (42 U.S.C. 300h-7)—State Programs to Establish Wellhead Protection Area

States are given the authority to develop wellhead protection programs that address the kinds of activities that might be conducted in proximity to wells, particularly seeking to limit activities that might pose threats to the well fields.

§1431 (42 U.S.C. 300i)—Emergency Powers

If there is an imminent and substantial endangerment to public health through drinking water, EPA is given emergency powers to act against contamination.

§1445 (42 U.S.C. 300j-4)—Records and Inspection

The EPA is to promulgate regulations requiring drinking water monitoring. This section also mandates that EPA establish record keeping requirements.

§1448 (42 U.S.C. 300j-7)—Judicial Review

The D.C. Circuit is given the jurisdiction of judicial review of national primary drinking water regulations. The U.C. Courts of Appeal are given jurisdiction for any other EPA action when the petitioner resides or transacts business in that particular region.

§1449 (42 U.S.C. 300j-8)—Citizen's Civil Action

This section gives citizens the right to file suit to enforce any mandatory provisions of the SDWA.

Part F—Additional Requirements to Regulate the Safety of Drinking Water

§1461–1465 (42 U.S.C. 300j-21 to 26)

These particular sections codify the Lead Contamination Act of 1988. This requires that any lead-lined tanks of drinking water coolers be recalled.

PERTINENT REGULATIONS

- **40 CFR Part 141—National Primary Drinking Water Regulations, EPA**
  One of the most significant provisions of the SDWA is the establishment of National Primary Drinking Water Regulations. The regulations are given life here, with the establishment of maximum contaminant levels (MCLs) for inorganic and organic constituents, Subpart B and maximum contaminant level goals (MCLGs), Subpart F. Much attention is given to the monitoring and analytical requirements for the regulated water quality parameters, Subpart C. Filtration and disinfection are given much the same kind of thorough treatment. Also included are sections covering control of lead and copper, muse of non-centralized treatment devices and treatment techniques.
- **40 CFR Part 142—National Primary Drinking Water Regulations Implementation, EPA**
  With only a few exceptions, this part applies to the public water system in each state. States are given primary enforcement responsibility, given that the state has an EPA-approved program. If a state wishes to revise its program, it may do so if it follows the requirements given in 142.12. All states with enforcement responsibility are required to submit to the EPA Administrator a report containing various components dealing with national primary drinking water regulations. Variances and exemptions to the primary regulations may be granted by either the states of the Administrator. Another section (142.60) discusses best available technologies for a list of contaminants as they pertain to national primary drinking water regulations.
- **40 CFR Part 143—National Secondary Drinking Water Regulations, EPA**
  While Part 141 of this title is concerned with primary drinking water regulations, this part deals with national secondary drinking water regulations. Secondary regulations control contaminants that primarily affect the aesthetic qualities of drinking water. Levels are given for selected contaminants and it is explained that states may establish higher or lower levels depending on special conditions with that state, given that public health and welfare are not adversely affected. Section 143.4 gives monitoring requirements, stating that monitoring should occur no less frequently than the schedule used for the National Interim Primary Drinking Water Regulations.
- **40 CFR Part 144—Underground Injection Control Program, EPA**
  This regulation stems from Part C of the SDWA which allows for the establishment of an Underground Injection Control (UIC) program in each state. General program requirements are given in Subpart B. These requirements address prohibition of unauthorized injection, prohibition of movement of fluid into underground sources of drinking water, prohibition of Class IV wells and requirements for wells injecting hazardous waste. In 144.16 of this subpart, direction is given for cases when injection does not occur into, through, or above an underground source of drinking water. Subpart C takes a look at authorization of underground injection by rule where each class of wells is given particular authorization requirements. Authorization is also available by permit, and this is the topic of Subpart D. Going hand-in-hand with Subpart D is Subpart E, which discusses permit conditions. Finally, in Subpart F, financial responsibility for hazardous waste injection wells is described.
- **40 CFR Part 146—Underground Injection Control Program: Criteria and Standards, EPA**
  By working closely with Part 144 of this title, this part sets forth the technical criteria and standards for the Underground Injection Control Programs. Criteria are given for exempted aquifers, which include aquifers used as an "underground source of drinking water". Injection wells are then classified as Class I–V, based on the material injected. The following subparts (B–G) then outline, in detail, the criteria for each class of injection wells.
- **40 CFR Part 149—Sole Source Aquifers, EPA**
  Pursuant to 1427 of the SDWA, this regulation was written to provide criteria for identifying critical aquifer protection areas. Section 149.3 refers to a Critical Aquifer Protection Area as one which was designated as a sole or principal source aquifer prior to June 19, 1986 for which an area wide ground-water

quality protection plan was approved. The second definition entails major recharge areas of a sole or principal source aquifer designated before June 19, 1988. Edwards Underground Reservoir, of the San Antonio area, is given attention in Subpart B.

- **Environmental Statement Memorandum No. ESM94-5—Environmental Impacts on Groundwater, Department of Interior**
  Implements CEQ memorandum issued on November 19, 1976, concerning 1424(e) of the Safe Drinking Water Act. EISs will analyze impacts on the quantity and quality of ground water with specific emphasis on drinking-water sources. For proposed actions which may affect ground water, early consultation is to be initiated with the appropriate District Hydrologist, Water Resources Division (WRD). If it is possible that the proposed action may affect an aquifer which has been designated or is being petitioned for designation for special protection, early consultation should be initiated with the Regional EPA Administrator.
- **516 DM 2 Appendix 2 (2.2), Department of Interior**
  Environmental documents (EA, EIS, FONSI) must be prepared for actions which may adversely affect such unique geographic characteristics as sole or principal source drinking-water aquifers.
- **518 DM 1—Comprehensive Waste Management, Department of the Interior**
  This chapter defines waste to include solid and hazardous waste, hazardous materials, and hazardous substances. Departmental policies, responsibilities, and functions regarding waste management are presented, with an aim towards prevention of hazardous waste generation. The use of sound waste management practices is mandated.

DEFINITIONS

Best Management Practices (BMP)

Schedules of activities, prohibitions of practices, maintenance procedures, and other management practices to prevent or reduce the pollution of "waters of the United States". (40 CFR Part 122.2)

Contaminant

Any physical, chemical, biological, or radiological substance or matter in water.(SDWA, 1401(6))

Discharge of a Pollutant

Any addition of any pollutant to navigable waters from any point source. (CWA, 502(12)(A))

Exemption

A document for water systems having technical and financial difficulty meeting national primary drinking water regulations effective for one year granted by EPA "due to compelling factors".

Maximum Contaminant Level (MCL)

The maximum permissible level of a contaminant in water which is delivered to any user of a public water system. (SDWA, 1401(3))

Maximum Contaminant Level Goal (MCLG)

The level at which no known or anticipated adverse effects on the health of persons occur and which allows an adequate margin of safety.

National Pollutant Discharge Elimination System (NPDES)

The national program for issuing, modifying, revoking and reissuing, terminating, monitoring and enforcing permits, and imposing and enforcing pretreatment requirements, under sections 307, 402, 318, and 405 of CWA. (40 CFR Part 122.2)

Navigable Waters

Waters of the United States, including the territorial seas. (CWA, 502(7))

Person

An individual, corporation, partnership, association, state, municipality, commission, or political subdivision of a State, or any interstate body. (CWA, 502(5))

Point Source

Any discernible, confined and discrete conveyance, including but not limited to any pipe, ditch, channel, tunnel, conduit, well, discrete fissure, container, rolling stock, concentrated animal feeding operation, or vessel, or other floating craft, from which pollutants are or may be discharged. This term does not include agricultural stormwater discharges and return flows from irrigated agriculture. (CWA, 502(14))

Pollutant

Dredged spoil, solid waste, incinerator residue, filter backwash, sewage, garbage, sewage sludge, munitions, chemical wastes, biological materials, radioactive materials (except those regulated under the Atomic Energy Act of 1954), heat, wrecked or discarded equipment, rock, sand, cellar dirt and industrial, municipal, and agricultural waste discharged into water. It does not mean: (a) sewage from vessels; or (b) water, gas, or other material which is injected into a well to facilitate production of oil or gas, or water derived in association with oil and gas production and disposed of in a well, if the well used either to facilitate production or for disposal purposes is approved by authority of the State in which the well is located, and if the State determines that the injection or disposal will not result in the degradation of ground or surface water sources. (CWA, 502(6))

Public Water System

A system for the provision to the public of piped water for human consumption, if such system has at least fifteen service connections or regularly serves at least twenty-five individuals. (SDWA, 1401(4))

Publicly Owned Treatment Works (POTW)

Any device or system used in the treatment of municipal sewage or industrial wastes of a liquid nature which is owned by a "State" or "municipality". This definition includes sewer, pipes, or other conveyances only if they convey wastewater to a POTW providing treatment. (40 CFR Part 122.2)

Recharge Zone

The area through which water enters a sole or principal source aquifer.

Significant Hazard to Public Health

Any level of contaminant which causes or may cause the aquifer to exceed any maximum contaminant level set forth in any promulgated National Primary Drinking Water Standard at any point where the water may be used for drinking purposes or which may otherwise adversely affect the health of persons, or which may require a public water system to install additional treatment to prevent such adverse effect.

Sole or Principal Source Aquifer

An aquifer which supplies 50 percent or more of the drinking water for an area.

Streamflow Source Zone

The upstream headwaters area which drains into an aquifer recharge zone.

Toxic Pollutants

Those pollutants...which after discharge and upon exposure, ingestion, inhalation or assimilation into any organism...,will, on the basis of the information available to the Administrator, cause death, disease, behavioral abnormalities, cancer, genetic mutations, physiological malfunctions or physical deformations, in such organisms or their offspring. (CWA, 502(13))

Variance

A document for water systems having technical and financial difficulty meeting national primary drinking water regulations which postpones compliance when the issuing of which "will not result in an unreasonable risk to health".

Waters of the United States

a) all waters which are currently used, were used in the past, or may be susceptible to use in interstate or foreign commerce, including all waters which are subject to the ebb and flow of the tide; b) all interstate waters, including interstate "wetlands"; c) all other waters such as interstate lakes, rivers, streams..., mudflats, sandflats, "wetlands", sloughs, prarie potholes, wet meadows, playa lakes, or natural ponds the use, degradation, or destruction of which would affect...interstate or foreign commerce...(40 CFR Part 122.2)

Wetlands

Those areas that are inundated or saturated by surface or groundwater at a frequency and duration sufficient to support...a prevalence of vegetation typically adapted for life in saturated soil conditions. Wetlands generally include swamps, marshes, bogs, and similar areas. (40 CFR Part 122.2)

APPLICABLE ACTIONS

Clean Water Act

NPDES permits are required for the discharge of pollutants from any point source into waters of the U.S. [40 CFR 122.51(c)]

Section 404 permits are required for the discharge of dredged or fill material into waters of the U.S. (33 CFR 323.3)

Safe Drinking Water Act

Any action which may result in degradation of groundwater quality and/or pose a hazard to public health.

SUMMARY OF IMPLEMENTATION PROCEDURES

Clean Water Act

The procedures for determination of potential water quality impact from USGS activities and the need for various permits are extremely complicated and require close coordination with the Environmental Affairs Office, USGS, with regional offices of the EPA and the Army Corps of Engineers, and with state water pollution control agencies.

Safe Drinking Water Act (Consult ESM 94-5 for details)

1. In all cases where a project may involve ground water, one or more aquifers, or a recharge area, consultation with the appropriate District Hydrologist, WRD, should be initiated early in the planning stages.
2. Where a project may have any possibility of direct or indirect contamination of a sole or principal source aquifer, a recharge zone for such an aquifer, or a related streamflow source zone that has been designated or for which a petition for designation is being processed, early consultation should be initiated with the Regional Administrator, EPA.
3. If a project-related ground water impact is possible, an environmental assessment should be prepared. If an EIS is to be prepared, a copy of the Notice of Intent should be sent to the appropriate EPA Regional Administrator and he/she should review the draft EIS.

4. Final plans for projects having a potential for adverse impacts should incorporate appropriate mitigation measures.

INTERFACE BETWEEN FEDERAL WATER QUALITY REGULATION AND STATE ALLOCATION OF WATER QUANTITY

Melinda R. Kassen
Trout Unlimited, Colorado Water Project
Boulder, Colorado

Healthy rivers and lakes require both clean and flowing water. In the United States, however, there are two separate and distinct systems for regulating water quality and water allocation, i.e., the amount of water taken out of a waterbody for beneficial use and, as a result, the amount of water left in the waterbody for environmental, aesthetic, or recreational purposes. States have the primary authority to allocate quantities of water for various users. Thus, state allocation decisions affect how much water flows in rivers and streams. At the same time, federal law—the Clean Water Act—regulates water quality, although the Act gives the states certain responsibilities and allows the Federal Government to delegate other responsibilities to the states, albeit in both cases with federal oversight. Because a river has aspects of both flow and quality; however, actions taken pursuant to either quality or quantity regulation can affect regulatory actions in the other system. Such instances of overlap create substantial friction both between levels of government and between the governments and those using or enjoying the nation's waters.

WATER QUANTITY

The 50 states each have developed their own systems of water allocation. However, the states in the western half of the country generally follow the system of prior appropriation, whereas the states in the wetter eastern half generally follow the riparian system of water allocation. Even federal agencies must procure their water rights through the state allocation systems.

In the temperate east of the country, states follow the riparian doctrine for water allocation. This common-law doctrine, imported from England with the first European settlers to North America, allows landowners bordering a river to take what water they need for use on their land—reasonable use—so long as such use does not impair others' use on the same river. Increasingly, riparian states are refining their doctrines and adopted statutory frameworks, including water use permits. Both common-law riparianism and the statutory variants provide that one owner's use should not adversely affect the water quality to the point where another riparian user cannot make reasonable use of the river. If such a situation develops, the injured riparian user can seek to enjoin the offending action and/or compensation for damages.

West of the 100th Meridian where natural precipitation is half or less than what is available further east (as well as in Alaska and Hawaii), states follow the doctrine of prior appropriation. Prior appropriation allows anyone to divert water from a river for beneficial use in priority. The priority system assigns water users seniority based on when they first started using the water; the "first in time, first in right" precept means that the most senior appropriator gets to fulfill his or her entire right before the next appropriator can take any water out of the stream.

Many western states do not have any public interest, environmental, or other balancing tests that prevent appropriators from drying up streams entirely. Although most western states do have some type of "instream flow" protection programs, these programs were typically added into the system within the past few decades and do not disturb existing diversionary rights.

WATER QUALITY

Before 1965, states had exclusive authority to regulate water quality except to the extent that the U.S. Army Corps of Engineers may have affected water quality through its regulation of navigable waters under the 1899 Rivers and Harbor Act. Even the 1965 federal legislation did not create federal legislative authority. So, it was not until 1972 that Congress passed the Federal Water Pollution Control Act, which, for the first time, created a federal floor for water quality protection and two mandatory permit programs. The goals of the Act were no less than to "restore and maintain the chemical, physical and biological integrity of the Nation's waters" and eliminate all discharges of pollutants into the nation's waters by 1985. [33 U.S.C. §1251 (2002)].

The Act, later amended and retitled the Clean Water Act, retained the preexisting scheme, albeit substantially strengthened, for states to adopt their own water quality standards—composed of the uses and chemical quality for all their water bodies—however, now subject to approval by the federal Environmental Protection Agency (EPA). In addition, federal agencies, or delegated state agencies, issued discharge permits to ensure compliance with the water quality standards. Finally, where a waterbody does not meet the water quality standards that the state has adopted for it, the Clean Water Act requires the state to list this waterbody as impaired and then to establish a "Total Maximum Daily Load" (TMDL) of pollutants that, once achieved, will result in the waterbody meeting its standards. This latter program is similar to the standards-setting program in that the states have primary authority for listing and setting TMDLs, but the EPA must ultimately approve the state's actions, and if the state's actions are insufficient to meet federal law, then EPA establishes the TMDLs.

The Clean Water Act creates two different permitting programs. The first, the National Pollutant Discharge Elimination System (NPDES), requires entities, including many industries but also municipal wastewater treatment facilities, discharging pollutants to a waterbody through a discrete conveyance or point source, often a pipe, to obtain a permit. This permit imposes "end of the pipe" effluent limits to ensure compliance with two distinct sets of requirements—the state-adopted water quality standards for the receiving water and a federally determined set of best-available technologies, established on an industry-by-industry basis. The second permit program regulates those who discharge dredged or fill material into waters of the United States, including wetlands tributary to rivers, lakes, and estuaries. Such permits must again protect the receiving water's water quality standards and must ensure that the discharge of dredged or fill materials does not have an "unacceptable adverse effect" on either municipal water supplies or the environment. Although Congress gave federal agencies the initial authority to administer both permit programs, the Act also encouraged states to seek delegation of the programs. Most states administer their own NPDES permit programs, rather than the EPA; only a handful of states have chosen to seek delegation of the permit program for dredged and fill materials from the U.S. Army Corps of Engineers. With regard to both types of permits, states must certify that the discharge, as ultimately regulated, will meet the state's water quality standards. In fact, states must certify compliance with water quality standards for all federal permits that affect water quality, even those not issued pursuant to the Clean Water Act, for example, hydropower licenses that the Federal Energy Regulatory Commission issues.

The Clean Water Act creates many additional programs and includes significant sources of federal funding to enable the states to meet their regulatory responsibility, and to enable municipalities to achieve the necessary levels of sewage treatment. In addition, 30 years after passage of the Act, most waters that remain impaired are so as a result of "non-point sources" of pollutants, for which the Clean Water Act does not require permits. These sources include runoff from farms, construction sites, and abandoned mines. Although the Clean Water Act did not mandate controls for these sources of contamination, it includes programs to encourage their voluntary cleanup.

INTEGRATION

In water rights adjudications and permit proceedings, in most western states, the decision maker has only a limited ability to consider water quality. At one extreme, California has fully integrated water quality protection into water rights decisions, such that the state agency responsible for granting water rights has in a few instances required a diverter to leave water instream for water quality protection. In addition, since 1983, California implements its common-law public trust doctrine to require water quality protection in the context of water diversions. Thus, the California Supreme Court found that the California Water Resources Board had an affirmative duty to protect Mono Lake's water quality and environment from adverse effects caused by diversions by the City of Los Angeles, although the court still recognized a need to balance water rights with environmental protection and did not limit existing diversions [*National Audubon Society v. Superior Court*, 658 P.2d 709 (Cal. 1983)]. Other states such as Colorado and Idaho have explicitly rejected this approach, whereas still others, e.g., Oklahoma and Nebraska, simply have not articulated a position in the decades since the Mono Lake ruling.

In fact, in most other states, the water quality protections embedded in water rights and permit proceedings are limited. Only 7 of the 17 western states have either statutory or public interest requirements to protect a senior diverter's water quality from injury as a result of changes to existing water rights.

Most states require a new diverter to protect water quality at least to the point that the quality is sufficient for a senior diverter's purposes. However, in Colorado, a senior water user cannot seek protection from a lessening of water quality because of new diversions, or changes to existing diversions, although that user is entitled to have water quality maintained when there is an exchange of water on the river, i.e., when the senior is receiving water from a different source because of changes elsewhere in the system. Thus, senior diverters on the Arkansas River were denied relief as a result of a change to water rights that resulted in water formerly used for irrigation on the plains being diverted in the headwaters out of the basin entirely, even though the change substantially reduced the water quality of the flows remaining the basin.

Although 1970s era slogan proclaimed that, "Dilution is not the solution to pollution," entities discharging pollutants to a stream pursuant to a Clean Water Act permit have also sought protection from changes to water rights that would remove dilution flows. In Colorado, at least, this effort has not been successful. Thus, the courts denied relief to a discharging industry that objected to a change of water rights that would have removed stream flow such that the discharger would have faced significant additional water treatment costs to be able to meet its permit limits. The court remarked on the subservient nature of water quality regulation to the water rights allocation system, which did not allow consideration of water quality, or injury to a discharger, as opposed to another water user, in determining whether to grant a water right [*City of Thornton v. Bijou Irrigation Co.*, 926 P.2d 1, 90–94 (Colo. 1996)].

On the water quality front, even as it adopted a federal regulatory scheme for water quality protection, Congress was mindful of the states' role in water allocation. Thus, the 1972 law includes a section that its provisions not "be construed as impairing or in any manner affecting any right or jurisdiction of the States with respect to the waters ... of such States" [33 U.S.C. §1370(2) (2002)]. Yet, by 1977, after 5 years of Clean Water Act implementation, some western water user interests had become alarmed as to the potential for Clean Water Act regulation to "interfere" with the exercise of water rights. Congress added an additional provision to the Act, as part of a broad package of amendments, that no regulatory actions it authorized should "supercede, abrogate or impair" the exercise of state water rights [33 U.S.C. §1251(g)(2002)]. However, for several reasons, including that this language is a goal and that legitimate water quality regulation may affect water rights, the Courts have upheld water quality regulation to occur that appears at least to affect individual water users. For example, a farmer could not discharge dredged and fill material without a permit, even though he was doing so to better use his water rights [*United States v. Akers*, 785 F.2d 814 (9th Cir. 1986)]. And, a state could impose conditions in a water certification for a federal permit that had the effect of limiting the amount of water the permittee could use [*PUD No. 1 of Jefferson County v. Wash. Dep't of Ecology*, 114 S.Ct. 1900 (1994); *see also Public Util. Dist. No.1 of Pend Oreille County v. Wash. Dep't of Ecology*, 51 P.3d 744 (Wash. 2002)]. Finally, an entity seeking to build a new dam must obtain a Clean Water Act dredge and fill permit even if that entity already has a state water right to build the dam [*Riverside Irrigation Dist. v. Andrews*, 758 F.2d 508 (10th Cir. 1985)]. In making these findings, the courts repeatedly have held that legitimate regulation under the Clean Water Act may proceed even if there are indirect effects on the exercise of a water right.

In contrast to the courts' rulings, state legislatures have often taken action expressly to exempt water rights holders from regulation under state Clean Water Act programs. Thus, Colorado adopted the federal Act's language regarding not to "supercede, abrogate or impair," but not in the goals section of the Act; rather the provision is part of the state Water Quality Control Act's implementation [Colo. Rev. Stat. §25-8-104 (2004)]. Similarly, after the U.S. and state Supreme Court rulings in Washington confirming the state water quality agency authority to impose minimum stream flows that might affect the yield of water rights through water quality certifications, the state legislature adopted statutory amendments forbidding the state agency from adversely affecting water rights in water quality regulation [Wash. Rev. Code §90-48-422 (2004)].

A recent set of cases that illustrate the potential tension between water quality regulation and state water allocation laws involve artificial transfers of water, i.e., the man-induced movement of water from one waterbody to another. In each case, an entity was moving water from a contaminated watershed to a relatively cleaner waterbody through a conveyance—via a snowmaking system at a ski area, a municipal drinking water collection system, and a flood control system. Three U.S. courts of Appeal ruled that the discharger had to obtain an NPDES permit, because there was a discharge of pollutants from a point source to waters of the United States. The U.S. Supreme Court considered one of these cases in 2004 but did not reach the ultimate issue, although the Court did hold that the transferring water from one distinct water to another through a pipe did constitute a point source subject to permitting, even though the discharger was not, itself, "adding" pollutants to the water being transported [*South Florida Water Mgmt. Dist. v. Miccosukee Tribe of Indians*, 124 S.Ct. 1537, 1543 (2004)]. In other words, it was enough that the water being discharged was polluted already and was being discharged to a cleaner system.

As all three of the cases developed east of the Mississippi, the dischargers did not argue that they had water rights that allowed them to operate their systems free from Clean Water Act regulation, but many western states and water users weighed in to make that point. In fact, diverters have argued that requiring all transbasin diversions of water to operate only under an NPDES permit would overwhelm the states' permitting systems and adversely affect the use of state water rights. Those in support of requiring such permits have suggested that many of these diversions with discharges could be accomplished under states' general permitting programs in a way that would not be onerous. Only one state, Pennsylvania, has historically required permits for these kinds of discharges.

There are other places in the Clean Water Act program where water rights issues collide with water quality regulation. The Clean Water Act requires states to establish water quality standards, including use designations and numeric or narrative criteria for individual pollutants (33 U.S.C. §1313). To the extent that the exercise of water rights has affected what uses a waterbody can attain, because of the loss of dilution flow or fundamental changes in the natural system because of a dam, diversion, or artificial increase in the quantity of water coursing through the channel, states wrestle with how to accommodate human activity while encouraging operations that will maintain the Clean Water Act's "fishable/swimmable" goal for water quality.

In the program that requires states to list waters that remain impaired after all permitted discharges have permits [33 U.S.C. 1313(d)] EPA has interpreted the Clean Water Act not to require listing for impairment because of pollution, the definition of which includes hydrologic modifications such as dams and diversions that allow parties to exercise their water rights [64 Fed. Reg. 46011 (August 23, 1999); 65 Fed. Reg. 43586, 43592 (July 13, 2000)]. The problem is that the exercise of water rights that result in changes to the flow regime can lead both to impairment and to certain pollutant levels above the adopted standards. Examples include excessive temperatures, sediment, and nutrient levels. As temperature, sediment, and nutrients are pollutants, there have been listings, for example, in the Yakima River Basin in Washington State, where the underlying issue is most likely a change in the flow regime as a result of the exercise of water rights. So as not to challenge the exercise of water rights, other states, including California, New Mexico, and Washington, try to solve high temperatures with requirements to plant shade trees along a river's banks. See, e.g., http://www.ecy.wa.gov/programs/wq/tmdl/-watershed/index.html. However, EPA has rejected state proposals to solve impairment with flow enhancement programs. See http://www.-deq.state.mt.us/wqinfo/TMDL/approvalLtrs/Approval_BigCreekYelBasin.pdf.

Thus, state resistance to integrating water quality regulation, a federal mandate, with state water allocation systems has hampered protection of the environment as well as protection of the water quality necessary to sustain competing water uses. However, fears regarding the costs associated with water users having to protect the environment have kept most state legislatures from adopting statutes that integrate water quality protection into their water allocation systems in a way that would impose additional burdens on water development.

REGULATORY ISSUES AND REMEDIATION: RISK, COSTS, AND BENEFITS

David B. Vance
ARCADIS
Midland, Texas

James A. Jacobs
Environmental Bio-Systems, Inc.
Mill Valley, California

Major regulatory changes have occurred over the past decade regarding enforcement and cleanup standards. These changes have occurred in part, because of the realization that enormous financial and human resources were being spent on the remediation of properties with low concentrations of residual fuel hydrocarbon (primarily gasoline and diesel) contamination with low impact to human receptors. A major report from a national laboratory in 1995 changed the regulatory approach to focus attention on the risk benefits and exposure pathways associated with subsurface fuel contamination. Risked based decision making takes into account the risk, costs, and benefits when evaluating regulatory approaches to remediation

INTRODUCTION

A Lawerence Livermore National Laboratory (LLNL) report (1) found that once fuel leak sources have been removed, petroleum hydrocarbon contamination generally does not spread far from the leak sites. Given time, naturally occurring indigenous microbes in the soil and water usually break down most of the contamination before drinking water sources are impacted. The report noted that only 136 sites of the 28,051 evaluated (<0.5 of 1%) affected drinking water supply wells. The LLNL study concluded that at many low-risk sites, active cleanup could be stopped and natural attenuation by naturally occurring biological and geochemical methods would reduce the contamination over time (1).

The LLNL study was important at the time. About 1200 out of 20,000 leaking underground fuel sites were examined in detail. It appeared that in most cases, the length of the hydrocarbon plumes was limited to several hundred feet and less than 50 feet vertically. Natural bioattenuation, abiotic degradation, and diffusion processes stabilized or attenuated these plumes over time (2). LLNL concluded that the threat to human health and long-term effects on the environment caused by petroleum hydrocarbon contamination may have been vastly overestimated. Water supplies can generally be purchased for about \$50 to \$200 per acre foot, or even \$1000 per acre foot at the most expensive price. Cleanups of petroleum contamination of groundwater were costing up to \$600,000 per acre feet to reclaim groundwater and to meet cleanup goals (2).

Critics to the LLNL report noted that LLNL had huge environmental cleanups to contend with, so their recommendations may have been politically motivated, none of the actual data discussed in the LLNL report were included in the report for examination, the report relies on written and verbal communications and papers supplied by the petroleum industry, and no public health professionals from CAL EPA, the Department of Health Services (DHS), or other agencies were solicited for their recommendations or review (3).

Even with the controversy, the LLNL report did focus public policy discussion on how clean is clean and what financial and human resources are worth using to clean up hydrocarbon contamination. These discussions led to consideration of risked based cleanup standards.

Changing from enforcement driven by arbitrary cleanup standards to action (or no action) driven by health-based risk assessments in conjunction with a cost/benefit analysis. The current debate includes the application of this concept at the highest level for the promulgation of new regulations or modification of existing regulations. However, the more powerful application of the new regulatory paradigm is on a site-specific basis. The purpose of risked based regulatory decision making is to examine the components of the site-specific risk/benefit/cost process. The following discussion is predicated on two basic assumptions: First, if a release of contaminants has occurred at a site, the source of that release will be abated. Second, free product hydrocarbons are not allowed to remain on or near the water table.

These actions quickly limit the potential size of a contaminant plume, minimize further migration of the core of that plume, and recover contaminants as free product versus dissolved phase. All of which offer significant cost savings in the remediation effort. A health-based risk assessment determines what concentration of contaminant represents an acceptable risk level. The cost side of the process is driven by the effort required to achieve that risk-determined contaminant concentration. Health-based risk assessment can be a complex process. However, it has two key technical components: first, is the dose/response relationship of a chemical and second, is the pathway through which that chemical can ultimately enter a human body. A third nontechnical issue is how accepting the public is of the process, particularly those exposed to the risk. Dose/response relationships are categorized into two broad areas: noncarcinogenic and carcinogenic.

Noncarcinogenic responses are generally characterized by a threshold dose. The threshold represents a level of exposure to a contaminant that can be tolerated by a person with no adverse effects until the protective mechanisms of the person are overwhelmed. As individual levels of tolerance vary, threshold determination can be controversial.

Carcinogenic responses are assumed to have no threshold. This assumption means that some finite cancer risk exists no matter how small the dose.

The primary routes of exposure to groundwater are ingestion, inhalation of vapors during bathing or showering, and dermal exposure during bathing or showering. Direct ingestion is straightforward, and inhalation and dermal adsorption are dependent on physical properties of the contaminant such as vapor pressure and partition coefficients.

The public perception of risk is contradictory. Significant risk to which a person will voluntarily expose themselves, such as driving, improper diet, or smoking are of concern, but deemed acceptable. In contrast, much smaller risks associated with environmental impairment that is imposed, and not voluntarily engaged, often generate alarm at any level above zero risk. This issue can only be resolved by education to each person's aggregate risk during their lifetime from many sources with a view toward the benefits each receives from our industrial society. This should be an important part of the current risk versus cost and benefit debate.

In a practical sense, the most effective method of controlling the risk from contaminated groundwater is simply to not use it, which in many situations may be the most viable alternative. However, our groundwater resources are finite and are diminishing in quality with time. One solution that is likely to become more common is surface treatment to remove contaminants. The most cost-effective goals are to maintain as much of our groundwater as possible in pristine condition and the timely restoration of contaminated aquifers to acceptable conditions.

Cost effectiveness enters from establishing a health-based contaminant concentration that is considered acceptable and expending the minimum effort sufficient to ensure that level is maintained. Whenever possible, natural attenuation should be used for the final clean up of the core and the dilute distal portions of the plume.

Groundwater recovery or *in situ* treatment should be designed with three goals in mind:

1. Control and treat the core of the plume in which the dissolved concentrations exceed the capacity of the aquifer to naturally attenuate in a timely fashion.
2. If required, recover groundwater at a rate just sufficient to achieve the above goal.
3. Design a system that takes into account the heterogeneity of the aquifer matrix.

A fine-grained matrix will in general take a significant amount of time to remediate because the overall mass transport rates the matrix will support are low and the amount of adsorbed contaminant is high. However, groundwater recovery rates or *in situ* remediation systems need only be sufficient to meet the low mass transfer requirements. On the other hand, the migration distances are likely to be minimal.

The worst case exists in aquifers that contain strata of fine-grained material mixed with highly permeable continuous layers. Pumping rates must be high to obtain hydraulic control, but treatment durations are extended because of the slow leaching of adsorbed contaminants from fine-grained material. In aquifers of this type with pronounced hydraulic gradients and subsequent high groundwater flow rates, it may be necessary to allow much of the contamination to spread through the aquifer such that concentrations will be at levels amenable to natural attenuation. A health-based risk assessment will be key in this scenario, which in essence uses an extensive portion of the aquifer matrix as a chemical/biological reactor for contaminant destruction.

For permeable heterogenetic aquifers with small hydraulic gradients and low groundwater flow rates, it will be more economical to run a groundwater recovery system or an *in situ* treatment system for short durations on a periodic basis. The core of the process is the diffusion of contaminants from fine-grained units into the more highly permeable portions of the aquifer. This contaminated pore volume is then treated to restore the contaminant concentrations to low levels providing a concentration gradient that drives the diffusion process. Initial capital costs must be met, but operation and maintenance costs will be significantly decreased over the life of the project.

If natural attenuation processes stop on a site because of a lack of a suitable electron acceptor, such as oxygen for aerobic bioremediation, other methods are available to enhance site conditions and to continue with the natural attenuation approach. The consumption of any available oxygen within the core of a hydrocarbon plume is quite common. In these cases, even the nitrate and sulfate concentrations may be low within the core of the plume because of the microbial consumption of these alternative electron acceptors. For these projects that are too high in concentration to close, but too low for active and aggressive remediation methods, enhanced bioremediation holds promise of lowering concentrations to regulatory acceptable levels, wherein natural attenuation can be used. More detailed information about natural and enhanced bioremediation methods can be found in Suthersan (4).

Risk-based decision making and natural attenuation has undergone changes over the past several years. Risk-based corrective action (RBCA) is a U.S. Environmental Protection Agency (EPA) cleanup process designed to protect human health and the environment while allowing for practical and cost-effective site-specific remedial measures. RBCA has an expanded use of natural attenuation as a cleanup option.

Some regulatory agencies are moving toward the use of risk-based screening levels and decision making. This approach uses lookup tables of conservative, risk-based screening levels (RBSLs) for over 100 chemicals commonly found in impacted soil and groundwater at sites where releases of hazardous substances have occurred. The approach is intended to help expedite the preparation of environmental risk assessments at sites where impacted soil and groundwater has been identified. As an alternative to preparing a formal risk assessment, soil and groundwater data collected at a site can be directly compared with the RBSLs and the need for additional work can be evaluated. It is anticipated that documents like these will be especially beneficial for use at small- to medium-sized sites, where the preparation of a more formal risk assessment may not be warranted or feasible because of time and cost constraints (5).

This new regulatory approach will rationalize the process of environmental restoration. Regulators, industry, and the public will benefit. For it to succeed, we all must understand and accept reasonable levels of risk. Cost effectiveness will hinge on the understanding and exploitation of site-specific aquifer characteristics, and acceptance of cleanup time frames that allow for the use of natural attenuation.

BIBLIOGRAPHY

1. Rice, D.W., Gose, R.D., Michaelsen, J.C., Dooher, B.P., MacQueen, D.H., Cullen, S.J., Kastenberg, W.E., Everett, L.G., and Marino, M.A. (1995). *California Leaking Underground Fuel Tank (LUFT) Historical Case Analyses*. Submitted to the California State Water Resources Control Board, Underground Storage Tank Program and the Senate Bill 1764 Leaking Underground Fuel Tank Advisory Committee. Livermore, CA, November 16.
2. Weiss, R.B. (1996). State Regulatory Policy Takes A Giant Step to Close Low Risk Sites. *California Environmental Business Council News*.
3. Elliott, J. (1995). State Loosens Gasoline Spill Cleanup Rules. *Albion Monitor*. December 21.
4. Suthersan, S. (2002). *Natural and Enhanced Remediation Systems*. CRC Press, Boca Raton, FL.
5. Brewer, R. (2003). Screening For Environmental Concerns at Sites With Contaminated Soil and Groundwater; regional Water Quality Control Board. San Francisco Region. California. Available: http://www.swrcb.ca.gov/rwqcb2/esl.htm.

THE SAFE DRINKING WATER ACT

Rolf R. von Oppenfeld
The TESTLaw Practice Group
Phoenix, Arizona

OVERVIEW OF THE SAFE DRINKING WATER ACT

The Safe Drinking Water Act (SDWA) was enacted in 1974 and subsequently amended in 1986 and 1996 to protect the public from risks posed by drinking water contamination. The Act required the Environmental Protection Agency (EPA) to establish standards or treatment techniques for contaminants that could adversely affect human health, establish requirements for monitoring the quality of drinking water supplies and for ensuring the proper operation and maintenance of water systems, and protect groundwater from subsurface injection of polluting liquids. EPA posted a general guide to the SDWA provisions at http://www.epa.gov/OGWDW/SDWAsumm.html and posted its general approach for implementing the provisions at http://www.OGWDW/sdwa/sdwa.html.

The 1996 Amendments are important because Congress gave the EPA greater flexibility under the SDWA and emphasized the cost/benefit analytical approach to regulation. For an overview of the program, see the Office of Groundwater and Drinkingwater at http://www.epa.gov/safewater.html. The SDWA regulations can be found at 40 C.F.R. Parts 141–149 and may be accessed on the internet at http://www.water.access.gpo.gov/nara/cfr or http://epa.gov/OGWDW.regs. html. Inquiries about the regulations can also be directed to the Safe Drinking Water Hotline at 1-800-426-4791.

The EPA has delegated primary compliance and enforcement responsibility for the SDWA program to qualifying states. A detailed description of state primacy requirements can be found at http://www.epa.gov/safewater/pws/primacy.html. If states fail to ensure proper standards, the EPA may bring independent enforcement against violators through the Act's federal enforcement procedures. With respect to the regulated community, the SWDA focuses on operators of public water systems (PWSs) and on persons engaged in underground injections of fluids into wells.

Regulation of Public Water Systems

The SDWA regulates public water systems (PWSs) that provide water for human consumption through pipes or other constructed conveyances and regularly provide service to more than 25 persons or have 15 or more service connections [42 U.S.C. § 300f(4)(A)]. The EPA separates PWSs into the following categories:

1. community water systems and
2. noncommunity water systems that may be either
   a. nontransient, or
   b. transient.

Community water systems (CWSs) supply at least 15 service connections used by year-round residents or regularly serve at least 25 year-round residents [42 U.S.C. § 300f(15)]. Of approximately 180,000 water systems in the United States, only 55,000 are considered CWSs. They range from large municipal systems that serve millions of people to small systems that serve fewer than 100 people. CWSs can be publicly owned systems, including systems owned and operated by municipalities, townships, counties, water districts, and water authorities; privately owned systems that may be owned and operated by groups ranging from investor-owned water companies to homeowners associations; or ancillary systems that provide water as an ancillary function of their principal business or enterprise. Ancillary systems primarily service mobile home parks and a variety of institutional water providers.

Noncommunity PWSs are those systems that are not community water systems [42 U.S.C. § 300f(16)]. Examples of noncommunity PWSs include small systems in recreational areas and transient situations. Noncommunity systems may not have year-round residents but would have at least 15 service connections used by travelers or intermittent users for at least 60 days a year or serve an average of 25 individuals for at least 60 days a year. Specific examples of noncommunity systems include restaurants, campgrounds, motels, schools, and factories with their own wells or other water supply.

Nontransient, noncommunity water systems (NTNCWS) regularly serve at least 25 of the same people for 6 months of the year (Id). Schools and factories that serve water to 25 or more of the same people for 6 or more months of the year are nontransient, noncommunity water systems. All remaining water systems are classified as transient noncommunity systems (TNCWS). TNCWSs do not serve the same people on a day-to-day basis. In fiscal year 2000, there were more than 54,000 CWSs, more than 20,000 NTNCWS, and more than 93,000 TNCWS.

Certain PWSs are excluded from coverage and need not comply with SDWA regulations (42 U.S.C. § 300g). Excluded PWSs

1. consist only of distribution and storage facilities (have no collection and treatment facilities);
2. obtain all their water from a regulated PWS (but are not owned or operated by that PWS);
3. do not sell water; and
4. are not interstate carriers of passengers (Id).

This exclusion means that facilities such as hotels, factories, schools, and other businesses are not subject to SDWA regulation merely because they operate a storage tank and act as conduits from a PWS to consumers. However, the exclusion does not apply to private facilities that maintain their own wells or water supply. The EPA provides guidance for its definition of PWSs at http://www.epa/gov/OGWDW/guide/define.html.

SDWA DRINKING WATER STANDARDS AND GOALS

The SDWA authorizes the EPA to set drinking water standards called National Primary Drinking Water Regulations (NPDWRs). The EPA's NPDWRs are located at 40 C.F.R. Parts 141 and 142. NPDWRs are designed to protect consumers against the adverse effects caused by contaminants in drinking water. The SDWA requires that the EPA publish NPDWRs for contaminants that may have an adverse affect on human health, occur, or are likely to occur in PWSs at harmful levels and where the health risk can be reduced by regulation.

The 1996 Amendments made dramatic changes in the SDWA implementation process. Congress required the EPA to consider the following criteria when setting future drinking water standards and goals:

1. use of best available, peer-reviewed science [42 U.S.C. § 300g-1(b)(3)(A)(i)];
2. the assessment of health effects of populations at greater risk [42 U.S.C. § 300g-1(b)(3)(C)(i)(V)];
3. a list of treatment technologies for small systems, considering the quality of source water [42 U.S.C. § 300g-1(b)(4)(E)(ii)];
4. an assessment of the incremental costs and benefits of each alternative MCL considered [42 U.S.C. § 300g-1(b)(3)(C)(i)(IV)].

Maximum Contaminant Levels (MCLs)

A NPDWR must specify either a maximum contaminant level (MCL) or a treatment technique [42 U.S.C. §§ 300f(1), 300g-1(a)(3)]. The EPA is authorized to promulgate a NPDWR "that requires the use of a treatment technique in lieu of establishing a MCL," if the EPA finds that "it is not economically or technologically feasible to ascertain the level of contaminant" [42 U.S.C. § 300f(1)(C)]. Two significant NPDWRs that rely on treatment techniques are the Surface Water Treatment Rule and the Lead and Copper Rule.

MCLs are enforceable standards and represent the maximum permissible level of a contaminant in water delivered to any use of a PWS (40 CFR § 141.2). MCLs must be set as close to the maximum containment level goals as feasible. "Feasible" means using the best technology, treatment techniques, or other means which the administrator finds available after considering costs [42 U.S.C. § 300g-1(b)(4)(D)]. There are limits, however, to the use of cost considerations. First, the costs considered must be limited to those likely to be incurred [42 U.S.C. § 300g-1(b)(3)(C)]. Second, no consideration of costs is allowed to establish a MCL for "contaminants that are disinfectants or disinfection byproducts, or to establish a maximum contaminant level or treatment technique requirement for controlling *Cryptosporidium*" [42 U.S.C. § 300g-1(b)(6)(C)].

Under certain conditions, MCLs may be set at a level different from the feasible level. A variation is allowed where the means employed to achieve the feasible level of one contaminant would increase concentrations of other contaminants that pose a threat to human health. A level other than the feasible level may also be set when the means of reaching the feasible level for one contaminant interferes with the treatment processes necessary to comply with other NPDWRs. In these situations, the MCL or treatment process specified must "minimize the overall risk of adverse health effects" [42 U.S.C. § 300g-1(b)(5)(B)(i)].

The EPA must publish findings based on a scientific risk assessment of the probable health benefits and probable costs of each alternative MCL under consideration [42 U.S.C. § 300g-1(b)(3)(C)]. The published findings should be based on all of the following:

1. The EPA must use the best available, peer-reviewed science and studies.
2. Data must be collected by accepted methods.
3. Information on health effects must be presented by the EPA in an objective, comprehensive, and understandable manner.
4. Analysis for MCLs must show the quantifiable and nonquantifiable health risk reduction benefits, costs, and any increased health risk of compliance with the MCL.
5. For treatment techniques, analysis must show the health risk benefits and costs likely to result from compliance and alternative treatment techniques.
6. Valid approaches for measuring benefits must be identified.

If the EPA determines that the costs of a promulgated MCL outweigh the benefits, the EPA may promulgate a new MCL [42 U.S.C. § 300g-1(b)(6)]. The choice of treatment technique is left to the PWS, but the EPA must list available technology and methods to meet the MCL [42 U.S.C. § 300g-1(b)(4)(E)]. In addition, this list must include affordable methods for small

PWSs (Id). The EPA maintains a list of MCLs at http://www.epa.gov/safewater/mcl.html.

Judicial Review of MCLs

Section 1448(a) of the SDWA provides for judicial review of EPA's MCLGs and MCLs [42 U.S.C. § 300j-7(a)]. Only the U.S. Court of Appeals for the District of Columbia may review MCL determinations. A petition for review must be filed within 45 days, starting on the date of the promulgation of the regulation, unless solely based on grounds occurring after expiration of the 45-day period. There is no judicial review of a regulation in any enforcement proceeding (Id). Judicial review of MCLs is conducted under a narrow standard of review. The Court of Appeals will reverse only if an EPA action is arbitrary, capricious, an abuse of discretion, or not in accordance with the law. The EPA is given great deference because of the level of technical expertise it uses to evaluate complex scientific data.

The Drinking Water Contaminant Candidate List

In 1996, Congress set its most recent schedule for setting standards for drinking water contaminants. See 42 U.S.C. § 300g-1(b)(1)(B). First, the EPA was required to publish a list of contaminants that are not subject to any proposed or promulgated NPDWR, that are known or anticipated to occur in PWSs, and that *may* require regulation under the SDWA. The EPA was required to publish a Drinking Water Contaminant Candidate List (CCL) in February 1998 and every 5 years thereafter. The list must be published only after consultation with the scientific community and public comment. Following publication of the list, the EPA must determine which of the listed contaminants to regulate. The determination must be completed every 5 years for at least five listed contaminants.

Unregulated Contaminant Monitoring Regulations

The 1996 amendments also required EPA to develop a program to monitor unregulated contaminants and publish a list of monitored contaminants [42 U.S.C. § 300j-4(a)(2)]. In response, the EPA promulgated the Unregulated Contaminant Monitoring Regulations (UCMR) [64 Fed. Reg. 50556 (Sept. 17, 1999)]. The UCMR works in conjunction with the CCL. The EPA can use the data generated through the UCMR to prioritize contaminants on the CCL and help determine which contaminants to regulate. Several states may also petition to have certain contaminants monitored under the UCMR [42 U.S.C. § 300j-4(a)(2)(B)]. The UCMR covers all CWS and NTNCWS that serve more than 10,000 customers. An additional, 800 systems serving under 10,000 customers were randomly selected to test for unregulated contaminants.

The UCMR created three lists of contaminants for monitoring. Substances were placed on lists depending on their known environmental occurrence and the ability to test analytically for them. List 1, the "Assessment Monitoring" list, contains those substances known to occur in more than two states or more than 10 PWSs, and for which monitoring methods were available. See 40 C.F.R. § 141.40 Table 1. EPA has little monitoring data for List 2, the "Screening Survey," and List 3, the "Pre-Screen Testing List."

Chemical Monitoring Reform (CMR)

To reduce unnecessary monitoring and reporting requirements, the 1996 Amendments contained a provision for modifying NPDWR monitoring requirements. This CMR provision streamlined drinking water monitoring requirements for 64 chemicals that may occur in the *source water* of PWSs. For a system to qualify for alternative monitoring, it must demonstrate that a contaminant is not present in its water supply or if it is present, that it is reliably and consistently below the MCL. If the contaminant is later detected at or above the MCL, then the system must eliminate the contamination, or it must monitor as required under the NPDWR.

Secondary Regulations

Section 1412(c) of the SDWA also authorizes the EPA to promulgate National Secondary Drinking Water Regulations (NSDWRs) [42 U.S.C. § 300g-1(c)]. A NSDWR is "a regulation which applies to PWSs and which specifies the maximum contaminant levels which, in the judgment of the Administrator, are requisite to protect the public welfare" [42 U.S.C. § 300f(2)]. The NSDWR "may apply to any contaminant in drinking water which may adversely affect the odor or appearance of such water and consequently may cause a substantial number of persons served by the PWSs providing such water to discontinue its use, or which may otherwise adversely affect the public welfare" (Id). NSDWRs are not federally enforceable but offer additional guidance based on odor, aesthetics, and appearance to water systems and states. NSDWRs may vary according to geography and other circumstances. To date, NSDWRs have been established for several contaminants, including copper, fluoride, pH, and total dissolved solids. These regulations are in 40 C.F.R. Part 143.

Consumer Information

A major objective of the SDWA is to increase communication between PWSs and their customers. Consequently, Congress required an annual "Consumer Confidence Report" for PWSs. EPA regulations require each CWS to mail to each customer, at least once annually, a report on the level of contaminants in the drinking water purveyed by that system. These reports also indicate the source of the water supplied to the consumer and notify customers that a source water assessment is available.

VARIANCES AND EXEMPTIONS

General Variance

A state that has primacy over the SDWA may issue a variance if it determines that a system cannot comply with a MCL, despite application of the best available technology (BAT). The EPA determines the BAT and treatment method for each NPDWR after examination for efficacy under field conditions, rather than solely under laboratory conditions, and after considering cost. (40 C.F.R. § 141.2). A state may not issue a variance where an unreasonable risk to health exists or where alternative sources of water are reasonably available to the system. Under Section 1415, EPA must propose and promulgate its BAT finding at the same time it proposes and promulgates a MCL for each contaminant (42 U.S.C. § 300g-4). BATs may vary among systems, depending upon the number of persons served by the system or for other physical conditions related to engineering feasibility and costs of complying with MCLs.

The 1996 amendments allow the EPA to review and object to any variance granted by a state [42 U.S.C. § 300g-4(e)(10)]. If the state does not resolve these objections, the EPA may overturn the variance when it violates SDWA requirements. Consumers served by a PWS that has been granted a variance may petition EPA to begin the objection and review process.

A variance will be granted for a maximum of 5 years with the possibility of a 3-year extension. At the time a variance is granted, it will contain a schedule of steps necessary to achieve ultimate compliance with the particular NPDWR applicable in the variance. The variance may also contain additional control measures for protecting public health during the period that the variance is in effect. The EPA's final rule, issued in August 1998, states that 5 years is a reasonable time frame for a PWS to come into compliance; however, the state has the flexibility to set a feasible schedule if it determines that more than 5 years are required [63 Fed. Reg. 43841 (Aug. 14, 1998)].

Small System Variance

Congress responded to concerns that NPDWR compliance costs unfairly burden small PWSs. The 1996 Amendments added a special variance applicable only to small systems. See 42 U.S.C. § 300g-4(e). A small system variance may be available to systems that serve fewer than 3,300 persons (or up to 10,000 with special EPA approval). The variance should be of special interest to small community systems as well as the schools, factories, and other institutions that qualify as noncommunity water systems under the Act. The small system variance is available only where EPA has identified a less expensive version of the best available technology, called "variance technology," and the water system cannot afford to comply with the regular NPDWR. All variances must continue to provide adequate protection of human health. The proposed maximum period for a small system variance is 3 years with the possibility of a 2-year extension to complete improvements, obtain an alternate water source, or obtain federal or state financial assistance.

Exemptions

Under Section 1416(a), a state may exempt a PWS from any MCL or treatment technique requirement if it finds that

1. due to compelling factors (which may include economic factors), the system is unable to comply;
2. the system was in operation on the effective date of the MCL or treatment technique, or, for a newer system, that no reasonable alternative source of drinking water is available to that system;
3. the exemption will not result in an unreasonable risk to health; and
4. restructuring and management changes cannot reasonably be made that will result in compliance or otherwise improve the quality of the water (42 U.S.C. § 300g-5).

Under section 1418(b), a state is to prescribe a compliance schedule and a schedule for implementing any required interim control measures at the same time it grants an exemption [42 U.S.C. § 300g-7(b)].

In a major change in the exemption provisions, the 1996 Amendments deleted provisions which limited an exemption to 12 months, subject to a 3-year extension [42 U.S.C. § 300g-5(b)(2)(A)]. The new provisions require that the schedule for an exemption include compliance with each contaminant level and treatment technique for which the exemption was granted, as soon as practicable, but not later than 3 years after the otherwise applicable compliance date. The only exception to this time period is for small systems that serve less than 3300 persons, under certain specified conditions, for which extensions may be renewed for one or more additional 2-year periods, but not to exceed a total of 6 years [42 U.S.C. § 300g-5(b)(2)(C)].

The amendments now require that the state also consider whether the PWS is a "disadvantaged community" and whether management or restructuring changes can be made that will result in compliance or, if compliance cannot be achieved, would improve the quality of the drinking water [42 U.S.C. § 300g-5(b)(4)]. Measures to develop an alternative source of water supply must also be considered.

The 1996 Amendments also gave states more flexibility to adopt alternative protective strategies. For example, a state exercising primary enforcement responsibility for a PWS may, case-by-case and after notice and opportunity for public comment, establish watershed management programs as an alternative to filtration for systems that have uninhabited, undeveloped watersheds. See 42 U.S.C. § 300g-1(b)(7)(C)(v).

UNDERGROUND INJECTION CONTROL (UIC) UNDER THE SDWA

Scope of the UIC Program

SDWA established the UIC program to protect existing and future underground sources of drinking water from contamination caused by the subsurface disposal of waste fluids into wells. See 42 U.S.C. §§ 300h-300h-5. Improper operation of an injection well that allows movement of contaminants into an underground source of drinking water can lead to substantial civil penalties. The SDWA allows states to administer their own UIC programs if the state regulations and statutes meet the EPA's minimum requirements specified in 40 C.F.R. Parts 144–146. See http://www.epa.gov/safewater/states.html.

The overriding standard applicable to all underground injection activities is that

> No owner or operator shall construct, operate, maintain, convert, plug, abandon, or conduct any other injection activity in a manner that allows the movement of fluid containing any contaminant into underground sources of drinking water, if the presence of that contaminant may cause a violation of any primary drinking water regulation under 40 C.F.R. Part 142 or may otherwise adversely affect the health of persons . . . [40 C.F.R. § 144.12(a)].

Underground injection is defined as well injection, or the subsurface emplacement of fluids through a bored, drilled or driven well, or through a dug well where the depth of the dug well is greater than the largest surface dimension" (40 C.F.R. § 144.3). Fluid is broadly defined as "any material or substance which flows or moves whether in a semisolid, liquid, sludge, gas, or any other former state" (Id).

EPA has defined five classes of underground injection systems:

Class I: Wells injecting hazardous or nonhazardous waste below the lower most underground source of drinking water

Class II: Wells injecting fluids associated with oil and gas production

Class III: Wells injecting fluids for mineral extraction

Class IV: Wells injecting hazardous waste into or above an underground source of drinking water

Class V: All injection wells not covered by Classes I–IV (40 C.F.R. § 144.6).

All Class IV wells are banned by SDWA (40 C.F.R. § 144.13). Class V wells are "authorized by rule," and no individual permit is required. There are currently few regulatory requirements for Class V wells other than reporting under 40 C.F.R. § 144.26. Class V wells are the systems most commonly used by industry and small commercial operations. Common Class V wells include air conditioning return flow wells, cesspools, and cooling water return flow wells. See EPA's Class V websites at http://www.epa.gov.sefewater/uic/classv.html and http://www.epa.gov/safewater/uic/c5fin-fs.html. The EPA has announced that it is discontinuing its permission to operate Class V wells for motor vehicle waste disposal and large capacity cesspools. [64 Fed.Reg. 68545 (Dec. 7, 1999)].

Certain Class I, II, and III wells are not required to obtain individual permits (40 C.F.R. § 144.21). These wells may be authorized by rule if the operator injects within 1 year of a new UIC rule or performs an inventory under 40 C.F.R. § 144.26 within 1 year. All other Class I, II, and III wells require an UIC permit. An application for a UIC permit for a new injection well must be submitted by a "reasonable time before construction is expected to begin" [40 C.F.R. § 144.31(c)(2)]. If a facility or activity is owned by one person but is operated by another person, it is the operator's duty to obtain an UIC permit [40 C.F.R. § 144.31(b)]. Applicants for a UIC permit must use the application form provided by the EPA [40 C.F.R. § 144.31(e)]. The EPA may issue a permit for an area, rather than for each well individually, if the wells are within the same well field, facility site, reservoir, project, or similar unit in the same state; the wells are operated by a single owner or operator; and if the wells are used to inject something other than hazardous waste [40 C.F.R. § 144.33(a)].

If cause exists, the Administrator may modify or revoke and reissue a UIC permit. The Administrator may terminate a UIC permit during its term, or deny a permit renewal application if there is noncompliance by the permittee, failure to disclose relevant facts, or the permitted activity endangers human health or the environment (40 CFR § 144.40). A detailed description of the UIC program is available at http://www.epa.gov/safewater/uic.html.

OTHER SDWA PROVISIONS

The SDWA gives the EPA Administrator the power to classify an aquifer as a "sole source aquifer" (SSA) if the aquifer is the sole source of drinking water for a community. The SSA designation gives the EPA greater ability to scrutinize federal projects in the designated area to ensure that the aquifer is not contaminated [42 U.S.C. § 300h-3(e)]. The SSA regulations are located at 40 C.F.R. Part 149. The SDWA also requires that states with primacy to implement a Source Water Assessment Program to protect source waters for PWSs better from contamination (42 U.S.C. § 300j-13). SDWA also provides federal funding under the Drinking Water State Revolving Loan Fund to states for drinking water infrastructure projects (42 U.S.C. § 300j-12).

REPRESENTING GEOPOLITICS OF (HYDRO) BORDERS IN SOUTH ASIA

Eva Saroch
Panjab University
Chandigarh, India

INTRODUCTION

In the tradition of political geography, the concept of boundary/border has been of major importance where researchers have understood it mainly as an empirical manifestation of political processes and particularly of the sovereignty of the state. But borders not only represent a 'fixed line' in space or time, rather, they need to be understood as being produced and reproduced in the *socio-spatial consciousness* of society (a broader form of consciousness that reflects the ideological and hegemonic structures). Therefore, it is argued here that boundaries, apart from being located on terra-firma as a physical line, also exist metaphorically, symbolically in many ways as a manifestation of socio-spatial consciousness and as 'plays of power' that generate and exert various codes, rules and procedures of 'distinction' in the process of Othering such as us/them, sacred/profane, inclusive/exclusive, to impose specific understanding through the construction of knowledge in sociocultural practices and discourses in politics, administration, economics, or the organization of ethnic relations on various scales.

In the context of traditional geopolitics of water, 'mastering space' by deploying strategies of spatial domination and control and acquiring and guarding access to freshwater resources has always been a highly contested geopolitical issue both among and within sovereign states. A critical geopolitical approach, focusing the intricate but intimate interplay between space and its natural endowments such as water, and the socio-spatial consciousness of its various stakeholders, seems to suggest that rivers, for example, are not simply 'located' physically on ground as water flows or as static lines on maps but are situated/implicated in the very process of 'Othering' or the construction of symbolic/cultural boundaries between 'us' and 'them' (see Campbell, 1992, in Ref. 1, p. 3). Attention is thus drawn to how growing disagreements and rivalries over who gets, when, how and how much transforms seemingly innocent water flows into highly resilient hydro-borders. A critical geopolitical perspective motivates us to look for the sites where hydroborders are produced and reproduced, both in the form of various discourses (administrative, political, economic, cultural, and religious) and practices on various—spatial scales local, provincial, and national. At yet another analytical level, it is worth exploring how the 'criticality' of water resources is inscribed on the mental maps of various stakeholders (at local, national, and regional levels) as symbols of socio-cultural–religious identification or differentiation, identity formation, and power-political pursuit of primacy and security.

Even though the states and societies of South Asia are hydrographically united—they share the Indus River Basin (India and Pakistan), the Ganges–Brahmaputra–Meghna Basin (India and Bangladesh), and the Ganges–Mahakali Basin (India and Nepal), which the brief profile of the South Asian waterways suggests—geopolitically they stand divided. Hydro-borders in the form of mental borders or barriers continue to appear and reappear in different guises within an existing asymmetry of wealth, knowledge, and information, and as part of conflict and struggle among various actors for spatial control or mastering space (2, p. 115).

Thus, it is argued here that in the context of traditional, state-centric geopolitics of water resources, which seems to persist in post-colonial, post-partition South Asia, the 'hydroborders' in South Asia continue to be constructed, reinforced, and legitimized through competing notions of domestic stability/instability, security/insecurity, and legitimacy/illegitimacy. Constructed through 'nationalist' metaphors, state-sponsored security narratives and even politically motivated religious speeches, these mental borders, deceptively hydrographic in argument as well as in appearance, remain integral to the politics of place-making and group differentiation in the region. Drawing upon insights from the ongoing tensions/conflicts between India–Pakistan, India–Bangladesh, and India–Nepal (as well as within these countries) over the so-called 'sharing' of water resources, this article critically examines the interplay between the boundary-defying/unifying nature of waterways and the bounding/excluding logic of geopolitical discourses, imaginations, and representations.

A PREFACE TO SOUTH ASIAN WATERWAYS

Ecologically speaking, a river knows no boundaries. Flowing through farms, cities, and nations, rivers join rather than divide places. Any river that flows through two or more nations is supported by an ecosystem that cuts across political boundaries and joins the basin states. In similar perspective, South Asian waterways are the epitomes of sociocultural synergies, binding the South Asian landscape into one composite whole. The ancient 'Sapt Sindu Valley' and the 'Gangetic Basin,' together known as the Indo–Gangetic Basin, encompass South Asian waterways. The River Indus and the River Ganga are the principal rivers of two separate systems, but their separation is overridden by the unity of the Indo–Gangetic Plain, which spreads between the Himalayas and the Deccan. The Great Plains consist of two river basins, those of the 'River Indus' and the 'Ganga–Brahmaputra,' and the network of these rivers constitutes the 'Himalayan river system,' which has not only maintained the geographical contiguity on the subcontinent but has also been a fulcrum of different cultural streams. Having sustained inhabitants, the aquatic communities, and the ecosystem over centuries, these vast river systems have been the unifying–binding force for the South Asian geo-economic and geo-cultural landscape.

PARTITION, OTHERNESS, AND HYDRO-BORDERS

A critical geopolitical approach to the study of partition suggests that the partition of 'British India' was the convergence of two different phenomena—ethno-politics and geopolitics. According to Denis Retaille, ethnopolitics takes identity not territory as its starting point (3, p. 34), whereas geopolitics is a question of rivalries between powers for land and for those who live there or rivalries for territory (4, p. 122). Each complements the other in producing the other's 'excess' or excessive geopolitics. Excessive geopolitics "What", as aptly put forth by Chaturvedi, "does" is transform ideological–ethnopolitical conflicts relating to the boundaries of identity groups into geopolitical rivalries over 'our' territory versus 'their' territory. Consequently, the cold-blooded logic of 'territory for the sake of territory,' or Lebensraum, subsumes the more humane logics of common habitat or shared spaces" (5, p. 150). What sustains excessive geopolitics is the 'reflexive Otherness.' The image of the 'Other' is significant in producing borders that are both physical and symbolic. The permanent presence of 'Other' in discourse over water conflicts in South Asia has produced a manifold character of hydro-borders that are manifest in various foreign policy pronouncements, in defining national identity, and in so-called development projects. Even the language of dispute or the dispute over language suggests the deadly hold of 'nationalist language' not only in conceptualizing or thematizing the problem but also strategizing the resolution of the dispute.

Sir Cyril Radcliffe, a British lawyer who was given just six weeks to draw boundaries in Punjab in the west and Bengal in the east, was aware that the network of canals, River Indus, and its five tributaries were a lifeline for united Punjab and also the continued integration of the irrigation system, at least on a functional basis, was fundamental for the basin as a whole. Such a joint venture run by both countries would not only safeguard the interests of both peoples but also form a basis of cooperation which might prove fruitful in the years to come (as cited in Ref. 6, p. 164). But Radcliffe's hope of joint management of the basin for the optimal benefit of the people and ecology was rebuffed by both Jinnah and Nehru. According to Jinnah, "he would rather have Pakistan deserts than fertile fields watered by courtesy of Hindus." Nehru curtly informed that "what India did with the Indian rivers was India's affair" (6). This is a classic example that shows how a common past is repudiated as a time of oppression in which 'we' were simply the victims of oppressive 'others.' It also shows how partition has always been a disciplining event that shuts out other possibilities of resolving the conundrum whose one form is partitioning country, homelands, or the network of rivers, the arteries of the basin.

Hence, the boundary line in Punjab was drawn between Lahore and Amritsar, cutting across the well-integrated canal system that draws its water from the Indus and its five tributaries, Jehlum, Chenab, Ravi, Sutluj, and Beas. The enforced geography of partition was such that the headwaters of the rivers of the Indus Basin remained in India, thereby giving India the status of upper riparian. Pakistan felt that its livelihood was threatened by the prospect of unilateral Indian control over the tributaries that fed water into the Pakistani portion of the basin. Thus, Pakistan apprehended that a 'hostile' India might, if it so desired, using tunnels, divert water within Kashmir from the Western to the Eastern rivers, affecting a transfer before the water even reached Pakistan or the fringes of 'Azad Kashmir' (which India chooses to call Pakistan-occupied Kashmir) (7, p. 234). Whereas India certainly had its own ambitions for the 'profitable development' of the basin, Pakistani media and the politicians launched a large-scale national campaign over a canal water dispute. Pakistan even used the Indus water dispute as a political tool in the battle over Kashmir being debated at the United Nations. Former Foreign Secretary of India, J.N. Dixit, while addressing a plenary session in a seminar, is reported to have said that, "it was well accepted in our strategic studies circles, that much of the passion on Kashmir was generated because India controls the headwaters of the Indus" (8, p. 161).

A greater insight into the deliberations of the pre-Indus Waters Treaty period reveals that the Treaty signed between India and Pakistan in 1960 was dictated and driven by the logic of 'Othering' that jettisoned and abandoned the more humane logic and vision of joint development and administration of the integrated Indus Basin for the optimal benefit of the people of the two countries. The Treaty led to the fissuring/partitioning of the Indus and its five tributaries (the three eastern rivers went to India and the three western rivers to Pakistan) and of the 'homelands' that dwelled in the canal colonies, but the Treaty also carried, within itself, seeds of bypassed, unresolved, old, and unanticipated new water conflicts in the basin.

Even after partition became an accomplished reality, dialogues to overcome the bitter divides created in the wake of partition became prey to the partitioned times. Nothing can be more illustrative in this context than the water disputes at provincial and local levels. There are intrastate water disputes, for example, between Punjab and Sindh (Pakistan), and Punjab, Haryana, and Rajasthan (India), among others. Within India, in the wake of the partition, the issue that agitated the Sikhs even more than the territorial dispute was the unjust and illegal allocation of Punjab's river waters and hydroelectric power generated from them to non-riparian neighbor states. More recently, Mr. Praksah Singh Badal, former Chief Minister of Punjab in India, reportedly said that every drop of water that flows through Punjab belongs to Punjab. In Pakistan, the water conflicts between Punjab and Sindh have gained momentum. According to Sindh, the upper riparian province has been progressively increasing its illegal share of the waters of the Indus River since 1889 (9). Apart from this, in 1960, the military regime in Pakistan surrendered, under the Indus Waters Treaty, the three eastern rivers to India along with Sindh's acknowledged share without prior consent or compensation to Sindh, leading to water scarcity in the province.

It is possible to argue that what lies at the very core of the matter is the entrenched notion of 'us' versus 'them', as well as a nagging sense of insecurity on the part of Islamabad that a 'permanently hostile' India, which also happens to be the upper riparian state, might use water resources as a politico-strategic weapon to paralyze Pakistan's economy completely. Such imaginations and suspicions continue to mar India–Pakistani relations, and they also keep alive the bitter memories of the bloody historical circumstances, in which the two 'Midnight Children' were born. Needless to say, the persistence of such hydro-borders makes any solutions of relatively fewer complex issues such as the Tulbul Navigation Project/Wular Barrage virtually impossible.

Similarly, the Ganges water dispute between India and Pakistan (and after 1971 between India and Bangladesh) also had its genesis in partition. The Farakka dispute dates back to 1951 when India decided to construct a barrage across the Ganges at Farakka to divert water to Bhagirathi to maintain its navigability, which was being hampered due to siltation. However, a deeper analysis suggests that Farakka is not only a technical problem involving the issue of water sharing but, as put forth by Begum Khurshida, a 'techno-political tangle' (10, p. 196). Technical debates have intermingled with political rhetoric, narratives, and nationalist myths; Farakka is not only a symbol of Indian domination (10, p. 202) but also projects an idiosyncratic 'upper riparian psyche' that intimidates lower riparian Bangladesh and jeopardizes its water security. The entire history of the dispute and the tortuous course of negotiations over Farakka suggests that what was being negotiated was not the flowing waters, but the strategic resource for nationalist politics between India and Bangladesh.

The Mahakali Treaty of 1996, signed between India and Nepal (concerning the integrated development of the Mahakali river), is yet another example where the ethics of water sharing could not bypass the deeply entrenched issues of sovereignty and (in)security or for that matter ideological contestations. In Nepal, the treaty is considered another sellout of water resources. The treaty is said to be 'the blatant manifestation of domination of Indian expansionism, on one hand, and total surrender to the Indian big brother on the other.' Thus, so far, the dominant trend among the two nation-states has been to perceive water issues both as a major problem area and as a valuable resource to enhance their respective 'national' power and defense (8, p. 167). In the process, the water issues have been (mis)used to sustain conventional geopolitical considerations.

It can also be argued that behind these self-locked positions are the deeply seated hydro-borders, created through various kind of geopolitical perceptions, metaphors/rhetorics/narratives by various protagonists both within and among the sovereign states involved at various levels in the dispute over water issues. As a consequence, water has become a crucial item of resource politics, nationalism, provincialism, and sub-regionalism (10, p. 196). All such linkages have shaped and are central to the elite politics of water and the so-called 'sustainable development' of South Asian waterways.

MAPPING OF RIVERS IN THE 'CARTOGRAPHIC ANXIETY' OF NATION

South Asian states continue to suffer from, what Sankaran Krishna (11, p. 82) calls, 'cartographic anxiety' surrounding the questions of national identity and survival. As meaning is invested in all aspects of cartography, so defined cartography becomes nothing less than the 'social and political production of nationality' itself. And these cartographic representations not only become alive during the election campaign, in the everyday life along the 'border' (see 12), but interestingly enough, different imaginations/mapping of the river that produce hydro-borders are invoked in the construction/representation of 'imagined landscape' or 'national identity' or securing 'territorial unity.'

Pakistan, argues Ian Talbot, has been a state searching for its national identity, and due to persistent failure to find such an identity, ethnic and tribal identities remain strong. Even five decades after its creation, the question remains whether Pakistan is a land for the Muslims or a nation of Muslims moving towards its destiny as a nation? (as cited in Ref. 12, p. 153). In an attempt to define 'national identity,' sometimes 'pasts' are invoked and 'histories are constructed' (12). For instance, in a major restatement (rather 'rewriting of history of the Indian sub-continent') Pakistani historian Aitzaz Ahsan, argues in his work *The Indus Saga and the Making of Pakistan,* that the Indus region (Pakistan) has a primordial existence outside India and centrifugal forces of its own (13). The essential purpose of this work is, in the words of its author, "to discover and define 'Indus person,' the Pakistani citizen." According to Ahsan, history states that Pakistan is said to have existed "for almost five and half thousand of the last six thousand years" and emphasizes that the "Indus has seldom been part of India."

Subsequently, his work contends that the Indian subcontinent has always been at least two different worlds. The divide between India and Indus is civilizational and cultural. There has existed a 'primordial divide' between the two: 'a palpable divide between two lands, two peoples and two civilisations: Indus and India. In fact the twain has seldom, if ever, truly met. Arguing that these differences are deep rooted, primordial, and many, he observes that

> The sub-continent is made up of two civilisations, Indus and Indic or Gangetic and that Indus has been one large, independent, politico-economic zone for the past countless centuries. The truly 'Indic region' comprising the Gangetic plain and peninsular India, on the one hand, and the Indus basin, consisting of Indus river and its tributaries (i.e., Pakistan) on the other. Further, more than the giant mountain range that separate the Indus region from the Central Asia, that discernible hump between Indus and India along a line roughly connecting Gurdaspur, in Eastern Punjab, to Kaithawar on the Arabian sea, has been the critical dividing line of history and culture. It approximates the division between India and Pakistan, thus giving to that border the sanction and strength of history (13, p. 289).

This kind of rewriting of history is an attempt to evade the conspicuous hydrographic oneness/unity of the Indus–Gangetic basin. By ingeniously selecting the facts that serve his 'partitioning and Othering' analyses and fitting his mapping of national identity based on the re-presentation and apt distortion of the 'Indus Basin' hydrographically, historically, culturally, and civilizationally, he is able to suitably construct a 'Indus Person' or Pakistani identity. Further, in reconstructing a narrative of a distinct cultural heritage, Ahsan has traced the civilizational roots of a nation-state that came into being only in 1947, into the distant past (14, p. 19). According to David Ludden, such an attempt which constructs beleaguered historical narratives around partition and sought to inscribe the past with a new meaning, is a feature of the politics of the cultural anxieties that have surfaced lately across contemporary South Asia (14).

In the traditional sense, river has been used as a boundary to define the territorial limits of a state, but to engage in rituals to purify 'frontier rivers' is not only to revive, the sacredness of the river but also to deploy new meaning(s) to the process of territorialization. The two-day Sindhu Darshan festival (the brainchild of Home Minister L.K. Advani), in Ladakh (on the Indian side of Kashmir), where the river Sindhu or Sindh flows before entering Pakistan territory, was at one level a welcome gesture toward integrating far-flung parts of the country into the mainland, whereas, on the other hand, a symbolic attempt to revive the link between 'Sindhu and Hindu', or a ceremony affirming the sacredness of the River Indus. What has followed 'Sindhu Darshan' is the so-called 'Brahmaputra Darshan,' a similar kind of festival in the northeast part of India, 50 km from the IndoChina border (15).

It is important to note that in imagination about India's national identity, there has always been a conceptual space for Hindu nationalism. Further, Hindu nationalism has two simultaneous impulses: a commitment to the territorial integrity of India as well as a political commitment to Hinduism. And in the geopolitical imagination of Hindu nationalists, 'national unity' and 'identity' are based on territory and religion (16, p. 235). Hence, it is argued that such festivals/practices, that are held at the banks of frontier rivers, are an attempt to engage in the process of sacralization and re-territorialization of the 'frontiers' of the nation-state and also to promote and consolidate the notion of 'Akhand Bharat' (undivided, greater India). These examples also reveal how different geopolitical imaginations and representations of the river have provided meaning(s) and a rationale to different cartographic anxieties, a facet of a larger post-colonial 'anxiety syndrome' or as Samaddar says, 'insecurity syndrome' (17, p. 32), surrounding the question of territorial unity and national identity. Needless to say, through representation of a river in a certain way, the state generates a representation of self and in the process the boundary-defying river also becomes a boundary-producing entity.

CONCLUSION

The ethics of sharing water resources posits that water, as a common natural heritage—especially transboundary watercourses, including aquifers—must be managed in an ecologically sustainable manner. But this ethical imperative is undermined as and when domestic geo-political rivalries, conflicting national ideologies, and geopolitical discourses conspire together to produce hydro-borders that are deeply entrenched and frequently (mis)used by the states as instruments or media of social distinction, domination, or control (18, p. 65). In their quest for domestic stability, political legitimacy, regime survival, and identity building, in both religious-communal and secular-national terms, South Asian states continue to articulate and defend a 'national vision,' that has little space to accommodate the concerns of other state or non-state actors. The process of Otherization that continues to make its place, that is, production and re-production of hydroborders in the bilateral or multilateral dialogue between India/Pakistan/or India/Nepal/Bangladesh over the water disputes does not reflect a dialogue of accommodation/agreement or cooperation but a dialogue of disagreement, separation, and partition itself. This also suggests that a new kind of resource geopolitics, in which different power structures are sustained through the discourse and practices, is in the making in South Asia. As long as the logic of Othering, as manifested in the form of hydroborders, continues to make its place in the South Asian geopolitics of water, the prospects for an ecologically sustainable, gender-sensitive, culturally appropriate, and socially just development and management of South Asian waterways is difficult to achieve. The prospects for peace in South Asian region also require that the construction and reconstruction of hydroborders be undermined new, normative and humane geopolitics of sharing and sustaining a common habitat in place of state-centric, power-political geopolitics of mastering space and resources.

BIBLIOGRAPHY

1. Paasi, A. (2000). *Space, Boundaries and The Territorial Construction of Identities*. Presented at the BRIT-IV, Conference *Geopolitics, Identities and Sustainability*, Feburary 20–26, Chandigarh.
2. Ahmed, I., Dixit, A., and Nandy, A. (1999). Water, power and people: A South Asian manifesto on the politics and knowledge of water. *Water Nepal* **7**: 113–121.
3. Retaille, D. (2000). Geopolitics in history. *Geopolitics* **5**(2), Autumn: 35–51.
4. Lacoste, Y. (2000). Rivalries for territories. *Geopolitics* **5**(2), Autumn: 120–158.
5. Chaturvedi, S. (2003). Towards a critical geography of partition(s): Some reflections on and from South Asia. A guest editorial in *Environ. Plann.: Soc. Space* **21**(2): 148–154.
6. Michel, A.A. (1967). *The Indus Rivers: A Study of the Effects of Partition*. Yale University Press, New Haven, CT.
7. Chapman, G.P. (2000). *The Geopolitics of South Asia: From Early Empires to India, Pakistan and Bangladesh*. Ashgate, Aldershot.
8. Gyawali, D. (2001). Water accords in South Asia: Pluralist politics under monistic design. In: *Peace as a Process: Reconciliation and Conflict Resolution in South Asia*. R. Samaddar and H. Reifeld (Eds.). Manohar, New Delhi.
9. Tarren, I. (2001). Sindhi Association of North America: Letters and petition. March 21. Available: http://www.sanalist.org/letters.htm.
10. Samaddar, R. (1997). Flowing waters and the nationalist metaphors. *Stud. Conflict Terrorism* **20**: 195–206.
11. Krishna, S. (1997). Cartographic anxiety: Mapping the body politic in India. In: *Political Geography: A Reader*. J. Agnew (Ed.). Arnold, London.
12. Chaturvedi, S. (2002). Process of othering in case of India and Pakistan. *J. Econ. Soc. Geogr.* **93**(2): 149–159.
13. Ahsan, A. (1996). *The Indus Saga and the Making of Pakistan*. Oxford University Press, Karachi.
14. Tan, T.Y. and Kudaisya, G. (2000). *The Aftermath of Partition in South Asia*. Routledge, London.
15. The India Express. (2001). After Sindhu, it's Brahmaputra Darshan in Arunachal. February 18.
16. Varshney, A. (1993). Contested meanings: India's national identity, Hindu nationalism, and politics of anxiety. *Daedelus* **122**(June): 227–261.
17. Samaddar, R. (2000–2001). The last hurrah that continues. *Transeuropeannes* **19/20**: 31. 48.
18. Chaturvedi, S. (2001). Mental borders and the prospects of peace in South Asia. In: *Peace as a Process: Reconciliation and Conflict Resolution in South Asia*. R. Samaddar and H. Reifeld (Eds.). Manohar, New Delhi, pp. 61–80.

WATER TRANSFERS

David L. Feldman
University of Tennessee
Knoxville, Tennessee

Water transfers are the diversion, reallocation, or appropriation of surface or groundwater from one watershed, drainage basin, or aquifer to another. All transfers entail water exchange between catchments. They may also move water from one *political jurisdiction* to another, exacerbating their impacts and complicating efforts to resolve stakeholder conflicts that they generate. In addition, transfers may be endeavored through various *means* including pipelines, tunnels, aqueducts, pumping facilities, or even seaborne tankers—an idea contemplated under proposals to transfer water from the Great Lakes to the Middle East.

MOTIVES AND MAJOR ISSUES

Water transfers are nearly as old as civilization. The first recorded large-scale transfers occurred in the ancient Near East. About 4000 years ago, the Sumerians discovered how to divert the Euphrates River (in what is present-day Iraq) to irrigate crops and reduce reliance on the vagaries of seasonal rainfall. Other nearby societies followed suit and built administrative infrastructures to support transfers, leading to speculation that the need to maintain such infrastructures contributed to the growth of political authoritarianism (1,2).

One motive for transfers is to provide sufficient water to satisfy agricultural, industrial, or other needs to regions perceived to have inadequate water supplies. A secondary reason is to promote regional development. This was a major motive for building diversion projects in the southwest United States in the nineteenth century, for example. For the exporting region, the benefits of transfers include economic compensation and better relations with neighbors. However, exporting regions are not always willing partners in transfers, in part because they may have unintended environmental consequences or may be undertaken without adequate consultation (3). Regardless of motive, *five* major issues tend to be associated with transfers:

- *Adequacy of Return Flow.* Most water transfers involve some return flow to the basin of origin. Issues raised by this return flow include how much and when water will be returned. If the return flow is relatively prompt and abundant, harm to ecological resources from low river levels, changes in hydrology, and/or losses to the local economy due to foregone benefits are likely to be less than if return flows are small in volume and/or late.
- *Loss of Water Rights.* Transfers impose impacts on those who have access to water through legal guarantee. These may be comparable from place-to-place, but remedies available under common law differ among regions. In states that have prior appropriation systems, there has tended to be less recourse to deterring transfers thought to be harmful than in states that have riparian rights systems—although water rights issues arise in *both* systems.
- *Adverse Ecological Impacts.* In general, water transfers generate two distinct forms of ecological impact: to in-stream flow (the volume, flow, and quality of water naturally found in rivers and streams prior to the building of impoundments or diversion projects) and biota transfer (the consignment, usually unintended, of exotic fish or other aquatic species from one basin to another, that adversely affect native species' welfare).
- *Water Agency Objectives.* Agencies that manage natural or cultural resources are affected by water transfer proposals in several ways. Often, transfer decisions are undertaken by an agency to meet obligations to promote regional development through water delivery (e.g., the Bureau of Reclamation). Conversely, an agency charged with protecting natural/cultural areas (e.g., National Park Service) may be reluctant to permit transfers.
- *Adverse Social Impacts.* Although difficult to measure, transfers generate significant, often adverse social impacts comparable to those engendered by other resource allocation decisions involving unequal power and authority, representation, and resource access. Specific impacts include changes in lifestyle, reductions in regional income, fewer opportunities for economic development, and abridgment of recreational amenities.

ADEQUATE RETURN FLOW

Water transfers, especially if large, can have profound effects on the quality of water in the basin from whence water is taken, *and* on the quality of water in the receiving basin. Here, we consider the first of these. Many water transfers involve what amount to temporary shifts of water between basins. Used, often untreated water is rerouted back to the basin of origin. Such return flows are often negotiated in agreements to contract transfers and, in riparian rights systems, may be *required* to ensure that natural stream flow is unimpeded (in general, this is one reason why riparian rights systems discourage transfers).

When transfers are permitted in riparian rights systems, return flows are supposed to occur in the vicinity of the same riparian landholding from which water was originally diverted, and its quality is supposed to be *unimpaired*. Historically, such return flows are rarely monitored in many states, do not occur with regularity, are usually easier to ensure when the transfer is between two adjoining watersheds *in* the same state, and are difficult to measure precisely when they occur. Moreover, if they do not occur during periods of low flow, when most needed, the ecological resources of the exporting basin may be harmed.

WATER LAW AND WATER TRANSFERS

All water law systems predicate certain principles regarding the judicious transfer of water from one basin to another. At the outset, two characteristics of American water law, significant for water transfers, must be noted. First, far more attention has been paid to regulating *surface*, as opposed to *groundwater*, transfers because it is difficult to determine exactly what constitutes a groundwater transfer. Underground aquifers are connected to one another and to surface waters in enigmatic ways often hard to prove in courts of law. Negotiating allocation formulas for surface waters usually entails measuring the amount of water to be apportioned at gauging stations at different seasons of the year and, then, limiting how much water an "upper" state or user can draw, *or* guaranteeing a certain amount of water to the "lower" state or user. For aquifers, however, there is no upper and lower user in the same sense as for surface waters (5).

Second, states may not discriminate in the way they manage out-of-state, as opposed to in-state, transfers. Article 1, Section 9 of the Constitution prohibits states from giving preference, in their regulation of enterprises, to businesses operating within their borders over those based in other states. Originally applied to regulation of ports and vessels traversing interstate waters, the clause has been construed by federal courts to mean that Congress has broad plenary power to regulate interstate commerce, manage and use water for navigation, and regulate nonnavigable streams. It has also been construed to mean that water is a commodity in interstate commerce and states cannot discriminate against one another's ability to engage in such commerce. This is an important consideration in efforts by states to regulate interbasin transfers (see "Conclusions").

Riparian Doctrine and Water Transfers

Under riparian law, the predominant legal system in the eastern United States, a landowner has the right to "reasonable" use of the water flowing past her/his property, subject to the rights of other downstream riparian landowners. Water used but not consumed must be returned to the watercourse unimpaired inquality. In general, diverting water from the drainage area of a watercourse *without* return flow is deemed "unreasonable" and, if a downstream riparian complains, can be expressly forbidden (6,7).

In practice, courts have ruled that "reasonable" diversions are allowable under riparian rights systems and typically define such transfers as those that are socially beneficial and impose no harm. Included under this definition are transfers undertaken for the use and benefit of public supply systems. In practice, definitions of "reasonableness" vary from case-to-case and court-to-court, and riparian doctrine and court precedents are unclear on the conditions for mitigating their adverse impacts. How to ensure adequately timed return flow is also uncertain (8,9). Moreover, a "reasonable use" transfer right can be superseded if a subsequent user can demonstrate a need that is more reasonable (10).

Transfers Under Prior Appropriation Doctrine

Appropriation law, the predominant water rights doctrine in 19 western and southwestern states lying west of the 100th meridian, quantifies and prioritizes water rights according to the "first in time, first in right" principle: the first to put water to a beneficial and reasonable use acquires a right superior to later appropriators. This principle grants use of a specified amount of water for a stated beneficial use each year, subject only to the rights of other appropriators (11,12). Land ownership is not the basis for this right: one can acquire it through purchase or other means. Moreover, water rights exist only when water is appropriated (i.e., diverted and used) for a *beneficial* purpose. Historically, this encouraged transfers.

Traditionally, off-stream uses (e.g., irrigation) were considered beneficial, whereas in-stream uses by fish and wildlife were treated as inherently wasteful (7). Senior appropriators could take their full allotment even if insufficient water remained for junior appropriators' needs. This practice served to bias the concept of "beneficial" to mean off-stream, consumptive uses. More important, appropriated waters could be used virtually anywhere, regardless of the distance from a stream. Most appropriation states now define beneficial use by statute. Preferred uses are explicitly defined, and reforms intended to reduce environmental damage through unregulated transfers, such as prescription of minimum stream flows, have been adopted (13,14).

ECOLOGICAL IMPACTS OF TRANSFERS

Dams, impoundments, pipelines, and other "hydrologic alterations" that enable water transfers may lead to severe changes in in-stream flow volumes, velocity, and quality in both exporting *and* destination watersheds, as well as fragmentation and loss of aquatic and riparian habitat (15,16). An important, but often overlooked impact is *biota transfer*, which can also result in both basins of origin and destination.

The Garrison Diversion Unit (GDU), a large transfer project in the Upper Missouri basin authorized by Congress in 1965, exemplifies these problems. The project has been criticized by Canadians due to concerns that return flows from its irrigation subunits could have negative impacts on fisheries and flood stage levels in the Red River flowing north into Manitoba. Canada claims that Missouri River biota would be transferred into Canadian waters in violation of the 1909 *Boundary Waters Treaty* designed to prevent contamination, diversion, or degradation of the transboundary waters shared by both nations. The project's size and scope have been significantly curtailed since the 1990s, but the biota transfer issue, especially, exemplifies two general problems of water transfers applicable elsewhere: (1) the benefits and costs of transfers may differ across affected political jurisdictions, and (2) uncertainty regarding impacts may be exacerbated by international disagreement. Canadian scientists were much more adamant about the risks of biota transfer than U.S. researchers whose work was sponsored by the Bureau of Reclamation (3).

AGENCY MANAGEMENT IMPACTS

Proposals for transferring water from one basin to another may challenge the responsibilities of an agency charged with protecting environmental quality or cultural amenities (e.g., historically significant sites) located within lands embraced by its mandate. By contrast, the provision of water supply may be viewed as so central to the mandated responsibility of another resource-development agency as to justify transfers and even override other considerations.

An example of how these divergent views may play out is offered by the experience of the National Park Service (NPS) and Bureau of Reclamation (BOR), agencies which have, on occasion, seen their interests conflict over water transfers. Under the National Park Service Act of 1916, NPS is responsible for protecting the natural, unimpaired flow of rivers flowing through its parks and other protected areas. By contrast, since 1902, the BOR has been responsible for developing irrigation and other water development projects in the West.

Generally, the NPS tries to balance its claims to needed in-stream flow against the reasonable desire of states (both eastern and western) to protect appropriative and riparian rights. To achieve this balance, the NPS asserts rights "for quantities determined to be the minimum amounts needed to protect the primary purposes" of a given park (12). According to one Interior Department solicitor, "appropriation for authorized federal purposes cannot be strictly limited by what state water law says is a 'diversion' of water or a 'beneficial use' for which water can be appropriated" (17).

In practice, when agencies cannot agree on conditions for transfers or when the economic and political stakes are high, Congress may intervene as it did in 1947 in requiring the NPS to allow the BOR to construct a water diversion project (the Colorado-Big Thompson Project) through a portion of Rocky Mountain National Park. This project's purpose was transferring water from the uppermost reaches of the Colorado River basin, which rises on the state's western slope, to the Platte River Basin northeast of Denver. The project's water storage capacity exceeds 710,000 acre-feet, and it now irrigates 1,000,000 acres (18). To compensate for potential harm, storage reservoirs were constructed for the project, in part, to ensure adequate in-stream flow to Upper Colorado River users and to permit reliable transfers to the eastern slope.

ADVERSE SOCIAL IMPACTS

All water transfers generate social impacts. In the western United States, where water marketing has gained widespread acceptance as a means of diverting water from low- to high-value uses, these impacts can be considerable. Studies show that water marketing in conjunction with transfer can be unpopular, especially in an importing region. In the San Joaquin Valley of California and the Grand Valley of western Colorado, two potential water-exporting areas, residents it was found, were more likely to oppose water transfers than residents of water-importing areas. Concerns included potential impacts to residents' economic welfare and way of life and whether allocations between exporting and importing regions provide for reciprocal benefits such as protecting future needs and preventing adverse environmental impacts (19). Generally, agricultural interests are more likely than other groups to approve of water transfers and to favor marketing as a tool for enabling such transfers to occur.

Despite such concerns, other studies suggest that water transfers may serve as an effective means of diverting water from low- to higher valued needs and even as a means of *restoring* in-stream flow (20). In parts of the American West, for example, including California, Montana, Nevada, Oregon, Texas, and Washington, special markets negotiated by environmental groups have had some success in reducing volumes of water diverted off-stream for irrigation and maintaining stream flows for fish and other environmental and recreational benefits (20: 14). In at least one instance, governmental entities have actually purchased the rights to in-stream water, establishing, in effect, a "reverse" transfer (21: pp. 5–11).

Potential barriers to this practice include the lack of a physical infrastructure for moving water from place-to-place; ambiguities in water law permitting markets, especially under riparian systems; and 'aligning' the price of water between buyer and seller. An underlying investment assumption in such deals is that, over time, transfer of water for restoring in-stream flow will yield a greater benefit than diversion for off-stream uses (22).

CONCLUSIONS—WATER TRANSFER DISPUTES AND THEIR IMPLICATIONS

Since the 1980s, disputes over water supply and allocation have arisen in many parts of the southeast United States, underscoring the fact that *no* region is immune from conflicts over competing demands for water. Sources of these disputes include population growth, urbanization, and competing demands, especially for agricultural irrigation, power, and public uses, as well as debate over how to balance off-stream and in-stream needs (23). One result of these disputes has been plans, real or contemplated, to transfer water from one basin to another.

One of the most contentious of such plans began in 1983 when the city of Virginia Beach, Virginia, indicated its intent to divert water from the Roanoke River to satisfy its growing needs. This led North Carolina to protest the project's potentially adverse impacts on ecological resources and the local economy of that part of the Roanoke Basin in the latter. After a nearly 15-year battle costing upward of $12 million, the project was approved after court challenges failed to demonstrate convincingly adverse effects from plans to divert approximately 60 MGD. A continuing legacy of this dispute, however, is the precedent which it may have established for *additional* diversions, as well as *how* to measure adverse impacts: as present-day losses or as potential future ones (24,25)?

Recently, conflict between Alabama, Florida, and Georgia over water allocation in the Apalachicola–Chattahoochee–Flint basin led some Atlanta-area officials—contemplating an eventual ceiling on withdrawals from a tri-state water compact—to look elsewhere for new supply. Among options explored, beginning in 1998, was the possibility of diverting water from the Tennessee River in Southeast Tennessee via pipeline. The mere consideration of this alternative partly convinced officials in Tennessee of that state's vulnerability to outside demands and led the legislature to pass an *Interbasin Water Transfer Act* in 2000 to regulate transfers of water from a basin originating in Tennessee to any other basin—in or outside the state (26–28).

These and other disputes over water transfers are instructive for one major reason: they underscore that decisions to undertake, or to deny, transfers, though partly based on perceived needs and environmental concerns, often result from the distribution of power and authority. Worldwide, a significant impediment to different jurisdictions reaching common accord on water allocation, and coming into conflict over transfers, is the perception by protagonists that they lack adequate influence over the outcome of negotiations or the ability to leverage decisions.

Oftentimes, this lack of influence is manifested in a lack of legal or economic tools (e.g., equitable water markets, effective regulatory frameworks, infrastructure to allocate water and manage competing demands) to resolve disputes. Just as inequalities of power and resources have been recognized as inhibiting durable solutions to water conflicts between, for example, Israel and Jordan, and Israel and the Palestinian West Bank (especially when compounded by secrecy and fundamental value differences), they may also inhibit compromise and accord among states within a nation (29–32). Likewise, when countries—like states within a country—find the means to cooperate, it is often the result of developing conjoint, equitable cooperative management schemes that are adequately funded, as in the Mekong River basin states of Thailand, Cambodia, and Vietnam, for example (33). The challenges facing transfer decisions are formidable, but opportunities for working toward common solutions are plentiful.

BIBLIOGRAPHY

1. Fradkin, P.L. (1996). *A River No More: The Colorado River and the West*. University of California Press, Berkeley, CA.
2. Bright, J. (2000). *A History of Israel*. John Knox Press, Westminster.
3. Feldman, D.L. (1991). The great plains garrison diversion unit and the search for an environmental ethic. *Policy Sci.* **24**(1): 41–64.
4. Jones, J.D., Sikora, V.A., and Woodward, J. (1983). *Study of Tennessee Water Resources Law: Legal Considerations for Effective Water Management Under Conditions of Shortage*. Research Report No. 97, Water Resources Research Center, The University of Tennessee, Knoxville, TN.
5. Grant, D.L. (1998). Introduction to interstate allocation problems. *Waters and Water Rights*. R.E. Beck (Ed.). Michie, Charlottesville, VA.
6. Laitos, J.G. and Tomain, J.P. (1992). *Energy and Natural Resources Law in a Nutshell*. West Publishing, St. Paul, MN.
7. Butler, L.L. (1990). Environmental water rights: an evolving concept of public property. *Va Environ. Law J.* **9**: 323–327.
8. Tarlock, A.D. (1997 and 1998 update). *Law of Water Rights and Resources*. Clark, Boardman and Callaghan, Deerfield, IL.
9. Wright, K.R. (Ed.). (1998). *Water Rights of the Eastern United States*. American Water Works Association, Denver, CO.
10. Samson, S.A. and Bacchus, S. (2000). Point—water marketing: the other side of the coin. *Water Resour. Impact* **2**(6) November: 15–16.
11. Pearson, R.P. (2000). Managing water scarcity—southwestern style. *Troubled Waters—Managing a Vital Resource—Global Issues Electron J.* March.
12. Gillilan, D.M. and Brown, T.C. (1997). *Instream Flow Protection: Seeking a Balance in Western Water Use*. Island Press, Washington, DC.
13. Tarlock, A.D., Corbridge, J.N., Jr., and Getches, D.H. (1993). *Water Resource Management: A Casebook in Law and Public Policy*, 4th Edn. The Foundation Press, Westbury, NY.
14. Kenney, D.S., McAllister, S.T., Caile, W.H., and Peckham, J.S. (2000). *The New Watershed Source Book—A Directory and Review of Watershed Initiatives in the Western United States*. Natural Resources Law Center, University of Colorado School of Law, Boulder, CO, April.
15. Pringle, C.M. (2000). Threats to U.S. public lands from cumulative hydrologic alterations outside of their boundaries. *Ecol. Appl.* **10**(4) August: 971–989.
16. Rosenberg, D.M., McCully, P., and Pringle, C.M. (2000). Global-scale environmental effects of hydrological alterations: Introduction. *BioScience* **50**(9) September: 746–751.
17. *Decisions of the Department of the Interior—Federal Water Rights of the National Park Service, Fish and Wildlife Service, Bureau of Reclamation and the Bureau of Land Management*. (1979). Office of the Solicitor, U.S. Department of the Interior, Washington, DC, June 25.
18. U.S. Bureau of Reclamation. (1957). *Colorado-Big Thompson Project,* Vol. 1. *Planning, Legislation, and General Description*. U.S. Department of the Interior, Denver, CO.
19. Keenan, S.P., Krannich, R.S., and Walker, M.S. (1999). Public perceptions of water transfers and markets: describing differences in water use communities. *Soc. Nat. Resour.* **12**: 279–292.
20. Landry, C.J. (2000). Agriculture and water markets in the new millennium. *Water Resour. Impact.* **2**(3) May: 13–15.
21. *Water in the West: Challenge for the Next Century*. (1998). Report of the Western Water Policy Review Advisory Commission. Published by NTIS: Springfield, VA, June.
22. *Restoring the Waters*. (1997). Natural Resources Law Center, the University of Colorado School of Law, Boulder, CO, May.
23. Arrandale, T. (1999). The eastern water wars *Governing* (August): 30–34.
24. Leahy, T. (1998). Lake Gaston project resolves water supply shortage. *Public Works: Eng. Constr. Maint.* **129**(7) June: 44–48.
25. (1995). Hopes dimming for pact to settle Gaston dispute. *Waterweek* June 19: 7.
26. (1997). *Atlanta Regional Water Supply Plan Update*. Adopted December 3, Atlanta Regional Commission—unpublished report.
27. Feldman, D.L. and Elmendorf, J. (2000). *Water Supply Challenges Facing Tennessee: Case Study Analyses and*

the Need for Long-Term Planning. Prepared for the Environmental Policy Office, Tennessee Department of Environment and Conservation, by the Energy, Environment and Resources Center, University of Tennessee, Nashville, TN, June.

28. (2000). *Inter-Basin Water Transfer Act*. H.B. 3002 & S.B. 3074. General Assembly of Tennessee.
29. Hassoun, R. (1998). Water between Arabs and Israelis: reaching twice-promised resources. In: *Water, Culture, and Power: Local Struggles in a Global Context*. J.M. Donahue and B.R. Johnston (Eds.). Island Press, Washington, DC, pp. 313–338.
30. Elmusa, S.S. (1995). Dividing common water resources according to international water law: the case of the Palestinian Israeli waters. *Nat. Resour. J.* **35**(Spring): 223.
31. Morris, M.E. (1993). *Dividing the Waters: Reaching Equitable Water Solutions in the Middle East*. RAND/P-7840. RAND Corporation, Santa Monica, CA.
32. Morris, M.E. (1992). *Poisoned Wells: The Politics of Water in the Middle East. Reprinted from Middle East Magazine*, RAND/RP-139. RAND Corporation, Santa Monica, CA.
33. Browder, G. and Ortolano, L. (2000). The evolution of an international water resources management regime in the Mekong River Basin. *Nat. Resour. J.* **40**(3): 499–532.

RESERVED WATER RIGHTS FOR INDIAN AND FEDERAL LANDS

Eve Woods
Denver, Colorado

Most rights to the use of water are created under state law. In general, states east of the Missouri River follow riparian law and the nineteen contiguous western states follow the prior appropriation doctrine or a hybrid of the two. The federal government has a strong policy of deference to state water law. However, some federal lands and tribal lands have a unique type of water right that occurs under federal law. These "reserved water rights" attach to federal lands that the Government kept or set aside for particular purposes and to Indian reservation lands that are held for the benefit of a federally recognized American Indian tribe.[1]

Historically, the United States set aside land from the public domain for federal purposes such as parks and national forest without taking the actions necessary to establish a water right under state law. It also negotiated treaties with American Indian Tribes that failed to address the question of a water supply for the lands set aside for the tribe. The doctrine of reserved water rights is based on the contract law theory that the federal government intended to reserve adequate water for itself when it set aside land for federal purposes, even though it did not say so expressly in the document setting aside the land (1). Likewise, when an Indian tribe reserved some of its land for itself while ceding the rest to the United States by treaty, this doctrine recognizes that the tribe must have intended to reserve some waters for itself, to make the land it reserved useable.

[1]A list of federally-recognized tribes is published annually by the Department of the Interior in the Federal Register.

The United States Supreme Court first recognized a reserved water right in the 1908 case of "Winters v. United States." The Court held that the Fort Belknap Tribe in Montana and the United States had impliedly reserved water for the Fort Belknap Reservation via their treaty of 1888. The Tribe was awarded 5000 acre feet of water from the Milk River for irrigation use on the Reservation, with a priority date of 1888, which gave the tribal water right priority over the rights of the upstream settlers who had begun farming the area in 1898. Reserved water rights are also called "Winters rights" after the name of that case (2).

From 1908 to 1963, the reserved rights doctrine was applied only to tribal water rights, and rarely at that. In 1963, in *Arizona v. California*, the reserved rights doctrine was extended to provide water rights for nontribal federal lands as well.

Reserved water rights share some characteristics with both riparian water rights and prior appropriation water rights. Reserved rights are riparian in the sense that they extend to the waters adjacent to or beneath the lands that they serve (3). They are riparian because they exist independent of whether the water has been diverted or applied to beneficial use. The full quantity of the reserved right develops in lieu of use and cannot be lost by nonuse. However, unlike riparian rights, a reserved right has a priority date. It is administered within the prior appropriation system—"first in time is first in right"—and so has the power to call out junior users in times of water shortage.

QUANTITY OF THE RIGHT

The quantity of a reserved water right is "that amount necessary to fulfill the purposes of the reservation" (4). It encompasses both the current and the future needs for water on the reserved lands (5).

This amorphous quantification standard has been difficult to implement. For federal lands set aside under an Act of Congress, the Congressional purpose stated or implied in that Act is the "purpose of the reservation" used for measuring the water right (6). The court uses expert testimony to determine the amount of water necessary to preserve wildlife or habitat, or to meet whatever purpose applies. If the land was set aside for more than one purpose, there can be only one reserved water right, and it will only be enough water to meet that primary purpose of the reservation (7).

Until recently, quantification of tribal reserved water rights was usually based on the PIA standard, i.e., enough water to grow crops on the "practicably irrigable acreage" of the Indian reservation, which was because many treaties stated that the purpose of the reservation was to develop a viable agricultural economy, intending that the tribal members would assimilate to the lifestyle of the nonIndian farmers who were then settling the West. To determine the number of practicably irrigable acres, courts look to economic and engineering experts to determine for how many acres the economic benefits of growing crops will at least equal the costs of transporting and applying the water to the land (8).

More recently, an Arizona court rejected the PIA standard in favor of a "homeland" standard for quantifying tribal reserved rights (9). The purpose of the reservation in that case was not agriculture, but was to provide a permanent homeland for the tribe. The court held that determining the minimal amount of water necessary to accomplish the homeland purpose—both now and in the future—would require analysis of six diverse factors concerning the tribe's cultural practices and the potential for economic development on the reservation. That new test has not yet been applied by any court.

The quantities of reserved rights vary greatly. They range from one thousand acre feet to hundreds of thousands of acre feet. Many tribal and federal reserved rights have yet to be quantified. The eventual use of those reserved rights will have a potentially enormous effect on the amount of water available to junior users on those streams.

McCARREN AMENDMENT

Although they are created under federal law, reserved water rights can be adjudicated by state courts (43 U.S.C. § 666). In the McCarran Act, the United States Congress waived federal sovereign immunity from suit to allow state courts to join federal claims into a comprehensive adjudication of a river system or other source. This 1952 Act prompted many western states to revise their statutes to allow for general stream adjudications, thereby forcing the quantification of reserved rights claims. Such adjudications are complex, requiring inter alia, a field analysis of each water claim in the basin to determine the quantity and priority date of each right.

PRIORITY DATE

The priority date of a reserved right is the date of the federal act or tribal treaty reserving the land, which is usually a relatively early right that will have priority over all water users who settled the area after that date. Under certain circumstances, a tribe may have a "time immemorial" priority date for its reserved right. This water right is, by definition, the earliest right on the stream (10).

The Pueblos of New Mexico also have time immemorial water rights. However, Pueblo water rights are not reserved rights. The quantity of a Pueblo water right is limited to past beneficial use and cannot be expanded to meet current or future Pueblo needs (11). Like reserved rights, Pueblo rights cannot be lost for nonuse.

BIBLIOGRAPHY

1. *United States v. Winans*. (1905). 198 U.S. 371.
2. *Winters v. United States*. (1908). 207 U.S. 564.
3. *In re General Adjudication of Gila River*. (1999). 989 P.2d 739 (Ariz).
4. *In re Big Horn*. (1988). 753 P.2d 76 (Wyo), *aff'd sub nom, Wyoming v. U.S.* (1989). 492 U.S. 406.
5. *Arizona v. California*. (1963). 373 U.S. 546.
6. *United States v. New Mexico*. (1978). 438 U.S. 696.
7. *Cappaert v. United States*. (1976). 426 U.S. 128.
8. *Arizona v. California*. (1963). 373 U.S. 546.
9. *In Re General Adjudication of Gila River*. (2001). 35 P.2d 68 (Ariz.).
10. *United States v. Adair*. (1984). 723 F.2d 1394 (9th Cir. 1983), *cert. denied*, 467 U.S. 1252.
11. *New Mexico v. Aamodt*. (1976). 537 F.2d 1102 (10th Cir.).

READING LIST

Tarlock, Corbridge, and Getches. (2002). *Water Resource Management: A Casebook in Law and Public Policy*, 5th Edn. Foundation Press, New York, pp. 839–909.

Colby, B. (2004). *Tribal Reserved Water Rights*.

WETLANDS POLICY IN THE UNITED STATES: FROM DRAINAGE TO RESTORATION

ANDREA K. GERLAK
Columbia University
New York, New York

Historically, swamps have been regarded as wastelands, evoking reactions of "disgust at their sight and smell, fear of malaria and yellow fever, and unease about rich resources running to waste within them." George Washington's leadership in the Great Dismal Swamp Company characterized early notions of swamps. The Company, designed to drain Virginia's swamp for timber production, promoted the notion that swamps were useless lands (1). The Swamp Land Acts of the 1850s are a further illustration as the federal government encouraged states to "reclaim" nearly 6.5 million acres of wetlands.

By the beginning of the twentieth century, perceptions began to change. Sportsmen and hunters called attention to declining number of waterfowl and initial federal efforts regarding wetlands focused almost exclusively on waterfowl habitat. In 1903, President Theodore Roosevelt established the first National Wildlife Refuge by executive order to protect endangered species on Pelican Island, Florida. With the Migratory Bird Hunting Stamp Act of 1934, Congress established a special fund to finance wetland acquisitions for duck habitat.

Despite such efforts, however, water resources continued to be managed largely for economic development purposes. The U.S. Army Corps of Engineers (Corps) became a key instrument in "modernization" during America's era of industrialization (2), and exerted broad discretion by undertaking improvements in the navigability of the nation's waterways under the 1880 Rivers and Harbors Act and the 1899 Refuse Act. Many federal programs provided incentives for wetland conversion, including subsidies for agriculture, grazing and forestry, reservoir construction, and flood relief and insurance.

By the 1960s, environmental values emerged to challenge the old paradigm. Public alarm over declines in many animal species, including the symbolic bald eagle, led to endangered species legislation in the 1960s and 1970s. The term "wetland" replaced "swamp." The Army Corps inched toward reform as well. In response to an emerging concern for water pollution, the Corps began to revise its permit regulations to take into account "public interest review," beyond pure economic impacts. Based on *Zabel* v. *Tabb* [430 F.2d 199 (5th Cir. 1970), cert. denied, 401 U.S. 910 (1971)] in 1971, the first federal court case over wetlands, the Corps was granted "ecological authority" to weigh noneconomic factors in reviewing and denying permit applications under its original 1899 authority. Real change occurred from passage of the Federal Water Pollution Control Act of 1972. Known as the Clean Water Act (CWA), this legislation marked the beginning of broad-based wetlands protection in the United States. Wetlands came to be valued as biologically diverse ecosystems providing great public benefits such as improved water quality and protection from floods. This marked the beginning of federal efforts at wetlands protection. During the next 30 years, wetlands management would become highly politicized (3).

WETLANDS, LEGISLATION, AND LITIGATION

By the Federal Water Pollution Control Act amendments of 1972, Congress established a comprehensive program of regulations and permits to control water pollution. Section 404 of the CWA marked the first federal policy designed to preserve wetlands. Section 404(b)(1) states that "the guiding principle should be that degradation or destruction of special sites may represent an irreversible loss of valuable aquatic resources." In less than two pages, Congress established a shared authority between the Army Corps and the newly created U.S. Environmental Protection Agency (EPA). As a political compromise, the Corps was designated as the lead agency to administer the permit system for discharging dredged and fill materials into navigable waters. The Corps' preliminary permit decisions would then be reviewed by the EPA that had authority to veto the Corps' decision. This compromise, although appearing innocuous, would complicate and burden Section 404 implementation.

A historic tradition in water resource management, the Corps embodied the older distributive/allocative model of policy-making, referred to as the "pork barrel." Such an approach allowed a high degree of local and state agenda-setting autonomy. The newly created EPA, meanwhile, embraced the new, centralized regulatory model, referred to as "command and control," and focused primarily on water quality. Given the distinct organizational culture of the agencies (4) implementation of Section 404 immediately was characterized by conflict between the EPA and Corps. Generally, the Corps preferred a limited approach, whereas the EPA pushed for expansive authority to regulate all wetlands in the United States.

The federal judiciary would play a powerful role as environmental groups challenged the Corps' implementation and regulatory practices. By *Natural Resources Defense Council* v. *Calloway* (392 F. Supp. 685) in 1975, the U.S. Supreme Court pushed the Corps towards broader wetlands protection and forced the Corps to revise its regulations to include significant intrastate waters, in addition to interstate waterbodies. This case sparked debate in Congress and by the 1977 amendments to the CWA, Congress refined and expanded its definition of wetlands demonstrating a clear congressional commitment to wetlands protection. In response to industry concerns, however, the amendments allowed exemptions for ongoing normal ranching, farming, and silviculture.

President Carter attempted to create a more unified wetland policy by issuing Executive Order 11990 (42 Fed. Reg. 26 (1977). In calling on all federal agencies to minimize wetland loss on land managed by the federal government, this action represents the highest administrative action ever taken to protect wetlands. Two year later, Carter's Attorney General Benjamin Civiletti gave greater weight to the EPA in wetlands implementation [43 Op. Attorney General 15 (1979)]. In attempting to resolve interagency disputes, he determined that the EPA enjoyed ultimate authority to decide the scope of wetlands regulation because that agency carried the responsibility for implementing the other sections of the 1972 CWA.

The courts pushed wetlands protection even further with *U.S.* v. *Riverside Bayview Homes* (474 U.S. 121) in 1985. Here, the U.S. Supreme Court issued a landmark decision in wetlands policy. By defining the waters covered broadly, the Court ruled that the Corps and EPA could regulate wetlands even when no surface water or other apparent hydrologic connection exists between the wetlands and an adjacent waterbody.

THE POLITICS OF WETLAND MANAGEMENT

At about the same time that it enacted the CWA, Congress passed legislation that brought still more federal agencies and departments into the wetlands policy arena. A 1970 statute, The Water Bank Act, created a federal program that paid farmers to preserve wetlands habitat for waterfowl. Administered by the Department of Agriculture's Soil Conservation Service (now the Natural Resources and Conservation Service), this program has grown significantly during the past 30 years. In 1973, still another dimension was added when Congress passed the Endangered Species Act (ESA). Administered by the U.S. Fish and Wildlife Service (FWS), this highly controversial statute added another federal player to the list of agencies implementing the wetlands protection policy. Consultation with the FWS became the norm because so many wetlands are home to endangered and threatened species. In some situations, the FWS has been given the authority to manage sensitive wetland areas. For example, in 1973, more than 107,000 acres of the Great Dismal Swamp was deeded to the Department of the Interior for the creation of the Great Dismal Swamp National Wildlife Refuge. How far we have come from the days of George Washington's Great Dismal Swamp Company. Thus, during the last 30 years, the FWS has become an integral component of national wetlands policy.

By the end of the "environmental decade" of the 1970s, at least four federal agencies had major responsibilities for protecting wetlands. A number of other federal agencies, including the Forest Service, the Bureau of Reclamation, and the Bureau of Land Management, had ancillary responsibilities. Table 1 provides a list of the primary federal agencies involved in wetlands protection today.

Other legislation added several more bureaucratic participants to an already crowded landscape. For example, the Farm Service Agency in the Department of Agriculture came to administer and enforce the Swampbuster provision contained in the 1985 Farm Bill. This disincentive program denies federal farm program benefits to producers who plant an agricultural commodity on converted wetlands. Officials of the Corps believe that Swampbuster has significantly reduced agriculture-related Section 404 permit applications (6). Environmentalists, however, have been critical of the program and the Corps' inadequate enforcement of it (7). Nonetheless, beginning in the 1980s, falling agricultural prices coupled with this shift in federal wetland policies helped reduce agriculture-related wetland loss. According to the U.S. Department of Agriculture analysis, wetland conversions on agricultural lands nationwide slowed to less than 30,000 acres per year between 1987 and 1991, down from 51,000 acres per year between 1982 and 1987 (8, p. 234). However, the building boom of the 1980s increased urban wetlands conversion to 89,000 acres per year, or 57% of the total gross conversion (9, p. 21).

Table 1. Primary Federal Agencies Involved in Wetland Protection[a]

Federal Agency	Role in Wetland Protection
Department of Defense, Army Corps of Engineers	Administers wetland permitting and enforcement under Section 404 of the Clean Water Act
Environmental Protection Agency	Performs oversight over Corps' permitting process and enforcement
Department of the Interior, Fish and Wildlife Service	Reviews Section 404 permit applications, makes recommendations to the Corps, and promotes voluntary restoration programs
Department of Agriculture, Farm Service Agency	Manages voluntary programs to help protect and restore wetlands associated with farmland
Department of Agriculture, Natural Resources Conservation Service	Delineates wetlands under farm bill legislation and provides technical and financial assistance to landowners to restore and enhance wetlands
Department of Commerce, National Oceanic and Atmospheric Administration	Initiates wetland restoration activities associated with marine ecosystems

[a]Reference 5.

In the modern era, implementation of Section 404 has been characterized by heightened polarization of competing interests and politicization of science. The polarization of interests can be traced to the wetlands manual debate of the late 1980s and early 1990s. In 1987, agencies moved to adopt a single manual to delineate wetlands during the permit process. Economic interests, led by the Farm Bureau and the National Wetlands Coalition (NWC), vehemently opposed such efforts. Mobilized under the NWC, this coalition of agriculture, oil, mining, utilities, and real estate developers, strove to "work with Congress and the Administration for legislative reform of regulatory improvements to the federal wetlands permitting program" (10). By drafting legislation and lobbying members of Congress, NWC seeks to protect private property rights and streamline the regulatory process to offer greater regulatory flexibility.

The White House responded with a proposal that restricted the Corps' and EPA's jurisdiction by narrowing the definition of a wetland. Environmental critics charged that the Bush Administration's 1991 manual would write off some 20 to 40 million acres of wetlands (11, p. 5). Field tests of the manual indicated that 30–80% of land delineated as wetlands in the 1989 manual were excluded by the 1991 manual (12). The debate over a politically acceptable definition of wetlands was never resolved. Implementing agencies returned to the 1987 wetlands manual for the wetlands delineation process. Congress eventually moved the wetland delineation issue out of the political arena and into the scientific one. In 1992, Congress authorized a study of wetlands by the National Research Council. Released in 1995, the National Research Council's report, titled "Wetlands: Characteristics and Boundaries," contained a scientifically grounded wetland definition and more than 80 recommendations for improving federal wetlands regulation (13). The report still awaits serious attention and implementation.

Most recently, the debate has focused on the 2001 Supreme Court decision, *Solid Waste Agency of Northern Cook County v. U.S. Army Corps of Engineers* [121 S.Cf. 675]. In this case, a county solid waste agency sought permission to develop a disposal site on 17 acres of land that contained isolated, seasonal ponds used by migratory birds. By a 5–4 decision, the Court denied CWA protection to isolated, nonnavigable ponds and wetlands contained within a single state under the Migratory Bird Rule. In doing so, the Court "reversed elements of more than 15 years of federal jurisdiction when it ruled that the federal government overstepped its authority in regulating certain isolated wetlands" (14, p. 2). Regional offices of the Corps and EPA were then instructed to withhold clean water protection from those types of isolated waters and to seek guidance from headquarters in determining whether to protect other small intrastate streams and waterways that enjoy federal protection (15). Both environmental and industry representatives agree that there is a need for congressional guidance following this ruling. Of course, both sides have radically different notions of what that response should entail.

THE MODERN DEBATE: WETLANDS CONSERVATION AND RESTORATION

Today, the wetlands debate has been framed by a "no-net-loss" goal. Adopted from the National Wetlands Policy Forum, a task force convened by the Conservation Foundation at the request of the EPA and headed by New Jersey Governor Thomas H. Kean in 1988, "no-net-loss" means conservation of wetlands whenever possible and that acres converted should be offset through restoration and creation, thereby, maintaining or increasing the wetland resource base (16). Every president since George H. Bush has adopted this approach, and in doing so, the debate has been framed around acreage. As such, it is the numbers that matter. Wetland values and functions become less important. Recently, the National Academy of Science's (NAS) Committee on Mitigating Wetlands Loss reported that it is "not convinced that the goal of no net loss for permitted wetlands is being met for wetland functions" (17, p. 3). According to the NAS, insufficient data exists to determine if the goal was even being met.

What has been determined is that wetland conversion rates have dropped during from 458,000 acres per year during the 1950s–1970s to 58,500 acres per year in the 1980s–1990s. This amounts to an 80% reduction in average annual loss from the prior decade (18, p. 9). Figure 1 displays wetlands loss from the 1950s to the 1990s. Conservation programs have helped stem wetland loss nationally. Recent reauthorization of the North American Wetlands Conservation Act where the federal government matches donations for state wildlife agencies, conservationists, landowners, and sportsmen demonstrates this. So too does reauthorization of the Wetlands Reserve Program, which provides financial incentives to farmers to restore and protect wetlands through long-term easements. Congress has elected the conservation approach as opposed to amending the Section 404 permit process. Employing incentive programs like these, as opposed to regulation, are understandably more popular with the public than the Section 404 permit process.

By focusing on wetlands acreage framed by the "no-net-loss" goal, restoration strategies have emerged as the preferred policy solution, attempting to reconcile economic growth and wetland protection. Restoration efforts are becoming increasingly popular as a strategy to mitigate

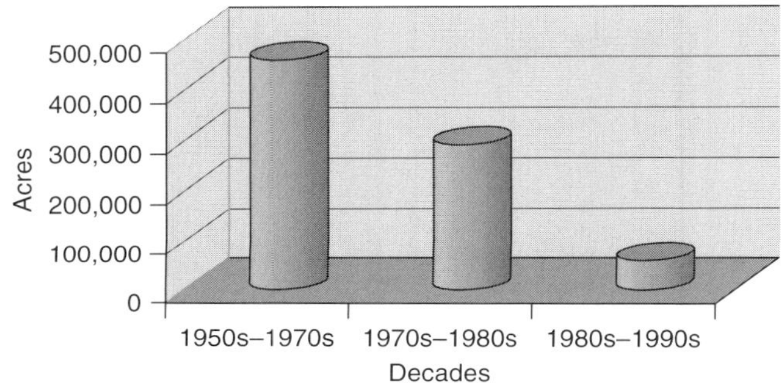

Figure 1. Wetlands loss figures for the mid-1950s to mid-1970s, mid-1970s to mid-1980s, and 1986–1997 (18).

wetland losses and counter past development. Restoration of the Florida Everglades, a joint federal–state effort at an estimated cost of $8 billion over the next two decades, represents the largest ecological restoration project ever attempted. Despite restorative successes of coastal areas (19), however, some scientists are increasingly concerned that hard to create wetlands like bogs and fens are being replaced with common, easy to create wetland types such as cattail marshes (20). Moreover, the National Research Council reports that restored wetlands often cannot re-create natural levels of biodiversity and sustainability (21). In the coming years, the debate over wetlands protection is likely to focus as much on the *quality* of wetlands as it does now on the *quantity* of acres restored.

CONCLUSION

In summary, by the 1960s and 1970s, environmental values challenged wetland drainage and destruction practices. Passage of the Clean Water Act marked the first significant federal efforts at wetland protection. Wetland management during the past 30 years has been characterized by litigation, an expansion of federal agency involvement, polarization of interest, and a politicization of science. We have not yet achieved the "no-net-loss" goal, but conservation programs and restoration efforts have helped to stem wetland loss nationally. Will they be enough to protect the wetlands that remain in the United States? Only time will tell.

BIBLIOGRAPHY

1. O'Hagan, K. (1991). Pumping with the intent to kill: Evading wetlands jurisdiction under Section 404 of the clean water act through draining. *DePaul Law Rev.* **40**: 1059–1080.
2. Feldman, D.L. (1991). *Water Resources Management: In Search of an Environmental Ethic*. Johns Hopkins University Press, Baltimore, MD.
3. Gerlak, A.K. (1996). Federal wetland policy: Is this policy all wet? In: *At the Nexus: Science Policy*. D.L. Soden (Ed.). Nova Science, Comack, NY.
4. Clarke, J.N. and McCool, D.C. (1996). *Staking Out the Terrain: Power and Performance Among Natural Resource Agencies*. State University of New York Press, Albany, NY.
5. U.S. General Accounting Office. (1998). *Wetlands Overview: Problems with Acreage Data Persist*. General Accounting Office, Washington, DC.
6. Zinn, J. and Copeland, C. (1996). *Agricultural Wetlands: Current Program and Legislative Proposals*. CRS Report for Congress 96-35 ENR, The Library of Congress, Congressional Research Service, Washington, DC.
7. Environmental Working Group. (1995). *"Taxing" the Taxpayer: Wetlands and Farm Subsidies in Perspective*. Environmental Working Group Policy Memorandum, Washington, DC.
8. Heimlich, R.E. (1994). Costs of an agricultural wetland reserve. *Land Econ.* **70**(2): May.
9. Heimlich, R.E., Wiebe, K.D., Claassen, R., Gadsby, D., and House, R.M. (1998). *Wetlands and Agriculture: Private Interests and Public Benefits*. United States Department of Agriculture, Economic Research Service, Washington, DC.

10. National Wetlands Coalition. *Mission Statement* at http://www.thenwc.org/home.htm, accessed September 5, 2003.
11. Keefer, M. (1991). *Wetlands: Going, Going, Going*. National Wildlife Federation Leader, Reston, VA.
12. Environmental Defense Fund and World Wildlife Fund. (1992). *How Wet is a Wetland? The Impacts of the Proposed Revisions to the Federal Wetlands Delineation Manual*. Environmental Defense Fund and World Wildlife Fund, New York and Washington, DC.
13. National Research Council. (1995). *Wetlands: Characteristics and Boundaries*. National Academy Press, Washington, DC.
14. National Wetlands Newsletter. (2001). *The SWANCC Decision*. Environmental Law Institute, Washington, DC, January–February.
15. U.S. Environmental Protection Agency. (2003). *Administration to Reaffirm Commitment to No Net Loss of Wetlands and Address Approach for Protecting Isolated Waters in Light of Supreme Court Ruling on Jurisdictional Issues*. Environmental Protection Agency, Washington, DC.
16. The Conservation Foundation. (1988). *Protecting America's Wetlands: An Action Agenda—The Final Report of the National Wetlands Policy Forum*. Conservation Foundation, Washington, DC.
17. National Research Council. (2001). *Compensating for Wetland Losses Under the Clean Water Act*. National Academy Press, Washington, DC.
18. Dahl, T.E. (2000). *Status and Trends of Wetlands in the Conterminous United States, 1986–1997*. United States Department of the Interior, Fish and Wildlife Service, Washington, DC.
19. Kukoy, S.J. and Canter, L.W. (1995). *Mitigation Banking as a Tool for Improving Wetland Preservation via Section 404 of the Clean Water Act*. The Environmental Professional 17, Gainesville, FL.
20. Hunt, R.J. (1996). *Do Created Wetlands Replace the Wetlands that are Destroyed*. U.S. Department of the Interior, U.S. Geological Survey, Madison, WI.
21. National Research Council. (2001). *Compensating for Wetland Losses under the Clean Water Act*. National Academy Press, Washington, DC.

WATER HISTORY, ART, AND CULTURE

CURIOUS USES OF AGRICULTURAL WATER IN THE WORLD

Eric P. Mollard
Institut de Recherche pour le Développement
France

The agricultural use of regularly flooded zones along large eastern rivers, from the Nile to the Yangtze, goes back to the births of cities and almost to the birth of agriculture. Elsewhere, wetland farming has been heterogeneous. Initially, marshes, flood plains, and coastal areas were sparsely reclaimed, whereas nearby identical wetlands were forsaken and are still unchanged today. At a later time, public intervention based on resource and technology resulting from the Industrial Revolution undertook a lot of total draining. The territorial gain for modern agriculture was accompanied by a dramatic loss of specific advantages such as incomparable aquatic and land biodiversity which can also serve agriculture.

Wetlands show two contradictory characteristics: they discourage and they attract. They are inhospitable because of the fear of marsh fevers, wizards, and legendary animals. More prosaically, great land fertility and moisture are beneficial at the cost of hard labor, not in the direction of general draining, but in that of the work for horticulture. The promise of such lands that have high productivity and are healthier when water is under control, could overcome the fears, real or imaginary, of the wilderness.

The physical duality of high productivity and unhealthiness was translated over time by a social duality, especially between natives and foreigners. Natives supplemented their livelihood by fishing and collecting reeds in a common resource land. The cases of fever were undoubtedly used to keep young people or free riders from defying the elders' authority and to discourage foreigners from seeking free lands. Due to a spooky atmosphere on foggy days, the local populations ended up believing in evil spells launched by their ancestors. Where the foreigners succeeded in settling, one often noted a dramatic social reversal. Indeed, through intense labor and the great fertility of the marsh soils, the last ones to settle became the first. The Aztecs in old Mexico epitomized such history. In the twelfth century A.D., a tribe, originating from the arid north and vassals of the kingdoms occupying the high plateau, could settle only on the immense lake that occupied most of the Valley of Mexico and had a few small islands full of snakes. Tenaciously, the Aztec people created a lake-based agriculture, the chinampas, that enabled them to build one of the largest cities in the world and to subjugate the neighboring kingdoms later.

The relation between wetlands, foreigners, a power that can become an empire, old legends, and current speeches from many stakeholders confers an integral social history on this environment. However, it would be erroneous to believe that the curious uses of water result today only in some anecdotes and relics. Two examples suffice to show how this history is not finished. The mulberry tree dikes on the Pearl River in China are a contemporary and extremely long-lived peculiar use of water. The second example is even more dynamic. The delta of the Chao Phraya River, which used to be a rice bowl in Thailand, is becoming a second Holland; it has horticulture for export, thanks to the technique of raised gardens.

The peculiar uses of water in agriculture prove to be invaluable elements in understanding the relation between water and societies, as well as in designing alternatives to commonplace agriculture instead of the common thought of large polder and irrigation schemes. In this article, we distinguish two types of uses: "raised" style and "hollow" style. The first refer to peasant reclamation, on an individual basis but as a large extension, whereas the second deal with some types of irrigation. The most curious feature of the story is that a local peculiarity is not so peculiar on a planetary level. Inventive men have found similar solutions on every continent.

THE AGRICULTURE OF HOLLOWS

The agriculture of hollows is not a simple adaptation to marshes and lowlands. It embraces the entire range of intensification from floating gardens to cultivated ponds.

Floating Agriculture

Nowadays, vegetables are still grown on natural, floating vegetation rafts or rafts man-made of reed bundles. They are used sometimes as seedbeds before firm ground transplanting. Floating gardens are located on some large lakes in Asia, Latin America, and Africa. Those of Lake Inlé were rediscovered when the border opened in Myamar, whereas those of Lake Dahl, in Kashmir, are less known. Several Chinese lakes, such as Lake Tien close to Kunming in Yunnan, would have witnessed this type of water farming. On Lake Titicaca in the Andes, the rafts of reed are platforms with fishing holes, as in an ice barrier, several thatched houses, some animals, and a small garden. It is not exactly a floating garden because agricultural activities take place on firm ground. On Lake Chad, the natural floating islands are used by fishermen who must continually move under penalty of sinking. In Yap, in Micronesia, the taro is sometimes cultivated on floating gardens cut in the marsh.

Former Chinese authors distinguished two types of rafts: wooden platforms and rafts made of reeds. Both were used for growing rice on the marshy edges of rivers, where a dike would have been difficult to build. Nowadays, the wooden rafts have disappeared, and the floating gardens

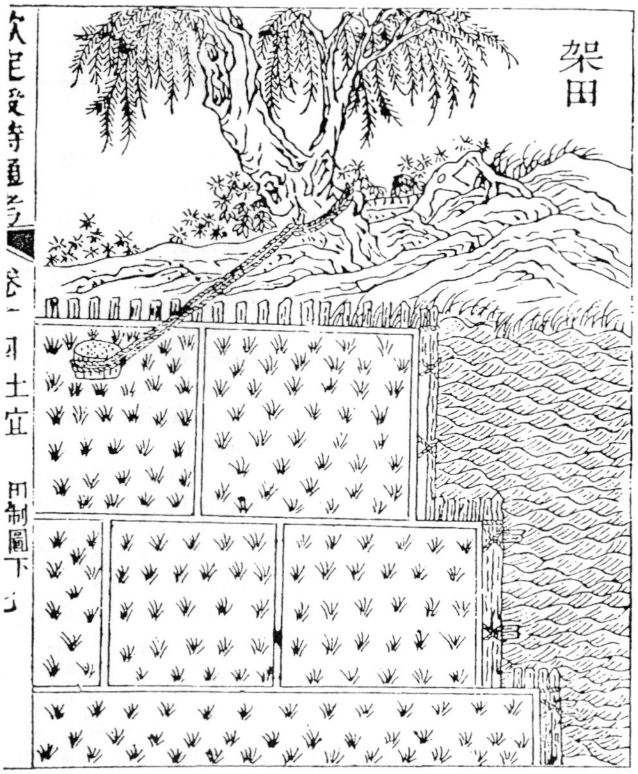

Figure 1. Floating fields.

on reeds are used for more lucrative market gardening (see Fig. 1).

Floating rice growing is a type of flood farming. When the water depth does not exceed 1 meter, one speaks of deep water rice growing with stand bearing and fast lengthening varieties; some of them survive short immersion. Beyond 1 meter, rice varieties have floating stems that lengthen in more than 5 meters of water and lie down when the water recedes. Such farming still exits in the extensive basins along the unprotected rivers in Asia (Chao Phraya, Ganges, Irrawaddy, and Mekong) and in Africa (the inland delta of Niger in Mali). If the water lasts, the harvest takes place in boats. The light work required during the season enabled a family to cultivate up to 20 ha. The great surplus fed the clergy, the warriors, and the rulers of the Angkor Empire in Kampuchea.

The Qochas

In the Quechua language, qocha means lake and generally, any artificial or natural basin. The qochas form connected sets of basins. The stretches of land that separate them are dedicated to animal breeding. Qochas are found in a few areas of the Bolivian and Peruvian altiplano at an altitude of 3850 meters, in particular in the department of Puno. Qochas have three basic forms: round, oblong, and rectangular (see Figs. 2 and 3). Most frequent are round ones that measure between 30 and 200 meters in diameter and have an average surface of 6,000 m^2. The rectangular ones are 150 m by 30 m. Whatever the form, the base is horizontal and the depth varies from 1 to 6 meters (on average, 2 m). Its diameter is crossed by a deep furrow through which water arrives and leaves; the base is surrounded by one or two circular furrows. Perpendicular furrows join the soft sloped edges. The qocha is used as a cultivated field or water tank. Its size is generally expressed in *masas*, the surface plowed by a team of two men and one woman in 1 day. Each family has six or seven *masas*. The crops are mainly potato, qaniwa, and kinuwa, rotating with oats, barley, corn, or a grazed fallow of variable duration.

The Diked Fields

Bordo and caja, dike and basin field, respectively, are words indiscriminately used for many techniques of counterfields or diked fields in Mexico. In tune with farmers, it is preferable to use a neutral terminology because, behind the appearance of a similar technique, the functions and practices vary according to site: flooded banks of a lake or river, rivers of a torrent, or mountain valley. Among the agricultural effects of the use of the technique are irrigation, protection against flooding, fertilization, control of weeds and pests, and water storage.

The Mexican midwest is a string of basins partitioned by volcanic cones of all sizes; many basins do not have natural discharge systems. Before the reclamation of many lakes, flooding was inescapable, and not very deep lakes retracted in dry season. Individual or organized diked fields were both used. Earth embankments measured between 60 centimeters and 2 meters high. They enclosed a surface ranging from several to more than 50 ha. A master diked field was filled by tapping a river or a source. In rainy years, the secondary diked fields were filled,

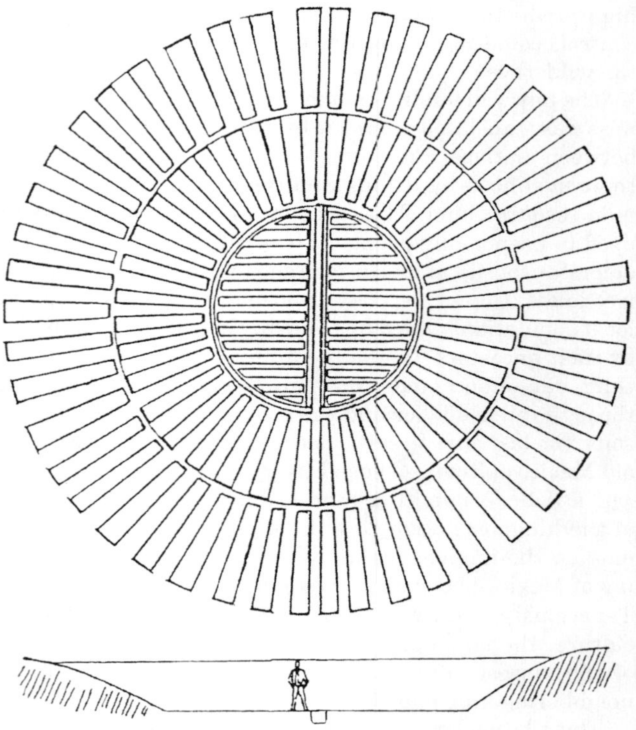

Figure 2. Diagram of a circular qocha.

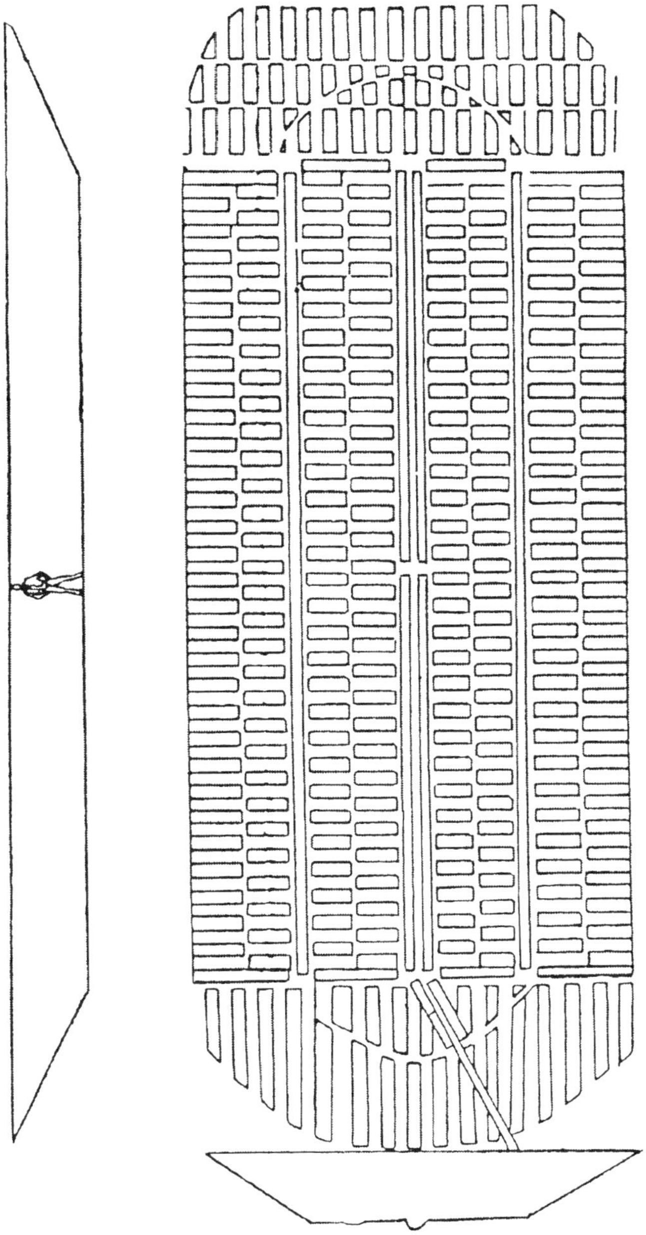

Figure 3. Diagram of a rectangular qocha.

too, to irrigate tens of hectares of wheat and sugar cane inside and outside the diked fields. In dry years, the fields downstream of the only master diked field were sown in chickpea, which received only one irrigation. Owing to the damp soil, the diked fields were farmed themselves with the same chickpea. The dry years provided a considerable surplus of quality fodder which made it possible to feed the cattle of the haciendas and also large pigsties. This region of Mexico is still the greatest industrial pig-breeding area in spite of the disappearance of most bordos during the twentieth century.

The haciendas built the systems of diked fields. After the agrarian reform, many of them were shared with many peasants who were unable to maintain such extensive systems. However, it is possible nowadays to find some diked fields used for lentil and chickpea farming.

Cultivated Ponds: Dombes and Pearl River

Dombes constitutes a undulating basin northeast of Lyons, France. The singularity of the area rests on 1 year of the crop, the assec (or dry pond), which alternates with two years of ponding, the évolage (the root eve comes from the Latin aqua). Agrifishing activity is closely connected with rights on water and land. These rights are contradictory to the French law that does not allow a land to be flooded by a dam built by a neighbor. Only a combination of circumstances and powerful events have perennialized this agriculture.

Dombes is a relatively wet area where clay soils are difficult to work and prevent agriculture from developing. As an independent principality until 1762, it was used as refuge against invaders. The proximity of Lyons, a gastronomical center requiring fresh fish of high quality, and the political independence of the close Dombes, which made it possible to maintain a special law there, are the causes of this agriculture. Indeed, each pond is a particular case combining technical adjustments and contractual agreements between landowners, the owners of the water and dam, and the sharecroppers growing assec, as well as the stakeholders upstream and downstream who can benefit or be affected by runoff, flooding, or water scarcity.

The technique of mulberry tree dikes associated with a fishing pond draws its peculiarity from the narrow association of farming and fishing. These joint activities are more productive than as separate activities. If integrated agriculture is well represented in China and in Vietnam, it seems that the delta of the Pearl River has world exclusivity in the association of the mulberry tree and fishing ponds. In the northern deltas, the frosts damage the mulberry trees and, in the south, heat and extensive flooding limit fish farming. Other situations in the world share this context, but they lack population and markets, which govern integrated agriculture, and the skills required in silkworm farming that the mulberry leaves supplied.

THE AGRICULTURE OF RAISED STYLE

The Gardens of the War

In the 1970s, the remote valleys in the center of New Guinea revealed a Stone Age society. The agriculture of the Dani tribes there is curious. Tuber crops are grown on raised fields built in marshes with spatula-ended sticks and worked by hand. A few-year fallow is characteristic of semi-intensive agriculture. Furthermore, they are "gardens of war" insofar as watchers prevent the raids of enemy tribes that seek to steal the working women. The Dani supplement their production with tubers produced in slash and burn farming. In such a society, where production is not for marketing, unlimited labor is first a social activity, as well as war.

Chinampas and Hortillonnages

Hortillonnages are raised beds from 0.80–1.50 m high. They are separated by ditches and right-angle channels;

both are full of water all year. The drainage makes the roots of the plants safe and gives them the possibility of benefiting from the soil fertility and permanent irrigation. The variety of raised fields in the world provides many local adaptations. The term "Chinese beds" must be reserved for raised beds constructed each season in paddy fields. As for the denominations of raised beds, drained fields, ridged agriculture, poldered fields, campos elevados, camellones, and chinampas, they apply to simple banked-up beds, more or less permanent traditional beds, furrow irrigation, and the various market gardening types. The French language has the specific term of hortillonnages. An unverified anecdote credits Julius Caesar with the use of the term hortilani (perhaps for garden) when he crossed the Somme river near Amiens. Dani from New Guinea show that this type of wetland use is accessible by technology from the Stone Age and that it can be old.

Classification must distinguish seasonal and permanent uses. Seasonal cropping is related to the drying up of channels, flooding, or winter. Permanent cropping is possible in tropical countries; irrigation is ensured, and flooding is controlled. In this case, one distinguishes individual flood control by poldered raised beds on the Thai pattern or by a collective dike on the lakes as was the case at the time of the Aztecs in Mexico. The first pattern includes two alternatives: on the one hand, those crafted for vegetable farming on lower beds (80 cm) that have bent surfaces and, on the other hand, orchard beds $1\frac{1}{2}$ meters high and flat surfaces. For vegetables, construction is by hand to avoid compacted soil, whereas the second option is mechanized. In Thailand, one also finds raised beds for coconut groves where water freely circulates from tides and flooding. In Mexico, Mayas built channeled market gardens where the ditches are dug on the edges of marshes; they are lengthened by traditional embanked raised beds.

In Europe, raised gardens are nowadays located under the cathedrals at the centers of cities, such as in Amiens and Bourges. Some are found in the marshes near Saint-Omer to produce cauliflower. The Atlantic marshes have preserved their raised meadows, but horticulture appeared only between the two wars; the fields were used mainly by cows, which are taken there in small boats. In Latin America, Mexico City nowadays preserves the chinampas in the district of Xochimilco wedged into the urban fabric. In Asia, they do not seem to be more than three centuries old, even in China. In Sumatra and Thailand they are promising a extension for export horticulture. In Burma and Vietnam, their development is still limited. In Java, at the beginning of the century, the surjan technique was developed, which associates rice growing in channels and a mixed crop on raised beds.

Finally, the many archaeological ruins gathered under the term camellon all over the American continent, which are vast undulations, do not necessarily proceed from raised bed or chinampa patterns. They would be closer to the semi-intensive agriculture practiced nowadays in Africa and in the valley bottoms of Rwanda, designed for producing sweet potatoes. Intensive, permanent horticulture requires a market and, as long as no close cities have been identified, camellon ruins bear witness to a more seasonal, low-input horticulture.

READING LIST

Altieri, M.A. Traditional farming in Latin America. *The Ecologist* **21**(2): 93–96.

Armillas, P. (1971). Gardens on swamps. Archeological research verifies historical data on Aztec land reclamation in the valley of Mexico. *Science* **174**(4010): 653–661.

Bérard, L. (1989). *Structures foncieres et organization sociale*. La Dombes au début du XXe siecle, Monde Alpin et Rhodanien (Le), No. 1–2, pp. 39–56.

Billaud, J-P. (1984). *Marais poitevin: rencontres de la terre et de l'eau*. L'Harmattan, Paris, p. 265.

Bruneau, M. and Bernot, L. (1972). Une population lacustre: les Intha du lac Inlé. *Jatba* **XIX**(10–11).

Chabert, J.P. (1984). Terres et eaux en Dombes. In: *Les hommes et l'eau*. Etudes Rurales Paris, No 93–94, pp. 87–89.

Coe, M.D. (1964). The chinampas of Mexico. *Sci. Am.* **211**(1): 90–98.

Dubois, P. (1913). La vallée de la Somme. Les hortillonnages, les pêcheries, les tourbières. *Bull. Soc. Géo. Lille* **59**: 324–338.

Jeffery, D. (1974). Burma's Leg Rowers and Floating Farms. *National Geography*: June 826–840.

Jimenez Osornio, J. (1990). Famous floating gardens threatened. *Ileia Newsletter*: October 26.

Jimenez-Osornio, J.J. and del Amo, R.S. (1988). *Intensive Mexican Traditional Agroecosystem: The Chinampa*. In: Allen and Van Dusen (Eds.). pp. 451–452.

Lamballe, P. (1993). Floating rice systems of South Prey Veng. *Gret* **20**.

Lambert, J.D.H., Siemens, A.H., and Arnason, J.T. (1984). Ancient Maya drained field agriculture: its possible application today in the new river floodplain, Belize. *Agric. Ecosystems Environ.* **11**: 67–84.

Leng, T.S. (1987). *Cultivation of deepwater and floating rice in Kampuchea*. Proc. Int. Deepwater Rice Workshop, IRRI, pp. 159–167.

Limouzin, P. (1994). Une agriculture périurbaine en voie de banalisation: l'exemple amiénois. *Bull. Assoc. Géogr. Franç.* **2**: 156–159.

Micha, J-C. (1987). L'aménagement des marais tropicaux en écosystème agro-piscicole. *Bull. Séanc. Acad. R. Sci.* Outre-Mer **31**(1985-3): 371–389.

Moriarty, J.R. (1968). Floating gardens (chinampas) agriculture in the old lakes of Mexico. *America Indigena 28(2) et Katunob* **7**(1) 1969: 59–75.

Perceveaux, P. (1972). Les étangs dombistes dans la première moitié du XIX siècle. *Bull. Soc. Naturalistes Archéologues Ain.* **86**: 83–107.

Salas, H. (1988). *Ecological reclamation of the chinampa area of Xochimilco, Mexico City*. In: A. Patricia and D. Van Dusen (Eds.). pp. 469–474.

Sceau, R. (1980). Les étangs de la Dombes: Fondements socio-économiques d'un système hydraulique. *Revue de géographie de Lyon* **55**(2): 129–159.

Serpenti, L.M. (1965). *Cultivators in the Swamps*. Amsterdam van Gorcum, Assen, p. 308.

Siemens, A.H. *Reducing the risk: some indications regarding pre-hispanic wetland agricultural intensification from contemporary use of a wetland/terra firma boundary zone in*

Central Veracruz, University of British Columbia, dep. geography/Vancouver, Ecological Studies.

Siemens, A.H. and Puleston, D.E. (1972). Ridged fields and associated features in Southern Campeche: new perspectives of the lowland Maya. *Am. Antiquity* **37**: 228–239.

Sluyter, A. (1994). Intensive wetland agriculture in Mesoamerica: space, time, and form. *Annals. Assoc. Am. Geographers* **84**(4): 557–584.

Song, X.F. (1997). Rice cultivation on floating-beds in different natural waters. *Chin. Rice Res. Newsl.* **5**(2): 4–5.

Thom, D.J. and Wells, J.C. (1987). Farming systems in the Niger inland delta, Mali. *Geographical Rev.* **77**(3): 328–342.

Warren, D.M. Slikkerveer, L.J., and Brokensha, D. (1995). *The Cultural Dimension of Development. Indigenous Knowledge Systems*. Intermediate Technology Publications, London.

Wilken, G.C. (1983). *A Note on Buoyency and Other Dubious Characteristics of the "Floating" Chinampas of Mexico*. In: Prehistoric Intensive agriculture in the tropics. Farrington (Ed.). British archeological Reports, Int. series 232, Oxford, pp. 31–48.

Wilken, G.C. (1997). Food-producing systems available to the ancient Maya. *Amer. Antiquity* **36**: 432–448.

You, X.L. (1991). Growing rice on the floating fields in ancient China. *Chin. J. Rice Sci.* **5**(2): 79–82.

WATER BETWEEN ARABS AND ISRAELIS: RESEARCHING TWICE-PROMISED RESOURCES

Clive D. Lipchin
Ann Arbor, Michigan

INTRODUCTION

On November 29, 1947, the General Assembly of the United Nations passed Resolution 181 in favor of partitioning Palestine into a Jewish state and an Arab state. The Zionist movement, which had been preparing itself for statehood accepted the plan, but the Palestine Arabs rejected it. On May 14, 1948, the state of Israel was proclaimed. The borders of the newly proclaimed state of Israel were a strategic nightmare representing two fighting serpents, one the new Jewish state, the other the West Bank under the autonomy of Jordan and the Gaza Strip under Egypt (1). The problems with the borders were many: lack of military strategic depth in the center of the country, an arid south, and a fractured north. No less problematic were access to water resources.

The majority of Israel's water resources are transboundary and are thus, in a sense, shared with her neighbors. The most contentious are the underground aquifers of the West Bank shared between Israel and the Palestinians and the tributaries to the Jordan River that originate in the occupied Golan Heights, which belonged to Syria prior to the 1967 Six-Day War. Since 1948, Israel has embarked on an ambitious development program for the country. The drive behind this program was the Zionist ethos of developing the land to create an attachment between the Jew and the soil. Massive agricultural and water projects ensued in the mid-1950s and 1960s in realizing the Zionist dream. The most significant was the National Water Carrier completed in 1964. The National Water Carrier is a complex system of pipes, pumps, and canals that transfers water from the Sea of Galilee in the north to the arid Negev Desert in the south for irrigation. The repercussions of this interbasin transfer have been the decline of the flow of the Jordan River that flows south from the Sea of Galilee to the Dead Sea. Today, the Jordan River is a trickle of its former self, and the Dead Sea is at its lowest level ever.

Since the mid-1980s, agricultural expansion has declined in Israel and has been surpassed by rapid industrial and urban development. Although agriculture is still the primary consumer of water in the country, the growing industrial and urban sectors require more and more water. In 1948, the Jewish population was approximately 500,000; today, it is approximately 5 million. At least 30% of Israel's water budget now comes from the aquifer systems that underlie the West Bank. These aquifers are recharged on the eastern slopes of the West Bank within autonomous Palestinian regions but flow westward to the Mediterranean across the 1967 "Green Line" into Israel. Both Palestinians and Israelis claim sole sovereignty over these resources due to this hydrologic reality.

Water has featured prominently in Israel's continuous peace negotiations with her Arab neighbors. Two significant opportunities have occurred in equitably managing these transboundary waters. The first was the shuttle diplomacy of U.S. special ambassador Eric Johnston who was sent to the region by the Eisenhower administration in 1953 and 1954. The second were the Oslo Interim Accords signed between Israel and the Palestinians in 1993 and the Israel–Jordan Peace Treaty of 1994.

THE FIRST PROMISE

During the early 1950s and mid-1960s, Israel embarked on two major public works projects to harness water resources for agricultural development. Both of these projects, the draining of the Hula Swamps north of the Sea of Galilee and the creation of the National Water Carrier on the northwestern shore of the Sea of Galilee, entailed diversion of waters from the Jordan River watershed without any formal agreement between the other riparians of the time, Syria and Jordan. Syria's objections to the projects were focused on the confiscation of Arab-owned land in the region of the swamps and Israel's activities in the demilitarized zones set up after the 1948 War of Independence (2). The rattling of sabers over the Jordan at this time prompted the Eisenhower administration to send special ambassador Eric Johnston to try to resolve the water dispute.

Eric Johnston came to the region with a proposal for a joint water development project to provide hydroelectricity and an irrigation network for the benefit of all the states in the region and, in addition, to create fertile land capable of supporting up to 900,000 Palestinian refugees on the West Bank of the Jordan (1). The plan was based on the Tennessee Valley Authority (TVA), which provides hydroelectric power to seven states in the Tennessee Valley. The TVA is the largest public utility in the United States.

Johnston visited the region in 1953 and 1954. The final agreement concerned allocation of the waters of the Jordan River. The final allocations were 45% to Israel and 55% to the Arabs (1). The Johnston Accords, as the agreement became known, was the first serious attempt at cooperation between Israel and her neighbors. It also vindicated third-party mediation (this time, on behalf of the United States) to bring the parties together. The role of external mediators in the Arab–Israeli conflict is now a requisite condition expressed by all parties. Nonetheless, the Arab leaders refused to sign the agreement because, as a matter of principle, they were opposed to the formal recognition of Israel (1). But both Jordan and Israel informed the Americans that they would treat the agreement as if it had been signed, and it served as the basis for Israel's diversion of water via the National Water Carrier and Jordan's diversion of the Yarmouk River via the East Ghor canal (2).

Israel's relationship with Syria, however, increased in belligerency. After completion of the National Water Carrier in 1964, Syria sought to deprive Israel of the headwaters of the Jordan by diverting the waters around Israel and began construction of a canal in 1965 for this purpose. According to Sherman (3), Syria had little to gain from the canal either economically or strategically, and Israel responded to the canal's construction as a personal affront to her national aspirations. In a series of military reprisals, Israel destroyed the bulldozers being used to dig the canal and, in effect, caused the Syrians to abandon the project. It is widely believed that these skirmishes contributed to the atmosphere of confrontation that finally led to the Six-Day War in 1967.

In the interim period between the Johnston Accords of 1954 and the signing of the Oslo Declaration of Principles in 1993, Israel and her neighbors undertook an independent approach to water development. Increasing populations and growing demands for more freshwater in the region coupled with a lack of a joint vision for shared management have created a serious deficit in the water budget of all the countries. The urgency of solving the water allocation problem, something that all the parties agree to, is now seen by many as the elusive catalyst for an overall final peace agreement.

THE SECOND PROMISE

At the conclusion of the Six-Day War in 1967, the hydrostrategic map of the region changed in favor of Israel. By the occupation of the West Bank and Gaza, Israel gained access to the mountain and coastal aquifers and to the headwaters of the Jordan in the Golan Heights. Since 1967, Israel's rapid economic growth has deepened her reliance on the aquifer systems of the West Bank. As the quality of life among Israel's citizens has risen, the gap between Israeli and Palestinian water consumption has grown. Before the peace process began, average annual Palestinian domestic consumption was less than 35 cubic meters, whereas in Israel, per capita consumption was about 100 cubic meters (2).

The Oslo process that began in the mid-1990s was considered the means to redress this balance as well as to lead to an overall peace settlement between Israel and the Palestinians. The principal protocol for implementing an interim accord between Israel and the Palestinians on water allocation is Article 40, together with its appendices, signed in Washington by the two parties in 1995 (4). This agreement is commonly known as Oslo Two.

The agreement created a Joint Water Committee (JWC) to deal with all water- and sewage-related issues during the interim period and led to the creation of the Palestinian Water Authority, which gradually took over control of some of the wells that were initially operated and maintained by the Israeli water company, Mekorot. The most significant aspect of Article 40 is Israel's recognition of Palestinian water rights that are to be determined in the Permanent Status Agreement. This agreement was to be completed within five years of the Oslo process.

The Oslo process and open negotiation between Israel and the Palestinians have suffered an irrevocable setback due to the current violence. Many consider the Oslo process now dead. Criticism among the Palestinians for the failure of Oslo is a lack of visible movement on the ground. Israeli settlements in the Occupied Territories continue to expand, the transfer of water resources to the Palestinian Authority by Israel has been delayed, and the freedom to transfer water and expand water and sewage networks among Palestinian urban centers has been curtailed. Part of this problem is the continued effort in Israeli political and military spheres to link water to security concerns. Nonetheless, even at the height of the recent violence, some channels remain open between Israeli and Palestinian scientists for cooperative water management. It remains to be seen if this cooperation can include the political and military sphere. Whether or not the Oslo process is dead, autonomous Palestinian institutions do exist to manage water resources for the majority of its citizens. Coordinated action, however, still remains a necessity for an equitable solution between Israel and the future Palestinian state.

The multilateral activities between Israel and her neighbors that began in Madrid in 1991 paved the way for the Israel–Jordan Treaty of Peace signed in 1994. Out of all of Israel's neighbors, Jordan has always been the most open to negotiations with the Jewish state. Tacit agreements between the two parties on strategic and water issues existed prior to the signing of the peace treaty. Having a formal peace treaty in place, both parties looked forward to peaceful cooperation and exchange in economics, culture, and security.

The Israel–Jordan peace treaty legally spells out mutually recognized water allocations for the Yarmouk and Jordan Rivers, as well as groundwater in the Arava/Araba valley, and calls for joint efforts to prevent water pollution (5). The treaty also acknowledges the water deficit facing both countries and calls for joint efforts to alleviate water shortages through cooperative efforts on both a regional and international scale (5). In the treaty, Israel agrees to provide storage services of 20 million cubic meters of water to Jordan during the winter in the Sea of Galilee and to work with Jordan to find an additional 50 million cubic meters (2). In return, Israel can continue to use wells in the Arava/Araba valley for agricultural purposes despite the fact that, according to the treaty, these wells now lie on the Jordanian side of the border. The region of the Arava/Araba valley is one of the few regions where open cooperation between Israel and an Arab state continues despite the current violence. Israeli agricultural expertise is being used to develop agricultural projects on the Jordanian side of the valley, and cooperation on combating and preventing oil spills in the Gulf of Aqaba continues between the two cities on the gulf, Eilat in Israel and Aqaba in Jordan.

RENEGOTIATING PROMISES MADE

Israel and her neighbors are inching toward a system of integrated watershed management. Today, agreements exist among three riparians of the Jordan River watershed: Israel, the Palestinian Authority, and Jordan. A fourth riparian has so far been excluded from the process. Movement on the Israeli–Syrian track is stalled in the starting gates. Syria insists on the precondition that Israel withdraw from all of the Golan Heights before any negotiations can begin. Israel rejects any form of preconditions as a basis for negotiation. The hydrostrategic importance of the Golan Heights as the headwaters of the Jordan are considered vital to Israel's water needs. For Syria, the issue of the Golan is not water, but pride in regaining the Golan and its position of influence in the Arab world; most of its water comes from the Tigris–Euphrates.

Accommodation with Syria will most likely require a "land for peace" deal. At the moment, the parties are far from agreement on what such a deal will constitute, but whatever deal is finally agreed on will affect the current agreements between Israel and the Palestinian Authority and Jordan. Both of these agreements discuss water allocations independently of Syria, the fourth riparian. Therefore, the allocation systems currently in place will have to be renegotiated once all the riparians are included ultimately which may seem impossible in today's climate, but it cannot be ignored.

Opportunities for agreement now exist to augment any allocation agreement that may be signed. Desalination is soon to become a reality in the region. Israel, as the dominant riparian, has a lot to gain politically and economically in cooperating with her neighbors in setting up and managing such plants. Trading in desalinated water may be one path toward economic cooperation among the riparians to improve the quality of life for all citizens. Desalination may also help to relieve the pressure on the existing freshwater in the region. Demand management policies also need to be considered, such as a move from agricultural hegemony over water, especially in Israel, toward domestic and industrial uses.

It is likely that the need and the recognition of a sustainable water supply in the region will push the parties together in negotiating an equitable water allocation and management system. This system can then provide the foundation for political agreement by the parties, an agreement that so far has proven to be an elusive promise.

BIBLIOGRAPHY

1. Shlaim, A. (2000). *The Iron Wall: Israel and the Arab World*. Penguin Books, New York.
2. Feitelson, E. (2000). The Ebb and flow of Arab-Israeli water conflicts: are past confrontations likely to resurface? *Water Policy* **2**: 343–363.
3. Sherman, M. (1999). *The Politics of Water in the Middle East: An Israeli Perspective on the Hydro-Political Aspects of the Conflict*. St. Martin's Press, New York.
4. Scarpa, D.J. (2001). *Oslo Interim Accords: With Special Reference to Article 40 (1995)*. Unpublished Manuscript.
5. Wolf, A.T. (2001). *Transboundary Waters: Sharing Benefits, Lessons Learned*. Thematic Background Paper. International Conference on Freshwater, Bonn, Germany, 2001.

THE MYTH OF BAD CHOLESTEROL: WHY WATER IS A BETTER CHOLESTEROL-LOWERING MEDICATION

F. Batmanghelidj
Global Health Solutions
Falls Church, Virginia

We in the medical profession, totally oblivious to the vital roles of cholesterol in the body, have been duped into thinking that this substance causes arterial disease of the heart and brain. The pharmaceutical industry has capitalized on the slogan of "bad cholesterol" and produced toxic chemicals that minimally lower the level of cholesterol in the body in the process, cause liver damage to thousands of people, some of whom die as a result of using these medications.

It is surprising that none of the frequently quoted and media-popularized doctors has reflected on the fact that cholesterol levels are measured in blood taken from the veins, yet nowhere in medical literature is there a single case of cholesterol having caused obstruction of the veins. Venous blood moves far move slowly than arterial blood and thus would be more inclined to have cholesterol deposits if the assumption of "bad cholesterol" were accurate. This mistake by us in the medical community and its exploitation by the pharmaceutical industry has caused unnecessary mass hysteria and a multibillion-dollar annual drain of financial resources by those who can least afford it in these times of serious economic crisis.

In truth, the so-called bad cholesterol is actually far more beneficial than is appreciated. The reason for its rise in the body is complications caused by chronic unintentional dehydration and insufficient urine production. Dehydration produces concentrated, acidic blood that becomes even more dehydrated during its passage through the lungs before reaching the heart—because of evaporation of water in the lungs during breathing. The membranes of the blood vessels of the heart and main arteries going up to the brain become vulnerable to the shearing pressure produced by the concentrated, acidic blood.

These membranes begin to lose water to the concentrated blood and thus lose resistance and structural integrity. The shearing force of thick, toxic blood tends to produce abrasions and minute tears in the lining of the arteries. Sections of these membranes could easily peel off and become emboli reaching the brain, kidneys, and other organs if some protective measure were not in place. To prevent the damaged blood vessel walls from getting peeled off by blood working its way under the torn lip of the membrane, low-density (so-called bad) cholesterol immediately coats and covers up the abrasions and tears and protects the underlying tissue like a waterproof bandage until the tissue heals. The advantage of "greasy" cholesterol is that blood components could roll over the patch without sticking to it and produce thrombosis in these vital arteries.

Thus, the vital, lifesaving role of low-density cholesterol proves that this substance is of utmost importance. It saves the lives of those who do not adequately hydrate their bodies by enabling their blood to flow easily through the blood vessels without causing damage.

Cholesterol has infinite vital roles. If the physiology of the body places a greater demand on some aspects of cholesterol functions, the liver manufactures more of it. Not knowing any of these functions and interfering with its actions is foolish—yet it is the present state of medicine, as far as cholesterol is concerned. Here are some other of these actions.

Cholesterol is an element from which many of our hormones are made. The body also makes vitamin D from cholesterol, which is deposited in skin that is exposed to sunlight. Cholesterol is used in the insulating membranes that cover our nerve systems. Cholesterol is used in dehydrated cells to bond the cell membrane structure together. There is no such thing as bad cholesterol. If all the primary ingredients are available for its normal functions, the human body does not engage in making things that are bad for its survival. Until now, we did not know that water is a vital nutrient that the body needs at all times—and in sufficient quantity.

By a convoluted process, the rise in the cholesterol concentration of blood is also an indicator of developing osteoporosis. Briefly, water manufactures hydroelectric energy for many of the energy-dependent physiological functions of the body; neurotransmission is on top of the list. In dehydration and low hydroelectric energy output, ATP energy trapped in calcium bonds becomes an alternative source of energy supply. Each atom of calcium that bonds to another calcium traps one unit of ATP energy.

Thus, the bones of the body, the repositories of much stored energy, begin to "decalcify" to contribute to the emergency energy needs of the body. As an integrated biochemical process in the body and to offset the catastrophic outcome of severely decalcified and softened bones, the manufacture of more vitamin D to increase the rate of bone formation becomes a dominant drive. To achieve this, the body has no alternative but to manufacture more cholesterol to be converted to vitamin D by sunlight on exposed skin. This process was used at the turn of last century to cure rickets in children, and it should be used for adults suffering from osteoporosis. They also need to be better hydrated for their bones to become firm and better calcified.

Water itself—not caffeine-containing beverages that further dehydrate—is a better cholesterol-lowering medication than any chemical on the market. It is absolutely safe, unlike the dangerous medications so often used. Please share this information with those you care for.

For more information about my medical breakthrough on the topic of chronic unintentional dehydration and the diseases it causes, read as many of the following titles as you can. They are the products of more than 24 years of focused research on the topic of deep dehydration inside the cells of the body: *Your Body's Many Cries for Water*; *ABC of Asthma, Allergies and Lupus; Water for Health, for Healing, for Life; How to Deal with Back Pain and Rheumatoid Joint Pain; Water Cures: Drugs Kill*; and my 2004 book, *Obesity, Cancer, and Depression: The Deadly Diseases of Dehydration*. For more information on dehydration, visit www.watercure.com and www.nafhim.org.

EARLY CLOCKS

Physics Laboratory, National Institute of Standards and Technology

Not until somewhat recently (that is, in terms of human history) did people find a need for knowing the time of day. As best we know, 5000 to 6000 years ago great civilizations in the Middle East and North Africa began to make clocks to augment their calendars. With their attendant

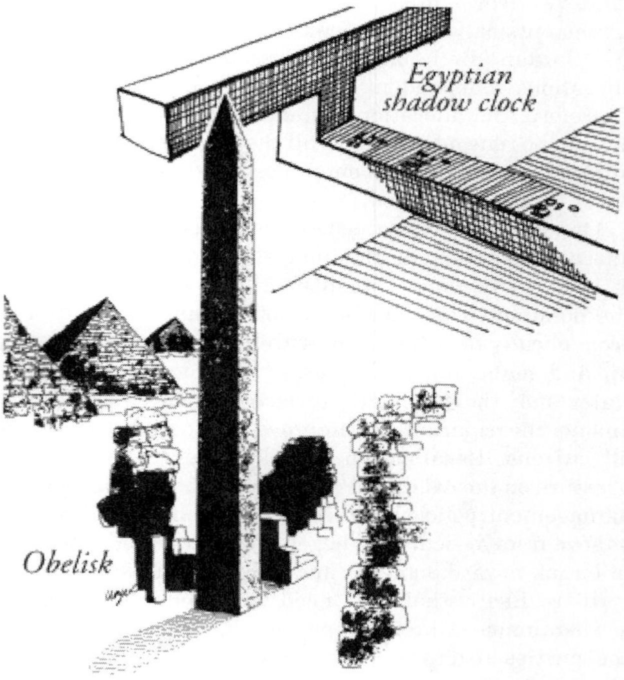

bureaucracies, formal religions, and other burgeoning societal activities, these cultures apparently found a need to organize their time more efficiently.

SUN CLOCKS

The Sumerian culture was lost without passing on its knowledge, but the Egyptians were apparently the next to formally divide their day into parts something like our hours. Obelisks (slender, tapering, four-sided monuments) were built as early as 3500 BCE. Their moving shadows formed a kind of sundial, enabling people to partition the day into morning and afternoon. Obelisks also showed the year's longest and shortest days when the shadow at noon was the shortest or longest of the year. Later, additional markers around the base of the monument would indicate further subdivisions of time.

Another Egyptian shadow clock or sundial, possibly the first portable timepiece, came into use around 1500 BCE. This device divided a sunlit day into 10 parts plus two "twilight hours" in the morning and evening. When the long stem with 5 variably spaced marks was oriented east and west in the morning, an elevated crossbar on the east end cast a moving shadow over the marks. At noon, the device was turned in the opposite direction to measure the afternoon "hours."

The *merkhet*, the oldest known astronomical tool, was an Egyptian development of around 600 BCE. A pair of merkhets was used to establish a north-south line (or meridian) by aligning them with the Pole Star. They could then be used to mark off nighttime hours by determining when certain other stars crossed the meridian.

In the quest for better year-round accuracy, sundials evolved from flat horizontal or vertical plates to more elaborate forms. One version was the hemispherical dial, a bowl-shaped depression cut into a block of stone, carrying a central vertical gnomon (pointer) and scribed with sets of hour lines for different seasons. The hemicycle, said to have been invented about 300 BCE, removed the useless half of the hemisphere to give an appearance of a half-bowl cut into the edge of a squared block. By 30 BCE, Vitruvius could describe 13 different sundial styles in use in Greece, Asia Minor, and Italy.

ELEMENTS OF A CLOCK

Before we continue describing the evolution of ways to mark the passage of time, perhaps we should broadly define what constitutes a clock. All clocks must have two basic components:

- *A regular, constant or repetitive process or action to mark off equal increments of time.* Early examples of such processes included the movement of the sun across the sky, candles marked in increments, oil lamps with marked reservoirs, sand glasses (hourglasses), and in the Orient, knotted cords and small stone or metal mazes filled with incense that would burn at a certain pace. Modern clocks use a balance wheel, pendulum, vibrating crystal, or electromagnetic waves associated with the internal workings of atoms as their regulators.
- *A means of keeping track of the increments of time and displaying the result.* Our ways of keeping track of the passage of time include the position of clock hands and digital time displays.

The history of timekeeping is the story of the search for ever more consistent actions or processes to regulate the rate of a clock.

WATER CLOCKS

Water clocks were among the earliest timekeepers that didn't depend on the observation of celestial bodies. One of the oldest was found in the tomb of the Egyptian pharaoh Amenhotep I, buried around 1500 BCE. Later named *clepsydras* ("water thieves") by the Greeks, who began using them about 325 BCE, these were stone vessels with sloping sides that allowed water to drip at a nearly constant rate from a small hole near the bottom. Other clepsydras were cylindrical or bowl-shaped containers designed to slowly fill with water coming in at a constant rate. Markings on the inside surfaces measured the passage of "hours" as the water level reached them. These clocks were used to determine hours at night, but may have been used in daylight as well. Another version consisted of a metal bowl with a hole in the bottom; when placed in a container of water the bowl would fill and sink in a certain time. These were still in use in North Africa in the 20th century.

More elaborate and impressive mechanized water clocks were developed between 100 BCE and 500 CE by Greek and Roman horologists and astronomers. The added complexity was aimed at making the flow more constant by regulating the pressure, and at providing fancier displays of the passage of time. Some water clocks rang bells and gongs; others opened doors and windows to show little figures of people, or moved pointers, dials, and astrological models of the universe.

A Macedonian astronomer, Andronikos, supervised the construction of his *Horologion*, known today as the Tower of the Winds, in the Athens marketplace in the first half

Early water clock

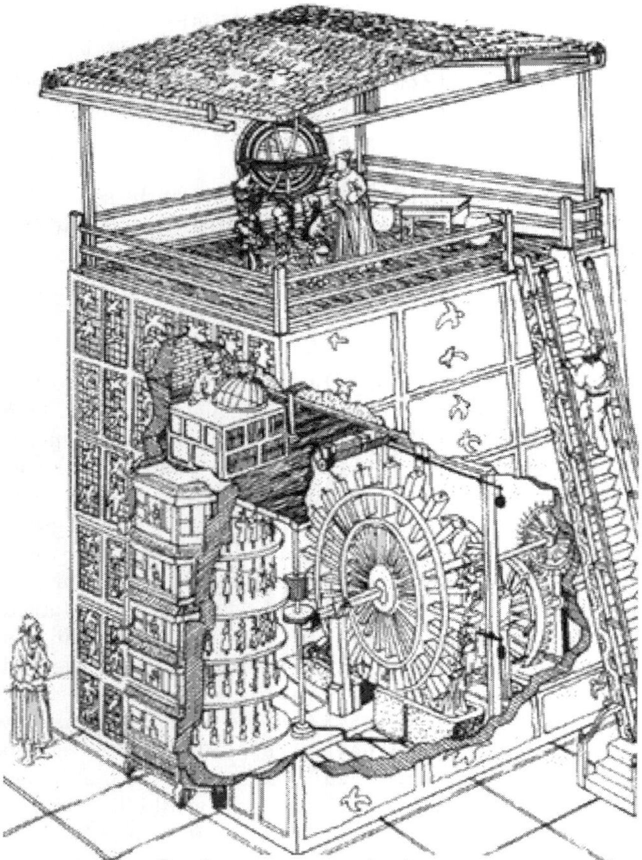

Su Sung water clock tower

of the first century BCE. This octagonal structure showed scholars and shoppers both sundials and mechanical hour indicators. It featured a 24 hour mechanized clepsydra and indicators for the eight winds from which the tower got its name, and it displayed the seasons of the year and astrological dates and periods. The Romans also developed mechanized clepsydras, though their complexity accomplished little improvement over simpler methods for determining the passage of time.

In the Far East, mechanized astronomical/astrological clock making developed from 200 to 1300 CE. Third-century Chinese clepsydras drove various mechanisms that illustrated astronomical phenomena. One of the most elaborate clock towers was built by Su Sung and his associates in 1088 CE. Su Sung's mechanism incorporated a water-driven escapement invented about 725 CE. The Su Sung clock tower, over 30 feet tall, possessed a bronze power-driven armillary sphere for observations, an automatically rotating celestial globe, and five front panels with doors that permitted the viewing of changing manikins which rang bells or gongs, and held tablets indicating the hour or other special times of the day.

Since the rate of flow of water is very difficult to control accurately, a clock based on that flow could never achieve excellent accuracy. People were naturally led to other approaches.

WATER CLOCKS

William E. Marks
Water Consciousness, Inc.
Martha's Vineyard,
Massachusetts

The science or art of measuring time is known as "horology," and, for the greater part of human history, time was calculated through the use of water clocks.

Ever since the dawn of humanity, we have enjoyed a relationship with water. It was this relationship with water that gave birth to the concept of measuring time.

It is believed that the first human-made timekeeping device was the water clock invented in Egypt around 1400 B.C. This was a very rudimentary timekeeping device based on the flow of water out of or into a bowl-type container. This filling or emptying of a measurable vessel with water provided a basis for telling time. The energy source that powered the first water clocks, and all others that followed, was the fluid properties of water responding to the influence of gravity.

The name given to early water clocks in Greece was "klepsydra" (pronounced KLEP-suh-druh). This name meant "thief of water" and was derived from two Greek words meaning "thief" and "water." These Greek root words can be traced to such present-day words as "kleptomaniac," for one who steals, and water words such as hydrant and hydrology.

The klepsydra migrated from Egypt to Greece, and then later into Rome. The Romans Latinized the Greek *klepsydra* to *clepsydra*, or the *horologium ex aqua*. In early history, there was probably one clepsydra for each village or town. Each clepsydra was set up in a central location and was watched over by responsible representatives, who took turns keeping watch over the device. This "watching" of the clepsydra was necessary to protect the time device and to have an attentive person standing by to fill the water basin constantly.

Some locations used a horn to sound the passage of a certain amount of time; others used a drum or the human voice. In this fashion, everyone within earshot had an idea of the time of day. This ancient system of sounding the passage of time remains in use today. Today, we find the sounding of bells at designated times in centrally located bell towers throughout the world, as well as the blowing of a horn. In some places, there are bell-keepers who ring the tower bell or bells manually at designated times. In other locations, the ringing of bells or blowing of a horn is mechanized and occurs on an automated basis.

In early England and later New England in the United States, there were so-called "walking clocks." A "walking clock" was a person who would use a loud voice to announce the time while walking the streets.

In Muslim countries, where Islam is widely practiced, prayer is timed at five different intervals during the day. In some Islamic areas, there are towers where a human occupant will loudly voice prayers to Allah at the five designated times. While praying aloud to Allah, the prayer-giver always faces in the direction of Mecca.

It is due to this keeping "watch" over the flow of water in early clocks and the watching of time in bell towers and by town criers that we have the etymology for the modern timepiece known today as the "watch."

THE FIRST CLOCK WAS A WATER CLOCK

Without doubt, the very first so-called "water clock" was humanity observing how water was connected to the timing of natural and cosmic events. The timing of moon tides, the timed event of spring floods when the earth turns on its axis and causes snow and icemelt in the mountains, and the calendar of humans watching celestial events to anticipate spring rains or monsoons are three examples.

To survive, humans require a steady source of freshwater. Therefore, the earliest humans naturally learned to observe, understand, and live within the water cycle of their respective regions. Each water cycle was connected to the sourcing of drinking water, the seasonal growth of various wild plant foods, and the migratory patterns of fish and animals. In this fashion, early humans became aware of a natural water rhythm that was connected to the timing cycle of rainfall, which in turn determined the seasons of life. Because of this intimate relationship of the cycle of water to the cycle of life, humans became aware of a water clock that was inherent in the scheme of nature.

For these reasons, early humans evolved in regions where freshwater was readily available constantly. This is why modern archaeologists find evidence of early civilizations evolving along river basins such as the Tigris–Euphrates Valley and along the Nile River. Besides using the river water for consumption and bathing, the wildlife within and around the rivers provided a steady and plentiful source of food. The rivers also provided a convenient way for evolving humans to travel.

As humans gained knowledge of their regional water cycles, they eventually figured out a methodology for raising livestock and growing crops. In other words, early humans became aware of the "big" water clock of nature, a water clock that had a timing sequence that coincided with certain celestial events. Consequently, we see the construction of structures, such as Stonehenge in England and New Grange in Ireland. These stone structures allowed early humans to forecast, or time, future events in nature based on the water cycle.

The flooding of the Nile River, for example, from the end of June into late October, was responsible for depositing rich sediment in Egypt's fertile Nile Valley. After the flooding subsided, various crops were planted from late October into late February. These crops were harvested from late February until the end of June, when the cycle would begin again.

The ancient Egyptian calendar was known as a "nilometer" because it was based on a vertical scale that recorded the annual flood levels of the Nile River. In essence, this was a "water clock" of huge dimension. Eventually, the Egyptians fine-tuned their nilometer water clock by breaking it down into twelve months of thirty days each; five days were added at the end. For the early Egyptians, this water calendar worked well with the seasonal flows of the Nile, which is why it became known as the "Nile year." The Nile year calendar, used by the Egyptians more than 6200 years ago, predated the invention of writing.

The use of the number "12" in creating the number of months is often related to the Egyptians observation of the moon cycle. There are about 12 moon cycles in a year. The division of the year into 12 months may also have provided a basis for the Egyptians and Babylonians to divide the day and night into 12 hour segments.

Originally, the Egyptians and Babylonians divided sunrise to sunset into 12 parts they called "hours," and sunset to sunrise into 12 parts. The lengths of hours for the day and night are not always the same through the year, so water clocks had to be adjusted every night and morning. Eventually they figured out that it was easier to keep time when they created the concept of the 24-hour day. The 24-hour day encompassed the passage of both day and night hours and therefore required little adjustment of water clocks.

Egyptian water clocks were among the earliest time-keeping devices that did not depend on the observation of celestial bodies. One of the oldest was found in the tomb of the Egyptian pharaoh, Amenhotep I, buried around 1500 B.C.E.

As explained by the National Institute of Standards and Technology Physics Laboratory, early water clocks were named clepsydras ("water thieves") by the Greeks. The Greeks began using clepsydras about 325 B.C.E. Clepsydras were stone vessels with sloping sides that allowed water to drip at a nearly constant rate from a small hole near the bottom. The drip-drip sound of water dropping from a water clock was an ancient precursor to the ticktock sound emitted by certain mechanical clocks invented in later times. Water clocks were replaced by the invention of mechanical clocks about 700 years ago.

Other clepsydras were cylindrical or bowl-shaped containers designed to fill slowly with water flowing in at a constant rate. Markings on the inside surfaces measured the passage of "hours" as the water level reached them. These clocks were especially used to determine the passing hours at night but may have been used in daylight as well.

Another version is known as the "sinking bowl" (see Fig. 1). This consisted of a metal or clay bowl with a hole in the bottom center. The larger the bowl, the longer it would take to fill with water and sink below the surface. When placed in a water container or calm body of water, the bowl would slowly fill and then sink after a certain amount of time. As soon as one bowl sank, the next bowl would be set on top of the water. Counting the number of bowls that sank during the daytime, or nighttime, created a crude basis for telling time. This method was often used to keep track of the "dark hours" through the night. The amount of time a bowl took to sink was sometimes measured during daylight so it could be calibrated with a sundial. This method helped ancient humans to compare the passage of time between daylight hours and dark hours, which allowed for knowledge of the coming change of seasons. As daylight hours grew shorter and nighttime hours longer, the earth naturally began to cool.

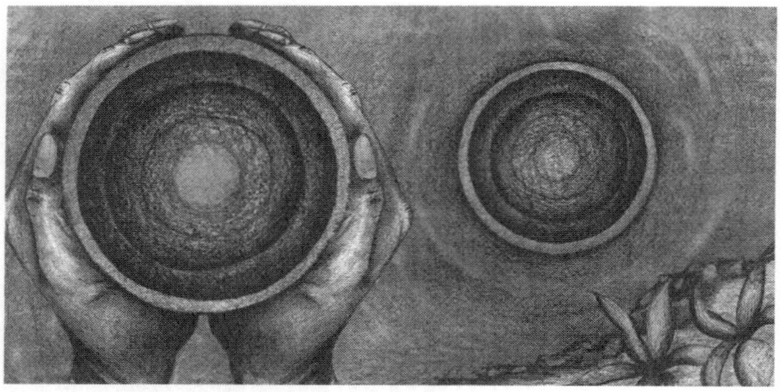

Figure 1. Sinking bowl water clock circa 1400 B.C. to present day. Drawing by: Sarah Adams.

The sinking bowl water clock was so low tech and functional that it remained in use for thousands of years in the Middle East, as well as in parts of North Africa into the twentieth century. The sinking bowl water clock remains in use in some parts of the world today.

More elaborate and impressive mechanized water clocks were developed between 100 B.C.E. and 500 C.E. by Greek and Roman horologists and astronomers. The added complexity was aimed at making the flow more constant by regulating the pressure and at providing fancier displays of the passage of time. Some water clocks rang bells and gongs; others opened doors and windows to show little figures of people, or moved pointers, dials, and astrological models of the universe.

The Greeks as well as the Romans used the water clock as a mechanism for regulating speaking time in courts. In Greece, the clepsydra was used to limit the time for pleading one's case in Athenian courts. Based on water clocks surviving from that time, the water would flow only for about 6 minutes. There are records of legal representatives asking for the flow of the water clock to be stopped so that certain laws or evidence may be presented to the court. On a more humorous level, there are stories where comments about how the orator should quench his thirst by drinking water from the water clock and thereby abbreviate his speaking time.

On a similar note, Romans also used the clepsydra to limit speaking time in Roman legal courts. This limitation of time on legal arguments by Roman orators was instituted by the command of Pompey and was alluded to by Cicero when he stated, "latrare ad clepsydram." When the clepsydra's water ran out, an orator's time ran out, and he was cut off from speaking further. Compared to the 6-minute klepsydras used by Greek courts, Roman courts used a 20-minute clepsydra.

During legal arguments, opposing parties were supposed to enjoy equal amounts of water. However, there are surviving records indicating that lawyers would sometimes request additional amounts of time by asking for a certain quantity of "clepsydrae." If a lawyer spoke too long, the opposing lawyers would argue that his water should be taken away by the court.

Martial, the Latin poet of Rome (c. 40–c. 102), once penned these words about a tiresome speaker who constantly moistened his throat by drinking from a glass of water:

> Seven water-clocks allowance you asked for in loud tones, Caecilianus, and the judge unwillingly granted them. But you speak much and long, and with backtilted head, swill tepid water out of glass flasks. That you may once for all sate your oratory and your thirst, we beg you, Caecilianus, now drink out of the water clock.

This legal limitation of time for presenting legal arguments remains in use in certain governing bodies throughout the world today. Depending on the governing rules of the land, an elected official or legal representative is allowed to hold the floor for only a designated amount of time.

It is also from the Roman legal system that we derived our modern-day A.M. and P.M. designations, representing before midday (*ante meridiem*, A.M.), and after midday (*post meridiem*, P.M.). This A.M. and P.M. split of the day was of legal import to the Romans because lawyers were required to appear in courts before noon.

On another note, the private ownership of a water clock in a home was considered a great status symbol during Roman times. Also, a person who gave the appearance of wasting time in life was referred to as "one who wastes water." Thus, we find an intimate connection between water and life, as perceived in the minds of ancient Greeks and Romans.

In Greece, a Macedonian astronomer, named Andronikos, supervised the construction of a complex Horologion. This mechanical device, known today as the Tower of the Winds, was located in the Athens marketplace during the first half of the first century B.C.E. This octagonal structure showed scholars and shoppers both sundials and mechanical hour indicators. It featured a 24-hour mechanized clepsydra and indicators for the eight winds from which the tower got its name, and it displayed the seasons of the year and astrological dates and periods.

The Romans also developed mechanized clepsydras (water clocks), though their complexity accomplished little improvement over simpler methods for determining the passage of time.

The Greeks divided the year into 12 parts they called "months." They then divided each month into 30 parts they called "days." The Greek year had a total of 360 days, which was equal to 12 times 30. The earth circles the Sun in 1 year, so the Greeks developed the mathematical concept of dividing the circle into 360°.

The division of the hour into 60 minutes and each minute into 60 seconds, probably finds its source in the ancient Sumerian sexagesimal system. This Sumerian system is based on the number 60 and was developed approximately 4000 years ago.

In the Far East, mechanized astronomical/astrological clock making developed from 200–1300 C.E. Third-century Chinese clepsydras drove various mechanisms that illustrated astronomical phenomena. One of the most elaborate clock towers was built by Su Sung and his associates in 1088 C.E. Su Sung's mechanism incorporated a water-driven escapement invented about 725 C.E. The Su Sung clock tower, over 30 feet tall, possessed a bronze, waterpower-driven armillary sphere for observations, an automatically rotating celestial globe, and five front panels with doors that permitted viewing changing manikins which rang bells or gongs and held tablets indicating the hour or other special times of the day. This and other mechanical efforts to use water for keeping time had their limitations. These limitations are inherent in the nature of water.

Water is a sensitive entity that is subject to linear and nonlinear influences. These influences include but are not limited to gravity; temperature; atmospheric pressure (weight of air above at any given moment); wind; alignment of the Moon and other planets; viscosity; sunspots (cause of ionic excitement in atmospheric water and related creation of aurora borealis and aurora australis); speed of the earth's rotation; the "Heisenberg principle" based on quantum mechanics and how that which is observed responds to being observed; sound (travels four times faster through water than through air); light; projection of intention, thought, and emotion as expressed in Masaru Emoto's pioneering water research; chemical and electrical composition of contact materials; suspended solids; the rate of molecular activity within water; the presence of vortical influences; and the spectrum of various wave influences.

Due to water's sensitivity and all of the variables that influence its character at any given second, the rate of flow of water is very difficult to control accurately. Therefore, a clock based on water flow has certain limitations and challenges for achieving high accuracy.

However, throughout time, there have been claims of water clocks that rivaled the accuracy of mechanical clocks. Once such example is an hour-glass type water clock, which had to be inverted as soon as the water from the upper glass chamber emptied into the lower glass chamber. One account of another accurate type of clepsydra, which was invented by a man named, Amontons, was published in the pamphlet, *Remarques & Experiences Physiques fur la Construction d'une nouvelle Clepsydre*, &c. Jombert, Paris, 1695.

Today, no matter what philosophical perspective a person may have about time, the basic fact is that the mechanics of keeping accurate time is a concept created for use as a linear tool.

Regardless of the variety and accuracy of human-made time devices, humankind remains subject to the vicissitudes and rhythm of flowing water, as well as its inherent cycle expressed in nature.

Understanding the rhythm and cycles of water is vital to the survival of life as well as civilization, so we cannot substitute mechanical devices for the water clock presented to us by our natural environment.

Our near future interaction with the earth's water cycle may well determine how much time our species has to survive. In other words, the available water in the earth's clepsydra is rapidly approaching empty at this time, so it behooves us to do more than just talk about altering the way we interact with water.

READING LIST

Beckman, J. (1814). *History of inventions and discoveries* [from old catalog]. Translated from the German by William Johnston, 2nd Edn. Printed for J. Walker [and others], London.

Hill, J.R. *Arabic Water-Clocks*. University of Aleppo, Institute for the History of Arabic Science, Syria, p. 1981.

Pollio, M.V. (1982). *De Architecture: The Como 1521 Edition in Exact Photographic Reproduction*. Ayer Co. Pub., North Stratford, NH.

Rees, A. *Ree's clocks, watches and chronometers, a selection from The Cyclopedia; or universal dictionary of the arts, sciences and literature*. pp. 1819–1820.

Rohr, R.R.J. (1996). *Sundials: History, Theory, and Practice*. Dover, Mineola, NY.

Waugh, A.E. (1993). *Sundials, Their Theory and Construction*. Dover, Mineola, NY.

National Institute of Standards and Technology Physics Laboratory.

www.ClocksWorld.com. *Water Clock*.

www.jim-diana-hart.home.att.net/clock.html. *Medieval Clocks*.

www.zetnet.co.uk/sea.jnp/earth.4/time.htm. *Time and the History of its Measurement, Reproduced from the Science and Technology Equipment News No. 6 by kind permission of the Scottish Schools Equipment Research Centre*.

www.inventors.about.com/library/weekly/aa071401a.htm. *The Invention of Clocks*.

www.cyrstalinks.com/clocks.html. *Clocks*.

www.rit.edu/-dansma/GoodenowOrrRossWaterClocks. *Mathematical Models of Water Clocks*.

www.sts.sunyit.edu/timetech/chinese.html. *Water Clocks*.

www.library.thinkquest.org/CO118228/water_clocks.htm. *Water Clocks*.

www.emie.cummings.net/clock.htm. *Clock - a History - Timekeepers*.

OUR EVOLVING WATER CONSCIOUSNESS

William E. Marks
Water Consciousness, Inc.
Martha's Vineyard,
Massachusetts

Since the dawn of our species, being conscious of water has been vital to our survival, health, and social behavior.

Hundreds of thousands of years before the written word, humans thought and communicated about water. Besides being essential for survival, water remains a mysterious fluid that science continues to research and study.

Prior to the invention of writing, communication about water found expression in human sign language and the spoken word. In early human writings, such as the Egyptian hieroglyphic system, the symbol for water is a ripple pattern that looks like this (〰). This early hieroglyphic symbol resembles almost exactly a purported water symbol carved into bone by a Neanderthal during the Paleolithic period.

Another ancient water symbol is the Chinese character "shui" (水). This character is of such antiquity that it predates Chinese recorded history and is of unknown origin. In *The Book of Changes* (I Ching), which is more than 3000 years old, there is the water trigram "Kan" (☵), whose symbol is obviously an evolution out of the more ancient water character of "shui."

The Chinese character for "water," it was recently discovered, exists as a natural water form. In Masaru Emoto's book, *Messages from Water*, he shows a photograph of crystallized water on page 12 that contains what appears to be a Chinese character. This photo is accompanied by the following text:

> "When water freezes, it becomes crystallized. At the moment right before it returns to its water form (with a rise in temperature, between −5 °C and 0 °C) it creates a crystal shape that is identical to the Chinese character for water. Did people in ancient times know this and make the Chinese character for "water" based on this information?"

Simply stated, the *Chinese Book of Changes* helps mankind deal with the idea of change. Or, as reported in the *Analects* (a book of conversations of Confucius, one of the authors of the I Ching), Confucius states, while standing by a river, "Everything flows on and on like this river, without pause, day and night."

THE TALE OF GILGAMESH

On another note, we find the incredible story of water expressed in the epic poem of Gilgamesh, a poem written in Sumer, an ancient civilization located in the lower Euphrates River Valley. Sumer is where writing is believed to have been invented. These first writings were written by Sumerians on clay tablets in the cuneiform script. Cuneiform script is composed of wedge-shaped characters imprinted into wet clay before it is heated (fired) for permanence.

The Sumerians believed that in the beginning only a vast swirling sea existed and that all things came from this sea. Because of this, it is no surprise to learn the Sumerian word mar meant "sea" as well as "womb." Archaeological evidence indicates that Sumer was the home of some of the earliest humans. The fact that humans lived continuously in Sumer since prehistoric times supports the theory of the way the region's ancient oral history found expression in the first writings of the Sumerians.

As can be expected, interpretations of the ancient texts found in Sumer vary. However, one thing stands without question—"water" served as the source of creation. The god Ea-Enki, the principal divinity of the liquid element—especially freshwater, is expressed as the Creator. Ea-Enki is presented as coming from the sea at the dawn of time and is said to be of supreme intelligence, skillful, ingenious, and the One who knows all things.

The name "Ea," which means "House of the Water," makes up the first two letters of "earth." Ea's domain was the Apsu—all the waters that surrounded the planet and kept it afloat, as well as the springs, rivers, and lakes. The freshwaters of the earth were considered the source of happiness, knowledge, and wisdom.

When writing was invented in Sumer, oral stories of old were rendered in clay. From these first writings came the Gilgamesh poem. However, the discovery of the Gilgamesh poem created concern in the minds of some religious leaders.

In 1872, members of the Society of Biblical Archaeology in London were surprised by the implications drawn by archaeologist, George Smith. Smith had recently deciphered the Gilgamesh poem. A poem that predated the Bible by thousands of years expressed a story about the "great flood," and included information about building a ship to save people and animals.

The discovery and translation of the Gilgamesh poem gave humanity a glimpse into water-related information and rituals passed down by oral history long before the invention of writing. Support for this can be found in the John Gardner and John Maier translation of Gilgamesh, where we are informed that the Gilgamesh "Creation Epic" was recited in Babylon each year during the "New Year's Festival," long before the creation of writing.

METAMORPHOSIS

From Gods to a Unifying Water Philosophy

More than 2500 years ago, in the Greek territory of Ionia, lived a man named Thales. Today, in philosophical circles, Thales is often described as "the world's first philosopher" and, according to Aristotle, "the founder of Ionian natural philosophy." For it is in Ionia where we find history's first naturalists recorded.

Thales, born in the rural Ionian town of Miletus, is renowned for his transformation of many of humankind's belief systems. The profound philosophical concept that Thales delivered to humankind was simply this—water is the first cause of all things seen and unseen.

Before Thales advanced his water philosophy, human thought was captured by the belief in polytheism. Under the concept of polytheism, people believed that all natural and imagined phenomena were created by a plethora of gods.

Besides contributing his water philosophy to Greek thought, Thales introduced other unique concepts to Greek thought. In fact, Thales' contributions were so profound and respected that during his time he was canonized as the wisest member of the Seven Sages of Greece.

Most of what we know about Thales flows from the writings of Plato, Aristotle, Plutarch, and other Greek philosophers. Thales was an original thinker who possessed a keen understanding of the myths and oral histories from hundreds of thousands of years in the past. At the same time, he was a scientist, astronomer, mathematician, teacher, author, politician, and a "seer" who could forecast future events.

For instance, Thales forecast a solar eclipse in the year 585 B.C. Modern scientists have checked on this event and confirm that a solar eclipse did take place on May 28, 585 B.C. and that it was visible in Ionia. How Thales was able to predict this event remains a mystery. According to information available about Thales' time in history, no mathematical formula existed for predicting an eclipse, and his feat is all the more impressive because Thales had to take into consideration the geographical latitude of Ionia.

Thales was born into a world filled with the so-called ancient mythologies and gods that were recorded in the first writings of humankind. By closely observing water and the natural world around him, Thales was able to synthesize his new philosophy into a form of natural science. This natural science helped to define the creation of life and humanity's connection to life and all creation through the entity of water.

Desiring to grasp the "first cause" of all things, Thales searched for a unifying hypothesis for the creation and sustenance of things both visible and invisible. This led him to the startling revelation that all things come from water, that the earth floats on water, that water is the constituent of all things, and that water is the cause of earthquakes. According to Plutarch's Miscellanies, Vol. III,

> Thales the Milesian doth affirm that water is the principle whence all things in the universe spring.... He pronounced, that all things had their original from water, and into water all things are resolved. His first reason was, that whatsoever was the prolific seed in all animals was a principle, and that is moist; so it is probable that all things receive their original from humidity. His second reason was, that all plants are nourished and fructified by that thing which is moist, of which being deprived they wither away. Thirdly, that the fire of which the sun and stars are made is nourished by water exhalations,—yea, and the world itself.

Today, we know that continental plates float and crash into each other and that there are chemical and other changes in groundwater prior to an earthquake. We also know from NASA planetary scientists that water was present in our part of the Universe before the creation of our solar system.

As far as Thales' theory "... that the fire of which the sun and stars are made is nourished by water exhalations,—yea, and the world itself," modern science has corroborated this fact. Recently, scientists discovered a direct relationship between the intensity and frequency of sunspot activity and weather patterns on the earth. Scientists also learned that sunspots contain water. As reported by Reuters on May 26, 1995,

> In research being released today in the journal *Science*, the experts report finding water not in the gaseous heart of the sun itself, but on somewhat cooler sunspots.
> "It's really sort of interesting," said Peter Bernath, a chemist who was part of the research team from the National Optical Astronomy Observatories in Tucson, Arizona, and Canada's University of Waterloo. "One is surprised that water can exist on the sun because it is so hot."
> Hydrogen and oxygen molecules both exist on the Sun, which is about 10,200 °F. They apparently combine to form water in sunspots, which are about 5200 °F. Because of the heat, the water molecules form vapor or steam, instead of pools or puddles.

Further scientific support of Thales' water theory was revealed by research data provided by a specially designed satellite called the Submillimeter-Wave Astronomy Satellite (SWAS). This satellite was launched in December 1998 to detect radio waves emitted by water molecules in space.

In an interview with the author of this article, Gary Melnick of the Harvard–Smithsonian Astrophysical Observatory in Cambridge provided several insights as to how modern science is providing support for Thales' water philosophy. Melnick is the leading scientist for the SWAS project. "We're seeing water everywhere in the universe. Every region we've looked at so far contains water," Melnick said.

Melnick explained that water found in interstellar space is crucial to the formation of stars. According to Melnick, when a big gas cloud begins to coalesce (form into a mass) under its own gravity, the gases begin to heat up. However, as the gases heat up, they begin to expand naturally. If water was not present to mitigate the heat generated by the condensing gas mixture, it would be difficult for the star to form. "It is because of this phenomenon," said Melnick, "that we see water helping to give birth to stars." Melnick also explained that one of the strangest discoveries of the SWAS project was finding a giant, water producing gas cloud in the heart of the Orion nebula. Based on SWAS data, Orion forms enough water every 24 minutes to fill all of the earth's oceans.

It is from our evolving scientific information, such as that expressed above, that we find corroboration of Thales' statements from more than 2500 years ago. However, along with our evolving water science and philosophies, there is also the wide practice of using water for ceremonial purposes. The use of water for the baptismal ceremony is very ancient and touches on the spiritual dimension of water.

Baptism by water

One of the most ancient and still widely used water rituals is that of baptism. From the second century forward, the value of the baptismal act was widely recognized and practiced in Christianity. Even though there are many meanings and beliefs relative to baptism, as practiced by Catholics, Protestants, and many others, water is always respected for its power to cleanse the body and regenerate the spirit.

For instance, the relationship of Moses to water harks back to his earliest beginnings. Besides his sprinkling of the covenant upon the people, we learn that the name Moses means "the water drawer." Also, not to be overlooked is Moses' origin from the Nile River. Exodus 2:1–10 tells us that as a baby, Moses was discovered floating in a basket by the Pharaoh's daughter. The word "tebah," used in the Old Testament for the basket carrying Moses, is also the same word used to describe Noah's Ark.

Besides Moses, there are other historic personages whose origins spring from water. The god Akki, "the water carrier," rescued Sargon, the founder of Babylon, from water. Karna, the icon of an early Indian epic, was found floating in the Ganges. And, just as there are many renowned people whose origins are linked with water, others find their way through various forms of water baptism.

The early "mystery" religions (Eleusinian, Orphic, Mithraic, Egyptian, Syrian) required baptism and other water rituals to wash away evil in preparation for the rites of communion with the god whose mastery of death assured immortality. These baptisms sometimes symbolized dying of the old life and rising again to the new. A similar idea underlies the baptism by immersion that is part of the initiation of proselytes to Judaism.

Today, many ancient water rituals can be found in almost all the world's religions.

Leonardo da Vinci (1452–1519)

About 500 years ago, Leonardo da Vinci was recording his thoughts about vortical energy as a result of his experiments with water. The subjects of water, hydrology, and hydraulics played a major part in Leonardo's lifetime studies. This fact was established when it was discovered that the largest number of Leonardo's notes were devoted to the subject of water than to any other subject.

The existence of water vortices intrigued Leonardo and prompted him to speculate about the existence and behavior of vortices in the air and in the cosmos. The power and meaning of vortices in water also led Leonardo to study closely the behavior of water under different conditions. In one of his experiments, Leonardo created elongated boxes with glass sidewalls so he could observe the motion of flowing water beneath the surface.

Based on his water experiments, Leonardo invented a double-gated dam. This dam used the pressure of water to make a watertight seal when the two gates met in the center with a projecting bias toward the high water side. This same principle is in evidence today at the Panama Canal and other places around the world.

Besides engineering and scientific inquiries into water, Leonardo's fascination with water dynamics and vortices is apparent in his art. The flowing motions of water vortices are expressed in his brush strokes and are obvious in his paintings, especially when depicting the hair of his subjects. Examples of this can be seen in his paintings "Leda, with the Swan;" the "Mona Lisa," and his "Self-Portrait" (see Fig. 1).

In the Mona Lisa, the hair emerges from the top of her head and flows downward like a fountain of water flowing from a spring, the breast area of her garment appears like a plunging waterfall, the flowing cloth over her left shoulder demonstrates the twisting vortex of water, and the dress sleeves on her arms appear as cascading ripples of darkness and reflected light, much like that of water overflowing from the base of a waterfall.

In Leonardo's "Self-Portrait," again we have the flow of hair springing from the head area. However, in this painting, the wavy flow of his head hair merges in dramatic fashion with his beard.

The existence of bubbles in water was also of interest to Leonardo; he noted from close observations that bubbles rise through water in a spiral motion. His written notes, as expressed in Leonardo's Codex Leicester Folio 23V (now owned by Bill Gates), depict how Leonardo documented that the "... motions of waters always move in a circle from surface to bottom."

Leonardo spent many years in his makeshift laboratory and in the field observing the movements of water and air. To see the fluid dynamics of water at work, Leonardo did experiments using glass so he could watch the motion of flowing water under various scenarios. During his field research, he maintained detailed notes and drawings about his experiences and observations. At times, Leonardo's mind would seemingly take off into other dimensions as his observations triggered ideas. Often, he would jot down or sketch these thoughts along the margins of his papers.

One such series of notes in the upper right hand corner of one of his papers was the outline for his proposed treatise on water. This treatise was divided into fifteen books, each book dealing with a different aspect of water. The classification of these books, as recorded by Leonardo, is as follows:

1. Of Water in Itself
2. Of The Sea
3. Of the Veins
4. Of Rivers

(a)

(b)

Figure 1. (**a**) Leonardo's "Self-Portrait," demonstrates water-like flow of hair springing from his head and merging in unifying waves with his beard. (**b**) Portrait of the Mona Lisa—depicts flow dynamics of water as her hair flows from her head, then cascades like a waterfall over her breasts down onto her arms, then continues flowing over her arms. Since da Vinci was careful in creating his compositions, it is important to note the mountains and "headwaters" in the background behind Mona Lisa's head, and the curving river meanders in the valleys behind both her shoulders that flow from left to right—in the same direction as the focus of Mona Lisa's eyes.

5. Of the Nature of Bottoms
6. Of Objects
7. Of Various Kinds of Gravel
8. Of the Surface of Water
9. Of Things Moving in It
10. Of River Repairs
11. Of Conduits
12. Of Canals
13. Of Machines Turned by Water
14. Of Raising Water
15. Of Things Worn Away by Water

Given the fullness of the life which Leonardo lived, he never found the time to complete the above books on water. His writings, especially those found in the Codex Leicester, contain many references and brief notes to be included in these books. An example of this can be found in Leonardo's notes dealing with precipitation:

> Write how clouds are formed and how they dissolve, and what it is that causes vapour to rise from the water of the earth into the air, and the cause of mists and of the air becoming thickened, and why it appears more blue or less blue at one time than another. Write in the same way of the regions of the air and the cause of snow and hail, and how water contracts and becomes hard in the form of ice, and of the new shapes that the snow forms in the air . . .

Another of Leonardo's field observations dealt with the physical expression of the vortex principle on a grand scale. His written observations about a water spout he saw along the seashore were recorded on a page entitled "Of wind twists and eddies involving water." In Leonardo's words,

> It often happens that, when one wind meets another at an obtuse angle, these two winds circle around together and twine themselves into the shape of a huge column, and becoming thus condensed, the air acquires weight. I once saw such winds, raging around together, produce a hollow in the sand of the seashore as deep as the height of a man, removing from it stones of considerable size, and carrying sand and seaweed though the air for the space of a mile and dropping them in the water, whirling them around and transforming them into a dense column, which formed dark thick clouds as its upper extremity. . . .

In spite of Leonardo's research and interest into various water phenomena, he was never able to provide a viable theory about the hydrologic cycle. The prevailing thinking in Leonardo's time was that the weight of the oceans waters forced water into the center of the earth. From the center of the earth, the water then flowed underground to where it eventually surfaced on land to appear in lakes, ponds, rivers, and swamps.

DISCOVERING WATER

Hydrologic Cycle

An important turning point in the correct human grasp of the hydrologic cycle came in 1580 with the publication of Bernard Palissy's book, *Discours admirables*. Palissy, a potter by trade, was a keen observer of nature and the workings of water. In *Discours admirables*, Palissy presented a concept of the hydrologic cycle that is accurate according to today's definition. Palissy's book also provided other new hydrologic concepts such as the origin of springs and rivers, how wells are recharged from nearby rivers, the hydrodynamics of artesian wells, the use of forestation for controlling erosion, and plans for constructing fountains for domestic water supply.

It was not until the late 1600s, however, that Edmund Halley, of Halley's Comet fame, actually proved that Palissy's theory was correct. Using his sense of observation and mathematical skills, Halley roughly figured the quantity of water draining from all the rivers of North Africa and Europe into the Mediterranean. From this, Halley determined that the combined flow was about equal to the water falling as rain or snow on the region's drainage area.

$H_2 + O$ = Water

In the late 1700s, Henry Cavendish, a chemist and physicist, experimented with sending electricity through water inside a sealed glass tube. To his surprise, the water disappeared. Many experiments later, Cavendish determined that water is made of two gases. Setting a match to different combinations of these two gases, Cavendish discovered that when he combined two parts of the "inflamable air" with one part of the "vital air," there was an explosion that left a residue of water drops inside the wall of the container. Because of these experiments, humans became aware of how the molecular structure of known matter could be split apart and recombined. Ironically, this experiment opened the door leading to the creation of the atom bomb.

Building upon the work of Cavendish, Antoine Lavoisier renamed "inflammable air" hydrogen—"the water producer"—and "vital air" as oxygen.

However, it was not until 1805 that Louis Joseph Gay-Lussac and Alexander von Humboldt proved that water was made up of two parts hydrogen and one part oxygen. This breakthrough proved that water is neither elemental (in the ancient use of the term), nor an element of chemistry, but a true compound unto itself. This new designation of water set it apart from air and earth (which are mixtures), and fire, which is a chemical process.

FIRST THERMOMETER

Before the invention of the thermometer, there was no way for people to qualify or quantify changes in temperature. The only way people could relate to changes in temperature was to comment how hot or cold something was. Comments like, "It's really a hot day today," or, "Brrrrr, the water is so cold, it gives me chills," were the only means of measuring temperature before the invention of the thermometer.

Galileo, along with others before his time, made the common sense observation that substances expanded when exposed to heat and contracted when chilled. In 1603, Galileo invented what is considered the world's first "thermometer" (Greek—"heat measure"). He did this by placing a heated glass tube upside down in a bowl of room-temperature water. As the hot air in the glass tube contracted, it pulled a column of water up into the tube. Through observation, Galileo made notes about how changes in room temperature caused the column of water in the tube to go up and down. This was one of the very first scientific instruments using glass, other than Leonardo da Vinci's earlier glass-sided boxes that were used for studying the vertical energy of flowing water.

In the mid-1600s, Ferdinand II, Grand Duke of Tuscany, discovered that placing water inside a round bulb with a narrow tube attached that was open at the top, provided a more sensitive way for observing temperature change. This allowed more accurate observations of water's response to temperature under different conditions, changes that, compared with Galileo's first thermometer, were more accurate because there was less influence from atmospheric air pressure.

Eventually, mercury replaced water or alcohol as the liquid of choice in thermometers. Mercury proved more stable because it does not freeze or evaporate. However, it was not until 1714, more than 100 years following Galileo's first thermometer, that Gabriel Daniel Fahrenheit invented a sealed mercury thermometer in combination with a graduated scale. History remains unclear how Fahrenheit selected the scale with which we are now familiar, a scale where water freezes at 32 °F, and boils at 212 °F. As is obvious, the study of water's response to temperature change determined the points on the thermometer. Determining the point on the scale where water turned to ice was considered a notable temperature event, as was the boiling point where water changed into a gas.

In 1742, 28 years after Fahrenheit's creation of a scaled thermometer, Anders Celsius of Sweden, fixed the freezing point of water at 0° and the boiling point at 100°. At first, this was recognized as the centigrade scale because of the 100-degree range where water remained in liquid form. However, in 1948 the centigrade scale became officially recognized by scientists as the "Celsius scale." This was to pay homage to the inventor and to be consistent with the naming of the Fahrenheit scale after its inventor.

The invention of the thermometer opened a whole new era of scientific and medical observation and inquiry. For instance, it was determined that the average human body temperature is 98.6 °F. A degree or two above or below this reading indicated of potential health problems.

Of course, the invention of the thermometer also gave humanity something more to talk about in its everyday life. Instead of just saying, "It's a really a hot day today," we can now say, "My goodness, it's over 90° today!"

SEAWATER BLOOD

About 100 years ago, the French scientist Rene' Quinton established the chemical similarities between seawater and blood plasma. This revelation created considerable media attention and opened the minds of many people to the possible origins of humankind and other life forms.

Today, science and medicine inform us that about 90% of blood plasma is water. This seawater-like blood water distributes necessary nutrients and oxygen to every cell in the human body through a network of approximately 60,000 miles of arteries and veins.

Because water is a universal solvent, seawater contains all elements known to humankind. However, scientists have determined that seawater contains only 11 elements in concentrations greater than 1/100%. Not surprisingly, these 11 elements are present in the human body at the same concentrations found in seawater.

OUR EVOLVING WATER CONSCIOUSNESS

In 1530, Nicolas Copernicus published his pivotal work, *De Revolutionibus*. This work theorized that the earth rotated on its axis every 24 hours and traveled around the Sun every 12 months, certainly, a startling theory for its time.

Today, there is an evolving water theory that may be as startling as Copernicus' discovery. This theory states that water is a conscious entity that responds to human thought and thereby creates our living reality. By doing so, water manifests the world within and around us in direct response to our thoughts and emotions. Think positive and loving thoughts—water responds by creating a healthy environment within and without. Think negative and destructive thoughts—water responds by creating a negative and destructive environment.

In her book, *The Entangled Universe*, Mae Wan Ho, Ph.D., establishes the recent discovery that we humans, as well as "... all living organisms are liquid crystalline." Because of our liquid crystalline makeup, we and other life forms are sensitive to various forms of energy, including the energy waves from sound, thoughts, and emotions. In this same book, Mae Wan Ho also writes:

> The visible body just happens to be where wave function of the organism is most dense. Invisible quantum waves are spreading out from each of us and permeating into all other organisms. At the same time, each of us has the waves of every other organism entangled within our own makeup.... We are participants in the creation drama that is constantly unfolding. We are constantly co-creating and re-creating ourselves and other organisms in the universe, shaping our common futures, making dreams come true, and realizing our potentials and ideals

In other words, we as liquid crystalline beings, are the creators of our living reality, the same reality that responds to our thoughts and emotions and, in return, influences our very own existence.

Crystals in Water

The fact that "all living organisms are liquid crystalline" and all living organisms contain water is consistent with the fact that water naturally contains crystals. The presence of crystals in water becomes obvious when water is frozen. Snowflakes are another form of frozen water crystals. And, just as no two human faces are exactly alike, the same is true for snowflakes and other forms of frozen water crystals.

In the book, *Messages from Water*, Masaru Emoto provides scientific proof from his research results and photographs, as to how the crystalline structures of various waters respond to environmental factors, pictures, thoughts, words, music, and other influences. Emoto's research shows how water is a sensitive, lifelike entity, that is responsive to a spectrum of human and other energies. For example, Emoto would take a sample of distilled water and freeze it to obtain a baseline of what the frozen crystals looked like. Emoto would then expose samples of the same distilled water to different influences, such as certain thoughts, music, and written words. He would then freeze the water from the samples to see what changes, if any, occurred to the water's crystals. For instance, the projection of loving and caring thoughts into a water sample resulted in fully formed hexagonal crystals, whereas the projection of negative and violent thoughts into a sample of the same water produced incomplete or malformed crystals.

A FUTURE BASED ON WATER

Due to present water shortages and dire forecasts, the subject of water will soon permeate the minds of all levels of people around the world. The building of dams, such as the Three Gorges project in China, the large-scale desalination of seawater in the Middle East, the overpumping of the world's groundwater aquifers, the proposed transoceanic transport of Canadian water via huge floating water bags, and the competition amongst the world's most powerful transnationals to control sources of water are examples of the way humans are attempting to manipulate and distribute water. Simply stated, whoever controls water has political and financial power, as well as power over which life forms live or die.

However, beyond the physical management of water for profit and/or power, there are hints of other dimensions of water that may soon lead humanity to higher levels of consciousness. The fact that water responds to human thoughts and emotions is opening new possibilities to human evolution. In the near future, we may find that the projection of positive and loving human energies into water will deliver us into a healthier world filled with abundant food. If so, such a conscious progression in our evolution will give humanity the opportunity to help restore life to a planet that has been most supportive of our earlier stages of evolution.

This evolution of our water intelligence also includes our gaining a better understanding of the DNA molecule. For we are now discovering that the configuration of DNA is based on hydrogen bonding—the same as water. DNA is the basis for passing along genetic codes, and the H-bond of DNA is intimately connected with surrounding water clusters, so we may find a new basis for understanding genetic intelligence.

Perhaps, in the final analysis, the evolutionary progression of humanity will read as follows: Stone Age; Bronze Age; Iron Age; Industrial Age; Space Age; Computer Age, and then finally, Age of Water.

Whether or not humanity survives the Age of Water, in my opinion, will be determined by humanity's ability to work consciously with the living values and energies existing within water.

EFFECTIVE WATER EDUCATION STRATEGIES IN A NONTRADITIONAL SETTING

Sharon L. Lien
The Groundwater Foundation
Lincoln, Nebraska

INTRODUCTION

The Groundwater Foundation (TGF) is a nonprofit organization whose mission is educating people to care about and for groundwater and encouraging groundwater stewardship in all citizens. TGF's belief, which has come out of 20 years of experience, is that once citizens understand the importance of groundwater, they act to protect it.

Because of this, TGF puts particular emphasis on education and programming for the general public, especially youth. However, these educational programs have not been in the formal classroom setting, but in nontraditional settings, such as a festival environment, in competitions, or in the field.

This article discusses three specific examples of effective water education in nontraditional settings—Children's Groundwater Festival, Groundwater University, and Awesome Aquifers for Science Olympiad—and the impact these efforts have had on educating youth on groundwater protection and conservation.

Children's Groundwater Festival

From its inception in 1989, the Children's Groundwater Festival has been a success. Formulated around the concept of making learning fun, the Festival is staged in a fair-like atmosphere that is stimulating and exciting for participants, very unlike traditional classroom education. Whether witnessing the water cycle or building a well in a cup, students learn about every aspect of water and the environment in a fun way that helps them retain concepts and change their behaviors.

The Festival was designed as an annual 1-day event for fourth- and fifth-grade students of Nebraska. It is a dynamic, energetic day filled with hands-on education. Students from all over the state travel to Grand Island to learn about groundwater and related resources through interactive presentations given by groundwater and natural resources professionals who represent government agencies, environmental organizations, higher education, private business, and interested citizens. At the Festival, students have access to many water-related, experiential activities. There is nowhere else that children can have access to more environmental experts in such a fun and energizing atmosphere. This is a field trip like no other!

What makes the festival concept so effective is the unique and innovative methods used to teach the students about water. They learn what an aquifer is by making one with ice cream, discover how water is delivered to their home by becoming a human water distribution system, and gain an appreciation of the value of water by operating a working model of a treatment plant. Students don't just listen; they experiment, create, participate, react and learn, empowering them to make a difference.

The Children's Groundwater Festival has become an international model of hands-on environmental education. Over the years, water educational festivals have sprung up across the United States and around the world. To meet the demand for specialized training for festival organizers, TGF has published several books and provides workshops on organizing a water festival.

Groundwater University

Groundwater University (GU) was created in 1994 to provide intensive, experience-based groundwater education for youth and to encourage them to pursue science and water-related studies or careers. This program, directed toward junior and senior high school students, provides 3 days of hands-on activities, field trips, and presentations from top water professionals. GU offers participants the opportunity to study groundwater in a summer camp setting. Student activities include

- constructing mini groundwater models
- hiking the Platte River and digging down to the water table
- Participating in field trips to monitor wells, streams, and lakes to collect and test water samples
- investigating the relationship between microorganisms and water quality under a microscope
- building leadership and group skills through individual and team projects on topics of interest

As part of this program, students have the opportunity to share information about many groundwater issues that are unique to their communities. Because GU faculty includes professional hydrologists, chemists, irrigation experts, well drillers, and geologists from government agencies, private water testing laboratories, universities and colleges, and municipal water systems, students gain insight into water and science careers.

All activities at GU are hands-on and require participation. Some activities are done by teams, others are individual. Students take a posttest, and results are compared with those on their pretests to measure the change in knowledge.

Several GU graduates have continued their groundwater-related activities. Through GU, TGF educates students about groundwater helps them understand the connection between safe water supplies and public health, and ignites in them an interest in pursuing science and water-related studies and careers.

Awesome Aquifers for Science Olympiad

Science Olympiad (SO) is an international program devoted to improving the quality of science education, increasing student interest in science, and providing recognition for outstanding achievement in science education by both students and teachers. Student teams in participating middle and high schools all across the United States learn about different scientific topics; build models for some events; and compete at the local, state, and national levels in dozens of different SO events. The program emphasizes science learning, participation, interaction, building team spirit, and having fun. More than 2 million students are involved annually in SO.

Because an event on groundwater did not exist in the SO program, TGF created Awesome Aquifers. This event is designed to increase student understanding of groundwater concepts. Prior to the tournament, students conduct groundwater research and design and build a groundwater flow model according to SO rules that are standardized among participating schools. Students are required to demonstrate certain concepts, but there are almost no restrictions on how they build their models, so the event encourages student creativity.

During competition, students use their models to demonstrate an understanding of the hydrologic cycle, the physical makeup of an aquifer, and changes in the groundwater system. Students can earn bonus points for demonstrating possible contamination remediation techniques, such as using an intervening well to intercept a contaminant plume, chemical or physical treatment, or pump and aeration treatment.

Awesome Aquifers provides an opportunity to teach some of the nation's brightest young scientists about a resource that is critical to human and environmental health at a time when earth science programs are being reduced in many schools. Other objectives for the event are to give students a chance to learn about groundwater-related careers from working scientists who serve as judges or mentors for the event, and to educate two other groups who often know little about groundwater: science teachers, who generally serve as Science Olympiad coaches, and parents, who are often quite involved with student teams.

IMPACT ON EDUCATING YOUTH

In 1994 and again in 2002, TGF invested the time and resources to determine the effectiveness of the Children's Groundwater Festival and Groundwater University in impacting students' knowledge and behaviors positively toward groundwater. Results from this third-party behavior impact study indicated:

- The formats of these programs are especially well geared to the desired learning styles and approaches of students, and students have few suggestions for ways to improve either the Festival or Groundwater University. Students who have participated in these programs are likely to remember them for years, and up to 6 years later are able to relate what they learned during the event.
- The Festival did have a definable impact on students both in terms of gain in knowledge and change in behavior. After students attended the Festival, they exhibited an increase in knowledge as demonstrated through pretest and posttest scores. They also demonstrated a willingness to change their

behaviors, and in some cases actually adopted water conservation and protective actions.

- For Groundwater University students, learning goes far beyond groundwater education; it encompasses the "whole child." Students stated that they attended to learn about groundwater, but equally important for them was the opportunity to learn socialization skills, which they considered just as valuable. Almost all students noted that they appreciated learning about the cultural differences among themselves and their peers from other areas of the country. Many stated that they appreciated opportunities simply to interact with "experts" in the field (noted speakers, college professors, groundwater professionals, etc.) who had been invited to lead parts of the program.
- Students see the connections between groundwater and their own academic and professional interests. Though much of the information presented at Groundwater University is science-related, students can quickly perceive the connection of groundwater to other areas of interest and study, including writing, art, education, etc. Most importantly, the study determined that Groundwater University translates into behavioral change for most of the students (1).

Evaluations conducted by TGF have shown that Awesome Aquifers is also an effective activity to educate students on groundwater concepts. Based on information provided by students who have participated in the event,

- 97 percent of the students who participated liked the event.
- 89 percent indicated that they would participate again.
- 87 percent indicated that the event influenced them to conserve or protect groundwater.

Additionally, students overwhelmingly stated that they learned about the importance of groundwater, the complexity of an aquifer, and that groundwater is an important water source. Comments from students about what they learned included

- "This event helped me learn about the connection between pollution and the environment."
- "I learned that it is very hard to remove pollutants from an aquifer."
- "I learned more about the water cycle and how pollution affects it."
- "What an aquifer is, how it works, and what can we do to help."
- "A lot more about groundwater and the wonderful resource it is!"
- "I learned more than I thought possible and loved every bit of it."
- "We learned a plethora of information about conservation and protection."
- "Human activities affect the groundwater system hugely."
- "I learned how complex the underground water system is."
- "How waste can affect the water."
- "I learned a lot about the groundwater system and had an experience in actually seeing how a real aquifer works."
- "All about wells and the art of well construction. It also taught me how to protect and conserve groundwater."
- "We learned about groundwater and land water systems, and their relationships."

Real changes usually emerge from educational strategies that give the student a sense of involvement and ownership. Hands-on activities are a reliable approach for this as is student-directed learning(2). Thinking out of the box, or in this case out of the classroom, can allow water educators to explore less traditional educational strategies. By building creative and diverse programs that engage and empower youth and measuring the results of these efforts, the goal of protecting our groundwater supplies now and for future generations can be achieved.

BIBLIOGRAPHY

1. Marsters, M. (2002). Follow-up Behavioral Impact Study on Groundwater Foundation Student Education Programs.
2. Coyle, K. (2004). *Understanding Environmental Literacy in America: And Making it a Reality*. National Environmental Education and Training Foundation, Washington, DC.

EVOLUTION

DAVID HALEY
Picacadilly
Ulverston, Cumbria, United Kingdom

> "Those stars are the fleshed forebears
> Of these dark hills, bowed like labourers,
> And of my blood." (1)

In the cosmic sense of evolution, water may be viewed as part of the design of a supernatural "creator," or a quirk of fate—that a certain condition occurred for a combination of atoms (H_2O) to coincide in a particular atmosphere and remain for a sustained period of time within the atmosphere of a planet. And as Theodore Schwenk writes in his analysis of flow forms, *Sensitive Chaos*: "It is not possible to speak of the beginning or end of a circulatory system; everything is inwardly connected and reciprocally related. Water is essentially the element of circulatory systems" (2). He continues: "The great rhythms of the tides are a response to forces that work in the interplay of earth and cosmos...and for which, through its great impressionability, the element of water is a receptive 'sense organ'" (3). Governed by the Sun and Moon, this earth-bound phenomenon is claimed by creation myths and scientific conjecture and is the source of inspiration for poets and artists.

> Fallen from heaven, lies across
> The lap of his mother, broken by world.
>
> But water will go on
> Issuing from heaven
>
> In dumbness uttering spirit brightness
> Through its broken mouth.
>
> Scattered in a million pieces and buried
> Its dry tombs will split, at a sign in the sky,
>
> At a rendering of veils.
> It will rise, in a time after times,
>
> After swallowing death and the pit
> It will return stainless
>
> For the delivery of this world.
> So the river is a god
>
> Knee-deep among reeds, watching men,
> Or hung by the heels down the door of a dam
>
> It is a god, and inviolable.
> Immortal. And will wash itself of all deaths (4).

Some cultures and communities believe in forms of divine creation with regard to evolution, but the understanding of much Western science continues to be based on a synthesis of Darwin's natural variation and selection ideas and Mendel's model of genetic inheritance. Central to this theory, known as the Neo-Darwinian synthesis, is the idea that natural selection is based on the transfer of genes through reproduction.

New knowledge of genetics and other evolutionary factors continue to fuel scientific debate, providing different interpretations with regard to the timescales and population patterns of evolutionary change and development. However, what seems to be emerging is that "... it may in general be inappropriate to use a theory of any kind as an ultimate framework for life as a whole" (5).

Lakoff and Johnson explore these concepts in their book, *Philosophy in the Flesh*: "In evolutionary theory, survival is keyed to the ability to fit ecological niches. Fitting a niche and hence surviving can occur for many reasons: coloration that hides one from predators, a large number of offspring, the availability of food ..." (6), and to these we may add the availability of freshwater and the ability to adapt to climate change. They then examine Western culture's understanding of evolution: "There is a common folk theory of evolution, that evolution is a competitive struggle to survive and reproduce. The folk theory has normative implications: Competitive struggle to survive and reproduce is natural. Moreover, it is good, because it got us where we are" (7).

In *The Web of Life: A New Synthesis of Matter and Mind*, Fritjof Capra uses Maturana and Varela's theory of autopoiesis and Lovelock and Margulis' Gaia theory to explore the unfolding of life differently from Darwinian theories. Capra sees evolution as a process of biological and environmental co-evolution—"an ongoing dance that proceeds through a subtle interplay of competition and cooperation, creation, and mutual adaptation" (8). Other developmental strategies are noted in the recombination of DNA among bacteria; Lynn Margulis' theory of "symbiogenesis" recognizes the vital importance of cooperation in evolutionary processes.

Another pattern of development to consider is known as convergence, when different organisms meet similar challenges by evolving in similar ways (e.g., the development of eyes in diverse species). Convergence has also come into play during periods of intense growth and innovation following planetary catastrophes. "Thus the disastrous depletion of hydrogen in the earth's atmosphere over two billion years ago led to one of the greatest evolutionary innovations, the use of water in photosynthesis" (9). However, this led to a catastrophic accumulation of oxygen, which in turn led to the evolution of oxygen-breathing bacteria and that recombined with the hydrogen that was escaping the atmosphere to form water, "...thus keeping the planet moist and preventing its oceans from escaping" (10). Other interactive living processes, like fermentation, also contribute to the balancing of the earth's systems, known as Gaia: "Gaia is just symbiosis as seen from space: all organisms are touching because all are bathed in the same air and the same flowing water" (11).

"The Russian mineralogist Vladimir Vernadsky (1863–1945) recognized life as a great geologic force. Anticipating Hypersea, he called living matter "animated water." Animated water is an excellent description of life" (12).

What is certain is that there is no life without water. Animals evolved from aquatic microorganisms about 700 million years ago and plants followed around 200 million years later. Then, between 400 and 450 million years ago, plants followed by animals started to leave water, to live on land. This major habitat transition called for major evolutionary innovations to cope with the absence of water. These included respiratory systems adapted to oxygen, skin that could cope with increased radiation from the Sun, and muscular and skeletal development to deal with increased gravity. To this day, however, all animals maintain their origins in the form of the

> ... womb that simulates the wetness, buoyancy and salinity of the ancient maritime environment. Moreover, the salt concentrations in the mammal blood and other bodily fluids are remarkably similar to those of the oceans. We came out of the ocean more than 400 million years ago, but we never completely left the seawater behind. We still find it in our blood, sweat and tears (13).

Capra brings the evolutionary story up to the present day:

> Far away from their African homeland, the early humans had to endure extremely harsh climatic conditions that had a strong impact on their further evolution. The entire evolutionary history of the human species, from the emergence of *Homo habilis* to the agricultural revolution almost two million years later, coincided with the famous ice ages (14).

CLIMATE CHANGE

It would seem that extreme and sudden changes in climatic conditions have given rise to great evolutionary innovations and further developments occur over relatively long periods of adaptation to prevailing conditions. However, as the dinosaurs found, not all species or groups of species survive such changes, and these concerns are among the important issues facing the Intergovernmental Panel on Climate Change (IPCC).

In recent history, an Irish scientist, John Tyndall (1820–1893), discovered the importance of water vapor in maintaining the earth's life-sustaining atmosphere.

> Tyndall's experiments also showed that molecules of water vapour, carbon dioxide and ozone are the best absorbers of heat radiation and that even in small quantities these gases absorb much more strongly than the atmosphere itself, a phenomenon of great meteorological importance. He concluded that among the constituents of the atmosphere, water vapour is the strongest absorber of radiant heat and is therefore the most important gas controlling the Earth's surface air temperature. He said that without water vapour the Earth's surface would be "held fast in the iron grip of frost." He later speculated how changes in water vapor and carbon dioxide could be related to climate change (15).

From the nineteenth century, Tyndall's grasp of the potential climatic problems facing the twentyfirst century are now the source of much speculation. In the publication Climate Change 2001: Synthesis Report, Summary for Policymakers, the IPCC was clear about certain consequences:

> Changes in sea level, snow cover, ice extent, and precipitation are consistent with a warming climate near the Earth's surface. Examples of these include a more active hydrological cycle with more heavy precipitation events and shifts in precipitation, widespread retreat of non-polar glaciers, increases in sea level and ocean-heat content, and decreases in snow cover and sea-ice extent and thickness. For instance, it is very likely that the twentieth century warming has contributed significantly to the observed sea-level rise, through thermal expansion of seawater and widespread loss of land ice (16).

A CONFLUENCE OF IDEAS

Uncertainty about how to model predictions of the future effects of global warming and climate change are still contentious issues among the world's top scientists. Different methods of selecting and presenting information produce different scenarios that favor opposing political, social, and commercial lobbying groups. Professor Peter Sammonds of University College London believes that "(t)here is a serious problem that the starting point for discussion on the environment has tended to not be considered scientific knowledge but alarmist interpretations of future climate set in a highly moralistic framework" (17). He ascribes most of the effects of climate change to naturally occurring phenomena, challenging the validity of available data and its misinterpretation. Sammonds criticizes the insurance industry for "(t)his talking up of the threat of global warming..." to meet their own ends and asserts that "... nothing should be done which would curtail global economic growth through misplaced strategies to curb carbon dioxide emission. Otherwise, we would be damaging our very ability to adapt to environmental changes."

However, the IPCC's 2001 Summary concludes,

> Significant progress has been made in ... many aspects of the knowledge required to understand climate change and the human response to it. However, there remain important areas where further work is required, in particular:

- The detection and attribution of climate change
- The understanding and prediction of regional changes in climate and climate extremes
- The quantification of climate change impacts at the global, regional, and local levels
- The analysis of adaptation and mitigation activities
- The integration of all aspects of the climate change issue into strategies for sustainable development
- Comprehensive and integrated investigations to support the judgement as to what constitutes dangerous anthropogenic interference with the climate system (18).

Writing from an explicitly "green" perspective, in her book, *Water Wars*, Vandana Shiva expresses her views on the relationship between water and climate change:

> Water is life, but too much or too little of it can become a threat to life. The stories of Noah and Vishnu Purana are tales of mythic floods that whipped out life on the planet. While floods and droughts have always occurred, they have become more intense and more frequent. These climatic extremes are linked to climate change, which is in turn linked to pollution of the atmosphere by fossil fuels ... The impact of climate crisis on all forms of life is mediated through water in the form of floods, cyclones, heat waves, and droughts (19).

This echoes the Renaissance concern for apocalyptic visions, and Leonardo da Vinci's (1452–1519) late drawings of the Deluge (Fig. 1) seem to express this Dantesque preoccupation—water becomes a frightening phenomenon in violent climatic conditions. But Leonardo's aesthetic was to envision these apparently catastrophic events as the spectacularly exciting stuff of life. This description illustrates his imagination and love of the dramatic, as much as his lifetime study of water that could equally find a lyrical likeness between a vortex in water and the curls of a young girl's hair:

> And the ruins of a mountain fall into the depth of a valley, creating a shore for the swollen waters of its river, which having already burst its banks, will rush onwards with monstrous waves, the greatest of which will strike and destroy the walls of the cities and farmhouses in the valley. And the ruins of the high buildings in these cities will throw up immense dust that rises like smoke or wreathed clouds against the falling rain. But the engorged waters will cascade around the pool that contains them, striking in eddying whirlpools

Figure 1. Leonardo da Vinci Deluge drawing No. 46 (146)—12383 Windsor collection.

> against different obstacles, throwing up muddy foam as they leap into the air before falling back and then again being dashed into the air. And the whirling waves that flow from the site of percussion are impelled across other eddies going in opposite direction, and afterward their recoil will be tossed up into the air, without dashing off from the surface. And at the outlet where the water flows from the pool, the spent waves will be seen spreading out toward the outlet, and afterward, falling or pouring through the air, and gaining weight and impetus movement, [this water] will hit the water below, pierce through it in a furious rush to reach bottom, and then being thrown back, the [water] returns to the surface of the lake, carrying with it the air that was submerged within it, and it, remains at the outlet, its foam mingled with logs of wood and other things lighter than water . . .The crests of the waves of the sea tumble down to their bases, whirling and falling with friction above the bubbles on their side (20).

Our evolutionary destiny and those of many other species depend on our ability to maintain adequate supplies of freshwater and adapt to the accelerating effects of climate change: "We must always bear in mind that the water we are using today is borrowed from future generations, who will require it for their survival" (21). Despite pollution, species depletion, and the uncertainties of the effects of global warming, the oceans still hold the greatest diversity of life on the planet and the potential for the continuum of evolution. As E. C. Pielou notes: "Living things depend on water, but water does not depend on living things. It has a life of its own" (22).

BIBLIOGRAPHY

1. Hughes, T. (1995). Fire-Eater. *Ted Hughes: New Selected Poems 1957–1994*. Faber and Faber, London, p. 30.
2. Schwenk, T. (1996) *Sensitive Chaos: The Creation of Flowing Forms in Water and Air*. Rudolf Steiner Press, London, p. 13.
3. Schwenk, T. (1996). Ibid.
4. Hughes, T. (1995). Ibid., *The River* p. 243.
5. Steen, F. (2001). *Evolutionary Theory: A Theory of Changes in Organic Design through Controlled Random Mutations and Contingent Selection*. Available: http://cogweb.ucla.edu/ep/Evolution.html.
6. Lakoff, G. and Johnson, M. (1999). *Philosophy in the Flesh: The Embodied Mind and its Challenge to Western Thought*. Basic Books, New York, p. 560.
7. Lakoff, G. and Johnson, M. (1999). Ibid., p. 557.
8. Capra, F. (1996). *The Web of Life: A New Synthesis of Mind and Matter*. HarperCollins, London, p. 222.
9. Capra, F. (1996). Ibid., p. 227.
10. Capra, F. (1996). Ibid., p. 232.
11. Margulis, L. (1998). *The Symbiotic Planet: A New Look at Evolution*. Weidenfrld & Nicholson, London, p. 2.
12. Margulis, L., (1998). Ibid.
13. Capra, F. (1996). Ibid., p. 244.
14. Capra, F. (1996). Ibid., p. 253.
15. The Tyndall Centre web site, *A brief biography of John Tyndall, FRS, DCL, LLD*. Available: www.tyn@uea.ac.uk.
16. The International Panel on Climate Change. (2001). *Climate Change 2001: Synthesis Report, Summary for Policymakers—IPCC Third Assessment Report*, p. 6.
17. Sammonds, P. (2002). Extreme climates. In: *Nature's Revenge? Hurricanes, Floods and Climate Change for Institute of Ideas*. Hodder & Stoughton, London, p. 21.
18. Ibid. (2001). *The International Panel on Climate Change*. p. 33.
19. Shiva, V. (2002). *Water Wars: Privatization, Pollution and Profit*. Pluto Press, London, p. 40.
20. Bambach, C.C. (Ed.). (2003). BAMBACH, C. C. quoting DA VINCI, L., Winsor RL 12665r; Richter 1970, Vol. 1, pp. 355–356, no. 609, *Leonardo Da Vinci Master Draftsman*. The Metropolitan Museum of Art, New York, p. 630.
21. Kyoto Excutive Committee for the 3rd World Water Forum. (2003). *Kyoto Water Declaration*.
22. Pielou, E.C. (1998). *Fresh Water*. University of Chicago Press, Chicago, IL.

HISTORY OF POND FISHERIES IN POLAND

Janusz Guziur
University of Warmia and
Mazury in Olsztyn
Olsztyn, Poland

The origins of carp breeding in Europe and Poland are sunk in the obscurity of bygone times. In Europe, the first historic record of the carp as a cure for ailments is in the writings of St. Hildegard (1098–1170), the prioress of a Benedictine convent. Some time later, carp breeding in ponds was mentioned by the monk, Albertus Magnus (1230–1310), in the archives of the Kladbrubkski monastery in Bohemia (1115), and in an investiture act by King Przemyslaw Otakar II granted in 1227 in Moravia.

Several hypotheses have been suggested concerning the introduction of carp breeding in Poland. It is believed that two routes were taken to bring the carp from its native Black Sea watershed to Poland, but whichever way it traveled, the fish was brought to Poland by the clergy (bishoprics). The oldest monastic order that first engaged in farming and fish breeding was the Order of the Cistercians, which was brought to Poland in twelfth century from the province of Brabant (on the border of Belgium and France).

The domestic mirror carp is known to have been bred in Poland since the mid-thirteenth century. This breed is not found in the wild in Europe or Asia, unlike the wild river carp (sazan), which has local forms that live free, for example, in the Danube. Because of its inferior body exterior and growth rate, wild sazan differs distinctly from domesticated mirror carp and fine-scale carp.

In the 1960s and 1970s, attempts were made to 'refresh' the blood of farm carp by breeding Danube sazan in fish ponds or crossing it with breeding carp (1–3).

According to Balon (3), mirror carp, also known as royal carp (*rex cyprinorium*—Johnston 1650) originated from mutants of the Danube sazan, which Roman legionnaires brought to Rome, where it made its home and from where it was later taken to monasteries around west Europe (Fig. 1). The monasteries propagated the breeding of the cultured (domestic) carp, also in Poland; this was encouraged by frequent and lengthy religious fasts (up to 180 days a year) (3,4).

The first known written document mentioning carp in Poland, dating back to the year 1466, was penned by the famous chronicler of the Battle of Grunwald, Jan Długosz (1415–1480). Długosz (5) describes three carp in the coat of arms of the Silesian nobles Korbczok (*tres pisces, qui carpiones vocantur*), which was later confirmed by Paprocki's Armorial (1584), or two trouts (*duo pisces, qui trutae appelantur*) in the coat of arms of the family Musznik de Dombrova. The carp also appears in the accounting books of the Order of Teutonic Knights (Deutscherdrensritter) of 1399 (6). This evidence proves that the carp was widespread in Poland as early as the thirteenth and fourteenth centuries.

Based on books by several authors (4,6–12), four periods of carp breeding development in Poland can be distinguished:

First Period from the Mid-Thirteenth to the Mid-Seventeenth Centuries (To the End of the Swedish Wars)

This was the time of dynamic development and flourishing of pond fish farming (Fig. 2). The first printed books on fish farming came out. The first pond fish farming textbook in Europe was written by Jan Dubrawski (*Dubravius*) from Ołomuniec (13), who also described fish farms in Polish lands (Cieszyn Silesia). Soon to follow was the first Polish textbook on fish ponds by Olbrycht Strumieński (14) "O sprawie, sypaniu, wymierzaniu i rybieniu stawów," and then a book by Stroynowski (15), reprinted in 1863. In 1568, the Polish historian Marcin Kromer included some information on carp farming in his description of the Polish state (4).

Figure 1. Fishing in a monastery pond (3).

Pond fish farming in Poland, and especially in Upper Silesia, the Polish cradle of pond management, was considered the most modern and strongest in Europe, on a par with fish breeding in Bohemia and Moravia, in terms of modern technologies of constructing fish ponds and management methods (including restocking) as well as the volume of production. Restocking, which involved moving growing fish to other ponds, prepared and filled with water in advance, developed in Poland in the second half of the sixteenth century and became strongest in the following centers: Oświęcim-Zatorze, estates of the archbishops of Gniezno and Krakow, and in the valley of the Barycz river (Milicz). Noteworthy is the fact that the first information about restocking can be found earlier in the works of Strumieński (14) and partly Dubravius (13). In the early eighteenth century, when pond fish breeding in Europe was in decline, this method was forsaken and remained forgotten until the end of the nineteenth century (4,10,16).

The most extensive information on fish management in the fifteenth–seventeenth centuries is available from the region of Silesia, especially Opawa-Cieszyn, Pszczyna, and Racibórz Silesia, described by the Czech historians of carp breeding, including Dubravius (13) and Hurt (17). The area of Silesia had favorable climatic (warm summers), soil (clays with enclaves of loess), and especially water conditions (the sources of the large Polish rivers, the Oder,

Figure 2. Pond fisherman from Milicz region, seventeenth century (9).

Vistula, Olza, and Soła). Politically, Silesia was loosely tied to the Czech Kingdom, which allowed the region to enjoy great economic independence. The total estimated area of fish ponds in upper Silesia in the fifteenth–sixteenth centuries was about 25–30 thousand ha, which contained all categories of ponds used presently (12). Those ponds were used for breeding other fish species apart from the carp: the so-called black fish, including all fish that have a dark skin and scale color (crucian carp, burbot, tench, mud loach, wels) and white fish (common bream, roach, perch, ide, chub, ruffe, bleak, sunbleak) as well as other less valuable species (9). Both groups of fish equaled 15–20% of the total amount of carp, pike, and walleye fished from ponds. From the sixteenth century on, brown trout were farmed in the mountainous parts of Silesia (Cieszyn, Opawa). The fish was carried by highlanders (traveling on foot) over the mountains and then transported to the court in Vienna.

In those days in Poland, as in neighboring countries (Bohemia, Moravia, Hungary, and Bavaria), some nobles and dukes commissioned construction of huge fish ponds built as 'monuments to glory,' such as the ponds *Stary* (700 ha) and *Grabownica* (500 ha) near Wrocław and Milicz, *Kaliski* (625 ha) near Opole, *Bieruński* (500 ha) near Katowice, and *Sigmund August's* (480 ha) near Knyszyn (Białystok). The latter, also called '*Czechowizna*,' was built in 1560 by Czechs brought to Poland by Queen Bona (12,18). Many ponds and whole fish farms have survived and are still in use, for example, **Gołysz, Ochaby**, and **Landek** near Cieszyn; **Zator, Grojec, Osiek**, and **Czernichów** near Kraków; **Łyszkowice** near Łowicz; **Knyszyn** near Białystok; and **Przygodzice** and **Milicz** in the Barycz river valley, together with several smaller ponds in the valleys of the upper Oder and Vistula rivers. The high status that fish and fish pond farming attained then is confirmed by the emblems of many Polish towns that bear images of carp, pike, trout, and other fish.

Second Period from the Mid-Seventeenth Century to the 1860s

This was the time when fish pond breeding declined in all of Europe. Carp breeding became unprofitable and was nearly discontinued. In Poland, more than half of the existing ponds were liquidated (dried) and used for more profitable production of cereals, predominantly wheat (9,17). It was then, however, that Revised Krzysztof Kluk published his excellent work on farming (19), which remained popular until the mid-nineteenth century. The third volume of this book was devoted to fish and fish ponds (9,20,21).

Third Period from 1868 to 1939

This was another period of dynamic development, especially in modern methods of fish pond management. The impulse for this renewed development came from the area of Cieszyn Silesia, a fish pond 'reservoir,' located in the upper Vistula river watershed. It was initiated by the precursor of modern carp breeding methods, Tomas Dubisz (*Dubisch*). Tomasz Dubisz (1813–1888), a Slovak who was born near the Danube and lived in Vienna, was an illiterate fish breeder brought to Ruda Rożaniecka (1868) by Baron Wattman. Later, he worked as a master fisherman in Landek located on the land estates belonging to Cieszyn Manors (property of Archduke Albrecht Habsburg), which today forms Poland's largest experimental fish farm of the Polish Academy of Sciences in Gołysz, Cieszyn Silesia. The year 1869 marked an important event in carp pond breeding, when Tomasz Dubish, master fisherman at a complex of ponds in Iłownica (PAN Gołysz) introduced a new (or actually forgotten) carp breeding method of restocking. That is the system of moving 1-year-old fingerlings to a succession of ponds, prepared earlier and filled with water, reducing the density of stocking at the same time (10,12,22). The creation and implementation of this system made it possible to select the breeding material, shorten the breeding cycle from 4–5 to 2–3 years, and achieve dynamic growth in annual profit (Dubisz himself attained an increase in profit from 3000 to 40,000 zlotys per pond). Soon this system of fish farming was adopted all over Europe (16,18,21) and, with small modifications, is practically used until the present (11,23).

The breakthrough in carp breeding achieved by Tomasz Dubish was continued by his pupils and successors, including Adolf Gash (1839–1915) from Czechowice, a creator of a new breed of Polish carp (known then as the Galician carp) characterized by a strong back; Michał Naimski (1842–1915); Paweł Morcinek (1961–1938) from Skoczów, a known fish breeder active in Silesia and Moravia and also the first biographer of Tomasz Dubisz; Andrzej Gostkowski (1839–1889) from Wadowice; and later the fish breeder Oskar Rudziński (1850–1919) from Osiek near Oświęcim, together with his son, Edward Rudziński (1892–1980), the senior of Polish carp selection breeders in the years between the First and Second World Wars.

Owing to the system developed by Dubisz and the introduction of the new Polish carp breed, the productivity of ponds immediately rose by more than 30% (18). All these achievements by Polish fish farmers and breeders at the turn of the nineteenth and twentieth centuries guaranteed Polish fish farming a leading position in Europe.

The 20 Years Between World Wars I and II

This was the time of a dynamic increase in the area of ponds in Poland. In 1919, the Parliament of the Polish Republic passed an act on leasing unused farmland properties, and a year later (1920), on their liquidation and lease and management free of charge. The two legal acts made it possible to diminish the area of fallow lands from 4.6 million ha (1918) to 0.37 million ha (1923). The total area of fish ponds increased from 37.6 thousand ha in 1923 to 88.8 thousand ha in 1938 (Table 1), mainly in the Lakelands; fewer ponds were in the regions of Pemarania, Wilno, Wołyń, and Podlasie. New fish ponds appeared on numerous water and unused land areas (9,10).

In the years prior to the outbreak of World War I, Poland produced more than 13,000 tons of fish (Table 2) and held first position as a European fish producer in the volume of output (30% of the European fish production) and the amount of fish produced per capita (0.37 kg). The average yield of Polish fish ponds was 163 kg/ha, with

Table 1. Fish Culture in Poland (1923–1938)[a]

Feature	Year 1923	1928	1938
Total area of ponds, ha	37,550	58,750	88,775
(%)	100	156	236
Number of pond farms, n	1,350	2,925	7,562
[%]	100	217	560
Pond production, tons	6,000	9,000	13,000
[%]	100	150	217

[a] References 4 and 10.

strong regional variations. The highest yields were still produced by fish breeders of Silesia and the areas near Kraków (to 800 kg/ha) and in Masovia (300–500 kg/ha), the lowest output was recorded in the provinces of Wilno, Polesie, and in Pomerania (to 123 kg/ha).

The first Polish Committee of Carp Selection was established in those years, headed by Prof. Franciszek Staff (1889–1966) and Edward Rudziński. After World War II, the Committee was presided over by Prof. Kaziemierz Stegman (1911–1978) of the SGGW in Warszawa.

Higher yields were obtained from fish ponds by widespread use of lupine to nourish fish (76% of the pond area) and mechanical and agronomic cultivation of the pond bottom, including fertilization. Pond operation was also aided by high unemployment and inexpensive labor in the countryside. There were also more than 50 centers of trout breeding (mountain and upland trout) of the area of 56 ha.

The FAO publications (Sarig 1966, cited by Ref. 11) indicate that carp from Poland and Germany were transported to Yugoslavia and further to Israel and the countries of the Far East, Indonesia (Java), North America, and Australia.

Postwar Period

After World War II and the subsequent change in the borders of the Polish state, the area of ponds in Poland diminished by more than 22 thousand ha and totaled 66,525 ha in 1945. Although about 9000 ha of fertile and well-managed ponds were taken over in the new lands in the west at the expense of 31.4 thousand ha of less fertile ponds in the former eastern territories of Poland (Table 3), the fish pond output attained at that time was comparable to that of the year 1938.

At present, the largest areas of carp ponds are in south and central Poland (three-quarters of the total area). In north Poland, in the territories of natural lakes (total area of 320 thousand ha) or in the mountainous regions, there are few ponds. The largest area of carp ponds is in the provinces of Silesia (Katowice)—11.4 thousand ha; Lower Silesia (Wrocław)—9.3 thousand ha; Masovia (Warsaw)—6.8 thousand ha; Pokarpackie (Rzeszów)—6.1

Table 3. Changes in the Structure of Inland Waters in Poland after World War II

Specification	Area of Waters Ha: Ponds	Lakes	Total
West Poland (new territories)[a]	9,150	202,850	212,000
—taken over by the USSR	31,400	103,536	134,936
—increase in total	—	+99,314	+77,064
Difference (−less)	−22,250	—	—

[a] Opole, Wrocław, Zielona Góra, Gorzów Wlkp., Szczecin.

Table 4. Fish Production in Ponds in Poland (1993–2001)[a]

Year	Eating Fish[b]: CARP (K_H)	Others	Stocking Fish[c]: K_1	K_2	Total Production [100.0%]	Mean Price of 1 kg of Carp: Zł	$ US
	Production (Thousand Tons)						
1993	20.4–61.4%	1.5	2.2	9.1	33.2	3.6	—
1995	19.6–64.5%	0.7	3.2	6.9	30.4	4.4	—
1996	21.4–63.5%	1.4	2.5	8.4	33.7	6.8	1.76
1999	21.4–64.1%	1.1	3.1	7.8	33.4	6.8	1.55
2000	22.7–60.2%	1.5	2.7	10.8	37.7	5.9	1.44
2001	20.3–68.1%	1.4	3.4	4.7	29.8	5.5	1.31

[a] Reference 24.
[b] K_h commercial, eating carp (2- and 3-year-old).
[c] K_1: 1-year-old carp (fingerling); K_2; 2-year-old carp.

Table 2. Total Area of Fish Ponds and Fish Production in Europe Between World Wars I and II[a]

Country	[year]	Area of Fish Ponds, ha: Total	Pond Area per 100,000 ha of Geographic Area	Carp Production, t	Fish Consumption, kg/person
Soviet Union[b]	[1935]	89,500	14.9	12,550	0.09
POLAND	[1938]	88,775	228.6	13,000	0.37
Germany	[1929]	75,000	131.6	7,500	0.11
Czechoslovakia	[1937]	52,000	359.4	2,700	0.16
Hungary	[1929]	9,800	93.3	2,600	0.25
Yugoslavia	[1931]	5,300	21.4	2,000	0.13
Latvia	[1930]	5,000	74.2	325	0.16
Lithuania	[1930]	3,000	35.7	300	0.08
Austria	[1932]	2,500	35.4	300	0.04
France	[1934]	1,200	21.5	.	.
Sweden	[1938]	1,600	3.5	.	0.03

[a] Reference 10.
[b] European part.

thousand ha; Wielkopolskie (Poznań)—5.8 thousand ha; and Lubelskie (Lublin)—5.7 thousand ha. In some parts of Poland, the total area of fish ponds reaches 6% of the geographical area of the locality (Pszczyna, Skoczów, Oświęcim) or up to 10% (Milicz near Wrocław). The total registered area of carp and trout ponds in Poland is 71 thousand ha (72% of which is the area used). This figure does not include several thousand home ponds built in the last 10 years by private individual farmers.

In the postwar years, inland fisheries in Poland, including pond fish farming, has been reorganized several times. As a result, until 1990 more than three-fourths of the pond area was managed by the State Fish Farms. The remaining area was exploited by fishermen's cooperatives, the Polish Angling Association, and, to a much smaller extent, by farmers and individual private fish farmers [about 16 thousand ha of ponds (23)].

Large and medium size trout farms are now located mainly in north Poland, in the area rich in clean, cold water. Fewer trout ponds are found in south Poland (the Karpaty and Beskidy mountains), predominantly smaller and older pounds (some dating back to prewar times), for example, łopuszna on the Dunajec River. The dynamic growth of this branch of the economy began in the 1960s and 1970s.

Currently there are more than 180 trout pounds in Poland, including very large fish farms whose output is 200–1,000 tons annually and produce mainly rainbow trout, for instance, Mylof (1150 t), Zabrodzie and Tarnowo (400–700 t). The output of trout in Poland was 10–12 thousand tons in the years 2000–2002 and of carp, more than 20,000 tons (Table 4).

Apart from carp farms, since the year 2001, the Polish Angling Association has been managing 16 trout breeding farms (producing mainly salmon fingerlings). These farms are located in the mountainous regions and in the lowlands. They produce fingerlings of salmonids, mainly salmon, and some species of river fish (idus, asp, vimba bream, barbel., chub, beaked carp undermouth, burbot).

The political and social transformation in Poland led to the liquidation of the State Fish Farms. A number of fish farms, some well functioning (both financially and technologically), have been divided into smaller units, not always organizationally efficient. Some of them, subject to restructuring, were taken over by the Agricultural Property Agency of the State Treasury (APA) as one-man companies of the State Treasury until their complete privatization. A considerable number of inland water reservoirs (lakes) has been leased and carp and trout ponds have been sold to private fish farmers and producers.

BIBLIOGRAPHY

1. Rudziński, E. (1961). Vergleichende Untesuschungen über den Wildkarpfen der Donau den Teichkarpfen. *Z. Fischer* **10**: 105–135.
2. Leszczyńska, W. and Biniakowski, L. (1967). Breeding of the wild carp in the sport fisheries stocking centre Otorowo. (in Polish). *Gospod. Rybna* **19**(2): 19–21.
3. Balon, E.K. (1974). *Domestication of the Carp (Cyprinus Carpio L).* R. Ont. Museum Life Sci. Misc. Publ, Ontario.
4. Włodek, J.M. (1980). Development of pond breeding of carp in south Poland. *Post. Nauk Roln.* **2/80**: 107–130.
5. Długosz, J. (1863–1887). *Chorografia Regni Poloniae. Historia Poloniae*. Opera omnia. Cracov.
6. Rudziński, E. (1962). Fragments of the history of the European domestic carp. (In Polish). *Gosp. Rybna* **9**: 3–5.
7. Rudziński, E. (1963). Carp in Poland. (In Polish). *Gosp. Rybna.* **1**: 6–8.
8. Górzyński, S. (1964). *Outline of the History of Fishing in Ancient Poland.* (In Polish). PWRiL, Warszawa.
9. Nyrek, A. (1966). *Fishery in Upper Silesia From Mid 16th to Mid 19th Century.* (In Polish, with French summaries). Travaux de la Societe des Scienc.et des Lett.de Wrocław, Vol. A-111.
10. Szczygielski, W. (1967). *A Brief History of Inland Fishery in Poland.* (In Polish, with Russian and English summaries). PWRiL, Warszawa.
11. Wolny, P. (1976). *Carp.* PWRiL, Warszawa, wyd. II.
12. Guziur, J. (1991). *Fisheries in Small Inland Reservoirs.* (In Polish). PWRiL, Warszawa.
13. Dubravius, J. (1543). *On Fish Ponds and Fish which Grow in Them* (Polish translation). Cracow, about 1600.
14. Strumieński, O. (1573). *On Management, Construction, Measuring and Stocking of Ponds.* (In ancient Polish). F. Kucharzewski, 1897, Cracov.
15. Stroynowski, S. (1609). *Description of Pond Order & Farm Management.* (In Old Polish). In: Gawarecki Z. and Kohn A. (1860). *Polskie stawowe gospodarstwo.* Warszawa.
16. Burda, V. (1935). *Grunlagen der Kapfenzucht.* Neudamm, Germany.
17. Hurt, R. (1960). *History of Pond Fisheries in Moravia and Silesia.* Vol. I–II. (In Czech). Brno, Opava.
18. Maylert, J. (1927). *Construction and Management of Polish Ponds.* (In Polish). Warszawa.
19. Kluk, K. (1780). *Natural History, Origins and Management of Domestic and Wild Animals ….* (In ancient Polish). Vol. III, Warszawa.
20. Leśniewski, P. (1837). *Polish Fishery.* (In Polish). Warszawa.
21. Morcinek, P. (1909). *Geschichte des Dubischverfahrens und des von Dubisch Herangezuchteten Schlesischen Karpfen.* Lands- und Fortswirtschafliche Gessellschaft, Troppau.
22. Guziur, J. (2001). *Polish and Czech Contacts in the Development of Pond Fishery in Cieszyn Silesia.* (In Czech). *Int. Fish Conf.* VURH-Ju Vodňany 2–5.05.
23. Guziur, J. and Turkowski, K. (1995). *Agriculture—Aquaculture Integrated Farming in North-East Poland—Present State and Prospects. Int. Aquacult. Conf. "Aquaculture in Eastern European Countries"*. St. Zagora (Bulg.), July 26–28, p. 15.
24. Krueger, A. (2000). *Pond Fisheries in Poland in 2001.* (In Polish). Int. Fish. Conf. PTR, Poznan, February 7-8, p. 8.

WATER: THE KEY TO NATURAL HEALTH AND HEALING

F. Batmanghelidj
Global Health Solutions
Falls Church, Virginia

It is true that the micromechanical side of medicine has advanced at dizzying rates, but our scientific understanding of the way the human body works is still in the Dark Ages. The inventive, resourceful, but commerce-motivated employees in science have managed to walk a tightrope between true science and illogical nonsense-science of medicine, to the point that the human body is now viewed as a badly designed product of nature in constant need of their interpretation and meddlesome interferences.

They have sold this false bill of goods to the gullible public to the point that a vast percentage of America's GNP is allocated to cope with their thought in our society. This trend started at the turn of the last century, when scientists were flexing their muscles to gain acceptance for their understanding of the human body. Some of these scientists did produce miracles—but a few did unimaginable damage.

Walter Bradford Cannon, for instance, a prominent physician of his time, did untold harm to the progress of medicine. Cannon asserted that thirst in the human body reveals itself by the presence of a dry mouth, brushing aside the earlier view that thirst is more of a general sensation that needs to be studied and defined. The simplistic but false dogma of "dry mouth portending thirst" gained a foothold in medical science and has been with us ever since.

This seemingly insignificant assertion changed the course of scientific inquiry into the way the human body works. It pushed science in the direction of the chemical composition of the body and how these solid components of the body interact with one another. It gave way to the dominance of "solutes" and application of the science of particle chemistry to the physiological functions of the body—an integrated and complex water-dependent multisystem system. Even now, the hydrolytic role of water in chemical reactions is not given any consideration when evaluating the outcome of chemical actions in the body.

This direction of totally erroneous thought has thrown medical science into ignorance about the complex, water-dependent natural design of the human body. Solute chemists took over and made a mess of medical science. Nowhere is this mistake more apparent than in AIDS research, when every ill that befalls a person with this diagnosis is blamed on a minute particle called a retrovirus—in complete disregard of the obvious.

To survive through the millennia, the human body has overcome deadly viruses such as polio, smallpox, measles, flu, and more, and has developed the ability to mount a very effective immune response within 9 days. In other words, if a virus has not been able to kill the person in the first 9 days of its attack, it will itself be killed, and the immune system will be all the wiser for the encounter. Surely if the body, generation after generation, has survived fast viruses, getting over slow viruses must be a cinch. With this logic installed within the human body, how could we blame a devastating "disease process" on a retrovirus?

Surely the presence of the retrovirus in people with AIDS has other, less commercial explanations that are being ignored for the obvious reason that a natural solution would emerge. If you read my book *Your Body's Many Cries for Water*, you will discover that the AIDS virus is a byproduct of a larger metabolic disturbance that has a drastic disruption of the amino acid pool composition of the body at the foundation of the problem. Naturally, from this direction of thought will emerge a noncommercial natural treatment protocol that will emancipate society from the burden of costly AIDS research.

THE ILLOGICAL FOUNDATIONS OF MEDICAL SCIENCE IN THE TWENTIETH CENTURY

In all, four errors in scientific assumptions have made a mess of our view of the human body—in itself an efficient, self-repairing, self-replicating design:

1. A dry mouth is the only signal indicating the body's need for water.
2. Water has no direct initiative in chemical reactions of the body and is only a subservient and passive element in the actions of the solid matter that it dissolves and circulates in the body.
3. The human body is efficient in regulating its water intake throughout the life span from childhood to old age.
4. Any fluid can satisfy the long-term water needs of the body.

WHY ARE THESE ASSUMPTIONS FALSE?

Dry mouth is not a reliable indicator of thirst because the body produces ample saliva even though there may be established and symptom-producing intracellular drought in different regions of the body. To facilitate the act of chewing and swallowing food, producing saliva, even in a dehydrated state, is an overriding primary function in the body. The salivary glands produce about 2 liters of saliva a day to supply the upper parts of the intestinal tract with water and some digestive enzymes, especially when the body is dehydrated. Oversalivation begins when the body is short of sodium, a contributing component of dehydration.

To think that water has no direct and dominant role in the chemical and physiological functions of the body reflects an earlier basic ignorance in molecular chemistry that has been applied to human physiology. It is now almost impossible to expunge this error from the way medical science is evolving. This thoughtless assumption is one of the main reasons we have a "sick-care system," not a healthcare system based on preventing disease.

Water itself must break down first before any chemical reaction can take place. It is the hydrolytic property of water that drives all chemical reactions. A chunk of meat has no energy value unless water acts on it and performs its hydrolytic act of breaking it down to small, energized particles that can then, and only then, be assimilated within the metabolic systems of the body.

Advising people to rely on the sharpness of their thirst sensation to hydrate their bodies optimally is yet another mistake of modern medical education. As we grow older, we gradually lose our thirst perception, to the point that the elderly can become obviously dehydrated yet not have the urge to reach for a glass of water sitting right next to them.

- Does this mean that the body can function efficiently in a comparatively drier state?
- Does this mean that fully hydrated plumlike cells can become prunelike without impacting most of the physiological events of the body that depend on their efficiency of function?

The answer to both questions is an emphatic no. In their earliest prunelike state, even minimally dehydrated cells begin to produce symptoms. Modern medicine has classified these manifestations as "diseases of unknown etiology." The dis-eases of dehydration have led to the creation of various fields of specialty medicine. Larger and larger hospitals and nursing homes have been formed to serve the medical profession and maintain a firm grip on a society that still does not know that water is a disease-preventing medication. It is essential for survival, and it is the master key to vibrant well-being of the human body.

The fourth mistake is the view that anything fluid can supply the long-term water needs of the body—hence the emergence of the soft drink industry, wines and beers included. Suffice it to say that both caffeine and alcohol are addictive and have cumulative adverse chemical effects on the body. Alcohol, even in small doses, negatively affects brain function by shutting down an important water-delivery system to the brain cells. In larger quantities, this effect becomes pronounced to the point of causing hangover headaches, indicators of brain thirst.

Caffeine is a chemical much like morphine and cocaine that some plants manufacture to protect themselves against animals that would feast on their nourishing leaves, seeds, and exotic poppies, the instruments of their survival as species on Mother Earth. Caffeine is a dehydrating substance. It also suppresses the brain tissue enzymes that are vital to the retentive powers of the nervous system. When the plant made caffeine in its process of development, it did so to make its feasting foes stupid enough to become easier prey to their own predators. The plant had no idea that humans would become its protectors and harvest its caffeine in greater and greater quantities. For the plant, caffeine is a chemical weapon.

If a child or a teenager is allowed to drink caffeinated sodas in place of water, attention deficit disorder and low scholastic results should be the early expected outcome. And all we need do is visit a few old age homes to see the state of the mind of people parked there by a society that does not understand the important difference between simple, tasteless natural water and taste-enhanced caffeine-laced water. The introduction of artificial sweeteners has been an added burden on our society. You can get a clear understanding of this problem by referring to my book, *Your Body's Many Cries for Water*.

About 100 years of fast-track travel on the train of these medical misconceptions has produced the commerce-driven quagmire of a sick-care system now known to be based on ignorance, if not fraud. The drug industry's hijacking of medical education has been so successful that even the shocking recent revelation that prescription medications—taken correctly and according to a physician's instructions—make over 2 million people sicker and kill more than 106,000 a year does not raise any eyebrows in official quarters. It is also known that another 98,000 die from mistaken prescriptions. These are only the *reported* cases; God knows how many die silently. These figures together—204,000 a year—put the use of prescription medications as the third most frequent cause of death in America, only after heart disease and cancer. Doctors are now fully aware that the chemicals they prescribe have been manufactured as palliatives.

WHY WATER IS A BETTER MEDICATION THAN ALL THE DRUGS ON THE MARKET

Water is a better medication than whatever is manufactured by the drug industry for any of the presently recognized painful degenerative diseases. These diseases are the products of an inefficient, water-deprived, prunelike state of cells that cannot keep up with the demands placed on them. Each year, more than 700,000 people die of heart disease, 500,000 die of cancer, and arterial disease of the brain kills over 160,000. All of these deaths might have been prevented if water was given a primary position in the daily diet.

THE NATURAL ELOQUENCE OF PAIN: IT SIGNIFIES THIRST

It took me many years of clinical and scientific research to uncover the significance of the pain mechanism as a primary indicator of water shortage in the body.

The human body is a satellite of water-dependent units of life—about a hundred trillion of them housed together in one capsule on their journey of life. The cells of the body, each endowed with the same genetic secrets of life, have accepted a division of labor for an orderly conduct of business in their collective land-based shelter—"the body." It has an intricate water-dependent design—a mobile chemical refinery—that has established a foothold on "dry land." The medical pundits have misunderstood the various ways the body deals with its inner cleansing processes. Nonetheless, they use commerce-suited treatment protocols for different pain sensations, without the slightest understanding why the body, at times, manifests diverse localized or general pains.

It is now clear that there is a direct link between overproduction of histamine, as the neurotransmitter that regulates the water balance of the body, and the induced pain-producing inflammatory effect of histamine that is its mechanism of directing increased circulation to a drought-stricken area. If the blood is concentrated because of water shortage, the pain threshold of the brain is automatically lowered and pain is felt until the drought-stricken area receives water. Many systems and mechanisms are engaged in this process, all of them regulated by the water level of the body.

The trigger mechanism is the balance between the acidity and the alkalinity inside cells. The interior of the cells of the body must be alkaline—a pH of about 7.4. Water and the element sodium from salt flow in and out of the cells and wash out the acid that is produced as a result of the metabolism of working cells. In dehydration, the rate of acid clearance is diminished, and the interiors of the cells in that area become more and more acidic. At a certain level of acidity—a level that obviously eats into and hurts the gene pool—an acid-sensitive substance in the nerve tissue called kininogen is converted to kinin. Kinin itself is a pain-producing substance. Thus, the nerve endings in the area report the local chemical environmental change to the brain.

The brain translates this information into a form of pain for the conscious mind and takes away the ability to use the area so that extra acid production is stopped until water comes and washes the excess acid away. This process causes a localized inflammatory reaction. The circulation opens up, and more blood is brought to cleanse the area and activate the repair processes to "undo" the damage. This damage is often in the joints of the hands and the legs; it could also be in the spinal column, in the heart muscle (causing angina), or in the intestinal tract (causing colitis).

There are about 10 frequently produced pains that should denote dehydration in the area in question: heartburn, joint pain, back pain, migraine headaches, angina, colitis pain, fibromyalgia, false appendicitis pain, peptic ulcer pains, and the eye pains of glaucoma. It is said that 110 million Americans suffer one or more of these pains at one time or another. The cost of pain management for these people is estimated at more than $100 billion a year.

RESOURCE MANAGEMENT: NOT DISEASE

Water is the universal solvent. It is also the solvent of what you might have been told is disease. When the body is short of water, the ingredients that it dissolves and carries around are also missing. Like any well-organized industrial complex, the body has a resource allocation program—a strict rationing plan. Water and the substances it distributes become subject to preplanned priority distribution protocols. Vital cells in vital organs are served first, even at the expense of the less critical parts of the body.

Water conservation, the emergency water-distribution and resource-management programs of the body, have been called allergies, asthma, hypertension, diabetes, and the autoimmune group of diseases.

Allergies

The body deals with foreign agents by producing antibodies that attach to the offending agents and chemically neutralize them, be they pollen, bacteria, or viruses. Histamine has the dual role of being the water- and resource-management regulator, as well as the immune system monitor and activator.

The design of the immune system is such that in dehydration and overactivity of histamine as the resource manager, the activity of the immune system is inhibited at—of all places—the very center of activity, the bone marrow. There are twice as many inhibitor cells responsive to histamine in the bone marrow as there are killer cells within the immune system. Thus, in dehydration and the added presence of histamine, the immune system becomes less and less active and leaves the body vulnerable to pollen, infections, and development of cancer cells. The relationship of allergies and the broader implications of the diminished immune system activity initiated by unintentional dehydration should be taken very seriously. Allergic reactions of the body in childhood translate to a variety of health problems in later years.

Asthma

Air that is exhaled in breathing carries with it lots of water—the winter steam in front of your face. In 24 hours, we lose about a liter of water through breathing. In its capacity as the body's primary water-management regulator, histamine brings about bronchial constrictions to restrict the free flow of air from the lungs, along with the water that would otherwise go with it. The restrictive breathing in asthma will totally disappear if the body becomes fully hydrated.

Seventeen million people in America, mostly children, suffer from asthma—never realizing that water is their lifesaving medication and is freely available in their own kitchens. I predict that in the near future, it will be revealed that the drug industry knew about the relationship of histamine and dehydration, but concealed this from doctors and the public to profit from innocent children who live in constant fear of suffocation and frequently die from asthma.

Hypertension

Sixty million people in America alone suffer from hypertension—yet these people are actually experiencing internal drought. The extra pressure is needed to filter water from an already concentrated blood and inject it through clusters of holes—as wide as only one water molecule—in the membranes of the cells in vital organs. This is how organs such as the brain, the liver, the lungs, the endocrine glands, the heart, and more, receive some water when the body is dehydrated. When the blood becomes concentrated, it tends to pull water osmotically out of the cells of the body. The process of injecting water into vital organs is a compensation mechanism. If adequate water and some minerals needed by the cell interior are taken daily to the point of making the blood more dilute, blood pressure levels will drop back to normal. For more information about hypertension, read *Your Body's Many Cries for Water*—and take a look at another book, *Water Cures: Drugs Kill*, to see how easily others have gotten over hypertension by adjusting their daily water intake.

Type II Diabetes

When food enters the stomach and intestines and is processed, insulin is produced to help feed the cells of the body. It facilitates the entry of circulating sugar and amino acids into the cell through its cell membrane receptor channels. Sugar and amino acids also pull the water that is osmotically attached to them into the cell through these channels. In a dehydrated body, this would be counterproductive for the needs of the brain. The brain is 85% water and does not need insulin to pick up its food components from blood circulation. To avoid the catastrophic outcome of delivering the vital needs of the brain to less vital organs, the process that regulates blood flow to the pancreas, when the body is suffering from internal drought, also inhibits insulin release from the pancreatic cells that produce it. The extra sugar in blood circulation also osmotically compensates for the salt that is missing in dehydration. Increased water intake and supplementation of minerals, with salt on top of the list, will correct the physiological imbalance in this condition, and the diabetes will clear up.

For more information about chronic unintentional dehydration and the diseases it causes, read as many of the following titles as you can. They are the products of more than 24 years of focused research on the topic of deep dehydration inside the cells of the body: *Your Body's Many Cries for Water; ABC of Asthma, Allergies and Lupus; Water for Health, for Healing, for Life; How to Deal with Back Pain and Rheumatoid Joint Pain; Water Cures: Drugs Kill*; and my 2004 book, *Obesity, Cancer, and Depression: The Deadly Diseases of Dehydration*. These books will leave no doubt in your mind that your body is truly a marvel of creation, designed to give your soul a perfect habitat for you to achieve the anxiety-free greatness you are capable of. When your brain is not preoccupied with the aches and pains of your body, you will have the freedom to soar through the limitless horizon of ideas and ideals—how the curative power of water came into light.

For more information on dehydration, visit www.watercure.com and www.nafhim.org.

WATER IN HISTORY

Geoff Petts
University of Birmingham
Birmingham, United Kingdom

The story of water in history is about human ambition to achieve food security, economic growth, and political power. It is both a social and a technological history. The history of civilization is about social organization and coordination, and for six millennia, successful water resource management was the basis of civilizations. Early cultures were built on the food surpluses produced by managing the floodwaters of large rivers for irrigation. In humid areas dominated by rain-fed agriculture, water was influential in other ways by enabling trade along inland waterways. Water also powered mills that fueled industrial and economic growth. But the successful use of riparian and floodplain sites was frequently interrupted by tragedies caused by flood and drought. The notorious Huang He was popularly known as 'China's Sorrow' because of its devastating floods and dramatic changes in its course, but it is also prone to severe droughts. In recent history, major floods in 1855 and 1938 took millions of lives and in destroying crops, brought widespread famine and disease. In the great drought of 1877–79, 13 million people died of starvation. Similarly, early Sumerian and Babylonian civilizations had to survive flood disasters. The Biblical famine described in Genesis (47: 13–21) may relate to a period of persistent drought when the Nile flood failed. In more recent times, the great Mississippi flood of 1927 was a disaster for millions of people, and it also changed the course of the American nation.

Throughout our history, the popular pioneering vision has seen the human conquest of nature as the route to greater wealth and security (1). In all long-settled regions of the world, modernization involved large-scale water projects to reclaim 'wastelands,' 'conserve' water, 'control' floods, and 'harvest' energy. But water is also fundamental to sustaining the earth's ecosystems upon which human societies depend. Two key events in history supported the establishment of the first civilizations: the advent of farming and the control of water. From that point on, humans gradually and then with increasing power, exerted control on the ecosystems in which they lived. During the past 6000 years, but especially since the Industrial Revolution, the human desire to control waters and to reclaim wetlands has led to widespread ecological degradation (2,3). From the mid-1960s and more widely during the past two decades, the role of water in history has changed to become a symbol for a new social philosophy (4), that focuses on *sustainable* environmental management, the protection of biodiversity, the restoration of river corridors and lakes, and the protection of the seas and oceans. The history of this relationship between humans and water is summarized in Table 1. This highlights four phases in the human conquest of water: (1) overcoming the threat of drought, (2) generating power, (3) protecting public health, and (4) protecting (and restoring) water environments.

WATER IN THE ANCIENT WORLD

The foundation of the earliest civilizations involved the efficient and equitable distribution of water through advancement of both technology and management systems. Goubert (5) considered our preoccupation with water one of the subdivisions of the religion of progress. In Judeo-Christian literature, water is a symbol of salvation demanding human vigilance and the imposition of order (6). However, the first revolution in water technology was possible only following the introduction of an indispensable new tool, script. This was necessary for organizing people, water, and soil in quantities that were too vast to be handled efficiently by the unrecorded memorizing of oral arrangements and instructions (7, p. 51). Thus, the Sumerian civilization depended for its survival on effective control and also on effective administration of the waters of the lower Tigris-Euphrates (7). Early, highly structured 'hydraulic civilizations' (8) developed throughout the Old World, in northern India, Southeast Asia, China, and MesoAmerica. These civilizations were founded on centralized state authorities. Water yielded

Table 1. Water in History

Years Ago	Date	
0	2000	
	1990	Emerging science of restoration ecology supports practical restoration schemes across Europe, North America, and Australia.
	1960	Concerns for public health and advances in water treatment begin to clean up water quality; and social concern for the environment prompts political action to control point-source pollution, protect wild areas, and assess environmental impacts of water developments.
50	1950	
	1930	Era of the megaproject begins with Tennessee Valley Authority (1933) and the Hoover Dam and Lake Mead on the Colorado River, U.S. (1935).
		Major rivers across Europe and in the U.S. become severely polluted.
100	1900	
	1890	Demonstration of the link between disease (e.g., cholera) and water.
	1880	First hydroelectric power schemes.
	1840	Extensive floodplain/wetland reclamation across Europe and the U.S.
		Urban growth, water pollution, and epidemics widespread.
200	1800	
		Water power drives the Industrial Revolution and supports growth of urban and industrial centres in upland valleys.
500	1500	Water power still chief provider of mechanical power for corn milling, textile manufacture, and iron works.
	1200	Earliest records of water pollution problems in Europe.
1000	1000	Water mills widespread.
	300	Romans built major, water-powered mills and engineered channels for inland transport.
2000	0	
	100 B.C.	The city of Rome has extensive and complex water-supply system.
		Large-scale land reclamation and flood control works in Egypt and China.
		Large-scale irrigation possible from introduction of Archimedes' screw pump and water wheel.
2500	500 B.C.	
	1500 B.C.	Introduction of water-lifting technology (shadouf) enabled localized perennial irrigation.
		1800–1550 B.C. famine in Egypt as Nile flood fails = ? Biblical famine (Genesis 47: 13–21)
		Regular Nile floods create stable and prosperous Egypt.
4000	2000 B.C.	
		Dry period (2200–2000 B.C.) caused Dark Age for many early civilizations from Greece to Mesopotamia.
		Ancient hydraulic civilizations appeared.
6000	4000 B.C.	

privileges for a nobility that generated considerable revenues from irrigated agriculture, mills, fisheries, and navigation. However, political power and social stability depended on advances in water management technology to secure food surpluses and water supplies for urban centers.

In these early civilizations, advances in water management were inspired by the desire to conquer drought and flood. Diversion weirs, storage reservoirs, and irrigation systems were developed along valley bottoms. In China, the Great Yu undertook complex drainage and flood control works on the Huang He about 4200 years ago (9). In mountain valleys, terracing was developed to conserve water, and soil conservation was an important by-product. However, two inventions had a particularly dramatic impact on the relationship between human civilizations and water. First, the invention of the arch by the Sumerians around 6000 years ago (10) overcame obstacles to transport and water conveyance. The bridge has been described as a symbol of the human conquest of nature, and the river crossing is prominent in social, economic, and military history (10). Worldwide, folklore indicates that bridges had an important place in the lives of early communities and a strong ritualistic character. River gods were made angry by the building of bridges that deprived them of their regular toll of drowning people and animals, and sacrifices were deliberately made during construction to appease the angry spirits (10). Second, the invention of the water wheel around 2300 years ago (11) made conservation and management of limited or fluctuating water supplies possible.

The Egyptian Nile spawned a great civilization some 5000 years ago which depended entirely on the river and its annual flood (11). The oldest agricultural sites in Egypt date from around 7200–6000 years ago. By the end of this period, the ancient Egyptian hydraulic civilization had evolved into strong state, kingship, and political capitals with temples, palaces, and royal cemeteries. Levees, cross dikes, dams, and canals were built to capture flood waters for cultivating a single winter crop. This farming method has been traditionally credited to King Menes, the founder of the 1st Dynasty in 5100 y B.P. However, a 200-year dry period between about 4200 and 4000 years ago had an adverse effect on the economic base of many early civilizations from Greece to Mesopotamia that entered a Dark Age dominated by social disorder, crop failures, famine, and political collapse. This pattern of economic prosperity and political stability during years of a reliable flood and periods of dry years of political impotence, famine, and civil war is repeated in the ancient history of these hydraulic civilizations.

Early basin agriculture was no more than an advanced stage of subsistence agriculture. The invention of lift irrigation about 3500 years ago sustained perennial irrigation. New technology (the shadouf or pale and bucket lever) and deep wells allowed the development of a second growing season during the summer months. This led to a new era of wealth. But this technology was effective only for small plots. Large-scale irrigation became possible only in Ptolemaic times, from about 2300 years ago, when the Archimedes screw pump and the water wheel were introduced. They could raise water to higher elevations and at a faster rate than the shadouf. Ptolemaic engineers were supported by a strong administration to supervise irrigation, land surveying, and dike, weir, and canal maintenance. Their efforts doubled the area of cultivable land and reclaimed some 325,000 acres of fertile land for agriculture by lowering the lake level in the Fayum basin. However, this was also the first era of major human impact upon river corridor ecosystems. Conversion of the floodplain to intensive agriculture destroyed the virgin pasture, wooded thickets, and pools.

ROME TO THE MIDDLE AGES

The Romans were the greatest engineers in ancient history. They built large water-powered mills and complex water-supply systems with long aqueducts. By about 2100 years ago, Rome had a large and complicated water distribution system capable of supplying an estimated 680,000–900,000 m^3 of water per day. The Roman Empire was in no sense a hydraulic civilization, but the monumental masonry aqueducts that led water to their cities endure as symbols of the power and glory of the empire.

The Romans channelized rivers to aid navigation, and important towns, such as York in the United Kingdom, became established at the head of navigable waterways (12). However, the most important legacy of the Roman Empire was the water mill, which Marx saw as a key force in the making of feudal civilizations. In England alone, 5000 water mills were listed in the Doomsday Survey of 1086 (13). Water power remained the chief provider of mechanical power for corn milling, textile manufacture and iron works until the nineteenth century.

However, another revolution in water engineering technology followed the change from feudal to early capitalist systems of land management. By the fifteenth century, the core areas of European market capitalism, upper Italy and Flanders, witnessed early innovation in the mechanical control of water flows and river regulation (6). From the late sixteenth century, the pace and scale of water engineering accelerated, underwritten legislatively and financially by the new centralized states and a capitalist land market. A primary driver was to reclaim marshlands and floodplains to increase the area for productive agriculture and the area that could contribute taxes, and to contribute to national pride in conquering nature (14). Thus, new engineering techniques were advanced to achieve more efficient land drainage. The Dutch developed dredging technology and designs for floodgates, retaining walls, and groins. In Italy, the problem of *La bonifica* (land reclamation in its broadest sense) stimulated advances in the systematic control of rivers. The 'fens' of England were drained between 1630 and 1653 by the Dutch engineer Cornelius Vermuyden using techniques previously employed in Holland and Italy. Once again, the management of the drainage required the transfer of administration from localized, decentralized, self-governing bodies to a more centralized authority, in this case, Parliament. But now there had evolved a new driver of progress, a scientific approach to water management. By the end of the seventeenth century the 'art' of water management had been replaced by a scientific methodology (15), not least as a result of the work of Castelli (1577–1643), the founder of modern hydraulics.

Prestigious projects were conceived. The Canal d'Entreroches, built between 1638 and 1664 and originally designed to link Lake Neuenberg and Lake Geneva, crossed the great European water divide between the catchments of the Rhine and Rhone rivers (16). This caught the public imagination and received funding from the Dutch and the citizens of Bern and Geneva. The scheme failed, but one section was completed and remained open for 180 years, transporting wine and salt. Other notable schemes include, in France, Vauban's great military works for Louis XIV in the Canal du Midi (6). Thus began an era of belief in the domination of nature by science, freeing societies from scarcity, want, and natural disasters. This belief grew into the nineteenth century when Europeans exhibited an extraordinary faith in the power of science to control, manipulate, and enhance the natural world. Scientific innovation and technical skill drove dramatic expansion of European commercial and industrial power. Heffernan (17) considers Roudaire's *la mer interieure*, a proposal to create a vast sea deep within the Saharan desert, a classical illustration of European imperial ambition.

The Postindustrial Revolution Era

In the Industrial Revolution, water assumed an ever-widening role for power, navigation, water supply, and waste disposal. Urban and economic growth brought wetland drainage, canalization of natural channels for boat traffic, and embanking and dredging of rivers for flood control. Freshwater springs dried up as groundwater abstraction increased to supply rapidly growing populations. Overcrowding resulted from the desire to house as many workers as possible in a small space near the industrial workplace. Contaminated ditches and streams became open sewers and breeding grounds for disease. Small streams and brooks were often lost, buried beneath expanding urban developments. Large lowland rivers stagnated, and important fisheries died. In London, chronic conditions had been reached on streams and rivers by the late eighteenth century, but even the tidal river was severely polluted by the 1840s (18). However, it was not until the link between water and health was established at the end of the nineteenth century that a revolution in water and wastewater management occurred (5,19).

Water and Health

Great epidemics of cholera and typhoid in nineteenth century Europe led to the establishment of Public Health Committees composed of public-spirited people and led to systematic sampling of water for laboratory analysis. From the 1830s, the notion of public health placed the issue of water supply in a wider context, including housing, cleanliness of towns, poverty, domestic habits, and disease. Cholera arrived in Britain in 1831 and in London, the most severe outbreak, in 1848–9, claimed 14,137 lives. Popular belief, known as the 'miasmic orthodoxy,' was that illness was spread by an infected atmosphere. The first connection between water and health was made by a London doctor, John Snow, who in 1849, suggested that water polluted by sewage might be the vehicle by which cholera was transmitted. In 1854, he traced the cause of a cholera epidemic in Soho to a single source, a polluted well. However, until the Pasteurian revolution (1880–1890), 'clean' water was synonymous with running water and 'dirty' water with stagnant water. Running waters were seen as self-purifying, capable of diluting wastes and destroying organic pollutants by oxidation. Public health management focused on draining ditches and marshes where mosquitoes bred, separating clean water supplies from sewage, and providing running water via complex systems of pipes and pumps to supply city dwellers. Paradoxically, public health problems were made worse by attempts to exploit the 'self-cleaning' capacity of running water by creating complex sewage collection and conveyance systems. The introduction of the water closet into common use in the 1840s (and the dramatic growth of London's population) coincided with a severe cholera outbreak in 1849 and appears to have promoted epidemics because the sewer systems to which they were connected could not support the volume of liquid waste. In London, the River Thames was so bad in the summer of 1858 that it became known as the 'Year of the Great Stink.'

Pasteur focused attention on the role of microbes in water pollution and established a new scientifically based definition of 'pure' water. In 1883, the German bacteriologist, Robert Koch, isolated the cholera bacillus and confirmed that cholera is usually spread by water contaminated by the feces of an infected person. Clean, running water became an indispensable element of the urban landscape; domestic space was restructured, the toilet became a separate room, personal hygiene became a private and solitary activity, and fountains gained utilitarian as well as aesthetic and religious significance.

The Technological Revolution in Water Storage and Power Generation

The twentieth century realized the control of water, both through the advancement of large-dam building technology (2) and of basinwide management programs. America's New Deal and the electrification of the Soviet Union were both as much social and ideological as they were symbols of national pride. Developments accelerated at the end of the nineteenth century following new advances in earth-moving and concrete technology. New schemes such as the Vyrnwy Dam in the River Severn Basin, United Kingdom, represented a new phase in both dam building (the dam was more than 40 m high and impounded a reservoir of 59.7 million m^3 (the largest in Europe at the time), the beginning of large-scale interbasin transfers across 100 km, and raised issues of multipurpose river regulation for water-supply, navigation, and fisheries (15).

Hydroelectric power was first demonstrated in 1882. The invention of electricity had a dramatic impact because it can be readily and exactly applied to a wide variety of functions and also because it can be easily delivered to a precise and remote point where it is wanted. The earliest operating combined water turbine and electricity generator was built in Northumberland, United Kingdom in 1879–82, and the first commercial hydroelectric power station was operated at Salmon Leap Falls on the River Bush in Northern Ireland in 1883 (20). However, the control of Niagara Falls between 1890 and 1910 inspired the utopian imagination through the capacity to control nature. The giant turbines allowed generating electricity on a new scale, and the scheme also demonstrated the potential for long-distance power transmission, on lines to New York (21).

During the 1930s in the United States, the wholesale transformation of river basins in the Appalachian valleys of the Tennessee Valley scheme and the construction of massive dams and reservoirs in the Colorado basin (22) opened the megaproject era. The 221-m high Hoover Dam and its 37,000-million m^3 reservoir, Lake Mead, dwarfed earlier schemes (2). In the Soviet Union, Lenin's vision of communism as socialism plus electrification of the whole country evolved in the great HEP schemes along the Volga, Dnieper and Don, and later along the Siberian River. All regions of the world experienced further acceleration in the rate of dam construction after 1950 to secure water supplies and generate hydroelectric power.

Harvey (23, p. 13) sees the twentieth century as one characterized by a new 'lust to dominate nature' and a fierce and uncompromising patriotism fed by reckless personal ambition. He suggests that the mammoth river regulation and HEP schemes built in Africa in the immediate postcolonial era, such as Kariba (1959) and Cabora Bassa (1974) on the Zambezi (1959) and Akosombo on the Volta (1965), are examples. By 1990, there were more than 40,000 large dams over 15 m high, and 400 megaprojects had been constructed or were planned. These megaprojects have dams over 150 m high, a dam volume of over 25 million m^3, a reservoir volume exceeding 25 billion m^3, and an installed capacity of more than 1000 megawatts (24). Irrigated agriculture produced 40% of the world's food but consumed some 2500 km^3 of water. By the mid-1970s, the great multipurpose dam, once a symbol of social advancement and technological prowess, was being attacked as environmentally and socially destructive (25). By 1984, there was a wealth of scientific research to demonstrate the ecological damage caused by dams and river regulation (2).

The most recent chapter in the history of the Nile valley (11) began in the nineteenth century when the desire to modernize Egypt focused on expanding the cultivation of cotton, a summer crop, to generate foreign exchange. Later rice and sugar cane, both large water users, became desirable commodities. The construction of dams sluices and canals began in 1861. The Aswan Dam was first completed in 1902 and then enlarged in 1912 and 1993. Finally, it created a storage reservoir of 5.2 billion cubic meters and supports an irrigated area five times that under early basin agriculture and a cropping intensity of nearly 170%. But this phase of development was not without its environmental consequences. The raising of soil moisture levels created favorable habitat for plant and animal parasites and caused waterlogged soils and salinization. Pesticide use exceeded 16,000 metric tons in the mid-1980s when the fertilizer application was nearly 150 kg per 1.038 acre. A high count of undesirable algae developed in the regulated flows causing odor and taste problems. Thirty of the 47 commercial fish species disappeared, and the sardine catch in the eastern Mediterranean was decimated. This stage in the regulation of the Nile changed the face of Egypt (11).

In 1990, Barbara Cummings published a book entitled *Dam the Rivers, Damn the People*, that focused attention on the wide environmental and social impacts of large hydroelectric schemes in Amazonia (26). In 1996, McCully in *Silenced Rivers. The Ecology and Politics of Large Dams*, wrote of 'pharaonic works' that mortgage an entire society to benefit the elite and to enhance national prestige, but which bring little long-term benefit to society at large (27). Yet megaprojects continued to be built, and new projects, such as the Three Gorges in China and Alqueva in Portugal, continue to fuel the sustainable development debate.

In the developed economies, problems of river regulation were subordinate to those of pollution. A typical industrial river in Europe, the Rhine (28), shows the growing severity of fecal and organic pollution until 1972. Since then, improved sewage and wastewater treatment have decreased the input of labile organics to rivers. However, following the impacts of fecal and organic pollution, other major water quality problems appeared (29). Three common problems have been accelerated eutrophication (30), resulting in nutrient input rates much higher those of natural environments; acidification (31) caused by 'acid rain' from long-range cross-frontier transport of air pollutants, especially sulphur dioxide; and salinization of rivers, lakes and soils in dry land areas (32). Pesticides, organic micropollutants, and metals have caused additional problems that often have slow, long-term effects where problems are magnified through food chains to impact higher animals severely, especially the top predators (33).

The change in the industrial base during the second half of the twentieth century led, finally, to the death of the waterfront. Shipping that had provided the transport system for the Industrial Revolution was replaced by road, rail and air transport, and a technological revolution caused the virtual disappearance of labor-intensive heavy industry. In the United Kingdom, the catastrophic decline of the powerful shipbuilding industry of the great deepwater cities, London, Liverpool, Newcastle, and Glasgow of the Thames, Mersey, Tyne, and Clyde, respectively, led to mass unemployment and decay of the urban fabric. The social response to urban living was one where people became increasingly alienated from, and disinterested in, the local natural environment. Nature and wildness became associated with remote areas, to be visited when desired, for recreation and leisure. Yet water has retained its value for societies, for supply and power production and also for its important amenity value. In aesthetic terms, water is strongly prospect oriented; the riverbank and lakeshore provide the classic compromise offering the advantage of good visibility and security at the same time. Indeed, the introduction of a reservoir or canal into an artificial landscape can greatly enhance landscape quality, and the waterfront has become an attractive and valuable asset for business and residential developments.

The New Age of Environmental Responsibility

The last 50 years of our history—since the quantitative revolution in science of the 1960s—has witnessed scientific advances that have demonstrated the full nature of human impacts on waters. Early classic works include *Man's Role in Changing the Face of the Earth* edited by William Thomas in 1956 and *Water, Earth and Man* edited by Richard Chorley in 1969 (34,35). However, from a political perspective, the dramatic impact of Rachel Carson's book, *Silent Spring* (36), aroused public concern about contamination of water and air and the threat to wildlife and human health and provided the catalyst for change in policy and practice.

By the last decade of the twentieth century, a new approach to water management emerged (4) that has five facets. The first focuses on wastewater treatment and pollution control. The second is concerned with allocating water to protect aquatic ecosystems from the impacts of abstraction and river flow regulation. The third is concerned with the 'nature-like' channel designs and environmentally sensitive civil engineering works. Fourth, there is increasing concern for establishing protected areas and habitat refuges isolated from human interference, including fishing and hunting as well as habitat disturbance. However, the fifth is particularly important: collaborative decision-making involving informed stakeholders and public participation, supported by public educational strategies promoting community ownership of waters and their associated ecosystems.

Advancements in the scientific understanding of large waterbodies (37) and complex water transfer schemes (38) have provided the context for these advances. New scientifically based tools for water management have evolved (39,40). However, legislation and public participation are key elements of the new approach. Legislation for managing commercial and recreational fisheries has a long history (41). In the late twentieth century, some governments passed legislation limiting the uses of waterbodies for a wide range of reasons, including nature conservation (42). At the Earth Summit of 1992, *The Rio Declaration* and *Agenda 21: Programme for Action for Sustainable Development* (43) placed biodiversity, ecosystems, and popular participation at the head of the 'development' agenda. We now need to close the debate about unsustainable exploitation or scientifically based stewardship.

BIBLIOGRAPHY

1. Cosgrove, D. and Petts, G. (Eds.). (1990). *Water, Engineering and Landscape*. Belhaven Press, London.
2. Petts, G.E. (1984). *Impounded Rivers*. John Wiley, Chichester, UK.
3. Brooks, A. (1988). Channelized Rivers: perspectives for environmental management. John Wiley and Sons, Chichester, UK.
4. Boon, P.J., Davies, B.R., and Petts, G.E. (Eds.). (2000). *Global Perspectives on River Conservation: Science, Policy and Practice*. John Wiley, Chichester, UK.
5. Goubert, J-P. (1986). *The Conquest of Water. The Advent of Health in the Industrial Age*. Princeton University Press, Princeton, NJ.
6. Cosgrove, D. (1990). *An Elemental Division: Water Control and Engineered Landscape*. In: Cosgrove and Petts (Eds.). *op cit.*, pp. 1–11.
7. Toynbee, A. (1976). *Mankind and Mother Earth*. Oxford University Press, Oxford.
8. Wittfogel, K.A. (1956). *The Hydraulic Civilisations*. In: Thomas, *op cit.*, pp. 152–164.
9. Changning, L. (1989). Problems in management of the Yellow River basin. *Regulated Rivers* **3**: 361–370.
10. Mare, E. (1954). *The Bridges of Britain*. B.T. Batsford, London.
11. Said, R. (1993). *The River Nile. Geology, Hydrology and Utilization*. Pergamon Press, Oxford.
12. Petts, G.E. (1990). Historical analysis of fluvial hydro systems. In: *Historical Change of Large Alluvial Rivers: Western Europe*. G.E. Petts, H. Moller, and A.L. Roux (Eds.). John Wiley, Chichester, UK, pp. 1–18.
13. Sheail, J. (1988). River regulation in the United Kingdom: An historical perspective. *Regulated Rivers* **2**: 221–232.
14. Butlin, R. (1990). *Drainage and Land-use in the Fenlands and Fen-Edge of Northeast Cambridgeshire in the Seventeenth and Eighteenth Centuries*. In: Cosgrove and Petts (Eds.). *op cit.*, pp. 54–76.
15. Petts, G.E. (1988). River regulation in the United Kingdom. *Regulated Rivers* **2**: 201–220.
16. Vischer, D. (1989). Impact of 18th and 19th century river training works: Three case studies from Switzerland. In: *Historical Change of Large Alluvial Rivers: Western Europe*. G.E. Petts, H. Moller, and A.L. Roux (Eds.). John Wiley, Chichester, UK, pp. 19–40.
17. Heffernan, M.J. (1990). *Bringing the Desert to Bloom: French Ambitions in the Sahara Desert During the Late Nineteenth Century—the Strange Case of 'La Mer Interieure'*. In: Cosgrove and Petts (Eds.). *op cit.*, pp. 94–114.
18. Wheeler, A. (1979). *The Tidal Thames: The History of a River and its Fishes*. Routledge and Kegan Paul, London.
19. Halliday, S. (1999). *The Great Stink of London. Sir Joseph Bazalgette and the Cleansing of the Victorian Capital*. Sutton, Stroud.
20. Gibbon, H.D. (1969). *The History of Water in Ulster*. Augustus M. Kelley, New York.
21. McKinsey, E. (1985). *Niagara Falls: Icon of the American Sublime*. Cambridge University Press, Cambridge.
22. Allen, F. (1952). *The Big Change: America Transforms Itself, 1900–1950*. Harper and Boss, New York.
23. Harvey, D. (1989). *The Condition of Postmodernity*. Blackwell, Oxford.
24. Gleick, P.H. (1998). *The World's Water. The Biennial Report on Freshwater Resources*. Island Press, Washington, DC.
25. White, G.F. (1977). *Environmental Effects of Complex River Development*. Westview, Boulder, CO.
26. Cummings, B.J. (1990). *Dam the Rivers, Damn the People*. Earthscan, London.
27. McCully, P. (1996). *Silenced Rivers: the Ecology and Politics of Large Dams*. Zed Books, London.
28. Kempe, S., Pettine, M., and Cauwet, G. (1991). Biogeochemistry of European rivers. In: *Biogeochemistry of Major World Rivers*. E.T. Degenes, S. Kempe, and J.E. Richey (Eds.). SCOPE 42, John Wiley, Chichester, UK, pp. 169–212.
29. Meybeck, M. and Helmer, R. (1989). The quality of rivers: from pristine stage to global pollution. *Palaeogeography, Palaeolimnology, Palaeoecology (Global and Planetary Change Section)* **75**: 283–309.
30. Harper, D.M. (1992). *Eutrophication of Freshwaters: Principles, Problems and Restoration*. Chapman and Hall, London.
31. Batterbee, R.W. (1994). Surface water acidification. In: *The Changing Environment*. N. Roberts (Ed.). Blackwell, Oxford, pp. 213–240.
32. Grainger, A. (1990). *The Threatening Desert*. Earlscar, London.
33. Carson, R. (1965). Silent Spring. Penguin, Bristol, UK.
34. Thomas, W.L., Jr. (Ed.). (1956). *Man's Role in Changing the Face of the Earth*. University of Chicago Press, Chicago, IL.
35. Chorley, R.J. (Ed.). (1969). *Water, Earth and Man*. Methuen, London.
36. Carson, R. (1966). *Silent Spring*. Penguin, Harmondsworth.
37. Lubinski, K., Wiener, J., and Bhowmik, N. (Eds.). (1995). Sustaining the ecological integrity of large floodplain rivers. *Regulated Rivers* **11**: 1–388.
38. Davies, B.R., Thoms, M.C., and Meador, M. (1992). The ecological impacts of inter-basin water transfers, and their threats to river basin integrity and conservation. *Aquatic Conservation: Mar. Freshwater Ecosystems* **2**: 325–349.
39. Calow, P. and Petts, G.E. (Eds.). (1992). *The Rivers Handbook*. Vol. 1. Blackwell Scientific, Oxford.
40. Calow, P. and Petts, G.E. (Eds.). (1994). *The Rivers Handbook*. Vol. 2. Blackwell Scientific, Oxford.
41. Welcomme, R.L. (1985). *River Fisheries*. Fish technical Paper 262. FAO, Rome.
42. Palmer, C.G., Peckham, B., and Soltan, F. (2000). *The Role of Legislation in River Conservation*. In: Boon, Davies, and Petts (Eds.). *op cit.*, pp. 475–492.
43. United Nations Conference on Environment and Development (UNCED). (1993). Agenda 21: Programme of Action for Sustainable Development. United Nations Department of Public Information, United Nations, New York.

READING LIST

Barry, J.M. (1997). *Rising Tide: The Great Mississippi Flood of 1927 and How It Changed America*. Simon and Schuster, New York.

Biswas, A.K. (1970). *History of Hydrology*. North Holland, Amsterdam, London.

Harris, L.E. (1957). Land drainage and reclamation. In: *A History of Technology*. C. Singer, E.J. Holmyard, A.R. Hall, and T.I. Williams (Eds.). Volume III From the *Renaissance to the Industrial Revolution c. 1500–1750*, Oxford.

Lambert, W.G. and Millard, A.R. (1969). *Atra-hasis—The Babylonian Story of the Flood*. Oxford University Press, London.

Showers, K.B. (2000). *Popular Participation in River Conservation*. In: Boon, Davies and Petts (Eds.). *op cit.*, pp. 459–474.

HOOVER DAM HISTORY

Bureau of Reclamation—U.S.
Department of the Interior

SPILLWAYS

The following article was put together by George, a former guide at Hoover Dam.

Many people who take the tour here at Hoover Dam want to know when they will get to see the water go over the top of the dam. Well, the water has never gone over the top of the dam and probably never will. We don't want the water to go over the top of the dam for a couple of reasons. First, the power house is located at the foot of the dam. The power house contains 17 large generators, each producing enough electricity to service about 100,000 people. All that water would be bad for the electrical generators (water and electricity don't play well together). Second, there are about 18,200 vehicles a day going across the top of the dam, and we don't want those vehicles to get swept away. The road across the top of the dam is a federal highway, and it is the shortest way to get from Las Vegas to points east.

Water will probably never go over the top of the dam due to the spillways. The spillways work just like the overflow hole in your bathtub or sink at home (if you don't remember seeing that hole, go look for it right now). If the water ever gets up that high, it will go in the hole and down the drain, not over the top and onto the bathroom floor (unless, you have children and they plugged up the hole). The spillways are located 27 feet below the top of the dam, one on each side of the dam. Any water getting up that high will go into the spillways then into tunnels 50 feet in diameter, and 600 feet long which are inclined at a steep angle and connect to two of the original diversion tunnels. Each spillway can handle 200,000 cubic feet per second (cfs) of water. The flow at Niagara Falls is about 200,000 cfs, so there is the potential for two Niagara Falls here.

Each spillway has four steel drum gates, each 100 feet long and 16 feet high. These gates can't stop the water going into the spillway, but they do allow an additional 16 feet of water to be stored in the reservoir. Each gate weighs approximately 5,000,000 pounds. Automatic control with optional manual operation is provided for raising and lowering the gates. When in raised position a gate may be held continuously in that position by the pressure of water against its bottom, until the water surface of the reservoir rises above a fixed point, when by action of a float the gate is automatically lowered. As the flood peak decreases, the gate can be operated manually so as to gradually empty the flood control portion of the reservoir without creation of flood conditions down stream. The spillways have been used twice. The first time, in 1941, was a test of the system. The second time, in 1983, was for a flood.

The Arizona spillway was placed in operation for the first time on August 6, 1941, soon after the reservoir level had reached a maximum elevation of 1220.44. The drum gates were raised for several hours on August 14, 1941, and a hurried inspection revealed that the tunnel lining was intact, and the inclined portion showed little or no signs of erosion at that time. Operations were then continued without interruption until the reservoir level had been lowered to elevation 1205.60 on December 1, 1942. The average discharge flow through the Arizona spillway during this period was approximately 13,500 cfs with a maximum flow of 38,000 cfs on October 28, 1941, when one of the drum gates dropped without warning.

That much water falling 600 feet down a very steep tunnel caused erosion of the tunnel lining. The eroded area was approximately 115 feet long and 30 feet wide, with a maximum depth of approximately 45 feet. The original volume of the cavity was 1069.6 cubic yards. Repair work was started almost immediately, but because it was believed that ordinary concrete was not suitable it was decided to utilize the Prepack and Intrusion process of concrete repair developed by the Dur-ite Company of Chicago, IL. After repair, the tunnel was polished smooth to help prevent future erosion.

During 1983, record flows into Lake Mead were recorded. The record surface elevation was recorded on July 24, with more than two feet of water spilling over the raised spillway gates of Nevada and Arizona. The record flows through the spillway tunnels again caused erosion in the concrete base, which had to be repaired. High water was responsible for wide spread damage throughout the project.

HYDROPSYCHOLOGY

Bellie Sivakumar
University of California
Davis, California

INTRODUCTION

Population growth and environmental degradation have become primary reasons for water scarcity on Earth. Added to this, inadequate water planning and management activities have resulted in numerous water conflicts around the world, whether in developed, developing, or underdeveloped areas. Although efforts to address these problems continue on one hand, one cannot help but also witness the role of regional, social, political, economic, and even religious and linguistic factors, among others, as an increasingly troubling menace in these conflicts.

This gathering threat is further evident from the major shift in human perception and expression of the water scarcity problem. If recent reports are any indication, we are faced with a situation where any or a combination of these factors can take precedence over "humanity" and, thus, we are left with no choice but to express our unhappiness/dissatisfaction about the consequences in the "most blunt and even antagonizing" way [e.g., "What a Lovely Water War!" (1); "Water shortages foster terrorism" (2)] rather than in a "diplomatic" manner.

Under these circumstances, what we, as water scientists, planners, and managers, can and must do to alleviate the violent and dangerous consequences, in addition to what we are already doing in terms of "water science" per se, is and will be under serious public scrutiny. Asking ourselves two relevant questions (and possibly others) would help address this:

1. How do our water planning and management activities affect human behavior?
2. How does human behavior, in turn, influence our water planning and management activities?

It is clear that answering these questions is not a simple and straightforward task, particularly in light of our traditional "rule" of conducting water research, i.e., quantity and quality (although not without "exceptions" involving also social, political, economic, and other factors). It is rather more complicated, warranting inclusion of yet another important dimension, the "human side."

In view of the significance of this "human dimension," Sivakumar (3) argues that studying the transactions between humans and water-related activities (e.g., planning and management) should be an essential part of water research, and also proposes the term "hydropsychology" for this field of research. Citing an ongoing interstate water conflict in south India (the Cauvery River water dispute) as an example, Sivakumar (3) discusses the contribution of "human behavior" to this conflict, in terms of the two questions raised above. He opines that the "linguistic" factor combined with politics, rather than social and economic factors, plays the key role in this conflict, which is only worsening with the passage of time. His question, particularly directed at water researchers and managers in these two states, is how do they balance their "science and management" on one hand against their "emotional and psychological interests" on the other. Specifically discussing their responsibilities and obligations in accurately reporting the ground realities on water data and disseminating research findings, Sivakumar (3) points out an important moral dilemma they (we) face: to be (more) faithful to science or to region, community, language, etc.?

With the increasing disorder, violence, and even hatred among various human sections around the world because of social, political, economic, and other factors, it should not be difficult to foresee a gloomy world order because of the continued worsening of the water situation as well. In fact, many observers even believe that (fresh)water *will* be the source of violent conflict among communities and nation-states in the not-too-distant future [e.g., (4,5)]. As a combination of "water scarcity" and "human behavior" can have profound consequences, the traditional "water quantity and quality" research must be supplemented and complemented with the new "hydropsychology" research. The purpose of this article is to present possible ways to make progress in hydropsychology research, which, in turn, could shed some light on studying the water crisis and the associated conflicts as a whole.

PROGRESS IN HYDROPSYCHOLOGY

The field of hydropsychology is new and, therefore, research is yet to take shape. In what follows, some ideas are presented to begin research in this field.

Problem Acknowledgment

Like in any other study, the first step in studying the role of human behavior in water conflicts is the recognition that it indeed exists and the acknowledgment that it is significant. It is often possible (and even considered wise) to "overlook" this role, because studying human behavior involves some "sensitive" aspects, such as social and economic inequality, political ideological differences, and cultural, racial, religious, and linguistic differences. Any effort toward making progress in hydropsychology, therefore, requires an open and gracious admission on our part of the need to study these sensitive issues and our willingness to do so.

Problem Identification

On acknowledging the problem, the second step is to investigate the "actual" relevance of human behaviors in water conflicts and vice-versa, which is extremely important, because some of the so-called "water" conflicts may not necessarily be related to water in itself, but water may be used merely as a "scapegoat" for advancing other interests. These interests may involve one or more of the sensitive aspects mentioned above. For example, Sivakumar (3) opines that linguistic interests and political compulsions in Karnataka and Tamil Nadu play more important roles than water itself in the Cauvery River dispute, either as sources of the dispute or as obstacles to find an amicable solution. Many "water" conflicts around the world exist that have "other" facets to them than just water, e.g., the Indus River water dispute between India and Pakistan (political and religious compulsions, national pride).

Although it is clear that accurate assessment of the "correlation" between human behaviors and water conflicts is important to move forward in hydropsychology research, this task is tremendously difficult, or at least can be tricky, at times. The only way to address this issue is by identifying all the relevant parties that, we feel, may have a role to play in the conflict, in the water sector, and/or in the human behavior sector. Depending on the conflict of interest, this may involve bringing the public (e.g., community leaders) and politicians together with researchers in various disciplines, such as water science, environmental science, political science, social science, psychology, theology, etc.

Problem Solving

With the above identification of the relevant parties, the next step is to conduct extensive individual and group consultations and discussions. Of particular importance among these parties are the water researchers and managers, politicians, and community leaders. Water researchers and managers play crucial roles in water planning and management, such as assessment of water availability, storage, and distribution; the existence of any conflict of interest to us may also motivate, and even define, our specific research agenda, as highlighted by Sivakumar (3), who cites his own (regional) interests in discussing the Cauvery River water dispute. Politicians make important policy decisions in regard to water and, therefore, have the authority to approve or reject our proposals for not only water storage and distribution but also water research itself; Sivakumar (3) extensively discusses the role of local and national politics, politicians, and political compulsions in the Cauvery water dispute. Community leaders (e.g., social, cultural, religious), by their own status and privilege, have a crucial responsibility and also the means to advise and guide the people (their followers) in ways that would negate conflicts and violence; in other words, they have a responsibility to keep a "check" on the people.

Although there is no question that these aspects have been studied in water science, in one way or another, such studies have been limited only to "quantity and quality" aspects. For example, farmer participation [e.g., (6)] and socio-economic analysis (7) have been increasingly studied. A host of scientific techniques have also been employed in these studies, such as Delphi Technique, Expert System, and Decision Support System. What has not been studied, however, is the specific role of human "behavioral or psychological" aspect in water science, which, unfortunately, seems to be a more important factor in the existing water conflicts around the world, and may become even more crucial in the future ones.

As studying the human behavioral aspect transcends our traditional water science expertise, developing ideas and techniques would indeed be challenging. It is obvious that we would benefit from, and even require, concepts and tools used in behavioral science. What efforts we will make to combine the knowledge gained in water science and in behavioral science and how (much) we will succeed in supplementing and complementing the ideas and concepts remains to be seen. In this regard, it is encouraging to note that recent advances made in the study of complex systems (e.g., nonlinear dynamics and chaos) and others have proven to be useful in water science [e.g., (8)] as well as in psychology (9). The existing techniques for studying other aspects in water science, such as farmer participation and socio-economic analysis, should also help us lay out a solid groundwork in this challenging task.

Research Dissemination

The outcomes of the hydropsychology research must be disseminated not just to researchers in water and other scientific fields but also to common people. Particular emphasis must be given to dissemination through media outlets, so that it would be easily accessible by all sections of the society. Also, the communication must be simple and effective for even the layman to understand, just like he also understands "What a Lovely Water War!" and "Water shortages foster terrorism." In other words, we, as water researchers and managers, must communicate with all sections of the society (particularly with the common people) with a clear message that we understand and care about their plight and that we are willing to work with everyone to solve their water crisis and the conflicts, which, of course, is easier said than done. Not only is this task extremely challenging, it may also result in "backlashes" and "negative effects." Although we must expect these, we must also keep in mind that these negative effects are often short-term ones. We must be willing, therefore, to continue to move forward with our efforts, only for the long-term benefits of "humanity as a whole" rather than the short-term benefits of "special interest groups."

CLOSING REMARKS

Recent trends in population growth and environmental degradation have brought us to a realization that fresh water *will* be the source of violent conflicts around the world in the near future. Although lack of water is certainly part of this problem, existing water conflicts around the world reveal that other facets also exist (e.g., socio-politico-economic and even regional, religious, and linguistic factors), which are reflections of human behavior.

This article is a follow-up of a recent "hydropsychology" proposal (3) that advocated studying the transactions between humans and water-related activities to improve our water resources planning and management. Some preliminary ideas to begin research in this field were presented, which range from acknowledging the problem in the first place to its identification and solution and, finally, to dissemination of outcomes.

We, as water researchers and managers, as members of the society, and as the more fortunate members of the scientific community dealing with the most precious ingredient for our life only next to air, have a moral responsibility and obligation to find amicable solutions to existing water conflicts as well as to avoid potential future ones. It is my hope that we will start making efforts on hydropsychology research sooner rather than later, and we will strive toward planning and managing water for humanity and peace on Earth.

BIBLIOGRAPHY

1. D'Souza, D. (2002). *What a Lovely Water War!* Rediff News. October 14. Available: http://www.rediff.com.
2. Sutherland, B. (2003). *Water Shortages 'Foster Terrorism'*. BBC News Online. March 18, Kyoto, Japan.
3. Sivakumar, B. (2004). Hydropsychology: the human side of water research. *Hydrol. Sci. J.* (in review).
4. Lorey, D.E. (2003). *Global Environmental Challenges of the Twenty-First Century*. Scholarly Resources Inc., Wilmington, DE, pp. 1–312.
5. El Kody, M. (2003). *The 3rd World Water Forum*, March 16–23, Kyoto, Japan.
6. Dick, M.R. (1997). Farmer participation in irrigation: 20 years of experience and lessons for the future. *Irri. Drain. Syst.* **11**(2): 103–118.
7. New South Wales Department of Land and Water Conservation. (1998). *Water Sharing the Way Forward: Community Based Socio-Economic Assessment*. Sydney, Australia.
8. Sivakumar, B. (2000). Chaos theory in hydrology: important issues and interpretations. *J. Hydrol.* **227**(1–4): 1–20.
9. Abraham, F.D. and Gilgen, A.R. (Eds.). (1995). *Chaos Theory in Psychology*. Praeger, New Haven, CT.

JACOB'S WELL

ARTHUR M. HOLST
Philadelphia Water Department
Philadelphia, Pennsylvania

Jacob's well was dug two miles southeast of the ancient city of Schechem, an area that is currently recognized as the city of Nablus in Israel. The generally accepted history of this well begins with Jacob, the original owner of the well and one of the Hebrew patriarchs. It is said that Jacob purchased the plot of land for grazing purposes "for a hundred pieces of money" (Gen. 33:18–19; Josh. 24:32; John 4:5–6). Jacob and his twelve sons were said to have dug the well together in the center of the property he purchased.

Jacob's well is famous for its religious connection to Jacob, and also to Jesus as well. It is believed that at this well, Jesus asked a Samaritan woman to retrieve water for him. Although its religious significance made it famous, there are other reasons that make this water source important. Limited rainfall and water supply plague the area. The story in the Bible (John 4:1–42) about Jesus and the Samaritan woman at Jacob's well, at a purely literal level, reveals the importance of fresh water sources to the people of this arid land.

> *"Jesus answered, 'Everyone who drinks this water will be thirsty again, but whoever drinks the water I give him will never thirst. Indeed, the water I give him will become a spring of water welling up to eternal life.*
>
> *The woman said to him, 'Sir, give me this water so that I won't get thirsty and have to keep coming here to draw water." (John 4:13–15)*

The Samaritan woman takes his offer literally as an end to her need for water, instead, Jesus attempted to draw a parallel to thing that she could relate to, her survival by means of water, his offer of salvation by drinking the "living water." Her response shows how important a water source, such as Jacob's well, was to their continued existence.

The well alone is an example how the people of this region and their livestock have sustained themselves since ancient times. The effort that was put into building a well of this magnitude illustrates the importance of this water source.

Water rights in this area have been contested for centuries. It was one of the conflicts between Samaritans and Jews in ancient times and part of the current conflict between the Israelis and the Arabs. The struggle is over a limited amount of land that is suitable for sustaining life in this area. It has become a more serious issue due to population growth over the years as a result of higher birth rates and immigration.

Jacob's well is approximately 100 feet deep and 9 feet in diameter, though it is very likely that when it was first dug, it exceeded this depth. Water gathers in the well from an underground spring. It has been called "living water" because the water comes from a moving, fresh stream. A well of these dimensions would have required a great deal of work and a large sum of money in ancient times. Today, Jacob's well is still being used; it has become a tourist attraction and religious shrine. A large, unfinished church has been erected over it. The land that surrounds the well is currently owned and tended by the Greek Orthodox Church.

READING LIST

Easton's Bible Dictionary, "Jacob's Well." www.studylight.org/dic/ebd/view.cgi?number=T1946 (July 12, 2001).

FFOZ Magazine, "Where There Water, There Is Life." www.ffoz/Magazine/Land/wateris.htm (July 12, 2001).

"Jacob's Well" www.furman.edu/mcknight/jerreg14.htm (July 12, 2001).

The Plain Truth, "The Woman at the Well," www.ptm.org/Mar WomanAtWell.htm (July 12, 2001).

"Shechem," www.worldmissions.org/Clipper/HistArch/Shechem.htm (July 12, 2001).

UCU Jacob's Well, "What Is Jacob's Well" http://www.canberra.edu.au/chaplaincy/catholic/cjwell.html (July 12, 2001).

White, Harold Dr. "Sermon Storehouse" http://members.nbci.com (July 12, 2001).

JAUBERT DE PASSA: THE FIRST WORLD HISTORY OF IRRIGATION IN 1846

Eric P. Mollard
Institut de Recherche pour le Développement
France

The first world history of irrigation appeared in 1846, and it has remained the only one to the present. The four volumes of the "Recherches sur les arrosages chez les peuples anciens" (Research on Watering among Ancients) describe hydraulic enterprizes from the most modest to the most monumental, and provide the political and social history of people who mobilized and organized themselves for this purpose. François Jaubert de Passa (1785–1856) was 61 years old when he wrote The First World History of Irrigation. This Catalan landowner, trained in law and after a political career, defended the conviction of a lifetime and intended to demonstrate it by systematically writing of irrigation from antiquity to the beginning of the nineteenth century. The lessons he drew from the comparisons in various cultural and political contexts are always topical, so much so that they match contemporary international issues. How is it possible that this early work, which had the potential for becoming an important foundation for organizing irrigation, has been all but ignored for one century and a half?

THE AUTHOR, HIS TIME, HIS CONVICTIONS

The book was published at a time when France experienced an institutional shift. The first half of the nineteenth century was under a political economy, and the second was marked by the reinforcement of technocracies and state intervention.

After the Napoleonic epic and the Congress of Vienna in 1815, France resumed military operations: the Spanish expedition, naval intervention in Greece, and the seizure of Algiers. The Geography Society of Paris has existed since 1822, the date when Champollion deciphered hieroglyphics. The military interventions continued (Rio de la Plata in 1838), and commercial establishments were set up in the Ivory Coast and Madagascar. Detailed accounts of travelers animated the discussions of many scientific societies under the reign of Louis-Philippe (1830–1848). In line with Montesquieu[1] and Toqueville,[2] political systems and their advantages for economic activity were discussed.[3]

In 1845, the Great Famine in Ireland began and lasted until 1867. Agronomists theorized around Liebig's discoveries in biology and, after the physiocrats, they strove to diffuse the intensive Norfolk rotation. At the end of the first half-century, the Research is directly influenced by the agronomy of notable landowners and the political economy. Jaubert de Passa gave an account of his observations in silkworm, mulberry tree, and oak cork farming, as well as in watering in the department of Pyrénées-Orientales. As a lawyer, he wrote the rules of the Têt users association, a river which crossed Perpignan and irrigated the plantations of Rousillon. Using the Research, he wanted to convince the politicians of the

[1]*De l'esprit des lois* was published in 1748.

[2]*De la démocratie en Amérique* was published in 1840.

[3]The Revue d'Économie Politique was created in 1833; the review Economie Politique Chrétienne of Villeneuve-Bargemont dates from 1834; the Société d'Économie Politique was created between Augustin Thierry et Michelet's books; the Annales de la Société d'Économie Politique were published in 1846.

productivity of irrigated agriculture:[4] "How much there are not, even in France, enlightened spirits persuaded that irrigation cannot have, in the fields of Gaul, results other than to multiply the meadows and to increase the desirable production of meat." He also tried to convince people of the role of institutions in guaranteeing freedom and authority. In addition to his own writings,[5] he gathered two sources, ancient authors' and travelers' descriptions. As a comparative history of civilizations covering the Old World, except sub-Saharan Africa, this treatise incorporated the sagacity of a field man sensitive to water government and social practices. In other words, Jaubert de Passa built an irrigation history on the basis of users' behavior, the real contracts that they have, and the local relationships that mediate the interaction between a regime and productive needs.

In 1851, the reign of Napoleon III began. Interventionism, which was legitimated by Auguste Comte's positivism,[6] resulted in great reclamation operations: the Suez Canal, development of the Gascogne Moors, and damming of the Rhone river. Upon the advent of state bodies of engineers, the debates gave up the ground of the political economy and the Research losts their interest at a time that emphasized technology. One of the possible reasons for the oblivion of the social component in irrigation was that local arrangements do not make the fame of the ruler sufficiently obvious, nor does homage to "progress." It is amusing to read, one decade in advance, an explanation of the exclusivity given to monumental technology. The next choices of Napoleon III echo with those of a pharaoh: "in his turn Menes raized dams and dug canals; moreover, he rectified the Nile riverbed and created artificial lakes to multiply jobs, to fertilize new soils, and to give royalty a character of beneficial greatness,"[7] as well as an Egyptian vice-pasha at the time: "The dam (at the apex of the delta) is not a modern idea ... Mehemet-Ali wanted to resume the project, to attach his name to a colossal work worthy of his genius."[8]

The interest in great irrigation under the aegis of the state was concretized one-half century later in Algeria and by general resident Lyautey in Morocco. It is possible that the Research influenced the French law of 1865 on user associations, which granted them legal status. Conversely, it is not certain whether small peasant irrigation was ever encouraged by the ruler as he had wished.

IRRIGATION AND GOVERNANCE IN THE "RESEARCH"

The Research is composed of six geographical parts (Table 1), which detail, region by region, the history of irrigation. Jaubert de Passa's objective is not theoretical, and only the final chapter advances some principles for an institutional framework to support the installation and durability of irrigation in France. However, one can highlight five great ideas through the combinations of situations and lessons that he draws through writing.

Irrigation and Civilization

Jaubert de Passa underlined the colossal aspect of antique hydraulic works, which are not unrelated to the Pyramids, the Hanging Gardens of Babylon, and the Great Wall of China: the construction of the Imperial canal spread out over five hundred years, "we will have an overall length of 440 leagues, the longest navigable canal in the world."[9] In a more pharaonic way, "... in the area that the Ionians were the first to indicate by the name Delta (Herodotus) ... with periodic floods on a flat ground without defense, it was required there to raize hillocks to build the villages and insulated dwellings. These hillocks required immense embankments because the delta had large cities, temples of extraordinary extension ...."[10]

Would Karl Wittfogel, one century later, have written "Oriental Despotism" (2) in the same manner if he had read Jaubert de Passa? Through the study of great hydraulic civilizations of the past, Wittfogel highlighted the need for a strong bureaucracy in creating and managing hydraulic projects, a bureaucracy that gave structure to the state apparatus and eventually to the whole society. The material and ideological bases of "Oriental Despotism" would reside in great irrigation. The two authors thus agreed on the institutional and political nature that underlies the construction of collective works.[11] On the other hand, Jaubert de Passa granted more credit to governance than to institutional structures. He showed that despotism or democracy can be either a brake on or a lever for irrigation.

Table 1. Composition of the Book

1st part	Assyrian Empire	Assyria, Babylon, Armenia ...
2nd part	Hindustan and Ganges valley	Hindustan, India, Thibet, Sindh, Nepal, Assam, Burma, Siam, Java ...
3rd part	China	Waterings, hydraulic works, agriculture, Japan, Siberia
4th part	Syria, Arabia, Egypt	Damascus, Antioch, Phonicia, Arabia, Yemen, Ethiopia ...
5th part	East	Cyprus, Ancient Greece, Thessalia, Asia Minor
6th part	Ancient Sicily, Italy, Roman Africa	Property among Romans, Republic, Empire, Tuscany, Algeria, the Sahara, Morocco
Chapter IV of the 6th part	About laws and water regime in relation to agriculture	Classification of water, concession, administration, competence, and jurisdiction
Index		

[4]"Irrigation is a practice as miraculous in its effects as common in its means." III 131 (Part III, p. 131 from reedition of 1981, Coll. Les Introuvables. Editions d'Aujourd'hui).
[5]Jaubert de Passa published observations about Spanish irrigation in Valencia and in Elche palm grove.
[6]1798–1857.
[7]IV325.
[8]IV394.
[9]III99.
[10]IV344.
[11]"In a hydraulic society, civil servants have privileged rents, which benefit owners in occidental society."

The Factors in the Origin of Hydraulic Works

A large hydraulic enterprize supposes the capacity to mobilize and organize an enormous labor force in a more or less coercive way. Wittfogel translates gigantism into social terms: "the delegates of power behave like tax masters of the national surplus and prestigious consumers." Unlike administrative terror and tax and legal submission in Wittfogel's book, Jaubert de Passa proposes a contract or, at least, sharing between leaders and people of social norms where religion would play an essential role, just like the confidence delegated to an administration that would animate a paternalist design of its role. Jaubert de Passa stated that tyranny within despotism is not durable.

The Research specified the factors in the origin of the canals, which do not necessarily have irrigation as their first goal. "From now on it was less a question of opening new canals than to improve, extend and especially to apply the discharge canals to the needs for agriculture."[12] The canals frequently have contradictory functions which require arbitration by the authority. The multifunctional character of the hydraulic technique applies besides to other sectors, in particular, the use of water in agriculture. Also, watering is used not only to irrigate, strictly speaking, to bring water to profit from additional production, but also to fertilize or improve the quality of production.[13] This plurality of functions, which results in distinguishing places and times, makes any attempt at generalization a delicate matter. Finally, irrigation is also related to land colonization and the possibilities of selling agricultural goods: "the multiplication of the canals accentuated the role of the Nile in transport."[14] Other factors took part in creating hydraulic works: trade, army, cities, tax department, and strengthening power,[15] and the ruler's fame. Water marketing is seldom practiced: "the public sale of derived water, in the only area of Ispahan, produced for the Shah, at the time of Chardin,[16] [...] 180,000 francs."[17] The comparison of tax pressures established by Jaubert de Passa displays the great disparity which prevailed between the countries and the times. Like ancient Greece, where the tax was at the sixteenth under Solon,[18] imperial China also had the wisdom to limit pressure: "At no time, the emperor could be unaware of the public sufferings ... China is thus the region in Asia where land is the least taxed."[19] In fact, the surpluses released by irrigated agriculture could lead to strong tax pressures. But it was not only the tax department which was to be taken into account, but all measures that encourage or discourage the farmer: "Not only the rates, but arbitrary taxes, in the East, harm agriculture."[20] Moreover, throughout his work, the author gave more emphasis to social practices and abuses than to formal government structures.

The Sharing of Social Standards

Social cement, which ensured the thousand-year-old sustainability of irrigated agriculture, rests on the standards shared by the various classes in a society: users, an attentive administration, and a fair ruler, whose richness comes from the farming community. Social submission and terror cannot function for long because farmers can unite in great masses and revolt against the establishment. Such revolts resulted in the overthrow of several Chinese dynasties by exasperated peasants. Civil disobedience, infrastructure destruction, revolts, abandonment of irrigated farming,[21] even emigration, as in the Nile delta under the Turks, are realities a government, tyrannical though it can be, authorizes only in the event of a major crisis. Consequently, the author suggested that communities placed their prerogatives and their confidence in the hands of the government to repair the infrastructure quickly, to protect rights, and to arbitrate conflicts. As the territory was extended by military invasion or land colonization, the more the state incorporated these prerogatives, the more it was able to mobilize labor and resources for increasingly monumental works. The surplus released by irrigated agriculture reinforced the class not depending on the land and the capacity to undertake increasingly colossal work, often at the request of the peasants themselves.

The contract was not always explicit because it was embedded in a total social fabric. In India, a religious reference is at the origin of irrigation: "There are even poor men who beg for several years to have a tarpa dug later in an uncultivated soil or on the edge of a commercial road. Bringing water in a deprived place is a commendable action in the eyes of Brahma;"[22] "In Egypt, in ancient times, the magistrates appointed for water supervision were taken from the caste of priests; their judgements were all the more respected because they were provided in front of the gates of the sanctuaries."[23] In an identical way, the civil water court in Valencia, Spain, always sits at the entrance to the cathedral.

Religion played a political role of countervailing power and a barrier to despotism: "As agents of this legislation, the priests always opposed a salutary brake on the sovereign authority and popularity; their

[12] III59.

[13] "Irrigation was mainly aimed at increasing the acreage and the return of meadows. In the rich grassland of Thessalia a horse breed adapted to warfare was reared" V57. "The Sibi district is crippled with rocky hills ... Man draws an admirable benefit from restricted resource of nature: with the streams and on poor and sandy soils, he created rich soils whose fertility seems inexhaustible" VI252.

[14] IV292.

[15] For instance, "The Persian king blocked a small valley, thereby holding the water and making a vast reservoir ... Suddenly deprived of irrigation, lowland farmers saw their millet and sesame crop perishing [...] Tired of so many pains, users came to implore at the palace gate, and they got, against high fees, that the water gates ... would be opened successively for the watering of the five small valleys" I191.

[16] 1643–1713. Author of "Voyage en Perse et aux Indes Orientales."

[17] I126.

[18] V45.

[19] III162.

[20] IV225.

[21] Except in Egypt, where rain-fed cropping is impossible.

[22] II204.

[23] VI351.

intervention imposed on the ruler circumspect functioning in conformity with the laws, at the same time as they gave to the people lessons of morals and more moderate feelings."[24] Jaubert de Passa added, "From absolute possession of the land to the control of the agricultural classes and the humiliation of the privileged caste, the distance is short and easy to cross: the priests understood it, and soon, advized or intimidated by them, Joseph returned to the owners of the lands which they had sold to the king, with the charge for them to pay annually to the royal treasure the fifth of all the products."[25] Not only is the purpose of one good administrative organization to intervene advizedly, but it must also share the social standard and be subject to control. The administration must justify its presence to the taxpayer. Tax and operational involvement of an administration are invariant and, for lack of countervailing power, it can easily justify itself in the eyes of the only ruler by an opinion which favors the monumental enterprize. This double function of the administration was particularly clear in Egypt: "Notifications transmitted by the inspector of the nilometer were going regularly to warn the governors of the provinces: the farmers and the officers of the king modeled themselves on these notifications, the ones, as Strabon says, to warn with the distribution of water and to do the necessary work on canals and dikes; others, to establish the tax, because they adjust it to the extent of the height of water."[26]

Conversely, the pragmatism that the management of water requires resulted sometimes in sacrificing the best principles. The Chinese administration thus chose the confusion of powers: "This apparent confusion in the performance of two quite distinct duties (delegate of water and judge) has as a result, in an empire where centralization and control are thorough until the last detail, to prevent conflicts between rival authorities and to facilitate the operations."[27]

The Decline of Civilizations

Nothing is ever gained, and the factors in the origin of great civilizations can be called into question. The ecological factors and the natural disasters were exploited little by Jaubert de Passa, who overlooked land salinization in Mesopotamia and the recurring silting up of tanks and canals.[28] Should we, therefore, assume that societies, at any given time, cannot mobilize themselves to rehabilitate drains, clean canals, and create new tanks?

His analysis of the decline of Egyptian irrigation under the Turkish empire led to detail governance resulting from the regime set up by the Divan.[29] First of all, he criticized the Ptolemaic kings: "Under Arcadius (395 A.D.), the tyranny of the tax department was extreme, but it respected watering, even by attacking ownership ... One of the laws stated that those who were accused of having opened a dam along the Nile were to be condemned to be burned at the stake ... Under Zenon in the year 474, the annual tribute of Egypt was suddenly raized from 50 pounds to 500 pounds of gold ... Under Phocas, the Egyptians were excluded from all public employment. Lastly, under Heraclius (year 641), the religious fights caused a civil turmoil which put an end to the domination of Greek emperors."[30] At the time of Arab domination, which was often favorable to the diffusion of irrigation, he showed that it resulted in weakening the pillars of traditional Egyptian religion. But it was especially the Turkish regime which caused the abandonment of a good part of the delta by farmers.[31]

The first problem occurred from new task sharing imposed by the Turkish government: "To reduce the burden on the treasury, the Divan maintained the old habits, and it declared the population responsible for maintenance, in spite of confiscation, which deprives them, mainly, of the benefit resulting from this work."[32] The negligence and incompetence of the high administration are added: "Since the drudgery was maintained, work from now on was exposed to the inexperience and arbitrariness of the governors: they forgot, rather generally, to carry out certain repairs in due course ... Often materials are missed, the time is short, and the unhappy fellah, curved under the whip, drives in posts, piles up corn canes."[33] The submission became physical and there remained nothing but terror to try, with dubious success, to preserve irrigated agriculture. The abuses and the privileges of big men resulted in appropriating water. Corruption increased in all levels of the administration. That will provide Bonaparte's words: "Better than Mamelucks, or let's retire."

Law and Institutions

Good legislation and freedom of undertakings are two principles on which, Jaubert de Passa claimed, irrigated agriculture needed to be developed in France. Watering must "be freed from the obstacles of the tax department, protected by the law and encouraged by the ruler."[34] He established a parallel between landed property and water rights. Private though it is, property requires an authority that is respected on the one hand; in addition, it is subject to the public interest which demands various constraints, in particular the right of easement (or land crossing). In a period when feudal survival had maintained local rights in many countries, Jaubert de Passa believed that water must belong in the public domain. In the East, such a principle is rooted in religion and is confirmed by legislation: "the laws of Zarathushtra, Manu and the

[24]IV331.
[25]IV400.
[26]IV300.
[27]VI306.
[28]"An earthquake would have ruined the Jordan valley, which Moises compared to the Garden of Eden" IV247.
[29]Turkish Council.
[30]IV417.
[31]"The whole Niphaiat was covered a long time ago with villages and fertile countryside ... Canals are nowadays silted; water is derived to the Damiette branch" (Champollion) IV350. Jaubert de Passa was aware of the progressive raising of the delta lands through history, but neither he nor others of his time ever made of it the explanation of its decline.
[32]IV386.
[33]IV386.
[34]II69.

followers of Buddha admit as private water only that which runs in an artificial bed."[35] In Italy, nationalization which appeared at the end of the Middle Ages authorized the reclamation of the Po Valley. In France, the phenomenon occurred later, but nationalization is essential not only due to the advantages for economic activity, but also because it trimmed feudal power down and reinforced the alliance between landowners and the crown. The state was able then to concede a right to a legal or physical entity for the benefit of the public interest. The administration must then supervise the use of this right and the application of the clauses of the concession.

JAUBERT DE PASSA BEFORE TECHNOCRACY

The "Research on Watering among Ancients" gives the primary role to local practices and governance. The Orient, where institutions favored irrigation, was at the time a source of inspiration, and this arrangement resulted in a model to be followed, even a legacy. After one and a half centuries of technical progress that came primarily from the Occident, the direction of moral debt was reversed. All the countries did not undergo colonial imposition, but all developed a qualified administration set on progress and technologies: it was gravity irrigation; it is now pressure irrigation. Jaubert de Passa had not envisaged this drift which marginalized social arrangements, nor that which, under cover of authority, led the French administration to create and manage directly the hydraulic equipment, exceeding the initial mandate of control, standardization, and sanction.

Jaubert de Passa attached less interest to the technique than to the social forms to which it owes its existence, its choice and its mode of management. In no case is the absence of irrigation ascribable to a defect in technological knowledge, and he rejects the exclusively ecological or technological explanations: "It is thus not the nature of the soil nor the defect of intelligence which deprived Sicily of the benefit of irrigation."[36] Jaubert de Passa supplemented the advantages of modern technology with a social analysis; that was the case during the debates caused by the construction of the Mehemet-Ali dam in Egypt: "Increasing the volume of water for irrigation is not only to add to the production of the land, but also to facilitate work and to assist the poor fellah; it is also to free the modest field workman from a fight, always disastrous, against the large usurpers of water; it is finally to bring back law and order in the villages, and to remove from the agents of power a means of oppression and a ceaseless cause of venality."[37]

Private initiative needs authority. It guarantees rights and secures against conflicts. However, Jaubert de Passa avoided setting up universal principles because he knew the price of flexibility when public and private interest change. His field experience led him to highlight the bad years, which can be catastrophic in irrigated agriculture: among Egyptians as among the Chinese, taxation was adjusted in bad years. Moreover, Egyptian irrigation functioned without concessions, and the great Roman landowners did their business without turning to the state for help. His demonstration applies to the French case where small holdings dominate. "Even under a successively despotic, feudal, monarchical and constitutional government one can make broad concessions with agriculture." [38] Egyptian and Chinese despotism conditioned the extraordinary epic of irrigation, whereas Turkish despotism undermined it. The Athenian and Roman democracies did not show great effectiveness in this field. It is the agreement between the type of farmers, the institutional context, and the sociocultural practices at which any reform must aim. From the remarkable diversity of situations, which is born from reading the Research, a last lesson arises: the importance of the details of arrangements for each one mirror obvious social participation. The latter cannot fit with bureaucratic monolithism, which does not lend itself either to the necessary flexibility for irrigated agriculture. This is the meaning we have to give to the paradigm of governance, which applies both in the time of Jaubert de Passa and the present.

[35]VI269.
[36]VI109.
[37]IV395.
[38]VI343.

BIBLIOGRAPHY

1. Jaubert de Passa, François-Jacques, Baron. (1846). Recherches sur les arrosages chez les peuples anciens.
2. Wittfogel, K. (1957). *Oriental Despotism. A Comparative Study of Total Power*. Yale University Press, 1967.

READING LIST

Jaubert de Paçà. (1824). Recherches historiques sur la langue catalane. Col. Fons Antic Mars 2001, p. 200.

Jaubert de Passa, François-Jacques, Baron. (1821). Mémoire sur les cours d'eau et les canaux d'arrosage des Pyrénées-Orientales.

Jaubert de Passa, François-Jacques, Baron. (1823). Notice historique sur la ville et le comté d'Empurias.

Jaubert de Passa, François-Jacques, Baron. (1823). Voyage en Espagne, dans les années 1816, 1817, 1818, 1819, ou Recherche sur les arrosages, sur les lois et coutumes qui les régissent [Extrait du Rapport sur le concours général de la pratique des arrosages, fait par M. Héricart de Thury.]

Jaubert de Passa, François-Jacques, Baron. (1833). Recherches historiques et géographiques sur la montagne de Roses et le cap de Creus.

Jaubert de Passa, François-Jacques, Baron. (1901). Journal de Voyage d'un Roussillonnais. In: Torreilles, Ph. De Barcelone à Valence en 1819, d'après le Journal de voyage d'un Roussillonnais, par Ph. Torreilles. (Extrait de la Revue des Pyrénées, tome XIII, 1901.)

Jaubert de Passa, n.d., Poesia eroticoburlesca Rossellonesa del segle XIX. Obres de Fra Pere i Poésies catalanes. Fons Antic 2001, p. 172.

Mattes, Jaubert. Notice biographique F. Jaubert de Passa.

Monglond, A. (1922). Au pays de la Vénus d'Ille. Mérimée et Jaubert de Passa. Revue d'histoire littéraire de la France.

BENJAMIN FRANKLIN: FROM KITE TO LIGHTNING ROD

William E. Marks
Water Consciousness, Inc.
Martha's Vineyard,
Massachusetts

Benjamin Franklin (1706–1790) is an unusual rags to riches story. At age 17, he left his home in Boston, after a falling out with his family, and found his way to Philadelphia. Arriving penniless, but rich in reading knowledge and from working as a writer and newspaper printer with his brother in Boston, he began his rise in life. He entered a common-law marriage in 1730 with Deborah Read, a Pennsylvania woman with whom he was long acquainted. They had one son who was lost to smallpox as a boy, but raised a daughter, Sally, along with Franklin's illegitimate son, William. The mother of William, with whom Franklin had an affair, has been lost to history.

Franklin was a founding father of the United States, the first Grandmaster of the Freemasons in North America, a family man, business man, a printer, inventor, author, publisher, scientist, musician, diplomat, and politician. Obviously, Franklin was many things to many people.

Franklin's relationship with water may be outlined as follows:

1. His kite flying stunt to experiment with understanding how storm clouds filled with water generated electricity. Lightning is a phenomenon that still perplexes science.
2. His invention of the lightning rod.
3. His invention of the Armonica water music instrument.
4. His study of the Gulf Stream and his epiphany for having it printed on a navigation chart.
5. His study of weather patterns.
6. His invention of swimming flippers.

FROM ELECTRICITY-PRODUCING WATER IN THE SKY TO LIGHTNING RODS

One stormy day in June of 1752, Benjamin Franklin took a walk into a field to go kite flying with his 21-year-old son, William. The field rested along the banks of the Schuylkill. Trusting in his fathers scientific know-how, William probably had no idea that he was participating in a potentially life-threatening experiment. At the time, little did William realize that Franklin had so much excitement and interest in studying storms that he could have been dubbed "the world's first official storm chaser."

As the story goes, Franklin attached a piece of metal wire to the fuselage of his homemade kite to help attract electrical charges flying among the overhead clouds. The hemp string played out through Franklin's fingers, and the kite twisted and turned in the turbulent winds as it soared higher and higher. After a considerable length of the string had played out, Franklin tied a key to the taut string. Then, with a dampened silk ribbon tied to the knuckles of his free hand, Franklin and William waited to see what would happen.

Whether it was a collection of static electricity gathered along the moist hemp string or a "gentle" lightning strike, we may never know. Regardless, Ben Franklin reported that when he placed his hand with the insulating silk ribbon near the key, he received a jolt of electricity as sparks leaped from the key. Franklin's grounded body acted as a grounding path for the electrical current built up in the electric field of the storm clouds overhead. In addition to proving that thunderclouds generate electricity, Franklin also extrapolated his negative/positive electrical research from his laboratory and theorized that the lower part of the thunderclouds was mostly negatively charged.

"The rest," as is often said, "is history." Franklin and his son, William, survived this experiment, But others who have tried to duplicate this experiment have paid dearly with their lives or their bodies were crippled.

Long before his famous kite-flying experiment, Franklin believed that lightning and static electricity shared similarities. While living in Boston in 1746, Franklin set up a laboratory to conduct his "electrical amusements." At the time, there were other scientists in Boston conducting electrical experiments. It was in 1746, during his first year of conducting electrical experiments, that Franklin suffered a not-too-amusing jolt of electricity. In a letter to a friend he expressed the shocking experience as, "...a universal blow throughout my whole body from head to foot, which seemed within as well as without; after which the first thing I took notice of was a violent quick shaking of my body..." As a result of this electrical shock, Franklin experienced numbness in his arms and the back of his neck for a few days.

As Franklin continued to conduct his electrical experiments, he wrote down his efforts and shared them in letters to Peter Collinson in London. Collinson was a friend and scientist who shared an interest in Franklin's experiments. In 1749, Franklin sent a letter to Collinson describing the concept of an electrical battery that could hold a charge for a brief period of time. In the same letter, Franklin expressed doubts as to how such an invention could be of service to humankind.

During Franklin's experiments, he observed that when glass was rubbed, it became "filled" with an electrical charge, making it "positively charged." He also observed that when other objects, such as amber, were rubbed, electricity flowed out from them, and therefore they became "negatively charged." This designation by Franklin of "positive" and "negative" electrical charges established a new base language for understanding electricity that remains with us to this day. Prior to Franklin's time the words "vitreous" and "resinous" were used to express the charges of electricity.

In his experiments, Franklin observed that when certain objects were "filled" with static electricity, they would discharge a "spark" of electricity when placed near a pointed metal needle. Franklin was not the first to speculate about a relationship between lightning and the harmless sparks created by static electricity. William

Wall, a British scholar, had suggested such a relationship in 1708.

However, Franklin was the first to propose an experiment using a lightning rod on top of a building to capture the "fire" from the clouds. The first lightning rod described by Franklin was an iron rod 8 to 10 feet long that came to a sharpened point at its uppermost tip. Figures 1–4 show a variety of lightning rods throughout the years.

Two years prior to his kite experiment, Franklin conducted experiments where he used a large sharp iron needle to draw electricity away from a charged metal sphere. Through his observations, Franklin extrapolated the results to his invention of the lightning rod to protect the tops of buildings.

In 1750 he wrote two letters to his friend, Collinson. In these letters, Franklin postulated:

> May not the knowledge of this power of points be of use to mankind, in preserving houses, churches, ships, etc., from the stroke of lightning, by directing us to fix, on the highest parts of those edifices, upright rods of iron made sharp as a needle...Would not these pointed rods probably draw the electrical fire silently out of a cloud before it came nigh enough to strike, and thereby secure us from that most sudden and terrible mischief!

Figure 1. Ventilator/Lightning Rod from Gay Head Lighthouse wick 1900, Martha's Vineyard, Massachusetts (photograph courtesy of William Waterway).

Figure 2. A weathervane and lightning rod combination. Location: Trinity Methodist Church, Oak Bluffs, Massachusetts (photograph courtesy of William Waterway).

> ...On top of some high tower or steeple, place a kind of sentry box big enough to contain a man and an electrical stand. From the middle of the stand, let an iron rod rise...upright 20 or 30 feet, pointed very sharp at the end. If the electrical stand be kept clean and dry, a man standing on it when such clouds are passing low might be electrified and afford sparks, the rod drawing fire to him from the cloud. If any danger to the man be apprehended (though there would be none) let him stand on the floor of his box, and now and then bring near to the rod the loop of a wire that has one end fastened to the leads; he holding it by a wax candle [i.e., insulating him from it]. So the sparks, if the rod is electrified, will strike from the rod to the wire and not affect him.

Even though Franklin's lightning rod proposal was first published in England in 1750 by *The Gentleman's Magazine* and then by the Royal Society of London in 1751, there was little interest in England in acting on Franklin's invention.

However, in 1752, Franklin's letters were translated into French. Soon after, the lightning rod concept was quickly embraced by King Louis XV and his two leading natural scientists, Comte de Buffon and Thomas-Francois D'Alibard.

On May 10, 1752 in the country village of Marly located near Paris, the French built a sentry box according to Franklin's published recommendations. The sentry box was equipped with a 40-foot lightning rod. A

Figure 3. A cross with lightning rod attached. Location: Town of Oak Bluffs on Martha's Vineyard, Massachusetts (photograph courtesy of William Waterway).

Figure 4. Four lightning rods on top of old Whaling Church, Edgartown, Massachusetts (photograph courtesy of William Waterway).

retired soldier, Thomas Francois D'Alibard, was somehow influenced to volunteer for the experiment. As fate would deliver, a storm passed over the sentry box in midafternoon. Much to the soldier's surprise, he and others witnessed the sparks leaping between the lightning rod and the insulated wire. Soon, the same experiment was conducted and verified in other areas of France.

When the French lightning rod experiment worked according to Franklin's theory, the King of France sent a message to Franklin expressing his compliments. However, due to the slow transit of mail in those days, Franklin did not receive the French King's message until August of 1752, two months after he had already proven his theory by successfully using the dangerous kite and key rig.

In June of 1752, Franklin was in Philadelphia watching the progress of a steeple being built on top of the Christ Church. Franklin was going to use this steeple for his first lightning rod experiment.

From his research and other "electrical amusements," Franklin speculated that the placement of pointed iron "rods" on top of buildings would prevent lightning from causing fires. He figured fires could be prevented because the pointed rod would capture the lightning's "fire sparks." On this note, Franklin was half right. It did not take Franklin long to figure out that lightning rods need to be "grounded." He accomplished this by using an insulated metal wire or cable to draw the charge from the lightning rod and to guide the electric charge harmlessly down into the ground away from the structure.

In 1753, Franklin published his lightning rod research in his Poor Richard's Almanac. He also provided a detailed description of his lightning rod invention. This helped to stimulate the installation of lightning rods throughout the American colonies.

The largest lightning rod installed in Franklin's lifetime was in 1788 at Maryland's State House. This rod was installed according to Franklin's specifications and has effectively protected the State House for more than 208 years.

Before Franklin's invention, lightning destroyed or damaged many buildings. Evidence of how rampant lightning damage was in the early 1700s can be found in various records. For instance, during a period of 33 years, in Germany alone, approximately 400 towers were damaged by lightning, and 120 bell ringers were killed. On a similar note, in France between 1753 to 1786, 386 church towers were struck, and 103 bell ringers were killed.

The installation of lightning rods in France and other countries immediately began to reduce the number of buildings struck by lightning and also reduced the deaths of bell ringers and other people in buildings with tall steeples. As the use of lightning rods slowly began to reduce the number of fires and deaths, the governments and the public began to appreciate the practical applications of Franklin's research into electricity and weather, especially, when one considers the antiquated fire fighting apparatus available in the 1700s. In Franklin's time, entire towns were known to burn down in fires ignited by lightning strikes.

Buildings, such as those with tall steeples (churches, tall mansions, and commercial buildings) were all susceptible to lightning strikes. Once outfitted with lightning rods, these tall structures were less likely to be struck by lightning. The tangible results of Franklin's invention in saving homes and buildings soon made him an international celebrity and a respected scientific genius.

However, when it came to installing lightning rods on churches or on public buildings where the church held influence, there was considerable resistance from many Catholic and Protestant church leaders. Relatively speaking, it was as though history was repeating itself; these same institutions also condemned the scientific theories of Copernicus and Galileo.

Both of these faiths taught and preached that being struck by lightning was a "sign" of God's punishment. When Franklin's experiments indicated that being struck by lightning was the unfortunate result of a random act in nature, he was immediately attacked by various church leaders.

Prior to Franklin's lightning rod, there was the age-old belief that the ringing of "consecrated" bells in a tower would help repel violent storms and lightning strikes. Therefore, as a thunder-and-lightning storm approached a town, the local bell ringer would do his religious duty to save the community. Little did he know that grabbing hold of the ringing rope and ringing the bells to ward off the evil storm was playing roulette with lightning. During thunderstorms, it was the accepted practice to ring the bells as violently as possible. This was done on the theory that the sound waves from the consecrated bells would break up the lightning in the overhead clouds. It is for this reason we find the words "Fulgaro Frango" inscribed on many medieval bells. "Fulgaro Frango" means "I break up the lightning flashes."

As people awakened to the relationship of lightning and the death of bell ringers, there soon became a struggle between the religious belief of ringing bells to ward off evil and the reality of being struck by lightning. Eventually, in 1786, 34 years after Franklin's lightning rod was first successfully tested in France, the Parliament of Paris signed an edict. The passing of this edict was "to make the custom of ringing church bells during storms illegal on account of the many deaths it caused to those pulling the ropes."

Due to this 1786 edict in France, we find one of the first examples of the forthcoming separation between church and state. A separation that was more or less prompted by Franklin's founding of the "scientific method," a method of inquiry into the nature of our world that caused some people to question the teachings of religion. This also helped to form the basis of fear from a religious perspective.

A well-documented attack against Franklin's lightning rod came in 1755 from a church leader in Massachusetts. The Reverend Thomas Prince, pastor of the Old South Church, published a sermon on the subject of lightning rods. In that sermon, he expressed the opinion that the recent big earthquake in Massachusetts was related to the installation of "iron points invented by the sagacious Mr. Franklin." His sermon also mentioned that when it came to installing lightning rods—"in Boston, are more erected than anywhere else in New England, and Boston seems to be more dreadfully shaken. Oh! there is no getting out of the mighty hand of God."

Due to the strong influence of religion in England, the first lightning rod installation on a church spire did not take place until 1762—10 years after the French began their installations. In the interim, London's Saint Bride's church was damaged by lightning several times.

In Austria's mountains of Carinthia, the Catholic church of Rosenberg was struck so frequently by lightning that peasants attending services were killed on several occasions. As a result, many of the peasants stopped attending services during stormy weather. The church steeple suffered lightning damage so severe, that it was completely rebuilt three times. When a lightning rod was installed in 1778, all lightning damage ceased.

In Italy, the tower of St. Mark's in Venice suffered lightning strike after strike. Records from 1417 onward show that the wooden spire was struck by lightning and consumed by fire on several occasions and suffered severe damage from many other lightning strikes. Due to the influence of the church, St. Mark's spire went unprotected. This is in spite of the fact that Beccaria, the respected Italian physicist, had introduced Franklin's invention to his country many years earlier. Even after St. Mark's spire was hit by lightning in 1761 and 1762, the church refused to install the "heretical rod." In time, as overwhelming evidence of the rod's effectiveness grew in neighboring countries, the church acquiesced, and a lightning rod was finally installed in 1766. Afterward, St. Mark's tower was never again damaged by lightning.

However, lessons such as this came too late to save the Church of St. Nazaire in Brescia, where the Republic of Venice had stored 100 tons of gunpowder. In those times, church vaults were often used for this purpose because they were dry. In 1767, a lightning strike hit the church spire and traveled down to the stored gunpowder. Over 3,000 people lost their lives, and a large portion of the city was destroyed.

Because of this common use of churches for storing gunpowder, there are similar stories well into the 1800s.

For example, in 1856, on the island of Rhodes, a lightning strike exploded the munition vaults and killed more than 4000 people.

Franklin's invention of the lightning rod proved to people the connection of lightning to storm clouds and thunder. This was the first proof that lightning was an electrical force generated in nature by the action of water molecules in the clouds. The mystery of how lightning is created by the motion of water molecules in storms was something that intrigued Franklin. However, he could only theorize as to the distribution of negative and positive charges in the clouds.

Franklin believed that his creation of the lightning rod was his greatest and most useful invention for the benefit of humankind. In spite of Franklin's apparent habit of keeping records and sharing information about his experiments through letters, it is of historic note that he never documented his kite experiment or wrote letters about the event. The only witness to the kite experiment was his son, William. And, the only written account of the kite and key experiment was by another person some 15 years after the event took place.

Franklin's lightning rod design was of an iron rod honed to a sharp point on its skyward tip. This design probably was developed as an extension of his laboratory experiments where he used an oversized needle with a sharpened point.

England was one of the last countries to begin installing lightning rods. The first lightning rod installation in England took place in 1760. The location was the Eddystone Lighthouse, an old wooden structure that had a history of damage by lightning. This lighthouse was vital to marine navigation safety and the commercial shipping interests of England.

As fate would deliver, nature further helped spur the use of Franklin's pointed lightning rod in England. In 1764, a bolt of lightning struck England's Saint Bride's steeple, already struck many times in the past. Saint Bride's church, a historic building in England, has a spire that stabs skyward prominently. The steeple's damage from lightning in 1764 prompted the Royal Society to recommend the use of Franklin's pointed lightning rods to protect the rest of the country's buildings.

As Franklin's pointed lightning rods were installed throughout Europe and America, King George III decided on a different approach. He decided to protect his palace with blunt-knobbed lightning rods, which he believed were superior to Franklin's. At the time, this also represented a political statement because King George viewed Franklin as a rebel. The use of blunt-knobbed lightning rods therefore represented a symbol of loyalty.

Regardless of King George III's, motivation to use blunt-tipped lightning rods in 1752, varying opinions and arguments have existed since that time, as to which lightning rod design is superior.

The controversy over which lightning rod was most effective was recently put to rest. In experiments conducted by the American Geophysical Union during the past decade, it was determined that the blunt lightning rod is more effective. Even though Franklin's design did not win this competition, he still deserves credit for bringing the lightning rod invention to the service of humankind. Besides the pointed and blunt versions of the lightning rod, there are many variations on the theme. Lightning rods have almost evolved into a folk art form, and many have considerable value to collectors.

On a closing note, as in the case of his inventions of the glass Armonica, bifocals, and the Franklin wood stove, Franklin never made money from his inventions, including the lightning rod. To quote Franklin, "As we enjoy great Advantages from the Inventions of others we should be glad of an Opportunity to serve others by any Invention of ours, and this we should do freely and generously."

This philosophy of Franklin's, about serving humanity, was also reflected in the words he wrote in 1790, the year of his death. From the tone of these words, it is evident that Franklin approached his earthly departure with philosophical humor:

> "I believe in one God, Creator of the universe....That the most acceptable service we render Him is doing good to His other children....As to Jesus...I have...some doubts as to his divinity; though it is a question I do not dogmatize upon, having never studied it, and think it needless to busy myself with it now, when I expect soon an opportunity of knowing the truth with less trouble."

WATER AND THE HISTORY OF MAN

Ann Azadpour-Keeley
National Risk Management Research Laboratory, ORD, U.S.EPA, Ada, Oklahoma

Jack W. Keeley
Environmental Engineer
Ada, Oklahoma

> And Isaac digged again the wells of water, which they had digged in the days of Abraham his father; for the Philistines had stopped them after the death of Abraham: and he called their names after the names by which his father had called them. (Genesis 26:18)

Before the earth was created, hydrogen, the oldest and most abundant material in the universe, and oxygen, created in the womb of stars, combined to form H_2O. Since the beginning of life on this planet and probably other planets scattered throughout infinite space, water has played a critical role in all facets of human existence, including the creation of life itself. We are made of water, and water produces the food we eat, from fish to fowl to fiber. It is the only substance on the earth that exists naturally in three phases, liquid, solid, and gas. As it serves us today in industry, agriculture, medicine, transportation, and the production of energy, it has served humans and all of the earth's creatures since the beginning of time.

In the days of Abraham, following family feuds and other conflicts, the victor often stopped wells by filling them with earth and stones. It was the equivalent of burning the village. Wells of water were so important that their owners gave them names, and communities which took root in the area assumed the names of the wells. In John 4:6, Jesus meets a woman from Samaria at Jacob's well and later reveals that He is the Messiah. The Christian Bible is replete with references to the importance of water to the fabric of life. It speaks of irrigation, of aqueducts, its creation, its purification, and even of it being turned into wine.

The Koran, the sacred book of the Moslems, was revealed to Mohammed by the angel Gabriel between A.D. 610 and 632. It mentions *water* 95 times in reference to such issues as springs pouring from rocks, hygiene, irrigation, pollution, prayer, thirst, and law. It even suggests that humans were made from water.

Hammurabi, the father of law, discussed the legal ramifications of water management in his Codes which were scripted 1700 years B.C. The Greeks constructed a 4200 foot tunnel or aqueduct to bring water to the city of Athens 2500 years ago, and the early Romans built 10 aqueducts which brought 220 million gallons of water to the city daily from mountain springs. The Aqua Claudia still stands in part around Rome, and the Pont du Gard in southern France was designed 20 centuries ago by the engineer Marcus Agrippa. The greatest water purveyor of all was the Zaghouan aqueduct in North Africa which crossed 130 miles of desert to deliver water to ancient Carthage. Even as the wheel was invented about 2000 B.C., the water clock (Clepsydra), fashioned by Plato, was first used to record time about 400 B.C., and the water mill came to be in the first century B.C.

Early civilizations took root near sources of abundant water: the Tigris–Euphrates Valley of Mesopotamia, the Nile Valley in Egypt, the Indus Valley in India, and the Hwang Ho Valley in China. These were the cradles of civilization which prospered, in great measure, because of the availability of water and the construction of extensive irrigation systems which made the land productive. In some cases, however, when the water systems failed, the civilization collapsed. It is believed, for example, that the Sumerian civilization of Mesopotamia failed because of poor irrigation practices which caused increased salinity in the soil and the eventual collapse of agriculture.

To an extent, the problems of managing a water system are mirrored today as major cities are forced to obtain water from greater and greater distances from the city. Often the search for available water as well as water rights litigation spans decades. The fall of groundwater levels increases pumping costs until irrigation is no longer cost-effective, and the effects of pollution increase both the cost of palatable water as well as the threat to human health.

Water has played an important as well as diverse role in many aspects of human history. It was the focus of religious rites, science, art, music, mythology, transportation, power, healing and architecture, to mention only the obvious. For example, the immersion into or the sprinkling of water have been forms of baptism long before the birth of Christianity. Ea or Enki had the distinction of being both the god of water and wisdom in several ancient religious peoples, including the Hittites and Babylonians and was often considered the creator of humanity. And, of course, the early scientists and philosophers such as Empedocles and Aristotle believed that everything was composed of only the four elements, earth, air, fire, and water. We all remember the story of the Greek mathematician Archimedes who formulated his fluid displacement law after stepping into a public bath and observed the overflowing water. He was so excited that he ran home naked crying Eureka! Eureka! He later used this water displacement knowledge to demonstrate that the king's goldsmith had adulterated a gold crown with less expensive silver.

For centuries, people around the world have bathed in mineral water to relax their souls and cure ailments. There are a number of religious shrines where the sick have been cured by holy water. One example is spring water near the site where Bernadette, a French saint, had a vision of the Virgin Mary in 1858. In music, we find Handel's Water Music (1717). Claude Monet produced a series of water-lily murals between 1900 and 1909, and in architecture Frank Lloyd Wright created the stunning Falling Water structure near Pittsburgh in 1936.

Even as water has given birth to all of the creatures of the earth and served them so well in many ways, so has it emulated the Apocalyptic horsemen of pestilence and death. Its overabundance has caused floods since the days of Noah, and its absence has resulted in drought and starvation for countless millions. Water from hurricanes and typhoons causes death and property damage of herculean proportions each year in all corners of the globe. Waterborne diseases (diarrhea is the leading cause of death in the world) cause 80% of the illness in Third World countries, and malaria from mosquitoes born in stagnant water affects 400 million people worldwide and results in 5 million deaths. The Industrial Revolution made sewers of streams and rivers.

The history of human association with water has been a union as blessed as it has been complicated, and on many occasions as wicked as Pandora's box. Perhaps it can best been capsulated in the words of John Dyer:

> And see the rivers how they run through woods and meads, in shade and sun, sometimes swift, sometimes slow, wave succeeding wave, they go a various journey to the deep like human life to endless sleep!

WATER, BACTERIA, LIFE ON MARS, AND MICROBIAL DIVERSITY

David B. Vance
ARCADIS
Midland, Texas

James A. Jacobs
Environmental Bio-Systems, Inc.
Mill Valley, California

Microbial life is extremely diverse with regard to ambient conditions, metabolic pathways, and geographic dispersion at depth in soils and sedimentary formations on Earth. Original bacterial life on Earth started in anaerobic conditions, without free oxygen in the atmosphere. Water chemistry in the primitive oceans would have been devoid of significant dissolved oxygen. The primitive atmosphere was chiefly carbon monoxide (CO), hydrogen sulfide (H_2S), nitrogen (N_2), hydrogen (H_2), and water. Some of the more volatile trace elements such as chromium, manganese, and sodium may also have been present (1). Since that time, almost every niche on Earth has been colonized by some type of microbial life. Given the robust nature of microbial life on Earth, it is possible that microbial life existed on other planets having similar elements and compounds. As the very early history of Mars may have been somewhat similar to that of Earth, Mars may have had the conditions for microbial life. In the oceans, the gases in the primitive atmosphere were present as dissolved constituents. In addition, given the reducing conditions, soluble forms of reduced metals were also present, with ferrous iron likely being a dominant species.

INTRODUCTION

From 3.5 to 1.5 billion years ago, the atmosphere of Earth was anoxic (primarily consisting of nitrogen and methane) and the dominant life forms were methanogenic, sulfate-reducing and iron-reducing anaerobic microbes. Near the end off this period, around 2 billion years ago, bacteria that could use nitrate to convert ferrous iron to ferric literally began to precipitate the soluble ferrous iron out of the ocean, forming the great Precambrian banded iron deposits and the later Paleozoic reworked redbeds.

Slowly starting about 3 billion years ago microbes that used the sun's light as an energy source became the dominant life form on the planet. These photosynthetic species began to become active and excrete oxygen as a waste product of their metabolism. Between 1.5 and 1.0 billion years ago, the "first" great environmental disaster occurred when photosynthetic life became abundant enough to elevate atmospheric oxygen to near current levels.

The type of bacteria that are indicated to have existed on Mars appear to be similar to those on Earth termed archaebacteria. These are bacteria with an ancient history, first appearing on Earth 3.5 billion years ago under the atmospheric conditions described above. Recent advances in genetic analysis indicate that these bacteria are a third form of life different from regular bacteria (Eubacteria) and all multicellular life (Eukaryotes). These archaebacteria survive through exotic metabolic processes that rely on a wide variety of chemical reactions to provide energy.

The archaebacteria fall into four major classes:

- Acidophiles that are able to grow in the pH range of 0.0 to 5.0.
- Thermophiles that can grow in a temperature range of +30 °C to +250 °C.
- Halophiles that survive in 10% to 27% sodium chloride solutions.
- Methanogens that are the best known archaebacteria and form methane in sediments and the intestinal tracts of animals.

MARTIAN ARCHAEACTERIA?

Also, modern (relatively speaking) Eubacteria are capable of adaptation to extreme environments, including:

- Alkalophiles that can grow in the pH range of 8.5 to 11.5.
- Barotolerant bacteria can exist under hydrostatic pressures up to 1100 atmospheres.
- Other chemoautotrophic bacteria such as denitrifiers, nitrate reducers, ferric iron reducers, sulfate reducers, hydrogen formers, sulfide oxidizers, and ferrous iron oxidizers.

The biochemical diversity is astounding. Just a few specifics are as follows:

- Bacteria have been found in aquifers at depths of 1.5 miles.
- They have been found in oil production formations at depths of almost 4 miles.
- Within the last year, bacteria have been found that survive using only surrounding rock and that generate hydrogen gas in the process; these are near 1 mile deep in basalt.
- Sulfate reduction is extremely common place in groundwater. Well water with the odor of hydrogen sulfide (rotten eggs) is a manifestation of bacteria-driven sulfate reduction.

Additional aquifer conditions under which it is common to find sulfate-reducing bacterial species are as follows:

- Eh conditions less than −200 mV.
- The presence of greater than 20 millimolar sulfate from geologic origins.

In the western United States, many aquifers are anaerobic and contain sulfate. Anthranilic acid, benzyl alcohol, and P-cresol have all been demonstrated to be biodegraded in the presence of sulfate-reducing bacteria. Research is currently under way evaluating the use of sulfate reducers for the degradation of chlorinated solvents and PCBs.

Two other dominant and effective anaerobic processes are denitrification and methanogenesis. In general, most aromatic hydrocarbons can be degraded anaerobically under methanogenic or denitrifying conditions.

Some thought should always be put into the exploitation of these alternative bacterial systems. The exact stoichiometrics should always be worked out with a complete assessment of all byproducts. For example, it is important to evaluate the overall electron flux in nitrate reduction systems:

For each mole of nitrate consumed, one mole of KOH or NaOH is generated, which in turn can lower aquifer pH and form limestone or other precipitates, that plug the infiltration or treatment zones. An acid must be added concurrent with the nitrate source in the appropriate amount.

Bacteria that produce hydrogen can sequentially dehalogenate chlorinated hydrocarbons to innocuous ethane or ethene gas.

Of course, all various metabolic systems are not equal with regard to energy yielded to the organism for each reaction that takes place.

On a scale of 1 to 10, with 10 having the highest energy yield:

— Aerobic oxygen based metabolism is 10.
— Nitrate reduction is 9.
— Iron reduction is 5
— Sulfate reduction is 4.
— Methanogenesis is 2.

Thus, the source of the dominance of aerobic life was precipitated by events 1 billion years ago (as described above). Given surface conditions, it is the most cost-effective metabolic system available. However, all other metabolic options still take precedent under the appropriate conditions. It has proven particularly difficult to drive hydrocarbon contaminated aquifers into oxic conditions mainly because of the poor transport properties of oxygen in water (the reason hemoglobin exists in our blood cells is to address that issue within our bodies). The power of many of the other metabolic systems is that they are driven by chemical species that are highly soluble in water (sulfate and nitrate, for example). The less-efficient energy yield is compensated by the dramatically improved transport properties.

In addition, these alternative metabolic systems are showing to have some capacity to degrade xenobiotic hydrocarbons that prove to be recalcitrant to aerobic attack.

The power of bioremediation lies in harnessing the vitality of bacterial life in the subsurface. The ubiquitous nature of bacterial life is constantly being revised upward. Archaebacteria exist under the extremes of physical/chemical conditions. Bacteria are found at such depths that it is almost certain they were laid down with the surrounding sediments and have been able to survive since. The potential presence of microbial life on Mars is an even more extreme example of the concept. *In situ* bioremediation is a process that appears to be viable almost under any subsurface conditions.

The history of Mars exploration has focused in a large part on the potential for past or current life on the planet. The first close-up picture of Mars was obtained in 1965. Since that time, a variety of satellites have traveled by Mars and a few spacecraft have landed on the surface of Mars. In 1976, the National Aeronautics and Space Adminstration's (NASA) Viking Mission to Mars had two spacecraft, *Viking 1* and *Viking 2*. Each craft consisted of an orbiter and a lander. NASA's Mars Global Surveyor began to orbit Mars in 1997 for the purpose of high-resolution imaging of the surface of Mars, studies of the topography and gravity, and the role of water and dust on the surface and in the atmosphere. This project also evaluated the weather and climate of Mars, the composition of the surface and atmosphere, and the existence and evolution of the Martian magnetic field (2). The Mars Pathfinder with the Sojourner Rover toured the surface of Mars in 1997. In 2001, the NASA Mars Surveyor 2001 lander was renamed the 2001 Mars Odyssey; the mission was to gather data to help determine whether the environment on Mars was ever conducive to life, to characterize the climate and geology of Mars, and to study potential radiation hazards on Mars.

The most recent information about Mars has come from NASA's two Rover Opportunity and Rover Spirit landing crafts, which inspected Mars during 2004. These remote studies indicate that the elements on Mars are the same elements found on Earth. On Earth, life started in water, and there is ample evidence of large amounts of water on Mars in the past (3).

Based on the similarities with Earth, the geologic conditions on Mars are still tantalizing scientists to examine the shallow Martian soil for any sign of past or current microbial life forms.

BIBLIOGRAPHY

1. Smith, D.G. (1981). *The Cambridge Encyclopedia of Earth Sciences*. Crown Publishers Inc./Cambridge University Press, New York.
2. National Aeronautics and Space Adminstration (NASA). (2004). Mars Global Surveyor. Available: http://nssdc.gsfc.nasa.gov/planetary/marsuv.html.
3. Baker, V. and Nummedal, D. (1978). *The Channeled Scabland: A Guide to the Geomorphology of the Columbia Basin, Washington*. Prepared for the Comparative Planetary Geology Field Conference held in the Columbia Basin, June 5–8, 1978. Sponsored by the Planetary Geology Program, Office of Space Science, National Aeronautics and Space Administration.

CANALS IN THE MEKONG DELTA: A HISTORICAL OVERVIEW FROM 200 C.E. TO THE PRESENT

David Biggs
University of Washington
Seattle, Washington

The Mekong Delta is one of the most productive agricultural regions in the world. It supports more than 20 million people living in a dense network of canals and creeks that today irrigate more than 2 million hectares of rice paddy. This landscape of dense settlement and intensive irrigated agriculture is relatively recent, although the earliest canal projects date to "Fu Nan" or Oc Eo culture circa 200 C.E. Today's water regime is not so much a single water system as it is a landscape built of multiple layers of canals in various stages of development and degradation. Some waterways continue to serve as primary transportation and irrigation works: others lie abandoned, their traces sometimes only visible through analysis of aerial photography or historical reports. This overview describes these historical layers of canalization and the context for some of the major projects in each period.

GEOLOGIC OVERVIEW

The Mekong Delta of today comprises roughly 6 million hectares (5 million in Vietnam and 1 million in Cambodia) of alluvial plains. It is bordered by the Gulf of Thailand in the west and the South China Sea in the east. The main channel of the Mekong River is believed to have formed sometime between 2 and 50 million years ago in the late Tertiary period. The older, higher terraces of its delta may have formed as early as 1 million years ago. Corings of sediment in the lower delta alluvium show alternating bands affected by freshwater and saltwater, suggesting that as recently as 6000 years ago, mean sea level may have reached a post-Ice Age high water point before declining to recent historical levels.

The slope of the delta is so flat that tidal surges affect water flow in canals and creeks up to 100 kilometers upstream. Two separate tidal regimes along the western and eastern coasts cause complicated periodic variations in water currents, depending on the proximity to both coasts. When opposing water currents meet, either at an intersection of waterways or along the path of a canal, they deposit sediment into bars that often block boat traffic. Vietnamese settlers, and probably earlier inhabitants, formed markets at these halfway points, convenient places for exchange. Choices for early modern settlements here were largely responses to water conditions and associated soil deposits.

From a soils perspective, the delta can be divided into subregions connected to geologic and ecological processes: a freshwater alluvial zone, the Dông Tháp floodplain, the Long Xuyên Quadrangle, the eastern coastal zone, and the Cà Mau Peninsula (Fig. 1). The Mekong River deposits fresh sediment along the banks of the channel, creating alluvial banks that rise 2–3 meters above mean sea level. This freshwater alluvium area fosters the development of the richest natural paddy in the delta, allowing two and three rice crops per year. The combination of high banks and good drainage allows farmers to irrigate easily and also to raise fruit trees without fear of flooding. Ancient alluvium is characterized by higher soils associated with a postulated high seas episode that occurred roughly 6000 years before the present. In the present era, this higher land forms a natural barrier to the Dông Tháp floodplain, a depressed area characterized by high soil acidity and annual floods. The other large flood zone is the Long Xuyên Quadrangle south of the main river channel. The Quadrangle is remarkable for granite and limestone

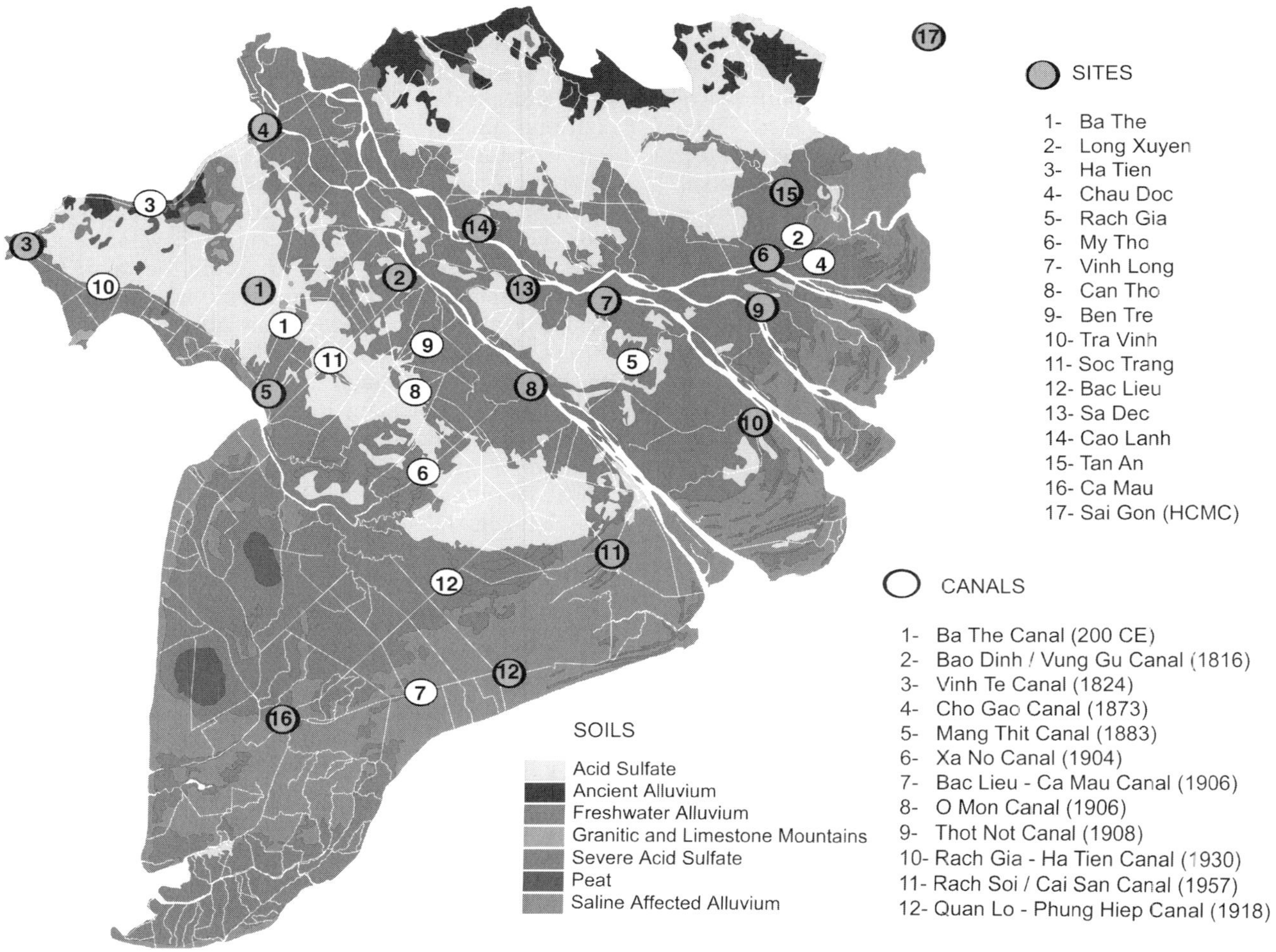

Figure 1. Mekong Delta divided into subregions.

mountains that rise up dramatically from the plains some 200–700 meters to form islands on the land. At the base of Ba Thê mountain in this Quadrangle lie some of the most substantial sites for what George Coedès described as the "Fu Nan Kingdom": the first example of an Indianized kingdom in Indochina.

Both of these two depressed floodplains, whose average altitudes are less than 1 meter above mean sea level, form two flood sinks that form virtual, inland seas each September as the Mekong River reaches its annual flood stage in the wet season monsoon. The annual flooding of the Mekong from August to October is important for far-reaching effects on almost all activity in the region. The Tonle Sap (Great Lake) in Cambodia grows to three times its dry season size in June, July, and August. As the lake reaches its limit, water flow reverses in the Tonle Sap River, flows downstream, and recombines with the mainstream rivers at Phnom Penh. The river then floods the Long Xuyên Quadrangle and the Dông Tháp Plain.

ANCIENT SETTLEMENT

The earliest traces of organized human settlement in the Mekong Delta date to the "Fu Nan" period (100 B.C.E.–550 C.E.). The types of evidence found in these settlements can be grouped roughly into wooden piers and structures, jewelry and statuary, and vestiges of infrastructure—brick foundations, brick towers, and canal traces. The largest brick foundations lie at the Ba Thê site in the Long Xuyên Quadrangle. Using aerial photography in 1931, an amateur archeologist, Pierre Paris, discovered traces of canals crossing the western delta, from Rạch Gía on the coast to Angkor Borei about 100 km to the northwest. What had previously been noted through ground surveys as straight creeks were actually the remaining traces of human-constructed waterways. Louis Malleret, a former director of the Ecole Francaise de l'Extrême-Orient, led digs in the 1940s and 1950s at canal sites discovered through aerial and ground surveys. Malleret's work and that of successive Vietnamese and international researchers has proven through comparison of pottery shards and other means that these canals dated to roughly 200–400 C.E. Styles in the pottery indicate containers for oils and other goods that resemble similar containers from eastern India in that time. The ancient canals also extended deep into the Cambodian portion of the Mekong Delta, connecting the gulf coast ports with inland river ports at Tak Ev and Angkor Borei.

In the ancient world, the Mekong Delta was an important halfway point for sailing ships travelling with the monsoons between India and China. As a possible warehousing center developed along these ancient canals, the Oc Eo site allowed ships from both destinations a useful trade interchange and provided them with important naval stores and valuable supplies brought downriver from the forest interior: rhinoceros horn, elephants, feathers, aromatic woods, and wild spices. Chinese records from both trade missions and accounts of Fu Nan monks visiting China note major settlements along the main course of the Mekong River below Phnom Penh, including the sites at Oc Eo/Ba Thê, Angkor Borei, and a site in Dông Tháp Muòi. Inscriptions described Indian merchant colonies at Southeast Asian ports, including those in the Mekong Delta. The situation of Indian merchants here enriched local rulers, allowing them to raise funds and craftsmen to build ships of their own.

Environmental factors may have played a role in the decline of pottery and evidence of population in the lower delta. Soil cores at the Oc Eo site reveal high concentrations of Rhizophera pollen spores in deeper layers, suggesting that saltwater mangrove trees populated a floodplain that is entirely freshwater today. Every year, the sea coast may have receded further from the Oc Eo site, and the canals may have slowly filled with sediment from the rivers, turning them into shallow creeks and ponds. This would in turn change the wetlands ecology to one favorable for freshwater tree species. This freshwater plant community would have also attracted mosquitoes that carry endemic malaria and parasites, thus putting dense populations at greater risk of disease outbreaks.

VIETNAMESE SETTLEMENT

There is no further evidence of major canal projects or dense settlement in the Mekong Delta for more than 1000 years after the decline of "Fu Nan." From the ninth to fourteenth centuries, the upstream kingdom of Angkor controlled the entire region and conducted its major trade routes by highways that stretched westward into present-day Thailand and to the west coast of the Isthmus of Kra. Archeological evidence near present-day coastal towns of Trà Vinh and Sóc Trang suggest that river ports here conducted small-scale local trade with Southeast Asian and Chinese merchant ships. One possible explanation for the lag in development of the Mekong Delta after Fu Nan could be that the very same waterways and flat landscape that facilitated easy transportation also allowed easy attack by enemy fleets. Steles at Angkor Thom (Siem Reap) depict naval battles with ships from Champa that attacked the Khmer capital by way of the Mekong River. The numerous islands in the Gulf of Thailand also provided easy harbor for all manner of pirates—Malay, Siamese, Chinese, European. Until Vietnamese settlers began migrating southward to the delta in the seventeenth century, island-like Khmer trading ports and administrative outposts of the Khmer kingdom characterized the area. Populations of Khmer villagers lived in some of the richest, naturally irrigated alluvial regions in areas that are today still ethnically Khmer: Sóc Trang, Trà Vinh, Trà Nóc, and Sa Déc.

One of the most important factors in Vietnamese settlement and early modern canalization was the increasing importance of rice as a trade commodity. Until the seventeenth century, rice was rarely traded by ship and in cities as were gold, spices, ivory, and aromatic woods. It was a local subsistence commodity and generally in greater abundance than the human labor necessary to cultivate it. In the seventeenth century, increasing population pressures in the Red River Delta brought on widespread famines and social unrest; to the present day, this population density in the north and the "southward progression" have played important roles in spurring development of waterways and infrastructure in the Mekong Delta. The upset of the ruling Minh Dynasty in 1644 also led to a mass exodus of Minh loyalists to Vietnam and other kingdoms that granted them asylum. In 1679, 3000 Fukien troops arrived in the Vietnamese kingdom. The Vietnamese ruler, anxious not to have them so close to his capital, sent them to the south where they formed the commercial towns of Biên Hòa, Sài Gòn, and Mỹ Tho. In 1681, another Chinese emigré, Mặc Cúu, immigrated to Cambodia and succeeded in winning a concession from the Khmer king to develop a trading port at Hà Tiên. From 1690 to 1780, Hà Tiên became a virtual city-state, trading most of the forest products of Cambodia by a Chinese-controlled shipping network centered at this Gulf Coast port. The more powerful kingdom of Siam now controlled the Isthmus of Kra, so Hà Tiên became an important trading port much in the same way as Oc Eo before it.

The Vietnamese kingdom established military and administrative posts in the Mekong Delta in the eighteenth century. In 1732, military campaigns led to the establishment of camps at Mesar (Mỹ Tho), Kampong Reussey (Bên Trè), Koh Gong (Gò Cong) and Peam Ba-rach (Long Xuyên). The court sent troops and new migrants to establish military plantations [dôn diên] in the new territories. These military plantations functioned as both garrisons and farming communities. At the same time, the Chinese developed mercantile posts and arranged the marketing of rice. The Tây Son civil war (1789–1802) and repeated Siamese invasions at Hà Tiên (1705, 1717, and 1771) led to the eventual abandonment of Hà Tiên and the establishment of Vietnamese control of the delta in 1802. For 60 years, the Court enticed settlers to establish garrisons and villages. In 1818, still worried about the threat of Siamese invasion, King Gia Long ordered his military governor in the delta, Thoại Ngọc Hâu, to assemble an army of conscript laborers to dig a canal 67 kilometers from a new river citadel at Châu Dôc to Hà Tiên. More than 50,000 laborers worked for 5 years and through several Khmer-led rebellions to complete the waterway. This strategic project was intended to allow greater Vietnamese control of the Gulf Coast and direct water routes to river garrisons. Two other major projects in this period connected the increasingly dense Sài Gòn watershed with the intermediate Vàm Co watershed and the Mekong river. The Bạo Dinh Canal in 1816 allowed more direct access to the Khmer territory east of Mỹ Tho and solidified trading routes and administrative communication in the eastern delta. The Thoại Hà (Long Xuyên–Rach Giá) Canal was also completed in 1816. The rapidity of this project (3 months) and its location in the heart of the former "Fu Nan" kingdom suggests that this waterway may have dated to a much earlier period and only been "cleaned" in 1816. Historical records in the nineteenth century do not specify this difference, and they often alternately describe canals as rivers.

Midcentury violence led to the abandonment of coastal outposts for periods of years, and the neglect of these new waterways led to their being choked off with silt. The natural processes of sediment accretion and growth of aquatic plants played an important role in canal projects, especially in the costs of their upkeep. The interplay of tidal pulses and opposing river currents led to formation of transverse sediment bars across the channel of a canal. In a matter of months, if not continually cleaned by hand, these bars would reduce the navigable channel to a depth of half a meter, even at high tide in the dry season. Besides this hydraulic challenge, the water hyacinth or luc bình could quickly reproduce and choke off an entire waterway with a thick, floating mat of roots on the water's surface. Thus, without the continuous application of human labor in large numbers, most artificial waterways could not remain navigable for long. With each social uprising, canals quickly deteriorated so that after a few years, the costs of redredging would approach the original cost of construction. Another theme of canal degradation that also continued through the colonial and postcolonial eras was its role in aiding rebel groups.

With the French colonial conquest beginning in 1858 and lasting until 1867, rebel groups could only survive in the deep swamps far away from deeper rivers and streams. Battles at the Plain of Reeds in 1873 proved that control of waterways was vital for the French colonial regime. For two decades, the French colony's ruling admirals focused their efforts on linking strategic areas with the colonial metropole at Sài Gòn. In 1875, Admiral Duperré organized a permanent committee to improve naval communications between Sài Gòn and the Mekong Delta. This commission surveyed existing and proposed routes until 1879. During this time, the colonial government mobilized armies of corvée [conscript] laborers into government service. These first canals included Trà On (1876), Chò Gạo (1877), Set Say (1878), Phụ Tục (1878), Mirador (1879), and Saintard (1879). Of these, the Chò Gao Canal was one of the most vital. Via several large waterways, it linked Sài Gòn with Mỹ Tho, the largest river port at the time. A gunboat or cannonière could now make the trip to the delta in just 8 hours.

The same problems of sediment accretion and controlling vegetation continued to dog French engineers as it had their Vietnamese predecessors. In just a few years, Chò Gạo was silted again, preventing passage of the larger boats and forcing them to make the much longer journey to the river port by entering the river from the sea. In 1883, the government paid the Eiffel company to construct steel railway bridges across the intermediate Vàm Co River to connect Sài Gòn with Mỹ Tho by train. As the colonial government grew stronger in the 1880s and the French succeeded in using mechanical dredges at Suez, the regime introduced mechanical dredges. The shallow and high-clay sediments here, however, forced the government to rely on labor crews until new dredges were built in the 1890s.

The period of colonial dredging contracts with a mechanical dredging firm lasted from 1894–1945. During the peak of colonial expansion from 1894–1930, the colonial regime dredged some of the major "transverse" waterways connecting the lower branch of the Mekong or Hậu Giang with the Gulf of Thailand. This lower part of the delta was the least inhabited, making it the easiest space for French colons to build plantations several thousand hectares in size. Major canals such as Xà No (1904), Lai Hiêu (1906), Thôt Nôt (1908), Ô

Môn (1906), Vinh Tê (redredged,1914), Quạn Lô–Phụng Hiêp (1918), Ba Thê (redredged,1923), and Tri Tôn (1928) were developed. They also, dredged important waterways into Dông Tháp (1890–1904), from Bac Liêu to Cà Mau (1906), and from Rạch Giá to Hà Tiên (1930). This massive effort in mechanical dredging was accompanied by the immigration of laborers from the north; the population increased from 500,000 to almost 4 million in 1930.

Some areas, especially the fields near Cân Tho, flourished, but many of the waterways proved to be both economic and ecological disasters. The Lai Hiêu Canal brought on higher flooding, and its embankments cut off existing water circulation that led both to the spread of cholera in 1908 and the failure of many preexisting fields. A series of typhoons from 1904–1908 caused abnormally high flooding in the Plain of Reeds and the abandonment of plantations and fields after several years of lost crops. Canals also led to drying of soils in floodplains that were high in sulfate content. This drying of soils produced acid sulfate that then leached into the waterways and in turn quickly reduced crop productivity. With the onset of the world depression in 1931 and the collapse in rice markets, the masses of migrant laborers employed as tenant farmers on the plantations began to protest. Hydraulic engineers also began to debate the future of projects in the area. One camp argued that it was necessary first to improve existing yields and to let some of the failed projects lie fallow. Another camp argued for more sophisticated mechanical approaches to drainage and irrigation, suggesting the redevelopment of casier or grids of irrigation canals where farmers could be resettled. Except for small settlement projects near Ba Thê, the colonial government did not develop any other large projects before it was ousted from power in 1945 and subsequently entered into a 9-year war with the Viêt Minh.

During the Second Indochina War (1945–1954), the Government of South Vietnam attempted to develop several of these casier styled resettlement projects in a fashion similar to those proposed by the French before them. Of these, the single permanent project was the Cái San Project, a series of "transverse" canals in the lower delta and a smaller network of intersecting canals numbered from 1 to 17 and letters A to H. From 1955–1962, the Government resettled approximately 50,000 settlers, mostly Catholic, from their original villages in North Vietnam to 3-hectare tracts at Cái Săn. During the 1960s, American military and private contractors further built up the area with a highway paralleling the Rạch Sọi Canal (Highway 80) and an airport at the southern end of the area (today Rạch Giá Airport). For the most part, both wars prevented steady extension of waterways due to continued guerilla attacks. Like preceding antistate rebels, preventing easy circulation of government troops on the waterways and (increasingly) roads was vital to survival. Most of the familiar scenes of combat during the "American War" were encounters in the swampy interiors of swamps where previous canals had either deteriorated or been abandoned because of their low agricultural utility.

Since the conclusion of the war in 1975, and especially with the market orientation stimulated by Dôi Mói Reform in 1986, the Government today has begun a series of new projects. Like projects enacted in the 1950s, many of these contemporary development projects were continuations of earlier plans developed in the 1960s and early 1970s. For example, in 1996, the Government began work on the Mỹ Thuận Bridge, a project first surveyed by the Nippon Koei Corporation in 1963. This bridge, completed in 2001, became the longest span bridge in Southeast Asia and for the first time, allowed continuous highway transport across the upper branch of the Mekong River. A second span is due to be completed near C`ân Tho in 2005. Now at peace, government leaders have refocused their attention on Dông Tháp and former guerilla strongholds for future agricultural development. The Vietnamese delta population now is well over 17 million persons, however, so decisions to build canals and flood dikes are becoming increasingly difficult. Flood control in one area often results in diversion of the floodwaters to another zone or province downstream, thus giving rise to interprovincial disputes over access to clean water. A related issue is the construction of dams on the upper and middle Mekong Basin. The Mekong Committee is an international advisory body made up of representatives of the six riparian countries. With the reunification of the region and all member states into both the Mekong Committee and ASEAN, downstream countries have begun to lobby against diversionary projects upstream.

Another future concern in the Mekong Delta is the impact of global warming. As sea level rises just a few feet, thousands of hectares along the coast would be permanently inundated with salt water. The Mekong Delta is one of the flattest deltas in the world, so fluctuations in both river levels and sea levels could require the construction of unprecedented sea dikes to protect areas of coastal rice paddy.

The apparent "system" of canals and water projects in the Mekong Delta reflects nearly 2000 years of experimentation. Some of those trials lasted into the present, and some others are barely visible now. The economic or environmental health of this system is largely dependent on regional social factors and the intensity of human activity here. Currently, the challenge for engineers and planners in this region is to reconcile these factors to distribute "clean" water equitably in a more predictable fashion.

READING LIST

Biggs, D. (2003). Problematic progress: Reading environmental and social change in the Mekong Delta. *J. Southeast Asian Stud* **34** (1): 77–96.

Coedes, G. (1967). *The Indianized States of Southeast Asia*. Translated by Susan Brown Cowing. W.F. Vella (Ed.). East-West Center Press, Honolulu, HI.

Paris, P. (1931). Anciens canaux reconnus sur photographies aériennes dans les provinces de Ta Kev et de Châu Dôc. *Bulletin de l'Ecole Française de l'Extrême-Orient* **XXXI**: 221–225.

Xuân, V.T. et al. (1998). *Development of Farming Systems in the Mekong Delta*. Hô Chí Minh City Publishing House, Hô Chí Minh City.

CONFLICT AND WATER USE IN THE MIDDLE EAST

CLIVE D. LIPCHIN
Ann Arbor, Michigan

INTRODUCTION

Freshwater is unevenly and patchily distributed across the surface of the earth, some regions have either ample supply or unlimited access, and other regions suffer from unpredictable rainfall patterns and long periods of drought. As human populations grow and their activities continue to alter environmental patterns, the demand for freshwater will increase as the availability and quality of freshwater decreases. These conditions make water-related conflicts more likely between nation states that are forced to share a common water resource under the strain of locally increasing populations and increasing regional demands for water. In the semiarid and arid regions of the globe, especially the Middle East, conflict over water seems a more and more likely prospect (Table 1).

Where water is scarce, competition for limited supplies can lead nations to seek access to water as a matter of national security. As recently as the mid-1980s, U.S. government intelligence services estimated that there were at least 10 places in the world where war could break out over dwindling shared water resources; the majority of these locations are in the Middle East. Jordan, Israel, Cyprus, Malta, and the countries of the Arabian Peninsula are sliding into the perilous zone where all available fresh surface and groundwater supplies will be fully used. In Israel, the Palestinian Authority, and Jordan, a water deficit already exists, and all available water resources are overexploited.

The conflicts in the Middle East are probably unique in the world and are many faceted and complex. As the birthplace of modern Western civilization and three of the world's largest religions, conflicts, whether political, racial, or religious, have been a part of the system since civilization first began between the twin rivers of the Tigris–Euphrates. The ongoing conflicts taking place on today's political stage can be traced back to Biblical times and beyond. In the face of such an historic record of conflict, solving disputes over shared water resources automatically involves cultural, economic, and political factors.

Table 1. Percapita Water Availability in 1990 and 2025[a]

Country	1990	2025, Cubic Meters Per Person Per Year
Kuwait	75	57
Saudia Arabia	306	113
United Arab Emirates	308	176
Jordan	327	121
Yemen	445	152
Israel	461	264
Qatar	1171	684
Oman	1266	410
Lebanon	1818	1113
Iran	2025	816
Syria	2914	1021
Iraq	5531	2162

[a]Reference 1.

The emergence of modern independent states since the decline of colonialism in the region has strengthened nationalistic ideals and exacerbated common property issues that transcend their borders. Given the importance of the region as the world's main supplier of fossil fuels, the international community has been quick to realize the delicate power balance in the region and how precarious and volatile it can be. In the current global political environment, it is in the world's best interest to foster cooperation among the states in equitably managing their shared waters.

Across the globe, many rivers, lakes, and groundwater aquifers transcend arbitrary political borders. This geographical fact has led to the geopolitical reality of disputes over shared waters; the situation in the Middle East (especially the Nile, Jordan, and Tigris–Euphrates) is just one such example. Other regions where water disputes have or are likely to develop are the Indus, Ganges, and Brahmaputra in southern Asia; and the Colorado, Rio Grande, and Parana in the Americas.

Water is a fundamental resource vital for survival, and it also provides a source of economic and political strength. Under these conditions, ensuring access to water provides a justification for going to war, and therefore water supply systems can become a goal of military conquest (2). Gleik (3) outlined the following characteristics that make water likely to be a source of strategic rivalry:

1. the degree of scarcity,
2. the extent to which the water supply is shared by more than one region or state,
3. the relative powers of the basin states, and
4. the ease of access to alternative freshwater sources.

The Middle East and its many ideological, religious, and geographical disputes fits within this framework.

SETTING

The current water crisis plaguing the region can be traced back to the arbitrary political division of the region into competing states by the withdrawal of the post-World War I colonial powers (4). The subsequent map of the region disregarded the issue of water in forming possible natural political boundaries. Except for Lebanon, none of the states that the colonial powers created was provided with independent water resources, and no mechanisms were established for coordinating the use of internationally shared water resources (4). The Tigris–Euphrates Basin was sectioned irregularly and placed in the domains of three competing states (Turkey, Syria, and Iraq); the Jordan River was divided up among four states (Israel, Jordan, the Palestinian Authority, and Syria) (4). The Nile River Basin is shared by no less than nine states, and the balance of power is heavily skewed toward the two

Table 2. International River Basins in the Middle East[a]

River Basin	Total Area of Basin, Square Kilometers	Countries in Basin	Area, Square Kilometers	Percentage of Total Area
Tigris	378,850	Iran	220,000	58
		Iraq	110,000	29
		Turkey	48,000	13
		Syria	850	<1
Euphrates	444,000	Iraq	177,000	40
		Turkey	125,000	28
		Syria	76,000	17
		Saudi Arabia	66,000	15
Jordan	19,580	Jordan	7,650	39
		Syria	7,150	36
		Israel	4,100	21
		Lebanon	950	5
Nile	3,031,000	Sudan	1,900,000	63
		Ethiopia	368,000	12
		Egypt	300,000	10
		Uganda	233,000	8
		Tanzania	116,000	4
		Kenya	55,000	2
		Zaire	23,000	1
		Rwanda	21,500	1
		Burundi	14,500	<1

[a]Adapted from Reference 1.

economically and militarily powerful downstream states, Egypt and the Sudan (Table 2).

An example to illustrate the tensions inherent in water resources development in the region is the conflict that took place between Israel and Syria. In the 1950s, Syria tried to prevent Israel from building its National Water Carrier, an extensive canal and piping system to provide water to the extremely arid areas of southern Israel. When Syria attempted to divert the headwaters of the Jordan in the mid-1960s, Israel used force, including air strikes against the diversion facilities, to prevent their construction and operation (5). These military actions, it is thought, contributed to the outbreak of the 1967 Six-Day War.

Having more mouths to feed, the allocation of freshwater becomes a pressing problem. More mouths means more food, and so, more water needs to be allocated to agricultural production. This issue is especially true for Turkey, Syria, and Iraq, which have all embarked independently on ambitious irrigation projects on the Tigris–Euphrates river system to increase their agricultural capacity (6). Water allocation, for the riparian states along the Jordan River, has become a balancing act between agriculture and domestic use as more of the population becomes urbanized. The allocation of water can lead immediately to issues of conflict as upstream states control the flow of a river course and inadvertently disadvantage those downstream that depend on the same flow.

The Tigris–Euphrates headwaters originate in Turkey but empty into the sea at the Persian Gulf in Iraq, flowing through Syria on the way to the sea. Egypt is entirely dependent on the Nile River for practically all of its water requirements but is at the mercy of the upstream states (especially Ethiopia) that contain the headwaters of the Nile, the Blue and White Nile, within their borders. Israel, Jordan, and the Palestinian Authority are in an even worse position. They possess few internal renewable water resources and all depend on one river system, the Jordan and associated aquifers, which has its headwaters originating outside of their territories in Lebanon and Syria.

Wolf (7) has provided a framework for analyzing transboundary waters and the potential for conflict. Within this framework, the problem, background, attempts at conflict management, and the potential for a positive outcome where conflict can be averted is provided. Wolf's framework is applied here in analyzing the three primary water systems in the Middle East: the Tigris–Euphrates, the Jordan, and the Nile.

THE TIGRIS–EUPHRATES RIVER BASIN

The Problem

In 1975, unilateral water developments came close to provoking warfare among the riparians of the Tigris–Euphrates (Turkey, Syria, and Iraq). Since the 1960s, population pressures drove the three states to pursue water development projects independently, primarily for agricultural expansion. Specifically, these were projects in southern Anatolia in Turkey, the Keban Dam (1965–1973), and in Syria, the Tabqa Dam (1968–1973) (7).

Background

From source to sea, the Euphrates is the longest river in the whole region of West Asia (8). It traverses a distance of 2,700 kilometers, of which some 40% is in the modern state of Turkey, 25% in Syria, and 35% in Iraq. Its twin, the Tigris, has a total length of 1,900 kilometers, of which about 20% lies in Turkey, 78% in Iraq, and only 2% along the pointed northeastern corner of Syria, known as the "Duck's Beak" (4).

Probably the most ambitious development project carried out in the region is Turkey's South East Anatolia Project known by the Turkish acronym GAP (for Guneydogu Anadolu Projesi, in Turkish). The project aims to develop the regions bordering Syria and Iraq that encompass the headwaters of the Tigris and Euphrates. The area is sparsely inhabited by a population of around 6 million; the majority consists of ethnic Kurds, who claim the region as part of their national homeland. By transforming this semiarid region into the country's breadbasket, Turkey hopes to encourage and promote the development of agriculture and industry and also to offset the Kurdish majority by attracting Turks into the area and so diluting Kurdish nationalistic claims (6).

The Turkish government has promoted the GAP project for its great eventual benefits for all, but recently, more than 250,000 people were displaced from their homes, which were to be inundated by dams. A conflict broke out between the Turkish government and the Kurdish Workers Party (PKK), which has repeatedly opposed GAP and regarded it as a Turkish theft of Kurdish waters (4).

The GAP plan calls for constructing 80 dams, 66 hydroelectric power stations that have a total capacity of 7700 megawatts, and 68 irrigation projects covering up to 2 million hectares. Among the principal dams are the Keban, the Karakaya, and the Ataturk, which is the linchpin of the entire GAP complex (4).

Despite its great agricultural potential, Iraq has, in the last decade, become a net importer of grain to feed its population of 26 million. Iraq's main irrigated lands are the regions of lower Mesopotamia. However, waterlogging and salinization plague this region and therefore an imperative need has developed for a coordinated scheme of rehabilitation and sustainable water management, especially to provide regional drainage.

In 1953, Iraq began work on a regional canal to provide comprehensive drainage. This canal has been dubbed "The Third River" because it flows midway between the Tigris and Euphrates. During the Gulf War, bombing caused great damage to Iraq's hydraulic works, but it managed to complete the project in December 1992.

The project was designed to drain 1.5 million hectares of land, allowing Iraq to increase its domestic food output significantly and thus help achieve its aims of economic independence. However, like Turkey's GAP project, Iraq's third river has displaced the indigenous Marsh Arabs, and the UN has criticized the canal as an environmental crime that threatens to destroy an entire ecosystem (6). An ulterior motive for draining the region has been the possible exposure of oil deposits that it is thought underlie the marshes and are considered as extensive as those in Kuwait.

More than Turkey and Iraq, Syria depends nearly entirely on the flow of the Euphrates for the development of its economy. Apart from the Euphrates, Syria has access to a few underground aquifers from which, through the sinking of thousands of wells, has already caused overdrawing and salinization of the groundwater. In 1974, Syria began its greatest engineering feat, the damming of the Euphrates and the creation of Lake Assad. The dam, built with Soviet engineering and financial aid, was supposed to irrigate some 400,000 hectares, generate electricity, and make the region prosperous (4). In this project, as for the projects described, 70,000 indigenous Bedouin were displaced. In addition, the project has turned out to be a disappointment; the imported Soviet water generators proved faulty, and the land originally designated for irrigation proved unsuitable because the soil contains large amounts of gypsum (4).

Attempts at Conflict Management

According to Shultz (8), the real problems that the three states currently face are management, apportionment, and development planning, which have led to disagreements among them. This joint dependency on the river waters clearly indicates that the national security of the three states are linked. Turkey, Syria, and Iraq may, therefore, be considered to form a hydropolitical security complex. A hydrosecurity complex is defined as those states that are geographically part owners and technically users of the rivers and, as a consequence, consider the rivers a major security issue (8).

Disputes over the Tigris–Euphrates waters consist of conflicts between upstream and downstream neighbors, but also between the state and ethnoreligious groups (Kurds, Marsh Arabs, and Bedouins). Both Syria and Iraq depend on Turkey for the continual flow of the Euphrates. The intensification in using the Euphrates for Turkey's GAP project has caused sporadic decreases in flow in both Syria and Iraq. Such a decrease in flow is especially important for Syria, as, unlike Iraq, it cannot rely on the flow of the Tigris. The issues of an amicable apportionment of the river's flow could probably be easily met because there is enough water for all three states (8). Surprisingly enough, however, no tripartite agreement has yet been signed by the three riparians on flow regulation, dams, and sustainable water management.

The issue in contention then is how to weigh historical rights against proportionate contributions to flow, taking into consideration the real needs of each country. These are the availability of energy, the need for hydroelectricity, the feasibility of developing economic alternatives to irrigation farming, the efficiency of water use, and the size of each country's population. In 1980, a Joint Economic Committee was established between Turkey and Iraq, which included meetings related to water resources. Syria began participating in 1983, but meetings have been sporadic at best (7).

Outcome

Turkish and Syrian leaders had agreed to resolve the issues of water allocation by the end of 1993, but no agreement has yet been negotiated. Thus far, it seems that the levels of mistrust among the countries makes any serious settlement unlikely in the near future.

THE JORDAN RIVER WATERSHED

The Problem

Ever since the mid-1950s, the issue facing the riparians of the Jordan River watershed has been one of recognizing rights and equitably apportioning surface and groundwaters. An agreement came close to being signed by the riparians in the early 1950s, which was based on the Johnson accords, named after U.S. special envoy Eric Johnson. Although Israel agreed to sign the accords, the Arab states officially rejected the agreement but have unofficially used the accords as a basis for ongoing negotiation. Since 1991, beginning with the Arab–Israeli peace accords, especially those between Israel and Jordan and between Israel and the Palestinian Authority, progress has been made on rights, use, and apportionment. The Oslo accords and the Israel–Jordan Peace Treaty of 1994 both included means for joint water resource management. Today, the ongoing Palestinian–Israeli violence has made the agreements signed at Oslo exceedingly difficult to implement.

Background

The Jordan Basin is an elongated valley in the central Middle East. Draining some 18,300 square kilometers, it extends from Mount Hermon in the north to the Dead Sea in the south and lies within the pre-June 1967 boundaries of Israel, Jordan, Lebanon, and Syria. Its waters, which originate in rainfall and in rivers and streams of Lebanon, Syria, and the Golan Heights, drain the lands to the east and to the west of the Jordan Valley. Precipitation in the basin ranges from more than 1000 mm/y in the north to less than 50 mm/y in the south but averages less than 200 mm/y on both sides of the Jordan River (9). Much of the basin is arid or semiarid and requires irrigation for agricultural development.

Critical to the region's water supply is access to the water resources of the Jordan and also to the underground aquifers of the region. Such access is especially crucial for Israel because 60% of its total potential water supplies are in the region's two largest aquifers, the coastal plain aquifer and the mountain aquifer (6).

As early as the mid-1970s, Israel has been overabstracting water from the coastal aquifer, causing saltwater intrusion. The Gaza Strip also depends on this aquifer, from which it is estimated that 65 MCM/y (million cubic meters per year) is available. Rapid population growth, however, has severely taxed the system and caused lowering of the water table and severe intrusion of seawater into wells. Consequently, most of the local drinking water in Gaza exceeds the salinity level of 500 ppm, considered the upper threshold for safe drinking water; in some areas, the salinity level has reached 1500 ppm and made drinking water next to intolerable (4).

The mountain aquifer or Yarkon–Taninim aquifer extends beyond the Green Line (Israel's pre-1967 borders) into the mountainous regions of the West Bank. The aquifer extends about 150 kilometers from north to south. The aquifer's sustainable yield of fresh water is estimated at about 300 MCM/y. It is tapped by more than 300 wells, whose total annual pumpage averages 375 MCM. Like the coastal aquifer, the mountain aquifer is suffering from overpumping and a decline in water quality.

Since the 1950s, Israel and Jordan have pursued unilateral plans for developing the Jordan River Basin. Israel drained the Huleh swamps to create arable land and constructed the National Water Carrier that pumps water from the Sea of Galilee to the central and southern regions of the country. Jordan created the King Abdullah canal on the eastern shores of the Jordan for irrigation and drinking purposes. Both of these projects have, in effect, dried up the natural flow of the Jordan to its terminus at the Dead Sea.

Attempts at Conflict Management

Although never signed, the Johnson Plan of 1955 offered a regional approach to the management of the Jordan, although groundwater was not considered. Today, groundwater is a key point in negotiations between Israelis and Palestinians because groundwater development will be integral to creating a Palestinian State.

Outcome

Since the Madrid talks of October 30, 1991, Israel and the Arab states agreed on the need for multilateral negotiations on water resources. This multilateral track runs in parallel with the bilateral negotiations on a political settlement between Israel and the Palestinian Authority (now stalled) and Israel and Syria (also stalled). The multilateral track on water serves as a venue for discussing problems of water supply and demand, and joint institutions have been created between Israelis and Palestinians and between Israel and Jordan on water issues. The idea of multilateral working groups is to create opportunity for progress on the bilateral track (7). In other words, agreement on water issues can become a catalyst for bilateral political agreements. The current Working Group on Water Resources has thus been more about fact-finding and sharing information than on the more contentious issues of water rights and allocations (7). It can be said, however, that Israel and Jordan have, to a large degree, agreed on these issues by signing the Israel–Jordan Peace Treaty in 1994. Nonetheless, a true accounting of rights and allocations in the region will need to include the Palestinian Authority as a fellow riparian to the Jordan, as well as Syria and Lebanon, that contain the river's headwaters.

THE NILE RIVER BASIN

The Problem

Management of the Nile waters is a case of historical rights (essentially claimed by Egypt) to the river and the political and economical power of the riparians. Egypt, as the power broker of the region, wields the greatest influence over the Nile, even though all of the Nile's water flow originates outside of Egypt. To date, only Egypt and the Sudan have negotiated on means for managing the Nile. However, nine sovereign states share the waters of the Nile. The core question is thus where best to control the river for the benefit of all riparians—upstream or down (7).

Background

The Nile is one of the great natural wonders of the world: It is the longest river and flows 6,825 km south to north across 35° latitude. The Nile's catchment basin covers approximately one-tenth of the African continent, an area of 3,007,000 square kilometers. Because of its size and variety of climates and topographies, the Nile Basin constitutes one of the most complex of all major river basins. Throughout the northern reaches of the Nile, human civilizations have depended on the river for their very survival; without the river, Egyptian civilization would not exist.

The Nile has two major tributaries, the White Nile and the Blue Nile. These two principal headwaters are very different from each other; they arise in contrasting climatic and physiographical areas and are characterized by disparate hydrologic regimes. The White Nile flows out of the tropical rainbelt of Central Africa and has relatively little interseasonal variation. The seasonal rainfall fluctuations in the Blue Nile headwater region of Ethiopia are much more marked than those prevailing at the source of the White Nile. In sum, approximately 86% of the Nile's flow originates in Ethiopia: The Blue

Nile contributes 73% and the Atbara 13%. The other 14% is contributed by the White Nile (10). The total volume of water available in any given year is subject to substantial hydrologic variability.

The Nile River Basin, along with all the water it carries, traverses the territories of nine separate nations: Burundi, Egypt, Ethiopia, Kenya, Rwanda, the Sudan, Tanzania, Uganda, and Zaire. Although these countries share one of the greatest rivers in the world, many inequities over this common resource exist and are especially difficult to redress. The Nile presents a classic riparian problem: A river does not recognize boundaries between states—it flows continuously. The ways in which it is diverted, collected, used, and returned to the stream determine the quality and quantity of water that each successive downstream nation will receive.

The Nile region is unique in economics and development. The pattern of water demand and use in the Nile Basin contrasts sharply with the pattern of supply: The upper valley states, in particular Ethiopia, are best placed geographically, but the lower valley states, especially Egypt, have a vital resource interest and can exercise decisive military and economic power. The paradox is that the countries that contribute the most water use the least, but the countries that use the most are those that have political and economic influence.

Egypt and the Sudan account for more than 90% of the water drawn from the Nile. However, there are seven other basin states, and their demand for water will inevitably grow. The upper riparian states, Tanzania, Uganda, Rwanda, Kenya, Burundi, and Zaire, use 0.05 km^3/yr collectively; Ethiopia uses only 0.6 km^3/yr. This is ironic because combined, these states contribute 72 km^3 of water per year, or about 86% of the Nile's flow (10).

Attempts at Conflict Management

Optimal development of the basin's water resources depends on agreement and cooperation among individual countries that share several common difficulties. Among them are population growth, insufficient agricultural production, and evaporative losses (10). These all directly impact the available water supply for all basin countries. An integrated approach to resolving these issues is most logical given that all countries have a stake in the outcome. In addition, these dilemmas transcend international boundaries, so any action taken by one nation will almost assuredly impact other nations in the region.

In 1929, the Nile Waters Agreement between Egypt and the Sudan was signed. This agreement guaranteed allocations to Sudan, but the entire seasonal flow of the river was reserved for Egypt (7). The agreement was heavily weighted in Egypt's favor by assurances that no works would be developed on the river that would threaten Egypt's interests. Intermittent negotiations occurred between Egypt and the Sudan on the building of the Aswan High Dam in Egypt in the late 1950s. No agreement on joint management of the dam was reached, in 1959, Sudan unilaterally built the Sennar Dam and by so doing, repudiated the 1929 agreement (7). However, when the military regime gained power in the Sudan in 1958, a Nile Waters Treaty was signed in 1959 called the Agreement for the Full Utilization of the Nile Waters.

The upstream states' reaction to the treaty was that the downstream states had simply divided the recognized average flow of the Nile between themselves and left nothing for any other countries (10).

Outcome

Egypt and the Sudan have agreed that any claims to Nile waters by the other riparians will be met with one unified Egyptian–Sudanese position. No other state riparian to the Nile has ever exercised a legal claim to the waters allocated in the 1959 Treaty (7). Ethiopia, which has an estimated 75–85% of the river's annual flow, has not been a major player in Nile hydropolitics. So far, it has pursued unilateral projects to meet its growing need for irrigation. Recently, however, a new Nile Initiative created dialogue among all the riparians on joint planning for the basin, and a new treaty, inclusive of all the riparians, may be in the offing.

CLOSING

Wars over water have not occurred. The last 50 years have seen only 37 acute disputes that involved violence. During the same period, 157 treaties have been negotiated and signed (7). But the growing lack of clean freshwater coupled with the rise in demand has led to intense political instability, and acute violence has occasionally resulted (7). Many of the world's water conflicts are also in the developing world, where political instability is coupled with socioeconomic ills such as poverty. Within the Middle East, the legacy of colonialism, mistrust between riparians, disparities in economic, infrastructural, and social capacity, and political orientation further complicate an integrated and regional approach to water resources development.

Middle Eastern water development has followed a pattern of unilateral development first due to perceived claims of sovereignty or to avoid political intricacies. At some point, a regional power may implement a project that will impact other riparians, such as the Aswan High Dam on the Nile and the GAP project in Turkey. The absence of relations or institutions for resolving these developments can be a flashpoint for violence. General legal principles for transboundary water management are, however, currently being defined and improved on, such as the Convention on the Non-Navigational Uses of International Watercourse ratified by the UN Assembly in 1997 (7). These principles are beginning to provide a framework to which countries can turn in formulating cooperative water regimes.

As described before, although the issues are highly contentious and complex, all-out war over water has not occurred in the Middle East. Positive outcomes to conflict resolution have occurred in some cases, for example, the Israel–Jordan peace treaty. In other conflicts, the countries are moving slowly toward resolution, as in the case of the Nile riparians.

BIBLIOGRAPHY

1. Gleik, P.H. (1994). Water, war and peace in the middle east. *Environment* **36**: 6–15.
2. Flakenmark, M. (1986). *Fresh Waters as a Factor in Strategic Policy and Action*. In: Westing (Ed.). Global Resources and International Conflict, pp. 85–113.
3. Gleik, P.H. (1993). Water and conflict. fresh water resources and international security. *International Security* **18**(1): 79–112.
4. Hillel, D. (1994). *Rivers of Eden. The Struggle for Water and the Quest for Peace in the Middle East*. Oxford University Press, New York.
5. Lowi, M.R. (1993). Bridging the divide. Transboundary resource disputes and the case of West Bank Water. *Int. Security* **18**(1): 113–138.
6. Lipchin, C. (1997). Water scarcity, international security and resource disputes-the case of the tigris-euphrates and jordan basin river systems. In: *Population-Environment Dynamics: Ten Case Studies*. S.L. Arlinghaus, W.D. Drake, and D. Jacobs (Eds.). University of Michigan, Ann Arbor, MI. Available: http://www-personal.umich.edu/wddrake/lipchin.html.
7. Wolf, A.T. (2001). *Transboundary Waters: Sharing Benefits, Lessons Learned*. Thematic Background Paper, International Conference on Freshwater, Bonn, Germany.
8. Shultz, M. (1995). Turkey, Syria and Iraq: A Hydropolitical Complex. In: *Hydropolitics. Conflict over Water as a Development Constraint*. L. Ohlson (Ed.). University Press, Dhaka.
9. Inbar, M. and Maos, J. (1984). Water Resource Planning and Development in the Northern Jordan Valley. *Water Int.* **9**(1): 19.
10. Smith, J. (1997). Nine Nations, One Nile. In: *Population-Environment Dynamics: Ten Case Studies*. S.L. Arlinghaus, W.D. Drake, and D. Jacobs (Eds.). University of Michigan, Ann Arbor, MI. Available: http://www-personal.umich.edu/wddrake/smith.html.

READING LIST

Starr, J.R. (1991). Water Wars. Financial Policy.

BENJAMIN FRANKLIN'S ARMONICA: A WATER MUSIC INSTRUMENT

William E. Marks
Water Consciousness, Inc.
Martha's Vineyard,
Massachusetts

The art of creating music from water-filled glass receptacles is probably as ancient as glassmaking. Some music scholars date the phenomenon of water music back to 2500 B.C. Because this time predates recorded history, all we have as documentation are references in early folklore and tales.

In the ancient days of Pythagoras, a variety of music was composed by musicians using water-filled bowls that held different levels of water. This use of creating water music according to a mathematical scale, resonated with Pythagorean philosophy, which believed in transmigration of the soul and a related belief in a numerical scale.

In Persia, where some methods of their glassblowing techniques remain a mystery to this day, we find accounts, as early as the fourteenth century, of water music being played from glasses. Largely because glass music was often played by amateur performers who created their own folk music, glass music has a history that is obscure and mysterious.

For instance, in Europe, the earliest reference to musical glasses dates back to 1492. Shortly thereafter, we find the visionary Nostradamus (1503–1566) using a water-filled bowl to see into the affairs of humankind across the span of time. This method of using a water bowl for divining the future is known as "scrying."

As expressed in the book, *The Holy Order of Water*, Nostradamus would sit alone at night on a "tripod stool of brass" and look deeply into a large bowl of water. Using a special branch that was wetted with water on both ends, he would rub one end of the wet branch in a circular motion around the rim of the water-filled bowl. This would produce tones, or ethereal voices, that according to Nostradamus, gave him information and insight about the future.

It is curious to note that there is little if any elaboration on the source of the water used by Nostradamus to see into other dimensions, a fact that may have been a hidden secret—just as it was hidden in writings concerning the creation of the "Philosopher's Stone," which is often referred to in ancient lore. Certainly, with what modern science is revealing at this time, we know that the quality and source of waters used for healing and for other "higher" purposes has to be of unique quality and handled with care.

When Galileo (1564–1642) was writing about pendulums, he wrote the following about water music being played with wine glasses filled with water:

> If one bows the bass string on a viola rather smartly and brings near it a goblet of fine, thin glass having the same tone [tuono] as that of the string, this goblet will vibrate and audibly resound. That the undulations of the medium are widely dispersed about the sounding body is evinced by the fact that a glass of water may be made to emit a tone merely by the friction of the fingertip upon the rim of the glass; for in this water is produced a series of regular waves. The same phenomenon is observed to better advantage by fixing the base of the goblet upon the bottom of a rather large vessel of water filled nearly to the edge of the goblet; for if, as before, we sound the glass by friction of the finger, we shall see ripples spreading with the utmost regularity and with high speed to large distances about the glass. I have often remarked, in thus sounding a rather large glass nearly full of water, that at first the waves are spaced with great uniformity, and when, as sometimes happens, the tone of the glass jumps an octave higher I have noted that at this moment each of the aforesaid waves divides into two; a phenomenon which shows clearly that the ratio involved in the octave [forma dell' ottava] is two (1,2).

For whatever reason, this ancient connection of the soul to water glass music plays itself again and again in various countries throughout history, including the life and times of Benjamin Franklin. Water is the source of life, and life depends on an ongoing healthy source of water, so the connection of creation and the soul to water cannot be taken for granted. As expressed above by Galileo, the water in glasses near the vibrating glass of water will vibrate as well. It takes only a small leap of imagination to grasp the concept that the cellular water within the bodies of people near a musician playing water glasses will also vibrate.

Throughout the seventeenth century, amateur performances are documented that featured sets of wine glasses arranged on a table, played with moistened fingers and tuned by the addition of small amounts of water. This "glass music" became popular throughout Europe and was played in churches and parlor circles (known as 'Gay Wine Music' in 1667 London). The unique qualities of this water glass music captured the imagination of many composers and was well entrenched in Europe during the height of Romanticism in the 1700s.

During 1743, Handel's Water Music was played on an 'Angelic Organ' (a set of tuned wine glasses) by the Irish musician Richard Pockridge. Franklin may have been exposed to this rendition of Handel's Water Music on water glasses because he attended many music concerts while in Europe. In 1759, Franklin wrote the following in his diary after watching Handel's last concert, "... the sublime old Handel was led to the organ and conducted the Messiah for the last time eight days before his death."

Also of musical record, we have the composer Gluck, who in 1746 impressed audiences throughout Europe with his "Verrillon." In doing so, Gluck used a set of wine glasses to produce his water music ('verres' is the French word for 'glass').

In 1757, the Pennsylvania assembly directed Franklin "to go home to England" and negotiate a fairer system of taxation for the colonies. That same year, while living in London he heard E. H. Delaval of the Royal Society play music on water-filled glasses. Afterward, Franklin wrote about the Delaval water music concert he attended:

> He collected a number of glasses of different sizes, fixed them near each other on a table, and tuned them by putting into them water, more or less as each note required. The tones were brought out by passing his fingers round their brims. He was unfortunately burnt here, with his instrument, in a fire which consumed the house he lived in (1759). Mr. E. (Edmund Hussey) Delaval, a most ingenious member of our Royal Society, made one in imitation of it, with a better choice and form of glasses, which was the first I saw or heard (3, p. 297).

Upon hearing the water glasses played, Franklin was moved to refer to it as, "That charming science." Having inherited an interest in music from his father, Franklin already had a strong foundation in the history, theory, and artistry of playing and composing music. Wanting to hear more harmonies with his melody, he set about inventing a way to mimic the sound of water glass music.

> Being charmed by the sweetness of its tones and the music he produced from it, I wished to see the glasses disposed in a more convenient form, and brought together in a narrower compass, so as to admit a greater number of tunes and all within reach of hand to a person sitting before the instrument (3, p. 298).

Enlisting the assistance of a London glassblower named Charles James, Franklin had 37 glass bowls (hemispheres) blown and ground to different sizes and thicknesses to produce 37 distinct tones. Each of these was then painted to help identify the note it produced. In the center of each glass bowl was a hole into which Franklin inserted a piece of cork. He then lined up the bowls chromatically, from lowest tone to highest, on a horizontal spindle (iron rod) according to size. The largest glass bowl was mounted to the left and then each subsequent bowl installed to the right was progressively smaller. The larger the bowl, the lower the tone, and the smaller the bowl, the higher the tone.

The iron spindle was attached to an apparatus similar to a spinning wheel, which was then turned by a foot treadle, a configuration common in those days on foot-operated sewing machines.

As the foot treadle turned the wheel, the spindle rotated the glass bowls. This allowed Franklin to sit at the Armonica, use his feet to turn the bowls, and by wetting his fingers from a bowl of water, produce music that rivaled that of the musical glasses containing water.

To make it easier to identify the notes of the bowls, Franklin painted the bowls using a combination of the seven colors of the rainbow along with the color white. Franklin's "Armonica" included two of the three octaves of the musical scale, giving musicians a greater creative range for their compositions (See Fig. 1).

The vibrations produced from the spinning glasses in the Armonica produced sounds with different frequencies when touched with moistened fingers. Therefore, a variety of musical tones could be created. The handmade glass used in Franklin's time was mostly soda-lime glass (containing sodium) or lead glass (containing lead).

When playing the Armonica, a moistened finger held to the rim of the glass as it rotates, produces a pure and constant tone. The pitch was altered by changing the pressure of the finger touching the rim of the glass, by adding more water to one's fingers, or by decreasing the size of the glass being touched. Therefore, Franklin's use of glasses of varying sizes arranged on the Armonica allowed a musician to create a range of melodies on the instrument. Because Franklin's Armonica instrument included two to three octaves of the musical scale, it permitted a composer greater personal freedom in creating musical arrangements.

When and how Franklin decided to coin the word "Armonica" for his musical invention is not known. What we do know is that on July 13, 1762, Franklin sent a letter to Padre Giambastista Beccaria of Italy, describing the attributes of this new musical invention:

> The advantages of this instrument are that its tones are incomparable sweet beyond those of any other; that they can be swelled and softened at pleasure by stronger or weaker pressures on the fingers, and continued at any length, and that the instrument, being once well tuned, never again wants tuning (3, p. 298).

Figure 1. 1920s painting of Franklin playing the Armonica. "Of all my inventions, the glass armonica has given me the greatest personal satisfaction"—Benjamin Franklin. (Photograph courtesy William Waterway).

In the same letter, Franklin writes, "In honour of your musical language, I have borrowed from it the name of this instrument, calling it the Armonica."

Considering this letter, it appears appropriate that Franklin chose "Armonica" as the name his musical instrument invention because "armonica" is the Italian word for "harmony."

Shortly after its introduction in England and Europe, the Armonica became widely accepted as a new musical instrument. As greater numbers of musicians began playing the Armonica, it soon became a fad. To meet the demand, Armonica factories were built. One of them reportedly employed more than 100 workers. It is estimated that more than 6000 Armonicas were manufactured during a period of about 40 years.

It is of historic note that many of the musicians who played the Armonica were women, which was unusual for that time. The French queen Marie Antoinette took lessons from a famed European Armonica player named Marianne Davies.

Always the charming statesman, Franklin would often take his Armonica to social gatherings, where he would set about entertaining the social elite. In short order, famous composers such as Schmittbauer, Richard Straus, J. A. Schultz, Mozart, Haydn, Galuppi, Martini, Beethoven, Donizetti, J. G. Naumann, Jommelli, J. F. Reichardt, Saint-Saens, and Leopold Rollig began composing music to fit Franklin's Armonica. In 1787, Rollig wrote, "...the sensation the harmonica [Armonica] produced after its first appearance and the unanimous applause of all who heard it, make the instrument...the most satisfying and most beautiful mankind has ever possessed."

Rollig's statement seems to echo what Franklin wrote 24 years earlier on December 11, 1763,

> I play some of the softest Tunes on my armonica, with which Entertainment our people here are quite charmed and conceive the Scottish Tunes to be the finest in the World. And indeed, there is so much simple Beauty in many of them, that in my Opinion they will never die, but in all ages find a Number of Admirers.

To say the least, Franklin was already renowned in England and Europe for his invention of the lightning rod. With the invention of the Armonica, Franklin's overseas popularity made him one of America's first international superstars.

Even though Franklin invented the Armonica, it is historical fact that Germany's society of musicians and fine glassmakers embraced this instrument more than musicians in any other nation. Most of the periods' literature on the Glass Armonica was available only in the German language. The only method book for playing the Armonica was authored by J. C. Müller in 1788 and published in Leipzig, Germany.

ARMONICA AS A HEALING WATER INSTRUMENT

Today, as we have evolved to understand the relationship of music to water for healing, it was something that Franklin seemed to understand in his day as well. We have the following account of Franklin using his Armonica for healing Princess Izabella Czartoryska of Poland from "melancholia."

As expressed in her own writings of 1772,

> I was ill, in a state of melancholia, and writing my testament and farewell letters. Wishing to distract me, my husband explained to me who Franklin was and to what he owed his fame...Franklin had a noble face with an expression of engaging kindness. Surprised by my immobility, he took my hands and gazed at me saying: pauvre jeune femme ["poor young lady"]. He then opened an armonica, sat down and played long. The music made a strong impression on me and tears began flowing from my eyes. Then Franklin sat by my side and looking with compassion said, "Madam, you are cured." Indeed that moment was a reaction to my melancholia. Franklin offered to teach me how to play the armonica—I accepted without hesitation, hence he gave me twelve lessons (4).

Another documented case of the Armonica used for healing involves Franz Mesmer (1734–1815). Mesmer was a Viennese psychiatrist who is known for his theory on "animal magnetism." Today, many people believe that Mesmer invented hypnosis. This misconception is due to the use of his name in the word "mesmerize." However, Mesmer did not invent or use hypnosis. One of Mesmer's water healing methods was to use a wooden tub filled with magnetized water. Besides using tubs with treated water, Mesmer also used the Armonica in his treatments.

The following is an account given by a Dr. Le Roux sometime between 1778 and 1779. Dr. Le Roux had taken an army surgeon to Mesmer's clinic for treatment of gout:

> After several turns around the room, Mr. Mesmer unbuttoned the patient's shirt and, moving back somewhat, placed his finger against the part affected. My friend felt a tickling pain. Mr. Mesmer then moved his finger perpendicularly across his abdomen and chest, and the pain followed the finger exactly. He then asked the patient to extend his index finger and pointed his own finger toward it at a distance of three or four steps, whereupon my friend felt an electric tingling at the tip of his finger, which penetrated the whole finger toward the palm. Mr. Mesmer then seated him near the harmonica [Armonica]; he had hardly begun to play when my friend was affected emotionally, trembled, lost his breath, changed color, and felt pulled toward the floor (5).

Being ahead of his time, Mesmer was using a water treatment method that is now becoming quite popular in Europe. Water, light, and sound are now used to stimulate healing while a patient is submerged in a tub of mineral water.

Mesmer, Mozart, and Franklin were Freemasons and were thus connected in their appreciation and use of the Armonica for helping humankind.

As noted in history, all three were highly regarded when it came to playing the Armonica. In 1773, Leopold Mozart met Marianne Davies in Vienna and started playing the Armonica shortly afterward. In a letter to his wife, Mozart wrote,

> Do you know that Herr von Mesmer plays Miss Davies' armonica unusually well? He is the only person in Vienna who has learned it, and he possesses a much finer instrument than Miss Davies does. Wolfgang too has played upon it. How I should love to have one!

Mesmer believed so much in the power of the Armonica that his dying request was to have a priest play the Armonica for him.

Besides the use of the Armonica for healing in the time of Benjamin Franklin, the harp was also used for healing. The ethereal connection of healing harp and Armonica music can be found in the adage, "If the harp is the instrument of the Angels, the Armonica is the voice of the Angels."

The use of the harp and Armonica for healing is based on the vibrational tones generated by these instruments. In theory, these tones cause the intracellular and intercellular water of the human body to be energized in a way that is conducive to healing. Testimony to this can be found in one of today's musical healing societies known as "Healing Harps (www.healingharps.org).

On another note, just as Franklin's lightning rod invention quickly found its way from America to all of Europe, in reverse fashion, his Armonica quickly found its way from Europe to all of America.

Needless to say, when Franklin returned home to America, he set up a room specifically for music and entertaining. This became known as the Franklins' "Blue Room." Even though this room was complemented with an assortment of musical instruments, the Armonica was always Franklin's favorite. In testimony to Franklin's appreciation for his musical invention, he leaves us the following quote, "Of all my inventions, the glass Armonica has given me the greatest personal satisfaction."

Ben Franklin collected no money from his glass Armonica invention. He refused to patent any of his inventions, saying: "As we enjoy great Advantages from the Inventions of others we should be glad of an Opportunity to serve others by any Invention of ours, and this we should do freely and generously."

THE DEATH AND RESURRECTION OF THE ARMONICA

After about 40 years as a popular musical instrument, the Armonica suddenly began to fall into disfavor during the early 1900s. There are various theories as to why this occurred. Some theorists say people were fearful of contracting lead poisoning from the bowls which were made of lead crystal. Others say that the Armonica was responsible for causing nervous disorders and possibly insanity and that the toning of the Armonica opens doorways to ethereal realms that caused maladies and possibly death. In Germany, where the Armonica was most popular, there was the unfortunate coincidence of a young child dying while an Armonica was being played in a performance hall. Rumors became so rampant that the Armonica became banned in some towns.

As controversy raged throughout Europe over the theorized negative health impacts of the Armonica, Franklin ignored these rumors and continued playing the Armonica his entire life without suffering any illness. However, the Armonica's popularity never really returned to what it was during the height of its popularity.

From a more practical perspective, it was believed that the growing popularity and versatility of the piano may have contributed to the Armonica's downfall. In about 1720, Christofori finished his real pianoforte. Christofori constructed a case superior to those used for harpsichords, which allowed the piano to withstand the increased strain of heavier strings. Christofori also added other improvements to the piano, which allowed the performer to produce a delicate pianissimo as well as a strong fortissimo, not possible on either the clavichord or harpsichord. These improvements in the piano also produced tones of music which could fill a concert hall and were easily heard by the entire audience.

Regardless of what caused the demise of the Armonica, there is now a renewed interest in its use. In the 1980s, American glassmaker, Gerhard Finkenbeiner of Waltham, Massachusetts, researched and resurrected Franklin's Armonica from obscurity. His glassmaking company, G. Finkenbeiner, Inc., makes a quality Quartz Glass Harmonica (Armonica) that contains no lead. According to Finkenbeiner, the use of pure crystal glass also gives a superior sound to the reinvented Armonica.

Finkenbeiner's version of the Armonica is finding interest and popularity as a musical instrument. It has been used at the Metropolitan Opera; in radio commercials, and in various movie tracks.

Today, a handful of musicians play the Armonica. Noted musician, William Wilde Zeitler, has composed several new musical compositions for the Armonica, which are available as CDs.

Another musician, Cecilia Brauer, is world renowned for her Armonica performances. Ms. Brauer gives lecture/demonstration programs called "Ben Franklin and the Armonica" at museums, historical sites, libraries, and social institutions such as the Carnegie Museum; the Franklin Institute, and the National Portrait Gallery in Washington, DC. Besides composing original music for several CDs, Ms. Brauer is also known for playing the Armonica at the Metropolitan Opera House, for the sound track of the movie "Interview with the Vampire," and for several PBS specials.

On an ethereal note, Gerhard Finkenbeiner, the glassmaker from Massachusetts responsible for resurrecting interest in the Armonica, disappeared under strange circumstances in the late 1990s. The mystery of his sudden disappearance has yet to be solved. However, company employees have rallied to keep the company as an ongoing enterprise. In 2003, the employees of G. Finkenbeiner, Inc., issued the following statement:

> The Quartz Glass Harmonica, Gerhard's invention and musical triumph, will also be continued faithfully by G. Finkenbeiner Inc., as Gerhard would have wanted it that way. This unique musical apparatus and its wonderfully ethereal tones will continue to delight listeners around the world.
> If you have any questions or comments, we would love to hear from you. Your satisfaction is our priority! Your continued support has been an invaluable asset during these somber times, and we thank you sincerely for your patronage.

THE ARMONICA'S FUTURE

To have a grasp of the Armonica's future, the author of this article contacted two the world's most prominent players, William Wilde Zeitler and Cecilia Brauer. Their responses speak for themselves:

William Wilde Zeitler (info@glassarmonica.com)

> In spite of the publicity the Armonica has received of late on the History Channel, NPR, etc., I think its future is quite precarious. At present I'm the ONLY full-time composer/performer for the instrument; I know of one other full-time player, and maybe a half-dozen part time players. That definitely puts it on "the endangered musical instrument list.

Cecilia Brauer (brauerarmonica@webtv.net)

> It is very difficult to predict the Armonica's future. So far as I know, the only manufacturer of the Armonica in the U.S. was Gerhard Finkenbeiner in Waltham, Massachusetts. He "disappeared" in a private airplane flight almost five years ago. And, unless his descendants decide to keep the Armonica segment of his company, G. Finkenbeiner Inc., as a going concern, I do not know what the prognosis is. I would hate to see it disappear again as it is such a beautiful, unique sounding instrument and also such an important part of Franklin's history. Franklin felt that of all the things he accomplished, the Armonica gave him his greatest personal joy.
> There is a man, Sacha Reckert in Germany, who makes them, but they are somewhat different than Gerhard's. Perhaps the people at Finkenbeiner's might be able to help. The contact is DHession@finkenbeiner.com.

Based on the opinions of William Wilde Zeitler and Cecilia Brauer, it certainly does sound as though the future of the Armonica is somewhat dubious. However, this is one of the few glass water instruments in the world, and due to its history in the cultures of Europe and America, the Armonica will hopefully continue to provide enjoyment for many future generations.

BIBLIOGRAPHY

1. Galileo, G. (1638). *Two New Sciences.*
2. Galileo, G. (1638). *Dialogues Concerning Two New Sciences.* H. Crew and A. de Salvio (Eds.). Dover Publications, New York, ISBN 0-486-60099-8.
3. Van Doren, C. (1991). *Benjamin Franklin,* New York.
4. Translated in Lipowski, Z.J. (1984). Benjamin Franklin as a psychotherapist: A forerunner of brief psychotherapy. *Perspect. Biol. Med.* **27**: 361–366.
5. Pattie, F.A. (1994). *Mesmer and Animal Magnetism.* Edmonsten Publishing Co., Hamilton, NY, p. 73.

WATER TRANSPORTATION OCCUPATIONS

Bureau of Labor Statistics, U.S.
Department of Labor

SIGNIFICANT POINTS

- Many jobs in water transportation occupations require a merchant mariner's document or a license from the U.S. Coast Guard.
- Merchant mariners on oceangoing ships are hired for periods ranging from a single voyage to several continuous voyages and may be away from home continuously for months.
- Jobs aboard oceangoing vessels have high pay, but competition for them remains keen, and merchant mariners might have to wait months between work opportunities.

NATURE OF THE WORK

The movement of huge amounts of cargo, as well as passengers, between nations and within our Nation depends on workers in water transportation occupations, also known on commercial ships as merchant mariners. They operate and maintain deep-sea merchant ships, tugboats, towboats, ferries, dredges, excursion vessels, and other waterborne craft on the oceans, the Great Lakes, rivers, canals, and other waterways, as well as in harbors. (Workers who operate watercraft used in commercial fishing are described in the section on **fishers and fishing vessel operators** elsewhere in the *Handbook*.)

Captains, mates, and pilots of water vessels command or supervise the operations of ships and water vessels, both within domestic waterways and on the deep sea. *Captains* or *masters* are in overall command of the operation of a vessel, and they supervise the work of all other officers and crew. They determine the course and speed of the vessel, maneuver to avoid hazards, and continuously monitor the vessel's position with charts and navigational aides. Captains either direct or oversee crew members who steer the vessel, determine its location, operate engines, communicate with other vessels, perform maintenance, handle lines, or operate equipment on the vessel. Captains and their department heads ensure that proper procedures and safety practices are followed, check to make sure that machinery and equipment are in good working order, and oversee the loading and discharging of cargo or passengers. They also maintain logs and other records tracking the ships' movements, efforts at controlling pollution, and cargo and passengers carried.

Deck officers or *mates* direct the routine operation of the vessel for the captain during the shifts when they are on watch. All mates stand watch for specified periods, usually 4 hours on and 8 hours off. However, on smaller vessels, there may be only one mate (called a *pilot* on some inland towing vessels), who alternates watches with the captain. The mate would assume command of the ship if the captain became incapacitated. When more than one mate is necessary aboard a ship, they typically are designated chief mate or first mate, second mate, third mate, etc. Mates also supervise and coordinate activities of the crew aboard the ship. They inspect the cargo holds during loading to ensure that the load is stowed according to specifications and regulations. Mates supervise crew members engaged in maintenance and the primary upkeep of the vessel.

Pilots guide ships in and out of harbors, through straits, and on rivers and other confined waterways where a familiarity with local water depths, winds, tides, currents, and hazards such as reefs and shoals are of prime importance. Pilots on river and canal vessels usually are regular crew members, like mates. Harbor pilots are generally independent contractors who accompany vessels while they enter or leave port. Harbor pilots may pilot many ships in a single day. *Motorboat operators* operate small, motor-driven boats that carry six of fewer passengers on fishing charters. They also take depth soundings in turning basins and serve as liaisons between ships, between ship and shore, between harbors and beaches, or on area patrol.

Ship engineers operate, maintain, and repair propulsion engines, boilers, generators, pumps, and other machinery. Merchant marine vessels usually have four engineering officers: A chief engineer and a first, second, and third assistant engineer. Assistant engineers stand periodic watches, overseeing the safe operation of engines and machinery.

Marine oilers and more experienced *qualified members of the engine department*, or QMEDs, maintain the vessel in proper running order in the engine spaces below decks, under the direction of the ship's engineering officers. These workers lubricate gears, shafts, bearings, and other moving parts of engines and motors; read pressure and temperature gauges; record data; and sometimes assist with repairs and adjust machinery.

Sailors operate the vessel and its deck equipment under the direction of the ship's officers and keep the nonengineering areas in good condition. They stand watch, looking out for other vessels and obstructions in the ship's path, as well as for navigational aids such as buoys and lighthouses. They also steer the ship, measure water depth in shallow water, and maintain and operate deck equipment such as lifeboats, anchors, and cargo-handling gear. On vessels handling liquid cargo, mariners designated as *pumpmen* hook up hoses, operate pumps, and clean tanks; on tugboats or tow vessels, they tie barges together into tow units, inspect them periodically, and disconnect them when the destination is reached. When docking or departing, they handle lines. They also perform routine maintenance chores, such as repairing lines, chipping rust, and painting and cleaning decks or other areas. Experienced sailors are designated *able seamen* on oceangoing vessels, but may be called simply *deckhands* on inland waters; larger vessels usually have a *boatswain*, or head seaman.

A typical deep-sea merchant ship has a captain, three deck officers or mates, a chief engineer and three assistant engineers, a radio operator, plus six or more unlicensed seamen, such as able seamen, oilers, QMEDs, and cooks or food handlers. The size and service of the ship determine the number of crewmembers for a particular voyage. Small vessels operating in harbors, on rivers, or along the coast may have a crew comprising only a captain and one deckhand. The cooking responsibilities usually fall under the deckhands' duties.

On larger coastal ships, the crew may include a captain, a mate or pilot, an engineer, and seven or eight seamen. Some ships may have special unlicensed positions for entry level apprentice trainees. Unlicensed positions on a large ship may include a full-time cook, an electrician, and machinery mechanics. On cruise ships, *bedroom stewards* keep passengers' quarters clean and comfortable.

WORKING CONDITIONS

Merchant mariners spend extended periods at sea. Most deep-sea mariners are hired for one or more voyages that last for several months; there is no job security after that. The length of time between voyages varies depending on job availability and personal preference.

The rate of unionization for these workers is about 24 percent, much higher than the average for all occupations. Consequently, merchant marine officers and seamen, both veterans and beginners, are hired for voyages through union hiring halls or directly by shipping companies. Hiring halls rank the candidates by the length of time the person has been out of work and fill open slots accordingly. Hiring halls typically are found in major seaports.

At sea, these workers usually stand watch for 4 hours and are off for 8 hours, 7 days a week. Those employed on Great Lakes ships work 60 days and have 30 days off, but do not work in the winter when the lakes are frozen. Workers on rivers, on canals, and in harbors are more likely to have year-round work. Some work 8- or 12-hour shifts and go home every day. Others work steadily for a week or a month and then have an extended period off. When working, they usually are on duty for 6 or 12 hours and off for 6 or 12 hours. Those on smaller vessels are normally assigned to one vessel and have steady employment.

People in water transportation occupations work in all weather conditions. Although merchant mariners try to avoid severe storms while at sea, working in damp and cold conditions often is inevitable. While it is uncommon nowadays for vessels to suffer disasters such as fire, explosion, or a sinking, workers face the possibility that they may have to abandon their craft on short notice if it collides with other vessels or runs aground. They also risk injury or death from falling overboard and hazards associated with working with machinery, heavy loads, and dangerous cargo. However, modern safety management procedures, advanced emergency communications, and effective international rescue systems place modern mariners in a much safer position.

Most newer vessels are air conditioned, soundproofed from noisy machinery, and equipped with comfortable living quarters. For some mariners, these amenities have helped ease the sometimes difficult circumstances of long periods away from home. Also, modern communications, especially email, link modern mariners to their families. Nevertheless, some mariners dislike the long periods away from home and the confinement aboard ship and consequently leave the occupation.

EMPLOYMENT

Water transportation workers held about 68,000 jobs in 2002. The total number that worked at some point in the year was perhaps twice as large because many merchant marine officers and seamen worked only part of the year. The following tabulation shows employment in the occupations that make up this group:

Sailors and marine oilers	27,000
Captains, mates, and pilots of water vessels	25,000
Ship engineers	8,200
Motorboat operators	4,100
All other water transportation workers	3,600

About 30 percent of all workers were employed in water transportation services. About one half worked in inland water transportation—primarily the Mississippi River system—while the other half were employed in water transportation on the deep seas, along the coasts, and on the Great Lakes. About another 28 percent worked in establishments related to port and harbor operations, marine cargo handling, or navigational services to shipping. The Federal government employed approximately 5 percent of all water transportation workers, most of whom worked on supply ships and are Civilian Mariners of the Department Navy's Military Sealift Command.

TRAINING, OTHER QUALIFICATIONS, AND ADVANCEMENT

Entry, training, and educational requirements for most water transportation occupations are established and regulated by the U.S. Coast Guard, an agency of the U.S. Department of Homeland Security. All officers and operators of commercially operated vessels must be licensed by the Coast Guard, which offers various kinds of licenses, depending on the position and type of vessel.

There are two ways to qualify for a deck or engineering officer's license: applicants either must accumulate sea time and meet regulatory requirements or must graduate from the U.S. Merchant Marine Academy or one of the six State maritime academies. In both cases, applicants must pass a written examination. Federal regulations also require that an applicant pass a physical examination, a drug screening, and a National Driver Register Check before being considered. Persons without formal training can be licensed if they pass the written exam and possess sea service appropriate to the license for which they are applying. However, it is difficult to pass the examination without substantial formal schooling or independent study. Also, because seamen may work 6 or fewer months a year, it can take 5 to 8 years to accumulate the necessary experience. The academies offer a 4-year academic program leading to a bachelor-of-science degree, a license (issued only by the Coast Guard) as a third mate (deck officer) or third assistant engineer (engineering officer), and, if the person is qualified, a commission as ensign in the U.S. Naval Reserve, Merchant Marine Reserve, or Coast Guard Reserve. With experience and additional training, third officers may qualify for higher rank.

Sailors and unlicensed engineers working on U.S. flagged deep-sea and Great Lakes vessels must hold a Coast Guard-issued document. In addition, they must hold certification when working aboard liquid-carrying vessels. Able seamen also must hold government-issued certification. For employment in the merchant marine as an unlicensed seaman, a merchant mariner's document issued by the Coast Guard is needed. Most of the jobs must be filled by U.S. citizens; however, a small percentage of applicants for merchant mariner documents do not need to be U.S. citizens, but must at least be aliens legally admitted into the United States and holding a green card. A medical certificate of excellent health attesting to vision, color perception, and general physical condition is required for higher level deckhands and unlicensed engineers. While no experience or formal schooling is required, training at a union-operated school is the best source. Beginners are classified as ordinary seamen and may be assigned to any of the three unlicensed departments: Deck, engine, or steward. With experience at sea and perhaps union-sponsored training, an ordinary seaman can pass the able-seaman exam and move up with 3 years of service.

No special training or experience is needed to become a seaman or deckhand on vessels operating in harbors or on rivers or other waterways. Newly hired workers generally are given a short introductory course and then learn skills on the job. After sufficient experience, they are eligible to take a Coast Guard exam to qualify as a mate, pilot, or captain. Substantial knowledge gained through experience, courses taught at approved schools, and independent study is needed to pass the exam.

Harbor pilot training usually consists of an extended apprenticeship with a towing company or a pilots' association. Entrants may be able seamen or licensed officers.

JOB OUTLOOK

Keen competition is expected to continue for jobs in water transportation occupations. Overall, employment in these occupations is projected to **grow more slowly than the average** for all occupations through the year 2012. Opportunities will vary by sector, and some of the best opportunities will be in scenic transportation and sightseeing and deep sea, coastal, and Great Lakes transportation.

Employment in deep-sea shipping for American mariners is expected to stabilize after several years of decline. New international regulations have raised shipping standards with respect to safety, training, and working conditions. Consequently, competition from ships that sail under foreign *flags of convenience* should lessen as insurance rates rise for ships that do not meet the new standards. Insuring ships under industrialized countries' flags, including that of the United States, should become less expensive, increasing the amount of international cargo carried by U.S. ships. A fleet of deep-sea U.S.-flagged ships is considered to be vital to the Nation's defense, so some receive Federal support through a maritime security subsidy and other provisions in laws that limit certain Federal cargoes to ships that fly the U.S. flag. Possible future developments include "fast ships"—oceangoing cargo vessels that use jet propulsion—which would decrease ocean-crossing times significantly. If such plans are successful, the industry will benefit in terms of increased business and employment.

Vessels on rivers and canals and on the Great Lakes carry mostly bulk products, such as coal, iron ore, petroleum, sand and gravel, grain, and chemicals. Although shipments of these products are expected to grow through the year 2012, current imports of steel are dampening employment on the Lakes, but actually leading to greater chances for overall employment for transport up the Mississippi River system. Employment in water transportation services is likely to rise, and efforts are underway at the Federal level that could lead to significantly greater use of ferries to handle commuter traffic around major metropolitan areas.

Employment growth also is expected in passenger cruise ships within U.S. waters. Vessels that operate between U.S. ports are required by law to be U.S.-flagged vessels. The building and staffing of several new cruise ships that will travel around the Hawaiian Islands will create new opportunities for employment at sea in the cruise line industry, which is composed mostly of foreign-flagged ships.

Openings within the traditional water transportation sector for mariners, although expanding only slightly, should be quite numerous because of the sizable need to replace those leaving the occupation Some experienced merchant mariners may continue to go without work for varying periods. However, this situation appears to be changing, with demand for licensed and unlicensed personnel rising. Maritime academy graduates who have not found licensed shipboard jobs in the U.S. merchant marine find jobs in related industries. Because they are commissioned as ensigns in the Naval or Coast Guard Reserve, some are selected for active duty in those branches of the Service. Some find jobs as seamen on U.S.-flagged or foreign-flagged vessels, tugboats, and other watercraft or enter civilian jobs with the U.S. Navy or Coast Guard. Some take land-based jobs with shipping companies, marine insurance companies, manufacturers of boilers or related machinery, or other related jobs.

EARNINGS

Earnings vary widely with the particular water transportation position and the worker's experience, ranging from the minimum wage for some beginning seamen or mate positions to more than $37.37 an hour for some experienced ship engineers. Median hourly earnings of water transportation occupations in 2002 were as follows:

Ship engineers	$24.61
Captains, mates, and pilots of water vessels	23.97
All other water transportation workers	14.67
Sailors and marine oilers	13.64
Motorboat operators	12.71

Annual pay for captains of larger vessels, such as container ships, oil tankers, or passenger ships, may exceed $100,000, but only after many years of experience. Similarly, captains of tugboats often earn more than the median reported here, with earnings dependent on the port and the nature of the cargo.

RELATED OCCUPATIONS

Workers in other occupations who make their living on the seas and coastal waters include **fishers and fishing vessel operators** and some **members of branches of the Armed Forces**.

SOURCES OF ADDITIONAL INFORMATION

Information on a program called "Careers Afloat", which includes a substantial listing of training and employment descriptive information and contacts in the U.S., may be obtained through:

- Maritime Administration, U.S. Department of Transportation, 400 7th St. SW., Room 7302, Washington, DC 20590. Internet: http://www.marad.dot.gov

Information on merchant marine careers, training, and licensing requirements is available from any of the following organizations:

- Military Sealift Command, APMC, PO Box 120, Camp Pendleton, Virginia Beach, VA 23548-0120.
- Seafarers' International Union, 5201 Auth Way, Camp Springs, MD 20746.
- Paul Hall Center for Maritime Training and Education, P.O. Box 75, Piney Point, MD 20674-0075. Internet: http://www.seafarers.org/phc
- International Organization of Masters, Mates, and Pilots, 700 Maritime Boulevard, Linthicum Heights, MD 21090-1941.
- U.S. Coast Guard National Maritime Center, 4200 Wilson Boulevard, Suite 630, Arlington, VA 22203-1804. Internet: http://www.uscg.mil/stcw/index.htm

OOH ONET CODES

53-5011.01, 53-5011.02, 53-5021.01, 53-5021.02, 53-5021.03, 53-5022.00, 53-5031.00

Suggested citation: Bureau of Labor Statistics, U.S. Department of Labor, *Occupational Outlook Handbook, 2004-05 Edition*, Water Transportation Occupations, on the Internet at http://www.bls.gov/oco/ocos247.htm (visited August 14, 2004).

A CONCISE GLIMPSE OF WATER IN THE HISTORY OF PHOTOGRAPHY

Basia Irland
University of New Mexico
Albuquerque, New Mexico

"#125. When you photograph water you strip it of its form: of its restless, liquid reality."
"#126. When you photograph water you give it an image that in essential ways bears no likeness to water."
"#127. When you photograph water you speak in graphic terms about something that has no graphic aspect."
Rebecca Horn, *Dictionary of Water* (1)

Water is universal and indispensable to life. Therefore, it has always been a central metaphor, both sprightly and exacting, for artists. In this brief article, we examine how a few photographers have looked to water for inspiration and also used it as the very substance of their work. There have, of course, been hundreds of photographers who have tested water as their subject matter or used it in their work in various ways at some stage of their career; some of the more notable and interesting are presented here. A few examples have been chosen, beginning with the earliest photographers of the nineteenth century.

Not just images of water caught on photographic paper, but the intricate process of developing and fixing these images depended on water. We tend to take the process for granted today as we have moved into an era of digital images, but it took many fumbling experiments before it was possible to capture the world "out there" and put it onto an object we could hold in our hands and gaze upon for generations (or at least as long as the chemical process would allow). The English scientist, Sir John Herschel (born 1792 and buried in 1871 next to Sir Isaac Newton at Westminster Abbey), explained in his letters to William Henry Fox Talbot (1800–1877), one of the earliest photographic artists, about the important role that clean water played in rinsing images. Although Herschel is best remembered for his discovery of hyposulfate as a fixative, he also discovered that one could fix photographs with plain water. After several unsuccessful attempts, it was found that the water must be as pure as possible and the rinsing process has to be repeated three times. Melting snow worked, as did distilled water. Salt traces in well water and the many forms of pollution finding their way into England's water supply would ruin a print. In March of 1839, Herschel communicated to his scientific colleagues at the Royal Society about the chemical compounds used in his new discoveries to aid ways of developing a print and how the photograph needed to be immersed two or three times in clean spring water and dried between blotting papers.

Most of Talbot's photographs of water were incidental fluid flowing under the more important human-made constructions of bridges (*Lattice Iron Bridge*, 1841; *The Hungerford Suspension Bridge*, 1845; and *The Suspension Bridge at Rouen*, 1843), or doubling the impact of architecture (*Lacock Abbey, Reflected in the Avon*, 1840–41). But in 1844, he took a series, *Sun Pictures of Scotland*, that leave the city and focus on the water itself with such titles as, *A Mountain Rivulet which flows at the foot of Doune Castle, Scenery of Loch Katrine*, and *A Waterfall*.

Colonel Stuart Wortley (1832–1890) presents an almost operatic view of the sea, complete with reflected moonlight, dark suspended cumulus clouds, and titles borrowed from poems such as "How calm, how beautiful comes on / The stilly hour when storms are gone" (1869) and "All the air was white with moonlight / All the water black with shadow" (1869). Wortley, along with Gustave Le Gray, was among the first photographers to capture a sense of stopped movement in an enlarged image. *A Wave Rolling In* (1861) shows a small, single breaking wave caught on film along with two tiny black figures in a rowboat off in the distance.

Gustave Le Gray (born in France, 1820), spent a 5-year period from 1855–1860 primarily involved in photographing sailing ships and ports, but mostly ocean waves. With a taste for the dramatic similar to Wortley's, Le Gray's seascapes (albumen prints from wet collodion on glass negatives) show dark, brooding bodies of water barely lit from the sun's dim rays hidden behind black ominous clouds. Most of these ocean views were taken on the Normandy coast and often included breaking surf in the lower right-hand corner. Le Gray was experimenting with ways to capture the movement of water and the atmospheric effects of light and cloud by using two overlaid negatives. Often, he would use the same seascape negative in combination with a different sky negative. *Grande vague—Sete* (1856/59) predates Wortley's wave photograph by 2 years. It shows a somber jetty jutting halfway across the print with a not so large wave breaking against shadowed rocks in the foreground. On October 27, 1999 this particular print broke the record for the sale of a single photograph at Sotheby's London auction, selling for $760,150. Le Gray spent the last 20 years of his life primarily photographing antiquities in Cairo, Egypt and died in 1882.

Anna Atkins (1799–1871), one of the earliest photographers, incorporated Herschel's cyanotype method, which was inexpensive and easily worked, to produce camera-less photograms using flowing pieces of sea algae. The first part of her series of these photograms was produced in 1843. *Photographs of British Algae; Cyanotype Impressions*, a handmade portfolio of images, was the first use of photography for scientific illustrations. Photograms, which use neither camera nor negative, are produced when objects, placed on a photosensitive surface and exposed to light briefly to create a shadow, are then developed. "Dispensing with lens, aperture, depth of field, and perspective, the photogram assumes the guise of a real trace, a visual footprint on the scale of the original, a memory of the essential contour devoid of redundant information" (2).

The ocean-blue tint of the cyanotype lent itself to the aquatic pictures of algae. Used until recently in the form of the architectural blueprint, cyanotypes involve only two chemical compounds, ferric ammonium citrate and potassium ferricyanide, which are applied to paper using a sponge or brush. After the paper is dried in the dark, an object such as Atkin's algae is placed on the paper and exposed to the sun. Rinsing the exposed cyanotype in water makes the print permanent and brings out the blue color. Halstead Place, where Atkins worked, was known for the quality of its well water.

Atkins was already an accomplished scientific illustrator, especially of seashells. In the handwritten cyanotype introduction to Part I of her book, Atkins explains that the need to make accurate drawings of such intricate algae led her to use the photographic process developed by Sir John Herschel to explore her studies of botany and make them available to a wider audience through publication. Atkins made numerous trips to the ocean to collect specimens of algae, or her friends would provide her with examples. Seaweed would be washed in water and pressed between blotting paper to dry within a few days. The silhouettes of algae hover on her photographic paper as if still in the sea. *Bangia fusco-purpurea* is as delicate as winter breath. *Polysiphonia violacea* is a soft, dendritic wisp. *Himanthalia lorea* begins with two bulbs and spreads out across the paper as if painted with a brush, exiting the confines of the page. In addition to her 10-year period of work with algae (Fig. 1), Atkins made prints of feathers (peacock, parrot, emu, partridge), grasses, lace, flowering plants, and an extensive study of ferns.

Two contemporary British photographers who have also created photographic images without using a camera are Susan Durges (born 1955) and Adam Fuss (1961). A baby of only a few weeks old floats in a liquid universe outside the mother's womb. A snake slithers silently through turquoise fluid leaving a trail of chain-link ripples. Perfectly formed concentric circles radiate outward from a still center. Adam Fuss fills a shallow glass tray with several inches of water over photosensitive paper and then places the baby or snake directly into this warm bath. As Eugenia Parry writes in her essay for the book, *Adam Fuss*, "He wasn't guided by Talbot or Atkins. He'd barely heard of them. But he shared their desires: To let what is mute speak; to capture and keep an object's mysterious self-inscription" (3).

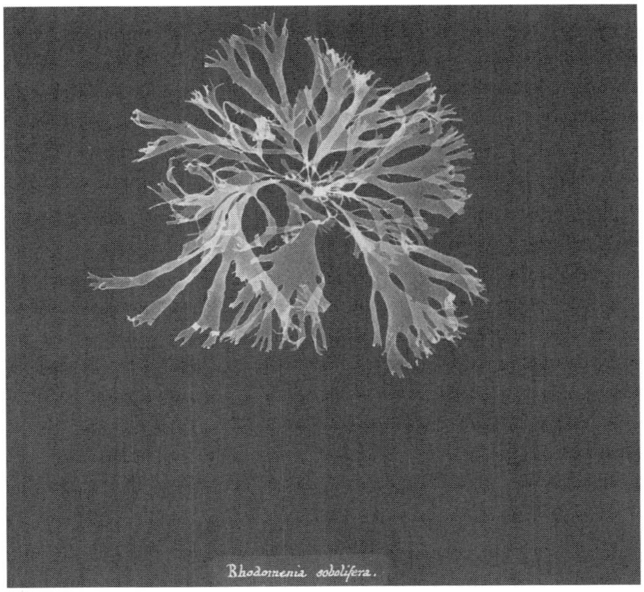

Figure 1. Artist: Anna Atkins. Title: Rhodomenia Sobolifera. Cyanotype negative, bound into a copy of Volume 1 of British Algae, 1850. 26.4 × 20.4 cm. Courtesy: Hans P. Kraus, Jr., Inc., New York.

Whereas most photographers have created photograms within the confines of the studio, Susan Derges (resides in Devon, England) in her *River Taw* series (1997–99), submerges photographic paper directly into a river at night. Enormous sheets of photosensitive paper (66″ × 24″) resting in an aluminum slide are placed just beneath the water surface and exposed to a bright, diffuse flash that captures images of water patterns sometimes overlaid with tree shadows. These are then processed in the darkroom. "Ambient light in the sky adds a colour cast to the Cibachrome (colour positive) images, which ranges from deep blue at full moon to dark green at new moon," writes Martin Kemp in an essay about the artist, in *Susan Derges Liquid Form 1985–99* (2).

Her *Waterfall* series (1997–98, 7′ × 3 1/2′) was created by holding the light-sensitive paper at the top of a water-mill chute using the same nighttime exposures. To create the *Shoreline* series, 1997–99, Derges put the photographic paper on the shoreline at night, waiting for the exact moment a wave would break and exposing it to a microsecond of flash. The ambient light from the moon or a nearby town affected the color of the print, often turning it magenta or dusky pink. Small bits of sand deposited by the wave created vortices and spiral forms.

Another British photographer who eschews preconceived notions of making a photograph is Steven Pippin (born 1960, Surrey, England). In 1993, Pippin began his performance/project, *The Continued Saga of an Amateur Photographer*, which took place in the bathroom on a British train. Ingrid Schaffner, in her essay, "Dirty Hole: Steven Pippin's Obscure Routes," describes how he "...goes about his business converting the toilet into a camera, taking a picture, and developing it. All the while the train rattles and hurls over the tracks" (4). The semicircular prints fit neatly into the commode bowl.

A toilet is not the only object Pippin converted into a camera. He also used washing machines in 1997 for *Laundromat-Locomotion.* In this project, he trip-wired the washers to take photographs and then developed them within the same machine. The wash and spin cycles ripped, scratched, and cracked the emulsion, creating prints that appear Victorian in age.

Across the ocean in the United States and back in the 1930s, we find Dr. Harold Edgerton, Center for Advanced Visual Studies, Massachusetts Institute of Technology, explaining in his book entitled *flash! Seeing the Unseen by Ultra High-Speed Photography* (1939), the process for taking photos that are now part of our iconic language of modernist images (5). These include a bullet passing through an apple, a child popping a balloon, and a drop of milk forming a crown-shape. In one photograph, using the stroboscope, a device that uses a flash to illuminate a moving object intermittently, what appears to be ice or translucent plastic is actually tap water coming out of a faucet caught in Edgerton's camera at 1/50,000 of a second. He also stopped the fluid action of water used to cut pulp and showed the figure eight emitting from a lawn sprinkler. He recorded spiral patterns revolving around a ship's propeller. These studies of motion are reminiscent of Leonardo da Vinci's investigative drawings in pen, ink, and red chalk done in the early 1500s. In *Studies of the Formation of Water in Motion* and *Studies of Water Formation* we see the same vortical shapes that Edgerton caught with his stop-action method 500 years later. The text written on da Vinci's 1510 *Studies of an Old Man and of Swirling Water* reads: "...thus the water forms eddying whirlpools, one part of which is due to the impetus of the principal current and the other to the incidental motion and return flow."

Theodor Schwenk was obsessed with these same vortices formed by fluids and in *Sensitive Chaos*, 1965, he compared liquid spiral forms with other natural phenomenon in his photographs of the bark of a cypress tree, descending hail clouds, unfurling fern leaves, and a drawing of horns of the African kudu antelope.

A few examples of Schwenk's descriptions in his index to the photographs of various vortices are "25. Trains of vortices also arise if a solid object is drawn in a straight line through stationary liquid. 26. The vortices push into the surrounding liquid like the ball part of a joint into its socket. A suitable length of exposure reveals a delicate structure. 33. A suitable length of exposure has revealed something of the structure of vortices. A meandering stream winds its way between the separate vortices. 41. A photograph of a vortex taken under water reveals the spiraling surface between the water and the air which is being sucked in" (6).

In *River Sutra*, Gita Metha describes how every river should have its own minstrel—someone to sing the praises of that body of flowing water (7). Perhaps every river, stream, brook and creek should also have its own photographer. Even though Roni Horn (8) describes water as too elusive an element to hold within the confines of a photograph's rectilinear frame, that certainly has not stopped artists throughout history from trying to capture something of the nature of a particular river. The 1875-mile length of the Rio Grande/Rio Bravo flows from its source in the San Juan Mountains of Southern Colorado, through New Mexico, and provides the border between Texas and Mexico. Photographer Laura Gilpin's (9) documentation in the 1940s of this muddy liquid ribbon (published in 1949 as *The Rio Grande, River of Destiny*) began in 1945 at the end of a war that left a scarcity of photographic materials and difficulties in traveling by car. She also experienced 2 years of drought in her three expeditions along the river. The book's simple chapter headings, "The Source," "Midstream," and "The Border," lead the viewer along a journey witnessing the flow of the river and also the surrounding towns and people living along the banks.

Photographer Jim Bones, in *Rio Grande, Mountains to the Sea*, however, focuses only on the plants and moving water within the riparian zone. "I fell in love with the Rio Grande because of a dream that repeatedly told me if I wanted to know the secrets of my own being, I must go first to the mountains and then to the sea. For the past twenty years, as I walked, rode, and floated throughout the Rio Grande's home country, I discovered that rivers cut to the very heart of things, of rock and life and even dreams" (10). Bones understands that the "end" of a river, as it empties into the sea, is also a beginning because it is here that the hydrologic cycle of evaporation, cloud formation, and rain continue the process of creating rivers all over again.

One of the chapters in Hans Silvester's book, H_2O *The Beauty and Mystery of Water* (11) is dedicated to "The Water Cycle, Disrupted" and mirrors back to the viewer the devastation humans can cause to bodies of water. It is a wake-up call to change our ways. A few of the disturbing images include five children playing in untreated sewage from Mexico City in the Tula River; a dead egret with a fly on its head floating in the polluted Coto Donana, Spain; and an irrigation canal, Bouches-du-Rhone, France, near Marseille's waste discharge where the trash, particularly plastic bags, is piled so high and wide that it is difficult to see the water.

The photographs of the River Thames for Roni Horn's book, *Another Water*, were taken in the Central London area in January and May of 1999 (8). Filling the page, with no reference points or shoreline, the water images could have been taken anywhere. Running continuously along the bottom of each page are 832 comments about the Thames or water in general. "#193. When you go down to the river you're killing two birds with one stone: you stand there and you go places." "#142. Water is sexy. I want to swim in it—to move, song-like, through it. I want to swim in it—push, kick, thrash, slide myself through it. I want to swim in it. I want to be held by it."

In among the references to water are brief descriptions of those who have chosen to take their life by drowning. "#465. An old man was found in the river last Christmas Eve. He was wearing so many layers of clothing (including two wool coats and a jacket) police couldn't lift him out of the water. (They towed it to the pier.)." "#586. A young Parisian woman came to London recently to drown herself in the river. It's curious how the Thames attracts people from far away. I've never heard of any other river doing this. I mean people don't travel from Canada to kill themselves in the Hudson—or even from Ohio." "#446. A young man was found in the river yesterday with a pair of Walkman headphones wrapped around his neck." And every so often there will be a whole white page with an autopsy or police report of someone who was found in the Thames, including details about their lives, whether they wore dentures and exactly where and how their body was located.

Another Water was followed the next year in 2001 by a second book, *Dictionary of Water,* using similar photographs of the Thames taken during the same months. In fact, Horn did not even take the photographs. Dusseldorf artist Nic Tenwiggenhorn made these images assisted by Uwe Schmidt and Helena Blaker. There is no mention within either of Horn's books about the Thames of the photographs being changed by computer, but on a web site providing information about *Dictionary of Water*, it states that this is a collection of digitally manipulated photographic images.

Industrial photographer Wim Cox (born Netherlands, 1938, moved to Germany, 1961) created likenesses of water to those similar by Horn, but they are accompanied by a very different tone of writing done by his brother, filmmaker Paul Cox (born 1940, Netherlands, resides, Melbourne, Australia) (12). The brief text is a poem—a kind of love poem—to water. It begins, "I am. I am water." And so, throughout history, water has pulled photographers to its side, embraced them, and helped in creating memorable images.

BIBLIOGRAPHY

1. Horn, R. (2001). *Dictionary of Water*, Edition' 7L, Steidl Publishers, Göttingen, Germany.
2. Kemp, M. (1999). *Susan Derges Liquid Form 1985–99*. Michael Hue Williams Fine Art, London.
3. Parry, E. (1997). *Adam Fuss*. Arena Editions, Santa Fe, NM.
4. Hirsch, F. "International Auction Review," and Schaffner, I. "Dirty Hole: Steven Pippin's Obscure Routes," *Art on Paper*. **4**: 3, (New York, January–February 2000), Fanning New York, 96, pp. 34–39
5. Edgerton, H.E. and Killian, J.R., Jr. (1939). *Flash! Seeing the Unseen by Ultra High-Speed Photography*. Hale, Cushman and Flint, Boston, MA.
6. Schwenk, T. (1965). *Sensitive Chaos*. Rudolf Steiner Press, Stuttgart, Germany.
7. Mehta, G. (1993). *A River Sutra*. Doubleday, Bantam, New York.
8. Horn, R. (2000). *Another Water*. Scalo, Zurich.
9. Gilpin, L. (1949). *The Rio Grande, River of Destiny*. Duell, Sloan and Pearce, New York.
10. Bones, J. (1985). *Rio Grande, Mountains to the Sea*. Texas Monthly Press, Austin, TX.
11. Silvester, H. (2001). *H_2O—The Beauty and Mystery of Water*. Harry N. Abrams, New York.
12. Cox, W. and Cox, P. (1997). *I Am*. Schuffelen, Cologne.

READING LIST

Buckland, G. (1980). *Fox Talbot and the Invention of Photography*. David R. Godine, Boston, MA.

Janis, E.P. (1987). *The Photography of Gustave Le Gray*. The Art Institute of Chicago and the University of Chicago Press, Chicago, IL.

Popham, A.E. (1945). (Deputy Keeper of Prints and Drawings, British Museum). *The Drawings of Leonardo da Vinci*. Reynal and Hitchcock, New York.

Schaaf, L.J. (1985). *Sun Gardens*. Victorian Photograms by Anna Atkins. An Aperture Book, Boston, MA.

Schaaf, L.J. (1992). *Out of the Shadows; Herschel, Talbot and the Invention of Photography*. Yale University Press, New Haven, CT.

WATER AS A HUMAN RIGHT

Laurel Phoenix
Green Bay, Wisconsin

"The human right to drinking water is fundamental to life and health. Sufficient and safe drinking water is a precondition for the realization of human rights... the human right to water entitles everyone to sufficient; affordable; physically accessible; safe and acceptable water for personal and domestic uses" (1).

The human right to physical and economic access to water has recently been declared by nongovernmental institutions and civil society groups worldwide. The movement to establish this right has been driven by two related issues. First, global populations are currently experiencing a water crisis predicted to become much worse. Second, the growing global trend of water privatization has created additional concerns regarding access to potable water.

GLOBAL WATER CRISIS

The United Nations, Earth Policy Institute, Pacific Institute, and other organizations have been following concurrent global trends of rising population and decreasing access to potable water. Currently, there are over 1 billion people lacking access to safe water and by 2010 this number will grow to 2.5 billion (2). Associated with this problem is lack of sanitation. There are currently 2.4 billion people lacking sanitation facilities. As this sanitation problem is exacerbated by growing populations, even more water becomes too polluted to safely drink.

PRIVATIZATION

Privatization of water supplies in developing countries has been met with ambivalence. Rather than increasing the number of people served with potable water, the higher water bills attendant with privatization have made poorer citizens lose access. In several instances, the very poor cannot afford to pay the higher rates without forfeiting some other necessity, like heating or food (3). Privatized facilities in many cases have shut off the water to those who cannot pay, forcing them to use polluted supplies, resulting in sickness or death. For example, Nkobongo, South Africa privatized their water system and rates were increased. As water was shut off to those who could not pay, people turned to polluted lakes and streams, creating a spectacular cholera epidemic killing almost 300 people and sickening over 250,000 over the course of 300 miles and several years (4). This scenario was repeated in many townships, causing the government to spend enormous amounts of money to deal with the health crisis. Moreover, when utilities are privatized in developing countries, there is little oversight regarding how often or to what degree rates are raised. Consequently, water shut-offs or the use of prepaid water meter cards can externalize the problems of poverty in one region to a wider area, while private water companies are insulated from the public health crisis that ensues.

Opponents of privatization demand that the minimum amount of water needed for basic needs ought to be provided for free, and then fees for water used per household can be charged for water used in addition to this subsistence requirement. This minimum amount varies from 20 liters per person per day for drinking and cooking to 40 liters per person per day to include water for hygiene. Opponents also contend that once a profit motive is introduced into provision of water, companies may demonstrate more concern for their profit margin than for the quality of the water, either through delaying needed maintenance, reducing quality control, or hiking prices even further and cutting off more citizens from a potable supply.

DECLARATIONS OF WATER AS A HUMAN RIGHT

In response to the global water crisis and privatization concerns, numerous groups have issued statements delineating the right to water. The National Forum on Water Privatization meeting in Ghana declared in May of 2001 that "water is a fundamental human right, essential to human life to which every person, rich or poor, man or woman, child or adult is entitled." The Blue Planet Project created a Treaty Initiative to Share and Protect the Global Water Commons, where they stated: "The global fresh water supply is a shared legacy, a public trust and a fundamental human right and, therefore, a collective responsibility" (5). By "collective" they mean every nation. The public trust theme is repeated in the 3rd World Water Forum: "Water, as a public trust and an inalienable human right, must be controlled by the peoples and communities that rely on it for their lives and livelihoods" (6).

IS THIS A NEW HUMAN RIGHT?

Human rights are not established in a vacuum. Citizens agitate for rights only when clear injustices or inequities occur. New rights are incorporated into law when those currently holding those rights agree to share those rights with the disenfranchised. Now that so many people do not have access to clean water globally, the call for human rights to water focuses world governments' attention on the distribution and access to water worldwide.

Human rights to critical natural resources spring from ancient laws regarding use of the commons. The Roman Emperor Justinian declared in 533–534 B.C.E. that running water, seas, air, and wildlife were things that could not be owned but could be used under regulation to ensure vital common resources were not exploited (7). Perhaps because of its ubiquity, governments never thought to declare a right to something so obvious. Even so, a right to water naturally comes before other rights for equality, freedom, and so on, because there would be no need to establish these other rights if you could not live to enjoy them. Scanlon et al. (8) surveyed documents written by international organizations such as the UN Charter or the Universal Declaration of Human Rights that support a "right to life" or "right to higher standards of living," noting they implied a right to water but did not state it. The Geneva Conventions and Protocols declare a right to drinking water only. Two human rights treaties, the Convention on the Elimination of All Forms of Discrimination against Women (CEDAW) and the Convention on the Rights of the Child, both specifically mention a right to water. For women, they declare a right to a water supply, and clean drinking water is mandated for children. Although mention of rights to water can be found in international law documents, they are not clearly defined, and recognized nation-states still need to incorporate them into national law.

RELATED ISSUES

Other issues interrelated with human rights to water are international influences [e.g., globalization and intergovernmental organizations (IGOs)] and governmental obligations to protect marginalized populations and manage water wisely.

Globalization

Globalization has hastened and consolidated the ability of multinational water corporations to search the world for new contracts. Globalization has also injected a new level of haste in developing countries wanting to industrialize, but first needing adequate water infrastructure to attract foreign investment. Countries wanting improved infrastructure but not having the means to pay for it often will enter into agreements for privatization that are not in their citizens' best interests. Mechanisms for institutionalizing globalization, such as the General Agreement on Tariffs and Trade (GATT), North American Free Trade Agreement (NAFTA), and many bilateral investment treaties (BITs), all assist in establishing free trade, deregulation at the state level, and the promotion of privatization. The negative effects on countries and their citizens with regard to democratic control over their water supplies or their ability to disagree with corporate objectives are analyzed in detail in *Thirst for Control* (9) and *Blue Gold: The Fight to Stop the Corporate Theft of the World's Water* (10).

IGOs

Related to globalization is the influence that IGOs and related multilateral financial institutions such as the World Bank and International Monetary Fund (IMF) have. Loans to developing countries are often preconditioned on the privatization of the country's water and/or sewer facilities (11). At the same time, guarantees regarding percentage of profit or length of contract are given to corporations. Consequently, countries can be assuming risks while corporations receive security from risk. The terms of these agreements can put citizens at risk and increase the likelihood that the poorest in cities will lose access to potable water.

Governmental Obligations

The human right to water is related to the public trust, a doctrine that recognizes the government as trustee for its citizens regarding common resources. Essentially, the government will hold ownership over common properties like water bodies, wildlife, and air; this then gives government the ability to regulate use or access to these common resources so that current and future populations will be able to benefit from them. This doctrine also presumes that a government will give equal access and treatment to all of its citizens, rich or poor, and uphold their right to access. Consequently, the human right to water is dependent on a government that takes its public trust doctrine seriously to declare a human right to safe water, and then to follow through with laws and policy that ensures each person will be able to physically and economically access that water.

Water for the Poor

The rural poor comprise 80% of the world population without access to safe water. The other 20% are the poor in city slums and squatter settlements on the city periphery. Rural people must walk hours to access dirty and open water sources contaminated by animal and human waste. The poor in the city either walk to an open stream or pay water vendors 6 to 12 times the cost of piped water (12). Money spent on expensive water will be money unavailable for food or housing or education.

Women's and Children's Rights

Declaring a human right to water benefits most women and children around the world, because it is commonly their job to search out and carry water back to their homes (13). Rural African women spend 26% of their day getting water from open sources such as rivers, muddy holes, or small lakes. Walking to distant water not only exposes women and children to rape, other physical attacks, schistosomiasis, or more mosquito bites but also takes time away from growing food, earning income, and consistent attendance at school. Children exposed to unsafe water become vulnerable to disease and death. Chemically polluted water endangers children more than adults because children consume more liters of water per kilogram of body weight. A study in Karachi, India concluded that people without water and sanitation spent six times more on medical bills than people linked to those services (14). Health is also compromised if there is not enough water to grow or cook a variety of food. As unsafe water translates into poor health, little money for other necessities, and little time to become educated, it predisposes the poor to a life of weak constitutions and unskilled, part-time labor. They remain unable to move up out of poverty.

Indigenous People

These people have often not enjoyed the same civil rights as other citizens of their country, so it is common for their waters to be diverted for use elsewhere (e.g., a large dam project) or to be polluted from mineral mining or drilling for oil. Civil governments rarely come to their aid and sometimes are the instigators of the water theft or pollution.

Intergenerational Water Resources

Because governments are assumed to be entities existing far beyond any human lifespan, the public trust implies that a government manage its natural resources for current as well as future citizens. In countries with growing populations, this requires a government to plan for future increased demand while managing current resources, which necessitates sustainable use of current resources as well as conservation measures to "produce" more water for future demand. Future generations are a marginalized population because their needs are barely acknowledged in current decision-making and highly discounted in cost/benefit calculations.

Irrigation and Water Management

Agriculture uses over 75% of the available fresh water in the world; therefore increases in efficiency would free up more water available for domestic use. Irrigation that "wastes" water has different driving forces in poor and rich countries. Irrigation in low or no technology countries tends to be the simplest forms, such as flooding the fields, diverting water into irrigation channels for too long, or using unlined irrigation ditches. These farmers generally cannot afford to install high-tech, pinpointed delivery systems, so much water is lost to evaporation or percolation.

In wealthier countries, subsidies for particular crops (e.g., cotton and rice) encourage the growth of water-intensive crops rather than other crops that the region's rainfall can support. If the farms rely primarily on groundwater, choosing a more water-intensive crop essentially exports additional water out of the area. This combination of agricultural subsidies and outdated water laws result in no monetary or legal incentives to conserve water.

CONCLUSION

The global water crisis and growing privatization have increased calls for human rights to water. Organizations interested in the establishment of this right argue that individual governments must protect their citizens from the negative effects of global forces, and that in securing and working toward full establishment of this right, they will fulfill their public trust obligations to their disenfranchised citizens. Creating "new" water by increasing global irrigation efficiency will require either increased expenditures on the part of developing countries, or a transfer of technology and financial assistance to them from the developed countries.

BIBLIOGRAPHY

1. United Nations. (2002). *General Comment on the Right to Water*. Adopted by the United Nations Committee on Economic, Social and Cultural Rights, November 27.
2. Intergovernmental Council of the IHP. (2004). *IHP Contribution to the World Water Assessment Programme (WWAP)*. Document # SC.2004/CONF.203/CLD.32; IHP/IC-XVI/INF.19.
3. Public Citizen's Critical Mass Energy and Environment Program. (2001). *Water Privatization Case Study: Cochabamba, Bolivia*. Public Citizen, Washington, DC.
4. International Consortium of Investigative Journalists. (2003). Metered to death: how a water experiment caused riots and a cholera epidemic. Available: http://cw.groupstone.net/Scripts/WebObj. . .jectid=D1D1364E000000F33188ACAE95C0CBF8.
5. This treaty can be downloaded at: http://www.blueplanetproject.net/english/treaty/.
6. This report can be downloaded at: http://www.unesco.org/water/wwap/wwdr/index.shtml. The most recent version of this report is: UN/WWAP (United Nations/World Water Assessment Programme). (2003). *UN World Water Development Report: Water for People, Water for Life*. UNESCO (United Nations Educational, Scientific and Cultural Organization) and Berghahn Books, Paris, New York and Oxford.
7. Getches, D.H. (1997). *Water Law in a Nutshell*, 3rd Edn. West Publishing, St. Paul, MN.
8. Scanlon, J., Cassar, A., and Nemes, N. (2004). *Water as a Human Right?* IUCN Environmental Policy and Law Paper # 51. International Union for Conservation of Nature and Natural Resources (IUCN).
9. Available: http://www.blueplanetproject.net/english/resources/publications/.
10. Barlow, M. and Clarke, T. (2002). *Blue Gold: The Fight to Stop the Corporate Theft of the World's Water*. The New Press, New York.
11. Public Citizen. (2004). Inter-American Development Bank water and sanitation sector loans promoting privatization. Available: http://www.citizen.org/print_article.cfm?ID=11376; Grusky, Sara. June 29, 2004. IMF forces water privatization on poor countries. Globalization challenge initiative. Available: http://www.waterobservatory.org/library.cfm?refID=33640; Eaton, J.M. July 20, 2002. Critical Perspectives on Water Privatization: Referenced Index and Digest of Major Studies, Reports, and Analyses. corporate accountability committee, Sierra Club. Available: http://www.waterobservatory.org/library.cfm?refID=33775; Grusksy, S. May, 2004. The IMF, the World Bank and the global water companies: a shared agenda. International Water Working Group; Grusky, S. and M. Fiil-Flynn. April 2004. Will the World Bank back down?: water privatization in a climate of global protest. Public Citizen. Available: http://www.worldbankwatch.org.
12. World Health Organization (WHO). (2003). *The Right to Water*. Health and Human Rights Publication Series #3. WHO, Paris, pp. 16, 23.
13. World Health Organization (WHO). (2003). *The Right to Water*. Health and Human Rights Publication Series #3. WHO, Paris.
14. WHO. (2003). *The Right to Water*. Health and Human Rights Publication Series #3. WHO, Paris, p. 24.

ANCIENT WATER AND SOIL CONSERVATION ECOSYSTEMS OF SRI LANKA

Lucius O. Mendis
Colombo, Sri Lanka

INTRODUCTION

Sri Lanka, an island of approximately 65,000 sq km, in the Indian ocean, lying between 6° and 8° North latitude, at the southern tip of the Indian subcontinent, experiences two monsoons, the northeast between October and March and the southwest between April and September, with occasional intermonsoon rains. Northeast monsoon rainfall is spread over much of the island, but the southwest monsoon is largely intercepted by a south-central hill massif, and a dry southwest wind blows over the north and east. On account of topography and rainfall, 103 rivers rising in the central highlands flow in a radial pattern to the sea (Fig. 1). Perennial rivers are called *ganga*, and nonperennial rivers and streams are called *oya, ara, dola*, or *ela* in Sinhala, and *aru* in Tamil. (Sinhalese constitute 74% of the population, Tamils about 12%.) A wet zone and dry zone are defined today; the criterion is 75 inches of annual rainfall. Humid-tropical conditions exist, ideal for year-round cultivation of rice, a water-intensive crop.

ANCIENT DEVELOPMENT

There were three political regions, Rajarata (King's country), Ruhunurata, and Mayarata in ancient Sri Lanka whose written history goes back to the mid-first millennium B.C. Actual writing began early in the Christian era on ola (palm) leaf manuscripts, many still preserved in temples and archives. The Mahavamsa, or great chronicle, and Culavamsa, little chronicle, translated by modern scholars, conform with contemporaneous rock inscriptions.

A famous king, Parakrama Bahu I, said *"Let not a drop of rain water flow to the sea without being made useful to man!"* (Mahavamsa, quoted in Reference 1). Human intervention in nature's hydrologic cycle is seen in the plains of ancient Rajarata and Ruhunurata in the form of river diversion systems and small, medium,

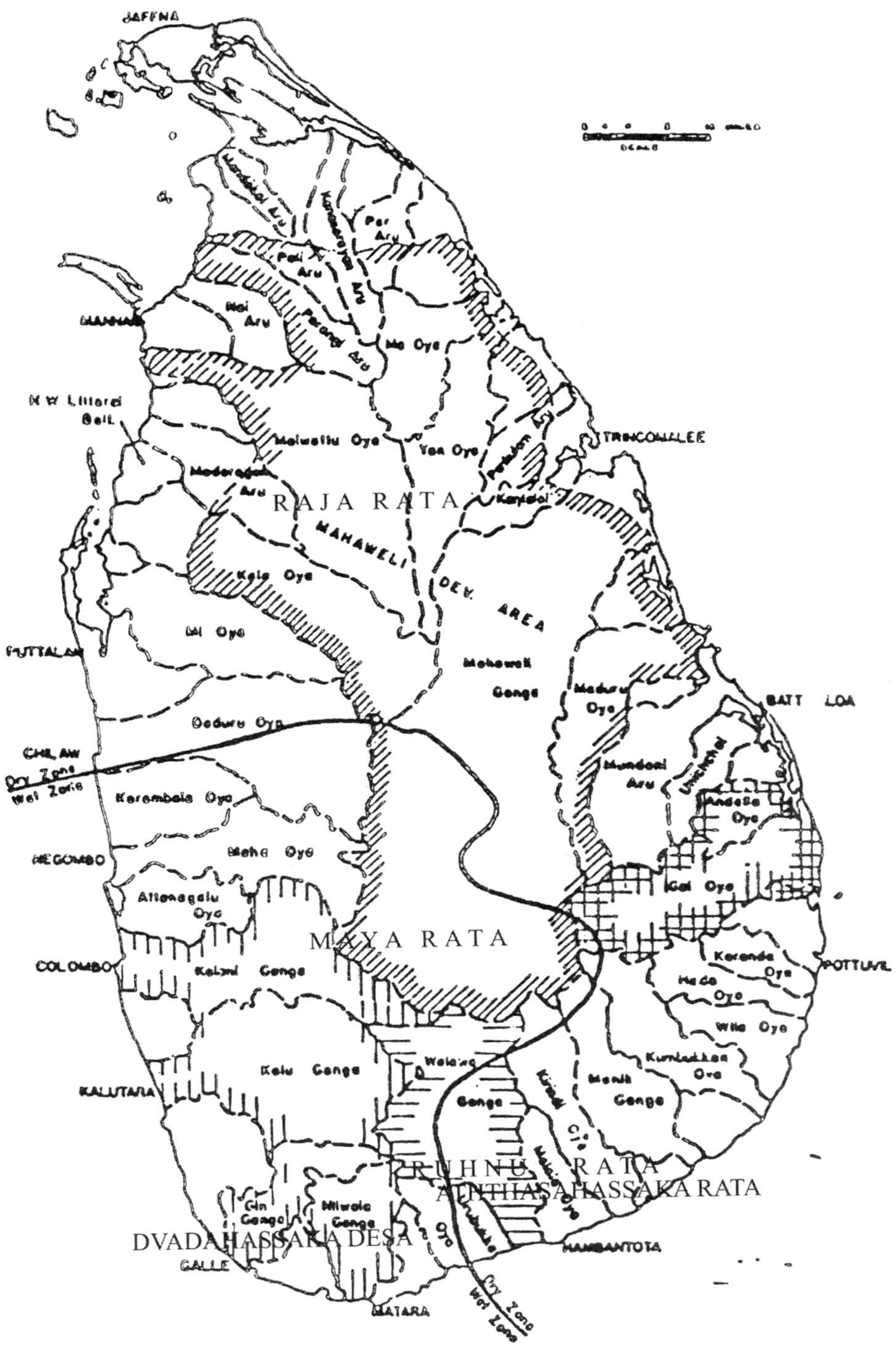

Figure 1. Rivers of Sri Lanka.

and large reservoirs. In wet zone uplands, much later, terraced rice fields saw reuse of irrigation water under gravity, benefiting from forest cover induced fertility. In colonial times, destruction of forests for planting coffee, and later tea, led to adverse impacts (2). Ancient diversion systems include stone anicuts (derived from the Tamil language) or weirs, called *amuna* in Sinhala and *tekkam* in Tamil, channels diverting streams for both irrigated agriculture, and storage in reservoirs. Large reservoirs with massive earth dams were built after the invention of the sluice with its *bisokotuwa* or access tower, forerunner of the valve tower (3, p. 379). Small tanks (Portuguese *tanque*) were synonymous with villages. Small earth dams without sluices, locally known as *vetiyas,* are incorrectly described on topographical survey sheets as *abandoned small tanks*. New metric-scale topo sheets have omitted many of these vetiyas, which checked surface runoff in nonperennial streams for storage in the earth itself, a water harvesting technique understood in India today (4) but not in Sri Lanka. Conservation of fauna began when

Table 1. Contrasting Perspectives on Ancient and Modern Projects

	Hydraulic Engineering Perspective—Hard Technology Design of Modern Projects	Ecosystems Perspective—Soft Technology Design seen in Ancient Projects
1. Water	Inanimate, active	animate, passive
2. Small tank	Inefficient, early stage in evolution and development of irrigation systems, to be submerged by a large reservoir built later	microwater and soil conservation ecosystem in downstream development area below a large reservoir; name of tank often identical to name of village
3. Large reservoir	Efficient irrigation system in combination with downstream channel distribution irrigation system without any small tanks in its command area	Main item in a macrowater and soil conservation ecosystem with microwater and soil conservation ecosystems in its command area
4. Diversion structure, dam or weir, and diversion channel	Built to augment a large reservoir—last stage in development of ancient irrigation systems in Sri Lanka (5,6)	Earliest stage in irrigated agriculture and evolution and development of ancient water and soil conservation ecosystems in Sri Lanka (7)
5. Vetiya, breached small earth embankment or bund, found in dry zone jungles	Embankment or bund of abandoned small tank	Deflection structure—microwater and soil conservation ecosystem; maintains water table
6. Downstream development area in modern projects	Cleared of all vegetation to lay out channel irrigation systems	Should be designed as a series of microwater and soil conservation ecosystems, including forest areas, as in ancient systems
7. Forest areas	Limited to catchment areas in modern projects	Not only in catchment areas in ancient projects, but interspersed with fields in downstream development areas for better nutrient flow

King Devanampiya Tissa, on a hunting trip about 223 B.C., was converted to Buddhism. Arahant Mahinda, son of Emperor Asoka of India, preached a sermon at Mihintale, a mount near the capital city Anuradhapura: *"The birds of the air and the beasts of the jungle have as much right to life as thou, O great King!"* (4). Hence, the ancient systems involved conservation of water and *interactions* of water, soil, flora, and fauna, with *functions* including irrigation, drainage, flood control, and domestic water supply. They should be called *water and soil conservation ecosystems*, as described in Table 1. Three aspects merit study:

- —their evolution and development beginning in about the mid-first millennium B.C.;
- —their stability and sustainability for about 17 centuries until about the thirteenth century A.D.; and
- —their apparently irreversible decline after the Parakrama Bahu I (1153–1186) era.

These are being addressed in a multidisciplinary project, *Science and Civilisation in Sri Lanka,* started recently following Joseph Needham's *Science and Civilisation in China* (8).

ANCIENT RAJARATA

Anuradhapura in Rajarata was the capital city of Sri Lanka from about the fourth century B.C. to about the tenth century. Polonnaruwa, further east, became an alternative capital in about the eighth century on account of invasions from southern India and remained the capital until the thirteenth century, after which drift to the southwest began (9). Water and soil conservation ecosystems in Rajarata were highly developed in the Anuradhapura period (5,10–13). Only restoration was necessary in modern times, until the advent of new projects, Gal oya (1947–1949) and the aid funded Mahaweli (1978–1986). However, the Sri Lanka National Atlas does not have a chapter on the ancient systems.

In western Rajarata, Anuradhapura city tanks were built in stages: Basawakkulama (fourth century B.C.), Nuwara weva (reservoir) (first century), Tissa weva (third century) and Nachchaduwa weva (fifth century). The $54\frac{1}{2}$-mile-long channel, Kalaweva Jayaganga (River of Victory), starts from the right bank sluice of the great twin reservoirs Kalaweva–Balaluweva built by King Dhatusena (459–477), feeding Tissa weva directly, with a branch channel to Nachchaduwa. Brohier (10) documented the *Interrelation of Large Reservoirs and Channels* in western and eastern Rajarata. Nicholas and Paranavitana (12) further expanded this (Fig. 2).

The drought-resisting capability of this interconnected system augmented *cascades of small tanks* whose names were synonymous with village names [Fig. 3 (14)] [Fig. 4 (15)].

In eastern Rajarata, ancient ecosystems were developed through more than 11 centuries from the first century to the twelfth century (5,13,16,17). The Elahera stone diversion anicut and the first 24 miles of the Elahera canal were built by King Vasabha (65–109); King Mahasena (276–303) built Minneriya weva at the tail end of this canal and started the extension of the

Elahera–Minneriya canal beyond Minneriya weva. King Agga Bodhi I (575–608) completed this extension, and King Agga Bodhi II (608–618) built Gantala weva, the modern Kantalai reservoir, at the tail end of this canal, and an extension to the sea near Trincomalee. Finally, King Parakrama Bahu I (1153–1186) improved the whole system and constructed three large reservoirs, called Seas of Parakrama: Parakrama Samudra, restored in modern times; Parakrama Talaka still unidentified; and Parakrama Sagara, created by raising the Elahera anicut and strengthening the earthen canal embankment between Elahera and Konduruweva. This formed a series of reservoirs because there is a fall of about 60 feet in this reach (5,15) (Fig. 5). Parakrama Sagara was also called *Koththabadhdhanijjara*, translated as *the weir furnished with a reservoir* and also as *the reservoir whose flood escape was walled up* (18,19).

ANCIENT RUHUNURATA

In ancient Ruhunurata, remains of thousands of ancient irrigation works are found in the southeastern area, once described as Aththasahasaka rata, the region of 8000 villages, and the southwestern area called Dvadahassaka desa, the region of 12,000 villages (Fig. 1). But they have not been studied and understood as those in ancient Rajarata, and modern development has impacted the ancient projects adversely.

JAFFNA PENINSULA

The climate and topography of the northern Jaffna peninsula are quite distinctive. Rainfall is comparatively sparse, and underlying limestone makes groundwater the

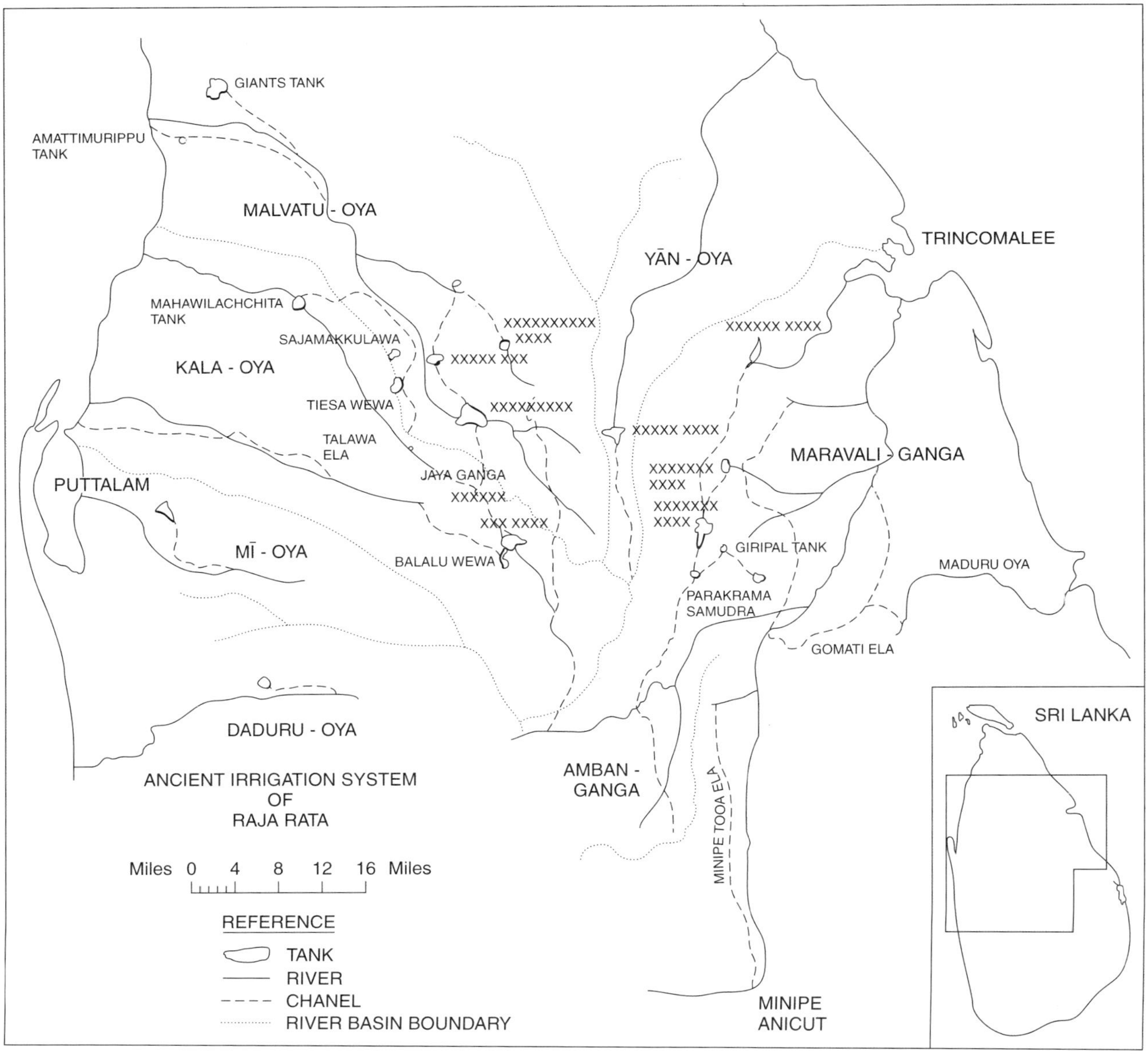

Figure 2. Interrelation of large reservoirs and channels.

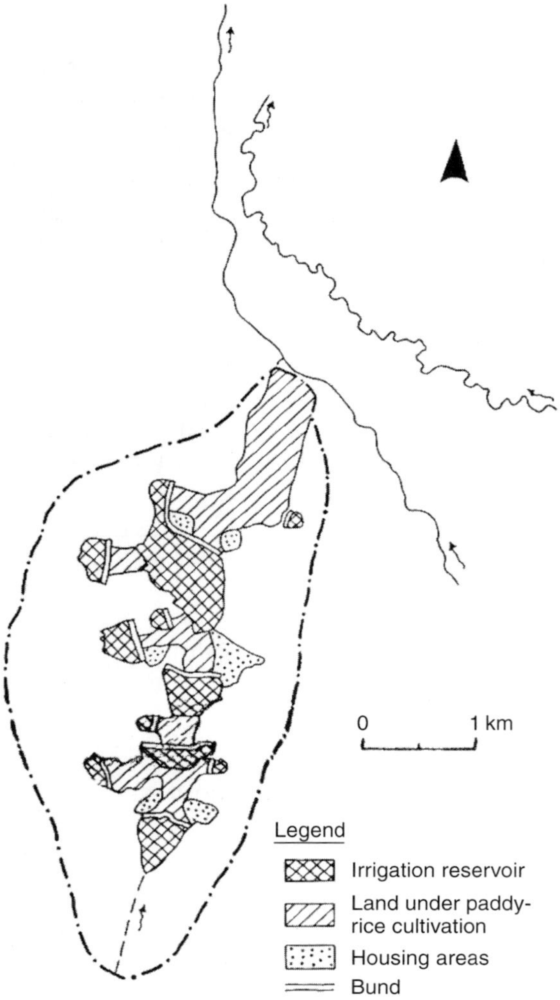

Figure 3. General layout of the cascade of four villages located in a local catchment.

principal source of water, extracted from dug wells, using the well sweep from ancient times (20).

MODERN DEVELOPMENT

Irrigation engineers today use a hydraulic engineering approach to design new projects, almost diametrically opposite to an ecosystems approach. In the hydraulic engineering or hard technology perspective, water is treated as inanimate and active, in the ecosystems or soft technology perspective, water is animate and passive. Consequences are shown in Table 1.

A four-stage hypothesis for the evolution and development of ancient irrigation systems follows the hydraulic engineering perspective (6,21):

—rainwater tanks from which water was bailed
—small storage reservoirs, called village tanks
—large storage reservoirs that submerged a number of small tanks
—augmentation of large reservoirs by diverting a river

Here, the evolution and development of storage reservoirs, in terms of storage capacity is followed by river diversion as a later development. But river diversion irrigation systems or water management in space existed before storage reservoirs or water management in time (which depended on the invention of the sluice) in all ancient civilizations.

Therefore, an alternative seven-stage hypothesis was proposed (7):

1. Rain-fed agriculture
2. Temporary or seasonal river diversion and inundation irrigation on river banks
3. Permanent river diversion and development of canal irrigation systems
4. Development of weirs and spillways on contour channels
5. Invention of the sorowwa (sluice) with its bisokotuwa (access tower)
6. Construction of small, medium, and large storage reservoirs
7. Damming a perennial river using a sluice for temporary river diversion, or the twin-tank technique, as in Kalaweva–Balaluweva in the fifth century

Meanwhile, a map entitled the Water Resources Development Plan (1959) showed proposed new large reservoirs, submerging ancient small village tanks, following the third stage of the four-stage hypothesis. Of ancient water and soil conservation ecosystems, only very large reservoirs and channels were shown.

In the southern area, location of the gigantic new Uda Walawe and Lunuganvehera reservoirs, follows the third stage in the four-stage hypothesis. Both Uda Walawe (1965–67) (Fig. 6) and Lunuganvehera (1978–1986) were built without considering alternative upstream sites. Environmental problems have made both these new projects unstable and unsustainable, contributing to insurrections against the government in 1971 and 1987–1989 that were suppressed with much loss of life and damage to property (22–25).

A similar possibility exists if Moragahakande, selected from the Water Resources Development Plan, 1959, for cost feasibility studies, is built (Fig. 5), compounded by the ethnic discord between separatist Tamils and the government. This large new reservoir will also destroy any possibility of restoring the potentially stable and sustainable ancient Parakrama Sagara.

A RIVER FOR JAFFNA—THE ARUMUGAM PLAN

A proposal called A River for Jaffna, prepared 50 years ago and partly implemented, will avoid all potential problems of Moragahakande. It involves enhancing both the quality and quantity of Jaffna groundwater by infinite dilution over the years by floodwater from the mainland Iranamadu reservoir (Fig. 7).

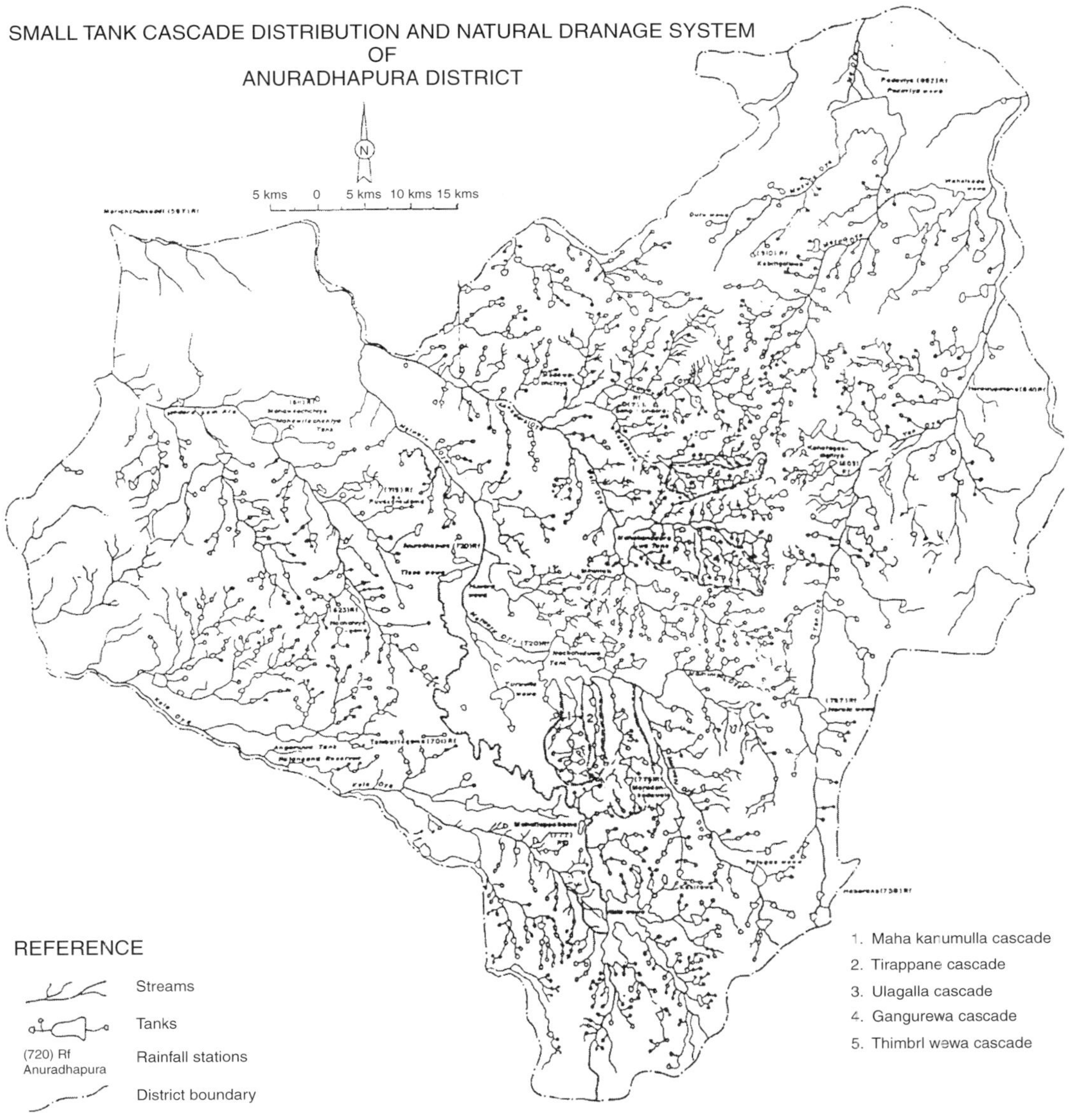

Figure 4. Cascades of small tanks.

BIBLIOGRAPHY

1. Weeramantry, C.G. (2000). *Separate Opinion* in the Gabcikovo Nagymaros case in the World Court, 1997. Republished for the Institution of Engineers, Sri Lanka. Vishwa Lekha.
2. Senanayake, D.S. (1934). *Agriculture and Patriotism*. Government Press, pp. 39–40.
3. Parker, H.W. (1909). *Ancient Ceylon*. Lusacs, London.
4. Center for Science and Environment. (1997). *Dying Wisdom*. State of India's Environment 4, p. 354.
5. Brohier, R.L. (1934). *Ancient Irrigation Works in Ceylon*. Government Press, Colombo.
6. Needham, J. (1971). *Science and Civilization in China*. Vol. 4, Part 3, Cambridge, p. 369.
7. Mendis, D.L.O. (1986). Evolution and development of irrigation ecosystems and social formations in ancient Sri Lanka. *Trans. Inst. Eng. Sri Lanka* 13–29.
8. Mendis, D.L.O. and Kamaladasa, B. (Eds.). (2003). *Water for People and Nature*. Arumugam Commemoration volume. Sri Lanka Water Heritage. Sarvodaya Vishwa Lekha, Ratmalana.
9. Indrapala, K. (1971). *Collapse of the Rajarata Civilisation and the Drift to the Southwest*. Ceylon Studies Seminar, University of Peradeniya.
10. Brohier, R.L. (1937). Inter-relation of groups of large reservoirs and channels in Rajarata. *J. R. Asiatic Soc. Ceylon* **34**(90): 70–89.
11. Brohier. (1991).
12. Nicholas, C.W. and Paranavitana, S. (1960). *Concise History of Ceylon*. Gunasena.
13. Arumugam, S. (1969). *Water Resources of Ceylon*. Water Resources Board, Colombo.
14. Sandell, K. (1988). Ecosystems in theory and practice. Farmers perspectives on water, nutrients and sustainability in low-resource agriculture in the dry zone of Sri Lanka. *Linkoping*, p. 94.

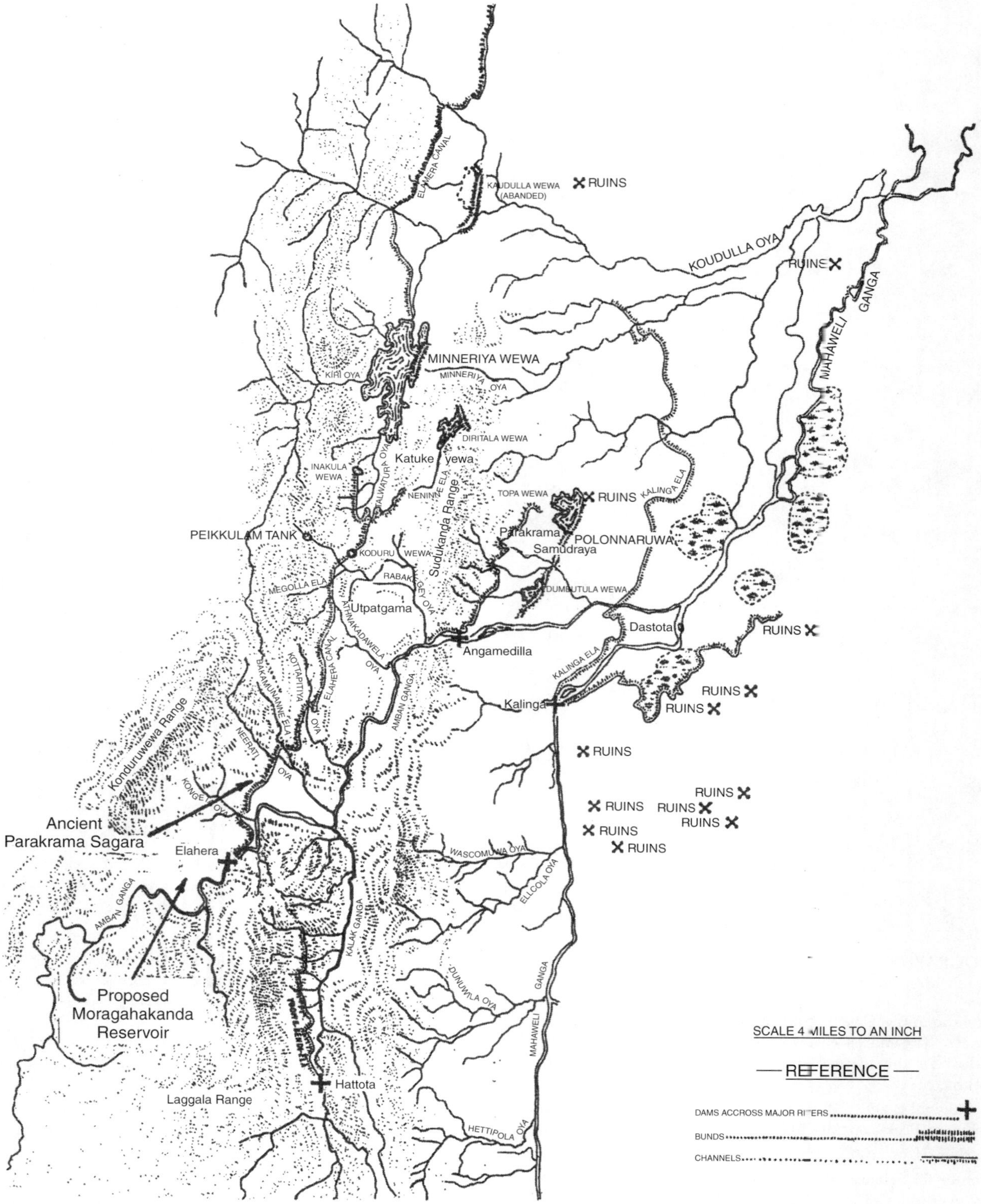

Reference : The Tamankaduwa District and the Elahera - Minneriya Canal R L Brohier, 1941
Diagram No.2 facing page 4, (Ancient Parakrama Sagara and proposed Moragahakanda reservoir, added)

Figure 5. Diagram showing the hydrography of Tamankaduwa and the ancient irrigation schemes.

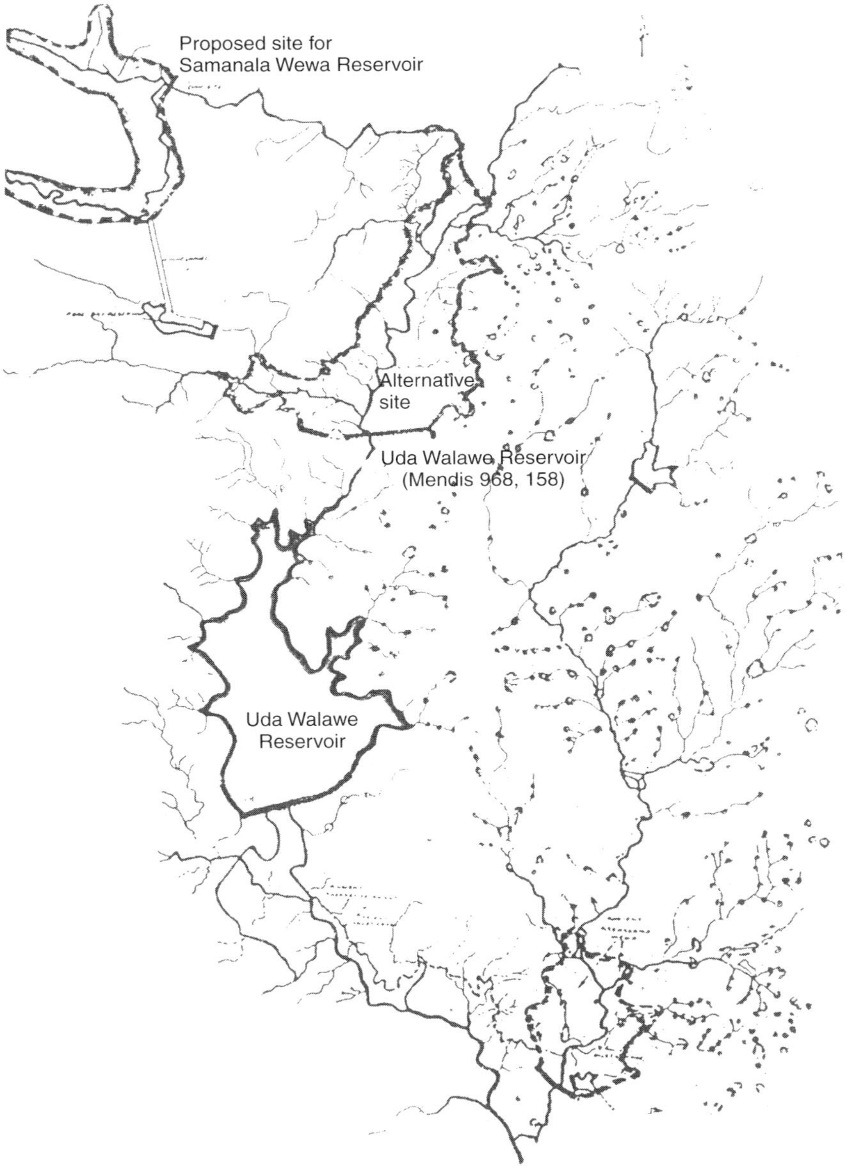

Figure 6. Walawe basin development ancient dispersed small-scale system vs modern centralized large-scale system.

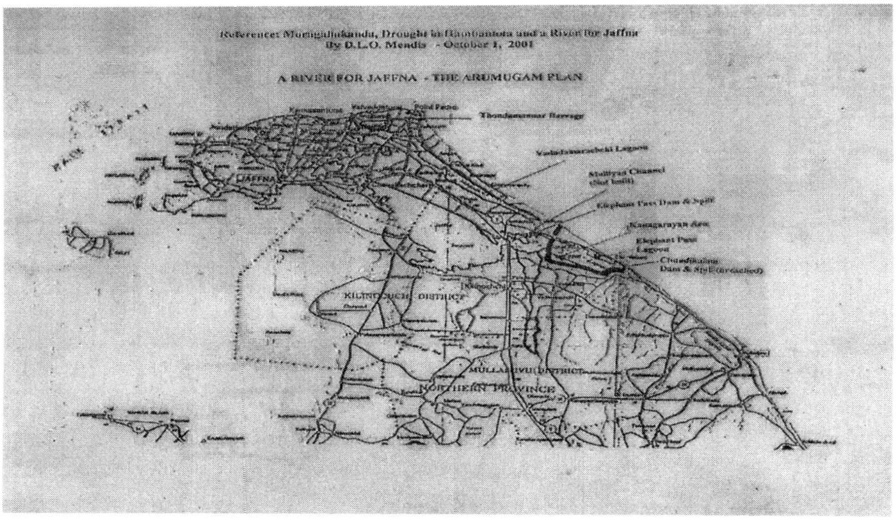

Figure 7. The Arumugam plan.

15. Panabokke, C.R. *Small Tank Cascades in the Rajarata*. In: Arumugam Commemoration, pp. 123–132.
16. Ward, SirH. (1857). Minutes of inspection. *Ceylon Almanac*. quoted in Brohier, 1934, Vol. 1, pp. 28–33.
17. Mendis, D.L.O. (2002). *Water Heritage of Sri Lanka*. History of Water Conservation series. Sarvodaya Vishwa Lekha, Ratmalana.
18. Geiger, W. (1953). *Mahavamsa (Translation)*. Government Press, Colombo.
19. Geiger, W. (1960). *Culture of Ceylon in Medieval Times*. Otto-Harrosowitz, Weisbaaden.
20. Tennant, J.E. (1859). *Ceylon*. Longmans, London. Reprinted Asian Educational Services, 1996, Vol. II, p. 552.
21. Brohier. (1956).
22. Mendis, D.L.O. (1989). *Hydraulic Civilisations, Irrigation Ecosystems and the Modern State*. Second Professor E O E Pereira Commemoration Lecture, Institution of Engineers, Sri Lanka.
23. Mendis, D.L.O. (1989). Development of underdevelopment in southern Sri Lanka: Destabilisation of ancient irrigation ecosystems by the impact of hydraulic engineering. *Trans. Inst. Eng. Sri Lanka:* 31–50.
24. Mendis, D.L.O. (1991). Non-ethnic causes of violent conflict in Sri Lanka: Environmental degradation on account of destruction of ancient irrigation ecosystems by the impact of hydraulic engineering. *Proc. 41st Int. Pugwash Conf. Sci. World Affairs*, Continental Press, Singapore, pp. 698–709.
25. Mendis, D.L.O. (1995). *Past, present and future conflict over water in Sri Lanka. Proc. 45th Int. Pugwash* Conf., Hiroshima. World Scientific, Singapore, pp. 791–800.

BEN FRANKLIN'S GULF STREAM WEATHER AND SWIM FINS

William E. Marks
Water Consciousness, Inc.
Martha's Vineyard,
Massachusetts

THE DISCOVERY OF THE GULF STREAM

In 1950, renowned nature writer Rachel Carson's book, *The Sea Around Us*, was published by Oxford University Press (1). In this book, Carson discusses at length, Benjamin Franklin's chart of the Gulf Stream (Fig. 1). A copy of Ben Franklin's chart is reproduced in Carson's book as an endpaper. In her writings, Carson credits the accuracy of Franklin's chart that was first created in 1769 and uses it and other information to help explain the flow of ocean currents.

In all fairness to the historic record, Franklin did not discover the Gulf Stream. The pre-Columbian Florida Seminole Indians, who traveled local waters, probably were aware of the Gulf Stream's current and its influence on their local sea travel and fisheries. As well, in 1492, Christopher Columbus pioneered Gulf Stream observations while sailing the equatorial current near the West Indies. While taking soundings, Columbus observed that the warmer surface waters were moving faster and in a different direction than the colder deeper waters (2).

In addition to the above, and thanks to research of Midshipman Pamela Phillips of the U.S. Naval Academy and Richard Gasparovic of the Applied Physics laboratory of John Hopkins University, we have the following time line:

1513: Ponce de Leon describes the Florida Current.

1519: Existence of the Gulf Stream is widely known to sailors.

1575: Andre Thevet associates the Gulf Stream with the rivers of the Gulf of Mexico.

1590: John White becomes the first person to mention the counter-currents on the coastward side of the Gulf Stream.

1606: Lescabot describes the meeting of the Labrador and Gulf Stream currents.

1660: Varenius classifies the Gulf Stream as a "perpetual special" motion of the sea and describes it as a giant stream.

1665: Kircher publishes showing the Gulf Stream.

1685: Happelius publishes second chart showing the Gulf Stream.

1769: Benjamin Franklin and Timothy Folger design and publish their chart of the Gulf Stream with notes on how to avoid it.

As expressed above, the first actual record of the Gulf Stream was on April 22, 1513, as related in the Chronicle of Ponce de Leon's Expedition, which eventually led to the so-called discovery of Florida. This 1513 document states:

> A current such that, although they had great wind, they could not proceed forward, but backward and it seemed that they were proceeding well; and at the end it was known that in such wise current was more powerful than the wind (3).

Further maritime observations of the Gulf Stream were recorded on an ongoing basis after its initial discovery. However, these observations remained the "secret" of the ship's captains and pilots. This fact was noted in an article by Franklin Bache, published by the American Philosophical Society in 1936 (4):

> Apparently the Spaniards soon learned that this northerly flowing current was not merely a local current but one of wide extent; for six years later [after 1513] when Antonio de Alaminos set out for Spain from Vera Cruz, he sailed northward with the Gulf Stream for a number of days before turning east toward Europe. This same Aliminos was pilot of Ponce de Leon's expedition of 1513, when the Gulf stream was first noted. It is, therefore, quite proper to credit the discovery of the Gulf Stream to Alaminos (Marmer H. A. *The Gulf Stream and its Problems*).
> For two and half centuries following its discovery the growth of knowledge regarding the Gulf Stream was slow. The story is told in detail by Kohl and more briefly by Pillsbury. During this period, to be sure, the mariner and more especially the whaler, became acquainted with the Gulf Stream throughout the greater part of its course. Much of this information, however, was kept as a professional secret,. . . (Marmer, 1929).

Figure 1. A chart of the Gulf Stream.

In 1513, after Ponce de Leon landed in Florida in the area known today as St. Augustine, he attempted to return to Puerto Rico after failing to locate the Fountain of Youth. However, as documented above, the current was so strong that his ships floated northward, even though they were sailing with a fair wind that should have taken them southward.

Six years later, Anton de Alaminos, found himself guiding Cortes's treasure ships from Mexico. With his prior knowledge as a pilot for Ponce de Leon, Alaminos used the northerly current of the Gulf Stream to his advantage.

Even though the discovery of the Gulf Stream by Alaminos was lost to history, there was another sailor who documented the current in 1735. Captain Walter Haxton, was a shipowner who also had a financial interest in understanding and using the currents of the ocean to his advantage. Besides being a captain, Haxton also owned several tobacco farms in the Maryland area and often shipped his produce to buyers in London.

In 1735, Haxton created the first detailed chart of the Chesapeake Bay area, which included bottom soundings, prevailing winds, and water current velocities and direction. As an addendum to this chart, Haxton plotted the latitude and longitude of the Gulf Stream (Northeast Current), as well as its direction and velocity.

Some 64 years before Franklin's chart in 1769, Haxton penned the following:

> It is generally known by those who trade to the northern parts of America that the current which comes out of the Gulph of Florida runs constantly along the coast of Carolina and Virginia and considerably further to the Northward, varying its course as it is obstructed by shores. Now if said current always runs nearly in the same part or space of the ocean (as from a great number of Tryals and observations which I have made in 23 voyages to Maryland, I have reason to think it does), the knowledge of its Limits Course and Strength may be very useful to those who have occasion to sail in it.

For some reason, Haxton's chart and related information about the Gulf Stream never received wide distribution or application. It was due to such oversight, as well as centuries of secrecy by various captains over the existence of the Gulf Stream, that the stage was set for the Franklin–Folger chart to make a splash in the annals of oceanic history. So much so, that most people today still harbor the belief that Benjamin Franklin discovered the Gulf Stream.

THE FRANKLIN/FOLGER GULF STREAM CHART

When Franklin had his and Folger's version of the Gulf Stream printed on a navigational chart in 1769, there was no such chart available. Franklin could only create the chart after receiving the benefit of Timothy Folger's drawings and advice. At the time, Folger was an active Nantucket whaling captain. Folger is credited for depicting the width and breadth of the Gulf Stream's current by drawing it on an Atlantic Ocean chart. Folger's chart included notes on how the Gulf Stream's current influenced the speed of ships crossing the Atlantic between America and Europe. Folger's drawing of the shape and speed of the Gulf Stream's current were extremely accurate for his time and in general, hold true to this day. Folger drew the Gulf Stream on a chart in answer to Franklin's inquiry about the forces of nature that may be impacting the speed of ships crossing the Atlantic Ocean. On another note, Folger was Franklin's cousin, twice removed (Nantucket Historical Society).

Today, it is generally recognized that the Franklin–Folger chart represents an accurate summary of the average width, length, and direction of the Gulf Stream, as well as the average speed of its current.

The name "Gulph Stream" was first suggested in 1762 by Benjamin Franklin because of the stream's sourcing from the Gulf of Mexico. Prior to 1762 it was generally known as the "Northeast Current."

The initial concept of creating Franklin's famous chart of the Gulf Stream was prompted in 1768 when Franklin was in London serving as Deputy Postmaster General for North America (Van Doren, 1938). While in London, Franklin received complaints from America as to how westbound mail packets [English ships] took up to two weeks longer than American merchant ships (5) *Am. Philos. Soc.* lxxvi (5). (1936). At about the same time, the English postal authorities had written to Franklin asking similar questions. These inquiries from Americans and the English about the speed of ships crossing the Atlantic prompted Franklin to do some research.

While researching the answers to these complaints in 1768, Franklin made inquiries of Timothy Folger, who happened to be in London at the time. Folger informed Franklin that based on experience, Nantucket whalers had developed a working knowledge of the Gulf Stream between Europe and America. This knowledge came from the whalers following the migration patterns of whales, which were consistently found north or south beyond the edges of the Gulf Stream, but never inside the Gulf Stream. Whaling captains from Nantucket and Martha's Vineyard knew that whales could be found feeding along the Gulf Streams plankton-rich boundaries. They also had learned that sailing back to New England whaling ports could be achieved in less time by staying north of the Gulf Stream and thereby avoiding its current.

Over time, Nantucket whalers learned to use the current of the Gulf Stream to their advantage when sailing in a northerly direction to England, but avoiding it when sailing in a southerly direction back to America. Folger told Franklin that American captains often told British captains about avoiding the current, but their hubris caused them to ignore the Americans advice. "They were too wise to be counselled by simple American fishermen," Folger told Franklin.

A letter from Franklin dated October 29, 1768, tells of Folger's role in the creation of the Gulf Stream chart:

> Discoursing with Captain Folger a very intelligent Mariner of the Island of Nantucket in New England concerning the long passages made by some Ships bound from England to New York I received from him the following information. Vizt
> That the Island in which he lives is Inhabited Chiefly by people concerned in the Whale Fishery, in which they employ near 150 Sail Vessels, that the whales are found generally near the Edges of the Gulph Stream, a strong current so called which comes out of the Gulph of Florida, passing Northeasterly along the Coast of America, and the turning off most Easterly running at the rate of 4, 3 1/2, 3 and 2 1/2 Miles an Hour; that the Whaling Business leading these people to Cruise along the Edges of the Stream in quest of Whales,. . ."

When the chart was created in 1769, Franklin wrote the following notes based in large part on the advice he received from Nantucket's Timothy Folger:

> By observing these directions and keeping between the stream and the shoals [that lie to the south of Nantucket and St. George's Banks], the passage from the Banks of Newfoundland to New York, Delaware, or Virginia, may be considerably shortened; for so you will have the advantage of the eddy current, which moves contrary to the Gulph Stream. Whereas if to avoid the shoals you keep too far to the southward, and get into that stream, you will be retarded by it at the rate of 60 or 70 miles a day.
> The Nantucket whale-men being extremely well acquainted with the Gulph Stream, its course, strength and extent, by their constant practice of whaling on the edges of it, from their island quite down to the Bahamas, this draft of that stream was obtained from one of them, Capt. Folger, and caused to be engraved on the old chart in London, for the benefit of navigators
> . . . Note, the Nantucket captains who are acquainted with this stream, make their voyages from England to Boston in as short a time generally as others take in going from Boston to England, viz. from 20 to 30 days.
> A stranger may know when he is in the Gulph Stream, by the warmth of the water, which is much greater than that of the water on each side of it. If then he is bound to the westward, he should cross the stream to get out of it as soon as possible (6).

Besides the above two sources referring to Timothy Folger's role in creating the Gulf Stream chart, we also have the following quote from Franklin's *Maritime Observations*. These observations were written at sea in August, 1785, and expressed in his letter to David Le Roy:

> I then observed it was a pity no notice was taken of this current upon the charts, and requested him [Timothy Folger]

> to mark it out for me, which he readily complied with, adding directions for avoiding it in sailing from Europe to North America. I procured it to be engraved by order from the general post-office, on the old chart of the Atlantic, at Mount and Page's, Tower Hill; and copies were sent down to Falmouth for the [English] captains of the packets [ships], who slighted it however; but is since printed in France, of which edition I annex a copy.

It is in Franklin's *Maritime Observations* that we find mention of his two charts of the Gulf Stream. Both of which he published—the first chart in England in 1769, and the second in France circa 1778. It was also during this 1785 voyage that Franklin expressed his theory about how the Trade Winds create the Gulf Stream. Franklin's theory was that the winds created the Gulf Stream current by the friction of the moving air on the surface of the water. This theory, in general has remained correct to this day. As expressed in Franklin's words:

> This Stream is probably generated by the great accumulation of water on the eastern coast of America between the tropics, by the Trade Winds which constantly blow there. It is known that a large piece of water ten miles broad and generally only three feet deep, has by a strong wind had its waters driven to one side and sustained so as to become six feet deep, while the windward side is laid dry.
> ... The power of the wind to raise water above its common level in the sea is known to us in America, by the high tides occasioned in all our seaports when a strong northeaster blows against the Gulf Stream.

After having the chart of the Gulf Stream printed, Franklin made copies available to British captains. However, Franklin noted that the British captains refused to believe in information about the Gulf Stream proffered by Nantucket whalers. On similar note, the Franklin–Folger chart, and its related information about the Gulf Stream were also rejected by English captains. For this reason, as well as the revolt of the colonies against England in the early 1770 s, Franklin's 1769 chart of the Gulf Stream became extremely rare.

It is believed by some historians that Franklin and other Americans in England may have suppressed the distribution of this 1769 chart. The reason that Franklin suppressed his 1769 chart of the Gulf Stream may have been to prevent the English Navy from using it to its advantage against the colonies. This has merit because it was in 1786 and after the Revolutionary War, that Franklin again published Folger's sketch of the Gulf Stream as part of an article in the *Proceedings of the American Philosophical Society of Philadelphia* (7).

When Franklin was returning to America from London in March 1775, he took temperature measurements of the Gulf Stream four times each day. During this voyage he discovered that the Gulf Stream was warmer than surrounding waters and that it had a color different from the colder waters on either side. These observations were consistent with those of Nantucket whalers.

In 1776, when Franklin was sent by the colonies as envoy to Paris to negotiate a treaty, he again took temperature measurements of the Gulf Stream and kept a record of his observations. These observations rekindled Franklin's interest in understanding the Gulf Stream. Upon arrival in France, he had the original Folger chart of the Gulf Stream copied and printed by Le Rouge. Soon after, the Le Rouge chart was made available to French captains sailing to America with arms and supplies. Even though the Le Rouge version of the Gulf Stream chart is undated, the copy in the Bibliotheque National has the following note inscribed on its underside, "communique' par M. de Franklin en 1778."

In 1785, at age 79, and five years before his death, Franklin made his last transatlantic voyage when he sailed home from France on the London packet. Continuing with his research in to the Gulf Stream, Franklin directed his grandnephew, Jonathan Williams, Jr., when and how to take temperature readings of the Gulf Stream. It was during this final voyage that Franklin wrote *Maritime Observations*. This paper was published in 1786 and includes the best known versions of the making of the Franklin–Folger chart of the Gulf Stream, as well as its related genesis.

The original Franklin–Folger chart of 1769 was lost to history for close to 200 years and was assumed forever lost. During that time, the only surviving record of the original 1769 chart was expressed in the writings of Franklin and others. Concern for finding a copy of the 1769 chart surfaced from time to time. For example, on April 24, 1936, The American Philosophical Society printed an article entitled, "WHERE IS FRANKLIN'S FIRST CHART OF THE GULF STREAM?." The article was authored by Franklin Bache, and was an international plea for people to search for a copy of the chart. It was not until 52 years later that this plea was answered. And, most appropriately, by a scientist who had interest in studying ocean currents.

In 1978, Dr. Philip L. Richardson of Woods Hole Oceanographic Institute found the original 1769 Franklin–Folger chart while doing research in France. As expressed by Richardson (8) *Science* (1980). **207**; 8.

> In September 1978, I found two prints of the Franklin–Folger chart in the Bibliotheque Nationale in Paris (Fig. 1) (7). It occurred to me that a copy of the chart might have been saved by the French because Franklin was envoy to France from 1776 to 1785 and both Franklin and his ideas were highly regarded by the French.
> Richardson would later publish his in-depth Gulf Stream research in *The Geographical Magazine*, London, LII (8), pgs. 575–58, in an article entitled "Progress on the Gulf Stream (9)."

When the author of this article contacted Dr. Richardson by email at Woods Hole Oceanographic Institute, he answered:

> I discovered the 1769 chart while I was working on Gulf Stream Rings, which are eddies that pinch off from Gulf Stream meanders. We were tracking them, measuring their paths in the ocean and their decay. I was spending a year in Paris analyzing some data and writing papers.
> I was thrilled by my discovery and spent a lot of time in further research of the Gulf Stream and Ben Franklin, and other early charts of the Gulf Stream. Based on that, I published the 1980 paper in Science vol 207, 643–645 and one in the book

Oceanography: The Past, (1980). Springer Verlag, New York. I also worked on derelict ship trajectories in the Stream, and compared them to drifting buoy trajectories and with Walter, Haxton's 1735 description of the Stream (10) (1982, *J. Mar. Res. Supp.*, **40**: 597–603).

Today, Dr. Philip L. Richardson has the status of Scientist Emeritus at Woods Hole Oceanographic Institute, where he continues to pursue his passion of oceanographic research.

FRANKLIN AND WEATHER

"Some are weatherwise, some are otherwise," wrote Franklin under his pen name of "Poor Richard" in 1735.

It was also in Poor Richard's Almanac that Franklin dared to be so weatherwise that he published some of the first recorded weather forecasts. Poor Richard's Almanac was first published in 1732 under Franklin's nom de plume of Richard Saunders.

Franklin recorded some of his first theories on weather phenomena while in Philadelphia in 1743. On the evening of October 21, he had plans to observe a lunar eclipse which was scheduled to occur at 8:30 P.M. As Franklin waited with anticipation to watch the eclipse, his plans were suddenly interrupted when a fast moving violent storm blocked his view of the moon.

As the following weeks passed, Franklin read in various newspapers that the violent storm tracked northeasterly, causing considerable damage from Virginia all the way to Boston. However, much to Franklin's surprise, the newspapers from Boston reported accounts of the lunar eclipse, and that the storm arrived in Boston long after the eclipse had come and gone.

This series of events intrigued Franklin; he became curious why Boston, which was located hundreds of miles northeast of Philadelphia, would experience the delayed arrival of the storm. Another factor that intrigued Franklin was that the winds and the direction of the storm seemed contrary. After much inquiry and observation of other storms, Franklin told his friend Jared Eliot that his research had led him to the "very singular opinion that, though the course of the wind is from the northeast to the southwest, yet the course of the storm is from the southwest to the northeast." This observation by Franklins established him as the first to observe that storm winds can rotate counterclockwise, even though the entire storm is traveling in a clockwise direction.

In his attempt to decipher how such weather patterns could occur, Franklin theorized that heated air rising skyward in the south created a vacuum effect (low pressure), which in turn attracted winds from the north (high pressure). This theory of low and high pressure systems was the first known explanation of how storms moved in the Northern Hemisphere.

Today, weather forecasters recognize and respect the accuracy of Franklin's theory as it contributes to our understanding of the hydrologic cycle and forecasting the movement of hurricanes. "With this began the science of weather prediction," said the renowned scholar, William Morris Davis, 150 years later. Perhaps this distinction qualifies Franklin as America's first meteorological scientist.

Franklin's penchant for weather phenomena surfaced again in 1755 when he had the occasion to see a dust devil, or whirlwind funnel, bouncing across the landscape, while he was on horseback in Maryland. Franklin was so excited that he chased the dust devil and rode beside it for a considerable time before it disappeared in the forest. He wrote of this experience at length, on August 25, to his longtime friend and correspondent, Peter Collinson.

This lifelong interest in weather surfaced again 6 years before his death when he published *Meteorological Imaginations and Conjectures*. In this publication, Franklin explores such weather phenomena as heat radiation, insulating clouds, the shifting direction of winds, and creation of fog. In his own fashion, Franklin correctly defines circulation patterns of the hydrologic cycle. As well, he is correct in his attempt to deduce the rare manifestation of hail and ice during the summer months. He correctly figures out that warm moist air reaches high into the atmosphere where it cools and freezes into ice. This heavy ice then falls to earth before it is melted by the summer sun.

Also, toward the twilight of his life, Franklin again made some brilliant deductions. Before he died in 1790, he researched the effects that volcanic eruptions had on weather patterns and cloud electrification. In doing so, he correctly hypothesized that the severe winter of 1783–84 was exacerbated by the summer 1783 volcanic eruption in Iceland. Franklin theorized that dust from volcanic eruptions was instrumental in reducing the amount of solar energy reaching the earth's surface.

SWIM FINS

Since boyhood, Franklin enjoyed the experience of swimming and had a reputation for his strength and agility at plying the waters. His innate ingenuity expressed itself early when, still a boy, Franklin invented swimming fins, which he secured to his hands to help him swim faster and with less fatigue. According to Franklin, he developed the swim fins as a way to reduce what he referred to as "a laborious and fatiguing operation."

No matter where he traveled, Franklin always found time to enjoy his favorite exercise of swimming. When in Boston, he swam in the Atlantic Ocean and the Charles River. And when in England, he swam the Thames River. Benjamin Franklin "was considered to be an expert swimmer and while in England attracted attention by swimming the two to three miles from Chelsea to Blackfriars" (Franklin Bache, 1936).

On his first trip to London in 1724, Franklin was known for entertaining observers with his "ornamental" maneuvers while splashing around in the Thames. Franklin enjoyed swimming so much that he once considered taking a full-time job as a swimming instructor while in England.

Out of respect for Franklin's lifelong interest in swimming, he was inducted posthumously into the International Swimming Hall of Fame.

On a closing note, it should be recognized that even though Benjamin Franklin was affectionately referred to and often addressed as Dr. Franklin, Benjamin Franklin never received a Ph.D. from any academic institution. However, this in no way detracts from the historic contributions Franklin made to the worlds of science, politics, music, invention, social studies, publishing, printing, meteorology, navigation, optometry, swimming, oceanography, and God knows what else.

In Franklin, we find a human being who had an unsatiable curiosity about the world and a strong belief that he could make a difference through his contributions.

BIBLIOGRAPHY

1. Carson, R.L. (1950). *The Sea Around Us*. Oxford University Press, London.
2. Franklin, B. (1793). *The Private Life of the Late Benjamin Franklin, LL.D.* (Originally written by Himself, and now translated from the French). London, Printed For J. Parsons, No. 21.
3. Scisco, L.D. (1913), The track of Ponce de Leon in 1513. *Bull. Am. Geogr. Soc.* **45**: 721–735.
4. Tucker, T. (2003). *Bolt of Fate: Benjamin Franklin and His Electric Kite Hoax*. Public Affairs Press, Springfield, IL.
5. *The Old Print Shop Portfolio*. (1952). Vol. XI, January, Number 5.
6. Brown, L.A. (1951). The River in the Ocean. In: *Essays Honoring Lawrence C. Wroth*. The Anthoensen Press, Portland, OR.
7. Clarence, S. B., Frederick, R. G., Curt, F. B., Goldschmidt, E.P., William, A. J., Rollo, G. S., Henry, R. W. et al. (1951). Bibliographers—including an introduction by Wilmarth Lewis. *Essays Honoring Lawrence C. Wroth*. The Anthoensen Press, Portland, OR.
8. Bache, F. (1936). Where is Franklin's First Chart of the Gulf Stream?, *Proceedings of the American Philosophical Society* **LXXVI**: 731–741.
9. Pillsbury, J. (1891). *The Gulf Stream: Methods of the Investigation and Results of the Research*. (Appendix No. 10—Report for 1890) United States Coast and Geodetic Survey (Washington 1891), pp. 461–620.
10. Cohen, I.B. (1972). Benjamin Franklin. In: *Dictionary of Scientific Biography*. C.C. Gillispie (Ed.). Charles Scribner's Sons, New York.
11. Translated in Lipowski, Z.J. (1984). Benjamin Franklin as a Psychotherapist: A Forerunner of brief Psychotherapy. *Perspectives in Biology and Medicine*, **27**: 361–366.
12. Tucker, T. (2003). *Bolt of Fate: Benjamin Franklin and His Electric Kite Hoax*. Public Affairs Press, Springfield, IL.
13. TIME Graphic by Gabel, E., and Dykman, J. (2003). Ben Franklin the Great Inventor. *Time Magazine* (July 7), pp. 60–61.
14. Golden, F. (2003). When Sparks Flew. *Time Magazine* (July 7), pp. 55, 58.
15. Hall, A.J. (1975). Philosopher of Dissent Benj. Franklin. *National Geographic* **148**(1): 93–122.
16. Marmer, H.A. (1929). The Gulf Stream and its Problems. Disbound copy published in 1929 Annual Report by Smithsonian Institution, Washington, DC.
17. Available: http://www.colonialhall.com/franklin.asp
18. Available: http://www.earlyamerica.com/lives/franklin
19. Available: http://www.educationalsynthesis.org/famamer/Franklin.html
20. Available: http://www.elywelyn.net/docs/quotes/franklin.html
21. Available: http://www.english.udel.edu/lemay/franklin
22. Available: http://www.famouspeople.com/famamer/Franklin.html
23. Available: http://www.geocities.com/peterroberts.geo/BFrWr.html
24. Available: http://www.ideafinder.com/history/inventors/franklin.htm
25. Available: http://www.inventors/about.com/library/inventors/blfranklin.htm
26. Available: http://www.pbs.org/franklin
27. Available: http://www.quotationspage.com/quotes/Benjamin_Franklin

WATER SYMBOLISM

D.L. Marrin
Hanalei, Hawaii

Since the beginning of recorded history, water has held a special place in the spiritual and physical lives of people who represent a wide range of cultures and civilizations. Though rituals, languages, and beliefs among ancient cultures differed greatly, water was routinely used to symbolize a myriad of places and events in both the observable and unobservable realms.

MESOPOTAMIA

The first generally recognized written history comes from the Sumerians, who inhabited the Mesopotamian region more than 5000 years ago. In the Sumerian pantheon, various gods personified heaven, earth, air, and water. *Enki*, the primary water god, represented wisdom and evolved from a great underworld ocean that also gave rise to heaven (1). The underworld was sometimes referred to as a river and apparently existed beneath the surface of the earth as a freshwater (not saline) body of water. Enki (later known as *Ea* by the Babylonians) ruled over this watery underworld and symbolized the primordial waters, which were associated with the "chaos" that existed before creation. Myths from around the world recognize that creation was preceded by an original state of chaos, which defined a state of formlessness and was often identified as a watery abyss or primordial sea (2). Such a sea apparently reflected the fact that ancient people recognized creation as the emergence of form from formlessness.

EGYPT

The Egyptians, similar to the Sumerians and Babylonians, believed that the heavens and the earth (including all the inhabitants of the earth) were created from the primordial or celestial waters. *Ra*, the great Egyptian sun god, was intimately involved in creating the manifested world as his Sun disk appeared above the waters to signal the beginning of time (3). Ra made his symbolic journey over the water in a boat because the Sun was made of fire and, therefore, could not have risen out of the waters of chaos on its own. According to the Egyptian pantheon, everything is created from the waters of chaos by a being that, similar to God in the Bible's Book of Genesis, appears above or within the primordial waters. The primordial waters or seas often symbolized the formless chaos and undifferentiated matter of the underworld waters that comprised the cosmos before its division into the realms of heaven and earth. Hence, heaven, earth, and all life emerged from "the waters."

CHINA

The Chinese culture was one of only a few that did not associate the primordial chaos with water in the form of a sea. Instead, the ancient Chinese apparently portrayed chaos as a misty vapor that embodied the cosmic energy governing space, time, and matter (4). The formless misty vapor, from which the duality of *yin* and *yang* were born, gave rise to all earthly life forms. Some historians believe that the ancient Chinese people divided even the spirit world into a Ministry of Waters, whereby seawater is overseen by four dragon-kings representing the cardinal directions and freshwater is ruled by four dragon-kings representing each of the major rivers of China.

OCEANIA

The ancient Maori culture of New Zealand referred to water as *wai*, which also means remembrance or the recollection of something that has been. In the Maori culture, water symbolizes the memory of all that has ever been and will be (5). Like so many ancient and indigenous cultures, the Maori consider that water's memory serves as the spark of life and is intimately associated with the process of creation. How water is able to retain or access memory is not a component of ancient or indigenous myths; however, the almost universal insight that such memory is constrained neither by space nor time is one that has stirred considerable controversy in the modern world.

GREECE

Born from the primordial chaos (water) was the Greek goddess *Gaia*, who represents the manifested world and the first pantheistic being to inhabit our planet after it was formed. Gaia (as the earth mother) and Uranus (as the sky) gave birth to *Oceanus*, who is generally considered the Greek personification of water. The god known as Oceanus is, of course, at the root of our English word "ocean," where most of the water on the earth is contained. The ancient Greeks believed that water was everywhere and also constituted everything, such that all things manifest in this world represented different transformations of water.

In addition to their pantheistic characters, the ancient Greeks were one of the first cultures to begin incorporating their intuitive or mythical views of water into a more philosophical and, from a modern perspective, comprehensible format. Though such early philosophies often echoed the message of ancient myths, they began to explain water's relationship to other aspects of the physical world. Losing little of its mystique as a symbol of the primordial chaos, water was associated with geometries, numbers, vibrations, and physical structures.

ELEMENTS AND ENERGIES

A sixth century B.C. philosopher, named Thales, hypothesized that water is the primary substance of all being, out of which everything is created and to which everything returns (6). He was perhaps the first to recognize water's unusual physical properties and its apparent universality, both as a solvent and as a so-called creational medium. During the fifth century B.C., a Hermetic philosopher, named Empedocles, expanded on Thales' notion by proposing that all matter in the universe is composed of differing combinations of four original substances and two moving forces. Empedocles referred to these four substances as the *elements* of fire, air, water, and earth, and he identified love and strife as the *moving forces*. The moving forces energized the combining or dissociating of elements, such that matter could not be created without love nor uncreated (destroyed) without its opposite moving force of strife.

In addition to the four elements, there is a fifth substance that, according to Greek philosopher Plato, was essential to creating the universe (7). This mysterious fifth substance was referred to as *aether*, a word coined by the Greeks to identify the substance filling all space beyond the earth. Similar to the four elements, aether is an integral component of most ancient beliefs, where it is generally associated with a life energy or vital force. The ancient Hindus also recognized the four elements and aether, or *akasha*, as fundamental components of the natural world.

THE PLATONIC SOLIDS

The four elements and aether were symbolized by three-dimensional geometries known as Platonic solids (see Fig. 1). These angular solids were believed to connect the etheric and worldly realms to the primordial waters of chaos (usually represented by a sphere). The solid that most often symbolizes water is the icosahedron, which represents the creative principle connecting the etheric and worldly realms. The reciprocal geometric relationship between the icosahedron (representing water) and the dodecahedron (representing aether) symbolized the intimate relationship between the two. Because matter or the material world (represented by a cube) was supposedly manifested from the aether via the water, it follows that the icosahedron and dodecahedron are mathematically related to the cube according to the golden ratio (as ϕ and $1/\phi$, respectively). The golden ratio ($\phi = 1.618\ldots$) was believed by some ancient cultures to represent the mathematical code underlying the creation of the manifested world.

BIBLICAL PERSPECTIVES

Water is frequently referenced in the Bible in a variety of contexts designed to communicate very different kinds of messages. Biblical scholars believe that the significance of water was heightened in a desert region (i.e., the Holy Land) where water was scarce and drought was a constant threat to life. Beyond water's more practical

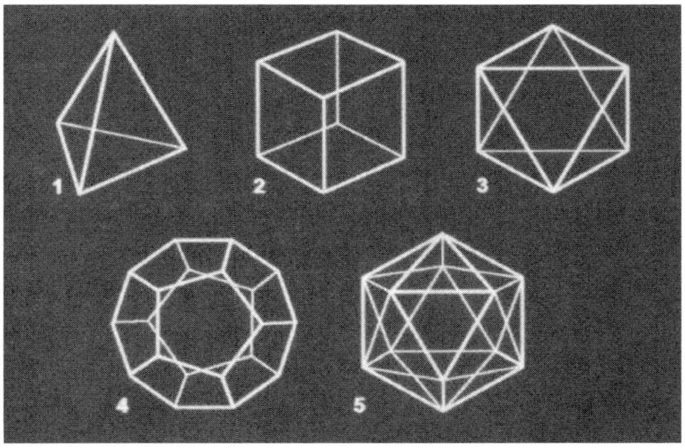

Figure 1. The five regular Platonic solids include (1) the tetrahedron symbolizing fire, (2) the cube symbolizing the earth, (3) the octahedron symbolizing air, (4) the dodecahedron symbolizing aether, and (5) the icosahedron symbolizing water.

considerations, the Bible portrays the spiritual aspects of water under several commonly recognized labels (8).

- *Cosmic water* describes a primordial force that only God can govern. The cosmic sea symbolizes the undifferentiated realm from which God created heaven and the earth.
- *Living water* symbolizes a source of life and a gift from God. Living water was considered distinct from ordinary water and symbolized both wisdom and knowledge.
- *Ceremonial water* symbolizes both the cleansing process and the passage between life and death. Ceremonial water is portrayed as an agent for removing or dissolving impurities.

If the biblical account of creation is consistent with that of other ancient accounts, heaven and earth are manifested from the symbolic, rather than literal, waters of which God is one. In this context, cosmic water is analogous to the waters of chaos rather than to literal water present in interstellar space (i.e., the modern scientific reference to cosmic water). Living water appears to be a designation for a special type of literal water, such that there is a recognizable difference between living and ordinary water. Living water is very rare, satisfies one's thirst, and also imparts life to the world. Most of the water used for religious ceremonies is ordinary water (sometimes symbolizing living water) that is blessed by an appropriate person.

ART AND NATURE

If one studies water in its natural state, it is difficult not to notice the variety of *flow forms* (e.g., whirlpools, eddies, ripples) that constantly appear and disappear. Many ancient cultures placed considerable import on these motions and believed that they held great powers for both creation and destruction. Renaissance artist Leonardo da Vinci is best known for his paintings and sculptures, but he was also a great student of water and its flow forms. He apparently believed that the vortical motion of fluids (particularly water) was a key to understanding and using the power of the universe. Two of the twentieth century's most renowned students of water flow forms were the European naturalists Theodor Schwenk and Viktor Schauberger, who expanded on ancient understandings and symbolisms of water.

According to Schauberger, water is the earth's blood and the essence of the *first substance*, whose boundaries and banks are the capillaries that guide it and in which it circulates (9). His view of water is similar to ancient perceptions inasmuch as water serves as the primordial substance of the universe and the fundamental component of both an organism's circulatory system and the planet's hydrologic cycle. Schwenk hypothesized that water's relationship to time was inherent in its rhythmical movement, which spanned cycles that ranged from seconds to years (10). He referred to different wave patterns and flow forms within a body of water as possessing distinct frequencies, overtones, and resonances—not unlike a musical instrument. Others have referred to this rhythm as "water's song," which often symbolizes the sound-shapes or music contained within the ripples and waves of water bodies.

POSTMODERN METAPHORS

The term "living water" is used today as a label or symbol for various aqueous concoctions in which ordinary water is treated with a combination of electrolytes, crystals, metals, colloids, and polymers, or exposed to various fields. The objective is to produce an optimal fluid that supports biological processes and, hence, retards both aging and disease. The exact mechanisms underlying a wide range of health claims have yet to be elucidated, but the modern practice of *water structuring* is based on the ancient understanding that water is able to retain or access a type of information that is colloquially referred to as "memory."

Perhaps the most recent water-related metaphor is that of *virtual water*, which was originally used to describe and, in some instances, quantify the influence of policy decisions on water resources in arid regions (11). Virtual water is a term used primarily in international trade for estimating the amount of water that is implicitly exchanged in goods and services (e.g., a ton of wheat requires 1000 tons of water) and the net gain for nations that import such goods and services. Although no actual water is traded during such transactions, the virtual water metaphor illustrates the extent to which real water is currently perceived as an economic commodity and is routinely transported across both geographic and political boundaries.

BIBLIOGRAPHY

1. Abraham, R. (1994). *Chaos, Gaia, Eros.* HarperCollins, San Francisco, CA, p. 126.
2. Andrews, T. (1998). *A Dictionary of Nature Myths.* Oxford University Press, New York, p. 36.

3. Mercatante, A. (1978). *Who's Who in Egyptian Mythology*. Clarkson N. Potter, New York, pp. 129–130.
4. McLeish, K. (1996). *Myth*. Facts on File, New York.
5. Brailsford, B. (1997). *Song of the Whale*. StonePrint Press, Hamilton, New Zealand, p. 174.
6. Lugt, H. (1983). *Vortex Flow in Nature and Technology*. John Wiley & Sons, New York, p. 6.
7. Lawlor, R. (1982). *Sacred Geometry*. Thames & Hudson, London, p. 96.
8. Ryken, L., Wilhoit, J., and Longman, T. (Eds.). (1998). *Dictionary of Biblical Imagery*. InterVarsity Press, Downers Grove, IL, pp. 929–932.
9. Coats, C. (1996). *Living Energies*. Gateway Books, Bath, UK, pp. 107–117.
10. Schwenk, T. (1965). *Sensitive Chaos*. Rudolf Steiner Press, London, pp. 18–20.
11. Wichelns, D. (2004). The policy relevance of virtual water can be enhanced by considering comparative advantages. *Agric. Water Manage.* **66**: 49–63.

GORDON AND FRANKLIN RIVERS AND THE TASMANIAN WILDERNESS WORLD HERITAGE AREA

ARTHUR M. HOLST
Philadelphia Water Department
Philadelphia, Pennsylvania

The Tasmanian Wilderness World Heritage Area (TWWHA) is one of the largest temperate wildernesses in the world. Covering more than 8500 square miles, or more than 20% of the island of Tasmania, it contains three state reserves, four national parks, four conservation areas, two protected archaeological sites, and three forest reserves. Mountain ranges, rivers, Australia's deepest caves, waterfalls, and many endangered species are also present, as well as three lakes: Lake Gordon, Lake Peddler, and Lake St. Clair. The TWWHA was adjudged worthy of environmental protection in 1982 by the World Heritage Committee.

There are many reasons that the Tasmanian Wilderness was chosen as a World Heritage Area. It contains the largest peatlands in the Southern Hemisphere. It has great geological significance; it contains rocks from all but one geological period, and its oldest rocks are from the Precambrian period, more than a billion years ago. Also, sandstones and siltstones contain fossils of early marine life, fossilized when Tasmania was still underwater (Fig. 1). Much of the Wilderness' mountain landscape is the result of three glacial periods that occurred during the Pleistocene Epoch.

One of TWWHA's attractions is the Cradle Mountain–Lake St. Clair National Park. Lake St. Clair, the source of the Derwent River, is Australia's deepest lake and is enclosed by mountains. The Park contains myrtle, pine, and beech trees. Another attraction is the Central Plateau and Walls of Jerusalem National Park. This plateau was formed by glaciers during the last Ice Age; the area's lakes and tarns were created when the Ice Age's glaciers melted.

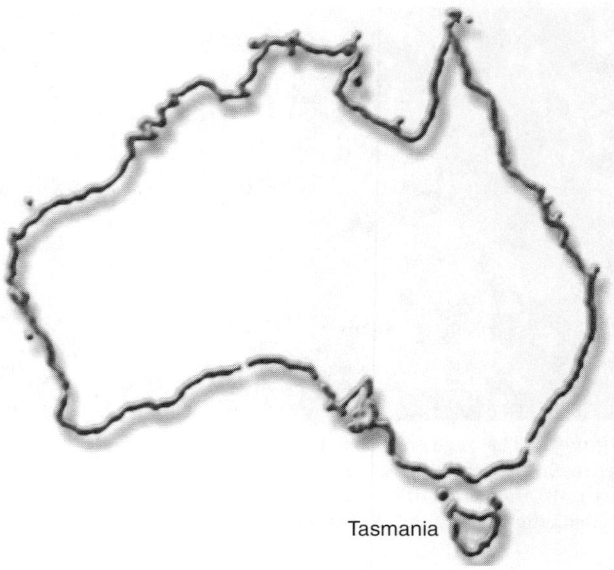

Figure 1. The island of Tasmania.

Many people visit the Tasmanian Wilderness World Heritage Area for its various activities. One of the most popular is known as bushwalking. This is basically hiking. There are advanced bushwalkers, and there are amateur bushwalkers, who take walks ranging from 15 minutes to an entire day. Advanced bushwalkers make longer and more arduous journeys over paths known as tracks. Tasmania's best-known track is called the Overland Track. About 80 km long, this track takes bushwalkers from 5–10 days to traverse. It features an alpine backdrop and passes the highest peak in Tasmania, Mount Ossa. This track is not exceptionally difficult; it is rated T1 by the Walking Track Management Strategy, the lowest rating, based on track hardening and track facilities.

Another favorite track of experienced bushwalkers is Frenchman's Cap. This track leads to the peak of the white quartzite Frenchman's Cap, the most prominent mountain in Franklin–Gordon Wild Rivers National Park. This track is considerably more difficult than the Overland Track; it has been assigned a T2 rating by the Walking Track Management Strategy. It usually takes bushwalkers 4 or 5 days to traverse this track.

A third popular track is the South Coast Track. It also has a T2 rating. This track has magnificent coastline views and spectacular beaches. It is located in Southwest National Park, and the main reason it is T2 is that the track is rather remote and far from any possible assistance. It is more than 80 km long and takes from 6–10 days to complete. It is often combined with the Port Davey Track to form a T3-grade track that takes from 10–16 days.

Another popular TWWHA activity is camping. Huts are provided along many common paths for general use, although tents are often used when the huts are full or where huts are unavailable.

The second most popular recreational activity in the TWWHA, after walking, is fishing. Local lakes, of which there are many, are often stocked with trout. These lakes, as well as rivers, are also sometimes used for boating, rafting, and kayaking. Some people opt to climb the peak of Frenchman's Cap. And the long, deep caves can be explored by experienced spelunkers. Marakoopa Cave is open to the general public.

Franklin–Gordon Rivers Wild Park (FGRWP), instated as a national park on May 13, 1981, is located in the southwest corner of Tasmania, between Southwest National Park, Cradle Mountain–Lake St. Clair National Park, and Walls of Jerusalem National Park. Not only the Franklin and Gordon Rivers, but also Jane River and Denison River are in FGRWP.

The Tasmanian World Wilderness Heritage Area is run by the Convention for the Protection of the World's Cultural and Natural Heritage, or World Heritage Convention. It was formed by UNESCO, the United Nations Educational, Scientific, and Cultural Organization, and it is the most extensive environmental convention in the world today. The area of the TWWHA is also world-renowned for its animal population. An unusually large number of species are endemic to Tasmania due to its remoteness from the rest of the world. Among these are the duck-billed platypus, the orange-bellied parrot, and the Tasmanian devil.

Along with all of these natural factors, the TWWHA possesses cultural significance as well. Thousands of years ago, Tasmania was detached from mainland Australia by Bass Strait, and the local Aborigines remained unaffected by outside influences until "discovered" by Abel Tasman in 1642. Many cave paintings have been found in Southwest Tasmania, including probably the southernmost cave paintings in the world.

The TWWHA's pristine quality is threatened by many factors. A major one is the soil erosion caused by bushwalkers hiking through the area. Many regulations are in place to attempt to lessen its impact on the environment. Another factor is cave degradation. Tasmania's many caves can be damaged easily by excessive visitors, so guides are available in many instances. Also, nonnative species have been introduced to the area, and they can have negative effects on local ecology. Fire can also devastate the environment; much of the TWWHA is designated fuel stove use only; the illegal use of campfires carries heavy fines.

There are many World Heritage Areas, but few can compare to the TWWHA in natural and cultural significance. It is one of only 13 World Heritage properties to satisfy all natural criteria. To this day, only one other World Heritage site has fulfilled as many criteria as the TWWHA.

READING LIST

Available: http://www.tasmaniatravel.com.au/misctext/hilite4.html.

Available: http://svc004.bne151p.server-web.com/spotlight/isncribed/taswilder.htm.

Available: http://www.dpiwe.tas.gov.au/inter.nsf/WebPages/UTAR-52SAD6?open.

Available: http://www.dpiwe.tas.gov.au/inter.nsf/Webpages/UTAR-52V8EK?open.

Available: http://www.dpiwe.tas.gov.au/inter.nsf/WebPages/UTAR-52W7QQ?open.

THE MISTAKE OF WAITING TO GET THIRSTY

F. Batmanghelidj
Global Health Solutions
Falls Church, Virginia

At the start of the twenty-first century, medicine is still muddled about the way the human body deals with a water shortage in its complex water-dependent inner environment. It is not yet clear even to physiologists of long standing in the field how local or general water shortages in the body manifest themselves. The following example is particularly dangerous: A respected scientist made an announcement that was quickly and widely publicized—unacceptable to me because its adoption will undoubtedly hurt people. As a means of airing the reasons why we need to prevent thirst rather that wait to get thirsty, I am responding here.

Heinz Valtin, M.D., an emeritus professor at Dartmouth Medical School, has ventured the opinion that there is no scientific merit in drinking eight 8-ounce glasses of water a day and not waiting to get thirsty before correcting dehydration. This view—published in the *American Journal of Physiology*, August 2002, and reiterated in the 2003 report of the Food and Nutrition Panel of the National Academy of Sciences—is the very foundation of all that is wrong with modern medicine, which is costing this nation $1.7 trillion a year (a figure rising at the rate of 12% annually). Dr. Valtin's view is as absurd as waiting for the final stages of a killer infection before giving a patient the appropriate antibiotics. His views are based on the erroneous inherited assumption that a dry mouth is an accurate sign of dehydration.

Dr. Valtin does not seem to be aware of an important paradigmatic shift in medicine. In the past, it was wrongly assumed that solutes in the body regulate all functions and the solvent has no direct role in any physiological functions. In medical schools, it is taught that water is only a solvent, a packing material and a means of transport and has no metabolic function of its own. I have come across this limited view at another Ivy League medical school from another eminent professor of physiology who, like Dr. Valtin, researched and taught the water-regulatory mechanisms of the kidney to medical students and doctors. Only when I asked him what "hydrolysis" is did the penny drop: He admitted that water is a nutrient and does indeed possess a dominant metabolic role in all physiological functions of the body.

Dr. Valtin's emphasis on the water-regulatory role of the kidneys limits his knowledge to the body's mechanisms of water deficit management. He seems to base his views of thirst management on the vital roles of vasopressin, the antidiuretic hormone, and the renin-angiotensin system, the elements that are engaged in the drought-management programs of the body when the body has already become dehydrated. He defines *dehydration* as a state of the body when it loses 5% of its water content. Yet he believes that we should wait until, at some such level of water loss, the urge to drink will correct the water deficit in the body. This view might have seemed plausible 25 years ago. Today, it exposes the tragic limitations of knowledge of human physiology available at a prestigious medical school in America.

Dr. Valtin does not take into consideration the fact that water is a nutrient. Its vital hydrolytic role would be lost to all the physiological functions affected by its shortage in its osmotically "free state." Another oversight is the fact that it is the interior of the cells of the body that would become drastically dehydrated. In dehydration, 66% of the water loss is from the interior of the cells, 26% of the loss is from extracellular fluid volume, and only 8% is borne by the blood tissue in the vascular system, which constricts within its network of capillaries and maintains the integrity of the circulation system.

Philippa M. Wiggin has shown that the mechanism that controls or brings about the effective function of cation pumps uses the energy-transforming property of water, the solvent: "The source of energy for cation transport or ATP synthesis lies in increases in chemical potentials with increasing hydration of small cations and polyphosphate anions in the highly structured interfacial aqueous phase of the two phosphorylated intermediates" (1). Waiting to get thirsty, when body fluids become concentrated, we lose the energy-generating properties of water in the dehydrated cells of the body. This is a major reason why we should prevent dehydration, rather than wait to correct it. This new understanding of the role of water in cation exchange is enough justification to let the body engage in prudent surplus water management rather than forcing it into drought and deficit water management, which is what Dr. Valtin is recommending.

In his research on "conformational change in biological macromolecules," Ephraim Katchalski-Katzir of the Weizmann Institute of Science has shown that the "proteins and enzymes of the body function more efficiently in solutions of lower viscosity"(2). Thus water loss from the interior of the cells would adversely affect their efficiency of function. This finding alone negates Dr. Valtin's view that we should let dehydration get established before drinking water. It is desirable that all cells of the body function efficiently within their physiological roles, so it would be more prudent to hydrate the body optimally rather than wait for the drought-management programs of the body to induce thirst. Furthermore, it is much easier for the body to deal with a slight surplus of water than to suffer from its shortfall and have to ration and allocate water to vital organs at the expense of less vital body functions. The circulation of concentrated blood in the vascular system is truly an invitation to catastrophe.

The tragedy of waiting to get thirsty hits home when it is realized that the sharpness of thirst perception is gradually lost as we get older. Phillips and associates have shown that even after 24 hours of water deprivation, the elderly still do not recognize that they are thirsty: "The important finding is that despite their obvious physiologic need, the elderly subjects were not markedly thirsty"(3). Bruce and associates have shown that, between the ages of 20 and 70, the ratio of water inside the cells to the amount of water outside the cells changes drastically—from 1.1 to 0.8.(4) Undoubtedly, this marked change in the intracellular water balance would not take place if the osmotic push and pull of life could favor water diffusion through the cell membranes everywhere in the body at the rate of 10^{-3} centimeters per second. Only by relying on the reverse osmotic process of expanding the extracellular water content of the body to filter and inject "load-free" water into vital cells by the actions of vasopressin and the renin-angiotensin-aldosterone systems—when the body physiology is constantly forced to rely on its drought-management programs—could such a drastic change in the water balance of the body result.

Two other scientific discoveries are disregarded when Dr. Valtin recommends that people should wait until they get thirsty before they drink water. One, the initiation of the thirst mechanisms is not triggered by vasopressin and the renin-angiotensin systems—these systems are involved only in water conservation and forced hydration of the cells. Thirst is initiated when the Na^+-K^+-ATPase pump is inadequately hydrated. It is water that generates voltage gradient by adequately hydrating the pump proteins in the neurotransmission systems of the body(1). This is the reason that brain tissue is 85% water(5) and cannot endure the level of "thirst-inducing" dehydration that is considered safe in the article published by Dr. Valtin.

Two, the missing piece of the scientific puzzle in the water-regulatory mechanisms of the body—which has been exposed since 1987, and Dr. Valtin and his colleagues need to know about it—is the coupled activity of the neurotransmitter histamine to the efficiency of the cation exchange, its role in initiating the drought-management programs, and its role in catabolic processes when the body is becoming more and more dehydrated(5). Based on the primary water-regulatory functions of histamine and the active role of water in all physiological and metabolic functions of the body—as the hydrolytic initiator of all solute functions—the symptoms of thirst are those produced by excess histamine activity and its subordinate mechanisms, which get engaged in the drought-management programs of the body. They include asthma, allergies, and the major pains of the body, such as heartburn, colitis pain, rheumatoid joint pain, back pain, migraine headaches, fibromyalgic pains, and even anginal pain(5,6). And because vasopressin and the renin-angiotensin-aldosterone activity in the body are subordinate to the activation of histamine, their role in raising blood pressure is a part of the drought-management programs of the body(6). The forced delivery of water into vital cells demands a greater injection pressure to counteract the direction of osmotic pull of water from inside the cells of the body, when it is dehydrated.

From the new perspective of my clinical and scientific research into the molecular physiology of dehydration and the peer-reviewed introduction of a paradigmatic shift in medical science recognizing histamine as a neurotransmitter in charge of the water regulation of the body, I can safely say that the 60 million Americans with hypertension, the 110 million with chronic pain, the

15 million with diabetes, the 17 million with asthma, the 50 million with allergies, and more, all did exactly as Dr. Valtin recommends. They all waited to get thirsty. Had they realized that water is a natural antihistamine(5,7,8) and a more effective diuretic, these people could have been saved the agony of their health problems.

For more information about the medical breakthrough on chronic unintentional dehydration and the diseases it causes, read as many of the following titles as you can. They are the products of more than 24 years of focused research on the topic of deep dehydration inside the cells of the body: *Your Body's Many Cries for Water; ABC of Asthma, Allergies and Lupus; Water for Health, for Healing, for Life; How to Deal with Back Pain and Rheumatoid Joint Pain; Water Cures: Drugs Kill*; and my 2004 book, *Obesity, Cancer, and Depression: The Deadly Diseases of Dehydration*. For more information on dehydration, visit www.watercure.com and www.nafhim.org.

BIBLIOGRAPHY

1. Wiggins, P.M. (1982). A Mechanism of ATP-driven cation pumps. In: *Biophysics of Water*. F. Franks and S.F. Mathis (Eds.). John Wiley & Sons, New York, pp. 266–269.
2. Katchalski-Katzir, E. (1984). Conformational changes in biological macromolecules. *Biorheology* **21**: 57–74.
3. Phillips, P.A., Rolls, B.J., Ledingham, J.G.G., Forsling, M.L., Morton, J.J., Crowe, M.J., and Wollner, L. (1985). Reduced thirst after water deprivation in healthy elderly men. *N. Engl. J. Med.* **311**(12): 753–759.
4. Bruce, A., Anderson, M., Arvidsson, B., and Isacksson, B. (1980). Body composition, predictions of normal body potassium, body water and body fat in adults on the basis of body height, body weight and age. *Scand. J. Clin. Lab. Invest.* **40** 461–473.
5. Batmanghelidj, F.M.D. (1987). Pain: A need for paradigm change. *Anticancer Res.* **7** (5B): 971–990. Available: http://www.watercure.com.
6. Batmanghelidj, F.M.D. *Your Body's Many Cries for Water*. Global Health Solutions.
7. Batmanghelidj, F.M.D. (1989). Neurotransmitter histamine: An alternative view. *3rd Interscience World Conf. Inflammation Analgesics Immunomodulators*, Monte Carlo, p. 37. Abstract and full article Available: http://www.watercure.com.
8. Batmanghelidj, F.M.D. *ABC of Asthma, Allergies and Lupus*. Global Health Solutions.

WATER AND WELL-BEING

David Haley
Picacadilly
Ulverston, Cumbria, United Kingdom

"Connected in water
Where the seas touch the sky
Tears bring forth oceans." (1)

If well-being is defined as "... the maximisation of the pleasure available to us" (2) and is related to health, safety, welfare, comfort, security, happiness, goodness, and contentment, then water must play an important part in maintaining this desirable state. For without water, none of these qualities of life can be achieved. Water connects all life and water connects all aspects of well-being, including social, economic, environmental, health, and aesthetic well-being, as a basic human right.

Article 25, Universal Declaration of Human Rights (1948) states: "Everyone has the right to a standard of living adequate for the health and well-being of himself and his family" (3). Kofi Annan, United Nations Secretary-General confirmed that "(a)ccess to safe water is a fundamental human need and therefore, a basic human right.

Contaminated water jeopardizes both the physical and social health of all people. It is an affront to human dignity" (4). But Dr. Gro Harlem Brundtland, Director-General World Health Organisation recognized that the right to water and well-being is not available to all: "Yes we have entered the new millennium with one of our most fundamental conditions of human development unmet: universal access to water. Of the world's 6 billion people, at least 1.1 billion lack access to safe drinking water. The lives of these people who are among the poorest on our planet are often devastated by this deprivation, which impedes the enjoyment of health and other human rights such as the right to food and adequate housing" (5).

NEEDS FOR WELL-BEING

Apart from the supply of safe drinking water and irrigation for food production, water is also essential to food preparation and personal hygiene. Environmental hygiene through adequate sanitation, safe housing, and water-based cultural practices are just as important to psychological well-being as they are to physical health. And water for the survival of ecosystems is critical to supporting biodiversity and also to maintaining aesthetic beauty. This, again is confirmed by the World Health Organisation: "Human livelihoods are intimately connected to natural life on the planet, and that, in turn, depends on water for survival. Natural ecosystems are often found in water or are heavily dependent on it. Without water, soil would parch, forests would wither, and species would die out" (6).

However, uncertainty—fear of the unknown—undermines well-being, particularly when it is related to an essential life-sustaining resource. As seen through history to the present day, uncertain access to adequate, safe water, provokes hostility and even war between communities competing for this vital element of survival. Likewise, the threat of flood has great physical and psychological impacts, threatening people's ability to house themselves and attachment to place.

CULTURAL DIVERSITY

"By means of water we give life to everything." (7)—Koran 21:30

Every culture has a particular and very important relationship with water. In Jewish, Christian, and Muslim religions, water plays a fundamental role in creation, sacred dwelling, blessings, affirmation of faith, and divine retribution. Vandana Shiva points out that "... the root of the Urdu word *abadi*, or human settlement, is *ab*, or water, reflecting the foundation of human settlements and civilization along water sources" (8).

From China, the Taoist yin/yang symbol depicts the dynamic balance and unity between opposite forces, representing our potential well-being with water and symbolizing the flow of life—water is by its very nature a dynamic element. However, too much or too little brings disaster, so our ability to adapt to change and the flow of water is a measure of our well-being. The ancient Chinese art of "geomancy," Feng Shui, translates into English as wind and water and pays particular attention to achieving well-being through the relationship with water. This important cultural concern influences the harmonious use of water within the home, office, restaurant, and other places of business because the water is believed to draw in money. Inappropriate use or placement of water could bring misfortune. Likewise, the beneficial design and positioning of buildings in relation to water is taken very seriously—supposedly, the tender for the construction of the Headquarters of the Hong Kong and Shanghi Bank was won by the architect, Sir Norman Foster, because his design incorporated Feng Shui principles of the building's relationship to the sea.

An old Chinese saying, "he who controls water, governs the empire," (9) refers to the emperor's ability to govern a contented population by controlling floods, rivers, and canals; a factor still understood, if contentiously applied in modern China, as plans for vast dam developments threaten the well-being of communities and ecosystems in return for hydroelectric power and cash crop irrigation. Like many societies, now, water is becoming valued as a commercial commodity, a finite resource that may be controlled and sold for private, industrial, agricultural, or leisure use. Reverence for water as a heavenly gift, spiritual source, or common resource seems to be diminishing.

In her book *Water Wars*, Vandana Shiva strongly maintains that "(w)ater is a common because it is the ecological basis of all life and because its sustainability and equitable allocation depend on cooperation among community members" (10). And she asserts that "(c)ommunity control avoids ecological breakdown and prevents social conflict. Over the centuries, indigenous water management systems have relied on ancient knowledge and evolved into complex systems that ensure the equitable distribution of water" (11). Water was, and still is recognized as a prime factor in well-being. This condition is beautifully expressed in the ancient Rig Veda hymn, *Water of Life:*

> Waters, you are the ones that bring us the life force.
> Help us to find nourishment,
> So that we may look upon great joy.
> Let us share in the most delicious sap that you have,
> As if you are loving mothers,
> Let us go straight to the house of the one,
> For whom you waters give us life and give us birth.
> For our well-being, let the goddesses be an aid to us,
> The waters be for us to drink.
> Let them cause well-being and health to flow over us.
> Mistresses of all the things that are chosen,
> Rulers over all peoples,
> The waters are the ones I beg for a cure.
> Waters—yield your cure as an armour for my body,
> So that I may see the sun for a long time.
> Waters—carry away all of this that has gone bad in me.
> Either that I have done in malicious deceit,
> Or whatever lie I have sworn to,
> I here sort the waters today.
> We have joined with their sap,
> Oh Agni, full of moisture,
> Come and flood me with splendour! (12)

THE FLOW OF LIFE—WATER AND WELL-BEING

Let us consider water and well-being as a flowing life system. "Water in one form or another circulates around us, all the time and everywhere" (13). Blockages, floods, and droughts may be thought of as disruptions to this flow. Evolution may be considered a process of morphing forms through structural diversity within an aquatic planetary system, rather than a linear progression. And climate change is one form of that evolutionary process. We may then see water as a dynamic major determinant in the structure, pattern, and process of evolution, life, and well-being. "The waters on the surface of the earth flow together wherever they can, as for example in the ocean, where all rivers come together" (14).

In his book, *Sensitive Chaos*: *The Creation of Flowing Forms in Water and Air*, Theodore Schwenk develops this idea:

> It is in the nature of water that it possesses those characteristics fundamental to all organisms for the maintenance of life; yet all is purely functional and without fixed form. These living qualities are nevertheless so clearly present and so evidently at work that the picture of the human organism is conjured up before us in three-fold nature—metabolic, nerve-sense and rhythmic organisations—revealing in harmony the basic characteristics of all living organisms.

And this notion is reinforced by architect philosopher, Christopher Alexander: "We came from the water; our bodies are largely water; and water plays a fundamental role in our psychology. We need constant access to water, all around us; and we cannot have it without reverence for water in all its forms." (15) Alexander continues: "Jung and the Jungian analysts take great bodies of water as representing the dreamer's unconscious. We even speculate, in light of the psychoanalytic evidence, that going into the water may bring a person closer to the unconscious processes in his life" (16).

And finally, marine biologist Commodore Jacques Cousteau expresses his love for water and the profound sense of well-being it can evokes:

> My body floated weightlessly through space, the water took possession of my skin, the clear outlines of marine creatures had something almost provocative, and economy of movement acquired moral significance. Gravity—I saw it in a flash—was

the original sin, committed by the first living beings who left the sea. Redemption could come only when we returned to the ocean as already the sea mammals have done. (17)

BIBLIOGRAPHY

1. Haley, D (2000). *Seeking an Unacceptable Profile: Species Nova [to look anew].* Performed at Between Nat. Int. Conf. Lancaster University, UK.
2. Lakoff, G. and Johnson, M. (1999). *Philosophy in the Flesh: The Embodied Mind and Its Challenge to Western Thought*. Basic Books, New York.
3. Article 25, Universal Declaration of Human Rights. (1948). *The Right to Water*. World Health Organization, France, p. 7.
4. Annan, K. (2003). *The Right to Water*. World Health Organization, France, p. 6.
5. Brutland, G.H. (2003). *The Right to Water*. World Health Organization, France, p. 3.
6. *The Right To Water*. (2003). World Health Organization, France, p. 19.
7. Koran 21:30. (2003). *The Right to Water*. World Health Organization, France, p. 3.
8. Shiva, V. (2002). *Water Wars: Privatization, Pollution and Profit*. Pluto Press, London, p. 20.
9. Rossbach, S. (1995). Feng Shui: An ancient theory of village siting. In: *The Village as Solar Ecology*. (East Falmouth, MA.: The New Alchemy Institute, 1980), *Feng Shui: Ancient Chinese wisdom on arranging a harmonious living environment*. Rider, London. p. 8.
10. Ibid., Shiva, V. (2002). p. 24.
11. Ibid., Shiva, V. (2002). p. 127.
12. Ibid., Shiva, V. (2002). p. VII.
13. Pielou, E.C. (1998). *Fresh Water*. University of Chicago Press, Chicago, IL, p. 246.
14. Trans. Wilhelm, R. (Eng. Trans. Baynes, C.F.). (1980). *The I Ching or Book of Changes*. Routledge & Kegan Paul, London, p. 35.
15. Alexander, C., Ishikawa, S., Silverstine, M., et al. (1977). *A Pattern Language: Towns, Buildings, Construction*. Oxford University Press, New York.
16. Ibid., Alexander, C., Ishikawa, S., Silverstine, M., et al. (1977). p. 324.
17. Schwenk, T. (1996). *Preface—Sensitive Chaos: The Creation of Flowing Forms in Water and Air*. J.Y. Cousteau (Ed.). Rudolf Steiner Press, London, p. 7.

FREE FLOWING WATER: A SOURCE OF WISDOM

MAGGIE LEE
Santa Fe, New Mexico

"Indeed it is because life is wisdom and water wisdom's element, that there can be such a thing as the water of life" (1).

In early August, while visiting gardens in the Boulder area of the Rocky Mountains, I discovered El Dorado Canyon, 10 minutes south. A brochure on the area relates. . .the tilted layers carry groundwater from the Rockies, down and eastward to a depth of 8,000′, before it is pushed to the surface as an artesian spring. Nearby, this tenacious spring feeds a 20 × 80′ pool. Swimming in this cool, turquoise, silky water, I moved with noticeable ease and buoyancy. . .refreshed in beneficent living water.

This article's title and beginning sentence, so eloquently expressed by Theodor Schwenk, continues with his experience of living water. "Water constantly pouring out of a spring in the middle of a flower-filled mountain meadow, sparkling in the pure cold air, with light reflected from the shimmering snowy peaks around." Generous, effervescent, living water.

Clearly, the beauty in the looking and the beauty in the seen, connected.

Sensing a similar mystery, David Abrams writes (when our perceptive practices reveal). . ."a reciprocal phenomenon organized as much by the surrounding worlds as by (the perceiver)" (2).

Recently I discovered a passage by Rudolf Steiner saying that springs are the eyes of the earth.

In his extraordinary, rich, and comprehensive lectures and writings of the early 1960s and into the 1970s, Theodor Schwenk (1910–1986) proposed a new 'water consciousness,' a transformative way of thinking about this universal element. He described and illustrated fluid circulation, movement, and rhythms of the earth's water cycle and resulting organic forms. His experiments demonstrated the rhythmical, metamorphic potential hidden within fluid processes. Included here are a few of his insights and discoveries.

Figure 1. Waterscapes p. 8 edited by Herbert Dreiseitl, Dieter Grau, and Karl H.C. Luduig, 2001, Birkhauser, Switzerland.

In *Water, The Element of Life*, Schwenk develops a picture.

> Water as mediator between the centric (earthly) and the universal forces, setting them in balance and interweaving them with its very substance, reveals the activity of the heavens on earth. It is the port of entry through which cosmic-peripheral forces pass into the earth realm, directing the shaping, 'formative forces', organizing and patterning organic forms (Fig. 1).
>
> Renouncing any form of its own, it becomes the creative matrix for form in everything else.
>
> Contemplating this vast living organism, Earth, one's attention is drawn again and again to the layer-structure of its great enveloping mantles and to the rhythms that play in and through them.
>
> A glance at the surface configuration of the earth reminds us that 71% of it is covered by water. This watery surface in its immense extension provides a plane of contact with the atmosphere. . .where it works as the great regulator in matters of climate, meteorological processes and their rhythms.

Water's ability to regulate the climate is due to its thermal properties, in its tremendous capacity for storing warmth. . .evident too, in the origin of ocean currents and biological processes.

> Water's ability to absorb gases allows life on continents to come together with deep seas and thus these seas become great regulators of the respiratory processes of earth. In autumn, the plant world withers away and no longer inhales carbon dioxide and in the breaking down processes of plant substance, in the disintegrating foliage, much carbon dioxide is released into the air and is absorbed by the seas. When, however, in spring, the plant world grows anew and forms the substance for its foliage out of the carbon dioxide in the air, the waters of the earth once again release some of the necessary carbon dioxide. By absorbing carbon dioxide, it increases its ability to dissolve solid substance.
>
> Almost every property of water shows itself, in its relationship to warmth, as maximally suited to support the life of earth and its inhabitants. In this vein, water is unequaled, as L.J. Henderson has shown; the temperature anomaly, as a result of which water reaches its greatest density at +4 °C and grows lighter again at the freezing point, is the reason solid ice does not sink, rather floats, thus keeping earth from becoming a totally lifeless block of ice (3).

"Water is an element that brings a state of balance everywhere. Rhythm is its life element and the more it can be active rhythmically, the more it remains alive in its innermost nature" (4).

Linking similarities between water and thinking, Schwenk offers,

> The capacity of water in the realm of substances to dissolve and bind together, reappears in thinking as a spiritual activity.
>
> Like water, thought can create forms, can unite and relate the forms to one another as ideas. We speak of a capacity to think fluently when someone is skillfully able to carry out creation of form in thought, harmoniously coordinating the stream of thoughts and progressing from one idea to another.
>
> In thinking there prevails the etheric life of the water forces; through water flows the wisdom of the universe. Is it not this wisdom itself which has created the element of water, a tool for its own activity.
>
> Everything in Nature forms one indivisible fabric woven of living interchange. An all encompassing world of life comes into being from the interplay of cosmic-peripheral forces, meteorological forces, forces of the elements, the earth and all its living organisms (5).

Viktor Schauberger was a pioneering Austrian forester and brilliant inventor (1885–1958). His inventions were linked to water supply, natural river regulation, agriculture, propulsion and energy generation. They were all based on his observations of an underlying principle governing all natural processes, implosion.

> Implosion is an inward, curving (centripetal) suctional motion which follows a spiraling, whirling path as a vortex. Modern technology is primarily based on explosion, that is the outward moving, centrifugal motion of heating, burning, pushing or exploding. Inefficient, most of the energy is lost during an explosive process through frictional resistance, which produces useless waste heat. A characteristic feature of the life-generating vortex, is that the outside moves slowly, while the inside moves rapidly. As water is imploded in a vortex, suspended particles, which are denser than water, are sucked into the center of flow, so the frictional resistance is reduced and the speed of flow increased (6).

In 1961, Theodor Schwenk joined with George Adams, a mathematician, who for decades, with the assistance of Olive Whicher, had been researching projective geometry connected with organic forms, to found the Institute of Flow Sciences in the Black Forest in southern Germany. Just then his classic book on water, *Sensitive Chaos*, was published. Describing his experiences in seas worldwide, Jacques Cousteau wrote in the Preface,. . . "all around us there arose from the living sea a hymn to the 'sensitive chaos.' These memories have now taken for me a new meaning suggested by the book of Theodor Schwenk" (Fig. 2).

John Wilkes, a sculptor, had also joined the staff and was responsible for technical production of geometric models. In 1970, Wilkes, now at Emerson College in Sussex, England, began developing the Flowform Method and founded the Flow Design Research Group, currently based at the Virbela Rhythm Research Institute at Emerson College.

Concerned with enhancing water's quality, he posed his initial question, "can we create a form for water that will enable it to manifest its potential for metamorphosis and order?"

> This was inspired by observing controlled experiments revealing paths of vortices in contrast to the complexity of water movements in outer nature. Streaming water generates asymmetrical forms, but studying the results of living processes, it became clear that these forms tend strongly toward symmetry. Such forms have to do with thrust processes. Wilkes created simple symmetrical channels with mirrored meandering walls and changing proportions. He was looking at symmetry in this moment as of a higher order, with the idea of lifting water towards the living. The fortuitous event

Figure 2. Tentacles of the sea curl inland, eroding the Dutch coastline along the Waddenzee. In this view from 14,000′, a very low tide has revealed pale plains of mud and sand, giving the tidelands the look of a river delta. The darker veins are channels cut deep by the land-stealing sea; when the tide is high, they serve fisherman as navigation lanes (5).

of proportion variation led to pulsing process in one particular place. This opened up for him a wide spectrum of investigation which led to the Flowform Method.

Water's capacity to support life is actually related to its ability to move within itself...to generate membranes (surfaces) within its own volume. These form sensitive 'organs' through which water mediates between individual organisms and its total surroundings, carrying information. Movement, especially vortical movement, sensitizes water to this task (Fig. 3).

The Flowform is a vessel, generally ovoid in shape, with a narrow entrance and exit, which create a resistance, inducing an oscillating figure of eight (lemniscate) movement. Flowing into alternate sides, whorls or vortices form, folding the water upon itself, oxygenating, enlivening the inner mobility of its many surfaces, the basis of its vitality, its freshness. Oxygenation fosters biological processes necessary to effect purification. The frequency of the rhythm varies with the size of Flowform, while the character of movement depends on the shape (7).

Special restorative value derives from water features designed in water's own terms...shaped surfaces caressed intimately by flowing, pulsing water.

Since 1973, in England and in collaboration with associates worldwide, well over 1000 Flowform installations have spanned more than 30 countries.

Also at that time, Arne Klingborg, director of the Rudolf Steiner Seminariet in Jarna, Sweden, invited Wilkes to test the beautifying and ecological possibilities of Flowforms in a lagoon purification system. Three

Figure 3. Such meander channels can be found below glaciers where streams have caused erosion over thousands of years. Photo supplied by John Wilkes.

Flowform cascades are operating through seven ponds with progressive wetland ecosystems from an algae pool to reed bed. Wilkes tells us, "not only is the coliform bacterial concentration in the treated water significantly below official levels permissible for public swimming, but the entire site serves as a community park and bird sanctuary" (Fig. 4).

Other applications of Flowforms include water purification, desalinated water treatment, stored-water treatment, swimming pools; air-conditioning and humidity; agricultural research centers, farms, homes, public gardens, parks and offices; aquaculture systems such as fish ladders for seasonally migrating species; pharmaceutics; and therapeutic and medicinal application.

In 1997, Flowforms America was founded. Its CEO, Sven Schunemann, worked with John Wilkes for a short period before this. Sven relates..."When water's movement is optimized for circularity, the liquid itself is able to be more effective as a natural element. Naturally flowing water is better aerated as more surface area is exposed."

John Wilkes added,

Figure 4. Malmö Flowform installed by a mountain stream at Sundet, Mösvatn, Norway. Water from the stream higher up is directed through the vessels which generate an ordered rhythmic swinging movement, like a living pulse, in contrast to the multitude of rhythmic movements in the streambed. If mounted below a hydroelectric turbine at a much larger scale, cast in situ, these more life-related rhythms, would improve the vital qualities of the water for irrigation, in a shorter distance! (11).

> It is necessary for Flowform creators to learn, in working with water's circular movements, to find the very special proportions which will finally generate pulsing processes.
> "Naturally flowing water is also more bioavailable...by allowing the nutrients it carries to be accessed more easily by plants. A plant, in turn, plays its part by taking nutrients out of such circulating water; thus fulfilling two roles, that of cleansing the water and of feeding itself" (8).

Wolfram Schwenk, a hydrobiologist and leading member of the Institute of Flow Sciences, writes about water and the creative process.

> If a creative artist wants to bring about cosmic and organic Flowforms in water, this cannot be done in the same way as a sculpture would be created. They can be evoked by handling water as an instrument. The creative process then takes place in the water itself. The artist learns how water twists and turns as it flows, eddies, trickles and spurts, surges, rests and reflects; enhancing and emphasizing by creative design, the theme develops as a joint work of art. Inspired, others may discover and experience special features of water. I know of no better, more attractive and sustainable ecological study course than this (9).

From his website, Herbert Dreiseitl informs, when referring to technology and infrastructure, of the lack of respect for the intrinsic aesthetics of water. He is actively participating in solving this dilemma.

During some months of 1978, he came to study the Flowform process with John Wilkes. In 1980, he established Atelier Dreiseitl in Uberlingen on Lake Constance in southern Germany. In his recently published book, *Waterscapes* (10), he writes, "to do justice to water, we have to go into the waterworld ourselves and experiment with it and learn to think in an integrated and interdisciplinary manner, about its flow and flexibility."

With regard to integrated planning, he states,

> water always creates a relationship between the detail and the whole. Each individual drop contributes to the balance of the earth's climate. Water projects become valuable when they help this process and can show that the place is being addressed and how it is connected with the world around it (10).

Further, he brings attention " to the special inner flexibility necessary to work with water successfully. The designs improve gradually and start to carry the signature of the water itself" (10).

Over the years, Atelier Dreiseitl has had opportunity through the scope of its commissions, which range from open-space planning in parks, water playgrounds, town squares to housing developments and urban hydrology, of linking water art to other complex themes, such as rainwater and surface water management, restoration and creation of waterbodies, incorporating Flowforms, storage and circulation cisterns, light and sound technologies, inlaid waterworks, and air-conditioning systems. Herbert relates, "demonstrating sustainable environmental technologies succeeds when everyone involved in the planning process, really does use interdisciplinary practices" (10) (Fig. 5).

Again, from his beautifully presented website, "water resources can only become sustainable when social values grant water the necessary priority. Water deserves everyones awareness" (10).

John Wilkes' new book, *Flowforms, The Rhythmic Power of Water*, is now available through Steiner Books (11).

Through our inventive, progressive and collaborative efforts, demonstrated and practiced with care, in our art and technology, may we allow water to be vigorous and free to the depths of its innermost nature.

> Choose to be changed. Oh be infused with the rapture of a fire that shows not what's changing in it as it burns for the

Figure 5. A water playground by the river—here everyone can join in. Water can be raised, played with and experienced! Water playground in Pforzheim, Germany (from 10).

directing spirit of the earth entire loves the figure of flight, not so much as the point where it turns

Sonnets to Orpheus II, 12 by Rilke
John Lash, translation after Walter Kaufman (12).

Maggie Lee
Santa Fe, New Mexico
January 28, 2004

BIBLIOGRAPHY

1. Schwenk, T. and Schwenk, W. (1989). *Water, The Element of Life*. Anthroposophic Press, New York.
2. Abrams, D. (1985). The perceptual implications of Gaia. *The Ecologist* **15**(3).
3. Henderson, L.J. (1914). *Die Umwelt des Lebens, Wiesbaden*. Cited in Water, The Element of Life, Schwenk.
4. Schwenk, Water, The Element of Life.
5. Schwenk, T. (1965). *Sensitive Chaos*. Rudolf Steiner Press. London, UK.
6. Coates, C. (1966). *Living Energies*. Gateway Books, UK Also see Alexandersson, O. (1996). *Living Water*. Gateway Books, UK, Bath.
7. Wilkes, J. Virbela Rhythm Research Institute, jwilkes 30@aol.com.
8. Schunemann, S. Available: http://www.flowformsamerica.com
9. Schwenk, W. (2001). Water as an open space. *Waterscapes*. H. Dreiseitl (Ed.). Birkhauser, Basel, Switzerland, pp. 106–107.
10. Dreiseitl, H. (2001). *Waterscapes*. Birkhauser, pp. 40–41.
11. Available: http://www.steinerbooks.org.
12. Website for John Lash is www.metahistory.org.
13. Baumgartner, W. (1989). *The Artesian Spring*. Energy Unlimited, Magdalena, NM.
14. Knight, D. and Stromber, J. (1998). *Nature's Miracles*. Centre for Implosion Research. Plymouth, UK www.ImplosionResearch.com. . .as advised by Dr. Hal Williams, Anthroposophic Physician, October 2002.

Sidebars

1. Why does water rise to such great heights in mountain springs or artesian wells?
   Artesian springs come bubbling up vertically out of the ground. The source of the water comes welling up in pulses from the center of a column. . .this movement is similar to a telescopic extension.
   We can observe that in every stream, the water at the edges flow more slowly than in the middle, where the water is colder. . .faster layers flow past slower layers. . .building layer upon layer and creating extensive inner surfaces. These layers have different temperatures and densities. The inside layer is most dense, coolest and has the highest electrical potential, resulting in increased centripetal, spiraling motion. The outside layers warm up and expand in that warming up.
   Within a closed channel, i.e., underground, the outer layers exert pressure on the inside layers. . .tension builds from the temperature difference of the layers, disturbing the equilibrium inviting water to move! Just as a plum seed is shot forward from its fruit if you squeeze it. . .the denser body shoots forward as relief from pressure. Water always seeks the path of least resistance (adapted from Reference 13).
2. The vortex with its different speeds, is closely akin to the great movements of the planetary system. A given planet circles round the Sun as though in a vortex, in as much as it moves faster when near the Sun and slowly when further away. The vortex in its law of movement is thus a miniature image of the solar system and its planets.
   Also, if a very small floating object with a fixed pointer is allowed to circle in a vortex, the pointer always points in the direction in which it was originally placed, that is, it always remains parallel to itself!. . .it is always directed to the same point at infinity. This illustrates how a vortex is oriented, as though by invisible threads, with respect to the entire firmament of fixed stars (5, pp. 44–45).
3. Water as an information carrier.
   Various researchers such as Professor Benveniste, Dr. W. Ludwig (Treven and Talkenhammer, ed., Umweltmedizin Move-Verlag-Idstein, 1991), and Professor David Schweitzer have provided clear proof that water acts as a liquid tape recorder and is able to receive, store, and transmit electromagnetic vibrations. Because water molecules have positive and negative poles, they behave like little magnets. They attach themselves to their neighboring molecules and form clusters of several hundred molecules. These clusters are very sensitive structures and are impressionable by vibrational influences. This is what give water the ability to store information.
   This is closely linked to homeopathy, where a substance is diluted so many times, that eventually there is no molecule of the original substance left. . .yet it still has an effect. Homeopathy works because of the cluster ability to store vibrational imprints.
   Every substance and element has its own individual vibrational pattern, a bit like an energy blueprint. If you carry out the homeopathic process of diluting and succussing, the vibrational pattern of the remedy material becomes locked into the cluster structure of the carrier water. When you take this homeopathically prepared remedy, the cluster structure is transferred into the body and you react to the vibrational pattern of the original substance from which the remedy was prepared (14).

THE MEDICINAL PROPERTIES OF THE WATERS OF SARATOGA SPRINGS

A. Trent Millet
Newark, Vermont

My personal experience with American waters led me to investigate the widespread use of water for medical purposes in Europe, the United States, and other countries using Western medical systems similar or identical to our own.

It is important to note initially, however, that a number of Europe's great medical universities were established on or around waters that for centuries have been known to have medicinal properties. The European health systems and insurance companies have a long-standing practice of sending their clientele to spa for three weeks a year, both prescriptively as a treatment and as a preventive medium. The most commonly stated reason, throughout countries such as France, Poland, Germany, Czechoslovakia, the former USSR, and Hungary, is an expected ten to fifteen year increase in vital life expectancy: "vital" meaning not only the possible increase of number of years alive, but also, most importantly, the number of years individuals continue to be active and contributing members of society.

Czechoslovakia, Hungary, and Germany as of the late 1950s and early 1960s all had over 300 state-run spas. The former USSR had over 3500 (1). At that time in Europe alone, over 3,700,000 people per year were attending spas.

Ninety percent of patients sent to Bad Neuenahr are directed there by their own doctors (Ref. 1, p. 506) 50% of patients there suffering from diabetes. Health insurance companies, physicians representing those companies, social security agencies, unions, and corporations and municipal governments refer 50% of those sent to Bad Hall (Ref. 1, p. 507). "Thus, the state insurance fund of Hungary declared that its computed outlay for therapy and compensation in the rheumatoid group of diseases diminished 23% percent by the use of spa therapy" (Ref. 1, p. 507).

Of 2 million people sent to German spas by insurance companies, 25% suffered from rheumatism, 21% from diseases of the heart and blood vessels, 21% from respiratory diseases, 12% from digestive track disorders, and 8% from conditions causing exhaustion (Ref. 1, p. 507). Similar statistics continue in present-day Europe (A. Trent Miller, personal observation).

I began investigating the reported healing properties and medical use of American waters in 1998. A large number of American waters were used medicinally up until the 1960s. I concentrated my research around the Saratoga, New York area due to the well cataloged medicinal and historic use of Saratoga's many waters.

The anecdotal evidence, personal interviews, and personal experience lie outside the scope of this work. We are concerned here with patient statistics, prescriptive use, and the qualitative medical properties of Saratoga's spa waters.

People have been coming to use the waters for health since the 1700s. The 300-year use of Saratoga's waters for healing reached its pinnacle with the building of Sartoga's Roosevelt Spa by President Franklin D. Roosevelt in the 1930s. The famous Lincoln and Washington Baths had already been operating for almost half a century with 107,299 treatments given by the two combined baths in 1927 (2). By 1930, Lincoln Bath had the ability to give 4500 treatments a day.

Roosevelt Spa and the surrounding Saratoga State Park contained bathhouses, a drinking hall, inhalation rooms, and a bottling plant. Saratoga had always been known for its multitude of drinking halls, several bottling plants, and large number of bathhouses. Due to the fact that Roosevelt Spa was under the direction of three eminent doctors, Simon Baruch, MD, Walter S. McClellan, MD, and Oskar Baudisch, Ph.D., who directed the Research Lab, a great deal of the medical evidence, studies, and resulting statistics exist today. Hence my focus on Roosevelt Spa. By the 1930s only 18 of Saratoga's original 145 waters remained available for medicinal and public use. (About 127 springs were shut down because of the exhaustive pumping of carbon dioxide for use in the increasingly popular soda fountains, lowering the water level of the entire area.) The spa waters were commonly used in conjunction with other treatments, such as electrotherapy, ultraviolet and interred therapy, massage, and standard allopathic medicines. Because of this, statistics pertaining to the results of water treatments alone are difficult to find. Of the comprehensive number of waters, we will be looking at the treatment rationale of several. Table 1 defines the specific, subtle, and very individual ingredients contained in 17 of the 70 plus waters once available (3).

In consideration of this, the following is an example of how doctors used the waters prescriptively. "The choice of springs should be guided by the nature of the local manifestations, and the preponderance of constitutional plethora or anemia. For the uncomplicated dyscrasia, notably alkaline waters, with comparatively slight saline admixture—such as the Kissengen and Vichy—may answer well. With abdominal torpidity, a greater proportion of chloride of sodium is indicated, and then the Patterson, Hathorn, Congress or Excelsior may be recommended; or, if plethoric engorgement, is still more conspicuous, the powerfully saline Carlsbad, Champion or Geyser. In atonic cases, with well marked debility, often resulting from repeated attacks, or too prolonged depressing treatment, the saline-chalybeates may prove most beneficial."

Of the springs mentioned above, only Hathorn, Congress, and Geyser still flow. The spring called "Old Red," which is an iron water, has been used for all kinds of eruptive skin diseases and inflammation of the eyes. Dr. Oskar Baudisch, an authority on magnetochemistry, worked on the iron molecule. "In the course of his research he investigated the iron contained in Saratoga mineral water. He demonstrated that the iron contained in the water in spite of the relatively small amount in solution exceeded much larger amounts in therapeutic value as a remedy for anemia. This effect he showed to be due to the fact that the peculiar internal structure of the specific iron molecule contained in the water was similar to that of the iron molecule contained in the hemoglobin of blood. For this reason the iron in the water can be readily assimilated in the body and is, therefore, more readily available to the body than are other forms of iron" (Ref. 2, p. 37).

In the case of Old Red mentioned above, the water also contains two human cell salts (2). It is explained by Dr. Grace Swagger in a video interview contained in "Spa City, The History of Saratoga Spas," that the iron and cell salts were originally dissolved from the flesh of dinosaurs, animals, and plants. Kenneth Cohen, in his book *Honoring the Medicine* (4), backs this up with a statement that "the

Table 1. Analyses of the Waters of the Saratoga Spa Ions and Radicals as Presented in Solution

Element or Radical	Geyser, ppm	Polaris, ppm	Hayes, ppm	Orenda, ppm	Hathorn # 3, ppm	Lincoln, ppm	State Seal, ppm
Sodium	850	560	3025	2420	3,820	1,150	2.0
Potassium	83	80	333	266	340	219	0.15
Lithium	2.1	5	11.9	8	10	6.3	0.01
Calcium	375	370	724	672	872	348	32.2
Magnesium	75	95	277	224	353	171	4.4
Barium	12.5	2.1	12	15.6	25	8.1	0.038
Strontium	1.0	—	10.5	10	12	9.9	0.11
Ammonium	12.0	1.2	13	13	14	6	Trace
Iron	4.0	1.1	1.8	1.7	2.2	8.2	0.40
Aluminum	4.1	6	9.0	8	4.0	3	0.034
Manganese	Trace	—	0.3	0.3	0.4	None	0.002
Bicarbonate	2,562	2130	4,550	3,600	4,850	2,608	104
Chloride	820	1000	4,500	3,800	6,030	1,538	4.8
Bromide	16	5	51	46	53	34	None
Iodide	0.9	—	2.1	2	2.9	1.3	None
Metaborate	6.0	0.9	7.0	7.5	7.0	4.1	0.25
Silica	12	17	11	11	12	51	5.0
Sulfate	None	None	None	None	None	None	22
Total solids dried at 110 °C	4,836	3,260	13,539	11,105	16,407	6,166	175
Radium 226, PCi/L		102	284	232	430	48	

same water that flowed through the bodies of the dinosaurs has been recycled for millions of years to flow through our own bodies today."

Hathorn Spring, a saline–alkaline carbon dioxide charged water (which has a taste similar to the commercial product Alka Seltzer) has been used for stomach and digestive problems for over a century. Hathorn water was listed as one of the four most important medicinal waters bottled in Saratoga. Apparently, the very specific type of minerals and their very refined particle size and frequency allow absorption into the system regardless of the patient's age. Hence, having the proper mineralization to interact with vitamins allows continued enzyme production throughout a person's life. The common belief that people approaching a half century or more cease to make digestive enzymes is not the case when ingesting this type of water.

One of the waters considered the most important is the carbon dioxide water. The CO_2 water of the Lincoln group of wells at Saratoga is considered to have the highest content of carbon dioxide in the world. Its most common use is in the treatment of heart-related ailments. "The natural CO_2 is considered to have special merit in the treatment of the cardiac patient because of its unique ability to cause vasodilatation of the cutaneous blood vessels without any increase in the work of the heart, such as would occur in a warm bath causing an equal vasodilatation. On the contrary, the CO_2 bath reduces the work of the heart. Though administered at a neutral temperature, the CO_2 bath gives the sensation of warmth because of the layer of insulating gas bubbles next to the skin. It has been found that the carbonic acid dissolved in a natural CO_2 bath diffuses through the skin at a rate of about 30 mL per minute. The hyperemia is due to released histamine and acetylcholine. As a result, the pressure in the afferent capillary loop is increased about four times; the venous capillary pressure increases; the deep venous plexuses of the skin are emptied; and the increased vascularity of the skin drains the splanchnic pool. The diastolic pressure and the heart rate are lowered. In addition, the hydrostatic pressure exerted by the bath improves the venous return flow through compression of the lymph channels and veins (5).

Although Saratoga utilized the CO_2 waters for heart ailments, it was largely in conjunction with other medical treatments. There are no records detailing patient results from just the CO_2 water treatments alone. However, Tables 2–7 show the use of the water by disease classification and the patient results, illustrating the effectiveness in diseases other than the heart.

Table 3 shows the distribution of all the patients in this group on the basis of the usual classification of rheumatic diseases. The largest group was diagnosed as osteoarthritis in its various forms and included 438 of the patients treated. Rheumatoid arthritis was second on the list with a total of 331 patients, nearly all of whom were classified as chronic. The remainder of the patients represented smaller groups as indicated in the table. From the study of this table it is clear that the two large groups of patients with osteoarthritis and rheumatoid arthritis accounted for a large majority of the patients.

The results of treatment recorded by the hospital physicians on discharge have been tabulated for each of the groups of rheumatic disease in Table 4. They show that 262 patients or 26.3% were slightly improved, 568 patients or 57.2% were moderately improved, and 86 patients or 8.7% were markedly improved. This gives a total improvement of 92.2%. Only 77 patients or 7.8% were unimproved or were transferred for other treatment. No deaths occurred in the series.

It is obvious immediately that any statistical analysis dealing with the number of patients treated in groups 3 to 8 would have little value. Because of the larger number of

Table 2. Total Number of Patients by Disease Classification

Primary Condition	Percent	July 1932 to June 1942	July 1942 to March 1952	Total
Heart and circulatory disorders including variations of blood pressure	30.8%	38,500	47,000	85,500
Rheumatic conditions including arthritis, myositis, fibrositis, and neuritis	23.7%	29,625	36,166	65,791
Gastrointestinal ailments including liver and gallbladder	17.5%	22,000	26,858	48,858
Nervous conditions including both functional and organic disorders	8.4%	10,500	12,818	23,318
Metabolic diseases including diabetes, obesity, and glandular disorders	4.1%	5,125	6,257	11,382
Skin disorders (noninfectious)	2.1%	2,625	3,255	5,880
Miscellaneous	3.2%	4,000	4,883	8,883
No disease including general debility	10.1%	12,625	15,428	28,053
Total	*100%*	*125,000*	*152,600*	*277,600*

Compiled by Dr. McClellan.

Table 3. Results of Treatment of 331 Patients with Rheumatoid Arthritis at the Saratoga Springs Veteran's Hospital

Number of Baths	Sight Improvement	Percent	Moderate Improvement	Percent	Marked Improvement	Percent	Unimproved	Percent
1–10	8	44.6	3	16.6	0	1	7	38.6
11–21	22	27.9	53	63.4	3	3.8	4	4.9
22–42	40	23	117	67.3	14	8.1	3	1.6
Over 42	11	19.3	40	70.2	0	0	6	10.5
Totals[a]	*81*	*24.5*	*213*	*64.3*	*17*	*5.1*	*20*	*6.1*

[a]Total percent of improvement is 93.9%.

Table 4. Results of Treatment

Type of Arthritis	Total Number	Slight Improvement	Percent	Moderate Improvement	Percent	Marked Improvement	Percent	Unimproved	Percent
1. Osteoarthritis	438	101	23.4	260	59.3	38	8.6	39	8.7
2. Rheumatoid arthritis	331	81	24.5	213	64.3	17	5.1	20	6.1
3. Traumatic arthritis	94	38	40.4	31	33.0	19	20.2	6	6.4
4. Mixed (rheumatoid and osteoarthritis)	41	12	29.0	21	51.2	4	9.9	4	9.9
5. Rheumatoid spondylitis	36	10	27.8	2.0	55.6	4	11.1	2	5.5
6. Specific infectious (Gc, Tb, etc.)	31	10	32.3	17	54.8	1	3.2	3	9.7
7. Gout	15	7	46.7	4	26.6	3	20.0	1	6.7
8. Unknown	7	3	42.8	2	28.6	0	0	2	28.6
Total	*993*	*262*	*26.3*	*568*	*57.2*	*86*	*8.7*	*77*	*7.8*

Compiled by Dr. McClellen and Dr. Comstock for Lincoln Water (CO_2), Saratoga, New York, July 1932 to March 1952.

Table 5. Osteoarthritis Results for a Total of 438 Patients

Number of Baths	Slight Improvement	Percent	Moderate Improvement	Percent	Marked Improvement	Percent	Unimproved	Percent
1–10	4	12.5	16	50	4	12.5	8	25.0
11–21	35	27.2	65	52	12	10.2	12	10.2
22–42	48	20.6	149	63	19	8.1	17	7.6
Over 42	14	23.5	30	61	3	6.1	2	4.2
Total[a]	*101*	*23.4*	*260*	*59*	*38*	*8.6*	*39*	*8.7*

[a]Total percent of improvement is 93.1%.

patients with osteoarthritis and with rheumatoid arthritis, a more detailed study was made.

In another series of observations, McClellan and Comstock found that radon added to the bath was also absorbed through the skin, both from baths of plain water and, to a greater degree, from baths in the naturally carbonated mineral water. Both carbon dioxide and radon are important substances in many spas about the world. The demonstration that they are actually absorbed from the water of the bath supports the conclusion that the absorbed carbon dioxide or radon plays a role in the physiologic changes observed.

Table 5 presents the results of treatment for patients with osteoarthritis analyzed in relation to the total amount of treatment received. The total amount of treatment was gauged by the number of mineral water baths, because for practically all patients it represented the major feature of the program. Even with 10 baths or less, some patients showed improvement, but the proportion of those who were not improved was much higher than in the other groups. When the patients received 11–21 baths, there was some increase in the percentage of patients who obtained benefit. When the baths were continued longer, over 21 in number, there was considerable increase in those who showed moderate improvement. When the treatment program extended beyond what is usually considered two average courses of treatment, little additional benefit was evident.

It is interesting to note that the variation of improvement was essentially in keeping with the figures, noted in the larger groups of patients. With the variety of conditions in the remaining 20% of the patients, it was impossible to apply this detailed analysis because of the relatively small numbers.

Moving southwest 45 miles west of Albany is Sharon Springs, New York. Although now open only a few weeks per year, at one time the sulfur magnesium and calcium waters were waters utilized in over 82,000 treatments a year during the 12-week season. After WWII and up to the present day, many of the victims of the German concentration camps, predominantly Hasidic Jews, and currently the Russian victims of similar camps use the water of Sharon Springs for recovery and rejuvenation. Most are paid for by the medical services of their countries of origin. The sulfur water is particularly beneficial for rheumatoid and osteoarthritis, skin problems, and general revitalization of the body.

Table 6. Rheumatoid Spondylitis

Number of Baths	Number of Patients	Improvement			
		Slight	Moderate	Marked	None
1–10	2	1	1	0	0
11–21	13	2	9	1	1
22–42	12	4	6	2	0
Over 42	9	3	4	1	1
Total	*36*	*10*	*20*	*4*	*2*

Table 7. Gout

Number of Baths	Number of Patients	Improvement			
		Slight	Moderate	Marked	None
1–10	0	0	0	0	0
11–21	2	1	1	0	0
22–42	7	2	2	3	0
Over 42	1	1	0	0	0
Total	*10*	*4*	*3*	*3*	*0*
No bath	5	3	1	0	1
Total	*15*				

It is common to find calcium and magnesium waters in close proximity to the sulfur springs. The calcium–magnesium balance is one of the most important balances in the body, along with the sodium chloride and acid alkalinity (pH, also adjustable with the use of water). The type of magnesium and calcium found in the water is ideally suited to the human body. Several qualified sources (6) have stated that the use of natural calcium and magnesium water maintains that ratio at ideal levels.

In closing, although this entry is certainly not a complete and comprehensive discussion of all the medicinal uses and statistics, it is an example of water's inherent natural medicinal qualities.

BIBLIOGRAPHY

1. American Congress of Physical Medicine and Rehabilitation. (1960). *Archives of Physical Medicine and Rehabilitation*. Vol. 41, Nov., pp. 504–513.
2. Swanner, G.M. (1988). *Saratoga Queen of Spas*. North Country Books.
3. Irwin, J.A. (1892). *Hydrotherapy at Saratoga*. Cassell Publishing, New York.
4. Cohen, K. (2003). *Honoring the Medicine*. Ballentine Books, New York.
5. European spas, the medicinal properties of the water. (1960). *Arch. Phys. Med. Rehab.* **41**: 9510.
6. Lithium Report.

A HISTORY OF HAWAIIAN FRESHWATER RESOURCES

M. Eric Benbow
Mollie D. McIntosh
Michigan State University
East Lansing, Michigan

Albert J. Burky
University of Dayton
Dayton, Ohio

INTRODUCTION

The present economy of the Hawaiian Islands is primarily dependent on tourism. Visitors spend millions of dollars at hotels, in shopping malls and shops, at restaurants, and travel to and from islands. This money provides the economic foundation for many local Hawaiian families, and it serves to drive the state's economy. However, the economy has not always been based on tourism, but rather on the agricultural strength of sugar cane from the late nineteenth century through most of the twentieth century. Historically, the rise of agriculture was based on freshwater, the resource that drives tourism and expanding commercial development today. However, the surface waters and aquifers are limited, and a balanced concern exists for the effects of continued freshwater withdrawal from these sources. In particular, the streams of Hawaii are home to many endemic organisms (found only in Hawaii) that are threatened by continued surface water withdrawal. Many of the aquifers are being overpumped, increasing the salinity of groundwater resources and sometimes influencing surface flows.

For this topic, we provide a brief historical description of the geological origin and predominate climate patterns of the Hawaiian Islands, an introduction to endemism and the freshwater ecosystems, early Hawaiian cultural use and contemporary freshwater management, the expansion of sugar cane as a catalyst for freshwater exploitation, the cultural consequences of sugar cane expansion, and existing political struggles. We conclude with a short discussion about the future of Hawaiian freshwater resources, one of delicate balance and immediate concern.

GEOLOGICAL ORIGIN

The Hawaiian Islands have grown from a permanent "hot spot" located at an unknown depth beneath the Pacific Ocean seafloor. As it has been difficult to determine detailed information on this hot spot, the geological formation of Hawaii remains partly theory (1). The hot spot is a huge lake of melted rock under tremendous pressure in the Earth's mantle where molten rock erupts, accumulates on the ocean floor, and finally emerges from the ocean surface after millions of years (2). The hot spot is believed to remain stationary, but the outer crustal plate has been shifting (1). The shifting plates, timing of eruptions, and sea level changes have resulted in a chain of submerged and remnant volcanic islands to the northwest (Emperor Seamount Chain) of Midway Atoll, and the existing main Hawaiian Islands to the southeast (2).

The largest Hawaiian Islands are the most recently formed. The two oldest are Kauai and Niihau, with Oahu and the islands of Maui Nui (Maui, Molokai, Lanai, and Kahoolawe) sequentially younger moving southeast, and the appropriately named Big Island (or Hawaii) that is volcanically active (Fig. 1) (3,4). A new island about 1000 m (called Loihi) below the ocean surface is developing off the southern coast of the Big Island (5). As one travels from the northwest to the southeast, the existing islands become younger, larger, taller, and less eroded.

CLIMATE PATTERNS

Remotely located in the Pacific Ocean, the climate and precipitation patterns of Hawaii are dependent on predominant wind patterns and ocean currents. The Trade Winds (Prevailing Westerly Winds) blow across the Hawaiian Islands in a northeasterly to southwesterly direction, with stronger intensity occurring during the drier summer months (Fig. 1) (6). As these winds move across the ocean collecting water vapor, they move up, around, and sometimes over the mountains. As the air moves to higher altitudes, it cools, clouds form, and it rains, causing a rain shadow effect responsible for perennially wet northeastern shores (windward) opposite dry southwestern shores (leeward) (6). High rainfall, lush tropical rainforest vegetation, rushing streams, and abundant cascades and waterfalls characterize the windward sides, whereas the leeward sides have little rainfall and desert lands of dry-tolerant shrubs and cacti. The spatial distribution of precipitation combined with steep elevational gradients and ocean currents create climate patterns conducive for biome and ecosystem variety in a relatively small geographical area, ecosystems where many organisms evolved and are found nowhere else on Earth.

HAWAIIAN ECOSYSTEMS AND ENDEMISM

The variety of ecosystems and biomes in Hawaii area a result of five general characteristics: (a) geographic

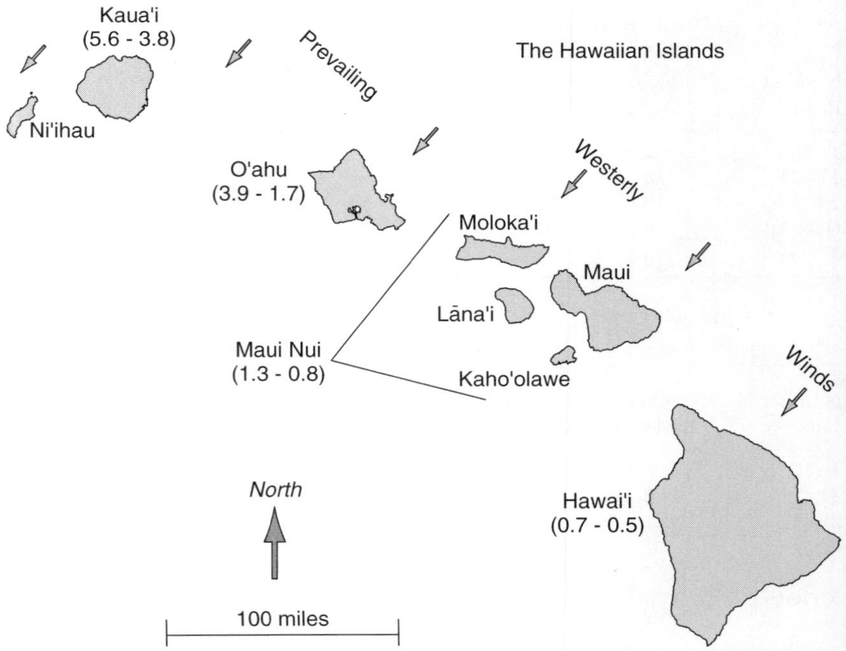

Figure 1. A map of the main Hawaiian Islands. Approximate island (or island group) ages are given in parentheses below the island name. Island ages are in millions of years and were estimated from two sources (3,4). Maui, Moloka'i, Lana'i, and Kaho'olawe were once a single island called Maui Nui until about 0.4–0.3 million years ago.

isolation, (b) small relative size and area, (c) recent age, (d) unique evolution, and (e) climate patterns (7). The Hawaiian Islands are some of the most isolated land masses in the world, which kept original immigrant species dispersal minimal (8). Relatively few species became established, but those that did underwent rapid speciation in the absence of many variables typical of mainland ecosystems (7). For example, some birds and many insects are flightless in response to evolving in predator absence (8). Many species never evolved effective defenses against predators or aggressive alien competitors, making the endemic flora and fauna easy prey, or out competed by introduced species that evolved with aggressive predators and competitors. Hawaii has approximately 10,000–15,000 endemic and indigenous species with over 30% of the total threatened or endangered species in the United States (4,7). In light of this, appropriate conservation, preservation, and management are needed to protect these rare habitats and endemic communities. The freshwater communities are threatened by continued and expanding riparian (land adjacent to a stream or river) habitat destruction, surface water removal, and introduced and invasive species.

HAWAIIAN STREAM ANIMAL COMMUNITIES

Hawaiian streams respond to unpredictable rainfall events that determine flow volume changes over short- and long-term time periods. The geology is commonly porous and precipitation quickly saturates riparian soils, creating rapid and high-volume surface runoff. Flash floods are common throughout the year. The local name for a flood is "freshet", indicative of the historical description, and contemporary observations of scoured and rearranged stream channels moved by tremendous hydrodynamic forces. It is thought that many Hawaiian stream organisms have evolved specialized life cycles (e.g., amphidromy, see below) to accommodate these unpredictable and frequently extreme events (9–11).

AMPHIDROMY

The native Hawaiian fish, shrimp, prawn, and snail species have a diadromous life cycle, meaning they migrate between freshwater and seawater (1). These native species are amphidromous, a subcategory of diadromy (Fig. 2) (12,13). Adults reproduce in streams, eggs hatch, and embryos (now called larvae) drift to the ocean for growth and development into postlarvae (from 10–30 mm body length) over several months (14–16), and then migrate back into the streams. In amphidromy, the purpose of migration is not for immediate breeding as in diadromy (e.g., salmon); instead, postlarvae, not adults, migrate into freshwater to feed, grow, and remain for the rest of their existence through multiple reproductive events (12,17,18).

There are four endemic and one indigenous amphidromous fish species in Hawaiian streams (Table 1) (16,19,20). The pelvic fins of the gobies are fused and form a ventral sucker used to cling to rocks and climb waterfalls as postlarvae and early juveniles (16,21). One fish species has been reported to climb single waterfalls >100 m high and combined waterfalls totaling a height of 300 m (19). The native shrimp, snails, and prawns (Table 1) also have the ability to climb waterfalls (20). Postlarval migration has been found to be directly related to stream discharge (22,23).

OTHER ENDEMIC STREAM ANIMALS

The native stream organisms have closely related marine relatives, and it is evident that the freshwater species

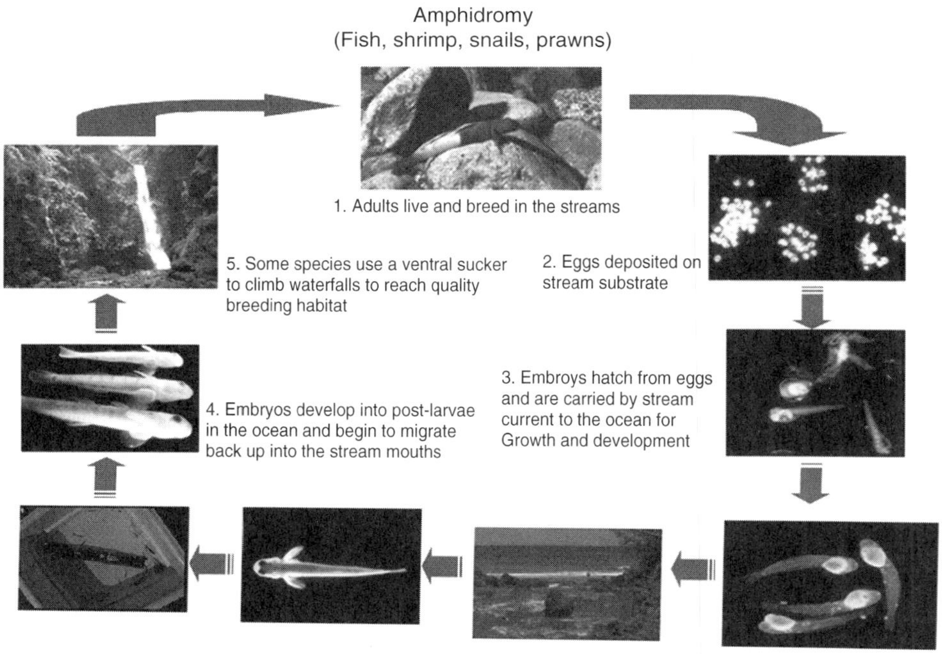

Figure 2. A schematic depicting the amphidromous life cycle of the native fish, snails, shrimp, and prawns found in Hawaiian streams. The life cycle requires postlarval development in the ocean and a migration back into the streams that is not immediately for breeding.

Table 1. The Freshwater Animals of Hawaiian Streams Including the Common Name, Scientific Name, Hawaiian Name, and Geographic Range

Taxa Type	Common Name	Scientific Name	Hawaiian Name	Range
Fish	general term for freshwater fishes		'o'opu	
	freshwater amphidromous gobies	*Lentipes concolor*	'o'opu 'alamo'o	endemic
			'o'opu hi'u kole	
			'o'opu hi'u'ula	
			'o'opu nu'ukole	
		Sicyopterus stimpsoni	'o'opu nopili	endemic
		Awaous guamensis	'o'opu nakea	indigenous
		Stenogobius hawaiiensis	'o'opu naniha	endemic
	freshwater amphidromous eleotrid	*Eleotris sandwicensis*	'o'opu 'akupa	endemic
	euryhaline-Hawaiian flag-tail fish	*Kuhlia sandvicensis*	aholehole	endemic
	euryhaline-striped or gray mullet	*Mugil cephalus*	'ama'ama	worldwide
Snails	freshwater amphidromous snail	*Neritina granosa*	hihiwai	endemic
			wi	
	estuarine amphidromous snail	*Neritina vespertina*	hapawai	endemic
			hapakai	
	brackish/marine neritid snail	*Theodoxus cariosus*	pipiwai	endemic
	lymnaeid snails	*Erinna newcombi*	*	endemic
Sponge	freshwater sponge	*Heteromyenia baileyi*	*	indigenous
Crustaceans	general term for freshwater shrimp		'opae	
	freshwater, mountain srhimp	*Atyoida bisulcata*	'opaekala'ole	endemic
			'opae kuahiwi	
			'opae kolo	
	freshwater prawn	*Macrobrachium grandimanus*	'opae 'oeha'a	endemic
	anchialine pond shrimp	*Halocaridina rubra*	'opae 'ula	endemic
Insects	adult dragonfly	*Anas* spp.	pinao	endemic
	adult damselfly	*Megalagrion* spp.	pinao 'ula	endemic
	immature dragonfly or damselfly	*Megalagrion* & *Anax* spp.	lohelohe	endemic
	immature dragonfly or damselfly	*Megalagrion* & *Anax* spp.	lohaloha	endemic
	immature dragonfly or damselfly	*Megalagrion* & *Anax* spp.	pua'alohehole	endemic
	immature dragonfly or damselfly	*Megalagrion* & *Anax* spp.	'olopelope	endemic
	many other endemic insects	various species	*	endemic

*Indicates that a Hawaiian name is not available.

evolved from marine habitats (Table 1) (10,16,24,25). Some examples of this can be found in stream insect communities consisting of midges, *Telmatogeton* (Chironomidae); beach flies, *Procanace* (Canacidae); and shore flies, *Scatella* (Ephydridae), which are thought to have radiated into streams from ancestral intertidal habitats (26–28). The Hawaiian *Telmatogeton* complex is unique with all five endemic species restricted to freshwater, the only representatives of an exclusively marine intertidal subfamily worldwide (26–29).

Another important insect group is the endemic damselflies of the *Megalagrion* complex. This genus has shown rapid speciation in concert with specific freshwater and terrestrial habitats (30–32). Contrary to many other aquatic insects of Hawaii, the endemic damselflies are thought to have evolved from one freshwater ancestor to Hawaii (31,32). As a result of recent extinctions, numerous research initiatives have been carried out to understand the evolution, biology, and conservation of this unique group of insects (31,32). Other freshwater organisms are reported in Table 1.

SURFACE WATER REMOVAL

The streams of Hawaii are habitat to relatively recently evolved communities with close marine ancestors (e.g., fish, snails, shrimp, and some insects) and/or isolated populations that have undergone rapid speciation (e.g., the damselflies). For millennia, these systems connected the mountains to the ocean along what is called in Hawaiian the mauka—makai continuum. The physical connection was the flowing streams supporting the biological connections of the amphidromous species. However, many streams were heavily diverted in the late 1800s for irrigating an expanding sugar cane industry. Stream flow removal and riparian degradation has had the following effects: (a) destroying and eliminating breeding habitat for all stream animals, (b) preventing amphidromous egg and larval drift to the ocean, (c) obstructing postlarval recruitment back into the streams, and (d) facilitating invasive species establishment (9,22,33–35). Water is a critical and limited resource, not only to humans but also to the native stream organisms. Balancing these will take compromise and scientifically based freshwater resource management.

FRESHWATER RESOURCE USE AND MANAGEMENT

Historic Management

The ahupua'a system was the first freshwater management system in Hawaii. Developed by the ali'i (chiefs) of the first Hawaiians 1500–1600 years ago, it maintained adequate freshwater resources for the entire population of each island. The ahupua'a was what now is considered a watershed and extended from the headwater springs into the ocean. Along this mauka—makai continuum, the land along the stream was divided into pie-like slices called 'ili kuponos, that were designated to maka'ainana (commoners) and overseen by a chief-appointed konohiki. The commoners did not own the land or water, for they only maintained them for the chiefs. Each konohiki was appointed to a single ahupua'a, and each 'ili kupono was divided into kalo lo'i (taro patches). Stream water was diverted by pani wai (diversions) and carried through small 'auwai (ditches) to the lo'i. The lo'i were connected in a stair-step manner by additional ditches maintained by the responsible commoners. If the ditches were not maintained upstream, all downstream lo'i would be affected. Contrary to contemporary freshwater management, the stream water was always returned to the stream before it entered the ocean, and was never carried out of the watershed. Therefore, the water continually connected the highest mountain reaches to the lowland settlements near the sea, maintaining the mauka—makai continuum.

The early Hawaiians understood the connectivity of the land, freshwater, and sea. The ahupua'a system maintained adequate freshwater resources for human use while balancing ecological function. Hawaiians also understood the power and importance of water to their culture. For example, the Hawaiian word for water, wai, is the root for wealth (waiwai) and law (kanawai) (36). Over the next 60 years, native Hawaiians would find just how important water could be to foreigners, or haoles—those without breathe, or mana (spirit). The sugar cane industry completely changed freshwater resource management: water would be removed from the ahupua'a, along with much of the native Hawaiian culture.

Contemporary Management

The influence of western missionaries and merchants shifted the Hawaiian perception of water from a shared resource to a commodity that provided maximum use and greatest reward to plantation owners (37). After the arrival of Captain James Cook in 1778, freshwater management changed little until the The Great Mahele of 1848, enacted by King Kamehameha III under heavy influence by the decedents of original missionaries, the new owners of the sugar cane industry (and pineapple later). The Great Mahele was the first act to privatize land and water in Hawaii. For nearly 25 years the kings of Hawaii attempted to negotiate reciprocity treaties with the United States In 1876, King Kalakaua finally established a reciprocity treaty assuring duty-free exchange between Hawaii and the United States, giving sugar companies a competitive edge in the world market and giving Pearl Harbor to the United States, which was the same year of the first water licenses, the creation of the first private water company, and the construction of the diversion-ditch systems, which relocated millions of gallons of water from the windward watersheds to leeward sugar cane fields. According to Wilcox (36), one pound of sugar takes approximately 4000 pounds, or 500 gallons, of water.

The growth of sugar cane and pineapple took over and "transformed Hawaii from a traditional, insular, agrarian, and debt-ridden society into a multicultural, cosmopolitan, and prosperous one" (36). During this time, the number of sugar plantations rose from 5 in 1857 to 90 in 1884 (36). From the late 1870s to the mid-1900s, the Hawaiian sugar cane industry continued to grow, in part because of political access to surface and groundwater resources. By the early 1900s, almost every stream on the windward sides of each island was at least partially diverted, and in 1920, the sugar cane industry was diverting >800 mgd and pumping an additional 400 mgd from aquifers (36). Almost 20 years later, the entire city of Boston was only consuming 80 mdg (36). The privately developed surface water of Hawaii was quite different from publicly controlled surface water development of the western United States, which put control of water into the hands of western businessmen, those who would eventually contribute to the overthrow of the last Hawaiian Monarchy. (Several books are available that describe the sequence of events that led to the overthrow (38–40).)

With the developing agricultural industry, the population and ethnic diversity expanded as well. A larger labor force was needed to construct the ditches and diversions and work the agricultural fields. For cheap labor, the agricultural companies looked to Japan, China, and other Asian nations. This immigration resulted in several ethnic populations living and working on plantation property, creating mixed nonHawaiian communities that are the roots of the ethnic diversity found in Hawaii today. As the population and successful industries increased, the popularity of the islands grew in the mid-1900s, whereas the sugar cane industry met fierce competition from other nations, and has resulted in the slow demise of the former economic strength. With the buildup and maintained military presence of Pearl Harber from World War II, the economic focus of Hawaii saw a change from agriculture to military establishment to tourism that is now the economic base of the state. Along the southwestern coast of each island, numerous hotels, shopping malls, golf courses, and restaurants have been developed because most people prefer to vacation on the sunny, dry side of the island. However, similar to the sugar cane scenario, water is not readily available on the leeward side. Therefore, water remains diverted from streams, or pumped from dwindling freshwater aquifers, in order to accommodate tourists, water the green lawns, and fill hotel swimming pools and spas.

A BRIEF INTRODUCTION TO HAWAIIAN WATER LAW

Water Rights

The expansion of the sugar industry resulted in many water disputes between sugar companies, various landowners, and native Hawaiians. These disputes helped recognize the following water rights: (a) appurtenant, (b) riparian, (c) prescriptive, and (d) surplus (41). Appurtenant rights developed from ancient rights and were officially declared in the earliest Hawaiian water case,

Peck vs. Baily (1867). This case determined that if land was entitled by a landowner to cultivate taro, than some stream water was also. Riparian rights permit water use to those who own land next to a stream; however, use must be reasonable and not harmful to other landowners (42). In riparian law, the focus is on maintaining stream integrity while allowing landowner priority uses (e.g., domestic drinking and washing) over nonpriority uses (e.g., irrigation, mining) (41). Both appurtenant and riparian water rights consider water as a common good. Prescriptive and surplus water rights address private ownership of water sources.

Prescriptive water rights are granted if a party has proven using water "belonging" to another party for an extended and continuous period of time (e.g., 20 years) (37). Surplus water is defined as the amount of water in a stream not covered by any prior water right (37). In court, cases of surplus water were heard independently, which precluded the establishment of defined standards for future disputes. For example, in the case of Hawaiian Commercial & Sugar Co. v. Wailuku Sugar Co. (1904), the court ruled that surplus water traveling through an ahupua'a belonged to the private land owner, but in Carter v. Territory (1917), the court concluded that surplus water was to be used reasonably according to riparian principles, thus shared among landowners. The inconsistent rulings and absence of an established in-stream flow standard left the status of surplus waters questionable until the state code in 1987.

Summary of Case Law

From the 1850s to the 1980s, the water rights of Hawaii were evolving according to case law, yet no standard criteria for the allocation, transport, and quantity of freshwater was resolved. There were several political, economical, and social reasons for no established criteria. First, the rise of the sugar cane industry boosted the economy, and with it came political power to the foreign owners. Second, during this period, the Hawaiian Islands shifted from a monarchy to a republic to a U.S. Territory, and finally became a U.S. state, which created an atmosphere of economic and political change that left many native Hawaiians in a state of social and political confusion (36), where they did not resist changes in water law, as it was not in their culture to protest actions of the King. As a result of this, there was difficulty in establishing a cause-effect relationship, and many of the villages most affected were too small to fight the wealthy plantation owners (36,37).

In 1978, the Hawaii State Constitution mandated that a water code and commission be established to manage water resources. And in the late 1980s, a Hawaii State Water Code and the Commission on Water Resource Management was passed by the Hawaii State Legislature (37). The State Water Code became a regulatory mechanism for state agencies and counties and required the Commission to monitor and guide water allocation by establishing minimum in-stream flow standards that were to enhance, protect, and re-establish beneficial in-stream uses. Current trends have been developing during the 1990s that are testing the utility of the Hawaii State Water Code. The first of several cases, and a landmark case for future petitions, was the Waiahole Combined Contested Case Hearing. In this case, the Hawaii Supreme Court ruled for the establishment of minimum in-stream flow standards and than any additional flow secondarily allocated for other uses. In-stream flow standards are currently being investigated. The Waiahole Ditch Case was the first test of the State Water Code, yet many similar cases are being petitioned at the time of this writing. The diminishing sugar industry is opening a new chapter in Hawaiian case law, a chapter that must not only consider economics but also "traditional and customary Hawaiian rights, protection and procreation of fish and wildlife, ecological balance, scenic beauty, public recreation, beneficial in-stream uses, and public interest" (36).

SUMMARY AND SYNTHESIS

The history of Hawaiian water resource issues is complex. The system of surface water management went through a dramatic change with the explosion of agriculture and the associated political power of missionary descendents; however, this explosion was inherently dependent on privately developed surface waters—a feedback proliferation of each other. The surface water development from the late 1800s through the mid-1900s drained watersheds of both water and native Hawaiian culture. The efficient removal of water high in the watersheds left a trickle downstream for native taro farmers. With the construction of massive diversion-ditch systems, native Hawaiians emigrated from the ahupua'a to build ditches and work sugar cane fields. At the same time, thousands of immigrants from China, Japan, and the Philippines melted into the sugar cane plantation farms, changing the demographics of Hawaii. The push by sugar businessmen to get an extended reciprocity treaty secured rights of Pearl Harbor to the United States; it also led to the coup of the Hawaiian Monarchy. At the same time, streams were being impacted by water removal and rapid riparian destruction, negatively affecting the native biological communities of the mauka–makai continuum.

The degradation of stream habitats in Hawaii has become an increasingly important issue to many native Hawaiians, scientists, private organizations, and individuals. Water resources have been, and will continue to be, a highly debated political issue. Water is necessary for agricultural, industrial, and municipal uses throughout Hawaii; however, in order to maintain the unique biodiversity of the streams, a compromise must be made between continued development and conservation. Thus, the future of the Hawaii's freshwater ecosystems and economic security is dependent on wise freshwater management and allocation, which should not be made in a scientific vacuum. Objective and sustainable answers can only be achieved with a solid scientific base, one free from political bias that has been the history of Hawaiian freshwater resources.

BIBLIOGRAPHY

1. Culliney, J.L. (1988). *Islands in a Far Sea: Nature and Man in Hawaii*. University of Hawaii Press, Honolulu, HI.
2. Hazlett, R.W. and Hyndman, D.W. (1996). *Roadside Geology of Hawaii*. Mountain Press, Missoula, MT.
3. Stearns, H.T. (1985). *Geology of the State of Hawaii*, 2nd Edn. Pacific books, Palo Alto, CA.
4. Juvik, S.P. and Juvik, J.O. (1998). *Atlas of Hawaii*, 3rd Edn. University of Hawaii Press, Honolulu, HI.
5. U.S.G.S. (1997). *U.S. Geological Survey Programs in Hawaii and the Pacific*. Department of Interior, United States Geological Survey.
6. Carlquist, S. (1980). *Hawaii: A Natural History*. Pacific Tropical Botanical Gardens, Honolulu, HI.
7. Mueller-Dombois, D., Bridges, K.W., and Carson, H.L. (1981). *Island Ecosystems: Biological Organization in Selected Hawaiian Communities*. Hutchinson Ross, Stroudsburg, PA.
8. Kay, E.A. (Ed.). (1994). *A Natural History of the Hawaiian Islands, Selected Readings II*. University of Hawaii Press, Honolulu, HI.
9. Way, C.M., Burky, A.J., Harding, J.M., Hau, S., and Puleloa, W.K.L.C. (1998). Reproductive biology of the endemic goby, *Lentipes concolor*, from Makamaka'ole Stream, Maui and Waikolu Stream, Moloka'i. *Environ. Biol. Fish.* **51**: 53–65.
10. Benbow, M.E. (1999). *Natural History and Bioenergetics of an Endemic Hawaiian Chironomid: Fluctuating Stability in a Stochastic Environment*. Ph.D. Dissertation. Department of Biology, University of Dayton, Dayton, OH.
11. Benbow, M.E., Burky, A.J., and Way, C.M. (1997). Larval habitat preference of the endemic Hawaiian midge, *Telmatogeton torrenticola* Terry (Telmatogetoninae). *Hydrobiologia* **346**: 129–136.
12. McDowall, R.M. (1996). Is there such a thing as amphidromy? *Micronesica* **30**(1): 3–14.
13. McDowall, R.M. (1997). The evolution of diadromy in fishes (revisited) and its place in phylogenetic analysis. *Rev. Fish Biol. Fish* **7**(4): 443–462.
14. Radtke, R.L. and Kinzie, R.A. (1996). Evidence of a marine larval stage in endemic Hawaiian stream gobies from isolated high-elevation locations. *T. Am. Fish Soc.* **125**: 613–621.
15. Radtke, R.L., Kinzie, R.A., and Folsom, S.D. (1988). Age at recruitment of Hawaiian freshwater gobies. *Environ. Biol. Fish* **23**(3): 205–213.
16. Benbow, M.E., Burky, A.J., and Way, C.M. (2004). Morphological characteristics and species separation of Hawaiian postlarval amphidromous fishes. *Micronesica* **37**(1): 127–143.
17. Lindstrom, D.P. (1998). *Reproduction, Early Development and Larval Transport Dynamics of Amphidromous Hawaiian Gobioids*. Ph.D. Dissertation, University of Hawaii, Honolulu, HI.
18. McDowall, R.M. (1992). Diadromy: origins and definitions of terminology. *Copeia* **1**: 248–251.
19. Kinzie, R.A., III (1990). *Species Profiles: Life Histories and Environmental Requirements of Coastal Vertebrates and Invertebrates Pacific Ocean Region. Amphidromous Macrofauna of Hawaiian Island Streams*. US Army Engineer Waterways Experiment Station, Vicksburg, MS.
20. Kinzie, R.A., III and Ford, J.I. (1982). *Population Biology in Small Hawaiian Streams*. Water Resources Research Center, University of Hawaii at Manoa, Honolulu, HI, p. 147.
21. Benbow, M.E., Orzetti, L.L., McIntosh, M.D., and Burky, A.J. (2002). A note on cascade climbing of migrating goby and shrimp postlarvae in two Maui streams. *Micronesica* **34**(2): 243–248.
22. Benbow, M.E., Burky, A.J. and Way, C.M. (2004). The use of two modified Breder traps to quantitatively study amphidromous upstream migration. *Hydrobiologia* **527**: 139–151.
23. Burky, A.J., Benbow, M.E., and Way, C.M. (2004). *Two modified Breder traps for quantitative studies of postlarval amphidromy, Hydrobiologia.*
24. Kinzie, R.A.I. (1997). Evolution and life history patterns in freshwater gobies. *Micronesica* **30**(1): 27–40.
25. McDowall, R.M. (2003). Hawaiian biogeography and the islands' freshwater fish fauna. *J. Biogeogr.* **30**(5): 703–710.
26. Hardy, D.E. and Delfinado, M.D. (1980). *Diptera: Cyclorrhapha III, Series Schizophora, Section Acalypterae, Exclusive of Family Drosophilidae*. Vol. 13. The University Press of Hawaii, Honolulu, HI.
27. Newman, L.J. (1988). Evolutionary relationships of the Hawaiian and North American *Telmatogeton* (Insecta; Diptera: Chironomidae). *Pacific. Sci.* **42**(1–2): 56–64.
28. Newman, L.J. (1977). Chromosomal evolution of the Hawaiian *Telmatogeton* (Chironomidae, Diptera). *Chromosoma* **64**: 349–369.
29. Wirth, W.W. (1947). A review of the genus *Telmatogeton* Schiner, with descriptions of three new Hawaiian species (Diptera: Tendipedidae). *Proceedings of the Hawaiian Entomological Society* **13**: 143–191.
30. Jordan, S., Simon, C., and Polhemus, D.A. (2003). Molecular systematics and adaptive radiation of Hawaii's endemic damselfly genus *Megalagrion* (Odonata: Coenagrionidae). *Syst. Biol.* **52**(1): 89–109.
31. Polhemus, D.A. (1997). Phylogenetic analysis of the Hawaiian damselfly genus *Megalagrion* (Odonata: Coenagrionidae): implications for biogeography, ecology, and conservation biology. *Pacific Science* **51**(4): 395–412.
32. Polhemus, D.A. and Asquith, A. (1996). *Hawaiian Damselflies: A Field Identification Guide*. Bishop Museum Press, Honolulu, HI.
33. Englund, R.A. and Filbert, R.B. (1999). Flow restoration and persistence of introduced species in Waikele Stream, O'ahu. *Micronesica* **32**(1): 143–154.
34. Benbow, M.E., Burky, A.J., and Way, C.M. (2003). Life cycle of a torrenticolous Hawaiian chironomid (*Telmatogeton torrenticola*): stream flow and microhabitat effects. *Ann. Limnol. Int. J. Lim.* **39**(2): 103–114.
35. McIntosh, M.D., Benbow, M.E., and Burky, A.J. (2002). Effects of stream diversion on riffle macroinvertebrate communities in a Maui, Hawaii, stream. *River Res. Applic.* **18**: 569–581.
36. Wilcox, C. (1996). *Sugar Water: Hawaii's Plantation Ditches*. University of Hawaii Press, Honolulu, HI.
37. Mackenzie, M.K. (1991). *Native Hawaiian Rights Handbook*. Native Hawaiian Legal Corporation, Honolulu, HI.
38. Liliuokalani. (1990). *Hawaii's Story by Hawaii's Queen*. Mutual Publishing, Honolulu, HI.
39. Daws, G. (1968). *Shoal of Time: A History of the Hawaiian Islands*. University of Hawaii Press, Honolulu, HI.
40. Cooper, G. and Daws, G. (1990). *Land and Power in Hawaii*. University of Hawaii Press, Honolulu, HI.
41. MacKenzie, M.K. (1991). Water rights, In: *Native Hawaiian Rights Handbook*. M.K. MacKenzie (Ed.). Native Hawaiian Legal Corporation, Honolulu, HI, pp. 149–172.
42. Schoenbaum, T.J. and Rosenburg, R.H. (1996). *Environmental Policy Law: Problems, Cases and Readings*. University Casebook Series, Westbury, NY.

INDEX

NOTE: Page numbers followed by f refer to figures, page numbers followed by t refer to tables.